DIGITAL
COMMUNICATIONS

17

DIGITAL COMMUNICATIONS
Fundamentals and Applications

BERNARD SKLAR

*The Aerospace Corporation, El Segundo, California
and
University of California, Los Angeles*

Prentice-Hall International, Inc.

Library of Congress Cataloging-in-Publication Data

SKLAR, BERNARD (date)
 Digital communications.

 Bibliography: p.
 Includes index.
 1. Digital communications. I. Title.
TK5103.7.S55 1988   621.38'0413   87-1316
ISBN 0-13-212713-X

Editorial/production supervision and
 interior design: Reynold Rieger
Manufacturing buyers: Gordon Osbourne and Paula Benevento

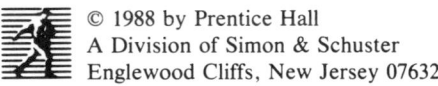

© 1988 by Prentice Hall
A Division of Simon & Schuster
Englewood Cliffs, New Jersey 07632

Printed in the United States of America

16   15   14

ISBN 0-13-212713-X 025

Prentice-Hall International (UK) Limited, *London*
Prentice-Hall of Australia Pty. Limited, *Sydney*
Prentice-Hall Canada Inc., *Toronto*
Prentice-Hall Hispanoamericana, S.A., *Mexico*
Prentice-Hall of India Private Limited, *New Delhi*
Prentice-Hall of Japan, Inc., *Tokyo*
Simon & Schuster Asia Pte. Ltd., *Singapore*
Editora Prentice-Hall do Brasil, Ltda., *Rio de Janeiro*
Prentice-Hall, *Englewood Cliffs, New Jersey*

To my mother, Ruth Sklar,
the memory of my father, Julius Sklar,
my wife, Gwen, and our children,
Debra, Sharon, and Dean

Contents

3 BANDPASS MODULATION AND DEMODULATION   117

4   COMMUNICATIONS LINK ANALYSIS

187

7 MODULATION AND CODING TRADE-OFFS   381

8 SYNCHRONIZATION

429

Maurice A. King, Jr.

9 MULTIPLEXING AND MULTIPLE ACCESS

475

11 SOURCE CODING

595

Fredric J. Harris

12   ENCRYPTION AND DECRYPTION   668

A A REVIEW OF FOURIER TECHNIQUES   710

B FUNDAMENTALS OF STATISTICAL DECISION THEORY   733

C RESPONSE OF CORRELATORS TO WHITE NOISE

D OFTEN USED IDENTITIES

E A CONVOLUTIONAL ENCODER/DECODER COMPUTER PROGRAM

F LIST OF SYMBOLS

INDEX

Preface

This book is intended to provide a comprehensive coverage of digital communication systems for senior-level undergraduates, first-year graduate students, and practicing engineers. Even though the emphasis of the book is on digital communications, necessary analog fundamentals are included, since analog waveforms are used for the radio transmission of digital signals.

The key feature of a digital communication system is that it deals with a finite set of discrete messages, in contrast to an analog communication system in which messages are defined on a continuum. The objective at the receiver of the digital system is *not* to reproduce a waveform with precision; it is, instead, to determine from a noise-perturbed signal which of the finite set of waveforms had been sent by the transmitter. In fulfillment of this objective, an impressive assortment of signal processing techniques has arisen over the past two decades.

The book develops these important techniques in the context of a unified structure. The structure, in block diagram form, appears at the beginning of each chapter; blocks in the diagram are emphasized, as appropriate, to correspond to the subject of that chapter. Major purposes of the book are (1) to add organization and structure to a field that has grown rapidly in the last two decades, and (2) to ensure awareness of the "big picture" even while delving into the details. The signals and key processing steps are traced from the information source through the transmitter, channel, receiver, and ultimately to the information sink. Signal transformations are organized according to functional classes: formatting and source coding, modulation, channel coding, multiplexing and multiple access, spreading, encryption, and synchronization. Throughout the book, emphasis is placed on system goals and the need to trade off basic system parameters such as signal-to-noise ratio, probability of error, and bandwidth (spectral) expenditure.

ORGANIZATION OF THE BOOK

It is assumed that the reader is familiar with Fourier methods and convolution. Appendix A reviews these techniques, emphasizing those properties that are particularly useful in the study of communication theory. It is also assumed that the reader has a knowledge of basic probability and has some familiarity with random variables. Appendix B builds on these disciplines for a short treatment on statistical decision theory with emphasis on hypothesis testing—so important in the understanding of detection theory. Chapter 1 introduces the overall digital communication system and the basic signal transformations that are highlighted in subsequent chapters. Some basic ideas of random variables and the additive white Gaussian noise (AWGN) model are reviewed. Also, the relationship between power spectral density and autocorrelation, and the basics of signal transmission through linear systems, are established. Chapter 2 covers the signal processing step, known as formatting, the step that renders an information signal compatible with a digital system. Chapter 2 also emphasizes the *transmission* of baseband signals. Chapter 3 deals with bandpass modulation and demodulation techniques. The detection of digital signals in Gaussian noise is stressed, and receiver optimization is examined. Chapter 4 deals with link analysis, an important subject for providing overall system insight; it considers some subtleties usually neglected at the college level. Chapters 5 and 6 deal with channel coding—a cost-effective way of providing improvement in system error performance. Chapter 5 emphasizes linear block coding, and Chapter 6 emphasizes convolutional coding.

Chapter 7 considers various modulation/coding system trade-offs dealing with probability of bit error performance, bandwidth efficiency, and signal-to-noise ratio. Chapter 8 deals with synchronization for digital systems. It covers phase-locked-loop implementation for achieving carrier synchronization; bit synchronization, frame synchronization, and network synchronization; and some fundamentals of synchronization as applied to satellite links.

Chapter 9 treats multiplexing and multiple access. It explores techniques that are available for utilizing the communication resource efficiently. Chapter 10 introduces spread-spectrum techniques and their application in such areas as multiple access, ranging, and interference rejection. This technology is particularly important for most military communication systems. The subject of source coding in Chapter 11 deals with data formatting, as is done in Chapter 2; the main difference between formatting and source coding is that source coding additionally involves data redundancy reduction. Rather than considering source coding immediately after formatting, source coding has purposely been treated in a later chapter. It is felt that the reader should be involved with the fundamental processing steps, such as modulation and channel coding, early in the book, before examining some of the special considerations of source coding. Chapter 12 covers some basic encryption/decryption ideas. It includes some classical encryption concepts, as well as some of the proposals for a class of encryption systems called public key cryptosystems.

If the book is used for a two-term course, a simple partitioning is suggested: the first six chapters to be taught in the first term, and the last six chapters in the second term. If the book is used for a one-term only course, it is suggested that the course material be selected from the following chapters: 1, 2, 3, 4, 5, 6, 8, and 10.

ACKNOWLEDGMENTS

This book is an outgrowth of my teaching activities at the University of California, Los Angeles, and my work in the Communications Division at The Aerospace Corporation. A number of people have contributed in many ways and it is a pleasure to acknowledge them. Dr. Maurice King, my colleague at Aerospace, carefully reviewed and made important contributions to each chapter. His continual assistance has been invaluable. He also contributed Chapter 8, Synchronization. Professor Fred Harris of San Diego State University suggested many improvements and contributed Chapter 11, Source Coding. I want to pay special thanks to Dr. Marvin Simon of the Jet Propulsion Laboratory for providing me with much encouragement and many valuable suggestions.

I also want to thank Professor Jim Omura of UCLA for sharing with me his considerable knowledge of encryption and thereby helping me improve Chapter 12. Professor Raymond Pickholtz of George Washington University gave me lots of beneficial advice throughout the writing process. Professors William Lindsey and Andreas Polydoros of the University of Southern California suggested important improvements. Professor James Modestino of Rensselaer Polytechnic Institute, Dr. Adam Lender of Lockheed Palo Alto Research Laboratory, and Professor Ron Iltis of the University of California, Santa Barbara, each provided valuable reviews. Dr. Todd Citron of Hughes Aircraft, Dr. Joe Odenwalder of MA/COM Linkabit, and Dr. Unjeng Cheng of Axiomatics were extremely helpful in the chapters on channel coding. Mr. Don Martin and Mr. Ned Feldman of The Aerospace Corporation made numerous suggestions and contributions. I also want to pay special thanks to Professor Wayne Stark of the University of Michigan, whose unique critical talents enhanced the manuscript's continuity.

The block diagrams in Figures 1.2 and 1.3, at each chapter opening, and on the cover of the book, first appeared in the two part paper: © 1983 IEEE; B. Sklar, ''A Structured Overview of Digital Communications—A Tutorial Review,'' *IEEE Communications Magazine*, August and October, 1983. Permission from IEEE to reprint these figures throughout the book is gratefully acknowledged.

My students at UCLA and those at Aerospace used early versions of chapters of this book and made many helpful contributions. I am indebted to all those students who have taken my courses and thus helped me with this project. I also want to express my appreciation to my management at Aerospace, Mr. Hal

McDonnell and Mr. Fred Jones, for their indulgence and moral support. I want to acknowledge and thank Ms. Cynthia Dickson for her diligence and speed in typing the entire manuscript.

Finally, I want to thank my wife, Gwen, for her very unselfish support, her understanding, and her endurance of the many months I had time for only *one* devotion—the writing of this book.

BERNARD SKLAR

Tarzana, California

CHAPTER 1

Signals and Spectra

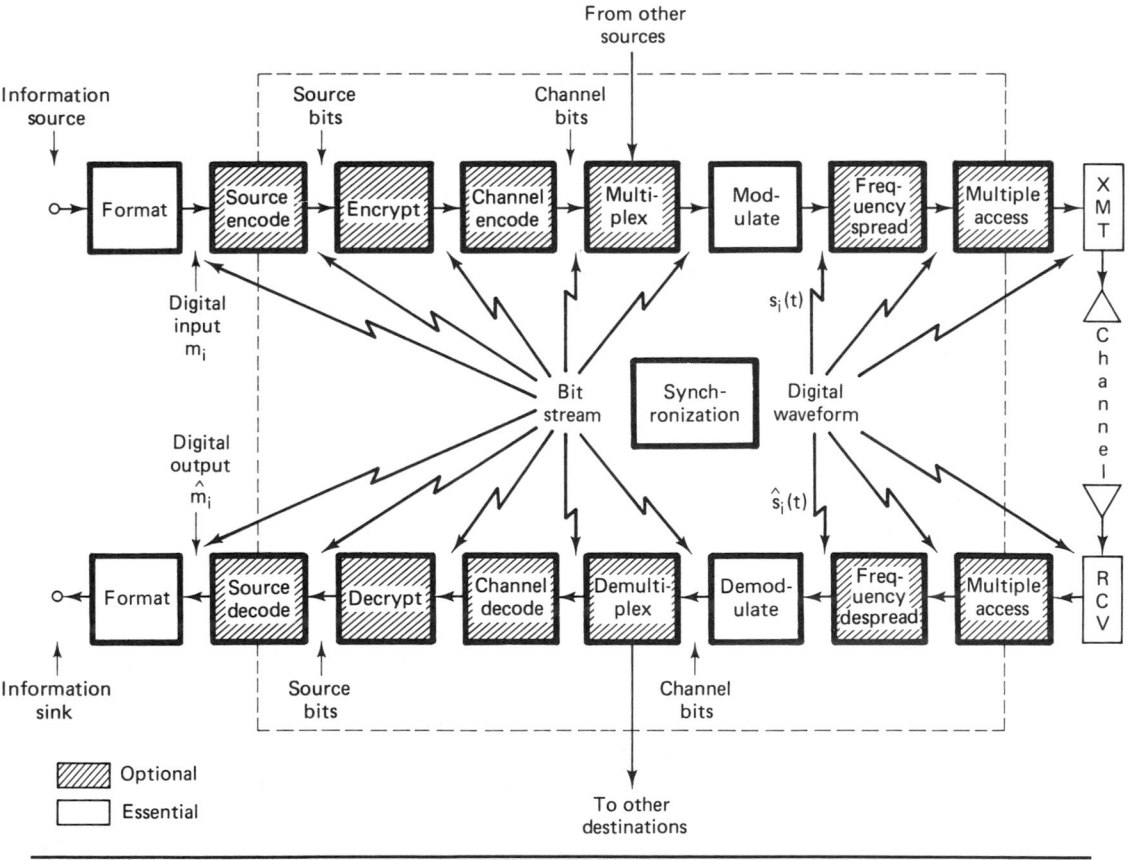

This book presents the ideas and techniques fundamental to digital communication systems. Emphasis is placed on system design goals and on the need for trade-offs among basic system parameters such as signal-to-noise ratio (SNR), probability of error, and bandwidth expenditure. Transmission bandwidth is a finite resource; there is a growing awareness that bandwidth must be conserved, shared, and used efficiently. In general, we shall see that system performance can often be improved through the use of increased transmission bandwidth. However, such an increase is not always possible, because of physical limitations or the constraint of government regulations concerning the allocation and conservation of the usable electromagnetic spectrum.

We shall deal with the transmission of information (voice, video, or data) over a path (channel) that may consist of wires, waveguides, or free space. Frequently, the treatment will be in the context of a satellite communications link. Communication via satellites has two unique characteristics: (1) the ability to cover the globe with a flexibility that cannot be duplicated with terrestrial links, and (2) the availability of bandwidth exceeding anything previously available for intercontinental communications. Until recently, most satellite communication systems have been analog in nature. However, digital communication is becoming increasingly attractive because of the ever-growing demand for data communication and because digital transmission offers data processing options and flexibilities not available with analog transmission.

The principal feature of a digital communication system (DCS) is that during a finite interval of time, it sends a waveform from a finite set of possible waveforms, in contrast to an analog communication system, which sends a waveform from an infinite variety of waveform shapes with theoretically infinite resolution. In a DCS, the objective at the receiver is *not* to reproduce a transmitted waveform with precision; it is, instead, to determine from a noise-perturbed signal which waveform from the finite set of waveforms had been sent by the transmitter. An important measure of system performance in a DCS is the probability of error (P_E).

1.1 DIGITAL COMMUNICATION SIGNAL PROCESSING

1.1.1 Why Digital?

Why are communication systems, military and commercial alike, "going digital"? There are many reasons. The primary advantage is the ease with which digital signals, compared to analog signals, are regenerated. Figure 1.1 illustrates an ideal binary digital pulse propagating along a transmission line. The shape of the waveform is affected by two basic mechanisms: (1) as all transmission lines and circuits have some nonideal transfer function, there is a distorting effect on the ideal pulse; and (2) unwanted electrical noise or other interference further distorts the pulse waveform. Both of these mechanisms cause the pulse shape to degrade as a function of line length, as shown in Figure 1.1. During the time that the transmitted pulse can still be reliably identified (before it is degraded to an ambiguous state by the transmission line), the pulse is amplified by a digital amplifier that recovers its original ideal shape. The pulse is thus "reborn" or regenerated. Circuits that perform this function at regular intervals along a transmission system are called *regenerative repeaters*.

Digital circuits are less subject to distortion and interference than are analog circuits. Since binary digital circuits operate in one of two states, fully on or fully off, to be meaningful a disturbance must be large enough to change the circuit operating point from one state to the other. Such two-state operation facilitates signal regeneration and thus prevents noise and other disturbances from accu-

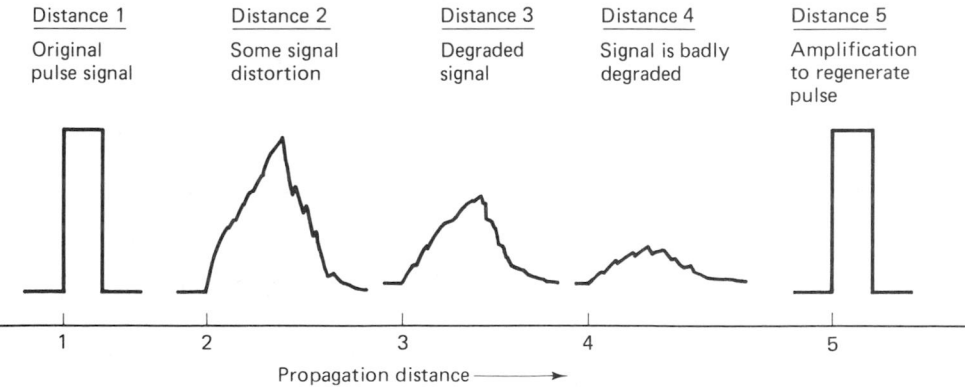

Figure 1.1 Pulse degradation and regeneration.

mulating in transmission. Analog signals, however, are *not* two-state signals; they can take an *infinite variety* of shapes. With analog circuits, even a small disturbance can render the reproduced waveform unacceptably distorted. Once the analog signal is distorted, the distortion cannot be removed by amplification. Since, with analog signals, accumulated noise is irrevocably bound to the signal, analog signals cannot be completely regenerated. Extremely low error rates producing high signal fidelity are possible through error detection and correction with digital techniques, but similar procedures are not available with analog.

There are other important advantages to digital communications. Digital circuits are *more reliable* and can be produced at lower cost than analog circuits. Also, digital hardware lends itself to *more flexible* implementation than analog hardware [e.g., microprocessors, digital switching, and large-scale integrated (LSI) circuits]. The combining of digital signals using time-division multiplexing (TDM) is *simpler* than the combining of analog signals using frequency-division multiplexing (FDM). Different types of digital signals (data, telegraph, telephone, television) can be treated as identical signals in transmission and switching—*a bit is a bit*. Also, for convenient switching, digital messages can be handled in autonomous groups called *packets*. Digital techniques lend themselves naturally to signal processing functions that protect against interference and jamming, or that provide encryption and privacy; such techniques are discussed in Chapters 10 and 12, respectively. Also, much data communication is computer to computer, or digital instrument or terminal to computer. Such digital terminations are naturally best served by digital communication links.

Most system choices entail trade-offs; system options are rarely all good or all bad. Thus far we have discussed only the *benefits* of digital transmission. What do you suppose are the *costs* or *liabilities*? A major disadvantage of digital transmission is that it typically requires a *greater system bandwidth* to communicate the same information in a digital format as compared to an analog format. Throughout this book we emphasize that bandwidth is a valuable resource, not always available. Bandwidth-efficient signaling techniques are discussed in Chapters 2 and 7. Another cost of digital transmission is that digital detection requires system synchronization (Chapter 8), whereas analog signals generally have no such requirement.

1.1.2 Typical Block Diagram and Transformations

The functional block diagram shown in Figure 1.2 illustrates the signal flow through a typical DCS. The upper blocks—format, source encode, encrypt, channel encode, multiplex, modulate, frequency spread, and multiple access—indicate the signal transformations from the source to the transmitter. The lower blocks indicate the signal transformations from the receiver to the sink; the lower blocks essentially reverse the signal processing steps performed by the upper blocks. It used to be that the only blocks within the dashed lines were the *modulator* and *demodulator,* together called a *modem*. During the past two decades, other signal processing functions were frequently incorporated within the same assembly as the modulator and demodulator. Consequently, at present, the term "modem"

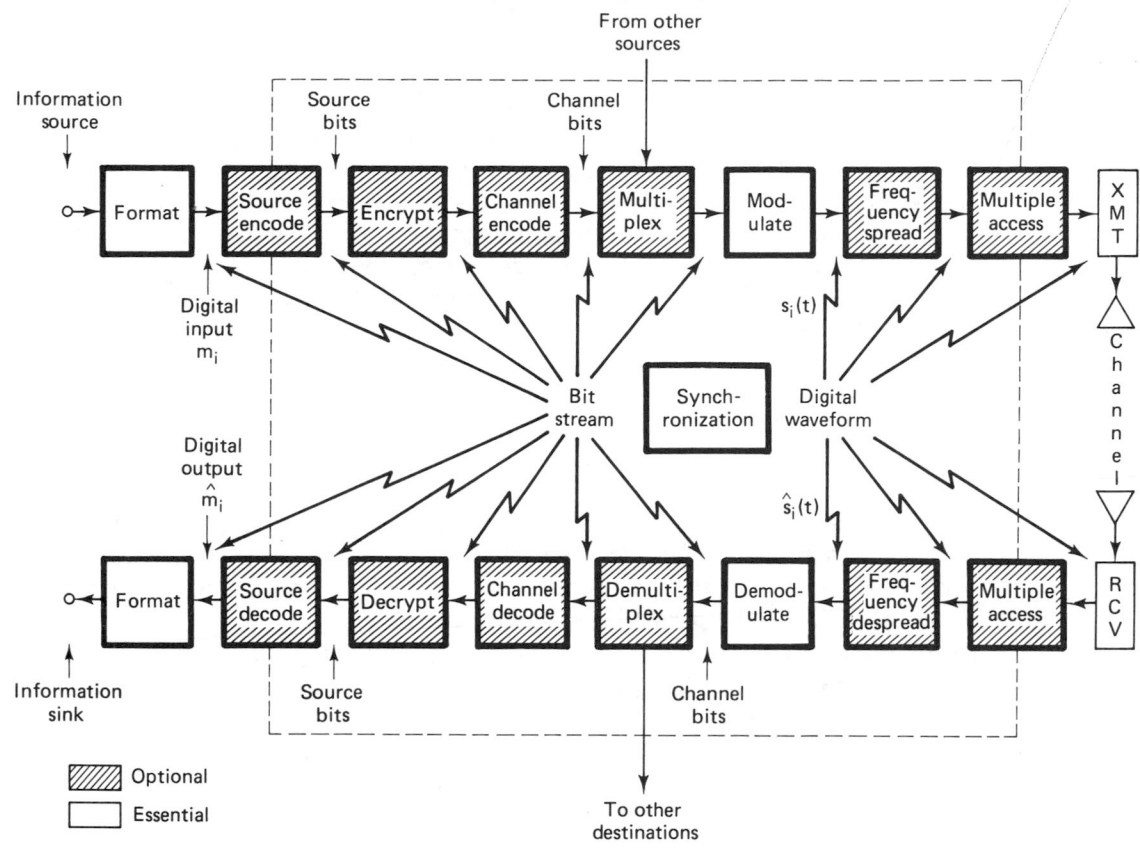

Figure 1.2 Block diagram of a typical digital communication system. (Reprinted with permission from B. Sklar, "A Structured Overview of Digital Communications," *IEEE Commun. Mag.*, August 1983, Fig. 1, p. 5. © 1983 IEEE.)

often encompasses all the processing steps shown within the dashed lines of Figure 1.2; when this is the case, the modem can be thought of as the "brains" of the system. Note that what constitutes a modem is not a precise concept; some of the blocks have purposely been shown *on* the dashed line rather than either inside or outside the modem. The transmitter and receiver can be thought of as the "muscles" of the system. The transmitter usually consists of a frequency up-conversion stage, a high-power amplifier, and an antenna. The receiver portion usually consists of an antenna, a low-noise amplifier (LNA), and a down-converter stage, typically to an intermediate frequency (IF).

Of all the signal processing steps, only formatting, modulation, and demodulation are essential for a DCS; the other processing steps within the modem are design options for specific system needs. *Formatting* transforms the source information into *digital symbols*; it makes the information compatible with the signal processing within a digital communication system. *Modulation* is the process by which the symbols are converted to *waveforms* that are compatible with the transmission channel.

The source encoding step produces analog-to-digital (A/D) conversion (for analog sources) *and* removes *redundant or unneeded* information. Encryption prevents unauthorized users from understanding messages and from injecting false messages into the system. Channel coding, for a given data rate, can reduce the probability of error (P_E), or reduce the signal-to-noise ratio (SNR) requirement, at the expense of bandwidth or decoder complexity. Channel coding can also reduce the system bandwidth requirement at the expense of SNR or P_E performance. Frequency spreading can produce a signal that is less vulnerable to interference (both natural and intentional) and can be used to enhance the privacy of the communicators. Multiplexing and multiple access procedures combine signals that might have different characteristics or might originate from different sources, so that they can share a portion of the communications resource.

The flow of the signal processing steps shown in Figure 1.2 represents a typical arrangement; however, the blocks are sometimes implemented in a different order. For example, multiplexing can take place prior to channel encoding, or prior to modulation, or—with a two-step modulation process (subcarrier and carrier)—it can be performed between the two modulation steps. Similarly, spreading can take place anywhere along the transmission chain; its precise location depends on the particular technique used. Figure 1.2 illustrates the reciprocal aspect of the procedure; any signal processing step that takes place in the transmitting chain must be reversed in the receiving chain. The figure also indicates that from the source to the modulator a message, also called a *baseband signal* or a *bit stream,* is characterized by a sequence of digital symbols. After modulation, the message takes the form of a digitally encoded waveform or *digital waveform*. Similarly, in the reverse direction, a received message appears as a digital waveform until it is demodulated. Thereafter it takes the form of a bit stream for all further signal processing steps. At various points along the signal route, noise corrupts the waveform $s(t)$ so that its reception must be termed an estimate $\hat{s}(t)$. Such noise and its deleterious effects on system performance are considered in Chapter 4.

Figure 1.3 shows the basic signal processing functions, which may be viewed as transformations from one signal space to another. The transformations are classified into seven basic groups:

1. Formatting and source coding
2. Modulation/demodulation
3. Channel coding
4. Multiplexing and multiple access
5. Spreading
6. Encryption
7. Synchronization

Although this organization has some inherent overlap, it provides a useful structure for the book. Beginning with Chapter 2, the seven basic transformations are considered individually. In Chapter 2 we discuss the basic formatting techniques for transforming the source information into digital symbols, as well as

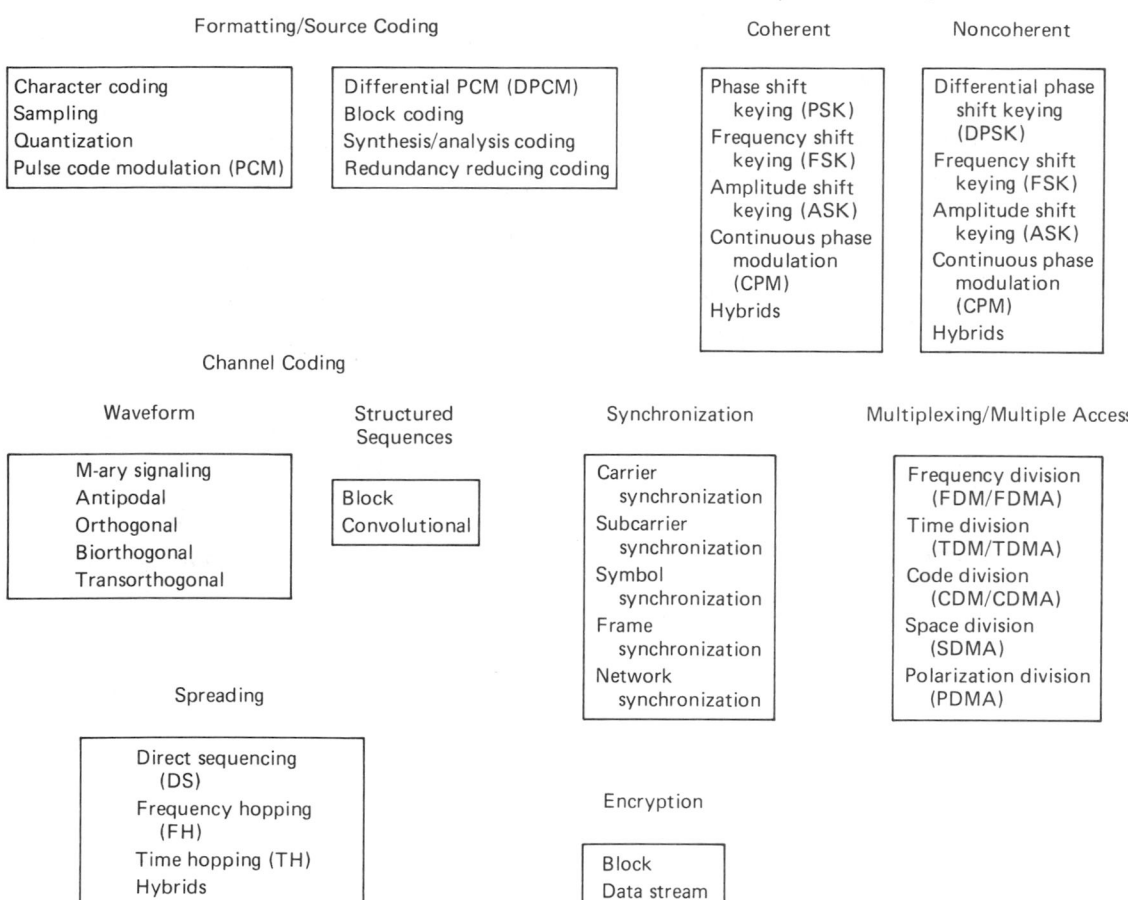

Figure 1.3 Basic digital communication transformations. (Reprinted with permission from B. Sklar, "A Structured Overview of Digital Communications," *IEEE Commun. Mag.*, August 1983, Fig. 2, p. 6. © 1983 IEEE.)

the selection of waveforms for making the symbols compatible with baseband transmission. As seen in Figure 1.3, formatting and source coding are grouped together; they are similar in that they involve data digitization. Since the term "source coding" has taken on the connotation of data redundancy reduction in addition to digitization, it is treated later, as a special formatting case, in Chapter 11.

In Figure 1.3, bandpass modulation/demodulation is partitioned into two basic categories, coherent and noncoherent. The process of *demodulation* involves the detection of the baseband information. Digital demodulation is typically accomplished with the aid of reference waveforms. When the references contain all the signal attributes, particularly phase information, the process is termed *coherent*; when phase information is not used, the process is termed *noncoherent*. Both techniques are detailed in Chapter 3.

Chapter 4 is devoted to link analysis. In the past, this area has received little attention in colleges or in textbooks, probably because it was considered straightforward and not worthy of detailed discussion. However, of the many specifications, analyses, and tabulations that support a developing communication system, link analysis stands out in its ability to provide overall system insight. In Chapter 4 we bring together all the link fundamentals that are essential for the analysis of most communication systems.

Channel coding deals with the techniques used to enhance digital signals so that they are less vulnerable to such channel impairments as noise, fading, and jamming. In Figure 1.3 channel coding is partitioned into two basic categories, waveform coding and structured sequences. *Waveform coding* involves the use of new waveforms, yielding improved detection performance over that of the original waveforms. *Structured sequences* involve the use of redundant bits to determine whether or not an error has occurred due to noise on the channel. One of these techniques, known as automatic repeat request (ARQ), simply recognizes the occurrence of an error and requests that the sender retransmit the message; other techniques, called forward error correction (FEC), are capable of automatically correcting the errors (within specified limitations). Under the heading of structured sequences, we shall discuss the two prevalent techniques, block coding and convolutional coding. In Chapter 5 we consider waveform coding and linear block coding. In Chapter 6 we consider convolutional coding, Viterbi decoding (and other decoding algorithms), hard versus soft decoding procedures, and interleaving and deinterleaving.

In Chapter 7 we summarize the design goals for a communication system and present various modulation and coding trade-offs that need to be considered in the design of a system. We discuss theoretical limitations such as the Nyquist criterion and the Shannon limit. We also examine bandwidth-efficient modulation schemes.

Chapter 8 deals with synchronization. In digital communications, synchronization involves the estimation of both time and frequency. The subject is partitioned as shown in Figure 1.3. Coherent systems need to synchronize their frequency reference with the carrier (and possibly subcarrier) in both frequency and phase. For noncoherent systems, phase synchronization is not needed. The fundamental time-synchronization process is symbol synchronization. The demodulator needs to know when to start and end the symbol detection procedure; a timing error will degrade detection performance. The next time-synchronization level, frame synchronization, allows the reconstruction of the message; and network synchronization allows coordination with other users in order to use the resource efficiently. In Chapter 8 we are concerned with the alignment of the timing of spatially separated periodic processes; the alignment is illustrated for the case of a satellite communications link.

Chapter 9 deals with multiplexing and multiple access. The two terms mean very similar things. Both involve the idea of resource sharing. The main difference between the two is that *multiplexing* takes place locally (e.g., on a printed circuit board, within an assembly, or even within a facility), and *multiple access* takes place remotely (e.g., multiple users share the use of a satellite transponder). Multiplexing involves an algorithm that is known a priori; usually, it is hard-wired into the system. Multiple access, on the other hand, is generally adaptive and may require overhead to enable the algorithm to operate. In Chapter 9 we discuss the classical ways of sharing the resource: frequency division, time division, and code division. We also consider some of the multiple access techniques that have emerged as a result of satellite communications.

Chapter 10 introduces a transformation of primary importance in military communications called spreading. The chapter deals with the spread-spectrum techniques that are emerging as important for achieving interference protection, privacy, or flexible access of the communications resource.

Chapter 11 treats source coding—techniques that deal with the task of forming efficient descriptions of source information. Source coding can be applied to digital data and to waveform signals; it can reduce data redundancy and thus reduce data rates. We will see that the advantage of source coding is a reduction of the system resources (i.e., bandwidth) required to describe the information.

The final chapter of the book, Chapter 12, deals with encryption and decryption, whose basic goals are privacy and authentication. *Privacy* refers to preventing unauthorized persons from extracting information (eavesdropping) from the channel. *Authentication* refers to preventing unauthorized persons from injecting spurious signals (spoofing) into the channel. In this chapter we highlight the data encryption standard (DES) and some current ideas for a class of encryption systems called public key cryptosystems.

1.1.3 Basic Digital Communication Nomenclature

Some of the basic digital signal nomenclature that frequently appears in digital communication literature is as follows:

Information source: the device producing information to be communicated by means of the DCS. Information sources can be analog or discrete. The output of an analog source can have any value in a continuous range of amplitudes, whereas the output of a discrete information source takes its value from a finite set. Analog information sources can be transformed into digital sources through the use of sampling and quantization. Sampling and quantization techniques called formatting and source coding (see Figure 1.3) are described in Chapters 2 and 11.

Textual message: a sequence of characters (see Figure 1.4a). For digital transmission, the message will be a sequence of digits or symbols from a finite symbol set or alphabet.

Character: a member of an alphabet or set of symbols (see Figure 1.4b). Characters may be mapped into a sequence of binary digits. There are several standardized codes used for character encoding, including the American Standard Code for Information Interchange (ASCII), Extended Binary Coded Decimal Interchange Code (EBCDIC), Hollerith, Baudot, Murray, and Morse.

(a) HOW ARE YOU?
OK
$9, 567, 216.73

(b) A
9
&

(c)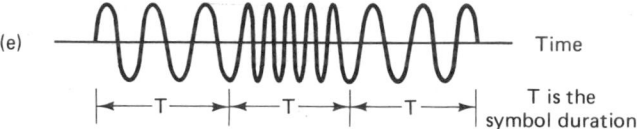

```
    H                    O                    W
```

0 0 0 1 0 0 1 1 1 1 1 0 0 1 1 1 1 0 1 0 1

(d)
1   Binary symbol ($k = 1$, $M = 2$)
10   Quaternary symbol ($k = 2$, $M = 4$)
011   8-ary symbol ($k = 3$, $M = 8$)

(e)   Time

$\leftarrow$ T $\rightarrow$ $\leftarrow$ T $\rightarrow$ $\leftarrow$ T $\rightarrow$ T is the
symbol duration

Figure 1.4 Nomenclature examples. (a) Textual messages. (b) Characters. (c) Bit stream (7-bit ASCII). (d) Symbols m_i, $i = 1, \ldots, M$, $M = 2^k$. (e) Bandpass digital waveform $s_i(t)$, $i = 1, \ldots, M$.

Binary digit (bit): the fundamental information unit for all digital systems. The term *bit* is also used as a unit of information content; this second usage is described in Chapter 7.

Bit stream: a sequence of binary digits (ones and zeros). Sometimes, a sequence of two-level pulses is used as a convenient illustration of the bit stream. The bit stream in Figure 1.4c uses a 7-bit ASCII character code for representing the message "HOW." A bit stream is often termed a *baseband signal,* which implies that its spectral content extends from (or near) dc up to some finite value, usually less than a few megahertz.

Symbol (digital message): groups of k bits considered as a unit or character m_i, from a finite symbol set or alphabet (see Figure 1.4d). The size of the alphabet, M, is $M = 2^k$ (i.e., k is the number of bits in the symbol). For transmission, each m_i symbol ($i = 1, \ldots, M$) will be represented by a corresponding waveform $s_1(t), s_2(t), \ldots, s_M(t)$. The symbol, m_i, is sent by transmitting the digital waveform, $s_i(t)$, for T seconds, the symbol time duration. The next symbol is sent during the next time interval, T. The fact that the symbol set transmitted by the DCS is finite is a primary difference between a DCS and an analog system. The DCS receiver need only decide which of the M waveforms was transmitted; however, an analog receiver must be capable of accurately estimating a continuous range of waveforms.

Digital waveform: a voltage or current waveform (a pulse for baseband transmission, or a sinusoid for bandpass transmission) that represents a digital symbol. The waveform characteristics (amplitude, width, position for pulses, or amplitude, frequency, phase for sinusoids) allow its identification as one of the symbols in the finite symbol alphabet. Figure 1.4e shows an example of a bandpass digital waveform. Even though the waveform is sinusoidal, and consequently has an analog appearance, it is called a *digital waveform* because it is encoded with digital information. In the figure, during each time interval, T, a preassigned frequency indicates the value of a digit.

Data rate: data rate in bits per second (bits/s) is given by $R = k/T = (1/T) \log_2 M$ bits/s, where k bits identify a symbol from an $M = 2^k$-symbol alphabet, and T is the k-bit symbol duration.

1.1.4 Digital versus Analog Performance Criteria

A principal difference between analog and digital communication systems has to do with the way in which we evaluate their performance. Analog systems draw their waveforms from a continuum, which therefore forms an infinite set; that is, a receiver must deal with an infinite number of possible waveshapes. The figure of merit for the performance of analog communication systems is a fidelity criterion, such as signal-to-noise ratio, percent distortion, or expected mean-square error between the transmitted and received waveforms.

By contrast, a digital communication system transmits signals that represent digits. These digits form a finite set or alphabet, and the set is known a priori to the receiver. A figure of merit for digital communication systems is the probability of incorrectly detecting a digit, or the probability of error (P_E).

1.2 CLASSIFICATION OF SIGNALS

1.2.1 Deterministic and Random Signals

A signal can be classified as *deterministic,* meaning that there is no uncertainty with respect to its value at any time, or as *random*, meaning that there is some degree of uncertainty before the signal actually occurs. Deterministic signals or waveforms are modeled by explicit mathematical expressions, such as $x(t) = 5 \cos 10t$. For a random waveform it is *not* possible to write such an explicit expression. However, when examined over a long period, a random waveform, also referred to as a *random process*, may exhibit certain regularities that can be described in terms of probabilities and statistical averages. Such a model, in the form of a probabilistic description of the random process, is particularly useful for characterizing signals and noise in communication systems.

1.2.2 Periodic and Nonperiodic Signals

A signal $x(t)$ is called *periodic in time* if there exists a constant $T_0 > 0$ such that

$$x(t) = x(t + T_0) \qquad \text{for } -\infty < t < \infty \tag{1.1}$$

where t denotes time. The smallest value of T_0 that satisfies this condition is called the *period* of $x(t)$. The period T_0 defines the duration of one complete cycle of $x(t)$. A signal for which there is no value of T_0 that satisfies Equation (1.1) is called a *nonperiodic signal*.

1.2.3 Analog and Discrete Signals

An *analog signal,* $x(t)$, is a continuous function of time; that is, $x(t)$ is uniquely defined for all t. An electrical analog signal arises when a physical waveform (e.g., speech) is converted into an electrical signal by means of a transducer. By comparison, a *discrete signal,* $x(kT)$, is one that exists only at discrete times; it is characterized by a sequence of numbers defined for each time, kT, where k is an integer and T is a fixed time interval.

1.2.4 Energy and Power Signals

An electrical signal can be represented as a voltage, $v(t)$, or a current, $i(t)$, with instantaneous power $p(t)$ across a resistor $\mathscr{R}$ defined by

$$p(t) = \frac{v^2(t)}{\mathscr{R}} \tag{1.2}$$

or

$$p(t) = i^2(t)\mathscr{R} \tag{1.3}$$

In communication systems, power is often normalized by assuming $\mathscr{R}$ to be $1\ \Omega$, although $\mathscr{R}$ may be another value in the actual circuit. If the actual value of the power is needed, it is obtained by "denormalization" of the normalized value. For the normalized case, Equations (1.2) and (1.3) have the same form. Therefore, regardless of whether the signal is a voltage or current waveform, the normalization convention allows us to express the instantaneous power as

$$p(t) = x^2(t) \tag{1.4}$$

where $x(t)$ is either a voltage or a current signal. The energy dissipated during the time interval $(-T/2,\ T/2)$ by a real signal with instantaneous power expressed by Equation (1.4) can then be written as

$$E_x^T = \int_{-T/2}^{T/2} x^2(t)\, dt \tag{1.5}$$

and the average power dissipated by the signal during the interval is

$$P_x^T = \frac{1}{T} \int_{-T/2}^{T/2} x^2(t) \, dt \qquad (1.6)$$

The performance of a communication system depends on the detected signal *energy*; higher-energy signals are detected more reliably (with fewer errors) than are lower-energy signals—the transmitted *energy does the work*. On the other hand, power is the *rate* at which energy is delivered. It is important for different reasons. The power determines the voltages that must be applied to a transmitter and the intensities of the electromagnetic fields that one must contend with in radio systems (i.e., fields in waveguides that connect the transmitter to the antenna, and fields around the radiating elements of the antenna).

In analyzing communication signals it is often desirable to deal with the *waveform energy*. We classify $x(t)$ as an *energy signal* if, and only if, it has nonzero but finite energy ($0 < E_x < \infty$) for all time, where

$$\begin{aligned} E_x &= \lim_{T \to \infty} \int_{-T/2}^{T/2} x^2(t) \, dt \\ &= \int_{-\infty}^{\infty} x^2(t) \, dt \end{aligned} \qquad (1.7)$$

In the real world we always transmit signals having finite energy ($0 < E_x < \infty$). However, in order to describe *periodic signals,* which by definition [Equation (1.1)] exist for all time and thus have infinite energy, and in order to deal with random signals that have infinite energy, it is convenient to define a class of signals called *power signals*. A signal is defined to be a power signal if, and only if, it has finite but nonzero power ($0 < P_x < \infty$) for all time, where

$$P_x = \lim_{T \to \infty} \frac{1}{T} \int_{-T/2}^{T/2} x^2(t) \, dt \qquad (1.8)$$

The energy and power classifications are mutually exclusive. An energy signal has finite energy but *zero average power,* whereas a power signal has finite average power but *infinite energy*. A waveform in a system may be constrained in either its power or energy values. As a general rule, periodic signals and random signals are classified as power signals, while signals that are both deterministic and non-periodic are classified as energy signals [1, 2].

As mentioned earlier, signal energy and power are both important parameters in specifying a communication system. The classification of a signal as either an energy signal or a power signal is a convenient model to facilitate the mathematical treatment of various signals and noise.

1.2.5 The Unit Impulse Function

A useful function in communication theory is the unit impulse or *Dirac delta function,* $\delta(t)$. The impulse function is an abstraction—an infinitely large amplitude pulse, with zero pulse width, and unity weight (area under the pulse), concentrated at the point where its argument is zero. The unit impulse is characterized by the following relationships:

$$\int_{-\infty}^{\infty} \delta(t)\, dt = 1 \tag{1.9}$$

$$\delta(t) = 0 \qquad \text{for } t \neq 0 \tag{1.10}$$

$$\delta(t) \text{ is unbounded at } t = 0 \tag{1.11}$$

$$\int_{-\infty}^{\infty} x(t)\delta(t - t_0)\, dt = x(t_0) \tag{1.12}$$

The unit impulse function, $\delta(t)$, is not a function in the usual sense. When operations involve $\delta(t)$, the convention is to interpret $\delta(t)$ as a unit-area pulse of finite amplitude and nonzero duration, after which the limit is considered as the pulse duration approaches zero. $\delta(t - t_0)$ can be depicted graphically as a spike located at $t = t_0$ with height equal to its integral or area. Thus $A\delta(t - t_0)$ with A constant represents an impulse function whose area or weight is equal to A, that is zero everywhere except at $t = t_0$.

Equation (1.12) is known as the *sifting* or *sampling property* of the unit impulse function; the unit impulse multiplier selects a sample of the function $x(t)$ evaluated at $t = t_0$.

1.3 SPECTRAL DENSITY

The *spectral density* of a signal characterizes the distribution of the signal's energy or power in the frequency domain. This concept is particularly important when considering filtering in communication systems. We need to be able to evaluate the signal and noise at the filter output. The energy spectral density (ESD) or the power spectral density (PSD) is used in the evaluation.

1.3.1 Energy Spectral Density

The total energy of a real-valued energy signal $x(t)$, defined over the interval $(-\infty, \infty)$, is described by Equation (1.7). Using Parseval's theorem [1], we can relate the energy of such a signal expressed in the time domain to the energy expressed in the frequency domain, as follows:

$$E_x = \int_{-\infty}^{\infty} x^2(t)\, dt = \int_{-\infty}^{\infty} |X(f)|^2\, df \tag{1.13}$$

where $X(f)$ is the Fourier transform of the nonperiodic signal $x(t)$ (for a review of Fourier techniques, see Appendix A). Let $\Psi_x(f)$ denote the squared magnitude spectrum, defined as

$$\Psi_x(f) = |X(f)|^2 \tag{1.14}$$

The quantity $\Psi_x(f)$ is the waveform *energy spectral density* (ESD) of the signal $x(t)$. Therefore, from Equation (1.13), we can express the total energy of the signal $x(t)$ by integrating the spectral density with respect to frequency, as follows:

$$E_x = \int_{-\infty}^{\infty} \Psi_x(f)\, df \qquad (1.15)$$

This equation states that the energy of a signal is equal to the area under the $\Psi_x(f)$ versus frequency curve. Energy spectral density describes the signal energy per unit bandwidth measured in joules/hertz. There are equal energy contributions from both positive and negative frequency components, since for a real signal, $x(t)$, $|X(f)|$ is an even function of frequency. Therefore, the energy spectral density is symmetrical in frequency about the origin, and thus the total energy of the signal $x(t)$ can be expressed as

$$E_x = 2 \int_{0}^{\infty} \Psi_x(f)\, df \qquad (1.16)$$

1.3.2 Power Spectral Density

The average power, P_x, of a real-valued power signal, $x(t)$, is defined in Equation (1.8). If $x(t)$ is a *periodic signal* with period T_0, it is classified as a power signal. The expression for the average power of a periodic signal takes the form of Equation (1.6), where the time average is taken over the signal period T_0, as follows:

$$P_x = \frac{1}{T_0} \int_{-T_0/2}^{T_0/2} x^2(t)\, dt \qquad (1.17a)$$

Parseval's theorem for a real-valued periodic signal [1] takes the form

$$P_x = \frac{1}{T_0} \int_{-T_0/2}^{T_0/2} x^2(t)\, dt = \sum_{n=-\infty}^{\infty} |c_n|^2 \qquad (1.17b)$$

where the $|c_n|$ terms are the complex Fourier series coefficients of the periodic signal (see Appendix A).

To apply Equation (1.17b), we need only know the magnitude of the coefficients, $|c_n|$. The *power spectral density* (PSD) function, $G_x(f)$, of the periodic signal, $x(t)$, is a real, even, and nonnegative function of frequency that gives the distribution of the power of $x(t)$ in the frequency domain, defined as

$$G_x(f) = \sum_{n=-\infty}^{\infty} |c_n|^2 \, \delta(f - nf_0) \qquad (1.18)$$

Equation (1.18) defines the power spectral density of a periodic signal, $x(t)$, as a succession of the weighted delta functions. Therefore, the PSD of a periodic signal is a discrete function of frequency. Using the PSD defined in Equation (1.18), we can now write the average normalized power of a real-valued signal, as follows:

$$P_x = \int_{-\infty}^{\infty} G_x(f) \, df = 2 \int_0^{\infty} G_x(f) \, df \qquad (1.19)$$

Equation (1.18) describes the PSD of periodic (power) signals only. If $x(t)$ is a nonperiodic signal it *cannot* be expressed by a Fourier series, and if it is a nonperiodic power signal (having infinite energy) it *may not* have a Fourier transform. However, we may still express the power spectral density of such signals in the *limiting sense*. If we form a *truncated version*, $x_T(t)$, of the nonperiodic power signal, $x(t)$, by observing it only in the interval $(-T/2, T/2)$, then $x_T(t)$ has finite energy, and has a proper Fourier transform, $X_T(f)$. It can be shown [2] that the power spectral density of the nonperiodic $x(t)$ can then be defined in the limit as

$$G_x(f) = \lim_{T \to \infty} \frac{1}{T} |X_T(f)|^2 \qquad (1.20)$$

Example 1.1 Average Normalized Power

(a) Find the average normalized power in the waveform, $x(t) = A \cos 2\pi f_0 t$, using time averaging.
(b) Repeat part (a) using the summation of spectral coefficients.

Solution

(a) Using Equation (1.17a), we have

$$P_x = \frac{1}{T_0} \int_{-T_0/2}^{T_0/2} A^2 \cos^2 2\pi f_0 t \, dt$$

$$= \frac{A^2}{2T_0} \int_{-T_0/2}^{T_0/2} (1 + \cos 4\pi f_0 t) \, dt$$

$$= \frac{A^2}{2T_0} (T_0) = \frac{A^2}{2}$$

(b) Using Equations (1.18) and (1.19) gives us

$$G_x(f) = \sum_{n=-\infty}^{\infty} |c_n|^2 \delta(f - nf_0)$$

$$\left.\begin{array}{l} c_1 = c_{-1} = \dfrac{A}{2} \\[2ex] c_n = 0 \quad \text{for } n = 0, \pm 2, \pm 3, \ldots \end{array}\right\} \quad \text{(see Appendix A)}$$

$$G_x(f) = \left(\frac{A}{2}\right)^2 \delta(f - f_0) + \left(\frac{A}{2}\right)^2 \delta(f + f_0)$$

$$P_x = \int_{-\infty}^{\infty} G_x(f) \, df = \frac{A^2}{2}$$

1.4 AUTOCORRELATION

1.4.1 Autocorrelation of an Energy Signal

Correlation is a matching process; *autocorrelation* refers to the matching of a signal with a delayed version of itself. The autocorrelation function, $R_x(\tau)$, of a real-valued energy signal, $x(t)$, is defined as

$$R_x(\tau) = \int_{-\infty}^{\infty} x(t)x(t + \tau) \, dt \qquad \text{for } -\infty < \tau < \infty \qquad (1.21)$$

The autocorrelation function, $R_x(\tau)$, provides a measure of how closely the signal matches a copy of itself as the copy is shifted τ units in time. The variable τ plays the role of a scanning or searching parameter. $R_x(\tau)$ is not a function of time; it is only a function of the time difference, τ, between the waveform and its shifted copy.

The autocorrelative function of a real-valued *energy* signal has the following properties:

1. $R_x(\tau) = R_x(-\tau)$ symmetrical in τ about zero

2. $R_x(\tau) \le R_x(0)$ for all τ maximum value occurs at the origin

3. $R_x(\tau) \leftrightarrow \Psi_x(f)$ autocorrelation and ESD form a Fourier transform pair, as designated by the double-headed arrows

4. $R_x(0) = \int_{-\infty}^{\infty} x^2(t) \, dt$ value at the origin is equal to the energy of the signal

If items 1 through 3 are satisfied, $R_x(\tau)$ satisfies the properties of an autocorrelation function. Property 4 can be derived from property 3 and thus need not be included as a basic test.

1.4.2 Autocorrelation of a Periodic (Power) Signal

The autocorrelation function of a real-valued power signal $x(t)$ is defined as

$$R_x(\tau) = \lim_{T \to \infty} \frac{1}{T} \int_{-T/2}^{T/2} x(t)x(t + \tau) \, dt \qquad \text{for } -\infty < \tau < \infty \qquad (1.22)$$

When the power signal, $x(t)$, is periodic with period T_0, the time average in Equation (1.22) may be taken over a *single period, T_0,* and the autocorrelation function can be expressed as follows:

$$R_x(\tau) = \frac{1}{T_0} \int_{-T_0/2}^{T_0/2} x(t)x(t + \tau) \, dt \qquad \text{for } -\infty < \tau < \infty \qquad (1.23)$$

The autocorrelation function of a real-valued *periodic* signal has properties similar to those of an energy signal, as follows:

1. $R_x(\tau) = R_x(-\tau)$ symmetrical in τ about zero

2. $R_x(\tau) \leq R_x(0)$ for all τ maximum value occurs at the origin

3. $R_x(\tau) \leftrightarrow G_x(f)$ autocorrelation and PSD form a Fourier transform pair

4. $R_x(0) = \dfrac{1}{T_0} \displaystyle\int_{-T_0/2}^{T_0/2} x^2(t)\, dt$ value at the origin is equal to the average power of the signal

1.5 RANDOM SIGNALS

The main objective of a communication system is the transfer of information over a channel. All useful message signals appear random; that is, the receiver does not know, a priori, which of the possible message waveforms will be transmitted. Also, the noise that accompanies the message signals is due to random electrical signals. Therefore, we need to be able to form efficient descriptions of random signals.

1.5.1 Random Variables

Let a *random variable, X(A)*, represent the functional relationship between a random event, A, and a real number. For notational convenience we shall designate the random variable by X, and let the functional dependence upon A be implicit. The random variable may be discrete or continuous. The *distribution function, $F_X(x)$*, of the random variable, X, is given by

$$F_X(x) = P(X \leq x) \tag{1.24}$$

where $P(X \leq x)$ is the probability that the value taken by the random variable, X, is less than or equal to a real number, x. The distribution function, $F_X(x)$, has the following properties:

1. $0 \leq F_X(x) \leq 1$

2. $F_X(x_1) \leq F_X(x_2)$ if $x_1 \leq x_2$

3. $F_X(-\infty) = 0$

4. $F_X(+\infty) = 1$

Another useful function relating to the random variable, X, is the *probability density function* (pdf), denoted $p_X(x)$, where

$$p_X(x) = \frac{dF_X(x)}{dx} \tag{1.25}$$

As in the case of the distribution function, the pdf is a function of a real number, x. The name "density function" arises from the fact that the probability of the event $x_1 \leq X \leq x_2$ equals

$$P(x_1 \leq X \leq x_2) = P(X \leq x_2) - P(X \leq x_1)$$

$$= F_X(x_2) - F_X(x_1)$$

$$= \int_{x_1}^{x_2} p_X(x)\, dx$$

The probability density function has the following properties

1. $p_X(x) \geq 0$

2. $\int_{-\infty}^{\infty} p_X(x)\, dx = F_X(+\infty) - F_X(-\infty) = 1$

Thus, a probability density function is always a nonnegative function with a total area of one. Throughout the book we use the designation, $p_X(x)$, for the probability density function of a *continuous* random variable. For ease of notation, we will often omit the subscript, X, and write simply, $p(x)$. We will use the designation $P(X = x_i)$ for the probability of a random variable, X, where X can take on *discrete* values only.

1.5.1.1 Ensemble Averages

The *mean value, m_X,* or *expected value* of a random variable, X, is defined by

$$m_X = \mathbf{E}\{X\} = \int_{-\infty}^{\infty} x p_X(x)\, dx \tag{1.26}$$

where $\mathbf{E}\{\cdot\}$ is called the *expected value operator*. The *nth moment* of a probability distribution of a random variable, X, is defined by

$$\mathbf{E}\{X^n\} = \int_{-\infty}^{\infty} x^n p_X(x)\, dx \tag{1.27}$$

For the purposes of communication system analysis, the most important moments of X are the first two moments. Thus, $n = 1$ in Equation (1.27) gives m_X as discussed above, whereas $n = 2$ gives the mean-square value of X, as follows:

$$\mathbf{E}\{X^2\} = \int_{-\infty}^{\infty} x^2 p_X(x)\, dx \tag{1.28}$$

We can also define *central moments,* which are the moments of the difference between X and m_X. The second central moment, called the *variance of X,* is defined as

$$\text{var}(X) = \mathbf{E}\{(X - m_X)^2\} = \int_{-\infty}^{\infty} (x - m_X)^2 p_X(x)\, dx \tag{1.29}$$

The variance of X is also denoted as σ_X^2, and its square root, σ_X, is called the *standard deviation* of X. Variance is a measure of the "randomness" of the random variable X. By specifying the variance of a random variable, we are constraining the width of its probability density function. The variance and the mean-square value are related by

$$\sigma_X^2 = \mathbf{E}\{X^2 - 2m_X X + m_X^2\}$$

$$= \mathbf{E}\{X^2\} - 2m_X \mathbf{E}\{X\} + m_X^2$$

$$= \mathbf{E}\{X^2\} - m_X^2$$

Thus, the variance is equal to the difference between the mean-square value and the square of the mean.

1.5.2 Random Processes

A random process, $X(A, t)$, can be viewed as a function of two variables, *an event A*, and *time*. Figure 1.5 illustrates a random process. In the figure there are *N sample functions* of time, $\{X_j(t)\}$. Each of the sample functions can be regarded as the output of a different noise generator. For a specific event A_j, we have a

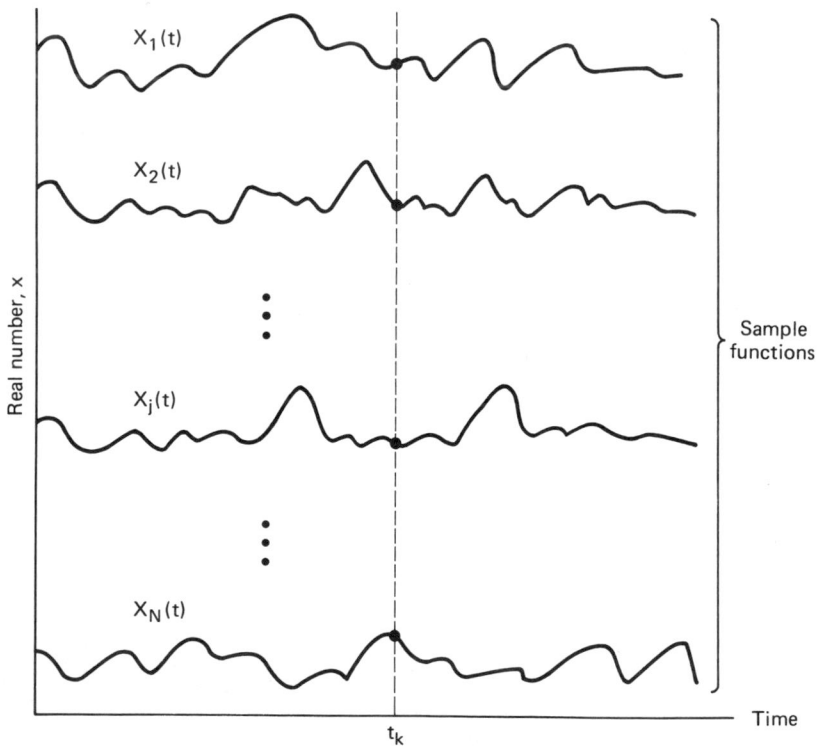

Figure 1.5 Random noise process.

single time function, $X(A_j, t) = X_j(t)$ (i.e., a sample function). The totality of all sample functions is called an *ensemble*. For a specific time t_k, $X(A, t_k)$ is a *random variable* $X(t_k)$, whose value depends on the event. Finally, for a specific event, $A = A_j$ and a specific time $t = t_k$, $X(A_j, t_k)$ is simply a *number*. For notational convenience we shall designate the random process by $X(t)$, and let the functional dependence upon A be implicit.

1.5.2.1 Statistical Averages of a Random Process

Because the value of a random process at any future time is unknown (since the identity of the event A is unknown), a random process whose distribution functions are continuous can be described statistically with a probability density function (pdf). In general the form of the pdf of a random process will be different for different times. In most situations it is not practical to determine empirically the probability distribution of a random process. However, a partial description consisting of the mean and autocorrelation function are often adequate for the needs of communication systems. We define the mean of the random process, $X(t)$, as

$$\mathbf{E}\{X(t_k)\} = \int_{-\infty}^{\infty} x p_{X_k}(x) \, dx = m_X(t_k) \tag{1.30}$$

where $X(t_k)$ is the random variable obtained by observing the random process at time t_k, and the pdf of $X(t_k)$, the density over the ensemble of events at time t_k, is designated $p_{X_k}(x)$.

We define the autocorrelation function of the random process, $X(t)$, to be a function of two variables, t_1 and t_2, as shown by

$$R_X(t_1, t_2) = \mathbf{E}\{X(t_1)X(t_2)\} \tag{1.31}$$

where $X(t_1)$ and $X(t_2)$ are random variables obtained by observing $X(t)$ at times t_1 and t_2. respectively. The autocorrelation function is a measure of the degree to which two time samples of the same random process are related.

1.5.2.2 Stationarity

A random process $X(t)$ is said to be *stationary* in the *strict sense* if none of its statistics are affected by a shift in the time origin. A random process is said to be *wide-sense stationary* (WSS) if two of its statistics, its mean and autocorrelation function, do not vary with a shift in the time origin. Thus, a process is WSS if

$$\mathbf{E}\{X(t)\} = m_X = \text{a constant} \tag{1.32}$$

and

$$R_X(t_1, t_2) = R_X(t_1 - t_2) \tag{1.33}$$

Strict-sense stationary implies wide-sense stationary, but not vice versa. Most of the useful results in communication theory are predicated on random information signals and noise being wide-sense stationary. From a practical point of view it is not necessary for a random process to be stationary for all time, but only for some observation interval of interest.

For stationary processes, the autocorrelation function in Equation (1.33) does not depend on time but only on the difference between t_1 and t_2. That is, all pairs of values of $X(t)$ at points in time separated by $\tau = t_1 - t_2$ have the same correlation value. Thus, for stationary systems, we can denote $R_X(t_1, t_2)$ simply as $R_X(\tau)$.

1.5.2.3 Autocorrelation of a Wide-Sense Stationary Random Process

Just as the variance provides a measure of randomness for random variables, the autocorrelation function provides a similar measure for random processes. For a wide-sense stationary process, the autocorrelation function is only a function of the *time difference* $\tau = t_1 - t_2$, that is,

$$R_X(\tau) = \mathbf{E}\{X(t)X(t + \tau)\} \qquad \text{for } -\infty < \tau < \infty \qquad (1.34)$$

For a zero mean WSS processes, $R_X(\tau)$ indicates the extent to which the random values of the process separated by τ seconds in time are statistically correlated. In other words, $R_X(\tau)$ gives us an idea of the frequency response that is associated with a random process. If $R_X(\tau)$ changes slowly as τ increases from zero to some value, it indicates that, on the average, sample values of $X(t)$ taken at $t = t_1$ and $t = t_1 + \tau$ are nearly the same. Thus, we would expect a frequency domain representation of $X(t)$ to contain a preponderance of low frequencies. On the other hand if $R_X(\tau)$ decreases rapidly as τ is increased, we would expect $X(t)$ to change rapidly with time and thereby contain mostly high frequencies.

Properties of the autocorrelation function of a real-valued wide-sense stationary process are:

1. $R_X(\tau) = R_X(-\tau)$ symmetrical in τ about zero

2. $R_X(\tau) \leq R_X(0)$ for all τ maximum value occurs at the origin

3. $R_X(\tau) \leftrightarrow G_X(f)$ autocorrelation and power spectral density form a Fourier transform pair

4. $R_X(0) = \mathbf{E}\{X^2(t)\}$ value at the origin is equal to the average power of the signal

1.5.3 Time Averaging and Ergodicity

To compute m_X and $R_X(\tau)$ by ensemble averaging, we would have to average across all the sample functions of the process and would need to have complete knowledge of the first- and second-order joint probability density functions. Such knowledge is generally not available.

When a random process belongs to a special class, known as an *ergodic process,* its time averages equal its ensemble averages, and the statistical properties of the process can be determined by *time averaging over a single sample function* of the process. For a random process to be ergodic it must be stationary in the strict sense. (The converse is not necessary.) However, for communication systems, where we are satisfied to meet the conditions of wide-sense stationarity, we are interested only in the mean and autocorrelation functions.

We can say that a random process is *ergodic in the mean* if

$$m_X = \lim_{T \to \infty} 1/T \int_{-T/2}^{T/2} X(t) \, dt \tag{1.35}$$

and it is *ergodic in the autocorrelation function* if

$$R_X(\tau) = \lim_{T \to \infty} 1/T \int_{-T/2}^{T/2} X(t)X(t + \tau) \, dt \tag{1.36}$$

Testing for the ergodicity of a random process is usually very difficult. In practice one makes an intuitive judgment as to whether it is reasonable to interchange the time and ensemble averages. A reasonable assumption in the analysis of most communication signals (in the absence of transient effects) is that the random waveforms are ergodic in the mean and the autocorrelation function. Since time averages equal ensemble averages for ergodic processes, fundamental electrical engineering parameters, such as dc value, rms value, and average power can be related to the moments of an ergodic random process. A summary of these relationships is:

1. The quantity $m_X = E\{X(t)\}$ is equal to the dc level of the signal.
2. The quantity m_X^2 is equal to the normalized power in the dc component.
3. The second moment of $X(t)$, $E\{X^2(t)\}$, is equal to the total average normalized power.
4. The quantity $\sqrt{E\{X^2(t)\}}$ is equal to the root-mean-square (rms) value of the voltage or current signal.
5. The variance, σ_X^2, is equal to the average normalized power in the time-varying or ac component of the signal.
6. If the process has zero mean (i.e., $m_X = m_X^2 = 0$), then $\sigma_X^2 = E\{X^2\}$, and the variance is the same as the mean-square value, or the variance represents the total power in the normalized load.
7. The standard deviation, σ_X, is the rms value of the ac component of the signal.
8. If $m_X = 0$, then σ_X is the rms value of the signal.

1.5.4 Power Spectral Density of a Random Process

A random process, $X(t)$, can generally be classified as a power signal having a power spectral density (PSD), $G_X(f)$, of the form shown in Equation (1.20). $G_X(f)$ is particularly useful in communications systems, because it describes the distribution of a signal's power in the frequency domain. The PSD enables us to evaluate the signal power that will pass through a network having known frequency characteristics. We summarize the principal features of PSD functions as follows:

1. $G_X(f) \geq 0$ and is always real valued

2. $G_X(f) = G_X(-f)$ for $X(t)$ real-valued

3. $G_X(f) \leftrightarrow R_X(\tau)$ PSD and autocorrelation form a Fourier transform pair

4. $P_X = \int_{-\infty}^{\infty} G_X(f)\, df$ relationship between average normalized power and PSD

Figure 1.6a illustrates a single sample waveform from a WSS random process, $X(t)$. The waveform is a binary random sequence with unit-amplitude positive and negative (bipolar) pulses. The positive and negative pulses occur with equal probability. The duration of each binary digit is T seconds, and the average or dc value of the random sequence is zero. Figure 1.6b shows the same sequence displaced τ_1 seconds in time; this sequence is therefore denoted $X(t - \tau_1)$. Let us assume that $X(t)$ is ergodic in the autocorrelation function so that we can use time averaging instead of ensemble averaging to find $R_X(\tau)$. The value of $R_X(\tau_1)$ is obtained by taking the product of the two sequences $X(t)$ and $X(t - \tau_1)$ and finding the average value using Equation (1.36). Equation (1.36) is accurate for ergodic processes *only in the limit*. However, integration over an integer number of periods can provide us with an estimate of $R_X(\tau)$. Notice that $R_X(\tau_1)$ can be obtained by a positive or negative shift of $X(t)$. Figure 1.6c illustrates such a calculation, using the single sample sequence (Figure 1.6a) and its shifted replica (Figure 1.6b). The cross-hatched areas under the product curve $X(t)X(t - \tau_1)$ contribute to positive values of the product, and the dotted areas contribute to negative values. The sequences can be further shifted by $\tau_2, \tau_3, \ldots$, each shift yielding a point on the overall autocorrelation function $R_X(\tau)$ shown in Figure 1.6d. Every random bit stream has an autocorrelation plot of the general shape shown in Figure 1.6d. The plot peaks at $R_X(0)$ [the best match occurs when τ equals zero, since $R(\tau) \leq R(0)$ for all τ], and it declines as τ increases. Figure 1.6d shows points corresponding to $R_X(0)$ and $R_X(\tau_1)$.

The analytical expression for the autocorrelation function $R_X(\tau)$ shown in Figure 1.6d, is [1]

$$R_X(\tau) = \begin{cases} 1 - \dfrac{|\tau|}{T} & \text{for } |\tau| \leq T \\[2mm] 0 & \text{for } |\tau| > T \end{cases} \tag{1.37}$$

The autocorrelation function allows us to express a random signal's power spectral density directly. Since the PSD and the autocorrelation function are Fourier transforms of each other, the PSD, $G_X(f)$, of the random binary sequence can be found,

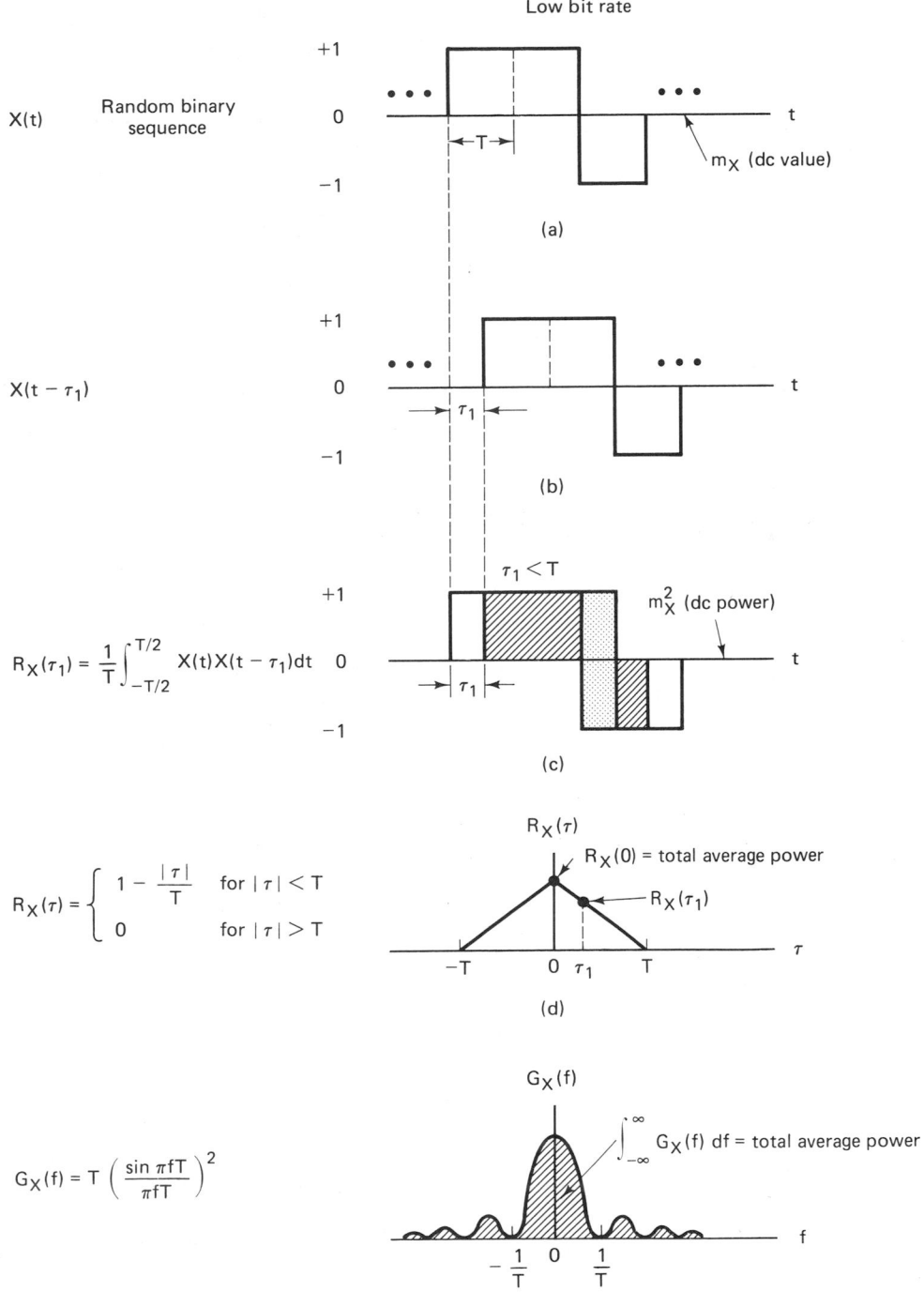

Figure 1.6 Autocorrelation and power spectral density.

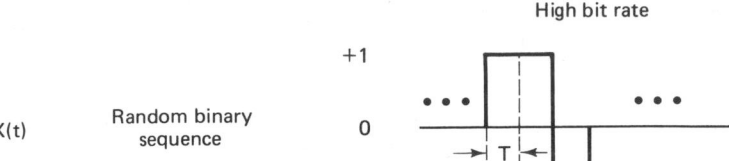

High bit rate

$X(t)$ Random binary
sequence

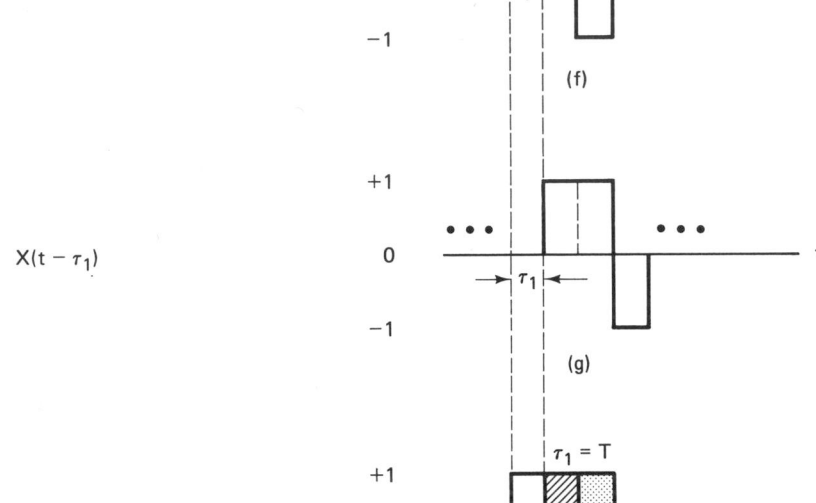

(f)

$X(t - \tau_1)$

(g)

$R_X(\tau_1) = \dfrac{1}{T} \displaystyle\int_{-T/2}^{T/2} X(t)X(t - \tau_1)dt$

(h)

$R_X(\tau) = \begin{cases} 1 - \dfrac{|\tau|}{T} & \text{for } |\tau| < T \\ 0 & \text{for } |\tau| > T \end{cases}$

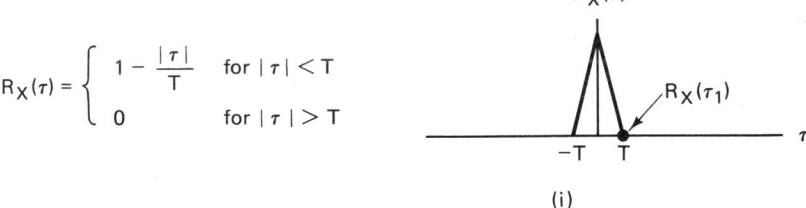

(i)

$G_X(f) = T\left(\dfrac{\sin \pi fT}{\pi fT}\right)^2$

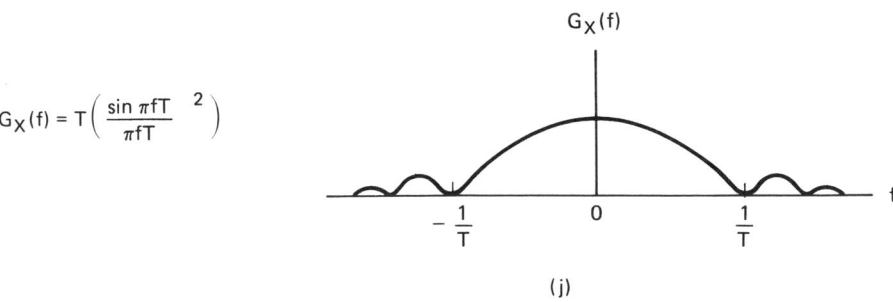

(j)

Figure 1.6 (*Continued*)

using Table A.1, as the transform of $R_X(\tau)$ in Equation (1.37). $G_X(f)$ is shown below, and its general shape is illustrated in Figure 1.6e.

$$G_X(f) = T \left(\frac{\sin \pi f T}{\pi f T} \right)^2 = T \operatorname{sinc}^2 fT \qquad (1.38)$$

where

$$\operatorname{sinc} y = \frac{\sin \pi y}{\pi y} \qquad (1.39)$$

Notice that the area under the PSD curve represents the average power in the signal. One convenient measure of *bandwidth* is the width of the main spectral lobe. Figure 1.6e illustrates that the bandwidth of a signal is inversely related to the symbol duration or pulse width. Figures 1.6f–j repeat the steps shown in Figures 1.6a–e, except that the bit duration is shorter. Notice that the shape of the shorter-bit-duration $R_X(\tau)$ is narrower, shown in Figure 1.6i, than it is for the longer-bit-duration $R_X(\tau)$, shown in Figure 1.6d. In Figure 1.6i, $R_X(\tau_1) = 0$; in other words, a shift of τ_1 in the case of the shorter-bit-duration example is enough to produce a zero match, or a complete decorrelation between the shifted sequences. Since the pulse duration, T, is shorter in Figure 1.6f, and the bit rate is higher than in Figure 1.6a, the bandwidth occupancy in Figure 1.6j is greater than the lower-bit-rate bandwidth occupancy shown in Figure 1.6e.

1.5.5 Noise in Communication Systems

The term *noise* refers to *unwanted* electrical signals that are always present in electrical systems. The presence of noise superimposed on a signal tends to obscure or mask the signal; it limits the receiver's ability to make correct symbol decisions, and thereby limits the rate of information transmission. Noise arises from a variety of sources, both man-made and natural. *Man-made noise* includes such sources as spark-plug ignition noise, switching transients, and other radiating electromagnetic signals. *Natural noise* includes electrical circuit and component noise, atmospheric disturbances, and galactic sources.

Good engineering design can eliminate much of the noise or its undesirable effect through filtering, shielding, the choice of modulation, and the selection of an optimum receiver site. For example, sensitive radio astronomy measurements are typically located at remote desert locations, far from man-made noise sources. However, there is one natural source of noise, called *thermal* or *Johnson noise,* that cannot be eliminated. Thermal noise [4, 5] is caused by the thermal motion of electrons in all dissipative components—resistors, wires, and so on. The same electrons that are responsible for electrical conduction are also responsible for thermal noise.

We can describe thermal noise as a zero-mean *Gaussian* random process. A Gaussian process, $n(t)$, is a random function whose value, n, at any arbitrary time, t, is statistically characterized by the Gaussian probability density function, $p(n)$:

$$p(n) = \frac{1}{\sigma\sqrt{2\pi}} \exp\left[-\frac{1}{2}\left(\frac{n}{\sigma}\right)^2\right] \tag{1.40}$$

where σ^2 is the variance of n. The *normalized* or *standardized Gaussian density function* of a zero-mean process is obtained by assuming that $\sigma = 1$. This normalized pdf is shown sketched in Figure 1.7.

    We will often represent a random signal as the sum of a Gaussian noise random variable and a dc signal:

$$z = a + n$$

where z is the random signal, a the dc component, and n the Gaussian noise random variable. The pdf $p(z)$ is then expressed as

$$p(z) = \frac{1}{\sigma\sqrt{2\pi}} \exp\left[-\frac{1}{2}\left(\frac{z-a}{\sigma}\right)^2\right] \tag{1.41}$$

where, as before, σ^2 is the variance of n. The Gaussian distribution is often used as the system noise model because of a theorem, called the *central limit theorem* [3], which states that under very general conditions the probability distribution of the sum of j statistically independent random variables approaches the Gaussian distribution as $j \to \infty$, no matter what the individual distribution functions may be. Therefore, even though individual noise mechanisms might have other than

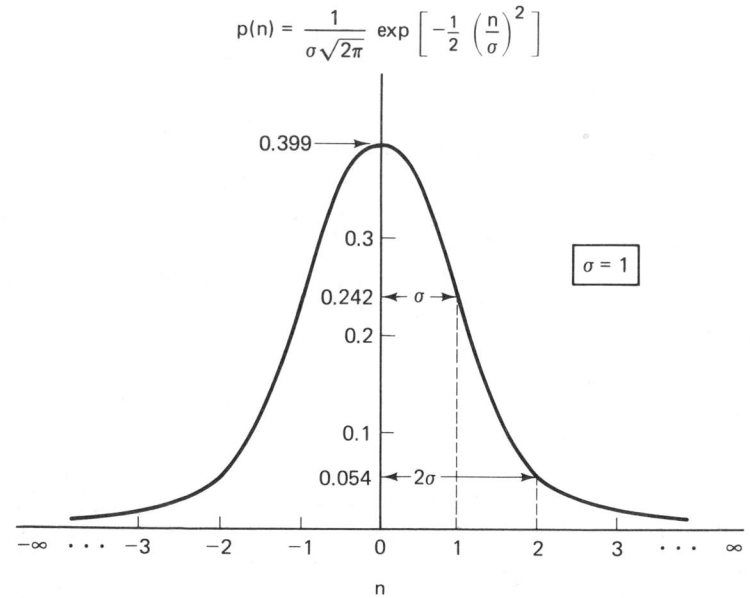

Figure 1.7   Normalized ($\sigma = 1$) Gaussian probability density function.

Gaussian distributions, the aggregate of many such mechanisms will tend toward the Gaussian distribution.

1.5.5.1 White Noise

The primary spectral characteristic of thermal noise is that its power spectral density is *the same* for all frequencies of interest in most communication systems; in other words, a thermal noise source emanates an equal amount of noise power per unit bandwidth at all frequencies—from dc to about 10^{12} Hz. Therefore, a simple model for thermal noise assumes that its power spectral density $G_n(f)$ is flat for all frequencies, as shown in Figure 1.8a, and is denoted as follows:

$$G_n(f) = \frac{N_0}{2} \quad \text{watts/hertz} \tag{1.42}$$

where the factor of 2 is included to indicate that $G_n(f)$ is a *two-sided* power spectral density. When the noise power has such a uniform spectral density, we refer to it as *white noise*. The adjective "white" is used in the sense that white light contains equal amounts of all frequencies within the visible band of electromagnetic radiation.

The autocorrelation function of white noise is given by the inverse Fourier transform of the noise power spectral density (see Table A.1) denoted as follows:

$$R_n(\tau) = \mathcal{F}^{-1}\{G_n(f)\} = \frac{N_0}{2} \delta(\tau) \tag{1.43}$$

Thus the autocorrelation of white noise is a delta function weighted by the factor $N_0/2$ and occurring at $\tau = 0$, as seen in Figure 1.8b. Note that $R_n(\tau)$ is zero for $\tau \neq 0$; that is, any two different samples of white noise, no matter how close together in time they are taken, are uncorrelated.

The average power, P_n, of white noise is *infinite* because its bandwidth is infinite. This can be seen by combining Equations (1.19) and (1.42) to yield.

$$P_n = \int_{-\infty}^{\infty} \frac{N_0}{2} df = \infty \tag{1.44}$$

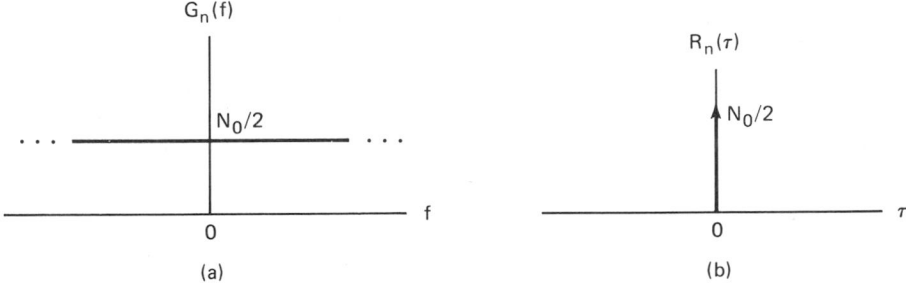

Figure 1.8 (a) Power spectral density of white noise. (b) Autocorrelation function of white noise.

Although white noise is a useful abstraction, no noise process can truly be white; however, the noise encountered in many real systems can be assumed to be approximately white. We can only observe such noise after it has passed through a real system which will have a finite bandwidth. Thus, as long as the bandwidth of the noise is appreciably larger than that of the system, the noise can be considered to have an infinite bandwidth.

The delta function in Equation (1.43) means that the noise signal, $n(t)$, is totally decorrelated from its time-shifted version, for any $\tau > 0$. Equation (1.43) indicates that *any* two different samples of a white noise process are uncorrelated. Since thermal noise is a Gaussian process and the samples are uncorrelated, the noise samples are also independent [3]. Therefore, the effect on the detection process of a channel with *additive white Gaussian noise* (AWGN) is that the noise affects each transmitted symbol *independently*. Such a channel is called a *memoryless channel*. The term "additive" means that the noise is simply superimposed or added to the signal—that there are no multiplicative mechanisms at work.

Since thermal noise is present in all communication systems and is the prominent noise source for most systems, the thermal noise characteristics—additive, white, and Gaussian—are most often used to model the noise in communication systems. Since zero-mean Gaussian noise is completely characterized by its *variance,* this model is particularly simple to use in the detection of signals and in the design of optimum receivers. In this book we shall assume, unless otherwise stated, that the system is corrupted by *additive zero-mean white Gaussian noise,* even though this is sometimes an oversimplification.

1.6 SIGNAL TRANSMISSION THROUGH LINEAR SYSTEMS

Having developed a set of models for signals and noise, we now consider the characterization of systems and their effects on such signals and noise. Since a system can be characterized equally well in the time domain or the frequency domain, techniques will be developed in both domains to analyze the response of a linear system to an arbitrary input signal. The signal, applied to the input of the system, as shown in Figure 1.9, can be described either as a time-domain signal, $x(t)$, or by its Fourier transform, $X(f)$. The use of time-domain analysis yields the time-domain output, $y(t)$, and in the process, $h(t)$, the characteristic or *impulse response* of the network, will be defined. When the input is considered in the frequency domain, we shall define a *frequency transfer function, $H(f)$,* for the system, which will determine the frequency-domain output, $Y(f)$. The system is assumed to be linear and time invariant. It is also assumed that there is no stored energy in the system at the time the input is applied.

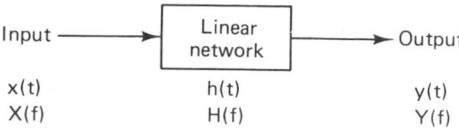

Input	Linear network	Output
x(t)	h(t)	y(t)
X(f)	H(f)	Y(f)

Figure 1.9 Linear system and its key parameters.

1.6.1 Impulse Response

The linear time-invariant system or network illustrated in Figure 1.9 is characterized in the time domain by an impulse response, $h(t)$, which is the response when the input is equal to a unit impulse $\delta(t)$; that is,

$$h(t) = y(t) \qquad \text{when } x(t) = \delta(t) \tag{1.45}$$

The response of the network to an arbitrary input $x(t)$ is then found by the convolution of $x(t)$ with $h(t)$, where * denotes the convolution operation (see Section A.5):

$$y(t) = x(t) * h(t) = \int_{-\infty}^{\infty} x(\tau)h(t - \tau)\, d\tau \tag{1.46}$$

The system is assumed to be *causal*, which means that there can be *no* output prior to the time, $t = 0$, when the input is applied. Therefore, the lower limit of integration can be changed to zero, and we can express the output $y(t)$ as

$$y(t) = \int_{0}^{\infty} x(\tau)h(t - \tau)\, d\tau \tag{1.47}$$

Equations (1.46) and (1.47) are called the *superposition integral* or the *convolution integral*.

1.6.2 Frequency Transfer Function

The frequency-domain output signal, $Y(f)$, is obtained by taking the Fourier transform of both sides of Equation (1.46). Since convolution in the time-domain transforms to multiplication in the frequency domain (and vice versa), Equation (1.46) yields

$$Y(f) = X(f)H(f) \tag{1.48}$$

or

$$H(f) = \frac{Y(f)}{X(f)} \tag{1.49}$$

provided, of course, that $X(f) \neq 0$ for all f. Here $H(f) = \mathscr{F}\{h(t)\}$, the Fourier transform of the impulse response function, is called the *frequency transfer function* or the *frequency response* of the network. In general, $H(f)$ is complex and can be written as

$$H(f) = |H(f)|\, e^{j\theta(f)} \tag{1.50}$$

where $|H(f)|$ is the magnitude response. The phase response, $\theta(f)$, is defined as

$$\theta(f) = \tan^{-1} \frac{\text{Im }\{H(f)\}}{\text{Re }\{H(f)\}} \tag{1.51}$$

where the terms "Re" and "Im" denote "the real part of" and "the imaginary part of," respectively.

The frequency transfer function of a linear time-invariant network can easily be measured in the laboratory with a sinusoidal generator at the input of the network and an oscilloscope at the output. When the input waveform $x(t)$ is expressed as

$$x(t) = A \cos 2\pi f_0 t$$

the output of the network will be

$$y(t) = A |H(f_0)| \cos [2\pi f_0 t + \theta(f_0)] \qquad (1.52)$$

The input frequency, f_0, is stepped through the values of interest; at each step, the amplitude and phase at the output are measured.

1.6.2.1 Random Processes and Linear Systems

If a random process forms the input to a time-invariant linear system, the output will also be a random process. That is, each sample function of the input process yields a sample function of the output process. The input power spectral density, $G_X(f)$, and the output power spectral density, $G_Y(f)$, are related as follows:

$$G_Y(f) = G_X(f) |H(f)|^2 \qquad (1.53)$$

Equation (1.53) provides a simple way of finding the power spectral density out of a time-invariant linear system when the input is a random process.

In Chapters 2 and 3 we consider the detection of signals in Gaussian noise. We will utilize a fundamental property of a Gaussian process applied to a linear system, stated as follows: It can be shown that if a Gaussian process, $X(t)$, is applied to a time-invariant linear filter, the random process, $Y(t)$, developed at the output of the filter is also Gaussian [6].

1.6.3 Distortionless Transmission

What is required of a network for it to behave like an *ideal* transmission line? The output signal from an ideal transmission line may have some time delay compared to the input, and it may have a different amplitude than the input (just a scale change), but otherwise it must have no distortion—it must have the same shape as the input. Therefore, for ideal distortionless transmission, we can describe the output signal as

$$y(t) = Kx(t - t_0) \qquad (1.54)$$

where K and t_0 are constants. Taking the Fourier transform of both sides (see Section A.3.1), we write

$$Y(f) = KX(f)e^{-j2\pi f t_0} \qquad (1.55)$$

Substituting the expression (1.55) for $Y(f)$ into Equation (1.49), we see that the required system transfer function for distortionless transmission is

$$H(f) = Ke^{-j2\pi f t_0} \qquad (1.56)$$

Therefore, to achieve *ideal distortionless transmission,* the overall system response must have a constant magnitude response, and its phase shift must be linear with frequency. It is not enough that the system amplify or attenuate all frequency components equally. All of the signal's frequency components must also arrive with identical time delay in order to add up correctly. Since time delay, t_0, is related to phase shift, θ, and radian frequency, $\omega = 2\pi f$, as follows,

$$t_0 \text{ (seconds)} = \frac{\theta \text{ (radians)}}{2\pi f \text{ (radians/second)}} \tag{1.57}$$

it is clear that phase shift must be proportional to frequency in order for the time delay of all components to be identical. In practice, a signal will be distorted in passing through some parts of a system. Phase or amplitude correction (*equalization*) networks may be introduced elsewhere in the system to correct for this distortion. It is the overall input–output characteristic of the system that determines its performance.

1.6.3.1 Ideal Filter

One cannot build the ideal network described in Equation (1.56). The problem is that Equation (1.56) implies an infinite bandwidth capability, where the bandwidth of a system is defined as the interval of positive frequencies over which the magnitude $|H(f)|$ remains within a specified value. In Section 1.7 various measures of bandwidth are enumerated. As an approximation to the ideal infinite-bandwidth network, let us choose a truncated network that passes, without distortion, all frequency components between f_ℓ and f_u, where f_ℓ is the lower cutoff frequency and f_u is the upper cutoff frequency, as shown in Figure 1.10. Each of these networks is called an *ideal filter.* Outside the range $f_\ell < f < f_u$, which is called the *passband,* the ideal filter is assumed to have a response of zero magnitude. The effective width of the passband is specified by the filter bandwidth $W_f = (f_u - f_\ell)$ hertz.

When $f_\ell \neq 0$ and $f_u \neq \infty$, the filter is called a *bandpass filter* (BPF), shown in Figure 1.10a. When $f_\ell = 0$ and f_u has a finite value, the filter is called a *low-pass filter* (LPF), shown in Figure 1.10b. When f_ℓ has a nonzero value and when $f_u \rightarrow \infty$, the filter is called a *high-pass filter* (HPF), shown in Figure 1.10c.

Following Equation (1.56), for the ideal low-pass filter transfer function with bandwidth $W_f = f_u$ hertz, shown in Figure 1.10b, we can write the transfer function as follows (letting $K = 1$):

$$H(f) = |H(f)| \, e^{-j\theta(f)} \tag{1.58}$$

where

$$|H(f)| = \begin{cases} 1 & \text{for } |f| < f_u \\ 0 & \text{for } |f| \geq f_u \end{cases} \tag{1.59}$$

and

$$e^{-j\theta(f)} = e^{-j2\pi f t_0} \tag{1.60}$$

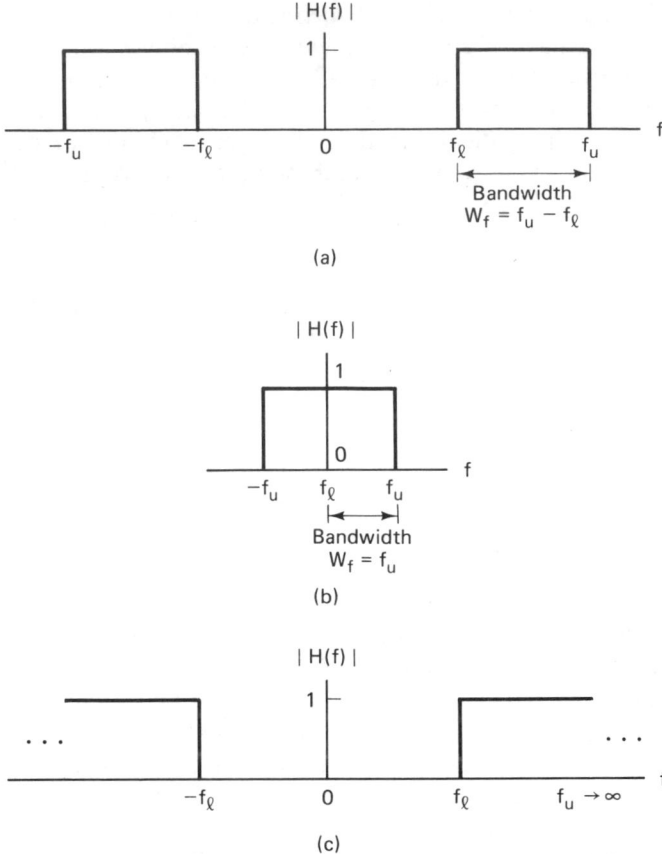

(a)

(b)

(c)

Figure 1.10 Ideal filter transfer function. (a) Ideal bandpass filter. (b) Ideal low-pass filter. (c) Ideal high-pass filter.

The impulse response $h(t)$ of the ideal low-pass filter, illustrated in Figure 1.11, is

$$h(t) = \mathcal{F}^{-1}\{H(f)\} = \int_{-\infty}^{\infty} H(f)e^{j2\pi ft} \, df \tag{1.61}$$

$$= \int_{-f_u}^{f_u} e^{-j2\pi ft_0} e^{j2\pi ft} \, df$$

$$= \int_{-f_u}^{f_u} e^{j2\pi f(t-t_0)} \, df$$

$$= 2f_u \frac{\sin 2\pi f_u(t - t_0)}{2\pi f_u(t - t_0)}$$

$$= 2f_u \text{ sinc } 2f_u(t - t_0) \tag{1.62}$$

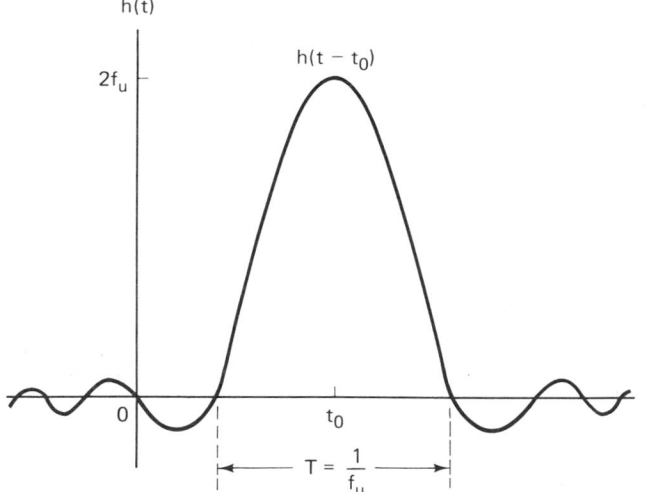

Figure 1.11   Impulse response of the ideal low-pass filter.

where sinc x is as defined in Equation (1.39). The impulse response shown in Figure 1.11 is noncausal, which means that it has a nonzero output prior to the application of an input at time $t = 0$. Therefore, it should be clear that the ideal filter described in Equation (1.58) is not realizable.

Example 1.2   Effect of an Ideal Filter on White Noise

White noise with power spectral density $G_n(f) = N_0/2$, shown in Figure 1.8a, forms the input to the ideal low-pass filter shown in Figure 1.10b. Find the power spectral density, $G_Y(f)$, and the autocorrelation function, $R_Y(\tau)$, of the output signal.

Solution

$$G_Y(f) = G_n(f)\,|H(f)|^2$$

$$= \begin{cases} \dfrac{N_0}{2} & \text{for } |f| < f_u \\ 0 & \text{otherwise} \end{cases}$$

The autocorrelation is the inverse Fourier transform of the power spectral density and is given by (see Table A.1)

$$R_Y(\tau) = N_0 f_u \frac{\sin 2\pi f_u \tau}{2\pi f_u \tau}$$

$$= N_0 f_u \, \text{sinc } 2f_u \tau$$

Comparing this result with Equation (1.62), we see that $R_Y(\tau)$ has the same shape as the impulse response of the ideal low-pass filter shown in Figure 1.11. In this example the ideal low-pass filter transforms the autocorrelation function of white noise (defined by the delta function) into a sinc function. After filtering, we no longer have white noise. The output noise signal will have zero correlation with shifted copies of itself, only at shifts of $\tau = n/2f_u$, where n is any integer other than zero.

1.6.3.2 Realizable Filters

The very simplest example of a realizable low-pass filter is made up of resistance ($\mathcal{R}$) and capacitance (C), as shown in Figure 1.12a; it is called an $\mathcal{R}C$ *filter,* and its transfer function can be expressed as [7]

$$H(f) = \frac{1}{1 + j2\pi f \mathcal{R}C} = \frac{1}{\sqrt{1 + (2\pi f \mathcal{R}C)^2}} \, e^{-j\theta(f)} \qquad (1.63)$$

where $\theta(f) = \tan^{-1} 2\pi f \mathcal{R}C$. The magnitude characteristic, $|H(f)|$, and the phase characteristic, $\theta(f)$ are plotted in Figures 1.12b and c, respectively. The low-pass filter bandwidth is defined to be its half-power point; this point is the frequency at which the output signal power has fallen to one-half of its peak value, or the frequency at which the magnitude of the output voltage has fallen to $1/\sqrt{2}$ of its peak value.

The half-power point is generally expressed in decibel (dB) units as the -3-dB point, or the point which is 3 dB down from the peak, where the decibel

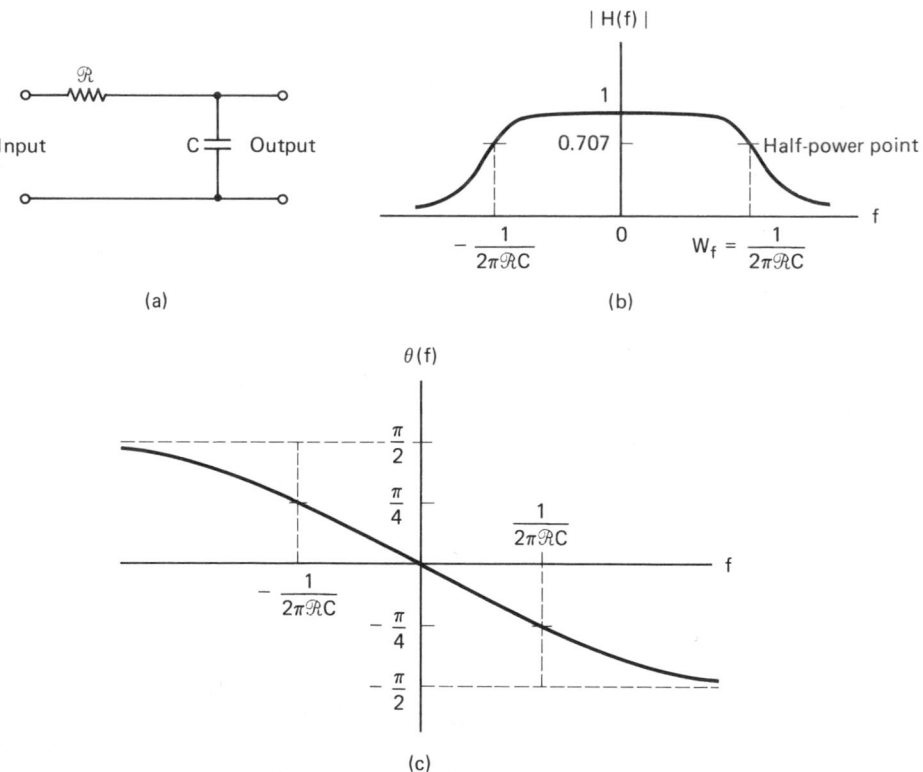

(a)

(b)

(c)

Figure 1.12 $\mathcal{R}C$ filter and its transfer function. (a) $\mathcal{R}C$ filter. (b) Magnitude characteristic of the $\mathcal{R}C$ filter. (c) Phase characteristic of the $\mathcal{R}C$ filter.

is defined as the ratio of two amounts of power, P_1 and P_2, existing at two points. By definition

$$\text{number of dB} = 10 \log_{10} \frac{P_2}{P_1} = 10 \log_{10} \frac{V_2^2/\mathcal{R}_2}{V_1^2/\mathcal{R}_1} \qquad (1.64a)$$

where V_1 and V_2 are voltages and $\mathcal{R}_1$ and $\mathcal{R}_2$ are resistances. For communication systems, *normalized power* is generally used for analysis; in this case, $\mathcal{R}_1$ and $\mathcal{R}_2$ are set equal to 1 Ω, so that

$$\text{number of dB} = 10 \log_{10} \frac{P_2}{P_1} = 10 \log_{10} \frac{V_2^2}{V_1^2} \qquad (1.64b)$$

The amplitude response, $|H(f)|$, can be expressed in decibels by

$$|H(f)|_{dB} = 20 \log_{10} \frac{V_2}{V_1} = 20 \log_{10} |H(f)| \qquad (1.64c)$$

where V_1 and V_2 are the input and output voltages, respectively, and where the input and output resistances have been assumed equal.

From Equation (1.63) it is easy to verify that the half-power point of the low-pass $\mathcal{R}C$ filter corresponds to $\omega = 1/\mathcal{R}C$ radians per second or $f = 1/(2\pi\mathcal{R}C)$ hertz. Thus the bandwidth W_f in hertz is $1/(2\pi\mathcal{R}C)$. The filter *shape factor* is a measure of how well a realizable filter approximates the ideal filter. It is typically defined as the ratio of the filter bandwidths at the -60-dB and -6-dB amplitude response points. A sharp-cutoff bandpass filter can be made with a shape factor as low as about 2. By comparison, the shape factor of the simple $\mathcal{R}C$ low-pass filter is almost 600.

There are several useful approximations to the ideal low-pass filter characteristic. One of these, the *Butterworth filter,* approximates the ideal low-pass filter with the following function:

$$|H_n(f)| = \frac{1}{\sqrt{1 + (f/f_u)^{2n}}} \qquad n \geq 1 \qquad (1.65)$$

where f_u is the upper -3-dB cutoff frequency. The magnitude function, $|H(f)|$, is sketched (single sided) for several values of n in Figure 1.13. Note that as n gets larger, the magnitude characteristics approach that of the ideal filter. Butterworth filters are popular because they are the best approximation to the ideal, in the sense of *maximal flatness* in the filter passband.

Example 1.3   Effect of an $\mathcal{R}C$ Filter on White Noise

White noise with spectral density, $G_n(f) = N_0/2$, shown in Figure 1.8a, forms the input to the $\mathcal{R}C$ filter shown in Figure 1.12a. Find the power spectral density, $G_Y(f)$, and the autocorrelation function, $R_Y(\tau)$, of the output signal.

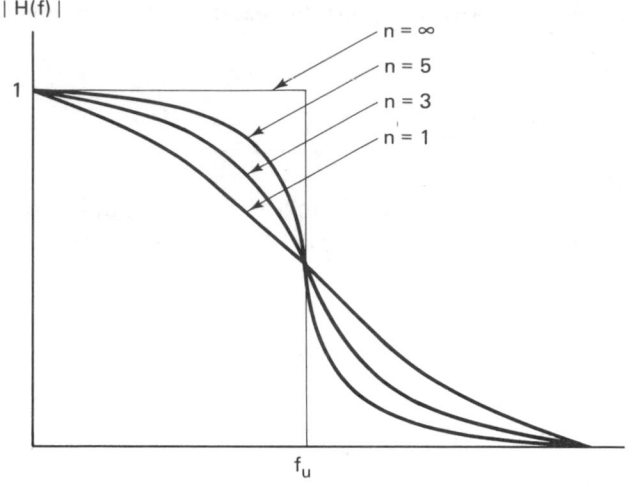

| H(f) |

n = ∞
n = 5
n = 3
n = 1

1

f_u

f   **Figure 1.13** Butterworth filter magnitude response.

Solution

$$G_Y(f) = G_n(f)\,|H(f)|^2$$

$$= \frac{N_0}{2}\,\frac{1}{1 + (2\pi f \mathcal{R}C)^2}$$

$$R_Y(\tau) = \mathcal{F}^{-1}\{G_Y(f)\}$$

Using Table A.1, the inverse Fourier transform of $G_Y(f)$ is

$$R_Y(\tau) = \frac{N_0}{4\mathcal{R}C}\,\exp\left(-\frac{|\tau|}{\mathcal{R}C}\right)$$

As might have been predicted, we no longer have white noise after filtering. The $\mathcal{R}C$ filter transforms the input autocorrelation function of white noise (defined by the delta function) into an exponential function. For a narrowband filter (a large $\mathcal{R}C$ product), the output noise will exhibit higher correlation between noise samples of a fixed time shift than will the output noise from a wideband filter.

1.6.4 Signals, Circuits, and Spectra

Signals have been described in terms of their spectra. Similarly, networks or circuits have been described in terms of their spectral characteristics or frequency transfer functions. How is a signal's bandwidth affected as a result of the signal passing through a filter circuit? Figure 1.14 illustrates two cases of interest. In Figure 1.14a (case 1), the input signal has a narrowband spectrum, and the filter transfer function is a wideband function. From Equation (1.48) we see that the output signal spectrum is simply the product of these two spectra. In Figure 1.14a we can verify that multiplication of the two spectral functions will result in a spectrum with a bandwidth approximately equal to the smaller of the two bandwidths (when one of the two spectral functions goes to zero, the multiplication yields zero). Therefore, for case 1, the output signal spectrum is constrained by

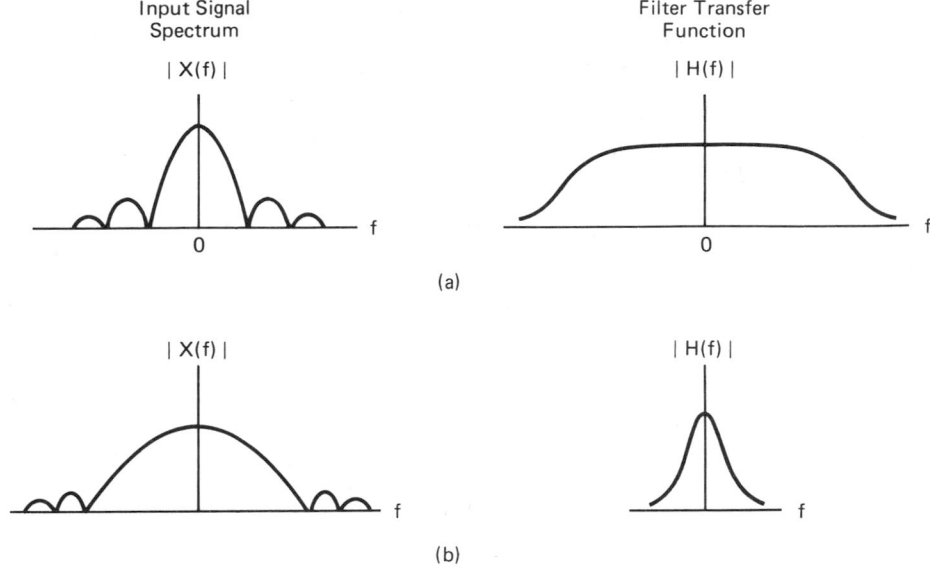

Figure 1.14 Spectral characteristics of the input signal and the circuit contribute to the spectral characteristics of the output signal. (a) Case 1: Output bandwidth is constrained by input signal bandwidth. (b) Case 2: Output bandwidth is constrained by filter bandwidth.

the input signal spectrum alone. Similarly, we see that for case 2, in Figure 1.14b, where the input signal is a wideband signal but the filter has a narrowband transfer function, the bandwidth of the output signal is constrained by the filter bandwidth; the output signal will be a filtered (distorted) rendition of the input signal.

The effect of a filter on a waveform can also be viewed in the time domain. The output, $y(t)$, resulting from convolving an ideal input pulse, $x(t)$ (having amplitude V_m and pulse width T), with the impulse response of a low-pass $\mathcal{R}C$ filter can be written as [8]

$$y(t) = \begin{cases} V_m(1 - e^{-t/\mathcal{R}C}) & \text{for } 0 \leq t \leq T \\ V'_m e^{-(t-T)/\mathcal{R}C} & \text{for } t > T \end{cases} \tag{1.66}$$

where

$$V'_m = V_m(1 - e^{-T/\mathcal{R}C}) \tag{1.67}$$

Let us define the pulse bandwidth, W_p, and the $\mathcal{R}C$ filter bandwidth, W_f, as

$$W_p = \frac{1}{T} \tag{1.68}$$

and

$$W_f = \frac{1}{2\pi\mathcal{R}C} \tag{1.69}$$

Sec. 1.6   Signal Transmission Through Linear Systems

39

The ideal input pulse, $x(t)$, and its magnitude spectrum $|X(f)|$, are shown in Figure 1.15. The $\mathcal{R}C$ filter and its magnitude characteristic, $|H(f)|$, are shown in Figures 1.12a and b, respectively. Following Equations (1.66) to (1.69), three cases are illustrated in Figure 1.16. Example 1 illustrates the case where $W_p \ll W_f$. Notice that the output response, $y(t)$, is a reasonably good approximation of the input pulse, $x(t)$, shown in dashed lines. This represents an example of *good fidelity*. In example 2, where $W_p \simeq W_f$, we can still recognize that a pulse had been transmitted from the output, $y(t)$. Finally, example 3 illustrates the case where $W_p \gg W_f$. Here the presence of the pulse is barely perceptible from the output, $y(t)$. Can you think of an application where the large filter bandwidth or good fidelity of example 1 is called for? A *precise ranging application,* perhaps, where the pulse time of arrival translates into distance, necessitates a pulse with a steep rise time. Which example characterizes the binary digital communications application? *It is example 2.* As we pointed out earlier regarding Figure 1.1, one of the principal features of binary digital communications is that each received pulse

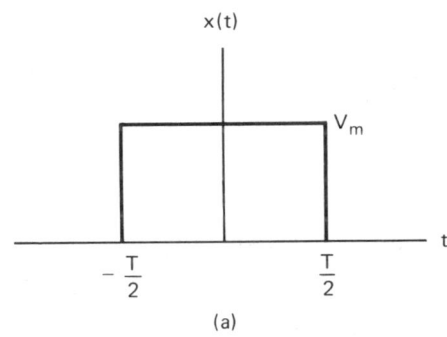

(a)

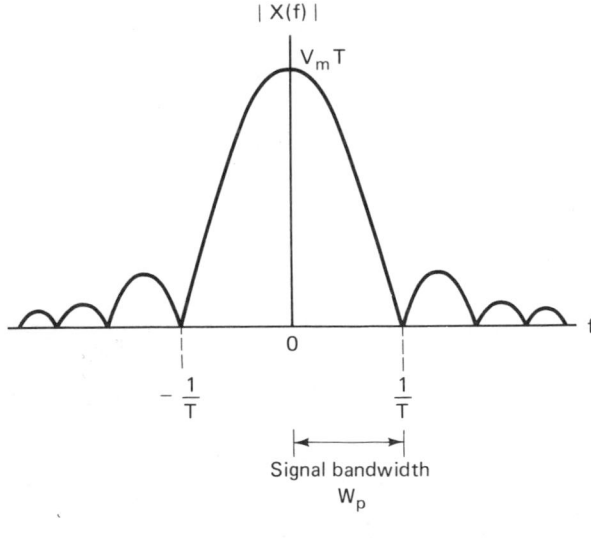

Signal bandwidth
W_p

(b)

Figure 1.15 (a) Ideal pulse. (b) Magnitude spectrum of the ideal pulse.

Signals and Spectra   Chap. 1

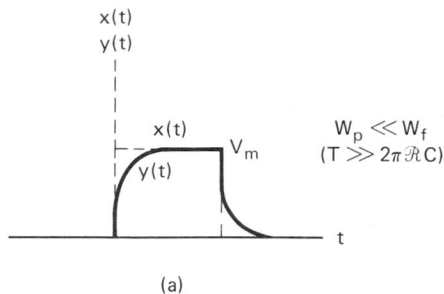

(a)

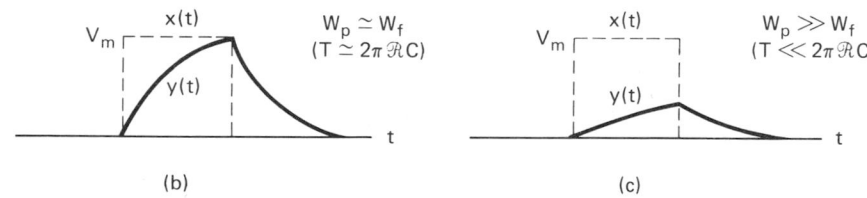

(b)   (c)

Figure 1.16 Three examples of filtering an ideal pulse. (a) Example 1: Good-fidelity output. (b) Example 2: Good-recognition output. (c) Example 3: Poor-recognition output.

need only be accurately *perceived* as being in one of its two states; a high-fidelity signal need not be maintained. Example 3 has been included for completeness; it would not be used as a design criterion for a practical system.

1.7 BANDWIDTH OF DIGITAL DATA

1.7.1 Baseband versus Bandpass

An easy way to translate the spectrum of a low-pass or baseband signal, $x(t)$, to a higher frequency is to multiply or *heterodyne* the baseband signal with a carrier wave, $\cos 2\pi f_c t$, as shown in Figure 1.17a. The resulting waveform, $x_c(t)$, is called a *double-sideband* (DSB) *modulated signal* and is expressed as

$$x_c(t) = x(t) \cos 2\pi f_c t \qquad (1.70)$$

From the frequency shifting theorem (see Section A.3.2) the spectrum of the DSB signal, $x_c(t)$, is given by $X_c(f)$:

$$X_c(f) = \tfrac{1}{2}[X(f - f_c) + X(f + f_c)] \qquad (1.71)$$

The magnitude spectrum $|X(f)|$ of the baseband signal, $x(t)$, having a bandwidth f_m, and the magnitude spectrum, $|X_c(f)|$, of the DSB signal, $x_c(t)$, having a bandwidth W_{DSB}, are shown in Figure 1.17b and c, respectively. In the plot of $|X_c(f)|$,

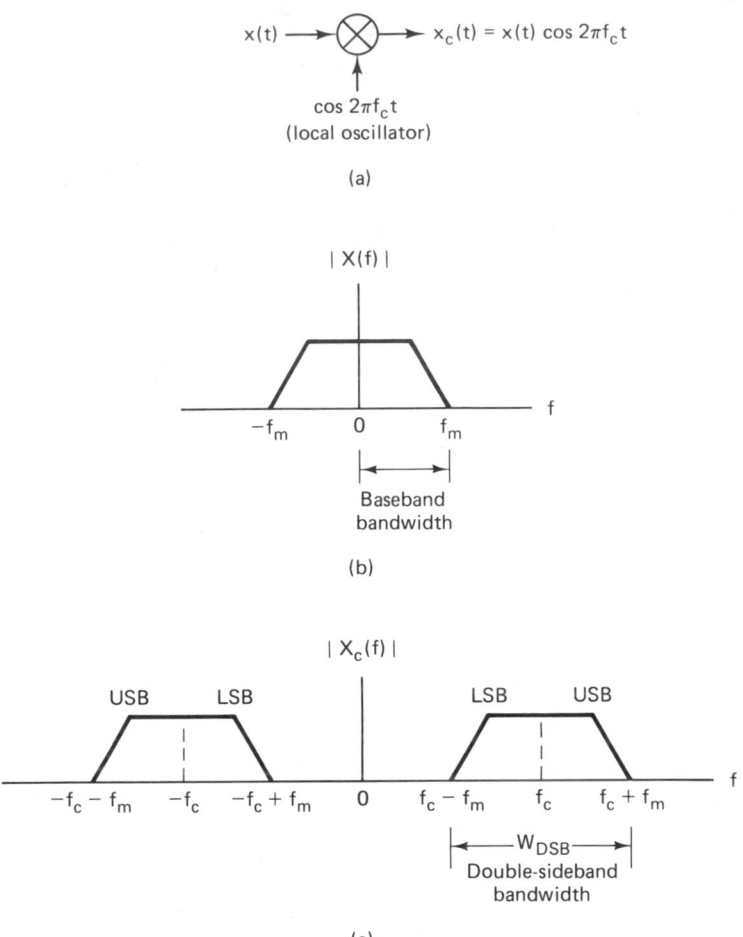

Figure 1.17 Comparison of baseband and double-sideband spectra. (a) Heterodyning. (b) Baseband spectrum. (c) Double-sideband spectrum.

spectral components corresponding to positive baseband frequencies, appear in the range f_c to $(f_c + f_m)$. This part of the DSB spectrum is called the *upper sideband* (USB). Spectral components corresponding to negative baseband frequencies appear in the range $(f_c - f_m)$ to f_c. This part of the DSB spectrum is called the *lower sideband* (LSB). Mirror images of the USB and LSB spectra appear in the negative-frequency half of the plot. The *carrier wave* is sometimes referred to as a *local oscillator* (LO) *signal,* a *mixing signal,* or a *heterodyne signal.* Generally, the carrier wave frequency is much higher than the bandwidth of the baseband signal; that is,

$$f_c \gg f_m$$

From Figure 1.17 we can readily compare the bandwidth f_m, required to transmit the baseband signal, with the bandwidth W_{DSB}, required to transmit the DSB signal; we see that

$$W_{\text{DSB}} = 2f_m \qquad (1.72)$$

That is, we need twice as much transmission bandwidth to transmit a DSB version of the signal than we do to transmit its baseband counterpart.

1.7.2 The Bandwidth Dilemma

Many important theorems of communication and information theory are based on the assumption of *strictly bandlimited* channels, which means that no signal power whatever is allowed outside the defined band. We are faced with the dilemma that strictly bandlimited signals are not realizable since they imply signals with infinite duration; nonbandlimited signals, having energy at arbitrarily high frequencies, appear just as unreasonable. It is no wonder that there is no single universal definition of bandwidth.

All bandwidth criteria have in common the attempt to specify a measure of the width, W, of a nonnegative real-valued power spectral density defined for all frequencies $|f| < \infty$. Figure 1.18 illustrates some of the most common definitions of bandwidth; in general, the various criteria are not interchangeable. The single-

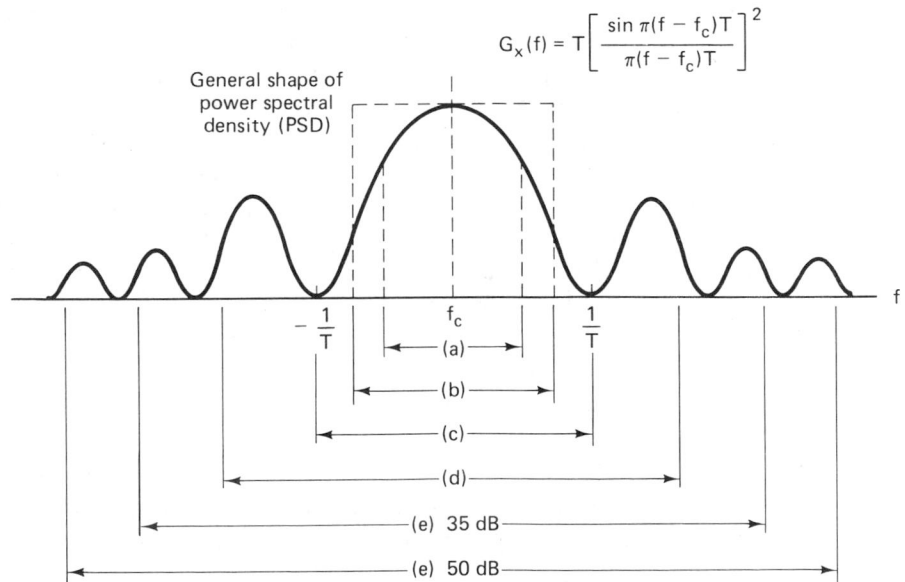

Figure 1.18 Bandwidth of digital data. (a) Half-power. (b) Noise equivalent. (c) Null to null. (d) 99% of power. (e) Bounded PSD (defines attenuation outside bandwidth) at 35 and 50 dB.

sided power spectral density, $G_x(f)$, for a single heterodyned pulse, $x_c(t)$, takes the analytical form

$$G_x(f) = T \left[\frac{\sin \pi(f - f_c)T}{\pi(f - f_c)T} \right]^2 \tag{1.73}$$

where f_c is the carrier wave frequency and T is the pulse duration. This power spectral density, whose general appearance is sketched in Figure 1.18, also characterizes a *random pulse sequence,* assuming that the averaging time is long relative to the pulse duration. The plot consists of a main lobe and smaller symmetrical sidelobes. The general shape of the plot is valid for most digital modulation formats; some formats, however, do not have well-defined lobes. The bandwidth criteria depicted in Figure 1.18 are as follows:

(a) *Half-power bandwidth.* This is the interval between frequencies at which $G_x(f)$ has dropped to half-power, or 3 dB below the peak value.

(b) *Equivalent rectangular or noise equivalent bandwidth.* The noise equivalent bandwidth was originally conceived to permit rapid computation of output noise power from an amplifier with a wideband noise input; the concept can similarly be applied to a signal bandwidth. The noise equivalent bandwidth W_N of a signal is defined by the relationship $W_N = P_x/G_x(f_c)$, where P_x is the total signal power over all frequencies and $G_x(f_c)$ is the value of $G_x(f)$ at the band center (assumed to be the maximum value over all frequencies).

(c) *Null-to-null bandwidth.* The most popular measure of bandwidth for digital communications is the width of the main spectral lobe, where most of the signal power is contained. This criterion lacks complete generality since some modulation formats lack well-defined lobes.

(d) *Fractional power containment bandwidth.* This bandwidth criterion has been adopted by the Federal Communications Commission (FCC Rules and Regulations Section 2.202) and states that the occupied bandwidth is the band that leaves exactly 0.5% of the signal power above the upper band limit and exactly 0.5% of the signal power below the lower band limit. Thus 99% of the signal power is inside the occupied band.

(e) *Bounded power spectral density.* A popular method of specifying bandwidth is to state that everywhere outside the specified band, $G_x(f)$ must have fallen at least to a certain stated level below that found at the band center. Typical attenuation levels might be 35 or 50 dB.

(f) *Absolute bandwidth.* This is the interval between frequencies, outside of which the spectrum is zero. This is a useful abstraction. However, for all realizable waveforms, the absolute bandwidth is infinite.

Example 1.4 Strictly Bandlimited Signals

The concept of a signal that is strictly limited to a band of frequencies is not realizable. Prove this by showing that a *strictly bandlimited* signal must also be a signal of *infinite time duration.*

Let $x(t)$ be a signal, with Fourier transform $X(f)$, that is strictly limited to the band of frequencies centered at $\pm f_c$ and of width $2W$. We may express $X(f)$ in terms of an ideal filter transfer function, $H(f)$, illustrated in Figure 1.19a, as follows:

$$X(f) = X'(f)H(f) \tag{1.74}$$

where, $X'(f)$ is the Fourier transform of a signal $x'(t)$, not necessarily bandlimited, where

$$H(f) = \text{rect}\left(\frac{f - f_c}{2W}\right) + \text{rect}\left(\frac{f + f_c}{2W}\right) \tag{1.75}$$

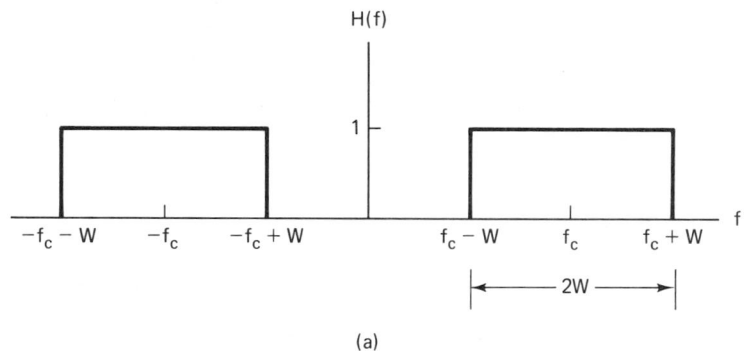

(a)

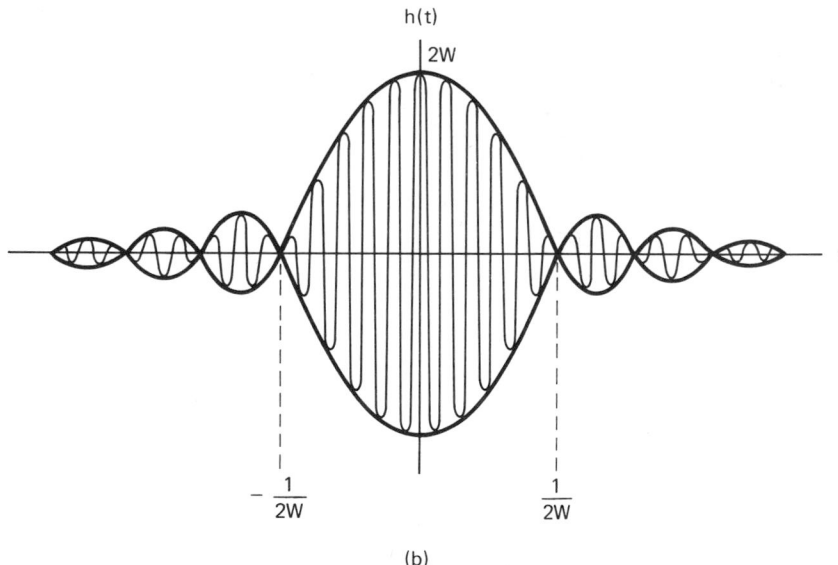

(b)

Figure 1.19 Transfer function and impulse response for a strictly bandlimited signal. (a) Ideal bandpass filter. (b) Ideal bandpass impulse response.

and where

$$\text{rect}\left(\frac{f}{2W}\right) = \begin{cases} 1 & \text{for } -W < f < W \\ 0 & \text{for } |f| > W \end{cases}$$

We can express $X(f)$ in terms of $X'(f)$ as

$$X(f) = \begin{cases} X'(f) & \text{for } (f_c - W) \leq |f_c| \leq (f_c + W) \\ 0 & \text{otherwise} \end{cases}$$

Multiplication in the frequency domain, as seen in Equation (1.74), transforms to convolution in the time domain as follows:

$$x(t) = x'(t) * h(t) \tag{1.76}$$

where $h(t)$, the inverse Fourier transform of $H(f)$, can be written as (see Tables A.1 and A.2)

$$h(t) = 2W \, (\text{sinc } 2Wt) \, \cos 2\pi f_c t$$

and is illustrated in Figure 1.19b. We note that $h(t)$ is of *infinite duration*. It follows, therefore, that $x(t)$ obtained in Equation (1.76) by convolving $x'(t)$ with $h(t)$ is also of infinite duration and therefore is *not realizable*.

1.8 CONCLUSION

In this chapter, the goals of the book have been outlined and the basic nomenclature has been defined. The fundamental concepts of time-varying signals, such as classification, spectral density, and autocorrelation, have been reviewed. Also, random signals have been considered, and white Gaussian noise, the primary noise model in most communication systems, has been characterized, statistically and spectrally. Finally, we have treated the important area of signal transmission through linear systems and have examined some of the realizable approximations to the ideal case. We have also established that the concept of an absolute bandwidth is an abstraction, and that in the real world we are faced with the need to choose a definition of bandwidth that is useful for our particular application. In the remainder of the book, each of the signal processing steps introduced in this chapter will be explored in the context of the typical system block diagram appearing at the beginning of each chapter.

REFERENCES

1. Haykin, S., *Communication Systems,* John Wiley & Sons, Inc., New York, 1983.
2. Shanmugam, K. S., *Digital and Analog Communication Systems,* John Wiley & Sons, Inc., New York, 1979.
3. Papoulis, A., *Probability, Random Variables, and Stochastic Processes,* McGraw-Hill Book Company, New York, 1965.
4. Johnson, J. B., "Thermal Agitation of Electricity in Conductors," *Phys. Rev.,* vol. 32, July 1928, pp. 97–109.

5. Nyquist, H., "Thermal Agitation of Electric Charge in Conductors," *Phys. Rev.,* vol. 32, July 1928, pp. 110–113.

6. Van Trees, H. L., *Detection, Estimation, and Modulation Theory,* Part 1, John Wiley & Sons, New York, 1968.

7. Schwartz, M., *Information Transmission, Modulation, and Noise,* McGraw-Hill Book Company, New York, 1970.

8. Millman, J., and Taub, H., *Pulse, Digital, and Switching Waveforms,* McGraw-Hill Book Company, New York, 1965.

PROBLEMS

1.1. Classify the following signals as energy signals or power signals. Find the normalized energy or normalized power of each.

(a) $x(t) = A \cos 2\pi f_0 t$ for $-\infty < t < \infty$

(b) $x(t) = \begin{cases} A \cos 2\pi f_0 t \\ 0 \end{cases}$ for $-T_0/2 \leq t \leq T_0/2$, where $T_0 = 1/f_0$ elsewhere

(c) $x(t) = \begin{cases} A \exp(-at) \\ 0 \end{cases}$ for $t > 0$, $a > 0$ elsewhere

(d) $x(t) = \cos t + 5 \cos 2t$ for $-\infty < t < \infty$

1.2. Determine the energy spectral density of a square pulse $x(t) = \text{rect}(t/T)$, where rect (t/T) equals 1, for $-T/2 \leq t \leq T/2$, and equals 0, elsewhere. Calculate the normalized energy E_x in the pulse.

1.3. Find an expression for the average normalized power in a periodic signal in terms of its complex Fourier series coefficients.

1.4. Using time averaging, find the average normalized power in the waveform $x(t) = 10 \cos 10t + 20 \cos 20t$.

1.5. Repeat Problem 1.4 using the summation of spectral coefficients.

1.6. Determine which, if any, of the following functions have the properties of autocorrelation functions. Justify your determination. [*Note:* $\mathcal{F}\{R(\tau)\}$ must be a nonnegative function. Why?]

(a) $x(\tau) = \begin{cases} 1 & \text{for } -1 \leq \tau \leq 1 \\ 0 & \text{otherwise} \end{cases}$

(b) $x(\tau) = \delta(\tau) + \sin 2\pi f_0 \tau$

(c) $x(\tau) = \exp(|\tau|)$

(d) $x(\tau) = 1 - |\tau|$ for $-1 \leq \tau \leq 1$

1.7. Determine which, if any, of the following functions have the properties of power spectral density functions. Justify your determination.

(a) $X(f) = \delta(f) + \cos^2 2\pi f$

(b) $X(f) = 10 + \delta(f - 10)$

(c) $X(f) = \exp(-2\pi |f - 10|)$

(d) $X(f) = \exp[-2\pi(f^2 - 10)]$

1.8. Find the autocorrelation function of $x(t) = A \cos(2\pi f_0 t + \phi)$ in terms of its period, $T_0 = 1/f_0$. Find the average normalized power of $x(t)$, using $P_x = R(0)$.

1.9. **(a)** Use the results of Problem 1.8 to find the autocorrelation function, $R(\tau)$, of waveform $x(t) = 10 \cos 10t + 20 \cos 20t$.

 (b) Use the relationship $P_x = R(0)$ to find the average normalized power in $x(t)$. Compare the answer with the answers to Problems 1.4 and 1.5.

1.10. For the function $x(t) = 1 + \cos 2\pi f_0 t$, calculate **(a)** the average value of $x(t)$; **(b)** the ac power of $x(t)$; **(c)** the rms value of $x(t)$.

1.11. Consider a random process given by $X(t) = A \cos(2\pi f_0 t + \phi)$, where A and f_0 are constants and ϕ is a random variable that is uniformly distributed over $(0, 2\pi)$. If $X(t)$ is an ergodic process, the time averages of $X(t)$ in the limit as $t \to \infty$ are equal to the corresponding ensemble averages of $X(t)$.

 (a) Use time averaging over an integer number of periods to calculate the approximations to the first and second moments of $X(t)$.

 (b) Use Equations (1.26) and (1.28) to calculate the ensemble-average approximations to the first and second moments of $X(t)$. Compare the results with your answers in part (a).

1.12. The Fourier transform of a signal, $x(t)$ is defined by $X(f) = \text{sinc } f$, where the sinc function is as defined in Equation (1.39). Find the autocorrelation function, $R_x(\tau)$, of the signal $x(t)$.

1.13. Use the sampling property of the unit impulse function to evaluate the following integrals.

 (a) $\displaystyle\int_{-\infty}^{\infty} \cos 6t \delta(t - 3)\, dt$

 (b) $\displaystyle\int_{-\infty}^{\infty} 10\delta(t)(1 + t)^{-1}\, dt$

 (c) $\displaystyle\int_{-\infty}^{\infty} \delta(t + 4)(t^2 + 6t + 1)\, dt$

 (d) $\displaystyle\int_{-\infty}^{\infty} \exp(-t^2)\delta(t - 2)\, dt$

1.14. Find $X_1(f) * X_2(f)$ for the spectra shown in Figure P1.1.

1.15. The two-sided power spectral density, $G_x(f) = 10^{-6} f^2$, of a waveform $x(t)$ is shown in Figure P1.2.

 (a) Find the normalized average power in $x(t)$ over the frequency band from 0 to 10 kHz.

 (b) Find the normalized average power contained in the frequency band from 5 to 6 kHz.

1.16. Decibels are logarithmic measures of *power ratios*, as described in Equation (1.64a). Sometimes, a similar formulation is used to express nonpower measurements in decibels (referenced to some designated unit). As an example, calculate how many decibels of hamburger meat you would buy to feed 2 hamburgers each to a group of 100 people. Assume that you and the butcher have agreed on the unit of "$\frac{1}{2}$ pound of meat" (the amount in one hamburger) as a reference unit.

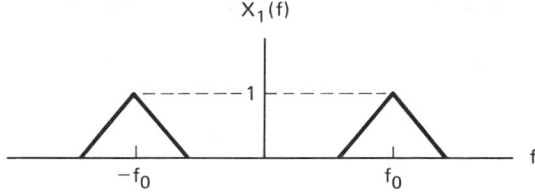

$X_1(f)$

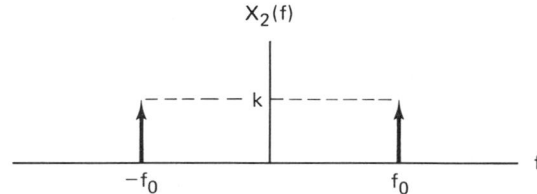

$X_2(f)$

Figure P1.1

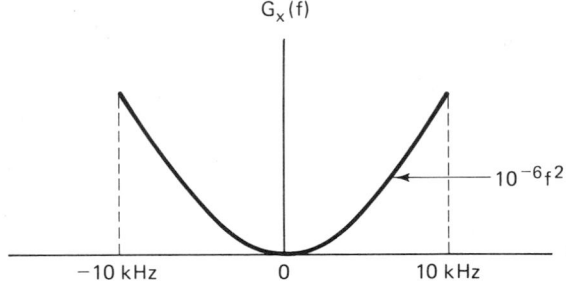

$G_x(f)$

$10^{-6}f^2$

−10 kHz   0   10 kHz

Figure P1.2

1.17. Consider the Butterworth low-pass amplitude response given in Equation (1.65).
   (a) Find the value of n so that $|H(f)|^2$ is constant to within ± 1 dB over the range $|f| \leq 0.9f_u$.
   (b) Show that as n approaches infinity, the amplitude response approaches that of an ideal low-pass filter.

1.18. Consider the network in Figure 1.9, whose frequency transfer function is $H(f)$. An impulse $\delta(t)$ is applied at the input. Show that the response $y(t)$ at the output is the inverse Fourier transform of $H(f)$.

1.19. An example of a *holding circuit,* commonly used in pulse systems, is shown in Figure P1.3. Determine the impulse response of this circuit.

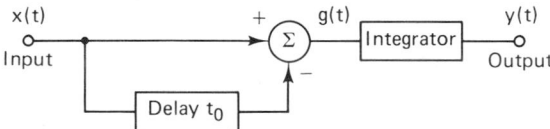

Figure P1.3

1.20. Given the spectrum

$$G_x(f) = 10^{-4} \left\{ \frac{\sin\left[\pi(f - 10^6)10^{-4}\right]}{\pi(f - 10^6)10^{-4}} \right\}^2$$

Find the value of the signal bandwidth using the following bandwidth definitions:

(a) Half-power bandwidth.

(b) Noise equivalent bandwidth.

(c) Null-to-null bandwidth.

(d) 99% of power bandwidth.

(e) Bandwidth beyond which the attenuation is 35 dB.

(f) Absolute bandwidth.

CHAPTER 2

Formatting
and
Baseband Transmission

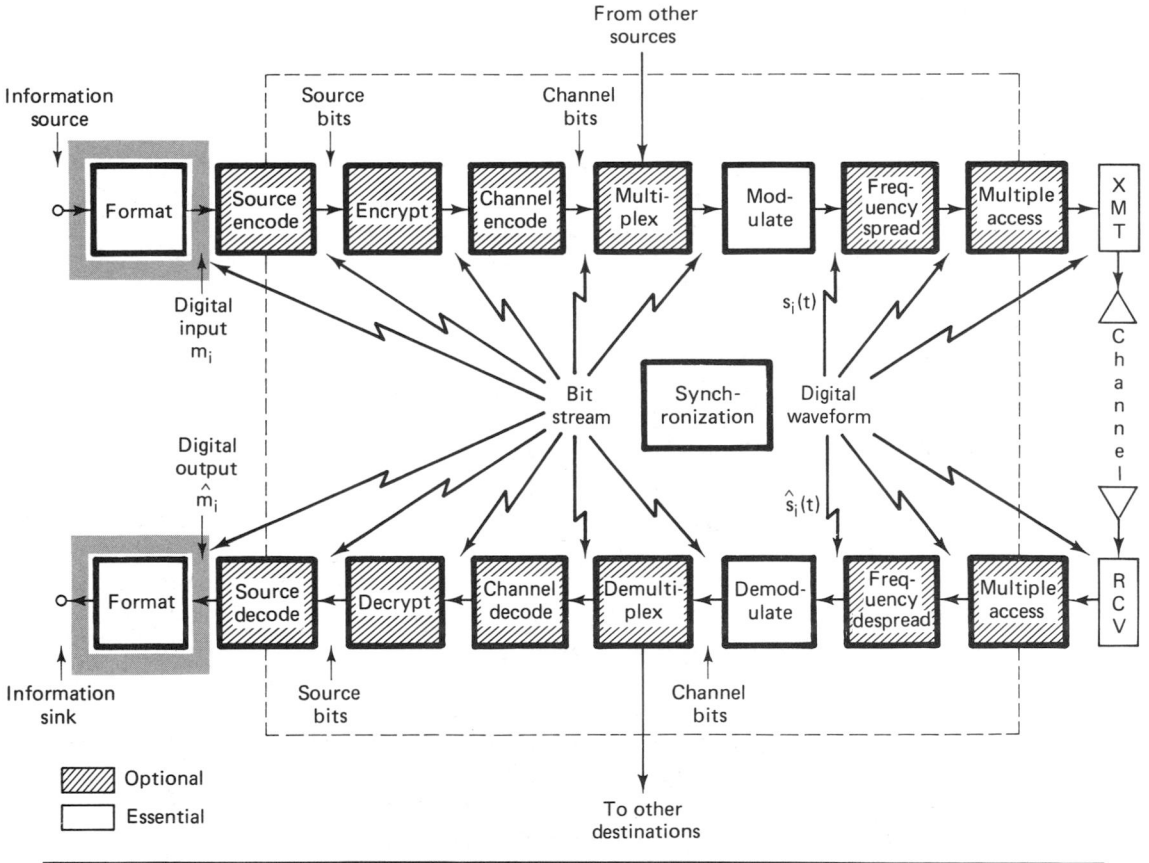

The first essential signal processing step, *formatting*, makes the source signal compatible with digital processing. *Transmit formatting* is a transformation from source information to digital symbols (in the receive chain, formatting is the reverse transformation). When there is data redundancy reduction or data compression, in addition to formatting, the process is termed *source coding*. Some authors consider formatting to be a special case of source coding. We treat formatting (and baseband transmission) in this chapter, and treat source coding as a special case of the *efficient description* of source information in Chapter 11. In Figure 2.1 the main formatting topics are highlighted—character coding, sampling, quantization, and pulse code modulation (PCM).

A signal whose spectrum extends from (or near) dc up to some finite value, usually less than a few megahertz, is called a *baseband* or *low-pass* signal. Such a signal is implied whenever we use the term "information," "message," or "data." For the transmission of baseband signals by a digital communication system, the information is *formatted* so that it is represented by digital symbols. Then, pulse waveforms are assigned that represent these symbols; this step is referred to as *pulse modulation* or *baseband modulation*. These waveforms can then be transmitted over a cable.

Baseband signals are not appropriate for propagation through many transmission media. Baseband signals whose spectrum has been shifted to a frequency band that is more appropriate for propagation through a transmission medium are called *bandpass modulation signals* or simply *bandpass signals*. Bandpass signals have their spectral content clustered in a band of frequencies near a value called

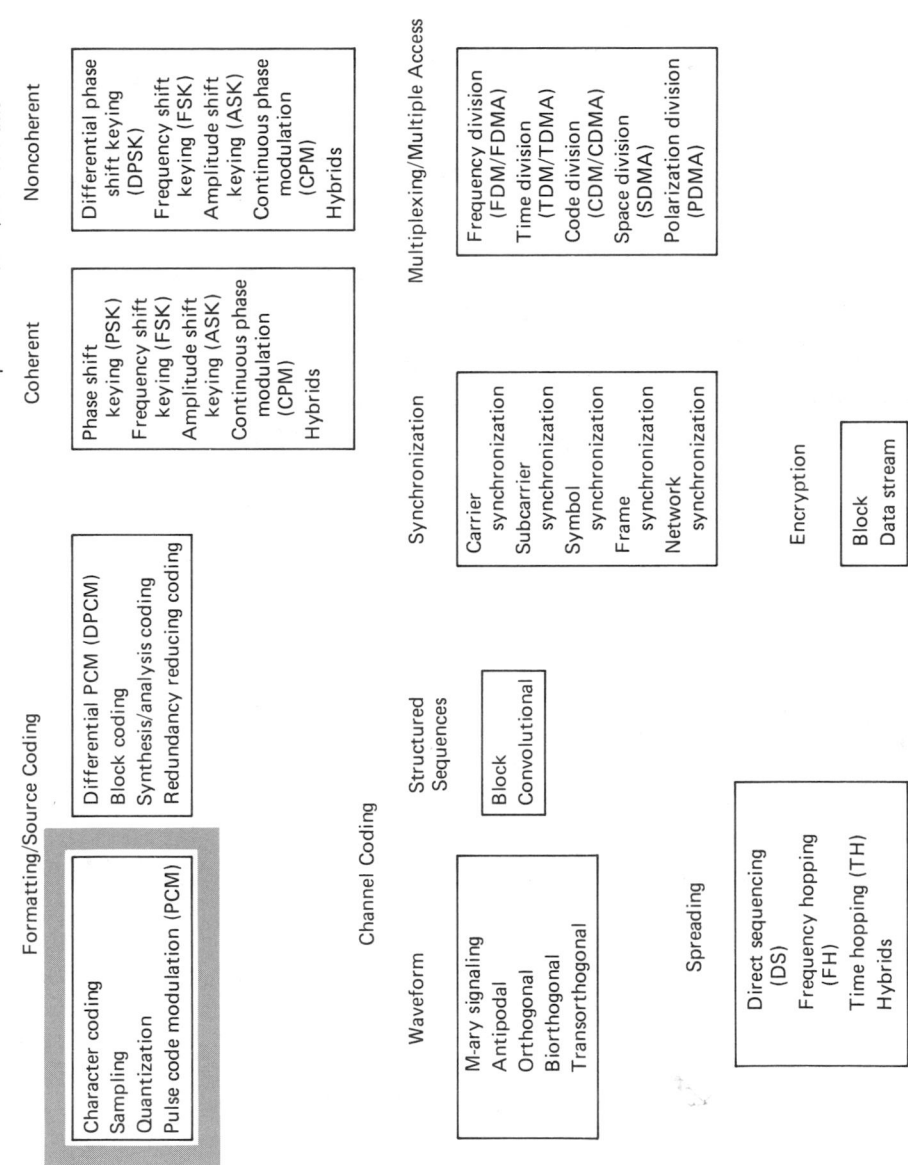

Figure 2.1 Basic digital communication transformations.

the *carrier frequency*. In Chapter 3 we deal with the modulation and demodulation of these bandpass signals.

2.1 BASEBAND SYSTEMS

In Figure 1.2 we presented a block diagram of a typical digital communication system. A version of this functional diagram, focusing primarily on the formatting and transmission of *baseband* signals, is shown in Figure 2.2. Data already in a digital format would bypass the formatting function. Textual information is transformed into binary digits by use of a coder. Analog information is formatted using three separate processes: sampling, quantization, and coding. In all cases, the formatting step results in a sequence of binary digits.

These digits are to be transmitted through a *baseband channel*, such as a pair of wires or a coaxial cable. However, no channel can be used for the transmission of binary digits without first transforming the digits to *waveforms* that are compatible with the channel. For baseband channels, compatible waveforms are pulses.

In Figure 2.2, the conversion from binary digits to pulse waveforms takes place in the block labeled *waveform encoder*, also called a *baseband modulator*. The output of the waveform encoder is typically a sequence of pulses with characteristics that correspond to the binary digits being sent. After transmission through the channel, the received waveforms are detected to produce an estimate of the transmitted digits, and then the final step, (reverse) formatting, recovers an estimate of the source information.

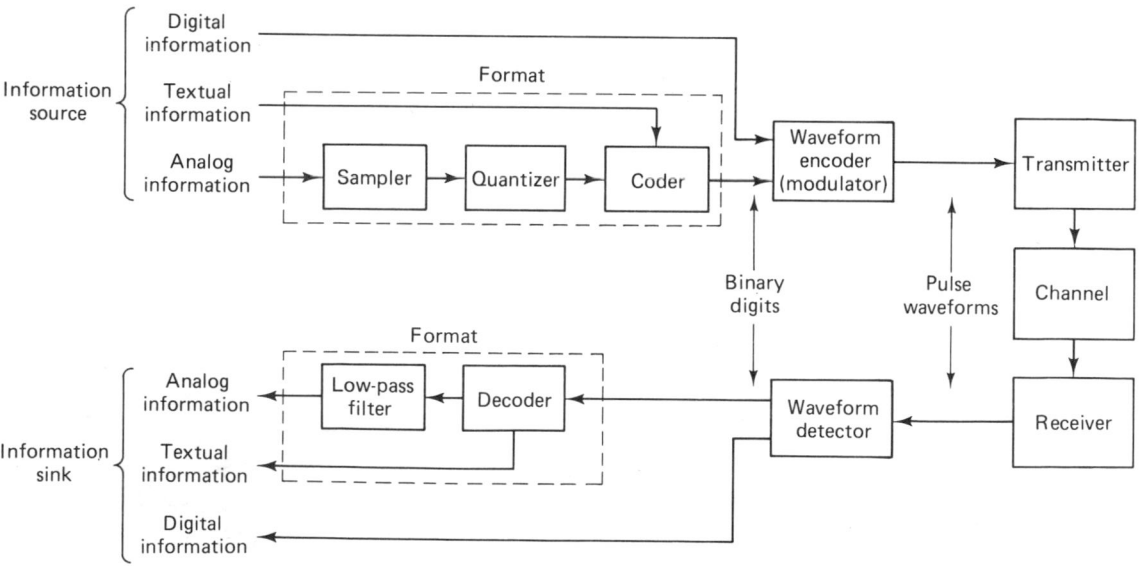

Figure 2.2 Formatting and transmission of baseband signals.

2.2 FORMATTING TEXTUAL DATA (CHARACTER CODING)

The original form of most communicated data (except for computer-to-computer transmissions) is either textual or analog. If the data consist of alphanumeric text, they will be character encoded with one of several standard formats, examples of which are, the American Standard Code for Information Interchange (ASCII), the Extended Binary Coded Decimal Interchange Code (EBCDIC), Baudot, and Hollerith. The textual material is thereby transformed into a digital format. The ASCII format is shown in Figure 2.3; the EBCDIC format is shown in Figure 2.4. The bit numbers signify the order of serial transmission, where bit number 1 is the first signaling element. Character coding, then, is the step that transforms text into binary digits (bits). Sometimes, existing character codes are modified to meet specialized needs. For example, the 7-bit ASCII code (Figure 2.3) can be modified to include an added bit for error detection purposes (see Chapter 5). On the other hand, sometimes the code is truncated to a 6-bit ASCII version, which provides capability for only 64 characters instead of the 128 characters allowed by 7-bit ASCII.

2.3 MESSAGES, CHARACTERS, AND SYMBOLS

Textual messages are comprised of a sequence of alphanumeric characters. When digitally transmitted the characters are first encoded into a sequence of bits, called a *bit stream* or *baseband signal*. Groups of k bits can then be combined to form new digits, or *symbols*, from a finite symbol set or alphabet of $M = 2^k$ such symbols. A system using a symbol set size of M is referred to as an *M-ary system*. The value of k or M represents an important initial choice in the design of any digital communication system. For $k = 1$, the system is termed *binary*, the size of the symbol set is $M = 2$, and the modulator uses one of the two different waveforms to represent the binary "one" and the other to represent the binary "zero." For this special case, the symbol and the bit are the same. For $k = 2$, the system is termed *quaternary* or *4-ary* ($M = 4$). At each symbol time, the modulator uses one of the four different waveforms that represents the symbol. The partitioning of the sequence of message bits is determined by the specification of the symbol set size, M. The following example should help clarify the relationship between the terms "message," "character," "symbol," "bit," and "digital waveform."

2.3.1 Example of Messages, Characters, and Symbols

Figure 2.5 shows examples of bit stream partitioning, based on the system specification for the values of k and M. The textual message in the figure is the word "THINK." Using 6-bit ASCII character coding (bit numbers 1 to 6 from Figure 2.3) yields a bit stream comprised of 30 bits. In Figure 2.5a, the symbol set size, M, has been chosen to be 8 (each symbol represents an 8-ary digit). The bits are therefore partitioned into groups of three ($k = \log_2 8$); the resulting 10 numbers

Bits			5	0	0	0	0	1	1	1	1
			6	0	0	1	1	0	0	1	1
			7	0	1	0	1	0	1	0	1
1	2	3	4								
0	0	0	0	NUL	DLE	SP	0	@	P	`	p
1	0	0	0	SOH	DC1	!	1	A	Q	a	q
0	1	0	0	STX	DC2	"	2	B	R	b	r
1	1	0	0	ETX	DC3	#	3	C	S	c	s
0	0	1	0	EOT	DC4	$	4	D	T	d	t
1	0	1	0	ENQ	NAK	%	5	E	U	e	u
0	1	1	0	ACK	SYN	&	6	F	V	f	v
1	1	1	0	BEL	ETB	'	7	G	W	g	w
0	0	0	1	BS	CAN	(	8	H	X	h	x
1	0	0	1	HT	EM	)	9	I	Y	i	y
0	1	0	1	LF	SUB	*	:	J	Z	j	z
1	1	0	1	VT	ESC	+	;	K	[	k	{
0	0	1	1	FF	FS	,	<	L	\	l	|
1	0	1	1	CR	GS	-	=	M	]	m	}
0	1	1	1	SO	RS	.	>	N	^	n	~
1	1	1	1	SI	US	/	?	O	_	o	DEL

NUL Null, or all zeros
SOH Start of heading
STX Start of text
ETX End of text
EOT End of transmission
ENQ Enquiry
ACK Acknowledge
BEL Bell, or alarm
BS   Backspace
HT   Horizontal tabulation
LF   Line feed
VT   Vertical tabulation
FF   Form feed
CR   Carriage return
SO   Shift out
SI   Shift in
DLE Data link escape

DC1 Device control 1
DC2 Device control 2
DC3 Device control 3
DC4 Device control 4
NAK Negative acknowledge
SYN Synchronous idle
ETB End of transmission block
CAN Cancel
EM   End of medium
SUB Substitute
ESC Escape
FS   File separator
GS   Group separator
RS   Record separator
US   Unit separator
SP   Space
DEL Delete

Figure 2.3 Seven-bit American standard code for information interchange (ASCII).

Figure 2.4 EBCDIC character code set.

Bits 1234 \ Bits 5678	0000	0001	0010	0011	0100	0101	0110	0111	1000	1001	1010	1011	1100	1101	1110	1111
0000	NUL	SOH	STX	ETX	PF	HT	LC	DEL			SMM	VT	FF	CR	SO	SI
0001	DLE	DC1	DC2	DC3	RES	NL	BS	IL	CAN	EM	CC		IFS	IGS	IRS	IUS
0010	DS	SOS	FS		BYP	LF	EOB	PRE			SM			ENQ	ACK	BEL
0011			SYN		PN	RS	US	EOT					DC4	NAK		SUB
0100	SP										¢	.	<	(	+	|
0101	&										!	$	*	)	;	¬
0110	-	/									¦	,	%	_	>	?
0111											:	#	@	'	=	"
1000		a	b	c	d	e	f	g	h	i						
1001		j	k	l	m	n	o	p	q	r						
1010			s	t	u	v	w	x	y	z						
1011																
1100		A	B	C	D	E	F	G	H	I						
1101		J	K	L	M	N	O	P	Q	R						
1110			S	T	U	V	W	X	Y	Z						
1111	0	1	2	3	4	5	6	7	8	9						

PF	Punch off
HT	Horizontal tab
LC	Lower case
DEL	Delete
SP	Space
UC	Upper case
RES	Restore
NL	New line
BS	Backspace
IL	Idle
PN	Punch on
EOT	End of transmission
BYP	Bypass
LF	Line feed
EOB	End of block
PRE	Prefix (ESC)
RS	Reader stop
SM	Start message
DS	Digit select
SOS	Start of significance
IFS	Interchange file separator
IGS	Interchange group separator
IRS	Interchange record separator
IUS	Interchange unit separator
Others	Same as ASCII

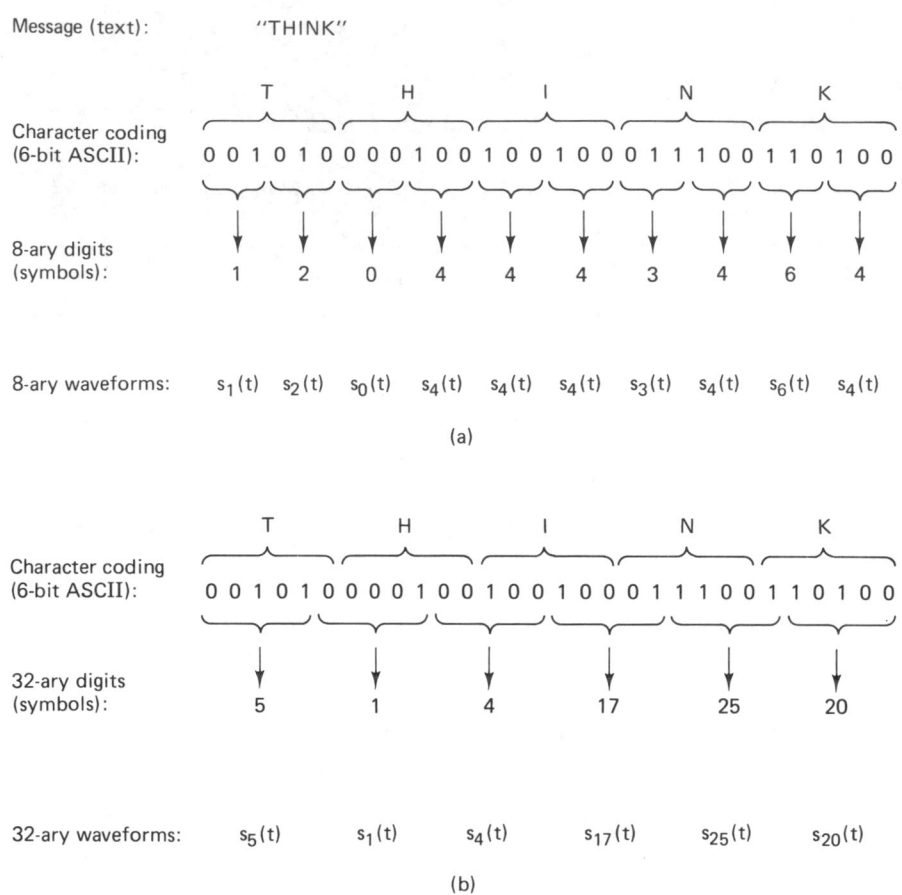

Message (text):   "THINK"

Character coding (6-bit ASCII):

T   H   I   N   K

0 0 1 0 1 0 0 0 0 1 0 0 1 0 0 1 0 0 0 1 1 1 0 0 1 1 0 1 0 0

8-ary digits (symbols):   1 2 0 4 4 4 3 4 6 4

8-ary waveforms: $s_1(t)$ $s_2(t)$ $s_0(t)$ $s_4(t)$ $s_4(t)$ $s_4(t)$ $s_3(t)$ $s_4(t)$ $s_6(t)$ $s_4(t)$

(a)

Character coding (6-bit ASCII):

T   H   I   N   K

0 0 1 0 1 0 0 0 0 1 0 0 1 0 0 1 0 0 0 1 1 1 0 0 1 1 0 1 0 0

32-ary digits (symbols):   5   1   4   17   25   20

32-ary waveforms: $s_5(t)$ $s_1(t)$ $s_4(t)$ $s_{17}(t)$ $s_{25}(t)$ $s_{20}(t)$

(b)

Figure 2.5 Messages, characters, and symbols. (a) 8-ary example. (b) 32-ary example.

represent the 10 octal symbols to be transmitted. The transmitter must have a repertoire of eight waveforms, $s_i(t)$, where $i = 1, \ldots, 8$, to represent the possible symbols, any one of which may be transmitted during a symbol time. The final row of Figure 2.5a lists the 10 waveforms that an 8-ary modulating system transmits to represent the textual message "THINK."

In Figure 2.5b, the symbol set size, M, has been chosen to be 32 (each symbol represents a 32-ary digit). The bits are therefore taken five at a time, and the resulting group of six numbers represent the six 32-ary symbols to be transmitted. Notice that there is no need for the symbol boundaries and the character boundaries to coincide. The first symbol represents $\frac{5}{6}$ of the first character, "T." The second symbol represents the remaining $\frac{1}{6}$ of the character "T" and $\frac{4}{6}$ of the next character, "H," and so on. It is not necessary that the characters be partitioned more aesthetically. The system sees the characters as a string of digits to be transmitted; only the end user (or the user's teleprinter machine) ascribes textual meaning to the final delivered sequence of bits. In this 32-ary case, a transmitter needs a repertoire of 32 waveforms, $s_i(t)$, where $i = 1, \ldots, 32$, one for each possible symbol that may be transmitted. The final row of the figure lists the six waveforms that a 32-ary modulating system transmits to represent the textual message "THINK."

2.4 FORMATTING ANALOG INFORMATION

If the information is analog, it cannot be character encoded as in the case of textual data; the information must first be transformed into a digital format. The process of transforming an analog waveform into a form that is compatible with a digital communication system starts with sampling the waveform to produce a discrete pulse-amplitude-modulated waveform, as described below.

2.4.1 The Sampling Theorem

The link between an analog waveform and its sampled version is provided by what is known as the *sampling process*. This process can be implemented in several ways, the most popular being the *sample-and-hold* operation. In this operation, a switch and storage mechanism (such as a transistor and a capacitor, or a shutter and a filmstrip) form a sequence of samples of the continuous input waveform. The output of the sampling process is called *pulse amplitude modulation* (PAM) because the successive output intervals can be described as a sequence of pulses with amplitudes derived from the input waveform samples. The analog waveform can be approximately retrieved from a PAM waveform by simple low-pass filtering. An important question is: How closely can a filtered PAM waveform approximate the original input waveform? This question can be answered by reviewing the *sampling theorem*, which states [1]: A bandlimited signal having no spectral components above f_m hertz can be determined uniquely by values sampled at uniform intervals of T_s seconds, where

$$T_s \le \frac{1}{2f_m} \tag{2.1}$$

This particular statement is also known as the *uniform sampling theorem*. Stated another way, the upper limit on T_s can be expressed in terms of the sampling rate, denoted $f_s = 1/T_s$. The restriction, stated in terms of the sampling rate, is known as the *Nyquist criterion*. The statement is

$$f_s \ge 2f_m \tag{2.2}$$

The sampling rate $f_s = 2f_m$ is also called the *Nyquist rate*. The Nyquist criterion is a theoretically sufficient condition to allow an analog signal to be *reconstructed completely* from a set of uniformly spaced discrete-time samples. In the sections that follow, the validity of the sampling theorem is demonstrated using different sampling approaches.

2.4.1.1 Impulse Sampling

Here we demonstrate the validity of the sampling theorem using the frequency convolution property of the Fourier transform. Let us first examine the case of *ideal sampling* with a sequence of unit impulse functions. Assume an analog waveform, $x(t)$, as shown in Figure 2.6a, with a Fourier transform, $X(f)$, which is zero outside the interval ($-f_m < f < f_m$), as shown in Figure 2.6b. The sampling of $x(t)$ can be viewed as the product of $x(t)$ with a periodic train of unit impulse functions, $x_\delta(t)$, shown in Figure 2.6c and defined as follows:

$$x_\delta(t) = \sum_{n=-\infty}^{\infty} \delta(t - nT_s) \tag{2.3}$$

where T_s is the sampling period and $\delta(t)$ is the unit impulse or Dirac delta function defined in Section 1.2.5. Let us choose $T_s = 1/2f_m$, so that the Nyquist criterion is just satisfied.

The *sifting property* of the impulse function (see Section A.4.1) states that

$$x(t)\delta(t - t_0) = x(t_0)\delta(t - t_0) \tag{2.4}$$

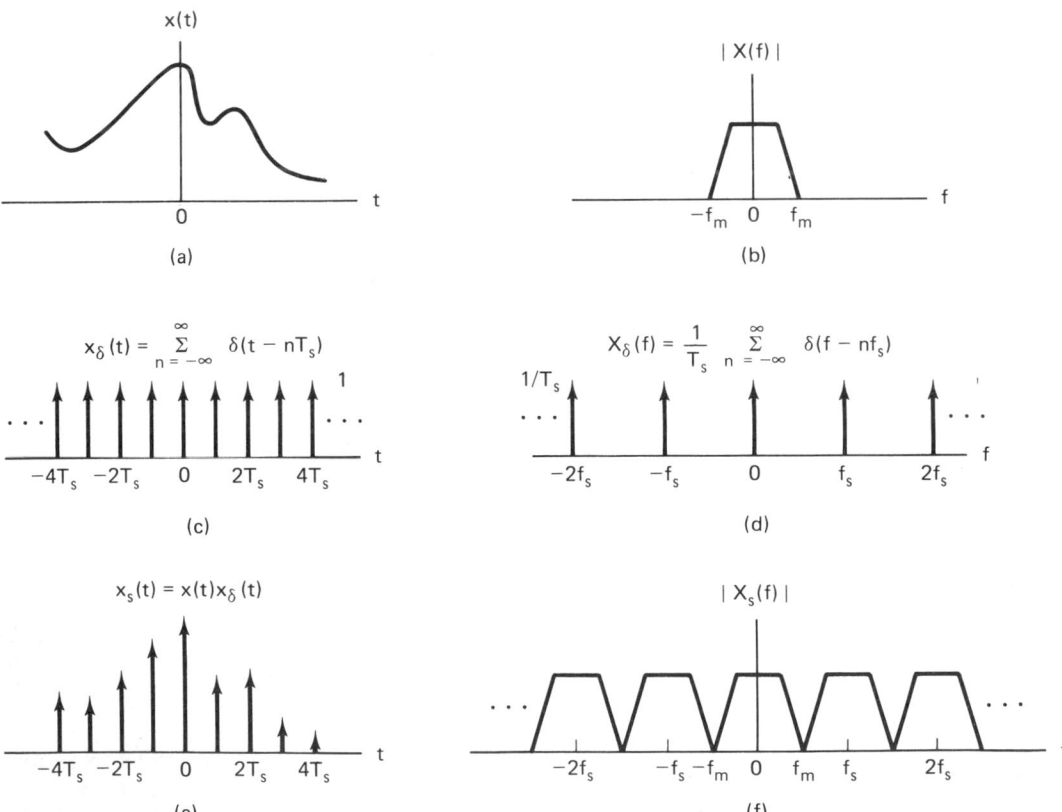

Figure 2.6 Sampling theorem using the frequency convolution property of the Fourier transform.

Using this property, we can see that $x_s(t)$, the sampled version of $x(t)$, shown in Figure 2.6e, is given by

$$x_s(t) = x(t)x_\delta(t) = \sum_{n=-\infty}^{\infty} x(t)\delta(t - nT_s)$$

$$= \sum_{n=-\infty}^{\infty} x(nT_s)\delta(t - nT_s) \tag{2.5}$$

Using the *frequency convolution property* of the Fourier transform (see Section A.5.3), the time-domain product $x(t)x_\delta(t)$ of Equation (2.5) transforms to the frequency-domain convolution $X(f) * X_\delta(f)$, where $X_\delta(f)$ is the Fourier transform of the impulse train $x_\delta(t)$,

$$X_\delta(f) = \frac{1}{T_s} \sum_{n=-\infty}^{\infty} \delta(f - nf_s) \tag{2.6}$$

and where $f_s = 1/T_s$ is the sampling frequency. Notice that the Fourier transform of an impulse train is another impulse train; the values of the periods of the two trains are reciprocally related to one another. Figures 2.6c and d illustrate the impulse train $x_\delta(t)$ and its Fourier transform $X_\delta(f)$, respectively.

Convolution with an impulse function simply shifts the original function, as follows:

$$X(f) * \delta(f - nf_s) = X(f - nf_s) \tag{2.7}$$

We can solve for the transform, $X_s(f)$, of the sampled waveform as follows:

$$X_s(f) = X(f) * X_\delta(f) = X(f) * \left[\frac{1}{T_s} \sum_{n=-\infty}^{\infty} \delta(f - nf_s) \right]$$

$$= \frac{1}{T_s} \sum_{n=-\infty}^{\infty} X(f - nf_s) \tag{2.8}$$

We therefore conclude that within the original bandwidth, the spectrum $X_s(f)$ of the sampled signal $x_s(t)$ is, to within a constant factor $(1/T_s)$, exactly the same as that of $x(t)$. In addition, the spectrum repeats itself periodically in frequency every f_s hertz. The sifting property of an impulse function makes the convolving of an impulse train with another function easy to visualize. The impulses act as sampling functions. Hence, convolution can be performed graphically by sweeping the impulse train, $X_\delta(f)$, in Figure 2.6d past the transform, $|X(f)|$, in Figure 2.6b. This sampling of $|X(f)|$ at each step in the sweep replicates $|X(f)|$ at each of the frequency positions of the impulse train, resulting in $|X_s(f)|$, shown in Figure 2.6f.

When the sampling rate is chosen, as it has been here, such that $f_s = 2f_m$, each spectral replicate is separated from each of its neighbors by a frequency band exactly equal to f_s hertz, and the analog waveform can theoretically be completely recovered from the samples, by the use of filtering. However, a filter with infinitely steep sides would be required. It should be clear that if $f_s > 2f_m$, the replications will move farther apart in frequency, as shown in Figure 2.7a, making it easier to perform the filtering operation. A typical low-pass filter characteristic that might be used to separate the baseband spectrum from those at higher frequencies is shown in the figure. When the sampling rate is reduced, such that $f_s < 2f_m$, the replications will overlap, as shown in Figure 2.7b, and some information will be lost. This phenomenon, the result of undersampling (sampling at too low a rate), is called *aliasing*. The Nyquist rate, $f_s = 2f_m$, is the sampling rate below which aliasing occurs; to avoid aliasing, the Nyquist criterion, $f_s \geq 2f_m$, must be satisfied.

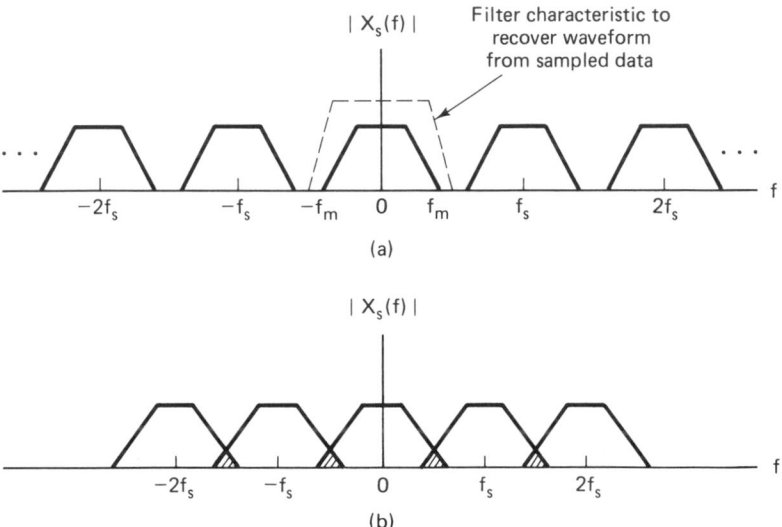

Figure 2.7 Spectra for various sampling rates. (a) Sampled spectrum ($f_s > 2f_m$). (b) Sampled spectrum ($f_s < 2f_m$).

As a matter of practical consideration, neither waveforms of engineering interest nor realizable bandlimiting filters are strictly bandlimited. These signals and filters can, however, be considered to be "essentially" bandlimited. By this we mean that a bandwidth can be determined beyond which the spectral components are attenuated to a level that is considered negligible.

2.4.1.2 Natural Sampling

Here we demonstrate the validity of the sampling theorem using the frequency shifting property of the Fourier transform. Although instantaneous sampling is a convenient model, a more practical way of accomplishing the sampling of a bandlimited analog signal, $x(t)$, is to multiply $x(t)$, shown in Figure 2.8a, by the pulse train or switching waveform, $x_p(t)$, shown in Figure 2.8c. Each pulse in $x_p(t)$ has width T and amplitude $1/T$. Multiplication by $x_p(t)$ can be viewed as the opening and closing of a switch. As before, the sampling frequency is designated f_s, and its reciprocal, the time period between samples, is designated T_s.

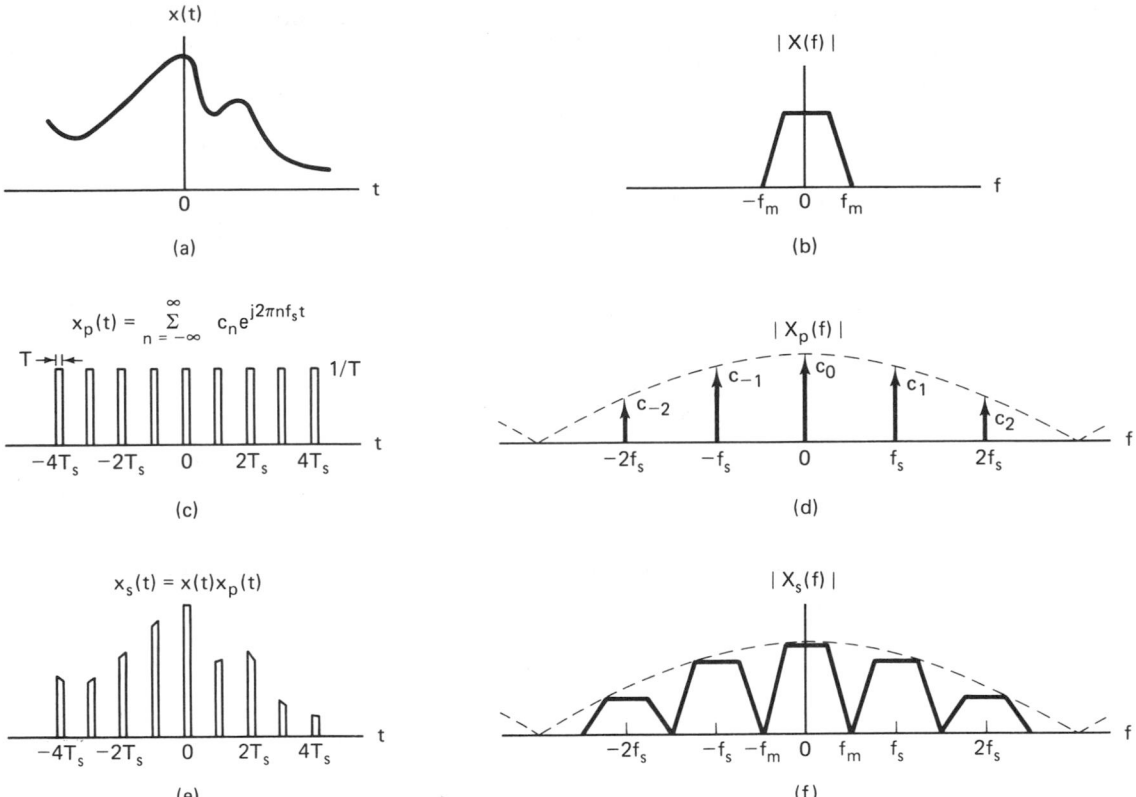

Figure 2.8 Sampling theorem using the frequency shifting property of the Fourier transform.

The resulting sampled-data sequence, $x_s(t)$, is illustrated in Figure 2.8e and is expressed as

$$x_s(t) = x(t)x_p(t) \tag{2.9}$$

The sampling here is termed *natural sampling*, since the top of each pulse in the $x_s(t)$ sequence retains the shape of its corresponding analog segment during the pulse interval. Using Equation (A.13), we can express the periodic pulse train $x_p(t)$ as a Fourier series in the form

$$x_p(t) = \sum_{n=-\infty}^{\infty} c_n e^{j2\pi n f_s t} \tag{2.10}$$

where the sampling rate, $f_s = 1/T_s$, is chosen equal to $2f_m$, so that the Nyquist criterion is just satisfied. From Equation (A.24), $c_n = (1/T_s)$ sinc (nT/T_s), where T is the pulse width, $1/T$ is the pulse amplitude, and

$$\text{sinc } y = \frac{\sin \pi y}{\pi y}$$

The envelope of the magnitude spectrum of the pulse train, seen as a dashed line in Figure 2.8d, has the characteristic sinc shape. Combining Equations (2.9) and (2.10), we can express $x_s(t)$ as

$$x_s(t) = x(t) \sum_{n=-\infty}^{\infty} c_n e^{j2\pi n f_s t} \qquad (2.11)$$

The transform, $X_s(f)$, of the sampled waveform is found as follows:

$$X_s(f) = \mathcal{F} \left\{ x(t) \sum_{n=-\infty}^{\infty} c_n e^{j2\pi n f_s t} \right\} \qquad (2.12)$$

For linear systems, we can interchange the operations of summation and Fourier transformation. Therefore, we can write

$$X_s(f) = \sum_{n=-\infty}^{\infty} c_n \mathcal{F}\{x(t)e^{j2\pi n f_s t}\} \qquad (2.13)$$

Using the *frequency translation* property of the Fourier transform (see Section A.3.2), we solve for $X_s(f)$ as follows:

$$X_s(f) = \sum_{n=-\infty}^{\infty} c_n X(f - nf_s) \qquad (2.14)$$

Similar to the unit impulse sampling case, Equation (2.14) and Figure 2.8f illustrate that $X_s(f)$ is a replication of $X(f)$, periodically repeated in frequency every f_s hertz. In this natural-sampled case, however, we see that $X_s(f)$ is weighted by the Fourier series coefficients of the pulse train, compared to a constant value in the impulse-sampled case. It is satisfying to note that *in the limit*, as the pulse width, T, approaches zero, c_n approaches $1/T_s$ for all n (see the example that follows), and Equation (2.14) converges to Equation (2.8).

Example 2.1 Comparison of Impulse Sampling and Natural Sampling

Consider a given waveform, $x(t)$, with Fourier transform, $X(f)$. Let $X_{s1}(f)$ be the spectrum of $x_{s1}(t)$, which is the result of sampling $x(t)$ with a unit impulse train $x_\delta(t)$. Let $X_{s2}(f)$ be the spectrum of $x_{s2}(t)$, the result of sampling $x(t)$ with a pulse train, $x_p(t)$, with pulse width, T, amplitude $1/T$ and period, T_s. Show that in the limit, as T approaches zero, $X_{s1}(f) = X_{s2}(f)$.

Solution

From Equation (2.8),

$$X_{s1}(f) = \frac{1}{T_s} \sum_{n=-\infty}^{\infty} X(f - nf_s)$$

and from Equation (2.14),

$$X_{s2}(f) = \sum_{n=-\infty}^{\infty} c_n X(f - nf_s)$$

As the pulse width $T \to 0$, and the pulse amplitude approaches infinity (the area of the pulse remains unity), $x_p(t) \to x_\delta(t)$. Using Equation (A.14), we can solve for c_n in the limit as follows:

$$c_n = \lim_{T \to 0} \frac{1}{T_s} \int_{-T_s/2}^{T_s/2} x_p(t) e^{-j2\pi n f_s t} \, dt$$

$$= \frac{1}{T_s} \int_{-T_s/2}^{T_s/2} x_\delta(t) e^{-j2\pi n f_s t} \, dt$$

Since, within the range of integration, $-T_s/2$ to $T_s/2$, the only contribution of $x_\delta(t)$ is that due to the impulse at the origin, we can write

$$c_n = \frac{1}{T_s} \int_{-T_s/2}^{T_s/2} \delta(t) e^{-j2\pi n f_s t} \, dt = \frac{1}{T_s}$$

Therefore, in the limit, $X_{s1}(f) = X_{s2}(f)$ for all n.

2.4.1.3 Sample-and-Hold Operation

The simplest and thus most popular sampling method, *sample and hold*, can be described by the convolution of the sampled pulse train, $[x(t)x_\delta(t)]$, shown in Figure 2.6e, with a unity amplitude rectangular pulse, $p(t)$, of pulse width T_s. This time convolution results in the *flat-top* sampled sequence, $x_s(t)$:

$$x_s(t) = p(t) * [x(t)x_\delta(t)]$$

$$= p(t) * \left[x(t) \sum_{n=-\infty}^{\infty} \delta(t - nT_s) \right] \tag{2.15}$$

The Fourier transform, $X_s(f)$, of the time convolution in Equation (2.15) is the frequency-domain product between the transform $P(f)$ of the rectangular pulse and the periodic spectrum, shown in Figure 2.6f, of the impulse-sampled data:

$$X_s(f) = P(f)\mathscr{F}\left\{ x(t) \sum_{n=-\infty}^{\infty} \delta(t - nT_s) \right\}$$

$$= P(f) \left\{ X(f) * \left[\frac{1}{T_s} \sum_{n=-\infty}^{\infty} \delta(f - nf_s) \right] \right\} \tag{2.16}$$

$$= P(f) \frac{1}{T_s} \sum_{n=-\infty}^{\infty} X(f - nf_s)$$

where $P(f)$ is of the form $T_s \, \text{sinc} \, fT_s$. The effect of this product operation results in a spectrum similar in appearance to the natural-sampled example presented in Figure 2.8f. The most obvious effect of the hold operation is the significant attenuation of the higher-frequency spectral replicates (compare Figure 2.8f to Figure 2.6f), which is a desired effect. Additional analog postfiltering is usually required to finish the filtering process by further attenuating the residual spectral components located at the multiples of the sample rate. A secondary effect of the hold operation is the nonuniform spectral gain, $P(f)$, applied to the desired baseband spectrum shown in Equation (2.16). The postfiltering operation can compensate for this attenuation by incorporating the inverse of $P(f)$ over the signal passband.

2.4.2 Aliasing

Figure 2.9 is a detailed view of the positive half of the baseband spectrum and one of the replicates from Figure 2.7b. It illustrates aliasing in the frequency domain. The overlapped region, shown in Figure 2.9b, contains that part of the spectrum which is aliased due to *undersampling*. The aliased spectral components represent ambiguous data that can be retrieved only under special conditions (see Section 11.4.4, on subband coding). In general, the ambiguity is not resolved and the ambiguous data appear in the frequency band between $(f_s - f_m)$ and f_m. Figure 2.10 illustrates that a higher sampling rate, f'_s, can eliminate the aliasing by separating the spectral replicates; the resulting spectrum in Figure 2.10b corresponds to the case in Figure 2.7a. Figures 2.11 and 2.12 illustrate two ways of eliminating aliasing using *antialiasing filters*. In Figure 2.11 the analog signal is *prefiltered* so that the new maximum frequency, f'_m, is reduced to $f_s/2$ or less. Thus there are no aliased components seen in Figure 2.11b, since $f_s > 2f'_m$. Eliminating the aliasing terms prior to sampling is good engineering practice. When the signal structure is well known, the aliased terms can be eliminated after sampling, with a low-pass filter operating on the sampled data [2]. In Figure 2.12 the aliased components are removed by *postfiltering* after sampling; the filter cutoff frequency, f''_m, removes the aliased components; f''_m needs to be less than

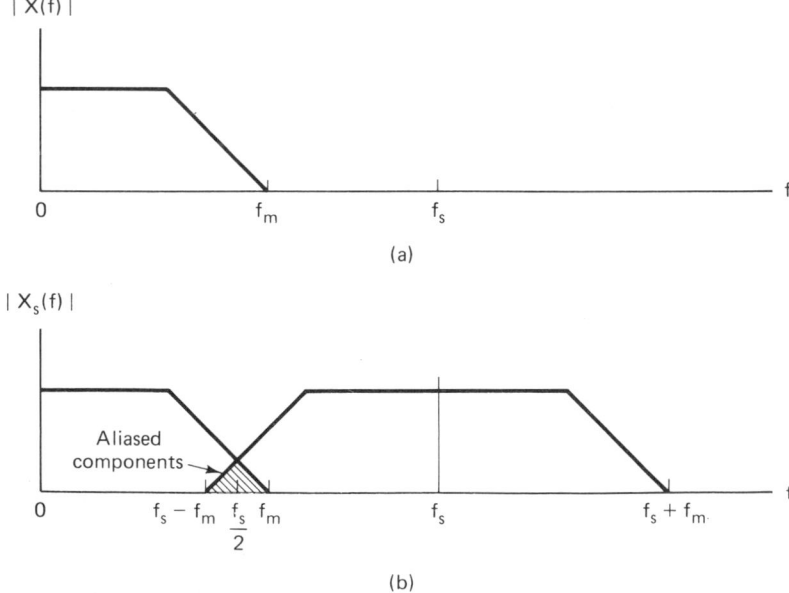

(a)

(b)

Figure 2.9 Aliasing in the frequency domain. (a) Continuous signal spectrum. (b) Sampled signal spectrum.

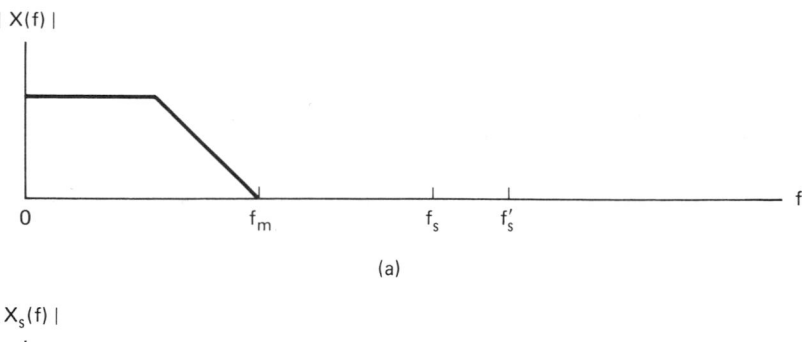

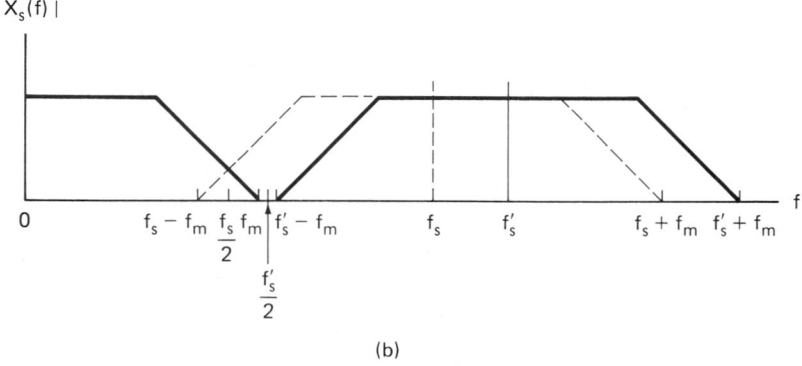

Figure 2.10 Higher sampling rate eliminates aliasing. (a) Continuous signal spectrum. (b) Sampled signal spectrum.

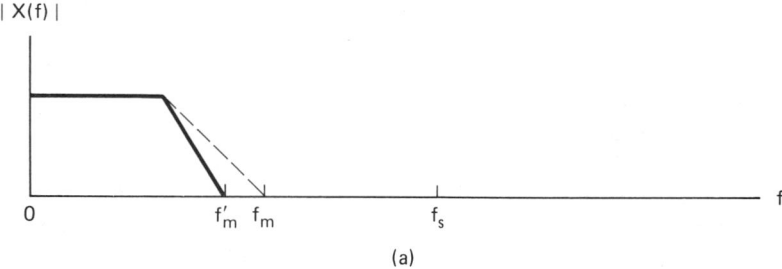

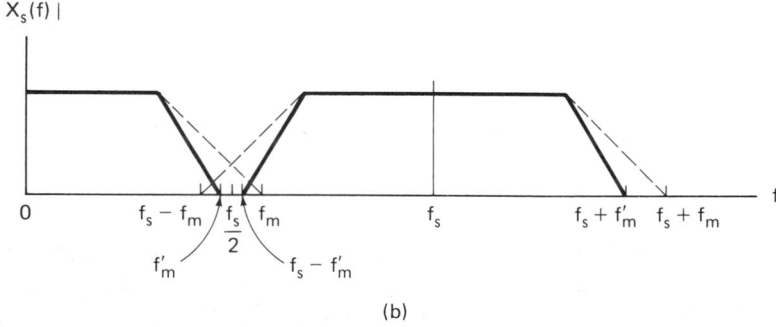

Figure 2.11 Sharper-cutoff filters eliminate aliasing. (a) Continous signal spectrum. (b) Sampled signal spectrum.

Sec. 2.4   Formatting Analog Information

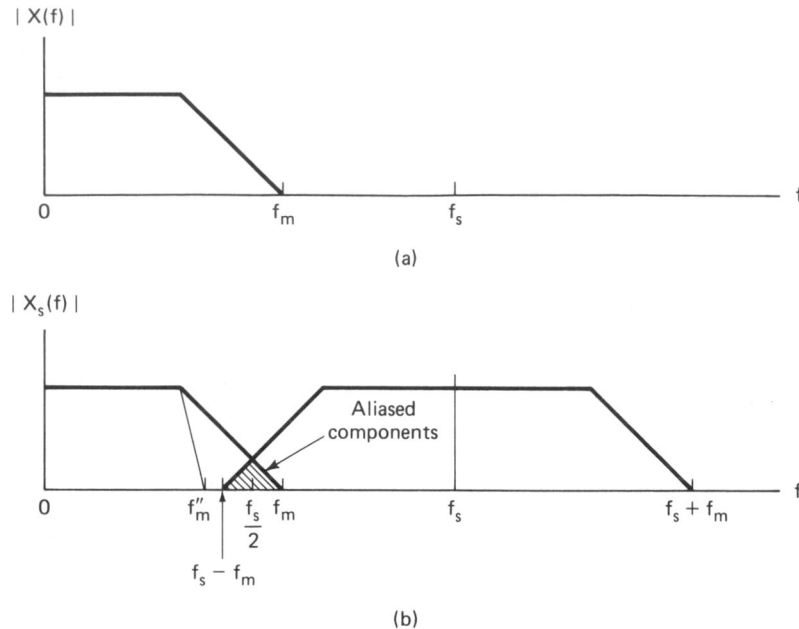

Figure 2.12 Postfilter eliminates aliased portion of spectrum. (a) Continuous signal spectrum. (b) Sampled signal spectrum.

$(f_s - f_m)$. Notice that the filtering techniques for eliminating the aliased portion of the spectrum in Figures 2.11 and 2.12 *will result in a loss* of some of the signal information. For this reason, the sample rate, cutoff bandwidth, and filter type selected for a particular signal bandwidth are all interrelated.

Realizable filters require a nonzero bandwidth for the transition between the passband and the required out-of-band attenuation. This is called the *transition bandwidth*. To minimize the system sample rate, we desire that the antialiasing filter have a small transition bandwidth. Filter complexity and cost rise sharply with narrower transition bandwidth, so a trade-off is required between the cost of a small transition bandwidth and the costs of the higher sampling rate, which are those of more storage and higher transmission rates. In many systems the answer has been to make the transition bandwidth between 10 and 20% of the signal bandwidth. If we account for the 20% transition bandwidth of the antialiasing filter, we have an *engineer's version* of the Nyquist sampling rate:

$$f_s \geq 2.2f_m \tag{2.17}$$

Figure 2.13 provides some insight into aliasing as seen in the time domain. The sampling instants of the solid-line sinusoid have been chosen so that the sinusoidal signal is undersampled. Notice that the resulting ambiguity allows one to draw a totally different (dashed-line) sinusoid, following the undersampled points.

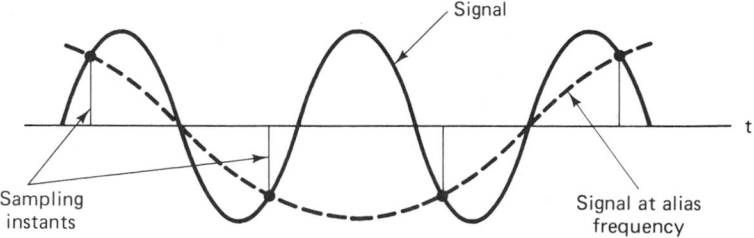

Figure 2.13 Alias frequency generated by sub-Nyquist sampling rate.

Example 2.2   Sampling Rate for a High-Quality Music System

We wish to produce a high-quality digitization of a 20-kHz bandwidth music source. We are to determine a reasonable sample rate for this source. By the engineer's version of the Nyquist rate, in Equation (2.17), the sampling rate should be greater than 44.0 ksamples/s. As a matter of comparison, the standard sampling rate for the compact disc digital audio player is 44.1 ksamples/s, and the standard sampling rate for studio-quality audio is 48.0 ksamples/s.

2.4.3 Signal Interface for a Digital System

Let us examine four ways in which analog source information can be described. Figure 2.14 illustrates the choices. Let us refer to the waveform in Figure 2.14a as the *original analog waveform*. Figure 2.14b represents a sampled version of the original waveform, typically referred to as *natural-sampled data* or PAM

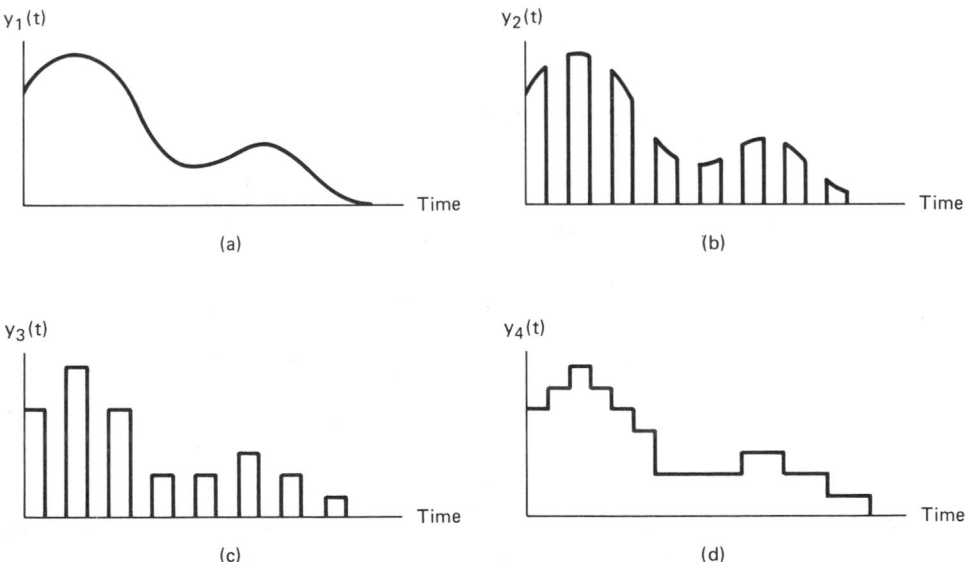

Figure 2.14   Amplitude and time coordinates of source data. (a) Original analog waveform. (b) Natural-sampled data. (c) Quantized samples. (d) Sample and hold.

(pulse amplitude modulation). Do you suppose that the sampled data in Figure 2.14b are compatible with a digital system? No, they are not, because the amplitude of each natural sample still has an infinite number of possible values; a digital system deals with a finite number of symbols. Even if the sampling is flat-top sampling, the possible pulse values form an infinite set, since they reflect all the possible values of the continuous analog waveform. Figure 2.14c illustrates the original waveform represented by discrete pulses. Here the pulses have flat tops *and* the pulse amplitude values are limited to a finite set. Each pulse is expressed as a level from a finite number of predetermined levels; each such level can be represented by a symbol from a finite alphabet. The pulses in Figure 2.14c are referred to as *quantized samples*; such a format is the obvious choice for interfacing with a digital system. The format in Figure 2.14d may be construed as the output of a sample-and-hold circuit. When the sample values are quantized to a finite set, this format can also interface with a digital system. After quantization, the analog waveform can still be recovered, but not precisely; improved reconstruction fidelity of the analog waveform can be achieved by increasing the number of quantization levels (requiring increased system bandwidth). Signal distortion due to quantization is treated in the following sections (and in Chapter 11).

2.5 SOURCES OF CORRUPTION

The analog signal recovered from the sampled, quantized, and transmitted pulses will contain corruption from several sources. The sources of corruption are related to (1) sampling and quantizing effects, and (2) channel effects. These effects are considered in the sections that follow.

2.5.1 Sampling and Quantizing Effects

2.5.1.1 Quantization Noise

The distortion inherent in quantization is a round-off or truncation error. The process of encoding the PAM waveform into a quantized waveform involves discarding some of the original analog information. This distortion, introduced by the need to approximate the analog waveform with quantized samples, is referred to as *quantization noise*; the amount of such noise is inversely proportional to the number of levels employed in the quantization process. The signal-to-noise ratio of quantized pulses is treated in Section 2.5.3.

2.5.1.2 Quantizer Saturation

The quantizer (or analog-to-digital converter) allocates L levels to the task of approximating the continuous range of inputs with a finite set of outputs. The range of inputs for which the difference between the input and output is small is called the *operating range* of the converter. If the input exceeds this range, the difference between the input and the output becomes large, and we say that the converter is operating in *saturation*. Saturation errors, being large, are more objectionable than quantizing noise. Generally, saturation is avoided by the use of automatic gain control (AGC), which effectively extends the operating range of the converter. Chapter 11 covers quantizer saturation in greater detail.

2.5.1.3 Timing Jitter

Our analysis of the sampling theorem predicted precise reconstruction of the signal based on uniformly spaced samples of the signal. If there is a slight jitter in the position of the sample, the sampling is no longer uniform. Although exact reconstruction is still possible if the sample positions are accurately known, the jitter is usually a random process and thus the sample positions are not accurately known. The effect of the jitter is equivalent to frequency modulation (FM) of the baseband signal. If the jitter is random, a low-level wideband spectral contribution is induced whose properties are very close to those of the quantizing noise. If the jitter exhibits periodic components, as might be found in data extracted from a tape recorder, the periodic FM will induce low-level spectral lines in the data. Timing jitter can be controlled with very good power supply isolation and stable clock references.

2.5.2 Channel Effects

2.5.2.1 Channel Noise

Thermal noise, interference from other users, and interference from circuit switching transients can cause errors in detecting the pulses carrying the digitized samples. Channel-induced errors can degrade the reconstructed signal quality quite quickly. This rapid degradation of output signal quality with channel-induced errors is called a *threshold effect*. If the channel noise is small, there will be no problem detecting the presence of the waveforms. Thus small noise does not corrupt the reconstructed signals. In this case, the only noise present in the reconstruction is the quantization noise. On the other hand, if the channel noise is large enough to affect our ability to detect the waveforms, the resultant detection error causes reconstruction errors. A large difference in behavior can occur for very small changes in channel noise level.

2.5.2.2 Intersymbol Interference

The channel is always bandlimited. A bandlimited channel disperses or spreads a pulse waveform passing through it (see Section 1.6.4). When the channel bandwidth is much greater than the pulse bandwidth, the spreading of the pulse will be slight. When the channel bandwidth is close to the signal bandwidth, the spreading will exceed a symbol duration and cause signal pulses to overlap. This overlapping is called *intersymbol interference* (ISI). Like any other source of interference, ISI causes system degradation (higher error rates); it is a particularly insidious form of interference because raising the signal power to overcome the interference will not improve the error performance. Details of how ISI is handled are presented in Section 2.11.

2.5.3 Signal-to-Noise Ratio for Quantized Pulses

Figure 2.15 illustrates an L-level linear quantizer for an analog signal with a peak-to-peak voltage range of $V_{pp} = V_p - (-V_p) = 2V_p$ volts. The quantized pulses assume positive and negative values, as shown in the figure. The step size between quantization levels, called the *quantile interval*, is denoted q volts. When the quantization levels are uniformly distributed over the full range, the quantizer is called a *uniform or linear quantizer*. Each sample value of the analog waveform is approximated with a quantized pulse; the approximation will result in an error no larger than $q/2$ in the positive direction or $-q/2$ in the negative direction. The degradation of the signal due to quantization is therefore limited to half a quantile interval, $\pm q/2$ volts.

A useful figure of merit for the uniform quantizer is the quantizer variance (mean-square error assuming zero mean). If we assume that the quantization error, e, is uniformly distributed over a single quantile interval q-wide (i.e., the analog input takes on all values with equal probability), the quantizer error variance is found to be

$$\sigma^2 = \int_{-q/2}^{+q/2} e^2 p(e)\ de \qquad (2.18a)$$

$$= \int_{-q/2}^{+q/2} e^2 \frac{1}{q}\ de = \frac{q^2}{12} \qquad (2.18b)$$

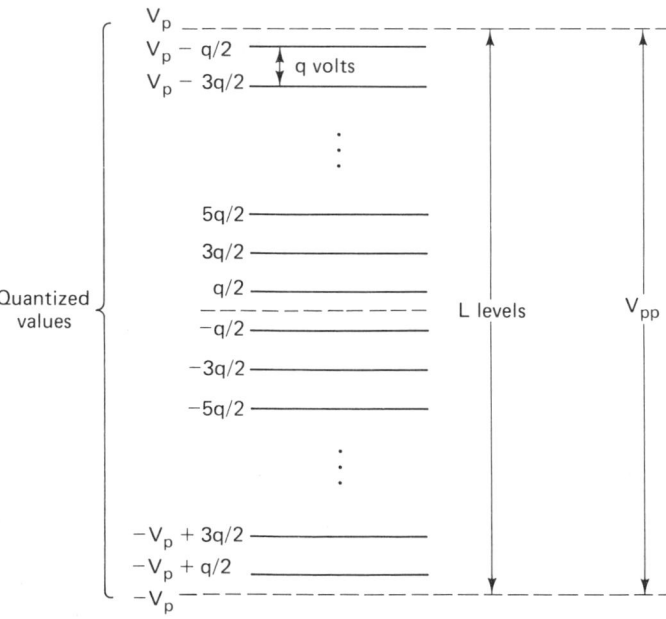

Figure 2.15   Quantization levels.

where $p(e) = 1/q$ is the (uniform) probability density function of the quantization error. The variance, σ^2, corresponds to the *average quantization noise power*. The peak power of the analog signal (normalized to 1 Ω) can be expressed as

$$V_p^2 = \left(\frac{V_{pp}}{2}\right)^2 = \left(\frac{Lq}{2}\right)^2 = \frac{L^2 q^2}{4} \qquad (2.19)$$

where L is the number of quantization levels. Equations (2.18) and (2.19) combined yield the ratio of peak signal power to average quantization noise power $(S/N)_q$, assuming that there are no errors due to ISI or channel noise:

$$\left(\frac{S}{N}\right)_q = \frac{L^2 q^2/4}{q^2/12} = 3L^2 \qquad (2.20)$$

It is intuitively satisfying to see that $(S/N)_q$ improves as a function of the number of quantization levels squared. In the limit (as $L \to \infty$), the signal approaches the PAM format (with no quantization), and the signal-to-quantization noise ratio is infinite; in other words, with an infinite number of quantization levels, there is zero quantization noise.

2.6 PULSE CODE MODULATION

Pulse code modulation (PCM) is the name given to the class of baseband signals obtained from the quantized PAM signals by encoding each quantized sample into a *digital word* [3]. The source information is sampled and quantized to one of L levels; then each quantized sample is digitally encoded into an ℓ-bit ($\ell = \log_2 L$) codeword. For baseband transmission, the codeword bits will then be transformed to pulse waveforms. The essential features of binary PCM are shown in Figure 2.16. Assume that an analog signal, $x(t)$, is limited in its excursions to the range -4 to $+4$ V. The step size between quantization levels has been set at 1 V. Thus eight quantization levels are employed; these are located at -3.5, -2.5, . . . , $+3.5$ V. We assign the code number 0 to the level at -3.5 V, the code number 1 to the level at -2.5 V, and so on, until the level at 3.5 V, which is assigned the code number 7. Each code number has its representation in binary arithmetic, ranging from 000 for code number 0 to 111 for code number 7.

The ordinate in Figure 2.16 is labeled with quantization levels and their code numbers. Each sample of the analog signal is assigned to the quantization level closest to the value of the sample. Beneath the analog waveform, $x(t)$, are seen four representations of $x(t)$, as follows: the natural sample values, the quantized sample values, the code numbers, and the PCM sequence.

Note that in the example of Figure 2.16, each sample is represented by a 3-bit codeword. If the signal, $x(t)$, had been quantized to 16 levels, a 4-bit codeword would be needed to characterize each sample, or if $x(t)$ had been quantized to four levels, a 2-bit codeword would be needed. From Equation (2.20) it can be seen that the greater the number of quantization levels, the lower will be the quantization noise. Hence quantization noise performance can be traded off versus data rate.

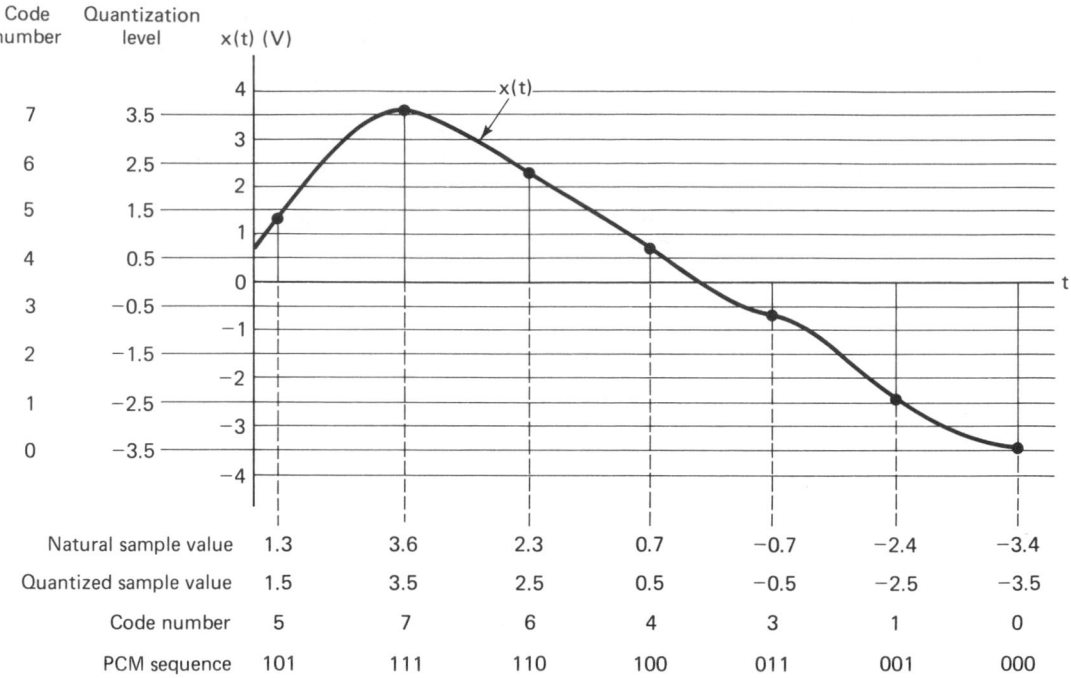

Natural sample value	1.3	3.6	2.3	0.7	−0.7	−2.4	−3.4
Quantized sample value	1.5	3.5	2.5	0.5	−0.5	−2.5	−3.5
Code number	5	7	6	4	3	1	0
PCM sequence	101	111	110	100	011	001	000

Figure 2.16 Natural samples, quantized samples, and pulse code modulation. (Reprinted with permission from Taub and Schilling, *Principles of Communication Systems*, McGraw-Hill Book Company, New York, 1971, Fig. 6.5-1, p. 205.)

2.7 UNIFORM AND NONUNIFORM QUANTIZATION

2.7.1 Statistics of Speech Amplitudes

Speech communication is a very important and specialized area of digital communications. Human speech is characterized by unique statistical properties; one such property is illustrated in Figure 2.17. The abscissa represents speech signal magnitudes, normalized to the root-mean-square (rms) value of such magnitudes through a typical communication channel, and the ordinate is probability. For most voice communication channels, very low speech volumes predominate; 50% of the time, the voltage characterizing detected speech energy is less than one-fourth of the rms value. Large amplitude values are relatively rare; only 15% of the time does the voltage exceed the rms value. We see from Equation (2.18b) that the quantization noise depends on the step size (size of the quantile interval). When the steps are uniform in size the quantization is known as *uniform quantization*. Such a system would be wasteful for speech signals; many of the quantizing steps would rarely be used. In a system that uses equally spaced quantization levels, the quantization noise is the same for all signal magnitudes. Therefore, with uniform quantization, the signal-to-noise ratio (SNR) is worse for low-level signals than for high-level signals. *Nonuniform quantization* can provide fine quantization of the weak signals and coarse quantization of the strong signals. Thus in the case of nonuniform quantization, quantization noise can be made

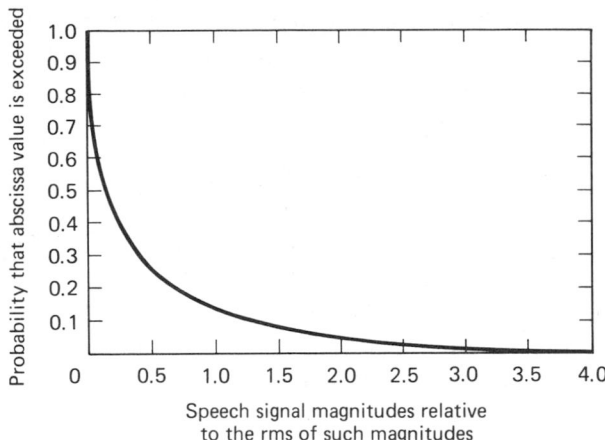

Figure 2.17 Statistical distribution of single-talker speech signal magnitudes.

proportional to signal size. The effect is to improve the overall SNR by reducing the noise for the predominant weak signals, at the expense of an increase in noise for the rarely occurring strong signals. Figure 2.18 compares the quantization of a strong versus a weak signal for uniform and nonuniform quantization. The staircase-like waveforms represent the approximations to the analog waveforms (after quantization distortion has been introduced). The SNR improvement that nonuniform quantization provides for the weak signal should be apparent. Nonuni-

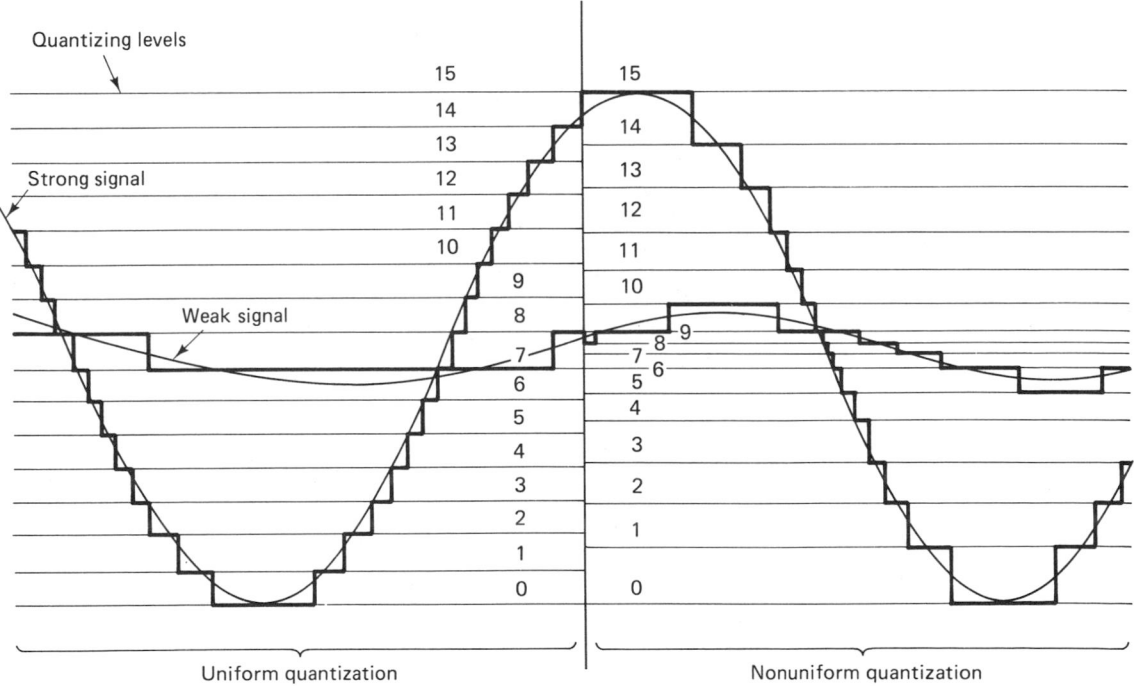

Figure 2.18 Uniform and nonuniform quantization of signals.

form quantization can be used to make the SNR a constant for all signals within the input range. For voice signals, the typical input signal dynamic range is 40 decibels (dB), where a decibel is defined in terms of the ratio of power P_2 to power P_1:

$$\text{number of dB} = 10 \log_{10} \frac{P_2}{P_1} \tag{2.21}$$

With a uniform quantizer, weak signals would experience a 40-dB-poorer SNR than that of strong signals. The standard telephone technique of handling the large range of possible input signal levels is to use a *logarithmic-compressed* quantizer instead of a uniform one. With such a nonuniform compressor the output SNR is independent of the distribution of input signal levels.

2.7.2 Nonuniform Quantization

One way of achieving nonuniform quantization is to use a nonuniform quantizer characteristic, shown in Figure 2.19a. More often, nonuniform quantization is achieved by first distorting the original signal with a logarithmic compression characteristic, as shown in Figure 2.19b, and then using a uniform quantizer. For small magnitude signals the compression characteristic has a much steeper slope than for large magnitude signals. Thus a given signal change at small magnitudes will carry the uniform quantizer through more steps than the same change at large

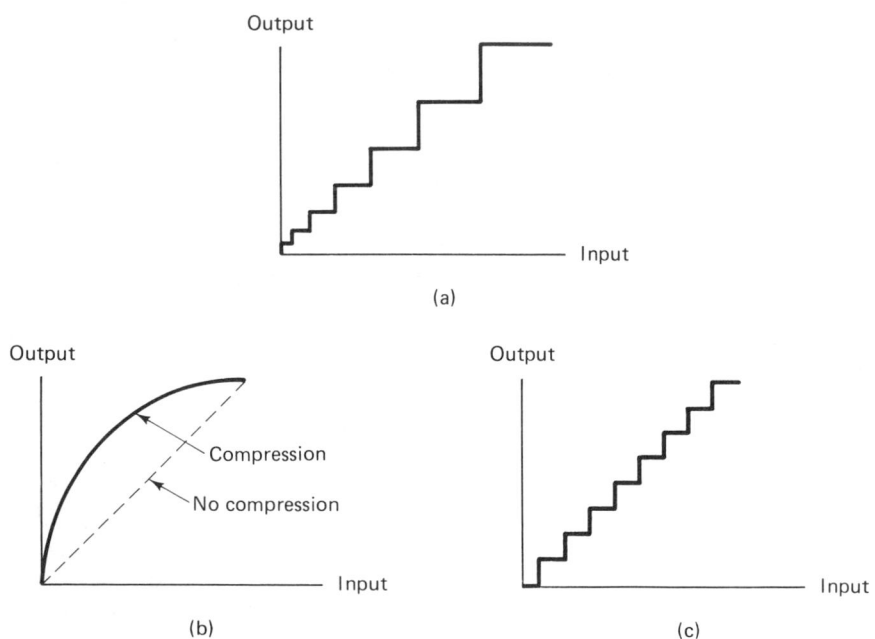

Figure 2.19 (a) Nonuniform quantizer characteristic. (b) Compression characteristic. (c) Uniform quantizer characteristic.

Formatting and Baseband Transmission   Chap. 2

magnitudes. The compression characteristic effectively changes the distribution of the input signal magnitudes so that there is not a preponderance of *low* magnitude signals at the output of the compressor. After compression, the distorted signal is used as the input to a uniform (linear) quantizer characteristic, shown in Figure 2.19c. At the receiver, an inverse compression characteristic, called *expansion*, is applied so that the overall transmission is not distorted. The processing pair (compression and expansion) is usually referred to as *companding*.

2.7.3 Companding Characteristics

The early PCM systems implemented a smooth logarithmic compression function. Today, most PCM systems use a piecewise linear approximation to the logarithmic compression characteristic. In North America a μ-law compression characteristic is used:

$$y = y_{max} \frac{\log_e[1 + \mu(|x|/x_{max})]}{\log_e(1 + \mu)} \text{ sgn } x \tag{2.22}$$

where

$$\text{sgn } x = \begin{cases} +1 & \text{for } x \geq 0 \\ -1 & \text{for } x < 0 \end{cases}$$

and where μ is a positive constant, x and y represent input and output voltages, and x_{max} and y_{max} are the maximum positive excursions of the input and output voltages, respectively. The compression characteristic is shown in Figure 2.20a for several values of μ. The standard value for μ is 255. Notice that $\mu = 0$ corresponds to linear amplification (uniform quantization).

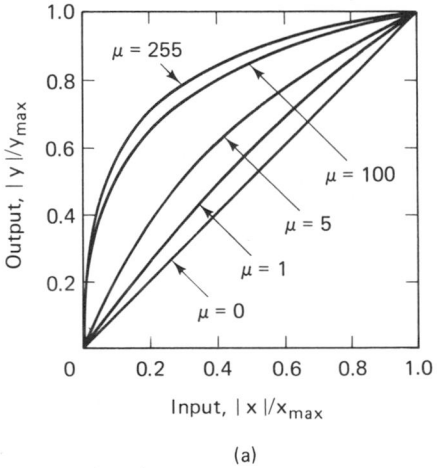

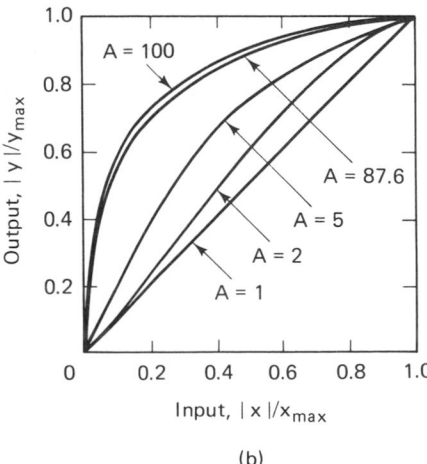

(a)   (b)

Figure 2.20 Compression characteristics. (a) μ-law characteristic. (b) A-law characteristic.

Sec. 2.7   Uniform and Nonuniform Quantization   77

Another compression characteristic, used mainly in Europe, is the A-law characteristic, defined as

$$
y = \begin{cases} y_{\max} \dfrac{A(|x|/x_{\max})}{1 + \log_e A} \, \mathrm{sgn}\, x & 0 < \dfrac{|x|}{x_{\max}} \le \dfrac{1}{A} \\[3ex] y_{\max} \dfrac{1 + \log_e[A(|x|/x_{\max})]}{1 + \log_e A} \, \mathrm{sgn}\, x & \dfrac{1}{A} < \dfrac{|x|}{x_{\max}} < 1 \end{cases} \tag{2.23}
$$

where A is a positive constant and x and y are as defined in Equation (2.22). The A-law compression characteristic is shown in Figure 2.20b for several values of A. A standard value for A is 87.6. See Chapter 11 for a more detailed treatment of μ-law and A-law companding characteristics.

2.8 BASEBAND TRANSMISSION

2.8.1 Waveform Representation of Binary Digits

We need to represent PCM binary digits by electrical pulses in order to transmit them through a baseband channel. Such a representation is shown in Figure 2.21. Codeword time slots are shown in Figure 2.21a, where the codeword is a 4-bit representation of each quantized sample. In Figure 2.21b, each binary one is represented by a pulse and each binary zero is represented by the absence of a pulse. Thus a sequence of electrical pulses having the pattern shown in Figure 2.21b can be used to transmit the information in the PCM bit stream, and hence the information in the quantized samples of a message.

At the receiver, a determination must be made as to the presence or absence of a pulse in each bit time slot. It will be shown in Section 2.9 that the likelihood of correctly detecting the presence of a pulse is a function of the pulse energy (or area under the pulse). Thus there is an advantage in making the pulse width, T', in Figure 2.21b as wide as possible. If we increase the pulse width to the maximum possible (equal to the bit time duration, T), we have the waveform shown in Figure 2.21c. Rather than describe this waveform as a sequence of present or absent pulses, we can describe it as a sequence of transitions between two levels. When the waveform occupies the upper voltage level it represents a binary one; when it occupies the lower voltage level it represents a binary zero.

2.8.2 PCM Waveform Types

Figure 2.22 illustrates the most commonly used PCM waveforms. The various waveforms are classified into the following groups:

1. Nonreturn-to-zero (NRZ)
2. Return-to-zero (RZ)
3. Phase encoded
4. Multilevel binary

Formatting and Baseband Transmission   Chap. 2

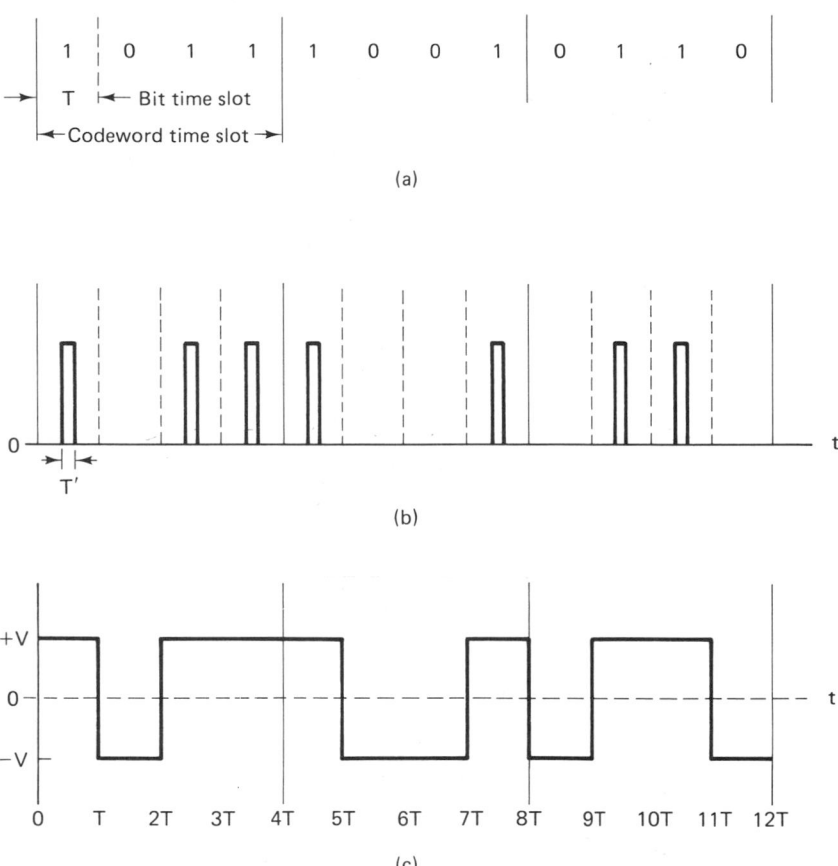

Figure 2.21 Example of waveform representation of binary digits. (a) PCM sequence. (b) Pulse representation of PCM. (c) Pulse waveform (transition between two levels).

The NRZ group is probably the most commonly used PCM waveform. It can be partitioned into the following subgroups: NRZ-L (L for level), NRZ-M (M for mark), and NRZ-S (S for space). NRZ-L is used extensively in digital logic. A binary one is represented by one level and a binary zero is represented by another level. There is a change in level whenever the data change from a one to a zero or from a zero to a one. With NRZ-M, the one, or *mark*, is represented by a change in level, and the zero, or *space*, is represented by no change in level. This is often referred to as *differential encoding*. NRZ-M is used primarily in magnetic tape recording. NRZ-S is the complement of NRZ-M: A one is represented by no change in level, and a zero is represented by a change in level.

The RZ waveforms consist of unipolar-RZ, bipolar-RZ, and RZ-AMI. These codes find application in baseband data transmission and in magnetic recording. With unipolar-RZ, a one is represented by a half-bit-wide pulse, and a zero is represented by the absence of a pulse. With bipolar-RZ, the ones and zeros are represented by opposite-level pulses that are one-half-bit wide. There is a pulse

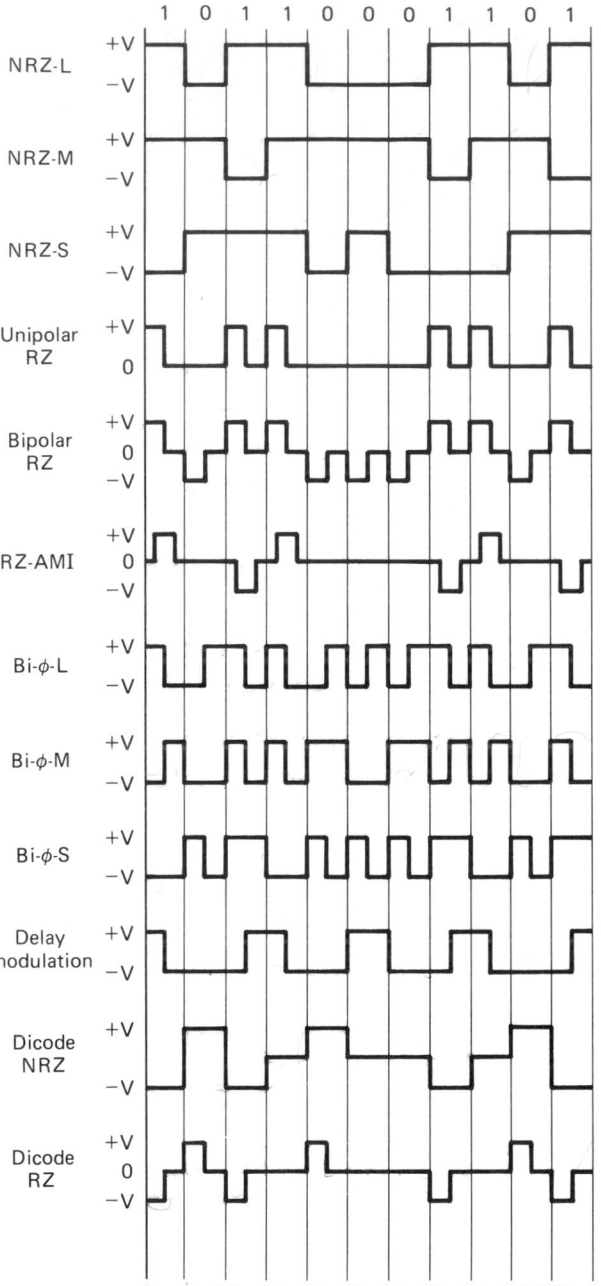

Figure 2.22 Various PCM waveforms.

present in each bit interval. RZ-AMI (AMI for "alternate mark inversion") is the coding scheme most often used in telemetry systems. The ones are represented by equal-amplitude alternating pulses. The zeros are represented by the absence of pulses.

The phase-encoded group consists of bi-φ-L (bi-phase-level), better known as *Manchester coding*; bi-φ-M (bi-phase-mark); bi-φ-S (bi-phase-space); and *delay modulation* (DM), or *Miller coding*. The phase-encoding schemes are used in magnetic recording systems and optical communications and in some satellite telemetry links. With bi-φ-L, a one is represented by a half-bit-wide pulse positioned during the first half of the bit interval; a zero is represented by a half-bit-wide pulse positioned during the second half of the bit interval. With bi-φ-M, a transition occurs at the beginning of every bit interval. A one is represented by a second transition one-half bit interval later; a zero is represented by no second transition. With bi-φ-S, a transition also occurs at the beginning of every bit interval. A one is represented by no second transition; a zero is represented by a second transition one-half bit interval later. With delay modulation [4], a one is represented by a transition at the midpoint of the bit interval. A zero is represented by no transition, unless it is followed by another zero. In this case, a transition is placed at the end of the bit interval of the first zero. Reference to the illustration in Figure 2.22 should help to make these descriptions clear.

Many binary waveforms use three levels, instead of two, to encode the binary data. Bipolar RZ and RZ-AMI belong to this group. The group also contains formats called *dicode* and *duobinary*. With dicode-NRZ, the one-to-zero or zero-to-one data transition changes the pulse polarity; without a data transition, the zero level is sent. With dicode-RZ, the one-to-zero or zero-to-one transition produces a half-duration polarity change; otherwise, a zero level is sent. The three-level duobinary signaling scheme is treated in Section 2.12.

One might ask why there are so many PCM waveforms. Are there really so many unique applications necessitating such a variety of waveforms to represent digits? The reason for the large selection relates to the differences in performance that characterize each waveform [5]. In choosing a coding scheme for a particular application, some of the parameters worth examining are the following:

1. *Dc component.* Eliminating the dc energy from the signal's power spectrum enables the system to be ac coupled. Magnetic recording systems, or systems using transformer coupling, have little sensitivity to very low frequency signal components. Thus low-frequency information could be lost.

2. *Self-Clocking.* Symbol or bit synchronization is required for any digital communication system. Some PCM coding schemes have inherent synchronizing or clocking features that aid in the recovery of the clock signal. For example, the Manchester code has a transition in the middle of every bit interval whether a one or a zero is being sent. This guaranteed transition provides a clocking signal.

3. *Error detection.* Some schemes, such as duobinary, provide the means of detecting data errors without introducing additional error-detection bits into the data sequence.

4. *Bandwidth compression.* Some schemes, such as multilevel codes, increase the efficiency of bandwidth utilization by allowing a reduction in required bandwidth for a given data rate; thus there is more information transmitted per unit bandwidth.

5. *Differential encoding.* This technique is useful because it allows the polarity of differentially encoded waveforms to be inverted without affecting the data detection. In communication systems where waveforms sometimes experience inversion, this is a great advantage. Differential encoding is treated in greater detail in Section 3.6.2.

6. *Noise immunity.* The various PCM waveform types can be further characterized by probability of bit error versus signal-to-noise ratio. Some of the schemes are more immune than others to noise. For example, the NRZ waveforms have better error performance than does the unipolar RZ waveform.

2.8.3 Spectral Attributes of PCM Waveforms

The most common criteria used for comparing PCM waveforms and for selecting one waveform type from the many available are: spectral characteristics, bit synchronization capabilities, error-detecting capabilities, interference and noise immunity, and cost and complexity of implementation. Figure 2.23 shows the spectral characteristics of some of the most popular PCM waveforms. The figure plots power spectral density in watts/hertz versus normalized bandwidth (frequency times pulse width). The spectral characteristic of a PCM waveform establishes the required system bandwidth and indicates how efficiently the bandwidth is being used. Bandwidth efficiency is addressed in detail in Chapter 7. The features that are easily observed in Figure 2.23 are the energy content at low frequency and the bandwidth requirements. Notice that the NRZ and duobinary schemes

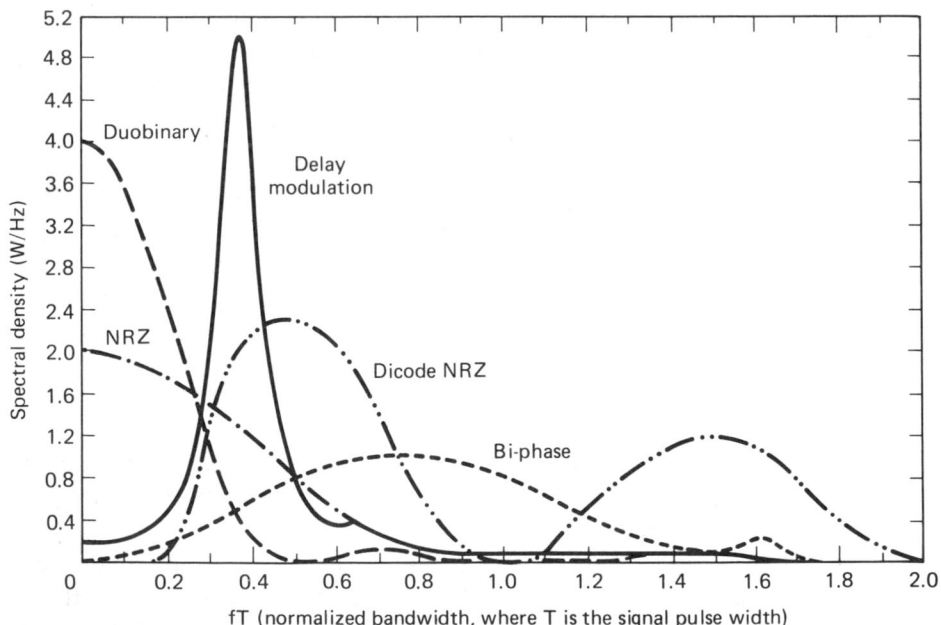

Figure 2.23 Spectral densities of various PCM waveforms.

have large spectral components at low frequency. Notice also that the bi-phase schemes have no energy at dc. However, bi-phase requires a relatively large system bandwidth, as does the dicode scheme. The methods that are particularly bandwidth efficient are the duobinary and delay modulation. Duobinary signaling is treated in Section 2.12.

2.9 DETECTION OF BINARY SIGNALS IN GAUSSIAN NOISE

Once the digital symbols are transformed into electrical waveforms, they can then be transmitted through the channel. During a given signaling interval, T, a binary system will transmit one of two waveforms, denoted $s_1(t)$ and $s_2(t)$. The transmitted signal over a symbol interval $(0, T)$ is represented by

$$s_i(t) = \begin{cases} s_1(t) & 0 \le t \le T & \text{for a binary 1} \\ s_2(t) & 0 \le t \le T & \text{for a binary 0} \end{cases}$$

The signal, $r(t)$, received by the receiver is represented by

$$r(t) = s_i(t) + n(t) \qquad i = 1, 2; \qquad 0 \le t \le T \tag{2.24}$$

where $n(t)$ is a zero-mean additive white Gaussian noise (AWGN) process.

Figure 2.24 highlights the *two separate* steps involved in signal detection. The *first step* consists of reducing the received waveform, $r(t)$ (whether baseband or bandpass), to a *single number*, $z(t = T)$. This operation can be performed by a linear filter followed by a sampler, as shown in block 1 of Figure 2.24, or optimally by a matched filter or correlator, which will be treated in later sections. The initial conditions of the filter or correlator are set to zero just prior to the arrival of each new symbol. At the end of a symbol duration, T, the output of block 1 yields the sample, $z(T)$, sometimes called the *test statistic*. We have assumed that the input noise is a Gaussian random process, and we have stated that the input filter is linear. A linear operation on a Gaussian random process will produce a second Gaussian random process [6]. Thus the filter output noise is Gaussian. If a nonlinear detector is used, the output noise will not be Gaussian

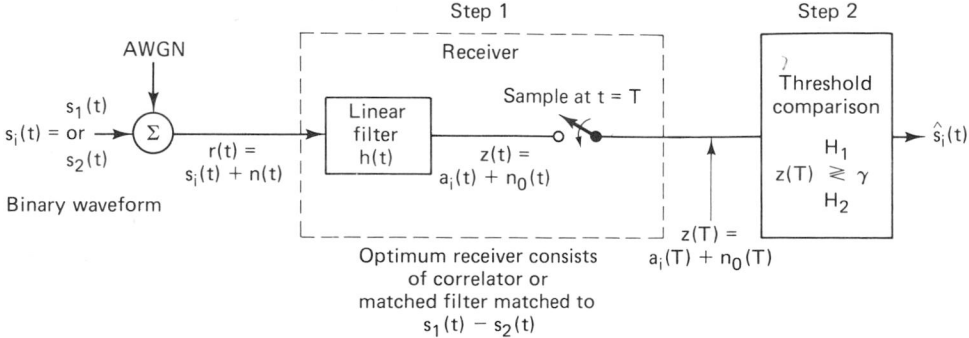

Figure 2.24 Two basic steps in digital signal detection.

and the following analysis will not apply. The output of block 1, sampled at $t = T$, yields

$$z(T) = a_i(T) + n_0(T) \qquad i = 1, 2 \tag{2.25}$$

where $a_i(T)$ is the signal component of $z(T)$ and $n_0(T)$ is the noise component. To shorten the notation, we sometimes write Equation (2.25) as $z = a_i + n_0$. The noise component, n_0, is a zero-mean *Gaussian random variable*, and thus $z(T)$ is a *Gaussian random variable* with a mean of either a_1 or a_2 depending on whether a binary one or binary zero was sent. The probability density function (pdf) of the Gaussian random noise, n_0, can be expressed as

$$p(n_0) = \frac{1}{\sigma_0 \sqrt{2\pi}} \exp\left[-\frac{1}{2} \left(\frac{n_0}{\sigma_0} \right)^2 \right] \tag{2.26}$$

where σ_0^2 is the noise variance. Thus it follows from Equations (2.25) and (2.26) that the conditional probability density functions (pdfs), $p(z|s_1)$ and $p(z|s_2)$ can be expressed as

$$p(z|s_1) = \frac{1}{\sigma_0 \sqrt{2\pi}} \exp\left[-\frac{1}{2} \left(\frac{z - a_1}{\sigma_0} \right)^2 \right] \tag{2.27}$$

$$p(z|s_2) = \frac{1}{\sigma_0 \sqrt{2\pi}} \exp\left[-\frac{1}{2} \left(\frac{z - a_2}{\sigma_0} \right)^2 \right] \tag{2.28}$$

These conditional pdfs are illustrated in Figure 2.25. The rightmost conditional pdf, $p(z|s_1)$, illustrates the probability density of the detector output, $z(T)$, given that $s_1(t)$ was transmitted. Similarly, the leftmost conditional pdf, $p(z|s_2)$, illustrates the probability density of $z(T)$ given that $s_2(t)$ was transmitted. The abscissa, $z(T)$, represents the full range of possible sample output values from block 1 of Figure 2.24.

The *second step* of the signal detection process consists of comparing the test statistic, $z(T)$, to a threshold level, γ, in block 2 of Figure 2.24, in order to estimate which signal, $s_1(t)$ or $s_2(t)$, has been transmitted. The filtering operation in block 1 does not depend on the decision criterion in block 2. Thus the choice of how best to implement block 1 can be independent of the particular decision strategy (choice of the threshold setting, γ).

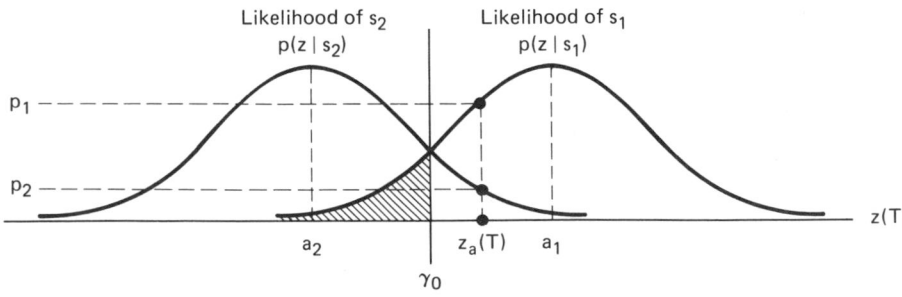

Figure 2.25 Conditional probability density functions: $p(z|s_1)$ and $p(z|s_2)$.

Once a received waveform, $r(t)$, is transformed to a number $z(T)$, the actual shape of the waveform is no longer important; all waveform types that are transformed to the same value of $z(T)$ are identical for detection purposes. We will see in Section 2.9.2 that a *matched filter* receiver in block 1 of Figure 2.24 is one that maps all signals of equal energy into the same point, $z(T)$. Therefore, the *signal energy* (not its shape) is the important parameter in the detection process. Thus the detection analysis for *baseband signals* is the same as that for *bandpass signals*. The final step in block 2 is to make the decision

$$z(T) \underset{H_2}{\overset{H_1}{\gtrless}} \gamma \qquad (2.29)$$

where H_1 and H_2 are the two possible (binary) hypotheses. Choosing H_1 is equivalent to deciding that signal $s_1(t)$ was sent, and choosing H_2 is equivalent to deciding that signal $s_2(t)$ was sent. The inequality relationship indicates that hypothesis H_1 is chosen if $z(T) > \gamma$, and hypothesis H_2 is chosen if $z(T) < \gamma$. If $z(T) = \gamma$, the decision can be an arbitrary one.

2.9.1 Maximum Likelihood Receiver Structure

A popular criterion for choosing the threshold level, γ, for the binary decision is based on minimizing the probability of error. The computation for this *minimum error* value of $\gamma = \gamma_0$ starts with forming an inequality expression between the ratio of conditional probability density functions and the signal a priori probabilities. The conditional density function, $p(z|s_i)$, is also called the *likelihood* of s_i. Thus the formulation as shown below is called the *likelihood ratio test* (see Appendix B).

$$\frac{p(z|s_1)}{p(z|s_2)} \underset{H_2}{\overset{H_1}{\gtrless}} \frac{P(s_2)}{P(s_1)} \qquad (2.30)$$

where $P(s_1)$ and $P(s_2)$ are the a priori probabilities that $s_1(t)$ and $s_2(t)$, respectively, are transmitted, and H_1 and H_2 are the two possible hypotheses. The rule for minimizing the error probability in Equation (2.30) states that we should choose hypothesis H_1 if the ratio of likelihoods is greater than the ratio of a priori probabilities.

It is shown in Section B.3.1 that if $P(s_1) = P(s_2)$, and if the likelihoods, $p(z|s_i)$ ($i = 1, 2$), are symmetrical, the substitution of Equations (2.27) and (2.28) into (2.30) yields

$$z(T) \underset{H_2}{\overset{H_1}{\gtrless}} \frac{a_1 + a_2}{2} = \gamma_0 \qquad (2.31)$$

where a_1 is the signal component of $z(T)$ when $s_1(t)$ is transmitted, and a_2 is the signal component of $z(T)$ when $s_2(t)$ is transmitted. The threshold level, γ_0, represented by $(a_1 + a_2)/2$, is the *optimum threshold* for minimizing the probability of making an incorrect decision for this important special case. This strategy is known as the *minimum error criterion*.

For equally likely signals, the optimum threshold, γ_0, passes through the intersection of the likelihood functions, as shown in Figure 2.25. Thus by following Equation (2.31), the decision stage effectively selects the hypothesis that corresponds to the signal with the *maximum likelihood*. For example, given an arbitrary detector output value, $z_a(T)$, for which there is a nonzero likelihood that $z_a(T)$ belongs to either signal class $s_1(t)$ or $s_2(t)$, one can think of the likelihood test as a comparison of the likelihood values $p(z_a|s_1)$ and $p(z_a|s_2)$. The signal corresponding to the maximum pdf is chosen as the most likely to have been transmitted. In other words, the detector chooses $s_1(t)$ if

$$p(z_a|s_1) > p(z_a|s_2) \tag{2.32}$$

Otherwise, the detector chooses $s_2(t)$. A detector that minimizes the error probability (for the case where the signal classes are equally likely) is also known as a *maximum likelihood detector*.

Figure 2.25 illustrates that Equation (2.32) is just a "common sense" way to make a decision when there exists statistical knowledge of the classes. Given the detector output value, $z_a(T)$, we see in Figure 2.25 that $z_a(T)$ intersects the likelihood of $s_1(t)$ at a value p_1, and it intersects the likelihood of $s_2(t)$ at a value p_2. What is the most reasonable decision for the detector to make? For this example, choosing class $s_1(t)$, which has the greater likelihood, is the most sensible choice. If this was an M-ary instead of a binary example, there would be a total of M likelihood functions representing the M signal classes to which a received signal might belong. The maximum likelihood decision would then be to choose the class that had the greatest likelihood of all M likelihoods. Refer to Appendix B for a review of decision theory fundamentals.

2.9.1.1 Error Probability

For the binary example in Figure 2.25, there are two ways in which errors can occur. An error, e, will occur when $s_1(t)$ is sent, and channel noise results in the receiver output signal, $z(T)$, being less than γ_0. The probability of such an occurrence is

$$P(e|s_1) = P(H_2|s_1) = \int_{-\infty}^{\gamma_0} p(z|s_1)\,dz \tag{2.33}$$

This is illustrated by the shaded area to the left of γ_0 in Figure 2.25. Similarly, an error occurs when $s_2(t)$ is sent, and the channel noise results in $z(T)$ being greater than γ_0. The probability of this occurrence is

$$P(e|s_2) = P(H_1|s_2) = \int_{\gamma_0}^{\infty} p(z|s_2)\,dz \tag{2.34}$$

The probability of an error is the sum of the probabilities of all the ways that an error can occur. For the binary case, we can express the probability of bit error, P_B, as follows:

$$P_B = \sum_{i=1}^{2} P(e, s_i) \tag{2.35}$$

Combining Equations (2.33) to (2.35), we can write

$$P_B = P(e|s_1)P(s_1) + P(e|s_2)P(s_2) \qquad (2.36a)$$

or equivalently,

$$P_B = P(H_2|s_1)P(s_1) + P(H_1|s_2)P(s_2) \qquad (2.36b)$$

That is, given that signal $s_1(t)$ was transmitted, an error results if hypothesis H_2 is chosen; or given that signal $s_2(t)$ was transmitted, an error results if hypothesis H_1 is chosen. For the case where the a priori probabilities are equal, that is, $P(s_1) = P(s_2) = \frac{1}{2}$,

$$P_B = \frac{1}{2}P(H_2|s_1) + \frac{1}{2}P(H_1|s_2) \qquad (2.37)$$

and because of the symmetry of the probability density functions

$$P_B = P(H_2|s_1) = P(H_1|s_2) \qquad (2.38)$$

The probability of a bit error, P_B, is numerically equal to the area under the "tail" of either likelihood function, $p(z|s_1)$ or $p(z|s_2)$, falling on the "incorrect" side of the threshold. We can therefore compute P_B by integrating $p(z|s_1)$ between the limits $-\infty$ and γ_0, or as shown below, by integrating $p(z|s_2)$ between the limits γ_0 and ∞:

$$P_B = \int_{\gamma_0 = (a_1 + a_2)/2}^{\infty} p(z|s_2) \, dz \qquad (2.39)$$

where $\gamma_0 = (a_1 + a_2)/2$ is the optimum threshold from Equation (2.31). Replacing the likelihood $p(z|s_2)$ with its Gaussian equivalent from Equation (2.28), we have

$$P_B = \int_{\gamma_0 = (a_1 + a_2)/2}^{\infty} \frac{1}{\sigma_0 \sqrt{2\pi}} \exp\left[-\frac{1}{2}\left(\frac{z - a_2}{\sigma_0}\right)^2 \right] dz \qquad (2.40)$$

where σ_0^2 is the variance of the noise out of the correlator.

Let $u = (z - a_2)/\sigma_0$. Then $\sigma_0 \, du = dz$ and

$$P_B = \int_{u = (a_1 - a_2)/2\sigma_0}^{u = \infty} \frac{1}{\sqrt{2\pi}} \exp\left(-\frac{u^2}{2}\right) du = Q\left(\frac{a_1 - a_2}{2\sigma_0}\right) \qquad (2.41)$$

where $Q(x)$, called the *complementary error function* or *co-error function*, is a commonly used symbol for the probability under the tail of the Gaussian distribution. It is defined as

$$Q(x) = \frac{1}{\sqrt{2\pi}} \int_x^{\infty} \exp\left(-\frac{u^2}{2}\right) du \qquad (2.42)$$

Note that the co-error function is defined in several ways (see Appendix B); however, all definitions are essentially equivalent. $Q(x)$ cannot be evaluated in closed form. It is presented in tabular form in Table B.1. Good approximations to $Q(x)$ by simpler functions can be found in Reference [7]. One such approximation, valid for $x > 3$, is

$$Q(x) \simeq \frac{1}{x\sqrt{2\pi}} \exp\left(-\frac{x^2}{2}\right) \qquad (2.43)$$

We have optimized (in the sense of minimizing P_B) the threshold level, γ, but have not optimized the filter in block 1 of Figure 2.24; we next consider optimizing this filter by maximizing the argument of $Q(x)$ in Equation (2.41).

2.9.2 The Matched Filter

A matched filter is a linear filter designed to provide the maximum signal-to-noise power ratio at its output for a given transmitted symbol waveform. Consider that a known signal $s(t)$ plus AWGN, $n(t)$, is the input to a linear, time-invariant filter followed by a sampler, as shown in Figure 2.24. At time $t = T$, the receiver output, $z(T)$, consists of a signal component, a_i, and a noise component, n_0. The variance of the output noise (average noise power) is denoted by σ_0^2, so that the ratio of the instantaneous signal power to average noise power, $(S/N)_T$, at time $t = T$, out of the receiver in block 1, is

$$\left(\frac{S}{N}\right)_T = \frac{a_i^2}{\sigma_0^2} \qquad (2.44)$$

We wish to find the filter transfer function, $H_0(f)$, that *maximizes* Equation (2.44). We can express the signal, $a(t)$, at the filter output, in terms of the filter transfer function, $H(f)$ (before optimization), and the Fourier transform of the input signal, as follows:

$$a(t) = \int_{-\infty}^{\infty} H(f)S(f)e^{j2\pi ft}\, df \qquad (2.45)$$

where $S(f)$ is the Fourier transform of the input signal, $s(t)$. If the two-sided power spectral density of the input noise is $N_0/2$ watts/hertz, then using Equations (1.19) and (1.53), we can express the output noise power, σ_0^2, as

$$\sigma_0^2 = \frac{N_0}{2} \int_{-\infty}^{\infty} |H(f)|^2\, df \qquad (2.46)$$

We then combine Equations (2.44) to (2.46) to express $(S/N)_T$, as follows:

$$\left(\frac{S}{N}\right)_T = \frac{\left|\int_{-\infty}^{\infty} H(f)S(f)e^{j2\pi fT}\, df\right|^2}{N_0/2 \int_{-\infty}^{\infty} |H(f)|^2\, df} \qquad (2.47)$$

We next find that value of $H(f) = H_0(f)$ for which the maximum $(S/N)_T$ is achieved, by using *Schwarz's inequality*. One form of the inequality can be stated as

$$\left| \int_{-\infty}^{\infty} f_1(x) f_2(x) \, dx \right|^2 \le \int_{-\infty}^{\infty} |f_1(x)|^2 \, dx \int_{-\infty}^{\infty} |f_2(x)|^2 \, dx \qquad (2.48)$$

The equality holds if $f_1(x) = k f_2^*(x)$, where k is an arbitrary constant and * indicates complex conjugate. If we identify $H(f)$ with $f_1(x)$ and $S(f) \, e^{j2\pi fT}$ with $f_2(x)$, we can write

$$\left| \int_{-\infty}^{\infty} H(f) S(f) e^{j2\pi fT} \, df \right|^2 \le \int_{-\infty}^{\infty} |H(f)|^2 \, df \int_{-\infty}^{\infty} |S(f)|^2 \, df \qquad (2.49)$$

Substituting into Equation (2.47) yields

$$\left(\frac{S}{N} \right)_T \le \frac{2}{N_0} \int_{-\infty}^{\infty} |S(f)|^2 \, df \qquad (2.50)$$

or

$$\max \left(\frac{S}{N} \right)_T = \frac{2E}{N_0} \qquad (2.51)$$

where the energy, E, of the input signal $s(t)$ is

$$E = \int_{-\infty}^{\infty} |S(f)|^2 \, df \qquad (2.52)$$

Thus the maximum output $(S/N)_T$ depends on the input *signal energy* and the power spectral density of the noise, *not on the particular shape* of the waveform that is used.

The equality in Equation (2.51) holds only if the optimum filter transfer function, $H_0(f)$, is employed, such that

$$H(f) = H_0(f) = kS^*(f) e^{-j2\pi fT} \qquad (2.53)$$

or

$$h(t) = \mathscr{F}^{-1}\{kS^*(f) e^{-j2\pi fT}\} \qquad (2.54)$$

Since $s(t)$ is a real-valued signal, we can write from Equations (A.29) and (A.31),

$$h(t) = \begin{cases} ks(T - t) & 0 \le t \le T \\ 0 & \text{elsewhere} \end{cases} \qquad (2.55)$$

Thus the impulse response of a filter that produces the maximum output signal-to-noise ratio is the mirror image of the message signal, $s(t)$, *delayed* by the symbol time duration, T. Note that the delay of T seconds makes Equation (2.55) *causal*; that is, the delay of T seconds makes $h(t)$ a function of positive time in the interval $0 \le t \le T$. Without the delay of T seconds, the response, $s(-t)$, is unrealizable because it describes a response as a function of negative time.

2.9.3 Correlation Realization of the Matched Filter

The term *matched filter* is often used synonymously with *product integrator* or *correlator*. Equation (2.55) and Figure 2.26a illustrate the matched filter's basic property: The impulse response of the filter is a delayed version of the mirror image (rotated on the $t = 0$ axis) of the signal waveform. Therefore, if the signal waveform is $s(t)$, its mirror image is $s(-t)$, and the mirror image delayed by T seconds is $s(T - t)$. The output, $z(t)$, of a causal filter can be described in the time domain as the convolution of a received input waveform, $r(t)$, with the impulse response of the filter (see Section A.5):

$$z(t) = r(t) * h(t) = \int_0^t r(\tau)h(t - \tau) \, d\tau \tag{2.56}$$

Substituting $h(t)$ of Equation (2.55) into $h(t - \tau)$ of Equation (2.56) and arbitrarily setting the constant k equal to unity, we get

$$z(t) = \int_0^t r(\tau)s[T - (t - \tau)] \, d\tau$$
$$= \int_0^t r(\tau)s(T - t + \tau) \, d\tau \tag{2.57}$$

When $t = T$, we can write Equation (2.57) as

$$z(T) = \int_0^T r(\tau)s(\tau) \, d\tau \tag{2.58}$$

The operation of Equation (2.58), the product integration of the received signal, $r(t)$, with a replica of the transmitted waveform, $s(t)$, over one symbol interval is known as the *correlation* of $r(t)$ with $s(t)$. Consider that a received signal, $r(t)$, is correlated with each prototype signal, $s_i(t)$ ($i = 1, \ldots, M$), using a bank of M correlators. The signal $s_i(t)$ whose product integration or correlation with $r(t)$ yields the maximum output $z_i(T)$ is the signal that matches $r(t)$ better than all the other $s_j(t)$, $j \neq i$. We will subsequently use this correlation characteristic for the optimum detection of signals.

2.9.3.1 Comparison of Convolution and Correlation

It is important to note that the correlator output and the matched filter output are the same *only at time $t = T$*. For a sine-wave input, the output of the correlator, $z(t)$, is approximately a linear ramp for $0 \leq t \leq T$. However, the matched filter output is approximately a sine-wave amplitude modulated by a linear ramp for $0 \leq t \leq T$. The comparison is shown in Figure 2.26b. To understand the similarities and differences between a matched filter and a product integrator, one might first ask: What are the similarities between *convolution* as expressed in Equation (2.56) and *correlation* as expressed in Equation (2.58)? With correlation, we simply multiply two functions together and integrate (compute the area under their product curve). We are calculating how closely two waveforms *match each other* in a given time period. With convolution, we sweep (step) two functions past one

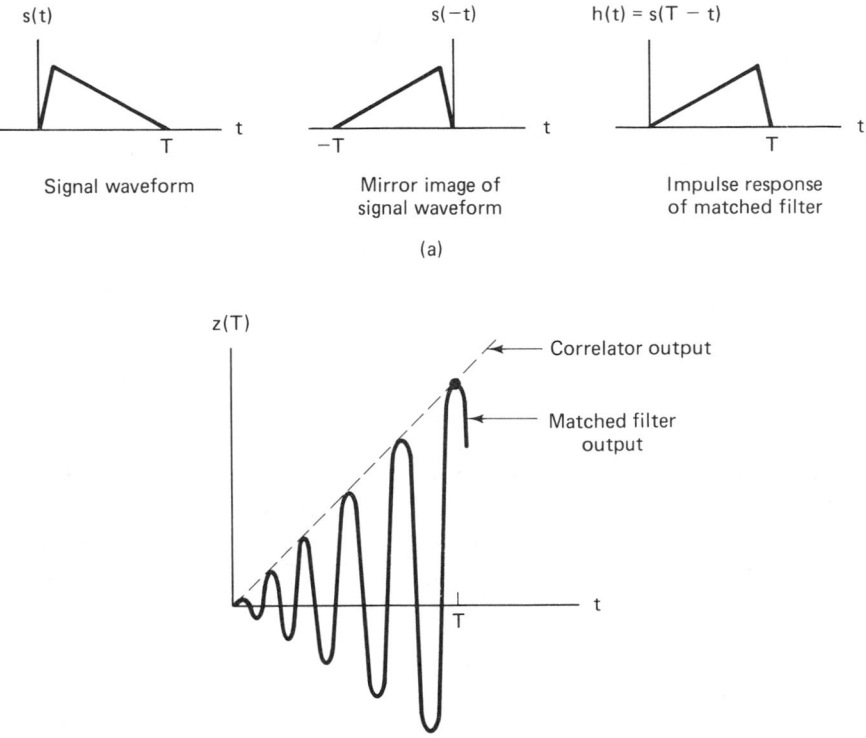

Figure 2.26 Correlator and matched filter. (a) Matched filter characteristic. (b) Comparison of correlator and matched filter outputs.

another and calculate a sequence of correlations (one for each step). The matched filter, used as a demodulator, only utilizes the correlation made at the symbol duration, T. Since the matched filter output and the correlator output are identical at the sampling time $t = T$, the matched filter and correlator functions, pictured in Figure 2.27, are used interchangeably.

2.9.4 Application of the Matched Filter

In Equation (2.41) we found that the optimum decision threshold resulted in $P_B = Q[(a_1 - a_2)/2\sigma_0]$. Finding the optimum threshold alone is not sufficient to optimize the detection process. To minimize P_B, we also need to select an optimum filter to maximize the argument of $Q(x)$. Thus we need to determine the linear filter that maximizes $(a_1 - a_2)/2\sigma_0$, or equivalently, that maximizes

$$\frac{(a_1 - a_2)^2}{\sigma_0^2} \tag{2.59}$$

where $(a_1 - a_2)$ is the difference of the signal components at the filter output, at time $t = T$, and the square of this difference signal is the instantaneous power

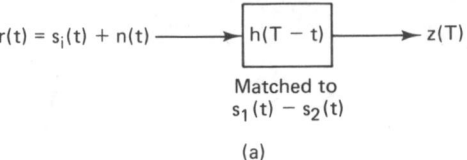

Matched to
$s_1(t) - s_2(t)$

(a)

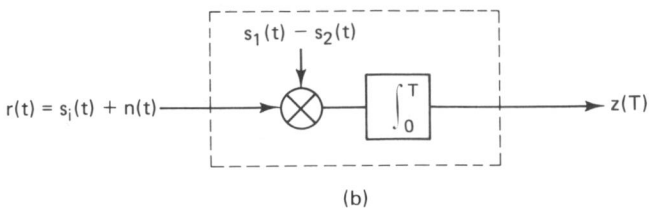

(b)

Figure 2.27 Equivalence of matched filter and correlator. (a) Matched filter. (b) Correlator.

of the difference signal. In Section 2.9.2 we described a filter that maximizes the output signal-to-noise ratio—the matched filter. Consider a filter that is *matched* to the input difference signal $[s_1(t) - s_2(t)]$. From Equations (2.44) and (2.51), the ratio of the instantaneous signal power to average noise power, $(S/N)_T$, at time $t = T$ out of this matched filter can be expressed as

$$\left(\frac{S}{N}\right)_T = \frac{(a_1 - a_2)^2}{\sigma_0^2} = \frac{2E_d}{N_0} \tag{2.60}$$

where $N_0/2$ is the two-sided power spectral density of the noise at the filter input, and E_d is the energy of the difference signal at the filter input:

$$E_d = \int_0^T [s_1(t) - s_2(t)]^2 \, dt \tag{2.61}$$

Thus, using Equations (2.41) and (2.60), we have

$$P_B = Q\left(\sqrt{\frac{E_d}{2N_0}}\right) \tag{2.62}$$

2.9.5 Error Probability Performance of Binary Signaling

2.9.5.1 Unipolar Signaling

Figure 2.28a illustrates an example of a baseband waveform used for unipolar signaling where

$$s_1(t) = A \quad 0 \le t \le T \quad \text{for binary 1}$$
$$s_2(t) = 0 \quad 0 \le t \le T \quad \text{for binary 0} \tag{2.63}$$

where $A > 0$ is the amplitude of signal $s_1(t)$. Assume that the unipolar signal plus white Gaussian noise is present at the input of a matched filter, with sampling time $t = T$. The correlator detector for such a signal type is shown in Figure 2.28b. The correlator multiplies and integrates the incoming signal, $r(t)$, with the

Formatting and Baseband Transmission   Chap. 2

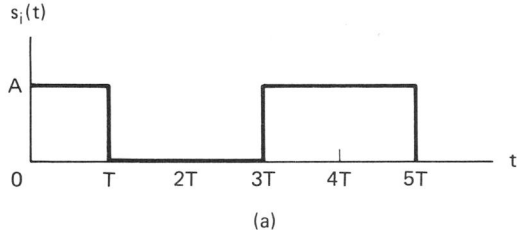

(a)

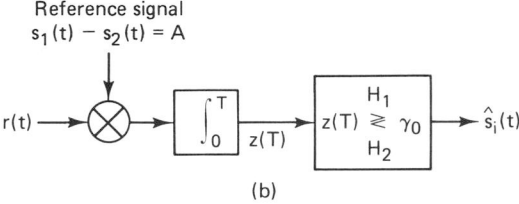

(b)

Figure 2.28 Detection of unipolar baseband signaling. (a) Unipolar signaling example. (b) Correlator detector.

difference of the prototype signals $[s_1(t) - s_2(t)] = A$, and after a symbol duration, T, compares the result, $z(T)$, with the threshold, γ_0. When $r(t) = s_1(t) + n(t)$, the signal component, $a_1(T)$, of $z(T)$ is found, using Equation (2.58), to be

$$a_1(T) = \mathbf{E}\{z(T)\} = \mathbf{E}\left\{ \int_0^T A^2 + An(t)\, dt \right\} = A^2 T$$

where $\mathbf{E}\{\cdot\}$ is the *expected value operator*. This follows since $\mathbf{E}\{n(t)\} = 0$. Similarly, when $r(t) = s_2(t) + n(t)$, then $a_2(T) = 0$. Thus the optimum threshold is $\gamma_0 = (a_1 + a_2)/2 = \frac{1}{2}A^2 T$. If the correlator output, $z(T)$, is greater than γ_0, the signal is declared to be $s_1(t)$; otherwise, it is declared to be $s_2(t)$.

The energy difference signal, from Equation (2.61), is $E_d = A^2 T$. Then the bit error performance at the output is obtained from Equation (2.62) as follows:

$$P_B = Q\left(\sqrt{\frac{E_d}{2N_0}} \right) = Q\left(\sqrt{\frac{A^2 T}{2N_0}} \right) = Q\left(\sqrt{\frac{E_b}{N_0}} \right) \qquad (2.64)$$

where the average energy per bit is $E_b = A^2 T/2$.

2.9.5.2 Bipolar Signaling

Figure (2.29a) illustrates an example of a bipolar baseband waveform, where

$$s_1(t) = +A \qquad 0 \le t \le T \qquad \text{for binary 1}$$
$$s_2(t) = -A \qquad 0 \le t \le T \qquad \text{for binary 0}$$
$$(2.65)$$

Binary waveforms that are the negative of one another, such as the bipolar pair above, where $s_1(t) = -s_2(t)$, are called *antipodal signals*. A correlator receiver for this antipodal type of waveform can be configured as shown in Figure 2.29b. One correlator multiplies and integrates the incoming signal $r(t)$ with the prototype signal, $s_1(t)$; the second correlator multiplies and integrates $r(t)$ with $s_2(t)$. The

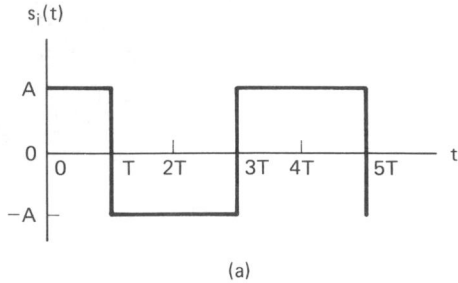

(a)

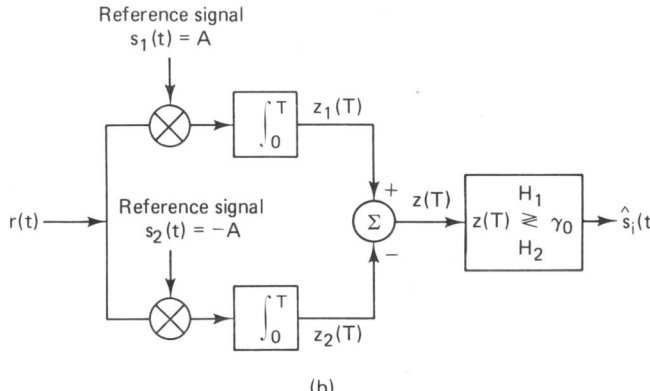

(b)

Figure 2.29 Detection of bipolar baseband signaling. (a) Bipolar signaling example. (b) Correlator detector.

correlator outputs are designated $z_i(T)$ ($i = 1, 2$). The point in the decision space, $z(T)$, is formed from the difference of the correlator outputs, as follows:

$$z(T) = z_1(T) - z_2(T) \tag{2.66}$$

and the decision is made according to Equation (2.31). For antipodal signals, $a_1 = -a_2$; therefore, $\gamma_0 = 0$. Thus if the *test statistic*, $z(T)$, is positive, the signal is declared to be $s_1(t)$, and if it is negative, it is declared to be $s_2(t)$.

The energy difference signal, $E_d = (2A)^2 T$. Then the bit error performance from Equation (2.62) is

$$P_B = Q\left(\sqrt{\frac{2A^2 T}{N_0}}\right) = Q\left(\sqrt{\frac{2E_b}{N_0}}\right) \tag{2.67}$$

where the average energy per bit is $E_b = A^2 T$. Figure 2.30 illustrates curves of P_B versus E_b/N_0 for unipolar and bipolar signaling. In examining the two curves, we can see a 3-dB error performance improvement for bipolar compared to unipolar signaling. This difference could have been predicted by the factor-of-2 difference in the coefficient of E_b in Equation (2.67) compared with Equation (2.64). In Chapter 3 we shall see that the error performance of *bandpass antipodal signaling* (e.g., coherently detected binary phase shift keying) is the same as that for *baseband antipodal signaling* (matched filter reception). Also, we shall see that the error performance of *bandpass orthogonal signaling* (e.g., coherently detected

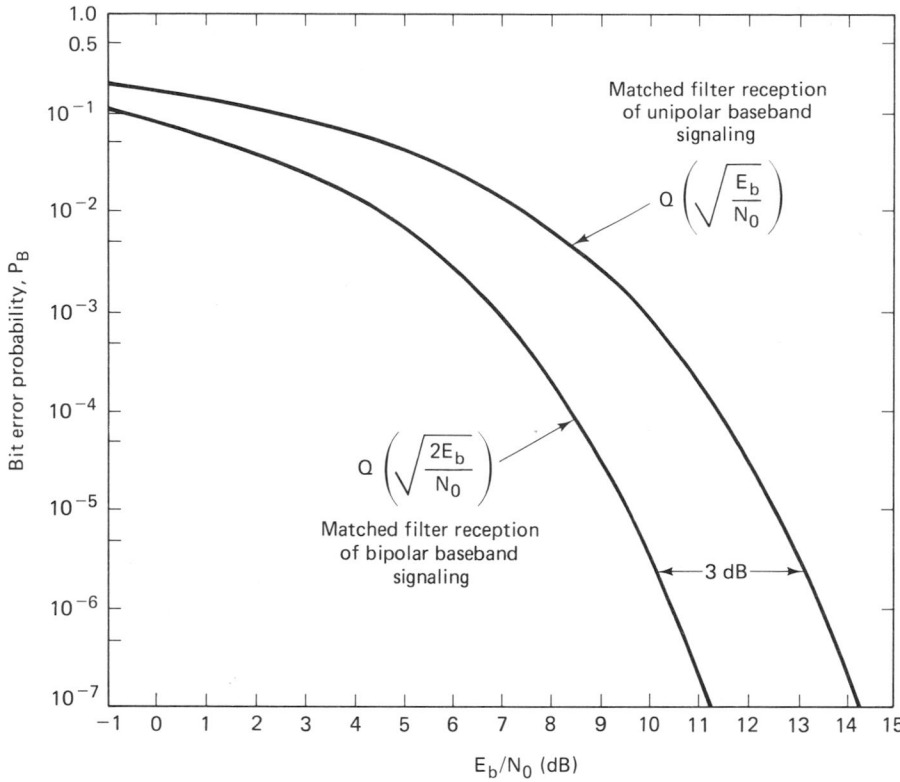

Figure 2.30 Bit error performance of unipolar and bipolar signaling.

frequency shift keying) is the same as that for *baseband unipolar signaling* (matched filter reception).

2.10 MULTILEVEL BASEBAND TRANSMISSION

The system bandwidth required for binary PCM signaling may be very large. What might we do to reduce the required bandwidth? One possibility is to use *multilevel signaling*. Consider a binary PCM bit stream with data rate R bits per second. Instead of transmitting a pulse waveform for each bit, we first partition the data into k-bit groups. We then use $M = 2^k$-level pulses for transmission. Each pulse waveform can now represent a k-bit symbol in a symbol stream of rate R/k symbols per second. Thus multilevel signaling, where $M > 2$, can be used to reduce the number of symbols transmitted per second, or thus to reduce the bandwidth requirements of the channel. Is there a price to be paid for such bandwidth reduction? Of course there is; it is discussed below.

Consider the task that the pulse receiver must perform; it needs to distinguish between the possible levels of each pulse. Can the receiver distinguish among the eight possible levels of each octal pulse in Figure 2.31a as easily as it can distin-

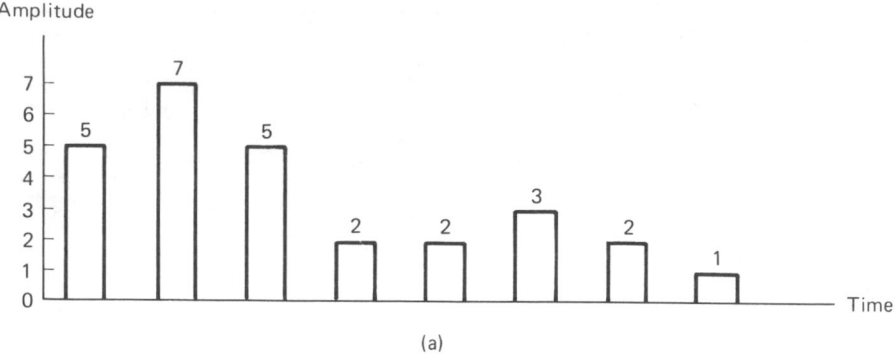

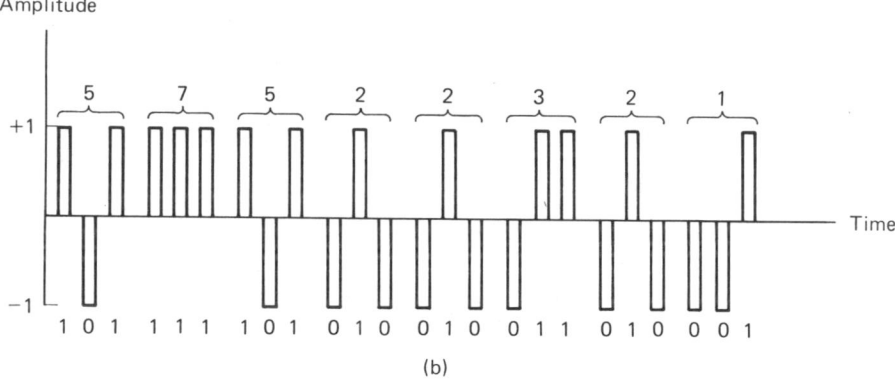

Figure 2.31 Pulse code modulation signaling. (a) Eight-level signaling. (b) Two-level signaling.

guish between the two possible levels of each binary pulse in Figure 2.31b? The transmission of an 8-level (compared to a 2-level) pulse requires a greater amount of energy for equivalent detection performance. (It is the amount of signal energy that determines how reliably a signal will be detected.) For equal average power in the binary and the octal pulses, it is easier to detect the binary pulses because the detector has more signal energy per level for making a binary decision than an 8-level decision. What price does a system designer pay if he or she chooses the transmission waveform to be the easier-to-detect binary PCM, rather than eight-level PCM? The engineer pays the price of needing three times as much system bandwidth for a given data rate, compared to the octal pulses, since each octal pulse must be replaced with three binary pulses (each one-third as wide as the octal pulses). One might ask: Why not use binary pulses with the same pulse duration as the original octal pulses, and suffer the information delay? For some cases this might be appropriate, but for most communication systems, such an increase in delay cannot be tolerated; the six o'clock news *must* be received at six o'clock. In Chapter 7 we examine in detail the trade-off between signal power and system bandwidth.

2.10.1 PCM Word Size

How many bits shall we assign to each analog sample? For digital telephone channels, each speech sample is PCM encoded using 8 bits, yielding 2^8 or 256 levels per sample. The choice of the number of levels, or bits per sample, depends on how much distortion we are willing to tolerate with the PCM format. It is useful to develop a general relationship between the required number of bits per analog sample (the PCM word size) and the allowable quantization distortion. Let the magnitude of the quantization distortion error, $|e|$, be specified not to exceed a fraction, p, of the peak-to-peak analog voltage, V_{pp}, as follows:

$$|e| \leq pV_{pp} \tag{2.68}$$

Since the quantization error can be no larger than $q/2$, where q is the quantile interval, we can write

$$|e|_{max} = \frac{q}{2} = \frac{V_{pp}}{2L} \tag{2.69}$$

where L is the number of quantization levels. Then

$$\frac{V_{pp}}{2L} \leq pV_{pp} \tag{2.70}$$

$$2^\ell = L \geq \frac{1}{2p} \text{ levels} \tag{2.71}$$

$$\ell \geq \log_2 \frac{1}{2p} \text{ bits} \tag{2.72}$$

It is important that we do not confuse the idea of bits per PCM word, denoted by ℓ in Equation (2.72), with the M-level transmission concept of k data bits per symbol. The following example should clarify the distinction.

Example 2.3   Quantization Levels and Multilevel Signaling

The information in an analog waveform, with maximum frequency $f_m = 3$ kHz, is to be transmitted over an M-level PCM system, where the number of pulse levels is $M = 16$. The quantization distortion is specified not to exceed $\pm 1\%$ of the peak-to-peak analog signal.

(a) What is the minimum number of bits/sample, or bits/PCM word, that should be used in this PCM system?

(b) What is the minimum required sampling rate, and what is the resulting bit transmission rate?

(c) What is the PCM pulse or symbol transmission rate?

In this example we are concerned with two types of *levels*: the number of quantization levels for fulfilling the distortion requirement, and the 16 levels of the multilevel PCM pulses.

Solution

(a) Using Equation (2.72), we calculate

$$\ell \geq \log_2 \frac{1}{0.02} = \log_2 50 \simeq 5.6$$

Therefore, use $\ell = 6$ bits/sample to meet the distortion requirement.

(b) Using the Nyquist sampling criterion, the minimum sampling rate $f_s = 2f_m = 6000$ samples/second (samples/s). From part (a), each sample will give rise to a PCM word composed of 6 bits. Therefore, the bit transmission rate $R = \ell f_s = 36,000$ bits/s.

(c) Since multilevel pulses are to be used with $M = 2^k = 16$ levels, $k = \log_2 16 = 4$ bits/symbol. Therefore, the bit stream will be partitioned into groups of 4 bits to form the new 16-level PCM digits, and the resulting symbol transmission rate R_s is $R/k = 36,000/4 = 9000$ symbols/s.

2.11 INTERSYMBOL INTERFERENCE

Figure 2.32a highlights the major filtering aspects of a typical baseband digital system; there are circuit reactances throughout the system—in the transmitter, in the receiver, and in the channel. The pulses at the input might be impulse-like samples, or flat-top samples. In either case, they are low-pass filtered at the transmitter to confine them to some desired bandwidth. Channel reactances can cause amplitude and phase variations that distort the pulses. The receiving filter, called the *equalizing filter*, should be configured to compensate for the distortion

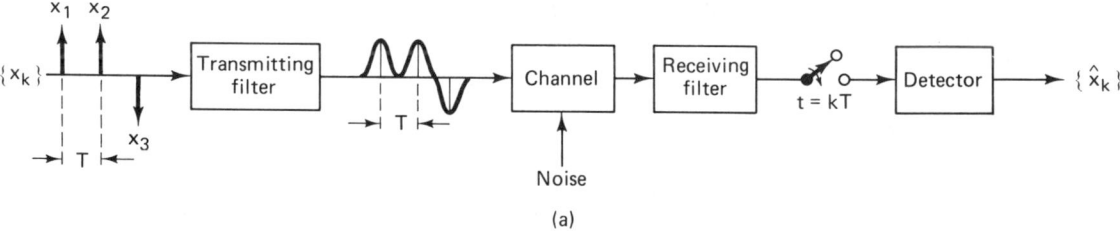

(a)

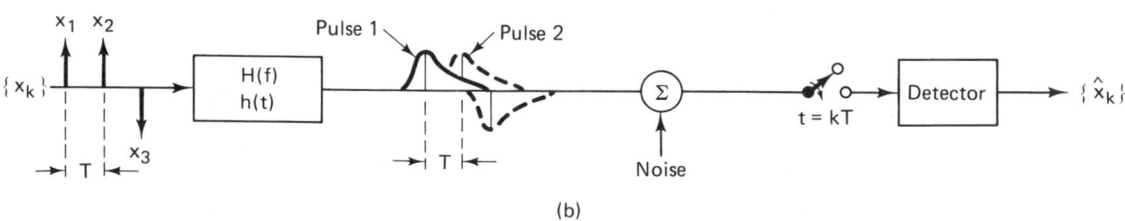

(b)

Figure 2.32 Intersymbol interference in the detection process. (a) Typical baseband digital system. (b) Equivalent model.

Formatting and Baseband Transmission Chap. 2

caused by the transmitter and the channel [8]. In a binary system with a commonly used PCM format, such as NRZ-L, the detector makes symbol decisions by comparing the received bipolar pulses to a threshold; for example, the detector decides that a binary one was sent if the received pulse is positive, and that a binary zero was sent if the received pulse is negative. Figure 2.32b illustrates a convenient model for the system, lumping all the filtering effects into one overall equivalent system transfer function, $H(f)$:

$$H(f) = H_t(f)H_c(f)H_r(f) \tag{2.73}$$

where $H_t(f)$ characterizes the transmitting filter, $H_c(f)$ the filtering within the channel, and $H_r(f)$ the receiving or equalizing filter. The characteristic $H(f)$, then, represents the composite system transfer function due to all of the filtering at various locations throughout the transmitter/channel/receiver chain. Due to the effects of system filtering, the received pulses overlap one another as shown in Figure 2.32b; the tail of one pulse "smears" into adjacent symbol intervals so as to interfere with the detection process; such interference is termed *intersymbol interference* (ISI). Even in the absence of noise, imperfect filtering and system bandwidth constraints lead to ISI. In practice, $H_c(f)$ is usually specified, and the problem remains to determine $H_t(f)$ and $H_r(f)$ such that the ISI of the pulses are minimized at the output of $H_r(f)$.

Nyquist [9] investigated the problem of specifying a received pulse shape so that no ISI occurs at the detector. He showed that the theoretical minimum system bandwidth needed to detect R_s symbols/s, without ISI, is $R_s/2$ hertz. This occurs when the system transfer function, $H(f)$, is made rectangular, as shown in Figure 2.33a. When $H(f)$ is such an ideal filter with bandwidth $1/2T$, its impulse response, the inverse Fourier transform of $H(f)$ (from Table A.1) is $h(t) = $ sinc (t/T), shown in Figure 2.33b. Thus $h(t)$ is the received pulse shape resulting from the application of an impulse at the input of such an ideal system. Nyquist established that if each pulse of a received sequence is of the form $h(t)$, the pulses can be detected without ISI. The bandwidth required to detect $1/T$ such pulses (symbols) per second is equal to $1/2T$; in other words, a system with bandwidth $W = 1/2T = R_s/2$ hertz can support a maximum transmission rate of $2W = 1/T = R_s$ symbols/s (*Nyquist bandwidth constraint*) without ISI. Figure

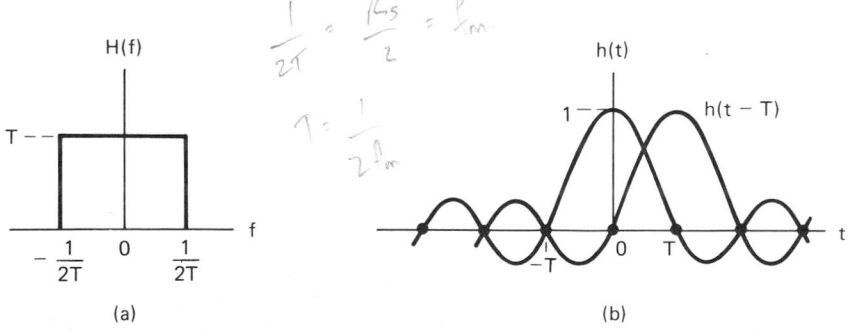

Figure 2.33 Nyquist channels for zero ISI. (a) Rectangular system transfer function $H(f)$. (b) Received pulse shape $h(t) = $ sinc (t/T).

2.33b illustrates how ISI is avoided. The figure shows two successive received pulses, $h(t)$ and $h(t - T)$. Even though $h(t)$ has a long tail, it passes through zero at the instant that $h(t - T)$ is sampled (at $t = T$) and therefore causes no degradation to the detection process. With such an ideal received pulse shape, the maximum possible symbol transmission rate per hertz, called the *symbol-rate packing*, is 2 symbols/s/Hz, without ISI.

What does the Nyquist bandwidth constraint say about the maximum number of *bits*/s/Hz that can be received without ISI? It says nothing about bits, directly. The constraint deals only with pulses or symbols, and the ability to detect their amplitude values without distortion from other pulses. The assignment of how many bits each symbol represents is a separate issue. In theory, each symbol can represent M levels or k bits ($M = 2^k$); as k or M increases in value, so does the complexity of the system. For example, when $k = 6$ bits/symbol, each symbol represents $M = 64$ levels. The number of bits/s/Hz that a system can support is referred to as the *bandwidth efficiency* of the system; this subject is treated separately in Chapter 7.

For most communication systems (with the exception of spread-spectrum systems, covered in Chapter 10), our goal is to reduce the required system bandwidth as much as possible; Nyquist has provided us with a basic limitation to such bandwidth reduction. What would happen if we tried to force a system to operate at smaller bandwidths than the constraint dictates? We would find that restricting the bandwidth would spread the pulses in time; this would degrade the system's error performance, due to the increase in ISI.

2.11.1 Pulse Shaping to Reduce ISI

The Nyquist requirement for a sinc (t/T) received pulse shape is not physically realizable since it dictates a rectangular bandwidth characteristic and an infinite time delay. Also, with such a characteristic, the detection process would be very sensitive to small timing errors. In Figure 2.33b the pulse $h(t)$ has zero value in adjacent pulse times *only* when the sampling is performed at exactly the correct sampling time; timing errors will produce ISI. Therefore, we cannot implement systems using the Nyquist bandwidth; we need to provide some "excess bandwidth" beyond the theoretical minimum. One frequently used system transfer function, $H(f)$, is called the *raised cosine filter*. It can be expressed as

$$H(f) = \begin{cases} 1 & \text{for } |f| < 2W_0 - W \\ \cos^2\left(\dfrac{\pi}{4}\dfrac{|f| + W - 2W_0}{W - W_0}\right) & \text{for } 2W_0 - W < |f| < W \\ 0 & \text{for } |f| > W \end{cases} \quad (2.74)$$

where W is the absolute bandwidth, and $W_0 = 1/2T$ represents the minimum Nyquist bandwidth for the rectangular spectrum and the -6-dB bandwidth (or half-amplitude point) for the raised cosine spectrum. The difference $(W - W_0)$ is termed the *excess bandwidth*; notice that $W = W_0$ for the rectangular spectrum.

The *roll-off factor* is defined to be $r = (W - W_0)/W_0$. It represents the excess bandwidth divided by the filter -6-dB bandwidth (i.e., the fractional excess bandwidth). For a given W_0, r specifies the required excess bandwidth (as a fraction of W_0) and characterizes the steepness of the filter roll-off. The raised cosine characteristic is illustrated in Figure 2.34a for roll-off values of $r = 0$, $r = 0.5$, and $r = 1.0$. The $r = 0$ roll-off is the Nyquist minimum-bandwidth case. Notice that when $r = 1.0$, the required excess bandwidth is 100%; a system with such an overall spectral characteristic can provide a symbol rate of R_s symbols/s using a bandwidth of R_s hertz (twice the Nyquist bandwidth), thus yielding a symbol-

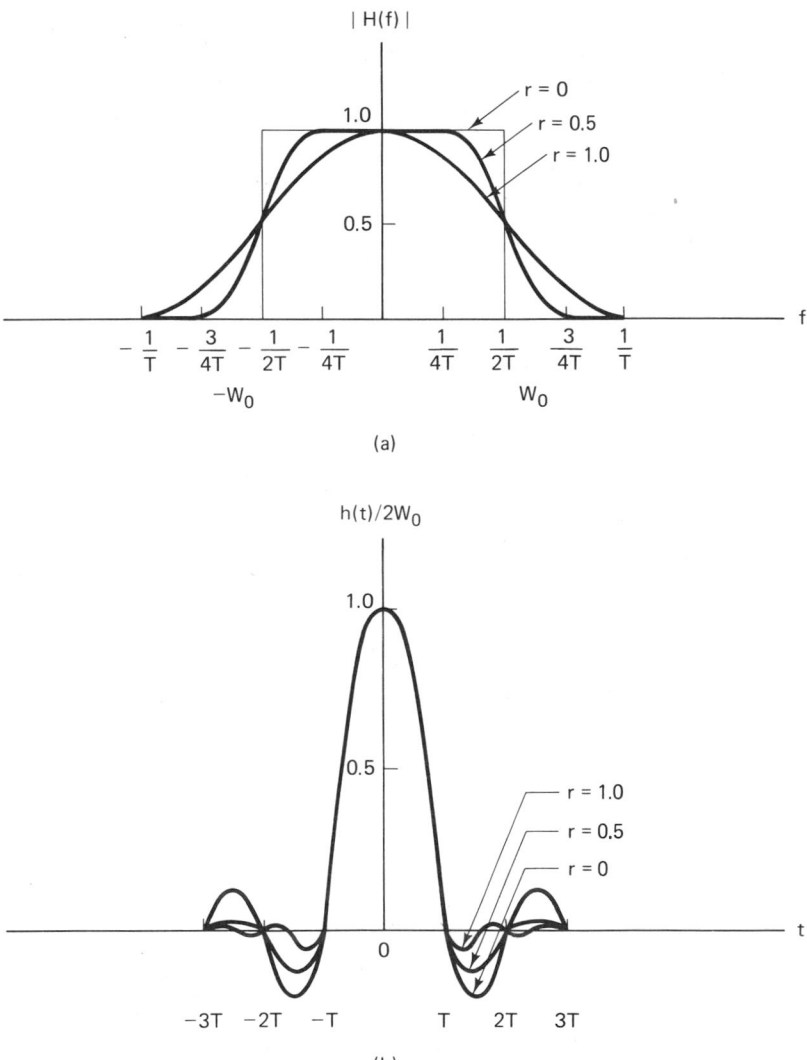

Figure 2.34 Raised cosine filter characteristics. (a) System transfer function. (b) System impulse response.

rate packing of 1 symbol/s/Hz. The corresponding impulse response for the $H(f)$ of Equation (2.74) is

$$h(t) = 2W_0(\text{sinc } 2W_0 t) \frac{\cos [2\pi(W - W_0)t]}{1 - [4(W - W_0)t]^2} \qquad (2.75)$$

The impulse response is shown in Figure 2.34b for $r = 0$, $r = 0.5$, and $r = 1.0$.

Recall that for zero ISI, we shall choose the system received pulse shape to be equal to $h(t)$; we can only do this approximately, since strictly speaking, the raised cosine pulse spectrum is not precisely physically realizable. A realizable frequency characteristic must have a time response that is zero prior to the pulse turn-on time, which is not the case for the family of raised cosine characteristics. These unrealizable filters are *noncausal* (the filter impulse response begins at time $t = -\infty$). However, a delayed version of $h(t)$, say $h(t - t_0)$, may be approximately generated by real filters if the delay t_0 is chosen such that $h(t - t_0) \simeq 0$, for $t < 0$. Notice in Figure 2.34b that timing errors will still result in some ISI degradation for $r = 1$. However, the problem is not as serious as it is for $r = 0$, because the tails of the $h(t)$ waveform are of much smaller amplitude for $r = 1$ than they are for $r = 0$.

The Nyquist bandwidth constraint states that the theoretical minimum required system bandwidth, W, for a symbol rate of R_s symbols/s without ISI, is $R_s/2$ hertz. A more general relationship between required bandwidth and symbol transmission rate involves the filter roll-off factor r, and can be stated as

$$W = \tfrac{1}{2}(1 + r)R_s \qquad \qquad W_0 = 1/2T \qquad (2.76)$$

Thus with $r = 0$, Equation (2.76) describes the required bandwidth for ideal rectangular filtering, also referred to as *Nyquist filtering*. Bandpass-modulated signals (baseband signals that have been shifted in frequency), such as amplitude shift keying (ASK) and phase shift keying (PSK), require twice the transmission bandwidth of the equivalent baseband signals (see Section 1.7.1). Such frequency-translated signals, occupying twice their baseband bandwidth, are often called double-sideband (DSB) signals. Therefore, for ASK- and PSK-modulated signals, the relationship between the required DSB bandwidth, W_{DSB}, and the symbol transmission rate, R_s, is

$$W_{\text{DSB}} = (1 + r)R_s \qquad (2.77)$$

Example 2.4   Bandwidth Requirements

(a) Find the minimum required bandwidth for the baseband transmission of a four-level PCM pulse sequence having a data rate of $R = 2400$ bits/s if the system transfer characteristic consists of a raised cosine spectrum with 100% excess bandwidth ($r = 1$).

(b) The same PCM sequence is modulated onto a carrier wave, so that the baseband spectrum is shifted and centered at frequency f_0. Find the minimum required DSB bandwidth for transmitting the modulated PCM sequence. Assume that the system transfer characteristic is the same as in part (a).

Solution

(a) $M = 2^k$; since $M = 4$ levels, then $k = 2$.

$$\text{Symbol or pulse rate } R_s = \frac{R}{k} = \frac{2400}{2} = 1200 \text{ symbols/s}$$

$$\text{Minimum bandwidth } W = \tfrac{1}{2}(1 + r)R_s = \tfrac{1}{2}(2)(1200) = 1200 \text{ Hz}$$

Figure 2.35a illustrates the baseband PCM received pulse in the time domain—an approximation to the $h(t)$ in Equation (2.75). Figure 2.35b illustrates the Fourier transform of $h(t)$—the raised cosine spectrum. Notice that the required bandwidth, W, starts at zero frequency and extends to $f = 1/T$; it is twice the size of the Nyquist theoretical minimum bandwidth.

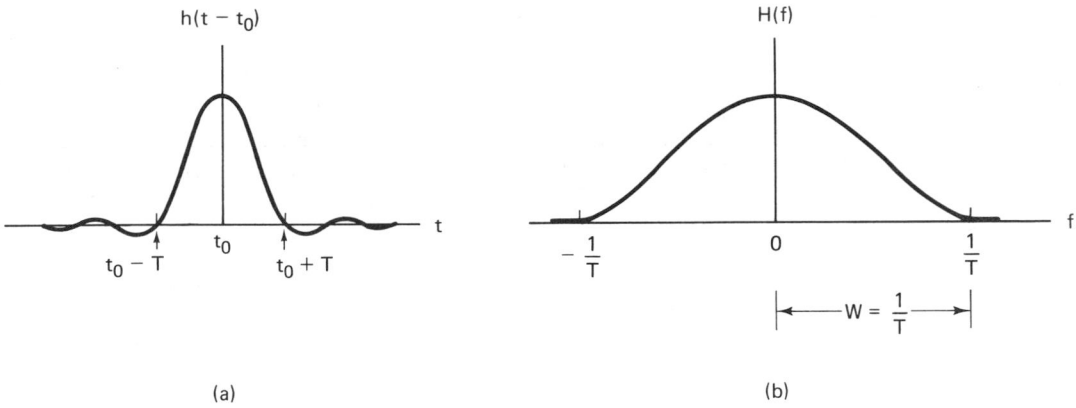

(a)   (b)

Figure 2.35   (a) Shaped pulse. (b) Baseband raised cosine spectrum.

(b) As in part (a),

$$R_s = 1200 \text{ symbols/s}$$

$$W_{\text{DSB}} = (1 + r)R_s = 2(1200) = 2400 \text{ Hz}$$

Figure 2.36a illustrates the modulated PCM received pulse. This waveform can be viewed as the product of a high-frequency sinusoidal carrier wave and a waveform with the pulse shape of Figure 2.35a. The single-sided spectral plot in Figure 2.36b illustrates that the modulated bandwidth, W_{DSB}, is

$$W_{\text{DSB}} = \left(f_0 + \frac{1}{T}\right) - \left(f_0 - \frac{1}{T}\right) = \frac{2}{T}$$

When the spectrum of Figure 2.35b is shifted up in frequency, the negative and positive halves of the baseband spectrum are shifted up in frequency, thereby doubling the required transmission bandwidth. As the name implies, the DSB signal has two sidebands: the upper sideband (USB), derived from the baseband positive half, and the lower sideband (LSB), derived from the baseband negative half.

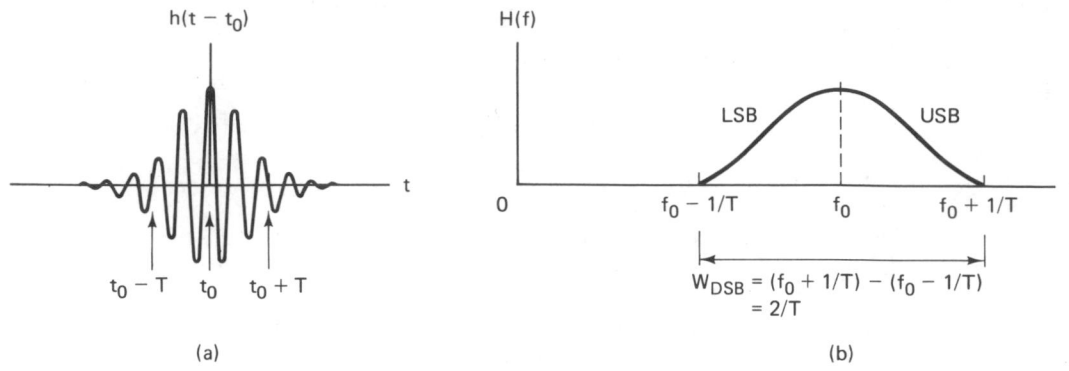

Figure 2.36 (a) Modulated shaped pulse. (b) DSB-modulated raised cosine spectrum.

Example 2.5 Digital Telephone Circuits

Compare the system bandwidth requirements for a 3-kHz analog telephone voice circuit versus a PCM voice circuit. Assume that the sampling rate for the analog-to-digital (A/D) conversion is 8000 samples/s. Also, assume that each voice sample is quantized to one of 256 levels (8-bit quantization).

Solution

The result of the sampling and quantization process yields a PAM signal such that each pulse (symbol) has one of 256 different levels. From Equation (2.76) we can write that the required system bandwidth (without ISI) for R_s symbols/s is

$$W \geq \frac{R_s}{2} \qquad \text{hertz}$$

where the equality sign holds true only for Nyquist filtering. For binary PCM, having $L = 256$ levels, each sample is converted to $\ell = \log_2 L = 8$ bits. Therefore, the system bandwidth required to transmit voice using PCM with 8-bit words is

$$W_{\text{PCM}} \geq (\log_2 L)\, \frac{R_s}{2} \qquad \text{hertz}$$

$$\geq \tfrac{1}{2}(8 \text{ bits/symbol})(8000 \text{ symbols/s}) = 32 \text{ kHz}$$

The 3-kHz analog voice circuit will generally require approximately 4 kHz of bandwidth (including some bandwidth separation between channels, called *guard bands*). Therefore, the PCM format using 8-bit quantization requires *at least* eight times the bandwidth required by the analog format.

2.11.2 Equalization

In practical systems, the frequency response of the channel is not known with sufficient precision to allow for a receiver design that will compensate for the intersymbol interference (ISI) for all time. In practice, the filter for handling ISI at the receiver contains various parameters that are adjusted on the basis of measurements of the channel characteristics. The process of thus correcting the channel-induced distortion is called *equalization*. A *transversal filter*—a delay line with T-second taps (where T is the symbol duration)—is a common choice for the *equalizer filter*. The outputs of the taps are amplified, summed, and fed to a decision device. The tap coefficients, c_n, are set to subtract the effects of interference from symbols that are adjacent in time to the desired symbol. Consider that there are $(2N + 1)$ taps with coefficients $c_{-N}, c_{-N+1}, \ldots, c_N$ as shown in Figure 2.37. Output samples, $\{y_k\}$, of the equalizer are then expressed in terms of the input samples, $\{x_k\}$, and tap coefficients as

$$y_k = \sum_{n=-N}^{N} c_n x_{k-n} \qquad k = -2N, \ldots, 2N \tag{2.78}$$

By defining the matrices $\mathbf{y}$, $\mathbf{c}$, and $\mathbf{x}$ as

$$\mathbf{y} = \begin{bmatrix} y_{-2N} \\ \vdots \\ y_0 \\ \vdots \\ y_{2N} \end{bmatrix} \qquad \mathbf{c} = \begin{bmatrix} c_{-N} \\ \vdots \\ c_0 \\ \vdots \\ c_N \end{bmatrix} \tag{2.79}$$

$$\mathbf{x} = \begin{bmatrix} x_{-N} & 0 & 0 & \cdots & 0 & 0 \\ x_{-N+1} & x_{-N} & 0 & \cdots & \cdots & \cdots \\ \vdots & & & & \vdots & \vdots \\ x_N & x_{N-1} & x_{N-2} & \cdots & x_{-N+1} & x_{-N} \\ \vdots & & & & \vdots & \vdots \\ 0 & 0 & 0 & \cdots & x_N & x_{N-1} \\ 0 & 0 & 0 & \cdots & 0 & x_N \end{bmatrix} \tag{2.80}$$

we can simplify the computation for $\{y_k\}$ as follows:

$$\mathbf{y} = \mathbf{xc} \tag{2.81}$$

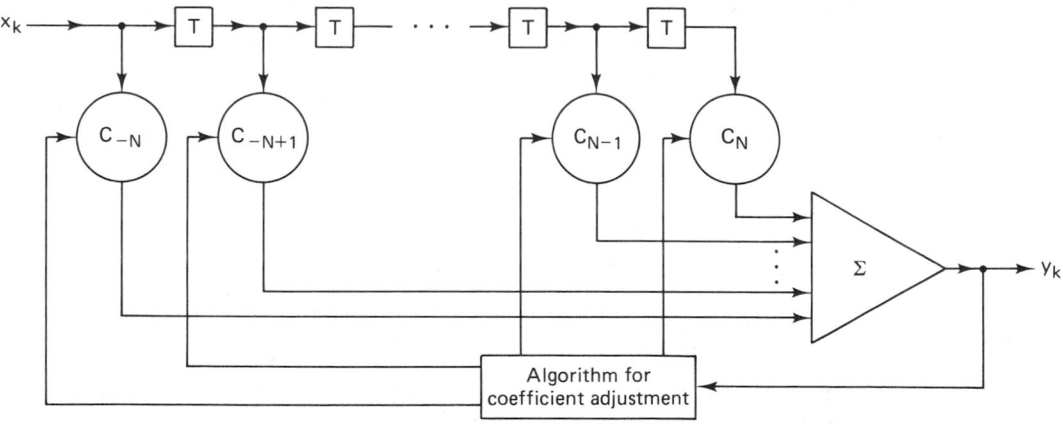

Figure 2.37 Transversal filter.

The criterion for selecting the c_n coefficients is typically based on the minimization of either peak distortion or mean-square distortion. Minimizing peak distortion can be accomplished by selecting the c_n coefficients so that the equalizer output is forced to zero at N sample points on either side of the desired pulse. That is,

$$y_k = \begin{cases} 1 & \text{for } k = 0 \\ 0 & \text{for } k = \pm 1, \pm 2, \ldots, \pm N \end{cases} \quad (2.82)$$

We then solve for c_n by combining Equations (2.79) to (2.81) and solving $2N + 1$ simultaneous equations. Minimizing the mean-square distortion similarly results in $2N + 1$ simultaneous equations.

There are two general types of automatic equalization. The first, *preset equalization*, transmits a training sequence that is compared at the receiver with a locally generated sequence. The differences between the two sequences are used to set the coefficients c_n. With the second method, *adaptive equalization*, the coefficients are continually and automatically adjusted directly from the transmitted data. A disadvantage of preset equalization is that it requires an initial training session, which must be repeated after any break in transmission. Also, a time-varying channel can degrade in ISI since the coefficients are fixed. Adaptive equalization can perform well if the channel error performance is satisfactory. However, if the error performance is poor, received channel errors may not allow the algorithm to converge. A common solution employs preset equalization initially to provide good channel error performance; once normal transmission begins, the system switches to an adaptive algorithm. A significant amount of research and development has taken place in the area of equalization during the past two decades [8, 10, 11].

2.12 PARTIAL RESPONSE SIGNALING

In 1963, Adam Lender [12, 13] showed that it is possible to transmit $2W$ symbols/s with zero ISI, using the theoretical minimum bandwidth of W hertz, without infinitely sharp filters. Lender used a technique called *duobinary signaling*, also referred to by the names *partial response signaling* and *correlative coding*. The basic idea behind the duobinary technique is to introduce some controlled amount of ISI into the data stream rather than trying to eliminate it completely. By introducing correlated interference between the pulses, and by changing the detection procedure, Lender, in effect, "canceled out" the interference at the detector, and thereby achieved the ideal symbol-rate packing of 2 symbols/s/Hz, an amount that had been considered unrealizable.

2.12.1 Duobinary Signaling

To understand how duobinary signaling introduces controlled ISI, let us look at a model of the process. We can think of the duobinary coding operation as if it were implemented as shown in Figure 2.38. Assume that a sequence of binary

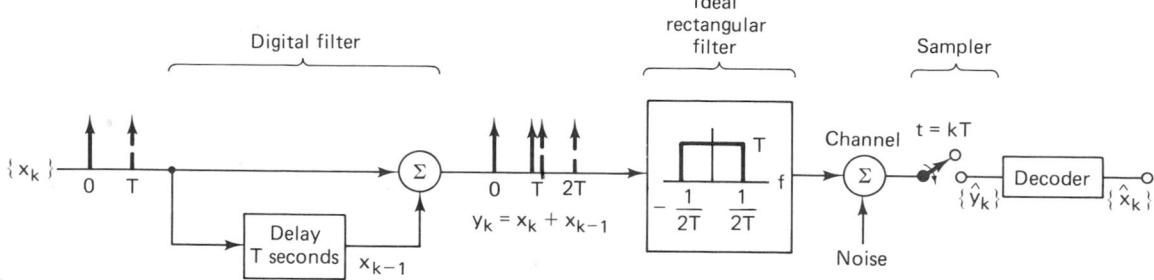

Figure 2.38 Duobinary signaling.

symbols $\{x_k\}$ is to be transmitted at the rate of R symbols/s over a system having an ideal rectangular spectrum of bandwidth $W = R/2 = 1/2T$ hertz. You might ask: How is this rectangular spectrum, in Figure 2.38, different from the unrealizable Nyquist characteristic? It has the same ideal characteristic; but we are not trying to implement the ideal rectangular filter. It is only the part of our equivalent model that is used for developing a filter that is easier to approximate. Before being shaped by the ideal filter, the pulses pass through a simple digital filter, as shown in the figure. The digital filter incorporates a one-digit delay; to each incoming pulse, the filter adds the value of the previous pulse. In other words, for every pulse into the digital filter, we get the summation of two pulses out. Each pulse of the sequence $\{y_k\}$ out of the digital filter can be expressed as

$$y_k = x_k + x_{k-1} \tag{2.83}$$

Hence the $\{y_k\}$ amplitudes are not independent; each y_k digit carries with it the *memory* of the prior digit. The ISI introduced to each y_k digit comes only from the preceding x_{k-1} digit. This correlation between the pulse amplitudes of $\{y_k\}$ can be thought of as the controlled ISI introduced by the duobinary coding. Controlled interference is the essence of this novel technique, because at the detector, such controlled interference can be removed as easily as it was added. The $\{y_k\}$ sequence is followed by the ideal Nyquist filter that does not introduce any ISI. At the receiver sampler, in Figure 2.38, we would expect to recover the sequence $\{y_k\}$, exactly in the absence of noise. Since all systems experience noise contamination, we shall refer to the *received* $\{y_k\}$ as the estimate of $\{y_k\}$ and denote it $\{\hat{y}_k\}$. Removing the controlled interference with the duobinary decoder yields an estimate of $\{x_k\}$ which we shall denote as $\{\hat{x}_k\}$.

2.12.2 Duobinary Decoding

If the binary digit x_k is equal to ± 1, then using Equation (2.83), y_k has one of three possible values: $+2$, 0, or -2. The duobinary code results in a three-level output: in general for M-ary transmission, partial response signaling results in $2M - 1$ output levels. The decoding procedure involves the inverse of the coding procedure, namely, subtracting the x_{k-1} decision from the y_k digit. Consider the following coding/decoding example.

Example 2.6   Duobinary Coding and Decoding

Use Equation (2.83) to demonstrate duobinary coding and decoding for the following sequence: $\{x_k\} = 0\ 0\ 1\ 0\ 1\ 1\ 0$. Consider the first bit of the sequence to be a startup digit, not part of the data.

Solution

Binary digit sequence $\{x_k\}$:	0	0	1	0	1	1	0
Bipolar amplitudes $\{x_k\}$:	-1	-1	$+1$	-1	$+1$	$+1$	-1
Coding rule: $y_k = x_k + x_{k-1}$:		-2	0	0	0	2	0

Decoding decision rule: If $\hat{y}_k = \quad 2$, decide that $\hat{x}_k = +1$ (or binary one)

If $\hat{y}_k = -2$, decide that $\hat{x}_k = -1$ (or binary zero).

If $\hat{y}_k = \quad 0$, decide opposite of the previous decision.

Decoded bipolar sequence $\{\hat{x}_k\}$:	-1	$+1$	-1	$+1$	$+1$	-1
Decoded binary sequence $\{\hat{x}_k\}$:	0	1	0	1	1	0

The decision rule simply implements the subtraction of each $\hat{x}_{k-1}$ decision from each $\hat{y}_k$. One drawback of this detection technique is that once an error is made, it tends to propagate, causing further errors, since present decisions depend on prior decisions. A means of avoiding this error propagation is known as *precoding*.

2.12.3 Precoding

Precoding is accomplished by first differentially encoding the $\{x_k\}$ binary sequence into a new $\{w_k\}$ binary sequence as follows:

$$w_k = x_k \oplus w_{k-1} \qquad (2.84)$$

where the symbol $\oplus$ represents modulo-2 addition (equivalent to the logical *exclusive-or* operation) of the binary digits. The rules of modulo-2 addition are as follows:

$$0 \oplus 0 = 0$$
$$0 \oplus 1 = 1$$
$$1 \oplus 0 = 1$$
$$1 \oplus 1 = 0$$

The $\{w_k\}$ binary sequence is then converted to a bipolar pulse sequence, and the coding operation proceeds in the same way as it did in Example 2.6. However, with precoding, the detection process is quite different from the detection of ordinary duobinary, as shown below in Example 2.7. The precoding model is shown in Figure 2.39; in this figure it is implicit that the modulo-2 addition producing the precoded $\{w_k\}$ sequence is performed on the *binary* digits, while the digital filtering producing the $\{y_k\}$ sequence is performed on the *bipolar* pulses.

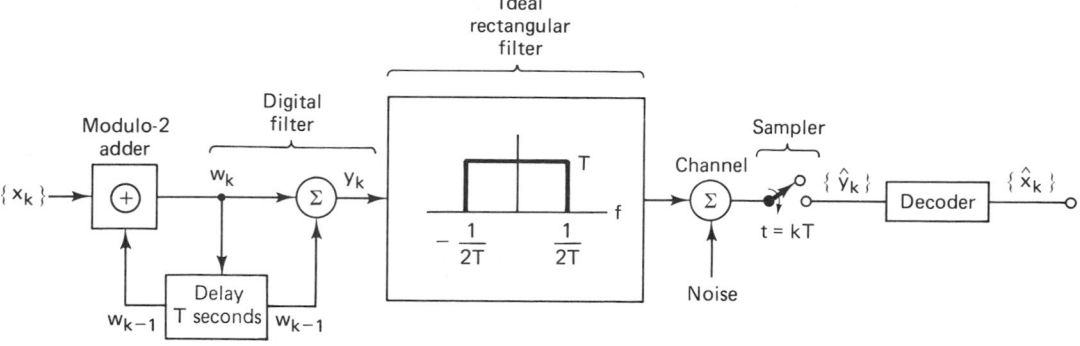

Figure 2.39 Precoded duobinary signaling.

Example 2.7 Duobinary Precoding

Illustrate the duobinary coding and decoding rules when using the differential precoding of Equation (2.84). Assume the same $\{x_k\}$ sequence as that given in Example 2.6.

Solution

Binary digit sequence $\{x_k\}$:			0	0	1	0	1	1	0
Precoded sequence $w_k = x_k \oplus w_{k-1}$:			0	0	1	1	0	1	1
Bipolar sequence $\{w_k\}$:			-1	-1	$+1$	$+1$	-1	$+1$	$+1$
Coding rule: $y_k = w_k + w_{k-1}$:				-2	0	$+2$	0	0	$+2$

Decoding decision rule: If $\hat{y}_k = \pm 2$, decide that $\hat{x}_k = $ binary zero.

If $\hat{y}_k = \quad 0$, decide that $\hat{x}_k = $ binary one.

Decoded binary sequence $\{\hat{x}_k\}$:				0	1	0	1	1	0

The differential precoding enables us to decode the $\{\hat{y}_k\}$ sequence by making a decision on each received sample singly, without resorting to prior decisions which could be in error. The major advantage is that in the event of a digit error due to noise, such an error does not propagate to other digits. Notice that the first bit in the differentially precoded binary sequence $\{w_k\}$ is an arbitrary choice. If the startup bit in $\{w_k\}$ had been chosen to be a binary one instead of a binary zero, the decoded result would have been the same.

2.12.4 Duobinary Equivalent Transfer Function

In Section 2.12.1 we described the duobinary transfer function as a digital filter incorporating a one-digit delay, followed by an ideal rectangular transfer function. Let us now examine an equivalent model. The Fourier transform of a delay can be described as $e^{-j2\pi fT}$ (see Section A.3.1); therefore, the input digital filter of

Figure 2.38 can be characterized with the frequency transfer function, $H_1(f)$, as follows:

$$H_1(f) = 1 + e^{-j2\pi fT} \qquad (2.85)$$

The transfer function of the ideal rectangular filter, designated $H_2(f)$, is shown below.

$$H_2(f) = \begin{cases} T & \text{for } |f| < \dfrac{1}{2T} \\ 0 & \text{elsewhere} \end{cases} \qquad (2.86)$$

The overall equivalent transfer function $H_e(f)$, of the digital filter cascaded with the ideal rectangular filter is then given by

$$
\begin{aligned}
H_e(f) &= H_1(f)H_2(f) \qquad \text{for } |f| < \frac{1}{2T} \\
&= (1 + e^{-j2\pi fT})T \qquad (2.87) \\
&= T(e^{j\pi fT} + e^{-j\pi fT})e^{-j\pi fT}
\end{aligned}
$$

$$|H_e(f)| = \begin{cases} 2T \cos \pi fT & \text{for } |f| < \dfrac{1}{2T} \\ 0 & \text{elsewhere} \end{cases} \qquad (2.88)$$

Thus $H_e(f)$, the composite transfer function for the cascaded digital and rectangular filters, has a gradual roll-off to the band edge, as can be seen in Figure 2.40a. The transfer function can be approximated by using realizable analog filtering; a separate digital filter is not needed. The duobinary equivalent $H_e(f)$ is called a *cosine filter* [14] (not to be confused with the raised cosine filter described in Section 2.11.1). The corresponding impulse response, $h_e(t)$, found by taking the inverse Fourier transform of $H_e(f)$ in Equation (2.87), is

$$h_e(t) = \text{sinc}\left(\frac{t}{T}\right) + \text{sinc}\left(\frac{t - T}{T}\right) \qquad (2.89)$$

and is plotted in Figure 2.40b. For every impulse, $\delta(t)$, at the input of Figure 2.38, the output is $h_e(t)$ with an appropriate polarity. Notice that there are only two nonzero samples, at T-second intervals, giving rise to controlled ISI from the adjacent bit. The introduced ISI is eliminated by use of the decoding procedure discussed in Section 2.12.2. Although the cosine filter is noncausal and therefore nonrealizable, it can be easily approximated. The implementation of the precoded duobinary technique described in Section 2.12.3 can be accomplished by first differentially encoding the binary sequence $\{x_k\}$ into the sequence $\{w_k\}$ (see Example 2.7). The pulse sequence $\{w_k\}$ is then filtered by the equivalent cosine characteristic described in Equation (2.88).

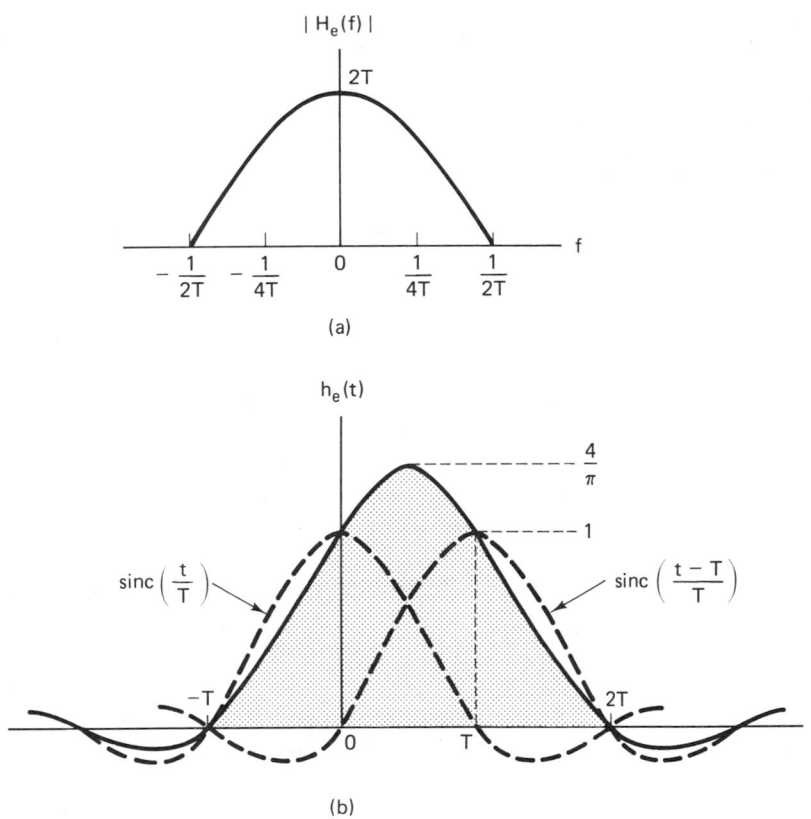

Figure 2.40 Duobinary transfer function and pulse shape. (a) Cosine filter. (b) Impulse response of the cosine filter.

2.12.5 Comparison of Binary with Duobinary Signaling

The duobinary technique introduces correlation between pulse amplitudes, whereas the more restrictive Nyquist criterion assumes that the transmitted pulse amplitudes are independent of one another. We have shown that duobinary signaling can exploit this introduced correlation to achieve zero ISI signal transmission, using a smaller system bandwidth than is otherwise possible. Do we get this performance improvement without paying a price? Such is hardly ever the case with engineering design options; there is almost always a trade-off involved. We saw that duobinary coding requires three levels, compared to the usual two levels for binary coding. Recall our discussion in Section 2.10, where we compared the performance and the required signal power for making eight-level PCM decisions versus two-level PCM decisions. For a fixed amount of signal power, the ease of making reliable decisions is inversely related to the number of levels that must be distinguished in each waveform. Therefore, it should be no surprise that although duobinary signaling accomplishes the zero ISI requirement with minimum bandwidth, duobinary also requires more power than binary signaling, for equivalent performance against noise. For a given probability of bit error (P_B), duobinary signaling requires approximately 2.5 dB greater SNR than binary signaling, while using only $1/(1 + r)$ the bandwidth that binary signaling requires [13], where r is the filter roll-off.

2.12.6 Polybinary Signaling

Duobinary signaling can be extended to more than three digits or levels, resulting in greater bandwidth efficiency; such systems are called *polybinary* [13, 15]. Consider that a binary message with two signaling levels is transformed into a signal with j signaling levels, numbered consecutively from zero to ($j - 1$). The transformation from binary to polybinary takes place in two steps. First, the original sequence $\{x_k\}$, consisting of binary ones and zeros, is converted into another binary sequence $\{y_k\}$, as follows: The present binary digit of sequence $\{y_k\}$ is formed from the modulo-2 addition of the ($j - 2$) immediately preceding digits of sequence $\{y_k\}$ and the present digit x_k. For example, let

$$y_k = x_k \oplus y_{k-1} \oplus y_{k-2} \oplus y_{k-3} \tag{2.90}$$

Here x_k represents the input binary digit and y_k the kth encoded digit. Since the expression involves ($j - 2$) = 3 bits preceding y_k, there are $j = 5$ signaling levels. Next, the binary sequence $\{y_k\}$ is transformed into a polybinary pulse train $\{z_k\}$ by adding *algebraically* the present bit of sequence $\{y_k\}$ to the ($j - 2$) preceding bits of $\{y_k\}$. Therefore, z_k modulo-2 = x_k, and the binary elements one and zero are mapped into even- and odd-valued pulses in the sequence $\{z_k\}$. Note that each digit in $\{z_k\}$ can be independently detected despite the strong correlation between bits. The primary advantage of such a signaling scheme is the redistribution of the spectral density of the original sequence $\{x_k\}$, so as to favor the low frequencies, thus improving system bandwidth efficiency.

2.13 CONCLUSION

In this chapter we have considered the first important step in any digital communication system, transforming the source information (both textual and analog) to a form that is compatible with a digital system. We treated various aspects of sampling, quantization (both uniform and nonuniform), and pulse code modulation (PCM). We also considered the selection of PCM waveforms for the transmission of baseband signals through the channel.

We described the detection of binary signals plus Gaussian noise in terms of two basic steps. In the first step the received waveform is reduced to a single number, $z(T)$, and in the second step a decision is made as to which signal was transmitted, on the basis of comparing $z(T)$ to a threshold. We discussed how to best choose this threshold. We also showed that a linear filter known as a matched filter or correlator is the optimum choice for maximizing the output signal-to-noise ratio, and thus minimizing the probability of error.

We defined intersymbol interference (ISI) and explained the importance of Nyquist's work in establishing a theoretical minimum bandwidth for symbol detection without ISI. We also introduced the duobinary concept of adding a controlled amount of ISI to achieve an improvement in bandwidth efficiency at the expense of an increase in power.

REFERENCES

1. Black, H. S., *Modulation Theory*, D. Van Nostrand Company, Princeton, N.J., 1953.

2. Oppenheim, A. V., *Applications of Digital Signal Processing*, Prentice-Hall, Inc., Englewood Cliffs, N.J., 1978.

3. Stiltz, H., ed., *Aerospace Telemetry*, Vol. 1, Prentice-Hall, Inc., Englewood Cliffs, N.J., 1961. p. 179.

4. Hecht, M., and Guida, A., "Delay Modulation," *Proc. IEEE*, vol. 57, no. 7, July 1969, pp. 1314–1316.

5. Deffebach, H. L., and Frost, W. O., "A Survey of Digital Baseband Signaling Techniques," *NASA Technical Memorandum NASATM X-64615*, June 30, 1971.

6. Van Trees, H. L., *Detection, Estimation, and Modulation Theory*, Part 1, John Wiley & Sons, Inc., New York, 1968.

7. Borjesson, P. O., and Sundberg, C. E., "Simple Approximations of the Error Function $Q(x)$ for Communications Applications," *IEEE Trans. Commun.*, vol. COM27, Mar. 1979, pp. 639–642.

8. Proakis, J. G., *Digital Communications*, McGraw-Hill Book Company, New York, 1983.

9. Nyquist, H., "Certain Topics of Telegraph Transmission Theory," *Trans. Am. Inst. Electr. Eng.*, vol. 47, Apr. 1928, pp. 617–644.

10. Korn, I., *Digital Communications*, Van Nostrand Reinhold Company, Inc., New York, 1985.

11. Wu, W. W., *Elements of Digital Satellite Communication*, Computer Science Press, Inc., Rockville, Md., 1984.

12. Lender, A., "The Duobinary Technique for High Speed Data Transmission," *IEEE Trans. Commun. Electron.*, vol. 82, May 1963, pp. 214–218.

13. Lender, A., "Correlative (Partial Response) Techniques and Applications to Digital Radio Systems," in K. Feher, *Digital Communications: Microwave Applications*, Prentice-Hall, Inc., Englewood Cliffs, N.J., 1981, Chap. 7.

14. Couch, L. W., II, *Digital and Analog Communication Systems*, Macmillan Publishing Company, New York, 1982.

15. Lender, A., "Correlative Digital Communication Techniques," *IEEE Trans. Commun. Technol.*, Dec. 1964, pp. 128–135.

PROBLEMS

2.1. You want to transmit the word "HOW" using an 8-ary system.

    (a) Encode the word "HOW" into a sequence of bits, using 7-bit ASCII coding, followed by an eighth bit for error detection, per character. The eighth bit is chosen so that the number of ones in the 8 bits is an even number. How many total bits are there in the message?

(b) Partition the bit stream into $k = 3$ bit segments. Represent each of the 3-bit segments as an octal number (symbol). How many octal symbols are there in the message?

(c) If the system were designed with 16-ary modulation, how many symbols would be used to represent the word "HOW"?

(d) If the system were designed with 256-ary modulation, how many symbols would be used to represent the word "HOW"?

2.2. We want to transmit 800 characters/s, where each character is represented by its 7-bit ASCII codeword, followed by an eighth bit for error detection, per character, as in Problem 2.1. A multilevel PCM format with $M = 16$ levels is used.
   (a) What is the effective transmitted bit rate?
   (b) What is the PCM symbol rate?

2.3. We wish to transmit a 100-character alphanumeric message in 2 s, using 7-bit ASCII coding, followed by an eighth bit for error detection, per character, as in Problem 2.1. A multilevel PCM format with $M = 32$ levels is used.
   (a) Calculate the effective transmitted bit rate and the PCM symbol rate.
   (b) Repeat part (a) for 16-level PCM, eight-level PCM, four-level PCM, and binary PCM.

2.4. Given an analog waveform that has been sampled at its Nyquist rate, f_s, using natural sampling, prove that a waveform (proportional to the original waveform) can be recovered from the samples, using the recovery techniques shown in Figure P2.1. The parameter mf_s is the frequency of the local oscillator, where m is an integer.

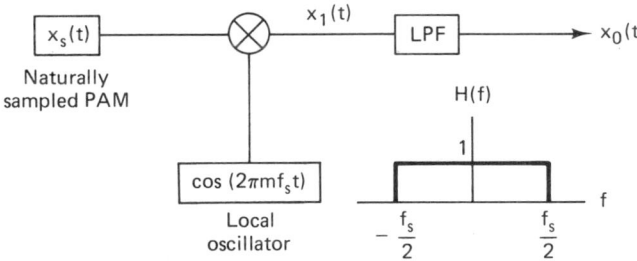

Figure P2.1

2.5. An analog signal is sampled at its Nyquist rate $1/T_s$, and quantized using L quantization levels. The derived digital signal is then transmitted on some channel.
   (a) Show that the time duration, T, of one bit of the transmitted binary encoded signal must satisfy $T \le T_s/(\log_2 L)$.
   (b) When is the equality sign valid?

2.6. Determine the number of quantization levels that are implied if the number of bits per sample in a given PCM code is (a) 5; (b) 8; (c) x.

2.7. Determine the minimum sampling rate necessary to sample and perfectly reconstruct the signal $x(t) = \sin(6280t)/(6280t)$.

2.8. Consider an audio signal with spectral components limited to the frequency band 300 to 3300 Hz. Assume that a sampling rate of 8000 samples/s will be used to generate a PCM signal. Assume that the ratio of peak signal power to average quantization noise power at the output needs to be 30 dB.
   (a) What is the minimum number of uniform quantization levels needed, and what is the minimum number of bits per sample needed?

(b) Calculate the system bandwidth (as specified by the main spectral lobe of the signal) required for the detection of such a PCM signal.

2.9. A waveform, $x(t) = 10 \cos (1000t + \pi/3) + 20 \cos (2000t + \pi/6)$ is to be uniformly sampled for digital transmission.

(a) What is the maximum allowable time interval between sample values that will ensure perfect signal reproduction?

(b) If we want to reproduce 1 hour of this waveform, how many sample values need to be stored?

2.10. (a) A waveform that is bandlimited to 50 kHz is sampled every 10 μs. Show graphically that these samples uniquely characterize the waveform. (Use a sinusoidal example for simplicity. Avoid sampling at points where the waveform equals zero.)

(b) If samples are taken 30 μs apart instead of 10 μs, show graphically that waveforms other than the original can be characterized by the samples.

2.11. Use the method of convolution to illustrate the effect of undersampling the waveform $x(t) = \cos 2\pi f_0 t$ for a sampling rate of $f_s = \frac{3}{2} f_0$.

2.12. (a) Sketch the complete $\mu = 10$ compression characteristic that will handle input voltages in the range -5 to $+5$ V.

(b) Plot the corresponding expansion characteristic.

(c) Draw a 16-level nonuniform quantizer characteristic that corresponds to the $\mu = 10$ compression characteristic.

2.13. Assume a binary sequence with equally likely binary levels. The sequence can be represented by either a bipolar or a unipolar signal set. Show that if the corresponding bipolar signal and unipolar signal have the same peak-to-peak amplitude separation, the bipolar signal uses less average power than the unipolar signal.

2.14. Assume that in a binary digital communication system, the signal component out of the correlator receiver is $a_i(T) = +1$ or -1 V with equal probability. If the Gaussian noise at the correlator output has unit variance, find the probability of a bit error.

2.15. A bipolar binary signal, $s_i(t)$, is a $+1$- or -1-V pulse during the interval $(0, T)$. Additive white Gaussian noise having two-sided power spectral density of 10^{-3} W/Hz is added to the signal. If the received signal is detected with a matched filter, determine the maximum bit rate that can be sent with a bit error probability of $P_B \le 10^{-3}$.

2.16. Bipolar pulse signals, $s_i(t)$ $(i = 1, 2)$, of amplitude ± 1 V are received in the presence of Gaussian noise with $\sigma^2 = 0.1$ V^2. Determine the optimum (minimum probability of error) detection threshold, γ_0, for matched filter detection if the a priori probabilities are: (a) $P(s_1) = 0.5$; (b) $P(s_1) = 0.7$; (c) $P(s_1) = 0.2$. (d) Explain the effect of the a priori probabilities on the value of γ_0. [*Hint:* Refer to Equations (B.10) to (B.12).]

2.17. A binary communication system transmits signals $s_i(t)$ $(i = 1, 2)$. The receiver test statistic, $z(T) = a_i + n_0$, where the signal component, a_i, is either $a_1 = +1$ or $a_2 = -1$, and the noise component, n_0 is uniformly distributed, yielding the conditional density functions $p(z|s_i)$ given by

$$p(z|s_1) = \begin{cases} \frac{1}{2} & \text{for } -0.2 \le z \le 1.8 \\ 0 & \text{otherwise} \end{cases}$$

$$p(z|s_2) = \begin{cases} \frac{1}{2} & \text{for } -1.8 \le z \le 0.2 \\ 0 & \text{otherwise} \end{cases}$$

Find the probability of a bit error, P_B, for the case of equally likely signaling and the use of an optimum decision threshold.

2.18. The information in an analog waveform, whose maximum frequency $f_m = 4000$ Hz, is to be transmitted using a 16-level PCM system. The quantization distortion must not exceed $\pm 1\%$ of the peak-to-peak analog signal.

(a) What is the minimum number of bits per sample or bits per PCM word that should be used in this PCM system?

(b) What is the minimum required sampling rate, and what is the resulting bit rate?

(c) What is the PCM pulse or symbol transmission rate?

2.19. (a) What is the theoretical minimum system bandwidth needed for a 10-Mbits/s signal using 16-level PCM without ISI?

(b) How large can the filter roll-off factor be if the allowable system bandwidth is 1.375 MHz?

2.20. A voice signal (300 to 3300 Hz) is digitized such that the quantization distortion $\leq \pm 0.1\%$ of the peak-to-peak signal voltage. Assume a sampling rate of 8000 samples/s and a multilevel PCM format with $M = 32$ levels. Find the theoretical minimum system bandwidth that avoids ISI.

2.21. A binary waveform of 9600 bits/s is converted to an octal waveform that is transmitted over a system having a raised cosine roll-off filter characteristic. The system has a conditioned (equalized) response out to 2.4 kHz.

(a) What is the octal symbol rate?

(b) What is the roll-off factor of the filter characteristic?

2.22. A voice signal in the range 300 to 3300 Hz is sampled at 8000 samples/s. We may transmit these samples directly as PAM, or we may first convert them into codewords using PCM.

(a) What is the minimum system bandwidth required for the detection of PAM with no ISI and with a filter roll-off characteristic of $r = 1$?

(b) Using the same filter roll-off characteristic, what is the minimum bandwidth required for the detection of binary PCM if the samples are quantized to eight levels?

(c) Repeat part (b) using 128 quantization levels.

2.23. A signal in the frequency range 300 to 3300 Hz is limited to a peak-to-peak swing of 10 V. It is sampled at 8000 samples/s and the samples are quantized to 64 evenly spaced levels. Calculate and compare the bandwidths and ratio of peak signal power to rms quantization noise if the quantized samples are transmitted either as binary pulses or as four-level pulses. Assume that the system bandwidth is defined by the main spectral lobe of the signal.

2.24. An analog signal is to be converted to a binary PCM signal and transmitted over a channel that is bandlimited to 100 kHz. Assume that 32 quantization levels are used and that the overall equivalent transfer function is of the raised cosine type with roll-off $r = 0.6$.

(a) Find the maximum PCM bit rate that can be used by this system without introducing ISI.

(b) Find the maximum signal bandwidth that can be accommodated for the analog signal.

(c) Repeat parts (a) and (b) for an eight-level PCM signal.

CHAPTER 3

Bandpass Modulation
and
Demodulation

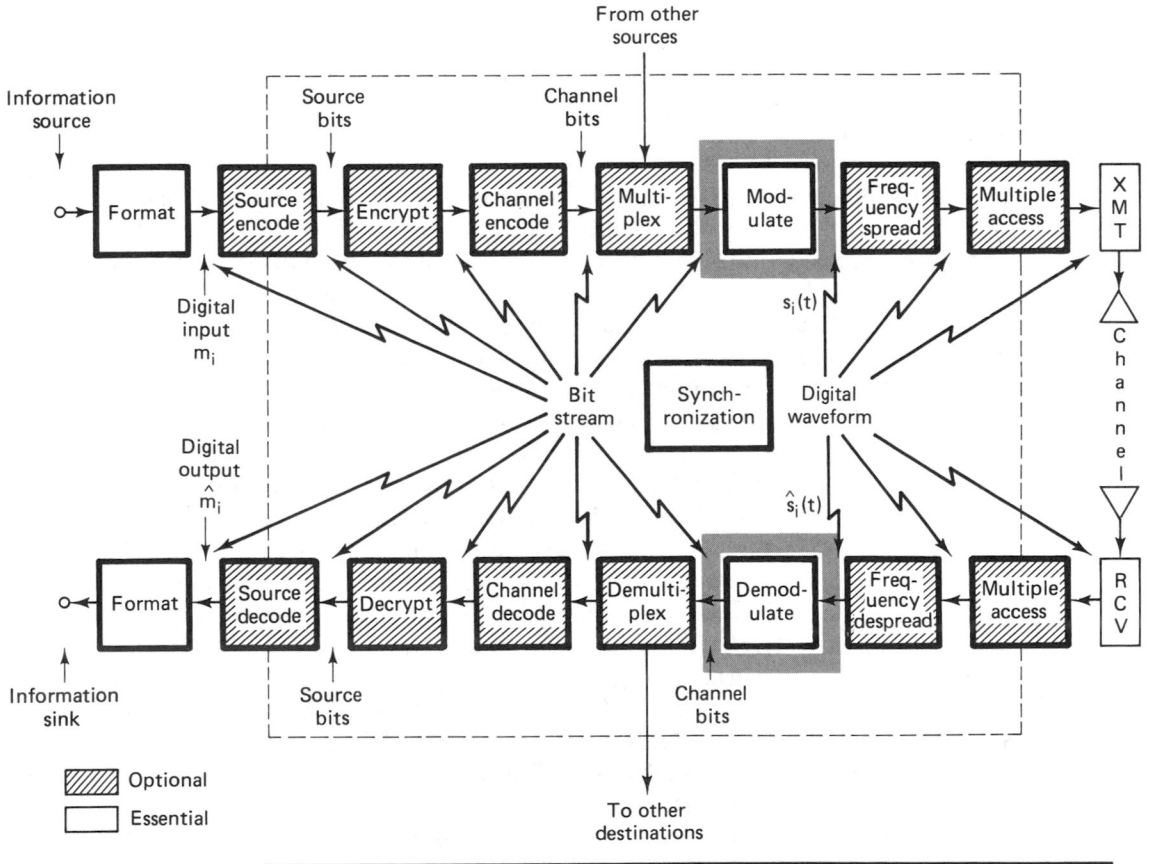

3.1 WHY MODULATE?

Digital modulation is the process by which digital symbols are transformed into waveforms that are compatible with the characteristics of the channel. In the case of baseband modulation, these waveforms are pulses, but in the case of *bandpass modulation* the desired information signal modulates a sinusoid called a *carrier wave,* or simply a *carrier*; for radio transmission the carrier is converted to an electromagnetic (EM) field for propagation to the desired destination. One might ask why it is necessary to use a carrier for the radio transmission of baseband signals. The answer is as follows. The transmission of EM fields through space is accomplished with the use of antennas. To efficiently couple the transmitted EM energy into space, the dimensions of the antenna aperture should be at least as large as the wavelength being transmitted. Wavelength, λ, is equal to c/f, where c, the speed of light, is 3×10^8 m/s. For a baseband signal with frequency $f = 3000$ Hz, $\lambda = 10^5$ m $\simeq 60$ miles. To efficiently transmit a 3000-Hz signal through space *without carrier-wave modulation*, an antenna that spans at least 60 miles would be required. Even if we were willing to inefficiently transmit the EM energy with an antenna measuring one-tenth of a wavelength, we are faced with an impossible antenna size. However, if the information to be transmitted is first modulated on a higher frequency carrier, for example a 30-GHz carrier, the equivalent antenna diameter is then less than $\frac{1}{2}$ in. For this reason, carrier-wave or bandpass modulation is an essential step for all systems involving radio transmission.

Bandpass modulation can provide other important benefits in signal transmission. If more than one signal utilizes a single channel, modulation may be used to separate the different signals. Such a technique, known as *frequency-division multiplexing,* is discussed in Chapter 9. Modulation can be used to minimize the effects of interference. A class of such modulation schemes, known as *spread-spectrum modulation*, requires a system bandwidth much larger than the minimum bandwidth that would be required by the message. The trade-off of bandwidth for interference rejection is considered in Chapter 10. Modulation can also be used to place a signal in a frequency band where design requirements, such as filtering and amplification, can be easily met. This is the case when radio-frequency (RF) signals are converted to an intermediate frequency (IF) in a receiver.

3.2 SIGNALS AND NOISE

3.2.1 Noise in Radio Communication Systems

The task of the demodulator or detector is to retrieve the bit stream from the received waveform, as nearly error free as possible, notwithstanding the distortion to which the signal may have been subjected. There are two primary causes for signal distortion. The first is the filtering effects of the transmitter, channel, and receiver discussed in Section 2.11. As described there, a nonideal system transfer function causes symbol "smearing," which can produce *intersymbol interference*.

The second cause for signal distortion is the noise that is produced by a variety of sources, such as galaxy noise, terrestrial noise, amplifier noise, and unwanted signals from other sources. An unavoidable cause of noise is the thermal motion of electrons in any conducting media. This motion produces *thermal noise* in amplifiers and circuits which corrupts the signal in an additive fashion; that is, the received signal, $r(t)$, is the sum of the transmitted signal, $s(t)$, and the thermal noise, $n(t)$. The statistics of thermal noise have been developed using quantum mechanics and are well known [1].

The primary statistical characteristic of thermal noise is that the noise amplitudes are distributed according to a normal or Gaussian distribution, discussed in Section 1.5.5 and shown in Figure 1.7. The probability density function (pdf), $p(n)$, of the zero-mean noise voltage is expressed as

$$p(n) = \frac{1}{\sigma\sqrt{2\pi}} \exp\left[-\frac{1}{2}\left(\frac{n}{\sigma}\right)^2 \right] \tag{3.1}$$

where σ^2 is the noise variance. In Figure 1.7 it can be seen that the most probable noise amplitudes are those with small positive or negative values. In theory, the noise can be infinitely large, but very large noise amplitudes are rare.

The primary spectral characteristic of thermal noise is that its two-sided power spectral density, $G_n(f) = N_0/2$, is flat for all frequencies of interest for radio communication systems. In other words, thermal noise, on the average, has just as much power per hertz in low-frequency fluctuations as in high-frequency fluctuations—up to a frequency of about 10^{12} hertz. When the noise power is characterized by a constant power spectral density, as shown in Figure 1.8a, we refer to it as *white noise*. Since thermal noise is present in all communication systems and is the predominant noise source for most systems, the thermal noise characteristics (additive, white, and Gaussian) are most often used to model the noise in the detection process and in the design of optimum receivers.

3.2.2 A Geometric View of Signals and Noise

Let us define an *N*-dimensional *orthogonal space* as one characterized by a set of *N* linearly independent functions, $\{\psi_j(t)\}$, called *basis functions*. Any arbitrary function in the space can be generated by a linear combination of these basis functions. The basis functions must satisfy the following conditions:

$$\int_0^T \psi_j(t)\psi_k(t)\,dt = K_j\delta_{jk} \qquad 0 \le t \le T; \quad j, k = 1, \ldots, N \qquad (3.2a)$$

$$\delta_{jk} = \begin{cases} 1 & \text{for } j = k \\ 0 & \text{otherwise} \end{cases} \qquad (3.2b)$$

where the operator δ_{jk} is called the *Kronecker delta function* and is defined by Equation (3.2b). When the K_j constants are nonzero, the signal space is called *orthogonal*. When the basis functions are normalized so that each $K_j = 1$, the space is called an *orthonormal* space. The principal requirement for orthogonality can be stated as follows: Each $\psi_j(t)$ function of the set of basis functions must be independent of the other members of the set. Each $\psi_j(t)$ must not interfere with any other members of the set in the detection process. From a geometric point of view, each $\psi_j(t)$ is mutually perpendicular to each of the other $\psi_k(t)$ for $j \ne k$. An example of such a space with $N = 3$ is shown in Figure 3.1, where the mutually perpendicular axes are designated $\psi_1(t)$, $\psi_2(t)$, and $\psi_3(t)$. If $\psi_j(t)$ corresponds to a real-valued voltage or current waveform component, associated with a 1-Ω

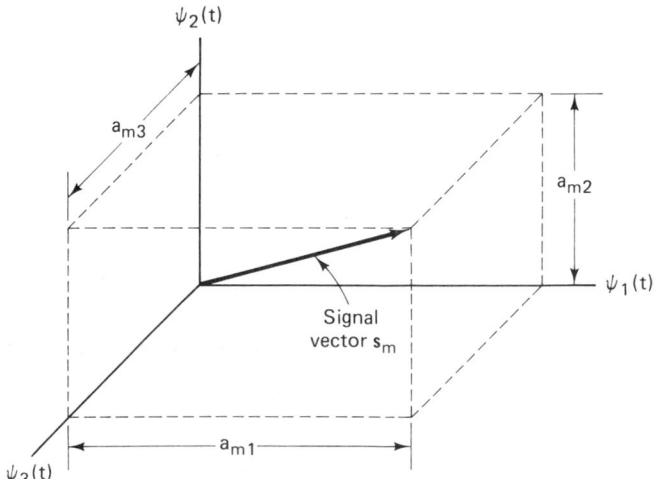

Figure 3.1 Vectorial representation of the signal waveform $s_m(t)$.

Bandpass Modulation and Demodulation   Chap. 3

resistive load, then using Equations (1.5) and (3.2), the normalized energy in joules dissipated in the load in T seconds, due to ψ_j, is

$$E_j = \int_0^T \psi_j^2(t)\, dt = K_j \tag{3.3}$$

One reason we focus on an *orthogonal signal space* is that Euclidean distance measurements, fundamental to the detection process, are easily formulated in such a space. However, even if the signaling waveforms do not comprise such an orthogonal set, they can be transformed into linear combinations of orthogonal waveforms. It can be shown [2] that *any arbitrary* finite set of waveforms $\{s_i(t)\}$ ($i = 1, \ldots, M$), where each member of the set is physically realizable and of duration T, can be expressed as a linear combination of N orthogonal waveforms $\psi_1(t), \psi_2(t), \ldots, \psi_N(t)$, where $N \leq M$, such that

$$s_1(t) = a_{11}\psi_1(t) + a_{12}\psi_2(t) + \cdots + a_{1N}\psi_N(t)$$

$$s_2(t) = a_{21}\psi_1(t) + a_{22}\psi_2(t) + \cdots + a_{2N}\psi_N(t)$$

$$\vdots \qquad\qquad\qquad\qquad \vdots$$

$$s_M(t) = a_{M1}\psi_1(t) + a_{M2}\psi_2(t) + \cdots + a_{MN}\psi_N(t)$$

These relationships are expressed in more compact notation as follows:

$$s_i(t) = \sum_{j=1}^N a_{ij}\psi_j(t) \qquad \begin{array}{l} i = 1, \ldots, M \\ N \leq M \end{array} \tag{3.4}$$

where

$$a_{ij} = \frac{1}{K_j} \int_0^T s_i(t)\psi_j(t)\, dt \qquad \begin{array}{l} i = 1, \ldots, M; \quad 0 \leq t \leq T \\ j = 1, \ldots, N \end{array} \tag{3.5}$$

The coefficient a_{ij} is the value of the $\psi_j(t)$ component of signal, $s_i(t)$. The form of the $\{\psi_j(t)\}$ is not specified; it is chosen for convenience and will depend on the form of the signal waveforms. The set of signal waveforms, $\{s_i(t)\}$, can be viewed as a set of vectors, $\{s_i\} = \{a_{i1}, a_{i2}, \ldots, a_{iN}\}$. If, for example, $N = 3$, we may plot the vector, s_m, corresponding to the waveform

$$s_m(t) = a_{m1}\psi_1(t) + a_{m2}\psi_2(t) + a_{m3}\psi_3(t)$$

as a point in a three-dimensional Euclidean space with coordinates (a_{m1}, a_{m2}, a_{m3}), as shown in Figure 3.1. The orientation among the signal vectors describes the relation of the signals to one another (with respect to phase or frequency), and the amplitude of each vector in the set $\{s_i\}$ is a measure of the signal energy transmitted during a symbol duration. In general, once a set of N orthogonal functions has been adopted, each of the transmitted signal waveforms, $s_i(t)$, is completely determined by the vector of its coefficients

$$s_i = (a_{i1}, a_{i2}, \ldots, a_{iN}) \qquad i = 1, \ldots, M \tag{3.6}$$

We shall employ the notation of signal vectors, {**s**}, or signal waveforms, {**s**(*t*)}, as best suits the discussion. A typical detection problem, conveniently viewed in terms of signal vectors, is illustrated in Figure 3.2. Vectors $\mathbf{s}_j$ and $\mathbf{s}_k$ represent *prototype* or *reference signals* belonging to the set of *M* waveforms, {$s_i(t)$}. The receiver knows, a priori, the location in the signal space of each prototype vector belonging to the *M*-ary set. During the transmission of any signal, the signal is perturbed by noise so that the resultant vector that is actually received is a perturbed version (e.g., $\mathbf{s}_j$ + **n** or $\mathbf{s}_k$ + **n**) of the original one, where **n** represents a noise vector. The noise is additive and has a Gaussian distribution; therefore, the resulting distribution of possible received signals is a cluster or cloud of points around $\mathbf{s}_j$ and $\mathbf{s}_k$. The cluster is dense in the center and becomes sparse with increasing distance from the prototype. The arrow marked **r** represents a signal vector that might arrive at the receiver during some symbol interval. The task of the receiver is to decide whether **r** has a close "resemblance" to the prototype $\mathbf{s}_j$, whether it more closely resembles $\mathbf{s}_k$, or whether it is closer to some other prototype signal in the *M*-ary set. The measurement can be thought of as a *distance* measurement. The question that the receiver or detector must resolve is: Which of the prototypes within the signal space is *closest* in distance to the received vector, **r**? The analysis of all demodulation or detection schemes involves this concept of *distance* between a received waveform and a set of possible transmitted waveforms. A simple rule for the detector to follow is to decide that **r** belongs to the same class as its nearest neighbor (nearest prototype vector).

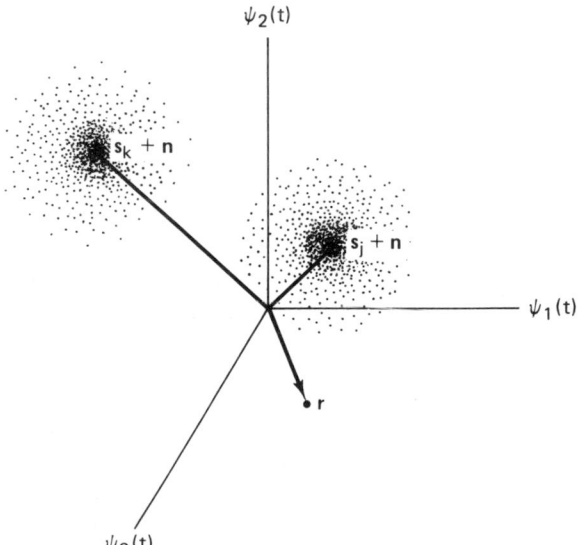

Figure 3.2 Signals and noise in a three-dimensional vector space.

3.2.2.1 Waveform Energy

Using Equations (1.5), (3.4), and (3.2), the normalized energy, E_i, associated with the waveform, $s_i(t)$, over a symbol interval, *T*, can be expressed in terms of the orthogonal components of $s_i(t)$ as follows:

$$E_i = \int_0^T s_i^2(t)\, dt = \int_0^T \left[\sum_j a_{ij}\psi_j(t) \right]^2 dt \tag{3.7}$$

$$= \int_0^T \sum_j a_{ij}\psi_j(t) \sum_k a_{ik}\psi_k(t)\, dt \tag{3.8}$$

$$= \sum_j \sum_k a_{ij}a_{ik} \int_0^T \psi_j(t)\psi_k(t)\, dt \tag{3.9}$$

$$= \sum_j \sum_k a_{ij}a_{ik}K_j\delta_{jk} \tag{3.10}$$

$$= \sum_{j=1}^N a_{ij}^2 K_j \qquad i = 1, \ldots, M \tag{3.11}$$

Equation (3.11) is a special case of Parseval's theorem relating the integral of the square of the waveform, $s_i(t)$, to the sum of the square of the orthogonal series coefficients. If orthonormal functions are used (i.e., $K_j = 1$), the normalized energy over a symbol duration T is given by

$$E_i = \sum_{j=1}^N a_{ij}^2 \tag{3.12}$$

If there is equal energy, E, in each of the $s_i(t)$ waveforms, we can write Equation (3.12) in the form

$$E = \sum_{j=1}^N a_{ij}^2 \qquad \text{for all } i \tag{3.13}$$

3.2.2.2 Generalized Fourier Transforms

The transformation described by Equations (3.2), (3.4), and (3.5) is referred to as the *generalized Fourier transformation*. In the case of ordinary Fourier transforms, the $\{\psi_j(t)\}$ set is comprised of sine and cosine harmonic functions. But in the case of generalized Fourier transforms, the $\{\psi_j(t)\}$ set is not constrained to any specific form; it must only satisfy the orthogonality statement of Equation (3.2). *Any* arbitrary integrable waveform set, as well as noise, can be represented as a linear combination of orthogonal waveforms through such a generalized Fourier transformation [2]. Therefore, in such an orthogonal space, we are justified in using distance (Euclidean distance) as a decision criterion for the detection of *any* signal set in the presence of AWGN. The most important application of this orthogonal transformation has to do with the way in which signals are actually transmitted and received. The transmission of a nonorthogonal signal set is generally accomplished by the appropriate weighting of the orthogonal carrier components. For example, in Section 3.5.3 we show that multiple phase shift keying (MPSK) signals are fully characterized by weighted sine and cosine components of the carrier.

Example 3.1 Orthogonal Representation of Waveforms

Figure 3.3 illustrates the statement that any arbitrary integrable waveform set can be represented as a linear combination of orthogonal waveforms. Figure 3.3a shows a set of three waveforms, $s_1(t)$, $s_2(t)$, $s_3(t)$.

(a) Demonstrate that these waveforms *do not* form an orthogonal set.

(b) Figure 3.3b shows a set of two waveforms, $\psi_1(t)$ and $\psi_2(t)$. Verify that these waveforms form an orthogonal set.

(c) Show how the nonorthogonal waveform set in part (a) can be expressed as a linear combination of the orthogonal set in part (b).

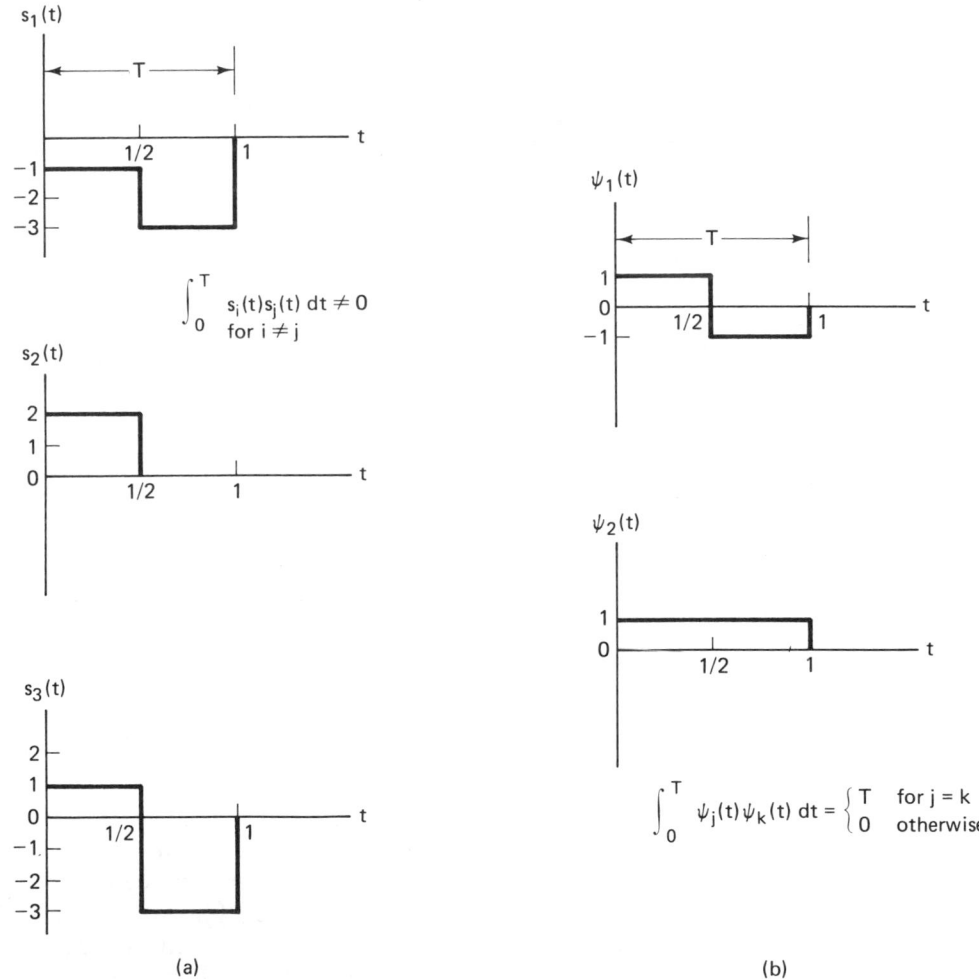

(a) (b)

Figure 3.3 Example of an arbitrary signal set in terms of an orthogonal set. (a) Arbitrary signal set. (b) Orthogonal basis functions.

Solution

(a) $s_1(t)$, $s_2(t)$, and $s_3(t)$ are clearly not orthogonal, since they do not meet the requirements of Equation (3.2); that is, the time integrated value (over a symbol duration) of the cross-product of any two of the three waveforms is not zero. Let us verify this for $s_1(t)$ and $s_2(t)$.

$$\int_0^T s_1(t)s_2(t)\, dt = \int_0^{T/2} s_1(t)s_2(t)\, dt + \int_{T/2}^T s_1(t)s_2(t)\, dt$$

$$= \int_0^{T/2} (-1)(2)\, dt + \int_{T/2}^T (-3)(0)\, dt = -T$$

Similarly, the integral over the interval, T, of each of the cross-products $s_1(t)s_3(t)$ and $s_2(t)s_3(t)$ results in nonzero values. Hence the waveform set $\{s_i(t)\}$ ($i = 1$, 2, 3) in Figure 3.3a is not an orthogonal set.

(b) Using Equation (3.2), we verify that $\psi_1(t)$ and $\psi_2(t)$ form an orthogonal set as follows:

$$\int_0^T \psi_1(t)\psi_2(t)\, dt = \int_0^{T/2} (1)(1)\, dt + \int_{T/2}^T (-1)(1)\, dt = 0$$

(c) We can express the nonorthogonal set $\{s_i(t)\}$ ($i = 1, 2, 3$) as a linear combination of the orthogonal basis waveforms $\{\psi_j(t)\}$ ($j = 1, 2$), as follows, by using Equation (3.5), where $K_j = T$:

$$s_1(t) = \psi_1(t) - 2\psi_2(t)$$

$$s_2(t) = \psi_1(t) + \psi_2(t)$$

$$s_3(t) = 2\psi_1(t) - \psi_2(t)$$

These relationships illustrate how an arbitrary waveform set $\{s_i(t)\}$ can be expressed as a linear combination of an orthogonal set $\{\psi_j(t)\}$, as described in Equations (3.4) and (3.5). What are the practical applications for being able to describe $s_1(t)$, $s_2(t)$, and $s_3(t)$, in terms of $\psi_1(t)$, $\psi_2(t)$, and the appropriate coefficients? If we want a system for transmitting waveforms $s_1(t)$, $s_2(t)$, and $s_3(t)$, the transmitter and the receiver need only be implemented using the two basis functions $\psi_1(t)$ and $\psi_2(t)$ instead of the three original waveforms. A convenient way in which an appropriate choice of a basis function set, $\{\psi_j(t)\}$, can be obtained for any given signal set, $\{s_i(t)\}$, is called the *Gram–Schmidt orthogonalization procedure*. It is described in Appendix 4A of Reference [3].

3.2.2.3 Representing White Noise with Orthogonal Waveforms

Additive white Gaussian noise (AWGN) can be expressed as a linear combination of orthogonal waveforms in the same way as signals. For the signal detection problem, the noise can be partitioned into two components,

$$n(t) = \hat{n}(t) + \tilde{n}(t) \tag{3.14}$$

where

$$\hat{n}(t) = \sum_{j=1}^N n_j\psi_j(t) \tag{3.15}$$

is taken to be the noise within the signal space, or the projection of the noise components on the signal coordinates $\psi_1(t), \ldots, \psi_N(t)$, and

$$\tilde{n}(t) = n(t) - \hat{n}(t) \qquad (3.16)$$

is defined as the noise outside the signal space. In other words, $\tilde{n}(t)$ may be thought of as the noise that is effectively tuned out by the detector. The symbol $\hat{n}(t)$ represents the noise that will interfere with the detection process. We can express the noise waveform, $n(t)$, as follows:

$$n(t) = \sum_{j=1}^{N} n_j \psi_j(t) + \tilde{n}(t) \qquad (3.17)$$

where

$$n_j = \frac{1}{K_j} \int_0^T n(t) \psi_j(t) \, dt \qquad \text{for all } j \qquad (3.18)$$

and

$$0 = \int_0^T \tilde{n}(t) \psi_j(t) \, dt \qquad (3.19)$$

The interfering portion of the noise, $\hat{n}(t)$, expressed in Equation (3.15) will henceforth be referred to simply as $n(t)$. We can express $n(t)$ by a vector of its coefficients similar to the way we did for signals in Equation (3.6).

$$\mathbf{n} = (n_1, n_2, \ldots, n_N) \qquad (3.20)$$

where $\mathbf{n}$ is a random vector with zero mean and Gaussian distribution, and where the noise components n_i $(i = 1, \ldots, N)$ are independent.

3.2.2.4 Variance of White Noise

White noise is an *idealized process* with two-sided power spectral density equal to a constant, $N_0/2$, for all frequencies from $-\infty$ to $+\infty$. Hence the noise variance (average noise power, since the noise has zero mean) is

$$\sigma^2 = \text{var}\,[n(t)] = \int_{-\infty}^{\infty} \left(\frac{N_0}{2}\right) df = \infty \qquad (3.21)$$

Although the variance for AWGN is infinite, the variance for *filtered* AWGN is finite. For example, if AWGN is correlated with one of a set of orthonormal functions $\psi_j(t)$, the variance of the correlator output is given by

$$\sigma^2 = \text{var}\,(n_j) = \mathbf{E}\left\{\left[\int_0^T n(t)\psi_j(t)\,dt\right]^2\right\} = \frac{N_0}{2} \qquad (3.22)$$

The proof of Equation (3.22) is given in Appendix C. Henceforth we shall assume that the noise of interest in the detection process is the output noise of a correlator or matched filter with variance $\sigma^2 = N_0/2$ as expressed in Equation (3.22).

3.3 DIGITAL BANDPASS MODULATION TECHNIQUES

Bandpass modulation (either analog or digital) is the process by which an information signal is converted to a sinusoidal waveform; for digital modulation, such a sinusoid of duration T is referred to as a digital symbol. The sinusoid has just three features that can be used to distinguish it from other sinusoids: amplitude, frequency, and phase. Thus bandpass modulation can be defined as the process whereby the amplitude, frequency, or phase of an RF carrier, or a combination of them, is varied in accordance with the information to be transmitted. The general form of the carrier wave, $s(t)$, is as follows:

$$s(t) = A(t) \cos \theta(t) \qquad (3.23)$$

where $A(t)$ is the time-varying amplitude and $\theta(t)$ is the time-varying angle. It is convenient to write

$$\theta(t) = \omega_0 t + \phi(t) \qquad (3.24)$$

so that

$$s(t) = A(t) \cos [\omega_0 t + \phi(t)] \qquad (3.25)$$

where ω_0 is the *radian frequency* of the carrier and $\phi(t)$ is the *phase*. The terms f and ω will each be used to denote frequency. When f is used, frequency in hertz is intended; when ω is used, frequency in radians per second is intended. The two frequency parameters are related by $\omega = 2\pi f$.

The basic *digital modulation/demodulation* types are listed in Figure 3.4. When the receiver exploits knowledge of the carrier's phase to detect the signals, the process is called *coherent detection*; when the receiver does not utilize such phase reference information, the process is called *noncoherent detection*. In digital communications, the terms *demodulation* and *detection* are used somewhat interchangeably, although demodulation emphasizes removal of the carrier, and detection includes the process of symbol decision. In ideal coherent detection, there is available at the receiver a prototype of each possible arriving signal. These prototype waveforms attempt to duplicate the transmitted signal set in every respect, even RF phase. The receiver is then said to be *phase locked* to the incoming signal. During detection, the receiver multiplies and integrates (correlates) the incoming signal with each of its prototype replicas. Under the heading of coherent modulation/demodulation in Figure 3.4 are listed phase shift keying (PSK), frequency shift keying (FSK), amplitude shift keying (ASK), continuous phase modulation (CPM), and hybrid combinations. The basic bandpass modulation formats are discussed in this chapter. Some specialized formats, such as offset quadrature PSK (OQPSK), minimum shift keying (MSK) belonging to the CPM class, and quadrature amplitude modulation (QAM), are treated in Chapter 7.

Noncoherent demodulation refers to systems employing demodulators that are designed to operate without knowledge of the absolute value of the incoming signal's phase; therefore, phase estimation is not required. Thus the advantage of noncoherent over coherent systems is reduced complexity, and the price paid is increased probability of error (P_E). In Figure 3.4 the modulation/demodulation

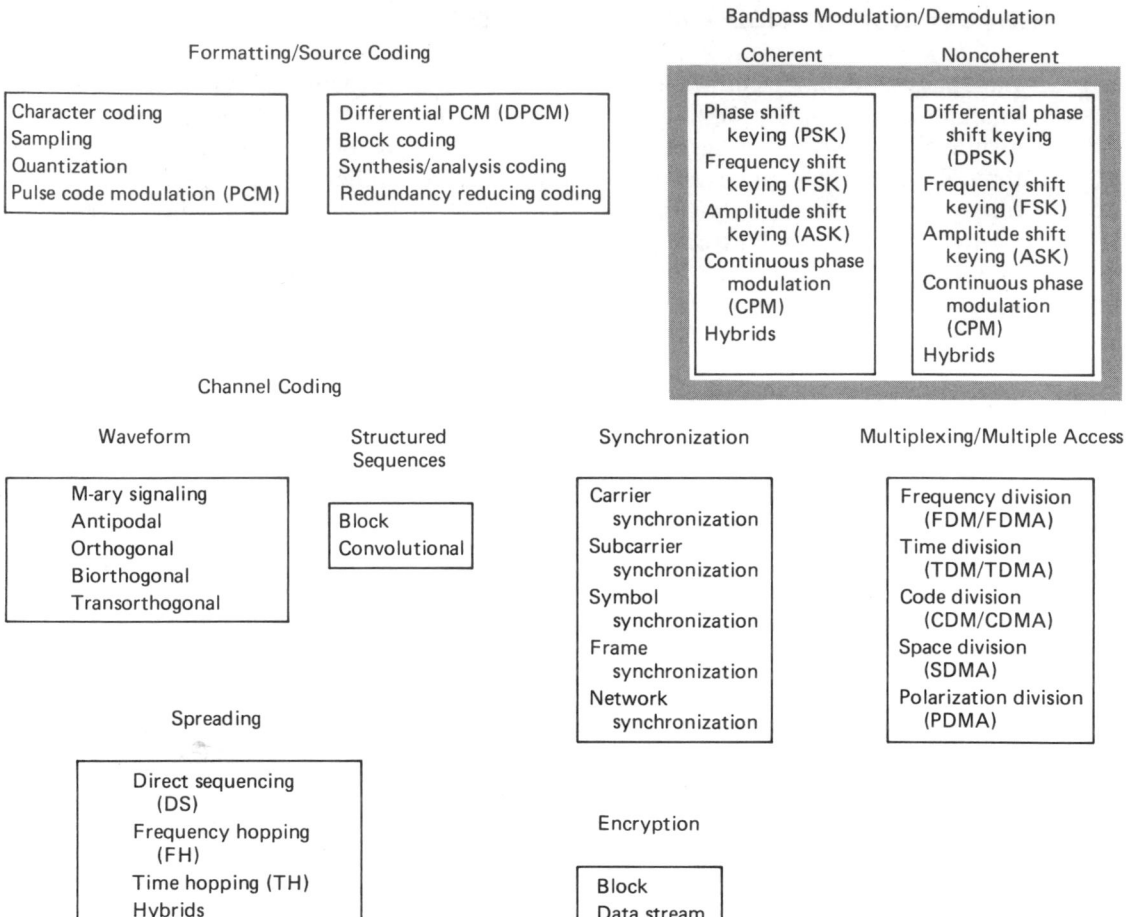

Figure 3.4 Basic digital communication transformations.

types that are listed in the noncoherent column, DPSK, FSK, ASK, CPM, and hybrids, are similar to those listed in the coherent column. We had implied that phase information is not used for noncoherent reception; how do you account for the fact that there is a form of phase shift keying under the noncoherent heading? It turns out that an important form of PSK can be classified as noncoherent (or differentially coherent) since it does not require a reference in phase with the received carrier. This "pseudo-PSK," termed *differential PSK* (DPSK), utilizes phase information of the prior symbol as a phase reference for detecting the current symbol. This is described in Sections 3.6.1 and 3.6.2.

Figure 3.5 illustrates examples of the most common digital modulation formats: PSK, FSK, ASK, and a hybrid combination of ASK and PSK (ASK/PSK or APK). The first column lists the analytic expression, the second is a typical pictorial of the waveform versus time, and the third is a vectorial schematic, with the orthogonal axes labeled $\{\psi_j(t)\}$. In the general *M*-ary signaling case, the pro-

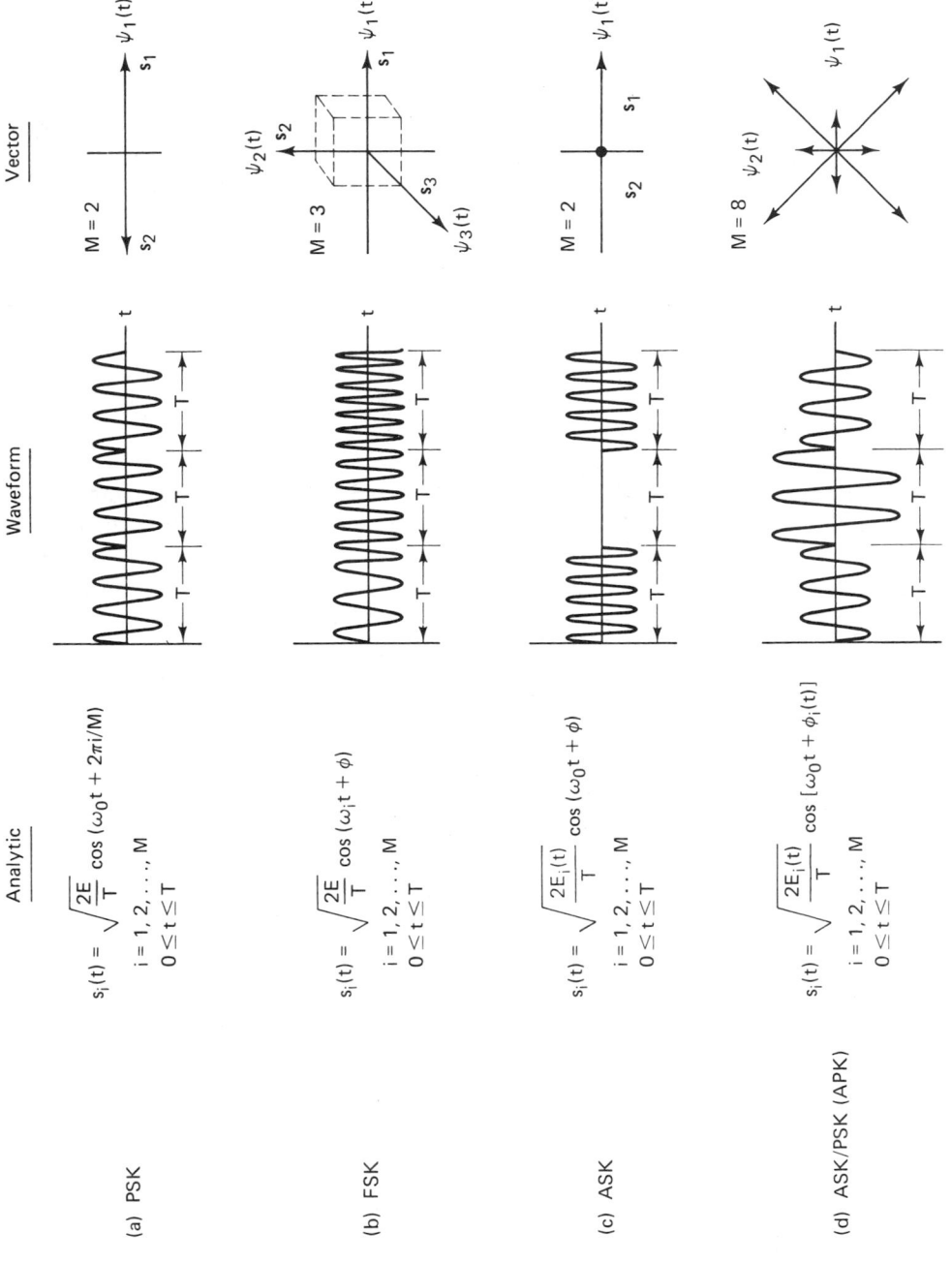

Figure 3.5 Digital modulations. (a) PSK. (b) FSK. (c) ASK. (d) ASK/PSK (APK).

cessor accepts k source bits at a time and instructs the modulator to produce one of an available set of $M = 2^k$ waveform types. Binary modulation, where $k = 1$, is just a special case of M-ary modulation. Each example shown in Figure 3.5 illustrates the set of signal waveforms with a particular value chosen for M.

3.3.1 Phase Shift Keying

Phase shift keying (PSK) was developed during the early days of the deep-space program; PSK is now widely used in both military and commercial communications systems. The general analytic expression for PSK is

$$s_i(t) = \sqrt{\frac{2E}{T}} \cos \left[\omega_0 t + \phi_i(t) \right] \quad \begin{array}{l} 0 \leq t \leq T \\ i = 1, \ldots, M \end{array} \quad (3.26)$$

where the phase term, $\phi_i(t)$, will have M discrete values, typically given by

$$\phi_i(t) = \frac{2\pi i}{M} \quad i = 1, \ldots, M$$

For the binary PSK (BPSK) example in Figure 3.5a, M is 2. The parameter E is symbol energy, T is symbol time duration, and $0 \leq t \leq T$. In BPSK modulation, the modulating data signal shifts the phase of the waveform, $s_i(t)$, to one of two states, either zero or π (180°). The waveform sketch in Figure 3.5a shows a typical BPSK waveform with its abrupt phase changes at the symbol transitions; if the modulating data stream were to consist of alternating ones and zeros, there would be such an abrupt change at each transition. The signal waveforms can be represented as vectors on a polar plot; the vector length corresponds to the signal amplitude, and the vector direction, for the general M-ary case, corresponds to the signal phase relative to the other $M - 1$ signals in the set. For the BPSK example, the vectorial picture illustrates the two 180° opposing vectors. Signal sets that can be depicted with such opposing vectors are called *antipodal signal sets*.

3.3.2 Frequency Shift Keying

The general analytic expression for FSK modulation is

$$s_i(t) = \sqrt{\frac{2E}{T}} \cos \left(\omega_i t + \phi \right) \quad \begin{array}{l} 0 \leq t \leq T \\ i = 1, \ldots, M \end{array} \quad (3.27)$$

where the frequency term, ω_i, will have M discrete values, and the phase term, ϕ, is an arbitrary constant. The FSK waveform sketch in Figure 3.5b illustrates the typical abrupt frequency changes at the symbol transitions. In this example, M has been chosen equal to 3, corresponding to the same number of waveform types (3-ary); note that this $M = 3$ choice for FSK has been selected to emphasize the mutually perpendicular axes. In practice, M is usually a nonzero power of 2 (2, 4, 8, 16, . . .). The signal set is characterized by Cartesian coordinates, such that each of the mutually perpendicular axes represents a sinusoid with a different frequency. As described earlier, signal sets that can be characterized with such mutually perpendicular vectors are called *orthogonal* signals. The required frequency spacing between the orthogonal tones is discussed in Section 3.6.4.

3.3.3 Amplitude Shift Keying

For the ASK example in Figure 3.5c, the general analytic expression is

$$s_i(t) = \sqrt{\frac{2E_i(t)}{T}} \cos{(\omega_0 t + \phi)} \qquad \begin{matrix} 0 \le t \le T \\ i = 1, \ldots, M \end{matrix} \qquad (3.28)$$

where the amplitude term, $\sqrt{2E_i(t)/T}$, will have M discrete values, and the phase term, ϕ, is an arbitrary constant. In Figure 3.5c, M has been chosen equal to 2, corresponding to two waveform types. The ASK waveform sketch in the figure can describe a radar transmission example, where the two signal amplitude states would be $\sqrt{2E/T}$ and zero. The vectorial picture utilizes the same phase–amplitude polar coordinates as the PSK example. Here we see a vector corresponding to the maximum-amplitude state, and a point at the origin corresponding to the zero-amplitude state. Binary ASK signaling (also called on–off keying) was one of the earliest forms of digital modulation used in radio telegraphy at the beginning of this century. Simple ASK is no longer widely used in digital communication systems; therefore, it will not be treated in detail.

3.3.4 Amplitude Phase Keying

For the combination of ASK and PSK (APK) example in Figure 3.5d, the general analytic expression

$$s_i(t) = \sqrt{\frac{2E_i(t)}{T}} \cos{[\omega_0 t + \phi_i(t)]} \qquad \begin{matrix} 0 \le t \le T \\ i = 1, \ldots, M \end{matrix} \qquad (3.29)$$

illustrates the indexing of both the signal amplitude term and the phase term. The APK waveform picture in Figure 3.5d illustrates some typical simultaneous phase and amplitude changes at the symbol transition times. For this example, M has been chosen equal to 8, corresponding to eight waveforms (8-ary). The figure illustrates a hypothetical eight-vector signal set on the phase–amplitude plane. Four of the vectors are at one amplitude; the other four vectors are at a different amplitude; and each of the vectors is separated by 45°. When the set of M symbols in the two-dimensional signal space are arranged in a rectangular constellation, the signaling is referred to as quadrature amplitude modulation (QAM); examples of QAM are considered in Chapter 7.

The vectorial picture for each of the modulation types described in Figure 3.5 (except the FSK case) is characterized on a plane whose *polar* coordinates represent signal *amplitude* and *phase*. The FSK case is characterized in a *Cartesian* coordinate space, with each axis representing a *frequency tone* ($\cos \omega_i t$) from the M-ary set of orthogonal tones.

3.3.5 Waveform Amplitude Coefficient

The waveform amplitude coefficient appearing in Equations (3.26) to (3.29) has the same general form, $\sqrt{2E/T}$, for all modulation formats. This expression is derived as follows:

$$s(t) = A \cos \omega t \tag{3.30}$$

where A is the peak value of the waveform. Since the peak value of a sinusoidal waveform equals $\sqrt{2}$ times the root-mean-square (rms) value, we can write

$$s(t) = \sqrt{2} A_{rms} \cos \omega t$$

$$= \sqrt{2 A_{rms}^2} \cos \omega t$$

Assuming the signal to be a voltage or a current waveform, A_{rms}^2 represents average power P (normalized to $1\ \Omega$). Therefore, we can write

$$s(t) = \sqrt{2P} \cos \omega t \tag{3.31}$$

Replacing P watts by E joules/T seconds, we get

$$s(t) = \sqrt{\frac{2E}{T}} \cos \omega t \tag{3.32}$$

We shall use either the amplitude notation, A, in Equation (3.30) or the designation $\sqrt{2E/T}$ in Equation (3.32). Since the *energy* in a signal is the key parameter in determining the error performance of the detection process, it is often more convenient to use the amplitude notation in Equation (3.32) because it facilitates solving directly for the probability of error, P_E, as a function of signal energy.

3.4 DETECTION OF SIGNALS IN GAUSSIAN NOISE

3.4.1 Decision Regions

Consider that the two-dimensional signal space in Figure 3.6 is the locus of the noise-perturbed prototype binary vectors $(\mathbf{s}_1 + \mathbf{n})$ and $(\mathbf{s}_2 + \mathbf{n})$. The noise vector, $\mathbf{n}$, is a zero-mean random vector; hence the received signal vector, $\mathbf{r}$, is a random vector with mean $\mathbf{s}_1$ or $\mathbf{s}_2$. The detector's task after receiving $\mathbf{r}$ is to decide which of the signals, $\mathbf{s}_1$ or $\mathbf{s}_2$, was actually transmitted. The method is usually to decide on the signal classification that yields the minimum expected P_E, although other strategies are possible [4]. For the case where M equals 2, with $\mathbf{s}_1$ and $\mathbf{s}_2$ being equally likely and with the noise being an additive white Gaussian noise (AWGN) process, we will see that the minimum-error decision rule is equivalent to choosing the signal class such that the distance $d(\mathbf{r}, \mathbf{s}_i) = \| \mathbf{r} - \mathbf{s}_i \|$ is minimized, where $\| \mathbf{x} \|$ is called the *norm* or *magnitude* of vector $\mathbf{x}$. This rule is often stated in terms of decision regions. In Figure 3.6, let us construct decision regions in the following

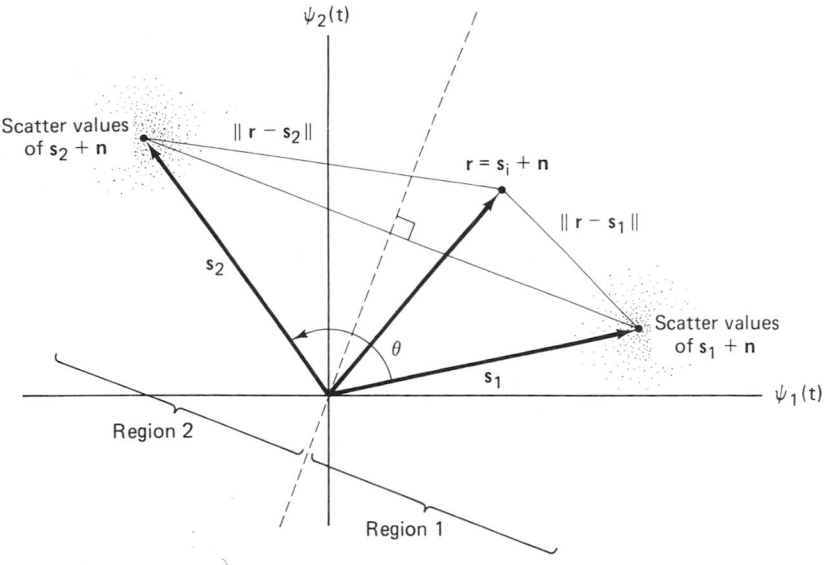

Figure 3.6 Two-dimensional signal space, with arbitrary equal-amplitude vectors s_1 and s_2.

way. Draw a line connecting the tips of the prototype vectors, s_1 and s_2. Next, construct the perpendicular bisector of the connecting line. Notice that this bisector passes through the origin of the space if s_1 and s_2 are equal in amplitude. For this $M = 2$ example in Figure 3.6, the constructed perpendicular bisector represents the locus of points equidistant between s_1 and s_2; hence the bisector describes the boundary between decision region 1 and decision region 2. The *decision rule* for the detector, stated in terms of *decision regions,* is: Whenever the received signal **r** is located in region 1, choose signal s_1; when it is located in region 2, choose signal s_2.

3.4.2 Correlation Receiver

In Section 2.9 we treated the detection of *baseband* binary signals in Gaussian noise. Since the detection of *bandpass* signals employs the same concepts, we shall summarize the key findings of that section. We focus particularly on that realization of a matched filter known as a *correlator*. In addition to binary detection, we also consider the more general case of *M*-ary detection. We assume that the only performance degradation is due to AWGN. The received signal, $r(t)$, is the sum of the transmitted prototype signal plus the random noise:

$$r(t) = s_i(t) + n(t) \quad \begin{matrix} 0 \le t \le T \\ i = 1, \ldots, M \end{matrix} \quad (3.33)$$

Given such a received signal, the detection process consists of *two basic steps.* In the first step, the received waveform, $r(t)$, is reduced to a *single random variable, $z(T)$,* or a *set of random variables, $z_i(T)$* $(i = 1, \ldots, M)$, formed at the output of the correlator(s) at time $t = T$, where T is the symbol duration. In the second step, a symbol decision is made, on the basis of comparing $z(T)$ to a threshold or on the basis of choosing the maximum $z_i(T)$. Step 1 can be thought of as transforming the waveform into a point in the decision space. Step 2 can be thought of as determining *in which decision region* the point is located. For the detector to be optimized (in the sense of minimizing the error probability), it is necessary to optimize the waveform-to-random-variable transformation, by using matched filters or correlators in step 1, and by also optimizing the decision criterion in step 2.

In Sections 2.9.2 and 2.9.3 we found that the matched filter provides the maximum signal-to-noise ratio at the filter output at time $t = T$. We described a correlator as one realization of a matched filter. We can define a *correlation receiver* comprised of M correlators, as shown in Figure 3.7a, that transforms a received waveform, $r(t)$, to a sequence of M numbers or correlator outputs, $z_i(T)$ $(i = 1, \ldots, M)$. Each correlator output is characterized by the following product integration or correlation with the received signal.

$$z_i(T) = \int_0^T r(t)s_i(t) \, dt \qquad i = 1, \ldots, M \tag{3.34}$$

The verb "to correlate" means "to match." The correlators attempt to match the incoming received signal, $r(t)$, with each of the candidate prototype waveforms, $s_i(t)$, known a priori to the receiver. A reasonable decision rule is to choose the waveform, $s_i(t)$, that *matches best* or has the *largest correlation* with $r(t)$. In other words, the decision rule is:

$$\text{Choose the } s_i(t) \text{ whose index} \atop \text{corresponds to the max } z_i(T) \tag{3.35}$$

Following Equation (3.4), any signal set, $\{s_i(t)\}$ $(i = 1, \ldots, M)$, can be expressed in terms of some set of basis functions, $\{\psi_j(t)\}$ $(j = 1, \ldots, N)$, where $N \leq M$. Then the bank of M correlators in Figure 3.7a may be replaced with a bank of N correlators, shown in Figure 3.7b, where the set of basis functions $\{\psi_j(t)\}$ form *reference signals*. The decision stage of this receiver consists of logic circuitry for choosing the signal, $s_i(t)$. The choice of $s_i(t)$ is made according to the best match of the coefficients, a_{ij}, seen in Equation (3.4), with the set of outputs $\{z_j(T)\}$. When the prototype waveform set, $\{s_i(t)\}$, is an orthogonal set, the receiver implementation in Figure 3.7a is identical to that in Figure 3.7b (differing perhaps by a scale factor). However, when $\{s_i(t)\}$ is *not* an orthogonal set, the receiver in Figure 3.7b, using N correlators instead of M, with reference signals $\{\psi_j(t)\}$, can represent a cost-effective implementation. We examine such an application for the detection of multiple phase shift keying (MPSK) in Section 3.5.3. For the other applications in this chapter, we shall assume a correlator receiver with reference signals $\{s_i(t)\}$.

In the case of *binary detection,* the correlation receiver can be configured as a single matched filter or product integrator, as shown in Figure 3.8a, with the reference signal being the difference between the binary prototype signals, $s_1(t) - s_2(t)$. The output of the correlator, $z(T)$, is fed directly to the decision stage.

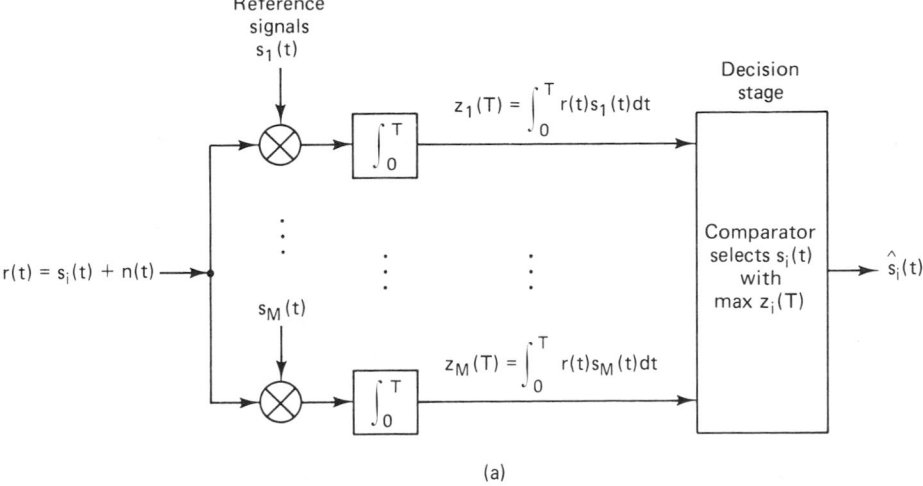

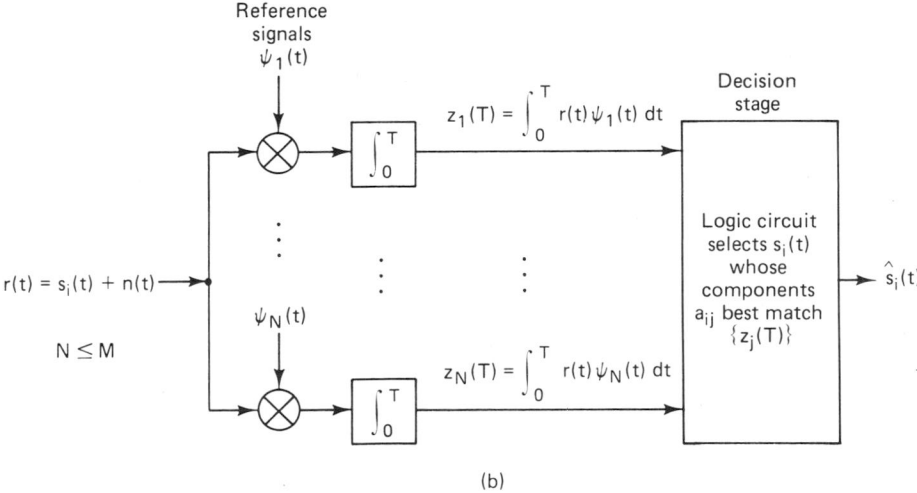

Figure 3.7 (a) Correlator receiver with reference signals $\{s_i(t)\}$. (b) Correlator receiver with reference signals $\{\psi_j(t)\}$.

For binary detection, the correlation receiver can also be drawn, as shown in Figure 3.8b, as two matched filters or product integrators, each of which is matched to one of the prototype reference signals, $s_1(t)$ or $s_2(t)$. The decision stage can then be configured to follow the rule in Equation (3.35), or the correlator outputs, $z_i(T)$ ($i = 1, 2$), can be differenced to form

$$z(T) = z_1(T) - z_2(T) \tag{3.36}$$

as shown in Figure 3.8b. Then, $z(T)$, called the *test statistic*, is fed to the decision stage, as in the case of the single correlator. In the *absence of noise*, an input

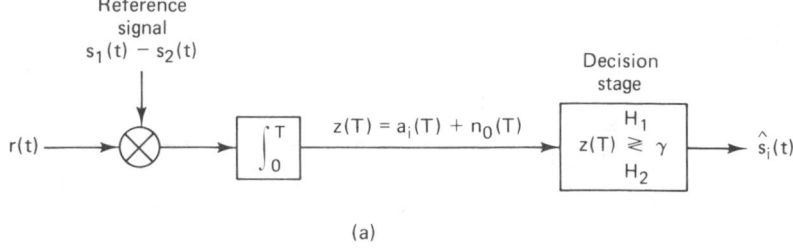

(a)

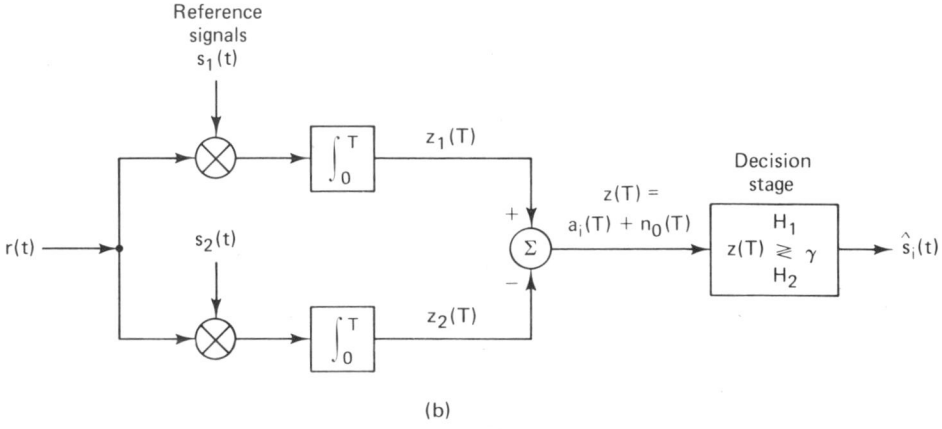

(b)

Figure 3.8 Binary correlator receiver. (a) Using a single correlator. (b) Using two correlators.

waveform, $s_i(t)$, yields the output, $z(T) = a_i(T)$, a signal-only component. The input noise, $n(t)$, is a Gaussian random process. Since the correlator is a *linear* device, the output noise is also a Gaussian random process [4]. Thus the output of the correlator, sampled at $t = T$, yields

$$z(T) = a_i(T) + n_0(T) \quad i = 1, 2$$

where $n_0(T)$ is the noise component. To shorten the notation we sometimes express $z(T)$ as $a_i + n_0$. The noise component, n_0, is a zero-mean *Gaussian random variable,* and thus $z(T)$ is a *Gaussian random variable* with a mean of either a_1 or a_2 depending on whether a binary one or binary zero was sent.

3.4.2.1 Binary Decision Threshold

For the random variable, $z(T)$, Figure 3.9 illustrates the two conditional probability density functions (pdfs), $p(z|s_1)$ and $p(z|s_2)$, with mean value of a_1 and a_2, respectively (these pdfs are also called the *likelihood* of s_1 and the *likelihood*

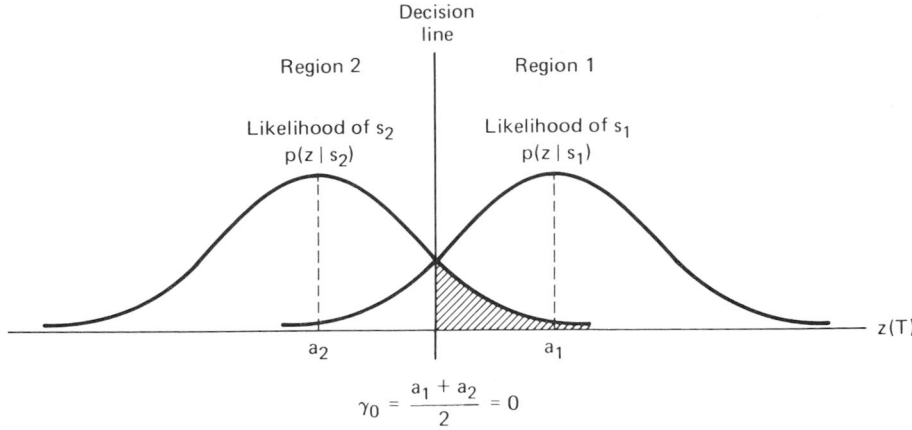

Figure 3.9 Conditional probability density functions: $p(z|s_1)$, $p(z|s_2)$.

of s_2, respectively):

$$p(z|s_1) = \frac{1}{\sigma_0\sqrt{2\pi}} \exp\left[-\frac{1}{2}\left(\frac{z - a_1}{\sigma_0}\right)^2\right] \qquad (3.37a)$$

$$p(z|s_2) = \frac{1}{\sigma_0\sqrt{2\pi}} \exp\left[-\frac{1}{2}\left(\frac{z - a_2}{\sigma_0}\right)^2\right] \qquad (3.37b)$$

where σ_0^2 is the noise variance. In Figure 3.9 the rightmost likelihood, $p(z|s_1)$, illustrates the probability density of the detector output, $z(T)$, given that $s_1(t)$ was transmitted. Similarly, the leftmost likelihood $p(z|s_2)$, illustrates the probability density of $z(T)$ given that $s_2(t)$ was transmitted. The abscissa, $z(T)$, represents the full range of possible sample output values from the correlation receiver in Figure 3.8.

With regard to optimizing the binary decision threshold for deciding in which region a received signal is located, we found in Section 2.9.1 that the *minimum error* criterion for equally likely binary signals corrupted by Gaussian noise can be stated as follows:

$$z(T) \underset{H_2}{\overset{H_1}{\gtrless}} \frac{a_1 + a_2}{2} = \gamma_0 \qquad (3.38)$$

where a_1 is the signal component of $z(T)$ when $s_1(t)$ is transmitted, and a_2 is the signal component of $z(T)$ when $s_2(t)$ is transmitted. The threshold level, γ_0, represented by $(a_1 + a_2)/2$, is the *optimum threshold* for minimizing the probability of making an incorrect decision given equally likely signals and symmetrical likelihoods. The decision rule in Equation (3.38) states that hypothesis H_1 should be selected [equivalent to deciding that signal $s_1(t)$ was sent] if $z(T) > \gamma_0$, and hypothesis H_2 should be selected [equivalent to deciding that $s_2(t)$ was sent] if $z(T) < \gamma_0$. If $z(T) = \gamma_0$, the decision can be an arbitrary one. For equal-energy, equally likely antipodal signals, where $s_1(t) = -s_2(t)$ and $a_1 = -a_2$, the optimum decision rule becomes

$$z(T) \underset{H_2}{\overset{H_1}{\gtrless}} \gamma_0 = 0 \tag{3.39a}$$

or

$$\begin{array}{ll} \text{decide } s_1(t) & \text{if } z_1(T) > z_2(T) \\ \text{decide } s_2(t) & \text{otherwise} \end{array} \tag{3.39b}$$

In the next section we illustrate the use of correlators and matched filters for the coherent detection of PSK and FSK modulation. In later sections we consider noncoherent detection, and we treat the error performance of various modulation types.

3.5 COHERENT DETECTION

3.5.1 Coherent Detection of PSK

The detector shown in Figure 3.7 can be used for the coherent detection of any digital waveforms. Such a correlating detector is often referred to as a *maximum likelihood detector*. Consider the following binary PSK (BPSK) example. Let

$$s_1(t) = \sqrt{\frac{2E}{T}} \cos (\omega_0 t + \phi) \qquad 0 \le t \le T \tag{3.40a}$$

$$s_2(t) = \sqrt{\frac{2E}{T}} \cos (\omega_0 t + \phi + \pi)$$

$$= -\sqrt{\frac{2E}{T}} \cos (\omega_0 t + \phi) \qquad 0 \le t \le T \tag{3.40b}$$

$n(t) =$ zero-mean white Gaussian random process

where the phase term, ϕ, is an arbitrary constant, so that the analysis is unaffected by setting $\phi = 0$. The parameter, E, is the signal energy per symbol, and T is the symbol duration. For this antipodal case, only a single basis function is needed. If an orthonormal signal space is assumed in Equations (3.4) and (3.5) (i.e., $K_j = 1$), we can express a basis function, $\psi_1(t)$, as follows:

$$\psi_1(t) = \sqrt{\frac{2}{T}} \cos \omega_0 t \qquad \text{for } 0 \le t \le T \tag{3.41}$$

Thus we may express the transmitted signals $s_i(t)$ in terms of $\psi_1(t)$ and the coefficients $a_{i1}(t)$:

$$s_i(t) = a_{i1}\psi_1(t) \tag{3.42a}$$

$$s_1(t) = a_{11}\psi_1(t) = \sqrt{E}\psi_1(t) \tag{3.42b}$$

$$s_2(t) = a_{21}\psi_1(t) = -\sqrt{E}\psi_1(t) \tag{3.42c}$$

Assume that $s_1(t)$ was transmitted. Then the expected values of the product integrators in Figure 3.7b, with reference signals $\psi_1(t)$ and $-\psi_1(t)$, are found as follows:

$$\mathbf{E}\{z_1|s_1\} = \mathbf{E}\left\{\int_0^T \sqrt{E}\psi_1^2(t) + n(t)\psi_1(t)\ dt\right\} \tag{3.43a}$$

$$\mathbf{E}\{z_2|s_1\} = \mathbf{E}\left\{\int_0^T -\sqrt{E}\psi_1^2(t) - n(t)\psi_1(t)\ dt\right\} \tag{3.43b}$$

$$\mathbf{E}\{z_1|s_1\} = \mathbf{E}\left\{\int_0^T \frac{2}{T}\sqrt{E}\cos^2\omega_0 t + n(t)\sqrt{\frac{2}{T}}\cos\omega_0 t\ dt\right\} = \sqrt{E} \tag{3.44a}$$

$$\mathbf{E}\{z_2|s_1\} = \mathbf{E}\left\{\int_0^T -\frac{2}{T}\sqrt{E}\cos^2\omega_0 t - n(t)\sqrt{\frac{2}{T}}\cos\omega_0 t\ dt\right\} = -\sqrt{E} \tag{3.44b}$$

where $\mathbf{E}\{\cdot\}$ denotes the ensemble average, referred to as the *expected value*. Equation (3.44) follows because $\mathbf{E}\{n(t)\} = 0$. The decision stage must decide which signal was transmitted by determining its location within the signal space. For this example, the choice of $\psi_1(t) = \sqrt{2/T}\cos\omega_0 t$ normalizes $\mathbf{E}\{z_i(T)\}$ to be $\pm\sqrt{E}$. The prototype signals $\{s_i(t)\}$ are the same as the reference signals $\{\psi_j(t)\}$ except for the normalizing scale factor. The decision stage chooses the signal with the largest value of $z_i(T)$. Thus, the received signal in this example is judged to be $s_1(t)$. The error performance for such coherently detected BPSK systems is treated in Section 3.7.1.

3.5.2 Sampled Matched Filter

In Section 2.9.2 we discussed the basic characteristic of the matched filter—namely, that its impulse response is a delayed version of the mirror image (rotated on the $t = 0$ axis) of the input signal waveform. Therefore, if the signal waveform is $s(t)$, its mirror image is $s(-t)$ and the mirror image delayed by T seconds is $s(T - t)$. The impulse response, $h(t)$, of a filter matched to $s(t)$ is then described by

$$h(t) = \begin{cases} s(T - t) & 0 \le t \le T \\ 0 & \text{elsewhere} \end{cases} \tag{3.45}$$

Figure 3.10a illustrates how a matched filter can be implemented using digital hardware. The input signal, $r(t)$, is comprised of the prototype signal, $s(t)$, plus noise, $n(t)$. The bandwidth of the signal is $W = 1/2T_s$, where the Nyquist sampling

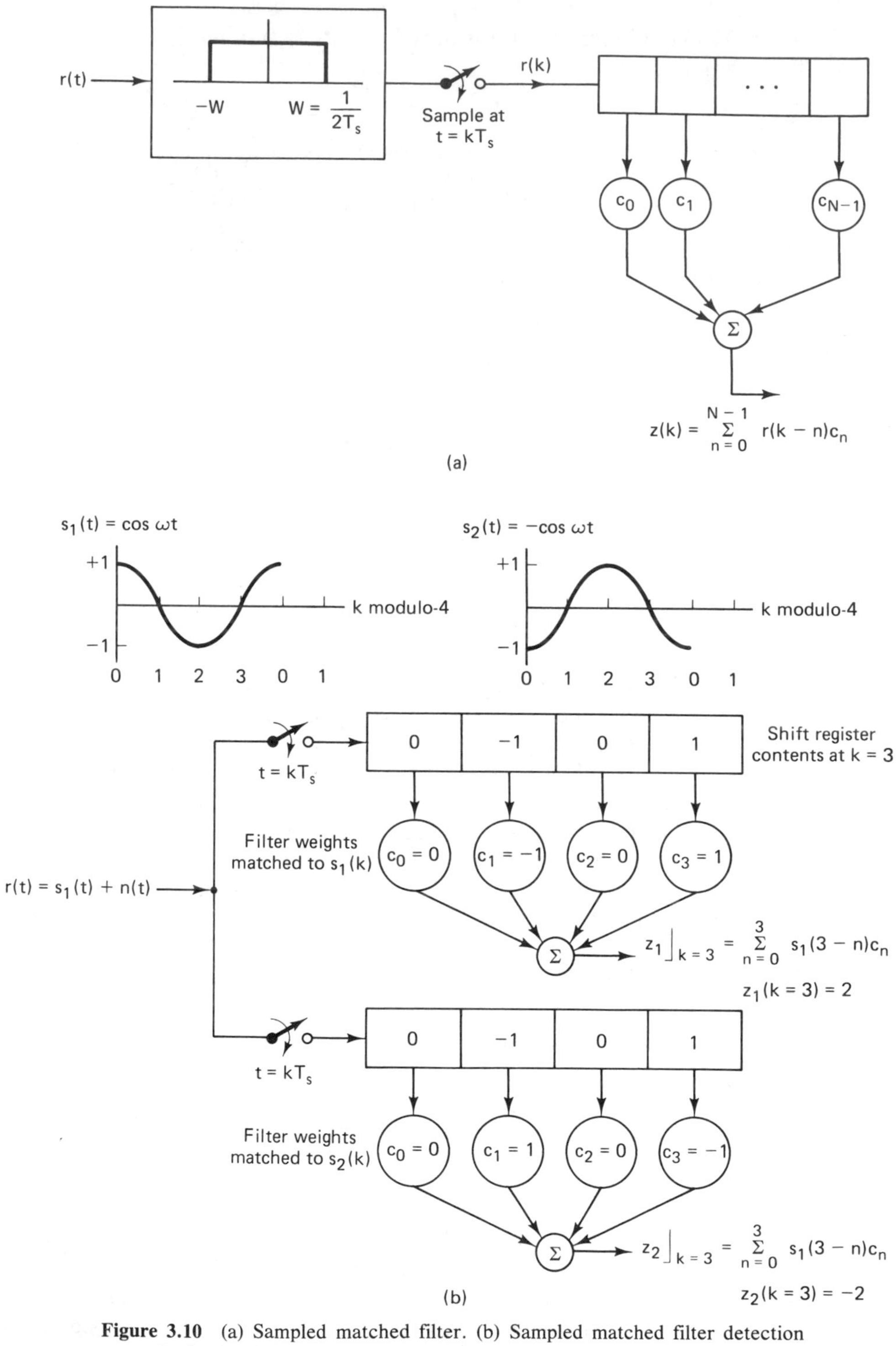

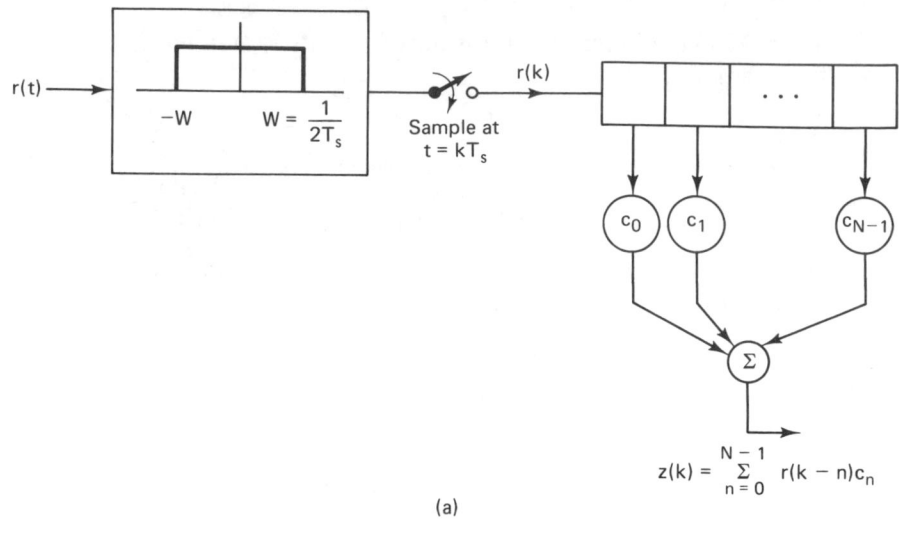

r(t)
−W
W = 1/2T_s
Sample at
t = kT_s
r(k)
. . .
c_0
c_1
c_N−1
Σ
z(k) = Σ_{n=0}^{N−1} r(k − n)c_n
(a)
s_1(t) = cos ωt
+1
−1
k modulo-4
0 1 2 3 0 1
s_2(t) = −cos ωt
+1
−1
k modulo-4
0 1 2 3 0 1
r(t) = s_1(t) + n(t)
t = kT_s
0
−1
0
1
Shift register
contents at k = 3
Filter weights
matched to s_1(k)
c_0 = 0
c_1 = −1
c_2 = 0
c_3 = 1
Σ
z_1 |_{k=3} = Σ_{n=0}^{3} s_1(3 − n)c_n
z_1(k = 3) = 2
t = kT_s
0
−1
0
1
Filter weights
matched to s_2(k)
c_0 = 0
c_1 = 1
c_2 = 0
c_3 = −1
Σ
z_2 |_{k=3} = Σ_{n=0}^{3} s_1(3 − n)c_n
(b)
z_2(k = 3) = −2

Figure 3.10 (a) Sampled matched filter. (b) Sampled matched filter detection example, in the absence of noise.

rate $f_s = 2W = 1/T_s$; hence the sampling interval is equal to T_s. At the clock times of $t = kT_s$, the analog signal is sampled and the samples are shifted into the register of Figure 3.10a from left to right. The shift register with its coefficients c_0 to c_{N-1} approximate a matched filter. Once the received signal has been sampled, the continuous time notation t is changed to kT_s or simply k to reflect the sampled notation

$$r(k) = s(k) + n(k) \qquad k = 0, 1, \ldots$$

where k represents a sample index. The output, $z(k)$, of the sampled matched filter, at a time corresponding to the kth sample is

$$z(k) = \sum_{n=0}^{N-1} r(k - n)c_n \qquad k = 0, 1, \ldots, \text{modulo-}N \qquad (3.46)$$

where x modulo-y is defined as the remainder of dividing x by y. For the binary demodulation application, $z_i(k)(i = 1, 2)$ outputs are compared to a threshold at each value of $k = N - 1$ corresponding to the end of a symbol. The c_n values are the filter weights constituting the filter impulse response that is matched to the signal, where n is the index of the weights and the register stages (from left to right) and k is the index of the samples as they are produced by the sampler. One can see the similarity between the convolution integral of Equation (2.56) and the summation of Equation (3.46), especially with regard to the mirror-image rotation of one of the functions prior to multiplication. Since we assume the noise to have zero mean, the expected value of a received sample for the binary case is expressed as

$$\mathbf{E}\{r(k)\} = s_i(k) \qquad i = 1, 2 \qquad (3.47)$$

If $s_1(t)$ had been transmitted, the expected matched filter outputs would be

$$\mathbf{E}\{z_i(k)\} = \sum_{n=0}^{N-1} s_1(k - n)c_n \qquad (3.48)$$

where the filter weights, c_n, are matched to the corresponding $s_i(k)$ for each branch.

Example 3.2 Sampled Matched Filter

Consider the BPSK waveform set

$$s_1(t) = \cos \omega t$$

and

$$s_2(t) = -\cos \omega t$$

Illustrate how a *sampled* matched filter or correlator, as shown in Figure 3.10a, can be used to detect a received signal, say $s_1(t)$, from the BPSK waveform set, in the absence of noise.

Solution

First, the waveform is sampled so that $s_1(t)$ is transformed into the set of samples, $\{s_1(k)\}$. The sampled matched filter receiver will be shown with two branches, following the analog implementation in Figure 3.8b. The top branch is made up of shift registers and coefficients matched to the $\{s_1(k)\}$ sample points. The bottom branch is similarly matched to the $\{s_2(k)\}$ sample points. The four equally spaced sample points ($k = 0, 1, 2, 3$) for each of the $\{s_i(k)\}$ are as follows (see Figure 3.10b):

$$s_1(k = 0) = 1, \qquad s_1(k = 1) = 0, \qquad s_1(k = 2) = -1, \qquad s_1(k = 3) = 0$$

$$s_2(k = 0) = -1, \qquad s_2(k = 1) = 0, \qquad s_2(k = 2) = 1, \qquad s_2(k = 3) = 0$$

The c_n coefficients represent the delayed mirror-image rotation of the signal to which the filter is matched. Therefore, $c_n = s_i(N - 1 - n)$, where $n = 0, \ldots, N - 1$, and we can write $c_0 = s_i(3)$, $c_1 = s_i(2)$, $c_2 = s_i(1)$, $c_3 = s_i(0)$. It is here that the reader can gain some insight as to why the convolution operation (with its mirror-image rotation) results in the appropriate lining up of the received signal samples with the weights (reference signal).

Consider the top branch in Figure 3.10b. At the $k = 0$ clock time, the first sample, $s_1(k = 0) = 1$, enters the leftmost stage of each register. At the next clock time, the second sample, $s_1(k = 1) = 0$, enters the leftmost stage of each register; at this same time the first sample, $s_1(k = 0) = 1$, has been shifted to the next right stage in each register, and so on. At the $k = 3$ clock time the sample, $s_1(k = 3) = 0$, enters the leftmost stage; by this time the first sample, $s_1(k = 0) = 1$, has been shifted into the rightmost stage. The four signal samples are now located in the registers in mirror-image arrangement compared to the way the prototype waveform, $s_1(t)$, is drawn in Figure 3.10b. The task of the demodulator is to find the best match to the incoming signal; the demodulator matches the reference coefficients of each branch with the incoming signal samples, in the order in which the samples arrive. Hence the convolution operation is an appropriate expression for describing the alignment of the incoming waveform samples with the reference coefficients, to maximize the correlation in the proper branch.

3.5.3 Coherent Detection of Multiple Phase Shift Keying

Figure 3.11 illustrates the signal space for a multiple phase shift keying (MPSK) signal set; the figure describes a four-level (4-ary) PSK or quadriphase shift keying (QPSK) example ($M = 4$). Binary source digits are collected two at a time, and for each symbol interval the two sequential digits instruct the modulator as to which of the four waveforms to produce. For typical coherent M-ary PSK (MPSK) systems, $s_i(t)$ can be expressed as

$$s_i(t) = \sqrt{\frac{2E}{T}} \cos \left(\omega_0 t - \frac{2\pi i}{M} \right) \qquad \begin{matrix} 0 \le t \le T \\ i = 1, \ldots, M \end{matrix} \qquad (3.49)$$

where E is the energy content of $s_i(t)$ over each symbol duration T, and ω_0 is the carrier frequency. If an orthonormal signal space is assumed in Equations (3.4) and (3.5), we can choose a convenient set of axes, as follows:

$$\psi_1(t) = \sqrt{\frac{2}{T}} \cos \omega_0 t \qquad (3.50a)$$

$$\psi_2(t) = \sqrt{\frac{2}{T}} \sin \omega_0 t \qquad (3.50b)$$

Bandpass Modulation and Demodulation   Chap. 3

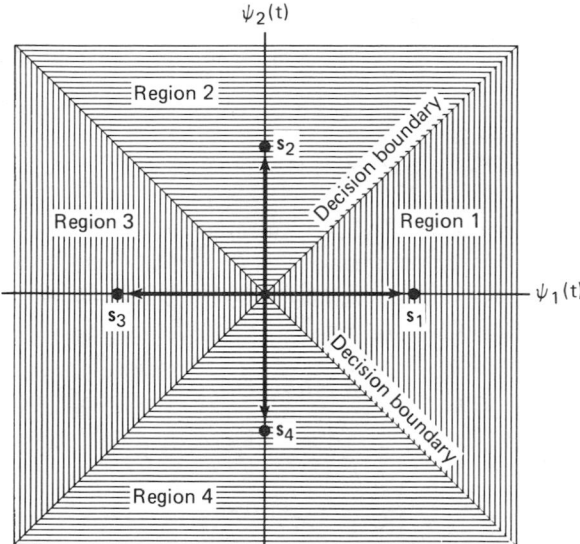

Figure 3.11 Signal space and decision regions for a QPSK system.

where the amplitude $\sqrt{2/T}$ has been chosen to normalize the expected output of the detector, as was done in Section 3.5.1. Now $s_i(t)$ can be written in terms of these orthonormal coordinates, giving

$$s_i(t) = a_{i1}\psi_1(t) + a_{i2}\psi_2(t) \qquad \begin{matrix} 0 \le t \le T \\ i = 1, \ldots, M \end{matrix} \qquad (3.51a)$$

$$= \sqrt{E} \cos\left(\frac{2\pi i}{M}\right) \psi_1(t) + \sqrt{E} \sin\left(\frac{2\pi i}{M}\right) \psi_2(t) \qquad (3.51b)$$

Notice that Equation (3.51) describes a set of M multiple phase waveforms (intrinsically nonorthogonal) in terms of only two orthogonal carrier-wave components. The $M = 4$ (QPSK) case is unique among MPSK signal sets in the sense that the QPSK waveform set is represented by a combination of antipodal and orthogonal members. The decision boundaries partition the signal space into $M = 4$ regions; the construction is similar to the procedure outlined in Section 3.4.1 and Figure 3.6 for $M = 2$. The decision rule for the detector (see Figure 3.11) is to decide that $s_1(t)$ was transmitted if the received signal vector falls in region 1, that $s_2(t)$ was transmitted if the received signal vector falls in region 2, and so on. In other words, the decision rule is to choose the ith waveform if $z_i(T)$ is the largest of the correlator outputs (seen in Figure 3.7).

The form of the correlator shown in Figure 3.7a implies that there are always M product correlators used for the demodulation of MPSK signals. The figure infers that for each of the M branches, a reference signal with the appropriate phase shift is configured. In practice, the implementation of an MPSK demodulator follows Figure 3.7b, requiring only $N = 2$ product integrators regardless of the size of the signal set M. The savings in implementation is possible because any arbitrary integrable waveform set can be expressed as a linear combination of orthogonal waveforms, as shown in Section 3.2.2. Figure 3.12 illustrates such a demodulator. The received signal, $r(t)$, can be expressed by combining Equations (3.50) and (3.51) as follows:

$$r(t) = \sqrt{\frac{2E}{T}}(\cos \phi_i \cos \omega_0 t + \sin \phi_i \sin \omega_0 t) + n(t) \quad \begin{array}{l} 0 \leq t \leq T \\ i = 1, \ldots, M \end{array} \quad (3.52)$$

where $\phi_i = 2\pi i/M$, and $n(t)$ is a zero-mean white Gaussian noise process. Notice in Figure 3.12 that there are only two reference waveforms or basis functions, $\psi_1(t) = \sqrt{2/T} \cos \omega_0 t$ for the upper correlator and $\psi_2(t) = \sqrt{2/T} \sin \omega_0 t$ for the lower correlator. The upper correlator computes

$$X = \int_0^T r(t)\psi_1(t) \, dt \quad (3.53)$$

and the lower correlator computes

$$Y = \int_0^T r(t)\psi_2(t) \, dt \quad (3.54)$$

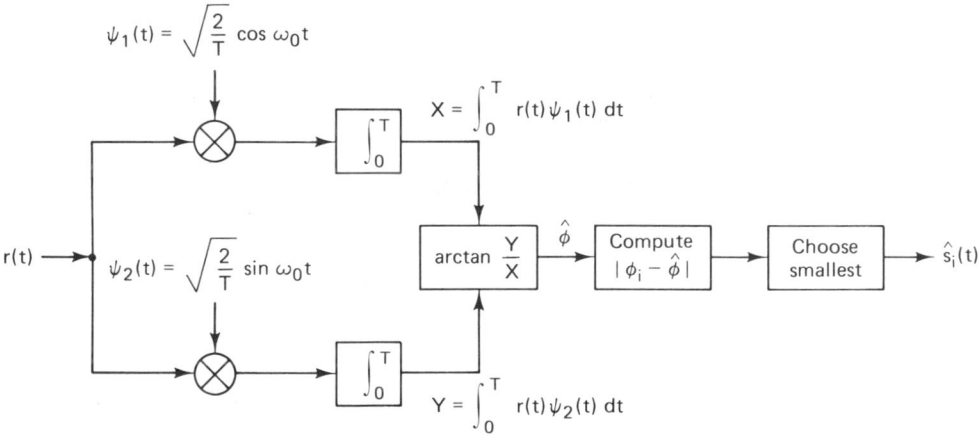

Figure 3.12 Demodulator for MPSK signals.

Figure 3.13 illustrates that the computation of the received phase angle $\hat{\phi}$ can be accomplished by computing the arctan of Y/X, where X can be thought of as the in-phase component of the received signal, Y is the quadrature component, and $\hat{\phi}$ is a noisy estimate of the transmitted ϕ_i. In other words, the upper correlator of Figure 3.12 produces an output X, the magnitude of the in-phase projection of the vector $\mathbf{r}$, and the lower correlator produces an output Y, the magnitude of the quadrature projection of the vector $\mathbf{r}$. The X and Y outputs of the correlators feed into the block marked arctan (Y/X). The resulting value of the angle $\hat{\phi}$ is compared with each of the stored prototype phase angles, ϕ_i. The demodulator selects the ϕ_i that is closest to the angle $\hat{\phi}$. In other words, the demodulator computes $|\phi_i - \hat{\phi}|$ for each of the ϕ_i prototypes and chooses the ϕ_i yielding the smallest output.

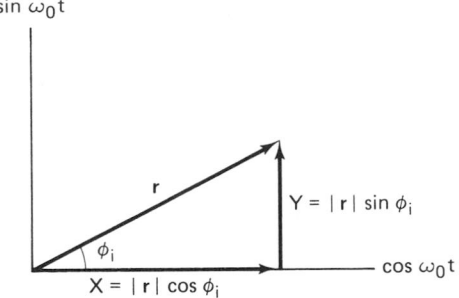

$$\hat{\phi} = \arctan\,(Y/X) \begin{cases} \text{Noisy estimate} \\ \text{of transmitted } \phi_i \end{cases}$$

Figure 3.13 In-phase and quadrature components of the received signal vector **r**.

3.5.4 Coherent Detection of FSK

FSK modulation is characterized by the information being contained in the frequency of the carrier. A typical set of FSK signal waveforms was described in Equation (3.27) as

$$s_i(t) = \sqrt{\frac{2E}{T}}\,\cos\,(\omega_i t\,+\,\phi) \qquad \begin{array}{l} 0 \le t \le T \\ i = 1,\ldots,M \end{array}$$

where E is the energy content of $s_i(t)$ over each symbol duration T, and $(\omega_{i+1} - \omega_i)$ is typically assumed to be an integral multiple of π/T. The phase term, ϕ, is an arbitrary constant and can be set equal to zero. Assuming that the basis functions $\psi_1(t), \psi_2(t), \ldots, \psi_N(t)$ form an orthonormal set, the most useful form for $\{\psi_j(t)\}$ is shown below.

$$\psi_j(t) = \sqrt{\frac{2}{T}}\,\cos\,\omega_j t \qquad j = 1,\ldots,N \tag{3.55}$$

where, as before, the amplitude $\sqrt{2/T}$ normalizes the expected output of the detector. From Equation (3.5) we can write

$$a_{ij} = \int_0^T \sqrt{\frac{2E}{T}}\,\cos\,(\omega_i t)\,\sqrt{\frac{2}{T}}\,\cos\,\omega_j t\,dt \tag{3.56}$$

Therefore,

$$a_{ij} = \begin{cases} \sqrt{E} & \text{for } i = j \\ 0 & \text{otherwise} \end{cases} \tag{3.57}$$

In other words, the ith prototype signal vector is located on the ith coordinate axis at a displacement $\sqrt{E}$ from the origin of the signal space. In this scheme, for the general M-ary case, the distance between any two prototype signal vectors s_i and s_j is constant:

$$d(s_i,\,s_j) = \|\,s_i\,-\,s_j\,\| = \sqrt{2E} \qquad \text{for } i \ne j \tag{3.58}$$

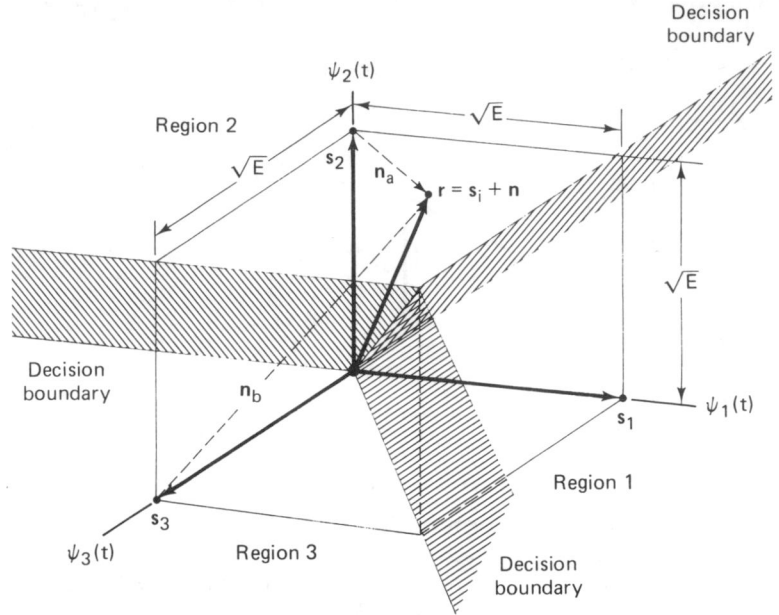

Figure 3.14 Partitioning the signal space for a 3-ary FSK signal.

Figure 3.14 illustrates the prototype signal vectors and the decision regions for a 3-ary ($M = 3$) coherently detected FSK system. As in the PSK case, the signal space is partitioned into M distinct regions, each containing one prototype signal vector; here, because the decision region is three-dimensional, the decision boundaries are planes instead of lines. The optimum decision rule is to decide that the transmitted signal belongs to the class whose index corresponds to the region where the received signal is found. In Figure 3.14, a received signal vector **r** is shown in region 2. Using the decision rule stated above, the detector classifies **r** as signal s_2. Since the noise is a Gaussian random vector, there is a probability greater than zero that **r** could have been produced by some signal other than s_2. For example, if the transmitter had sent s_2, then **r** would be the sum of signal plus noise, $s_2 + \mathbf{n}_a$, and the decision to choose s_2 is correct; however, if the transmitter had actually sent s_3, then **r** would be the sum of signal plus noise, $s_3 + \mathbf{n}_b$ and the decision to select s_2 is an error. The error performance of coherently detected FSK systems is treated in Section 3.7.3.

3.6 NONCOHERENT DETECTION

3.6.1 Detection of Differential PSK

The name *differential PSK* (DPSK) sometimes needs clarification because two separate aspects of the modulation/demodulation format are being referred to: the encoding procedure and the detection procedure. The term *differential encoding* refers to the procedure of encoding the data differentially; that is, the presence of a binary one or zero is manifested by the symbol's similarity or difference when compared to the preceding symbol. The term *differentially coherent detection* of differentially encoded PSK, the usual meaning of DPSK, refers to a detection scheme often classified as noncoherent because it does not require a reference in phase with the received carrier. Sometimes, differentially encoded PSK is *coherently* detected. This will be discussed in Section 3.7.2.

With noncoherent systems, no attempt is made to determine the actual value of the phase of the incoming signal. Therefore, if the transmitted waveform is

$$s_i(t) = \sqrt{\frac{2E}{T}} \cos [\omega_0 t + \theta_i(t)] \qquad \begin{matrix} 0 \leq t \leq T \\ i = 1, \dots, M \end{matrix}$$

the received signal can be characterized by

$$r(t) = \sqrt{\frac{2E}{T}} \cos [\omega_0 t + \theta_i(t) + \alpha] + n(t) \qquad \begin{matrix} 0 \leq t \leq T \\ i = 1, \dots, M \end{matrix} \qquad (3.59)$$

where α is an arbitrary constant and is typically assumed to be a random variable uniformly distributed between zero and 2π, and $n(t)$ is an AWGN process.

For coherent detection, matched filters (or their equivalents) are used; for noncoherent detection, this is not possible because the matched filter output is a function of the unknown angle α. However, if we assume that α varies slowly relative to two period times $(2T)$, the phase difference between two successive waveforms, $\theta_j(T_1)$ and $\theta_k(T_2)$ is independent of α, that is,

$$[\theta_k(T_2) + \alpha] - [\theta_j(T_1) + \alpha] = \theta_k(T_2) - \theta_j(T_1) = \phi_i(T_2) \qquad (3.60)$$

The basis for *differentially coherent detection* of differentially encoded PSK (DPSK) is as follows. The carrier phase of the previous signaling interval can be used as a phase reference for demodulation. Its use requires *differential encoding* of the message sequence at the transmitter since the information is carried by the difference in phase between two successive waveforms. To send the ith message $(i = 1, 2, \dots, M)$, the present signal waveform must have its phase advanced by $\phi_i = 2\pi i/M$ radians over the previous waveform. The detector, in general, calculates the coordinates of the incoming signal by correlating it with locally generated waveforms such as $\sqrt{2/T} \cos \omega_0 t$ and $\sqrt{2/T} \sin \omega_0 t$. The detector then measures the angle between the currently received signal vector and the previously received signal vector, as illustrated in Figure 3.15.

In general, DPSK signaling performs less efficiently than PSK, because the errors in DPSK tend to propagate (to adjacent symbol times) due to the correlation between signaling waveforms. One way of viewing the difference between PSK and DPSK is that the former compares the received signal with a clean reference; in the latter, however, two noisy signals are compared with each other. We might say that there is twice as much noise associated with DPSK signaling compared to PSK signaling. Consequently, as a first guess, we might estimate that DPSK manifests a degradation of approximately 3 dB when compared with PSK; this degradation decreases rapidly with increasing signal-to-noise ratio. The trade-off

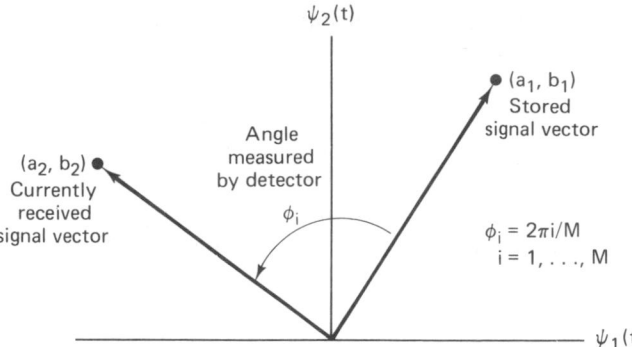

Figure 3.15 Signal space for DPSK.

for this performance loss is reduced system complexity. The error performance for the detection of DPSK is treated in Section 3.7.5.

3.6.2 Binary Differential PSK Example

The essence of differentially coherent detection in DPSK is that the identity of the data is inferred from the changes in phase from symbol to symbol. Therefore, since the data are detected by differentially examining the waveform, the transmitted waveform would first be encoded in a differential fashion. Figure 3.16a illustrates a differential encoding of a binary message data stream, $m(k)$, where k is the sample time index. The differential encoding starts (third row in the figure) with the first bit of the code bit sequence, $c(k = 0)$, chosen arbitrarily (here taken to be a one). Then the sequence of encoded bits, $c(k)$, can, in general, be encoded in one of two ways:

$$c(k) = c(k - 1) \oplus m(k) \tag{3.61}$$

or

$$c(k) = \overline{c(k - 1) \oplus m(k)} \tag{3.62}$$

where the symbol $\oplus$ represents modulo-2 addition (defined in Section 2.12.3) and the overbar denotes complement. In Figure 3.16a the differentially encoded message was obtained by using Equation (3.62). In other words, the present code bit, $c(k)$, is a one if the message bit, $m(k)$, and the prior coded bit, $c(k - 1)$, are the same, otherwise, $c(k)$ is a zero. The fourth row translates the coded bit sequence, $c(k)$, into the phase shift sequence, $\theta(k)$, where a one is characterized by a $180°$ phase shift, and a zero is characterized by a $0°$ phase shift.

Figure 3.16b illustrates the binary DPSK detection scheme in block diagram form. Notice that the basic product integrator of Figure 3.7 is the essence of this detection process; as with coherent PSK, we are still attempting to correlate a received signal with a reference. The interesting difference here is that the reference signal is simply a delayed version of the received signal. In other words, during each symbol time, we are matching a received symbol with the prior symbol and looking for a correlation or an anticorrelation (180° out of phase).

Consider the received signal with phase shift sequence, $\theta(k)$, entering the

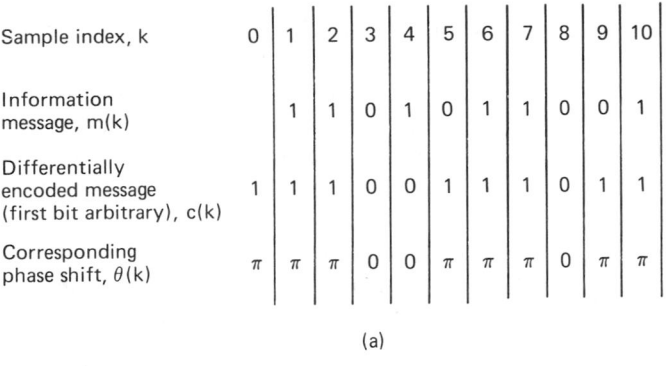

Sample index, k	0	1	2	3	4	5	6	7	8	9	10
Information message, m(k)		1	1	0	1	0	1	1	0	0	1
Differentially encoded message (first bit arbitrary), c(k)	1	1	1	0	0	1	1	1	0	1	1
Corresponding phase shift, $\theta(k)$	π	π	π	0	0	π	π	π	0	π	π

(a)

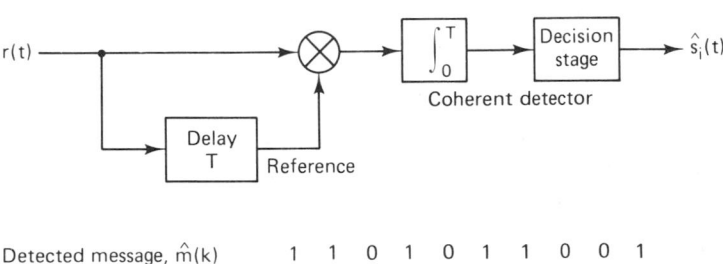

Detected message, $\hat{m}(k)$ 1   1   0   1   0   1   1   0   0   1

(b)

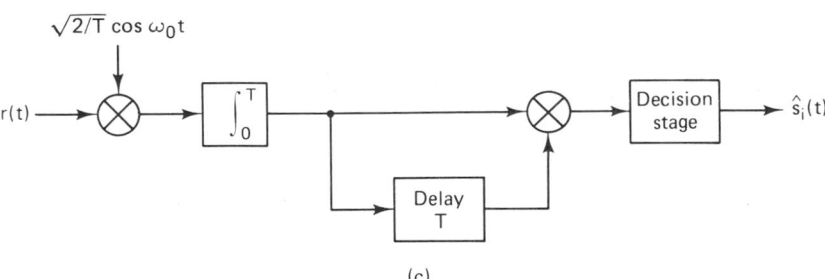

(c)

Figure 3.16 Differential PSK (DPSK). (a) Differential encoding. (b) Differentially coherent detection. (c) Optimum differentially coherent detection.

detector of Figure 3.16b, in the absence of noise. The phase, $\theta(k = 1)$, is matched with $\theta(k = 0)$; they have the same value, π; hence the first bit of the detected output is $\hat{m}(k = 1) = 1$. Then $\theta(k = 2)$ is matched with $\theta(k = 1)$; again they have the same value, and $\hat{m}(k = 2) = 1$. Then $\theta(k = 3)$ is matched with $\theta(k = 2)$; they are different, so that $\hat{m}(k = 3) = 0$, and so on.

It must be pointed out that the detector in Figure 3.16b is suboptimum [5] in the sense of error performance. The optimum differential detector for DPSK requires a reference carrier in frequency but not necessarily in phase with the received carrier. Hence the optimum differential detector is shown in Figure 3.16c [6]. Its performance is treated in Section 3.7.5.

3.6.3 Noncoherent Detection of FSK

A detector for the noncoherent detection of FSK waveforms described by Equation (3.27) can be implemented with correlators similar to those shown in Figure 3.7. However, the hardware must be configured as an *energy detector*, without exploiting phase measurements. For this reason, the noncoherent detector typically requires twice as many channel branches as the coherent detector. Figure 3.17 illustrates the in-phase (I) and quadrature (Q) channels used to detect a binary FSK (BFSK) signal set noncoherently. Notice that the upper two branches are configured to detect the signal with frequency ω_1; the reference signals are $\sqrt{2/T} \cos \omega_1 t$ for the I branch and $\sqrt{2/T} \sin \omega_1 t$ for the Q branch. Similarly, the lower two branches are configured to detect the signal with frequency ω_2; the reference signals are $\sqrt{2/T} \cos \omega_2 t$ for the I branch and $\sqrt{2/T} \sin \omega_2 t$ for the Q branch. Imagine that the received signal $r(t)$, by chance alone, is exactly of the form $\cos \omega_1 t + n(t)$; that is, the phase is exactly zero, and thus the signal component of the received signal exactly matches the top-branch reference signal with regard to frequency and phase. In that event, the product integrator of the top branch should yield the maximum output. The second branch should yield a near-zero output (integrated zero-mean noise) since its reference signal $\sqrt{2/T} \sin \omega_1 t$

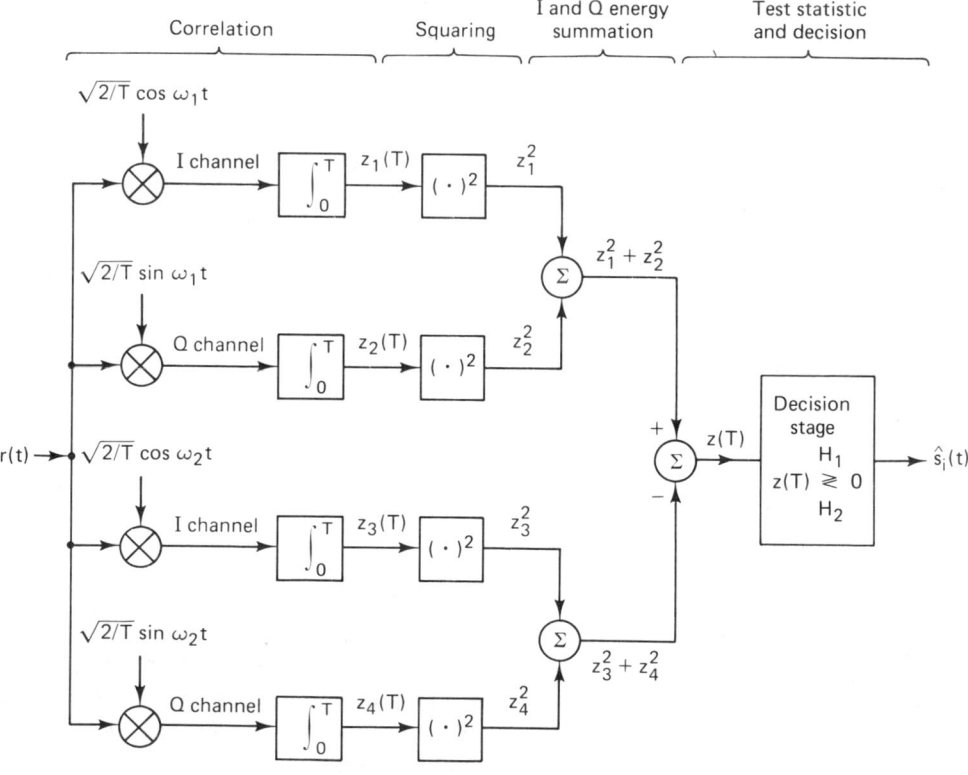

Figure 3.17   Quadrature receiver.

is orthogonal to the signal component of $r(t)$. The third and fourth branches should also yield near-zero outputs since their ω_2 reference signals are also orthogonal to the signal component of $r(t)$.

Now, imagine a different scenario; suppose that by chance alone, the received signal, $r(t)$, is of the form $\sin \omega_1 t + n(t)$. In that event, the second branch in Figure 3.17 should yield the maximum output, while the others should yield near-zero outputs. In actual practice, the most likely scenario is that $r(t)$ is of the form $\cos (\omega_1 t + \phi) + n(t)$; that is, the incoming signal will *partially* correlate with the $\cos \omega_1 t$ reference and *partially* correlate with the $\sin \omega_1 t$ reference. Now it should be obvious why a noncoherent quadrature receiver uses twice as many branches as a coherent one; the receiver knows nothing about the incoming signal's phase. The receiver essentially resolves the signal into an I component and a Q (90° out of phase) component. In Figure 3.17 the blocks following the product integrators perform a squaring operation to prevent the appearance of any negative values. Then for each of the signal types in the set (two in this binary example) the energy from the I and Q channels is added. The final stage forms the test statistic, $z(T)$, and chooses the signal with frequency ω_1 or the signal with frequency ω_2 depending on which pair of energy detectors yielded the maximum output.

Another possible implementation for noncoherent FSK detection uses bandpass filters, centered at $f_i = \omega_i/2\pi$, with bandwidth, $W_f = 1/T$, followed by *envelope detectors*, as shown in Figure 3.18. An envelope detector consists of a rectifier and a low-pass filter. The detectors are matched to the *signal envelopes* and not to the signals themselves. The phase of the carrier is of no importance in defining the envelope; hence no phase information is used. In the case of binary FSK, the decision as to whether a one or a zero was transmitted is made on the basis of which of two envelope detectors has the largest amplitude at the moment of measurement. Similarly, for a multiple frequency shift keying (MFSK) system, the decision as to which of M signals was transmitted is made on the basis of which of the M envelope detectors has the maximum output.

Even though the envelope detector block diagram of Figure 3.18 looks func-

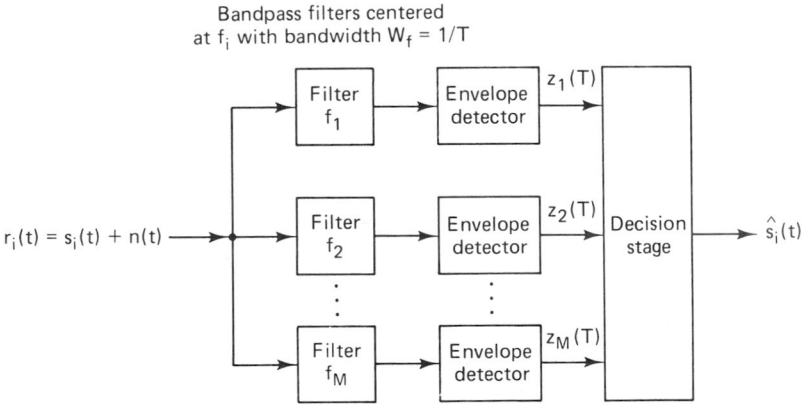

Figure 3.18 Noncoherent detection of FSK using envelope detectors.

tionally simpler than the quadrature receiver of Figure 3.17, the use of filters usually results in the envelope detector design having greater weight and cost than the quadrature receiver. Quadrature receivers can be implemented digitally; thus, with the advent of large-scale integrated (LSI) circuits, they are often the preferred choice for noncoherent detectors. The detector in Figure 3.18 can also be implemented digitally by performing discrete Fourier transformations instead of using analog filters, but such a design is usually more complex than a digital implementation of the quadrature receiver.

3.6.4 Minimum Required Tone Spacing for Noncoherent Orthogonal FSK Signaling

Frequency shift keying is usually implemented as orthogonal signaling where each tone (sinusoid) in the signal set cannot interfere with any of the other tones. In order for the signal set to be orthogonal, any pair of adjacent tones must have a frequency separation of a multiple of $1/T$ hertz. A tone with frequency f_i, that is switched on for a symbol duration of T seconds and then switched off, such as the FSK tone described in Equation (3.27), can be analytically described by

$$s_i(t) = (\cos 2\pi f_i t) \; \text{rect} \; (t/T)$$

$$\text{where rect} \; (t/T) = \begin{cases} 1 & \text{for } -T/2 \le t \le T/2 \\ 0 & \text{for } |t| > T/2 \end{cases}$$

The Fourier transform of $s_i(t)$, from Table A.1, is

$$\mathscr{F} \{s_i(t)\} = T \; \text{sinc} \; (f - f_i)T$$

where the sinc function is as defined in Equation (1.39). The spectra of two such adjacent tones, tone 1 with frequency f_1 and tone 2 with frequency f_2, are plotted in Figure 3.19.

In order that the two tones not interfere with each other during detection, the peak of the spectrum of tone 1 must coincide with one of the zero crossings of the spectrum of tone 2 and similarly, the peak of the tone 2 spectrum must coincide with one of the zero crossings of the tone 1 spectrum. The frequency difference between the center of the spectral main lobe and the first zero crossing represents the *minimum required spacing*. This corresponds to a minimum tone separation of $1/T$ hertz.

Example 3.3   Minimum Tone Spacing for Noncoherent Orthogonal FSK

Consider two waveforms $\cos (2\pi f_1 t + \phi)$ and $\cos 2\pi f_2 t$ to be used for *noncoherent* FSK signaling, where $f_1 > f_2$. The symbol rate is equal to $1/T$ symbols/s, where T is the symbol duration and ϕ is a constant arbitrary angle from 0 to 2π.

(a) Prove that the minimum tone spacing for *noncoherently detected* orthogonal FSK signaling is $1/T$.

(b) What is the minimum tone spacing for *coherently detected* orthogonal FSK signaling?

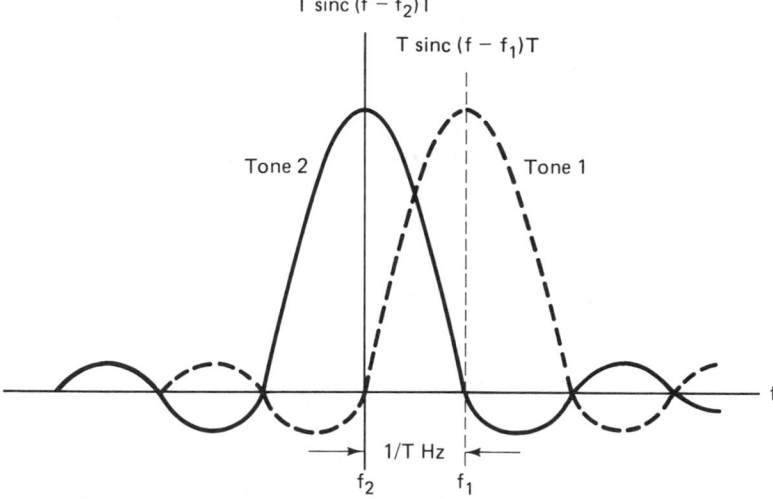

Figure 3.19 Minimum tone spacing for noncoherently detected orthogonal FSK signaling.

Solution

(a) For the two waveforms to be orthogonal, they must fulfill the orthogonality constraint of Equation (3.2):

$$\int_0^T \cos(2\pi f_1 t + \phi) \cos 2\pi f_2 t \, dt = 0 \tag{3.63}$$

Using the basic trigonometric identities shown in Equations (D.6) and (D.1) to (D.3), we can write Equation (3.63) as

$$\cos\phi \int_0^T \cos 2\pi f_1 t \cos 2\pi f_2 t \, dt$$

$$- \sin\phi \int_0^T \sin 2\pi f_1 t \cos 2\pi f_2 t \, dt = 0 \tag{3.64}$$

$$\cos\phi \int_0^T [\cos 2\pi(f_1 + f_2)t + \cos 2\pi(f_1 - f_2)t] \, dt$$

$$- \sin\phi \int_0^T [\sin 2\pi(f_1 + f_2)t + \sin 2\pi(f_1 - f_2)t] \, dt = 0 \tag{3.65}$$

$$\cos\phi \left[\frac{\sin 2\pi(f_1 + f_2)t}{2\pi(f_1 + f_2)} + \frac{\sin 2\pi(f_1 - f_2)t}{2\pi(f_1 - f_2)}\right]_0^T$$

$$+ \sin\phi \left[\frac{\cos 2\pi(f_1 + f_2)t}{2\pi(f_1 + f_2)} + \frac{\cos 2\pi(f_1 - f_2)t}{2\pi(f_1 - f_2)}\right]_0^T = 0 \tag{3.66}$$

$$\cos\phi \left[\frac{\sin 2\pi(f_1 + f_2)T}{2\pi(f_1 + f_2)} + \frac{\sin 2\pi(f_1 - f_2)T}{2\pi(f_1 - f_2)}\right]$$

$$+ \sin\phi \left[\frac{\cos 2\pi(f_1 + f_2)T - 1}{2\pi(f_1 + f_2)} + \frac{\cos 2\pi(f_1 - f_2)T - 1}{2\pi(f_1 - f_2)}\right] = 0 \tag{3.67}$$

We can assume that $f_1 + f_2 \gg 1$ and can thus make the following approximation:

$$\frac{\sin 2\pi(f_1 + f_2)T}{2\pi(f_1 + f_2)} \cong \frac{\cos 2\pi(f_1 + f_2)T}{2\pi(f_1 + f_2)} \cong 0 \qquad (3.68)$$

Then, combining Equations (3.67) and (3.68), we can write

$$\cos \phi \sin 2\pi(f_1 - f_2)T + \sin \phi \, [\cos 2\pi(f_1 - f_2)T - 1] \simeq 0 \qquad (3.69)$$

Note that for arbitrary ϕ, the terms in Equation (3.69) can sum to zero only when $\sin 2\pi(f_1 - f_2)T = 0$, and simultaneously $\cos 2\pi(f_1 - f_2)T = 1$.

Since

$$\sin x = 0 \qquad \text{for } x = n\pi$$

and

$$\cos x = 1 \qquad \text{for } x = 2k\pi$$

where n and k are integers, then both $\sin x = 0$ and $\cos x = 1$ occur simultaneously when $n = 2k$. From Equation (3.69), for arbitrary ϕ, we can therefore write:

$$2\pi(f_1 - f_2)T = 2k\pi$$
$$f_1 - f_2 = \frac{k}{T} \qquad (3.70)$$

Thus the minimum tone spacing for *noncoherent* FSK signaling occurs for $k = 1$:

$$f_1 - f_2 = \frac{1}{T} \qquad (3.71)$$

(b) To find the minimum tone spacing for *coherent* FSK, where the angle ϕ is zero, we simply rewrite Equation (3.69) with $\phi = 0$, which gives

$$\sin 2\pi(f_1 - f_2)T = 0 \qquad (3.72)$$

$$f_1 - f_2 = \frac{n}{2T} \qquad (3.73)$$

Thus the minimum tone spacing for *coherent* FSK signaling occurs for $n = 1$ as follows:

$$f_1 - f_2 = \frac{1}{2T} \qquad (3.74)$$

Therefore, for the same symbol rate, coherently detected FSK can occupy less bandwidth than noncoherently detected FSK and still retain orthogonal signaling. We can say that coherent FSK is more *bandwidth efficient*. The subject of bandwidth efficiency is addressed in greater detail in Chapter 7.

3.7 ERROR PERFORMANCE FOR BINARY SYSTEMS

3.7.1 Probability of Bit Error
for Coherently Detected BPSK

An important measure of performance used for comparing digital modulation schemes is the probability of error, P_E. For the correlator or matched filter detector, the calculations for obtaining P_E can be viewed geometrically (see Figure 3.6). They involve finding the probability that given a particular transmitted signal vector, say s_1, the noise vector, n, will give rise to a received signal falling outside region 1. The probability of the detector making an incorrect decision is termed the *probability of symbol error* (P_E). It is often convenient to specify system performance by the probability of bit error (P_B), even when decisions are made on the basis of symbols for which $M > 2$. The relationship between P_B and P_E is treated in Section 3.9.3 for orthogonal signaling and in Section 3.9.4 for multiple phase signaling.

For convenience, this section is restricted to the coherent detection of BPSK modulation. For this case the symbol error probability is the bit error probability. Assume that the signals are equally likely. Also assume that when signal, $s_i(t)$ ($i = 1, 2$), is transmitted, the received signal, $r(t)$, is equal to $s_i(t) + n(t)$, where $n(t)$ is an AWGN process. The antipodal signals, $s_1(t)$ and $s_2(t)$, can be characterized in a one-dimensional signal space as described in Section 3.5.1, where

$$s_1(t) = \sqrt{E}\psi_1(t)$$
$$0 \le t \le T \qquad (3.75)$$
$$s_2(t) = -\sqrt{E}\psi_1(t)$$

The decision stage of the detector will choose the $s_i(t)$ with the largest correlator output $z_i(T)$, or in this case of equal-energy antipodal signals, the detector, using the decision rule in Equation (3.39a), decides

$$
\begin{array}{ll}
s_1(t) & \text{if } z(T) > \gamma_0 = 0 \\
s_2(t) & \text{otherwise}
\end{array}
\qquad (3.76)
$$

Two types of errors can be made, as shown in Figure 3.9: The first type of error takes place if signal $s_1(t)$ is transmitted but the noise is such that the detector measures a negative value for $z(T)$ and chooses hypothesis H_2 [the hypothesis that signal $s_2(t)$ was sent]. The second type of error takes place if signal $s_2(t)$ is transmitted but the detector measures a positive value for $z(T)$ and chooses hypothesis H_1 [the hypothesis that signal $s_1(t)$ was sent].

To calculate the probability of a bit error, P_B, for this binary *minimum error* detector, we use the relationships developed in Section 2.9, starting with Equation (2.36b):

$$P_B = P(H_2|s_1)P(s_1) + P(H_1|s_2)P(s_2) \qquad (3.77)$$

For the case when the a priori probabilities are equal, that is, $P(s_1) = P(s_2) = \frac{1}{2}$, we can write

$$P_B = \tfrac{1}{2}P(H_2|s_1) + \tfrac{1}{2}P(H_1|s_2) \tag{3.78}$$

Because of the symmetry of the probability density functions in Figure 3.9, we can also write

$$P_B = P(H_2|s_1) = P(H_1|s_2) \tag{3.79}$$

Thus the probability of a bit error, P_B, is numerically equal to the area under the "tail" of either pdf, $p(z|s_1)$ or $p(z|s_2)$, that falls on the "incorrect" side of the threshold. We can therefore compute P_B by integrating $p(z|s_1)$ between the limits $-\infty$ and γ_0, or as shown below, by integrating $p(z|s_2)$ between the limits γ_0 and ∞.

$$P_B = \int_{\gamma_0 = (a_1 + a_2)/2}^{\infty} p(z|s_2)\, dz \tag{3.80}$$

where the likelihoods, $p(z|s_i)$ $(i = 1, 2)$, are Gaussian functions with mean value, a_i, and the optimum threshold, γ_0, as shown in Section B.3.1, is equal to $(a_1 + a_2)/2$. The area-related probability of bit error, P_B, is seen to be the shaded area in Figure 3.9. It is shown in Section B.3.2 that Equation (3.80) reduces to

$$P_B = \int_{(a_1 - a_2)/2\sigma_0}^{\infty} \frac{1}{\sqrt{2\pi}} \exp\left(-\frac{u^2}{2}\right) du = Q\left(\frac{a_1 - a_2}{2\sigma_0}\right) \tag{3.81}$$

where σ_0 is the standard deviation of the noise out of the correlator. The function, $Q(x)$, called the *complementary error function* or *co-error function*, is defined as

$$Q(X) = \frac{1}{\sqrt{2\pi}} \int_x^{\infty} \exp\left(-\frac{u^2}{2}\right) du \tag{3.82}$$

and is described in greater detail in Sections 2.9 and B.3.2.

For equal-energy antipodal signaling, such as the BPSK format in Equation (3.75), the receiver output signal components are $a_1 = \sqrt{E_b}$ when $s_1(t)$ is sent and $a_2 = -\sqrt{E_b}$ when $s_2(t)$ is sent, where E_b is the signal energy per binary symbol. For AWGN we can replace the noise variance, σ_0^2, out of the correlator with $N_0/2$ (see Appendix C), so that we can rewrite Equation (3.81) as follows:

$$P_B = \int_{\sqrt{2E_b/N_0}}^{\infty} \frac{1}{\sqrt{2\pi}} \exp\left(-\frac{u^2}{2}\right) du \tag{3.83}$$

$$= Q\left(\sqrt{\frac{2E_b}{N_0}}\right) \tag{3.84}$$

This result could also have been obtained by noting that the energy difference, E_d, between the *antipodal signal vectors*, s_1 and s_2, with amplitudes of $\pm\sqrt{E_b}$, as seen in Figure 3.20a, can be computed as the square of the distance between

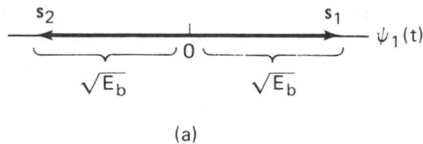

(a)

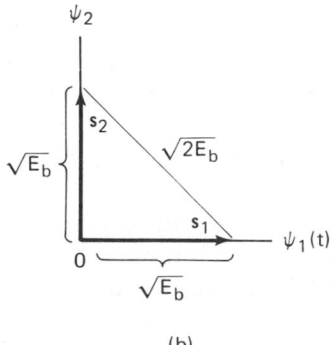

(b)

Figure 3.20 Binary signal vectors. (a) Antipodal. (b) Orthogonal.

the heads of the antipodal vectors, or in terms of the waveforms

$$E_d = \int_0^T [s_1(t) - s_2(t)]^2 \, dt \tag{3.85}$$

$$= \int_0^T s_1^2(t) \, dt + \int_0^T s_2^2(t) \, dt - 2 \int_0^T s_1(t)s_2(t) \, dt \tag{3.86}$$

Assuming equal energy signals,

$$E_b = \int_0^T s_1^2(t) \, dt = \int_0^T s_2^2(t) \, dt \tag{3.87}$$

$$E_d = 2E_b - 2E_b\rho = 2E_b(1 - \rho) \tag{3.88}$$

where

$$\rho = \frac{1}{E_b} \int_0^T s_1(t)s_2(t) \, dt \tag{3.89}$$

is the time cross-correlation coefficient and E_b is the average energy of the binary signals, $s_1(t)$ and $s_2(t)$. The correlation coefficient, ρ, is a measure of similarity between the two signals, $s_1(t)$ and $s_2(t)$, such that

$$-1 \leq \rho \leq 1 \tag{3.90}$$

In terms of signal vectors, the cross-correlation coefficient can be written

$$\rho = \cos \theta \tag{3.91}$$

where θ is the angle between the two signal vectors s_1 and s_2 (see Figure 3.6). In Equation (2.62), we developed an expression for the probability of bit error in terms of the energy difference between the two binary signals, as follows:

$$P_B = Q\left(\sqrt{\frac{E_d}{2N_0}}\right) \tag{3.92}$$

Substituting Equation (3.88) into Equation (3.92), we get

$$P_B = Q\left[\sqrt{\frac{E_b(1 - \rho)}{N_0}}\right] \tag{3.93}$$

For $\rho = 1$ (or $\theta = 0$), the signals are perfectly correlated (identical). For $\rho = -1$ (or $\theta = \pi$), the signals are anticorrelated (antipodal). Since the binary PSK signals are antipodal, we can set $\rho = -1$, and Equation (3.93) is then identical to Equation (3.84).

Note that the bit error probability, P_B, for the coherent detection of bandpass antipodal signals, as seen in Equation (3.84), is the same as the P_B for the matched filter detection of baseband antipodal (bipolar) signals in Equation (2.67).

3.7.1.1 The Basic SNR Parameter for Digital Communication Systems

The parameter E_b/N_0 in Equation (3.84) can be expressed as the ratio of average signal power to average noise power, S/N (or SNR). By introducing the signal bandwidth W, we can write the following identities, showing the relationship between E_b/N_0 and SNR for binary signals.

$$\frac{E_b}{N_0} = \frac{ST}{N_0} = \frac{S}{RN_0} = \frac{SW}{RN_0W} = \frac{S}{N}\left(\frac{W}{R}\right) \tag{3.94}$$

where

$$S = \text{average modulating signal power}$$

$$T = \text{bit time duration}$$

$$R = 1/T = \text{bit rate}$$

$$N = N_0W$$

Analysis similar to that used for developing P_B in Equations (3.84) and (3.93) is used in finding the P_B expressions for other types of modulation. Figure 3.21 illustrates the "waterfall-like" shape of most probability of error curves in the field of digital communications. The curve describes a system's error probability performance in terms of available E_b/N_0. For $E_b/N_0 \geq x_0$, $P_E \leq P_0$. The dimensionless ratio E_b/N_0 is a standard quality measure for digital communications system performance. Note that optimum digital signal detection implies a correlator (or matched filter) implementation, in which case the signal bandwidth is equal to the noise bandwidth. Often we are faced with a system model for which this is not the case; in practice, we include a factor in the required E_b/N_0 that accounts for such suboptimal detection performance. Required E_b/N_0 can be considered a metric that characterizes the performance of one system versus another;

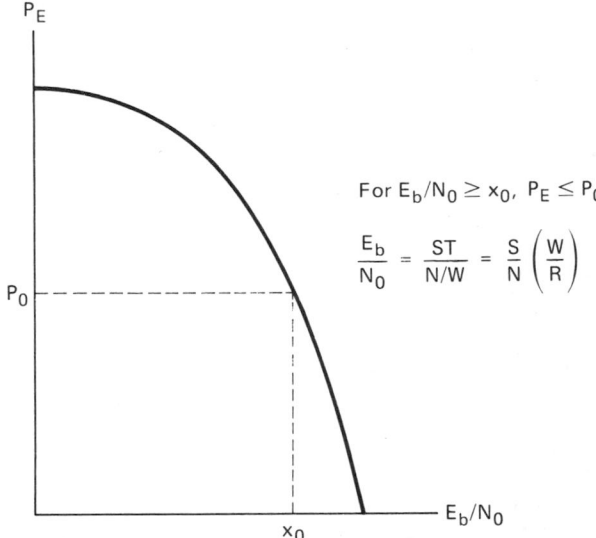

For $E_b/N_0 \geq x_0$, $P_E \leq P_0$

$$\frac{E_b}{N_0} = \frac{ST}{N/W} = \frac{S}{N}\left(\frac{W}{R}\right)$$

Figure 3.21 General shape of the P_E versus E_b/N_0 curve.

the *smaller* the required E_b/N_0, the *more efficient* is the system modulation and detection process for a given probability of error. Figure 3.22 is a plot comparing the bit error probability, P_B, of several binary modulation/demodulation types. The P_B for coherent detection of PSK, as shown in Equation (3.84), is plotted as the leftmost P_B curve.

Example 3.4 Bit Error Probability for BPSK Signaling

Find the bit error probability for a BPSK system with a bit rate of 1 Mbit/s. The received waveforms, $s_1(t) = A \cos \omega_0 t$ and $s_2(t) = -A \cos \omega_0 t$, are coherently detected with a matched filter. The value of A is 10 mV. Assume that the single-sided noise power spectral density is $N_0 = 10^{-11}$ W/Hz and that signal power and energy per bit are normalized relative to a 1-Ω load.

Solution

$$A = \sqrt{\frac{2E_b}{T}} = 10^{-2} \text{ V} \qquad T = \frac{1}{R} = 10^{-6} \text{ s}$$

Thus

$$E_b = \frac{A^2}{2} T = 5 \times 10^{-11} \text{ J} \quad \text{and} \quad \sqrt{\frac{2E_b}{N_0}} = 3.16$$

$$P_B = Q\left(\sqrt{\frac{2E_b}{N_0}}\right) = Q(3.16)$$

Using Table B.1 or Equation (2.43), we obtain

$$P_B = 8 \times 10^{-4}$$

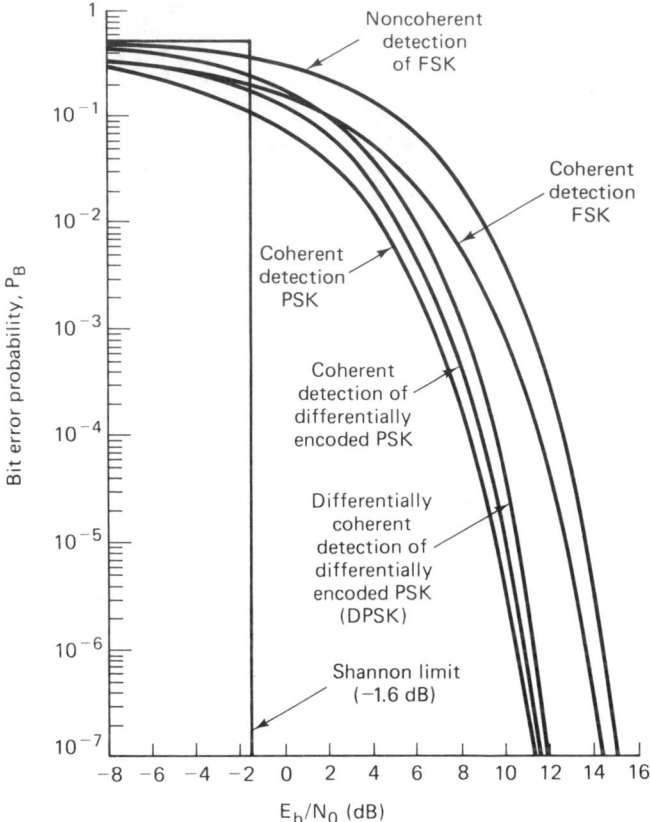

Figure 3.22 Bit error probability for several types of binary systems.

3.7.2 Probability of Bit Error for Coherently Detected Differentially Encoded PSK

Channel waveforms sometimes experience inversion; for example, when using a coherent reference generated by a phase-locked loop (see Chapter 8), one may have phase ambiguity. If the carrier phase were reversed in a DPSK modulation application, what would be the effect on the message? The only effect would be an error in the bit during which inversion occurred or the bit just after inversion, since the message information is encoded in the similarity or difference between adjacent symbols. The similarity or difference quality remains unchanged if the carrier is inverted. Sometimes, systems are *differentially encoded* and *coherently detected*, simply to avoid these phase ambiguities.

The probability of bit error for coherently detected, differentially encoded PSK is given by [7]

$$P_B = 2Q\left(\sqrt{\frac{2E_b}{N_0}}\right)\left[1 - Q\left(\sqrt{\frac{2E_b}{N_0}}\right)\right] \tag{3.95}$$

This relationship is plotted in Figure 3.22. Notice that there is a slight degradation of error performance compared to the coherent detection of PSK. This is due to the differential encoding since any single detection error results in two decision errors. Error performance for the more popular differentially coherent detection (DPSK) is covered in Section 3.7.5.

3.7.3 Probability of Bit Error for Coherently Detected FSK

Equations (3.83) and (3.84) describe the probability of bit error for coherent antipodal signals. A more general treatment for binary coherent signals (not limited to antipodal signals) yields the following equation for P_B [8]:

$$P_B = \frac{1}{\sqrt{2\pi}} \int_{\sqrt{(1-\rho)E_b/N_0}}^{\infty} \exp\left(-\frac{u^2}{2}\right) du \qquad (3.96)$$

From Equation (3.91), $\rho = \cos\theta$ is the time cross-correlation coefficient between signal $s_1(t)$ and $s_2(t)$, where θ is the angle between signal vectors s_1 and s_2 (see Figure 3.6). For antipodal signals such as BPSK, $\theta = \pi$, thus $\rho = -1$.

For orthogonal signals such as binary FSK (BFSK), $\theta = \pi/2$, since the s_1 and s_2 vectors are perpendicular to each other; thus $\rho = 0$, as can be verified with Equation (3.89), and Equation (3.96) can then be written

$$P_B = \frac{1}{\sqrt{2\pi}} \int_{\sqrt{E_b/N_0}}^{\infty} \exp\left(-\frac{u^2}{2}\right) du = Q\left(\sqrt{\frac{E_b}{N_0}}\right) \qquad (3.97)$$

where the *co-error function, $Q(x)$,* is defined in Equation (3.82). The result could also have been obtained by noting that the energy difference between the orthogonal signal vectors, s_1 and s_2, with amplitudes of $\sqrt{E_b}$, as shown in Figure 3.20b, can be computed as the square of the distance between the heads of the orthogonal vectors, to be $E_d = 2E_b$. Using this result in Equation (3.92) yields the same result as in Equation (3.97). Equation (3.97) is plotted in Figure 3.22 (coherent detection of FSK). If we compare Equation (3.97) with Equation (3.84), we can see that 3 dB (a factor of 2) more E_b/N_0 is required for BFSK to provide the same performance as BPSK. It should not be surprising that the performance of BFSK signaling is worse than BPSK signaling, since for a given signal power, orthogonal vectors are spaced closer to one another than antipodal vectors.

The bit error probability, P_B, for the coherent detection of orthogonal bandpass signals as seen in Equation (3.97) is the same as the P_B for the matched filter detection of baseband unipolar signals in Equation (2.64). As mentioned earlier, the details of on–off keying (OOK) are not treated in this book. However, it is worth noting that the P_B, described in Equation (3.97), is also identical to the error performance for the coherent detection of OOK signaling (matched filter reception).

3.7.4 Probability of Bit Error for Noncoherently Detected FSK

Consider the equally likely binary FSK signal set, $\{s_i(t)\}$, defined in Equation (3.27) as follows:

$$s_i(t) = \sqrt{\frac{2E}{T}} \cos(\omega_i t + \phi) \qquad 0 \leq t \leq T, \qquad i = 1, 2$$

The phase term, ϕ, is unknown and assumed constant. The detector is characterized by $M = 2$ channels of bandpass filters and envelope detectors, as shown in Figure 3.18. The input to the detector consists of the received signal, $r(t) = s_i(t) + n(t)$, where $n(t)$ is a white Gaussian noise process with two-sided power spectral density, $N_0/2$. Assume that $s_1(t)$ and $s_2(t)$ are separated in frequency sufficiently that they have negligible overlap. We start the probability of error, P_B, computation the same way that we did for coherently detected PSK, with Equation (3.78).

$$P_B = \tfrac{1}{2}P(H_2|s_1) + \tfrac{1}{2}P(H_1|s_2) \tag{3.98}$$

$$= \frac{1}{2}\int_0^T p(z|s_1)\,dz + \frac{1}{2}\int_0^T p(z|s_2)\,dz$$

For the binary case, the *test statistic*, $z(T)$, is defined by $z_1(T) - z_2(T)$. Assume that the bandwidth of the filter, W_f, is $1/T$, so that the envelope of the FSK signal is (approximately) preserved at the filter output. If there was no noise at the receiver, the value of $z(T) = \sqrt{2E/T}$ when $s_1(t)$ is sent, and $z(T) = -\sqrt{2E/T}$ when $s_2(t)$ is sent. Because of this symmetry, the optimum threshold is $\gamma_0 = 0$. The pdf $p(z|s_1)$ is similar to $p(z|s_2)$; that is,

$$p(z|s_1) = p(-z|s_2) \tag{3.99}$$

Therefore, we can write

$$P_B = \int_0^T p(z|s_2)\,dz \tag{3.100}$$

or

$$P_B = P(z_1 > z_2|s_2) \tag{3.101}$$

where z_1 and z_2 denote the outputs $z_1(T)$ and $z_2(T)$ from the envelope detectors shown in Figure 3.18. For the case where the tone $s_2(t) = \cos \omega_2 t$ is sent, such that $r(t) = s_2(t) + n(t)$, the output, $z_1(T)$, is a *Gaussian noise random variable only*; it has no signal component. A Gaussian distribution into the *nonlinear envelope detector* yields a Rayleigh distribution at the output [8], so that

$$p(z_1|s_2) = \begin{cases} \dfrac{z_1}{\sigma_0^2} \exp\left(-\dfrac{z_1^2}{2\sigma_0^2}\right) & z_1 \geq 0 \\[2mm] 0 & z_1 < 0 \end{cases} \tag{3.102}$$

where σ_0^2 is the noise at the filter output. On the other hand, $z_2(T)$ has a Rician distribution, since the input to the lower envelope detector is a sinusoid plus noise [8]. The pdf, $p(z_2|s_2)$, is written as

$$p(z_2|s_2) = \begin{cases} \dfrac{z_2}{\sigma_0^2} \exp\left[-\dfrac{(z_2^2 + A^2)}{2\sigma_0^2}\right] I_0\left(\dfrac{z_2 A}{\sigma_0^2}\right) & z_2 \geq 0 \\ \\ 0 & z_2 < 0 \end{cases} \tag{3.103}$$

where $A = \sqrt{2E/T}$, and as before, σ_0^2 is the noise at the filter output. The function $I_0(x)$, known as the modified zero-order Bessel function of the first kind [9], is defined as

$$I_0(x) = \frac{1}{2\pi} \int_0^{2\pi} \exp\left(x \cos \theta\right) d\theta \tag{3.104}$$

When $s_2(t)$ is transmitted, the receiver makes an error whenever the envelope sample $z_1(T)$ obtained from the upper channel (due to noise alone) exceeds the envelope sample $z_2(T)$ obtained from the lower channel (due to signal plus noise). Thus the probability of this error can be obtained by integrating $p(z_1|s_2)$ with respect to z_1 from z_2 to infinity, and then averaging over all possible values of z_2. That is,

$$P_B = P(z_1 > z_2|s_2)$$

$$= \int_0^\infty p(z_2|s_2)\left[\int_{z_2}^\infty p(z_1|s_2)\, dz_1\right] dz_2 \tag{3.105}$$

$$= \int_0^\infty \frac{z_2}{\sigma_0^2} \exp\left[-\frac{(z_2^2 + A^2)}{2\sigma_0^2}\right] I_0\left(\frac{z_2 A}{\sigma_0^2}\right)\left[\int_{z_2}^\infty \frac{z_1}{\sigma_0^2} \exp\left(-\frac{z_1^2}{2\sigma_0^2}\right) dz_1\right] dz_2 \tag{3.106}$$

where $A = \sqrt{2E/T}$ and where the inner integral is the conditional probability of an error for a fixed value of z_2, given that $s_2(t)$ was sent, and the outer integral averages this conditional probability over all possible values of z_2. This integral can be evaluated [10], to yield

$$P_B = \frac{1}{2} \exp\left(-\frac{A^2}{4\sigma_0^2}\right) \tag{3.107}$$

Using Equation (1.19), we can express the filter output noise, σ_0^2, as

$$\sigma_0^2 = 2\left(\frac{N_0}{2}\right) W_f \tag{3.108}$$

where $G_n(f) = N_0/2$ and W_f is the filter bandwidth. Thus Equation (3.107) becomes

$$P_B = \frac{1}{2} \exp\left(-\frac{A^2}{4N_0 W_f}\right) \tag{3.109}$$

Equation (3.109) indicates that the error performance depends on the bandpass filter bandwidth, and that P_B becomes smaller as W_f is decreased. The result is valid only when the intersymbol interference (ISI) is negligible. The minimum W_f allowed (i.e., for no ISI) is obtained from Equation (2.77) with the filter roll-off factor $r = 0$. Thus $W_f = R$ bits/s $= 1/T$, and we can write Equation (3.109) as

$$P_B = \frac{1}{2} \exp\left(-\frac{A^2 T}{4N_0}\right) \tag{3.110}$$

$$= \frac{1}{2} \exp\left(-\frac{E_b}{2N_0}\right) \tag{3.111}$$

where $E_b = (1/2)A^2 T$ is the energy per bit. When comparing the error performance of noncoherent FSK with coherent FSK (see Figure 3.22), it is seen that for the same P_B, noncoherent FSK requires approximately 1 dB more E_b/N_0 than that for coherent FSK (for $P_B \leq 10^{-4}$). The noncoherent receiver is easier to implement, since coherent reference signals need not be generated. Therefore, almost all FSK receivers use noncoherent detection. It can be seen in the following section that when comparing noncoherent FSK to noncoherent DPSK, the same 3-dB difference occurs as for the comparison between coherent FSK and coherent PSK.

As mentioned earlier, the details of on–off keying (OOK) are not treated in this book. However, it is worth noting that the bit error probability, P_B, described in Equation (3.111) is identical to the P_B for the noncoherent detection of OOK signaling.

3.7.5 Probability of Bit Error for DPSK

Let us define a BPSK signal set

$$x_1(t) = \sqrt{\frac{2E}{T}} \cos(\omega_0 t + \phi) \qquad 0 \leq t \leq T$$

$$\tag{3.112}$$

$$x_2(t) = \sqrt{\frac{2E}{T}} \cos(\omega_0 t + \phi \pm \pi) \qquad 0 \leq t \leq T$$

A characteristic of DPSK is that there are no fixed decision regions in the signal space. Instead, the decision is based on the phase difference between successively received signals. Then for DPSK signaling we are really transmitting each bit with the binary signal pair

$$s_1(t) = (x_1, x_1) \quad \text{or} \quad (x_2, x_2) \qquad 0 \leq t \leq 2T$$
$$s_2(t) = (x_1, x_2) \quad \text{or} \quad (x_2, x_1) \qquad 0 \leq t \leq 2T \tag{3.113}$$

where (x_i, x_j) $(i, j = 1, 2)$ denotes $x_i(t)$ followed by $x_j(t)$ defined in Equation (3.112). The first T seconds of each waveform are actually the last T seconds of the previous waveform. Note that $s_1(t)$ and $s_2(t)$ can each have either of two possible forms and that $x_1(t)$ and $x_2(t)$ are antipodal signals. Thus the correlation between $s_1(t)$ and $s_2(t)$ for *any combination* of forms can be written as

$$z(2T) = \int_0^{2T} s_1(t)s_2(t)\, dt$$

$$= \int_0^T [x_1(t)]^2\, dt - \int_0^T [x_1(t)]^2\, dt = 0$$

(3.114)

Therefore, pairs of DPSK signals can be represented as orthogonal signals $2T$ seconds long. Detection could correspond to noncoherent envelope detection with four channels matched to each of the possible envelope outputs, as shown in Figure 3.23a. Since the two envelope detectors representing each symbol are negatives of each other, the envelope sample of each will be the same. Hence we can implement the detector as a single channel for $s_1(t)$ matched to either (x_1, x_1) or (x_2, x_2), and a single channel for $s_2(t)$ matched to either (x_1, x_2) or (x_2, x_1), as shown in Figure 3.23b. The DPSK detector is therefore reduced to a standard two-channel noncoherent detector. In reality, the filter can be matched to the difference signal so that only one channel is necessary. For orthogonal signals, this operates with the bit error probability in Equation (3.111). Since the DPSK signals have a bit interval of $2T$, the $s_i(t)$ signals defined in Equation (3.113) have

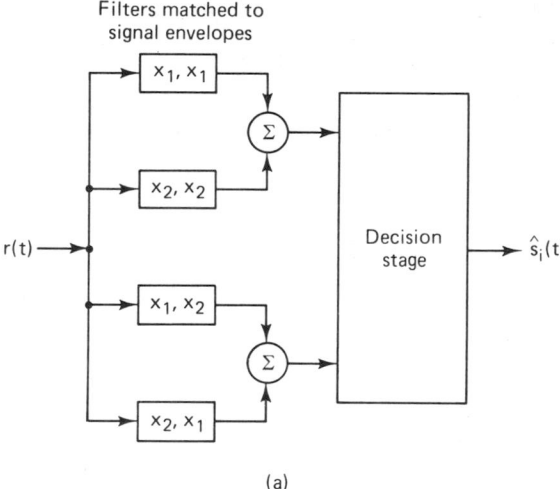

(a)

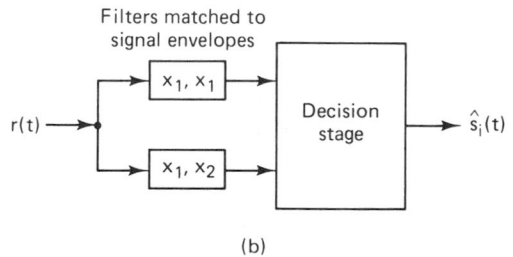

(b)

Figure 3.23 DPSK detection. (a) Four-channel differentially coherent detection of binary DPSK. (b) Equivalent two-channel detector for binary DPSK.

twice the energy of a signal defined over a single-symbol duration. Thus we may write P_B as

$$P_B = \frac{1}{2} \exp \left(-\frac{E_b}{N_0} \right) \tag{3.115}$$

Equation (3.115) is seen plotted in Figure 3.22, designated as differentially coherent detection of differentially encoded PSK, or simply DPSK. This expression is valid for the optimum DPSK detector shown in Figure 3.16c. For the detector shown in Figure 3.16b, the error probability will be slightly inferior to that given in Equation (3.115) [5]. When comparing the error performance of Equation (3.115) with that of coherent PSK (see Figure 3.22), it is seen that for the same P_B, DPSK requires approximately 1 dB more E_b/N_0 than does BPSK (for $P_B \leq 10^{-4}$). It is easier to implement a DPSK system than a PSK system, since the DPSK receiver does not need phase synchronization. For this reason, DPSK, although less efficient than PSK, is sometimes the preferred choice between the two.

3.7.6 Comparison of Bit Error Performance for Various Modulation Types

The P_B expressions for the best known of the binary modulation schemes discussed above are listed in Table 3.1 and are illustrated in Figure 3.22. For $P_B = 10^{-4}$, it can be seen that there is approximately a 4-dB difference between the best (coherent PSK) and the worst (noncoherent FSK) that were discussed here. In some cases, 4 dB is a small price to pay for the implementation simplicity gained in going from coherent PSK to noncoherent FSK; however, for other cases, even a 1-dB saving is worthwhile. There are other considerations besides P_B and system complexity; for example, in some cases (such as a randomly fading channel), a noncoherent system is more desirable because there may be difficulty in establishing and maintaining a coherent reference. Signals that can withstand significant degradation before their ability to be detected is affected are clearly desirable in military and space applications.

TABLE 3.1 Probability of Error for Selected Binary Modulation Schemes

Modulation	P_B
Coherent PSK	$Q\left(\sqrt{\dfrac{2E_b}{N_0}} \right)$
Noncoherent DPSK	$\dfrac{1}{2} \exp\left(-\dfrac{E_b}{N_0} \right)$
Coherent FSK	$Q\left(\sqrt{\dfrac{E_b}{N_0}} \right)$
Noncoherent FSK	$\dfrac{1}{2} \exp\left(-\dfrac{1}{2}\dfrac{E_b}{N_0} \right)$

3.8 *M*-ARY SIGNALING AND PERFORMANCE

3.8.1 Ideal Probability of Bit Error Performance

The typical probability of error versus E_b/N_0 curve was shown to have a waterfall-like shape in Figure 3.21. The probability of bit error (P_B) characteristics of various binary modulation schemes in AWGN also display this shape, as shown in Figure 3.22. What should an *ideal* P_B versus E_b/N_0 curve look like? Figure 3.24 displays the ideal characteristic as the *Shannon limit*. The limit represents the threshold E_b/N_0 below which reliable communication cannot be maintained. Shannon's work is described in greater detail in Chapter 7.

We can describe the ideal curve in Figure 3.24 as follows. For all values of E_b/N_0 above the Shannon limit, P_B is zero. Once E_b/N_0 is reduced below the Shannon limit, P_B degrades to the worst-case value of $\frac{1}{2}$. (Note that $P_B = 1$ is not the worst case for binary signaling, since that value is just as good as $P_B = 0$; if the probability of making a bit error is 100%, the bit stream could simply be inverted to retrieve the correct data.) It should be clear, by comparing the typical P_B curve with the ideal one in Figure 3.24 that the large arrow in the figure describes the desired direction of movement to achieve improved P_B performance.

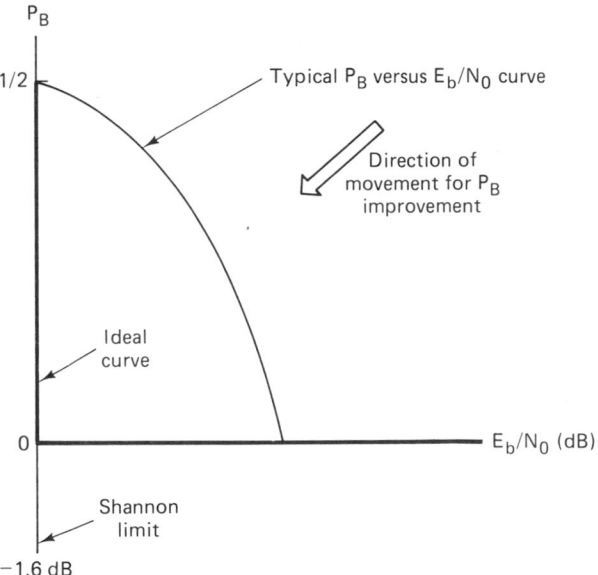

Figure 3.24 Ideal P_B versus E_b/N_0 curve.

3.8.2 *M*-ary Signaling

Let us review *M*-ary signaling. The processor considers *k* bits at a time. It instructs the modulator to produce one of $M = 2^k$ waveforms; binary signaling is the special case where $k = 1$. Does *M*-ary signaling improve or degrade performance? Be careful with your answer—the question is a loaded one. Figure 3.25 illustrates the probability of bit error, $P_B(M)$, versus E_b/N_0 for coherently detected *orthog-*

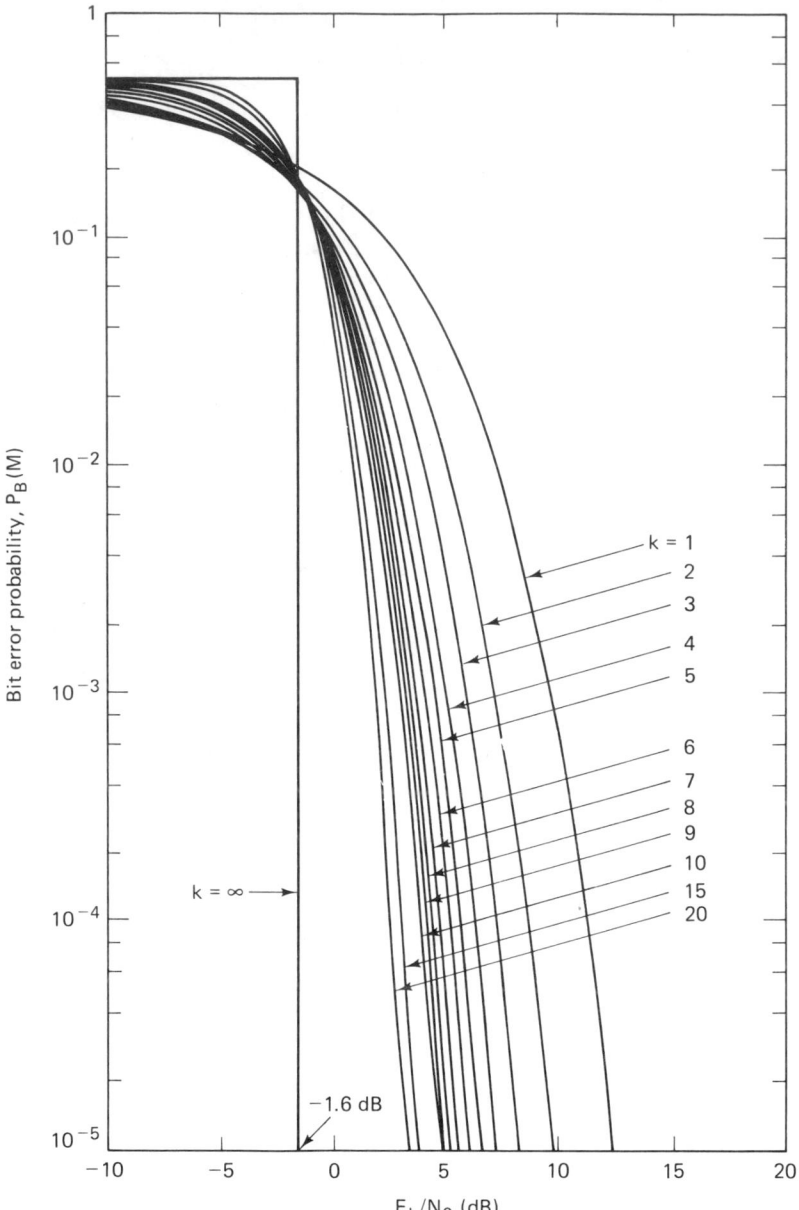

Figure 3.25 Bit error probability for coherently detected M-ary orthogonal signaling. (Reprinted from W. C. Lindsey and M. K. Simon, *Telecommunication Systems Engineering*, Prentice-Hall, Inc., Englewood Cliffs, N.J., 1973, courtesy of W. C. Lindsey and Marvin K. Simon.)

onal M-ary signaling over a Gaussian channel. Figure 3.26 similarly illustrates $P_B(M)$ versus E_b/N_0 for coherently detected *multiple phase M*-ary signaling over a Gaussian channel. In which direction do the curves move as the value of k (or M) increases? From Figure 3.24 we know the directions of curve movement for

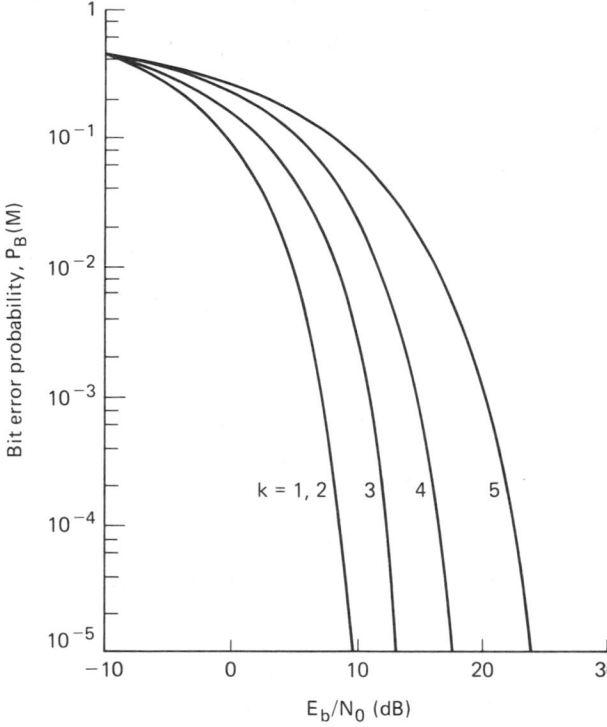

Figure 3.26 Bit error probability for coherently detected multiple phase signaling.

improved and degraded error performance. In Figure 3.25, as k increases, the curves move in the direction of improved error performance. In Figure 3.26, as k increases, the curves move in the direction of degraded error performance. Such movement tells us that M-ary signaling produces improved error performance with orthogonal signaling and degraded error performance with multiple phase signaling. Can that be true? Why would anyone ever use multiple phase PSK signaling if it provides degraded error performance compared to binary PSK signaling? It *is* true, and many systems do use multiple phase signaling. The question, as stated, is loaded because it implies that error probability versus E_b/N_0 is the *only* performance criterion; there are many others (e.g., bandwidth, power, throughput, complexity), but in Figures 3.25 and 3.26, error performance is the characteristic that stands out explicitly.

A performance characteristic that is not explicitly seen in Figures 3.25 and 3.26 is the required system bandwidth. For the curves characterizing M-ary orthogonal signals in Figure 3.25, as k increases, the required bandwidth also increases. For the M-ary multiple phase curves in Figure 3.26, as k increases, a larger bit rate can be transmitted within the same bandwidth. In other words, for a fixed data rate, the required bandwidth is decreased. Therefore, *both* the orthogonal and multiple phase error performance curves tell us that M-ary signaling represents a vehicle for performing a system trade-off. In the case of orthogonal signaling, error performance improvement can be achieved at the expense of bandwidth. In the case of multiple phase signaling, bandwidth performance can be achieved at the expense of error performance. Error performance versus bandwidth performance, a fundamental communications trade-off, is treated in greater detail in Chapter 7.

3.8.3 Vectorial View of MPSK Signaling

Figure 3.27 illustrates MPSK signal sets for $M = 2, 4, 8$, and 16. In Figure 3.27a we see the binary ($k = 1, M = 2$) antipodal vectors s_1 and s_2 positioned 180° apart. The decision boundary is drawn so as to partition the signal space into two regions. On the figure is also shown a noise vector **n** equal in magnitude to s_1. The figure establishes the magnitude and orientation of the minimum energy noise vector that would cause the detector to make a symbol error.

In Figure 3.27b we see the 4-ary ($k = 2, M = 4$) vectors positioned 90° apart. The decision boundaries (only one line is drawn) divide the signal space into four regions. Again a noise vector **n** is drawn (from the head of a signal vector, normal to the closest decision boundary) to illustrate the minimum energy noise vector that would cause the detector to make a symbol error. Notice that the minimum energy noise vector of Figure 3.27b is smaller than that of Figure 3.27a, illustrating that the 4-ary system is more vulnerable to noise than the 2-ary system (signal energy being equal for each case). As we move on to Figure 3.27c for the 8-ary case and Figure 3.27d for the 16-ary case, it should be clear that for multiple phase signaling, as M increases, we are crowding more signal vectors into the signal plane. As the vectors are moved closer together, a smaller amount of noise energy is required to cause an error.

Figure 3.27 adds some insight as to why the curves of Figure 3.26 behave as they do as k is increased. Figure 3.27 also provides some insight into a basic trade-off in multiple phase signaling. Crowding more signal vectors into the signal space is tantamount to increasing the data rate without increasing the system bandwidth (the vectors are all confined to the same plane). In other words, we have increased the bandwidth utilization at the expense of error performance.

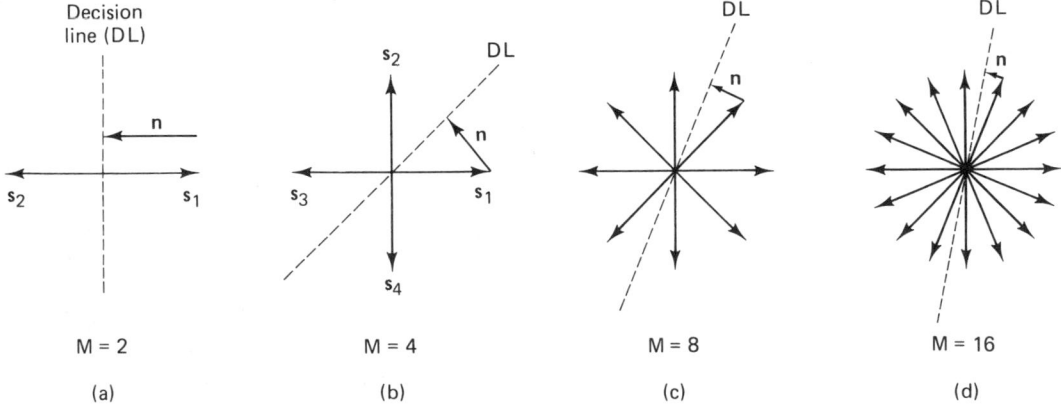

Figure 3.27 MPSK signal sets for $M = 2, 4, 8, 16$.

Look at Figure 3.27d, where the error performance is worse than any of the other examples in Figure 3.27. How might we "buy back" the degraded error performance; that is, what can we trade-off so that the distance between neighboring signal vectors in Figure 3.27d is increased to that in Figure 3.27a? We can increase the signal strength (make the signal vectors larger) until the minimum distance from the head of a signal vector to a decision line equals the length of the noise vector in Figure 3.27a. Therefore, in a multiple phase system, as M is increased, we can either achieve improved bandwidth performance at the expense of error performance, or if we increase the E_b/N_0 so that the error probability is not degraded, we can achieve improved bandwidth performance at the expense of increasing E_b/N_0.

3.8.4 BPSK and QPSK Have the Same Bit Error Probability

In Equation (3.94) we stated the general relationship between E_b/N_0 and S/N_0 for binary transmission, as follows:

$$\frac{E_b}{N_0} = \frac{S}{N_0} \cdot \left(\frac{1}{R}\right) \tag{3.116}$$

where S is the average signal power and R is the bit rate. A BPSK signal with the available E_b/N_0 found from Equation (3.116) will perform with a P_B that can be read from the $k = 1$ curve in Figure 3.26. QPSK can be characterized as two orthogonal BPSK channels. The QPSK bit stream is usually partitioned into an even and odd (I and Q) stream; each new stream modulates an orthogonal component of the carrier at half the bit rate of the original stream. The I stream modulates the cos $\omega_0 t$ term and the Q stream modulates the sin $\omega_0 t$ term. If the magnitude of the original QPSK vector has the value A, the magnitude of the I and Q component vectors each has a value of $A/\sqrt{2}$, as shown in Figure 3.28. Thus, each of the quadrature BPSK signals has half of the average power of the original QPSK signal. Hence if the original QPSK waveform has a bit rate of R bits/s and an average power of S watts, the quadrature partitioning results in each

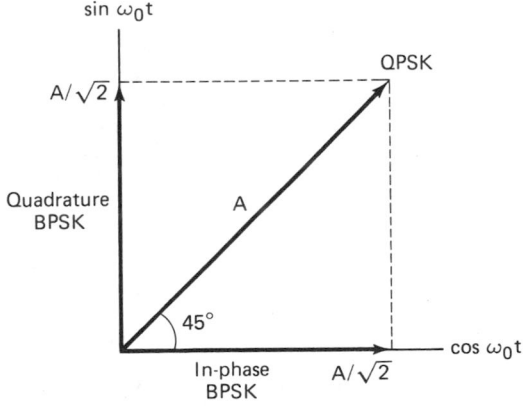

Figure 3.28 In-phase and quadrature BPSK components of QPSK signaling.

of the BPSK waveforms having a bit rate of $R/2$ bits/s and an average power of $S/2$ watts.

Therefore, the E_b/N_0 characterizing each of the orthogonal BPSK channels, comprising the QPSK signal, is equivalent to the E_b/N_0 in Equation (3.116) since it can be written as

$$\frac{E_b}{N_0} = \frac{S/2}{N_0}\left(\frac{W}{R/2}\right) = \frac{S}{N_0}\left(\frac{1}{R}\right) \tag{3.117}$$

Thus each of the orthogonal BPSK channels, and hence the composite QPSK signal, is characterized by the same E_b/N_0 and hence the same P_B performance as a BPSK signal. The natural orthogonality of the 90° phase shifts between adjacent QPSK symbols results in the *bit error probabilities* being equal for both BPSK and QPSK signaling. It is important to note that the *symbol error probabilities* are *not* equal for BPSK and QPSK signaling. The relationship between bit error probability and symbol error probability is treated in Sections 3.9.3 and 3.9.4.

3.8.5 Vectorial View of MFSK Signaling

In Section 3.8.3, Figure 3.27 provides some insight as to why the error performance of MPSK signaling degrades as k (or M) increases. It would be useful to have a similar vectorial illustration for the error performance of MFSK signaling as seen in the curves of Figure 3.25. Since the MFSK signal space is characterized by M mutually perpendicular axes, we can only conveniently illustrate the cases, $M = 2$ and $M = 3$. In Figure 3.29a we see the binary orthogonal vectors s_1 and s_2 positioned 90° apart. The decision boundary is drawn so as to partition the signal space into two regions. On the figure is also shown a noise vector $\mathbf{n}$, which represents the minimum noise vector that would cause the detector to make an error.

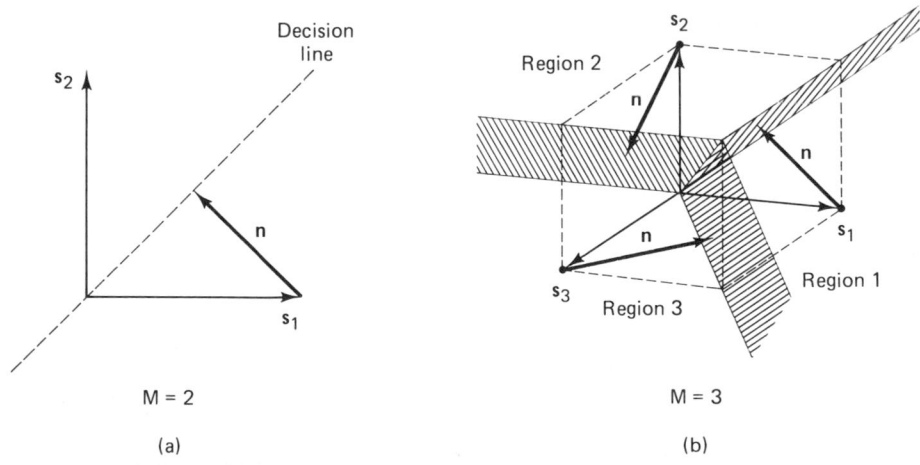

Figure 3.29 MFSK signal sets for $M = 2, 3$.

In Figure 3.29b we see a 3-ary signal space with axes positioned 90° apart. Here decision planes partition the signal space into three regions. Noise vectors **n** are shown added to each of the prototype signal vectors s_1, s_2, and s_3; each noise vector illustrates an example of the minimum noise energy that would cause the detector to make a symbol error. The minimum noise energy vectors in Figure 3.29b are the same length as the noise vector in Figure 3.29a. In Section 3.5.4 we stated that the distance between any two prototype signal vectors s_i and s_j in an M-ary orthogonal space is constant. It follows that the minimum distance between a prototype signal vector and any of the decision boundaries remains fixed as M increases. Unlike the case of MPSK signaling, where adding new signals to the signal set makes the signals vulnerable to smaller noise vectors, here in the case of MFSK signaling, adding new signals to the signal set does *not* make the signals vulnerable to smaller noise vectors.

It would be convenient to illustrate the point by drawing higher-dimensional orthogonal spaces, but of course this is not possible. We can only use our "mind's eye" to understand that increasing the signal set, M, by adding additional axes, where each new axis is mutually perpendicular to all the others, does not crowd the signal set more closely together; thus a transmitted signal from an orthogonal set is *not* more vulnerable to a noise vector when the set is increased in size. In fact, we see from Figure 3.25 that as k increases, the bit error performance improves.

Understanding the error performance improvement of orthogonal signaling, as illustrated in Figure 3.25, is facilitated by comparing the probability of symbol error (P_E) versus unnormalized SNR, with P_E versus E_b/N_0. Figure 3.30 represents a set of P_E performance curves plotted against unnormalized SNR for coherent FSK signaling. Here we see that P_E degrades as M is increased. Didn't we say that an orthogonal signal is *not* made more vulnerable to a given noise vector, as the orthogonal set is increased in size? It is correct that for orthogonal signaling, with a given SNR it takes a fixed-size noise vector to perturb a transmitted signal into an error region; the signals do not become vulnerable to smaller noise vectors as M increases. However, as M increases, more neighboring decision regions are introduced; thus the number of ways in which a symbol error can be made increases. Figure 3.30 reflects the degradation in P_E versus unnormalized SNR as M is increased; there are $(M - 1)$ ways to make an error. Examining performance under the condition of a fixed SNR (as M increases) is not very useful for digital communications. A fixed SNR means a fixed amount of energy per symbol; thus as M increases, there is a fixed amount of energy to be apportioned over a larger number of bits, or there is less energy per bit. The most useful way of comparing one digital system with another is on the basis of *bit-normalized SNR* or E_b/N_0. The error performance improvement with increasing M, seen in Figure 3.25, manifests itself only when error probability is plotted against E_b/N_0. For this case, as M increases, the required E_b/N_0 (to meet a given error probability) is reduced for a fixed SNR; therefore, we need to map the Figure 3.30 plot into a new plot, similar to Figure 3.25, where the abscissa represents E_b/N_0 instead of SNR. Figure 3.31 illustrates such a mapping; it demonstrates that curves manifesting degraded P_E with increasing M (such as Figure 3.30) are

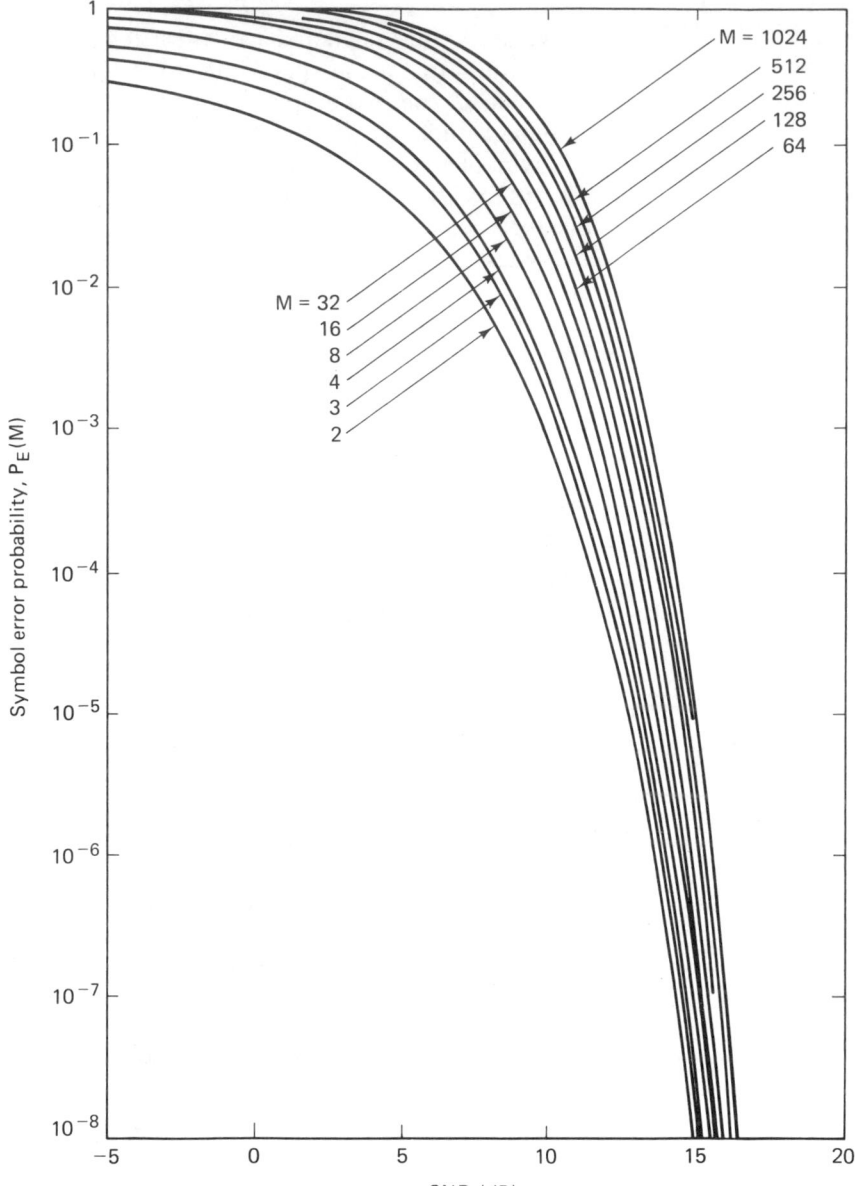

Figure 3.30 Symbol error probability versus SNR for coherent FSK signaling. (From Bureau of Standards, *Technical Note 167*, March 1963.) (Reprinted from *Central Radio Propagation Laboratory Technical Note 167*, March 25, 1963, Fig. 1, p. 5, courtesy of National Bureau of Standards.)

transformed into curves manifesting improved P_E with increasing M. The basic mapping relationship is expressed in Equation (3.94):

$$\frac{E_b}{N_0} = \frac{S}{N}\left(\frac{W}{R}\right)$$

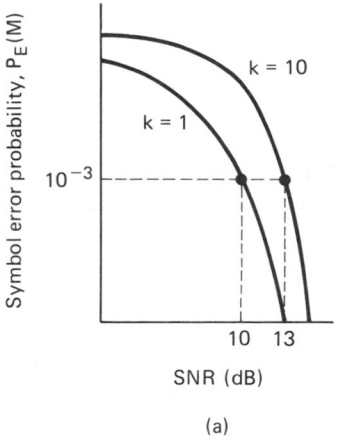

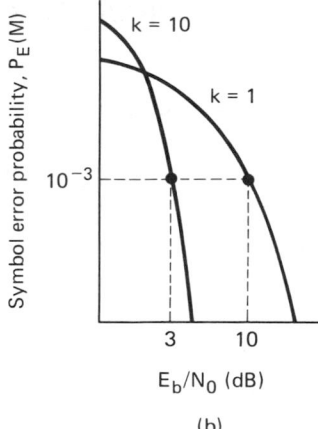

Figure 3.31 Mapping P_E versus SNR into P_E versus E_b/N_0 for orthogonal signaling. (a) Unnormalized. (b) Normalized.

where W is the detection bandwidth. Since

$$R = \frac{\log_2 M}{T} = \frac{k}{T}$$

where T is the symbol duration, we can then write

$$\frac{E_b}{N_0} = \frac{S}{N}\left(\frac{WT}{\log_2 M}\right) = \frac{S}{N}\left(\frac{WT}{k}\right) \tag{3.118}$$

For FSK signaling the detection bandwidth, W in hertz, is typically equal in value to the symbol rate $1/T$, in other words, $WT \simeq 1$. Therefore,

$$\frac{E_b}{N_0} \simeq \frac{S}{N}\left(\frac{1}{k}\right) \tag{3.119}$$

Figure 3.31 illustrates the mapping from P_E versus SNR to P_E versus E_b/N_0 for coherently detected M-ary orthogonal signaling. In Figure 3.31a, on the $k = 1$ curve is shown an operating point corresponding to $P_E = 10^{-3}$ and SNR = 10 dB. On the $k = 10$ curve is shown an operating point at the same $P_E = 10^{-3}$ but with SNR = 13 dB (approximate values taken from Figure 3.30). Here we clearly see the degradation in error performance as k increases. Consider the same $k = 1$ and $k = 10$ cases mapped onto the Figure 3.31b plane, where the abscissa is E_b/N_0. The $k = 1$ case looks exactly the same as it does in Figure 3.31a. But for the $k = 10$ case, the required E_b/N_0 is obtained from Equation (3.119) as follows: $E_b/N_0 = 20(\frac{1}{10}) = 2$ (3 dB), thus showing the error performance improvement as k is increased. In digital communication systems, error performance is almost always considered in terms of E_b/N_0, since such a measurement makes for a meaningful comparison between one system's performance and another. Therefore, the curves of Figures 3.30 and 3.31a are hardly ever seen.

3.9 SYMBOL ERROR PERFORMANCE FOR M-ARY SYSTEMS ($M > 2$)

3.9.1 Probability of Symbol Error for MPSK

For large energy-to-noise ratios, the symbol error performance, $P_E(M)$, for equally likely coherently detected M-ary PSK signaling can be expressed [9] as follows:

$$P_E(M) \simeq 2Q \left(\sqrt{\frac{2E_s}{N_0}} \sin \frac{\pi}{M} \right) \tag{3.120}$$

where $P_E(M)$ is the probability of symbol error, $E_s = E_b(\log_2 M)$ is the energy per symbol, and $M = 2^k$ is the size of the symbol set. The $P_E(M)$ performance curves for coherently detected MPSK signaling are plotted versus E_b/N_0 in Figure 3.32.

The symbol error performance for differentially coherent detection of M-

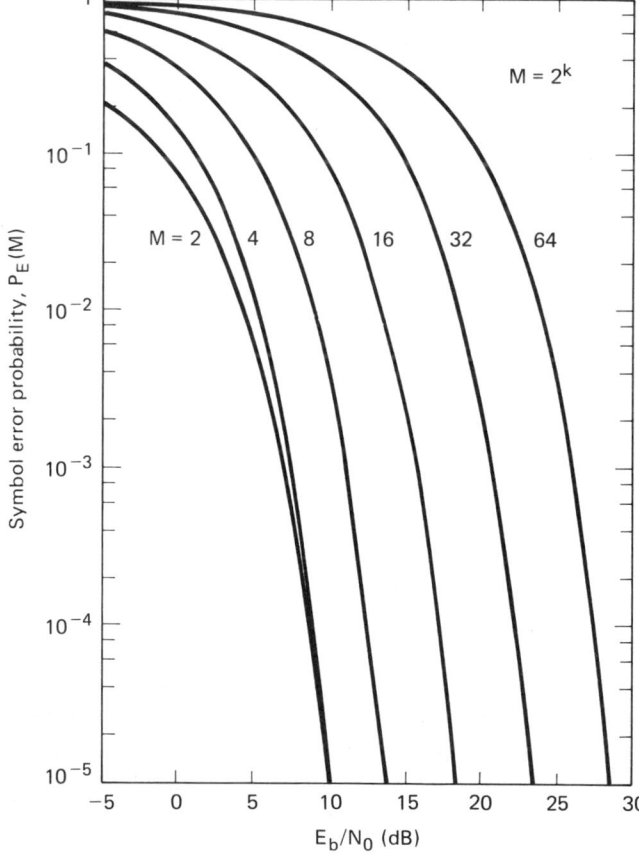

Figure 3.32 Symbol error probability for coherently detected multiple phase signaling. (Reprinted from W. C. Lindsey and M. K. Simon, *Telecommunication Systems Engineering*, Prentice-Hall, Inc., Englewood Cliffs, N.J., 1973, courtesy of W. C. Lindsey and Marvin K. Simon.)

ary DPSK (for large E_s/N_0) is similarly expressed [9] as

$$P_E(M) \simeq 2Q\left(\sqrt{\frac{2E_s}{N_0}}\, \sin\frac{\pi}{\sqrt{2}M}\right) \tag{3.121}$$

3.9.2 Probability of Symbol Error for MFSK

The symbol error performance $P_E(M)$, for equally likely *coherently* detected M-ary orthogonal signaling can be upper bounded [7] as follows:

$$P_E(M) \le (M-1)Q\left(\sqrt{\frac{E_s}{N_0}}\right) \tag{3.122}$$

where $E_s = E_b(\log_2 M)$ is the energy per symbol and M is the size of the symbol set. The $P_E(M)$ performance curves for coherently detected M-ary orthogonal signaling are plotted versus E_b/N_0 in Figure 3.33.

The symbol error performance for equally likely *noncoherently* detected M-ary orthogonal signaling is [11]

$$P_E(M) = \frac{1}{M}\exp\left(-\frac{E_s}{N_0}\right)\sum_{j=2}^{M}(-1)^j\binom{M}{j}\exp\left(\frac{E_s}{jN_0}\right) \tag{3.123}$$

where

$$\binom{M}{j} = \frac{M!}{j!\,(M-j)!} \tag{3.124}$$

is the standard binomial coefficient yielding the number of ways in which j symbols out of M may be in error. Note that for the binary case, Equation (3.123) reduces to

$$P_B = \frac{1}{2}\exp\left(-\frac{E_b}{2N_0}\right) \tag{3.125}$$

which is the same result as that described by Equation (3.111). The $P_E(M)$ performance curves for noncoherently detected M-ary orthogonal signaling are plotted versus E_b/N_0 in Figure 3.34. If we compare this noncoherent orthogonal $P_E(M)$ performance with the corresponding $P_E(M)$ results for the coherent detection of orthogonal signals in Figure 3.33, it can be seen that for $k > 7$, there is a negligible difference. An upper bound for coherent as well as noncoherent reception of orthogonal signals is [11]

$$P_E(M) < \frac{M-1}{2}\exp\left(-\frac{E_s}{2N_0}\right) \tag{3.126}$$

where E_s is the energy per symbol and M is the size of the symbol set.

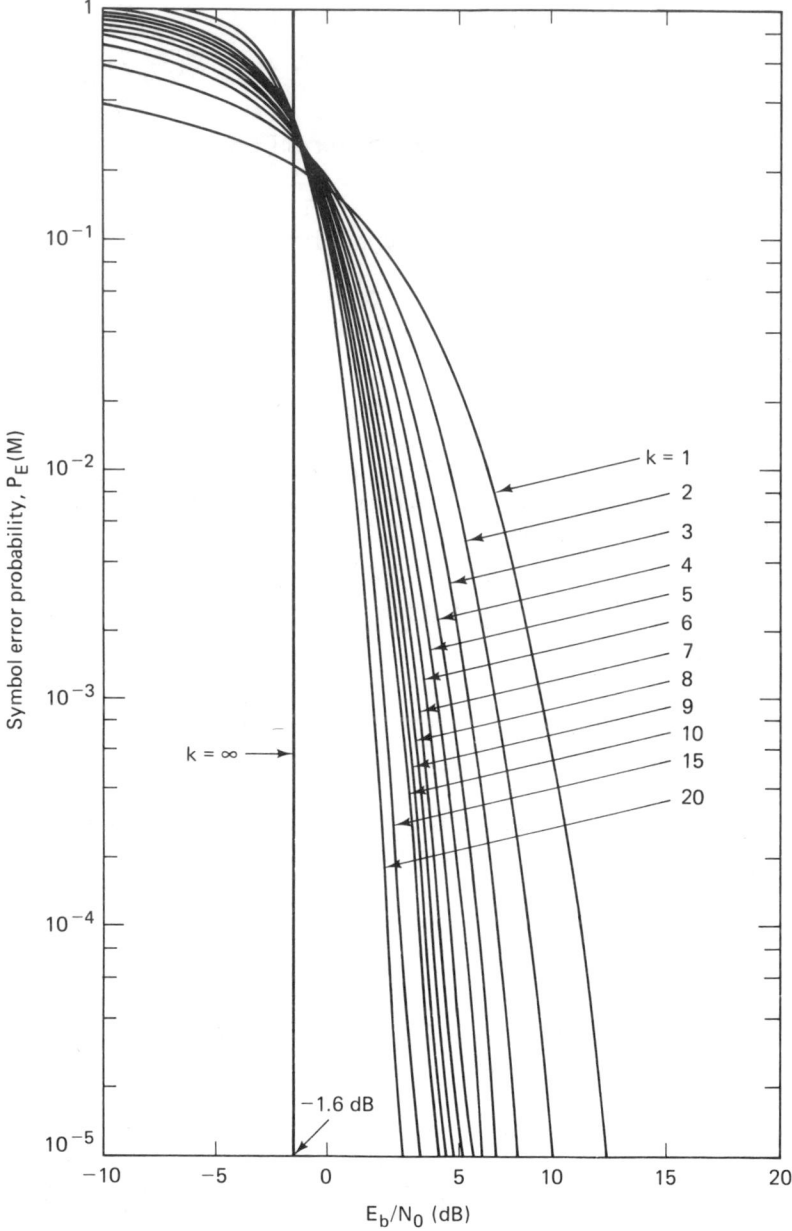

Figure 3.33 Symbol error probability for coherently detected *M*-ary orthogonal signaling. (Reprinted from W. C. Lindsey and M. K. Simon, *Telecommunication Systems Engineering*, Prentice-Hall, Inc., Englewood Cliffs, N.J., 1973, courtesy of W. C. Lindsey and Marvin K. Simon.)

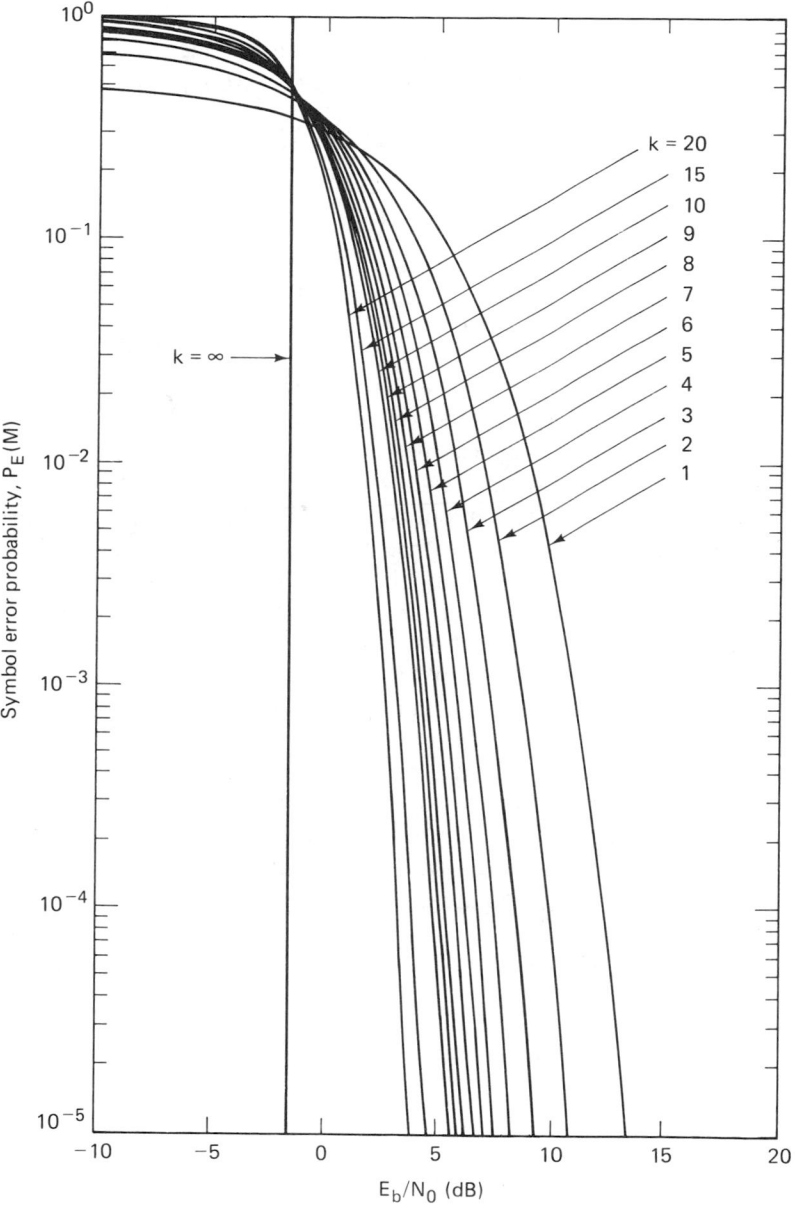

Figure 3.34 Symbol error probability for noncoherently detected *M*-ary orthogonal signaling. (Reprinted from W. C. Lindsey and M. K. Simon, *Telecommunication Systems Engineering*, Prentice-Hall, Inc., Englewood Cliffs, N.J., 1973, courtesy of W. C. Lindsey and Marvin K. Simon.)

3.9.3 Bit Error Probability versus Symbol Error Probability for Orthogonal Signals

It can be shown [11] that the relationship between probability of bit error (P_B) and probability of symbol error (P_E) for an M-ary orthogonal signal set is

$$\frac{P_B}{P_E} = \frac{2^{k-1}}{2^k - 1} = \frac{M/2}{M-1} \tag{3.127}$$

In the limit as k increases we get

$$\lim_{k \to \infty} \frac{P_B}{P_E} = \frac{1}{2}$$

A simple example will make Equation (3.127) intuitively acceptable. Figure 3.35 describes an octal message set. The message symbols (assumed equally likely) are to be transmitted on orthogonal waveforms such as FSK. With orthogonal signaling, a decision error willl transform the correct signal into any one of the $(M - 1)$ incorrect signals with equal probability. The example in Figure 3.35 indicates that the symbol comprised of bits 0 1 1 was transmitted. An error might occur in any one of the other $2^k - 1 = 7$ symbols, with equal probability. Notice that just because a symbol error is made does not mean that all the bits within the symbol will be in error. In Figure 3.35, if the receiver decides that the transmitted symbol is the bottom one listed, comprised of bits 1 1 1, two of the three transmitted symbol bits will be correct; only one bit will be in error. It should be apparent that P_B will be less than or equal to P_E.

Consider any of the bit-position columns in Figure 3.35. For each bit position, the digit occupancy consists of 50% ones and 50% zeros. In the context of the first bit position (rightmost column) and the transmitted symbol, how many ways are there to cause an error to the binary one? There are $2^{k-1} = 4$ ways (four places where zeros appear in the column) that a bit error can be made; it is the same for each of the columns. The final relationship, P_B/P_E, for orthogonal sig-

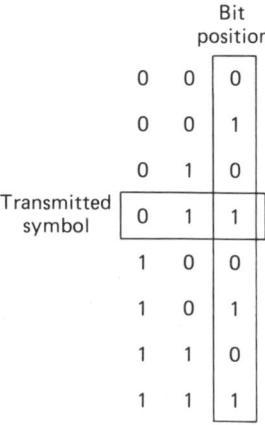

Figure 3.35   Example of P_B versus P_E.

naling, in Equation (3.127), is obtained by forming the following ratio: the number of ways that a bit error can be made (2^{k-1}) divided by the number of ways that a symbol error can be made ($2^k - 1$). For the Figure 3.35 example, $P_B/P_E = 4/7$.

3.9.4 Bit Error Probability versus Symbol Error Probability for Multiple Phase Signaling

For the case of MPSK signaling, P_B is less than or equal to P_E, just as in the case of MFSK signaling. However, there is an important difference. For orthogonal signaling, selecting any one of the $(M - 1)$ erroneous symbols is equally likely. In the case of MPSK signaling, each signal vector is not equidistant from all of the others. Figure 3.36a illustrates an 8-ary decision space with the pie-shaped regions denoted by the 8-ary symbols in binary notation. If symbol (0 1 1) is transmitted, it is clear that should an error occur, the transmitted signal will most likely be mistaken for one of its closest neighbors, (0 1 0) or (1 0 0). The likelihood that (0 1 1) would get mistaken for (1 1 1) is relatively remote. If the assignment of bits to symbols follows the binary sequence shown in the symbol decision regions of Figure 3.36a, some symbol errors will usually result in two or more bit errors, even with a large signal-to-noise ratio.

For nonorthogonal schemes, such as MPSK signaling, one often uses a binary-to-M-ary code such that binary sequences corresponding to adjacent symbols (phase shifts) differ in only one bit position; thus when an M-ary symbol error occurs, it is more likely that only one of the k input bits will be in error. A code that provides this desirable feature is the Gray code [9]; Figure 3.36b illustrates the bit-to-symbol assignment using a Gray code for 8-ary PSK. Here it can be seen that neighboring symbols differ from one another in only one bit position. Therefore, the occurrence of a multibit error, for a given symbol error, is much

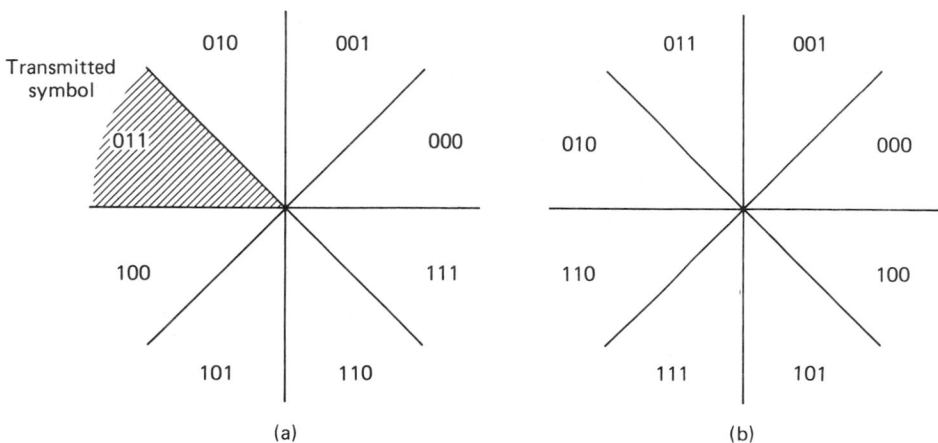

Figure 3.36 Binary-coded versus Gray-coded decision regions in an MPSK signal space. (a) Binary coded. (b) Gray coded.

reduced compared to the uncoded binary assignment seen in Figure 3.36a. Utilizing the Gray code assignment, it can be shown [7] that

$$P_B \simeq \frac{P_E}{\log_2 M} = \frac{P_E}{k} \qquad \text{(for } P_E \ll 1\text{)} \qquad (3.128)$$

Recall from Section 3.8.4 that BPSK and QPSK signaling have the same bit error probability. Here in Equation (3.128) we verify that they do not have the same symbol error probability. For BPSK, $P_E = P_B$. However, for QPSK, $P_E \simeq 2P_B$.

An exact closed-form expression for the bit error probability, P_B, of 8-ary PSK, together with tight upper and lower bounds on P_B for M-ary PSK with larger M, may be found in Lee [12].

3.9.5 Effects of Intersymbol Interference

In the previous sections and in Chapter 2 we have treated the detection of signals in the presence of AWGN under the assumption that there is no intersymbol interference (ISI). Thus the analysis has been straightforward, since the zero-mean AWGN process is characterized by its variance alone. In practice we find that ISI is often a second source of interference which must be accounted for. As explained in Section 2.11, ISI can be generated by the use of bandlimiting filters at the transmitter output, in the channel, or at the receiver input. The result of this additional interference is to degrade the error probabilities for coherent as well as for noncoherent reception. Analysis involving ISI in addition to AWGN is much more complicated since it involves the impulse response of the channel. The subject will not be treated here; however, for those readers interested in the details of the analysis, References [13–18] should prove interesting.

3.10 CONCLUSION

We have catalogued some basic bandpass digital modulation formats, particularly phase shift keying (PSK) and frequency shift keying (FSK). We have considered a geometric view of signal vectors and noise vectors, particularly antipodal and orthogonal signal sets. This geometric view allows us to consider the detection problem in the light of an orthogonal signal space and signal regions. This view of the space, and the effect of noise vectors causing transmitted signals to be received in the incorrect region, facilitates the understanding of the detection problem and the performance of various modulation and demodulation techniques. In Chapter 7 we reconsider the subjects of modulation and demodulation, and we investigate some bandwidth-efficient modulation techniques.

REFERENCES

1. Nyquist, H., "Thermal Agitation of Electric Charge in Conductors," *Phys. Rev.*, vol. 32, July 1928, pp. 110–113.

2. Arthurs, E., and Dym, H., "On the Optimum Detection of Digital Signals in the Presence of White Gaussian Noise—A Geometric Interpretation of Three Basic Data Transmission Systems," *IRE Trans. Commun. Syst.*, December 1962.

3. Wozencraft, J. M., and Jacobs, I. M., *Principles of Communication Engineering,* John Wiley & Sons, Inc., New York, 1965.

4. Van Trees, H. L., *Detection, Estimation, and Modulation Theory,* Part 1, John Wiley & Sons, Inc., New York, 1968.

5. Park, J. H., Jr., "On Binary DPSK Detection," *IEEE Trans. Commun.,* vol. COM26, no. 4, Apr. 1978, pp. 484–486.

6. Ziemer, R. E., and Peterson, R. L., *Digital Communications and Spread Spectrum Systems,* Macmillan Publishing Company, Inc., New York, 1985.

7. Lindsey, W. C., and Simon, M. K., *Telecommunication Systems Engineering,* Prentice-Hall, Inc., Englewood Cliffs, N.J., 1973.

8. Whalen, A. D., *Detection of Signals in Noise,* Academic Press, Inc., New York, 1971.

9. Korn, I., *Digital Communications,* Van Nostrand Reinhold Company, Inc., New York, 1985.

10. Couch, L. W. II, *Digital and Analog Communication Systems,* Macmillan Publishing Company, New York, 1983.

11. Viterbi, A. J., *Principles of Coherent Communications,* McGraw-Hill Book Company, New York, 1966.

12. Lee, P. J., "Computation of the Bit Error Rate of Coherent *M*-ary PSK with Gray Code Bit Mapping," *IEEE Trans. Commun.,* vol. COM34, no. 5, May 1986, pp. 488–491.

13. Hoo, E. Y., and Yeh, Y. S., "A New Approach for Evaluating the Error Probability in the Presence of Intersymbol Interference and Additive Gaussian Noise," *Bell Syst. Tech. J.,* vol. 49, Nov. 1970, pp. 2249–2266.

14. Shimbo, O., Fang, R. J., and Celebiler, M., "Performance of *M*-ary PSK Systems in Gaussian Noise and Intersymbol Interference," *IEEE Trans. Inf. Theory,* vol. IT19, Jan. 1973, pp. 44–58.

15. Prabhu, V. K., "Error Probability Performance of *M*-ary CPSK Systems with Intersymbol Interference," *IEEE Trans. Commun.,* vol. COM21, Feb. 1973, pp. 97–109.

16. Yao, K., and Tobin, R. M., "Moment Space Upper and Lower Error Bounds for Digital Systems with Intersymbol Interference," *IEEE Trans. Inf. Theory,* vol. IT22, Jan. 1976, pp. 65–74.

17. King, M. A., Jr., "Three Dimensional Geometric Moment Bounding Techniques," *J. Franklin Inst.,* vol. 309, no. 4, Apr. 1980, pp. 195–213.

18. Prabhu, V. K., and Salz, J., "On the Performance of Phase-Shift Keying Systems," *Bell Syst. Tech. J.,* vol. 60, Dec. 1981, pp. 2307–2343.

PROBLEMS

3.1. Determine whether or not $s_1(t)$ and $s_2(t)$ are orthogonal over the interval $(-1.5T_2 < t < 1.5T_2)$, where $s_1(t) = \cos(2\pi f_1 t + \phi_1)$, $s_2(t) = \cos(2\pi f_2 t + \phi_2)$, and $f_2 = 1/T_2$ for the following cases.
   (a) $f_1 = f_2$ and $\phi_1 = \phi_2$
   (b) $f_1 = \frac{1}{3}f_2$ and $\phi_1 = \phi_2$

(c) $f_1 = 2f_2$ and $\phi_1 = \phi_2$

(d) $f_1 = \pi f_2$ and $\phi_1 = \phi_2$

(e) $f_1 = f_2$ and $\phi_1 = \phi_2 + \pi/2$

(f) $f_1 = f_2$ and $\phi_1 = \phi_2 + \pi$

3.2. (a) Show that the three functions illustrated in Figure P3.1 are pairwise orthogonal over the interval $(-2, 2)$.

(b) Determine the value of the constant, A, that makes the set of functions in part (a) an orthonormal set.

(c) Express the following waveform, $x(t)$, in terms of the orthonormal set of part (b).

$$x(t) = \begin{cases} 1 & \text{for } 0 \leq t \leq 2 \\ 0 & \text{otherwise} \end{cases}$$

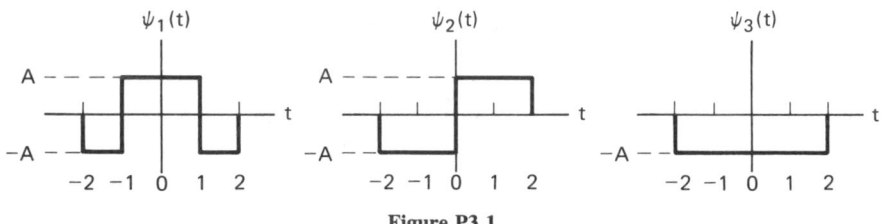

Figure P3.1

3.3. Consider the functions

$$\psi_1(t) = \exp(-|t|) \quad \text{and} \quad \psi_2 = 1 - A \exp(-2|t|)$$

Determine the constant, A, such that $\psi_1(t)$ and $\psi_2(t)$ are orthogonal over the interval $(-\infty, \infty)$.

3.4. Find the expected number of bit errors made in one day by the following continuously operating coherent BPSK receiver. The data rate is 5000 bits/s. The input digital waveforms are $s_1(t) = A \cos \omega_0 t$ and $s_2(t) = -A \cos \omega_0 t$, where $A = 1$ mV and the single-sided noise power spectral density is $N_0 = 10^{-11}$ W/Hz. Assume that signal power and energy per bit are normalized relative to a 1-Ω resistive load.

3.5. A continuously operating coherent BPSK system makes errors at the average rate of 100 errors per day. The data rate is 1000 bits/s. The single-sided noise power spectral density is $N_0 = 10^{-10}$ W/Hz.

(a) If the system is ergodic, what is the average bit error probability?

(b) If the value of received average signal power per bit is adjusted to be 10^{-6} W, will this received power be adequate to maintain the error probability found in part (a)?

3.6. If a system's main performance criterion is bit error probability, which of the following two modulation schemes would be selected for an AWGN channel? Show computations.

Binary coherent orthogonal FSK with $E_b/N_0 = 12$ dB

Binary noncoherent orthogonal FSK with $E_b/N_0 = 14$ dB

3.7. If a system's main performance criterion is bit error probability, which of the following two modulation schemes would be selected for an AWGN channel? Show computations.

Binary noncoherent orthogonal FSK with $E_b/N_0 = 13$ dB

Binary coherent PSK with $E_b/N_0 = 8$ dB

3.8. The bit stream

$$1\ 0\ 1\ 0\ 1\ 0\ 1\ 1\ 1\ 1\ 0\ 1\ 0\ 1\ 0\ 1\ 0\ 0\ 0\ 0\ 1\ 1\ 1\ 1$$

is to be transmitted using DPSK modulation. Show four different differentially encoded sequences that can represent the data sequence above, and explain the algorithm that generated each.

3.9. **(a)** Calculate the minimum required bandwidth for a noncoherently detected orthogonal binary FSK system. The higher-frequency signaling tone is 1 MHz and the symbol duration is 1 ms.

(b) What is the minimum required bandwidth for a noncoherent MFSK system having the same symbol duration?

3.10. Consider a BPSK system with equally likely waveforms $s_1(t) = \cos \omega_0 t$ and $s_2(t) = -\cos \omega_0 t$. At the matched filter detector, the $s_1(t)$ reference is $\cos (\omega_0 t + \phi)$, where ϕ is a phase error. Calculate the value of the phase error that would increase the probability of bit error from 2.0×10^{-3} to 2.5×10^{-3} relative to no phase error for an AWGN channel.

3.11. Find the probability of bit error, P_B, for the coherent matched filter detection of the equally likely binary FSK signals

$$s_1(t) = 0.5 \cos 2000\pi t$$

$$s_2(t) = 0.5 \cos 2020\pi t$$

where the two-sided AWGN power spectral density is $N_0/2 = 0.0001$. Assume that the symbol duration is $T = 0.01$ s.

3.12. Find the optimum (minimum probability of error) threshold, γ_0, for detecting the equally likely signals $s_1(t) = \sqrt{2E/T} \cos \omega_0 t$ and $s_2(t) = \sqrt{\frac{1}{2}E/T} \cos (\omega_0 t + \pi)$ in AWGN, using a correlator receiver as shown in Figure 3.7b. Assume a reference signal of $\psi_1(t) = \sqrt{2/T} \cos \omega_0 t$.

3.13. A system using matched filter detection of equally likely BPSK signals, $s_1(t) = \sqrt{2E/T} \cos \omega_0 t$ and $s_2(t) = \sqrt{2E/T} \cos (\omega_0 t + \pi)$, operates in AWGN with a received E_b/N_0 of 6.8 dB. Assume that $E\{z(T)\} = \pm\sqrt{E}$.

- **(a)** Find the minimum probability of bit error, P_B, for this signal set and E_b/N_0.
- **(b)** If the decision threshold is $\gamma = 0.1\sqrt{E}$, find P_B.
- **(c)** The threshold of $\gamma = 0.1\sqrt{E}$ is optimum for a particular set of a priori probabilities, $P(s_1)$ and $P(s_2)$. Find the values of these probabilities (refer to Section B.2).

3.14. A binary source with equally likely symbols controls the switch position in a transmitter operating over an AWGN channel, as shown in Figure P3.2. The noise has two-sided spectral density $N_0/2$. Assume antipodal signals of time duration T seconds and energy E joules. The system clock produces a clock pulse every T seconds, and the binary source rate is $1/T$ bits/s. Under *normal* operation, the switch is up when the source produces a binary zero, and it is down when the source produces a binary one. However, the switch is *faulty*. With probability, p, it will be thrown in the wrong direction during a given T-second interval. The presence of a switch error during any interval is independent of the presence of a switch error at any other time. Assume that $E\{z(T)\} = \pm\sqrt{E}$.

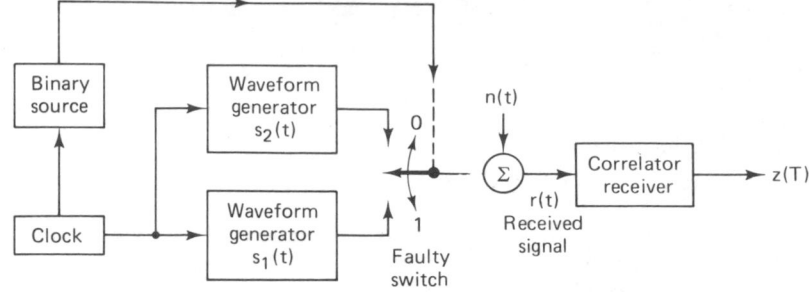

Figure P3.2

(a) Sketch the conditional probability functions, $p(z|s_1)$ and $p(z|s_2)$.

(b) The correlator receiver observes $r(t)$ in the interval $(0, T)$. Sketch the block diagram of an optimum receiver for minimizing the bit error probability when it is known that the switch is faulty with probability, p.

(c) Which one of the following two systems would you prefer to have?

$$p = 0.1 \quad \text{and} \quad \frac{E_b}{N_0} = \infty$$

$$p = 0 \quad \text{and} \quad \frac{E_b}{N_0} = 7 \text{ dB}$$

3.15. (a) Consider a 16-ary PSK system with symbol error probability, $P_E = 10^{-5}$. A Gray code is used for the symbol to bit assignment. What is the approximate bit error probability?

(b) Repeat part (a) for a 16-ary orthogonal FSK system.

3.16. Consider a coherent orthogonal MFSK system with $M = 8$ having the equally likely waveforms $s_i(t) = A \cos 2\pi f_i t$, $i = 1, \ldots, M$, $0 \le t \le T$, where $T = 0.2$ ms. The received carrier amplitude, A, is 1 mV, and the two-sided AWGN spectral density, $N_0/2$, is 10^{-11} W/Hz. Calculate the probability of bit error, P_B.

3.17. A bit error probability of $P_B = 10^{-3}$ is required for a system with a data rate of 100 kbits/s to be transmitted over an AWGN channel using coherently detected MPSK modulation. The system handwidth is 50 kHz. Assume that the filter has a roll-off characteristic of $r = 1$ and that a Gray code is used for the symbol to bit assignment.
(a) What E_s/N_0 is required for the specified P_B?
(b) What E_b/N_0 is required?

3.18. A differentially coherent MPSK system operates over an AWGN channel with an E_b/N_0 of 10 dB. What is the symbol error probability for $M = 8$ and equally likely symbols?

3.19. If a system's main performance criterion is bit error probability, which of the following two modulation schemes would be selected for transmission over an AWGN channel? Show computations.

$$\text{coherent 8-ary orthogonal FSK with } \frac{E_b}{N_0} = 8 \text{ dB}$$

$$\text{coherent 8-ary PSK with } \frac{E_b}{N_0} = 13 \text{ dB}$$

(Assume that a Gray code is used for the MPSK symbol-to-bit assignment.)

CHAPTER 4

Communications Link Analysis

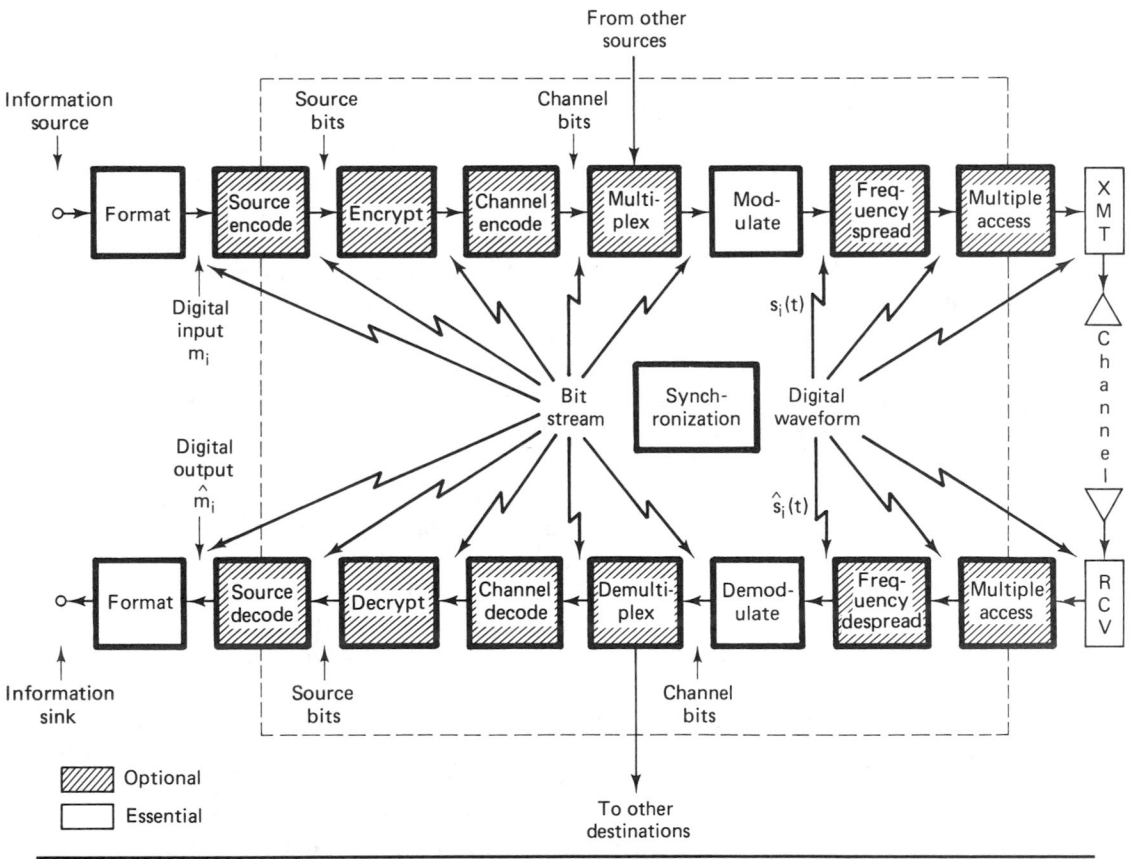

4.1 WHAT THE SYSTEM LINK BUDGET TELLS THE SYSTEM ENGINEER

When we talk about a communications *link,* to what part of the system are we referring? Is it simply the channel or region between the transmitter and receiver? No, it is far more than that. The link encompasses the entire communications path, from the information source, through all the encoding and modulation steps, through the transmitter and the channel, up to and including the receiver with all its signal processing steps, and terminating at the information sink.

What is a link analysis, and what purpose does it serve in the development of a communication system? The link analysis, and its output, the *link budget,* consist of the calculations and tabulation of the useful signal power and the interfering noise power available at the receiver. The link budget is a balance sheet of gains and losses; it outlines the detailed apportionment of transmission and reception resources, noise sources, signal attenuators, and effects of processes throughout the link. Some of the budget parameters are statistical (e.g., allowances for the fading of signals due to meteorological events); the budget is therefore an *estimation* technique for evaluating communication system performance. In Chapter 3 we examined probability of error versus E_b/N_0 curves having a "waterfall-like" shape, such as the one shown in Figure 3.21. We thereby related error probability to E_b/N_0 for various modulation types in Gaussian noise. Once a modulation scheme has been chosen, the requirement to meet a particular error probability dictates a particular operating point on the curve; in other words, the required error performance dictates the value of E_b/N_0 that must be made available at the receiver in order to meet that performance. The primary purpose of a link analysis is to determine the *actual* system operating point in Figure 3.21 and to establish that the error probability associated with that point is less than or equal to the system requirement. Of the many specifications, analyses, and tabulations that are used in the development of a communication system, the link budget stands out as a basic tool for providing the system engineer with overall system insight.

By examining the link budget, one can learn many things about overall system design and performance. For example, from the link margin, one learns whether the system will meet its requirements comfortably, marginally, or not at all. It will be evident if there are any hardware constraints, and whether such constraints can be compensated for in other parts of the link. The link budget is often used as a "score sheet" in considering system trade-offs and configuration changes, and in understanding subsystem nuances and interdependencies. From a quick examination of the link budget and its supporting documentation, one can judge whether the analysis was done precisely or if it represents a rough estimate. Together with other modeling techniques, the link budget can help predict equipment weight, size, prime power requirements, technical risk, and cost. The link budget is one of the system manager's most useful documents; it represents the "bottom-line" tally in the search for optimimum system performance.

4.2 THE CHANNEL

The propagating medium or electromagnetic path connecting the transmitter and receiver is called the *channel*. In general, a communications channel might consist of wires, coaxial cables, fiber optic cables, and in the case of radio-frequency (RF) links, waveguides, the atmosphere, or empty space. For most terrestrial communication links, the channel space is occupied by the atmosphere and partially bounded by the earth's surface. For satellite links, the channel is occupied mostly by empty space. Although some atmospheric effects occur at altitudes up to 100 km, the *bulk* of the atmosphere extends to an altitude of 20 km. Therefore, only a small part (0.05%) of the total synchronous altitude (35,800 km) path is occupied by significant amounts of atmosphere. Most of this chapter is presented in the context of such a satellite communications link.

4.2.1 The Concept of Free Space

The concept of *free space* assumes a channel free of all hindrances to RF propagation, such as absorption, reflection, refraction, or diffraction. If there is any atmosphere in the channel, it must be perfectly uniform and meet all these conditions. Also, we assume that the earth is infinitely far away or that its reflection coefficient is negligible. The RF energy arriving at the receiver is assumed to be a function only of distance from the transmitter (following the inverse-square law of optics). A free-space channel characterizes an ideal RF propagation path; in practice, propagation through the atmosphere and near the ground results in absorption, reflection, refraction, and diffraction, which modify the free-space transmission. Atmospheric absorption is treated in later sections. Reflection, refraction, and diffraction, which play an important role in determining terrestrial communications performance, are not treated here; Panter [1] provides a comprehensive treatment of these mechanisms.

4.2.2 Signal-to-Noise Ratio Degradation

The signal-to-noise power ratio (SNR) defined below is a convenient measure of performance at various points in the link.

$$SNR = \frac{\text{signal power}}{\text{noise power}}$$

Unless otherwise stated, SNR refers to *average* signal power and *average* noise power. The signal can be an information signal, a baseband waveform, or a modulated carrier. The SNR can degrade in two ways: (1) through the decrease of the desired signal power, and (2) through the increase of noise power, or the increase of interfering signal power. Let us refer to these degradations as *loss* and *noise* (or *interference*), respectively. Losses occur when a portion of the signal is absorbed, diverted, scattered, or reflected along its route to the intended receiver; thus a portion of the transmitted energy does not arrive at the receiver. There are four primary noise sources: (1) thermal noise can be generated within the link, (2) sky noise (e.g., galaxy noise, atmospheric noise) can be introduced into the link, (3) system nonlinearities can cause spurious signals to be created within the link, and (4) interfering signals from other users of the same frequency can be introduced into the link. Industry usage of the terms *loss* and *noise* frequently confuses the underlying degradation mechanism; however, the net effect on the SNR is the same.

4.2.3 Sources of Signal Loss and Noise

Figure 4.1 is a block diagram of a satellite communications link, emphasizing the sources of signal loss and noise. In the figure a signal loss is distinguished from a noise source by a dot pattern or line pattern, respectively. The contributors of *both* signal loss *and* noise are identified by a crosshatched line pattern. The following list of 21 sources of degradation represents a partial catalog of the major contributors to SNR degradation. The numbers correspond to the numbered circles in Figure 4.1.

1. *Bandlimiting loss.* All systems use filters in the transmitter to ensure that the transmitted energy is confined to the allocated or assigned bandwidth. This is to avoid interfering with other channels or users and to meet the requirements of regulatory agencies. Such filtering reduces the total amount of energy that would otherwise have been transmitted; the result is a *loss* in signal.

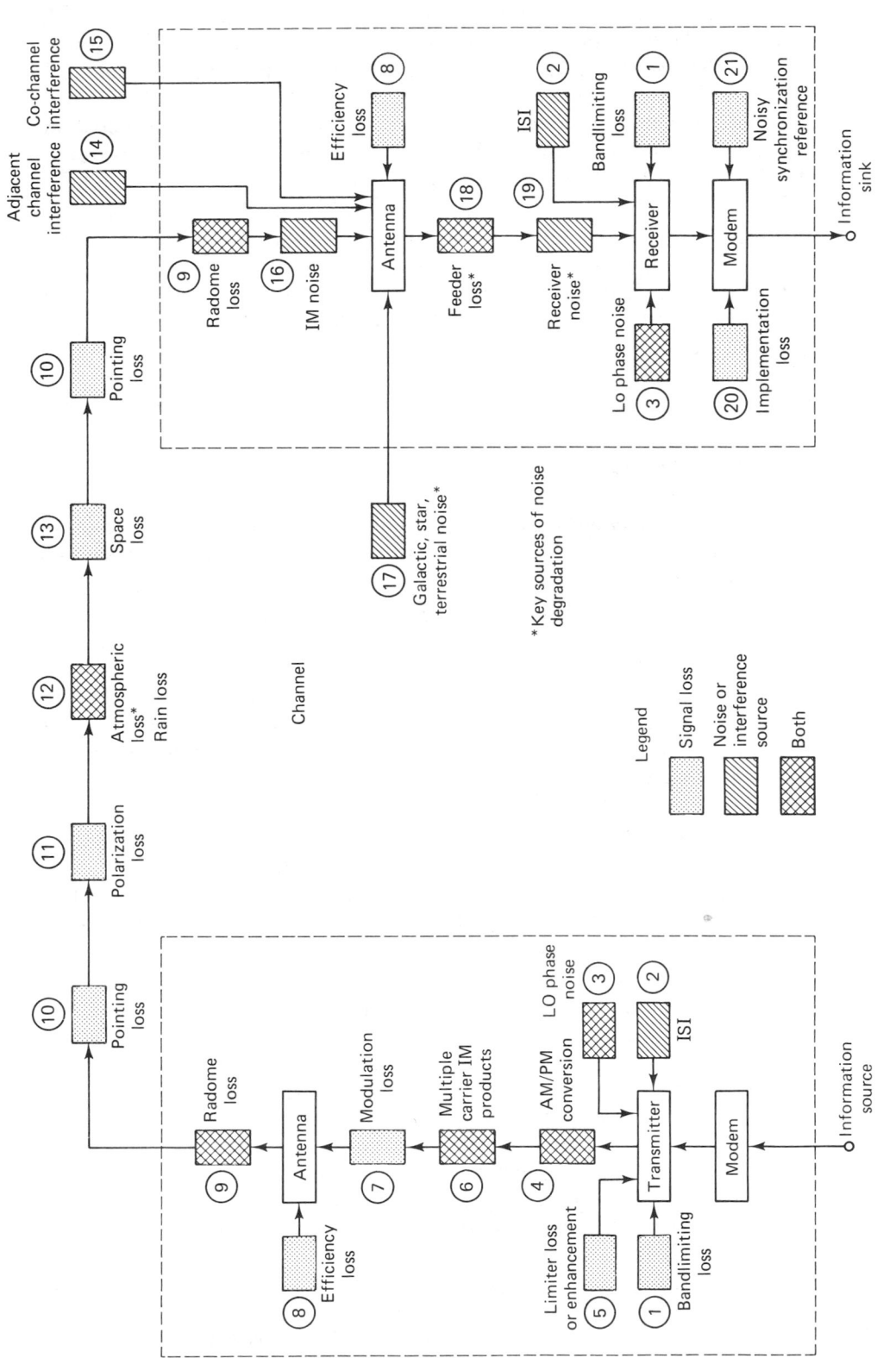

Figure 4.1 Satellite transmitter-to-receiver link with typical loss and noise sources.

2. *Intersymbol interference (ISI)*. As discussed in Chapter 2, filtering throughout the system—in the transmitter, in the receiver, and in the channel—can result in ISI. The received pulses overlap one another; the tail of one pulse "smears" into adjacent symbol intervals so as to *interfere* with the detection process. Even in the absence of thermal noise, imperfect filtering and system bandwidth constraints lead to ISI degradation.

3. *Local oscillator (LO) phase noise*. When an LO is used in signal mixing, phase fluctuations or jitter adds phase *noise* to the signal. When used as the reference signal in a receiver correlator, phase jitter can cause detector degradation and hence signal *loss*. At the transmitter, phase jitter can cause out-of-band signal spreading, which, in turn, will be filtered out and cause a *loss* in signal.

4. *AM/PM conversion*. AM-to-PM conversion is a phase *noise* phenomenon occurring in nonlinear devices such as traveling-wave tubes (TWT). Signal amplitude fluctuations (amplitude modulation) produce phase variations that contribute phase *noise* to signals that will be coherently detected. AM-to-PM conversion can also cause extraneous sidebands, resulting in signal *loss*.

5. *Limiter loss or enhancement*. A hard limiter can enhance the stronger of two signals, and suppress the weaker; this can result in either a signal *loss* or a signal *gain* [2].

6. *Multiple-carrier intermodulation (IM) products*. When several signals having different carrier frequencies are simultaneously present in a nonlinear device, such as a TWT, the result is a multiplicative interaction between the carrier frequencies which can produce signals at all combinations of sum and difference frequencies. The energy apportioned to these spurious signals (intermodulation or IM products) represents a *loss* in signal energy. In addition, if these IM products appear within the bandwidth region of these or other signals, the effect is that of added *noise* for those signals.

7. *Modulation loss*. The link budget is a calculation of received useful power (or energy). Only the power associated with information-bearing signals is useful. Error performance is a function of energy per transmitted symbol. Any power used for transmitting the carrier rather than the modulating signal (symbols) is a modulation *loss*. (However, energy in the carrier may be useful in aiding synchronization.)

8. *Antenna efficiency*. Antennas are transducers that convert electronic signals into electromagnetic fields, and vice versa. They are also used to focus the electromagnetic energy in a desired direction. The larger the antenna aperture (area), the larger is the resulting signal power density in the desired direction. An antenna's efficiency is described by the ratio of its effective aperture to its physical aperture. Mechanisms contributing to a reduction in efficiency (*loss* in signal strength) are known as amplitude tapering, aperture blockage, scattering, re-radiation, spillover, edge diffraction, and dissipative loss [3]. Typical efficiencies due to the combined effects of these mechanisms range between 50 and 80%.

9. *Radome loss and noise*. A radome is a protective cover, used with some

antennas, for shielding against weather effects. The radome, being in the path of the signal, will scatter and absorb some of the signal energy, thus resulting in a signal *loss*. A basic law of physics holds that a body capable of absorbing energy also radiates energy (at temperatures above 0 K). Some of this energy falls in the bandwidth of the receiver and constitutes injected *noise*.

10. *Pointing loss.* There is a *loss* of signal when either the transmitting antenna or the receiving antenna is imperfectly pointed.

11. *Polarization loss.* The polarization of an electromagnetic (EM) field is defined as the direction in space along which the field lines point, and the polarization of an antenna is described by the polarization of its radiated field. There is a *loss* of signal due to any polarization mismatch between the transmitting and receiving antennas.

12. *Atmospheric loss and noise.* The atmosphere is responsible for signal loss and is also a contributor of unwanted noise. The bulk of the atmosphere extends to an altitude of approximately 20 km; yet within that relatively short path, important loss and noise mechanisms are at work. Figure 4.2 is a plot of the theoretical one-way attenuation from a specified height to the top of the atmosphere. The calculations were made for several heights (0 km is sea level) and for a water vapor content of 7.5 g/m^3 at the earth's surface. The plot qualifies the magnitude of signal *loss* by indicating where it occurs, due to oxygen (O_2) and water vapor absorption as a function of carrier frequency. Local maxima of attenuation occur in the vicinities of 22 GHz (water vapor), and 60 and 120 GHz (O_2). The atmosphere also contributes *noise* energy into the link. As in the case of the radome, molecules that absorb energy also radiate energy. The oxygen and water vapor molecules radiate noise throughout the RF spectrum. The portion of this noise that falls within the bandwidth of a given communication system will degrade its SNR. A primary atmospheric cause of signal *loss* and contributor of *noise* is rainfall. The more intense the rainfall, the more signal energy it will absorb. Also, on a day when rain passes through the antenna beam, there is a larger amount of atmospheric noise radiated into the system receiver than there is on a clear day. More will be said about atmospheric noise in later sections.

13. *Space loss.* There is a decrease in the electric field strength, and thus in signal strength (power density or flux density), as a function of distance. For a satellite communications link, the space loss is the largest single *loss* in the system. It is a loss in the sense that all the radiated energy is not focused on the intended receiving antenna.

14. *Adjacent channel interference.* This *interference* is characterized by unwanted signals from other frequency channels "spilling over" or injecting energy into the channel of interest. The proximity with which channels can be located in frequency is determined by the modulation spectral roll-off and the width and shape of the main spectral lobe.

15. *Co-channel interference.* This *interference* refers to the degradation caused by an interfering waveform appearing within the signal bandwidth. It can be

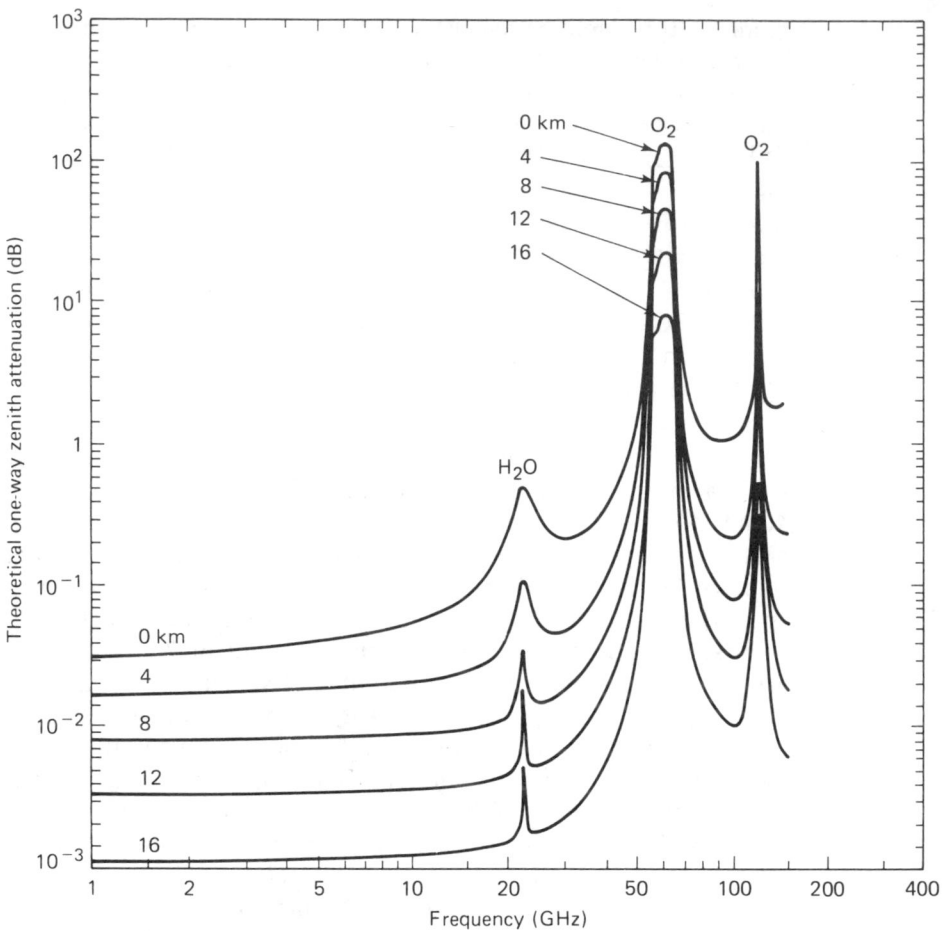

Figure 4.2 Theoretical vertical one-way attenuation from specified height to top of atmosphere for 7.5 g/m³ of water vapor at the surface. (Does not include effect of rain or cloud attenuation.) (Reprinted from NASA Reference Publication 1082(03), "Propagation Effects Handbook for Satellite Systems Design," June 1983, Fig. 6.2-1, p. 218, courtesy of the National Aeronautics and Space Administration.)

introduced by a variety of ways, such as accidental transmissions, insufficient vertical and horizontal polarization discrimination, or by radiation spillover from an antenna sidelobe (low-energy beam surrounding the main antenna beam). It can be brought about by other authorized users of the same spectrum.

16. *Intermodulation (IM) noise.* The IM products described in item 6 result from multiple-carrier signals interacting in a nonlinear device. Such IM products are sometimes called *active intermods*; as described in item 6, they can either cause a loss in signal energy or be responsible for noise injected into a link. Here we consider *passive intermods*; these are caused by multiple-carrier

transmission signals interacting with nonlinear components at the transmitter output. These nonlinearities generally occur at the junction of waveguide coupling joints, at corroded surfaces, and at surfaces having poor electrical contact. When large EM fields impinge on surfaces that have a diode-like transfer function (work potential), they cause multiplicative products, and hence *noise*. If such noise radiates into a closely located receiving antenna, it can seriously degrade the receiver performance.

17. *Galactic or cosmic, star, and terrestrial noise.* All the celestial bodies, such as the stars and the planets, radiate energy. Such *noise* energy in the field of view of the antenna will degrade the SNR.

18. *Feeder line loss.* The level of the received signal might be very small (e.g., 10^{-12} W), and thus will be particularly susceptible to noise degradation. The receiver front end, therefore, is a region where great care is taken to keep the noise as small as possible until the signal has been suitably amplified. The waveguide or cable (feeder line) between the receiving antenna and the receiver front end contributes both signal *attenuation* and thermal *noise*; the details are treated in Section 4.5.3.

19. *Receiver noise.* This is the thermal *noise* generated within the receiver; the details are treated in Sections 4.5.1 to 4.5.4.

20. *Implementation loss.* This *loss* in performance is the difference between theoretical detection performance and the actual performance due to imperfections such as timing errors, frequency offsets, finite rise and fall times of waveforms, and finite-value arithmetic.

21. *Imperfect synchronization reference.* When the carrier phase, the subcarrier phase, and the symbol timing references are all derived perfectly, the error probability is a well-defined function of E_b/N_0 discussed in Chapter 3. In general, they are not derived perfectly, resulting in a system *loss*.

4.3 RECEIVED SIGNAL POWER AND NOISE POWER

4.3.1 The Range Equation

In radio communication systems, the carrier wave is propagated from the transmitter by the use of a transmitting antenna. The transmitting antenna is a transducer that converts electronic signals into electromagnetic (EM) fields. At the receiver, a receiving antenna performs the reverse function; it converts EM fields into electronic signals. The development of the fundamental power relationship between the receiver and transmitter usually begins with the assumption of an omnidirectional RF source, transmitting uniformly over 4π steradians. Such an ideal source, called an *isotropic radiator,* is illustrated in Figure 4.3. The power density, $p(d)$, on a hypothetical sphere at a distance, d, from the source is related to the transmitted power, P_t, by

$$p(d) = \frac{P_t}{4\pi d^2} \qquad \text{watts/m}^2 \qquad (4.1)$$

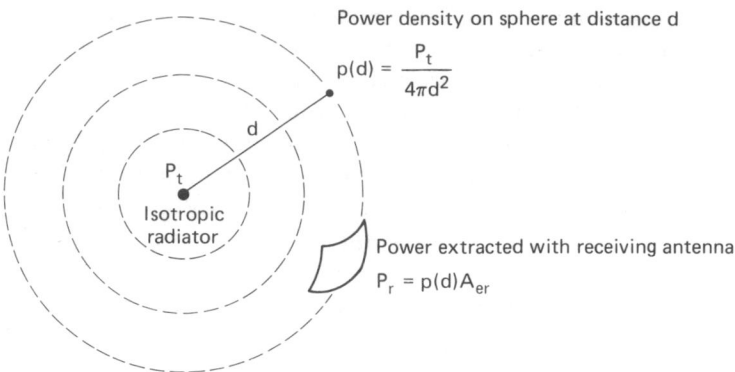

Figure 4.3 Range equation. Expresses received power in terms of distance.

since $4\pi d^2$ is the area of the sphere. The power, P_r, extracted with the receiving antenna can be written

$$P_r = p(d)A_{er} = \frac{P_t A_{er}}{4\pi d^2} \tag{4.2}$$

where the parameter, A_{er}, is the absorption cross section (effective area) of the receiving antenna, defined by

$$A_{er} = \frac{\text{total power extracted}}{\text{incident power flux density}} \tag{4.3}$$

If the antenna under consideration is a transmitting antenna, its effective area is designated by A_{et}. If the antenna in question is unspecified as to its receiving or transmitting function, its effective area is designated simply by A_e.

An antenna's effective area, A_e, and physical area, A_p, are related by an efficiency parameter, η, as

$$A_e = \eta A_p \tag{4.4}$$

which accounts for the fact that the total incident power is not extracted; it is lost through various mechanisms [3]. Nominal values for η are 0.55 for a dish (parabolic-shaped reflector) and 0.75 for a horn-shaped antenna.

The antenna parameter that relates the power output (or input) to that of an isotropic radiator as a purely geometric ratio is the antenna directivity or *directive gain, G,* where

$$G = \frac{\text{maximum power intensity}}{\text{average power intensity over } 4\pi \text{ steradians}} \tag{4.5}$$

In the absence of any dissipative loss or impedance mismatch loss, the antenna *gain* (in the direction of maximum intensity) is defined simply as the directive gain in Equation (4.5). However, in the event that there exists some dissipative or impedance mismatch loss, the antenna gain is then equal to the directive gain times a loss factor to account for these losses [4]. In this chapter we shall assume that the dissipative loss is zero and that the impedances are perfectly matched. Therefore, Equation (4.5) describes the *peak antenna gain*; it can be viewed as the result of concentrating the RF flux in some restricted region less than 4π steradians, as shown in Figure 4.4. Now we can define an *effective radiated power,* with respect to an isotropic radiator (EIRP), as the product of the transmitted power, P_t, and the gain of the transmitting antenna, G_t, as follows:

$$\text{EIRP} = P_t G_t \qquad (4.6)$$

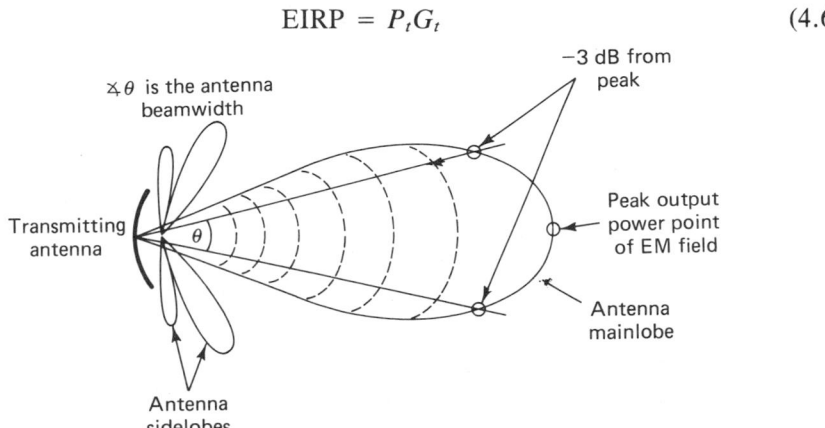

Figure 4.4 Antenna gain is the result of concentrating the isotropic RF flux.

Example 4.1   Effective Isotropic Radiated Power

Show that the same value of EIRP can be produced equally well by using a transmitter with $P_t = 100$ W or with $P_t = 0.1$ W, by employing the appropriate antenna in each case.

Solution

Figure 4.5a depicts a 100-W transmitter coupled to an isotropic antenna; the EIRP $= P_t G_t = 100 \times 1 = 100$ W. Figure 4.5b depicts a 0.1-W transmitter coupled to an antenna with gain $G_t = 1000$; the EIRP $= P_t G_t = 0.1 \times 1000 = 100$ W. If field-strength meters were positioned, as shown, to measure the effective power, the measurements could not distinguish between the two cases.

4.3.1.1 Back to the Range Equation

For the more general case in which the transmitter has some antenna gain relative to an isotropic antenna, we replace P_t with EIRP in Equation (4.2) to yield

$$P_r = \text{EIRP}\, \frac{A_{er}}{4\pi d^2} \qquad (4.7)$$

The relationship between antenna gain, G, and antenna effective area, A_e, is [4]

$$G = \frac{4\pi A_e}{\lambda^2} \qquad \text{(for } A_e \gg \lambda^2 \text{)} \qquad (4.8)$$

where λ is the wavelength of the carrier. Wavelength λ and frequency f are

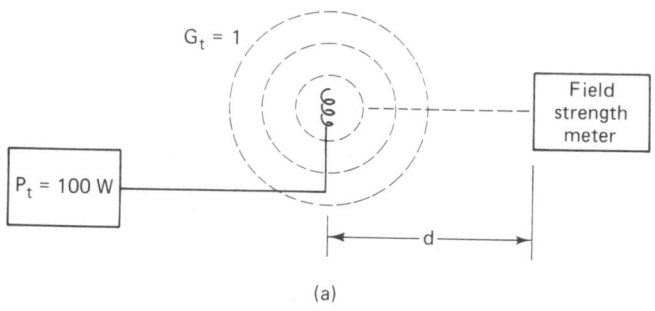

(a)

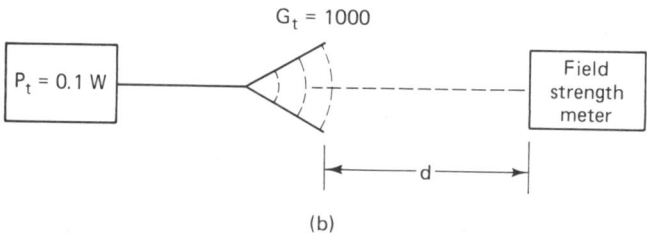

(b)

Figure 4.5 The same value of EIRP produced two different ways.

reciprocally related by $\lambda = c/f$, where c is the speed of light ($\approx 3 \times 10^8$ m/s). Similar expressions apply for both the transmitting and receiving antennas. The *reciprocity theorem* states that for a given antenna and carrier wavelength, the transmitting and receiving gains are identical [4].

The antenna field of view is a measure of the solid angle into which most of the field power is concentrated. Field of view is a measure of the directional properties of the antenna; it is inversely related to antenna gain—high-gain antennas are commensurate with narrow fields of view. Instead of using the solid-angle field of view, we often deal with the planar angle *beamwidth* measured in radians or degrees. Figure 4.4 pictures a directive antenna pattern and illustrates the common definition of the antenna beamwidth. The beamwidth is the angle that subtends the points at which the peak field power is reduced by 3 dB. How does the antenna beamwidth vary with signal frequency? How does the beamwidth vary with antenna size? As can be seen from Equation (4.8), the antenna gain increases with decreased wavelength (increased frequency); antenna gain also increases with increased effective area. Increasing antenna gain is tantamount to focusing the flux density into a more restricted cone angle; hence, increasing either the signal frequency or the antenna size results in a *narrower beamwidth*.

We can calculate the effective area of an isotropic antenna by setting $G = 1$ in Equation (4.8), and solving for A_e as follows:

$$A_e = \frac{\lambda^2}{4\pi} \tag{4.9}$$

Then to find the power received, P_r, when the receiving antenna is isotropic, we substitute Equation (4.9) into Equation (4.7) to get

$$P_r = \frac{\text{EIRP}}{(4\pi d/\lambda)^2} = \frac{\text{EIRP}}{L_s} \tag{4.10}$$

where the collection of terms $(4\pi d/\lambda)^2$, called the *path loss* or *free-space loss,* is designated by L_s. Notice that Equation (4.10) states that the power received by an isotropic antenna is equal to the effective transmitted power, reduced only by the path loss. When the receiving antenna is not isotropic, replacing A_{er} in Equation (4.7) with $G_r\lambda^2/4\pi$ from Equation (4.8) yields the more general expresion for P_r.

$$P_r = \frac{\text{EIRP } G_r\lambda^2}{(4\pi d)^2} = \frac{\text{EIRP } G_r}{L_s} \tag{4.11}$$

where G_r is the receiving antenna gain.

4.3.2 Received Signal Power as a Function of Frequency

Since the transmitting antenna and the receiving antenna can each be expressed as a gain or an area, P_r can be expressed four different ways:

$$P_r = \frac{P_t G_t A_{er}}{4\pi d^2} \tag{4.12}$$

$$P_r = \frac{P_t A_{et} A_{er}}{\lambda^2 d^2} \tag{4.13}$$

$$P_r = \frac{P_t A_{et} G_r}{4\pi d^2} \tag{4.14}$$

$$P_r = \frac{P_t G_t G_r\lambda^2}{(4\pi d)^2} \tag{4.15}$$

where A_{er} and A_{et} are the effective areas of the receiving and transmitting antennas, respectively.

In Equations (4.12) to (4.15) the dependent variable is received signal power, P_r, and the independent variables involve parameters such as transmitted power, antenna gain, antenna area, wavelength, and range. Suppose that we ask the question: How does received power vary as wavelength is decreased (or as frequency is increased), all other independent variables remaining constant? From Equations (4.12) and (4.14) it appears that P_r and wavelength are not related at all. From Equation (4.13), P_r appears to be inversely proportional to wavelength squared, and from Equation (4.15), P_r appears to be directly proportional to wavelength squared. Is there a paradox here? Of course there is not; Equations (4.12) to (4.15) seem to conflict only because antenna gain and antenna area are wavelength related, as stated in Equation (4.8). When should one use each of the Equations (4.12) to (4.15) for determining P_r as a function of wavelength? Consider a system that is already configured; that is, the antennas have already been built or their dimensions are fixed (A_{et} and A_{er} are fixed). Then Equation (4.13) is the appropriate choice for calculating the P_r performance. Equation (4.13) states that for fixed-size antennas, the received power increases as the wavelength is decreased.

Consider the use of Equation (4.12), where G_t and A_{er} are independent variables. We want G_t and A_{er} held fixed over the range of P_r versus wavelength calculations. What happens to the gain of a fixed-dimension transmitting antenna as the independent variable, λ, is decreased? G_t increases [see Equation (4.8)]. But we cannot have G_t increasing in Equation (4.12)—we want G_t held fixed. In other words, to ensure that G_t remains fixed, we would need to reduce the transmitting antenna size as wavelength decreases. It should be apparent that Equation (4.12) is the appropriate equation when starting with a *fixed transmitting antenna gain* (or beamwidth) requirement and the parameter, A_{et}, is not fixed. For similar reasons, Equation (4.14) is used when A_{et} and G_r are fixed, and Equation (4.15) is used when both the transmitting and receiving antenna gains (or beamwidths) are fixed.

Figure 4.6 illustrates a satellite application where the downlink antenna beam is required to provide earth coverage (a beamwidth of approximately 17° from synchronous altitude). Since the satellite antenna gain, G_t, must be fixed, the resulting P_r is independent of wavelength, as shown in Equation (4.12). If the transmission at some frequency f_1 ($= c/\lambda_1$) provides earth coverage, then a frequency change to f_2, where $f_2 > f_1$, will result in reduced coverage (since for a given antenna, G_t will increase), hence the antenna size must be reduced to maintain the required earth coverage or beamwidth. Thus earth coverage antennas become smaller as the carrier frequency is increased.

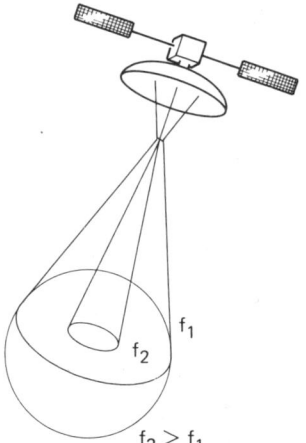

f_1

f_2

$f_2 > f_1$

Figure 4.6 Received power as a function of frequency.

4.3.3 Path Loss Is Frequency Dependent

From Equation (4.10) it can be seen that path loss, L_s, is wavelength (frequency) dependent. The question is often asked: Why should path loss, which is just a geometric inverse-square loss, be a function of frequency? The answer is that path loss, as characterized in Equation (4.10), is a *definition* predicated on the use of an isotropic receiving antenna ($G_r = 1$). Hence path loss is a convenient tool; it represents a hypothetical received-power loss that *would occur if the receiving antenna were isotropic*. Figure 4.3 and Equation (4.1) have established that power density, $p(d)$, is a function of distance—a purely geometric consideration; $p(d)$ is *not* a function of frequency. However, since path loss is predicated on $G_r = 1$, when we attempt to collect some P_r with an *isotropic antenna,* the result is characterized by Equation (4.10). Again let us emphasize that L_s can be viewed as a convenient collection of terms that have been assigned the unfortunate name *path loss*. The name conjures up an image of a purely geometric effect and fails to emphasize the requirement that $G_r = 1$. A better choice of a name would have been *unity-gain propagation loss*. In a radio communication system, path loss accounts for the largest loss in signal power. In satellite systems, the path loss for a C-band (6-GHz) link to a synchronous altitude satellite is typically 200 dB.

Example 4.2   Antenna Design for Measuring Path Loss

Design a hypothetical experiment to measure path loss L_s, at frequencies $f_1 = 30$ MHz and $f_2 = 60$ MHz, when the distance between the transmitter and receiver is 100 km. Find the effective area of the receiving antenna, and calculate the path loss in decibels for each case.

Solution

Figure 4.7 illustrates the two links for measuring L_s at frequencies f_1 and f_2, respectively. The power density, $p(d)$, at each receiver is identical and equal to

$$p(d) = \frac{\text{EIRP}}{4\pi d^2}$$

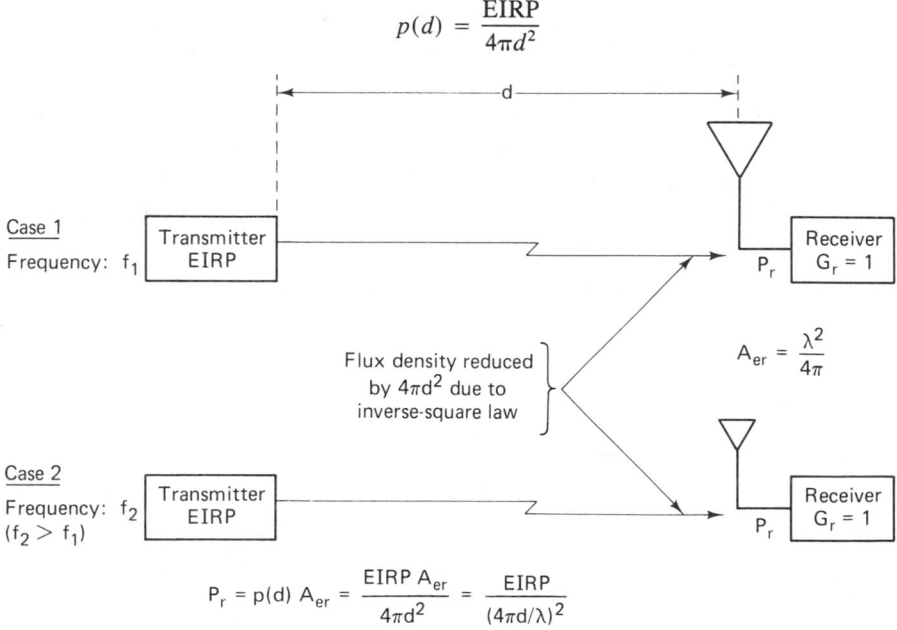

$$P_r = p(d) \, A_{er} = \frac{\text{EIRP} \, A_{er}}{4\pi d^2} = \frac{\text{EIRP}}{(4\pi d/\lambda)^2}$$

Figure 4.7   Path loss versus frequency. Hypothetical experiment to measure path loss at two different frequencies.

This reduction in power density is due *only* to the inverse-square law. The actual power received at each receiver is found by multiplying the power density $p(d)$ at the receiver by the effective area, A_{er}, of the collecting antenna, as shown in Equation (4.7). Since path loss is predicated on $G_r = 1$, we compute the effective area, A_{er1}, at frequency f_1, and A_{er2} at frequency f_2, using Equation (4.9):

$$A_{er} = \frac{\lambda^2}{4\pi} = \frac{(c/f)^2}{4\pi}$$

$$A_{er1} = \frac{(3 \times 10^8/30 \times 10^6)^2}{4\pi} \simeq 8 \text{ m}^2$$

$$A_{er2} = \frac{(3 \times 10^8/60 \times 10^6)^2}{4\pi} \simeq 2 \text{ m}^2$$

The path loss for each case in decibels is

$$L_{s1} = 10 \times \log_{10}\left(\frac{4\pi d}{\lambda_1}\right)^2 = 10 \times \log_{10}\left(\frac{4\pi \times 10^5}{3 \times 10^8/30 \times 10^6}\right)^2$$

$$\simeq 102 \text{ dB}$$

$$L_{s2} = 10 \times \log_{10}\left(\frac{4\pi d}{\lambda_2}\right)^2 = 10 \times \log_{10}\left(\frac{4\pi \times 10^5}{3 \times 10^8/60 \times 10^6}\right)^2$$

$$= 108 \text{ dB}$$

4.3.4 Thermal Noise Power

Thermal noise is caused by the thermal motion of electrons in all conductors. It is generated in the lossy coupling between an antenna and receiver and in the first stages of the receiver. The noise power spectral density is constant at all frequencies up to about 10^{12} Hz, giving rise to the name *white noise*. The thermal noise process in communication receivers is modeled as an additive white Gaussian noise (AWGN) process, as described in Section 1.5.5. The physical model [5, 6] for thermal or Johnson noise is a noise generator with an open-circuit mean-square voltage of $4\kappa T°W\mathcal{R}$, where

$$\kappa = \text{Boltzmann's constant} = 1.38 \times 10^{-23} \text{ J/K or W/K-Hz}$$

$$= -228.6 \text{ dBW/K-Hz}$$

$T° = $ temperature, kelvin

$W = $ bandwidth, hertz

$\mathcal{R} = $ resistance, ohms

The maximum thermal noise power, N, that could be coupled from the noise generator into the front end of an amplifier, is

$$N = \kappa T°W \qquad \text{watts} \qquad (4.16)$$

Thus the maximum single-sided noise power spectral density, N_0 (noise power in a 1-Hz bandwidth), available at the amplifier input is

$$N_0 = \frac{N}{W} = \kappa T° \qquad \text{watts/hertz} \qquad (4.17)$$

It might seem that the noise power should depend on the magnitude of the resistance—but it does not. Consider an intuitive argument to verify this. Electrically connect a large resistance to a small one, such that they form a closed path and such that their physical temperatures are the same. If noise power were a function of resistance, there would be a net power flow from the large resistance to the small one; the large resistance would become cooler and the small one would become warmer. This violates our experience, not to mention the second law of thermodynamics. Therefore, the power delivered from the large resistance to the small one must be equal to the power it receives.

The available power from a thermal noise source is dependent on the ambient temperature of the source (the *noise temperature*), as is seen in Equation (4.16). This leads to the useful concept of an *effective noise temperature* for noise sources that are not necessarily thermal in origin (e.g., galactic, atmospheric, interfering signals) that can be introduced into the receiving antenna. The effective noise temperature of such a noise source is defined as the temperature of a hypothetical thermal noise source that would give rise to an equivalent amount of interfering power. The subject of noise temperature is treated in greater detail in Section 4.5.

Example 4.3   Maximum Available Noise Power

Using a noise generator with mean-square voltage equal to $4\kappa T° W\mathscr{R}$, demonstrate that the maximum amount of noise power that can be coupled from this source into an amplifier is $N_i = \kappa T° W$.

Solution

A theorem from network theory states that maximum power is delivered to a load when the value of the load impedance is made equal to the complex conjugate of the generator impedance [7]. In this case the generator is a pure resistance, $\mathscr{R}$; therefore, the condition for maximum power transfer is fulfilled when the input resistance of the amplifier equals $\mathscr{R}$. Figure 4.8 illustrates such a network. The input thermal noise source is represented by an electrically equivalent model consisting of a noise-

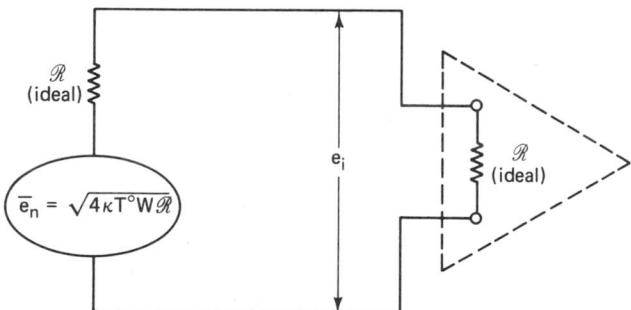

Figure 4.8 Electrical model of maximum available thermal noise power at amplifier input.

less source resistor in series with an ideal voltage generator whose rms noise voltage is $\sqrt{4\kappa T^{\circ}W\mathcal{R}}$. The input resistance of the amplifier is made equal to $\mathcal{R}$. The noise voltage delivered to the amplifier input is just one-half the generator voltage, following basic circuit principles. The noise power, N_i, delivered to the amplifier input can accordingly be expressed as follows:

$$N_i = \frac{(\sqrt{4\kappa T^{\circ}W\mathcal{R}}/2)^2}{\mathcal{R}} = \frac{4\kappa T^{\circ}W\mathcal{R}}{4\mathcal{R}}$$

$$= \kappa T^{\circ}W$$

4.4 LINK BUDGET ANALYSIS

In evaluating system performance, the quantity of greatest interest is the signal-to-noise ratio (SNR). This is because the basic system design centers on our ability to detect the signal, with an acceptable error probability, in the presence of noise. Since the desired signal here is a modulated carrier waveform, we often speak of the average carrier power-to-noise ratio (C/N) or (P_r/N) as the SNR of particular interest. We obtain P_r/N by dividing Equation (4.11) by noise power, N:

$$\frac{P_r}{N} = \frac{\text{EIRP } G_r/N}{L_s} \tag{4.18}$$

Equation (4.18) applies to any one-way RF link. With *analog receivers,* the noise bandwidth (generally referred to as the effective or equivalent noise bandwidth) seen by the demodulator is usually greater than the signal bandwidth, and P_r/N is the main parameter for measuring signal detectability and performance quality. With *digital receivers,* however, correlators or matched filters are usually implemented, and signal bandwidth is taken to be equal to noise bandwidth. Rather than consider input noise power, a common formulation for digital links is to replace noise power with *noise power spectral density*. We can use Equation (4.17) to rewrite Equation (4.18) as

$$\frac{P_r}{N_0} = \frac{\text{EIRP } G_r/T^{\circ}}{\kappa L_s L_o} \tag{4.19}$$

where the system effective temperature, T°, is a function of the noise radiated into the antenna and the thermal noise generated within the first stages of the receiver. Note that the receiving antenna gain, G_r, and system temperature, T°, are grouped together. The grouping, G_r/T°, is sometimes called the *receiver sensitivity*. The reason for treating these terms in this way is explained in Section 4.6.2.

It is important to emphasize that the effective temperature, T°, is a parameter that *models* the effect of various noise sources; the subject is treated in greater detail in Section 4.5. In Equation (4.19) we have introduced a term, L_o, to represent all other losses and degradation factors not specifically addressed by the other terms of Equation (4.18). The factor L_o allows for the large assortment of different losses and noise sources cataloged earlier. Equation (4.19) summarizes the key parameters of any link analysis; they are the received signal power-to-noise power spectral density (P_r/N_0), the effective transmitted power (EIRP), the receiver sensitivity ($G_r/T°$), and the losses (L_s, L_o).

If we assume that all the received power is in the modulating (information bearing) signal, we can write from Equation (3.94),

$$\frac{P_r}{N_0} = \frac{S}{N_0} = \frac{E_b}{N_0} R \tag{4.20}$$

recalling that S is the average modulating signal power, E_b/N_0 the bit energy per noise power spectral density, and R the bit rate. If some of the received power is unmodulated carrier power (a signal power loss), we can still employ Equation (4.20), except that the carrier power is accounted for as a loss factor [within the parameter L_o of Equation (4.19)].

4.4.1 Two E_b/N_0 Values of Interest

We have referred to E_b/N_0 as that value of bit energy per noise power spectral density required to yield a specified error probability. To facilitate calculating a margin or safety factor M, we need to differentiate between the *required* E_b/N_0 and the actual or *received* E_b/N_0. From this point on we will refer to the former as $(E_b/N_0)_{\text{reqd}}$ and to the latter as $(E_b/N_0)_r$. Figure 4.9 depicts an example with two operating points. The first is associated with $P_B = 10^{-3}$; let us call this operating point the system required error performance. Let us assume that an $(E_b/N_0)_{\text{reqd}}$ value of 10 dB will yield this required performance. Do you suppose we would build this system so that the demodulator received this 10-dB value *exactly*? Of course not; we would specify and design the system to have a safety margin, so that the $(E_b/N_0)_r$ actually received would be somewhat larger than the $(E_b/N_0)_{\text{reqd}}$. Thus we might design the system to operate at the second operating point on Figure 4.9; here $(E_b/N_0)_r = 12$ dB and $P_B = 10^{-5}$. For this example we can describe the safety margin or *link margin,* as providing a two-order-of-magnitude improved P_B, or as is more usual, we can describe the link margin in terms of providing 2 dB more E_b/N_0 than is required. We can rewrite Equation (4.20), introducing the link margin parameter, M, as

$$\frac{P_r}{N_0} = \left(\frac{E_b}{N_0}\right)_r R = M \left(\frac{E_b}{N_0}\right)_{\text{reqd}} R \tag{4.21}$$

The difference in decibels between $(E_b/N_0)_r$ and $(E_b/N_0)_{\text{reqd}}$ yields the link margin:

$$M \text{ (dB)} = \left(\frac{E_b}{N_0}\right)_r \text{ (dB)} - \left(\frac{E_b}{N_0}\right)_{\text{reqd}} \text{ (dB)} \tag{4.22}$$

The parameter $(E_b/N_0)_{\text{reqd}}$ reflects the differences from one system design to another; these might be due to differences in modulation or coding schemes. A larger than expected $(E_b/N_0)_{\text{reqd}}$ may be due to a suboptimal RF system, which manifests large timing errors or which allows more noise into the detection process than does an ideal matched filter.

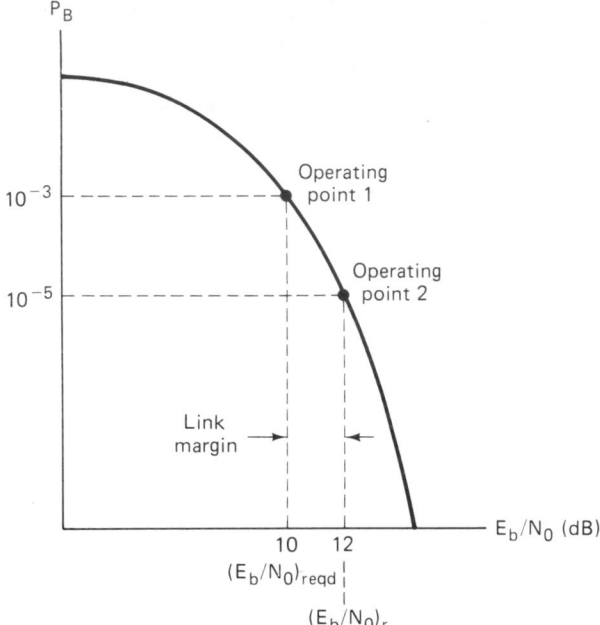

Figure 4.9 Two E_b/N_0 values of interest.

Combining Equations (4.19) and (4.21) and solving for the link margin M yields

$$M = \frac{\text{EIRP } G_r/T^\circ}{(E_b/N_0)_{\text{reqd}}R\kappa L_s L_o}$$ (4.23)

4.4.2 Link Budgets Are Typically Calculated in Decibels

Since link budget analysis is typically calculated in decibels, we can express Equation (4.23) as

$$M \text{ (dB)} = \text{EIRP (dBW)} + G_r \text{ (dBi)} - \left(\frac{E_b}{N_0}\right)_{\text{reqd}} \text{(dB)} - R \text{ (dB-bit/s)}$$

$$- \kappa T^\circ \text{ (dBW/Hz)} - L_s \text{ (dB)} - L_o \text{ (dB)} \quad (4.24)$$

Transmitted signal power, EIRP, is expressed in decibel-watts (dBW); noise power spectral density, N_0, is in decibel-watts per hertz (dBW/Hz); antenna gain, G_r, is in decibels referenced to isotropic gain (dBi); data rate, R, is in decibels referenced to 1 bit/s (dB-bit/s); and all other terms are in decibels (dB). The numerical values of the Equation (4.24) parameters constitute the link budget, a useful tool for allocating communications resources. In an effort to maintain a positive margin, we might trade off any parameter with any other parameter; we might choose to reduce transmitter power by giving up excess margin, or we might elect to increase the data rate by reducing $(E_b/N_0)_{\text{reqd}}$ (through the selection of improved modulation and coding). Any one of the Equation (4.24) decibels, regardless of the parameter from which it stems, is just as good as any other decibel. It should be noted, however, that as requirements become more constrained, it may not be possible to trade or yield on some items. For example, even though binary PSK modulation outperforms binary FSK (in the P_B sense), requirements to operate on a fading channel may dictate the avoidance of PSK and the choice of the more robust FSK. Also, certain beamwidth requirements may constrain antenna dimensions, so that one might *not* have the freedom of trading off or selecting any antenna gain that one desires.

4.4.3 How Much Link Margin Is Enough?

The question of how much link margin should be designed into a system is asked frequently. The answer is that if all sources of gain, loss, and noise have been rigorously detailed (worst case), and if the link parameters with large variances (e.g., fades due to weather) match the statistical requirements for link availability, very little additional margin is needed. The margin needed depends on how much confidence one has in each of the link budget entries. For systems employing new technology or new operating frequencies, one needs more margin than for systems that have been repeatedly built and tested. Sometimes the link budget provides an allowance for fades due to weather directly, as a line item. Other times, however, the required value of margin reflects the link requirements for a given rain degradation. For satellite communications at C-band (uplink at 6 GHz, downlink at 4 GHz), where the parameters are well known and fairly well behaved, it should be possible to design a system with only 1 dB of link margin. Receive-only television stations operating with 16-ft-diameter dishes at C-band are frequently designed with only a fraction of a decibel of margin. However, telephone communications via satellite using standards of 99.9% availability require considerably more margin; some of the INTELSAT systems have 4 to 5 dB of margin. When nominal rather than worst-case computations are performed, allowances are usually made for unit-to-unit equipment variations over the operating temperature range, line voltage variations, and mission duration. Also, for space communications, there may be an allowance for errors in tracking a satellite's location.

Designs using higher frequencies (e.g., 14/12 GHz) generally call for larger (weather) margins because atmospheric losses increase with frequency and are highly variable. It should be noted that a by-product of the attenuation due to atmospheric loss is greater antenna noise. With low-noise amplifiers, small weather changes can result in increases of 40 to 50 K in antenna temperature. Table 4.1 represents a link analysis proposed to the Federal Communications Commission (FCC) by Satellite Television Corporation for the Direct Broadcast Satellite (DBS) service. Notice that the downlink budget is tabulated for two alternative weather conditions: clear, and 5-dB loss due to rain. The signal loss due to atmospheric attenuation is only a small fraction of a decibel for clear weather and is the full 5 dB during rain. The next item in the downlink tabulation, home receiver $G/T°$, illustrates the additional degradation caused by the rain; additional thermal noise irradiates the receiving antenna, making the effective system noise temperature, $T°$, increase, and the home receiver $G/T°$ decrease

TABLE 4.1 Proposed Direct Broadcast Satellite (DBS) from Satellite Television Corp.

Uplink	
Earth station EIRP	86.6 dBW
Free-space loss (17.6 GHz, 48° elevation)	208.9 dB
Assumed rain attenuation	12.0 dB
Satellite $G/T°$	7.7 dB/K
Uplink $C/\kappa T°$	102.0 dB-Hz

Downlink	Atmospheric condition	
	Clear	5-dB rain attenuation
Satellite EIRP	57.0 dBW	57.0 dBW
Free-space loss (12.5 GHz, 30° elevation)	206.1 dB	206.1 dB
Atmospheric attenuation	0.14 dB	5.0 dB
Home receiver $G/T°$ (0.75 m)	9.4 dB/K	8.1 dB/K
Receiver pointing loss (0.5° error)	0.6 dB	0.6 dB
Polarization mismatch loss (average)	0.04 dB	0.04 dB
Downlink $C/\kappa T°$	88.1 dB-Hz	82.0 dB-Hz
Overall $C/\kappa T°$	87.9 dB-Hz	82.0 dB-Hz
Overall C/N (in 16 MHz)	15.9 dB	10.0 dB
Reference threshold C/N	10.0 dB	10.0 dB
Margin over threshold	5.9 dB	0.0 dB

(from 9.4 dB/K to 8.1 dB/K). Therefore, when extra margin is allowed for weather loss, additional margin should simultaneously be added to compensate for the increase in antenna noise temperature.

With regard to satellite links, in industry one often hears such expressions as "the link *can* be closed," meaning that the margin, in decibels, has a positive value and the required error performance will be satisfied, or "the link *cannot* be closed," meaning that the margin has a negative value and the required error performance will *not* be satisfied. Even though the words "the link closes" or "the link does not close" give the impression of an "on–off" condition, it is worth emphasizing that lack of link closure, or a negative margin, means that the error performance falls short of the system requirement; it does not necessarily mean that communications cease. For example, consider a system whose $(E_b/N_0)_{\text{reqd}}$ = 10 dB, as shown in Figure 4.9, but whose $(E_b/N_0)_r$ = 8 dB. Assume that 8 dB corresponds to $P_B = 10^{-2}$. Thus there is a margin of -2 dB, and a bit error probability of 10 times the specified error probability. The link may still be useful, though degraded.

4.4.4 Link Availability

Link availability is usually a measure of long-term link utility stated on an average annual basis; for a given geographical location, the link availability measures the percentage of time the link can be closed. For example, for a particular link between Washington, D.C., and a satellite repeater, the long-term weather pattern may be such that a 10-dB weather margin is adequate for link closure 98% of the time; for 2% of the time, heavy rains result in greater than 10 dB SNR degradation, so that the link does not close. Since the effect of rain on SNR degradation is a function of signal frequency, link availability and required margin must be examined in the context of a particular transmission frequency.

Figure 4.10 summarizes worldwide satellite link availability at a frequency of 44 GHz. The plot illustrates percentage of the earth visible (the link closes, and a prescribed probability of error is met) as a function of margin for the case of three equispaced geostationary satellites. A *geostationary satellite* is located in a circular orbit in the same plane as the earth's equatorial plane and at the synchronous altitude of 35,800 km. The satellite's orbital period is identical with that of the earth's rotational period, and therefore the satellite appears stationary when viewed from the earth. Figure 4.10 shows a family of visibility curves with different required link availabilities, ranging from benign (95% availability) to fairly stringent (99% availability). In general, for a fixed link margin, visibility is inversely proportional to required availability, and for a fixed availability, visibility increases monotonically with margin [8]. Figures 4.11 to 4.13 illustrate, by unshaded and shaded areas, the parts of the earth from which the 44-GHz link can and cannot be closed 99% of the time for three different values of link margin. Figure 4.11 illustrates the link coverage of such locations for a margin value of 14 dB. Notice that this figure can be used to pinpoint the regions of heaviest

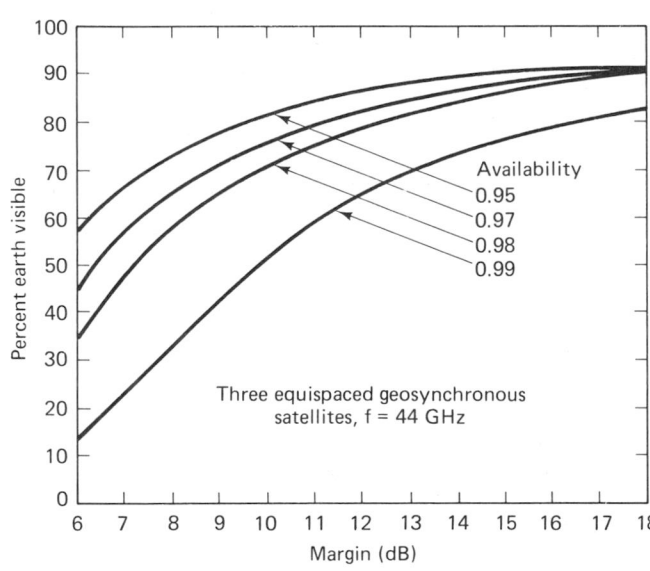

Figure 4.10 Earth coverage versus link margin for various values of link availability. (Reprinted from L. M. Schwab, "World-Wide Link Availability for Geostationary and Critically Inclined Orbits Including Rain Effects," *Lincoln Laboratory*, *Rep. DCA-9*, Jan. 27, 1981, Fig. 14, p. 38, courtesy of Lincoln Laboratory.)

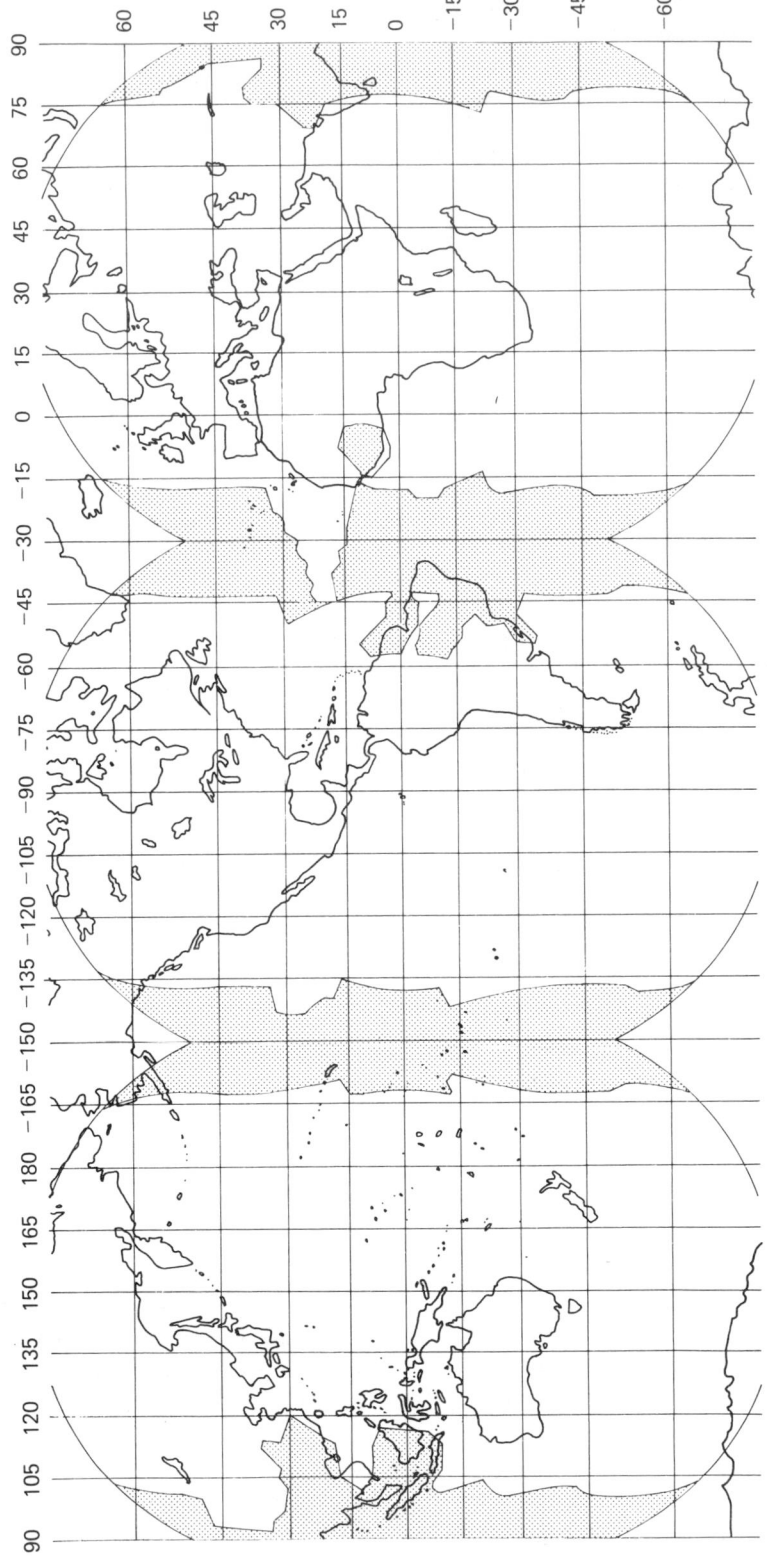

Figure 4.11 Earth coverage (unshaded) for 0.99 link availability for three equispaced geostationary satellites, f = 44 GHz, link margin = 14 dB. (Reprinted from L. M. Schwab, "World-Wide Link Availability for Geostationary and Critically Inclined Orbits Including Rain Effects," *Lincoln Laboratory, Rep. DCA-9,* Jan. 27, 1981, Fig. 17, p. 42, courtesy of Lincoln Laboratory.)

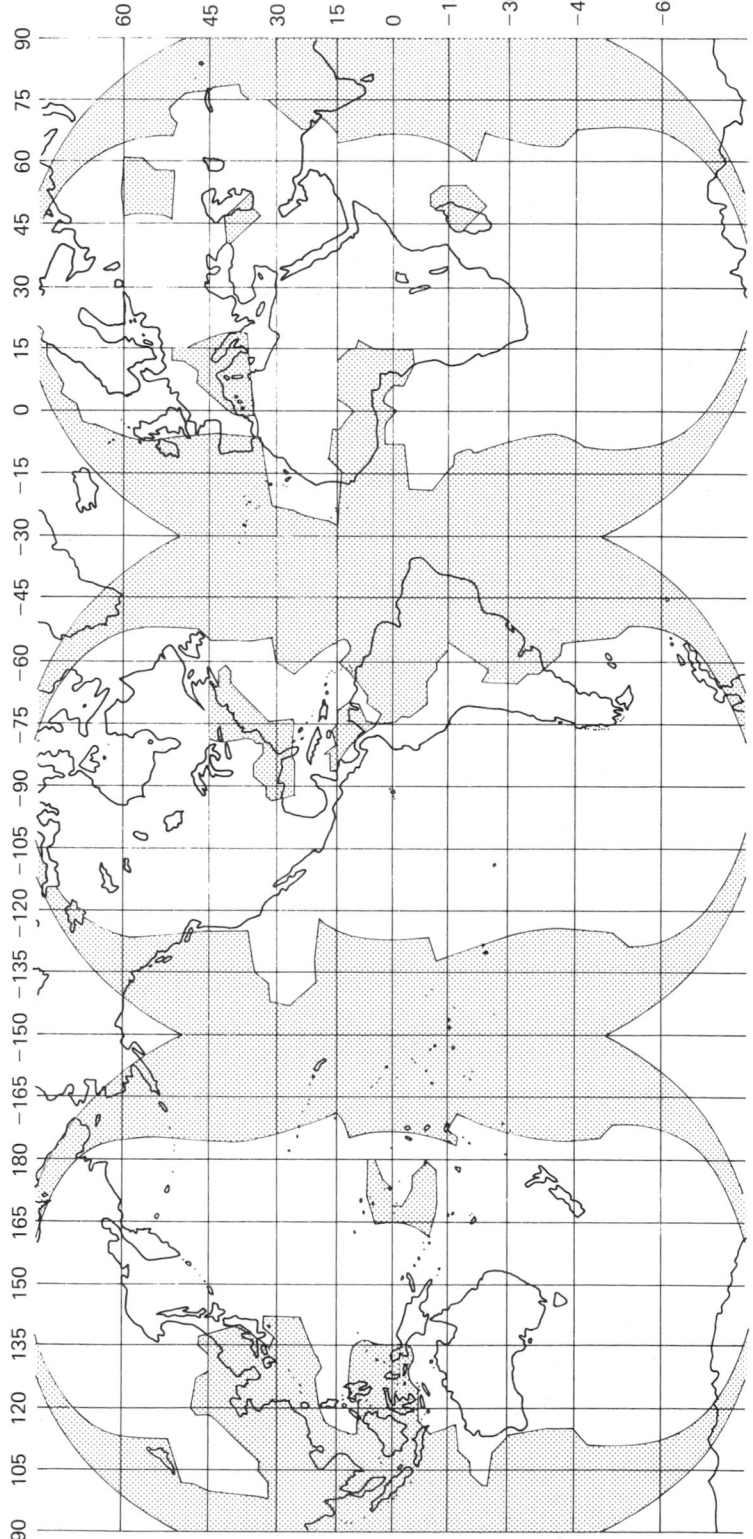

Figure 4.12 Earth coverage (unshaded) for 0.99 link availability for three equispaced geostationary satellites, $f = 44$ GHz, link margin $= 10$ dB. (Reprinted from L. M. Schwab, "World-Wide Link Availability for Geostationary and Critically Inclined Orbits Including Rain Effects," *Lincoln Laboratory, Rep.-DCA-9*, Jan. 27, 1981, Fig. 18, p. 43, courtesy of Lincoln Laboratory.)

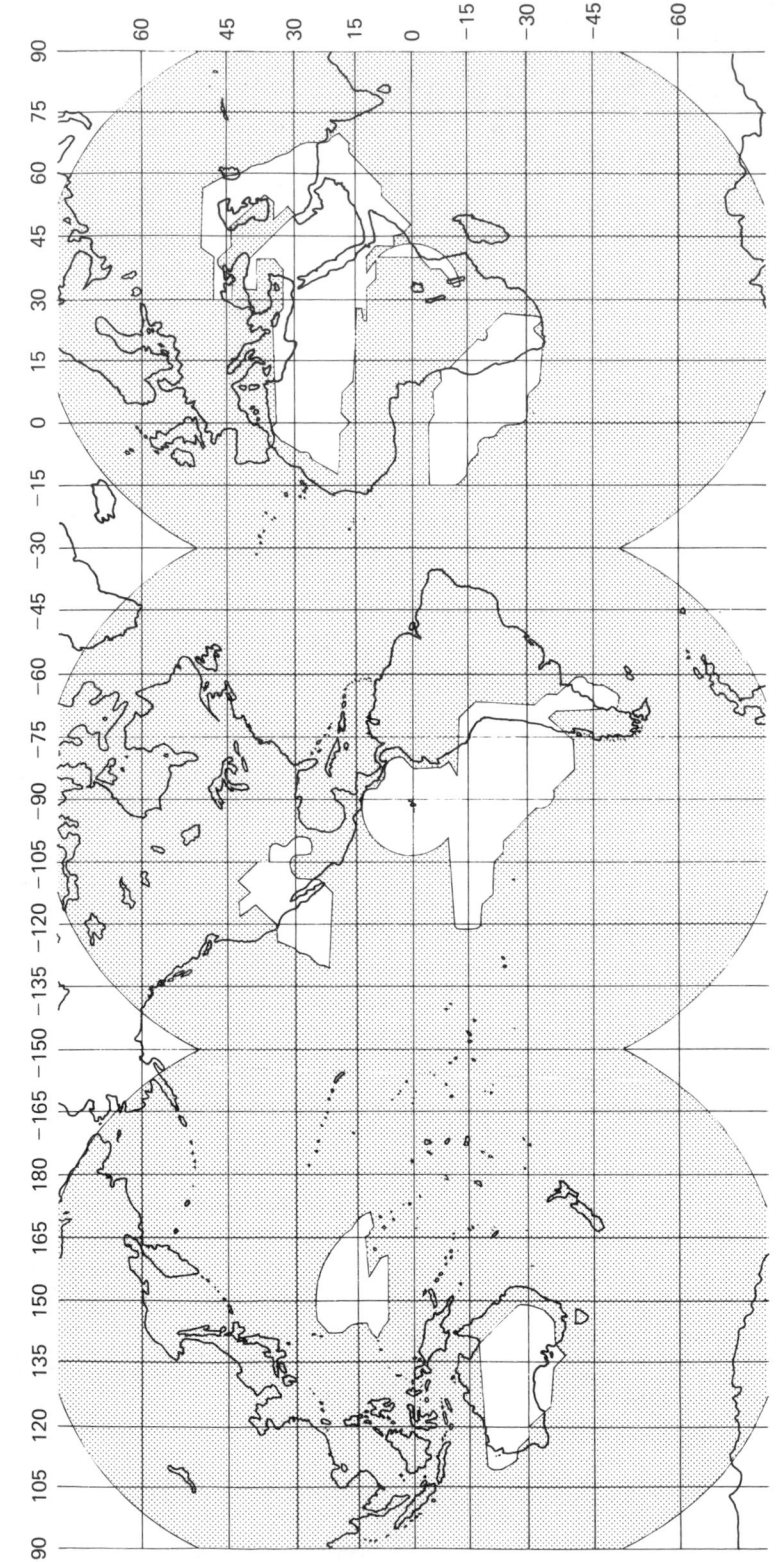

Figure 4.13 Earth coverage (unshaded) for 0.99 link availability for three equispaced geostationary satellites, f = 44 GHz, link margin = 6 dB. (Reprinted from L. M. Schwab, "World-Wide Link Availability for Geostationary and Critically Inclined Orbits Including Rain Effects," *Lincoln Laboratory, Rep. DCA-9*, Jan. 27, 1981, Fig. 19, p. 44, courtesy of Lincoln Laboratory.)

rainfall, such as Brazil and Indonesia. The figure represents the result of a link calculation performed in concert with a weather model of the earth.

In Figure 4.11 there are shaded strips on the east and west boundaries of each satellite's field of view. Why do you suppose the link availability is not met in these regions? At the edge of the earth the propagation path between the satellite and ground is longer than the path directly beneath the satellite. Degradation occurs in three ways: (1) the longer path results in reduced power density at the receiving antenna; (2) the edge of coverage sites will experience reduced satellite antenna gain, unless the satellite antenna pattern is designed to be uniform over its entire field of view (typically the pattern is -3 dB at the beam edge compared to the peak gain at the beam center); and (3) propagation to the edge of the earth traverses a thicker atmospheric layer because of the oblique path and the earth's curvature. The third item is of prime importance at those signal frequencies that are most attenuated by the atmosphere. Why do you suppose you do not see the same shaded areas near the north and south poles in Figure 4.11? Snowfall does not have the same deleterious effect on signal propagation as does rainfall; the phenomenon is known as the *freeze effect*.

Figure 4.12 illustrates the parts of the earth that can and cannot close the 44-GHz link 99% of the time with 10-dB link margin. Notice that the shaded areas have grown considerably compared to the 14-dB margin case; now, the east coast of the United States, the Mediterranean, and most of Japan cannot close the link 99% of the time. Figure 4.13 illustrates similar link performance for a margin of 6 dB. Whereas Figure 4.11 could be used to locate the regions of greatest rainfall, Figure 4.13 can be used to locate the driest weather regions on the earth; such areas are seen to be the southwestern part of the United States, most of Australia, the coast of Peru and Chile, and the Sahara desert in Africa.

4.5 NOISE FIGURE, NOISE TEMPERATURE, AND SYSTEM TEMPERATURE

4.5.1 Noise Figure

Noise figure, F, relates the SNR at the input of a network to the SNR at the output of the network. Thus noise figure measures the SNR degradation caused by the network. Figure 4.14 illustrates such an example. Figure 4.14a depicts the SNR at an *amplifier input,* $(SNR)_{in}$, as a function of frequency. At its peak, the signal is 40 dB above the noise floor. Figure 4.14b depicts the SNR at the *amplifier output* $(SNR)_{out}$. The amplifier gain has increased the signal by 20 dB; however, the amplifier has added its own additional noise. The output signal, at its peak, is only 30 dB above the noise floor. Since the SNR degradation from input to output is 10 dB, this is tantamount to describing the amplifier as having a 10-dB noise figure. Noise figure is a parameter that expresses the noisiness of a two-port network or device, such as an amplifier, compared to a reference noise source at the input port. It can be written as follows:

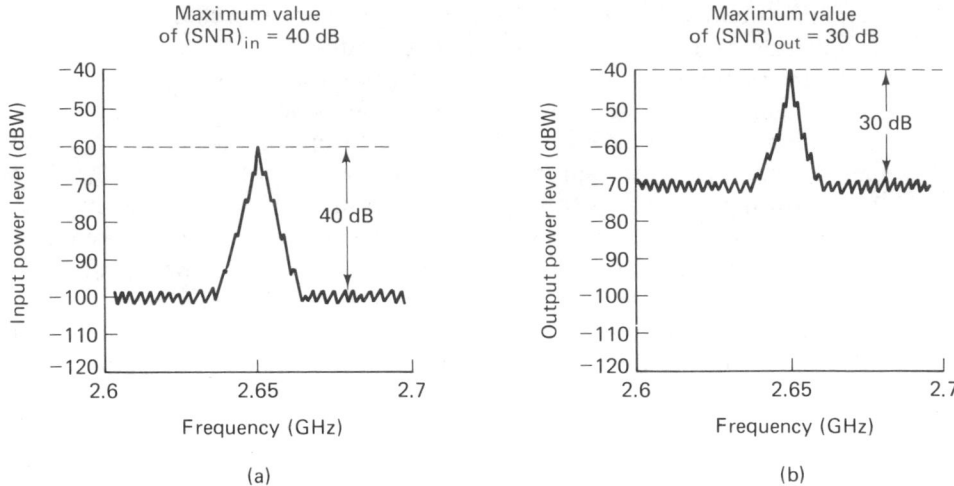

Figure 4.14 Amplifier signal and noise levels as a function of frequency. (a) Amplifier input. (b) Amplifier output.

$$F = \frac{(\text{SNR})_{\text{in}}}{(\text{SNR})_{\text{out}}} = \frac{S_i/N_i}{GS_i/G(N_i + N_{ai})} \tag{4.25}$$

where

$$S_i = \text{signal power at the amplifier input port}$$

$$N_i = \text{noise power at the amplifier input port}$$

$$N_{ai} = \text{amplifier noise referred to the input port}$$

$$G = \text{amplifier gain}$$

Figure 4.15 is an example illustrating Equation (4.25). Figure 4.15a represents a *realizable amplifier* example with a gain, $G = 100$, and internal noise power, $N_a = 10\ \mu\text{W}$. The source noise, external to the amplifier, is $N_i = 1\ \mu\text{W}$. In Figure 4.15b we assume that the *amplifier is ideal,* and we ascribe the noisiness of the real amplifier, from part (a) of the figure, to an external source, N_{ai}, in series with the original source, N_i. The value of N_{ai} is obtained by reducing N_a by the amplifier gain. As shown in Figure 4.15b, Equation (4.25) references all noise to the amplifier input, whether the noise is actually present at the input or is internal to the device. As can be seen in Figure 4.15, the noise power output from the real amplifier is identical to that of its electrically equivalent model.

Equation (4.25) reduces to the following:

$$F = \frac{N_i + N_{ai}}{N_i} = 1 + \frac{N_{ai}}{N_i} \tag{4.26}$$

Notice from Equation (4.26) that the noise figure expresses the noisiness of a

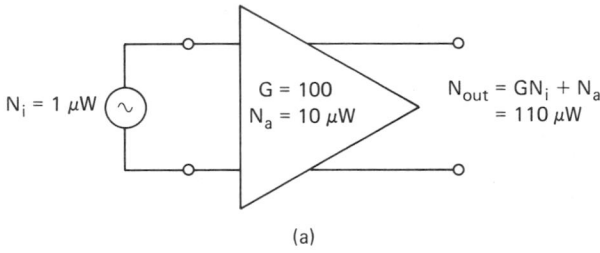

(a)

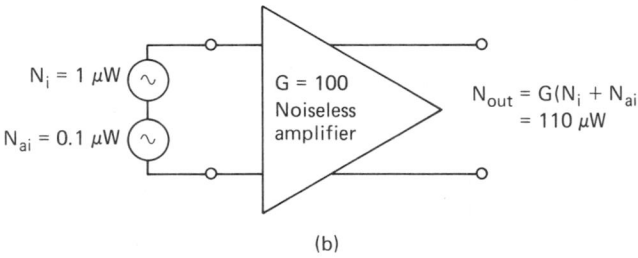

(b)

Figure 4.15 Example of noise treatment in amplifiers.

network relative to an input source noise; noise figure is *not* an absolute measure of noise. An ideal amplifier or network, one that contributes no noise ($N_{ai} = 0$), has a noise figure equal to unity (0 dB).

For the concept of noise figure to have utility, we need to be able to make equitable comparisons among devices on the basis of Equation (4.26). We must, therefore, choose a value of N_i as a *reference*. The noise figure of any device will then represent a measure of how much noisier the device is than the reference. In 1944, Friis [9] suggested that noise figure be defined for a noise source at a reference temperature of $T_0^\circ = 290$ K. That suggestion was subsequently adopted by the IEEE as part of its standard definition for noise figure [10]. From Equation (4.17) we see that the maximum available noise power spectral density from any source resistance is established by specifying its temperature. The value of 290 K was selected as the reference because it is a reasonable approximation of the source temperature for many links. Also with T_0° chosen to be 290 K, the value of noise spectral density N_0 at T_0° results in an aesthetically pleasing number, as shown below:

$$N_0 = \kappa T_0^\circ = 1.38 \times 10^{-23} \times 290 = 4.00 \times 10^{-21} \text{ W/Hz}$$

Expressed in decibels,

$$N_0 = -204 \text{ dBW/Hz}$$

4.5.2 Noise Temperature

Rearranging Equation (4.26), we can write

$$N_{ai} = (F - 1)N_i \tag{4.27}$$

From Equation (4.16) we can replace N_i with $\kappa T_0^\circ W$ and N_{ai} with $\kappa T_R^\circ W$, where

T_0° is the reference temperature of the source and T_R° is called the *effective noise temperature* of the receiver (or network). We can then write

$$\kappa T_R^\circ W = (F - 1)\kappa T_0^\circ W$$
$$T_R^\circ = (F - 1)\, T_0^\circ$$

or since T_0° has been chosen to be 290 K,

$$T_R^\circ = (F - 1)290 \text{ K} \tag{4.28}$$

Equation (4.26) uses the concept of noise figure to characterize the noisiness of an amplifier. Equation (4.28) represents an alternative but equivalent characterization known as *effective noise temperature*. Note that the noise figure is a measurement relative to a reference. However, noise temperature has no such constraint.

We can think of available noise power spectral density and effective noise temperature, in the context of Equation (4.17), as equivalent ways of characterizing noise sources. Equation (4.28) tells us that the noisiness of an amplifier can be modeled as if it were caused by an additional noise source, as seen in Figure 4.15b, operating at some effective temperature called T_R°. For a purely resistive termination, T_R° is never less than ambient temperature unless it is cooled. It is important to note that for reactive terminations, such as uncooled parametric amplifiers or other low-noise devices, T_R° can be much less than 290 K, even though the ambient temperature is higher [11]. For the output noise of an amplifier, N_{out}, as a function of its effective temperature, we can use Equations (4.25) and (4.16) to write

$$N_{\text{out}} = GN_i + GN_{ai} \tag{4.29}$$
$$= G\kappa T_g^\circ W + G\kappa T_R^\circ W = G\kappa(T_g^\circ + T_R^\circ)W$$

where T_g° is the temperature of the source.

4.5.3 Line Loss

The difference between amplifier networks and lossy line networks can be viewed in the context of the degradation mechanisms *loss* and *noise*, described earlier. Noisy networks in Sections 4.5.1 and 4.5.2 were discussed with amplifiers in mind. We saw that SNR degradation resulted from injecting additional (amplifier) noise into the link, as shown in Figure 4.15. However, in the case of a lossy line, we shall show that the SNR degradation results from the signal being attenuated while the noise remains fixed (for the case where the line temperature is equal to or less than the source temperature). The degradation effect will nonetheless be measured as an increase in noise figure or effective noise temperature.

Consider the lossy line or network shown in Figure 4.16. Assume the line is matched with its characteristic impedance at the source and at the load. We shall define power loss, L, as

$$L = \frac{\text{input power}}{\text{output power}}$$

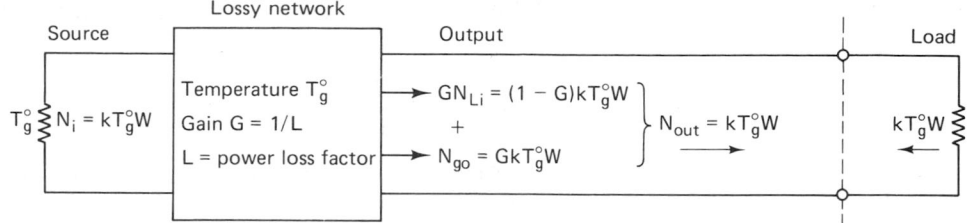

Figure 4.16 Lossy line: impedance matched and temperature matched at both ends.

Then the network gain, G, equals $1/L$ (less than unity for a lossy line). Let all components be at temperature T_g°. The total output noise power flowing from the network into the load is

$$N_{\text{out}} = \kappa T_g^\circ W$$

since the network output appears as a pure resistance at the temperature T_g°. The total power flowing from the load back into the network must also equal N_{out}, to ensure thermal equilibrium. Recall that available noise power, $\kappa T^\circ W$, is dependent only on temperature, bandwidth, and impedance matching; it is not dependent on the resistance value. N_{out} can be considered as being comprised of two components, N_{go} and GN_{Li}, such that

$$N_{\text{out}} = \kappa T_g^\circ W = N_{go} + GN_{Li} \qquad (4.30)$$

where

$$N_{go} = G\kappa T_g^\circ W \qquad (4.31)$$

is the component of output noise power due to the source and GN_{Li} is the component of output noise power due to the lossy network, where N_{Li} is the network noise relative to its input. Combining Equations (4.30) and (4.31), we can write

$$\kappa T_g^\circ W = G\kappa T_g^\circ W + GN_{Li} \qquad (4.32)$$

Solving for N_{Li} yields

$$N_{Li} = \frac{1 - G}{G} \kappa T_g^\circ W = \kappa T_L^\circ W \qquad (4.33)$$

Therefore, the effective noise temperature of the line is

$$T_L^\circ = \frac{1 - G}{G} T_g^\circ \qquad (4.34)$$

and since $G = 1/L$,

$$T_L^\circ = (L - 1)T_g^\circ \qquad (4.35)$$

Choosing $T_g^\circ = 290$ K as the reference temperature, we can write

$$T_L^\circ = (L - 1)290 \text{ K} \qquad (4.36)$$

Using Equations (4.28) and (4.36), the *noise figure for a lossy line* can be expressed as

$$F = 1 + \frac{T_L^\circ}{290} = L \tag{4.37}$$

Note that some authors use the parameter L to mean the reciprocal of the loss factor defined here. In such cases, noise figure $F = 1/L$.

Example 4.4   Lossy Line

A line at temperature $T^\circ = 290$ K is fed from a source whose noise temperature is $T_g^\circ = 290$ K. The input signal power, S_i, is 100 picowatts (pW) and the signal bandwidth is 1 GHz. The line has a loss factor $L = 2$. Calculate the $(SNR)_{in}$, the effective line temperature, T_L°, the output signal power, S_{out}, the $(SNR)_{out}$, and the noise figure, F.

Solution

$$N_i = \kappa T_g^\circ W$$

$$= 1.38 \times 10^{-23} \text{ W/K-Hz} \times 290 \text{ K} \times 10^9 \text{ Hz}$$

$$= 4 \times 10^{-12} \text{ W} = 4 \text{ pW}$$

$$(SNR)_{in} = \frac{100 \text{ pW}}{4 \text{ pW}} = 25 \text{ (14 dB)}$$

$$T_L^\circ = (L - 1)290 \text{ K} = 290 \text{ K}$$

$$S_{out} = \frac{S_i}{L} = \frac{100 \text{ pW}}{2} = 50 \text{ pW}$$

$$N_{out} = \kappa T^\circ W = 4 \text{ pW}$$

$$(SNR)_{out} = \frac{50 \text{ pW}}{4 \text{ pW}} = 12.5 \text{ (11 dB)}$$

$$F = L = 2 \text{ (3 dB)}$$

4.5.4 Composite Noise Figure and Composite Noise Temperature

When two networks are connected in series, as shown in Figure 4.17a, their composite noise figure can be written as

$$F_{comp} = F_1 + \frac{F_2 - 1}{G_1} \tag{4.38}$$

where G_1 is the gain associated with network 1. When n networks are connected in series the relationship between stages expressed in Equation (4.38) continues, so that the *composite noise figure* for a sequence of n stages is written as

$$F_{comp} = F_1 + \frac{F_2 - 1}{G_1} + \frac{F_3 - 1}{G_1 G_2} + \cdots + \frac{F_n - 1}{G_1 G_2 \cdots G_{n-1}} \tag{4.39}$$

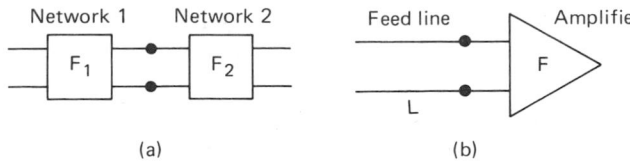

Figure 4.17 Networks connected in series.

(a) (b)

Can you guess from Equation (4.39) what the design goals for the front end of the receiver (especially the first stage or the first couple of stages) should be? At the front end of the receiver, the signal is most susceptible to added noise; therefore, the first stage should have as low a noise figure, F_1, as possible. Also, because the noise figure of each subsequent stage is reduced by the gains of the prior stages, it behooves us to strive for as high a gain, G_1, as possible. Simultaneously achieving the lowest F_1 and the highest G_1 represents conflicting goals; therefore, compromises are always necessary.

Equations (4.39) and (4.28) can be combined to express the composite effective noise temperature of a sequence of n stages:

$$T^\circ_{\text{comp}} = T^\circ_1 + \frac{T^\circ_2}{G_1} + \frac{T^\circ_3}{G_1 G_2} + \cdots + \frac{T^\circ_n}{G_1 G_2 \cdots G_{n-1}} \qquad (4.40)$$

Figure 4.17b illustrates a feed line in series with an amplifier; this is a typical arrangement following a receiving antenna. Using Equation (4.38) to find F_{comp} for such a lossy line and amplifier arrangement, we can write

$$F_{\text{comp}} = L + L(F - 1) = LF \qquad (4.41)$$

since the noise figure of the lossy line is L and the gain of the line is $1/L$. By analogy with Equation (4.36), we can write the composite temperature as

$$T^\circ_{\text{comp}} = (LF - 1)290 \text{ K} \qquad (4.42)$$

We can also write the composite temperature of line and amplifier as follows:

$$T^\circ_{\text{comp}} = (LF - 1 + L - L)290 \text{ K}$$

$$= [(L - 1) + L(F - 1)]290 \text{ K} \qquad (4.43)$$

$$= T^\circ_L + LT^\circ_R$$

4.5.4.1 Comparison of Noise Figure and Noise Temperature

Since noise figure, F, and effective noise temperature, T°, characterize the noise performance of devices, some engineers feel compelled to select one of these measures as the more useful. However, they each have their place. For terrestrial applications, F is almost universally used; the concept of SNR degradation for a 290 K source temperature makes sense, because terrestrial source temperatures are typically close to 290 K. Terrestrial noise figure values typically fall in the convenient range 1 to 10 dB.

For space applications, T° is the more common figure of merit. The range of values for commercial systems is typically between 30 and 150 K, giving adequate resolution for comparing performance between systems. A disadvantage of using noise figures for such low-noise networks is that the values obtained are all close to unity (0.5 to 1.5 dB), which makes it difficult to compare devices. For low-noise applications, F (in decibels) would need to be expressed to two decimal places to provide the same resolution or precision as does $T°$. For space applications, a reference temperature of 290 K is not as appropriate as it is for terrestrial applications. When using effective temperature, no reference temperature (other than absolute zero K) is needed for judging degradation. The effective input noise temperature is simply compared to the effective source noise temperature. In general, applications involving very low noise devices seem to favor the effective temperature measure over the noise figure.

4.5.5 System Effective Temperature

Figure 4.18 represents a simplified schematic of a receiving system, identifying those areas—the antenna, the line, and the preamplifier—that play a primary role in SNR degradation. We have already discussed the degradation role of the preamplifier—additional noise is injected into the link. And we have discussed line loss—the signal is attenuated, while the noise is held fixed (for the case where the line temperature is less than or equal to the source temperature). The remaining source of degradation is via the receiving antenna. An antenna is like a lens. Its noise contributions are dictated by what the antenna is "looking at." If the antenna is pointed at a cool portion of the sky, very little thermal noise is introduced, but if the antenna is pointed at a warm body, a greater amount of thermal noise is introduced. The *antenna temperature* is a measure of the effective temperature integrated over the entire antenna pattern.

We now find *system temperature*, $T_S°$, by adding together all the system noise contributors (in terms of effective temperature). The summation is expressed as

$$T_S° = T_A° + T_{comp}° \tag{4.44}$$

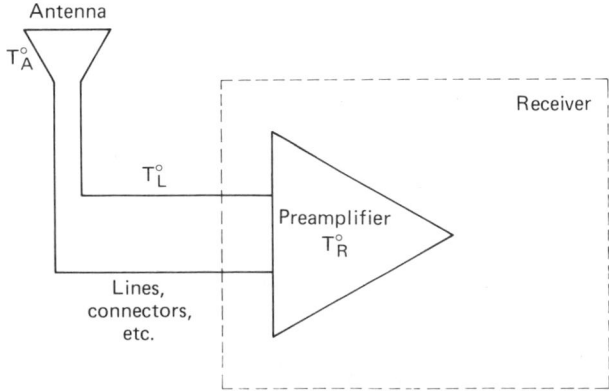

Figure 4.18 Major noise contributors of a receiving system.

where T_A° is the antenna temperature and T_{comp}° is the composite temperature of the line and the preamplifier. Since the system temperature, T_S°, is a new composite, made up of T_A° and the composite effective temperature of the line and preamp, one might ask: Why doesn't Equation (4.44) appear to have the same sequential gain reduction factors as those in Equation (4.40)? We have assumed that the antenna has *no dissipative parts*; its gain, unlike an amplifier or attenuator, can be thought of as a processing gain. Whatever effective temperature is introduced at the antenna comes through, unaltered by the antenna; the antenna represents the source noise, or source temperature, at the input to the line.

Using Equation (4.43), we can modify Equation (4.44) as follows:

$$T_S^\circ = T_A^\circ + T_L^\circ + LT_R^\circ \tag{4.45}$$

$$= T_A^\circ + (L - 1)290 \text{ K} + L(F - 1)290 \text{ K}$$

$$= T_A^\circ + (LF - 1)290 \text{ K} \tag{4.46}$$

If LF is provided in units of decibels, we must first convert LF to a ratio, so that T_S° takes the form

$$T_S^\circ = T_A^\circ + (10^{LF/10} - 1)290 \text{ K} \tag{4.47}$$

Example 4.5   Noise Figure and Noise Temperature

A receiver front end, shown in Figure 4.19a, has a noise figure of 10 dB, a gain of 80 dB, and a bandwidth of 6 MHz. The input signal power, S_i, is 10^{-11} W. Assume that the line loss is zero and the antenna temperature is 150 K. Find T_R°, T_S°, N_{out}, $(\text{SNR})_{\text{in}}$, and $(\text{SNR})_{\text{out}}$.

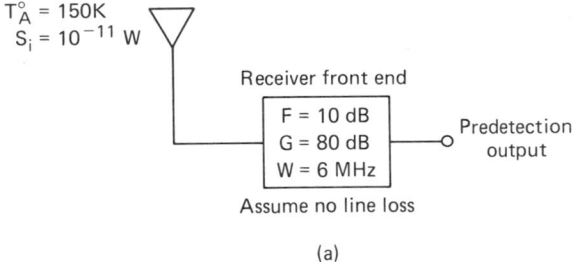

(a)

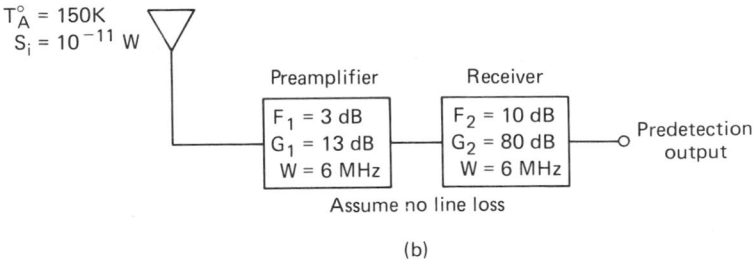

(b)

Figure 4.19   Improving a receiver front end with a low-noise preamplifier.

Solution

First convert all decibel values to ratios.

$$T_R^\circ = (F - 1)290 \text{ K} = 2610 \text{ K}$$

$$T_S^\circ = T_A^\circ + T_R^\circ = 150 \text{ K} + 2610 \text{ K} = 2760 \text{ K}$$

$$N_{\text{out}} = G\kappa T_A^\circ W + G\kappa T_R^\circ W = G\kappa T_S^\circ W$$

$$= 10^8 \times 1.38 \times 10^{-23} \times 6 \times 10^6 (150 \text{ K} + 2610 \text{ K})$$

$$= \underbrace{1.2 \text{ } \mu\text{W}}_{\substack{\text{source} \\ \text{contribution}}} + \underbrace{21.6 \text{ } \mu\text{W}}_{\substack{\text{front-end} \\ \text{contribution}}} = 22.8 \text{ } \mu\text{W}$$

$$(\text{SNR})_{\text{in}} = \frac{S_i}{\kappa T_A^\circ W} = \frac{10^{-11}}{1.24 \times 10^{-14}} = 806.5 \text{ (29.1 dB)}$$

$$(\text{SNR})_{\text{out}} = \frac{S_{\text{out}}}{N_{\text{out}}} = \frac{10^8 \times 10^{-11}}{22.8 \times 10^{-6}} = 43.9 \text{ (16.4 dB)}$$

Notice in this example that the amplifier noise is significantly larger than the source noise and represents the major cause of SNR degradation.

Example 4.6 Improving SNR with a Low-Noise Preamplifier

Use a preamplifier, as shown in Figure 4.19b, with a noise figure of 3 dB, a gain of 13 dB, and a bandwidth of 6 MHz to improve the SNR of the receiver in Example 4.5. Find T_{comp}° for the composite preamplifier and receiver. Find T_S°, F_{comp}, N_{out}, and $(\text{SNR})_{\text{out}}$. Assume zero line loss.

Solution

Again, convert all decibel values to ratios before proceeding.

$$T_{R1}^\circ = (F_1 - 1)290 \text{ K} = 290 \text{ K}$$

$$T_{R2}^\circ = (F_2 - 1)290 \text{ K} = 2610 \text{ K}$$

$$T_{\text{comp}}^\circ = T_{R1}^\circ + \frac{T_{R2}^\circ}{G_1} = 290 \text{ K} + \frac{2610 \text{ K}}{20} = 420.5 \text{ K}$$

$$T_S^\circ = T_A^\circ + T_{\text{comp}}^\circ = 150 \text{ K} + 420.5 \text{ K} = 570.5 \text{ K}$$

$$F_{\text{comp}} = F_1 + \frac{F_2 - 1}{G_1} = 2 + \frac{9}{20} = 2.5 \text{ (4 dB)}$$

$$N_{\text{out}} = G\kappa T_A^\circ W + G\kappa T_{\text{comp}}^\circ W = G\kappa T_S^\circ W$$

$$= 20 \times 10^8 \times 1.38 \times 10^{-23} \times 6 \times 10^6 (150 \text{ K} + 420.5 \text{ K})$$

$$= \underbrace{24.8 \text{ } \mu\text{W}}_{\substack{\text{source} \\ \text{contribution}}} + \underbrace{69.6 \text{ } \mu\text{W}}_{\substack{\text{front-end} \\ \text{contribution}}} = 94.4 \text{ } \mu\text{W}$$

$$(\text{SNR})_{\text{out}} = \frac{S_{\text{out}}}{N_{\text{out}}} = \frac{10^{-11} \times 20 \times 10^8}{94.4 \times 10^{-6}} = 212.0 \text{ (23.3 dB)}$$

With the added preamplifier the (predetection) output noise has increased (from 22.8 μW in Example 4.5) to 94.4 μW. Even though the noise power has increased, the lower system temperature has resulted in a 6.9-dB improvement in SNR (from 16.4 dB in Example 4.5, to 23.3 dB here). The price we pay for this improvement is the need to provide an F_{comp} improvement of 6 dB (from 10 dB in Example 4.5, to 4 dB in this example).

The unwanted noise is, in part, *injected via the antenna* ($\kappa T_A^\circ W$), and in part, *generated internally* in the receiver front end ($\kappa T_{comp}^\circ W$). The amount of system improvement that can be rendered via front-end design depends on what portion of the total noise the front end contributes. We saw in Example 4.5 that the front end contributed the major portion of the noise. Therefore, in Example 4.6, providing a low-noise preamplifier improved the system SNR significantly. In the next example, we show the case where the major portion of the noise is injected via the antenna; we shall see that introducing a low-noise preamplifier in such a case will not help the SNR appreciably.

Example 4.7 Attempting SNR Improvement When the Value of T_A° Is Large

Repeat Examples 4.5 and 4.6 with one change: let $T_A^\circ = 8000$ K. In other words, the preponderant amount of noise is being injected via the antenna; the antenna might have a very hot body (the sun) fully occupying its field of view. Calculate the SNR improvement that would be provided by the preamplifier used in Example 4.6 and Figure 4.19b, and compare the result with that of Example 4.6.

Solution

Without preamplifier

$$N_{out} = G\kappa W(T_A^\circ + T_R^\circ)$$

$$= 10^8 \times 1.38 \times 10^{-23} \times 6 \times 10^6(8000 \text{ K} + 2610 \text{ K})$$

$$= \underbrace{66.2 \text{ μW}}_{\substack{\text{source} \\ \text{contribution}}} + \underbrace{21.6 \text{ μW}}_{\substack{\text{front-end} \\ \text{contribution}}} = 87.8 \text{ μW}$$

$$(\text{SNR})_{out} = \frac{S_{out}}{N_{out}} = \frac{10^8 \times 10^{-11}}{87.8 \times 10^{-6}} = 11.4 \ (10.6 \text{ dB})$$

With preamplifier

$$N_{out} = 20 \times 10^8 \times 1.38 \times 10^{-23} \times 6 \times 10^6(8000 \text{ K} + 420.5 \text{ K})$$

$$= \underbrace{1324.8 \text{ μW}}_{\substack{\text{source} \\ \text{contribution}}} + \underbrace{69.6 \text{ μW}}_{\substack{\text{front-end} \\ \text{contribution}}} = 1394.4 \text{ μW}$$

$$(\text{SNR})_{out} = \frac{20 \times 10^8 \times 10^{-11}}{1.39 \times 10^{-3}} = 14.4 \ (11.6 \text{ dB})$$

Therefore, for this case, the SNR improvement is only 1 dB, a far cry from the 6.9 dB accomplished earlier. When the noise is mostly due to devices within the receiver, it is possible to improve the SNR by introducing low-noise devices. However, when the noise is mostly due to external causes, improving the receiver front end will not help much.

Noise figure is a definition, predicated on a reference temperature of 290 K. When the source temperature is other than 290 K, as is the case in Examples 4.5, 4.6, and 4.7, it is necessary to define a *working* or *effective noise figure* that describes the actual $(\text{SNR})_{in}$ versus $(\text{SNR})_{out}$ relationship shown in Equation (4.25). Such an operational noise figure can be found as follows:

$$F_{op} = \frac{T_S^\circ}{T_A^\circ} = \frac{T_A^\circ + T_R^\circ}{T_A^\circ}$$

since

$$F = 1 + \frac{T_R^\circ}{T_0^\circ}$$

then

$$F_{op} = (F-1)\,\frac{T_0^\circ}{T_A^\circ} + 1 \qquad (4.48)$$

4.5.6 Sky Noise Temperature

The receiving antenna collects random noise emissions from galactic, solar, and terrestrial sources, constituting the sky background noise. The sky background appears as a combination of galactic effects that decrease with frequency, and atmospheric effects that start becoming significant at 10 GHz and increase with frequency. Figure 4.20 illustrates the sky temperature, as measured from the earth, due to both these effects. Notice that there is a region, between 1 and 10 GHz, where the temperature is lowest; the galaxy noise has become quite small at 1 GHz, and for satellite communications the blackbody radiation noise due to the absorbing atmosphere is not significant below 10 GHz. (For other applications, e.g., passive radiometry, it is still a problem.) This region, known as the *microwave window* or *space window,* is particularly useful for satellite or deep-space communication. The low sky noise is the principal reason that such systems primarily use carrier frequencies in this part of the spectrum. The galaxy and atmospheric noise curves in Figure 4.20 are comprised of a family of curves each at a different elevation angle θ. When θ is zero, the receiving antenna points at the horizon and the propagation path encompasses the longest possible atmospheric layer; when θ is 90°, the receiving antenna points to the zenith, and the resulting propagation path contains the shortest possible atmospheric layer. Thus the upper curve of the family represents the near worst-case (clear-weather) sky noise temperature versus frequency, and the lower curve represents the most benign case. Also shown in Figure 4.20 is a plot of noise temperature versus frequency *due to rain.* Since the intensity of any rainstorm can only be expressed

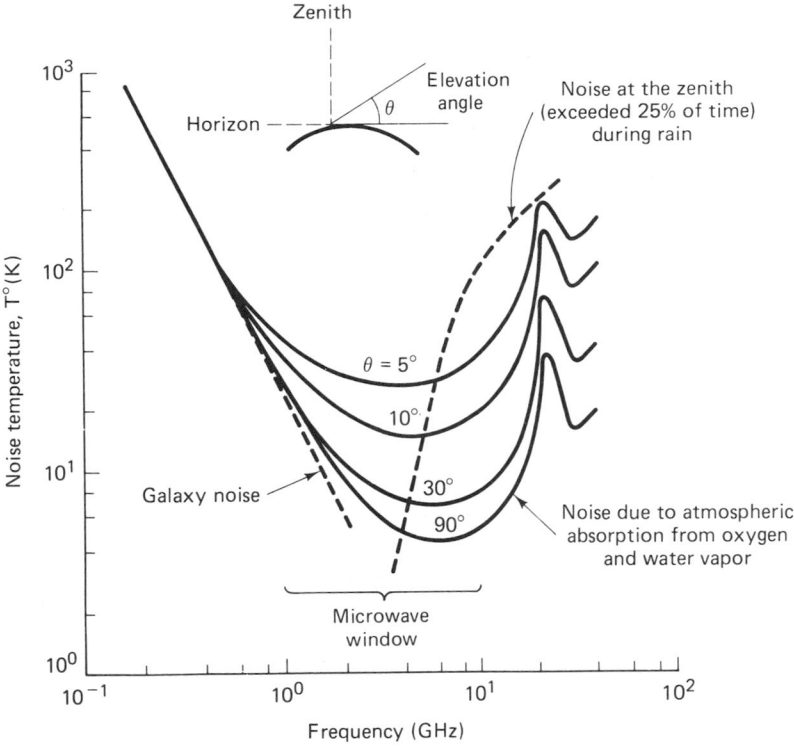

Figure 4.20 Sky noise temperature.

statistically, the noise temperatures shown are values that are exceeded 25% of the time (at the zenith). Which spectral region appears the most benign for space communications when rainfall is taken into account? It is the region at the low end of the space window. For this reason, systems such as the Space Ground Link Subsystem or SGLS (military) and the Unified S-Band Telemetry, Tracking, and Control System (NASA) are located in the 1.8 to 2.4-GHz band.

4.5.6.1 Radio Maps of the Sky

Various researchers have mapped the galactic noise radiation as a function of frequency. Figure 4.21 is such a radio temperature map, after Ko and Kraus [12], indicating the temperature contours of the sky in the region of 250 MHz when viewed from the earth. In general, the sky is composed of localized galactic sources (sun, moon, planets, etc.), each having its own temperature. The map is effectively a weighted sum of the individual galactic source temperatures plus a constant sky background. The coordinates of the map, *declination and right ascension,* can be thought of as celestial latitude and longitude with an earth reference (right ascension is calibrated in units of hour-angle, where 24 hours corresponds to a complete rotation of the earth). On Figure 4.21, the temperature contours range from a low of 90 K to a high of 1000 K. The measurements were

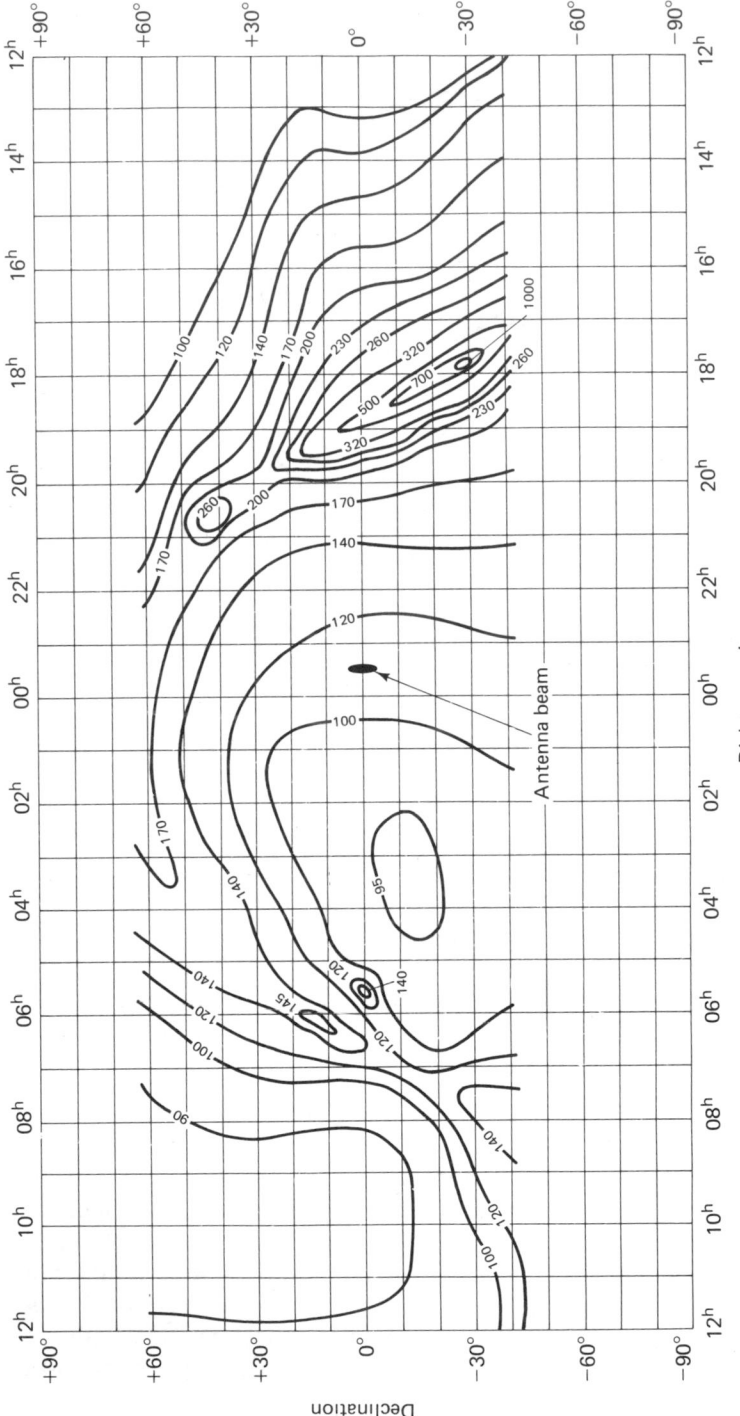

Figure 4.21 Radio map of the sky background at 250 MHz. (Reprinted from H. C. Ko and J. D. Kraus, "A Radio Map of the Sky at 1.2 Meters," *Sky Telesc.*, vol. 16, Feb. 1957, p. 160, with permission from *Sky and Telescope* astronomy magazine, Cambridge, Mass.)

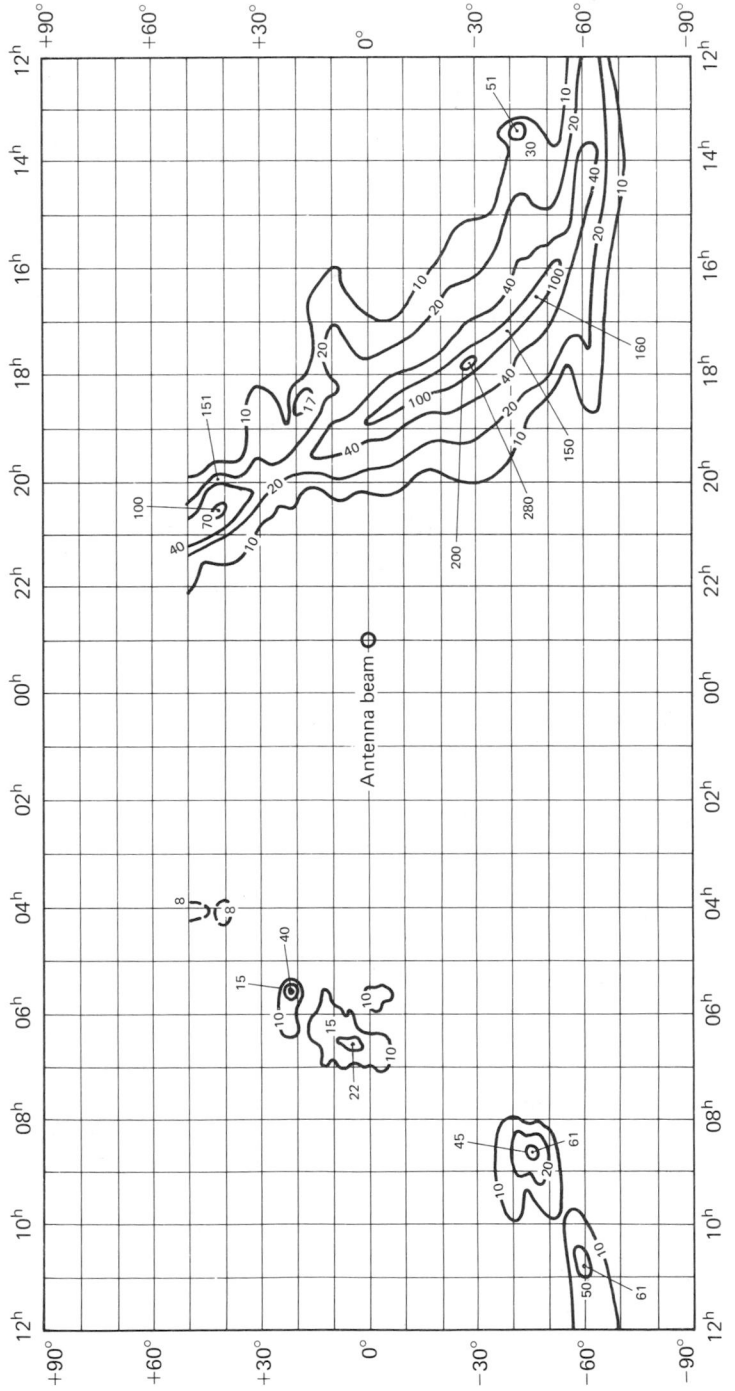

Figure 4.22 Radio map of the sky background at 600 MHz. (Reprinted with permission from J. H. Piddington and G. H. Trent, "A Survey of Cosmic Radio Emission at 600 Mc/s," *Aust. J. Phys.*, vol. 9, Dec. 1956, Fig. 1, pp. 483–486.)

made so as to exclude the sun (night sky). The antenna beam in the center of the map indicates the size of the sky area over which measurements are made (each measurement is an average over that beam area). The narrower the beam, the finer the resolution of the temperature contours; the wider the beam, the coarser the resolution.

Figure 4.22 is another such radio map at 600 MHz, after Piddington and Trent [13]. At this frequency the galaxy noise is reduced compared to Figure 4.21, as predicted in Figure 4.20; the low is 8 K and the high is 280 K. If you examine Figures 4.21 and 4.22 for the region of greatest noise radiation, where on the map do you see the most activity, and what is its significance? It is seen as an elongated region in the right-hand midsection of each map; the longitudinal axis of the elongation designates the location of our *galactic plane,* where such cosmic noise radiation is most intense.

4.6 SAMPLE LINK ANALYSIS

In Section 4.4 we developed the basic link parameter relationships. In this section we use these relationships to calculate a sample link budget, shown in Table 4.2. The table may appear to house a formidable listing of terms; one can get the false impression that the link budget represents a complex compilation. Just the opposite is true, and we introduce Figure 4.23 to underscore this assertion. In this figure we have reduced the set of line items from Table 4.2 to a few key parameters. The goal of a link analysis is to determine whether or not the required error performance is met, by examining the E_b/N_0 actually received and comparing it to the E_b/N_0 required to meet the system specification. The principal items needed for this determination are the EIRP (how much effective power is transmitted), the G/T° figure of merit (how much sensitivity the receiver has for collecting this power), L_s (the largest single loss, the space loss), and L_o (other contributing losses and degradations). That is *all* there is to it!

4.6.1 Link Budget Details

The link budget example in Table 4.2 consists of three columns of numbers. Only the middle column represents the link budget. The other columns consist of ancillary information, such as antenna beamwidth, or computations to support the main tabulation. Losses are bracketed in the usual bookkeeping way. Subtotals are shown enclosed in a box; the final link margin is shown in a double box. The computations are performed as in Equation (4.24), which is repeated below, with the exception that the terms G_r and T° are grouped together as G_r/T° instead of being listed separately.

$$M \text{ (dB)} = \text{EIRP (dBW)} + \frac{G_r}{T^\circ} \text{ (dB/K)} - \left(\frac{E_b}{N_0}\right)_{\text{reqd}} \text{ (dB)} - R \text{ (dB-bits/s)}$$
$$- \kappa \text{ (dBW/K-Hz)} - L_s \text{ (dB)} - L_o \text{ (dB)}$$

Let us examine the 21 line items listed in Table 4.2.

TABLE 4.2 Earth Terminal to Satellite Link Budget Example: Frequency = 8 GHz, Range = 21,915 Nautical Miles.

1.	Transmitter power (dBW)	(100.00 W)	20.0	P_t
2.	Transmitter circuit loss (dB)		⟨2.0⟩	L_o
3.	Transmitter antenna gain (peak dBi)		51.6	G_t
	Dish diameter (ft)	20.00		
	Half-power beamwidth (degrees)	0.45		
4.	Terminal EIRP (dBW)		69.6	EIRP
5.	Path loss (dB)	(10° elev.)	⟨202.7⟩	L_s
6.	Fade allowance (dB)		⟨4.0⟩	L_o
7.	Other losses (dB)		⟨6.0⟩	L_o
8.	Received isotropic power (dBW)		−143.1	
9.	Receiver antenna gain (peak dBi)		35.1	G_r
	Dish diameter (ft)	3.00		
	Half-power beamwidth (degrees)	2.99		
10.	Edge of coverage loss (dB)		⟨2.0⟩	L_o
11.	Received signal power (dBW)		−110.0	P_r
	Receiver noise figure at antenna port (dB)			11.5
	Receiver temperature (dB-K)			35.8 (3806 K)
	Receiver antenna temperature (dB-K)			24.8 (300 K)
12.	System temperature (dB-K)			36.1 (4106 K)
13.	System $G/T°$ (dB-K)		−1.0	$G/T°$
14.	Boltzmann's constant (dBW/K-Hz)			−228.60
15.	Noise spectral density (dBW/Hz)		⟨−192.5⟩	$N_0 = kT°$
16.	Received P_r/N_0 (dB-Hz)		82.5	$(P_r/N_0)_r$
17.	Data rate (dB-bit/s)	(2 Mbits/s)	⟨63.0⟩	R
18.	Received E_b/N_0 (dB)		19.5	$(E_b/N_0)_r$
19.	Implementation loss (dB)		⟨1.5⟩	L_o
20.	Required E_b/N_0 (dB)		⟨10.0⟩	$(E_b/N_0)_{reqd}$
21.	Margin (dB)		8.0	M

1. Transmitter power is 100 W (20 dBW).
2. Circuit loss between the transmitter and antenna is 2 dB.
3. Transmitting antenna gain is 51.6 dBi.
4. The net tally of items 1 to 3 yields the EIRP = 69.6 dBW.
5. The path loss has been calculated for the range shown in the table title, corresponding to a 10° elevation angle at the earth terminal.
6 **and 7.** Here are allowances made for weather fades and a variety of other, unspecified losses.

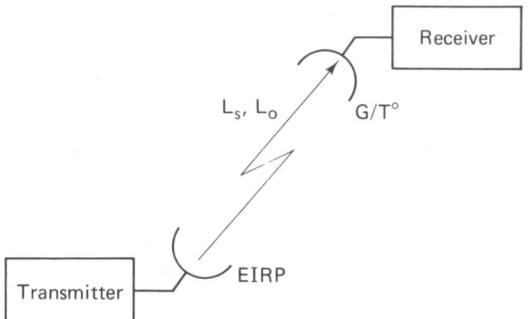

Figure 4.23 Key parameters of a link analysis.

8. Received isotropic power refers to the power that would be received, -143.1 dBW, if the receiving antenna were isotropic.

9. The peak gain of the receiving antenna is 35.1 dBi.

10. Edge of coverage loss is due to the off-axis antenna gain (compared to peak gain) and to the increased range for users at the extreme edge of communication coverage (a nominal 2-dB loss is shown here.)

11. The input power to the receiver, tallied from items 8, 9, and 10, is -110 dBW.

12. Off to the side, in column 3, we compute the system temperature T_S°, from the receiver noise figure and the antenna temperature.

13. We then form G/T° by combining the G_r in item 9, with T_S°.

14. Boltzmann's constant is -228.6 dBW/K-Hz.

15. Boltzmann's constant in decibels (item 14), plus system temperature in decibels (item 12), yields noise power spectral density.

16. Finally, we can form the received signal-to-noise spectral density, 82.5 dB-Hz, by subtracting noise spectral density in decibels (item 15), from received signal power in decibels (item 11).

17. The data rate is listed in dB-bit/s.

18. Since $E_b/N_0 = (1/R)(P_r/N_0)$, we need to subtract R in decibels (item 17), from P_r/N_0 in decibels (item 16), yielding $(E_b/N_0)_r = 19.5$ dB.

19. An implementation loss, here taken to be 1.5 dB, accounts for the difference between theoretically predicted detection performance, and the performance of the actual detector.

20. This is our required E_b/N_0, a result of the modulation and coding chosen, and the probability of error specified.

21. The difference between the received and the required E_b/N_0 in decibels (taking implementation loss into account), yields the final margin.

4.6.2 Receiver Figure of Merit

An explanation of why receiving antenna gain and system temperature have been grouped together as G/T° rather than kept separately, is as follows. In the early days of satellite communications development, the G_r and the T_S° were specified separately. A contractor who agreed to meet these specifications would need to allow himself some safety margin for meeting each specification. Even though the user was generally only interested in the "bottom-line" performance, and not in the explicit value of G_r or T_S°, the contractor would not be able to exploit potential trade-offs. The net result was an overspecified (more costly) system than was necessary. Recognition of such overspecification resulted in specifying the antenna and receiver front end as a single figure-of-merit parameter, G/T°, sometimes called the *receiver sensitivity,* such that cost-effective trade-offs between the antenna design and the receiver design might be employed.

4.6.3 Received Isotropic Power

Another recognized area of overspecification in receiver design is in the separate specification of the required P_R/N_0 (or E_b/N_0) and receiver G/T°. If P_r/N_0 and G/T° are specified separately, the system contractor is forced to meet each value. The contractor will plan for a margin in both places. As in the G/T° case of the preceding section, there are advantages in specifying P_R/N_0 and G/T° as one parameter; this new parameter, called the *received isotropic power* (RIP), can be written as follows:

$$\text{RIP (dBW)} = \frac{P_r}{N_0} \text{ (dB-Hz)} - \frac{G}{T^\circ} \text{ (dB/K)} + \kappa \text{ (dBW/K-Hz)} \qquad (4.49)$$

or, in terms of ratios,

$$\text{RIP} = \frac{P_r}{\kappa T^\circ} \left(\frac{\kappa T^\circ}{G_r} \right) = \frac{P_r}{G_r} \qquad (4.50)$$

It is important to note that P_r/N_0 refers to the predetection signal-to-noise spectral density ratio (SNR) *required* for a particular error probability when using a particular modulation scheme (it usually includes an allowance for *detector implementation losses*). Let us designate the theoretically required SNR to yield a particular P_B as $(P_r/N_0)_{\text{th-rq}}$. We can therefore write

$$\frac{P_r}{N_0} = L_o' \left(\frac{P_r}{N_0} \right)_{\text{th-rq}} \qquad (4.51)$$

where L_o' is called the implementation loss and accounts for the hardware and operational losses in the detection process. Combining Equations (4.50) and (4.51), we can write

$$\text{RIP} = L_o' \left(\frac{P_r}{\kappa T^\circ} \right)_{\text{th-rq}} \frac{\kappa T^\circ}{G_r} \qquad (4.52)$$

Specifying the RIP required to meet the system error performance allows the contractor to commit to meeting a single parameter value. The contractor is allowed to trade off P_r/N_0 versus G/T° and L_o' performance. As G/T° is improved, the detector performance can be degraded, and vice versa.

4.7 SATELLITE REPEATERS

Satellite repeaters retransmit the messages they receive (with a translation in carrier frequency). A *regenerative* (digital) repeater regenerates, that is, demodulates and reconstitutes the digital information embedded in the received waveforms before retransmission; however, a *nonregenerative* repeater only amplifies and retransmits. A nonregenerative repeater, therefore, can be used with many different modulation formats (simultaneously or sequentially without any switching), but a regenerative repeater is usually designed to operate with only one, or a very few, modulation formats. A link analysis for a regenerative satellite repeater treats the uplink and downlink as two separate point-to-point analyses. To calculate the overall bit error performance of a regenerative repeater link, it is necessary to determine separately the bit error probability on the uplink and downlink. Let P_u and P_d be the probability of a bit being in error on the uplink and downlink, respectively. A bit will be correct in the end-to-end link if either the bit is correct on both the up- and downlink, or if it is in error on both the up- and downlink. Therefore, the overall probability that a bit is correct, P_c, is

$$P_c = (1 - P_u)(1 - P_d) + P_u P_d \tag{4.53}$$

and the overall probability that a bit is in error, P_B, is

$$P_B = 1 - P_c = P_u + P_d - 2P_u P_d \tag{4.54}$$

For low values of P_u and P_d, the overall bit error performance is approximated simply by summing the individual uplink and downlink bit error probabilities:

$$P_B \simeq P_u + P_d \tag{4.55}$$

4.7.1 Nonregenerative Repeaters

Link analysis for a nonregenerative repeater treats the entire "round trip" (uplink transmission to the satellite and downlink retransmission to an earth terminal) as a single analysis. Features that are unique to nonregenerative repeaters, are the dependence of the overall SNR on the uplink SNR and the sharing of the repeater downlink power in proportion to the uplink power from each of the various uplink signals and noise. Henceforth, reference to a repeater or transponder will mean a *nonregenerative repeater,* and for simplicity, we will assume that the transponder is operating in its linear range.

A satellite transponder is limited in transmission capability by its downlink power, the earth terminal's uplink power, satellite and earth terminal noise, and channel bandwidth. One of these usually is a dominant performance constraint; most often the downlink power or the channel bandwidth proves to be the major system limitation. Figure 4.24 illustrates the important link parameters of a linear satellite repeater channel. The repeater transmits all uplink signals (or noise, in the absence of signal) without any processing beyond amplification and frequency translation. Let us assume that there are multiple simultaneous uplinks within the receiver's bandwidth W and that they are separated from one another through

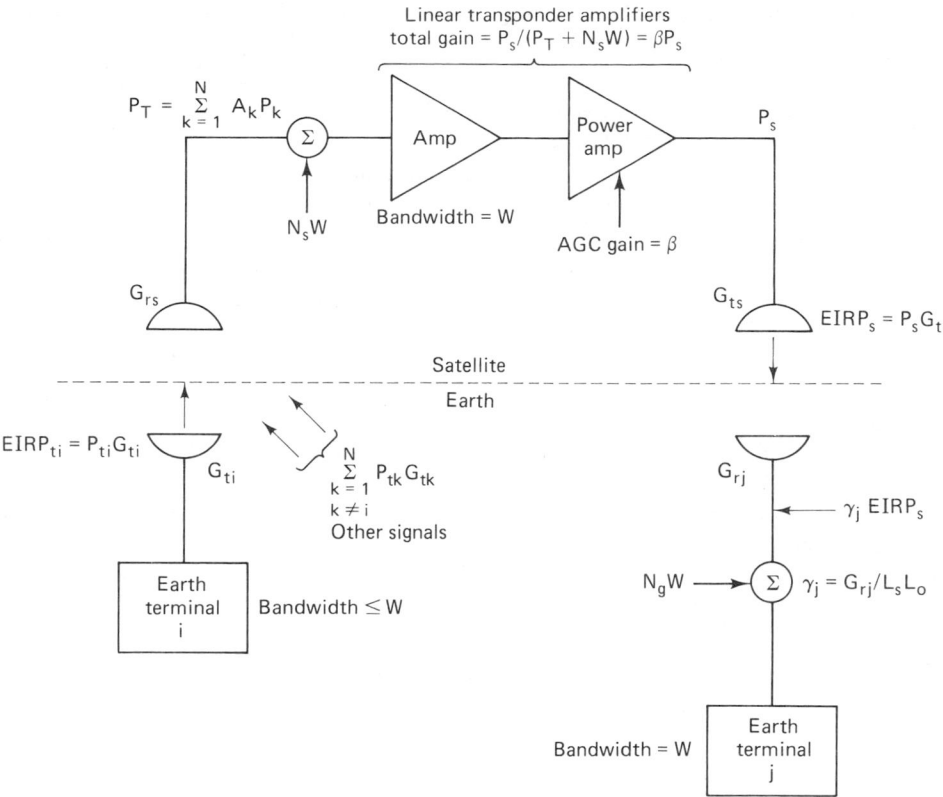

Figure 4.24 Nonregenerative satellite repeater.

the use of a technique called *frequency-division multiple access* (FDMA). FDMA is a communications resource-sharing technique whereby different users occupy disjoint portions of the transponder bandwidth; FDMA is treated in Chapter 9. The satellite effective downlink power $EIRP_s$ is constant and since we are assuming a linear transponder, $EIRP_s$ is shared among the multiple uplink signals (and noise) in proportion to their respective input power levels.

The transmission starts from a ground station (bandwidth $\leq W$), say terminal i, with a terminal $EIRP_{ti} = P_{ti}G_{ti}$. Simultaneously, other signals are being transmitted to the satellite (from other terminals). The EIRP from the kth terminal will henceforth be referred to simply as P_k. At the satellite, a total signal power $P_T = \sum A_k P_k$ is received, where A_k reflects the uplink propagation loss and the satellite receive antenna gain for each terminal. $N_s W$ is the satellite uplink noise power, where N_s is the composite noise power spectral density due to noise radiated into the satellite antenna *and* generated in the satellite receiver. The total satellite down-link $EIRP_s = P_s G_{ts}$, where P_s is the satellite transponder output power and G_{ts} is the satellite transmitting antenna gain, can be expressed by the following identity [14]:

$$EIRP_s = EIRP_s \beta [A_i P_i + (P_T - A_i P_i) + N_s W] \qquad (4.56)$$

Both the left and right sides of Equation (4.56) express the total satellite EIRP. On the right side, the term $\beta[A_iP_i + (P_T - A_iP_i) + N_sW]$ constitutes the fractional apportionment of EIRP_s for the various users and uplink noise, such that the composite expression equals unity. The usefulness of this identity should become clear shortly. The total power gain of the transponder can be expressed as βP_s. Since P_s is fixed and the input signals can vary, $\beta = 1/(P_T + N_sW)$ represents an automatic gain control (AGC) term. The total received uplink signal power, P_T, has purposely been written as $A_iP_i + (P_T - A_iP_i)$ to separate signal i power from the remainder of the simultaneous signals in the transponder. The total power received at the jth earth terminal, with bandwidth W, can be written as

$$P_{rj} = \text{EIRP}_s\gamma_j\beta[A_iP_i + (P_T - A_iP_i) + N_sW] + N_gW \qquad (4.57)$$

where $\gamma_j = G_{rj}/L_sL_o$ accounts for downlink losses and receiving antenna gain for the jth earth terminal. $\text{EIRP}_s\gamma_j$ represents the portion of EIRP_s that is received by the jth earth terminal, and N_g is the downlink noise power spectral density generated and introduced into that terminal receiver. Equation (4.57) describes the essence of downlink power apportionment among the various users and noise in a repeater. Let us rewrite Equation (4.57) by replacing β with its equivalent $1/(P_T + N_sW)$, as follows:

$$P_{rj} = \text{EIRP}_s\gamma_j\left(\frac{A_iP_i}{P_T + N_sW} + \frac{P_T - A_iP_i}{P_T + N_sW} + \frac{N_sW}{P_T + N_sW}\right) + N_gW \quad (4.58)$$

To facilitate our discussion, let us amplify Equation (4.58) with words

$$P_{rj} = \text{EIRP}_s\gamma_j\left(\frac{S_i \text{ U/L power}}{\text{total } (S + N) \text{ U/L power}} + \frac{\text{balance } S \text{ U/L power}}{\text{total } (S + N) \text{ U/L power}}\right.$$
$$\left. + \frac{\text{U/L noise power}}{\text{total } (S + N) \text{ U/L power}}\right) + N_gW$$

where S stands for signal power, N for noise power, and U/L for uplink.

From Equation (4.58), can you recognize an important relationship that must exist among multiple users sharing a nonregenerative transponder? The users must *cooperate with one another,* by not exceeding agreed-upon uplink transmission power levels. Equation (4.58) states that the portion of the downlink EIRP dedicated to any one user (or to uplink noise) is determined by the ratio of that user's uplink power to the total uplink signal plus noise power. Hence if one of the sharing users should choose to "cheat" by increasing his or her uplink power, the effect would be an enhancement of this user's downlink signal level, at the expense of the other users' downlink signal levels. Notice from Equation (4.58) that the uplink noise shares the downlink resource along with the other users. This coupling of uplink noise onto the downlink is a feature unique to nonregenerative repeaters.

From Equation (4.58) we can express the P_r/N for signal i received at the jth terminal as follows:

$$\left(\frac{P_r}{N}\right)_{ij} = \frac{\text{EIRP}_s\gamma_j[A_iP_i/(P_T + N_sW)]}{\text{EIRP}_s\gamma_j[N_sW/(P_T + N_sW)] + N_gW} \tag{4.59}$$

and we can write the overall P_r/N_0 for signal i received at the jth terminal as [14]

$$\left(\frac{P_r}{N_0}\right)_{ij} = \frac{\text{EIRP}_s\gamma_j\beta A_iP_i}{\text{EIRP}_s\gamma_j\beta N_s + N_g} \tag{4.60}$$

Equations (4.58) to (4.60) illustrate that the uplink repeater noise degrades the overall SNR in two ways—it "steals" downlink EIRP, and it contributes to the total system noise. When the satellite uplink noise dominates, that is, when $P_T \ll N_sW$, the link is said to be *uplink limited*, and most of the downlink EIRP$_s$ is wastefully allocated to uplink noise power. When this is the case and when $\text{EIRP}_s\gamma_j \gg N_gW$, we can rewrite Equation (4.60) as

$$\left(\frac{P_r}{N_0}\right)_{ij} \simeq \frac{\text{EIRP}_s\gamma_jA_iP_i/N_sW}{(\text{EIRP}_s\gamma_j/W) + N_g} \simeq \frac{A_iP_i}{N_s} \tag{4.61}$$

Equation (4.61) illustrates that in the case of an uplink limited channel, the overall P_r/N_0 ratio essentially follows the uplink SNR. The more common situation is the *downlink limited* channel, in which case $P_T \gg N_sW$, and the satellite EIRP is limited. In this case Equation (4.60) can be rewritten as

$$\left(\frac{P_r}{N_0}\right)_{ij} \simeq \frac{\text{EIRP}_s\gamma_jA_iP_i/P_T}{N_g} \tag{4.62}$$

The power of the transponder is then shared primarily among the various uplink transmitted signals; very little uplink noise is transmitted on the downlink. The performance of the repeater, in this case, is constrained only by its downlink parameters.

Table 4.3 illustrates a link analysis example (full round trip) for a nonregenerative repeater. The uplink portion by itself does not constitute a link budget since the transmission is not demodulated at the satellite. Without demodulation, *there are no bits* and therefore there is no way to measure the bit-error performance. After the full round trip, the signal is demodulated at the earth terminal; only then does the link analysis yield the margin. The example in Table 4.3 represents a case where the satellite transponder is servicing 10 simultaneous users. In the block marked "A" is shown the ratio $P_r/(P_T + N_sW)$, which dictates the apportionment of the downlink EIRP for the signal of interest. In this example, with all users transmitting the same power level, each of the signals is allocated 9.8% of the downlink EIRP. In the block marked "B" we see the apportionment of the downlink EIRP. The total is 1514.7 W; the user of interest gets 148.5 W; the other nine users get a total of 1336.1 W; and the uplink noise is apportioned 30.1 W.

An estimate of the performance described in Equation (4.60) can be obtained by using the uplink and downlink values of E_b/N_0 (or P_r/N_0), combined as follows, in the *absence of intermodulation noise* [15]:

$$\left(\frac{E_b}{N_0}\right)_{ov}^{-1} = \left(\frac{E_b}{N_0}\right)_{u}^{-1} + \left(\frac{E_b}{N_0}\right)_{d}^{-1} \qquad (4.63)$$

where the subscripts ov, u, and d, indicate overall, uplink, and downlink values of E_b/N_0, respectively.

Most commercial satellite transponder designs are nonregenerative. However, it seems clear that future commercial systems will require on-board processing, switching, or selective message addressing, and will use regenerative repeaters to transform the received waveforms to message bits. Besides the potential for sophisticated data processing, one of the principal advantages of regenerative compared to nonregenerative repeaters is that the uplink is decoupled from the downlink so that the uplink noise is not retransmitted on the downlink. There are significant performance improvements possible with regenerative satellite repeaters in terms of the E_b/N_0 values needed on the uplinks and downlinks, relative to the values needed for the conventional nonregenerative designs in use today. Improvements of as much as 5 dB on the uplink and 6.8 dB on the downlink (using coherent QPSK modulation, with $P_B = 10^{-4}$) have been demonstrated [16].

4.7.2 Nonlinear Repeater Amplifiers

Power is severely limited in most satellite communication systems, and the inefficiencies associated with linear power amplification stages are expensive to bear. For this reason, many satellite repeaters employ nonlinear power amplifiers. Efficient power amplification is obtained at the cost of signal distortion due to nonlinear operation. The major undesirable effects of the repeater nonlinearities are:

1. Intermodulation (IM) noise due to the interaction of different carriers. The harm is twofold; useful power can be lost from the channel as IM energy (typically 1 to 2 dB), and spurious IM products can be introduced into the channel as interference. The latter problem can be quite serious.

2. AM-to-PM conversion is a phenomenon common to nonlinear devices such as traveling-wave tubes (TWT). Fluctuations in the signal envelope (amplitude modulation) produce phase variations that can affect the error performance, for systems using a coherent or differentially coherent modulation format.

3. In hard limiters, weak signals can be suppressed, relative to stronger signals, by as much as 6 dB [2]. In saturated TWTs, the suppression of weak signals is due not only to limiting, but also to the fact that the signal coupling mechanism of the tube is optimized in favor of the stronger signals. The effect can cause weak signals to be suppressed by as much as 18 dB [17].

Conventional nonregenerative repeaters are generally operated *backed-off* from their highly nonlinear saturated region; this is done to avoid appreciable IM noise and thus to allow efficient utilization of the system's entire bandwidth. However, backing off to the linear region is a compromise; some level of IM noise must be accepted to achieve a useful level of output power.

TABLE 4.3 Link Budget Example For a Nonregenerative Satellite Repeater with 10 Users: Uplink Frequency = 375 MHz, Downlink Frequency = 275 MHz, Range = 22,000 Nautical Miles

	Uplink		Downlink	
Transmitter power (dBW)		27.0 (500.0 W)		13.0 (20.0 W)
Transmitter circuit losses (dB)		1.0		1.0
Transmitter antenna gain (peak-dBi)		19.0		19.8
Dish diameter (ft)	10.00		15.00	
Half-power beamwidth (degrees)	19.16		17.42	
EIRP (dBW)		45.0		31.8 (1514.7 W)
Path loss (dB)		176.1		173.4
Transmitted signal power (dBW)				21.7 (148.5 W)
Transmitted other signal power (dBW)				31.3 (1336.1 W)
Transmitted U/L noise power (dBW)				14.8 (30.1 W) **B**
Other losses (dB)		2.0		2.0
Received isotropic signal power (dBW)		−133.1		−153.7
Received isotropic U/L noise power (dBW)				−160.6
Receiver antenna gain (peak dBi)		22.5		16.3
Dish diameter (ft)	15.00		10.00	
Half-power beamwidth (degrees)	12.77		26.13	
Received signal power (dBW)		−110.6		−137.4
Received U/L noise power (dBW)				−144.3
Receiver antenna temperature (dB/K)		24.6 (290 K)		20.0 (100 K)
Receiver noise figure at antenna port (dB)		10.8		2.0
Receiver temperature (dB/K)		35.1 (3197 K)		22.3 (170 K)
System temperature (dB/K)		35.4 (3487 K)		24.3 (270 K)
System $G/T°$ (dB/K)		−12.9		−8.0
Boltzmann's constant (dBW/K-Hz)		−228.6		−228.6
Noise spectral density (dBW/Hz)		−193.2		−204.3
System bandwidth (dB/Hz)		75.6 (36.0 MHz)		75.6 (36.0 MHz)
Noise power (dBW)		−117.6		−128.7
U/L noise + D/L noise power (dBW)				−128.6
Simultaneous accesses	10			
Received other signal power (dBW)		−101.1		
Other signals + noise (dBW)		−101.0		
$P_r/(P_T + N_sW)$ (dB)		−10.1 (0.098) **A**		
P_r/N (dB)		7.0		−8.7
Overall P_r/N (dB)				−8.8
P_r/N_0 (dB-Hz)		82.6		66.9
Overall P_r/N_0 (dB-Hz)				66.8
Data rate (dB-bit/s)				50.0 (100,000 bit/s)
Available E_b/N_0 (dB)				16.8
Required E_b/N_0 (dB)				10.0
Margin (dB)				6.8

4.8 SYSTEM TRADE-OFFS

The link budget example in Table 4.3 is a resource allocation document. With such a link tabulation, one can examine potential system trade-offs and attempt to optimize system performance. A decibel is a decibel is a decibel; any decibel, from wherever it comes, is just as good as any other decibel in contributing to the overall performance requirements. The link budget is a natural starting point for considering all sorts of potential trade-offs: margin versus noise figure, antenna size versus transmitter power, and so on. Table 4.4 represents an example of a computer exercise for examining a possible trade-off between the earth station transmitting power and the system noise margin at the receiving terminal. The first row in the table is taken from the Table 4.3 link budget. Suppose a system engineer is concerned that a 500-W transmitter is not practical because of some physical constraints within the transmitting earth terminal; the engineer might then consider a trade-off of transmitter power versus thermal noise margin. The listing of candidate trade-offs is a trivial task for a computer. Table 4.4 was generated by repeating the link budget computation multiple times, and at each iteration, reducing P_t by one-half.

The result is a selection of transmitters (in steps of 3 dB) and uplink, downlink, and overall SNRs, and margin, associated with each transmitter value. The system engineer need only peruse the list to find a likely candidate. For example, if the engineer were satisfied with a margin of 3 to 4 dB, it appears he could reduce the transmitter from 500 W to 20 to 30 W. Or, he might be willing to provide a transmitter with, say, $P_t = 100$ W, since he may want to consider additional trade-offs (perhaps because of having misgivings about one of the other subsystems, say the antenna size). The engineer would then start a new tabulation with $P_t = 100$ W, and again perform a succession of link budget computations, to produce a similar enumeration of other possible trade-offs.

Notice from Table 4.4 that one can recognize the uplink-limited and downlink-limited regions, discussed earlier. In the first few rows, where the uplink SNR is high, a 3-dB degradation in uplink SNR results in only a few tenths of a

TABLE 4.4 Potential Trade-Off: P_t versus Margin

P_t (W)	$(P_r/N_0)_u$ (dB-Hz)	$(P_r/N_0)_d$ (dB-Hz)	$(P_r/N_0)_{ov}$ (dB-Hz)	Margin (dB)
500.0	82.6	66.9	66.8	6.8
250.0	79.6	66.8	66.6	6.6
125.0	76.6	66.6	66.2	6.2
62.5	73.6	66.3	65.5	5.5
31.3	70.5	65.7	64.5	4.5
15.6	67.5	64.8	62.9	2.9
7.8	64.5	63.3	60.8	0.8
3.9	61.5	61.4	58.4	-1.6
2.0	58.4	59.0	55.7	-4.3
1.0	55.4	56.4	52.9	-7.2
0.5	52.4	53.6	49.9	-10.1

decibel degradation to the overall SNR. Here the system is *downlink limited*; that is, the system is constrained primarily by its downlink parameters and is hardly affected by the uplink parameters. In the bottom few rows of the table, we see that a 3-dB degradation to the uplink affects the overall SNR by almost 3 dB. Here the system is *uplink limited*; that is, the system is constrained primarily by the uplink parameters.

4.9 CONCLUSION

Of the many analyses that support a developing communication system, the link budget stands out in its ability to provide overall system insight. By examining the link budget, one can learn many things about the overall system design and performance. For example, from the link margin, one learns whether the system will meet its requirements comfortably, marginally, or not at all. It will be evident if there are any hardware constraints, and whether such constraints can be compensated for in other parts of the link. The link budget is often used for considering system trade-offs and configuration changes, and in understanding subsystem nuances and interdependencies. Together with other modeling techniques, the link budget can help predict weight, size, and cost. We have considered how to formulate this budget and how it might be used for system trade-offs. The link budget is one of the system manager's most useful documents; it represents a "bottom-line" tally in the search for optimum system performance.

REFERENCES

1. Panter, P. F., *Communication Systems Design: Line-of-Sight and Tropo-Scatter Systems,* R. E. Krieger Publishing Co., Inc., Melbourne, Fla., 1982.
2. Jones, J. J., "Hard Limiting of Two Signals in Random Noise," *IEEE Trans. Inf. Theory,* vol. IT9, January 1963.
3. Silver, S., *Microwave Antenna Theory and Design,* MIT Radiation Laboratory Series, Vol. 12, McGraw-Hill Book Company, New York, 1949.
4. Kraus, J. D., *Antennas,* McGraw-Hill Book Company, New York, 1950.
5. Johnson, J. B., "Thermal Agitation of Electricity in Conductors," *Phys. Rev.,* vol. 32, July 1928, pp. 97–109.
6. Nyquist, H., "Thermal Agitation of Electric Charge in Conductors," *Phys. Rev.,* vol. 32, July 1928, pp. 110–113.
7. Desoer, C. A., and Kuh, E. S., *Basic Circuit Theory,* McGraw-Hill Book Company, New York, 1969.
8. Schwab, L. M., "World-Wide Link Availability for Geostationary and Critically Inclined Orbits Including Rain Attenuation Effects," *Lincoln Laboratory, Rep. DCA-9,* Jan. 27, 1981.
9. Friis, H. T., "Noise Figure of Radio Receivers," *Proc. IRE,* July 1944, pp. 419–422.

10. IRE Subcommittee 7.9 on Noise, "Description of the Noise Performance of Amplifiers and Receiving Systems," *Proc. IEEE,* Mar. 1963, pp. 436–442.

11. Blackwell, L. A., and Kotzebue, K. L., *Semiconductor Diode Parametric Amplifiers,* Prentice-Hall, Inc., Englewood Cliffs, N.J., 1961.

12. Ko, H. C., and Kraus, J. D., "A Radio Map of the Sky at 1.2 Meters," *Sky Telesc.,* vol. 16, Feb. 1957, pp. 160–161.

13. Piddington, J. H., and Trent, G. H., "A Survey of Cosmic Radio Emission at 600 Mc/s," *Aust. J. Phys.,* vol. 9, Dec. 1956, pp. 481–493.

14. Spilker, J. J., *Digital Communications by Satellite,* Prentice-Hall, Inc., Englewood Cliffs, N.J., 1977.

15. Pritchard, W. L., and Sciulli, J. A., *Satellite Communication Systems Engineering,* Prentice-Hall, Inc., Englewood Cliffs, N.J., 1986.

16. Campanella, S. J., Assal, F., and Berman, A., "Onboard Regenerative Repeaters," *Int. Conf. Commun.,* Chicago, vol. 1, 1977, pp. 6.2-121–66.2-125.

17. Wolkstein, H. J., "Suppression and Limiting of Undesired Signals in Travelling-Wave-Tube Amplifiers," Publication ST-1583, *RCA Rev.,* vol. 22, no. 2, June 1961, pp. 280–291.

PROBLEMS

4.1. (a) What is the value in decibels of the free-space loss for a carrier frequency of 100 MHz and a range of 3 miles?

(b) The transmitter output power is 10 W. Assume that both the transmitting and receiving antennas are isotropic and that there are no other losses. Calculate the received power in dBW.

(c) If in part (b) the EIRP is equal to 20 W, calculate the received power in dBW.

(d) If the diameter of a dish antenna is doubled, calculate the antenna gain increase in decibels.

(e) For the system of part (a), what must the diameter of a dish antenna be in order for the antenna gain to be 10 dB? Assume an antenna efficiency of 0.55.

4.2. A transmitter has an output of 2 W at a carrier frequency of 2 GHz. Assume that the transmitting and receiving antennas are parabolic dishes each 3 ft in diameter. Assume that the efficiency of each antenna is 0.55.

(a) Evaluate the gain of each antenna.

(b) Calculate the EIRP of the transmitted signal in units of dBW.

(c) If the receiving antenna is located 25 miles from the transmitting antenna over a free-space path, find the available signal power out of the receiving antenna in units of dBW.

4.3. From Table 4.1 we see that the proposal from Satellite Television Corporation called for a direct broadcast satellite (DBS) EIRP of 57 dBW and a downlink transmission frequency of 12.5 GHz. Assume that the only loss is the downlink space loss shown. Suppose that the downlink information consists of a digital signal with a data rate of 5×10^7 bits/s. Assume that the required E_b/N_0 is 10 dB, the system temperature at your home receiver is 600 K, and that your rooftop dish has an efficiency of 0.55. What is the minimum dish diameter that you can use in order to close the link? Do you think the neighbors will object?

4.4. An amplifier has an input and output resistance of 50 Ω, a 60-dB gain, and a bandwidth of 10 kHz. When a 50-Ω resistor at 290 K is connected to the input, the output rms noise voltage is 100 μV. Determine the effective noise temperature of the amplifier.

4.5. An amplifier has a noise figure of 4 dB, a bandwidth of 500 kHz, and an input resistance of 50 Ω. Calculate the input signal voltage needed to yield an output SNR = 1 when the amplifier is connected to a signal source of 50 Ω at 290 K.

4.6. Consider a communication system with the following specifications: transmission frequency = 3 GHz, modulation format is BPSK, bit-error probability = 10^{-3}, data rate = 100 bits/s, link margin = 3 dB, EIRP = 100 W, receiver antenna gain = 10 dB, distance between transmitter and receiver = 40,000 km. Assume that the line loss between the receiving antenna and the receiver is negligible.
(a) Calculate the maximum permissible noise power spectral density in watts/hertz referenced to the receiver input.
(b) What is the maximum permissible effective noise temperature in kelvin for the receiver if the antenna temperature is 290 K?
(c) What is the maximum permissible noise figure in dB for the receiver?

4.7. A receiver preamplifier has a noise figure of 13 dB, a gain of 60 dB, and a bandwidth of 2 MHz. The antenna temperature is 490 K, and the input signal power is 10^{-12} W.
(a) Find the effective temperature, in kelvin, of the preamplifier.
(b) Find the system temperature in kelvin.
(c) Find the output SNR in decibels.

4.8. Assume that a receiver has the following parameters: gain = 50 dB, noise figure = 10 dB, bandwidth = 500 MHz, input signal power = 50×10^{-12} W, source temperature, $T_A^\circ = 10$ K, line loss = 0 dB. You are asked to insert a preamplifier between the antenna and the receiver. The preamplifier is to have a gain of 20 dB and a bandwidth of 500 MHz. Find the preamplifier noise figure that would be required to provide a 10-dB improvement in overall system SNR.

4.9. Find the maximum allowable effective system temperature, T_s°, required to *just close* a particular link with a bit error probability of 10^{-5} for a data rate of $R = 10$ kbits/s. The link parameters are as follows: transmission frequency = 12 GHz, EIRP = 10 dBW, receiver antenna gain = 0 dB, modulation type is noncoherently detected BFSK, other losses = 0 dB, and the distance between transmitter and receiver = 100 km.

4.10. Consider a receiver made up of the following three steps: The input stage is a preamplifier with a gain of 20 dB and a noise figure of 6 dB. The second stage is a 3-dB lossy network. The output stage is an amplifier with a gain of 60 dB and a noise figure of 16 dB.
(a) Find the composite noise figure for the receiver.
(b) Repeat part (a) with the preamplifier removed.

4.11. (a) Find the effective input noise temperature, T_R°, of a receiver comprised of three amplifier stages connected in series with power gains, from input to output, of 10, 16, and 20 dB, and effective noise temperatures, from input to output, of 1800, 2700, and 4800 K.
(b) What would the gain of the first stage have to be to reduce the contribution to T_R° of all stages after the first to 10% of the first-stage contribution?

4.12. The effective temperature of a particular multiple-stage receiver is required to be 300 K. Assume that the effective temperatures and gains of stages 2 through 4 are as follows: $T_2^\circ = 600$ K, $T_3^\circ = T_4^\circ = 2000$ K, $G_2 = 13$ dB, and $G_3 = G_4 = 20$ dB.

(a) Compute the required gain, G_1, for the first stage, under the conditions that $T_1^\circ = 200, 230, 265, 290, 295,$ and 300 K.

(b) Plot the G_1 versus T_1° trade-off.

(c) Regarding contributions to the effective temperature of the receiver, why is it reasonable in this case to ignore all stages beyond the fourth stage?

(d) In a practical engineering trade-off between T_1° and G_1, what range of T_1° values do you think should be considered?

4.13. A receiver consists of a preamplifier followd by multiple amplifier stages. The composite effective temperature of all the amplifier stages is 1000 K, referenced to the preamplifier output.

(a) Compute the receiver effective noise temperature, referenced to the preamplifier input, for a single-stage preamplifier with a noise temperature of 400 K and gains of 3, 6, 10, 16, and 20 dB.

(b) Repeat part (a) for a two-stage preamplifier with 400-K noise per stage and gains of 3, 6, 10, and 13 dB per stage.

(c) Plot the receiver effective temperature versus the gain of the first stage for parts (a) and (b).

4.14. Consider increasing the capacity of a working radio link by replacing the antennas with larger antennas. List three reasons why this might not be practical.

4.15. A receiver with 80-dB gain and an effective noise temperature of 3000 K is connected to an antenna that has a noise temperature of 600 K.

(a) Find the noise power that is available from the source over a 40-MHz band.

(b) Find the receiver noise power referenced to the receiver input.

(c) Find the receiver output noise power over a 40-MHz band.

4.16. An antenna is pointed in a direction such that it has a noise temperature of 50 K. It is connected to a preamplifier that has a noise figure of 2 dB and an available gain of 30 dB over an effective bandwidth of 20 MHz. The input signal to the preamplifier has a value of 10^{-12} W.

(a) Find the effective input noise temperature of the preamplifier.

(b) Find the SNR out of the preamplifier.

4.17. A receiver with a noise figure of 13 dB is connected to an antenna through 75 ft of 300-Ω transmission line that has a loss of 3 dB per 100 ft.

(a) Evaluate the composite noise figure of the line and the receiver.

(b) If a 20-dB preamplifier with a 3-dB noise figure is inserted between the line and the receiver, evaluate the composite noise figure of the line, the preamplifier, and the receiver.

(c) Evaluate the composite noise figure if the preamplifier is inserted between the antenna and the transmission line.

4.18. A satellite communication system uses a transmitter that produces 20 W of RF power at a carrier frequency of 8 GHz that is fed into a 2-ft parabolic antenna. The distance to the receiving earth station is 20,000 nautical miles. The receiving system uses an 8-ft parabolic antenna and has a 100-K system noise temperature. Assume that each antenna has an efficiency of 0.55. Also assume that the incidental losses amount to 2 dB.

(a) Calculate the maximum data rate that can be used if the modulation is differentially coherent PSK (DPSK) and the bit error probability is not to exceed 10^{-5}.

(b) Repeat part (a) assuming that the downlink transmission is at a carrier frequency of 2 GHz.

4.19. Consider that an unmanned spacecraft with a carrier frequency of 2 GHz and a 10-W transponder is in the vicinity of the planet Saturn (a distance of 7.9×10^8 miles from the earth). The receiving earth station has a 75-ft antenna and a system noise temperature of 20 K. Calculate the size of the spacecraft antenna that would be required to just close a 100-bits/s data link. Assume that the required E_b/N_0 is 10 dB and that there are incidental losses amounting to 3 dB. Also assume that each antenna has an efficiency of 0.55.

4.20. **(a)** Assume a receiver front end with the following parameters: gain = 60 dB, bandwidth = 500 MHz, noise figure = 6 dB, input signal power = 6.4×10^{-11} W, source temperature, T_A° = 290 K, line loss = 0 dB. A preamplifier with the following characteristics is inserted between the antenna and the receiver: gain = 10 dB, noise figure = 1 dB. Find the composite receiver noise figure, in decibels. How much noise figure improvement, in decibels, has been realized?

(b) Find the output SNR improvement, in decibels, as a result of the improved noise figure.

(c) Repeat part (b) for T_A° = 6000 K. What is the output SNR improvement in decibels?

(d) Repeat part (b) for T_A° = 15 K. What is the output SNR improvement in decibels?

(e) What conclusions can you draw from your answers with regard to how the improvement in output SNR tracks the improvement in noise figure? Explain.

4.21. **(a)** Given the following link parameters, find the maximum allowable receiver noise figure. The modulation is coherent BPSK with a bit-error probability of 10^{-5} for a data rate of 10 Mbits/s. The transmission frequency is 12 GHz. The EIRP is 0 dBW. The receiving antenna diameter is 0.1 m (assume an efficiency of 0.55), and the antenna temperature is 800 K. The distance between the transmitter and receiver is 10 km. The margin is 0 dB and the incidental losses are assumed to be 0 dB.

(b) If the data rate is doubled, how will that affect the value of the noise figure in part (a)?

(c) If the antenna diameter is doubled, how will that affect the value of the noise figure in part (a)?

4.22. **(a)** Ten users simultaneously access a nonregenerative satellite repeater with a 50-MHz bandwidth using an FDMA access scheme. Assume that each user's EIRP is 10 dBW; also assume that each user's coefficient, $A_i = G_{rs}/L_s L_o = -140$ dB. What is the total power, P_T, received by the satellite receiver?

(b) Assume that the satellite system noise temperature is 2000 K. What is the value of the satellite receiver noise power in watts, referenced to the receiver input?

(c) What is the uplink SNR at the satellite receiver for each user's signal?

(d) Assuming the received power at the satellite from each user is the same, what fraction of the satellite EIRP is allocated to each of the 10 users' signals? If the satellite downlink $EIRP_s$ = 1000 W, how many watts per user is downlinked?

(e) How much of the satellite EIRP is allocated to the transmission of uplink thermal noise?

(f) Is the satellite uplink limited or downlink limited? Explain.

(g) At the earth station, the receiver noise temperature is 800 K. What is the resultant (overall) average signal-to-noise power spectral density (P_r/N_0) for a single user's transmission across a 50-MHz band? Assume that the coefficient $\gamma = G_r/L_s L_o$ = -140 dB.

(h) Recalculate P_r/N_0 for a single user's transmission, using an approximation resulting from your answer to part (f).

(i) In the absence of intermodulation noise, the following repeater relationship is often used.

$$\text{overall} \left(\frac{P_r}{N_0}\right)^{-1} = \text{uplink} \left(\frac{P_r}{N_0}\right)^{-1} + \text{downlink} \left(\frac{P_r}{N_0}\right)^{-1}$$

Recalculate P_r/N_0 using this relationship, and compare the result with your answers to parts (g) and (h):

4.23. How many users can simultaneously access a nonregenerative satellite repeater with a 100-MHz bandwidth, such that each user is allocated 50 W of the satellite's EIRP of 5000 W? At the satellite, the effective system temperature, $T_s^\circ = 3500$ K. Assume that each user's uplink EIRP is 10 dBW and that the G_r/L_sL_o term that reduces this EIRP at the satellite receiver is equal to -140 dB for each user.

CHAPTER 5

Channel Coding:
Part 1

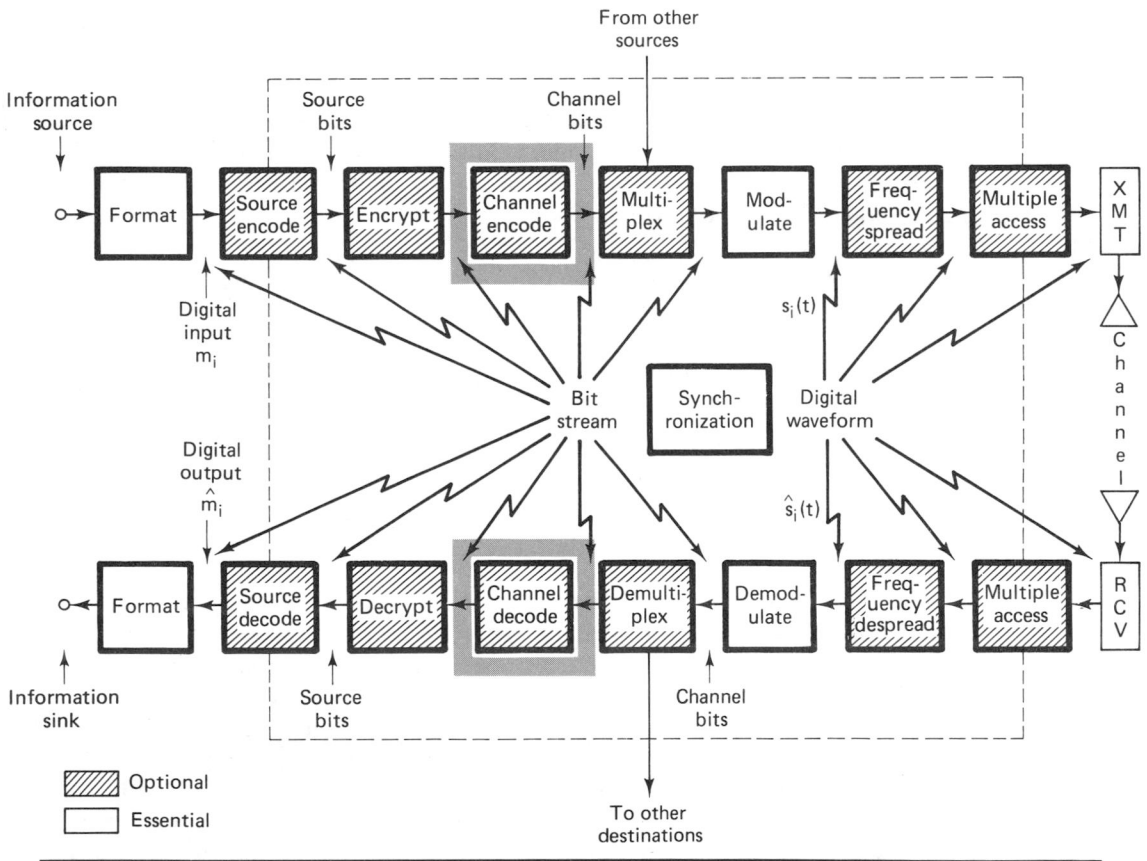

Channel coding refers to the class of signal transformations designed to improve communications performance by enabling the transmitted signals to better withstand the effects of various channel impairments, such as noise, fading, and jamming. Usually, the goal of channel coding is to reduce the probability of bit error (P_B), or to reduce the required E_b/N_0, at the cost of expending more bandwidth than would otherwise be necessary. The exceptions to this are the combined modulation and coding techniques for bandlimited channels described in Chapter 7. Why do you suppose channel coding has become such a popular way to provide performance improvement? The use of large-scale integrated (LSI) circuits has made it possible to provide as much as an 8-dB performance improvement through coding, at much less cost than through the use of other methods such as higher-power transmitters or larger antennas.

5.1 WAVEFORM CODING

Channel coding can be partitioned into two study areas, waveform (or signal design) coding and structured sequences (or structured redundancy), as shown in Figure 5.1. *Waveform coding* deals with transforming waveforms into "better waveforms," to make the detection process less subject to errors. *Structured sequences* deals with transforming data sequences into "better sequences," having structured redundancy (redundant bits). The redundant bits can then be used for the detection and correction of errors. The encoding procedure provides the coded signal (whether waveforms or structured sequences) with better distance

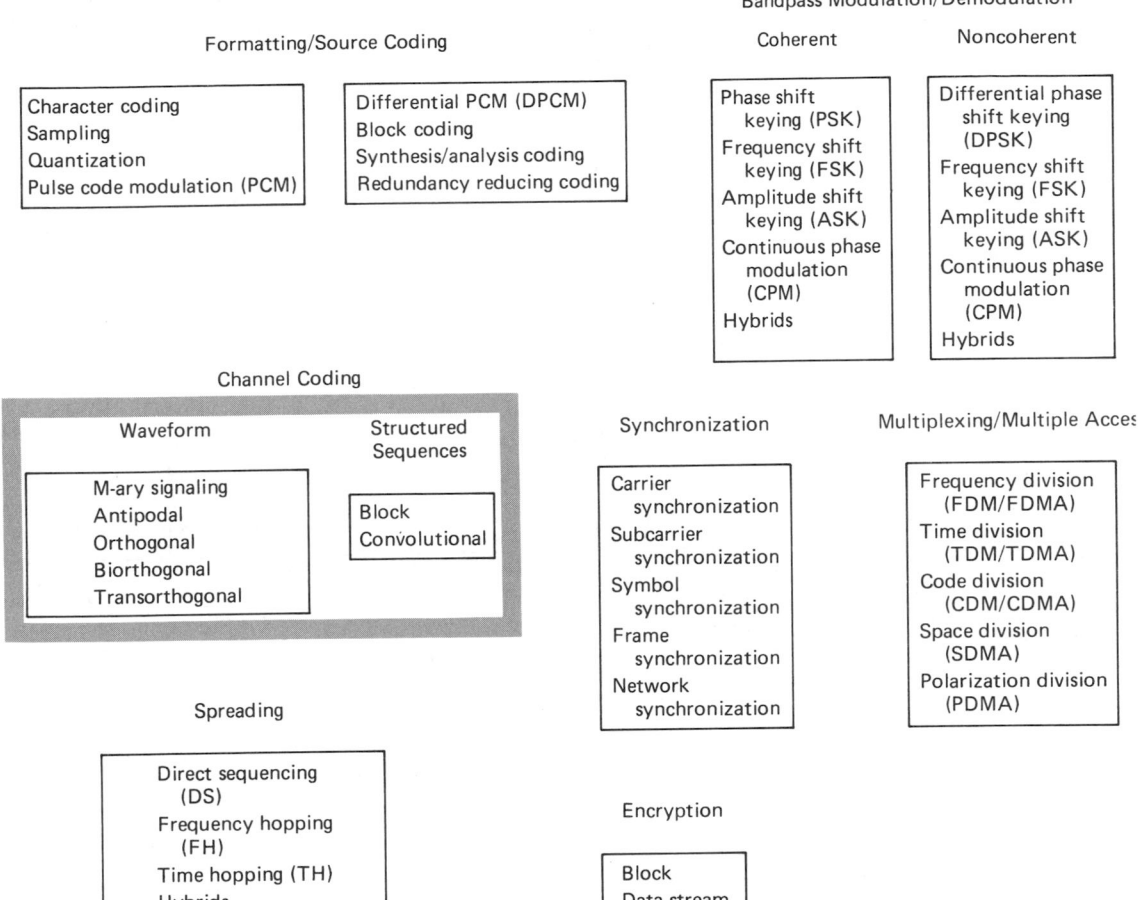

Figure 5.1 Basic digital communication transformations.

properties than those of their uncoded counterparts. First, we consider some waveform coding techniques. Then, starting with Section 5.3, we treat the more popular subject of structured sequences.

5.1.1 Antipodal and Orthogonal Signals

Antipodal and orthogonal signals have been discussed in Chapter 3; we shall repeat the paramount features of these signal classes. The example shown in Figure 5.2 illustrates the analytical representation, $s_1(t) = -s_2(t) = \sin \omega_0 t$, $0 \le t \le T$, of an antipodal signal set, as well as its waveform representation and its vectorial representation. What are some synonyms or analogies that are used to describe *antipodal signals*? We can say that such signals are mirror images, or that one signal is the negative of the other, or that the signals are 180° apart.

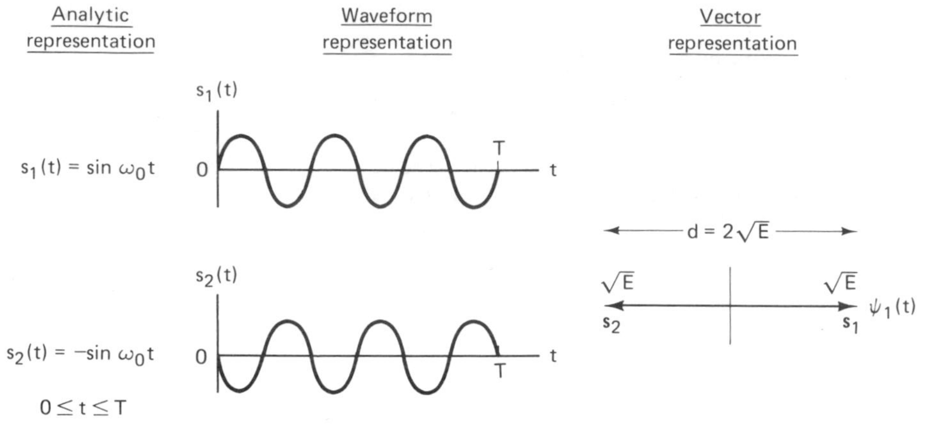

Figure 5.2 Example of an antipodal signal set.

The example shown in Figure 5.3 illustrates an orthogonal signal set. We know that sin x and cos x are orthogonal functions; similarly, sin mx and sin nx, where m and n are integers and $m \neq n$, are also orthogonal functions (see Section A.2.1). In Figure 5.3 we have chosen a pulse waveform example because it provides a clearer picture of orthogonality. The pulse waveform is described by

$$\begin{aligned} s_1(t) &= p(t) \qquad & 0 \leq t \leq T \\ s_2(t) &= p\left(t - \frac{T}{2}\right) \qquad & 0 \leq t \leq T \end{aligned} \tag{5.1}$$

where $p(t)$ is a pulse with duration $\tau = T/2$, and T is the symbol duration. In general, a set of equal energy signals $s_i(t)$, where $i = 1, 2, \ldots, M$, constitutes

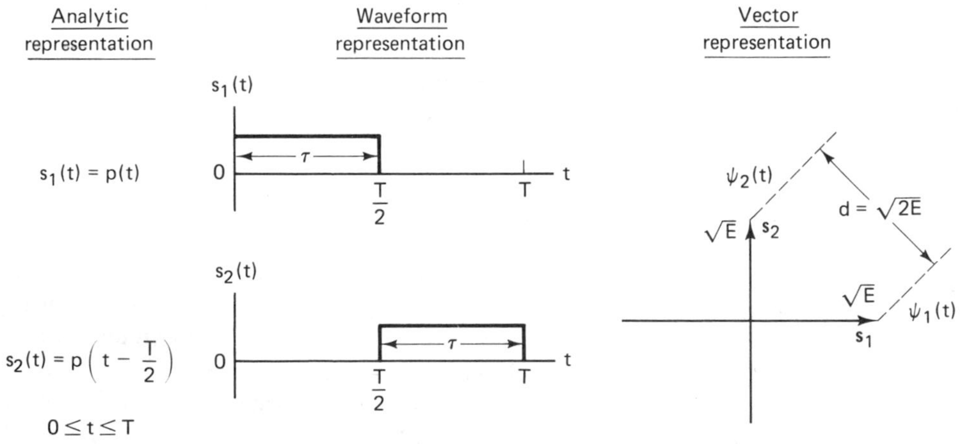

Figure 5.3 Example of a binary orthogonal signal set.

an orthogonal set, if, and only if,

$$z_{ij} = \frac{1}{E} \int_0^T s_i(t)s_j(t) \, dt = \begin{cases} 1 & \text{for } i = j \\ 0 & \text{otherwise} \end{cases} \qquad (5.2)$$

where z_{ij} is called the *cross-correlation coefficient*, and where E is the signal energy expressed as

$$E = \int_0^T s_i^2(t) \, dt \qquad (5.3)$$

The waveform representation in Figure 5.3 illustrates that $s_1(t)$ and $s_2(t)$ cannot interfere with one another because they are disjoint in time. The vectorial representation illustrates the perpendicular relationship between orthogonal signals. Let us consider some alternative descriptions of orthogonal signals or vectors. We can say that the inner or dot product of two different vectors in the orthogonal set must equal zero. In a two- or three-dimensional Cartesian coordinate space, we can describe the signal vectors, geometrically, as being mutually perpendicular to one another. We can say that one vector has zero projection on the other, or that one signal cannot interfere with the other, since they do not share the same *signal space*.

5.1.2 *M*-ary Signaling

With M-ary signaling, the processor accepts k data bits at a time. It then instructs the modulator to produce one of $M = 2^k$ waveforms; binary signaling is the special case where $k = 1$. For $k > 1$, M-ary signaling, as described in Chapter 3, can be regarded as a *waveform coding* procedure. For orthogonal signaling (e.g., MFSK), as k increases there will be an improved error performance or a reduction in required E_b/N_0, at the expense of bandwidth; nonorthogonal signaling (e.g., MPSK) can manifest improved bandwidth efficiency, at the expense of degraded error performance or an increase in required E_b/N_0. By the appropriate choice of signal waveforms, one can trade off error performance versus E_b/N_0 performance, versus bandwidth efficiency. Such trade-offs are treated in greater detail in Chapter 7.

5.1.3 Waveform Coding with Correlation Detection

Waveform coding procedures transform a waveform set into an improved waveform set. The improved waveform set can then be used to provide improved P_B compared to the original set. The most popular of such *waveform codes* are referred to as *orthogonal* and *biorthogonal codes*. The encoding procedure endeavors to make each of the waveforms in the coded signal set as unalike as possible; the goal is to render the cross-correlation coefficient, z_{ij}, among all pairs of signals, as described in Equation (5.2), as small as possible. The smallest possible value of the cross-correlation coefficient occurs when the signals are anticorrelated ($z_{ij} = -1$); however, this can be achieved only when the number of symbols in the set is two ($M = 2$) and the symbols are *antipodal*. In general, it is possible to make all the cross-correlation coefficients equal to zero [1]. The set is then said to be *orthogonal*. Antipodal signal sets are optimum in the sense that each signal is most distant from the other signal in the set; this is seen in Figure 5.2 where the distance, d, between signal vectors is seen to be $d = 2\sqrt{E}$, where E represents the signal energy during a symbol duration T, as expressed in Equation (5.3). Compared to antipodal signals, the distance properties of orthogonal signal sets can be thought of as "second best" (for a given level of waveform energy). In Figure 5.3 the distance between the orthogonal signal vectors is seen to be $d = \sqrt{2E}$.

The *cross-correlation* between two signals is a measure of the *distance* between the signal vectors. The smaller the cross-correlation, the more distant are the vectors from each other. This can be verified in Figure 5.2, where the antipodal signals (whose $z_{ij} = -1$) are represented by vectors that are most distant from each other, and in Figure 5.3, where the orthogonal signals (whose $z_{ij} = 0$) are represented by vectors that are closer to one another than the antipodal vectors. It should be obvious that the distance between two identical waveforms (whose $z_{ij} = 1$) is zero.

Figure 5.4 illustrates the replacement of a 2-bit data set with an improved (orthogonal) codeword set. Both the original data set and the codeword replacement set are comprised of the binary digits (1, 0). Also shown in the figure is the

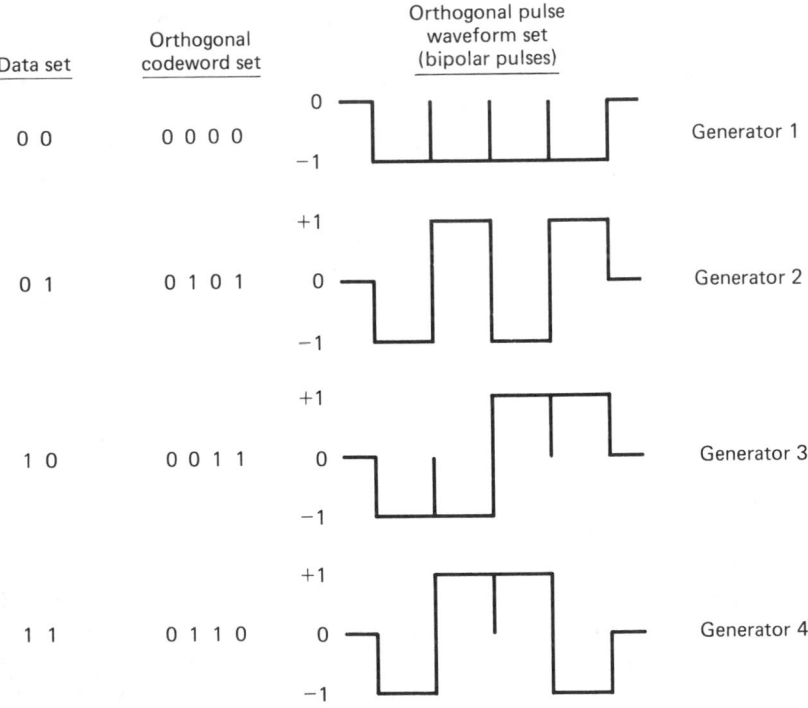

Figure 5.4 Replacement of data set with orthogonal codeword set and waveform set.

waveform set comprised of bipolar pulses $(+1, -1)$ that represents the codeword set. Equation (5.2) is stated in terms of waveforms. However, when the waveform set, $\{s_i(t)\}$, is represented by binary digits, it is easy to show that Equation (5.2) can be simplified as follows:

$$z_{ij} = \frac{\text{number of digit agreements } - \text{ number of digit disagreements}}{\text{total number of digits}}$$

$$z_{ij} = \begin{cases} 1 & \text{for } i = j \\ 0 & \text{otherwise} \end{cases}$$

(5.4)

where $i, j = 1, \ldots, M$, and M is the size of the codeword set. Using Equation (5.4), one can quickly verify that the codeword set in Figure 5.4 is orthogonal. Transmitting data with such an orthogonal set in place of the original data set results in larger distances among signaling waveforms, and thus yields better error performance for a given SNR.

Consider a set of $M = 2^k$ messages that are to be transmitted, using PSK modulation, over a channel disturbed by additive white Gaussian noise (AWGN). The transmitter shown in Figure 5.5, stores or generates the M pulse waveforms of the type shown in Figure 5.4. A message is transmitted by selecting one of the M waveform generators to *phase modulate* the carrier, such that the phase $(\phi_j = 0$ or $\pi)$ of the carrier during each bit time, $0 \le t \le T_b$, corresponds to the amplitudes $(j = -1$ or $1)$ of the generating pulse waveform. At the receiver in Figure 5.6 the noisy signal is demodulated to baseband and fed to the M correlators (or matched filters). Correlation is performed over a codeword duration, $0 \le t \le T$, where $T = (\log_2 M)T_b = kT_b$. With orthogonally coded waveforms, in the absence of noise, the outputs of all correlators, except the one corresponding to the transmitted codeword, are zero.

5.1.4 Orthogonal Codes

A 1-bit data set can be transformed, using *orthogonal* codewords of two-digits each, described by the matrix $\mathbf{H}_1$ as follows:

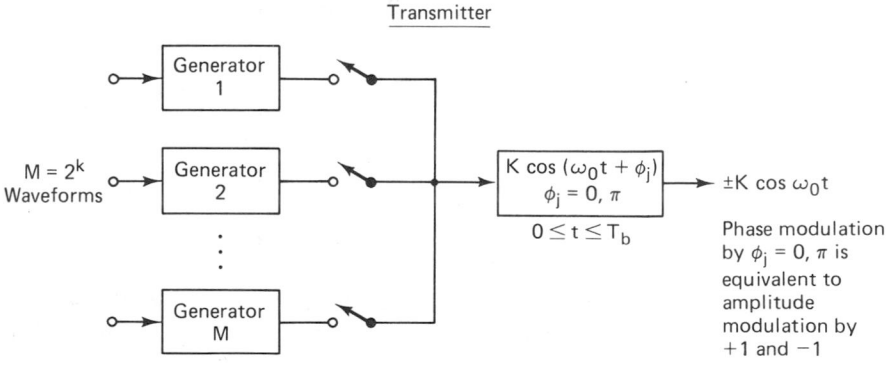

Figure 5.5 Waveform-encoded phase coherent system (transmitter).

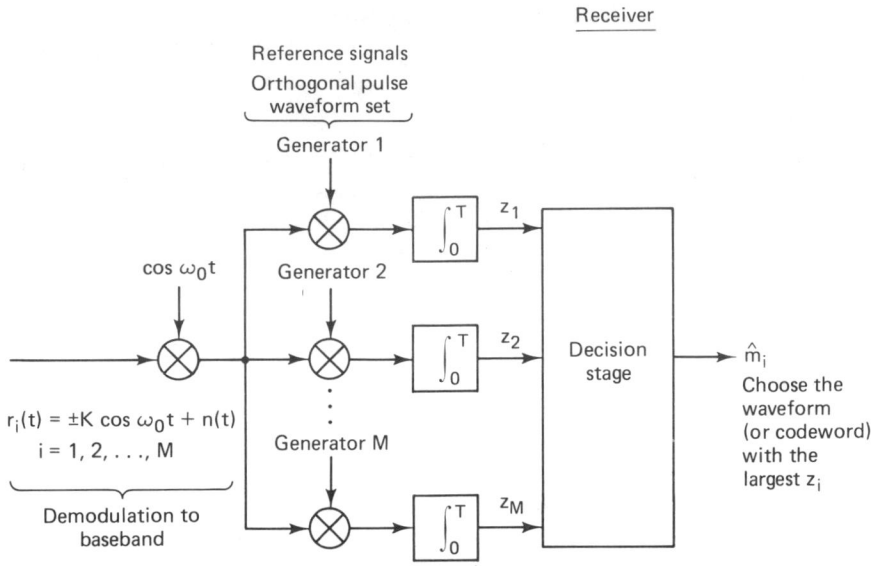

Figure 5.6 Waveform coding with correlation detection.

Data set	Orthogonal codeword set

$$
\begin{array}{cc}
0 \\
1
\end{array}
\qquad
\mathbf{H}_1 = \begin{bmatrix} 0 & 0 \\ 0 & 1 \end{bmatrix}
$$

For this, and the following examples, use Equation (5.4) to verify the orthogonality of the codeword set. To encode a 2-bit data set, we extend the foregoing set both horizontally and vertically, creating matrix $\mathbf{H}_2$.

Data set	Orthogonal codeword set

$$
\begin{array}{cc}
0\ 0 \\
0\ 1 \\
1\ 0 \\
1\ 1
\end{array}
\qquad
\mathbf{H}_2 =
\left[
\begin{array}{cc:cc}
0 & 0 & 0 & 0 \\
0 & 1 & 0 & 1 \\
\hdashline
0 & 0 & 1 & 1 \\
0 & 1 & 1 & 0
\end{array}
\right]
=
\begin{bmatrix}
\mathbf{H}_1 & \mathbf{H}_1 \\
\mathbf{H}_1 & \overline{\mathbf{H}_1}
\end{bmatrix}
$$

The lower right quadrant is the complement of the prior codeword set. We continue the same construction rule to obtain an orthogonal set $\mathbf{H}_3$ for a 3-bit data set.

Data set	Orthogonal codeword set

$$
\mathbf{H}_3 =
\begin{array}{ccc|ccc}
\end{array}
$$

Data set:

0 0 0

0 0 1

0 1 0

0 1 1

1 0 0

1 0 1

1 1 0

1 1 1

$$
\mathbf{H}_3 = \left[
\begin{array}{cccc:cccc}
0 & 0 & 0 & 0 & 0 & 0 & 0 & 0 \\
0 & 1 & 0 & 1 & 0 & 1 & 0 & 1 \\
0 & 0 & 1 & 1 & 0 & 0 & 1 & 1 \\
0 & 1 & 1 & 0 & 0 & 1 & 1 & 0 \\
\hdashline
0 & 0 & 0 & 0 & 1 & 1 & 1 & 1 \\
0 & 1 & 0 & 1 & 1 & 0 & 1 & 0 \\
0 & 0 & 1 & 1 & 1 & 1 & 0 & 0 \\
0 & 1 & 1 & 0 & 1 & 0 & 0 & 1
\end{array}
\right] = \begin{bmatrix} \mathbf{H}_2 & \mathbf{H}_2 \\ \mathbf{H}_2 & \overline{\mathbf{H}_2} \end{bmatrix}
$$

In general, we can construct a codeword set, $\mathbf{H}_k$, of dimension $2^k \times 2^k$, called a *Hadamard matrix*, for a k-bit data set from the $\mathbf{H}_{k-1}$ matrix, as follows:

$$
\mathbf{H}_k = \begin{bmatrix} \mathbf{H}_{k-1} & \mathbf{H}_{k-1} \\ \mathbf{H}_{k-1} & \overline{\mathbf{H}_{k-1}} \end{bmatrix}
$$

Each pair of words in each codeword set, $\mathbf{H}_1, \mathbf{H}_2, \mathbf{H}_3, \ldots, \mathbf{H}_k, \ldots$, has as many digit agreements as disagreements [2]. Hence, in accordance with Equation (5.4), $z_{ij} = 0$ (for $i \neq j$), and each of the sets is orthogonal.

Just as M-ary signaling with an orthogonal modulation format (such as MFSK) improves the P_B performance, waveform coding with an orthogonally constructed signal set, in combination with correlation detection, produces *exactly the same* improvement. For equally likely, equal-energy orthogonal signals, the probability of codeword (symbol) error can be upper bounded, as follows [2]:

$$
P_E(k) \leq (2^k - 1)Q\left(\sqrt{\frac{kE_b}{N_0}}\right) \tag{5.5}
$$

where $Q(x)$ is defined in Equation (2.42). For fixed k, as E_b/N_0 is increased, the bound becomes increasingly tight. For $P_E(k) \leq 10^{-3}$, Equation (5.5) is a good approximation of the error probability. The relationship between $P_B(k)$ and $P_E(k)$ given in Equation (3.127) is repeated here:

$$\frac{P_B(k)}{P_E(k)} = \frac{2^{k-1}}{2^k - 1} \tag{5.6}$$

Combining Equations (5.5) and (5.6), the probability of bit error can be bounded as follows:

$$P_B(k) \leq (2^{k-1})Q\left(\sqrt{\frac{kE_b}{N_0}}\right) \tag{5.7}$$

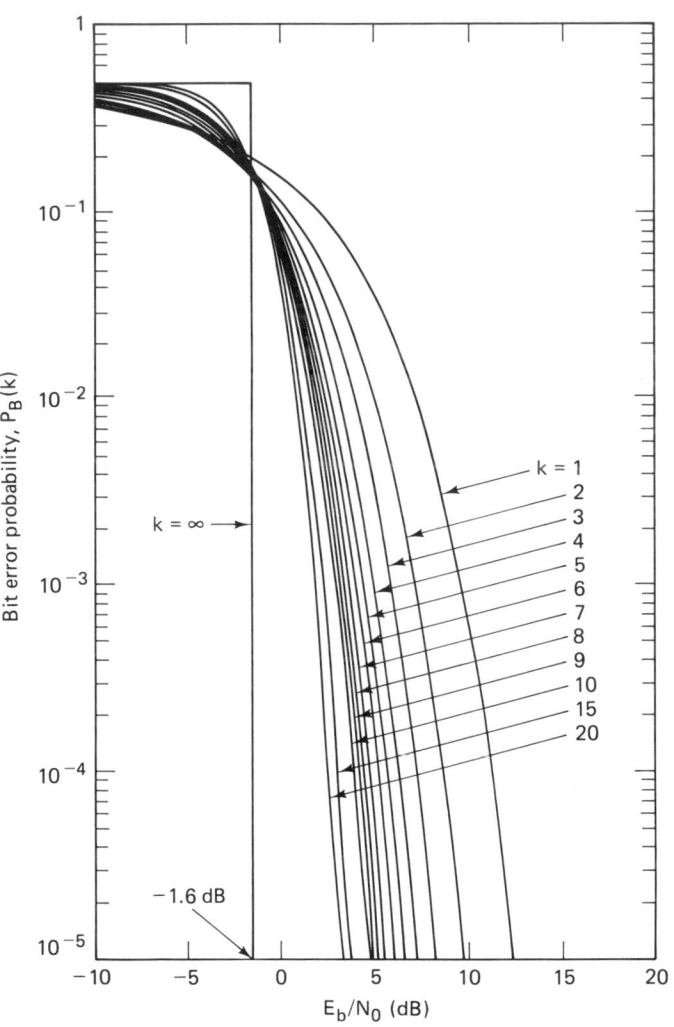

Figure 5.7 Coherent detection of orthogonally coded transmission. (Reprinted from W. C. Lindsey and M. K. Simon, *Telecommunication Systems Engineering*, Prentice-Hall, Inc., Englewood Cliffs, N.J., 1973, courtesy of W. C. Lindsey and Marvin K. Simon.)

$P_B(k)$ is plotted in Figure 5.7 for various values of k; the uncoded case corresponds to the $k = 1$ curve. The performance improvement for $k > 1$ should be obvious. The curves are identical to the orthogonal signaling performance (such as FSK) of Figure 3.25. What price do we pay for this improvement? We need to expend more transmission bandwidth. The orthogonal codes can be described as having $(2^k - k)$ redundant digits. For example, the orthogonal H_3 matrix above reassigns 3-bit messages into 8-bit codewords, resulting in five redundant digits. Therefore, the bandwidth is increased by $\frac{8}{3}$ or, in general, by $2^k/k$. For orthogonal codes, the required transmission bandwidth increases exponentially with k. Compared to structured sequences, this type of coding *does not utilize bandwidth efficiently*.

5.1.5 Biorthogonal Codes

A *biorthogonal* signal set of M total signals or codewords can be obtained from an orthogonal set of $M/2$ signals by augmenting it with the negative of each signal, as follows:

$$\mathbf{B}_k = \begin{bmatrix} \mathbf{H}_{k-1} \\ \hline \mathbf{H}_{k-1} \end{bmatrix}$$

For example, a 3-bit data set can be transformed into a biorthogonal codeword set as follows:

Data set			Biorthogonal codeword set				
0	0	0		0	0	0	0
0	0	1		0	1	0	1
0	1	0		0	0	1	1
0	1	1		0	1	1	0
1	0	0		1	1	1	1
1	0	1		1	0	1	0
1	1	0		1	1	0	0
1	1	1		1	0	0	1

$$\mathbf{B}_3 = \begin{bmatrix} 0 & 0 & 0 & 0 \\ 0 & 1 & 0 & 1 \\ 0 & 0 & 1 & 1 \\ 0 & 1 & 1 & 0 \\ \hline 1 & 1 & 1 & 1 \\ 1 & 0 & 1 & 0 \\ 1 & 1 & 0 & 0 \\ 1 & 0 & 0 & 1 \end{bmatrix}$$

The biorthogonal set is really two sets of orthogonal codes such that each codeword in one set has its antipodal codeword in the other set. The biorthogonal set consists of a *combination of orthogonal and antipodal* signals. With respect to z_{ij} of Equations (5.2) or (5.4), biorthogonal codes can be characterized as

$$z_{ij} = \begin{cases} 1 & \text{for } i = j \\ -1 & \text{for } i \neq j, \, |i - j| = \dfrac{M}{2} \\ 0 & \text{for } i \neq j, \, |i - j| \neq \dfrac{M}{2} \end{cases} \tag{5.8}$$

One advantage of a biorthogonal code over an orthogonal one for the same data set, is that the biorthogonal code requires *one-half* as many bits per codeword (compare the $\mathbf{B}_3$ matrix with the $\mathbf{H}_3$ matrix). Thus the bandwidth requirements for

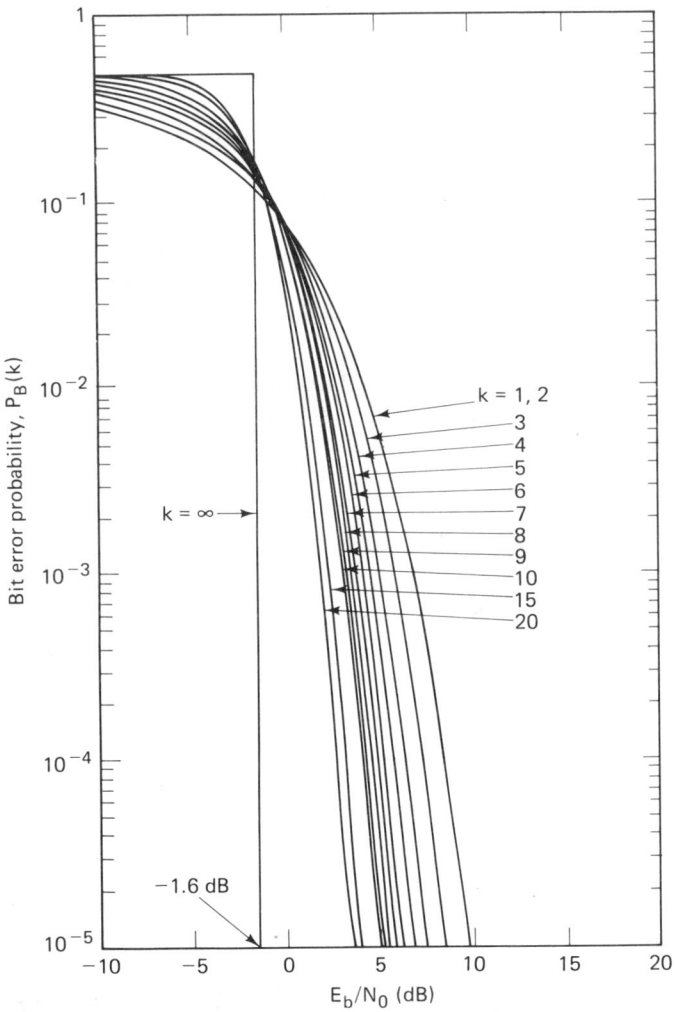

Figure 5.8 Coherent detection of biorthogonally coded transmission. (Reprinted from W. C. Lindsey and M. K. Simon, *Telecommunication Systems Engineering*, Prentice-Hall, Inc., Englewood Cliffs, N.J., 1973, courtesy of W. C. Lindsey and Marvin K. Simon.)

Channel Coding: Part 1 Chap. 5

biorthogonal codes are one-half the requirements for comparable orthogonal ones. Since antipodal signal vectors have better distance properties than orthogonal ones, it should come as no surprise that biorthogonal codes perform slightly better than orthogonal ones. For equally likely, equal-energy biorthogonal signals, the probability of codeword (symbol) error can be upper bounded, as follows [2]:

$$P_E(k) \leq (2^k - 2)Q\left(\sqrt{\frac{kE_b}{N_0}}\right) + Q\left(\sqrt{\frac{2kE_b}{N_0}}\right) \tag{5.9}$$

which becomes increasingly tight for fixed k as E_b/N_0 is increased. $P_B(k)$ is a complicated function of $P_E(k)$; we can approximate it with the relationship [2]

$$P_B(k) \simeq \frac{P_E(k)}{2}$$

The approximation is quite good for $k > 3$. Therefore, we can write

$$P_B(k) \leq \frac{1}{2}\left[(2^k - 2)Q\left(\sqrt{\frac{kE_b}{N_0}}\right) + Q\left(\sqrt{\frac{2kE_b}{N_0}}\right)\right] \tag{5.10}$$

The P_B performance of these biorthogonal codes, shown in Figure 5.8, offers improved performance, compared to the performance of the orthogonal codes shown in Figure 5.7, and requires only *half the bandwidth* of orthogonal codes.

5.1.6 Transorthogonal (Simplex) Codes

A code generated from an orthogonal set by deleting the first digit of each codeword is called a *transorthogonal* or *simplex code*. Such a code is characterized by

$$z_{ij} = \begin{cases} 1 & \text{for } i = j \\ \dfrac{-1}{M-1} & \text{for } i \neq j \end{cases} \tag{5.11}$$

A simplex code represents the *minimum energy* equivalent (in the error probability sense) of the equally likely orthogonal set. In comparing the error performance of orthogonal, biorthogonal, and simplex codes, we can state that simplex coding requires the minimum E_b/N_0 for a specified symbol error rate. However, for a *large value of k*, all three schemes are *essentially identical* in error performance. Biorthogonal coding requires half the bandwidth of the others. However, for each of these codes, bandwidth requirements (and system complexity) grow exponentially with the value of k; therefore, such coding schemes are attractive only when large bandwidths are available. When bandwidth is not plentiful, the structured redundancy techniques (see Section 5.3 through Chapter 6) are more attractive [3]. When bandwidth is *very scarce*, the so-called combined modulation and coding techniques for bandlimited channels are most promising (see Sections 7.10.6 and 7.10.7).

5.2 TYPES OF ERROR CONTROL

Before we discuss the details of structured redundancy, let us describe the two basic ways such redundancy is used for controlling errors. The first, *error detection and retransmission*, utilizes *parity bits* (redundant bits added to the data) to detect that an error has been made. The receiving terminal does not attempt to correct the error; it simply requests the transmitter to retransmit the data. Notice that a two-way link is required for such dialogue between the transmitter and receiver. The second type of error control, *forward error correction* (FEC), requires a one-way link only, since in this case the parity bits are designed for both the detection and correction of errors. We shall see that not all error patterns can be corrected; error-correcting codes are classified according to their error-correcting capabilities.

5.2.1 Terminal Connectivity

Communication terminals are often classified according to their connectivity with other terminals. The possible connections, shown in Figure 5.9, are termed *simplex* (not to be confused with the simplex or transorthogonal codes), *half-duplex*, and *full-duplex*. The simplex connection, in Figure 5.9a, is a one-way link. Transmissions are made from terminal A to terminal B only, never in the reverse direction. The half-duplex connection, in Figure 5.9b, is a link whereby transmissions may be made in either direction but not simultaneously. Finally, the full-duplex connection, in Figure 5.9c, is a two-way link, where transmissions may proceed in both directions simultaneously.

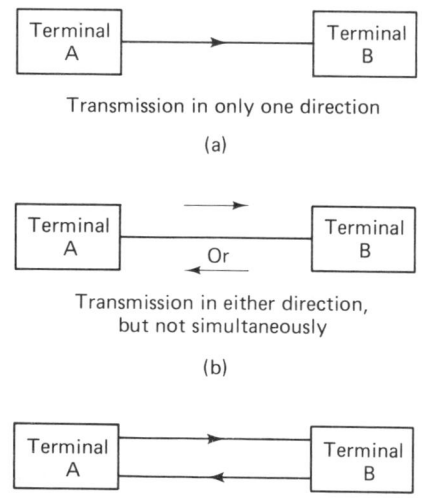

Transmission in only one direction

(a)

Transmission in either direction, but not simultaneously

(b)

Transmission in both directions simultaneously

(c)

Figure 5.9 Terminal connectivity classifications. (a) Simplex. (b) Half-duplex. (c) Full-duplex.

5.2.2 Automatic Repeat Request

When the error control consists of error detection only, the communication system generally needs to provide a means of alerting the transmitter that an error has been detected and that a retransmission is necessary. Such error control procedures are known as *automatic repeat request* or automatic retransmission query (ARQ) methods. Figure 5.10 illustrates three of the most popular ARQ procedures. In each of the diagrams, time is advancing from left to right. The first procedure, called *stop-and-wait ARQ*, is shown in Figure 5.10a. It requires a half-duplex connection only, since the transmitter waits for an acknowledgment (ACK) of each transmission before it proceeds with the next transmission. In the figure, the third transmission block is received in error; therefore, the receiver responds with a negative acknowledgment (NAK), and the transmitter retransmits this third

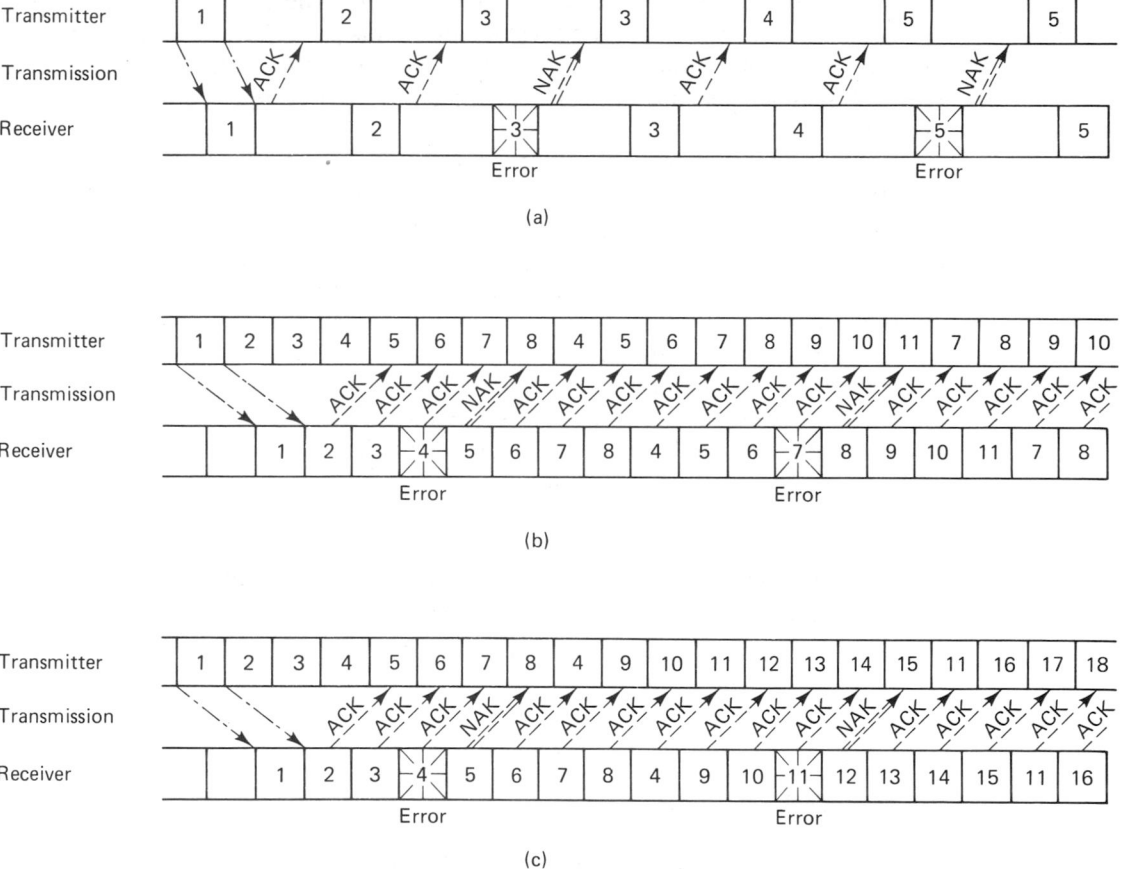

Figure 5.10 Automatic repeat request (ARQ). (a) Stop-and-wait ARQ (half-duplex). (b) Continuous ARQ with pullback (full-duplex). (c) Continuous ARQ with selective repeat (full-duplex).

message block before transmitting the next in the sequence. The second ARQ procedure, called *continuous ARQ with pullback*, is shown in Figure 5.10b. Here a full-duplex connection is necessary. Both terminals are transmitting simultaneously; the transmitter is sending message data and the receiver is sending acknowledgment data. Notice that a sequence number has to be assigned to each block of data. Also, the ACKs and NAKs need to reference such numbers, or else there needs to be a priori knowledge of the propagation delays so that the transmitter knows which messages are associated with which acknowledgments. In the example of Figure 5.10b there is a fixed separation of four blocks between the message being transmitted and the acknowledgment being simultaneously received. For example, when message 8 is being sent, a NAK corresponding to the corrupted message 4 is being received. In this ARQ procedure, the transmitter "pulls back" to the message in error and retransmits all message data, starting with the corrupted message. The final method, called *continuous ARQ with selective repeat*, is shown in Figure 5.10c. Here, as with the second ARQ procedure, a full-duplex connection is needed. However, in this procedure, only the corrupted message is repeated; then the transmitter continues the transmission sequence where it had left off instead of repeating any subsequent correctly received messages.

The choice of which ARQ procedure to choose is a trade-off between the requirements for efficient utilization of the communications resource and the need to provide full-duplex connectivity. The half-duplex connectivity required in Figure 5.10a is less costly than full-duplex; the associated inefficiency can be measured by the blank time slots. The more efficient utilization illustrated in Figures 5.10b and c requires the more costly full-duplex connectivity.

The major advantage of ARQ over forward error correction (FEC) is that error detection requires much simpler decoding equipment and much less redundancy than does error correction. Also, ARQ is adaptive in the sense that information is retransmitted only when errors occur. On the other hand, FEC may be desirable in place of, or in addition to, error detection, for any of the following reasons:

1. A reverse channel is not available or the delay with ARQ would be excessive.
2. The retransmission strategy is not conveniently implemented.
3. The expected number of errors, without corrections, would require excessive retransmissions.

5.3 STRUCTURED SEQUENCES

In Section 3.8 we considered digital signaling by means of $M = 2^k$ signal waveforms (*M*-ary signaling), where each waveform contains k bits of information. We saw that in the case of orthogonal *M*-ary signaling, we can decrease P_B by increasing M (expanding the bandwidth). Similarly, in Section 5.1 we showed that it is possible to decrease P_B by encoding k binary digits into one of M orthogonal codewords. The major disadvantage with such orthogonal coding techniques is the associated inefficient use of bandwidth. The required transmission bandwidth grows exponentially with k for an orthogonal set of $M = 2^k$ waveforms. In this and subsequent sections we abandon the need for antipodal or orthogonal properties and focus on a class of encoding procedures known as *parity-check codes*. Such channel coding procedures are classified as *structured sequences* because they represent methods of inserting structured redundancy into the source data so that the presence of errors can be detected or the errors corrected. Structured sequences are partitioned into two important subcategories as shown in Figure 5.1: *block coding* and *convolutional coding*. Block coding (primarily) is treated in this chapter, and convolutional coding is treated in Chapter 6. These techniques allow us to attain a P_B performance comparable to waveform encoding techniques but with lower bandwidth requirements. The codewords of these codes (structured sequences) are usually *nonorthogonal* [3].

5.3.1 Channel Models

5.3.1.1 Discrete Memoryless Channel

A *discrete memoryless channel* (DMC) is characterized by a discrete input alphabet, a discrete output alphabet, and a set of conditional probabilities, $P(j|i)$ $(1 \leq i \leq M, 1 \leq j \leq Q)$, where i represents a modulator M-ary input symbol, j represents a demodulator Q-ary output symbol, and $P(j|i)$ is the probability of receiving j given that i was transmitted. Each output symbol of the channel depends only on the corresponding input, so that for a given input sequence $\mathbf{U} = u_1, u_2, \ldots, u_m, \ldots, u_N$ the conditional probability of a corresponding output sequence $\mathbf{Z} = z_1, z_2, \ldots, z_m, \ldots, z_N$ may be expressed as

$$P(\mathbf{Z}|\mathbf{U}) = \prod_{m=1}^{N} P(z_m|u_m) \tag{5.12}$$

In the event that the channel *has memory* (i.e., noise or fading that occurs in bursts), the conditional probability of the sequence $\mathbf{Z}$ would need to be expressed as the *joint* probability of all the elements of the sequence. Equation (5.12) expresses the *memoryless* condition of the channel. Since the channel noise in a memoryless channel is defined to affect each symbol independently of all the other symbols, the conditional probability of $\mathbf{Z}$ is seen as the product of the independent element probabilities.

5.3.1.2 Binary Symmetric Channel

A *binary symmetric channel* (BSC) is a special case of a DMC; the input and output alphabet sets consist of the binary elements (0 and 1). The conditional probabilities are symmetric:

$$\begin{aligned} P(0|1) &= P(1|0) = p \\ P(1|1) &= P(0|0) = 1 - p \end{aligned} \tag{5.13}$$

Equation (5.13) states the channel *transition probabilities*. That is, given that a channel symbol was transmitted, the probability that it is received in error is p (related to the symbol energy), and the probability that it is received correctly is $(1 - p)$. Since the demodulator output consists of the discrete elements 0 and 1, the demodulator is said to make a firm or *hard decision* on each symbol. A commonly used code system consists of BPSK modulated coded data, hard decision demodulated. Then the channel symbol error probability is found using the methods discussed in Section 3.7.1 and Equation (3.84) to be

$$p = Q\left(\sqrt{\frac{2E_c}{N_0}} \right)$$

where E_c/N_0 is the channel symbol energy per noise density, and $Q(x)$ is defined in Equation (2.42).

When such hard decisions are used in a binary coded system, the demodulator feeds the two-valued *code symbols* or *channel bits* to the decoder. Since the decoder then operates on the hard decisions made by the demodulator, decoding with a BSC channel is called *hard-decision decoding*.

5.3.1.3 Gaussian Channel

We can generalize our definition of the DMC to channels with alphabets that are not discrete. An example is the *Gaussian channel* with a discrete input alphabet and a continuous output alphabet over the range $(-\infty, \infty)$. The channel adds noise to the symbols. Since the noise is a Gaussian random variable, with zero mean and variance σ^2, the resulting probability density function (pdf) of the received random variable z, conditioned on the symbol u_k (the likelihood of u_k), can be written

$$p(z|u_k) = \frac{1}{\sigma\sqrt{2\pi}} \exp\left[\frac{-(z - u_k)^2}{2\sigma^2} \right] \qquad (5.14)$$

for all z, where $k = 1, 2, \ldots, M$. For this case, *memoryless* has the same meaning as it does in Section 5.3.1.1, and Equation (5.12) can be used to obtain the conditional probability for the sequence, **Z**.

When the demodulator output consists of a continuous alphabet or its quantized approximation (with greater than two quantization levels), the demodulator is said to make *soft decisions*. In the case of a coded system, the demodulator feeds such quantized code symbols to the decoder. Since the decoder then operates on the soft decisions made by the demodulator, decoding with a Gaussian channel is called *soft-decision decoding*.

In the case of a hard-decision channel, we are able to characterize the detection process with a channel symbol error probability. However, in the case of a soft-decision channel, the detector makes the kind of decisions (soft decisions) that cannot be labeled as correct or incorrect. Thus, since there are no firm decisions, there cannot be a probability of making an error; the detector can only formulate a family of conditional probabilities or likelihoods of the different symbol types.

It is possible to design decoders using soft decisions, but block code soft-decision decorders are substantially more complex than hard-decision decoders; therefore, block codes are usually implemented with hard-decision decoders. For convolutional codes, both hard- and soft-decision implementations are equally popular. In this chapter we consider that the channel is a binary symmetric channel (BSC), and hence the decoder employs hard decisions. In Chapter 6 we further discuss channel models, as well as hard- versus soft-decision decoding for convolutional codes.

5.3.2 Code Rate and Redundancy

In the case of block codes, the source data are segmented into blocks of k data bits, also called information bits or message bits; each block can represent any one of 2^k distinct messages. The encoder transforms each k-bit data block into a larger block of n bits, called code bits or channel symbols. The $(n - k)$ bits, which the encoder adds to each data block, are called *redundant bits*, *parity bits*, or *check bits*; they carry no new information. The code is referred to as an (n, k) code. The ratio of redundant bits to data bits, $(n - k)/k$, within a block is called the *redundancy* of the code, and the ratio of data bits to total bits, k/n, is called the *code rate*. The code rate can be thought of as the portion of a code bit that constitutes information. For example, in a rate $\frac{1}{2}$ code, each code bit carries $\frac{1}{2}$ bit of information.

In this chapter and Chapter 6 we consider those coding techniques that provide redundancy by increasing the required transmission bandwidth. For example, an error control technique that employs a rate 1/2 code (100% redundancy) will require double the bandwidth of an uncoded system. However, if a rate 3/4 code is used, the redundancy is 33% and the bandwidth expansion is only 4/3. In Chapter 7 we consider modulation/coding techniques for bandlimited channels where complexity, instead of bandwidth, is traded for error performance improvement.

5.3.3 Parity-Check Codes

5.3.3.1 Single-Parity-Check Code

Parity-check codes use linear sums of the information bits, called *parity symbols* or *parity bits*, for error detection or correction. A single-parity check code is constructed by adding a single-parity bit to a block of data bits. The parity bit takes on the value of one or zero as needed to ensure that the summation of all the bits in the codeword yields an even (or odd) result. The summation operation is performed using modulo-2 arithmetic (exclusive-or logic), as described in Section 2.12.3. If the added parity is designed to yield an even result, the method is termed *even parity*, and if designed to yield an odd result, it is termed *odd*

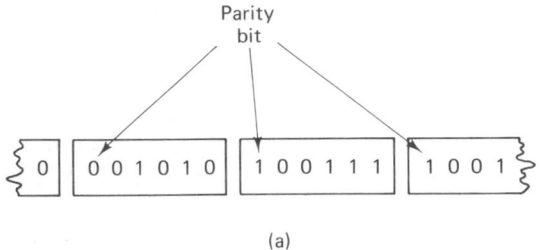

(a)

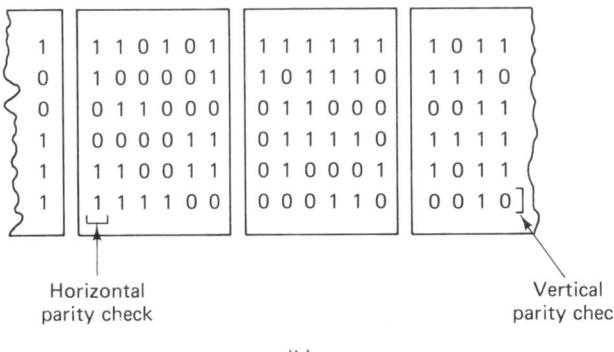

Horizontal
parity check

Vertical
parity check

Figure 5.11 Parity checks for serial
and parallel transmission. (a) Serial
transmission. (b) Parallel transmission.

(b)

parity. Figure 5.11a illustrates a serial data transmission (the rightmost bit is the earliest bit). A single-parity bit is added (the leftmost bit in each block) to yield even parity.

At the receiving terminal, the decoding procedure consists of testing that the modulo-2 sum of the codeword bits yields a zero result (even parity). If the result is found to be one instead of zero, the codeword is known to contain errors. The rate of the code can be expressed as $k/(k + 1)$. Do you suppose the decoder can automatically *correct* a digit that is received in error? No, it cannot. It can only *detect* the presence of an odd number of bit errors (if an even number of bits are inverted, the parity test will appear correct; this represents the case of an *undetected error*). Assuming that all bit errors are equally likely and occur independently, we can write the probability of *j* errors occurring in a block of *n* symbols as

$$P(j, n) = \binom{n}{j} p^j (1 - p)^{n-j} \tag{5.15}$$

where *p* is the probability that a *channel symbol* is received in error, and where

$$\binom{n}{j} = \frac{n!}{j!(n - j)!}$$

is the number of various ways in which *j* bits out of *n* may be in error. Thus for a single-parity error-detection code, the probability of an undetected error, P_{nd}, within a block of n bits is computed, as follows:

$$P_{nd} = \sum_{j=1}^{\substack{n/2 \text{ (for } n \text{ even)} \\ (n-1)/2 \text{ (for } n \text{ odd)}}} \binom{n}{2j} p^{2j}(1 - p)^{n-2j} \qquad (5.16)$$

Example 5.1   Even-Parity Code

Configure a (4, 3) even-parity error-detection code such that the parity symbol appears as the leftmost symbol of the codeword. Which error patterns can the code detect? Compute the probability of an undetected message error, assuming that all symbol errors are independent events and that the probability of a channel symbol error is $p = 10^{-3}$.

Solution

Message	Parity	Codeword
000	0	0 000
100	1	1 100
010	1	1 010
110	0	0 110
001	1	1 001
101	0	0 101
011	0	0 011
111	1	1 111

parity   message

The code is capable of detecting all single- and triple-error patterns. The probability of an undetected error is equal to the probability that two or four errors occur anywhere in a codeword.

$$P_{nd} = \binom{4}{2} p^2 (1 - p)^2 + \binom{4}{4} p^4$$

$$= 6p^2 (1 - p)^2 + p^4$$

$$= 6p^2 - 12p^3 + 7p^4$$

$$= 6(10^{-3})^2 - 12(10^{-3})^3 + 7(10^{-3})^4 \approx 6 \times 10^{-6}$$

5.3.3.2 Rectangular Code

A *rectangular code*, also called a *product code*, can be thought of as a parallel data transmission, depicted in Figure 5.11b. First we form a rectangle of message bits comprised of M rows and N columns; then a horizontal parity check is appended to each row and a vertical parity check is appended to each column, resulting in an augmented array of dimensions $(M + 1) \times (N + 1)$. The rate of the rectangular code, k/n, can then be written as

$$\frac{k}{n} = \frac{MN}{(M + 1)(N + 1)} \qquad (5.17)$$

How much more powerful is the rectangular code than the single-parity code, which is only capable of error detection? Notice that any single bit error will cause a parity check failure in one of the array columns *and* in one of the array rows. Therefore, the rectangular code can correct a single error pattern since the error is uniquely located at the intersection of the error-detecting row and the error-detecting column. For the example shown in Figure 5.11b, the array dimensions are $M = N = 5$; therefore, the figure depicts a (36, 25) code that can correct a single error located anywhere in the 36 bit positions. For an error-correcting block code, we compute the probability that the decoded block has an uncorrected error by accounting for all the ways in which a *message error* can be made. Starting with the probability of j errors in a block of n symbols, expressed in Equation (5.15), we can write the probability of a message error, also called a *block error* or *word error*, P_M, for a code that can correct all t and fewer error patterns:

$$P_M = \sum_{j=t+1}^{n} \binom{n}{j} p^j (1 - p)^{n-j} \qquad (5.18)$$

where p is the probability that a *channel symbol* is received in error. For the example in Figure 5.11b, the code can correct all single error patterns ($t = 1$) within the rectangular block of $n = 36$ bits. Hence the summation in Equation (5.18) starts with $j = 2$:

$$P_M = \sum_{j=2}^{36} \binom{36}{j} p^j (1 - p)^{36-j} \qquad (5.19)$$

When p is reasonably small, the first term in the summation is the dominant one; we can therefore write for this (36, 25) rectangular code example

$$P_M \simeq \binom{36}{2} p^2 (1 - p)^{34}$$

The *bit error probability*, P_B, depends on the particular code and decoder. An approximation for P_B is given in Section 5.5.3.

5.3.4 Coding Gain

Figure 5.12 illustrates the probability of bit error, P_B, versus E_b/N_0 for coherent binary PSK modulation in combination with examples of various (n, k) codes over a Gaussian channel. The (1, 1) curve illustrates the uncoded PSK performance, while the (24, 12) and (127, 92) curves illustrate coded PSK performance using block codes with $(n - k) = 12$ parity bits and 35 parity bits, respectively. From Figure 3.24 we know in which direction the waterfall-like curves move, corresponding to P_B performance improvement. Look at the various curves in Figure 5.12. Can you explain why the coded curves (to which we attribute P_B performance improvement) appear to be moving in the wrong direction when compared with the uncoded curve? Where does the strength of the code manifest itself? The curves in Figure 5.12 indicate that the strength of a code is seen only after an E_b/N_0 threshold has been exceeded (approximately 5.5 dB in this example). For values of E_b/N_0 less than the threshold, the coding manifests itself only as over-

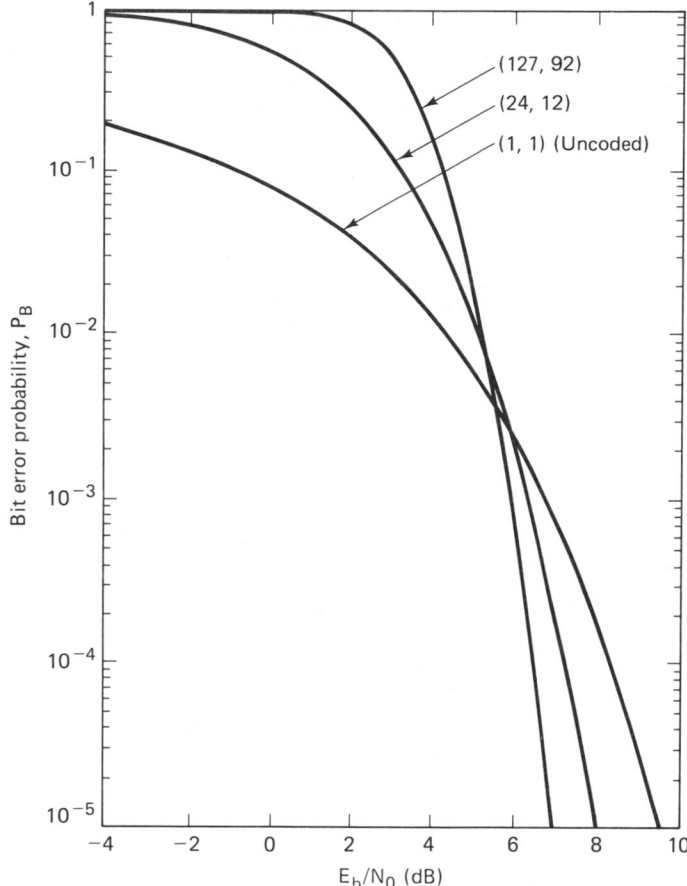

Figure 5.12 Coded versus uncoded bit error performance for coherent PSK with various (n, k) codes.

head bits resulting in *reduced energy per bit*, compared to the uncoded case; before the threshold is exceeded, the redundant bits are simply "excess baggage" without the ability to improve performance. Once the threshold is exceeded, the performance improvement of the code more than compensates for the reduction in energy per coded bit. Therefore, in Figure 5.12, once the threshold value of $E_b/N_0 = 5.5$ dB is exceeded, the relative positions of the curves reverse themselves compared to their positions at less-than-threshold E_b/N_0. *Coding gain* is defined as the reduction, expressed in decibels, in the required E_b/N_0 to achieve a specified error performance of an error-correcting coded system over an uncoded one with the same modulation. For example, in Figure 5.12, for $P_B = 10^{-5}$, the (24, 12) code has a coding gain of about 1.5 dB.

Example 5.2   Coded versus Uncoded Performance

Compare the message error probability for a communications link with and without the use of error-correction coding. Assume that the uncoded transmission characteristics are: BPSK modulation, Gaussian noise, $S/N_0 = 43,776$, data rate $R = 4800$ bits/s. For the coded case, also assume the use of a (15, 11) error-correcting code that is capable of correcting any single-error pattern within a block of 15 bits. Consider that the demodulator makes hard decisions and thus feeds the demodulated code bits directly to the decoder, which in turn outputs an estimate of the original message.

Solution

Following Equation (3.84), let $p_u = Q\sqrt{2E_b/N_0}$ and $p_c = Q\sqrt{2E_c/N_0}$ be the uncoded and coded channel symbol error probabilities, respectively, where E_b/N_0 is uncoded bit energy per noise spectral density and E_c/N_0 is the coded bit energy per noise spectral density.

Without coding

$$\frac{E_b}{N_0} = \frac{S}{RN_0} = 9.12 \text{ (9.6 dB)}$$

$$p_u = Q\left(\sqrt{\frac{2E_b}{N_0}}\right) = Q(\sqrt{18.24}) = 1.02 \times 10^{-5} \qquad (5.20)$$

where the following approximation of $Q(x)$ from Equation (2.43) was used:

$$Q(x) \simeq \frac{1}{x\sqrt{2\pi}} \exp\left(\frac{-x^2}{2}\right) \qquad \text{for } x > 3$$

The probability that the uncoded message block, P_M^u, will be received in error is 1 minus the product of the probabilities that each bit will be detected correctly. Thus

$$P_M^u = 1 - (1 - p_u)^k$$

$$= 1 - \underbrace{(1 - p_u)^{11}}_{\substack{\text{probability that all} \\ \text{11 bits in uncoded} \\ \text{block are correct}}} \qquad = \underbrace{1.12 \times 10^{-4}}_{\substack{\text{probability that at} \\ \text{least 1 bit out of} \\ \text{11 is in error}}} \qquad (5.21)$$

With coding:

The channel symbol rate, sometimes called the coded bit rate, R_c is 15/11 times the data bit rate.

$$R_c = 4800 \times \tfrac{15}{11} \approx 6545 \text{ bps}$$

$$\frac{E_c}{N_0} = \frac{S}{R_cN_0} = 6.688 \text{ (8.25 dB)}$$

The E_c/N_0 for each code bit is less than that for the uncoded bit because the channel bit rate has increased but the transmitter power is assumed to be fixed.

$$p_c = Q\left(\sqrt{\frac{2E_c}{N_0}}\right) = Q(\sqrt{13.38}) = 1.36 \times 10^{-4} \qquad (5.22)$$

It can be seen by comparing the results of Equation (5.20) with (5.22) that the channel bit error probability has degraded. More bits must be detected during the same time interval, and with the same available power; the performance improvement due to the coding *is not yet apparent*. We now compute the coded message error rate, P_M^c, using Equation (5.18).

$$P_M^c = \sum_{j=2}^{n=15} \binom{15}{j}(p_c)^j(1 - p_c)^{15-j}$$

The summation is started with $j = 2$ since the code corrects all single errors within a block of $n = 15$ bits. A good approximation is obtained by using only the first term of the summation. For p_c we use the value calculated in Equation (5.22):

$$P_M^c = \binom{15}{2}(p_c)^2(1 - p_c)^{13} = 1.94 \times 10^{-6} \tag{5.23}$$

By comparing the results of Equation (5.21) with (5.23), it is seen that the probability of message error has improved by a factor of 58 due to the error-correcting code used in this example.

5.4 LINEAR BLOCK CODES

Linear block codes (such as the one in Example 5.2) are a class of parity check codes that can be characterized by the (n, k) notation described earlier. The encoder transforms a block of k message digits (a message vector) into a longer block of n codeword digits (a code vector), constructed from a given alphabet of elements. When the alphabet consists of two elements (0 and 1), the code is a binary code comprised of binary digits (bits). Our discussion of linear block codes is restricted to binary codes, unless otherwise noted.

The k-bit messages form 2^k distinct message sequences referred to as *k-tuples* (sequences of k digits). The n-bit blocks can form as many as 2^n distinct sequences, referred to as *n-tuples*. The encoding procedure assigns to each of the 2^k message k-tuples *one* of the 2^n n-tuples. A block code represents a one-to-one assignment, whereby the 2^k message k-tuples are *uniquely* mapped into a new set of 2^k codeword n-tuples; the mapping can be accomplished via a look-up table. For *linear codes*, the mapping transformation is, of course, *linear*.

5.4.1 Vector Spaces

The set of all binary n-tuples, V_n, is called a *vector space* over the binary field of two elements (0 and 1). The binary field has two operations, addition and multiplication, such that the results of all operations are in the same set of two elements. The arithmetic operations of addition and multiplication are defined by the conventions of the algebraic field [4]. For example, in a binary field, the rules of addition and multiplication are as follows:

Addition	Multiplication
$0 \oplus 0 = 0$	$0 \cdot 0 = 0$
$0 \oplus 1 = 1$	$0 \cdot 1 = 0$
$1 \oplus 0 = 1$	$1 \cdot 0 = 0$
$1 \oplus 1 = 0$	$1 \cdot 1 = 1$

The addition operation, designated with the symbol $\oplus$, is the same modulo-2 operation described in Section 2.12.3.

5.4.2 Vector Subspaces

A subset S of the vector space V_n is called a *subspace* if the following two conditions are met:

1. The all-zeros vector is in S.
2. The sum of any two vectors in S is also in S (known as the *closure property*).

These properties are fundamental for the algebraic characterization of *linear block codes*. Suppose that $\mathbf{V}_i$ and $\mathbf{V}_j$ are two codewords (also called code vectors) in an

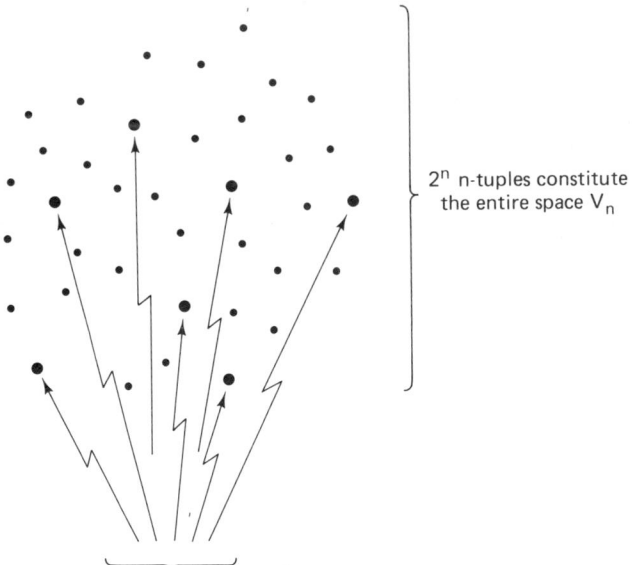

2^n n-tuples constitute the entire space V_n

2^k n-tuples constitute the subspace of codewords

Figure 5.13 Linear block-code structure.

(n, k) binary block code. The code is said to be *linear* if, and only if, $(\mathbf{V}_i \oplus \mathbf{V}_j)$ is also a code vector. A linear block code, then, is one in which vectors outside the subspace cannot be created by the addition of legitimate code vectors (members of the subspace).

For example, the vector space V_4 is totally populated by the following 2^4 = sixteen 4-tuples:

$$0000 \quad 0001 \quad 0010 \quad 0011 \quad 0100 \quad 0101 \quad 0110 \quad 0111$$

$$1000 \quad 1001 \quad 1010 \quad 1011 \quad 1100 \quad 1101 \quad 1110 \quad 1111$$

An example of a subset of V_4 that forms a subspace is

$$0000 \quad 0101 \quad 1010 \quad 1111$$

It is easy to verify that the addition of any two vectors in the subspace can only yield one of the other members of the subspace. A set of 2^k n-tuples is called a *linear block code* if, and only if, it is a subspace of the vector space V_n of all n-tuples. Figure 5.13 illustrates, with a simple geometric analogy, the structure behind linear block codes. We can imagine the vector space V_n comprised of 2^n n-tuples. Within this vector space there exists a subset of 2^k n-tuples comprising a subspace. These 2^k vectors or points, shown "sprinkled" among the more numerous 2^n points, represent the legitimate or allowable codeword assignments. A message is encoded into one of the 2^k allowable code vectors and then transmitted. Because of noise in the channel, a perturbed version of the code vector (one of the other 2^n vectors in the n-tuple space) may be received. If the perturbed vector is not too unlike (not too distant from) the valid code vector, the decoder can decode the message correctly. The basic goals in choosing a particular code, similar to the goals in selecting a set of modulation waveforms, can be stated in the context of Figure 5.13 as follows:

1. We want to strive for coding efficiency by packing the V_n space with as many code vectors as possible. This is tantamount to saying that we only want to expend a *small amount of redundancy* (excess bandwidth).
2. We want the code vectors to be as *far apart from one another* as possible, so that even if the vectors experience some corruption during transmission, they may still be correctly decoded, with a high probability.

5.4.3 A (6, 3) Linear Block Code Example

Examine the following coding assignment that describes a (6, 3) code. There are $2^k = 2^3 = 8$ message vectors, and therefore eight code vectors. There are $2^n = 2^6 = $ sixty-four 6-tuples in the V_6 vector space.

Message vector	Code vector
0 0 0	0 0 0 0 0 0
1 0 0	1 1 0 1 0 0
0 1 0	0 1 1 0 1 0
1 1 0	1 0 1 1 1 0
0 0 1	1 0 1 0 0 1
1 0 1	0 1 1 1 0 1
0 1 1	1 1 0 0 1 1
1 1 1	0 0 0 1 1 1

It is easy to check that the eight code vectors shown above form a subspace of V_6 (the all-zeros vector is present, and the sum of any two code vectors yields another code vector member of the subspace). Therefore, these code vectors represent a *linear block code*, as defined in Section 5.4.2.

5.4.4 Generator Matrix

If k is large, a *table look-up* implementation of the encoder becomes prohibitive. For a (127, 92) code there are 2^{92} or approximately 5×10^{27} code vectors. If the encoding procedure consists of a simple look-up table, imagine the size of the memory necessary to contain such a large number of code vectors. Fortunately, it is possible to reduce complexity by generating the required code vectors as needed, instead of storing them.

Since a set of code vectors that forms a linear block code is a k-dimensional subspace of the n-dimensional binary vector space ($k < n$), it is always possible to find a set of n-tuples, fewer than 2^k, that can generate all the 2^k member vectors of the subspace. The generating set of vectors is said to *span* the subspace. The smallest *linearly independent* set that spans the subspace is called a *basis* of the subspace, and the number of vectors in this basis set is the dimension of the subspace. Any basis set of k linearly independent n-tuples $\mathbf{V}_1, \mathbf{V}_2, \ldots, \mathbf{V}_k$ can be used to generate the required linear block code vectors, since each code vector is a linear combination of $\mathbf{V}_1, \mathbf{V}_2, \ldots, \mathbf{V}_k$. That is, each of the set of 2^k code vectors $\mathbf{U}$ can be described by

$$\mathbf{U} = m_1\mathbf{V}_1 + m_2\mathbf{V}_2 + \cdots + m_k\mathbf{V}_k$$

where $m_i = (0 \text{ or } 1)$ are the message digits and $i = 1, \ldots, k$.

In general, we can define a *generator matrix* by the following $k \times n$ array:

$$\mathbf{G} = \begin{bmatrix} \mathbf{V}_1 \\ \mathbf{V}_2 \\ \vdots \\ \mathbf{V}_k \end{bmatrix} = \begin{bmatrix} v_{11} & v_{12} & \cdots & v_{1n} \\ v_{21} & v_{22} & \cdots & v_{2n} \\ \vdots & & & \\ v_{k1} & v_{k2} & \cdots & v_{kn} \end{bmatrix} \tag{5.24}$$

Code vectors, by convention, are usually designated as row vectors. Thus, the message $\mathbf{m}$, a sequence of k message bits, is shown below as a row vector ($1 \times k$ matrix having one row and k columns).

$$\mathbf{m} = m_1, m_2, \ldots, m_k$$

The generation of the code vector, $\mathbf{U}$, is written in matrix notation as the product of $\mathbf{m}$ and $\mathbf{G}$, as follows:

$$\mathbf{U} = \mathbf{m}\mathbf{G} \qquad (5.25)$$

where, in general, the matrix multiplication $\mathbf{C} = \mathbf{AB}$ is performed in the usual way by using the rule

$$c_{ij} = \sum_{k}^{n} a_{ik}b_{kj} \qquad i = 1, \ldots, l \quad j = 1, \ldots, m$$

where $\mathbf{A}$ is an $l \times n$ matrix, $\mathbf{B}$ is an $n \times m$ matrix, and the result $\mathbf{C}$ is an $l \times m$ matrix. For the example introduced in the preceding section, we can fashion a generator matrix as follows:

$$\mathbf{G} = \begin{bmatrix} \mathbf{V}_1 \\ \mathbf{V}_2 \\ \mathbf{V}_3 \end{bmatrix} = \begin{bmatrix} 1 & 1 & 0 & 1 & 0 & 0 \\ 0 & 1 & 1 & 0 & 1 & 0 \\ 1 & 0 & 1 & 0 & 0 & 1 \end{bmatrix} \qquad (5.26)$$

where $\mathbf{V}_1$, $\mathbf{V}_2$, and $\mathbf{V}_3$ are three *linearly independent vectors* (a subset of the eight code vectors) that can generate all the code vectors. Notice that the sum of any two generating vectors does not yield any of the other generating vectors (opposite of closure). Let us generate the code vector for the message vector 1 1 0, using the generator matrix of Equation (5.26).

$$\mathbf{U} = \begin{bmatrix} 1 & 1 & 0 \end{bmatrix} \begin{bmatrix} \mathbf{V}_1 \\ \mathbf{V}_2 \\ \mathbf{V}_3 \end{bmatrix} = 1 \cdot \mathbf{V}_1 + 1 \cdot \mathbf{V}_2 + 0 \cdot \mathbf{V}_3$$

$$= 1\ 1\ 0\ 1\ 0\ 0 + 0\ 1\ 1\ 0\ 1\ 0 + 0\ 0\ 0\ 0\ 0\ 0$$

$$= 1\ 0\ 1\ 1\ 1\ 0 \quad \text{(code vector for the message vector 1 1 0)}$$

Thus the code vector corresponding to a message vector is a linear combination of the rows of $\mathbf{G}$. Since the code is totally defined by $\mathbf{G}$, the encoder need only store the k rows of G instead of the total 2^k vectors of the code. For this example notice that the generator array of dimension 3×6 replaces the original code vector array of dimension 8×6, representing a reduction in system complexity.

5.4.5 Systematic Linear Block Codes

A systematic (n, k) linear block code is a mapping from a k-dimensional message vector to an n-dimensional code vector in such a way that part of the sequence generated coincides with the k message digits. The remaining $(n - k)$ digits are parity digits. A systematic linear block code will have a generator matrix of the form

$$\mathbf{G} = \left[\begin{array}{c|c} \mathbf{P} & \mathbf{I}_k \end{array}\right]$$

$$= \begin{bmatrix} p_{11} & p_{12} & \cdots & p_{1,(n-k)} & 1 & 0 & \cdots & 0 \\ p_{21} & p_{22} & \cdots & p_{2,(n-k)} & 0 & 1 & \cdots & 0 \\ \vdots & & & & & & \vdots \\ p_{k1} & p_{k2} & \cdots & p_{k,(n-k)} & 0 & 0 & \cdots & 1 \end{bmatrix} \qquad (5.27)$$

where **P** is the parity array portion of the generator matrix, $p_{ij} = (0 \text{ or } 1)$, and $\mathbf{I}_k$ is the $k \times k$ identity matrix (ones on the main diagonal and zeros elsewhere). Notice that with this systematic generator, the encoding complexity is further reduced since it is not necessary to store the identity matrix portion of the array. By combining Equations (5.25) and (5.27), each code vector is expressed as follows:

$$u_1, u_2, \ldots, u_n = [m_1, m_2, \ldots, m_k]$$

$$\times \begin{bmatrix} p_{11} & p_{12} & \cdots & p_{1,(n-k)} & 1 & 0 & \cdots & 0 \\ p_{21} & p_{22} & \cdots & p_{2,(n-k)} & 0 & 1 & \cdots & 0 \\ \vdots & & & & & & \vdots \\ p_{k1} & p_{k2} & \cdots & p_{k,(n-k)} & 0 & 0 & \cdots & 1 \end{bmatrix}$$

where

$$u_i = m_1 p_{1i} + m_2 p_{2i} + \cdots + m_k p_{ki} \qquad \text{for } i = 1, \ldots, (n-k)$$

$$= m_{i-n+k} \qquad \text{for } i = (n-k+1), \ldots, n$$

Given the message k-tuple

$$\mathbf{m} = m_1, m_2, \ldots, m_k$$

and the general code vector n-tuple

$$\mathbf{U} = u_1, u_2, \ldots, u_n$$

the systematic code vector can be expressed as

$$\mathbf{U} = \underbrace{p_1, p_2, \ldots, p_{n-k}}_{\text{parity bits}}, \underbrace{m_1, m_2, \ldots, m_k}_{\text{message bits}} \qquad (5.28)$$

where

$$p_1 = m_1 p_{11} + m_2 p_{21} + \cdots + m_k p_{k1}$$

$$p_2 = m_1 p_{12} + m_2 p_{22} + \cdots + m_k p_{k2} \qquad (5.29)$$

$$p_{n-k} = m_1 p_{1,(n-k)} + m_2 p_{2,(n-k)} + \cdots + m_k p_{k,(n-k)}$$

Systematic code vectors are sometimes written so that the message bits occupy the left-hand portion of the code vector and the parity bits occupy the right-hand portion. This reordering has no effect on the error detection or error correction properties of the code, and will not be considered further.

For the (6, 3) code example in Section 5.4.3, the code vectors are described as follows:

$$U = [m_1, m_2, m_3] \underbrace{\begin{bmatrix} 1 & 1 & 0 & \vdots & 1 & 0 & 0 \\ 0 & 1 & 1 & \vdots & 0 & 1 & 0 \\ 1 & 0 & 1 & \vdots & 0 & 0 & 1 \end{bmatrix}}_{\substack{P \qquad I_3}} \qquad (5.30)$$

$$U = \underbrace{m_1 + m_3}_{u_1}, \underbrace{m_1 + m_2}_{u_2}, \underbrace{m_2 + m_3}_{u_3}, \underbrace{m_1}_{u_4}, \underbrace{m_2}_{u_5}, \underbrace{m_3}_{u_6}, \qquad (5.31)$$

Equation (5.31) gives us some insight regarding the structure of linear block codes. We see that the redundant digits are produced in a variety of ways. The first parity bit is the sum of the first and third message bits; the second parity bit is the sum of the first and second message bits, and the third parity bit is the sum of the second and third message bits. Intuition tells us that such structure, compared to single-parity checks or simple digit-repeat procedures, may provide greater ability to detect and correct errors.

5.4.6 Parity-Check Matrix

Let us define a matrix, H, called the *parity-check matrix*, that will enable us to decode the received vectors. For each $(k \times n)$ generator matrix, G, there exists an $(n - k) \times n$ matrix, H, such that the rows of G are orthogonal to the rows of H; that is $GH^T = 0$, where H^T is the *transpose* of H, and 0 is a $k \times (n - k)$ all-zeros matrix. H^T is an $n \times (n - k)$ matrix whose rows are the columns of H and whose columns are the rows of H. To fulfill the orthogonality requirements, the components of the H matrix are written

$$H = [I_{n-k} \vdots P^T] \qquad (5.32)$$

Hence, the H^T matrix is written

$$H^T = \begin{bmatrix} I_{n-k} \\ \text{------} \\ P \end{bmatrix} \qquad (5.33a)$$

$$= \begin{bmatrix} 1 & 0 & \cdots & 0 \\ 0 & 1 & \cdots & 0 \\ \vdots & & & \\ 0 & 0 & \cdots & 1 \\ p_{11} & p_{12} & \cdots & p_{1,(n-k)} \\ p_{21} & p_{22} & \cdots & p_{2,(n-k)} \\ \vdots & & & \\ p_{k1} & p_{k2} & \cdots & p_{k,(n-k)} \end{bmatrix} \qquad (5.33b)$$

It is easy to verify that the product $\mathbf{UH}^T$ of each code vector $\mathbf{U}$ generated by $\mathbf{G}$, and the $\mathbf{H}^T$ matrix, yields the zero vector:

$$\mathbf{UH}^T = p_1 + p_1, p_2 + p_2, \ldots, p_{n-k} + p_{n-k} = \mathbf{0}$$

where the parity bits $p_1, p_2, \ldots, p_{n-k}$ are defined in Equation (5.29). Thus once the *parity-check matrix*, $\mathbf{H}$, is constructed to fulfill the foregoing orthogonality requirements, we can use it to test whether a received vector is a valid member of the codeword set. $\mathbf{U}$ is a code vector generated by matrix $\mathbf{G}$ if, and only if, $\mathbf{UH}^T = \mathbf{0}$.

5.4.7 Syndrome Testing

Let $\mathbf{r} = r_1, r_2, \ldots, r_n$ be the received code vector (one of 2^n n-tuples) resulting from the transmission of $\mathbf{U} = u_1, u_2, \ldots, u_n$ (one of 2^k n-tuples). We can therefore describe $\mathbf{r}$ as

$$\mathbf{r} = \mathbf{U} + \mathbf{e} \tag{5.34}$$

where $\mathbf{e} = e_1, e_2, \ldots, e_n$ is the error vector or error pattern introduced by the channel. There are a total of $2^n - 1$ potential nonzero error patterns in the space of 2^n n-tuples. The *syndrome* of $\mathbf{r}$ is defined as

$$\mathbf{S} = \mathbf{rH}^T \tag{5.35}$$

The syndrome is the result of a parity check performed on $\mathbf{r}$ to determine whether $\mathbf{r}$ is a valid member of the codeword set. If, in fact, $\mathbf{r}$ is a member, the syndrome $\mathbf{S}$ has a value $\mathbf{0}$. If $\mathbf{r}$ contains detectable errors, the syndrome has some nonzero value. If $\mathbf{r}$ contains correctable errors, the syndrome (like the symptom of an illness) has some nonzero value which can earmark the particular error pattern. The decoder, depending upon whether it has been implemented to perform FEC or ARQ, will then take actions to locate the errors and correct them (FEC), or will request a retransmission (ARQ). Combining Equations (5.34) and (5.35), the syndrome of $\mathbf{r}$ is seen to be

$$\mathbf{S} = (\mathbf{U} + \mathbf{e})\mathbf{H}^T$$

$$= \mathbf{UH}^T + \mathbf{eH}^T \tag{5.36}$$

However, $\mathbf{UH}^T = \mathbf{0}$ for all members of the codeword set. Therefore,

$$\mathbf{S} = \mathbf{eH}^T \tag{5.37}$$

The foregoing development, starting with Equation (5.34) and terminating with Equation (5.37), is evidence that the syndrome test, whether performed on either a corrupted code vector or on the error pattern that caused it, yields the same syndrome. An important property of linear block codes, fundamental to the decoding process, is that the mapping between correctable error patterns and syndromes is one to one.

It is interesting to note the following two required properties of the parity-check matrix.

1. No column of **H** can be all zeros, or else an error in the corresponding code vector position would not affect the syndrome and would be undetectable.
2. All columns of **H** must be unique. If two columns of **H** were identical, errors in these two corresponding code vector positions would be indistinguishable.

Example 5.3   Syndrome Test

Suppose that code vector **U** = 1 0 1 1 1 0 from the example in Section 5.4.3 is transmitted and the vector **r** = 0 0 1 1 1 0 is received; that is, the leftmost bit is received in error. Find the syndrome vector value $\mathbf{S} = \mathbf{rH}^T$ and verify that it is equal to $\mathbf{eH}^T$.

Solution

$$\mathbf{S} = \mathbf{rH}^T$$

$$= [0 \ \ 0 \ \ 1 \ \ 1 \ \ 1 \ \ 0] \begin{bmatrix} 1 & 0 & 0 \\ 0 & 1 & 0 \\ 0 & 0 & 1 \\ 1 & 1 & 0 \\ 0 & 1 & 1 \\ 1 & 0 & 1 \end{bmatrix}$$

$$= [1, \ \ 1 + 1, \ \ 1 + 1] = [1 \ \ 0 \ \ 0] \quad \text{(syndrome of corrupted code vector)}$$

Next, we verify that the syndrome of the corrupted code vector is the same as the syndrome of the error pattern that caused the error.

$$\mathbf{S} = \mathbf{eH}^T = [1 \ \ 0 \ \ 0 \ \ 0 \ \ 0 \ \ 0]\mathbf{H}^T = [1 \ \ 0 \ \ 0] \quad \text{(syndrome of error pattern)}$$

5.4.8 Error Correction

We have detected a single error and have shown that the syndrome test performed on either the corrupted code vector, or on the error pattern that caused it, yields the same syndrome. This should be a clue that we not only can detect the error, but since there is a one-to-one correspondence between correctable error patterns and syndromes, we can correct such error patterns. Let us arrange the 2^n n-tuples that represent possible received vectors in an array, called the *standard array*, such that the first row contains all the code vectors, starting with the all-zeros vector, and the first column contains all the correctable error patterns. Recall from the basic properties of linear codes (see Section 5.4.2) that the all-zeros vector must be a member of the codeword set. Each row, called a *coset*, consists of an error pattern in the first column, called the *coset leader*, followed by the code vectors perturbed by that error pattern. The standard array format for an (n, k) code is as follows:

$$
\begin{array}{ccccccc}
\mathbf{U}_1 & \mathbf{U}_2 & \cdots & \mathbf{U}_i & \cdots & \mathbf{U}_{2^k} \\
\mathbf{e}_2 & \mathbf{U}_2 + \mathbf{e}_2 & \cdots & \mathbf{U}_i + \mathbf{e}_2 & \cdots & \mathbf{U}_{2^k} + \mathbf{e}_2 \\
\mathbf{e}_3 & \mathbf{U}_2 + \mathbf{e}_3 & \cdots & \mathbf{U}_i + \mathbf{e}_3 & \cdots & \mathbf{U}_{2^k} + \mathbf{e}_3 \\
\vdots & \vdots & & \vdots & & \vdots \\
\mathbf{e}_j & \mathbf{U}_2 + \mathbf{e}_j & \cdots & \mathbf{U}_i + \mathbf{e}_j & \cdots & \mathbf{U}_{2^k} + \mathbf{e}_j \\
\vdots & \vdots & & \vdots & & \\
\mathbf{e}_{2^{n-k}} & \mathbf{U}_2 + \mathbf{e}_{2^{n-k}} & \cdots & \mathbf{U}_i + \mathbf{e}_{2^{n-k}} & \cdots & \mathbf{U}_{2^k} + \mathbf{e}_{2^{n-k}}
\end{array}
\tag{5.38}
$$

The array contains all 2^n n-tuples in the space V_n (each n-tuple appears in *only one* location). Each coset consists of 2^k n-tuples. Therefore, there are $(2^n/2^k) = 2^{n-k}$ cosets. Suppose that a code vector $\mathbf{U}_i$ is transmitted over a noisy channel. If the error pattern caused by the channel is a coset leader, the received vector will be decoded correctly into the transmitted code vector $\mathbf{U}_i$. If the error pattern is not a coset leader, an erroneous decoding will result.

5.4.8.1 The Syndrome of a Coset

If $\mathbf{e}_j$ is the coset leader or error pattern of the jth coset, then $\mathbf{U}_i + \mathbf{e}_j$ is an n-tuple in this coset. The syndrome of this n-tuple can be written

$$\mathbf{S} = (\mathbf{U}_i + \mathbf{e}_j)\mathbf{H}^T = \mathbf{U}_i\mathbf{H}^T + \mathbf{e}_j\mathbf{H}^T$$

Since $\mathbf{U}_i$ is a code vector, $\mathbf{U}_i\mathbf{H}^T = \mathbf{0}$, and we can write, as in Equation (5.37)

$$\mathbf{S} = (\mathbf{U}_i + \mathbf{e}_j)\mathbf{H}^T = \mathbf{e}_j\mathbf{H}^T \qquad (5.39)$$

From Equation (5.39) it is clear that all members of a coset have the *same syndrome*, and in fact, the syndrome is used to estimate the error pattern. The syndrome for every coset is different.

5.4.8.2 Error Correction Decoding

The procedure for error correction decoding proceeds as follows:

1. Calculate the syndrome of $\mathbf{r}$ using $\mathbf{S} = \mathbf{r}\mathbf{H}^T$.
2. Locate the coset leader (error pattern), $\mathbf{e}_j$, whose syndrome equals $\mathbf{r}\mathbf{H}^T$.
3. This error pattern is assumed to be the corruption caused by the channel.
4. The corrected received vector, or code vector, is identified as $\mathbf{U} = \mathbf{r} + \mathbf{e}_j$. We can say that we retrieve the valid code vector by subtracting out the identified error; in modulo-2 arithmetic the operation of subtraction is identical to that of addition.

5.4.8.3 Locating the Error Pattern

Returning to the example of Section 5.4.3, we arrange the $2^6 =$ sixty-four 6-tuples in a standard array as shown in Figure 5.14. The valid code vectors are the eight vectors in the first row, and the *correctable error patterns* are the eight *coset leaders* in the first column. Notice that all 1-bit error patterns are correctable. Also notice that after exhausting all 1-bit error patterns, there remains some error-correcting capability since we have not yet accounted for all sixty-four 6-tuples. There is one unassigned coset leader; therefore, there remains the capability of correcting one additional error pattern. We have the flexibility of choosing this error pattern to be any of the n-tuples in the remaining coset. In Figure 5.14 this final correctable error pattern is chosen, somewhat arbitrarily, to be the 2-bit error pattern 0 1 0 0 0 1. Decoding will be correct if, and only if, the error pattern caused by the channel is one of the coset leaders.

We now determine the syndrome corresponding to each of the correctable

000000	110100	011010	101110	101001	011101	110011	000111
000001	110101	011011	101111	101000	011100	110010	000110
000010	110110	011000	101100	101011	011111	110001	000101
000100	110000	011110	101010	101101	011001	110111	000011
001000	111100	010010	100110	100001	010101	111011	001111
010000	100100	001010	111110	111001	001101	100011	010111
100000	010100	111010	001110	001001	111101	010011	100111
010001	100101	001011	111111	111000	001100	100010	010110

Figure 5.14 Example of a standard array for a (6, 3) code.

error sequences by computing $\mathbf{e}_j \mathbf{H}^T$ for each coset leader, as follows:

$$
\mathbf{S} = \mathbf{e}_j \begin{bmatrix} 1 & 0 & 0 \\ 0 & 1 & 0 \\ 0 & 0 & 1 \\ 1 & 1 & 0 \\ 0 & 1 & 1 \\ 1 & 0 & 1 \end{bmatrix}
$$

The results are listed in Table 5.1. Since each syndrome in the table is unique, the decoder can identify the error pattern $\mathbf{e}$ to which it corresponds.

TABLE 5.1 Syndrome Look-Up Table

Error pattern	Syndrome
0 0 0 0 0 0	0 0 0
0 0 0 0 0 1	1 0 1
0 0 0 0 1 0	0 1 1
0 0 0 1 0 0	1 1 0
0 0 1 0 0 0	0 0 1
0 1 0 0 0 0	0 1 0
1 0 0 0 0 0	1 0 0
0 1 0 0 0 1	1 1 1

5.4.8.4 Error Correction Example

As outlined in Section 5.4.8.2, we receive the vector $\mathbf{r}$ and calculate its syndrome using $\mathbf{S} = \mathbf{r}\mathbf{H}^T$. We then use the syndrome look-up table (Table 5.1), developed in the preceding section, to find the corresponding error pattern. This error pattern is an estimate of the error, and we denote it $\hat{\mathbf{e}}$. The decoder then adds $\hat{\mathbf{e}}$ to $\mathbf{r}$ to obtain an estimate of the transmitted code vector $\hat{\mathbf{U}}$.

$$
\hat{\mathbf{U}} = \mathbf{r} + \hat{\mathbf{e}} = (\mathbf{U} + \mathbf{e}) + \hat{\mathbf{e}} = \mathbf{U} + (\mathbf{e} + \hat{\mathbf{e}}) \tag{5.40}
$$

If the estimated error pattern is the same as the actual error pattern, that is, if $\hat{e}$ = e, then the estimate $\hat{U}$ is equal to the transmitted code vector U. On the other hand, if the error estimate is incorrect, the decoder will estimate a code vector that was not transmitted, and we have an *undetectable decoding error*.

Example 5.4   Error Correction

Assume that code vector U = 1 0 1 1 1 0, from the Section 5.4.3 example, is transmitted, and the vector r = 0 0 1 1 1 0 is received. Show how a decoder, using the Table 5.1 syndrome look-up table, can correct the error.

Solution

The syndrome of r is computed:

$$S = [0 \ \ 0 \ \ 1 \ \ 1 \ \ 1 \ \ 0]H^T = [1 \ \ 0 \ \ 0]$$

Using Table 5.1, the error pattern corresponding to the syndrome above is estimated to be

$$\hat{e} = 1 \ 0 \ 0 \ 0 \ 0 \ 0$$

The corrected vector is then estimated by

$$\hat{U} = r + \hat{e}$$
$$= 0 \ 0 \ 1 \ 1 \ 1 \ 0 + 1 \ 0 \ 0 \ 0 \ 0 \ 0$$
$$= 1 \ 0 \ 1 \ 1 \ 1 \ 0$$

Since the estimated error pattern is the actual error pattern in this example, the error correction procedure yields $\hat{U} = U$.

5.5 CODING STRENGTH

5.5.1 Weight and Distance of Binary Vectors

It should be clear that not all error patterns can be correctly decoded. The error correction capability of a code will be investigated by first defining its structure. The *Hamming weight*, $w(U)$, of a vector U is defined to be the number of nonzero elements in U. For a binary vector this is equivalent to the number of ones in the vector. For example, if U = 1 0 0 1 0 1 1 0 1, then $w(U)$ = 5. The *Hamming distance* between two code vectors U and V, denoted $d(U, V)$, is defined to be the number of elements in which they differ: for example,

$$U = 1 \ 0 \ 0 \ 1 \ 0 \ 1 \ 1 \ 0 \ 1$$
$$V = 0 \ 1 \ 1 \ 1 \ 1 \ 0 \ 1 \ 0 \ 0$$
$$d(U, V) = 6$$

By the properties of modulo-2 addition, we note that the sum of two binary vectors is another vector whose binary ones are located in those positions in which the two vectors differ: for example,

$$U + V = 1\ 1\ 1\ 0\ 1\ 1\ 0\ 0\ 1$$

Thus we observe that the Hamming distance between two code vectors is equal to the Hamming weight of their sum: that is, $d(U, V) = w(U + V)$. Also, we see that the Hamming weight of a code vector is equal to its Hamming distance from the all-zeros vector.

5.5.2 Minimum Distance of a Linear Code

Consider the set of distances between all pairs of code vectors in the space V_n. The smallest member of the set is the *minimum distance* of the code and is denoted d_{min}. Why do you suppose we have an interest in the minimum distance; why not the maximum distance? The minimum distance, like the weakest link in a chain, gives us a measure of the code's minimum capability and therefore characterizes the code's strength.

   As discussed earlier, the sum of any two code vectors yields another code vector member of the subspace. This property of linear codes is stated simply as: If U and V are code vectors, then $W = U + V$ must also be a code vector. Hence the distance between two code vectors is equal to the weight of a third code vector; that is, $d(U, V) = w(U + V) = w(W)$. Thus the minimum distance of a linear code can be ascertained without examining the distance between all combinations of code vector pairs. We only need to examine the weight of each code vector (excluding the all-zeros vector) in the subspace; the minimum weight corresponds to the minimum distance, d_{min}. Equivalently, d_{min} corresponds to the smallest of the set of distances between the all-zeros code vector and all the other code vectors.

5.5.3 Error Detection and Correction

The task of the decoder, having received the vector r, is to estimate the transmitted code vector U_i. The optimal decoder strategy can be expressed in terms of the *maximum likelihood* algorithm (see Appendix B) as follows: Decide in favor of U_i if

$$P(r|U_i) = \max_{\text{over all } U_j} P(r|U_j) \tag{5.41}$$

Since for the binary symmetric channel (BSC), the likelihood of U_i with respect to r is inversely proportional to the distance between r and U_i, we can write: Decide in favor of U_i if

$$d(r, U_i) = \min_{\text{over all } U_j} d(r, U_j) \tag{5.42}$$

In other words, the decoder determines the distance between **r** and each of the possible transmitted code vectors $\mathbf{U}_i$, and selects as most likely a $\mathbf{U}_i$ for which

$$d(\mathbf{r}, \mathbf{U}_i) \le d(\mathbf{r}, \mathbf{U}_j) \qquad \text{for } i, j = 1, \ldots, M \quad \text{and} \quad i \ne j \qquad (5.43)$$

where $M = 2^k$ is the size of the code vector set. If the minimum is not unique, the choice between minimum distance codewords is arbitrary. Distance metrics are treated further in Chapter 6.

In Figure 5.15 the distance between two code vectors **U** and **V** is shown using a number line calibrated in *Hamming distance*. Each black dot represents a corrupted code vector. Figure 5.15a illustrates the reception of vector $\mathbf{r}_1$, which is distance 1 from **U** and distance 4 from **V**. An error-correcting decoder, following the maximum likelihood strategy, will select **U** upon receiving $\mathbf{r}_1$. If $\mathbf{r}_1$ had been the result of a 1-bit corruption to the transmitted code vector **U**, the decoder has successfully corrected the error. But if $\mathbf{r}_1$ had been the result of a 4-bit corruption to the transmitted code vector **V**, the result is a decoding error. Similarly, a double error in transmission of **U** might result in the received vector $\mathbf{r}_2$, which is distance 2 from **U** and distance 3 from **V**, as shown in Figure 5.15b. Here, too, the decoder will select **U** upon receiving $\mathbf{r}_2$. A triple error in transmission of **U** might result in a received vector $\mathbf{r}_3$ which is distance 3 from **U** and distance 2 from **V**, as shown in Figure 5.15c. Here the decoder will select **V** upon receiving $\mathbf{r}_3$, and will have made an error in decoding. From Figure 5.15 it should be clear that if error

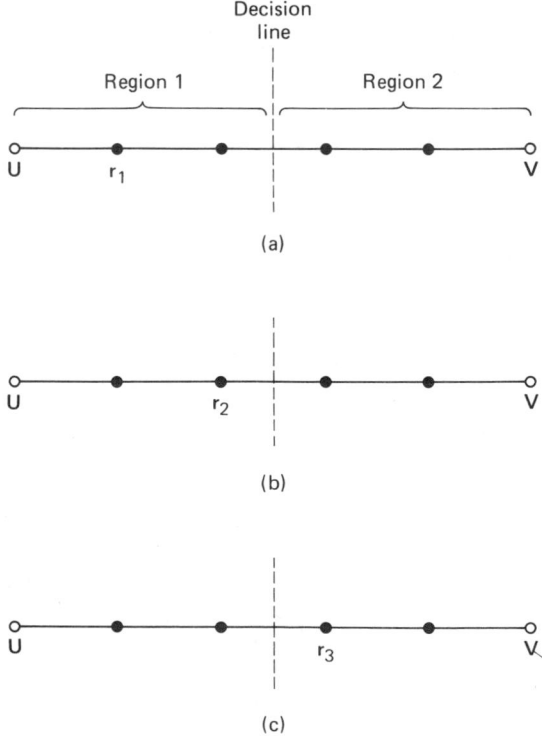

(a)

(b)

(c)

Figure 5.15 Error correction and detection strength. (a) Received vector $\mathbf{r}_1$. (b) Received vector $\mathbf{r}_2$. (c) Received vector $\mathbf{r}_3$.

Channel Coding: Part 1   Chap. 5

detection and not correction is the task, a corrupted vector, characterized by a black dot and representing a 1-bit, 2-bit, 3-bit, or 4-bit error, can be detected. However, five errors in transmission might result in code vector **V** being received when code vector **U** was actually transmitted; such an error would be *undetectable*.

From Figure 5.15 we can see that the error-detecting and error-correcting capabilities of a code are related to the *minimum distance* between code vectors. The decision line in the figure serves the same purpose in the process of decoding as it does in demodulation, to define the decision regions. In the Figure 5.15 example, the decision criterion of choosing **U** if **r** falls in region 1, and choosing **V** if **r** falls in region 2, illustrates that such a code, with $d_{min} = 5$, can correct two errors. In general, the *error-correcting capability*, t, of a code is defined as the maximum number of guaranteed correctable errors per codeword, and is written [4]

$$ t = \left\lfloor \frac{d_{min} - 1}{2} \right\rfloor $$

(5.44)

where $\lfloor x \rfloor$ means the largest integer not to exceed x. Often, a code that corrects all possible sequences of t or fewer errors can also correct certain sequences of $t + 1$ errors. This can be seen in Figure 5.14. In this example $d_{min} = 3$, and thus from Equation (5.44), we can see that *all* $t = 1$ bit-error patterns are correctable. Also, *a single* $t + 1$ or 2-bit error pattern is correctable. In general, a t-error-correcting (n, k) linear code is capable of correcting a total of 2^{n-k} error patterns. If a t-error-correcting block code is used strictly for error correction on a binary symmetric channel (BSC) with transition probability p, the probability that the decoder commits an erroneous decoding, and that the n-bit block is in error, can be calculated by using Equation (5.18) as an upper bound:

$$ P_M \leq \sum_{j=t+1}^{n} \binom{n}{j} p^j (1 - p)^{n-j} $$

(5.45)

The bound becomes an equality when the decoder corrects all combinations of errors up to and including t errors, but no combinations of errors greater than t. Such decoders are called *bounded distance decoders*. The decoded bit-error probability depends on the particular code and decoder. It can be expressed [5] by the following approximation:

$$ P_B \simeq \frac{1}{n} \sum_{j=t+1}^{n} j \binom{n}{j} p^j (1 - p)^{n-j} $$

(5.46)

A code can be used to detect errors prior to, or instead of, correcting them. It should be clear from Figure 5.15 that any received vector characterized by a black dot (a corrupted code vector) can be identified as an error. Therefore, the error-detecting capability, e, is defined in terms of d_{min} as

$$ e = d_{min} - 1 $$

(5.47)

A block code with minimum distance d_{min} guarantees that all error patterns of

$d_{min} - 1$ or fewer errors can be detected. Such a code is also capable of detecting a large fraction of error patterns with d_{min} or more errors. In fact, an (n, k) code is capable of detecting $2^n - 2^k$ error patterns of length n. The reasoning is as follows. There are a total of $2^n - 1$ possible nonzero error patterns in the space of 2^n n-tuples. Even the bit pattern of a valid codeword represents a potential error pattern. Thus there are $2^k - 1$ error patterns that are identical to the $2^k - 1$ nonzero codewords. If any of these $2^k - 1$ error patterns occurs, it alters the transmitted codeword $\mathbf{U}_i$ into another codeword $\mathbf{U}_j$. Thus $\mathbf{U}_j$ will be received and its syndrome is zero. The decoder accepts $\mathbf{U}_j$ as the transmitted codeword and thereby commits an incorrect decoding. Therefore, there are $2^k - 1$ undetectable error patterns. If the error pattern is not identical to one of the 2^k codewords, the syndrome test on the received vector $\mathbf{r}$ yields a nonzero syndrome, and the error is detected. Therefore, there are exactly $2^n - 2^k$ detectable error patterns. For large n, where $2^k \ll 2^n$, only a small fraction of error patterns are undetected.

5.5.3.1 Code Vector Weight Distribution

Let A_j be the number of code vectors of weight j within an (n, k) linear code. The numbers $A_0, A_1, \ldots, A_n$ are called the *weight distribution* of the code. If the code is used only for error detection, on a BSC, the probability, P_{nd}, that the decoder does not detect an error can be computed from the weight distribution of the code [5] as follows:

$$P_{nd} = \sum_{j=1}^{n} A_j p^j (1 - p)^{n-j} \tag{5.48}$$

where p is the transition probability of the BSC. If the minimum distance of the code is d_{min}, the values of A_1 to $A_{d_{min}-1}$ are zero.

Example 5.5 Probability of an Undetected Error in an Error Detecting Code

Consider that the $(6, 3)$ code, given in Section 5.4.3, is used only for error detection. Calculate the probability of an undetected error if the channel is a BSC and the transition probability is 10^{-2}.

Solution

The weight distribution of this code is $A_0 = 1$, $A_1 = A_2 = 0$, $A_3 = 4$, $A_4 = 3$, $A_5 = 0$, $A_6 = 0$. Therefore, we can write, using Equation (5.48),

$$P_{nd} = 4p^3(1 - p)^3 + 3p^4(1 - p)^2$$

For $p = 10^{-2}$, the probability of an undetected error is 3.9×10^{-6}.

5.5.3.2 Simultaneous Error Correction and Detection

It is possible to trade correction capability from the maximum guaranteed (t), where t is defined in Equation (5.44), for the ability to simultaneously detect a class of errors. A code can be used for the simultaneous correction of α errors and detection of β errors where $\beta \geq \alpha$, provided that its minimum distance is [4]

$$d_{min} \geq \alpha + \beta + 1 \tag{5.49}$$

When t or fewer errors occur, the code is capable of detecting and correcting them. When more than t but fewer than $e + 1$ errors occur, where e is defined in Equation (5.47), the code is capable of detecting their presence but correcting only a subset of them. For example, a code with $d_{min} = 7$ can be used to simultaneously detect and correct in any one of the following ways:

Detect (β)	Correct (α)
3	3
4	2
5	1
6	0

Note that correction implies prior detection. For the above example, when there are three errors, all of them can be detected and corrected. When there are five errors, all of them can be detected but only a subset of them (one) can be corrected.

5.5.4 Visualization of a 6-Tuple Space

Figure 5.16 is a visualization of the eight codewords from the example of Section 5.4.3. The codewords are generated from linear combinations of the three independent 6-tuples in Equation (5.26); the codewords form a three-dimensional subspace. The figure shows such a subspace completely occupied by the eight codewords (large black circles); the coordinates of the subspace have purposely been drawn to emphasize their nonorthogonality. Figure 5.16 is an attempt to illustrate the entire space, containing sixty-four 6-tuples, even though there is no precise way to draw or construct such a model. Spherical layers or shells are shown around each codeword. Each of the nonintersecting inner layers is a Hamming distance of 1 from its associated codeword; each outer layer is a Hamming distance of 2 from its codeword. Larger distances are not useful in this example. For each codeword, the two layers shown are occupied by perturbed codewords. There are six such points on each inner sphere (a total of 48 points), representing the six possible 1-bit error-perturbed vectors associated with each codeword. These 1-bit perturbed codewords are distinct in the sense that they can best be associated with only one codeword, and therefore can be corrected. As is seen from the standard array of Figure 5.14, there is also one 2-bit error pattern that can be corrected. There is a total of $\binom{6}{2} = 15$ different 2-bit error patterns that can be inflicted on each codeword, but only one of them, in our example the 0 1 0 0 0 1 error pattern, can be corrected. The other fourteen 2-bit error patterns yield vectors that cannot be uniquely identified with just one codeword; these noncorrectable error patterns yield vectors that are equivalent to the error-perturbed vectors of two or more codewords. In the figure, all correctable (fifty-six) 1- and 2-bit error-perturbed codewords are shown as small black circles. Perturbed codewords that cannot be corrected are shown as small clear circles.

Figure 5.16 is useful for visualizing the properties of a class of codes known as *perfect codes*. A t-error-correcting code is called a perfect code if its standard array has all the error patterns of t or fewer errors and no others as coset leaders.

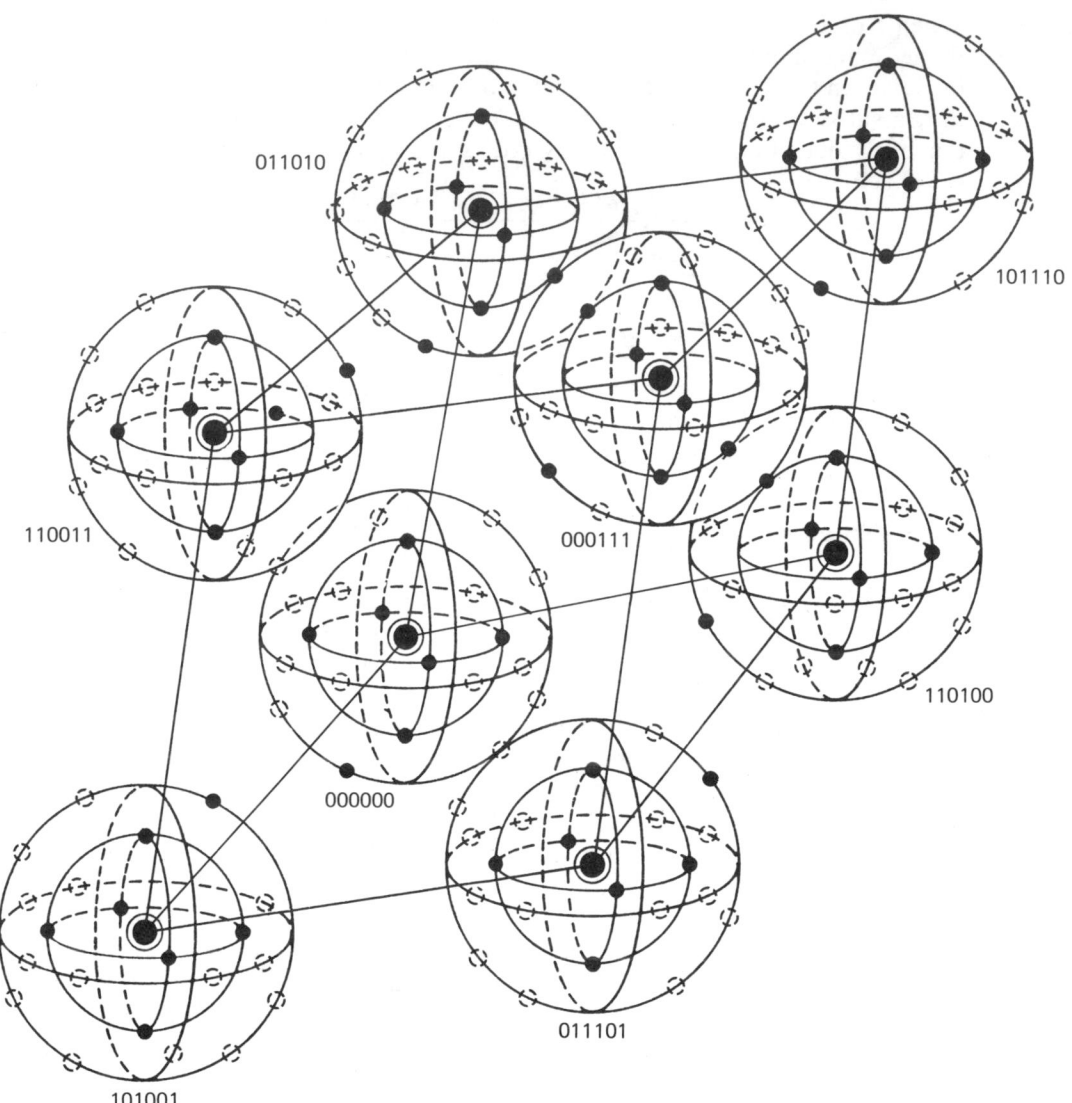

Figure 5.16 Example of eight codewords in a 6-tuple space.

In terms of Figure 5.16, a t-error-correcting perfect code is one that can, with maximum likelihood decoding, correct all perturbed code vectors occupying a shell at Hamming distance t or less from its originating codeword, and cannot correct any perturbed vectors occupying shells at distances greater than t.

Figure 5.16 is also useful for understanding the basic goal in the search for good codes. We would like for the space to be filled with as many codewords as possible (efficient utilization of the added redundancy), and we would also like these codewords to be as far away from one another as possible. Obviously, these goals conflict.

5.5.5 Erasure Correction

A receiver may be designed to declare a symbol *erased* when it is received ambiguously or when the receiver recognizes the presence of interference or a transient malfunction. Such a channel has an input alphabet of size Q and an output alphabet of size $Q + 1$; the extra output symbol is called an *erasure flag*, or simply an *erasure*. When a demodulator makes a symbol error, two parameters are needed to correct that error, its *location* and its *correct* symbol value. In the case of binary symbols, this reduces to needing only the error location. However, if the demodulator declares a symbol *erased*, although the correct symbol value is not known, the symbol location *is* known, and for this reason, the decoding of erased codewords can be simpler than error correcting. An error control code can be used to correct erasures or to correct errors and erasures simultaneously. If the code has minimum distance d_{min}, any pattern of ρ or fewer erasures can be corrected if [6]

$$d_{min} \geq \rho + 1 \tag{5.50}$$

Assume for the moment that no errors occur outside the erasure positions. The advantage of correcting by means of erasures is expressed quantitatively as follows: If a code has a minimum distance d_{min}, then from Equation (5.50), $d_{min} - 1$ erasures can be reconstituted. Since the number of errors that can be corrected without erasure information is $(d_{min} - 1)/2$ at most, from Equation (5.44), the advantage of correcting by means of erasures is clear. Further, any pattern of α errors and γ erasures can be corrected simultaneously if [6]

$$d_{min} \geq 2\alpha + \gamma + 1 \tag{5.51}$$

Simultaneous erasure correction and error correction can be accomplished in the following way. First, the γ-erased positions are replaced with zeros and the resulting codeword is decoded normally. Next, the γ-erased positions are replaced with ones, and the decoding operation is repeated on this version of the codeword. Of the two codewords obtained (one with erasures replaced by zeros, and the other with erasures replaced by ones) the one corresponding to the smallest number of errors corrected outside the γ-erased positions is selected. This technique will always result in correct decoding if Equation (5.51) is satisfied.

Example 5.6 Erasure Correction

Consider the codeword set presented in Section 5.4.3:

000000   110100   011010   101110   101001   011101   110011   000111

Suppose that the codeword 110011 was transmitted and that the two leftmost digits were declared by the receiver to be erasures. Verify that the received flawed sequence xx0011 can be corrected.

Solution

Since $d_{min} = \rho + 1 = 3$, the code can correct as many as $\rho = 2$ erasures. This is easily verified above or with Figure 5.14 by comparing the rightmost four digits of xx0011 with each of the allowable codewords. The codeword that was actually transmitted is closest in Hamming distance to the flawed sequence.

5.6 CYCLIC CODES

Binary cyclic codes are an important subclass of linear block codes. The codes are easily implemented with feedback shift registers; the syndrome calculation is easily accomplished with similar feedback shift registers; and the underlying algebraic structure of a cyclic code lends itself to efficient decoding methods. An (n, k) linear code is called a *cyclic code* if it can be described by the following property. If the n-tuple $\mathbf{U} = (u_0, u_1, u_2, \ldots, u_{n-1})$ is a code vector in the subspace S, then $\mathbf{U}^{(1)} = (u_{n-1}, u_0, u_1, \ldots, u_{n-2})$ obtained by an end-around shift, is also a code vector in S. Or in general, $\mathbf{U}^{(i)} = (u_{n-i}, u_{n-i+1}, \ldots, u_{n-1}, u_0, u_1, \ldots, u_{n-i-1})$, obtained by i end-around or cyclic shifts, is also a code vector in S.

The components of a code vector $\mathbf{U} = (u_0, u_1, u_2, \ldots, u_{n-1})$ can be treated as the coefficients of a polynomial $\mathbf{U}(X)$ as follows:

$$\mathbf{U}(X) = u_0 + u_1 X + u_2 X^2 + \cdots + u_{n-1}X^{n-1} \tag{5.52}$$

The polynomial function $\mathbf{U}(X)$ can be thought of as a "placeholder" for the digits of the code vector $\mathbf{U}$; that is, an n-tuple vector is described by a polynomial of degree $n - 1$ or less. The presence or absence of each term in the polynomial indicates the presence of a 1 or 0 in the corresponding location of the n-tuple. If the u_{n-1} component is nonzero, the polynomial is of degree $n - 1$. The usefulness of this polynomial description of a codeword will become clear as we discuss the algebraic structure of the cyclic codes.

5.6.1 Algebraic Structure of Cyclic Codes

Expressing the code vectors in polynomial form, the cyclic nature of the code manifests itself in the following way. If $\mathbf{U}(X)$ is an $(n - 1)$-degree codeword polynomial, then $\mathbf{U}^{(i)}(X)$, the remainder resulting from dividing $X^i\mathbf{U}(X)$ by $X^n + 1$, is also a codeword; that is,

$$\frac{X^i\mathbf{U}(X)}{X^n + 1} = \mathbf{q}(X) + \frac{\mathbf{U}^{(i)}(X)}{X^n + 1} \tag{5.53}$$

or, multiplying through by $X^n + 1$,

$$X^i\mathbf{U}(X) = \mathbf{q}(X)(X^n + 1) \underbrace{+ \mathbf{U}^{(i)}(X)}_{\text{remainder}} \tag{5.54}$$

which can also be described in terms of modulo arithmetic as follows:

$$U^{(i)}(X) = X^i U(X) \text{ modulo } (X^n + 1) \qquad (5.55)$$

where x modulo y is defined as the remainder obtained from dividing x by y. Let us demonstrate the validity of Equation (5.55) for the case of $i = 1$.

$$U(X) = u_0 + u_1 X + u_2 X^2 + \cdots + u_{n-2} X^{n-2} + u_{n-1} X^{n-1}$$

$$XU(X) = u_0 X + u_1 X^2 + u_2 X^3 + \cdots + u_{n-2} X^{n-1} + u_{n-1} X^n$$

We now add and subtract u_{n-1}, or since we are using modulo-2 arithmetic, we add u_{n-1} twice, as follows:

$$XU(X) = \underbrace{u_{n-1} + u_0 X + u_1 X^2 + u_2 X^3 + \cdots + u_{n-2} X^{n-1}}_{U^{(1)}(X)} + u_{n-1} X^n + u_{n-1}$$

$$= U^{(1)}(X) + u_{n-1}(X^n + 1)$$

Since $U^{(1)}(X)$ is of degree $n - 1$, it cannot be divided by $X^n + 1$. Thus we can write from Equation (5.53)

$$U^{(1)}(X) = XU(X) \text{ modulo } (X^n + 1)$$

By extension we can write

$$U^{(i)}(X) = X^i U(X) \text{ modulo } (X^n + 1) \qquad (5.56)$$

Example 5.7   Cyclic Shift of a Code Vector

Let $\mathbf{U} = 1 \; 1 \; 0 \; 1$, for $n = 4$. Express the code vector in polynomial form, and using Equation (5.54), solve for the third end-around shift of the code vector.

Solution

$$U(X) = 1 + X + X^3 \qquad \text{(polynomial is written low order to high order)}$$

$$X^i U(X) = X^3 + X^4 + X^6 \qquad \text{where } i = 3$$

Divide $X^3 U(X)$ by $X^4 + 1$, and solve for the remainder using polynomial division.

$$
\begin{array}{r}
X^2 + 1 \\
X^4 + 1 \overline{)X^6 + X^4 + X^3} \\
\underline{X^6 + X^2} \\
X^4 + X^3 + X^2 \\
\underline{X^4 + 1} \\
X^3 + X^2 + 1 \qquad \text{remainder } U^{(3)}(X)
\end{array}
$$

Writing the remainder low order to high order: $1 + X^2 + X^3$, the codeword $\mathbf{U}^{(3)} = 1 \; 0 \; 1 \; 1$ is three cyclic shifts of $\mathbf{U} = 1 \; 1 \; 0 \; 1$. Remember that for binary codes, the addition operation is performed modulo-2, so that $+1 = -1$, and we consequently do not show any minus signs in the computation.

5.6.2 Binary Cyclic Code Properties

We can generate a cyclic code using a *generator polynomial* in much the way that we generated a block code using a generator matrix. The generator polynomial $\mathbf{g}(X)$ for an (n, k) cyclic code is unique and is of the form

$$\mathbf{g}(X) = g_0 + g_1 X + g_2 X^2 + \cdots + g_r X^r \qquad (5.57)$$

where g_0 and g_r must equal 1. Every codeword polynomial in the subspace is of the form $\mathbf{U}(X) = \mathbf{m}(X)\mathbf{g}(X)$, where $\mathbf{U}(X)$ is a polynomial of degree $n - 1$ or less. Therefore, the message polynomial $\mathbf{m}(X)$ is written

$$\mathbf{m}(X) = m_0 + m_1 X + m_2 X^2 + \cdots + m_{n-r-1} X^{n-r-1} \qquad (5.58)$$

There are 2^{n-r} codeword polynomials, and there are 2^k code vectors in an (n, k) code. Since there must be one codeword polynomial for each code vector

$$n - r = k$$

or

$$r = n - k$$

Hence $\mathbf{g}(X)$, as shown in Equation (5.57), must be of degree $n - k$, and every codeword polynomial in the (n, k) cyclic code can be expressed as

$$\mathbf{U}(X) = (m_0 + m_1 X + m_2 X^2 + \cdots + m_{k-1} X^{k-1})\mathbf{g}(X) \qquad (5.59)$$

U is said to be a valid code vector of the subspace S if, and only if, $\mathbf{g}(X)$ divides into $\mathbf{U}(X)$ without a remainder.

A generator polynomial $\mathbf{g}(X)$ of an (n, k) cyclic code is a factor of $X^n + 1$; that is, $X^n + 1 = \mathbf{g}(X)\mathbf{h}(X)$. For example,

$$X^7 + 1 = (1 + X + X^3)(1 + X + X^2 + X^4)$$

Using $\mathbf{g}(X) = 1 + X + X^3$ as a generator polynomial of degree $n - k = 3$, we can generate an $(n, k) = (7, 4)$ cyclic code. Or, using $\mathbf{g}(X) = 1 + X + X^2 + X^4$ where $n - k = 4$ we can generate a (7, 3) cyclic code. In summary, if $\mathbf{g}(X)$ is a polynomial of degree $n - k$ and is a factor of $X^n + 1$, then $\mathbf{g}(X)$ uniquely generates an (n, k) cyclic code.

5.6.3 Encoding in Systematic Form

In Section 5.4.5 we introduced the *systematic* form and discussed the reduction in complexity that makes this encoding form attractive. Let us use some of the algebraic properties of the cyclic code to establish a systematic encoding procedure. We can express the message vector in polynomial form, as follows:

$$\mathbf{m}(X) = m_0 + m_1 X + m_2 X^2 + \cdots + m_{k-1} X^{k-1} \qquad (5.60)$$

In systematic form, the message digits are utilized as part of the code vector. We can think of shifting the message digits into the rightmost k stages of a codeword register, and then appending the parity digits by placing them in the leftmost $n - k$ stages. Therefore, we want to manipulate the message polynomial algebraically so that it is right-shifted $n - k$ positions. If we multiply $\mathbf{m}(X)$ by X^{n-k} we get the right-shifted message polynomial:

$$X^{n-k}\mathbf{m}(X) = m_0 X^{n-k} + m_1 X^{n-k+1} + \cdots + m_{k-1} X^{n-1} \qquad (5.61)$$

If we next divide Equation (5.61) by $\mathbf{g}(X)$, the result can be expressed as

$$X^{n-k}\mathbf{m}(X) = \mathbf{q}(X)\mathbf{g}(X) + \mathbf{r}(X) \qquad (5.62)$$

where

$$\mathbf{r}(X) = r_0 + r_1 X + r_2 X^2 + \cdots + r_{n-k-1} X^{n-k-1}$$

We can also say that

$$\mathbf{r}(X) = X^{n-k}\mathbf{m}(X) \text{ modulo } \mathbf{g}(X) \qquad (5.63)$$

Adding $\mathbf{r}(X)$ to both sides of Equation (5.62), using modulo-2 arithmetic, we get

$$\mathbf{r}(X) + X^{n-k}\mathbf{m}(X) = \mathbf{q}(X)\mathbf{g}(X) = U(X) \qquad (5.64)$$

The left-hand side of Equation (5.64) is recognized as a valid codeword polynomial, since it is a polynomial of degree $n - 1$ or less, and when divided by $\mathbf{g}(X)$ there is a zero remainder. This codeword can be expanded into its polynomial terms as follows:

$$\mathbf{r}(X) + X^{n-k}\mathbf{m}(X) = r_0 + r_1 X + \cdots + r_{n-k-1} X^{n-k-1}$$
$$+ m_0 X^{n-k} + m_1 X^{n-k+1} + \cdots + m_{k-1} X^{n-1}$$

The codeword polynomial corresponds to the code vector

$$\mathbf{U} = \underbrace{(r_0, r_1, \ldots, r_{n-k-1},}_{(n-k) \text{ parity bits}} \underbrace{m_0, m_1, \ldots, m_{k-1})}_{k \text{ message bits}} \qquad (5.65)$$

Example 5.8   Cyclic Code in Systematic Form

Using the generator polynomial $\mathbf{g}(X) = 1 + X + X^3$, generate a systematic code vector from the (7, 4) codeword set for the message vector $\mathbf{m} = 1\ 0\ 1\ 1$.

Solution

$$\mathbf{m}(X) = 1 + X^2 + X^3, \quad n = 7, \quad k = 4, \quad n - k = 3$$

$$X^{n-k}\mathbf{m}(X) = X^3(1 + X^2 + X^3) = X^3 + X^5 + X^6$$

Dividing $X^{n-k}\mathbf{m}(X)$ by $\mathbf{g}(X)$ using polynomial division, we can write

$$X^3 + X^5 + X^6 = \underbrace{(1 + X + X^2 + X^3)}_{\substack{\text{quotient} \\ \mathbf{q}(X)}} \underbrace{(1 + X + X^3)}_{\substack{\text{generator} \\ \mathbf{g}(X)}} + \underbrace{1}_{\substack{\text{remainder} \\ \mathbf{r}(X)}}$$

Using Equation (5.64) yields

$$\mathbf{U}(X) = \mathbf{r}(X) + X^3\mathbf{m}(X) = 1 + X^3 + X^5 + X^6$$

$$\mathbf{U} = \underbrace{1\ 0\ 0}_{\substack{\text{parity} \\ \text{bits}}} \quad \underbrace{1\ 0\ 1\ 1}_{\substack{\text{message} \\ \text{bits}}}$$

5.6.4 Circuit for Dividing Polynomials

We have seen that the cyclic shift of a codeword polynomial and that the encoding of a message polynomial involves the division of one polynomial by another. Such an operation is readily accomplished by a *dividing circuit* (feedback shift register). Given two polynomials $\mathbf{V}(X)$ and $\mathbf{g}(X)$, where

$$\mathbf{V}(X) = v_0 + v_1X + v_2X^2 + \cdots + v_mX^m$$

and

$$\mathbf{g}(X) = g_0 + g_1X + g_2X^2 + \cdots + g_rX^r$$

such that $m \geq r$, the divider circuit of Figure 5.17 performs the polynomial division steps of dividing $\mathbf{V}(X)$ by $\mathbf{g}(X)$, thereby determining the quotient and remainder terms:

$$\frac{\mathbf{V}(X)}{\mathbf{g}(X)} = \mathbf{q}(X) + \frac{\mathbf{r}(X)}{\mathbf{g}(X)}$$

The stages of the register are first initialized by being filled with zeros. The first r shifts enter the most significant (higher-order) coefficients of $\mathbf{V}(X)$. After the rth shift, the quotient output is $g_r^{-1}v_m$; this is the highest-order term in the quotient. For each quotient coefficient q_i the polynomial $q_i\mathbf{g}(X)$ must be subtracted from the dividend. The feedback connections in Figure 5.17 perform this subtraction. The difference between the leftmost r terms remaining in the dividend

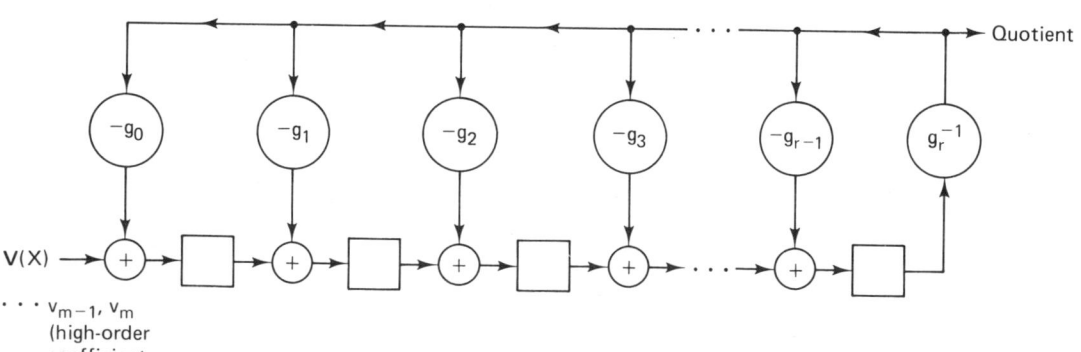

Figure 5.17 Circuit for dividing polynomials.

Channel Coding: Part 1   Chap. 5

and the feedback terms $q_i \mathbf{g}(X)$ is formed on each shift of the circuit and appears as the contents of the register. At each shift of the register, the difference is shifted one stage; the highest-order term (which by construction is zero) is shifted out, while the next significant coefficient of $\mathbf{V}(X)$ is shifted in. After $m + 1$ total shifts into the register, the quotient has been serially presented at the output and the remainder resides in the register.

Example 5.9 Dividing Circuit

Use a dividing circuit of the form shown in Figure 5.17 to divide $\mathbf{V}(X) = X^3 + X^5 + X^6$ ($\mathbf{V} = 0\ 0\ 0\ 1\ 0\ 1\ 1$) by $\mathbf{g}(X) = (1 + X + X^3)$. Find the quotient and remainder terms. Compare the circuit implementation to the polynomial division steps performed by hand.

Solution

The dividing circuit needs to perform the following operation:

$$\frac{X^3 + X^5 + X^6}{1 + X + X^3} = \mathbf{q}(X) + \frac{\mathbf{r}(X)}{1 + X + X^3}$$

The required feedback shift register, following the general form of Figure 5.17, is shown in Figure 5.18. Assume that the register contents are initially zero. The operational steps of the circuit are as follows:

Input queue	Shift number	Register contents	Output
0 0 0 1 0 1 1	0	0 0 0	–
0 0 0 1 0 1	1	1 0 0	0
0 0 0 1 0	2	1 1 0	0
0 0 0 1	3	0 1 1	0
0 0 0	4	0 1 1	1
0 0	5	1 1 1	1
0	6	1 0 1	1
–	7	1 0 0	1

After the fourth shift, the quotient coefficients $\{q_i\}$ serially presented at the output are seen to be 1 1 1 1, or the quotient polynomial is $\mathbf{q}(X) = 1 + X + X^2 + X^3$. The remainder coefficients $\{r_i\}$ are 1 0 0, or the remainder polynomial $\mathbf{r}(X) = 1$. In

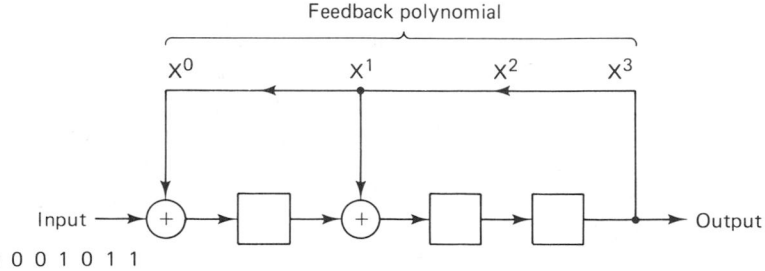

Figure 5.18   Dividing circuit for Example 5.9.

summary, the circuit computation $V(X)/g(X)$ is seen to be

$$\frac{X^3 + X^5 + X^6}{1 + X + X^3} = 1 + X + X^2 + X^3 + \frac{1}{1 + X + X^3}$$

The polynomial division steps are as follows:

Output after shift number:

```
            4     5     6     7
            ↓     ↓     ↓     ↓
            X³ +  X² +  X  +  1
X³ + X + 1 )X⁶ + X⁵      + X³
            X⁶      + X⁴ + X³ ←──────────────── feedback after 4th shift
            ─────────────────
                 X⁵ + X⁴ ←──────────────────── register after 4th shift
                 X⁵      + X³ + X² ←─────────── feedback after 5th shift
                 ────────────────
                      X⁴ + X³ + X² ←────────── register after 5th shift
                      X⁴      + X² + X ←─────── feedback after 6th shift
                      ──────────────
                           X³      + X ←────── register after 6th shift
                           X³      + X + 1 ←── feedback after 7th shift
                           ──────────────
                                     1 ←────── register after 7th shift
                                              (remainder)
```

5.6.5 Systematic Encoding with an ($n - k$)-Stage Shift Register

The encoding of a cyclic code in systematic form has been shown, in Section 5.6.3, to involve the computation of parity bits as the result of the formation of $X^{n-k}m(X)$ modulo $g(X)$, in other words, the *division* of an *upshifted* (right shifted) message polynomial by a generator polynomial $g(X)$. The need for upshifting is to make room for the parity bits, which are appended to the message bits, yielding the code vector in systematic form. Upshifting the message bits by $n - k$ positions is a trivial operation and is not really performed as part of the dividing circuit. Instead, only the parity bits are computed; they are then placed in the appropriate location alongside the message bits. The parity polynomial is the *remainder* after dividing by the generator polynomial; it is available in the register after n shifts through the ($n - k$)-stage feedback register shown in Figure 5.18. Notice that the first $n - k$ shifts through the register are simply filling the register. We cannot have any feedback until the rightmost stage has been filled; we therefore can shorten the shifting cycle by loading the input data to the output of the last stage, as shown in Figure 5.19. Further, the feedback term into the leftmost stage is the sum of the input and the rightmost stage. We guarantee that this sum is generated by ensuring that $g_0 = g_{n-k} = 1$ for any generator polynomial $g(X)$. The circuit feedback connections correspond to the coefficients of the generator polynomial, which is written

$$g(X) = 1 + g_1 X + g_2 X^2 + \cdots + g_{n-k-1}X^{n-k-1} + X^{n-k} \qquad (5.66)$$

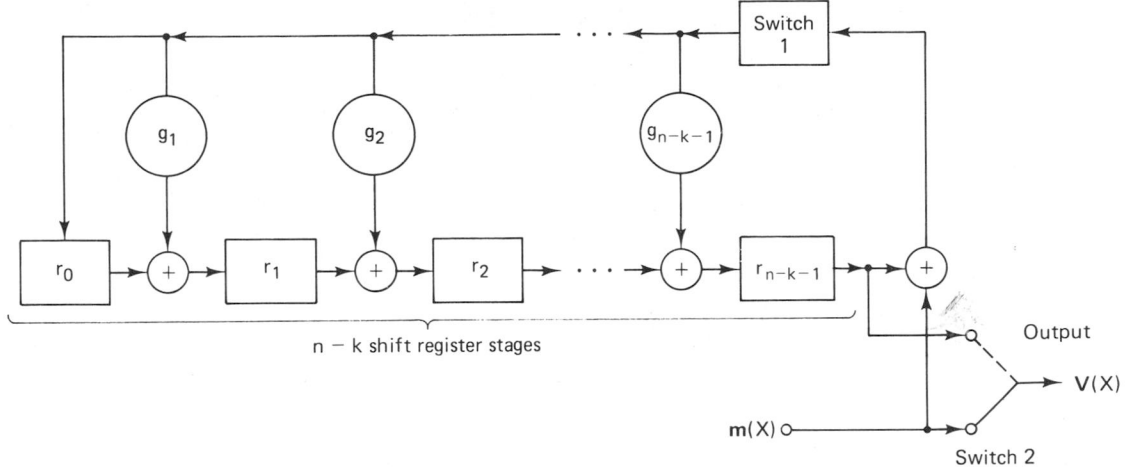

Figure 5.19 Encoding with an $(n - k)$-stage shift register.

The following steps describe the encoding procedure used with the Figure 5.19 encoder.

1. **Switch 1 is closed during the first k shifts, to allow transmission of the message bits into the $n - k$ stage encoding shift register.**

2. **Switch 2 is in the down position to allow transmission of the message bits directly to an output register during the first k shifts.**

3. **After transmission of the kth message bit, switch 1 is opened and switch 2 is moved to the up position.**

4. **The remaining $n - k$ shifts clear the encoding register by moving the parity bits to the output register.**

5. **The total number of shifts is equal to n, and the contents of the output register is the codeword polynomial $r(X) + X^{n-k}m(X)$.**

Example 5.10   Systematic Encoding of a Cyclic Code

Use a feedback shift register of the form shown in Figure 5.19 to encode the message vector $\mathbf{m} = 1\ 0\ 1\ 1$ into a $(7, 4)$ code vector using the generator polynomial $\mathbf{g}(X) = 1 + X + X^3$.

Solution

$$\mathbf{m} = 1\ 0\ 1\ 1$$

$$\mathbf{m}(X) = 1 + X^2 + X^3$$

$$X^{n-k}\mathbf{m}(X) = X^3\mathbf{m}(X) = X^3 + X^5 + X^6$$

$$X^{n-k}\mathbf{m}(X) = \mathbf{q}(X)\mathbf{g}(X) + \mathbf{r}(X)$$

$$\mathbf{r}(X) = X^3 + X^5 + X^6 \text{ modulo } (1 + X + X^3)$$

For the $(n - k) = 3$-stage encoding shift register shown in Figure 5.20, the operational steps are as follows:

Input queue	Shift number	Register contents	Output
1 0 1 1	0	0 0 0	–
1 0 1	1	1 1 0	1
1 0	2	1 0 1	1
1	3	1 0 0	0
–	4	1 0 0	1

After the fourth shift, switch 1 is opened, switch 2 is moved to the up position, and the parity bits contained in the register are shifted to the output. The output code vector is $\mathbf{U} = 1\ 0\ 0\ 1\ 0\ 1\ 1$, or in polynomial form, $\mathbf{U}(X) = 1 + X^3 + X^5 + X^6$.

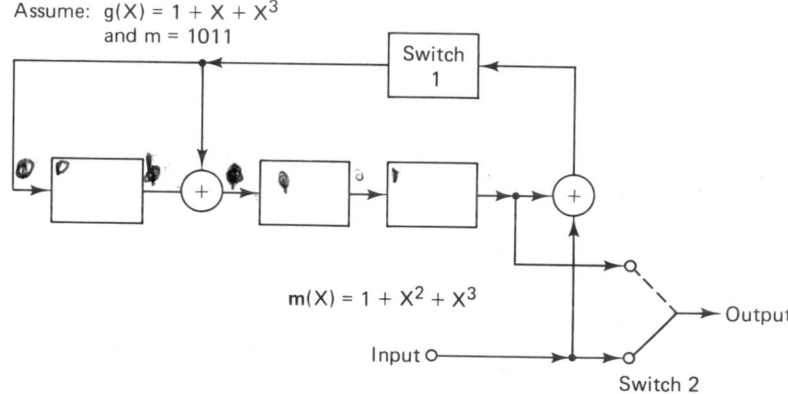

Assume: $g(X) = 1 + X + X^3$
and $m = 1011$

Switch 1

$m(X) = 1 + X^2 + X^3$

Input

Switch 2

Output

Figure 5.20 Example of encoding a (7, 4) cyclic code with an $(n - k)$-stage shift register.

5.6.6 Error Detection with an $(n - k)$-Stage Shift Register

A transmitted code vector may be perturbed by noise, and hence the vector received may be a corrupted version of the transmitted code vector. Let us assume that a codeword with polynomial representation $\mathbf{U}(X)$ is transmitted and that a vector with polynomial representation $\mathbf{Z}(X)$ is received. Since $\mathbf{U}(X)$ is a code polynomial, it must be a multiple of the generator polynomial $\mathbf{g}(X)$, that is,

$$\mathbf{U}(X) = \mathbf{m}(X)\mathbf{g}(X) \tag{5.67}$$

and $\mathbf{Z}(X)$, the corrupted version of $\mathbf{U}(X)$, can be written

$$\mathbf{Z}(X) = \mathbf{U}(X) + \mathbf{e}(X) \tag{5.68}$$

where $\mathbf{e}(X)$ is the error pattern polynomial. The decoder tests whether $\mathbf{Z}(X)$ is a codeword polynomial, that is, whether it is divisible by $\mathbf{g}(X)$, with a zero re-

mainder. This is accomplished by *calculating the syndrome* of the received polynomial. The syndrome $S(X)$ is equal to the remainder resulting from dividing $Z(X)$
by $g(X)$, that is,

$$Z(X) = q(X)g(X) + S(X) \qquad (5.69)$$

where $S(X)$ is a polynomial of degree $n - k - 1$ or less. Thus the syndrome is
an $(n - k)$-tuple. By combining Equations (5.67) to (5.69), we obtain

$$e(X) = [m(X) + q(X)]g(X) + S(X) \qquad (5.70)$$

By comparing Equations (5.69) and (5.70), we see that the syndrome $S(X)$, obtained as the remainder of $Z(X)$ modulo $g(X)$, is exactly the same polynomial
obtained as the remainder of $e(X)$ modulo $g(X)$. Thus the syndrome of the received
polynomial $Z(X)$ contains the information needed for correction of the error pattern. The syndrome calculation is accomplished by a division circuit, almost identical to the encoding circuit used at the transmitter. An example of syndrome
calculation with an $n - k$ shift register is shown in Figure 5.21 using the code
vector generated in Example 5.10. Switch 1 is initially closed, and switch 2 is
open. The received vector is shifted into the register input, with all stages initially
set to zero. After the entire received vector has been entered into the shift register,
the contents of the register is the syndrome. Switch 1 is then opened and switch
2 is closed, so that the syndrome vector can be shifted out of the register. The
operational steps of the decoder are as follows:

Input queue	Shift number	Register contents
1 0 0 1 0 1 1	0	0 0 0
1 0 0 1 0 1	1	1 0 0
1 0 0 1 0	2	1 1 0
1 0 0 1	3	0 1 1
1 0 0	4	0 1 1
1 0	5	1 1 1
1	6	1 0 1
–	7	0 0 0 Syndrome

If the syndrome is an all-zeros vector, the received vector is assumed to be a
valid code vector. If the syndrome is a nonzero vector, the received vector is a

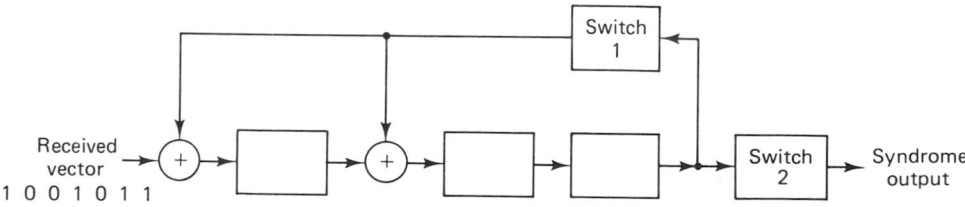

Figure 5.21 Example of syndrome calculation with an $(n - k)$-stage shift register.

perturbed code vector and errors have been detected; such errors can be corrected by adding the error vector (indicated by the syndrome) to the received vector, similar to the procedure described in Section 5.4.8. This method of decoding is useful for simple codes. More complex codes require the use of algebraic techniques to obtain practical decoders [6, 7].

5.7 WELL-KNOWN BLOCK CODES

5.7.1 Hamming Codes

Hamming codes are a simple class of block codes characterized by the following (n, k) structure:

$$(n, k) = (2^m - 1, 2^m - 1 - m) \tag{5.71}$$

where $m = 2, 3, \ldots$. These codes have a minimum distance of 3 and thus, from Equations (5.44) and (5.47), they are capable of correcting all single errors or detecting all combinations of two or fewer errors within a block. Syndrome decoding is especially suited for Hamming codes. In fact, the syndrome can be formed to act as a binary pointer to identify the error location [5]. Although Hamming codes are not very powerful, they belong to a very limited class of block codes known as *perfect* codes, described in Section 5.5.4.

Assuming hard decision decoding the bit error probability can be written, from Equation (5.46), as follows:

$$P_B \simeq \frac{1}{n} \sum_{j=2}^{n} j \binom{n}{j} p^j (1 - p)^{n-j} \tag{5.72}$$

where p is the channel symbol error probability (transition probability on the binary symmetric channel). In place of Equation (5.72) we can use the following equivalent equation. Its identity with Equation (5.72) is proven in Appendix D, Equation (D.16).

$$P_B \cong p - p(1 - p)^{n-1} \tag{5.73}$$

Figure 5.22 is a plot of P_B versus channel symbol error probability, illustrating the comparative performance for different types of block codes. For the Hamming codes, the plots are shown for $m = 3, 4,$ and 5, or $(n, k) = (7, 4), (15, 11),$ and $(31, 26)$. For performance over a Gaussian channel using coherently demodulated BPSK, we can express the channel symbol error probability in terms of E_c/N_0, similar to Equation (3.84), as follows:

$$p = Q\left(\sqrt{\frac{2E_c}{N_0}}\right) \tag{5.74}$$

where E_c/N_0 is the code symbol energy per noise spectral density, and where $Q(x)$ is as defined in Equation (2.42). To relate E_c/N_0 to information bit energy per noise spectral density (E_b/N_0), we use

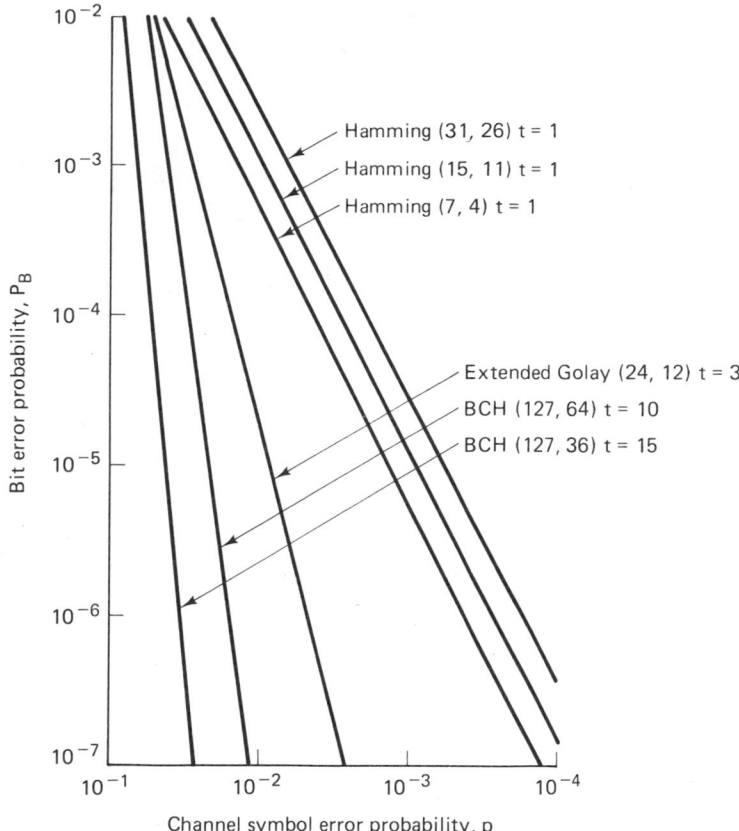

Figure 5.22 Bit error probability versus channel symbol error probability for several block codes.

$$\frac{E_c}{N_0} = \left(\frac{k}{n}\right)\frac{E_b}{N_0} \tag{5.75}$$

For Hamming codes, Equation (5.75) becomes

$$\frac{E_c}{N_0} = \frac{2^m - 1 - m}{2^m - 1}\frac{E_b}{N_0} \tag{5.76}$$

Combining Equations (5.73), (5.74), and (5.76), P_B can be expressed as a function of E_b/N_0 for coherently demodulated BPSK over a Gaussian channel. The results are plotted in Figure 5.23 for different types of block codes. For the Hamming codes, plots are shown for $(n, k) = (7, 4)$, $(15, 11)$, and $(31, 26)$.

Example 5.11   Error Probability for Modulated and Coded Signals

A coded BFSK modulated signal is transmitted over a Gaussian channel. The signal is noncoherently detected and hard-decision decoded. Find the decoded bit error probability if the coding is a Hamming (7, 4) block code and the received E_b/N_0 is equal to 20.

Solution

First we need to find E_c/N_0 using Equation (5.75):

$$\frac{E_c}{N_0} = \frac{4}{7}(20) = 11.43$$

Then, for coded noncoherent BFSK, we can relate the probability of a channel symbol error to E_c/N_0, similar to Equation (3.111), as follows:

$$p = \frac{1}{2}\exp\left(-\frac{E_c}{2N_0}\right)$$

$$= \frac{1}{2}\exp\left(-\frac{11.43}{2}\right) = 1.6 \times 10^{-3}$$

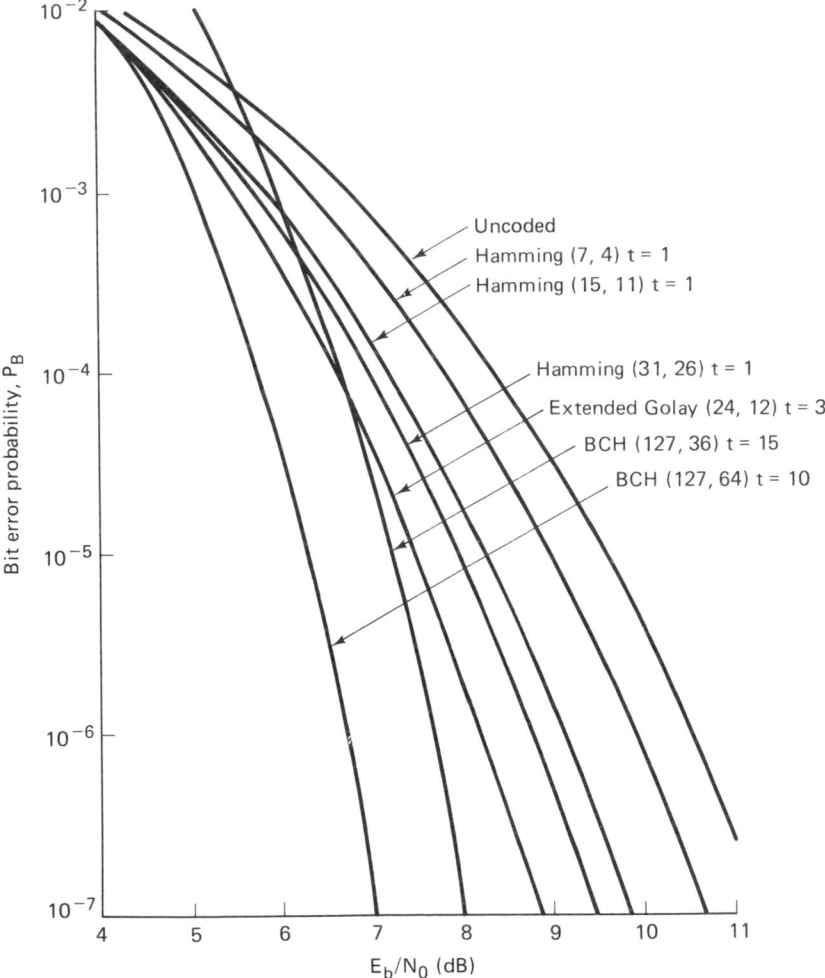

Figure 5.23 P_B versus E_b/N_0 for coherently demodulated BPSK over a Gaussian channel for several block codes.

Using this result in Equation (5.73), we solve for the probability of a decoded bit error, as follows:

$$P_B \simeq p - p(1 - p)^6 \simeq 1.6 \times 10^{-5}$$

5.7.2 Extended Golay Code

One of the more useful block codes is the binary (24, 12) *extended Golay code*, which is formed by adding an overall parity bit to the perfect (23, 12) code, known as the *Golay code*. This added parity bit increases the minimum distance d_{min} from 7 to 8 and produces a rate $\frac{1}{2}$ code, which is easier to implement (with regard to system clocks) than the rate 12/23 original Golay code. Extended Golay codes are considerably more powerful than the Hamming codes described in the preceding section. The price paid for the improved performance is a more complex decoder, a lower code rate, and hence a larger bandwidth expansion.

Since $d_{min} = 8$ for the extended Golay code, we see from Equation (5.44) that the code is guaranteed to correct all triple errors. The decoder can additionally be designed to correct *some but not all* four-error patterns. Since only 19% of the four-error patterns can be corrected, the decoder, for the sake of simplicity, is usually designed to only correct three-error patterns [5]. Assuming hard decision decoding, the bit error probability for the extended Golay code can be written as a function of the channel symbol error probability, p, from Equation (5.46), as follows:

$$P_B \simeq \frac{1}{24} \sum_{j=4}^{24} j \binom{24}{j} p^j (1 - p)^{24-j} \qquad (5.77)$$

The plot of Equation (5.77) is shown in Figure 5.22; the error performance of the extended Golay code is seen to be significantly better than that of the Hamming codes. Combining Equations (5.77), (5.74), and (5.75), we can relate P_B versus E_b/N_0 for coherently demodulated BPSK with extended Golay coding over a Gaussian channel. The result is plotted in Figure 5.23.

5.7.3 BCH Codes

Bose–Chadhuri–Hocquenghem (BCH) codes are a generalization of Hamming codes that allow multiple error correction. They are a *powerful class of cyclic codes* that provide a large selection of block lengths, code rates, alphabet sizes, and error-correcting capability. Table 5.2 lists some commonly used code generators, $\mathbf{g}(x)$, for the construction of BCH codes [8], for various values of n, k, and t, up to a block length of 255. The coefficients of $\mathbf{g}(x)$ are presented as octal numbers arranged so that when they are converted to binary digits the rightmost digit corresponds to the zero-degree coefficient of $\mathbf{g}(x)$. BCH codes are important, because at block lengths of a few hundred, the BCH codes outperform all other block codes with the same block length and code rate. The most commonly used BCH codes employ a binary alphabet and a codeword block length of $n = 2^m - 1$, where $m = 3, 4, \ldots$.

TABLE 5.2 Generators of Primitive BCH Codes

n	k	t	g(x)	n	k	t	g(x)
7	4	1	13	255	171	11	15416214212342356077061630637
15	11	1	23		163	12	7500415510075602551574724514601
	7	2	721		155	13	37575130054076650157225064646477633
	5	3	2467		147	14	1642130173537165525304165305441011711
31	26	1	45		139	15	461401732060175561570722730247453567445
	21	2	3551		131	18	2157133314715101512612502774421420241 65471
	16	3	107657		123	19	1206140522420660037172103265164122622 72506267
	11	5	5423325		115	21	6052666557210024726363640460027635255 6313472737
	6	7	313365047		107	22	2220577232206626563124173002353474201 7 6574750154441
63	57	1	103		99	23	106566672534731742227441162015743322 5 2411076432303431
	51	2	12471		91	25	6750265030327444172723631724732511075 5507627207243445 61
	45	3	1701317		87	26	1101367634147432364352316343071720462 0672254527311721317
	39	4	166623567		79	27	66700035663765750000202703442073661746 2 1015326711766541342355
	36	5	1033500423				
	30	6	157464165547				
	24	7	17323260404441				
	18	10	1363026512351725				
	16	11	6331141367235453				
	10	13	472622305527250155				
	7	15	5231045543503271737				

n	k	t	g(x) (octal)
127	120	1	211
	113	2	41567
	106	3	11554743
	99	4	3447023271
	92	5	624730022327
	85	6	130704476322273
	78	7	26230002166130115
	71	9	6255010713253127753
	64	10	1206534025570773100045
	57	11	335265252505705053517721
	50	13	54446512523314012421501421
	43	14	17721772213651227521220574343
	36	15	3146074666522075044764574721735
	29	21	4031144613676700603667301441176155
	22	23	123376070404722522435445626637647043
	15	27	22057042445604554770523013762217604353
	8	31	7047264052751030651476224271567733130217
255	247	1	435
	239	2	267543
	231	3	156720665
	223	4	75626641375
	215	5	23157564726421
	207	6	16176560567636227
	199	7	7633031270420722341
	191	8	2663470176115333714567
	187	9	52755313540001322236351
	179	10	22624710717340432416300455
	71	29	24024710520644321515554172112331163205444250362557643221706035
	63	30	10754475055163544432531517357707003666111726455267613656702543301
	55	31	731542520350110013301527530603205432541432675501055704442603473617
	47	42	2533542017062646563030413774062331751233341454446045005066024552543173
	45	43	152020560552341611311013463764237015636702447076237303320157025051541
	37	45	5136330255067007414177447245437530420735706174323432347644354737403044003
	29	47	30257155366730714655270640123613771153422423242011741140602547574110403565037
	21	55	1256215257060332656001773153607612103227341405653074542521153121614466513473725
	13	59	464173200505256454442657371425006600433067744547656140317467721357026134460500547
	9	63	1572602521747246320103104325535513461416367212044074545112766115547705561677516057

Source: Reprinted with permission from "Table of Generators for BCH Codes," *IEEE Trans. Inf. Theory,* vol. IT10, no. 4, Oct. 1964, p. 391. © 1964 IEEE.

The title of Table 5.2 indicates that the generators shown are for those BCH codes known as *primitive codes*. The term "primitive" is a number-theoretic concept requiring an elaborate algebraic development [9–11], which will not be presented here. In Figures 5.22 and 5.23 are plotted error performance curves of two BCH codes (127, 64) and (127, 36), to illustrate comparative performance. Assuming hard decision decoding, the P_B versus channel error probability is shown in Figure 5.22. The P_B versus E_b/N_0 for coherently demodulated BPSK over a Gaussian channel is shown in Figure 5.23. The curves in Figure 5.23 seem to depart from our expectations. They each have the same block size, yet the more redundant (127, 36) code does not exhibit as much coding gain as does the less redundant (127, 64) code. It has been shown that a relatively broad maximum of coding gain versus code rate for fixed n occurs roughly between coding rates of $\frac{1}{3}$ and $\frac{3}{4}$ for BCH codes [12]. Performance over a Gaussian channel degrades substantially at very high or very low rates [11].

Figure 5.24 represents computed performance of BCH codes [13] using coherently demodulated BPSK with both *hard-* and *soft-decision decoding*. Soft-decision decoding is not usually used with block codes because of its complexity. However, whenever it is implemented, it offers an approximate 2-dB coding gain over hard-decision decoding. For a given code rate, the decoded error probability is known to improve with increasing block length n [4]. Thus for a given code rate, it is interesting to compare the block length that would be required for the hard-decision-decoding performance to be comparable to the soft-decision-decoding performance. In Figure 5.24, the BCH codes shown all have code rates of approximately $\frac{1}{2}$. From the figure [13] it appears that for a fixed code rate, the hard-decision-decoded BCH code of length 8 times n or longer has a better performance than that of a soft-decision-decoded BCH code of length n.

5.7.4 Reed–Solomon Codes

One special subclass of the BCH codes (the discovery of which preceded the BCH codes) is the particularly useful *nonbinary* set called Reed–Solomon codes. Reed–Solomon codes achieve the *largest possible code minimum distance* for any linear code with the same encoder input and output block lengths. For nonbinary codes, the distance between two code words is defined as the number of nonbinary symbols in which the sequences differ. For Reed–Solomon codes the code minimum distance is given by [14]

$$d_{\min} = n - k + 1 \tag{5.78}$$

where k is the number of data symbols being encoded, and n is the total number of code symbols in the encoded block. Following Equation (5.44), the code is capable of correcting any combination of t or fewer symbol errors, as follows:

$$t = \frac{d_{\min} - 1}{2} = \frac{n - k}{2} \tag{5.79}$$

and thus requires no more than $2t$ parity check symbols.

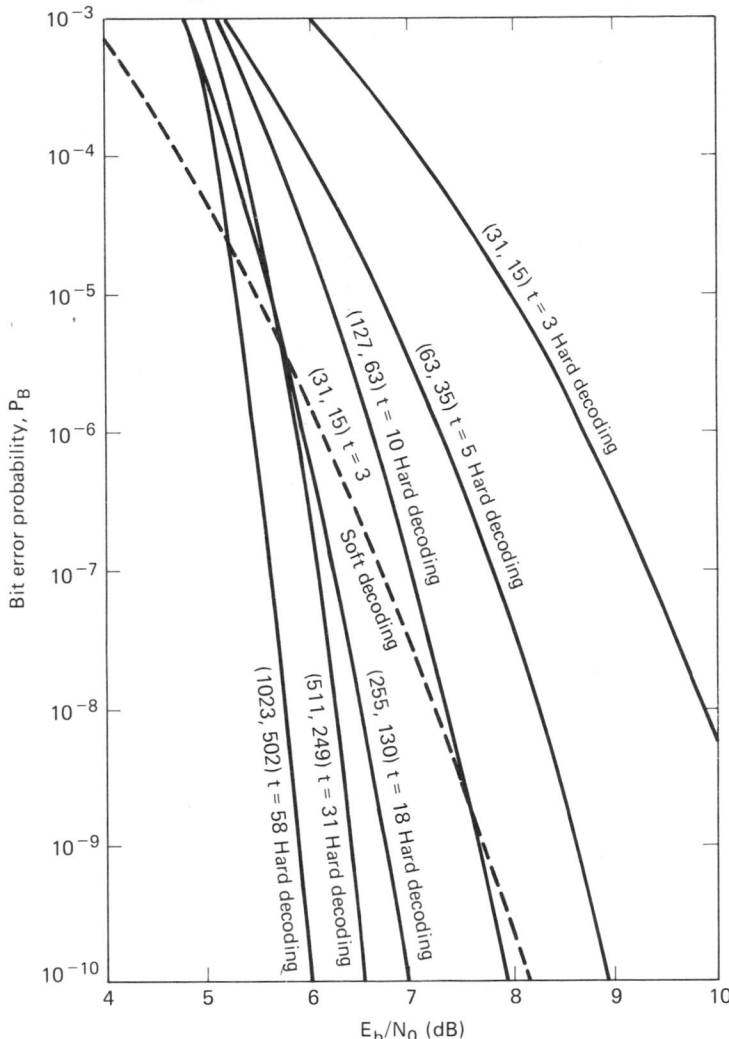

Figure 5.24 P_B versus E_b/N_0 for coherently demodulated BPSK over a Gaussian channel using BCH codes. (Reprinted with permission from L. J. Weng, "Soft and Hard Decoding Performance Comparisons for BCH Codes," *Proc. Int. Conf. Commun.*, 1979, Fig. 3, p. 25.5.5. © 1979 IEEE.)

A *t*-error-correcting Reed–Solomon code with an alphabet of 2^m symbols has $n = 2^m - 1$ and $k = 2^m - 1 - 2t$, where $m = 2, 3, \ldots$. An advantage of nonbinary codes such as a Reed–Solomon code can be seen by the following comparison. Consider a binary $(n, k) = (7, 3)$ code. The entire *n*-tuple space amounts to $2^n = 2^7 = 128$ binary words, of which $2^k = 2^3 = 8$ (or $\frac{1}{16}$ of the *n*-tuples) are codewords. Next, consider a nonbinary $(n, k) = (7, 3)$ code where each symbol is comprised of $m = 3$ bits. The *n*-tuple space amounts to $2^{nm} = 2^{21} = 2,097,152$ binary words, of which $2^{km} = 2^9 = 512$ (or $1/4096$ of the *n*-tuples) are codewords. With symbols, each made up of *m* bits, only a small fraction (i.e., 2^{km} of the large number 2^{nm}) of possible different words of *n* symbols become codewords. This fraction decreases with increasing values of m. When a small fraction of the n-tuple space is used for codewords, a large $d_{\min}$ can be created.

The Reed–Solomon (R–S) codes are particularly useful for *burst-error correction*; that is, they are effective for channels that have memory. Also, they can be used efficiently on channels where the set of input symbols is large. An interesting feature of the R–S code is that as many as two information symbols can be added to an R–S code of length n without reducing its minimum distance. This extended R–S code has length $n + 2$ and the same number of parity check symbols as the original code. From Equation (5.46) the R–S decoded symbol error probability, P_E, can be written in terms of the channel symbol error probability, p, as follows [5]:

$$P_E \simeq \frac{1}{2^m - 1} \sum_{j=t+1}^{2^m - 1} j \binom{2^m - 1}{j} p^j (1 - p)^{2^m - 1 - j} \qquad (5.80)$$

The bit error probability can be upper bounded by the symbol error probability

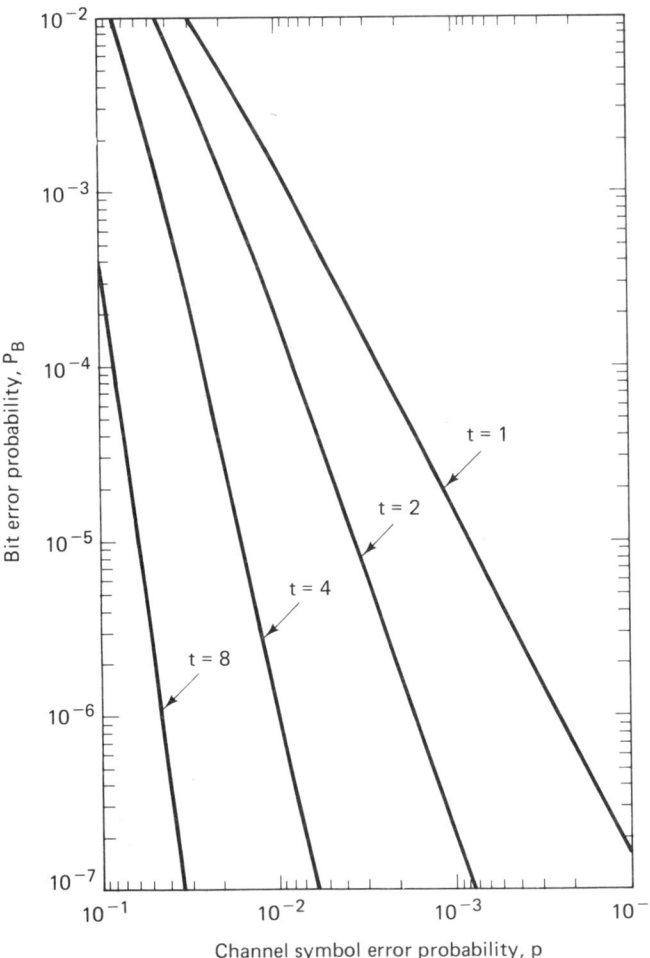

Figure 5.25 P_B versus p for 32-ary orthogonal signaling and $n = 31$, t-error-correcting Reed–Solomon coding. (Reprinted with permission from *Data Communications, Networks, and Systems*, ed. Thomas C. Bartee, Howard W. Sams Company, Indianapolis, Ind., 1985, p. 311. Originally published in J. P. Odenwalder, *Error Control Coding Handbook*, M/A-COM LINKABIT, Inc., San Diego, Calif., July 15, 1976, p. 91.)

Channel Coding: Part 1   Chap. 5

for specific modulation types. For MFSK modulation with $M = 2^m$, the relationship between P_B and P_E as given in Equation (3.127) is repeated here:

$$\frac{P_B}{P_E} = \frac{2^{m-1}}{2^m - 1} \qquad (5.81)$$

Figure 5.25 shows P_B versus the channel symbol error probability, p, plotted from Equations (5.80) and (5.81) for various t-error-correcting 32-ary orthogonal Reed–

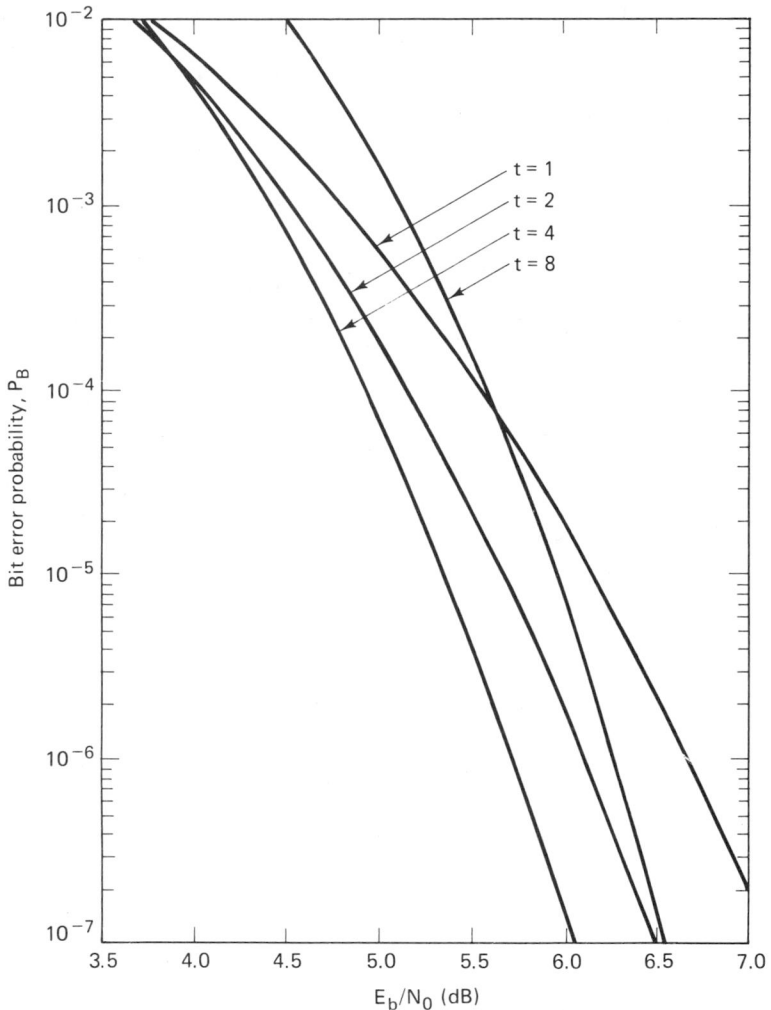

Figure 5.26 Bit error probability versus E_b/N_0 performance of several $n = 31$, t-error-correcting Reed–Solomon coding systems with 32-ary MFSK modulation over an AWGN channel. (Reprinted with permission from *Data Communications, Networks, and Systems*, ed. Thomas C. Bartee, Howard W. Sams Company, Indianapolis, Ind., 1985, p. 312. Originally published in J. P. Odenwalder, *Error Control Coding Handbook*, M/A-COM LINKABIT, Inc., San Diego, Calif., July 15, 1976, p. 92.)

Solomon codes with $n = 31$ (thirty-one 5-bit symbols per code block). Figure 5.26 shows P_B versus E_b/N_0 for such a coded system using 32-ary MFSK modulation and noncoherent demodulation over an AWGN channel [5]. For R–S codes, error probability is an exponentially decreasing function of block length, n, and decoding complexity is proportional to a small power of the block length [14]. The R–S codes are sometimes used in a concatenated arrangement. In such a system, an inner convolutional decoder first provides some error control by operating on soft-decision demodulator outputs; the convolutional decoder then presents hard-decision data to the outer Reed–Solomon decoder, which further reduces the probability of error. In Chapter 6 we discuss further the use of concatenated and R–S coding as applied to the compact disc (CD) digital audio system.

5.8 CONCLUSION

In this chapter we have explored the general goals of channel coding, all leading to improve probability of error performance at a minimum cost in bandwidth. We partitioned channel coding into two study groups: waveform coding and structured sequences. Waveform coding represents a transformation of waveforms into improved waveforms, such that the distance properties are improved compared to the original waveforms. Structured sequences involve the addition of parity digits to the data such that the parity digits can then be employed for detecting and/or correcting specific error patterns. The main advantage of structured sequence coding over waveform coding is that it can accomplish improved P_B performance using less bandwidth.

We particularly examined linear block codes. Geometric analogies can be drawn between the coding and modulation disciplines. They both seek to pack the signal space efficiently and to maximize the distance between signals in the signaling set. Within block codes we looked at cyclic codes, which are relatively easy to implement using modern integrated circuit techniques. We considered the polynomial representation of codes and the correspondence between the polynomial structure, the necessary algebraic operations, and the hardware implementation. We have also looked at performance details of some of the well-known block codes. A large class of codes, the convolutional codes, have been left for consideration in Chapter 6.

REFERENCES

1. Viterbi, A. J., "On Coded Phase-Coherent Communications," *IRE Trans. Space Electron. Telem.,* vol. SET7, Mar. 1961, pp. 3–14.
2. Lindsey, W. C., and Simon, M. K., *Telecommunication Systems Engineering,* Prentice-Hall, Inc., Englewood Cliffs, N.J., 1973.

3. Proakis, J. G., *Digital Communications,* McGraw-Hill Book Company, New York, 1983.

4. Lin, S., and Costello, D. J., Jr., *Error Control Coding: Fundamentals and Applications,* Prentice-Hall, Inc., Englewood Cliffs, N.J., 1983.

5. Odenwalder, J. P., *Error Control Coding Handbook,* Linkabit Corporation, San Diego, Calif., July 15, 1976.

6. Blahut, R. E., *Theory and Practice of Error Control Codes,* Addison-Wesley Publishing Company, Inc., Reading, Mass., 1983.

7. Blahut, R. E., "Algebraic Fields, Signal Processing, and Error Control," *Proc. IEEE,* vol. 73, May 1985, pp. 874–893.

8. Stenbit, J. P., "Tables of Generators for Bose–Chadhuri Codes, *IEEE Trans. Inf. Theory,* vol. IT10, no. 4, Oct. 1964, pp. 390–391.

9. Berlekamp, E. R., *Algebraic Coding Theory,* McGraw-Hill Book Company, New York, 1968.

10. Peterson, W. W., and Weldon, E. J., *Error Correcting Codes,* 2nd ed., The MIT Press, Cambridge, Mass., 1972.

11. Clark, G. C., Jr., and Cain, J. B., *Error-Correction Coding for Digital Communications,* Plenum Press, New York, 1981.

12. Wozencraft, J. M., and Jacobs, I. M., *Principles of Communication Engineering,* John Wiley & Sons, Inc., New York, 1965.

13. Weng, L. J., "Soft and Hard Decoding Performance Comparisons for BCH Codes," *Proc. Int. Conf. Commun.,* 1979, pp. 25.5.1–25.5.5.

14. Gallager, R. G., *Information Theory and Reliable Communication,* John Wiley & Sons, Inc., New York, 1968.

PROBLEMS

5.1. Design an (n, k) single-parity code that will detect all 1-, 3-, 5-, and 7-error patterns in a block. Show the values of n and k, and find the probability of an undetected block error if the probability of channel symbol error is 10^{-2}.

5.2. Calculate the probability of message error for a 12-bit data sequence encoded with a (24, 12) linear block code. Assume that the code corrects all 1-bit and 2-bit error patterns and assume that it corrects no error patterns with more than two errors. Also, assume that the probability of a channel symbol error is 10^{-3}.

5.3. Consider a (127, 92) linear block code capable of triple error corrections.
  (a) What is the probability of message error for an uncoded block of 92 bits if the channel symbol error probability is 10^{-3}?
  (b) What is the probability of message error when using the (127, 92) block code if the channel symbol error probability is 10^{-3}?

5.4. Calculate the improvement in probability of message error relative to an uncoded transmission for a (24, 12) double-error-correcting linear block code. Assume that coherent BPSK modulation is used and that the received $E_b/N_0 = 10$ dB.

5.5. Consider a (24, 12) linear block code capable of double-error corrections. Assume that a noncoherently detected BFSK modulation format is used and that the received $E_b/N_0 = 14$ dB.

(a) Does the code provide any improvement in probability of message error? If it does, how much? If it does not, explain why not.

(b) Repeat part (a) with $E_b/N_0 = 10$ dB.

5.6. The telephone company uses a "best-of-five" encoder for some of its digital data channels. In this system every data bit is repeated five times, and at the receiver, a majority vote decides the value of each data bit. If the uncoded probability of bit error is 10^{-3}, calculate the coded bit error probability when using such a best-of-five code.

5.7. The minimum distance for a particular linear block code is 11. Find the maximum error-correcting capability, the maximum error-detecting capability, and the maximum erasure-correcting capability in a block length.

5.8. Consider a (7, 4) code whose generator matrix is

$$
\mathbf{G} = \begin{bmatrix} 1 & 1 & 1 & 1 & 0 & 0 & 0 \\ 1 & 0 & 1 & 0 & 1 & 0 & 0 \\ 0 & 1 & 1 & 0 & 0 & 1 & 0 \\ 1 & 1 & 0 & 0 & 0 & 0 & 1 \end{bmatrix}
$$

(a) Find all the code vectors of the code.

(b) Find $\mathbf{H}$, the parity-check matrix of the code.

(c) Compute the syndrome for the received vector 1 1 0 1 1 0 1. Is this a valid code vector?

(d) What is the error-correcting capability of the code?

(e) What is the error-detecting capability of the code?

5.9. Consider a systematic block code whose parity-check equations are

$$p_1 = m_1 + m_2 + m_4$$

$$p_2 = m_1 + m_3 + m_4$$

$$p_3 = m_1 + m_2 + m_3$$

$$p_4 = m_2 + m_3 + m_4$$

where m_i are message digits and p_i are check digits.

(a) Find the generator matrix and the parity-check matrix for this code.

(b) How many errors can the code correct?

(c) Is the vector 10101010 a codeword?

(d) Is the vector 01011100 a codeword?

5.10. Consider the linear block code with the codeword defined by

$$\mathbf{U} = m_1 + m_2 + m_4 + m_5, \ m_1 + m_3 + m_4 + m_5, \ m_1 + m_2 + m_3 + m_5,$$
$$m_1 + m_2 + m_3 + m_4, \ m_1, m_2, m_3, m_4, m_5$$

(a) Show the generator matrix.

(b) Show the parity-check matrix.

(c) Find n, k, and d_{min}.

5.11. Design a (4, 2) linear block code.
  (a) Choose the codewords to be in systematic form, and choose them with the goal of maximizing d_{min}.
  (b) Find the generator matrix for the codeword set.
  (c) Calculate the parity-check matrix.
  (d) Enter the sixteen 4-tuples into a standard array.
  (e) What are the error-correcting and error-detecting capabilities of the code?
  (f) Make a syndrome table for the correctable error patterns.

5.12. Consider the (5, 1) repetition code, which consists of the two codewords 00000 and 11111, corresponding to messages 0 and 1, respectively. Derive the standard array for this code. Is this a perfect code?

5.13. Design a (3, 1) code that will correct all single-error patterns. Choose the codeword set and show the standard array.

5.14. Is a (7, 3) code a perfect code? Is a (7, 4) code a perfect code? Is a (15, 11) code a perfect code? Justify your answers.

5.15. A (15, 11) linear block code can be defined by the following parity array:

$$P = \begin{bmatrix} 0 & 0 & 1 & 1 \\ 0 & 1 & 0 & 1 \\ 1 & 0 & 0 & 1 \\ 0 & 1 & 1 & 0 \\ 1 & 0 & 1 & 0 \\ 1 & 1 & 0 & 0 \\ 0 & 1 & 1 & 1 \\ 1 & 1 & 1 & 0 \\ 1 & 1 & 0 & 1 \\ 1 & 0 & 1 & 1 \\ 1 & 1 & 1 & 1 \end{bmatrix}$$

  (a) Show the parity-check matrix for this code.
  (b) List the coset leaders from the standard array. Is this code a perfect code? Justify your answer.
  (c) A received vector is V = 0 1 1 1 1 1 0 0 1 0 1 1 0 1 1. Compute the syndrome. Assuming that a single bit error has been made, find the correct codeword.
  (d) How many erasures can this code correct? Explain.

5.16. Is it possible that a nonzero error pattern can produce a syndrome of $S = 0$? If yes, how many such error patterns can give this result for an (n, k) code? Use Figure 5.14 to justify your answer.

5.17. Determine which, if any, of the following polynomials can generate a cyclic code with codeword length $n \leq 7$. Find the (n, k) values of any such codes that can be generated.
  (a) $1 + X^3 + X^4$
  (b) $1 + X^2 + X^4$
  (c) $1 + X + X^3 + X^4$
  (d) $1 + X + X^2 + X^4$
  (e) $1 + X^3 + X^5$

5.18. Encode the message 1 0 1 in systematic form using polynomial division and the generator $g(X) = 1 + X + X^2 + X^4$.

5.19. Design a feedback shift register encoder for an (8, 5) cyclic code with a generator $g(x) = 1 + X + X^2 + X^3$. Use the encoder to find the codeword for the message 1 0 1 0 1 in systematic form.

5.20. In Figure P5.1 the signal is differentially coherent PSK (DPSK), the encoded symbol rate is 10,000 code symbols per second, and the decoder is a single-error-correcting (7, 4) decoder. Is a predetection signal-to-noise spectral density ratio of $P_r/N_0 = 48$ dBW sufficient to provide a probability of message error of 10^{-3} at the output? Justify your answer. Assume that a message block contains 4 data bits and that any single-error pattern in a block length of 7 bits can be corrected.

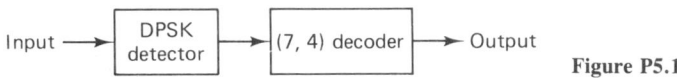

Figure P5.1

5.21. A (15, 5) cyclic code has a generator polynomial as follows:

$$g(X) = 1 + X + X^2 + X^5 + X^8 + X^{10}$$

(a) Draw a diagram of an encoder for this code.

(b) Find the code polynomial (in systematic form) for the message $m(X) = 1 + X^2 + X^4$.

(c) Is $V(X) = 1 + X^4 + X^6 + X^8 + X^{14}$ a code polynomial in this system? Justify your answer.

5.22. Consider the (15, 11) cyclic code generated by $g(X) = 1 + X + X^4$.

(a) Devise a feedback register encoder and decoder for this code.

(b) Illustrate the encoding procedure with the message vector 11001101011 by listing the states of the register (the rightmost bit is the earliest bit).

(c) Repeat part (b) for the decoding procedure.

5.23. For a fixed probability of channel symbol error, the probability of bit error for a Hamming (15, 11) code is worse than that for a Hamming (7, 4) code. Explain why. What, then, is the advantage of the (15, 11) code? What basic trade-off is involved?

5.24. A (63, 36) BCH code can correct five errors. Nine blocks of a (7, 4) code can correct nine errors. Both codes have the same code rate.

(a) The (7, 4) code can correct more errors. Is it more powerful? Explain.

(b) Compare the two codes when five errors occur randomly in 63 bits.

5.25. Information from a source is organized in 36-bit messages that are to be transmitted over an AWGN channel using noncoherently detected BFSK modulation.

(a) If no error control coding is used, compute the E_b/N_0 required to provide a message error probability of 10^{-3}.

(b) Consider the use of a (127, 36) linear block code (minimum distance is 31) in the transmission of these messages. Compute the coding gain for this code for a message error probability of 10^{-3}. (*Hint:* The coding gain is defined as the difference between the E_b/N_0 required without coding and the E_b/N_0 required with coding.)

5.26. (a) Consider a data sequence encoded with a (127, 64) BCH code and then modulated using coherent 16-ary PSK. If the received E_b/N_0 is 10 dB, find the MPSK probability of symbol error, the probability of coded bit error (assuming that a Gray code is used for symbol-to-bit assignment), and the probability of information bit error.

(b) For the same probability of information bit error found in part (a), determine the value of E_b/N_0 required if the modulation in part (a) is changed to coherent 16-ary FSK. Explain the difference.

5.27. A message consists of English text (assume that each word in the message contains six letters). Each letter is encoded using the 7-bit ASCII character code. Thus, each word of text consists of a 42-bit sequence. The message is to be transmitted over a channel having a symbol error probability of 10^{-3}.

(a) What is the probability that a word will be received in error?

(b) If a repetition code is used such that each letter in each word is repeated three times, and at the receiver, majority voting is used to decode the message, what is the probability that a decoded word will be in error?

(c) If a (126, 42) BCH code with error-correcting capability of $t = 14$ is used to encode each 42-bit word, what is the probability that a decoded word will be in error?

(d) For a real system, it is not fair to compare uncoded versus coded message error performance on the basis of a fixed probability of channel symbol error, since this implies a fixed level of received E_c/N_0 for all choices of coding (or lack of coding). Therefore, repeat parts (a), (b), and (c) under the condition that the channel symbol error probability is determined by a received E_b/N_0 of 12 dB, where E_b/N_0 is the information bit energy per noise spectral density. Assume that the information rate must be the same for all choices of coding or lack of coding. Also assume that noncoherent binary FSK modulation is used over an AWGN channel.

(e) Discuss the relative error performance capabilities of the above coding schemes under the two postulated conditions—fixed channel symbol error probability, and fixed E_b/N_0. Under what circumstances can a repetition code offer error performance improvement? When will it cause performance degradation?

CHAPTER 6

Channel Coding:
Part 2

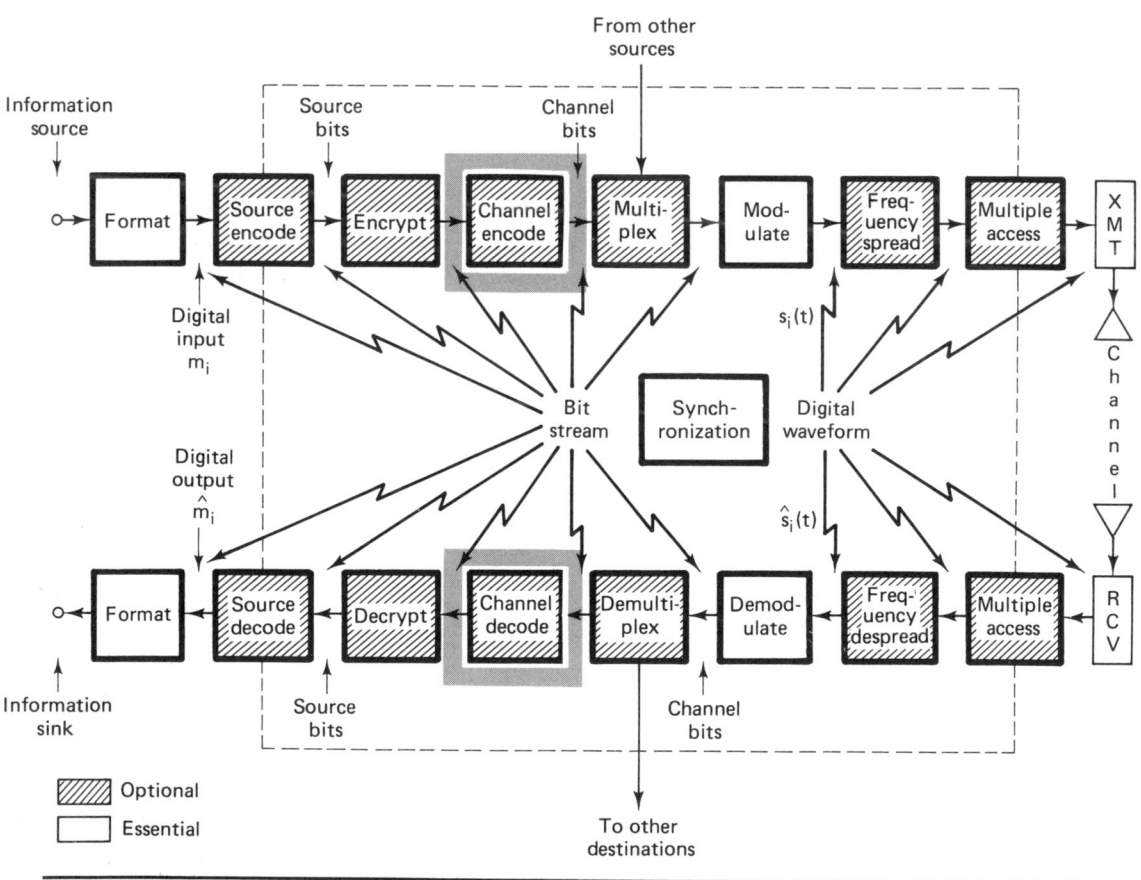

There are two major categories of channel codes: block and convolutional. Chapter 5 deals mainly with block coding. This chapter deals mainly with convolutional coding. A *linear block code* is described by two integers, n and k, and a generator matrix or polynomial. The integer k is the number of data bits that form an input to a block encoder. The integer n is the total number of bits in the associated codeword out of the encoder. A characteristic of linear block codes is that each codeword n-tuple is uniquely determined by the input message k-tuple. The ratio k/n is called the *rate* of the code—a measure of the amount of added redundancy. A *convolutional code* is described by three integers, n, k, and K, where the ratio k/n has the same code rate significance (information per coded bit) that it has for block codes; however, n does *not* define a block or codeword length as it does for block codes. The integer K is a parameter known as the *constraint length*; it represents the number of k-tuple stages in the encoding shift register. An important characteristic of convolutional codes, different from block codes, is that the encoder has memory—the n-tuple emitted by the convolutional encoding procedure is not only a function of an input k-tuple, but is also a function of the previous $K - 1$ input k-tuples. In practice, n and k are small integers and K is varied to control the redundancy.

6.1 CONVOLUTIONAL ENCODING

In Figure 1.2 we presented a typical block diagram of a digital communication system. A version of this functional diagram, focusing primarily on the convolutional encode/decode and modulate/demodulate portions of the communication link, is shown in Figure 6.1. The input message source is denoted by the sequence $\mathbf{m} = m_1, m_2, \ldots, m_i, \ldots$, where each m_i represents a binary digit (bit). We shall assume that each m_i is equally likely to be a one or a zero, and independent from digit to digit. Being independent, the bit sequence lacks any redundancy; that is, knowledge about bit m_i gives no information about m_j $(i \neq j)$. The encoder transforms each sequence $\mathbf{m}$ into a unique codeword sequence $\mathbf{U} = G(\mathbf{m})$. Even though the sequence $\mathbf{m}$ uniquely defines the sequence $\mathbf{U}$, a key feature of convolutional codes is that a given k-tuple within $\mathbf{m}$ does *not* uniquely define its associated n-tuple within $\mathbf{U}$ since the encoding of each k-tuple is *not only* a function of that k-tuple but is also a function of the $K - 1$ input k-tuples that precede it. The sequence $\mathbf{U}$ can be partitioned into a sequence of branch words: $\mathbf{U} = U_1$, $U_2, \ldots, U_i, \ldots$. Each branch word U_i is made up of binary *code symbols*, often called *channel symbols*, *channel bits*, or *coded bits*; unlike the input message bits the code symbols are not independent.

In a typical communication application, the codeword sequence $\mathbf{U}$ modulates a waveform $s(t)$. During transmission, the waveform $s(t)$ is corrupted by noise, resulting in a received waveform $\hat{s}(t)$ and a demodulated sequence $\mathbf{Z} = Z_1, Z_2,$ $\ldots, Z_i, \ldots$, as indicated in Figure 6.1. The task of the decoder is to produce an estimate $\hat{\mathbf{m}} = \hat{m}_1, \hat{m}_2, \ldots, \hat{m}_i, \ldots$, of the original message sequence, using the received sequence $\mathbf{Z}$ together with a priori knowledge of the encoding procedure.

A general convolutional encoder, shown in Figure 6.2, is mechanized with a kK-stage shift register and n modulo-2 adders, where K is the constraint length. The constraint length represents the number of k-bit shifts over which a single information bit can influence the encoder output. At each unit of time, k bits are shifted into the first k stages of the register; all bits in the register are shifted k stages to the right, and the outputs of the n adders are sequentially sampled to

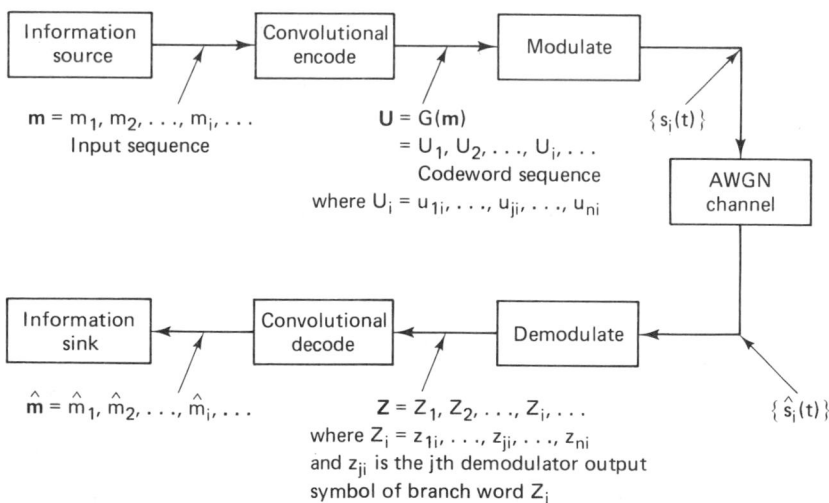

Figure 6.1 Encode/decode and modulate/demodulate portions of a communication link.

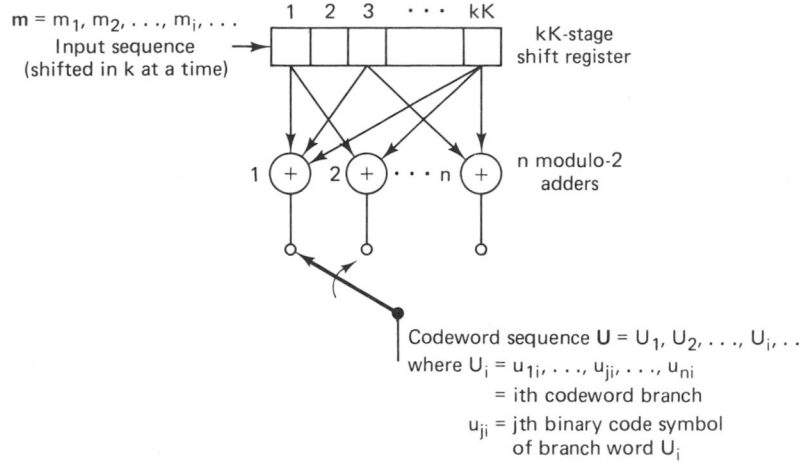

Figure 6.2 Convolutional encoder with constraint length K and rate k/n.

yield the binary code symbols or coded bits. These code symbols are then used by the modulator to specify the waveforms to be transmitted over the channel. Since there are n coded bits for each input group of k message bits, the code rate is k/n message bit per coded bit, where $k < n$.

We shall consider only the most commonly used binary convolutional encoders for which $k = 1$, that is, those encoders in which the message bits are shifted into the encoder one bit at a time, although generalization to higher-order alphabets is straightforward [1, 2]. For the $k = 1$ encoder, at the ith unit of time, message bit m_i is shifted into the first shift register stage; all previous bits in the register are shifted one stage to the right, and as in the more general case, the outputs of the n adders are sequentially sampled and transmitted. Since there are n coded bits for each message bit, the code rate is $1/n$. The n code symbols occurring at time t_i comprise the ith branch word, $U_i = u_{1i}, u_{2i}, \ldots, u_{ni}$, where u_{ji} ($j = 1, 2, \ldots, n$) is the jth code symbol belonging to the ith branch word. Note that for the rate $1/n$ encoder, the kK-stage shift register can be referred to simply as a K-stage register, and the constraint length K, which was expressed in units of k-tuple stages, can be referred to as constraint length in units of bits.

6.2 CONVOLUTIONAL ENCODER REPRESENTATION

To describe a convolutional code, one needs to characterize the encoding function $G(\mathbf{m})$, so that given an input sequence $\mathbf{m}$, one can readily compute the output sequence $\mathbf{U}$. Several methods are used for representing a convolutional encoder, the most popular being the *connection pictorial*, *connection vectors or polynomials*, the *state diagram*, the *tree diagram*, and the *trellis diagram*. They are each described below.

6.2.1 Connection Representation

We shall use the convolutional encoder, shown in Figure 6.3, as a model for discussing convolutional encoders. The figure illustrates a (2, 1) convolutional encoder with constraint length $K = 3$. There are $n = 2$ modulo-2 adders; thus the code rate k/n is $\frac{1}{2}$. At each input bit time, a bit is shifted into the leftmost stage and the bits in the register are shifted one position to the right. Next, the output switch samples the output of each modulo-2 adder (i.e., first the upper adder, then the lower adder), thus forming the code symbol pair making up the branch word associated with the bit just inputted. The sampling is repeated for each inputted bit. The choice of connections between the adders and the stages of the register gives rise to the characteristics of the code. Any change in the choice of connections results in a different code. The connections are, of course, *not* chosen or changed arbitrarily. The problem of choosing connections to yield good distance properties is complicated and has not been solved in general; however, good codes have been found by computer search for all constraint lengths less than about 20 [3–5].

Unlike a block code that has a fixed word length n, a convolutional code has no particular block size. However, convolutional codes are often forced into a block structure by *periodic truncation*. This requires a number of zero bits to be appended to the end of the input data sequence, for the purpose of clearing or *flushing* the encoding shift register of the data bits. Since the added zeros carry no information, the \effective code rate falls below k/n. To keep the code rate close to k/n, the truncation period is generally made as long as practical.

One way to represent the encoder is to specify a set of n *connection vectors*, one for each of the n modulo-2 adders. Each vector has dimension K and describes the connection of the encoding shift register to that modulo-2 adder. A one in the ith position of the vector indicates that the corresponding stage in the shift register is connected to the modulo-2 adder, and a zero in a given position indicates that no connection exists between the stage and the modulo-2 adder. For the encoder example in Figure 6.3, we can write the connection vector $\mathbf{g}_1$ for the upper connections and $\mathbf{g}_2$ for the lower connections as follows:

$$\mathbf{g}_1 = 1\ 1\ 1$$

$$\mathbf{g}_2 = 1\ 0\ 1$$

Consider that a message vector $\mathbf{m} = 1\ 0\ 1$ is convolutionally encoded with the encoder shown in Figure 6.3. The three message bits are inputted, one at a time, at times t_1, t_2, and t_3, as shown in Figure 6.4. Subsequently, $(K - 1) = 2$ zeros are inputted at times t_4 and t_5 to flush the register and thus ensure that the tail end of the message is shifted the full length of the register. The output sequence is seen to be 1 1 1 0 0 0 1 0 1 1, where the leftmost symbol represents the earliest transmission. The entire output sequence, including the code symbols as a result of flushing, are needed to decode the message. To flush the message from the encoder requires one less zero than the number of stages in the register, or $K - 1$ flush bits. Another zero input is shown at time t_6, for the reader to verify that the corresponding branch word output is then 00.

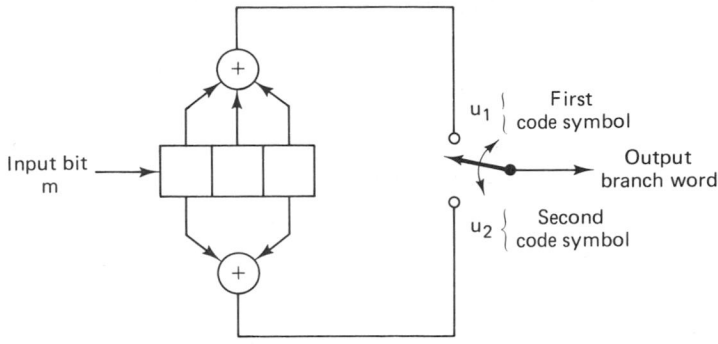

Figure 6.3 Convolutional encoder (rate $\frac{1}{2}$, $K = 3$).

6.2.1.1 Impulse Response of the Encoder

We can approach the encoder in terms of its *impulse response*—that is, the response of the encoder to a single "one" bit that moves through it. Consider the contents of the register in Figure 6.3 as a one moves through it.

Register contents	Branch word	
	u_1	u_2
1 0 0	1	1
0 1 0	1	0
0 0 1	1	1

Input sequence: 1   0   0
Output sequence: 1 1 1 0 1 1

The output sequence for the input "one" is called the impulse response of the encoder. Then for the input sequence **m** = 1 0 1, the output may be found by the *superposition* or the *linear addition* of the time-shifted input "impulses" as follows:

Input **m**	Output				
1	1 1	1 0	1 1		
0		0 0	0 0	0 0	
1			1 1	1 0	1 1
Modulo-2 sum:	1 1	1 0	0 0	1 0	1 1

Observe that this is the same output as that obtained in Figure 6.4, demonstrating that *convolutional codes are linear*—just as the linear block codes of Chapter 5. It is from this property of generating the output by the linear addition of time-shifted impulses, or the convolution of the input sequence with the impulse response of the encoder, that we derive the name *convolutional encoder*. Often,

m = 101 ⟶ [Encoder] ⟶ U

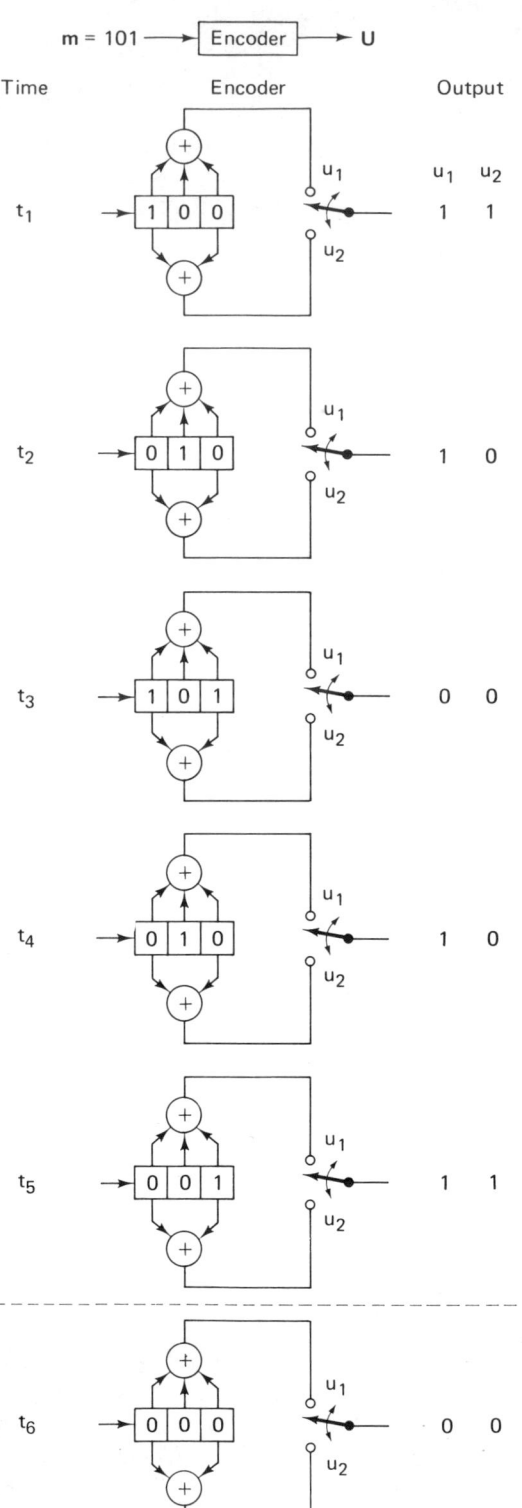

Figure 6.4 Convolutionally encoding a message sequence with a rate $\frac{1}{2}$, $K = 3$ encoder.

this encoder characterization is presented in terms of an infinite-order generator matrix [6].

Notice that the *effective code rate* for the foregoing example with 3-bit input sequence and 10-bit output sequence is $k/n = \frac{3}{10}$—quite a bit less than the rate $\frac{1}{2}$ that might have been expected from the knowledge that each input data bit yields a pair of output channel bits. The reason for the disparity is that the final data bit into the encoder needs to be shifted through the encoder. All of the output channel bits are needed in the decoding process. If the message had been longer, say 300 bits, the output codeword sequence would contain 604 bits, resulting in a code rate of 300/604—much closer to $\frac{1}{2}$.

6.2.1.2 Polynomial Representation

Sometimes, the encoder connections are characterized by *generator polynomials*, similar to those used in Chapter 5 for describing the feedback shift register implementation of cyclic codes. We can represent a convolutional encoder with a set of n generator polynomials, one for each of the n modulo-2 adders. Each polynomial is of degree $K - 1$ or less and describes the connection of the encoding shift register to that modulo-2 adder, much the same way that a connection vector does. The coefficient of each term in the $(K - 1)$-degree polynomial is either 1 or 0, depending on whether a connection exists or does not exist between the shift register and the modulo-2 adder in question. For the encoder example in Figure 6.3, we can write the generator polynomial $g_1(X)$ for the upper connections and $g_2(X)$ for the lower connections as follows:

$$g_1(X) = 1 + X + X^2$$

$$g_2(X) = 1 + X^2$$

where the lowest-order term in the polynomial corresponds to the input stage of the register. The output sequence is found as follows:

$$U(X) = m(X)g_1(X) \text{ interlaced with } m(X)g_2(X)$$

First, express the message vector $\mathbf{m} = 1\ 0\ 1$ as a polynomial—that is, $m(X) = 1 + X^2$. We shall again assume the use of zeros following the message bits, to flush the register. Then the output polynomial, $U(X)$, or the output sequence, U, of the Figure 6.3 encoder can be found for the input message $\mathbf{m}$ as follows:

$$m(X)g_1(X) = (1 + X^2)(1 + X + X^2) = 1 + X + X^3 + X^4$$

$$m(X)g_2(X) = (1 + X^2)(1 + X^2) = 1 + X^4$$

$$\overline{m(X)g_1(X) = 1 + \ \ X + 0X^2 + \ \ X^3 + X^4}$$

$$m(X)g_2(X) = 1 + 0X + 0X^2 + 0X^3 + X^4$$

$$\overline{U(X) = (1, 1) + (1, 0)X + (0, 0)X^2 + (1, 0)X^3 + (1, 1)X^4}$$

$$U = 1\ 1 \quad\quad 1\ 0 \quad\quad 0\ 0 \quad\quad 1\ 0 \quad\quad 1\ 1$$

In this example we started with another point of view—that the convolutional encoder can be treated as a set of *cyclic code shift registers*. We represented the encoder with *polynomial generators* as used for describing cyclic codes. However, we arrived at the same output sequence as in Figure 6.4, and the same output sequence as the impulse response treatment of the preceding section. For a good presentation of convolutional code structure in the context of linear sequential circuits, see Reference [7].

6.2.2 State Representation and the State Diagram

The state of a rate $1/n$ convolutional encoder is defined as the contents of the rightmost $K - 1$ stages (see Figure 6.3). Knowledge of the state together with knowledge of the next input is necessary and sufficient to determine the next output. Let the state of the encoder at time, t_i, be defined as $X_i = m_{i-1}, m_{i-2}, \ldots, m_{i-K+1}$. The ith codeword branch, U_i, is completely determined by state X_i and the present input bit m_i; thus the state X_i represents the past history of the encoder in determining the encoder output. The encoder state is said to be *Markov*, in the sense that the probability, $P(X_{i+1}|X_i, X_{i-1}, \ldots, X_0)$, of being in state X_{i+1}, given all previous states, depends only on the most recent state, X_i; that is, the probability is equal to $P(X_{i+1}|X_i)$.

One way to represent simple encoders is with a *state diagram*; such a representation for the encoder in Figure 6.3 is shown in Figure 6.5. The states, shown in the boxes of the diagram, represent the possible contents of the rightmost $K - 1$ stages of the register, and the paths between the states represent the output branch words resulting from such state transitions. The states of the register are designated $a = 00$, $b = 10$, $c = 01$, and $d = 11$; the diagram shown in Figure 6.5 illustrates all the state transitions that are possible for the encoder in Figure

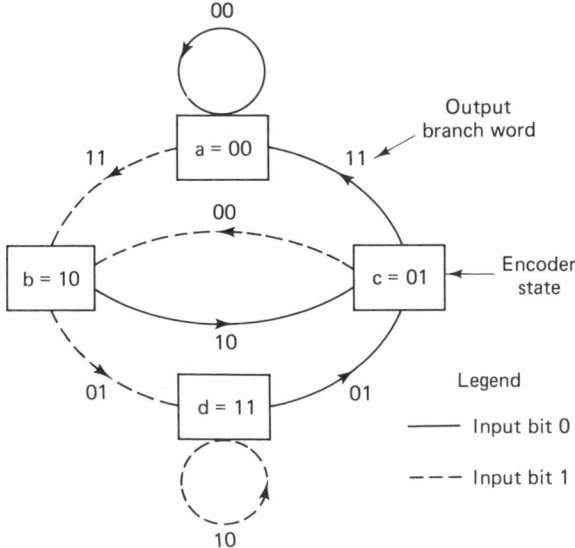

Figure 6.5 Encoder state diagram (rate $\frac{1}{2}$, $K = 3$).

Channel Coding: Part 2   Chap. 6

6.3. There are *only two transitions* emanating from each state, corresponding to the two possible input bits. Next to each path between states is written the output branch word associated with the state transition. In drawing the path, we use the convention that a solid line denotes a path associated with an input bit, zero, and a dashed line denotes a path associated with an input bit, one. Notice that it is *not possible* in a single transition to move from a given state to *any arbitrary state*. As a consequence of shifting-in one bit at a time, there are only two possible state transitions that the register can make at each bit time. For example, if the present encoder state is 00, the *only possibilities* for the state at the next shift are 00 or 10.

Example 6.1   Convolutional Encoding

For the encoder shown in Figure 6.3, show the state changes and the resulting output codeword sequence U for the message sequence m = 1 1 0 1 1, followed by $K - 1 = 2$ zeros to flush the register. Assume that the initial contents of the register are all zeros.

Solution

Input bit m_i	Register contents	State at time t_i	State at time t_{i+1}	Branch word at time t_i u_1	u_2
–	0 0 0	0 0	0 0	–	
1	1 0 0	0 0	1 0	1	1
1	1 1 0	1 0	1 1	0	1
0	0 1 1	1 1	0 1	0	1
1	1 0 1	0 1	1 0	0	0
1	1 1 0	1 0	1 1	0	1
0	0 1 1	1 1	0 1	0	1
0	0 0 1	0 1	0 0	1	1

state t_i

state t_{i+1}

Output sequence: U = 1 1   0 1   0 1   0 0   0 1   0 1   1 1

Example 6.2   Convolutional Encoding

In Example 6.1 the initial contents of the register are all zeros. This is equivalent to the condition that the given input sequence is preceded by two zero bits (the encoding is a function of the present bit and the $K - 1$ prior bits). Repeat Example 6.1 with the assumption that the given input sequence is preceded by two one bits, and verify that now the codeword sequence U for input sequence m = 1 1 0 1 1 is different than the codeword found in Example 6.1.

Solution

The entry "×" signifies "don't know."

Input bit m_i	Register contents	State at time t_i	State at time t_{i+1}	Branch word at time t_i	
				u_1	u_2
–	1 1 ×	1 ×	1 1	–	
1	1 1 1	1 1	1 1	1	0
1	1 1 1	1 1	1 1	1	0
0	0 1 1	1 1	0 1	0	1
1	1 0 1	0 1	1 0	0	0
1	1 1 0	1 0	1 1	0	1
0	0 1 1	1 1	0 1	0	1
0	0 0 1	0 1	0 0	1	1

state t_i

state t_{i+1}

Output sequence: U = 1 0   1 0   0 1   0 0   0 1   0 1   1 1

By comparing this result with that of Example 6.1, we can see that each branch word of the output sequence **U** is *not only* a function of the input bit, but is also a function of the $K - 1$ prior bits.

6.2.3 The Tree Diagram

Although the state diagram completely characterizes the encoder, one cannot easily use it for tracking the encoder transitions as a function of time since the diagram cannot represent time history. The tree diagram adds the *dimension of time* to the state diagram. The tree diagram for the covolutional encoder shown in Figure 6.3 is illustrated in Figure 6.6. At each successive input bit time the encoding procedure can be described by traversing the diagram from left to right, each tree branch describing an output branch word. The branching rule for finding a codeword sequence is as follows: If the input bit is a zero, its associated branch word is found by moving to the next rightmost branch in the upward direction. If the input bit is a one, its branch word is found by moving to the next rightmost branch in the downward direction. Assuming that the initial contents of the encoder is all zeros, the diagram shows that if the first input bit is a zero, the output branch word is 00 and, if the first input bit is a one, the output branch word is 11. Similarly, if the first input bit is a one and the second input bit is a zero, the second output branch word is 10. Or, if the first input bit is a one and the second input bit is a one, the second output branch word is 01. Following this procedure we see that the input sequence 1 1 0 1 1 traces the heavy line drawn on the tree

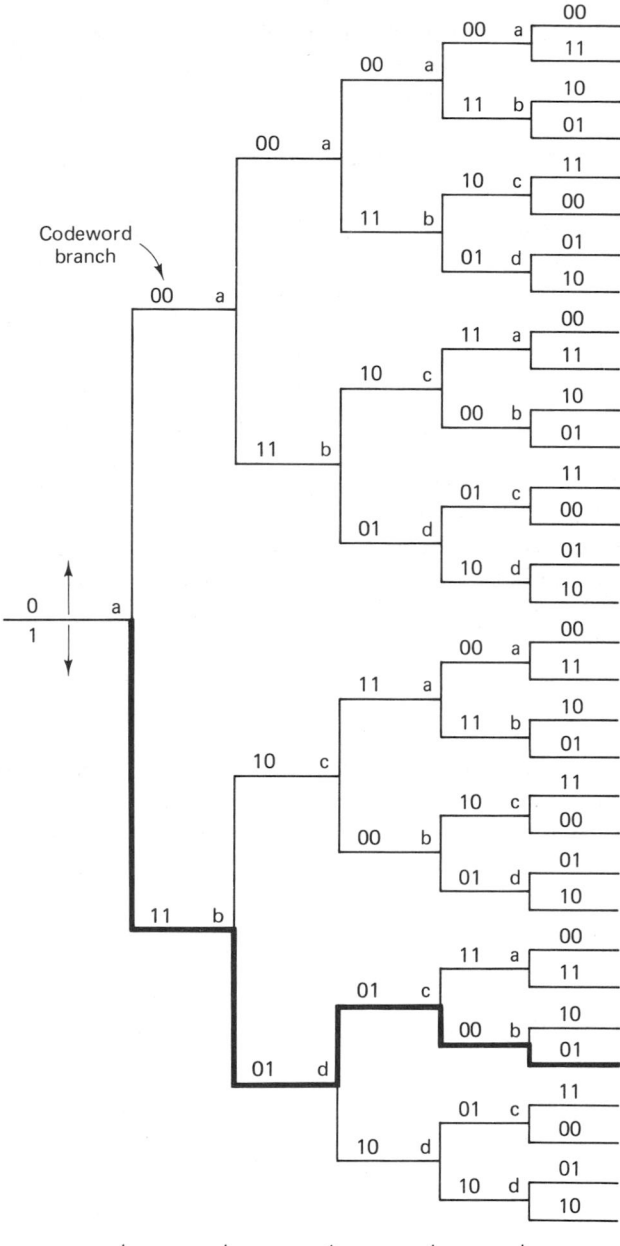

Codeword branch

Figure 6.6 Tree representation of encoder (rate $\frac{1}{2}$, $K = 3$).

t_1 t_2 t_3 t_4 t_5

diagram in Figure 6.6. This path corresponds to the following output codeword sequence: 1 1 0 1 0 1 0 0 0 1.

The added dimension of time in the tree diagram (compared to the state diagram) allows one to dynamically describe the encoder as a function of a particular input sequence. However, can you see one problem in trying to use a tree diagram for describing a sequence of any length? The number of branches increase as a function of 2^L, where L is the number of bits in the input sequence. You would quickly run out of paper, and patience.

6.2.4 The Trellis Diagram

Observation of the Figure 6.6 tree diagram shows that for this example, the structure repeats itself at time t_4, after the third branching (in general, the tree structure *repeats after K branchings*, where K is the constraint length). We label each node in the tree of Figure 6.6 to correspond to the four possible states in the shift register, as follows: $a = 00$, $b = 10$, $c = 01$, and $d = 11$. The first branching of the tree structure, at time t_1, produces a pair of nodes labeled a and b. At each successive branching the number of nodes double. The second branching, at time t_2, results in four nodes labeled a, b, c, and d. After the *third* branching there are a total of eight nodes; two of them are labeled a, two are labeled b, two are labeled c, and two are labeled d. We can see that all branches emanating from two nodes of the same state generate identical branch word sequences. From this point on, the upper and the lower halves of the tree are identical. The reason for this should be obvious from examination of the encoder in Figure 6.3. As the fourth input bit enters the encoder on the left, the first input bit is ejected on the right and no longer influences the output branch words. Consequently, the input sequences 1 0 0 x y . . . and 0 0 0 x y . . . , where the leftmost bit is the earliest bit, generate the same branch words after the ($K = 3$)rd branching. This means that any two nodes having the same state label, at the same time t_i, can be merged since all succeeding paths will be indistinguishable. If we do this to the tree structure of Figure 6.6, we obtain another diagram, called the trellis. The *trellis dia-*

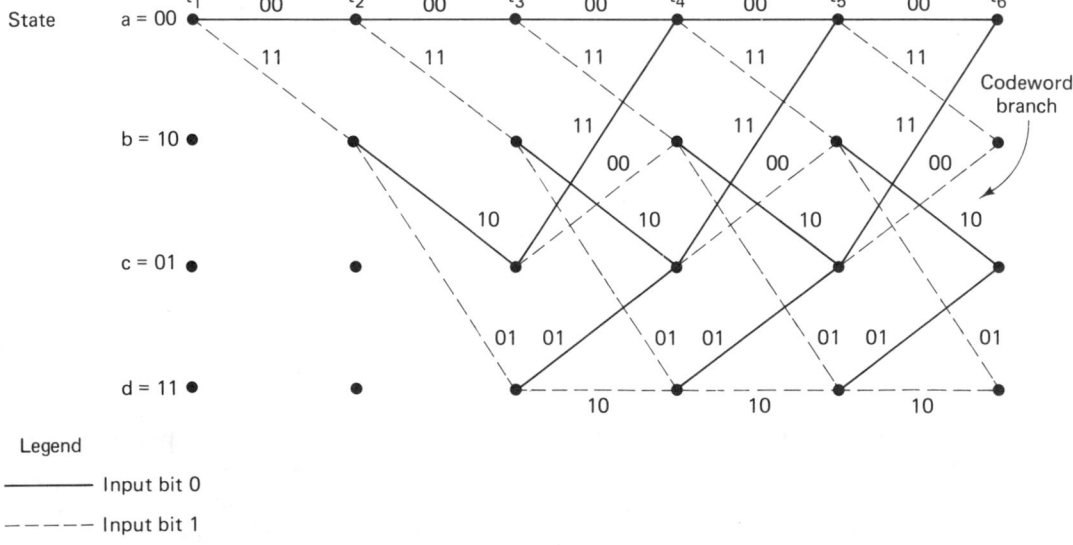

Figure 6.7 Encoder trellis diagram (rate $\frac{1}{2}$, $K = 3$).

gram, by exploiting the repetitive structure, provides a more manageable encoder description than does the tree diagram. The trellis diagram for the convolutional encoder of Figure 6.3 is shown in Figure 6.7.

In drawing the trellis diagram, we use the same convention that we introduced with the state diagram—that a solid line denotes the output generated by an input bit, zero, and a dashed line denotes the output generated by an input bit, one. The nodes of the trellis characterize the encoder states; the first row nodes correspond to the state $a = 00$, the second and subsequent rows correspond to the states $b = 10$, $c = 01$, and $d = 11$. At each unit of time the trellis requires 2^{K-1} nodes to represent the 2^{K-1} possible encoder states. The trellis in our example assumes a fixed periodic structure after trellis depth 3 is reached (at time t_4). In the general case, the fixed structure prevails after depth K is reached. After this point, each of the states can be entered from either of two preceding states. Also, each of the states can transition to one of two states. Of the two outgoing branches, one corresponds to an input bit zero and the other corresponds to an input bit one. On Figure 6.7 the output branch words corresponding to the state transitions appear as labels on the trellis branches.

6.3 FORMULATION OF THE CONVOLUTIONAL DECODING PROBLEM

6.3.1 Maximum Likelihood Decoding

If all input message sequences are equally likely, a decoder that achieves the minimum probability of error is one that compares the conditional probabilities, also called the *likelihood functions*, $P(\mathbf{Z}|\mathbf{U}^{(m)})$, where $\mathbf{Z}$ is the received sequence and $\mathbf{U}^{(m)}$ is one of the possible transmitted sequences, and chooses the maximum. The decoder chooses $\mathbf{U}^{(m')}$ if

$$P(\mathbf{Z}|\mathbf{U}^{(m')}) = \max_{\text{all } \mathbf{U}^{(m)}} P(\mathbf{Z}|\mathbf{U}^{(m)}) \tag{6.1}$$

The *maximum likelihood* concept, as stated in Equation (6.1), is a fundamental development of decision theory (see Appendix B); it is the formalization of a "common-sense" way to make decisions when there is statistical knowledge of the possibilities. In the binary demodulation treatment in Chapters 2 and 3 there were *only two* equally likely possible signals, $s_1(t)$ or $s_2(t)$, that might have been transmitted. Therefore, to make the binary maximum likelihood decision, given a received signal, meant only to decide that $s_1(t)$ was transmitted if

$$p(z|s_1) > p(z|s_2)$$

otherwise, to decide that $s_2(t)$ was transmitted. The parameter z represents $z(T)$, the receiver output at a symbol duration time $t = T$. However, when applying maximum likelihood to the convolutional decoding problem, there are typically a *multitude* of possible codeword sequences that might have been transmitted. To be specific, an L-bit codeword sequence is a member of a set of 2^L possible sequences. Therefore, in the maximum likelihood context, we can say that the decoder chooses a particular $\mathbf{U}^{(m')}$ as the transmitted sequence if the likelihood $P(\mathbf{Z}|\mathbf{U}^{(m')})$ is greater than the likelihoods of all the other possible transmitted sequences. Such an optimal decoder, which minimizes the error probability (for the case where all transmitted sequences are equally likely), is known as a *maximum likelihood decoder*. The likelihood functions are given or computed from the specifications of the channel.

We will assume that the noise is additive white Gaussian with zero mean and the channel is *memoryless*, which means that the noise affects each code symbol *independently* of all the other symbols. For a convolutional code of rate $1/n$, we can therefore express the likelihood, $P(\mathbf{Z}|\mathbf{U}^{(m)})$ as follows:

$$P(\mathbf{Z}|\mathbf{U}^{(m)}) = \prod_{i=1}^{\infty} P(Z_i|U_i^{(m)}) = \prod_{i=1}^{\infty} \prod_{j=1}^{n} P(z_{ji}|u_{ji}^{(m)}) \qquad (6.2)$$

where Z_i is the ith branch of the received sequence $\mathbf{Z}$, $U_i^{(m)}$ the ith branch of a particular codeword sequence $\mathbf{U}^{(m)}$, z_{ji} the jth code symbol of Z_i, and $u_{ji}^{(m)}$ the jth code symbol of $U_i^{(m)}$, each branch comprising n code symbols. The decoder problem consists of choosing a path through the trellis of Figure 6.7 (each possible path defines a codeword) such that

$$\prod_{i=1}^{\infty} \prod_{j=1}^{n} P(z_{ji}|u_{ji}^{(m)}) \text{ is maximized} \qquad (6.3)$$

Generally, it is computationally more convenient to use the logarithm of the likelihood function since this permits the summation, instead of the multiplication, of terms. We are able to use this transformation because the logarithm is a monotonically increasing function and thus will not alter the final result in our codeword selection. We can define the log-likelihood function $\gamma_{\mathbf{U}}(m)$ as

$$\gamma_{\mathbf{U}}(m) = \log P(\mathbf{Z}|\mathbf{U}^{(m)}) = \sum_{i=1}^{\infty} \log P(Z_i|U_i^{(m)}) = \sum_{i=1}^{\infty} \sum_{j=1}^{n} \log P(z_{ji}|u_{ji}^{(m)}) \qquad (6.4)$$

The decoder problem now consists of choosing a path through the tree of Figure 6.6 or the trellis of Figure 6.7 such that $\gamma_{\mathbf{U}}(m)$ is maximized. For the decoding of convolutional codes, either the tree or the trellis structure can be used. In the tree representation of the code, the fact that the paths remerge is ignored. Since the number of possible sequences for an L-symbol-long sequence is 2^L, maximum likelihood decoding of an L-bit-long received sequence, using a tree diagram, requires the "brute force" or exhaustive comparison of 2^L accumulated log-likelihood metrics, representing all the possible different codewords that could have been transmitted. Hence it is not practical to consider maximum likelihood decoding with a tree structure. It is shown in a later section that with the use of the trellis representation of the code, it is possible to configure a decoder which can discard the paths that could not possibly be candidates for the maximum likelihood sequence. The decoded path is chosen from some reduced set of *surviving paths*. Such a decoder is still optimum in the sense that the decoded path is the same as the decoded path obtained from a "brute force" maximum likelihood decoder, but the early rejection of unlikely paths reduces the decoding complexity.

For an excellent tutorial on the structure of convolutional codes, maximum likelihood decoding, and code performance, see Reference [8]. There are several algorithms that yield *approximate* solutions to the maximum likelihood decoding problem, including sequential [9, 10] and threshold [11]. Each of these algorithms is suited to certain special applications, but are all suboptimal. In contrast, the *Viterbi decoding algorithm* performs maximum likelihood decoding and is therefore optimal. This does not imply that the Viterbi algorithm is best for every application; there are severe constraints imposed by hardware complexity. The Viterbi algorithm is considered in Sections 6.3.3 and 6.3.4.

6.3.2 Channel Models: Hard versus Soft Decisions

Before specifying an algorithm that will determine the maximum likelihood decision, let us describe the channel. The codeword sequence $\mathbf{U}^{(m)}$, made up of branch words, with each branch word comprised of n code symbols, can be considered to be an endless stream, as opposed to a block code, in which the source data and their codewords are partitioned into precise block sizes. The codeword sequence shown in Figure 6.1 emanates from the convolutional encoder and enters the modulator, where the code symbols are transformed into signal waveforms. The modulation may be baseband (e.g., pulse waveforms) or bandpass (e.g., PSK or FSK). In general, ℓ symbols at a time, where ℓ is an integer, are mapped into signal waveforms $s_i(t)$, where $i = 1, 2, \ldots, M = 2^\ell$. When $\ell = 1$, the modulator maps each code symbol into a binary waveform. The channel over which the waveform is transmitted is assumed to corrupt the signal with Gaussian noise. When the corrupted signal is received, it is first processed by the demodulator and then by the decoder.

Consider that a binary signal, transmitted over a symbol interval $(0, T)$, is represented by $s_1(t)$ for a binary one and $s_2(t)$ for a binary zero. The received signal is $r(t) = s_i(t) + n(t)$, where $n(t)$ is a zero-mean Gaussian noise process. In Sections 2.9 and 3.4 we described the detection of $r(t)$ in terms of two basic steps. In the first step, the received waveform is reduced to a single number, $z(T) = a_i + n_0$, where a_i is the signal component of $z(T)$ and n_0 is the noise component. The noise component, n_0, is a zero-mean *Gaussian random variable*, and thus $z(T)$ is a *Gaussian random variable* with a mean of either a_1 or a_2 depending on whether a binary one or binary zero was sent. In the second step of the detection process a decision was made as to which signal was transmitted, on the basis of comparing $z(T)$ to a threshold. The conditional probabilities of $z(T)$, $p(z|s_1)$, and $p(z|s_2)$ are shown in Figure 6.8, labeled likelihood of s_1 and likelihood of s_2. The demodulator in Figure 6.1, converts the set of time-ordered random variables, $\{z(T)\}$, into a code sequence, $\mathbf{Z}$, and passes it on to the decoder. The demodulator output can be configured in a variety of ways. It can be implemented to make a *firm or hard decision* as to whether $z(T)$ represents a zero or a one. In this case, the output of the demodulator is quantized to two levels, zero and one, and fed

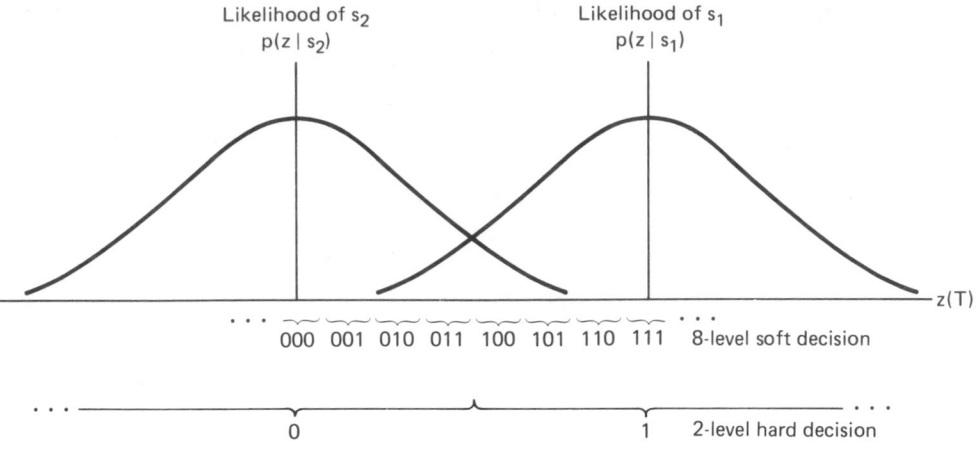

Figure 6.8 Hard and soft decoding decisions.

into the decoder (this is exactly the same threshold decision that was made in Chapters 2 and 3). Since the decoder operates on the hard decisions made by the demodulator, the decoding is called *hard-decision decoding*.

The demodulator can also be configured to feed the decoder with a *quantized value* of $z(T)$ *greater than two levels*, or with an unquantized or analog value of $z(T)$. Such an implementation furnishes the decoder with more information than is provided in the hard-decision case. When the quantization level of the demodulator output is greater than two, the decoding is called *soft-decision decoding*. Eight levels (3-bits) of quantization are illustrated on the abscissa of Figure 6.8. When the demodulator sends a hard binary decision to the decoder, it sends it a single binary symbol. When the demodulator sends a soft binary decision, quantized to eight levels, it sends the decoder a 3-bit word describing an interval along $z(T)$. In effect, sending such a 3-bit word in place of a single binary symbol is equivalent to sending the decoder a *measure of confidence* along with the code symbol. Referring to Figure 6.8, if the demodulator sends 1 1 1 to the decoder, this is tantamount to declaring the code symbol to be a one with very high confidence, while sending a 1 0 0 is tantamount to declaring the code symbol to be a one with very low confidence. It should be clear that ultimately, every message decision out of the decoder must be a hard decision; otherwise, one might see computer printouts that read: "think it's a 1," "think it's a 0," and so on. The idea behind the demodulator *not making hard decisions* and sending more data (soft decisions) to the decoder can be thought of as an interim step to provide the decoder with more information, which the decoder then uses for recovering the message sequence (with better error performance than it could in the case of hard-decision decoding).

For a Gaussian channel, eight-level quantization results in a performance improvement of approximately 2 dB in required signal-to-noise ratio compared to two-level quantization. This means that eight-level soft-decision decoding can provide the same probability of bit error as that of hard-decision decoding, but requires 2 dB *less* E_b/N_0 for the same performance. Analog (or infinite-level quantization) results in a 2.2-dB performance improvement over two-level quantization; therefore, *eight-level quantization* results in a loss of approximately 0.2 dB compared to infinitely fine quantization. For this reason, quantization to more than eight levels can yield little performance improvement [12]. What price is paid for such improved soft-decision-decoder performance? In the case of hard-decision decoding, a single bit is used to describe each code symbol, while for eight-level quantized soft-decision decoding 3 bits are used to describe each code symbol; therefore, three times the amount of data must be handled during the decoding process. Hence the price paid for soft-decision decoding is an increase in required memory size at the decoder (and possibly a speed penalty).

Block decoding algorithms and convolutional decoding algorithms have been devised to operate with hard *or* soft decisions. However, soft-decision decoding is generally not used with block codes because it is considerably more difficult than hard-decision decoding to implement. The most prevalent use of soft-decision decoding is with the *Viterbi convolutional decoding algorithm*, since with Viterbi decoding, soft decisions represent only a trivial increase in computation.

6.3.2.1 Binary Symmetric Channel

A binary symmetric channel (BSC) is a discrete memoryless channel (see Section 5.3.1) that has binary input and output alphabets and symmetric transition probabilities. It can be described by the conditional probabilities

$$P(0|1) = P(1|0) = p$$
$$P(1|1) = P(0|0) = 1 - p$$

(6.5)

as illustrated in Figure 6.9. The probability that an output symbol will differ from the input symbol is p, and the probability that the output symbol will be identical to the input symbol is $(1 - p)$. The BSC is an example of a *hard-decision channel*, which means that, even though continuous-valued signals may be received by the demodulator, a BSC allows only firm decisions such that each demodulator output symbol, z_{ji}, as shown in Figure 6.1, consists of one of two binary values. The indexing of z_{ji} pertains to the jth code symbol of the ith branch word, Z_i. The demodulator then feeds the sequence $\mathbf{Z} = \{Z_i\}$ to the decoder.

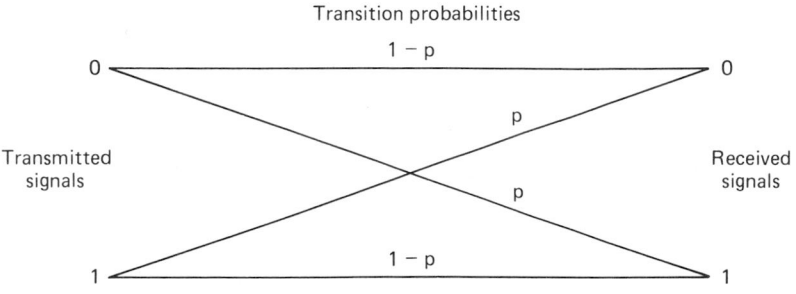

Figure 6.9 Binary symmetric channel (hard-decision channel).

Let $U^{(m)}$ be a transmitted codeword over a BSC with symbol error probability p, and let Z be the corresponding received decoder sequence. As noted previously, a maximum likelihood decoder chooses the codeword $U^{(m')}$ which maximizes the likelihood, $P(Z|U^{(m)})$, or its logarithm. For a BSC, this is equivalent to choosing the codeword, $U^{(m')}$, that is closest in *Hamming distance* to Z [8]. Thus Hamming distance is an appropriate metric to describe the distance or closeness of fit between $U^{(m)}$ and Z. From all the possible transmitted sequences, $U^{(m)}$, the decoder chooses the $U^{(m')}$ sequence for which the distance to Z is minimum.

Suppose that $U^{(m)}$ and Z are each L-bit-long sequences and that they differ in d_m positions [i.e., the Hamming distance between $U^{(m)}$ and Z is d_m]. Then, since the channel is assumed memoryless, the probability that this $U^{(m)}$ was transformed to the specific received Z at distance d_m from it can be written

$$P(Z|U^{(m)}) = p^{d_m}(1-p)^{L-d_m} \tag{6.6}$$

and the log-likelihood function is

$$\log P(Z|U^{(m)}) = -d_m \log\left(\frac{1-p}{p}\right) + L\log(1-p) \tag{6.7}$$

If we compute this quantity for each possible transmitted sequence, the second term will be constant in each case. Assuming that $p < 0.5$, we can express Equation (6.7) as

$$\log P(Z|U^{(m)}) = -Ad_m - B \tag{6.8}$$

where A and B are positive constants. Therefore, choosing the codeword $U^{(m')}$ such that the Hamming distance, d_m, to the received sequence Z is minimized corresponds to *maximizing the likelihood or log-likelihood metric*. Consequently, over a BSC, the log-likelihood metric is conveniently replaced by the Hamming distance, and a maximum likelihood decoder will choose, in the tree or trellis diagram, the path whose corresponding sequence, $U^{(m')}$, is at the *minimum Hamming distance* to the received sequence Z.

6.3.2.2 Gaussian Channel

For a Gaussian channel, each demodulator output symbol, z_{ji}, as shown in Figure 6.1, is a value from a continuous alphabet. The symbol z_{ji} cannot be labeled as a correct or incorrect detection decision. Sending the decoder such soft decisions can be viewed as sending a family of conditional probabilities of the different symbols (see Section 5.3.1). It can be shown [8] that maximizing $P(Z|U^{(m)})$ is equivalent to maximizing the inner product between the codeword sequence, $U^{(m)}$ (consisting of binary symbols), and the analog-valued received sequence, Z. Thus the decoder chooses the codeword $U^{(m')}$ if it maximizes

$$\sum_{i=1}^{\infty}\sum_{j=1}^{n} z_{ji}u_{ji}^{(m)} \tag{6.9}$$

This is equivalent to choosing the codeword $U^{(m')}$ that is closest in *Euclidean distance* to Z. Even though the hard- and soft-decision channels require different metrics, the concept of choosing the codeword $\mathbf{U}^{(m')}$ that is closest to the received sequence, $\mathbf{Z}$, is the same in both cases. To implement the maximization of Equation (6.9) exactly, the decoder would have to be able to handle analog-valued arithmetic operations. This is impractical because the decoder is generally implemented digitally. Thus it is necessary to quantize the received symbols z_{ji}. Does Equation (6.9) remind you of the demodulation treatment in Chapter 3? Equation (6.9) is the discrete version of correlating an input received waveform, $r(t)$, with a reference waveform, $s_i(t)$, as expressed in Equation (3.34). The quantized Gaussian channel, typically referred to as a *soft-decision channel*, is the channel model assumed for the soft-decision decoding described earlier.

6.3.3 The Viterbi Convolutional Decoding Algorithm

The Viterbi decoding algorithm was discovered and analyzed by Viterbi [13] in 1967. The Viterbi algorithm essentially performs maximum likelihood decoding; however, it reduces the computational load by taking advantage of the special structure in the code trellis. The advantage of Viterbi decoding, compared with brute-force decoding, is that the complexity of a Viterbi decoder is not a function of the number of symbols in the codeword sequence. The algorithm involves calculating a *measure of similarity*, *or distance*, between the received signal, at time t_i, and all the trellis paths entering each state at time t_i. The Viterbi algorithm removes from consideration those trellis paths that could not possibly be candidates for the maximum likelihood choice. When two paths enter the same state, the one having the best metric is chosen; this path is called the *surviving path*. This selection of surviving paths is performed for all the states. The decoder continues in this way to advance deeper into the trellis, making decisions by eliminating the least likely paths. The early rejection of the unlikely paths reduces the decoding complexity. In 1969, Omura [14] demonstrated that the Viterbi algorithm is, in fact, maximum likelihood. Note that the goal of selecting the optimum path can be expressed, equivalently, as choosing the codeword with the *maximum likelihood metric*, or as choosing the codeword with the *minimum distance metric*.

6.3.4 An Example of Viterbi Convolutional Decoding

For simplicity, a BSC is assumed; thus Hamming distance is a proper distance measure. The encoder for this example is shown in Figure 6.3, and the encoder trellis diagram is shown in Figure 6.7. A similar trellis can be used to represent the decoder, as shown in Figure 6.10. The basic idea behind the decoding procedure can best be understood by examining the Figure 6.7 encoder trellis in concert with the Figure 6.10 decoder trellis. For the decoder trellis it is convenient to label each trellis branch at time t_i with the *Hamming distance* between the received code symbols and the corresponding branch word from the encoder trellis. The example in Figure 6.10, shows a message sequence, $\mathbf{m}$, the corresponding codeword sequence, $\mathbf{U}$, and a noise corrupted received sequence, $\mathbf{Z} =$ 11 01 01 10 01 . . . . The branch words seen on the *encoder trellis* branches

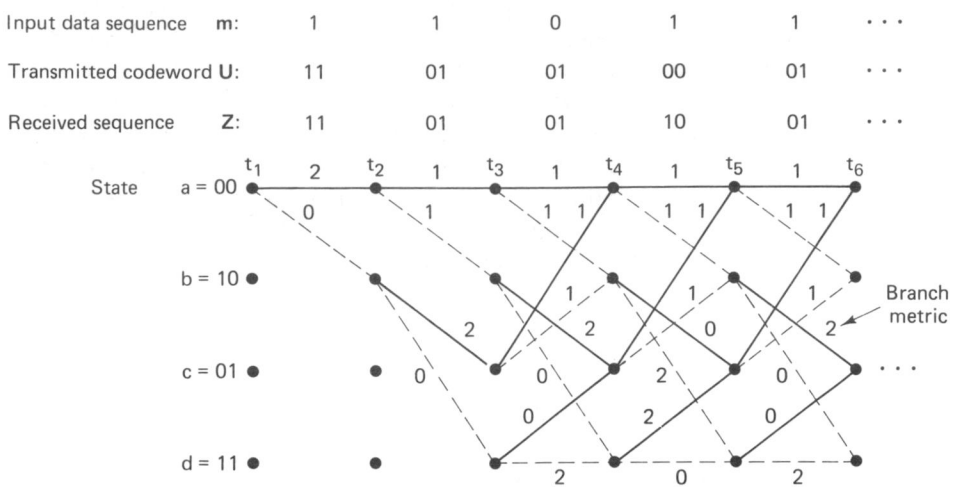

Figure 6.10 Decoder trellis diagram (rate $\frac{1}{2}$, $K = 3$).

characterize the encoder in Figure 6.3, and are known a priori to both the encoder and the decoder. These encoder branch words are the code symbols that would be expected to come from the encoder output as a result of each of the state transitions. The labels on the *decoder trellis* branches are accumulated by the decoder *on the fly*. That is, as the code symbols are received, each branch of the decoder trellis is labeled with a metric of similarity (Hamming distance) between the received code symbols and each of the branch words for that time interval. From the received sequence, **Z**, in Figure 6.10, we see that the code symbols received at time t_1 are 11. In order to label the decoder branches at time t_1 with the appropriate Hamming distance metric, we look at the Figure 6.7 encoder trellis. Here we see that a state 00 → 00 transition yields an output branch word of 00. But we received 11. Therefore, on the decoder trellis we label the state 00 → 00 transition with the Hamming distance between them, namely 2. Looking at the encoder trellis again, we see that a state 00 → 10 transition yields an output branch word of 11, which corresponds exactly with the code symbols we received at time t_1. Therefore, on the decoder trellis, we label the state 00 → 10 transition with a Hamming distance of 0. We continue labeling the decoder trellis branches in this way as the symbols are received at each time t_i. The decoding algorithm uses these Hamming distance metrics to find the *most likely* (minimum distance) path through the trellis.

The basis of *Viterbi decoding* is the following observation: If any two paths in the trellis merge to a single state, one of them can always be eliminated in the search for an optimum path. For example, Figure 6.11 shows two paths merging at time t_5 to state 00. Let us define the *cumulative Hamming path metric* of a given path at time t_i as the sum of the branch Hamming distance metrics along that path up to time t_i. In Figure 6.11 the upper path has metric 4; the lower has metric 1. The upper path cannot be a portion of the optimum path because the lower path, which enters the same state, has a lower metric. This observation holds because of the Markov nature of the encoder state: The present state summarizes the encoder history in the sense that previous states cannot affect future states or future output branches.

At each time t_i there are 2^{K-1} states in the trellis, where K is the constraint length, and each state can be *entered by means of two paths*. Viterbi decoding consists of computing the metrics for the two paths entering each state and *eliminating one of them*. This computation is done for each of the 2^{K-1} nodes at time t_i; then the decoder moves to time t_{i+1} and repeats the process. The first few steps in our decoding example are as follows (see Figure 6.12). Assume that the input data sequence **m**, codeword **U**, and received sequence **Z** are as shown in Figure 6.10. Assume that the decoder knows the correct initial state of the trellis. (This assumption is not necessary in practice, but simplifies the explanation.) At time t_1 the received code symbols are 11. From state 00 the only possible transitions are to state 00 or state 10, as shown in Figure 6.12a. State $00 \rightarrow 00$ transition has branch metric 2; state $00 \rightarrow 10$ transition has branch metric 0. At time t_2 there are two possible branches leaving each state, as shown in Figure 6.12b. The cumulative path metrics of these branches are labeled λ_a, λ_b, λ_c, and λ_d, corresponding to the terminating state. At time t_3 in Figure 6.12c there are again two branches diverging from each state. As a result, there are two paths entering each state at time t_4. As noted previously, one path entering each state can be eliminated, namely, the one having the larger cumulative path metric. Should metrics of the two entering paths be of equal value, one path is chosen for elimination by using an arbitrary rule. The surviving path into each state is shown in Figure 6.12d. At this point in the decoding process, there is only a single surviving path between times t_1 and t_2. Therefore, the decoder can now decide that the state transition which occurred between t_1 and t_2 was $00 \rightarrow 10$. Since this transition is produced by an input bit one, the decoder outputs a one as the first decoded bit. Here we can see how the decoding of the surviving branch is facilitated by having drawn the lattice branches with solid lines for input zeros and dashed lines for input ones. Note that the first bit was not decoded until the path metric computation had proceeded to a much greater depth into the trellis. For a typical

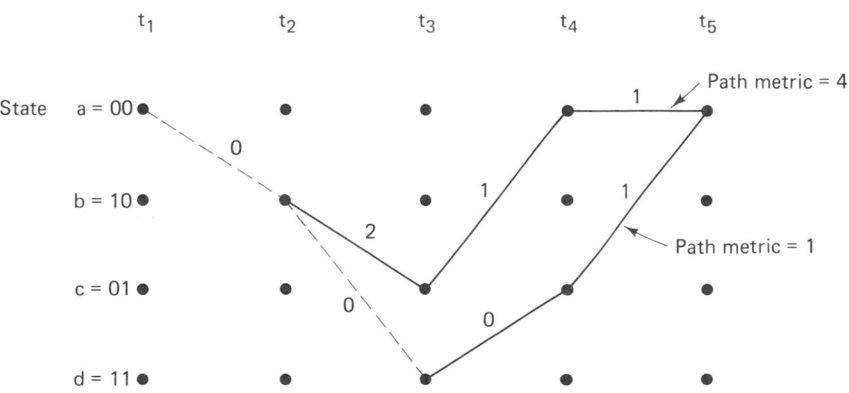

Figure 6.11   Path metrics for two merging paths.

decoder implementation, this represents a decoding delay which can be as much as five times the constraint length in bits.

At each succeeding step in the decoding process, there will always be two possible paths entering each state; one of the two will be eliminated by comparing the path metrics. Figure 6.12e shows the next step in the decoding process. Again, at time t_5 there are two paths entering each state, and one of each pair can be eliminated. Figure 6.12f shows the survivors at time t_5. Notice that in our example we cannot yet make a decision on the second input data bit because there still are two paths leaving the state 10 node at time t_2. At time t_6 in Figure 6.12g we again see the pattern of remerging paths, and in Figure 6.12h we see the survivors at time t_6. Also, in Figure 6.12h the decoder outputs one as the second decoded bit, corresponding to the single surviving path between t_2 and t_3. The decoder continues in this way to advance deeper into the trellis and to make decisions on the input data bits by eliminating all paths but one.

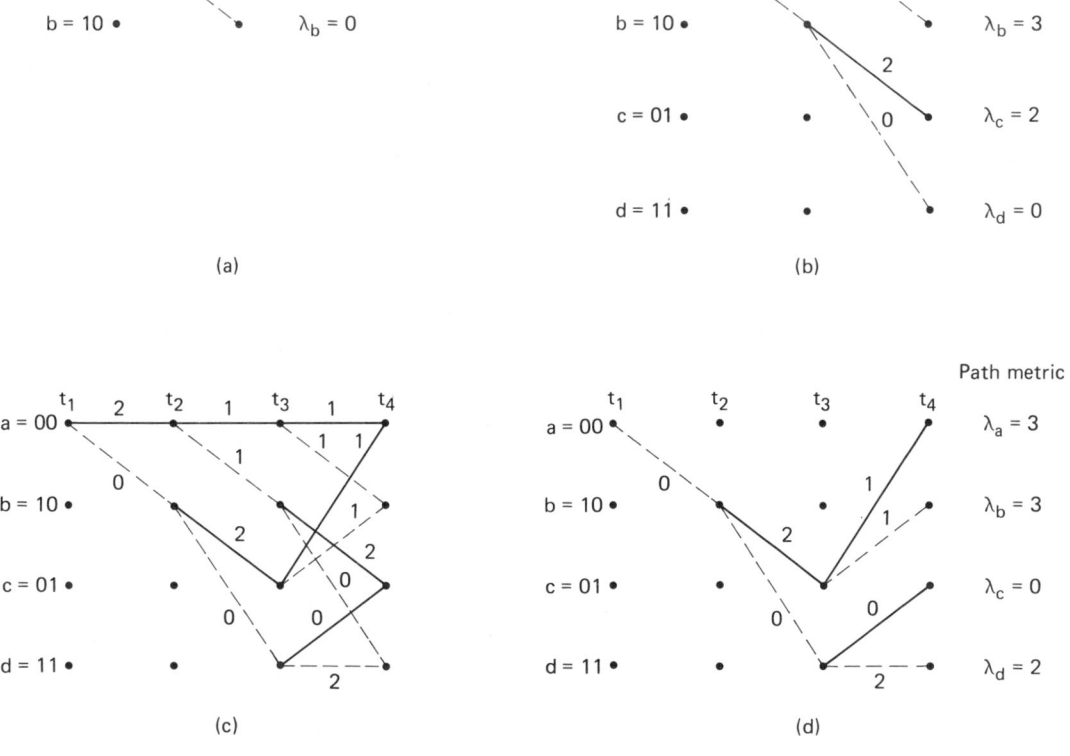

Figure 6.12 Selection of survivor paths. (a) Survivors at t_2. (b) Survivors at t_3. (c) Metric comparisons at t_4. (d) Survivors at t_4. (e) Metric comparisons at t_5. (f) Survivors at t_5. (g) Metric comparisons at t_6. (h) Survivors at t_6.

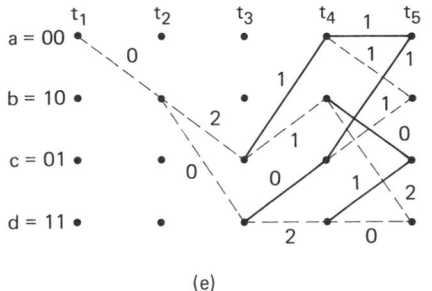

(e)

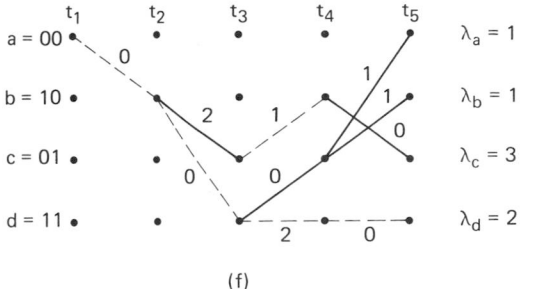

(f)

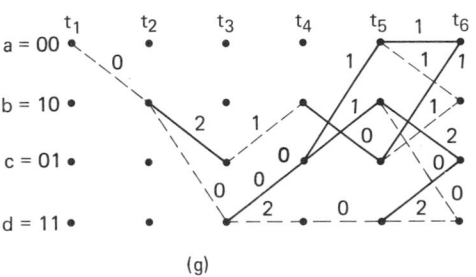

(g)

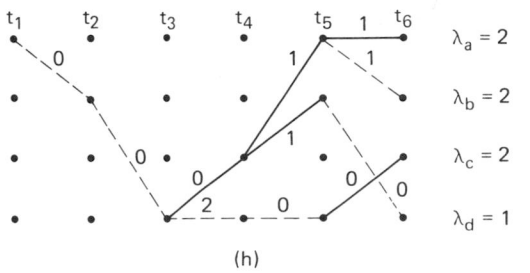

(h)

Figure 6.12 (*Continued*)

6.3.5 Path Memory and Synchronization

The storage requirements of the Viterbi decoder grow exponentially with constraint length K. For a code with rate $1/n$, the decoder retains a set of 2^{K-1} paths after each decoding step. With high probability, these paths will not be mutually disjoint very far back from the present decoding depth [12]. All of the 2^{K-1} paths tend to have a common stem which eventually branches to the various states. Thus if the decoder stores enough of the history of the 2^{K-1} paths, the oldest bits on all paths will be the same. A simple decoder implementation, then, contains a *fixed amount of path history* and outputs the oldest bit on an arbitrary path each time it steps one level deeper into the trellis. The amount of path storage required, u, is [12]

$$u = h2^{K-1} \tag{6.10}$$

where h is the length of the information bit path history per state. A refinement, which minimizes the value of h, uses the oldest bit on the most likely path as the decoder output, instead of the oldest bit on an arbitrary path. It has been demonstrated [12] that a value of h of 4 or 5 times the code constraint length is sufficient for near-optimum decoder performance. The storage requirement, u, is the basic limitation on the implementation of Viterbi decoders. The current state of the art limits decoders to a constraint length of about $K = 10$. Efforts to increase coding gain by further increasing constraint length are met by the exponential increase in memory requirements (and complexity) that follows from Equation (6.10).

Branch word synchronization is the process of determining the beginning of a branch word in the received sequence. Such synchronization can take place without new information being added to the transmitted symbol stream because the received data appear to have an excessive error rate when not synchronized. Therefore, a simple way of accomplishing synchronization is to monitor some concomitant indication of this large error rate, that is, the rate at which the path metrics are increasing or the rate at which the surviving paths in the trellis merge. The monitored parameters are compared to a threshold, and synchronization is then adjusted accordingly.

6.4 PROPERTIES OF CONVOLUTIONAL CODES

6.4.1 Distance Properties of Convolutional Codes

Let us consider the distance properties of convolutional codes in the context of our simple encoder in Figure 6.3 and its trellis diagram in Figure 6.7. We want to evaluate the distance between all possible pairs of codeword sequences. As in the case of block codes (see Section 5.5.2), we are interested in the *minimum distance* between all pairs of such codeword sequences in the code, since the minimum distance is related to the error-correcting capability of the code. Because a convolutional code is a group or *linear code* [6], there is no loss in generality in simply finding the minimum distance between each of the codeword sequences and the all-zeros sequence. Assuming that the all-zeros input sequence was transmitted, the paths of interest are those that start and end in the 00 state and do not return to the 00 state anywhere in between. An error will occur whenever the distance of any other path that merges with the $a = 00$ state at time t_i is less than that of the all-zeros path up to time t_i, causing the all-zeros path to be discarded in the decoding process. In other words, given the all-zeros transmission, an error occurs whenever the *all-zeros path does not survive*. The minimum distance for making such an error can be found by exhaustively examining every path from the 00 state to the 00 state. First, let us redraw the trellis diagram, shown in Figure 6.13, labeling each branch with its Hamming distance from the all-zeros codeword instead of with its branch word symbols. The Hamming distance between two unequal-length sequences will be found by first appending the necessary number of zeros to the shorter sequence to make the two sequences equal in length. Consider all the paths that diverge from the all-zeros path and then remerge for the first time at some arbitrary node. From Figure 6.13 we can compute the distances of these paths from the all-zeros path. There is one path at distance 5 from the all-zeros path; this path departs from the all-zeros path at time t_1 and merges with it at time t_4. Similarly, there are two paths at distance 6, one which departs at time t_1 and merges at time t_5, and the other which departs at time t_1 and merges at time t_6, and so on. We can also see from the dashed and solid lines

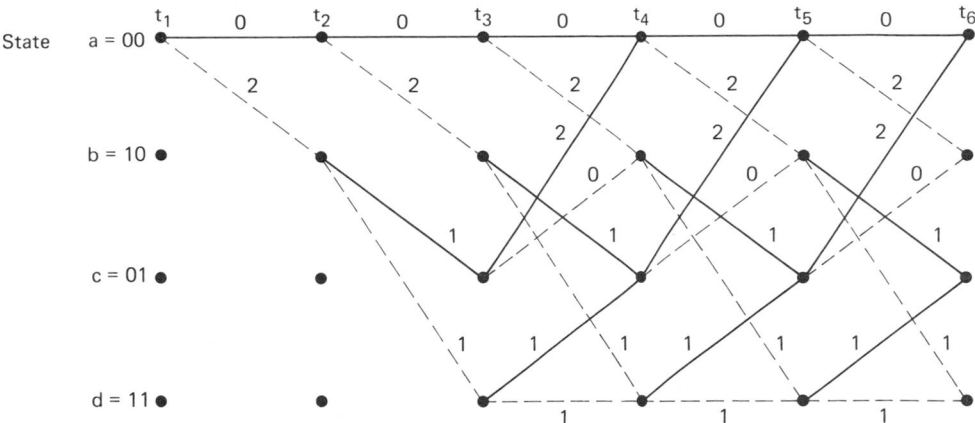

Figure 6.13 Trellis diagram, labeled with distances from the all-zeros path.

of the diagram that the input bits for the distance 5 path are 1 0 0; it differs in only one input bit from the all-zeros input sequence. Similarly, the input bits for the distance 6 paths are 1 1 0 0 and 1 0 1 0 0; each differs in two positions from the all-zeros path. The minimum distance in the set of all arbitrarily long paths that diverge and remerge, called the *minimum free distance* or simply the *free distance*, is seen to be 5 in this example. For calculating the error-correcting capability of the code, we repeat Equation (5.44) with the minimum distance, $d_{\min}$, replaced by the free distance, d_f.

$$t = \left\lfloor \frac{d_f - 1}{2} \right\rfloor \qquad (6.11)$$

where $\lfloor x \rfloor$ means the largest integer no greater than x. Setting $d_f = 5$, we see that the code, characterized by the Figure 6.3 encoder, can correct any two channel errors.

Although Figure 6.13 presents the computation of free distance in a straightforward way, a more direct closed-form expression can be obtained by starting with the state diagram in Figure 6.5. First, we label the branches of the state diagram as either $D^0 = 1$, D^1, or D^2, shown in Figure 6.14, where the exponent of D denotes the Hamming distance from the branch word of that branch to the all-zeros branch. The self-loop at node a can be eliminated since it contributes nothing to the distance properties of a codeword sequence relative to the all-zeros sequence. Furthermore, node a can be split into two nodes (labeled a and e), one of which represents the input and the other the output of the state diagram. All paths originating at $a = 00$ and terminating at $e = 00$ can be traced on the modified state diagram of Figure 6.14. We can calculate the transfer function of path $a\ b\ c\ e$ (starting and ending at state 00) in terms of the indeterminate "placeholder" D, as $D^2\ D\ D^2 = D^5$. The exponent of D represents the cumulative tally of the number of ones in the path, and hence the Hamming distance from the all-zeros path. Similarly, the paths $a\ b\ d\ c\ e$ and $a\ b\ c\ b\ c\ e$ each have the transfer

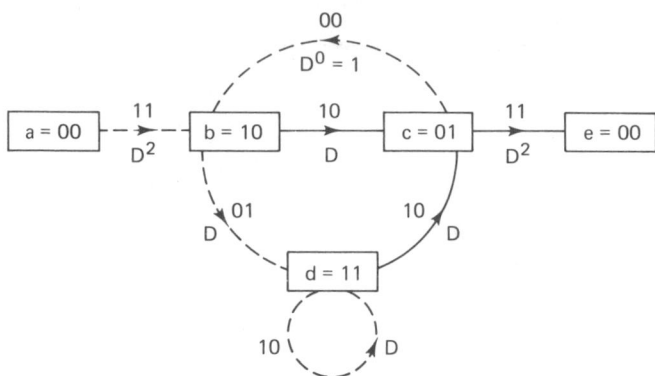

Figure 6.14 State diagram, labeled according to distance from the all-zeros path.

function D^6 and thus a Hamming distance of 6 from the all-zeros path. We now write the following state equations:

$$X_b = D^2 X_a + X_c$$

$$X_c = DX_b + DX_d$$

$$X_d = DX_b + DX_d$$ (6.12)

$$X_e = D^2 X_c$$

where $X_a, \ldots, X_e$ are dummy variables for the partial paths to the intermediate nodes. The *transfer function*, $T(D)$, sometimes called the *generating function* of the code can be expressed as $T(D) = X_e/X_a$. By solving the state equations shown in Equation (6.12), we obtain [15, 16]

$$T(D) = \frac{D^5}{1 - 2D}$$ (6.13)

$$= D^5 + 2D^6 + 4D^7 + \cdots + 2^\ell D^{\ell+5} + \cdots$$

The transfer function for this code indicates that there is a single path of distance 5 from the all-zeros path, two of distance 6, four of distance 7, and in general, there are 2^ℓ paths of distance $\ell + 5$ from the all-zeros path, where $\ell = 0, 1, 2, \ldots$. The free distance d_f of the code is the Hamming weight of the lowest-order term in the expansion of $T(D)$. In this example $d_f = 5$. In evaluating distance properties, the transfer function, $T(D)$, cannot be used for long constraint lengths since the complexity of $T(D)$ increases exponentially with constraint length.

The transfer function can be used to provide more detailed information than just the distance of the various paths. Let us introduce a factor L into each branch of the state diagram so that the exponent of L can serve as a counter to indicate the number of branches in any given path from state $a = 00$ to state $e = 00$. Furthermore, we can introduce a factor N into all branch transitions caused by the input bit one. Thus, as each branch is traversed, the cumulative exponent on N increases by one, only if that branch transition is due to an input bit one. For the convolutional code characterized in our Figure 6.3 example, the additional factors L and N are shown on the modified state diagram of Figure 6.15. We can now modify Equations (6.12) as follows:

$$X_b = D^2LNX_a + LNX_c$$

$$X_c = DLX_b + DLX_d$$

$$X_d = DLNX_b + DLNX_d \qquad (6.14)$$

$$X_e = D^2LX_c$$

The transfer function of this augmented state diagram is

$$T(D, L, N) = \frac{D^5L^3N}{1 - DL(1 + L)N}$$

$$= D^5L^3N + D^6L^4(1 + L)N^2 + D^7L^5(1 + L)^2N^3 \qquad (6.15)$$

$$+ \cdots + D^{\ell+5}L^{\ell+3}N^{\ell+1} + \cdots$$

Thus we can verify some of the path properties displayed in Figure 6.13. There is one path of distance 5, length 3, which differs in one input bit from the all-zeros path. There are two paths of distance 6, one of which is length 4, the other length 5, and both differ in two input bits from the all-zeros path. Also, of the distance 7 paths, one is of length 5, two are of length 6, and one is of length 7; all four paths correspond to input sequences that differ in three input bits from the all-zeros path. Thus if the all-zeros path is the correct path and the noise causes us to choose one of the incorrect paths of distance 7, three bit errors will be made.

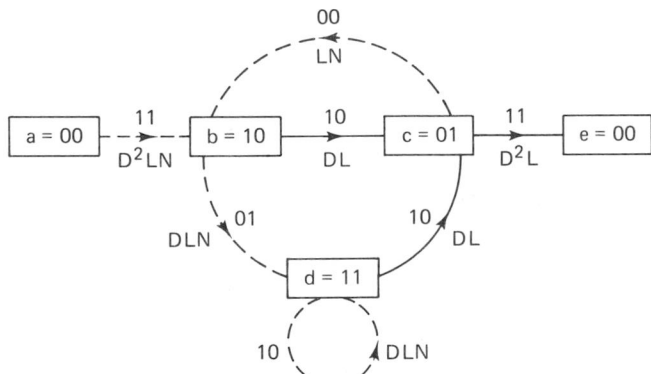

Figure 6.15 State diagram, labeled according to distance, length, and number of input ones.

6.4.1.1 Error-Correcting Capability of Convolutional Codes

In the study of block codes in Chapter 5, we saw that the error-correcting capability, t, represented the number of code symbol errors that could, with maximum likelihood decoding, be corrected in each block length of the code. However, when decoding convolutional codes, the error-correcting capability cannot be stated so succinctly. With regard to Equation (6.11), we can say that the code can, with maximum likelihood decoding, correct t errors within a few constraint lengths, where "few" here means 3 to 5. The exact length depends on how the errors are distributed. For a particular code and error pattern, the length can be bounded using transfer function methods. A computer program for convolutional decoding with the Viterbi algorithm, called VITALG, is provided in Appendix E. The interested reader can use this tool for verifying the capability of Viterbi decoding of convolutional codes with various choices of code generators, code rates, constraint lengths, and path memory lengths.

6.4.2 Systematic and Nonsystematic Convolutional Codes

A *systematic* convolutional code is one in which the input k-tuple appears as part of the output branch word n-tuple associated with that k-tuple. Figure 6.16 shows a binary, rate $\frac{1}{2}$, $K = 3$ systematic encoder. For linear block codes, any nonsystematic code can be transformed into a systematic code with the same block distance properties. This is not the case for convolutional codes. The reason for this is that convolutional codes depend largely on *free distance*; making the convolutional code systematic, in general, *reduces* the maximum possible free distance for a given constraint length and rate.

Table 6.1 shows the maximum free distance for rate $\frac{1}{2}$ systematic and nonsystematic codes for $K = 2$ through 8. For large constraint lengths the results are even more widely separated [17].

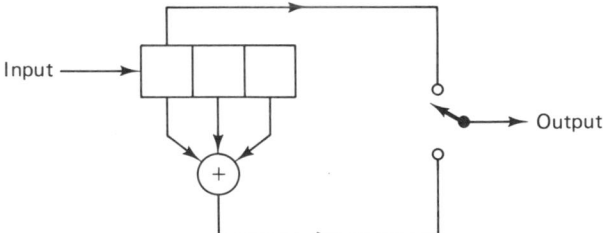

Figure 6.16 Systematic convolutional encoder, rate $\frac{1}{2}$, $K = 3$.

6.4.3 Catastrophic Error Propagation in Convolutional Codes

A *catastrophic error* is defined as an event whereby a finite number of code symbol errors cause an infinite number of decoded data bit errors. Massey and Sain [18] have derived a necessary and sufficient condition for convolutional codes to display catastrophic error propagation. For rate $1/n$ codes with register taps designated by polynomial generators, as described in Section 6.2.1, the condition for catastrophic error propagation is that the generators have a *common polynomial factor* (of degree at least one). For example, Figure 6.17a illustrates a rate $\frac{1}{2}$, $K = 3$ encoder with upper polynomial $\mathbf{g}_1(X)$ and lower polynomial $\mathbf{g}_2(X)$, as follows:

$$\mathbf{g}_1(X) = 1 + X$$
$$\mathbf{g}_2(X) = 1 + X^2 \qquad (6.16)$$

Constraint length	Free distance systematic	Free distance nonsystematic
2	3	3
3	4	5
4	4	6
5	5	7
6	6	8
7	6	10
8	7	10

Source: A. J. Viterbi and J. K. Omura, *Principles of Digital Communication and Coding*, McGraw-Hill Book Company, New York, 1979, p. 251.

The generators $\mathbf{g}_1(X)$ and $\mathbf{g}_2(X)$ have in common the polynomial factor, $1 + X$, since

$$1 + X^2 = (1 + X)(1 + X)$$

Therefore, the encoder in Figure 6.17a can manifest *catastrophic error propagation*.

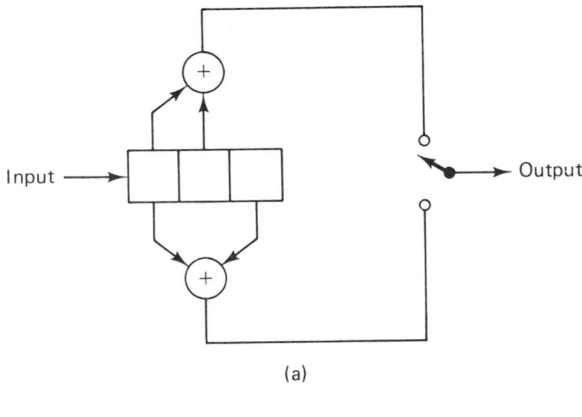

(a)

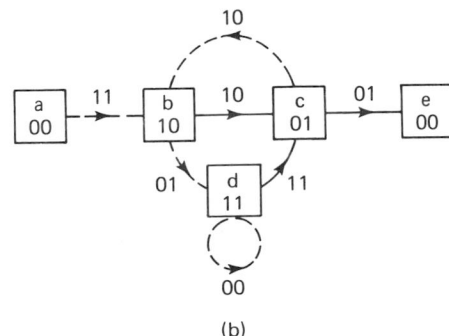

(b)

Figure 6.17 Encoder displaying catastrophic error propagation. (a) Encoder. (b) State diagram.

In terms of the state diagram for any-rate code, catastrophic errors can occur if, and only if, any closed-loop path in the diagram has zero weight (zero distance from the all-zeros path). To illustrate this, consider the example of Figure 6.17. The state diagram in Figure 6.17b is drawn with the state $a = 00$ node split into two nodes, a and e, as before. Assuming that the all-zeros path is the correct path, the incorrect path $a\ b\ d\ d\ \ldots\ d\ c\ e$ has exactly 6 ones, no matter how many times we go around the self-loop at node d. Thus for a BSC, for example, three channel errors may cause us to choose this incorrect path. An arbitrarily large number of errors (two plus the number of times the self-loop is traversed) can be made on such a path. We observe that for rate $1/n$ codes, if each adder in the encoder has an even number of connections, the self-loop corresponding to the all-ones data state will have zero weight, and consequently, *the code will be catastrophic*.

The only advantage of a systematic code, described earlier, is that it can never be catastrophic, since each closed loop must contain at least one branch generated by a nonzero input bit, and thus each closed loop must have a nonzero code symbol. However, it can be shown [19] that only a small fraction of non-systematic codes (excluding those where all adders have an even number of taps) are catastrophic.

6.4.4 Performance Bounds for Convolutional Codes

The probability of bit error, P_B, for a binary convolutional code using hard-decision decoding can be shown [8] to be upper bounded as follows:

$$P_B \leq \frac{dT(D, N)}{dN}\bigg|_{N=1, D=2\sqrt{p(1-p)}} \tag{6.17}$$

where p is the probability of channel symbol error. For the example of Figure 6.3, $T(D, N)$ is obtained from $T(D, L, N)$ by setting $L = 1$ in Equation (6.15).

$$T(D, N) = \frac{D^5 N}{1 - 2DN} \tag{6.18}$$

and

$$\frac{dT(D, N)}{dN}\bigg|_{N=1} = \frac{D^5}{(1 - 2D)^2} \tag{6.19}$$

Combining Equations (6.17) and (6.19), we can write

$$P_B \leq \frac{\{2[p(1 - p)]^{1/2}\}^5}{\{1 - 4[p(1 - p)]^{1/2}\}^2} \tag{6.20}$$

For coherent BPSK modulation over an additive white Gaussian noise (AWGN) channel, it can be shown [8] that the bit error probability is bounded by

$$P_B \leq Q\left(\sqrt{2d_f \frac{E_c}{N_0}}\right) \exp\left(d_f \frac{E_c}{N_0}\right) \frac{dT(D, N)}{dN}\bigg|_{N=1, D=\exp(-E_c/N_0)} \quad (6.21)$$

where

$$E_c/N_0 = rE_b/N_0$$

E_b/N_0 = ratio of information bit energy to noise power spectral density

E_c/N_0 = ratio of channel symbol energy to noise power spectral density

$r = k/n$ = rate of the code

and where $Q(x)$ is defined in Equations (2.42) and (2.43) and tabulated in Table B.1. Therefore, for the rate $\frac{1}{2}$ code with free distance $d_f = 5$, in conjunction with coherent BPSK and hard-decision decoding, we can write

$$P_B \leq Q\left(\sqrt{\frac{5E_b}{N_0}}\right) \exp\left(\frac{5E_b}{2N_0}\right) \frac{\exp(-5E_b/2N_0)}{[1 - 2\exp(-E_b/2N_0)]^2}$$

$$\leq \frac{Q(\sqrt{5E_b/N_0})}{[1 - 2\exp(-E_b/2N_0)]^2} \quad (6.22)$$

6.4.5 Coding Gain

Coding gain is defined as the reduction, usually expressed in decibels, in the required E_b/N_0 to achieve a specified error probability of the coded system over an uncoded system with the same modulation and channel characteristics. Table 6.2 lists an upper bound on the coding gains, compared to uncoded coherent BPSK, for several maximum free distance convolutional codes with constraint lengths varying from 3 to 9 over a Gaussian channel with hard-decision decoding. The table illustrates that it is possible to achieve significant coding gain even with

TABLE 6.2 Coding Gain Upper Bounds for Some Convolutional Codes

Rate $\frac{1}{2}$ codes			Rate $\frac{1}{3}$ codes		
K	d_f	Upper bound (dB)	K	d_f	Upper bound (dB)
3	5	3.97	3	8	4.26
4	6	4.76	4	10	5.23
5	7	5.43	5	12	6.02
6	8	6.00	6	13	6.37
7	10	6.99	7	15	6.99
8	10	6.99	8	16	7.27
9	12	7.78	9	18	7.78

Source: V. K. Bhargava, D. Haccoun, R. Matyas, and P. Nuspl, *Digital Communications by Satellite*, John Wiley & Sons, Inc., New York, 1981.

a simple convolutional code. The actual coding gain will vary with the required bit error probability [20].

Table 6.3 lists the measured coding gains, compared to uncoded coherent BPSK, achieved with hardware implementation or computer simulation over a Gaussian channel with soft-decision decoding [21]. The uncoded E_b/N_0 is given in the leftmost column. From Table 6.3 we can see that coding gain increases as the bit error probability is decreased. However, the coding gain cannot increase indefinitely; it has an upper bound as shown in the table. This bound in decibels can be shown [21] to be

$$\text{coding gain} \leq 10 \log_{10} (rd_f) \tag{6.23}$$

where r is the code rate and d_f is the free distance. Examination of Table 6.3 also reveals that at $P_B = 10^{-7}$, for code rates of $\frac{1}{2}$ and $\frac{2}{3}$, the weaker codes tend to be closer to the upper bound than are the more powerful codes.

Typically, Viterbi decoding is used over binary input channels with either hard or 3-bit soft quantized outputs. The constraint lengths vary between 3 and 9, the code rate is rarely smaller than $\frac{1}{3}$, and the path memory is usually a few constraint lengths [12]. The path memory refers to the depth of the input bit history stored by the decoder. From the Viterbi decoding example in Section 6.3.4, one might question the notion of a fixed path memory. It seems from the example that the decoding of a branch word, at any arbitrary node, can take place as soon as there is only a single surviving branch at that node. That is true; however, to actually implement the decoder in this way would entail an extensive amount of processing to continually check when the branch word can be decoded. Instead, *a fixed delay is provided*, after which the branch word is decoded. It has been shown [12, 22] that a fixed amount of path history, namely 4 or 5 times the constraint length, is sufficient to limit the degradation from the optimum decoder performance to about 0.1 dB for the BSC and Gaussian channels. Typical error performance curves are shown in Figure 6.18 for rate $\frac{1}{2}$ codes using coherent BPSK over a soft (8-level) quantized channel, with Viterbi decoding, and a 32-bit path memory. Also plotted are the transfer function bounds for infinitely fine quantized received data [12]. Figure 6.19 gives the simulation results for Viterbi decoding with hard decision quantization [12]. Notice that each increment in constraint

TABLE 6.3   Basic Coding Gain (dB) for Soft Decision Viterbi Decoding

Uncoded E_b/N_0 (dB)	Code rate		$\frac{1}{3}$		$\frac{1}{2}$			$\frac{2}{3}$		$\frac{3}{4}$	
	P_B	K	7	8	5	6	7	6	8	6	9
6.8	10^{-3}		4.2	4.4	3.3	3.5	3.8	2.9	3.1	2.6	2.6
9.6	10^{-5}		5.7	5.9	4.3	4.6	5.1	4.2	4.6	3.6	4.2
11.3	10^{-7}		6.2	6.5	4.9	5.3	5.8	4.7	5.2	3.9	4.8
	Upper bound		7.0	7.3	5.4	6.0	7.0	5.2	6.7	4.8	5.7

Source: I. M. Jacobs, "Practical Applications of Coding," *IEEE Trans. Inf. Theory*, vol. IT20, May 1974, pp. 305–310.

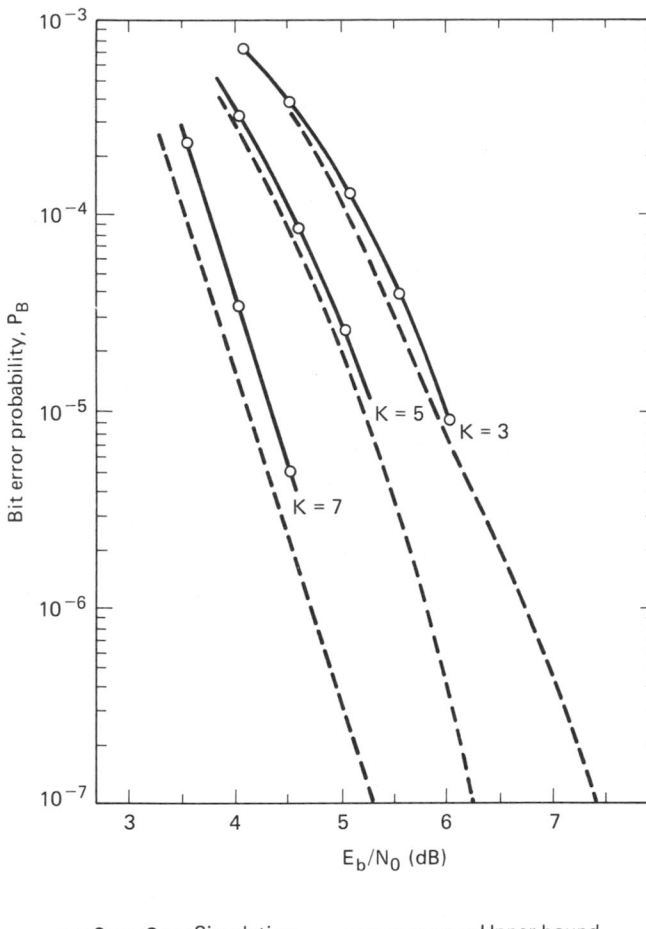

Bit error probability, P_B

E_b/N_0 (dB)

K = 5
K = 3
K = 7

⊸—⊸— Simulation   ▬ ▬ ▬ ▬ Upper bound

Figure 6.18 Bit error probability versus E_b/N_0 for rate $\frac{1}{2}$ codes using coherent BPSK over a soft quantized channel, Viterbi decoding, and a 32-bit path memory. (Reprinted with permission from J. A. Heller and I. M. Jacobs, "Viterbi Decoding for Satellite and Space Communication," *IEEE Trans. Commun. Technol.*, vol. COM19, no. 5, October 1971, Fig. 5, p. 84. © 1971 IEEE.)

length improves the required E_b/N_0 by a factor of approximately 0.5 dB at $P_B = 10^{-5}$. Also, as expected, the 3-bit soft decisions of the channel output result in approximately a 2-dB gain over the hard quantized BSC.

6.4.6 Best Known Convolutional Codes

The connection vectors or polynomial generators of a convolutional code are usually selected based on the code's free distance properties. The first criterion is to select a code that does not have catastrophic error propagation and that has the maximum free distance for the given rate and constraint length. Then the number of paths at the free distance d_f, or the number of data bit errors the paths represent, should be minimized. The selection procedure can be further refined by considering the number of paths or bit errors at $d_f + 1$, at $d_f + 2$, and so on, until only one code or class of codes remains. A list of the best known codes of rate $\frac{1}{2}$, $K = 3$ to 9, and rate $\frac{1}{3}$, $K = 3$ to 8, based on this criterion was compiled by Odenwalder [3, 23] and is given in Table 6.4. The connection vectors in this

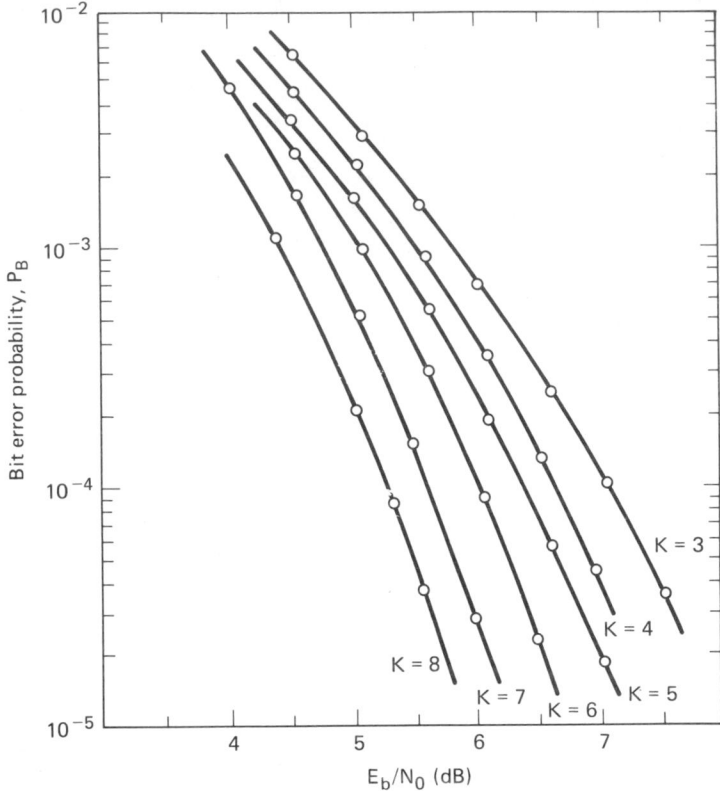

Figure 6.19 Bit error probability versus E_b/N_0 for rate $\frac{1}{2}$ codes using coherent BPSK over a BSC, Viterbi decoding, and a 32-bit path memory. (Reprinted with permission from J. A. Heller and I. M. Jacobs, "Viterbi Decoding for Satellite and Space Communication," *IEEE Trans. Commun. Technol.*, vol. COM19, no. 5, October 1971, Fig. 7, p. 84. © 1971 IEEE.)

table represent the presence or absence (1 or 0) of a tap connection on the corresponding stage of the convolutional encoder. The leftmost term corresponds to the leftmost stage of the encoder register, and the rightmost term corresponds to the rightmost stage, following the notation established in Figure 6.3. It is interesting to note that these connections can be inverted (leftmost and rightmost can be interchanged in the above description). Under the condition of Viterbi decoding, the inverted connections give rise to codes with identical distance properties, and hence identical performance, as those in Table 6.4.

6.4.7 Convolutional Code Rate Trade-Off

6.4.7.1 Performance with Coherent PSK Signaling

The error-correcting capability of a coding scheme increases as the number of channel symbols n per information bit k increases or the rate, k/n, decreases. However, the channel bandwidth and the decoder complexity both increase with n. The advantage of lower code rates when using convolutional codes with co-

TABLE 6.4 Optimum Short Constraint Length Convolutional Codes (Rate $\frac{1}{2}$ and Rate $\frac{1}{3}$)

Rate	Constraint length	Free distance	Code vector
$\frac{1}{2}$	3	5	111 101
$\frac{1}{2}$	4	6	1111 1011
$\frac{1}{2}$	5	7	10111 11001
$\frac{1}{2}$	6	8	101111 110101
$\frac{1}{2}$	7	10	1001111 1101101
$\frac{1}{2}$	8	10	10011111 11100101
$\frac{1}{2}$	9	12	110101111 100011101
$\frac{1}{3}$	3	8	111 111 101
$\frac{1}{3}$	4	10	1111 1011 1101
$\frac{1}{3}$	5	12	11111 11011 10101
$\frac{1}{3}$	6	13	101111 110101 111001
$\frac{1}{3}$	7	15	1001111 1010111 1101101
$\frac{1}{3}$	8	16	11101111 10011011 10101001

Source: J. P. Odenwalder, *Error Control Coding Handbook*, Linkabit Corp., San Diego, Calif., July 15, 1976.

herent PSK, is that the required E_b/N_0 is decreased (for a large range of code rates), permitting the transmission of higher data rates for a given amount of power, or permitting reduced power for a given data rate. Simulation studies have shown [16, 22] that for a fixed constraint length, a decrease in the code rate from $\frac{1}{2}$ to $\frac{1}{3}$ results in a reduction of the required E_b/N_0 of roughly 0.4 dB. However, the corresponding increase in decoder complexity is about 17%. For smaller values of code rate, the improvement in performance relative to the increased decoding complexity diminishes rapidly [22]. Eventually, a point is reached where further decrease in code rate is characterized by a reduction in coding gain.

6.4.7.2 Performance with Noncoherent Orthogonal Signaling

In contrast to PSK, there is an optimum code rate of about $\frac{1}{2}$ for noncoherent orthogonal signaling. Error performance at rates of $\frac{1}{3}$, $\frac{2}{3}$, and $\frac{3}{4}$ are each worse than those for rate $\frac{1}{2}$. For a fixed constraint length, the rate $\frac{1}{3}$, $\frac{2}{3}$, and $\frac{3}{4}$ codes typically degrade by about 0.25, 0.5, and 0.3 dB, respectively, relative to the rate $\frac{1}{2}$ performance [16].

6.5 OTHER CONVOLUTIONAL DECODING ALGORITHMS

6.5.1 Sequential Decoding

Prior to the discovery of an optimum algorithm by Viterbi, other algorithms had been proposed for decoding convolutional codes. The earliest was the *sequential decoding algorithm*, originally proposed by Wozencraft [24, 25] and subsequently modified by Fano [2]. A sequential decoder works by generating hypotheses about the transmitted codeword sequence; it computes a metric between these hypotheses and the received signal. It goes forward as long as the metric indicates that its choices are likely; otherwise, it goes backward, changing hypotheses until, through a systematic trial-and-error search, it finds a likely hypothesis. Sequential decoders can be implemented to work with hard or soft decisions, but soft decisions are usually avoided because they greatly increase the amount of the required storage and the complexity of the computations.

Consider that using the encoder shown in Figure 6.3, a sequence **m** = 1 1 0 1 1 is encoded into the codeword sequence **U** = 1 1 0 1 0 1 0 0 0 1, as shown in Example 6.1. Assume that the received sequence **Z** is, in fact, a *correct* rendition of **U**. The decoder has available a replica of the encoder code tree, shown in Figure 6.6, and can use the received sequence **Z** to penetrate the tree. The decoder starts at the time t_1 node of the tree and generates both paths leaving that node. The decoder follows that path which agrees with the received n code symbols. At the next level in the tree, the decoder again generates both paths leaving that node, and follows the path agreeing with the second group of n code symbols. Proceeding in this manner, the decoder quickly penetrates the tree.

Suppose, however, that the received sequence **Z** is a *corrupted* version of **U**. The decoder starts at the time t_1 node of the code tree and generates both paths leading from that node. If the received n code symbols coincide with one of the generated paths, the decoder follows that path. If there is not agreement, the decoder follows the *most likely path* but keeps a cumulative count on the number of disagreements between the received symbols and the branch words on the path being followed. If two branches appear equally likely, the receiver uses an arbitrary rule, such as following the zero input path. At each new level in the tree, the decoder generates new branches and compares them with the next set of n received code symbols. The search continues to penetrate the tree along the most likely path and maintains the cumulative disagreement count.

If the disagreement count exceeds a certain number (which may increase as we penetrate the tree), the decoder decides that it is on an incorrect path, backs out of the path, and tries another. The decoder keeps track of the discarded pathways to avoid repeating any path excursions. For example, assume that the encoder in Figure 6.3 is used to encode the message sequence $\mathbf{m} = 1\ 1\ 0\ 1\ 1$ into the codeword sequence $\mathbf{U}$ as shown in Example 6.1. Suppose that the fourth and seventh bits of the transmitted sequence $\mathbf{U}$ are received in error, such that:

Time:		t_1	t_2	t_3	t_4	t_5
Message sequence:	$\mathbf{m} =$	1	1	0	1	1
Transmitted sequence:	$\mathbf{U} =$	1 1	0 1	0 1	0 0	0 1
Received sequence:	$\mathbf{Z} =$	1 1	0 0	0 1	1 0	0 1

Let us follow the decoder path trajectory with the aid of Figure 6.20. Assume that a cumulative path disagreement count of 3 is the criterion for backing up and trying an alternative path. On Figure 6.20 the numbers along the path trajectory represent the current disagreement count.

1. At time t_1 we receive symbols 11 and compare them with the branch words leaving the first node.
2. The most likely branch is the one with branch word 11 (corresponding to an input bit one or downward branching), so the decoder decides that input bit one is the correct decoding, and moves to the next level.
3. At time t_2, the decoder receives symbols 00 and compares them with the available branch words 10 and 01 at this second level.
4. There is no "best" path, so the decoder arbitrarily takes the input bit zero (or branch word 10) path, and the disagreement count registers a disagreement of 1.
5. At time t_3, the decoder receives symbols 01 and compares them with the available branch words 11 and 00 at this third level.
6. Again, there is no best path, so the decoder arbitrarily takes the input zero (or branch word 11) path, and the disagreement count is increased to 2.
7. At time t_4, the decoder receives symbols 10 and compares them with the available branch words 00 and 11 at this fourth level.
8. Again, there is no best path, so the decoder takes the input bit zero (or branch word 00) path, and the disagreement count is increased to 3.
9. But a disagreement count of 3 is the turnaround criterion, so the decoder "backs out" and tries the alternative path. The disagreement counter is reset to 2.
10. The alternative path is the input bit one (or branch word 11) path at the t_4

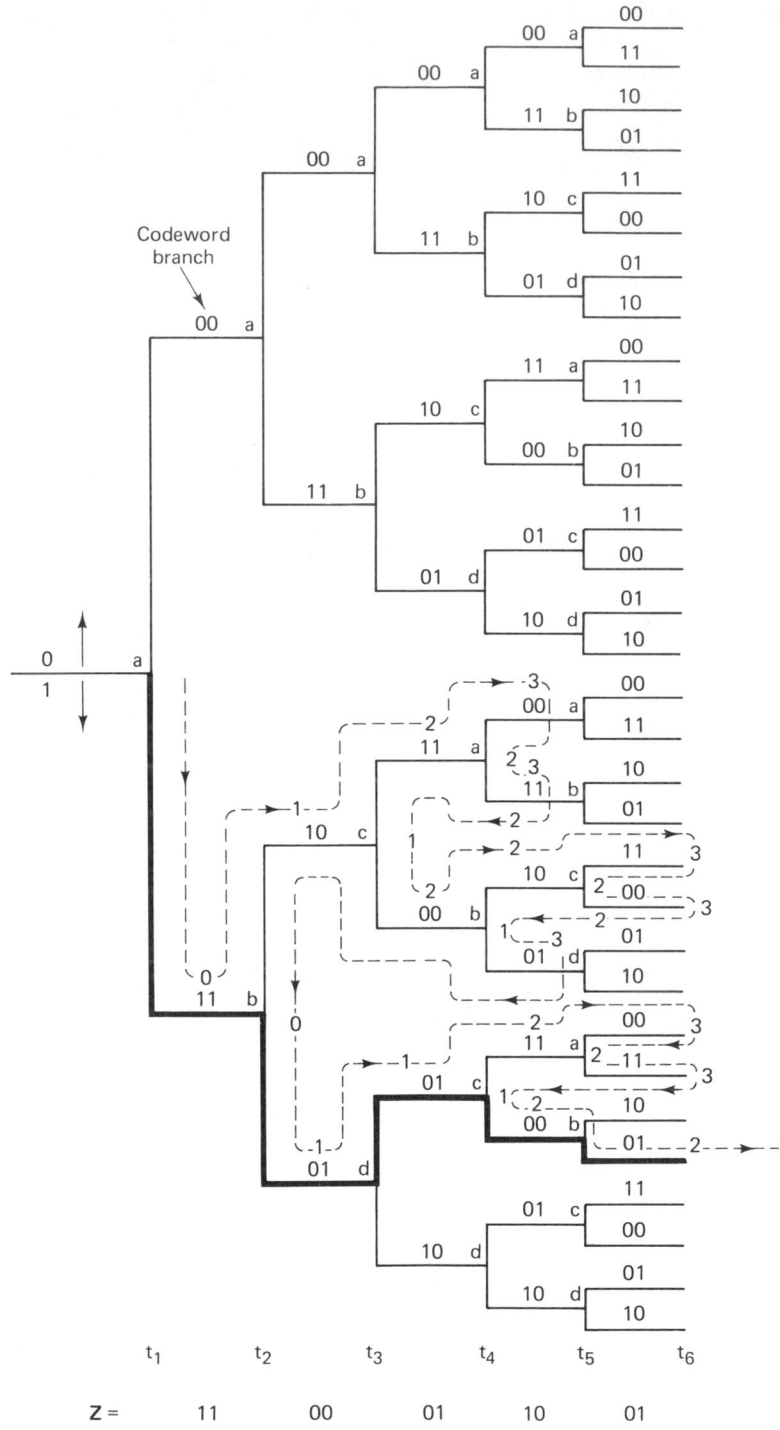

Figure 6.20 Sequential decoding example.

level. The decoder tries this, but compared to the received symbols 10, there is still a disagreement of 1, and the counter is reset to 3.

11. But, 3 being the turnaround criterion, the decoder backs out of this path, and the counter is reset to 2. All of the alternatives have now been traversed at this t_4 level, so the decoder returns to the node at t_3, and resets the counter to 1.

12. At the t_3 node, the decoder compares the symbols received at time t_3, namely 01, with the untried 00 path. There is a disagreement of 1, and the counter is increased to 2.

13. At the t_4 node, the decoder follows the branch word 10 that matches its t_4 code symbols of 10. The counter remains unchanged at 2.

14. At the t_5 node, there is no best path, so the decoder follows the upper branch, as is the rule, and the counter is increased to 3.

15. At this count, the decoder backs up, resets the counter to 2, and tries the alternative path at node t_5. Since the alternate branch word is 00, there is a disagreement of 1 with the received code symbols 01 at time t_5, and the counter is again increased to 3.

16. The decoder backs out of this path, and the counter is reset to 2. All of the alternatives have now been traversed at this t_5 level, so the decoder returns to the node at t_4 and resets the counter to 1.

17. The decoder tries the alternative path at t_4, which raises the metric to 3 since there is a disagreement in two positions of the branch word. This time the decoder must back up all the way to the time t_2 node because all of the other paths at higher levels have been tried. The counter is now decremented to zero.

18. At the t_2 node, the decoder now follows the branch word 01, and because there is a disagreement of 1 with the received code symbols 00 at time t_2, the counter is increased to 1.

The decoder continues in this way. As shown in Figure 6.20, the final path, which has not increased the counter to its turnaround criterion, yields the correctly decoded message sequence, 1 1 0 1 1. Sequential decoding can be viewed as a trial-and-error technique for searching out the correct path in the code tree. It performs the search in a sequential manner, always operating on just a single path at a time. If an incorrect decision is made, subsequent extensions of the path will be wrong. The decoder can eventually recognize its error by monitoring the path metric. The algorithm is similar to the case of an automobile traveler following a road map. As long as the traveler recognizes that the passing landmarks correspond to those on the map, he continues on the path. When he notices strange landmarks (an increase in his dissimilarity metric) the traveler eventually assumes that he is on an incorrect road, and he backs up to a point where he can now recognize the landmarks (his metric returns to an acceptable range). He then tries an alternative road.

6.5.2 Comparisons and Limitations of Viterbi and Sequential Decoding

The major drawback of the Viterbi algorithm is that while error probability decreases exponentially with constraint length, the number of code states, and consequently decoder complexity, *grows exponentially with constraint length*. On the other hand, the computational complexity of the Viterbi algorithm is independent of channel characteristics (compared to hard-decision decoding, soft-decision decoding requires only a trivial increase in the number of computations). Sequential decoding achieves asymptotically the same error probability as maximum likelihood decoding but without searching all possible states. In fact, with sequential decoding the number of states searched is essentially *independent of constraint length*, thus making it possible to use very large ($K = 41$) constraint

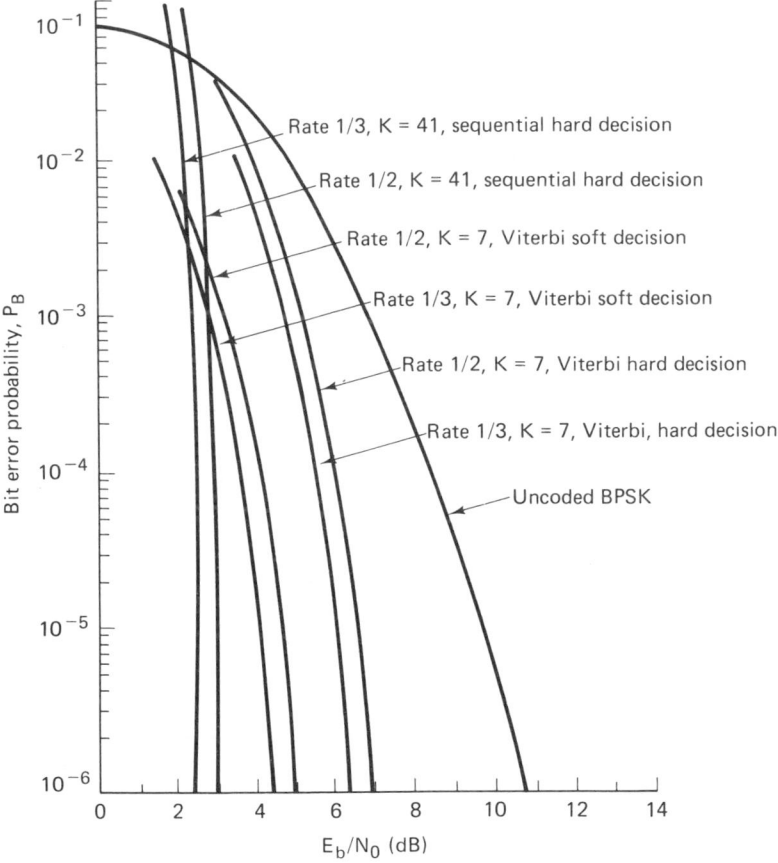

Figure 6.21 Bit error performance for various Viterbi and sequential decoding schemes using coherent BPSK over an AWGN channel. (Reprinted with permission from J. K. Omura and B. K. Levitt, "Coded Error Probability Evaluation for Antijam Communication Systems," *IEEE Trans. Commun.*, vol. COM30, no. 5, May 1982, Fig. 4, p. 900. © 1982 IEEE.)

lengths. This is an important factor in providing such low error probabilities. The major drawback of sequential decoding is that the number of state metrics searched is a random variable. For sequential decoding, the expected number of poor hypotheses and backward searches is a function of the channel SNR. With a low SNR, more hypotheses must be tried than with a high SNR. Because of this variability in computational load, buffers must be provided to store the arriving sequences. Under low SNR, the received sequences must be buffered while the decoder is laboring to find a likely hypothesis. If the average symbol arrival rate exceeds the average symbol decode rate, the buffer will overflow, no matter how large it is, causing a loss of data. The sequential decoder typically puts out error-free data until the buffer overflows, at which time the decoder has to go through a recovery procedure. The buffer overflow threshold is a very sensitive function of SNR. Therefore, an important part of a sequential decoder specification is the *probability of buffer overflow*.

In Figure 6.21, some typical P_B versus E_b/N_0 curves for these two popular solutions to the convolutional decoding problem, Viterbi decoding and sequential decoding, illustrate their comparative performance using coherent BPSK over an AWGN channel. The curves compare Viterbi decoding (rates $\frac{1}{2}$ and $\frac{1}{3}$ hard decision, $K = 7$) versus Viterbi decoding (rates $\frac{1}{2}$ and $\frac{1}{3}$ soft decision, $K = 7$) versus sequential decoding (rates $\frac{1}{2}$ and $\frac{1}{3}$ hard decision, $K = 41$). One can see from Figure 6.21 that coding gains of approximately 8 dB at $P_B = 10^{-6}$ can be achieved with sequential decoders. Since the work of Shannon [26] foretold the potential of approximately 11 dB of coding gain compared to uncoded BPSK, it appears that the major portion of what is theoretically possible can already be accomplished.

6.5.3 Feedback Decoding

A *feedback decoder* makes a hard decision on the data bit at stage j based on metrics computed from stages $j, j + 1, \ldots, j + m$, where m is a preselected positive integer. *Look-ahead length*, L, is defined as $L = m + 1$, the number of received code symbols, expressed in terms of the corresponding number of encoder input bits that are used to decode an information bit. The decision of whether the data bit is zero or one depends on which branch the minimum Hamming distance path traverses in the *look-ahead window* from stage j to stage $j + m$. The detailed operation is best understood in terms of a specific example. Let us consider the use of a feedback decoder for the rate $\frac{1}{2}$ convolutional code shown in Figure 6.3. Figure 6.22 illustrates the tree diagram and the operation of the feedback decoder for $L = 3$. That is, in decoding the bit at branch j, the decoder considers the paths at branches $j, j + 1$, and $j + 2$.

Beginning with the first branch, the decoder computes 2^L or eight cumulative Hamming path metrics and decides that the bit for the first branch is zero if the minimum distance path is contained in the upper part of the tree, and decides one if the minimum distance path is in the lower part of the tree. Assume that the received sequence is $Z = 1\ 1\ 0\ 0\ 0\ 1\ 0\ 0\ 0\ 1$. We now examine the eight paths from time t_1 through time t_3 in the block marked A in Figure 6.22, and compute

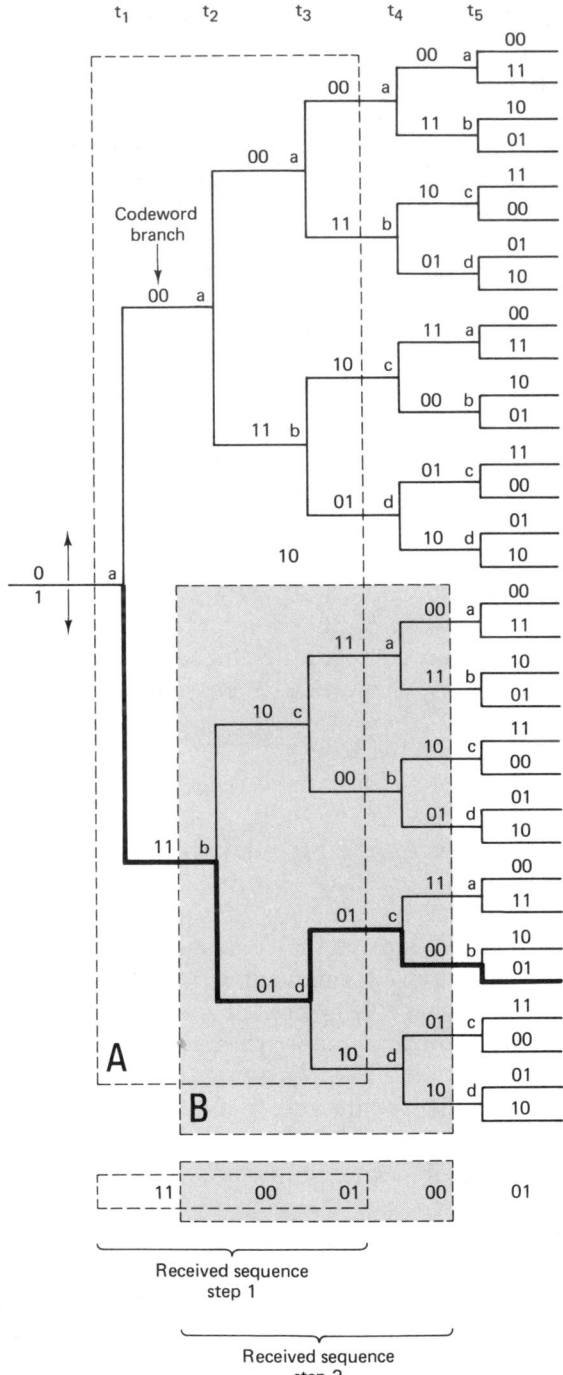

Figure 6.22 Feedback decoding example.

metrics comparing these eight paths with the first six received code symbols (three branches deep times two symbols per branch). Listing the Hamming cumulative path metrics (starting from the top path), they are:

Upper-half metrics: 3, 3, 6, 4

Lower-half metrics: 2, 2, 1, 3

We see that the minimum metric is contained in the lower part of the tree. Therefore, the first decoded bit is one (characterized by a downward movement on the tree). The next step is to extend the lower part of the tree (the part that survived) one stage deeper, and again compute eight metrics, this time from t_2 through t_4. Having decoded the first two code symbols, we now slide over two code symbols to the right and again compute the path metrics for six code symbols. This takes place in the block marked B in Figure 6.22. Again, listing the metrics from top path to bottom path, they are:

Upper-half metrics: 2, 4, 3, 3

Lower-half metrics: 3, 1, 4, 4

For the assumed received sequence, the minimum metric is found in the lower half of block B. Therefore, the second decoded bit is one.

    The same procedure continues until the entire message is decoded. The decoder is called a *feedback decoder* because the detection decisions are *fed back* to the decoder in determining the subset of code paths that are to be considered next. On the BSC, the feedback decoder can perform nearly as well as the Viterbi decoder [17] in that it can correct all the more probable error patterns, namely all those of weight $(d_f - 1)/2$ or less, where d_f is the free distance of the code. An important design parameter for feedback convolutional decoders is L, the look-ahead length. Increasing L increases the coding gain but also increases the decoder implementation complexity.

6.6 INTERLEAVING AND CONCATENATED CODES

Throughout this chapter and Chapter 5 we have assumed that the channel is *memoryless,* since we have considered codes that are designed to combat random independent errors. A channel that has *memory* is one that exhibits mutually dependent signal transmission impairments. An example of such a channel is a fading channel, particularly when the fading varies slowly compared to one symbol time. Another type of impairment, called *multipath,* involves signal arrivals at the receiver over two or more paths of different lengths. The effect is that the signals *arrive out of phase* with each other, and the cumulative received signal is distorted. High-frequency (HF) and tropospheric propagation radio channels suffer from such phenomena. Also, some channels suffer from switching noise and other burst noise (e.g., telephone channels or channels disturbed by pulse jamming). All of these time-correlated impairments result in statistical dependence among successive symbol transmissions. That is, the disturbances tend to cause errors that occur in bursts, instead of as isolated events.

Under the assumption that the channel has memory, the errors no longer can be characterized as single randomly distributed bit errors whose occurrence is independent from bit to bit. Most block or convolutional codes are designed to combat random independent errors. The result of a channel having memory on such coded signals is to cause degradation in error performance. Coding techniques for channels with memory have been proposed [27, 28], but the greatest problem with such coding is the difficulty in obtaining accurate models of the often time-varying statistics of such channels. One technique, which only requires a knowledge of the *duration or span* of the channel memory, *not* its exact statistical characterization, is the use of time diversity or *interleaving*.

Interleaving the coded message before transmission and deinterleaving after reception causes bursts of channel errors to be spread out in time and thus to be handled by the decoder as if they were random errors. Since, in all practical cases, the channel memory decreases with time separation, the idea behind interleaving is to separate the codeword symbols in time. The intervening times are similarly filled by the symbols of other codewords. Separating the symbols in time effectively transforms a channel with memory to a *memoryless* one, and thereby enables the random-error-correcting codes to be useful in a burst-noise channel.

The interleaver shuffles the code symbols over a span of several block lengths (for block codes) or several constraint lengths (for convolutional codes). The span required is determined by the burst duration. The details of the bit redistribution pattern must be known to the receiver in order for the symbol stream to be deinterleaved before being decoded. Figure 6.23 illustrates a simple interleaving example. In Figure 6.23a we see seven uninterleaved codewords, A through G. Each codeword is comprised of seven code symbols. Let us assume that the code has a single-error-correcting capability within each seven-symbol sequence. If the memory span of the channel is one codeword in duration, such a seven-symbol-time noise burst could destroy the information contained in one or two codewords. However, suppose that, after having encoded the data, the code symbols were then *interleaved* or shuffled, as shown in Figure 6.23b. That is, each code symbol of each codeword is separated from its preinterleaved neighbors by a span of seven symbol times. The interleaved stream is then used to modulate a waveform that is transmitted over the channel. A contiguous channel noise burst occupying seven symbol times is seen in Figure 6.23b, to affect one code symbol from each of the original seven codewords. Upon reception, the stream is first deinterleaved so that it resembles the original coded sequence in Figure 6.23a. Then the stream is decoded. Since each codeword possesses a single-error-correcting capability, the burst noise has no degrading effect on the final sequence.

Interleaving techniques have proven useful for all the convolutional and block codes described here and in Chapter 5. Two types of interleavers are commonly used, *block interleavers* and *convolutional interleavers*. They are each described below.

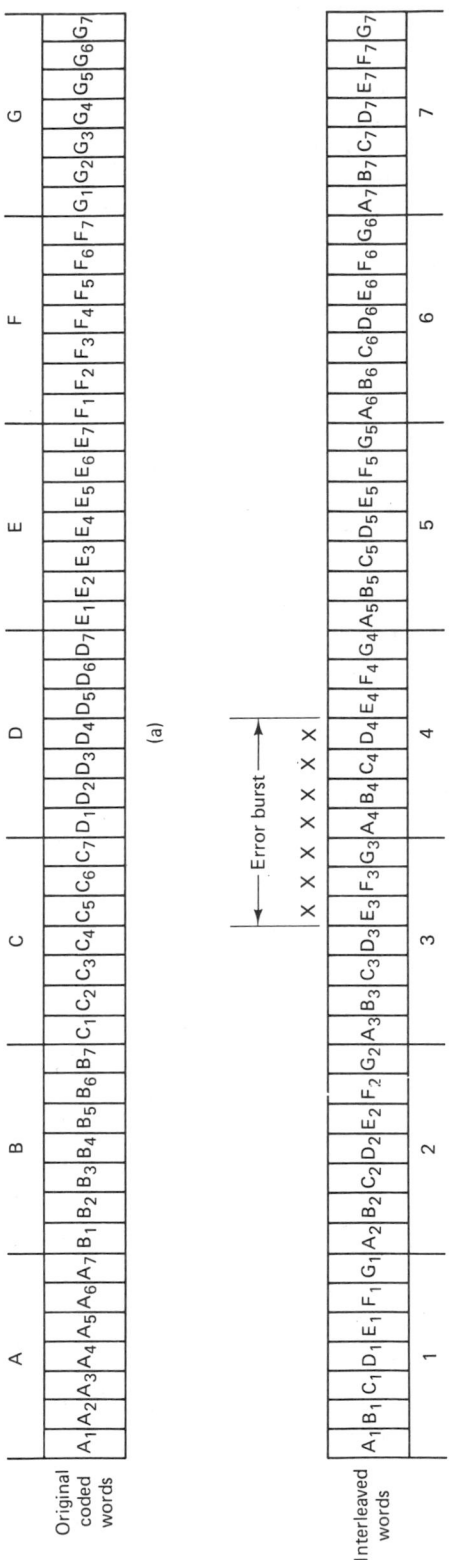

Figure 6.23 Interleaving example. (a) Original uninterleaved codewords, each comprised of seven code symbols. (b) Interleaved code symbols.

6.6.1 Block Interleaving

A block interleaver accepts the coded symbols in blocks from the encoder, permutes the symbols, and then feeds the rearranged symbols to the modulator. The usual permutation of the block is accomplished by *filling the columns* of an M-row-by N-column ($M \times N$) array with the encoded sequence. After the array is completely filled, the symbols are then fed to the modulator *one row at a time* and transmitted over the channel. At the receiver, the deinterleaver performs the inverse operation; it accepts the symbols from the demodulator, deinterleaves them, and feeds them to the decoder. Symbols are entered into the deinterleaver array by rows, and removed by columns. Figure 6.24a illustrates an example of an interleaver with $M = 4$ rows and $N = 6$ columns. The entries in the array illustrate the order in which the 24 code symbols are placed into the interleaver. The output sequence to the transmitter consists of code symbols removed from the array by rows, as shown in the figure. The most important characteristics of such a block interleaver are as follows:

1. Any burst of less than N contiguous channel symbol errors results in isolated errors at the deinterleaver output that are separated from each other by at least M symbols.
2. Any bN burst of errors, where $b > 1$, results in output bursts from the deinterleaver of no more than $\lceil b \rceil$ symbol errors. Each output burst is separated from the other bursts by no less than $M - \lfloor b \rfloor$ symbols. The notation $\lceil x \rceil$ means the smallest integer no less than x, and $\lfloor x \rfloor$ means the largest integer no greater than x.
3. A periodic sequence of single errors spaced N symbols apart results in a single burst of errors of length M at the deinterleaver output.
4. The interleaver/deinterleaver end-to-end delay is approximately $2MN$ symbol times. To be precise, only $M(N - 1) + 1$ memory cells need to be filled before transmission can begin (as soon as the first symbol of the last column of the $M \times N$ array is filled). A corresponding number needs to be filled at the receiver before decoding begins. Thus the minimum end-to-end delay is $(2MN - 2M + 2)$ symbol times, not including any channel propagation delay.
5. The memory requirement is MN symbols for each location (interleaver and deinterleaver). However, since the $M \times N$ array needs to be (mostly) filled before it can be read out, a memory of $2MN$ symbols is generally implemented at each location to allow the emptying of one $M \times N$ array while the other is being filled, and vice versa.

Example 6.3   Interleaver Characteristics

Using the $M = 4$, $N = 6$ interleaver structure of Figure 6.24a, verify each of the block interleaver characteristics described above.

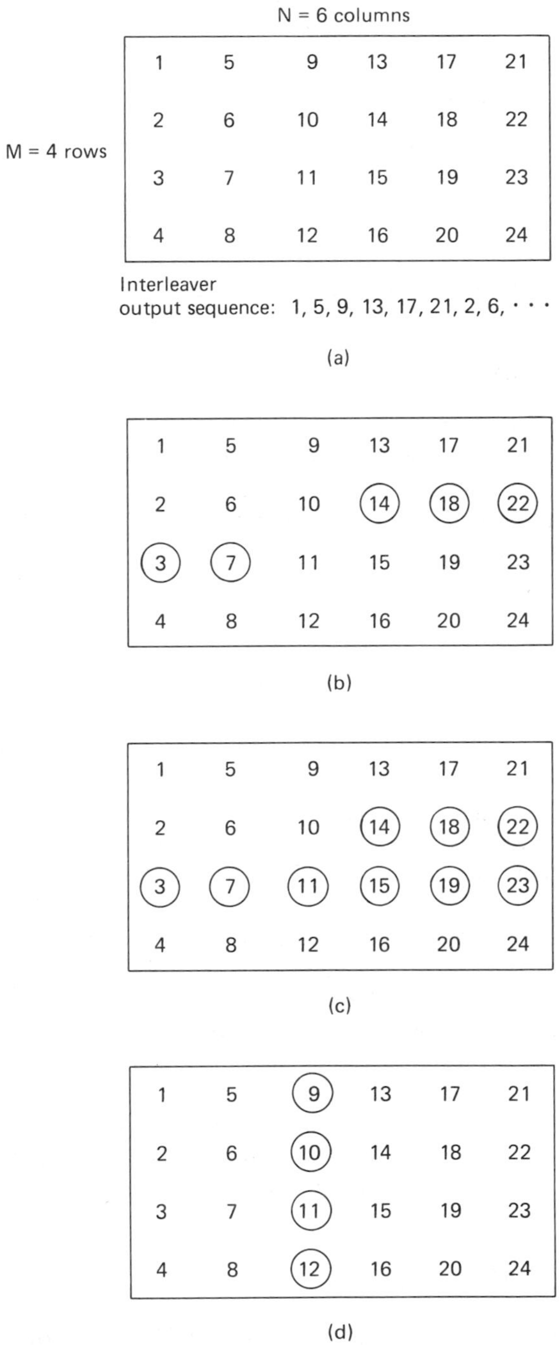

N = 6 columns

1	5	9	13	17	21
2	6	10	14	18	22
3	7	11	15	19	23
4	8	12	16	20	24

M = 4 rows

Interleaver
output sequence: 1, 5, 9, 13, 17, 21, 2, 6, · · ·

(a)

1	5	9	13	17	21
2	6	10	(14)	(18)	(22)
(3)	(7)	11	15	19	23
4	8	12	16	20	24

(b)

1	5	9	13	17	21
2	6	10	(14)	(18)	(22)
(3)	(7)	(11)	(15)	(19)	(23)
4	8	12	16	20	24

(c)

1	5	(9)	13	17	21
2	6	(10)	14	18	22
3	7	(11)	15	19	23
4	8	(12)	16	20	24

(d)

Figure 6.24 Block interleaver example. (a) $M \times N$ block interleaver. (b) Five-symbol error burst. (c) Nine-symbol error burst. (d) Periodic single-error sequence spaced $N = 6$ symbols apart.

Solution

1. Let there be a noise burst of five symbol times, such that the symbols shown encircled in Figure 6.24b experience errors in transmission. After deinterleaving at the receiver, the sequence is

    1 2 ③ 4 5 6 ⑦ 8 9 10 11 12

    13 ⑭ 15 16 17 ⑱ 19 20 21 ㉒ 23 24

    where the encircled symbols are in error. It is seen that the smallest separation between symbols in error is $M = 4$.

2. Let $b = 1.5$ so that $bN = 9$. Figure 6.24c illustrates an example of a nine-symbol error burst. After deinterleaving at the receiver, the sequence is

    1 2 ③ 4 5 6 ⑦ 8 9 10 ⑪ 12

    13 ⑭ ⑮ 16 17 ⑱ ⑲ 20 21 ㉒ ㉓ 24

    Again, the encircled symbols are in error. It is seen that the bursts consist of no more than $\lceil 1.5 \rceil = 2$ contiguous symbols and that they are separated by at least $M - \lfloor 1.5 \rfloor = 4 - 1 = 3$ symbols.

3. Figure 6.24d illustrates a sequence of single errors spaced by $N = 6$ symbols apart. After deinterleaving at the receiver, the sequence is

    1 2 3 4 5 6 7 8 ⑨ ⑩ ⑪ ⑫

    13 14 15 16 17 18 19 20 21 22 23 24

    It is seen that the deinterleaved sequence has a single error burst of length $M = 4$ symbols.

4. End-to-end delay: The minimum end-to-end delay due to the interleaver and deinterleaver is $(2MN - 2M + 2) = 42$ symbol times.

5. Memory requirement: The interleaver and the deinterleaver arrays are each of size $M \times N$. Therefore, storage for $MN = 24$ symbols is required at each end of the channel. As mentioned earlier, storage for $2MN = 48$ symbols would generally be implemented.

Typically, for use with a single-error-correcting code the interleaver parameters are selected such that the number of columns N overbounds the *expected burst length*. The choice of the number of rows M is dependent on the coding scheme used. For block codes, M should be larger than the code block length, while for convolutional codes, M should be larger than the constraint length. Thus a burst of length N can cause at most a single error in any block codeword; similarly, with convolutional codes, there will be at most a single error in any decoding constraint length. For t-error-correcting codes, the choice of N need only overbound the expected burst length divided by t.

6.6.2 Convolutional Interleaving

Convolutional interleavers have been proposed by Ramsey [29] and Forney [30]. The structure proposed by Forney appears in Figure 6.25. The code symbols are

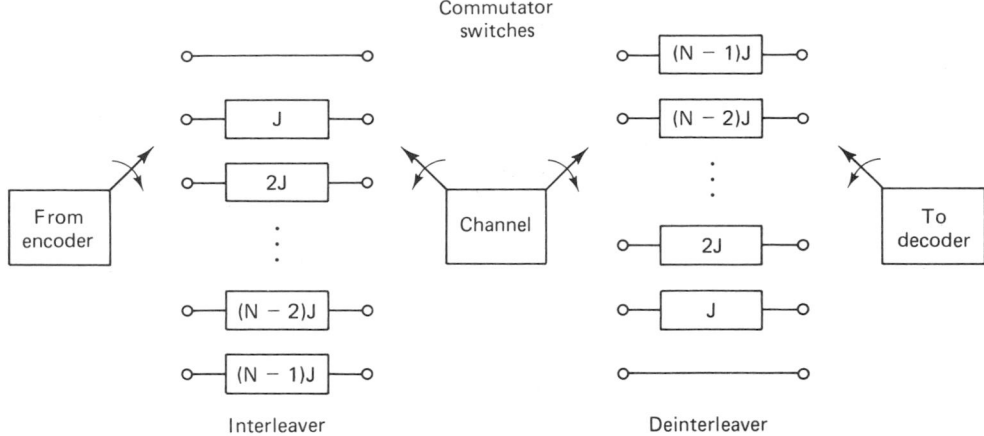

From
encoder

J

2J

(N − 2)J

(N − 1)J

Interleaver

Channel

(N − 1)J

(N − 2)J

2J

J

Deinterleaver

To
decoder

Figure 6.25 Shift register implementation of a convolutional interleaver/
deinterleaver.

sequentially shifted into the bank of N registers; each successive register provides
J symbols more storage than did the preceding one. The zeroth register provides
no storage (the symbol is transmitted immediately). With each new code symbol
the commutator switches to a new register, and the new code symbol is shifted
in while the oldest code symbol in that register is shifted out to the modulator/
transmitter. After the $(N − 1)$th register, the commutator returns to the zeroth
register and starts again. The deinterleaver performs the inverse operation, and
the input and output commutators for both interleaving and deinterleaving must
be synchronized.

Figure 6.26 illustrates an example of a simple convolutional four-register
($J = 1$) interleaver being loaded by a sequence of code symbols. The synchronized
deinterleaver is shown simultaneously feeding the deinterleaved symbols to the
decoder. Figure 6.26a shows symbols 1 to 4 being loaded; the ×s represent unknown states. Figure 6.26b shows the first four symbols shifted within the registers
and the entry of symbols 5 to 8 to the interleaver input. Figure 6.6c shows symbols
9 to 12 entering the interleaver. The deinterleaver is now filled with message
symbols, but nothing useful is being fed to the decoder yet. Finally, Figure 6.6d
shows symbols 13 to 16 entering the interleaver, and at the output of the deinterleaver, symbols 1 to 4 are being passed to the decoder. The process continues
in this way until the entire codeword sequence, in its original preinterleaved form,
is presented to the decoder.

The performance of a convolutional interleaver is very similar to that of a
block interleaver. The important advantage of convolutional over block interleaving is that with convolutional interleaving the end-to-end delay is $M(N − 1)$
symbols, where $M = NJ$, and the memory required is $M(N − 1)/2$ at both ends
of the channel. Therefore, there is a reduction of one-half in delay and memory
over the block interleaving requirements [16].

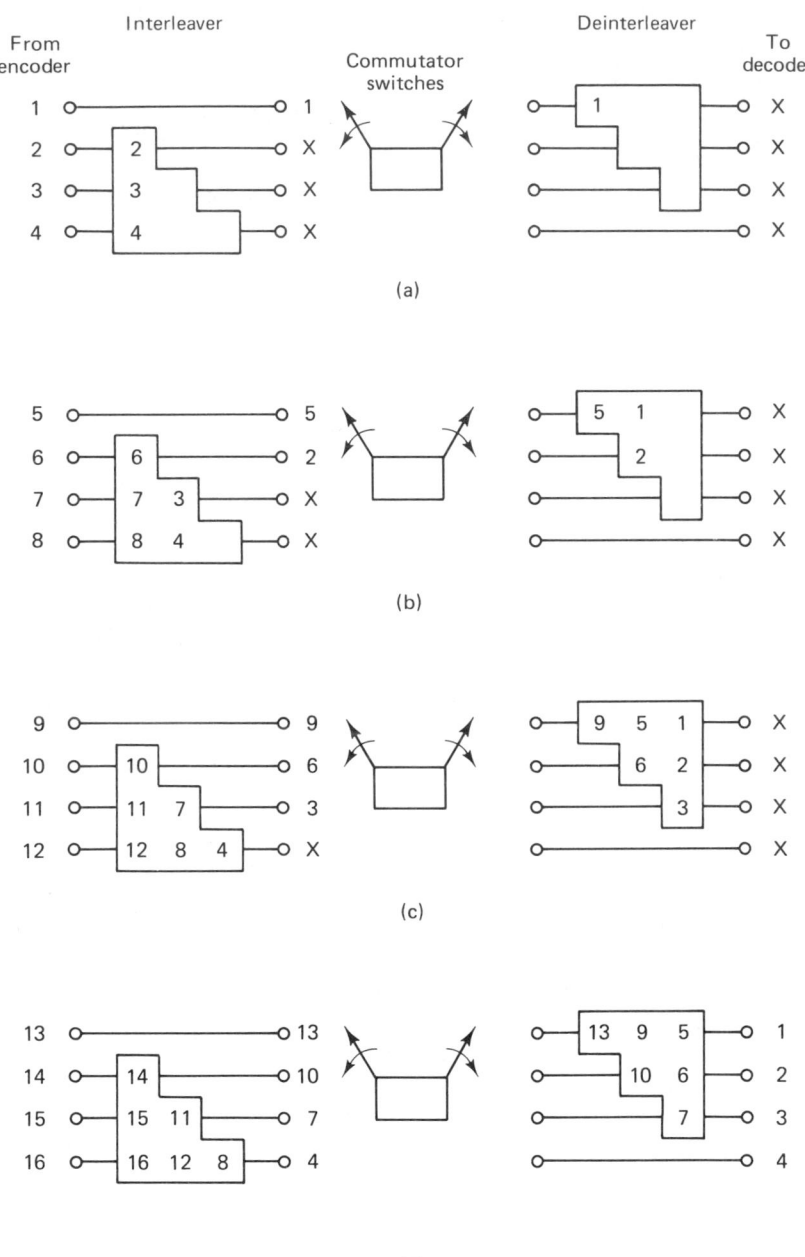

Figure 6.26 Convolutional interleaver/deinterleaver example.

6.6.3 Concatenated Codes

A concatenated code is one that uses two levels of coding, an inner code and an outer code, to achieve the desired error performance. Figure 6.27 illustrates the order of encoding and decoding. The inner code, the one that interfaces with the modulator/demodulator and channel, is usually configured to correct most of the channel errors. The outer code, usually a higher-rate (lower-redundancy) code, then reduces the probability of error to the specified level. The primary reason for using a concatenated code is to achieve a low error rate with an overall implementation complexity which is less than that which would be required by a single coding operation. In Figure 6.27 an interleaver is shown between the two coding steps. This is usually required to spread any error bursts that may appear at the output of the inner coding operation.

One of the most popular concatenated coding systems uses a Viterbi-decoded convolutional inner code and a Reed–Solomon (R–S) outer code, with interleaving between the two coding steps [23]. Operation of such systems with E_b/N_0 in the range 2.0 to 2.5 dB to achieve $P_B = 10^{-5}$ (only about 4 dB away from the Shannon limit) is now feasible with practical hardware [16]. In this system, the demodulator outputs soft quantized code symbols to the inner convolutional decoder, which in turn outputs hard quantized code symbols with bursty errors to the R–S decoder. (In a Viterbi-decoded system, the output errors tend to occur in bursts.) The outer R–S code is formed from m-bit segments of the binary data stream (see Section 5.7.4). The performance of such a (nonbinary) R–S code depends only on the number of *symbol errors* in the block. The code is undisturbed by burst errors within an m-bit symbol. That is, for a given symbol error, the R–S code performance is the same whether the symbol error is due to one bit being in error or m bits being in error. However, the concatenated system performance is severely degraded by correlated errors among successive symbols. Hence the interleaving between codes needs to take place at the symbol level (not at the bit level). In the next section we consider a popular consumer application of such symbol interleaving in a concatenated system.

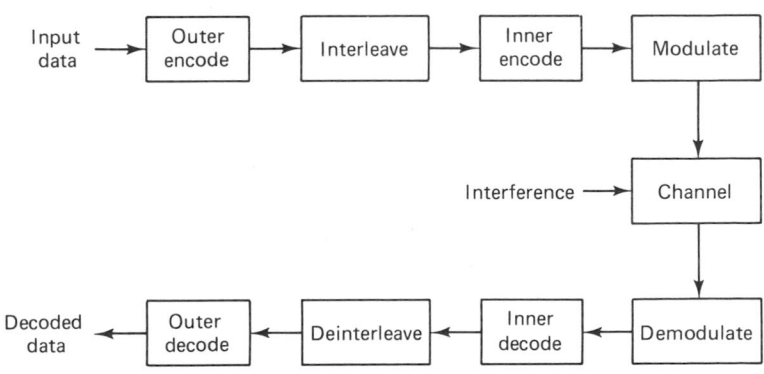

Figure 6.27 Block diagram of a concatenated coding system.

6.7 CODING AND INTERLEAVING APPLIED TO THE COMPACT DISC DIGITAL AUDIO SYSTEM

In 1979, Philips Corp. of the Netherlands and Sony Corp. of Japan defined a standard for the digital storage and reproduction of audio signals, known as the *compact disc (CD) digital audio system*. This CD system has become the world standard for achieving fidelity of sound reproduction that far surpasses any other available technique. A plastic disc 120 mm in diameter is used to store the digitized audio waveform. The waveform is sampled at 44.1 kilosamples/s to provide a recorded bandwidth of 20 kHz; each audio sample is uniformly quantized to one of 2^{16} levels (16 bits/sample), resulting in a dynamic range of 96 dB and a total harmonic distortion of 0.005%. A single disc (playing time approximately 70 minutes) stores about 10^{10} bits in the form of minute *pits* that are optically scanned by a laser.

There are several sources of channel errors: (1) small unwanted particles or air bubbles in the plastic material or pit inaccuracies arising in manufacturing, and (2) fingerprints or scratches during handling. It is difficult to predict how, on the average, a CD will get damaged; but in the absence of an accurate channel model, it is safe to assume that the channel mainly has a *burstlike* error behavior, since a scratch or fingerprint will cause *several* consecutive data samples to be in error. An important aspect of the system design contributing to the high-fidelity performance is a concatenated control scheme called the *cross-interleave Reed–Solomon* code (CIRC). The data are rearranged in time so that digits stemming from contiguous samples of the waveform are *spread out in time*. In this way, error bursts are made to appear as single random events (see the earlier sections on interleaving). The digital information is protected by adding parity bytes derived in two Reed–Solomon (R–S) encoders (see Section 5.7.4). Error control applied to the compact disc depends mostly on R–S coding and multiple layers of interleaving. Material on the CD is treated in this chapter rather than in Chapter 5 with R–S coding because it follows naturally after the subject of interleaving and concatenated codes in the previous sections.

In digital audio applications, an undetected decoding error is very serious since it results in clicks, while occasional *detected* failures are not so serious because they can be concealed. The CIRC error-control scheme in the CD system involves both *correction* and *concealment* of errors. The performance specifications for the CIRC are given in Table 6.5. From the specifications in the table it would appear that the CD can endure much damage (e.g., 8-mm holes punched in the disc) without any noticeable effect on the sound quality.

The CIRC system achieves its error control by a hierarchy of the following techniques:

1. The decoder provides a level of error correction.
2. If the error correction capability is exceeded, the decoder provides a level of erasure correction (see Section 5.5.5).
3. If the erasure correction capability is exceeded, the decoder attempts to

TABLE 6.5 Specifications for the CD Cross-Interleave Reed–Solomon Code

Maximum correctable burst length	≈ 4000 bits (2.5-mm track length on the disc)
Maximum interpolatable burst length	$\approx 12{,}000$ bits (8 mm)
Sample interpolation rate	One sample every 10 hours at $P_B = 10^{-4}$
	1000 samples/min at $P_B = 10^{-3}$
Undetected error samples (clicks)	Less than one every 750 hours at $P_B = 10^{-3}$
	Negligible at $P_B \leq 10^{-4}$
New discs are characterized by	$P_B \approx 10^{-4}$

conceal unreliable data samples by *interpolating* between reliable neighboring samples.

4. If the interpolation capability is exceeded, the decoder blanks out or *mutes* the system for the duration of the unreliable samples.

6.7.1 CIRC Encoding

Figure 6.28 illustrates the basic CIRC encoder block diagram (within the CD recording equipment) and the decoder block diagram (within the CD player equipment). Encoding consists of the encoding and interleaving steps designated as: Δ interleave, C_2 encode, D^* interleave, C_1 encode, and D interleave. The decoder steps, consisting of deinterleaving and decoding, are performed in the *reverse* order of the encoding steps and are designated as: D deinterleave, C_1 decode, D^* deinterleave, C_2 decode, and Δ deinterleave.

Figure 6.29 illustrates the basic system frame time, comprised of six sampling periods, each made up of a stereo sample pair (16-bit left sample and 16-bit right sample). The bits are organized into symbols or bytes of 8 bits each. Therefore, each sample pair contains 4 bytes, and the uncoded frame contains $k = 24$ bytes. Figure 6.29a–e summarizes the *five encoding steps* that characterize the CIRC system. The function of each of these steps will best be understood when we consider the decoding operation.

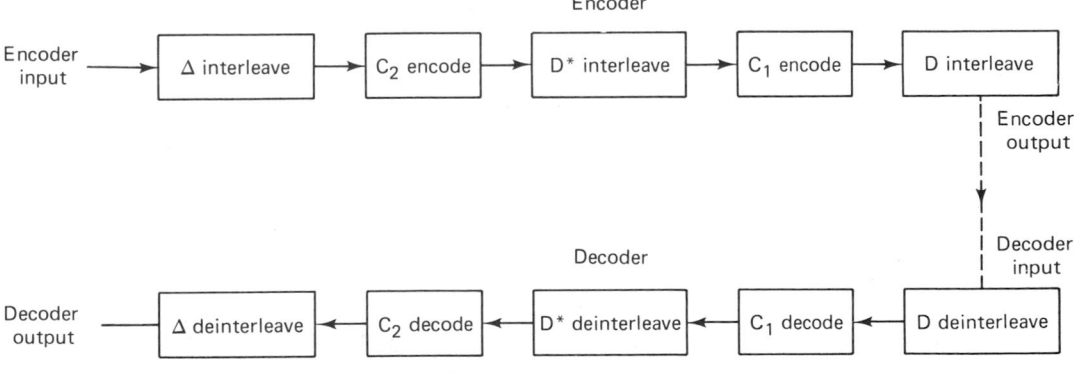

Figure 6.28 CIRC encoder and decoder.

Sec. 6.7   Coding and Interleaving Used in the CD Digital Audio System   **367**

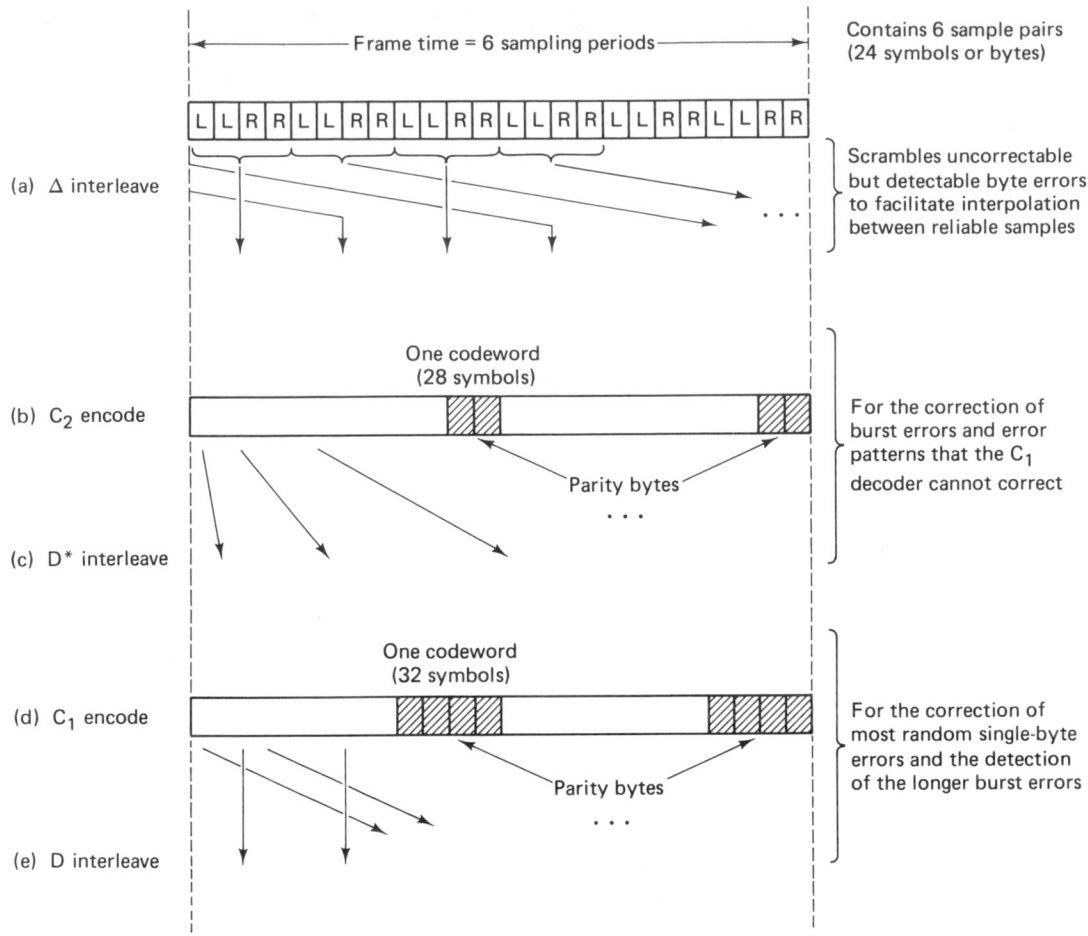

Figure 6.29 Compact disc encoder. (a) Δ interleave. (b) C_2 encode. (c) D^* interleave. (d) C_1 encode. (e) D interleave.

(a) Δ *interleave*. Even-numbered samples are separated from odd-numbered samples by two frame times in order to scramble uncorrectable but detectable byte errors. This facilitates the interpolation process.

(b) C_2 *encode*. Four Reed–Solomon (R–S) parity bytes are added to the Δ-interleaved 24-byte frame, resulting in a total of $n = 28$ bytes. This (28, 24) code is called the *inner code*.

(c) D^* *interleave*. Here each byte is delayed a different length, thereby spreading errors over several codewords. C_2 encoding together with D^* interleaving have the function of providing for the correction of burst errors and error patterns that the C_1 decoder cannot correct.

(d) C_1 *encode*. Four R–S parity bytes are added to the $k = 28$ bytes of the D^*-interleaved frame, resulting in a total of $n = 32$ bytes. This (32, 28) code is called the *outer code*.

(e) *D interleave.* The purpose is to *cross-interleave* the *even bytes* of a frame with the *odd bytes* of the next frame. By this procedure, two consecutive bytes on the disc will always end up in two different codewords. Upon decoding, this interleaving, together with the C_1 decoding, results in the correction of most random single errors and the detection of longer burst errors.

6.7.1.1 Shortening the R–S Code

In Section 5.7.4 an (n, k) R–S code is expressed in terms of $n = 2^m - 1$ total symbols and $k = 2^m - 1 - 2t$ data symbols, where m is the number of bits per symbol and t is the error-correcting capability of the code in symbols. For the CD system, where a symbol is made up of 8 bits, a 2-symbol error-correcting code can be configured as a (255, 251) code. However, the CD system uses a considerably shorter block length. Any block code (in systematic form) can be shortened without affecting the number of errors that can be corrected within a block length. In terms of the (255, 251) R–S code, imagine that 227 of the 251 data symbols are a set of all-zero symbols (which are not actually transmitted and hence are not subject to any errors). Then the code is really a (28, 24) code with the same 2-symbol error-correcting capability. This is what is done in the C_2 encoder of the CD system.

We can think of the 28 total symbols out of the C_2 encoder as the data symbols into the C_1 encoder. Again, we can configure a shortened 2-symbol error-correcting (255, 251) code by throwing away 223 data symbols—the result being a (32, 28) code.

6.7.2 CIRC Decoding

The inner and outer R–S codes with (n, k) values (32, 28) and (28, 24) each use four parity bytes. The code rate of the CIRC is $(k_1/n_1)(k_2/n_2) = 24/32 = 3/4$. From Equation (5.78) the minimum distance of the C_1 and C_2 R–S codes is $d_{min} = n - k + 1 = 5$. From Equations (5.79) and (5.50),

$$t \le \frac{d_{min} - 1}{2} \tag{6.24}$$

$$\rho \le d_{min} - 1 \tag{6.25}$$

where t is the error-correcting capability and ρ is the erasure-correcting capability, it is seen that the C_1 or C_2 decoder can correct a maximum of 2 symbol errors or 4 symbol erasures per codeword. Or, as described by Equation (5.51), it is possible to correct any pattern of α errors and γ erasures simultaneously provided that

$$d_{min} \ge 2\alpha + \gamma + 1 \tag{6.26}$$

There is a trade-off between error correction and erasure correction; the larger the error correcting capability used, the smaller will be the erasure correcting capability.

The benefits of CIRC are best seen at the *decoder*, where the processing steps, shown in Figure 6.30, are in the reverse order of the encoder steps. The decoder steps are as follows:

1. *D deinterleave*. This function is performed by the alternating delay lines marked D. The 32 bytes $(B_{i1}, \ldots, B_{i32})$ of an encoded frame are applied in parallel to the 32 inputs of the D deinterleaver. Each delay is equal to the duration of 1 byte, so that the information of the *even bytes* of a frame is cross-deinterleaved with that of the *odd bytes* of the next frame.

2. C_1 *decode*. The D deinterleaver and the C_1 decoder are designed to correct a single byte error in the block of 32 bytes and to detect larger burst errors. If multiple errors occur, the C_1 decoder passes them on unchanged, attaching to all 28 remaining bytes an erasure flag, sent via the dashed lines (the four parity bytes used in the C_1 decoder are no longer retained).

3. *D* deinterleave*. Due to the different lengths of the deinterleaving delay lines $D^*(1, \ldots, 27)$, errors that occur in one word at the output of the C_1 decoder are *spread over a number of words* at the input of the C_2 decoder. This results in reducing the number of errors per input word of the C_2 decoder, enabling the C_2 decoder to correct these errors.

4. C_2 *decode*. The C_2 decoder is intended for the correction of burst errors that the C_1 decoder could not correct. If the C_2 decoder cannot correct these errors, the 24-byte codeword is passed on unchanged to the Δ deinterleaver

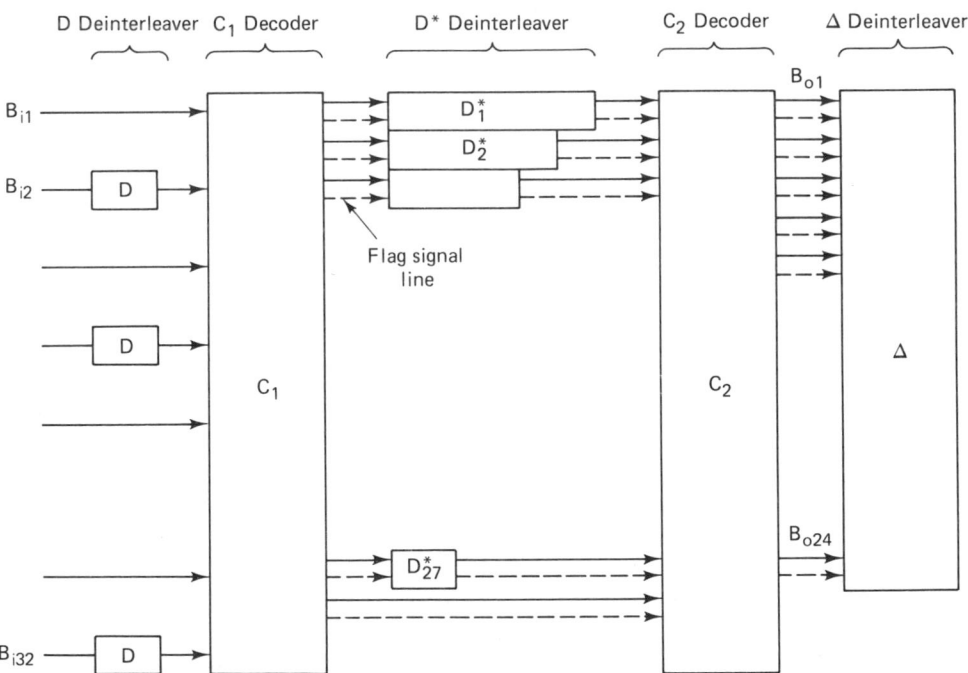

Figure 6.30 Compact disc decoder.

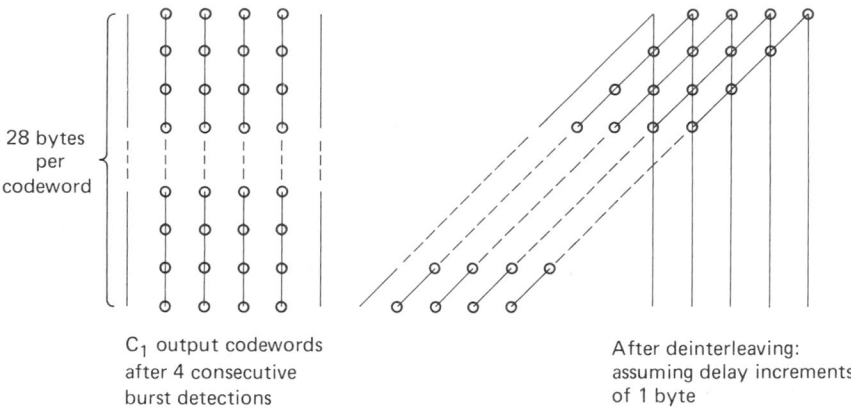

28 bytes
per
codeword

C_1 output codewords
after 4 consecutive
burst detections

After deinterleaving:
assuming delay increments
of 1 byte

Figure 6.31 Example of 4-byte erasure capability. (Rightmost event is at the earliest time.)

and the associated positions are given an *erasure flag* via the dashed output lines, $B_{o1}, \ldots, B_{o24}$.

5. Δ *deinterleave*. The final operation deinterleaves uncorrectable but detected byte errors in such a way that *interpolation* can be used between reliable neighboring samples.

Figure 6.31 highlights the decoder steps 2, 3, and 4. At the output of the C_1 decoder is seen a sequence of four 28-byte codewords that have exceeded the 1 byte per codeword error correction design. Therefore, each of the symbols in these codewords is tagged with an erasure flag (shown with circles). The D* deinterleaver provides a staggered delay for each byte of a codeword, so that the bytes of a given codeword arrive in different codewords at the input to the C_2 decoder. If we assume that the delay increments of the D* deinterleaver in Figure 6.31 are 1 byte, it would be possible to correct error bursts of as many as four consecutive C_1 codewords (since the C_2 decoder is capable of four erasure corrections per codeword). In the actual CD system, the delay increments are 4 bytes; therefore, the maximum burst error correction capability consists of 16 consecutive uncorrectable C_1 words.

6.7.3 Interpolation and Muting

Samples that cannot be corrected by the C_2 decoder could cause audible disturbances. The function of the *interpolation* process is to insert new samples, estimated from reliable neighbors, in place of the unreliable ones. If an entire C_2 word is detected as unreliable, this would make it impossible to apply interpolation without additional interleaving, since both even- and odd-numbered samples are unreliable. This can happen if the C_1 decoder fails to detect an error but the C_2 decoder detects it. It is the purpose of Δ deinterleaving (over a span of two frame

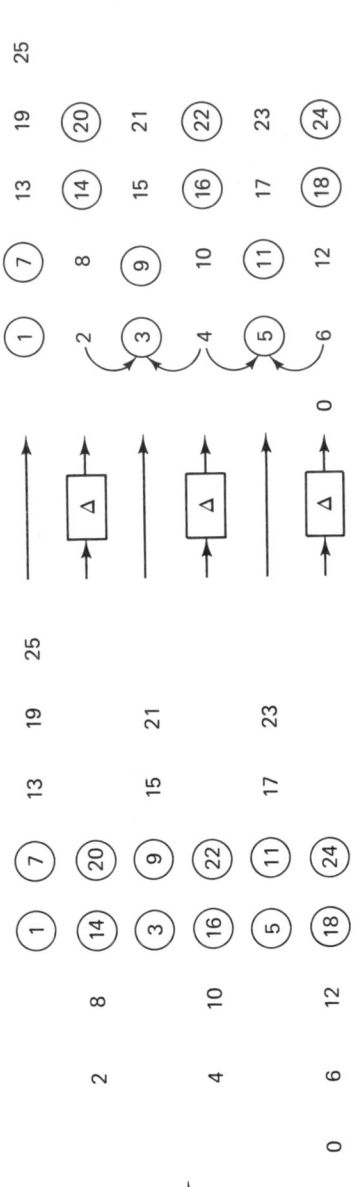

Figure 6.32 Effect of interleaving. (Rightmost event is at the earliest time.)

372

times) to obtain a pattern where even-numbered samples can be interpolated from reliable odd-numbered samples, or vice versa.

Two successive unreliable words consisting of 12 sample pairs are shown in Figure 6.32. A sample pair consists of a sample (2 bytes) from the right audio channel and a sample from the left audio channel. The numbers indicate the ordering of the sets of samples. An encircled sample set denotes an *erasure flag*. After Δ deinterleaving, the unreliable samples shown in the figure are estimated by a first-order linear interpolation between neighboring samples that stem from a different location on the disc.

In CD players, another level of error control is provided in case a burst length of 48 frames is exceeded and 2 or more consecutive unreliable samples result. In this case the system is *muted* (audio is softly blanked out), which is not discernible to the human ear if the muting time does not exceed a few milliseconds. For a more detailed treatment of the CIRC coding scheme in the CD system, see References [31–34].

6.8 CONCLUSION

In the last decade, coding emphasis has been in the area of convolutional codes since in almost every application, convolutional codes outperform block codes for the same implementation complexity of the encoder–decoder. For satellite communication channels, forward error correction techniques can easily reduce the required SNR for a specified error performance by 5 to 6 dB. This coding gain can translate directly into an equivalent reduction in required satellite effective radiated power (EIRP), with consequently reduced satellite weight and cost.

In this chapter we have outlined the essential structural difference between block codes and convolutional codes—the fact that rate $1/n$ convolutional codes have a memory of the prior $K - 1$ bits, where K is the encoder constraint length. With such memory, the encoding of each input data bit not only depends on the value of that bit but on the values of the $K - 1$ input bits that precede it. We presented the decoding problem in the context of the maximum likelihood algorithm, examining all the candidate codeword sequences which could possibly be created by the encoder, and selecting the one that appears statistically most likely; the decision is based on a distance metric for the received code symbols. The error performance analysis of convolutional codes is more complicated than the simple binomial expansion describing the error performance of many block codes. We laid out the concept of free distance, and we presented the relationship between free distance and error performance in terms of bounds. We also described the basic idea behind sequential decoding and feedback decoding and showed some comparative performance curves and tables for various coding schemes.

Finally, we described a technique, interleaving, that allows the popular block and convolutional coding schemes to be used over channels that exhibit bursty noise or periodic fading, without suffering degradation. We used the CD digital audio system as an example of how interleaving plays an important role in ameliorating the effects of burst noise.

Appendix E consists of a FORTRAN program called VITALG for the convolutional encoding and Viterbi decoding of messages. The messages can be in the form of binary sequences or ASCII characters. The user has a choice of code rate, constraint length, connection vectors, and path memory length. The program can be used to insert errors into a bit stream after it has been encoded. From the program output, the user sees the error correcting that results from the use of hard-decision Viterbi decoding of his chosen message. It should prove interesting to use the VITALG program for verifying the performance of the optimum Odenwalder codes shown in Table 6.4.

REFERENCES

1. Gallager, R. G., *Information Theory and Reliable Communication,* John Wiley & Sons, Inc., New York, 1968.

2. Fano, R. M., "A Heuristic Discussion of Probabilistic Decoding," IRE *Trans. Inf. Theory,* vol. IT9, no. 2, 1963, pp. 64–74.

3. Odenwalder, J. P., *Optimal Decoding of Convolutional Codes,* Ph.D. dissertation, University of California, Los Angeles, 1970.

4. Curry, S. J., *Selection of Convolutional Codes Having Large Free Distance,* Ph.D. dissertation, University of California, Los Angeles, 1971.

5. Larsen, K. J., "Short Convolutional Codes with Maximal Free Distance for Rates $\frac{1}{2}$, $\frac{1}{3}$, and $\frac{1}{4}$," *IEEE Trans. Inf. Theory,* vol. IT19, no. 3, 1973, pp. 371–372.

6. Lin, S., and Costello, D. J., Jr., *Error Control Coding: Fundamentals and Applications,* Prentice-Hall, Inc., Englewood Cliffs, N.J., 1983.

7. Forney, G. D., Jr., "Convolutional Codes: I. Algebraic Structure," *IEEE Trans. Inf. Theory,* vol. IT16, no. 6, Nov. 1970, pp. 720–738.

8. Viterbi, A., "Convolutional Codes and Their Performance in Communication Systems," *IEEE Trans. Commun. Technol.,* vol. COM19, no. 5, Oct. 1971, pp. 751–772.

9. Forney, G. D., Jr., and Bower, E. K., "A High Speed Sequential Decoder: Prototype Design and Test," *IEEE Trans. Commun. Technol.,* vol. COM19, no. 5, Oct. 1971, pp. 821–835.

10. Jelinek, F., "Fast Sequential Decoding Algorithm Using a Stack," *IBM J. Res. Dev.,* vol. 13, Nov. 1969, pp. 675–685.

11. Massey, J. L., *Threshold Decoding,* The MIT Press, Cambridge, Mass., 1963.

12. Heller, J. A., and Jacobs, I. W., "Viterbi Decoding for Satellite and Space Communication," *IEEE Trans. Commun. Technol.,* vol. COM19, no. 5, October 1971, pp. 835–848.

13. Viterbi, A. J., "Error Bounds for Convolutional Codes and an Asymptotically Optimum Decoding Algorithm," *IEEE Trans. Inf. Theory,* vol. IT13, April 1967, pp. 260–269.

14. Omura, J. K., "On the Viterbi Decoding Algorithm" (correspondence), *IEEE Trans. Inf. Theory,* vol. IT15, Jan. 1969, pp. 177–179.

15. Mason, S. J., and Zimmerman, H. J., *Electronic Circuits, Signals, and Systems,* John Wiley & Sons, Inc., New York, 1960.

16. Clark, G. C., Jr., and Cain, J. B., *Error-Correction Coding for Digital Communications,* Plenum Press, New York, 1981.

17. Viterbi, A. J., and Omura, J. K., *Principles of Digital Communication and Coding,* McGraw-Hill Book Company, New York, 1979.

18. Massey, J. L., and Sain, M. K., "Inverse of Linear Sequential Circuits," *IEEE Trans. Comput.,* vol. C17, Apr. 1968, pp. 330–337.

19. Rosenberg, W. J., *Structural Properties of Convolutional Codes,* Ph.D. dissertation, University of California, Los Angeles, 1971.

20. Bhargava, V. K., Haccoun, D., Matyas, R., and Nuspl, P., *Digital Communications by Satellite,* John Wiley & Sons, Inc., New York, 1981.

21. Jacobs, I. M., "Practical Applications of Coding," *IEEE Trans. Inf. Theory,* vol. IT20, May 1974, pp. 305–310.

22. Linkabit Corporation, "Coding Systems Study for High Data Rate Telemetry Links," *NASA Ames Res. Center, Final Rep. CR-114278,* Contract NAS-2-6-24, Moffett Field, Calif., 1970.

23. Odenwalder, J. P., *Error Control Coding Handbook,* Linkabit Corporation, San Diego, Calif., July 15, 1976.

24. Wozencraft, J. M., "Sequential Decoding for Reliable Communication," *IRE Natl. Conv. Rec.,* vol. 5, pt. 2, 1957, pp. 11–25.

25. Wozencraft, J. M., and Reiffen, B., *Sequential Decoding,* The MIT Press, Cambridge, Mass., 1961.

26. Shannon, C. E., "A Mathematical Theory of Communication," *Bell Syst. Tech. J.,* vol. 27, 1948, pp. 379–423, 623–656.

27. Brayer, K., "Error Correcting Code Performance on HF, Troposcatter, and Satellite Channels," *IEEE Trans. Commun. Technol.,* vol. COM19, 1971, pp. 835–848.

28. Kohlenberg, A., and Forney, G. D., "Convolutional Coding for Channels with Memory," *IEEE Trans. Inf. Theory,* vol. IT2, 1968, pp. 618–626.

29. Ramsey, J. L., "Realization of Optimum Interleavers, *IEEE Trans. Inf. Theory,* vol. IT16, no. 3, May 1970, pp. 338–345.

30. Forney, G. D., "Burst-Correcting Codes for the Classic Bursty Channel," *IEEE Trans. Commun. Technol.,* vol. COM19, Oct. 1971, pp. 772–781.

31. Peek, J. B. H., "Communications Aspects of the Compact Disc Digital Audio System," *IEEE Commun. Mag.,* vol. 23, no. 2, Feb. 1985, pp. 7–20.

32. Berkhout, P. J., and Eggermont, L. D. J., "Digital Audio Systems," *IEEE ASSP Mag.,* Oct. 1985, pp. 45–67.

33. Driessen, L. M. H. E., and Vries, L. B., "Performance Calculations of the Compact Disc Error Correcting Code on a Memoryless Channel," *Fourth Int. Conf. Video and Data Record.,* Southampton, England, Apr. 20–23, 1982, *IERE Conf. Proc.,* vol. 54, pp. 385–395.

34. Hoeve, H., Timmermans, J., and Vries, L. B., "Error Correction in the Compact Disc System," *Philips Tech. Rev.,* vol. 40, no. 6, 1982, pp. 166–172.

PROBLEMS

6.1. Draw the state diagram, tree diagram, and trellis diagram for the $K = 3$, rate $\frac{1}{3}$ code generated by

$$g_1(X) = X + X^2$$

$$g_2(X) = 1 + X$$

$$g_3(X) = 1 + X + X^2$$

6.2. Given a $K = 3$, rate $\frac{1}{2}$, binary convolutional code with the partially completed state diagram shown in Figure P6.1, find the complete state diagram and sketch a diagram for the encoder.

6.3. Draw the state diagram, tree diagram, and trellis diagram for the convolutional encoder characterized by the block diagram in Figure P6.2.

6.4. Suppose that you were trying to find the quickest way to get from London to Vienna by boat or train. The diagram in Figure P6.3 was constructed from various schedules. The labels on each path are travel times. Using the Viterbi algorithm, find the fastest route from London to Vienna. In a general sense, explain how the algorithm works, what calculations must be made, and what information must be retained in the memory used by the algorithm.

6.5. Consider the convolutional encoder shown in Figure P6.4.
(a) Write the connection vectors and polynomials for this encoder.
(b) Draw the state diagram, tree diagram, and trellis diagram.

6.6. What is the impulse response of the encoder of Problem 6.5? Using the impulse response, determine the output sequence when the input is 1 0 1. Verify by using the generator polynomials.

6.7. Does the encoder of Problem 6.5 allow catastrophic error propagation? Justify your answer with an example.

6.8. Find the free distance of the encoder of Problem 6.3 by the transfer function method.

6.9. Let the codewords of a coding scheme be

$$a = 0\ 0\ 0\ 0\ 0\ 0$$

$$b = 1\ 0\ 1\ 0\ 1\ 0$$

$$c = 0\ 1\ 0\ 1\ 0\ 1$$

$$d = 1\ 1\ 1\ 1\ 1\ 1$$

If the received sequence over a binary symmetric channel is 1 1 1 0 1 0 and a maximum likelihood decoder is used, what will be the decoded symbol?

6.10. Consider that the $K = 3$, rate $\frac{1}{2}$ encoder of Figure 6.3 is used over a binary symmetric channel (BSC). Assume that the initial encoder state is the 00 state. At the output of the BSC, the sequence $Z = (1\ 1\ 0\ 0\ 0\ 0\ 1\ 0\ 1\ 1$ rest all "0") is received.
(a) Find the maximum likelihood path through the trellis diagram, and determine the first 5 decoded information bits. If a tie occurs between any two merged paths, choose the upper branch entering the particular state.
(b) Identify any channel bits in Z that were inverted by the channel during transmission.

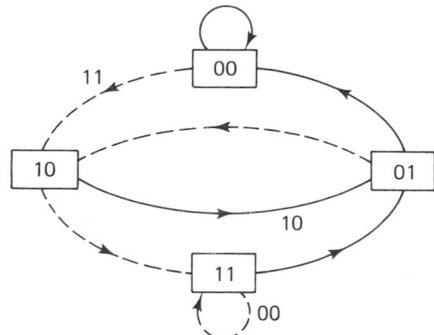

Figure P6.1

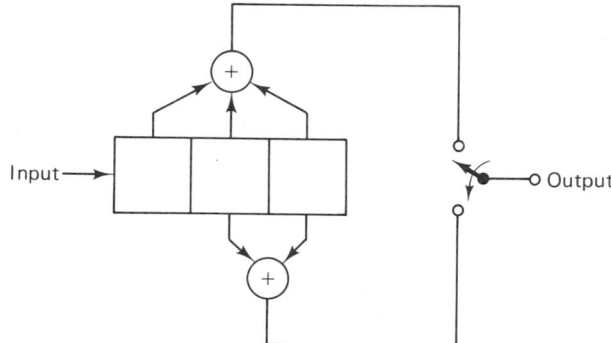

Figure P6.2

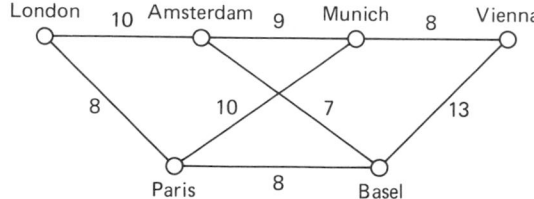

Figure P6.3

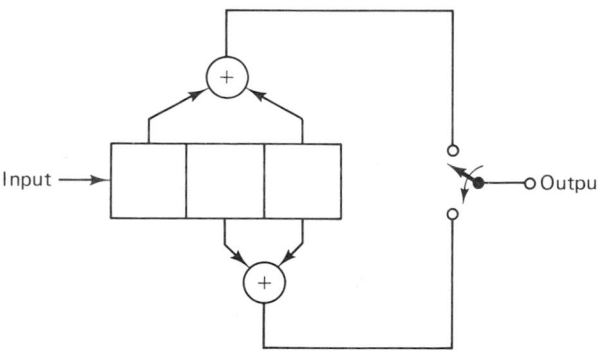

Figure P6.4

6.11. Determine which of the following rate $\frac{1}{2}$ codes are catastrophic.

(a) $g_1(X) = X^2$, $\quad g_2(X) = 1 + X + X^3$

(b) $g_1(X) = 1 + X^2$, $\quad g_2(X) = 1 + X^3$

(c) $g_1(X) = 1 + X + X^2$, $\quad g_2(X) = 1 + X + X^3 + X^4$

(d) $g_1(X) = 1 + X + X^3 + X^4$, $\quad g_2(X) = 1 + X^2 + X^4$

(e) $g_1(X) = 1 + X^4 + X^6 + X^7$, $\quad g_2(X) = 1 + X^3 + X^4$

(f) $g_1(X) = 1 + X^3 + X^4$, $\quad g_2(X) = 1 + X + X^2 + X^4$

6.12. (a) Consider a coherently detected BPSK signal encoded with the encoder shown in Figure 6.3. Find an upper bound on the bit error probability, P_B, if the available E_b/N_0 is 6 dB. Assume hard decision decoding.

(b) Compare P_B with the uncoded case and calculate the improvement factor.

6.13. Using sequential decoding, illustrate the path along the tree diagram shown in Figure 6.20 when the received sequence is 0 1 1 1 0 0 0 1 1 1. The backup criterion is three disagreements.

6.14. Repeat the decoding example of Problem 6.13 using feedback decoding, with a look-ahead length of 3. In the event of a tie, select the upper half of the tree.

6.15. Figure P6.5 depicts a constraint length 2 convolutional encoder.

(a) Draw the state diagram, tree diagram, and trellis diagram.

(b) Assume that a received message from this encoder is 1 1 0 0 1 0. Use a feed-back decoding algorithm with a look-ahead length of 2 to decode the coded message sequence.

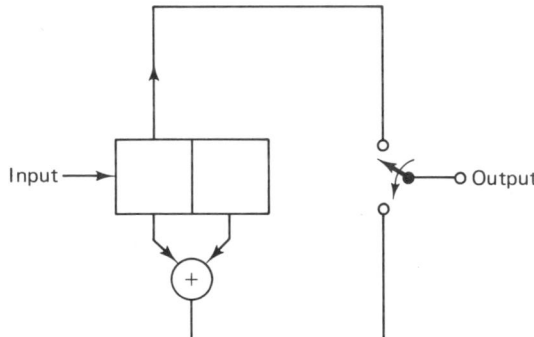

Figure P6.5

6.16. Using the branch word information on the encoder trellis of Figure 6.7, decode the sequence $\mathbf{Z} = (01\ 11\ 00\ 01\ 11$ rest all "0"), using hard-decision Viterbi decoding.

6.17. Consider the rate $\frac{2}{3}$ convolutional encoder shown in Figure P6.6. In this encoder, $k = 2$ bits at a time are shifted into the encoder and $n = 3$ bits are generated at the encoder output. There are $kK = 4$ stages in the register, and the constraint length is $K = 2$ in units of 2-bit bytes. The state of the encoder is defined as the contents of the rightmost $K - 1$ k-tuple stages. Draw the state diagram, the tree diagram, and the trellis diagram.

6.18. Find the ratio of the predetection signal-to-noise spectral density, P_r/N_0, in decibels, required to yield a decoded data rate of 1 Mbit/s with a bit error probability of 10^{-5}. Assume binary noncoherent FSK modulation. Also, assume convolutional encoding with the following decoder relationship:

$$P_b = 2000\ P_B^4$$

where P_B and P_b are bit error probabilities into and out of the decoder, respectively.

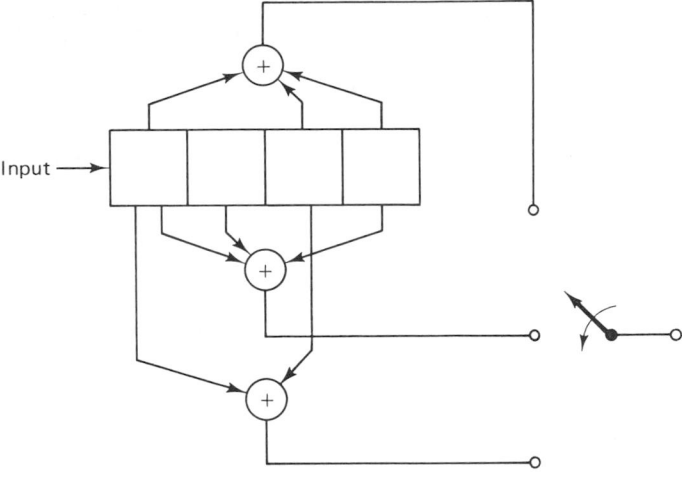

Figure P6.6

6.19. The sequence

1 0 1 1 0 1 1 0 0 0 1 0 1 1 0 0

is the input to a 4 × 4 block interleaver. What is the output sequence? The same input sequence is applied to the convolutional interleaver of Figure 6.26. What is the output sequence?

6.20. Using the computer program VITALG listed in Appendix E, perform the following calculations. Let the uncoded message consist of a sequence of binary zeros, where the number of zeros is 10 times the constraint length of the code being used. Convolutionally encode the message and emulate a memoryless AWGN channel by inserting random transmission errors into the coded sequence (space the errors at approximately a uniform distance from one another). Decode the corrupted coded message using the Viterbi algorithm with the path memory chosen to be five times the constraint length. Use the code generators described as optimum by Odenwalder in Table 6.4. Record the errors corrected by the code, and tabulate the maximum number of errors that can be corrected using each of the following codes:

(a) Rate $\frac{1}{2}$, constraint length 3
(b) Rate $\frac{1}{3}$, constraint length 3
(c) Rate $\frac{1}{2}$, constraint length 5
(d) Rate $\frac{1}{3}$, constraint length 5
(e) Rate $\frac{1}{2}$, constraint length 7
(f) Rate $\frac{1}{3}$, constraint length 7
Explain the results.

6.21. Repeat Problem 6.20. However, instead of a memoryless AWGN channel, emulate a channel that has memory by inserting error bursts into the coded message. Let a burst consist of an uninterrupted sequence of errors placed approximately in the middle of the message. Tabulate the maximum number of errors that can be corrected using each of the code types (a) through (f) listed in Problem 6.20, and compare the error-correcting capabilities of the codes with these two different channel environments. Explain the results.

6.22. Repeat Problem 6.20, parts (e) and (f), for both the uniformly spaced error pattern and the burst error pattern (described in Problems 6.20 and 6.21). In each case compare the performance of the Odenwalder generators in Table 6.4 to other generators of your own choosing. Tabulate the results. Do your findings support the premise that the Table 6.4 generators are optimum?

6.23. For each of the following conditions, design an interleaver for a communication system operating over a bursty noise channel at a transmission rate of 19,200 coded symbols/s.

    (a) A contiguous noise burst typically lasts for 250 ms. The system code consists of a (127, 36) BCH code with $d_{min} = 31$. The end-to-end delay is not to exceed 5 s.

    (b) A contiguous noise burst typically lasts for 20 ms. The system code consists of a rate $\frac{1}{2}$ convolutional code with a feedback decoding algorithm that corrects an average of 3 symbols in a sequence of 21 symbols. The end-to-end delay is not to exceed 160 ms.

6.24. **(a)** Calculate the probability of a byte (symbol) error after decoding the data stored on a compact disc (CD) as described in Section 6.7. Assume that the probability of a channel symbol error for the disc is 10^{-3}. Also assume that the inner and outer R–S decoders are each configured to correct all 2-symbol errors, and that the interleaving process results in channel symbol errors being uncorrelated from one another.

    (b) Repeat part (a) for a disc that has a probability of channel symbol error equal to 10^{-2}.

CHAPTER 7

Modulation and Coding Trade-Offs

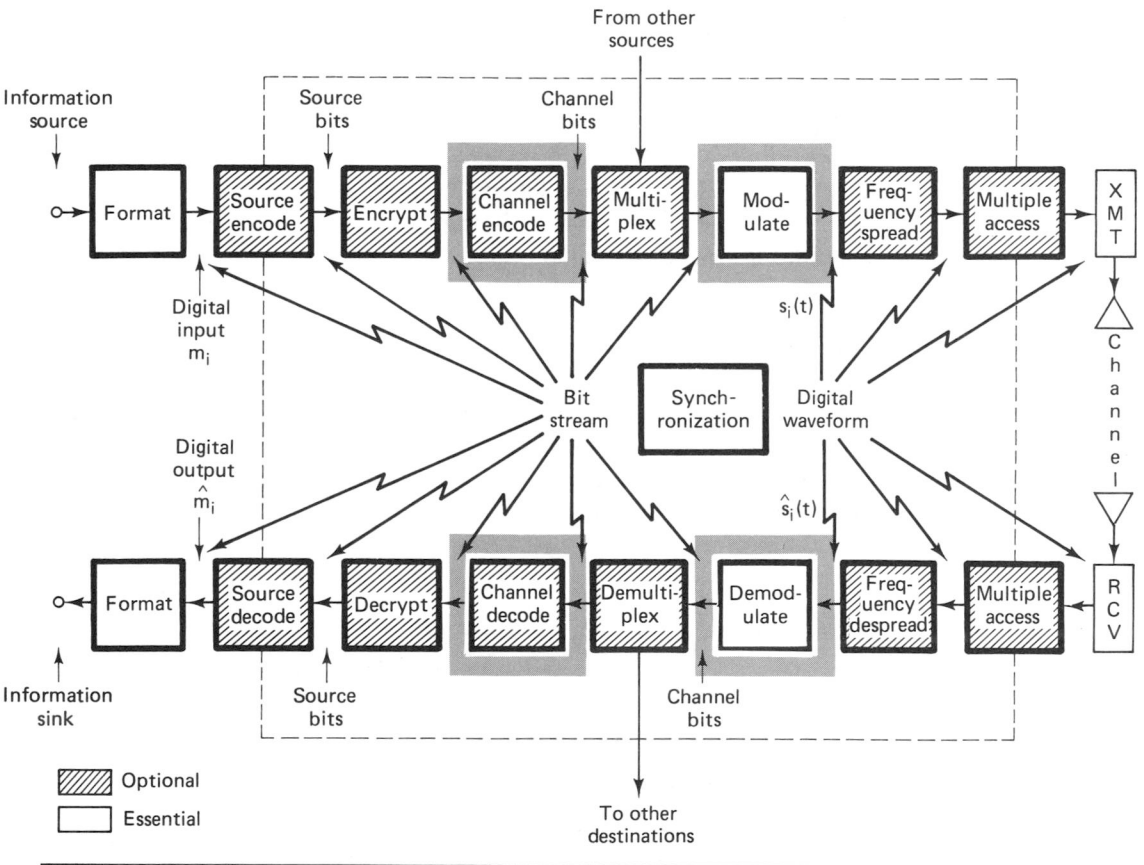

7.1 GOALS OF THE COMMUNICATIONS SYSTEM DESIGNER

System trade-offs are fundamental to all digital communication designs. The goals of the designer are (1) to maximize transmission bit rate, R; (2) to minimize probability of bit error, P_B; (3) to minimize required power, or equivalently, to minimize required bit energy to noise power spectral density, E_b/N_0; (4) to minimize required system bandwidth, W; (5) to maximize system utilization, that is, to provide reliable service for a maximum number of users with minimum delay and with maximum resistance to interference; and (6) to minimize system complexity, computational load, and system cost. A good system designer seeks to achieve all these goals simultaneously. However, goals 1 and 2 are clearly in conflict with goals 3 and 4; they call for simultaneously maximizing R, while minimizing P_B, E_b/N_0, and W. There are several constraints and theoretical limitations that necessitate the trading off of any one system requirement with each of the others. Some of the constraints are:

The Nyquist theoretical minimum bandwidth requirement
The Shannon–Hartley capacity theorem (and the Shannon limit)
Government regulations (e.g., frequency allocations)
Technological limitations (e.g., state-of-the art components)
Other system requirements (e.g., satellite orbits)

Some of the realizable modulation and coding trade-offs can best be viewed as a change in operating point on one of two performance planes. These planes will be referred to as the error probability plane and the bandwidth efficiency plane; they are described in the following sections.

7.2 ERROR PROBABILITY PLANE

Figure 7.1 illustrates the family of P_B versus E_b/N_0 curves for the coherent detection of orthogonal signaling (Figure 7.1a) and multiple phase signaling (Figure 7.1b). For signaling schemes that process k bits at a time, the signaling is called M-ary (see Section 3.8). The modulator uses one of its $M = 2^k$ waveforms to represent each k-bit sequence, where M is the size of the symbol set. Figure 7.1a illustrates the potential bit error improvement with orthogonal signaling as k (or M) is increased. For orthogonal signal sets, such as frequency shift keying (FSK) modulation, increasing the size of the symbol set can provide an improvement in P_B, or a reduction in the E_b/N_0 required, at the cost of increased bandwidth. Figure 7.1b illustrates potential bit error degradation with nonorthogonal signaling as k (or M) increases. For nonorthogonal signal sets, such as multiple phase shift keying (MPSK) modulation, increasing the size of the symbol set can reduce the bandwidth requirement, but at the cost of a degraded P_B, or an increased E_b/N_0 requirement. We shall refer to these families of curves (Figure 7.1a or b) as *error probability performance curves,* and to the plane on which they are plotted as an *error probability plane.* Such a plane describes the locus of operating points available for a particular type of modulation and coding. For a given system information rate, each curve in the plane can be associated with a different fixed minimum required bandwidth; therefore, the set of curves can be termed *equibandwidth curves.* As the curves move in the direction of the ordinate, the required transmission bandwidth increases; as the curves move in the opposite direction, the required bandwidth decreases. Once a modulation and coding scheme and an available E_b/N_0 are determined, system operation is characterized by a particular point in the error probability plane. Possible trade-offs can be viewed as changes in the operating point on one of the curves or as changes in the operating point from one curve to another curve of the family. These trade-offs are seen in Figure 7.1a and b as changes in the system operating point in the direction shown by the arrows. Movement of the operating point along line 1, between points a and b, can be viewed as trading off P_B for E_b/N_0 performance (with W fixed). Similarly, movement along line 2, between points c and d, is seen as trading P_B for W performance (with E_b/N_0 fixed). Finally, movement along line 3, between points e and f, illustrates trading W for E_b/N_0 performance (with P_B fixed). Movement along line 1 is effected by increasing or decreasing the available E_b/N_0. This can be achieved, for example, by increasing transmitter power, which means that the trade-off might be accomplished simply by "turning a knob," even after the system is configured. However, the other trade-offs (movement along line 2 or line 3) involve some change in the system modulation or coding scheme, and therefore need to be accomplished during the system design phase.

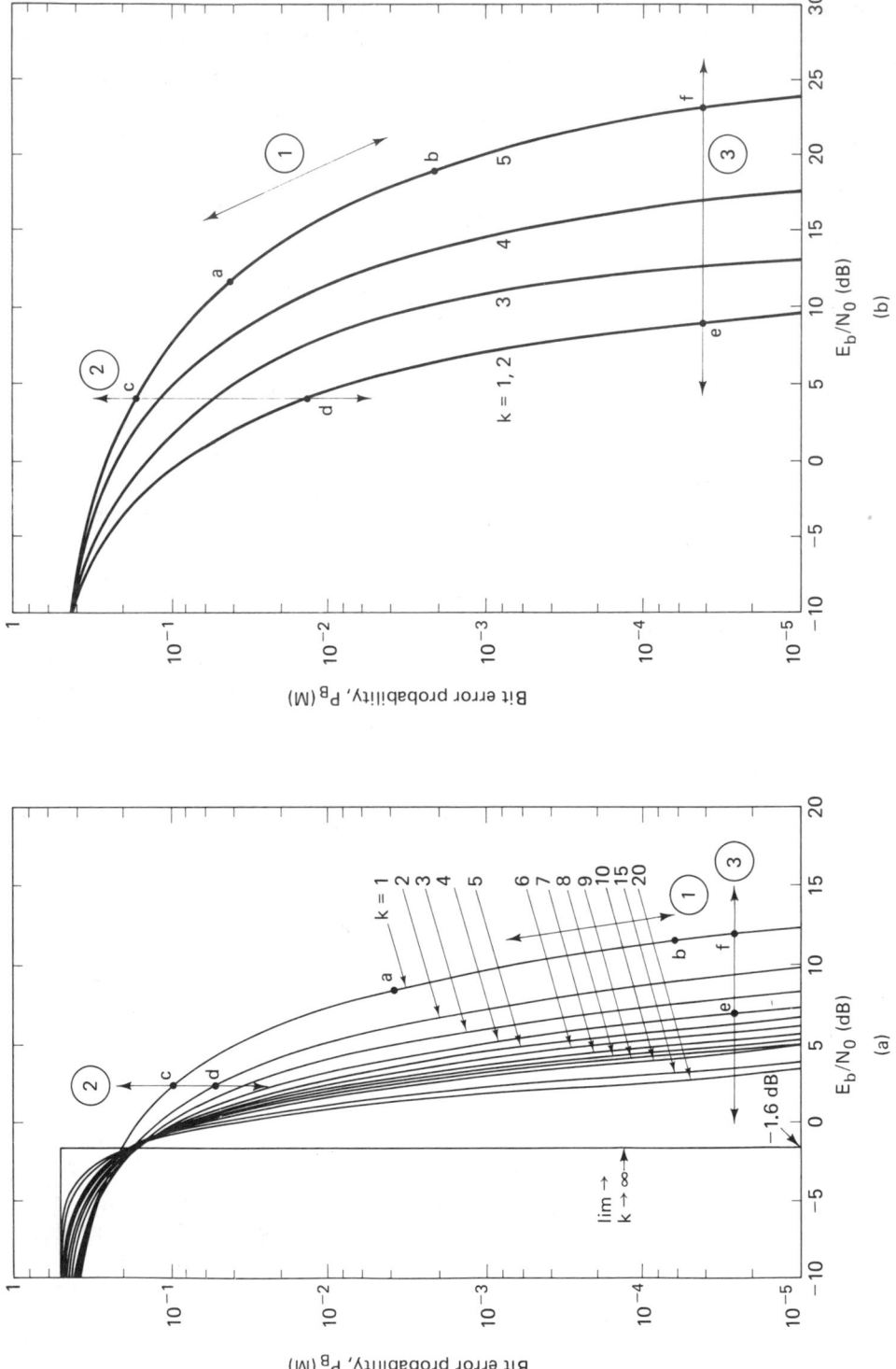

Figure 7.1 Bit error probability versus E_b/N_0 for coherently detected M-ary signaling. (a) Orthogonal signaling. (b) Multiple phase signaling.

7.3 NYQUIST MINIMUM BANDWIDTH

Every realizable system having some nonideal filtering will suffer from intersymbol interference (ISI)—the tail of one pulse spilling over into adjacent symbol intervals so as to interfere with correct detection. Nyquist [1] showed that, in theory, R_s symbols per second could be detected without ISI in an $R_s/2$ hertz minimum bandwidth (Nyquist bandwidth); this is a basic theoretical constraint, limiting the designer's goal to expend as little bandwidth as possible (see Section 2.11). In practice, R_s hertz is typically required for the transmission of R_s symbols per second. In other words, *typical* digital communication throughput, without ISI, is limited to 1 symbol/s per hertz. The modulation or coding system assigns to each symbol, of its set of M symbols, a k-bit meaning, where $M = 2^k$. For a signaling scheme with a fixed bandwidth, such as MPSK, as k increases, the allowable data rate, R, increases, and hence the bandwidth efficiency, R/W, measured in bits per second per hertz, also increases. For example, movement along line 3, from point e to point f in Figure 7.1b represents trading E_b/N_0 for a reduced bandwidth requirement. In other words, with the same system bandwidth one can transmit at an increased data rate, hence at an increased R/W.

7.4 SHANNON–HARTLEY CAPACITY THEOREM

Shannon [2] showed that the system capacity, C, of a channel perturbed by additive white Gaussian noise (AWGN) is a function of the average received signal power, S, the average noise power, N, and the bandwidth, W. The capacity relationship (Shannon–Hartley theorem) can be stated as

$$C = W \log_2 \left(1 + \frac{S}{N} \right) \tag{7.1}$$

When W is in hertz and the logarithm is taken to the base 2, as shown, the capacity is given in bits/s. It is theoretically possible to transmit information over such a channel at any rate, R, where $R \leq C$, with an *arbitrarily small* error probability by using a sufficiently complicated coding scheme. For an information rate $R > C$, it is not possible to find a code that can achieve an arbitrarily small error probability. Shannon's work showed that the values of S, N, and W *set a limit on transmission rate, not on error probability*. Shannon [3] used Equation (7.1) to graphically exhibit a bound for the achievable performance of practical systems. This plot, shown in Figure 7.2, gives the normalized channel capacity C/W in bits/s/Hz as a function of the channel signal-to-noise ratio (SNR). A related plot, shown in Figure 7.3, indicates the normalized channel bandwidth W/C in Hz/bits/s as a function of SNR in the channel. Figure 7.3 is sometimes used to illustrate the power–bandwidth trade-off inherent in the ideal channel. However, it is not a pure trade-off [4] because the detected noise power is proportional to bandwidth.

$$N = N_0 W \tag{7.2}$$

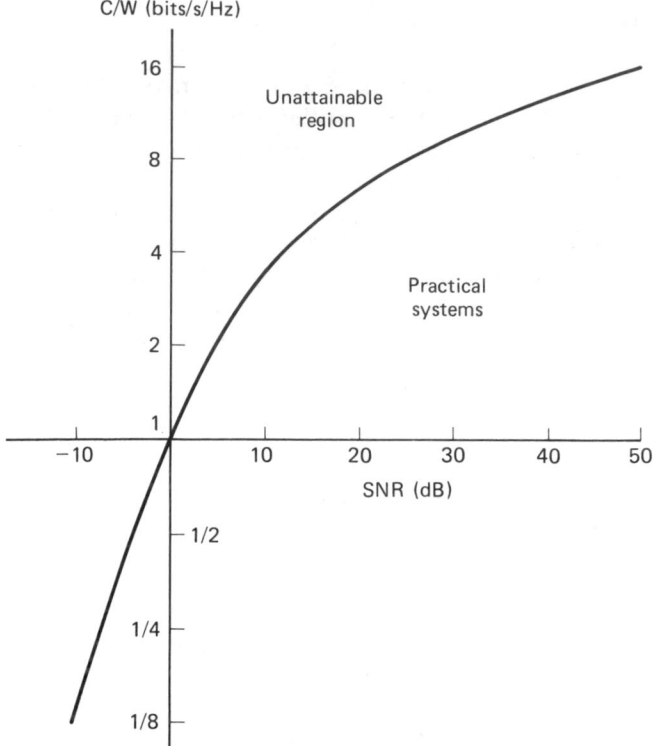

Figure 7.2 Normalized channel capacity versus channel SNR.

Substituting Equation (7.2) into Equation (7.1) and rearranging terms yields

$$\frac{C}{W} = \log_2 \left(1 + \frac{S}{N_0 W} \right) \tag{7.3}$$

For the case where transmission bit rate is equal to channel capacity, $R = C$, we can use the identity presented in Equation (3.94) to write

$$\frac{S}{N_0 C} = \frac{E_b}{N_0} \tag{7.4}$$

Hence we can modify Equation (7.3) as follows:

$$\frac{C}{W} = \log_2 \left[1 + \frac{E_b}{N_0} \left(\frac{C}{W} \right) \right] \tag{7.5}$$

$$2^{C/W} = 1 + \frac{E_b}{N_0} \left(\frac{C}{W} \right)$$

$$\frac{E_b}{N_0} = \frac{W}{C} (2^{C/W} - 1) \tag{7.6}$$

Figure 7.4 is a plot of W/C versus E_b/N_0 in accordance with Equation (7.6).

Modulation and Coding Trade-Offs   Chap. 7

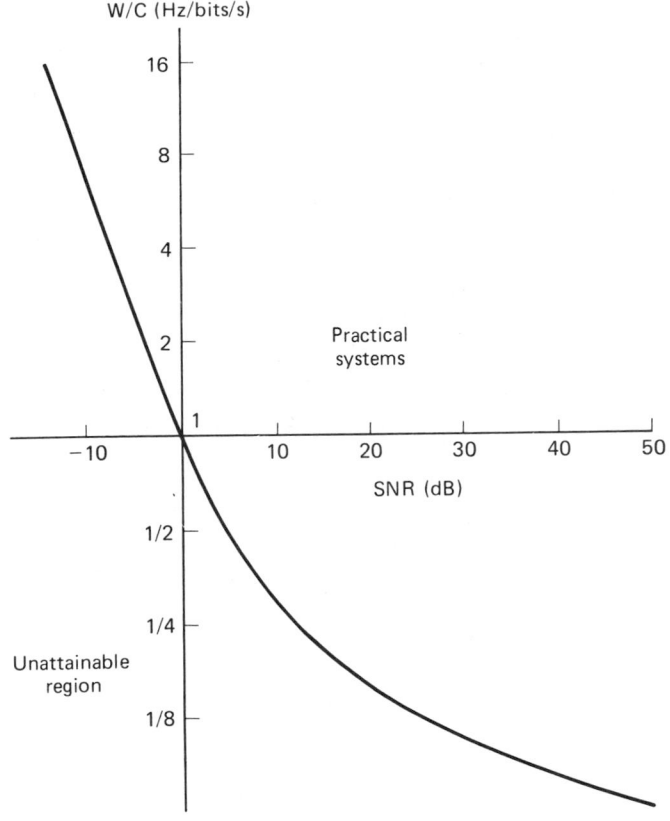

W/C (Hz/bits/s)

Practical systems

SNR (dB)

Unattainable region

Figure 7.3 Normalized channel bandwidth versus channel SNR.

The asymptotic behavior of this curve as $C/W \to 0$ (or $W/C \to \infty$) is discussed in the next section.

7.4.1 Shannon Limit

There exists a limiting value of E_b/N_0 below which there can be no error-free communication at any information rate. Using the identity

$$\lim_{x \to 0} (1 + x)^{1/x} = e$$

we can calculate the limiting value of E_b/N_0 as follows. Let

$$x = \frac{E_b}{N_0} \left(\frac{C}{W}\right)$$

Then from Equation (7.5),

$$\frac{C}{W} = x \log_2 (1 + x)^{1/x}$$

$$1 = \frac{E_b}{N_0} \log_2 (1 + x)^{1/x}$$

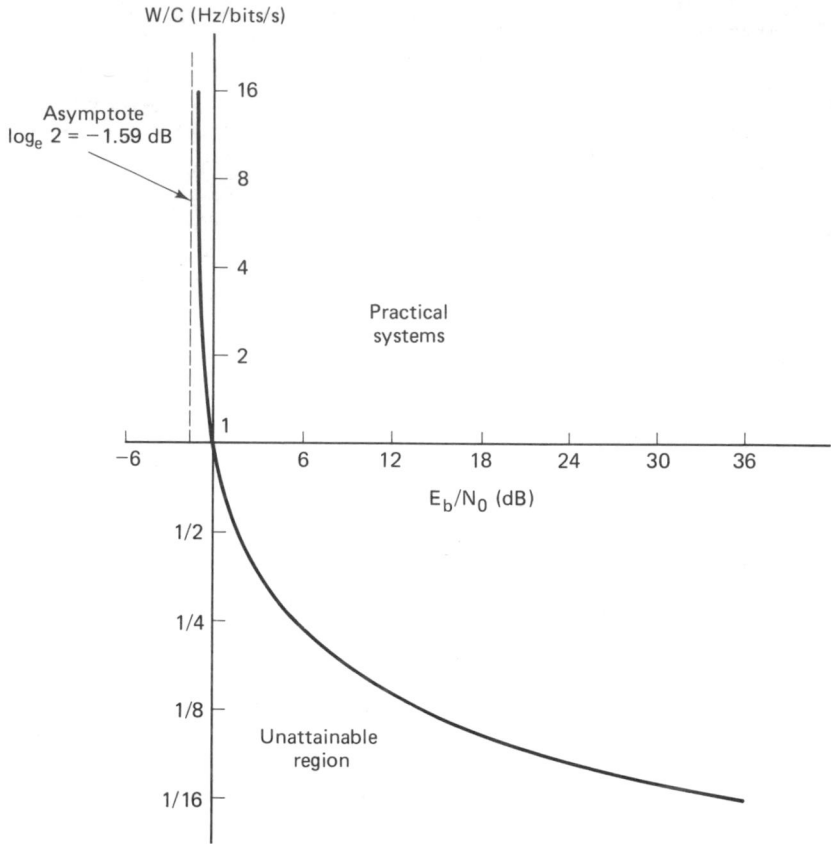

Figure 7.4 Normalized channel bandwidth versus channel E_b/N_0.

In the limit, as $C/W \rightarrow 0$, we get

$$\frac{E_b}{N_0} = \frac{1}{\log_2 e} = 0.693 \tag{7.7}$$

or, in decibels $= -1.59$ dB

This value of E_b/N_0 is called the *Shannon limit*. On Figure 7.1a the Shannon limit is the P_B versus E_b/N_0 curve corresponding to $k \rightarrow \infty$. The curve is discontinuous, going from a value of $P_B = \frac{1}{2}$ to $P_B = 0$ at $E_b/N_0 = -1.59$ dB. It is not possible in practice to reach the Shannon limit, because as k increases without bound, the bandwidth requirement and the implementation complexity increase without bound. Shannon's work provided a theoretical proof for the existence of codes that could improve the P_B performance, or reduce the E_b/N_0 required, from the levels of the uncoded binary modulation schemes to levels approaching the limiting curve. For a bit error probability of 10^{-5}, binary phase shift keying (BPSK) modulation requires an E_b/N_0 of 9.6 dB (the optimum uncoded binary modulation). Therefore, Shannon's work promised the existence of a theoretical performance improvement of 11.2 dB over the performance of optimum uncoded binary modulation, through the use of coding techniques. Today, most of that promised improvement (approximately 7 dB) is realizable [5]. Optimum system design can best be described as a search for rational compromises or trade-offs among the various constraints and conflicting goals. The modulation and coding trade-off, that is, the selection of modulation and coding techniques to make the best use of transmitter power and channel bandwidth, is important, since there are strong incentives to reduce the cost of generating power and to conserve the radio spectrum.

7.4.2 Entropy

To design a communications system with a specified message handling capability, we need a metric for measuring the information content to be transmitted. Shannon [2] developed such a metric, H, called the entropy of the message source (having n possible outputs). *Entropy* is defined as the average amount of information per source output and is expressed by

$$H = - \sum_{i=1}^{n} p_i \log_2 p_i \qquad \text{bits/source output} \qquad (7.8)$$

where p_i is the probability of the ith output and $\sum p_i = 1$. In the case of a binary message or a source having only two possible outputs, with probabilities p and $q = (1 - p)$, the entropy is written

$$H = -(p \log_2 p + q \log_2 q) \qquad (7.9)$$

and is plotted versus p in Figure 7.5.

The quantity H has a number of interesting properties, including the following:

1. When the logarithm in Equation (7.8) is taken to the base 2, as shown, the unit for H is average bits per event. The unit *bit*, here, is a measure of *information content* and is not to be confused with the term "bit," meaning "binary digit."

2. The term "entropy" has the same uncertainty connotation as it does in certain formulations of statistical mechanics. For the information source with two equally likely possibilities (e.g., the flipping of a fair coin), it can be seen from Figure 7.5 that the uncertainty in the event, and hence the average information content, is maximum. As the probabilities depart from the equally likely case, the average information content decreases. In the limit, when one of the probabilities goes to zero, H also goes to zero. We know the result before the event happens, so the result conveys no additional information.

3. To illustrate that information content is related to a priori probability (if the a priori message probability at the receiver is zero or one, we need not send the message), consider the following example: At the end of her nine-month

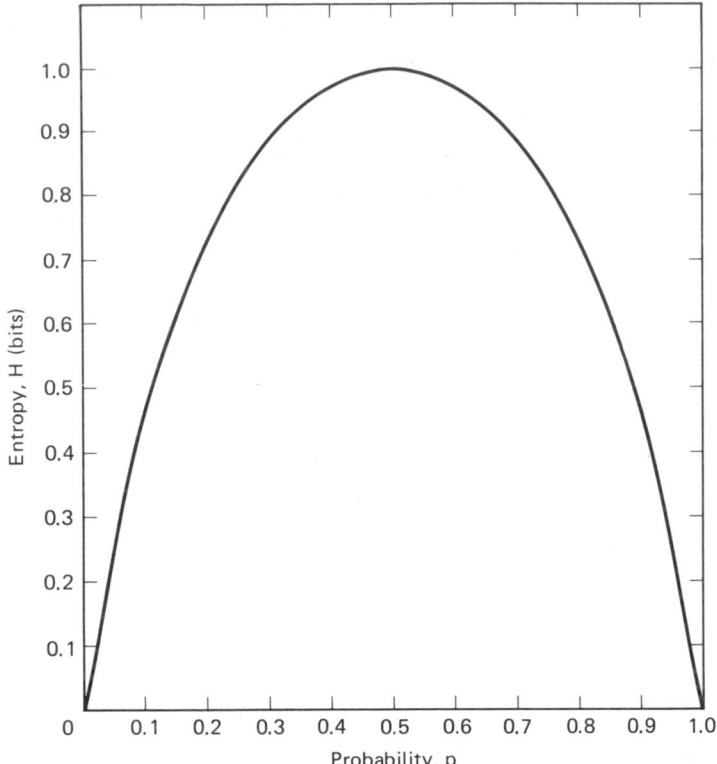

Figure 7.5 Entropy versus probability (two events).

pregnancy, a woman enters the delivery room of a local hospital to give birth. Her husband waits anxiously in the waiting room. After some time, a physician approaches the husband and says: "Congratulations, you are the father of a child." How much information has the physician given the father *beyond the medical outcome*? Almost none; the father has known with virtual certainty that a child was forthcoming. Had the physician said, "you are the father of a boy" or "you are the father of a girl," he would have transmitted 1 bit of information, since there was a 50% chance that the child could have been a boy or a girl.

Example 7.1 Average Information Content in the English Language

(a) Calculate the average information in bits/character for the English language, assuming that each of the 26 characters in the alphabet occurs with equal likelihood. Neglect spaces and punctuation.

(b) Since the alphabetic characters do not appear with equal frequency in the English language (or any other language), the answer to part (a) will represent an upper bound on average information content per character. Repeat part (a) under the assumption that the alphabetic characters occur with the following probabilities:

$$p = 0.10: \quad \text{for the letters a, e, o, t}$$

$$p = 0.07: \quad \text{for the letters h, i, n, r, s}$$

$$p = 0.02: \quad \text{for the letters c, d, f, l, m, p, u, y}$$

$$p = 0.01: \quad \text{for the letters b, g, j, k, q, v, w, x, z}$$

Solution

(a) $H = - \sum\limits_{i=1}^{26} \dfrac{1}{26} \log_2 \left(\dfrac{1}{26} \right)$

$\quad\quad = 4.7 \text{ bits/character}$

(b) $H = -(4 \times 0.1 \log_2 0.1 + 5 \times 0.07 \log_2 0.07$
$\quad\quad\quad + 8 \times 0.02 \log_2 0.02 + 9 \times 0.01 \log_2 0.01)$
$\quad\quad = 4.17 \text{ bits/character}$

If we want to express the 26 letters of the alphabet with some binary-digit coding scheme, we generally need five binary digits for each character. Example 7.1 demonstrates that there may be a way to encode the English language with a fewer number of binary digits per character, *on the average,* by exploiting the fact that the average amount of information contained within each character is less than 5 bits. The subject of source coding, which deals with this exploitation, is treated in Chapter 11.

7.4.3 Equivocation and Effective Transmission Rate

Suppose that we are transmitting information at a rate of 1000 binary symbols/s over a binary symmetric channel (defined in Section 5.3.1), and that the a priori probability of transmitting either a one or a zero is equally likely. Suppose also that the noise in the channel is so great that the probability of receiving a one is $\frac{1}{2}$, whatever was transmitted, and similarly for receiving a zero. In such a case, half the received symbols would be correct *due to chance alone,* and the system might appear to be providing 500 bits/s while actually no information is being received at all. Equally "good" reception could be obtained by dispensing with the channel entirely and "flipping a coin" within the receiver. The proper correction to apply to the amount of information transmitted is the amount of information that is lost in the channel. Shannon [2] uses a correction factor called *equivocation* to account for the uncertainty in the received signal. Equivocation is defined as the *conditional entropy* of the message X, given Y, as shown below:

$$H(X|Y) = - \sum\limits_{X,Y} P(X, Y) \log_2 P(X|Y)$$
$$= - \sum\limits_{Y} P(Y) \sum\limits_{X} P(X|Y) \log_2 P(X|Y) \tag{7.10}$$

where X is the transmitted source message, Y is the received signal, $P(X, Y)$ is the joint probability of X and Y, and $P(X|Y)$ is the conditional probability of X given Y. Equivocation can be thought of as the uncertainty that message X was sent, having received Y. For an *error-free channel*, $H(X|Y) = 0$, because having received Y, there is complete certainty about the message X. However, for a channel with a nonzero probability of symbol error, $H(X|Y) > 0$, because the channel introduces uncertainty. Consider a binary sequence, X, where the a priori source probabilities are $P(X = 1) = P(X = 0) = \frac{1}{2}$, and where, on the average, the channel produces one error in a received sequence of 100 bits ($P_B = 0.01$). Using Equation (7.10), the equivocation $H(X|Y)$ is expressed as

$$H(X|Y) = -[(1 - P_B) \log_2 (1 - P_B) + P_B \log_2 P_B]$$

$$= -(0.99 \log_2 0.99 + 0.01 \log_2 0.01)$$

$$= 0.081 \text{ bit/received symbol}$$

Thus, the channel introduces 0.081 bit of uncertainty to each received symbol.

Shannon showed that the average effective information content, H_{eff}, at the receiver, is obtained by subtracting the equivocation from the entropy of the source. Therefore,

$$H_{eff} = H(X) - H(X|Y) \tag{7.11}$$

For a system transmitting equally likely binary symbols, the entropy, $H(X)$, is 1 bit/symbol. When the symbols are received with $P_B = 0.01$ the equivocation is 0.081 bit/received symbol as was calculated above. Then using Equation (7.11), the effective entropy of the received signal, H_{eff}, is

$$H_{eff} = 1 - 0.081 = 0.919 \text{ bit/received symbol}$$

Thus, if $R = 1000$ binary symbols transmitted per second, for example, the effective information bit rate, R_{eff}, can be expressed as

$$R_{eff} = R H_{eff} \tag{7.12}$$

$$= 1000 \text{ symbols/s} \times 0.919 \text{ bit/symbol} = 919 \text{ bits/s}$$

Notice that in the extreme case where $P_B = 0.5$,

$$H(X|Y) = -(0.5 \log_2 0.5 + 0.5 \log_2 0.5)$$

$$= 1 \text{ bit/symbol}$$

and, applying Equations (7.12) and (7.11) to the $R = 1000$ symbols/s example, yields

$$R_{eff} = 1000 \text{ symbols/s} (1 - 1) = 0 \text{ bit/s}$$

as should be expected.

Example 7.2 Apparent Contradiction in the Shannon Limit

Plots of P_B versus E_b/N_0 typically display a smooth increase of P_B as E_b/N_0 is decreased. For example, the bit error probability for the curves in Figure 7.1 shows P_B *tending* to 0.5 in the limit as E_b/N_0 approaches zero. Thus there is apparently always a nonvanishing information rate, regardless of how small E_b/N_0 becomes. This *appears to contradict* the Shannon limit of $E_b/N_0 = -1.59$ dB, below which no error-free information rate can be supported per unit bandwidth, or below which even an infinite bandwidth cannot support a finite information rate (see Figure 7.4).

(a) Suggest a way of resolving the apparent contradiction.
(b) Show how Shannon's equivocation correction can resolve it for a binary PSK system where the source has an entropy of 1 bit/symbol. Consider that the operating point on Figure 7.1b corresponds to $E_b/N_0 = 0.1$ (-10 dB).

Solution

(a) The value of E_b, traditionally used in link calculations for practical systems, is invariably the received signal energy per *transmitted symbol*. However, the meaning of E_b in Equation (7.6) is the signal energy per bit of *received information*. The information loss caused by the noisy channel must be taken into account to resolve the apparent contradiction.
(b) Following Equation (3.84) for BPSK,

$$P_B = Q(\sqrt{2E_b/N_0}) = Q(0.447)$$

where Q is defined in Equation (2.42) and tabulated in Table B.1. From the tabulation, P_B is found to be 0.33. Next, we solve for the equivocation and effective entropy:

$$\begin{aligned}
H(X|Y) &= -[(1 - P_B) \log_2 (1 - P_B) + P_B \log_2 P_B] \\
&= -(0.67 \log_2 0.67 + 0.33 \log_2 0.33) \\
&= 0.915 \text{ bit/symbol}
\end{aligned}$$

$$\begin{aligned}
H_{\text{eff}} &= H(X) - H(X|Y) \\
&= 1 - 0.915 \\
&= 0.085 \text{ bit/symbol}
\end{aligned}$$

Hence

$$\begin{aligned}
\left(\frac{E_b}{N_0}\right)_{\text{eff}} &= \frac{(E_b/N_0)}{H_{\text{eff}}} \frac{\text{joules per symbol/watts per hertz}}{\text{bits/symbol}} \\
&= \frac{0.1}{0.085} = 1.176 \frac{\text{joules per bit}}{\text{watts/Hz}} \\
&= 0.7 \text{ dB}
\end{aligned}$$

Thus the effective value of E_b/N_0 is equal to 0.7 dB per received information bit, which is well above Shannon's limit of -1.59 dB.

7.5 BANDWIDTH-EFFICIENCY PLANE

Using Equation (7.6), we can plot normalized channel bandwidth W/C in Hz/bits/s versus E_b/N_0, as shown in Figure 7.4. Here, with the abscissa taken as E_b/N_0, we see the *true power–bandwidth trade-off* at work. It can be shown [4] that well-designed systems tend to operate near the "knee" of this power–bandwidth trade-off curve for the ideal ($R = C$) channel. Actual systems are frequently within 10 dB or less of the performance of the ideal. The existence of the knee means that systems seeking to reduce the channel bandwidth they occupy or to reduce the signal power they require must make an increasingly unfavorable exchange in the other parameter. For example, from Figure 7.4, an ideal system operating at an E_b/N_0 of 1.8 dB and using a normalized bandwidth of 0.5 Hz/bits/s would have to increase E_b/N_0 to 20 dB to reduce the bandwidth occupancy to 0.1 Hz/bits/s. Trade-offs in the other direction are similarly inequitable.

Using Equation (7.6), we can also plot C/W versus E_b/N_0. This relationship is shown plotted on the R/W versus E_b/N_0 plane in Figure 7.6. We shall denote

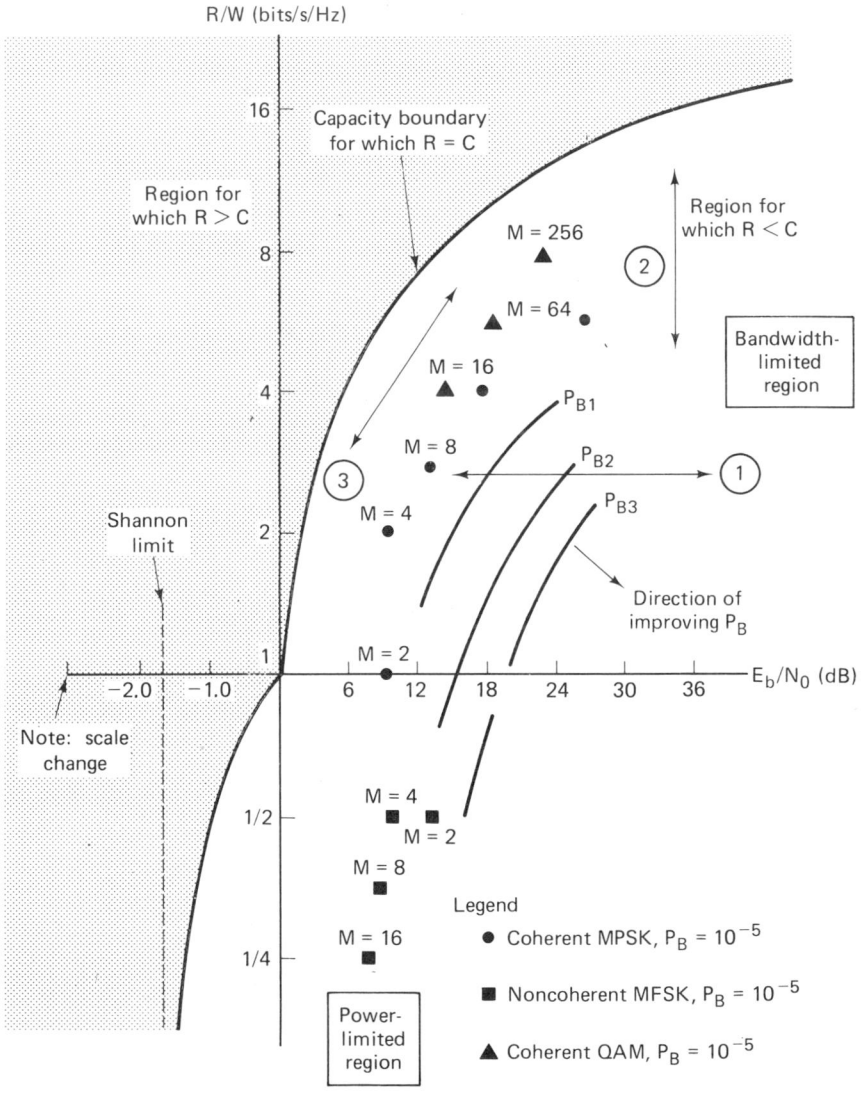

Figure 7.6 Bandwidth-efficiency plane.

this plane as the *bandwidth-efficiency plane*. The ordinate, R/W, is a measure of how much data can be communicated in a specified bandwidth within a given time; it therefore reflects how efficiently the bandwidth resource is utilized. The abscissa is E_b/N_0 in units of decibels. For the case where $R = C$ in Figure 7.6, the curve represents a boundary that separates a region characterizing practical communication systems from a region where such communication systems are not theoretically possible. Like Figure 7.2, the bandwidth-efficiency plane in Figure 7.6 sets the limiting performance that can be achieved by practical systems. Since the abscissa in Figure 7.6 is E_b/N_0 rather than SNR, Figure 7.6 is more useful for comparing digital communication modulation and coding trade-offs than is Figure 7.2.

7.5.1 Bandwidth Efficiency of MPSK and MFSK Modulation

On the bandwidth-efficiency plane of Figure 7.6 are plotted the operating points for coherent MPSK modulation at a bit error probability of 10^{-5}. We assume Nyquist (ideal rectangular) filtering at baseband, so that the minimum double-sideband (DSB) bandwidth at an intermediate frequency (IF) is $W_{\mathrm{IF}} = 1/T$, where T is the symbol duration. Thus the bandwidth efficiency is $R/W = \log_2 M$, where M is the symbol set size. For realistic channels and waveforms, the performance must be reduced to account for the bandwidth increase required to implement realizable filters. Notice that for MPSK modulation, R/W increases with increasing M. Notice also that the location of the MPSK points indicates that BPSK ($M = 2$) and quaternary PSK or QPSK ($M = 4$) require the same E_b/N_0. That is, for the same value of E_b/N_0, QPSK has a bandwidth efficiency of 2 bits/s/Hz, compared to 1 bit/s/Hz for BPSK. This unique feature stems from the fact that QPSK is effectively a composite of two BPSK signals transmitted on orthogonal components of the carrier.

Also plotted on the bandwidth-efficiency plane of Figure 7.6 are the operating points for noncoherent MFSK modulation at a bit error probability of 10^{-5}. We assume that the IF transmission bandwidth is $W_{\mathrm{IF}} = M/T$, and thus the bandwidth efficiency is $R/W = k/M$. Notice that for MFSK modulation, R/W decreases with increasing M. Notice also that the position of the MFSK points indicates that BFSK ($M = 2$) and quaternary FSK ($M = 4$) have the same bandwidth efficiency, even though the former requires greater E_b/N_0 for the same error probability. The bandwidth efficiency varies with the modulation index (tone spacing in hertz divided by bit rate). Under the assumption that an equal increment of bandwidth is required for each MFSK tone the system uses, it can be seen that for $M = 2$, the bandwidth efficiency is 1 bit/s/2 Hz or $\frac{1}{2}$, and for $M = 4$, similarly, the R/W is 2 bits/s/4 Hz or $\frac{1}{2}$.

Operating points for coherent quadrature amplitude modulation (QAM) are also plotted in Figure 7.6. Of the modulations shown, QAM is clearly the most bandwidth efficient; it is treated in greater detail in Section 7.9.3.

7.5.2 Analogies between Bandwidth-Efficiency and Error Probability Planes

The bandwidth-efficiency plane in Figure 7.6 is analogous to the error probability plane in Figure 7.1. The Shannon limit of the Figure 7.1 plane is analogous to the capacity boundary of the Figure 7.6 plane. The curves in Figure 7.1 were referred to as equibandwidth curves. In Figure 7.6, we can analogously describe equi-error-probability curves for various modulation and coding schemes. The curves, labeled P_{B1}, P_{B2}, and P_{B3}, are hypothetical constructions for some arbitrary modulation and coding scheme; the P_{B1} curve represents the largest error probability of the three curves, and the P_{B3} curve represents the smallest. The general direction in which the curves move for improved P_B is indicated on the figure.

Just as potential trade-offs among P_B, E_b/N_0, and W were considered for the error probability plane, the same trade-offs can be considered on the bandwidth efficiency plane. The potential trade-offs are seen in Figure 7.6 as changes in operating point in the direction shown by the arrows. Movement of the operating point along line 1 can be viewed as trading P_B for E_b/N_0, with R/W fixed. Similarly, movement along line 2 is seen as trading P_B for W (or R/W), with E_b/N_0 fixed. Finally, movement along line 3 illustrates trading W (or R/W) for E_b/N_0, with P_B fixed. In Figure 7.6, as in Figure 7.1, movement along line 1 can be effected by increasing or decreasing the available E_b/N_0. However, movement along line 2 or line 3 requires changes in the system modulation or coding scheme.

The two primary communications resources are the transmitted power and the channel bandwidth. In many communication systems, one of these resources may be more precious than the other, and hence most systems can be classified as either power limited or bandwidth limited. In *power-limited systems,* coding schemes can be used to save power at the expense of bandwidth, whereas in *bandwidth-limited systems,* spectrally efficient modulation techniques can be used to save bandwidth at the expense of power.

7.6 POWER-LIMITED SYSTEMS

For the case of power-limited systems, systems in which power is scarce but system bandwidth is available (e.g., a space communication link), the following trade-offs might be made: (1) improved P_B can be achieved by expending bandwidth (for a given E_b/N_0); or (2) required E_b/N_0 can be reduced by expending bandwidth (for a given P_B). The error probability plane of Figure 7.1a can be very useful for examining these potential trade-offs. It is on such a plane that we can verify whether or not a candidate modulation or code offers improvement in required E_b/N_0 for a particular channel and for a specified P_B (or whether the modulation or code offers improvement in P_B for a given E_b/N_0).

7.7 BANDWIDTH-LIMITED SYSTEMS

Any digital scheme that transmits $\log_2 M$ bits in T seconds using a bandwidth of W hertz operates at a bandwidth efficiency of $R/W = (\log_2 M)/WT$ bits/s/Hz. From this expression it can be seen that the smaller the WT product, the more bandwidth efficient will be the system. Signals with small WT products are more often used with bandwidth-limited systems—systems in which channel bandwidth is constrained but power is available. For this case the usual objective is to design the link so as to maximize the transmitted information rate over the bandlimited channel, at the expense of E_b/N_0 (while maintaining a specified value of P_B). For bandlimited operation, bandwidth efficiency is a useful criterion of system performance, and the bandwidth-efficiency plane of Figure 7.6 is useful for examining potential trade-offs.

Two regions, the bandwidth-limited region and the power-limited region, are shown on the bandwidth efficiency plane of Figure 7.6. Notice that the desirable trade-offs associated with each of these regions are not equitable. For the bandwidth-limited region, large R/W is desired; however, as E_b/N_0 is increased, the capacity boundary curve flattens out and ever-increasing amounts of additional E_b/N_0 are required to achieve improvement in R/W. A similar relationship is at work in the power-limited region. Here a savings in E_b/N_0 is desired, but the capacity boundary curve is steep; to achieve a small reduction in required E_b/N_0 requires a large reduction in R/W.

7.8 MODULATION AND CODING TRADE-OFFS

Figure 7.7 is useful in pointing out analogies between the two performance planes, the error probability plane of Figure 7.1 and the bandwidth efficiency plane of Figure 7.6. Figure 7.7a and b represent the same planes as Figures 7.1 and 7.6, respectively. They have been redrawn as symmetrical, by choosing appropriate scales. In each case the arrows and their labels describe the general effect of moving an operating point in the direction of the arrow by means of appropriate modulation and coding techniques. The notations G, C, and F stand for the trade-off considerations "*G*ained or achieved," "*C*ost or expended," and "*F*ixed or unchanged," respectively. The parameters being traded are P_B, W, R/W, and P (power or S/N). Just as the movement of an operating point toward the Shannon limit in Figure 7.7a can achieve improved P_B or reduced required transmitter power at the cost of bandwidth, so too movement toward the capacity boundary in Figure 7.7b can improve bandwidth efficiency at the cost of increased required power or degraded P_B.

Most often, these trade-offs are examined with a fixed P_B (constrained by the system requirement) in mind. Therefore, the most interesting arrows are those having bit error probability (marked F: P_B). There are four such arrows on Figure

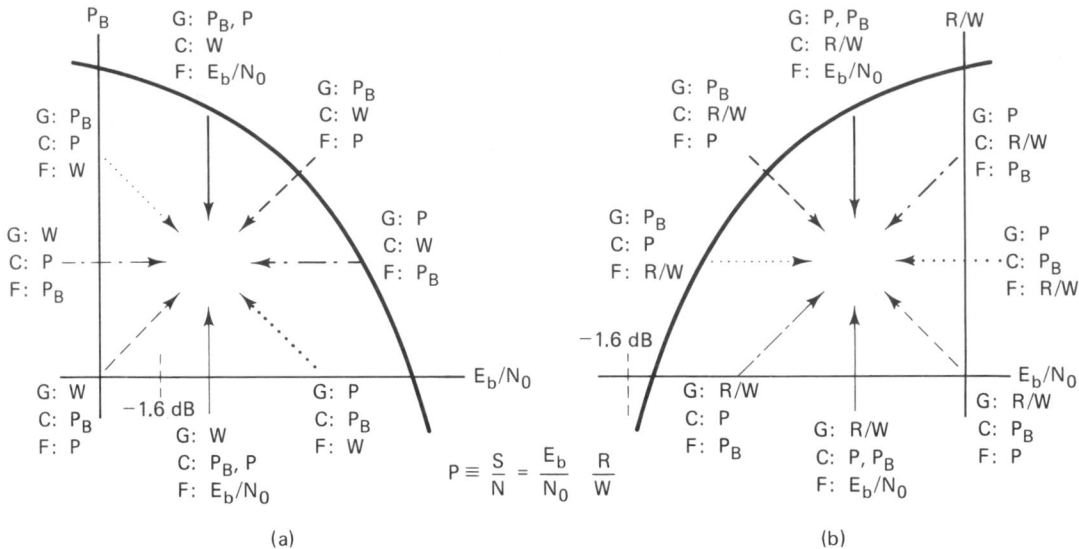

Figure 7.7 Modulation/coding trade-offs. (a) Error probability plane. (b) Bandwidth-efficiency plane.

7.7, two on the error probability plane and two on the bandwidth-efficiency plane. Arrows marked with the same pattern indicate correspondence between the two planes. System operation can be characterized by either of these two planes. The planes represent two ways of looking at some of the key system parameters; each plane highlights slightly different aspects of the overall design problem. The error probability plane tends to be most useful with *power-limited systems,* where as we move from curve to curve, the bandwidth requirements are only inferred, while the bit error probability is clearly displayed. The bandwidth efficiency plane is generally more useful for examining *bandwidth-limited systems*; here as we move from curve to curve, the bit error probability is only inferred, but the bandwidth requirements are explicit.

The two system trade-off planes, error probability and bandwidth efficiency, have been presented *heuristically* with simple examples (orthogonal and multiple phase signaling) to provide some insight into the design issues of trading-off error probability, bandwidth, and power. The ideas are useful for *most modulation and coding schemes,* with the following caveat. For *some* codes or combined modulation and coding schemes, the performance curves *do not move as predictably* as those for the examples chosen here. The reason has to do with the strength and bandwidth expansion features of the particular code. For example, the performance of coherent PSK combined with several codes was illustrated in Figure 5.23. Examine the curves characterizing the two BCH codes, (127, 64) and (127, 36). It should be clear from their relative positions that the (127, 64) code manifests *greater coding gain* than the (127, 36) code. This violates our expectations since, within the same block size, the latter code has greater redundancy (requires more bandwidth expansion) than the former. Also, in the area of trellis-coded modulation covered in Section 7.10.6, we consider codes that provide coding gain without any bandwidth expansion. Performance curves for such coding schemes will also behave differently from the curves of most modulation and coding schemes discussed so far.

7.9 BANDWIDTH-EFFICIENT MODULATION

The primary objective of spectrally efficient modulation techniques is to maximize bandwidth efficiency. The increasing demand for digital transmission channels has led to the investigation of spectrally efficient modulation techniques [6] to maximize bandwidth efficiency and thus help ameliorate the spectral congestion problem.

Some systems have additional modulation requirements besides spectral efficiency. For example, satellite systems with highly nonlinear transponders require a constant envelope modulation. This is because the nonlinear transponder produces extraneous sidebands when passing a signal with amplitude fluctuations (due to a mechanism called AM-to-PM conversion). These sidebands deprive the information signals of some of their portion of transponder power, and also can interfere with nearby channels (adjacent channel interference) or with other communication systems (co-channel interference). *Offset QPSK* (OQPSK) and *Minimum shift keying* (MSK) are two examples of constant envelope modulation schemes that are attractive for systems using nonlinear transponders.

7.9.1 QPSK and Offset QPSK Signaling

Figure 7.8 illustrates the partitioning of a typical pulse stream for QPSK modulation. Figure 7.8a shows the original data stream $d_k(t) = d_0, d_1, d_2, \ldots$ consisting of bipolar pulses; that is, the values of $d_k(t)$ are $+1$ or -1, representing binary one and zero, respectively. This pulse stream is divided into an in-phase stream, $d_I(t)$, and a quadrature stream, $d_Q(t)$, illustrated in Figure 7.8b, as follows:

$$d_I(t) = d_0, d_2, d_4, \ldots \quad \text{(even bits)}$$
$$d_Q(t) = d_1, d_3, d_5, \ldots \text{(odd bits)}$$
(7.13)

Note that $d_I(t)$ and $d_Q(t)$ each have half the bit rate of $d_k(t)$. A convenient orthogonal realization of a QPSK waveform, $s(t)$, is achieved by amplitude modulating the in-phase and quadrature data streams onto the cosine and sine functions of a carrier wave, as follows:

$$s(t) = \frac{1}{\sqrt{2}} d_I(t) \cos \left(2\pi f_0 t + \frac{\pi}{4} \right) + \frac{1}{\sqrt{2}} d_Q(t) \sin \left(2\pi f_0 t + \frac{\pi}{4} \right) \quad (7.14)$$

Using the trigonometric identities shown in Equations (D.5) and (D.6), Equation

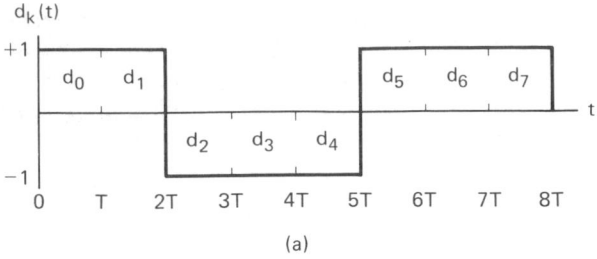

(a)

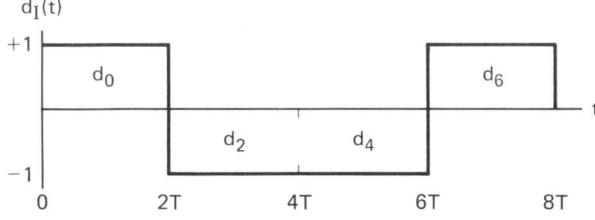

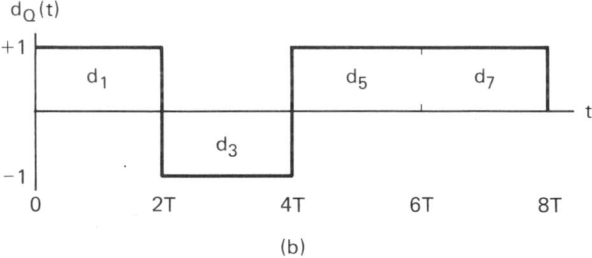

(b)

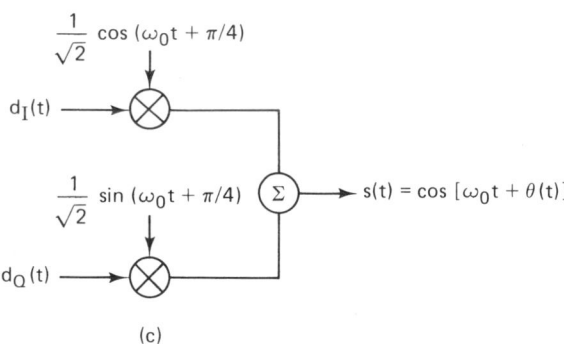

(c)

Figure 7.8 QPSK modulation.

(7.14) can also be written as

$$s(t) = \cos\left[2\pi f_0 t + \theta(t)\right] \tag{7.15}$$

The QPSK modulator is shown in the block diagram of Figure 7.8c. The pulse stream $d_I(t)$ amplitude-modulates the cosine function with an amplitude of $+1$ or -1. This is equivalent to shifting the phase of the cosine function by 0 or π; consequently, this produces a BPSK waveform. Similarly, the pulse stream $d_Q(t)$ modulates the sine function, yielding a BPSK waveform orthogonal to the cosine function. The summation of these two orthogonal components of the carrier yields the QPSK waveform. The value of $\theta(t)$ will correspond to one of the four possible combinations of $d_I(t)$ and $d_Q(t)$ in Equation (7.14). These values are: $\theta(t) = 0°$, $\pm 90°$, or $180°$, and the resulting signal vectors are seen in the signal space illustrated in Figure 7.9. Because $\cos (2\pi f_0 t + \pi/4)$ and $\sin (2\pi f_0 t + \pi/4)$ are orthogonal, the two BPSK signals can be detected separately.

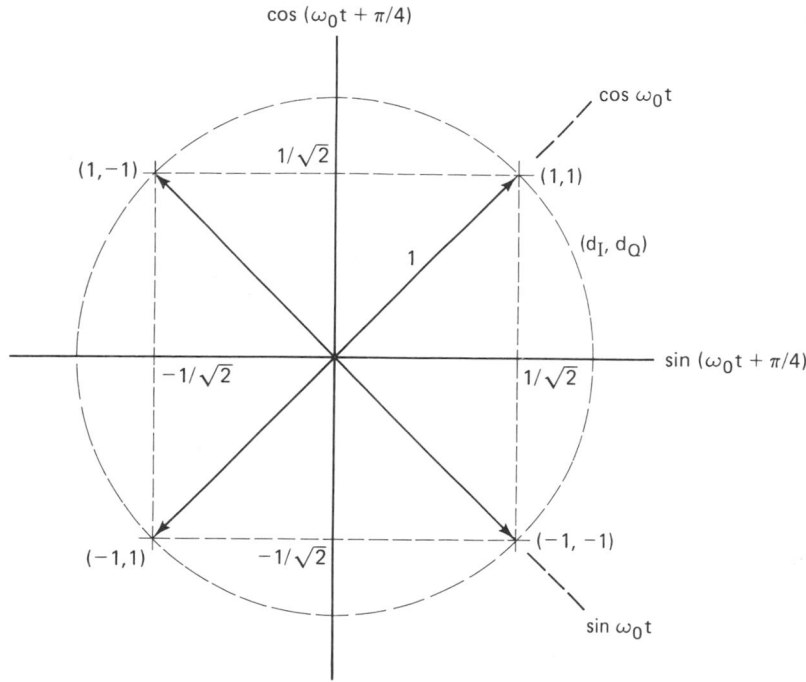

Figure 7.9 Signal space for QPSK and OQPSK.

Offset QPSK (OQPSK) signaling can also be represented by Equations (7.14) and (7.15); the difference between the two modulation schemes, QPSK and OQPSK, is only in the *alignment* of the two baseband waveforms. As shown in Figure 7.8, the duration of each original pulse is T (Figure 7.8a), and hence in the partitioned streams of Figure 7.8b, the duration of each pulse is $2T$. In standard QPSK, the odd and even pulse streams are both transmitted at the rate of $1/2T$ bits/s and are synchronously aligned, such that their transitions coincide, as shown in Figure 7.8b. In OQPSK, sometimes called *staggered QPSK* (SQPSK), there is the same data stream partitioning and orthogonal transmission; the difference is that the timing of the pulse stream $d_I(t)$ and $d_Q(t)$ is shifted such that the alignment of the two streams is offset by T. Figure 7.10 illustrates this offset.

In standard QPSK, due to the coincident alignment of $d_I(t)$ and $d_Q(t)$, the carrier phase can change only once every $2T$. The carrier phase during any $2T$ interval can be any one of the four phases shown in Figure 7.9, depending on the values of $d_I(t)$ and $d_Q(t)$ during that interval. During the next $2T$ interval, if neither pulse stream changes sign, the carrier phase remains the same. If only one of the

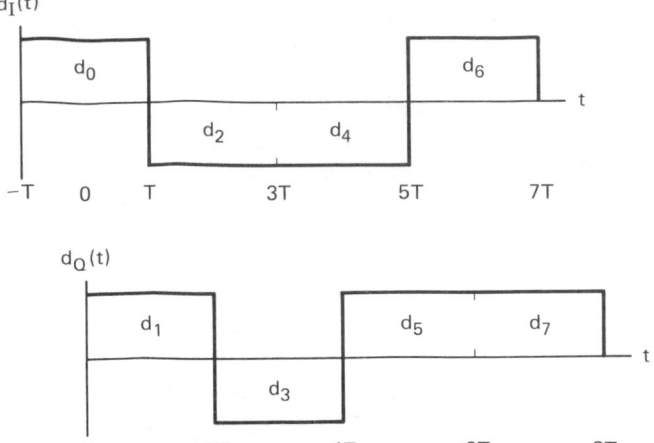

Figure 7.10 Offset QPSK (OQPSK) data streams.

pulse streams changes sign, a phase shift of $\pm 90°$ occurs. A change in both streams results in a carrier phase shift of 180°. Figure 7.11a shows a typical QPSK waveform for the sample sequence $d_I(t)$ and $d_Q(t)$ shown in Figure 7.8.

If a QPSK modulated signal undergoes filtering to reduce the spectral sidelobes, the resulting waveform will no longer have a constant envelope and in fact,

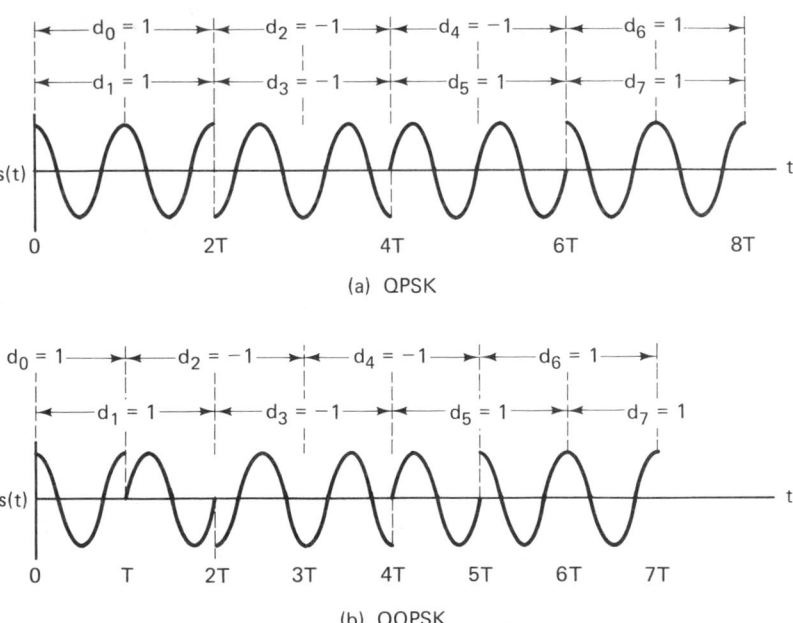

Figure 7.11 (a) QPSK and (b) OQPSK waveforms. (Reprinted with permission from S. Pasupathy, "Minimum Shift Keying: A Spectrally Efficient Modulation," *IEEE Commun. Mag.*, July 1979, Fig. 4, p. 17. © 1979 IEEE.)

the occasional 180° phase shifts will cause the envelope to go to zero momentarily (see Figure 7.11a). When these signals are used in satellite channels employing highly nonlinear amplifiers, the constant envelope will tend to be restored. However, at the same time, all of the *undesirable* frequency side-lobes, which can interfere with nearby channels and other communication systems, are also restored.

In OQPSK, the pulse streams $d_I(t)$ and $d_Q(t)$ are staggered and thus do not change states simultaneously. The possibility of the carrier changing phase by 180° is eliminated, since only one component can make a transition at one time. Changes are limited to 0° and $\pm 90°$ every T seconds. Figure 7.11b shows a typical OQPSK waveform for the sample sequence in Figure 7.10. When an OQPSK signal undergoes bandlimiting, the resulting intersymbol interference causes the envelope to droop slightly to the region of $\pm 90°$ phase transition, but since the phase transitions of 180° have been avoided in OQPSK, the envelope will not go to zero as it does with QPSK. When the bandlimited OQPSK goes through a nonlinear transponder, the envelope droop is removed; however, the high-frequency components associated with the collapse of the envelope are not reinforced. Thus out-of-band interference is avoided [7].

7.9.2 Minimum Shift Keying

The main advantage of OQPSK over QPSK, that of suppressing out-of-band interference, suggests that further improvement is possible if the OQPSK format is modified to avoid discontinuous phase transitions. This was the motivation for designing continuous phase modulation (CPM) schemes. *Minimum shift keying* (MSK) is one such scheme [7–9]. MSK can be viewed as either a special case of *continuous-phase frequency shift keying* (CPFSK), or a special case of OQPSK with sinusoidal symbol weighting. When viewed as CPFSK, the MSK waveform can be expressed as [8]

$$s(t) = \cos\left[2\pi\left(f_0 + \frac{d_k}{4T}\right)t + x_k\right] \qquad kT < t < (k + 1)T \qquad (7.16)$$

where f_0 is the carrier frequency, $d_k = \pm 1$ represents the bipolar data being transmitted at a rate $R = 1/T$, and x_k is a phase constant which is valid over the kth binary data interval. Notice that for $d_k = 1$, the frequency transmitted is $f_0 + 1/4T$, and for $d_k = -1$, the frequency transmitted is $f_0 - 1/4T$. The tone spacing in MSK is thus one-half that employed for noncoherently demodulated orthogonal FSK, giving rise to the name *minimum* shift keying. During each T-second data interval, the value of x_k is a constant, that is, $x_k = 0$ or π, determined by the requirement that the phase of the waveform be continuous at $t = kT$. This requirement results in the following recursive phase constraint for x_k:

$$x_k = \left[x_{k-1} + \frac{\pi k}{2}(d_{k-1} - d_k)\right] \text{ modulo } 2\pi \qquad (7.17)$$

Equation (7.16) can be expressed in a quadrature representation, as follows, using the identities in Equations (D.5) and (D.6):

$$s(t) = a_k \cos \frac{\pi t}{2T} \cos 2\pi f_0 t - b_k \sin \frac{\pi t}{2T} \sin 2\pi f_0 t$$

$$kT < t < (k+1)T \quad (7.18)$$

where

$$a_k = \cos x_k = \pm 1$$
$$b_k = d_k \cos x_k = \pm 1 \quad (7.19)$$

The in-phase (I) component is identified as $a_k \cos (\pi t/2T) \cos 2\pi f_0 t$, where cos $2\pi f_0 t$ is the carrier, $\cos (\pi t/2T)$ can be regarded as a *sinusoidal symbol weighting,* and a_k is a data-dependent term. Similarly, the quadrature (Q) component is identified as $b_k \sin (\pi t/2T) \sin 2\pi f_0 t$, where $\sin 2\pi f_0 t$ is the quadrature carrier term, $\sin (\pi t/2T)$ can be regarded as a sinusoidal symbol weighting, and b_k is a data-dependent term. It might appear that the a_k and b_k terms can change every T seconds, since the source data, d_k, can change every T seconds. However, because of the continuous phase constraint, the a_k term can only change value at the zero crossings of $\cos (\pi t/2T)$ and the b_k term can only change value at the zero crossings of $\sin (\pi t/2T)$. Thus the symbol weighting in either the I- or Q-channel is a half-cycle sinusoidal pulse of duration $2T$ seconds with alternating sign. As in the case of OQPSK, the I and Q components are offset T seconds with respect to one another.

Notice that x_k in Equation (7.17) is a function of the difference between the prior data bit and the present data bit (differential encoding). Hence the a_k and b_k terms in Equation (7.18) can be viewed as *differentially encoded* components of the d_k source data. However, for bit-to-bit independent data d_k, the signs of successive I- or Q-channel pulses are also random from one $2T$-second pulse interval to the next. Thus when viewed as a special case of OQPSK, Equation (7.18) can be rewritten with more straightforward (nondifferential) data encoding [8] as follows:

$$s(t) = d_I(t) \cos \frac{\pi t}{2T} \cos 2\pi f_0 t + d_Q(t) \sin \frac{\pi t}{2T} \sin 2\pi f_0 t \quad (7.20)$$

where $d_I(t)$ and $d_Q(t)$ have the same in-phase and quadrature data stream interpretation as in Equation (7.13). This MSK format in Equation (7.20) is sometimes referred to as *precoded MSK.* Figure 7.12 illustrates Equation (7.20) pictorially. Figure 7.12a and c show the sinusoidal weighting of the I- and Q-channel pulses. These sequences represent the same data sequences as in Figure 7.10, but here, multiplication by a sinusoid results in more gradual phase transitions compared to those of the original data representation. Figure 7.12b and d illustrate the modulation of the orthogonal components $\cos 2\pi f_0 t$ and $\sin 2\pi f_0 t$, respectively, by the sinusoidally shaped data streams. Figure 7.12e illustrates the summation of the orthogonal components from Figure 7.12b and d. In summary, the following properties of MSK modulation can be deduced from Equation (7.20) and Figure

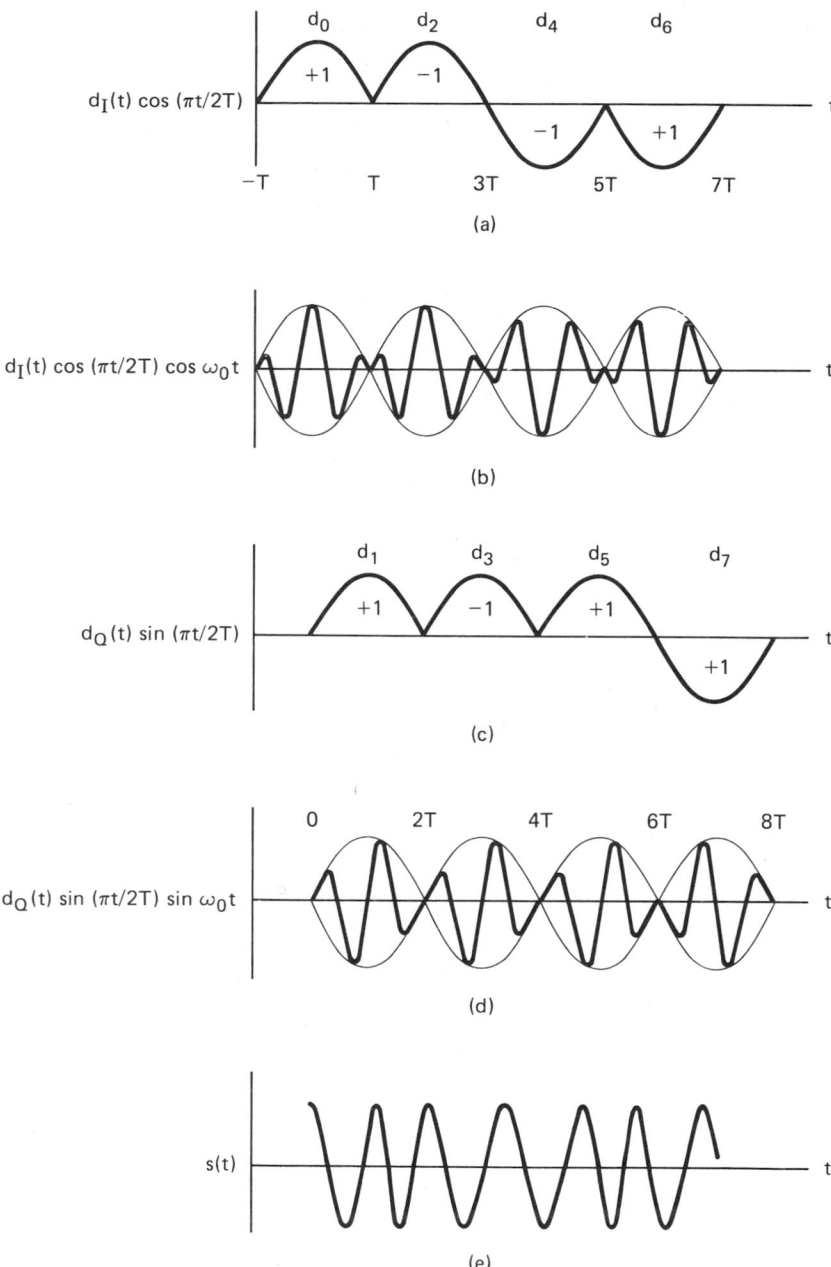

Figure 7.12 Minimum shift keying (MSK). (a) Modified I bit stream. (b) I bit stream times carrier. (c) Modified Q bit stream. (d) Q bit stream times carrier. (e) MSK waveform. (Reprinted with permission from S. Pasupathy, "Minimum Shift Keying: A Spectrally Efficient Modulation," *IEEE Commun. Mag.*, July 1979, Fig. 5, p. 18. © 1979 IEEE.)

7.12: (1) the waveform $s(t)$ has constant envelope; (2) there is phase continuity in the RF carrier at the bit transitions; and (3) the waveform $s(t)$ can be regarded as an FSK waveform with signaling frequencies $f_0 + 1/4T$ and $f_0 - 1/4T$. Therefore, the minimum tone separation required for MSK modulation is

$$\left(f_0 + \frac{1}{4T} \right) - \left(f_0 - \frac{1}{4T} \right) = \frac{1}{2T} \tag{7.21}$$

which is equal to half the bit rate. Notice that the required tone spacing for MSK is one-half the spacing, $1/T$, required for the noncoherent detection of FSK signals (see Section 3.6.4). That is because it is being coherently demodulated.

The power spectral density $G(f)$ for QPSK and OQPSK is given by [8]

$$G(f) = 2PT \left(\frac{\sin 2\pi fT}{2\pi fT} \right)^2 \tag{7.22}$$

where P is the average power in the modulated waveform. For MSK, $G(f)$ is given by [8]

$$G(f) = \frac{16PT}{\pi^2} \left(\frac{\cos 2\pi fT}{1 - 16f^2T^2} \right)^2 \tag{7.23}$$

The normalized power spectral density ($P = 1$ W) for QPSK, OQPSK, and MSK are sketched in Figure 7.13. A spectral plot of BPSK is included for comparison. The fact that BPSK requires more bandwidth than the others for a given level of spectral density should come as no surprise. In Section 7.5.1 and Figure 7.6 we saw that the theoretical bandwidth efficiency of BPSK is half that of QPSK. It is seen from Figure 7.13 that MSK has lower sidelobes than QPSK or OQPSK. This is a consequence of multiplying the data stream with a sinusoid, yielding more *gradual phase transitions*. The more gradual the transition, the faster the spectral tails drop to zero. MSK is *spectrally more efficient* than QPSK or OQPSK; however, as can be seen from Figure 7.13, the MSK spectrum has a wider mainlobe than QPSK and OQPSK. Therefore, MSK may not be the preferred method for narrowband links.

7.9.2.1 Error Performance of OQPSK and MSK

We have seen that BPSK and QPSK have the same bit error probability because QPSK is configured as two BPSK signals modulating orthogonal components of the carrier. Since staggering the bit streams does not change the orthogonality of the carriers, OQPSK has the same theoretical bit error performance as BPSK and QPSK.

Minimum shift keying uses antipodal symbol shapes, $\pm \cos (\pi t/2T)$ and $\pm \sin (\pi t/2T)$, over $2T$ to modulate the two quadrature components of the carrier. Thus when a matched filter is used to recover the data from each of the quadrature components independently, MSK, as defined in Equation (7.20), has the same error performance properties as BPSK, QPSK, and OQPSK [7]. However, if MSK is coherently detected as an FSK signal over an observation interval of T seconds, it would be poorer than BPSK by 3 dB [7]. MSK, with differentially encoded

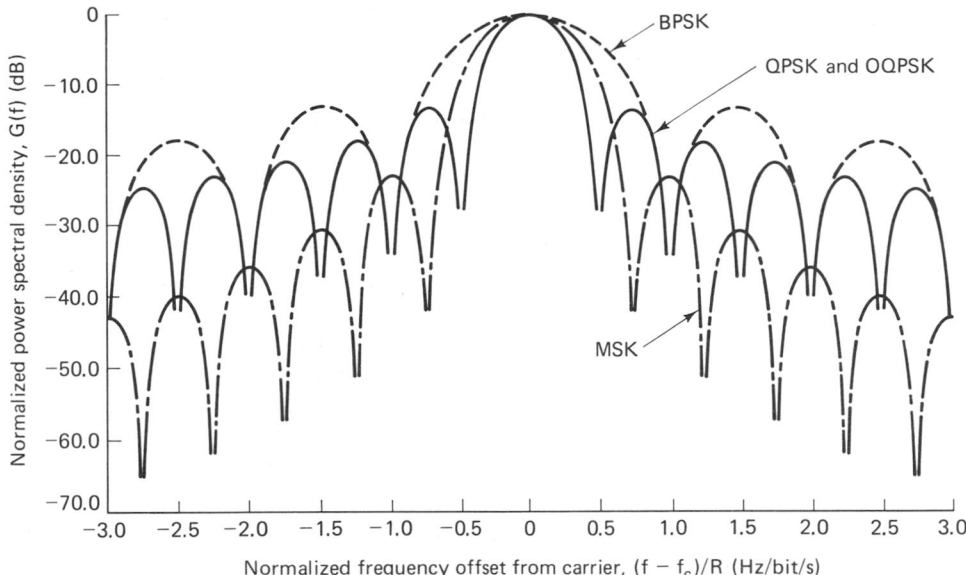

Figure 7.13 Normalized power spectral density for BPSK, QPSK, OQPSK, and MSK. (Reprinted with permission from F. Amoroso, "The Bandwidth of Digital Data Signals," *IEEE Commun. Mag.*, vol. 18, no. 6, Nov. 1980, Fig. 2A, p. 16. © 1980 IEEE.)

data, as defined in Equation (7.16), has the same error probability performance as the coherent detection of differentially encoded PSK.

QPSK systems require a fully coherent or differentially coherent detection scheme. However, since MSK is a type of FSK, it can also be noncoherently detected. This permits inexpensive demodulation of MSK when the value of received E_b/N_0 permits.

7.9.3 Quadrature Amplitude Modulation

Coherent *M*-ary phase shift keying (MPSK) modulation is a well-known technique for achieving bandwidth reduction. Instead of using a binary alphabet with 1 bit of information per channel symbol period, an alphabet with M symbols is used, permitting the transmission of $k = \log_2 M$ bits during each symbol period. Since the use of *M*-ary symbols allows a k-fold increase in the data rate within the same bandwidth, then for a fixed data rate, use of *M*-ary PSK reduces the required bandwidth by a factor k (see Section 3.8.3).

From Equation (7.14) it can be seen that QPSK modulation consists of two independent streams. One stream amplitude-modulates the cosine function of a carrier wave with levels $+1$ and -1, and the other stream similarly amplitude-modulates the sine function. The resultant waveform is termed a double-sideband suppressed-carrier (DSB-SC) wave, since the RF bandwidth is twice the baseband bandwidth (see Section 1.7.1) and there is no isolated carrier term. *Quadrature amplitude modulation* (QAM) can be considered a logical extension of QPSK, since QAM also consists of two independently amplitude-modulated carriers in quadrature. Each block of k bits (k assumed even) can be split into two ($k/2$)-bit blocks which use ($k/2$)-bit digital-to-analog (D/A) converters to provide the required modulating voltages for the carriers. At the receiver, each of the two signals is independently detected using matched filters. QAM signaling can also be viewed as a combination of amplitude shift keying (ASK) and phase shift keying (PSK), giving rise to the alternative name, *amplitude phase keying* (APK). Finally, it can also be viewed as amplitude shift keying in two dimensions, giving rise to the name *quadrature amplitude shift keying* (QASK).

Figure 7.14a illustrates a two-dimensional signal space and a set of 16-ary QAM signal vectors or points arranged in a rectangular constellation. A canonical QAM modulator is shown in Figure 7.14b. Assuming that Gaussian noise is the only channel disturbance, the simple channel model of Figure 7.14c applies. Signals are sent in pairs (x, y). The model indicates that the signal point coordinates (x, y) are transmitted over separate channels and independently perturbed by Gaussian noise variables (n_x, n_y), each with zero mean and variance N. Or we can say that the two-dimensional signal point is perturbed by a two-dimensional Gaussian noise variable. If the average signal energy (mean-square value of the signal coordinates) is S, then the signal-to-noise ratio is S/N. The simplest method of digital signaling through such a system is to use one-dimensional pulse amplitude modulation (PAM) independently for each signal coordinate. In PAM, to

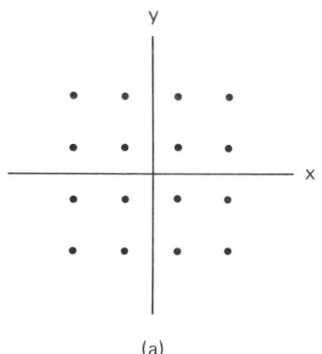

(a)

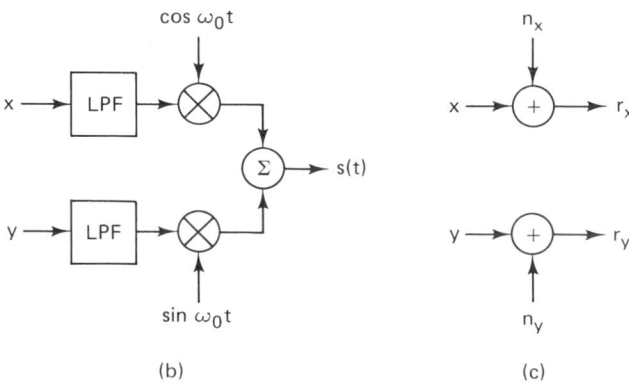

(b)   (c)

Figure 7.14 QAM modulation. (a) 16-ary signal space. (b) Canonical QAM modulator. (c) QAM channel model.

send k bits/dimension over a Gaussian channel, each signal point coordinate takes on one of 2^k equally likely equispaced amplitudes. By convention, the signal points are grouped about the center of the space at amplitudes $\pm 1, \pm 3, \ldots, \pm(2^k - 1)$.

7.9.3.1 QAM Probability of Bit Error

For a rectangular constellation, a Gaussian channel, and matched filter reception, the probability of bit error is expressed [10] by

$$P_B \simeq \frac{2(1 - L^{-1})}{\log_2 L} Q \left[\sqrt{\left(\frac{3 \log_2 L}{L^2 - 1}\right) \frac{2E_b}{N_0}} \right] \qquad (7.24)$$

where $Q(x)$ is as defined in Equation (2.42) and L represents the number of amplitude levels in one dimension. We assume that a sequence of $\log_2 L$ bits are assigned to an L-ary symbol using a Gray code (defined in Section 3.9.4).

7.9.3.2 Bandwidth–Power Trade-Off

The bandwidth–power trade-off of M-ary QAM at a bit error probability of 10^{-5} is displayed on the bandwidth-efficiency plane in Figure 7.6, with the abscissa measured in average E_b/N_0. We assume Nyquist filtering of the baseband pulses so that the DSB transmission bandwidth at IF is $W_{IF} = 1/T$, where T is the symbol duration. Thus the bandwidth efficiency is $R/W = \log_2 M$, where M is the symbol set size. For realistic channels and waveforms, the performance must be reduced to account for the increased bandwidth necessary to implement realizable filters. From Figure 7.6 it can be seen that QAM represents a method of reducing the bandwidth required for the transmission of digital data. As with M-ary PSK, bandwidth efficiency can be exchanged for power or E_b/N_0; however, in the case of QAM, a *much more efficient exchange* is possible than in the case of M-ary PSK.

For a comparative treatment of digital modulation techniques in general, Reference [11] contains useful performance data and an extensive list of other references.

Example 7.3   Waveform Design

Assume that a data stream with data rate $R = 144$ Mbits/s is to be transmitted on an RF channel using a DSB modulation scheme. Assume Nyquist filtering and an allowable DSB bandwidth of 36 MHz. Which modulation technique would you choose for this requirement? If the available E_b/N_0 is 20, what would be the resulting probability of bit error?

Solution

The required spectral efficiency is

$$\frac{R}{W} = \frac{144 \text{ Mbits/s}}{36 \text{ MHz}} = 4 \text{ bits/s/Hz}$$

From Figure 7.6 we note that 16-ary QAM, with a theoretical spectral efficiency of 4 bits/s/Hz, requires a lower E_b/N_0 than that of 16-ary PSK for the same P_B. Based on these considerations we choose a 16-ary QAM modem.

With the available E_b/N_0 given as 20, we use Equation (7.24) to calculate the expected bit error probability as

$$P_B \simeq \frac{3}{4} Q\left(\sqrt{\frac{4}{5}\frac{E_b}{N_0}}\right) = 2.5 \times 10^{-5}$$

Example 7.4 Spectral Efficiency

(a) Explain the computation of the QAM spectral efficiency in Example 7.3, considering that QAM is transmitted on orthogonal components of a carrier wave.

(b) Since the DSB bandwidth is 36 MHz in Example 7.3, consider using half that amount at baseband to transmit the 144-Mbits/s data stream, using multilevel PAM. What is the spectral efficiency needed to accomplish this, and how many levels of PAM would be required? Assume Nyquist filtering.

Solution

(a) *Bandpass channel using QAM:* The 144-Mbits/s data stream is partitioned into a 72-Mbits/s in-phase and a 72-Mbits/s quadrature stream; one stream amplitude-modulates the cosine component of a carrier over a bandwidth of 36 MHz, and the other stream amplitude-modulates the sine component of the carrier wave over the same 36-MHz bandwidth. Since each 72-Mbits/s stream modulates an orthogonal component of the carrier, the 36 MHz suffices for both streams, or for the full 144 Mbits/s. Thus the spectral efficiency is (144 Mbits/s)/36 MHz = 4 bits/s/Hz.

(b) *Required spectral efficiency at baseband*

$$\frac{R}{W} = \frac{144 \text{ Mbits/s}}{18 \text{ MHz}} = 8 \text{ bits/s/Hz}$$

Assuming Nyquist filtering, a bandwidth of 18 MHz can support a maximum symbol rate of $R_s = 2W = 36$ megasymbols/s [see Equation (2.76)]. Each PAM pulse must therefore have an ℓ-bit meaning, such that

$$R = \ell R_s$$

Hence

$$\ell = \frac{144 \text{ Mbits/s}}{36 \text{ megapulses/s}} = 4 \text{ bits/pulse}$$

where $\ell = \log_2 L$, and $L = 16$ levels.

7.10 MODULATION AND CODING FOR BANDLIMITED CHANNELS

The channel coding techniques of Chapters 5 and 6 have generally *not* been associated with voice-grade telephone channels (although the first field test of sequential decoding of convolutional codes was on a telephone line). Recently, however, there has been considerable interest in techniques that can provide coding gain for bandlimited channels. The motivation is to enable the reliable transmission of *higher data rates* over voice-grade channels. The potential gain is about 3 bits/symbol (for a given signal-to-noise ratio) [12] or, alternatively, a given error performance could be achieved with a power savings of 9 dB [12].

The greatest interest is in the following three separate coding research areas:

1. Optimum signal constellation boundaries (choosing a closely packed signal subset from any regular array or lattice of candidate points)
2. Higher-density lattice structures (adding improvement to the signal subset choice by starting with the densest possible lattice for the space)
3. Trellis-coded modulation (combined modulation and coding techniques for obtaining coding gain for bandlimited channels)

The first two areas are not "true" error control coding schemes. By "true error control coding" we refer to those techniques that employ some structured redundancy to improve the error performance. Only the third technique, trellis-coded modulation, involves redundancy. Each of these coding research areas and their expected performance improvements are discussed below.

7.10.1 Commercial Telephone Modems

The use of efficient modulation techniques has traditionally been spearheaded by the telecommunications industry, since the telephone company's foremost resource consists of sharply bandlimited voice-grade channels. The typical telephone channel is characterized by a high signal-to-noise ratio (SNR) of approximately 30 dB and a bandwidth of approximately 3 kHz. Table 7.1 lists the evolution of high-speed telephone modems with bandwidth efficiencies (R/W) ranging from 2 to 8 bits/s/Hz. The list starts with the Bell 201, introduced in about 1962, which used QPSK in a nominal 1200-Hz bandwidth to achieve 2400 bits/s on private lines. The first commercially important 4800-bits/s modem was the Milgo 4400/48, introduced in about 1967. It utilized a nominal 1600-Hz bandwidth in conjunction with 8-ary PSK to achieve a bandwidth efficiency of 3 bits/s/Hz. In 1971 the Codex 9600C was introduced. It provided 9600 bits/s in a 2400-Hz

TABLE 7.1 Modem Milestones

Year	Model	Speed (bits/s)	Bandwidth (Hz)	Modulation	R/W (bits/s/Hz)
1962	Bell 201	2,400	1200	4-PSK	2
1967	Milgo 4400/48	4,800	1600	8-PSK	3
1971	Codex 9600C	9,600	2400	16-QAM	4
1980	Paradyne MP14400	14,400	2400	64-QAM	6
1981	Codex SP14.4	14,400	2400	64-QAM	6
1984	Codex 2660	14,400	2400	Trellis-coded QAM	6
1985	Codex 2680	19,200	2400	Trellis-coded QAM	8

bandwidth ($R/W = 4$ bits/s/Hz) using 16-ary QAM. Note that as channel equalization techniques (see Section 2.11.2) improved, a larger bandwidth portion of the voice-grade channel became usable. Whereas in 1962, only 1200 Hz could be reliably employed, that value doubled by 1971.

In 1980, first-generation 14,400-bits/s modems were introduced by Paradyne (MP14400), followed in 1981 by Codex (SP14.4). These modems improved the bandwidth efficiency by utilizing 64-ary QAM with an R/W of 6 bits/s/Hz. In a second generation, appearing in 1984, trellis-coded QAM modulation (treated in Section 7.10.6) was introduced to provide better error performance. In 1985, Codex introduced a modem with $R/W = 8$ bits/s/Hz, thereby achieving a data rate of 19,200 bits/s in a nominal bandwidth of 2400 Hz. Without any major upgrading of the telephone network, 19,200 bits/s is considered the maximum achievable data rate for an unconditioned voice-grade telephone channel [12].

7.10.2 Signal Constellation Boundaries

Several researchers [13–17] have examined large numbers of possible QAM signal constellations in a search for designs that result in the best error performance for a given average signal-to-noise ratio. Figure 7.15 illustrates some examples of symbol constellations for $M = 4$, 8, and 16 that have been considered [13]. The circular sets are designated by the notation $(a, b, \ldots)$, where there are a quantity of a signals on the inner circle, b signals on the next circle, and so on. In general, the constellation rule, known as the Campopiano–Glazer construction rule [15], that yields optimum signal set performance can be summarized as follows: From an infinite array of points closely packed in a *regular array or lattice,* select a closely packed subset of 2^k points as a signal constellation. In this case "optimum" means minimum average or peak power for a given error probability. In a two-dimensional signal space the optimum boundary surrounding an array of points tends toward a circle. Figure 7.16 illustrates examples of 64-ary ($k = 6$) and 128-ary ($k = 7$) signal sets from a rectangular array. The cross-shaped boundaries are a compromise to the optimum circle. The $k = 6$ constellation was used in the Paradyne 14.4-kbits/s modem. Compared to a square, the performance improvement resulting from a circular boundary is only a modest 0.2 dB [12].

7.10.3 Higher-Dimensional Signal Constellations

For any particular information rate and a channel noise process that is independent and identically distributed in the two dimensions, signaling in a two-dimensional QAM space can provide the same error performance with less average (or peak) power than signaling in a one-dimensional pulse amplitude (PAM) space. We stated earlier that this is accomplished by choosing signal points on a two-dimensional lattice from within a circular rather than a rectangular boundary. In the same way, by going to a higher number, N, of dimensions and choosing points on an N-dimensional lattice from within an N-sphere rather than an N-cube, further energy savings are possible. Several researchers [18–21] have studied multidimensional signal constellations. Consider the four-dimensional configuration

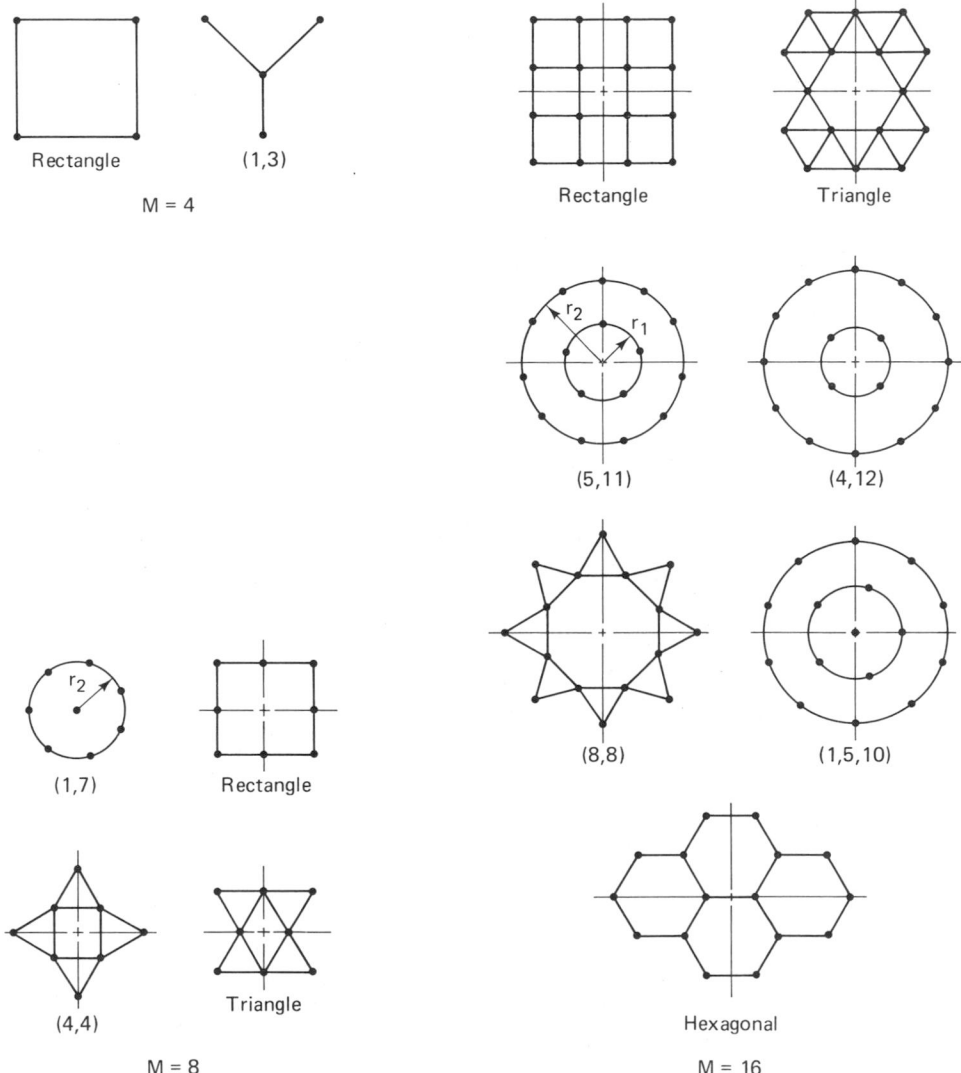

Figure 7.15 *M*-ary symbol constellations. (Reprinted with permission from C. M. Thomas, M. Y. Weidner, and S. H. Durrani, "Digital Amplitude-Phase Keying with *M*-ary Alphabets," *IEEE Trans. Commun.*, vol. COM22, no. 2, Feb. 1974, Figs. 2 and 3, p. 170. © 1974 IEEE.)

illustrated in Figure 7.17. The transmitter transmits four simultaneous sequences of pulses over four bandlimited Gaussian channels. We assume that the source produces one of *M* symbols, $m_i = 1, 2, \ldots, M$, every *T* seconds. A given symbol m_i causes four pulses to be emitted—$a_i s(t)$, $b_i s(t)$, $c_i s(t)$, $d_i s(t)$—as shown in Figure 7.17. These are transmitted on separate noninterfering channels. The pulses are distorted by independent AWGN in each channel, and at the receiver they are detected separately with matched filters. The four independent channels can

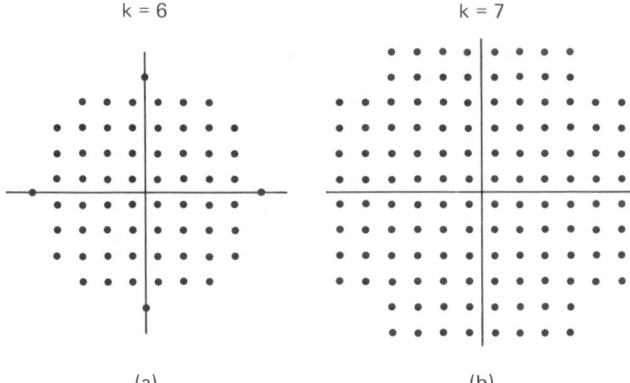

k = 6

k = 7

(a)

(b)

Figure 7.16 Examples of *M*-ary constellations using a rectangular array.

be implemented in a number of ways:

1. Two bandpass channels can be used, each with separately modulated in-phase and quadrature components (QAM or MPSK modulation on each channel).
2. The two bandpass channels can be time- or frequency-division multiplexed and carried on a common transmission line.
3. Orthogonal electromagnetic wave polarization can be used.

Let us compare a two-dimensional 16-ary QAM system with a four-dimensional alternative. In the two-dimensional modulation case, during each *T*-second

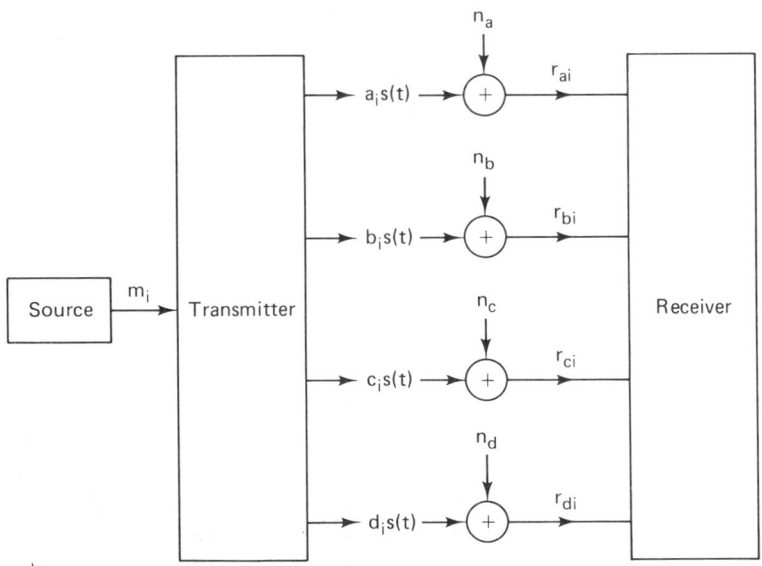

Figure 7.17 Four-dimensional system configuration.

interval, a symbol (4 bits) is transmitted by the modulation of the in-phase and quadrature components of a carrier. In other words, two 4-ary numbers can be transmitted to describe a signal vector in the 16-ary two-dimensional signaling plane. In the case of four-dimensional modulation, two successive symbols (8 bits) are sent each $2T$ seconds by transmitting four 4-ary numbers representing a point in a 256-ary signal space. It can be shown that increasing the dimensionality of the signal space offers a potential savings in average signal energy for a given level of error performance. That is, in going to a higher-dimensional space, one can effect an energy savings based on the selection of signal points from an N-sphere versus an N-cube of the same volume—the average energy of the signal points from the N-sphere is less than that from the N-cube. Table 7.2 gives the energy savings possible in N dimensions. Of course, the implementation of such a scheme involves added complexity. To send n bits per symbol in N dimensions (assuming N even), incoming bits must be grouped in blocks of $nN/2$. A mapping must then be performed into the space of $2^{(nN/2)}$ N-dimensional vectors which have the least energy among all such vectors. A corresponding inverse mapping must be made at the receiver. The added complexity may, of course, outweigh the performance gain. As N goes to infinity, the gain goes to 1.53 dB [12].

TABLE 7.2 Energy Savings from N-Sphere Mapping versus N-Cube Mapping

Dimensions (N)	N-sphere mapping gain (dB)
2	0.20
4	0.45
8	0.73
16	0.98
24	1.10
32	1.17
48	1.26
64	1.31

Source: G. D. Forney, Jr., et al., "Efficient Modulation for Bandlimited Channels," *IEEE J. Sel. Areas Commun.*, vol. SAC2, no. 5, September 1984, pp. 632–647.

7.10.4 Higher-Density Lattice Structures

In Section 7.10.3 we discussed the selection of a closely packed subset of points from any regular array or lattice. Here we consider the added improvement by starting with the *densest possible lattice* in the space. In a two-dimensional signal space, the densest lattice is the hexagonal lattice (try penny packing). The result of employing a hexagonal lattice instead of a rectangular one, such as those shown in Figure 7.16, can be a 0.6-dB savings in average energy. Figure 7.18 illustrates some examples of hexagonal packing. The strange-looking $k = 4$ constellation in Figure 7.18a was discovered by Foschini et al. [17] and is still the best 16-ary

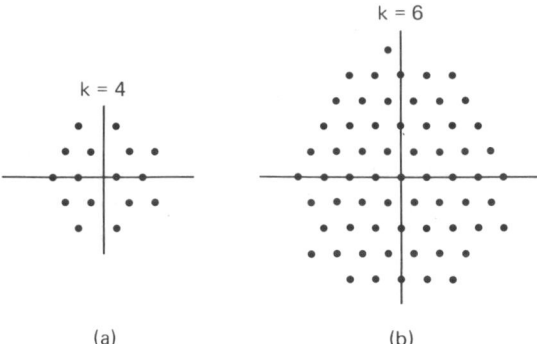

k = 4

k = 6

(a)

(b)

Figure 7.18 Examples of *M*-ary constellations using a hexagonal array.

constellation known. The $k = 6$ constellation in Figure 7.18b is used in the Codex SP14.4 modem.

The hexagonal lattice is optimum for two dimensions. For higher dimensions there are other lattice structures that provide the densest packing. Table 7.3 gives the gain over the rectangular lattice, in decibels, due to the densest packings currently known for various dimensions.

TABLE 7.3 Energy Savings from Dense Lattices versus the Rectangular Lattice

Dimensions (N)	Dense lattice gain (dB)
2	0.62
4	1.51
8	3.01
16	4.52
24	6.02
32	6.02
48	7.78
64	8.09

Source: G. D. Forney, Jr., et al., "Efficient Modulation for Bandlimited Channels," *IEEE J. Sel. Areas Commun.*, vol. SAC2, no. 5, September 1984, pp. 632–647.

7.10.5 Combined Gain: *N*-Sphere Mapping and Dense Lattice

It is possible to combine the benefits of the Campopiano–Glazer boundary construction in N dimensions with the gain from the densest lattice in N-space. The resulting gain is a combination of N-sphere versus N-cube boundary gain of Table 7.2 and the lattice packing density gain of Table 7.3. The combined energy savings are shown in Table 7.4.

TABLE 7.4 Combined Energy Savings
from N-Sphere Mapping and Dense Lattices

Dimensions (N)	Combined savings gain (dB)
2	0.82
4	1.96
8	3.74
16	5.50
24	7.12
32	7.19
48	9.04
64	9.40

Source: G. D. Forney, Jr., et al., ''Efficient Modulation for Bandlimited Channels,'' *IEEE J. Sel. Areas Commun.*, vol. SAC2, no. 5, September 1984, pp. 632–647.

7.10.6 Trellis-Coded Modulation

The codes described in Chapters 5 and 6 achieve an improvement in bit error probability (P_B) by *bandwidth expansion.* In the case of both block codes and convolutional codes, bandwidth is increased by replacing each k-tuple message with an n-tuple codeword, where $n > k$. In the case of bandlimited channels, *bandwidth expansion is not possible.* In the past, therefore, coding has never been popular for bandlimited channels such as telephone channels. Recently, however, there has been increasing interest in some types of combined modulation and coding schemes, called *trellis-coded modulation,* that achieve coding gain without any bandwidth expansion. At first it may seem that this statement violates some basic power–bandwidth–error probability trade-off principle. However, there is still a trade-off at work; trellis-coded modulation achieves coding gain at the expense of *decoder complexity.*

Trellis-coded modulation combines a *multilevel/phase* modulation signaling set with a state-oriented trellis coding scheme. Multilevel/phase signal sets are signal constellations having multiple amplitudes, multiple phases, or a combination of multiple amplitudes and multiple phases. A trellis code is one that can be characterized with a trellis diagram. The convolutional codes described in Chapter 6 are linear trellis codes, but trellis codes are *not constrained to be linear.* Coding gains can be realized with block codes or trellis codes, but we shall consider only trellis codes because the availability of the Viterbi decoding algorithm makes trellis decoding attractive. Coding for bandlimited channels still requires controlled introduction of redundancy. However, in this case, the redundancy is due to an increased signal alphabet, achieved through multilevel/phase signaling, so that channel bandwidth is not increased. Ungerboeck [22] investigated the design of multilevel/phase trellis codes that provide *coding gain without band-*

width expansion. He showed that in the presence of AWGN, net coding gains of 3 to 6 dB, relative to the uncoded case, could be achieved.

7.10.6.1 The Idea behind Trellis-Coded Modulation

The error performance of an uncoded nonorthogonal *M*-ary modulation (such as PAM, PSK, or QAM) depends on the distance between the closest pair of signal points. This minimum distance is determined by the average transmitter power and the number and position of the signal points. For a constant average power, the minimum distance between points decreases as the number of points increases. Therefore, assuming a constant channel symbol rate and constant average power, the error performance degrades for systems that attempt to increase the transmission bit rate by increasing the size of the symbol set. The objective of trellis coding is to increase the minimum distance between the signals that are the *most likely to be confused,* without increasing the average power.

Trellis coding may be implemented with a convolutional encoder (see Chapter 6) wherein k current bits and $K - 1$ prior bits are used to produce $n = k + p$ coded bits, where K is the encoder constraint length and where p is the number of parity bits. The $n = k + p$ coded bits require 2^n binary channel symbols for transmission. Notice that encoding increases the signal set size from 2^k to 2^{k+p}. Figure 7.19a illustrates an uncoded 4-ary PAM signal set, before and after being rate $\frac{2}{3}$ encoded into an 8-ary PAM signal set. Similarly, Figure 7.19b illustrates an uncoded 4-ary PSK (QPSK) signal set before and after being rate $\frac{2}{3}$ encoded into an 8-ary PSK signal set. Similarly, Figure 7.19c illustrates an uncoded 16-

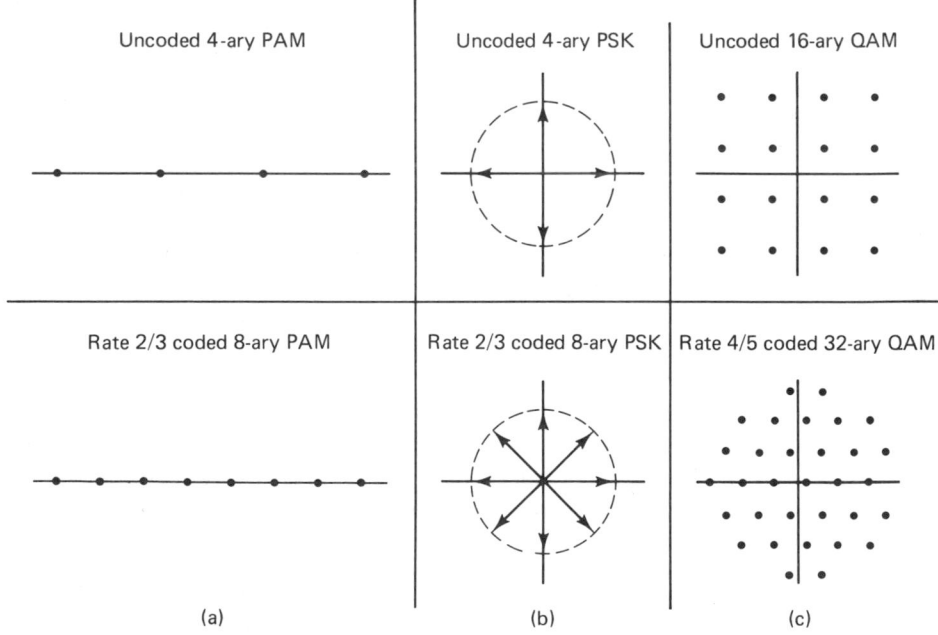

Figure 7.19 Increase of signal set size for trellis-coded modulation.

ary QAM signal set before and after being rate $\frac{4}{5}$ encoded into a 32-ary QAM signal set. In each of these three cases the system is configured to use the same average signal power before and after coding. The examples in Figure 7.19 illustrate the basic idea behind trellis-coded modulation. In each case the symbol set size is increased from 2^k to 2^{k+1} (there are twice as many coded symbols as uncoded ones) to provide the needed coding redundancy; however, in each case, the increase in the number of signals *does not* result in an increase in required bandwidth. The expanded signal set does result in a *reduced distance* between adjacent symbol points for a given average power. However, because of the redundancy introduced by the code, this reduced distance no longer determines the error performance. Instead, the *free distance* (see Section 6.4.1), which is the minimum distance between members of the set of *allowable code symbol sequences,* determines the error performance. Ungerboeck [22] investigated the increase in channel capacity achievable by signal set expansion and concluded that by *doubling* the number of channel signals ($p = 1$), it is possible to gain almost all the channel capacity that can be gained. This can be accomplished by encoding with a rate $k/(k + 1)$ code, and subsequently mapping groups of $k + 1$ bits into the larger set of 2^{k+1} channel symbols.

7.10.6.2 An Error Event

Figure 7.20 illustrates an error event in a trellis code; that is, the figure illustrates a transmitted sequence marked $\mathbf{U} = \ldots, U_1, U_2, U_3, \ldots$ and an alternative sequence marked $\mathbf{V} = \ldots, V_1, V_2, V_3, \ldots$. The alternative sequence is

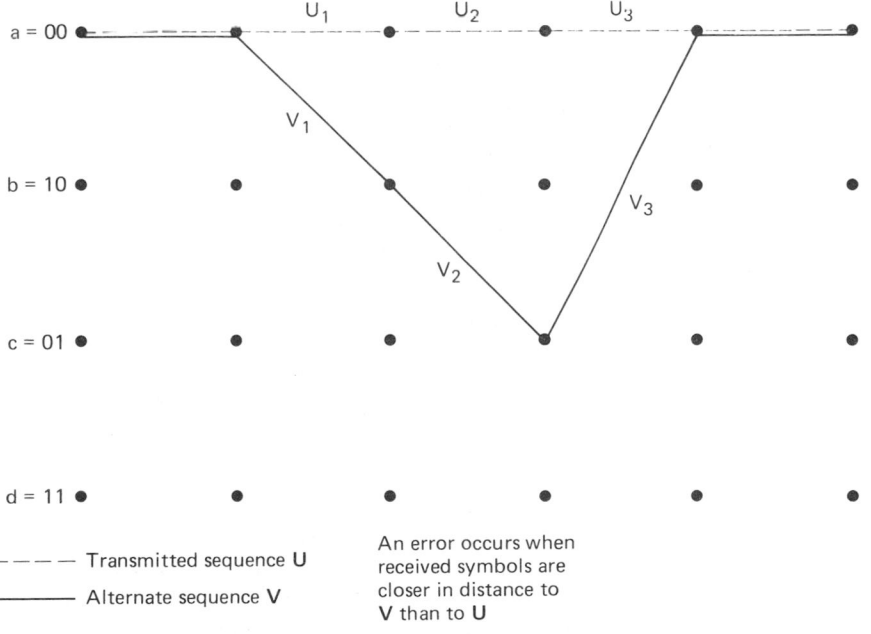

Figure 7.20 Illustration of an error event.

seen to diverge and then remerge with the transmitted sequence. Assuming soft-decision decoding, an *error event occurs* whenever the received symbols are closer in Euclidean distance to some alternative sequence, **V**, than to the actually transmitted sequence, **U**. Thus the separation of **U** and **V** describes an error event. This implies that codes for multilevel/phase signals should be designed to achieve *maximum free Euclidean distance* rather than maximum free Hamming distance; for soft-decision decoding, the larger the Euclidean distance, the lower the probability of error. Therefore, assigning signal points to the coded bits in a way that maximizes Euclidean distance is the key to optimizing the trellis codes. Ungerboeck [22–24] investigated this bit-to-symbol mapping problem and devised an assignment procedure, called the *method of set partitioning,* which will always provide coding gain, given an adequate choice of trellis states. The rules for this bit-to-symbol mapping are based on the method of set partitioning, which can be summarized as follows:

1. All parallel transitions in the trellis structure are separated by the maximum possible Euclidean distance. *Parallel transitions* refer to the branch words resulting from the transmission of uncoded bits together with coded bits (see the example in the following section). The reasoning behind this is based on the fact that parallel transitions imply that single signal-error events can occur. This limits the achievable free Euclidean distance to the minimum distance in the subsets of signals assigned to parallel transitions.

2. All transitions diverging from or merging into a trellis state are assigned the next maximum possible Euclidean distance separation.

In summary, trellis coding for bandlimited channels employs larger signal alphabets achieved through multilevel/phase signaling, such that channel bandwidth is not increased (e.g., M-ary PAM, MPSK, or QAM). Even though the increase in signal set size *reduces* the minimum distance between symbols, the free Euclidean distance between trellis code sequences *more than compensates* for the signal points being crowded together. The result is a net error-performance gain of 3 to 6 dB without any bandwidth expansion [22]. We illustrate these ideas by considering a rate $\frac{2}{3}$ convolutional encoder in the following section.

7.10.7 Trellis-Coding Example

A rate $\frac{2}{3}$ convolutional encoder with constraint length $K = 3$ is shown in Figure 7.21. The rate $\frac{2}{3}$ encoding is accomplished by transmitting one bit from each pair of bits in the input sequence unmodified, and encoding the other bit into two channel bits using a rate $\frac{1}{2}$ encoder. The resulting trellis diagram is shown in Figure 7.22, where the parallel transitions are due to the uncoded bit m_1 shown as the leftmost bit on each trellis branch. The two upper branches emerging from each state represent transitions due to $m_1 m_2$ being 00 and 10, respectively; the two lower branches represent transitions due to $m_1 m_2$ being 01 and 11, respectively. The Viterbi decoding technique for finding the maximum likelihood path through the trellis proceeds in exactly the same way as in the example of Section 6.3.4.

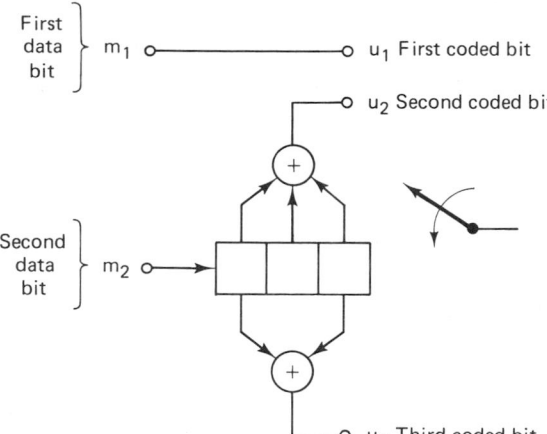

First data bit m_1 o——————o u_1 First coded bit

u_2 Second coded bit

Second data bit m_2

u_3 Third coded bit

Figure 7.21 Rate $\frac{2}{3}$ convolutional encoder.

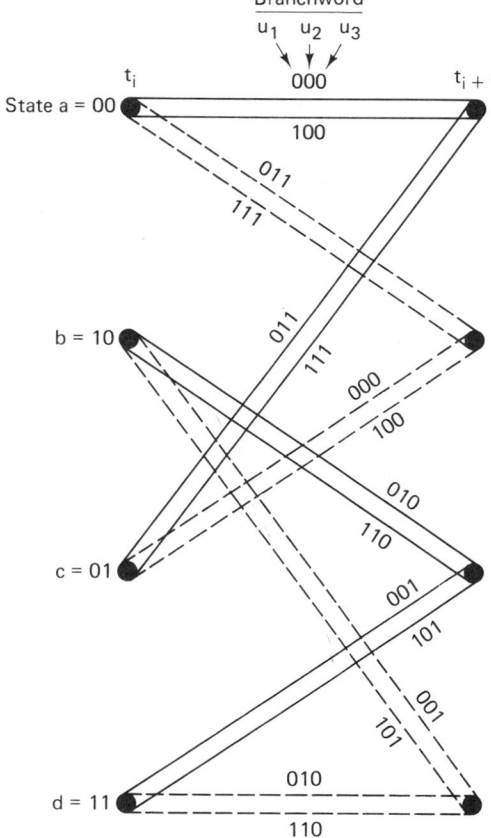

Branchword
u_1 u_2 u_3

t_i 000 t_{i+1}

State a = 00

100

011

111

b = 10

011

111

000

100

010

110

c = 01

001

101

001

101

010

d = 11

110

Figure 7.22 Trellis diagram (rate $\frac{2}{3}$ code).

The only (operational) differences are: (1) In this rate $\frac{2}{3}$ example, there are twice as many branches to consider than in the ordinary $K = 3$ convolutional code; and (2) in choosing the two decoded bits for a surviving branch, the first decoded bit of the pair is the same as the first bit u_1 of that branch word since u_1 is the same as the *uncoded* bit m_1. The second decoded bit of the pair corresponds to the input bit m_2 that produced the state transition of the branch being decoded. In Figure 7.22 a branch having a solid line corresponds to $m_2 = 0$, and a branch having a dashed line corresponds to $m_2 = 1$.

7.10.7.1 Coding Gain for Trellis Coding

Consider the coding gain of the rate $\frac{2}{3}$ trellis-coding example described in the preceding section. Let us assume a simple one-dimensional signal space with multilevel pulse amplitude modulation (PAM), as shown in Figure 7.23. In Figure 7.23a is an 8-ary PAM signal set. Since soft decisions are assumed, the appropriate distance metric is the Euclidean distance. The Euclidean distance of each signal, from the center of the signal space, is shown in arbitrary units. Also shown in Figure 7.23a is the bit-to-symbol assignment according to the set partitioning rules outlined earlier. Notice the adherence to these rules, by comparing Figure 7.23a with Figure 7.22. All parallel transitions are separated by a distance of eight units, and all branches diverging from a given state are separated by at least four units.

The average signal power, S_{av}, is computed as follows:

$$S_{av} = \frac{d_1^2 + d_2^2 + \cdots + d_M^2}{M} \tag{7.25}$$

where d_i is the Euclidean distance of the ith signal from the center of the space, and M is the number of codeword symbols in the set. For the signal set shown in Figure 7.23a, where $M = 8$, use of Equation (7.25) yields $S_{av} = 21$. Figure 7.23b illustrates a 4-ary PAM signal set which is the uncoded equivalent of the rate $\frac{2}{3}$ codeword set; the Euclidean distances have been chosen to yield the same average signal power as in the coded case in Figure 7.23a.

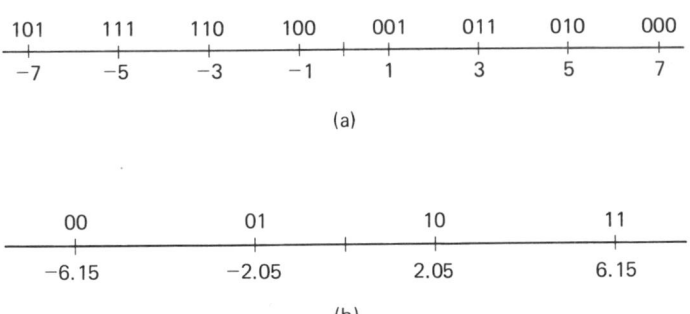

(a)

(b)

Figure 7.23 8-ary and 4-ary PAM signal sets.

Figure 7.24 illustrates the minimum distance error event for the rate $\frac{2}{3}$ encoder shown in Figure 7.21. The transmitted sequence is assumed to correspond to the all-zeros path. Each of the branch words on this path has a Euclidean distance of 7 units from the center of the space. The error event diverges from the all-zeros path by first transitioning to state 10, then state 01, and finally re-

Modulation and Coding Trade-Offs   Chap. 7

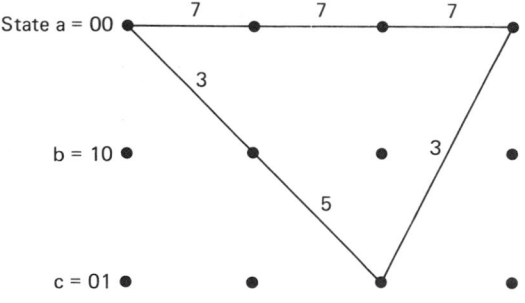

State a = 00

7　　　7　　　7

3

b = 10

3

5

c = 01

Figure 7.24 Minimum distance error event for rate $\frac{2}{3}$ convolutional code.

d = 11

merging to state 00. On each branch of the error event is written its Euclidean distance from the center of the space, assuming the uncoded bit $m_1 = 0$; this assumption assures that the distance between the all-zeros transmission and the error event is minimum. To verify the Euclidean distances for the error event in Figure 7.24, first note the state from which and to which each branch transitions. From Figure 7.22 determine the branch word sequence for each transition (with $m_1 = 0$), and then from Figure 7.23a determine the Euclidean distance from the center of the space for each branch in the error event.

Signal amplitude corresponds to the distance of the signal point from the center of the signal space; signal power corresponds to the square of this distance. In comparing the relative performance of the uncoded 4-ary PAM modulation with the trellis-coded 8-ary modulation, it is therefore appropriate to compare the square of the minimum distance d_{min}^2 for an error event in each system, given that the average power is the same in both cases. In general, allowing the average signal power to be different in each case, we can solve for the coding gain G, as follows [25]:

$$G = \frac{(d_{min}^2/S_{av})_{coded}}{(d_{min}^2/S_{av})_{uncoded}} \qquad (7.26)$$

For the rate $\frac{2}{3}$ trellis-coding example, the d_{min}^2 value for each error event is calculated as follows:

$$(d_{min}^2)_{coded} = (7 - 3)^2 + (7 - 5)^2 + (7 - 3)^2 = 36$$

$$(d_{min}^2)_{uncoded} = (6.15 - 2.05)^2 = 16.8$$

Then the coding gain is calculated by using Equation (7.26) as follows:

$$G = \frac{36/21}{16.8/21} = 2.14$$

or, in decibels $= 3.31$ dB

Therefore, even for this simple $K = 3$ example, a significant amount of coding gain has been provided *without any bandwidth expansion*. Larger coding gains can be achieved with an increased number of trellis states (larger constraint length) at the expense of increased decoding complexity. Table 7.5 lists the average power coding gain as a function of number of trellis states [26] for the rate $\frac{2}{3}$ coded 8-ary PAM example discussed here. Gain is computed relative to an uncoded 4-ary PAM signal set.

TABLE 7.5 Coding Gain Obtained for 8-ary PAM with Rate 2/3 Trellis Coding

Number of trellis states	Constraint length (K)	Average power gain (dB)
4	3	3.31
8	4	3.77
16	5	4.18
32	6	4.56
64	7	5.23
128	8	5.23
256	9	5.83

Source: G. C. Clark, Jr. and J. B. Cain, *Error Correction Coding for Digital Communications*, Plenum Press, New York, 1981, p. 388.

At the transmitter, there is only a slight increase in complexity due to the trellis coding. However, the decoding problem at the receiver is made much more complex [12], so that the trade-off consists of evaluating the *coding gain versus the decoding complexity*. The increased availability of large-scale integrated (LSI) circuits and very high speed integrated circuits (VHSIC) can ameliorate this problem and make these coding techniques extremely attractive for achieving coding gain for bandlimited channels. Before concluding, we point out that further coding gain, without bandwidth expansion, is possible by introducing asymmetry into the signal point constellation [27–28].

7.11 CONCLUSION

In this chapter we have integrated some of the ideas in Chapters 3, 5, and 6 dealing with modulation and coding. We have reviewed the basic system design goals: to maximize data rate while simultaneously minimizing error probability, bandwidth, E_b/N_0, and complexity. We examined the trade-offs heuristically on two performance planes: the error probability plane and the bandwidth efficiency plane. The former explicitly illustrates the P_B versus E_b/N_0 trade-offs while only implicitly displaying the bandwidth expenditure. The latter explicitly illustrates the R/W versus E_b/N_0 trade-offs while only implicitly displaying the P_B performance. We discussed some of the basic constraints to improvement without limit. The Nyquist criterion establishes that we cannot continue to reduce system bandwidth indefinitely. There is a theoretical limitation; in order to transmit R_s symbols/second without intersymbol interference, we must utilize a minimum of $R_s/2$ hertz of bandwidth. The Shannon–Hartley theorem relates to the power–bandwidth trade-off and results in another important limitation, the Shannon limit. The Shannon limit of -1.59 dB is the minimum amount of E_b/N_0 that is necessary (in concert with channel coding) to achieve an arbitrarily low error probability over an AWGN channel. The more general limitation is the channel capacity, above which there cannot be error-free signaling. We have also examined some of the bandwidth-efficient modulation schemes, such as minimum shift keying (MSK), quadrature amplitude modulation (QAM), and trellis-coded modulation. The latter technique offers an attractive way to obtain coding gain without paying the price of additional bandwidth.

REFERENCES

1. Nyquist, H., "Certain Topics on Telegraph Transmission Theory," *Trans. Am. Inst. Electr. Eng.,* vol. 47, Apr. 1928, pp. 617–644.

2. Shannon, C. E., "A Mathematical Theory of Communication," *Bell Syst. Tech. J.,* vol. 27, 1948, pp. 379–423, 623–657.

3. Shannon, C. E., "Communication in the Presence of Noise," *Proc. IRE,* vol. 37, no. 1, Jan. 1949, pp. 10–21.

4. Bedrosian, E., "Spectrum Conservation by Efficient Channel Utilization," *Rand Corp., Rep. WN-9275-ARPA,* Contract DAHC-15-73-C-0181, Santa Monica, Calif., Oct. 1975.

5. Odenwalder, J. P., *Error Control Coding Handbook,* Linkabit Corporation, San Diego, Calif., July 15, 1977.

6. Smith, J. G., "Spectrally Efficient Modulation," *Proc. IEEE Int. Conv. Commun. (ICC '77),* June 1977, pp. 3.1-37–3.1-41.

7. Pasupathy, S., "Minimum Shift Keying: A Spectrally Efficient Modulation," *IEEE Commun. Mag.,* July 1979, pp. 14–22.

8. Gronemeyer, S. A., and McBride, A. L., "MSK and Offset QPSK Modulation," *IEEE Trans. Commun.,* vol. COM-24, Aug. 1976, pp. 809–820.

9. M. K. Simon, "A Generalization of Minimum Shift Keying (MSK) Type Signaling Based upon Input Data Symbol Pulse Shaping," *IEEE Trans. Commun.,* vol. COM24, Aug. 1976, pp. 845–857.

10. Korn, I., *Digital Communications,* Van Nostrand Reinhold Company, Inc., New York, 1985.

11. Oetting, J. D., "A Comparison of Modulation Techniques for Digital Radio," *IEEE Trans. Commun.,* vol. COM27, no. 12, Dec. 1979, pp. 1752–1762.

12. Forney, G. D., Jr. et al., "Efficient Modulation for Bandlimited Channels," *IEEE J. Sel. Areas Commun.,* vol. SAC2, no. 5, Sept. 1984, pp. 632–647.

13. Thomas, C. M., Weidner, M. Y., and Durrani, S. H., "Digital Amplitude-Phase Keying with M-ary Alphabets," *IEEE Trans. Commun.,* vol. COM22, no. 2, Feb. 1974, pp. 168–180.

14. Lucky, R. W., and Hancock, J. C., "On the Optimum Performance of N-ary Systems Having Two Degrees of Freedom," *IRE Trans. Commun. Syst.,* vol. CS10, June 1962, pp. 185–192.

15. Campopiano, C. N., and Glazer, B. G., "A Coherent Digital Amplitude and Phase Modulation Scheme," *IRE Trans. Commun. Syst.,* vol. CS10, June 1962, pp. 90–95.

16. Cahn, C. R., "Combined Digital Phase and Amplitude Modulation Communication Systems," *IRE Trans. Commun. Technol.,* Sept. 1960.

17. Foschini, G. J., Gitlin, R. D., and Weinstein, S. B., "Optimization of Two Dimensional Signal Constellations in the Presence of Gaussian Noise," *IEEE Trans. Commun.,* vol. COM22, no. 1, Jan. 1974, pp. 28–38.

18. Welti, G. R., and Jhong, S. L., "Digital Transmission with Coherent Four-Dimensional Modulation," *IEEE Trans. Inf. Theory,* vol. IT20, no. 4, July 1974, pp. 497–502.

19. Gersho, A., and Lawrence, V. B., "Multidimensional Signal Constellations for Voiceband Data Transmission," *IEEE J. Sel. Areas Commun.,* vol. SAC2, no. 5, Sept. 1984, pp. 687–702.

20. Zetterberg, L. H., and Brandstrom, H., "Codes for Combined Phase and Amplitude Modulated Signals in a Four-Dimensional Space," *IEEE Trans. Commun.,* vol. COM25, no. 9, Sept. 1977, pp. 943–950.

21. Wilson, S. G., Sleeper, H. A., and Srinath, N. K., "Four-Dimensional Modulation and Coding: An Alternative to Frequency Reuse," *IEEE 1984 Intl. Commun. Conf.,* pp. 919–923.

22. Ungerboeck, G., "Channel Coding with Multilevel/Phase Signals," *IEEE Trans. Inf. Theory,* vol. IT28, Jan. 1982, pp. 55–67.

23. Ungerboeck, G., "Trellis-Coded Modulation with Redundant Signal Sets, Part I. Introduction," *IEEE Commun. Mag.,* vol. 25, no. 2, Feb. 1987, pp. 5–11.

24. Ungerboeck, G., "Trellis-Coded Modulation with Redundant Signal Sets, Part II; State of the Art," *IEEE Commun. Mag.,* vol. 25, no. 2, Feb. 1987, pp. 12–21.

25. Thapar, H. K., "Real-Time Application of Trellis Coding to High-Speed Voiceband Data Transmission," *IEEE J. Sel. Areas Commun.,* vol. SAC2, no. 5, Sept. 1984, pp. 648–658.

26. Clark, G. C., Jr., and Cain, J. B., *Error Correction Coding for Digital Communications,* Plenum Press, New York, 1981.

27. Divsalar, D., and Yuen, J. H., "Asymmetric MPSK for Trellis Codes," *GLOBECOM '84,* Nov. 26–29, 1984.

28. Divsalar, D., Simon, M. K., and Yuen, J. H., "Trellis Coding with Asymmetric Modulations," *IEEE Trans. Commun.,* vol. COM35, no. 2, Feb. 1987.

PROBLEMS

7.1. Consider a voice-grade telephone circuit with a bandwidth of 3 kHz. Assume that the circuit can be modeled as an AWGN channel.
   (a) What is the capacity of such a circuit if the SNR is 30 dB?
   (b) What is the minimum SNR required for a data rate of 4800 bits/s on such a voice-grade circuit?
   (c) Repeat part (b) for a data rate of 19,200 bits/s.

7.2. Consider that a 100-kbits/s data stream is to be transmitted on a voice-grade telephone circuit (with a bandwidth of 3 kHz). Is it possible to achieve error-free transmission with a SNR of 10 dB? Justify your answer. If it is not possible, suggest system modifications that might be made.

7.3. Consider a source that produces six messages with probabilities $\frac{1}{2}, \frac{1}{4}, \frac{1}{8}, \frac{1}{16}, \frac{1}{32}$, and $\frac{1}{32}$. Determine the average information content in bits, of a message.

7.4. A given source alphabet consists of 300 words, of which 15 occur with probability 0.06 each and the remaining 285 words occur with probability 0.00035 each. If 1000 words are transmitted each second, what is the average rate of information transmission?

7.5. (a) Find the average capacity in bits per second that would be required to transmit a high-resolution black-and-white TV signal at the rate of 32 pictures per second if each picture is made up of 2×10^6 picture elements and 16 different brightness levels. All picture elements are assumed to be independent and all levels have equal likelihood of occurrence.

(b) For color TV, this system additionally provides for 64 different shades of color. How much more system capacity is required for a color system compared to the black and white system?

(c) Find the required capacity if 100 of the possible brightness–color combinations occur with a probability of 0.003 each, 300 of the combinations occur with a probability of 0.001, and 624 of the combinations occur with a probability of 0.00064.

7.6. Prove that entropy is maximized when all source outputs have equal probability.

7.7. Compute the equivocation or message uncertainty in bits per character for a textual transmission using 7-bit ASCII coding. Assume that each character is equally likely and that the noise on the channel results in a bit error probability of 0.01.

7.8. Suppose a binary noncoherent FSK link has a maximum data rate of 2.4 kbits/s without ISI over a channel whose nominal bandwidth is 2.4 kHz. Suggest ways of increasing the data rate under the following system constraints.

(a) The system is power limited.

(b) The system is bandwidth limited.

(c) The system is both power and bandwidth limited.

7.9. Table P7.1 characterizes four different satellite-to-earth-terminal links. For each link assume that the space loss is 196 dB, the margin is 0 dB, and there are no other incidental losses. For each link, plot an operating point on the bandwidth efficiency plane, R/W versus E_b/N_0, and characterize the link according to one of the following descriptions: bandwidth limited, severely bandwidth limited, power limited, and severely power limited. Justify your answers.

7.10. In designing a communication system for a 9600-bits/s data stream with a bit error

TABLE P7.1 Downlink Capacity for Four Satellite Links

Satellite	Receive terminal	Maximum data rate
INTELSAT IV EIRP = 22.5 dBW Bandwidth = 36 MHz	Large fixed antenna diameter = 30 m G/T = 40.7 dB/K	165 Mbits/s
DSCS II EIRP = 28 dBW Bandwidth = 50 MHz	Shipboard antenna diameter = 4 ft G/T = 10 dB/K	100 kbits/s
DSCS II EIRP = 28 dBW Bandwidth = 50 MHz	Large fixed antenna diameter = 60 ft G/T = 39 dB/K	72 Mbits/s
GAPSAT/MARISAT EIRP = 28 dBW Bandwidth = 500 kHz	Aircraft antenna gain = 0 dB G/T = −30 dB/K	500 bits/s

probability of 10^{-5} or better, assume that you are required to choose the modulation, coding, and interleaving from the schemes described below.

8-ary noncoherent FSK

16-ary QAM (matched filter detection)

(127, 92) BCH code, $d_{min} = 11$

Rate $\frac{1}{2}$, feedback-decoded convolutional code, corrects an average of three symbol errors out of a sequence of 21 symbols

Block interleaver (16 × 32)

Convolutional interleaver (150 × 300)

Choose a modulation technique and if deemed necessary, a coding scheme or a coding/interleaving scheme for the following applications. Justify your choices.

(a) Voice-grade telephone channel with 2400 Hz of usable bandwidth and available $E_b/N_0 = 14$ dB.

(b) Satellite channel with 40 kHz of usable bandwidth and with an available E_b/N_0 of 7.3 dB.

(c) Voice-grade link over a bursty noise channel. A noise burst typically lasts for 100 ms. The usable bandwidth is 3400 Hz and the available $E_b/N_0 = 10$ dB.

7.11. (a) For a fixed error probability, show that the relationship between alphabet size, M, and required average power for MPSK versus QAM can be expressed as

$$\frac{\text{average power for MPSK}}{\text{average power for QAM}} \simeq \frac{3M^2}{2(M - 1)\pi^2}$$

(b) Discuss the advantage of one type of signaling over the other.

7.12. Telephone modems operating at 19.2 kbits/s are now available using trellis-coded QAM modulation.

(a) Calculate the bandwidth efficiency of such modems, assuming that the usable channel bandwidth is 2400 Hz.

(b) Assuming AWGN and an available $E_b/N_0 = 10$ dB, calculate the theoretically available capacity in the 2400-Hz bandwidth.

(c) What is the required E_b/N_0 that will enable a 2400-Hz bandwidth to have a capacity of 19.2 kbits/s?

7.13. Figure 7.15 shows several 16-ary symbol constellations.

(a) For the (5, 11) circular constellations, compute the minimum radial distances r_1 and r_2 if the minimum distance between each symbol must be 1 unit.

(b) Compute the average signal power for the (5, 11) circular constellation, and compare it to the average signal power for the 4 × 4 ($M = 16$) square constellation (with the same minimum distance between symbols).

(c) Why might the square constellation be more practical?

7.14. Consider that the rate $\frac{2}{3}$ trellis-coded system of Section 7.10.7 is used over a binary symmetric channel (BSC). Assume that the initial encoder state is the 00 state. At the output of the BSC, the sequence $\mathbf{Z} = (1\ 1\ 1\ 0\ 0\ 1\ 1\ 0\ 1\ 0\ 1\ 1$ rest all "0") is received.

(a) Find the maximum likelihood path through the trellis diagram and determine the first 6 decoded information bits. If a tie occurs between any two merged paths, choose the upper branch entering the particular state.

(b) Determine if any channel bits in $\mathbf{Z}$ had been inverted by the channel during transmission, and if so, identify them.

(c) Explain how you would proceed with the problem if the channel were specified as a Gaussian channel instead of a BSC.

CHAPTER 8

Synchronization

Maurice A. King, Jr.
The Aerospace Corporation
El Segundo, California

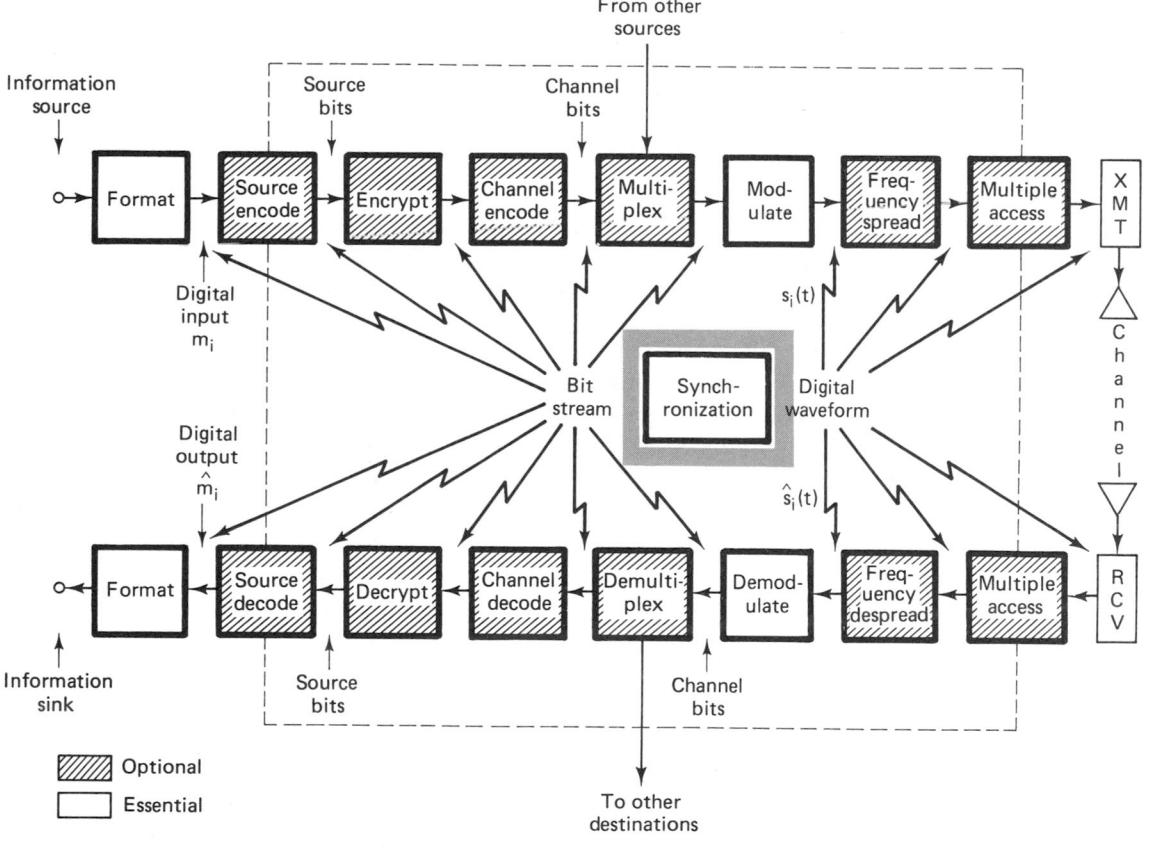

8.1 SYNCHRONIZATION IN THE CONTEXT OF DIGITAL COMMUNICATIONS

8.1.1 What It Means to Be Synchronized

In most of this book some level of signal synchronization is assumed, although perhaps not explicitly. In the discussions of demodulation and channel decoding it is assumed that many things about the incoming signal are accurately known. For example, in the case of coherent phase modulation (PSK), the receiver is assumed to be able to generate a set of reference signals whose phases are identical (except perhaps for a constant offset) to the phases of the signaling alphabet in use at the transmitter. These reference signals are compared to the incoming signals in the process of making maximum likelihood symbol decisions.

In order to be able to generate these reference signals, the receiver has to be in synchronization with the received carrier. This means that there has to be phase concurrence between the incoming carrier sinusoid and a replica of it in the receiver. In other words, if there were no information modulated on the incoming carrier, the incoming sinusoid and the replica in the receiver would pass through zero simultaneously. This is what is known as being in *phase lock* and is a condition that must be closely approximated if coherently modulated signals are going to be accurately demodulated at the receiver. If the information-bearing signal is not modulated directly on the carrier but indirectly through the use of a subcarrier, both the phase of the carrier and that of the subcarrier must be determined. If the carrier and subcarrier are not kept in phase synchronism by the transmitter (they typically are not), this will require the generation of a replica of the subcarrier by the receiver, where the phase of the subcarrier replica is controlled separately from that of the carrier replica. This will enable the receiver to achieve phase lock on both the carrier and subcarrier.

It is also assumed that the receiver has accurate knowledge of when an incoming symbol started and when it is over. This knowledge is required in order to know the proper symbol integration interval—the interval over which energy is integrated prior to making symbol decisions. Clearly, if the receiver integrates over an interval of an inappropriate length, or over an interval that spans two symbols, the ability to make accurate symbol decisions will be degraded.

It can be seen that symbol synchronization and phase synchronization are similar in that they both involve producing in the receiver a replica of a portion of the transmitted signal. For phase synchronization it was the sinusoidal carrier. For symbol synchronization the replica is a square wave at the symbol transition rate. The receiver must be able to produce a square wave that will transition through zero simultaneously with the incoming signal's transitions between symbols. A receiver that is able to do this can be said to have symbol synchronization, or be in *symbol lock*. Since there are typically a very large number of carrier cycles per symbol period, this second level of synchronization is much coarser than phase synchronization and is usually done with different circuitry than that used for phase synchronization.

In many communication systems an even higher level of synchronization is required. This is usually called *frame synchronization*. Frame synchronization is required when the information is organized in blocks, or messages of some uniform number of symbols. This will occur, for example, if a block code is used for forward error control, or if the communications channel is being time shared, on a regular basis, by several users (TDMA). In the case of block coding, the decoder needs to know where the boundaries between codewords are in order to decode the message correctly. In the case of a time-shared channel, it is necessary to know where the boundaries between channel users are, in order to know where the information is coming from and to where it should be routed. Similar to symbol synchronization, frame synchronization is equivalent to being able to generate a square wave at the frame rate, with the zero crossings coincident with the transitions from one frame to the next.

Most digital communications systems using coherent modulation require all three of these levels of synchronization: phase, symbol, and frame. Systems using noncoherent modulation techniques will typically require symbol and frame synchronization, but since the modulation is not coherent, phase lock is not required. Instead, noncoherent systems require *frequency synchronization*. Frequency synchronization differs from phase synchronization in that with frequency synchronization, the replica of the carrier that is generated by the receiver is allowed to have an arbitrary constant phase offset from the received carrier. Receiver designs can be simplified by removing the requirement to determine the exact value of the incoming carrier phase. Unfortunately, as is shown in the discussion of modulation techniques, this simplification carries a penalty in terms of degraded performance versus signal-to-noise ratio. The relative trade-offs of synchronization levels versus performance and system versatility are discussed further in the next section.

All of the discussion thus far has been oriented toward the receiving end of a communications link. There are instances, however, when the transmitter assumes the more active role in synchronization, by varying the timing and frequency of its transmissions to correspond to the expectations of the receiver. An example of this situation is a satellite communication network, where many terrestrial terminals are beaming signals toward a single satellite receiver. In most of these cases the transmitter relies on a return path from the receiver to determine the accuracy of its synchronization. Thus transmitter synchronization often implies two-way communications, or a network, in order to be successful. Thus transmitter synchronization is often called *network synchronization*. Transmitter or network synchronization is discussed later in this chapter.

8.1.2 Costs versus Benefits of Synchronization Levels

The need for a receiver to be synchronized with a communication system implies costs to the system. Each level of synchronization that is added implies more cost. The most obvious cost is in the need for additional hardware or software in the receiver for acquisition and tracking loops. Possibly less obvious costs lie in the extra time required to achieve synchronization before commencing communications, or in the energy expended by the transmitter on signals to be used at the receiver as acquisition or tracking aids. In the face of these costs to the system, one might question why a communications system designer would consider a system design requiring a high degree of synchronization. The answers are improved performance and versatility.

Consider a standard commercial AM radio. This radio may be considered part of a broadcast communication system involving a central transmitter and many receivers. This communication system involves no synchronization. However, the receiver passband must be wide enough to accommodate not only the information-bearing signal, but also any fluctuations in the carrier, due perhaps to Doppler shift* or drift in the transmitter's frequency reference. This requirement on the receiver passband means that additional noise energy is passed to the detector, over and above the amount theoretically required by the bandwidth of the information. A somewhat more complicated receiver that includes a carrier-frequency tracking loop would be able to keep a narrow passband filter centered about the carrier, thereby substantiallly reducing the detected noise energy and improving the received signal-to-noise ratio. Thus although a standard radio may be perfectly adequate for reception of signals from large transmitters a few tens of kilometers distant, it may prove totally inadequate under less benign conditions.

---

* An offset in frequency as perceived by a receiver, from the nominally transmitted frequency, caused by relative motion of the transmitter and receiver. Ignoring second- and higher-order effects, the value of the frequency offset, Δf, is given by $v f_0 / c$, where v is the relative velocity (positive when the relative distance between transmitter and receiver is being reduced), f_0 the nominal frequency, and c the speed of light.

For digital communications, another example of a trade-off between performance and receiver complexity is in the choice of modulation. Among the simplest digital receivers are those designed to be used with noncoherently detected binary FSK. The only synchronization requirements are bit timing and frequency tracking. However, the same bit error probability could be achieved with approximately 4 dB less signal-to-noise ratio if the modulation is coherent BPSK. The disadvantage of BPSK is that the receiver requires a phase tracking loop, which can be a complex design problem for signals that may experience high Doppler rates* or deep signal fades.†

A third instance of a cost versus performance trade-off involving synchronization is in the use of error control coding. As was established in Chapters 5 and 6, there are substantial performance advantages in the use of appropriate error control coding techniques. The cost, measured in receiver complexity, however, can be high. For a block decoder to operate properly requires the receiver to achieve block, message, or frame synchronism. This is a procedure over and above the usual decoding procedure, although some error-correcting codes have been designed with block synchronization aids built in [1]. Convolutional codes also require some degree of additional synchronization in order to provide optimum performance. Although the performance analysis of convolutional codes often makes the assumption that the input data sequence is infinitely long, in practice it is not. In order to provide the minimum error probability, the decoder must know the beginning state (usually all zeros) when the data sequence will begin, the eventual ending state, and when the ending state is to be reached. Knowing when the beginning state was left and when the ending state is to be reached, however, is equivalent to having frame synchronization. In addition, the decoder will have to know how to group the channel symbols in order to make branch decisions. This is also a synchronization requirement.

The trade-offs discussed thus far have been in terms of the performance versus complexity of individual links and receivers. The ability to synchronize has a large potential consequence in terms of system efficiency and versatility as well. Frame synchronization allows the use of advanced, versatile multiple access techniques, such as the variety of demand assignment multiple access (DAMA) schemes, which are becoming increasingly popular as communication channel resources become increasingly scarce. In addition, the use of spread-spectrum techniques, both as multiple access schemes and for interference rejection, requires a high level of system synchronization. Spread-spectrum techniques are treated in depth in following chapters. It will be seen that these techniques provide the potential for a great deal of system versatility, which is a very valuable feature if the system will encounter changing or unstable conditions, such as the effects of intentional and unintentional interference from external sources.

---

* The rate of change of the Doppler shift. This rate sets requirements on the tracking ability of the phase tracking loop.

† Variations in the received signal strength caused by random variations in channel parameters.

8.2 RECEIVER SYNCHRONIZATION

All digital communication systems require some degree of synchronization to incoming signals by the receivers. In this section the fundamentals of the various levels of receiver synchronization are discussed. The discussion will begin with the basic level of synchronization required for coherent reception—phase synchronization—and a brief discussion of the principles of phase-locked-loop (PLL) operation and design. The discussion will then broaden into the topics of frequency and symbol synchronization. Frequency and symbol synchronization are required for all digital communications reception, either coherent or noncoherent. The final topics in the section are receiver frame synchronization and techniques for achieving and maintaining it.

8.2.1 Coherent Systems: Phase-Locked Loops

At the heart of all phase synchronization circuits is some version of a phase-locked loop (PLL). A schematic diagram of the basic PLL is given in Figure 8.1. Phase-locked loops are servo-control loops, whose controlled parameter is the phase of a locally generated replica of the incoming carrier signal. Phase-locked loops have three basic components: a phase detector, a loop filter, and a voltage-controlled oscillator (VCO). The phase detector is a device that produces a measure of the difference in phase between an incoming signal and the local replica. As the incoming signal and the local replica change with respect to each other, the phase difference (or phase error) becomes a time-varying signal into the loop filter. The loop filter governs the PLL's response to these variations in the error signal. A well-designed loop should be able to track changes in the incoming signal's phase but not be overly responsive to receiver noise. The VCO is the device that produces the carrier replica. The VCO, as the name implies, is a sinusoidal oscillator whose frequency is controlled by a voltage level at the device input. In Figure 8.1 the phase detector is shown as a multiplier, the loop filter is described by its impulse response function, $f(t)$, with Fourier transform, $F(\omega)$, and the VCO is so indicated.

A VCO is an oscillator whose output frequency is a linear function of its input voltage over some range of input and output. A positive input voltage will cause the VCO output frequency to be greater than its uncontrolled value, ω_0,

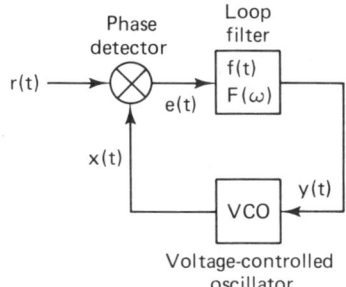

Figure 8.1 Schematic of the basic phase-locked loop.

while a negative voltage will cause it to be less. Phase lock is achieved by feeding a filtered version of the phase difference (i.e., the phase error) between the incoming signal, $r(t)$, and the output of the VCO, $x(t)$, back to the input of the VCO, $y(t)$.

Consider a normalized input signal of the form

$$r(t) = \sin [\omega_0 t + \theta(t)] \tag{8.1}$$

where ω_0 is the nominal carrier frequency and $\theta(t)$ is a slowly varying phase. Similarly, consider a normalized VCO output of the form

$$x(t) = 2 \cos [\omega_0 t + \hat{\theta}(t)] \tag{8.2}$$

These signals will produce an output error signal at the phase detector output of the form

$$e(t) = x(t)r(t) = 2 \cos [\omega_0 t + \hat{\theta}(t)] \sin [\omega_0 t + \theta(t)]$$

$$= \sin [\theta(t) - \hat{\theta}(t)] + \sin [2\omega_0 t + \theta(t) + \hat{\theta}(t)] \tag{8.3}$$

Assuming that the loop filter is low pass, the second term on the right hand side of Equation (8.3) will be filtered out and can be ignored. This low-pass assumption is a reasonable loop design decision. A low-pass filter provides an error signal that is solely a function of the difference in phases between the input [Equation (8.1)] and the VCO output [Equation (8.2)]. This is exactly the error signal that is needed. The VCO output frequency is the time derivative of the argument of the sine function in Equation (8.2). If we make the assumption that ω_0 is the uncontrolled frequency of the VCO (the output frequency when the input voltage is zero), we can express the difference in the VCO output frequency from ω_0 as the time differential of the phase term $\hat{\theta}(t)$. The output frequency of the VCO is a linear function of the input voltage. Therefore, since an input voltage of zero produces an output frequency of ω_0, the difference in the output frequency from ω_0 will be proportional to the value of the input voltage, $y(t)$, or

$$\Delta\omega(t) = \frac{d}{dt} [\hat{\theta}(t)] = K_0 y(t)$$

$$= K_0 e(t) * f(t)$$

$$\simeq K_0 [\theta(t) - \hat{\theta}(t)] * f(t) \tag{8.4}$$

where $\Delta\omega(t)$ denotes the frequency difference, the notation $*$ indicates the convolution operation (see Appendix A), and the small-angle approximation [i.e., $e(t) = \sin [\theta(t) - \hat{\theta}(t)] \simeq \theta(t) - \hat{\theta}(t)$] has been used in the last line of Equation (8.4). The small-angle approximation will be accurate when the output phase error is small (the loop is close to phase lock). This will be the situation when the loop is operating normally. The factor K_0 is the gain of the VCO, and $f(t)$ is the loop filter impulse response. This linear differential equation in $\hat{\theta}(t)$ (utilizing the small-angle approximation) is known as the linearized loop equation. It is the single most useful relationship in determining loop behavior during normal operation (where the phase error is small).

Example 8.1   Linearized Loop Equation

Show that for appropriately chosen K_0 and $f(t)$ the linearized loop equation [Equation (8.4)] demonstrates a tendency toward phase lock—that is, the phase difference between the incoming signal and the VCO output tends to decrease.

Solution

Consider the case where the phase of the input signal, $\theta(t)$, is slowly varying with time. It can be seen that if the phase difference on the right-hand side of Equation (8.4) is positive [i.e., $\theta(t) > \hat{\theta}(t)$], then by appropriate choice of K_0 and $f(t)$, the time derivative of $\hat{\theta}(t)$ will be positive, so that $\hat{\theta}(t)$ will increase with time, which will tend to reduce the magnitude of the difference $|\theta(t) - \hat{\theta}(t)|$. On the other hand, if the phase difference is negative, $\hat{\theta}(t)$ will decrease with time, which will also reduce the magnitude of the phase difference. Finally, if $\theta(t) = \hat{\theta}(t)$, then Equation (8.4) indicates that $\hat{\theta}(t)$ will not change with time, and the equality will be maintained.

Consider the Fourier transform of Equation (8.4),

$$j\omega\hat{\Theta}(\omega) = K_0[\Theta(\omega) - \hat{\Theta}(\omega)]F(\omega) \tag{8.5}$$

where the capitalized functions of ω are the Fourier transforms of the lowercase functions of t in Equation (8.4). That is, $\hat{\Theta}(\omega) \leftrightarrow \hat{\theta}(t)$, $\Theta(\omega) \leftrightarrow \theta(t)$, and $F(\omega) \leftrightarrow f(t)$. Reorganizing Equation (8.5) provides

$$\frac{\hat{\Theta}(\omega)}{\Theta(\omega)} = \frac{K_0 F(\omega)}{j\omega + K_0 F(\omega)} = H(\omega) \tag{8.6}$$

The term $H(\omega)$ is known as the closed-loop transfer function of the PLL. This term is very useful in characterizing the transient response of a PLL. The order of a PLL is defined to be the order of the highest-order term in $j\omega$ in the denominator of $H(\omega)$. Equation (8.6) indicates that this is always one more than the order of the loop filter, $F(\omega)$. This is because when $F(\omega)$ is expressed analytically as $F(\omega) = N(\omega)/D(\omega)$, the denominator of $H(\omega)$ when expressed as a polynomial in $j\omega$ will have the term $j\omega D(\omega)$, which must have a term in $j\omega$ that is one order higher than the highest-order term in $D(\omega)$ alone. The order of a PLL is critical for determining the loop's steady-state response to a steady-state input. This is discussed in the next section.

8.2.1.1   Steady-State Errors and Loop Filter Characteristics

By reorganizing Equation (8.6) we can obtain an expression for the Fourier transform of the phase error.

$$
\begin{aligned}
E(\omega) &= \mathcal{F}\{e(t)\} \\
&= \Theta(\omega) - \hat{\Theta}(\omega) \\
&= [1 - H(\omega)]\Theta(\omega) \\
&= \frac{j\omega\Theta(\omega)}{j\omega + K_0 F(\omega)}
\end{aligned} \tag{8.7}
$$

Equation (8.7) can be used in conjunction with the final value theorem of Fourier transforms to determine the steady-state error response of a loop to a variety of possible input characteristics. The steady-state error is the residual error after all transients have died away, and thus provides a measure of a loop's ability to cope with various types of changes in the input. The final value theorem states that

$$\lim_{t \to \infty} e(t) = \lim_{j\omega \to 0} j\omega E(\omega) \tag{8.8}$$

Combining Equations (8.7) and (8.8) yields

$$\lim_{t \to \infty} e(t) = \lim_{j\omega \to 0} \frac{(j\omega)^2 \Theta(\omega)}{j\omega + K_0 F(\omega)} \tag{8.9}$$

Example 8.2   Response to a Phase Step

Consider a loop's steady-state response to a phase step at the loop input.

Solution

Assuming that the PLL was originally in phase lock, a phase step will throw the loop out of lock. Having abruptly changed, however, the input phase again becomes stable. This should be the easiest type of phase disturbance for a PLL to deal with. The Fourier transform of a phase step will be taken to be

$$\Theta(\omega) = \mathcal{F}\{\Delta\phi \, u(t)\}$$

$$= \frac{\Delta\phi}{j\omega} \tag{8.10}$$

where $\Delta\phi$ is the magnitude of the step and $u(t)$ is the unit step function,

$$u(t) = \begin{cases} 1 & \text{for } t > 0 \\ 0 & \text{for } t < 0 \end{cases}$$

$$= \int_{-\infty}^{t} \delta(\tau) \, d\tau$$

where $\delta(\tau)$ is the Dirac delta function. From Equations (8.9) and (8.10),

$$\lim_{t \to \infty} e(t) = \lim_{j\omega \to 0} \frac{j\omega \, \Delta\phi}{j\omega + K_0 F(\omega)} = 0$$

assuming that $F(0) \neq 0$. Thus the loop will eventually track out any phase step that appears at the input if the loop filter has a nonzero dc response. This means that for any loop filter with the property that $F(\omega) = N(\omega)/D(\omega)$ and $N(0) \neq 0$, the PLL will automatically tend to recover phase lock if the input is displaced by a constant phase. This is clearly a very desirable loop characteristic.

Example 8.3   Response to a Frequency Step

Next, consider a loop's steady-state response to a frequency step at the input.

Solution

A frequency step can approximate the effect of a Doppler shift in the incoming signal frequency due to relative motion between the transmitter and the receiver. Thus, this is an important example for systems with mobile terminals. Since phase is the integral of frequency, the input phase will change linearly as a function of time for a constant input frequency offset. The Fourier transform of the phase characteristic will be the transform of the integral of the frequency characteristic. Since the frequency characteristic is a step, and the transform of an integral is the transform of the integrand divided by the parameter $j\omega$,

$$\Theta(\omega) = \frac{\Delta\omega}{(j\omega)^2} \tag{8.11}$$

where $\Delta\omega$ is the magnitude of the frequency step. Substituting Equation (8.11) into Equation (8.9) yields

$$\lim_{t\to\infty} e(t) = \lim_{j\omega\to0} \frac{\Delta\omega}{j\omega + K_0F(\omega)} = \frac{\Delta\omega}{K_0F(0)} \tag{8.12}$$

The steady-state result in this case depends on more properties of the loop filter than merely a nonzero dc response. If the filter is an all-pass,

$$F_{ap}(\omega) = 1 \tag{8.13}$$

a low-pass,

$$F_{\ell p}(\omega) = \frac{\omega_1}{j\omega + \omega_1} \tag{8.14}$$

or a lead-lag,

$$F_{\ell\ell}(\omega) = \left(\frac{\omega_1}{\omega_2}\right) \frac{j\omega + \omega_2}{j\omega + \omega_1} \tag{8.15}$$

Equation (8.12) indicates that the loop will track the input phase ramp with a constant steady-state error whose value will depend on the gain term, K_0, and the magnitude of the frequency step. Using any of $F_{ap}(\omega)$, $F_{\ell p}(\omega)$, or $F_{\ell\ell}(\omega)$ for $F(\omega)$ in Equation (8.12) yields

$$\lim_{t\to\infty} e(t) = \frac{\Delta\omega}{K_0}$$

Notice that a product of several filters with filter characteristics of the form of Equation (8.13), (8.14), or (8.15) would still produce this result. This steady-state error, which is called the *velocity error*, will exist regardless of the order of the filter, unless the denominator of $F(\omega)$, contains $j\omega$ as a factor [$\omega_1 = 0$ in the denominator of Equation (8.14) or (8.15) with the appropriate renormalization in the numerators]. Having $j\omega$ as a factor of $D(\omega)$ is equivalent to having a perfect integrator in the loop filter. It is not possible to build a perfect integrator, but one may be closely approximated either digitally or by using active integrated circuits [2]. Thus if the system design requires the tracking of Doppler shifts with zero steady-state error, the loop filter design must contain an approximation to a perfect integrator. It should be noted that even with a nonzero velocity error, the frequency is still being tracked; there are important applications where tracking to zero phase error is not important. Noncoherent signaling, such as the standard use of FSK modulation, is an example. For noncoherent signaling it is actually frequency tracking that is required, and the absolute value of phase is unimportant.

Example 8.4   Response to a Frequency Ramp

Consider a loop's steady-state response when the input frequency is changing linearly with time (a frequency ramp function).

Solution

This example corresponds to the effect of a step change in the time derivative of the input frequency. This would approximate a change in the Doppler rate, which could model acceleration in the motion between a satellite or an aircraft and a ground receiver. In this case the Fourier transform of the phase characteristic is given by

$$\Theta(\omega) = \frac{\Delta\dot{\omega}}{(j\omega)^3} \tag{8.16}$$

where $\Delta\dot{\omega}$ is the magnitude of the rate of frequency change. In this case, Equation (8.9) yields

$$\lim_{t\to\infty} e(t) = \lim_{j\omega\to0} \frac{\Delta\dot{\omega}/j\omega}{j\omega + K_0F(\omega)} = \lim_{j\omega\to0} \frac{\Delta\dot{\omega}}{j\omega K_0F(\omega)} \tag{8.17}$$

If the loop has a nonzero velocity error, that is, if the right-hand side of Equation (8.12) is not equal to zero, Equation (8.17) shows the steady-state phase error due to a frequency ramp to be unbounded. This says that a PLL with loop filters given by any of Equations (8.13) to (8.15) will not be able to track a frequency ramp. In order to track a frequency ramp, the denominator of the loop filter transform, $D(\omega)$, must have $j\omega$ as a factor. From Equation (8.17) it can be seen that a loop filter with this type of transfer function, $F(\omega) = N(\omega)/j\omega D_1(\omega)$, will allow the PLL to track a frequency ramp with a constant phase error. This implies that in order to track a signal with a linearly changing Doppler shift (constant relative acceleration), the receiver must have a PLL that is second order or higher. To track a frequency ramp with zero phase error, the loop filter would be required to have a transfer function with $(j\omega)^2$ as a factor of the denominator, $F(\omega) = N(\omega)/(j\omega)^2D_2(\omega)$. This implies a PLL that is third order or higher. Thus high-performance aircraft that need to track phase accurately through violent maneuvers may require third- or higher-order PLLs. In all cases, frequency lock is available with a loop of one order less than that required for phase lock. Steady-state error analysis is therefore a useful indicator of the required complexity of the loop filters.

In practice, the vast majority of PLL designs are second order. This is because a second-order loop can be made to be unconditionally stable [2] (i.e., the loop will always try to track the input—no set of input conditions, regardless of how extreme, will cause the loop to become unable to respond in the appropriate direction to changes in the input). Second-order loops will track out the effect of a frequency step (Doppler shift), and they are relatively easy to analyze, since the closed-form results obtained for first-order loops are good approximations for second-order loop performance. Third-order loops are used for some special applications [e.g., some Global Positioning System (GPS) navigation receivers have a third-order PLL], but loop performance for third-order loops is relatively difficult to determine, and third- and higher-order loops are only conditionally stable. A typical communication system design decision is that if the signal dynamics are expected to be such that high-order loops would be required for coherent demodulation, noncoherent demodulation is used instead.

8.2.1.2 Performance in Noise

The steady-state analysis of the preceding section tacitly assumed that the input signal was noise free. In some situations this may be approximately correct, but as in other parts of communication analysis, the more general case would include the effects of noise.

Reconsider the normalized loop input signal of Equation (8.1) and Figure 8.1. With the inclusion of normalized narrowband additive Gaussian noise, $n(t)$, the expression for the input becomes

$$r(t) = \sin (\omega_0 t + \theta) + n(t) \qquad (8.18)$$

where, for the moment, we consider the input phase offset, θ, to be a constant. The noise process, $n(t)$, assumed to be a zero-mean narrowband Gaussian process, can be expanded into quadrature components about the carrier frequency as [3]

$$n(t) = n_c(t) \cos \omega_0 t + n_s(t) \sin \omega_0 t \qquad (8.19)$$

where both $n_c(t)$ and $n_s(t)$ are zero-mean Gaussian random processes and are statistically independent. Now the output of the phase detector can be written as [see Equation (8.3)]

$$e(t) = x(t)r(t)$$

$$= \sin (\theta - \hat{\theta}) + n_c(t) \cos \hat{\theta} - n_s(t) \sin \hat{\theta}$$

$$+ \text{(terms at twice the carrier frequency)} \qquad (8.20)$$

As before, the twice-carrier-frequency terms are assumed to be eliminated by the loop filter. Denoting the second and third terms of Equation (8.20) as

$$n'(t) = n_c(t) \cos \hat{\theta} - n_s(t) \sin \hat{\theta} \qquad (8.21)$$

it is easy to verify that the variance of $n'(t)$ is identical to the variance of $n(t)$. This variance will be denoted by σ_n^2.

Consider the autocorrelation function of $n'(t)$,

$$R(t_1, t_2) = \mathbf{E}\{n'(t_1)n'(t_2)\}$$

$$= \mathbf{E}\{n_c(t_1)n_c(t_2)\} \cos^2 \hat{\theta} + \mathbf{E}\{n_s(t_1)n_s(t_2)\} \sin^2 \hat{\theta}$$

$$- [\mathbf{E}\{n_c(t_1)n_s(t_2)\} + \mathbf{E}\{n_s(t_1)n_c(t_2)\}] \sin \hat{\theta} \cos \hat{\theta} \qquad (8.22)$$

where $\mathbf{E}\{\cdot\}$ denotes the expected value. The cross-terms on the right-hand side of Equation (8.22) are equal to zero because n_c and n_s are mutually independent and have zero means [3]. With the assumption of wide-sense stationarity [4] we have

$$R(\tau) = R_c(\tau) \cos^2 \hat{\theta} + R_s(\tau) \sin^2 \hat{\theta} \qquad (8.23)$$

where $\tau = t_1 - t_2$. Taking Fourier transforms, the power spectral density of $n'(t)$ is seen to be

$$G(\omega) = \mathscr{F}[R(\tau)]$$

$$= G_c(\omega) \cos^2 \hat{\theta} + G_s(\omega) \sin^2 \hat{\theta} \qquad (8.24)$$

where G_c and G_s are the Fourier transforms of R_c and R_s, respectively. But from Equation (8.19), it can be seen that the spectra G_c and G_s are made of shifted versions of the spectra of the original noise process $n(t)$. Therefore, because of our construction [5],

$$G_s(\omega) = G_c(\omega) = G_n(\omega_0 - \omega) + G_n(\omega_0 + \omega)$$

where $G_n(\omega)$ is the spectral density of the original bandpass noise process $n(t)$. Equation (8.24) can be rewritten as

$$G(\omega) = G_n(\omega_0 - \omega) + G_n(\omega_0 + \omega) \qquad (8.25)$$

For the special case of white noise, we have $G_n(\omega) = N_0/2$ watts/hertz, where N_0 is the single-sided spectral density of the white noise. Thus, from Equation (8.25), for this important special case,

$$G(\omega) = N_0 \qquad (8.26)$$

The value in this development is that for the same small-angle approximations that were made in the preceding section, the spectral density of the VCO phase, $G_{\hat{\theta}}$, is related to the spectral density of the noise process through the loop transfer function [Equation (8.6)]. That is,

$$G_{\hat{\theta}}(\omega) = G(\omega) |H(\omega)|^2 \qquad (8.27)$$

where $G(\omega)$ is as given in Equation (8.25) and $H(\omega)$ as defined in Equation (8.6). The variance of the output phase is then

$$\sigma_{\hat{\theta}}^2 = \frac{1}{2\pi} \int_{-\infty}^{\infty} G(\omega) |H(\omega)|^2 \, d\omega \qquad (8.28)$$

For the special case of white noise,

$$\sigma_{\hat{\theta}}^2 = \frac{N_0}{2\pi} \int_{-\infty}^{\infty} |H(\omega)|^2 \, d\omega \qquad (8.29)$$

The integral in Equation (8.29) (renormalized to natural frequency) is called the *two-sided loop bandwidth*, W_L. The *single-sided loop bandwidth* is termed B_L. The definitions of these terms are

$$W_L = 2B_L = \frac{1}{2\pi} \int_{-\infty}^{\infty} |H(\omega)|^2 \, d\omega \qquad \text{hertz} \qquad (8.30)$$

Thus, if the noise process is white and the small-angle approximation holds (in other words, the loop is successfully tracking the input phase), the phase variance is given by

$$\sigma_{\hat{\theta}}^2 = 2N_0 B_L \tag{8.31}$$

The phase variance is a measure of the amount of jitter or wobble in the VCO output due to noise at the input. Equations (8.31) and (8.7) highlight one of the many trade-offs in communication theory. Clearly, one would wish $\sigma_{\hat{\theta}}^2$ to be small, which for a given noise level implies a small loop bandwidth, B_L, which from Equation (8.30) implies a narrow $H(\omega)$. However, it can be inferred from Equation (8.7) that the narrower the effective bandwidth of $H(\omega)$, the poorer will be the loop's ability to track incoming signal phase changes, $\Theta(\omega)$. Thus a loop design must balance noise response with desired input phase response. The designer's dilemma is to design a loop that responds appropriately to the changes in the input signal, while not being overly responsive to the apparent changes, which are actually only artifacts of the noise process.

8.2.1.3 Nonlinear Loop Analysis

All of the PLL discussion in the previous sections has utilized what is called the linearized PLL model. This model is shown schematically in Figure 8.2. The model makes use of the small-angle approximation

$$\sin(\theta - \hat{\theta}) \simeq \theta - \hat{\theta} \tag{8.32}$$

which is accurate when the loop is "in lock" and performing as desired (i.e., with small phase errors). Clearly, these conditions form only part of the picture. A complete analysis of PLL performance must allow for the times when Equation (8.32) is not accurate. When the small-angle approximation is inaccurate, an appropriate model is the one shown schematically in Figure 8.3. From Equations (8.4), (8.20), and (8.21) and Figure 8.3, the model can be described by the differential equation

$$\frac{d}{dt}[\hat{\theta}(t)] = K_0 f(t) * \sin[\theta(t) - \hat{\theta}(t)] + f(t) * n'(t) \tag{8.33}$$

where, as before, $*$ denotes the convolution operation. In spite of the best efforts of many researchers, this differential equation has resisted general solution for many years. However, Viterbi [5] has derived a closed-form solution for an important special case.

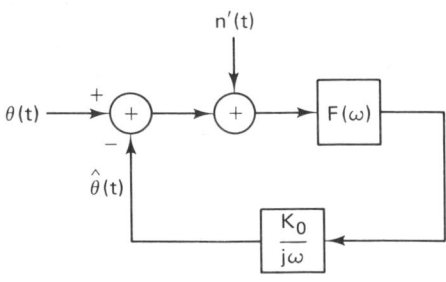

Figure 8.2 Schematic of linearized PLL model.

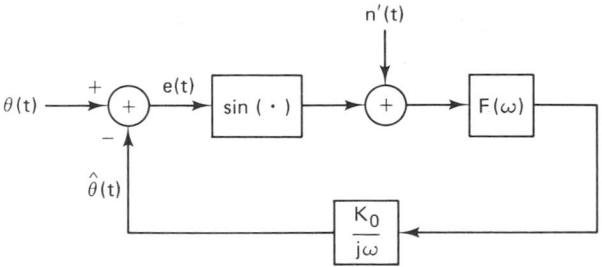

Figure 8.3 Schematic of nonlinearized PLL model.

Consider the case where $\theta(t)$, the input phase as a function of time, is a constant, θ. We can now define a new phase variable

$$\phi(t) = [\theta - \hat{\theta}(t)] \text{ modulo } 2\pi \qquad (8.34)$$

Since θ is constant, Equation (8.33) can be rewritten as

$$\frac{d}{dt}[\phi(t)] = K_0 f(t) * \sin \phi(t) + f(t) * n'(t) \qquad (8.35)$$

Since from Equation (8.35), $\phi(t)$ is a function of the random process $n'(t)$, $\phi(t)$ itself is a random process. Because $\phi(t)$ is defined modulo 2π, it can be shown [2] that $\phi(t)$ is stationary in the limit when all transient effects have died down (i.e., θ is a constant). Viterbi [5] has determined that for a first-order PLL [i.e., the loop filter is a short circuit, or equivalently $f(t) = \delta(t)$], the probability density function of ϕ is of the form

$$p(\phi) = \frac{\exp(\rho \cos \phi)}{2\pi I_0(\rho)} \qquad \text{for } |\phi| \leq \pi \qquad (8.36)$$

where $\rho = 1/\sigma_{\hat{\theta}}^2$ [see Equation (8.31)] is the normalized (to unit signal energy) loop signal-to-noise ratio, and $I_0(\rho)$ is the zeroth-order modified Bessel function of the first kind, evaluated at ρ. The phase variance, modulo 2π, can now be computed using Equation (8.36). The resulting value of the phase variance will be exact for first-order loops, and is an extremely useful approximation for the behavior of many second-order loops [2]. It has also been shown to be an exact form for higher-order loops under a modified definition of ρ [6].

The change of variable from a phase that can take any real value to a phase that is modulo 2π results in the concept of loop cycle slips. A cycle slip occurs when the magnitude of the original phase error, $|\theta - \hat{\theta}(t)|$, exceeds 2π radians. This will cause the value of ϕ [Equation (8.34)] to abruptly change from about 2π to about 0. This event can be thought of as a momentary loss of lock with an almost immediate reacquisition. The statistics of cycle slips can be as important an indicator of PLL performance as phase variance—especially at low loop signal-to-noise ratios when cycle slips may occur frequently.

By manipulating his phase distribution results, Viterbi [5] derived an expression for the mean time to the first cycle slip, T_m, beginning at some arbitrary reference time,

$$T_m = \frac{\pi^2 \rho I_0^2(\rho)}{2B_L} \tag{8.37}$$

For large ρ this expression can be approximated by

$$T_m \simeq \frac{\pi \exp (2\rho)}{4B_L} \tag{8.38}$$

As was true with the probability density function of Equation (8.36), these results were derived for first-order loops, but they are useful approximations for the behavior of second-order loops, and provide an upper bound to second-order loop performance at medium and large loop signal-to-noise ratios. In addition, computer simulations and laboratory measurements [2] indicate that the time between cycle slips, T, is exponentially distributed,

$$P(T) = 1 - \exp \left(-\frac{T}{T_m} \right) \tag{8.39}$$

This is to say that the probability that a loop will cycle-slip within time T, starting from zero phase error, is given by Equation (8.39).

8.2.1.4 Suppressed Carrier Loops

The discussion of PLLs to this point has presumed that the carrier input is a fairly stable sinusoid with some known positive average energy. In the case of a phase modulated communication system, if the carrier phase variation due to the modulation is less than $\pi/2$ radians, there will be positive energy at the carrier frequency. This is called a system design that has a residual carrier component, and all of the discussion of PLL development to this point would apply directly to this residual component. A diagram of the signal space for a binary phase modulated system with a residual carrier component is given in Figure 8.4, for a modulating angle of $\gamma \leq \pi/2$. At one time, most phase modulated systems were designed in this way. However, the residual carrier component is, in a sense, wasted energy—in the sense that the energy in the residual carrier is not being used to transmit the information, only to transmit the carrier. Thus most modern phase modulated systems are suppressed carrier systems. This means that there is no average energy transmitted at the carrier frequency. All of the transmitted energy goes into the modulation. Unfortunately, this means that there is no longer any signal for the basic PLL of Figure 8.1 to track.

Consider, as an example, a BPSK signal

$$r(t) = m(t) \sin (\omega_0 t + \theta) + n(t) \tag{8.40}$$

where $m(t) = \pm 1$ with equal probability. This is a suppressed carrier transmission—the average energy at radian frequency ω_0 is zero. In Figure 8.4 this signal is the situation when $\gamma = \pi/2$. The figure indicates that for this case the vertical carrier component will vanish. To acquire and track the phase of the carrier, the effects of the modulation must be eliminated. One way to eliminate the modulation

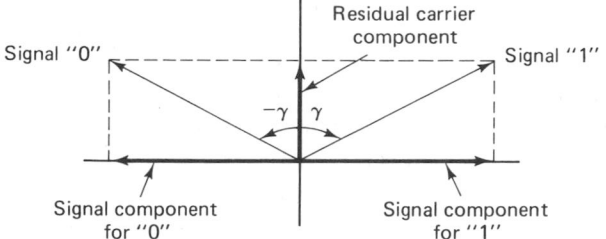

Residual carrier
component

Signal "0"

Signal "1"

$-\gamma$ γ

Signal component
for "0"

Signal component
for "1"

Figure 8.4 Residual carrier binary
phase modulation.

is to square the signal;

$$r^2(t) = m^2(t) \sin^2 (\omega_0 t + \theta) + n^2(t) + 2n(t)m(t) \sin (\omega_0 t + \theta)$$

$$= \tfrac{1}{2} - \tfrac{1}{2} \cos (2\omega_0 t + 2\theta) + n^2(t) + 2n(t)m(t) \sin (\omega_0 t + \theta) \quad (8.41)$$

where use has been made of the fact that $m^2(t) = 1$. The second term on the right-hand side of Equation (8.41) is a carrier-related term (at twice the original carrier frequency) that can be acquired and tracked with a basic PLL of the type illustrated in Figure 8.1. Such an arrangement is illustrated in Figure 8.5. When the incoming suppressed-carrier waveform is squared, the resulting twice-carrier component can be acquired and tracked by a PLL of standard design. Some of the problems with this procedure can be inferred from Equation (8.41). The first problem is simply that all phase angles have been doubled. Thus the phase noise and phase jitter has been doubled, and the phase error variance (related to the phase noise squared) is larger by a factor of 4 than that of the original signal. This angle doubling is offset by the divide-by-2 circuit at the VCO output, and therefore, does not directly affect the accuracy of the loop's output signal that is used by the data demodulator, but this larger internal variation will cause the PLL to require a 6-dB-larger carrier signal-to-noise ratio than a residual carrier system in order to maintain phase lock. In addition, there are now two effective noise

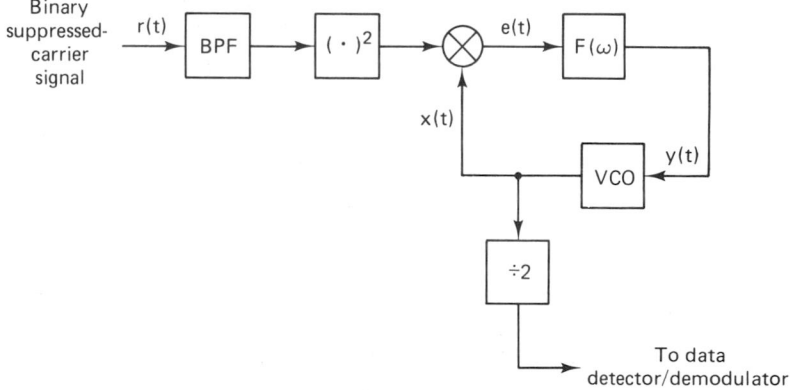

Binary
suppressed-
carrier
signal

$r(t)$

BPF

$(\cdot)^2$

$e(t)$

$F(\omega)$

$x(t)$

VCO

$y(t)$

$\div 2$

To data
detector/demodulator

Figure 8.5 Basic squaring loop schematic.

terms interfering with loop operation, because of the cross-correlation term between noise and signal in Equation (8.41). For cases of medium or low loop signal-to-noise ratio, these two noise terms will reduce the available signal-to-noise ratio even further, relative to the original unmodulated carrier. This additional loss due to signal-times-noise and noise-times-noise terms is called the *loop squaring loss*, S_L. Gardner [2] shows that if the input noise process, $n(t)$, is a narrowband Gaussian noise of bandwidth B_i, the squaring loss is upper bounded by

$$S_L = 1 + N_0 B_i \qquad (8.42)$$

where, as before, N_0 is the single-sided power spectral density of the prefiltered, normalized white Gaussian noise process. Equation (8.42) is an upper bound because the filter bandwidth B_i is tacitly assumed to be wide enough to pass the signal undistorted. In an actual design, signal distortion can be traded for squaring loss, as is shown in [7].

Since the normalization in Equation (8.42) is with respect to the signal power, the second term is proportional to a signal-to-noise ratio

$$\rho_i = \frac{1}{2 N_0 B_i} \qquad (8.43)$$

where ρ_i is the signal-to-noise ratio in the input filter bandwidth. For large loop signal-to-noise ratios, the output phase variance can now be expressed as

$$\sigma_\theta^2 = 2 N_0 B_L S_L = 2 N_0 B_L \left(1 + \frac{1}{2 \rho_i} \right) \qquad (8.44)$$

The leading term on the right-hand side of Equation (8.44) can be seen to be identical to that of Equation (8.31), the phase variance of the standard PLL. It can also be seen that for large input signal-to-noise ratios, the second term in the squaring loss will vanish, and we are left with the phase variance of the standard PLL.

Another potentially serious problem, associated mainly with suppressed carrier loops, is that of *false lock* [2, 8–10]. This can be a problem especially during acquisition or reacquisition of carrier phase. The interaction of the data stream with the loop nonlinearities (especially the squaring circuit) and loop filters will produce sidebands in the spectrum that is input to the phase detector. These sidebands can contain stable frequency components. Care must be taken that these stable components are not allowed to capture the tracking loop. If the loop is captured, it will appear to be operating correctly; the VCO control signal, $y(t)$, will be small but the VCO output will be offset in frequency from the correct carrier component. This is false lock. The loop is tracking a sideband frequency component, and the real carrier is being filtered out by the loop filter. False lock is a hardware implementation problem that typically sets an effective lower limit on the bandwidth of the loop filters. Residual carrier loops, having fewer nonlinear elements, are not usually bothered by false locking.

8.2.1.5 Costas Loops

An important form of a suppressed carrier loop is the Costas loop, shown schematically in Figure 8.6. This loop design is important because it eliminates the square-law device, which can be difficult to implement at carrier frequencies, and replaces it with a multiplier and relatively simple low-pass filters. Although the appearance of the circuits in Figures 8.5 and 8.6 is quite different, their theoretical performance can be shown to be the same [2]. The main remaining implementation problem with Costas loops is that to achieve the theoretically optimum performance, the two low-pass arm filters must be perfectly matched. This can only be approximated in any hardware implementation. Thus the decision as to whether to implement a Costas loop or the classical design of Figure 8.5 amounts to a design decision between the difficulty of implementing the squaring device and the difficulty of keeping the arm filters nearly matched. This design decision will depend on the parameters and requirements of the particular receiving system, and cannot be generalized here.

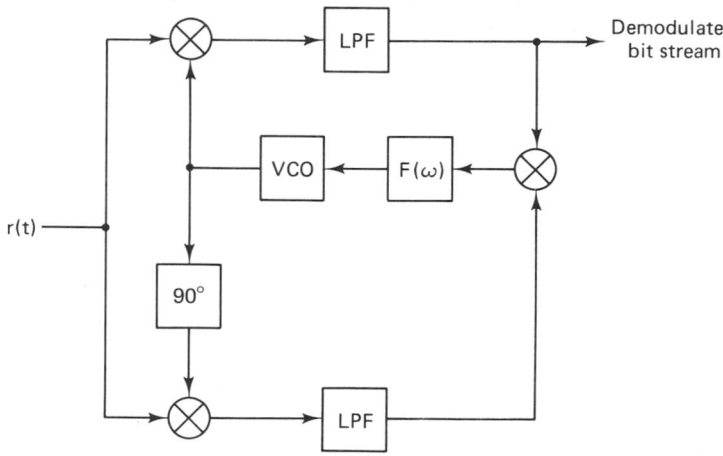

Figure 8.6 Costas loop.

8.2.1.6 High-Order Suppressed-Carrier Loops

Binary phase shift keying is not the only type of suppressed-carrier modulation. In fact, assuming that all signals are equally likely a priori, any modulation scheme whose average amplitude, averaged over the signal set, is zero will have no average energy in the transmitted carrier. Perhaps the most common nonbinary suppressed-carrier modulation is quadrature phase shift keying, or QPSK (4-ary PSK). If a QPSK signal is squared, the result "looks like" a BPSK signal. Thus for equally likely QPSK signals, the carrier is still suppressed. However, squaring the signal a second time, equivalent to taking the original signal to the fourth power, can be seen to produce a term with a carrier component, at four times the transmitted carrier's frequency. As in the binary case, operating on the incoming signal with a power-law device produces cross products among the noise terms and signal terms, and introduces the equivalent of a "squaring loss." Under the assumption that the noise bandwidth will pass the signal undistorted, the loss for fourth-power loops is upper bounded by [2]

$$S_L = 1 + \frac{9}{\rho_i} + \frac{6}{\rho_i^2} + \frac{3}{2\rho_i^3} \qquad (8.45)$$

As was the case with the squaring loop, for sufficiently high input signal-to-noise ratios, ρ_i, Equation (8.45) indicates that the additional loss terms vanish, and the loop performance approaches that of the basic loop. As was also the case for the squaring loop, there are Costas loop designs equivalent to fourth-order loops [2, 11, 12] that may exhibit hardware implementation advantages. Their theoretical performance, however, is the same as that of the straightforward fourth-power design.

Example 8.5   Squaring Loss Bounds

Compare the upper bounds on squaring loss, S_L, given by Equations (8.42) and (8.45) for second- and fourth-power loops, respectively, for an input loop signal-to-noise ratio, ρ_i, of 10 dB.

Solution

A 10 dB signal-to-noise ratio is also 10 in terms of its power ratio. Therefore, from Equations (8.42)) to (8.44), for the squaring loop,

$$S_L = 1 + \frac{1}{2\rho_i} = 1.05 = 0.2 \text{ dB}$$

From Equation (8.45), for the fourth-power loop

$$S_L = 1 + 0.9 + 0.06 + 0.0015 = 1.9615 = 2.9 \text{ dB}$$

Thus, while an input signal-to-noise ratio of 10 dB is adequate to keep losses small for the squaring loop, the same signal-to-noise ratio may allow significant losses for the fourth-power loop.

8.2.1.7 Acquisition

In most of the discussion thus far, the assumption has been that the PLL is in lock. This was the justification for assuming that the phase error, $|\theta - \hat{\theta}|$, was small. At one time or another, however, every loop must acquire lock; that is, it must be brought into lock. Acquisition can be accomplished with the aid of external circuits or signals (aided acquisition) or in some cases by an unaided PLL (self-acquisition) [2].

Acqusition is an inherently nonlinear operation, and therefore is difficult to analyze in general. However, some intuition may be obtained by considering a noise-free first-order loop. Such a loop is shown schematically in Figure 8.3, where $n'(t) = 0$ (noise-free) and $F(\omega) = 1$ (first-order). Denote the input phase as

$$\theta(t) = \omega_i t$$

and the output phase as

$$\hat{\theta}(t) = \omega_0 t + \int_0^t K_0 \sin e(\tau) \, d\tau + \hat{\theta}(0) \tag{8.46}$$

where ω_i and ω_0 are the radian frequencies of the input and output signals, respectively. Thus the phase error is given by

$$e(t) = \theta(t) - \hat{\theta}(t)$$

$$= (\omega_i - \omega_0)t - K_0 \int_0^t \sin e(\tau) \, d\tau - \hat{\theta}(0) \tag{8.47}$$

Differentiating both sides and letting $\Delta \omega = \omega_i - \omega_0$ provides

$$\frac{de}{dt} = \Delta \omega - K_0 \sin e \tag{8.48}$$

where the time dependence of the function $e(t)$ has been suppressed to ease notation. This differential equation describes the behavior of the first-order noise-free PLL. The loop being in lock requires that

$$\frac{de}{dt} = 0 \tag{8.49}$$

Equation (8.49) is a necessary but not a sufficient condition for phase lock. This can be verified by observing the phase plane diagram of Figure 8.7. This figure is obtained by dividing both sides of Equation (8.48) by the gain term K_0, and plotting the results. First observe point a. If the phase error is displaced a little to the left or right of point a, the sign of the derivative term is such that the phase

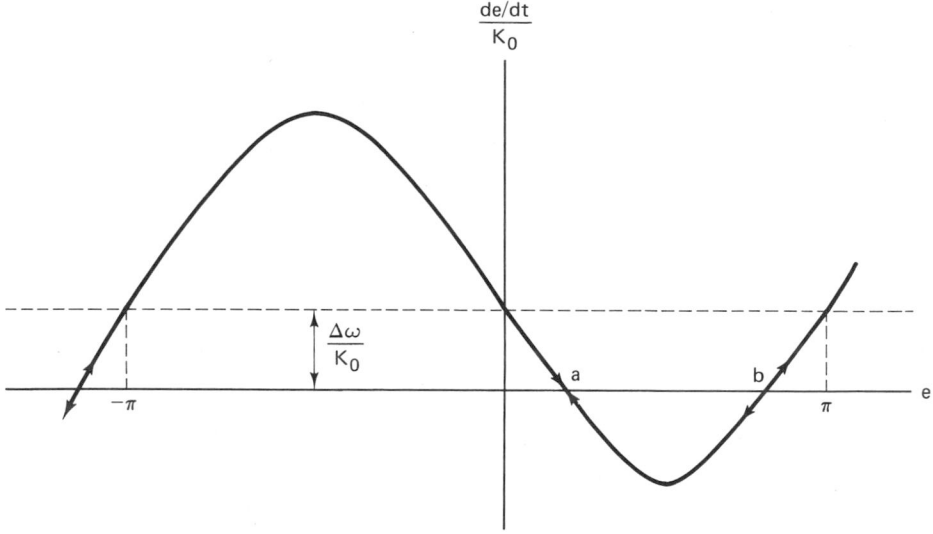

Figure 8.7 Phase-plane plot of first-order loop.

error e, will be driven back toward a. Thus point a is a stable point of the system, a point where phase lock can be obtained and will be maintained. For the case of point b, however, if the phase error is exactly at b, Equation (8.49) will be satisfied, but if there is any slight offset from b, the sign of the derivative term will be such that the error will be driven away from b. Thus b is a point of marginal stability for the loop, a point where Equation (8.49) is satisfied, but not a stable lock point.

The amount of time required for a loop to come into lock can be a very important system design consideration. By observing Equation (8.48) we can see that the requirement of Equation (8.49) for phase lock cannot be met unless

$$\frac{|\Delta\omega|}{|K_0|} \leq 1 \qquad (8.50)$$

This is because sinusoidal functions have a maximum amplitude of unity. This range of frequency difference, $-K_0 < \Delta\omega < K_0$, is sometimes called the lock-in range of the loop. Assuming that Equation (8.50) holds, Gardner [2] gives a rule of thumb of $3/K_0$ seconds for the time required for loop acquisition. Actual values can be obtained from Equation (8.47) for well-defined sets of initial conditions. It can be seen from the phase plot of Figure 8.7 that the time required will vary widely as a function of the initial phase error. For phase errors very close to point b, the driving force $(de/dt)/K_0$ will be very small. Thus, for this worst-case phase error, the error could "linger" in the vicinity of b for a long time. This phenomenon is called *terminal loop hang-up* [13] and can be a serious problem for system designs that depend on self-acquisition.

Perhaps the most important operational difference between first-order and higher-order loops is the higher-order loop's ability to "pull in" from frequency differences that are larger than the lock-in range. A first-order loop with a frequency error larger than the lock-in range will drift toward lock but never quite lock in (why?). Second- and higher-order loops can pull in and achieve phase lock because of their more complicated phase-plane characteristics. Interested readers should consult Viterbi [5] and other texts on PLLs for more details [2, 6, 14–16].

The study of self-acquisition for phase-locked loops is mostly of academic interest. Gardner [2] states that loops using self-acquisition can be guaranteed to acquire in reasonable time only under very benign circumstances. This, unfortunately, is rarely the case in practice.

Acquisition aiding drives the loop through the region of phase space expected to contain the lock-in region by means of some external driving signal. This is the most common means of achieving acquisition. Aiding can be implemented by simply applying a voltage ramp to the input of the VCO. This driving signal will cause the VCO output frequency to vary linearly with time. As was shown earlier [Equation (8.17)], loops with loop filters that do not contain $j\omega$ as a factor of their transfer function's denominator cannot track a frequency ramp with finite phase error. Therefore, if frequency sweeping is to be employed with a first-order loop, or a second-order loop without this transfer function characteristic, the rate of frequency sweep must be slow enough so that when the loop achieves lock, the presence of phase lock can be detected and the sweeping signal removed before it drives the loop back out of lock. With loops that contain $j\omega$ as a factor of $D(\omega)$, it may not be necessary to remove the sweeping signal at all, because, at least in theory, the loop will be able to track out the frequency ramp. In any case, the sweep rate must not be too large, or the loop will be driven through the lock point so fast that it will fail to acquire. For a second-order loop with loop transfer function [see Equation (8.6)]

$$H(\omega) = \frac{1}{-(j\omega/\omega_n)^2 + 2\zeta(j\omega/\omega_n) + 1} \tag{8.51}$$

Gardner [2] indicates that the maximum sweep rate, $\Delta\dot\omega$, must be in the vicinity of

$$\Delta\dot\omega \simeq \tfrac{1}{2}\omega_n^2(1 - 2\sigma_{\hat\theta}) \tag{8.52}$$

where $\sigma_{\hat\theta}$ is as defined in Equation (8.31), and ω_n, implicitly defined in Equation (8.51), is called the *natural frequency* of a second-order PLL and is related to the loop bandwidth, B_L, and loop damping factor, ζ [6], by

$$\omega_n = \frac{8\zeta}{4\zeta^2 + 1} B_L$$

Blanchard [14] gives more detailed results for aided phase acquisition.

8.2.1.8 Phase Tracking Errors and Link Performance

If a loop is unable to track out all phase errors, the received symbol error probability will be degraded relative to what is theoretically achievable. The analysis required to determine the amount of the degradation is very involved, but for most of the standard coherent signaling systems, curves are available [11, 12, 17]. Figure 8.8 is an example of such a performance curve for a residual carrier-phase tracking loop operating on a signal with BPSK modulation in additive Gaussian noise. It can be seen that for signal-to-noise ratios of moderate value, small phase errors produce very little degradation. It is only when the standard deviation of phase error exceeds 0.3 that the degradations become significant. This means that the inherent degradation in performance caused by a well-designed loop operating in benign conditions can generally be ignored. The curve also indicates that if conditions are such that the phase variance is large, increasing the data signal-to-Gaussian-noise ratio may not be effective in reducing the detected error probability. It should be noted that the presence of an irreducible error in these situations is a characteristic of residual carrier designs with constant loop signal-to-noise ratios, ρ_i. Suppressed carrier tracking loops tend not to have irreducible errors, because an increase in the data signal-to-noise ratio will increase the signal-to-noise ratio of the suppressed carrier tracking loop, reducing the tracking error.

Example 8.6 PLL Signal-to-Noise Ratio

Develop an integral expression for the effect on link bit error probability of slowly varying phase tracking errors for a residual carrier BPSK link. Compare the effect

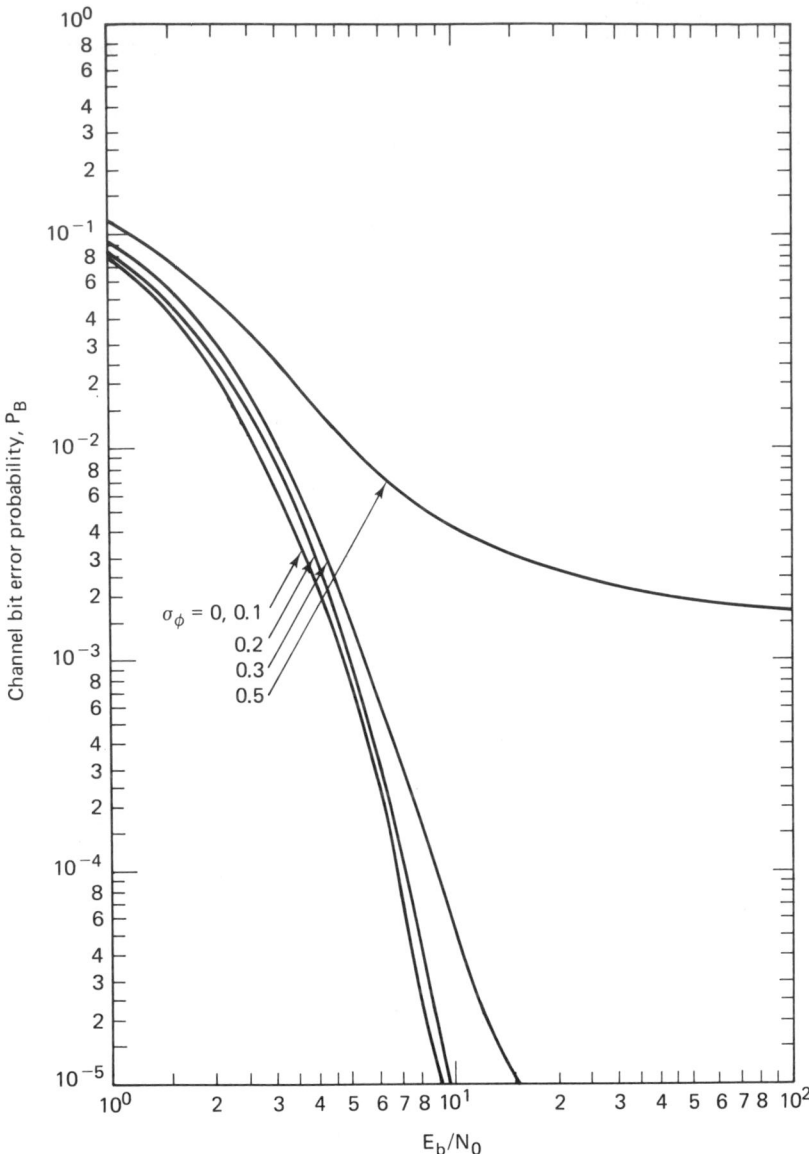

Figure 8.8 Channel bit error probability versus E_b/N_0 for BPSK with imperfect carrier synchronization. (Reprinted with permission from J. J. Stiffler, *Theory of Synchronous Communications*, Prentice-Hall, Inc., Englewood Cliffs, N.J., 1971, Fig. 9.1, p. 270.)

of a normalized loop signal-to-noise ratio ($\rho = 1/\sigma_\theta^2$) of 20 dB with one of 10 dB on error performance at a desired bit error probability of 10^{-5} using Figure 8.8.

Solution

From Chapter 3 we have that the theoretically possible bit error probability for a BPSK link in additive white Gaussian noise of single-sided spectral density N_0

watts/hertz is given by

$$P_B = Q\left(\sqrt{\frac{2E_b}{N_0}}\right)$$

where E_b is the received energy per bit time. From the derivation of this expression for error probability, it can be shown that if there is a slowly varying (with respect to the data rate) phase tracking error of β radians, the resulting probability of error will be given by

$$P_B(\beta) = Q\left(\sqrt{\frac{2E_b \cos \beta}{N_0}}\right)$$

Now if the phase error β is the result of tracking errors caused by system noise, β will be described stochastically by some probability density function $p(\beta)$. Then the expected bit error probability is given by

$$P_B = \int_0^{2\pi} P_B(\beta)p(\beta)\,d\beta$$

For the special case of a first-order loop, the probability density function is given by Equation (8.36). Then the final expression for expected bit error probability is given by

$$P_B = \int_0^{2\pi} Q\left(\sqrt{\frac{2E_b \cos \beta}{N_0}}\right) \frac{\exp(\rho \cos \beta)}{2\pi I_0(\rho)}\,d\beta$$

A loop signal-to-noise ratio (ρ_i) of 20 dB will correspond to a standard deviation of phase noise of $\sigma_{\hat{\theta}} = 0.1$ rad. From Figure 8.8, this small amount of phase noise produces no appreciable degradation in the bit error probability. A loop ρ_i of 10 dB, however, corresponds to a phase noise standard deviation of $\sigma_{\hat{\theta}} = 0.32$ rad. It can be seen from Figure 8.8 that for a bit error probability of 10^{-5}, this phase noise standard deviation will require a data SNR of somewhat more than 11 (10.4 dB) rather than a data SNR of 9.1 (9.6 dB) for perfect phase tracking. Thus this loop signal-to-noise will cause an error performance degradation of somewhat more than 0.8 dB at an error probability of 10^{-5}. It should be noted that for loop SNRs less than about 10 dB, the degradation in performance increases very rapidly. Thus 10 dB is something of a threshold for reasonable system performance for residual carrier designs. Suppressed carrier designs, having no problem with irreducible error, may do better.

8.2.2 Symbol Synchronization

All digital receivers need to have their demodulators synchronized to the incoming digital symbol transitions in order to achieve optimum demodulation. In the discussion that follows, we will consider several of the basic types of designs of symbol or data synchronizers. The discussion will center on a random binary baseband signal, for ease of terminology and notation, but it is expected that the extension to nonbinary baseband signals will be apparent.

The symbol synchronizers that will be considered here can be classified into two basic groups. The first group consists of the *open-loop synchronizers*. These circuits recover a replica of the transmitter data clock output directly from operations on the incoming data stream. The second group comprises the *closed-loop synchronizers*. Closed-loop data synchronizers attempt to lock a local data clock to the incoming signal by use of comparative measurements on the local and incoming signals. Closed-loop methods tend to be more accurate, but are much more costly and complex.

8.2.2.1 Open-Loop Symbol Synchronizers

Open-loop symbol synchronizers are also occasionally called *nonlinear filter synchronizers* [17], a very descriptive title. This class of synchronizers generate a frequency component at the symbol rate by operating on the incoming baseband sequence with a combination of filtering and a nonlinear device. The operation is analogous to carrier recovery in a suppressed carrier tracking loop. In the present case, the desired frequency component, at the data symbol rate, is isolated with a bandpass filter, and "shaped" with a high-gain saturating amplifier. The shaping recovers the square-wave appearance of the data clock signal.

Three examples of open-loop bit synchronizers are shown in Figure 8.9. In the first example (Figure 8.9a), the incoming signal, $s(t)$, is filtered with a matched filter. The output of this filter will be the autocorrelation function of the input signal shape. For square-wave signaling, for example, the output will be the familiar isosceles-triangular waveshape. The sequence of bit autocorrelation wave-

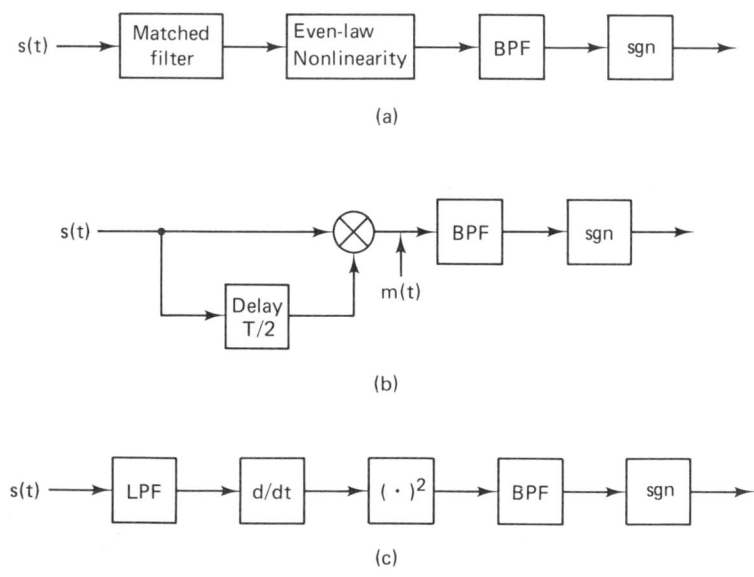

(a)

(b)

(c)

Figure 8.9 Three types of open-loop bit synchronizers.

shapes is then "rectified" by some type of memoryless even-law nonlinearity, a square-law device, for example. The resulting waveform will have positive amplitude peaks that correspond, to within a time delay, with the input symbol transitions. This sequence of processes is illustrated in Figure 8.10. Thus the output waveform from the even-law device will contain a Fourier component at the fundamental frequency of the data clock. This frequency component is isolated from its harmonics with a bandpass filter (BPF), and shaped with an ideal saturating amplifier, with transfer function sgn x, defined as follows:

$$\text{sgn } x = \begin{cases} 1 & \text{for } x > 0 \\ -1 & \text{otherwise} \end{cases} \tag{8.53}$$

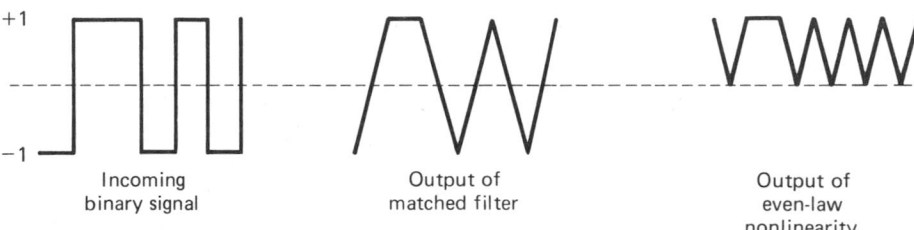

Figure 8.10 Open-loop bit synchronizer illustration.

The second example in Figure 8.9 produces a Fourier component at the data clock frequency by means of a delay and multiply. The delay shown in Figure 8.9b is half a bit period, which is the best value because it provides the strongest Fourier component [17]. The waveform $m(t)$ will always be positive in the second half of every bit period, but will have a negative first half if there has been a state change in the incoming bit stream, $s(t)$. This produces a square-wave signal with spectral components at the data rate and all harmonics, as in Figure 8.9a. As before, the appropriate spectral component can be isolated with a BPF and shaped.

The final example (Figure 8.9c) amounts to an edge detector. The main operations are those of differentiation and rectification (by use of a square-law device). For a square-wave input, the differentiator will produce positive or negative spikes at all symbol transitions. When rectified, the resulting sequence of positive spikes will have a Fourier component at the data symbol rate. A potential problem with this particular scheme is that differentiators are typically very sensitive to wideband noise. This necessitates the low-pass filter (LPF) that precedes the differentiator in Figure 8.9c. The LPF, however, will also remove the high-frequency components of the data symbols, causing them to lose their original rectangular waveshape. This will cause the resulting differential signal to have some finite rise and fall time, rather than being a set of impulses.

Clearly, there will be some hardware delay associated with the signal processing steps illustrated in Figure 8.9. Wintz and Luecke [18] have shown that for a BPF that effectively averages K input symbols (bandwidth $\simeq 1/KT$), the magnitude of the fractional mean time error (delay) is approximated by

$$\frac{|\bar{\epsilon}|}{T} \simeq \frac{0.33}{\sqrt{KE_b/N_0}} \qquad \text{for } \frac{E_b}{N_0} > 5, \quad K \geq 18 \qquad (8.54)$$

where T is the bit period, E_b the detected energy per bit, and N_0 the single-sided received noise spectral density. Wintz and Luecke have also shown that at high signal-to-noise ratios the fractional standard deviation of the fractional timing error is given by

$$\frac{\sigma_\epsilon}{T} \simeq \frac{0.411}{\sqrt{KE_b/N_0}} \qquad \text{for } \frac{E_b}{N_0} > 1 \qquad (8.55)$$

Thus, for a given BPF, when the received signal-to-noise ratio is sufficiently large, all of the techniques shown in Figure 8.9 will provide accurate bit timing.

8.2.2.2 Closed-Loop Symbol Synchronizers

The primary disadvantage of open-loop symbol synchronization methods is that there is an unavoidable non-zero-mean tracking error. This error can be made small for large signal-to-noise ratios, but since the synchronization signal waveform depends directly on the incoming signal, the error will never vanish.

Closed-loop symbol data synchronizers use comparative measurements on the incoming signal and a locally generated data clock signal to bring the locally generated signal into synchronism with the incoming data transitions. The procedure is essentially the same as that used for closed-loop carrier tracking.

Among the most popular of the closed-loop symbol synchronizers is the early/late-gate synchronizer. An example of such a synchronizer is shown schematically in Figure 8.11. The synchronizer operates by performing two separate integrations of the incoming signal energy over two different $(T - d)$ second portions of a symbol interval. The first integration (the early gate) begins integration at the loop's best estimate of the beginning of a symbol period (the nominal time zero), and integrates for the next $(T - d)$ seconds. The second integral (the late gate) delays the start of its integration for d seconds, and then integrates to the end of the symbol period (the nominal time T). The difference in the absolute values of the outputs of these two integrations, y_1 and y_2, is a measure of the receiver's symbol timing error, and can be fed back to the loop's timing reference to correct loop timing.

The action of the early/late-gate synchronizer can be understood by referring to Figure 8.12. In the case of perfect synchronization, Figure 8.12a shows that both gates are entirely within a single symbol interval. In this case, both integrators will accumulate the same amount of signal, and their difference (the error signal, e, in Figure 8.11) is zero. Thus, when the device is synchronized, it is stable—there is no tendency to drive itself away from synchronization. The case shown in Figure 8.12b is for a receiver whose data clock is early relative to the incoming data. In this case the first portion of the early gate falls in the previous bit interval, while the late gate is still entirely inside the current symbol. The late-gate integrator will accumulate signal over its entire $(T - d)$ integration interval, as in the

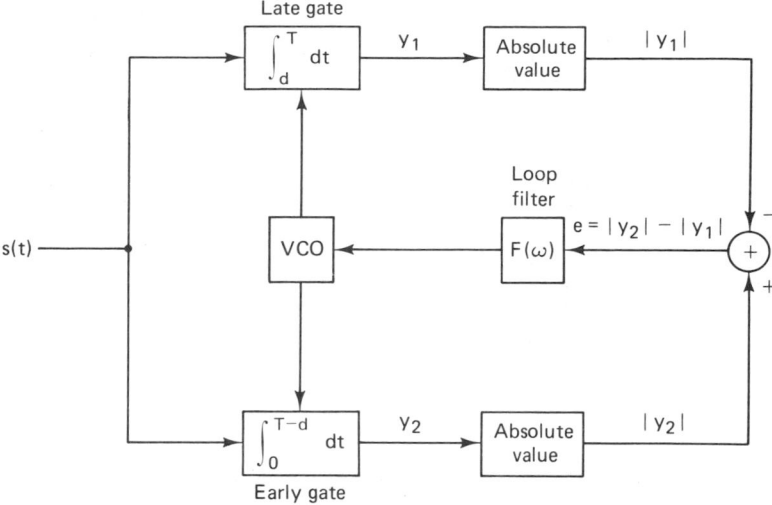

Figure 8.11 Early/late-gate data synchronizer.

case in Figure 8.12a, but the early-gate integrator will end up with energy accumulated only over $[(T - d) - 2\Delta)]$, where Δ is the portion of the early-gate interval falling in the previous bit interval. Thus, for this case the error signal will be $e = -2\Delta$, which will lower the input voltage to the VCO in Figure 8.11. This will reduce the VCO output frequency and retard the receiver's timing to bring it back toward the incoming signal's bit timing. Using Figure 8.12 as a guide, it can be seen that if the receiver's timing had been late, the amounts of energy integrated in the early gate and late gate would be reversed, as would the sign of the error signal. Thus late receiver timing produces an increase in the VCO input voltage, increasing the output frequency and advancing the receiver's timing toward that of the incoming signal.

The example illustrated in Figure 8.12 tacitly assumes that there will be data state changes before and after the channel symbol of interest. If there are no transitions, it can be seen that the early gate and late gate will have the same integrated energy. Thus there will be no error signal generated for cases where there is no data state change. This is a practical implementation consideration in the use of all symbol synchronizers. Reconsider Figure 8.11. It is not possible to build two integrators that are exactly the same. Thus the signals from the two arms of the early/late-gate loop will contain an offset with respect to each other, even when they should be identical. This offset will be small for well-designed integrators but will cause the loop to drift out of synchronism if there are long sequences of identical data symbols. There are two common responses to this problem. The first, and perhaps most obvious, is to format the data in a manner which ensures that there will be no transitionless intervals that are long enough to allow the loop to break lock. The second response is to modify the loop design so that it contains a single integrator. An example of this type of modified design

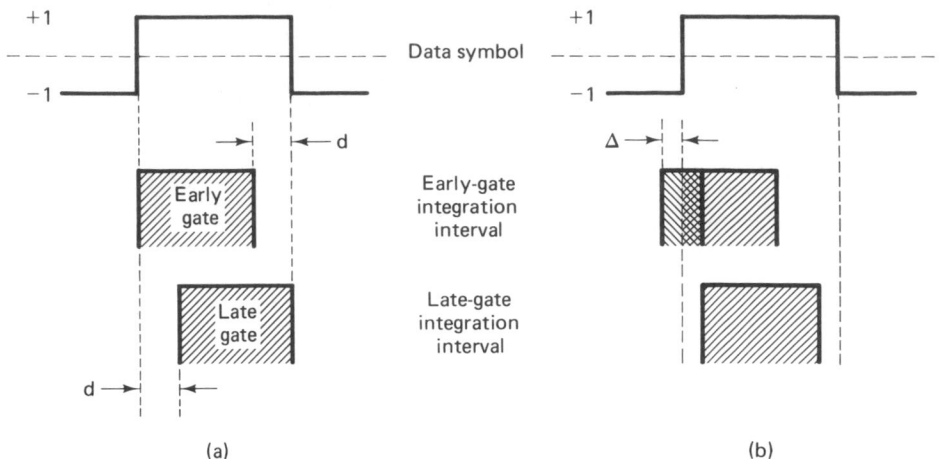

Figure 8.12 (a) Correct receiver timing. (b) Early receiver timing.

is the tau-dither loop, considered in conjunction with the synchronization of spread-spectrum systems in Chapter 10.

Another loop design issue is the integration interval of the two gates. The example illustrated in Figure 8.12 shows the gates to occupy about three-fourths of a symbol period. Actually, this interval can vary from half a symbol interval to nearly a whole symbol interval (why not less than half?). The trade-off is between the amount of integrated noise and interference in a gate versus the amount of signal. As was true with the nonlinear model of phase-locked loops, loops of this type are difficult to analyze; the determination of the best design may require computer simulation. This will be especially true for overlapping gates, as in Figure 8.12, because the noise samples in the two gates will be correlated. Gardner [2] has shown that for a normalized incoming signal of one volt, additive white Gaussian noise, random data (the probability of a transition is $\frac{1}{2}$), and early and late gates that are half a bit interval in duration, for large loop signal-to-noise ratios the fractional timing jitter is approximated by

$$\frac{\sigma_e^2}{T^2} = 2N_0 B_L \tag{8.56}$$

where N_0 is the (normalized) noise power spectral density, T the symbol interval, and B_L the loop bandwidth.

8.2.2.3 Symbol Synchronization Errors and Symbol Error Performance

The effect of symbol synchronization error on bit error probability for a BPSK signal in additive white Gaussian noise is shown in Figure 8.13. It can be seen from the figure that the degradation is less than about 1 dB in signal-to-noise ratio for a fractional timing jitter of less than 5%. Comparing symbol timing error effects to the effect of phase noise (see Figure 8.8), it is seen that the symbol

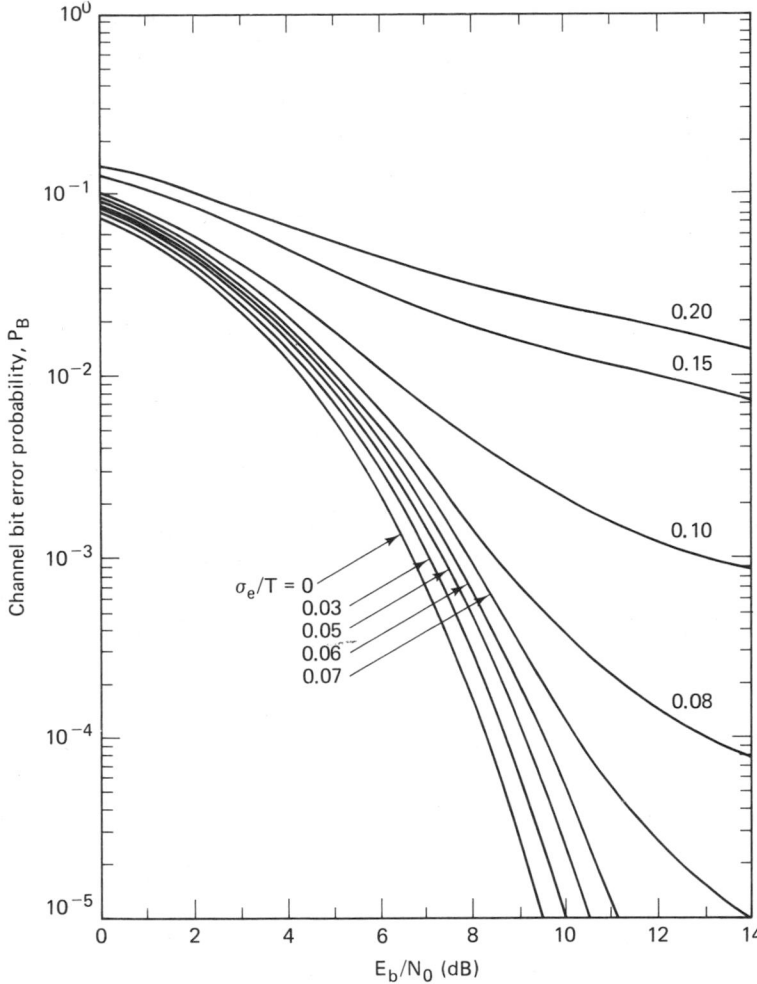

Figure 8.13 Channel bit error probability versus E_b/N_0 with the standard deviation of the symbol sync error σ_e as a parameter. (Reprinted from W. C. Lindsey and M. K. Simon, *Telecommunication Systems Engineering*, Prentice-Hall, Inc., Englewood Cliffs, N.J., 1973, courtesy of W. C. Lindsey and Marvin K. Simon.)

synchronization error, taken as a fraction of the symbol interval, does not affect system performance as strongly as does phase noise taken as a fraction of a cycle. In both cases, however, the degradation increases with increases in error.

Example 8.7   Effect of Timing Jitter

Through the use of Figure 8.13, determine the effect of a 10% symbol fractional timing jitter on a system required to maintain a 10^{-3} bit error probability.

Solution

It can be seen from Figure 8.13 that a 10^{-3}-bit error probability will require a SNR of about 6.7 dB in the absence of all timing jitter. The same figure indicates that for a fractional timing jitter of 10% ($\sigma_e/T = 0.1$) a SNR of about 12.9 dB is required. Thus the ability to accommodate this large timing jitter would require a 6.2-dB higher signal-to-noise ratio than that needed to maintain a 10^{-3}-bit error probability without jitter. This illustrates a use to which Figure 8.13 can be put; however, this example is clearly extreme. No communication system would be designed with over four times the nominally required power level in order to accommodate a large symbol synchronization error. Some other answer would be found, such as redesigning the system filtering to increase the value of K in Equation (8.55), which will reduce the symbol timing jitter.

8.2.3 Frame Synchronization

Almost all digital data streams have some sort of frame structure. This is to say that the data stream is organized into uniformly sized groups of bits. If the data stream is digitized TV, example groups might correspond to horizontal raster scans, which would be further organized in terms of vertical raster scans. Computer data are typically organized into words in one of the several standard digital formats (ASCII, for example) and these, in turn, are organized into card images or files. Any system that uses block error control coding must be organized around the codeword length. Even digital speech is typically transmitted in packets of bits with a constant number of bits in a packet.

For a receiver to make sense of the incoming data stream, the receiver needs to be synchronized with the data stream's frame structure. Frame synchronization is usually accomplished with the aid of some special signaling procedure from the transmitter. This procedure may be very simple, or fairly involved, depending on the environment in which the system is required to operate.

Probably the simplest frame synchronization aid is the *frame marker,* illustrated in Figure 8.14. The frame marker is a single bit, or a short pattern of bits that the transmitter injects periodically into the data stream. The receiver must know the pattern and the injection interval. The receiver, having achieved data synchronization, correlates the known pattern with the incoming data stream at the known injection interval. If the receiver is not in synchronization with the framing pattern, the accumulated correlation will be low. When the receiver comes into frame synch, however, the correlation should be nearly perfect, blemished only by an occasional detection error.

The advantage of the frame marker is its simplicity. Even a single bit can suffice as a frame marker if a sufficient number of correlations are accumulated before deciding whether or not the system has achieved synchronization. The major drawback is that the sufficient number may be very large, and thus the expected time required to acquire synchronization would be long. Therefore, frame markers are most useful in systems that transmit data continuously, like some computer links, and would be inappropriate for systems that transmit in isolated bursts or systems that require rapid frame acquisition.

An approach for systems with inconsistent or bursty transmissions, or systems with rapid acquisition requirements, is a *synchronization codeword.* A synchronization codeword would typically be sent as part of a message header. The

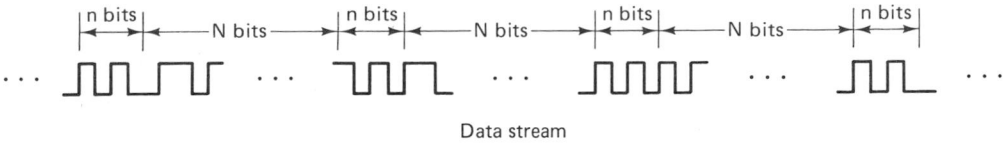

Data stream

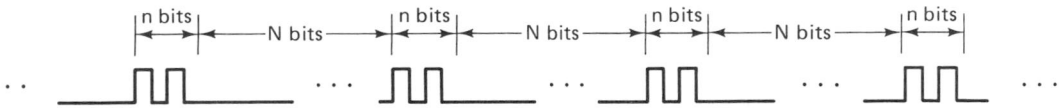

Receiver generated frame marker replica

Figure 8.14 Frame marker illustration.

receiver must know the codeword and be constantly searching for it in the data stream, possibly with a matched filter correlator. Detection of the codeword would indicate a known position (typically the beginning) in the data frame. The advantage of this system is that frame acquisition can be essentially immediate. The only delay would be that required to process the incoming codeword. The disadvantage is that the codeword must be relatively long, relative to the frame marker, to keep the probability of false detections low. The complexity of the correlation operation is proportional to the length of the sequence, so the correlator must be relatively complicated.

A good synchronization codeword is one that has the property that the absolute value of its "correlation sidelobes" is small. A correlation sidelobe is the value of the correlation of a codeword with a time-shifted version of itself. Thus the correlation sidelobe value, C_k, for a k-symbol shift of an N-bit code sequence, $\{X_j\}$, is given by

$$C_k = \sum_{j=1}^{N-k} X_j X_{j+k} \tag{8.57}$$

where X_i ($1 \leq i \leq N$) is an individual code symbol taking values ± 1, and the adjacent data symbols (associated with index values $i > N$) are assumed to be zero. An example of correlation sidelobe computation is shown in Figure 8.15. The 5-bit sequence in the example is seen to have good correlation properties, in that the largest sidelobe is one-fifth of the main lobe, C_0. Sequences like the example in Figure 8.15, with the property that their largest sidelobe has a magnitude of unity, are known as Barker sequences or Barker words [19]. There is no known constructive method for finding Barker words, and only 10 unique words are known, the longest of which has 13 symbols. The known unique Barker words are given in Table 8.1. Some thought should make it clear that a completely exhaustive list of known Barker sequences would include those sequences produced by inverting the sign of the symbols and those produced by reversing the time ordering of the symbols in the sequences of Table 8.1.

The sidelobe correlation properties of Barker codes are based on the as-

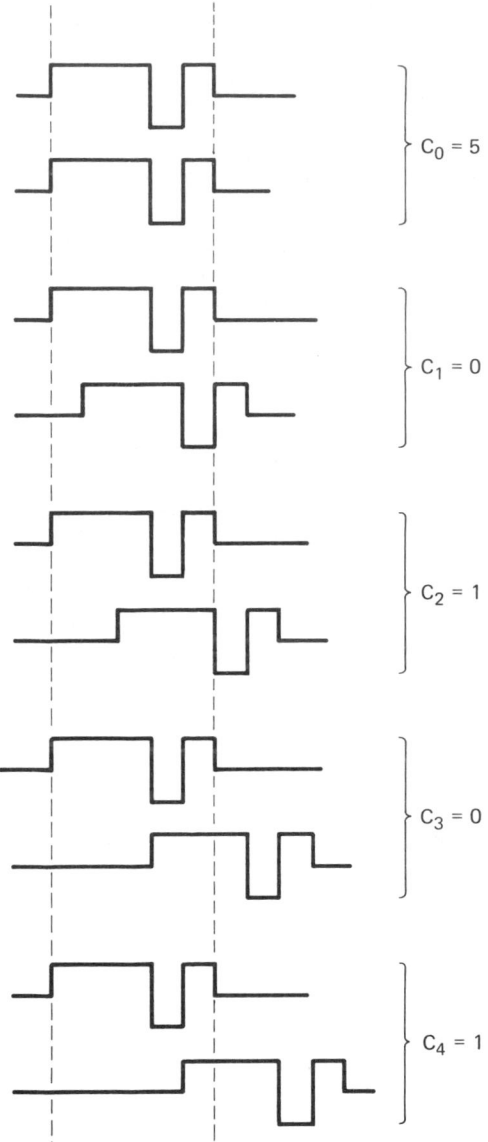

Figure 8.15 Correlation sidelobe example.

sumption that the adjacent symbols have zero value. This is an approximation to the effect of equally likely random binary data adjacent to the Barker word, taking the values ±1. Unfortunately, the Barker sequences are too short for this approximation to provide the best codeword in random binary data in all cases. Willard [20] found the best sequences, in terms of the minimum probability of false synchronization, by the use of computer simulation, for random adjacent symbols for the Barker word lengths. The Willard sequences are shown in Table 8.2.

TABLE 8.1 Barker Synchronization
Codewords

N	Barker sequence
1	+
2	+ + or + −
3	+ + −
4	+ + + − or + + − +
5	+ + + − +
7	+ + + − − + −
11	+ + + − − − + − − + −
13	+ + + + + − − + + − + − +

Two probabilities characterize the performance of a system using a synchronization word. These are the probability of a missed detection and the probability of false alarm. Clearly, the system designer would wish both probabilities to be as small as possible. These are conflicting desires. In order to decrease the probability of a miss, the system design may allow less than perfect correlation of an incoming synchronization word. That is, a word may be accepted even if it contains a small number of errors. This, however, enlarges the number of symbol patterns that will be accepted and thereby increases the probability of a false alarm. The probability of a miss for an N-bit word where k or fewer errors are accepted is given by

$$P_m = \sum_{j=k+1}^{N} \binom{N}{j} p^j (1 - p)^{N-j} \tag{8.58}$$

where p is the probability of a detector bit error. The probability of a false alarm generated by N bits of random data is given by

$$P_{FA} = \sum_{j=0}^{k} \frac{\binom{N}{j}}{2^N} \tag{8.59}$$

It can be seen that for small p, P_m will decrease roughly exponentially with in-

TABLE 8.2 Willard Synchronization
Codewords

N	Willard sequences
1	+
2	+ −
3	+ + −
4	+ + − −
5	+ + − + −
7	+ + + − + − −
11	+ + + − + + − + − − −
13	+ + + + + − − + − + − − −

creasing k. Unfortunately, P_{FA} increases roughly exponentially with increasing k. To obtain acceptable values of both P_m and P_{FA} for a given value of p, the system designer often needs values of N larger than those provided by the Barker and Willard sequences. Fortunately, there is a fairly large and growing body of literature dealing with long sequences. Most of these sequences have been discovered through exhaustive computer searches. Spilker [17] lists sequences of up to $N = 24$ found by Newman and Hofman [21] and mentions that their original paper has sequences to $N = 100$. Wu [22] provides a list of Maury-Styles sequences to length $N = 30$, a list of Linder sequences to length 40, and a fairly complete discussion of the topic of synchronization sequences, including constructive techniques for reasonable but nonoptimum sequences, and insight into the frame synchronization procedures of some operational satellite digital communication systems.

8.3 NETWORK SYNCHRONIZATION

For systems using coherent modulation techniques, one-direction communications such as broadcast channels, or single-link communications, such as most microwave links, land-line links, or fiber optics links, the synchronization architecture that makes the most sense is to make synchronization totally a receiver function. For communications systems using noncoherent modulation techniques, or that involve many users accessing a central communication node, such as many satellite communication systems, it often makes sense for synchronization to be mostly or entirely a terminal function. This means that the terminal transmitter parameters are modified to achieve synchronization, rather than modifying the central node's receiver parameters. This must be the approach if the system uses time-division multiple access (TDMA). In TDMA each user is allotted a segment of time in which to transmit its information. The terminal transmitter must be synchronized with the system in order for its transmitted burst of data to arrive at the central node at the time when the node is prepared to receive the data. Synchronization of the terminal transmitter also makes sense with systems that combine signal processing at the central node with frequency-division multiple access (FDMA). If the terminals precorrect their transmission to be synchronized with the central node, the node can use a fixed set of channel filters and a single timing reference for the processing of all channels. Otherwise, the node would require a separate time and frequency acquisition and tracking capability for each incoming channel, and would need to deal with the possibility of varying amounts of adjacent channel interference. It seems clear that terminal transmitter synchronization is often the cleaner, more reasonable system approach to synchronizing a network.

Transmitter synchronization procedures may be classified as being either open loop or closed loop. Open-loop techniques do not rely on any measurement of the arriving signal parameters at the central node. The terminal precorrects its transmission based on stored knowledge of link parameters that have been provided by some external authority but may possibly be modified by observations of a return signal from the central node. Open-loop techniques rely on link parameters being accurately known and predictable. They work best when link geometry is nominally fixed, and the links themselves operate continually for relatively long periods, once established. They tend to be difficult to use efficiently when the link geometry is not static or when the terminals access the system sporadically. The main advantages of the open-loop methods are that acquisition is fast, the procedure can work without a return link, and the amount of real-time computation that is required is small. The disadvantages of open-loop methods are that they require the existence of the external authority that provides knowledge of the required link parameters and that they are relatively inflexible. The lack of any direct real-time measure of system characteristics means that the system cannot adjust quickly to any unplanned change in conditions.

Closed-loop techniques, on the other hand, require little in the way of a priori knowledge of link parameters. Knowledge would be useful in reducing the time required for acquisition, but need not be precise as is required by open-loop methods. Closed-loop methods involve measurements of the synchronization accuracy of the incoming transmissions from the terminal upon their arrival at the central node, and the return of the results of these measurements to the terminal via a return path. Thus closed-loop methods require a return path that provides a response to the terminal's transmission, the ability in the terminal to recognize the response for what it is, and the ability in the terminal to modify the transmitter characteristics appropriately, based on the response. This amounts to a requirement for a relatively large amount of real-time processing in the terminal, and two-way links between every terminal and the central node. The disadvantages of closed-loop methods are that they require a relatively large amount of real-time processing, require two-way links to every terminal, and that acquisition can take a relatively long time. The advantages are that no external source of knowledge is required for the system to work, and the responses on the return link allow the system to adapt easily and quickly to changing geometries and link conditions.

8.3.1 Open-Loop Transmitter Synchronization

Open-loop systems can be further subdivided into systems that employ information gained by observing a return link, and those that do not. Those that do not are the simplest of all, in terms of real-time processing requirements, but communication performance for these simple terminals is clearly very dependent on stable link characteristics.

All transmitter synchronization schemes attempt to precorrect the timing and transmission frequency of the signal in such a manner that the signal will arrive at a receiver with the expected frequency and at the expected time. Thus, to precorrect time, a transmitter would divide the distance between itself and the receiver by the speed of light to get the transmission transit time, and then shift the message transmission timing that much ahead. By transmitting the signal early, it will arrive at the receiver at the appropriate time. The time of arrival at the node, T_A, is given by

$$T_A = T_t + \frac{d}{c} \tag{8.60}$$

where T_t is the actual transmission start time, d the transmit distance, and c the speed of light. Similarly, to precorrect the transmission frequency, the transmitter must allow for the Doppler shift caused by relative motion between the transmitter and the intended receiver. To be received correctly, the required *transmission radian frequency* is

$$\omega \simeq \left(1 - \frac{v}{c}\right) \omega_0 \tag{8.61}$$

where c is the speed of light, v the relative velocity (positive for decreasing transmission distance), and ω_0 the nominal transmission radian frequency.

Unfortunately, in practice neither the time nor the frequency precorrection can be done exactly. Even satellites in nominally geostationary orbits move slightly with respect to a point on the earth, and the behavior of the time and frequency references in the terminal and the central node are never entirely predictable. Thus there will always be some time and frequency precorrection error. The time error may be expressed as

$$T_e = \frac{r_e}{c} + \Delta t \tag{8.62}$$

where r_e is the error in the range estimate and Δt is the difference between the time reference at the terminal and the reference at the receiver. The frequency error may be expressed as

$$\omega_e = \frac{v_e \omega_0}{c} + \Delta \omega \tag{8.63}$$

where v_e is the error in the measured or predicted relative velocity of the transmitter and receiver—the Doppler error, and $\Delta \omega$ is the frequency difference between the transmitter and the receiver frequency references. There are many other sources of time and frequency error in addition to those mentioned here, but they are typically much less important. Spilker [17] gives a reasonably complete accounting of time and frequency error sources for satellite systems.

The error terms Δt and $\Delta \omega$ are typically due to random fluctuations in frequency references. The time reference for a transmitter or receiver is generally obtained by counting cycles of the frequency reference, so errors in the accuracy of the time and frequency references are related. The fluctuation in a frequency reference are very difficult to characterize statistically, although the power spectral density of the fluctuations is approximated by a sequence of power-law segments [12]. Frequency references are often specified in terms of a maximum allowable fractional frequency change per day:

$$\delta = \frac{\Delta \omega}{\omega_0} \qquad \text{hertz/hertz/day} \tag{8.64}$$

Typical values for δ range from 10^{-5} to 10^{-6} for inexpensive crystal oscillators, to 10^{-9} to 10^{-11} for high-quality crystal oscillators, to 10^{-12} for rubidium standards, to 10^{-13} for cesium standards. An effect of specifying system frequency references by the maximum fractional frequency is that if there is no intervention, the offset from the nominal frequency, ω_0, can grow linearly with time.

$$\Delta\omega(T) = \omega_0 \int_0^T \delta\, dt + \Delta\omega(0) = \omega_0\, \delta T + \Delta\omega(0) \qquad \text{hertz} \qquad (8.65)$$

For a cycle-counting time reference, however, the cumulative time offset is related to the cumulative phase error of the reference:

$$\Delta t(T) = \int_0^T \frac{\Delta\omega(t)}{\omega_0}\, dt + \Delta t(0)$$

$$= \int_0^T \delta t\, dt + \int_0^T \frac{\Delta\omega(0)}{\omega}\, dt + \Delta t(0)$$

$$= \frac{1}{2}\, \delta T^2 + \frac{\Delta\omega(0)\, T}{\omega_0} + \Delta t(0) \qquad (8.66)$$

Thus, without intervention, a time reference error can grow quadratically with time. For open-loop transmitter synchronization systems this quadratic growth in time error often sets limits on how often the exernal authority must intervene, either to update the terminal's knowledge of receiver timing, or to reset both the receiver's and the transmitter's time references to nominal. The quadratic error growth usually means that timing errors are more of an operational problem than are frequency errors, although this will depend on the system design.

If the transmitter does not have information from measurements on a return link, the time and frequency offsets as modeled by Equations (8.62) to (8.66) will allow a system designer the ability to determine the maximum interval between interventions, based on a probability-of-error criterion. Time and frequency reference recalibration is often a burdensome procedure, to be done as rarely as possible.

If a terminal has access to a return link from the central node and the ability to make comparative measurements between the local reference and incoming signal parameters, the interval between recalibrations can be made much longer. Large satellite control stations can measure and model the orbital parameters of nominally geostationary satellites to an accuracy of a few tens of feet in range and a few feet/second in velocity relative to the ground terminal. Thus, for the important special case of a synchronous satellite as the central node, the first terms on the right-hand side of Equations (8.62) and (8.63) are usually negligible. When this is true, the differences between the incoming signal parameters and those generated by the terminal's time and frequency references will approximate the error terms Δt and $\Delta\omega$. These error terms measured on the downlink can be used to compute appropriate corrections to the uplink transmissions. On the other hand, if the time and frequency references are known to be accurate but the link geometry is somewhat in question, perhaps because the terminal is mobile or the satellite is nongeostationary, the same sort of return link measurement could be used to resolve range or velocity uncertainties. These measures of range or relative velocity can then be used to precorrect uplink timing and frequency.

The case where a terminal is able to utilize measurements made on a return link signal is sometimes called quasi-closed-loop transmitter synchronization. The quasi-closed technique is clearly more adaptable to uncertainties in the communication system than is the purely open-loop system. The purely open-loop system requires complete a priori knowledge of all important link parameters in order to operate successfully. Unanticipated changes in the links cannot be tolerated. The quasi-closed-loop system, on the other hand, requires a priori knowledge of all but one of the important parameters in each of time and frequency, but the remaining term can be determined from observations of the return link. This adds complexity to the terminal, but also adds the ability to adapt to certain types of unplanned link changes. This degree of adaptability can greatly reduce the frequency of required system calibration.

8.3.2 Closed-Loop Transmitter Synchronization

Closed-loop transmitter synchronization involves the transmission of special synchronization signals that are used to determine the signal's time or frequency error relative to the desired timing or frequency when the signal arrives at the receiver. The results of this determination are then fed back to the transmitter on a return link. The determination of synchronization errors can be either implicit or explicit. If the central node has sufficient processing capacity, the central node may make an actual error measurement. Such a measurement might be the amount and direction of offset, or perhaps simply the direction alone. This information would be formatted and returned to the transmitter on a return link. If the central node has little processing capability, the special synchronization signal may simply be turned around and returned to the transmitter on the return link. In this case it becomes part of the transmitter's task to interpret the returned signal for itself. The design of a special synchronization signal that lends itself to easy unambiguous interpretation can be a challenge.

The relative advantages and disadvantages of the two types of closed-loop systems have to do with the location of the signal processing capability and the efficiency of channel usage. A major advantage of having the processing at the central node is that results of the error measurements that are transmitted on the return link can be a short digital sequence. This efficient use of the return link can be important if a single return link is time-division multiplexed between a large number of terminals. A second potential advantage is that the error-measuring capability in the central node can be shared by all terminals communicating through the node. This can amount to a large savings in system processing capability. The principal potential advantage in having the processing at the terminal is that the central node may not be easily accessible, and reliability considerations may dictate a simple design. This has typically been the case when the central node is a space satellite. With continuing improvements in satellite technology, simplicity requirements can be expected to be less dominant in the future than in the past. Another potential advantage to having the processing in the terminal is that the response can be quicker because there is little processing delay in the central node. This may be important if link parameters are changing very rapidly. The primary disadvantages are the inefficient use of the return channel and that the return signals may be difficult to interpret. This difficulty would arise when the central node is not just a simple repeater, but makes symbol decisions and transmits these decisions on the return link. This symbol decision capability can greatly improve the terminal-to-terminal error performance, but complicates the synchronization procedure. This is because the effects of a time or frequency offset are resident in the return signal indirectly—only as they have affected the symbol decisions. Consider the example of a BFSK transmission to a central node that makes noncoherent bit decisions. The decisions will be dependent on the detected signal energy in the mark and space detectors. If the transmitted signal is an alternating sequence of marks and spaces, the signal at the central node can be modeled as

$$r(t) = \begin{cases} \sin\left[(\omega_0 + \omega_s + \Delta\omega)t + \theta\right] & 0 \le t \le \Delta t \\ \sin\left[(\omega_0 + \Delta\omega)t + \theta\right] & \Delta t < t \le T \end{cases} \tag{8.67}$$

where T is the symbol interval, ω_0 one symbol frequency, $(\omega_0 + \omega_s)$ the other symbol frequency, $\Delta\omega$ the frequency error at the central node, Δt the signal arrival time error at the central node, and θ an arbitrary phase angle. Now if x and y represent the detector quadrature components,

$$x = \frac{1}{T} \int_0^T r(t) \cos \omega_0 t \, dt \tag{8.68}$$

$$y = \frac{1}{T} \int_0^T r(t) \sin \omega_0 t \, dt \tag{8.69}$$

the detected signal energy can be expressed as

$$z^2 = x^2 + y^2$$

$$= \left(\frac{\sin\left[(\omega_s + \Delta\omega)\,\Delta t/2\right]}{(\omega_s + \Delta\omega)T}\right)^2 + \left(\frac{\sin\left[\Delta\omega(T - \Delta t)/2\right]}{\Delta\omega T}\right)^2 \tag{8.70}$$

$$+ \frac{\cos(\Delta\omega\,\Delta t) + \cos\left[\Delta\omega\,T - (\omega_s + \Delta\omega)\,\Delta t\right] - \cos(\Delta\omega\,T) - \cos(\omega_s\,\Delta t)}{2\Delta\omega(\omega_s + \Delta\omega)T^2}$$

For the special case where the time error, Δt, is zero, Equation (8.70) simplifies to

$$z^2 = \left[\frac{\sin(\Delta\omega\,T/2)}{\Delta\omega\,T}\right]^2 \tag{8.71}$$

For the case where the frequency offset is zero,

$$z^2 = \left(\frac{T - \Delta t}{2T}\right)^2 + \left[\frac{\sin(\omega_s\,\Delta t/2)}{\omega_s T}\right]^2 \tag{8.72}$$

Sec. 8.3   Network Synchronization

469

The important thing to notice in Equations (8.70) to (8.72) is that any time error or frequency offset or combination of both will decrease the detected signal energy in the correct symbol detector and introduce signal energy into the incorrect signal detector. This will reduce the effective distance between signals in signal space and degrade error performance. A measurement of error performance, however, which is all that is available on the return link, gives no insight into whether the problem is a frequency offset, a time error, or a combination of both. Thus the transmission of standard signals is not likely to provide a useful response for synchronization.

A useful technique for determining the correct frequency precorrection for our example of BFSK signaling is to transmit a constant tone whose frequency is the average of the two symbol frequencies. Such a tone should produce a random binary sequence on the return link with equal numbers of marks and spaces. A frequency offset from the average would produce predominately marks or spaces. Finding the center frequency in this way allows accurate frequency precorrection of the signals. Once the correct frequency is found, the transmitter can transmit an alternating sequence of marks and spaces in order to discover correct timing. By varying the timing of the transmission through a range of half a symbol interval, the transmitter can look for the timing that provides the worst error performance. When the transmission arrival at the central node is displaced from correct timing by half a symbol interval, the two detectors will detect equal amounts of energy, and the binary sequence on the return link will be random. Determining the time when the transmitted and return signal are decorrelated will allow the transmitter to compute the correct transmission timing. Notice that this procedure works better than attempting to find the point at which error performance is the best. Any well-designed system will have sufficient transmission energy to allow for slight timing offsets, so an error-free return signal could be achieved with less than perfect timing. In fact, the larger the signal-to-noise ratio, the worse a best-finding procedure works. A worst-finding system, however, will work well for any well-designed system and will improve in potential accuracy with increasing signal-to-noise ratio. This can be seen intuitively, because increased signal-to-noise ratios will allow the system to tolerate larger timing errors, so the improvement in error performance as the timing error decreases from half a symbol time will be more rapid in the large signal-to-noise case than in the smaller signal-to-noise ratio case. This will allow a more precise determination of the half-symbol timing position.

8.4 CONCLUSION

This chapter has outlined the fundamental problems and issues associated with synchronization in digital communications. The trade-offs are generally between expense and complexity, on the one hand, and error performance on the other. We have discussed receiver synchronization and phase-locked loops (PLL) in particular. Typically, it is the receiver that takes the most active role in the synchronization of a communications link. Even in cases where a terminal's transmitter assumes the more active role, as in some satellite links, the process is often aided by a return path that has been acquired by the terminal's receiver. Thus receiver synchronization is more fundamental. Phase-locked loops, and their variations, are the primary control circuits used to track variations in phase of an incoming signal. The mathematics needed to describe the response of a PLL to a given input involves the solution to a nonlinear differential equation. It was shown, however, that under steady-state conditions, a linearized model provides a useful approximation to system performance. In circumstances where the linearized model cannot be accurately applied, results by Viterbi [5] for first-order loops were introduced. Although exact for first-order loops only, these results have been shown to be useful approximations to the performance of higher-order loops as well [2].

The extremely important special case of suppressed-carrier loops was discussed. Suppressed-carrier loops are required to track the phase of an incoming signal that has no average energy at the carrier frequency. The common example of such a signal is one that has been modulated with standard antipodal BPSK. In this situation, a harmonic of the suppressed carrier is produced through the use of a nonlinearity, and the harmonic is tracked.

The next higher level of synchronization treated here was symbol synchronization. Two primary classes of symbol synchronization were discussed. Open-loop synchronizers operate directly on the modulated signal to produce a symbol transition indication. Closed-loop synchronizers use a closed-cycle control loop to acquire and track the symbol transitions.

The highest level of synchronization considered was frame synchronization. To receive the data in a useful form, the receiver must determine which symbols belong to which frames. This knowledge is equivalent to having frame sync, which is usually accomplished by including some recognizable pattern, known to the receiver, with the data symbols. The receiver scans the incoming data until it recognizes the pattern. Synchronization can be checked by looking for periodic repetitions of the pattern.

This chapter has necessarily been only an outline of the important problems, issues, and results relating to the synchronization of digital communication systems. The interested reader will find that the references listed are worthy works that will provide much greater depth of coverage than space has allowed here.

REFERENCES

1. Peterson, W. W., and Weldon, E. J., *Error-Correcting Codes,* The MIT Press, Cambridge, Mass., 1972.

2. Gardner, F. M., *Phaselock Techniques,* 2nd ed., John Wiley & Sons, Inc., New York, 1979.

3. Davenport, W. B., and Root, W. L., *Random Signals and Noise,* McGraw-Hill Book Company, New York, 1958.

4. Papoulis, A., *Probability, Random Variables, and Stochastic Processes,* McGraw-Hill Book Company, New York, 1965.

5. Viterbi, A. J., *Principles of Coherent Communications,* McGraw-Hill Book Company, New York, 1966.

6. Lindsey, W. C., *Synchronization Systems in Communication and Control,* Prentice-Hall, Inc., Englewood Cliffs, N.J., 1972.

7. Lindsey, W. C., and Simon, M. K., "Detection of Digital FSK and PSK Using a First-Order Phase-Locked Loop," *IEEE Trans. Commun.,* vol. COM25, no. 2, Feb. 1977, pp. 200–214.

8. Develet, J. A., Jr., "The Influence of Time Delay on Second-Order Phase Lock Loop Acquisition Range," *Int. Telem. Conf.,* London, 1963.

9. Johnson, W. A., "A General Analysis of the False-Lock Problem Associated with the Phase-Lock Loop," *The Aerospace Corp., Rep. TOR-269(4250-45)-1,* NASA Accession N64-13776, 1963.

10. Tausworthe, R. C., "Acquisition and False-Lock Behavior of Phase-Locked Loops with Noisy Inputs," *Jet Propulsion Laboratory, JPL SPS 37-46,* vol. 4, 1967.

11. Franks, L. E., "Synchronization Subsystems: Analysis and Design," in K. Feher, *Digital Communications, Satellite/Earth Station Engineering,* Prentice-Hall, Inc., Englewood Cliffs, N.J., 1981, Chap. 7.

12. Simon, M. K., and Yuen, J. H., "Receiver Design and Performance Characteristics," in J. H. Yuen, ed., *Deep Space Telecommunications Systems Engineering,* Plenum Press, New York, 1983.

13. Gardner, F. M., "Hangup in Phase-Lock Loops," *IEEE Trans. Commun.,* COM25, October 1977.

14. Blanchard, A., *Phase-Locked Loops,* John Wiley & Sons, Inc., New York, 1976.

15. Holmes, J. K., *Coherent Spread Spectrum Systems,* John Wiley & Sons, Inc., New York, 1982.

16. Lindsey, W. C., and Simon, M. K., eds., *Phase Locked Loops and Their Applications,* IEEE Press, New York, 1977.

17. Spilker, J. J., Jr., *Digital Communications by Satellite,* Prentice-Hall, Inc., Englewood Cliffs, N.J., 1977.

18. Wintz, P. A., and Luecke, E. J., "Performance of Optimum and Suboptimum Synchronizers," *IEEE Trans. Commun. Technol.,* June 1969, pp. 380–389.

19. Barker, R. H., "Group Synchronization of Binary Digital Systems," in W. Jackson, ed., *Communication Theory,* Academic Press, Inc., New York, 1953.

20. Willard, M. W., "Optimum Code Patterns for PCM Synchronization," *Proc. Natl. Telem. Conf.,* 1962, paper 5-5.

21. Newman, F., and Hofman, L., "New Pulse Sequences with Desirable Correlation Properties," *Proc. Natl. Telem. Conf.,* 1971, pp. 272–282.

22. Wu, W. W., *Elements of Digital Satellite Communication,* Vol. 1, Computer Science Press, Inc., Rockville, Md., 1984.

PROBLEMS

8.1. A transmitter is sending an unmodulated tone of constant energy (a beacon) to a distant receiver. The receiver and transmitter are in motion with respect to each other such that $d(t) = D[1 - \sin(mt)] + D_0$, where $d(t)$ is the distance between the transmitter and receiver (possibly this represents an aircraft doing "figure-eight" maneuvers over a ground station), and D, m, and D_0 are constants. This relative motion will cause a Doppler shift in the received transmitter frequency of

$$\Delta\omega_D(t) = \frac{\omega_0 v(t)}{c}$$

where $\Delta\omega_D$ is the Doppler shift, ω_0 the nominal carrier frequency, $v(t) = \dot{d}(t)$ the relative velocity between the transmitter and receiver, and c the speed of light. Assuming that the linearized loop equations hold and that the receiver's PLL is in lock (zero phase error) at $t = 0$, show that an appropriately designed first-order loop can maintain frequency lock.

8.2. Consider a transmitter and receiver that are in relative motion as in Problem 8.1. Once again assume that the linearized loop equations hold. Under this assumption determine the PLL phase error as a function of time for the all-pass and low-pass loop filters of Equations (8.13) and (8.14). Demonstrate that the validity of the assumption of the linearized loop equations depends on the value of the gain K_0.

8.3. A high-performance aircraft is transmitting an unmodulated carrier signal to a ground terminal. The ground terminal is initially in phase lock with the signal. The aircraft performs a maneuver whose dynamics are described by the equation for acceleration, $a(t) = At^2$, where A is a constant. Assuming that the linearized equations apply, determine the minimum order of the phase-locked loop required to track the signal from this aircraft.

8.4. Show that the loop bandwidth of a first-order phase-locked loop is given by $B_L = K_0/4$, where K_0 is the loop gain.

8.5. A second-order phase-locked loop has a low-pass loop filter given by

$$F(\omega) = \frac{\omega_1}{j\omega + \omega_1}$$

and a loop gain of K_0. Under the assumption that $K_0 \geq \omega_1/4$, show that the loop bandwidth of this phase-locked loop is given by $B_L = K_0/8$. [*Hint:**

$$\int_{-\infty}^{\infty} \frac{dx}{R} = \frac{\pi \cos(h/2)}{2cq^3 \sin h} \qquad \text{for } 4ac > b^2$$

where $R = a + bx^2 + cx^4$, $q = \sqrt[4]{a/c}$, and $\cos h = -b/2\sqrt{ac}$.]

8.6. A first-order phase-locked loop with loop gain K_0 is disturbed by additive white Gaussian noise of normalized (to unit signal energy) two-sided power spectral density of $N_0/2$ watts/hertz. Determine the necessary relationship between noise power spectral density and loop gain if the loop is designed to cycle slip no more often than once per day.

8.7. Viterbi [5] determined that the probability density function of the output phase of a first-order phase-locked loop disturbed by white Gaussian noise is given by

$$p(\phi) = \frac{\exp(\rho \cos \phi)}{2\pi I_0(\rho)}, \qquad |\phi| \leq \pi, \quad \rho \geq 0$$

Demonstrate that $p(\phi)$ given above is in fact a probability density function, and compute the mean and variance of ϕ.

8.8. Computer simulations and laboratory measurements have indicated that the time

* I. S. Gradshteyn and I. M. Ryzhik, *Table of Integrals, Series and Products* (New York: Academic Press, 1965), 2.161.1.

between cycle slips is exponentially distributed; that is, the distribution function of the time between cycle slips, T, is given by

$$p(T) = 1 - \exp\left(-\frac{T}{T_m}\right)$$

Given this distribution function, find the mean time between cycle slips and the variance about this mean as functions of T_m. If the mean time between cycle slips is 1 day, what is the probability of cycle slips less than 1 hour apart? More than 3 days apart?

8.9. Consider a second-order phase-locked loop with a low-pass loop filter

$$F(\omega) = \frac{\omega_1}{j\omega + \omega_1}$$

During aided acquisition it is desired that this loop be scanned throughout a 1000-radian uncertainty region in 1 s. If the relationship between loop gain and filter constant is $K_0 = 2\omega_1$, what is the required relationship between loop gain and the single-sided additive white Gaussian noise power spectral density, N_0? Determine the largest value of N_0 that can be accommodated.

8.10. Consider the operation of an open-loop symbol synchronizer whose band pass filter (BPF) has a bandwidth of $0.1/T$ hertz, where T is the symbol period. For a bit energy-to-noise power spectral density ratio (E_b/N_0) of 10 dB, determine the magnitude of the approximate mean and variance of the fractional tracking error, and compute an upper bound on the probability that the tracking error exceeds three times its approximate fractional mean. (*Hint:* Consider the Chebyshev inequality [4].)

8.11. A communication system is used to transmit commands to a payload at a data rate of 100 bits/s. Each command is preceded by an N-bit header that identifies it in the data stream. Assuming that except possibly for the header, the bits appear to be random [$P(1) = P(0) = \frac{1}{2}$], what is the minimum-length header that would provide an expected frequency of false alarms of one per year? For a channel bit error probability of 10^{-5}, what is the probability of missing this header? What is the miss probability if the channel error probability is 2×10^{-2}? If the system is redesigned to accept the header with up to two errors, what is the minimum required length for an expected false alarm rate of one per year? What is the miss probability with this new system and a channel error probability of 2×10^{-2}?

8.12. A deep-space probe is moving away from the earth at a nominal velocity of 15,000 m/s with a velocity uncertainty of ± 3 m/s. The probe frequency reference is specified to have a drift rate of no more than 10^{-9} Hz/Hz/day. The nominal downlink transmission frequency is 8 GHz. After a 1-month (30-day) silence the probe begins a scheduled transmission toward an earth terminal. The earth terminal contains a cesium standard. What center frequency and frequency search bandwidth should be used by the ground station? Assuming that the range to the probe was accurately known at the beginning of the month, and that the uncertainty in the probe's time and frequency references were zero [$\Delta t(0) = 0$, $\Delta\omega(0) = 0$], what is the uncertainty in the time of arrival of the downlink transmission?

8.13. A communications link operates at a nominal center frequency of 10 GHz for a single brief period once a day. The receiver operates with a second-order PLL that has a pull-in range of ± 1 kHz. Assuming that the loop acquires using self-acquisition and that both the transmitter and receiver use the same type of frequency reference, what type of frequency reference must this be?

CHAPTER 9

Multiplexing
and
Multiple Access

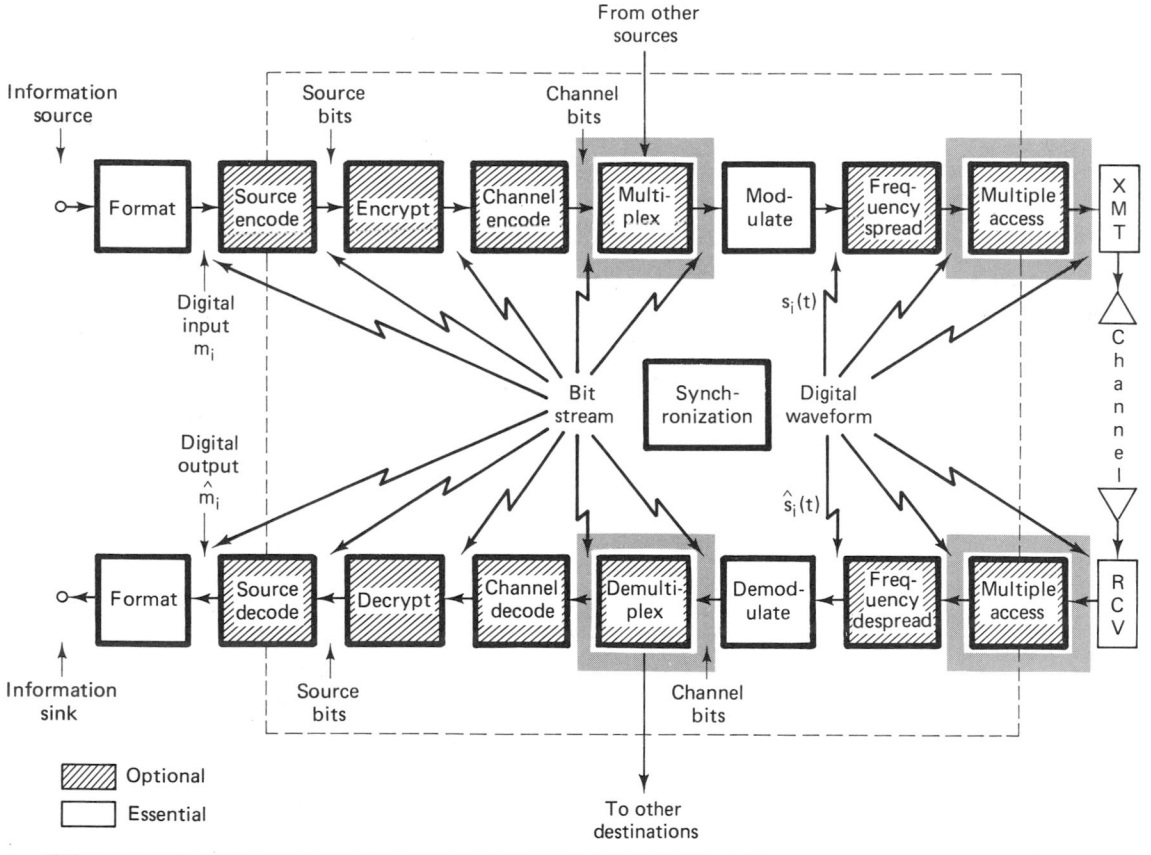

The terms "multiplexing" and "multiple access" refer to the sharing of a fixed communications resource (CR). There is a subtle difference between multiplexing and multiple access. With *multiplexing*, users' requirements or plans for CR sharing are fixed, or at most, slowly changing. The resource allocation is assigned a priori, and the sharing is usually a process that takes place within the confines of a *local site* (e.g., a circuit board). *Multiple access*, however, usually involves the *remote sharing* of a resource, such as a satellite. With a dynamically changing multiple access scheme, a system controller must become aware of each user's CR needs; the amount of time required for this information transfer constitutes an overhead and sets an upper limit on the efficiency of the utilization of the CR.

9.1 ALLOCATION OF THE COMMUNICATIONS RESOURCE

There are three basic ways to increase the throughput (total data rate) of a communications resource (CR). The first way is either to increase the transmitter's effective isotropic radiated power (EIRP) or to reduce system losses so that the received E_b/N_0 is increased. The second way is to provide more channel bandwidth. The third approach is to make the allocation of the CR more efficient. This third approach is the domain of communications multiple access. The problem, in the context of a satellite transponder, is to efficiently allocate portions of the transponder's fixed CR to a large number of users who seek to communicate digital information to each other at a variety of bit rates and duty cycles. The

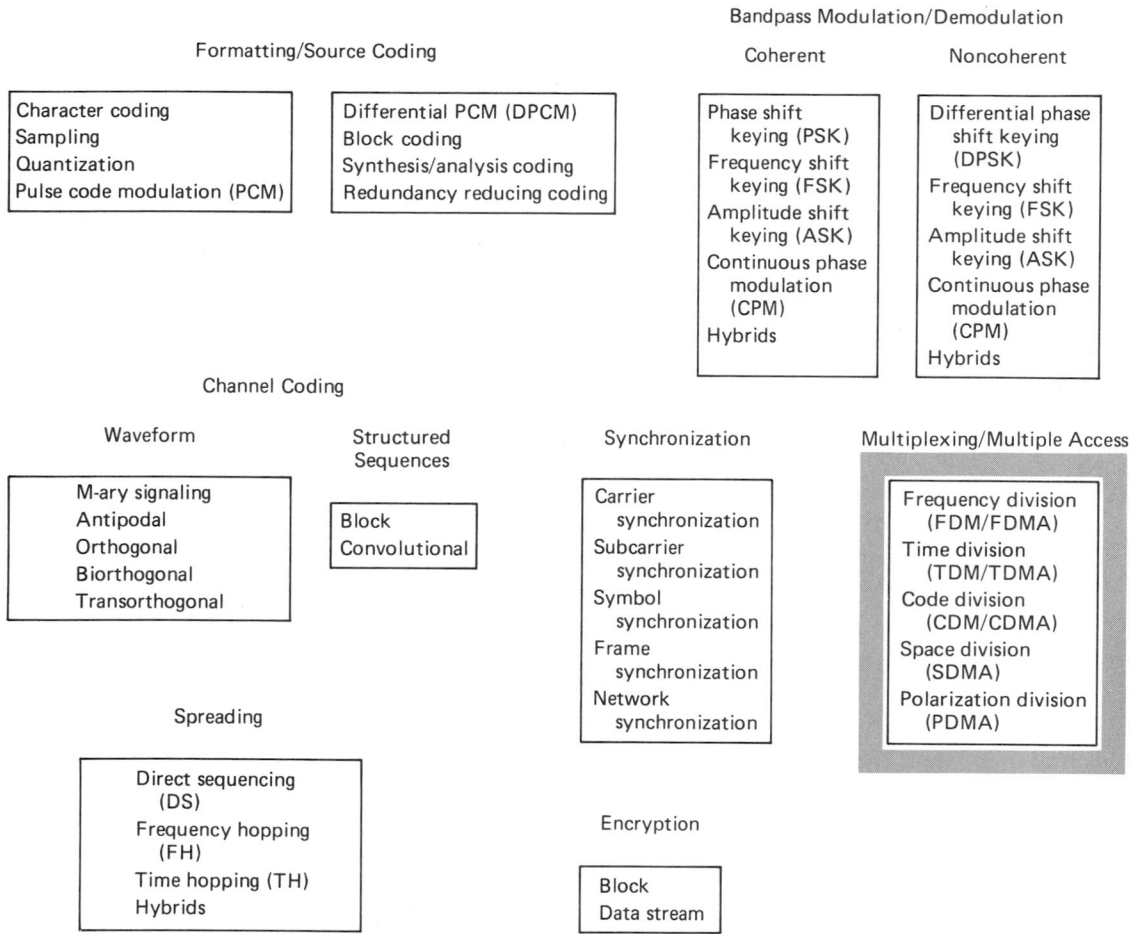

Figure 9.1 Basic digital communication transformations.

basic ways of distributing the communications resource. listed under the heading "multiplexing/multiple access" in Figure 9.1, are:

1. *Frequency division (FD).* Specified subbands of frequency are allocated.
2. *Time division (TD).* Periodically recurring time slots are identified. With some systems, users are provided a fixed assignment in time. With others, users may access the resource at random times.
3. *Code division (CD).* Specified members of a set of orthogonal or nearly orthogonal spread spectrum codes (each using the full channel bandwidth) are allocated.
4. *Space division (SD) or multiple beam frequency reuse.* Spot beam antennas

are used to separate radio signals by pointing in different directions. It allows for reuse of the same frequency band.

5. *Polarization division (PD) or dual polarization frequency reuse.* Orthogonal polarizations are used to separate signals, allowing for reuse of the same frequency band.

The key to *all* multiplexing and multiple access schemes is that various signals share a CR without creating unmanageable interference to each other in the detection process. The allowable limit of such interference is that signals on one CR channel should not significantly increase the probability of error in another channel. Orthogonal signals on separate channels will avoid interference between users. Signal waveforms $x_i(t)$, where $i = 1, 2, \ldots$, are defined to be orthogonal if they can be described in the time domain by

$$\int_{-\infty}^{\infty} x_i(t)x_j(t)\, dt = \begin{cases} K & \text{for } i = j \\ 0 & \text{otherwise} \end{cases} \qquad (9.1)$$

where K is a nonzero constant. Similarly, the signals are orthogonal if they can be described in the frequency domain by

$$\int_{-\infty}^{\infty} X_i(f)X_j(f)\, df = \begin{cases} K & \text{for } i = j \\ 0 & \text{otherwise} \end{cases} \qquad (9.2)$$

where the functions $X_i(f)$ are the Fourier transforms of the signal waveforms $x_i(t)$. Channelization characterized by orthogonal waveforms, as shown in Equation (9.1), is called time-division multiplexing or time-division multiple access (TDM/TDMA), and that characterized by orthogonal spectra, as shown in Equation (9.2), is called frequency-division multiplexing or frequency-division multiple access (FDM/FDMA).

9.1.1 Frequency-Division Multiplexing/Multiple Access

9.1.1.1 Frequency-Division Multiplex Telephony

In the early days of telephony, a separate pair of wires was needed for each telephone trunk circuit (trunk circuits interconnect intercity switching centers). As illustrated in Figure 9.2, the skies of all the major cities in the world grew dark with overhead wires as the demand for telephone service grew. A major development in the early 1900s, frequency-division multiplex (FDM) telephony, made it possible to transmit several telephone signals simultaneously on a single wire, and thereby transformed the methods of telephone transmission.

The communications resource (CR) is illustrated in Figure 9.3 as the frequency–time plane. The channelized spectrum shown here is an example of FDM or FDMA. The assignment of a signal or user to a frequency band is *long term* or *permanent*; the CR can simultaneously contain several spectrally separate signals. The first frequency band contains signals that operate between frequencies f_0 and f_1, the second between frequencies f_2 and f_3, and so on. The spectral regions between assignments, called *guard bands*, act as buffer zones to reduce

Figure 9.2 In the early days of telephony a pair of wires was needed for each trunk circuit.

interference between adjacent frequency channels. We might ask: How does one transform a baseband signal so that it occupies a higher frequency band? The answer is, by *heterodyning* or *mixing*, also called *modulating* the signal with a fixed frequency from a sine-wave oscillator.

If two input signals to a mixer are sinusoids with frequencies f_A and f_B, the mixing or multiplication will yield new sum and difference frequencies at f_{A+B} and f_{A-B}. The trigonometric identity

$$\cos A \cos B = \tfrac{1}{2}[\cos (A + B) + \cos (A - B)] \tag{9.3}$$

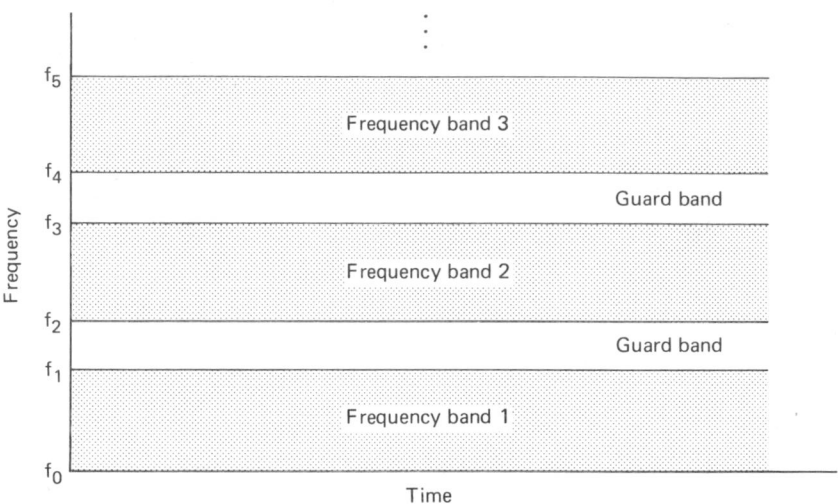

Figure 9.3 Frequency-division multiplexing.

describes the effect of the mixer. Figure 9.4a illustrates the mixing of a typical voice-grade telephone signal, $x(t)$ (baseband frequency range is 300 to 3400 Hz) with a sinusoid from a 20-kHz oscillator. The baseband two-sided magnitude spectrum, $|X(f)|$, is shown in Figure 9.4a. Can the mixer be a linear device? *No.* The output signal of a linear device will only consist of the *same* component frequencies as the input signal, differing only in amplitude and/or phase.

Figure 9.4b illustrates the one-sided magnitude spectrum, $|X(f - f_0)|$, at the mixer output. As a result of the mixing described by Equation (9.3), the output spectrum is a frequency-upshifted version of the baseband spectrum, centered at the oscillator frequency of 20 kHz. This spectrum is called a *double-sideband* (DSB) *spectrum* because the information appears in two different bands of the positive frequency domain. Figure 9.4c shows the lower sideband (LSB), whose frequency range is 16,600 to 19,700 Hz, the result of filtering the DSB spectrum. This sideband is sometimes referred to as the *inverted sideband* because the order of low-to-high frequency components is the reverse of that of the baseband components. Filtering can similarly be used to separate the upper sideband (USB), whose frequency range is 20,300 to 23,400 Hz, as shown in Figure 9.4d. This sideband is sometimes referred to as the *erect sideband* because the order of the low-to-high frequency components corresponds to that of the baseband components. Each sideband of the DSB spectrum contains the same information. Thus, only one sideband, either the USB or the LSB, is needed in order to retrieve the original baseband data.

A simple FDM example with three translated voice channels is seen in Figure 9.5. In channel 1, the 300- to 3400-Hz voice signal is mixed with a 20-kHz oscillator. In channels 2 and 3, a similar type of voice signal is mixed with a 16-kHz and 12-kHz oscillator, respectively. Only the lower sidebands are retained; the result of the mixing and filtering (to remove the upper sidebands) yields the frequency-shifted voice channels shown in Figure 9.5. The composite output waveform is just the sum of the three signals, having a total bandwidth in the range 8.6 to 19.7 kHz.

Figure 9.6 illustrates the two lowest levels of the FDM multiplex hierarchy for telephone channels. The first level consists of a *group* of 12 channels modulated onto subcarriers shown in the range 60 to 108 kHz. The second level is made up of five groups (60 channels) called a *supergroup* modulated onto the subcarriers shown in the range 312 to 552 kHz. The multiplexed channels are now treated as a composite signal that can be transmitted over cables or can be further modulated onto a carrier wave for radio transmission.

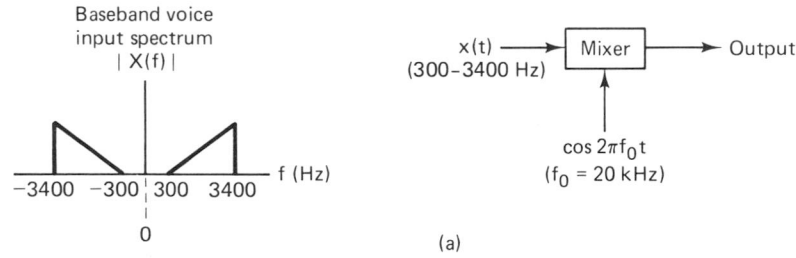

(a)

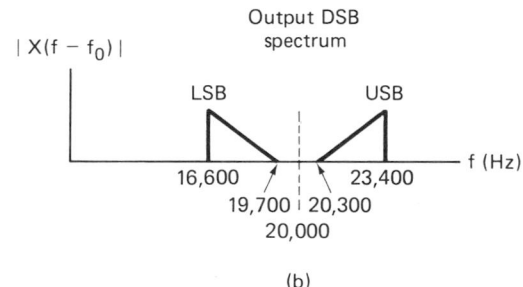

(b)

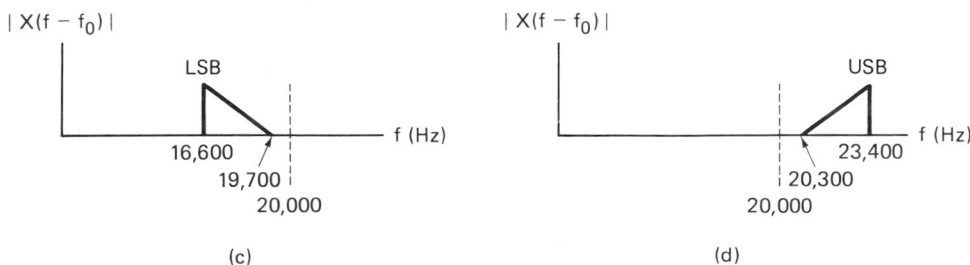

(c)   (d)

Figure 9.4 Heterodyning (mixing). (a) Mixing operation. (b) Mixer output spectrum. (c) Lower sideband. (d) Upper sideband.

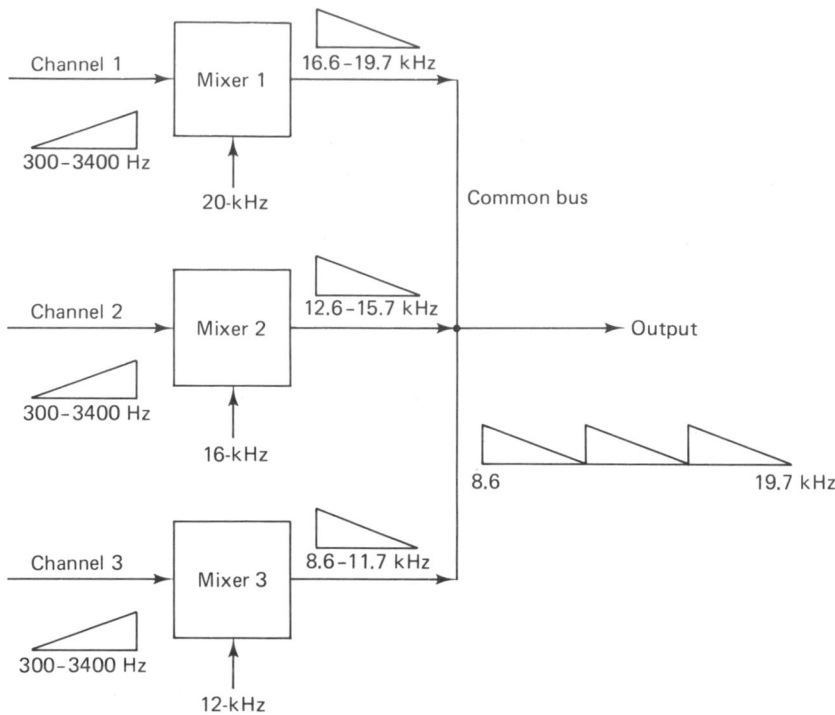

Figure 9.5 Simple FDM example. Three frequency-shifted voice channels.

9.1.1.2 Frequency-Division Multiple Access of Satellite Systems

Most of the free world's communication satellites are positioned in a *geostationary* or *geosynchronous* orbit. This means that the satellite is in a circular orbit, in the same plane as the earth's equatorial plane, and at such an altitude (approximately 19,330 nautical miles) that the orbital period is identical with the earth's rotational period. Since such satellites appear stationary when viewed from the earth, three of them spaced 120° apart can provide worldwide coverage (except for the polar regions). Most communication satellite systems are made up of nonregenerative repeaters or transponders. *Nonregenerative* means that the uplink (earth-to-satellite) transmissions are simply amplified, frequency shifted, and retransmitted on the downlink (satellite-to-earth) without any demodulation/remodulation or signal processing (see Section 4.7.1). The most popular frequency band for commercial satellite communications, called *C-band*, uses a 6-GHz carrier for the uplink and a 4-GHz carrier for the downlink. For C-band satellite systems, *each satellite* is permitted, by international agreement, to use a 500-MHz-wide spectral assignment. Typically, each satellite has 12 transponders with a bandwidth of 36 MHz each. The most common 36-MHz transponders operate in an FDM/FM/FDMA (frequency-division multiplex, frequency-modulated, frequency-division multiple access) multidestination mode. Let us consider each component of this name:

1. *FDM*. Signals such as telephone signals, each one having a single-sideband 4-kHz spectrum (including guard bands) are FDM'd to form a multichannel composite signal.
2. *FM*. The composite signal is frequency-modulated (FM) onto a carrier and transmitted to the satellite.
3. *FDMA*. Subdivisions of the 36-MHz transponder bandwidth may be assigned to different users. Each user receives a specific bandwidth allocation whereby he or she can access the transponder.

Thus, composite FDM channels are FM modulated and transmitted to the satellite within the bandwidth allocation of an FDMA plan. The major advantage of FDMA (compared to TDMA) is its simplicity. The FDMA channels require no

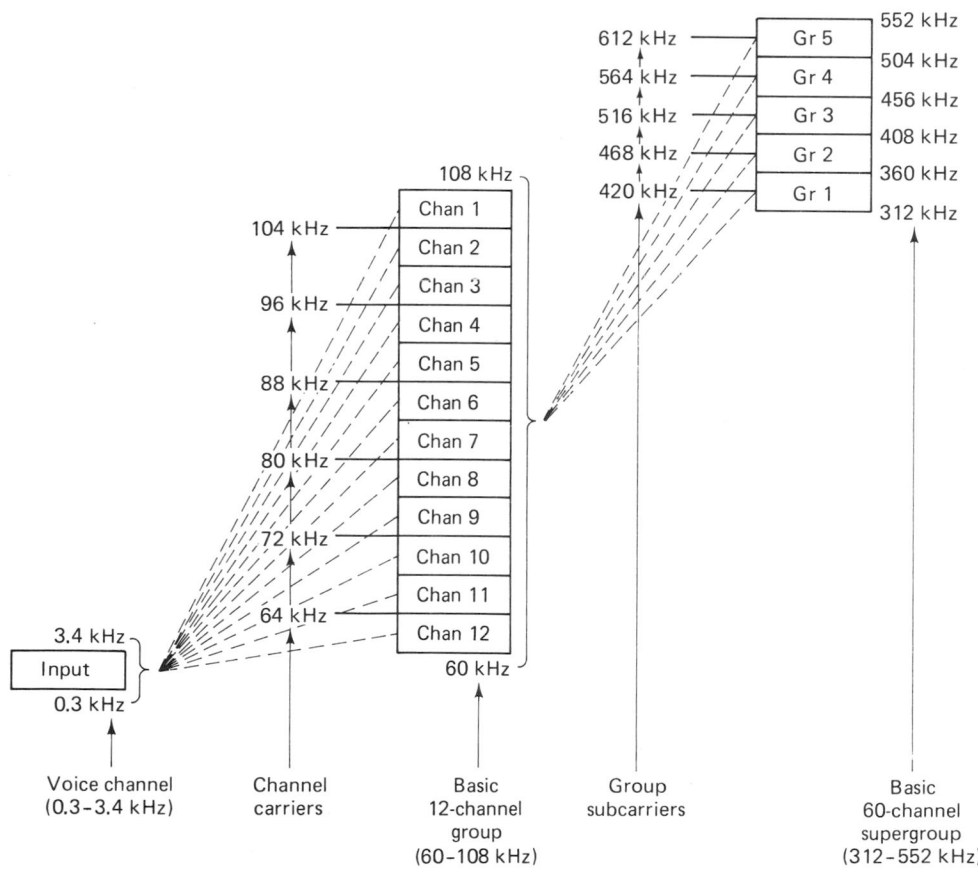

Figure 9.6 Modulation plan of a typical frequency-division multiplex system.

synchronization or central timing; each channel is almost independent of all other channels. Later we discuss some advantages of TDMA compared to FDMA.

9.1.2 Time-Division Multiplexing/Multiple Access

In Figure 9.3, sharing of the communications resource (CR) is accomplished by allocating frequency bands. In Figure 9.7, the same CR is shared by assigning each of M signals or users the full spectral occupancy of the system for a short duration of time called a *time slot*. The unused time regions between slot assignments, called *guard times*, allow for some time uncertainty between signals in adjacent time slots, and thus act as buffer zones to reduce interference. Figure 9.8 is an illustration of a typical TDMA satellite application. Time is segmented into intervals called frames. Each frame is further partitioned into assignable user time slots. The frame structure repeats, so that a fixed TDMA assignment constitutes one or more slots that periodically appear during each frame time. Each earth station transmits its data in bursts, timed so as to arrive at the satellite coincident with its designated time slot(s). When the bursts are received by the satellite transponder, they are retransmitted on the downlink, together with the bursts from other stations. A receiving station detects and demultiplexes the appropriate bursts and feeds the information to the intended user.

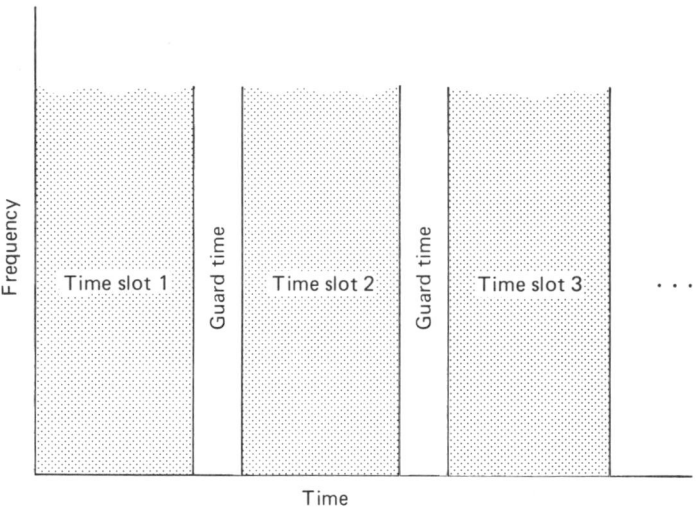

Figure 9.7 Time-division multiplexing.

9.1.2.1 Fixed-Assignment TDM/TDMA

The simplest TDM/TDMA scheme, called *fixed-assignment TDM/TDMA*, is so named because the M time slots that make up each frame are preassigned to signal sources, long term. Figure 9.9 illustrates, in block diagram form, the operation of such a system. The multiplexing operation consists of providing each

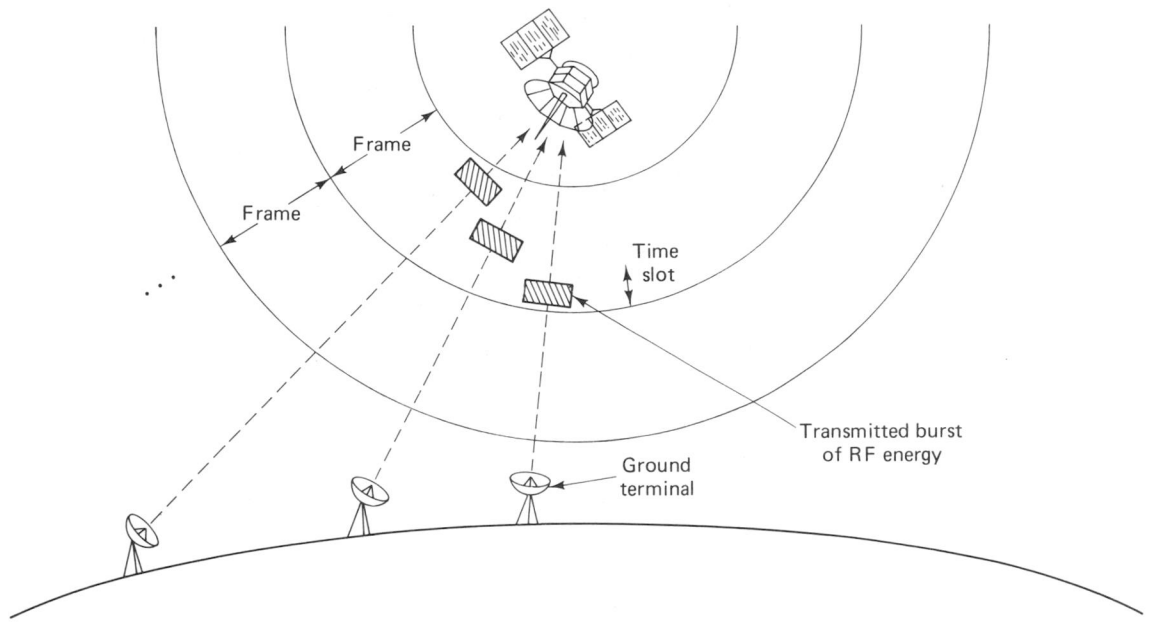

Figure 9.8 Typical TDMA configuration.

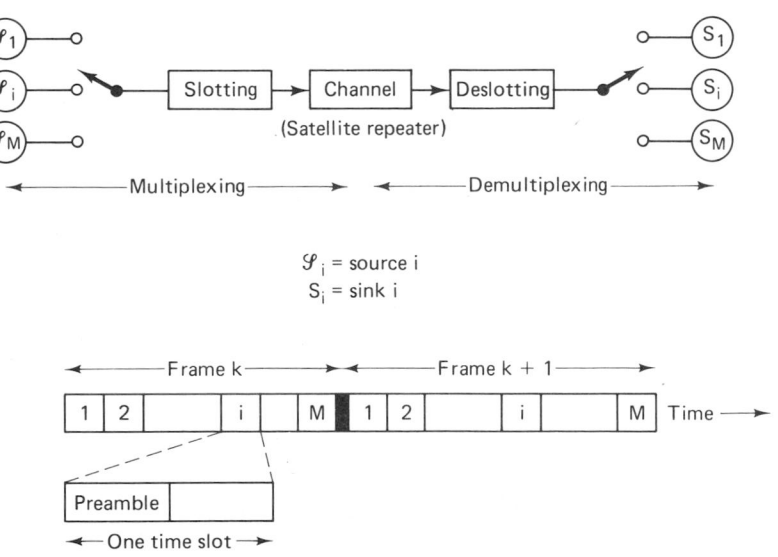

$\mathscr{S}_i$ = source i
S_i = sink i

Figure 9.9 Fixed-assignment TDM.

source with an opportunity to occupy one or more slots. The demultiplexing operation consists of deslotting the information and delivering the data to the intended sink. The two commutating switches in Figure 9.9 have to be synchronized so that the message corresponding to source 1, for example, appears on the channel 1 output, and so on. The message itself is generally comprised of a preamble portion and a data portion. The preamble portion usually contains synchronization, addressing, and error control sequences.

A fixed-assignment TDM/TDMA scheme is extremely efficient when the source requirements are predictable, and the traffic is heavy (the time slots are most always filled). However, for bursty or sporadic traffic, the fixed-assignment scheme is wasteful. Consider the simple example shown in Figure 9.10. In this example there are four time slots per frame; each slot is preassigned to users A, B, C, and D, respectively. In Figure 9.10a we see a typical activity profile of the four users. During the first frame time, user C has no data to transmit; during the second frame time, user B has none, and during the third frame time, user A has none. In a fixed-assignment TDMA scheme, all of the slots within a frame are preassigned. If the "owner" of a slot has *no* data to send during a particular frame, that slot is wasted. The data stream, shown in Figure 9.10b, illustrates the wasted time slots in this example. When source requirements are unpredictable, as in this example, there can be more efficient schemes, involving the dynamic assignment of the slots rather than a fixed assignment. Such schemes are variously known as packet-switched systems, statistical multiplexers, or concentrators; the effect, shown in Figure 9.10c, is to use all the slots in a frame in such a way that capacity is conserved. In later sections we discuss the TDMA systems used in INTELSAT V and VI.

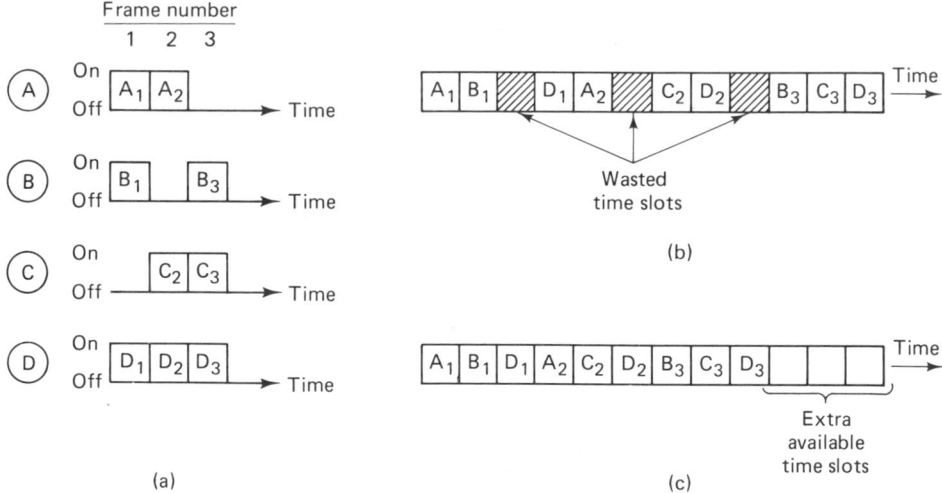

Figure 9.10 Fixed-assignment TDM versus packet switching. (a) Data source activity profiles. (b) Fixed-assignment time-division multiplexing. (c) Time-division packet switching (concentration).

9.1.3 Communications Resource Channelization

In Figure 9.3 we considered that the CR is partitioned into spectral bands, and in Figure 9.7 we viewed the same CR as being partitioned into time slots. Figure 9.11 represents a more general organization of the CR allowing for the assignment of a frequency band for a prescribed period of time. Such a multiple access scheme is· referred to as *combined FDMA/TDMA*. For the assignments of frequency bands, let us assume an equal apportionment of the total bandwidth, W, among M user groups or classes, so that M disjoint frequency bands of width W/M hertz are continuously available to· their assigned group. Similarly, for the assignment of time slots, the time axis is partitioned into time frames, each of duration T, and the frames are partitioned into N slot times, each of duration T/N. We assume that the users are time synchronized and that the assigned slots are located periodically within the frames. Each user in each frequency band is permitted to transmit during each periodic appearance of the user's assigned slot, and is permitted to use the assigned channel bandwidth for the slot duration. A slot is uniquely determined as the mth slot within the nth frame. Referring to Figure 9.11, we can describe the time of a particular slot (n, m) with reference to time zero as follows:

$$\text{time of slot } (n, m) = nT + \frac{(m-1)T}{N} \leq t \leq nT + \frac{mT}{N}$$

$$n = 0, 1, \ldots; m = 1, 2, \ldots, N. \quad (9.4)$$

The nth frame time, T, is denoted by the time interval $[nT, (n+1)T]$. As can be seen in Figure 9.11, the domain of the unit signal is the intersection of the time slot (n, m) and the frequency band (j). Assume that a modulation/coding system is chosen so that the full bandwidth W of the CR can support R bits/s. In any

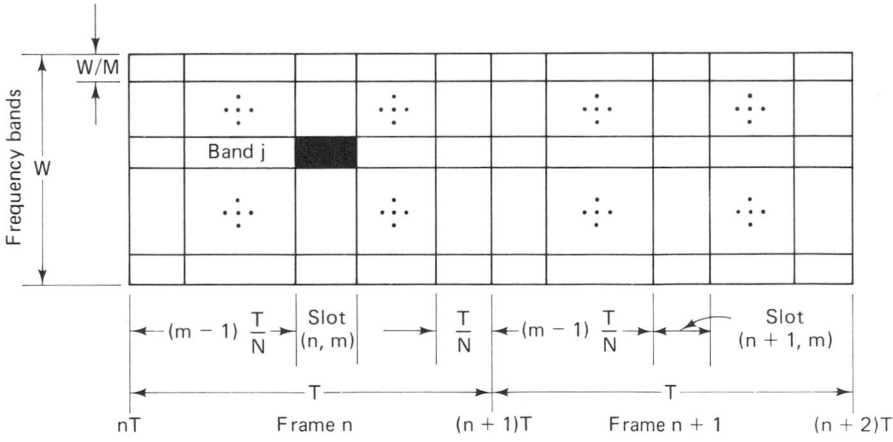

Figure 9.11 Communications resource: time/frequency channelization.

frequency band having a bandwidth of W/M hertz, the associated bit rate will be R/M bits/s. FDMA alone would provide M bands each with a bandwidth of $1/M$ of the full bandwidth of the CR. TDMA alone would provide the full system bandwidth for each of the N slots, where the duration of each slot is $1/N$ of the frame time.

9.1.4 Performance Comparison of FDMA and TDMA

9.1.4.1 Bit Rate Equivalence of FDMA and TDMA

Figure 9.12 highlights the basic differences between an FDMA and TDMA system in a communications resource capable of supporting a total of R bits/s. In Figure 9.12a the system bandwidth is divided into M orthogonal frequency bands. Hence each of the M sources, $\mathscr{S}_m (1 \leq m \leq M)$ can simultaneously transmit at a bit rate of R/M bits/s. In Figure 9.12b the frame is divided into M orthogonal time slots. Hence each of the M sources bursts its transmission at R bits/s, M times faster than the equivalent FDMA user for $(1/M)$th the time. In both cases, the source $\mathscr{S}_m$ transmits information at an average rate of R/M bits/s.

Let the information generated by each of the sources in Figure 9.12 be organized into b-bit groups, or *packets*. In the case of FDMA, the b-bit packets are transmitted in T seconds over each of the M disjoint channels. Therefore, the

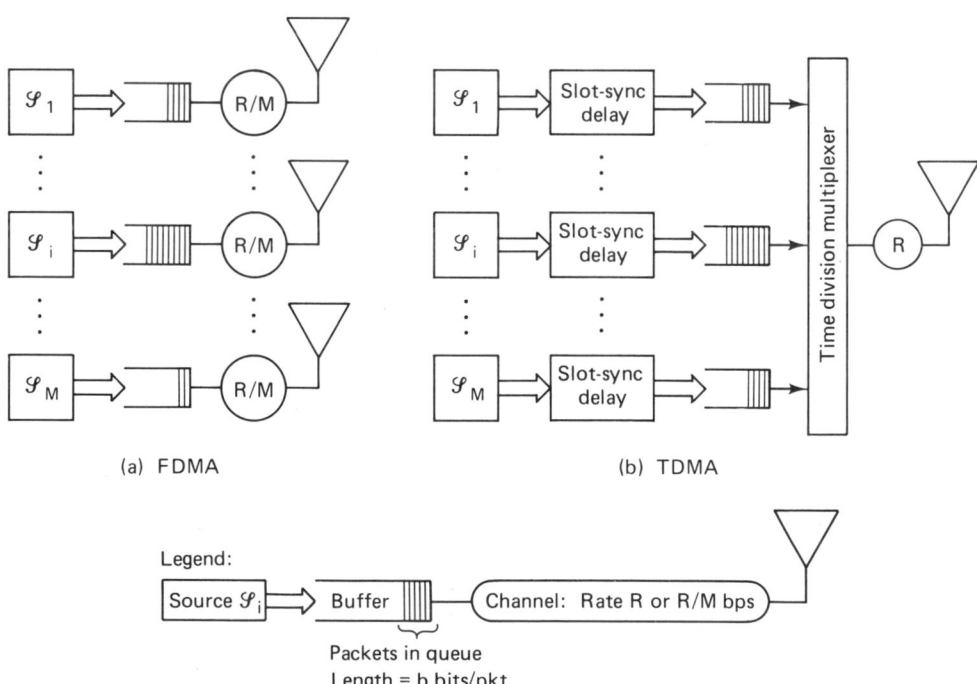

(a) FDMA   (b) TDMA

Figure 9.12 (a) FDMA: frequency divided into M orthogonal frequency bands. (b) TDMA: time divided into M orthogonal time slots (one packet per time slot).

total bit rate, R_{FD}, required is

$$R_{FD} = M \frac{b}{T} \quad \text{bits/s} \tag{9.5}$$

In the case of TDMA, the b bits are transmitted in T/M seconds from each source. Therefore, the bit rate, R_{TD}, required is

$$R_{TD} = \frac{b}{T/M} \quad \text{bits/s} \tag{9.6}$$

Since Equations (9.5) and (9.6) yield identical results, we can conclude that

$$R_{FD} = R_{TD} = R = \frac{Mb}{T} \quad \text{bits/s} \tag{9.7}$$

Thus both systems require the same full CR data rate, R bits/s.

9.1.4.2 Message Delays in FDMA and TDMA

From the previous sections it might appear that the duality between FDMA and TDMA will result in equivalent performance. This is not the case when the metric of performance is the average packet *delay*. It can be shown [1, 2] that TDMA is inherently superior to FDMA in the sense that the average packet delay using TDMA is less than the delay using FDMA.

As before, we assume that in the case of FDMA the system bandwidth is divided into M orthogonal frequency bands, and in the case of TDMA the frame is divided into M orthogonal time slots. For the analysis of message delay, the simplest case is that of deterministic data sources. It is assumed that the CR is 100% utilized, so that all frequency bands in the case of FDMA, and all time slots in the case of TDMA, are filled with data packets. For simplicity, it is also assumed that there are *no* overhead costs such as guard bands or guard times.

The message delay, D, can be defined as

$$D = w + \tau \tag{9.8}$$

where w is the average packet waiting time (prior to transmission) and τ is the packet transmission time. In the FDMA case, each packet is sent over a T-second interval, so the packet transmission time for FDMA, τ_{FD}, is simply

$$\tau_{FD} = T \tag{9.9}$$

In the TDMA case, each packet is sent in slots of T/M seconds. We can thus write the TDMA packet transmission time, τ_{TD}, with the use of Equation (9.7), as

$$\tau_{TD} = \frac{T}{M} = \frac{b}{R} \tag{9.10}$$

Since the FDMA channel is continuously available and packets are sent as soon as they are generated, the waiting time, w_{FD}, for FDMA is

$$w_{FD} = 0 \qquad (9.11)$$

FDMA and TDMA bit streams are compared in Figure 9.13. For TDMA, Figure 9.13a illustrates that each user's slot begins at a different point in the T-second frame; that is, packet S_{mk} will start at $(m - 1)T/M$ seconds $(1 \leq m \leq M)$ after the packet generation instant. Therefore, the average waiting time, w_{TD}, that a TDMA packet sustains before transmission begins is

$$w_{TD} = \frac{1}{M} \sum_{m=1}^{M} (m - 1) \frac{T}{M} = \frac{T}{M^2} \sum_{n=0}^{M-1} n = \frac{T}{M^2} \frac{(M - 1)(M)}{2}$$

$$= \frac{T}{2} \left(1 - \frac{1}{M} \right) \qquad (9.12)$$

The maximum waiting time before transmission of a packet is $(M - 1)T/M$ seconds, and on the average a packet will wait $\frac{1}{2}(M - 1)(T/M) = (T/2)(1 - 1/M)$ seconds, as given by Equation (9.12).

To compare the average delay times, D_{FD} and D_{TD}, for FDMA and TDMA, respectively, we combine Equations (9.9) and (9.11) into Equation (9.8), and sim-

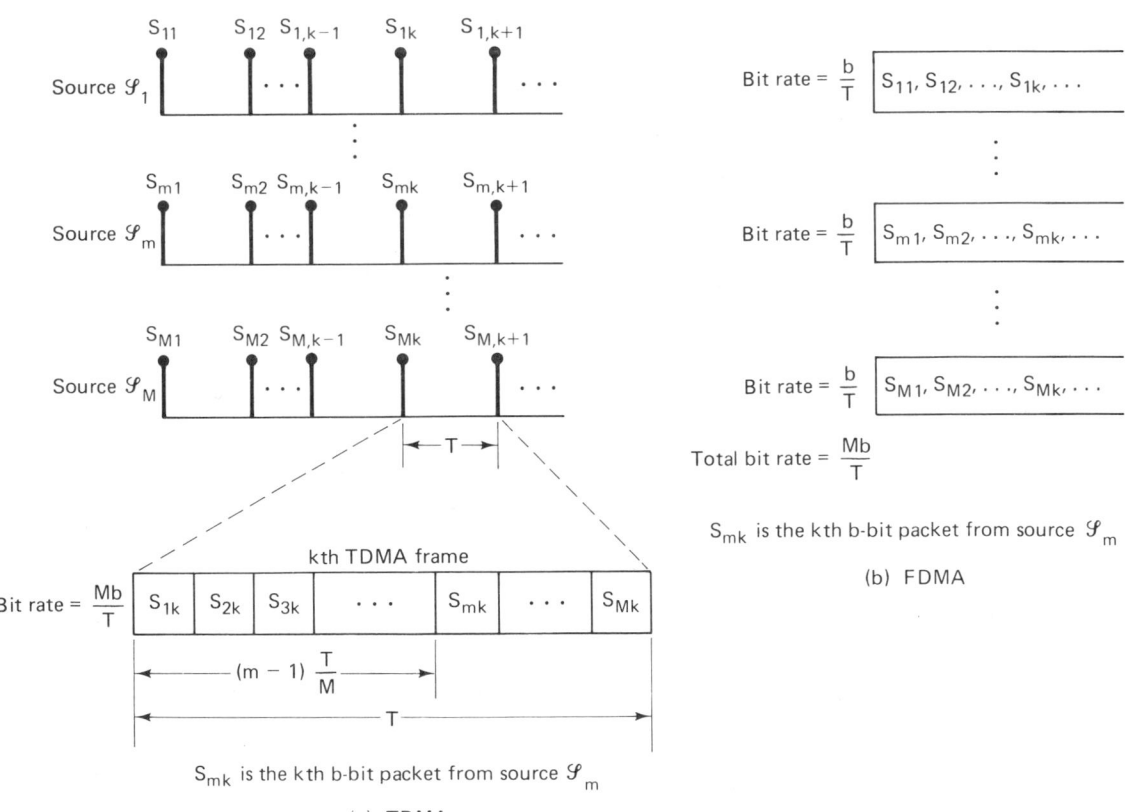

Figure 9.13   (a) TDMA and (b) FDMA channelization.

ilarly combine Equations (9.10) and (9.12) into Equation (9.8), yielding

$$D_{\text{FD}} = T \tag{9.13}$$

$$D_{\text{TD}} = \frac{T}{2}\left(1 - \frac{1}{M}\right) + \frac{T}{M} = D_{\text{FD}} - \frac{T}{2}\left(1 - \frac{1}{M}\right) \tag{9.14}$$

Using Equation (9.7), Equation (9.14) can be written as

$$D_{\text{TD}} = D_{\text{FD}} - \frac{b}{2R}\left(M - 1\right) \tag{9.15}$$

The result indicates that TDMA is inherently superior to FDMA, from a message delay point of view. Although Equation (9.15) assumed that the data source is deterministic, the smaller average message delays for TDMA schemes hold up for any independent message arrival process [1, 2].

9.1.5 Code-Division Multiple Access

In Figure 9.3 the CR plane was illustrated as being shared by slicing it horizontally to form FDMA frequency bands, and in Figure 9.7 the same CR plane was illustrated as being shared by slicing it vertically to form TDMA time slots. These two techniques are the most common choices for multiple access applications. Figure 9.14 illustrates the CR being partitioned by the use of a hybrid combination of FDMA and TDMA known as *code-division multiple access* (CDMA). CDMA

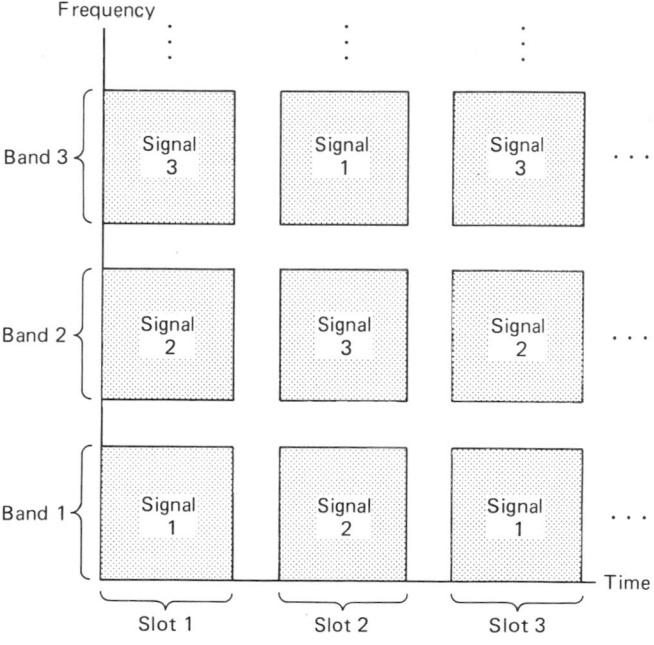

Figure 9.14 Code-division multiplexing.

is an application of spread-spectrum (SS) techniques. Spread-spectrum techniques can be classified into two major categories: *direct-sequence* SS and *frequency hopping* SS. We introduce frequency hopping CDMA (FH-CDMA) in this chapter, and we treat direct-sequence CDMA together with the overall subject of spread-spectrum techniques in Chapter 10.

It is easiest to visualize *frequency hopping* CDMA, illustrated in Figure 9.14, as the short-term assignment of a frequency band to various signal sources. At each successive time slot, whose duration is usually brief, the frequency band assignments are reordered. In Figure 9.14, during time slot 1, signal 1 occupies band 1, signal 2 occupies band 2, and signal 3 occupies band 3. During time slot 2, signal 1 hops to band 3, signal 2 hops to band 1, and signal 3 hops to band 2, and so on. The CR can thus be fully utilized, but the participants, having their frequency bands reassigned at each time slot, appear to be playing "musical chairs." Each user employs a pseudonoise (PN) code, orthogonal (or nearly orthogonal) to all the other user codes, that dictates the frequency hopping band assignments. Details of PN code sequences are treated in Section 10.2. Figure 9.14 is an oversimplified view of the way the CR is shared in frequency hopping CDMA, since the symmetry implies that each frequency hopping signal is in time synchronism with each of the other signals. This is *not the case*. In fact, one of the attractions of CDMA compared to TDMA is that there is no need for synchronization among user groups (only between a transmitter and a receiver within a group).

The block diagram in Figure 9.15 illustrates the frequency hopping modulation process. At each frequency hop time the PN generator feeds a code sequence to a device called a *frequency hopper*. The frequency hopper synthesizes one of the allowable hop frequencies. Assume that the data modulation has an *M*-ary frequency shift keying (MFSK) format. The essential difference between a conventional MFSK system and a frequency hopping (FH) MFSK system is that in the conventional system, a data symbol modulates a carrier wave that is *fixed* in frequency, but in the hopping system, the data symbol modulates a carrier

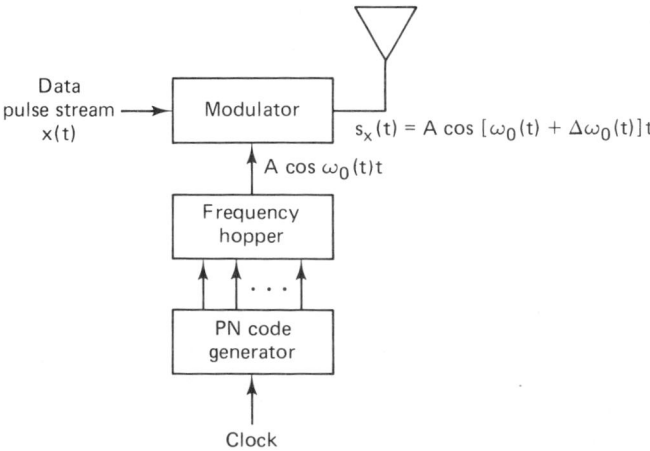

Figure 9.15 CDMA frequency hopping modulation process.

wave that *hops* across the total CR bandwidth. The FH modulation in Figure 9.15 can be thought of as a two-step process—data modulation and frequency hopping modulation—even though it can be implemented in a single step, where the modulator produces a transmission tone based on the simultaneous dictates of the PN code and the data. Frequency hopping systems are covered in detail in Section 10.4.

One might ask: Don't the FDMA and TDMA options provide sufficient multiple access flexibility? FDMA and TDMA methods can surely be relied on to apportion the communications resource equitably. Of what use is this hybrid technique? CDMA offers some unique advantages, as follows:

1. *Privacy.* When the code for a particular user group is only distributed among authorized users, the CDMA process provides communications privacy, since the transmissions cannot easily be intercepted by unauthorized users without the code.
2. *Fading channels.* If a particular portion of the spectrum is characterized by fading, signals in that frequency range are attenuated. In an FDMA scheme, a user who was unfortunate enough to be assigned to the fading position of the spectrum might experience highly degraded communications for as long as the fading persists. However, in a FH-CDMA scheme, only during the time a user hops into the affected portion of the spectrum will the user experience degradation. Therefore, with CDMA, such degradation is shared among all the users.
3. *Jam resistance.* During a given CDMA hop, the signal bandwidth is identical to the bandwidth of conventional MFSK, which is typically equal to the minimum bandwidth necessary to transmit the MFSK symbol. However, over a duration of many time slots, the system will hop over a frequency band which is much wider than the data bandwidth. We refer to this utilization of bandwidth as spread spectrum. In Chapter 10 we develop, in detail, the resistance to jamming that spread spectrum affords a user.
4. *Flexibility.* The most important advantage of CDMA schemes, compared to TDMA, is that there need be no precise time coordination among the various simultaneous transmitters. The orthogonality between user transmissions on different codes is not affected by transmission-time variations. This will become clear upon closer examination of the autocorrelation and cross-correlation properties of the codes, considered in Chapter 10.

9.1.6 Space-Division and Polarization-Division Multiple Access

Figure 9.16a depicts the INTELSAT IVA application of space-division multiple access (SDMA), also called *multiple-beam frequency reuse*. INTELSAT IVA used a dual-beam receive antenna feeding two receivers to allow simultaneous access of the satellite from two different regions of the earth. The frequency band

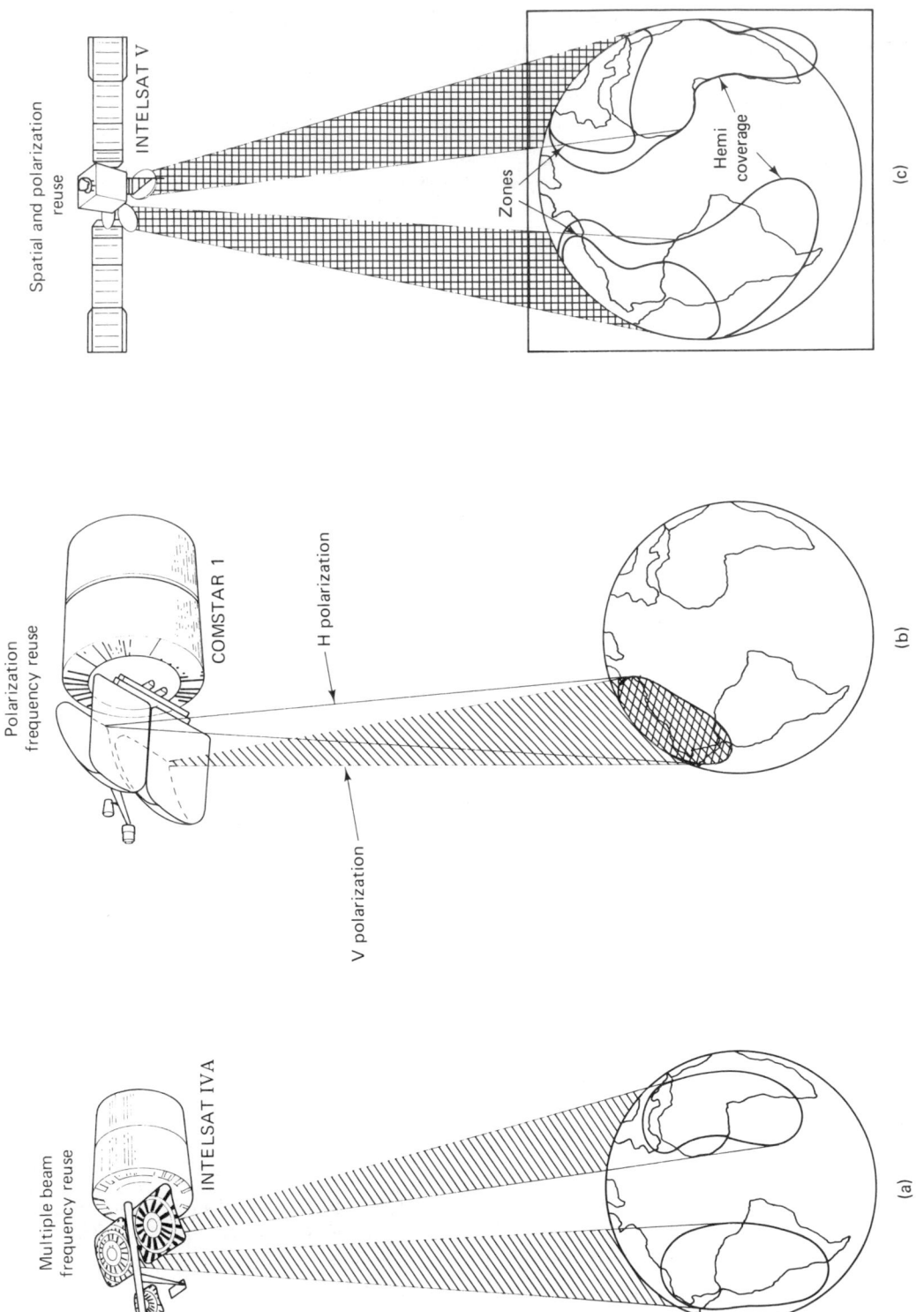

Figure 9.16 SDMA and PDMA. (a) INTELSAT IVA. (b) COMSTAR 1. (c) INTELSAT V (Atlantic coverage).

allocated to each receive beam was identical because the uplink signals were spatially separated. In such cases, the frequency band is said to be *reused*.

Figure 9.16b depicts an application of polarization-division multiple access (PDMA), also called *dual-polarization frequency reuse*, from COMSTAR 1. Here separate antennas are used, each with different polarization and followed by separate receivers, allowing simultaneous access of the satellite from the same region of the earth. Each corresponding earth station antenna needs to be polarized in the same way as its counterpart in the satellite. (This is generally accomplished by providing each participating earth station with an antenna that has dual polarization.) The frequency band allocated to each antenna beam can be identical because the uplink signals are orthogonal in polarization. As with SDMA, the frequency band in PDMA is said to be reused. Figure 9.16c depicts an application of the simultaneous use of SDMA and PDMA in INTELSAT V. There are two separate hemispheric coverages, west and east. There are also two smaller zone beams; each zone beam overlaps a portion of one of the hemispheric beams and is separated from it by orthogonal polarization. Thus there is a fourfold reuse of the spectrum.

9.2 MULTIPLE ACCESS COMMUNICATIONS SYSTEM AND ARCHITECTURE

A *multiple access protocol* or *multiple access algorithm* (MAA) is that rule by which a user knows how to use time, frequency, and code functions to communicate through a satellite to other users. A multiple access system is a com-

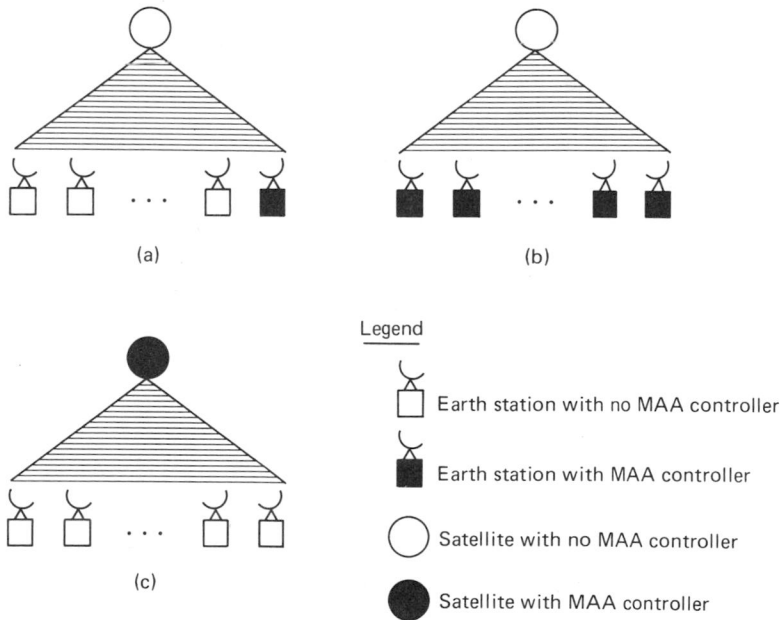

Figure 9.17 Satellite multiple access architecture. (a) Single earth station control. (b) Distributed earth station control. (c) Satellite control.

bination of hardware and software that supports the MAA. The general goal of a multiple access system is to provide communications service in a timely, orderly, and efficient way.

Figure 9.17 illustrates some basic choices for the architecture of a satellite multiple access system. The legend indicates the symbols used for an earth station with and without an MAA controller, and a satellite with and without an MAA controller. Figure 9.17a illustrates the case where one earth station is designated as the master, or the controller. This earth station possesses an MAA computer and responds to the service requests of all other users. Notice that a user's request entails a transmission through the satellite and back down to the controller. The controller's response entails another transmission through the satellite; hence there are two up- and downlink transmissions required for each service assignment. Figure 9.17b illustrates the case where the MAA control is distributed among all the earth stations; there is no single controller. Each earth station uses the same algorithm and they each have identical knowledge regarding access requests and assignments; therefore, only one round trip is required for each service assignment. Figure 9.17c illlustrates the case where the MAA controller is in the satellite. A service request goes from user to satellite, and the response from the satellite can follow immediately; therefore, only one round trip is required for each service assignment.

9.2.1 Multiple Access Information Flow

Figure 9.18 is a flow diagram describing the basic flow of information between the multiple access algorithm (MAA) or controller and an earth station; the numbers below correspond to those on the figure. Recall from the preceding section that the control may be lodged in the satellite, in a master station, or distributed among all the earth stations.

1. *Channelization.* This term refers to the most general allocation information [e.g., channels 1 to N may be allocated for the Army and channels $(N + 1)$

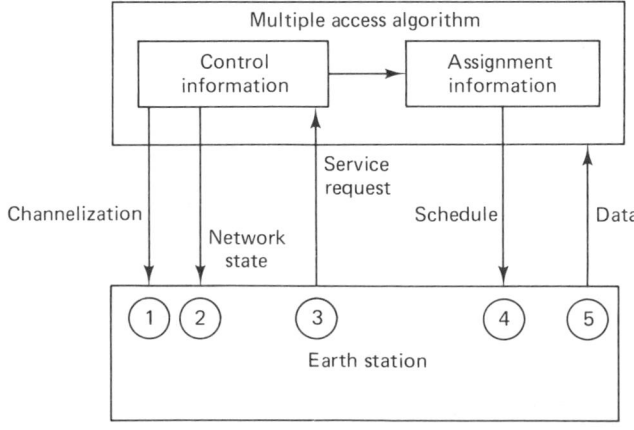

Figure 9.18 Multiple access information flow.

to *M* for the Navy]. This information seldom changes, and may be distributed to the earth stations by the use of a newsletter rather than via the communication system.

2. *Network state* (*NS*). This term refers to the state of the CR. A station is advised regarding the availability of the communications resource and where in the resource (e.g., time, frequency, code position) to transmit its service request(s).

3. *Service request.* Then the station makes its request(s) for service (e.g., allocation for *m* message slots).

4. Upon receipt of the service request(s), the controller sends the station a schedule regarding where and when to position its data in the CR.

5. The station transmits its data according to its assigned schedule.

9.2.2 Demand-Assignment Multiple Access

Multiple access schemes are termed *fixed assignment* when a station has periodic access to the channel independent of its actual need. By comparison, dynamic assignment schemes, sometimes called *demand-assignment multiple access* (DAMA), give the station access to the channel only when it requests access. If the traffic from a station tends to be burst-like or intermittent, DAMA procedures can be much more efficient than fixed-assignment procedures. A DAMA scheme capitalizes on the fact that actual demand *rarely* equals the peak demand. If a system's capacity is equal to the total peak demand and if the traffic is bursty, the system will be underutilized most of the time. However, by using buffers and DAMA, a system with reduced average capacity can handle bursty traffic, at the cost of some queueing delay. Figure 9.19 summarizes the difference between a fixed system, whose capacity is equal to the sum of the user requirements, and a dynamic system, whose capacity is equal to the average of the user requirements.

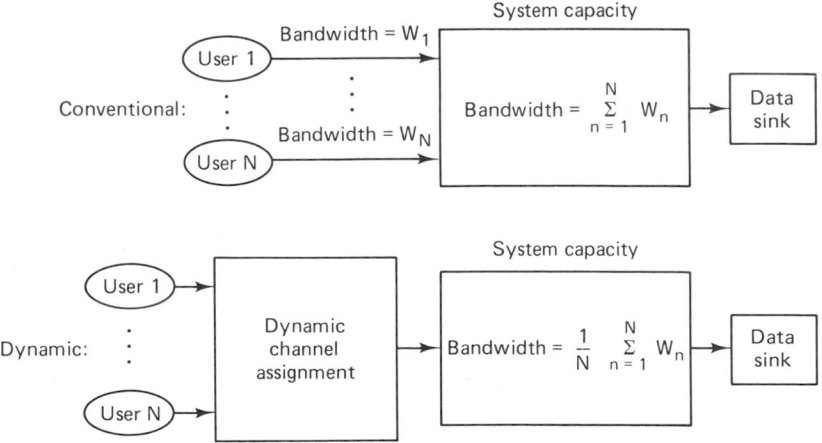

Figure 9.19   Bandwidth reduction for systems using dynamic channel assignment.

9.3 ACCESS ALGORITHMS

9.3.1 ALOHA

In 1971, the University of Hawaii began operation of its ALOHA system. A communication satellite was used to interconnect the several university computers by use of a random access protocol [3–7]. The system concept was extremely simple, consisting of the following modes:

1. *Transmission mode.* Users transmit at any time they desire, encoding their transmissions with an error detection code.
2. *Listening mode.* After a message transmission, a user listens for an acknowledgment (ACK) from the receiver. Transmissions from different users will sometimes overlap in time, causing reception errors in the data in each of the contending messages. We say that the messages have *collided*. In such cases, the errors are detected, and the users receive a negative acknowledgment (NAK).
3. *Retransmission mode.* When a NAK is received, the messages are simply retransmitted. Of course, if the colloding users were to retransmit immediately, they would collide again. Therefore, the users retransmit after a *random* delay.
4. *Timeout mode.* If, after a transmission, the user does not receive either an ACK or NAK within a specified time, the user retransmits the message.

9.3.1.1 Message Arrival Statistics

Assume that the total system demand requires an average message or packet arrival rate of λ successful or accepted messages per second. Because of the presence of collisions, some of the messages will be unsuccessful or rejected. Therefore, we define the total traffic arrival rate, λ_t, as the acceptance rate, λ, plus the rejection rate, λ_r, as follows:

$$\lambda_t = \lambda + \lambda_r \tag{9.16}$$

Let us denote the length of each message or packet as b bits. Then we can define the average amount of successful traffic or *throughput*, ρ', on the channel in units of bits per second, as

$$\rho' = b\lambda \tag{9.17}$$

We can also define the *total traffic*, G', on the channel in units of bits per second, as

$$G' = b\lambda_t \tag{9.18}$$

With the channel capacity (maximum bit rate) designated as R bits per second, let us further define a *normalized throughput*, ρ, and a *normalized total traffic*, G, as

$$\rho = \frac{b\lambda}{R} \qquad (9.19)$$

$$G = \frac{b\lambda_t}{R} \qquad (9.20)$$

Normalized throughput, ρ, expresses throughput as a fraction ($0 \le \rho \le 1$) of channel capacity. Normalized total traffic, G, expresses total traffic as a fraction ($0 \le G \le \infty$) of the channel capacity. Notice that G can take on values greater than unity.

We can also define the transmission time of each packet as folllows:

$$\tau = \frac{b}{R} \quad \text{seconds/packet} \qquad (9.21)$$

By substituting equation (9.21) into Equations (9.19) and (9.20), we can write

$$\rho = \lambda\tau \qquad (9.22)$$

$$G = \lambda_t\tau \qquad (9.23)$$

A user can successfully transmit a message as long as no other user began one within the previous τ seconds or starts one within the next τ seconds. If another user began a message within the previous τ seconds, its tail end will collide with the current message. If another user begins a message within the next τ seconds, it will collide with the tail end of the current message. Thus a space of 2τ seconds is needed for each message.

The message arrival statistics for unrelated users of a communication system is often modeled as a Poisson process. The probability of having K new messages arrive during a time interval of τ seconds is given by the Poisson distribution [8] to be

$$P(K) = \frac{(\lambda\tau)^K e^{-\lambda\tau}}{K!} \qquad K \ge 0 \qquad (9.24)$$

where λ is the average message arrival rate. Because the users transmit without regard for each other in the ALOHA system, this expression is useful for calculating the probability that exactly $K = 0$ other messages are transmitted during a time interval 2τ. This is the probability, P_s, that a user's message transmission was successful (experienced no collisions). To compute P_s, assuming that all traffic is Poisson, we use λ_t and 2τ in Equation (9.24). Thus

$$P_s = P(K = 0) = \frac{(2\tau\lambda_t)^0 e^{-2\tau\lambda_t}}{0!} = e^{-2\tau\lambda_t} \qquad (9.25)$$

In Equation (9.16) we defined total traffic arrival rate λ_t, in terms of the successful portion, λ, and the repetition or unsuccessful portion, λ_r; then, by definition, the probability of a successful packet can be expressed as

$$P_s = \frac{\lambda}{\lambda_t} \qquad (9.26)$$

By combining Equations (9.25) and (9.26), we have

$$\lambda = \lambda_t e^{-2\tau\lambda_t} \qquad (9.27)$$

By combining Equation (9.27) with Equations (9.22) and (9.23), we can write

$$\rho = Ge^{-2G} \qquad (9.28)$$

Equation (9.28) relates the normalized throughput, ρ, to the normalized total traffic, G, on the channel for the ALOHA system. A plot of this relationship labeled "pure ALOHA" is shown in Figure 9.20. As G increases, ρ increases until a point is reached where further traffic increases create a large enough collision rate to cause a reduction in the throughput. The maximum ρ, equal to $1/2e$ = 0.18, occurs at a value of $G = 0.5$. Therefore, for a pure ALOHA channel, only 18% of the CR can be utilized. Simplicity of control is achieved at the expense of channel capacity [7, 9].

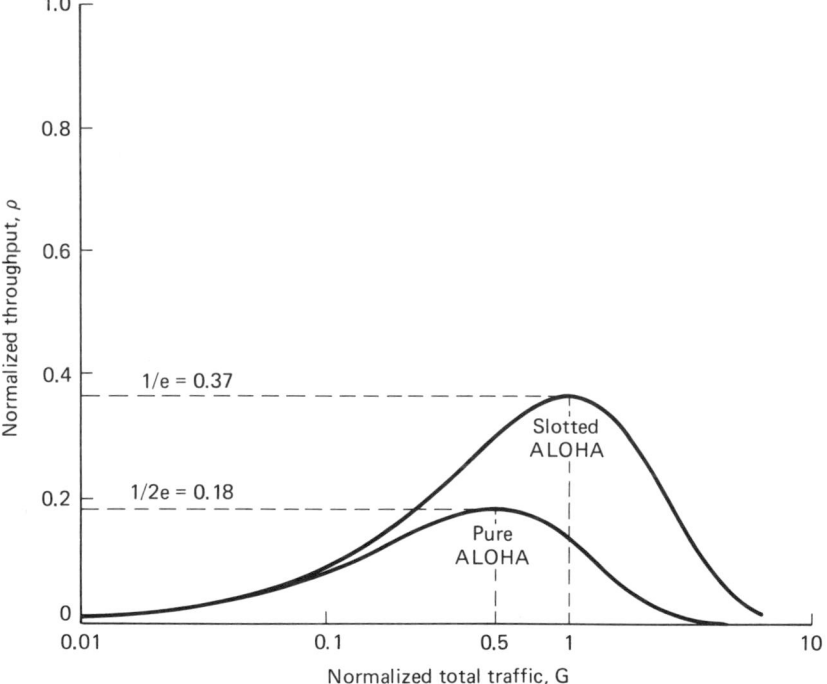

Figure 9.20 Throughput in ALOHA channels (successful transmissions versus total transmissions).

9.3.2 Slotted ALOHA

The pure ALOHA scheme can be improved by requiring a small amount of co-ordination among the stations. The slotted ALOHA (S-ALOHA) is such a system. A sequence of synchronization pulses is broadcast to all stations. As with pure

ALOHA, packet lengths are constant. Messages are required to be sent in the slot time between synchronization pulses, and can be started only at the *beginning* of a time slot. This simple change reduces the rate of collisions by half, since only messages transmitted in the same slot can interfere with one another. It can be shown [9, 10] that for S-ALOHA, the reduction in the *collision window* from 2τ to τ results in the following relationship between normalized throughput, ρ, and normalized total traffic, G.

$$\rho = Ge^{-G} \tag{9.29}$$

The plot of Equation (9.29) is shown in Figure 9.20 labeled "slotted ALOHA." Here the maximum value of ρ is $1/e = 0.37$, or an improvement of two times the pure ALOHA protocol.

The retransmission mode described for the pure ALOHA system was modified for S-ALOHA so that if a negative acknowledgment (NAK) occurs, the user retransmits after a *random* delay of an integer number of slot times. Figure 9.21 illustrates the S-ALOHA operation. A packet of data bits is shown transmitted by user k followed by the satellite acknowledgment (ACK). Also shown are users m and n simultaneously transmitting packets, which results in a collision; a NAK is returned. Each using station employs a random-number generator to select its retransmission time. The figure illustrates an example of the m and n retransmission at their respective randomly selected times. Of course, there is some probability that users m and n will recollide. However, in that case, they simply repeat the retransmission, using another random delay.

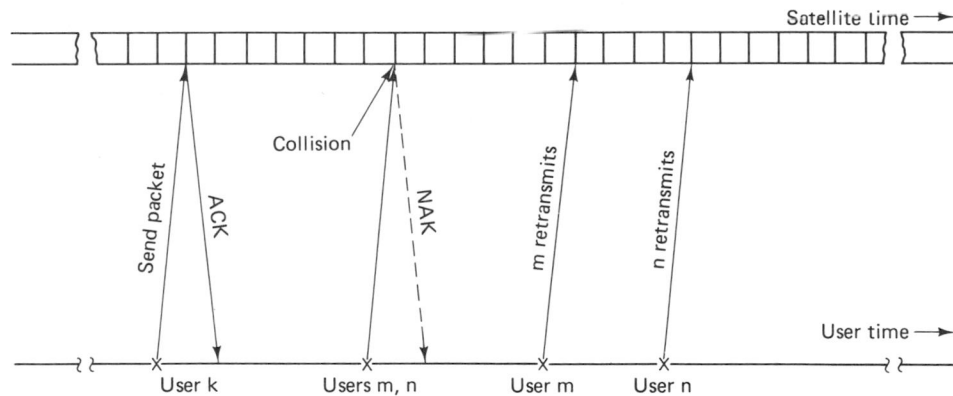

Figure 9.21 Random access scheme: slotted ALOHA operation.

Example 9.1 Poisson Process

Assuming that packet transmissions and retransmission can both be described as a Poisson process, calculate the *probability* that a data packet transmission in an S-ALOHA system will experience a collision with *one other user*. Assume that the total traffic rate $\lambda_t = 10$ packets/s and the packet duration $\tau = 10$ ms.

Solution

$$P(K = 1) = \left.\frac{(\tau\lambda_t)^K e^{-\tau\lambda_t}}{K!}\right|_{K=1}$$

$$= (10 \times 0.01)^1 e^{-0.1} = 0.1e^{-0.1}$$

$$= 0.09$$

9.3.3 Reservation-ALOHA

A significant improvement was made to the ALOHA system with the introduction of the reservation-ALOHA (R-ALOHA) [11] scheme. The R-ALOHA system has two basic modes: an unreserved mode and a reserved mode; each is described below.

Unreserved Mode (Quiescent State)

1. A time frame is established and divided into a number of small reservation subslots.
2. Users use these small subslots to reserve message slots.
3. After requesting a reservation, the user listens for an acknowledgment and a slot assignment.

Reserved Mode

1. The time frame is divided into $M + 1$ slots whenever a reservation is made.
2. The first M slots are used for message transmissions.
3. The last slot is subdivided into subslots to be used for reservation/requests.
4. Users send message packets only in their assigned portions of the M slots.

Consider the R-ALOHA example shown in Figure 9.22. In the quiescent state, with no reservations, time is partitioned into short subslots for making reservations. Once a reservation is made, the system is configured so that $M = 5$ message slots followed by $V = 6$ reservation subslots becomes the timing format. The figure illustrates a request and an acknowledgment in progress. In this example the station seeks to reserve three message slots. The reservation acknowledgment advises the using station where to locate its first data packet. Since the control is distributed so that all participants receive the downlink transmissions and are thus aware of the reservations and time format, the acknowledgment need not disclose any more than the location of the first slot. As shown in Figure 9.22, the station sends its second packet in the slot following the first packet. The user further knows that the next slot is comprised of six subslots for reservations, so *no* packets are transmitted during this time. The third and final packet is sent in the following slot. When there are no reservations taking place, the system reverts back to its quiescent format of subslots only. Since the control is distributed, all the participants are made aware of the quiescent format by receiving appropriate

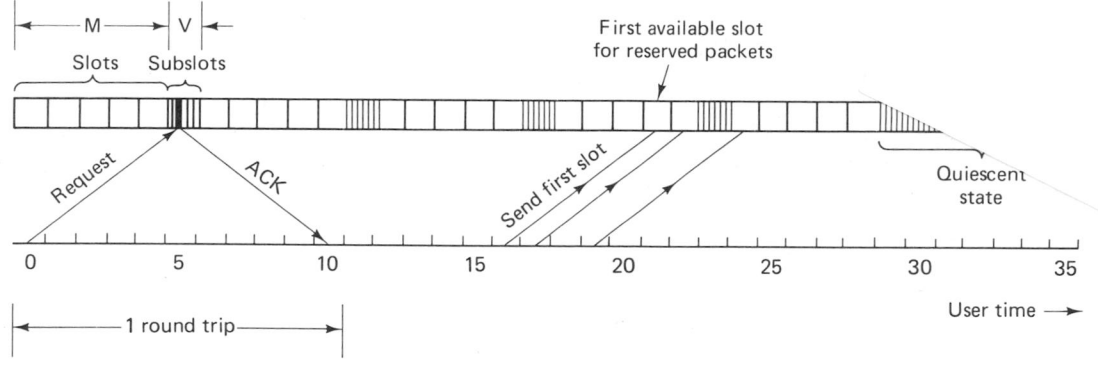

Figure 9.22 Example of reservation ALOHA. Station seeks to reserve three slots ($M = 5$ slots, $V = 6$ subslots).

synchronizing pulses on the downlink. Other interesting reservation schemes are discussed in References [12, 13].

9.3.4 Performance Comparison of S-ALOHA and R-ALOHA

From Chapter 3 the basic quality measure of a digital modulation scheme is its P_B versus E_b/N_0 curve. This measure is particularly useful because E_b/N_0 is a *normalized signal-to-noise ratio*; being normalized, the curves allow us to compare the performance of various modulation schemes. There is a similar performance measure for multiple access schemes. Here we are interested in the average delay versus normalized throughput. What would an *ideal delay–throughput curve* look like? Figure 9.23 illustrates such a curve. For normalized throughput values

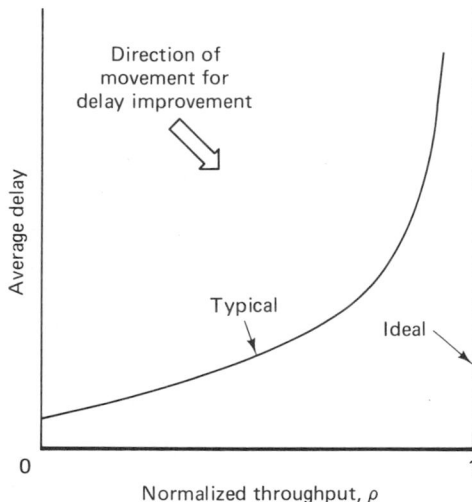

Figure 9.23 Delay–throughput characteristic.

of, $0 \leq \rho < 1$, the delay equals zero until $\rho = 1$; then the delay increases without bound. Figure 9.23 also shows a *typical* delay–throughput curve and the direction in which the curve will move as delay performance improves.

Figure 9.24 compares the delay–throughput performance of S-ALOHA with that of R-ALOHA (formatted with two message slots and six reservation subslots). Knowing the location of the *ideal* curve it is easy to compare the delay performance of these two systems. For a throughput of less than approximately 0.20, the S-ALOHA manifests less average delay than does R-ALOHA. But for values of ρ between 0.20 and 0.67, it is apparent that R-ALOHA is superior, since the average delay is less. Why does the S-ALOHA perform better at low traffic intensity? The S-ALOHA algorithm does not require the overhead of the reservation subslots as does R-ALOHA. Therefore, at low values of ρ, R-ALOHA pays the price of greater delay due to the greater overhead. For $\rho > 0.2$, the collisions and retransmissions inherent in the S-ALOHA system cause it to incur greater delay (unbounded at $\rho = 0.37$), more quickly than the R-ALOHA system. At higher throughput ($0.2 < \rho < 0.67$), the overhead structure of R-ALOHA ensures that its delay degradation grows in a more orderly manner than S-ALOHA. For R-ALOHA, an unbounded delay is not reached until $\rho = 0.67$.

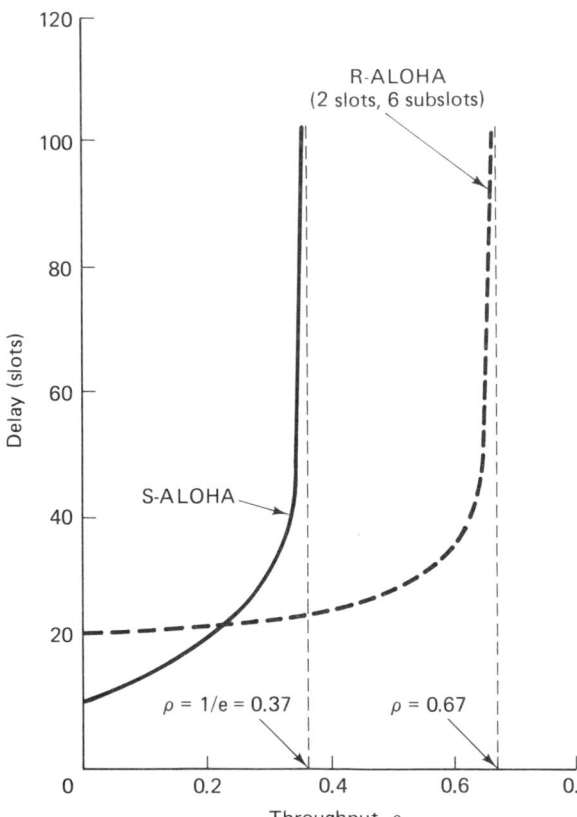

Figure 9.24 Delay–throughput comparison: S-ALOHA versus R-ALOHA on a satellite channel.

Example 9.2   Channel Utilization

    (a) Normalized throughput, ρ, is a measure of channel utilization. It can be found by forming the ratio of the successfully transmitted message traffic, in bits per second to the total message traffic, including rejected messages, in bits per second. Calculate the normalized throughput of a channel that has a maximum data rate $R = 50$ kbits/s and operates with $M = 10$ ground stations, each station transmitting at the average rate of $\lambda = 2$ packets/second. The system format provides for $b = 1350$ bits/packet.

    (b) Which of the three ALOHA schemes discussed—pure, slotted, and reservation—could be successfully used with this channel?

Solution

    (a) Generalizing Equation (9.19) to allow for traffic from multiple stations, we have

$$\rho = \frac{Mb\lambda}{R}$$

$$= \frac{10(1350)(2)}{50,000}$$

$$= 0.54$$

    (b) Only the R-ALOHA scheme could be used for this system, since with each of the other schemes, 54% of the resource cannot be utilized.

9.3.5 Polling Techniques

One way to impose order on a system with multiple users having random access requirements is to institute a controller that periodically polls the user population to determine their service requests. If the user population is large (e.g., thousands of terminals) and the traffic is bursty, the time required to poll the population can be an excessive overhead burden. One technique for rapidly polling a user population [4, 14] is called a *binary tree search*. Figure 9.25 illustrates a satellite example of such a tree search to resolve contention among users. In this example, assume that the total user population is eight terminals; let them be identified by the binary numbers 000 to 111 as shown in Figure 9.25. Assume that terminals 001, 100, and 110 are contending for the service of a single channel. The tree search operates by continually partitioning the population until there is just a single branch remaining. The terminal corresponding to that branch is the "winner" and hence the first terminal to access the channel. The operation is repeated and again yields a single terminal that may next use the channel. The algorithm proceeds according to the following steps (see Figure 9.25):

1. The satellite requests the transmission of the contending terminals' first (leftmost) bit of their identification (ID) numbers.

2. Terminal 001 transmits a zero, and terminals 100 and 110 each transmit a one. The satellite, on the basis of received signal strength, selects one or zero as the bit it "heard." In this example the satellite chooses binary one

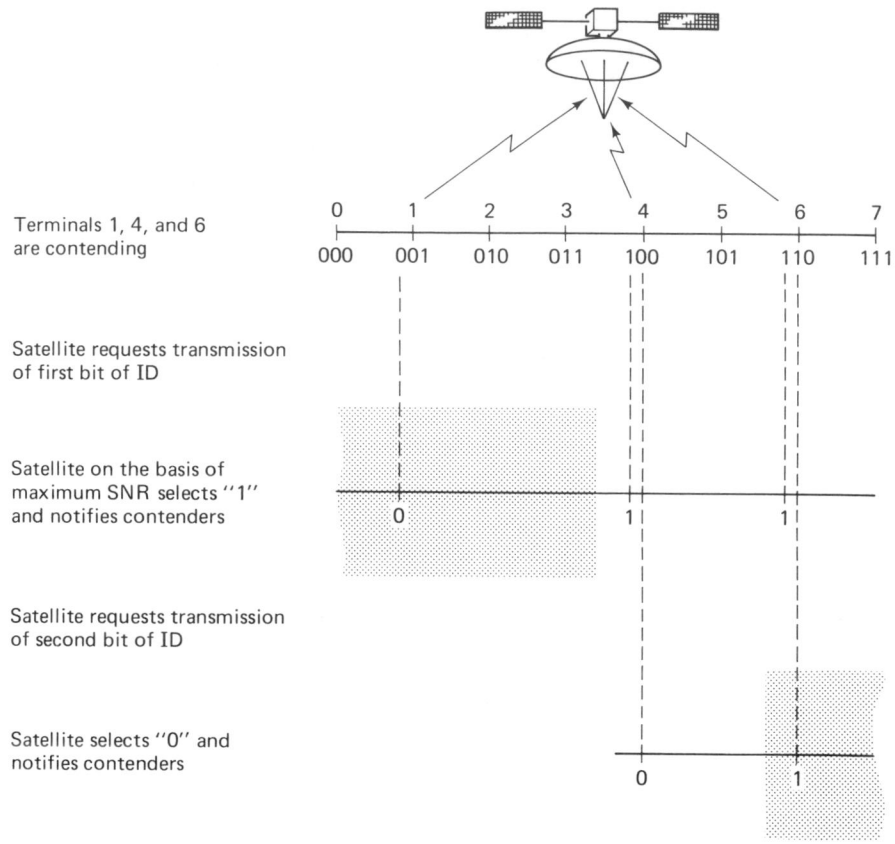

Figure 9.25 Tree search to resolve contention (eight terminal example).

and informs the users accordingly. Half the user population now knows that it has not been selected. The terminals in the "losing" half "bow out" of contention during this pass through the tree. In this example terminal 001 bows out.

3. The satellite requests the transmission of the second identifying bit from the remaining contending terminals.

4. Terminal 100 transmits a zero, and terminal 110 transmits a one.

5. Assume that the satellite selects the zero and notifies the contenders accordingly. Terminal 110 bows out. The process continues until it is clear that terminal 100 is free to access the satellite.

6. When the channel becomes available, steps 1–5 are repeated.

**Example 9.3 Comparison between Binary Tree Search and
Straight Polling**

(a) A binary tree search requires $n = \log_2 Q$ decisions for each pass through a population of Q terminals. A savings in time is possible with a tree search if the population is large and the average demand for service is small. Calculate the time needed for the straight polling of a population of 4096 terminals, to provide channel availability to 100 terminals requesting service. Compare the result with the time needed to perform a binary tree search 100 times, over the same population. Assume that the time required to poll one terminal and the time required for one decision of a binary tree search are each equal to 1 s.

(b) Develop an expression for Q', the largest number of terminals that results in the same (or less) time expended for binary tree searching as compared to straight polling.

(c) Compute Q' for part (a).

Solution

(a) Straight polling of 4096 terminals:

$$T = 4096 \times 1 \text{ s} = 4096 \text{ s}$$

Binary tree search for 100 terminals requires 100 passes through the binary tree:

$$T' = (100 \times \log_2 4096) \times 1 \text{ s} = 1200 \text{ s}$$

(b) Q' is the maximum number of terminals that will result in $T' \leq T$ in part (a). This will occur when

$$Q'' \log_2 Q \times 1 \text{ s/decision} = Q \times 1 \text{ s/poll}$$

$$Q' = \lfloor Q'' \rfloor = \left\lfloor \frac{Q}{\log_2 Q} \right\rfloor \tag{9.30}$$

where $\lfloor x \rfloor$ is the largest integer no greater than x.

(c) Q' for part (a)

$$Q' = \left\lfloor \frac{4096}{\log_2 4096} \right\rfloor = 341 \text{ terminals}$$

A binary tree search for 341 terminals entails a search time of 4092 s.

9.4 MULTIPLE ACCESS TECHNIQUES EMPLOYED WITH INTELSAT

The first commercial, geostationary communication satellite (INTELSAT I, or Early Bird) launched in 1965, represented the start of a new telecommunications era. Its 240 voice circuits provided more capacity than the undersea cables laid between the United States and Europe during the previous 10 years [15].

Early Bird featured a hard-limiting nonlinear transponder using FDMA. When several signals having different carrier frequencies simultaneously occupy a nonlinear device, the result is the production of intermodulation products which are signals at all combinations of sum and difference frequencies [16–18]. The energy apportioned to these intermodulation or IM products represents a *loss* in the useful signal energy. In addition, if these IM products appear within the bandwidth of other signals, the effect is that of added *noise* for the other signals.

The nonlinear transponder in Early Bird allowed for only two earth stations (one in the United States and one in Europe) to simultaneously access the satellite.

Figure 9.26 illustrates this satellite's operation between the United States and Europe. Three European earth stations were interconnected via a terrestrial network. Each month a different European station accessed the satellite and distributed the traffic to the other two stations.

9.4.1 Preassigned FDM/FM/FDMA or MCPC Operation

INTELSAT II and III improved multiple access capability by operating their travelling-wave tube amplifiers (TWTA) in the linear region instead of the hard-limiting region. This kept the IM products at an acceptable level, allowing more than two simultaneous accesses. (The price paid was a reduction in power amplifier efficiency.) Thus many FM carriers from various earth stations could simultaneously access these satellites. The operation is designated preassigned multidestination FDM/FM/FDMA or simply FDM/FM, or multichannel per carrier (MCPC), and is illustrated in Figure 9.27. Long-distance calls originating in country A enter the telephone exchange and are multiplexed into a supergroup (five groups of 12 voice circuits each). Country A transmits the supergroup on a single FM carrier at frequency f_A. Each group within the supergroup has been preassigned to an earth station in country A for telephone traffic destined to countries

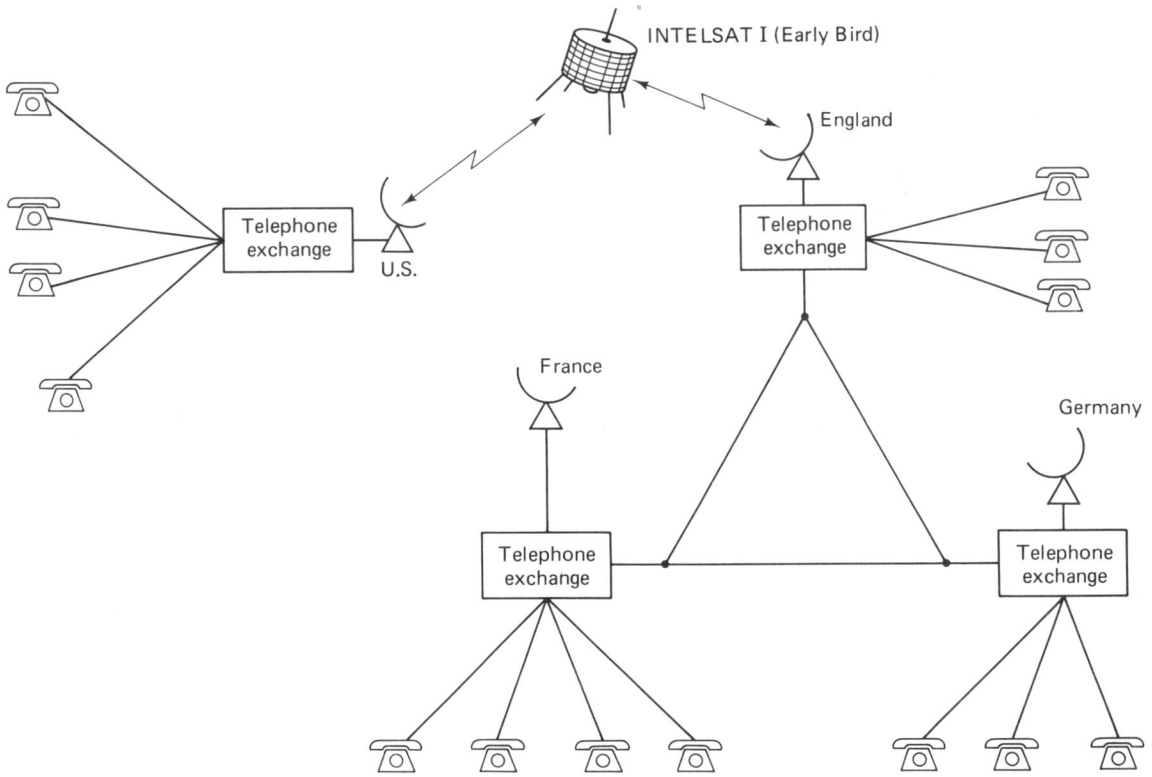

Figure 9.26 Early satellite operation.

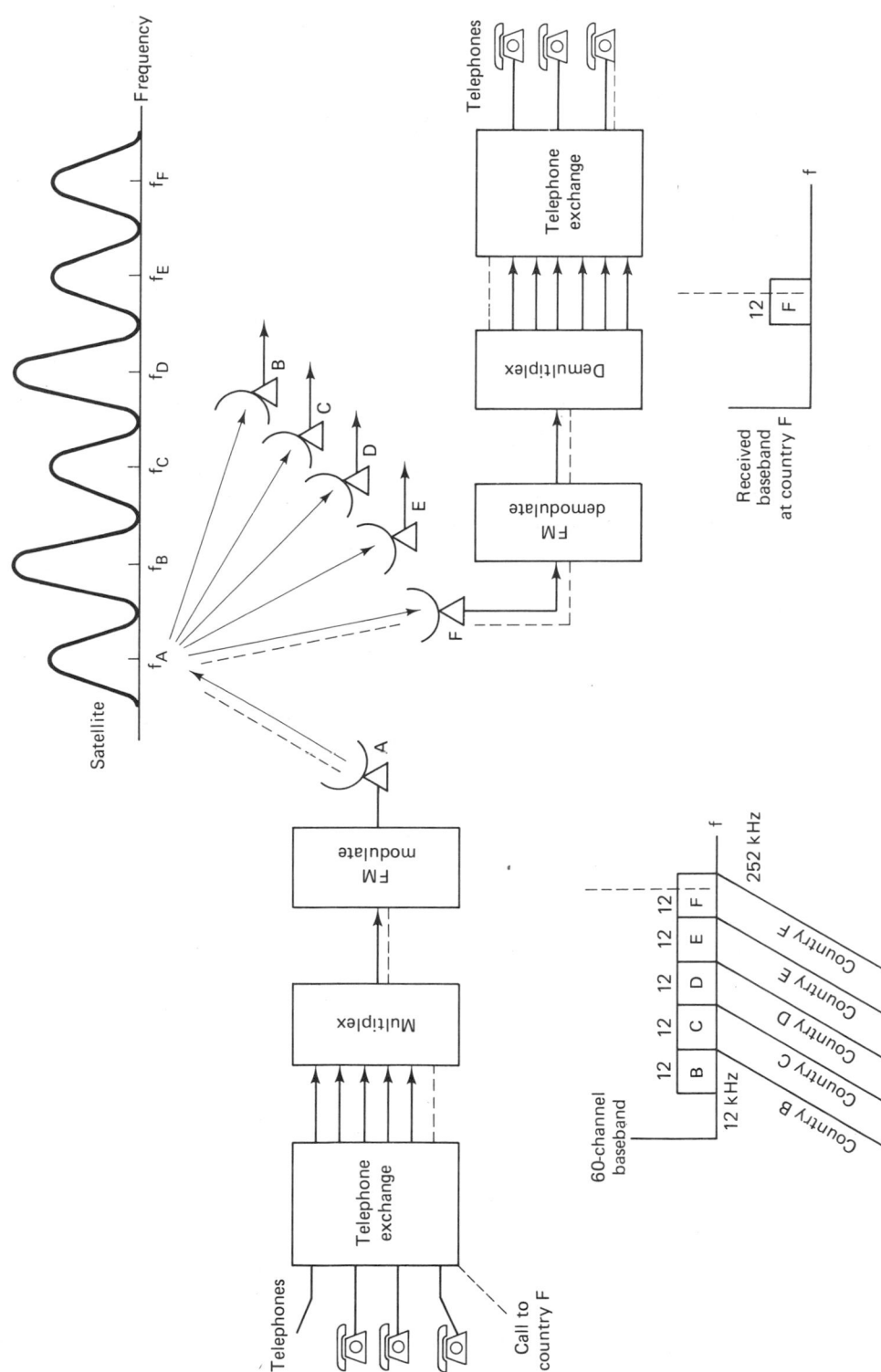

Figure 9.27 Preassigned multidestination FDM/FM carriers. (Reprinted with permission from J. G. Puente and A. M. Werth, "Demand-Assigned Service for the INTELSAT Global Network," *IEEE Spectrum*, Jan. 1971. © 1971 IEEE.)

509

B through *F*. These countries each receive the signal on frequency f_A. The received signal is demodulated and demultiplexed at the destination country, selecting only those 12 channels preassigned to it.

9.4.2 MCPC Modes of Accessing an INTELSAT Satellite

INTELSAT has standardized the ways in which each 36-MHz transponder may be shared by specifying the occupied RF bandwidth and the number of 4-kHz channels per user. Some of these standard channels are shown in Table 9.1. Notice that the capacity of the transponder (last column in Table 9.1) drops as the number of carriers increases. The reasons are as follows:

1. Guard bands are needed between carrier bands; the more carriers there are, the more guard bands are needed. Hence capacity is reduced.
2. Multiple carriers in the nonlinear TWTA cause intermodulation (IM) products. If the TWTA is backed off into the linear region to reduce interference, the TWTA can provide less overall power. The channel becomes power limited and can service fewer carriers.

TABLE 9.1 Standard INTELSAT MCPC Accessing Modes

Number of carriers per transponder	Carrier bandwidth	Number of 4-kHz channels per carrier	Number of 4-kHz channels per transponder
1	36 MHz	900	900
4	3 at 10 MHz	132	456
	1 at 5 MHz	60	
7	5 MHz	60	420
14	2.5 MHz	24	336

Table 9.1 indicates that a single carrier provides the most efficient use of the transponder. Why doesn't INTELSAT always operate its transponders in this mode? The answer is that not all earth stations have enough traffic to justify the assignment of an entire 36-MHz transponder. The other modes are needed so that various combinations of stations having less traffic will be able to share a transponder.

9.4.2.1 Bandwidth-Limited versus Power-Limited Conditions

In the preceding section it was stated that the backed-off transponder cannot support as many channels as the fully saturated transponder. It is useful to examine the two extreme transponder conditions, bandwidth limited and power limited, in the context of a satellite transponder. In Figure 9.28 we assume a 36-MHz transponder with a maximum power output of 20 W. Figure 9.28a illustrates an MCPC mode of operation whereby four carriers share the 36-MHz bandwidth. Assume that each carrier requires 4 W. The total output power is 16 W (less than the maximum capability of the amplifier); therefore, there is still power to spare.

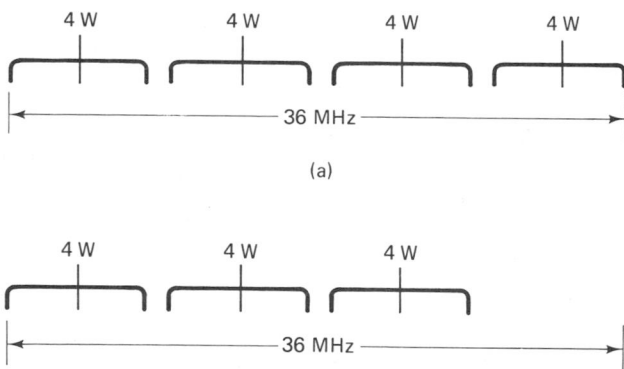

Figure 9.28 Bandwidth-limited versus power-limited configurations. (a) Bandwidth-limited example. (b) Power-limited example.

However, should another user want to access the transponder, the total 36-MHz bandwidth has already been allocated to the existing four carriers; there is no additional bandwidth to spare. Figure 9.28a illustrates this bandwidth-limited case.

Suppose that the previous example results in the production of serious IM products at the transponder. Assume that it is necessary to linearize the transponder by operating it at a reduced maximum power output of 12 W. With only a 12-W capability, the transponder can no longer support four users with 4 W each. One of the users must be "thrown off," as illustrated in Figure 9.28b. Therefore, we have bandwidth to support another user, but not sufficient power. Figure 9.28b illustrates this power-limited case.

9.4.3 SPADE Operation

The preassigned MCPC multiple access scheme is very efficient when the traffic is heavy enough so that the channels are most always filled. However, if out of a 12-channel group, only one channel is active, the other 11 cannot be turned off. The FDM/FM transmission is made with or without actual telephone traffic on the channels. Therefore, the long-term preassignment of carriers to stations having light traffic is wasteful. Since there are many light traffic links, a flexible method to service them was needed. Also, an efficient way to handle overflow traffic from medium-capacity preassigned links was needed. Such was the motivation for a novel DAMA scheme known as SPADE, first used with INTELSAT IV. The acronym SPADE stands for "single-channel-per-carrier PCM multiple access demand assignment equipment." The principal features characterizing SPADE operation [15] are:

1. A single voice-grade channel is analog-to-digital (A/D) converted at a bit rate of 64 kbits/s.
2. This baseband digital signal modulates a carrier using quadrature phase shift keying (QPSK). Unlike the MCPC case, there is *only one* voice channel per carrier.

3. The channel spacing is 45 kHz. Within a transponder, there is bandwidth available for 800 channel carriers. Six carrier positions are vacant by design; thus there are 794 usable carriers.

4. The carrier is dynamically assigned, *upon demand.*

5. The dynamic assignment is accomplished over a 160-kHz common signaling channel (CSC) used as an "order-wire" or control circuit. The bit rate on the CSC is 128 kbits/s, and the modulation is binary phase shift keying (BPSK).

Figure 9.29 illustrates the frequency allocations for the CSC and the 800 carriers in the SPADE system. The SPADE operation can best be understood with the aid of Figure 9.30. The CSC operates in a fixed-assignment TDMA broadcast mode; that is, all earth stations monitor the CSC and are aware of the current state of channel assignments. Each earth station has a 1-ms time slot on the CSC (once every 50 ms) for requesting or releasing a channel. When an earth station needs a channel, it "seizes" a free one by requesting a frequency pair at random and transmitting its selection on the CSC. Random selection makes it unlikely that two stations will simultaneously request the same channel unless there are very few remaining. As soon as the channel is allocated, each of the other earth station processors deletes it from its list of available channels. The list is continually kept updated via the CSC. Thus control of the SPADE access scheme is *distributed* among all the participating earth stations.

When the station finishes with the channel, the station indicates the channel's release by transmitting a signal in its time slot on the CSC. Each station receives this signal and designates the released channel as available. If two stations simultaneously seize the same channel, they each get a "busy" indication. They try again, selecting at random from the pool of available channels.

9.4.3.1 Transponder Capacity Utilization with SPADE

Table 9.2 is a continuation of Table 9.1. We see that the transponder bandwidth utilization with SPADE results in a total capacity of 800 voice channels per transponder. Compare Table 9.2 with Table 9.1. In Table 9.1, as the number of carriers increases from 1 to 14, the total number of channels decreases from 900 to 336. Why doesn't the SPADE system in Table 9.2 exhibit less capacity than the 336 channels associated with 14 carriers? The improved utilization comes about as follows. When there is only one voice channel per carrier, the carrier can be switched off when no speech is detected. Even with all channels operating, they can be switched off approximately 60% of the time. The transponders are power limited; power savings means that more channels can be transmitted. Also, SPADE uses digital voice transmission (QPSK); the bandwidth efficiency of the system is commensurate with the single-carrier FDM/FM case.

9.4.3.2 SPADE Efficiency

With MCPC, capacity is preassigned; a station's unused channels cannot be reallocated to other stations. SPADE is a DAMA system where all channels are shared. The channels are allocated to users as needed. An important telephone

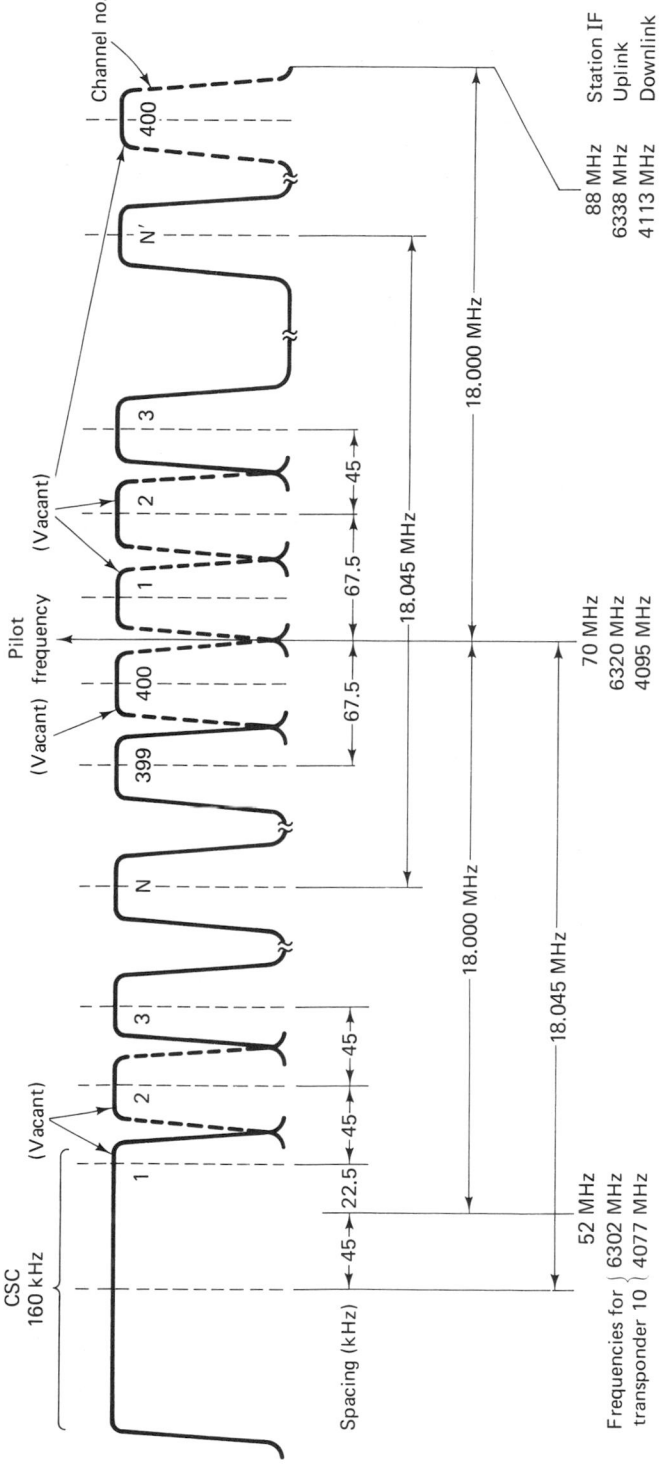

Figure 9.29 SPADE frequency allocations. (Reprinted with permission from J. G. Puente and A. M. Werth, "Demand-Assigned Service for the INTELSAT Global Network," *IEEE Spectrum*, Jan. 1971. © 1971 IEEE.)

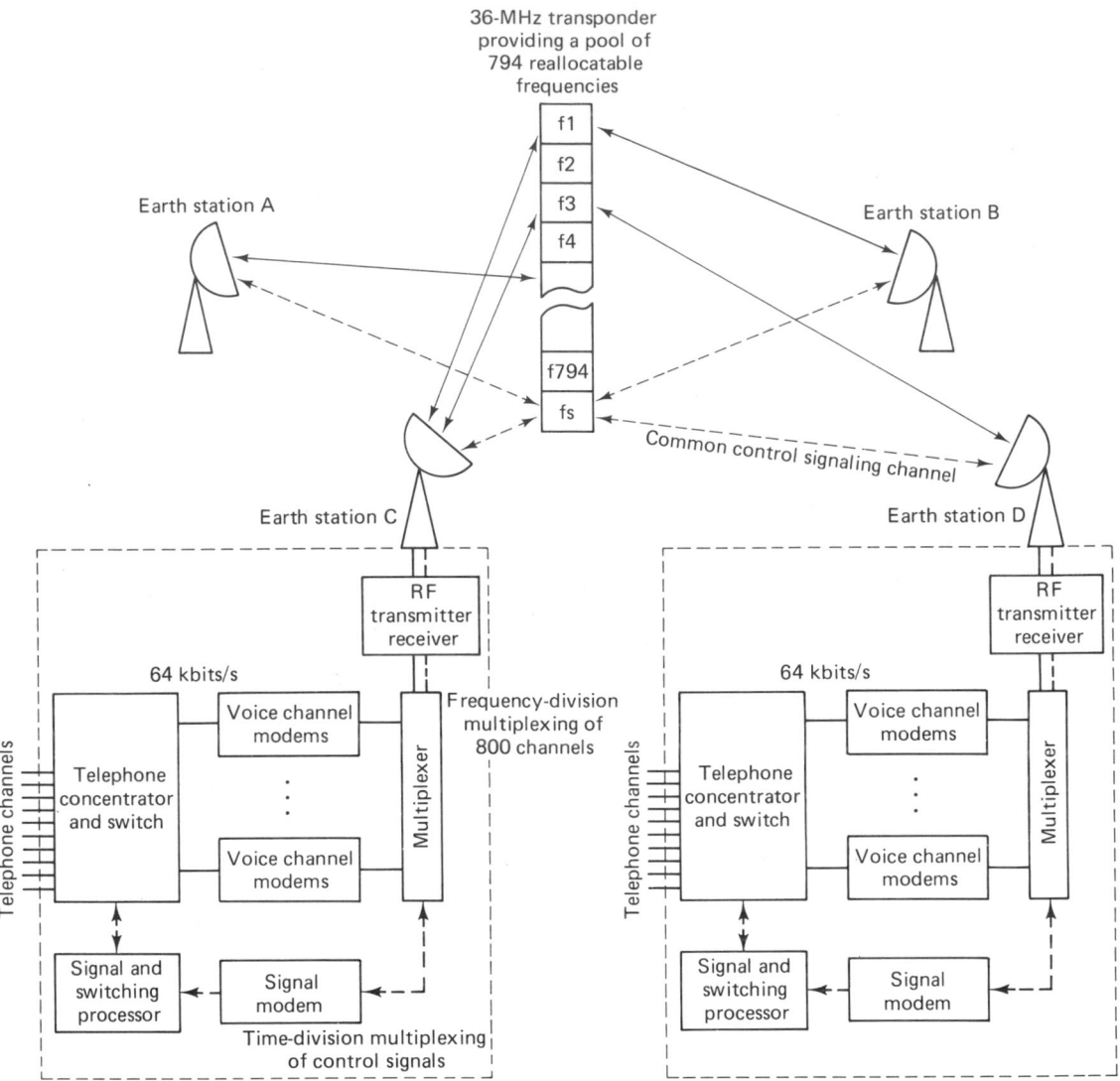

Figure 9.30 SPADE operation. (From James Martin, *Communications Satellite Systems*, © 1978, Fig. 15.2, p. 236. Reprinted by permission of Prentice-Hall, Englewood Cliffs, N.J.)

TABLE 9.2 SPADE Accessing

Number of carriers per transponder	Carrier bandwidth	Number of 4-kHz channels per carrier	Number of 4-kHz channels per transponder
800	45 kHz	1	800

system quality measure, called the probability of blocking, is the probability that a requested circuit is not available. To achieve 1% probability of blocking requires four times as many MCPC channels as SPADE channels. A SPADE transponder with 800 channels is equivalent to 3200 MCPC channels [15].

9.4.3.3 Mixed-Size Earth Station Network Using SPADE

A standard-size INTELSAT earth station has a receiver sensitivity $G/T° = 40.7$ dB/K, whereas the smaller size stations have a $G/T° = 35$ dB/K. If 125 SPADE channels are destined for small stations, the total transponder capacity of 800 standard channels is reduced to 525 channels. This is the point at which half the available power is used to service the standard stations. The relationship between transponder capacity and channels allocated to small stations is shown in Figure 9.31. An explanation of this relationship can best be seen in Figure 9.32. When the total TWTA power provides service to large stations, Figure 9.32a illustrates that the 36-MHz bandwidth transponder is occupied by approximately 800 carriers each at a power level of x dBW (the bandwidth-limited case). When half the power is required to service small stations, Figure 9.32b illustrates that 400 carriers (half of the original 800) each at a power level of x dBW are reserved for the standard stations. Consider what happens to the remaining 400 carrier positions. From Chapter 4 we know that the error performance of a link is directly related to the product of EIRP and $G/T°$. For any link, one can trade off these two parameters, thereby maintaining a fixed level of performance. Since the small station has a $G/T°$ of 5.7 dB less than that of the standard station, it is necessary to supply the small station with 5.7 dB more EIRP for equivalent performance. The carrier power is *increased* by approximately 5.7 dB for each small station, thus the quan-

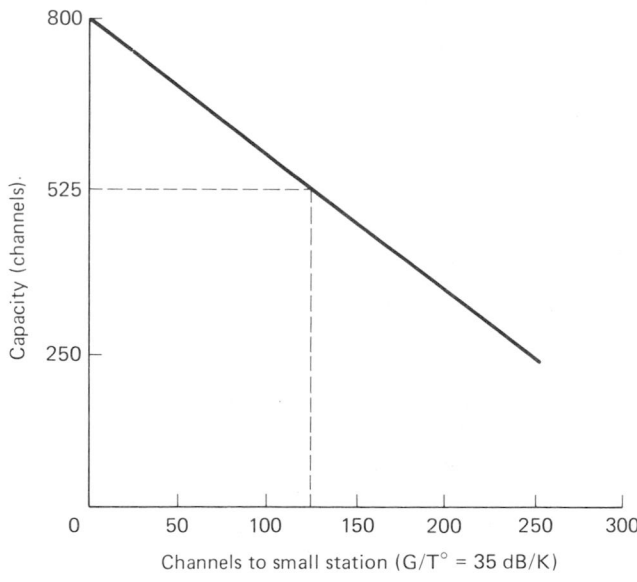

Figure 9.31 SPADE transponder capacity in a mixed-size earth station network.

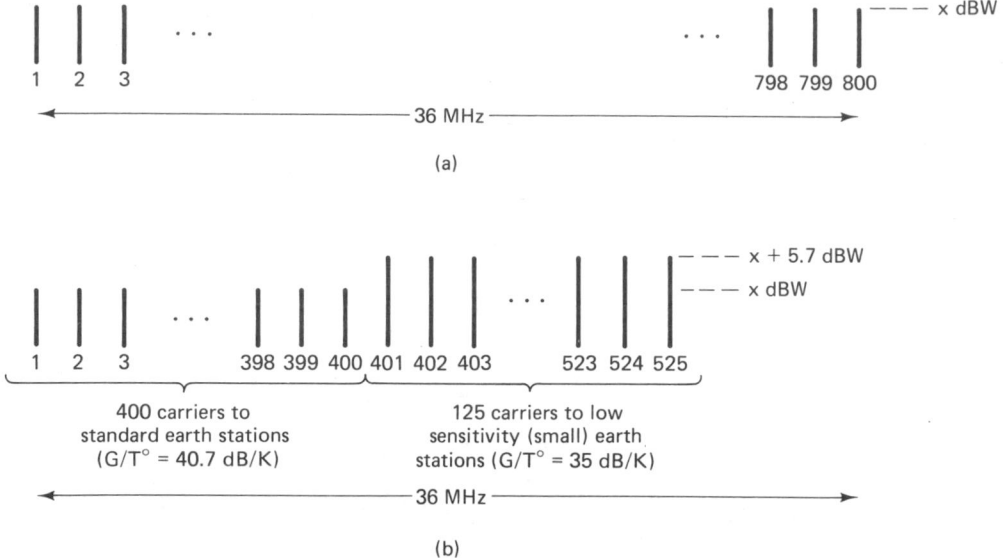

Figure 9.32 Mixed-size earth station network. (a) When the total TWTA power services large stations: bandwidth limited (800 carriers). (b) When half the TWTA power services small stations: power limited (525 carriers).

tity of the remaining carriers serving these small stations is *decreased* by a similar amount. Therefore instead of 400 carriers, 125 (a reduction of 5.1 dB) are used to serve the small stations; the transponder is now power limited.

At the time a channel is assigned to a call, the transmitting station is apprised of the size of the destination station. Recall that these satellites are nonregenerative so that the apportionment of downlink EIRP is established by the transmitting station (see Section 4.7.1). The transmitting station sets its power level according to the needs of the receiving station.

9.4.4 TDMA in INTELSAT

The first generation of multiple access communication systems has been dominated by FDMA systems. The trend, however, is now in favor of TDMA systems, made possible by the availability of precise clocks and high-speed switching elements [19–24]. INTELSAT IV used a 128-kbits/s TDMA scheme for the common signaling channel that controls the SPADE network. Intelsat V introduced a 120-Mbits/s TDMA scheme for multiple-beam international digital service. One disadvantage or cost in implementing a TDMA scheme is the need for providing precise *synchronization* among the participating earth stations and the satellite. FDMA systems, not having such requirements, are less complex from a networking point of view. Comparisons of TDMA versus FDMA operation are summarized as follows:

1. FDMA can cause IM products. This can be avoided by operating the TWTA in its linear region, thereby reducing the available power output.

2. With TDMA, there is only one carrier present at a time in the TWTA. Thus IM distortion cannot occur.

3. TDMA earth station equipment is more sophisticated and hence more costly than FDMA equipment. However, for earth stations providing multiple point-to-point channels, FDMA stations require separate radio-frequency (RF) up-conversion and down-conversion signal processing stages. Thus with FDMA, the amount of equipment grows with the amount of simultaneous connectivity. With TDMA, such growth does not take place since channel selectivity is accomplished in time rather than frequency. Therefore, for a large multiply connected earth station, TDMA can be more cost-effective than FDMA.

4. In multiple-beam systems, each beam may need to communicate with every other beam. TDMA lends itself to conveniently forming connections sequentially as in satellite-switched TDMA (SS/TDMA). INTELSAT VI uses such satellite-switched TDMA (SS/TDMA), described in Section 9.4.5.

An example of the comparative performance of TDMA, FDM/FM, and SPADE is shown for an INTELSAT IV transponder as a plot of channel capacity versus earth station $G/T°$ in Figure 9.33. Figure 9.33a is for an earth coverage antenna, and Figure 9.33b is for a spot-beam antenna. From synchronous altitudes these antennas have half-power beamwidths of $17°$ and $4.5°$, respectively. From these plots it is seen that single-carrier FDM/FM is as efficient as TDMA when the system is operated with standard earth stations ($G/T° = 40.7$ dB/K). For smaller earth stations ($G/T° \leq 31$ dB/K) working through earth-coverage transponders, SPADE is more efficient than TDMA and multicarrier FDM/FM (MCPC); only the four-carrier case is plotted. For earth stations having $G/T°$ in the range 19 to 40.7 dB/K working through a spot-beam transponder, TDMA is superior to SPADE and MCPC. For smaller earth stations having $G/T°$ in the range 6 to 19 dB/K working through a spot-beam transponder, SPADE is superior to TDMA and MCPC. In general, when working through *standard* earth stations it is seen [19] that TDMA is the most efficient multiple access scheme for INTELSAT IV.

9.4.4.1 PCM Multiplex Frame Structures

There are two digital telephony standards for PCM frame structures in operation. The North American standard is called *T-Carrier*; it is built around the 193-bit frame shown in Figure 9.34a. There are 24 channels; each channel contains an 8-bit voice sample. Also, there is one bit per frame with alternating value 1 0 1 0 . . . from frame to frame, used for frame alignment. Since a voice-grade telephone channel has a bandwidth of $W = 4$ kHz (including guard bands), the Nyquist sampling rate for recovering the analog information within 4 kHz is $f_s = 2W = 8000$ samples/s. Therefore, the basic PCM frame, called the *Nyquist*

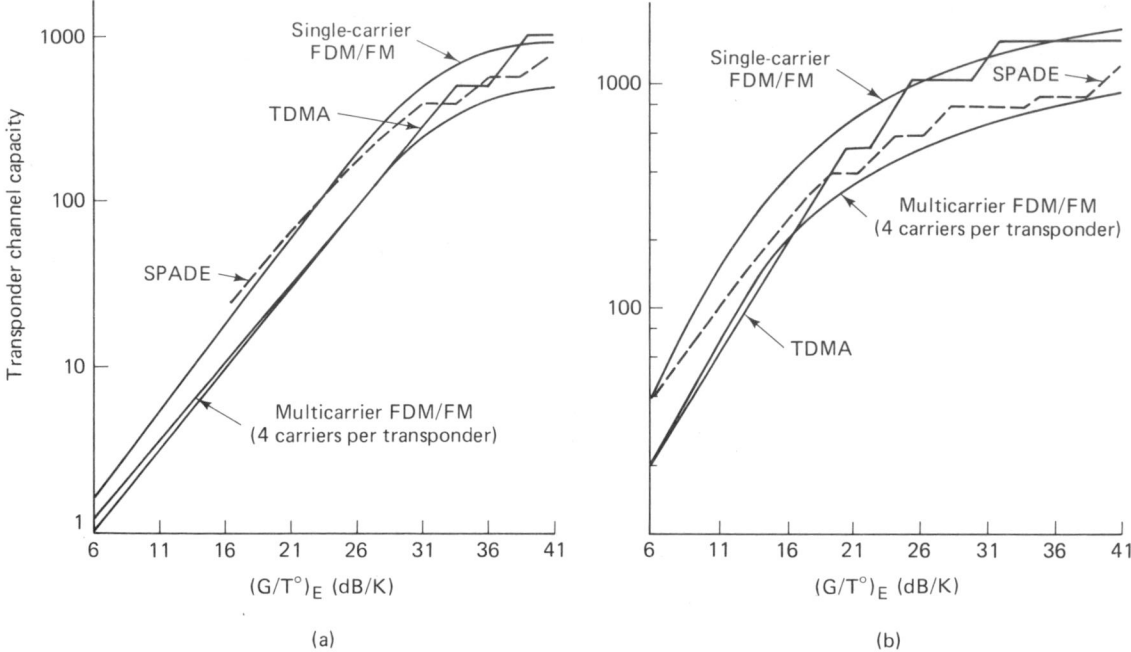

Figure 9.33 Channel capacity versus earth station $G/T°$ for FDMA, TDMA, and SPADE. (a) Global-beam transponder channel capacity as function of $(G/T°)_E$, where $(G/T°)_E$ means earth station G/T. (b) Spot-beam transponder channel capacity as function of $(G/T°)_E$. [From D. Chakraborty, "INTELSAT IV Satellite System (Voice) Channel Capacity versus Earth Station Performance," *IEEE Trans. Commun. Tech.*, vol. COM19, no. 3, June 1971, pp. 355–362. © 1971 IEEE.]

frame, which contains 24 voice samples from 24 different message sources, has a frame rate of 8000 frames/s (duration of 125 µs). Thus the basic T-Carrier bit rate is 193 bits/frame × 8000 frames/s = 1.544 Mbits/s.

The European standard is built around a 256-bit frame shown in Figure 9.34b. There are 30 message channels, each containing an 8-bit voice sample. Also, one 8-bit time slot is used for frame alignment and another 8-bit time slot is used for signaling (addressing) information. The European frame rate is the same as that of the T-Carrier. Therefore, the basic European bit rate is 256 bits/frame × 8000 frames/s = 2.048 Mbits/s.

9.4.4.2 The High-Rate TDMA Frame for Europe

Sixteen Nyquist frames of the European PCM Multiplex format are shown in Figure 9.35a. Each frame contains an 8-bit sample from each of 30 terrestrial channels, plus 8 bits of framing and 8 bits of signaling information. The TDMA frame duration is

$$16 \text{ Nyquist frames} \times 125 \text{ µs/Nyquist frame} = 2 \text{ ms}$$

Within this 2-ms frame are contained

16 Nyquist frames × 256 bits/Nyquist frame = 4096 bits

The basic idea behind TDMA is that a user's low-rate data stream can share the CR with similar streams from other users by *bursting* the transmission at a much faster rate than the rate at which it is generated. Figure 9.35b illustrates a 2-ms high-rate TDMA frame. The frame begins with a reference burst, RB1, emitted by a reference station. The burst contains information necessary to enable other stations to precisely position their message traffic bursts in the frame. There may be a second burst, RB2, for reliability, followed by a sequence of traffic slots. The traffic slots may be preassigned, or they may be assigned according to a DAMA protocol [20].

The PCM multiplex signal with a bit rate of $R_0 = 2.048$ Mbits/s and a frame duration of $T = 2$ ms is compressed (by a factor of 59) and transmitted using QPSK modulation at a burst rate of $R_T = 120.832$ Mbits/s (symbol rate of 60.416 megasymbols/s). The duration of the traffic data field T_{tr} in the high rate TDMA

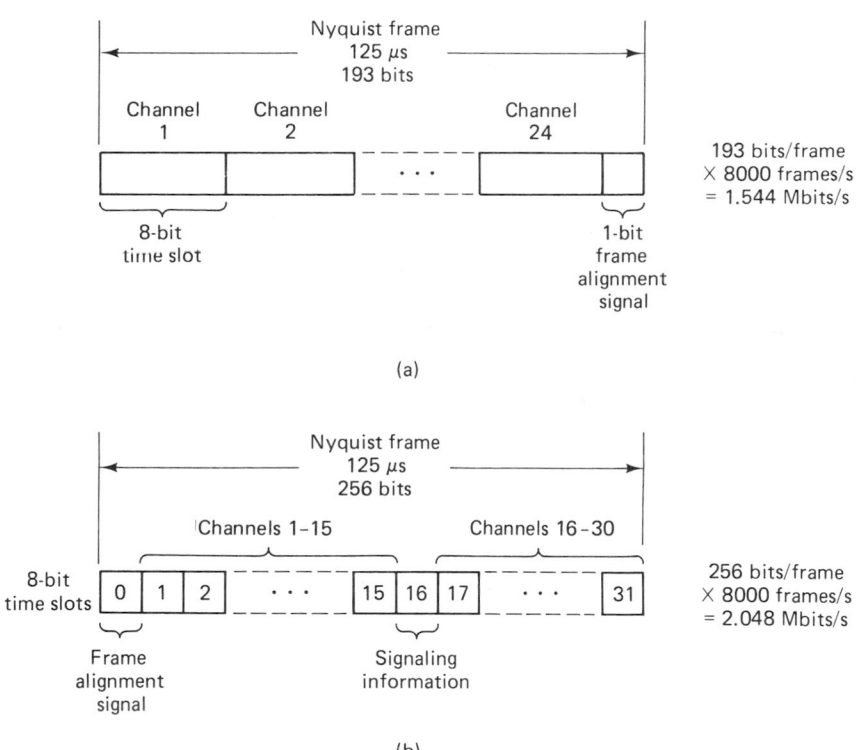

(a)

(b)

Figure 9.34 PCM multiplex frame structure. (a) Frame structure for T-Carrier (North American) PCM multiplex. (b) Frame structure for the European PCM multiplex.

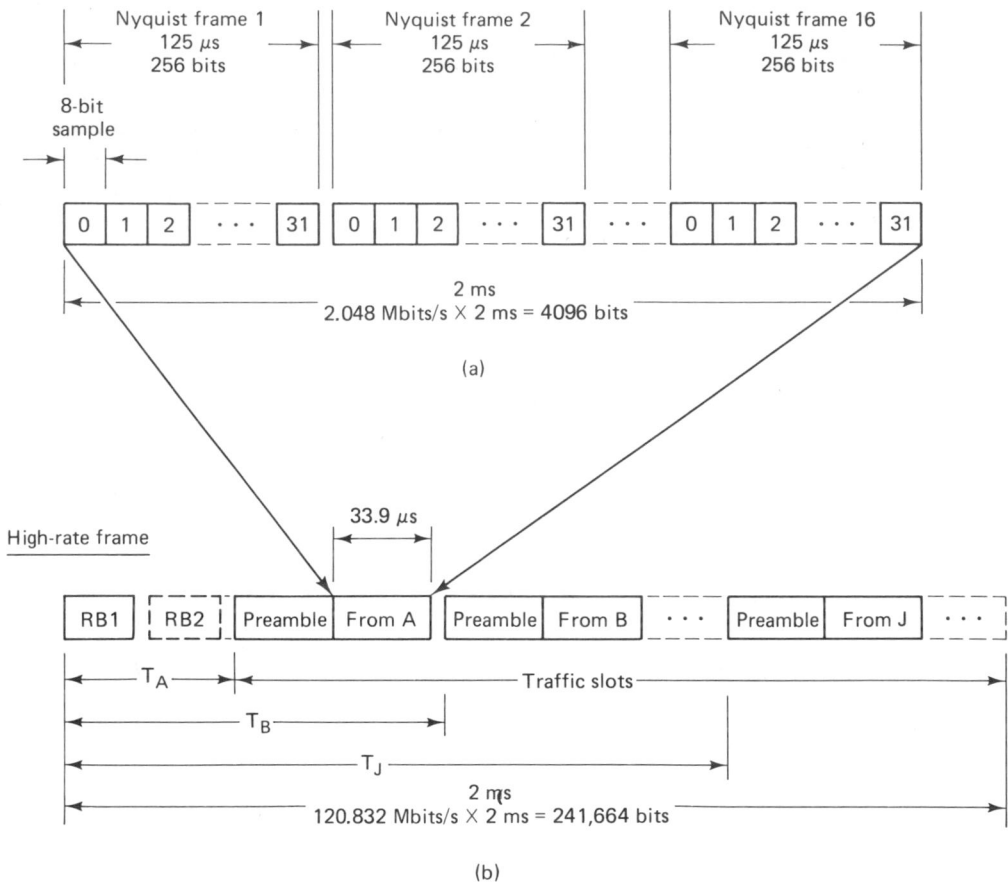

Figure 9.35 INTELSAT digital transmission standards for Europe. (a) Terrestrial PCM multiplex. (b) High-rate frame.

frame is calculated as follows:

$$T_{\text{tr}} = \frac{R_0 T}{R_T} \tag{9.31}$$

$$= \frac{2.048 \times 10^6 \times 2 \times 10^{-3}}{120.832 \times 10^6}$$

$$= 33.9 \ \mu\text{s}$$

To obtain the total duration of a traffic burst, the time used for the preamble must be added. If the preamble contains S_P symbols, then assuming QPSK modulation, the total length of the traffic burst measured in number of symbols, S_T, is

$$S_T = \frac{R_0 T}{2} + S_P \tag{9.32}$$

and the burst-time duration is

$$T_T = \frac{2S_T}{R_T} \tag{9.33}$$

If the preamble contains 300 symbols, then

$$S_T = \frac{2.048 \times 10^6 \times 2 \times 10^{-3}}{2} + 300$$

$$= 2348 \text{ symbols}$$

Using this in Equation (9.33), we obtain

$$T_T = \frac{2 \times 2348}{120.832 \times 10^6} = 38.9 \ \mu s$$

9.4.4.3 The High-Rate TDMA Frame for North America

The INTELSAT TDMA burst (bit) rate of $R_T = 120.832$ Mbits/s was chosen to be compatible with both the European and North American standards. Figure 9.36 is similar to Figure 9.35 except that the PCM multiplex signal is the 24-channel T-Carrier instead of the 30-channel European standard. The essential T-Carrier features that are different from the European standard are listed below and are shown on the figure.

1. Each Nyquist frame is comprised of 24 channels or samples $\times$ 8 bits $+$ 1 frame alignment bit $= 193$ bits.
2. The 16 Nyquist frames contain $16 \times 193 = 3088$ bits.
3. The T-Carrier data rate is 1.544 Mbits/s.
4. The duration of the traffic data field in the high-rate TDMA frame is calculated from Equation (9.31).

$$T_{tr} = \frac{1.544 \times 10^6 \times 2 \times 10^{-3}}{120.832 \times 10^6}$$

$$= 25.6 \ \mu s$$

9.4.4.4 INTELSAT TDMA Operation

At the transmitting earth station, the continuous low-rate data stream enters one of a pair of buffers illustrated in Figure 9.37a. When one buffer is filling at the low rate (1.544 Mbits/s or 2.048 Mbits/s), the other is emptying at the burst rate (120.832 Mbits/s). The buffers alternate functions at each TDMA frame. The time of application of the high-rate clock is controlled so that the traffic burst is transmitted in the proper interval to arrive at the satellite in its assigned position in the TDMA frame.

At the receiving station, the received traffic burst is routed to one of a pair of expansion buffers, shown in Figure 9.37b, that have the inverse function of

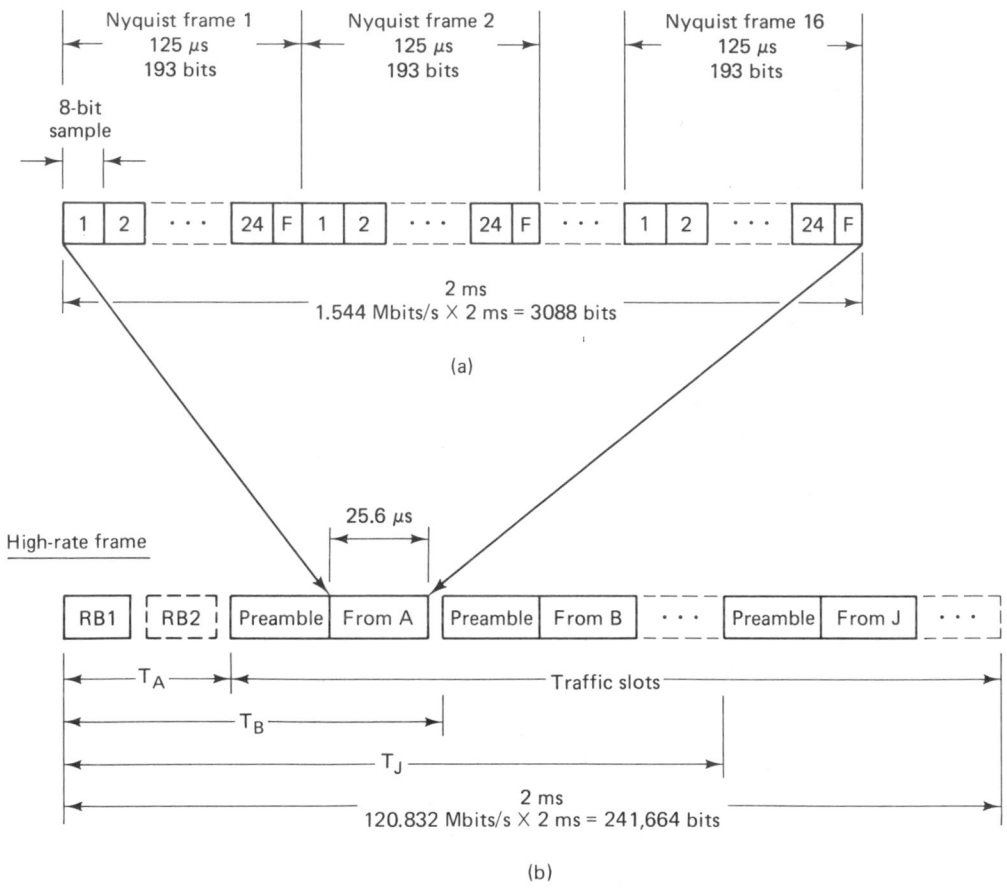

Figure 9.36 INTELSAT digital transmissions standards for T-Carrier. (a) Terrestrial PCM multiplex. (b) High-rate frame.

the compression buffers in Figure 9.37a. When one buffer is filling at the high rate, the other is emptying at the desired output rate.

   The most critical aspect of TDMA operation is the precise synchronization needed to assure orthogonality of the time slots [20]. Figure 9.38 illustrates the general idea behind most commercial satellite synchronization schemes. One station is designated as the master or control station. This station transmits periodic bursts of reference timing pulses. User stations also transmit their timing pulses, designated as slave pulses in Figure 9.38. On the downlink, the using station receives the master or reference pulses in addition to its own slave pulses. The time difference between the master and slave pulses corresponds to the timing error. The station adjusts its clock so as to reduce this timing error.

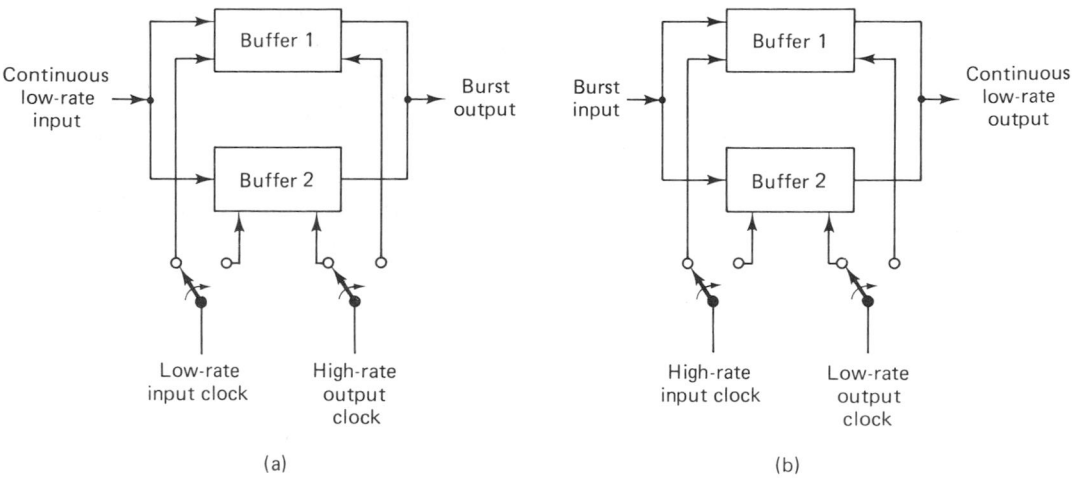

Figure 9.37 Burst compression and expansion buffers. (a) Compression buffers at transmitter. (b) Expansion buffers at receiver.

9.4.5 Satellite-Switched TDMA in INTELSAT

Modern communication satellites often employ several regional antenna beams. For a satellite based over the Atlantic Ocean, separate beams might be aimed at North America, Europe, South America, and Africa. Switches are used to allow the interconnection of stations in one region to communicate with stations in another region. The basic goal of a satellite-switched TDMA (SS/TDMA) scheme is to provide an efficient way of cyclically providing interconnection of TDMA data among various coverage regions.

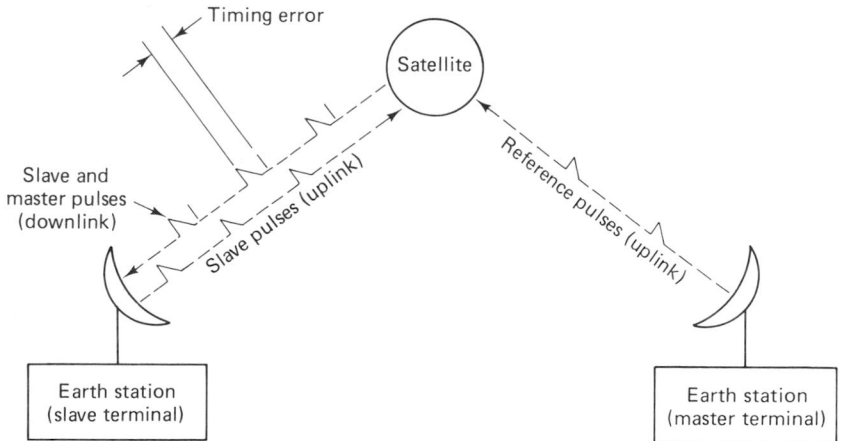

Figure 9.38 TDMA synchronization concept.

The heart of the system consists of a microwave switch matrix, located in the satellite, that is programmed via ground control to change states cyclically in rapid sequence, thus interconnecting distinct uplink beams to distinct downlink beams at each switching time. An earth station in the network communicates with those in other beams by transmitting TDMA bursts in the proper timing positions in the sequence. The pattern of switch states is selected so as to maximize the usable system capacity under the constraints of the traffic demands [21]. For complete interconnectivity between N beams, a total of $N!$ different satellite switch states or *modes* are required. Table 9.3 illustrates the six modes required for the full interconnectivity of a three-beam system.

In mode 1, the satellite receivers in beams A, B, and C are connected to the satellite transmitters for beams A, B, and C, respectively. An earth station in one of these beams can then communicate with other earth stations in the same beam. The beam is said to be *looped back* on itself.

Figure 9.39 illustrates a three-beam (beams A, B, and C) example of a SS/TDMA system. The satellite microwave switch matrix is configured in a *crossbar* design. This design can be thought of as being made up of row and column lines; when one row and one column are energized, contact is made at the intersection. A crossbar design only permits a single row to communicate with a single column at a time. If uplink A_U is connected to downlink B_D, *neither* A_U *nor* B_D can be simultaneously connected to any other beam.

In Figure 9.39, three different traffic patterns during time slot intervals T_1, T_2, and T_3, with three different switch states S_1, S_2, and S_3 are shown. During interval T_1, switch state S_1 interconnects the beams in a loop-back fashion which permits the uplink messages in slot T_1 to be delivered to their correct destinations. During time interval T_2, switch state S_2 interconnects uplink beam A_U to downlink beam B_D, uplink beam B_U to downlink beam C_D, and uplink beam C_U to downlink beam A_D. This connection pattern assures that the uplink messages in slot T_2 are delivered to their correct destinations. During time interval T_3, switch state S_3 similarly connects uplink transmissions to downlink beams to assure correct delivery of the data.

The traffic patterns and their durations are programmed to optimize the resource capacity and to serve the users as efficiently as possible. The cyclic pattern can be reprogrammed by ground command to meet changing traffic requirements.

9.4.5.1 Traffic Matrix

Figure 9.40 is a matrix describing the communication traffic among N spot-beam coverages. In this figure, t_{ij} is the traffic volume from the ith beam to the jth beam. The subtotal S_i is the total traffic originating from the ith uplink beam, expressed as

$$S_i = \sum_{j=1}^{N} t_{ij} \tag{9.34}$$

TABLE 9.3 Three-Beam Satellite Switch Modes

Input	Output					
	Mode 1	Mode 2	Mode 3	Mode 4	Mode 5	Mode 6
A	A	A	B	B	C	C
B	B	C	A	C	A	B
C	C	B	C	A	B	A

and R_j is the total traffic received in the jth downlink beam, expressed as

$$R_j = \sum_{i=1}^{N} t_{ij} \tag{9.35}$$

When the traffic in a SS/TDMA system is controlled by a nonblocking switch (one that allows for the transmission of *all* messages, without any "busy" signals)

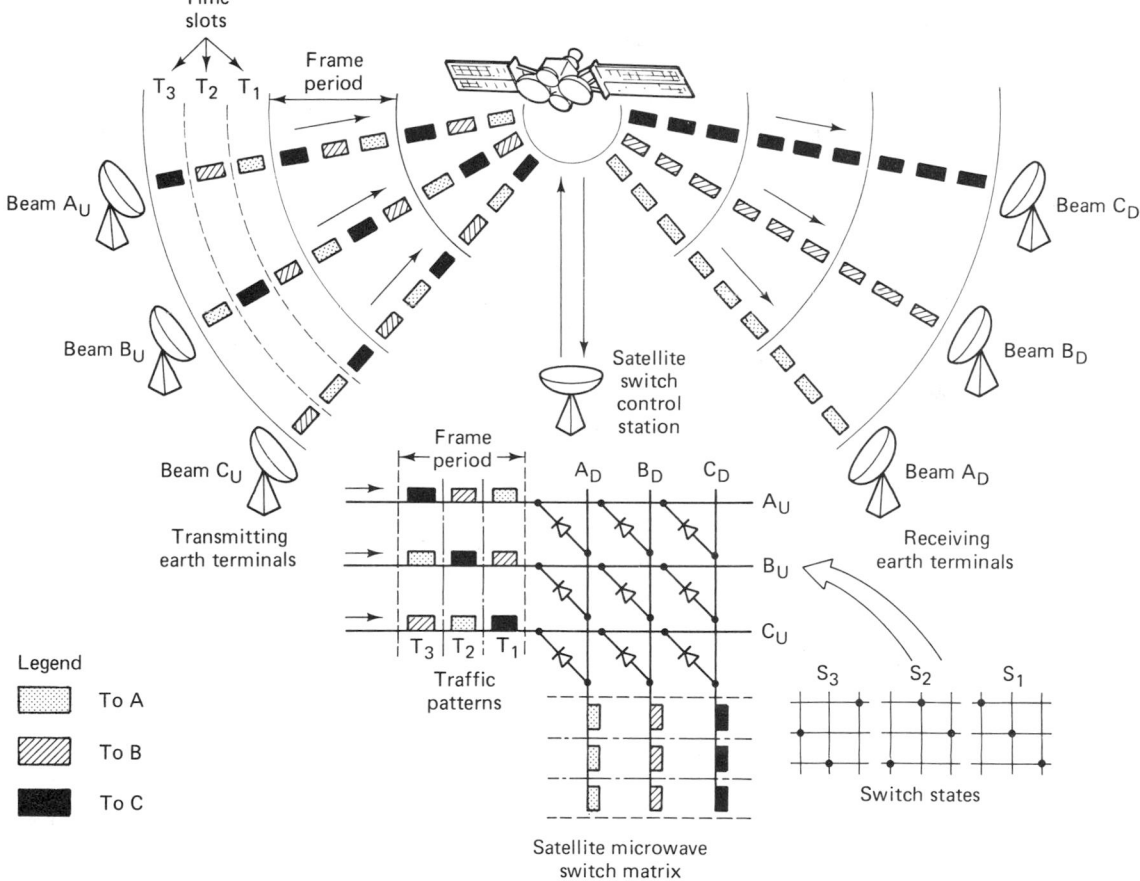

Figure 9.39 Satellite-switched TDMA (SS/TDMA).

Origin \ Destination	1	2	· · ·	j	· · ·	N	Originating traffic subtotals
1	t_{11}	t_{12}		t_{1j}		t_{1N}	S_1
2	t_{21}	t_{22}		t_{2j}		t_{2N}	S_2
⋮							
i	t_{i1}	t_{i2}		t_{ij}		t_{iN}	S_i
⋮							
N	t_{N1}	t_{N2}		t_{Nj}		t_{NN}	S_N
Received traffic subtotals	R_1	R_2		R_j		R_N	Total

Figure 9.40 Traffic matrix.

a k-second time slot will be assigned to each channel in the TDMA frame. For efficient utilization of the CR, the total traffic in Figure 9.40 should be transmitted within a frame time T which should be made as short as possible. The minimum frame time, T_{min}, for providing such nonblocking connectivity can be expressed [22] as follows:

$$T_{\text{min}} = k \max(\{S_i\}, \{R_j\}) \tag{9.36}$$

where max $(\{S_i\}, \{R_j\})$ is the maximum value taken over the set of all $\{S_i\}$ and $\{R_j\}$. Equation (9.36) describes the minimum time to communicate *all* of the traffic in the traffic matrix, for equal bandwidth per channel.

9.5 MULTIPLE ACCESS TECHNIQUES FOR LOCAL AREA NETWORKS

A local area network (LAN) can be used to interconnect computers, terminals, printers, and so on, located within a building or a small set of buildings. While long-haul networks use the public telephone network for economic reasons, LAN designers usually lay their own high-bandwidth cables. Bandwidth is not as scarce as it is in the long-haul cases. Not being forced to optimize bandwidth, a LAN can use simple access algorithms [6, 25–27].

9.5.1 Carrier-Sense Multiple Access Networks

Ethernet is a LAN access scheme developed by the Xerox Corporation. The Ethernet scheme is based on the assumption that each local machine can sense the state of a common broadcast channel before attempting to use it. The technique is known as *carrier-sense multiple access with collision detection* (CSMA/CD). The word "carrier," here, means *any* electrical activity on the cable. Figure 9.41a

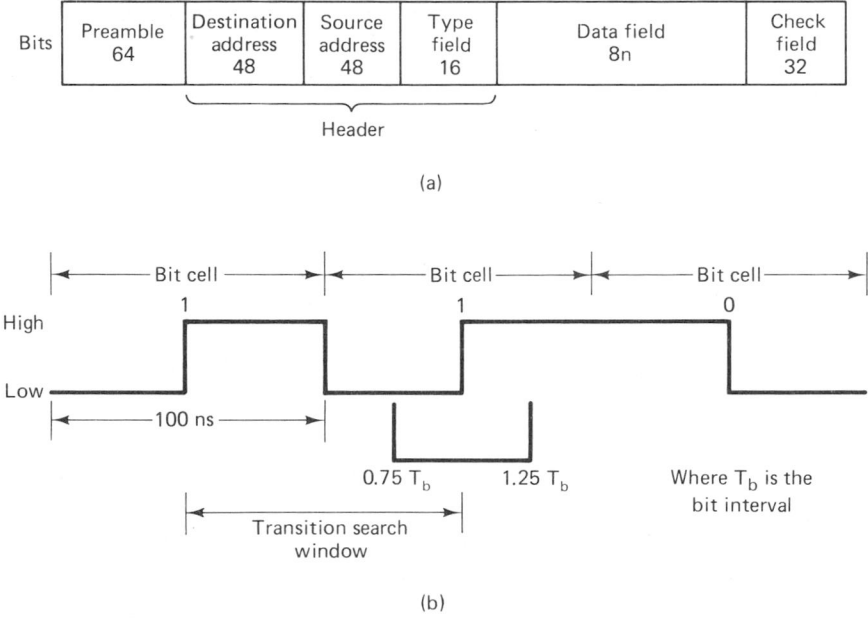

Bits

| Preamble 64 | Destination address 48 | Source address 48 | Type field 16 | Data field 8n | Check field 32 |

Header

(a)

High

Low

Bit cell 1   Bit cell 1   Bit cell 0

100 ns

$0.75\ T_b$ $1.25\ T_b$

Where T_b is the bit interval

Transition search window

(b)

Figure 9.41 Ethernet bit field and PCM format. (a) Ethernet specification. (b) Manchester PCM format.

illustrates the bit field format for the Ethernet specification; the details are listed below.

1. The maximum packet size is 1526 bytes, where a byte is 8 bits. The packet breakdown is 8-byte preamble + 14-byte header + 1500-byte data + 4-byte parity.
2. The minimum packet size is 72 bytes, consisting of an 8-byte preamble + 14-byte header + 46-byte data + 4-byte parity.
3. The minimum spacing between packets is 9.6 μs.
4. The preamble contains a 64-bit synchronization pattern of alternating ones and zeros, ending with two consecutive ones: (1 0 1 0 1 0 . . . 1 0 1 0 1 1).
5. The receiving station examines a destination address field in the header to see if it should accept a particular packet. The first bit indicates the type of address (0 = single address, 1 = group address); an entire field of ones means an all-station broadcast.
6. The source address is the unique address of the transmitting machine.
7. The type field determines how the data field is to be interpreted. For example, bits in the type field can be used to describe such things as data encoding, encryption, message priority, and so on.
8. The data field is an integer number of bytes from a minimum of 46 to a maximum of 1500.

9. The parity check field houses the parity bits which are generated by the following generating polynomial (see Section 5.6):

$$X^{32} + X^{26} + X^{23} + X^{22} + X^{14} + X^{12} + X^{11} + X^{10}$$
$$+ X^8 + X^7 + X^5 + X^4 + X^2 + X + 1$$

The Ethernet multiple access algorithm defines the following user action or response:

1. *Defer*. The user must not transmit when the carrier is present or within the minimum packet spacing time.
2. *Transmit*. The user may transmit if not deferring until the end of the packet or until a collision is detected.
3. *Abort*. If a collision is detected, the user must terminate packet transmission and transmit a short jamming signal to ensure that all collision participants are aware of the collision.
4. *Retransmit*. The user must wait a random delay time (similar to the ALOHA system) and then attempt retransmission.
5. *Backoff*. The delay before the nth attempt is a uniformly distributed random number from 0 to $2^n - 1$, for $(0 < n \le 10)$. For $n > 10$, the interval remains 0 to 1023. The unit of time for the retransmission delay is 512 bits (51.2 μs).

Figure 9.41b illustrates a 10-Mbits/s data stream with Manchester PCM formatting from the Ethernet specification. Notice that with such formatting, each bit cell or bit position contains a transition. A binary one is characterized by transitioning from a low level to a high level, while a binary zero has the opposite transition. Therefore, the presence of data transitions denotes to all "listeners" that the carrier is present. If a transition is not seen between 0.75 and 1.25 bit times since the last transition, the carrier has been lost, indicating the end of a packet.

9.5.2 Token-Ring Networks

A carrier-sense network consists of a cable onto which all stations are passively connected. A *ring network,* by comparison, consists of a series of point-to-point cables between consecutive stations. The interfaces between the ring and the stations are active rather than passive. Figure 9.42a illustrates a typical unidirectional ring with interface connections to several stations. Figure 9.42b illustrates the state of the interface for the listen mode and the transmit mode. In the *listen mode* the input bits are copied to the output with a delay of one bit time. In the *transmit mode,* the connection is broken so that the station can enter its own data onto the ring. The token is defined as a special bit pattern (e.g., 1 1 1 1 1 1 1 1) which circulates on the ring whenever all stations are idle. How does the system ensure that message data do not contain a tokenlike sequence? *Bit stuffing* is used to prevent this pattern from occurring in the data. For the

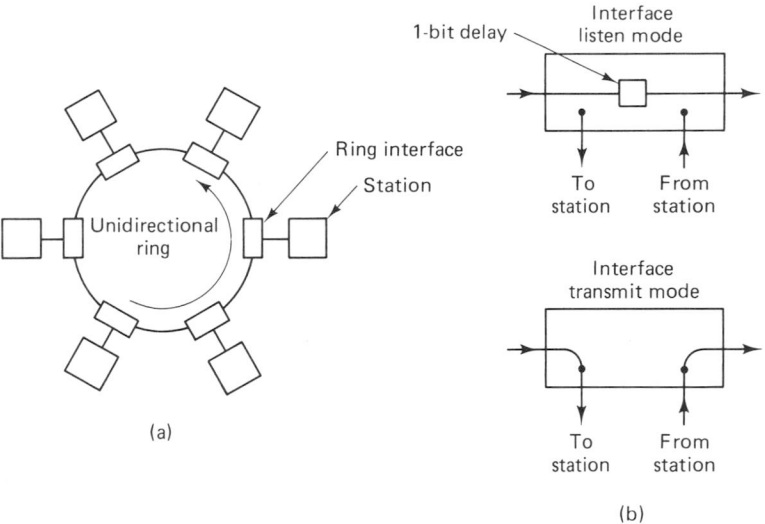

Figure 9.42 Token-ring network. (a) Network. (b) Listen and transmit modes.

8-bit token example shown, a bit-stuffing algorithm would insert a zero into the data stream after each sequence of seven consecutive ones. The data receiver would use a similar algorithm to dispose of the inserted bit following any sequence of seven consecutive ones. The token-ring access scheme works as follows:

1. A station wanting to send a message monitors the token appearing at the interface. When the last bit of the token appears, the station inverts it (e.g., 1 1 1 1 1 1 1 0). The station then breaks the interface connection and enters its own data onto the ring.
2. As bits come back around the ring, they are removed by the sender. There is no limit on the size of the packets, because the entire packet never appears on the ring at one instant.
3. After transmitting the last bit of its message, the station must regenerate the token. After the last data bit has circled the ring and has been removed, the interface is switched back to the listen mode.
4. Contention is not possible with a token-ring system. During heavy traffic, as soon as a token is regenerated, the next downstream station requiring service will see and remove the token. Thereby, permission to transmit rotates smoothly around the ring. Since there is only one token, there is no contention.

The ring itself must have sufficient delay to enable a complete token to circulate when all stations are idle. A major issue in ring network design is the propagation distance or "length" of a bit. If the data rate is R Mbits/s, a bit is emitted every $(1/R)$ microseconds. Since the propagation rate along a typical coaxial cable is 200 m/μs, each bit occupies $200/R$ meters on the ring.

Example 9.4   Minimum Ring Size

If an 8-bit token is to be used on a 5-Mbits/s token-ring network, calculate the minimum *propagation distance, d_p,* needed for the ring circumference. Assume that the propagation velocity v_p is 200 m/μs.

Solution

$$R = 5 \text{ Mbits/s}$$

Time to emit one bit, t_b:

$$t_b = \frac{1}{5 \times 10^6} \text{ s}$$

Time to emit the 8-bit token, t_t:

$$t_t = \frac{8}{5 \times 10^6} \text{ s}$$

Propagation distance for the 8-bit token:

$$d_p = t_t \times v_p$$

$$= \tfrac{8}{5} \, \mu\text{s} \times 200 \text{ m/}\mu\text{s}$$

$$= 320 \text{ m}$$

9.5.3 Performance Comparison of CSMA/CD and Token-Ring Networks

Figure 9.43 compares the delay-throughput characteristics of a CSMA/CD network with a token-ring network. In each case, the cable length is 2 km, there are 50 stations on the network, the average packet length is 1000 bits, and the header size is 24 bits. Figure 9.43a, the case where the transmission rate is 1 Mbits/s, illustrates that under these assumptions, CSMA/CD and token ring perform almost equally well. In Figure 9.43b, only one parameter has been changed as compared to Figure 9.43a; the transmission rate was increased to 10 Mbits/s. The difference for CSMA/CD is considerable; for normalized throughput, $\rho < 0.22$, CSMA/CD performs better than token ring. However, for $\rho > 0.22$, token ring clearly manifests better delay-throughput characteristics. To understand the reason for the poor CSMA/CD performance in Figure 9.43b, let us review the definition of ρ, described in Equations (9.17) and (9.19) and shown as

$$\rho = \frac{b\lambda}{R} = \frac{\rho'}{R}$$

where $\rho' = b\lambda$ is channel throughput in bits per second and R is the channel capacity (maximum transmission bit rate). As R increases, channel throughput must increase accordingly for a given value of ρ. At higher channel throughput rates a significant portion of the CSMA/CD transmission attempts ends in collision [26].

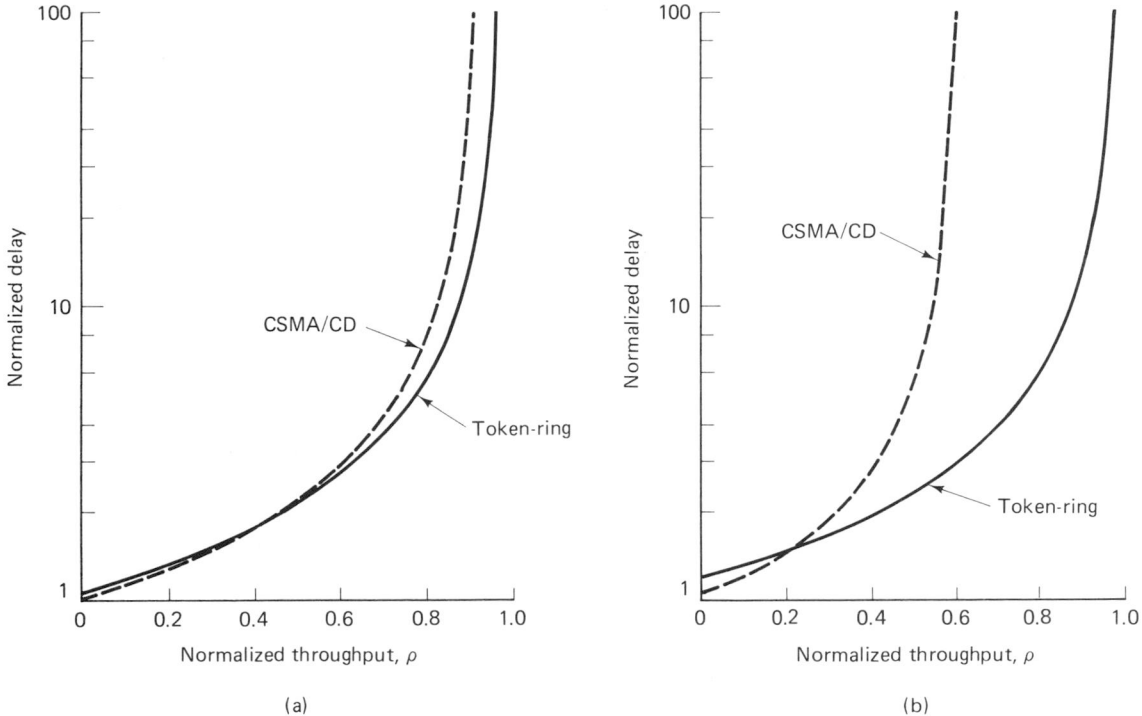

Figure 9.43 Delay versus throughput performance for CSMA/CD and token-ring networks. (a) Transmission rate = 1 Mbits/s. (b) Transmission rate = 10 Mbits/s. (Reprinted with permission from W. Bux, "Local-Area Subnetworks: A Performance Comparison," *IEEE Trans. Commun.,* vol. COM29, no. 10, Oct. 1981, pp. 1465–1473. © 1981 IEEE.)

9.6 CONCLUSION

In this chapter we have outlined the concepts of resource sharing. The classical approaches of FDM/FDMA and TDM/TDMA were discussed in some detail. We also described a hybrid multiple access technique called CDMA, and introduced some of the satellite multiple access techniques that became popular in the 1970s and 1980s, known as multiple-beam frequency reuse and dual-polarization frequency reuse.

We described the demand-assignment (DAMA) techniques in the context of several versions of the ALOHA algorithm, and we considered several of the multiple-access techniques employed with INTELSAT, such as FDM/FM, SPADE, TDMA, and SS/TDMA. Finally, we examined two popular algorithms used for local area networks: carrier-sense multiple access with collision detection (CSMA/CD) and a token-ring network. The goals of the chapter were to introduce an assortment of multiple access techniques rather than attempting a rigorous treatment of any of them.

REFERENCES

1. Rubin, I., "Message Delays in FDMA and TDMA Communication Channels," *IEEE Trans. Commun., vol.* COM27, no. 5, May 1979, pp. 769–777.

2. Nirenberg, L. M., and Rubin, I., "Multiple Access System Engineering—A Tutorial," *IEEE WESCON/78 Professional Program,* Modern Communication Techniques and Applications, session 21, Los Angeles, Sept. 13, 1978.

3. Abramson, N., "The ALOHA System—Another Alternative for Computer Communications," *Proc. Fall Joint Comput. Conf. AFIPS,* vol. 37, 1970, pp. 281–285.

4. Hayes, J. F., "Local Distribution in Computer Communications," *IEEE Commun. Mag.,* Mar. 1981, pp. 6–14.

5. Schwartz, M., *Computer-Communication Network Design and Analysis,* Prentice-Hall, Inc., Englewood Cliffs, N.J., 1977.

6. Tanenbaum, A. S., *Computer Networks,* Prentice-Hall, Inc., Englewood Cliffs, N.J., 1981.

7. Abramson, N., "The ALOHA System," in N. Abramson and F. F. Kuo, eds., *Computer Communication Networks,* Prentice-Hall, Inc., Englewood Cliffs, N.J., 1973.

8. Kleinrock, L., *Queueing Systems,* Vol. 1, *Theory,* John Wiley & Sons, Inc., New York, 1975.

9. Abramson, N., "Packet Switching with Satellites," *AFIPS Conf. Proc.,* vol. 42, June 1973, pp. 695–702.

10. Rosner, R. D., *Packet Switching,* Lifelong Learning Publications, Wadsworth Publishing Company, Inc., Belmont, Calif., 1982.

11. Crowther, W., Rettberg, R., Walden, D., Ornstein, S., and Heart, F., "A System for Broadcast Communication: Reservation ALOHA," *Proc. Sixth Hawaii Int. Conf. Syst. Sci.,* Jan. 1973, pp. 371–374.

12. Roberts, L., "Dynamic Allocation of Satellite Capacity through Packet Reservation," *AFIPS Conf. Proc.,* vol. 42, June 1973, p. 711.

13. Binder, R., "A Dynamic Packet-Switching System for Satellite Broadcast Channels," *Proc. Int. Conf. Commun.,* June 1975, pp. 41-1–41-5.

14. Capetanakis, J., "Tree Algorithms for Packet Broadcast Channels," *IEEE Trans. Inf. Theory,* vol. IT25, Sept. 1979, pp. 505–515.

15. Puente, J. G., and Werth, A. M., "Demand-Assigned Service for the INTELSAT Global Network," *IEEE Spectrum,* Jan. 1971, pp. 59–69.

16. Jones, J. J., "Hard Limiting of Two Signals in Random Noise," *IEEE Trans. Inf. Theory,* vol. IT9, Jan. 1963, pp. 34–42.

17. Bond, F. E., and Meyer, H. F., "Intermodulation Effects in Limiter Amplifier Repeaters," *IEEE Trans. Commun. Technol.,* vol. COM18, no. 2, Apr. 1970, pp. 127–135.

18. Shimbo, O., "Effects of Intermodulation, AM-PM Conversion, and Additive Noise in Multicarrier TWT Systems," *Proc. IEEE,* vol. 59, Feb. 1971, pp. 230–238.

19. Chakraborty, D. "INTELSAT IV Satellite System (Voice) Channel Capacity versus Earth-Station Performance," *IEEE Trans. Commun. Technol.,* vol. COM19, no. 3, June 1971, 355–362.

20. Campanella, S., and Schaefer, D., "Time Division Multiple Access Systems

(TDMA)," in K. Feher, *Digital Communications, Satellite/Earth Station Engineering*, Prentice-Hall, Inc., Englewood Cliffs, N.J., 1983.

21. Scarcella, T., and Abbott, R. V., "Orbital Efficiency Through Satellite Digital Switching," *IEEE Commun. Mag.*, May 1983, pp. 38–46.

22. Muratani, T., Satellite-Switched Time-Domain Multiple Access," *Proc. IEEE Electron. and Aerosp. Conf. (EASCON)*, 1974, pp. 189–196.

23. Dill, G. D., "TDMA, The State-of-the-Art," *Rec. IEEE Electron. Aerosp. Syst. Conv. (EASCON)*, Sept. 26–28, 1977, pp. 31-5A–31-5I.

24. Jarett, K., "Operational Aspects of Intelsat VI Satellite-Switched TDMA Communication System," *AIAA Tenth Commun. Satell. Syst. Conf.* Mar. 1984, pp. 107–111.

25. Stallings, W., "Local Network Performance," *IEEE Commun. Mag.*, vol. 22, No. 2, Feb. 1984, pp. 27–36.

26. Bux, W., Local-Area Subnetworks: A Performance Comparison," *IEEE Trans. Commun.*, vol. COM29, no. 10, Oct. 1981, pp. 1465–1473.

27. Dixon, R. C., Strole, N. C., and Markov, J. D., "A Token-Ring Network for Local Data Communications," *IBM Syst. J.*, vol. 22, no. 1–2, 1983, pp. 47–62.

PROBLEMS

9.1. Design an FDM signal set consisting of five voice channels, each in the frequency range 300 to 3400 Hz. The multiplexed composite is to be made up of inverted sidebands and is to occupy the spectral region from 30 to 50 kHz.
   (a) Draw the composite spectrum, indicating individual spectrum and guard band frequency locations.
   (b) Draw a block diagram showing the heterodyning and filtering details and the required local oscillator values.

9.2. A receiver is tuned to receive the lower sideband (LSB) of a radio-frequency (RF) carrier wave with frequency, $f_c = 8$ MHz. The bandwidth of the LSB signal is 100 kHz. The receiver employs a local oscillator (LO) with frequency, f_{LO}, for heterodyning the received signal down to a lower intermediate frequency (IF). Assume that $f_{LO} > f_c$, and that the IF amplifier is centered at 2 MHz. Draw a block diagram of the heterodyning conversion, including the RF filter, the LO, and the IF filter. Indicate the center frequency of each filter and typical spectra of the signals at various points in the diagram.

9.3. Equations (9.13) to (9.15) demonstrate that the average message delay time for TDMA is less than that for FDMA. Discuss the practical benefits of such reduced delay in TDMA, as a function of frame time, for a satellite link with a one-way range of 36,000 km. For what values of frame time can there be a significant advantage of TDMA over FDMA?

9.4. A group of stations share a 56-kbits/s pure ALOHA channel. Each station outputs a packet on the average of once every 10 s, even if the previous one has not yet been sent (i.e., the stations buffer the packets). Each packet is comprised of 3000 bits. What is the maximum number of stations that can share this channel, assuming that the arrival process is Poisson?

9.5. A group of three stations share a 56-kbits/s pure ALOHA channel. The average bit rate transmitted from each of the three stations is $R_1 = 7.5$ kbits/s, $R_2 = 10$ kbits/s, and $R_3 = 20$ kbits/s. The size of each packet is 100 bits/packet. Find the normalized total traffic on the channel, the normalized throughput, the probability of successful transmission, and the arrival rate of successful packets. Assume that the arrival process is Poisson.

9.6. Verify that for a pure ALOHA access scheme, the normalized throughput is bounded by $1/2e$ and that this maximum occurs when the normalized total traffic is equal to 0.5.

9.7. (a) Verify that Equation (9.24) is a valid probability density function (pdf) for a discrete random variable.
   (b) Calculate the mean of a discrete random variable having a pdf like the one given in Equation (9.24).
   (c) Show that your result in part (b) is consistent with the claim that λ is the average packet arrival rate.

9.8. Consider the pure ALOHA arrival scenario shown in Figure P9.1. The vertical arrows indicate packet arrival times. N_n is the number of arriving packets in the time interval $(T_{n-1}, T_n]$, where $(t_x, t_y]$ indicates the interval $t_x < t \leq t_y$. N_{n+1} is the number of arriving packets in $(T_n, T_{n+1}]$, and τ is the time duration per packet in seconds. The average arrival rate is λ_t. Assume the arrivals are independent of each other.
   (a) Write an expression for the joint pdf of N_n and N_{n+1}.
   (b) Let T_n define the time at which user A's packet arrives. Express, in terms of the joint pdf of N_n and N_{n+1}, the probability that user A's transmission will be successful.

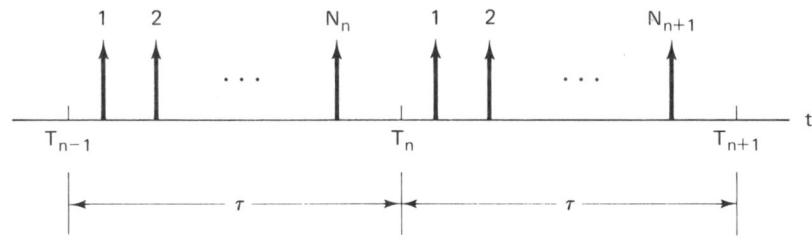

Figure P9.1

9.9. Let $N = N_n + N_{n+1}$, where N_n and N_{n+1} are as defined in Problem 9.8. Write an expression for the pdf of N, and give an interpretation for N.

9.10. Six thousand stations are competing for the use of a single slotted ALOHA channel. The average station makes 30 requests per hour, where each request is for one slot of 500-μs duration. Calculate the normalized total traffic on the channel.

9.11. Consider the arrival scenario of Figure P9.1; the location of the packet arrival times are permissible as shown under pure ALOHA, but not under slotted ALOHA, where arrivals are permitted only at the discrete times T_i, where $i = 0, 1, \ldots$. Assume that the average arrival rate is λ_t.
   (a) How would Figure P9.1 need to be modified if slotted ALOHA is used? How would the pdfs of N_n and N_{n+1} change?
   (b) If user A's packet arrives at time T_n, what is the probability of successful transmission?

9.12. A group of slotted-ALOHA stations generate a total of 120 requests per second, including both original and retransmissions. Each request is for a 12.5-ms duration slot.

(a) What is the normalized total traffic on the channel?

(b) What is the probability of a successful transmission on the first attempt?

(c) What is the probability of exactly two collisions before a successful transmission?

9.13. Measurements of a slotted-ALOHA channel show that 20% of the slots are idle.

(a) What is the normalized total traffic on the channel?

(b) What is the normalized throughput?

(c) Is the channel underloaded or overloaded?

9.14. Show that the sum of two Poisson processes, with rates λ_1 and λ_2, is also a Poisson process, with rate $\lambda_t = \lambda_1 + \lambda_2$. Generalize your result for the sum of n Poisson processes.

9.15. A 10-MHz transponder is occupied by 200 identical carriers, half servicing stations with $G/T = 40$ dB/K, the other half servicing stations with $G/T = 37$ dB/K. All stations have a requirement to operate with a bit error probability of 10^{-5}. The transponder is power limited under this configuration.

(a) What is the maximum possible bandwidth for each carrier?

(b) Suppose that each carrier has a bandwidth of 40 kHz, and the transponder is required to service a group of larger ($G/T = 40$ dB/K) stations only. How many stations can the transponder handle? Will the transponder be power or bandwidth limited?

(c) Repeat part (b) for the case where the transponder is to service a group of small ($G/T = 37$ dB/K) stations only.

9.16. A TDMA system operates at 100 Mbits/s with a 2-ms frame time. Assume that all slots are of equal length and that a guard time of 1 μs is required between slots.

(a) Compute the efficiency of the communications resource (CR) for the case of 1, 2, 5, 10, 20, 50, and 100 slots per frame.

(b) Repeat part (a) assuming that a 100-bit preamble is required at the start of each slot. Compute the efficiency of the CR in terms of the desired information transmission.

(c) Graph the results of parts (a) and (b).

9.17. With reference to Equation (9.36):

(a) Discuss the efficiency of the CR use if all S_i and R_j are equal.

(b) Discuss the effect of a few S_i or R_j being much larger than the majority. How can the efficiency of the CR be improved?

(c) When are the distributions of S_i and R_j likely to be similar? Dissimilar?

9.18. (a) Consider a token-ring network operating at a transmission rate of 10 Mbits/s over a cable having a propagation velocity of 200 m/μs. How many meters of cable is equal to a delay of 1 bit at each ring interface?

(b) If the token is 10 bits long, and all but three stations are switched off during evening hours, what is the minimum cable length needed for the ring?

CHAPTER 10

Spread-Spectrum
Techniques

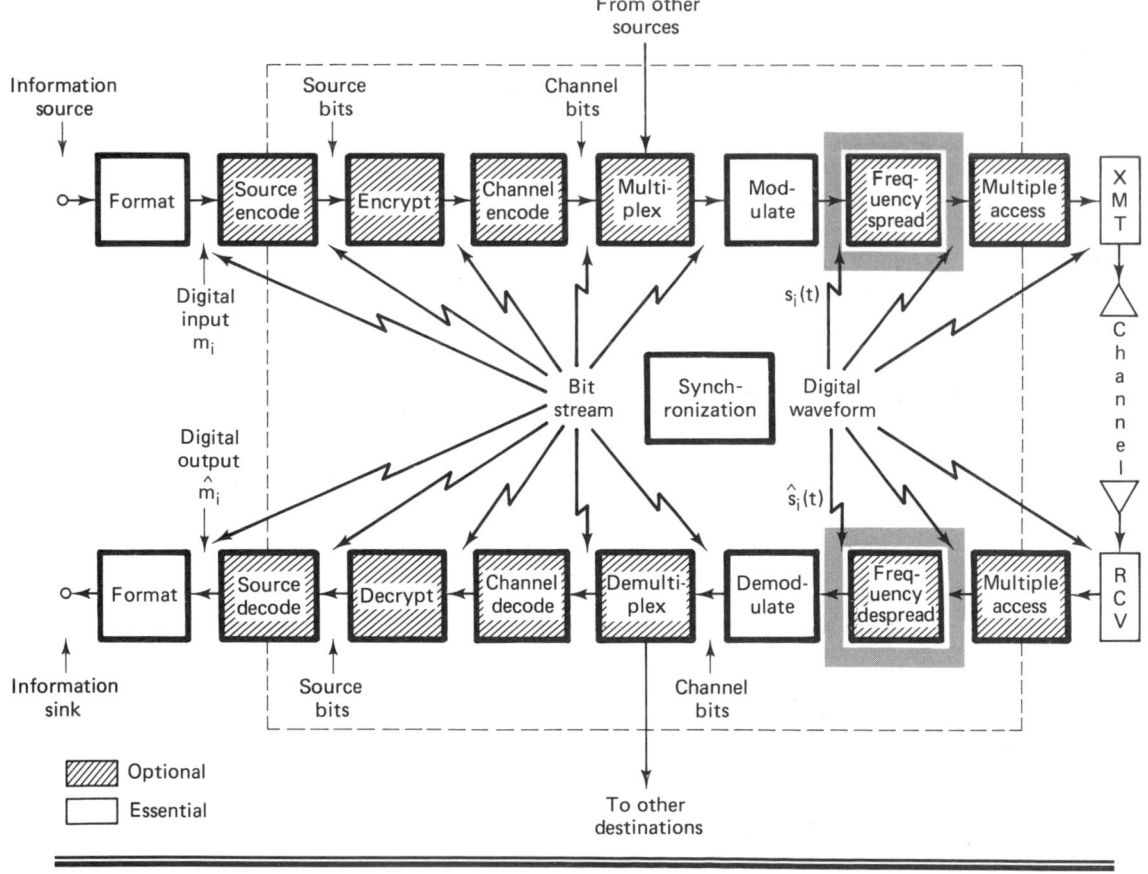

10.1 SPREAD-SPECTRUM OVERVIEW

The initial application of spread-spectrum (SS) techniques was in the development of military guidance and communication systems. By the end of World War II, spectrum spreading for jamming resistance was already a familiar concept to radar engineers [1], and during subsequent years, SS investigation was motivated primarily by the desire to achieve highly jam-resistant communication systems. As a result of this research, there emerged an assortment of other applications in such areas as energy density reduction, high-resolution ranging, and multiple access, which will be discussed in later sections. The techniques considered in this chapter are called *spread spectrum* because the transmission bandwidth employed is much greater than the minimum bandwidth required to transmit the information. A system is defined to be a spread-spectrum system if it fulfills the following requirements:

1. The signal occupies a bandwidth much in excess of the minimum bandwidth necessary to send the information.
2. Spreading is accomplished by means of a *spreading signal*, often called a *code signal*, which is independent of the data. The details of some spreading signals are described in later sections.
3. At the receiver, despreading (recovering the original data) is accomplished by the correlation of the received spread signal with a synchronized replica of the spreading signal used to spread the information.

Standard modulation schemes such as frequency modulation and pulse code modulation also spread the spectrum of an information signal, but they do not qualify as spread-spectrum systems since they do not satisfy all the conditions outlined above.

10.1.1 The Beneficial Attributes of Spread-Spectrum Systems

10.1.1.1 Interference Suppression Benefits

White Gaussian noise is a mathematical model that, by definition, has infinite power spread uniformly over all frequencies. Effective communication is possible with this interfering noise of infinite power because only the finite-power noise components that are present within the signal space (in other words, share the *same coordinates* as the signal components) can interfere with the signal. The balance of the noise power may be thought of as noise that is effectively tuned out by the detector (see Section 3.2.2). For a typical narrowband signal, this means that only the noise in the signal bandwidth can degrade performance. The idea behind a spread-spectrum anti-jam (AJ) system is as follows. Consider that many orthogonal signal coordinates or dimensions are available to a communication link and that only a small subset of these signal coordinates are used at any time. We assume that the jammer cannot determine the signal subset that is currently in use. For signals of bandwidth W and duration T, the number of signaling dimensions can be shown [2] to be approximately $2WT$. Given a specific signal design, the error performance of such a system is only a function of E_b/N_0. Against white Gaussian noise, with *infinite* power, the use of spreading (large $2WT$) offers no performance improvement. However, when the noise stems from a jammer with a *fixed finite* power and with uncertainty as to where in the signal space the signal coordinates are located, the jammer's choices are limited to those shown below.

1. Jam *all* the signal coordinates of the system, with an *equal* amount of power in each one, with the result that *little* power is available for each coordinate.
2. Jam a *few* signal coordinates with *increased* power in each of the jammed coordinates (or more generally, jam all the coordinates with various amounts of power in each).

Figure 10.1 compares the effect of spectrum spreading in the presence of white noise with spreading in the presence of an intentional jammer. The power spectral density of the signal is denoted $G(f)$ before spreading, and $G_{ss}(f)$ after spreading. For simplicity, the figure treats the frequency dimension only. In Figure 10.1a it can be seen that the single-sided power spectral density of white noise, N_0, is unchanged as a result of expanding the signal bandwidth from W to W_{ss}. The average power of white noise (area under the spectral density curve) is infinite. Hence, the use of spreading offers no performance improvement here. Figure 10.1b (upper diagram) illustrates the case of received (fixed finite) jammer power, J, and power spectral density, $J_0' = J/W$, where W is the unspread band-

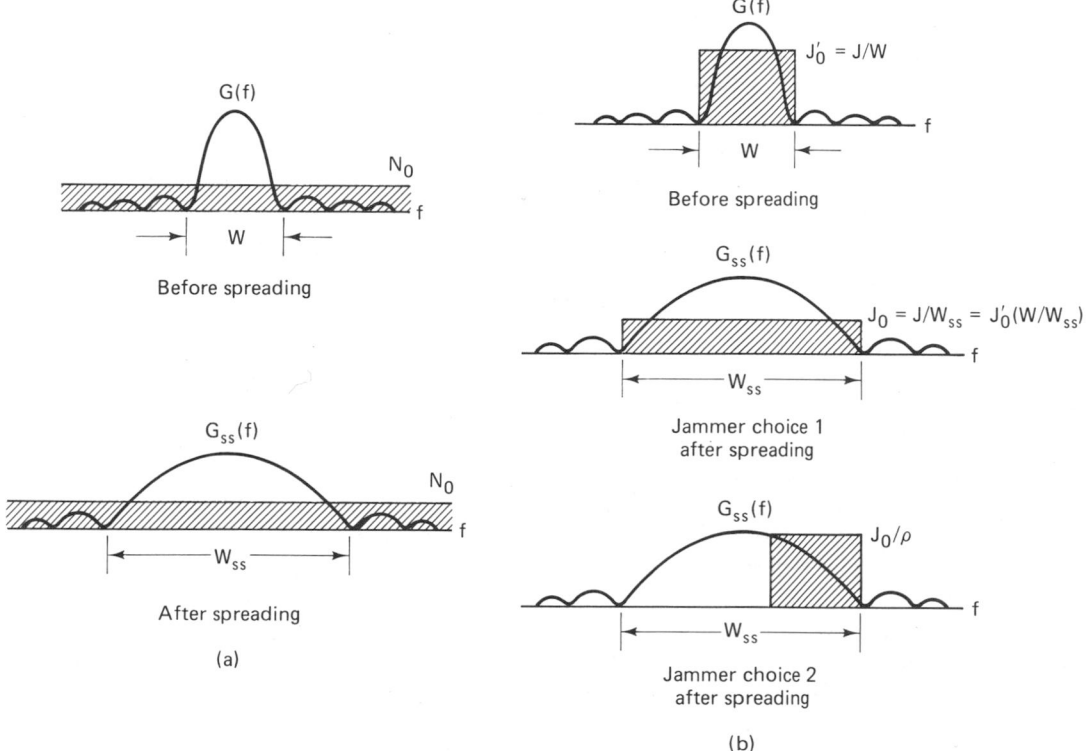

Figure 10.1 Effect of spectrum spreading. (a) Spectrum spreading in the presence of white noise. (b) Spectrum spreading in the presence of an intentional jammer.

width being jammed. Once the signal bandwidth is spread, the jammer can make one of the two choices listed earlier—choice 1 results in a reduction in jammer noise spectral density, J_0', by a factor (W/W_{ss}) across the spread spectrum. The resulting noise spectral density, $J_0 = J/W_{ss}$, is referred to as the *broadband jammer noise spectral density*. Choice 2 results in a reduction in the number of signal coordinates that the jammer occupies. However, with choice 2 the jammer can increase its noise spectral density from J_0 to J_0/ρ ($0 < \rho \leq 1$), where ρ is the portion of the spread-spectrum band the jammer elects to jam. If the jammer makes a poor choice in the coordinates to be jammed, the average effect of jamming will be less than if it makes a good choice. The larger the dimensionality of the signal set or the more signal coordinates the communicator can choose from, the greater is the jammer's uncertainty regarding the effectiveness of the jamming technique, and the better will be the protection against jamming.

Jamming is not always the result of an intentional act. Sometimes, the jamming signal is caused by natural phenomena, and sometimes it is the result of self-interference caused by *multipath*, in which delayed versions of the signal, arriving via alternative paths, interfere with the direct path transmission.

10.1.1.2 Energy Density Reduction

One can imagine situations where it is desired that a communications link be operated without being detected by anyone other than the intended receiver. Systems designed for this special task are known as *low probability of detection* (LPD) or *low probability of intercept* (LPI) communication systems. These systems are designed to make the detection of their signals as difficult as possible by anyone but the intended receiver. The goal of such a system is to use the minimum signal power and the optimum signaling scheme that results in the minimum probability of being detected. Since, in spread-spectrum systems, the signal is spread over many more signaling coordinates than in conventional modulation schemes, the resulting signal power is, on the average, spread thinly and uniformly in the spread domain. Therefore, not only can the spread-spectrum signal be made difficult to jam, but additionally, the signal's very existence may be rendered difficult to perceive. To anyone who does not possess a synchronized replica of the spreading signal, the spread-spectrum signal will seem "buried in the noise."

A *radiometer* is a simple power measuring instrument that can be used by an adversary to detect the presence of spread-spectrum signals within some bandwidth W. The radiometer, illustrated in Figure 10.2, consists of a bandpass filter (BPF) with bandwidth W, a squaring circuit to ensure a positive output value, since the presence of *signal energy* is being detected, and an integrating circuit. At time $t = T$, the output of the integrator is compared to a preset threshold. If the output of the integrator is larger than the threshold, a signal is declared present; otherwise, the signal is declared absent. References [3, 4] provide details on the detectability of spread-spectrum signals, using radiometers and other more complicated instruments that make use of the features of the SS signal itself.

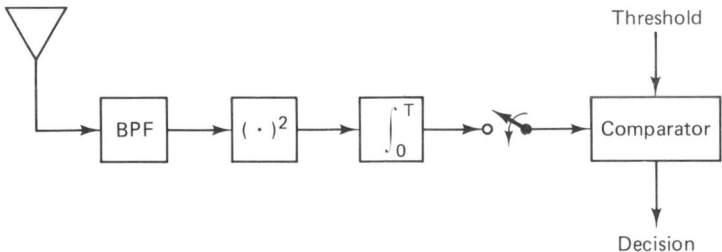

Figure 10.2 Radiometer.

Spread-spectrum systems that are designed to exhibit LPI may also exhibit a *low probability of position fix* (LPPF), which means that even if the presence of the signal is perceived, the direction of the transmitter is difficult to pinpoint. Some spread-spectrum systems also exhibit a *low probability of signal exploitation* (LPSE), which means that the identification of the source is difficult to ascertain.

Another, unrelated application of spread-spectrum signaling deals with the fact that in some cases energy density reduction may be required to meet national allocation regulations. Downlink transmissions from satellites must meet international regulations on the spectral density that impinges on the earth. By spreading the downlink energy over a wider bandwidth, the total transmitted power can be increased and hence performance improved, while the energy density regulations are followed.

10.1.1.3 Fine Time Resolution

Spread-spectrum signals can be used for ranging or determination of position location. Distance can be determined by measuring the time delay of a pulse as it traverses the channel. Uncertainty in the delay measurement is inversely proportional to the bandwidth of the signal pulse. This can be seen by the illustration in Figure 10.3. The uncertainty of the measurement, Δt, is proportional to the rise time of the pulse, which is inversely proportional to the bandwidth of the pulse signal; that is,

$$\Delta t \simeq \frac{1}{W} \tag{10.1}$$

The larger the bandwidth, the more precisely one can measure range. Over a Gaussian channel, a one-shot measurement on a single pulse is not very reliable. The spread-spectrum technique, however, uses a code signal consisting of a long sequence of polarity changes (e.g., a binary PSK-modulated signal) in place of the single pulse. Upon reception, the received sequence is correlated against a local replica and the results of the correlation are used to perform an accurate time-delay or range measurement.

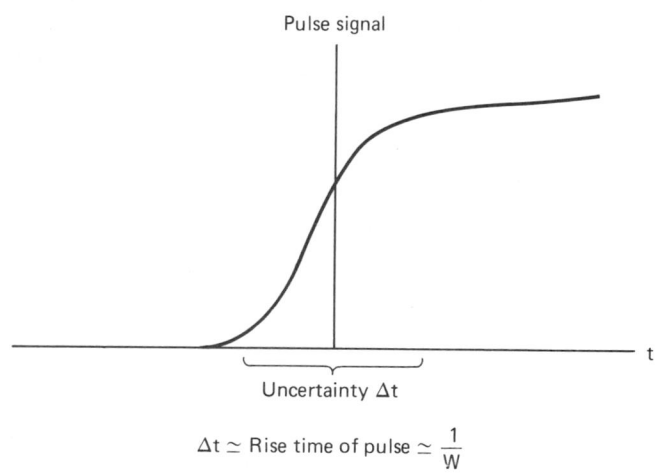

$$\Delta t \simeq \text{Rise time of pulse} \simeq \frac{1}{W}$$

Figure 10.3   Time-delay measurement.

10.1.1.4 Multiple Access

Spread-spectrum methods can be used as a multiple access technique, in order to share a communications resource among numerous users in a coordinated manner. The technique, termed *code-division multiple access* (CDMA), since each simultaneous user employs a unique spread-spectrum signaling code, was discussed briefly in Chapter 9. One of the by-products of this type of multiple access is the ability to provide communication privacy between users with different spreading signals. An unauthorized user (a user not having access to a spreading signal) cannot easily monitor the communications of the authorized users. A more detailed treatment is presented in a later section.

10.1.2 Model for Spread-Spectrum Interference Rejection

Figure 10.4 illustrates a model for spread-spectrum interference rejection. At the modulator, the information signal $x(t)$, with a data rate of R bits/s, is multiplied by a spreading code signal, $g(t)$, having a code symbol rate, usually called the code *chip rate*, R_p chips/second. Assume that the transmission bandwidths for $x(t)$ and $g(t)$ are R hertz and R_p hertz, respectively. Multiplication in the time domain transforms to convolution in the frequency domain:

$$x(t)g(t) \leftrightarrow X(\omega) * G(\omega) \tag{10.2}$$

Therefore, if the data signal is narrowband compared to the spreading signal, the resulting product signal $x(t)g(t)$ will have approximately the bandwidth of the spreading signal (see Section A.5).

At the demodulator, the received signal is ideally multiplied by a synchronized replica of the spreading code signal, $g(t)$, which results in the despreading of the signal. A filter with bandwidth R is used to remove any spurious higher-frequency components. If there is any undesired signal at the receiver, the multiplication by $g(t)$ will spread this undesired signal, in the same way that the multiplication by $g(t)$ at the transmitter spread the desired signal originally. Consider the effect on a jammer that attempts to position a narrowband jamming signal within the information bandwidth. The first operation at the receiver input is multiplication by the spreading signal. Hence the jamming tone is spread to the bandwidth of the spreading signal.

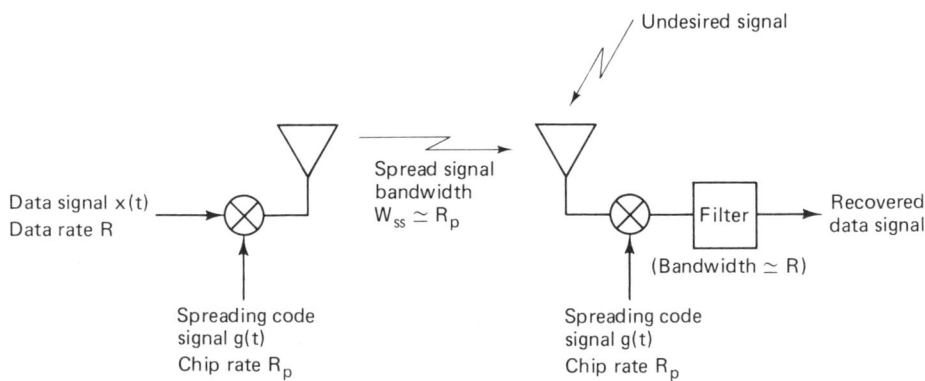

Figure 10.4 Basic spread-spectrum technique.

The essence behind the interference rejection capability of a spread-spectrum system can be summarized as follows:

1. Multiplication by the spreading signal *once* spreads the signal bandwidth.
2. Multiplication by the spreading signal *twice*, followed by filtering, recovers the original signal.
3. The desired signal gets multiplied *twice*, but the interference signal gets multiplied only *once*.

10.1.3 A Catalog of Spreading Techniques

Figure 10.5 highlights the popular techniques for spreading the information signal over a large number of signal coordinates or dimensions. For signals of bandwidth W and duration T, the dimensionality of the signaling space is approximately $2WT$.

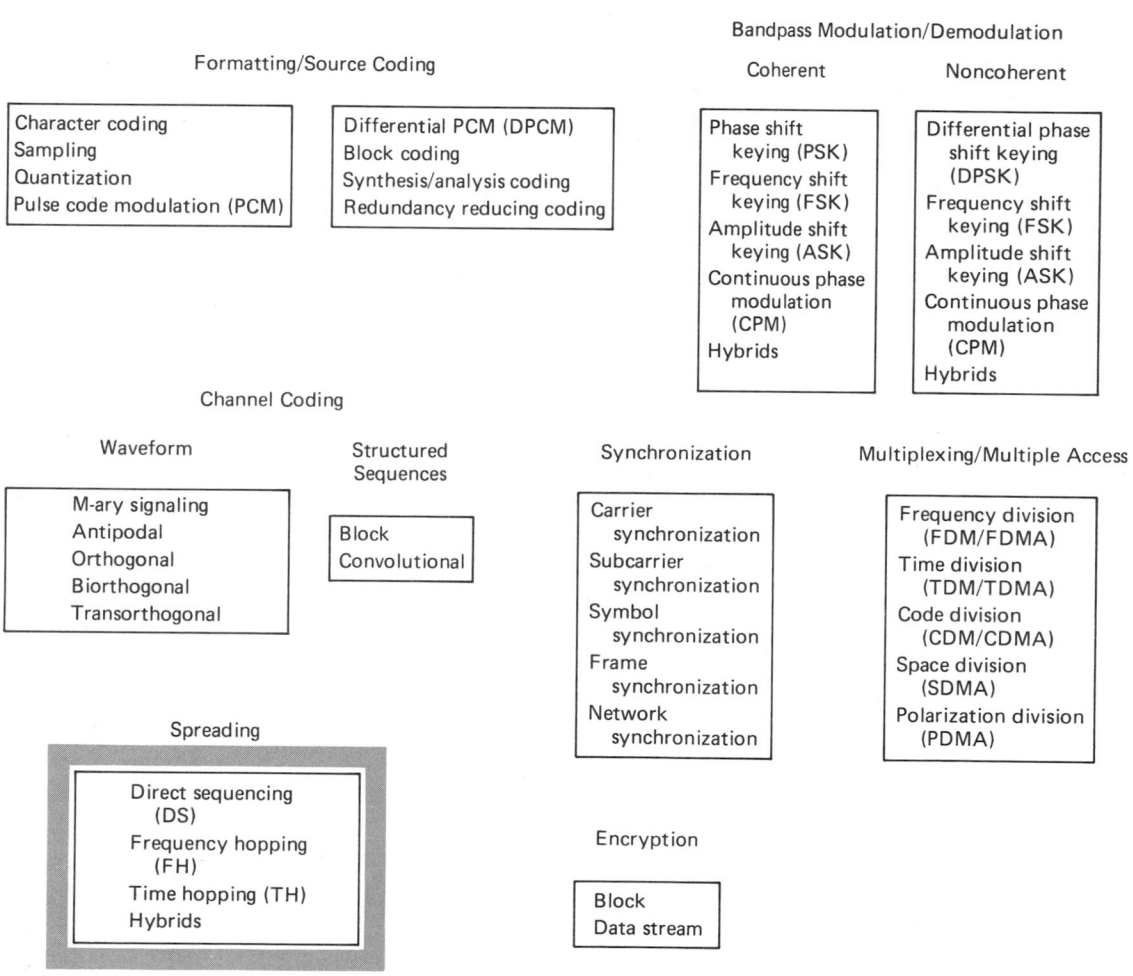

Figure 10.5 Basic digital communication transformations.

To increase the dimensionality, we can either increase W by spectrum spreading, or increase T by time spreading or time hopping (TH). With spectrum spreading the signal is spread in the frequency domain. With time hopping, a message with data rate R is allocated a longer transmission-time duration than would be used with a conventional modulation scheme. During this longer time the data are sent in bursts according to the dictates of a code. We can say that with time hopping the signal is spread in the time domain. For both cases, frequency spreading and time spreading, a jammer will be uncertain regarding the signaling subset that is currently in use.

In Figure 10.5, the first two items listed under the category of spreading, *direct sequencing* (DS) and *frequency hopping* (FH), are the most commonly used techniques for spectrum spreading. As a jamming-rejection technique, *time hopping* (TH), the third item in the list, is similar to spread spectrum, in that the location of the signal coordinates is hidden from potential adversaries. Also, there are hybrid combinations of the spreading techniques, for example, DS/FH, FH/TH, and DS/FH/TH; however, these techniques can be viewed as simple extensions of the material presented here and will not be elaborated on. In this chapter we focus only on the two major spread-spectrum techniques, direct sequencing and frequency hopping.

10.1.4 Historical Background

10.1.4.1 Transmitted Reference versus Stored Reference

During the early years of spread-spectrum investigation one technique that was considered for operating a transmitter and receiver synchronously with a *truly random* spreading signal such as wideband noise, was called a *transmitted reference* (TR) system. In a TR system the transmitter would send two versions of an unpredictable wideband carrier, one modulated by data and the other unmodulated. These two signals were transmitted on separate channels. The receiver used the unmodulated carrier as the reference signal for despreading (correlating) the data-modulated carrier. The principal advantage of a TR system was that there were no significant synchronization problems at the receiver, since the data-modulated signal and the spreading signal used for despreading were transmitted simultaneously. The principal disadvantages of TR systems were that (1) the spreading code was sent in the clear and thus was available to any listener; (2) the system could be easily spoofed by a jammer sending a pair of waveforms acceptable to the receiver; (3) performance degraded at low signal levels since noise was present on both signals; and (4) twice the bandwidth and transmitted power were required because of the need to transmit the reference.

Modern spread-spectrum systems all use a technique called *stored reference* (SR), whereby the spreading code signal is independently generated at both the transmitter and the receiver. The main advantage of an SR system is that a well-designed code signal cannot be predicted by monitoring the transmission. Note that the noiselike code signal in an SR system cannot be truly random as it could in the case of a TR system. Since the same code must be generated independently at two or more sites, the code sequence must be deterministic, even though it should appear random to unauthorized listeners. Such random-appearing deterministic signals are called pseudonoise (PN) or pseudorandom signals; their generation is treated later in greater detail.

10.1.4.2 Noise Wheels

In the late 1940s and early 1950s, Mortimer Rogoff, working at ITT, demonstrated the fundamental operation of spectrum spreading systems with a novel experiment [5]. Using photographic techniques, Rogoff built a "noise wheel" for storing a noiselike signal. He randomly selected 1440 numbers not ending in 00 from the Manhattan telephone directory, and radially plotted the middle two of the last four digits so that the radius at every $\frac{1}{4}°$ represented a new random number. The drawing was transferred to the wheel-shaped film shown in Figure 10.6. When the wheel was rotated past a slit of light, the resulting intensity-modulated light beam provided a stored noiselike spreading signal to be sensed by a photocell.

Rogoff mounted two such identical wheels on a single axis driven by a 900-rpm synchronous motor. One wheel's noiselike spreading signal was modulated

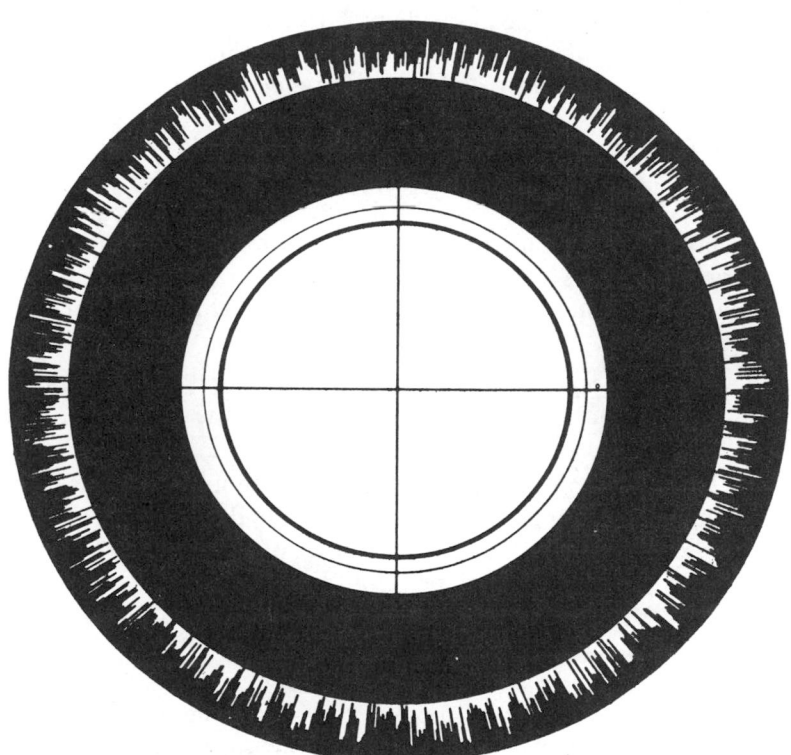

Figure 10.6 Rogoff's noise wheel. [Reprinted from Section I (Communications) of "Application of Statistical Methods to Secrecy Communication Systems," Proposal 946, Fed. Telecomm. Lab., August 28, 1950, Fig. 6, courtesy of ITT.]

with data (and interference) to provide one input to the receiving correlator, while the other wheel's unmodulated spreading signal provided the other input to the correlator. These baseband experiments, performed with data rates of 1 bit/s, demonstrated the feasibility of conveying information hidden in noiselike signals [6].

10.2 PSEUDONOISE SEQUENCES

The spread-spectrum approach called *transmitted reference* (TR) can utilize a *truly* random code signal for spreading and despreading, since the code signal and the data-modulated code signal are simultaneously transmitted over different regions of the spectrum. The *stored reference* (SR) approach *cannot* use a truly random code signal since the code needs to be stored or generated at the receiver. For the SR system a *pseudonoise* or *pseudorandom* code signal must be used.

How does a pseudorandom signal differ from a random one? A random signal *cannot* be predicted; its future variations can only be described in a statistical sense. However, a pseudorandom signal is not random at all; it is a deterministic, periodic signal that is known to both the transmitter and receiver. Why the name "pseudonoise" or "pseudorandom"? Even though the signal is deterministic, it appears to have the statistical properties of sampled white noise. It appears, to an unauthorized listener, to be a truly random signal.

10.2.1 Randomness Properties

What are these randomness properties that make a pseudorandom signal appear truly random? There are three basic properties that can be applied to any periodic binary sequence as a test for the appearance of randomness. The properties, called *balance, run,* and *correlation,* are described below for binary signals:

1. *Balance property.* Good balance requires that in each period of the sequence, the number of binary ones differs from the number of binary zeros by at most one digit.
2. *Run property.* A *run* is defined as a sequence of a single type of binary digit(s). The appearance of the alternate digit in a sequence starts a new run. The length of the run is the number of digits in the run. Among the runs of ones and zeros in each period, it is desirable that about one-half the runs of each type are of length 1, about one-fourth are of length 2, one-eighth are of length 3, and so on.
3. *Correlation property.* If a period of the sequence is compared term by term with any cyclic shift of itself, it is best if the number of agreements differs from the number of disagreements by not more than one count.

In the next section, a PN sequence is generated to test these properties.

10.2.2 Shift Register Sequences

Consider the linear feedback shift register illustrated in Figure 10.7. It is made up of a four-stage register for storage and shifting, a modulo-2 adder, and a feedback path from the adder to the input of the register (modulo-2 addition has been defined in Section 2.12.3). The shift register operation is controlled by a sequence of clock pulses (not shown). At each clock pulse the contents of each stage in the register is shifted one stage to the right. Also, at each clock pulse the contents of stages X_3 and X_4 are modulo-2 added (a linear operation), and the result is fed back to stage X_1. The shift register sequence is defined to be the output of the last stage—stage X_4 in this example.

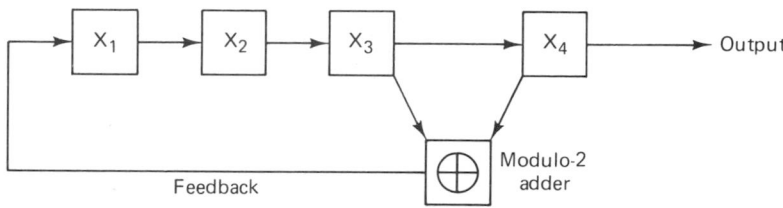

Figure 10.7   Linear feedback shift register example.

Assume that stage X_1 is initially filled with a one and the remaining stages are filled with zeros, that is, the initial state of the register is 1 0 0 0. From Figure 10.7 we can see that the succession of register states will be as follows:

1 0 0 0   0 1 0 0   0 0 1 0   1 0 0 1   1 1 0 0   0 1 1 0   1 0 1 1   0 1 0 1

1 0 1 0   1 1 0 1   1 1 1 0   1 1 1 1   0 1 1 1   0 0 1 1   0 0 0 1   1 0 0 0

Since the last state, 1 0 0 0, corresponds to the initial state, we see that the register repeats the foregoing sequence after 15 clock pulses. The output sequence is obtained by noting the contents of stage X_4 at each clock pulse. The output sequence is seen to be

$$0 \ 0 \ 0 \ 1 \ 0 \ 0 \ 1 \ 1 \ 0 \ 1 \ 0 \ 1 \ 1 \ 1 \ 1$$

where the leftmost bit is the earliest bit. Let us test the sequence above for the randomness properties outlined in the preceding section. First, the balance property; there are seven zeros and eight ones in the sequence—therefore, the sequence meets the balance condition. Next, the run property; consider the zero runs—there are four of them. One-half are of length 1, and one-fourth are of length 2. The same is true for the one runs. The sequence is too short to go further, but we can see that the run condition is met. The correlation property is treated in Section 10.2.3.

The shift register generator produces sequences that depend on the number of stages, the feedback tap connections, and initial conditions. The output sequences can be classified as either *maximal length* or *nonmaximal length*. Maximal length sequences have the property that for an n-stage linear feedback shift register the sequence repetition period in clock pulses p is

$$p = 2^n - 1 \qquad (10.3)$$

Thus it can be seen that the sequence generated by the shift register generator of Figure 10.7 is an example of a maximal length sequence. If the sequence length is less than $(2^n - 1)$, the sequence is classified as a nonmaximal length sequence.

10.2.3 PN Autocorrelation Function

The autocorrelation function $R_x(\tau)$ of a periodic waveform $x(t)$, with period T_0, was given in Equation (1.23) and is shown below in normalized form.

$$R_x(\tau) = \frac{1}{K}\left(\frac{1}{T_0}\right) \int_{-T_0/2}^{T_0/2} x(t)x(t + \tau)\, dt \qquad \text{for } -\infty < \tau < \infty \qquad (10.4)$$

where

$$K = \frac{1}{T_0} \int_{-T_0/2}^{T_0/2} x^2(t)\, dt \qquad (10.5)$$

When $x(t)$ is a periodic pulse waveform representing a PN code, we refer to each fundamental pulse as a *PN code symbol* or a *chip*. For such a PN waveform of unit chip duration and period p chips, the normalized autocorrelation function may be expressed as

$$R_x(\tau) = \frac{1}{p} \cdot \begin{pmatrix} \text{number of agreements less number of disagreements} \\ \text{in a comparison of one full period of the sequence} \\ \text{with a } \tau \text{ position cyclic shift of the sequence} \end{pmatrix} \qquad (10.6)$$

The normalized autocorrelation function for a maximal length sequence, $R_x(\tau)$, is shown plotted in Figure 10.8. It is clear that for $\tau = 0$, that is, when $x(t)$ and its replica are perfectly matched, $R(\tau) = 1$. However, for any cyclic shift between $x(t)$ and $x(t + \tau)$ with $(1 \leq \tau < p)$, the autocorrelation function is equal to $-1/p$ (for large p, the sequences are virtually decorrelated for a shift of a *single chip*).

It is now easy to test the output PN sequence of the shift register in Figure 10.7 for the third randomness property—correlation. Below is shown the output sequence; also shown is the same sequence with a single end-around shift:

$$0\ 0\ 0\ 1\ 0\ 0\ 1\ 1\ 0\ 1\ 0\ 1\ 1\ 1\ 1$$
$$1\ 0\ 0\ 0\ 1\ 0\ 0\ 1\ 1\ 0\ 1\ 0\ 1\ 1\ 1$$

$$d\ a\ a\ d\ d\ a\ d\ a\ d\ d\ d\ d\ a\ a\ a$$

The digits that agree are labeled a and those that disagree are labeled d. Following Equation (10.6), the value of the autocorrelation function for this single one-chip shift is seen to be

$$R(\tau = 1) = \tfrac{1}{15}\,(7 - 8) = -\tfrac{1}{15}$$

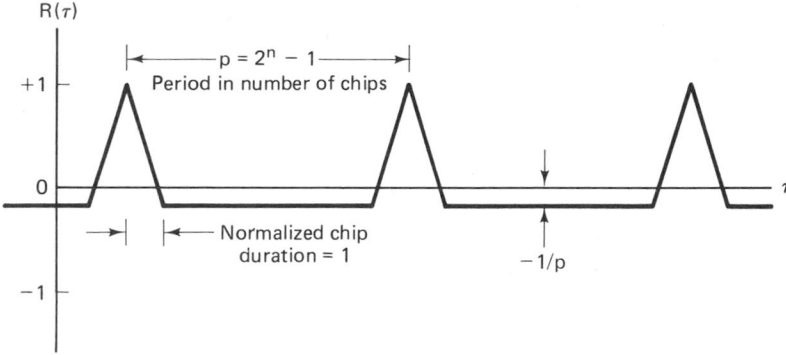

Figure 10.8 PN autocorrelation function.

Any cyclic shift yielding a mismatch from perfect synchronization results in the same autocorrelation value, $-1/p$. Hence the sequence meets the third randomness property.

10.3 DIRECT-SEQUENCE SPREAD-SPECTRUM SYSTEMS

The block diagram in Figure 10.9a depicts a *direct-sequence* (DS) modulator. "Direct sequence" is the name given to the spectrum spreading technique whereby a carrier wave is first modulated with a data signal $x(t)$, then the data-modulated signal is again modulated with a high-speed (wideband) spreading signal $g(t)$. Consider a constant-envelope data-modulated carrier having power P, radian frequency ω_0, and data phase modulation $\theta_x(t)$, given by

$$s_x(t) = \sqrt{2P} \cos [\omega_0 t + \theta_x(t)] \qquad (10.7)$$

Upon further constant-envelope modulation by the spreading signal, $g(t)$, the transmitted waveform can be expressed as

$$s(t) = \sqrt{2P} \cos [\omega_0(t) + \theta_x(t) + \theta_g(t)] \qquad (10.8)$$

where the phase of the carrier is now seen to have two components: $\theta_x(t)$ due to the data and $\theta_g(t)$ due to the spreading sequence.

In Chapter 3 it was shown that ideal suppressed carrier binary phase shift keying (BPSK) modulation results in instantaneous changes of π radians to the phase of the carrier, according to the dictates of the data. We can equivalently express Equation (10.7) as the multiplication of the carrier wave by $x(t)$, an antipodal pulse stream with pulse values of $+1$ or -1:

$$s_x(t) = \sqrt{2P} \, x(t) \cos \omega_0 t \qquad (10.9)$$

If, like the data, the spreading sequence modulation is also BPSK, and $g(t)$ is an antipodal pulse stream with pulse values of $+1$ or -1, Equation (10.8) can be written as

$$s(t) = \sqrt{2P}\, x(t)g(t)\cos\omega_0 t \qquad (10.10)$$

A modulator based on Equation (10.10) is illustrated in Figure 10.9b. The data pulse stream and the spreading pulse stream are first multiplied, and then the composite $x(t)g(t)$ modulates the carrier. If the assignment of pulse value to binary value is

Pulse value	Binary value
1	0
-1	1

then the initial step in the DS/BPSK modulation can be accomplished by the modulo-2 addition of the binary data sequence with the binary spreading sequence.

Demodulation of the DS/BPSK signal is accomplished by correlating or re-modulating the received signal with a synchronized replica of the spreading signal $g(t - \hat{T}_d)$ as seen in Figure 10.9c, where $\hat{T}_d$ is the receiver's estimate of the propagation delay T_d from the transmitter to the receiver. In the absence of noise and interference, the output signal from the correlator can be written as

$$A\sqrt{2P}\, x(t - T_d)g(t - T_d)g(t - \hat{T}_d)\cos[\omega_0(t - T_d) + \phi] \qquad (10.11)$$

where the constant A is a system gain parameter and ϕ is a random phase angle in the range $(0, 2\pi)$. Since $g(t) = \pm 1$, the product $g(t - T_d)g(t - \hat{T}_d)$ will be unity if $\hat{T}_d = T_d$, that is, if the code signal at the receiver is exactly synchronized with the code signal at the transmitter. When it is synchronized, the output of the receiver correlator is the despread data-modulated signal (except for a random phase ϕ and delay T_d). The despreading correlator is then followed by a conventional demodulator for recovering the data.

10.3.1 Example of Direct Sequencing

Figure 10.10 is an example of DS/BPSK modulation and demodulation following the block diagrams of Figure 10.9b and c. In Figure 10.10a are shown the binary data sequence (1, 0) and its bipolar pulse waveform equivalent $x(t)$, where the binary to pulse value assignments are the same as those described in the preceding section. Examples of a binary spreading sequence and its bipolar pulse waveform equivalent $g(t)$ are shown in Figure 10.10b. The modulo-2 addition of the data sequence and the code sequence, and the equivalent waveform of the product $x(t)g(t)$, is shown in Figure 10.10c.

For the BPSK modulation described by Equations (10.8) and (10.10), it is shown in Figure 10.10d that the phase of the carrier, $\theta_x(t) + \theta_g(t)$, equals π when the value of the product waveform $x(t)g(t)$ equals -1 (or the modulo-2 sum of data and code is binary 1). Similarly, the phase of the carrier is zero when the value of $x(t)g(t)$ equals $+1$ (or the modulo-2 sum of data and code is binary 0). One can appreciate the *signal hiding* property of spread-spectrum signals by comparing the code waveform in Figure 10.10b with the composite waveform in Figure 10.10c. The latter has the signal $x(t)$ "hidden" within it. Just as your eye has

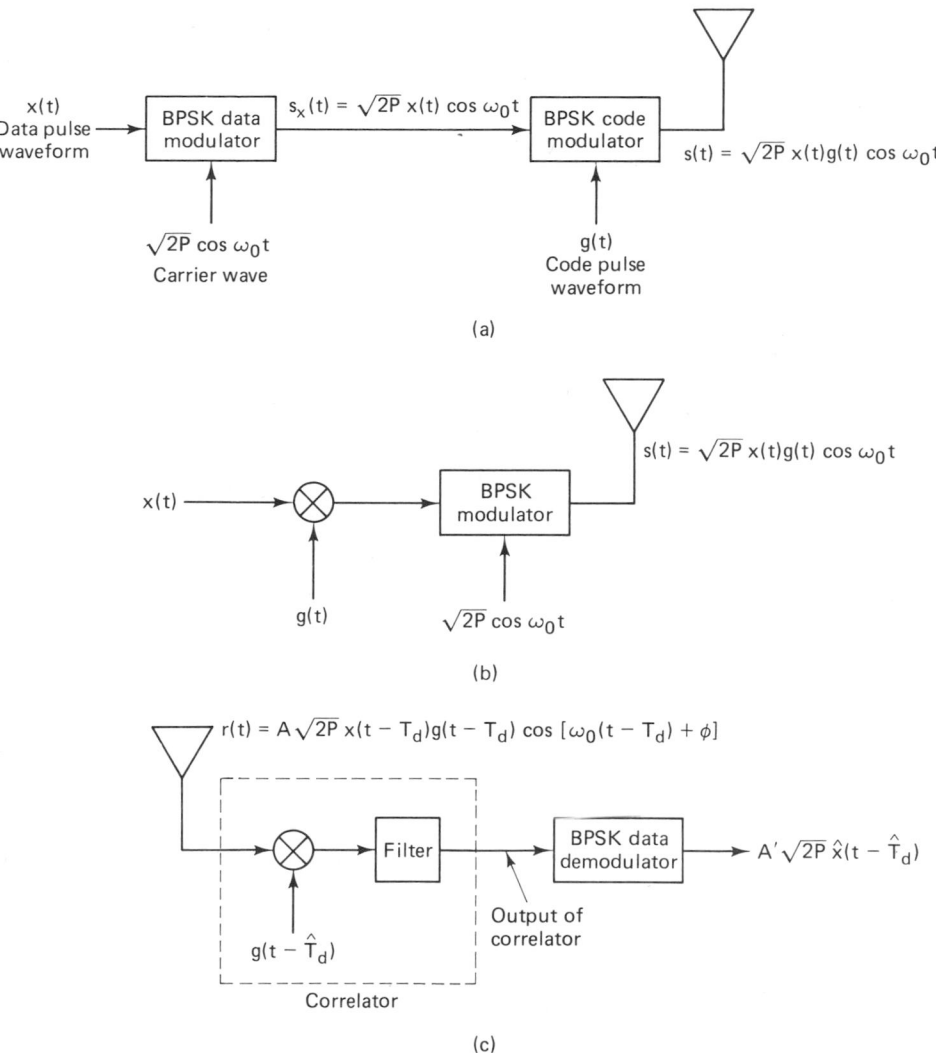

Figure 10.9 Direct-sequence spread-spectrum system. (a) BPSK direct-sequence transmitter. (b) Simplified BPSK direct-sequence transmitter. (c) BPSK direct-sequence receiver.

difficulty finding the slowly moving data signal in the rapidly moving code signal, it is similarly difficult for a receiver to recover a slowly moving signal from a rapidly moving code without having an exact replica of the code.

As shown in Figure 10.9c, DS/BPSK demodulation is a two-step process. The first step, despreading, is accomplished by correlating the received signal with a synchronized replica of the code. The second step, data demodulation, is accomplished with a conventional demodulator. In the example of Figure 10.10

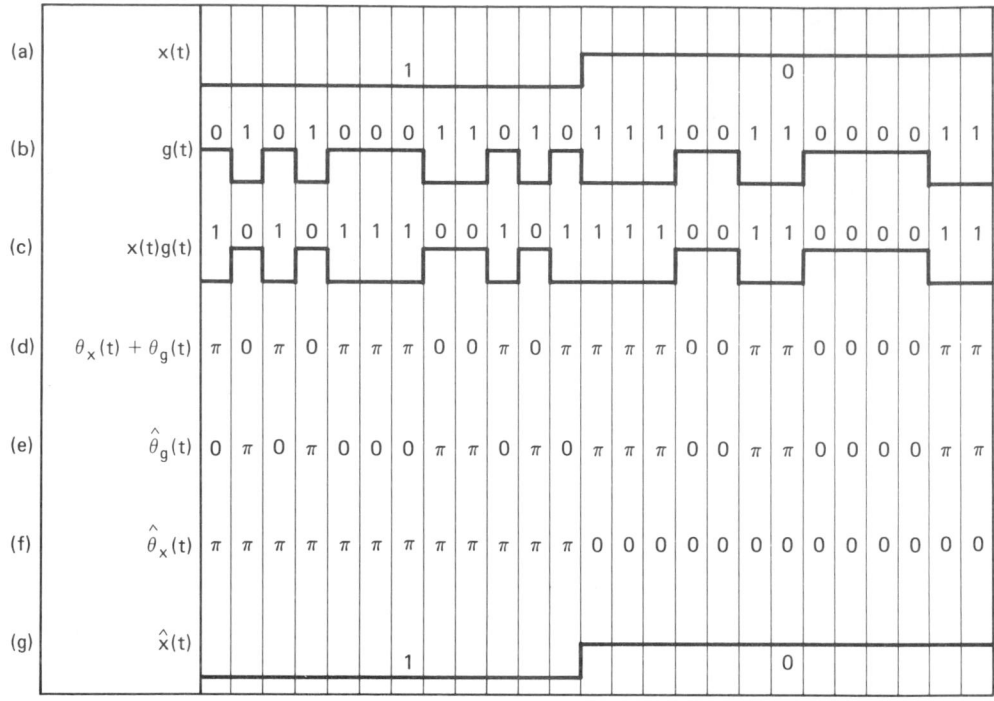

Figure 10.10 Spread-spectrum example using direct sequencing. (a) Binary data waveform to be transmitted. (b) Code sequence. (c) Transmitted sequence. (d) Phase of transmitted carrier. (e) Phase shift produced by receiver code. (f) Phase of received carrier after phase shift by receiver code. (g) Demodulated data waveform.

we see the code replica $\hat{\theta}_g(t)$, in Figure 10.10e, as the phase shift (either 0 or π) that is produced at the receiver by the despreading code. Figure 10.10f illustrates the resulting estimate of the carrier phase, $\hat{\theta}_x(t)$, after despreading or after $\hat{\theta}_g(t)$ has been added to $\theta_x(t) + \theta_g(t)$. At this point one can recognize the original data pattern in the phase terms of the carrier wave. The final step, shown in Figure 10.10g, is to recover an estimate of the data waveform, $\hat{x}(t)$, by the use of a BPSK demodulator.

10.3.2 Processing Gain and Performance

A fundamental issue in spread-spectrum systems is *how much* protection spreading can provide against interfering signals with finite power. Spread-spectrum techniques distribute a relatively low-dimensional signal in a large-dimensional signal space. The signal is "hidden" within the signal space, since we assume that a jammer does not know which signal coordinates are being transmitted at any time. The only recourse for the jammer, intent upon communication disruption, is to jam the entire space with its fixed total power, thus inducing a limited amount of interference in each signal coordinate, or to jam a portion of the signal space with its total power, thus leaving the remainder of the signal space free of interference.

Consider a set of D orthogonal signals, $s_i(t)$, $1 \leq i \leq D$, in an N-dimensional space, where in general, $D \ll N$. Following the development in Section 3.2.2, we can write

$$s_i(t) = \sum_{j=1}^{N} a_{ij}\psi_j(t) \qquad \begin{array}{l} i = 1, 2, \ldots, D; \quad 0 \leq t \leq T \\ D \ll N \end{array} \tag{10.12}$$

where

$$a_{ij} = \int_0^T s_i(t)\psi_j(t)\, dt \tag{10.13}$$

and

$$\int_0^T \psi_j(t)\psi_k(t)\, dt = \begin{cases} 1 & \text{for } j = k \\ 0 & \text{otherwise} \end{cases} \tag{10.14}$$

The $\{\psi_j(t)\}$ are linearly independent functions that *span* or characterize the N-dimensional orthonormal space and are called *basis* functions of the space. For every information symbol that is transmitted, a set of coefficients $\{a_{ij}\}$ is chosen independently, using a pseudorandom spreading code, in order to hide the D-dimensional signal set in the larger N-dimensional space. The set of random variables $\{a_{ij}\}$ assume the values $\pm a$, each with a probability of $\frac{1}{2}$. The receiver, of course, has access to each set of coefficients chosen in order to perform the necessary correlation despreading. Even if the same ith symbol is sent repeatedly, the set $\{a_{ij}\}$ used to transmit it is newly selected from symbol to symbol. The energy in each signal waveform of the D signal set will be assumed equal, so that we can write the average energy for each signal as follows:

$$E_s = \int_0^T \overline{s_i^2(t)}\, dt = \sum_{j=1}^{N} \overline{a_{ij}^2} \qquad i = 1, 2, \ldots, D \tag{10.15}$$

where the overbar means the expected value over the ensemble of many symbol transmissions. The independent coefficients have zero mean and correlation:

$$\overline{a_{ij}a_{ik}} = \begin{cases} \dfrac{E_s}{N} & \text{for } j = k \\[2mm] 0 & \text{otherwise} \end{cases} \tag{10.16}$$

The standard assumption is that the jammer has no a priori knowledge regarding the selection of the signaling coefficients $\{a_{ij}\}$. As far as the jammer is concerned, the coefficients are uniformly distributed over the N basis coordinates. If the jammer chooses to distribute its power uniformly over the total signal space, the jammer waveform $w(t)$ can be written

$$w(t) = \sum_{j=1}^{N} b_j\psi_j(t) \tag{10.17}$$

with total energy

$$E_w = \int_0^T w^2(t)\, dt = \sum_{j=1}^N b_j^2 \tag{10.18}$$

A reasonable goal for a jammer would be to devise a strategy for selecting the portions b_j^2, of its fixed total energy E_w so as to minimize the desired signal-to-noise ratio (SNR) at the receiver after demodulation.

At the receiver, the detector output (ignoring receiver noise),

$$r(t) = s_i(t) + w(t) \tag{10.19}$$

is correlated with the set of possible transmitted signals, so that the output of the ith correlator z_i is

$$z_i = \int_0^T r(t)s_i(t)\, dt = \sum_{j=1}^N \left(a_{ij}^2 + b_j a_{ij}\right) \tag{10.20}$$

The second term on the right side of Equation (10.20) averages to zero over the ensemble of all possible pseudorandom code sequences, since the set of random variables $\{a_{ij}\}$ assume the values $\pm a$, each with probability $\frac{1}{2}$. Therefore, given that $s_m(t)$ was transmitted, the expected value of the output of the ith correlator, $\mathbf{E}(z_i|s_m)$, can be written, following the development in References [7, 8],

$$\mathbf{E}(z_i|s_m) = \sum_{j=1}^N \overline{a_{ij}^2} = \begin{cases} E_s & \text{for } i = m \\ 0 & \text{otherwise} \end{cases} \tag{10.21}$$

In Equation (10.21), the term $\mathbf{E}(z_i|s_m)$ for $i = m$ is to be interpreted as follows. Given that $s_i(t)$ is to be transmitted, N coefficients a_{ij} ($1 \le j \le N$) are chosen pseudorandomly (the receiver is assumed to have access to each choice of the a_{ij} for correlation despreading). Hence, in computing $\mathbf{E}(z_i|s_i)$, even though the ith information symbol is specified at the transmitter, the pattern of coefficients used to send it appears random (to the unauthorized receiver) for each transmission. Equation (10.21) presumes that the jammer has not been successful in its attempt to employ some clever tactics (described in Section 10.7).

Let us assume that all D signals are equally likely. Then the expected value at the output of any of the D correlators is

$$\mathbf{E}(z_i) = \frac{E_s}{D} \tag{10.22}$$

Similarly, using Equations (10.15) to (10.21), we compute $\text{var}\,(z_i|s_i)$, the variance at the output of the ith correlator, given that the ith signal was transmitted.

$$\text{var}\,(z_i|s_i) = \sum_{j,k} b_j b_k \overline{a_{ij} a_{ik}}$$

$$= \sum_{j=1}^N b_j^2 \overline{a_{ij}^2} \tag{10.23}$$

$$= \sum_{j=1}^{N} b_j^2 \frac{E_s}{N}$$

$$= \frac{E_w E_s}{N} \tag{10.24}$$

For completeness, the variance at the output of the ith correlator, var $(z_i|s_m)$, given that the mth signal was transmitted, where $i \neq m$, can similarly be computed to be

$$\text{var } (z_i|s_m) = \frac{E_w E_s}{N} + \frac{E_s^2}{N} \tag{10.25}$$

The signal-to-jammer ratio (SJR) at the output of the ith correlator can be defined as

$$\text{SJR} = \sum_{m=1}^{D} \frac{\mathbf{E}^2(z_i|s_m)}{\text{var } (z_i|s_m)} P(s_m) = \frac{E_s^2/D}{E_w E_s/N} = \frac{E_s N}{E_w D} \tag{10.26}$$

where the probability of the mth signal $P(s_m) = 1/D$, since the signals are assumed to occur with equal probability, and where the signal energy and the jammer energy in the ith correlator are denoted by $\mathbf{E}^2(z_i)$ and var (z_i), respectively. Because of Equation (10.21), the only terms in the summation of Equation (10.26) not equal to zero are those for which $i = m$. The result is independent of the way in which the jammer chooses to distribute its energy. Therefore, regardless of how b_j is chosen, subject to $\sum_j b_j^2 = E_w$, the SJR in Equation (10.26) indicates that spreading gives the signal an advantage of a factor of N/D over the jammer. The ratio N/D is known as the *processing gain* G_p.

Since the approximate dimensionality of a signal with bandwidth W and duration T is $2WT$, we can express the processing gain as

$$G_p = \frac{N}{D} \simeq \frac{2W_{ss}T}{2W_{min}T} = \frac{W_{ss}}{R} \tag{10.27}$$

where W_{ss} is the spread-spectrum bandwidth (the total bandwidth used by the spreading technique) and W_{min} is the minimum bandwidth of the data (taken to be the data rate, R). For direct sequence systems, W_{ss} is approximately the code chip rate R_p, and W_{min} is similarly the data rate R, giving

$$G_p = \frac{R_p}{R} \tag{10.28}$$

10.4 FREQUENCY HOPPING SYSTEMS

We now consider a spread-spectrum technique called frequency hopping (FH). The modulation most commonly used with this technique is M-ary frequency shift keying (MFSK), where $k = \log_2 M$ information bits are used to determine which one of M frequencies is to be transmitted. The position of the M-ary signal set is shifted pseudorandomly by the frequency synthesizer over a hopping bandwidth W_{ss}. A typical FH/MFSK system block diagram is shown in Figure 10.11. In a conventional MFSK system, the data symbol modulates a *fixed frequency* carrier; in an FH/MFSK system, the data symbol modulates a carrier whose frequency is *pseudorandomly* determined. In either case, a single tone is transmitted. The FH system in Figure 10.11 can be thought of as a two-step modulation process—data modulation and frequency hopping modulation—even though it can be implemented as a single step whereby the frequency synthesizer produces a transmission tone based on the simultaneous dictates of the PN code and the data. At each frequency hop time a PN generator feeds the frequency synthesizer a frequency word (a sequence of ℓ chips) which dictates one of 2^{ℓ} symbol-set positions. The frequency hopping bandwidth, W_{ss}, and the minimum frequency spacing between consecutive hop positions, Δf, dictate the minimum number of chips necessary in the frequency word.

For a given hop, the occupied transmission bandwidth is identical to the bandwidth of conventional MFSK, which is typically much smaller than W_{ss}. However, averaged over many hops, the FH/MFSK spectrum occupies the entire spread-spectrum bandwidth. Current technology permits FH bandwidths of the order of several gigahertz, which is an order of magnitude larger than implementable DS bandwidths [9], thus allowing for larger processing gains in FH compared to DS systems. Since frequency hopping techniques operate over such wide bandwidths, it is difficult to maintain phase coherence from hop to hop. Therefore, such schemes are usually configured using noncoherent demodulation. Nevertheless, consideration has been given to coherent FH in Reference [10].

In Figure 10.11 we see that the receiver reverses the signal processing steps of the transmitter. The received signal is first FH demodulated (dehopped) by mixing it with the same sequence of pseudorandomly selected frequency tones that was used for hopping. Then the dehopped signal is applied to a conventional bank of M noncoherent energy detectors to select the most likely symbol.

Example 10.1   Frequency Word Size

A hopping bandwidth W_{ss} of 400 MHz and a frequency step size Δf of 100 Hz are specified. What is the minimum number of PN chips that are required for each frequency word?

Solution

$$\text{Number of tones contained in } W_{ss} = \frac{W_{ss}}{\Delta f} = \frac{400 \text{ MHz}}{100 \text{ Hz}}$$

$$= 4 \times 10^6$$

$$\text{Minimum number of chips} = \lceil \log_2 (4 \times 10^6) \rceil$$

$$= 22 \text{ chips}$$

where $\lceil x \rceil$ indicates the smallest integer value not less than x.

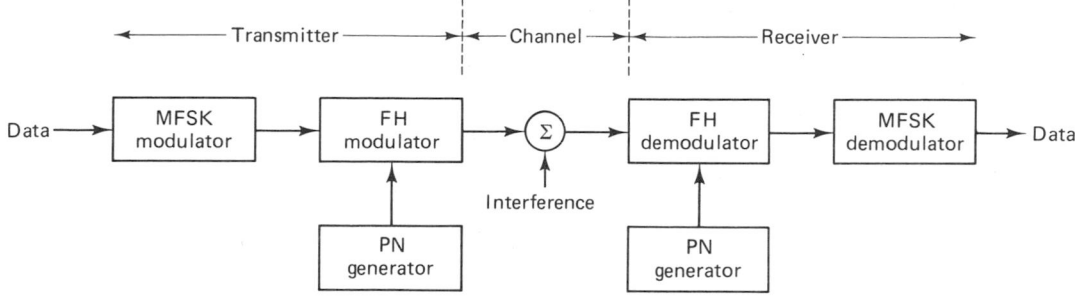

Figure 10.11 FH/MFSK system.

10.4.1 Frequency Hopping Example

Consider the frequency hopping example illustrated in Figure 10.12. The input data consist of a binary sequence with a data rate of $R = 150$ bits/s. The modulation is 8-ary FSK. Therefore, the symbol rate is $R_s = R/(\log_2 8) = 50$ symbols/s (the symbol duration $T = 1/50 = 20$ ms). The frequency is hopped once per symbol, and the hopping is time-synchronous with the symbol boundaries. Thus the hopping rate is 50 hops/s. Figure 10.12 depicts the time–bandwidth plane of the communication resource; the abscissa represents time, and the ordinate represents

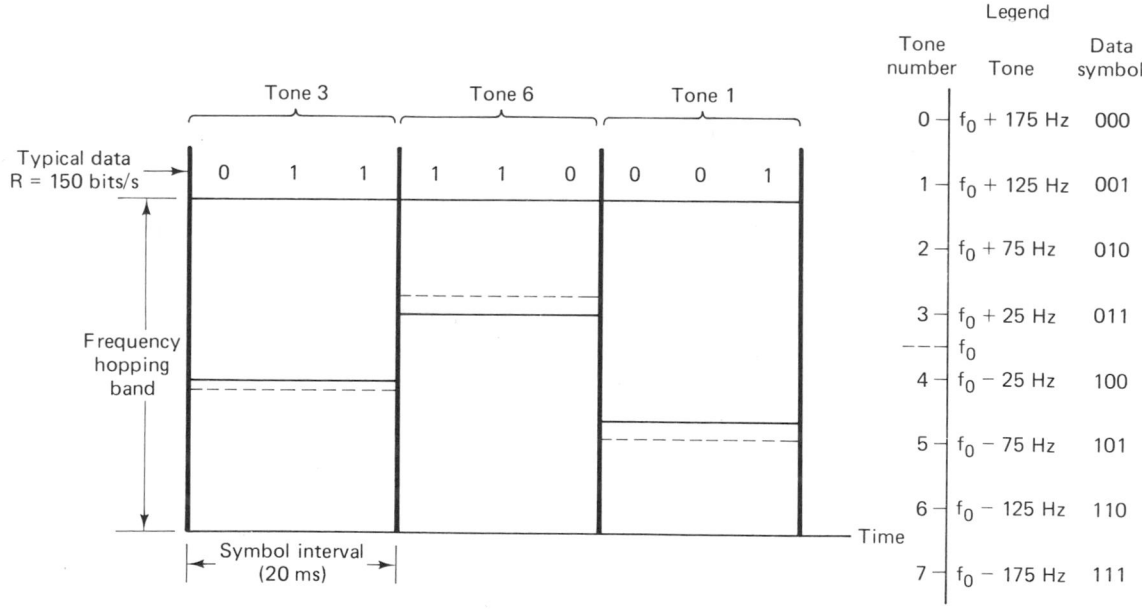

Figure 10.12 Frequency hopping example using 8-ary FSK modulation.

the hopping bandwidth, W_{ss}. The legend on the right side of the figure illustrates a set of 8-ary FSK symbol-to-tone assignments. Notice that the tone separation specified is $1/T = 50\,\text{Hz}$, which corresponds to the minimum required tone spacing for the orthogonal signaling of this noncoherent FSK example (see Section 3.6.4).

A typical binary data sequence is shown at the top of Figure 10.12. Since the modulation is 8-ary FSK, the bits are grouped three at a time to form symbols. In a *conventional* 8-ary FSK scheme, a single-sideband tone, offset from f_0, the *fixed* center frequency of the data band, would be transmitted (according to an assignment like the one shown in the legend). The only difference in this FH/MFSK example is that the center frequency of the data band, f_0, is *not fixed*. For each new symbol, f_0 hops to a new position in the hop bandwidth, and the entire data-band structure moves with it. In the example of Figure 10.12, the first symbol in the data sequence, 0 1 1, yields a tone 25 Hz above f_0. The diagram depicts f_0 with a dashed line and the symbol tone with a solid line. During the second symbol interval, f_0 has hopped to a new spectral location, as indicated by the dashed line. The second symbol, 1 1 0, dictates that a tone indicated by the solid line, 125 Hz below f_0, shall be transmitted. Similarly, the final symbol in this example, 0 0 1, calls for a tone 125 Hz above f_0. Again, the center frequency has moved, but the relative positions of the symbol tones remain fixed.

10.4.2 Robustness

A common dictionary definition describes the term *robustness* as the state of being strong and healthy; full of vigor; hardy. In the context of communications, the usage is not too different. Robustness characterizes a signal's ability to withstand impairments from the channel, such as noise, jamming, fading, and so on. A signal configured with multiple replicate copies, each transmitted on a different frequency, has a greater likelihood of survival than does a single such signal with equal total power. The greater the diversity (multiple transmissions, at different frequencies, spread in time), the more robust the signal against random interference.

The following example should clarify the concept. Consider a message consisting of four symbols s_1, s_2, s_3, s_4. The introduction of diversity starts by repeating the message N times. Let us choose $N = 8$. Then, the repeated symbols called *chips* can be written

$$s_1 s_1 s_1 s_1 s_1 s_1 s_1 s_1 s_2 s_2 s_2 s_2 s_2 s_2 s_2 s_2 s_3 s_3 s_3 s_3 s_3 s_3 s_3 s_3 s_4 s_4 s_4 s_4 s_4 s_4 s_4 s_4$$

Each chip is transmitted at a different hopping frequency (the center of the data bandwidth is changed for each chip). The resulting transmissions at frequencies f_i, f_j, f_k, . . . yield a more robust signal than without such divesity. A target-shooting analogy is that a pellet from a barrage of shotgun pellets has a better chance of hitting a target, compared to the action of a single bullet.

10.4.3 Frequency Hopping with Diversity

In Figure 10.13 we extend the example illustrated in Figure 10.12, with the additional feature of a chip repeat factor of $N = 4$. During each 20-ms symbol interval, there are now four columns, corresponding to the four separate chips to be transmitted for each symbol. At the top of the figure we see the same data sequence, with $R = 150$ bps, as in the earlier example; and we see the same 3-bit partitioning to form the 8-ary symbols. Each symbol is transmitted four times, and for each transmission the center frequency of the data band is hopped to a new region of the hopping band, under the control of a PN code generator. Therefore, for this example, each chip interval, T_c, is equal to $T/N = 20$ ms/4 = 5 ms in duration, and the hopping rate is now

$$\frac{NR}{\log_2 8} = 200 \text{ hops/s}$$

Notice that the spacing between frequency tones must change to meet the changed requirement for orthogonality. Since the duration of each FSK tone is now equal to the chip duration, that is, $T_c = T/N$, the minimum separation between tones is $1/T_c = N/T = 200$ Hz. As in the earlier example, Figure 10.13 illustrates that the center of the data band (plus the modulation structure) is shifted at each new

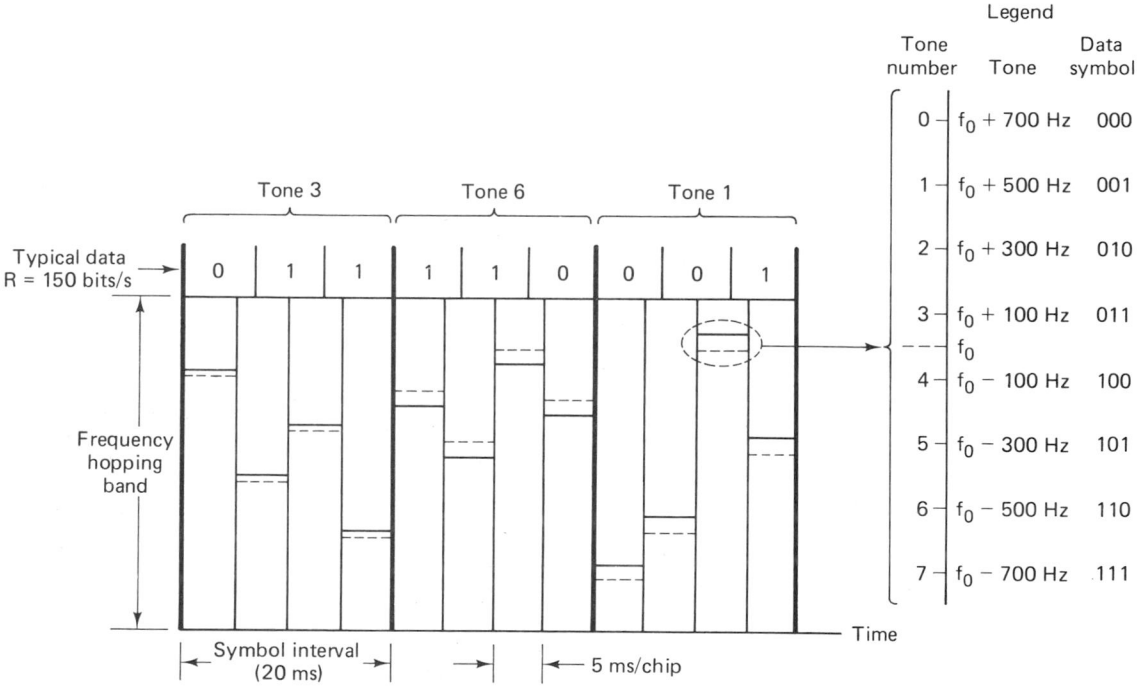

Figure 10.13 Frequency hopping example with diversity ($N = 4$).

chip time. The position of the solid line (transmission frequency) has the same relationship to the dashed line (center of the data band) for each of the chips associated with a given symbol.

10.4.4 Fast Hopping versus Slow Hopping

In the case of direct-sequence spread-spectrum systems, the term "chip" refers to the PN code symbol (the symbol of shortest duration in a DS system). In a similar sense for frequency hopping systems, the term "chip" is used to characterize the shortest uninterrupted waveform in the system. Frequency hopping systems are classified as *slow frequency hopping* (SFH), which means there are several modulation symbols per hop, or as *fast frequency hopping* (FFH), which means that there are several frequency hops per modulation symbol. For SFH, the shortest uninterrupted waveform in the system is that of the data symbol; however, for FFH, the shortest uninterrupted waveform is that of the hop. Figure 10.14a illustrates an example of FFH; the data symbol rate is 30 symbols/s and the frequency hopping rate is 60 hops/s. The figure illustrates the waveform $s(t)$ over one symbol duration ($\frac{1}{30}$ s). The waveform change in (the middle of) $s(t)$ is

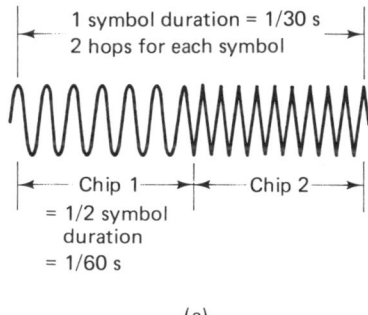

(a)

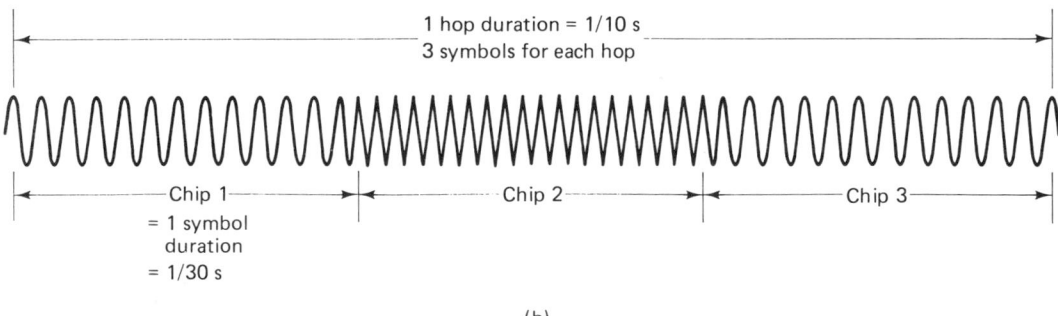

(b)

Figure 10.14 Chip—in the context of an FH/MFSK system. (a) Example 1: Frequency hopping MFSK system with symbol rate = 30 symbols/s and hopping rate = 60 hops/s. 1 chip = 1 hop. (b) Example 2: Same as part (a) except hopping rate = 10 hops/s. 1 chip = 1 symbol.

due to a new frequency hop. In this example, a chip corresponds to a hop since the hop duration is shorter than the symbol duration. Each chip corresponds to half a symbol. Figure 10.14b illustrates an example of SFH; the data symbol rate is still 30 symbols/s, but the frequency hopping rate has been reduced to 10 hops/s. The waveform $s(t)$ is shown over a duration of three symbols ($\frac{1}{10}$ s). In this example, the hopping boundaries appear only at the beginning and end of the three-symbol duration. Here, the changes in the waveform are due to the modulation state changes; therefore, in this example a chip corresponds to a data symbol, since the data symbol is shorter than the hop duration.

Figure 10.15a illustrates an FFH example of a binary FSK system. The diversity is $N = 4$. There are 4 chips transmitted per bit. As in Figure 10.13, the dashed line in each column corresponds to the center of the data band and the solid line corresponds to the symbol frequency. Here, for FFH, the chip duration is the hop duration. Figure 10.15b illustrates an example of an SFH binary FSK system. In this case, there are 3 bits transmitted during the time duration of a single hop. Here, for SFH, the chip duration is the bit duration. If this SFH example were changed from a binary system to an 8-ary system, what would the

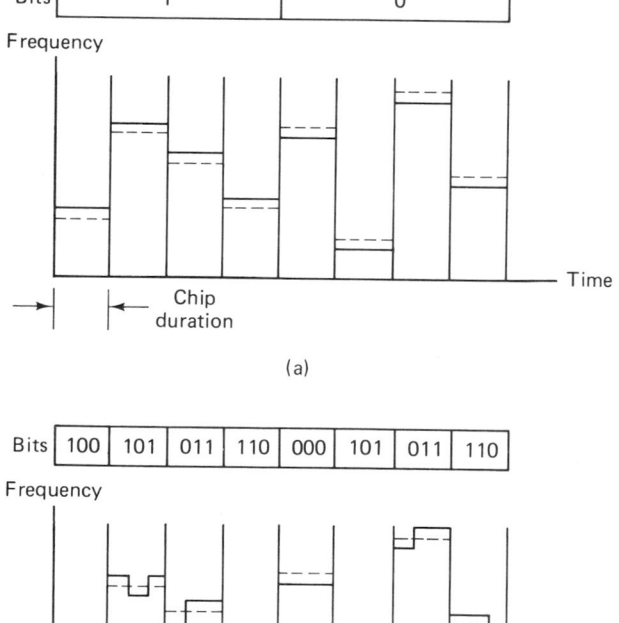

Figure 10.15 Fast hopping versus slow hopping in a binary system. (a) Fast-hopping example: 4 hops/bit. (b) Slow-hopping example: 3 bits/hop.

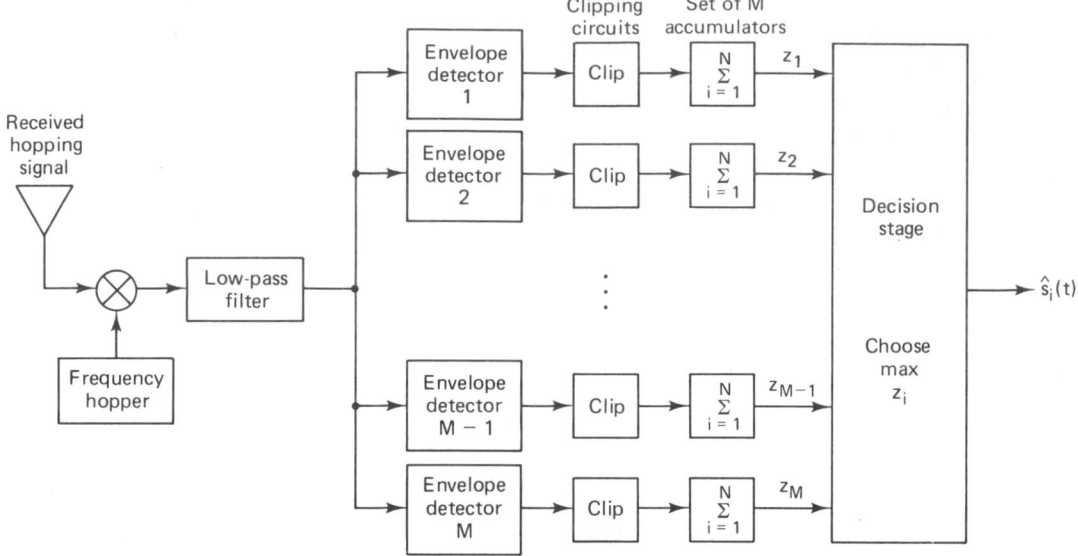

Figure 10.16 FFH/MFSK demodulator.

chip duration then correspond to? If the system were implemented as an 8-ary scheme, each 3 bits would be transmitted as a single data symbol. The symbol boundaries and the hop boundaries would then be the same, and the chip duration, the hop duration, and the symbol duration would all be the same.

10.4.5 FFH/MFSK Demodulator

Figure 10.16 illustrates the schematic for a typical fast frequency hopping MFSK (FFH/MFSK) demodulator. First, the signal is dehopped using a PN generator identical to the one used for hopping. Then, after filtering with a low-pass filter that has a bandwidth equal to the data bandwidth, the signal is demodulated using a bank of M envelope or energy detectors. Each envelope detector is followed by a clipping circuit and an accumulator. The clipping circuit serves an important function in the presence of an intentional jammer or other strong unpredictable interference; it is treated in a later section. The demodulator does *not* make symbol decisions on a chip-by-chip basis. Instead, the energy from the N chips are accumulated, and after the energy from the Nth chip is added to the $N - 1$ earlier ones, the demodulator makes a symbol decision by choosing the symbol that corresponds to the accumulator, z_i $(i = 1, 2, \ldots, M)$, with maximum energy.

10.5 SYNCHRONIZATION

For both DS and FH spread-spectrum systems, a receiver must employ a *synchronized* replica of the spreading or code signal to demodulate the received signal successfully. The process of synchronizing the locally generated spreading signal with the received spread-spectrum signal is usually accomplished in two steps.

The first step, called *acquisition,* consists of bringing the two spreading signals into *coarse* alignment with one another. Once the received spread-spectrum signal has been acquired, the second step, called *tracking,* takes over and continuously maintains the best possible waveform *fine* alignment by means of a feedback loop.

10.5.1 Acquisition

The acquisition problem is one of searching throughout a region of time and frequency uncertainty in order to synchronize the received spread-spectrum signal with the locally generated spreading signal. Acquisition schemes can be classified as coherent or noncoherent. Since the despreading process typically takes place before carrier synchronization, and therefore the carrier phase is unknown at this point, most acquisition schemes utilize noncoherent detection. When determining the limits of the uncertainty in time and frequency, the following items must be considered:

1. Uncertainty in the distance between the transmitter and the receiver translates into uncertainty in the amount of propagation delay.
2. Relative clock instabilities between the transmitter and the receiver result in phase differences between the transmitter and receiver spreading signals that will tend to grow as a function of elapsed time between synchronization.
3. Uncertainty of the receiver's relative velocity with respect to the transmitter translates into uncertainty in the value of Doppler frequency offset of the incoming signal.
4. Relative oscillator instabilities between the transmitter and the receiver result in frequency offsets between the two signals.

10.5.1.1 Correlator Structures

A common feature of all acquisition methods is that the received signal and the locally generated signal are first correlated to produce a measure of similarity between the two. This measure is then compared to a threshold to decide if the two signals are in synchronism. If they are, the tracking loop takes over.* If they are not, the acquisition procedure provides for a phase or frequency change in the locally generated code as a part of a systematic search through the receiver's phase and frequency uncertainty region, and another correlation is attempted.

Consider the direct-sequence *parallel-search* acquisition system shown in Figure 10.17. The locally generated code $g(t)$ is available with delays that are spaced one-half chip ($T_c/2$) apart. If the time uncertainty between the local code and the received code is N_c chips and a complete parallel search of the entire time uncertainty region is to be accomplished in a single search time, $2N_c$ correlators are used. Each correlator simultaneously examines a sequence of λ chips, after which the $2N_c$ correlator outputs are compared. The locally generated code, corresponding to the correlator with the largest output is chosen. Conceptually, this is the simplest of the search techniques; it considers all possible code positions

---

* Quite often to maintain a small false alarm probability, the threshold crossing must be further verified by a suitable verification algorithm before the tracking loop takes over [4].

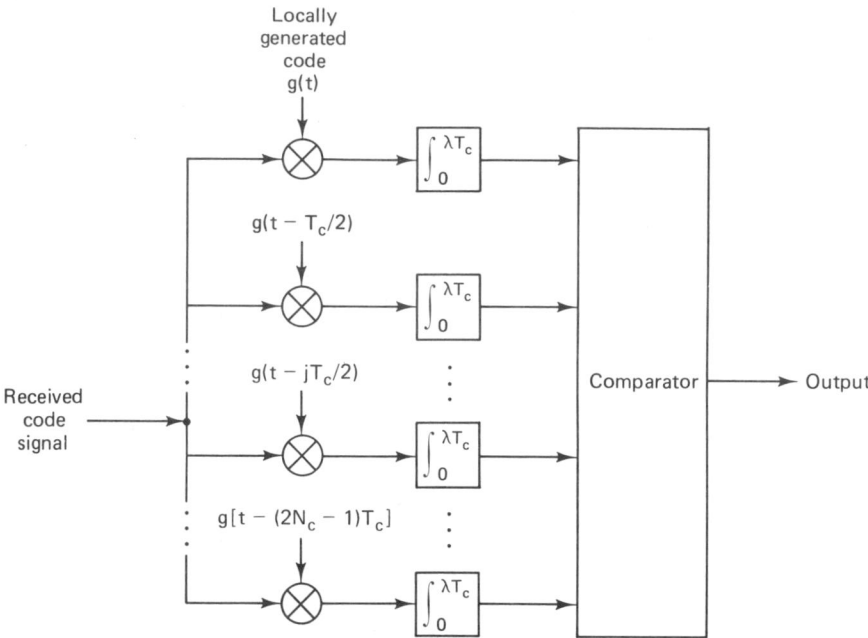

Figure 10.17 Direct-sequence parallel search acquisition.

(or fractional code positions) in parallel and uses a maximum likelihood algorithm for acquiring the code. Each detector output pertains to the identical observation of received signal plus noise. As λ increases, the synchronization error probability (i.e., the probability of choosing the incorrect code alignment) decreases. Thus λ is chosen as a compromise between minimizing the probability of a synchronization error and minimizing the time to acquire.

Figure 10.18 illustrates a simple acquisition scheme for a frequency hopping system. Assume that a sequence of N consecutive frequencies from the hop sequence is chosen as a synchronization pattern (without data modulation). The N noncoherent matched filters each consists of a mixer followed by a bandpass filter (BPF) and a square-law envelope detector (an envelope detector followed by a square-law device). If the frequency hopping sequence is $f_1, f_2, \ldots, f_N$, delays are inserted into the matched filters so that when the correct frequency hopping sequence appears, the system produces a large output, indicating detection of the synchronization sequence. Acquisition can be accomplished rapidly because all possible code offsets are examined simultaneously.

If, during each correlation, λ chips are examined, the maximum time required, $(T_{\text{acq}})_{\text{max}}$, for a fully parallel search is

$$(T_{\text{acq}})_{\text{max}} = \lambda T_c \qquad (10.29)$$

The mean acquisition time of a parallel search system can be approximated by noting that after integrating over λ chips, a correct decision will be made with probability P_D, called the *probability of detection*. If an incorrect output is chosen,

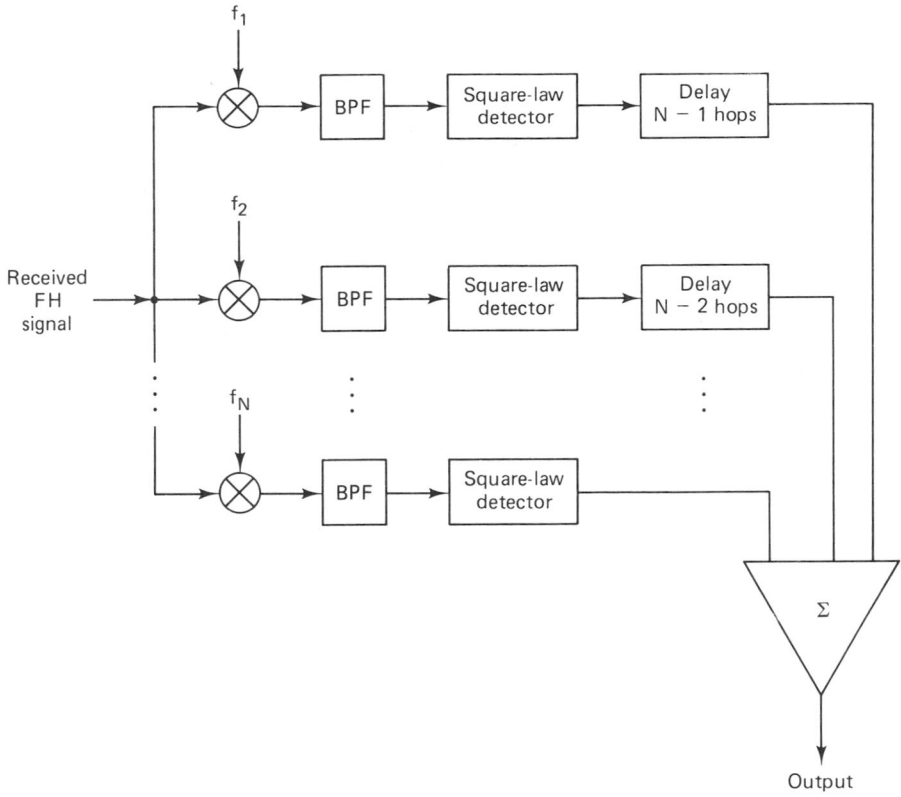

Figure 10.18 Frequency hopping acquisition scheme.

an additional λ chips are again examined to make a determination of the correct output. Therefore, on the average, the acquisition time is [6]

$$\overline{T}_{\text{acq}} = \lambda T_c P_D + 2\lambda T_c P_D (1 - P_D) + 3\lambda T_c P_D (1 - P_D)^2 + \cdots$$

$$= \frac{\lambda T_c}{P_D} \tag{10.30}$$

Since the required number of correlators or matched filters can be prohibitively large, fully parallel acquisition techniques are not usually used. In place of Figures 10.17 and 10.18, a single correlator or matched filter can be implemented that will *serially search* until synchronization is achieved. Naturally, trade-offs between fully parallel, fully serial, and combinations of the two involve hardware complexity versus time to acquire for the same uncertainty and chip rate.

10.5.1.2 Serial Search

A popular strategy for the acquisition of spread-spectrum signals is to use a single correlator or matched filter to serially search for the correct phase of the DS code signal or the correct hopping pattern of the FH signal. A considerable reduction in complexity, size, and cost can be achieved by a serial implementation that repeats the correlation procedure for each possible sequence shift. Figures 10.19 and 10.20 illustrate the basic configuration for DS and FH spread-spectrum schemes, respectively. In a stepped serial acquisition scheme for a DS system, the timing epoch of the local PN code is set, and the locally generated PN signal is correlated with the incoming PN signal. At fixed examination intervals of λT_c (search dwell time), where $\lambda \gg 1$, the output signal is compared to a preset threshold. If the output is below the threshold, the phase of the locally generated code signal is incremented by a fraction (usually one-half) of a chip and the correlation is reexamined. When the threshold is exceeded, the PN code is assumed to have been acquired, the phase-incrementing process of the local code is inhibited, and the code tracking procedure will be initiated. In a similar scheme for FH systems, shown in Figure 10.20, the PN code generator controls the frequency hopper. Acquisition is accomplished when the local hopping is aligned with that of the received signal.

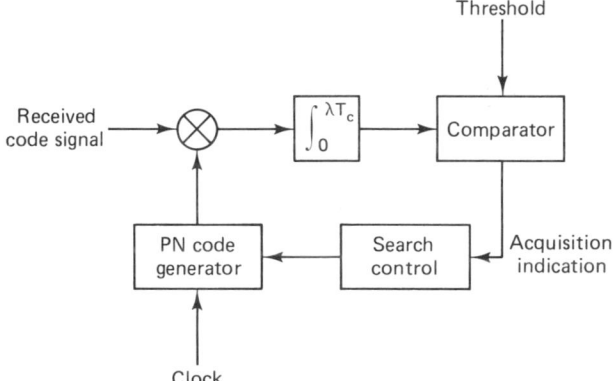

Figure 10.19 Direct-sequence serial search acquisition.

The maximum time required for a fully serial DS search, assuming that the search proceeds in half-chip increments, is

$$(T_{\text{acq}})_{\max} = 2N_c\lambda T_c \tag{10.31}$$

where the uncertainty region to be searched is N_c chips long. The mean acquisition time of a serial DS search system can be shown, for $N_c \gg \frac{1}{2}$ chip, to be [4]

$$\overline{T}_{\text{acq}} = \frac{(2 - P_D)(1 + KP_{\text{FA}})}{P_D}(N_c\lambda T_c) \tag{10.32}$$

where λT_c is the search dwell time, P_D the probability of correct detection, and P_{FA} the probability of false alarm. We can regard the time interval $K\lambda T_c$, where $K \gg 1$, as the time needed to verify a detection. Therefore, in the event of a false alarm, $K\lambda T_c$ seconds is the time penalty incurred. For $N_c \gg \frac{1}{2}$ chip and $K \ll 2N_c$, the variance of the acquisition time is

$$(\text{var})_{\text{acq}} = (2N_c\lambda T_c)^2(1 + KP_{\text{FA}})\left(\frac{1}{12} + \frac{1}{P_D^2} - \frac{1}{P_D}\right) \tag{10.33}$$

Spread-Spectrum Techniques   Chap. 10

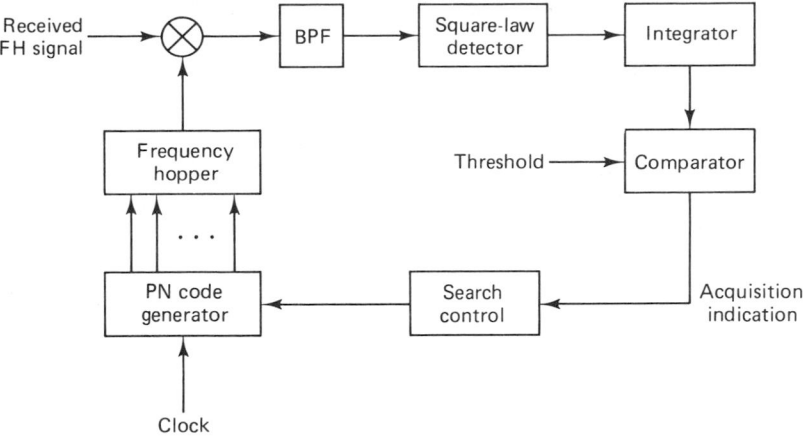

Figure 10.20 Frequency hopping serial search acquisition.

10.5.1.3 Sequential Estimation

Another search technique, called *rapid acquisition by sequential estimation* (RASE), proposed by Ward [11], is illustrated in Figure 10.21. The switch is initially in position 1. The RASE system enters its best estimate of the first n received code chips into the n stages of its local PN generator. The fully loaded register defines a starting state from which the generator begins its operation. A PN sequence has the property that the next combination of register states depends only on the present combination of states. Therefore, if the first n received chips are correctly estimated, all the following chips from the local PN generator will

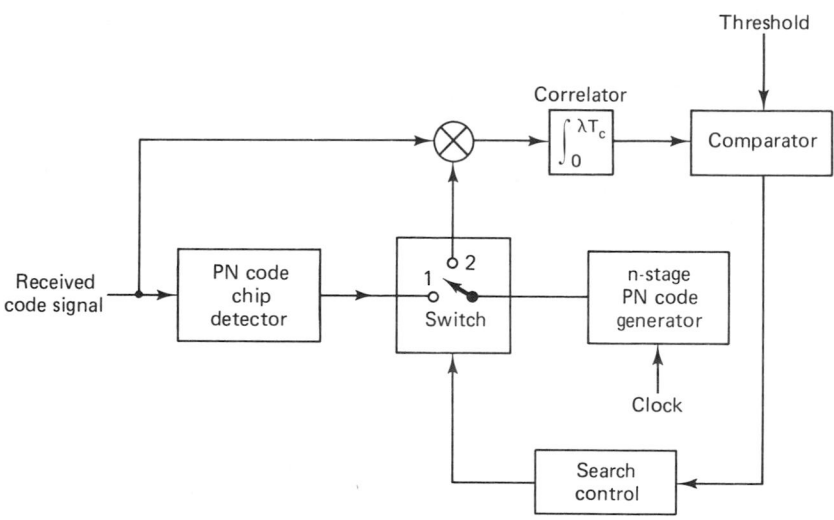

Figure 10.21 Rapid acquisition by sequential estimation.

Sec. 10.5   Synchronization

567

be correctly generated. The switch is next thrown to position 2. If the starting state had been correctly estimated, the local generator generates the same sequence as the incoming waveform, in the absence of noise. If the correlator output after λT_c exceeds a preset threshold level, we assume that synchronization has occurred. If the output is less than the threshold, the switch is returned to position 1, the register is reloaded with estimates of the next n received chips, and the procedure is repeated. Once synchronization has occurred, the system no longer needs estimates of the input code chips. We can calculate the *minimum* acquisition time for the case when no noise is present. The first n chips will be correctly loaded into the register, and therefore the acquisition time is

$$T_{\text{acq}} = nT_c \tag{10.34}$$

While the RASE system has a rapid acquisition capability it has the drawback of being highly vulnerable to noise and interference signals. The reason for this is that the estimation process consists of a simple chip-by-chip hard-decision demodulation, without using the interference rejection benefits of the PN code.

For an extensive treatment of sequential estimation, see Reference [4].

10.5.2 Tracking

Once acquisition or coarse synchronization is completed, tracking or fine synchronization takes place. Tracking code loops can be classified as coherent or noncoherent. A coherent loop is one in which the carrier frequency and phase are known exactly so that the loop can operate on a baseband signal. A noncoherent loop is one in which the carrier frequency is not known exactly (due to Doppler effects, for example), nor is the phase. In most instances, since the carrier frequency and phase are not known exactly, a priori, a noncoherent code loop is used to track the received PN code. Tracking loops are further classified as a *full-time* early-late tracking loop, often referred to as a *delay-locked loop* (DLL), or as a *time-shared* early-late tracking loop, frequently referred to as a *tau-dither loop* (TDL). A basic noncoherent DLL loop for a direct-sequence spread-spectrum system using binary phase shift keying (BPSK) is shown in Figure 10.22. The data $x(t)$ and the code $g(t)$ each modulate the carrier wave using BPSK, and as before in the absence of noise and interference, the received waveform can be expressed as

$$r(t) = A\sqrt{2P}\, x(t)g(t) \cos{(\omega_0 t + \phi)} \tag{10.35}$$

where the constant A is a system gain parameter and ϕ is a random phase angle in the range $(0, 2\pi)$. The locally generated code of the tracking loop is offset in phase from the incoming $g(t)$ by a time τ, where $\tau < T_c/2$. The loop provides *fine* synchronization by first generating two PN sequences $g(t + T_c/2 + \tau)$ and $g(t - T_c/2 + \tau)$ delayed from each other by one chip. The two bandpass filters are designed to pass the data and to average the product of $g(t)$ and the two PN sequences $g(t \pm T_c/2 + \tau)$. (See Reference [4] for the optimum filter bandwidth for a given filter type.) The square-law envelope detector eliminates the data since $|x(t)| = 1$. The output of each envelope detector is given approximately by

$$E_D \simeq \mathbf{E} \left\{ \left| g(t)g\left(t \pm \frac{T_c}{2} + \tau\right) \right| \right\} = \left| R_g\left(\tau \pm \frac{T_c}{2}\right) \right| \qquad (10.36)$$

where the operator $\mathbf{E}\{\cdot\}$ means *expected value* and $R_g(x)$ is the autocorrelation function of the PN waveform as shown in Figure 10.8. The feedback signal $Y(\tau)$ is shown in Figure 10.23. When τ is positive, the feedback signal $Y(\tau)$ instructs the voltage-controlled oscillator (VCO) to increase its frequency, thereby forcing τ to decrease, and when τ is negative, $Y(\tau)$ instructs the VCO to decrease, thereby forcing τ to increase. When τ is a suitably small number, $g(t)g(t + \tau) \simeq 1$, yielding the despread signal $Z(t)$, which is then applied to the input of a conventional data demodulator. Detailed analysis of the DLL can be found in References [4, 12–14].

A problem with the DLL is that the early and late arms must be precisely gain balanced or else the feedback signal $Y(\tau)$ will be offset and will not produce a zero signal when the error is zero. This problem is solved by using a time-shared tracking loop in place of the full-time delay-locked loop. The time-shared loop time shares the use of the early-late correlators. The main advantages are that only one correlator need be used in the design of the loop, and further, that dc offset problems are reduced.

An offshoot of the time-shared tracking loop is called the *tau-dither loop* (TDL), shown in Figure 10.24. This design has the advantage that only one correlator is needed to provide the code *tracking* function *and* the *despreading* function. Just as in the case of a DLL, the received signal is correlated with an early and a late version of the locally generated PN code. As shown in Figure 10.24, the PN code generator is driven by a clock signal whose phase is *dithered* back

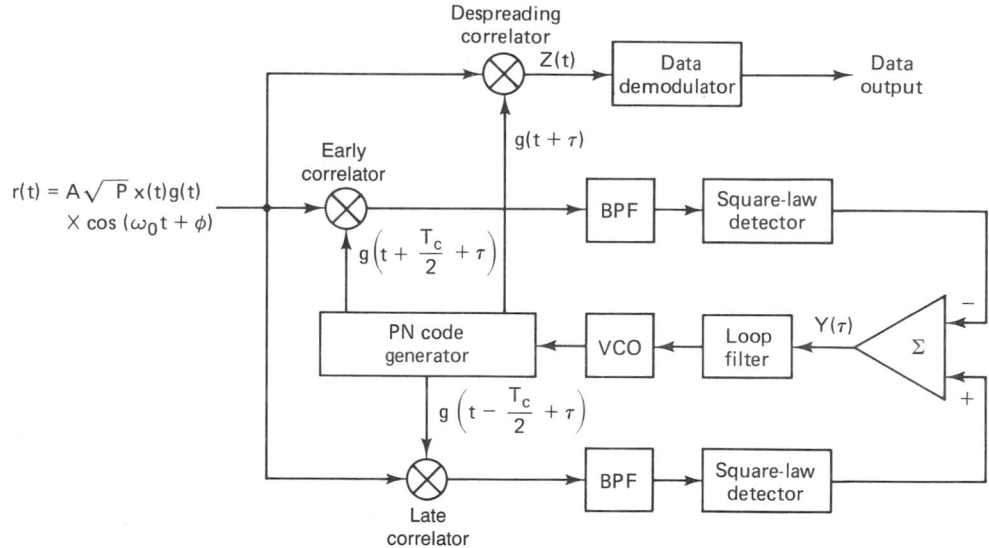

Figure 10.22 Delay-locked loop for tracking direct-sequence signals.

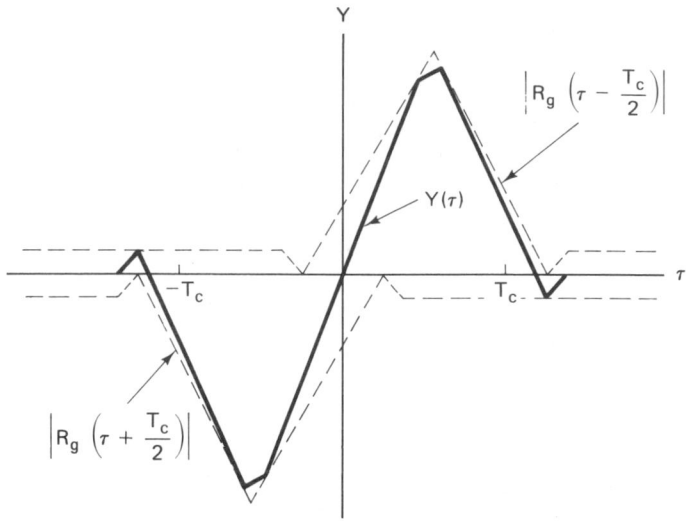

Figure 10.23 DLL feedback signal $Y(\tau)$.

and forth with a square-wave switching function; this eliminates the necessity of ensuring identical transfer functions of the early and late paths. The signal-to-noise performance of the TDL is only about 1.1 dB worse than that of the DLL if the arm filters are designed properly [4]. For a comprehensive treatment of synchronization of PN codes, see References [4, 15,16].

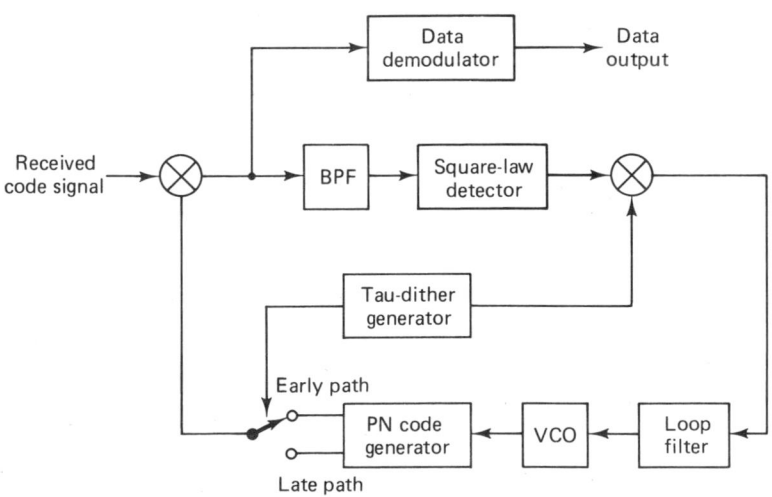

Figure 10.24 Tau-dither tracking loop.

10.6 SPREAD-SPECTRUM APPLICATIONS

10.6.1 Code-Division Multiple Access

Spread-spectrum multiple access techniques allow multiple signals occupying the same RF bandwidth to be transmitted simultaneously without interfering with one another. The application of spread-spectrum techniques to the problem of multiple access was discussed in Chapter 9 for a frequency hopped code-division multiple access (FH/CDMA) scheme. Here we consider CDMA using direct sequence (DS/CDMA). In these schemes, each of N user groups is given its own code, $g_i(t)$, where $i = 1, 2, \ldots, N$. The user codes are approximately orthogonal, so that the cross-correlation of two different codes is near zero. The main advantage of a CDMA system is that all the participants can share the full spectrum of the resource asynchronously; that is, the transition times of the different users' symbols do not have to coincide.

A typical DS/CDMA block diagram is shown in Figure 10.25. The first block illustrates the data modulation of a carrier, $A \cos \omega_0 t$. The output of the data modulator belonging to a user from group 1, $s_1(t)$, is shown below. The waveform is very general in form; no restriction has been placed on the type of modulation that can be used.

$$s_1(t) = A_1(t) \cos [\omega_0 t + \phi_1(t)] \qquad (10.37)$$

Next, the data-modulated signal is multiplied by the spreading signal $g_1(t)$ belonging to user group 1, and the resulting signal, $g_1(t)s_1(t)$, is transmitted over the channel. Simultaneously, users from group 2 through N multiply their signals by their own code functions. Frequently, each code function is kept secret, and its use is restricted to the community of authorized users. The signal present at the receiver is the linear combination of the emanations from each of the users. Neglecting signal delays, we show this linear combination below.

$$g_1(t)s_1(t) + g_2(t)s_2(t) + \cdots + g_N(t)s_N(t) \qquad (10.38)$$

As mentioned earlier, multiplication of $s_1(t)$ by $g_1(t)$ produces a signal whose spectrum is the convolution of the spectrum of $s_1(t)$ with the spectrum of $g_1(t)$. Thus, assuming that the signal $s_1(t)$ is relatively narrowband compared with the code or spreading signal $g_1(t)$, the product signal $g_1(t)s_1(t)$ will have approximately the bandwidth of $g_1(t)$. Assume that the receiver is configured to receive messages from user group 1. Assume, too, that the $g_1(t)$ code, generated at the receiver, is perfectly synchronized with the received signal from a group 1 user. The first stage of the receiver multiplies the incoming signal of Equation (10.38) by $g_1(t)$. The output of the multiplier will yield the following terms:

Desired signal: $\qquad g_1^2(t)s_1(t)$

Plus a composite of

undesired signals: $\qquad g_1(t)g_2(t)s_2(t) + g_1(t)g_3(t)s_3(t)$

$$+ \cdots + g_1(t)g_N(t)s_N(t) \qquad (10.39)$$

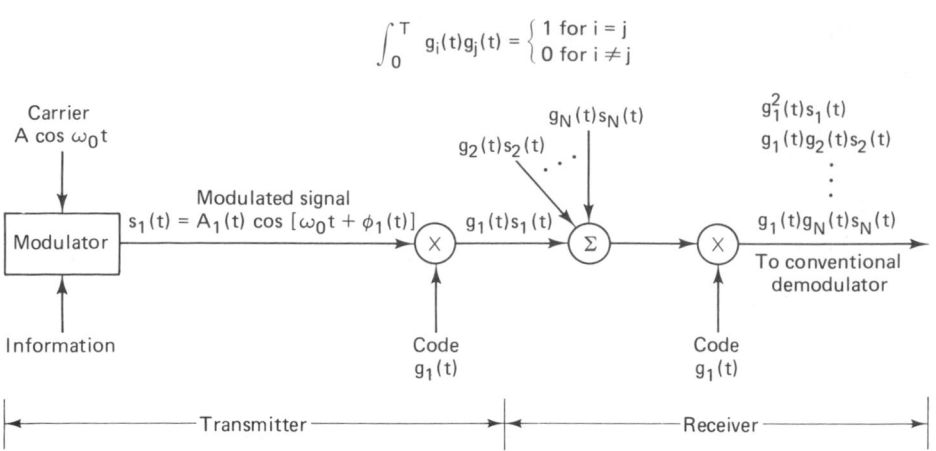

$$\int_0^T g_i(t)g_j(t) = \begin{cases} 1 \text{ for } i = j \\ 0 \text{ for } i \neq j \end{cases}$$

Figure 10.25 Code-division multiple access.

If the code functions, $\{g_i(t)\}$, are chosen with orthogonal properties, similar to Equation (10.14), the desired signal can be extracted perfectly in the absence of noise since $\int_0^T g_i^2(t) = 1$, and the undesired signals are easily rejected, since $\int_0^T g_i(t)g_j(t)\,dt = 0$ for $i \neq j$. In practice, the codes are not perfectly orthogonal; hence the cross-correlation between user codes introduces performance degradation, which limits the maximum number of simultaneous users.

Consider the frequency-domain view of the DS/CDMA receiver. Figure 10.26a illustrates the wideband input to the receiver; it consists of wanted and unwanted signals, each spread by its own code with code rate R_p, and each having a power spectral density of the form $\text{sinc}^2(f/R_p)$. Receiver thermal noise is also shown as having a flat spectrum across the band. The combined waveform of Equation (10.39) (desired plus undesired signals) is applied to the input of the receiver correlator driven by a synchronous replica of $g_1(t)$. Figure 10.26b illustrates the spectrum after correlation with the code $g_1(t)$ (despreading). The desired

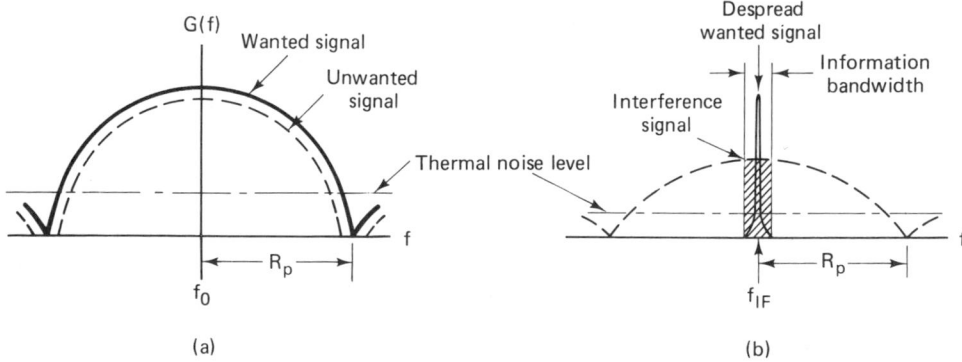

Figure 10.26 Spread-spectrum signal detection. (a) Spectrum at the input to receiver. (b) Spectrum after correlation with the correct and synchronized PN code.

signal, occupying the information bandwidth centered at an intermediate frequency (IF), is then applied to a conventional demodulator, with bandwidth just wide enough to accommodate the despread signal. The undesired signals of Equation (10.39) remain effectively spread by $g_1(t)g_i(t)$. Only that portion of the spectrum of the unwanted signals falling in the information bandwidth of the receiver will cause interference with the desired signal.

Pursley [17] presents an excellent treatment on the performance of SSMA using DS, taking correlation properties of the code sequences into account. Also, Geraniotis [18] and Geraniotis and Pursley [19, 20] evaluate the performance of FH and DS multiple access systems subject to interference.

10.6.2 Multipath Channels

Consider a DS binary PSK communication system operating over a multipath channel that has more than one path from the transmitter to the receiver. Such multiple paths may be due to atmospheric reflection or refraction, or reflections from buildings or other objects, and may result in fluctuations in the received signal level. The different paths may consist of several discrete paths each with a different attentuation and time delay, or they might consist of a continuum of paths. Figure 10.27 illustrates a communication link with two discrete paths. The multipath wave is delayed by some time, τ, compared to the direct wave. In television receivers, signals such as these cause "ghosts," or under extreme conditions, complete loss of picture synchronization.

In a direct-sequence spread-spectrum system, if we assume that the receiver is synchronized to the time delay and RF phase of the direct path, the received signal can be expressed as

$$r(t) = Ax(t)g(t) \cos \omega_0 t + \alpha Ax(t - \tau)g(t - \tau) \cos (\omega_0 t + \theta) + n(t) \quad (10.40)$$

where $x(t)$ is the data signal, $g(t)$ the code signal, $n(t)$ a zero-mean Gaussian noise process, and τ the differential time delay between the two paths, assumed to be in the interval $0 < \tau < T$. The angle θ is a random phase, assumed to be uniformly distributed in the range $(0, 2\pi)$, and α is the attenuation of the multipath signal

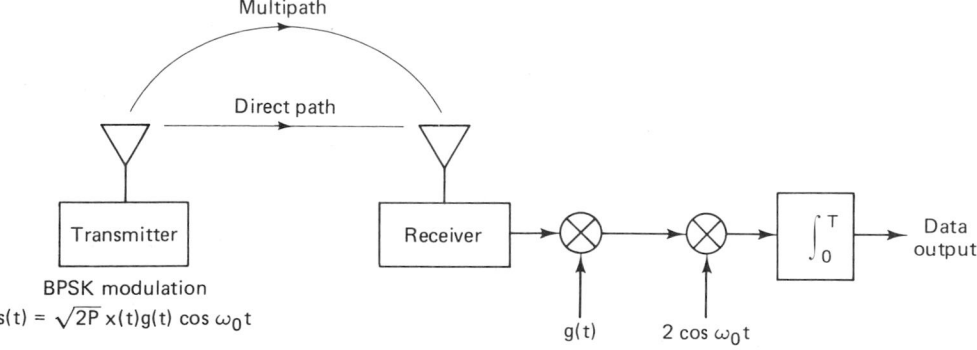

Figure 10.27 Direct-sequence BPSK system operating over a multipath channel.

relative to the direct path signal. For the receiver, synchronized to the direct path signal, the output of the correlator, $z(t = T)$, can be written as

$$z(t = T) = \int_0^T [Ax(t)g^2(t) \cos \omega_0 t$$

$$+ \alpha Ax(t - \tau)g(t)g(t - \tau) \cos (\omega_0 t + \theta) + n(t)g(t)]2 \cos \omega_0 t \, dt \quad (10.41)$$

where $g^2(t) = 1$. Also, for $\tau > T_c$, $g(t)g(t - \tau) \simeq 0$ (for codes with long periods), where T_c is the chip duration. Therefore, if T_c is less than the differential time delay between the multipath and direct path signals, we can write

$$z(t = T) = \int_0^T 2Ax(t) \cos^2 \omega_0 t + 2n(t)g(t) \cos \omega_0 t \, dt = Ax(T) + n_0(T) \quad (10.42)$$

where $n_0(T)$ is a zero-mean Gaussian random variable. We see that the spread-spectrum system, similar to the case of CDMA, effectively eliminates the multipath interference by virtue of its code-correlation receiver.

If frequency hopping (FH) is used against the multipath problem, improvement in system performance is also possible but through a different mechanism. FH receivers avoid multipath losses by rapid changes in the transmitter frequency band, thus avoiding the interference by changing the receiver band position before the arrival of the multipath signal.

10.6.3 The Jamming Game

The goals of a jammer are to deny reliable communications to his adversary and to accomplish this at minimum cost. The goals of the communicator are to develop a jam-resistant communication system under the following assumptions: (1) complete invulnerability is not possible; (2) the jammer has a priori knowledge of most system parameters, such as frequency bands, timing, traffic, and so on; (3) the jammer has *no* a priori knowledge of the PN spreading or hopping codes. The signaling waveform should be designed so that the jammer cannot gain any appreciable jamming advantage by choosing a jammer waveform and strategy other than wideband Gaussian noise (i.e., being clever should gain nothing for the jammer). The fundamental design rule in specifying a jam-resistant system is to make it as costly as possible for the jammer to succeed in jamming the system.

10.6.3.1 Jammer Waveforms

There are many different waveforms that can be used for jamming communication systems. The most appropriate choice depends on the targeted system. Figure 10.28 shows power spectral density plots of examples of jammer waveforms versus a communicator's frequency hopped M-ary FSK (FH/MFSK) tone. The range of the abscissa represents the spread-spectrum bandwidth W_{ss}. The three columns in the figure represent three instances in time (three hop times) when symbols having spectra G_1, G_2, and G_3, respectively, are being transmitted. Figure 10.28a illustrates a relatively low-level noise jammer occupying the full spread-spectrum bandwidth. In Figure 10.28b the jammer strategy is to trade bandwidth

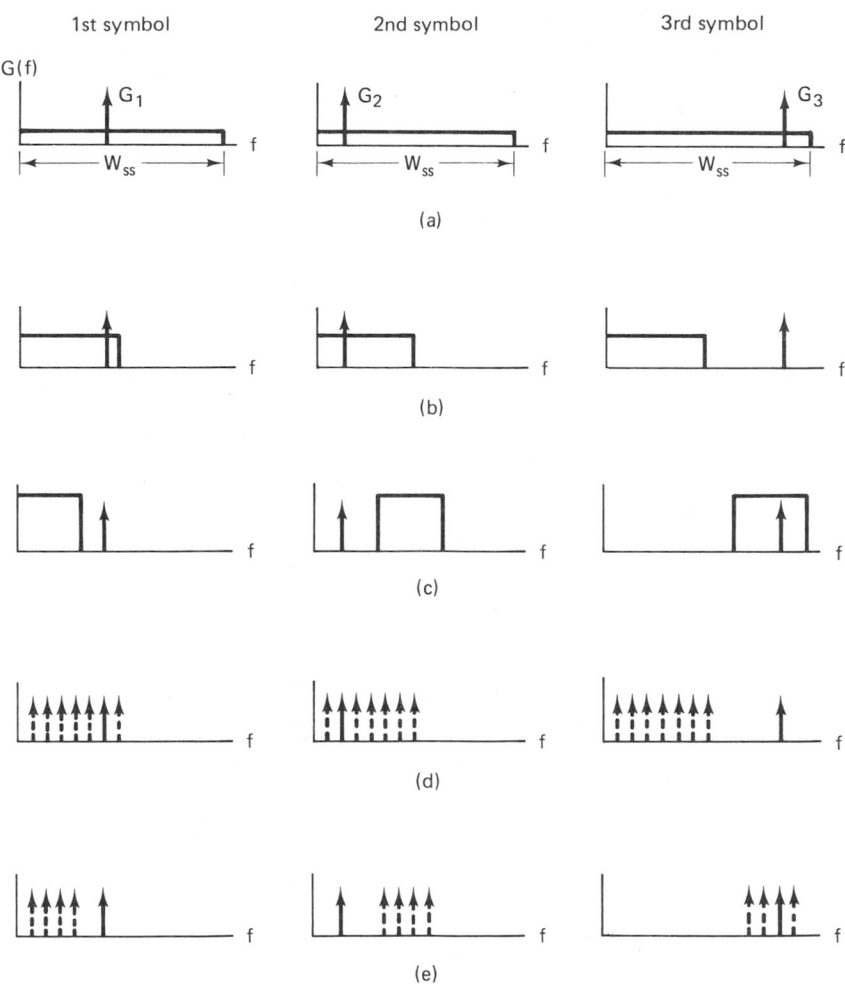

Figure 10.28 Jammer waveforms. (a) Full-band noise. (b) Partial-band noise. (c) Stepped noise. (d) Partial-band tones. (e) Stepped tones.

occupancy for greater power spectral density (the total power, or area under the curve, remains the same). The figure indicates that in this case, the jammer noise does not always share the same bandwidth region as the signal, but when it does, the effect can be destructive. In Figure 10.28c the noise jammer strategy is again to jam only part of the band, so that the jammer power spectral density can be increased, but in this case the jammer steps through different regions of the band at random times, thus preventing the communicator from using adaptive techniques to avoid the jamming. In Figure 10.28d and e the jammer uses a group of tones, instead of a continuous frequency band, in partial-band (Figure 10.28d) and stepped fashion (Figure 10.28e). This is a technique most often used against FH systems. Another jamming technique, not shown in Figure 10.28, is a pulse jammer, consisting of pulse-modulated bandlimited noise. Unless otherwise stated, we shall assume that the jammer waveform is wideband noise and that the jammer strategy is to jam the entire bandwidth W_{ss} continuously. The effects of partial band jamming and pulse jamming are considered later.

10.6.3.2 Tools of the Communicator

The usual design goal for an anti-jam (AJ) communication system is to force a jammer to expend its resources over (1) a wide-frequency band, (2) for a maximum time, and (3) from a diversity of sites. The most prevalent design options are (1) frequency diversity, by the use of direct-sequence and frequency hopping spread-spectrum techniques; (2) time diversity, by the use of time hopping; (3) spatial discrimination, by the use of a narrow-beam antenna which forces a jammer to enter the receiver via an antenna sidelobe and hence suffer, typically, a 20- to 25-dB disadvantage, and (4) combinations of the above.

10.6.3.3 *J/S Ratio*

In Chapter 4 we were concerned primarily with link error performance as a function of thermal noise interference. Emphasis was placed on the signal-to-noise ratio parameters—required E_b/N_0 and available E_b/N_0 for meeting a specified error performance. In this section we are similarly concerned with link error performance as a function of interference. However, here the source of interference is the noise power of a jammer in addition to thermal noise. Therefore, the SNR of interest is $E_b/(N_0 + J_0)$, where J_0 is the noise power spectral density due to the jammer. Unless otherwise specified, J_0 is assumed equal to J/W_{ss}, where J is the average received jammer power (jammer power referred to the receiver front end) and W_{ss} is the spread-spectrum bandwidth. Since the jammer power is generally much greater than the thermal noise power, the SNR of interest in a jammed environment is usually taken to be E_b/J_0. Therefore, similar to the thermal noise case, we define $(E_b/J_0)_{reqd}$ as the bit energy per jammer noise power spectral density *required* for maintaining the link at a specified error probability. The parameter E_b can be written as

$$E_b = ST_b = \frac{S}{R}$$

where S is the received signal power, T_b the bit duration, and R the data rate in bits/s. Then we can express $(E_b/J_0)_{reqd}$ as

$$\left(\frac{E_b}{J_0}\right)_{reqd} = \left(\frac{S/R}{J/W_{ss}}\right)_{reqd} = \frac{W_{ss}/R}{(J/S)_{reqd}} = \frac{G_p}{(J/S)_{reqd}} \qquad (10.43)$$

where $G_p = W_{ss}/R$ is denoted the *processing gain,* and $(J/S)_{reqd}$ can be written

$$\left(\frac{J}{S}\right)_{reqd} = \frac{G_p}{(E_b/J_0)_{reqd}} \qquad (10.44)$$

The ratio $(J/S)_{reqd}$ is a figure of merit that provides a measure of how *invulnerable* a system is to interference. Which system has better jammer-rejection capability: one with a larger $(J/S)_{reqd}$ or a smaller $(J/S)_{reqd}$? The *larger* the $(J/S)_{reqd}$, the *greater* is the system's noise rejection capability, since this figure of merit describes how much noise power relative to signal power is *required* in order to degrade the system's specified error performance. Of course, the communicator would like the communication system *not* to degrade at all.

Another way of describing the relationship in Equation (10.44) is as follows. An adversary would like to employ a jamming strategy that forces the effective $(E_b/J_0)_{reqd}$ to be as large as possible. The adversary may employ pulse, tone, or partial-band jamming rather than wideband noise jamming. A large $(E_b/J_0)_{reqd}$ implies a small $(J/S)_{reqd}$ ratio for a fixed processing gain. This may force the communicator to employ a larger processing gain to increase the $(J/S)_{reqd}$. The system designer strives to choose a signaling waveform such that the jammer can gain no special advantage by using a jamming strategy other than wideband Gaussian noise.

10.6.3.4 Anti-Jam Margin

Sometimes the $(J/S)_{reqd}$ ratio is referred to as the *anti-jam* (AJ) *margin* since it characterizes the system jammer-rejection capability. But this is not really a good use of the phrase since AJ margin usually means the safety margin against a *particular threat*. Using the same approach as in Chapter 4 (for calculating the margin against thermal noise), we can define the AJ margin M_{AJ}, as follows:

$$M_{AJ}(dB) = \left(\frac{E_b}{J_0}\right)_r (dB) - \left(\frac{E_b}{J_0}\right)_{reqd} (dB) \qquad (10.45)$$

where $(E_b/J_0)_r$ is the E_b/J_0 *actually received*. Following the same format as Equation (10.43), we can express $(E_b/J_0)_r$ as

$$\left(\frac{E_b}{J_0}\right)_r = \frac{G_p}{(J/S)_r} \qquad (10.46)$$

where $(J/S)_r$, or simply J/S, is the ratio of the actually received jammer power to signal power. We can now combine Equations (10.43), (10.45), and (10.46), as follows:

$$M_{AJ} (dB) = \frac{G_p}{(J/S)_r} (dB) - \frac{G_p}{(J/S)_{reqd}} (dB) \qquad (10.47)$$

$$= \left(\frac{J}{S}\right)_{reqd} (dB) - \left(\frac{J}{S}\right)_r (dB) \qquad (10.48)$$

Example 10.2   Satellite Jamming

Figure 10.29 illustrates a satellite jamming scenario. The airplane terminal is equipped with a frequency hopping (FH) spread-spectrum system transmitting with an $EIRP_T$ = 20 dBW. The data rate is R = 100 bits/s. The jammer is transmitting wideband Gaussian noise, continually, with an $EIRP_J$ = 60 dBW. Assume that $(E_b/J_0)_{reqd}$ = 10 dB and that the path loss is identical for both the airplane terminal and the jammer.

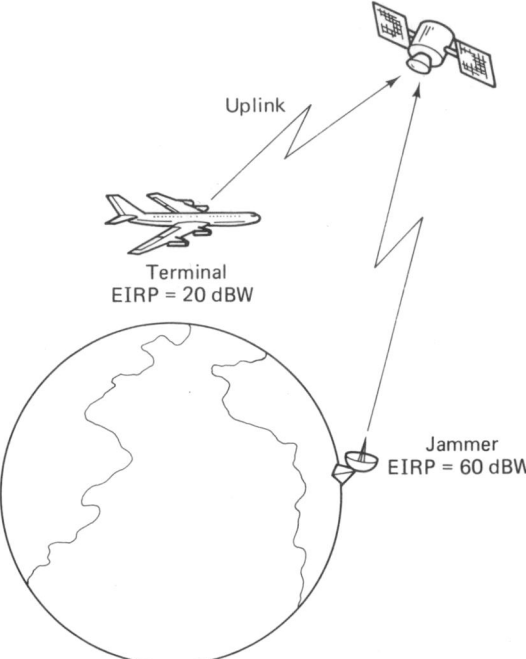

Uplink

Terminal
EIRP = 20 dBW

Jammer
EIRP = 60 dBW

Figure 10.29   Satellite jamming scenario.

(a) Should the communicators be concerned more with the jamming of the uplink or with that of the downlink?
(b) If it is desired to have an AJ margin of 20 dB, what should be the value of the hopping bandwidth W_{ss}?

Solution

(a) Jamming the uplink is of much greater concern, since such single-point interference could degrade the communications of a multitude of terminals that are simultaneously using the satellite transponder. To achieve an equivalent degradation by jamming the downlink, the jammer would have to jam each of the receiving terminals. Downlink jamming is of some concern for critical military missions, but of less concern than uplink jamming.
(b) With the assumption that the path loss is the same for both the communicator and the jammer, we can replace $(J/S)_r$ in Equation (10.48) with the ratio of *transmitted* jammer-to-signal power, $\text{EIRP}_J/\text{EIRP}_T$. Therefore, we can write

$$M_{\text{AJ}}\,(\text{dB}) = (J/S)_{\text{reqd}}\,(\text{dB}) + \text{EIRP}_T\,(\text{dBW}) - \text{EIRP}_J\,(\text{dBW})$$

$$= G_p\,(\text{dB}) - \left(\frac{E_b}{J_0}\right)_{\text{reqd}}\,(\text{dB}) + \text{EIRP}_T\,(\text{dBW}) - \text{EIRP}_J\,(\text{dBW})$$

$$G_p = 20\,\text{dB} + 10\,\text{dB} - 20\,\text{dBW} + 60\,\text{dBW} = 70\,\text{dB}$$

$$W_{ss} = G_p\,(\text{dB}) + R\,(\text{dB-Hz}) = 70\,\text{dB} + 20\,\text{dB-Hz}$$

$$= 90\,\text{dB-Hz} = 1\,\text{GHz}$$

Example 10.3   Satellite Downlink Jamming

In Example 10.2 the distance from the transmitting airplane to the receiving satellite and the distance from the jammer to the satellite were assumed identical. Certainly, the closer the jammer gets to the receiver, the greater will be the jamming interference. Consider a downlink jamming scenario where the satellite $EIRP_s = 35$ dBW, the jammer $EIRP_J = 60$ dBW, the space loss from the satellite to the receiving terminal is $L_s = 200$ dB, and the space loss from the jammer to the receiving terminal is $L'_s = 160$ dB. How much processing gain is needed to close the link with an AJ margin of 0 dB? Assume that $(E_b/J_0)_{reqd} = 10$ dB.

Solution

For the downlink jamming scenario the proximity of the jammer to the receiving airplane is much closer than that of the satellite to the airplane. These distances show up as the space losses in the $(J/S)_r$ term of Equation (10.48), as follows:

$$M_{AJ} \text{ (dB)} = \left(\frac{J}{S}\right)_{reqd} \text{ (dB)} - \left(\frac{J}{S}\right)_r \text{ (dB)}$$

where

$$\left(\frac{J}{S}\right)_r \text{ (dB)} = EIRP_J \text{ (dBW)} - L'_s \text{ (dB)} - EIRP_s \text{ (dBW)} + L_s \text{ (dB)}$$

and

$$\left(\frac{J}{S}\right)_{reqd} \text{ (dB)} = \frac{W_{ss}}{R} \text{ (dB)} - \left(\frac{E_b}{J_0}\right)_{reqd} \text{ (dB)}$$

Combining the above equations, and solving for processing gain, $G_p = W_{ss}/R$, yields

$$G_p \text{ (dB)} = 75 \text{ dB}$$

10.7 FURTHER JAMMING CONSIDERATIONS

10.7.1 Broadband Noise Jamming

If the jamming signal is modeled as a zero-mean wide-sense-stationary Gaussian noise process with a flat power spectral density over the frequency range of interest, then for a fixed jammer received power, J, the jammer power spectral density J_0 is equal to J/W, where W is the bandwidth that the jammer chooses to occupy. If the jammer strategy is to jam the entire spread-spectrum bandwidth, W_{ss}, with its fixed power, the jammer is referred to as a wideband or *broadband jammer,* and the jammer power spectral density is

$$J_0 = \frac{J}{W_{ss}} \tag{10.49}$$

In Chapter 3 it was shown that the bit error probability P_B for a coherently demodulated BPSK system (without channel coding) is

$$P_B = Q\left(\sqrt{\frac{2E_b}{N_0}}\right) \tag{10.50}$$

where $Q(x)$ is defined in Equations (2.42) and (2.43) and tabulated in Table B.1. The single-sided noise power spectral density N_0 represents thermal noise at the front end of the receiver. The presence of the jammer increases this noise power spectral density from N_0 to $(N_0 + J_0)$. Thus the average bit error probability for a coherent BPSK system in the presence of broadband jamming is

$$P_B = Q\left(\sqrt{\frac{2E_b}{N_0 + J_0}}\right) = Q\left[\sqrt{\frac{2E_b/N_0}{1 + (E_b/N_0)(J/S)/G_p}}\right] \tag{10.51}$$

When P_B is plotted versus E_b/N_0 for a given J/S ratio, the resulting curves are such as those in Figure 10.30, [7, 21]. The curves in Figure 10.30, shown for two different values of processing gain, *tend to flatten out* as E_b/N_0 increases, indicating that for a given ratio of jammer power to signal power, the jammer will cause some irreducible error probability. The only way to reduce this error probability is to increase the processing gain.

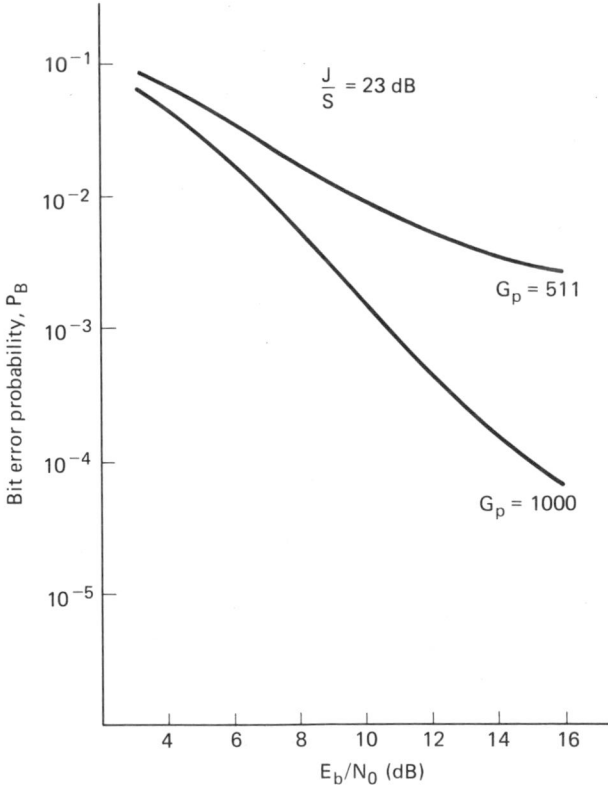

Figure 10.30 Bit error probability versus E_b/N_0 for a given J/S ratio. (Reprinted with permission from R. L. Pickholtz, D. L. Schilling, and L. B. Milstein, "Theory of Spread-Spectrum Communications—A Tutorial," *IEEE Trans. Commun.*, vol. COM30, no. 5, May 1982, Fig. 11, p. 866. © 1982 IEEE.)

10.7.2 Partial-Band Noise Jamming

A jammer can often increase the degradation to a FH system by employing *partial-band* jamming. Assuming that the frequency hopped modulation format is non-coherently detected binary FSK, the probability of a bit error, from Equation (3.111) is

$$P_B = \frac{1}{2} \exp\left(-\frac{E_b}{2N_0}\right) \qquad (10.52)$$

Let us define a parameter, ρ, where $0 < \rho \leq 1$, representing the fraction of the band being jammed. The jammer can trade bandwidth jammed for in-band jammer power, such that by jamming a band $W = \rho W_{ss}$, the jammer noise power spectral density can be concentrated to a level J_0/ρ, thus maintaining a constant average jamming received power J where $J = J_0 W_{ss}$.

In the case of partial-band jamming, a specific transmitted symbol will be received unjammed, with probability $(1 - \rho)$, and will be perturbed by jammer power with spectral density J_0/ρ, with probability ρ. Therefore, the average bit error probability can be written from Equation (10.52), as follows:

$$P_B = \frac{1 - \rho}{2} \exp\left(-\frac{E_b}{2N_0}\right) + \frac{\rho}{2} \exp\left[-\frac{E_b}{2(N_0 + J_0/\rho)}\right] \qquad (10.53)$$

Since, in a jamming environment, it is often the case that $J_0 \gg N_0$, we can simplify Equation (10.53) to the form

$$P_B = \frac{\rho}{2} \exp\left(-\frac{\rho E_b}{2J_0}\right) \qquad (10.54)$$

Figure 10.31 illustrates the probability of bit error versus E_b/J_0 for various values of the fraction, ρ. Clearly, the jammer would choose the fraction $\rho = \rho_0$ that maximizes P_B. Notice that ρ_0 decreases with increasing values of E_b/J_0 (see the ρ_0 locus in Figure 10.31). An expression for ρ_0 is easily found by differentiation (setting $dP_B/d\rho = 0$ and solving for ρ). This yields

$$\rho_0 = \begin{cases} \dfrac{2}{E_b/J_0} & \text{for } \dfrac{E_b}{J_0} > 2 \\[3mm] 1 & \text{for } \dfrac{E_b}{J_0} \leq 2 \end{cases} \qquad (10.55)$$

In this case, $(P_B)_{\max}$ is given by

$$(P_B)_{\max} = \begin{cases} \dfrac{e^{-1}}{E_b/J_0} & \text{for } \dfrac{E_b}{J_0} > 2 \\[3mm] \dfrac{1}{2} \exp\left(-\dfrac{E_b}{2J_0}\right) & \text{for } \dfrac{E_b}{J_0} \leq 2 \end{cases} \qquad (10.56)$$

where e is the base of the natural logarithm ($e = 2.7183$). This result is dramatic; the effect of a worst-case partial-band jammer on a system with spread spectrum *but without coding* changes the exponential relationship of Equation (10.54) into the inverse linear one of Equation (10.56). The ρ_0 locus in Figure 10.31 illustrates the P_B versus E_b/J_0 performance for the worst-case partial-band jammer. Here at 10^{-6} bit error probability there is over 40-dB difference between broadband noise jamming and the worst-case partial-band jamming for the same jamming

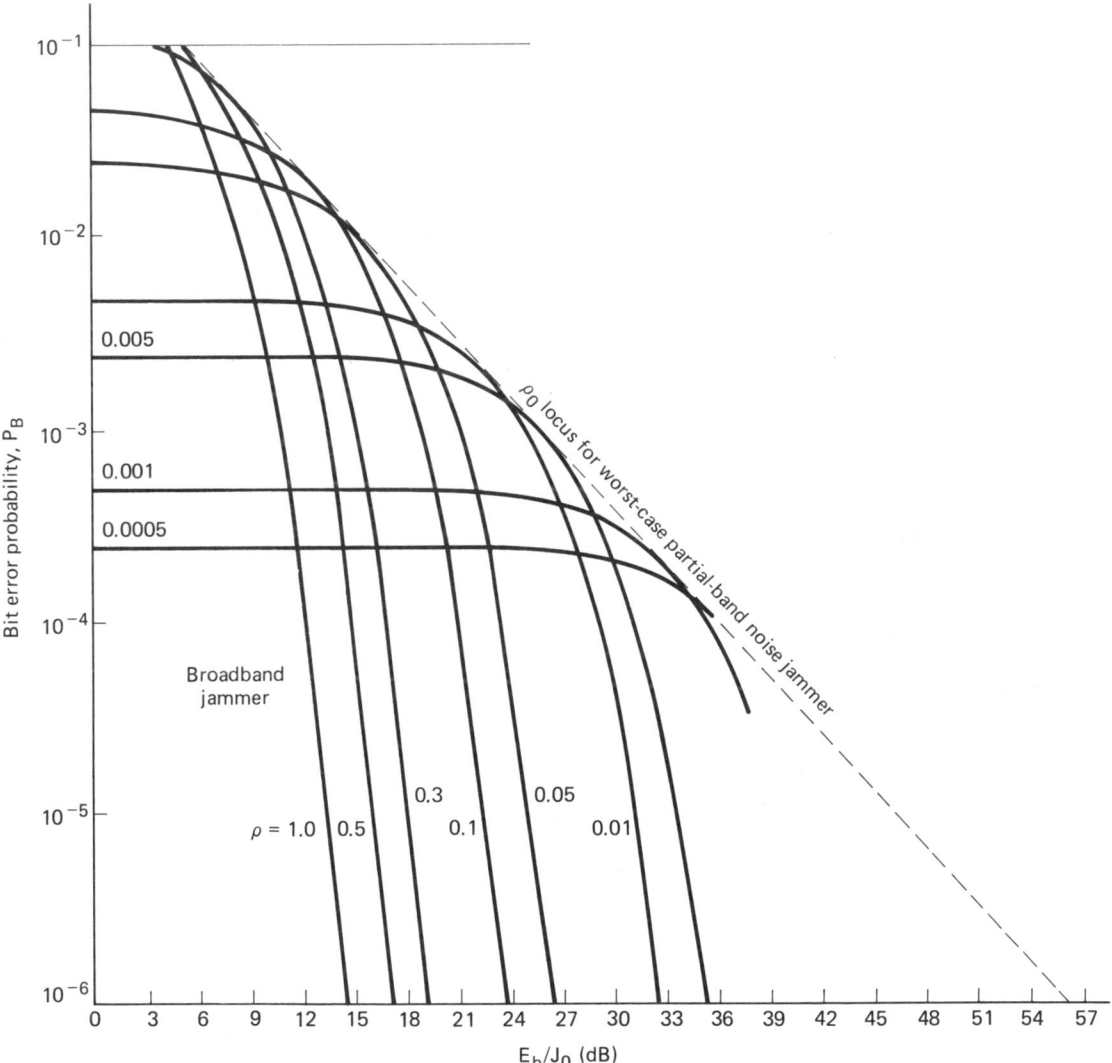

Figure 10.31 Partial-band noise jammer (FH/BFSK signaling). (Reprinted from M. K. Simon, J. K. Omura, R. A. Scholtz, and B. K. Levitt, *Spread Spectrum Communications*, Vol. 1, Fig. 3.24, p. 173. © 1985, with permission of the publisher, Computer Science Press, Inc., 1803 Research Blvd., Rockville, Md. 20850 USA.)

power [6, 22]. Hence, an intelligent jammer, with fixed finite power, can produce significantly greater degradation with partial-band jamming than is possible with broadband jamming. Forward error correction (FEC) coding with appropriate interleaving can mitigate this degradation [9]. In fact, for codes with low-enough rates, FEC can *force* a partial-band jammer to be a worst-case jammer only when operating as a broadband jammer [23, 24].

10.7.3 Multiple-Tone Jamming

In the case of *multiple-tone jamming,* the jammer divides its total received power, J, into distinct, equal-power, random-phase CW tones. These are distributed over the spread-spectrum bandwidth, W_{ss}, according to some strategy [9]. The analysis of the effects of tone jamming is more complicated than that of noise jamming, especially for DS systems. Therefore, the effect of a despread tone is often approximated as Gaussian noise. Reference [25] provides analysis of the performance of DS systems in the presence of multiple-tone interference. For a noncoherent FH/FSK system operating in the presence of partial-band tone jamming, the performance is often assumed the same as that of partial-band noise jamming [26]. However, multiple-CW-tone jamming can be more effective than partial-

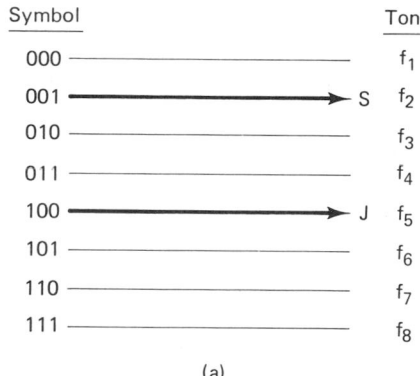

(a)

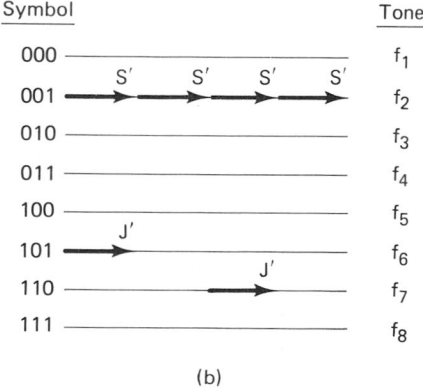

(b)

Figure 10.32 Fast hopping symbol repeat versus tone jamming. (a) One frequency hop. (b) Four frequency hops.

band noise against FH/MFSK signals because CW tones are the most efficient way for a jammer to inject energy into noncoherent detectors [9]. References [9, 10, 26, 27] provide extensive treatment and analysis of the performance of various communication systems in the presence of various types of jammers.

In the FFH/MFSK demodulator of Figure 10.16, a chip clipping circuit is shown between each envelope detector and accumulator. The function of such a circuit in a tone jamming environment can best be understood with the aid of the example shown in Figure 10.32. An 8-ary FSK frequency hopping system with no diversity, indicated in Figure 10.32a, is compared with a *fast*-frequency hopping system that combines chip repeating ($N = 4$ in this example) with the clipping of each chip, indicated in Figure 10.32b. Each row in the figures reprsents one of the $M = 8$ accumulators shown in Figure 10.16. The presence of a signal in the accumulator is indicated by a vector. In Figure 10.32a we see that, for a particular frequency hop, the data band is occupied by a received message symbol with received signal power, S. If, by chance, a jamming tone with received power J, where $J \geq S$, falls on a different tone within this data band during the same hop, the detector would not be able to decide reliably on the correct symbol.

In Figure 10.32b, the communicator's four chips (the length of each vector is a measure of the clipped signal power, S') sum to the maximum capacity of the accumulator. If the jammer tones, by chance, fall in the same spectral region as that of the signal, they will not confuse the detector, since the jamming tones are also clipped to the same level, $J' = S'$, as the signal chips. In Figure 10.32b, two of the jamming tones fall in the data band, but because they are clipped, there is no confusion about the correct symbol decision.

10.7.4 Pulse Jamming

Consider a spread-spectrum DS/BPSK communication system in the presence of a pulse-noise jammer. A pulse-noise jammer transmits pulses of bandlimited white Gaussian noise having a time-averaged received power, J, although the actual power during a jamming pulse duration is larger. Assume that the jammer can choose the center frequency and bandwidth of the noise to be the same as the receiver's center frequency and bandwidth. Assume also that the jammer can trade duty cycle for increased (concentrated) jammer power, such that if the jamming is present for a fraction $0 < \rho < 1$ of the time, then during this time, the jammer power spectral density is increased to a level J_0/ρ, thus maintaining a constant time-averaged power J (where $J = J_0 W_{ss}$ and W_{ss} is the system spread-spectrum bandwidth).

The bit error probability P_B for a coherently demodulated BPSK system (without channel coding) was given in Equation (10.50):

$$P_B = Q\left(\sqrt{\frac{2E_b}{N_0}} \right)$$

The single-sided noise power spectral density N_0 represents thermal noise at the front end of the receiver. The presence of the jammer increases this noise power spectral density from N_0 to $(N_0 + J_0/\rho)$. Since the jammer transmits with duty cycle ρ, the average bit error probability is

$$P_B = (1 - \rho)Q\left(\sqrt{\frac{2E_b}{N_0}}\right) + \rho Q\left(\sqrt{\frac{2E_b}{N_0 + J_0/\rho}}\right) \qquad (10.57)$$

We can generally assume that in a jamming environment, N_0 can be neglected. Therefore, we can write

$$P_B \simeq \rho Q\left(\sqrt{\frac{2E_b\rho}{J_0}}\right) \qquad (10.58)$$

The jammer will, of course, attempt to choose the duty cycle ρ that maximizes P_B. Figure 10.33 illustrates P_B for various values of ρ. The value of $\rho = \rho_0$ that maximizes P_B decreases with increasing values of E_b/J_0, as was the case with partial-band jamming. This is seen by differentiating Equation (10.58) to obtain [6]

$$\rho_0 = \begin{cases} \dfrac{0.709}{E_b/J_0} & \text{for } \dfrac{E_b}{J_0} > 0.709 \\[3mm] 1 & \text{for } \dfrac{E_b}{J_0} \leq 0.709 \end{cases} \qquad (10.59)$$

which results in the maximum bit error probability

$$(P_B)_{\max} = \begin{cases} \dfrac{0.083}{E_b/J_0} & \text{for } \dfrac{E_b}{J_0} > 0.709 \\[3mm] Q\left(\sqrt{\dfrac{2E_b}{J_0}}\right) & \text{for } \dfrac{E_b}{J_0} \leq 0.709 \end{cases} \qquad (10.60)$$

The effect of a worst-case pulse jammer upon a system with spread spectrum *but without coding* changes the complementary error function relationship of Equation (10.58) into the inverse linear one of Equation (10.60). As a result, at an error probability of 10^{-6}, there is almost a 40-dB difference in E_b/J_0 between the broadband jammer and the worst-case pulse jammer (see Figure 10.33). For the same jammer power, the jammer can do considerably more harm to an uncoded DS/BPSK system with pulse jamming than with constant power jamming. The effect of a pulse-noise jammer on uncoded DS/BPSK is similar to the effect of a partial-band noise jammer on uncoded FH/BFSK, treated in Section 10.7.2. In both cases considerable degradation is brought about by concentrating more jammer power on a fraction of the transmitted uncoded symbols. Forward error correction coding with appropriate interleaving can almost fully restore this degraded performance [9, 23–25, 28].

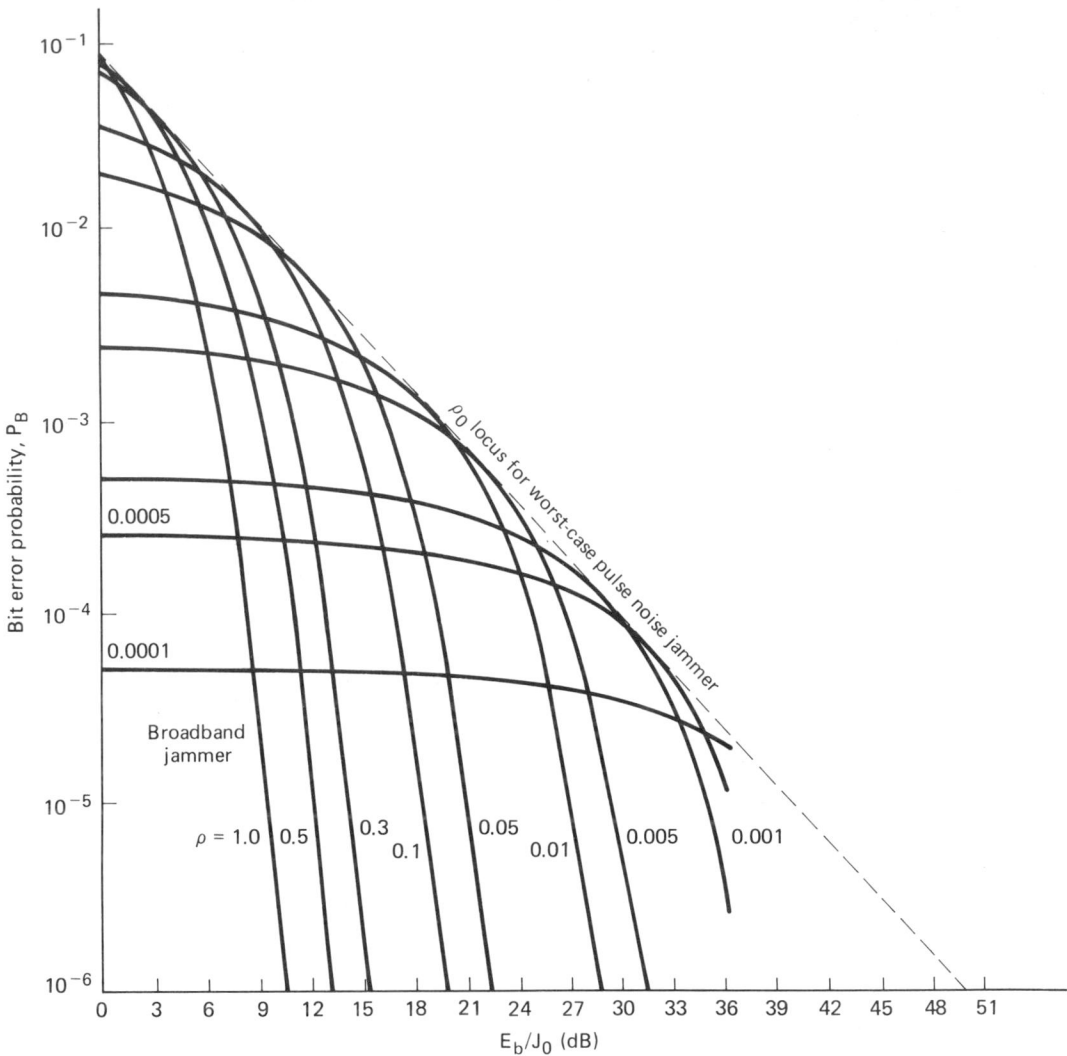

Figure 10.33 Pulse noise jammer (DS/BPSK signaling). (Reprinted from M. K. Simon, J. K. Omura, R. A. Scholtz, and B. K. Levitt, *Spread Spectrum Communications*, Vol. 1, Fig. 3.7, p. 150. © 1985, with permission of the publisher, Computer Science Press, Inc., 1803 Research Blvd., Rockville, Md. 20850 USA.)

10.7.5 Repeat-Back Jamming

In Examples 10.2 and 10.3 we considered an FH spread-spectrum system performance against a broadband Gaussian noise jammer. Notice that the frequency hopping rate did not enter into the margin computations. Isn't this disturbing? Intuitively, it would seem that the faster the frequency hops, the easier it is to "hide" the signal from the jammer. If the hopping rate truly does not enter into the computations, why not hop only once a day or once a week? The answer is that the meaure of jammer-rejection capability, namely processing gain, G_p, is based on the assumption that the jammer is a "dumb" jammer; that is, the jammer knows the extent of the spread-spectrum bandwidth, W_{ss}, but does *not* know the exact spectral location of the signal at any moment in time. We assume that the hopping rate is *fast enough* to preclude the jammer from monitoring the transmitted signal so as to usefully change this jamming strategy. Under what condition is this assumption questionable? There are "smart" jammers that are known as *repeat-back jammers* or *frequency-follower (FF) jammers*. These jammers monitor a communicator's signal (usually via a sidelobe beam from the transmitting antenna). They possess wideband receivers and high-speed signal processing capability that enable them to rapidly concentrate their jamming signal power in the spectral vicinity of a communicator's FH/FSK signal. By so doing, the smart jammer can increase the jamming power in the communicator's instantaneous bandwidth, thereby gaining an advantage over a wideband jammer. Notice that this strategy is useful only against frequency hopping signals. In direct-sequence systems, there is no instantaneous narrowband signal for the jammer to detect.

What can be done to defeat the repeat-back jammer? One method is to simply hop so fast that by the time the jammer receives, detects, and transmits the jamming signal, the communicator is already transmitting at a *new* hop (which of course will be unaffected by jamming at the frequency of the prior hop). The following example should make this point clear.

Example 10.4   Fast Hopping to Evade the Repeat-Back Jammer

Assume that a repeat-back jammer is located $d = 30$ km away from the communicator. Assume further that the jammer can monitor any uplink transmission from the communicator to a nearby satellite, as shown in Figure 10.34. How fast must the communicator hop his frequency to evade the repeat-back jammer? Assume that the jammer can change its jamming frequency in zero time, and that the only differential delay between the communicator's uplink signal and the jamming uplink signal is the propagation delay from the communicator to the jammer.

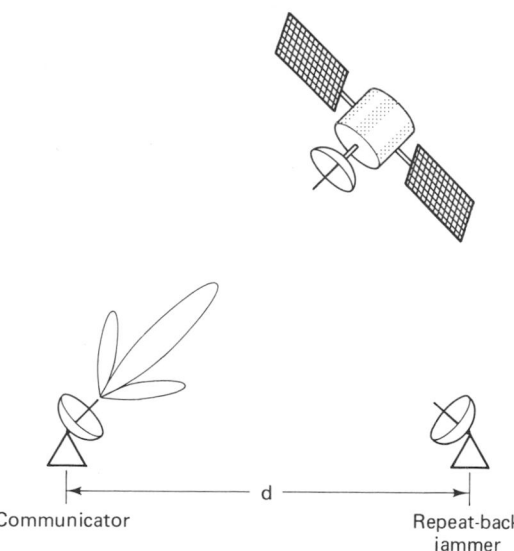

|← d →|
Communicator   Repeat-back   **Figure 10.34**   Example of fast hopping
                                jammer   to evade the repeat-back jammer.

Solution

To ensure that the communicator's tone transmission and the jammer's attempt to disrupt that tone do not overlap in time, it is necessary that the duration of each hop, T_{hop}, have the value

$$T_{hop} \leq \frac{d}{c} = \frac{3 \times 10^4 \text{ m}}{3 \times 10^8 \text{ m/s}} = 10^{-4} \text{ s}$$

where c is the speed of light. Then $R_{hop} \geq 10,000$ hops/s.

10.7.6 BLADES System

Another technique capable of defeating the repeat-back jammer dates back to the mid-1950s when Sylvania engineers developed a system named the Buffalo Laboratories Application of Digitally Exact Spectra, or BLADES. The system used its code generator to independently select two new frequencies for each bit; the *final choice* of the frequency tone actually transmitted was dictated by the data bit about to be transmitted. Figure 10.35 illustrates a typical data stream of binary ones and zeros, called *marks* and *spaces,* respectively, and a sequence of frequency pairs f_1 and f'_1, f_2 and f'_2, ... The appearance of a mark dictates the choice of frequency f_i, while the appearance of a space dictates the choice of frequency f'_i. As shown in the figure, the data stream in this example gives rise to the sequence of transmitted tones, f'_1, f_2, f'_3, f'_4, f_5, and so on. How can such a system defeat a repeat-back jammer? The jammer monitors the transmissions and sends up energy in the neighborhood of the frequencies it perceives. The modulation of the BLADES system has no structure in the usual sense; either there *is* energy present or there is *no* energy present at a given frequency. The jammer sending narrowband energy in the same spectral neighborhood as the signal, is not destroying any modulation structure. For a noncoherent system, the jammer is only enhancing the communicator's signal. The only recourse for the repeat-back jammer is to change strategy by becoming a broadband jammer, and to jam the entire spread-spectrum bandwidth.

Notice that it is not really necessary to have a *pair* of frequencies for each

Figure 10.35 BLADES system.

bit. A *single* frequency will do. The communicator then transmits the pseudo-random frequency for a binary one and sends nothing for a binary zero. The receiver has the same code generator and therefore monitors the same pseudo-random frequencies. A binary one is detected by virtue of energy at the monitored frequency, and a binary zero is known by a lack of energy at the monitored frequency. Of course, the system is not as robust as when the marks and spaces are each transmitted on independently selected frequencies.

10.8 CONCLUSION

Spread-spectrum (SS) technology has only emerged since the 1950s. Yet this novel approach to applications such as multiple access, ranging, and interference rejection has rendered SS techniques extremely important to most current NASA and military communication systems. In this chapter we presented an overview enumerating the benefits and types of spread-spectrum techniques, as well as some historical background.

Pseudorandom sequences are at the heart of all present-day SS systems; we therefore treated PN generation and properties. Emphasis was placed on the two major spread-spectrum techniques: direct sequence and frequency hopping. Also, consideration was given to synchronization, a crucial aspect of spread-spectrum operation. Some application examples were considered, such as code-division multiple access and communications with multipath conditions. Also, attention was devoted to the subject of jamming and jam-resistant systems, since this area represents one of the primary uses for spread-spectrum systems.

REFERENCES

1. Scholtz, R. A., "The Origins of Spread Spectrum Communications," *IEEE Trans. Commun.*, vol. COM30, no. 5, May 1982, pp. 822–854.

2. Shannon, C. E., "Communication in the Presence of Noise," *Proc. IRE,* Jan. 1949, pp. 10–21.

3. Dillard, R. A., "Detectability of Spread Spectrum Signals," *IEEE Trans. Aerosp. Electron. Syst.*, July 1979.

4. Simon, M. K., Omura, J. K., Scholtz, R. A., and Levitt, B. K., *Spread Spectrum Communications,* Vol. 3, Computer Science Press, Inc., Rockville, Md., 1985.

5. de Rosa, L. A., and Rogoff, M., Sec. I (Communications) of *Application of Statistical Methods to Secrecy Communication Systems,* Proposal 946, Fed. Telecommun. Lab., Nutley, N.J., Aug. 28, 1950.

6. Simon, M. K., Omura, J. K., Scholtz, R. A., and Levitt, B. K., *Spread Spectrum Communications,* Vol. 1, Computer Science Press, Inc., Rockville, Md., 1985.

7. Pickholtz, R. L., Schilling, D. L., and Milstein, L. B., "Theory of Spread-Spectrum Communications—A Tutorial," *IEEE Trans. Commun.,* vol. COM30, no. 5, May 1982, pp. 855–884.

8. Pickholtz, R. L., Schilling, D. L., and Milstein, L. B., Revisions to "Theory of Spread-Spectrum Communications—A Tutorial," *IEEE Trans. Commun.,* vol. COM32, no. 2, Feb. 1984, pp. 211–212.

9. Simon, M. K., Omura, J. K., Scholtz, R. A., and Levitt, B. K., *Spread Spectrum Communications,* Vol. 2, Computer Science Press, Inc., Rockville, Md., 1985.

10. Simon, M. K., and Polydoros, A., "Coherent Detection of Frequency-Hopped Quadrature Modulations in the Presence of Jamming: Part I. QPSK and QASK; Part II. QPR class I Modulation," *IEEE Trans. Commun.,* vol. COM29, Nov. 1981, pp. 1644–1668.

11. Ward, R. B., "Acquisition of Pseudonoise Signals by Sequential Estimation," *IEEE Trans. Commun.,* COM13, Dec. 1965, pp. 475–483.

12. Spilker, J. J., and Magill, D. T., "The Delay-Lock Discriminator—An Optimum Tracking Device," *Proc. IRE,* Sept. 1961.

13. Spilker, J. J., "Delay-Lock Tracking of Binary Signals," *IEEE Trans. Space Electron. Telem.,* Mar. 1963.

14. Simon, M. K., "Noncoherent Psuedonoise Code Tracking Performance of Spread Spectrum Receivers," *Commun.,* vol. COM25, Mar. 1977.

15. Ziemer, R. E., and Peterson, R. L., *Digital Communications and Spread Spectrum Systems,* Macmillan Publishing Company, New York, 1985.

16. Holmes, J. K., *Coherent Spread Spectrum Systems,* John Wiley & Sons, Inc., New York, 1982.

17. Pursley, M. B., "Performance Evaluation for Phase-Coded Spread-Spectrum Multiple-Access Communication: Part I. System Analysis," *IEEE Trans. Commun.,* vol. COM25, no. 8, Aug. 1977, pp. 795–799.

18. Geraniotis, E., "Noncoherent Hybrid DS-SFH Spread-Spectrum Multiple-Access Communications," *IEEE Trans. Commun.,* vol. COM34, no. 9, Sept. 1986, pp. 862–872.

19. Geraniotis, E., and Pursley, M. B., "Error Probability for Direct-Sequence Spread-Spectrum Multiple-Access Communications: Part I. Upper and Lower Bounds," *IEEE Trans. Commun.,* vol. COM30, no. 5, May 1982, pp. 985–995.

20. Geraniotis, E., and Pursley, M. B., "Error Probabilities for Direct-Sequence Spread-Spectrum Multiple-Access Communications: Part II. Approximations," *IEEE Trans. Commun.,* vol. COM30, no. 5, May 1982, pp. 996–1009.

21. Schilling, D. L., Milstein, L. B., Pickholtz, R. L., and Brown, R. W., "Optimization of the Processing Gain of an *M*-ary Direct Sequence Spread Spectrum Communication System," *IEEE Trans. Commun.,* vol. COM28, no. 8, Aug. 1980, pp. 1389–1398.

22. Viterbi, A. J., and Jacobs, I. M., "Advances in Coding and Modulation for Noncoherent Channels Affected by Fading, Partial Band, and Multiple Access Interference," in A. S. Viterbi, ed., *Advances in Communication Systems,* Vol. 4, Academic Press, Inc., New York, 1975.

23. Stark, W. E., "Coding for Frequency-Hopped Spread-Spectrum Communication with Partial-Band Interference: Part I. Capacity and Cutoff Rate," *IEEE Trans. Commun.,* vol. COM33, no. 10, Oct. 1985, pp. 1036–1044.

24. Stark, W. E., "Coding for Frequency-Hopped Spread-Spectrum Communication with Partial-Band Interference: Part II. Coded Performance," *IEEE Trans. Commun.,* vol. COM33, no. 10, Oct. 1985, pp. 1045–1057.

25. Milstein, L. B., Davidovici, S., and Schilling, D. L., "The Effect of Multiple-Tone

Interfering Signals on a Direct Sequence Spread Spectrum Communication System,'' *IEEE Trans. Commun.*, vol. COM30, Mar. 1982, pp. 436–446.

26. Milstein, L. B., Pickholtz, R. L., and Schilling, D. L., "Optimization of the Processing Gain of an FSK-FH system," *IEEE Trans. Commun.*, vol. COM28, July 1980, pp. 1062–1079.

27. Huth, G. K., "Optimization of Coded Spread Spectrum Systems Performance," *IEEE Trans. Commun.*, vol. COM25, Aug. 1977, pp. 763–770.

28. Viterbi, A. J., "Spread Spectrum Communications—Myths and Realities," *IEEE Commun. Mag.*, May 1979, pp. 11–18.

PROBLEMS

10.1. Explain why a maximal-length n-stage linear feedback shift register can produce a sequence with a period no greater than $2^n - 1$.

10.2. Show that in a maximal-length n-stage linear feedback shift register the output stage must always be an input to the feedback network.

10.3. Consider the DS/BPSK spread-spectrum transmitter of Figure 10.9a or b. Let $x(t)$ be the sequence 1 0 0 1 1 0 0 0 1, arriving at a rate of 75 bits/s, where the leftmost bit is the earliest bit. Let $g(t)$ be generated by the shift register of Figure 10.7, with an initial state of 1 1 1 1 and a clock rate of 225 Hz.
   (a) Sketch the final transmitted sequence $x(t)g(t)$.
   (b) What is the bandwidth of the transmitted (spread) signal?
   (c) What is the processing gain?
   (d) Suppose that the estimated delay, $\hat{T}_d$, of Figure 10.9c is too large by one chip time. Sketch the despread chip sequence.
   (e) Choose a decision rule for deciding on $\hat{x}(t)$ and identify the errors.

10.4. A total of 24 equal-power terminals are to share a frequency band through a code-division multiple access (CDMA) system. Each terminal transmits information at 9.6 kbits/s with a direct-sequence spread-spectrum BPSK modulated signal. Calculate the minimum chip rate of the PN code in order to maintain a bit error probability of 10^{-3}. Assume that the receiver noise is negligible with respect to the interference from the other users.

10.5. A feedback shift register PN generator produces a 31-bit PN sequence at a clock rate of 10 MHz. What are the equation and graphical form of the autocorrelation function and power spectral density of the sequence? Assume that the pulses have values of ± 1.

10.6. Consider an FH/MFSK system such as the one shown in Figure 10.11. Let the PN generator be defined by a 20-stage linear feedback shift register with a maximal length sequence. Each state of the register dictates a new center frequency within the hopping band. The minimum step size between center frequencies (hop to hop) is 200 Hz. The register clock rate is 2 kHz. Assume that 8-ary FSK modulation is used and that the data rate is 1.2 kbits/s.
   (a) What is the hopping bandwidth?
   (b) What is the chip rate?
   (c) How many chips are there in each data symbol?
   (d) What is the processing gain?

10.7. The block diagram of Figure 10.16 is described in Section 10.4.5 for a fast frequency hopping (FFH) demodulator. Draw a similar block diagram for a slow frequency hopping (SFH) demodulator, and explain how it would work.

10.8. Find the mean and the standard deviation of the time needed to acquire a 10-megachip/s BPSK modulated PN code sequence using a serial search where 100 chips are examined at a time. Assume that a correct detection results when all 100 received chips match the locally generated ones. The ratio of received chip energy to noise power spectral density is 9.6 dB, and the uncertainty time between the received and local code sequences is 1 ms. Assume that the probability of false lock (false alarm) is negligible.

10.9. There are 11 equal-power terminals in a CDMA communication system, transmitting signals toward a central node. Each terminal transmits information at 1 kbit/s on a 100-kbits/s direct-sequence spreading signal using BPSK modulation.
   (a) If receiver noise is negligible with respect to the interference from other users, what is the received ratio of bit energy to interference power spectral density (E_b/I_0) experienced by each user?
   (b) What is the effect on E_b/I_0 if all users double their output power?
   (c) If the users wish to expand their service to 101 equal-power users, what must be done to the spreading codes to maintain the original E_b/I_0 ratio?

10.10. A CDMA system uses direct-sequence modulation with a data bandwidth of 10 kHz and a spread bandwidth of 10 MHz. With only one signal being transmitted, the received E_b/N_0 is 16 dB.
   (a) If the required E_b/N_0 is 10 dB, how many equal-power users can share the band?
   (b) If each user's transmitted power is reduced by 3 dB, how many equal-power users can share the band?
   (c) What is the maximum number of users that can share the band?
   (d) How many equal-power users could share the band if they switch to TDMA with 98% efficient use of the communications resource?
   (e) Why is the answer to part (d) so much greater than the answer to part (c)? What is the disadvantage of TDMA compared to CDMA (i.e., what penalty is paid to accommodate more users)?

10.11. A DS/SS system is used to combat multipath. If the path length of the multipath wave is 100 m longer than that of the direct wave, what is the minimum chip rate necessary to reject the multipath interference?

10.12. A ground-to-synchronous satellite link must be closed in a jamming environment. The data rate is 1 kbit/s and the ground station has a 60-ft antenna. Antijam protection is provided by a 10-Mbits/s direct-sequence spread-spectrum code. The jammer has a 150-ft antenna and a transmitter with 400 kW of power. Assume equal space and propagation losses. How much power is required of the earth station transmitter to achieve an E_b/J_0 of 16 dB at the satellite receiver? Assume that the receiver noise is negligible.

10.13. Input data at 75 bits/s are channel encoded using a rate $\frac{1}{2}$ encoder. The coded bits are then modulated using 8-ary FSK. The FSK symbols are then spread by frequency hopping at a rate of 2000 hops/s.
   (a) What is the chip rate?
   (b) What is the order of diversity?
   (c) If there are two such signals, time-division multiplexed (TDM'd) on the channel

at the same hopping rate, how would this affect the chip rate, symbol rate, and order of diversity?

(d) If there are 80 such signals TDM'd on the channel, how would this effect the chip rate, symbol rate, and order of diversity?

10.14. A frequency hopping noncoherent binary FSK system operates at an E_b/N_0 of 30 dB with a hopping bandwidth of 2 GHz. Assume that no channel coding is used. A jammer operating over the same broadband bandwidth yields a received $J_0 = 100N_0$.

(a) What is the bit error probability, P_B?

(b) If the jammer becomes a partial-band jammer, what bandwidth should it occupy to be most effective?

(c) What is P_B as a result of such optimum partial-band jamming?

(d) What is the unjammed P_B?

10.15. A noncoherent frequency hopping 8-ary FSK system hops at 12,000 hops/s over a bandwidth of 1 MHz. The symbol rate is 3000 symbols/s. Assume that channel coding is not used. The signal power at the input of the receiver is 10^{-12} W. A partial-band noise jammer occupies 50 kHz (assumed to be entirely within the hopping bandwidth of the signal). The received jammer power is 10^{-11} W. Assume that the system temperature is 290 K. What is the probability of bit error?

10.16. A coherent DS/BPSK system is transmitting at a data rate of 10 kbits/s in the presence of a broadband jammer. Assume that the system does not use channel coding. Also assume that the propagation losses are the same for the system and the jammer.

(a) If the EIRP of the communicator is 20 kW and the EIRP of the jammer is 60 kW, calculate the required spread-spectrum bandwidth to achieve a bit error probability of $P_B = 10^{-5}$.

(b) If the jammer is a pulse jammer, calculate the pulse duty cycle that results in worst-case jamming. What is the value of P_B at this duty cycle?

10.17. A communicator intends to use frequency hopping at a hop rate of 10,000 hops/s to avoid a threat of repeat-back jamming.

(a) Ignoring the curvature of the earth, and assuming that the communicator is transmitting to a satellite at geosynchronous altitude (approximately 36,000 km) that is directly overhead, compute the *radius of vulnerability*, which is the radius outside of which the communicator is unconditionally safe from repeat-back jamming by a ground-based jammer.

(b) If the communicator knows that the jammer requires a minimum of 10 μs to identify the transmission frequency and tune the jammer output, compute the radius of vulnerability conditioned on this information.

10.18. Consider an airborne repeat-back jammer as shown in Figure P10.1. The communicator is using a FH/SS system. What is the minimum hop rate required in order that the repeat-back jamming does not degrade the message? What would be the minimum required hopping rate if the communicator and jammer switched positions (i.e., fixed land jammer and airborne communicator).

10.19. Spread-spectrum techniques can be used to meet government regulations regarding flux (power) density radiating the surface of the earth. If a satellite at synchronous altitude (36,000 km) transmits 4-kbits/s data using 100 W of EIRP, what spreading bandwidth is required to maintain a flux density on the earth's surface no greater than -151 dBW/m^2 in any 4-kHz band?

10.20. A communicator uses noncoherent BFSK modulation and frequency hopping to combat the effects of a jammer. The power of the communicator's signal at the receiver input is 10 μW. The SNR in the absence of jamming is assumed to be very large. The power of the jamming signal at the receiver input is 1 W.

(a) If the jammer jams the entire hopping bandwidth with equal amounts of Gaussian noise (the noise will be white within the band), what bandwidth expansion factor will allow the communicator to maintain a bit error probability of 10^{-4}?

(b) Assume that the jammer decides to "color" its jamming noise by reducing its energy by a fraction, α ($0 \leq \alpha \leq 1$), in half the hop bandwidth, and increasing it by a like amount in the other half (thereby keeping its transmitted energy constant). Assuming that the communicator does not modify his hopping pattern to avoid the jammer strategy, develop an expression for the bit error probability for this case of colored jamming.

(c) Determine the fraction, α, that is optimum from the jammer standpoint for each of the limiting cases (i) when the effective SNR is large and, (ii) when it is small.

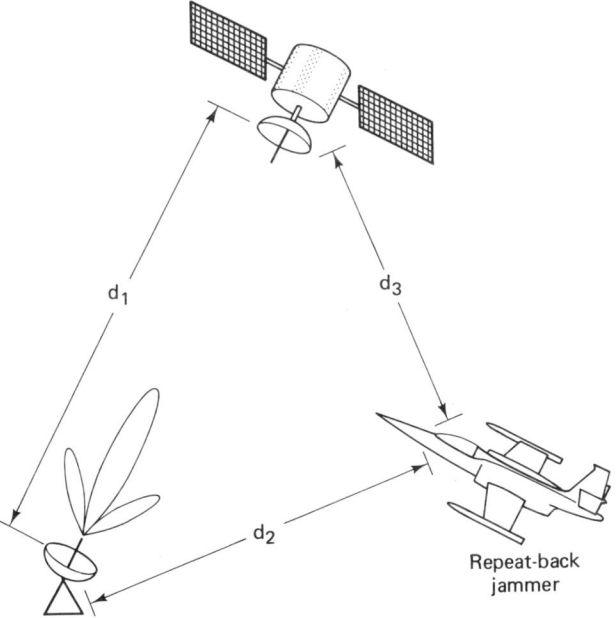

Communicator

Repeat-back jammer

Figure P10.1

CHAPTER 11

Source Coding

Fredric J. Harris
San Diego State University
San Diego, California

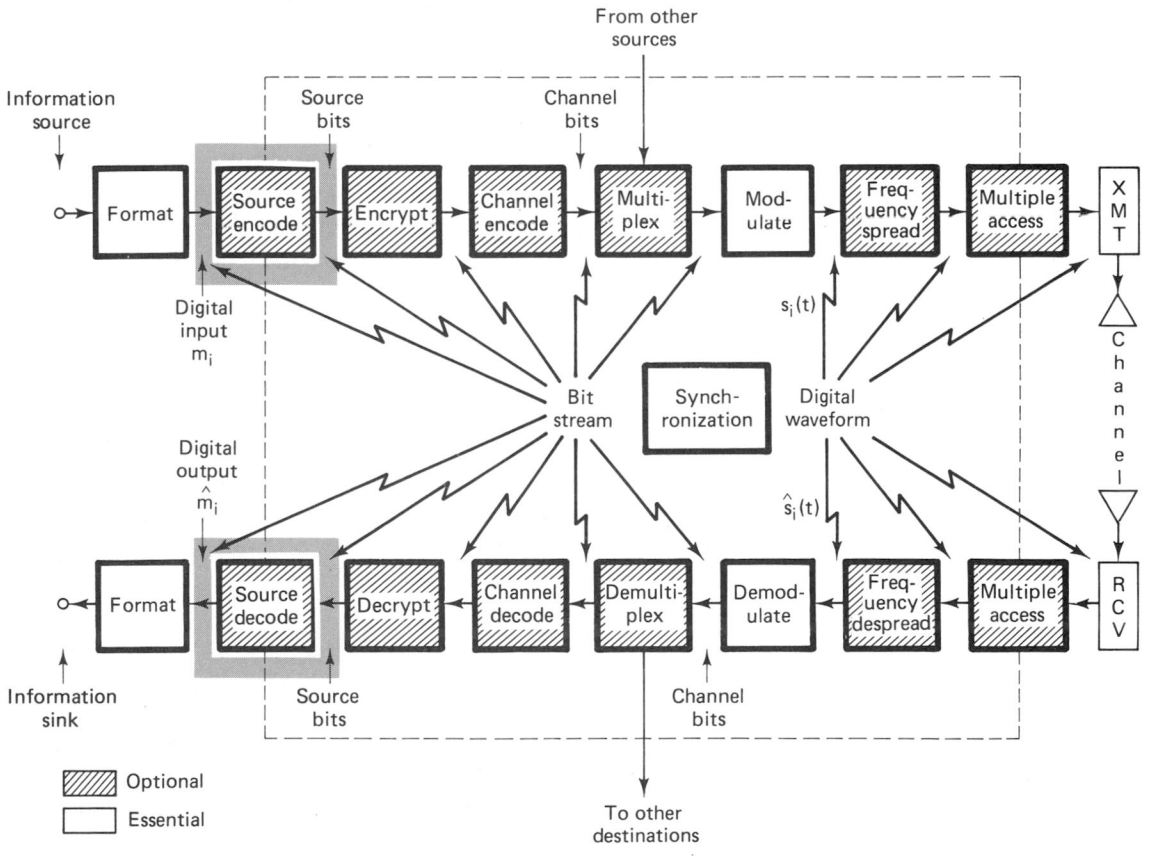

11.1 SOURCES

Source coding deals with the task of forming efficient descriptions of information sources. For discrete sources, the ability to form reduced data rate descriptions is related to the information content and the statistical correlation among the source symbols. For analog sources, the ability to form reduced data rate descriptions, subject to a fixed fidelity criterion, is related to the amplitude distribution and the temporal correlation of the source waveforms. The goal of source coding is either to improve the SNR for a given bit rate or to reduce the bit rate for a given SNR. To understand where the tools and techniques of source coding are effective, it is important to have common measures of source parameters. For this reason, in this section we examine simple models of discrete and analog sources and then describe how well real sources match the ideal models.

11.1.1 Discrete Sources

A discrete source generates (or emits) at a uniform rate a sequence of symbols $x(kT)$, selected from a source alphabet at discrete time intervals, kT, where $k = 1, 2, 3, \ldots$ is an index. If the alphabet contains a finite number of symbols, say N symbols, the source is said to be a *finite discrete source*. An example of such a source is the output of a 12-bit digital-to-analog converter (one of 4096 output levels) or the output of a 10-bit analog-to-digital converter (one of 1024 binary 10-tuples). Another example of a discrete source is the succession of 8-bit ASCII characters emitted by a remote computer terminal.

A finite discrete source is defined by the list of source symbols (sometimes called the alphabet) and the probability assignment to these symbols (or letters). We will assume that the source is short-term stationary, that is, that the probability assignment is fixed over the observation interval. An example in which the alphabet is fixed but the probability assignment changes is found in the sequence of symbols emitted by a terminal for someone typing English text followed by typing Spanish text and then again French text.

If we know that the probability of each symbol, X_j, is $P(X_j) = p_j$, we know the *self-information*, $I(X_j)$, for each symbol in the alphabet set.

$$I(X_j) = -\log_2 p_j \tag{11.1}$$

The average self-information for the symbols in an alphabet is denoted $H(X)$ and is called the *source entropy*.

$$H(X) = E\{I(X_j)\} = -\sum_{j=1}^{N} p_j \log_2 p_j \tag{11.2}$$

where $E\{X\}$ is the expected value of X. The source entropy is defined as the average amount of information per source output; it can be considered to be the average amount of uncertainty that is resolved by use of the alphabet. It is thus the average amount of information that must be moved through the communication channel to resolve that uncertainty. It can be shown that this amount of information in bits per symbol is bounded below by zero if there is no uncertainty, and is bounded above by $\log_2 N$ if there is maximum uncertainty.

$$0 \le H(X) \le \log_2 N \tag{11.3}$$

Example 11.1   Entropy of a Binary Source

Consider the binary source that generates independent symbols 0 and 1 with probabilities equal to p and $(1 - p)$, respectively. We described this source in Section 7.4.2 and presented its entropy function in Figure 7.4. If $p = 0.1$ and $(1 - p) = 0.9$, the source entropy is

$$H(X) = -[p \log p + (1 - p) \log (1 - p)] \tag{11.4}$$
$$= 0.47 \text{ bit/symbol}$$

Thus this source can be described (with the use of appropriate coding) with less than half a bit per symbol rather than with one bit per symbol in its present form.

We note that the first reason that source coding works is because the information content of an N-symbol alphabet used in real communication systems is always less than or equal to the upper bound of Equation (11.3). We know from experience, as we have noted in Example 7.1, that the symbols of English text are not all equally likely. For instance, we use the high probability of certain letters in text as part of the strategy to initialize the game of Hangman. (In this game a player must guess the letters, but not the positions, of a hidden word of known length. Penalties accrue to false guesses, and the letters of the entire word must be found prior to the occurrence of six false guesses.)

A discrete source is said to be *memoryless* if the sequence of symbols emitted by the source are statistically independent. In particular this means that for symbols taken two at a time, the joint probability of two elements is simply the product of their respective probabilities.

$$P(X_j, X_k) = P(X_j|X_k)P(X_k) = P(X_j)P(X_k) \tag{11.5}$$

A result of statistical independence is that the information required to transmit a sequence of M symbols (called an M-tuple) from a given alphabet is precisely M times the average information required to transmit a single symbol. This happens because the probability of a statistically independent M-tuple is given by

$$P(X_1, X_2, \ldots, X_M) = \prod_{m=1}^{M} P(X_m) \tag{11.6}$$

so that the average entropy per symbol, $H_M(X)$, of a statistically independent M-tuple is given by

$$H_M(X) = \frac{1}{M} \mathbf{E}\{-\log_2 P(X_1, X_2, \ldots, X_M)\}$$

$$= \frac{1}{M} \sum_{X_m} [-P(X_m) \log_2 P(X_m)] \tag{11.7}$$

$$= H(X)$$

A discrete source is said to have memory if the sequence of source elements are not independent. The dependency between symbols means that in a sequence of M symbols there is reduced uncertainty about the Mth symbol when we know the previous $(M - 1)$ symbols. For instance, is there much uncertainty in the next symbol for the 10-tuple CALIFORNI_? Thus the M-tuple with dependent symbols contains less information, or resolves less uncertainty, than does one with independent symbols. The entropy of a source with memory is the limit

$$H(X) = \lim_{M \to \infty} H_M(X) \tag{11.8}$$

We observe that the entropy of an M-tuple from a source with memory is always less than the entropy of a source with the same alphabet and symbol probability but without memory.

$$H_M(X)_{\text{with memory}} < H_M(X)_{\text{with no memory}} \tag{11.9}$$

For example, given a symbol (or letter) "q" in English text, we know that the next symbol will probably be a "u." Hence in a communication task, being told that the letter "u" follows a letter "q" adds little information to our knowledge of the word being transmitted. As another example, given the letters "th," the most likely symbols to follow are: a, e, i, o, u, r, and space. Thus adding the next symbol to the given set resolves some uncertainty but not much. A formal statement of this awareness is that the average entropy per symbol of an M-tuple from a source with memory *decreases as the length M increases*. Hence it is more efficient to encode symbols from a source with memory three at a time than it would be to encode them two at a time or one a time. Encoder complexity, memory constraints, and delay considerations require that practical source encoding be performed on finite-length sequences.

To help us understand the gains to be had in coding sources with memory, we form simple models of these sources. One such model is called a *first-order Markov source* [1]. This model identifies a number of states (or symbols in the context of information theory) and the conditional probabilities of transitioning to each next state. In the first-order model, the transition probabilities depend only on the present state. That is, $P(X_{i+1}|X_i, X_{i-1}, \ldots) = P(X_{i+1}|X_i)$. In the context of a binary sequence, this expression gives the probability of the state of the next digit conditioned on the state of the current digit.

Example 11.2   Entropy of a Binary Source with Memory

Consider the binary (i.e., two-symbol) first-order Markov source described by the state transition diagram shown in Figure 11.1. The source is defined by the state transition probabilities $P(0|1)$ and $P(1|0)$ of 0.45 and 0.05, respectively. The entropy of the source, X, is the weighted sum of the conditional entropies that correspond to the transition probabilities of the model.

$$H(X) = P(0)\,H(X|0) + P(1)\,H(X|1) \tag{11.10}$$

where

$$H(X|0) = -[P(0|0) \log_2 P(0|0) + P(1|0) \log_2 P(1|0)]$$

and

$$H(X|1) = -[P(1|1) \log_2 P(1|1) + P(0|1) \log_2 P(0|1)]$$

The a priori probability of each state is found by the total probability equations

$$P(0) = P(0)P(0|0) + P(1)P(0|1)$$

$$P(1) = P(1)P(1|1) + P(0)P(1|0)$$

$$P(0) + P(1) = 1$$

Solving for the a priori probabilities using the transition probabilities, we have

$$P(0) = 0.9 \quad \text{and} \quad P(1) = 0.1$$

Solving for the source entropy using Equation (11.10), we have

$$H(X) = P(0)H(X|0) + P(1)H(X|1)$$

$$= (0.9)(0.286) + (0.1)(0.993)$$

$$= 0.357 \text{ bit/symbol}$$

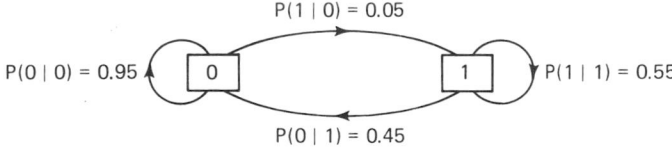

Figure 11.1   State transition diagram for a first-order Markov model.

Comparing this result to the result of Example 11.1, we see that the source with memory has a *lower entropy* than the source without memory even though the a priori symbol probabilities are the same.

Example 11.3   Extension Codes

The source alphabet for the binary Markov source of Example 11.2 consists of 0 and 1 and occur with probabilities of 0.9 and 0.1, respectively. Successive symbols are not independent and we can define a new set of code symbols as binary 2-tuples (*an extension code*) to take advantage of this dependency.

Binary 2-tuple	Extension symbol	Extension symbol probability
00	a	$P(a) = P(0|0)P(0) = (0.95)(0.9) = 0.855$
11	b	$P(b) = P(1|1)P(1) = (0.55)(0.1) = 0.055$
01	c	$P(c) = P(0|1)P(1) = (0.45)(0.1) = 0.045$
10	d	$P(d) = P(1|0)P(0) = (0.05)(0.9) = 0.045$

where the rightmost digit of the 2-tuple is the earliest digit. The entropy for this code extension alphabet is found by using an extension of Equation (11.10)) as follows:

$$H(\mathbf{X}_2) = P(a)H(\mathbf{X}_2|a) + P(b)H(\mathbf{X}_2|b) + P(c)H(\mathbf{X}_2|c) + P(d)H(\mathbf{X}_2|d)$$

$$H(\mathbf{X}_2) = 0.825 \text{ bit/output symbol}$$

$$= 0.412 \text{ bit/input symbol}$$

where $\mathbf{X}_k$ is the kth-order extension of the source, X. A longer extension code, which takes advantage of the adjacent symbol dependency, is of the form

Binary 3-tuple	Extension symbol	Extension symbol probability
000	a	$P(0|00)P(00) = (0.95)(0.855) = 0.8123$
100	b	$P(1|00)P(00) = (0.05)(0.855) = 0.0428$
001	c	$P(0|01)P(01) = (0.95)(0.045) = 0.0428$
111	d	$P(1|11)P(11) = (0.55)(0.055) = 0.0303$
110	e	$P(1|10)P(10) = (0.55)(0.045) = 0.0303$
011	f	$P(0|11)P(11) = (0.45)(0.055) = 0.0248$
010	g	$P(0|10)P(10) = (0.45)(0.045) = 0.0248$
101	h	$P(1|01)P(01) = (0.05)(0.045) = 0.0023$

Again, using an extension of Equation (11.10), the entropy for this extension code is found to be

$$H(\mathbf{X}_3) = 1.223 \text{ bits/output symbol}$$

$$= 0.408 \text{ bit/input symbol}$$

We note that the entropy of the one-symbol, two-symbol, and three-symbol descriptions of the source (0.470, 0.412, and 0.408 bit, respectively) are decreasing asymptotically toward the source entropy of 0.357 bit/input symbol. Remember that the source entropy is the lower bound in bits per input symbol for this (infinite memory) alphabet and this bound can only be approached asymptotically with finite-length coding.

11.1.2 Waveform Sources

A waveform source is a random process of some independent variable. We classically consider this variable to be time, so that the waveform of interest is a time-varying waveform. Important examples of time-varying waveforms are the outputs of transducers used in process control, such as temperature, pressure, velocity, displacement, and flow rates. Examples of particularly high interest include speech and music waveforms. The waveform can also be a function of one or more spatial variables (e.g., displacement). Important examples of spatial waveforms include single images such as a photograph, or moving images such as the successive images (at 24 frames/s) of moving picture film. Spatial images are often converted to time-varying functions by a simple scanning operation. This, for example, is done for facsimile transmission and with a slight modification (called interlacing) for standard broadcast television.

11.1.2.1 Amplitude Density Functions

Discrete sources were described by a list of their possible elements (called letters of an alphabet) and their multidimensional probability density functions (pdfs) of all orders. By analogy, waveform sources are similarly described in terms of their probability density functions as well as parameters and functions derived from these functions. We model many waveforms as random processes with classical probability distribution functions and with simple correlation properties. In the modeling process we distinguish between short-term or local (time) characteristics and long-term or global characteristics. This partition is necessary because many waveforms are nonstationary.

The probability density function of the actual process may not be available to the system designer. Sample density functions can, of course, be rapidly formed in real time during a short preceding interval and used as reasonable estimates over the present interval. A less ambitious task is simply to make estimates of short-term waveform-related averages. These include the sample mean (or time-average value), the sample variance (or mean-square value assuming zero mean), and correlation coefficients formed over the previous sample interval. In many applications of waveform analysis, the input waveform is converted to a zero-mean waveform by subtracting the estimates of the mean. This happens, for example, in a digital panel meter in which an auxiliary circuit measures the effects of the internal dc offset voltages and subtracts them in a process known as *auto-zero*. Further, the variance estimate is often used to scale the range of the input waveform to match the dynamic amplitude range of subsequent waveform-handling equipment. This process, performed in the digital panel meter, is called *autoranging* or *automatic gain control* (AGC). The function of these signal conditioning operations, mean removal and variance control (gain adjustment) shown in Figure 11.2, is to normalize the probability density functions of the input waveform. This normalization assures optimal utility of the limited dynamic range of subsequent recording, transmission, or processing subsystems.

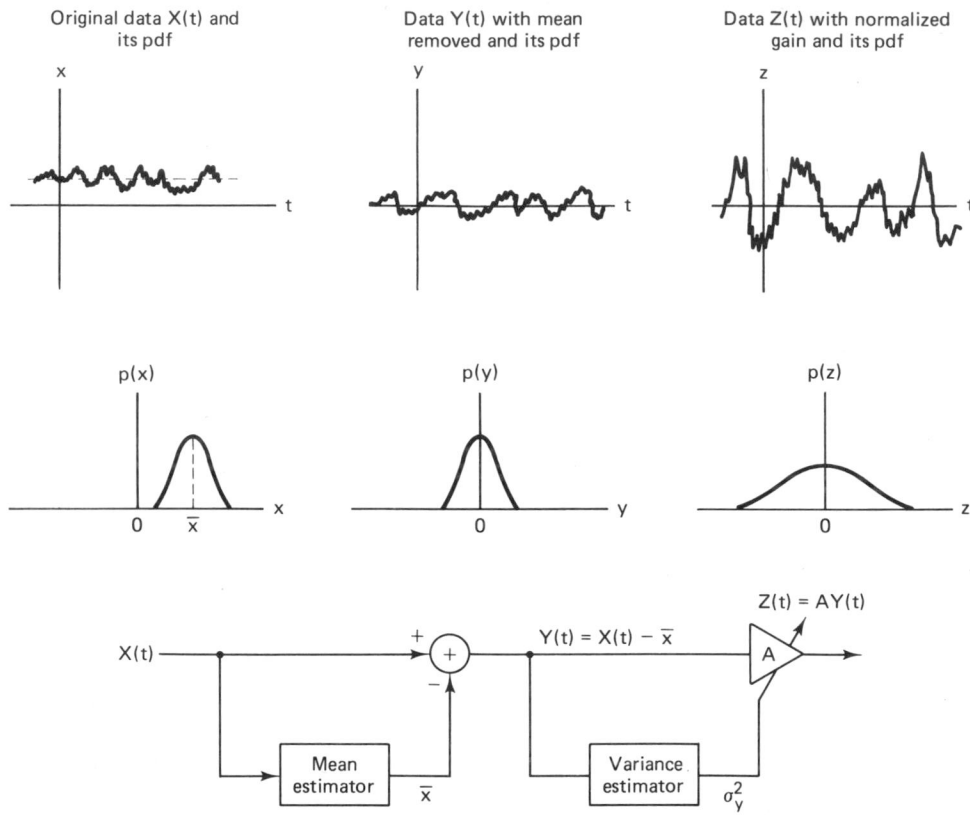

Figure 11.2 Mean removal and variance control (gain adjustment) for a data-dependent signal conditioning system.

11.1.2.2 Autocorrelation Function, Power Spectrum, and Models

There is significant correlation between the amplitudes of many waveform sources in successive time intervals. This correlation means that successive time samples are not independent. If the time sequence is truly independent, the autocorrelation function of the sequence would be an impulse function. The width of the autocorrelation function (in seconds) is called the correlation time of the process and is akin to the time constant of a filter. This time interval is an indication of how much shift along the time axis is necessary to find uncorrelated data samples. If the correlation time is large, we interpret this to mean that the waveform makes significant amplitude changes slowly. Conversely, if the correlation time is small, we infer that the waveform makes significant amplitude changes very quickly.

The Fourier transform of the autocorrelation function is the power spectral density of the waveform process. Thus an alternative description of the autocorrelation function, which reflects the amount of intersample dependence, is the degree of flatness in the waveform power spectrum. A flat spectrum, sometimes called a *white spectrum*, corresponds to source waveforms with independent values sample to sample. A power spectrum with a wide bandwidth implies a time function capable of rapid changes in envelope, while a power spectrum with a narrow bandwidth suggests a time function capable of only slow changes. In general, the larger the deviation from flatness, the more correlation will be found in the waveform samples. Very large changes from flatness in the power spectrum may warrant source descriptions which partition the spectrum, via filters, into subbands each of which is described and quantized separately.

11.2 AMPLITUDE QUANTIZING

Amplitude quantizing is the task of mapping samples of a continuous amplitude waveform to a finite set of amplitudes. The hardware that performs the mapping is the analog-to-digital converter (ADC or A-D). The amplitude quantizing occurs after the sample-and-hold operation. The simplest quantizer to visualize performs an instantaneous mapping from each continuous input sample level to one of the preassigned equally spaced output levels. Quantizers that exhibit equally spaced increments between possible quantized output levels are called *uniform quantizers* or sometimes *linear quantizers*. Possible instantaneous input–output characteristics are easily visualized by a simple staircase graph consisting of risers and treads of the types shown in Figure 11.3. Figure 11.3a, b, and d show quantizers with uniform quantizing steps, while Figure 11.3c is a quantizer with nonuniform quantizing steps. Figure 11.3a depicts a quantizer with *midtread* at the origin, while Figure 11.3b and d present quantizers with *midrisers* at the origin. A distinguishing property of midriser and midtread converters is related to the presence or absence, respectively, of output level changes when the input to the converter is idle noise. Further, Figure 11.3d presents a *biased* (i.e., truncation) quantizer, while the remaining quantizers in the figure are unbiased and are referred to as *rounding quantizers*. Most quantizers are truncation quantizers due to implementation considerations. The terms "midtread" and "midriser" are staircase terms used to describe whether the horizontal or vertical member of the staircase is at the origin. The unity-slope line passing through the origin represents the ideal nonquantized input–output characteristic we are trying to approximate with the staircase. The difference between the staircase and the unity-slope-line segment represents the approximation error made by the quantizer at each input level. Figure 11.4 illustrates the approximation error amplitude versus input amplitude function for each quantizer characteristic in Figure 11.3. Parts (a) through (d) of Figure 11.4 correspond to the same parts in Figure 11.3. This error is often modeled as quantizing noise because the error sequence obtained when quantizing a

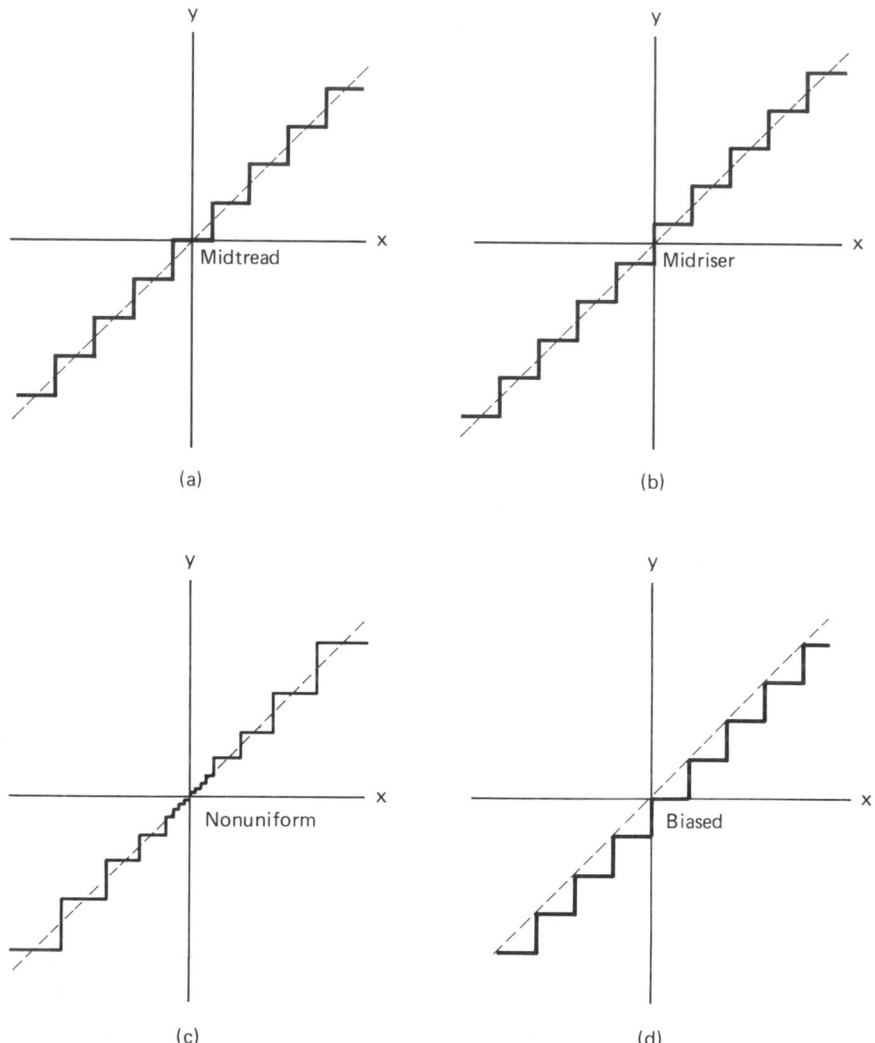

Figure 11.3 Various quantizer transfer functions.

wideband random process is reminiscent of an additive noise sequence. Unlike true additive noise sources, the quantizing errors are signal dependent and are highly structured. It is desirable to break up this structure; this can be accomplished by introducing an independent noise perturbation, known as *dither*, prior to the quantization step. This is discussed in Section 11.2.4.

The linear quantizer is simple to implement and is particularly easy to understand. It is the universal form of the quantizer in the sense that it makes no assumptions about the amplitude statistics and correlation properties of the input waveform, nor does it take advantage of user-related fidelity specifications. Quantizers that take advantage of these considerations are more efficient as source

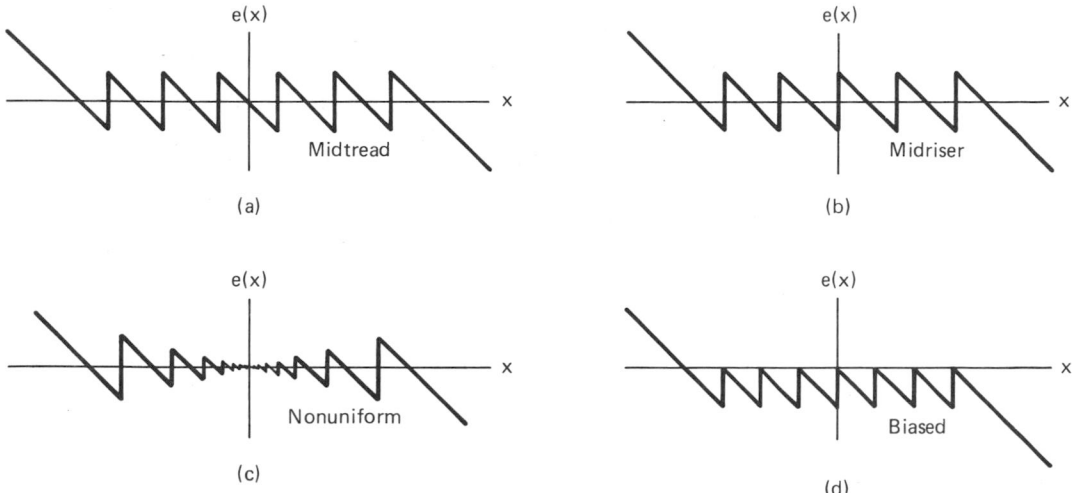

Figure 11.4 Instantaneous error voltage for various quantizer transfer functions.

coders and are more task specific then the general linear quantizer; these quantizers are often more complex and more expensive but are sometimes justified in terms of improved system performance. There are applications for which the uniform quantizer is the most desirable amplitude quantizer. These include signal processing applications, graphics and display applications, and process control applications. There are other applications for which nonuniform, adaptive quantizers are more desirable amplitude quantizers. These include waveform encoders for efficient storage and communication, contour encoders for images, vector encoders for speech, and analysis/synthesis encoders (such as the vocoder) for speech.

11.2.1 Quantizing Noise

The difference between the input and output of a quantizer is called the *quantizing error*. In Figure 11.5 we demonstrate the process of mapping the input sequence $X(t)$ to the quantized output sequence $\hat{X}(t)$. We can visualize forming $\hat{X}(t)$ by adding to each $X(t)$ an error sequence, $e(t)$.

$$\hat{X}(t) = X(t) + e(t)$$

The error sequence, $e(t)$, is deterministically defined by the input amplitude through the instantaneous error versus amplitude characteristic of the form in Figure 11.4. We note that the error sequence exhibits two distinct characteristics over different input operating regions.

The first operating interval is the *granular error region* corresponding to the sawtooth-shaped error characteristic. Within this interval the quantizer errors are confined by the size of the nearby staircase risers. The errors that occur in this region are called the granular errors or sometimes the quantizing errors. The input

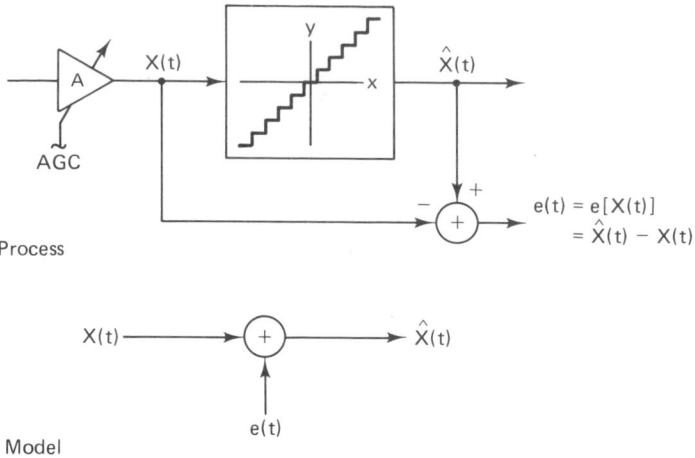

Process

Model

Figure 11.5 Process and model of quantizing noise corruption of input signal.

interval for which the quantizing errors are granular defines the *dynamic range of the quantizer*. This interval is sometimes called the *region of linear operation*. Proper use of the quantizer requires that the input signal conditioning somehow match the dynamic range of the input signal to the dynamic range of the quantizer. This is a function of the signal-dependent gain control system called automatic gain control (AGC), indicated in Figure 11.5.

The second operating interval is the nongranular error region corresponding to the linearly increasing (or decreasing) error characteristic. The errors that occur in this interval are called *saturation or overload errors*. When the quantizer operates in this region, we say that the quantizer is saturated. Saturation errors are larger than the granular errors and may have a more objectionable effect on reconstruction fidelity.

The quantization error corresponding to each input amplitude represents an error or noise term associated with that input amplitude. Under the assumptions that the quantization interval is small compared to the dynamic range of the input signal and that the input signal has a smooth probability density function over the quantization interval, we can assume that the quantization errors are uniformly distributed over that interval, as illustrated in Figure 11.6. The pdf with zero mean corresponds to a rounding quantizer, while the pdf with a mean of $-q/2$ corresponds to a truncation quantizer.

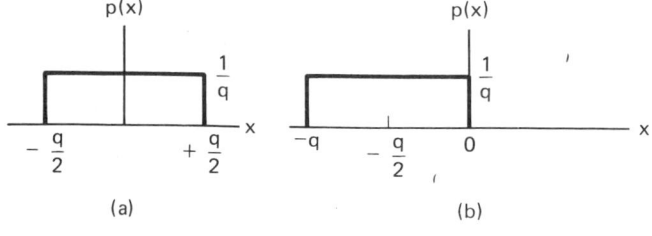

Figure 11.6 Probability density functions for quantizing error uniformly distributed over one quantile, q. (a) Probability density function for a rounding quantizer. (b) Probability density function for a truncating quantizer.

A quantizer or analog-to-digital converter (ADC) is defined by the number, size, and location of its quantizing step boundaries, and the corresponding step sizes. In a uniform quantizer, the step sizes are all equal and are equally spaced. The number of step sizes, N, is usually a power of 2 of the form $N = 2^b$, where b is the number of bits used in the conversion process. This number of levels is equally distributed over the dynamic range of the possible input levels. Normally, this range is defined as $\pm E_{max}$, such as ± 1.0 V or ± 5.0 V. Thus, accounting for the full range of $2E_{max}$, the size of a quantization step q is

$$q = \frac{2E_{max}}{2^b} \tag{11.11}$$

As an example, using Equation (11.11), the quantizing step hereafter called a *quantile*, for a 10-bit converter operating over the ± 1.0 V range is 1.953 mV. Occasionally, the operating range of a converter is altered so that the quantile is a "whole" number. For example, changing the operating range of the converter to ± 1.024 V results in a quantizing step size of 2.0 mV.

A useful figure of merit for the uniform quantizer is the quantizer output variance. If we assume that the quantization error is uniformly distributed over a single quantile interval q-wide, the quantizer variance, assuming zero-mean error, is found to be

$$\sigma^2 = \int_{-q/2}^{+q/2} e^2 p(e)\, de = \int_{-q/2}^{+q/2} e^2 \frac{1}{q}\, de = \frac{q^2}{12} \tag{11.12}$$

where $p(e) = 1/q$ is the probability density function of the quantization error. Thus the rms quantizer noise in a quantile interval of width q is found to be $q/\sqrt{12} = 0.29$ quantile. Equation (11.12) determines the quantizing noise power over one quantile, assuming that the errors are equiprobable over one quantile. In general, the quantiles are not equally sized over the range of input variables. We can account for this amplitude-dependent error by averaging the squared error over the amplitude variable and weighting by the probability of that amplitude. This is expressed by

$$\sigma_q^2 = E\{(x - q(x)]\}^2 = \int_{-\infty}^{+\infty} [e(x)]^2 p(x)\, dx \tag{11.13}$$

where x is the input variable, $q(x)$ is its quantized version, and $p(x)$ is the amplitude probability density function. We can partition the interval of integration in Equation (11.13) into two main intervals, one accounting for errors in the granular region of the quantizer and the second accounting for errors in the saturation region. We shall define the saturation amplitude of the quantizer as E_{max}. Also, we assume an odd symmetric transfer function for the quantizer, a symmetric pdf for the input signal, and redefine the error power to be the total error power, σ_D^2. We now write

$$\sigma_D^2 = 2 \int_0^{+\infty} [e(x)]^2 p(x)\, dx \tag{11.14a}$$

$$= 2 \int_0^{+E_{max}} [e(x)]^2 p(x)\, dx + 2 \int_{+E_{max}}^{+\infty} [e(x)]^2 p(x)\, dx \tag{11.14b}$$

or
$$\sigma_D^2 = \sigma_q^2 + \sigma_s^2 \tag{11.14c}$$

where σ_q^2 is the power in the linear region and σ_s^2 is the power in the saturation region. The noise power σ_q^2 can be further divided into subintervals corresponding to the successive discrete quantizer output levels (i.e., quantiles). If we assume that there are N such quantile levels, the integral becomes

$$\sigma_q^2 = \sum_{n=1}^{N} \int_{x_n}^{x_{n+1}} [e(x)]^2 p(x)\, dx \tag{11.15}$$

If we now assume that the density function is approximately uniform over each quantile interval, Equation (11.15) simplifies to

$$\sigma_q^2 = \frac{1}{12} \sum_{n=1}^{N} [x_{n+1} - x_n]^2 \int_{x_n}^{x_{n+1}} p(x)\, dx \tag{11.16}$$

11.2.2 Uniform Quantizing

If the quantizer has uniform quantiles equal to q, Equation (11.16) simplifies further to

$$\sigma_q^2 = \frac{1}{12} q^2\, 2 \int_0^{Nq/2} p(x)\, dx \tag{11.17}$$

where the limits of integration account for the $\pm N/2$ quantiles.

Noise power alone will not fully describe the noise performance of the quantizer. A more meaningful measure of quality is the ratio of output quantizing noise variance to input signal variance. Assuming that the input signal has zero mean, the signal variance is

$$\sigma_X^2 = \int_{-\infty}^{\infty} x^2 p(x)\, dx \tag{11.18}$$

Further insight into the average quantizer noise requires that we examine a specific density function and a specific quantizer.

Example 11.4   Uniform Quantizer

Determine the quantizer variance for a signal that is uniformly distributed over the full dynamic range of a uniform quantizer with 2^b equally spaced quantile levels. In this case there is no saturation noise and only the granular noise term must be computed. Each quantile interval, denoted by q is

$$q = (2E_{max})2^{-b} \tag{11.19}$$

where $2E_{max}$ is the input interval between the positive and negative boundaries of the linear quantizing range.

Solution

Substituting Equation (11.19) into Equation (11.12) or (11.17), we have the quantizing noise power (in the linear region):

$$\sigma_q^2 = \tfrac{1}{12}(2E_{\max}2^{-b})^2 = \tfrac{1}{3}(E_{\max})^2 2^{-2b} \tag{11.20}$$

The input signal power is found by performing the integration of Equation (11.18) for a uniform probability density function in the zero-mean interval spanning $2E_{\max}$. In this interval the density is $1/(2E_{\max})$, so the signal variance is found to be

$$\sigma_X^2 = \int_{-E_{\max}}^{+E_{\max}} \frac{1}{2E_{\max}} x^2 \, dx = \frac{1}{3}(E_{\max})^2 \tag{11.21}$$

Taking the ratio of noise power to signal power (NSR), we have

$$\text{NSR} = \frac{\sigma_q^2}{\sigma_X^2} = 2^{-2b} \tag{11.22}$$

Now converting the NSR to decibels, we have

$$\text{NSR}_{dB} = 10 \log_{10} \text{NSR} = 10 \log_{10} 2^{-2b} \tag{11.23a}$$

$$= -20b \log_{10} 2 = -6.02b \ (dB) \tag{11.23b}$$

For our example, Equation (11.23b) suggests that each bit used in the conversion process is worth -6.02 dB in noise-to-signal ratio. In fact, the NSR for any uniform quantizer, not operating in saturation, is of the form

$$\text{NSR}_{dB} = -6.02b + C \tag{11.24}$$

where the term C depends on the signal density function and the ratio of signal standard deviation to quantizer saturation level.

Figure 11.7 presents the discrete Fourier transform of two sinusoids that have been sampled by a linear 10-bit ADC. The two simusoids have relative amplitudes of 1.0 and 0.01 (i.e., one is reduced 40 dB with respect to the other). In Figure 11.7a the input signal is scaled to full dynamic range of the 10-bit converter. With the full-scale sinusoid and quantizer range set to 1.0, the 10-bit quantizer exhibits a quantile of amplitude 0.001953 and an rms quantization noise amplitude of 0.000564 for a peak signal-to-noise ratio (SNR) of 65 dB. The discrete Fourier transforms performed for Figure 11.7 were of length 256, and since the SNR of a transform increases proportional to integration time, there is an improvement in SNR of 21 dB due to the transform [2]. Thus, at the transform output, the average SNR due to quantizing is 86 dB. There are significant variations about the expected noise power level, typically on the order of 9 dB. Thus the minimum SNR at the transform is on the order of 77 dB, which is the range we observe in Figure 11.7. In Figure 11.7b and c, the input signal is attenuated by 20 and 40 dB, respectively, relative to the full-scale levels of Figure 11.7a. Note that the higher-frequency sinusoidal input signal in Figure 11.7c, now attenuated 80 dB relative to full scale, is at the noise level of the converter and is lost. The low-frequency sinusoid in Figure 11.7c is now attenuated 40 dB relative to full scale, and therefore exhibits a SNR that is 40 dB lower than the same signal in Figure 11.7a.

Given the task of minimizing the average quantizing noise-to-signal ratio in a quantizer, we are faced with a conflict of requirements. On one hand, we wish to keep the signals large with respect to the quantizing level, q, in order to achieve

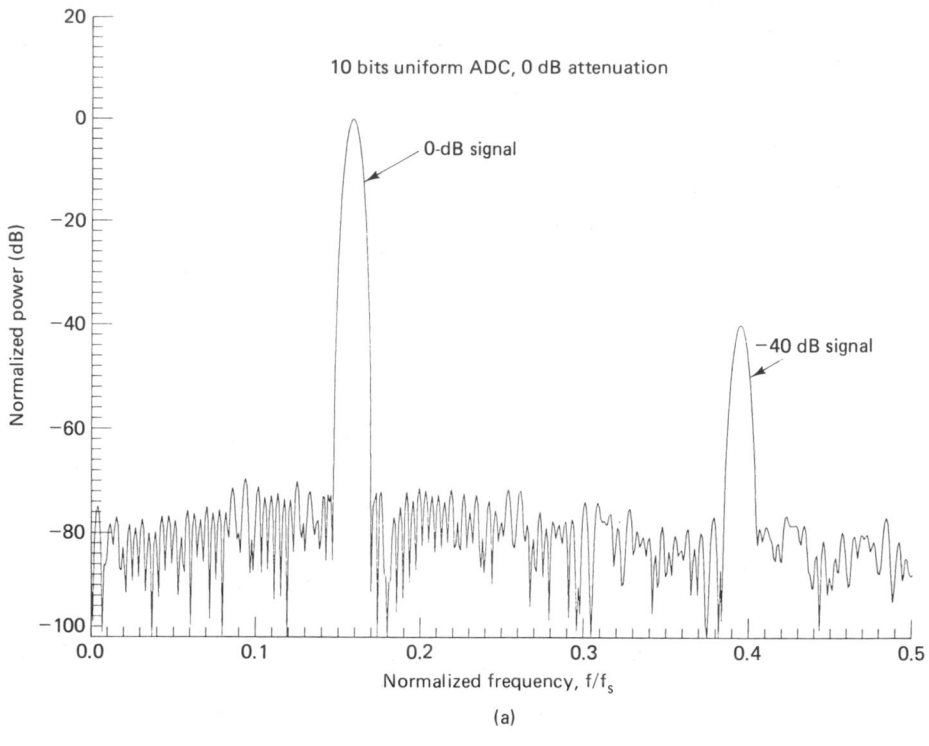

(a)

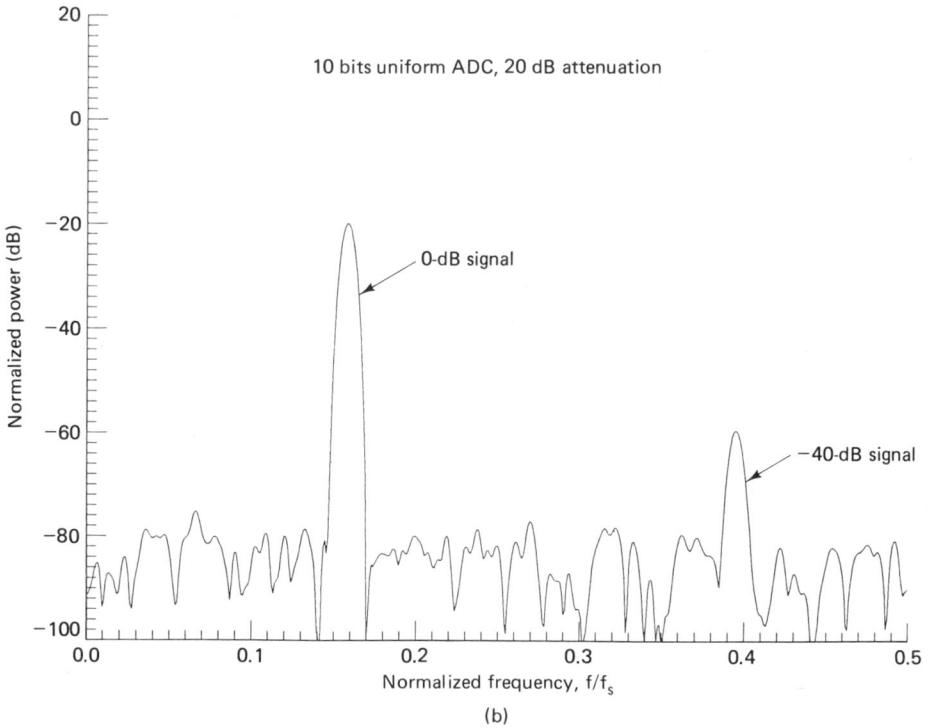

(b)

Figure 11.7 Power spectrum of uniformly quantized signals.

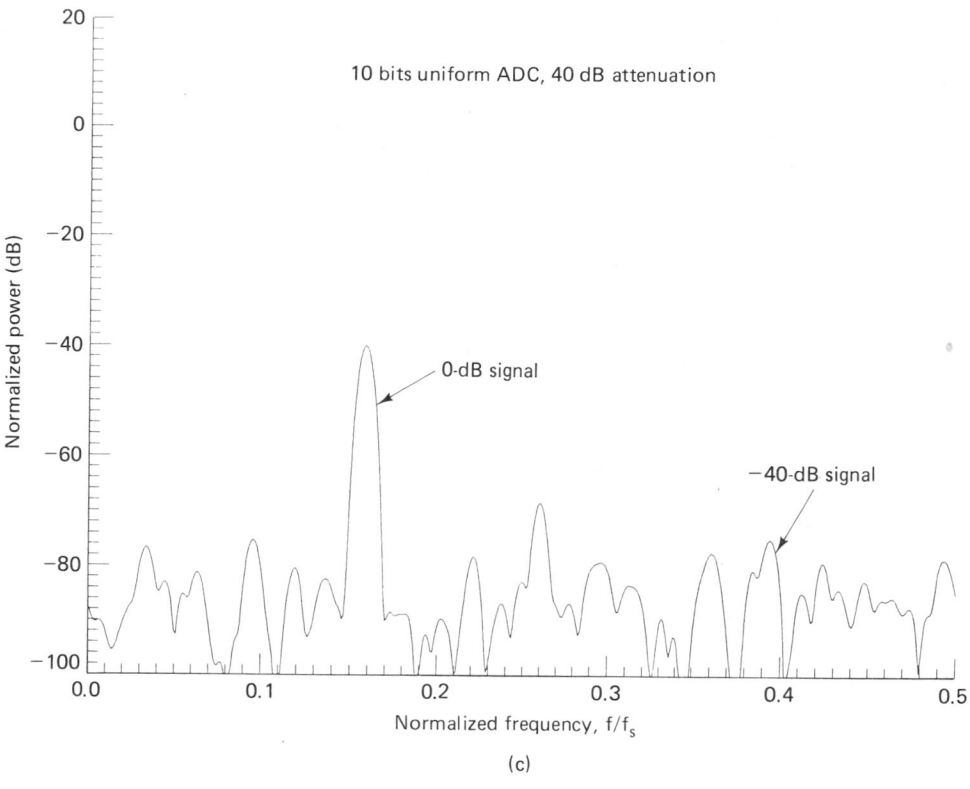

Figure 11.7 (*Continued*)

a high average signal-to-noise ratio. We also find it necessary to keep the signal small to avoid saturating the quantizer. We resolve the opposing requirements by scaling the input signal so that its rms value is a specified fraction of the full-scale quantizer range. The specified fraction is chosen to balance the saturation errors (weighted by their probability of occurrence) against the quantizing errors (which are similarly weighted) and thus achieve a minimum noise-to-signal ratio.

11.2.3 Saturation

Figure 11.8 presents the average NSR of a uniform quantizer as a function of the ratio of quantizer saturation level to rms value of input signal. The figure dramatically demonstrates that saturation noise is more severe than is quantizing noise. This can be simply explained by examining the instantaneous error characteristic, as shown in Figure 11.4, and noting that saturation errors are very large relative to the quantizing errors. Thus a small amount of saturation, even if it occurs infrequently, will make a large contribution to the average noise levels of the quantizer.

Saturation noise and quantization noise differ in another important way. *Quantization noise* tends to be white noise. Dither signals may be intentionally

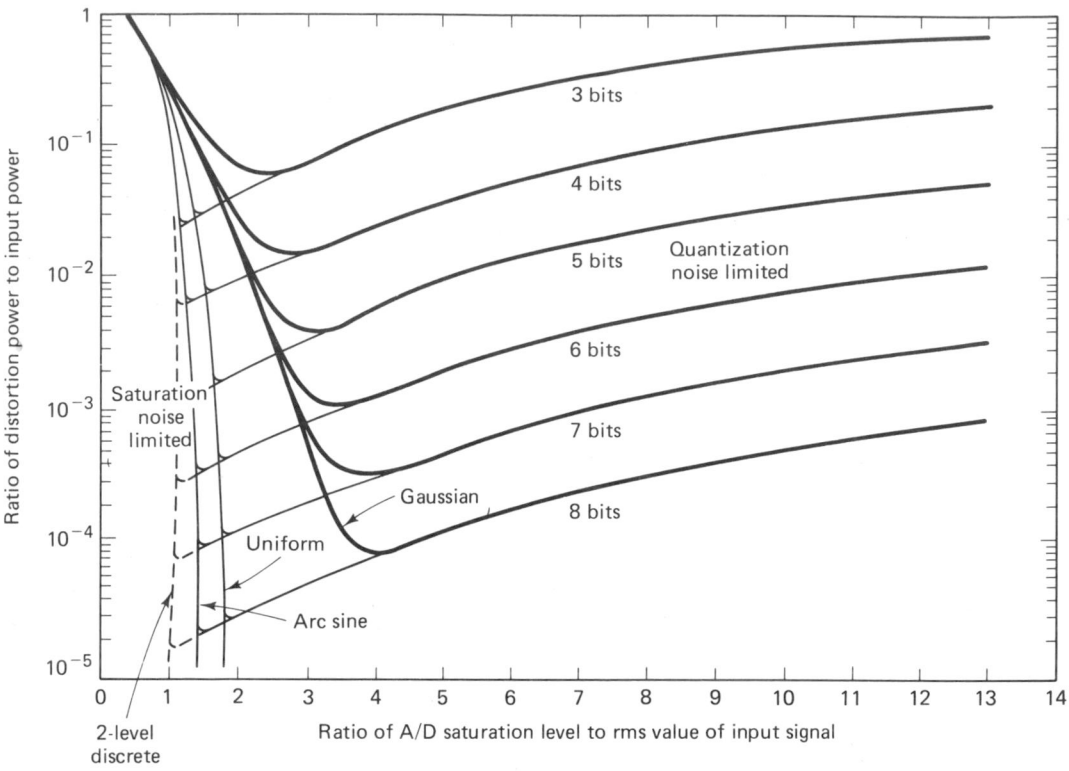

Figure 11.8 Quantizer NSR parameterized on number of bits in ADC for signals with different input probability densities.

added to the analog signal prior to the quantizer to assure this property. *Saturation noise*, on the other hand, tends to be white only when the input signal has a broad bandwidth and tends to be harmonically related to the input signal if it has a narrow bandwidth. Thus the effects of quantizing noise can be filtered or averaged because it truly has the characteristics of noise. Saturation noise, on the other hand, is indistinguishable from signal content and generally cannot be reduced by subsequent averaging or filtering techniques.

Figure 11.9 presents the discrete Fourier transforms of the same signal set presented in Section 11.2.2—two sinusoids of relative amplitude 1.0 and 0.01, quantized with a 10-bit ADC. In Figure 11.9a, b, and c, the peak signal amplitudes were adjusted to 5%, 10%, and 20% (0.42, 0.83, and 1.58 dB, respectively) above the ADC saturation level. Note the very many spectral artifacts caused by the saturation. The saturation noise grows larger as the signal excursion increases into saturation. Also note that some of these artifacts are only down 20 to 40 dB relative to full-scale signal. Compare these figures to Figure 11.7 to see the dramatic difference that too little signal attenuation, hence saturation, makes in the noise output of an ADC.

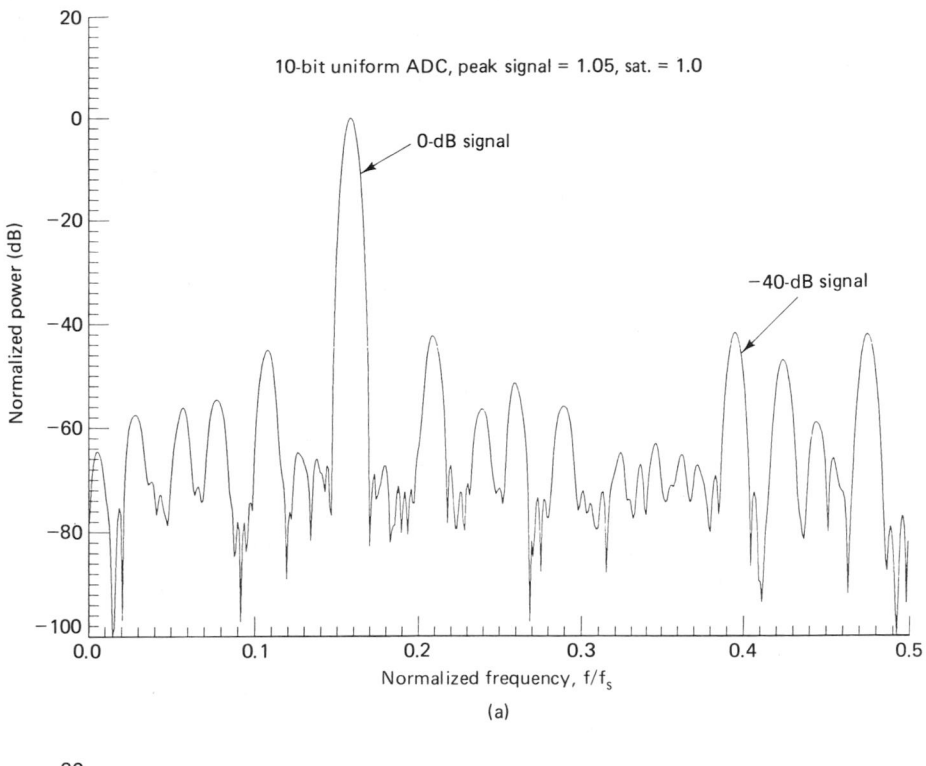

(a)

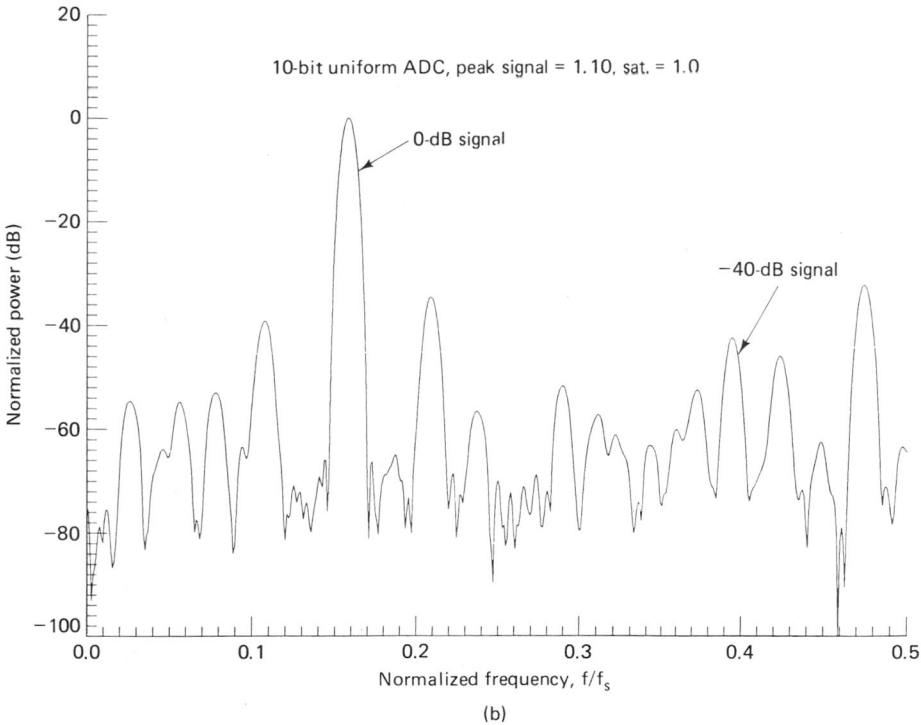

(b)

Figure 11.9 Power spectrum of uniformly quantized signals, with the quantizer saturating on the signal peaks.

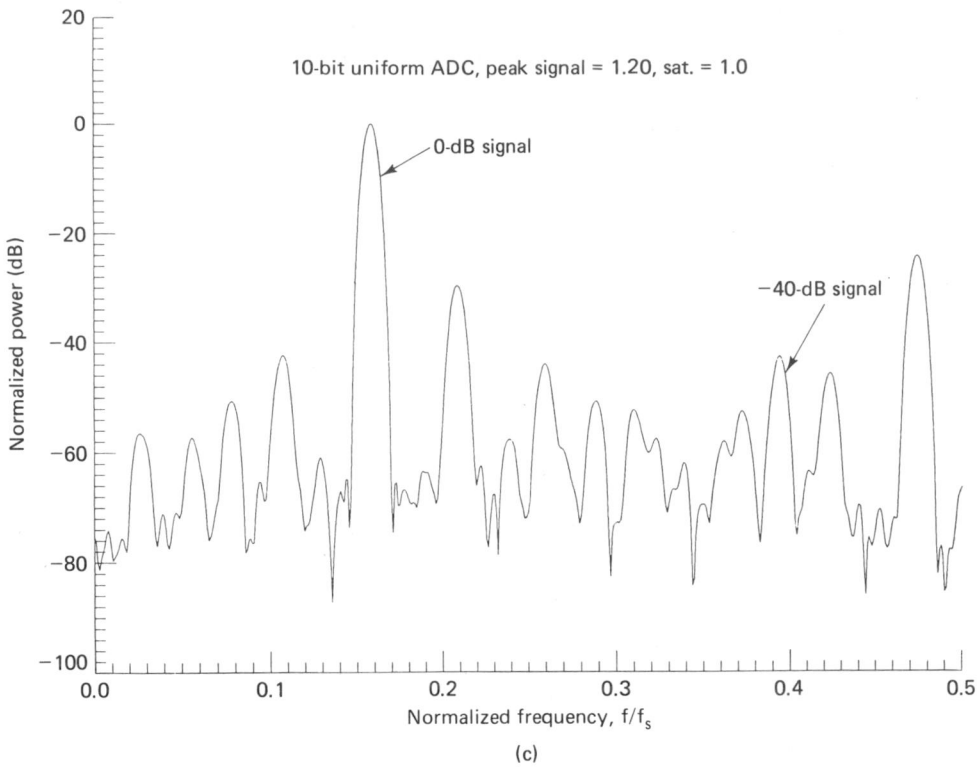

Figure 11.9 (*Continued*)

11.2.4 Dithering

Dithering is one of the most clever applications of noise as a useful engineering tool. To *dither*, according to *Webster's Ninth New Collegiate Dictionary*, is to "act indecisively, to vacillate". A *dither signal* is a small perturbation or disturbance added to a measurement process to reduce the effect of small local nonlinearities. The most familiar form of dither is the slight tapping we apply to the side of a d'Arsonval meter movement prior to taking the reading (before the days of digital meters). The tapping is a sequence of little impulses for displacing the needle movement beyond the local region which exhibits a nonlinear coefficient of friction at low velocities. A more sophisticated example of this same effect is the mechanical dither applied to the counterrotating laser beams of a laser beam gyro to break up low-level frequency entrapment known as deadband [3].

In the analog-to-digital converter application, the effect of the dither is to reduce or eliminate the local discontinuities (i.e., the risers and flats) of the instantaneous input–output transfer function. We can best visualize the effect of these discontinuities by listing the desired properties of the error sequence formed by the quantizer process and then examining the actual properties of the same sequence. The quantizer error sequence is modeled as additive noise. The desired properties of such a noise sequence, $e(n)$, are

1. *Zero mean:* $\qquad\qquad\qquad E[e(n)] = 0$
2. *White:* $\qquad\qquad\qquad\quad E[e(n)e(n + m)] = \sigma^2\delta(m)$
3. *Uncorrelated with data X(n):* $\quad E[e(n)X(n + m)] = 0$

where n and m are sample indices, and $\delta(m)$ is a Dirac delta function. In Figure 11.10, we examine a sequence of samples formed by a truncating ADC and make the following observations:

1. The error sequence is all of the same polarity; therefore, it is not zero mean.
2. The error sequence is not independent, sample to sample; therefore, it is not white.
3. The error sequence is correlated with the input; therefore, it is not independent.

Repeated measurements of the same signal would result in the same noise, and thus no amount of averaging could reduce the deviation from the true input signal. Paradoxically, we would like this noise to be "noisier." If the noise were independent on successive measurements, averaging would reduce the deviation from the true values. Thus, faced with the problem that the noise we get is not

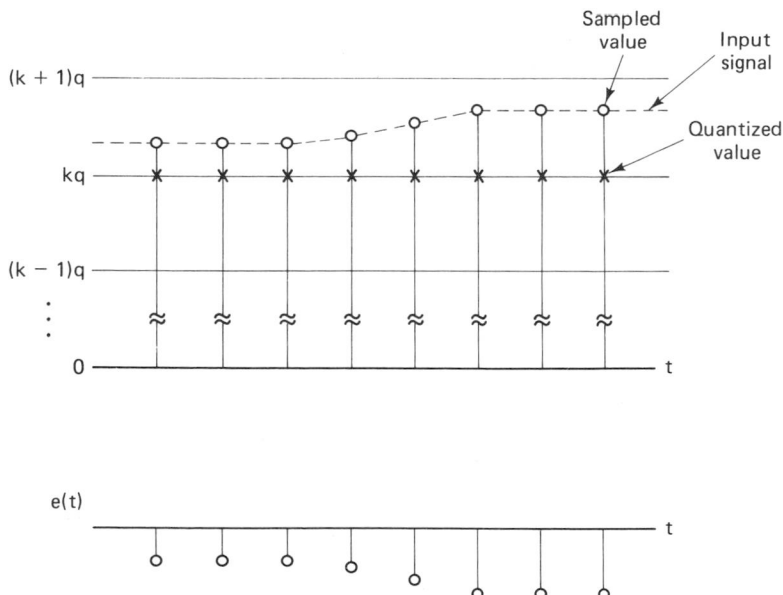

Figure 11.10 Sampled data sequence quantized to next-lowest quantile level and associated error sequence.

the noise we want, we choose to alter that noise by adding our own. We add a perturbation to the measurement to override the undesired low-level structure of the quantizer noise. The added perturbation, in a sense, converts *bad noise* to *good noise* [4].

Example 11.5 Dither Linearization

We hypothesize a quantizer that can only measure integers and converts input data into the next lowest integer (a process called truncation). We make 10 measurements of a signal, say of amplitude 3.7. In the absence of a dither we have readings all equal to 3.0. Now add a uniformly distributed (over 0 to 1) random number sequence to the input prior to performing the reading. The sequence of data has the form

Reading	Raw	Quantized	Dithered	Quantized
1	3.7	3.0	4.0485	4.0
2	3.7	3.0	4.5685	4.0
3	3.7	3.0	3.9789	3.0
4	3.7	3.0	4.0615	4.0
5	3.7	3.0	3.8074	3.0
6	3.7	3.0	3.9629	3.0
7	3.7	3.0	4.6252	4.0
8	3.7	3.0	4.2599	4.0
9	3.7	3.0	4.0408	4.0
10	3.7	3.0	4.2228	4.0
Averages =		3.0	4.1576	3.7
Dither mean =			0.5000	
(Average − Dither mean) =			3.6576	

Note that the average value of the dithered signal is the sum of the signal plus the mean value (0.5) of the dither. Here we have used a biased dither to remove the quantizer bias. The average of the dithered and quantized readings is (for this example) a correct reading and, in general, will be closer to the true signal than will the nondithered and quantized measurements [5, 6].

To help us understand the effect that dithering has on the quantization process, consider the following experiment. Let us apply 60 dB of attenuation to a sinusoidal signal of amplitude 1.0. The attenuated signal, then, has a full-scale amplitude of 0.001, which is approximately one half of the quantization interval 0.001953 of a ten-bit uniform quantizer. The sampled output of a rounding quantizer, with this signal as input, will be essentially all zeros except for an occasional count of ± 1, which occurs when the input crosses the $\pm q/2$ level of 0.000976 (corresponding to the least significant bit of the ADC). If the input signal were attenuated another 0.23 dB, the threshold levels of the least significant bit would never be crossed and the output sequence would be all zeros. Now, let us add a *dither signal* of amplitude 0.001 rms to the attenuated signal of 0.001 so that the signal plus dither regularly crosses the $\pm q/2$ levels of the ADC. Figure 11.11 shows the power spectra obtained by transforming and averaging 32 realizations of this dithered signal. Lo and behold, the 60-dB attenuated signal, at the edge of the ADC's resolvability, is indeed present and has been accurately measured.

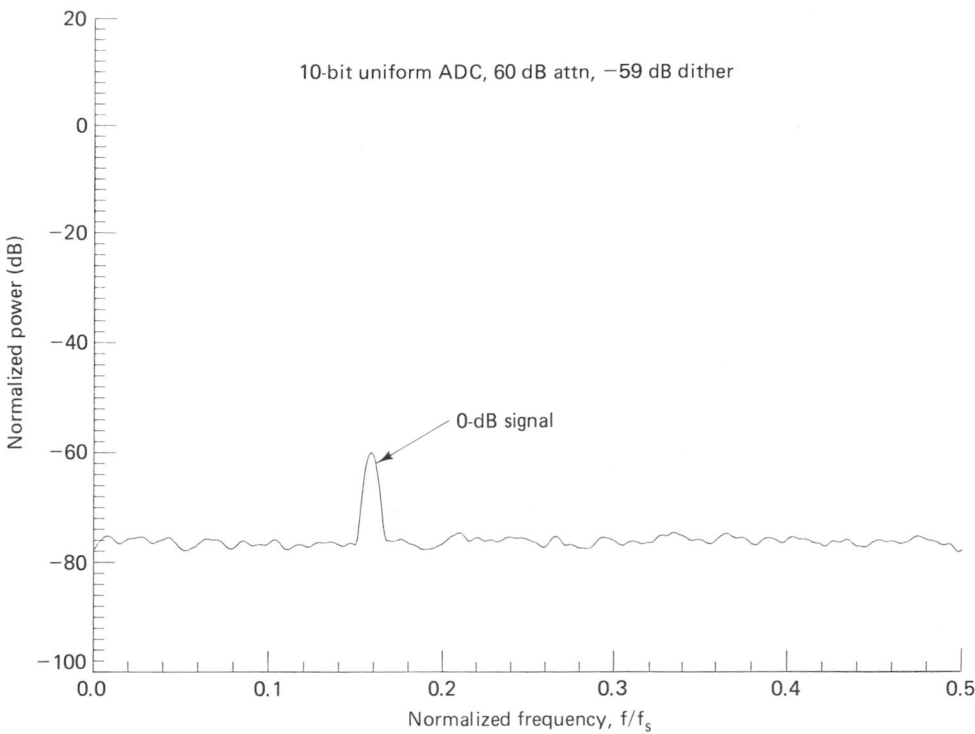

Figure 11.11 Power spectrum of uniformly quantized dithered signal.

The dither signal has had the effect of extending the dynamic range of the ADC (typically by 9 dB or 1.5 bits) and has improved the effective linearity of the low-level ADC staircase approximation.

11.2.5 Nonuniform Quantizing

Uniform quantizers are the common type of analog-to-digital converters because they are the most *robust*. By "robust" we mean that they are relatively insensitive to small changes in the input statistics. They achieve this robustness by not being finely tuned to one specific set of input parameters. This allows them to perform well even in the face of uncertain input parameters, and it means that small changes in input statistics will result in only small changes in output statistics.

When there is small uncertainty in the input signal statistics, it is possible to design a nonuniform quantizer which exhibits a smaller quantizer NSR than a uniform quantizer using the same number of bits. This is accomplished by partitioning the input dynamic range into nonuniform intervals such that the noise power, weighted by the probability of occurrence in each interval, is the same. Iterative solutions for the decision boundaries and step sizes for an optimal quantizer can be found for specific density functions and for a small number of bits. This task is simplified by modeling the nonuniform quantizer as a sequence of operators, as depicted in Figure 11.12. The input levels are first mapped, via a nonlinear function called a compressor, to an alternative range of levels. These levels are uniformly quantized and the quantized signal levels are then mapped, via a complementary nonlinear function called an expander, to the output range of levels. Borrowing part of the name from each of the operations COMpress and exPAND, we form the acronym by which this process is commonly identified, *companding*.

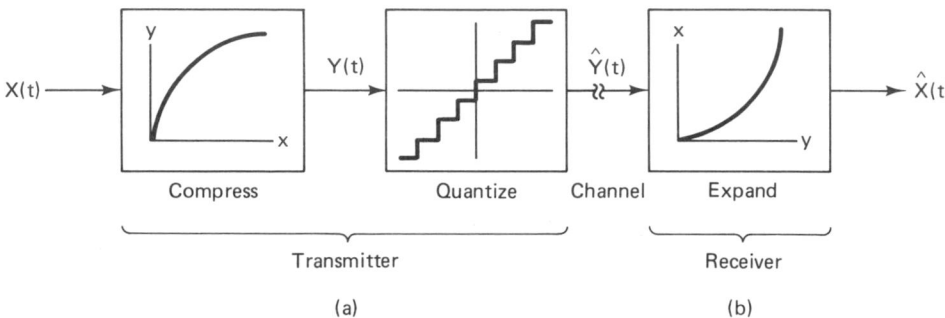

Figure 11.12 Nonuniform quantizer as a sequence of compression, uniform quantization, and expansion.

11.2.5.1 (Near) Optimal Nonuniform Quantizing

Examining the compressor characteristics, $\hat{y} = C(x)$, of Figure 11.13, we note that the quantizing step sizes for the output variable, y, are related to the quantizing step sizes for the input variable, x, through the slope $\dot{C}(x)$ [i.e., $\Delta y = \Delta x \dot{C}(x)$]. Under reasonable conditions, such as a large number of quantizing levels and a smooth pdf for the input variable, we can arrive at the output quantizing noise variance [7]

$$\sigma_q^2 = \frac{q^2}{12} \int_{-x_{max}}^{+x_{max}} \frac{p(x)}{|\dot{C}(x)|^2} \, dx \tag{11.25}$$

For a specific pdf, the compression characteristic, $C(x)$, can be found which minimizes σ_q^2. The optimal compressor law for a given pdf is [8]

$$C(x) = \int_0^x \sqrt[3]{Kp(z)} \, dz \tag{11.26}$$

We find that the optimal compressor characteristic is proportional to the integral of the cube root of the input probability density function. This is called *fine tuning*. If the compressor is designed to operate with one density function and it is used with some other density function (including scaled versions), the quantizer is said to be *mismatched* and there may be severe performance degradation due to the mismatch [6].

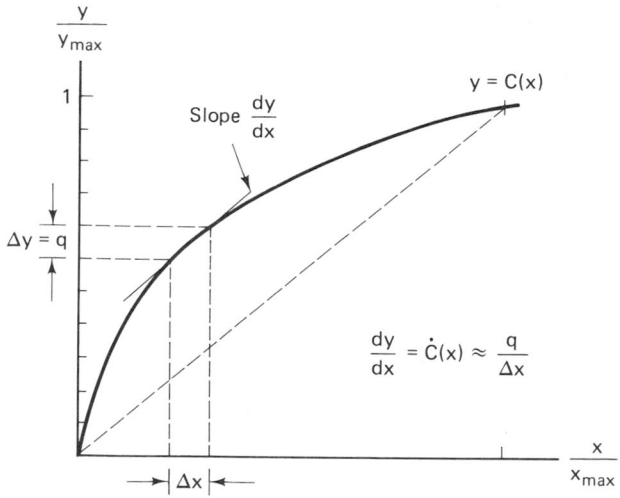

Figure 11.13 Compressor characteristics $y = C(x)$ and estimates to local slope $\dot{C}(x)$.

11.2.5.2 Logarithmic Compression

In the preceding section we presented the compression law for the case in which the input probability density function is well defined. We now address the case for which little is known about the density function. This case occurs, for instance, when the average power of the input signal is a *random variable*. As an example, the voice level of a randomly chosen telephone user may vary from one extreme of a barely audible whisper to the other extreme of a bellowing shout.

For the case of an *unknown density function*, the compressor characteristics of the nonuniform quantizer must be selected such that the resultant noise performance is independent of the specific density function. Although this is a worthy undertaking, it may not be possible to achieve this independence. We are willing to compromise, however, and we will settle for virtual independence over a large range of input variance and input density functions. An example of a quantizer that exhibits a SNR independent of the input density can be visualized with the aid of Figure 2.18. There we saw the very large difference in NSR ratio for different amplitude input signals when quantized with a uniform quantizer. By comparison, we saw that the nonuniform quantizer permits large errors for large signals. This makes intuitive sense. If the SNR is to be independent of the amplitude distribution, the quantizing noise must be proportional to the input level.

Equation (11.25) presented the quantizer noise variance for an arbitrary density function and compressor characteristics. The signal variance for any density function is

$$\sigma_X^2 = \int_{-\infty}^{\infty} x^2 p(x)\, dx \tag{11.27}$$

In the absence of saturation, the quantizer signal-to-noise ratio is of the form

$$\frac{\sigma_x^2}{\sigma_q^2} = \frac{\int_{-x_{max}}^{+x_{max}} x^2 p(x) \, d(x)}{(q^2/12) \int_{-x_{max}}^{+x_{max}} [p(x)/|\dot{C}(x)|^2] \, dx} \tag{11.28}$$

To have the SNR be independent of the specific density function, we require that the numerator be a scaled version of the denominator. This happens if the following is true:

$$|\dot{C}(x)|^2 = \left|\frac{K}{x}\right|^2 \tag{11.29}$$

or

$$\dot{C}(x) = \frac{K}{x} \tag{11.30}$$

from which we obtain by integration,

$$C(x) = \int_0^x \frac{K}{z} \, dz \tag{11.31}$$

or

$$C(x) = \log_e x + \text{constant} \tag{11.32}$$

This result is intuitively appealing. A *logarithmic compressor* allows large signals to have larger errors because of the log scale—equal distances (or errors) are, in fact, equal ratios, which is what we were seeking, a *constant signal-to-noise ratio*. Here the constant is present to match the boundary conditions between x_{max} and y_{max}. Accounting for this boundary condition, we have the logarithmic converter of the form

$$\frac{y}{y_{max}} = \frac{C(x)}{y_{max}} = \log_e \frac{x}{x_{max}} \tag{11.33}$$

The form of the compression suggested by the logarithm function is shown in Figure 11.14a. The first problem with this function is that it does not map the negative input signals. We account for the negative signals by adding a reflected version of the log to the negative axis. This modification results in Figure 11.14b and is of the form

$$\frac{y}{y_{max}} = \frac{C(x)}{y_{max}} = \log_e \frac{|x|}{x_{max}} \, \text{sgn} \, x \tag{11.34}$$

where

$$\text{sgn} \, x = \begin{cases} +1 & \text{for } x > 0 \\ -1 & \text{for } x < 0 \end{cases}$$

The remaining problem we face is that the resultant compression is not continuous through the origin; in fact, it completely misses the origin. We need

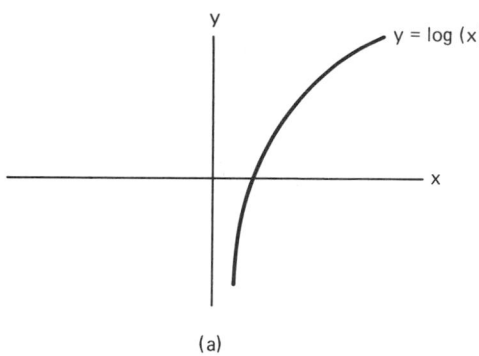

(a)

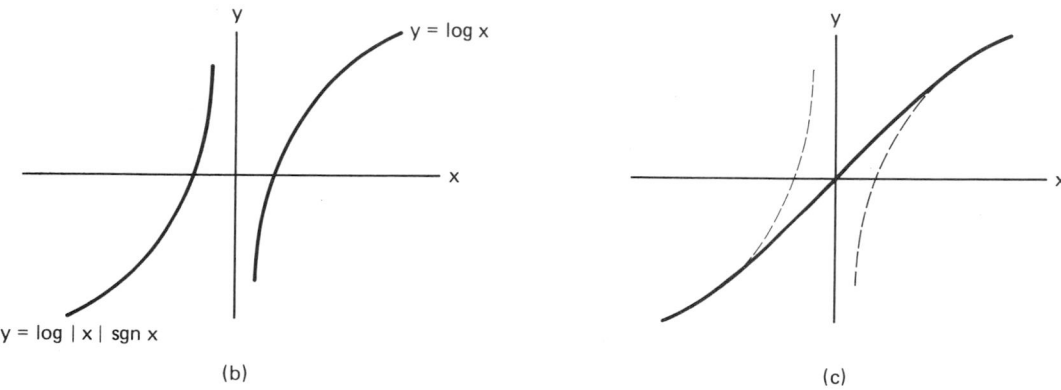

$y = \log x$

$y = \log |x| \operatorname{sgn} x$

(b)   (c)

Figure 11.14 (a) Log function prototype for compression law. (b) Log $|x|$ sgn x function prototype for compression law. (c) Log $|x|$ sgn x function with a smooth transition between segments.

to make a smooth transition between the logarithmic function and a linear segment passing through the origin. There are two standard compression functions that perform this transition: the μ-law and A-law companders.

μ-Law Compander. The μ-law compander is the Bell System (hence the North American standard) compression law. It is of the form

$$y = C(x) = y_{max} \frac{\log_e [1 + \mu(|x|/x_{max})]}{\log_e (1 + \mu)} \operatorname{sgn} x \qquad (11.35)$$

The approximate behavior of this compressor in the regions corresponding to small and large values of the argument are

$$y = C(x) = \begin{cases} y_{max} \dfrac{\mu(|x|/x_{max})}{\log_e \mu} & \mu\left(\dfrac{|x|}{x_{max}}\right) \ll 1 \\[3ex] y_{max} \dfrac{\log_e [\mu(|x|/x_{max})]}{\log_e \mu} & \mu\left(\dfrac{|x|}{x_{max}}\right) \gg 1 \end{cases} \qquad (11.36)$$

The parameter μ in the μ-law compander had originally been set to 100 for use with a 7-bit converter. It was later changed to 255 for use with an 8-bit converter. The 8-bit $\mu = 255$ μ-law converter has become the standard North American conversion law.

Example 11.6 Average SNR for μ-Law Compressor

The SNR for the μ-law compressor can be estimated by substituting the μ-law expression into Equation (11.28). For positive values of the input variable x, the compression law is

$$y = C(x) = y_{max} \frac{\log_e [1 + \mu(|x| /x_{max})]}{\log_e (1 + \mu)} \tag{11.37}$$

Then the derivative $\dot{C}(x)$ is

$$\dot{y} = \dot{C}(x) = y_{max} \frac{1}{\log_e (1 + \mu)} \frac{\mu(1/x_{max})}{1 + \mu(|x| /x_{max})} \tag{11.38}$$

For values of the input variable for which $\mu(x/x_{max})$ is large compared to unity, the derivative becomes

$$\dot{y} = \dot{C}(x) = \frac{1}{x} \frac{y_{max}}{\log_e \mu} \tag{11.39}$$

Substituting for $1/\dot{C}(x)$ in Equation (11.28), we find

$$\text{SNR} = \frac{\sigma_s^2}{\sigma_q^2} = \frac{1}{[(q^2/12)(\log_e \mu/y_{max})]^2} \tag{11.40}$$

$$= 3 \left(\frac{2y_{max}}{q} \right)^2 \left(\frac{1}{\log_e \mu} \right)^2 \tag{11.41}$$

The ratio $2y_{max}/q$ is the number of quantizing levels (2^b) of the compressed quantizer. For the 8-bit converter with $\mu = 255$, the SNR is found to be

$$\text{SNR} = 3 \left[\frac{2^8}{\log_e(255)} \right]^2 = 3(46.166)^2$$

$$= 38.1 \text{ dB} \tag{11.42}$$

For a comparison, the SNR of an actual μ-law quantizer is presented in Figure 11.15. There the SNR is plotted for input sinusoids of different amplitudes. The serration of the performance curve is due to the piecewise linear approximation to the continuous μ-law curve. This is described shortly.

Figure 11.16 presents the discrete Fourier transform of the pair of input sinusoids of relative amplitude 1.0 and 0.01. Here the input signal is quantized with a 10 bit μ-law ($\mu = 40$) converter and the signal levels are attenuated by 0, 20, and 40 dB, respectively, relative to full-scale input. Note that the quantizing noise levels for the full-scale signal are equivalent to that of the uniform quantizer (ADC) (see Figure 11.7). The difference in performance between the log compressed ADC and the uniform ADC is to be seen for the attenuated signals. We see that as the input signal levels are attenuated, the quantizing noise is also reduced. Thus the log compressed ADC has no problem "seeing" the low-level input signal even with 40-dB

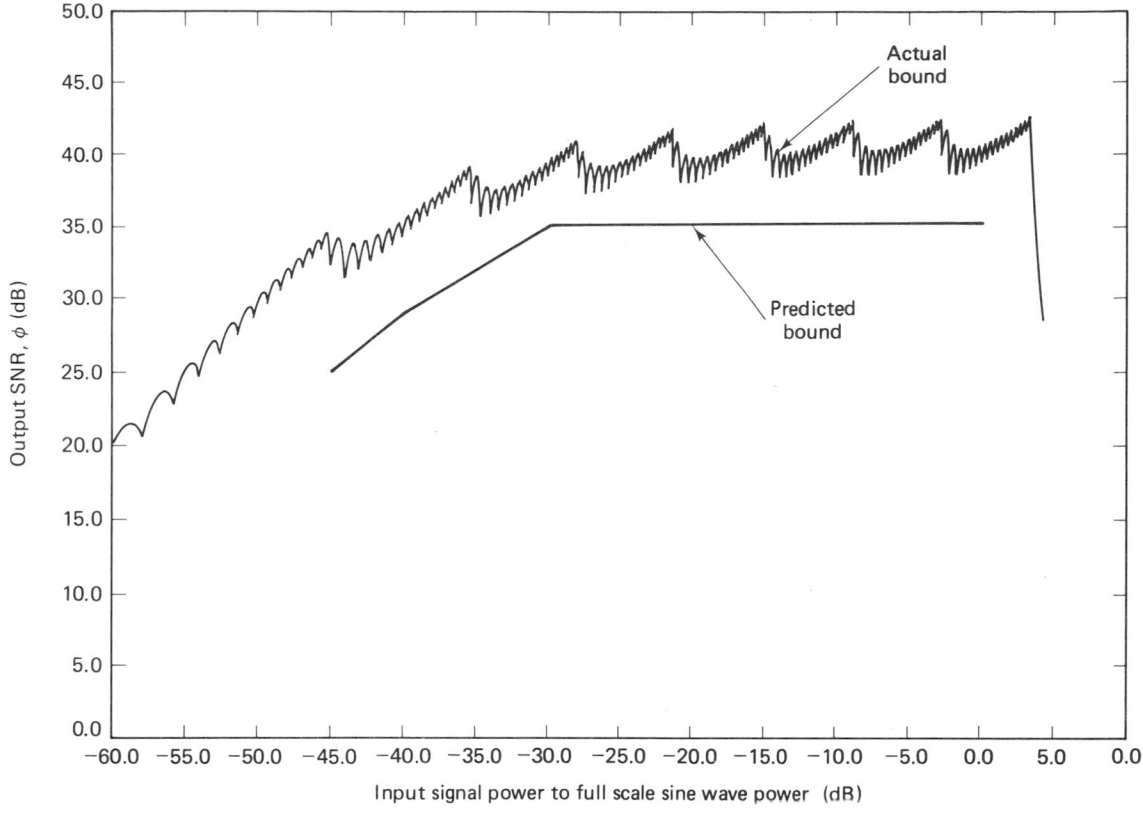

Figure 11.15 Predicted and measured SNR for a μ-law quantizer.

attenuation in Figure 11.16a, while the same signal is lost in the noise of the uniform converter (see Figure 11.7c).

The μ-law compressor realization differs from the expression presented in Equation (11.35) in a minor way. The functional expression is approximated by 16 linear chord segments over the possible 256 output levels, as shown in Figure 11.17. Eight of these segments are in the first quadrant, eight are in the third quadrant, and the "0" segment has the same slope in both quadrants. Over each chord segment the quantization is uniform in the four lower-order conversion bits. Thus the 8-bit compressed conversion format is of the form

$$\underbrace{b_7} \quad \underbrace{b_6 \; b_5 \; b_4} \quad \underbrace{b_3 \; b_2 \; b_1 \; b_0}$$

sign bit   segment   position on
(quadrant)   segment

It is the piecewise chord approximation to the smooth function and a staircase

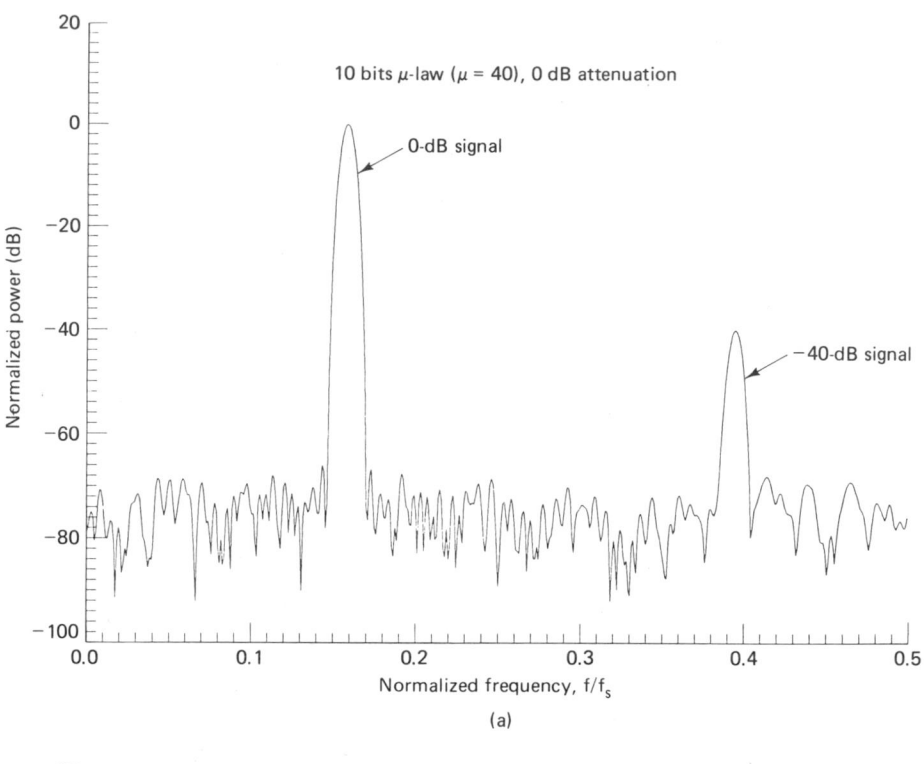

(a)

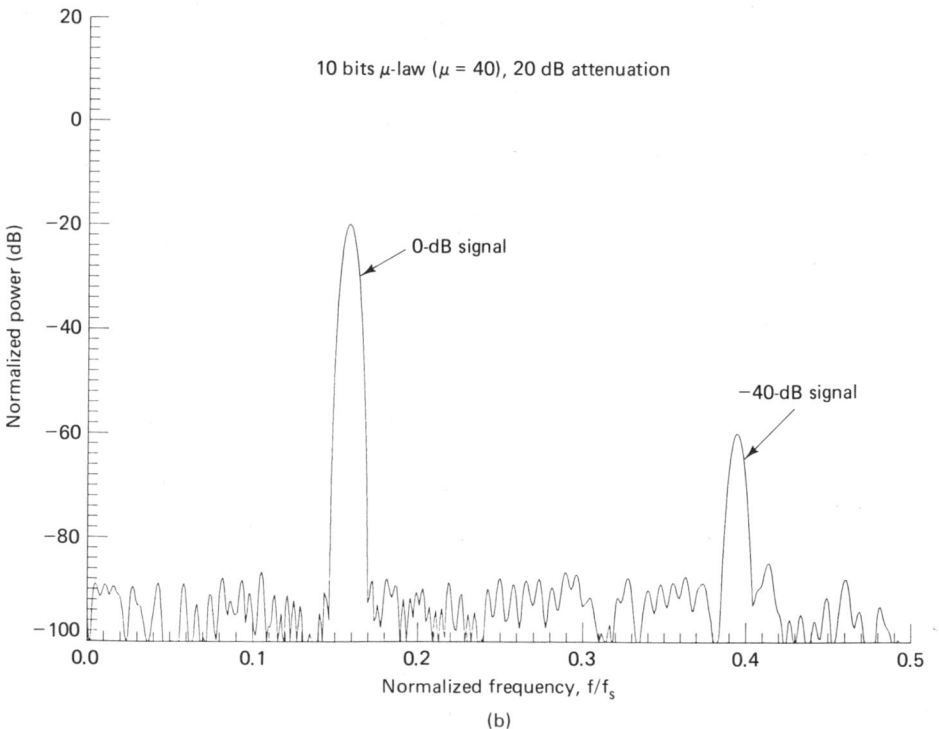

(b)

Figure 11.16 Power spectrum of μ-law quantized signals.

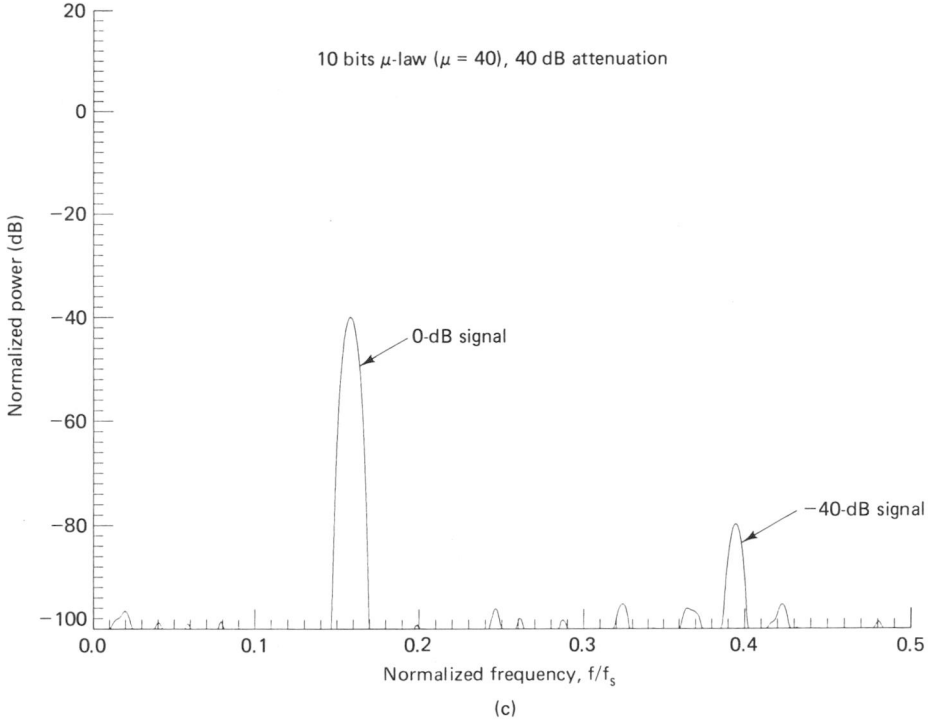

10 bits μ-law (μ = 40), 40 dB attenuation

0-dB signal

−40-dB signal

Normalized frequency, f/f_s

(c)

Figure 11.16 (*Continued*)

approximation of each chord that accounts for the cusps in the SNR shown in Figure 11.15.

A-Law Compander. The *A*-law compander is the CCITT (hence the European) standard approximation to the logarithmic compression. The form of the compressor is

$$
y = C(x) = \begin{cases} y_{max} \dfrac{A(\,|x|\,/x_{max})}{1 + \log_e A} \, \text{sgn } x & 0 < \dfrac{|x|}{x_{max}} < \dfrac{1}{A} \\[3mm] y_{max} \dfrac{1 + \log_e[A(\,|x|\,/x_{max})]}{1 + \log A} \, \text{sgn } (x) & \dfrac{1}{A} < \dfrac{|x|}{x_{max}} < 1 \end{cases}
\tag{11.43}
$$

The standard value of the parameter *A* is 87.56, and for this value, using an 8-bit conversion, the average SNR is 38.0 dB. The *A*-law compression characteristic is approximated, in a manner similar to the μ-law compressor, by a sequence of 16 linear chords spanning the output range. The lower two chords in each quadrant are in fact a single chord corresponding to the linear segment of the *A*-law compressor. One important difference between the *A*-law and the μ-law compression characteristics is that the *A*-law standard has a midriser at the origin, while the μ-law standard has a midtread at the origin. Thus the *A*-law compressor has no

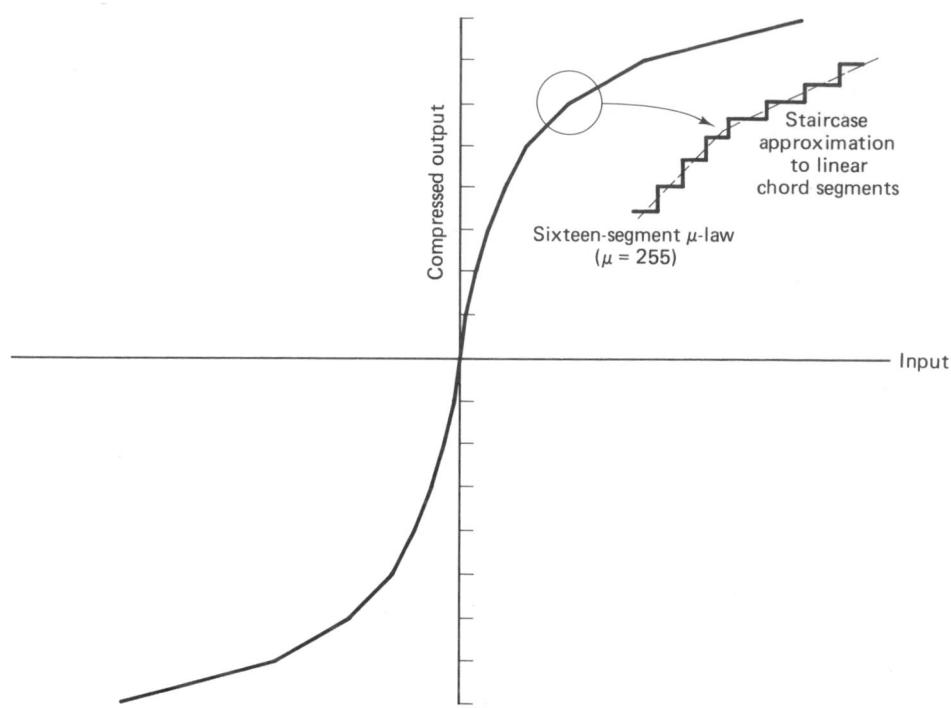

Figure 11.17 Seven-bit compressed quantization with 16-segment approximation to μ-law.

zero value and hence exhibits no interval for which data are not being transmitted for zero input.

There are direct mappings from the *A*-law 8-bit compressed conversion format to a 12-bit linear binary code and from the μ-law 8-bit compressed format to a 13-bit linear code [8]. This operation permits the A/D conversion to be performed with a uniform quantizer and then be mapped to the smaller number of bits in a code converter. This also permits the inverse mapping at the receiver (i.e., the expansion) to be performed on the digital sample.

Pulse Code Modulation. One of the tasks performed by a pulse code modulation (PCM) process is the conversion of a waveform source to a binary sequence discrete source. This task is performed in a three-step process—sampling, quantizing, and encoding—as indicated in Figure 2.2. We have addressed the sampling process in Chapter 2, and have addressed the quantizing process in this chapter as well as in Chapter 2. We note that the encoding process, which was shown in Figure 2.2 to follow quantization, is often embedded in the hardware that performs the quantization. It can be described as follows: successive approximation (SA) analog-to-digital (A/D) converters form the successive bits of the encoded data by a feedback, comparison, and decision process. In the feedback process, a binary search is conducted over the range of possible input levels by repeatedly asking: Is the input signal above or below the midpoint of the remaining uncertainty interval? By this technique the uncertainty interval is reduced by one-half for each comparison and decision step until the uncertainty range matches the allowable quantizing interval.

In the SA conversion, the results of each previous decision reduce the uncertainty to be resolved during the next decision. In a similar manner, the results of the previous A/D conversions can be used to reduce the uncertainty to be resolved during the next conversion. This reduction in uncertainty is achieved by carrying forward to the next sample auxiliary information from earlier samples. This information is called the redundant part of the signal, and by carrying it forward, we reduce the interval of uncertainty over which the quantizer and encoder must search for the next signal sample. Carrying data forward is one method of achieving *redundancy reduction*.

11.3 DIFFERENTIAL PULSE CODE MODULATION

By the use of past data to assist in measuring (i.e., quantizing) new data, we leave ordinary PCM and enter the realm of differential PCM (DPCM). In DPCM, a prediction of the next sample value is formed from past values. This prediction can be thought of as instructions for the quantizer to conduct its search for the next sample value in a particular interval. By using the redundancy in the signal to form a prediction, the region of uncertainty is reduced and the quantization can be performed with a reduced number of decisions (or bits) for a given quantization level or with reduced quantization levels for a given number of decisions (or bits). The reduction in redundancy is realized by subtracting the prediction from the next sample value. This difference is called the *prediction error*.

The quantizing methods described in Section 11.2 are called *instantaneous or zero memory quantizers* because the digital conversion is based on the single (current) input sample. In Section 11.1 we identified the properties of sources that permitted source rate reductions. These properties were non-equiprobable source levels and nonindependent sample values. Instantaneous quantizers achieve source coding gains by taking into account the probability density assignment for each sample. The quantizing methods that take account of sample-to-sample correlation are noninstantaneous quantizers. These quantizers reduce source redundancy by first converting the correlated input sequence into a related sequence with reduced correlation, reduced variance, or reduced bandwidth. This new sequence is then quantized with fewer bits.

The correlation characteristics of a source can be visualized in the time domain by samples of its autocorrelation function and in the frequency domain by its power spectrum. If we examine a power spectrum, $G_X(f)$, of a short-term speech signal, as shown in Figure 11.18, we find that the spectrum has a local peak in the neighborhood of 300 to 800 Hz and falls off at a rate of 6 to 12 dB/octave. By interpreting this power spectrum, we can infer certain properties of the time function from which it was derived. We observe that large changes in the signal occur slowly (low frequency) and that rapid changes in the signal (high frequency) must be of low amplitude. An equivalent interpretation can be found

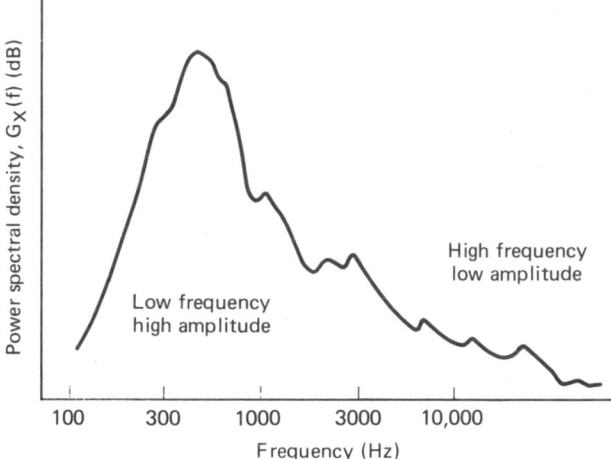

Figure 11.18 Typical power spectrum for speech signals.

in the autocorrelation function, $R_X(\tau)$, of the signal, as shown in Figure 11.19. Here a broad, slowly changing autocorrelation function suggests that there will be only slight change on a sample-to-sample basis, and that a time interval exceeding the correlation distance is required for a full amplitude change. In particular, correlation values for typical single-sample delay is on the order of 0.79 to 0.87 and the correlation distance to the first zero crossing is on the order of 4 to 6 sample intervals of T seconds per interval.

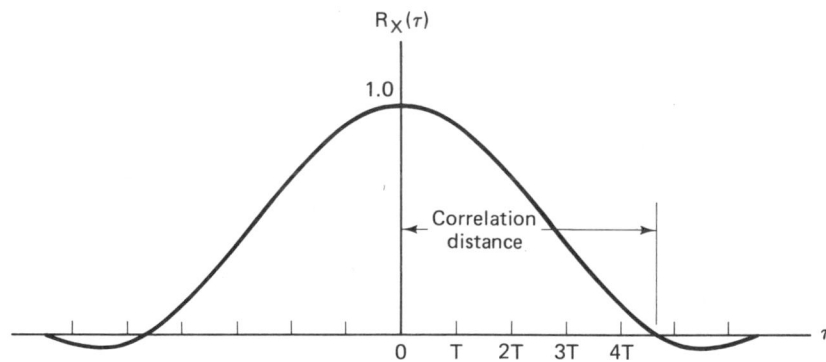

Figure 11.19 Autocorrelation function for typical speech signals.

Since the difference between adjacent time samples for speech is small, sampling techniques have evolved based on transmitting sample-to-sample differences rather than actual sample values. Successive differences are in fact a special case of a class of noninstantaneous converters called one-tap linear predictive coders. These are sometimes called *predictor-corrector coders*. This structure is shown in Figure 11.20. In this type of converter, the transmitter and the receiver have the same prediction model, which is derived from the signal's correlation characteristics. Each use the model to predict the next sample value based

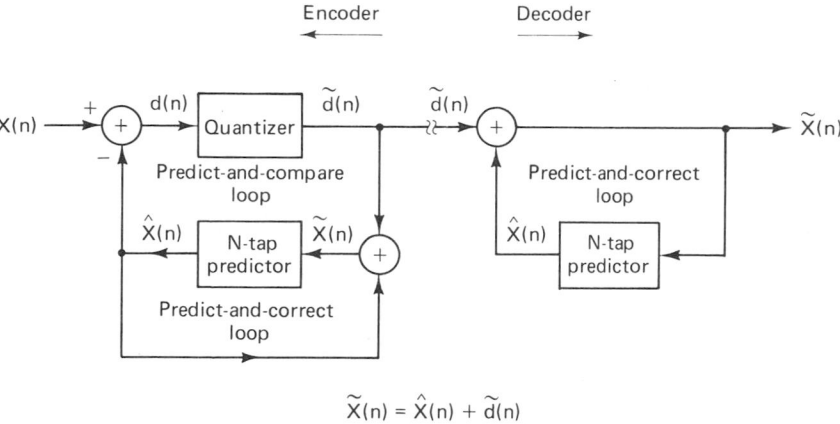

$$\widetilde{X}(n) = \hat{X}(n) + \widetilde{d}(n)$$

Figure 11.20 One-tap predictor differential pulse code modulator (DPCM).

on recent past sample values. The transmitter forms the *prediction error* (or the *residue*) as the difference between the next measured sample value and the predicted sample value. The equation for the prediction loop is

$$d(n) = X(n) - \hat{X}(n)$$

where $X(n)$ is the nth input sample, $\hat{X}(n)$ the predicted value of that sample, and $d(n)$ the associated prediction error. This is performed in the predict and compare loop, the upper loop of the transmitter in Figure 11.20. The transmitter corrects its prediction by forming the sum of its prediction and the prediction error. The equations for the correction loop are

$$\tilde{d}(n) = \text{quant}[d(n)]$$
$$\tilde{X}(n) = \hat{X}(n) + \tilde{d}(n)$$

where quant($\cdot$) represents the quantization operation, and $\tilde{d}(n)$ is the quantized version of the prediction error and $\tilde{X}(n)$ is the corrected (and quantized) version of the input sample. This is performed in the predict and correct loop, the lower loop of the encoder and the only loop of the decoder in Figure 11.20. The decoder must also be informed of the prediction error so that it can use its correction loop to correct its prediction. Thus the communication task is that of transmitting the difference (the error signal) between the predicted and the actual data sample. For this reason, this class of coder is often called a *differential pulse code modulator (DPCM)*. If the prediction model forms predictions which are close to the actual sample values, the residues will exhibit reduced variance (relative to the original signal). From Section 11.2 we know that the number of bits required to move data through the channel with a given fidelity is related to the signal variance. Hence the reduced variance sequence of residues can be moved through the channel with a reduced data rate.

The predictive converters must have a short-term memory that supports the fast-time operations required for the prediction algorithm. In addition, they will often have a long-term memory that supports the slow time, often data-dependent operations, such as automatic gain control and filter coefficient adjustments. Predictors that incorporate the slower, data-dependent, adjustment algorithms are called *adaptive predictors*.

11.3.1 One-Tap Prediction

The one-tap linear prediction coding (LPC) filter in the DPCM process predicts the next input sample value based on the previous input sample value. The prediction equation is of the form

$$X(n \mid n - 1) = aX(n - 1 \mid n - 1) \tag{11.44}$$

where $X(n \mid m)$ is the estimate of X at time n given all the samples collected up through time m, and where a is a parameter used to minimize the prediction error. The prediction error available after the measurement is of the form

$$d(n) = [X(n) - X(n \mid n - 1)] \tag{11.45a}$$

$$= [X(n) - aX(n - 1 \mid n - 1)] \tag{11.45b}$$

The mean-squared error is of the form

$$\mathbf{E}\{d^2(n)\} = \mathbf{E}\{X(n)X(n) - 2aX(n)X(n - 1 \mid n - 1)$$

$$+ a^2 X(n - 1 \mid n - 1)X(n - 1 \mid n - 1)\} \tag{11.46}$$

If $X(n - 1 \mid n - 1)$ is an unbiased estimate of $X(n - 1)$, Equation (11.46) can be written as

$$R_d(0) = R_X(0) - 2aR_X(1) + a^2 R_X(0) \tag{11.47a}$$

$$= R_X(0)[1 + a^2 - 2aC_X(1)] \tag{11.47b}$$

where $R_d(n)$ and $R_X(n)$ are the autocorrelation functions of the prediction error and the input signal, respectively, $R_d(0)$ is the power in the error, $R_X(0)$ the power in the signal, and $C_X(n) = R_X(n)/R_X(0)$ the normalized autocorrelation function. We can select the parameter a to minimize the prediction error power of Equation (11.47) by setting to zero the partial of $R_d(0)$ with respect to a.

$$\frac{\partial R_d(0)}{\partial a} = R_X(0)[2a - 2C_X(1)] \tag{11.48}$$

Setting to zero and solving for a^*, where the * implies the optimal solution, we have

$$a^* = C_X(1) \tag{11.49}$$

Substituting a^* back into Equation (11.47), we have

$$R_d^*(0) = R_X(0)[1 + a^*C_X(1) - 2a^*C_X(1)] \tag{11.50a}$$

$$= R_X(0)[1 - a^*C_X(1)] \tag{11.50b}$$

$$= R_X(0)[1 - C_X^2(1)] \tag{11.50c}$$

We can define the *prediction gain* of the encoder as the ratio of input to output variances, $R_X(0)/R_d(0)$. For a fixed bit rate this gain represents an increase in output SNR, while for a fixed output SNR this gain represents a reduced bit rate description. We note that the prediction gain for the optimal predictor is always greater than one for any value of signal correlation [Equation (11.50b)]. On the other hand, the prediction gain is greater than one for the unity-gain, one-tap predictor, only if the signal correlation exceeds 0.5 [Equation (11.47b)].

Example 11.7   Prediction Gain of a One-Tap LPC Filter

A signal with correlation coefficient $C_X(1)$ equal to 0.8 is to be quantized with a one-tap LPC filter. Determine the prediction gain when the prediction coefficient is (a) optimized with respect to the minimum prediction error, or (b) set to unity.

Solution

(a) From Equation (11.50c),

$$R_d^*(0) = R_X(0)(1 - 0.64) = 0.36 R_X(0) \tag{11.51a}$$

$$\text{Prediction gain} = \frac{1}{0.36} = 2.78 \text{ or } 4.44 \text{ dB} \tag{11.51b}$$

(b) From Equation (11.47b),

$$R_d(0) = 2R_X(0)(1 - 0.8) = 0.40 R_X(0) \tag{11.51c}$$

$$\text{Prediction gain} = \frac{1}{0.40} = 2.50 \text{ or } 3.98 \text{ dB} \tag{11.51d}$$

11.3.2 *N*-Tap Prediction

The N-tap LPC filter predicts the next sample value based on a linear combination of the previous N sample values. We will assume that the quantized estimates used by the prediction filters are unbiased and error free. With this assumption, we can drop the double indices (we used in Section 11.3.1) from the data in the filter but still use them for the predictions. Then the N-tap prediction equation takes the form

$$X(n \mid n - 1) = a_1 X(n - 1) + a_2 X(n - 2) + \cdots + a_N X(n - N) \tag{11.52}$$

The prediction error takes the form

$$d(n) = X(n) - X(n \mid n - 1) \tag{11.53a}$$

$$= X(n) - a_1 X(n - 1) - a_2 X(n - 2) - \cdots - a_N X(n - N) \tag{11.53b}$$

The mean-square prediction error is of the form

$$E\{d(n)d(n)\} = E\{[X(n) - X(n \mid n - 1)]^2\} \tag{11.54}$$

Clearly, the mean-square prediction error is quadratic in the filter coefficients a_i. As we did in Section 11.3.1, we can take the partial of the mean-squared error with respect to each coefficient and set those partials to zero. Formally, taking the partial with respect to the jth coefficient prior to expanding $X(n \mid n - 1)$, we have

$$\frac{\partial R_d(0)}{\partial a_j} = \mathrm{E}\left\{2[X(n) - X(n \mid n - 1)] \frac{\partial X(n \mid n - 1)}{\partial a_j} X(n \mid n - 1)\right\} \quad (11.55\mathrm{a})$$

$$= \mathrm{E}\{2[X(n) - X(n \mid n - 1)] [-X(n - j)]\} \quad (11.55\mathrm{b})$$

$$= 2\mathrm{E}\{[X(n) - a_1 X(n - 1) - a_2 X(n - 2)$$

$$- \cdots - a_N X(n - N)] X(n - j)\} \quad (11.55\mathrm{c})$$

$$= 2[R_X(j) - a_1 R_X(j - 1) - a_2 R_X(j - 2)$$

$$- \cdots - a_N R_X(j - N)] \quad (11.55\mathrm{d})$$

This collection of equations (one for each j) can be arranged in matrix form known as the *normal equations*. This form is

$$
\begin{bmatrix} R_X(1) \\ R_X(2) \\ R_X(3) \\ \vdots \\ R_X(N) \end{bmatrix}
$$

$$
= \begin{bmatrix} R_X(0) & R_X(-1) & R_X(-2) & \cdots & R_X(1 - N) \\ R_X(1) & R_X(0) & R_X(-1) & \cdots & R_X(2 - N) \\ R_X(2) & R_X(1) & R_X(0) & \cdots & R_X(3 - N) \\ & & & & \vdots \\ R_X(N - 1) & R_X(N - 2) & R_X(N - 3) & \cdots & R_X(0) \end{bmatrix}
\begin{bmatrix} a_1 \\ a_2 \\ a_3 \\ \vdots \\ a_N \end{bmatrix}^*
$$

$$(11.56\mathrm{a})$$

The normal equations can be written more compactly as

$$\mathbf{R}_X(1, N) = \mathbf{R}_X \mathbf{a}^* \quad (11.56\mathrm{b})$$

where the $\mathbf{R}_X(1, N)$ is the correlation vector of delays from 1 through N, $\mathbf{R}_X$ is the correlation matrix (assuming a zero-mean process), and $\mathbf{a}^*$ is the optimum filter weight vector.

To gain insight into the solution of the normal equations, we now recast the mean-square-error equation (11.54) in matrix form.

$$R_d(0) = \mathrm{E}\{[X(n) - \mathbf{a}^T \mathbf{X}(n - 1)][X(n) - \mathbf{X}^T(n - 1)\mathbf{a}]\} \quad (11.57\mathrm{a})$$

$$= R_X(0) - \mathbf{R}_X^T(1, N)\mathbf{a} - \mathbf{a}^T \mathbf{R}_X(-1, -N) + \mathbf{a}^T \mathbf{R}_X \mathbf{a} \quad (11.57\mathrm{b})$$

where $\mathbf{R}^T$ is the transpose of $\mathbf{R}$. Substituting the right-hand side of the optimal weight vector solution of Equation (11.56b) into (11.57b), we have

$$R_d(0) = R_X(0) - \mathbf{R}_X^T(1, N)\mathbf{a}^* - \mathbf{a}^T \mathbf{R}_X(-1, -N) + \mathbf{a}^{*T} \mathbf{R}_X(1, N) \quad (11.58\mathrm{a})$$

$$= R_X(0) - \mathbf{R}_X^T(-1, -N)\mathbf{a}^* \quad (11.58\mathrm{b})$$

We can now bring the right-hand side of Equation (11.56) over to the left-hand side, and use Equation (11.58b) to augment the top row of the matrix to obtain the *whitening form* of the optimal predictor. In this form, the only nonzero output of the matrix product occurs at time zero, which is akin to an output impulse.

$$
\begin{bmatrix}
R_X(0) & R_X(-1) & R_X(-2) & R_X(-3) & \cdots & R_X(-N) \\
R_X(1) & R_X(0) & R_X(-1) & R_X(-2) & \cdots & R_X(1-N) \\
R_X(2) & R_X(1) & R_X(0) & R_X(-1) & \cdots & R_X(2-N) \\
R_X(3) & R_X(2) & R_X(1) & R_X(0) & \cdots & R_X(3-N) \\
\vdots & & & & & \\
R_X(N) & R_X(N-1) & R_X(N-2) & R_X(N-3) & \cdots & R_X(0)
\end{bmatrix}
\begin{bmatrix}
1 \\ -a_1 \\ -a_2 \\ -a_3 \\ \vdots \\ -a_N
\end{bmatrix}^*
$$

$$
=
\begin{bmatrix}
R_d(0) \\ 0 \\ 0 \\ 0 \\ \vdots \\ 0
\end{bmatrix}
\tag{11.59}
$$

The top row of Equation (11.59) states that the power in the prediction error is of the form

$$
R_d(0) = R_X(0)[1 - a_1 C_X(1) - a_2 C_X(2) - \cdots - a_N C_X(N)] \tag{11.60}
$$

Compare this form to Equation (11.50b). An interesting property of the optimal N-tap predictor filter is this: The coefficient set that obtains the minimum mean-square prediction error also predicts, with zero error, the next $N - 1$ correlation samples from the previous $N - 1$ correlation samples.

For fixed filter coefficients, the DPCM coder can achieve a prediction gain, relative to linear quantizing, of 6 to 8 dB [9]. This prediction gain is essentially independent of filter length once the length exceeds three or four taps. Additional gain is available if the coder has slow adaptive capabilities. Adaptive coders are introduced in Section 11.3.3 and discussed in some detail in Section 11.3.4.

11.3.3 Delta Modulation

Delta modulation, often called *delta mod*, Δ-mod, or DM, is a particularly simple form of one-tap DPCM coding. Equation (11.50c) demonstrates that the prediction gain for a one-tap predictor can be large if the normalized correlation coefficient, $C_X(1)$, is close to unity. Working toward the goal of high sample-to-sample correlation, the predictive filter is generally operated at a rate that far exceeds the Nyquist rate. For example, the sample rate might be chosen at four times Nyquist. Then for a 3.3-kHz bandwidth with a nominal sample rate of 8 kHz, the high correlation prediction filter would operate at a 32-kHz sample rate. The justification for the high correlation, hence the high sample rate, is that the quantizer operating in the error loop can be very simple. The simplest form of the quantizer is a one-bit quantizer, which is in fact, only a comparator which detects and reports the sign of the difference signal. As a result, the prediction error signal is a 1-bit word which has the interesting advantage of not requiring word framing in a communication system. The original attraction of this technique was simplicity of hardware and algorithm structure, neither of which is a major driver in today's integrated circuit marketplace. Delta modulation is still an important option because of its simplicity, ease of analysis, relative ease of embedding adaptive options, and its ability to perform robustly in the presence of channel errors [10, 11].

The block diagram of a delta modulator version of the one-tap linear predictor (presented in Section 11.3.1) is shown in Figure 11.21. Note that the predictor-corrector loop at the receiver is followed by an analog postreconstruction filter. This filter removes the out-of-band quantizing noise and overload distortion which is generated by the two-level coding and which extends beyond the information bandwidth of this coding process. The coder is completely characterized by the sampling frequency, the quantizing step size, Δs, and the postreconstruction filter.

The equations for prediction and for the residual error of the delta modulator are of the form

$$X(n|n-1) = aX(n-1|n-1) \tag{11.61a}$$

$$d(n) = X(n) - X(n|n-1) \tag{11.61b}$$

where n is a sample index. The weighting term a in Equation (11.61a) is nominally set to the correlation coefficient $C_X(1)$, which in light of the high sample rate is very close to unity. Some authors distinguish between delta modulation and 1-bit DPCM on the basis of this parameter, a. When $a = 1$, the system is called delta modulation, and when a is not limited to 1, it is called 1-bit DPCM. The predictor-corrector loop is then characterized as an integrator. Robust recovery from channel-induced errors at the receiver requires that the weighting term not be set to unity, but rather to a smaller value, so that the errors have only a finite

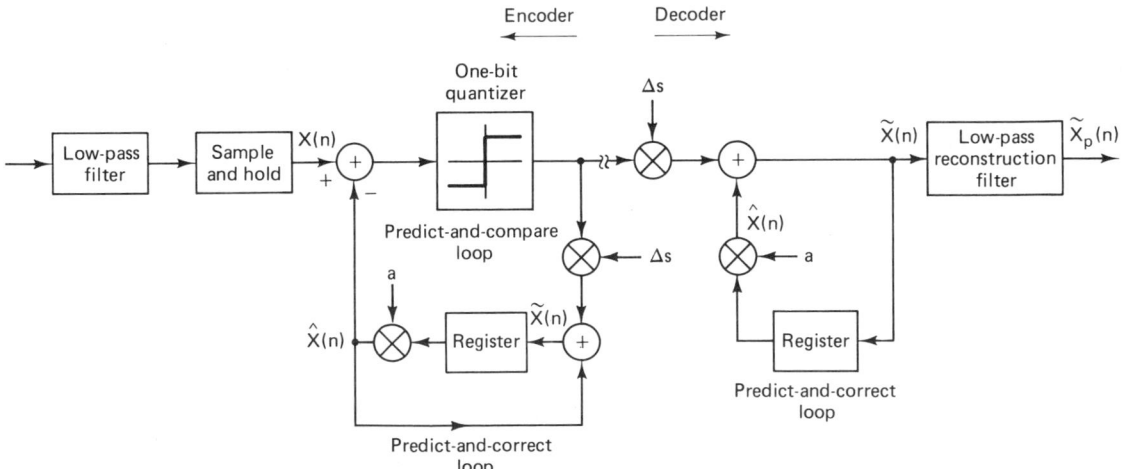

Figure 11.21　One-tap predictor delta modulation process.

persistence. This one-tap predictor is characterized as a leaky integrator; the optimum value of the coefficient to minimize the effects of channel-induced errors while simultaneously controlling prediction errors is [8]

$$a^* = \frac{1 - \sqrt{1 - C_X^2(1)}}{C_X(1)} \tag{11.62}$$

The 1-bit quantized error term is of the form

$$e(n) = \text{sgn } d(n) \tag{11.63a}$$

The correction equation derived from the 1-bit quantized error term is

$$X(n|n) = X(n|n - 1) + \Delta s(n) \text{ sgn } d(n) \tag{11.63b}$$

The step size, $\Delta s(n)$, in Equation (11.63b) is set to a constant for ordinary delta modulation, sometimes called *linear delta mod* (LDM), and is changed by an adaption algorithm in response to the error sequence in the adaptive forms of delta modulation.

An example of the performance of a delta modulator one-tap linear predictor-corrector to an input signal is shown in Figure 11.22. Shown are the input signal, the prediction signal, $\hat{X}(n)$, overlaid, the low-pass filtered version of the prediction signal, and the delta modulation sequence. In this example, the sampling fre-

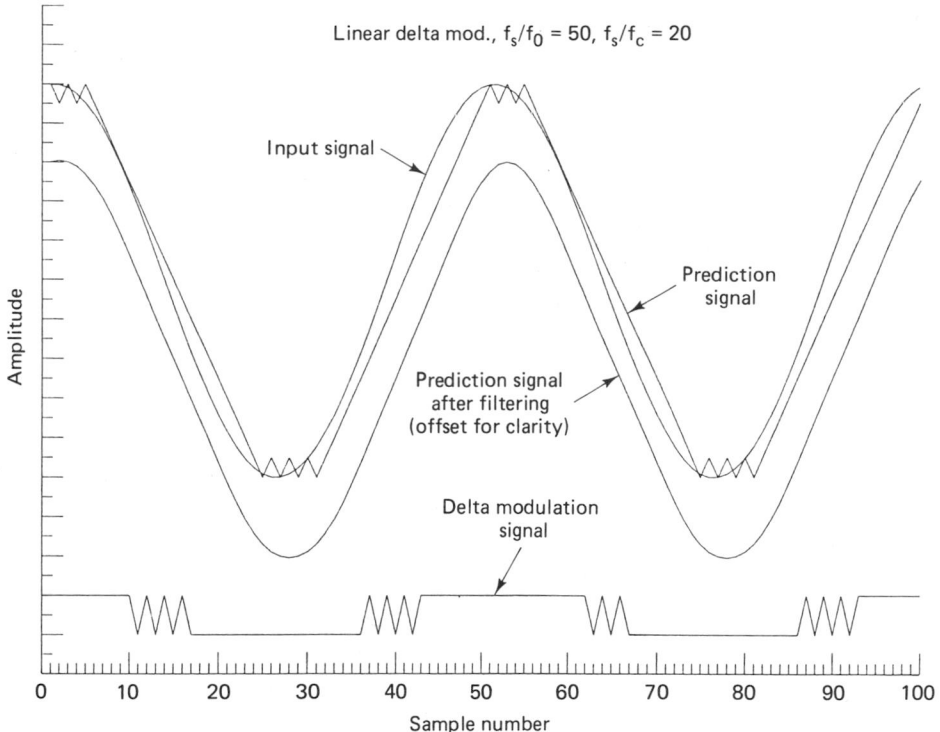

Figure 11.22 Example of waveshapes from a linear delta modulation process.

Sec. 11.3   Differential Pulse Code Modulation

quency, f_s, is 50 times the frequency of the input sinusoid (i.e., $f_s/f_0 = 50$), and f_s is 20 times the cut-off frequency, f_c, of the low-pass reconstruction filter (i.e., $f_s/f_c = 20$). Note that the difference between the input and the corrected prediction can be characterized by two essentially different types of tracking errors: slope overload and granular noise. *Slope overload* occurs when the slope of the input signal exceeds the delta mod's maximum slope of $\Delta s(1/T_s)$, where T_s is the sampling time. *Granular noise* occurs when the input signal slope falls within the slope capabilities of the delta modulator and the prediction loop attempts to match the local slope with a sequence of positive and negative Δs steps.

We note that the granular noise can be reduced by *decreasing* the size of the increment Δs, and that the slope overload noise can be reduced by *increasing* the size of the increment Δs. This, of course, is the justification for an adaptive step size. For a fixed step size, Δs must be selected to minimize the two sources of error. We first examine the performance of the delta modulator operating without slope overload and then examine a more general case.

Let us assume that the input bandwidth is defined as zero to f_m. This specifies the analog postfilter and the Nyquist sampling rate of $2f_m$. Let the actual sample rate of the LDM exceed the Nyquist rate by the factor F (i.e., $f_s = 2Ff_m$). Further, the step size of the LDM is Δs. The input to the LDM is a sinusoid of amplitude A and of frequency $2\pi f_0$, such as

$$x(t) = A \sin 2\pi f_0 t \tag{11.64}$$

The derivative of the input is

$$\dot{x}(t) = 2\pi f_0 A \cos 2\pi f_0 t \tag{11.65a}$$

and the maximum input slope is

$$\dot{x}_{\max} = 2\pi f_0 A \tag{11.65b}$$

The maximum slope capability of the delta mod is

$$\text{slope}_{\max} = \Delta s \left(\frac{1}{T_s}\right) = \Delta s f_s \tag{11.66}$$

Equating the maximum slope of the predictor to the maximum slope of the input, we have the maximum amplitude of the input signal as a function of step size, sample frequency, and input frequency.

$$A = \frac{\Delta s f_s}{2\pi f_0} \tag{11.67}$$

As expected, the amplitude of the input sinusoid which avoids slope overload is proportional to the step size Δs and the ratio of the sample rate to the input frequency. The signal power, σ_s^2 in the sinusoid is

$$\sigma_s^2 = \frac{A^2}{2} = \frac{1}{2} \left(\frac{\Delta s f_s}{2\pi f_0}\right)^2 \tag{11.68}$$

The only source of reconstruction error is the granular quantizing noise. We can reasonably assume that this noise is uniformly distributed between $+\Delta s$ and $-\Delta s$. As shown in Equation (11.21), the variance, σ_q^2, of this noise is

$$\sigma_q^2 = \tfrac{1}{3}(\Delta s)^2 \tag{11.69}$$

Thus, using Equations (11.68) and (11.69), the prefiltering signal-to-noise ratio is

$$\text{SNR}_{\text{pre}} = \frac{\sigma_s^2}{\sigma_q^2} \frac{\tfrac{1}{2}(\Delta s f_s/2\pi f_0)^2}{\tfrac{1}{3}(\Delta s)^2} = \frac{3}{8\pi^2} \left(\frac{f_s}{f_0}\right)^2 \tag{11.70}$$

We now estimate the filtering gain (improvement in SNR due to the postreconstruction filtering). Since the reconstruction error waveform is a pattern of random bipolar binary pulses, the autocorrelation function, $R_q(\tau)$, of the error pattern is a triangle wave of peak value $\tfrac{1}{3}(\Delta s)^2$, and the error power spectral density, $G_q(f)$, is

$$G_q(f) = \frac{T_s}{3}(\Delta s)^2 \left(\frac{\sin \pi f T_s}{\pi f T_s}\right)^2 \tag{11.71}$$

where T_s is the sampling time. This power spectral density has a main lobe width defined by zeros located at $-1/T_s$ and $+1/T_s$, as shown in Figure 11.23. The mainlobe amplitude can be approximated by a constant over a narrow bandwidth between $-f_m$ and $+f_m$, the bandwidth of the postprocessing filter. If we assume that the postprocessing low-pass filter following the reconstruction is an ideal filter with unity gain between $-f_m$ and $+f_m$, with $f_m \ll 1/T_s$, the integrated output noise power, $(\sigma_q^2)_{\text{out}}$, between these limits is approximately

$$(\sigma_q^2)_{\text{out}} = 2f_m \frac{T_s}{3}(\Delta s)^2 = \frac{2f_m}{f_s} \frac{1}{3}(\Delta s)^2 \tag{11.72a}$$

where $f_s = 1/T_s$. Equation (11.72a) is obtained by assuming that the input quantizing noise power, $(\sigma_q^2)_{\text{in}}$, from Equation (11.69) is

$$(\sigma_q^2)_{\text{in}} = \tfrac{1}{3}(\Delta s)^2 \tag{11.72b}$$

The ratio of the input to output noise powers is the *postfiltering gain*, which is seen from Equation (11.72) to be $f_s/2f_m$. Thus the postfiltering signal-to-noise ratio, SNR_{pst}, is obtained by multiplying Equation (11.70) by this filtering gain, yielding

$$\text{SNR}_{\text{pst}} = \frac{3}{16\pi^2} \frac{f_s^3}{f_0^2 f_m} \tag{11.73}$$

Note that the SNR for the LDM is proportional to the cube of the bit rate (same as sample rate for LDM).

Example 11.8 Sampled Sinusoid

Determine the SNR for a 1-kHz sinusoid, sampled at 32 kHz, without slope overload, and followed by a 4-kHz postreconstruction filter.

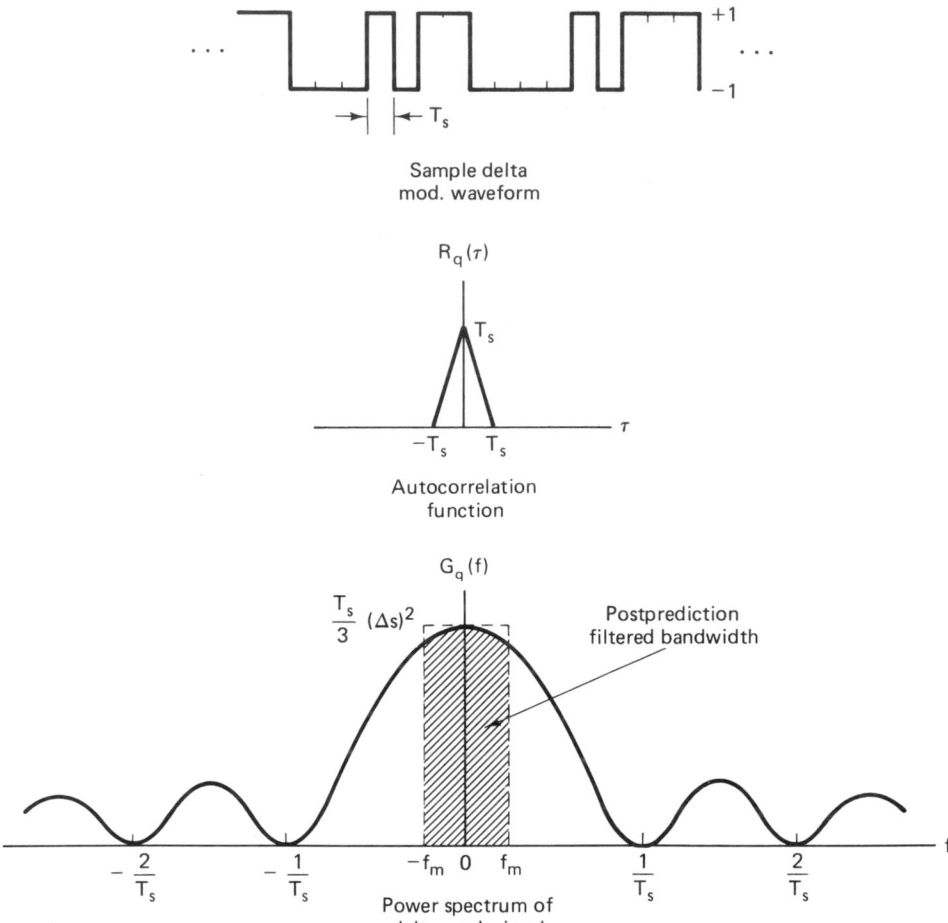

Figure 11.23 Sample time-series, autocorrelation function, and power spectrum for a delta modulation process.

Solution

Substituting in Equation (11.73), we find

$$\text{SNR}_{\text{pst}} = \frac{0.0190(32)^3}{1^2 \times 4} = 155.6$$

$$= 21.9 \text{ dB}$$

We note that the data rate for Example 11.8 is 32 kbits/s, which is the same bit rate obtained by sampling at 8 kHz (Nyquist) with 4 bits/sample. Applying the equations developed in Example 11.4 to the sampling of a full-scale sinusoid, a linear 4-bit PCM quantizer has an average SNR of

$$\text{SNR} = (1.5)2^{2b} = 1.7 + 6b \text{ (dB)} \tag{11.74a}$$

$$= 25.7 \text{ (dB)} \tag{11.74b}$$

For all the simplicity of the LDM, it does not perform as well as even a 4-bit linear PCM coder. This is true! The reason the linear PCM does so well is that its SNR increases exponentially with the bit rate while the LDM's SNR increases only as the cube of the bit rate. The gains to be realized with the delta modulator will be treated after incorporating an adaptive step size.

11.3.4 Adaptive Prediction

The prediction gain to be had in classical predictive coders are proportional to the ratio of the *signal variance* to *prediction error variance*. This is because for a fixed quantizing noise level, fewer bits are required to describe a signal with smaller energy. The utility of the predictive coder is limited by possible mismatches between the source signal and the predictor filter. The sources of mismatch are related to the time-varying behavior (i.e., nonstationarity) of the amplitude distribution and of the spectral or correlation properties of the signal. Adaptive encoders incorporate (slow time) auxiliary loops to estimate the parameters required to obtain locally optimal performance. These auxiliary loops periodically schedule modifications to the prediction loop parameters and thus avoid predictor mismatch. The International Telegraph and Telephone Consultative Committee (CCITT) has selected a 32-kbits/s adaptive differential pulse code modulation (ADPCM) coder as a standard for toll-quality speech. This achieves a 2:1 savings in bit rate relative to 64-kbits/s logarithmic compressed PCM.

11.3.4.1 Forward Adaption

In forward adaption algorithms, the input data to be encoded are buffered and processed in order to estimate the local statistics, such as the first N samples of the autocorrelation function. The zero-delay correlation sample, $R_X(0)$, is a short-term estimate of the local variance (zero mean). This estimate is used to adjust the automatic gain control (AGC) in order to obtain an optimal match of the scaled input signal to that of the quantizer dynamic range. This is denoted AQF, for adaptive quantization forward control. The remaining $N - 1$ correlation estimates are used to form new filter coefficients for the prediction filter. This adaption is called *adaptive prediction forward (APF) control*. Figure 11.24 shows this form of the adaptive algorithm. This is an extension of the structure presented in Figure 11.20. Here the predictor coefficients are derived from the input data, now called *side information*, and must be transmitted along with the prediction errors from the encoder to the decoder. The update rate of these adaptive coefficients is related to the length of time the input signal can be considered locally stationary. For example, speech caused by mechanical displacement of the speech articulators (tongue, lips, teeth, etc.) cannot change characteristics more rapidly than 10 or 20 times per second. This suggests an update interval of 50 to 100 ms. Using arithmetically simple, but suboptimal estimating algorithms to compute the local filter parameters makes a higher update rate necessary. An update every 20 ms to compute the parameters of a 10- to 12-tap filter has become a common rate. Prediction gains of 10 to 16 dB can be had with 10-tap filters when feedforward adaption is used with predictive coders [12].

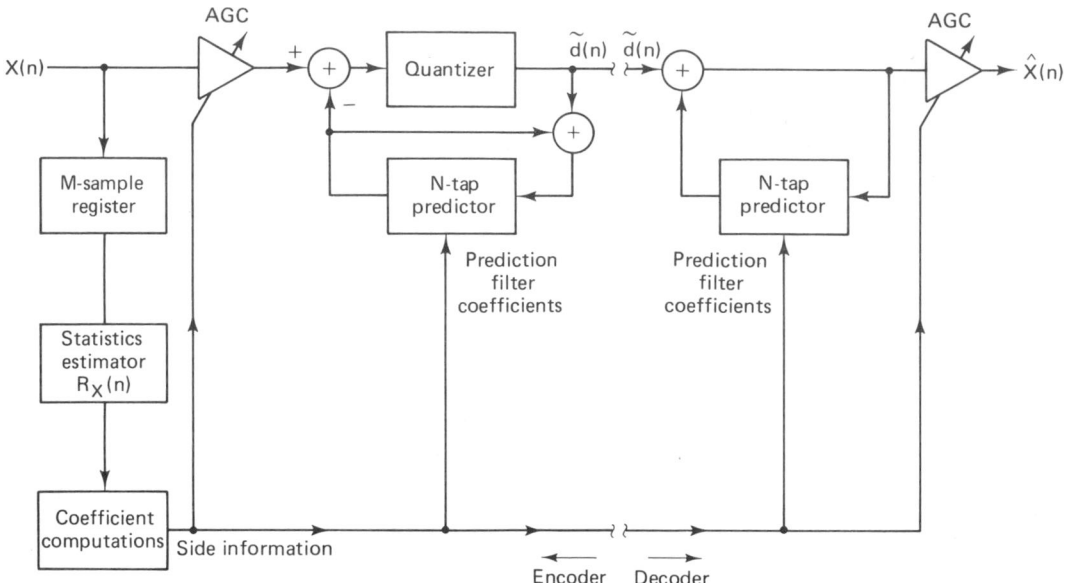

Figure 11.24 Forward adaptive prediction and quantization coding.

11.3.4.2 Backward Adaption

A primary disadvantage of forward adaption algorithms is the need to transmit the side information formed during each adaption cycle. An alternative to the forward adaption algorithms are those that form *backward adaption*. In these algorithms, processing is performed on the output data of the process, as is shown in Figure 11.25, as opposed to being performed on the input data. These same output data are transmitted to the receiver through the communication channel. Thus, applying the same algorithm as the transmitter, the receiver can reconstruct the adaption parameters. Backward adaption of quantizing size is denoted AQB, and backward adaption of predictor coefficients is called APB.

A particularly simple form of the backward adaption is the modification of linear delta modulation (LDM) to form adaptive delta modulation. As described earlier, there are conflicting requirements for large and small LDM step sizes to avoid slope overload and granular noise, respectively. In ADM, a decision algorithm changes the step size based on short run lengths of the successive prediction error polarities. The structure of all ADM algorithms is based on the following argument. When successive errors are of opposing polarity, the delta modulator is successfully tracking and the resultant errors are granular. In this mode, there may be an advantage to reducing the step size $\Delta s(n)$. On the other hand, if a run of successive errors are of the same polarity, the delta modulator may be operating in slope overload. The proper response to the possible slope overload is to increase the step size.

The change of step sizes can be performed by fixed additive scalars or by fixed multiplicative scalars, in which case the algorithm is called *constant factor*

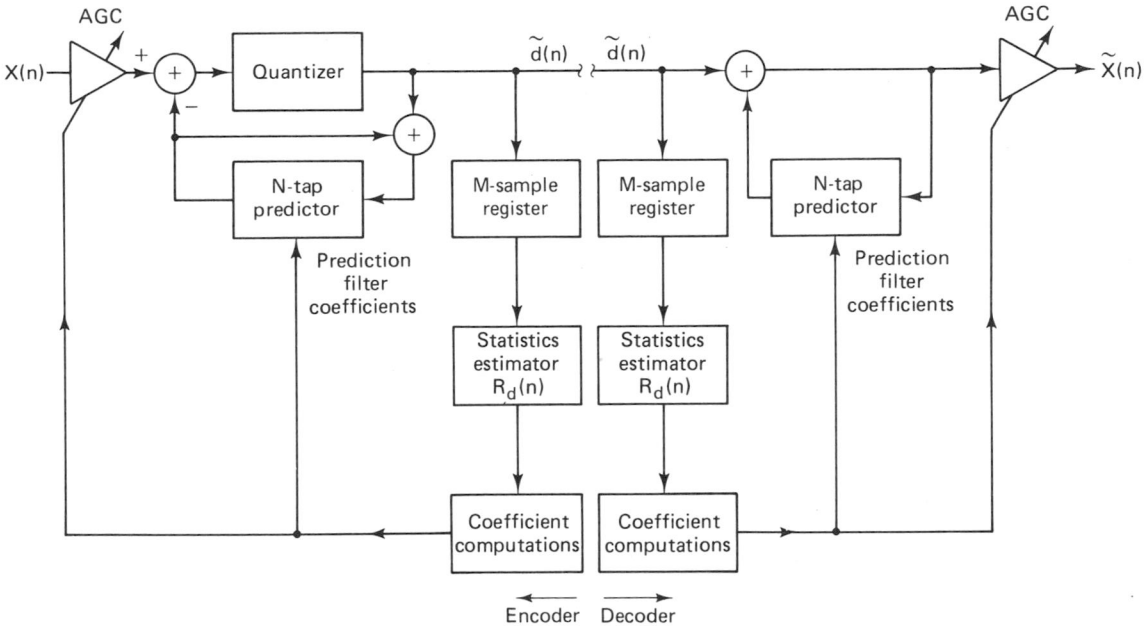

Figure 11.25 Backward adaptive prediction and quantization coding.

delta modulation (CFDM), or may be by a combination of exponential decay and a fixed additive scalar, in which case the algorithm is called *continuously variable slope delta* (CVSD) *modulation*.

Figure 11.26 presents an adaptive delta modulator based on increasing or decreasing the step size by a factor of 50% at each adaption iteration [13]. This is of the form

$$\Delta s(n) = \begin{cases} |\Delta s(n-1)| \, [d(n) + 0.5d(n-1)] & \text{if } |\Delta s(n-1)| > \Delta s_{\min} \\ \Delta s_{\min} \, d(n) & \text{if } |\Delta s(n-1)| < \Delta s_{\min} \end{cases}$$

(11.75)

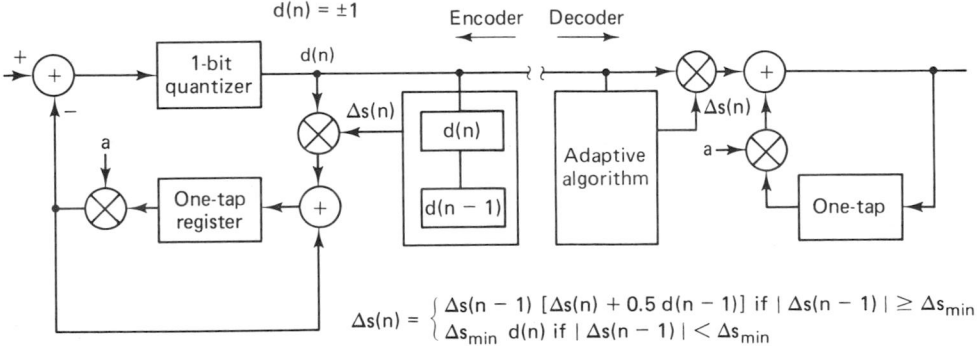

$$\Delta s(n) = \begin{cases} \Delta s(n-1) \, [\Delta s(n) + 0.5 \, d(n-1)] & \text{if } |\Delta s(n-1)| \geq \Delta s_{\min} \\ \Delta s_{\min} \, d(n) & \text{if } |\Delta s(n-1)| < \Delta s_{\min} \end{cases}$$

Figure 11.26 Adaptive delta modulation.

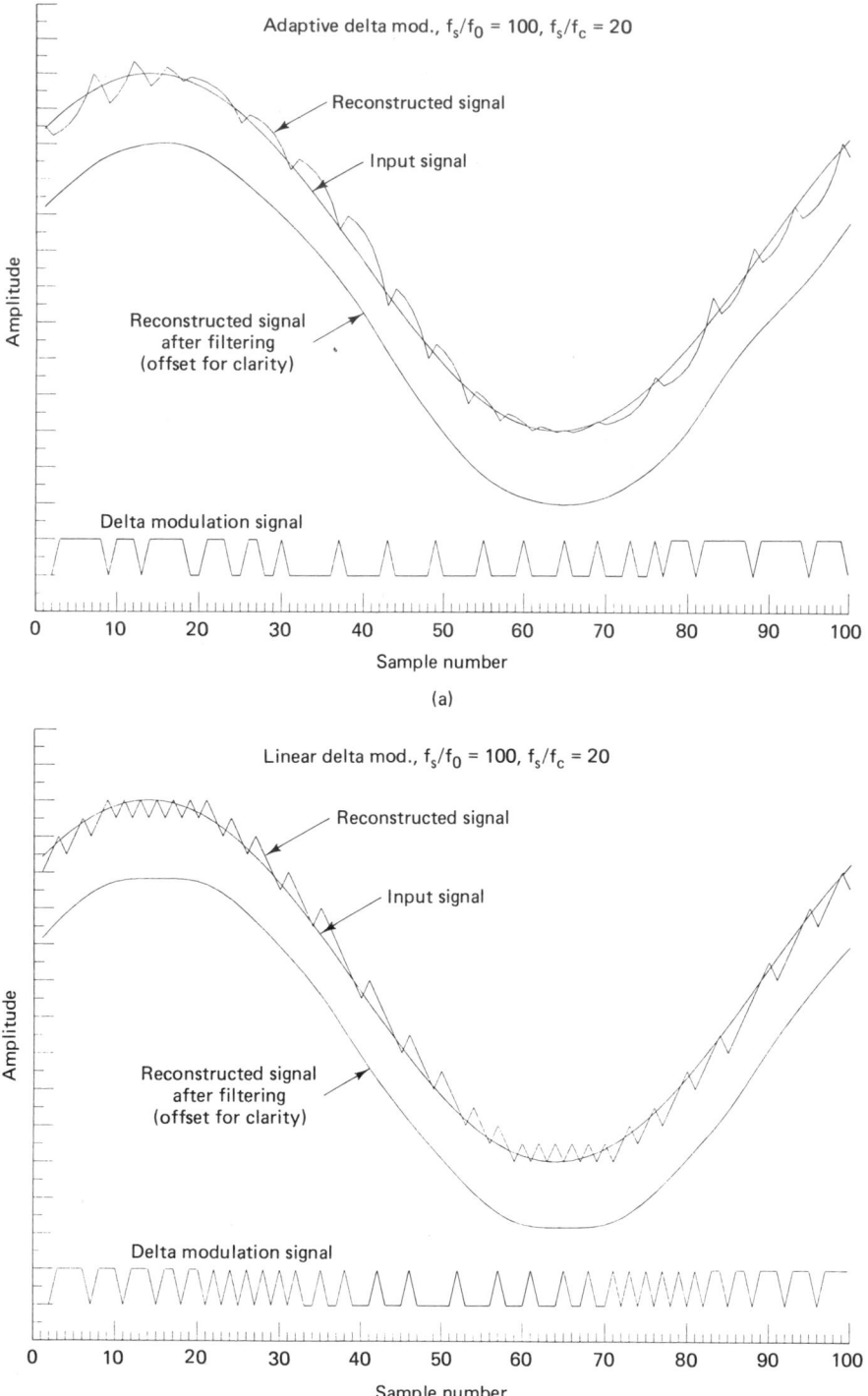

Figure 11.27 Reconstruction performance of an adaptive and linear delta modulation process. (a) Adaptive. (b) Linear.

Figure 11.27a demonstrates the reconstruction performance of this algorithm, and for comparison, Figure 11.27b presents the same signal using a linear delta modulator. The Δs_{min} for this example is $\frac{1}{8}$, the Δs for the LDM is 1, and the amplitude of the input sinusoid in both cases is 10. The ratio of sample frequency to sinusoid frequency is 100 and the cutoff frequency of the analog post filter is $\frac{1}{20}$ of the sample frequency. We see, in Figure 11.27a, that while tracking, the ADM adapts its step sizes, hence its slope, in successive clock intervals. We note that the reduced step size of the ADM results in smaller errors near the extremes of the input signal. We also note that the errors in the two systems are comparable in regions of moderate slope, and while not demonstrated, we can see that the increased slope of the ADM would avoid slope overload if the input signal were to exhibit high slope.

11.4 BLOCK CODING

The quantizers we have examined up to now have been *scalar quantizers*. Scalar quantizers form a *single output sample* based on the present input sample and (possibly) the N previous output samples. Block coders, on the other hand, form a *vector of output samples* based on the present and the N previous input samples. The *coding gain* of a waveform coder is the ratio of the input SNR to the output SNR. When the noise variances of the input and output are equal, this gain is simply the ratio of input-to-output signal variances. The ratio converts directly to 6 dB per bit for the difference between the number of input bits per sample and the average number of output bits per sample. Block coders can achieve impressive coding gains. On the average they can represent sequences quantized to 8 bits with only 1 or 2 bits per sample [8]. Block coding techniques are varied, but a common thread that runs through block coding techniques is the mapping of an input sequence to an alternative coordinate system. This mapping may be to a subspace of a complete space, so that the mapping may not be reversible [8]. Alternatively, a data-dependent editing scheme may be used to identify the subspace of the mapping from which the quantized data are extracted. Block coding techniques are often classified by their mapping techniques, which include, for example, vector quantizers, various orthogonal transform coders, and channelized coders such as the subband coder. Block coders are further described by their algorithmic structures, such as codebook coders, tree coders, trellis coders, discrete Fourier transform, discrete cosine transform, discrete Walsh–Hadamard transform, discrete Karhunen–Loeve transform, and quadrature mirror filter bank. We now examine examples of the various block coding schemes.

11.4.1 Vector Quantizing

Vector quantizers represent an extension of conventional scalar quantization. In scalar quantization a scalar value is selected from a finite list of possible values to represent an input sample. The value is selected to be close (in some sense) to the sample it is representing. The fidelity measures are various weighted mean-square measures which preserve our intuitive concept of distance in terms of ordinary vector lengths. By extension, in vector quantization, a vector is selected from a finite list of possible vectors to represent an input vector of samples. The selected vector is chosen to be close (in some sense) to the vector it is representing.

Each input vector can be visualized as a point in an N-dimensional space. The quantizer is defined by a partition of this space into a set of nonoverlapping volumes [14]. These volumes are called intervals, polygons, and polytopes, respectively, for one-, two-, and N-dimensional vector spaces. The task of the vector quantizer is to determine the volume in which an input vector is located. The output of the optimal quantizer is the vector identifying the centroid of that volume. As in the one-dimensional quantizer, the mean-square error is a function of the boundary locations for the partition and the multidimensional pdf of the input vector.

The description of a vector quantizer can be cast as two distinct tasks. The first is the code design task, which deals with the problem of performing the multidimensional volume quantization (or partition) and selecting the allowable output sequences. The second task is that of using the code and deals with searching for the particular volume in this partition which corresponds (according to some fidelity criterion) to the best description of the source. The two tasks, the partition and the search, may be coupled by the form of the algorithm selected to control the complexity of encoding and decoding. The standard vector coding methods are codebook, tree, and trellis coding algorithms [15].

11.4.1.1 Codebook, Tree, and Trellis Coders

The codebook coders are essentially table look-up algorithms. A list of candidate patterns (codewords) is stored in the codebook memory. Each pattern is identified by an address or pointer index. The coding routine searches through the list of patterns for the one that is closest to the input pattern and transmits to the receiver the address where that pattern can be found in its codebook.

The tree and trellis coders are sequential coders. As such, the allowable codewords of the code cannot be selected independently but must exhibit a node steering structure. This is similar to the structure of the sequential error detection and correction algorithms, which traverse the branches of a graph while forming the branch weight approximation to the input sequence (see Section 6.5.1). A tree graph suffers from exponential memory growth as the dimension or depth of the tree increases. The trellis graph reduces the dimensionality problem by tracking simultaneous contender paths, with an associated path weight metric called *intensity*, through a finite state trellis (see Section 6.3.3).

11.4.1.2 Code Population

The code vectors stored in the codebook, tree, or trellis are the likely or typical vectors. The first task, that of code design, in which the likely code vectors are identified, is called *populating* the code. The methods of determining the code population are classically *deterministic*, *stochastic*, and *iterative*. The deterministic population is a list of preassigned possible outputs based on a simple suboptimal or user perception fidelity criterion or based on a simple decoding algorithm. An example of the former is the coding of the samples in 3-space of the red, green, and blue (RGB) components of a color TV signal. The eye does not have the same resolution to each color and it would appear that the coding can be applied independently to each color to reflect this different sensitivity. The resulting quantizing volumes would be rectangular parallelepipeds. The problem with independent quantizing is that we do not see images in this coordinate system; rather, we see images in the coordinates of luminance, hue, and saturation. A black-and-white photo, for example, uses only the luminance coordinate. Thus quantizing RGB coordinates independently does not result in the smallest amount of user-perceived distortion for a given number of bits. To obtain improved distortion performance, the RGB quantizer should partition its space into regions that reflect the partitions in the alternate space. Alternatively, the quantization could be performed independently in the alternative space by the use of transform coding, treated in Section 11.4.2. Deterministic coding is the easiest to implement but leads to the smallest coding gain (smallest reduction in bit rate for a given SNR).

The stochastic population would be chosen based on an assumed underlying pdf of the input samples. Iterative solutions to the optimal partitions exist and can be determined for any assumed pdf. The overall samples are modeled by the assumed pdf. In the absence of an underlying pdf, iterative techniques based on a large population of training sequences can be used to form the partition and the output population. Training sequences may involve tens of thousands of representative input samples.

11.4.1.3 Searching

Given an input vector and a populated codebook, tree, or trellis, the coder algorithm must conduct a search to determine the best matching contender vector. An exhaustive search over all possible contenders will assure the best match. Coder performance improves for larger-dimensional spaces, but so does complexity. An exhaustive search over a large dimension may be prohibitively time consuming. An alternative is to conduct a nonexhaustive, suboptimal search scheme with acceptably small degradations from the optimal path. Memory requirements and computational complexity are often a driving consideration in the selection of search algorithms. Examples of search algorithms include single-path (best leaving branch) algorithms, multiple-path algorithms, and binary (successive approximation) codebook algorithms. Most of the search algorithms attempt to identify and discard unlikely patterns without having to test the entire pattern.

11.4.2 Transform Coding

In Section 11.4.1 we examined vector quantizers in terms of a set of likely patterns and techniques to determine the one pattern in the set closest to the input pattern. One measure of goodness of approximation is the weighted mean-square error of the form

$$d(\mathbf{X}, \hat{\mathbf{X}}) = (\mathbf{X} - \hat{\mathbf{X}})\mathbf{B}(\mathbf{X})(\mathbf{X} - \hat{\mathbf{X}})^T \qquad (11.76)$$

where $B(\mathbf{X})$ is a weight matrix and $\mathbf{X}^T$ is the transpose of $\mathbf{X}$. The minimization may be computationally simpler if the weighting matrix is a diagonal matrix. A diagonal weighting matrix implies a decoupled (or uncorrelated) coordinate set so that the error minimization due to quantization can be performed independently over each coordinate.

Thus transform coding entails the following set of operations, which are shown in Figure 11.28.

1. An invertible transform is applied to the input vector.
2. The coefficients of the transform are quantized.
3. The quantized coefficients are transmitted and received.
4. The transform is inverted with quantized coefficients.

Note that the transform does not perform any source encoding; it merely allows for a more convenient description of the signal vector to permit ease of source encoding. The task of the transform is to map a correlated input sequence into a different coordinate system in which the coordinates have reduced correlation. Recall that this is precisely the task performed by predictive coders. The source encoding occurs with the bit assignment to the various coefficients of the transform. As part of this assignment, the coefficients may be partitioned into subsets which are quantized with different number of bits but not with different quantizing step sizes. This assignment reflects the dynamic range (variance) of each coefficient and may be weighted by a measure that reflects the importance, relative to the human perception [16], of the basis element carried by each coefficient. A subset of the coefficients, for instance, may be set to zero amplitude, or may be quantized with 1 or 2 bits.

The transformation can be chosen to be independent of the data vector. Examples of such transforms are the discrete Fourier transform (DFT), discrete Walsh–Hadamard transform (DWHT), discrete cosine transform (DCT), and the

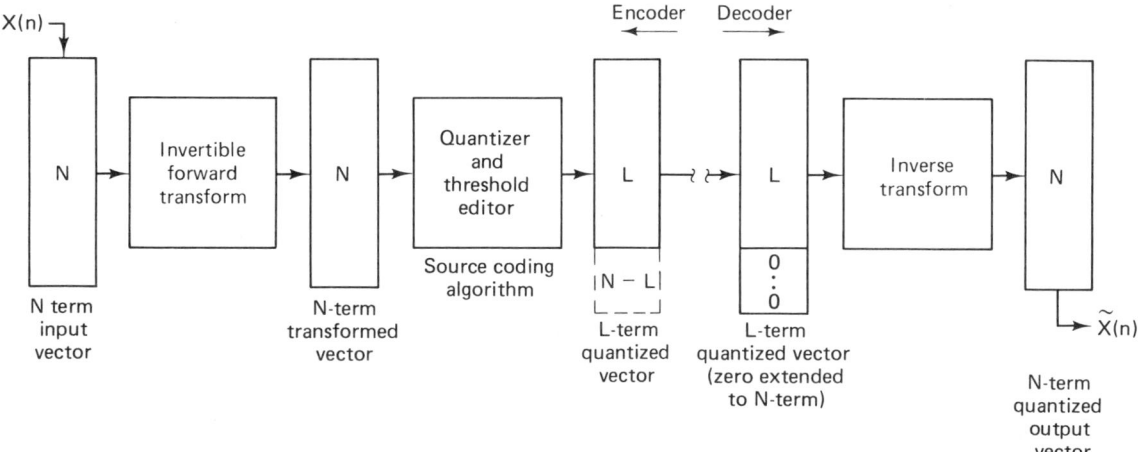

Figure 11.28 Transform coding.

discrete slant transform (DST). The transformation can also be derived from the data vector, as in the discrete Karhunen–Loeve transform (DKLT), sometimes called the principal component transform (PCT) [17].

The data-independent transforms are easiest to implement but do not perform as well as the data-dependent transforms. Often, the attraction of computational simplicity is sufficient justification for using the data-independent transformations. The coding gain penalty for using a good suboptimal transformation is small, typically less than 2 dB, and the degradation is usually cited when demonstrating performance characteristics.

11.4.3 Quantization for Transform Coding

Transform coders are called spectral encoders because the signal is described in terms of a spectral decomposition (in a selected basis set). The spectral terms are computed for nonoverlapped successive blocks of input data. Thus the output of a transform coder can be viewed as a set of time series, one series for each spectral term. The variance of each series can be determined and each can be quantized with a different number of bits. By permitting independent quantization of each transform coefficient we have the option to allocate a fixed number of bits among the transform coefficients to obtain a minimum quantizing error.

11.4.4 Subband Coding

The transform coders of Section 11.4.3 were described as a partition of an input signal into a collection of slowly varying time series, each of which is associated with a particular basis vector of the transform. The spectral terms, the inner product of the data with the basis vectors, are computed by a set of inner products. The set of inner products can be computed by a set of *finite impulse response* (FIR) filters [18]. With this perspective, the transform coder can be considered to be performing a channelization of the input data. By extension, a *subband coder*, which performs a spectral channelization by a bank of contiguous narrowband filters, can be considered a special case of a transform coder. A typical subband coder is shown in Figure 11.29.

Casting the spectral decomposition of the data as a filtering task affords us the option to form a class of custom basis sets (i.e., spectral filters): in particular, basis sets that reflect our user perception preferences and our source models. For example, the quantizing noise generated in a band with large variance will be confined to that band, not spilling into a nearby band with low variance and hence susceptible to low-level noise masking. We also have the option to form filters with equal or with unequal bandwidths (as seen in Figure 11.29). Thus we can independently assign to each subband the sample rate appropriate to its bandwidth and a number of quantizing bits appropriate to its variance. By comparison, in conventional transform coding, each basis vector amplitude is sampled at the same rate.

The subband coder can be designed as a conventional transmultiplexer. Here the input signal is considered to be composed of a number of independent narrow-bandwidth frequency-division-multiplexed subchannels. The encoder dechannelizes the input frequency-division-multiplexed (FDM) signal into a set of low-data-rate time-division-multiplexed (TDM) channels. After quantization and transmission, the decoder reverses the filtering process, converting the TDM channels back to the original FDM signal. In the classic approach to this process the input signal is demodulated by a bank of single-sideband narrowband filters. This is shown in Figure 11.30. These filters perform a complex (i.e., cosine and sine, or I and Q) heterodyne, thus basebanding the selected center frequency by the $e^{j\phi_k n}$ multiplication, where k is a frequency index and n is a time index. This is a filtering operation that reduces the input bandwidth to the selected channel bandwidth and resamples the signal to the lowest rate that avoids aliasing of the reduced bandwidth channelized data. This down-sampling is often called *decimation*, for reasons that escape logic, since nothing happens in tens and nothing is being destroyed. At the receiver, the reverse process is performed. The channelized signals are passed through interpolating filters to increase their sample rate to the desired output sample rate, are heterodyned back to their proper spectral position, and are combined to form the original composite signal.

For speech encoding, or more generally, for signals that are related to mechanical resonances, filter banks with nonequal center frequencies and nonequal bandwidths are desirable. Such filters are called constant-Q, or proportional, filter banks. These filters have logarithmically spaced center frequencies with bandwidths proportional to the center frequencies. This proportional spacing appears as uniform spacing and bandwidth when viewed on a log scale and reflects the spectral properties of many physical acoustic sources.

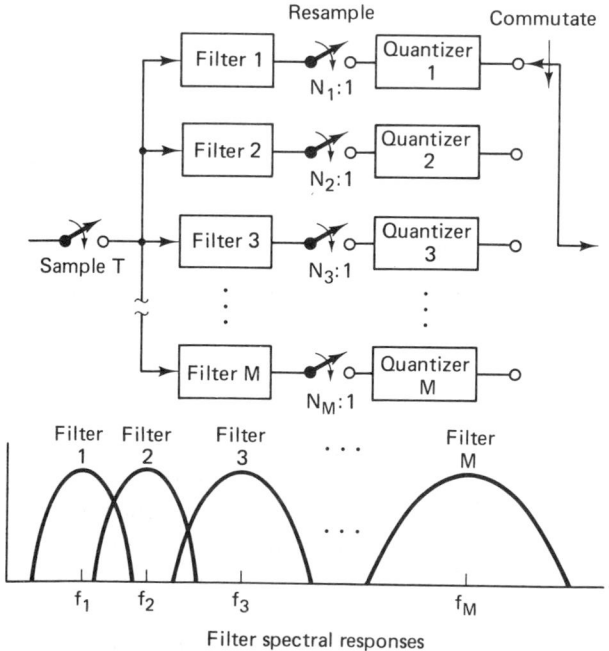

Figure 11.29 Subband coding performed by a channelized spectral decomposition.

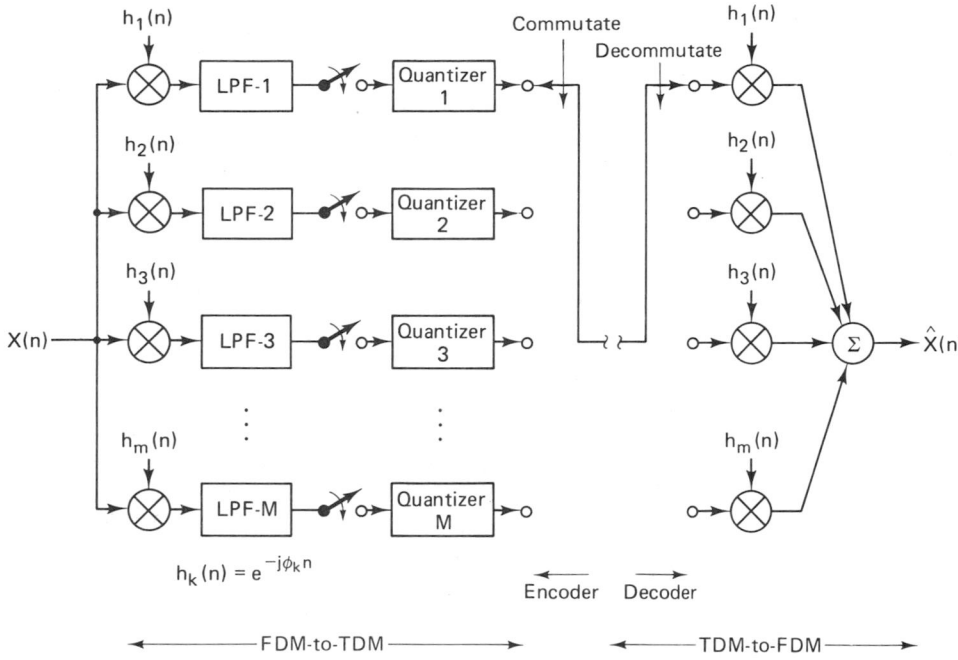

Figure 11.30 Transmultiplexer model of channel coding.

11.5 SYNTHESIS/ANALYSIS CODING

The encoding schemes we have examined until now can be classed as waveform encoders. They construct approximations to the input signals, which minimizes some distance measure between the signal and the approximation. These techniques are very general and can be applied to any signal source. Synthesis/analysis coders, on the other hand, are very signal specific; in particular, they are designed primarily for voice signals. These encoders take advantage of the fact that while the hearing mechanism responds to the amplitude content of a signal's short-term spectrum, it is fairly insensitive to its phase structure. Thus this class of encoder forms a reconstructed signal which approximates the magnitudes and time-varying characteristic of a sequence of the signal's short-term spectra, but makes no attempt to preserve its relative phase.

The spectral characteristics of voice appears to be stationary over periods between 20 and 50 ms. A number of techniques have evolved which analyze the spectral characteristics of voice every 20 ms and use the results of that analysis to synthesize a waveform which exhibits the same short-term power spectrum. Some techniques employ a model of the speech generation mechanism for which model parameters have to be estimated at the update rate. Examples of this type of encoder is the vocoder (voice coder), in its various forms, and the linear predictive coder (LPC). Other techniques manipulate the signal by combinations of spectral modifications and time partitions, which, with side information, reduces the number of time samples required to faithfully reconstruct the original spectrum. An example of this type of encoder is the Digitalker process of National Semiconductor [19].

The common thread that runs through all analysis/synthesis encoders is that the voice signal is not required to "look" like the original signal, but rather, to "sound" like it.

11.5.1 Vocoders

The vocoder is a model of the speech generation mechanism. Speech is a two-step mechanical process, one of excitation and one of response. The excitation consists of the acoustic emissions caused by the interaction of the diaphragm-

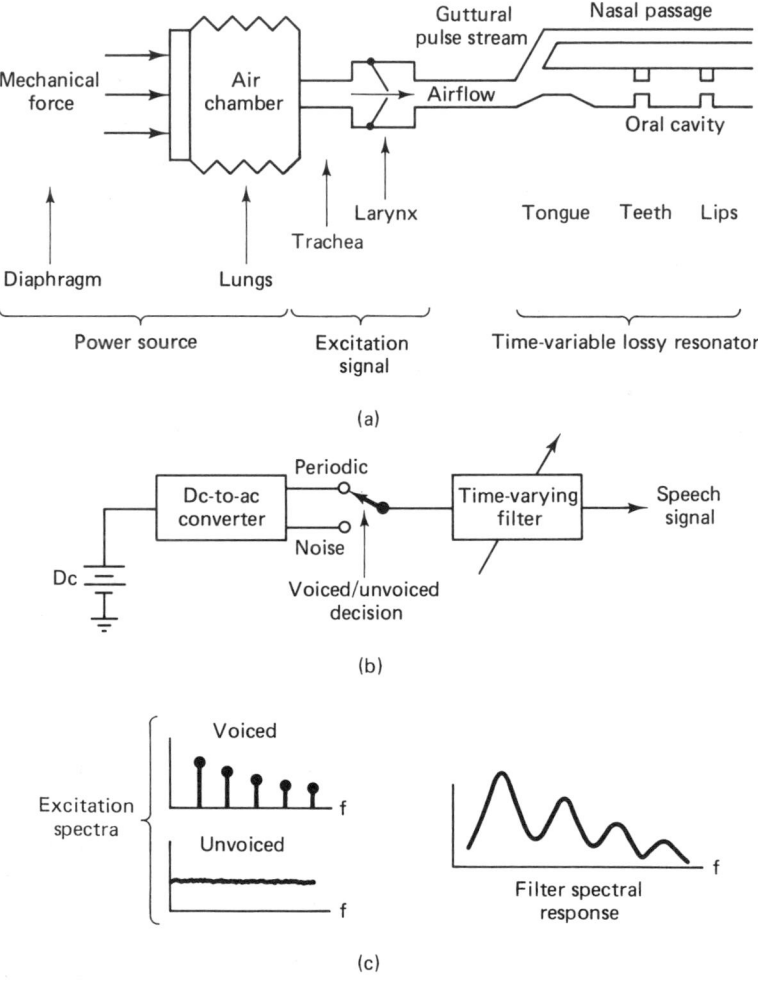

Figure 11.31 Mechanical and electrical modeling of the voice generation process. (a) Mechanical model. (b) Electrical model. (c) Excitation power spectrum and filter response.

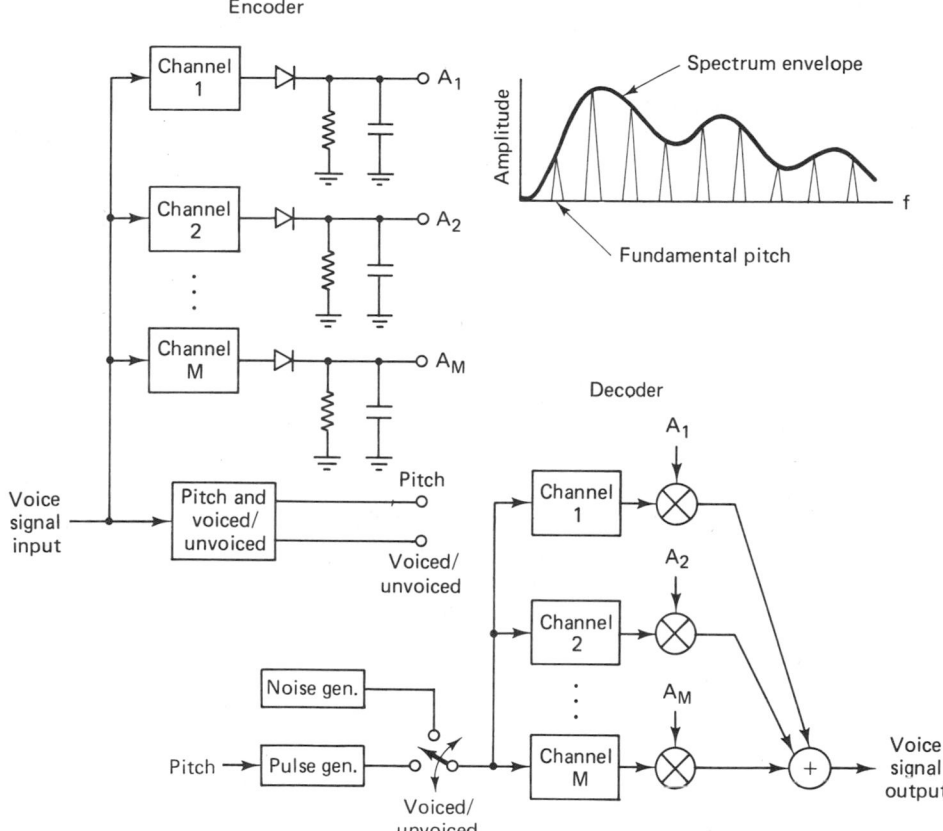

Figure 11.32 Channel vocoder.

supplied air pressure and the valvelike operation of the larynx (or vocal chords). Mechanical and electrical models of the process are shown in Figure 11.31a and b. By control over the vocal chords, the acoustic emissions can be made periodic or turbulent. The periodic emissions are called *voiced sounds*; the turbulent emissions are called *unvoiced*. The frequency of the voiced sound is called pitch. The oral and nasal cavities can be modeled as slowly time-varying lossy resonators which affect the spectral content of the excitation. The changes in the resonators are caused by mechanical displacement of the oral articulators: the lips, teeth, tongue, and so on. A typical excitation power spectrum and filter response are shown in Figure 11.31c. The vocoder models the speech mechanism in a two-step process: the excitation model and the time-varying lossy resonator or filter model.

The first vocoder was described by H. Dudley of Bell Labs and was demonstrated at the 1939 New York World's Fair [20]. The form of his vocoder, now called the *channel vocoder*, is shown in Figure 11.32. The channel vocoder analyzes the signal in what is essentially a channelized spectrum analyzer. This consists of a bank of narrowband filters, detectors, and averagers, which form a local estimate of the power distribution over the frequency band. Also embedded in the vocoder is the excitation estimator, which makes the voiced/unvoiced decision and estimates the pitch frequency if the excitation is voiced. These estimates are transmitted to the receiver, which forms the speech generation model. The spectral envelope estimates are used to alter the channel gains of a bank of narrowband filters, and the voiced/unvoiced decision selects a pulse or noise generator to drive the filter bank.

The bandwidths necessary to describe the parameters of the vocoder can be estimated as follows. Assume that a channel vocoder has 15 filters which span the bandwidth. Each filter is followed by a detector and a low-pass filter with a 20-Hz bandwidth. Thus the channelized bandwidth is 15 times 20 or 300 Hz. Additional bandwidth must be allocated for the voiced/unvoiced and pitch information. The most difficult part of a vocoder implementation is the voiced/unvoiced decision. To preserve speaker naturalness, the lower band of frequencies (200 to 600 Hz), which contains the pitch signal, is simply transmitted to the receiver. The receiver then generates harmonics of this lower band (with a nonlinearity) to excite the channel filters in the remaining spectral region. This implementation is called a *voice-excited vocoder*. Thus, the total vocoder bandwidth is 300 Hz for the channelized data and 400 Hz for the excitation, a total of 700 Hz.

There are many variants of the vocoder. One example is the *formant vocoder*, shown in Figure 11.33, which transmits to the receiver the amplitude and

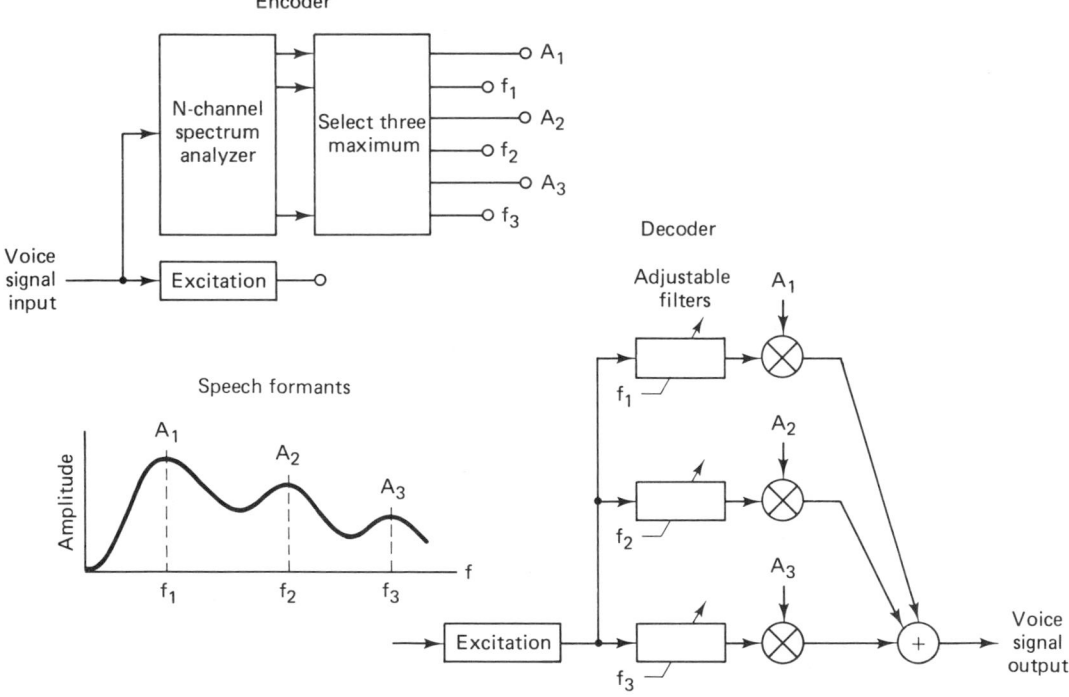

Figure 11.33 Formant vocoder.

spectral position of only the three dominant peaks of the spectral envelope. These dominant peaks, labeled A_1, A_2, and A_3, called formants, represent the dominant resonances of the oral cavity.

11.5.2 Linear Predictive Coding

The adaptive predictors, described in Section 11.3.4, were designed to predict or form good estimates of an input speech signal. In the adaptive form, the prediction coefficients are recomputed as side information from periodic examination of the input data. Then the difference between the input and the prediction is transmitted to the receiver to resolve the prediction error. *Linear predictive coders* (LPCs) are the natural extension of N-tap predictive coders. When the filter coefficients are periodically computed with an optimal algorithm, the prediction is so good that there is (essentially) no prediction error information worth transmitting to the receiver. Rather than transmit these low-level prediction errors, the LPC system transmits the filter coefficients and the voiced/unvoiced excitation decision for the model. Thus the only data sent in LPC is the high-quality side information of the classic adaptive algorithm. An LPC model for voice synthesis is shown in Figure 11.34. The Texas Instruments Speak and Spell learning games use a 12-tap LPC speech synthesizer implemented by a single microchip.

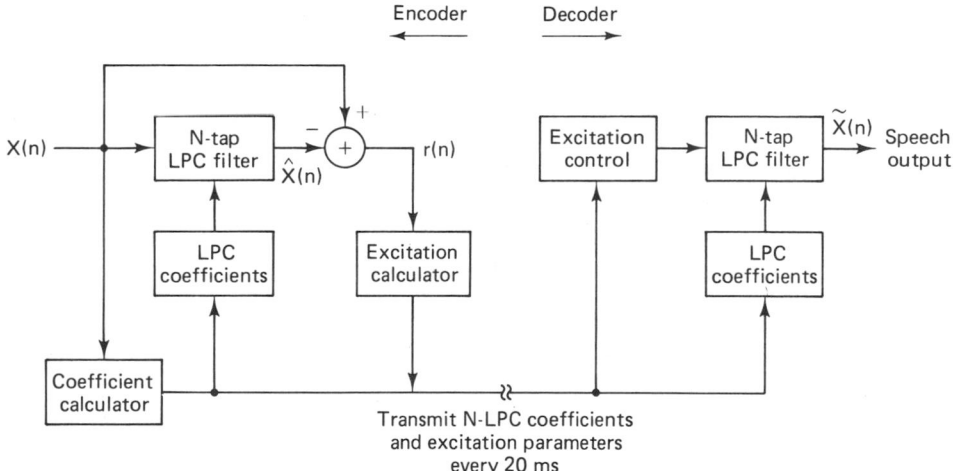

Figure 11.34 Linear predictive coefficient (LPC) speech modeling.

11.6 REDUNDANCY-REDUCING CODING

Coding to reduce the redundancy of a data source entails the selection of an efficient (usually) binary representation of that source. Often this requires the substitution of one binary representation of the source symbols with an alternative representation. The substitution is usually temporary and is performed to achieve an economy of storage or transmission of the discrete source symbols. The binary code assigned to each source symbol must satisfy certain constraints to permit reversal of the substitution. In addition, the code may be further constrained by system considerations such as memory limits or implementation ease.

We are so used to assigning binary codes to represent source symbols that we may lose sight of the arbitrariness of this assignment. The most common example of this is the binary assignments to the cardinal numbers (let's not even consider the negative numbers). We can count in straight binary, binary-coded octal, binary-coded decimal, binary-coded hexadecimal, two-out-of-five decimal, excess-three decimal, and on and on. In this example, ease of computation, error detection, ease of display, or convenience of coding are the considerations for selecting the assignment. For the specific task of data compression, *reduced number of bits* is the primary consideration.

Finite discrete sources are characterized by a set of distinct symbols, $X(n)$ ($n = 1, 2, 3, \ldots, N$), called the source alphabet, where n is a data index. A complete characterization requires the probability of each symbol and the joint probabilities of the symbols taken two at a time, three at a time, and so on. The symbols may represent a two-level (binary) source, such as the black-and-white levels of a facsimile scan, or a many-symbol source, such as the 49 common characters of Sanskrit. Another common many-symbol alphabet is the keyboard of a computer terminal. These nonbinary symbols are mapped, via a dictionary called a character code (see Figure 2.2 for the ASCII code and Figure 2.3 for the EBCDIC code), to a binary alphabet description.

The standard character codes are of fixed length, such as 5, 6, or 7 bits. The length is usually chosen so that there are enough binary characters to assign a unique sequence to each input alphabet character. These may include the upper- and lowercase letters of the alphabet (A, B, C, . . . , Z, a, b, c, . . . , z), numerals (0, 1, 2, . . . , 9), punctuation (!, ?, :, '', ', . . .), special characters (@, #, $, %, &, *, +, /, . . .), and control characters, such as backspace, return, and so on. Fixed-length codes have the property that character boundaries are separated by a fixed bit count. This allows the conversion of a serial data stream to a parallel data stream by a simple bit counter.

Two code standards may define the same symbol in different ways. For example, the ASCII (7-bit) code has enough bits to assign different binary sequences to the upper- and lowercase versions of each letter. On the other hand, the *Baudot* (5-bit) code with only 32 binary sequences cannot do the same. To account for the full character set, the Baudot code defines two control characters, called *letter shift* (LS) and *figure shift* (FS), to be used as prefixes. When used, these control characters reassign the binary-to-symbol mapping. This works very much like the *shift key* on a typewriter; the shift key reassigns a completely new character set to the keyboard. In a similar fashion, the keyboards on some calculators have two prefix character keys, so that each key stroke can have three possible meanings. Also, some word processor instruction codes use double- and triple-stroke command functions. In a very real sense, these two- and three-word instructions represent a variable-length code assignment. These longer code words are assigned to characters (or instructions) that do not occur as often as those assigned single codewords. What we receive in exchange for using the occasional longer words is more efficient storage (smaller keyboard) or transmission of the source.

Data compression codes are often variable-length codes. Intuitively, we would expect the length of a binary sequence assigned to each alphabet symbol to be inversely related to the probability of that symbol. After all, if a symbol occurs with high probability, it contains little information and should not be assigned much of the system resources. In a similar manner, it would not seem unreasonable to find that when all symbols are equally likely, the code should be of fixed length. Perhaps the best known variable-length code is the Morse code. Samuel Morse counted the quantity of letters in a printer's font drawer to determine the relative frequency of letters in normal text. The variable-length code assignment reflects this relative frequency.

A significant amount of *data compression* can be realized when there is a wide difference in the probabilities of the symbols. To achieve this compression there must also be a sufficiently large number of symbols. Sometimes, in order to have a large enough set of symbols, we form a new set of symbols derived from the original set called an *extension code*. We have already seen this trick in Example 11.3 and will examine the general technique in the next section.

11.6.1 Properties of Codes

Earlier we alluded to properties that a code must satisfy for it to be useful. Some of these properties are obvious, some are not. It is worth listing and demonstrating the *desired properties*. We will consider a three-symbol alphabet with the given probability assignment, shown below. Listed with the input alphabet are six binary code assignments. Scan these for a moment and try to determine which codes are practical.

X_i	$P(X_i)$
a	0.73
b	0.25
c	0.02

Symbol	Code 1	Code 2	Code 3	Code 4	Code 5	Code 6
a	00	00	0	1	1	1
b	00	01	1	10	00	01
c	11	10	11	100	01	11

Uniquely Decodeable Property. Uniquely decodeable codes are those that allow us to invert the mapping to the original symbol alphabet. Obviously, code 1 above is not uniquely decodeable because the symbol a and b are assigned the same binary sequence. Thus the first requirement of a useful code is that each symbol be assigned a unique binary sequence. By this condition, all the other codes appear satisfactory until we examine codes 3 and 6 carefully. These codes indeed have unique binary sequences assigned to each symbol. The problem occurs when these code sequences are strung together. For instance, try to decode the binary pattern 1 0 1 1 1 in code 3; is it b, a, b, b, b or b, a, b, c or b, a, c, b? Trying to decode the same sequence in code 6 gives similar difficulties. These codes are not uniquely decodeable even though the individual characters have unique code assignments.

Prefix-Free Property. A sufficient (but not necessary) condition to assure that a code is uniquely decodeable is that no codeword be the prefix of any other code word. Codes that satisfy this condition are called prefix-free codes. Note that code 4 is not prefix-free but is uniquely decodeable. Prefix-free codes also have the property that they are instantaneously decodeable. Code 4 has a property that may be undesirable; it is not instantaneously decodeable. An instantaneously decodeable code is one for which the boundary of the present codeword can be identified by the end of the present codeword rather than by the beginning of the next codeword. For instance, in transmitting the symbol b with the binary sequence 1 0 in code 4, the reciver cannot determine if this is the whole codeword for symbol b or the partial codeword for symbol c.

11.6.1.1 Code Length and Source Entropy

At the beginning of the chapter we described the formal concept of information content and source entropy. We identified the self-information, in bits, about the symbol X_n, denoted $I(X_n)$, as $\log_2 [1/P(X_n)]$. From the perspective that information resolves uncertainty we recognize that the information content of a symbol goes to zero as the probability of that symbol goes to unity. We also defined the *entropy* of a finite discrete source as the average information of that source. From the perspective that information resolves uncertainty, the entropy is the average amount of uncertainty resolved per use of the alphabet. It also represents the average number of bits per symbol required to describe the source. In this sense it is also the lower bound of what can be achieved with some variable-length data compression codes. A number of considerations prevent an actual code from achieving the entropy bound of the input alphabet. These include uncertainty in probability assignment and buffering constraints. The average bit length achieved by a given code is denoted by $\bar{n}$. This average length is computed as the sum of the binary code lengths n_i weighted by the probability of that code symbol $P(X_i)$.

$$\bar{n} = \sum_i n_i P(X_i)$$

A great deal is implied about the performance of a variable-length code when we say *average number of bits*. In a variable-length code assignment some symbols will have code lengths which exceed the average length, while some will have code lengths which are smaller than the average. It may occur that a long pattern of symbols with long codewords are delivered to the coder. The short-term bit rate required to transmit these symbols will exceed the average bit rate of the code. If a channel is expecting data at the average rate, the local excess rate must be buffered in a memory. By the same token, a long pattern of symbols with short codewords may be delivered to the coder. The short-term bit rate required to transmit these symbols will fall short of the average rate of the code. Here the channel will find itself waiting for bits that are not to be had. For this reason, *data buffering* is required to smooth the local statistical variations associated with the input alphabet.

The last caveat is that variable-length codes are designed to operate with a specified list of symbols and proabilities. If the data presented to the coder have a significantly different list of probabilities, the coder buffers may not be able to support the mismatch and underflow or overflow will occur.

11.6.2 Huffman Code

The Huffman code [21] is a prefix-free variable-length code which can achieve the shortest average code length $\bar{n}$ for a given input alphabet. The shortest average code length for a particular alphabet may be significantly greater than the entropy of the source alphabet. This inability to exploit the promised data compression is related to the alphabet, not to the coding technique. Often the alphabet can be modified to form an extension code, and the same coding technique is then reapplied to achieve better compression performance. Compression performance is measured by the *compression* ratio. This measure is equal to the ratio of the average number of bits per sample before compression to the average number of bits per sample after compression.

The Huffman coding procedure can be applied for transforming between any two alphabets. We will demonstrate the application of the procedure between an arbitrary input alphabet and a binary output alphabet. The Huffman code is generated as part of a tree-forming process. The process starts by listing the input alphabet symbols, along with their probabilities (or relative frequencies), in descending order of occurrence. These tabular entries correspond to the branch ends of a tree, as shown in Figure 11.35. Each branch is assigned a branch weight equal to the probability of that branch. The process now forms the tree which supports these branches. The two entries with the lowest relative frequency are merged (at a branch node) to form a new branch with their composite probability. After every merging, the new branch and the remaining branches are reordered (if necessary) to assure that the reduced table preserves the descending probability of occurrence. We call this reordering *bubbling* [22]. During the rearrangement after each merging, the new branch rises through the table until it can rise no further. Thus if we form a branch with a weight of 0.2 and during the bubbling process find two other branches already with the 0.2 weight, the new branch is bubbled to the top of the 0.2 group, as opposed to simply joining it. The bubbling to the top of the group results in a code with reduced code length variance but otherwise a code with the same average length as that obtained by simply joining the group. This reduced code length variance lowers the chance of buffer overflow.

As an example of this part of the code process, we will apply the Huffman procedure to the input alphabet shown in Figure 11.35. The tabulated alphabet and the associated probabilities are shown on the figure. After forming the tree,

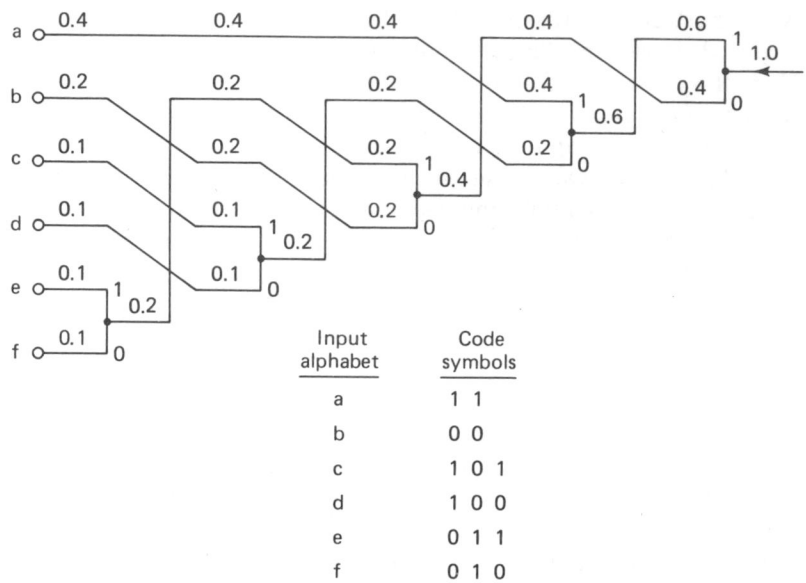

Input alphabet	Code symbols
a	1 1
b	0 0
c	1 0 1
d	1 0 0
e	0 1 1
f	0 1 0

Figure 11.35 Huffman coding tree for a six-character set.

each branch node is labeled with a binary 1/0 decision to distinguish the two branches. The labeling is arbitrary, but for consistency, at each node we will label the branch going up with a "1" and the branch going down with a "0." After labeling the branch nodes we trace the tree path from the base of the tree (far right) to each output branch (far left). The path contains the binary sequence to reach that branch. In the table below, we have listed at each end branch the path sequence corresponding to each path.

X_i	$P(X_i)$	Code	n_i	$n_i P(X_i)$
a	0.4	1 1	2	0.8
b	0.2	0 0	2	0.4
c	0.1	1 0 1	3	0.3
d	0.1	1 0 0	3	0.3
e	0.1	0 1 1	3	0.3
f	0.1	0 1 0	3	0.3

$$\bar{n} = 2.4$$

where $i = 1, \ldots, 6$. We find that the average code length, $\bar{n}$, for this alphabet is 2.4 bits per character. It does not mean that we have to find a way to transmit a noninteger number of bits. Rather, it means that on the average, 240 bits will have to be moved through the communication channel when transmitting 100 input symbols. For comparison, a fixed-length code required to span the six-character input alphabet would be of length 3 bits, and the entropy of the input alphabet, using Equation (11.2) is 2.32 bits. Thus this code offers a compression ratio of 1.25 (3.0/2.4) and achieves 96.7% (2.32/2.40) of the possible compression ratio.

As another example, one for which we can demonstrate the use of code extension, let us examine the three-character alphabet presented in Section 11.6.1.

X_i	$P(X_i)$
a	0.73
b	0.25
c	0.02

The Huffman code tree for this alphabet is shown in Figure 11.36, and the details are tabulated below.

X_i	$P(X_i)$	Code	n_i	$n_i P(X_i)$
a	0.73	1	1	0.73
b	0.25	01	2	0.54
c	0.02	00	2	0.04
			$\bar{n}$ =	1.31

where $i = 1, \ldots, 3$. The average code length for this Huffman code is 1.31 bits; it would be 2 bits for a fixed-length code. The compression ratio for this code is 1.53. Again, using Equation (11.2), the entropy for the alphabet is 0.9443 bit, so that the efficiency (0.944/1.31 = 72%) of the code is significantly smaller than for the preceding example.

To improve coding efficiency or to achieve greater compression gain, we have to redefine the source alphabet. A larger source alphabet holds the promise of increased variability, one requirement to realize a reduction in average code length, and an increased number of tree branches for assigning the variable-length code. We do this by selecting characters two at a time from the source alphabet to be new characters in the extension alphabet. If we assume that the symbols are independent, the probability of each new element is the product of the indi-

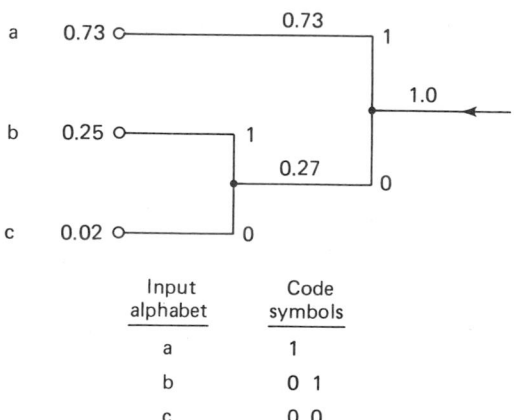

Input alphabet	Code symbols
a	1
b	0 1
c	0 0

Figure 11.36 Huffman coding tree for a three-character set.

vidual probabilities. The extension alphabet is as follows:

X_i	$P(X_i)$	Code	n_i	$n_i P(X_i)$
aa	0.5329	1	1	0.5329
ab	0.1825	00	2	0.3650
ba	0.1825	011	3	0.5475
bb	0.0625	0101	4	0.2500
ac	0.0146	01000	5	0.0730
ca	0.0146	010011	6	0.0876
bc	0.0050	0100100	7	0.0350
cb	0.0050	01001011	8	0.0400
cc	0.0002	01001010	8	0.0016

$$\bar{n} = 1.9326 \text{ bits/two symbols}$$

$$= 0.9663 \text{ bit/symbol}$$

where $i = 1, \ldots, 9$, and the code sequence for each X_i has been found by the use of the Huffman procedure described above. The compression ratio of this extension code is 2.07 and the coding efficiency is 97.7%.

Extension codes offer a very powerful technique to include the effects of nonindependent symbol sets. For example, in English text, adjacent letters are highly correlated; very common pairs include

th	re	in
sh	he	e_
de	ed	s_
ng	at	r_
te	es	d_

where the dash represents a space. Similarly, common English three tuples include

the	and	for
ing	ion	ess

Thus rather than perform Huffman coding on the individual letters, it is more efficient to extend the alphabet to include all 1-tuples plus common 2-tuples and 3-tuples, and then perform the coding on the extension code.

11.6.3 Run-Length Codes

In many applications, a sequence of symbols to be transmitted or stored is characterized by lengthy runs of specific symbols. Rather than code each symbol of a lengthy run, it makes sense to describe the run with an efficient substitution code. As an example, runs of spaces (the most common symbol in text) are encoded in many communication protocols by a control character followed by the character count. The IBM 3780 BISYNC protocol has an option to replace runs of spaces with an "IGS" character (if EBCDIC; or "GS" if ASCII) followed by a count of 2 to 63. Longer runs are partitioned into successive runs of 63 characters.

The run-length substitution coding can be applied to the original symbol alphabet or the binary representation of that alphabet. Run-length coding is particularly attractive for binary alphabets derived from specific sources. The most important commercial example is facsimile coding used for transmitting documents by instant electronic mail (i.e., U.S. Post Office, *U.S.A. Today*) [23].

11.6.3.1 Huffman Coding for Facsimile Transmission

Facsimile transmission is the process of transmitting a two-dimensional image as a sequence of successive line scans. The most common images are, in fact, documents containing text and figures. The position of the scan lines and the position along a scan line are quantized into spatial locations that define a two-dimensional grid of picture elements called *pixels*. The standard CCITT document is defined to be of width 8.27 in. (20.7 cm) and of length 11.7 in. (29.2 cm), almost 8.5 in. by 11.0 in. The spatial quantization for normal resolution is 1188 pixels/line and 1728 lines/document. The standard also defines a high-resolution quantization of 2376 pixels/line with the same 1728 lines/document. The total number of individual pixels for a normal-resolution facsimile transmission is 2,052,864 and is doubled for high resolution. For comparison, the number of pixels in NTSC (National Television Standard Committee) standard commercial television is 480 × 640 or 307,200. Thus facsimile has 6.7 or 13.4 times the resolution of a standard TV image.

The relative brightness or darkness of the scanned image at each position in the scan is quantized into two levels: B for black and W for white. Thus the signal observed during a scan line is a two-level pattern representing the B and W image intensity under the scan. It is easy to see that a horizontal scan line across this sheet of paper will exhibit a pattern consisting of long runs of B and W levels. The standard CCITT run-length coding scheme to compress the run of B and W levels is based on a modified variable-length Huffman code; it is listed in Table 11.1. Two types of patterns are identified, runs of W and runs of B. Each run length is described by a *partitioned codeword*. The first partition, called the *makeup codeword* or most significant bits (MSB), identifies runs with lengths that are multiples of 64. The second partition, called the *terminating codeword* or least significant bits (LSB), identifies the length of the remaining run. Each run of B (or W) of length from 0 through 63 is assigned a unique Huffman code word, as is each run of length $64 \times K, K = 1, 2, \ldots, 27$. A unique END OF LINE (EOL) is also defined in the code, which indicates that no black pixels follow, hence the next line should be started; this is akin to a carriage return on a typewriter.

Example 11.0 Run-Length Code

Use the modified Huffman code to compress the line consisting of 1188 pixel elements

200 W, 10 B, 10 W, 84 B, 884W

TABLE 11.1 Modified Huffman Code for CCITT Facsimile Standard

Run length	White	Black	Run length	White	Black
			Makeup codewords		
64	11011	0000001111	960	011010100	0000001110011
128	10010	000011001000	1024	011010101	0000001110100
192	010111	000011001001	1088	011010110	0000001110101
256	0110111	000001011011	1152	011010111	0000001110110
320	00110110	000000110011	1216	011011000	0000001110111
384	00110111	000000110100	1280	011011001	0000001010010
448	01100100	000000110101	1344	011011010	0000001010011
512	01100101	0000001101100	1408	011011011	0000001010100
576	01101000	0000001101101	1472	010011000	0000001010101
640	01100111	0000001001010	1536	010011001	0000001011010
704	011001100	0000001001011	1600	010011010	0000001011011
768	011001101	0000001001100	1664	011000	0000001100100
832	011010010	0000001001101	1728	010011011	0000001100101
896	011010011	0000001110010	EOL	000000000001	000000000001

Run length	White	Black	Run length	White	Black
			Terminating codewords		
0	00110101	000110111	32	00011011	000001101010
1	000111	010	33	00010010	000001101011
2	0111	11	34	00010011	000011010010
3	1000	10	35	00010100	000011010011
4	1011	011	36	00010101	000011010100
5	1100	0011	37	00010110	000011010101
6	1110	0010	38	00010111	000011010110
7	1111	00011	39	00101000	000011010111
8	10011	000101	40	00101001	000001101100
9	10100	000100	41	00101010	000001101101
10	00111	0000100	42	00101011	000011011010
11	01000	0000101	43	00101100	000011011011
12	001000	0000111	44	00101101	000001010100
13	000011	00000100	45	00000100	000001010101
14	110100	00000111	46	00000101	000001010110
15	110101	000011000	47	00001010	000001010111
16	101010	0000010111	48	00001011	000001100100
17	101011	0000011000	49	01010010	000001100101
18	0100111	0000001000	50	01010011	000001010010
19	0001100	00001100111	51	01010100	000001010011
20	0001000	00001101000	52	01010101	000001000100
21	0010111	00001101100	53	00100100	000000110111
22	0000011	00000110111	54	00100101	000000111000
23	0000100	00000101000	55	01011000	000000100111
24	0101000	00000010111	56	01011001	000000101000
25	0101011	00000011000	57	01011010	000001011000
26	0010011	000011001010	58	01011011	000001011001
27	0100100	000011001011	59	01001010	000000101011
28	0011000	000011001100	60	01001011	000000101100
29	00000010	000011001101	61	00110010	000001011010
30	00000011	000001101000	62	00110011	000001100110
31	00011010	000001101001	63	00110100	000001100111

Solution

Using Table 11.1, we determine the coding for this pattern to be (the spaces are for our reading benefit)

010111   10011   0000100   00111   0000001111   00001101000   000000000001

192 W   8 W   10 B   10 W   64 B   20 B   EOL

Only 56 bits are required to send this line containing a sequence of 1188 bits.

11.7 CONCLUSION

In this chapter we have presented some of the highlights of source coding. We saw that source coding can be applied to digital data and to waveform signals. Digital data can be reconstructed exactly from a reduced data description of a source if the source exhibits correlation between alphabet elements or if the elements are not equally likely. Waveform signals, in general, experience distortion when represented by a digital description. This distortion can be made arbitrarily small by an appropriate increase in bit rate required to describe the source. Source coding can also be applied to waveform sources to obtain reduced-data-rate descriptions if the source exhibits a long correlation interval or if the possible amplitudes are not equally probable.

The system advantage of source coding is the reduced need for the system resources of bandwidth and/or energy per bit required to deliver a description of the source. This advantage is available in exchange for a third system resource—computation and memory. With the cost of these latter resources continuing to fall as they have over the past decade, source coding promises to fill an ever-increasing role in future communication and storage systems. The interested reader is encouraged to examine References [8, 17, 24–26] dealing with source coding.

REFERENCES

1. Papoulis, A., *Probability, Random Variables, and Stochastic Processes,* McGraw-Hill Book Company, New York, 1965.
2. Harris, F. J., "Windows, Harmonic Analysis, and the Discrete Fourier Transform," *Proc. IEEE,* vol. 67, no., Jan. 1979.
3. Martin, G., "Gyroscopes May Cease Spinning," *IEEE Spectrum,* vol. 23, no. 2, Feb. 1986, pp. 48–53.
4. Vanderkooy, J., and Lipshitz, S. T., "Resolution beyond the Least Significant Bit with Dither," *J. Audio Eng. Soc.,* no. 3, Mar. 1984, pp. 106–112.
5. Blesser, B. A., "Digitalization of Audio: A Comprehensive Examination of Theory, Implementation, and Current Practice," *J. Audio Eng. Soc.,* vol. 26, no. 10, Oct. 1978, pp. 739–771.
6. Sluyter, R. J., "Digitalization of Speech," *Philips Tech. Rev.,* vol. 41, no. 7–8, 1983–84, pp. 201–221.

7. Bell Telephone Laboratories Staff, *Transmission Systems for Communications*, Western Electric Co. Technical Publications, Winston-Salem, N.C., 1971.

8. Jayant, N. S., and Noll, P., *Digital Coding of Waveforms*, Prentice-Hall, Inc., Englewood Cliffs, N.J., 1984.

9. Markel, J. D., and Gray, A. H., Jr., *Linear Prediction of Speech*, Springer-Verlag, New York, 1976.

10. Abate, J., "Linear and Adaptive Delta Modulation," *Proc. IEEE*, vol. 55, Mar. 1967, pp. 298–308.

11. Steele, R., *Delta Modulation Systems*, John Wiley & Sons, Inc., New York, 1975.

12. Cummisky, P., Jayant, N., and Flanagan, J., "Adaptive Quantization in Differential PCM Coding of Speech," *Bell Syst. Tech. J.*, vol. 52, 1973, pp. 115–118.

13. Song, C., Garodnick, J., and Schilling, D., "A Variable Step Size Robust Delta Modulator," *IEEE Trans. Commun.*, vol. 19, no. 6, 1971, pp. 1033–1044.

14. Gersho, A., "Asymptotically Optimal Block Quantization," *IEEE Trans. Inf. Theory*, vol. IT25, no. 4, July 1979, pp. 373–380.

15. Gersho, A., "On the Structure of Vector Quantizers," *IEEE Trans. Inf. Theory*, vol. IT28, no. 2, Mar. 1982, pp. 157–166.

16. Jefffress, L., "Masking," in J. Tobias, ed., *Foundations of Modern Auditory Theory*, Academic Press, Inc., New York, 1970.

17. Lynch, T. J., *Data Compression Techniques and Applications*, Lifetime Learning Publications, New York, 1985.

18. Schafer, R. W., and Rabiner, L. R., "Design of Digital Filter Banks for Speech Analysis," *Bell Syst. Tech. J.*, vol. 50, no. 10, Dec. 1971, pp. 3097–3115.

19. DIGITALKER, *Application Note*, National Semiconductor Linear Applications Manual, 1983.

20. Dudley, H., "The Vocoder," *Bell Lab. Rec.*, no. 18, 1919/40, pp. 122–126.

21. Huffman, D. A., "A Method for the Construction of Minimum Redundancy Codes," *Proc. IRE.*, vol. 40, Sept. 1952, pp. 1098–1101.

22. Hamming, R. W., *Coding and Information Theory*, Prentice-Hall, Inc., Englewood Cliffs, N.J., 1980.

23. Usubuchi, T., Omachi, T., and Iinuma, K., "Adaptive Predictive Coding for Newspaper Facsimile," *Proc. IEEE*, vol. 68, July 1980, pp. 807–812.

24. Viterbi, A. J., and J. K. Omura, *Principles of Digital Communication and Coding*, McGraw-Hill Book Company, New York, 1979.

25. Flanagan, J. L., Schroeder, M. R., Atal, B. S., Crochiere, R. E., Jayant, N. S., and Tribolet, J. M., "Speech Coding," *IEEE Trans. Commun.*, COM27, 1979, pp. 710–737.

26. Held, G., *Data Compression, Techniques and Applications, Hardware and Software Considerations*, John Wiley & Sons, Inc., New York, 1983.

PROBLEMS

11.1 A discrete source generates three independent symbols *A*, *B*, and *C* with probabilities 0.9, 0.08, and 0.02, respectively. Determine the entropy of the source.

11.2. A discrete source generates two dependent symbols *A* and *B* with conditional probabilities

$$P(A|A) = 0.8 \qquad P(B|A) = 0.2$$
$$P(A|B) = 0.6 \qquad P(B|B) = 0.4$$

(a) Determine the probabilities of symbols A and B.

(b) Determine the entropy of the source.

(c) Determine the entropy of the source if the symbols were independent with the same probabilities.

11.3. A 16-bit linear analog-to-digital converter operates over an input range of ± 5.0 V.

(a) Determine the size of a quantile.

(b) Determine the rms quantizing noise voltage.

(c) Determine the average SNR (due to quantizing) for a full-scale sinusoidal input signal.

(d) Consider that the distance traveled on a 100-mile automobile trip is measured to the same accuracy as that of the 16-bit converter. What is the rms error in feet?

11.4. Use the following Microsoft® FORTRAN program to demonstrate the effect of dithering as a linearization technique. The program prompts for an input value to which it adds independent, zero-mean, unit variance random noise prior to truncation and averaging. The subroutine "RAN" is a random number generator. The output is the result of the averaging. See Example 11.5.

```
C   PROGRAM TO DEMONSTRATE THE EFFECT OF DITHERING
         WRITE(*,*) ' '
         WRITE(*,'(A\)') ' ENTER DECIMAL NUMBER TO BE QUANTIZED   -> '
         READ(*,*) ANUMBR
         WRITE(*,'(A\)') ' ENTER NUMBER OF SAMPLES TO BE AVERAGED -> '
         READ(*,*) ICNT
         JSEED=2234
         AVGX=0.0
         VARX=0.0
         ADD=0.5
         IF(ANUMBR.LT.0.0)ADD=-ADD
         WRITE(*,4)
4        FORMAT(/,30X,'QUANTIZE',/,' INDEX',4X,'INPUT',4X,
     C   'DITHERED',3X,'DITHERED',4X,'RUNNING',4X,'RUNNING',/,9X,
     C   'SAMPLE',5X,'SAMPLE',5X,'SAMPLE',7X,'AVG',8X,'VAR',/)
         DO 15 I=1,ICNT
         XDITHR=ANUMBR+RAN(JSEED)-RAN(JSEED)+RAN(JSEED)-RAN(JSEED)
         IXQUANT=IFIX(XDITHR+ADD)
         X=FLOAT(IXQUANT)
         AVGX=AVGX+(X-AVGX)/FLOAT(I)
         VARX=VARX+((AVGX-ANUMBR)*(AVGX-ANUMBR)-VARX)/FLOAT(I)
         WRITE(*,10)I,ANUMBR,XDITHR,IXQUANT,AVGX,SQRT(VARX)
10       FORMAT(I4,4X,F7.3,4X,F7.3,5X,I4,6X,F7.3,5X,F6.4)
15       CONTINUE
         END
```

11.5. A 10-bit A-to-D converter (ADC) is designed to operate over a full-scale range of ± 5.0 V.

(a) Determine the size of a single quantile step.

(b) For a 5.0-V (full-scale) sinusoid, determine the output signal-to-quantizing noise ratio.

(c) For a 0.050-V ($\frac{1}{100}$ of full-scale) sinusoid, determine the output signal-to-quantizing noise ratio.

(d) For an input signal with a Gaussian-distributed amplitude, the probability of saturation is controlled by adjusting the input attenuator so that the saturation level corresponds to four standard deviations. Determine the output signal-to-quantizing noise ratio for this case.

(e) Determine the probability of signal saturation for the signal described in part (d).

11.6. Determine the optimal compression characteristic for the input density function shown in Figure P11.1 (an approximation to a continuous density function).

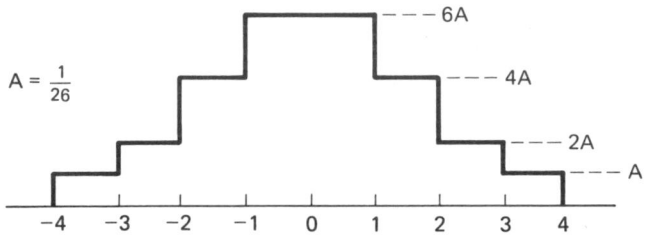

$A = \dfrac{1}{26}$

Figure P11.1

11.7. A 10-bit μ-law converter is designed to operate over a full-scale range of ±5.0 V.
   (a) If μ = 100, determine the output signal-to-quantizing noise ratio for a 5.0-V (full-scale) sinusoid.
   (b) If μ = 100, determine the output signal-to-quantizing noise ratio for a 0.050-V ($\frac{1}{100}$ of full scale) sinusoid.
   (c) Repeat parts (a) and (b) for μ = 250.

11.8. A compact disc (CD) recording system samples each of two stereo signals with a 16-bit A-to-D converter (ADC) at 44.1 kilosamples/s.
   (a) Determine the output signal-to-noise ratio for a full-scale sinusoid.
   (b) If the recorded music is designed to have a crest factor (peak-to-rms ratio) of 20, determine the average output signal-to-quantizing noise ratio.
   (c) The bit stream of digitized data is augmented by the addition of error-correcting bits, substitution bits to aid the clock extraction by a phase-locked loop (PLL), and display and control bit fields. These additional bits represent 100% overhead; that is, 2 bits are stored for each bit generated by the ADC. Determine the output bit rate of the CD recorder system.
   (d) The CD can record an hour's worth of music. Determine the number of bits recorded on a CD.
   (e) For a comparison, a good collegiate dictionary may contain 1500 pages, 2 columns/page, 100 lines/column, 7 words/line, 6 letters/word, and 6 bits/letter. Determine the number of bits required to describe the dictionary and estimate the number of comparable books that can be stored on a CD.

11.9. A 1-bit quantizer is being designed to sample an input sinusoid of amplitude A with uniformly distributed phase. Determine the amplitude X_0, the output level of the 1-bit quantizer, which minimizes the mean-square quantization error.

11.10. A one-step linear predictive filter is to be used to sample a constant-amplitude sinusoid. The ratio of sample frequency to sinusoid frequency is 10.0. Determine the prediction coefficient of the filter. Determine the ratio of output power to input power for the one-tap predictor.

11.11. A two-tap linear predictor filter is being designed to operate in a DPCM system.

The predictor is of the form

$$\hat{X}(n) = a_1 X(n - 1) + a_2 X(n - 2)$$

(a) Determine the values a_1^* and a_2^* which minimize the mean-square prediction error.

(b) Determine the expression for the mean-square prediction error.

(c) Determine the prediction error power if the correlation coefficient of the input signal is of the form

$$C(n) = \begin{cases} 1 - \dfrac{|n|}{4} & n = 0, 1, 2, 3, 4 \\ 0 & \text{otherwise} \end{cases}$$

(d) Determine the prediction error power if the correlation coefficient of the input signal is of the form

$$C(n) = \cos \theta_0 n$$

11.12. A linear delta modulator is designed to operate at six times the Nyquist rate for a signal with a 3-kHz bandwidth. The modulation step size is 250 mV.

(a) Determine the maximum amplitude of an 800-Hz input signal for which the delta modulator is not in the slope-overload condition.

(b) Determine the prefiltered output signal-to-noise ratio for the signal described in part (a).

(c) Determine the postfiltered output signal-to-noise ratio for the signal described in part (a).

11.13. Design a binary Huffman code for a discrete source of three independent symbols A, B, and C with probabilities 0.9, 0.08, and 0.02, respectively. Determine the average code length for the code.

11.14. Design a binary first-order extension code (two symbols at a time) for the discrete source described in Problem 11.13. Determine the average code length per symbol for this code.

11.15. An input alphabet (a keyboard on a word processor) consists of 100 characters.

(a) If the keystrokes are encoded by a fixed-length code, determine the required number of bits for the encoding.

(b) We make the simplifying assumption that 10 of the keystrokes are equally likely and that each occurs with probability 0.05. We also assume that the remaining 90 keystrokes are equally likely. Determine the average number of bits required to encode this alphabet using a variable-length Huffman code.

11.16. Use the CCITT-modified Huffman facsimile code to encode the following single-line sequence of 2047 black-and-white pixels. Determine the ratio of coded bits to input bits.

1W   1B   2W   2B   4W   4B   8W   8B   16W   16B   32W   32B

64W   64B   128W   128B   256W   256B   512W   512B   1W

CHAPTER 12

Encryption
and
Decryption

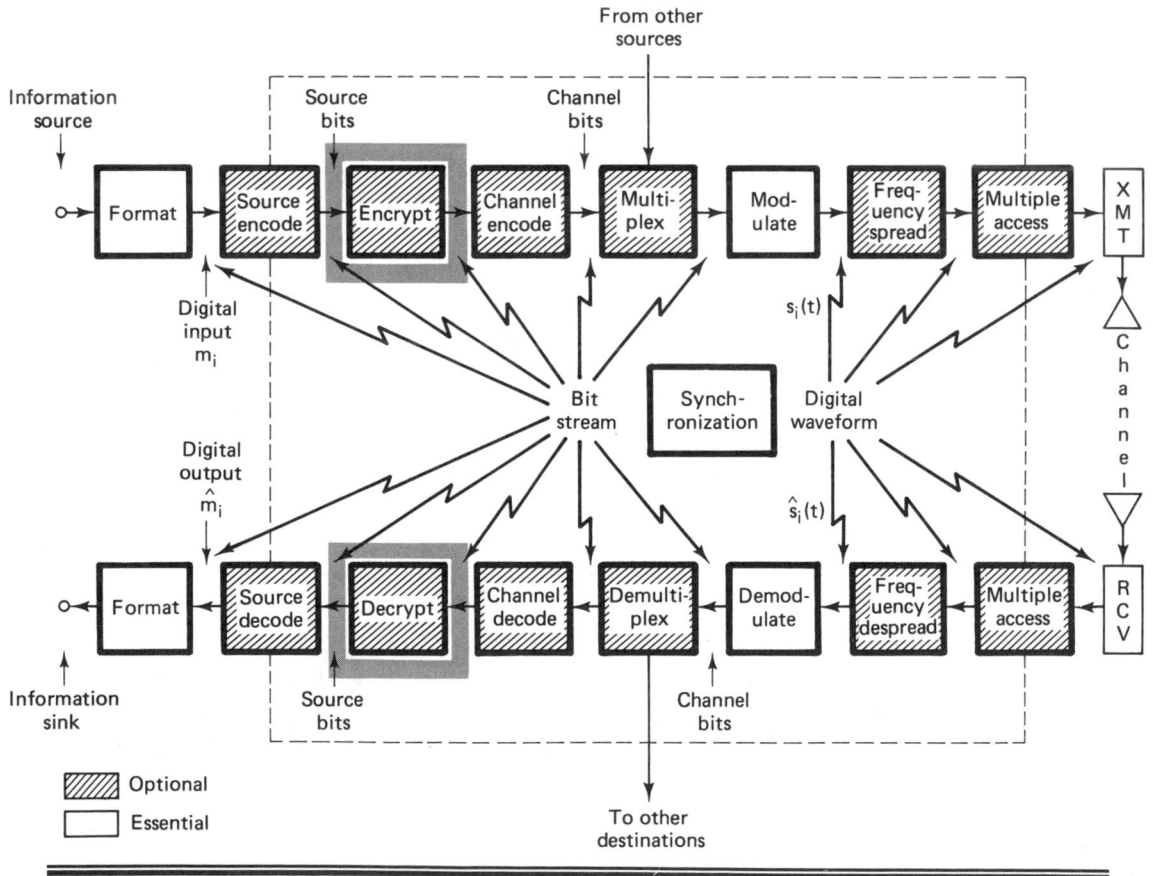

12.1 MODELS, GOALS, AND EARLY CIPHER SYSTEMS

12.1.1 A Model of the Encryption and Decryption Process

The desire to communicate privately is a human trait that dates back to earliest times. Hence the history of secret communications is rich with unique inventions and colorful anecdotes [1]. The study of ways to disguise messages so as to avert unauthorized interception is called *cryptography*. The terms *encipher* and *encrypt* refer to the message transformation performed at the transmitter, and the terms *decipher* and *decrypt* refer to the inverse transformation performed at the receiver. The two primary reasons for using cryptosystems in communications are (1) *privacy*, to prevent unauthorized persons from extracting information from the channel (eavesdropping); and (2) *authentication*, to prevent unauthorized persons from injecting information into the channel (spoofing). Sometimes, as in the case of electronic funds transfer or contract negotiations, it is important to provide the electronic equivalent of a *written signature* in order to avoid or settle any dispute between the sender and receiver as to what message, if any, was sent.

Figure 12.1 illustrates a model of a cryptographic channel. A message, or plaintext, M, is encrypted by the use of an invertible transformation, E_K, that produces a ciphertext, $C = E_K(M)$. The ciphertext is transmitted over an insecure or *public channel*. When an authorized receiver obtains C, he decrypts it with the inverse transformation, $D_K = E_K^{-1}$, to obtain the original plaintext message, as follows:

$$D_K(C) = E_K^{-1}[E_K(M)] = M \qquad (12.1)$$

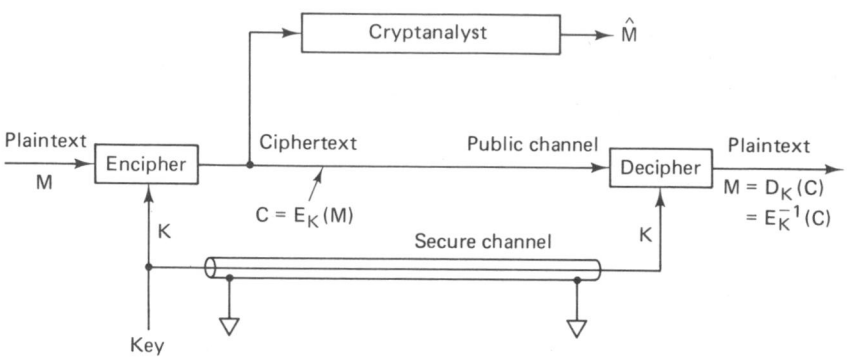

Figure 12.1 Model of a cryptographic channel.

The parameter K refers to a set of symbols or characters called a *key*, which dictates a specific encryption transformation, E_K, from a family of cryptographic transformations. Originally, the security of cryptosystems depended on the secrecy of the entire encryption process, but eventually systems were developed for which the general nature of the encryption transformation or algorithm could be publicly revealed, since the security of the system depended on the specific key. The key is supplied along with the plaintext message for encryption, and along with the ciphertext message for decryption. There is a close analogy here with a general-purpose computer and a computer program. The computer, like the cryptosystem, is capable of a large variety of transformations, from which the computer program, like the specific key, selects one. In most cryptosystems, anyone with access to the key can both encrypt and decrypt messages. The key is transmitted to the community of authorized users over a secure channel (as an example, a courier may be used to hand-carry the sensitive key information); the key usually remains unchanged for a considerable number of transmissions. The goal of the *cryptanalyst* (eavesdropper or adversary) is to produce an estimate of the plaintext, $\hat{M}$, by analyzing the ciphertext obtained from the public channel, without benefit of the key.

Encryption schemes fall into two generic categories: *block encryption*, and *data-stream* or simply *stream encryption*. With block encryption, the plaintext is segmented into blocks of fixed size; each block is encrypted independently from the others. For a given key, a particular plaintext block will therefore be carried into the same ciphertext block each time it appears (similar to block encoding). With data-stream encryption, similar to convolutional coding, there is no fixed block size. Each plaintext bit, m_i, is encrypted with the ith element, k_i, of a sequence of symbols (key stream) generated with the key. The encryption is *periodic* if the key stream repeats itself after p characters for some fixed p; otherwise, it is nonperiodic.

In general, the properties desired in an encryption scheme are quite different from those desired in a channel coding scheme. For example, with encryption, plaintext data should never appear directly in the ciphertext, but with channel coding, codes are often in *systematic form* comprised of unaltered message bits plus parity bits (see Section 5.4.5). Consider another example of the differences between encryption and channel coding. With block encryption, a single bit error at the input of the decryptor might change the value of many of the output bits in the block. This effect, known as *error propagation*, is often a desirable cryptographic property since it makes it difficult for unauthorized users to succeed in spoofing a system. However, in the case of channel coding, we would like the system to correct as many errors as possible, so that the output is relatively unaffected by input errors.

12.1.2 System Goals

The major requirements for a cryptosystem can be stated as follows:

1. To provide an *easy* and *inexpensive* means of encryption and decryption to all authorized users in possession of the appropriate key
2. To ensure that the cryptanalyst's task of producing an estimate of the plaintext without benefit of the key is made *difficult* and *expensive*

Successful cryptosystems are classified as being either *unconditionally secure* or *computationally secure*. A system is said to be *unconditionally secure* when the amount of information available to the cryptanalyst is insufficient to determine the encryption and decryption transformations, no matter how much computing power the cryptanalyst has available. One such system, called a *one-time pad*, involves encrypting a message with a random key that is used one time only. The key is never reused; hence the cryptanalyst is denied information that might be useful against subsequent transmissions with the same key. Although such a system is unconditionally secure (see Section 12.2.1), it has limited use in a conventional communication system, since a new key would have to be distributed for each new message—a great logistical burden. The distribution of keys to the authorized users is a major problem in the operation of any cryptosystem, even when a key is used for an extended period of time. Although some systems can be proven to be unconditionally secure, currently there is no known way to demonstrate security for an arbitrary cryptosystem. Hence the specifications for most cryptosystems rely on the less formal designation of *computational security* for x number of years, which means that under circumstances favorable to the cryptanalyst (i.e., using state-of-the-art computers) the system security could be broken in a period of x years, but could not be broken in less than x years.

12.1.3 Classic Threats

The weakest classification of cryptanalytic threat on a system is called a *ciphertext-only attack*. In this attack the cryptanalyst might have *some* knowledge of the general system and the language used in the message, but the only significant data available to him is the encrypted transmission intercepted from the public channel.

A more serious threat to a system is called a *known plaintext attack*; it involves knowledge of the plaintext *and* knowledge of its ciphertext counterpart. The rigid structure of most business forms and programming languages often provides an opponent with much a priori knowledge of the details of the plaintext message. Armed with such knowledge and with a ciphertext message, the cryptanalyst can mount a known plaintext attack. In the diplomatic arena, if an encrypted message directs a foreign minister to make a particular public statement, and if he does so without paraphrasing the message, the cryptanalyst may be privy to both the ciphertext *and* its exact plaintext translation. While a known plaintext attack is not always possible, its occurrence is frequent enough that a system is not considered secure unless it is designed to be secure against the plaintext attack [2].

When the cryptanalyst is in the position of *selecting* the plaintext, the threat is termed a *chosen plaintext attack*. Such an attack was used by the United States to learn more about the Japanese cryptosystem during World War II. On May 20, 1942, Admiral Yamamoto, Commander-in-Chief of the Imperial Japanese Navy, issued an order spelling out the detailed tactics to be used in the assault of Midway island. This order was intercepted by the Allied listening posts. By this time, the Americans had learned enough of the Japanese code to decrypt most of the message. Still in doubt, however, were some important parts, such as the *place* of the assault. They suspected that the characters ''AF'' meant Midway island, but to be sure, Joseph Rochefort, head of the Combat Intelligence Unit, decided to use a chosen plaintext attack to trick the Japanese into providing concrete proof. He had the Midway garrison broadcast a distinctive plaintext message in which Midway reported that its fresh-water distillation plant had broken down. The American cryptanalysts needed to wait only two days before they intercepted a Japanese ciphertext message stating that AF was short of fresh water [1].

12.1.4 Classic Ciphers

One of the earliest examples of a monoalphabetic cipher was the *Caesar Cipher*, used by Julius Caesar during the Gallic wars. Each plaintext letter is replaced with a new letter obtained by an *alphabetic shift*. Figure 12.2a illustrates such an encryption transformation, consisting of three end-around shifts of the alphabet. When using this Caesar's alphabet, the message, ''now is the time'' is encrypted as follows:

Plaintext: N O W I S T H E T I M E

Ciphertext: Q R Z L V W K H W L P H

The decryption key is simply the number of alphabetic shifts; the code is changed by choosing a new key. Another classic cipher system, illustrated in Figure 12.2b, is called the *Polybius square*. Letters I and J are first combined and treated as a single character since the final choice can easily be decided from the context of the message. The resulting 25 character alphabet is arranged in a 5 × 5 array.

Plaintext:	A B C D E F G H I J K L M N O P Q R S T U V W X Y Z
Chiphertext:	D E F G H I J K L M N O P Q R S T U V W X Y Z A B C

(a)

	1	2	3	4	5
1	A	B	C	D	E
2	F	G	H	IJ	K
3	L	M	N	O	P
4	Q	R	S	T	U
5	V	W	X	Y	Z

(b)

Figure 12.2 (a) Caesar's alphabet with a shift of 3. (b) Polybius square.

Encryption of any character is accomplished by choosing the appropriate row–column (or column-row) number pair. An example of encryption with the use of the Polybius square follows:

Plaintext: N O W I S T H E T I M E

Ciphertext: 33 43 25 42 34 44 32 51 44 42 23 51

The code is changed by a rearrangement of the letters in the 5 × 5 array.

The *Trithemius progressive key*, shown in Figure 12.3, is an example of a *polyalphabetic cipher*. The row labeled shift 0 is identical to the usual arrangement of the alphabet. The letters in the next row are shifted one character to the left with an end-around shift for the leftmost position. Each successive row follows the same pattern of shifting the alphabet one character to the left as compared to the prior row. This continues until the alphabet has been depicted in all possible arrangements of end-around shifts. One method of using such an alphabet is to select the first cipher character from the shift 1 row, the second cipher character from the shift 2 row, and so on. An example of such encryption is

Plaintext: N O W I S T H E T I M E

Ciphertext: O Q Z M X Z O M C S X Q

There are several interesting ways that the Trithemius progressive key can be used. One way, called the *Vigenere key method*, employs a keyword. The key dictates the row choices for encryption and decryption of each successive character in the message. For example, suppose that the word "TYPE" is selected as the key; then an example of the Vigenere encryption method is

Key: T Y P E T Y P E T Y P E

Plaintext: N O W I S T H E T I M E

Ciphertext: G M L M L R W I M G B I

Plaintext:

Shift:	a	b	c	d	e	f	g	h	i	j	k	l	m	n	o	p	q	r	s	t	u	v	w	x	y	z
0	A	B	C	D	E	F	G	H	I	J	K	L	M	N	O	P	Q	R	S	T	U	V	W	X	Y	Z
1	B	C	D	E	F	G	H	I	J	K	L	M	N	O	P	Q	R	S	T	U	V	W	X	Y	Z	A
2	C	D	E	F	G	H	I	J	K	L	M	N	O	P	Q	R	S	T	U	V	W	X	Y	Z	A	B
3	D	E	F	G	H	I	J	K	L	M	N	O	P	Q	R	S	T	U	V	W	X	Y	Z	A	B	C
4	E	F	G	H	I	J	K	L	M	N	O	P	Q	R	S	T	U	V	W	X	Y	Z	A	B	C	D
5	F	G	H	I	J	K	L	M	N	O	P	Q	R	S	T	U	V	W	X	Y	Z	A	B	C	D	E
6	G	H	I	J	K	L	M	N	O	P	Q	R	S	T	U	V	W	X	Y	Z	A	B	C	D	E	F
7	H	I	J	K	L	M	N	O	P	Q	R	S	T	U	V	W	X	Y	Z	A	B	C	D	E	F	G
8	I	J	K	L	M	N	O	P	Q	R	S	T	U	V	W	X	Y	Z	A	B	C	D	E	F	G	H
9	J	K	L	M	N	O	P	Q	R	S	T	U	V	W	X	Y	Z	A	B	C	D	E	F	G	H	I
10	K	L	M	N	O	P	Q	R	S	T	U	V	W	X	Y	Z	A	B	C	D	E	F	G	H	I	J
11	L	M	N	O	P	Q	R	S	T	U	V	W	X	Y	Z	A	B	C	D	E	F	G	H	I	J	K
12	M	N	O	P	Q	R	S	T	U	V	W	X	Y	Z	A	B	C	D	E	F	G	H	I	J	K	L
13	N	O	P	Q	R	S	T	U	V	W	X	Y	Z	A	B	C	D	E	F	G	H	I	J	K	L	M
14	O	P	Q	R	S	T	U	V	W	X	Y	Z	A	B	C	D	E	F	G	H	I	J	K	L	M	N
15	P	Q	R	S	T	U	V	W	X	Y	Z	A	B	C	D	E	F	G	H	I	J	K	L	M	N	O
16	Q	R	S	T	U	V	W	X	Y	Z	A	B	C	D	E	F	G	H	I	J	K	L	M	N	O	P
17	R	S	T	U	V	W	X	Y	Z	A	B	C	D	E	F	G	H	I	J	K	L	M	N	O	P	Q
18	S	T	U	V	W	X	Y	Z	A	B	C	D	E	F	G	H	I	J	K	L	M	N	O	P	Q	R
19	T	U	V	W	X	Y	Z	A	B	C	D	E	F	G	H	I	J	K	L	M	N	O	P	Q	R	S
20	U	V	W	X	Y	Z	A	B	C	D	E	F	G	H	I	J	K	L	M	N	O	P	Q	R	S	T
21	V	W	X	Y	Z	A	B	C	D	E	F	G	H	I	J	K	L	M	N	O	P	Q	R	S	T	U
22	W	X	Y	Z	A	B	C	D	E	F	G	H	I	J	K	L	M	N	O	P	Q	R	S	T	U	V
23	X	Y	Z	A	B	C	D	E	F	G	H	I	J	K	L	M	N	O	P	Q	R	S	T	U	V	W
24	Y	Z	A	B	C	D	E	F	G	H	I	J	K	L	M	N	O	P	Q	R	S	T	U	V	W	X
25	Z	A	B	C	D	E	F	G	H	I	J	K	L	M	N	O	P	Q	R	S	T	U	V	W	X	Y

Figure 12.3 Trithemius progressive key.

where the first letter, T, of the key indicates that the row choice for encrypting the first plaintext character is the row starting with T (shift 19). The next row choice starts with Y (shift 24), and so on. A variation of this key method, called the *Vigenere auto (plain) key method*, starts with a single letter or word used as a *priming key*. The priming key dictates the starting row or rows for encrypting the first or first few plaintext characters, as in the preceding example. Next, the *plaintext characters* themseleves are used as the key for choosing the rows for encryption. An example using the letter "F" as the priming key follows:

Key:	F N O W I S T H E T I M
Plaintext:	N O W I S T H E T I M E
Ciphertext:	S B K E A L A L X B U Q

With the auto key method, it should be clear that feedback has been introduced to the encryption process. With this feedback, the choice of the ciphertext is dictated by the contents of the message.

A final variation of the Vigenere method, called the *Vigenere auto (cipher) key method*, is similar to the plain key method in that a priming key and feedback are used. The difference is that after encryption with the priming key, each successive key character in the sequence is obtained from the prior *ciphertext character* instead of from the plaintext character. An example should make this clear. As before, the letter "F" is used as the priming key.

Key:	F S G C K C V C G Z H T
Plaintext:	N O W I S T H E T I M E
Ciphertext:	S G C K C V C G Z H T X

Although each key character can be found from its preceding ciphertext character, it is functionally dependent on *all* the preceding characters in the message plus the priming key. This has the effect of diffusing the statistical properties of the plaintext across the ciphertext, making statistical analysis very difficult for a cryptanalyst. One weakness of the cipher key example depicted here is that the ciphertext contains key characters which will be exposed on the public channel "for all to see." Variations of this method can be employed to prevent such overt exposure [3]. By today's standards Vigenere's encryption schemes are not very secure; his basic contribution was the discovery that nonrepeating key sequences could be generated by using the messages themselves or functions of the messages.

12.2 THE SECRECY OF A CIPHER SYSTEM

12.2.1 Perfect Secrecy

Consider a cipher system with a finite message space $\{M\} = M_0, M_1, \ldots, M_{N-1}$ and a finite ciphertext space $\{C\} = C_0, C_1, \ldots, C_{U-1}$. For any M_i, the a priori probability that M_i is transmitted is $P(M_i)$. Given that C_j is received, the a posteriori probability that M_i was transmitted is $P(M_i|C_j)$. A cipher system is said to have *perfect secrecy* if for every message M_i and every ciphertext C_j, the a posteriori probability is equal to the a priori probability:

$$P(M_i|C_j) = P(M_i) \tag{12.2}$$

Thus for a system with perfect secrecy, a cryptanalyst who intercepts C_j obtains no further information to enable him or her to determine which message was transmitted. A necessary and sufficient condition for perfect secrecy is that for every M_i and C_j,

$$P(C_j|M_i) = P(C_j) \tag{12.3}$$

The schematic in Figure 12.4 illustrates an example of perfect secrecy. In this example, $\{M\} = M_0, M_1, M_2, M_3, \{C\} = C_0, C_1, C_2, C_3, \{K\} = K_0, K_1, K_2, K_3, N = U = 4$, and $P(M_i) = P(C_j) = \frac{1}{4}$. The transformation from message

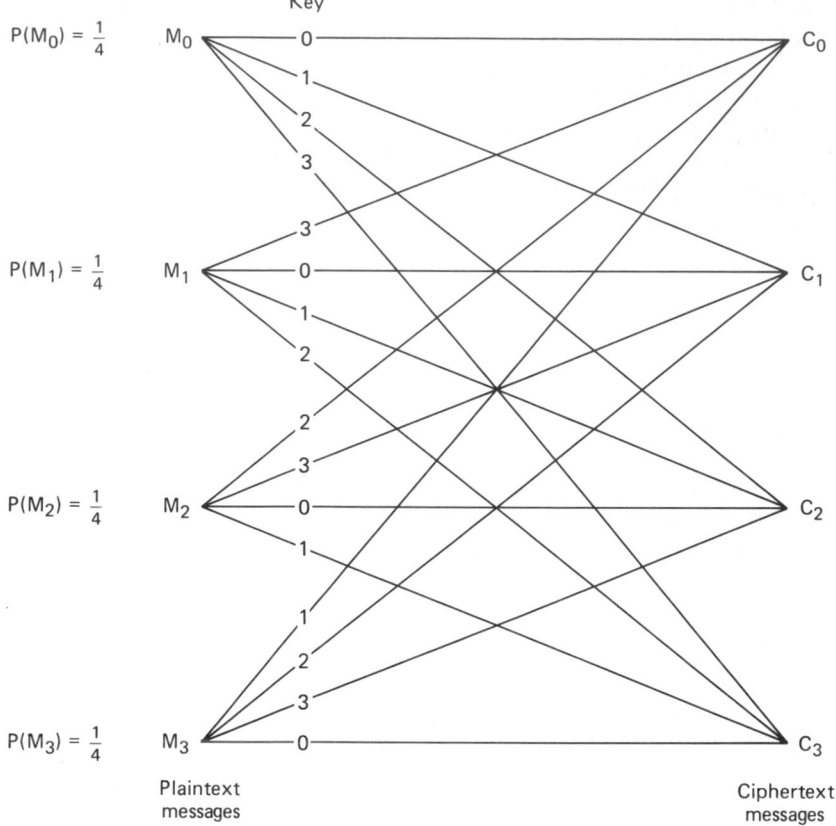

Figure 12.4 Example of perfect secrecy.

to ciphertext is obtained by

$$C_s = T_{K_j}(M_i)$$

$$s = (i + j) \text{ modulo-}N$$

(12.4)

where T_{K_j} indicates a transformation under the key, K_j, and x modulo-y is defined as the remainder of dividing x by y. Thus $s = 0, 1, 2, 3$. A cryptanalyst intercepting one of the ciphertext messages $C_s = C_0, C_1, C_2$, or C_3 would have no way of determining which of the four keys was used, and therefore whether the correct message is M_0, M_1, M_2, or M_3. A cipher system in which the number of messages, the number of keys, and the number of ciphertext transformations are all equal is said to have perfect secrecy if and only if the following two conditions are met:

1. There is only one key transforming each message to each ciphertext.
2. All keys are equally likely.

If these conditions are not met, there would be some message M_i such that for a given C_j, there is no key that can decipher C_j into M_i, implying that $P(M_i|C_j)$ = 0 for some i and j. The cryptanalyst could then eliminate certain plaintext messages from consideration, thereby simplifying the task. Perfect secrecy is a very desirable objective since it means that the cipher system is unconditionally secure. It should be apparent, however, that for systems which transmit a large number of messages, the amount of key that must be distributed for perfect secrecy can result in formidable management problems, making such systems impractical. Since in a system with perfect secrecy, the number of different keys is at least as great as the number of possible messages, if we allow messages of unlimited length, perfect secrecy requires an infinite amount of key.

Example 12.1   Breaking a Cipher System When the Key Space Is Smaller Than the Message Space

Consider that the 29-character ciphertext

G R O B O K B O D R O R O B Y O C Y P I O C D O B I O K B

was produced by a Caesar cipher (see Section 12.1.4) such that each letter has been shifted by K positions, where $1 \leq K \leq 25$. Show how a cryptanalyst can break this code.

Solution

Because the number of possible keys (there are 25) is smaller than the number of possible 29-character meaningful messages (there are a myriad), perfect secrecy cannot be achieved. In the original polyalphabetic cipher of Figure 12.3, a plaintext character is replaced by a letter of increasingly higher rank as the row number (K) increases. Hence, in analyzing the ciphertext, we reverse the process by creating rows such that each ciphertext letter is replaced by letters of decreasing rank. The cipher is easily broken by trying all the keys, from 1 to 25, as shown in Figure 12.5, yielding only one key ($K = 10$) that produces the meaningful message: WHERE ARE THE HEROES OF YESTERYEAR (The spaces have been added.)

Example 12.2   Perfect Secrecy

We can modify the key space of Example 12.1 to create a cipher having perfect secrecy. In this new cipher system each character in the message is encrypted using a *randomly selected* key value. The key, K, is now given by the sequence k_1, k_2, . . . , k_{29}, where each k_i is a random integer in the range (1, 25) dictating the shift used for the ith character; thus there are a total of $(25)^{29}$ different key sequences. Then the 29-character ciphertext in Example 12.1 could correspond to *any* meaningful 29-character message. For example, the ciphertext could correspond to the plaintext (the spaces have been added)

ENGLISH AND FRENCH ARE SPOKEN HERE

derived by the key: 2, 4, 8, 16, 6, 18, 20, ... Most of the 29-character possibilities can be ruled out because they are not meaningful messages (this much is known without the ciphertext). Perfect secrecy is achieved because interception of the ciphertext in this system reveals no additional information about the plaintext message.

Text

0	G R O B O K B O D R O R O B Y O C Y P I O C D O B I O K B
1	F Q N A N J A N C Q N Q N A X N B X O H N B C N A H N J A
2	E P M Z M I Z M B P M P M Z W M A W N G M A B M Z G M I Z
3	D O L Y L H Y L A O L O L Y V L Z V M F L Z A L Y F L H Y
4	C N K X K G X K Z N K N K X U K Y U L E K Y Z K X E K G X
5	B M J W J F W J Y M J M J W T J X T K D J X Y J W D J F W
6	A L I V I E V I X L I L I V S I W S J C I W X I V C I E V
7	Z K H U H D U H W K H K H U R H V R I B H V W H U B H D U
8	Y J G T G C T G V J G J G T Q G U Q H A G U V G T A G C T
9	X I F S F B S F U I F I F S P F T P G Z F T U F S Z F B S
10	W H E R E A R E T H E H E R O E S O F Y E S T E R Y E A R
11	V G D Q D Z Q D S G D G D Q N D R N E X D R S D Q X D Z Q
12	U F C P C Y P C R F C F C P M C Q M D W C Q R C P W C Y P
13	T E B O B X O B Q E B E B O L B P L C V B P Q B O V B X O
14	S D A N A W N A P D A D A N K A O K B U A O P A N U A W N
15	R C Z M Z V M Z O C Z C Z M J Z N J A T Z N O Z M T Z V M
16	Q B Y L Y U L Y N B Y B Y L I Y M I Z S Y M N Y L S Y U L
17	P A X K X T K X M A X A X K H X L H Y R X L M X K R X T K
18	O Z W J W S J W L Z W Z W J G W K G X Q W K L W J Q W S J
19	N Y V I V R I V K Y V Y V I F V J F W P V J K V I P V R I
20	M X U H U Q H U J X U X U H E U I E V O U I J U H O U Q H
21	L W T G T P G T I W T W T G D T H D U N T H I T G N T P G
22	K V S F S O F S H V S V S F C S G C T M S G H S F M S O F
23	J U R E R N E R G U R U R E B R F B S L R F G R E L R N E
24	I T Q D Q M D Q F T Q T Q D A Q E A R K Q E F Q D K Q M D
25	H S P C P L C P E S P S P C Z P D Z Q J P D E P C J P L C

Figure 12.5 Example of breaking a cipher system when the key space is smaller than the message space.

12.2.2 Entropy and Equivocation

As discussed in Chapter 7, the amount of information in a message is related to the probability of occurrence of the message. Messages with probability of either 0 or 1 contain no information, since we can be very confident concerning our prediction of their occurrence. The more uncertainty there is in predicting the occurrence of a message, the greater is the information content. Hence when each of the messages in a set is equally likely, we can have *no* confidence in our ability to predict the occurrence of a particular message, and the uncertainty or information content of the message is maximum.

Entropy, $H(X)$, is defined as the average amount of information per message. It can be considered a measure of how much *choice* is involved in the selection of a message, X. It is expressed by the following summation over all possible messages:

$$H(X) = - \sum_X P(X) \log_2 P(X) = \sum_X P(X) \log_2 \frac{1}{P(X)} \qquad (12.5)$$

When the logarithm is taken to the base 2, as shown, $H(X)$ is the *expected number of bits* in an *optimally encoded* message, X. This is not quite the measure that a cryptanalyst desires. He will have intercepted some ciphertext and will want to know how confidently he can predict a message (or key) given that this particular ciphertext was sent. *Equivocation*, $H(X|Y)$, defined as the conditional entropy of X given Y, is a more useful measure for the cryptanalyst in attempting to break the cipher.

$$H(X|Y) = - \sum_{X,Y} P(X, Y) \log_2 P(X|Y)$$

$$= \sum_Y P(Y) \sum_X P(X|Y) \log_2 \frac{1}{P(X|Y)} \qquad (12.6)$$

Equivocation can be thought of as the uncertainty that message X was sent, having received Y. The cryptanalyst would like $H(X|Y)$ to approach zero as the amount of intercepted ciphertext, Y, increases.

Example 12.3   Entropy and Equivocation

Consider a sample message set consisting of eight equally likely messages $\{X\} = X_1, X_2, \ldots, X_8$.

(a) Find the entropy associated with a message from the set $\{X\}$.
(b) Given another equally likely message set $\{Y\} = Y_1, Y_2$. Consider that the occurrence of each message Y narrows the possible choices of X in the following way:

If Y_1 is present: only $X_1, X_2, X_3,$ or X_4 is possible

If Y_2 is present: only $X_5, X_6, X_7,$ or X_8 is possible

Find the equivocation of message X conditioned on message Y.

Solution

(a) $P(X) = \frac{1}{8}$
   $H(X) = 8[(\frac{1}{8}) \log_2 8] = 3$ bits/message

(b) $P(Y) = \frac{1}{2}$. For each Y, $P(X|Y) = \frac{1}{4}$ for four of the X's and $P(X|Y) = 0$ for the remaining four X's. Using Equation (12.6), we obtain

$$H(X|Y) = 2[(\frac{1}{2})4(\frac{1}{4} \log_2 4)] = 2 \text{ bits/message}$$

We see that knowledge of Y has reduced the uncertainty of X from 3 bits/message to 2 bits/message.

12.2.3 Rate of a Language and Redundancy

The *true rate* of a language, r, is defined as the average number of *information bits* contained in each character and is expressed for messages of length N by

$$r = \frac{H(X)}{N} \tag{12.7}$$

where $H(X)$ is the message entropy, or the number of bits in the *optimally encoded* message. For large N, estimates of r for written English range between 1.0 and 1.5 bits/character [4]. The *absolute rate* or maximum entropy, r', of a language is defined as the maximum number of information bits contained in each character assuming that all possible sequences of characters are equally likely. The absolute rate is given by

$$r' = \log_2 L \tag{12.8}$$

where L is the number of characters in the language. For the English alphabet $r' = \log_2 26 = 4.7$ bits/character. The true rate of English is, or course, much less than its absolute rate since, like most languages, English is highly redundant and structured.

The *redundancy*, D, of a language is defined in terms of its true rate and absolute rate as follows:

$$D = r' - r \tag{12.9}$$

For the English language with $r' = 4.7$ bits/character and $r = 1.5$ bits/character, $D = 3.2$, and the ratio $D/r' = 0.68$ is a measure of the redundancy in the language.

12.2.4 Unicity Distance and Ideal Secrecy

We stated earlier that perfect secrecy requires an infinite amount of key if we allow messages of unlimited length. With a finite key size, the equivocation of the key $H(K|C)$ generally approaches zero, implying that the key can be uniquely determined and the cipher system can be broken. The *unicity distance* is defined as the smallest amount of ciphertext, N, such that the key equivocation $H(K|C)$ is close to zero. Therefore, the unicity distance is the amount of ciphertext needed to uniquely determine the key and thus break the cipher system. Shannon [5] described an *ideal secrecy* system as one in which $H(K|C)$ does not approach zero as the amount of ciphertext approaches infinity; that is, no matter how much ciphertext is intercepted, the key cannot be determined. The term "ideal secrecy" describes a system that does not achieve perfect secrecy but is nonetheless unbreakable (unconditionally secure) because it does not reveal enough information to determine the key.

Most cipher systems are too complex to determine the probabilities required to derive the unicity distance. However, it is sometimes possible to approximate unicity distance, as shown by Shannon [5] and Hellman [6]. Following Hellman, assume that each plaintext and ciphertext message comes from a finite alphabet of L symbols. Thus there are $2^{r'N}$ possible messages of length N, where r' is the absolute rate of the language. We can consider the total message space partitioned into two classes, meaningful messages, M_1, and meaningless messages M_2:

$$\text{number of meaningful messages} = 2^{rN} \tag{12.10}$$

$$\text{number of meaningless messages} = 2^{r'N} - 2^{rN} \tag{12.11}$$

where r is the true rate of the language, and where the a priori probabilities of the message classes are

$$P(M_1) = \frac{1}{2^{rN}} = 2^{-rN} \qquad M_1 \text{ meaningful} \tag{12.12}$$

$$P(M_2) = 0 \qquad M_2 \text{ meaningless} \tag{12.13}$$

Let us assume that there are $2^{H(K)}$ possible keys (size of the key alphabet), where $H(K)$ is the entropy of the key (number of bits in the key). Assume that all keys are equally likely, that is,

$$P(K) = \frac{1}{2^{H(K)}} = 2^{-H(K)} \tag{12.14}$$

The derivation of the unicity distance is based on a *random cipher* model, which states that for each key K and ciphertext C, the decryption operation $D_K(C)$ yields an independent random variable distributed over all the possible $2^{r'N}$ messages (both meaningful and meaningless). Therefore, for a given K and C, the $D_K(C)$ operation can produce any one of the plaintext messages with equal probability.

Given an encryption described by $C_i = E_{K_i}(M_i)$, a *false solution*, F, arises whenever encryption under another key K_j could also produce C_i either from the message M_i or from some other message M_j; that is,

$$C_i = E_{K_i}(M_i) = E_{K_j}(M_i) = E_{K_j}(M_j) \tag{12.15}$$

A cryptanalyst intercepting C_i would not be able to pick the correct key and hence could not break the cipher system. We are not concerned with the decryption operations that produce *meaningless* messages because these are easily rejected.

For every correct solution to a particular ciphertext there are $2^{H(K)} - 1$ incorrect keys, each of which has the same probability $P(F)$ of yielding a false solution. Because each meaningful plaintext message is assumed equally likely, the probability of a false solution, $P(F)$, is the same as the probability of getting a meaningful message.

$$P(F) = \frac{2^{rN}}{2^{r'N}} = 2^{(r-r')N} = 2^{-DN} \tag{12.16}$$

where $D = r' - r$ is the redundancy of the language. The expected number of false solutions $\overline{F}$ is then

$$\overline{F} = [2^{H(K)} - 1]P(F) = [2^{H(K)} - 1]2^{-DN} \tag{12.17}$$
$$\simeq 2^{H(K)-DN}$$

Because of the rapid decrease of $\overline{F}$ with increasing N,

$$\log_2 \overline{F} = H(K) - DN = 0 \qquad (12.18)$$

is defined as the point where the number of false solutions is sufficiently small so that the cipher can be broken. The resulting unicity distance is therefore

$$N = \frac{H(K)}{D} \qquad (12.19)$$

We can see from Equation (12.17) that if $H(K)$ is much larger than DN, there will be a large number of meaningful decryptions, and thus a small likelihood of a cryptanalyst distinguishing which meaningful message is the correct message. In a loose sense, DN represents the number of equations available for solving for the key, and $H(K)$ the number of unknowns. When the number of equations is smaller than the number of unknown key bits, a unique solution is not possible and the system is said to be unbreakable. When the number of equations is larger than the number of unknowns, a unique solution is possible and the system can no longer be characterized as unbreakable (although it may still be computationally secure).

It is the predominance of meaningless decryptions that enables cryptograms to be broken. Equation (12.19) indicates the value of using *data compression* techniques prior to encryption. Data compression removes redundancy, thereby increasing the unicity distance. Perfect data compression would result in $D = 0$ and $N = \infty$ for any key size.

Example 12.4   Unicity Distance

Calculate the unicity distance for a written English encryption system, where the key is given by the sequence $k_1, k_2, \ldots, k_{29}$, where each k_i is a random integer in the range $(1, 25)$ dictating the shift number (Figure 12.3) for the ith character. Assume that each of the possible key sequences is equally likely.

Solution

There are $(25)^{29}$ possible key sequences, each of which is equally likely. Therefore, using Equations (12.5), (12.8), and (12.19) we have:

Key entropy: $H(K) = \log_2 (25)^{29} \approx 135$ bits

Absolute rate for English: $r' = \log_2 26 = 4.7$ bits/character

Assumed true rate for English: $r = 1.5$ bits/character

Redundancy: $D = r' - r = 3.2$ bits/character

$$N = \frac{H(K)}{D} = \frac{135}{3.2} \cong 43 \text{ characters}$$

In Example 12.2, perfect secrecy was illustrated using the same type of key sequence described here, with a 29-character message. In this example we see that if the available ciphertext is 43 characters long (which implies that some portion of the key sequence must be used twice), a unique solution may be possible. However, there is no indication as to the computational difficulty in finding the solution. Even though we have estimated the theoretical amount of ciphertext required to break the cipher, it might be computationally infeasible to accomplish this.

12.3 PRACTICAL SECURITY

For ciphertext sequences greater than the unicity distance any system can be solved, in principle, merely by trying each possible key until the unique solution is obtained. This is completely impractical, however, except when the key is extremely small. For example, for a key configured as a permutation of the alphabet, there are $26! \simeq 4 \times 10^{26}$ possibilities (considered small in the cryptographic context). In an exhaustive search, one might expect to reach the right key at about halfway through the search. If we assume that each trial requires a computation time of 1 μs, the total search time exceeds 10^{12} years. Hence techniques other than a brute-force search (e.g., statistical analysis) must be employed if a cryptanalyst is to have any hope of success.

12.3.1 Confusion and Diffusion

A statistical analysis using the frequency of occurrence of individual characters and character combinations can be used to solve many cipher systems. Shannon [5] suggested two encryption concepts for frustrating the statistical endeavors of the cryptanalyst. He termed these encryption transformations confusion and diffusion. *Confusion* involves substitutions that render the final relationship between the key and ciphertext as complex as possible. This makes it difficult to utilize a statistical analysis to narrow the search to a particular subset of the key variable space. Confusion ensures that the majority of the key is needed to decrypt even very short sequences of ciphertext. *Diffusion* involves transformations that smooth out the statistical differences between characters and between character combinations. An example of diffusion with a 26-letter alphabet is to transform a message sequence $M = M_0, M_1, \ldots$ into a new message sequence $Y = Y_0, Y_1, \ldots$ as follows:

$$Y_n = \sum_{i=0}^{s-1} M_{n+i} \qquad \text{modulo-26} \qquad (12.20)$$

where each character in the sequence is regarded as an integer modulo-26, s is some chosen integer, and $n = 1, 2, \ldots$. The new message, Y, will have the same redundancy as the original message, M, but the letter frequencies of Y will be more uniform than in M. The effect is that the cryptanalyst needs to intercept a longer sequence of ciphertext before any statistical analysis can be useful.

12.3.2 Substitution

Substitution encryption techniques, such as the Caesar cipher and the Trithemius progressive key cipher, are widely used in puzzles. Such simple substitution ciphers offer little encryption protection. For a substitution technique to fulfill Shannon's concept of *confusion*, a more complex relationship is required. Figure 12.6 shows one example of providing greater substitution complexity through the use of a nonlinear transformation. In general, n input bits are first represented as one of 2^n different characters (binary-to-octal transformation in the example of Figure 12.6). The set of 2^n characters are then permuted so that each character is transposed to one of the others in the set. The character is then converted back to an n-bit output.

It can be easily shown that there are $(2^n)!$ different substitution or connection patterns possible. The cryptanalyst's task becomes computationally unfeasible as n gets large, say $n = 128$; then $2^n = 10^{38}$, and $(2^n)!$ is an astronomical number. We recognize that for $n = 128$, this substitution box (S-box) transformation is complex (confusion). However, although we can identify the S-box with $n = 128$ as ideal, its implementation is not feasible because it would require a unit with $2^n = 10^{38}$ wiring connections.

To verify that the S-box example in Figure 12.6 performs a *nonlinear transformation*, we need only use the superposition theorem stated below as a test. Let

$$C = Ta + Tb$$
$$C' = T(a + b)$$

$$(12.21)$$

where a and b are input terms, C and C' are output terms, and T is the transformation.

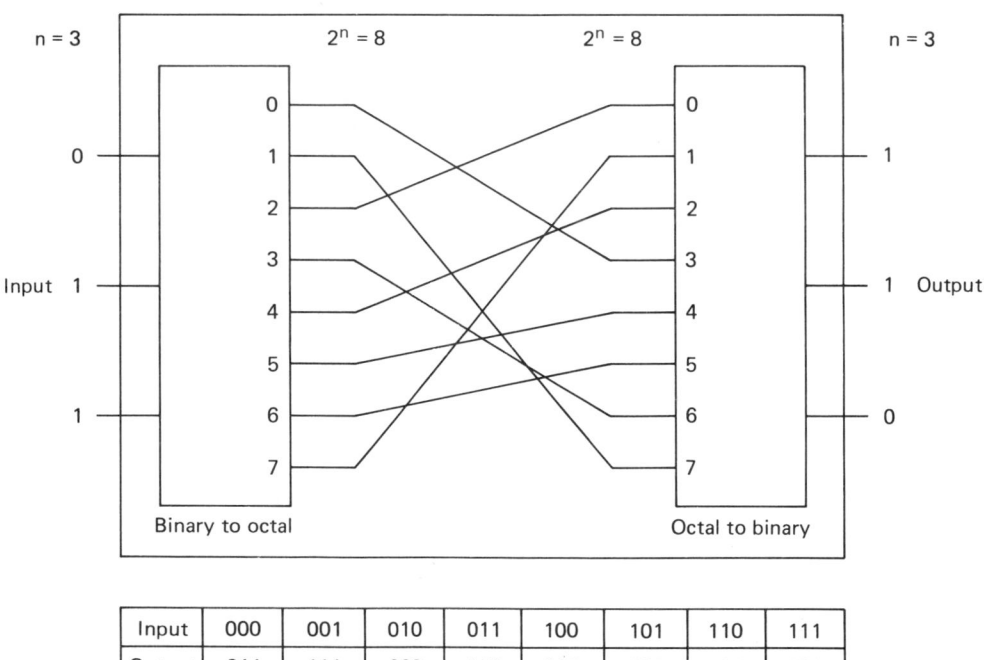

Input	000	001	010	011	100	101	110	111
Output	011	111	000	110	010	100	101	001

Figure 12.6 Substitution box.

$$\text{If } T \text{ is linear:} \quad C = C' \text{ for all inputs}$$

$$\text{If } T \text{ is nonlinear:} \quad C \neq C'$$

Suppose that $a = 001$ and $b = 010$; then using T as described in Figure 12.6,

$$C = T(001) \oplus T(010) = 111 \oplus 000 = 111$$

$$C' = T(001 \oplus 010) = T(011) = 110$$

where the symbol $\oplus$ represents modulo-2 addition. Since $C \neq C'$, the S-box is nonlinear.

12.3.3 Permutation

In permutation (transposition), the positions of the plaintext letters in the message are simply rearranged, rather than being substituted with other letters of the alphabet as in the classic ciphers. For example, the word THINK might appear, after permutation, as the ciphertext HKTNI. Figure 12.7 represents an example of binary data permutation (a linear operation). Here we see that the input data are simply rearranged or permuted (P-box). The technique has one major disadvantage when used alone; it is vulnerable to trick messages. A trick message is illustrated in Figure 12.7. A single 1 at the input and all the rest 0 quickly reveals one of the internal connections. If the cryptanalyst can subject the system to a

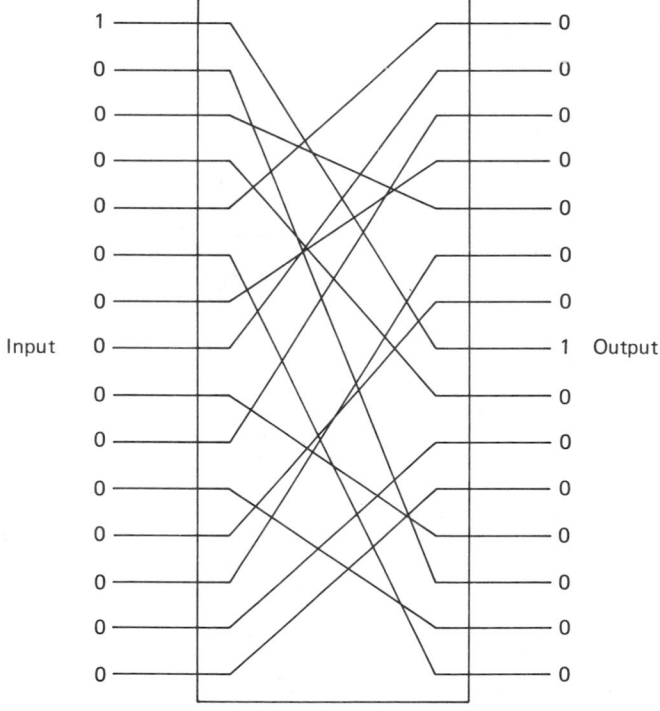

Figure 12.7 Permutation box.

plaintext attack, he will transmit a sequence of such trick messages, moving the single 1 one position for each transmission. In this way, each of the connections from input to output is revealed. This is an example of why a system's security should not depend on its architecture.

12.3.4 Product Cipher System

For transformations involving reasonable numbers of n-message symbols, both of the foregoing cipher systems (the S-box and the P-box) are by themselves wanting. Shannon [5] suggested using a *product cipher* or a combination of S-box and P-box transformations, which together could yield a cipher system more powerful than either one alone. This approach of alternately applying substitution and permutation transformations has been used by IBM in the LUCIFER system [7, 8], and has become the basis for the national Data Encryption Standard (DES) [9]. Figure 12.8 illustrates such a combination of P-boxes and S-boxes. Decryption is accomplished by running the data backward, using the inverse of each S-box. The system as pictured in Figure 12.8 is difficult to implement since each S-box is different, a randomly generated key is not usable, and the system does not lend itself to repeated use of the same circuitry. To avoid these difficulties, the LU-CIFER system [8] used two different types of S-boxes, S_1 and S_0, which could be publicly revealed. Figure 12.9 illustrates such a system. The input data are transformed by the sequence of S-boxes and P-boxes under the dictates of a key. The 25-bit key in this example designates, with a binary one or zero, the choice (S_1 or S_0) of each of the 25 S-boxes in the block. The details of the encryption devices can be revealed since security of the system is provided by the key.

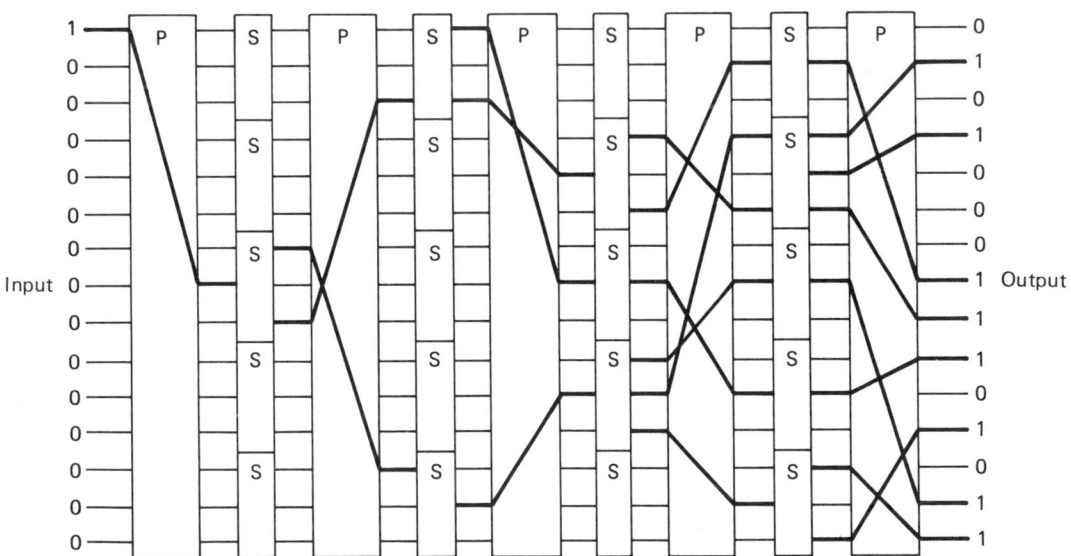

Figure 12.8 Product cipher system.

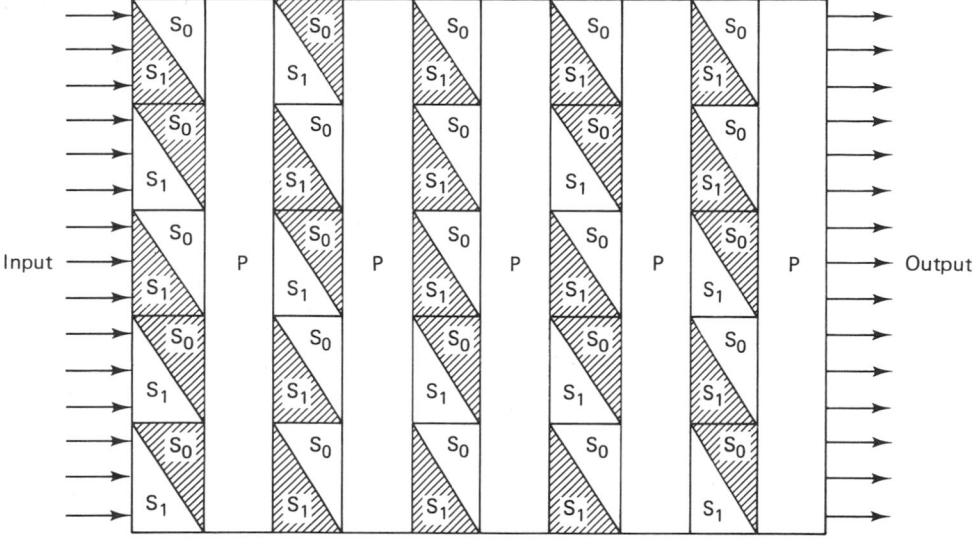

Shaded boxes correspond to the symbols of the binary key below.

Example of binary key

1 0 1 0 0 0 1 0 1 1 1 1 1 0 1 1 0 1 0 1 1 1 0 1 0

Figure 12.9 Individual keying capability.

The iterated structure of the product cipher system in Figure 12.9 is typical of most present-day block ciphers. The messages are partitioned into successive blocks of n bits, each of which is encrypted with the same key. The n-bit block represents one of 2^n different characters, allowing for $(2^n)!$ different substitution patterns. Consequently, for a reasonable implementation, the substitution part of the encryption scheme is performed in parallel on small segments of the block. An example of this is seen in the next section.

12.3.5 The Data Encryption Standard

In 1977, the National Bureau of Standards adopted a modified Lucifer system as the national Data Encryption Standard (DES) [9]. From a system input–output point of view, DES can be regarded as a block encryption system with an alphabet size of 2^{64} symbols, as shown in Figure 12.10. An input block of 64 bits, regarded

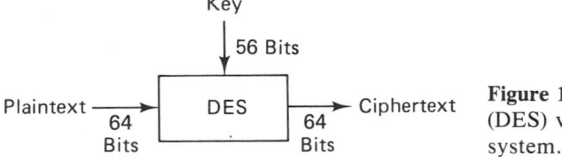

Figure 12.10 Data encryption standard (DES) viewed as a block encryption system.

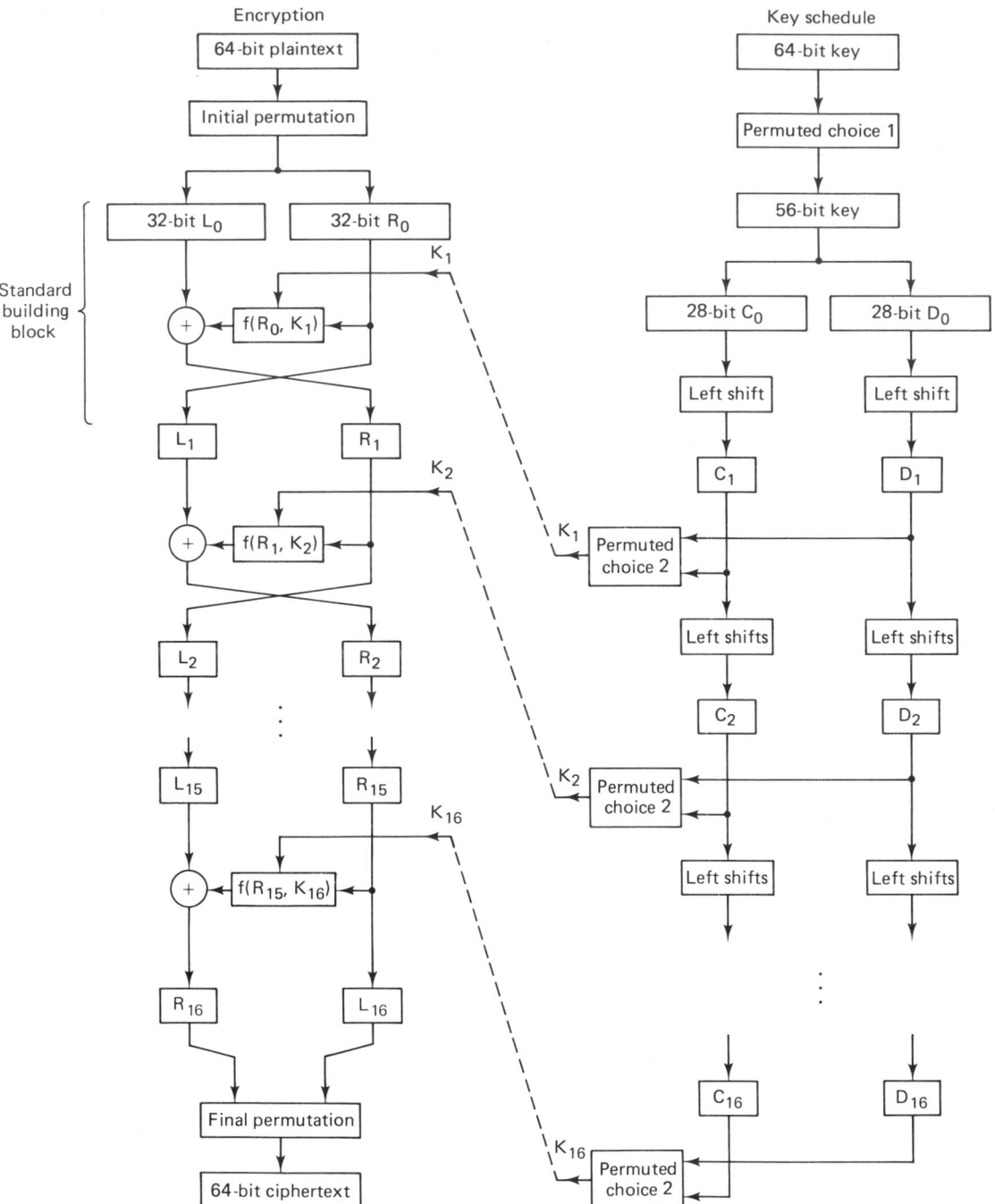

Figure 12.11 Data encryption standard.

as a plaintext symbol in this alphabet, is replaced with a new ciphertext symbol. Figure 12.11 illustrates the system functions in block diagram form. The encryption algorithm starts with an initial permutation (IP) of the 64 plaintext bits, described in the IP-table (Table 12.1). The IP-table is read from left to right and from top to bottom, so that bits $x_1, x_2, \ldots, x_{64}$ are permuted to $x_{58}, x_{50}, \ldots,$ x_7. After this initial permutation, the heart of the encryption algorithm consists of 16 iterations using the standard building block (SBB) shown in Figure 12.12. The standard building block uses 48 bits of key to transform the 64 input data bits into 64 output data bits, designated as 32 left-half bits and 32 right-half bits. The output of each building block becomes the input to the next building block. The input right-half 32 bits (R_{i-1}) are copied unchanged to become the output left-half 32 bits (L_i). The R_{i-1} bits are also *extended* and transformed into 48 bits with the E-table (Table 12.2), and then modulo-2 summed with the 48 bits of the key. As in the case of the IP-table, the E-table is read from left to right and from top to bottom. The table expands bits

$$R_{i-1} = x_1, x_2, \ldots, x_{32}$$

into

$$(R_{i-1})_E = x_{32}, x_1, x_2, \ldots, x_{32}, x_1 \tag{12.22}$$

Notice that the bits listed in the first and last columns of the E-table are those bit positions that are used twice to provide the 32 bit-to-48 bit expansion.

Next, $(R_{i-1})_E$ is modulo-2 summed with the ith key selection, explained later, and the result is segmented into eight 6-bit blocks

$$B_1, B_2, \ldots, B_8$$

that is,

$$(R_{i-1})_E \oplus K_i = B_1, B_2, \ldots, B_8 \tag{12.23}$$

Each of the eight 6-bit blocks, B_j, is then used as an input to an S-box function which returns a 4-bit block, $S_j(B_j)$. Thus the input 48 bits are transformed by the S-box to 32 bits. The S-box mapping function, S_j, is defined in Table 12.3. The transformation of $B_j = b_1, b_2, b_3, b_4, b_5, b_6$ is accomplished as follows. The integer corresponding to bits $b_1 \ b_6$ selects a row in the table, and the integer corresponding to bits $b_2 \ b_3 \ b_4 \ b_5$ selects a column in the table. For example, if

TABLE 12.1 Initial Permutation (IP)

58	50	42	34	26	18	10	2
60	52	44	36	28	20	12	4
62	54	46	38	30	22	14	6
64	56	48	40	32	24	16	8
57	49	41	33	25	17	9	1
59	51	43	35	27	19	11	3
61	53	45	37	29	21	13	5
63	55	47	39	31	23	15	7

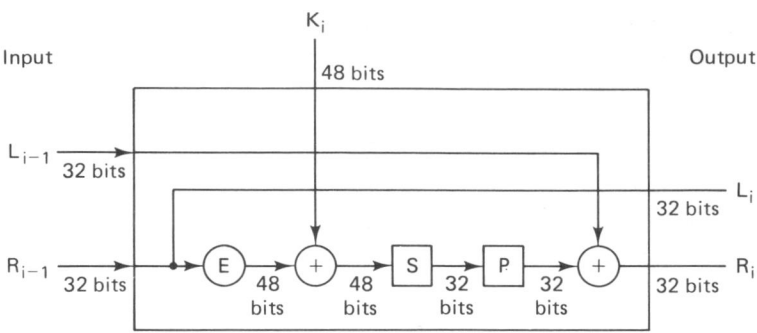

Figure 12.12 Standard building block (SBB).

$b_1 = 110001$, then S_1 returns the value in row 3, column 8, which is the integer 5 and is represented by the bit sequence 0101. The resulting 32-bit block out of the S-box is then permuted using the P-table (Table 12.4). As in the case of the other tables, the P-table is read from left to right and from top to bottom, so that bits $x_1, x_2, \ldots, x_{32}$ are permuted to $x_{16}, x_7, \ldots, x_{25}$. The 32-bit output of the P-table is modulo-2 summed with the input left-half 32 bits (L_{i-1}), forming the output right-half 32 bits (R_i).

The algorithm of the standard building block can be represented by

$$L_i = R_{i-1} \qquad\qquad\qquad (12.24)$$

$$R_i = L_{i-1} \oplus f(R_{i-1}, K_i) \qquad\qquad (12.25)$$

where $f(R_{i-1}, K_i)$ denotes the functional relationship comprised of the E-table, S-box, and P-table described above. After 16 iterations of the SBB, the data are transposed according to the final inverse permutation (IP^{-1}) described in the IP^{-1}-table (Table 12.5), where the output bits are read from left to right and from top to bottom, as before.

To decrypt, the same algorithm is used but the key sequence that is used in the standard building block is taken in the reverse order. Note that the value of $f(R_{i-1}, K_i)$ which can also be expressed in terms of the output of the ith block as $f(L_i, K_i)$, makes the decryption process possible.

TABLE 12.2 E-Table Bit Selection

32	1	2	3	4	5
4	5	6	7	8	9
8	9	10	11	12	13
12	13	14	15	16	17
16	17	18	19	20	21
20	21	22	23	24	25
24	25	26	27	28	29
28	29	30	31	32	1

TABLE 12.3 S-Box Selection Functions

Row	0	1	2	3	4	5	6	7	8	9	10	11	12	13	14	15	
								Column									
0	14	4	13	1	2	15	11	8	3	10	6	12	5	9	0	7	
1	0	15	7	4	14	2	13	1	10	6	12	11	9	5	3	8	S_1
2	4	1	14	8	13	6	2	11	15	12	9	7	3	10	5	0	
3	15	12	8	2	4	9	1	7	5	11	3	14	10	0	6	13	
0	15	1	8	14	6	11	3	4	9	7	2	13	12	0	5	10	
1	3	13	4	7	15	2	8	14	12	0	1	10	6	9	11	5	S_2
2	0	14	7	11	10	4	13	1	5	8	12	6	9	3	2	15	
3	13	8	10	1	3	15	4	2	11	6	7	12	0	5	14	9	
0	10	0	9	14	6	3	15	5	1	13	12	17	11	4	2	8	
1	13	7	0	9	3	4	6	10	2	8	5	14	12	11	15	1	S_3
2	13	6	4	9	8	15	3	0	11	1	2	12	5	10	14	7	
3	1	10	13	0	6	9	8	7	4	15	14	3	11	5	2	12	
0	7	13	14	3	0	6	9	10	1	2	8	5	11	12	4	15	
1	13	8	11	5	6	15	0	3	4	7	2	12	1	10	14	9	S_4
2	10	6	9	0	12	11	7	13	15	1	3	14	5	2	8	4	
3	3	15	0	6	10	1	13	8	9	4	5	11	12	7	2	14	
0	2	12	4	1	7	10	11	6	8	5	3	15	13	0	14	9	
1	14	11	2	12	4	7	13	1	5	0	15	10	3	9	8	6	S_5
2	4	2	1	11	10	13	7	8	15	9	12	5	6	3	0	14	
3	11	8	12	7	1	14	2	13	6	15	0	9	10	4	5	3	
0	12	1	10	15	9	2	6	8	0	13	3	4	14	7	5	11	
1	10	15	4	2	7	12	9	5	6	1	13	14	0	11	3	8	S_6
2	9	14	15	5	2	8	12	3	7	0	4	10	1	13	11	6	
3	4	3	2	12	9	5	15	0	11	14	1	7	6	0	8	13	
0	4	11	2	14	15	0	8	13	3	12	9	7	5	10	6	1	
1	13	0	11	7	4	9	1	10	14	3	5	12	2	15	8	6	S_7
2	1	4	11	13	12	3	7	14	10	15	6	8	0	5	9	2	
3	6	11	13	8	1	4	10	7	9	5	0	15	14	2	3	12	
0	13	2	8	4	6	15	11	1	10	9	3	14	5	0	12	7	
1	1	15	13	8	10	3	7	4	12	5	6	11	0	14	9	2	S_8
2	7	11	4	1	9	12	14	2	0	6	10	13	15	3	5	8	
3	2	1	14	7	4	10	8	13	15	12	9	0	3	5	6	11	

TABLE 12.4 P-Table Permutation

16	7	20	21
29	12	28	17
1	15	23	26
5	18	31	10
2	8	24	14
32	27	3	9
19	13	30	6
22	11	4	25

12.3.5.1 Key Selection

Key selection also proceeds in 16 iterations, as seen in the key schedule portion of Figure 12.11. The input key consists of a 64-bit block with 8 parity bits in positions 8, 16, . . . , 64. The permuted choice 1 (PC-1) discards the parity bits and permutes the remaining 56 bits as shown in Table 12.6. The output of PC-1 is split into two halves, C and D, of 28 bits each. Key selection proceeds in 16 iterations in order to provide a different set of 48 key bits to each SBB encryption iteration. The C and D blocks are successively shifted as follows:

$$C_i = \text{LS}_i(C_{i-1}) \quad \text{and} \quad D_i = \text{LS}_i(D_{i-1}) \tag{12.26}$$

where LS_i is a left circular shift by the number of positions shown in Table 12.7. The sequence C_i, D_i is then transposed according to the permuted choice 2 (PC-2) shown in Table 12.8. The result is the key sequence, K_i, which is used in the ith iteration of the encryption algorithm.

The DES can be implemented as a block encryption system (see Figure 12.11), which is sometimes referred to as a *codebook* method. A major disadvantage of this method is that a given block of input plaintext will always result in the same output ciphertext (under the same key). Another encryption mode, called the *cipher feedback* mode, encrypts single bits rather than characters, resulting in a stream encryption system [3]. With the cipher feedback scheme (described later), the encryption of a segment of plaintext not only depends on the key and the current data, but also on some of the earlier data.

Since the late 1970s, two points of contention have been widely publicized about the DES [10]. The first concerns the key variable length. Some researchers

TABLE 12.5 Final Permutation (IP^{-1})

40	8	48	16	56	24	64	32
39	7	47	15	55	23	63	31
38	6	46	14	54	22	62	30
37	5	45	13	53	21	61	29
36	4	44	12	52	20	60	28
35	3	43	11	51	19	59	27
34	2	42	10	50	18	58	26
33	1	41	9	49	17	57	25

TABLE 12.6 Key Permutation PC-1

57	49	41	33	25	17	9
1	58	50	42	34	26	18
10	2	59	51	43	35	27
19	11	3	60	52	44	36
63	55	47	39	31	23	15
7	62	54	46	38	30	22
14	6	61	53	45	37	29
21	13	5	28	20	12	4

TABLE 12.7 Key Schedule of Left Shifts

Iteration, i	Number of left shifts
1	1
2	1
3	2
4	2
5	2
6	2
7	2
8	2
9	1
10	2
11	2
12	2
13	2
14	2
15	2
16	1

TABLE 12.8 Key Permutation PC-2

14	17	11	24	1	5
3	28	15	6	21	10
23	19	12	4	26	8
16	7	27	20	13	2
41	52	31	37	47	55
30	40	51	45	33	48
44	49	39	56	34	53
46	42	50	36	29	32

felt that 56 bits are not adequate to preclude an exhaustive search. The second concerns the details of the internal structure of the S-boxes, which were never released by IBM. The National Security Agency (NSA), which had been involved in the testing of the DES algorithm, had requested that the information not be publicly discussed, since it was sensitive. The critics feared that NSA had been involved in design selections that would allow NSA to "tap into" any DES-encrypted messages [10].

12.4 STREAM ENCRYPTION

Earlier, we defined a *one-time pad* as an encryption system with a random key, used one time only, that exhibits unconditional security. One can conceptualize a stream encryption implementation of a one-time pad using a truly random key stream (the key sequence never repeats). Thus perfect secrecy can be achieved for an infinite number of messages, since each message would be encrypted with a different portion of the random key stream. The development of stream encryption schemes represents an attempt to emulate the one-time pad. Great emphasis was placed on generating key streams that appeared to be random, yet could easily be implemented for decryption, because they could be generated by algorithms. Such stream encryption techniques use pseudorandom (PN) sequences, which derive their name from the fact that they appear random to the casual observer; binary pseudorandom sequences have statistical properties similar to the random flipping of a fair coin. However, the sequences, of course, are deterministic (see Section 10.2). These techniques are popular because the encryption and decryption algorithms are readily implemented with feedback shift registers. At first glance it may appear that a PN key stream can provide the same security as the one-time pad, since the period of the sequence generated by a maximum-length linear shift register is $2^n - 1$ bits, where n is the number of stages in the register. If the PN sequence were implemented with a 50-stage register and a 1-MHz clock rate, the sequence would repeat every $2^{50} - 1$ microseconds, or every 35 years. In this era of large-scale integrated (LSI) circuits, it is just as easy to provide an implementation with 100 stages, in which case the sequence would repeat every 4×10^{16} years. Therefore, one might suppose that since the PN sequence does not repeat itself for such a long time, it would appear truly random and yield perfect secrecy. There is one important difference between the PN sequence and a truly random sequence used by a one-time pad. The PN sequence is generated by an algorithm; thus, knowing the algorithm, one knows the entire sequence. In Section 12.4.2 we will see that an encryption scheme that uses a linear feedback shift register in this way is very vulnerable to a *known plaintext attack*.

12.4.1 Example of Key Generation Using a Linear Feedback Shift Register

Stream encryption techniques generally employ shift registers for generating their PN key sequences. A shift register can be converted into a pseudorandom sequence generator by including a feedback loop that computes a new term for the first stage based on the previous n terms. The register is said to be linear if the numerical operation in the feedback path is linear. The PN generator example from Section 10.2 is repeated in Figure 12.13. For this example, it is convenient to number the stages as shown in Figure 12.13, where $n = 4$ and the outputs from stages 1 and 2 are modulo-2 added (linear operation) and fed back to stage 4. If the initial state of stages (x_4, x_3, x_2, x_1) is 1 0 0 0, the succession of states triggered by clock pulses would be 1 0 0 0, 0 1 0 0, 0 0 1 0, 1 0 0 1, 1 1 0 0, and so on. The output sequence is made up of the bits shifted out from the rightmost stage of the register, that is, 1 1 1 1 0 1 0 1 1 0 0 1 0 0 0, where the rightmost bit in this sequence is the earliest output and the leftmost bit is the most recent output. Given any linear feedback shift register of degree n, the output sequence is ultimately periodic.

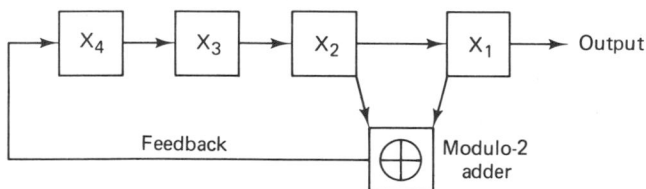

Figure 12.13 Linear feedback register example.

12.4.2 Vulnerabilities of Linear Feedback Shift Registers

An encryption scheme that uses a linear feedback shift register (LFSR) to generate the key stream is very vulnerable to attack. A cryptanalyst needs only $2n$ bits of plaintext and its corresponding ciphertext to determine the feedback taps, the initial state of the register, and the entire sequence of the code. In general, $2n$ is very small compared to the period $2^n - 1$. Let us illustrate this vulnerability with the LFSR example illustrated in Figure 12.13. Imagine that a cryptanalyst, who knows nothing about the internal connections of the LFSR, manages to obtain $2n = 8$ bits of ciphertext and its plaintext equivalent. These are shown below, where the rightmost bit is the earliest received and the leftmost bit is the most recent that was received.

Plaintext: 0 1 0 1 0 1 0 1

Ciphertext: 0 0 0 0 1 1 0 0

The cryptanalyst adds the two sequences together, modulo-2, to obtain the segment of the key stream, 0 1 0 1 1 0 0 1, illustrated in Figure 12.14. The key stream sequence shows the contents of the LFSR stages at various times. The rightmost border surrounding four of the key bits shows the contents of the shift register at time t_1. As we successively slide the "moving" border one digit to the left, we see the shift register contents at times $t_2, t_3, t_4, \ldots$. From the linear structure of the four-stage shift register, we can write

$$g_4 x_4 + g_3 x_3 + g_2 x_2 + g_1 x_1 = x_5 \qquad (12.27)$$

where x_5 is the digit fed back to the input and $g_i (= 1 \text{ or } 0)$ defines the ith feedback

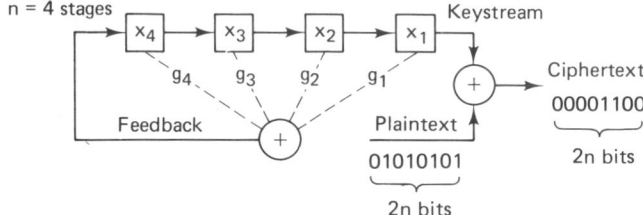

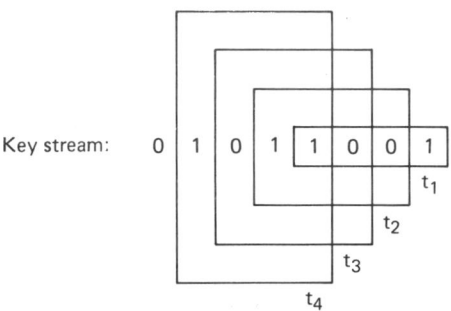

Key stream: 0 | 1 | 0 | 1 | 1 | 0 | 0 | 1

Figure 12.14 Example of vulnerability of a linear feedback shift register.

connection. For this example we can thus write the following four equations with four unknowns, by examining the contents of the shift register at the four times shown in Figure 12.14.

$$g_4(1) + g_3(0) + g_2(0) + g_1(1) = 1$$

$$g_4(1) + g_3(1) + g_2(0) + g_1(0) = 0$$

$$g_4(0) + g_3(1) + g_2(1) + g_1(0) = 1 \qquad (12.28)$$

$$g_4(1) + g_3(0) + g_2(1) + g_1(1) = 0$$

whose solution is $g_1 = 1$, $g_2 = 1$, $g_3 = 0$, $g_4 = 0$, corresponding to the LFSR shown in Figure 12.13. The cryptanalyst has thus learned the connections of the LFSR, together with the starting state of the register at time t_1. He can therefore know the sequence for all time [3]. To generalize this example for any n-stage LFSR, we rewrite Equation (12.27) as follows:

$$x_{n+1} = \sum_{i=1}^{n} g_i x_i \qquad (12.29)$$

We can write Equation (12.29) as the matrix equation

$$\mathbf{x} = \mathbf{X}\mathbf{g} \qquad (12.30)$$

where

$$\mathbf{x} = \begin{bmatrix} x_{n+1} \\ x_{n+2} \\ \vdots \\ x_{2n} \end{bmatrix} \qquad \mathbf{g} = \begin{bmatrix} g_1 \\ g_2 \\ \vdots \\ g_n \end{bmatrix}$$

and

$$\mathbf{X} = \begin{bmatrix} x_1 & x_2 & \cdots & x_n \\ x_2 & x_3 & \cdots & x_{n+1} \\ \vdots & \vdots & & \vdots \\ x_n & x_{n+1} & \cdots & x_{2n-1} \end{bmatrix}$$

It can be shown [3] that the columns of $\mathbf{X}$ are linearly independent; thus $\mathbf{X}$ is nonsingular (its determinant is nonzero) and has an inverse. Hence,

$$\mathbf{g} = \mathbf{X}^{-1}\mathbf{x} \qquad (12.31)$$

The matrix inversion requires at most on the order of n^3 operations and is thus easily accomplished by computer for any reasonable value of n. For example, if $n = 100$, $n^3 = 10^6$, and a computer with a 1-μs operation cycle would require 1 s for the inversion. The weakness of a LFSR is caused by the linearity of Equation (12.31). The use of *nonlinear feedback* in the shift register makes the cryptanalyst's task much more difficult, if not computationally intractable.

12.4.3 Synchronous and Self-Synchronous Stream Encryption Systems

We can categorize stream encryption systems as either *synchronous* or *self-synchronous*. In the former, the key stream is generated independently of the message, so that a lost character during transmission necessitates a resynchronization of the transmission and receiver key generators. A synchronous stream cipher is shown in Figure 12.15. The starting state of the kcy generator is initialized with a known input, I_0. The ciphertext is obtained by the modulo addition of the ith key character, k_i, with the ith message character, m_i. Such synchronous ciphers are generally designed to utilize *confusion* (see Section 12.3.1) but not *diffusion*. That is, the encryption of a character is not diffused over some block length of message. For this reason, synchronous stream ciphers do not exhibit *error propagation*.

In a *self-synchronous* stream cipher, each key character is derived from a fixed number, n, of the preceding ciphertext characters, giving rise to the name *cipher feedback*. In such a system, if a ciphertext character is lost during trans-

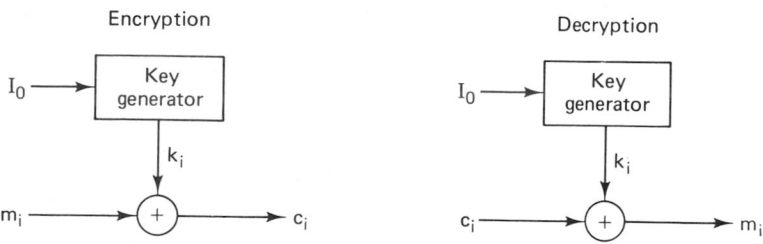

Figure 12.15 Synchronous stream cipher.

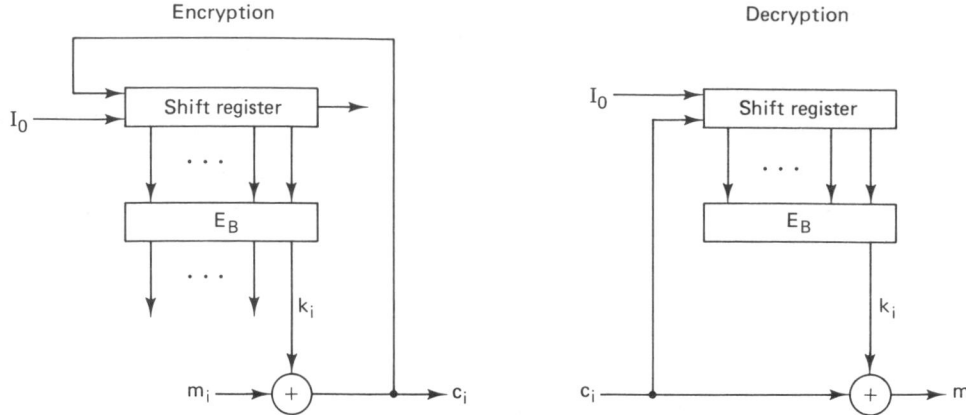

Figure 12.16 Cipher feedback mode.

mission, the error propagates forward for n characters, but the system resynchronizes itself after n correct ciphertext characters are received.

In Section 12.1.4 we looked at an example of cipher feedback in the Vigenere auto key cipher. We saw that the advantages of such a system are that (1) a nonrepeating key is generated, and (2) the statistics of the plaintext message are diffused throughout the ciphertext. However, the fact that the key was exposed in the ciphertext was a basic weakness. This problem can be eliminated by passing the ciphertext characters through a nonlinear block cipher to obtain the key characters. Figure 12.16 illustrates a shift register key generator operating in the cipher feedback mode. Each output ciphertext character, c_i (formed by the modulo addition of the message character, m_i, and the key character, k_i), is fed back to the input of the shift register. As before, initialization is provided by a known input, I_0. At each iteration, the output of the shift register is used as input to a (nonlinear) block encryption algorithm, E_B. The low-order output character from E_B becomes the next key character, k_{i+1}, to be used with the next message character, m_{i+1}. Since, after the first few iterations, the input to the algorithm depends only on the ciphertext, the system is self-synchronizing.

12.5 PUBLIC KEY CRYPTOSYSTEMS

The concept of public key cryptosystems was introduced in 1976 by Diffie and Hellman [11]. In conventional cryptosystems the encryption algorithm can be revealed since the security of the system depends on a safeguarded key. The same key is used for both encryption and decryption. Public key cryptosystems utilize *two different* keys, one for encryption and the other for decryption. In public key cryptosystems, not only the encryption algorithm but also the encryption key can be publicly revealed without compromising the security of the system. In fact, a public directory, much like a telephone directory, is envisioned, which contains the encryption keys of all the subscribers. Only the decryption keys are kept secret. Figure 12.17 illustrates such a system. The important features of a public key cryptosystem are as follows:

1. The encryption algorithm, E_K, and the decryption algorithm, D_K, are invertible transformations on the plaintext, M, or the ciphertext, C, defined by the key K. That is, for each K and M, if $C = E_K(M)$, then $M = D_K(C) = D_K[E_K(M)]$.
2. For each K, E_K and D_K are easy to compute.
3. For each K, the computation of D_K from E_K is computationally intractable.

Such a system would enable secure communication between subscribers who have never met or communicated before. For example, as seen in Figure 12.17, subscriber A can send a message, M, to subscriber B by looking up B's encryption key in the directory and applying the encryption algorithm, E_B, to obtain the ciphertext $C = E_B(M)$, which he transmits on the public channel. Subscriber B is the only party who can decrypt C by applying his decryption algorithm, D_B, to obtain $M = D_B(C)$.

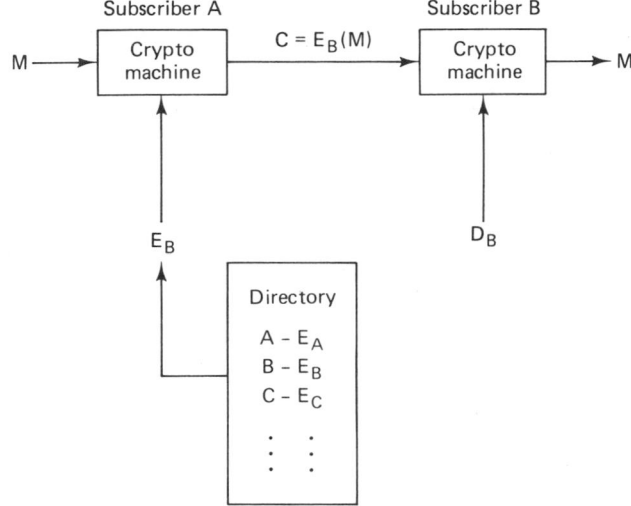

Figure 12.17 Public key cryptosystem.

12.5.1 Signature Authentication Using a Public Key Cryptosystem

Figure 12.18 illustrates the use of a public key cryptosystem for signature authentication. Subscriber A "signs" his message by first applying his decryption algorithm, D_A, to the message, yielding $S = D_A(M) = E_A^{-1}(M)$. Next, he uses the encryption algorithm, E_B, of subscriber B to encrypt S, yielding $C = E_B(S) = E_B[E_A^{-1}(M)]$, which he transmits on a public channel. When subscriber B receives C, he first decrypts it using his private decryption algorithm, D_B, yielding $D_B(C) = E_A^{-1}(M)$. Then he applies the encryption algorithm of subscriber A to produce $E_A[E_A^{-1}(M)] = M$.

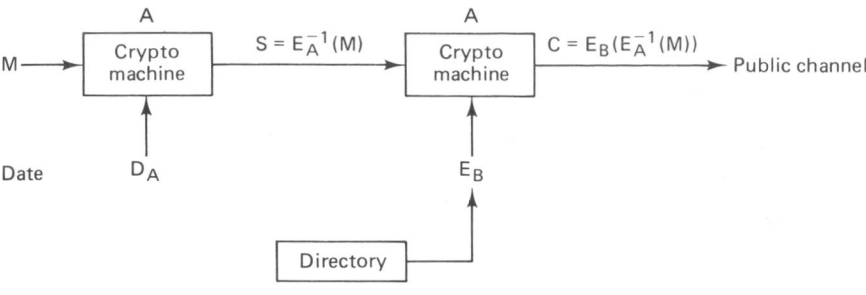

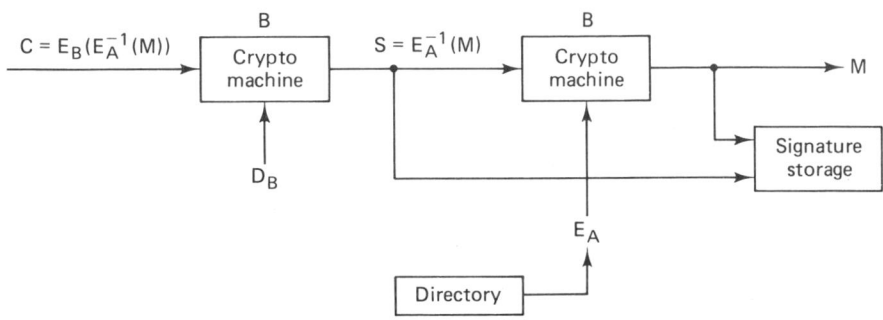

Figure 12.18 Signature authentication using a public key cryptosystem.

If the result is an intelligible message, it must have been initiated by subscriber A, since no one else could have known A's secret decryption key to form $S = D_A(M)$. Notice that S is both message dependent and signer dependent, which means that while B can be sure that the received message indeed came from A, at the same time A can be sure that no one can attribute any false messages to him.

12.5.2 A Trapdoor One-Way Function

Public key cryptosystems are based on the concept of trapdoor one-way functions. Let us first define a *one-way function* as an easily computed function whose inverse is computationally infeasible to find. For example, consider the function $y = x^5 + 12x^3 + 107x + 123$. It should be apparent that given x, y is easy to compute, but given y, x is relatively difficult to compute. A *trapdoor one-way function* is a one-way function whose inverse is easily computed if certain features, used to design the function, are known. Like a trapdoor, such functions are easy to go through in one direction. Without special information the reverse process takes an impossibly long time. We will apply the concept of a trapdoor in Section 12.5.5, when we discuss the Merkle–Hellman scheme.

12.5.3 The Rivest–Shamir–Adelman Scheme

In the Rivest–Shamir–Adelman (RSA) scheme messages are first represented as integers in the range $(0, n-1)$. Each user chooses his own value of n and another pair of positive integers, e and d, in a manner to be described below. The user places his encryption key, the number pair (n, e), in the public directory. The decryption key consists of the number pair (n, d), of which d is kept secret. Encryption of a message, M, and decryption of a ciphertext, C, are defined as follows:

$$\text{Encryption:} \quad C = E(M) = (M)^e \text{ modulo-}n \qquad (12.32)$$
$$\text{Decryption:} \quad M = D(C) = (C)^d \text{ modulo-}n$$

They are each easy to compute and the results of each operation are integers in the range $(0, n-1)$. In the RSA scheme, n is obtained by selecting *two large prime numbers*, p and q, and multiplying them together:

$$n = pq \qquad (12.33)$$

Although n is made public, p and q are kept hidden, due to the great difficulty in factoring n. Then

$$\phi(n) = (p-1)(q-1) \qquad (12.34)$$

called *Euler's totient function*, is formed. The parameter $\phi(n)$ has the interesting property [12] that for any integer X in the range $(0, n-1)$ and any integer k,

$$X = X^{k\phi(n)+1} \text{ modulo-}n \qquad (12.35)$$

Therefore, while all other arithmetic is done modulo-n, arithmetic in the exponent is done modulo-$\phi(n)$. A large integer, d, is randomly chosen so that it is relatively prime to $\phi(n)$, which means that $\phi(n)$ and d must have no common divisors other than 1, expressed as

$$\gcd[\phi(n), d] = 1 \qquad (12.36)$$

where gcd means "greatest common divisor." Any prime number greater than the larger of (p, q) will suffice. Then the integer e, where $0 < e < \phi(n)$, is found from the following relationship:

$$ed \text{ modulo-}\phi(n) = 1 \qquad (12.37)$$

which, from Equation (12.35), is tantamount to choosing e and d to satisfy

$$X = X^{ed} \text{ modulo-}n \qquad (12.38)$$

Therefore,

$$E[D(X)] = D[E(X)] = X \qquad (12.39)$$

and decryption works correctly. Given an encryption key (n, e), one way that a cryptanalyst might attempt to break the cipher is to factor n into p and q, compute

$\phi(n) = (p - 1)(q - 1)$, and compute d from Equation (12.37). This is all straight-forward except for the factoring of n.

The RSA scheme is based on the fact that it is easy to generate two large prime numbers, p and q, and multiply them together, but it is very much more difficult to factor the result. The product can therefore be made public as part of the encryption key, without compromising the factors that would reveal the decryption key corresponding to the encryption key. By making each of the factors roughly 100 digits long, the multiplication can be done in a fraction of a second, but the exhaustive factoring of the result should take billions of years [2].

12.5.3.1 Use of the RSA Scheme

Using the example in Reference [12], let $p = 47$, $q = 59$. Therefore, $n = pq = 2773$ and $\phi(n) = (p - 1)(q - 1) = 2668$. The parameter d is chosen to be relatively prime to $\phi(n)$. For example, choose $d = 157$. Next, the value of e is computed as follows (the details are shown in the next section):

$$ed \text{ modulo } \phi(n) = 1$$

$$157e \text{ modulo } 2668 = 1$$

Therefore, $e = 17$. Consider the plaintext example

ITS ALL GREEK TO ME

By replacing each letter with a two-digit number in the range (01, 26) corresponding to its position in the alphabet, and encoding a blank as 00, the plaintext message can be written as

0920 1900 0112 1200 0718 0505 1100 2015 0013 0500

Each message needs to be expressed as an integer in the range $(0, n - 1)$; therefore, for this example, encryption can be performed on blocks of four digits at a time since this is the maximum number of digits that will always yield a number less than $n - 1 = 2772$. The first four digits (0920) of the plaintext are encrypted as follows:

$$C = (M)^e \text{ modulo-}n = (920)^{17} \text{ modulo-}2773 = 948$$

Continuing this process for the remaining plaintext digits, we get

$$C = 0948\ 2342\ 1084\ 1444\ 2663\ 2390\ 0778\ 0774\ 0219\ 1655$$

The plaintext is returned by applying the decryption key, as follows:

$$M = (C)^{157} \text{ modulo-}2773$$

12.5.3.2 How to Compute e

A variation of Euclid's algorithm [13] for computing the gcd of $\phi(n)$ and d is used to compute e. First, compute a series $x_0, x_1, x_2, \ldots$, where $x_0 = \phi(n)$, $x_1 = d$, and $x_{i+1} = x_{i-1}$ modulo-x_i, until an $x_k = 0$ is found. Then the gcd $(x_0, x_1) = x_{k-1}$. For each x_i compute numbers a_i and b_i such that $x_i = a_i x_0 + b_i x_1$.

If $x_{k-1} = 1$, then b_{k-1} is the multiplicative inverse of x_1 modulo-x_0. If b_{k-1} is a negative number, the solution is $b_{k-1} + \phi(n)$.

Example 12.5 Computation of e from d and $\phi(n)$

For the previous example, with $p = 47$, $q = 59$, $n = 2773$, $\phi(n) = 2688$, and d chosen to be 157, use the Euclid algorithm to verify that $e = 17$.

Solution

i	x_i	a_i	b_i	y_i
0	2668	1	0	
1	157	0	1	16
2	156	1	-16	1
3	1	-1	17	

where

$$y_i = \left\lfloor \frac{x_{i-1}}{x_i} \right\rfloor$$

$$x_{i+1} = x_{i-1} - y_i x_i$$

$$a_{i+1} = a_{i-1} - y_i a_i$$

$$b_{i+1} = b_{i-1} - y_i b_i$$

Hence

$$e = b_3 = 17$$

12.5.4 The Knapsack Problem

The classic knapsack problem is illustrated in Figure 12.19. The knapsack is filled with a subset of the items shown with weights indicated in grams. Given the weight of the filled knapsack (the scale is calibrated to deduct the weight of the empty knapsack), determine which items are contained in the knapsack. For this simple example, the solution can easily be found by trial and error. However, if there are 100 possible items in the set, instead of 10, the problem may become computationally infeasible.

Let us express the knapsack problem in terms of a knapsack vector and a data vector. The knapsack vector is an n-tuple of distinct integers (analogous to the set of possible knapsack items)

$$\mathbf{a} = a_1, a_2, \ldots, a_n$$

The data vector is an n-tuple of binary symbols

$$\mathbf{x} = x_1, x_2, \ldots, x_n$$

The knapsack, S, is the sum of a subset of the components of the knapsack vector

$$S = \sum_{i=1}^{n} a_i x_i \qquad \text{where } x_i = 0, 1$$

$$= \mathbf{ax}$$

(12.40)

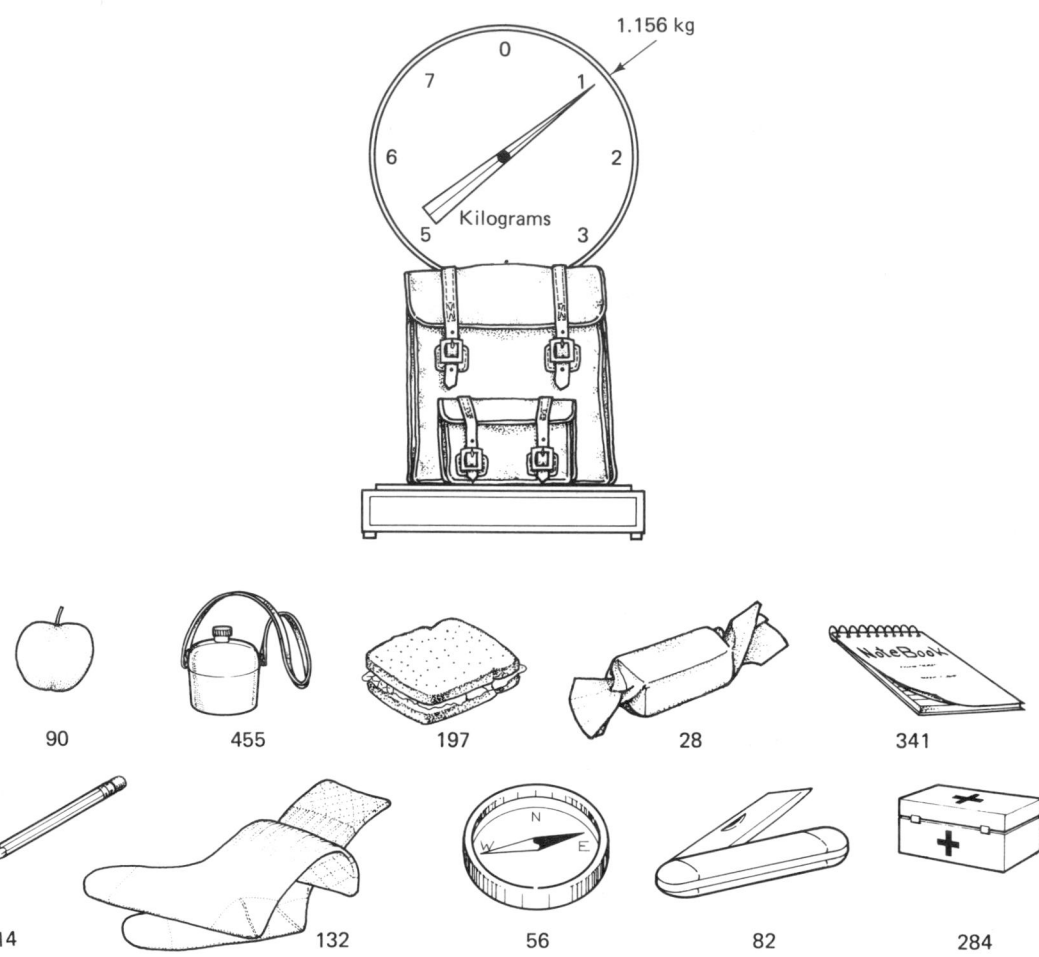

Figure 12.19 Knapsack problem.

The knapsack problem can be stated as follows: Given S and knowing $\mathbf{a}$, determine $\mathbf{x}$.

Example 12.6 Knapsack Example

Given $\mathbf{a} = 1, 2, 4, 8, 16, 32$ and $S = \mathbf{ax} = 26$, find $\mathbf{x}$.

Solution

In this example $\mathbf{x}$ is seen to be the *binary* representation of S. The decimal-to-binary conversion should appear more familiar with $\mathbf{a}$ expressed as $2^0, 2^1, 2^2, 2^3, 2^4, 2^5$. The data vector $\mathbf{x}$ is easily found since $\mathbf{a}$ in this example is *super-increasing*, which means that each component of the n-tuple $\mathbf{a}$ is larger than the sum of the preceding components. That is,

$$a_i > \sum_{j=1}^{i-1} a_j \qquad i = 2, 3, \ldots, n \tag{12.41}$$

When **a** is super-increasing, the solution of **x** is found by starting with $x_n = 1$ if $S \geq a_n$ (otherwise $x_n = 0$), and continuing, as follows:

$$
x_i =
\begin{cases}
1 & \text{if } S - \sum_{j=i+1}^{n} x_j a_j \geq a_i \\
\\
0 & \text{otherwise}
\end{cases}
\tag{12.42}
$$

where $i = n - 1, n - 2, \ldots, 1$. From Equation (12.42) it is easy to compute $\mathbf{x} = 0\ 1\ 0\ 1\ 1\ 0$.

Example 12.7   Knapsack Example

Given $\mathbf{a} = 171, 197, 459, 1191, 2410, 4517$ and $\mathbf{S} = \mathbf{ax} = 3798$, find **x**.

Solution

As in Example 12.6, **a** is super-increasing; therefore, we can compute **x** using Equation (12.42), which again yields

$$\mathbf{x} = 0\ 1\ 0\ 1\ 1\ 0$$

12.5.5 A Public Key Cryptosystem Based on a Trapdoor Knapsack

This scheme, also known as the Merkle–Hellman scheme [14], is based on the formation of a knapsack vector that is not super-increasing and is therefore not easy to solve. However, an essential part of this knapsack is a *trapdoor* that enables the authorized user to solve it.

First, we form a super-increasing n-tuple, $\mathbf{a}'$. Then we select a prime number M such that

$$
M > \sum_{i=1}^{n} a_i'
\tag{12.43}
$$

We also select a random number, W, where $1 < W < M$, and we form W^{-1} to satisfy the following relationship:

$$WW^{-1} \text{ modulo-}M = 1 \tag{12.44}$$

The vector $\mathbf{a}'$ and the numbers M, W, and W^{-1} are all kept hidden. Next, we form **a** with the elements from $\mathbf{a}'$, as follows:

$$a_i = Wa_i' \text{ modulo-}M \tag{12.45}$$

The formation of **a** using Equation (12.45) constitutes forming a knapsack vector with a *trapdoor*. When a data vector **x** is to be transmitted, we multiply **x** by **a**, yielding the number S, which is sent on the public channel. Using Equation (12.45), S can be written as follows:

$$
S = \mathbf{ax} = \sum_{i=1}^{n} a_i x_i = \sum_{i=i}^{n} (Wa_i' \text{ modulo-}M) x_i
\tag{12.46}
$$

The authorized user receives S and, using Equation (12.44), converts it to S':

$$S' = W^{-1}S \text{ modulo-}M = W^{-1} \sum_{i=1}^{n} (Wa_i' \text{ modulo-}M)x_i \text{ modulo-}M$$

$$= \sum_{i=1}^{n} (W^{-1}Wa_i' \text{ modulo-}M)x_i \text{ modulo-}M \tag{12.47}$$

$$= \sum_{i=1}^{n} a_i'x_i \text{ modulo-}M$$

$$= \sum_{i=1}^{n} a_i'x_i$$

Since the authorized user knows the secretly held super-increasing vector $\mathbf{a}'$, he or she can use S' to find $\mathbf{x}$.

12.5.5.1 Use of the Merkle–Hellman Scheme

Suppose that user A wants to construct public and private encryption functions. He first considers the super-increasing vector $\mathbf{a}' = (171, 197, 459, 1191, 2410, 4517)$

$$\sum_{i=1}^{6} a_i' = 8945$$

He then chooses a prime number M larger than 8945, a random number W, where $1 \le W < M$, and calculates W^{-1} to satisfy $WW^{-1} = 1$ modulo-M.

$$\left. \begin{array}{l} \text{Choose } M = 9109 \\ \text{choose } W = 2251 \\ \text{then } W^{-1} = 1388 \end{array} \right\} \text{kept hidden}$$

He then forms the trapdoor knapsack vector as follows:

$$a_i = a_i' \, 2251 \text{ modulo-}9109$$

$$\mathbf{a} = 2343, 6215, 3892, 2895, 5055, 2123$$

User A makes public the vector $\mathbf{a}$, which is clearly not super-increasing. Suppose that user B wants to send a message to user A.

If $\mathbf{x} = 0\ 1\ 0\ 1\ 1\ 0$ is the message to be transmitted, user B forms

$$S = \mathbf{ax} = 14{,}165 \text{ and transmits it to user } A$$

User A, who receives S, converts it to S':

$$S' = \mathbf{a'x} = W^{-1}S \text{ modulo-}M$$

$$= 1388 \cdot 14{,}165 \text{ modulo-}9109$$

$$= 3798$$

Using $S' = 3798$ and the super-increasing vector $\mathbf{a}'$, user A easily solves for $\mathbf{x}$.

Public key schemes are generally too slow for data encryption. With large encrypted data networks, the biggest problem is how to distribute and manage the keys; the public key systems appear to be very promising for use in key management.

12.6 CONCLUSION

In this chapter we have presented the basic models and goals of the cryptographic process. We looked at some early cipher systems and reviewed the mathematical theory of secret communications established by Shannon. We defined a system that can exhibit perfect secrecy and established that such systems can be implemented but that they are not practical for use where high-volume communications are required. We also considered practical security systems that employ Shannon's techniques known as confusion and diffusion to frustrate the statistical endeavors of a cryptanalyst.

The outgrowth of Shannon's work was utilized by IBM in the LUCIFER system, which later grew into the National Bureau of Standards' Data Encryption Standard (DES). We outlined the DES algorithm in detail. We also considered the use of linear feedback shift registers (LFSR) for stream encryption systems, and demonstrated the intrinsic vulnerability of an LFSR used as a key generator.

Finally, we looked at the novel area of public key cryptosystems and examined two schemes, the Rivest–Shamir–Adelman (RSA) scheme, based on the product of two large prime numbers, and the Merkle–Hellman scheme, based on the classical knapsack problem. The Merkle–Hellman scheme is now considered broken [15], so that today the RSA scheme seems to be the primary means of implementing public key cryptosystems.

REFERENCES

1. Kahn, D., *The Codebreakers*, Macmillan Publishing Company, New York, 1967.
2. Diffie, W., and Hellman, M. E., "Privacy and Authentication: An Introduction to Cryptography," *Proc. IEEE*, vol. 67, no. 3, Mar. 1979, pp. 397–427.
3. Beker, H., and Piper, F., *Cipher Systems*, John Wiley & Sons, Inc., New York, 1982.
4. Denning, D. E. R., *Cryptography and Data Security*, Addison-Wesley Publishing Company, Reading, Mass., 1982.
5. Shannon, C. E., "Communication Theory of Secrecy Systems," *Bell Syst. Tech. J.*, vol. 28, Oct. 1949, pp. 656–715.
6. Hellman, M. E., "An Extension of the Shannon Theory Approach to Cryptography," *IEEE Trans. Inf. Theory*, vol. IT23, May 1978, pp. 289–294.
7. Smith, J. L., "The Design of Lucifer, a Cryptographic Device for Data Communications," *IBM Research Rep. RC-3326*, 1971.
8. Feistel, H. "Cryptography and Computer Privacy," *Sci. Am.*, vol. 228, no. 5, May 1973, pp. 15–23.

9. National Bureau of Standards, "Data Encryption Standard," *Federal Information Processing Standard (FIPS)*, Publication no. 46, Jan. 1977.

10. United States Senate Select Committee on Intelligence, "Unclassified Summary: Involvement of NSA in the Development of the Data Encryption Standard," *IEEE Commun. Soc. Mag.*, vol. 16, no. 6, Nov. 1978, pp. 53–55.

11. Diffie, W., and Hellman, M. E., "New Directions in Cryptography," *IEEE Trans. Inf. Theory*, vol. IT22, Nov. 1976, pp. 644–654.

12. Rivest, R. L., Shamir, A., and Adelman, L., "On Digital Signatures and Public Key Cryptosystems," *Commun. ACM*, vol. 21, Feb. 1978, pp. 120–126.

13. Knuth, D. E., *The Art of Computer Programming*, Vol. 2, *Seminumerical Algorithms*, 2nd ed., Addison-Wesley Publishing Company, Reading, Mass., 1981.

14. Merkle, R. C., and Hellman, M. E., "Hiding Information and Signatures in Trap-Door Knapsacks," *IEEE Trans. Inf. Theory*, vol. IT24, Sept. 1978, pp. 525–530.

15. Shamir, A., "A Polynomial Time Algorithm for Breaking the Basic Merkle-Hellman Cryptosystem," *IEEE 23rd Ann. Symp. Found. Comput. Sci.*, 1982, pp. 145–153.

PROBLEMS

12.1. Let X be an integer variable represented with 64 bits. The probability is $\frac{1}{2}$ that X is in the range $(0, 2^{16} - 1)$, the probability is $\frac{1}{4}$ that X is in the range $(2^{16}, 2^{32} - 1)$, and the probability is $\frac{1}{4}$ that X is in the range $(2^{32}, 2^{64} - 1)$. Within each range the values are equally likely. Compute the entropy of X.

12.2. A set of equally likely weather messages are: sunny (S), cloudy (C), light rain (L), and heavy rain (H). Given the added information concerning the time of day (morning or afternoon), the probabilities change as follows:

$$\text{Morning:} \quad P(S) = \tfrac{1}{8}, P(C) = \tfrac{1}{8}, P(L) = \tfrac{3}{8}, P(H) = \tfrac{3}{8}$$

$$\text{Afternoon:} \quad P(S) = \tfrac{3}{8}, P(C) = \tfrac{3}{8}, P(L) = \tfrac{1}{8}, P(H) = \tfrac{1}{8}$$

(a) Find the entropy of the weather message.
(b) Find the entropy of the message conditioned on the time of day.

12.3. The Hawaiian alphabet has only 12 letters—the vowels, a, e, i, o, u, and the consonants, h, k, l, m, n, p, w. Assume that each vowel occurs with probability 0.116, and that each consonant occurs with probability 0.06. Also assume that the average number of *information bits* per letter is the same as that for the English language. Calculate the unicity distance for an encrypted Hawaiian message if the key sequence consists of a random permutation of the 12-letter alphabet.

12.4. Estimate the unicity distance for an English language encryption system that uses a key sequence made up of 10 random alphabetic characters:
(a) Where each key character can be any one of the 26 letters of the alphabet (duplicates are allowed).
(b) Where the key characters may not have any duplicates.

12.5. Repeat Problem 12.4 for the case where the key sequence is made up of ten integers randomly chosen from the set of numbers 0 to 999.

12.6. (a) Find the unicity distance for a DES system which encrypts 64-bit blocks (eight alphabetic characters) using a 56-bit key.

(b) What is the effect on the unicity distance in part (a) if the key is increased to 128 bits?

12.7. In Figures 12.8 and 12.9, P-boxes and S-boxes alternate. Is this arrangement any more secure than if all the P-boxes were first grouped together, followed by all the S-boxes similarly grouped together? Justify your answer.

12.8. What is the output of the first iteration of the DES algorithm when the plaintext and the key are each made up of zero sequences?

12.9. Consider the 10-bit plaintext sequence 0 1 0 1 1 0 1 0 0 1 and its corresponding ciphertext sequence 0 1 1 1 0 1 1 0 1 0, where the rightmost bit is the earliest bit. Describe the five-stage linear feedback shift register (LFSR) that produced the key sequence and show the initial state of the register. Is the output sequence of maximal length?

12.10. Following the RSA algorithm and parameters in Example 12.5, compute the encryption key, e, when the decryption key is chosen to be 151.

12.11. Given e and d that satisfy ed modulo-$\phi(n) = 1$, and a message that is encoded as an integer number, M, in the range $(0, n - 1)$ such that the gcd $(M, n) = 1$. Prove that $(M^e$ modulo-$n)^d$ modulo-$n = M$.

12.12. Use the RSA scheme to encrypt the message $M = 3$. Use the prime numbers $p = 5$ and $q = 7$. Choose the decryption key, d, to be 11, and calculate the value of the encryption key, e.

12.13. Consider the following for the RSA scheme.
(a) If the prime numbers are $p = 7$ and $q = 11$, list five allowable values for the decryption key, d.
(b) If the prime numbers are $p = 13$, $q = 31$, and the decryption key is $d = 37$, find the encryption key, e, and describe how you would use it to encrypt the word "DIGITAL."

12.14. Use the Merkle–Hellman public key scheme with the super-increasing vector, $\mathbf{a}'$ = 1, 3, 5, 10, 20. Use the following additional parameters: a large prime number $M = 51$ and a random number $W = 37$.
(a) Find the nonsuper-increasing vector, $\mathbf{a}$, to be made public, and encrypt the data vector 1 1 0 1 1.
(b) Show the steps by which an authorized receiver decrypts the ciphertext.

APPENDIX A

A Review
of Fourier Techniques

A.1 SIGNALS, SPECTRA, AND LINEAR SYSTEMS

Electrical communication signals consist of time-varying voltage or current waveforms, typically described in the time domain. It is also convenient to describe such signals in the frequency domain. A signal's frequency-domain description is called its *spectrum*. Spectral concepts are important in communication analysis and design; they can describe a signal by its average power or energy content at various frequencies, and they illustrate how much of the electromagnetic spectrum (bandwidth) the signal occupies. Broadcast stations are required by the Federal Communications Commission (FCC) to operate at their assigned frequency with very tight tolerances on the occupied bandwidth; for example, amplitude-modulated (AM) radio channels are spaced 10 kHz apart, and television channels are spaced 6 MHz apart. Our interest in spectra and Fourier techniques has to do with the real-world constraints of ensuring that our communication signals are confined to specified spectral boundaries.

Frequency spectral characteristics can be ascribed to both signal waveforms and to circuits. When we say that a particular spectrum describes a *signal*, we mean that one way of characterizing the signal waveform is to specify its amplitude and phase as a function of frequency. However, when we talk about the spectral attributes of a *circuit* we are referring to the output versus input frequency-domain transfer function of the circuit; in other words, we are characterizing the circuit by how much of a specific input signal spectrum is allowed to pass through it.

A.2 FOURIER TECHNIQUES FOR LINEAR SYSTEM ANALYSIS

Fourier techniques are often used for analyzing linear circuits or systems in the following ways: (1) by predicting the system response, (2) by determining the system dynamic specification (transfer function), and (3) by evaluating or interpreting test results. Item 1, predicting system response, is illustrated schematically in Figure A.1. Let the input be an arbitrary periodic waveform with period equal to T_0 seconds. Fourier techniques allow us to describe such an input as a sum of sinusoidal waveforms, as shown in the figure. The lowest-frequency sinusoid, or the *fundamental* frequency of the input periodic, has frequency $1/T_0$ hertz; the balance of the sinusoids have frequencies that are integral *harmonics* ($2/T_0$, $3/T_0$, . . .) of this fundamental frequency. An important attribute of a linear system is that *superposition* applies, which means that the response to the sum of excitations is the sum of the responses to the individually applied excitations. In fact, this is used as a definition of linearity. Specifically, if

$$y_1(t) = \text{system response to } x_1(t)$$

$$y_2(t) = \text{system response to } x_2(t)$$

and

$$ay_1(t) + by_2(t) = \text{system response to } ax_1(t) + bx_2(t)$$

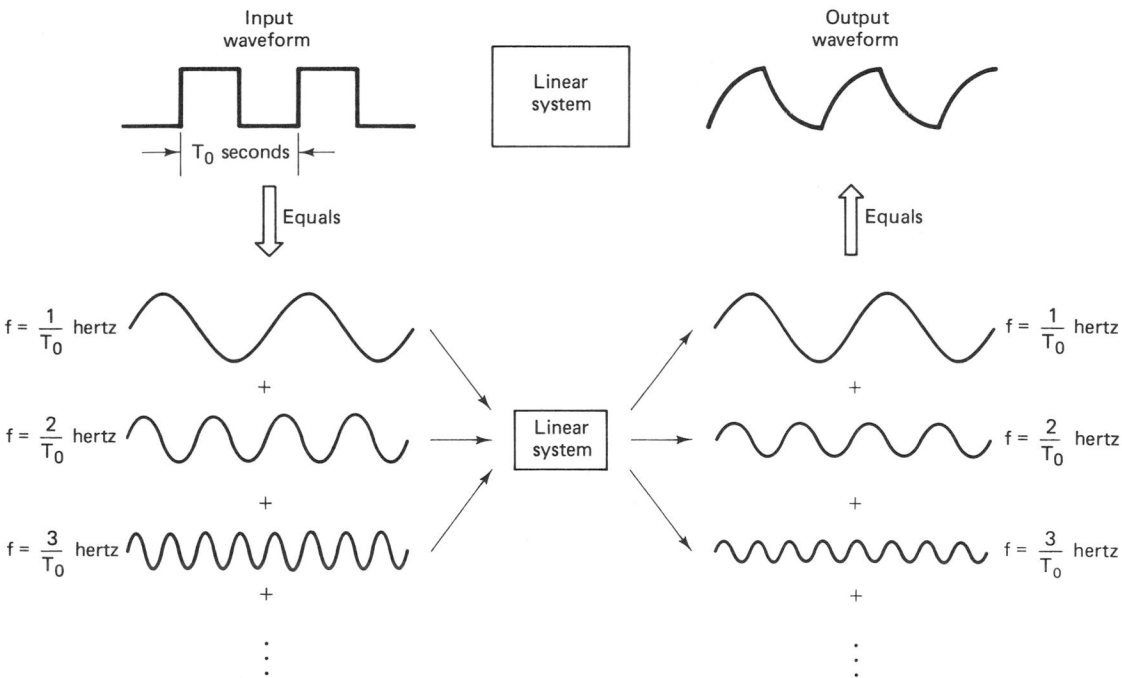

Figure A.1 Predicting system response.

for all a, b, $x_1(t)$, and $x_2(t)$, then the system is linear. A consequence of this definition is that the output response of a *linear* system with sinusoidal input waveforms must be comprised of sinusoidal waveforms having the *same frequencies* as the input waveforms; such a system is typically specified by an output versus input *frequency transfer function* (magnitude and phase versus frequency) as shown in Figure A.2. Figure A.2a illustrates a typical example of signal magnitude versus frequency; similarly, Figure A.2b illustrates a typical example of signal phase versus frequency.

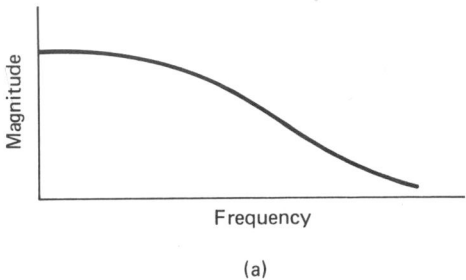

Frequency

(a)

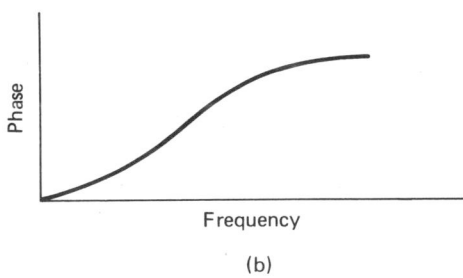

Frequency

(b)

Figure A.2 System transfer function. (a) Magnitude response. (b) Phase response.

The system transfer function serves as a performance specification; it describes the system response to each of the component sinusoids. Therefore, with the system transfer function in hand, one can predict each of the resulting output components. Using the principle of superposition, the final step of the analysis is to sum the individual output responses, thus forming the resulting overall response to the input periodic (see Figure A.1). In a similar manner, one can determine a system's transfer function, or evaluate a system's test results from knowledge of the input and output waveforms.

The development of Fourier methods had a major impact on the analysis of linear systems; it provided the translation between transient phenomena and sinusoidal techniques, and it simplified the analysis of linear systems under the excitation of any arbitrary input waveform. Just as logarithms allow the operation of multiplication to be treated as addition, so Fourier techniques allow the replacement of complex waveforms with sinusoidal components and sinusoidal methods.

A.2.1 Fourier Series Transform

Signals that are periodic with finite energy within each period can be represented by the *Fourier series*. Equation (A.1) describes such an arbitrary periodic waveform, $x(\lambda)$, in terms of an infinite number of increasing harmonic sine and cosine components.

$$x(\lambda) = \tfrac{1}{2}a_0 + a_1 \cos \lambda + a_2 \cos 2\lambda + a_3 \cos 3\lambda$$

$$+ \cdots + b_1 \sin \lambda + b_2 \sin 2\lambda + b_3 \sin 3\lambda + \cdots \quad (A.1)$$

The terms $\cos \lambda$ and $\sin \lambda$ are called the *fundamental terms*; the terms $\cos n\lambda$ and $\sin n\lambda$, for $n > 1$, are called *harmonic terms*, where n is an integer. The terms a_n and b_n represent the coefficients of the fundamental and harmonics, and $\tfrac{1}{2}a_0$ is the constant or dc term.

The function $x(\lambda)$ must have a period of 2π, or a submultiple thereof, and it must be single valued. The Fourier series can be thought of as a "recipe" for synthesizing any *arbitrary periodic* waveform using sinusoidal components. To be useful, the series must converge; that is, the sum of the series, as more and more of the higher harmonics are added, must approach a limit.

The process of synthesizing an arbitrary periodic waveform, from the coefficient values describing the mix of harmonics, is termed *synthesis*. The inverse process of calculating the coefficient values is termed *anaylsis*. Calculation of the coefficients is facilitated by the fact that the average of the sine and cosine cross-products is zero, as well as the average of any sinusoid. Equations (A.2) to (A.4), listed below, illustrate the basic averaging properties of the sine, cosine, their products and cross-products.

$$\left.\begin{array}{c} \displaystyle\int_{-\pi}^{\pi} \sin m\lambda \, d\lambda = 0 \\[1em] \displaystyle\int_{-\pi}^{\pi} \cos m\lambda \, d\lambda = 0 \\[1em] \displaystyle\int_{-\pi}^{\pi} \sin m\lambda \cos n\lambda \, d\lambda = 0 \end{array}\right\} \quad \text{where } m \text{ and } n \text{ are any integers} \quad (A.2)$$

$$\left.\begin{array}{c} \displaystyle\int_{-\pi}^{\pi} \sin m\lambda \sin n\lambda \, d\lambda = 0 \\[1em] \displaystyle\int_{-\pi}^{\pi} \cos m\lambda \cos n\lambda \, d\lambda = 0 \end{array}\right\} \quad \text{for } m \neq n \quad (A.3)$$

$$\left.\begin{array}{c} \displaystyle\int_{-\pi}^{\pi} (\sin m\lambda)^2 \, d\lambda = \pi \\[1em] \displaystyle\int_{-\pi}^{\pi} (\cos m\lambda)^2 \, d\lambda = \pi \end{array}\right\} \quad \text{for } m = n \quad (A.4)$$

Consider how one could go about finding the value of the coefficient, a_n or

b_n, in Equation (A.1). To find the coefficient a_3, for example, we can multiply both sides of Equation (A.1) by cos 3λ $d\lambda$ and integrate, as follows:

$$\int_{-\pi}^{\pi} x(\lambda) \cos 3\lambda \, d\lambda = \int_{-\pi}^{\pi} \tfrac{1}{2}a_0 \cos 3\lambda \, d\lambda + \int_{-\pi}^{\pi} a_1 \cos \lambda \cos 3\lambda \, d\lambda$$

$$+ \int_{-\pi}^{\pi} a_2 \cos 2\lambda \cos 3\lambda \, d\lambda + \int_{-\pi}^{\pi} a_3 (\cos 3\lambda)^2 \, d\lambda + \cdots$$

$$+ \int_{-\pi}^{\pi} b_1 \sin \lambda \cos 3\lambda \, d\lambda + \int_{-\pi}^{\pi} b_2 \sin 2\lambda \cos 3\lambda \, d\lambda$$

$$+ \int_{-\pi}^{\pi} b_3 \sin 3\lambda \cos 3\lambda \, d\lambda + \cdots$$

$$\int_{-\pi}^{\pi} x(\lambda) \cos 3\lambda \, d\lambda = \int_{-\pi}^{\pi} a_3(\cos 3\lambda)^2 \, d\lambda = a_3 \pi$$

$$a_3 = \frac{1}{\pi} \int_{-\pi}^{\pi} x(\lambda) \cos 3\lambda \, d\lambda$$

We can generalize the analysis above, to get

$$a_n = \frac{1}{\pi} \int_{-\pi}^{\pi} x(\lambda) \cos n\lambda \, d\lambda \tag{A.5}$$

$$b_n = \frac{1}{\pi} \int_{-\pi}^{\pi} x(\lambda) \sin n\lambda \, d\lambda \tag{A.6}$$

a_0 is found by solving Equation (A.5) with $n = 0$. This results in

$$\tfrac{1}{2}a_0 = \tfrac{1}{2}\pi \int_{-\pi}^{\pi} x(\lambda) \, d\lambda \tag{A.7}$$

which represents the zero-frequency term, or the average value of the periodic waveform. The synthesis process of Equation (A.1) can be expressed in more compact form as follows:

$$x(\lambda) = \tfrac{1}{2}a_0 + \sum_{n=1}^{\infty} (a_n \cos n\lambda + b_n \sin n\lambda) \tag{A.8}$$

There are several ways to express the *transform pair* (analysis and synthesis) of the Fourier series. The most common form makes use of the following identities to express the sine and cosine in exponential form:

$$\cos \lambda = \frac{e^{j\lambda} + e^{-j\lambda}}{2} \tag{A.9}$$

$$\sin \lambda = \frac{e^{j\lambda} - e^{-j\lambda}}{2j} \tag{A.10}$$

A periodic function with period T_0 seconds has frequency components of

f_0, $2f_0$, $3f_0$, . . . , where $f_0 = 1/T_0$ is called the *fundamental frequency*. We also refer to the frequency components as ω_0, $2\omega_0$, $3\omega_0$, . . . , where $\omega_0 = 2\pi/T_0$ is called the fundamental *radian* frequency. The terms f and ω are each used to denote frequency. When f is used, frequency in hertz is intended; when ω is used, frequency in radians/second is intended. Let us replace the $n\lambda$ terms of Equations (A.5) to (A.8) with $2\pi n f_0 t = 2\pi n t/T_0$ as the general argument of the sinusoidal components, where n is an integer. For $n = 1$, nf_0 represents the fundamental frequency; for $n > 1$, nf_0 represents harmonics of the fundamental frequency. Using Equations (A.8) to (A.10), we can express $x(t)$ in exponential form as follows:

$$x(t) = \frac{a_0}{2} + \frac{1}{2} \sum_{n=1}^{\infty} [(a_n - jb_n)e^{j2\pi nf_0 t} + (a_n + jb_n)e^{-j2\pi nf_0 t}] \qquad (A.11)$$

Let c_n denote the complex coefficients, or spectral components of $x(t)$, related to a_n and b_n by

$$c_n = \begin{cases} \frac{1}{2}(a_n - jb_n) & \text{for } n > 0 \\ \dfrac{a_0}{2} & \text{for } n = 0 \\ \frac{1}{2}(a_n + jb_n) & \text{for } n < 0 \end{cases} \qquad (A.12)$$

Then we can simplify Equation (A.11) as follows:

$$x(t) = \sum_{n=-\infty}^{\infty} c_n e^{j2\pi nf_0 t} \qquad (A.13)$$

where the coefficients of the exponential harmonics, c_n, are

$$c_n = \frac{1}{T_0} \int_{-T_0/2}^{T_0/2} x(t)e^{-j2\pi nf_0 t}\, dt \qquad (A.14)$$

To verify Equation (A.14) we multiply both sides of Equation (A.13) by $e^{-j2\pi mf_0 t}\, dt/T_0$, integrate over the interval $(-T_0/2, T_0/2)$, and use the following relationship:

$$\frac{1}{T_0} \int_{-T_0/2}^{T_0/2} e^{j(n-m)2\pi f_0 t}\, dt = \delta_{nm} = \begin{cases} 1 & \text{for } n = m \\ 0 & \text{for } n \neq m \end{cases} \qquad (A.15)$$

where δ_{nm} is known as the *Kronecker delta*. By multiplying and integrating in this way we obtain, for all integers m,

$$\frac{1}{T_0} \int_{-T_0/2}^{T_0/2} x(t)e^{-j2\pi mf_0 t}\, dt = \sum_{n=-\infty}^{\infty} c_n \delta_{nm} = c_m \qquad (A.16)$$

In general, the coefficient c_n is a complex number; it can be expressed in the form

$$c_n = |c_n|\, e^{j\theta_n} \qquad (A.17)$$

$$c_{-n} = |c_n|\, e^{-j\theta_n} \qquad (A.18)$$

where

$$|c_n| = \tfrac{1}{2}\sqrt{a_n^2 + b_n^2} \qquad (A.19)$$

$$\theta_n = \tan^{-1} -\frac{b_n}{a_n} \qquad (A.20)$$

$$b_0 = 0 \quad \text{and} \quad c_0 = \frac{a_0}{2}$$

The value of $|c_n|$ defines the magnitude of the nth harmonic component of the periodic waveform, so that a plot of $|c_n|$ versus frequency, called the *magnitude spectrum*, yields the magnitude of each of the n discrete harmonics in the signal. Similarly, a plot of θ_n versus frequency, called the *phase spectrum*, yields the phase of each harmonic component in the signal.

The Fourier coefficients of a real-valued periodic time function exhibit the following relationship:

$$c_{-n} = c_n^* \qquad (A.21)$$

where c_n^* is the complex conjugate of c_n. We therefore have

$$|c_{-n}| = |c_n| \qquad (A.22)$$

and the magnitude spectrum is an even function of frequency. Similarly, the phase spectrum θ_n is an odd function of frequency, because from Equation (A.20),

$$\theta_{-n} = -\theta_n \qquad (A.23)$$

The Fourier series is particularly useful in characterizing arbitrary periodic waveforms, with finite energy in each period, as presented above. The Fourier series can also be used to characterize nonperiodic signals having finite energy over a finite interval. However, a more convenient frequency-domain representation for such signals uses the Fourier integral transform (presented in Section A.2.3).

A.2.2 Spectrum of a Pulse Train

A signal of great interest in digital communications is an ideal periodic sequence of rectangular pulses, called a *pulse train*, illustrated in Figure A.3. For the pulse train, $x_p(t)$, with pulse amplitude A, pulse width T, and period T_0, the reader can verify, using Equations (A.14) and (A.10), the following expression for the Fourier series coefficients:

$$c_n = \frac{AT}{T_0} \frac{\sin(\pi nT/T_0)}{\pi nT/T_0} = \frac{AT}{T_0} \operatorname{sinc} \frac{nT}{T_0} \qquad (A.24)$$

where

$$\operatorname{sinc} y = \frac{\sin \pi y}{\pi y}$$

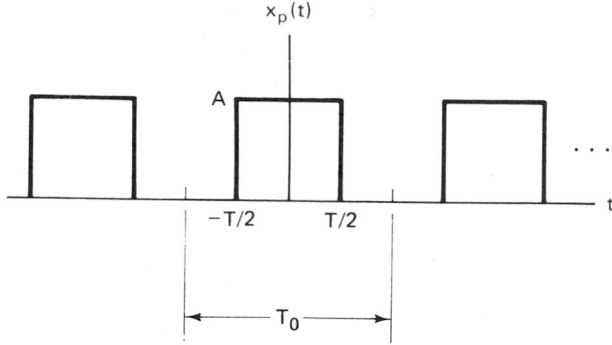

Figure A.3 Pulse train.

The sinc function, as shown in Figure A.4, has a maximum value of unity at $y = 0$ and approaches zero as y approaches infinity, oscillating through positive and negative values. It goes through zero at $y = \pm 1, \pm 2, \ldots$ The pulse train magnitude spectrum, $|c_n|$ as a function of n/T_0, is plotted in Figure A.5a, and the phase spectrum, θ_n, is plotted in Figure A.5b. The positive and negative frequencies of the two-sided spectrum represent a useful way of expressing the spectrum mathematically; of course, only the positive frequencies can be reproduced in a laboratory.

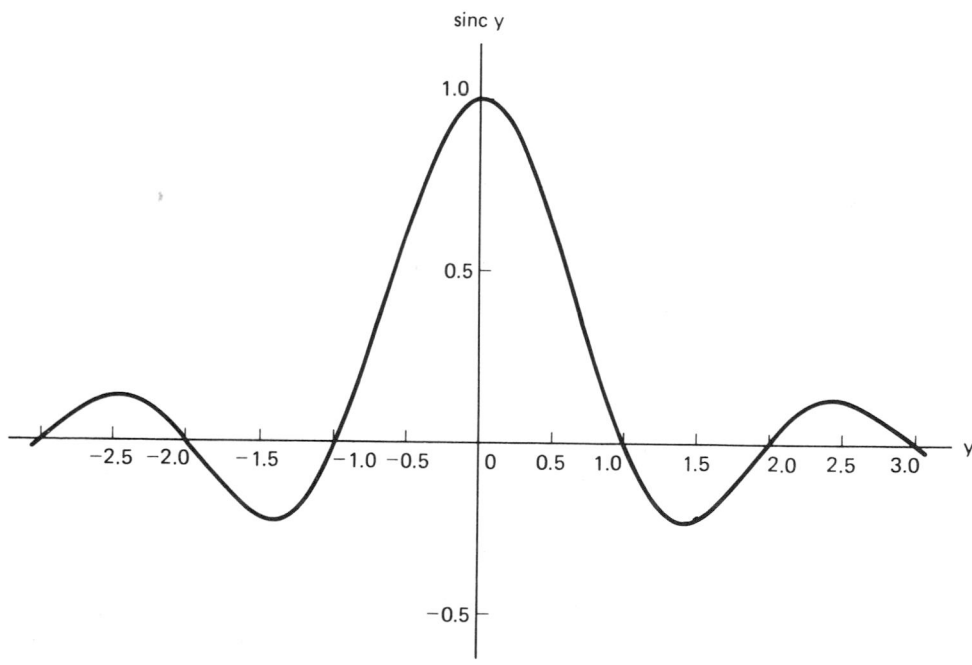

Figure A.4 Sinc function.

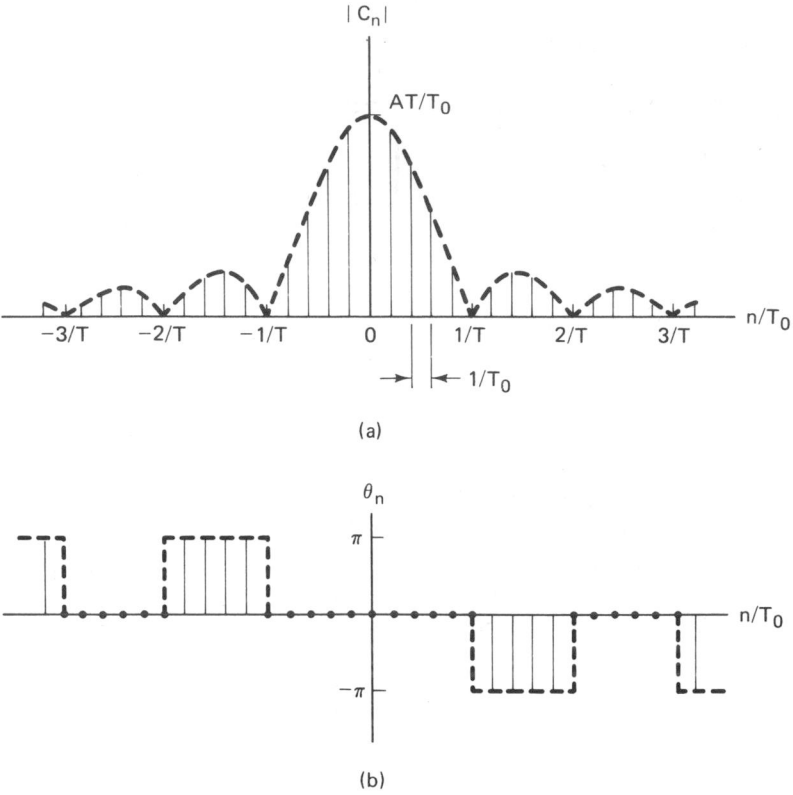

Figure A.5 Spectrum of a pulse train. (a) Magnitude spectrum. (b) Phase spectrum.

Synthesis is performed by substituting the coefficients of Equation (A.24) into Equation (A.13). The resulting series below yields the original ideal pulse train, $x_p(t)$, synthesized from its component parts.

$$x_p(t) = \frac{AT}{T_0} \sum_{n=-\infty}^{\infty} \text{sinc}\, \frac{nT}{T_0}\, e^{j2\pi n f_0 t} \qquad (A.25)$$

The ideal periodic pulse train contains frequency components at all integer multiples of the fundamental. In communication systems the significant portion of a baseband signal's power or energy is often assumed to be contained within the frequencies from zero to the first null of the magnitude spectrum (see Figure A.5a). Therefore, $1/T$ is often used as a measure of signal *bandwidth*, in hertz, for a pulse train with pulse width T. Note that bandwidth is inversely proportional to pulse width; the narrower are the pulses, the wider is the bandwidth associated with these pulses. Also, notice that the spacing between spectral lines $\Delta f = 1/T_0$ is inversely proportional to the pulse period; as the period increases, the lines move closer together.

A.2.3 Fourier Integral Transform

In communication systems we often encounter nonperiodic signals having finite energy in a finite interval, and having zero energy outside this interval. Such signals can be conveniently characterized using the Fourier integral transform, or simply the *Fourier transform*. We can describe the nonperiodic signal as a periodic one, in the limiting sense. For example, consider the pulse train shown in Figure A.3. As $T_0 \rightarrow \infty$ and the pulse train approaches a single pulse, $x(t)$, the number of spectral lines approaches infinity and the spectral plot approaches a smooth frequency spectrum $X(f)$. For this limiting case, we can define a Fourier integral transform pair. This pair, expressed below, can be used to describe the time–frequency relationship for nonperiodic signals.

$$X(f) = \int_{-\infty}^{\infty} x(t)e^{-j2\pi ft}\, dt \tag{A.26}$$

$$x(t) = \int_{-\infty}^{\infty} X(f)e^{j2\pi ft}\, df \tag{A.27}$$

where f is frequency measured in hertz. Henceforth, the Fourier integral transform operation will be designated by the notation $\mathscr{F}\{\cdot\}$, and the inverse Fourier integral transform will be designated by $\mathscr{F}^{-1}\{\cdot\}$. The relationship between the time and frequency domains will be indicated by using the double arrow as follows:

$$x(t) \leftrightarrow X(f)$$

This notation indicates that $X(f)$ is the Fourier transform of $x(t)$ and that $x(t)$ is the inverse Fourier transform of $X(f)$. In the typical communications context, $x(t)$ is a real-valued function and $X(f)$ is a complex function, having real and imaginary components; in polar form, shown below, the spectrum, $X(f)$, can be specified by a magnitude characteristic and a phase characteristic.

$$X(f) = |X(f)|\, e^{j\theta(f)} \tag{A.28}$$

The properties of $X(f)$, the spectrum of a nonperiodic waveform, are similar to those of the spectrum for a periodic waveform, presented in Equations (A.17) to (A.23); that is, when $x(t)$ is real valued

$$X(-f) = X^*(f) \tag{A.29}$$

$$= |X(f)|\, e^{-j\theta(f)} \tag{A.30}$$

where X^* is the complex conjugate of X. The magnitude spectrum $|X(f)|$ is an even function of f and the phase spectrum is an odd function of f. In many cases $X(f)$ is either purely real or purely imaginary, and only one plot suffices to describe it.

A.3 FOURIER TRANSFORM PROPERTIES

There are many excellent references dealing with the details of Fourier transforms and their properties [1–4]. In this appendix we will emphasize the properties that are fundamental to communication systems. Some of the key features affecting signal transmission in communication systems are time delay, phase shift, multiplication by other signals, frequency translation, waveform convolution, and spectral convolution. We shall focus on the Fourier properties (shifting and convolution) needed to describe these key communication features.

A.3.1 Time Shifting Property

If $x(t) \leftrightarrow X(f)$,

$$\mathcal{F}\{x(t - t_0)\} = \int_{-\infty}^{\infty} x(t - t_0)e^{-j2\pi ft} \, dt$$

Let $\mu = t - t_0$; then

$$\mathcal{F}\{x(t - t_0)\} = \int_{-\infty}^{\infty} x(\mu)e^{-j2\pi f(\mu + t_0)} \, d\mu$$
$$= X(f)e^{-j2\pi ft_0} \tag{A.31}$$

As a signal is delayed in time, the magnitude of its frequency spectrum remains unchanged, but its phase spectrum experiences a phase shift. A time shift of t_0 in the time domain is equivalent to multiplication by $e^{-j2\pi ft_0}$ (a phase shift of $-2\pi ft_0$) in the frequency domain.

A.3.2 Frequency Shifting Property

If $x(t) \leftrightarrow X(f)$,

$$\mathcal{F}\{x(t)e^{j2\pi f_0 t}\} = \int_{-\infty}^{\infty} x(t)e^{j2\pi f_0 t}e^{-j2\pi ft} \, dt$$

$$= \int_{-\infty}^{\infty} x(t)e^{-j2\pi(f - f_0)t} \, dt \tag{A.32}$$

$$= X(f - f_0)$$

This is the basic *frequency translating* property that describes the shifted spectrum resulting from multiplying a signal by $e^{j2\pi f_0 t}$. Equation (A.32) can be used in conjunction with Equation (A.9) to yield the Fourier transform of a waveform multiplied by a cosine wave, as follows:

$$x(t) \cos 2\pi f_0 t = \tfrac{1}{2}[x(t)e^{j2\pi f_0 t} + x(t)e^{-j2\pi f_0 t}]$$
$$x(t) \cos 2\pi f_0 t \leftrightarrow \tfrac{1}{2}[X(f - f_0) + X(f + f_0)] \tag{A.33}$$

This property is also called the *mixing* or *modulation* theorem. Multiplication of an arbitrary signal by a sinusoid of frequency f_0 translates the original signal spectrum by f_0, and also by $-f_0$.

A.4 USEFUL FUNCTIONS

A.4.1 Unit Impulse Function

A useful function in communication theory is the unit impulse or *Dirac delta* function, $\delta(t)$. The impulse function can be developed from any of several fundamental functions (e.g., a rectangular pulse or a triangular pulse). In each development, the impulse function is defined in the limiting sense (the pulse amplitude approaches infinity, the pulse width approaches zero, but the area under the pulse is constrained to be unity) [5]. The unit impulse function has the following important properties:

$$\int_{-\infty}^{\infty} \delta(t) \, dt = 1 \tag{A.34}$$

$$\delta(t) = 0 \quad \text{for } t \neq 0 \tag{A.35}$$

$$\delta(t) \text{ is unbounded at } t = 0 \tag{A.36}$$

$$\mathcal{F}\{\delta(t)\} = \mathcal{F}^{-1}\{\delta(f)\} = 1 \tag{A.37}$$

$$\int_{-\infty}^{\infty} x(t)\delta(t - t_0) \, dt = x(t_0) \tag{A.38}$$

Equation (A.38) is known as the *sifting* or *sampling property*; the unit impulse multiplier selects a sample of the function $x(t)$ evaluated at $t = t_0$.

In some problems it is useful to use the following equivalent integrals for an impulse function, defined in the time domain or the frequency domain [3]:

$$\delta(t) = \int_{-\infty}^{\infty} e^{j2\pi ft} \, df \tag{A.39}$$

$$\delta(f) = \int_{-\infty}^{\infty} e^{-j2\pi ft} \, dt \tag{A.40}$$

A.4.2 Spectrum of a Sinusoid

For the purpose of representing a sinusoidal waveform by a Fourier transform, the waveform may be assumed to exist only in the interval $(-T_0/2 < t < T_0/2)$. Under these conditions the function has a Fourier transform as long as T_0 is finite. In the limit, T_0 is made very large, but finite. The spectrum of the waveform $x(t) = A \cos 2\pi f_0 t$ can be found by using Equations (A.9) and (A.26):

$$X(f) = \int_{-\infty}^{\infty} \frac{A}{2} (e^{j2\pi f_0 t} + e^{-j2\pi f_0 t}) e^{-j2\pi ft} \, dt$$

$$= \frac{A}{2} \int_{-\infty}^{\infty} e^{-j2\pi(f - f_0)t} + e^{-j2\pi(f + f_0)t} \, dt$$

As described in Equation (A.40), the integral expression above can be equated

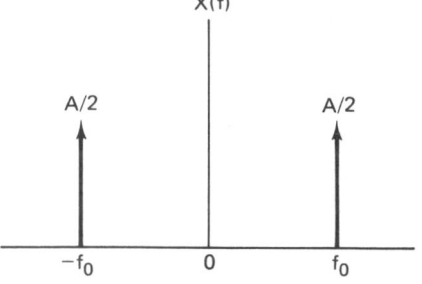

Figure A.6 Spectrum for $x(t) = A \cos 2\pi f_0 t$.

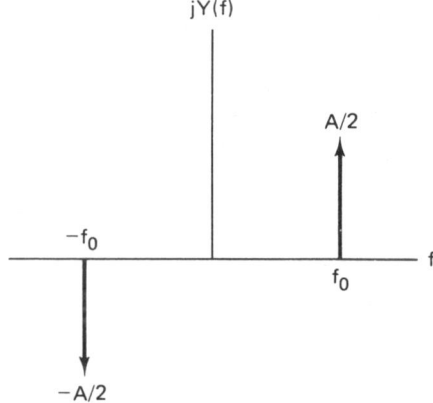

Figure A.7 Spectrum for $y(t) = A \sin 2\pi f_0 t$.

to unit impulse functions located at frequencies $\pm f_0$ as follows:

$$X(f) = \frac{A}{2}[\delta(f - f_0) + \delta(f + f_0)] \tag{A.41}$$

Similarly, the spectrum of a sine waveform $y(t) = A \sin 2\pi f_0 t$ can be shown to be equal to

$$Y(f) = \frac{A}{2j}[\delta(f - f_0) - \delta(f + f_0)] \tag{A.42}$$

The cosine waveform spectrum is shown in Figure A.6, and the sine waveform spectrum is shown in Figure A.7. Each of the impulse functions shown on these spectral plots is depicted as a spike with a weight of $A/2$ or $-A/2$.

A.5 CONVOLUTION

Convolution was used by Oliver Heaviside in the late nineteenth century to calculate electrical circuit output current when the input voltage waveform was more complicated than a simple battery source. The use of the methods of Heaviside

A Review of Fourier Techniques   App. A

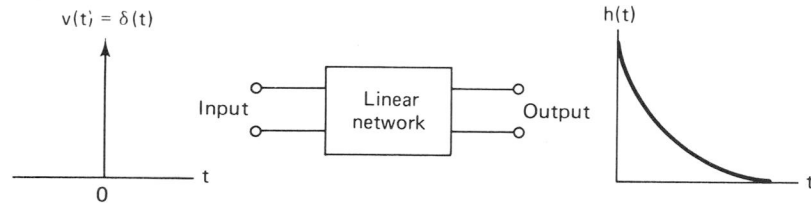

Figure A.8 Impulse response of a linear system.

predates the use of the analytical methods developed by Fourier and Laplace (even though publications by Fourier and Laplace came earlier).

The response of a circuit to an impulse voltage $v(t) = \delta(t)$ is called the *impulse response* and is denoted by $h(t)$, as shown in Figure A.8; it is simply the output voltage that would result if the input were a delta function. Heaviside approximated an arbitrary voltage waveform, like the one shown in Figure A.9a, by a set of equally spaced pulses. Such pulses of finite height and duration are shown in Figure A.9b. In the limit as the pulse width $\Delta\tau$ approaches zero, each pulse approaches an impulse function with weight equal to the area under that

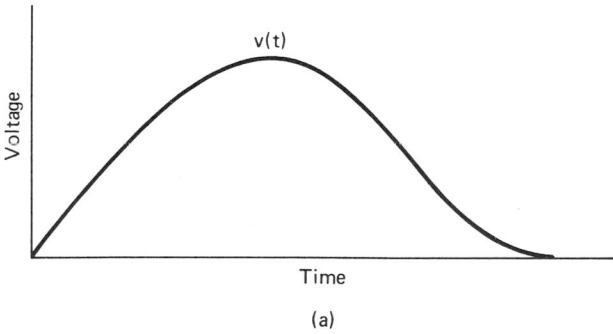

(a)

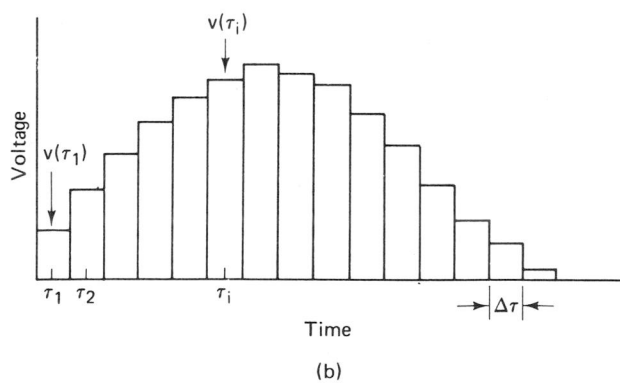

Time

(b)

Figure A.9 (a) Input voltage waveform. (b) Approximate input voltage waveform.

pulse. In the following discussion we shall refer to these equally spaced pulses as *impulses* even though they are impulses *only in the limit*.

Care needs to be taken with the notation of time, since we are interested in the times at which impulses are applied, and also the times at which their output responses are observed. We need to identify these two different time sequences; we shall use the following notation:

1. Time of the input application will be termed τ, so that the input voltage impulses are designated $v(\tau_1)$, $v(\tau_2)$, . . . , $v(\tau_N)$.

2. Time of the output response will be termed t, so that the output currents are designated $i(t_1)$, $i(t_2)$, . . . , $i(t_N)$.

Heaviside found the response or current produced by each input impulse independently; then he added the individual responses to get the total current. The weight of the impulse produced by the rectangular voltage at time τ_1 is the product $v(\tau_1) \, \Delta\tau$. The series of impulses can approximate the arbitrary input voltage as closely as desired by allowing $\Delta\tau$ to approach zero. Note again that the instant at which an impulse is applied is called τ_i, and the instance at which the system response is determined is called t_i, where τ is the input time variable, t is the output time variable, and $i = 1, \ldots, N$.

Figure A.10 illustrates the output response $i(t) = A_1 h(t - \tau_1)$ to an impulse with height $v(\tau_1)$. Since the input impulse at τ_1 is *not* a unit impulse, we weight it with its strength or area, $A_1 = v(\tau_1) \, \Delta\tau$. At some time t_1, where $t_1 > \tau_1$, the output response to the impulse $v(\tau_1)$ is expressed as

$$i(t_1) = A_1 h(t_1 - \tau_1) \qquad \text{for } t_1 > \tau_1$$

as shown in Figure A.10. When there are several input impulses, the total output response for a linear system is simply the sum of the individual responses. Figure A.11 illustrates the response of the network to two input impulses. For N impulses,

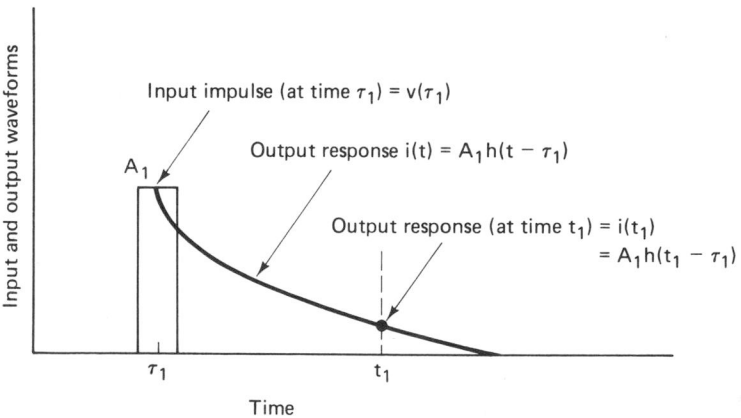

Figure A.10   Output response to an impulse at time τ_1.

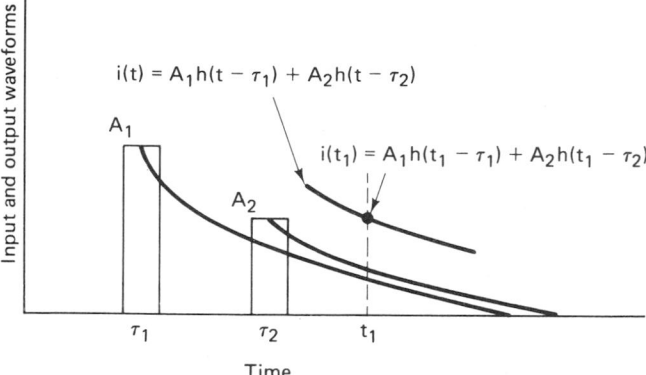

$i(t) = A_1h(t - \tau_1) + A_2h(t - \tau_2)$

A_1

$i(t_1) = A_1h(t_1 - \tau_1) + A_2h(t_1 - \tau_2)$

A_2

τ_1 τ_2 t_1

Time

Input and output waveforms

Figure A.11 Output response to two impulses.

the output current measured at time t_1 can be expressed as

$$i(t_1) = A_1h(t_1 - \tau_1) + A_2h(t_1 - \tau_2) + \cdots + A_N(t_1 - \tau_N)$$

where the impulses are applied at $\tau_1, \tau_2, \ldots, \tau_N$, and where $t_1 > \tau_N$.

Any impulses applied at times greater than t_1 are disregarded, for they contribute nothing to $i(t_1)$. This corresponds to the *causality* requirement for physically realizable systems, which states that the system response must be zero prior to the application of the excitation. By generalizing, we get the output current at any time t,

$$i(t) = A_1h(t - \tau_1) + A_2h(t - \tau_2) + \cdots + A_Nh(t - \tau_N)$$

or

$$i(t) = \sum_{i=1}^{N} v(\tau_i)\,\Delta\tau h(t - \tau_i) \tag{A.43}$$

since the height of the impulse at τ_i is equal to $v(\tau_i)$. As $\Delta\tau$ approaches zero, the sum of the input impulses approaches the actual applied voltage $v(\tau)$; we can replace $\Delta\tau$ with $d\tau$, and the summation becomes the *convolution integral*:

$$i(t) = \int_{-\infty}^{\infty} v(\tau)h(t - \tau)\,d\tau \tag{A.44}$$

In shorthand notation this is expressed as

$$i(t) = v(t) * h(t) \tag{A.45}$$

In summary, $i(t)$ is the sum of the individual impulse responses as a function of output time t. Each impulse response is due to an impulse applied at some input time τ and is weighted by the strength of that impulse.

Sec. A.5   Convolution   **725**

A.5.1 Graphical Illustration of Convolution

Consider that an input square pulse $v(t)$ is applied to a linear network whose impulse response is labeled $h(t)$ as shown in Figure A.12a. The output response is characterized by the convolution integral expressed in Equation (A.44).

The independent variable in the convolution integral is τ. The functions $v(\tau)$ and $h(-\tau)$ are shown in Figure A.12b. Note that $h(-\tau)$ is obtained by folding $h(\tau)$ about $\tau = 0$. The term $h(t - \tau)$ represents the function $h(-\tau)$ shifted by t seconds along the positive τ axis. Figure A.12c shows the function $h(t_1 - \tau)$. The value of the convolution integral at $t = t_1$ is given by Equation (A.44) evaluated at $t = t_1$. This is simply the area under the product curve of $v(\tau)$ and $h(t_1 - \tau)$, shown shaded in Figure A.12d. Similarly, the convolution integral evaluated at $t = t_2$ is equal to the shaded area in Figure A.12e. Figure A.12f is a plot of the output response as a result of the square pulse input to the circuit with impulse response shown in Figure A.12a. Each evaluation of the convolution integral, at some time t_i, yields one point, $i(t_i)$, on the plot of Figure A.12f.

A.5.2 Time Convolution Property

If $x_1(t) \leftrightarrow X_1(f)$ and $x_2(t) \leftrightarrow X_2(f)$,

$$x_1(t) * x_2(t) = \int_{-\infty}^{\infty} x_1(\tau)x_2(t - \tau) \, d\tau$$

$$\mathcal{F}\{x_1(t) * x_2(t)\} = \int_{-\infty}^{\infty} \int_{-\infty}^{\infty} x_1(\tau)x_2(t - \tau) \, d\tau \, e^{-j2\pi ft} \, dt$$

For linear systems, we may exchange the order of integration as follows:

$$\mathcal{F}\{x_1(t) * x_2(t)\} = \int_{-\infty}^{\infty} x_1(\tau) \, d\tau \int_{-\infty}^{\infty} x_2(t - \tau)e^{-j2\pi ft} \, dt \qquad \text{(A.46)}$$

By the Fourier *time shifting property*, the second integral expression of the right-hand side is equal to $X_2(f)e^{-j2\pi f\tau}$:

$$\mathcal{F}\{x_1(t) * x_2(t)\} = X_2(f) \int_{-\infty}^{\infty} x_1(\tau)e^{-j2\pi f\tau} \, d\tau$$
$$= X_1(f)X_2(f) \qquad \text{(A.47)}$$

Therefore, the operation of *convolution* in the time domain can be replaced by *multiplication* in the frequency domain.

A.5.3 Frequency Convolution Property

Because of the symmetry of the Fourier transform pair in Equations (A.26) and (A.27), it can be shown that multiplication in the time domain transforms to convolution in the frequency domain

$$x_1(t)x_2(t) \leftrightarrow X_1(f) * X_2(f) \qquad \text{(A.48)}$$

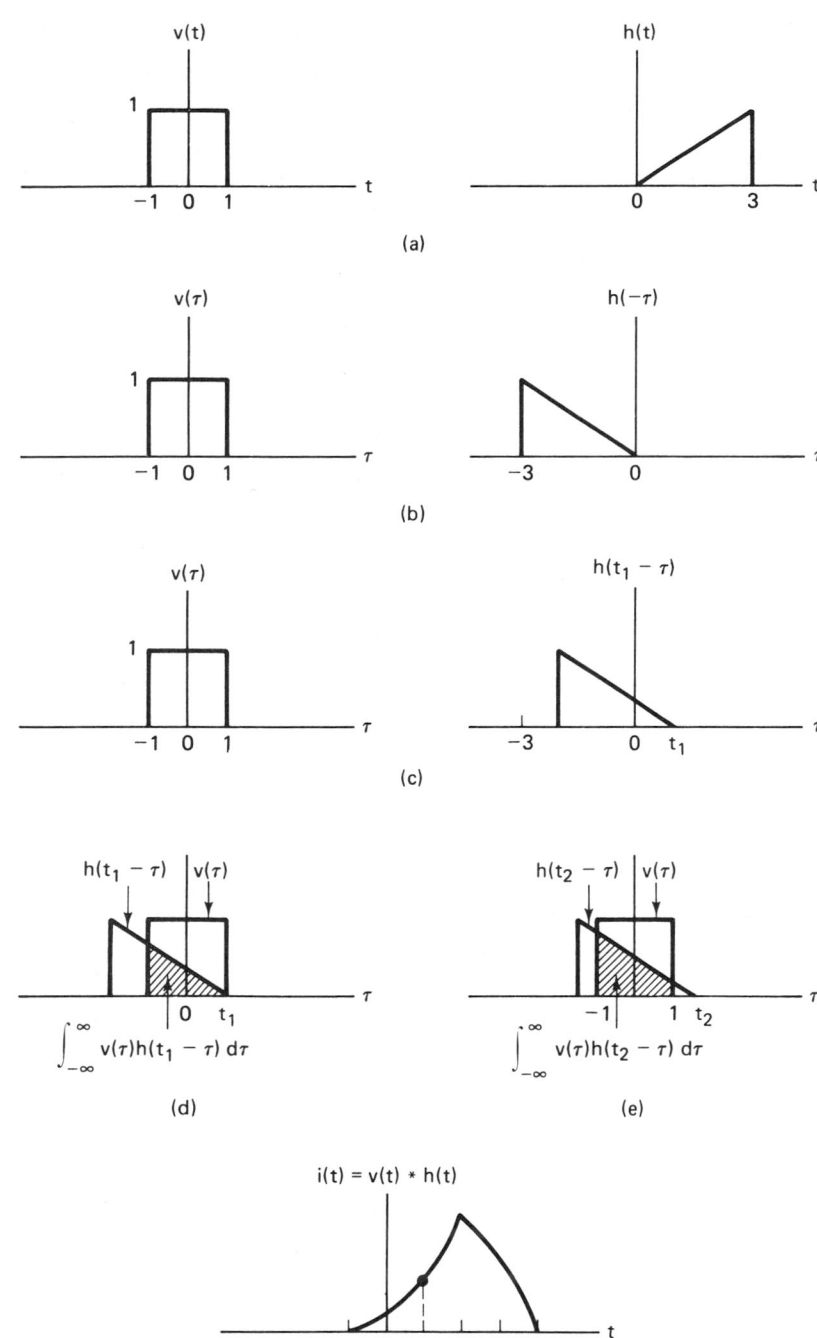

Figure A.12 Graphical example of convolution.

The properties that transform multiplication in one domain to convolution in the other domain are particularly useful, since one operation is often easier to perform than the other. For example, we discussed earlier that Heaviside used convolution to solve for the output current of a linear system when the input was excited by an arbitrary voltage waveform. Such methods involve the (sometimes tedious) convolution of an input waveform with the impulse response of a system. Since convolution in the time domain is transformed into multiplication in the frequency domain, as shown in Equation (A.47), for a linear system we can simply multiply the input waveform spectrum by the system transfer function. The output waveform is then found by taking the inverse Fourier transform of the product.

$$i(t) = \mathcal{F}^{-1}\{V(f)H(f)\} \qquad (A.49)$$

Solutions of the form shown in Equation (A.49) are often much easier to perform than those described by Equation (A.45). However, under certain circumstances, the operation of convolution is so simple that it can be performed graphically, by inspection. For example, suppose that we wished to multiply an arbitrary waveform by some fixed frequency cosine wave, such as a carrier wave, in the case of modulation. By applying Equation (A.48), we can convolve the spectrum of the arbitrary waveform with the spectrum of the cosine wave. This is easily accomplished, as is shown in the next section.

A.5.4 Convolution of a Function with a Unit Impulse

By the property shown in Equation (A.47), it should be clear that if

$$x(t) \leftrightarrow X(f)$$

and since

$$\delta(t) \leftrightarrow 1$$

then

$$x(t) * \delta(t) \leftrightarrow X(f) \qquad (A.50)$$

It should also be evident that

$$x(t) * \delta(t) = x(t) \qquad (A.51)$$

and

$$X(f) * \delta(f) = X(f) \qquad (A.52)$$

We therefore conclude that convolution of a function with a unit impulse function reproduces the original function. A simple extension of Equation (A.52) yields

$$X(f) * \delta(f - f_0) = X(f - f_0) \qquad (A.53)$$

Figure A.13 illustrates the ease of convolving the spectrum of an arbitrary waveform with the spectrum of a cosine wave. Figure A.13a shows an arbitrary baseband spectrum $X(f)$. Figure A.13b shows a spectrum, $Y(f) = \delta(f - f_0) +$

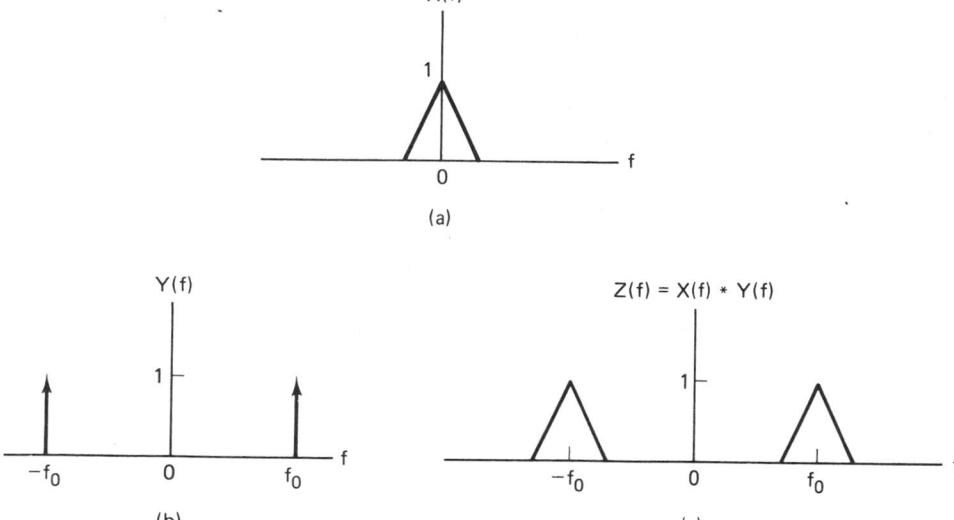

Figure A.13 Convolving a signal spectrum with a cosine-wave spectrum.

$\delta(f + f_0) = \mathcal{F}\{2 \cos 2\pi f_0 t\}$. The output, $Z(f) = X(f) * Y(f)$, in Figure A.13c is obtained by convolving the waveform spectrum with the impulse functions of $Y(f)$ according to Equation (A.53), where the impulses act as sampling functions. Hence, for this simple example, convolution can be performed graphically by sweeping the sampling impulses past the waveform spectrum. Multiplication by the impulse functions at each step in the sweep yields replications of the waveform spectrum. The result, shown in Figure A.13c, is a shifted version of the original spectrum $X(f)$ to the locations of the impulse functions in Figure A.13b.

A.5.5 Demodulation Application of Convolution

In Section A.5.4 we examined a waveform multiplied by $2 \cos 2\pi f_0 t$. We illustrated the frequency-domain view of convolving the waveform spectrum with a cosine-wave spectrum. In this section we look at the reverse process. A waveform that has been multiplied by $2 \cos 2\pi f_0 t$ is to be demodulated (the waveform is to be restored to its baseband frequency range).

Figure A.14a represents the spectrum, $Z(f)$, of the waveform that has been upshifted in frequency. We can demodulate this upshifted waveform and recover the baseband waveform, by multiplying it by $2 \cos 2\pi f_0 t$. Instead, we shall illustrate the detection process in the frequency domain by convolving $Z(f)$ with the spectrum of the carrier, $Y(f) = \delta(f - f_0) + \delta(f + f_0)$, shown in Figure A.14b.

A simple extension of Equations (A.52) and (A.53) yields

$$X(f - f_0) * \delta(f - f_1) = X(f - f_0 - f_1) \qquad \text{(A.54)}$$

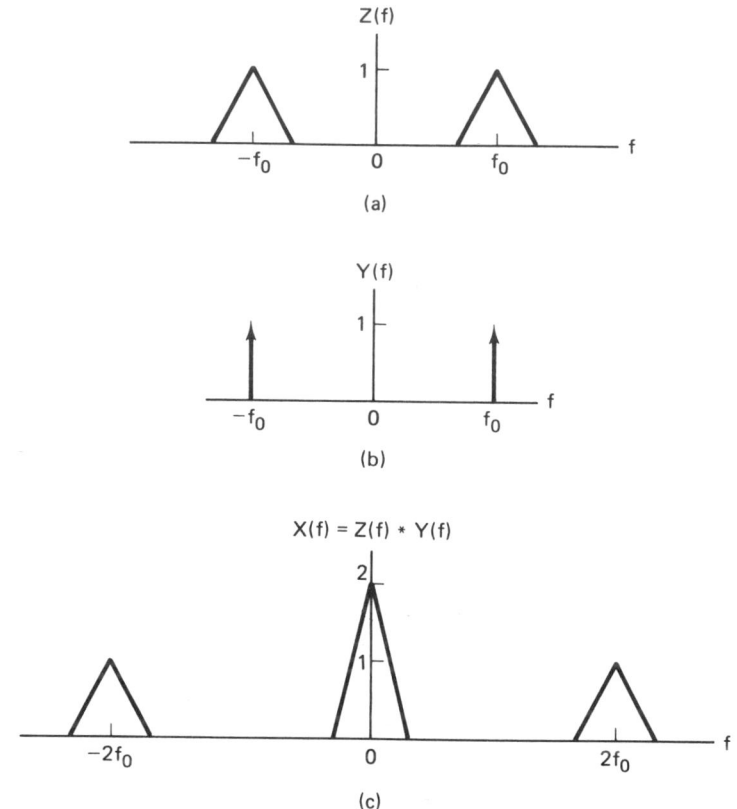

Figure A.14 Demodulation application.

Therefore, the result of demodulation, $X(f) = Z(f) * Y(f)$ is obtained by applying Equation (A.54). The resulting signal spectrum appears at baseband (detected) and also at frequencies $\pm 2f_0$, as shown in Figure A.14c. As in the previous section, the convolution can be performed graphically. The resulting Figure A.14c contains the following terms:

$$[Z(f - f_0) + Z(f + f_0)] * [\delta(f - f_0) + \delta(f + f_0)]$$

$$= Z(f - f_0) * \delta(f - f_0) + Z(f - f_0) * \delta(f + f_0)$$

$$+ Z(f + f_0) * \delta(f - f_0) + Z(f + f_0) * \delta(f + f_0) \qquad \text{(A.55)}$$

$$= 2Z(f) + Z(f - 2f_0) + Z(f + 2f_0)$$

Notice that the resulting terms consist of the baseband spectrum plus terms associated with higher-frequency components. The result is typical of the detection process; the higher-frequency terms are filtered and discarded, leaving the demodulated baseband spectrum.

Commonly used Fourier transforms and operations are tabulated below. The transforms are listed in Table A.1 and the operations in Table A.2.

TABLE A.1 Fourier Transforms

$x(t)$	$X(f)$
1. $\delta(t)$	1
2. 1	$\delta(f)$
3. $\cos 2\pi f_0 t$	$\dfrac{1}{2}[\delta(f - f_0) + \delta(f + f_0)]$
4. $\sin 2\pi f_0 t$	$\dfrac{1}{2j}[\delta(f + f_0) - \delta(f - f_0)]$
5. $\delta(t - t_0)$	$\exp(-j2\pi f t_0)$
6. $\exp(j2\pi f_0 t)$	$\delta(f - f_0)$
7. $\exp(-a|t|), \quad a > 0$	$\dfrac{2a}{a^2 + (2\pi f)^2}$
8. $\exp\left[-\pi \left(\dfrac{t}{T}\right)^2\right]$	$T \exp[-\pi(fT)^2]$
9. $u(t) = \begin{cases} 1 & \text{for } t > 0 \\ 0 & \text{for } t < 0 \end{cases}$	$\dfrac{1}{2}\delta(f) + \dfrac{1}{j2\pi f}$
10. $\exp(-at) u(t), \quad a > 0$	$\dfrac{1}{a + j2\pi f}$
11. $t \exp(-at) u(t), \quad a > 0$	$\dfrac{1}{(a + j2\pi f)^2}$
12. $\text{rect}\left(\dfrac{t}{T}\right)$	$T \text{ sinc } fT$
13. $\cos 2\pi f_0 t \left[\text{rect}\left(\dfrac{t}{T}\right)\right]$	$\dfrac{T}{2}[\text{sinc } (f - f_0)T + \text{sinc } (f + f_0)T]$
14. $W \text{ sinc } Wt$	$\text{rect}\left(\dfrac{f}{W}\right)$
15. $\begin{cases} 1 - \dfrac{|t|}{T} & \text{for } |t| \le T \\ 0 & \text{for } |t| > T \end{cases}$	$T \text{ sinc}^2 fT$
16. $\displaystyle\sum_{m=-\infty}^{\infty} \delta(t - mT_0)$	$\dfrac{1}{T_0}\displaystyle\sum_{n=-\infty}^{\infty} \delta\left(f - \dfrac{n}{T_0}\right)$

Note: $\text{rect}(f/2W) = 1$ for $-W < f < W$, 0 for $|f| > W$, and $\text{sinc } x = (\sin \pi x)/\pi x$.

TABLE A.2 Fourier Operations

Operation	$x(t)$	$X(f)$
1. Scaling	$x(at)$	$\dfrac{1}{\|a\|} X\left(\dfrac{f}{a}\right)$
2. Time shifting	$x(t - t_0)$	$X(f) \exp(-j2\pi f t_0)$
3. Frequency shifting	$x(t) \exp(j2\pi f_0 t)$	$X(f - f_0)$
4. Time differentiation	$\dfrac{d^n x}{dt^n}$	$(j2\pi f)^n X(f)$
5. Frequency differentiation	$(-jt)^n x(t)$	$\dfrac{d^n X}{df^n}$
6. Time integration	$\displaystyle\int_{-\infty}^{t} x(\tau)d\tau$	$\dfrac{1}{j2\pi f} X(f) + \dfrac{1}{2}X(0)\delta(f)$
7. Time convolution	$x_1(t) * x_2(t)$	$X_1(f)X_2(f)$
8. Frequency convolution	$x_1(t)x_2(t)$	$X_1(f) * X_2(f)$

REFERENCES

1. Papoulis, A., *Signal Analysis*, McGraw-Hill Book Company, New York, 1977.
2. Panter, P. F., *Modulation, Noise, and Spectral Analysis*, McGraw-Hill Book Company, New York, 1965.
3. Bracewell, R., *The Fourier Transform and Its Applications*, McGraw-Hill Book Company, New York, 1978.
4. Haykin, S., *Communication Systems*, John Wiley & Sons, Inc., New York, 1983.
5. Schwartz, M., *Information, Transmission, Modulation, and Noise*, McGraw-Hill Book Company, New York, 1980.

Fundamentals
of Statistical
Decision Theory

The basic elements of a statistical decision problem are (1) a set of hypotheses that characterize the possible true states of nature, (2) a test in which data are obtained from which we wish to infer the truth, (3) a decision rule that operates on the data to decide in an optimal fashion which hypothesis best describes the true state of nature, and (4) a criterion of optimality. These fundamental steps are treated in the material that follows. The *optimality criterion* we will choose for the decision rule is to minimize the probability of making an erroneous decision, although other criteria are possible [1].

The subject of statistical decision theory and hypothesis testing builds on the mathematical discipline of probability theory and random variables. It is assumed that the reader has a familiarity with these subjects; if not, Reference [2] is a suggested resource.

B.1 BAYES' THEOREM

The mathematical foundations of hypothesis testing rest on Bayes' theorem, which is derived from the definition of the relationship between the conditional and joint probability of the random variables A and B:

$$P(A|B)P(B) = P(B|A)P(A) = P(A, B) \qquad \text{(B.1)}$$

A statement of the theorem is

$$P(A|B) = \frac{P(B|A)P(A)}{P(B)} \qquad \text{(B.2)}$$

Bayes' theorem allows us to infer the conditional probability, $P(A|B)$, from the conditional probability $P(B|A)$.

B.1.1 Discrete Form of Bayes' Theorem

Bayes' theorem can be expressed in discrete form, as follows:

$$P(s_i|z_j) = \frac{P(z_j|s_i)P(s_i)}{P(z_j)} \qquad \begin{array}{l} i = 1, \ldots, M \\ j = 1, \ldots \end{array} \qquad \text{(B.3)}$$

where

$$P(z_j) = \sum_{i=1}^{M} P(z_j|s_i)P(s_i)$$

In a communications application, s_i is the ith signal class, from a set of M classes, and z_j is the jth sample of a received signal. Equation (B.3) can be thought of as the description of an experiment involving a received sample and some statistical knowledge of the signal classes to which the received sample may belong. The probability of occurrence of the ith signal class, $P(s_i)$, before the experiment, is called the *a priori probability*. As a result of examining a particular received sample, z_j, we can find a statistical measure of the *likelihood* that z_j belongs to class s_i from the conditional probability density function (pdf) $P(z_j|s_i)$. *After* the experiment, we can compute the *a posteriori probability*, $P(s_i|z_j)$, which can be thought of as a "refinement" of our prior knowledge. Thus we enter into the experiment with some a priori knowledge concerning the probability of the state of nature, and after examining a sample signal, we are provided with an "after-the-fact" a posteriori probability. The parameter $P(z_j)$ is the probability of the received sample, z_j, over the entire space of signal classes. This term, $P(z_j)$, can be thought of as a scaling factor, since its value is the same for *each* signal class.

Example B.1 Use of Bayes' Theorem (Discrete Form)

Given two boxes of parts. Box 1 contains 1000 parts, of which 10% are defective, and box 2 contains 2000 parts, of which 5% are defective. If a box is randomly chosen and then a part is randomly chosen from it, tested, and found to be good, what is the probability that the part came from box 1?

Solution

$$P(\text{box } 1|\text{GP}) = \frac{P(\text{GP}|\text{box } 1)P(\text{box } 1)}{P(\text{GP})}$$

where GP means "good part."

$$P(\text{GP}) = P(\text{GP}|\text{box } 1)P(\text{box } 1) + P(\text{GP}|\text{box } 2)P(\text{box } 2)$$

$$= (0.90)(0.5) + (0.95)(0.5)$$

$$= 0.450 + 0.475 = 0.925$$

$$P(\text{box } 1|\text{GP}) = \frac{0.450}{0.925} = 0.486$$

Before the experiment, the a priori probability of having chosen either box 1 or box

2 was equally likely. After obtaining a good part, the Bayesian computation can be regarded as a way of "fine tuning" our thinking that $P(\text{box } 1) = 0.5$ to yield the a posteriori probability of 0.486. The Bayes' theorem is simply a formalization of common sense. Having selected a good part from one of the two boxes, isn't it intuitively reasonable that there is a higher probability that the part came from the box with the larger concentration of good parts, and a lower probability that it came from the box with the smaller concentration of good parts? The Bayes' theorem has refined the a priori statistic into an a posteriori statistic for the probability of box selection.

Example B.2   Decision Theory Applied to a Betting Game

A box has three coins: a fair coin, a two-headed coin, and a two-tailed coin. You are asked to pick one coin at random, look at one side only, and guess head or tail for the other side. What is the optimum decision strategy for this game?

Solution

We can view this problem as a signal detection problem. A signal is transmitted, but because of the channel noise, the received signal is somewhat obscured. Not being able to look at the other side of the coin is tantamount to receiving a noise perturbed signal. Let H_i represent the hypotheses ($i = F, H, T$), where F, H, and T, stand for fair, head, and tail, respectively:

$$H_F: \quad H, T \text{ (fair coin)}$$

$$H_H: \quad H, H \text{ (two-headed coin)}$$

$$H_T: \quad T, T \text{ (two-tailed coin)}$$

Let z_j represent the received sample ($j = H, T$), where z_H is a head and z_T is a tail. Let the a priori probabilities of the hypotheses be equally likely, so that $P(H_F) = P(H_H) = P(H_T) = \frac{1}{3}$. Using Bayes' theorem,

$$P(H_i|z_j) = \frac{P(z_j|H_i)P(H_i)}{\sum_i P(z_j|H_i)P(H_i)}$$

we need to compute the probability for each hypothesis, given each signal class. Thus we need to examine the results of *six* computations before we can establish an optimum decision strategy. In each case the value of $P(z_j|H_i)$ can be obtained from the conditional probabilities drawn in Figure B.1. Consider that we choose a coin and view a head (z_H), we compute the following three a posteriori probabilities:

$$P(H_F|z_H) = \frac{(\frac{1}{2})(\frac{1}{3})}{(\frac{1}{2})(\frac{1}{3}) + (1)(\frac{1}{3}) + 0} = \frac{1}{3}$$

$$P(H_H|z_H) = \frac{(1)(\frac{1}{3})}{(\frac{1}{2})(\frac{1}{3}) + (1)(\frac{1}{3}) + 0} = \frac{2}{3}$$

$$P(H_T|z_H) = 0$$

If the received sample is a tail (z_T), we similarly compute

$$P(H_F|z_T) = \frac{1}{3}$$

$$P(H_H|z_T) = 0$$

$$P(H_T|z_T) = \frac{2}{3}$$

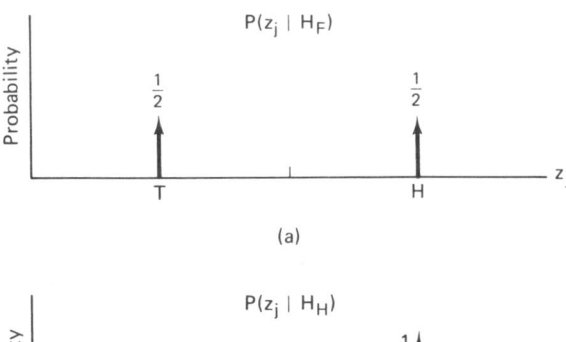

(a)

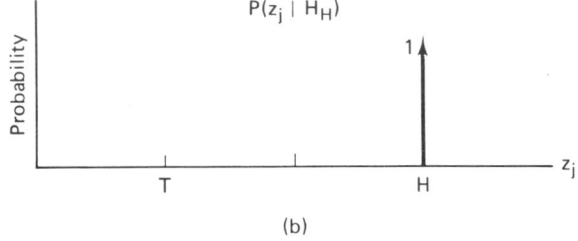

(b)

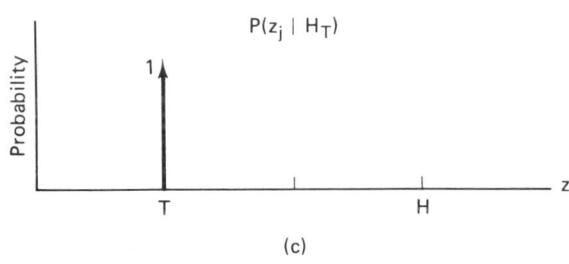

(c)

Figure B.1 Conditional probability $P(z_j \mid H_i)$. (a) Conditioned on the fair-coin hypothesis. (b) Conditioned on the two-headed-coin hypothesis. (c) Conditioned on the two-tailed-coin hypothesis.

The optimum decision strategy then is as follows: If a head, z_H, is received, choose hypothesis H_H (that the other side is also a head). If a tail, z_T, is received, choose hypothesis H_T (that the other side is also a tail).

B.1.2 Mixed Form of Bayes' Theorem

For most communication engineering applications of interest, the possible values of the received samples are *continuous* in range, because of the additive Gaussian noise in the channel. Therefore, the most useful form of Bayes' theorem contains a continuous- instead of a discrete-valued pdf. We shall rewrite Equation (B.3) to emphasize this change:

$$P(s_i|z) = \frac{p(z|s_i)P(s_i)}{p(z)} \qquad i = 1, \ldots, M$$

$$p(z) = \sum_{i=1}^{M} p(z|s_i)P(s_i)$$

(B.4)

where $p(z|s_i)$ is the conditional pdf of the received continuous-valued sample, z, conditioned on the signal class, s_i.

Example B.3 A Pictorial View of Bayes' Theorem

Consider two signal classes, s_1 and s_2, characterized by the triangular-shaped conditional pdfs, $p(z|s_1)$ and $p(z|s_2)$, illustrated in Figure B.2. A signal is received; it might have any value on the z-axis. If the pdfs did not overlap, we could classify the signal with certainty. For the example shown in Figure B.2, we need a rule to help us classify received signals, since some signals will fall in the region where the two pdfs overlap. Consider a received signal, z_a. Assume that the two signal classes, s_1 and s_2, are equally likely, and calculate the two alternative a posteriori probabilities. Suggest a decision rule that the receiver should use for deciding to which signal class z_a belongs. Repeat this for signal z_b.

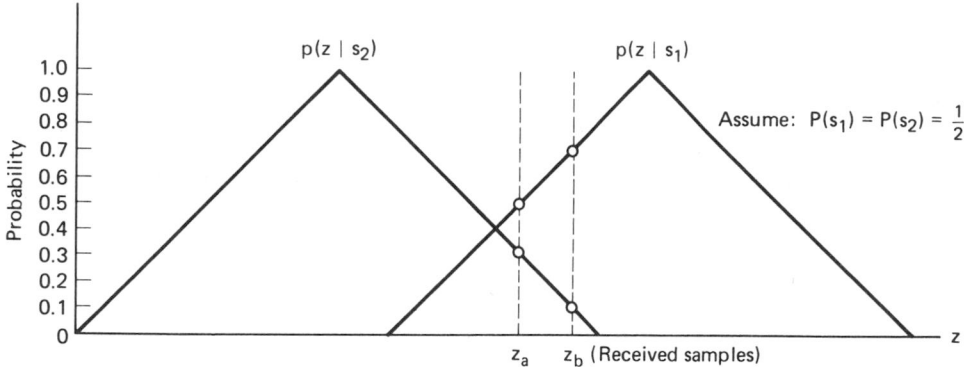

Figure B.2 Pictorial view of Bayes' theorem.

Solution

From Figure B.2 we can see that $p(z_a|s_1) = 0.5$ and $p(z_a|s_2) = 0.3$. Thus

$$P(s_1|z_a) = \frac{p(z_a|s_1)P(s_1)}{p(z_a|s_1)P(s_1) + p(z_a|s_2)P(s_2)}$$

$$= \frac{(0.5)(0.5)}{(0.5)(0.5) + (0.3)(0.5)} = \frac{5}{8}$$

and

$$P(s_2|z_a) = \frac{(0.3)(0.5)}{(0.5)(0.5) + (0.3)(0.5)} = \frac{3}{8}$$

One rule is to decide that the received signal belongs to the class with the maximum a posteriori probability (class s_1). An equivalent rule, for the case of equal a priori probabilities, is to examine the value of the pdf conditioned on each signal class (referred to as the likelihood of the signal class) and choose the class with the maximum. Examine Figure B.2 and notice that this *maximum likelihood rule* parallels our intuition. The likelihood that signal z_a belongs to each class corresponds to an encircled point on each pdf. The maximum likelihood rule is to choose the signal class that yields the largest conditional probability of all the alternatives. We repeat the computations for the received signal z_b, as follows:

$$P(s_1|z_b) = \frac{(0.7)(0.5)}{(0.7)(0.5) + (0.1)(0.5)} = \frac{7}{8}$$

$$P(s_2|z_b) = \frac{(0.1)(0.5)}{(0.7)(0.5) + (0.1)(0.5)} = \frac{1}{8}$$

As before, the maximum likelihood rule dictates that we choose signal class s_1. Notice that in the case of received sample z_b, we can have greater confidence in the correctness of our choice compared to the case of signal z_a. This is because the ratio of $p(z_b|s_1)$ to $p(z_b|s_2)$ is considerably larger than the ratio of $p(z_a|s_1)$ to $p(z_a|s_2)$.

B.2 DECISION THEORY

B.2.1 Components of the Decision Theory Problem

Having reviewed hypothesis testing based on Bayesian statistics, let us examine more carefully the components of the decision theory problem in the context of a communication system, as shown in Figure B.3. The signal source at the transmitter consists of a set $\{s_i(t)\}$, $i = 1, \ldots, M$, of waveforms (or hypotheses). A signal waveform $r(t) = s_i(t) + n(t)$ is received, where $n(t)$ is an additive white Gaussian noise (AWGN) process introduced in the channel. At the receiver, the waveform is reduced to a single number, $z(t = T)$, that may appear anywhere on the z-axis. Because the noise is a Gaussian process and the receiver is assumed linear, the output, $z(t)$, is also a Gaussian process [1], and the number, $z(T)$, is a *continuous-valued random variable*.

$$z(T) = a_i(T) + n_0(T) \tag{B.5}$$

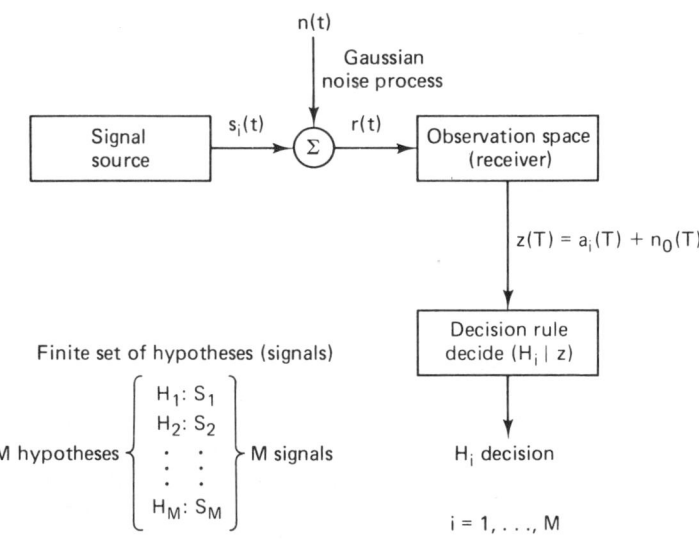

Figure B.3 Components of the decision theory problem in the context of a communication system.

The sample $z(T)$ is made up of a signal component, $a_i(T)$, and a noise component, $n_0(T)$. The time T is the symbol duration. At each kT, where k is an integer, the receiver uses a decision rule for deciding which signal class has been received. For ease of notation, Equation (B.5) is sometimes written simply as $z = a_i + n_0$, where the functional dependence on T is implicit.

B.2.2 The Likelihood Ratio Test and the Maximum A Posteriori Criterion

A reasonable starting point for establishing the receiver decision rule is shown below for the case of *two* signal classes.

$$P(s_1|z) \underset{H_2}{\overset{H_1}{\gtrless}} P(s_2|z) \tag{B.6}$$

Equation (B.6) states that we should choose hypothesis H_1 if the a posteriori probability $P(s_1|z)$ is greater than the a posteriori probability $P(s_2|z)$. Otherwise, we should choose hypothesis H_2.

We can replace the a posteriori probabilities of Equation (B.6) with their equivalent expressions from Bayes' theorem [Equation (B.4)], yielding

$$p(z|s_1) P(s_1) \underset{H_2}{\overset{H_1}{\gtrless}} P(z|s_2) P(s_2) \tag{B.7}$$

We now have a decision rule in terms of pdfs (likelihoods). If we rearrange Equation (B.7) as shown below

$$\frac{p(z|s_1)}{p(z|s_2)} \underset{H_2}{\overset{H_1}{\gtrless}} \frac{P(s_2)}{P(s_1)} \tag{B.8}$$

the left-hand ratio is known as the *likelihood ratio*, and the entire equation is often referred to as the *likelihood ratio test*. Equation (B.8) corresponds to making a decision based on a comparison of a measurement of a received signal to a threshold. Since the test is based on choosing the signal class with maximum a posteriori probability, the decision criterion is called the *maximum a posteriori* (MAP) criterion. It is also called the *minimum error criterion*, since on the average, this criterion yields the minimum number of incorrect decisions. It should be emphasized that this criterion is optimum only when each of the error types are equally harmful or costly. When some of the error types are more costly than others, a criterion that incorporates relative cost of the errors should best be employed [1].

B.2.3 The Maximum Likelihood Criterion

Very often there is no knowledge available about the a priori probabilities of the hypotheses or signal classes. Even when such information is available, its accuracy is sometimes mistrusted. In those instances, decisions are usually made by assuming the most conservative a priori probabilities possible; that is, the values of the a priori probabilities are selected so that the classes are *equally likely*. When this is done, the MAP criterion is known as the *maximum likelihood criterion*, and Equation (B.8) can be written as follows:

$$\frac{p(z|s_1)}{p(z|s_2)} \underset{H_2}{\overset{H_1}{\gtrless}} 1 \tag{B.9}$$

Notice that the maximum likelihood criterion of Equation (B.9) is the same as the maximum likelihood rule that was described in Example B.3.

B.3 SIGNAL DETECTION EXAMPLE

B.3.1 The Maximum Likelihood Binary Decision

The pictorial view of the decision process in Example B.3 dealt with triangular-shaped probability density functions as a convenient example. Figure B.4 illustrates the conditional pdfs for the binary noise-perturbed output signals, $z(T) = a_1 + n_0$ and $z(T) = a_2 + n_0$ from a typical receiver. The signals, a_1 and a_2, are mutually independent and are equally likely. The noise, n_0, is assumed to be an independent Gaussian random variable with zero mean, variance σ_0^2, and pdf $p(n_0)$ given by

$$p(n_0) = \frac{1}{\sigma_0 \sqrt{2\pi}} \exp\left[-\frac{1}{2}\left(\frac{n_0^2}{\sigma_0^2}\right)\right] \tag{B.10}$$

We can therefore write the likelihood ratio, $L(z)$, described in Equation (B.8) as follows:

$$L(z) = \frac{p(z|s_1)}{p(z|s_2)}$$

$$= \frac{\dfrac{1}{\sigma_0 \sqrt{2\pi}} \exp\left[-\dfrac{1}{2}\left(\dfrac{z - a_1}{\sigma_0}\right)^2\right]}{\dfrac{1}{\sigma_0 \sqrt{2\pi}} \exp\left[-\dfrac{1}{2}\left(\dfrac{z - a_2}{\sigma_0}\right)^2\right]} \underset{H_2}{\overset{H_1}{\gtrless}} \frac{P(s_2)}{P(s_1)}$$

$$= \frac{\exp\left(-\dfrac{z^2}{2\sigma_0^2}\right) \exp\left(-\dfrac{a_1^2}{2\sigma_0^2}\right) \exp\left(\dfrac{2za_1}{2\sigma_0^2}\right)}{\exp\left(-\dfrac{z^2}{2\sigma_0^2}\right) \exp\left(-\dfrac{a_2^2}{2\sigma_0^2}\right) \exp\left(\dfrac{2za_2}{2\sigma_0^2}\right)} \underset{H_2}{\overset{H_1}{\gtrless}} \frac{P(s_2)}{P(s_1)} \tag{B.11}$$

$$= \exp\left[\frac{z(a_1 - a_2)}{\sigma_0^2} - \frac{a_1^2 - a_2^2}{2\sigma_0^2}\right] \underset{H_2}{\overset{H_1}{\gtrless}} \frac{P(s_2)}{P(s_1)}$$

where a_1 is the receiver output signal component when $s_1(t)$ is sent, and a_2 is the output signal component when $s_2(t)$ is sent. The inequality relationship described by Equation (B.11) is preserved for any *monotonically* increasing (or decreasing)

Fundamentals of Statistical Decision Theory   App. B

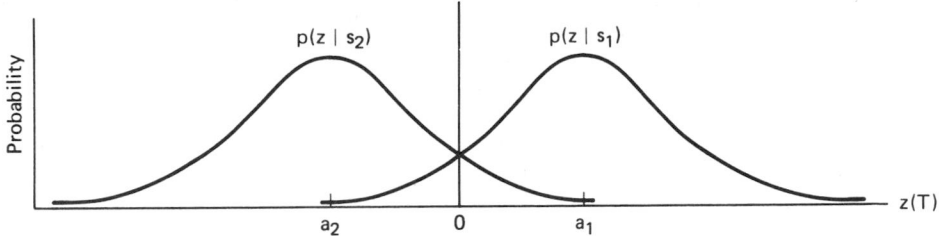

Figure B.4 Conditional pdfs for a typical binary receiver.

transformation. Therefore, to simplify Equation (B.11), we take the natural logarithm of both sides, resulting in the log-likelihood ratio, $l(z)$.

$$l(z) = \frac{z(a_1 - a_2)}{\sigma_0^2} - \frac{a_1^2 - a_2^2}{2\sigma_0^2} \underset{H_2}{\overset{H_1}{\gtrless}} \ln \frac{P(s_2)}{P(s_1)} \qquad \text{(B.12)}$$

When the classes are equally likely,

$$\ln \frac{P(s_2)}{P(s_1)} = 0$$

so that

$$z \underset{H_2}{\overset{H_1}{\gtrless}} \frac{a_1^2 - a_2^2}{2(a_1 - a_2)}$$

$$z \underset{H_2}{\overset{H_1}{\gtrless}} \frac{a_1 + a_2}{2} = \gamma_0 \qquad \text{(B.13)}$$

For *antipodal signals*, $s_1(t) = -s_2(t)$ and $a_1 = -a_2$; thus we can write

$$z \underset{H_2}{\overset{H_1}{\gtrless}} 0 \qquad \text{(B.14)}$$

Therefore, the maximum likelihood rule for the case of equally likely antipodal signals compares the received sample to a zero threshold, which is tantamount to deciding $s_1(t)$ if the sample is positive, and $s_2(t)$ if the signal is negative.

B.3.2 Probability of Bit Error

For the binary example in Section B.3.1, we want to compute the bit error probability, P_B, using the decision rule in Equation (B.13). The probability of an error is calculated by summing the probabilities of the various ways that an error can be made:

$$P_B = P(H_2|s_1)P(s_1) + P(H_1|s_2)P(s_2) \qquad \text{(B.15)}$$

That is, given that class $s_1(t)$ was transmitted, an error results if hypothesis H_2 is chosen; or, given that class $s_2(t)$ was transmitted, an error results if hypothesis

H_1 is chosen. For the special case of symmetric probability density functions, and for $P(s_1) = P(s_2) = 0.5$, we can write

$$P_B = P(H_2|s_1) = P(H_1|s_2) \tag{B.16}$$

The probability of an error, P_B, is equal to the probability that an incorrect hypothesis, H_1, will be decided when $s_2(t)$ is sent, or that H_2 will be decided when $s_1(t)$ is sent. Thus P_B is numerically equal to the area under the "tail" of either

TABLE B.1 Complementary Error Function $Q(x) = \int_x^\infty (1/\sqrt{2\pi}) \exp(-u^2/2)\, du$

					$Q(x)$					
x	0.00	0.01	0.02	0.03	0.04	0.05	0.06	0.07	0.08	0.09
0.0	0.5000	0.4960	0.4920	0.4880	0.4840	0.4801	0.4761	0.4721	0.4681	0.4641
0.1	0.4602	0.4562	0.4522	0.4483	0.4443	0.4404	0.4364	0.4325	0.4286	0.4247
0.2	0.4207	0.4168	0.4129	0.4090	0.4052	0.4013	0.3974	0.3936	0.3897	0.3859
0.3	0.3821	0.3783	0.3745	0.3707	0.3669	0.3632	0.3594	0.3557	0.3520	0.3483
0.4	0.3446	0.3409	0.3372	0.3336	0.3300	0.3264	0.3228	0.3192	0.3156	0.3121
0.5	0.3085	0.3050	0.3015	0.2981	0.2946	0.2912	0.2877	0.2843	0.2810	0.2776
0.6	0.2743	0.2709	0.2676	0.2643	0.2611	0.2578	0.2546	0.2514	0.2483	0.2451
0.7	0.2420	0.2389	0.2358	0.2327	0.2296	0.2266	0.2236	0.2206	0.2168	0.2148
0.8	0.2169	0.2090	0.2061	0.2033	0.2005	0.1977	0.1949	0.1922	0.1894	0.1867
0.9	0.1841	0.1814	0.1788	0.1762	0.1736	0.1711	0.1685	0.1660	0.1635	0.1611
1.0	0.1587	0.1562	0.1539	0.1515	0.1492	0.1469	0.1446	0.1423	0.1401	0.1379
1.1	0.1357	0.1335	0.1314	0.1292	0.1271	0.1251	0.1230	0.1210	0.1190	0.1170
1.2	0.1151	0.1131	0.1112	0.1093	0.1075	0.1056	0.1038	0.1020	0.1003	0.0985
1.3	0.0968	0.0951	0.0934	0.0918	0.0901	0.0885	0.0869	0.0853	0.0838	0.0823
1.4	0.0808	0.0793	0.0778	0.0764	0.0749	0.0735	0.0721	0.0708	0.0694	0.0681
1.5	0.0668	0.0655	0.0643	0.0630	0.0618	0.0606	0.0594	0.0582	0.0571	0.0559
1.6	0.0548	0.0537	0.0526	0.0516	0.0505	0.0495	0.0485	0.0475	0.0465	0.0455
1.7	0.0446	0.0436	0.0427	0.0418	0.0409	0.0401	0.0392	0.0384	0.0375	0.0367
1.8	0.0359	0.0351	0.0344	0.0336	0.0329	0.0322	0.0314	0.0307	0.0301	0.0294
1.9	0.0287	0.0281	0.0274	0.0268	0.0262	0.0256	0.0250	0.0244	0.0239	0.0233
2.0	0.0228	0.0222	0.0217	0.0212	0.0207	0.0202	0.0197	0.0192	0.0188	0.0183
2.1	0.0179	0.0174	0.0170	0.0166	0.0162	0.0158	0.0154	0.0150	0.0146	0.0143
2.2	0.0139	0.0136	0.0132	0.0129	0.0125	0.0122	0.0119	0.0116	0.0113	0.0110
2.3	0.0107	0.0104	0.0102	0.0099	0.0096	0.0094	0.0091	0.0089	0.0087	0.0084
2.4	0.0082	0.0080	0.0078	0.0075	0.0073	0.0071	0.0069	0.0068	0.0066	0.0064
2.5	0.0062	0.0060	0.0059	0.0057	0.0055	0.0054	0.0052	0.0051	0.0049	0.0048
2.6	0.0047	0.0045	0.0044	0.0043	0.0041	0.0040	0.0039	0.0038	0.0037	0.0036
2.7	0.0035	0.0034	0.0033	0.0032	0.0031	0.0030	0.0029	0.0028	0.0027	0.0026
2.8	0.0026	0.0025	0.0024	0.0023	0.0023	0.0022	0.0021	0.0021	0.0020	0.0019
2.9	0.0019	0.0018	0.0018	0.0017	0.0016	0.0016	0.0015	0.0015	0.0014	0.0014
3.0	0.0013	0.0013	0.0013	0.0012	0.0012	0.0011	0.0011	0.0011	0.0010	0.0010
3.1	0.0010	0.0009	0.0009	0.0009	0.0008	0.0008	0.0008	0.0008	0.0007	0.0007
3.2	0.0007	0.0007	0.0006	0.0006	0.0006	0.0006	0.0006	0.0005	0.0005	0.0005
3.3	0.0005	0.0005	0.0005	0.0004	0.0004	0.0004	0.0004	0.0004	0.0004	0.0003
3.4	0.0003	0.0003	0.0003	0.0003	0.0003	0.0003	0.0003	0.0003	0.0003	0.0002

pdf, $p(z|s_1)$ or $p(z|s_2)$, falling on the *incorrect side* of the threshold. We can therefore compute P_B by integrating $p(z|s_1)$ between the limits $-\infty$ and γ_0, or as shown below, by integrating $p(z|s_2)$ between the limits γ_0 and ∞.

$$P_B = \int_{\gamma_0 = (a_1 + a_2)/2}^{\infty} p(z|s_2)\, dz$$

$$= \int_{(a_1 + a_2)/2}^{\infty} \frac{1}{\sigma_0 \sqrt{2\pi}} \exp\left[-\frac{1}{2}\left(\frac{z - a_2}{\sigma_0}\right)^2 \right] dz \qquad \text{(B.17)}$$

Let

$$u = \frac{z - a_2}{\sigma_0}$$

Then $\sigma_0\, du = dz$, and

$$P_B = \int_{u = (a_1 - a_2)/2\sigma_0}^{u = \infty} \frac{1}{\sqrt{2\pi}} \exp\left(-\frac{u^2}{2} \right) du = Q\left(\frac{a_1 - a_2}{2\sigma_0}\right) \qquad \text{(B.18)}$$

where $Q(x)$, called the *complementary error function* or *co-error function*,* is tabulated in Table B.1.

Another form of the co-error function which is frequently used is

$$\text{erfc}\,(x) = \frac{2}{\sqrt{\pi}} \int_x^{\infty} \exp\,(-u^2)\, du \qquad \text{(B.19)}$$

The two co-error functions, $Q(x)$ and erfc (x), are related as follows:

$$\text{erfc}\,(x) = 2Q(x\sqrt{2}) \qquad \text{(B.20)}$$

$$Q(x) = \frac{1}{2}\,\text{erfc}\left(\frac{x}{\sqrt{2}}\right) \qquad \text{(B.21)}$$

REFERENCES

1. Van Trees, H. L., *Detection, Estimation, and Modulation Theory*, Part 1, John Wiley & Sons, Inc., New York, 1968.
2. Papoulis, A., *Probability, Random Variables, and Stochastic Processes*, McGraw-Hill Book Company, New York, 1965.

---

* Note that the co-error function is defined in several ways; however, all definitions are essentially equivalent.

APPENDIX C

Response of Correlators to White Noise

The inputs to a bank of N correlators represent a white Gaussian noise process, $n(t)$, with zero mean and two-sided power spectral density, $N_0/2$. The output of each correlator, n_j, at time $t = T$, is a *Gaussian random variable* defined by

$$n_j = \int_0^T n(t)\psi_j(t)\,dt \qquad j = 1, \ldots, N \tag{C.1}$$

where $\{\psi_j(t)\}$ forms an orthonormal set. Since n_j is Gaussian, it is characterized completely by its mean and variance. The mean, $\bar{n}_j$, is equal to

$$\bar{n}_j = \mathbf{E}\{n_j\} = \mathbf{E}\left\{\int_0^T n(t)\psi_j(t)\,dt\right\} \tag{C.2}$$

where $\mathbf{E}\{\cdot\}$ is the expected value operator. The variance, σ_j^2, of n_j is equal to

$$\sigma_j^2 = \mathbf{E}\{n_j^2\} - \bar{n}_j^2 \tag{C.3}$$

$$= \mathbf{E}\left\{\int_0^T n(t)\psi_j(t)\,dt \int_0^T n(s)\psi_j(s)\,ds\right\} - \bar{n}_j^2 \tag{C.4}$$

$$= \int_0^T \int_0^T \mathbf{E}\{n(t)n(s)\psi_j(t)\psi_j(s)\}\,dt\,ds - \bar{n}_j^2 \tag{C.5}$$

Since $n(t)$ is a zero-mean process, then

$$\mathbf{E}\{n(t)\} = 0 \tag{C.6}$$

which implies that

$$\bar{n}_j = \mathbf{E}\{n_j\} = 0 \tag{C.7}$$

The autocorrelation function $R_n(t, s)$ of the process $n(t)$ is equal to

$$R_n(t, s) = \mathbf{E}\{n(t)n(s)\} \tag{C.8}$$

If the noise, $n(t)$, is assumed stationary, then $R_n(t, s)$ is only a function of the time difference, $\tau = t - s$. From Equation (C.5) we have

$$\sigma_j^2 = \text{var} \{n_j\} = \int_0^T \int_0^T R_n(\tau)\psi_j(t)\psi_j(s) \, dt \, ds \tag{C.9}$$

For a stationary random process, the power spectral density, $G_n(f)$, and the autocorrelation function, $R_n(\tau)$, form a Fourier transform pair. Thus we can write

$$R_n(\tau) = \int_{-\infty}^{\infty} G_n(f)e^{j2\pi f\tau} \, df \tag{C.10}$$

Since $n(t)$ is white noise, its power spectral density, $G_n(f)$, is $N_0/2$ for all f, and we can write Equation (C.10) as

$$R_n(\tau) = \int_{-\infty}^{\infty} \frac{N_0}{2} e^{j2\pi f\tau} \, df = \frac{N_0}{2} \delta(\tau) \tag{C.11}$$

where $\delta(\tau)$ is the unit impulse function defined in Section A.4.1. Substituting Equation (C.11) into Equation (C.9), we get

$$\sigma_j^2 = \frac{N_0}{2} \int_0^T \int_0^T \delta(t - s)\psi_j(t)\psi_j(s) \, dt \, ds \tag{C.12}$$

$$= \frac{N_0}{2} \int_0^T \psi_j^2(t) \, dt = \frac{N_0}{2} \qquad j = 1, \ldots, N \tag{C.13}$$

where we have utilized the *sifting property* of the unit impulse function (see Section A.4.1) and the fact that $\{\psi_j(t)\}$, $j = 1, \ldots, N$, constitutes an orthonormal set. Thus for white Gaussian noise with two-sided power spectral density $N_0/2$ watts/hertz, the output noise power from each of the N correlators is equal to $N_0/2$ watts.

APPENDIX D

Often Used Identities

$$\cos x \cos y = \tfrac{1}{2} \cos (x + y) + \tfrac{1}{2} \cos (x - y) \qquad \text{(D.1)}$$

$$\sin x \sin y = -\tfrac{1}{2} \cos (x + y) + \tfrac{1}{2} \cos (x - y) \qquad \text{(D.2)}$$

$$\sin x \cos y = \tfrac{1}{2} \sin (x + y) + \tfrac{1}{2} \sin (x - y) \qquad \text{(D.3)}$$

$$\cos x \sin y = \tfrac{1}{2} \sin (x + y) - \tfrac{1}{2} \sin (x - y) \qquad \text{(D.4)}$$

$$\sin (x \pm y) = \sin x \cos y \pm \cos x \sin y \qquad \text{(D.5)}$$

$$\cos (x \pm y) = \cos x \cos y \mp \sin x \sin y \qquad \text{(D.6)}$$

$$\cos^2 x = \tfrac{1}{2}(1 + \cos 2x) \qquad \text{(D.7)}$$

$$\sin^2 x = \tfrac{1}{2}(1 - \cos 2x) \qquad \text{(D.8)}$$

$$\sin x \cos x = \tfrac{1}{2} \sin 2x \qquad \text{(D.9)}$$

$$\sin x + \sin y = 2 \sin \tfrac{1}{2}(x + y) \cos \tfrac{1}{2}(x - y) \qquad \text{(D.10)}$$

$$\sin x - \sin y = 2 \cos \tfrac{1}{2}(x + y) \sin \tfrac{1}{2}(x - y) \qquad \text{(D.11)}$$

$$\cos x + \cos y = 2 \cos \tfrac{1}{2}(x + y) \cos \tfrac{1}{2}(x - y) \qquad \text{(D.12)}$$

$$\cos x - \cos y = -2 \sin \tfrac{1}{2}(x + y) \sin \tfrac{1}{2}(x - y) \qquad \text{(D.13)}$$

$$\sin x = \frac{e^{jx} - e^{-jx}}{2j} \qquad \text{(D.14)}$$

$$\cos x = \frac{e^{jx} + e^{-jx}}{2} \qquad \text{(D.15)}$$

$$P_B = \frac{1}{n} \sum_{j=2}^{n} j \binom{n}{j} p^j (1-p)^{n-j} = p - p(1-p)^{n-1} \tag{D.16}$$

Proof:

$$j \binom{n}{j} = j \frac{n!}{j!\,(n-j)!} = \frac{n!}{(j-1)!\,(n-j)!} = n \frac{(n-1)!}{(j-1)!\,[(n-1)-(j-1)]!}$$

$$= n \binom{n-1}{j-1}$$

$$P_B = \sum_{j=2}^{n} \binom{n-1}{j-1} p^j (1-p)^{n-j} = p \sum_{j=2}^{n} \binom{n-1}{j-1} p^{j-1} (1-p)^{(n-1)-(j-1)}$$

Change of parameter: $\qquad i = (j-1)$

Therefore, $(j = 2)$ becomes $(i = 1)$, and $(j = n)$ becomes $(i = n - 1)$.

$$P_B = p \sum_{i=1}^{n-1} \binom{n-1}{i} p^i (1-p)^{(n-1)-i}$$

$$= p \sum_{i=0}^{n-1} \left[\binom{n-1}{i} p^i (1-p)^{(n-1)-i} - \binom{n-1}{0} p^0 (1-p)^{(n-1)-0} \right]$$

$$= p[1 - (1-p)^{n-1}]$$

$$= p - p(1-p)^{n-1}$$

APPENDIX E

A Convolutional Encoder/Decoder Computer Program

```
      PROGRAM VITALG
C REVISION H, JULY 6, 1989
C
C                    CONVOLUTIONAL ENCODER/DECODER PROGRAM
C                    WRITTEN IN LAHEY FORTRAN FOR IBM PC
C                    BY A. H. YAMADA AND M. A. ROLENZ
C                       THE AEROSPACE CORPORATION
C                       2350 EAST EL SEGUNDO BLVD.
C                       EL SEGUNDO, CA. 90245
C
      COMMON/ENCDE/NCODE(3,128),IPREV(128),NSEQ,IREG,NCRATE,
     +NSQHLF,KL,ITAP(3)
C COMMON BLOCK ENCODE DEFINES CODE STRUCTURES
C        NCODE(I,J)    CODE BIT I FOR PATH J
C        IPREV(I)      STARTING STATE IN TRELLIS FOR PATH I
C        NSEQ          NUMBER OF STATES IN TRELLIS (2**(KL-1))
C        IREG          ENCODER SHIFT REGISTER CONTENTS
C        NCRATE        NUMBER OF CODE GENERATORS (2 OR 3)
C        NSQHLF        HALF OF NSEQ  (2**(KL-2))
C        KL            CONSTRAINT LENGTH
C        ITAP(I)       DECIMAL EQUIVALENT OF CODE GENERATOR TAPS
      COMMON/DECDE/METRIC(64),MLSEQ(64,70),MEMLEN
C COMMON BLOCK DECODE IS FOR BUFFERING VITERBI DECODER TRELLIS HISTORY
C        METRIC(I)     CURRENT PATH METRIC FOR STATE I
C        MLSEQ(I,J)    TRACE OF SURVIVED LIKELIHOOD SEQUENCE I
C        MEMLEN        LENGTH OF MLSEQ(I,J), PATH MEMORY LENGTH
      COMMON/DECOD2/ILEVEL(3)
C COMMON BLOCK DECODE2 STORES CURRENT INPUTS TO VITERBI DECODER
C        ILEVEL(I)     CURRENT (ITIME) BITS TO BE DECODED
      COMMON/TIME/ITIME,ITIME1,INTIME,IBEST
C COMMON BLOCK HOLDS POINTERS FOR SUBROUTINES DECODE AND DECIDE
C        ITIME         TIMING POINTER FOR CURRENT TIME
C        ITIME1        ITIME-1
C        INTIME        TIMING POINTER FOR THE LAST DECODED BIT
C        IBEST         POINTER FOR THE BEST SURVIVING PATH
```

748

```
      COMMON/VITEST/IDCBIT(646),ICODED(1938),IBFFER(646),NDEC,LENIBF
C COMMON BLOCK VITEST HOLDS MESSAGE BUFFERS
C         IDCBIT          DECODED ASCII BITS
C         ICODED          ENCODED BIT STREAM
C         IBFFER          INPUT ASCII BITS
C         NDEC            COUNTER FOR DECODED BITS
C         LENIBF          NUMBER OF ASCII BITS AND KL-1 FLUSH BITS
      COMMON/FLUSH/NRECVD,NREP,NTAIL
C COMMON BLOCK FLUSH IS USED TO FLUSH DECODER AT THE END OF MESSAGE
C         NRECVD          COUNTER FOR NUMBER OF BITS INTO THE DECODER
C         NREP            NUMBER OF CODED BITS (LENIBF*NCRATE)
C         NTAIL           DESIGNATES THE BEGINNING OF THE FLUSH BITS
      COMMON/FLAG/IFLAG
      CHARACTER*80 BUFFER,BUFFR2
      CHARACTER*1 ANSWER
      CHARACTER*2 DRIVE
      CHARACTER*6 FILNAM
      CHARACTER*12 FILOUT
C RANDOMIZE THE INITIAL SEED
      SEED=RRAND()
      WRITE(6,1080)
1080  FORMAT(' This program encodes a binary or ASCII message into a',
     +' convolutionally encoded'/' bit stream.  The code parameters',
     +' are user inputs.  The program uses Viterbi'/' decoding (hard',
     +' or soft decision) and compares the transmitted sequence',
     +' against'/' the decoded sequence.'/)
C   INPUT FILE NAME AND OPEN FILE
      PRINT,' ENTER OUTPUT FILE NAME (6 CHARACTERS MAXIMUM) '
      READ(5,*) FILNAM
      PRINT,' ENTER DISK DRIVE FOR OUTPUT (e.g. A:,B:,C: etc.) '
      READ(5,*) DRIVE
      PRINT,' '
      FILOUT=DRIVE//CHARNB(FILNAM)//'.OUT'
      PRINT,' '
      PRINT,'    OUTPUT FILE WILL BE     -          ',FILOUT
      PRINT,' '
      OPEN(UNIT=2,FILE=FILOUT,STATUS='NEW')
C INITIAL DEFAULT CODE PARAMETERS
      NCRATE = 2
      MEMLEN = 15
      KL     = 3
      WRITE(6,1082)
 1082 FORMAT(' INPUT CODE PARAMETERS:'
     +/' THE INITIAL DEFAULT IS THE RATE 1/2, K=3 ODENWALDER CODE')
C   BEGINNING OF MAIN LOOP
      1 CONTINUE
C   INPUT OF CODING PARAMETERS
      CALL PARMIN(NCRATE,MEMLEN,KL,ITAP)
C   INITIALIZE CODE STRUCTURE
      CALL INTIAL
C   INPUT MESSAGE
      CALL MSGIN(BUFFER,LENBUF)
C   CONVERT MESSAGE TO ASCII BITS
      CALL ASCII1(BUFFER,LENBUF,IBFFER,LENIBF,KL)
      NREP = NCRATE*LENIBF
      NTAIL=NREP+(2-KL)*NCRATE
C   ENCODE BITS
      CALL ENCODE
C   INTRODUCE ERRORS INTO BIT STREAM
      PRINT, ' DO YOU WANT RANDOM ERRORS, ASSUMING BPSK MODULATION'
      PRINT, ' AND AN AWGN CHANNEL (Y/N) ? '
      INPUT, ANSWER
      IF((ANSWER.EQ.'Y').OR.(ANSWER.EQ.'y'))THEN
        CALL RANDOM(BUFFER,ICODED,NREP,NCRATE,KL,SNROBS)
      ELSE
        CALL ERRORS(BUFFER,ICODED,NREP,NCRATE,KL)
      ENDIF
```

```
C   DECODER TRELLIS TRACE OUTPUT OPTION
      PRINT,' '
      PRINT,' DO YOU WANT TO SEE THE TRELLIS TRACE (Y/N) ? '
      INPUT, ANSWER
      IF((ANSWER.EQ.'Y').OR.(ANSWER.EQ.'y'))THEN
        WRITE(6,1085)
        WRITE(2,1085)
 1085   FORMAT(/' TRELLIS TRACE'
     +/' The decimal numbers in the array, when converted to binary,'
     +/' represent the trellis states.')
        ITRS=1
      ELSE
        ITRS=0
      ENDIF
      DO 10 I=1,NSEQ
        DO 11 J=1,MEMLEN
          MLSEQ(I,J)=1000
   11   CONTINUE
        METRIC(I)=0
        MLSEQ(I,1)=I
   10 CONTINUE
      IREG=0
      METRIC(1)=1000
      ITIME1=1
      ITIME=1
      INTIME=1
      NRECVD=0
      NDEC=0
      PRINT,' '
      PRINT,' Simulating reception and decoding'
  100 CONTINUE
C   READ NCRATE RECEIVED BITS INTO THE DECODER
      DO 110 I=1,NCRATE
        NRECVD=NRECVD+1
        ILEVEL(I)=ICODED(NRECVD)
  110 CONTINUE
      CALL DECODE(ITRS)
      IF (NRECVD.LT.NREP) GO TO 100
C   CONVERT RECEIVED BITS BACK TO ASCII
      CALL ASCII2(BUFFR2,LENBUF,IDCBIT,LENIBF)
C   COUNT THE NUMBER OF ERRORS
      NBTERR = 0
      DO 200 INDEX = 1 , LENIBF
        IF( IBFFER(INDEX) .NE. IDCBIT(INDEX) ) NBTERR=NBTERR+1
  200 CONTINUE
      MSIZE=LENIBF-KL+1
      BEROBS=FLOAT(NBTERR)/FLOAT(MSIZE)
      WRITE(6,1000) MSIZE,NBTERR,BEROBS
      WRITE(2,1000) MSIZE,NBTERR,BEROBS
      NCHERR = 0
      DO 300 INDEX = 1 , LENBUF
        IF(BUFFER(INDEX:INDEX).NE.BUFFR2(INDEX:INDEX)) NCHERR=NCHERR+1
  300 CONTINUE
C   OUTPUT
      IF (IFLAG.EQ.1) GO TO 402
      WRITE(6,1001) LENBUF,NCHERR
      WRITE(2,1001) LENBUF,NCHERR
  402 WRITE(6,1025)
      WRITE(2,1025)
      NLINES = LENIBF/72 + 1
      DO 400 INDEX = 1 , NLINES
        NBEGIN = (INDEX -1)*72 + 1
        NEND   = NBEGIN + 71
        IF(NEND.GT.MSIZE) NEND=MSIZE
        WRITE(6,1030) (IBFFER(I),I=NBEGIN,NEND)
        WRITE(6,1031) (IDCBIT(I),I=NBEGIN,NEND)
        WRITE(2,1030) (IBFFER(I),I=NBEGIN,NEND)
        WRITE(2,1031) (IDCBIT(I),I=NBEGIN,NEND)
```

```
      400 CONTINUE
          IF (IFLAG.EQ.1) GO TO 401
          WRITE(6,1015)
          WRITE(2,1015)
          WRITE(6,1020)  (BUFFER(I:I),I=1,LENBUF)
          WRITE(6,1020)  (BUFFR2(I:I),I=1,LENBUF)
          WRITE(2,1020)  (BUFFER(I:I),I=1,LENBUF)
          WRITE(2,1020)  (BUFFR2(I:I),I=1,LENBUF)
      401 PRINT,' '
          PRINT,' START AGAIN (Y/N) ? '
          READ(5,1010) ANSWER
          IF((ANSWER.EQ.'Y').OR.(ANSWER.EQ.'y'))GO TO 1
     1000 FORMAT(//,' ',I4,' decoded bits ',/,
         +' ',I4,' bits in error ',/
         +' ',E10.4,' decoded bit-error rate '/)
     1001 FORMAT(//,' ',I4,' decoded ASCII characters',/,
         +' ',I4,' characters in error',/)
     1010 FORMAT(A1)
     1015 FORMAT(/,' Original vs. decoded ASCII message',/)
     1020 FORMAT(' ',80A1)
     1025 FORMAT(/,' Original vs. decoded bit stream',/)
     1030 FORMAT(' DATA: ',72I1)
     1031 FORMAT(' DEC''D: ',72I1)
          END
C
          SUBROUTINE INTIAL
          COMMON/ENCDE/NCODE(3,128),IPREV(128),NSEQ,IREG,NCRATE,
         +NSQHLF,KL,ITAP(3)
          COMMON/DECDE/METRIC(64),MLSEQ(64,70),MEMLEN
          NPATH=2**KL
          NSEQ=NPATH/2
          NSQHLF=NSEQ/2
C    LOOP TO GENERATE OUTPUTS FOR EACH PATH
          DO 100 IPATH=1,NPATH
C    IBIT = BIT PATTERN IN THE REGISTER FOR EACH PATH
          IBIT=IPATH-1
          IPREV(IPATH)=MOD(IBIT,NSEQ)+1
C    LOOP TO GENERATE EACH OUTPUT BIT
          DO 10 IOUTN=1,NCRATE
C    IREG=BIT PATTERN AFTER MASKED BY A G VECTOR
          IREGTAP=ITAP(IOUTN)
          IREGBIT=IBIT
C    ICNT COUNTS WEIGHTS
          ICNT=0
          IAMONT=NPATH
C    LOOP TO COUNT WEIGHTS STARTS HERE
          DO 30 IBTCNT=1,KL
          IAMONT=IAMONT/2
          IF(IREGBIT.LT.IAMONT) GO TO 31
          IREGBIT=IREGBIT-IAMONT
          IF(IREGTAP.LT.IAMONT) GO TO 30
          ICNT=ICNT+1
     31   CONTINUE
          IF(IREGTAP.LT.IAMONT) GO TO 30
          IREGTAP=IREGTAP-IAMONT
     30   CONTINUE
C    CHECK IF WEIGHT IS ZERO
          IF (ICNT.EQ.0) GO TO 43
C    CHECK IF WEIGHT IS  EVEN (TO 43)
          IF (ICNT/2.NE.(ICNT-1)/2) GO TO 43
          NCODE(IOUTN, IPATH)=1
          GO TO 10
     43   NCODE(IOUTN, IPATH)=-1
     10   CONTINUE
      100 CONTINUE
          RETURN
          END
```

App. E   A Convolutional Encoder/Decoder Computer Program   **751**

```
            SUBROUTINE ENCODE
            COMMON/VITEST/IDCBIT(646),ICODED(1938),IBFFER(646),NDEC,LENIBF
            COMMON/ENCDE/NCODE(3,128),IPREV(128),NSEQ,IREG,NCRATE,
           +NSQHLF,KL,ITAP(3)
            DO 200 INDEX = 1 , LENIBF
              INFO=IBFFER(INDEX)
              IREG=IREG/2+INFO*NSEQ
              IRGPTR=IREG+1
              DO 210 IOUTN=1,NCRATE
                ICODED(IOUTN+(INDEX-1)*NCRATE)=NCODE(IOUTN,IRGPTR)
      210     CONTINUE
      200   CONTINUE
            RETURN
            END
C
            SUBROUTINE DECODE(ITRS)
            DIMENSION IDATED(64),MLST(64)
            COMMON/ENCDE/NCODE(3,128),IPREV(128),NSEQ,IREG,NCRATE,
           +NSQHLF,KL,ITAP(3)
            COMMON/DECDE/METRIC(64),MLSEQ(64,70),MEMLEN
            COMMON/DECOD2/ILEVEL(3)
            COMMON/TIME/ITIME,ITIME1,INTIME,IBEST
            COMMON/FLUSH/NRECVD,NREP,NTAIL
C    METRIC CALCULATION
            DO 200 ISTATE=1,NSEQ
C    FOR EACH STATE, THERE ARE TWO PATHS (IPATH1 AND IPATH2).
            IPATH2=2*ISTATE
            IPATH1=IPATH2-1
            ITMP1=METRIC(IPREV(IPATH1))
            ITMP2=METRIC(IPREV(IPATH2))
            DO 210 IOUTN=1,NCRATE
              ITMP1=ITMP1+NCODE(IOUTN,IPATH1)*ILEVEL(IOUTN)
              ITMP2=ITMP2+NCODE(IOUTN,IPATH2)*ILEVEL(IOUTN)
      210   CONTINUE
C    METRICS:  THE LARGER THE BETTER
            IF(ITMP2-ITMP1) 225,225,223
      223   IDATED(ISTATE)=ITMP2
            MLST(ISTATE)=IPREV(IPATH2)
            GO TO 200
      225   IDATED(ISTATE)=ITMP1
            MLST(ISTATE)=IPREV(IPATH1)
C    MLST HOLDS THE STARTING NODE LOCATION FOR BRANCH TO ISTATE
      200 CONTINUE
C    MAX. LIKELIHOOD BRANCH IS SELECTED FOR EACH NODE AT THIS POINT
C    CHECK FOR BRANCHING NODES
            DO 240 ITPHLF=1,NSQHLF
            IBMHLF=NSQHLF+ITPHLF
            IBRNCH=MLST(ITPHLF)
            IF(IBRNCH.NE.MLST(IBMHLF))GOTO250
            IMISS=IBRNCH-1
            IF(MOD(IBRNCH,2).EQ.1) IMISS=IMISS+2
C    SEARCH FOR BRANCHING AND DROPPED SEQUENCES
            DO 241 ISEQ=1,NSEQ
              IF (MLSEQ(ISEQ,ITIME1).EQ.IMISS) GO TO 242
      241   CONTINUE
            STOP
      242   IMISSQ=ISEQ
            DO 243 ISEQ=1,NSEQ
              IF (MLSEQ(ISEQ,ITIME1).EQ.IBRNCH) GO TO 244
      243   CONTINUE
            STOP
      244   IBRNSQ=ISEQ
C    REPLACE A DROPPED SEQUENCE WITH A BRANCHING SEQUENCE UP TO ITIME1
            JTIME=INTIME
      245   MLSEQ(IMISSQ,JTIME)=MLSEQ(IBRNSQ,JTIME)
            IF(JTIME.EQ.ITIME1) GO TO 249
            JTIME=JTIME+1
```

```
            IF(JTIME.GT.MEMLEN) JTIME=1
            GO TO 245
C     NOW EXTEND THE ML SEQUENCES WITH CURRENT STATES
      250   CONTINUE
            DO 251 ISEQ=1,NSEQ
               IF (MLSEQ(ISEQ,ITIME1).EQ.MLST(ITPHLF)) GO TO 252
      251   CONTINUE
            STOP
      252   IMISSQ=ISEQ
            DO 256 ISEQ=1,NSEQ
               IF (MLSEQ(ISEQ,ITIME1).EQ.MLST(IBMHLF)) GO TO 258
      256   CONTINUE
            STOP
      258   IBRNSQ=ISEQ
      249   MLSEQ(IMISSQ,ITIME)=ITPHLF
            MLSEQ(IBRNSQ,ITIME)=IBMHLF
      240   CONTINUE
C     UPDATING METRICS ASSOCIATED WITH EACH STATE
            DO 391 ISTATE=1,NSEQ
      391 METRIC(ISTATE)=IDATED(ISTATE)
C     FORCE METRICS FOR TAIL BIT FLUSHING AT END OF MESSAGE
            IF(NRECVD.LT.NTAIL) GO TO 401
            MNSTATE=2**((NREP-NRECVD)/2)+1
            DO 392 ISTATE=MNSTATE,NSEQ
            METRIC(ISTATE)=-1000
      392 CONTINUE
C     SEARCH OF INPUT BIT STARTS HERE
      401 CONTINUE
C     CHECK IF ANY SEQUENCE MATCHES
            DO 402 ISEQ=1,NSEQ
            IF (MLSEQ(ISEQ,INTIME).NE.MLSEQ(1,INTIME)) GO TO 500
      402 CONTINUE
C     ALL PATHS ARE MERGING
            IBEST=1
            CALL DECIDE(1)
            GO TO 401
      500   CONTINUE
            ITIME1=ITIME
            ITIME=ITIME+1
            IF(ITIME.GT.MEMLEN) ITIME=1
            IF(ITRS.EQ.1) THEN
C THE FOLLOWING 5 COMMANDS ARE USED FOR OBSERVING THE TRELLIS TRACE
            WRITE(6,1110) INTIME,ITIME1
            WRITE(2,1110) INTIME,ITIME1
            DO 299 I=1,NSEQ
            WRITE(6,1111) (MLSEQ(I,J)-1,J=1,MEMLEN)
            WRITE(2,1111) (MLSEQ(I,J)-1,J=1,MEMLEN)
      1110  FORMAT(/' PATHS DIVERGE AT...',I2,8X,'CURRENT INPUT AT...',I2)
      1111  FORMAT(1X,32I2)
      299   CONTINUE
            ENDIF
            IF (ITIME.NE.INTIME) GO TO 599
C     OVERFLOW OF PATH MEMORY DETECTED
            IBEST=1
            DO 520 ISEQ=2,NSEQ
            IF(METRIC(MLSEQ(IBEST,ITIME1)).LT.
         +  METRIC(MLSEQ(ISEQ,ITIME1))) IBEST=ISEQ
      520 CONTINUE
            CALL DECIDE(2)
      599 CONTINUE
            IF(NRECVD.LT.NREP) GO TO 600
C     END OF MESSAGE DETECTED.  FLUSHING THE DECODER
            IBEST=1
            DO 590 ISEQ=2,NSEQ
            IF(METRIC(MLSEQ(IBEST,ITIME1)).LT.
         +METRIC(MLSEQ(ISEQ,ITIME1))) IBEST=ISEQ
      590 CONTINUE
```

```
          CALL DECIDE(3)
    600 CONTINUE
          RETURN
          END
C
          SUBROUTINE DECIDE(MODE)
          COMMON/VITEST/IDCBIT(646),ICODED(1938),IBFFER(646),NDEC,LENIBF
          COMMON/ENCDE/NCODE(3,128),IPREV(128),NSEQ,IREG,NCRATE,
         +NSQHLF,KL,ITAP(3)
          COMMON/DECDE/METRIC(64),MLSEQ(64,70),MEMLEN
          COMMON/TIME/ITIME,ITIME1,INTIME,IBEST
      1 NDEC=NDEC+1
          IDCBIT(NDEC)=1
          IF(MLSEQ(IBEST,INTIME).LE.NSQHLF) IDCBIT(NDEC)=0
          INTIME=INTIME+1
          IF(INTIME.GT.MEMLEN) INTIME=1
          IF(MODE.NE.3) GO TO 20
          IF(NDEC.LT.LENIBF) GO TO 1
     20 CONTINUE
          RETURN
          END
C
          SUBROUTINE MSGIN(BUFFER,LENBUF)
          COMMON/FLAG/IFLAG
C    THIS SUBROUTINE ASKS THE USER FOR AN INPUT MESSAGE
C    IN EITHER BINARY OR CHARACTER FORM
          CHARACTER*80 BUFFER
          CHARACTER*1 ANSWER,ONE
          CHARACTER*8 BITBUF,BLANK
          DATA ONE/'1'/
          DATA BLANK/'        '/
          IFLAG = 0
          PRINT,' BINARY MESSAGE (Y/N) ? '
          READ(5,1000) ANSWER
          IF((ANSWER.EQ.'Y').OR.(ANSWER.EQ.'y'))THEN
            IFLAG = 1
            PRINT,' '
            PRINT,' ENTER 8 BITS (0 OR 1) PER LINE.  FOR AN ALL-ZEROS'
            PRINT,' MESSAGE, HIT ENTER FOR EACH 8-BIT SEQUENCE OF ZEROS.'
            PRINT,' TYPE "END" WHEN DONE'
            PRINT,' '
            KOUNT = 0
    100     CONTINUE
            BITBUF = BLANK
            PRINT,'>'
            READ(5,1020) BITBUF
            IF((BITBUF.EQ.'END     ').OR.(BITBUF.EQ.'end     '))THEN
                GO TO 200
              ELSE
                NCHAR = 0
                DO 10 INDEX = 1 , 8
                  IF( BITBUF(INDEX:INDEX) .EQ. ONE ) THEN
                    NCHAR = NCHAR + 2**(INDEX-1)
                  ENDIF
     10         CONTINUE
                KOUNT = KOUNT + 1
                BUFFER(KOUNT:KOUNT) = CHAR( NCHAR )
            ENDIF
            GO TO 100
    200     CONTINUE
            LENBUF = KOUNT
          ELSE
            DO 1 INDEX = 1 , 80
              BUFFER(INDEX:INDEX) = ' '
      1     CONTINUE
            PRINT,' ENTER ASCII MESSAGE'
            PRINT,' '
```

```
          READ(5,1010) BUFFER
          DO 50 INDEX = 80 , 1 , -1
             IF( BUFFER(INDEX:INDEX) .NE. ' ') THEN
                LENBUF = INDEX
                GO TO 55
             ENDIF
   50     CONTINUE
   55     CONTINUE
        ENDIF
        RETURN
 1000 FORMAT(A1)
 1010 FORMAT(A80)
 1020 FORMAT(A8)
      END
C
      SUBROUTINE PARMIN(NCRATE,MEMLEN,KL,ITAP)
C    THIS SUBROUTINE ASKS THE USER FOR HIS/HER CHOICE OF CODING
C    PARAMETERS
      DIMENSION ITAP(3),ITAPB(3,2:3,3:7)
C TAP LOCATIONS FOR CODE GENERATORS ARE FROM TABLE 6.4 P.349
      DATA ITAPB/111,101,0,111,111,101,1111,1011,0,1111,1011,1101,
     +10111,11001,0,11111,11011,10101,101111,110101,0,101111,110101,
     +111001,1001111,1101101,0,1001111,1010111,1101101/
 2014 PRINT,' '
      WRITE(6,1000) KL,NCRATE,MEMLEN,(ITAPB(I,NCRATE,KL),I=1,NCRATE)
 1000 FORMAT(' THE DEFAULT CODES ARE THE ODENWALDER CODES LISTED IN',
     +' TABLE 6.4, PAGE 349'//
     +' 1. CONSTRAINT LENGTH (7 MAX.)             ',I1/
     +' 2. NUMBER OF CODE GENERATORS (3 MAX.)     ',I1/
     +' 3. PATH MEMORY LENGTH (70 MAX.)           ',I2//
     +' 4. FIRST GENERATOR                ',I7/
     +' 5. SECOND GENERATOR               ',I7/
     +' 6. THIRD GENERATOR                ',I7)
      WRITE(6,1001)
 1001 FORMAT(/' Note that the code generators are shown as a binary',
     +' sequence, where'/' each 1 represents a connected tap, and',
     +' each 0 represents an unconnected tap.'/)
      PRINT,' ENTER PARAMETER NUMBER TO CHANGE (ENTER 0 IF DONE) > '
      READ(5,*) NUMCNG
      IF ( NUMCNG .EQ. 0 ) THEN
          GO TO 2015
        ELSE IF ( NUMCNG .EQ. 1 ) THEN
          PRINT,' ENTER NEW CONSTRAINT LENGTH       > '
          READ(5,*) KL
          GO TO 2014
        ELSE IF ( NUMCNG .EQ. 2 ) THEN
          PRINT,' ENTER NUMBER OF CODE GENERATORS   > '
          READ(5,*) NCRATE
          GO TO 2014
        ELSE IF ( NUMCNG .EQ. 3 ) THEN
          PRINT,' ENTER NEW PATH MEMORY LENGTH      > '
          READ(5,*) MEMLEN
          GO TO 2014
        ELSE IF ( NUMCNG .EQ. 4 ) THEN
          PRINT,' ENTER FIRST GENERATOR             > '
          READ(5,*) ITAPB(1,NCRATE,KL)
          GO TO 2014
        ELSE IF ( NUMCNG .EQ. 5 ) THEN
          PRINT,' ENTER SECOND GENERATOR            > '
          READ(5,*) ITAPB(2,NCRATE,KL)
          GO TO 2014
        ELSE IF ( NUMCNG .EQ. 6 ) THEN
          PRINT,' ENTER THIRD GENERATOR             > '
          READ(5,*) ITAPB(3,NCRATE,KL)
          GO TO 2014
        ELSE
          PRINT,' NUMBER OF CHANGE OUT OF RANGE.  RE-ENTER'
```

```
            GO TO 2014
        ENDIF
 2015 CONTINUE
C   CALCULATE ITAP FROM ITAPB
        DO 100 ICRATE=1,NCRATE
          IACCUM=0
          IDIGIT=10
          IBNARY=1
          IWORD=ITAPB(ICRATE,NCRATE,KL)
          DO 110 IBIT=1,KL
            ITEST=MOD(IWORD,IDIGIT)
            IF(ITEST.NE.0) THEN
              IACCUM=IACCUM+IBNARY
              IWORD=IWORD-ITEST
            ENDIF
            IDIGIT=IDIGIT*10
            IBNARY=IBNARY*2
  110     CONTINUE
          ITAP(ICRATE)=IACCUM
  100   CONTINUE
        WRITE(2,1000) KL,NCRATE,MEMLEN,(ITAPB(I,NCRATE,KL),I=1,NCRATE)
  999 FORMAT(A1)
        RETURN
        END
C
        SUBROUTINE ASCII1(BUFFER,LENBUF,IBFFER,LENIBF,KL)
C   THIS SUBROUTINE CONVERTS THE CHARACTER MESSAGE TO A BIT STREAM
        CHARACTER*80 BUFFER
        DIMENSION IBFFER(646)
        KOUNT = 0
        DO 100 INDEX1 = 1 , LENBUF
          NCHAR = ICHAR(BUFFER(INDEX1:INDEX1))
          DO 200 INDEX2 = 1 , 8
            KOUNT = KOUNT + 1
            IBIT = MOD(NCHAR,2**INDEX2)
            IF (IBIT.NE.0) THEN
                IBFFER(KOUNT) = 1
              ELSE
                IBFFER(KOUNT) = 0
            ENDIF
            NCHAR=NCHAR-IBIT
  200     CONTINUE
  100   CONTINUE
        DO 300 INDEX1 = 1 , KL-1
          KOUNT = KOUNT + 1
          IBFFER(KOUNT) = 0
  300   CONTINUE
        LENIBF = KOUNT
        RETURN
        END
C
        SUBROUTINE ASCII2(BUFFER,LENBUF,IBFFER,LENIBF)
C   THIS SUBROUTINE CONVERTS A BIT STREAM TO CHARACTERS
        CHARACTER*80 BUFFER
        DIMENSION IBFFER(646)
        KOUNT = 0
        LENBUF = LENIBF/8
        DO 100 INDEX1 = 1 , LENBUF
          NCHAR = 0
          DO 200 INDEX2 = 1 , 8
            KOUNT = KOUNT + 1
            NCHAR = NCHAR + 2**(INDEX2-1) * (IBFFER(KOUNT))
  200     CONTINUE
          BUFFER(INDEX1:INDEX1) = CHAR(NCHAR)
  100   CONTINUE
        RETURN
        END
```

```
      SUBROUTINE ERRORS(BUFFER,ICODED,NREP,NCRATE,KL)
C    THIS SUBROUTINE QUERIES THE USER WHERE HE/SHE WOULD LIKE ERRORS
C    TO OCCUR IN THE TRANSMITTED BIT STREAM
      CHARACTER*80 BUFFER
      DIMENSION ICODED(NREP)
      CHARACTER*24 CURSOR
      LINESZ = 8 * NCRATE
      NERROR = 0
      NLINES = NREP/LINESZ
      DO 100 INDEX = 1 , NLINES
        DO 110 INDEX2 = 1 , 24
          CURSOR(INDEX2:INDEX2) = ' '
  110   CONTINUE
      PRINT,' '
      PRINT,' CREATE THE DESIRED ERROR PATTERN BY MOVING THE CURSOR'
      PRINT,' TO A POSITION DIRECTLY BELOW THE BIT WHERE AN ERROR'
      PRINT,' IS TO OCCUR, AND TYPING ANY CHARACTER'
      PRINT,' '
      PRINT,' '
      NBEGIN = MIN0( LINESZ * (INDEX-1) + 1 , NREP)
      NEND   = MIN0( NBEGIN + (LINESZ - 1) , NREP)
      WRITE(6,1005) BUFFER(INDEX:INDEX)
      WRITE(2,1005) BUFFER(INDEX:INDEX)
      PRINT,' ENCODED CHARACTER '
      WRITE(6,1010)  ((ICODED(I)+1)/2,I=NBEGIN,NEND)
      WRITE(2,1011)  ((ICODED(I)+1)/2,I=NBEGIN,NEND)
      WRITE(6,1020)
      READ(5,1000) CURSOR
      DO 120 INDEX2 = 1 , LINESZ
        IF( CURSOR(INDEX2:INDEX2) .NE. ' ' ) THEN
          INDEX3 = NBEGIN + (INDEX2-1)
          ICODED( INDEX3 ) = -ICODED( INDEX3 )
          NERROR=NERROR+1
        ENDIF
  120   CONTINUE
      PRINT,' RECEIVED CHARACTER BIT STREAM'
      WRITE(6,1012)  ((ICODED(I)+1)/2,I=NBEGIN,NEND)
      WRITE(2,1013)  ((ICODED(I)+1)/2,I=NBEGIN,NEND)
      WRITE(2,1014) CURSOR
      WRITE(2,1020)
  100 CONTINUE
      WRITE(2,1020)
      WRITE(6,1020)
      WRITE(2,1030) NREP-NCRATE*(KL-1),NERROR
      WRITE(6,1030) NREP-NCRATE*(KL-1),NERROR
      RETURN
 1000 FORMAT(A24)
 1005 FORMAT(' Transmitted character...',A1)
 1010 FORMAT(1X,24I1)
 1012 FORMAT(1X,24I1)
 1011 FORMAT(' Transmitted coded bits  ',24I1)
 1013 FORMAT('    Received coded bits  ',24I1)
 1014 FORMAT(' Coded bit error pattern ',A24)
 1020 FORMAT(' ')
 1030 FORMAT(I5,' coded bits sent',I7,' coded bits received in error')
      END
C
      SUBROUTINE RANDOM(BUFFER,ICODED,NREP,NCRATE,KL,SNROBS)
C    THIS SUBROUTINE GENERATES A DEMODULATED BPSK SIGNAL, CORRUPTED
C    BY AWGN, AND EXPRESSED AS A QUANTIZED CODED-SYMBOL STREAM
      CHARACTER*80 BUFFER
      DIMENSION ICODED(NREP),IHIST(32)
      LINESZ = 8 * NCRATE
      NERROR = 0
      VAR=0.
      NLINES = NREP/LINESZ
      PRINT,' ENTER PREDETECTION EB/N0 IN DECIBELS        > '
```

```
      INPUT,SNR
      SNR=10.**(SNR/10.)
      SIGMA=1./SQRT(2.*SNR/NCRATE)
      PRINT,' ENTER NUMBER OF SOFT-DECISION BITS (6 MAX.)  > '
      INPUT,NQ
      NBIN=2.**(NQ-1)
      NLEVEL=2.**NQ-1
      DO 10 K=1,NBIN
        IHIST(K)=0
   10 CONTINUE
      DO 100 INDEX = 1 , NLINES
        NBEGIN = MIN0( LINESZ * (INDEX-1) + 1 , NREP)
        NEND = MIN0( NBEGIN + (LINESZ - 1) , NREP)
        WRITE(6,1005) BUFFER(INDEX:INDEX)
        WRITE(2,1005) BUFFER(INDEX:INDEX)
        WRITE(6,1011) (ICODED(I),I=NBEGIN,NEND)
        WRITE(2,1011) (ICODED(I),I=NBEGIN,NEND)
        DO 120 INDEX2 = 1 , LINESZ
          INDEX3 = NBEGIN + (INDEX2-1)
          CALL NOISE(AWGN)
          VAR=VAR+AWGN*AWGN
          CODED=SIGMA*AWGN + ICODED(INDEX3)
          ERROR=CODED*ICODED(INDEX3)
          IF(ERROR.LE.0.) NERROR=NERROR+1
          Q=CODED*NLEVEL
          IF(Q.GT.NLEVEL) Q=NLEVEL
          IF(Q.LT.-NLEVEL) Q=-NLEVEL
          IQ=INT(Q)
          IQMOD=MOD(IQ,2)
          IF(IQMOD.EQ.0) IQ=IQ+SIGN(1.,Q)
          ICODED(INDEX3)=IQ
          IQP=IABS(IQ)/2+1
          IHIST(IQP)=IHIST(IQP)+1
  120   CONTINUE
        WRITE(6,1013) (ICODED(I),I=NBEGIN,NEND)
        WRITE(2,1013) (ICODED(I),I=NBEGIN,NEND)
  100 CONTINUE
      WRITE(2,1030) NREP-NCRATE*(KL-1),NERROR
      WRITE(6,1030) NREP-NCRATE*(KL-1),NERROR
      VAR=VAR/(NREP-1)
      SNROBS=10.*ALOG10(FLOAT(NCRATE)/2./SIGMA/SIGMA/VAR)
      WRITE(2,1040) SNROBS
      WRITE(6,1040) SNROBS
      WRITE(2,1041) (IHIST(I),I=1,NBIN)
      WRITE(6,1041) (IHIST(I),I=1,NBIN)
      RETURN
 1005 FORMAT(/' Xmitted character...',A1)
 1011 FORMAT(' Xmtd:   ',24I3)
 1013 FORMAT(' Rcvd:   ',24I3)
 1030 FORMAT(/,I5,' coded bits sent',I7,' coded bits received',
     +' in error')
 1040.FORMAT(/,' Observed SNR in dB ',F5.2,/)
 1041 FORMAT(' Quantized likelihood function histogram, where the'
     +/' rightmost entry represents the most assured decisions'//32
     +(1X,I4))
      END
C
      SUBROUTINE NOISE(AWGN)
C THIS SUBROUTINE GENERATES A ZERO-MEAN UNIT-VARIANCE GAUSSIAN RANDOM
C VARIABLE SAMPLE
      AMP=SQRT(-ALOG(RND())*2.)
      PHASE=RND()*6.283185308
      AWGN=AMP*COS(PHASE)
      RETURN
      END
```

APPENDIX F

List of Symbols

a_{ij}	Coefficient of jth basis function
a_j	Signal component output of jth correlator
A	Peak amplitude of a waveform
Λ_e	Effective area of an antenna
B_L	Single-sided loop bandwidth
c	Speed of light $\simeq 3 \times 10^8$ m/s
C	Channel capacity
C	Electrical capacitance
$C/\kappa T°$	Ratio of average carrier power to noise power spectral density
d	Distance
d_f	Free distance
$d_{\min}$	Minimum distance
D	Delay time of message
D	Redundancy of a language
D	Decryption transformation
e	The natural number 2.7183
$\mathbf{e}$	Error pattern vector
$e(t)$	Error signal
$\mathbf{e}(X)$	Error pattern polynomial
E	Encryption transformation
E_x	Energy of waveform $x(t)$
$\mathbf{E}\{X\}$	Expected value of the random variable X
EIRP	Effective radiated power with reference to an isotropic source
E_b/J_0	Ratio of bit energy to jammer power spectral density

E_b/N_0	Ratio of bit energy to noise power spectral density
E_c/N_0	Ratio of channel symbol energy to noise power spectral density
f	Frequency (hertz)
f_c	Carrier-wave frequency
f_m	Maximum frequency
f_s	Sampling frequency
f_ℓ	Lower cutoff filter frequency
f_u	Upper cutoff filter frequency
F	Noise figure
$\mathcal{F}\{x\}$	Fourier transform of the function $x(t)$
$\mathcal{F}^{-1}\{X\}$	Inverse Fourier transform of the function $X(f)$
$g(t)$	Pseudorandom code function
$\mathbf{g}(X)$	Generator polynomial for a cyclic code
G	Antenna gain
G	Coding again
$\mathbf{G}$	Generator matrix for a linear block code
G	Normalized total message traffic
G_p	Processing gain
$G_x(f)$	Power spectral density of waveform $x(t)$
$h(t)$	Impulse response of a network
$\mathbf{H}$	Parity-check matrix for a code
H_i	The ith hypothesis
$\mathbf{H}_k$	Hadamard matrix
$H(f)$	Frequency transfer function of a network
$H_0(f)$	Optimum frequency transfer function
$H(X)$	Entropy of information source X
$H(X|Y)$	Conditional entropy (entropy of X, given Y)
$i(t)$	Electrical current waveform
I	Electrical current
$I_0(x)$	Zero-order modified Bessel function of the first kind
$I(X)$	Self-information of information source X
J	Received average jammer power
J_0	Jammer power spectral density
J/S	Ratio of received average jammer power to average signal power
k	Number of bits per M-ary signal set
k/n	Code rate (ratio of number of data bits to total bits in codeword)
K	Constraint length of a convolutional encoder
K	Key, dictating a specific encryption or decryption transformation
ℓ	Number of quantization bits
L	Look-ahead length for convolutional feedback decoding
L	Number of bits in sequence
L	Number of quantization levels
L_s	Space loss
L_o	Other losses
$\mathbf{m}$	Message vector
$\mathbf{m}(X)$	Message polynomial

m_i	Data bit
M	Margin
M	Waveform or signal set size
(n, k)	Code designation by number of total bits (n) and data bits (k) in codeword
$\bar{n}$	Average number of bits per character
n_0	Noise random variable output of correlator at symbol time $t = T$
$n(t)$	Gaussian noise process
N	Noise power
N	Unicity distance
N_0	Level of single-sided power spectral density of white noise
NSR	Ratio of average noise power to average signal power
p	Probability of channel symbol error
p_i	Parity bit
$p(t)$	Instantaneous power
$p(x)$	Probability density function of a continuous random variable
$p(x\|y)$	Probability density function of x conditioned on y
$\mathbf{P}$	Parity array
P_B	Probability of bit error
P_E	Probability of symbol error
P_{FA}	Probability of false alarm
P_m	Probability of miss
P_M	Probability of message or block error
P_{nd}	Probability of undetected error
P_r/N_0	Ratio of received average signal power to noise power spectral density
$P(X)$	Probability of a discrete random variable
P_x	Average power in waveform $x(t)$
q	Quantization step size (quantile interval)
$\mathbf{q}(X)$	Quotient polynomial
$Q(x)$	Complementary error function (integral of the tail beyond x of the Gaussian density function)
r	Filter roll-off factor
r	True rate of a language
r'	Absolute rate of a language
$r(t)$	Received signal waveform
$r(X)$	Remainder polynomial
R	Data rate (bits/second)
R_c	Coded data rate (coded bits/second)
R_p	Code chip rate (chips/second)
R_s	Symbol rate (symbols/second)
$R_x(\tau)$	Autocorrelation function of waveform $x(t)$
$\mathscr{R}$	Electrical resistance
$s(t)$	Signal waveform
$\hat{s}(t)$	Estimate of signal waveform
$\mathbf{s}$	Signal vector
sgn x	Sign function of x

S	Signal power
$\mathbf{S}$	Syndrome vector
SJR	Ratio of average signal power to average jammer power
SNR	Ratio of average signal power to average noise power
S/N	Ratio of signal power to noise power
$S(f)$	Fourier transform of the waveform $s(t)$
$\mathbf{S}(X)$	Syndrome polynomial
t	Number of errors correctable in an error-correcting code
t	Independent time variable
t_0	Time delay
t_{ij}	Amount of message traffic from i to j
T	Pulse width
T	Symbol interval
$T(D)$	Transfer function or generating function of convolutional code
T_{hop}	Duration of a hop
T_s	Sampling interval
T°	Temperature
T_A°	Antenna temperature
T_L°	Effective line temperature
T_R°	Effective receiver temperature
T_S°	System temperature
T_{acq}	Time to acquire
u_i	Code symbol
$u(t)$	Unit step function
$\mathbf{U}$	Codeword vector
$\mathbf{U}(X)$	Codeword polynomial
v	Relative velocity
$v(t)$	Electrical voltage waveform
var (X)	Variance of random variable X
V	Electrical voltage
$w(t)$	Jammer waveform
W	Bandwidth
W_f	Filter bandwidth
W_{DSB}	Double-sideband bandwidth
W_N	Noise equivalent bandwidth
W_{ss}	Spread-spectrum bandwidth
$x(t)$	Normalized VCO output
$y(t)$	VCO input voltage function
$z(t)$	Output of matched filter or correlator
γ	Threshold level
γ_0	Optimum threshold level
δ	Fractional frequency drift per day
δ_{mn}	Kronecker delta function
$\delta(t)$	Impulse (Dirac delta) function
ϵ	Error

ζ	Loop damping characteristic (second-order loop)
η	Antenna efficiency
$\theta(t)$	Time-varying phase
$\Theta(\omega)$	Fourier transform of $\theta(t)$
λ	Wavelength
λ	Packet arrival rate
π	Pi, 3.14159
ρ	Fraction of the frequency band being jammed
ρ	Fraction of the time the jammer is ''on''
ρ	Normalized loop signal-to-noise ratio
ρ	Normalized message throughput
ρ	Number of erasures correctable in an error-correcting code
ρ	Time-correlation coefficient
ρ_0	Value of ρ that maximizes bit error probability (worst-case jamming)
σ_X	Standard deviation of random variable X
σ_X^2	Variance of random variable X
τ	Pulse width
τ	Time shift (independent variable of the autocorrelation function)
$\phi(t)$	Time-varying phase
$\psi_j(t)$	Basis function
$\Psi_x(f)$	Energy spectral density of waveform $x(t)$
ω	Radian frequency (radians per second)
κ	Boltzmann's constant, 1.38×10^{-23} J/K

Index

■ **TABLE 1.7**
Approximate Physical Properties of Some Common Gases at Standard Atmospheric Pressure (BG Units)

Gas	Temperature (°F)	Density, ρ (slugs/ft³)	Specific Weight, γ (lb/ft³)	Dynamic Viscosity, μ (lb·s/ft²)	Kinematic Viscosity, ν (ft²/s)	Gas Constant,[a] R (ft·lb/slug·°R)	Specific Heat Ratio,[b] k
Air (standard)	59	2.38 E − 3	7.65 E − 2	3.74 E − 7	1.57 E − 4	1.716 E + 3	1.40
Carbon dioxide	68	3.55 E − 3	1.14 E − 1	3.07 E − 7	8.65 E − 5	1.130 E + 3	1.30
Helium	68	3.23 E − 4	1.04 E − 2	4.09 E − 7	1.27 E − 3	1.242 E + 4	1.66
Hydrogen	68	1.63 E − 4	5.25 E − 3	1.85 E − 7	1.13 E − 3	2.466 E + 4	1.41
Methane (natural gas)	68	1.29 E − 3	4.15 E − 2	2.29 E − 7	1.78 E − 4	3.099 E + 3	1.31
Nitrogen	68	2.26 E − 3	7.28 E − 2	3.68 E − 7	1.63 E − 4	1.775 E + 3	1.40
Oxygen	68	2.58 E − 3	8.31 E − 2	4.25 E − 7	1.65 E − 4	1.554 E + 3	1.40

[a]Values of the gas constant are independent of temperature.
[b]Values of the specific heat ratio depend only slightly on temperature.

■ **TABLE 1.8**
Approximate Physical Properties of Some Common Gases at Standard Atmospheric Pressure (SI Units)

Gas	Temperature (°C)	Density, ρ (kg/m³)	Specific Weight, γ (N/m³)	Dynamic Viscosity, μ (N·s/m²)	Kinematic Viscosity, ν (m²/s)	Gas Constant,[a] R (J/kg·K)	Specific Heat Ratio,[b] k
Air (standard)	15	1.23 E + 0	1.20 E + 1	1.79 E − 5	1.46 E − 5	2.869 E + 2	1.40
Carbon dioxide	20	1.83 E + 0	1.80 E + 1	1.47 E − 5	8.03 E − 6	1.889 E + 2	1.30
Helium	20	1.66 E − 1	1.63 E + 0	1.94 E − 5	1.15 E − 4	2.077 E + 3	1.66
Hydrogen	20	8.38 E − 2	8.22 E − 1	8.84 E − 6	1.05 E − 4	4.124 E + 3	1.41
Methane (natural gas)	20	6.67 E − 1	6.54 E + 0	1.10 E − 5	1.65 E − 5	5.183 E + 2	1.31
Nitrogen	20	1.16 E + 0	1.14 E + 1	1.76 E − 5	1.52 E − 5	2.968 E + 2	1.40
Oxygen	20	1.33 E + 0	1.30 E + 1	2.04 E − 5	1.53 E − 5	2.598 E + 2	1.40

[a]Values of the gas constant are independent of temperature.
[b]Values of the specific heat ratio depend only slightly on temperature.

Why WileyPLUS for Engineering?

WileyPLUS offers today's Engineering students the interactive and visual learning materials they need to help them grasp difficult concepts—and apply what they've learned to solve problems in a dynamic environment.

⊕ *A robust variety of examples and exercises enable students to work problems, see their results, and obtain instant feedback including hints and reading references linked directly to the online text.*

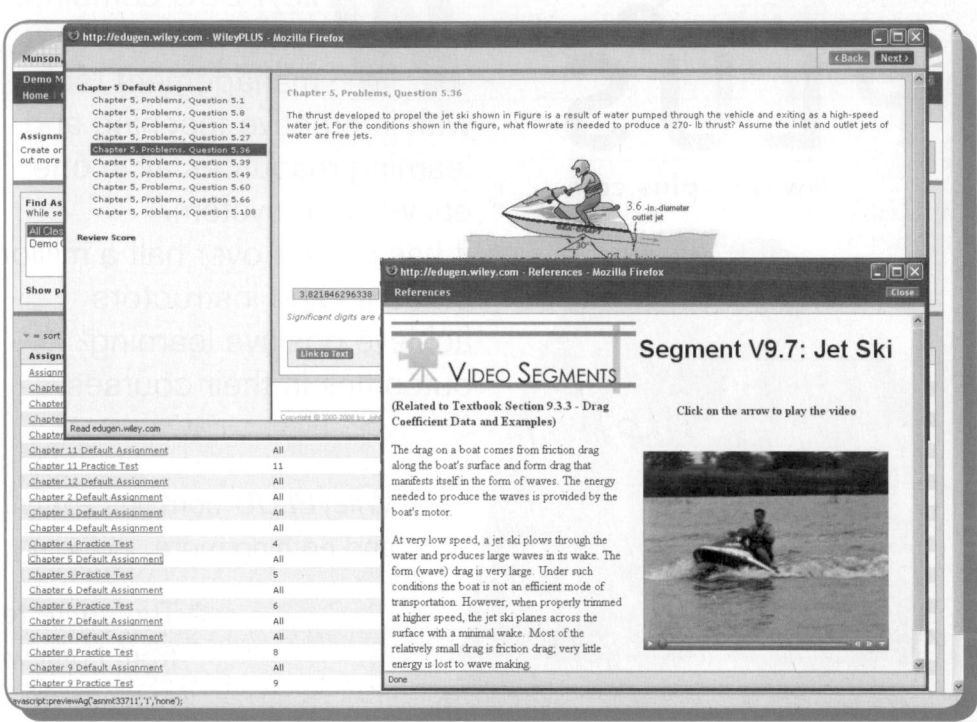

⊕ *Students can visualize concepts from the text by linking to dynamic resources such as animations, videos, and interactive LearningWare.*

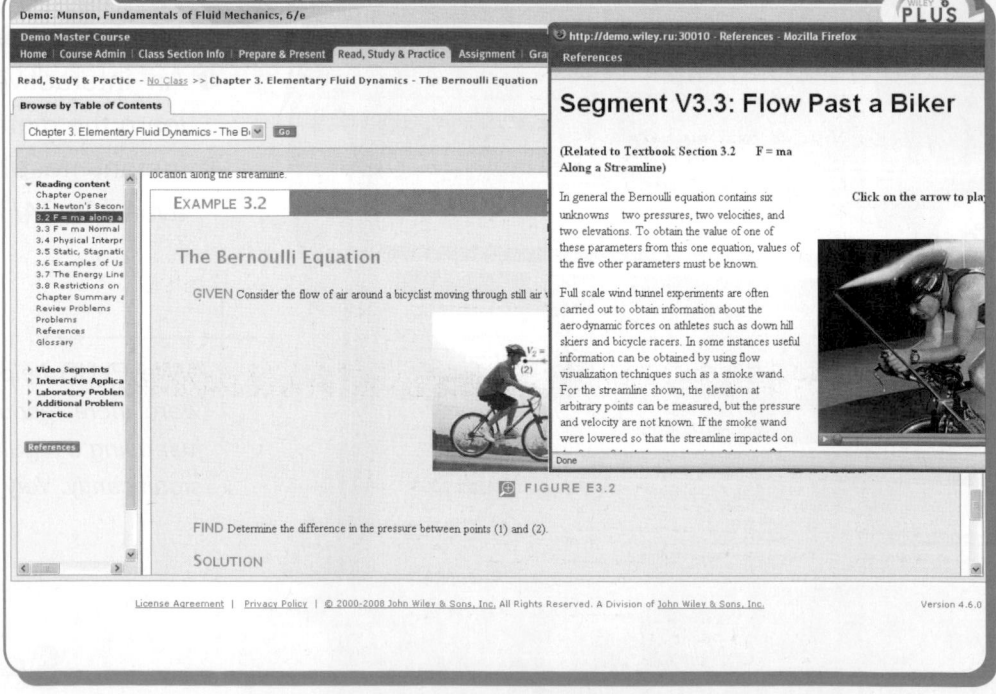

See and try WileyPLUS in action!
Details and Demo: www.wileyplus.com

WileyPLUS combines robust course management tools with the complete online text and all of the interactive teaching & learning resources you and your students need in one easy-to-use system.

"I loved this program [WileyPLUS] and I hope I can use it in the future." — Anthony Pastin, West Virginia University

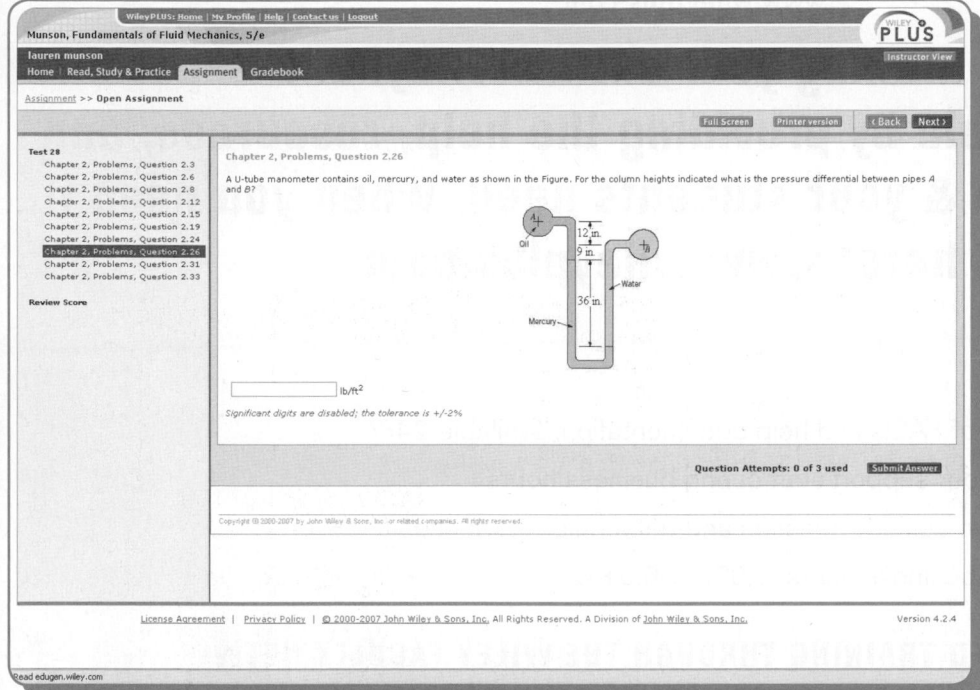

Algorithmic questions allow a group of students to work on the same problem with differing values. Students can also rework a problem with differing values for additional practice.

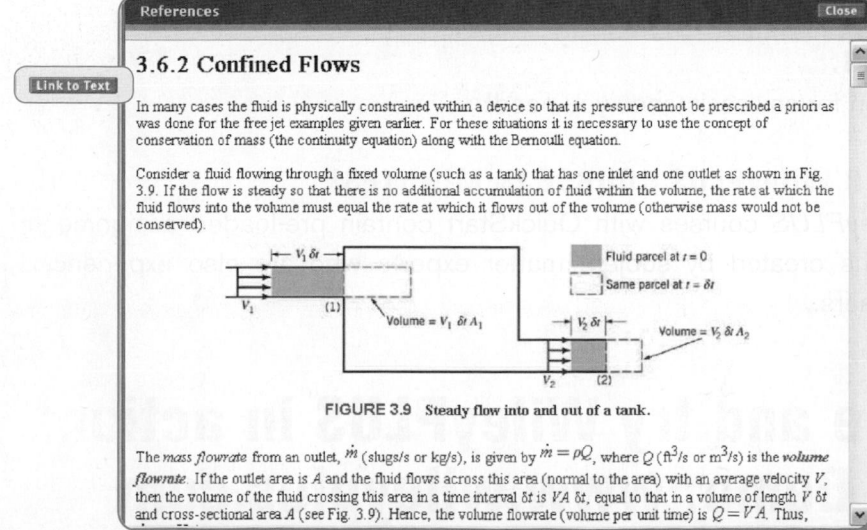

MultiPart Problems and **GoTutorials** lead students through a series of steps, providing instant feedback along the way, to help them develop a logical, structured approach to problem solving.

Or, they can link directly to the online text to read about this concept before attempting the problem again—with or without the same values.

www.wileyplus.com

Wiley is committed to making your entire WileyPLUS experience productive & enjoyable by providing the help, resources, and personal support you & your students need, when you need it. It's all here: www.wileyplus.com

TECHNICAL SUPPORT:

- ⊕ A fully searchable knowledge base of FAQs and help documentation, available 24/7
- ⊕ Live chat with a trained member of our support staff during business hours
- ⊕ A form to fill out and submit online to ask any question and get a quick response
- ⊕ **Instructor-only** phone line during business hours: 1.877.586.0192

FACULTY-LED TRAINING THROUGH THE WILEY FACULTY NETWORK:
Register online: www.wherefacultyconnect.com
Connect with your colleagues in a complimentary virtual seminar, with a personal mentor in your field, or at a live workshop to share best practices for teaching with technology.

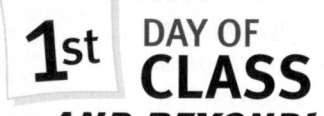

1ST DAY OF CLASS...AND BEYOND!
Resources You & Your Students Need to Get Started & Use WileyPLUS from the first day forward.

- ⊕ 2-Minute Tutorials on how to set up & maintain your *WileyPLUS* course
- ⊕ User guides, links to technical support & training options
- ⊕ *WileyPLUS for Dummies*: Instructors' quick reference guide to using *WileyPLUS*
- ⊕ Student tutorials & instruction on how to register, buy, and use *WileyPLUS*

YOUR WileyPLUS ACCOUNT MANAGER:
Your personal *WileyPLUS* connection for any assistance you need!

SET UP YOUR WileyPLUS COURSE IN MINUTES!
Selected *WileyPLUS* courses with QuickStart contain pre-loaded assignments & presentations created by subject matter experts who are also experienced *WileyPLUS* users.

Interested? See and try WileyPLUS in action!
Details and Demo: www.wileyplus.com

Sixth Edition

Fundamentals of Fluid Mechanics

International Student Version

BRUCE R. MUNSON
DONALD F. YOUNG

Department of Aerospace Engineering and Engineering Mechanics

THEODORE H. OKIISHI

Department of Mechanical Engineering
Iowa State University
Ames, Iowa, USA

WADE W. HUEBSCH

Department of Mechanical and Aerospace Engineering
West Virginia University
Morgantown, West Virginia, USA

WILEY

John Wiley & Sons, Inc.

Contributing Subject Matter Expert: Wade W. Huebsch, Associate Professor, West Virginia University, Dept. of Mechanical & Aerospace Engr.

ISBN: 978-0-470-39881-4

Printed in Asia

10 9 8 7 6 5 4 3 2 1

About the Authors

Bruce R. Munson, Professor Emeritus of Engineering Mechanics at Iowa State University, received his B.S. and M.S. degrees from Purdue University and his Ph.D. degree from the Aerospace Engineering and Mechanics Department of the University of Minnesota in 1970.

Prior to joining the Iowa State University faculty in 1974, Dr. Munson was on the mechanical engineering faculty of Duke University from 1970 to 1974. From 1964 to 1966, he worked as an engineer in the jet engine fuel control department of Bendix Aerospace Corporation, South Bend, Indiana.

Dr. Munson's main professional activity has been in the area of fluid mechanics education and research. He has been responsible for the development of many fluid mechanics courses for studies in civil engineering, mechanical engineering, engineering science, and agricultural engineering and is the recipient of an Iowa State University Superior Engineering Teacher Award and the Iowa State University Alumni Association Faculty Citation.

He has authored and coauthored many theoretical and experimental technical papers on hydrodynamic stability, low Reynolds number flow, secondary flow, and the applications of viscous incompressible flow. He is a member of The American Society of Mechanical Engineers.

Donald F. Young, Anson Marston Distinguished Professor Emeritus in Engineering, received his B.S. degree in mechanical engineering, his M.S. and Ph.D. degrees in theoretical and applied mechanics from Iowa State Uiversity, and has taught both undergraduate and graduate courses in fluid mechanics at Iowa State for many years. In addition to being named a Distinguished Professor in the College of Engineering, Dr. Young has also received the Standard Oil Foundation Outstanding Teacher Award and the Iowa State University Alumni Association Faculty Citation. He has been engaged in fluid mechanics research for more than 35 years, with special interests in similitude and modeling and the interdisciplinary field of biomedical fluid mechanics. Dr. Young has contributed to many technical publications and is the author or coauthor of two textbooks on applied mechanics. He is a Fellow of The American Society of Mechanical Engineers.

Ted H. Okiishi, Professor Emeritus of Mechanical Engineering at Iowa State University, joined the faculty there in 1967 after receiving his undergraduate and graduate degrees from that institution.

From 1965 to 1967, Dr. Okiishi served as a U.S. Army officer with duty assignments at the National Aeronautics and Space Administration Lewis Research Center, Cleveland, Ohio, where he participated in rocket nozzle heat transfer research, and at the Combined Intelligence Center, Saigon, Republic of South Vietnam, where he studied seasonal river flooding problems.

Professor Okiishi and his students have been active in research on turbomachinery fluid dynamics. Some of these projects have involved significant collaboration with government and industrial laboratory researchers with two of their papers winning the ASME Melville Medal (in 1989 and 1998).

Dr. Okiishi has received several awards for teaching. He has developed undergraduate and graduate courses in classical fluid dynamics as well as the fluid dynamics of turbomachines.

He is a licensed professional engineer. His professional society activities include having been a vice president of The American Society of Mechanical Engineers (ASME) and of the American Society for Engineering Education. He is a Life Fellow of The American Society of Mechanical Engineers and past editor of its *Journal of Turbomachinery*. He was recently honored with the ASME R. Tom Sawyer Award.

Wade W. Huebsch, Associate Professor in the Department of Mechanical and Aerospace Engineering at West Virginia University, received his B.S. degree in aerospace engineering from San Jose State University where he played college baseball. He received his M.S. degree in mechanical engineering and his Ph.D. in aerospace engineering from Iowa State University in 2000.

Dr. Huebsch specializes in computational fluid dynamics research and has authored multiple journal articles in the areas of aircraft icing, roughness-induced flow phenomena, and boundary layer flow control. He has taught both undergraduate and graduate courses in fluid mechanics and has developed a new undergraduate course in computational fluid dynamics. He has received multiple teaching awards such as Outstanding Teacher and Teacher of the Year from the College of Engineering and Mineral Resources at WVU as well as the Ralph R. Teetor Educational Award from SAE. He was also named as the Young Researcher of the Year from WVU. He is a member of the American Institute of Aeronautics and Astronautics, the Sigma Xi research society, the Society of Automotive Engineers, and the American Society of Engineering Education.

Preface

This book is intended for junior and senior engineering students who are interested in learning some fundamental aspects of fluid mechanics. We developed this text to be used as a first course. The principles considered are classical and have been well-established for many years. However, fluid mechanics education has improved with experience in the classroom, and we have brought to bear in this book our own ideas about the teaching of this interesting and important subject. This sixth edition has been prepared after several years of experience by the authors using the previous editions for introductory courses in fluid mechanics. On the basis of this experience, along with suggestions from reviewers, colleagues, and students, we have made a number of changes in this edition. The changes (listed below, and indicated by the word *New* in descriptions in this preface) are made to clarify, update, and expand certain ideas and concepts.

New to This Edition

In addition to the continual effort of updating the scope of the material presented and improving the presentation of all of the material, the following items are new to this edition.

With the wide-spread use of new technologies involving the web, DVDs, digital cameras and the like, there is an increasing use and appreciation of the variety of visual tools available for learning. This fact has been addressed in the new edition by the inclusion of numerous new illustrations, graphs, photographs, and videos.

Illustrations: The book contains more than 260 *new* illustrations and graphs. These illustrations range from simple ones that help illustrate a basic concept or equation to more complex ones that illustrate practical applications of fluid mechanics in our everyday lives.

Photographs: The book contains more than 256 *new* photographs. Some photos involve situations that are so common to us that we probably never stop to realize how fluids are involved in them. Others involve new and novel situations that are still baffling to us. The photos are also used to help the reader better understand the basic concepts and examples discussed.

Videos: The video library for the book has been significantly enhanced by the addition of 80 *new* video segments directly related to the text material. They illustrate many of the interesting and practical applications of real-world fluid phenomena. There are now 159 videos.

Examples: All of the examples are *newly* outlined and carried out with the problem solving method of "Given, Find, Solution, and Comment."

Learning objectives: Each chapter begins with a set of learning objectives. This *new* feature provides the student with a brief preview of the topics covered in the chapter.

List of equations: Each chapter ends with a *new* summary of the most important equations in the chapter.

Problems: Approximately 30% *new* homework problems have been added for this edition. They are all *newly* grouped and identified according to topic. Typically, the first few problems in each group are relatively easy ones. In many groups of problems there are one or two *new* problems in which the student is asked to find a photograph/image of a particular flow situation and write a paragraph describing it. Each chapter contains *new* Life Long Learning Problems (i.e., one aspect of the life long learning as interpreted by the authors) that ask the student to obtain information about a given, new flow concept and to write a brief report about it.

Fundamentals of Engineering Exam: A set of FE exam questions is *newly* available on the book web site.

Key Features

Illustrations, Photographs, and Videos

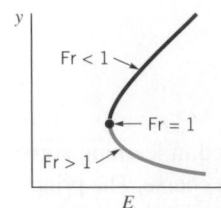

Fluid mechanics has always been a "visual" subject—much can be learned by viewing various aspects of fluid flow. In this new edition we have made several changes to reflect the fact that with new advances in technology, this visual component is becoming easier to incorporate into the learning environment, for both access and delivery, and is an important component to the learning of fluid mechanics. Thus, approximately 516 *new* photographs and illustrations have been added to the book. Some of these are within the text material; some are used to enhance the example problems; and some are included as margin figures of the type shown in the left margin to more clearly illustrate various points discussed in the text. In addition, 80 *new* video segments have been added, bringing the total number of video segments to 159. These video segments illustrate many interesting and practical applications of real-world fluid phenomena. Many involve *new* CFD (computational fluid dynamics) material. Each video segment is identified at the appropriate location in the text material by a video icon and thumbnail photograph of the type shown in the left margin. Each video segment has a separate associated text description of what is shown in the video. There are approximately 160 homework problems that are directly related to the topics in the videos.

V1.5 Floating Razor Blade

Examples

One of our aims is to represent fluid mechanics as it really is—an exciting and useful discipline. To this end, we include analyses of numerous everyday examples of fluid-flow phenomena to which students and faculty can easily relate. In the sixth edition 163 examples are presented that provide detailed solutions to a variety of problems. Many of the examples have been *newly* extended to illustrate what happens if one or more of the parameters is changed. This gives the user a better feel for some of the basic principles involved. In addition, many of the examples contain *new* photographs of the actual device or item involved in the example. Also, all of the examples are *newly* outlined and carried out with the problem solving methodology of "Given, Find, Solution, and Comment" as discussed on page 5 in the "Note to User" before Example 1.1.

Fluids in the News

The set of approximately 60 short "Fluids in the News" stories has been *newly* updated to reflect some of the latest important, and novel ways that fluid mechanics affects our lives. Many of these problems have homework problems associated with them.

Homework Problems

A set of more than 1330 homework problems (approximately 30% *new* to this edition) stresses the practical application of principles. The problems are *newly* grouped and identified according to topic. An effort has been made to include several *new*, easier problems at the start of each group. The following types of problems are included:

1) "standard" problems,
2) computer problems,
3) discussion problems,
4) supply-your-own-data problems,
5) review problems with solutions,
6) problems based on the "Fluids in the News" topics,
7) problems based on the fluid videos,
8) Excel-based lab problems,
9) *new* "Life long learning" problems,
10) *new* problems that require the user to obtain a photograph/image of a given flow situation and write a brief paragraph to describe it,
11) simple CFD problems to be solved using FlowLab,
12) *new* Fundamental of Engineering (FE) exam questions available on book web site.

Lab Problems—There are 30 extended, laboratory-type problems that involve actual experimental data for simple experiments of the type that are often found in the laboratory portion of many introductory fluid mechanics courses. The data for these problems are provided in Excel format.

Life Long Learning Problems—There are more than 40 *new* life long learning problems that involve obtaining additional information about various new state-of-the-art fluid mechanics topics and writing a brief report about this material.

Review Problems—There is a set of 186 review problems covering most of the main topics in the book.

Well-Paced Concept and Problem-Solving Development

Since this is an introductory text, we have designed the presentation of material to allow for the gradual development of student confidence in fluid problem solving. Each important concept or notion is considered in terms of simple and easy-to-understand circumstances before more complicated features are introduced. Each page contains a brief summary (a highlight) sentence that serves to prepare or remind the reader about an important concept discussed on that page.

Several brief components have been added to each chapter to help the user obtain the "big picture" idea of what key knowledge is to be gained from the chapter. A *new* brief Learning Objectives section is provided at the beginning of each chapter. It is helpful to read through this list prior to reading the chapter to gain a preview of the main concepts presented. Upon completion of the chapter, it is beneficial to look back at the original learning objectives to ensure that a satisfactory level of understanding has been acquired for each item. Additional reinforcement of these learning objectives is provided in the form of a Chapter Summary and Study Guide at the end of each chapter. In this section a brief summary of the key concepts and principles introduced in the chapter is included along with a listing of important terms with which the student should be familiar. These terms are highlighted in the text. A *new* list of the main equations in the chapter is included in the chapter summary.

Topical Organization

In the first four chapters the student is made aware of some fundamental aspects of fluid motion, including important fluid properties, regimes of flow, pressure variations in fluids at rest and in motion, fluid kinematics, and methods of flow description and analysis. The Bernoulli equation is introduced in Chapter 3 to draw attention, early on, to some of the interesting effects of fluid motion on the distribution of pressure in a flow field. We believe that this timely consideration of elementary fluid dynamics increases student enthusiasm for the more complicated material that follows. In Chapter 4 we convey the essential elements of kinematics, including Eulerian and Lagrangian mathematical descriptions of flow phenomena, and indicate the vital relationship between the two views. For teachers who wish to consider kinematics in detail before the material on elementary fluid dynamics, Chapters 3 and 4 can be interchanged without loss of continuity.

Chapters 5, 6, and 7 expand on the basic analysis methods generally used to solve or to begin solving fluid mechanics problems. Emphasis is placed on understanding how flow phenomena are described mathematically and on when and how to use infinitesimal and finite control volumes. The effects of fluid friction on pressure and velocity distributions are also considered in some detail. A formal course in thermodynamics is not required to understand the various portions of the text that consider some elementary aspects of the thermodynamics of fluid flow. Chapter 7 features the advantages of using dimensional analysis and similitude for organizing test data and for planning experiments and the basic techniques involved.

Owing to the growing importance of computational fluid dynamics (CFD) in engineering design and analysis, material on this subject is included in Appendix A. This material may be omitted without any loss of continuity to the rest of the text. This introductory CFD overview includes examples and problems of various interesting flow situations that are to be solved using FlowLab software.

Chapters 8 through 12 offer students opportunities for the further application of the principles learned early in the text. Also, where appropriate, additional important notions such as boundary layers, transition from laminar to turbulent flow, turbulence modeling, and flow separation are introduced. Practical concerns such as pipe flow, open-channel flow, flow measurement, drag and lift, the effects of compressibility, and the fluid mechanics fundamentals associated with turbomachines are included.

Students who study this text and who solve a representative set of the exercises provided should acquire a useful knowledge of the fundamentals of fluid mechanics. Faculty who use this text are provided with numerous topics to select from in order to meet the objectives of their own courses. More material is included than can be reasonably covered in one term. All are reminded of the fine collection of supplementary material. We have cited throughout the text various articles and books that are available for enrichment.

Student and Instructor Resources

Student Companion Site—The student section of the book website at www.wiley.com/go/global/munson contains the assets listed below. Access is free-of-charge with the registration code included in the front of every new book.

Video Library	CFD Driven Cavity Example
Review Problems with Answers	FlowLab Tutorial and User's Guide
Lab Problems	FlowLab Problems
Comprehensive Table of Conversion Factors	

Instructor Companion Site—The instructor section of the book website at www.wiley.com/go/global/munson contains the assets in the Student Companion Site, as well as the following, which are available only to professors who adopt this book for classroom use:

- Instructor Solutions Manual, containing complete, detailed solutions to all of the problems in the text.
- Figures from the text, appropriate for use in lecture slides.

These instructor materials are password-protected. Visit the Instructor Companion Site to register for a password.

Acknowledgments

We express our thanks to the many colleagues who have helped in the development of this text, including:

Donald Gray of West Virginia University for help with Chapter 10;

Bruce Reichert for help with Chapter 11;

Patrick Kavanagh of Iowa State University;

Dave Japiske of Concepts NREC for help with Chapter 12;

Bud Homsy for permission to use many of the new video segments.

We wish to express our gratitude to the many persons who supplied the photographs used throughout the text and to the many persons who provided suggestions for this and previous editions through reviews and surveys. In addition, we wish to express our thanks to the reviewers and contributors of the *WileyPLUS* course:

David Benson, Kettering University

Andrew Gerhart, Lawrence Technological University

Philip Gerhart, University of Evansville

Alison Griffin, University of Central Florida

Jay Martin, University of Wisconsin—Madison

John Mitchell, University of Wisconsin—Madison

Pierre Sullivan, University of Toronto

Mary Wolverton, Mississippi State University

Finally, we thank our families for their continued encouragement during the writing of this sixth edition.

Working with students over the years has taught us much about fluid mechanics education. We have tried in earnest to draw from this experience for the benefit of users of this book. Obviously we are still learning and we welcome any suggestions and comments from you.

BRUCE R. MUNSON

DONALD F. YOUNG

THEODORE H. OKIISHI

WADE W. HUEBSCH

*F*eatured in this Book

FLUIDS IN THE NEWS

Throughout the book are many brief news stories involving current, sometimes novel, applications of fluid phenomena. Many of these stories have homework problems associated with them.

F l u i d s   i n   t h e   N e w s

Weather, barometers, and bars One of the most important indicators of weather conditions is *atmospheric pressure*. In general, a falling or low pressure indicates bad weather; rising or high pressure, good weather. During the evening TV weather report in the United States, atmospheric pressure is given as so many inches (commonly around 30 in.). This value is actually the height of the mercury column in a mercury *barometer* adjusted to sea level. To determine the true atmospheric pressure at a particular location, the elevation relative to sea level must be known. Another unit used by meteorologists to indicate atmospheric pressure is the *bar*, first used in weather reporting in 1914; and defined as 10^5 N/m². The definition of a bar is probably related to the fact that standard sea-level pressure is 1.0133×10^5 N/m², that is, only slightly larger than one bar. For typical weather patterns, "sea-level equivalent" atmospheric pressure remains close to one bar. However, for extreme weather conditions associated with tornadoes, hurricanes, or typhoons, dramatic changes can occur. The lowest atmospheric sea-level pressure ever recorded was associated with a typhoon, Typhoon Tip, in the Pacific Ocean on October 12, 1979. The value was 0.870 bars (25.8 in. Hg). (See Problem 2.19.)

SUMMARY SENTENCES

A brief summary sentence is given on each page to prepare or remind the reader about an important concept discussed on that page.

2.6   Manometry

Manometers use vertical or inclined liquid columns to measure pressure.

A standard technique for measuring pressure involves the use of liquid columns in vertical or inclined tubes. Pressure measuring devices based on this technique are called *manometers*. The mercury barometer is an example of one type of manometer, but there are many other configurations possible, depending on the particular application. Three common types of manometers include the piezometer tube, the U-tube manometer, and the inclined-tube manometer.

PHOTOGRAPHS AMD ILLUSTRATIONS

More than 515 new photographs and illustrations have been added to help illustrate various concepts in the text.

(Photograph courtesy of Cameron Balloons.)

2.11.1 Archimedes' Principle

When a stationary body is completely submerged in a fluid (such as the hot air balloon shown in the figure in the margin), or floating so that it is only partially submerged, the resultant fluid force acting on the body is called the *buoyant force*. A net upward vertical force results because pressure increases with depth and the pressure forces acting from below are larger than the pressure forces acting from above. This force can be determined through an approach similar to that used in the previous section for forces on curved surfaces. Consider a body of arbitrary shape, having a volume $\forall$, that is immersed in a fluid as illustrated in Fig. 2.24*a*. We enclose the body in a parallelepiped and draw a free-body diagram of the parallelepiped with the body removed as shown in Fig. 2.24*b*. Note that the forces F_1, F_2, F_3, and F_4 are simply the forces exerted on the plane surfaces of the parallelepiped (for simplicity the forces in the *x* direction are not shown), $\mathcal{W}$ is the weight of the shaded fluid volume (parallelepiped minus body), and F_B is the force the body is exerting *on the fluid*. The forces on the vertical surfaces, such as F_3 and F_4, are all equal and cancel, so the equilibrium equation of interest is in the *z* direction and can be expressed as

$$F_B = F_2 - F_1 - \mathcal{W} \qquad (2.21)$$

If the specific weight of the fluid is constant, then

$$F_2 - F_1 = \gamma(h_2 - h_1)A$$

where A is the horizontal area of the upper (or lower) surface of the parallelepiped, and Eq. 2.21 can be written as

$$F_B = \gamma(h_2 - h_1)A - \gamma[(h_2 - h_1)A - \forall]$$

Simplifying, we arrive at the desired expression for the buoyant force

$$\boxed{F_B = \gamma \forall} \qquad (2.22)$$

FLUID VIDEOS

A set of 159 videos illustrating interesting and practical applications of fluid phenomena is provided on the book website. An icon in the margin identifies each video. Approximately 160 homework problems are tied to the videos.

V2.6 Atmospheric buoyancy

BOXED EQUATIONS

Important equations are boxed to help the user identify them.

2.13 Chapter Summary and Study Guide

In this chapter the pressure variation in a fluid at rest is considered, along with some important consequences of this type of pressure variation. It is shown that for incompressible fluids at rest the pressure varies linearly with depth. This type of variation is commonly referred to as hydrostatic pressure distribution. For compressible fluids at rest the pressure distribution will not generally be hydrostatic, but Eq. 2.4 remains valid and can be used to determine the pressure distribution if additional information about the variation of the specific weight is specified. The distinction between absolute and gage pressure is discussed along with a consideration of barometers for the measurement of atmospheric pressure.

Pressure measuring devices called manometers, which utilize static liquid columns, are analyzed in detail. A brief discussion of mechanical and electronic pressure gages is also included. Equations for determining the magnitude and location of the resultant fluid force acting on a plane surface in contact with a static fluid are developed. A general approach for determining the magnitude and location of the resultant fluid force acting on a curved surface in contact with a static fluid is described. For submerged or floating bodies the concept of the buoyant force and the use of Archimedes' principle are reviewed.

The following checklist provides a study guide for this chapter. When your study of the entire chapter and end-of-chapter exercises has been completed you should be able to

Pascal's law
surface force
body force
incompressible fluid
hydrostatic pressure
*   distribution*
pressure head
compressible fluid
U.S. standard
*   atmosphere*
absolute pressure
gage pressure
vacuum pressure
barometer
manometer
Bourdon pressure
*   gage*
center of pressure
buoyant force
Archimedes' principle
center of buoyancy

■ write out meanings of the terms listed here in the margin and understand each of the related concepts. These terms are particularly important and are set in *italic, bold, and color* type in the text.

■ calculate the pressure at various locations within an incompressible fluid at rest.

■ calculate the pressure at various locations within a compressible fluid at rest using Eq. 2.4 if the variation in the specific weight is specified.

■ use the concept of a hydrostatic pressure distribution to determine pressures from measurements using various types of manometers.

■ determine the magnitude, direction, and location of the resultant hydrostatic force acting on a plane surface.

CHAPTER SUMMARY AND STUDY GUIDE

At the end of each chapter is a brief summary of key concepts and principles introduced in the chapter along with key terms and a summary of key equations involved.

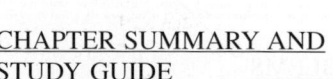

LEARNING OBJECTIVES

At the beginning of each chapter is a set of learning objectives that provides the student a preview of topics covered in the chapter.

EXAMPLE PROBLEMS

A set of example problems provides the student detailed solutions and comments for interesting, real-world situations.

2 Fluid Statics

CHAPTER OPENING PHOTO: *Floating iceberg:* An iceberg is a large piece of fresh water ice that originated as snow in a glacier or ice shelf and then broke off to float in the ocean. Although the fresh water ice is lighter than the salt water in the ocean, the difference in densities is relatively small. Hence, only about one ninth of the volume of an iceberg protrudes above the ocean's surface, so that what we see floating is literally "just the tip of the iceberg." (Photograph courtesy of Corbis Digital Stock/Corbis Images)

Learning Objectives

After completing this chapter, you should be able to:

- determine the pressure at various locations in a fluid at rest.
- explain the concept of manometers and apply appropriate equations to determine pressures.
- calculate the hydrostatic pressure force on a plane or curved submerged surface.
- calculate the buoyant force and discuss the stability of floating or submerged objects.

In this chapter we will consider an important class of problems in which the fluid is either at rest or moving in such a manner that there is no relative motion between adjacent particles. In both instances there will be no shearing stresses in the fluid, and the only forces that develop on the surfaces of the particles will be due to the pressure. Thus, our principal concern is to investigate pressure and its variation throughout a fluid and the effect of pressure on submerged surfaces. The absence of shearing stresses greatly simplifies the analysis and, as we will see, allows us to obtain relatively simple solutions to many important practical problems.

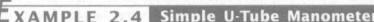

EXAMPLE 2.4 Simple U-Tube Manometer

GIVEN A closed tank contains compressed air and oil ($SG_{oil} = 0.90$) as is shown in Fig. E2.4. A U-tube manometer using mercury ($SG_{Hg} = 13.6$) is connected to the tank as shown. The column heights are $h_1 = 36$ in., $h_2 = 6$ in., and $h_3 = 9$ in.

FIND Determine the pressure reading (in psi) of the gage.

SOLUTION

Following the general procedure of starting at one end of the manometer system and working around to the other, we will start at the air–oil interface in the tank and proceed to the open end where the pressure is zero. The pressure at level (1) is

$$p_1 = p_{air} + \gamma_{oil}(h_1 + h_2)$$

This pressure is equal to the pressure at level (2), since these two points are at the same elevation in a homogeneous fluid at rest. As we move from level (2) to the open end, the pressure must decrease by $\gamma_{Hg}h_3$, and at the open end the pressure is zero. Thus, the manometer equation can be expressed as

$$p_{air} + \gamma_{oil}(h_1 + h_2) - \gamma_{Hg}h_3 = 0$$

or

$$p_{air} + (SG_{oil})(\gamma_{H_2O})(h_1 + h_2) - (SG_{Hg})(\gamma_{H_2O})h_3 = 0$$

For the values given

$$p_{air} = -(0.9)(62.4\ \text{lb/ft}^3)\left(\frac{36 + 6}{12}\ \text{ft}\right)$$
$$+ (13.6)(62.4\ \text{lb/ft}^3)\left(\frac{9}{12}\ \text{ft}\right)$$

so that

$$p_{air} = 440\ \text{lb/ft}^2$$

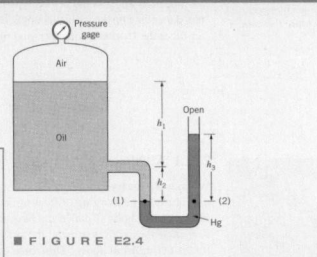

■ FIGURE E2.4

Since the specific weight of the air above the oil is much smaller than the specific weight of the oil, the gage should read the pressure we have calculated; that is,

$$p_{gage} = \frac{440\ \text{lb/ft}^2}{144\ \text{in.}^2/\text{ft}^2} = 3.06\ \text{psi} \qquad (\text{Ans})$$

COMMENTS Note that the air pressure is a function of the height of the mercury in the manometer and the depth of the oil (both in the tank and in the tube). It is not just the mercury in the manometer that is important.

Assume that the gage pressure remains at 3.06 psi, but the manometer is altered so that it contains only oil. That is, the mercury is replaced by oil. A simple calculation shows that in this case the vertical oil-filled tube would need to be $h_3 = 11.3$ ft tall, rather than the original $h_3 = 9$ in. There is an obvious advantage of using a heavy fluid such as mercury in manometers.

REVIEW PROBLEMS

On the book web site are nearly 200 Review Problems covering most of the main topics in the book.

2.111 An open container of oil rests on the flatbed of a truck that is traveling along a horizontal road at 55 mi/hr. As the truck slows uniformly to a complete stop in 5 s, what will be the slope of the oil surface during the period of constant deceleration?

2.112 A 5-gal, cylindrical open container with a bottom area of 120 in.^{2} is filled with glycerin and rests on the floor of an elevator. (a) Determine the fluid pressure at the bottom of the container when the elevator has an upward acceleration of 3 ft/s^2. (b) What resultant force does the container exert on the floor of the elevator during this acceleration? The weight of the container is negligible. (Note: 1 gal = 231 in.^{3})

2.113 An open rectangular tank 1 m wide and 2 m long contains gasoline to a depth of 1 m. If the height of the tank sides is 1.5 m, what is the maximum horizontal acceleration (along the long axis of the tank) that can develop before the gasoline would begin to spill?

2.114 If the tank of Problem 2.113 slides down a frictionless plane that is inclined at 30° with the horizontal, determine the angle the free surface makes with the horizontal.

2.115 A closed cylindrical tank that is 8 ft in diameter and 24 ft long is completely filled with gasoline. The tank, with its long axis horizontal, is pulled by a truck along a horizontal surface. Determine the pressure difference between the ends (along the long axis of the tank) when the truck undergoes an acceleration of 5 ft/s^2.

2.121 (See Fluids in the News article titled "Rotating mercury mirror telescope," Section 2.12.2.) The largest liquid mirror telescope uses a 6-ft-diameter tank of mercury rotating at 7 rpm to produce its parabolic-shaped mirror as shown in Fig. P2.121. Determine the difference in elevation of the mercury, Δh, between the edge and the center of the mirror.

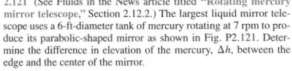

■ FIGURE P2.121

■ **Lab Problems**

2.122 This problem involves the force needed to open a gate that covers an opening in the side of a water-filled tank. To proceed with this problem, go to Appendix H which is located on the book's web site, www.wiley.com/college/munson.

LAB PROBLEMS

On the book website is a set of lab problems in Excel format involving actual data for experiments of the type found in many introductory fluid mechanics labs.

xiv

Review Problems

Go to Appendix G for a set of review problems with answers. Detailed solutions can be found in *Student Solution Manual and Study Guide for Fundamentals of Fluid Mechanics,* by Munson et al. (© 2009 John Wiley and Sons, Inc.).

Problems

Note: Unless otherwise indicated, use the values of fluid properties found in the tables on the inside of the front cover. Problems designated with an (*) are intended to be solved with the aid of a programmable calculator or a computer. Problems designated with a (†) are "open-ended" problems and require critical thinking in that to work them one must make various assumptions and provide the necessary data. There is not a unique answer to these problems.

Answers to the even-numbered problems are listed at the end of the book. Access to the videos that accompany problems can be obtained through the book's web site, www.wiley.com/college/munson. The lab-type problems can also be accessed on this web site.

Section 3.2 F = ma along a Streamline

3.1 Obtain a photograph/image of a situation which can be analyzed by use of the Bernoulli equation. Print this photo and write a brief paragraph that describes the situation involved.

3.2 Air flows steadily along a streamline from point (1) to point (2) with negligible viscous effects. The following conditions are measured: At point (1) $z_1 = 2$ m and $p_1 = 0$ kPa; at point (2) $z_2 = 10$ m, $p_2 = 20$ N/m^2, and $V_2 = 0$. Determine the velocity at point (1).

front of the object and V_0 is the upstream velocity. (a) Determine the pressure gradient along this streamline. (b) If the upstream pressure is p_0, integrate the pressure gradient to obtain the pressure $p(x)$ for $-\infty \le x \le -a$. (c) Show from the result of part (b) that the pressure at the stagnation point ($x = -a$) is $p_0 + \rho V_0^2/2$, as expected from the Bernoulli equation.

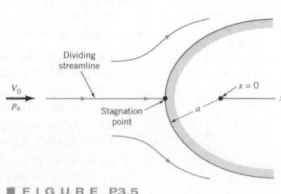

■ FIGURE P3.5

HOMEWORK PROBLEMS

Homework problems at the end of each chapter stress the practical applications of fluid mechanics principles. Over 1350 homework problems are included.

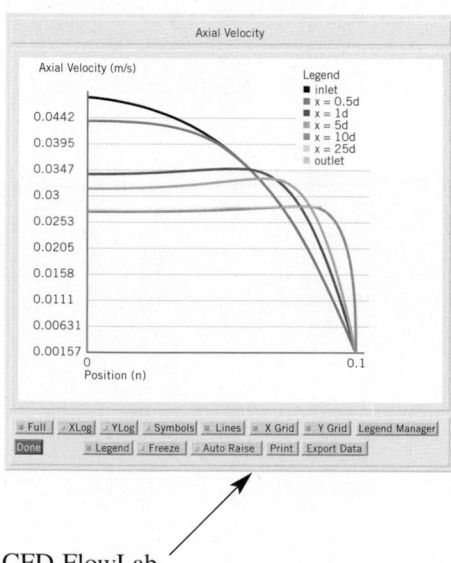

CFD FlowLab

For those who wish to become familiar with the basic concepts of computational fluid dynamics, a new overview to CFD is provided in Appendices A and I. In addition, the use of FlowLab software to solve interesting flow problems is described in Appendices J and K.

5.118   Water flows by gravity from one lake to another as sketched in Fig. P5.118 at the steady rate of 80 gpm. What is the loss in available energy associated with this flow? If this same amount of loss is associated with pumping the fluid from the lower lake to the higher one at the same flowrate, estimate the amount of pumping power required.

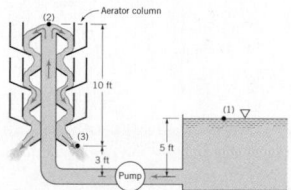

■ **F I G U R E   P5.118**

5.119   Water is pumped from a tank, point (1), to the top of a water plant aerator, point (2), as shown in Video V5.14 and Fig. P5.119 at a rate of 3.0 ft³/s. (a) Determine the power that the pump adds to the water if the head loss from (1) to (2) where $V_2 = 0$ is 4 ft. (b) Determine the head loss from (2) to the bottom of the aerator column, point (3), if the average velocity at (3) is $V_3 = 2$ ft/s.

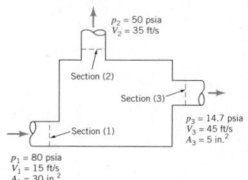

■ **F I G U R E   P5.119**

5.120   A liquid enters a fluid machine at section (1) and leaves at sections (2) and (3) as shown in Fig. P5.120. The density of the fluid is constant at 2 slugs/ft³. All of the flow occurs in a horizontal plane and is frictionless and adiabatic. For the above-mentioned and additional conditions indicated in Fig. P5.120, determine the amount of shaft power involved.

$p_2 = 50$ psia
$V_2 = 35$ ft/s

Section (2)

Section (3)

$p_3 = 14.7$ psia
$V_3 = 45$ ft/s
$A_3 = 5$ in.²

Section (1)

$p_1 = 80$ psia
$V_1 = 15$ ft/s
$A_1 = 30$ in.²

■ **F I G U R E   P5.120**

5.121   Water is to be moved from one large reservoir to another at a higher elevation as indicated in Fig. P5.121. The loss of available

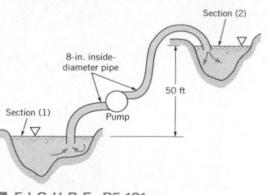

■ **F I G U R E   P5.121**

energy associated with 2.5 ft³/s being pumped from sections (1) to (2) is loss $= 61\bar{V}^2/2$ ft²/s², where $\bar{V}$ is the average velocity of water in the 8-in. inside diameter piping involved. Determine the amount of shaft power required.

5.122   Water is to be pumped from the large tank shown in Fig. P5.122 with an exit velocity of 6 m/s. It was determined that the original pump (pump 1) that supplies 1 kW of power to the water did not produce the desired velocity. Hence, it is proposed that an additional pump (pump 2) be installed as indicated to increase the flowrate to the desired value. How much power must pump 2 add to the water? The head loss for this flow is $h_L = 250Q^2$, where h_L is in m when Q is in m³/s.

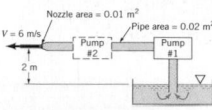

■ **F I G U R E   P5.122**

5.123   (See Fluids in the News article titled "Curtain of air," Section 5.3.3.) The fan shown in Fig. P5.123 produces an air curtain to separate a loading dock from a cold storage room. The air curtain is a jet of air 10 ft wide, 0.5 ft thick moving with speed $V = 30$ ft/s. The loss associated with this flow is loss $= K_L V^2/2$, where $K_L = 5$. How much power must the fan supply to the air to produce this flow?

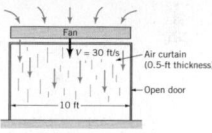

■ **F I G U R E   P5.123**

Section 5.3.2   Application of the Energy Equation— Combined with Linear momentum

5.124   If a $\frac{3}{4}$-hp motor is required by a ventilating fan to produce a 24-in. stream of air having a velocity of 40 ft/s as shown in Fig. P5.124, estimate (a) the efficiency of the fan and (b) the thrust of the supporting member on the conduit enclosing the fan.

5.125   Air flows past an object in a pipe of 2-m diameter and exits as a free jet as shown in Fig. P5.125. The velocity and pressure upstream are uniform at 10 m/s and 50 N/m², respectively. At the

Contents

4
FLUID KINEMATICS   147

5
FINITE CONTROL VOLUME
ANALYSIS   187

6
DIFFERENTIAL ANALYSIS OF
FLUID FLOW   263

A

COMPUTATIONAL FLUID DYNAMICS AND FLOWLAB

B

PHYSICAL PROPERTIES OF FLUIDS

C

PROPERTIES OF THE U.S. STANDARD ATMOSPHERE

D

COMPRESSIBLE FLOW DATA FOR AN IDEAL GAS

ONLINE APPENDIX LIST

E

COMPREHENSIVE TABLE OF CONVERSION FACTORS
See book web site, www.wiley.com/ go/global/munson, for this material.

F

VIDEO LIBRARY
See book web site, www.wiley.com/ go/global/munson, for this material.

G

REVIEW PROBLEMS
See book web site, www.wiley.com/ go/global/munson, for this material.

H

LABORATORY PROBLEMS
See book web site, www.wiley.com/ go/global/munson, for this material.

I

CFD DRIVEN CAVITY EXAMPLE
See book web site, www.wiley.com/ go/global/munson, for this material.

J

FLOWLAB TUTORIAL AND USER'S GUIDE
See book web site, www.wiley.com/ go/global/munson, for this material.

K

FLOWLAB PROBLEMS
See book web site, www.wiley.com/ go/global/munson, for this material.

ANSWERS

INDEX

VIDEO INDEX

1 Introduction

CHAPTER OPENING PHOTO: The nature of air bubbles rising in a liquid is a function of fluid properties such as density, viscosity, and surface tension. (Left: air in oil; right: air in soap.) (Photographs copyright 2007 by Andrew Davidhazy, Rochester Institute of Technology.)

Learning Objectives

After completing this chapter, you should be able to:

- determine the dimensions and units of physical quantities.
- identify the key fluid properties used in the analysis of fluid behavior.
- calculate common fluid properties given appropriate information.
- explain effects of fluid compressibility.
- use the concepts of viscosity, vapor pressure, and surface tension.

Fluid mechanics is that discipline within the broad field of applied mechanics that is concerned with the behavior of liquids and gases at rest or in motion. It covers a vast array of phenomena that occur in nature (with or without human intervention), in biology, and in numerous engineered, invented, or manufactured situations. There are few aspects of our lives that do not involve fluids, either directly or indirectly.

1

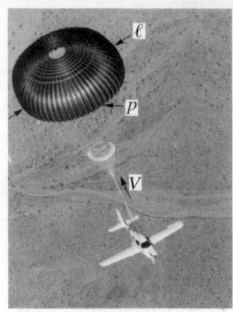

(Photo courtesy of CIR-RUS Design Corporation.)

V1.1 Mt. St. Helens Eruption

The immense range of different flow conditions is mind-boggling and strongly dependent on the value of the numerous parameters that describe fluid flow. Among the long list of parameters involved are (1) the physical size of the flow, ℓ; (2) the speed of the flow, V; and (3) the pressure, p, as indicated in the figure in the margin for a light aircraft parachute recovery system. These are just three of the important parameters which, along with many others, are discussed in detail in various sections of this book. To get an inkling of the range of some of the parameter values involved and the flow situations generated, consider the following.

■ Size, ℓ

Every flow has a characteristic (or typical) length associated with it. For example, for flow of fluid within pipes, the pipe diameter is a characteristic length. Pipe flows include the flow of water in the pipes in our homes, the blood flow in our arteries and veins, and the air flow in our bronchial tree. They also involve pipe sizes that are not within our everyday experiences. Such examples include the flow of oil across Alaska through a 1.2 m diameter, 1286 km-long pipe, and, at the other end of the size scale, the new area of interest involving flow in nano-scale pipes whose diameters are on the order of 10^{-8} m. Each of these pipe flows has important characteristics that are not found in the others.

Characteristic lengths of some other flows are shown in Fig. 1.1a.

■ Speed, V

As we note from The Weather Channel, on a given day the wind speed may cover what we think of as a wide range, from a gentle 8 km/h breeze to a 160 km/h hurricane or a

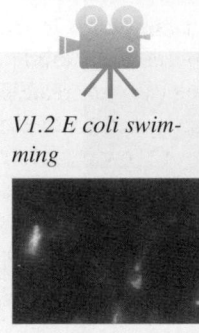

V1.2 E coli swimming

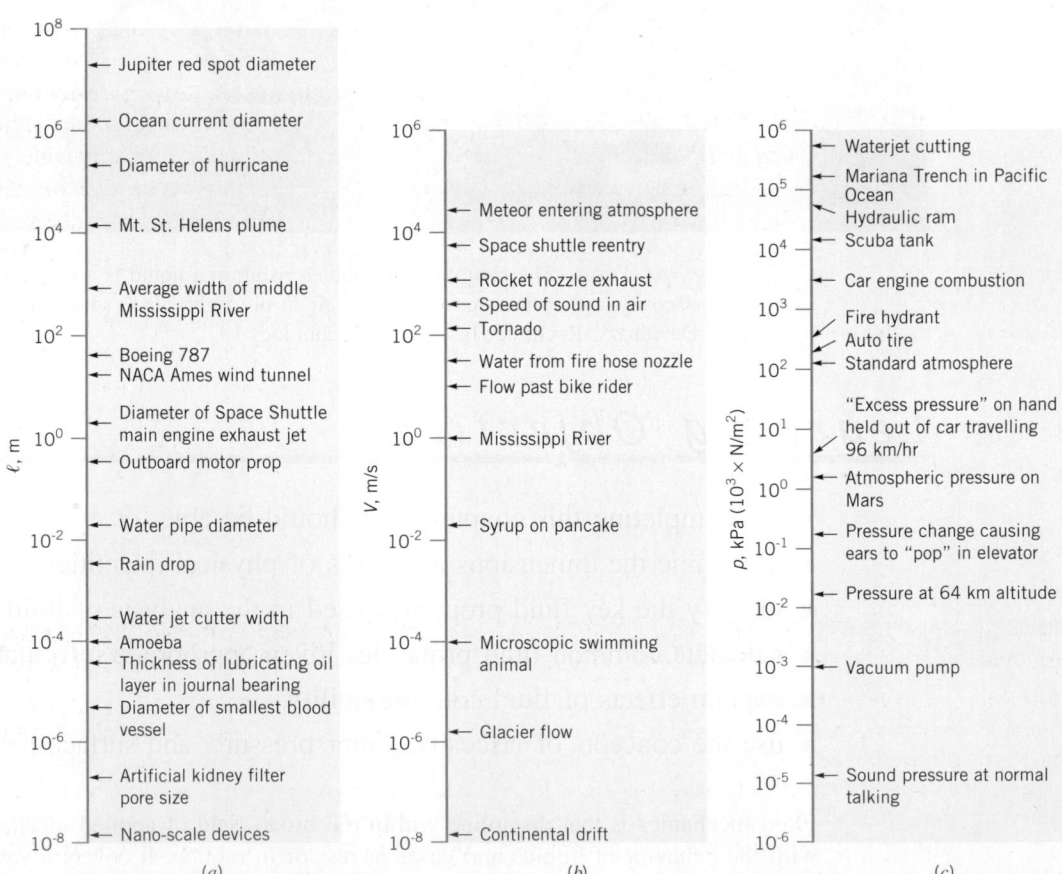

■ **FIGURE 1.1** **Characteristic values of some fluid flow parameters for a variety of flows. (a) Object size, (b) fluid speed, (c) fluid pressure.**

400 km/h tornado. However, this speed range is small compared to that of the almost imperceptible flow of the fluid-like magma below the earth's surface which drives the motion of the tectonic plates at a speed of about 2×10^{-8} m/s or the 3×10^4 m/s hypersonic air flow past a meteor as it streaks through the atmosphere.

Characteristic speeds of some other flows are shown in Fig. 1.1*b*.

■ Pressure, *p*

The pressure within fluids covers an extremely wide range of values. We are accustomed to the 241 kPa pressure within our car's tires, the "120 over 70" typical blood pressure reading, or the standard 101.3 kPa atmospheric pressure. However, the large 69 MPa pressure in the hydraulic ram of an earth mover or the tiny 1.4×10^{-5} kPa pressure of a sound wave generated at ordinary talking levels are not easy to comprehend.

Characteristic pressures of some other flows are shown in Fig. 1.1*c*.

The list of fluid mechanics applications goes on and on. But you get the point. Fluid mechanics is a very important, practical subject that encompasses a wide variety of situations. It is very likely that during your career as an engineer you will be involved in the analysis and design of systems that require a good understanding of fluid mechanics. Although it is not possible to adequately cover all of the important areas of fluid mechanics within one book, it is hoped that this introductory text will provide a sound foundation of the fundamental aspects of fluid mechanics.

1.1   Some Characteristics of Fluids

One of the first questions we need to explore is, What is a fluid? Or we might ask, What is the difference between a solid and a fluid? We have a general, vague idea of the difference. A solid is "hard" and not easily deformed, whereas a fluid is "soft" and is easily deformed (we can readily move through air). Although quite descriptive, these casual observations of the differences between solids and fluids are not very satisfactory from a scientific or engineering point of view. A closer look at the molecular structure of materials reveals that matter that we commonly think of as a solid (steel, concrete, etc.) has densely spaced molecules with large intermolecular cohesive forces that allow the solid to maintain its shape, and to not be easily deformed. However, for matter that we normally think of as a liquid (water, oil, etc.), the molecules are spaced farther apart, the intermolecular forces are smaller than for solids, and the molecules have more freedom of movement. Thus, liquids can be easily deformed (but not easily compressed) and can be poured into containers or forced through a tube. Gases (air, oxygen, etc.) have even greater molecular spacing and freedom of motion with negligible cohesive intermolecular forces and as a consequence are easily deformed (and compressed) and will completely fill the volume of any container in which they are placed. Both liquids and gases are fluids.

Both liquids and gases are fluids.

<hr>

F l u i d s   i n   t h e   N e w s

Will what works in air work in water? For the past few years a San Francisco company has been working on small, maneuverable submarines designed to travel through water using wings, controls, and thrusters that are similar to those on jet airplanes. After all, water (for submarines) and air (for airplanes) are both fluids, so it is expected that many of the principles governing the flight of airplanes should carry over to the "flight" of winged submarines. Of course, there are differences. For example, the submarine must be designed to withstand external pressures of nearly 700 pounds per square inch (4826 kPa) greater than that inside the vehicle. On the other hand, at high altitude where commercial jets fly, the exterior pressure is 3.5 psi (24 kPa) rather than standard sea level pressure of 14.7 psi (101.3 kPa), so the vehicle must be pressurized internally for passenger comfort. In both cases, however, the design of the craft for minimal drag, maximum lift, and efficient thrust is governed by the same fluid dynamic concepts.

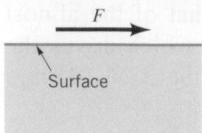

Surface

Although the differences between solids and fluids can be explained qualitatively on the basis of molecular structure, a more specific distinction is based on how they deform under the action of an external load. Specifically, *a fluid is defined as a substance that deforms continuously when acted on by a shearing stress of any magnitude.* A shearing stress (force per unit area) is created whenever a tangential force acts on a surface as shown by the figure in the margin. When common solids such as steel or other metals are acted on by a shearing stress, they will initially deform (usually a very small deformation), but they will not continuously deform (flow). However, common fluids such as water, oil, and air satisfy the definition of a fluid—that is, they will flow when acted on by a shearing stress. Some materials, such as slurries, tar, putty, toothpaste, and so on, are not easily classified since they will behave as a solid if the applied shearing stress is small, but if the stress exceeds some critical value, the substance will flow. The study of such materials is called *rheology* and does not fall within the province of classical fluid mechanics. Thus, all the fluids we will be concerned with in this text will conform to the definition of a fluid given previously.

Although the molecular structure of fluids is important in distinguishing one fluid from another, it is not yet practical to study the behavior of individual molecules when trying to describe the behavior of fluids at rest or in motion. Rather, we characterize the behavior by considering the average, or macroscopic, value of the quantity of interest, where the average is evaluated over a small volume containing a large number of molecules. Thus, when we say that the velocity at a certain point in a fluid is so much, we are really indicating the average velocity of the molecules in a small volume surrounding the point. The volume is small compared with the physical dimensions of the system of interest, but large compared with the average distance between molecules. Is this a reasonable way to describe the behavior of a fluid? The answer is generally yes, since the spacing between molecules is typically very small. For gases at normal pressures and temperatures, the spacing is on the order of 10^{-6} mm, and for liquids it is on the order of 10^{-7} mm. The number of molecules per cubic millimeter is on the order of 10^{18} for gases and 10^{21} for liquids. It is thus clear that the number of molecules in a very tiny volume is huge and the idea of using average values taken over this volume is certainly reasonable. We thus assume that all the fluid characteristics we are interested in (pressure, velocity, etc.) vary continuously throughout the fluid—that is, we treat the fluid as a *continuum.* This concept will certainly be valid for all the circumstances considered in this text. One area of fluid mechanics for which the continuum concept breaks down is in the study of rarefied gases such as would be encountered at very high altitudes. In this case the spacing between air molecules can become large and the continuum concept is no longer acceptable.

1.2 Dimensions, Dimensional Homogeneity, and Units

Since in our study of fluid mechanics we will be dealing with a variety of fluid characteristics, it is necessary to develop a system for describing these characteristics both *qualitatively* and *quantitatively.* The qualitative aspect serves to identify the nature, or type, of the characteristics (such as length, time, stress, and velocity), whereas the quantitative aspect provides a numerical measure of the characteristics. The quantitative description requires both a number and a standard by which various quantities can be compared. The standard for length is the meter, for time an hour or second, and for mass is the kilogram. Such standards are called *units,* and several systems of units are in common use as described in the following section. The qualitative description is conveniently given in terms of certain *primary quantities,* such as length, L, time, T, mass, M, and temperature, Θ. These primary quantities can then be used to provide a qualitative description of any other *secondary quantity*: for example, area $\doteq L^2$, velocity $\doteq LT^{-1}$, density $\doteq ML^{-3}$, and so on, where the symbol $\doteq$ is used to indicate the *dimensions* of the secondary quantity in terms of the primary quantities. Thus, to describe qualitatively a velocity, V, we would write

Fluid characteristics can be described qualitatively in terms of certain basic quantities such as length, time, and mass.

$$V \doteq LT^{-1}$$

■ **TABLE 1.1**
Dimensions Associated with Common Physical Quantities

	FLT System	MLT System		FLT System	MLT System
Acceleration	LT^{-2}	LT^{-2}	Power	FLT^{-1}	ML^2T^{-3}
Angle	$F^0L^0T^0$	$M^0L^0T^0$	Pressure	FL^{-2}	$ML^{-1}T^{-2}$
Angular acceleration	T^{-2}	T^{-2}	Specific heat	$L^2T^{-2}\Theta^{-1}$	$L^2T^{-2}\Theta^{-1}$
Angular velocity	T^{-1}	T^{-1}			
Area	L^2	L^2	Specific weight	FL^{-3}	$ML^{-2}T^{-2}$
			Strain	$F^0L^0T^0$	$M^0L^0T^0$
Density	$FL^{-4}T^2$	ML^{-3}	Stress	FL^{-2}	$ML^{-1}T^{-2}$
Energy	FL	ML^2T^{-2}	Surface tension	FL^{-1}	MT^{-2}
Force	F	MLT^{-2}	Temperature	Θ	Θ
Frequency	T^{-1}	T^{-1}			
Heat	FL	ML^2T^{-2}	Time	T	T
			Torque	FL	ML^2T^{-2}
Length	L	L	Velocity	LT^{-1}	LT^{-1}
Mass	$FL^{-1}T^2$	M	Viscosity (dynamic)	$FL^{-2}T$	$ML^{-1}T^{-1}$
Modulus of elasticity	FL^{-2}	$ML^{-1}T^{-2}$	Viscosity (kinematic)	L^2T^{-1}	L^2T^{-1}
Moment of a force	FL	ML^2T^{-2}			
Moment of inertia (area)	L^4	L^4	Volume	L^3	L^3
			Work	FL	ML^2T^{-2}
Moment of inertia (mass)	FLT^2	ML^2			
Momentum	FT	MLT^{-1}			

and say that "the dimensions of a velocity equal length divided by time." The primary quantities are also referred to as *basic dimensions.*

For a wide variety of problems involving fluid mechanics, only the three basic dimensions, L, T, and M are required. Alternatively, L, T, and F could be used, where F is the basic dimensions of force. Since Newton's law states that force is equal to mass times acceleration, it follows that $F \doteq MLT^{-2}$ or $M \doteq FL^{-1}T^2$. Thus, secondary quantities expressed in terms of M can be expressed in terms of F through the relationship above. For example, stress, σ, is a force per unit area, so that $\sigma \doteq FL^{-2}$, but an equivalent dimensional equation is $\sigma \doteq ML^{-1}T^{-2}$. Table 1.1 provides a list of dimensions for a number of common physical quantities.

All theoretically derived equations are *dimensionally homogeneous*—that is, the dimensions of the left side of the equation must be the same as those on the right side, and all additive separate terms must have the same dimensions. We accept as a fundamental premise that all equations describing physical phenomena must be dimensionally homogeneous. If this were not true, we would be attempting to equate or add unlike physical quantities, which would not make sense. For example, the equation for the velocity, V, of a uniformly accelerated body is

$$V = V_0 + at \tag{1.1}$$

where V_0 is the initial velocity, a the acceleration, and t the time interval. In terms of dimensions the equation is

$$LT^{-1} \doteq LT^{-1} + LT^{-1}$$

and thus Eq. 1.1 is dimensionally homogeneous.

Some equations that are known to be valid contain constants having dimensions. The equation for the distance, d, traveled by a freely falling body can be written as

$$d = 4.90t^2 \tag{1.2}$$

and a check of the dimensions reveals that the constant must have the dimensions of LT^{-2} if the equation is to be dimensionally homogeneous. Actually, Eq. 1.2 is a special form of the well-known equation from physics for freely falling bodies,

$$d = \frac{gt^2}{2} \tag{1.3}$$

in which g is the acceleration of gravity. Equation 1.3 is dimensionally homogeneous and valid in any system of units. For $g = 9.81 \text{ m/s}^2$ the equation reduces to Eq. 1.2 and thus Eq. 1.2 is valid only for the system of units using meter and seconds. Equations that are restricted to a particular system of units can be denoted as *restricted homogeneous equations,* as opposed to equations valid in any system of units, which are *general homogeneous equations.* The preceding discussion indicates one rather elementary, but important, use of the concept of dimensions: the determination of one aspect of the generality of a given equation simply based on a consideration of the dimensions of the various terms in the equation. The concept of dimensions also forms the basis for the powerful tool of *dimensional analysis,* which is considered in detail in Chapter 7.

General homogeneous equations are valid in any system of units.

Note to the users of this text. All of the examples in the text use a consistent problem-solving methodology which is similar to that in other engineering courses such as statics. Each example highlights the key elements of analysis: *Given, Find, Solution,* and *Comment.*

The *Given* and *Find* are steps that ensure the user understands what is being asked in the problem and explicitly list the items provided to help solve the problem.

The *Solution* step is where the equations needed to solve the problem are formulated and the problem is actually solved. In this step, there are typically several other tasks that help to set up the solution and are required to solve the problem. The first is a drawing of the problem; where appropriate, it is always helpful to draw a sketch of the problem. Here the relevant geometry and coordinate system to be used as well as features such as control volumes, forces and pressures, velocities, and mass flow rates are included. This helps in gaining a visual understanding of the problem. Making appropriate assumptions to solve the problem is the second task. In a realistic engineering problem-solving environment, the necessary assumptions are developed as an integral part of the solution process. Assumptions can provide appropriate simplifications or offer useful constraints, both of which can help in solving the problem. Throughout the examples in this text, the necessary assumptions are embedded within the *Solution* step, as they are in solving a real-world problem. This provides a realistic problem-solving experience.

The final element in the methodology is the *Comment.* For the examples in the text, this section is used to provide further insight into the problem or the solution. It can also be a point in the analysis at which certain questions are posed. For example: Is the answer reasonable, and does it make physical sense? Are the final units correct? If a certain parameter were changed, how would the answer change? Adopting the above type of methodology will aid in the development of problem-solving skills for fluid mechanics, as well as other engineering disciplines.

EXAMPLE 1.1   Restricted and General Homogeneous Equations

GIVEN   A liquid flows through an orifice located in the side of a tank as shown in Fig. E1.1. A commonly used equation for determining the volume rate of flow, Q, through the orifice is

$$Q = 0.61 \, A\sqrt{2gh}$$

where A is the area of the orifice, g is the acceleration of gravity, and h is the height of the liquid above the orifice.

FIND   Investigate the dimensional homogeneity of this formula.

SOLUTION

The dimensions of the various terms in the equation are $Q =$ volume/time $\doteq L^3T^{-1}$, $A =$ area $\doteq L^2$, $g =$ acceleration of gravity $\doteq LT^{-2}$, and $h =$ height $\doteq L$.

These terms, when substituted into the equation, yield the dimensional form:

$$(L^3T^{-1}) \doteq (0.61)(L^2)(\sqrt{2})(LT^{-2})^{1/2}(L)^{1/2}$$

or

$$(L^3T^{-1}) \doteq [(0.61)\sqrt{2}](L^3T^{-1})$$

It is clear from this result that the equation is dimensionally homogeneous (both sides of the formula have the same dimensions of L^3T^{-1}), and the numbers (0.61 and $\sqrt{2}$) are dimensionless.

If we were going to use this relationship repeatedly we might be tempted to simplify it by replacing g with its standard value of 9.81 m/s² and rewriting the formula as

$$Q = 2.70\,A\sqrt{h} \qquad (1)$$

A quick check of the dimensions reveals that

$$L^3T^{-1} \doteq (2.70)(L^{5/2})$$

and, therefore, the equation expressed as Eq. 1 can only be dimensionally correct if the number 2.70 has the dimensions of $L^{1/2}T^{-1}$. Whenever a number appearing in an equation or formula has dimensions, it means that the specific value of the number will depend on the system of units used. Thus, for the case being considered with meter and seconds used as units,

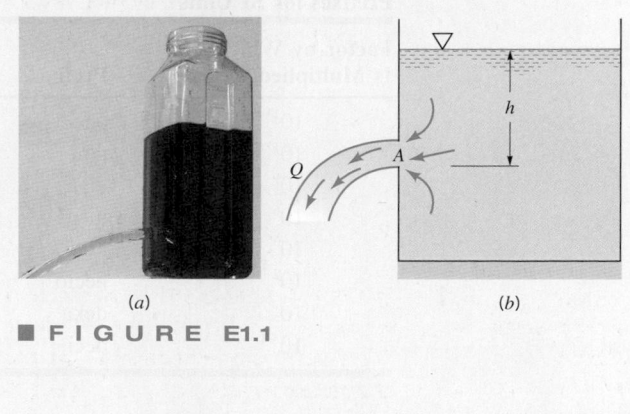

(a) (b)

■ **FIGURE E1.1**

the number 2.70 has units of m¹ᐟ²/s. Equation 1 will only give the correct value for Q (in m³/s) when A is expressed in square meter and h in meter. Thus, Eq. 1 is a *restricted* homogeneous equation, whereas the original equation is a *general* homogeneous equation that would be valid for any consistent system of units.

COMMENT   A quick check of the dimensions of the various terms in an equation is a useful practice and will often be helpful in eliminating errors—that is, as noted previously, all physically meaningful equations must be dimensionally homogeneous. We have briefly alluded to units in this example, and this important topic will be considered in more detail in the next section.

1.2.1 Systems of Units

In addition to the qualitative description of the various quantities of interest, it is generally necessary to have a quantitative measure of any given quantity. For example, if we measure the width of this page in the book and say that it is 10 units wide, the statement has no meaning until the unit of length is defined. If we indicate that the unit of length is a meter, and define the meter as some standard length, a unit system for length has been established (and a numerical value can be given to the page width). In addition to length, a unit must be established for each of the remaining basic quantities (force, mass, time, and temperature). There are several systems of units in use and we shall consider three systems that are commonly used in engineering.

International System (SI). In 1960 the Eleventh General Conference on Weights and Measures, the international organization responsible for maintaining precise uniform standards of measurements, formally adopted the *International System of Units* as the international standard. This system, commonly termed SI, has been widely adopted worldwide and is widely used (although certainly not exclusively) in the United States. It is expected that the long-term trend will be for all countries to accept SI as the accepted standard and it is imperative that engineering students become familiar with this system. In SI the unit of length is the meter (m), the time unit is the second (s), the mass unit is the kilogram (kg), and the temperature unit is the kelvin (K). Note that there is no degree symbol used when expressing a temperature in kelvin units. The kelvin

■ **TABLE 1.2**

Prefixes for SI Units

Factor by Which Unit Is Multiplied	Prefix	Symbol	Factor by Which Unit Is Multiplied	Prefix	Symbol
10^{15}	peta	P	10^{-2}	centi	c
10^{12}	tera	T	10^{-3}	milli	m
10^{9}	giga	G	10^{-6}	micro	μ
10^{6}	mega	M	10^{-9}	nano	n
10^{3}	kilo	k	10^{-12}	pico	p
10^{2}	hecto	h	10^{-15}	femto	f
10	deka	da	10^{-18}	atto	a
10^{-1}	deci	d			

temperature scale is an absolute scale and is related to the Celsius (centigrade) scale (°C) through the relationship

$$K = °C + 273.15$$

Although the Celsius scale is not in itself part of SI, it is common practice to specify temperatures in degrees Celsius when using SI units.

The force unit, called the newton (N), is defined from Newton's second law as

$$1 \text{ N} = (1 \text{ kg})(1 \text{ m/s}^2)$$

In mechanics it is very important to distinguish between weight and mass.

Thus, a 1-N force acting on a 1-kg mass will give the mass an acceleration of 1 m/s². Standard gravity in SI is 9.807 m/s² (commonly approximated as 9.81 m/s²) so that a 1-kg mass weighs 9.81 N under standard gravity. Note that weight and mass are different, both qualitatively and quantitatively! The unit of *work* in SI is the joule (J), which is the work done when the point of application of a 1-N force is displaced through a 1-m distance in the direction of a force. Thus,

$$1 \text{ J} = 1 \text{ N} \cdot \text{m}$$

The unit of *power* is the watt (W) defined as a joule per second. Thus,

$$1 \text{ W} = 1 \text{ J/s} = 1 \text{ N} \cdot \text{m/s}$$

Prefixes for forming multiples and fractions of SI units are given in Table 1.2. For example, the notation kN would be read as "kilonewtons" and stands for 10^3 N. Similarly, mm would be read as "millimeters" and stands for 10^{-3} m. The centimeter is not an accepted unit of length in the SI system, so for most problems in fluid mechanics in which SI units are used, lengths will be expressed in millimeters or meters.

How long is a foot? Today, in the United States, the common length *unit* is the *foot,* but throughout antiquity the unit used to measure length has quite a history. The first length units were based on the lengths of various body parts. One of the earliest units was the Egyptian cubit, first used around 3000 B.C. and defined as the length of the arm from elbow to extended fingertips. Other measures followed, with the foot simply taken as the length of a man's foot. Since this length obviously varies from person to person it was often "standardized" by using the length of the current reigning royalty's foot. In 1791 a special French commission proposed that a new universal length unit called a meter (metre) be defined as the distance of one-quarter of the earth's meridian (north pole to the equator) divided by 10 million. Although controversial, the meter was accepted in 1799 as the standard. With the development of advanced technology, the length of a meter was redefined in 1983 as the distance traveled by light in a vacuum during the time interval of 1/299,792,458 s. The foot is now defined as 0.3048 meters. Our simple rulers and yardsticks indeed have an intriguing history.

EXAMPLE 1.2 | Units

GIVEN A tank of liquid having a total mass of 36 kg rests on a support in the equipment bay of the Space Shuttle.

FIND Determine the force (in newtons) that the tank exerts on the support shortly after lift off when the shuttle is accelerating upward as shown in Fig. E1.2a at 4.5 m/s².

SOLUTION

A free-body diagram of the tank is shown in Fig. E1.2b, where $\mathcal{W}$ is the weight of the tank and liquid, and F_f is the reaction of the floor on the tank. Application of Newton's second law of motion to this body gives

$$\sum \mathbf{F} = m\,\mathbf{a}$$

or

$$F_f - \mathcal{W} = ma \qquad (1)$$

where we have taken upward as the positive direction. Since $\mathcal{W} = mg$, Eq. 1 can be written as

$$F_f = m(g + a) \qquad (2)$$

Before substituting any number into Eq. 2, we must decide on a system of units, and then be sure all of the data are expressed in these units. Since we want F_f in newtons, we will use SI units so that

$$F_f = 36 \text{ kg} \left[9.81 \text{ m/s}^2 + 4.5 \text{ m/s}^2 \right]$$
$$= 515 \text{ kg} \cdot \text{m/s}^2$$

■ **FIGURE E1.2a** (Photograph courtesy of NASA.)

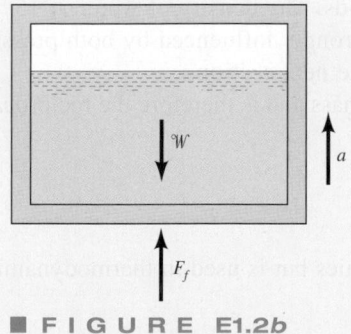

■ **FIGURE E1.2b**

Since $1 \text{ N} = 1 \text{ kg} \cdot \text{m/s}^2$, it follows that

$$F_f = 515 \text{ N} \qquad \text{(downward on floor)} \qquad \textbf{(Ans)}$$

The direction is downward since the force shown on the free-body diagram is the force of the support *on the tank* so that the force the tank exerts *on the support* is equal in magnitude but opposite in direction.

COMMENT Be careful not to interchange the physical properties of mass and weight.

1.3   Analysis of Fluid Behavior

The study of fluid mechanics involves the same fundamental laws you have encountered in physics and other mechanics courses. These laws include Newton's laws of motion, conservation of mass, and the first and second laws of thermodynamics. Thus, there are strong similarities between the general approach to fluid mechanics and to rigid-body and deformable-body solid mechanics. This is indeed helpful since many of the concepts and techniques of analysis used in fluid mechanics will be ones you have encountered before in other courses.

The broad subject of fluid mechanics can be generally subdivided into *fluid statics,* in which the fluid is at rest, and *fluid dynamics,* in which the fluid is moving. In the following chapters we will consider both of these areas in detail. Before we can proceed, however, it will be necessary to define and discuss certain fluid *properties* that are intimately related to fluid behavior. It is obvious that different fluids can have grossly different characteristics. For example, gases are light and compressible, whereas liquids are heavy (by comparison) and relatively incompressible. A syrup flows slowly from a container, but water flows rapidly when poured from the same container. To quantify these differences, certain fluid properties are used. In the following several sections the properties that play an important role in the analysis of fluid behavior are considered.

1.4   Measures of Fluid Mass and Weight

1.4.1 Density

The density of a fluid is defined as its mass per unit volume.

The *density* of a fluid, designated by the Greek symbol ρ (rho), is defined as its mass per unit volume. Density is typically used to characterize the mass of a fluid system. In the SI system, ρ has units of kg/m^3.

The value of density can vary widely between different fluids, but for liquids, variations in pressure and temperature generally have only a small effect on the value of ρ. The small change in the density of water with large variations in temperature is illustrated in Fig. 1.2. Table 1.3 lists values of density for several common liquids. The density of water at 15 °C (288 K) is 999 kg/m^3. Unlike liquids, the density of a gas is strongly influenced by both pressure and temperature, and this difference will be discussed in the next section.

The *specific volume, v,* is the *volume* per unit mass and is therefore the reciprocal of the density—that is,

$$v = \frac{1}{\rho} \tag{1.4}$$

This property is not commonly used in fluid mechanics but is used in thermodynamics.

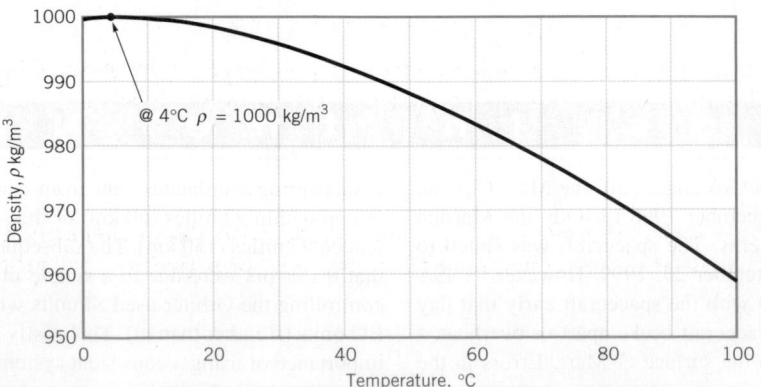

■ **FIGURE 1.2**   **Density of water as a function of temperature.**

■ **TABLE 1.3**
Approximate Physical Properties of Some Common Liquids (SI Units)

(See inside of front cover.)

1.4.2 Specific Weight

Specific weight is weight per unit volume; specific gravity is the ratio of fluid density to the density of water at a certain temperature.

The *specific weight* of a fluid, designated by the Greek symbol γ (gamma), is defined as its *weight* per unit volume. Thus, specific weight is related to density through the equation

$$\gamma = \rho g \tag{1.5}$$

where g is the local acceleration of gravity. Just as density is used to characterize the mass of a fluid system, the specific weight is used to characterize the weight of the system. In the SI system, γ has units of N/m^3. Under conditions of standard gravity ($g = 9.807$ m/s^2), water at 15 °C (288 K) has a specific weight of 9.80 kN/m^3. Table 1.3 lists values of specific weight for several common liquids (based on standard gravity). More complete tables for water can be found in Appendix B (Table B.1).

1.4.3 Specific Gravity

The *specific gravity* of a fluid, designated as *SG*, is defined as the ratio of the density of the fluid to the density of water at some specified temperature. Usually the specified temperature is taken as 4 °C (277 k), and at this temperature the density of water is 1000 kg/m^3. In equation form, specific gravity is expressed as

$$SG = \frac{\rho}{\rho_{H_2O@4\,°C}} \tag{1.6}$$

13.55

Water

Mercury

and since it is the *ratio* of densities, the value of *SG* does not depend on the system of units used. For example, the specific gravity of mercury at 20 °C (293 K) is 13.55. This is illustrated by the figure in the margin. Thus, the density of mercury can be readily calculated in SI units through the use of Eq. 1.6 as

$$\rho_{Hg} = (13.55)(1000 \text{ kg/m}^3) = 13.6 \times 10^3 \text{ kg/m}^3$$

It is clear that density, specific weight, and specific gravity are all interrelated, and from a knowledge of any one of the three the others can be calculated.

1.5   Ideal Gas Law

Gases are highly compressible in comparison to liquids, with changes in gas density directly related to changes in pressure and temperature through the equation

$$\rho = \frac{p}{RT} \tag{1.7}$$

where p is the absolute pressure, ρ the density, T the absolute temperature,[1] and R is a gas constant. Equation 1.7 is commonly termed the *ideal* or *perfect gas law,* or the *equation of state* for

---

[1]We will use T to represent temperature in thermodynamic relationships although T is also used to denote the basic dimension of time.

an ideal gas. It is known to closely approximate the behavior of real gases under normal conditions when the gases are not approaching liquefaction.

In the ideal gas law, absolute pressures and temperatures must be used.

Pressure in a fluid at rest is defined as the normal force per unit area exerted on a plane surface (real or imaginary) immersed in a fluid and is created by the bombardment of the surface with the fluid molecules. From the definition, pressure has the dimension of FL^{-2}, and in SI units is expressed as N/m^2. In SI, 1 N/m^2 is defined as a *pascal,* abbreviated as Pa, and pressures are commonly specified in pascals. The pressure in the ideal gas law must be expressed as an *absolute pressure,* denoted (abs), which means that it is measured relative to absolute zero pressure (a pressure that would only occur in a perfect vacuum). Standard sea-level atmospheric pressure (by international agreement) is 101.33 kPa (abs). For most calculations these pressures can be rounded to 101 kPa. In engineering it is common practice to measure pressure relative to the local atmospheric pressure, and when measured in this fashion it is called *gage pressure.* Thus, the absolute pressure can be obtained from the gage pressure by adding the value of the atmospheric pressure. For example, as shown by the figure in the margin on the next page, a pressure of 207 kPa (gage) in a tire is equal to 308 kPa (abs) at standard atmospheric pressure. Pressure is a particularly important fluid characteristic and it will be discussed more fully in the next chapter.

EXAMPLE 1.3 | Ideal Gas Law

GIVEN The compressed air tank shown in Fig. E1.3a has a volume of 0.024 m³. The temperature is 20 °C (293 K) and the atmospheric pressure is 101.3 kPa (abs).

FIND When the tank is filled with air at a gage pressure of 345 kPa, determine the density of the air and the weight of air in the tank.

SOLUTION

The air density can be obtained from the ideal gas law (Eq. 1.7)

$$\rho = \frac{p}{RT}$$

so that

$$\rho = \frac{(345 \text{ kPa} + 101.3 \text{ kPa})}{(286.9 \text{ J/kg} \cdot \text{K})[(20 + 273)\text{K}]}$$

$$= 5.30 \text{ kg/m}^3 \qquad \text{(Ans)}$$

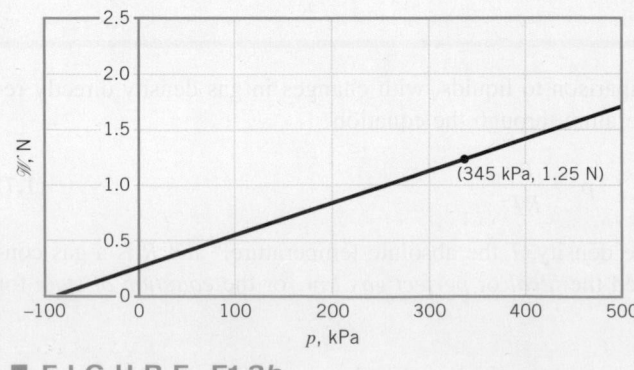

■ **FIGURE E1.3a** (Photograph courtesy of Jenny Products, Inc.)

Note that both the pressure and temperature were changed to absolute values.

The weight, $\mathcal{W}$, of the air is equal to

$$\mathcal{W} = \rho g \times \text{(volume)}$$
$$= (5.30 \text{ kg/m}^3)(9.81 \text{ m/s}^2)(0.024 \text{ m}^3)$$
$$= 1.25 \text{ kg} \cdot \text{m/s}^2$$

[graph: vertical axis $\mathcal{W}$, N from 0 to 2.5; horizontal axis p, kPa from −100 to 500; line with point labeled (345 kPa, 1.25 N)]

■ **FIGURE E1.3b**

so that since $1 \, \text{N} = 1 \, \text{kg} \cdot \text{m/s}^2$

$$\mathcal{W} = 1.25 \, \text{N} \qquad \text{(Ans)}$$

COMMENT By repeating the calculations for various values of the pressure, p, the results shown in Fig. E1.3b are

obtained. Note that doubling the gage pressure does not double the amount of air in the tank, but doubling the absolute pressure does. Thus, a scuba diving tank at a gage pressure of 690 kPa does not contain twice the amount of air as when the gage reads 345 kPa.

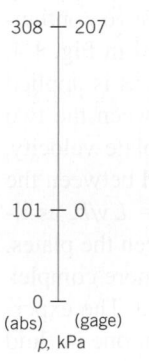

308 ┬ 207

101 ┬ 0

0 ┴
(abs)   (gage)
  p, kPa

■ **TABLE 1.4**

Approximate Physical Properties of Some Common Gases at Standard Atmospheric Pressure (SI Units)

(See inside of front cover.)

The gas constant, R, which appears in Eq. 1.7, depends on the particular gas and is related to the molecular weight of the gas. Values of the gas constant for several common gases are listed in Table 1.4. Also in these tables the gas density and specific weight are given for standard atmospheric pressure and gravity and for the temperature listed. More complete tables for air at standard atmospheric pressure can be found in Appendix B (Table B.2).

1.6   Viscosity

V1.3 Viscous fluids

V1.4 No-slip condition

The properties of density and specific weight are measures of the "heaviness" of a fluid. It is clear, however, that these properties are not sufficient to uniquely characterize how fluids behave since two fluids (such as water and oil) can have approximately the same value of density but behave quite differently when flowing. There is apparently some additional property that is needed to describe the "fluidity" of the fluid.

To determine this additional property, consider a hypothetical experiment in which a material is placed between two very wide parallel plates as shown in Fig. 1.3a. The bottom plate is rigidly fixed, but the upper plate is free to move. If a solid, such as steel, were placed between the two plates and loaded with the force P as shown, the top plate would be displaced through some small distance, δa (assuming the solid was mechanically attached to the plates). The vertical line AB would be rotated through the small angle, $\delta \beta$, to the new position AB'. We note that to resist the applied force, P, a shearing stress, τ, would be developed at the plate–material interface, and for equilibrium to occur, $P = \tau A$ where A is the effective upper plate area (Fig. 1.3b). It is well known that for elastic solids, such as steel, the small angular displacement, $\delta \beta$ (called the shearing strain), is proportional to the shearing stress, τ, that is developed in the material.

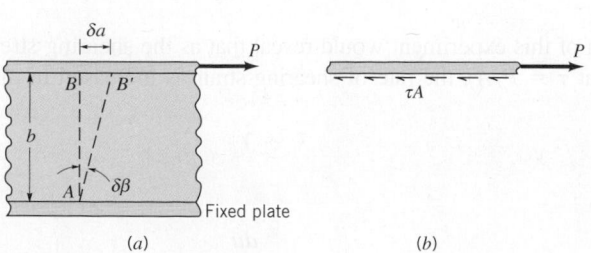

■ **FIGURE 1.3** (a) Deformation of material placed between two parallel plates. (b) Forces acting on upper plate.

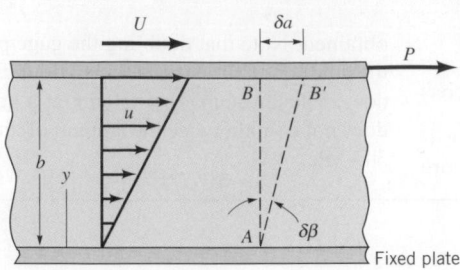

■ **FIGURE 1.4** **Behavior of a fluid placed between two parallel plates.**

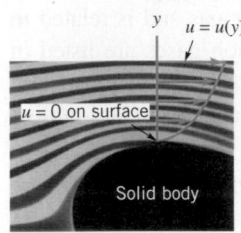

Real fluids, even though they may be moving, always "stick" to the solid boundaries that contain them.

What happens if the solid is replaced with a fluid such as water? We would immediately notice a major difference. When the force P is applied to the upper plate, it will move continuously with a velocity, U (after the initial transient motion has died out) as illustrated in Fig. 1.4. This behavior is consistent with the definition of a fluid—that is, if a shearing stress is applied to a fluid it will deform continuously. A closer inspection of the fluid motion between the two plates would reveal that the fluid in contact with the upper plate moves with the plate velocity, U, and the fluid in contact with the bottom fixed plate has a zero velocity. The fluid between the two plates moves with velocity $u = u(y)$ that would be found to vary linearly, $u = Uy/b$, as illustrated in Fig. 1.4. Thus, a *velocity gradient, du/dy*, is developed in the fluid between the plates. In this particular case the velocity gradient is a constant since $du/dy = U/b$, but in more complex flow situations, such as that shown by the photograph in the margin, this is not true. The experimental observation that the fluid "sticks" to the solid boundaries is a very important one in fluid mechanics and is usually referred to as the *no-slip condition.* All fluids, both liquids and gases, satisfy this condition.

In a small time increment, δt, an imaginary vertical line AB in the fluid would rotate through an angle, $\delta \beta$, so that

$$\tan \delta \beta \approx \delta \beta = \frac{\delta a}{b}$$

Since $\delta a = U\,\delta t$ it follows that

$$\delta \beta = \frac{U\,\delta t}{b}$$

We note that in this case, $\delta \beta$ is a function not only of the force P (which governs U) but also of time. Thus, it is not reasonable to attempt to relate the shearing stress, τ, to $\delta \beta$ as is done for solids. Rather, we consider the *rate* at which $\delta \beta$ is changing and define the *rate of shearing strain, $\dot{\gamma}$*, as

$$\dot{\gamma} = \lim_{\delta t \to 0} \frac{\delta \beta}{\delta t}$$

which in this instance is equal to

$$\dot{\gamma} = \frac{U}{b} = \frac{du}{dy}$$

A continuation of this experiment would reveal that as the shearing stress, τ, is increased by increasing P (recall that $\tau = P/A$), the rate of shearing strain is increased in direct proportion—that is,

$$\tau \propto \dot{\gamma}$$

or

$$\tau \propto \frac{du}{dy}$$

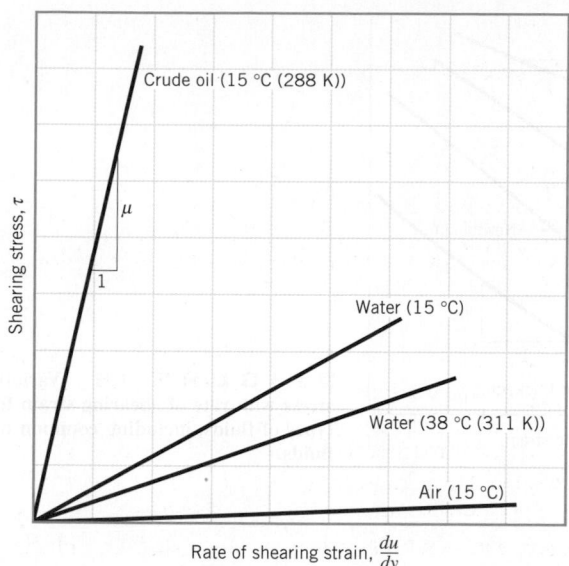

Shearing stress, τ

Crude oil (15 °C (288 K))

μ

1

Water (15 °C)

Water (38 °C (311 K))

Air (15 °C)

Rate of shearing strain, $\frac{du}{dy}$

■ **FIGURE 1.5**   **Linear
variation of shearing stress with rate of
shearing strain for common fluids.**

This result indicates that for common fluids such as water, oil, gasoline, and air the shearing
stress and rate of shearing strain (velocity gradient) can be related with a relationship of the
form

$$\tau = \mu \frac{du}{dy} \tag{1.8}$$

*V1.5 Capillary tube
viscometer*

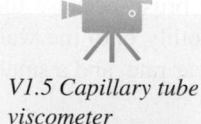

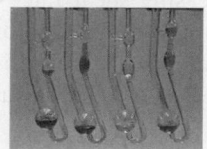

*Dynamic viscosity
is the fluid property
that relates shear-
ing stress and fluid
motion.*

where the constant of proportionality is designated by the Greek symbol μ (mu) and is called the
absolute viscosity, dynamic viscosity, or simply the *viscosity* of the fluid. In accordance with
Eq. 1.8, plots of τ versus du/dy should be linear with the slope equal to the viscosity as illustrated
in Fig. 1.5. The actual value of the viscosity depends on the particular fluid, and for a particular
fluid the viscosity is also highly dependent on temperature as illustrated in Fig. 1.5 with the two
curves for water. Fluids for which the shearing stress is *linearly* related to the rate of shearing strain
(also referred to as rate of angular deformation) are designated as *Newtonian fluids* after I. Newton
(1642–1727). Fortunately most common fluids, both liquids and gases, are Newtonian. A more general formulation of Eq. 1.8 which applies to more complex flows of Newtonian fluids is given in
Section 6.8.1.

*For non-Newtonian
fluids, the apparent
viscosity is a func-
tion of the shear
rate.*

Fluids for which the shearing stress is not linearly related to the rate of shearing strain are
designated as *non-Newtonian fluids.* Although there is a variety of types of non-Newtonian
fluids, the simplest and most common are shown in Fig. 1.6. The slope of the shearing stress

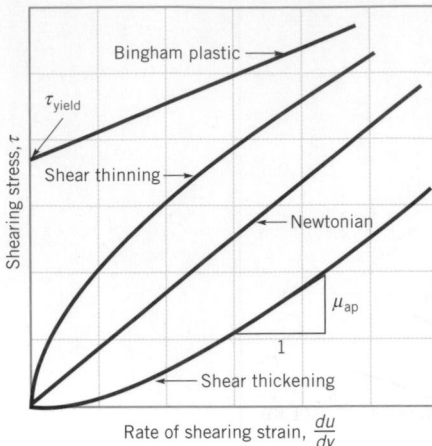

■ **F I G U R E 1.6** Variation of shearing stress with rate of shearing strain for several types of fluids, including common non-Newtonian fluids.

versus rate of shearing strain graph is denoted as the *apparent viscosity, μ_{ap}.* For Newtonian fluids the apparent viscosity is the same as the viscosity and is independent of shear rate.

For *shear thinning fluids* the apparent viscosity decreases with increasing shear rate—the harder the fluid is sheared, the less viscous it becomes. Many colloidal suspensions and polymer solutions are shear thinning. For example, latex paint does not drip from the brush because the shear rate is small and the apparent viscosity is large. However, it flows smoothly onto the wall because the thin layer of paint between the wall and the brush causes a large shear rate and a small apparent viscosity.

For *shear thickening fluids* the apparent viscosity increases with increasing shear rate—the harder the fluid is sheared, the more viscous it becomes. Common examples of this type of fluid include water–corn starch mixture and water–sand mixture ("quicksand"). Thus, the difficulty in removing an object from quicksand increases dramatically as the speed of removal increases.

The other type of behavior indicated in Fig. 1.6 is that of a *Bingham plastic,* which is neither a fluid nor a solid. Such material can withstand a finite, nonzero shear stress, τ_{yield}, the yield stress, without motion (therefore, it is not a fluid), but once the yield stress is exceeded it flows like a fluid (hence, it is not a solid). Toothpaste and mayonnaise are common examples of Bingham plastic materials. As indicated in the figure in the margin, mayonnaise can sit in a pile on a slice of bread (the shear stress less than the yield stress), but it flows smoothly into a thin layer when the knife increases the stress above the yield stress.

From Eq. 1.8 it can be readily deduced that the dimensions of viscosity are FTL^{-2}. Thus, in SI units viscosity is given as $N \cdot s/m^2$. Values of viscosity for several common liquids and gases are listed in Tables 1.3 and 1.4. A quick glance at these tables reveals the wide variation in viscosity among fluids. Viscosity is only mildly dependent on pressure and the effect of pressure is usually neglected. However, as previously mentioned, and as illustrated in Fig. 1.8, viscosity is very sensitive to temperature. For example, as the temperature of water changes from 20 to 38 °C (288–311 K) the density decreases by less than 1% but the viscosity decreases by about 40%. It is thus clear that particular attention must be given to temperature when determining viscosity.

Figure 1.7 shows in more detail how the viscosity varies from fluid to fluid and how for a given fluid it varies with temperature. It is to be noted from this figure that the viscosity of liquids decreases with an increase in temperature, whereas for gases an increase in temperature causes an increase in viscosity. This difference in the effect of temperature on the viscosity of liquids and gases can again be traced back to the difference in molecular structure. The liquid molecules are closely spaced, with strong cohesive forces between molecules, and the resistance to relative motion between adjacent layers of fluid is related to these intermolecular forces. As the temperature increases, these cohesive forces are reduced with a corresponding reduction

The various types of non-Newtonian fluids are distinguished by how their apparent viscosity changes with shear rate.

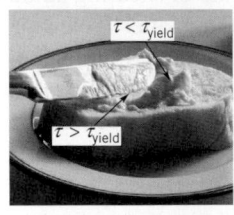

V1.6 Non-Newtonian behavior

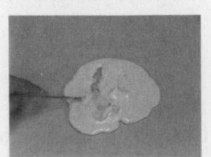

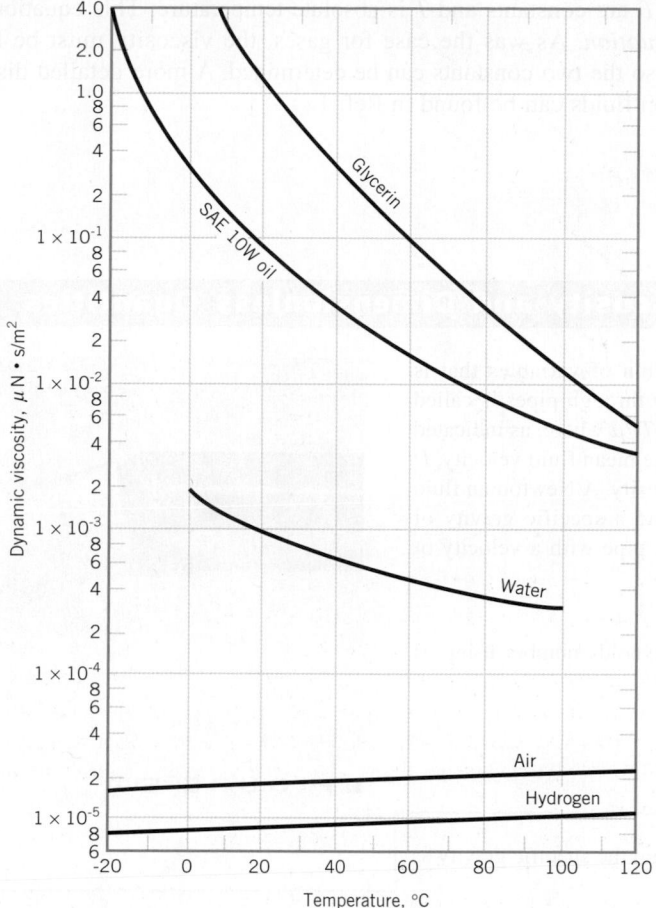

■ FIGURE 1.7 **Dynamic (absolute) viscosity of some common fluids as a function of temperature.**

in resistance to motion. Since viscosity is an index of this resistance, it follows that the viscosity is reduced by an increase in temperature. In gases, however, the molecules are widely spaced and intermolecular forces negligible. In this case, resistance to relative motion arises due to the exchange of momentum of gas molecules between adjacent layers. As molecules are transported by random motion from a region of low bulk velocity to mix with molecules in a region of higher bulk velocity (and vice versa), there is an effective momentum exchange which resists the relative motion between the layers. As the temperature of the gas increases, the random molecular activity increases with a corresponding increase in viscosity.

The effect of temperature on viscosity can be closely approximated using two empirical formulas. For gases the *Sutherland equation* can be expressed as

Viscosity is very sensitive to temperature.

$$\mu = \frac{CT^{3/2}}{T + S} \qquad (1.9)$$

where C and S are empirical constants, and T is absolute temperature. Thus, if the viscosity is known at two temperatures, C and S can be determined. Or, if more than two viscosities are known, the data can be correlated with Eq. 1.9 by using some type of curve-fitting scheme.

For liquids an empirical equation that has been used is

$$\mu = De^{B/T} \qquad (1.10)$$

where D and B are constants and T is absolute temperature. This equation is often referred to as *Andrade's equation.* As was the case for gases, the viscosity must be known at least for two temperatures so the two constants can be determined. A more detailed discussion of the effect of temperature on fluids can be found in Ref. 1.

EXAMPLE 1.4   Viscosity and Dimensionless Quantities

GIVEN   A dimensionless combination of variables that is important in the study of viscous flow through pipes is called the *Reynolds number,* Re, defined as $\rho V D / \mu$ where, as indicated in Fig. E1.4, ρ is the fluid density, V the mean fluid velocity, D the pipe diameter, and μ the fluid viscosity. A Newtonian fluid having a viscosity of $0.38 \text{ N} \cdot \text{s/m}^2$ and a specific gravity of 0.91 flows through a 25-mm-diameter pipe with a velocity of 2.6 m/s.

FIND   Determine the value of the Reynolds number using SI units.

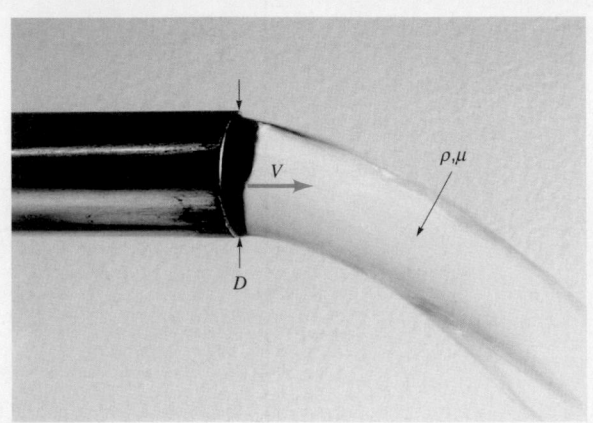

■ **FIGURE E1.4**

SOLUTION

(a)   The fluid density is calculated from the specific gravity as

$$\rho = SG\, \rho_{H_2O@4\,°C} = 0.91\,(1000 \text{ kg/m}^3) = 910 \text{ kg/m}^3$$

and from the definition of the Reynolds number

$$\text{Re} = \frac{\rho V D}{\mu} = \frac{(910 \text{ kg/m}^3)(2.6 \text{ m/s})(25 \text{ mm})(10^{-3} \text{ m/mm})}{0.38 \text{ N} \cdot \text{s/m}^2}$$

$$= 156\,(\text{kg} \cdot \text{m/s}^2)/\text{N}$$

However, since $1 \text{ N} = 1 \text{ kg} \cdot \text{m/s}^2$ it follows that the Reynolds number is unitless—that is,

$$\text{Re} = 156 \qquad \text{(Ans)}$$

The value of any dimensionless quantity does not depend on the system of units used if all variables that make up the quantity are expressed in a consistent set of units.

COMMENTS   Dimensionless quantities play an important role in fluid mechanics and the significance of the Reynolds number as well as other important dimensionless combinations will be discussed in detail in Chapter 7. It should be noted that in the Reynolds number it is actually the ratio μ/ρ that is important, and this is the property that is defined as the kinematic viscosity.

EXAMPLE 1.5   Newtonian Fluid Shear Stress

GIVEN   The velocity distribution for the flow of a Newtonian fluid between two wide, parallel plates (see Fig. E1.5a) is given by the equation

$$u = \frac{3V}{2}\left[1 - \left(\frac{y}{h}\right)^2\right]$$

where V is the mean velocity. The fluid has a viscosity of $2 \text{ N} \cdot \text{s/m}^2$. Also, $V = 0.6 \text{ m/s}$ and $h = 5 \text{ mm}$.

FIND   Determine: (a) the shearing stress acting on the bottom wall, and (b) the shearing stress acting on a plane parallel to the walls and passing through the centerline (midplane).

SOLUTION

For this type of parallel flow the shearing stress is obtained from Eq. 1.8,

$$\tau = \mu \frac{du}{dy} \quad (1)$$

Thus, if the velocity distribution $u = u(y)$ is known, the shearing stress can be determined at all points by evaluating the velocity gradient, du/dy. For the distribution given

$$\frac{du}{dy} = -\frac{3Vy}{h^2} \quad (2)$$

(a) Along the bottom wall $y = -h$ so that (from Eq. 2)

$$\frac{du}{dy} = \frac{3V}{h}$$

and therefore the shearing stress is

$$\tau_{\substack{\text{bottom} \\ \text{wall}}} = \mu\left(\frac{3V}{h}\right) = \frac{(2\,\text{N} \cdot \text{s/m}^2)(3)(0.6\,\text{m/s})}{(5\,\text{mm})(1\,\text{m/1000 mm})}$$

$$= 720\,\text{N/m}^2 \text{ (in direction of flow)} \quad \textbf{(Ans)}$$

This stress creates a drag on the wall. Since the velocity distribution is symmetrical, the shearing stress along the upper wall would have the same magnitude and direction.

(b) Along the midplane where $y = 0$ it follows from Eq. 2 that

$$\frac{du}{dy} = 0$$

and thus the shearing stress is

$$\tau_{\text{midplane}} = 0 \qquad \textbf{(Ans)}$$

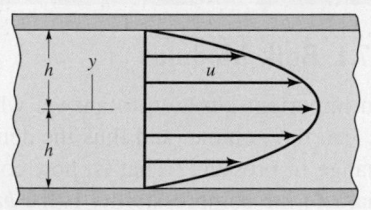

■ **F I G U R E E1.5a**

COMMENT From Eq. 2 we see that the velocity gradient (and therefore the shearing stress) varies linearly with y and in this particular example varies from 0 at the center of the channel to $720\,\text{N/m}^2$ at the walls. This is shown in Fig. E1.5b. For the more general case the actual variation will, of course, depend on the nature of the velocity distribution.

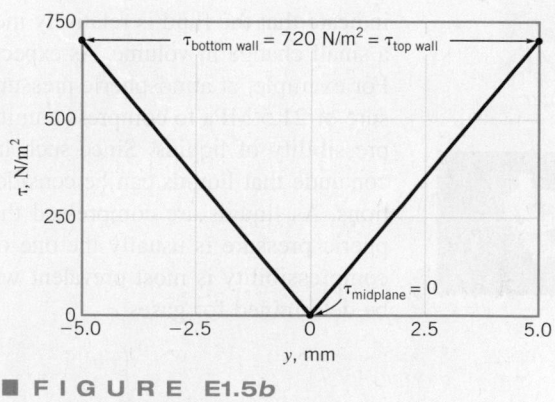

■ **F I G U R E E1.5b**

Quite often viscosity appears in fluid flow problems combined with the density in the form

$$\nu = \frac{\mu}{\rho}$$

Kinematic viscosity is defined as the ratio of the absolute viscosity to the fluid density.

This ratio is called the *kinematic viscosity* and is denoted with the Greek symbol ν (nu). The dimensions of kinematic viscosity are L^2/T, and the SI units are m^2/s. Values of kinematic viscosity for some common liquids and gases are given in Tables 1.3 and 1.4. More extensive tables giving both the dynamic and kinematic viscosities for water and air can be found in Appendix B (Tables B.1 and B.2), and graphs showing the variation in both dynamic and kinematic viscosity with temperature for a variety of fluids are also provided in Appendix B (Fig. B.1).

Although in this text we are primarily using SI units, dynamic viscosity is often expressed in the metric CGS (centimeter-gram-second) system with units of dyne · s/cm². This combination is called a *poise,* abbreviated P. In the CGS system, kinematic viscosity has units of cm²/s, and this combination is called a *stoke,* abbreviated St.

1.7 Compressibility of Fluids

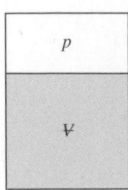

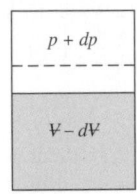

1.7.1 Bulk Modulus

An important question to answer when considering the behavior of a particular fluid is how easily can the volume (and thus the density) of a given mass of the fluid be changed when there is a change in pressure? That is, how compressible is the fluid? A property that is commonly used to characterize compressibility is the *bulk modulus, E_v,* defined as

$$E_v = -\frac{dp}{d\forall/\forall} \tag{1.11}$$

where dp is the differential change in pressure needed to create a differential change in volume, $d\forall$, of a volume $\forall$. This is illustrated by the figure in the margin. The negative sign is included since an increase in pressure will cause a decrease in volume. Since a decrease in volume of a given mass, $m = \rho\forall$, will result in an increase in density, Eq. 1.11 can also be expressed as

$$E_v = \frac{dp}{d\rho/\rho} \tag{1.12}$$

The bulk modulus (also referred to as the *bulk modulus of elasticity*) has dimensions of pressure, FL^{-2}. In SI units, values for E_v are usually given as N/m^2 (Pa). Large values for the bulk modulus indicate that the fluid is relatively incompressible—that is, it takes a large pressure change to create a small change in volume. As expected, values of E_v for common liquids are large (see Table 1.3). For example, at atmospheric pressure and a temperature of 15 °C (288 K) it would require a pressure of 21.5 MPa to compress a unit volume of water 1%. This result is representative of the compressibility of liquids. Since such large pressures are required to effect a change in volume, we conclude that liquids can be considered as *incompressible* for most practical engineering applications. As liquids are compressed the bulk modulus increases, but the bulk modulus near atmospheric pressure is usually the one of interest. The use of bulk modulus as a property describing compressibility is most prevalent when dealing with liquids, although the bulk modulus can also be determined for gases.

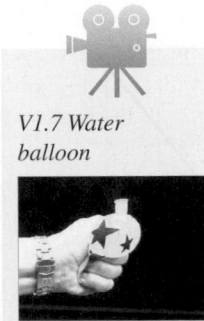

V1.7 Water balloon

F l u i d s   i n   t h e   N e w s

This water jet is a blast Usually liquids can be treated as incompressible fluids. However, in some applications the *compressibility* of a liquid can play a key role in the operation of a device. For example, a water pulse generator using compressed water has been developed for use in mining operations. It can fracture rock by producing an effect comparable to a conventional explosive such as gunpowder. The device uses the energy stored in a water-filled accumulator to generate an ultrahigh-pressure water pulse ejected through a 10- to 25-mm-diameter discharge valve. At the ultrahigh pressures used (300 to 400 MPa, or 3000 to 4000 atmospheres), the water is compressed (i.e., the volume reduced) by about 10 to 15%. When a fast-opening valve within the pressure vessel is opened, the water expands and produces a jet of water that upon impact with the target material produces an effect similar to the explosive force from conventional explosives. Mining with the water jet can eliminate various hazards that arise with the use of conventional chemical explosives, such as those associated with the storage and use of explosives and the generation of toxic gas by-products that require extensive ventilation. (See Problem 1.81.)

1.7.2 Compression and Expansion of Gases

When gases are compressed (or expanded), the relationship between pressure and density depends on the nature of the process. If the compression or expansion takes place under constant temperature conditions (*isothermal process*), then from Eq. 1.7

$$\frac{p}{\rho} = \text{constant} \tag{1.13}$$

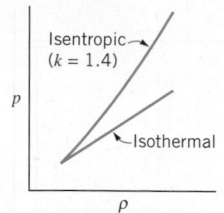

Isentropic (k = 1.4)

p

Isothermal

ρ

The value of the bulk modulus depends on the type of process involved.

If the compression or expansion is frictionless and no heat is exchanged with the surroundings (*isentropic process*), then

$$\frac{p}{\rho^k} = \text{constant} \tag{1.14}$$

where k is the ratio of the specific heat at constant pressure, c_p, to the specific heat at constant volume, c_v (i.e., $k = c_p/c_v$). The two specific heats are related to the gas constant, R, through the equation $R = c_p - c_v$. As was the case for the ideal gas law, the pressure in both Eqs. 1.13 and 1.14 must be expressed as an absolute pressure. Values of k for some common gases are given in Table 1.4, and for air over a range of temperatures, in Appendix B (B.2). The pressure–density variations for isothermal and isentropic conditions are illustrated in the margin figure.

With explicit equations relating pressure and density, the bulk modulus for gases can be determined by obtaining the derivative $dp/d\rho$ from Eq. 1.13 or 1.14 and substituting the results into Eq. 1.12. It follows that for an isothermal process

$$E_v = p \tag{1.15}$$

and for an isentropic process,

$$E_v = kp \tag{1.16}$$

Note that in both cases the bulk modulus varies directly with pressure. For air under standard atmospheric conditions with $p = 101.3$ kPa (abs) and $k = 1.40$, the isentropic bulk modulus is 142 kPa. A comparison of this figure with that for water under the same conditions ($E_v = 2150$ MPa) shows that air is approximately 15,000 times as compressible as water. It is thus clear that in dealing with gases, greater attention will need to be given to the effect of compressibility on fluid behavior. However, as will be discussed further in later sections, gases can often be treated as incompressible fluids if the changes in pressure are small.

EXAMPLE 1.6   Isentropic Compression of a Gas

GIVEN A 0.03 m³ of air at an absolute pressure of 101.3 kPa is compressed isentropically to 0.015 m³ by the tire pump shown in Fig. E1.6a.

FIND What is the final pressure?

SOLUTION

For an isentropic compression

$$\frac{p_i}{\rho_i^k} = \frac{p_f}{\rho_f^k}$$

where the subscripts i and f refer to initial and final states, respectively. Since we are interested in the final pressure, p_f, it follows that

$$p_f = \left(\frac{\rho_f}{\rho_i}\right)^k p_i$$

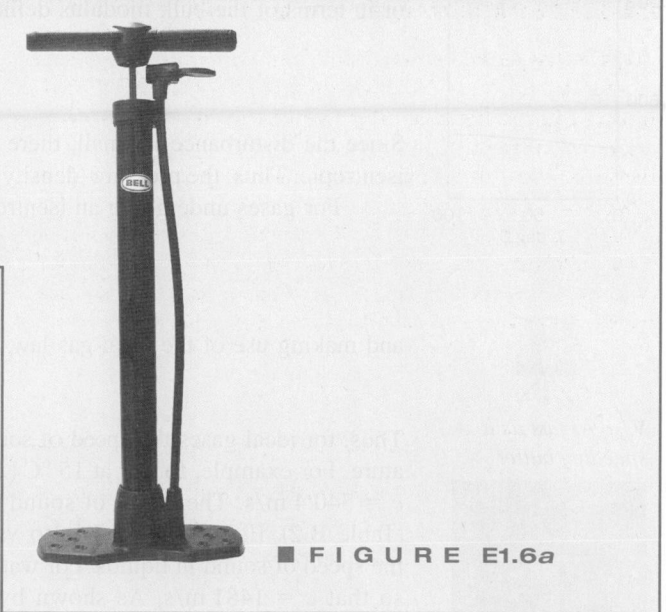

■ FIGURE E1.6a

As the volume, V, is reduced by one-half, the density must double, since the mass, $m = \rho V$, of the gas remains constant. Thus, with $k = 1.40$ for air

$$p_f = (2)^{1.40}(101.3 \text{ kPa}) = 267 \text{ kPa (abs)} \qquad \text{(Ans)}$$

COMMENT By repeating the calculations for various values of the ratio of the final volume to the initial volume, V_f/V_i, the results shown in Fig. E1.6b are obtained. Note that even though air is often considered to be easily compressed (at least compared to liquids), it takes considerable pressure to significantly reduce a given volume of air as is done in an automobile engine where the compression ratio is on the order of $V_f/V_i = 1/8 = 0.125$.

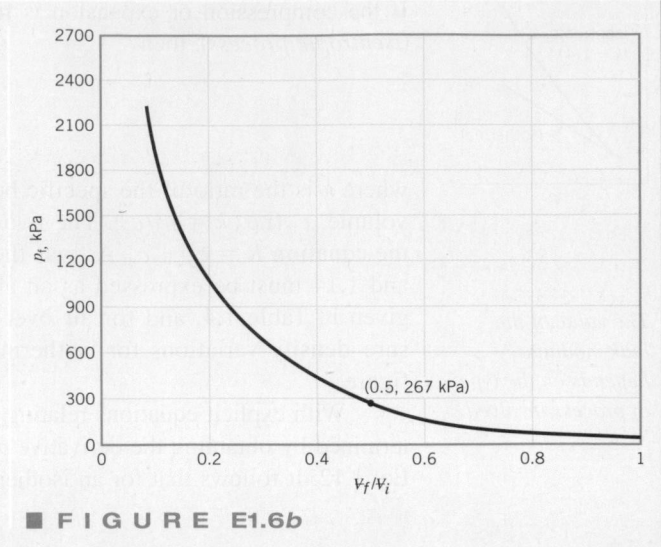

■ **FIGURE E1.6b**

1.7.3 Speed of Sound

The velocity at which small disturbances propagate in a fluid is called the speed of sound.

Another important consequence of the compressibility of fluids is that disturbances introduced at some point in the fluid propagate at a finite velocity. For example, if a fluid is flowing in a pipe and a valve at the outlet is suddenly closed (thereby creating a localized disturbance), the effect of the valve closure is not felt instantaneously upstream. It takes a finite time for the increased pressure created by the valve closure to propagate to an upstream location. Similarly, a loudspeaker diaphragm causes a localized disturbance as it vibrates, and the small change in pressure created by the motion of the diaphragm is propagated through the air with a finite velocity. The velocity at which these small disturbances propagate is called the *acoustic velocity* or the *speed of sound, c.* It will be shown in Chapter 11 that the speed of sound is related to changes in pressure and density of the fluid medium through the equation

$$c = \sqrt{\frac{dp}{d\rho}} \qquad (1.17)$$

or in terms of the bulk modulus defined by Eq. 1.12

$$c = \sqrt{\frac{E_v}{\rho}} \qquad (1.18)$$

Since the disturbance is small, there is negligible heat transfer and the process is assumed to be isentropic. Thus, the pressure–density relationship used in Eq. 1.17 is that for an isentropic process. For gases undergoing an isentropic process, $E_v = kp$ (Eq. 1.16) so that

$$c = \sqrt{\frac{kp}{\rho}}$$

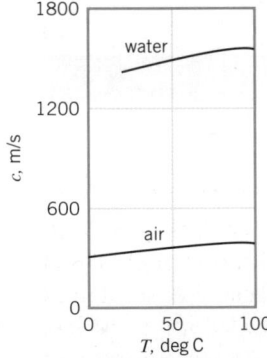

and making use of the ideal gas law, it follows that

$$c = \sqrt{kRT} \qquad (1.19)$$

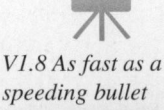

V1.8 As fast as a speeding bullet

Thus, for ideal gases the speed of sound is proportional to the square root of the absolute temperature. For example, for air at 15 °C (288 K) with $k = 1.40$ and $R = 286.9$ J/kg · K, it follows that $c = 340.4$ m/s. The speed of sound in air at various temperatures can be found in Appendix B (Table B.2). Equation 1.18 is also valid for liquids, and values of E_v can be used to determine the speed of sound in liquids. For water at 20 °C (293 K), $E_v = 2.19$ GN/m^2 and $\rho = 998.2$ kg/m^3 so that $c = 1481$ m/s. As shown by the figure in the margin, the speed of sound in water is much higher than in air. If a fluid were truly incompressible ($E_v = \infty$) the speed of sound would be infinite. The speed of sound in water for various temperatures can be found in Appendix B (Table B.1).

EXAMPLE 1.7 | Speed of Sound and Mach Number

GIVEN A jet aircraft flies at a speed of 885 km/h at an altitude of 10,500 m, where the temperature is $-54\ °C$ and the specific heat ratio is $k = 1.4$.

FIND Determine the ratio of the speed of the aircraft, V, to that of the speed of sound, c, at the specified altitude.

SOLUTION

From Eq. 1.19 the speed of sound can be calculated as

$$
\begin{aligned}
c &= \sqrt{kRT} \\
&= \sqrt{(1.40)(286.9\ \text{J/kg} \cdot \text{K})(-54 + 273.15)\ \text{K}} \\
&= 297\ \text{m/s}
\end{aligned}
$$

Since the air speed is

$$
V = \frac{(885\ \text{km/h})(1000\ \text{m/km})}{(3600\ \text{s/hr})} = 246\ \text{m/s}
$$

the ratio is

$$
\frac{V}{c} = \frac{246\ \text{m/s}}{297\ \text{m/s}} = 0.828 \qquad \textbf{(Ans)}
$$

COMMENT This ratio is called the *Mach number*, Ma. If Ma < 1.0 the aircraft is flying at *subsonic* speeds, whereas for Ma > 1.0 it is flying at *supersonic* speeds. The Mach number is an important dimensionless parameter used in the study of the flow of gases at high speeds and will be further discussed in Chapters 7 and 11.

By repeating the calculations for different temperatures, the results shown in Fig. E1.7 are obtained. Because the speed of

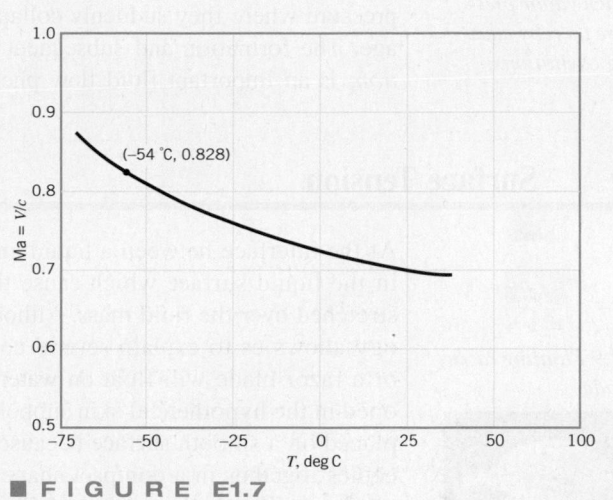

(−54 °C, 0.828)

■ FIGURE E1.7

sound increases with increasing temperature, for a constant airplane speed, the Mach number decreases as the temperature increases.

1.8 Vapor Pressure

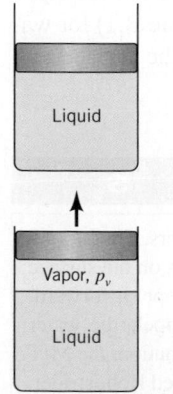

A liquid boils when the pressure is reduced to the vapor pressure.

It is a common observation that liquids such as water and gasoline will evaporate if they are simply placed in a container open to the atmosphere. Evaporation takes place because some liquid molecules at the surface have sufficient momentum to overcome the intermolecular cohesive forces and escape into the atmosphere. If the container is closed with a small air space left above the surface, and this space evacuated to form a vacuum, a pressure will develop in the space as a result of the vapor that is formed by the escaping molecules. When an equilibrium condition is reached so that the number of molecules leaving the surface is equal to the number entering, the vapor is said to be saturated and the pressure that the vapor exerts on the liquid surface is termed the *vapor pressure*, p_v. Similarly, if the end of a completely liquid-filled container is moved as shown in the figure in the margin without letting any air into the container, the space between the liquid and the end becomes filled with vapor at a pressure equal to the vapor pressure.

Since the development of a vapor pressure is closely associated with molecular activity, the value of vapor pressure for a particular liquid depends on temperature. Values of vapor pressure for water at various temperatures can be found in Appendix B (Table B.1), and the values of vapor pressure for several common liquids at room temperatures are given in Table 1.3.

Boiling, which is the formation of vapor bubbles within a fluid mass, is initiated when the absolute pressure in the fluid reaches the vapor pressure. As commonly observed in the kitchen, water

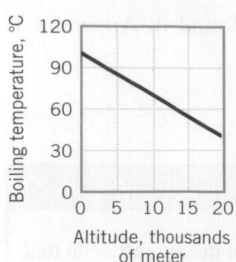

In flowing liquids it is possible for the pressure in localized regions to reach vapor pressure thereby causing cavitation.

at standard atmospheric pressure will boil when the temperature reaches 100 °C (373 K)—that is, the vapor pressure of water at 100 °C (373 K) is 101.3 kPa (abs). However, if we attempt to boil water at a higher elevation, say 9000 m above sea level (the approximate elevation of Mt. Everest), where the atmospheric pressure is 30 kPa (abs), we find that boiling will start when the temperature is about 69 °C (294 K). At this temperature the vapor pressure of water is 30 kPa (abs). For the U.S. Standard Atmosphere (see Section 2.4), the boiling temperature is a function of altitude as shown in the figure in the margin. Thus, boiling can be induced at a given pressure acting on the fluid by raising the temperature, or at a given fluid temperature by lowering the pressure.

An important reason for our interest in vapor pressure and boiling lies in the common observation that in flowing fluids it is possible to develop very low pressure due to the fluid motion, and if the pressure is lowered to the vapor pressure, boiling will occur. For example, this phenomenon may occur in flow through the irregular, narrowed passages of a valve or pump. When vapor bubbles are formed in a flowing fluid, they are swept along into regions of higher pressure where they suddenly collapse with sufficient intensity to actually cause structural damage. The formation and subsequent collapse of vapor bubbles in a flowing fluid, called *cavitation,* is an important fluid flow phenomenon to be given further attention in Chapters 3 and 7.

1.9   Surface Tension

V1.9 Floating razor blade

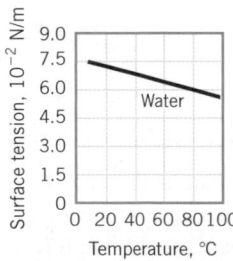

At the interface between a liquid and a gas, or between two immiscible liquids, forces develop in the liquid surface which cause the surface to behave as if it were a "skin" or "membrane" stretched over the fluid mass. Although such a skin is not actually present, this conceptual analogy allows us to explain several commonly observed phenomena. For example, a steel needle or a razor blade will float on water if placed gently on the surface because the tension developed in the hypothetical skin supports it. Small droplets of mercury will form into spheres when placed on a smooth surface because the cohesive forces in the surface tend to hold all the molecules together in a compact shape. Similarly, discrete bubbles will form in a liquid. (See the photograph at the beginning of Chapter 1.)

These various types of surface phenomena are due to the unbalanced cohesive forces acting on the liquid molecules at the fluid surface. Molecules in the interior of the fluid mass are surrounded by molecules that are attracted to each other equally. However, molecules along the surface are subjected to a net force toward the interior. The apparent physical consequence of this unbalanced force along the surface is to create the hypothetical skin or membrane. A tensile force may be considered to be acting in the plane of the surface along any line in the surface. The intensity of the molecular attraction per unit length along any line in the surface is called the *surface tension* and is designated by the Greek symbol σ (sigma). For a given liquid the surface tension depends on temperature as well as the other fluid it is in contact with at the interface. The dimensions of surface tension are FL^{-1} with SI units of N/m. Values of surface tension for some common liquids (in contact with air) are given in Table 1.3 and in Appendix B (Table B.1) for water at various temperatures. As indicated by the figure in the margin, the value of the surface tension decreases as the temperature increases.

F l u i d s   i n   t h e   N e w s

Walking on water Water striders are insects commonly found on ponds, rivers, and lakes that appear to "walk" on water. A typical length of a water strider is about 0.4 in. (1 cm), and they can cover 100 body lengths in one second. It has long been recognized that it is *surface tension* that keeps the water strider from sinking below the surface. What has been puzzling is how they propel themselves at such a high speed. They can't pierce the water surface or they would sink. A team of mathematicians and engineers from the Massachusetts Institute of Technology (MIT) applied conventional flow visualization techniques and high-speed video to examine in detail the movement of the water striders. They found that each stroke of the insect's legs creates dimples on the surface with underwater swirling vortices sufficient to propel it forward. It is the rearward motion of the vortices that propels the water strider forward. To further substantiate their explanation, the MIT team built a working model of a water strider, called Robostrider, which creates surface ripples and underwater vortices as it moves across a water surface. Waterborne creatures, such as the water strider, provide an interesting world dominated by surface tension. (See Problem 1.97.)

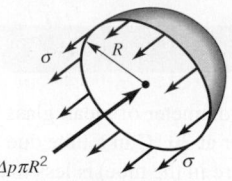

■ FIGURE 1.8   **Forces acting on one-half of a liquid drop.**

The pressure inside a drop of fluid can be calculated using the free-body diagram in Fig. 1.8. If the spherical drop is cut in half (as shown), the force developed around the edge due to surface tension is $2\pi R\sigma$. This force must be balanced by the pressure difference, Δp, between the internal pressure, p_i, and the external pressure, p_e, acting over the circular area, πR^2. Thus,

$$2\pi R\sigma = \Delta p\,\pi R^2$$

or

$$\Delta p = p_i - p_e = \frac{2\sigma}{R} \tag{1.20}$$

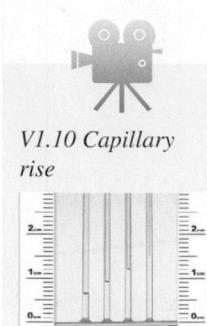

V1.10 Capillary rise

It is apparent from this result that the pressure inside the drop is greater than the pressure surrounding the drop. (Would the pressure on the inside of a bubble of water be the same as that on the inside of a drop of water of the same diameter and at the same temperature?)

Among common phenomena associated with surface tension is the rise (or fall) of a liquid in a capillary tube. If a small open tube is inserted into water, the water level in the tube will rise above the water level outside the tube, as is illustrated in Fig. 1.9a. In this situation we have a liquid–gas–solid interface. For the case illustrated there is an attraction (adhesion) between the wall of the tube and liquid molecules which is strong enough to overcome the mutual attraction (cohesion) of the molecules and pull them up the wall. Hence, the liquid is said to *wet* the solid surface.

The height, h, is governed by the value of the surface tension, σ, the tube radius, R, the specific weight of the liquid, γ, and the *angle of contact,* θ, between the fluid and tube. From the free-body diagram of Fig. 1.9b we see that the vertical force due to the surface tension is equal to $2\pi R\sigma\cos\theta$ and the weight is $\gamma\pi R^2 h$ and these two forces must balance for equilibrium. Thus,

$$\gamma\pi R^2 h = 2\pi R\sigma\cos\theta$$

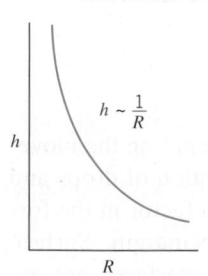

$h \sim \dfrac{1}{R}$

h

R

so that the height is given by the relationship

$$h = \frac{2\sigma\cos\theta}{\gamma R} \tag{1.21}$$

The angle of contact is a function of both the liquid and the surface. For water in contact with clean glass $\theta \approx 0°$. It is clear from Eq. 1.21 that the height is inversely proportional to the tube radius, and therefore, as indicated by the figure in the margin, the rise of a liquid in a tube as a result of capillary action becomes increasingly pronounced as the tube radius is decreased.

Capillary action in small tubes, which involves a liquid–gas–solid interface, is caused by surface tension.

If adhesion of molecules to the solid surface is weak compared to the cohesion between molecules, the liquid will not wet the surface and the level in a tube placed in a nonwetting liquid will actually be depressed, as shown in Fig. 1.9c. Mercury is a good example of a nonwetting liquid when it is in contact with a glass tube. For nonwetting liquids the angle of contact is greater than 90°, and for mercury in contact with clean glass $\theta \approx 130°$.

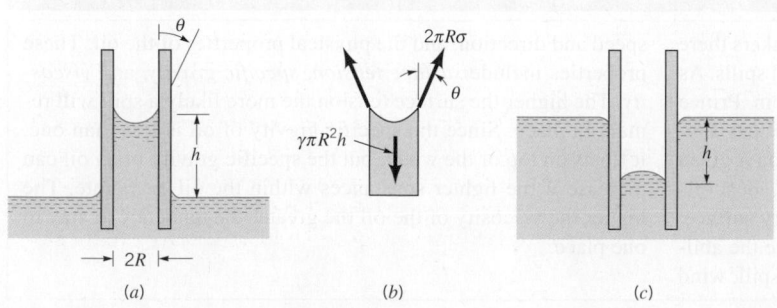

(a)   (b)   (c)

■ FIGURE 1.9   **Effect of capillary action in small tubes. (a) Rise of column for a liquid that wets the tube. (b) Free-body diagram for calculating column height. (c) Depression of column for a nonwetting liquid.**

EXAMPLE 1.8 Capillary Rise in a Tube

GIVEN Pressures are sometimes determined by measuring the height of a column of liquid in a vertical tube.

FIND What diameter of clean glass tubing is required so that the rise of water at 20 °C in a tube due to capillary action (as opposed to pressure in the tube) is less than $h = 1.0$ mm?

SOLUTION

From Eq. 1.22

$$h = \frac{2\sigma \cos\theta}{\gamma R}$$

so that

$$R = \frac{2\sigma \cos\theta}{\gamma h}$$

For water at 20 °C (from Table B.1), $\sigma = 0.0728$ N/m and $\gamma = 9.789$ kN/m³. Since $\theta \approx 0°$ it follows that for $h = 1.0$ mm,

$$R = \frac{2(0.0728 \text{ N/m})(1)}{(9.789 \times 10^3 \text{ N/m}^3)(1.0 \text{ mm})(10^{-3} \text{ m/mm})}$$

$$= 0.0149 \text{ m}$$

and the minimum required tube diameter, D, is

$$D = 2R = 0.0298 \text{ m} = 29.8 \text{ mm} \qquad \textbf{(Ans)}$$

COMMENT By repeating the calculations for various values of the capillary rise, h, the results shown in Fig. E1.8 are obtained.

Note that as the allowable capillary rise is decreased, the diameter of the tube must be significantly increased. There is always some capillarity effect, but it can be minimized by using a large enough diameter tube.

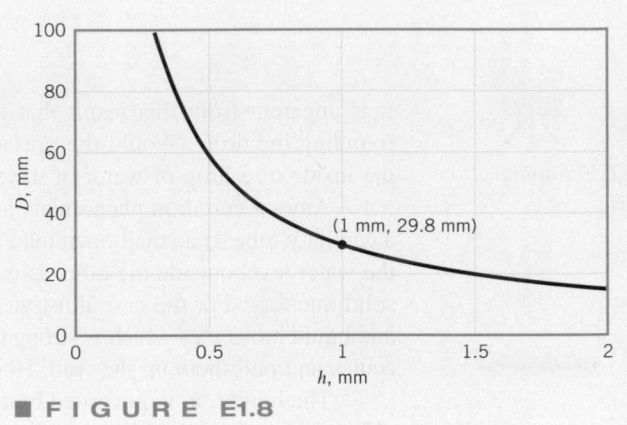

(1 mm, 29.8 mm)

■ **FIGURE E1.8**

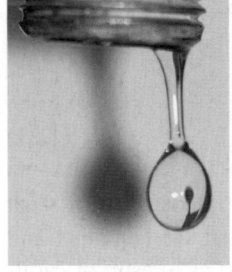

(Photograph copyright 2007 by Andrew Davidhazy, Rochester Institute of Technology.)

Surface tension effects play a role in many fluid mechanics problems, including the movement of liquids through soil and other porous media, flow of thin films, formation of drops and bubbles, and the breakup of liquid jets. For example, surface tension is a main factor in the formation of drops from a leaking faucet, as shown in the photograph in the margin. Surface phenomena associated with liquid–gas, liquid–liquid, and liquid–gas–solid interfaces are exceedingly complex, and a more detailed and rigorous discussion of them is beyond the scope of this text. Fortunately, in many fluid mechanics problems, surface phenomena, as characterized by surface tension, are not important, since inertial, gravitational, and viscous forces are much more dominant.

Fluids in the News

Spreading of oil spills With the large traffic in oil tankers there is great interest in the prevention of and response to oil spills. As evidenced by the famous *Exxon Valdez* oil spill in Prince William Sound in 1989, oil spills can create disastrous environmental problems. It is not surprising that much attention is given to the rate at which an oil spill spreads. When spilled, most oils tend to spread horizontally into a smooth and slippery surface, called a slick. There are many factors which influence the ability of an oil slick to spread, including the size of the spill, wind speed and direction, and the physical properties of the oil. These properties include *surface tension, specific gravity,* and *viscosity*. The higher the surface tension the more likely a spill will remain in place. Since the specific gravity of oil is less than one, it floats on top of the water, but the specific gravity of an oil can increase if the lighter substances within the oil evaporate. The higher the viscosity of the oil the greater the tendency to stay in one place.

1.10   A Brief Look Back in History

Before proceeding with our study of fluid mechanics, we should pause for a moment to consider the history of this important engineering science. As is true of all basic scientific and engineering disciplines, their actual beginnings are only faintly visible through the haze of early antiquity. But, we know that interest in fluid behavior dates back to the ancient civilizations. Through necessity there was a practical concern about the manner in which spears and arrows could be propelled through the air, in the development of water supply and irrigation systems, and in the design of boats and ships. These developments were of course based on trial and error procedures without any knowledge of mathematics or mechanics. However, it was the accumulation of such empirical knowledge that formed the basis for further development during the emergence of the ancient Greek civilization and the subsequent rise of the Roman Empire. Some of the earliest writings that pertain to modern fluid mechanics are those of Archimedes (287–212 B.C.), a Greek mathematician and inventor who first expressed the principles of hydrostatics and flotation. Elaborate water supply systems were built by the Romans during the period from the fourth century B.C. through the early Christian period, and Sextus Julius Frontinus (A.D. 40–103), a Roman engineer, described these systems in detail. However, for the next 1000 years during the Middle Ages (also referred to as the Dark Ages), there appears to have been little added to further understanding of fluid behavior.

As shown in Fig. 1.10, beginning with the Renaissance period (about the fifteenth century) a rather continuous series of contributions began that forms the basis of what we consider to be the science of fluid mechanics. Leonardo da Vinci (1452–1519) described through sketches and writings many different types of flow phenomena. The work of Galileo Galilei (1564–1642) marked the beginning of experimental mechanics. Following the early Renaissance period and during the seventeenth and eighteenth centuries, numerous significant contributions were made. These include theoretical and mathematical advances associated with the famous names of Newton, Bernoulli, Euler, and d'Alembert. Experimental aspects of fluid mechanics were also advanced during this period, but unfortunately the two different approaches, theoretical and experimental, developed along separate paths. *Hydrodynamics* was the term associated with the theoretical or mathematical study of idealized, frictionless fluid behavior, with the term *hydraulics* being used to describe the applied or experimental aspects of real fluid behavior, particularly the behavior of water. Further contributions and refinements were made to both theoretical hydrodynamics and experimental hydraulics during the nineteenth century, with the general differential equations describing fluid motions that are used in modern fluid mechanics being developed in this period. Experimental hydraulics became more of a science, and many of the results of experiments performed during the nineteenth century are still used today.

At the beginning of the twentieth century, both the fields of theoretical hydrodynamics and experimental hydraulics were highly developed, and attempts were being made to unify the two. In 1904 a classic paper was presented by a German professor, Ludwig Prandtl (1875–1953), who introduced the concept of a "fluid boundary layer," which laid the foundation for the unification of

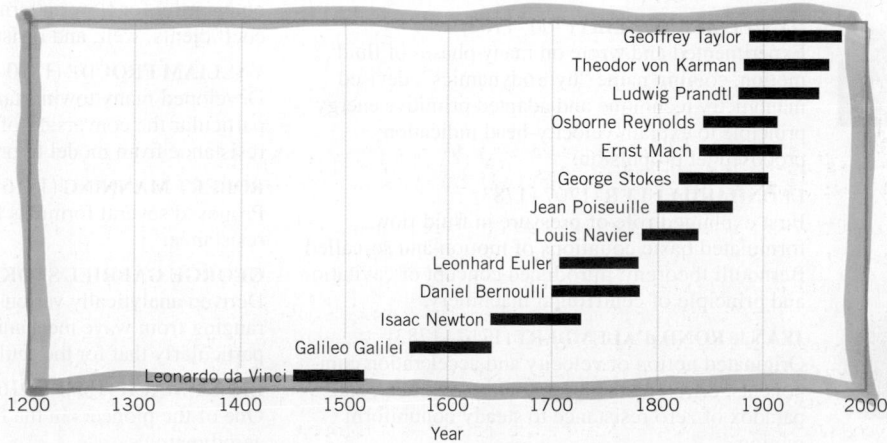

■ F I G U R E 1.10   **Time line of some contributors to the science of fluid mechanics.**

the theoretical and experimental aspects of fluid mechanics. Prandtl's idea was that for flow next to a solid boundary a thin fluid layer (boundary layer) develops in which friction is very important, but outside this layer the fluid behaves very much like a frictionless fluid. This relatively simple concept provided the necessary impetus for the resolution of the conflict between the hydrodynamicists and the hydraulicists. Prandtl is generally accepted as the founder of modern fluid mechanics.

Also, during the first decade of the twentieth century, powered flight was first successfully demonstrated with the subsequent vastly increased interest in *aerodynamics*. Because the design of aircraft required a degree of understanding of fluid flow and an ability to make accurate predictions of the effect of air flow on bodies, the field of aerodynamics provided a great stimulus for the many rapid developments in fluid mechanics that took place during the twentieth century.

As we proceed with our study of the fundamentals of fluid mechanics, we will continue to note the contributions of many of the pioneers in the field. Table 1.5 provides a chronological

The rich history of fluid mechanics is fascinating, and many of the contributions of the pioneers in the field are noted in the succeeding chapters.

Leonardo da Vinci

Isaac Newton

Daniel Bernoulli

Ernst Mach

■ TABLE 1.5
Chronological Listing of Some Contributors to the Science of Fluid Mechanics Noted in the Text[a]

ARCHIMEDES (287–212 B.C.)
Established elementary principles of buoyancy and flotation.

SEXTUS JULIUS FRONTINUS (A.D. 40–103)
Wrote treatise on Roman methods of water distribution.

LEONARDO da VINCI (1452–1519)
Expressed elementary principle of continuity; observed and sketched many basic flow phenomena; suggested designs for hydraulic machinery.

GALILEO GALILEI (1564–1642)
Indirectly stimulated experimental hydraulics; revised Aristotelian concept of vacuum.

EVANGELISTA TORRICELLI (1608–1647)
Related barometric height to weight of atmosphere, and form of liquid jet to trajectory of free fall.

BLAISE PASCAL (1623–1662)
Finally clarified principles of barometer, hydraulic press, and pressure transmissibility.

ISAAC NEWTON (1642–1727)
Explored various aspects of fluid resistance—inertial, viscous, and wave; discovered jet contraction.

HENRI de PITOT (1695–1771)
Constructed double-tube device to indicate water velocity through differential head.

DANIEL BERNOULLI (1700–1782)
Experimented and wrote on many phases of fluid motion, coining name "hydrodynamics"; devised manometry technique and adapted primitive energy principle to explain velocity-head indication; proposed jet propulsion.

LEONHARD EULER (1707–1783)
First explained role of pressure in fluid flow; formulated basic equations of motion and so-called Bernoulli theorem; introduced concept of cavitation and principle of centrifugal machinery.

JEAN le ROND d'ALEMBERT (1717–1783)
Originated notion of velocity and acceleration components, differential expression of continuity, and paradox of zero resistance to steady nonuniform motion.

ANTOINE CHEZY (1718–1798)
Formulated similarity parameter for predicting flow characteristics of one channel from measurements on another.

GIOVANNI BATTISTA VENTURI (1746–1822)
Performed tests on various forms of mouthpieces—in particular, conical contractions and expansions.

LOUIS MARIE HENRI NAVIER (1785–1836)
Extended equations of motion to include "molecular" forces.

AUGUSTIN LOUIS de CAUCHY (1789–1857)
Contributed to the general field of theoretical hydrodynamics and to the study of wave motion.

GOTTHILF HEINRICH LUDWIG HAGEN (1797–1884)
Conducted original studies of resistance in and transition between laminar and turbulent flow.

JEAN LOUIS POISEUILLE (1799–1869)
Performed meticulous tests on resistance of flow through capillary tubes.

HENRI PHILIBERT GASPARD DARCY (1803–1858)
Performed extensive tests on filtration and pipe resistance; initiated open-channel studies carried out by Bazin.

JULIUS WEISBACH (1806–1871)
Incorporated hydraulics in treatise on engineering mechanics, based on original experiments; noteworthy for flow patterns, nondimensional coefficients, weir, and resistance equations.

WILLIAM FROUDE (1810–1879)
Developed many towing-tank techniques, in particular the conversion of wave and boundary layer resistance from model to prototype scale.

ROBERT MANNING (1816–1897)
Proposed several formulas for open-channel resistance.

GEORGE GABRIEL STOKES (1819–1903)
Derived analytically various flow relationships ranging from wave mechanics to viscous resistance—particularly that for the settling of spheres.

ERNST MACH (1838–1916)
One of the pioneers in the field of supersonic aerodynamics.

Osborne Reynolds

Ludwig Prandtl

■ **TABLE 1.5** (continued)

OSBORNE REYNOLDS (1842–1912)
Described original experiments in many fields—cavitation, river model similarity, pipe resistance—and devised two parameters for viscous flow; adapted equations of motion of a viscous fluid to mean conditions of turbulent flow.

JOHN WILLIAM STRUTT, LORD RAYLEIGH (1842–1919)
Investigated hydrodynamics of bubble collapse, wave motion, jet instability, laminar flow analogies, and dynamic similarity.

VINCENZ STROUHAL (1850–1922)
Investigated the phenomenon of "singing wires."

EDGAR BUCKINGHAM (1867–1940)
Stimulated interest in the United States in the use of dimensional analysis.

MORITZ WEBER (1871–1951)
Emphasized the use of the principles of similitude in fluid flow studies and formulated a capillarity similarity parameter.

LUDWIG PRANDTL (1875–1953)
Introduced concept of the boundary layer and is generally considered to be the father of present-day fluid mechanics.

LEWIS FERRY MOODY (1880–1953)
Provided many innovations in the field of hydraulic machinery. Proposed a method of correlating pipe resistance data which is widely used.

THEODOR VON KÁRMÁN (1881–1963)
One of the recognized leaders of twentieth century fluid mechanics. Provided major contributions to our understanding of surface resistance, turbulence, and wake phenomena.

PAUL RICHARD HEINRICH BLASIUS (1883–1970)
One of Prandtl's students who provided an analytical solution to the boundary layer equations. Also, demonstrated that pipe resistance was related to the Reynolds number.

[a]Adapted from Ref. 2; used by permission of the Iowa Institute of Hydraulic Research, The University of Iowa.

listing of some of these contributors and reveals the long journey that makes up the history of fluid mechanics. This list is certainly not comprehensive with regard to all of the past contributors, but includes those who are mentioned in this text. As mention is made in succeeding chapters of the various individuals listed in Table 1.5, a quick glance at this table will reveal where they fit into the historical chain.

It is, of course, impossible to summarize the rich history of fluid mechanics in a few paragraphs. Only a brief glimpse is provided, and we hope it will stir your interest. References 2 to 5 are good starting points for further study, and in particular Ref. 2 provides an excellent, broad, easily read history. Try it—you might even enjoy it!

1.11   Chapter Summary and Study Guide

This introductory chapter discussed several fundamental aspects of fluid mechanics. Methods for describing fluid characteristics both quantitatively and qualitatively are considered. For a quantitative description, units are required, and in this text, International (SI) System units are used. For the qualitative description the concept of dimensions is introduced in which basic dimensions such as length, L, time, T, and mass, M, are used to provide a description of various quantities of interest. The use of dimensions is helpful in checking the generality of equations, as well as serving as the basis for the powerful tool of dimensional analysis discussed in detail in Chapter 7.

Various important fluid properties are defined, including fluid density, specific weight, specific gravity, viscosity, bulk modulus, speed of sound, vapor pressure, and surface tension. The ideal gas law is introduced to relate pressure, temperature, and density in common gases, along with a brief discussion of the compression and expansion of gases. The distinction between absolute and gage pressure is introduced and this important idea is explored more fully in Chapter 2.

The following checklist provides a study guide for this chapter. When your study of the entire chapter and end-of-chapter exercises has been completed you should be able to

■ write out meanings of the terms listed here in the margin and understand each of the related concepts. These terms are particularly important and are set in *italic, bold, and color* type in the text.

■ determine the dimensions of common physical quantities.

■ determine whether an equation is a general or restricted homogeneous equation.

■ use SI system of units.

■ calculate the density, specific weight, or specific gravity of a fluid from a knowledge of any two of the three.

■ calculate the density, pressure, or temperature of an ideal gas (with a given gas constant) from a knowledge of any two of the three.

■ relate the pressure and density of a gas as it is compressed or expanded using Eqs. 1.13 and 1.14.

■ use the concept of viscosity to calculate the shearing stress in simple fluid flows.

■ calculate the speed of sound in fluids using Eq. 1.18 for liquids and Eq. 1.19 for gases.

■ determine whether boiling or cavitation will occur in a liquid using the concept of vapor pressure.

■ use the concept of surface tension to solve simple problems involving liquid–gas or liquid–solid–gas interfaces.

Some of the important equations in this chapter are:

Specific weight	$\gamma = \rho g$	**(1.5)**
Specific gravity	$SG = \dfrac{\rho}{\rho_{H_2O@4\,°C}}$	**(1.6)**
Ideal gas law	$\rho = \dfrac{p}{RT}$	**(1.7)**
Newtonian fluid shear stress	$\tau = \mu\,\dfrac{du}{dy}$	**(1.8)**
Bulk modulus	$E_v = -\dfrac{dp}{d\forall/\forall}$	**(1.11)**
Speed of sound in an ideal gas	$c = \sqrt{kRT}$	**(1.19)**
Capillary rise in a tube	$h = \dfrac{2\sigma\cos\theta}{\gamma R}$	**(1.21)**

References

1. Reid, R. C., Prausnitz, J. M., and Sherwood, T. K., *The Properties of Gases and Liquids,* 3rd Ed., McGraw-Hill, New York, 1977.

2. Rouse, H. and Ince, S., *History of Hydraulics,* Iowa Institute of Hydraulic Research, Iowa City, 1957, Dover, New York, 1963.

3. Tokaty, G. A., *A History and Philosophy of Fluid Mechanics,* G. T. Foulis and Co., Ltd., Oxfordshire, Great Britain, 1971.

4. Rouse, H., *Hydraulics in the United States 1776–1976,* Iowa Institute of Hydraulic Research, Iowa City, Iowa, 1976.

5. Garbrecht, G., ed., *Hydraulics and Hydraulic Research—A Historical Review,* A. A. Balkema, Rotterdam, Netherlands, 1987.

6. Brenner, M. P., Shi, X. D., Eggens, J., and Nagel, S. R., *Physics of Fluids,* Vol. 7, No. 9, 1995.

7. Shi, X. D., Brenner, M. P., and Nagel, S. R., *Science,* Vol. 265, 1994.

Review Problems

Go to Appendix G for a set of review problems with answers. Detailed solutions can be found in *Student Solution Manual and Study* *Guide for Fundamentals of Fluid Mechanics,* by Munson, et al. (© 2009 John Wiley and Sons, Inc.).

Problems

Note: Unless specific values of required fluid properties are given in the statement of the problem, use the values found in the tables on the inside of the front cover. Problems designated with an (*) are intended to be solved with the aid of a programmable calculator or a computer. Problems designated with a (†) are "open-ended" problems and require critical thinking in that to work them one must make various assumptions and provide the necessary data. There is not a unique answer to these problems.

   Answers to the even-numbered problems are listed at the end of the book. Access to the videos that accompany problems can be obtained through the book's web site, www.wiley.com/college/munson. The lab-type problems can also be accessed on this web site.

Section 1.2 Dimensions, Dimensional Homogeneity, and Units

1.1 The force, F, of the wind blowing against a building is given by $F = C_D \rho V^2 A/2$, where V is the wind speed, ρ the density of the air, A the cross-sectional area of the building, and C_D is a constant termed the drag coefficient. Determine the dimensions of the drag coefficient.

1.2 Verify the dimensions, in both the *FLT* and *MLT* systems, of the following quantities which appear in Table 1.1: (a) volume, (b) acceleration, (c) mass, (d) moment of inertia (area), and (e) work.

1.3 Determine the dimensions, in both the *FLT* system and the *MLT* system, for (a) the product of force times acceleration, (b) the product of force times velocity divided by area, and (c) momentum divided by volume.

1.4 Verify the dimensions, in both the *FLT* system and the *MLT* system, of the following quantities which appear in Table 1.1: (a) frequency, (b) stress, (c) strain, (d) torque, and (e) work.

1.5 If u is a velocity, x a length, and t a time, what are the dimensions (in the *MLT* system) of (a) $\partial u/\partial t$, (b) $\partial^2 u/\partial x \partial t$, and (c) $\int (\partial u/\partial t)\, dx$?

1.6 If p is a pressure, V a velocity, and ρ a fluid density, what are the dimensions (in the *MLT* system) of (a) p/ρ, (b) $pV\rho$, and (c) $p/\rho V^2$?

1.7 If V is a velocity, ℓ a length, and ν a fluid property (the kinematic viscosity) having dimensions of $L^2 T^{-1}$, which of the following combinations are dimensionless: (a) $V\ell\nu$, (b) $V\ell/\nu$, (c) $V^2\nu$, (d) $V/\ell\nu$?

1.8 If V is a velocity, determine the dimensions of Z, α, and G, which appear in the dimensionally homogeneous equation

$$V = Z(\alpha - 1) + G$$

1.9 The volume rate of flow, Q, through a pipe containing a slowly moving liquid is given by the equation

$$Q = \frac{\pi R^4 \Delta p}{8\mu\ell}$$

where R is the pipe radius, Δp the pressure drop along the pipe, μ a fluid property called viscosity $(FL^{-2}T)$, and ℓ the length of pipe. What are the dimensions of the constant $\pi/8$? Would you classify this equation as a general homogeneous equation? Explain.

1.10 According to information found in an old hydraulics book, the energy loss per unit weight of fluid flowing through a nozzle connected to a hose can be estimated by the formula

$$h = (0.04 \text{ to } 0.09)(D/d)^4 V^2/2g$$

where h is the energy loss per unit weight, D the hose diameter, d the nozzle tip diameter, V the fluid velocity in the hose, and g the acceleration of gravity. Do you think this equation is valid in any system of units? Explain.

1.11 The pressure difference, Δp, across a partial blockage in an artery (called a *stenosis*) is approximated by the equation

$$\Delta p = K_v \frac{\mu V}{D} + K_u \left(\frac{A_0}{A_1} - 1\right)^2 \rho V^2$$

where V is the blood velocity, μ the blood viscosity $(FL^{-2}T)$, ρ the blood density (ML^{-3}), D the artery diameter, A_0 the area of the unobstructed artery, and A_1 the area of the stenosis. Determine the dimensions of the constants K_v and K_u. Would this equation be valid in any system of units?

1.12 Assume that the speed of sound, c, in a fluid depends on an elastic modulus, E_v, with dimensions FL^{-2}, and the fluid density, ρ, in the form $c = (E_v)^a (\rho)^b$. If this is to be a dimensionally homogeneous equation, what are the values for a and b? Is your result consistent with the standard formula for the speed of sound? (See Eq. 1.19b.)

1.13 A formula to estimate the volume rate of flow, Q, flowing over a dam of length, B, is given by the equation

$$Q = 1.70 \, BH^{3/2}$$

where H is the depth of the water above the top of the dam (called the head). This formula gives Q in m³/s when B and H are in meter. Is the constant, 1.70, dimensionless? Would this equation be valid if units other than meter and seconds were used?

†1.14 Cite an example of a restricted homogeneous equation contained in a technical article found in an engineering journal in your field of interest. Define all terms in the equation, explain why it is a restricted equation, and provide a complete journal citation (title, date, etc.).

1.15 Water flows from a large drainage pipe at a rate of 4500 L/min. What is this volume rate of flow in m³/s?

1.16 An important dimensionless parameter in certain types of fluid flow problems is the *Froude number* defined as $V/\sqrt{g\ell}$, where V is a velocity, g the acceleration of gravity, and ℓ a length. Determine the value of the Froude number for $V = 3$ m/s, $g = 9.81$ m/s², and $\ell = 0.6$ m.

Section 1.4   Measures of Fluid Mass and Weight

1.17 Obtain a photograph/image of a situation in which the density or specific weight of a fluid is important. Print this photo and write a brief paragraph that describes the situation involved.

1.18 A tank contains 500 kg of a liquid whose specific gravity is 2. Determine the volume of the liquid in the tank.

1.19 Clouds can weigh thousands of newtons due to their liquid water content. Often this content is measured in grams per cubic meter (g/m^3). Assume that a cumulus cloud occupies a volume of one cubic kilometer, and its liquid water content is $0.2 \ g/m^3$. **(a)** What is the volume of this cloud in cubic kilometers? **(b)** How much does the water in the cloud weigh in newtons?

1.20 A tank of oil has a mass of 365 kg. **(a)** Determine its weight in newtons at the earth's surface. **(b)** What would be its mass (in kg) and its weight (in newtons) if located on the moon's surface where the gravitational attraction is approximately one-sixth that at the earth's surface?

1.21 A certain object weighs 300 N at the earth's surface. Determine the mass of the object (in kilograms) and its weight (in newtons) when located on a planet with an acceleration of gravity equal to $1.2 \ m/s^2$.

1.22 The density of a certain type of jet fuel is $775 \ kg/m^3$. Determine its specific gravity and specific weight.

1.23 A *hydrometer* is used to measure the specific gravity of liquids. (See **Video V2.8**.) For a certain liquid, a hydrometer reading indicates a specific gravity of 1.15. What is the liquid's density and specific weight?

1.24 An open, rigid-walled, cylindrical tank contains $0.1 \ m^3$ of water at $4 \ °C$. Over a 24-hour period of time the water temperature varies from 4 to $32 \ °C$. Make use of the data in Appendix B to determine how much the volume of water will change. For a tank diameter of 0.6 m, would the corresponding change in water depth be very noticeable? Explain.

†1.25 Estimate the number of newtons of mercury it would take to fill your bathtub. List all assumptions and show all calculations.

1.26 A mountain climber's oxygen tank contains 4.45 N of oxygen when he begins his trip at sea level where the acceleration of gravity is $9.81 \ m/s^2$. What is the weight of the oxygen in the tank when he reaches the top of Mt. Everest where the acceleration of gravity is $9.78 \ m/s^2$? Assume that no oxygen has been removed from the tank; it will be used on the descent portion of the climb.

1.27 The information on a can of pop indicates that the can contains 355 mL. The mass of a full can of pop is 0.369 kg while an empty can weighs 0.153 N. Determine the specific weight, density, and specific gravity of the pop and compare your results with the corresponding values for water at $20 \ °C$.

*1.28 The variation in the density of water, ρ, with temperature, T, in the range $20 \ °C \leq T \leq 50 \ °C$, is given in the following table.

Density (kg/m^3)	998.2	997.1	995.7	994.1	992.2	990.2	988.1
Temperature $(°C)$	20	25	30	35	40	45	50

Use these data to determine an empirical equation of the form $\rho = c_1 + c_2 T + c_3 T^2$ which can be used to predict the density over the range indicated. Compare the predicted values with the data given. What is the density of water at $42.1 \ °C$?

1.29 If 1 cup of cream having a density of $1005 \ kg/m^3$ is turned into 3 cups of whipped cream, determine the specific gravity and specific weight of the whipped cream.

†1.30 The presence of raindrops in the air during a heavy rainstorm increases the average density of the air–water mixture. Estimate by what percent the average air–water density is greater than that of just still air. State all assumptions and show calculations.

Section 1.5   Ideal Gas Law

1.31 Determine the mass of air in a $2 \ m^3$ tank if the air is at room temperature, $20 \ °C$, and the absolute pressure within the tank is 200 kPa (abs).

1.32 Nitrogen is compressed to a density of $4 \ kg/m^3$ under an absolute pressure of 400 kPa. Determine the temperature in degrees Celsius.

1.33 The temperature and pressure at the surface of Mars during a Martian spring day were determined to be $-50 \ °C$ and 900 Pa, respectively. **(a)** Determine the density of the Martian atmosphere for these conditions if the gas constant for the Martian atmosphere is assumed to be equivalent to that of carbon dioxide. **(b)** Compare the answer from part (a) with the density of the earth's atmosphere during a spring day when the temperature is $18 \ °C$ and the pressure 101.6 kPa (abs).

1.34 A closed tank having a volume of $0.06 \ m^3$ is filled with 1.3 N of a gas. A pressure gage attached to the tank reads 83 kPa when the gas temperature is $27 \ °C$. There is some question as to whether the gas in the tank is oxygen or helium. Which do you think it is? Explain how you arrived at your answer.

1.35 A compressed air tank contains 5 kg of air at a temperature of $80 \ °C$. A gage on the tank reads 300 kPa. Determine the volume of the tank.

1.36 A rigid tank contains air at a pressure of 620 kPa (abs) and a temperature of $15 \ °C$. By how much will the pressure increase as the temperature is increased to $43 \ °C$?

1.37 The helium-filled blimp shown in Fig. P1.37 is used at various athletic events. Determine the number of newtons of helium within it if its volume is $1926 \ m^3$ and the temperature and pressure are $27 \ °C$ and 98 kPa (abs), respectively.

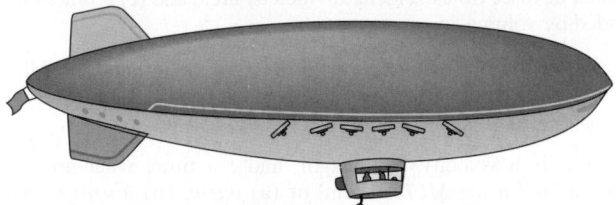

■ **F I G U R E   P1.37**

*1.38 Develop a computer program for calculating the density of an ideal gas when the gas pressure in pascals (abs), the temperature in degrees Celsius, and the gas constant in $J/kg \cdot K$ are specified. Plot the density of helium as a function of temperature from $0 \ °C$ to $200 \ °C$ and pressures of 50, 100, 150, and 200 kPa (abs).

Section 1.6   Viscosity (Also see Lab Problems 1.98 and 1.99.)

1.39 Obtain a photograph/image of a situation in which the viscosity of a fluid is important. Print this photo and write a brief paragraph that describes the situation involved.

1.40 For flowing water, what is the magnitude of the velocity gradient needed to produce a shear stress of $1.0 \ N/m^2$?

1.41 Make use of the data in Appendix B to determine the dynamic viscosity of glycerin at 29 °C.

1.42 One type of *capillary-tube viscometer* is shown in **Video V1.5** and in Fig. P1.42. For this device the liquid to be tested is drawn into the tube to a level above the top etched line. The time is then obtained for the liquid to drain to the bottom etched line. The kinematic viscosity, ν, in m^2/s is then obtained from the equation $\nu = KR^4t$ where K is a constant, R is the radius of the capillary tube in mm, and t is the drain time in seconds. When glycerin at 20 °C is used as a calibration fluid in a particular viscometer, the drain time is 1430 s. When a liquid having a density of 970 kg/m^3 is tested in the same viscometer the drain time is 900 s. What is the dynamic viscosity of this liquid?

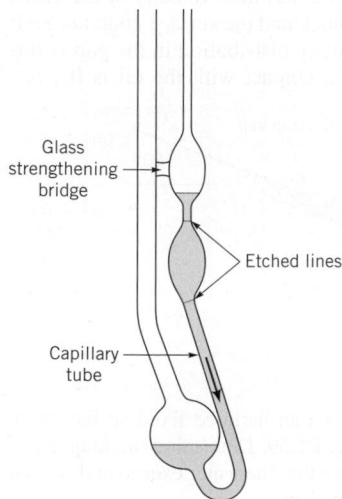

Glass strengthening bridge

Etched lines

Capillary tube

■ **F I G U R E   P1.42**

1.43 The viscosity of a soft drink was determined by using a capillary tube viscometer similar to that shown in Fig. P1.42 and **Video V1.5**. For this device the kinematic viscosity, ν, is directly proportional to the time, t, that it takes for a given amount of liquid to flow through a small capillary tube. That is, $\nu = Kt$. The following data were obtained from regular pop and diet pop. The corresponding measured specific gravities are also given. Based on these data, by what percent is the absolute viscosity, μ, of regular pop greater than that of diet pop?

	Regular pop	Diet pop
$t(s)$	377.8	300.3
SG	1.044	1.003

1.44 Determine the ratio of the dynamic viscosity of water to air at a temperature of 60 °C. Compare this value with the corresponding ratio of kinematic viscosities. Assume the air is at standard atmospheric pressure.

1.45 The kinematic viscosity of oxygen at 20 °C and a pressure of 150 kPa (abs) is 10.4 mm^2/s. Determine the dynamic viscosity of oxygen at this temperature and pressure.

*1.46 Fluids for which the shearing stress, τ, is not linearly related to the rate of shearing strain, $\dot{\gamma}$, are designated as non-Newtonian fluids. Such fluids are commonplace and can exhibit unusual behavior, as shown in **Video V1.6**. Some experimental data obtained for a particular non-Newtonian fluid at 27 °C are shown below.

τ(kPa)	0	0.1	0.37	0.89	1.5
$\dot{\gamma}$ (s^{-1})	0	50	100	150	200

Plot these data and fit a second-order polynomial to the data using a suitable graphing program. What is the apparent viscosity of this fluid when the rate of shearing strain is 70 s^{-1}? Is this apparent viscosity larger or smaller than that for water at the same temperature?

1.47 Water flows near a flat surface and some measurements of the water velocity, u, parallel to the surface, at different heights, y, above the surface are obtained. At the surface $y = 0$. After an analysis of the data, the lab technician reports that the velocity distribution in the range $0 < y < 0.1$ m is given by the equation

$$u = 0.81 + 9.2y + 4.1 \times 10^3 y^3$$

with u in m/s when y is in m. (**a**) Do you think that this equation would be valid in any system of units? Explain. (**b**) Do you think this equation is correct? Explain. You may want to look at **Video 1.4** to help you arrive at your answer.

1.48 Calculate the Reynolds numbers for the flow of water and for air through a 4-mm-diameter tube, if the mean velocity is 3 m/s and the temperature is 30 °C in both cases (see Example 1.4). Assume the air is at standard atmospheric pressure.

1.49 For air at standard atmospheric pressure the values of the constants that appear in the Sutherland equation (Eq. 1.9) are $C = 1.458 \times 10^{-6}$ $kg/(m \cdot s \cdot K^{1/2})$ and $S = 110.4$ K. Use these values to predict the viscosity of air at 10 °C and 90 °C and compare with values given in Table B.2 in Appendix B.

*1.50 Use the values of viscosity of air given in Table B.2 at temperatures of 0, 20, 40, 60, 80, and 100 °C to determine the constants C and S which appear in the Sutherland equation (Eq. 1.9). Compare your results with the values given in Problem 1.49. (*Hint:* Rewrite the equation in the form

$$\frac{T^{3/2}}{\mu} = \left(\frac{1}{C}\right)T + \frac{S}{C}$$

and plot $T^{3/2}/\mu$ versus T. From the slope and intercept of this curve, C and S can be obtained.)

1.51 The viscosity of a fluid plays a very important role in determining how a fluid flows. (See **Video V1.3**.) The value of the viscosity depends not only on the specific fluid but also on the fluid temperature. Some experiments show that when a liquid, under the action of a constant driving pressure, is forced with a low velocity, V, through a small horizontal tube, the velocity is given by the equation $V = K/\mu$. In this equation K is a constant for a given tube and pressure, and μ is the dynamic viscosity. For a particular liquid of interest, the viscosity is given by Andrade's equation (Eq. 1.10) with $D = 239 \times 10^{-7}$ $N \cdot s/m^2$ and $B = 2222$ K. By what percentage will the velocity increase as the liquid temperature is increased from 4 to 38 °C (277 to 311 K)? Assume all other factors remain constant.

*1.52 Use the value of the viscosity of water given in Table B.1 at temperatures of 0, 20, 40, 60, 80, and 100 °C to determine the constants D and B which appear in Andrade's equation (Eq. 1.10). Calculate the value of the viscosity at 50 °C and compare with the value given in Table B.1. (*Hint:* Rewrite the equation in the form

$$\ln \mu = (B)\frac{1}{T} + \ln D$$

and plot $\ln \mu$ versus $1/T$. From the slope and intercept of this curve, B and D can be obtained. If a nonlinear curve-fitting program is available the constants can be obtained directly from Eq. 1.10 without rewriting the equation.)

1.53 For a parallel plate arrangement of the type shown in Fig. 1.4 it is found that when the distance between plates is 2 mm, a shearing stress of 150 Pa develops at the upper plate when it is pulled at a velocity of 1 m/s. Determine the viscosity of the fluid between the plates.

1.54 Two flat plates are oriented parallel above a fixed lower plate as shown in Fig. P1.54. The top plate, located a distance b above the fixed plate, is pulled along with speed V. The other thin plate is located a distance cb, where $0 < c < 1$, above the fixed plate. This plate moves with speed V_1, which is determined by the viscous shear forces imposed on it by the fluids on its top and bottom. The fluid on the top is twice as viscous as that on the bottom. Plot the ratio V_1/V as a function of c for $0 < c < 1$.

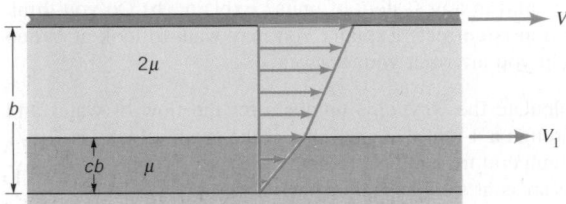

■ **F I G U R E   P1.54**

1.55 There are many fluids that exhibit non-Newtonian behavior (see, for example, Video V1.6). For a given fluid the distinction between Newtonian and non-Newtonian behavior is usually based on measurements of shear stress and rate of shearing strain. Assume that the viscosity of blood is to be determined by measurements of shear stress, τ, and rate of shearing strain, du/dy, obtained from a small blood sample tested in a suitable viscometer. Based on the data given below determine if the blood is a Newtonian or non-Newtonian fluid. Explain how you arrived at your answer.

τ(N/m^2)	0.04	0.06	0.12	0.18	0.30	0.52	1.12	2.10
du/dy (s^{-1})	2.25	4.50	11.25	22.5	45.0	90.0	225	450

1.56 The sled shown in Fig. P1.56 slides along on a thin horizontal layer of water between the ice and the runners. The horizontal force that the water puts on the runners is equal to 5.3 N when the sled's speed is 15 m/s. The total area of both runners in contact with the water is 0.007 m^2, and the viscosity of the water is 168×10^{-5} N·s/m^2. Determine the thickness of the water layer under the runners. Assume a linear velocity distribution in the water layer.

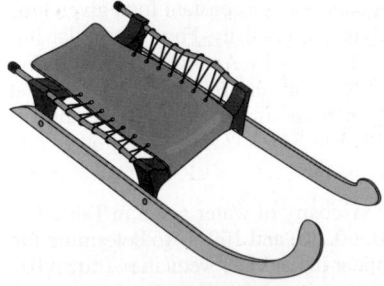

■ **F I G U R E   P1.56**

1.57 A 25-mm-diameter shaft is pulled through a cylindrical bearing as shown in Fig. P1.57. The lubricant that fills the 0.3-mm gap between the shaft and bearing is an oil having a kinematic viscosity of 8.0×10^{-4} m^2/s and a specific gravity of 0.91. Determine the force P required to pull the shaft at a velocity of 3 m/s. Assume the velocity distribution in the gap is linear.

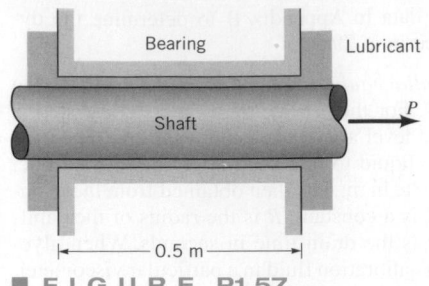

■ **F I G U R E   P1.57**

1.58 A 10-kg block slides down a smooth inclined surface as shown in Fig. P1.58. Determine the terminal velocity of the block if the 0.1-mm gap between the block and the surface contains SAE 30 oil at 15 °C. Assume the velocity distribution in the gap is linear, and the area of the block in contact with the oil is 0.1 m^2.

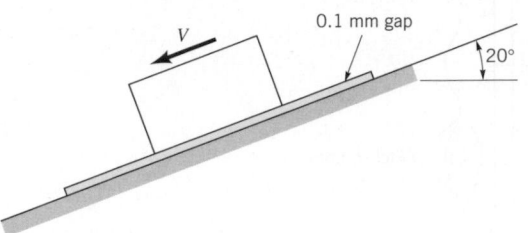

■ **F I G U R E   P1.58**

1.59 A layer of water flows down an inclined fixed surface with the velocity profile shown in Fig. P1.59. Determine the magnitude and direction of the shearing stress that the water exerts on the fixed surface for $U = 2$ m/s and $h = 0.1$ m.

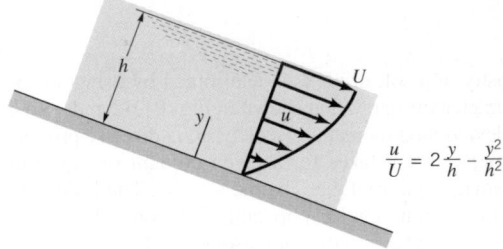

$$\frac{u}{U} = 2\frac{y}{h} - \frac{y^2}{h^2}$$

■ **F I G U R E   P1.59**

*1.60 Standard air flows past a flat surface and velocity measurements near the surface indicate the following distribution:

y (m)	15×10^{-4}	3×10^{-4}	6×10^{-4}	12×10^{-4}	18×10^{-4}	24×10^{-4}
u (m/s)	0.23	0.46	0.92	1.94	3.11	4.39

The coordinate y is measured normal to the surface and u is the velocity parallel to the surface. **(a)** Assume the velocity distribution is of the form

$$u = C_1 y + C_2 y^3$$

and use a standard curve-fitting technique to determine the constants C_1 and C_2. **(b)** Make use of the results of part (a) to determine the magnitude of the shearing stress at the wall ($y = 0$) and at $y = 0.015$ m.

1.61 A new computer drive is proposed to have a disc, as shown in Fig. P1.61. The disc is to rotate at 10,000 rpm, and the reader head is to be positioned 0.012 mm. above the surface of the disc. Estimate the shearing force on the reader head as a result of the air between the disc and the head.

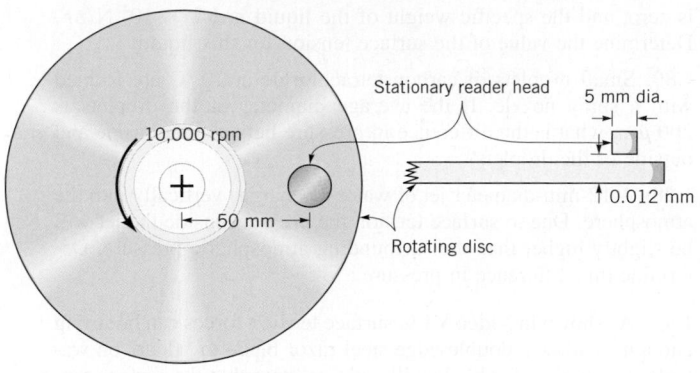

■ **F I G U R E   P1.61**

1.62 The space between two 15 cm-long concentric cylinders is filled with glycerin (viscosity = 407×10^{-3} N · s/m²). The inner cylinder has a radius of 7.6 cm and the gap width between cylinders is 0.25 cm. Determine the torque and the power required to rotate the inner cylinder at 180 rev/min. The outer cylinder is fixed. Assume the velocity distribution in the gap to be linear.

1.63 A pivot bearing used on the shaft of an electrical instrument is shown in Fig. P1.63. An oil with a viscosity of $\mu = 0.479$ N·s/m² fills the 0.025 mm gap between the rotating shaft and the stationary base. Determine the frictional torque on the shaft when it rotates at 5,000 rpm.

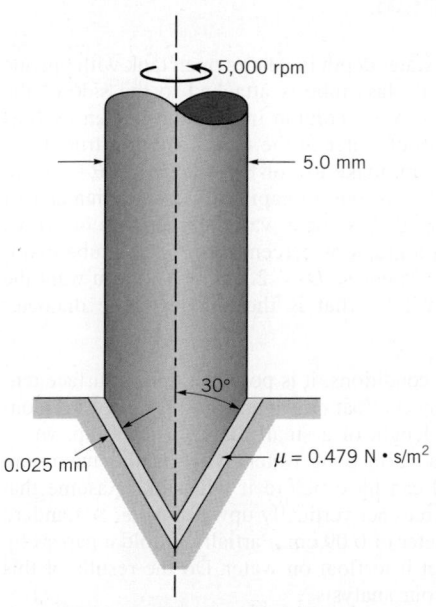

■ **F I G U R E   P1.63**

1.64 The viscosity of liquids can be measured through the use of a *rotating cylinder viscometer* of the type illustrated in Fig. P1.64. In this device the outer cylinder is fixed and the inner cylinder is rotated with an angular velocity, ω. The torque $\mathscr{T}$ required to develop ω is measured and the viscosity is calculated from these two measurements. **(a)** Develop an equation relating μ, ω, $\mathscr{T}$, ℓ, R_o, and R_i. Neglect end effects and assume the velocity distribution in the gap is linear. **(b)** The following torque-angular velocity data were obtained with a rotating cylinder viscometer of the type discussed in part (a).

Torque (N · m)	17.8	35.3	53.6	71.5	87.9	106.5
Angular velocity (rad/s)	1.0	2.0	3.0	4.0	5.0	6.0

For this viscometer $R_o = 6.35$ cm, $R_i = 6.22$ cm, and $\ell = 12.7$ cm Make use of these data and a standard curve-fitting program to determine the viscosity of the liquid contained in the viscometer.

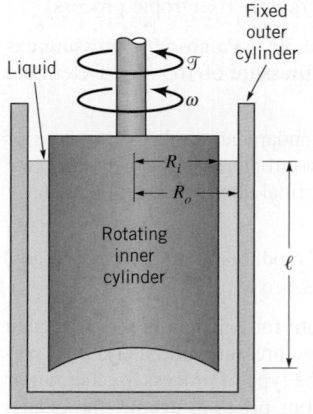

■ **F I G U R E   P1.64**

1.65 A 30 cm-diameter circular plate is placed over a fixed bottom plate with a 0.25 cm gap between the two plates filled with glycerin as shown in Fig. P1.65. Determine the torque required to rotate the circular plate slowly at 2 rpm. Assume that the velocity distribution in the gap is linear and that the shear stress on the edge of the rotating plate is negligible.

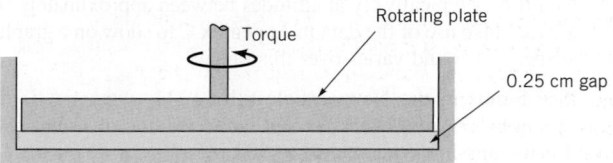

■ **F I G U R E   P1.65**

†**1.66** Vehicle shock absorbers damp out oscillations caused by road roughness. Describe how a temperature change may affect the operation of a shock absorber.

1.67 Some measurements on a blood sample at 37 °C indicate a shearing stress of 0.52 N/m² for a corresponding rate of shearing strain of 200 s⁻¹. Determine the apparent viscosity of the blood and compare it with the viscosity of water at the same temperature.

Section 1.7   Compressibility of Fluids

1.68 Obtain a photograph/image of a situation in which the compressibility of a fluid is important. Print this photo and write a brief paragraph that describes the situation involved.

1.69 A sound wave is observed to travel through a liquid with a speed of 1500 m/s. The specific gravity of the liquid is 1.5. Determine the bulk modulus for this fluid.

1.70 Estimate the increase in pressure (in kPa) required to decrease a unit volume of mercury by 0.1%.

1.71 A 1-m³ volume of water is contained in a rigid container. Estimate the change in the volume of the water when a piston applies a pressure of 35 MPa.

1.72 Determine the speed of sound at 20 °C in **(a)** air, **(b)** helium, and **(c)** natural gas (methane). Express your answer in m/s.

1.73 Air is enclosed by a rigid cylinder containing a piston. A pressure gage attached to the cylinder indicates an initial reading of 172 kPa. Determine the reading on the gage when the piston has compressed the air to one-third its original volume. Assume the compression process to be isothermal and the local atmospheric pressure to be 101.3 kPa.

1.74 Repeat Problem 1.73 if the compression process takes place without friction and without heat transfer (isentropic process).

1.75 Carbon dioxide at 30 °C and 300 kPa absolute pressure expands isothermally to an absolute pressure of 165 kPa. Determine the final density of the gas.

1.76 Natural gas at 21 °C and standard atmospheric pressure of 101.3 kPa (abs) is compressed isentropically to a new absolute pressure of 483 kPa. Determine the final density and temperature of the gas.

1.77 Compare the isentropic bulk modulus of air at 101 kPa (abs) with that of water at the same pressure.

*1.78 Develop a computer program for calculating the final gage pressure of gas when the initial gage pressure, initial and final volumes, atmospheric pressure, and the type of process (isothermal or isentropic) are specified. Check your program against the results obtained for Problem 1.73.

1.79 An important dimensionless parameter concerned with very high-speed flow is the *Mach number*, defined as V/c, where V is the speed of the object such as an airplane or projectile, and c is the speed of sound in the fluid surrounding the object. For a projectile traveling at 1290 km/h through air at 10 °C and standard atmospheric pressure, what is the value of the Mach number?

1.80 Jet airliners typically fly at altitudes between approximately 0 to 12,200 m. Make use of the data in Appendix C to show on a graph how the speed of sound varies over this range.

1.81 (See Fluids in the News article titled "This water jet is a blast," Section 1.7.1) By what percent is the volume of water decreased if its pressure is increased to 304 MPa?

Section 1.8 Vapor Pressure

1.82 During a mountain climbing trip it is observed that the water used to cook a meal boils at 90 °C rather than the standard 100 °C at sea level. At what altitude are the climbers preparing their meal? (See Tables B.1 and C.1 for data needed to solve this problem.)

1.83 When a fluid flows through a sharp bend, low pressures may develop in localized regions of the bend. Estimate the minimum absolute pressure (in kPa) that can develop without causing cavitation if the fluid is water at 70 °C.

1.84 Estimate the minimum absolute pressure (in pascals) that can be developed at the inlet of a pump to avoid cavitation if the fluid is carbon tetrachloride at 20 °C.

1.85 When water at 70 °C flows through a converging section of pipe, the pressure decreases in the direction of flow. Estimate the minimum absolute pressure that can develop without causing cavitation.

1.86 At what atmospheric pressure will water boil at 35 °C?

Section 1.9 Surface Tension

1.87 Obtain a photograph/image of a situation in which the surface tension of a fluid is important. Print this photo and write a brief paragraph that describes the situation involved.

1.88 When a 2-mm-diameter tube is inserted into a liquid in an open tank, the liquid is observed to rise 10 mm above the free surface of the liquid. The contact angle between the liquid and the tube is zero, and the specific weight of the liquid is 1.2×10^4 N/m^3. Determine the value of the surface tension for this liquid.

1.89 Small droplets of carbon tetrachloride at 20 °C are formed with a spray nozzle. If the average diameter of the droplets is 200 μm, what is the difference in pressure between the inside and outside of the droplets?

1.90 A 12-mm-diameter jet of water discharges vertically into the atmosphere. Due to surface tension the pressure inside the jet will be slightly higher than the surrounding atmospheric pressure. Determine this difference in pressure.

1.91 As shown in **Video V1.9**, surface tension forces can be strong enough to allow a double-edge steel razor blade to "float" on water, but a single-edge blade will sink. Assume that the surface tension forces act at an angle θ relative to the water surface as shown in Fig. P1.91. **(a)** The mass of the double-edge blade is 0.64×10^{-3} kg, and the total length of its sides is 206 mm. Determine the value of θ required to maintain equilibrium between the blade weight and the resultant surface tension force. **(b)** The mass of the single-edge blade is 2.61×10^{-3} kg, and the total length of its sides is 154 mm. Explain why this blade sinks. Support your answer with the necessary calculations.

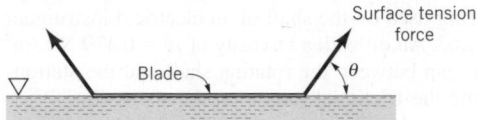

■ **FIGURE P1.91**

1.92 To measure the water depth in a large open tank with opaque walls, an open vertical glass tube is attached to the side of the tank. The height of the water column in the tube is then used as a measure of the depth of water in the tank. **(a)** For a true water depth in the tank of 1 m, make use of Eq. 1.21 (with $\theta \simeq 0°$) to determine the percent error due to capillarity as the diameter of the glass tube is changed. Assume a water temperature of 30 °C. Show your results on a graph of percent error versus tube diameter, D, in the range 0.25 cm $< D < 2.5$ cm. **(b)** If you want the error to be less than 1%, what is the smallest tube diameter allowed?

1.93 Under the right conditions, it is possible, due to surface tension, to have metal objects float on water. (See **Video V1.9**.) Consider placing a short length of a small diameter steel (sp. wt. = 77 kN/m^3) rod on a surface of water. What is the maximum diameter that the rod can have before it will sink? Assume that the surface tension forces act vertically upward. *Note:* A standard paper clip has a diameter of 0.09 cm. Partially unfold a paper clip and see if you can get it to float on water. Do the results of this experiment support your analysis?

1.94 An open, clean glass tube, having a diameter of 3 mm, is inserted vertically into a dish of mercury at 20 °C. How far will the column of mercury in the tube be depressed?

1.95 An open, clean glass tube ($\theta = 0°$) is inserted vertically into a pan of water. What tube diameter is needed if the water level in the tube is to rise one tube diameter (due to surface tension)?

1.96 Determine the height that water at 15 °C will rise due to capillary action in a clean, 0.6 cm-diameter tube. What will be the height if the diameter is reduced to 0.03 cm?

1.97 (See Fluids in the News article titled "Walking on water," Section 1.9.) **(a)** The water strider bug shown in Fig. P1.97 is supported on the surface of a pond by surface tension acting along the interface between the water and the bug's legs. Determine the minimum length of this interface needed to support the bug. Assume the bug weighs 10^{-4} N and the surface tension force acts vertically upwards. **(b)** Repeat part (a) if surface tension were to support a person weighing 750 N.

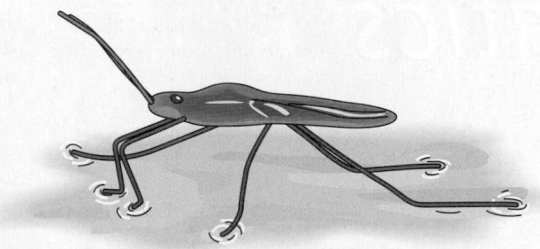

■ **F I G U R E   P1.97**

■ **Lab Problems**

1.98   This problem involves the use of a Stormer viscometer to determine whether a fluid is a Newtonian or a non-Newtonian fluid. To proceed with this problem, go to Appendix H, which is located on the book's web site, www.wiley.com/college/munson.

1.99   This problem involves the use of a capillary tube viscometer to determine the kinematic viscosity of water as a function of temperature. To proceed with this problem, go to Appendix H, which is located on the book's web site, www.wiley.com/college/munson.

■ **Life Long Learning Problems**

1.100   Although there are numerous non-Newtonian fluids that occur naturally (quick sand and blood among them), with the advent of modern chemistry and chemical processing, many new, manmade non-Newtonian fluids are now available for a variety of novel application. Obtain information about the discovery and use of newly developed non-Newtonian fluids. Summarize your findings in a brief report.

1.101   For years, lubricating oils and greases obtained by refining crude oil have been used to lubricate moving parts in a wide variety of machines, motors, and engines. With the increasing cost of crude oil and the potential for the reduced availability of it, the need for nonpetroleum based lubricants has increased considerably. Obtain information about non-petroleum based lubricants. Summarize your findings in a brief report.

1.102   It is predicted that nano-technology and the use of nano-sized objects will allow many processes, procedures, and products that, as of now, are difficult for us to comprehend. Among new nano-technology areas is that of nano-scale fluid mechanics. Fluid behavior at the nano-scale can be entirely different than that for the usual everyday flows with which we are familiar. Obtain information about various aspects of nano-fluid mechanics. Summarize your findings in a brief report.

■ **FE Exam Problems**

Sample FE (Fundamentals of Engineering) exam question for fluid mechanics are provided on the book's web site, www.wiley.com/college/munson.

2 Fluid Statics

Learning Objectives

After completing this chapter, you should be able to:

- determine the pressure at various locations in a fluid at rest.
- explain the concept of manometers and apply appropriate equations to determine pressures.
- calculate the hydrostatic pressure force on a plane or curved submerged surface.
- calculate the buoyant force and discuss the stability of floating or submerged objects.

In this chapter we will consider an important class of problems in which the fluid is either at rest or moving in such a manner that there is no relative motion between adjacent particles. In both instances there will be no shearing stresses in the fluid, and the only forces that develop on the surfaces of the particles will be due to the pressure. Thus, our principal concern is to investigate pressure and its variation throughout a fluid and the effect of pressure on submerged surfaces. The absence of shearing stresses greatly simplifies the analysis and, as we will see, allows us to obtain relatively simple solutions to many important practical problems.

2.1 Pressure at a Point

As we briefly discussed in Chapter 1, the term pressure is used to indicate the normal force per unit area at a given point acting on a given plane within the fluid mass of interest. A question that immediately arises is how the pressure at a point varies with the orientation of the plane passing

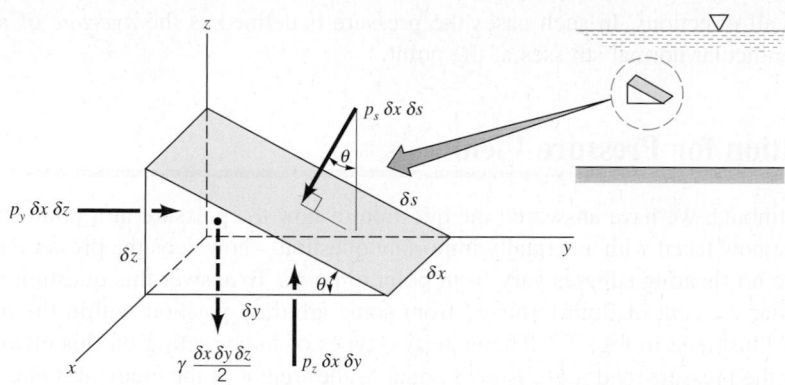

■ FIGURE 2.1   Forces on an arbitrary wedge-shaped element of fluid.

through the point. To answer this question, consider the free-body diagram, illustrated in Fig. 2.1, that was obtained by removing a small triangular wedge of fluid from some arbitrary location within a fluid mass. Since we are considering the situation in which there are no shearing stresses, the only external forces acting on the wedge are due to the pressure and the weight. For simplicity the forces in the x direction are not shown, and the z axis is taken as the vertical axis so the weight acts in the negative z direction. Although we are primarily interested in fluids at rest, to make the analysis as general as possible, we will allow the fluid element to have accelerated motion. The assumption of zero shearing stresses will still be valid so long as the fluid element moves as a rigid body; that is, there is no relative motion between adjacent elements.

The equations of motion (Newton's second law, $\mathbf{F} = m\mathbf{a}$) in the y and z directions are, respectively,

$$\sum F_y = p_y\, \delta x\, \delta z - p_s\, \delta x\, \delta s \sin\theta = \rho\, \frac{\delta x\, \delta y\, \delta z}{2}\, a_y$$

$$\sum F_z = p_z\, \delta x\, \delta y - p_s\, \delta x\, \delta s \cos\theta - \gamma\, \frac{\delta x\, \delta y\, \delta z}{2} = \rho\, \frac{\delta x\, \delta y\, \delta z}{2}\, a_z$$

where p_s, p_y, and p_z are the average pressures on the faces, γ and ρ are the fluid specific weight and density, respectively, and a_y, a_z the accelerations. Note that a pressure must be multiplied by an appropriate area to obtain the force generated by the pressure. It follows from the geometry that

$$\delta y = \delta s \cos\theta \qquad \delta z = \delta s \sin\theta$$

so that the equations of motion can be rewritten as

$$p_y - p_s = \rho a_y \frac{\delta y}{2}$$

$$p_z - p_s = (\rho a_z + \gamma)\frac{\delta z}{2}$$

The pressure at a point in a fluid at rest is independent of direction.

Since we are really interested in what is happening at a point, we take the limit as δx, δy, and δz approach zero (while maintaining the angle θ), and it follows that

$$p_y = p_s \qquad p_z = p_s$$

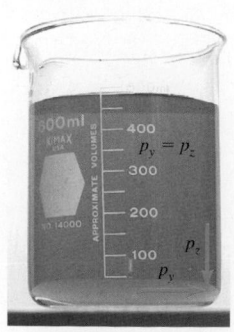

or $p_s = p_y = p_z$. The angle θ was arbitrarily chosen so we can conclude that *the pressure at a point in a fluid at rest, or in motion, is independent of direction as long as there are no shearing stresses present.* This important result is known as *Pascal's law,* named in honor of Blaise Pascal (1623–1662), a French mathematician who made important contributions in the field of hydrostatics. Thus, as shown by the photograph in the margin, at the junction of the side and bottom of the beaker, the pressure is the same on the side as it is on the bottom. In Chapter 6 it will be shown that for moving fluids in which there is relative motion between particles (so that shearing stresses develop), the normal stress at a point, which corresponds to pressure in fluids at rest, is not necessarily the same in all directions. In such cases the pressure is defined as the *average* of any three mutually perpendicular normal stresses at the point.

2.2 Basic Equation for Pressure Field

Although we have answered the question of how the pressure at a point varies with direction, we are now faced with an equally important question—how does the pressure in a fluid in which there are no shearing stresses vary from point to point? To answer this question consider a small rectangular element of fluid removed from some arbitrary position within the mass of fluid of interest as illustrated in Fig. 2.2. There are two types of forces acting on this element: *surface forces* due to the pressure, and a *body force* equal to the weight of the element. Other possible types of body forces, such as those due to magnetic fields, will not be considered in this text.

The pressure may vary across a fluid particle.

If we let the pressure at the center of the element be designated as p, then the average pressure on the various faces can be expressed in terms of p and its derivatives, as shown in Fig. 2.2. We are actually using a Taylor series expansion of the pressure at the element center to approximate the pressures a short distance away and neglecting higher order terms that will vanish as we let δx, δy, and δz approach zero. This is illustrated by the figure in the margin. For simplicity the surface forces in the x direction are not shown. The resultant surface force in the y direction is

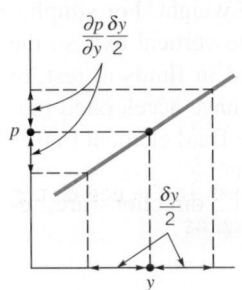

$$\delta F_y = \left(p - \frac{\partial p}{\partial y}\frac{\delta y}{2} \right)\delta x\,\delta z - \left(p + \frac{\partial p}{\partial y}\frac{\delta y}{2} \right)\delta x\,\delta z$$

or

$$\delta F_y = -\frac{\partial p}{\partial y}\,\delta x\,\delta y\,\delta z$$

Similarly, for the x and z directions the resultant surface forces are

$$\delta F_x = -\frac{\partial p}{\partial x}\,\delta x\,\delta y\,\delta z \qquad \delta F_z = -\frac{\partial p}{\partial z}\,\delta x\,\delta y\,\delta z$$

The resultant surface force acting on the element can be expressed in vector form as

$$\delta \mathbf{F}_s = \delta F_x \hat{\mathbf{i}} + \delta F_y \hat{\mathbf{j}} + \delta F_z \hat{\mathbf{k}}$$

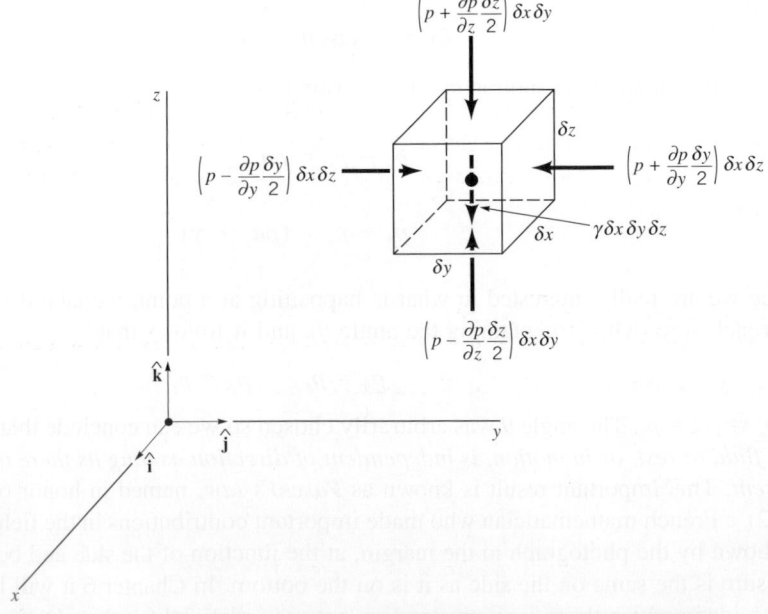

■ **FIGURE 2.2** Surface and body forces acting on small fluid element.

The resultant sur-face force acting on a small fluid ele-ment depends only on the pressure gradient if there are no shearing stresses present.

or

$$\delta \mathbf{F}_s = -\left(\frac{\partial p}{\partial x}\,\hat{\mathbf{i}} + \frac{\partial p}{\partial y}\,\hat{\mathbf{j}} + \frac{\partial p}{\partial z}\,\hat{\mathbf{k}}\right)\delta x\,\delta y\,\delta z \tag{2.1}$$

where $\hat{\mathbf{i}}, \hat{\mathbf{j}}$, and $\hat{\mathbf{k}}$ are the unit vectors along the coordinate axes shown in Fig. 2.2. The group of terms in parentheses in Eq. 2.1 represents in vector form the *pressure gradient* and can be written as

$$\frac{\partial p}{\partial x}\,\hat{\mathbf{i}} + \frac{\partial p}{\partial y}\,\hat{\mathbf{j}} + \frac{\partial p}{\partial z}\,\hat{\mathbf{k}} = \nabla p$$

where

$$\nabla(\ ) = \frac{\partial(\ )}{\partial x}\,\hat{\mathbf{i}} + \frac{\partial(\ )}{\partial y}\,\hat{\mathbf{j}} + \frac{\partial(\ )}{\partial z}\,\hat{\mathbf{k}}$$

and the symbol ∇ is the *gradient* or "del" vector operator. Thus, the resultant surface force per unit volume can be expressed as

$$\frac{\delta \mathbf{F}_s}{\delta x\,\delta y\,\delta z} = -\nabla p$$

Since the z axis is vertical, the weight of the element is

$$-\delta \mathscr{W}\hat{\mathbf{k}} = -\gamma\,\delta x\,\delta y\,\delta z\,\hat{\mathbf{k}}$$

where the negative sign indicates that the force due to the weight is downward (in the negative z direction). Newton's second law, applied to the fluid element, can be expressed as

$$\sum \delta \mathbf{F} = \delta m\,\mathbf{a}$$

where $\Sigma\,\delta\mathbf{F}$ represents the resultant force acting on the element, $\mathbf{a}$ is the acceleration of the element, and δm is the element mass, which can be written as $\rho\,\delta x\,\delta y\,\delta z$. It follows that

$$\sum \delta \mathbf{F} = \delta \mathbf{F}_s - \delta \mathscr{W}\hat{\mathbf{k}} = \delta m\,\mathbf{a}$$

or

$$-\nabla p\,\delta x\,\delta y\,\delta z - \gamma\,\delta x\,\delta y\,\delta z\,\hat{\mathbf{k}} = \rho\,\delta x\,\delta y\,\delta z\,\mathbf{a}$$

and, therefore,

$$-\nabla p - \gamma\hat{\mathbf{k}} = \rho\mathbf{a} \tag{2.2}$$

Equation 2.2 is the general equation of motion for a fluid in which there are no shearing stresses. We will use this equation in Section 2.12 when we consider the pressure distribution in a moving fluid. For the present, however, we will restrict our attention to the special case of a fluid at rest.

2.3 Pressure Variation in a Fluid at Rest

For a fluid at rest $\mathbf{a} = 0$ and Eq. 2.2 reduces to

$$\nabla p + \gamma\hat{\mathbf{k}} = 0$$

or in component form

$$\frac{\partial p}{\partial x} = 0 \qquad \frac{\partial p}{\partial y} = 0 \qquad \frac{\partial p}{\partial z} = -\gamma \tag{2.3}$$

These equations show that the pressure does not depend on x or y. Thus, as we move from point to point in a horizontal plane (any plane parallel to the x–y plane), the pressure does not

For liquids or gases at rest, the pressure gradient in the vertical direction at any point in a fluid depends only on the specific weight of the fluid at that point.

change. Since p depends only on z, the last of Eqs. 2.3 can be written as the ordinary differential equation

$$\frac{dp}{dz} = -\gamma \tag{2.4}$$

Equation 2.4 is the fundamental equation for fluids at rest and can be used to determine how pressure changes with elevation. This equation and the figure in the margin indicate that the pressure gradient in the vertical direction is negative; that is, the pressure decreases as we move upward in a fluid at rest. There is no requirement that γ be a constant. Thus, it is valid for fluids with constant specific weight, such as liquids, as well as fluids whose specific weight may vary with elevation, such as air or other gases. However, to proceed with the integration of Eq. 2.4 it is necessary to stipulate how the specific weight varies with z.

If the fluid is flowing (i.e., not at rest with $\mathbf{a} = 0$), then the pressure variation is much more complex than that given by Eq. 2.4. For example, the pressure distribution on your car as it is driven along the road varies in a complex manner with x, y, and z. This idea is covered in detail in Chapters 3, 6, and 9.

2.3.1 Incompressible Fluid

Since the specific weight is equal to the product of fluid density and acceleration of gravity ($\gamma = \rho g$), changes in γ are caused either by a change in ρ or g. For most engineering applications the variation in g is negligible, so our main concern is with the possible variation in the fluid density. In general, a fluid with constant density is called an *incompressible fluid.* For liquids the variation in density is usually negligible, even over large vertical distances, so that the assumption of constant specific weight when dealing with liquids is a good one. For this instance, Eq. 2.4 can be directly integrated

$$\int_{p_1}^{p_2} dp = -\gamma \int_{z_1}^{z_2} dz$$

to yield

$$p_2 - p_1 = -\gamma(z_2 - z_1)$$

or

$$p_1 - p_2 = \gamma(z_2 - z_1) \tag{2.5}$$

where p_1 and p_2 are pressures at the vertical elevations z_1 and z_2, as is illustrated in Fig. 2.3.

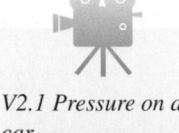

V2.1 Pressure on a car

Equation 2.5 can be written in the compact form

$$p_1 - p_2 = \gamma h \tag{2.6}$$

or

$$p_1 = \gamma h + p_2 \tag{2.7}$$

where h is the distance, $z_2 - z_1$, which is the depth of fluid measured downward from the location of p_2. This type of pressure distribution is commonly called a *hydrostatic distribution,* and Eq. 2.7

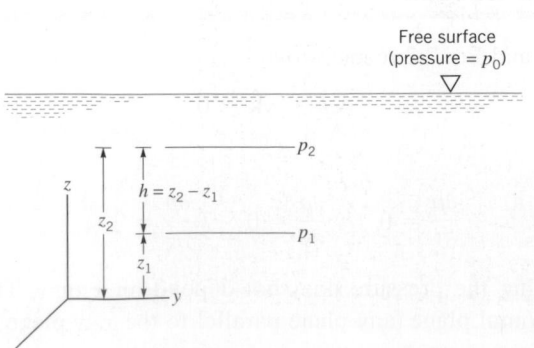

■ **F I G U R E   2.3**   Notation for pressure variation in a fluid at rest with a free surface.

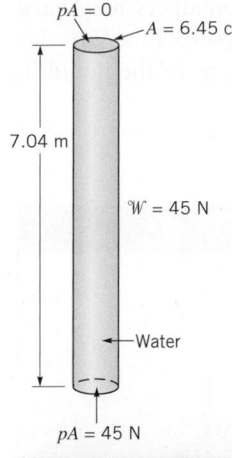

$pA = 0$

$A = 6.45$ cm^2

7.04 m

$W = 45$ N

Water

$pA = 45$ N

shows that in an incompressible fluid at rest the pressure varies linearly with depth. The pressure must increase with depth to "hold up" the fluid above it.

It can also be observed from Eq. 2.6 that the pressure difference between two points can be specified by the distance h since

$$h = \frac{p_1 - p_2}{\gamma}$$

In this case h is called the *pressure head* and is interpreted as the height of a column of fluid of specific weight γ required to give a pressure difference $p_1 - p_2$. For example, a pressure difference of 69 kPa can be specified in terms of pressure head as 7.04 m of water ($\gamma = 9800$ N/m^3), or 518 mm of Hg ($\gamma = 133$ kN/m^3). As illustrated by the figure in the margin, a 7.04-m-tall column of water with a cross-sectional area of 6.45 cm^2 weighs 45 N.

F l u i d s   i n   t h e   N e w s

Giraffe's blood pressure A giraffe's long neck allows it to graze up to 6 m above the ground. It can also lower its head to drink at ground level. Thus, in the circulatory system there is a significant *hydrostatic pressure* effect due to this elevation change. To maintain blood to its head throughout this change in elevation, the giraffe must maintain a relatively high blood pressure at heart level—approximately two and a half times that in humans. To prevent rupture of blood vessels in the high-pressure lower leg re-

gions, giraffes have a tight sheath of thick skin over their lower limbs which acts like an elastic bandage in exactly the same way as do the g-suits of fighter pilots. In addition, valves in the upper neck prevent backflow into the head when the giraffe lowers its head to ground level. It is also thought that blood vessels in the giraffe's kidney have a special mechanism to prevent large changes in filtration rate when blood pressure increases or decreases with its head movement. (See Problem 2.14.)

When one works with liquids there is often a free surface, as is illustrated in Fig. 2.3, and it is convenient to use this surface as a reference plane. The reference pressure p_0 would correspond to the pressure acting on the free surface (which would frequently be atmospheric pressure), and thus if we let $p_2 = p_0$ in Eq. 2.7 it follows that the pressure p at any depth h below the free surface is given by the equation:

$$p = \gamma h + p_0 \tag{2.8}$$

As is demonstrated by Eq. 2.7 or 2.8, the pressure in a homogeneous, incompressible fluid at rest depends on the depth of the fluid relative to some reference plane, and it is *not* influenced by the *size* or *shape* of the tank or container in which the fluid is held. Thus, in Fig. 2.4

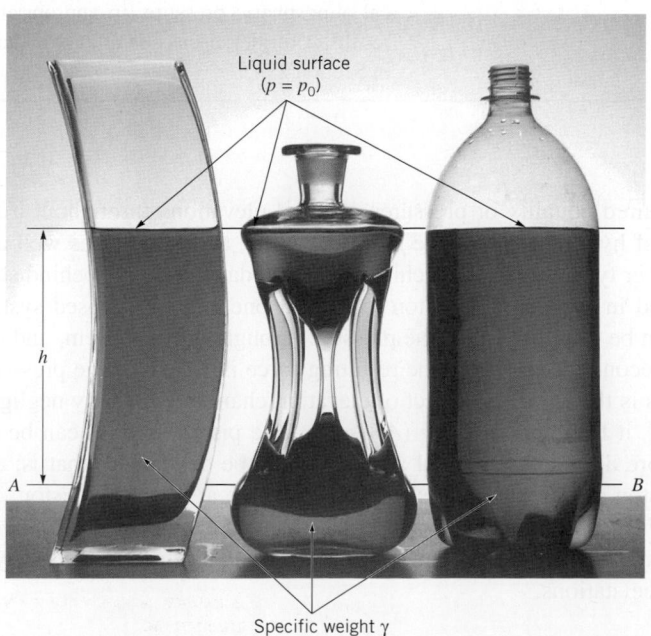

Liquid surface
($p = p_0$)

h

A

B

Specific weight γ

■ **F I G U R E   2.4   Fluid pressure in containers of arbitrary shape.**

the pressure is the same at all points along the line *AB* even though the containers may have the very irregular shapes shown in the figure. The actual value of the pressure along *AB* depends only on the depth, *h*, the surface pressure, p_0, and the specific weight, γ, of the liquid in the container.

EXAMPLE 2.1 Pressure–Depth Relationship

GIVEN   Because of a leak in a buried gasoline storage tank, water has seeped in to the depth shown in Fig. E2.1. The specific gravity of the gasoline is *SG* = 0.68.

FIND   Determine the pressure at the gasoline–water interface and at the bottom of the tank. Express the pressure in units of N/m^2, N/mm^2, and as a pressure head in meters of water.

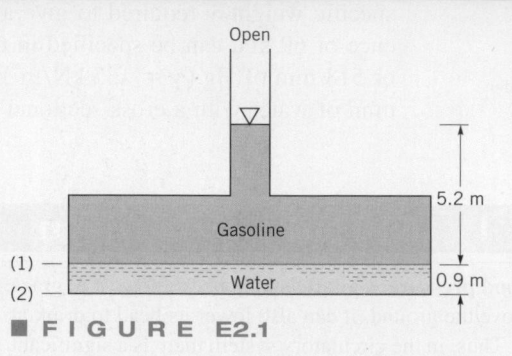

■ **FIGURE E2.1**

SOLUTION

Since we are dealing with liquids at rest, the pressure distribution will be hydrostatic, and therefore the pressure variation can be found from the equation:

$$p = \gamma h + p_0$$

With p_0 corresponding to the pressure at the free surface of the gasoline, then the pressure at the interface is

$$p_1 = SG\gamma_{H_2O}h + p_0$$
$$= (0.68)(9800 \text{ N/m}^3)(5.2 \text{ m}) + p_0$$
$$= 34.7 + p_0 \text{ (kN/m}^2)$$

If we measure the pressure relative to atmospheric pressure (gage pressure), it follows that $p_0 = 0$, and therefore

$$p_1 = 34.7 \text{ kN/m}^2 \qquad \text{(Ans)}$$

$$p_1 = \frac{34.7 \text{ kN/m}^2}{10^6 \text{ mm}^2/\text{m}^2} = 0.035 \text{ N/mm}^2 \qquad \text{(Ans)}$$

$$\frac{p_1}{\gamma_{H_2O}} = \frac{34.7 \text{ kN/m}^2}{9800 \text{ N/m}^3} = 3.54 \text{ m} \qquad \text{(Ans)}$$

It is noted that a rectangular column of water 3.54 m tall and 1 m² in cross section weighs 34.7 kN. A similar column with a 1-mm² cross section weighs 0.035 N.

We can now apply the same relationship to determine the pressure at the tank bottom; that is,

$$p_2 = \gamma_{H_2O} h_{H_2O} + p_1$$
$$= (9800 \text{ N/m}^3)(0.9 \text{ m}) + 34.7 \text{ kN/m} \quad \text{(Ans)}$$
$$= 43.5 \text{ kN/m}^2$$

$$p_2 = \frac{43.5 \text{ kN/m}^2}{10^6 \text{ mm}^2/\text{m}^2} = 0.044 \text{ N/mm}^2 \qquad \text{(Ans)}$$

$$\frac{p_2}{\gamma_{H_2O}} = \frac{43.5 \text{ kN/m}^2}{9800 \text{ N/m}^3} = 4.44 \text{ m} \qquad \text{(Ans)}$$

COMMENT   Observe that if we wish to express these pressures in terms of *absolute* pressure, we would have to add the local atmospheric pressure (in appropriate units) to the previous results. A further discussion of gage and absolute pressure is given in Section 2.5.

The transmission of pressure throughout a stationary fluid is the principle upon which many hydraulic devices are based.

The required equality of pressures at equal elevations throughout a system is important for the operation of hydraulic jacks (see Fig. 2.5*a*), lifts, and presses, as well as hydraulic controls on aircraft and other types of heavy machinery. The fundamental idea behind such devices and systems is demonstrated in Fig. 2.5*b*. A piston located at one end of a closed system filled with a liquid, such as oil, can be used to change the pressure throughout the system, and thus transmit an applied force F_1 to a second piston where the resulting force is F_2. Since the pressure *p* acting on the faces of both pistons is the same (the effect of elevation changes is usually negligible for this type of hydraulic device), it follows that $F_2 = (A_2/A_1)F_1$. The piston area A_2 can be made much larger than A_1 and therefore a large mechanical advantage can be developed; that is, a small force applied at the smaller piston can be used to develop a large force at the larger piston. The applied force could be created manually through some type of mechanical device, such as a hydraulic jack, or through compressed air acting directly on the surface of the liquid, as is done in hydraulic lifts commonly found in service stations.

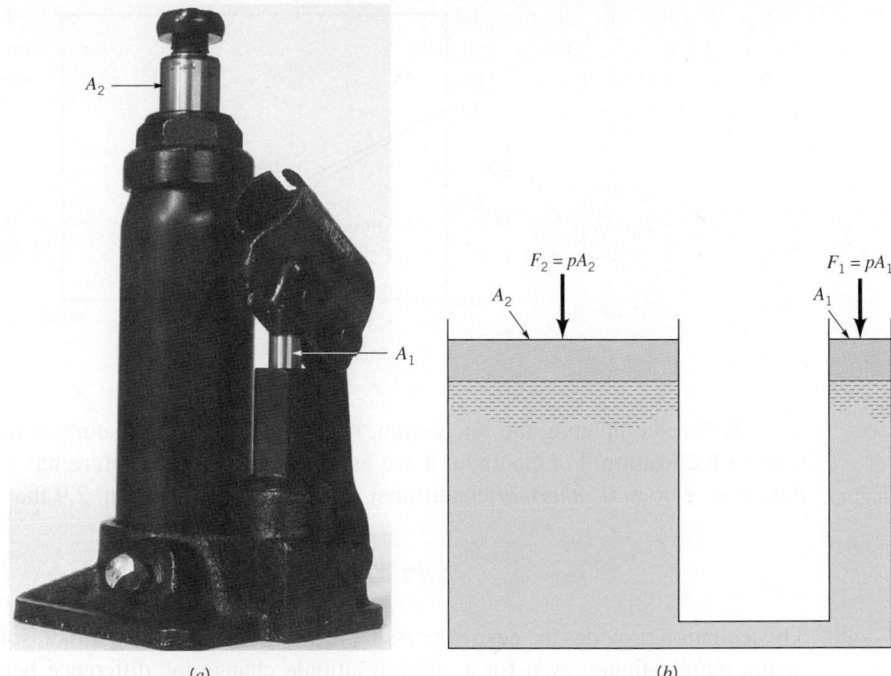

(a) (b)

■ **FIGURE 2.5**   (a) Hydraulic jack, (b) Transmission of fluid pressure.

2.3.2 Compressible Fluid

We normally think of gases such as air, oxygen, and nitrogen as being **compressible fluids** since the density of the gas can change significantly with changes in pressure and temperature. Thus, although Eq. 2.4 applies at a point in a gas, it is necessary to consider the possible variation in γ before the equation can be integrated. However, as was discussed in Chapter 1, the specific weights of common gases are small when compared with those of liquids. For example, the specific weight of air at sea level 15 °C (288 K) is 12 N/m^3, whereas the specific weight of water under the same conditions is 9800 N/m^3. Since the specific weights of gases are comparatively small, it follows from Eq. 2.4 that the pressure gradient in the vertical direction is correspondingly small, and even over distances of several hundred meters the pressure will remain essentially constant for a gas. This means we can neglect the effect of elevation changes on the pressure in gases in tanks, pipes, and so forth in which the distances involved are small.

For those situations in which the variations in heights are large, on the order of thousands of feet, attention must be given to the variation in the specific weight. As is described in Chapter 1, the equation of state for an ideal (or perfect) gas is

$$\rho = \frac{p}{RT}$$

If the specific weight of a fluid varies significantly as we move from point to point, the pressure will no longer vary linearly with depth.

where p is the absolute pressure, R is the gas constant, and T is the absolute temperature. This relationship can be combined with Eq. 2.4 to give

$$\frac{dp}{dz} = -\frac{gp}{RT}$$

and by separating variables

$$\int_{p_1}^{p_2} \frac{dp}{p} = \ln \frac{p_2}{p_1} = -\frac{g}{R} \int_{z_1}^{z_2} \frac{dz}{T} \tag{2.9}$$

where g and R are assumed to be constant over the elevation change from z_1 to z_2. Although the acceleration of gravity, g, does vary with elevation, the variation is very small (see Table C.1 in Appendix C), and g is usually assumed constant at some average value for the range of elevation involved.

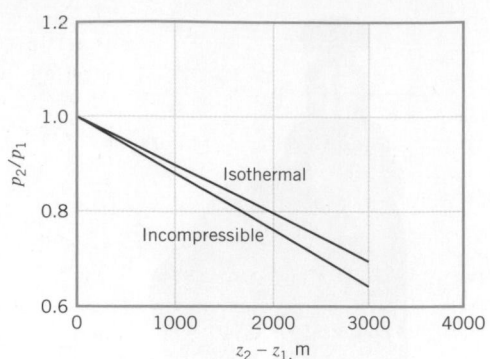

Before completing the integration, one must specify the nature of the variation of temperature with elevation. For example, if we assume that the temperature has a constant value T_0 over the range z_1 to z_2 (*isothermal* conditions), it then follows from Eq. 2.9 that

$$p_2 = p_1 \exp\left[-\frac{g(z_2 - z_1)}{RT_0}\right] \tag{2.10}$$

This equation provides the desired pressure–elevation relationship for an isothermal layer. As shown in the margin figure, even for a 3000-m altitude change the difference between the constant temperature (isothermal) and the constant density (incompressible) results are relatively minor. For nonisothermal conditions a similar procedure can be followed if the temperature–elevation relationship is known, as is discussed in the following section.

EXAMPLE 2.2   Incompressible and Isothermal Pressure–Depth Variations

GIVEN In 2007 the Burj Dubai skyscraper being built in the United Arab Emirates reached the stage in its construction where it became the world's tallest building. When completed it is expected to be at least 693.4 m tall, although its final height remains a secret.

FIND (a) Estimate the ratio of the pressure at the projected 693.4-m top of the building to the pressure at its base, assuming the air to be at a common temperature of 15 °C (288 K) (b) Compare the pressure calculated in part (a) with that obtained by assuming the air to be incompressible with $\gamma = 12.0 \text{ N/m}^3$ at 101.3 kPa (abs) (values for air at standard sea level conditions).

SOLUTION

For the assumed isothermal conditions, and treating air as a compressible fluid, Eq. 2.10 can be applied to yield

$$\frac{p_2}{p_1} = \exp\left[-\frac{g(z_2 - z_1)}{RT_0}\right]$$

$$= \exp\left\{\frac{-(9.81 \text{ m/s}^2)(693.4 \text{ m})}{(286.9 \text{ J/kg} - \text{K})(288 \text{ K})}\right\}$$

$$= 0.921 \tag{Ans}$$

If the air is treated as an incompressible fluid we can apply Eq. 2.5. In this case

$$p_2 = p_1 - \gamma(z_2 - z_1)$$

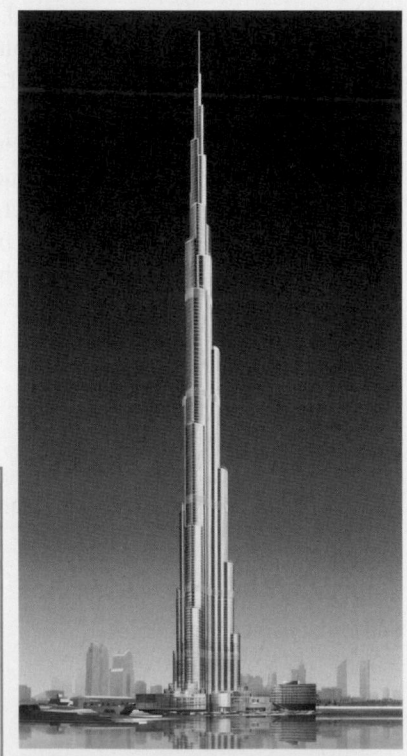

■ **FIGURE E2.2** **(Figure courtesy of Emaar Properties, Dubai, UAE.)**

or

$$\frac{p_2}{p_1} = 1 - \frac{\gamma(z_2 - z_1)}{p_1}$$

$$= 1 - \frac{(12.0 \text{ N/m}^3) \times (693.4 \text{ m})}{101.3 \text{ kPa}} = 0.918 \quad \textbf{(Ans)}$$

COMMENTS Note that there is little difference between the two results. Since the pressure difference between the bottom and top of the building is small, it follows that the variation in fluid

density is small and, therefore, the compressible fluid and incompressible fluid analyses yield essentially the same result.

We see that for both calculations the pressure decreases by approximately 8% as we go from ground level to the top of this tallest building. It does not require a very large pressure difference to support a 693.4-m-tall column of fluid as light as air. This result supports the earlier statement that the changes in pressures in air and other gases due to elevation changes are very small, even for distances of hundreds of meters. Thus, the pressure differences between the top and bottom of a horizontal pipe carrying a gas, or in a gas storage tank, are negligible since the distances involved are very small.

2.4 Standard Atmosphere

The standard atmosphere is an idealized representation of mean conditions in the earth's atmosphere.

An important application of Eq. 2.9 relates to the variation in pressure in the earth's atmosphere. Ideally, we would like to have measurements of pressure versus altitude over the specific range for the specific conditions (temperature, reference pressure) for which the pressure is to be determined. However, this type of information is usually not available. Thus, a "standard atmosphere" has been determined that can be used in the design of aircraft, missiles, and spacecraft, and in comparing their performance under standard conditions. The concept of a standard atmosphere was first developed in the 1920s, and since that time many national and international committees and organizations have pursued the development of such a standard. The currently accepted standard atmosphere is based on a report published in 1962 and updated in 1976 (see Refs. 1 and 2), defining the so-called *U.S. standard atmosphere,* which is an idealized representation of middle-latitude, year-round mean conditions of the earth's atmosphere. Several important properties for standard atmospheric conditions at *sea level* are listed in Table 2.1, and Fig. 2.6 shows the temperature profile for the U.S. standard atmosphere. As is shown in this figure the temperature decreases with altitude in the region nearest the earth's surface (*troposphere*), then becomes essentially constant in the next layer (*stratosphere*), and subsequently starts to increase in the next layer. Typical events that occur in the atmosphere are shown in the figure in the margin.

Since the temperature variation is represented by a series of linear segments, it is possible to integrate Eq. 2.9 to obtain the corresponding pressure variation. For example, in the troposphere, which extends to an altitude of about 11 km, the temperature variation is of the form

$$T = T_a - \beta z \qquad (2.11)$$

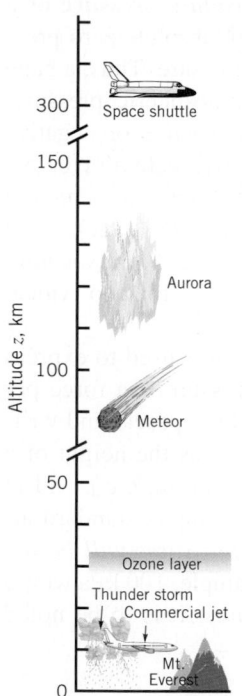

■ TABLE 2.1

Properties of U.S. Standard Atmosphere at Sea Level[a]

Property	SI Units
Temperature, T	288.15 K (15 °C)
Pressure, p	101.33 kPa (abs)
Density, ρ	1.225 kg/m^3
Specific weight, γ	12.014 N/m^3
Viscosity, μ	1.789 × 10^{-5} N · s/m^2

[a]Acceleration of gravity at sea level = 9.807 m/s^2.

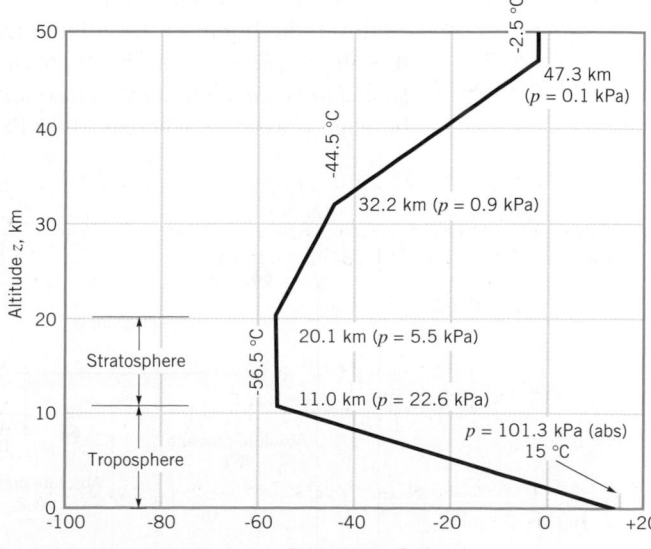

■ FIGURE 2.6 Variation of temperature with altitude in the U.S. standard atmosphere.

where T_a is the temperature at sea level ($z = 0$) and β is the *lapse rate* (the rate of change of temperature with elevation). For the standard atmosphere in the troposphere, $\beta = 0.00650$ K/m.

Equation 2.11 used with Eq. 2.9 yields

$$p = p_a\left(1 - \frac{\beta z}{T_a}\right)^{g/R\beta} \tag{2.12}$$

where p_a is the absolute pressure at $z = 0$. With p_a, T_a, and g obtained from Table 2.1, and with the gas constant $R = 286.9$ J/kg · K, the pressure variation throughout the troposphere can be determined from Eq. 2.12. This calculation shows that at the outer edge of the troposphere, where the temperature is -56.5 °C (217 K), the absolute pressure is about 23 kPa. It is to be noted that modern jetliners cruise at approximately this altitude. Pressures at other altitudes are shown in Fig. 2.6, and tabulated values for temperature, acceleration of gravity, pressure, density, and viscosity for the U.S. standard atmosphere are given in Table C.1 in Appendix C.

2.5   Measurement of Pressure

Pressure is designated as either absolute pressure or gage pressure.

Since pressure is a very important characteristic of a fluid field, it is not surprising that numerous devices and techniques are used in its measurement. As is noted briefly in Chapter 1, the pressure at a point within a fluid mass will be designated as either an *absolute* pressure or a *gage* pressure. Absolute pressure is measured relative to a perfect vacuum (absolute zero pressure), whereas gage pressure is measured relative to the local atmospheric pressure. Thus, a gage pressure of zero corresponds to a pressure that is equal to the local atmospheric pressure. Absolute pressures are always positive, but gage pressures can be either positive or negative depending on whether the pressure is above atmospheric pressure (a positive value) or below atmospheric pressure (a negative value). A negative gage pressure is also referred to as a *suction* or *vacuum* pressure. For example, 69 kPa (abs) could be expressed as -32.3 kPa (gage), if the local atmospheric pressure is 101.3 kPa, or alternatively 32.3 kPa suction or 32.3 kPa vacuum. The concept of gage and absolute pressure is illustrated graphically in Fig. 2.7 for two typical pressures located at points 1 and 2.

In addition to the reference used for the pressure measurement, the *units* used to express the value are obviously of importance. As is described in Section 1.5, pressure is a force per unit area, and the unit in the SI system is N/m²; this combination is called the pascal and written as Pa (1 N/m² = 1 Pa). As noted earlier, pressure can also be expressed as the height of a column of liquid. Then, the units will refer to the height of the column (mm, m, etc.), and in addition, the liquid in the column must be specified (H_2O, Hg, etc.). For example, standard atmospheric pressure can be expressed as 760 mm Hg (abs). *In this text, pressures will be assumed to be gage pressures unless specifically designated absolute.* For example, 100 kPa would be gage pressures, whereas 100 kPa (abs) would refer to absolute pressures. It is to be noted

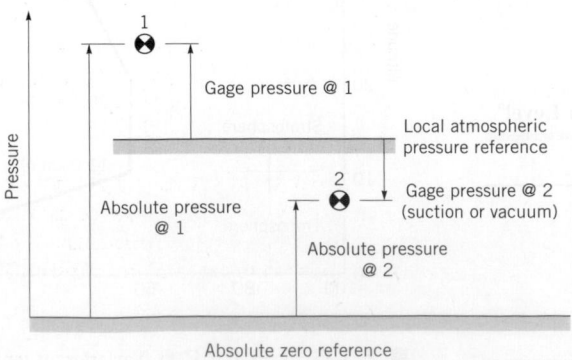

■ **FIGURE 2.7**   **Graphical representation of gage and absolute pressure.**

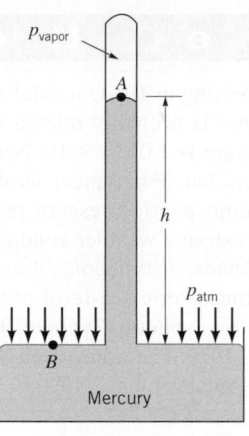

■ **F I G U R E   2.8**   Mercury barometer.

that *pressure differences* are independent of the reference, so that no special notation is required in this case.

The measurement of atmospheric pressure is usually accomplished with a mercury *barometer,* which in its simplest form consists of a glass tube closed at one end with the open end immersed in a container of mercury as shown in Fig. 2.8. The tube is initially filled with mercury (inverted with its open end up) and then turned upside down (open end down), with the open end in the container of mercury. The column of mercury will come to an equilibrium position where its weight plus the force due to the vapor pressure (which develops in the space above the column) balances the force due to the atmospheric pressure. Thus,

$$p_{atm} = \gamma h + p_{vapor} \tag{2.13}$$

where γ is the specific weight of mercury. For most practical purposes the contribution of the vapor pressure can be neglected since it is very small [for mercury, $p_{vapor} = 0.00016$ kPa (abs) at a temperature of 20 °C (293 K)], so that $p_{atm} \approx \gamma h$. It is conventional to specify atmospheric pressure in terms of the height, h, in millimeters of mercury. Note that if water were used instead of mercury, the height of the column would have to be approximately 10.4 m rather than 76 cm of mercury for an atmospheric pressure of 101.3 kPa. This is shown to scale in the figure in the margin. The concept of the mercury barometer is an old one, with the invention of this device attributed to Evangelista Torricelli in about 1644.

EXAMPLE   2.3   Barometric Pressure

GIVEN   A mountain lake has an average temperature of 10 °C and a maximum depth of 40 m. The barometric pressure is 598 mm Hg.

FIND   Determine the absolute pressure (in pascals) at the deepest part of the lake.

SOLUTION

The pressure in the lake at any depth, h, is given by the equation

$$p = \gamma h + p_0$$

where p_0 is the pressure at the surface. Since we want the absolute pressure, p_0 will be the local barometric pressure expressed in a consistent system of units; that is

$$\frac{p_{barometric}}{\gamma_{Hg}} = 598 \text{ mm} = 0.598 \text{ m}$$

and for $\gamma_{Hg} = 133$ kN/m^3

$$p_0 = (0.598 \text{ m})(133 \text{ kN/m}^3) = 79.5 \text{ kN/m}^2$$

From Table B.1, $\gamma_{H_2O} = 9.804$ kN/m^3 at 10 °C and therefore

$$p = (9.804 \text{ kN/m}^3)(40 \text{ m}) + 79.5 \text{ kN/m}^2$$
$$= 392 \text{ kN/m}^2 + 79.5 \text{ kN/m}^2$$
$$= 472 \text{ kPa (abs)} \qquad \text{(Ans)}$$

COMMENT   This simple example illustrates the need for close attention to the units used in the calculation of pressure; that is, be sure to use a *consistent* unit system, and be careful not to add a pressure head (m) to a pressure (Pa).

2.6   Manometry

Manometers use vertical or inclined liquid columns to measure pressure.

A standard technique for measuring pressure involves the use of liquid columns in vertical or inclined tubes. Pressure measuring devices based on this technique are called *manometers*. The mercury barometer is an example of one type of manometer, but there are many other configurations possible, depending on the particular application. Three common types of manometers include the piezometer tube, the U-tube manometer, and the inclined-tube manometer.

2.6.1 Piezometer Tube

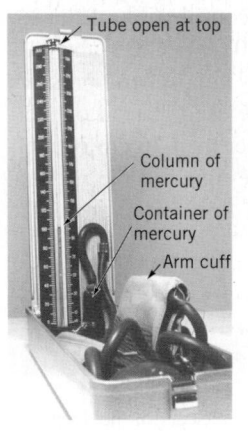

Tube open at top

Column of mercury

Container of mercury

Arm cuff

The simplest type of manometer consists of a vertical tube, open at the top, and attached to the container in which the pressure is desired, as illustrated in Fig. 2.9. The figure in the margin shows an important device whose operation is based upon this principle. It is a sphygmomanometer, the traditional instrument used to measure blood pressure.

Since manometers involve columns of fluids at rest, the fundamental equation describing their use is Eq. 2.8

$$p = \gamma h + p_0$$

which gives the pressure at any elevation within a homogeneous fluid in terms of a reference pressure p_0 and the vertical distance h between p and p_0. Remember that in a fluid at rest pressure will *increase* as we move *downward* and will decrease as we move *upward*. Application of this equation to the piezometer tube of Fig. 2.9 indicates that the pressure p_A can be determined by a measurement of h_1 through the relationship

$$p_A = \gamma_1 h_1$$

where γ_1 is the specific weight of the liquid in the container. Note that since the tube is open at the top, the pressure p_0 can be set equal to zero (we are now using gage pressure), with the height

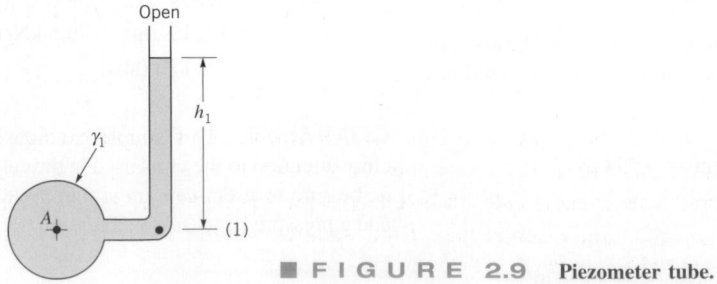

Open

h_1

γ_1

A

(1)

■ **F I G U R E   2.9**   **Piezometer tube.**

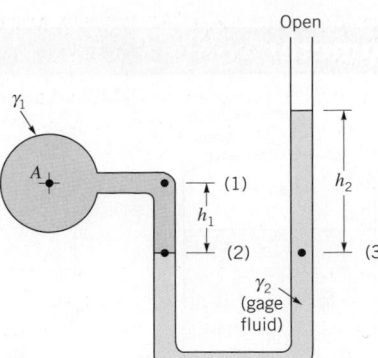

■ **F I G U R E 2.10**   Simple U-tube manometer.

h_1 measured from the meniscus at the upper surface to point (1). Since point (1) and point A within the container are at the same elevation, $p_A = p_1$.

Although the piezometer tube is a very simple and accurate pressure measuring device, it has several disadvantages. It is only suitable if the pressure in the container is greater than atmospheric pressure (otherwise air would be sucked into the system), and the pressure to be measured must be relatively small so the required height of the column is reasonable. Also, the fluid in the container in which the pressure is to be measured must be a liquid rather than a gas.

2.6.2 U-Tube Manometer

To overcome the difficulties noted previously, another type of manometer which is widely used consists of a tube formed into the shape of a U, as is shown in Fig. 2.10. The fluid in the manometer is called the *gage fluid*. To find the pressure p_A in terms of the various column heights, we start at one end of the system and work our way around to the other end, simply utilizing Eq. 2.8. Thus, for the U-tube manometer shown in Fig. 2.10, we will start at point A and work around to the open end. The pressure at points A and (1) are the same, and as we move from point (1) to (2) the pressure will increase by $\gamma_1 h_1$. The pressure at point (2) is equal to the pressure at point (3), since the pressures at equal elevations in a continuous mass of fluid at rest must be the same. Note that we could not simply "jump across" from point (1) to a point at the same elevation in the right-hand tube since these would not be points within the same continuous mass of fluid. With the pressure at point (3) specified, we now move to the open end where the pressure is zero. As we move vertically upward the pressure decreases by an amount $\gamma_2 h_2$. In equation form these various steps can

The contribution of gas columns in manometers is usually negligible since the weight of the gas is so small.

be expressed as

$$p_A + \gamma_1 h_1 - \gamma_2 h_2 = 0$$

and, therefore, the pressure p_A can be written in terms of the column heights as

$$p_A = \gamma_2 h_2 - \gamma_1 h_1 \qquad \textbf{(2.14)}$$

A major advantage of the U-tube manometer lies in the fact that the gage fluid can be different from the fluid in the container in which the pressure is to be determined. For example, the fluid in A in Fig. 2.10 can be either a liquid or a gas. If A does contain a gas, the contribution of the gas column, $\gamma_1 h_1$, is almost always negligible so that $p_A \approx p_2$, and in this instance Eq. 2.14 becomes

$$p_A = \gamma_2 h_2$$

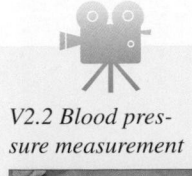

V2.2 Blood pressure measurement

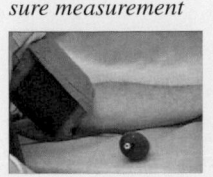

Thus, for a given pressure the height, h_2, is governed by the specific weight, γ_2, of the gage fluid used in the manometer. If the pressure p_A is large, then a heavy gage fluid, such as mercury, can be used and a reasonable column height (not too long) can still be maintained. Alternatively, if the pressure p_A is small, a lighter gage fluid, such as water, can be used so that a relatively large column height (which is easily read) can be achieved.

EXAMPLE 2.4 Simple U-Tube Manometer

GIVEN A closed tank contains compressed air and oil ($SG_{oil} = 0.90$) as shown in Fig. E2.4. A U-tube manometer using mercury ($SG_{Hg} = 13.6$) is connected to the tank as shown. The column heights are $h_1 = 91.4$ cm, $h_2 = 15.2$ cm, and $h_3 = 22.9$ cm.

FIND Determine the pressure reading (in N/cm²) of the gage.

SOLUTION

Following the general procedure of starting at one end of the manometer system and working around to the other, we will start at the air–oil interface in the tank and proceed to the open end where the pressure is zero. The pressure at level (1) is

$$p_1 = p_{air} + \gamma_{oil}(h_1 + h_2)$$

This pressure is equal to the pressure at level (2), since these two points are at the same elevation in a homogeneous fluid at rest. As we move from level (2) to the open end, the pressure must decrease by $\gamma_{Hg}h_3$, and at the open end the pressure is zero. Thus, the manometer equation can be expressed as

$$p_{air} + \gamma_{oil}(h_1 + h_2) - \gamma_{Hg}h_3 = 0$$

or

$$p_{air} + (SG_{oil})(\gamma_{H_2O})(h_1 + h_2) - (SG_{Hg})(\gamma_{H_2O})h_3 = 0$$

For the values given

$$p_{air} = -(0.9)(9800 \text{ N/m}^3)\left(\frac{91.4 + 15.2}{100} \text{ m}\right)$$
$$+ (13.6)(9800 \text{ N/m}^3)\left(\frac{22.9}{100} \text{ m}\right)$$

so that

$$p_{air} = 21 \text{ kPa}$$

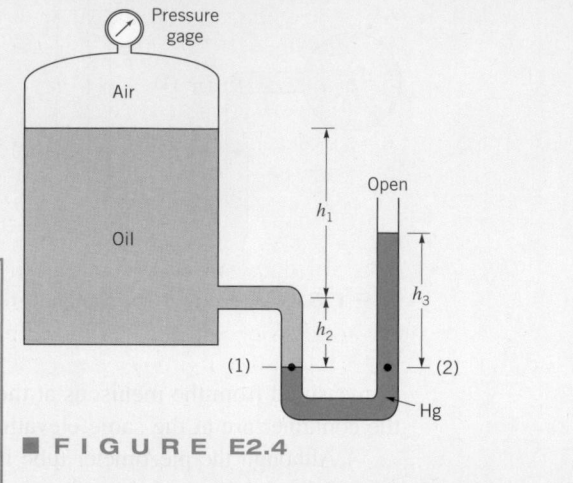

■ **FIGURE** E2.4

Since the specific weight of the air above the oil is much smaller than the specific weight of the oil, the gage should read the pressure we have calculated; that is,

$$p_{gage} = \frac{21 \text{ kPa}}{10^4 \text{ cm/m}^2} = 2.1 \text{ N/cm}^2 \qquad \textbf{(Ans)}$$

COMMENTS Note that the air pressure is a function of the height of the mercury in the manometer and the depth of the oil (both in the tank and in the tube). It is not just the mercury in the manometer that is important.

Assume that the gage pressure remains at 2.1 N/cm², but the manometer is altered so that it contains only oil. That is, the mercury is replaced by oil. A simple calculation shows that in this case the vertical oil-filled tube would need to be $h_3 = 3.44$ m tall, rather than the original $h_3 = 22.9$ cm. There is an obvious advantage of using a heavy fluid such as mercury in manometers.

Manometers are often used to measure the difference in pressure between two points.

The U-tube manometer is also widely used to measure the *difference* in pressure between two containers or two points in a given system. Consider a manometer connected between containers A and B as is shown in Fig. 2.11. The difference in pressure between A and B can be found

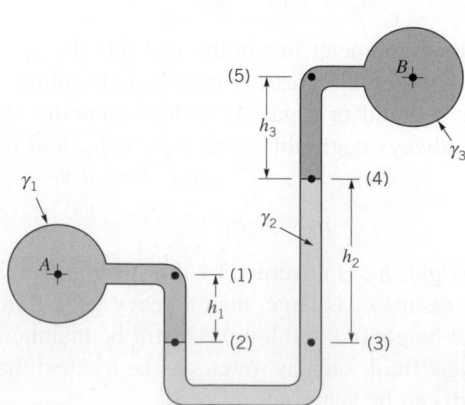

■ **FIGURE** 2.11 **Differential U-tube manometer.**

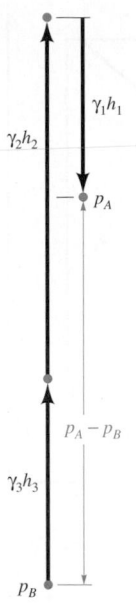

by again starting at one end of the system and working around to the other end. For example, at A the pressure is p_A, which is equal to p_1, and as we move to point (2) the pressure increases by $\gamma_1 h_1$. The pressure at p_2 is equal to p_3, and as we move upward to point (4) the pressure decreases by $\gamma_2 h_2$. Similarly, as we continue to move upward from point (4) to (5) the pressure decreases by $\gamma_3 h_3$. Finally, $p_5 = p_B$, since they are at equal elevations. Thus,

$$p_A + \gamma_1 h_1 - \gamma_2 h_2 - \gamma_3 h_3 = p_B$$

Or, as indicated in the figure in the margin, we could start at B and work our way around to A to obtain the same result. In either case, the pressure difference is

$$p_A - p_B = \gamma_2 h_2 + \gamma_3 h_3 - \gamma_1 h_1$$

When the time comes to substitute in numbers, be sure to use a consistent system of units!

Capillarity due to surface tension at the various fluid interfaces in the manometer is usually not considered, since for a simple U-tube with a meniscus in each leg, the capillary effects cancel (assuming the surface tensions and tube diameters are the same at each meniscus), or we can make the capillary rise negligible by using relatively large bore tubes (with diameters of about 1.3 cm or larger; see Section 1.9). Two common gage fluids are water and mercury. Both give a well-defined meniscus (a very important characteristic for a gage fluid) and have well-known properties. Of course, the gage fluid must be immiscible with respect to the other fluids in contact with it. For highly accurate measurements, special attention should be given to temperature since the various specific weights of the fluids in the manometer will vary with temperature.

EXAMPLE 2.5 | U-Tube Manometer

GIVEN As will be discussed in Chapter 3, the volume rate of flow, Q, through a pipe can be determined by means of a flow nozzle located in the pipe as illustrated in Fig. E2.5a. The nozzle creates a pressure drop, $p_A - p_B$, along the pipe which is related to the flow through the equation $Q = K\sqrt{p_A - p_B}$, where K is a constant depending on the pipe and nozzle size. The pressure drop is frequently measured with a differential U-tube manometer of the type illustrated.

FIND (a) Determine an equation for $p_A - p_B$ in terms of the specific weight of the flowing fluid, γ_1, the specific weight of the gage fluid, γ_2, and the various heights indicated. (b) For $\gamma_1 = 9.80 \text{ kN/m}^3$, $\gamma_2 = 15.6 \text{ kN/m}^3$, $h_1 = 1.0 \text{ m}$, and $h_2 = 0.5 \text{ m}$, what is the value of the pressure drop, $p_A - p_B$?

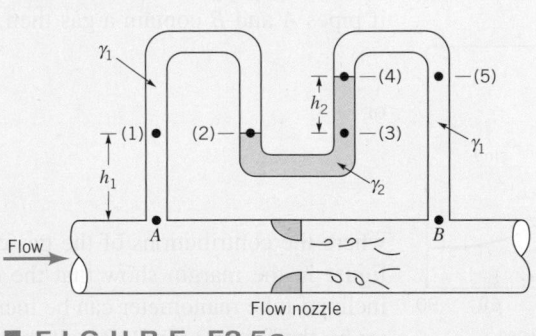

■ **FIGURE E2.5a**

SOLUTION

(a) Although the fluid in the pipe is moving, the fluids in the columns of the manometer are at rest so that the pressure variation in the manometer tubes is hydrostatic. If we start at point A and move vertically upward to level (1), the pressure will decrease by $\gamma_1 h_1$ and will be equal to the pressure at (2) and at (3). We can now move from (3) to (4) where the pressure has been further reduced by $\gamma_2 h_2$. The pressures at levels (4) and (5) are equal, and as we move from (5) to B the pressure will increase by $\gamma_1(h_1 + h_2)$. Thus, in equation form

$$p_A - \gamma_1 h_1 - \gamma_2 h_2 + \gamma_1(h_1 + h_2) = p_B$$

or

$$p_A - p_B = h_2(\gamma_2 - \gamma_1) \qquad \textbf{(Ans)}$$

COMMENT It is to be noted that the only column height of importance is the differential reading, h_2. The differential

manometer could be placed 0.5 or 5.0 m above the pipe ($h_1 = 0.5$ m or $h_1 = 5.0$ m), and the value of h_2 would remain the same.

(b) The specific value of the pressure drop for the data given is

$$p_A - p_B = (0.5 \text{ m})(15.6 \text{ kN/m}^3 - 9.80 \text{ kN/m}^3)$$
$$= 2.90 \text{ kPa} \qquad \textbf{(Ans)}$$

COMMENT By repeating the calculations for manometer fluids with different specific weights, γ_2, the results shown in Fig. E2.5b are obtained. Note that relatively small pressure differences can be measured if the manometer fluid has nearly the same specific weight as the flowing fluid. It is the difference in the specific weights, $\gamma_2 - \gamma_1$, that is important.

Hence, by rewriting the answer as $h_2 = (p_A - p_B)/(\gamma_2 - \gamma_1)$ it is seen that even if the value of $p_A - p_B$ is small, the value of h_2 can be large enough to provide an accurate reading provided the value of $\gamma_2 - \gamma_1$ is also small.

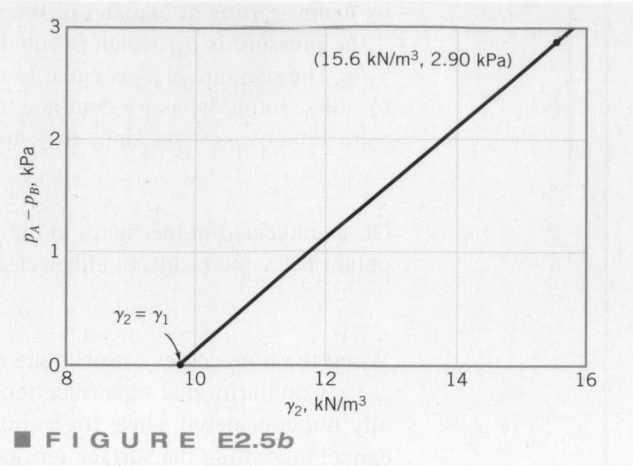

■ **F I G U R E   E2.5b**

2.6.3 Inclined-Tube Manometer

To measure small pressure changes, a manometer of the type shown in Fig. 2.12 is frequently used. One leg of the manometer is inclined at an angle θ, and the differential reading ℓ_2 is measured along the inclined tube. The difference in pressure $p_A - p_B$ can be expressed as

$$p_A + \gamma_1 h_1 - \gamma_2 \ell_2 \sin \theta - \gamma_3 h_3 = p_B$$

or

$$p_A - p_B = \gamma_2 \ell_2 \sin \theta + \gamma_3 h_3 - \gamma_1 h_1 \tag{2.15}$$

Inclined-tube manometers can be used to measure small pressure differences accurately.

where it is to be noted the pressure difference between points (1) and (2) is due to the *vertical* distance between the points, which can be expressed as $\ell_2 \sin \theta$. Thus, for relatively small angles the differential reading along the inclined tube can be made large even for small pressure differences. The inclined-tube manometer is often used to measure small differences in gas pressures so that if pipes A and B contain a gas then

$$p_A - p_B = \gamma_2 \ell_2 \sin \theta$$

or

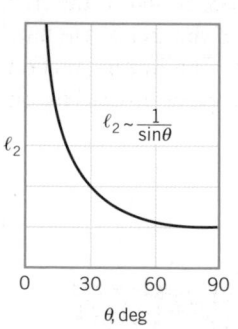

$$\ell_2 = \frac{p_A - p_B}{\gamma_2 \sin \theta} \tag{2.16}$$

where the contributions of the gas columns h_1 and h_3 have been neglected. Equation 2.16 and the figure in the margin show that the differential reading ℓ_2 (for a given pressure difference) of the inclined-tube manometer can be increased over that obtained with a conventional U-tube manometer by the factor $1/\sin \theta$. Recall that $\sin \theta \to 0$ as $\theta \to 0$.

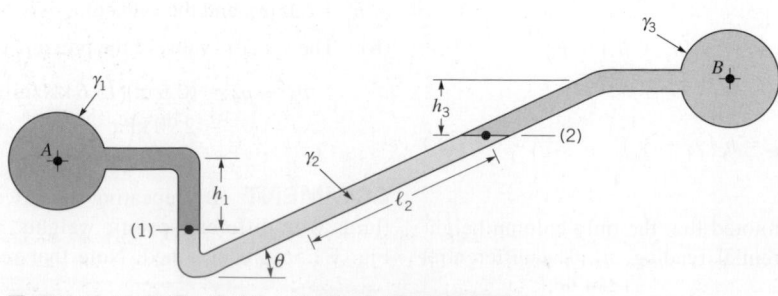

■ **F I G U R E   2.12**   Inclined-tube manometer.

2.7   Mechanical and Electronic Pressure Measuring Devices

A Bourdon tube pressure gage uses a hollow, elastic, and curved tube to measure pressure.

V2.3 Bourdon gage

Although manometers are widely used, they are not well suited for measuring very high pressures, or pressures that are changing rapidly with time. In addition, they require the measurement of one or more column heights, which, although not particularly difficult, can be time consuming. To overcome some of these problems numerous other types of pressure measuring instruments have been developed. Most of these make use of the idea that when a pressure acts on an elastic structure the structure will deform, and this deformation can be related to the magnitude of the pressure. Probably the most familiar device of this kind is the *Bourdon* **pressure gage,** which is shown in Fig. 2.13*a*. The essential mechanical element in this gage is the hollow, elastic curved tube (Bourdon tube) which is connected to the pressure source as shown in Fig. 2.13*b*. As the pressure within the tube increases the tube tends to straighten, and although the deformation is small, it can be translated into the motion of a pointer on a dial as illustrated. Since it is the difference in pressure between the outside of the tube (atmospheric pressure) and the inside of the tube that causes the movement of the tube, the indicated pressure is gage pressure. The Bourdon gage must be calibrated so that the dial reading can directly indicate the pressure in a suitable unit such as pascal. A zero reading on the gage indicates that the measured pressure is equal to the local atmospheric pressure. This type of gage can be used to measure a negative gage pressure (vacuum) as well as positive pressures.

The *aneroid* barometer is another type of mechanical gage that is used for measuring atmospheric pressure. Since atmospheric pressure is specified as an absolute pressure, the conventional Bourdon gage is not suitable for this measurement. The common aneroid barometer contains a hollow, closed, elastic element which is evacuated so that the pressure inside the element is near absolute zero. As the external atmospheric pressure changes, the element deflects, and this motion can be translated into the movement of an attached dial. As with the Bourdon gage, the dial can be calibrated to give atmospheric pressure directly, with the usual unit being millimeters of mercury.

For many applications in which pressure measurements are required, the pressure must be measured with a device that converts the pressure into an electrical output. For example, it may be desirable to continuously monitor a pressure that is changing with time. This type of pressure measuring device is called a *pressure transducer,* and many different designs are used. One possible type of transducer is one in which a Bourdon tube is connected to a linear variable differential transformer (LVDT), as is illustrated in Fig. 2.14. The core of the LVDT is connected to the free end of the Bourdon tube so that as a pressure is applied the resulting motion of the end of the tube moves the core through the coil and an output voltage develops. This voltage is a linear function of the pressure and could be recorded on an oscillograph or digitized for storage or processing on a computer.

(a)   (b)

■ **F I G U R E   2.13**   (*a*) **Liquid-filled Bourdon pressure gages for various pressure ranges.** (*b*) **Internal elements of Bourdon gages. The "C-shaped" Bourdon tube is shown on the left, and the "coiled spring" Bourdon tube for high pressures of 6895 kPa and above is shown on the right. (Photographs courtesy of Weiss Instruments, Inc.)**

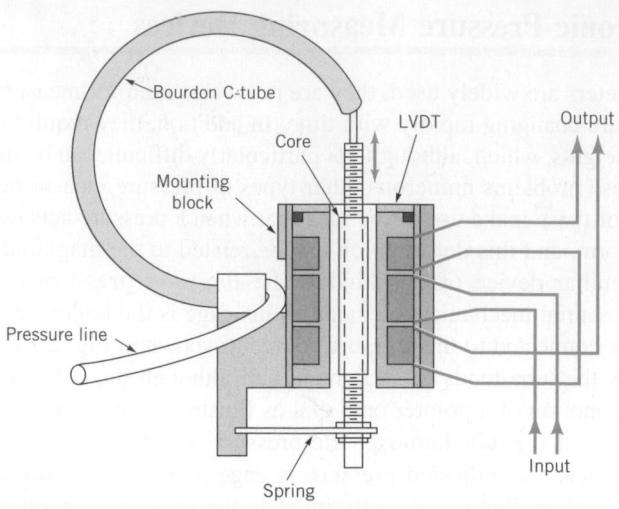

■ **F I G U R E 2.14** **Pressure transducer which combines a linear variable differential transformer (LVDT) with a Bourdon gage. (From Ref. 4, used by permission.)**

F l u i d s   i n   t h e   N e w s

Tire pressure warning Proper tire inflation on vehicles is important for more than ensuring long tread life. It is critical in preventing accidents such as rollover accidents caused by underinflation of tires. The National Highway Traffic Safety Administration is developing a regulation regarding four-tire tire-pressure monitoring systems that can warn a driver when a tire is more than 25 percent underinflated. Some of these devices are currently in operation on select vehicles; it is expected that they will soon be required on all vehicles. A typical tire-pressure monitoring system fits within the tire and contains a *pressure transducer* (usually either a piezo-resistive or a capacitive type transducer) and a transmitter that sends the information to an electronic control unit within the vehicle. Information about tire pressure and a warning when the tire is underinflated is displayed on the instrument panel. The environment (hot, cold, vibration) in which these devices must operate, their small size, and required low cost provide challenging constraints for the design engineer.

It is relatively complicated to make accurate pressure transducers for the measurement of pressures that vary rapidly with time.

One disadvantage of a pressure transducer using a Bourdon tube as the elastic sensing element is that it is limited to the measurement of pressures that are static or only changing slowly (quasistatic). Because of the relatively large mass of the Bourdon tube, it cannot respond to rapid changes in pressure. To overcome this difficulty, a different type of transducer is used in which the sensing element is a thin, elastic diaphragm which is in contact with the fluid. As the pressure changes, the diaphragm deflects, and this deflection can be sensed and converted into an electrical voltage. One way to accomplish this is to locate strain gages either on the surface of the diaphragm not in contact with the fluid, or on an element attached to the diaphragm. These gages can accurately sense the small strains induced in the diaphragm and provide an output voltage proportional to pressure. This type of transducer is capable of measuring accurately both small and large pressures, as well as both static and dynamic pressures. For example, strain-gage pressure transducers of the type shown in Fig. 2.15 are used to measure arterial blood pressure, which is a relatively small pressure that varies periodically with a fundamental frequency of about 1 Hz. The transducer is usually connected to the blood vessel by means of a liquid-filled, small diameter tube called a pressure catheter. Although the strain-gage type of transducers can be designed to have very good frequency response (up to approximately 10 kHz), they become less sensitive at the higher frequencies since the diaphragm must be made stiffer to achieve the higher frequency response. As an alternative, the diaphragm can be constructed of a piezoelectric crystal to be used as both the elastic element and the sensor. When a pressure is applied to the crystal, a voltage develops because of the deformation of the crystal. This voltage is directly related to the applied pressure. Depending on the design, this type of transducer can be used to measure both very low and high pressures (up to approximately 690 MPa) at high frequencies. Additional information on pressure transducers can be found in Refs. 3, 4, and 5.

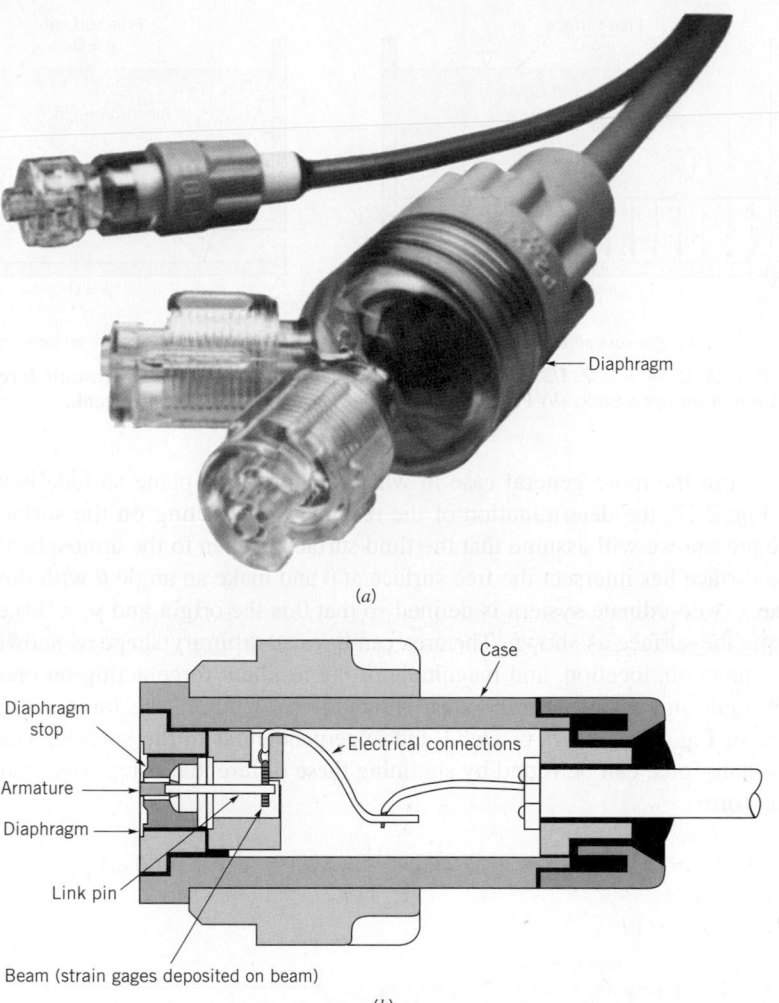

(a)

(b)

■ **FIGURE 2.15**   (a) **Two different sized strain-gage pressure transducers (Spectramed Models P10EZ and P23XL) commonly used to measure physiological pressures. Plastic domes are filled with fluid and connected to blood vessels through a needle or catheter. (Photograph courtesy of Spectramed, Inc.)** (b) **Schematic diagram of the P23XL transducer with the dome removed. Deflection of the diaphragm due to pressure is measured with a silicon beam on which strain gages and an associated bridge circuit have been deposited.**

2.8   Hydrostatic Force on a Plane Surface

V2.4 Hoover dam

When a surface is submerged in a fluid, forces develop on the surface due to the fluid. The determination of these forces is important in the design of storage tanks, ships, dams, and other hydraulic structures. For fluids at rest we know that the force must be *perpendicular* to the surface since there are no shearing stresses present. We also know that the pressure will vary linearly with depth as shown in Fig. 2.16 if the fluid is incompressible. For a horizontal surface, such as the bottom of a liquid-filled tank (Fig. 2.16a), the magnitude of the resultant force is simply $F_R = pA$, where p is the uniform pressure on the bottom and A is the area of the bottom. For the open tank shown, $p = \gamma h$. Note that if atmospheric pressure acts on both sides of the bottom, as is illustrated, the *resultant* force on the bottom is simply due to the liquid in the tank. Since the pressure is constant and uniformly distributed over the bottom, the resultant force acts through the centroid of the area as shown in Fig. 2.16a. As shown in Fig. 2.16b, the pressure on the ends of the tank is not uniformly distributed. Determination of the resultant force for situations such as this is presented below.

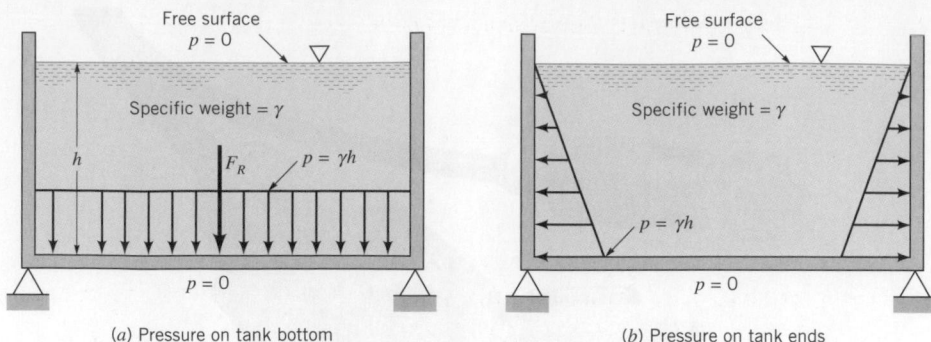

(a) Pressure on tank bottom

(b) Pressure on tank ends

■ **FIGURE 2.16** *(a)* **Pressure distribution and resultant hydrostatic force on the bottom of an open tank.** *(b)* **Pressure distribution on the ends of an open tank.**

The resultant force of a static fluid on a plane surface is due to the hydrostatic pressure distribution on the surface.

For the more general case in which a submerged plane surface is inclined, as is illustrated in Fig. 2.17, the determination of the resultant force acting on the surface is more involved. For the present we will assume that the fluid surface is open to the atmosphere. Let the plane in which the surface lies intersect the free surface at 0 and make an angle θ with this surface as in Fig. 2.17. The x–y coordinate system is defined so that 0 is the origin and $y = 0$ (i.e., the x-axis) is directed along the surface as shown. The area can have an arbitrary shape as shown. We wish to determine the direction, location, and magnitude of the resultant force acting on one side of this area due to the liquid in contact with the area. At any given depth, h, the force acting on dA (the differential area of Fig. 2.17) is $dF = \gamma h \, dA$ and is perpendicular to the surface. Thus, the magnitude of the resultant force can be found by summing these differential forces over the entire surface. In equation form

$$F_R = \int_A \gamma h \, dA = \int_A \gamma y \sin \theta \, dA$$

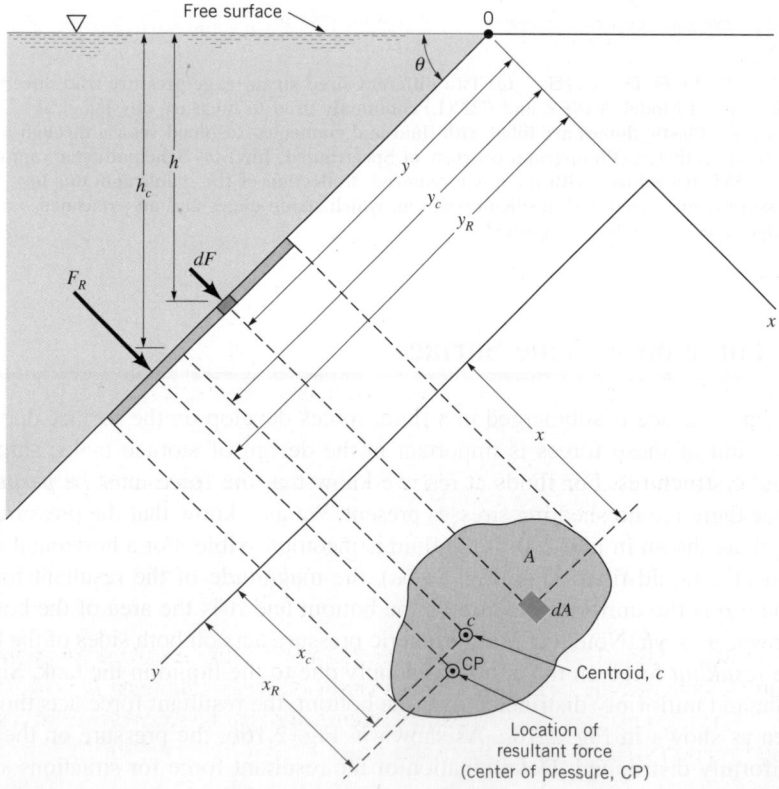

■ **FIGURE 2.17** **Notation for hydrostatic force on an inclined plane surface of arbitrary shape.**

where $h = y \sin \theta$. For constant γ and θ

$$F_R = \gamma \sin \theta \int_A y \, dA \qquad (2.17)$$

The integral appearing in Eq. 2.17 is the *first moment of the area* with respect to the x axis, so we can write

$$\int_A y \, dA = y_c A$$

where y_c is the y coordinate of the centroid of area A measured from the x axis which passes through 0. Equation 2.17 can thus be written as

$$F_R = \gamma A y_c \sin \theta$$

The magnitude of the resultant fluid force is equal to the pressure acting at the centroid of the area multiplied by the total area.

or more simply as

$$\boxed{F_R = \gamma h_c A} \qquad (2.18)$$

where h_c is the vertical distance from the fluid surface to the centroid of the area. Note that the magnitude of the force is independent of the angle θ. As indicated by the figure in the margin, it depends only on the specific weight of the fluid, the total area, and the depth of the centroid of the area below the surface. In effect, Eq. 2.18 indicates that the magnitude of the resultant force is equal to the pressure at the centroid of the area multiplied by the total area. Since all the differential forces that were summed to obtain F_R are perpendicular to the surface, the resultant F_R must also be perpendicular to the surface.

Although our intuition might suggest that the resultant force should pass through the centroid of the area, this is not actually the case. The y coordinate, y_R, of the resultant force can be determined by summation of moments around the x axis. That is, the moment of the resultant force must equal the moment of the distributed pressure force, or

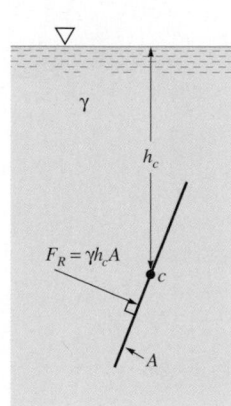

$$F_R y_R = \int_A y \, dF = \int_A \gamma \sin \theta \, y^2 \, dA$$

and, therefore, since $F_R = \gamma A y_c \sin \theta$

$$y_R = \frac{\displaystyle\int_A y^2 \, dA}{y_c A}$$

The integral in the numerator is the *second moment of the area (moment of inertia)*, I_x, with respect to an axis formed by the intersection of the plane containing the surface and the free surface (x axis). Thus, we can write

$$y_R = \frac{I_x}{y_c A}$$

Use can now be made of the parallel axis theorem to express I_x as

$$I_x = I_{xc} + A y_c^2$$

where I_{xc} is the second moment of the area with respect to an axis passing through its *centroid* and parallel to the x axis. Thus,

$$\boxed{y_R = \frac{I_{xc}}{y_c A} + y_c} \qquad (2.19)$$

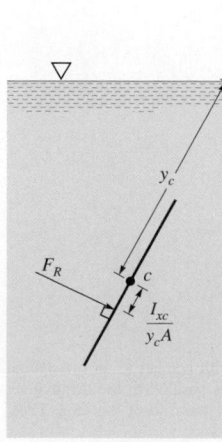

As shown by Eq. 2.19 and the figure in the margin, the resultant force does not pass through the centroid but for nonhorizontal surfaces is always *below* it, since $I_{xc}/y_c A > 0$.

The x coordinate, x_R, for the resultant force can be determined in a similar manner by summing moments about the y axis. Thus,

$$F_R x_R = \int_A \gamma \sin \theta \, xy \, dA$$

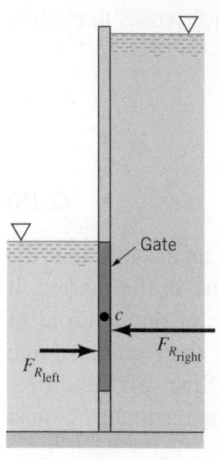

The resultant fluid force does not pass through the centroid of the area.

and, therefore,

$$x_R = \frac{\int_A xy\, dA}{y_c A} = \frac{I_{xy}}{y_c A}$$

where I_{xy} is the product of inertia with respect to the x and y axes. Again, using the parallel axis theorem,[1] we can write

$$x_R = \frac{I_{xyc}}{y_c A} + x_c \tag{2.20}$$

where I_{xyc} is the product of inertia with respect to an orthogonal coordinate system passing through the *centroid* of the area and formed by a translation of the x–y coordinate system. If the submerged area is symmetrical with respect to an axis passing through the centroid and parallel to either the x or y axes, the resultant force must lie along the line $x = x_c$, since I_{xyc} is identically zero in this case. The point through which the resultant force acts is called the *center of pressure*. It is to be noted from Eqs. 2.19 and 2.20 that as y_c increases the center of pressure moves closer to the centroid of the area. Since $y_c = h_c/\sin\theta$, the distance y_c will increase if the depth of submergence, h_c, increases, or, for a given depth, the area is rotated so that the angle, θ, decreases. Thus, the hydrostatic force on the right-hand side of the gate shown in the margin figure acts closer to the centroid of the gate than the force on the left-hand side. Centroidal coordinates and moments of inertia for some common areas are given in Fig. 2.18.

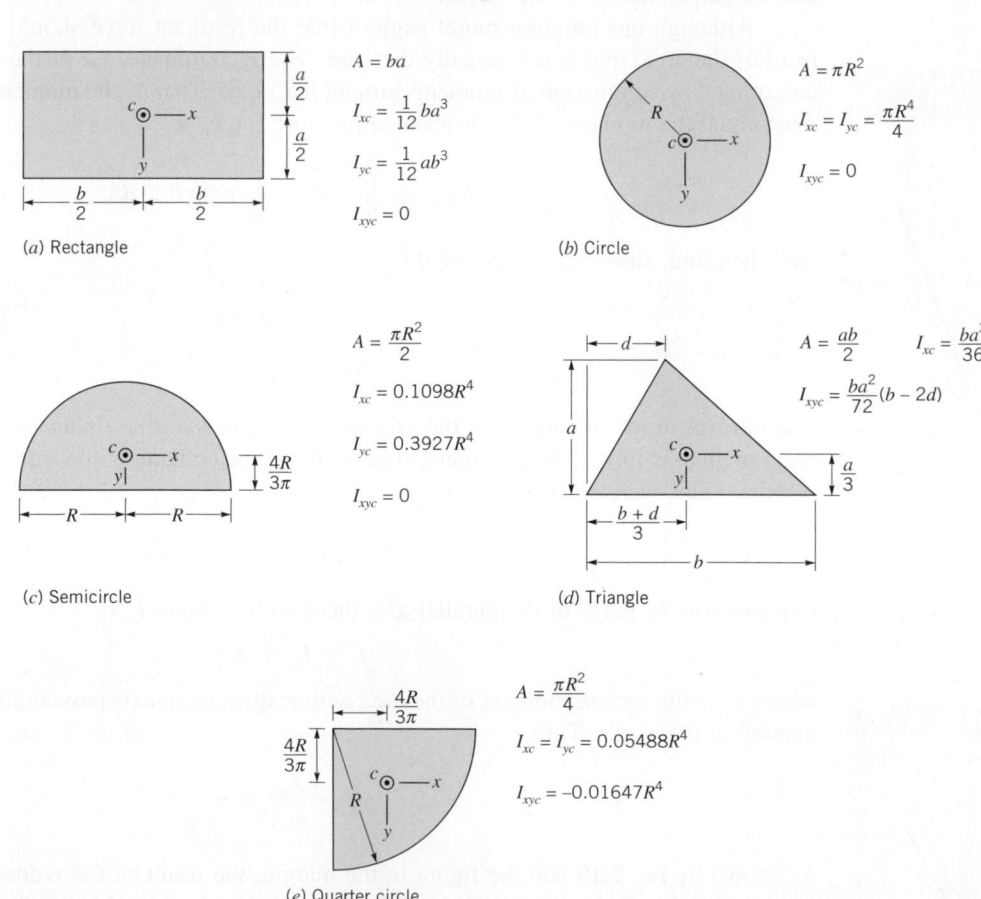

(a) Rectangle

$$A = ba$$
$$I_{xc} = \frac{1}{12}ba^3$$
$$I_{yc} = \frac{1}{12}ab^3$$
$$I_{xyc} = 0$$

(b) Circle

$$A = \pi R^2$$
$$I_{xc} = I_{yc} = \frac{\pi R^4}{4}$$
$$I_{xyc} = 0$$

(c) Semicircle

$$A = \frac{\pi R^2}{2}$$
$$I_{xc} = 0.1098 R^4$$
$$I_{yc} = 0.3927 R^4$$
$$I_{xyc} = 0$$

(d) Triangle

$$A = \frac{ab}{2} \qquad I_{xc} = \frac{ba^3}{36}$$
$$I_{xyc} = \frac{ba^2}{72}(b - 2d)$$

(e) Quarter circle

$$A = \frac{\pi R^2}{4}$$
$$I_{xc} = I_{yc} = 0.05488 R^4$$
$$I_{xyc} = -0.01647 R^4$$

■ **FIGURE 2.18** **Geometric properties of some common shapes.**

[1]Recall that the parallel axis theorem for the product of inertia of an area states that the product of inertia with respect to an orthogonal set of axes (x–y coordinate system) is equal to the product of inertia with respect to an orthogonal set of axes parallel to the original set and passing through the centroid of the area, plus the product of the area and the x and y coordinates of the centroid of the area. Thus, $I_{xy} = I_{xyc} + Ax_c y_c$.

The Three Gorges Dam The Three Gorges Dam being constructed on China's Yangtze River will contain the world's largest hydroelectric power plant when in full operation. The dam is of the concrete gravity type, having a length of 2309 meters with a height of 185 meters. The main elements of the project include the dam, two power plants, and navigation facilities consisting of a ship lock and lift. The power plants will contain 26 Francis type turbines, each with a capacity of 700 megawatts. The spillway section, which is the center section of the dam, is 483 meters long with 23 bottom outlets and 22 surface sluice

gates. The maximum discharge capacity is 102,500 cubic meters per second. After more than 10 years of construction, the dam gates were finally closed, and on June 10, 2003, the reservoir had been filled to its interim level of 135 meters. Due to the large depth of water at the dam and the huge extent of the storage pool, *hydrostatic pressure forces* have been a major factor considered by engineers. When filled to its normal pool level of 175 meters, the total reservoir storage capacity is 39.3 billion cubic meters. The project is scheduled for completion in 2009. (See Problem 2.79.)

EXAMPLE 2.6   Hydrostatic Force on a Plane Circular Surface

GIVEN The 4-m-diameter circular gate of Fig. E2.6a is located in the inclined wall of a large reservoir containing water ($\gamma = 9.80$ kN/m³). The gate is mounted on a shaft along its horizontal diameter, and the water depth is 10 m above the shaft.

FIND Determine

(a) the magnitude and location of the resultant force exerted on the gate by the water and

(b) the moment that would have to be applied to the shaft to open the gate.

SOLUTION

(a) To find the magnitude of the force of the water we can apply Eq. 2.18,

$$F_R = \gamma h_c A$$

and since the vertical distance from the fluid surface to the centroid of the area is 10 m, it follows that

$$F_R = (9.80 \times 10^3 \text{ N/m}^3)(10 \text{ m})(4\pi \text{ m}^2)$$
$$= 1230 \times 10^3 \text{ N} = 1.23 \text{ MN} \quad \text{(Ans)}$$

To locate the point (center of pressure) through which F_R acts, we use Eqs. 2.19 and 2.20,

$$x_R = \frac{I_{xyc}}{y_c A} + x_c \qquad y_R = \frac{I_{xc}}{y_c A} + y_c$$

For the coordinate system shown, $x_R = 0$ since the area is symmetrical, and the center of pressure must lie along the diameter A-A. To obtain y_R, we have from Fig. 2.18

$$I_{xc} = \frac{\pi R^4}{4}$$

and y_c is shown in Fig. E2.6b. Thus,

$$y_R = \frac{(\pi/4)(2 \text{ m})^4}{(10 \text{ m/sin } 60°)(4\pi \text{ m}^2)} + \frac{10 \text{ m}}{\sin 60°}$$
$$= 0.0866 \text{ m} + 11.55 \text{ m} = 11.6 \text{ m}$$

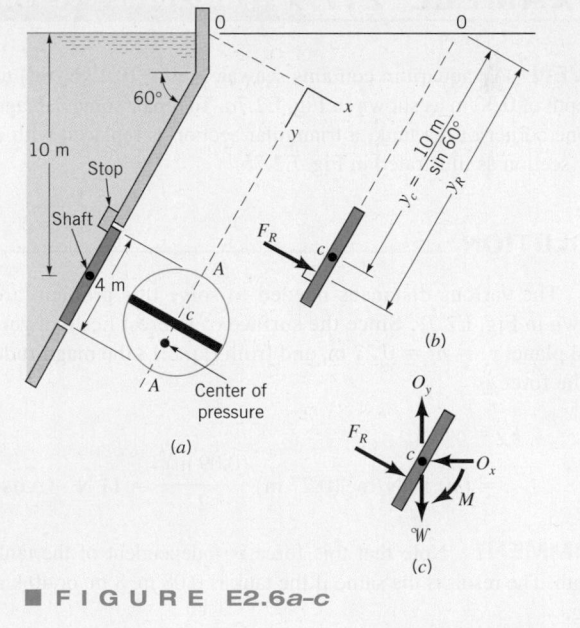

■ **FIGURE E2.6a–c**

and the distance (along the gate) below the shaft to the center of pressure is

$$y_R - y_c = 0.0866 \text{ m} \quad \text{(Ans)}$$

We can conclude from this analysis that the force on the gate due to the water has a magnitude of 1.23 MN and acts through a point along its diameter A-A at a distance of 0.0866 m (along the gate) below the shaft. The force is perpendicular to the gate surface as shown in Fig. E2.6b.

COMMENT By repeating the calculations for various values of the depth to the centroid, h_c, the results shown in Fig. E2.6d are obtained. Note that as the depth increases, the distance between the center of pressure and the centroid decreases.

(b) The moment required to open the gate can be obtained with the aid of the free-body diagram of Fig. E2.6c. In this diagram $\mathcal{W}$

is the weight of the gate and O_x and O_y are the horizontal and vertical reactions of the shaft on the gate. We can now sum moments about the shaft

$$\sum M_c = 0$$

and, therefore,

$$M = F_R(y_R - y_c)$$
$$= (1230 \times 10^3 \text{ N})(0.0866 \text{ m})$$
$$= 1.07 \times 10^5 \text{ N} \cdot \text{m} \qquad \text{(Ans)}$$

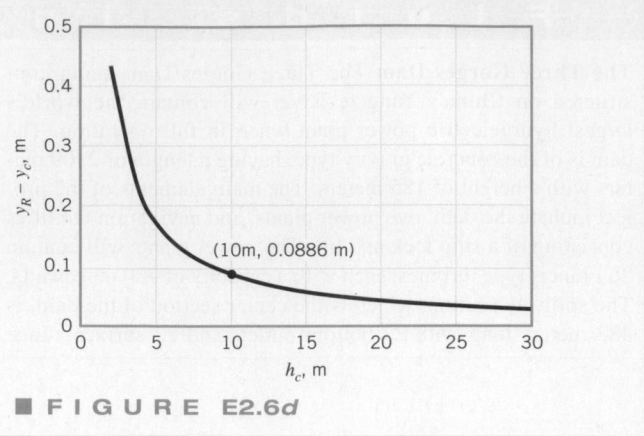

■ **FIGURE E2.6d**

EXAMPLE 2.7 ■ Hydrostatic Pressure Force on a Plane Triangular Surface

GIVEN An aquarium contains seawater ($\gamma = 10.1 \text{ kN/m}^3$) to a depth of 0.30 m as shown in Fig. E2.7a. To repair some damage to one corner of the tank, a triangular section is replaced with a new section as illustrated in Fig. E2.7b.

FIND Determine

(a) the magnitude of the force of the seawater on this triangular area, and

(b) the location of this force.

SOLUTION

(a) The various distances needed to solve this problem are shown in Fig. E2.7c. Since the surface of interest lies in a vertical plane, $y_c = h_c = 0.27$ m, and from Eq. 2.18 the magnitude of the force is

$$F_R = \gamma h_c A$$

$$= (10.1 \text{ kN/m}^3)(0.27 \text{ m}) \frac{(0.09 \text{ m})^2}{2} = 11 \text{ N} \quad \text{(Ans)}$$

COMMENT Note that this force is independent of the tank length. The result is the same if the tank is 0.08 m, 8 m, or 40 km long.

(b) The y coordinate of the center of pressure (CP) is found from Eq. 2.19,

$$y_R = \frac{I_{xc}}{y_c A} + y_c$$

and from Fig. 2.18

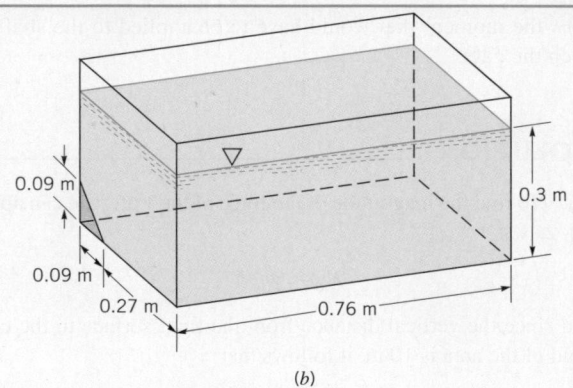

(b)

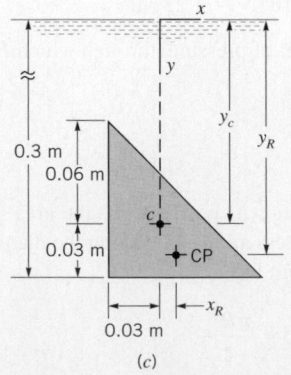

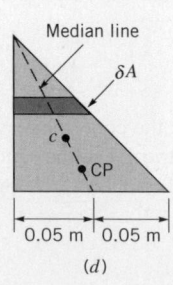

(c)   (d)

■ **FIGURE E2.7b–d**

■ **FIGURE E2.7a** (Photograph courtesy of Tenecor Tanks, Inc.)

$$I_{xc} = \frac{(0.09 \text{ m})(0.09 \text{ m})^3}{36} = \frac{6.6 \times 10^{-5}}{36} \text{ m}^4$$

so that

$$y_R = \frac{6.6 \times 10^{-5}/36 \text{ m}^4}{(0.27 \text{ m})(8.1 \times 10^{-3}/2 \text{ m}^2)} + 0.27 \text{ m}$$

$$= 0.0017 \text{ m} + 0.27 \text{ m} = 0.272 \text{ m} \qquad \text{(Ans)}$$

Similarly, from Eq. 2.20

$$x_R = \frac{I_{xyc}}{y_c A} + x_c$$

and from Fig. 2.18

$$I_{xyc} = \frac{(0.09 \text{ m})(0.09 \text{ m})^2(0.09 \text{ m})}{72} = \frac{6.6 \times 10^{-5}}{72} \text{ m}^4$$

so that

$$x_R = \frac{6.6 \times 10^{-5}/72 \text{ m}^4}{(0.27 \text{ m})(8.1 \times 10^{-3}/2 \text{ m}^2)} + 0 = 8.38 \times 10^{-4} \text{ m} \quad \text{(Ans)}$$

COMMENT   Thus, we conclude that the center of pressure is 0.84 mm to the right of and 1.7 mm below the centroid of the area. If this point is plotted, we find that it lies on the median line for the area as illustrated in Fig. E2.7d. Since we can think of the total area as consisting of a number of small rectangular strips of area δA (and the fluid force on each of these small areas acts through its center), it follows that the resultant of all these parallel forces must lie along the median.

2.9   Pressure Prism

An informative and useful graphical interpretation can be made for the force developed by a fluid acting on a plane rectangular area. Consider the pressure distribution along a vertical wall of a tank of constant width b, which contains a liquid having a specific weight γ. Since the pressure must vary linearly with depth, we can represent the variation as is shown in Fig. 2.19a, where the pressure is equal to zero at the upper surface and equal to γh at the bottom. It is apparent from this diagram that the average pressure occurs at the depth $h/2$, and therefore the resultant force acting on the rectangular area $A = bh$ is

$$F_R = p_{av} A = \gamma\left(\frac{h}{2}\right)A$$

which is the same result as obtained from Eq. 2.18. The pressure distribution shown in Fig. 2.19a applies across the vertical surface so we can draw the three-dimensional representation of the pressure distribution as shown in Fig. 2.19b. The base of this "volume" in pressure-area space is the plane surface of interest, and its altitude at each point is the pressure. This volume is called the *pressure prism*, and it is clear that the magnitude of the resultant force acting on the rectangular surface is equal to the volume of the pressure prism. Thus, for the prism of Fig. 2.19b the fluid force is

The magnitude of the resultant fluid force is equal to the volume of the pressure prism and passes through its centroid.

$$F_R = \text{volume} = \frac{1}{2}(\gamma h)(bh) = \gamma\left(\frac{h}{2}\right)A$$

where bh is the area of the rectangular surface, A.

The resultant force must pass through the *centroid* of the pressure prism. For the volume under consideration the centroid is located along the vertical axis of symmetry of the surface, and at a distance of $h/3$ above the base (since the centroid of a triangle is located at $h/3$ above its base). This result can readily be shown to be consistent with that obtained from Eqs. 2.19 and 2.20.

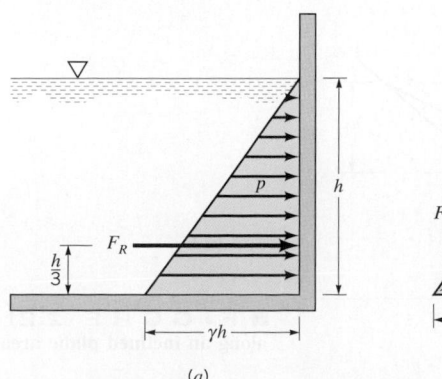

(a)

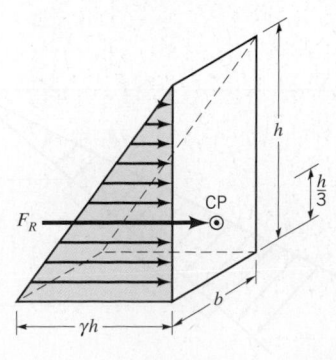

(b)

■ **FIGURE 2.19**
Pressure prism for vertical rectangular area.

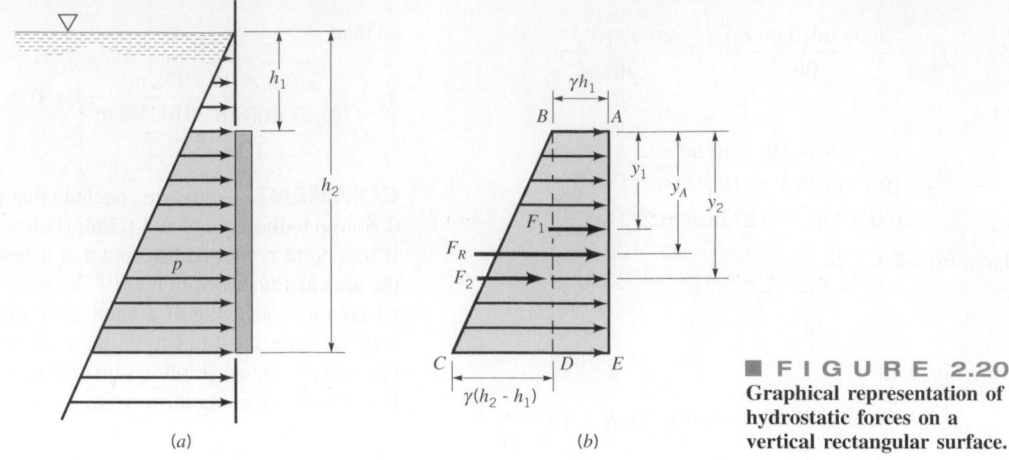

This same graphical approach can be used for plane rectangular surfaces that do not extend up to the fluid surface, as illustrated in Fig. 2.20a. In this instance, the cross section of the pressure prism is trapezoidal. However, the resultant force is still equal in magnitude to the volume of the pressure prism, and it passes through the centroid of the volume. Specific values can be obtained by decomposing the pressure prism into two parts, ABDE and BCD, as shown in Fig. 2.20b. Thus,

$$F_R = F_1 + F_2$$

where the components can readily be determined by inspection for rectangular surfaces. The location of F_R can be determined by summing moments about some convenient axis, such as one passing through A. In this instance

$$F_R y_A = F_1 y_1 + F_2 y_2$$

and y_1 and y_2 can be determined by inspection.

For inclined plane rectangular surfaces the pressure prism can still be developed, and the cross section of the prism will generally be trapezoidal, as is shown in Fig. 2.21. Although it is usually convenient to measure distances along the inclined surface, the pressures developed depend on the vertical distances as illustrated.

The use of pressure prisms for determining the force on submerged plane areas is convenient if the area is rectangular so the volume and centroid can be easily determined. However, for other nonrectangular shapes, integration would generally be needed to determine the volume and centroid. In these circumstances it is more convenient to use the equations developed in the previous section, in which the necessary integrations have been made and the results presented in a convenient and compact form that is applicable to submerged plane areas of any shape.

The effect of atmospheric pressure on a submerged area has not yet been considered, and we may ask how this pressure will influence the resultant force. If we again consider the pressure distribution on a plane vertical wall, as is shown in Fig. 2.22a, the pressure varies from zero at the surface to γh at the bottom. Since we are setting the surface pressure equal to zero, we are using

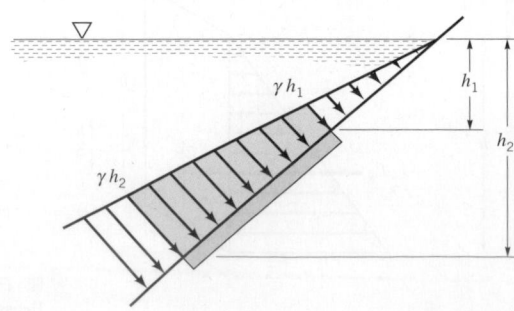

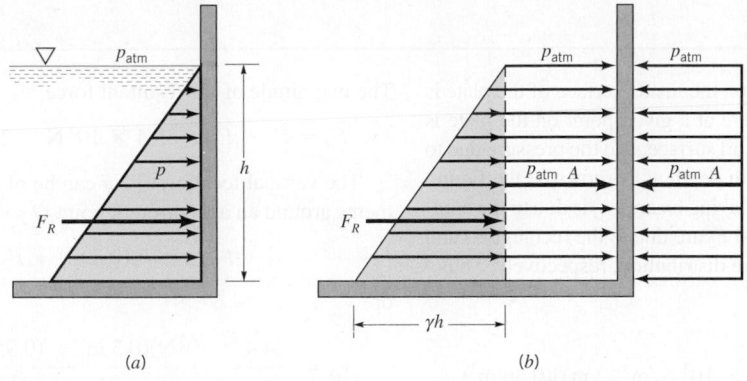

■ **F I G U R E   2.22**   Effect of atmospheric pressure on the resultant force acting on a plane vertical wall.

atmospheric pressure as our datum, and thus the pressure used in the determination of the fluid force is gage pressure. If we wish to include atmospheric pressure, the pressure distribution will be as is shown in Fig. 2.22b. We note that in this case the force on one side of the wall now consists of F_R as a result of the hydrostatic pressure distribution, plus the contribution of the atmospheric pressure, $p_{atm}A$, where A is the area of the surface. However, if we are going to include the effect of atmospheric pressure on one side of the wall, we must realize that this same pressure acts on the outside surface (assuming it is exposed to the atmosphere), so that an equal and opposite force will be developed as illustrated in the figure. Thus, we conclude that the *resultant* fluid force on the surface is that due only to the gage pressure contribution of the liquid in contact with the surface—the atmospheric pressure does not contribute to this resultant. Of course, if the surface pressure of the liquid is different from atmospheric pressure (such as might occur in a closed tank), the resultant force acting on a submerged area, A, will be changed in magnitude from that caused simply by hydrostatic pressure by an amount $p_s A$, where p_s is the gage pressure at the liquid surface (the outside surface is assumed to be exposed to atmospheric pressure).

The resultant fluid force acting on a submerged area is affected by the pressure at the free surface.

EXAMPLE 2.8   Use of the Pressure Prism Concept

GIVEN   A pressurized tank contains oil ($SG = 0.90$) and has a square, 0.6-m by 0.6-m plate bolted to its side, as is illustrated in Fig. E2.8a. The pressure gage on the top of the tank reads 50 kPa, and the outside of the tank is at atmospheric pressure.

FIND   What is the magnitude and location of the resultant force on the attached plate?

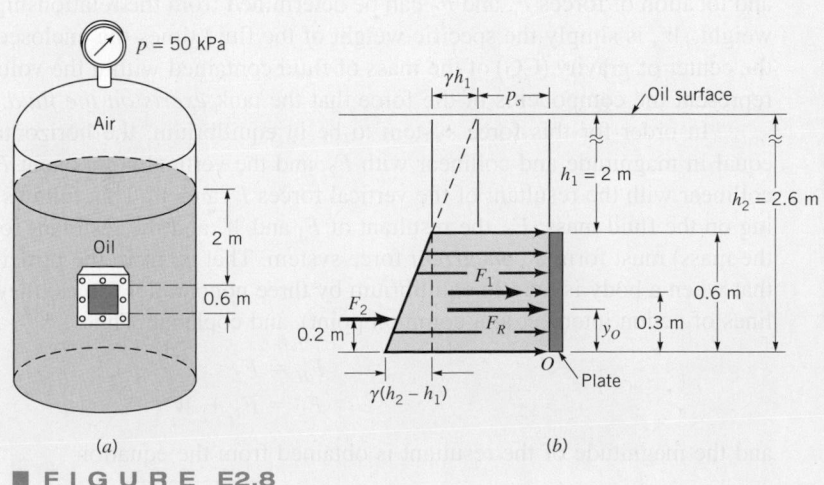

■ **F I G U R E   E2.8**

SOLUTION

The pressure distribution acting on the inside surface of the plate is shown in Fig. E2.8b. The pressure at a given point on the plate is due to the air pressure, p_s, at the oil surface, and the pressure due to the oil, which varies linearly with depth as is shown in the figure. The resultant force on the plate (having an area A) is due to the components, F_1 and F_2, where F_1 and F_2 are due to the rectangular and triangular portions of the pressure distribution, respectively. Thus,

$$F_1 = (p_s + \gamma h_1)A$$
$$= [50 \times 10^3 \text{ N/m}^2$$
$$+ (0.90)(9.81 \times 10^3 \text{ N/m}^3)(2 \text{ m})](0.36 \text{ m}^2)$$
$$= 24.4 \times 10^3 \text{ N}$$

and

$$F_2 = \gamma \left(\frac{h_2 - h_1}{2}\right)A$$
$$= (0.90)(9.81 \times 10^3 \text{ N/m}^3)\left(\frac{0.6 \text{ m}}{2}\right)(0.36 \text{ m}^2)$$
$$= 0.954 \times 10^3 \text{ N}$$

The magnitude of the resultant force, F_R, is therefore

$$F_R = F_1 + F_2 = 25.4 \times 10^3 \text{ N} = 25.4 \text{ kN} \qquad \text{(Ans)}$$

The vertical location of F_R can be obtained by summing moments around an axis through point O so that

$$F_R y_O = F_1(0.3 \text{ m}) + F_2(0.2 \text{ m})$$

or

$$y_O = \frac{(24.4 \times 10^3 \text{ N})(0.3 \text{ m}) + (0.954 \times 10^3 \text{ N})(0.2 \text{ m})}{25.4 \times 10^3 \text{ N}}$$
$$= 0.296 \text{ m} \qquad \text{(Ans)}$$

Thus, the force acts at a distance of 0.296 m above the bottom of the plate along the vertical axis of symmetry.

COMMENT Note that the air pressure used in the calculation of the force was gage pressure. Atmospheric pressure does not affect the resultant force (magnitude or location), since it acts on both sides of the plate, thereby canceling its effect.

2.10 Hydrostatic Force on a Curved Surface

V2.5 Pop bottle

The equations developed in Section 2.8 for the magnitude and location of the resultant force acting on a submerged surface only apply to plane surfaces. However, many surfaces of interest (such as those associated with dams, pipes, and tanks) are nonplanar. The domed bottom of the beverage bottle shown in the figure in the margin shows a typical curved surface example. Although the resultant fluid force can be determined by integration, as was done for the plane surfaces, this is generally a rather tedious process and no simple, general formulas can be developed. As an alternative approach we will consider the equilibrium of the fluid volume enclosed by the curved surface of interest and the horizontal and vertical projections of this surface.

For example, consider a curved portion of the swimming pool shown in Fig. 2.23a. We wish to find the resultant fluid force acting on section BC (which has a unit length perpendicular to the plane of the paper) shown in Fig. 2.23b. We first isolate a volume of fluid that is bounded by the surface of interest, in this instance section BC, the horizontal plane surface AB, and the vertical plane surface AC. The free-body diagram for this volume is shown in Fig. 2.23c. The magnitude and location of forces F_1 and F_2 can be determined from the relationships for planar surfaces. The weight, $\mathcal{W}$, is simply the specific weight of the fluid times the enclosed volume and acts through the center of gravity (CG) of the mass of fluid contained within the volume. The forces F_H and F_V represent the components of the force that the tank *exerts on the fluid*.

In order for this force system to be in equilibrium, the horizontal component F_H must be equal in magnitude and collinear with F_2, and the vertical component F_V equal in magnitude and collinear with the resultant of the vertical forces F_1 and $\mathcal{W}$. This follows since the three forces acting on the fluid mass (F_2, the resultant of F_1 and $\mathcal{W}$, and the resultant force that the tank exerts on the mass) must form a *concurrent* force system. That is, from the principles of statics, it is known that when a body is held in equilibrium by three nonparallel forces, they must be concurrent (their lines of action intersect at a common point), and coplanar. Thus,

$$F_H = F_2$$
$$F_V = F_1 + \mathcal{W}$$

and the magnitude of the resultant is obtained from the equation

$$F_R = \sqrt{(F_H)^2 + (F_V)^2}$$

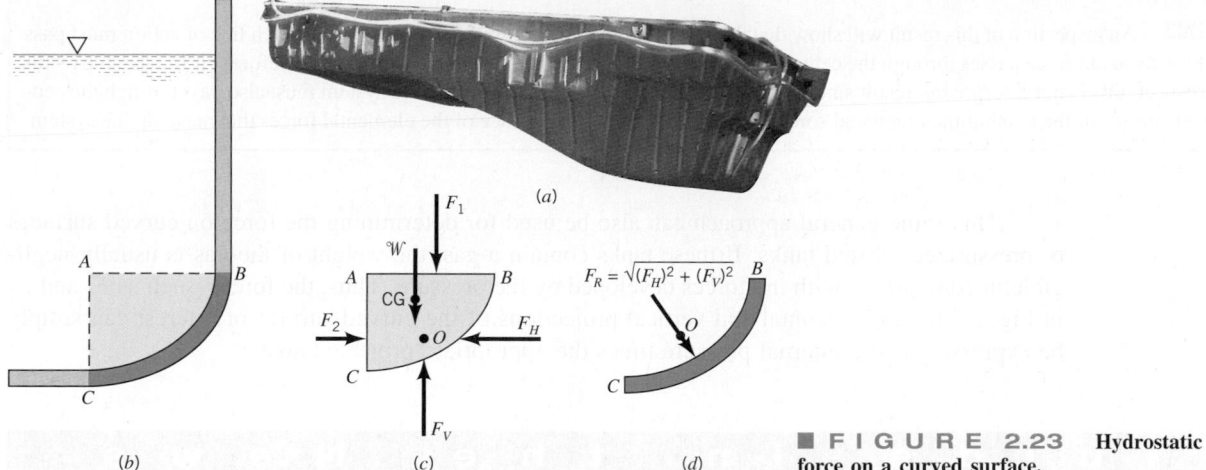

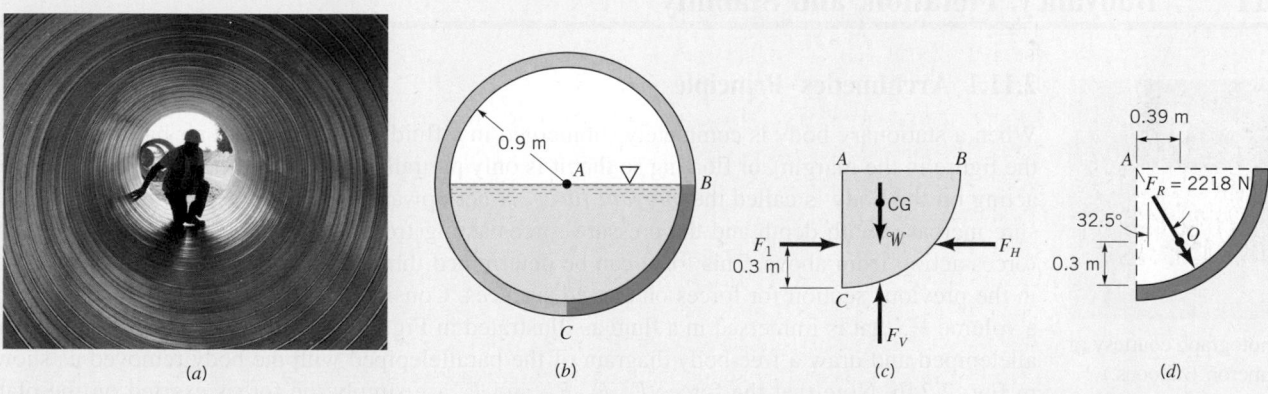

FIGURE 2.23   **Hydrostatic force on a curved surface.**

The resultant F_R passes through the point O, which can be located by summing moments about an appropriate axis. The resultant force of the fluid acting *on the curved surface BC* is equal and opposite in direction to that obtained from the free-body diagram of Fig. 2.23*c*. The desired fluid force is shown in Fig. 2.23*d*.

EXAMPLE 2.9   Hydrostatic Pressure Force on a Curved Surface

GIVEN   A 1.8-m-diameter drainage conduit of the type shown in Fig. E2.9*a* is half full of water at rest, as shown in Fig. E2.9*b*.

FIND   Determine the magnitude and line of action of the resultant force that the water exerts on a 0.3 m length of the curved section *BC* of the conduit wall.

FIGURE E2.9   **(Photograph courtesy of CONTECH Construction Products, Inc.)**

SOLUTION

We first isolate a volume of fluid bounded by the curved section *BC*, the horizontal surface *AB*, and the vertical surface *AC*, as shown in Fig. E2.9*c*. The volume has a length of 0.3 m. The forces acting on the volume are the horizontal force, F_1, which acts on the vertical surface *AC*, the weight, $\mathcal{W}$, of the fluid contained within the volume, and the horizontal and vertical components of the force of the conduit wall on the fluid, F_H and F_V, respectively.

The magnitude of F_1 is found from the equation

$$F_1 = \gamma h_c A = (9800 \text{ N/m}^3)(0.45 \text{ m})(0.27 \text{ m}^2) = 1191 \text{ N}$$

and this force acts 0.3 m above *C* as shown. The weight $\mathcal{W} = \gamma \mathcal{V}$, where $\mathcal{V}$ is the fluid volume, is

$$\mathcal{W} = \gamma \mathcal{V} = (9800 \text{ N/m}^3)\left(\frac{0.81\pi}{4}\text{m}^2\right)(0.3 \text{ m}) = 1870 \text{ N}$$

and acts through the center of gravity of the mass of fluid, which according to Fig. 2.18 is located 0.39 m to the right of *AC* as shown. Therefore, to satisfy equilibrium

$$F_H = F_1 = 1191 \text{ N} \qquad F_V = \mathcal{W} = 1870 \text{ N}$$

and the magnitude of the resultant force is

$$F_R = \sqrt{(F_H)^2 + (F_V)^2}$$
$$= \sqrt{(1191 \text{ N})^2 + (1870 \text{ N})^2} = 2218 \text{ N} \quad \textbf{(Ans)}$$

The force the water exerts *on* the conduit wall is equal, but *opposite in direction,* to the forces F_H and F_V shown in Fig. E2.9*c*. Thus, the resultant force *on the conduit wall* is shown in Fig. E2.9*d*. This force acts through the point *O* at the angle shown.

COMMENT An inspection of this result will show that the line of action of the resultant force passes through the center of the conduit. In retrospect, this is not a surprising result since at each point on the curved surface of the conduit the elemental force due to the pressure is normal to the surface, and each line of action must pass through the center of the conduit. It therefore follows that the resultant of this concurrent force system must also pass through the center of concurrence of the elemental forces that make up the system.

This same general approach can also be used for determining the force on curved surfaces of pressurized, closed tanks. If these tanks contain a gas, the weight of the gas is usually negligible in comparison with the forces developed by the pressure. Thus, the forces (such as F_1 and F_2 in Fig. 2.23c) on horizontal and vertical projections of the curved surface of interest can simply be expressed as the internal pressure times the appropriate projected area.

F l u i d s   i n   t h e   N e w s

Miniature, exploding pressure vessels Our daily lives are safer because of the effort put forth by engineers to design safe, lightweight pressure vessels such as boilers, propane tanks, and pop bottles. Without proper design, the large *hydrostatic pressure forces on the curved surfaces* of such containers could cause the vessel to explode with disastrous consequences. On the other hand, the world is a more friendly place because of miniature pressure vessels that are designed to explode under the proper conditions—popcorn kernels. Each grain of popcorn contains a small amount of water within the special, impervious hull (pressure vessel) which, when heated to a proper temperature, turns to steam, causing the kernel to explode and turn itself inside out. Not all popcorn kernels have the proper properties to make them pop well. First, the kernel must be quite close to 13.5% water. With too little moisture, not enough steam will build up to pop the kernel; too much moisture causes the kernel to pop into a dense sphere rather than the light fluffy delicacy expected. Second, to allow the pressure to build up, the kernels must not be cracked or damaged.

2.11   Buoyancy, Flotation, and Stability

2.11.1 Archimedes' Principle

(Photograph courtesy of Cameron Balloons.)

When a stationary body is completely submerged in a fluid (such as the hot air balloon shown in the figure in the margin), or floating so that it is only partially submerged, the resultant fluid force acting on the body is called the *buoyant force.* A net upward vertical force results because pressure increases with depth and the pressure forces acting from below are larger than the pressure forces acting from above. This force can be determined through an approach similar to that used in the previous section for forces on curved surfaces. Consider a body of arbitrary shape, having a volume $\mathcal{V}$, that is immersed in a fluid as illustrated in Fig. 2.24a. We enclose the body in a parallelepiped and draw a free-body diagram of the parallelepiped with the body removed as shown in Fig. 2.24b. Note that the forces F_1, F_2, F_3, and F_4 are simply the forces exerted on the plane surfaces of the parallelepiped (for simplicity the forces in the x direction are not shown), $\mathcal{W}$ is the weight of the shaded fluid volume (parallelepiped minus body), and F_B is the force the body is exerting *on the fluid.* The forces on the vertical surfaces, such as F_3 and F_4, are all equal and cancel, so the equilibrium equation of interest is in the z direction and can be expressed as

$$F_B = F_2 - F_1 - \mathcal{W} \tag{2.21}$$

If the specific weight of the fluid is constant, then

$$F_2 - F_1 = \gamma(h_2 - h_1)A$$

V2.6 Atmospheric buoyancy

where A is the horizontal area of the upper (or lower) surface of the parallelepiped, and Eq. 2.21 can be written as

$$F_B = \gamma(h_2 - h_1)A - \gamma[(h_2 - h_1)A - \mathcal{V}]$$

Simplifying, we arrive at the desired expression for the buoyant force

$$F_B = \gamma \mathcal{V} \tag{2.22}$$

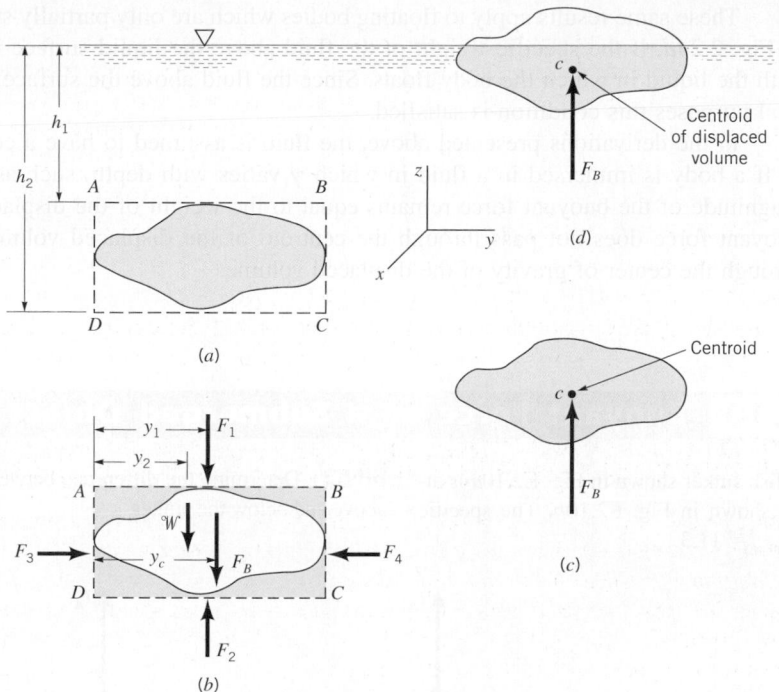

■ **F I G U R E 2.24** Buoyant force on submerged and floating bodies.

where γ is the specific weight of the fluid and $\forall$ is the volume of the body. The direction of the buoyant force, which is the force of the fluid *on the body,* is opposite to that shown on the free-body diagram. Therefore, the buoyant force has a magnitude equal to the weight of the fluid displaced by the body and is directed vertically upward. This result is commonly referred to as *Archimedes' principle* in honor of Archimedes (287–212 B.C.), a Greek mechanician and mathematician who first enunciated the basic ideas associated with hydrostatics.

The location of the line of action of the buoyant force can be determined by summing moments of the forces shown on the free-body diagram in Fig. 2.24*b* with respect to some convenient axis. For example, summing moments about an axis perpendicular to the paper through point *D* we have

$$F_B y_c = F_2 y_1 - F_1 y_1 - \mathcal{W} y_2$$

and on substitution for the various forces

$$\forall y_c = \forall_T y_1 - (\forall_T - \forall) y_2 \tag{2.23}$$

where $\forall_T$ is the total volume $(h_2 - h_1)A$. The right-hand side of Eq. 2.23 is the first moment of the displaced volume $\forall$ with respect to the *x–z* plane so that y_c is equal to the *y* coordinate of the centroid of the volume $\forall$. In a similar fashion it can be shown that the *x* coordinate of the buoyant force coincides with the *x* coordinate of the centroid. Thus, we conclude that the *buoyant force passes through the centroid of the displaced volume* as shown in Fig. 2.24*c*. The point through which the buoyant force acts is called the *center of buoyancy.*

F l u i d s   i n   t h e   N e w s

Concrete canoes A solid block of concrete thrown into a pond or lake will obviously sink. But, if the concrete is formed into the shape of a canoe it can be made to float. Of course the reason the canoe floats is the development of the *buoyant force* due to the displaced volume of water. With the proper design, this vertical force can be made to balance the weight of the canoe plus passengers—the canoe floats. Each year since 1988 there is a National Concrete Canoe Competition for university teams. It's jointly sponsored by the American Society of Civil Engineers and Master Builders Inc. The canoes must be 90% concrete and are typically designed with the aid of a computer by civil engineering students. Final scoring depends on four components: a design report, an oral presentation, the final product, and racing. For the 2007 competition the University of Wisconsin's team won for its fifth consecutive national championship with a 179-lb (796 N), 19.11-ft (5.8 m) canoe. (See Problem 2.107.)

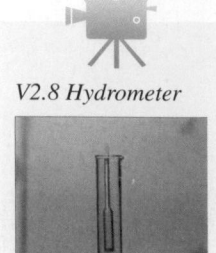

V2.8 Hydrometer

These same results apply to floating bodies which are only partially submerged, as illustrated in Fig. 2.24*d*, if the specific weight of the fluid above the liquid surface is very small compared with the liquid in which the body floats. Since the fluid above the surface is usually air, for practical purposes this condition is satisfied.

In the derivations presented above, the fluid is assumed to have a constant specific weight, γ. If a body is immersed in a fluid in which γ varies with depth, such as in a layered fluid, the magnitude of the buoyant force remains equal to the weight of the displaced fluid. However, the buoyant force does not pass through the centroid of the displaced volume, but rather, it passes through the center of gravity of the displaced volume.

EXAMPLE 2.10   Buoyant Force on a Submerged Object

GIVEN The 1.8-N lead fish sinker shown in Fig. E2.10*a* is attached to a fishing line as shown in Fig. E2.10*b*. The specific gravity of the sinker is $SG_{sinker} = 11.3$.

FIND Determine the difference between the tension in the line above and below the sinker.

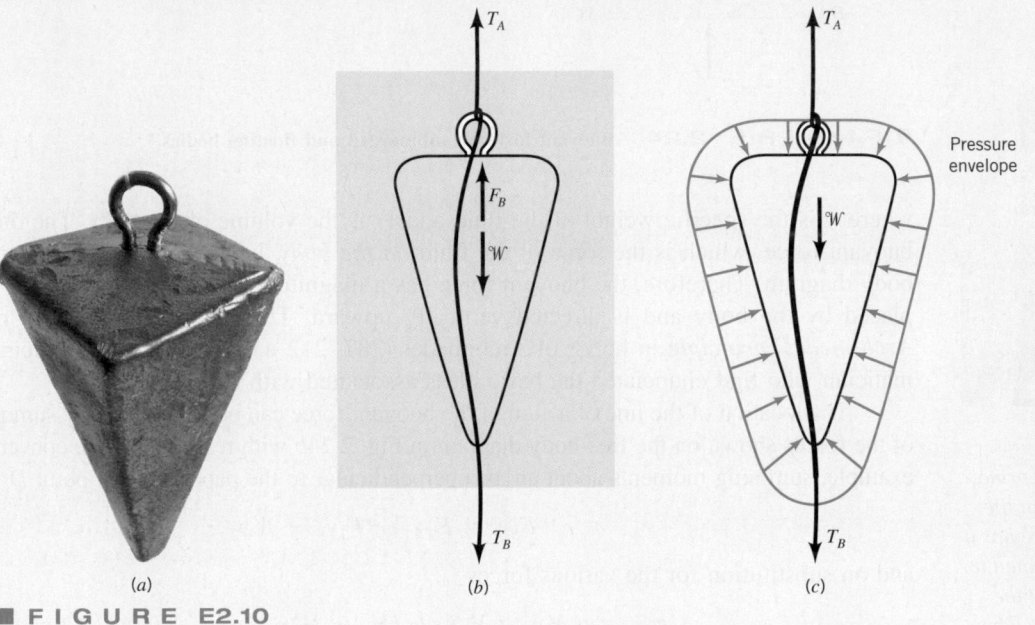

(a)   *(b)*   *(c)*

■ **FIGURE E2.10**

SOLUTION

A free body diagram of the sinker is shown in Fig. E.10*b*, where W is the weight of the sinker, F_B is the buoyant force acting on the sinker, and T_A and T_B are the tensions in the line above and below the sinker, respectively. For equilibrium it follows that

$$T_A - T_B = W - F_B \tag{1}$$

Also,

$$W = \gamma_{sinker} V = \gamma \, SG_{sinker} \, V \tag{2}$$

where γ is the specific weight of water and V is the volume of the sinker. From Eq. 2.22,

$$F_B = \gamma V \tag{3}$$

By combining Eqs. 2 and 3 we obtain

$$F_B = W/SG_{sinker} \tag{4}$$

Hence, from Eqs. 1 and 4 the difference in the tensions is

$$T_A - T_B = W - W/SG_{sinker} = W[1 - (1/SG_{sinker})] \tag{5}$$
$$= 1.8 \text{ N} [1 - (1/11.3)] = 1.64 \text{ N} \tag{Ans}$$

COMMENTS Note that if the sinker were raised out of the water, the difference in tension would equal the entire weight of the sinker ($T_A - T_B = 1.8$ N) rather than the 1.64 N when it is in the water. Thus, since the sinker material is significantly heavier than water, the buoyant force is relatively unimportant. As seen from Eq. 5, as SG_{sinker} becomes very large, the buoyant force becomes insignificant, and the tension difference becomes nearly equal to the weight of the sinker. On the other hand, if $SG_{sinker} = 1$, then $T_A - T_B = 0$ and the sinker is no longer a "sinker." It is neutrally buoyant and no external force from the line is required to hold it in place.

In this example we replaced the hydrostatic pressure force on the body by the buoyant force, F_B. Another correct free-body diagram of the sinker is shown in Fig. E2.20c. The net effect of the pressure forces on the surface of the sinker is equal to the upward force of magnitude F_B (the buoyant force). Do not include both the buoyant force and the hydrostatic pressure effects in your calculations—use one or the other.

F l u i d s   i n   t h e   N e w s

Explosive Lake In 1986 a tremendous explosion of carbon dioxide (CO_2) from Lake Nyos, west of Cameroon, killed more than 1700 people and livestock. The explosion resulted from a build up of CO_2 that seeped into the high pressure water at the bottom of the lake from warm springs of CO_2-bearing water. The CO_2-rich water is heavier than pure water and can hold a volume of CO_2 more than five times the water volume. As long as the gas remains dissolved in the water, the stratified lake (i.e., pure water on top, CO_2 water on the bottom) is stable. But if some mechanism causes the gas

bubbles to nucleate, they rise, grow, and cause other bubbles to form, feeding a chain reaction. A related phenomenon often occurs when a pop bottle is shaken and then opened. The pop shoots from the container rather violently. When this set of events occurred in Lake Nyos, the entire lake overturned through a column of rising and expanding *buoyant* bubbles. The heavier-than-air CO_2 then flowed through the long, deep valleys surrounding the lake and asphyxiated human and animal life caught in the gas cloud. One victim was 27 km downstream from the lake.

2.11.2 Stability

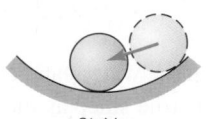

Stable

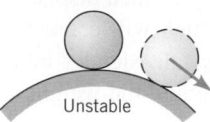

Unstable

The stability of a body can be determined by considering what happens when it is displaced from its equilibrium position.

Another interesting and important problem associated with submerged or floating bodies is concerned with the stability of the bodies. As illustrated by the figure in the margin, a body is said to be in a *stable equilibrium* position if, when displaced, it returns to its equilibrium position. Conversely, it is in an *unstable equilibrium* position if, when displaced (even slightly), it moves to a new equilibrium position. Stability considerations are particularly important for submerged or floating bodies since the centers of buoyancy and gravity do not necessarily coincide. A small rotation can result in either a restoring or overturning couple. For example, for the *completely* submerged body shown in Fig. 2.25, which has a center of gravity below the center of buoyancy, a rotation from its equilibrium position will create a restoring couple formed by the weight, W, and the buoyant force, F_B, which causes the body to rotate back to its original position. Thus, for this configuration the body is stable. It is to be noted that as long as the center of gravity falls *below* the center of buoyancy, this will always be true; that is, the body is in a *stable equilibrium* position with respect to small rotations. However, as is illustrated in Fig. 2.26, if the center of gravity of the completely submerged body is above the center of buoyancy, the resulting couple formed by the weight and the buoyant force will cause the body to overturn and move to a new equilibrium position. Thus, a completely submerged body with its center of gravity *above* its center of buoyancy is in an *unstable equilibrium* position.

For *floating* bodies the stability problem is more complicated, since as the body rotates the location of the center of buoyancy (which passes through the centroid of the displaced volume) may

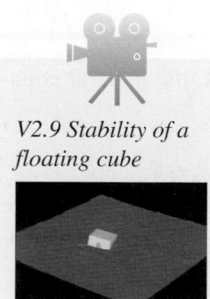

V2.9 Stability of a floating cube

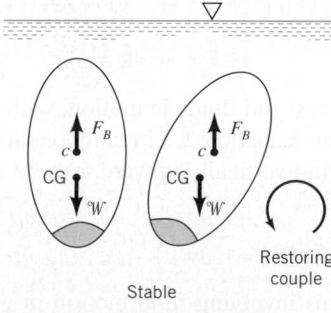

Stable

Restoring couple

■ F I G U R E   2.25
Stability of a completely immersed body—center of gravity below centroid.

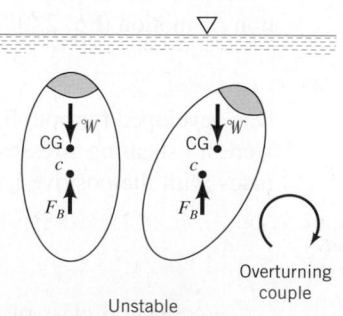

Unstable

Overturning couple

■ F I G U R E   2.26
Stability of a completely immersed body—center of gravity above centroid.

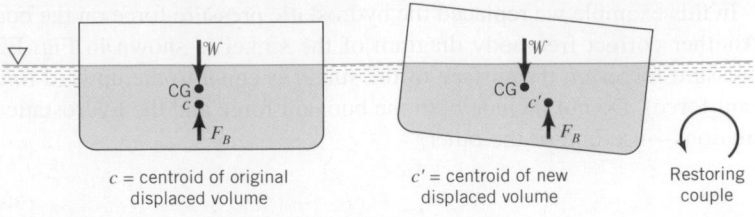

c = centroid of original   c' = centroid of new   Restoring
displaced volume   displaced volume   couple

Stable

■ **F I G U R E 2.27** **Stability of a floating body—stable configuration.**

Marginally stable

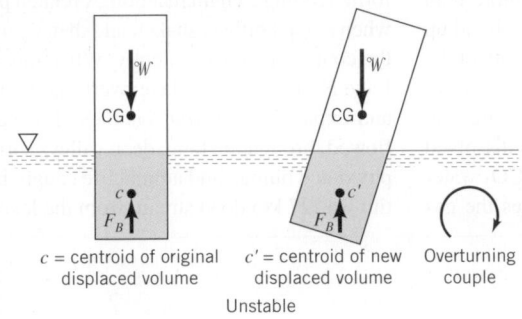

c = centroid of original   c' = centroid of new   Overturning
displaced volume   displaced volume   couple

Unstable

■ **F I G U R E 2.28** **Stability of a floating body—unstable configuration.**

Very stable

V2.10 Stability of a model barge

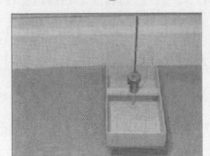

change. As is shown in Fig. 2.27, a floating body such as a barge that rides low in the water can be stable even though the center of gravity lies above the center of buoyancy. This is true since as the body rotates the buoyant force, F_B, shifts to pass through the centroid of the newly formed displaced volume and, as illustrated, combines with the weight, $\mathcal{W}$, to form a couple which will cause the body to return to its original equilibrium position. However, for the relatively tall, slender body shown in Fig. 2.28, a small rotational displacement can cause the buoyant force and the weight to form an overturning couple as illustrated.

It is clear from these simple examples that the determination of the stability of submerged or floating bodies can be difficult since the analysis depends in a complicated fashion on the particular geometry and weight distribution of the body. Thus, although both the relatively narrow kayak and the wide houseboat shown in the figures in the margin are stable, the kayak will overturn much more easily than the houseboat. The problem can be further complicated by the necessary inclusion of other types of external forces such as those induced by wind gusts or currents. Stability considerations are obviously of great importance in the design of ships, submarines, bathyscaphes, and so forth, and such considerations play a significant role in the work of naval architects (see, for example, Ref. 6).

2.12 Pressure Variation in a Fluid with Rigid-Body Motion

Although in this chapter we have been primarily concerned with fluids at rest, the general equation of motion (Eq. 2.2)

$$-\nabla p - \gamma \hat{\mathbf{k}} = \rho \mathbf{a}$$

was developed for both fluids at rest and fluids in motion, with the only stipulation being that there were no shearing stresses present. Equation 2.2 in component form, based on rectangular coordinates with the positive z axis being vertically upward, can be expressed as

$$-\frac{\partial p}{\partial x} = \rho a_x \qquad -\frac{\partial p}{\partial y} = \rho a_y \qquad -\frac{\partial p}{\partial z} = \gamma + \rho a_z \qquad \text{(2.24)}$$

Even though a fluid may be in motion, if it moves as a rigid body there will be no shearing stresses present.

A general class of problems involving fluid motion in which there are no shearing stresses occurs when a mass of fluid undergoes rigid-body motion. For example, if a container of fluid accelerates along a straight path, the fluid will move as a rigid mass (after the initial sloshing motion has died out) with each particle having the same acceleration. Since there is no deformation,

There is no shear stress in fluids that move with rigid-body motion or with rigid-body rotation.

there will be no shearing stresses and, therefore, Eq. 2.2 applies. Similarly, if a fluid is contained in a tank that rotates about a fixed axis, the fluid will simply rotate with the tank as a rigid body, and again Eq. 2.2 can be applied to obtain the pressure distribution throughout the moving fluid. Specific results for these two cases (rigid-body uniform motion and rigid-body rotation) are developed in the following two sections. Although problems relating to fluids having rigid-body motion are not, strictly speaking, "fluid statics" problems, they are included in this chapter because, as we will see, the analysis and resulting pressure relationships are similar to those for fluids at rest.

2.12.1 Linear Motion

We first consider an open container of a liquid that is translating along a straight path with a constant acceleration **a** as illustrated in Fig. 2.29. Since $a_x = 0$, it follows from the first of Eqs. 2.24 that the pressure gradient in the x direction is zero ($\partial p / \partial x = 0$). In the y and z directions

$$\frac{\partial p}{\partial y} = -\rho a_y \tag{2.25}$$

$$\frac{\partial p}{\partial z} = -\rho(g + a_z) \tag{2.26}$$

The change in pressure between two closely spaced points located at y, z, and $y + dy$, $z + dz$ can be expressed as

$$dp = \frac{\partial p}{\partial y} \, dy + \frac{\partial p}{\partial z} \, dz$$

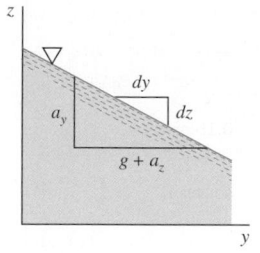

or in terms of the results from Eqs. 2.25 and 2.26

$$dp = -\rho a_y \, dy - \rho(g + a_z) \, dz \tag{2.27}$$

Along a line of *constant* pressure, $dp = 0$, and therefore from Eq. 2.27 it follows that the slope of this line is given by the relationship

$$\frac{dz}{dy} = -\frac{a_y}{g + a_z} \tag{2.28}$$

This relationship is illustrated by the figure in the margin. Along a free surface the pressure is constant, so that for the accelerating mass shown in Fig. 2.29 the free surface will be inclined if $a_y \neq 0$. In addition, all lines of constant pressure will be parallel to the free surface as illustrated.

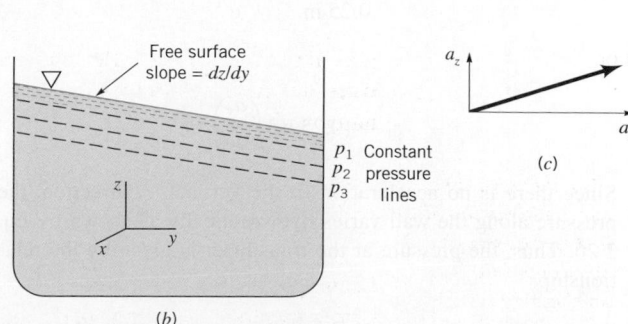

(a)   (b)

■ **F I G U R E   2.29**   **Linear acceleration of a liquid with a free surface.**

The pressure distribution in a fluid mass that is accelerating along a straight path is not hydrostatic.

For the special circumstance in which $a_y = 0$, $a_z \neq 0$, which corresponds to the mass of fluid accelerating in the vertical direction, Eq. 2.28 indicates that the fluid surface will be horizontal. However, from Eq. 2.26 we see that the pressure distribution is not hydrostatic, but is given by the equation

$$\frac{dp}{dz} = -\rho(g + a_z)$$

For fluids of constant density this equation shows that the pressure will vary linearly with depth, but the variation is due to the combined effects of gravity and the externally induced acceleration, $\rho(g + a_z)$, rather than simply the specific weight ρg. Thus, for example, the pressure along the bottom of a liquid-filled tank which is resting on the floor of an elevator that is accelerating upward will be increased over that which exists when the tank is at rest (or moving with a constant velocity). It is to be noted that for a *freely falling* fluid mass ($a_z = -g$), the pressure gradients in all three coordinate directions are zero, which means that if the pressure surrounding the mass is zero, the pressure throughout will be zero. The pressure throughout a "blob" of orange juice floating in an orbiting space shuttle (a form of free fall) is zero. The only force holding the liquid together is surface tension (see Section 1.9).

EXAMPLE 2.11 Pressure Variation in an Accelerating Tank

GIVEN The cross section for the fuel tank of an experimental vehicle is shown in Fig. E2.11. The rectangular tank is vented to the atmosphere and the specific gravity of the fuel is $SG = 0.65$. A pressure transducer is located in its side as illustrated. During testing of the vehicle, the tank is subjected to a constant linear acceleration, a_y.

FIND (a) Determine an expression that relates a_y and the pressure (in N/m^2) at the transducer. (b) What is the maximum acceleration that can occur before the fuel level drops below the transducer?

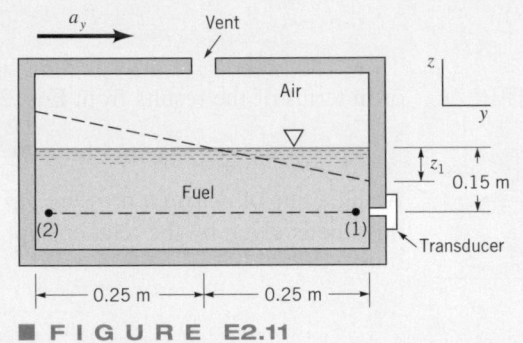

■ **FIGURE E2.11**

SOLUTION

(a) For a constant horizontal acceleration the fuel will move as a rigid body, and from Eq. 2.28 the slope of the fuel surface can be expressed as

$$\frac{dz}{dy} = -\frac{a_y}{g}$$

since $a_z = 0$. Thus, for some arbitrary a_y, the change in depth, z_1, of liquid on the right side of the tank can be found from the equation

$$-\frac{z_1}{0.25\ \text{m}} = -\frac{a_y}{g}$$

or

$$z_1 = (0.25\ \text{m})\left(\frac{a_y}{g}\right)$$

Since there is no acceleration in the vertical, z, direction, the pressure along the wall varies hydrostatically as shown by Eq. 2.26. Thus, the pressure at the transducer is given by the relationship

$$p = \gamma h$$

where h is the depth of fuel above the transducer, and therefore

$$p = (0.65)(9800\ \text{N/m}^3)[0.15\ \text{m} - (0.25\ \text{m})(a_y/g)]$$

$$= \left(955.5 - 1592.5\frac{a_y}{g}\right) \qquad \text{(Ans)}$$

for $z_1 \leq 0.15$ m As written, p would be given in N/m^2

(b) The limiting value for $(a_y)_{\text{max}}$ (when the fuel level reaches the transducer) can be found from the equation

$$0.15\ \text{m} = (0.25\ \text{m})\left[\frac{(a_y)_{\text{max}}}{g}\right]$$

or

$$(a_y)_{\text{max}} = \frac{3g}{5}$$

and for standard acceleration of gravity

$$(a_y)_{\text{max}} = \tfrac{3}{5}(9.81\ \text{m/s}^2) = 5.9\ \text{m/s}^2 \qquad \text{(Ans)}$$

COMMENT Note that the pressure in horizontal layers is not constant in this example since $\partial p/\partial y = -\rho a_y \neq 0$. Thus, for example, $p_1 \neq p_2$.

2.12.2 Rigid-Body Rotation

After an initial "start-up" transient, a fluid contained in a tank that rotates with a constant angular velocity ω about an axis as is shown in Fig. 2.30 will rotate with the tank as a rigid body. It is known from elementary particle dynamics that the acceleration of a fluid particle located at a distance r from the axis of rotation is equal in magnitude to $r\omega^2$, and the direction of the acceleration is toward the axis of rotation, as is illustrated in the figure. Since the paths of the fluid particles are circular, it is convenient to use cylindrical polar coordinates r, θ, and z, defined in the insert in Fig. 2.30. It will be shown in Chapter 6 that in terms of cylindrical coordinates the pressure gradient ∇p can be expressed as

$$\nabla p = \frac{\partial p}{\partial r}\,\hat{\mathbf{e}}_r + \frac{1}{r}\frac{\partial p}{\partial \theta}\,\hat{\mathbf{e}}_\theta + \frac{\partial p}{\partial z}\,\hat{\mathbf{e}}_z \tag{2.29}$$

Thus, in terms of this coordinate system

$$\mathbf{a}_r = -r\omega^2\,\hat{\mathbf{e}}_r \qquad \mathbf{a}_\theta = 0 \qquad \mathbf{a}_z = 0$$

and from Eq. 2.2

$$\frac{\partial p}{\partial r} = \rho r\omega^2 \qquad \frac{\partial p}{\partial \theta} = 0 \qquad \frac{\partial p}{\partial z} = -\gamma \tag{2.30}$$

These results show that for this type of rigid-body rotation, the pressure is a function of two variables r and z, and therefore the differential pressure is

$$dp = \frac{\partial p}{\partial r}\,dr + \frac{\partial p}{\partial z}\,dz$$

or

$$dp = \rho r\omega^2\,dr - \gamma\,dz \tag{2.31}$$

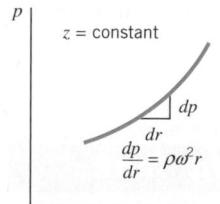

On a horizontal plane ($dz = 0$), it follows from Eq. 2.31 that $dp/dr = \rho\omega^2 r$, which is greater than zero. Hence, as illustrated in the figure in the margin, because of centrifugal acceleration, the pressure increases in the radial direction.

Along a surface of constant pressure, such as the free surface, $dp = 0$, so that from Eq. 2.31 (using $\gamma = \rho g$)

$$\frac{dz}{dr} = \frac{r\omega^2}{g}$$

Integration of this result gives the equation for surfaces of constant pressure as

$$z = \frac{\omega^2 r^2}{2g} + \text{constant} \tag{2.32}$$

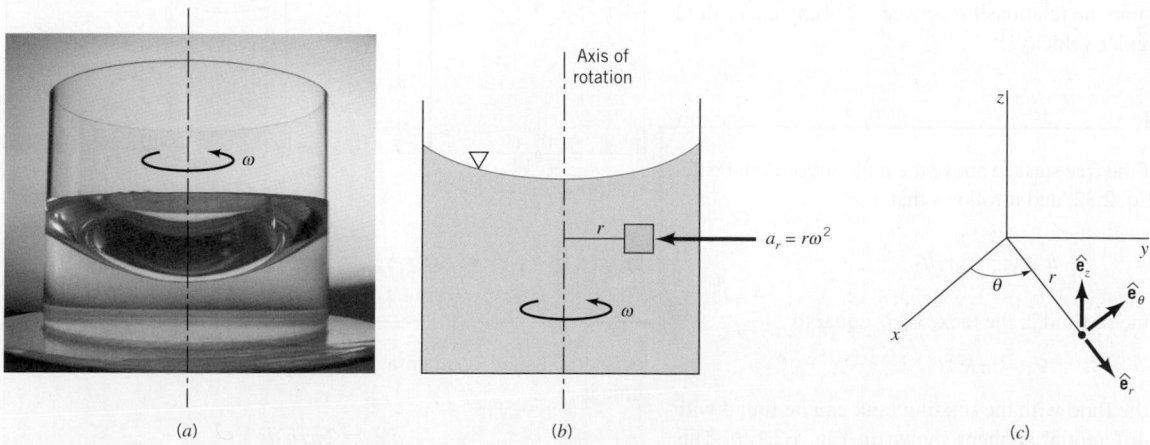

(a)   (b)   (c)

■ **F I G U R E   2.30**   **Rigid-body rotation of a liquid in a tank. (Photograph courtesy of Geno Pawlak.)**

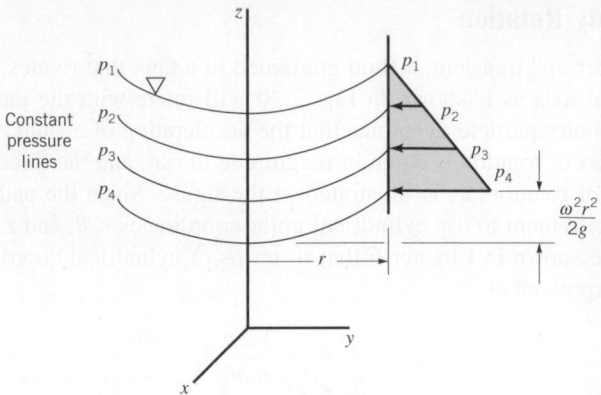

■ **FIGURE 2.31** Pressure distribution in a rotating liquid.

The free surface in a rotating liquid is curved rather than flat.

This equation reveals that these surfaces of constant pressure are parabolic, as illustrated in Fig. 2.31. Integration of Eq. 2.31 yields

$$\int dp = \rho\omega^2 \int r\,dr - \gamma \int dz$$

or

$$p = \frac{\rho\omega^2 r^2}{2} - \gamma z + \text{constant} \qquad (2.33)$$

where the constant of integration can be expressed in terms of a specified pressure at some arbitrary point r_0, z_0. This result shows that the pressure varies with the distance from the axis of rotation, but at a fixed radius, the pressure varies hydrostatically in the vertical direction as shown in Fig. 2.31.

EXAMPLE 2.12 Free Surface Shape of Liquid in a Rotating Tank

GIVEN It has been suggested that the angular velocity, ω, of a rotating body or shaft can be measured by attaching an open cylinder of liquid, as shown in Fig. E2.12a, and measuring with some type of depth gage the change in the fluid level, $H - h_0$, caused by the rotation of the fluid.

FIND Determine the relationship between this change in fluid level and the angular velocity.

SOLUTION

The height, h, of the free surface above the tank bottom can be determined from Eq. 2.32, and it follows that

$$h = \frac{\omega^2 r^2}{2g} + h_0$$

The initial volume of fluid in the tank, V_i, is equal to

$$V_i = \pi R^2 H$$

The volume of the fluid with the rotating tank can be found with the aid of the differential element shown in Fig. E2.12b. This

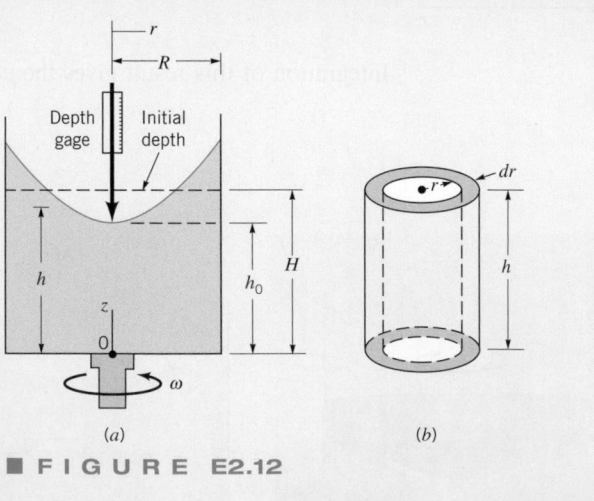

■ **FIGURE E2.12**

cylindrical shell is taken at some arbitrary radius, r, and its volume is

$$dV = 2\pi r h\,dr$$

The total volume is, therefore,

$$\mathcal{V} = 2\pi \int_0^R r\left(\frac{\omega^2 r^2}{2g} + h_0\right) dr = \frac{\pi\omega^2 R^4}{4g} + \pi R^2 h_0$$

Since the volume of the fluid in the tank must remain constant (assuming that none spills over the top), it follows that

$$\pi R^2 H = \frac{\pi\omega^2 R^4}{4g} + \pi R^2 h_0$$

or

$$H - h_0 = \frac{\omega^2 R^2}{4g} \qquad \text{(Ans)}$$

COMMENT   This is the relationship we were looking for. It shows that the change in depth could indeed be used to determine the rotational speed, although the relationship between the change in depth and speed is not a linear one.

F l u i d s   i n   t h e   N e w s

Rotating mercury mirror telescope A telescope mirror has the same shape as the parabolic free surface of a *liquid in a rotating tank*. The liquid mirror telescope (LMT) consists of a pan of liquid (normally mercury because of its excellent reflectivity) rotating to produce the required parabolic shape of the free surface mirror. With recent technological advances, it is possible to obtain the vibration-free rotation and the constant angular velocity necessary to produce a liquid mirror surface precise enough for astronomical use. Construction of the largest LMT, located at the University of British Columbia, has recently been completed. With a diameter of 6 ft (1.8 m) and a rotation rate of 7 rpm, this mirror uses 30 liters of mercury for its 1-mm thick, parabolic-shaped mirror. One of the major benefits of a LMT (compared to a normal glass mirror telescope) is its low cost. Perhaps the main disadvantage is that a LMT can look only straight up, although there are many galaxies, supernova explosions, and pieces of space junk to view in any part of the sky. The next generation LMTs may have movable secondary mirrors to allow a larger portion of the sky to be viewed. (See Problem 2.121.)

2.13 Chapter Summary and Study Guide

Pascal's law
surface force
body force
incompressible fluid
hydrostatic pressure
   distribution
pressure head
compressible fluid
U.S. standard
   atmosphere
absolute pressure
gage pressure
vacuum pressure
barometer
manometer
Bourdon pressure
   gage
center of pressure
buoyant force
Archimedes' principle
center of buoyancy

In this chapter the pressure variation in a fluid at rest is considered, along with some important consequences of this type of pressure variation. It is shown that for incompressible fluids at rest the pressure varies linearly with depth. This type of variation is commonly referred to as hydrostatic pressure distribution. For compressible fluids at rest the pressure distribution will not generally be hydrostatic, but Eq. 2.4 remains valid and can be used to determine the pressure distribution if additional information about the variation of the specific weight is specified. The distinction between absolute and gage pressure is discussed along with a consideration of barometers for the measurement of atmospheric pressure.

Pressure measuring devices called manometers, which utilize static liquid columns, are analyzed in detail. A brief discussion of mechanical and electronic pressure gages is also included. Equations for determining the magnitude and location of the resultant fluid force acting on a plane surface in contact with a static fluid are developed. A general approach for determining the magnitude and location of the resultant fluid force acting on a curved surface in contact with a static fluid is described. For submerged or floating bodies the concept of the buoyant force and the use of Archimedes' principle are reviewed.

The following checklist provides a study guide for this chapter. When your study of the entire chapter and end-of-chapter exercises has been completed you should be able to

■ write out meanings of the terms listed here in the margin and understand each of the related concepts. These terms are particularly important and are set in *italic, bold, and color* type in the text.

■ calculate the pressure at various locations within an incompressible fluid at rest.

■ calculate the pressure at various locations within a compressible fluid at rest using Eq. 2.4 if the variation in the specific weight is specified.

■ use the concept of a hydrostatic pressure distribution to determine pressures from measurements using various types of manometers.

■ determine the magnitude, direction, and location of the resultant hydrostatic force acting on a plane surface.

- determine the magnitude, direction, and location of the resultant hydrostatic force acting on a curved surface.
- use Archimedes' principle to calculate the resultant hydrostatic force acting on floating or submerged bodies.
- analyze, based on Eq. 2.2, the motion of fluids moving with simple rigid-body linear motion or simple rigid-body rotation.

Some of the important equations in this chapter are:

Pressure gradient in a stationary fluid	$\dfrac{dp}{dz} = -\gamma$	**(2.4)**
Pressure variation in a stationary incompressible fluid	$p_1 = \gamma h + p_2$	**(2.7)**
Hydrostatic force on a plane surface	$F_R = \gamma h_c A$	**(2.18)**
Location of hydrostatic force on a plane surface	$y_R = \dfrac{I_{xc}}{y_c A} + y_c$	**(2.19)**
	$x_R = \dfrac{I_{xyc}}{y_c A} + x_c$	**(2.20)**
Buoyant force	$F_B = \gamma \Psi$	**(2.22)**
Pressure gradient in rigid-body motion	$-\dfrac{\partial p}{\partial x} = \rho a_x, \quad -\dfrac{\partial p}{\partial y} = \rho a_y, \quad -\dfrac{\partial p}{\partial z} = \gamma + \rho a_z$	**(2.24)**
Pressure gradient in rigid-body rotation	$\dfrac{\partial p}{\partial r} = \rho r \omega^2, \quad \dfrac{\partial p}{\partial \theta} = 0, \quad \dfrac{\partial p}{\partial z} = -\gamma$	**(2.30)**

References

1. *The U.S. Standard Atmosphere, 1962*, U.S. Government Printing Office, Washington, D.C., 1962.
2. *The U.S. Standard Atmosphere, 1976*, U.S. Government Printing Office, Washington, D.C., 1976.
3. Benedict, R. P., *Fundamentals of Temperature, Pressure, and Flow Measurements*, 3rd Ed., Wiley, New York, 1984.
4. Dally, J. W., Riley, W. F., and McConnell, K. G., *Instrumentation for Engineering Measurements*, 2nd Ed., Wiley, New York, 1993.
5. Holman, J. P., *Experimental Methods for Engineers*, 4th Ed., McGraw-Hill, New York, 1983.
6. Comstock, J. P., ed., *Principles of Naval Architecture*, Society of Naval Architects and Marine Engineers, New York, 1967.
7. Hasler, A. F., Pierce, H., Morris, K. R., and Dodge, J., "Meteorological Data Fields 'In Perspective'," *Bulletin of the American Meteorological Society*, Vol. 66, No. 7, July 1985.

Review Problems

Go to Appendix G for a set of review problems with answers. Detailed solutions can be found in *Student Solution Manual and Study Guide for Fundamentals of Fluid Mechanics*, by Munson et al. (© 2009 John Wiley and Sons, Inc.).

Problems

Note: Unless otherwise indicated, use the values of fluid properties found in the tables on the inside of the front cover. Problems designated with an (*) are intended to be solved with the aid of a programmable calculator or a computer. Problems designated with a (†) are "open-ended" problems and require critical thinking in that to work them one must make various assumptions and provide the necessary data. There is not a unique answer to these problems.

Answers to the even-numbered problems are listed at the end of the book. Access to the videos that accompany problems can be obtained through the book's web site, www.wiley.com/college/munson. The lab-type problems can also be accessed on this web site.

Section 2.3 **Pressure Variation in a Fluid at Rest**

2.1 Obtain a photograph/image of a situation in which the fact that in a static fluid the pressure increases with depth is important. Print this photo and write a brief paragraph that describes the situation involved.

2.2 A closed, 5-m-tall tank is filled with water to a depth of 4 m. The top portion of the tank is filled with air which, as indicated by a pressure gage at the top of the tank, is at a pressure of 20 kPa. Determine the pressure that the water exerts on the bottom of the tank.

2.3 A closed tank is partially filled with glycerin. If the air pressure in the tank is 4.1 N/cm^2 and the depth of glycerin is 3.0 m, what is the pressure in N/m^2 at the bottom of the tank?

2.4 Blood pressure is usually given as a ratio of the maximum pressure (systolic pressure) to the minimum pressure (diastolic pressure). As shown in **Video V2.2**, such pressures are commonly measured with a mercury manometer. A typical value for this ratio for a human would be 120/70, where the pressures are in mm Hg. **(a)** What would these pressures be in pascals? **(b)** If your car tire was inflated to 120 mm Hg, would it be sufficient for normal driving?

2.5 An unknown immiscible liquid seeps into the bottom of an open oil tank. Some measurements indicate that the depth of the unknown liquid is 1.5 m and the depth of the oil (specific weight = 8.5 kN/m^3) floating on top is 5.0 m. A pressure gage connected to the bottom of the tank reads 65 kPa. What is the specific gravity of the unknown liquid?

2.6 Bathyscaphes are capable of submerging to great depths in the ocean. What is the pressure at a depth of 5 km, assuming that seawater has a constant specific weight of 10.1 kN/m^3? Express your answer in pascals.

2.7 For the great depths that may be encountered in the ocean the compressibility of seawater may become an important consideration. **(a)** Assume that the bulk modulus for seawater is constant and derive a relationship between pressure and depth which takes into account the change in fluid density with depth. **(b)** Make use of part **(a)** to determine the pressure at a depth of 6 km assuming seawater has a bulk modulus of 2.3×10^9 Pa and a density of 1030 kg/m^3 at the surface. Compare this result with that obtained by assuming a constant density of 1030 kg/m^3.

2.8 Sometimes when riding an elevator or driving up or down a hilly road a person's ears "pop" as the pressure difference between the inside and outside of the ear is equalized. Determine the pressure difference (in kPa) associated with this phenomenon if it occurs during a 46 m elevation change.

2.9 Develop an expression for the pressure variation in a liquid in which the specific weight increases with depth, h, as $\gamma = Kh + \gamma_0$, where K is a constant and γ_0 is the specific weight at the free surface.

*2.10 In a certain liquid at rest, measurements of the specific weight at various depths show the following variation:

h (m)	γ (N/m^3)
0	10.9
3	11.9
6	13.2
9	14.3
12	15.2
15	16.0
18	16.8
21	17.2
24	17.6
27	17.9
30	18.1

The depth $h = 0$ corresponds to a free surface at atmospheric pressure. Determine, through numerical integration of Eq. 2.4, the corresponding variation in pressure and show the results on a plot of pressure (in kPa) versus depth (in meters).

†2.11 Because of elevation differences, the water pressure in the second floor of your house is lower than it is in the first floor. For tall buildings this pressure difference can become unacceptable. Discuss possible ways to design the water distribution system in very tall buildings so that the hydrostatic pressure difference is within acceptable limits.

*2.12 Under normal conditions the temperature of the atmosphere decreases with increasing elevation. In some situations, however, a temperature inversion may exist so that the air temperature increases with elevation. A series of temperature probes on a mountain give the elevation–temperature data shown in the table below. If the barometric pressure at the base of the mountain is 83.4 kPa (abs), determine by means of numerical integration the pressure at the top of the mountain.

Elevation (m)	Temperature (°C)
1524	10.1 (base)
1676	12.9
1829	15.7
1951	17.0
2164	19.4
2256	20.2
2500	21.1
2621	20.8
2804	20.0
3018	19.5 (top)

†2.13 Although it is difficult to compress water, the density of water at the bottom of the ocean is greater than that at the surface because of the higher pressure at depth. Estimate how much higher the ocean's surface would be if the density of seawater were instantly changed to a uniform density equal to that at the surface.

2.14 (See Fluids in the News article titled **"Giraffe's blood pressure,"** Section 2.3.1.) **(a)** Determine the change in hydrostatic pressure in a giraffe's head as it lowers its head from eating leaves 6 m above the ground to getting a drink of water at ground level as shown in Fig. P2.14. Assume the specific gravity of blood is $SG = 1$. **(b)** Compare the pressure change calculated in part **(a)** to the normal 120 mm of mercury pressure in a human's heart.

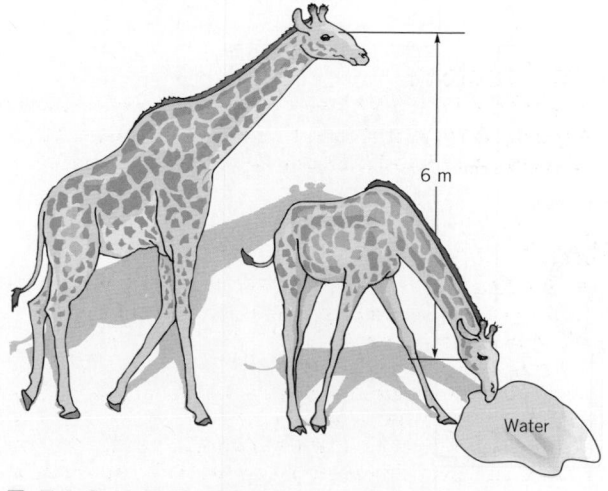

6 m

Water

■ **FIGURE P2.14**

Section 2.4 **Standard Atmosphere**

2.15 Assume that a person skiing high in the mountains at an altitude of 4500 m takes in the same volume of air with each breath as she does while walking at sea level. Determine the ratio of the mass of oxygen inhaled for each breath at this high altitude compared to that at sea level.

2.16 Pikes Peak near Denver, Colorado, has an elevation of 4300 m. **(a)** Determine the pressure at this elevation, based on Eq. 2.12. **(b)** If the air is assumed to have a constant specific weight of 12 N/m³, what would the pressure be at this altitude? **(c)** If the air is assumed to have a constant temperature of 15 °C, what would the pressure be at this elevation? For all three cases assume standard atmospheric conditions at sea level (see Table 2.1).

2.17 Equation 2.12 provides the relationship between pressure and elevation in the atmosphere for those regions in which the temperature varies linearly with elevation. Derive this equation and verify the value of the pressure given in Table C.1 in Appendix C for an elevation of 5 km.

2.18 As shown in Fig. 2.6 for the U.S. standard atmosphere, the troposphere extends to an altitude of 11 km where the pressure is 22.6 kPa (abs). In the next layer, called the stratosphere, the temperature remains constant at −56.5 °C. Determine the pressure and density in this layer at an altitude of 15 km. Assume $g = 9.77$ m/s² in your calculations. Compare your results with those given in Table C.1 in Appendix C.

2.19 (See Fluids in the News article titled **"Weather, barometers, and bars,"** Section 2.5.) The record low sea-level barometric pressure ever recorded is 65.5 cm of mercury. At what altitude in the standard atmosphere is the pressure equal to this value?

Section 2.5 **Measurement of Pressure**

2.20 On a given day, a barometer at the base of the Washington Monument reads 76 cm of mercury. What would the barometer reading be when you carry it up to the observation deck 152 m above the base of the monument?

2.21 Bourdon gages (see **Video V2.3** and Fig. 2.13) are commonly used to measure pressure. When such a gage is attached to the closed water tank of Fig. P2.21 the gage reads 34.5 kPa. What is the absolute air pressure in the tank? Assume standard atmospheric pressure of 101.3 kPa.

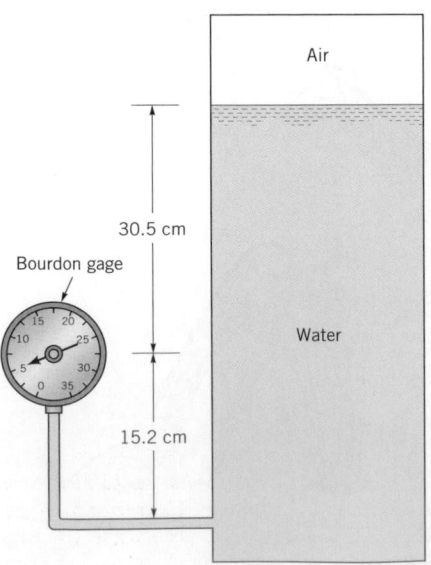

Air

30.5 cm

Bourdon gage

Water

15.2 cm

■ **F I G U R E P2.21**

2.22 On the suction side of a pump a Bourdon pressure gage reads 40 kPa vacuum. What is the corresponding absolute pressure if the local atmospheric pressure is 100 kPa (abs)?

Section 2.6 **Manometry**

2.23 Obtain a photograph/image of a situation in which the use of a manometer is important. Print this photo and write a brief paragraph that describes the situation involved.

2.24 A water-filled U-tube manometer is used to measure the pressure inside a tank that contains air. The water level in the U-tube on the side that connects to the tank is 1.5 m above the base of the tank. The water level in the other side of the U-tube (which is open to the atmosphere) is 0.6 m above the base. Determine the pressure within the tank.

2.25 A barometric pressure of 74.7 cm. Hg corresponds to what value of atmospheric pressure in pascals?

2.26 For an atmospheric pressure of 101 kPa (abs) determine the heights of the fluid columns in barometers containing one of the following liquids: **(a)** mercury, **(b)** water, and **(c)** ethyl alcohol. Calculate the heights including the effect of vapor pressure, and compare the results with those obtained neglecting vapor pressure. Do these results support the widespread use of mercury for barometers? Why?

2.27 A mercury manometer is connected to a large reservoir of water as shown in Fig. P2.27. Determine the ratio, h_w/h_m, of the distances h_w and h_m indicated in the figure.

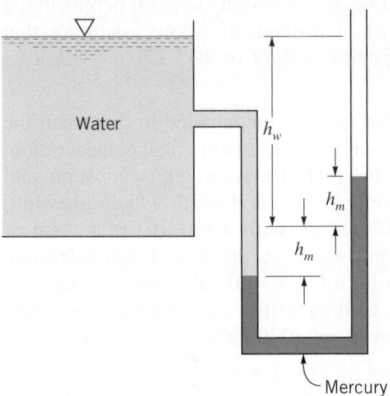

Water

h_w

h_m

h_m

Mercury

■ **F I G U R E P2.27**

2.28 A U-tube manometer is connected to a closed tank containing air and water as shown in Fig. P2.28. At the closed end of the manometer the air pressure is 110.3 kPa (abs). Determine the reading

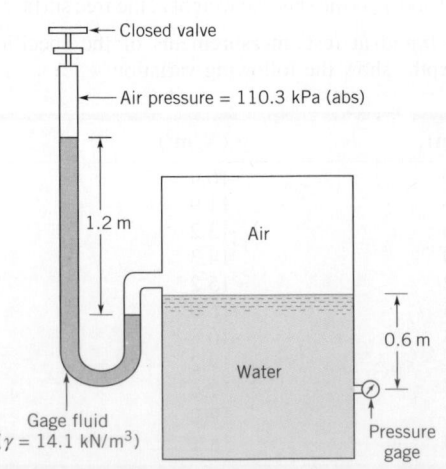

Closed valve

Air pressure = 110.3 kPa (abs)

1.2 m

Air

0.6 m

Water

Gage fluid
($\gamma = 14.1$ kN/m³)

Pressure gage

■ **F I G U R E P2.28**

on the pressure gage for a differential reading of 1.2 m on the manometer. Express your answer in kPa (gage). Assume standard atmospheric pressure and neglect the weight of the air columns in the manometer.

2.29 A closed cylindrical tank filled with water has a hemispherical dome and is connected to an inverted piping system as shown in Fig. P2.29. The liquid in the top part of the piping system has a specific gravity of 0.8, and the remaining parts of the system are filled with water. If the pressure gage reading at A is 60 kPa, determine: **(a)** the pressure in pipe B, and **(b)** the pressure head, in millimeters of mercury, at the top of the dome (point C).

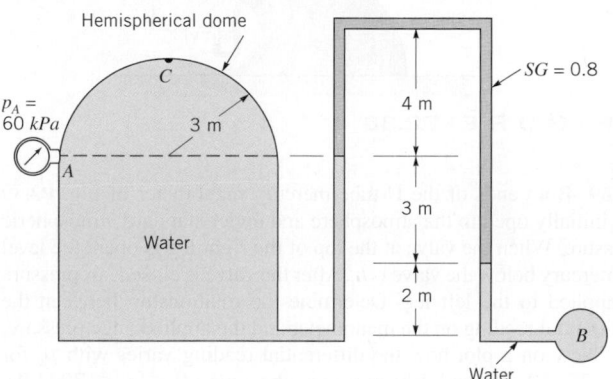

■ FIGURE P2.29

2.30 Two pipes are connected by a manometer as shown in Fig. P2.30. Determine the pressure difference, $p_A - p_B$, between the pipes.

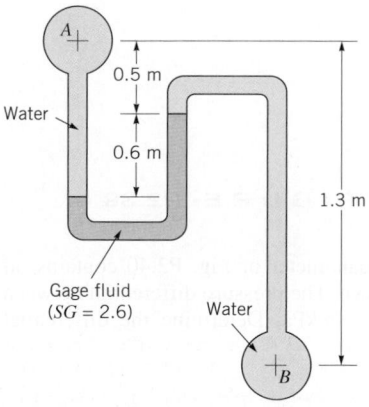

■ FIGURE P2.30

2.31 A U-tube manometer is connected to a closed tank as shown in Fig. P2.31. The air pressure in the tank is 3.4 kPa and the liquid in

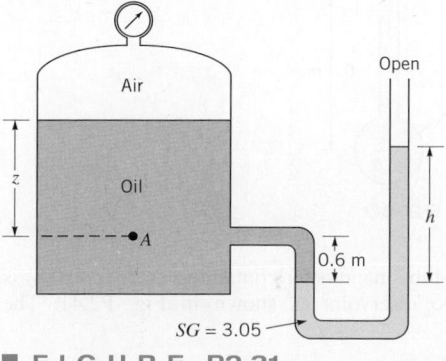

■ FIGURE P2.31

the tank is oil ($\gamma = 8.5$ kN/m³). The pressure at point A is 13.8 kPa. Determine: **(a)** the depth of oil, z, and **(b)** the differential reading, h, on the manometer.

2.32 For the inclined-tube manometer of Fig. P2.32 the pressure in pipe A is 4.1 kPa. The fluid in both pipes A and B is water, and the gage fluid in the manometer has a specific gravity of 2.6. What is the pressure in pipe B corresponding to the differential reading shown?

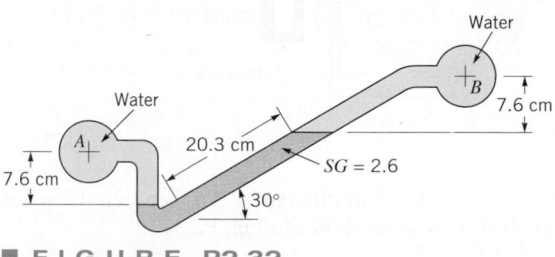

■ FIGURE P2.32

2.33 A flowrate measuring device is installed in a horizontal pipe through which water is flowing. A U-tube manometer is connected to the pipe through pressure taps located 7.6 cm on either side of the device. The gage fluid in the manometer has a specific weight of 17.6 kN/m³. Determine the differential reading of the manometer corresponding to a pressure drop between the taps of 0.3 N/cm².

2.34 Small differences in gas pressures are commonly measured with a *micromanometer* of the type illustrated in Fig. P2.34. This device consists of two large reservoirs each having a cross-sectional area A_r which are filled with a liquid having a specific weight γ_1 and connected by a U-tube of cross-sectional area A_t containing a liquid of specific weight γ_2. When a differential gas pressure, $p_1 - p_2$, is applied, a differential reading, h, develops. It is desired to have this reading sufficiently large (so that it can be easily read) for small pressure differentials. Determine the relationship between h and $p_1 - p_2$ when the area ratio A_t/A_r is small, and show that the differential reading, h, can be magnified by making the difference in specific weights, $\gamma_2 - \gamma_1$, small. Assume that initially (with $p_1 = p_2$) the fluid levels in the two reservoirs are equal.

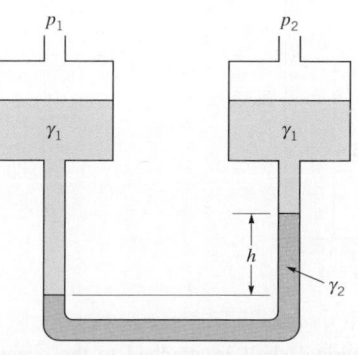

■ FIGURE P2.34

2.35 The cyclindrical tank with hemispherical ends shown in Fig. P2.35 contains a volatile liquid and its vapor. The liquid density is 800 kg/m³, and its vapor density is negligible. The pressure in the vapor is 120 kPa (abs), and the atmospheric pressure is 101 kPa (abs). Determine: **(a)** the gage pressure reading on the pressure gage; and **(b)** the height, h, of the mercury manometer.

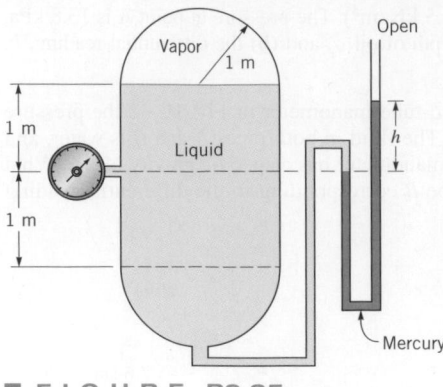

2.36 Determine the elevation difference, Δh, between the water levels in the two open tanks shown in Fig. P2.36.

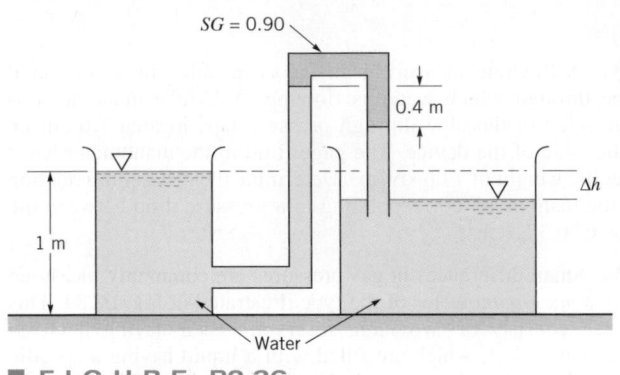

■ F I G U R E P2.36

2.37 For the configuration shown in Fig. P2.37 what must be the value of the specific weight of the unknown fluid? Express your answer in N/m^3.

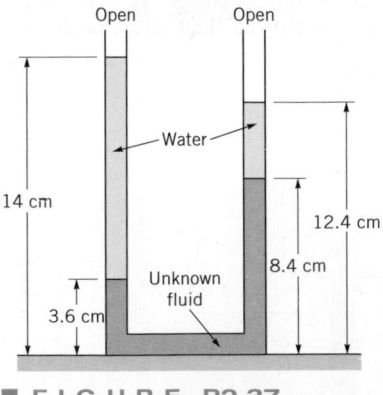

■ F I G U R E P2.37

2.38 An air-filled, hemispherical shell is attached to the ocean floor at a depth of 10 m as shown in Fig. P2.38. A mercury barometer located inside the shell reads 765 mm Hg, and a mercury U-tube manometer designed to give the outside water pressure indicates a differential reading of 735 mm Hg as illustrated. Based on these data what is the atmospheric pressure at the ocean surface?

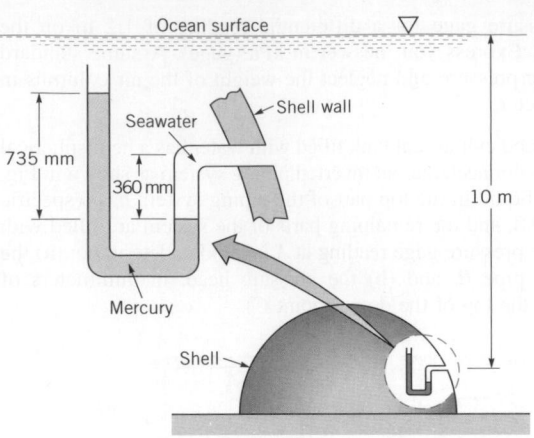

■ F I G U R E P2.38

***2.39** Both ends of the U-tube mercury manometer of Fig. P2.39 are initially open to the atmosphere and under standard atmospheric pressure. When the valve at the top of the right leg is open, the level of mercury below the valve is h_i. After the valve is closed, air pressure is applied to the left leg. Determine the relationship between the differential reading on the manometer and the applied gage pressure, p_g. Show on a plot how the differential reading varies with p_g for $h_i = 25, 50, 75,$ and 100 mm over the range $0 \leq p_g \leq 300$ kPa. Assume that the temperature of the trapped air remains constant.

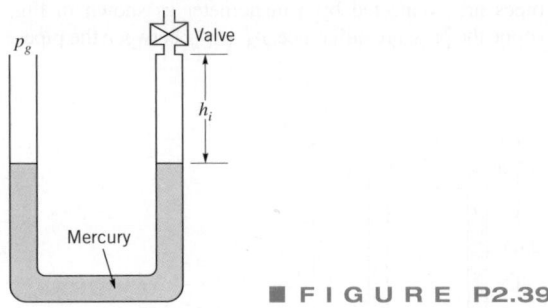

■ F I G U R E P2.39

2.40 The inverted U-tube manometer of Fig. P2.40 contains oil ($SG = 0.9$) and water as shown. The pressure differential between pipes A and B, $p_A - p_B$, is -5 kPa. Determine the differential reading, h.

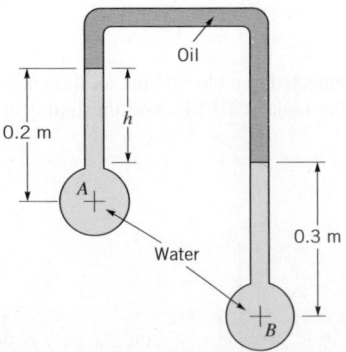

■ F I G U R E P2.40

2.41 An inverted U-tube manometer containing oil ($SG = 0.8$) is located between two reservoirs as shown in Fig. P2.41. The

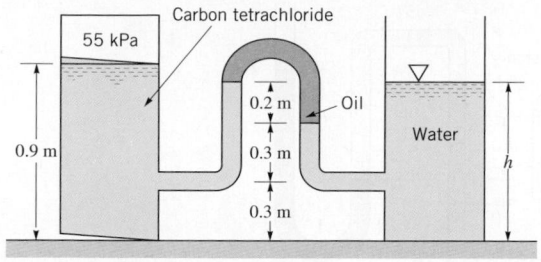

■ FIGURE P2.41

reservoir on the left, which contains carbon tetrachloride, is closed and pressurized to 55 kPa. The reservoir on the right contains water and is open to the atmosphere. With the given data, determine the depth of water, h, in the right reservoir.

2.42 Determine the pressure of the water in pipe A shown in Fig. P2.42 if the gage pressure of the air in the tank is 13.8 kPa.

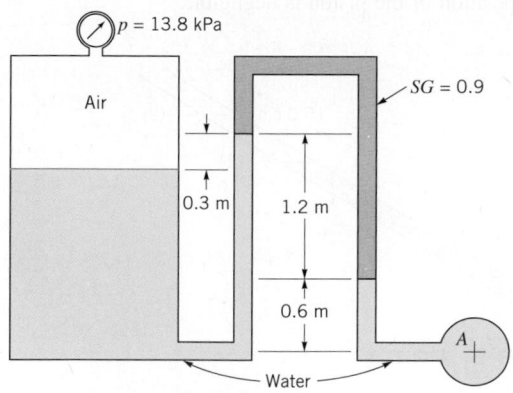

■ FIGURE P2.42

2.43 In Fig. P2.43 pipe A contains gasoline ($SG = 0.7$), pipe B contains oil ($SG = 0.9$), and the manometer fluid is mercury. Determine the new differential reading if the pressure in pipe A is decreased 25 kPa, and the pressure in pipe B remains constant. The initial differential reading is 0.30 m as shown.

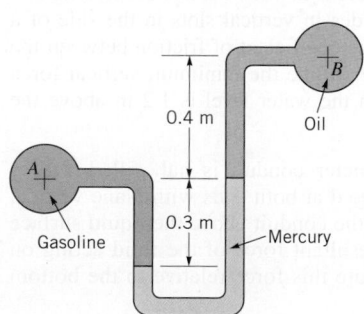

■ FIGURE P2.43

2.44 The inclined differential manometer of Fig. P2.44 contains carbon tetrachloride. Initially the pressure differential between pipes A and B, which contain a brine ($SG = 1.1$), is zero as illustrated in the figure. It is desired that the manometer give a differential reading of 30.5 cm (measured along the inclined tube) for a pressure differential of 0.7 kPa. Determine the required angle of inclination, θ.

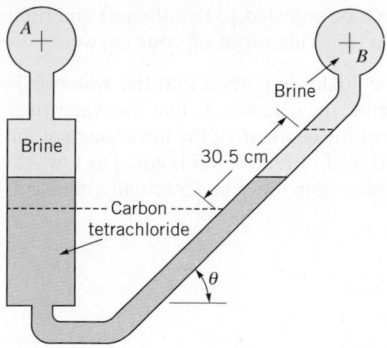

■ FIGURE P2.44

2.45 Determine the new differential reading along the inclined leg of the mercury manometer of Fig. P2.45, if the pressure in pipe A is decreased 10 kPa and the pressure in pipe B remains unchanged. The fluid in A has a specific gravity of 0.9 and the fluid in B is water.

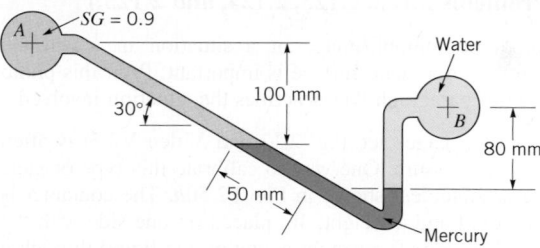

■ FIGURE P2.45

2.46 Determine the change in the elevation of the mercury in the left leg of the manometer of Fig. P2.46 as a result of an increase in pressure of 34.4 kPa in pipe A while the pressure in pipe B remains constant.

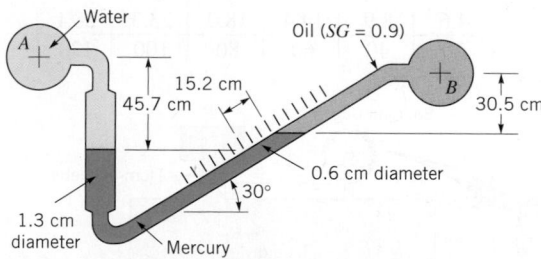

■ FIGURE P2.46

2.47 The U-shaped tube shown in Fig. P2.47 initially contains water only. A second liquid with specific weight, γ, less than water is placed on top of the water with no mixing occurring. Can the

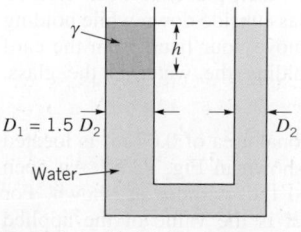

■ FIGURE P2.47

height, h, of the second liquid be adjusted so that the left and right levels are at the same height? Provide proof of your answer.

*2.48 An inverted hollow cylinder is pushed into the water as is shown in Fig. P2.48. Determine the distance, ℓ, that the water rises in the cylinder as a function of the depth, d, of the lower edge of the cylinder. Plot the results for $0 \leq d \leq H$, when H is equal to 1 m. Assume the temperature of the air within the cylinder remains constant.

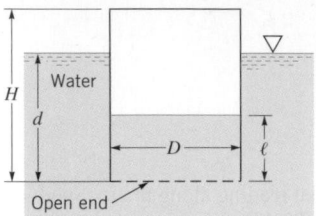

■ **FIGURE P2.48**

Section 2.8 Hydrostatic Force on a Plane Surface (Also see Lab Problems 2.122, 2.123, 2.124, and 2.125.)

2.49 Obtain a photograph/image of a situation in which the hydrostatic force on a plane surface is important. Print this photo and write a brief paragraph that describes the situation involved.

*2.50 A Bourdon gage (see Fig. 2.13 and **Video V2.3**) is often used to measure pressure. One way to calibrate this type of gage is to use the arrangement shown in Fig. P2.50a. The container is filled with a liquid and a weight, $\mathcal{W}$, placed on one side with the gage on the other side. The weight acting on the liquid through a 1-cm-diameter opening creates a pressure that is transmitted to the gage. This arrangement, with a series of weights, can be used to determine what a change in the dial movement, θ, in Fig. P2.50b, corresponds to in terms of a change in pressure. For a particular gage, some data are given below. Based on a plot of these data, determine the relationship between θ and the pressure, p, where p is measured in kPa.

$\mathcal{W}$ (N)	0	4.6	8.9	14.4	18.0	23.3	28.1
θ (deg.)	0	20	40	60	80	100	120

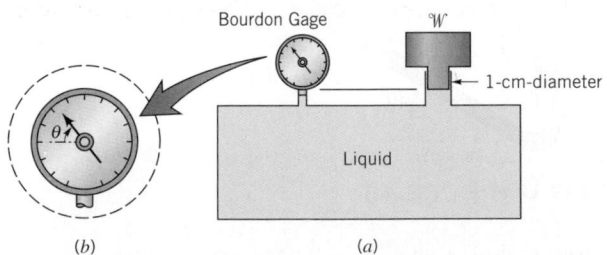

■ **FIGURE P2.50**

2.51 You partially fill a glass with water, place an index card on top of the glass, and then turn the glass upside down while holding the card in place. You can then remove your hand from the card and the card remains in place, holding the water in the glass. Explain how this works.

2.52 A piston having a cross-sectional area of 0.07 m² is located in a cylinder containing water as shown in Fig. P2.52. An open U-tube manometer is connected to the cylinder as shown. For $h_1 = 60$ mm and $h = 100$ mm, what is the value of the applied force, P, acting on the piston? The weight of the piston is negligible.

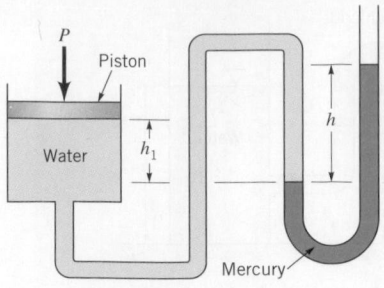

■ **FIGURE P2.52**

2.53 A 15-cm-diameter piston is located within a cylinder which is connected to a 1.3-cm-diameter inclined-tube manometer as shown in Fig. P2.53. The fluid in the cylinder and the manometer is oil (specific weight = 9.3 kN/m³). When a weight, $\mathcal{W}$, is placed on the top of the cylinder, the fluid level in the manometer tube rises from point (1) to (2). How heavy is the weight? Assume that the change in position of the piston is negligible.

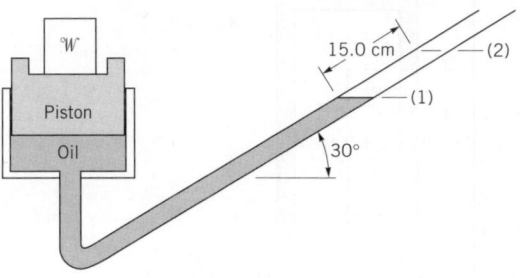

■ **FIGURE P2.53**

2.54 A circular 2-m-diameter gate is located on the sloping side of a swimming pool. The side of the pool is oriented 60° relative to the horizontal bottom, and the center of the gate is located 3 m below the water surface. Determine the magnitude of the water force acting on the gate and the point through which it acts.

2.55 A vertical rectangular gate is 2.4 m wide and 3.0 m long and weighs 26.7 kN. The gate slides in vertical slots in the side of a reservoir containing water. The coefficient of friction between the slots and the gate is 0.03. Determine the minimum vertical force required to lift the gate when the water level is 1.2 m above the top edge of the gate.

2.56 A horizontal 2-m-diameter conduit is half filled with a liquid ($SG = 1.6$) and is capped at both ends with plane vertical surfaces. The air pressure in the conduit above the liquid surface is 200 kPa. Determine the resultant force of the fluid acting on one of the end caps, and locate this force relative to the bottom of the conduit.

2.57 Forms used to make a concrete basement wall are shown in Fig. P2.57. Each 1.2-m-long form is held together by four ties— two at the top and two at the bottom as indicated. Determine the tension in the upper and lower ties. Assume concrete acts as a fluid with a weight of 23.6 kN/m³.

2.58 A structure is attached to the ocean floor as shown in Fig. P2.58. A 2-m-diameter hatch is located in an inclined wall and hinged on one edge. Determine the minimum air pressure, p_1, within the container that will open the hatch. Neglect the weight of the hatch and friction in the hinge.

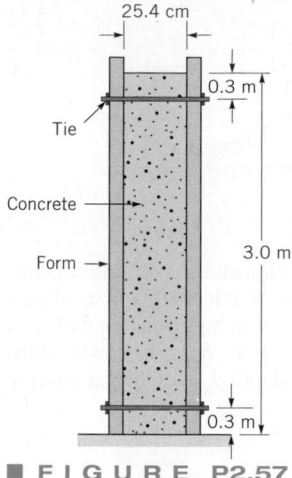

■ **FIGURE P2.57**

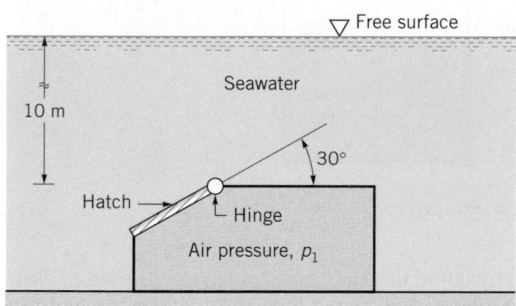

■ **FIGURE P2.58**

2.59 A long, vertical wall separates seawater from freshwater. If the seawater stands at a depth of 7 m, what depth of freshwater is required to give a zero resultant force on the wall? When the resultant force is zero will the moment due to the fluid forces be zero? Explain.

2.60 A pump supplies water under pressure to a large tank as shown in Fig. P2.60. The circular-plate valve fitted in the short discharge pipe on the tank pivots about its diameter A–A and is held shut against the water pressure by a latch at B. Show that the force on the latch is independent of the supply pressure, p, and the height of the tank, h.

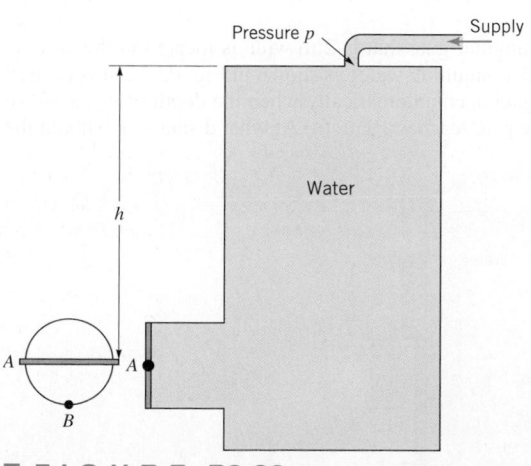

■ **FIGURE P2.60**

2.61 A homogeneous, 1.2-m-wide, 2.4-m-long rectangular gate weighing 3.6 kN is held in place by a horizontal flexible cable as

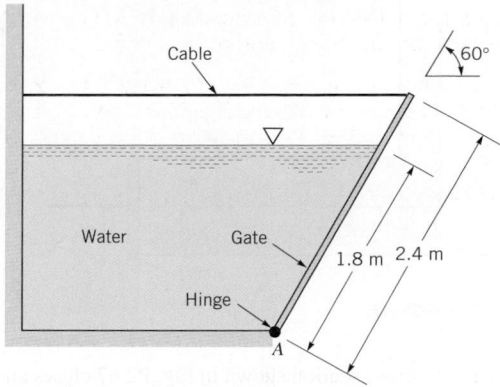

■ **FIGURE P2.61**

shown in Fig. P2.61. Water acts against the gate which is hinged at point A. Friction in the hinge is negligible. Determine the tension in the cable.

†2.62 Sometimes it is difficult to open an exterior door of a building because the air distribution system maintains a pressure difference between the inside and outside of the building. Estimate how big this pressure difference can be if it is "not too difficult" for an average person to open the door.

2.63 An area in the form of an isosceles triangle with a base width of 1.8 m and an altitude of 2.4 m lies in the plane forming one wall of a tank which contains a liquid having a specific weight of 12.5 kN/m³. The side slopes upward, making an angle of 60° with the horizontal. The base of the triangle is horizontal and the vertex is above the base. Determine the resultant force the fluid exerts on the area when the fluid depth is 6.0 m above the base of the triangular area. Show, with the aid of a sketch, where the center of pressure is located.

2.64 Solve Problem 2.63 if the isosceles triangle is replaced with a right triangle having the same base width and altitude as the isosceles triangle.

2.65 A vertical plane area having the shape shown in Fig. P2.65 is immersed in an oil bath (specific weight = 8.75 kN/m³). Determine the magnitude of the resultant force acting on one side of the area as a result of the oil.

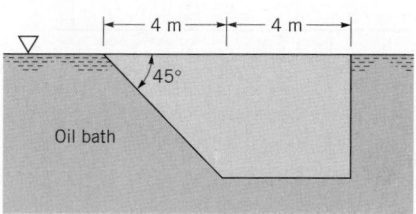

■ **FIGURE P2.65**

2.66 A 3-m-wide, 8-m-high rectangular gate is located at the end of a rectangular passage that is connected to a large open tank filled with water as shown in Fig. P2.66. The gate is hinged at its bottom and held closed by a horizontal force, F_H, located at the center of the gate. The maximum value for F_H is 3500 kN. **(a)** Determine the maximum water depth, h, above the center of the gate that can exist without the gate opening. **(b)** Is the answer the same if the gate is hinged at the top? Explain your answer.

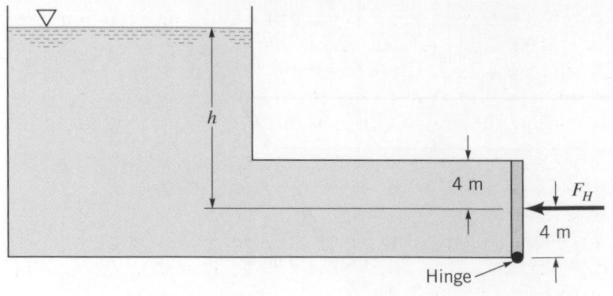

■ **FIGURE P2.66**

2.67 A gate having the cross section shown in Fig. P2.67 closes an opening 1.5 m wide and 1.2 m high in a water reservoir. The gate weighs 2.2 kN and its center of gravity is 0.3 m to the left of *AC* and 0.6 m above *BC*. Determine the horizontal reaction that is developed on the gate at *C*.

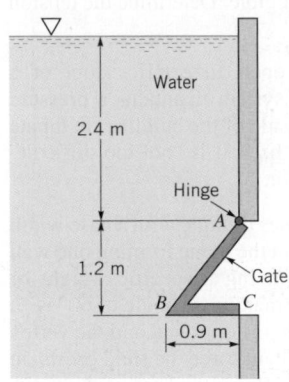

■ **FIGURE P2.67**

2.68 The massless, 1.2-m-wide gate shown in Fig. P2.68 pivots about the frictionless hinge O. It is held in place by the 8.9 kN counterweight, *W*. Determine the water depth, *h*.

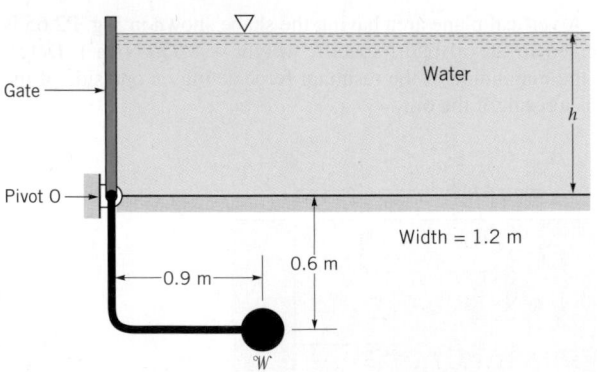

■ **FIGURE P2.68**

*2.69** A 890 N homogeneous gate of 3.0 m width and 1.5 m length is hinged at point *A* and held in place by a 3.7-m-long brace as shown in Fig. P2.69. As the bottom of the brace is moved to the right, the water level remains at the top of the gate. The line of action of the force that the brace exerts on the gate is along the brace. **(a)** Plot the magnitude of the force exerted on the gate by the brace as a function of the angle of the gate, θ, for $0 \leq \theta \leq 90°$. **(b)** Repeat the calculations for the case in which the weight of the gate is negligible. Comment on the results as $\theta \to 0$.

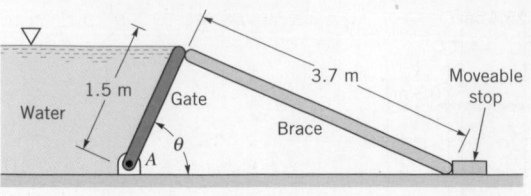

■ **FIGURE P2.69**

2.70 An open tank has a vertical partition and on one side contains gasoline with a density $\rho = 700 \text{ kg/m}^3$ at a depth of 4 m, as shown in Fig. P2.70. A rectangular gate that is 4 m high and 2 m wide and hinged at one end is located in the partition. Water is slowly added to the empty side of the tank. At what depth, *h*, will the gate start to open?

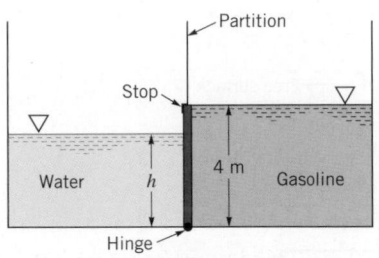

■ **FIGURE P2.70**

2.71 A 1.2 m by 0.9 m massless rectangular gate is used to close the end of the water tank shown in Fig. P2.71. A 890 N weight attached to the arm of the gate at a distance ℓ from the frictionless hinge is just sufficient to keep the gate closed when the water depth is 0.6 m, that is, when the water fills the semicircular lower portion of the tank. If the water were deeper the gate would open. Determine the distance ℓ.

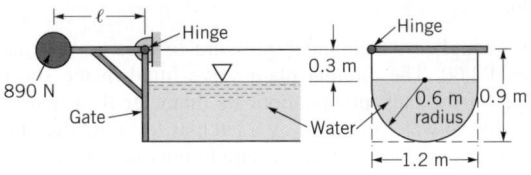

■ **FIGURE P2.71**

2.72 A rectangular gate that is 2 m wide is located in the vertical wall of a tank containing water as shown in Fig. P2.72. It is desired to have the gate open automatically when the depth of water above the top of the gate reaches 10 m. **(a)** At what distance, *d*, should the

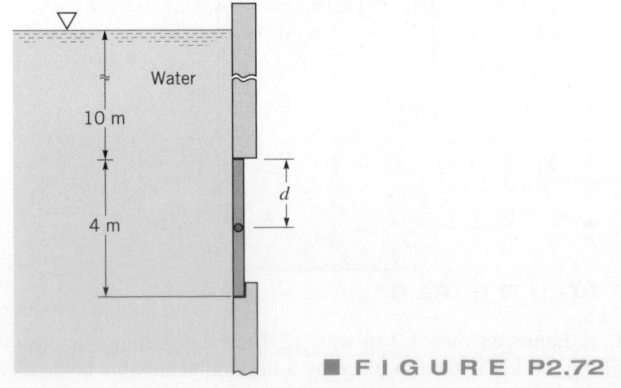

■ **FIGURE P2.72**

frictionless horizontal shaft be located? **(b)** What is the magnitude of the force on the gate when it opens?

2.73 A thin 1.2-m-wide, right-angle gate with negligible mass is free to pivot about a frictionless hinge at point O, as shown in Fig. P2.73. The horizontal portion of the gate covers a 0.3-m-diameter drain pipe which contains air at atmospheric pressure. Determine the minimum water depth, h, at which the gate will pivot to allow water to flow into the pipe.

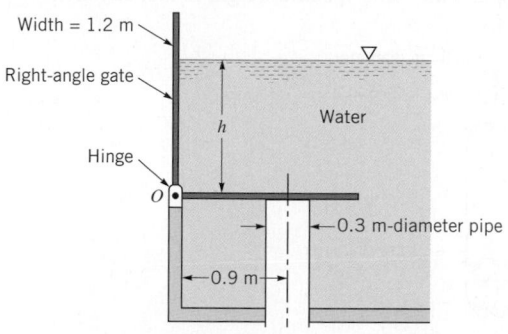

Width = 1.2 m

Right-angle gate

Water

h

Hinge

O

0.3 m-diameter pipe

0.9 m

■ **F I G U R E P2.73**

2.74 An open rectangular tank is 2 m wide and 4 m long. The tank contains water to a depth of 2 m and oil ($SG = 0.8$) on top of the water to a depth of 1 m. Determine the magnitude and location of the resultant fluid force acting on one end of the tank.

*__2.75__ An open rectangular settling tank contains a liquid suspension that at a given time has a specific weight that varies approximately with depth according to the following data:

h (m)	γ (kN/m³)
0	10.0
0.4	10.1
0.8	10.2
1.2	10.6
1.6	11.3
2.0	12.3
2.4	12.7
2.8	12.9
3.2	13.0
3.6	13.1

The depth $h = 0$ corresponds to the free surface. Determine, by means of numerical integration, the magnitude and location of the resultant force that the liquid suspension exerts on a vertical wall of the tank that is 6 m wide. The depth of fluid in the tank is 3.6 m.

2.76 The closed vessel of Fig. P2.76 contains water with an air pressure of 69 kPa at the water surface. One side of the vessel

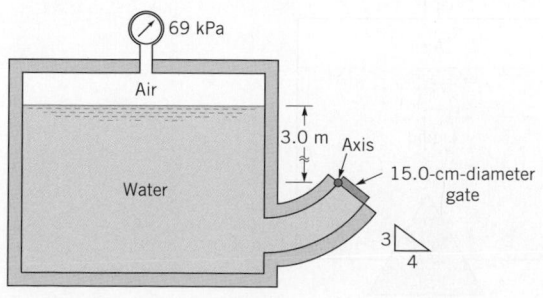

69 kPa

Air

3.0 m Axis

15.0-cm-diameter gate

Water

3

4

■ **F I G U R E P2.76**

contains a spout that is closed by a 15.0-cm-diameter circular gate that is hinged along one side as illustrated. The horizontal axis of the hinge is located 3.0 m below the water surface. Determine the minimum torque that must be applied at the hinge to hold the gate shut. Neglect the weight of the gate and friction at the hinge.

2.77 A 1.2-m-tall, 20.8-cm-wide concrete (23.6 kN/m³) retaining wall is built as shown in Fig. P2.77. During a heavy rain, water fills the space between the wall and the earth behind it to a depth h. Determine the maximum depth of water possible without the wall tipping over. The wall simply rests on the ground without being anchored to it.

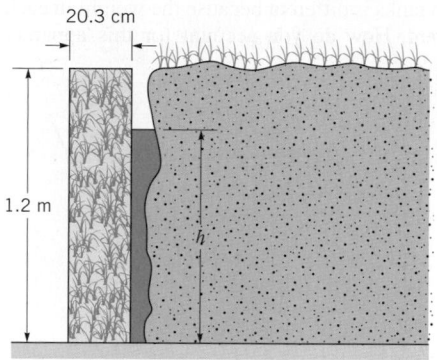

20.3 cm

1.2 m

h

■ **F I G U R E P2.77**

*__2.78__ Water backs up behind a concrete dam as shown in Fig. P2.78. Leakage under the foundation gives a pressure distribution under the dam as indicated. If the water depth, h, is too great, the dam will topple over about its toe (point A). For the dimensions given, determine the maximum water depth for the following widths of the dam: $\ell = 6, 9, 12, 15,$ and 18 m. Base your analysis on a unit length of the dam. The specific weight of the concrete is 23.6 kN/m³.

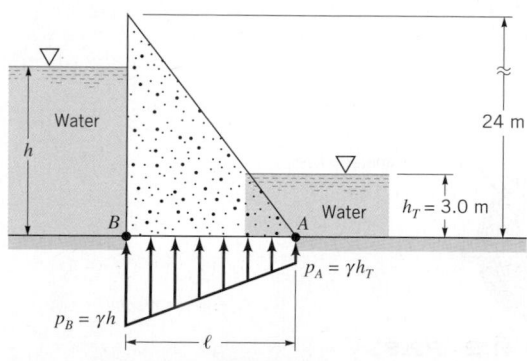

Water

h

24 m

Water

$h_T = 3.0$ m

B

A

$p_A = \gamma h_T$

$p_B = \gamma h$

ℓ

■ **F I G U R E P2.78**

2.79 (See Fluids in the News article titled "The Three Gorges Dam," Section 2.8.) **(a)** Determine the horizontal hydrostatic force on the 2309-m-long Three Gorges Dam when the average depth of the water against it is 175 m. **(b)** If all of the 6.4 billion people on Earth were to push horizontally against the Three Gorges Dam, could they generate enough force to hold it in place? Support your answer with appropriate calculations.

Section 2.10 Hydrostatic Force on a Curved Surface

2.80 Obtain a photograph/image of a situation in which the hydrostatic force on a curved surface is important. Print this photo and write a brief paragraph that describes the situation involved.

2.81 A 0.6-m-diameter hemispherical plexiglass "bubble" is to be used as a special window on the side of an above-ground swimming pool. The window is to be bolted onto the vertical wall of the pool and faces outward, covering a 0.6-m-diameter opening in the wall. The center of the opening is 1.2 m below the surface. Determine the horizontal and vertical components of the force of the water on the hemisphere.

2.82 Two round, open tanks containing the same type of fluid rest on a table top as shown in Fig. P2.82. They have the same bottom area, *A*, but different shapes. When the depth, *h*, of the liquid in the two tanks is the same, the pressure force of the liquids on the bottom of the two tanks is the same. However, the force that the table exerts on the two tanks is different because the weight in each of the tanks is different. How do you account for this apparent paradox?

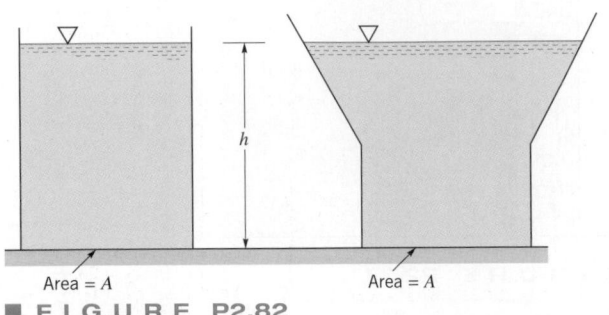

■ **FIGURE P2.82**

2.83 Two hemispherical shells are bolted together as shown in Fig. P2.83. The resulting spherical container, which weighs 1.3 kN is filled with mercury and supported by a cable as shown. The container is vented at the top. If eight bolts are symmetrically located around the circumference, what is the vertical force that each bolt must carry?

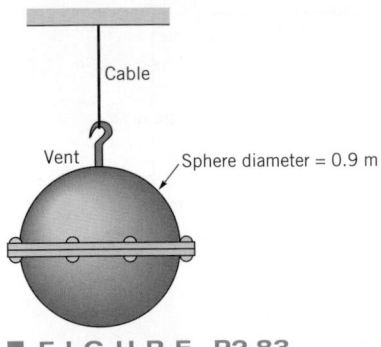

■ **FIGURE P2.83**

2.84 The 5.5-m-long gate of Fig. P2.84 is a quarter circle and is hinged at *H*. Determine the horizontal force, *P*, required to hold the gate in place. Neglect friction at the hinge and the weight of the gate.

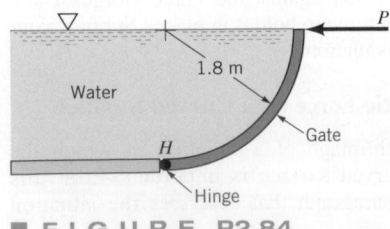

■ **FIGURE P2.84**

2.85 The air pressure in the top of the 2-liter pop bottle shown in **Video V2.5** and Fig. P2.85 is 276 kPa, and the pop depth is 25 cm. The bottom of the bottle has an irregular shape with a diameter of 11 cm **(a)** If the bottle cap has a diameter of 2.5 cm what is the magnitude of the axial force required to hold the cap in place? **(b)** Determine the force needed to secure the bottom 5 cm of the bottle to its cylindrical sides. For this calculation assume the effect of the weight of the pop is negligible. **(c)** By how much does the weight of the pop increase the pressure 5 cm above the bottom? Assume the pop has the same specific weight as that of water.

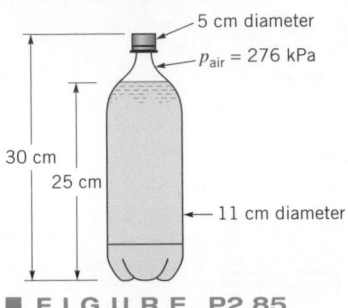

■ **FIGURE P2.85**

2.86 Hoover Dam (see **Video 2.4**) is the highest arch-gravity type of dam in the United States. A cross section of the dam is shown in Fig. P2.86(*a*). The walls of the canyon in which the dam is located are sloped, and just upstream of the dam the vertical plane shown in Figure P2.86(*b*) approximately represents the cross section of the water acting on the dam. Use this vertical cross section to estimate the resultant horizontal force of the water on the dam, and show where this force acts.

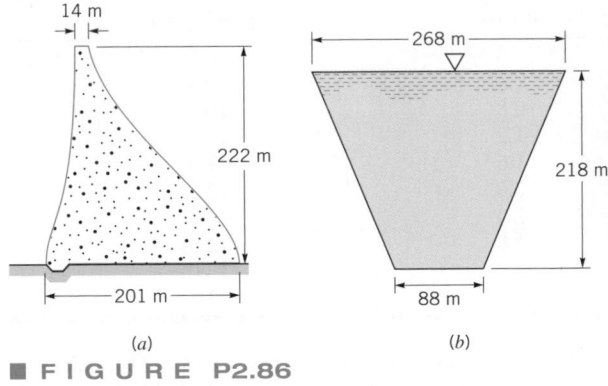

(*a*)   (*b*)

■ **FIGURE P2.86**

2.87 A plug in the bottom of a pressurized tank is conical in shape, as shown in Fig. P2.87. The air pressure is 40 kPa and the liquid in

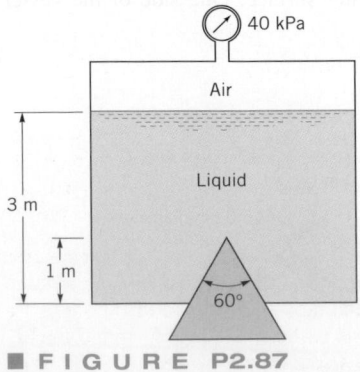

■ **FIGURE P2.87**

the tank has a specific weight of 27 kN/m³. Determine the magnitude, direction, and line of action of the force exerted on the curved surface of the cone within the tank due to the 40-kPa pressure and the liquid.

2.88 The homogeneous gate shown in Fig. P2.88 consists of one quarter of a circular cylinder and is used to maintain a water depth of 4 m. That is, when the water depth exceeds 4 m, the gate opens slightly and lets the water flow under it. Determine the weight of the gate per meter of length.

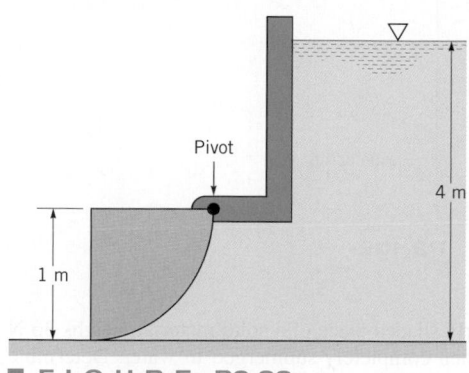

■ **FIGURE P2.88**

2.89 The concrete (specific weight = 23.6 kN/m³) seawall of Fig. P2.89 has a curved surface and restrains seawater at a depth of 7 m. The trace of the surface is a parabola as illustrated. Determine the moment of the fluid force (per unit length) with respect to an axis through the toe (point A).

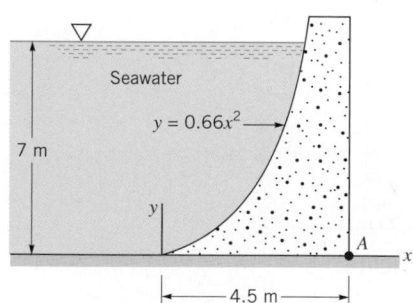

■ **FIGURE P2.89**

2.90 A cylindrical tank with its axis horizontal has a diameter of 2.0 m and a length of 4.0 m. The ends of the tank are vertical planes. A vertical, 0.1-m-diameter pipe is connected to the top of the tank. The tank and the pipe are filled with ethyl alcohol to a level of 1.5 m above the top of the tank. Determine the resultant force of the alcohol on one end of the tank and show where it acts.

2.91 If the tank ends in Problem 2.90 are hemispherical, what is the magnitude of the resultant horizontal force of the alcohol on one of the curved ends?

2.92 An open tank containing water has a bulge in its vertical side that is semicircular in shape as shown in Fig. P2.92. Determine the horizontal and vertical components of the force that the water exerts on the bulge. Base your analysis on a 0.3-m length of the bulge.

2.93 A closed tank is filled with water and has a 1.2 m-diameter hemispherical dome as shown in Fig. P2.93. A U-tube manometer is connected to the tank. Determine the vertical force of the water on the dome if the differential manometer reading is 2 m and the air pressure at the upper end of the manometer is 87 kPa.

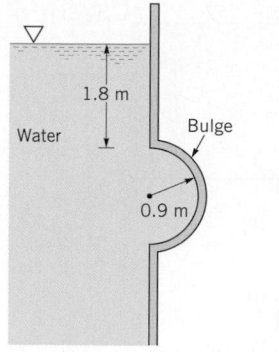

■ **FIGURE P2.92**

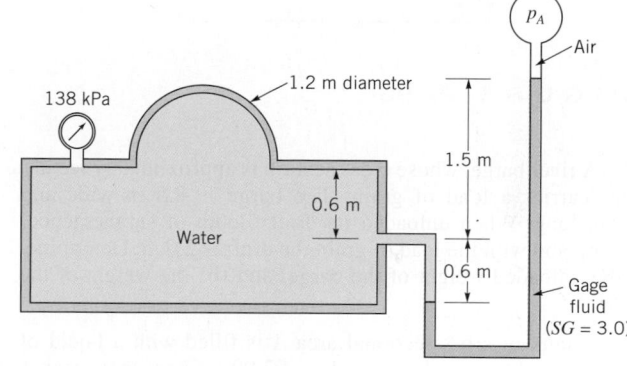

■ **FIGURE P2.93**

2.94 A 3-m-diameter open cylindrical tank contains water and has a hemispherical bottom as shown in Fig. P2.94. Determine the magnitude, line of action, and direction of the force of the water on the curved bottom.

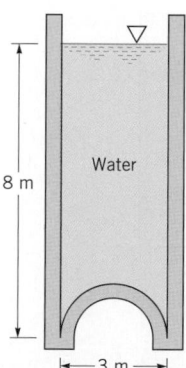

■ **FIGURE P2.94**

2.95 Three gates of negligible weight are used to hold back water in a channel of width b as shown in Fig. P2.95 on the next page. The force of the gate against the block for gate (b) is R. Determine (in terms of R) the force against the blocks for the other two gates.

Section 2.11 Buoyancy, Flotation, and Stability

2.96 Obtain a photograph/image of a situation in which Archimedes' principle is important. Print this photo and write a brief paragraph that describes the situation involved.

2.97 A freshly cut log floats with one fourth of its volume protruding above the water surface. Determine the specific weight of the log.

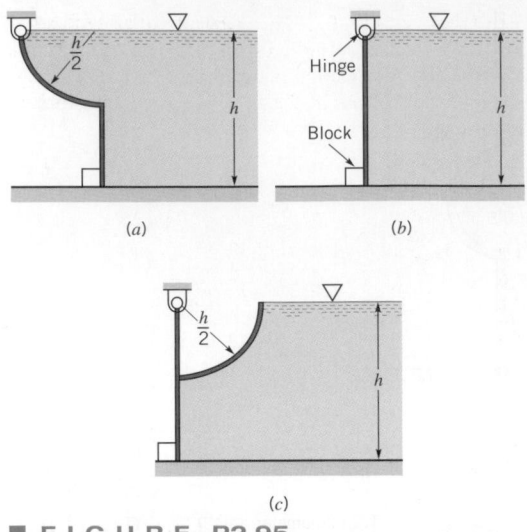

(a)   *(b)*

(c)

■ **F I G U R E P2.95**

2.98 A river barge, whose cross section is approximately rectangular, carries a load of grain. The barge is 8.5 m wide and 27.4 m long. When unloaded its draft (depth of submergence) is 1.5 m, and with the load of grain the draft is 2.0 m. Determine: **(a)** the unloaded weight of the barge, and **(b)** the weight of the grain.

2.99 A tank of cross-sectional area A is filled with a liquid of specific weight γ_1 as shown in Fig. P2.99*a*. Show that when a cylinder of specific weight γ_2 and volume Ψ is floated in the liquid (see Fig. P2.99*b*), the liquid level rises by an amount $\Delta h = (\gamma_2 / \gamma_1) \Psi / A$.

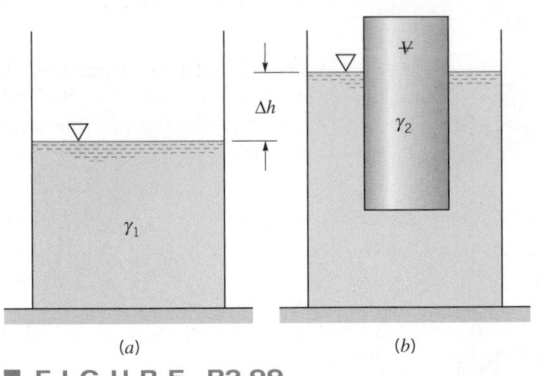

(a)   *(b)*

■ **F I G U R E P2.99**

2.100 When the Tucurui Dam was constructed in northern Brazil, the lake that was created covered a large forest of valuable hardwood trees. It was found that even after 15 years underwater the trees were perfectly preserved and underwater logging was started. During the logging process a tree is selected, trimmed, and anchored with ropes to prevent it from shooting to the surface like a missile when cut. Assume that a typical large tree can be approximated as a truncated cone with a base diameter of 2.4 m, a top diameter of 0.6 m, and a height of 30.0 m. Determine the resultant vertical force that the ropes must resist when the completely submerged tree is cut. The specific gravity of the wood is approximately 0.6.

†2.101 Estimate the minimum water depth needed to float a canoe carrying two people and their camping gear. List all assumptions and show all calculations.

2.102 An inverted test tube partially filled with air floats in a plastic water-filled soft drink bottle as shown in **Video V2.7** and Fig. P2.102. The amount of air in the tube has been adjusted so that it just floats. The bottle cap is securely fastened. A slight squeezing of the plastic bottle will cause the test tube to sink to the bottom of the bottle. Explain this phenomenon.

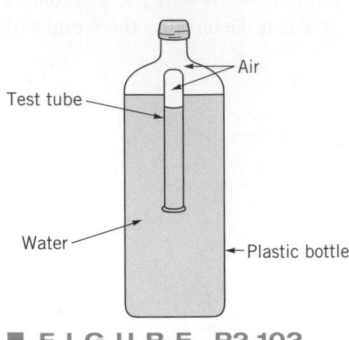

■ **F I G U R E P2.102**

2.103 An irregularly shaped piece of a solid material weighs 36 N in air and 23 N when completely submerged in water. Determine the density of the material.

2.104 A 1-m-diameter cylindrical mass, M, is connected to a 2-m-wide rectangular gate as shown in Fig. P2.104. The gate is to open when the water level, h, drops below 2.5 m. Determine the required value for M. Neglect friction at the gate hinge and the pulley.

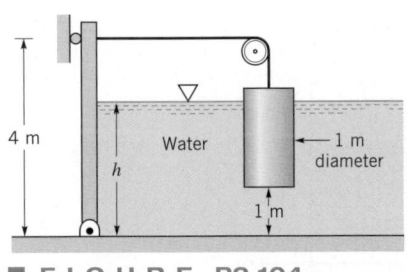

■ **F I G U R E P2.104**

2.105 When a hydrometer (see Fig. P2.105 and **Video V2.8**) having a stem diameter of 0.7 cm is placed in water, the stem protrudes 8 cm above the water surface. If the water is replaced with a liquid having a specific gravity of 1.10, how much of the stem would protrude above the liquid surface? The hydrometer weighs 0.2 N.

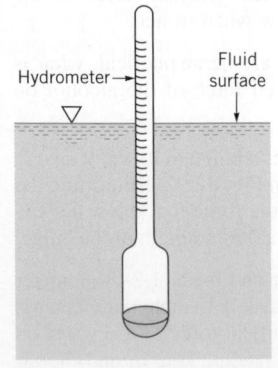

■ **F I G U R E P2.105**

2.106 A 0.6 m thick block constructed of wood ($SG = 0.6$) is submerged in oil ($SG = 0.8$), and has a 0.6-m-thick aluminum (specific weight = 26.3 kN/m³) plate attached to the bottom as indicated in Fig. P2.106. Determine completely the force required to hold the block in the position shown. Locate the force with respect to point A.

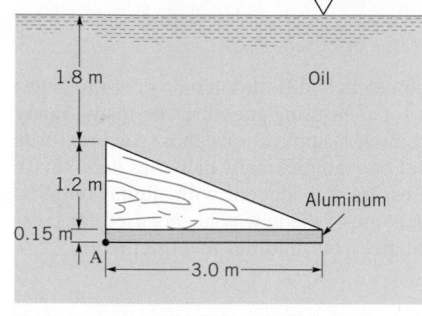

■ **FIGURE P2.106**

2.107 (See Fluids in the News article titled "Concrete canoe," Section 2.11.1.) How much extra water does a 654 kN concrete canoe displace compared to an ultralightweight 169 N Kevlar canoe of the same size carrying the same load?

2.108 An iceberg (specific gravity 0.917) floats in the ocean (specific gravity 1.025). What percent of the volume of the iceberg is under water?

Section 2.12 Pressure Variation in a Fluid with Rigid-Body Motion

2.109 Obtain a photograph/image of a situation in which the pressure variation in a fluid with rigid-body motion is involved. Print this photo and write a brief paragraph that describes the situation involved.

2.110 It is noted that while stopping, the water surface in a glass of water sitting in the cup holder of a car is slanted at an angle of 15° relative to the horizontal street. Determine the rate at which the car is decelerating.

2.111 An open container of oil rests on the flatbed of a truck that is traveling along a horizontal road at 89 km/hr. As the truck slows uniformly to a complete stop in 5 s, what will be the slope of the oil surface during the period of constant deceleration?

2.112 A 19 ℓ, cylindrical open container with a bottom area of 775 cm² is filled with glycerin and rests on the floor of an elevator. **(a)** Determine the fluid pressure at the bottom of the container when the elevator has an upward acceleration of 0.9 m/s². **(b)** What resultant force does the container exert on the floor of the elevator during this acceleration? The weight of the container is negligible. (Note: 1 ℓ = 1000 cm³)

2.113 An open rectangular tank 1 m wide and 2 m long contains gasoline to a depth of 1 m. If the height of the tank sides is 1.5 m, what is the maximum horizontal acceleration (along the long axis of the tank) that can develop before the gasoline would begin to spill?

2.114 If the tank of Problem 2.113 slides down a frictionless plane that is inclined at 30° with the horizontal, determine the angle the free surface makes with the horizontal.

2.115 A closed cylindrical tank that is 2.4 m in diameter and 7.3 m long is completely filled with gasoline. The tank, with its long axis horizontal, is pulled by a truck along a horizontal surface. Determine the pressure difference between the ends (along the long axis of the tank) when the truck undergoes an acceleration of 1.5 m/s².

2.116 The open U-tube of Fig. P2.116 is partially filled with a liquid. When this device is accelerated with a horizontal acceleration a, a differential reading h develops between the manometer legs which are spaced a distance ℓ apart. Determine the relationship between a, ℓ, and h.

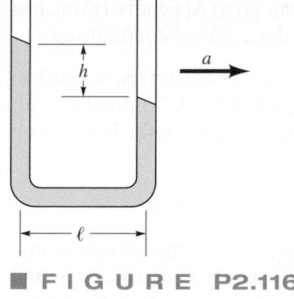

■ **FIGURE P2.116**

2.117 An open 1-m-diameter tank contains water at a depth of 0.7 m when at rest. As the tank is rotated about its vertical axis the center of the fluid surface is depressed. At what angular velocity will the bottom of the tank first be exposed? No water is spilled from the tank.

2.118 An open, 0.6-m-diameter tank contains water to a depth of 0.9 m when at rest. If the tank is rotated about its vertical axis with an angular velocity of 180 rev/min, what is the minimum height of the tank walls to prevent water from spilling over the sides?

2.119 A child riding in a car holds a string attached to a floating, helium-filled balloon. As the car decelerates to a stop, the balloon tilts backwards. As the car makes a right-hand turn, the balloon tilts to the right. On the other hand, the child tends to be forced forward as the car decelerates and to the left as the car makes a right-hand turn. Explain these observed effects on the balloon and child.

2.120 A closed, 0.4-m-diameter cylindrical tank is completely filled with oil ($SG = 0.9$) and rotates about its vertical longitudinal axis with an angular velocity of 40 rad/s. Determine the difference in pressure just under the vessel cover between a point on the circumference and a point on the axis.

2.121 (See Fluids in the News article titled "Rotating mercury mirror telescope," Section 2.12.2.) The largest liquid mirror telescope uses a 1.8-m-diameter tank of mercury rotating at 7 rpm to produce its parabolic-shaped mirror as shown in Fig. P2.121. Determine the difference in elevation of the mercury, Δh, between the edge and the center of the mirror.

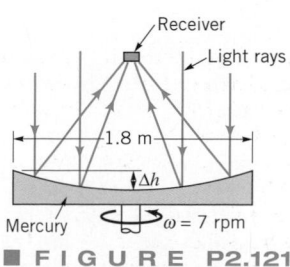

■ **FIGURE P2.121**

■ **Lab Problems**

2.122 This problem involves the force needed to open a gate that covers an opening in the side of a water-filled tank. To proceed with this problem, go to Appendix H which is located on the book's web site, www.wiley.com/college/munson.

2.123 This problem involves the use of a cleverly designed apparatus to investigate the hydrostatic pressure force on a submerged rectangle. To proceed with this problem, go to Appendix H which is located on the book's web site, www.wiley.com/college/munson.

2.124 This problem involves determining the weight needed to hold down an open-bottom box that has slanted sides when the box is filled with water. To proceed with this problem, go to Appendix H which is located on the book's web site, www.wiley.com/college/munson.

2.125 This problem involves the use of a pressurized air pad to provide the vertical force to support a given load. To proceed with this problem, go to Appendix H which is located on the book's web site, www.wiley.com/college/munson.

■ **Life Long Learning Problems**

2.126 Although it is relatively easy to calculate the net hydrostatic pressure force on a dam, it is not necessarily easy to design and construct an appropriate, long-lasting, inexpensive dam. In fact, inspection of older dams has revealed that many of them are in peril of collapse unless corrective action is soon taken. Obtain information about the severity of the poor conditions of older dams throughout the country. Summarize your findings in a brief report.

2.127 Over the years the demand for high-quality, first-growth timber has increased dramatically. Unfortunately, most of the trees that supply such lumber have already been harvested. Recently, however, several companies have started to reclaim the numerous high-quality logs that sank in lakes and oceans during the logging boom times many years ago. Many of these logs are still in excellent condition. Obtain information, particularly that associated with the use of fluid mechanics concepts, about harvesting sunken logs. Summarize your findings in a brief report.

2.128 Liquid-filled manometers and Bourdon tube pressure gages have been the mainstay for measuring pressure for many, many years. However, for many modern applications, these tried-and-true devices are not sufficient. For example, many new uses need small, accurate, inexpensive pressure transducers with digital outputs. Obtain information about some of the new concepts used for pressure measurement. Summarize your findings in a brief report.

■ **FE Exam Problems**

Sample FE (Fundamentals of Engineering) exam question for fluid mechanics are provided on the book's web site, www.wiley.com/college/munson.

3 Elementary Fluid Dynamics—The Bernoulli Equation

CHAPTER OPENING PHOTO: Flow past a blunt body: On any object placed in a moving fluid there is a stagnation point on the front of the object where the velocity is zero. This location has a relatively large pressure and divides the flow field into two portions—one flowing to the left, and one flowing to the right of the body. (Dye in water.) (Photograph by B. R. Munson.)

Learning Objectives

After completing this chapter, you should be able to:

- discuss the application of Newton's second law to fluid flows.
- explain the development, uses, and limitations of the Bernoulli equation.
- use the Bernoulli equation (stand-alone or in combination with the continuity equation) to solve simple flow problems.
- apply the concepts of static, stagnation, dynamic, and total pressures.
- calculate various flow properties using the energy and hydraulic grade lines.

The Bernoulli equation may be the most used and abused equation in fluid mechanics.

In this chapter we investigate some typical fluid motions (fluid dynamics) in an elementary way. We will discuss in some detail the use of Newton's second law ($\mathbf{F} = m\mathbf{a}$) as it is applied to fluid particle motion that is "ideal" in some sense. We will obtain the celebrated Bernoulli equation and apply it to various flows. Although this equation is one of the oldest in fluid mechanics and the assumptions involved in its derivation are numerous, it can be used effectively to predict and analyze a variety of flow situations. However, if the equation is applied without proper respect for its restrictions, serious errors can arise. Indeed, the Bernoulli equation is appropriately called "the most used and the most abused equation in fluid mechanics."

A thorough understanding of the elementary approach to fluid dynamics involved in this chapter will be useful on its own. It also provides a good foundation for the material in the following chapters where some of the present restrictions are removed and "more nearly exact" results are presented.

93

3.1 Newton's Second Law

As a fluid particle moves from one location to another, it usually experiences an acceleration or deceleration. According to Newton's second law of motion, the net force acting on the fluid particle under consideration must equal its mass times its acceleration,

$$\mathbf{F} = m\mathbf{a}$$

In this chapter we consider the motion of inviscid fluids. That is, the fluid is assumed to have zero viscosity. If the viscosity is zero, then the thermal conductivity of the fluid is also zero and there can be no heat transfer (except by radiation).

In practice there are no inviscid fluids, since every fluid supports shear stresses when it is subjected to a rate of strain displacement. For many flow situations the viscous effects are relatively small compared with other effects. As a first approximation for such cases it is often possible to ignore viscous effects. For example, often the viscous forces developed in flowing water may be several orders of magnitude smaller than forces due to other influences, such as gravity or pressure differences. For other water flow situations, however, the viscous effects may be the dominant ones. Similarly, the viscous effects associated with the flow of a gas are often negligible, although in some circumstances they are very important.

We assume that the fluid motion is governed by pressure and gravity forces only and examine Newton's second law as it applies to a fluid particle in the form:

> (Net pressure force on a particle) + (net gravity force on particle) =
> (particle mass) × (particle acceleration)

The results of the interaction between the pressure, gravity, and acceleration provide numerous useful applications in fluid mechanics.

To apply Newton's second law to a fluid (or any other object), we must define an appropriate coordinate system in which to describe the motion. In general the motion will be three-dimensional and unsteady so that three space coordinates and time are needed to describe it. There are numerous coordinate systems available, including the most often used rectangular (x, y, z) and cylindrical (r, θ, z) systems shown by the figure in the margin. Usually the specific flow geometry dictates which system would be most appropriate.

In this chapter we will be concerned with two-dimensional motion like that confined to the x–z plane as is shown in Fig. 3.1a. Clearly we could choose to describe the flow in terms of the components of acceleration and forces in the x and z coordinate directions. The resulting equations are frequently referred to as a two-dimensional form of the *Euler equations* of motion in rectangular Cartesian coordinates. This approach will be discussed in Chapter 6.

As is done in the study of dynamics (Ref. 1), the motion of each fluid particle is described in terms of its velocity vector, $\mathbf{V}$, which is defined as the time rate of change of the position of the particle. The particle's velocity is a vector quantity with a magnitude (the speed, $V = |\mathbf{V}|$) and direction. As the particle moves about, it follows a particular path, the shape of which is governed by the velocity of the particle. The location of the particle along the path is a function of where the particle started at the initial time and its velocity along the path. If it is *steady flow* (i.e., nothing changes with time at a given location in the flow field), each successive particle that passes through a given point [such as point (1) in Fig. 3.1a] will follow the same path. For such cases the

Inviscid fluid flow is governed by pressure and gravity forces.

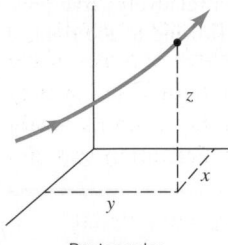

Rectangular

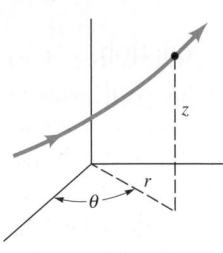

Cylindrical

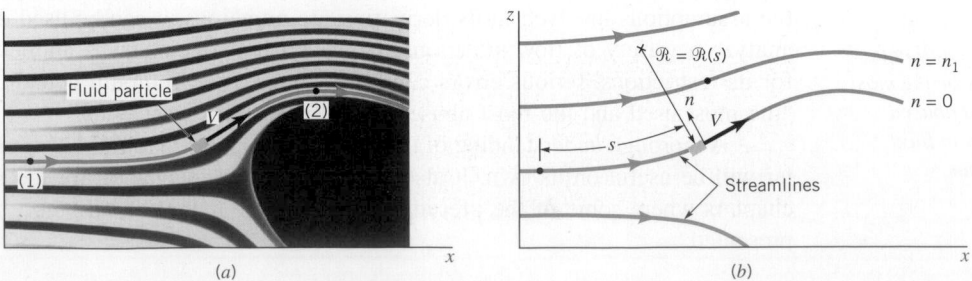

(a)   (b)

■ **FIGURE 3.1** (a) Flow in the x–z plane. (b) Flow in terms of streamline and normal coordinates.

path is a fixed line in the x–z plane. Neighboring particles that pass on either side of point (1) follow their own paths, which may be of a different shape than the one passing through (1). The entire x–z plane is filled with such paths.

For steady flows each particle slides along its path, and its velocity vector is everywhere tangent to the path. The lines that are tangent to the velocity vectors throughout the flow field are called *streamlines*. For many situations it is easiest to describe the flow in terms of the "streamline" coordinates based on the streamlines as are illustrated in Fig. 3.1*b*. The particle motion is described in terms of its distance, $s = s(t)$, along the streamline from some convenient origin and the local radius of curvature of the streamline, $\mathcal{R} = \mathcal{R}(s)$. The distance along the streamline is related to the particle's speed by $V = ds/dt$, and the radius of curvature is related to the shape of the streamline. In addition to the coordinate along the streamline, s, the coordinate normal to the streamline, n, as is shown in Fig. 3.1*b*, will be of use.

To apply Newton's second law to a particle flowing along its streamline, we must write the particle acceleration in terms of the streamline coordinates. By definition, the acceleration is the time rate of change of the velocity of the particle, $\mathbf{a} = d\mathbf{V}/dt$. For two-dimensional flow in the x–z plane, the acceleration has two components—one along the streamline, a_s, the streamwise acceleration, and one normal to the streamline, a_n, the normal acceleration.

The streamwise acceleration results from the fact that the speed of the particle generally varies along the streamline, $V = V(s)$. For example, in Fig. 3.1*a* the speed may be 30 m/s at point (1) and 15 m/s at point (2). Thus, by use of the chain rule of differentiation, the s component of the acceleration is given by $a_s = dV/dt = (\partial V/\partial s)(ds/dt) = (\partial V/\partial s)V$. We have used the fact that speed is the time rate of change of distance, $V = ds/dt$. Note that the streamwise acceleration is the product of the rate of change of speed with distance along the streamline, $\partial V/\partial s$, and the speed, V. Since $\partial V/\partial s$ can be positive, negative, or zero, the streamwise acceleration can, therefore, be positive (acceleration), negative (deceleration), or zero (constant speed).

The normal component of acceleration, the centrifugal acceleration, is given in terms of the particle speed and the radius of curvature of its path. Thus, $a_n = V^2/\mathcal{R}$, where both V and $\mathcal{R}$ may vary along the streamline. These equations for the acceleration should be familiar from the study of particle motion in physics (Ref. 2) or dynamics (Ref. 1). A more complete derivation and discussion of these topics can be found in Chapter 4.

Thus, the components of acceleration in the s and n directions, a_s and a_n, are given by

$$a_s = V \frac{\partial V}{\partial s}, \qquad a_n = \frac{V^2}{\mathcal{R}} \tag{3.1}$$

where $\mathcal{R}$ is the local radius of curvature of the streamline, and s is the distance measured along the streamline from some arbitrary initial point. In general there is acceleration along the streamline (because the particle speed changes along its path, $\partial V/\partial s \neq 0$) and acceleration normal to the streamline (because the particle does not flow in a straight line, $\mathcal{R} \neq \infty$). Various flows and the accelerations associated with them are shown in the figure in the margin. As discussed in Section 3.6.2, for incompressible flow the velocity is inversely proportional to the streamline spacing. Hence, converging streamlines produce positive streamwise acceleration. To produce this acceleration there must be a net, nonzero force on the fluid particle.

To determine the forces necessary to produce a given flow (or conversely, what flow results from a given set of forces), we consider the free-body diagram of a small fluid particle as is shown in Fig. 3.2. The particle of interest is removed from its surroundings, and the reactions of the

Fluid particles accelerate normal to and along streamlines.

V3.1 Streamlines past an airfoil

$a_s = a_n = 0$

$a_s > 0$

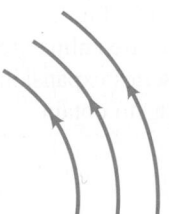

$a_n > 0$

$a_s > 0, a_n > 0$

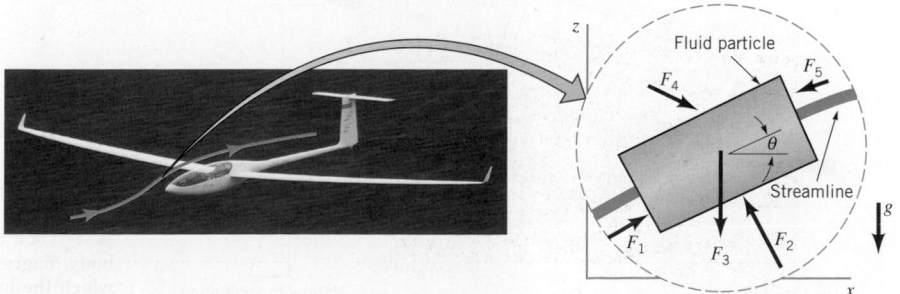

■ **FIGURE 3.2** Isolation of a small fluid particle in a flow field. (Photo courtesy of Diana Sailplanes.)

surroundings on the particle are indicated by the appropriate forces present, $\mathbf{F}_1$, $\mathbf{F}_2$, and so forth. For the present case, the important forces are assumed to be gravity and pressure. Other forces, such as viscous forces and surface tension effects, are assumed negligible. The acceleration of gravity, g, is assumed to be constant and acts vertically, in the negative z direction, at an angle θ relative to the normal to the streamline.

3.2   F = ma along a Streamline

Consider the small fluid particle of size δs by δn in the plane of the figure and δy normal to the figure as shown in the free-body diagram of Fig. 3.3. Unit vectors along and normal to the streamline are denoted by $\hat{\mathbf{s}}$ and $\hat{\mathbf{n}}$, respectively. For steady flow, the component of Newton's second law along the streamline direction, s, can be written as

$$\sum \delta F_s = \delta m \, a_s = \delta m \, V \frac{\partial V}{\partial s} = \rho \, \delta \mkern-2mu\Psi \, V \frac{\partial V}{\partial s} \tag{3.2}$$

where $\sum \delta F_s$ represents the sum of the s components of all the forces acting on the particle, which has mass $\delta m = \rho \, \delta \mkern-2mu\Psi$, and $V \, \partial V / \partial s$ is the acceleration in the s direction. Here, $\delta \mkern-2mu\Psi = \delta s \, \delta n \, \delta y$ is the particle volume. Equation 3.2 is valid for both compressible and incompressible fluids. That is, the density need not be constant throughout the flow field.

The gravity force (weight) on the particle can be written as $\delta \mkern-2mu\mathcal{W} = \gamma \, \delta \mkern-2mu\Psi$, where $\gamma = \rho g$ is the specific weight of the fluid ($\mathrm{N/m^3}$). Hence, the component of the weight force in the direction of the streamline is

$$\delta \mkern-2mu\mathcal{W}_s = -\delta \mkern-2mu\mathcal{W} \sin \theta = -\gamma \, \delta \mkern-2mu\Psi \sin \theta$$

In a flowing fluid the pressure varies from one location to another.

If the streamline is horizontal at the point of interest, then $\theta = 0$, and there is no component of particle weight along the streamline to contribute to its acceleration in that direction.

As is indicated in Chapter 2, the pressure is not constant throughout a stationary fluid ($\nabla p \neq 0$) because of the fluid weight. Likewise, in a flowing fluid the pressure is usually not constant. In general, for steady flow, $p = p(s, n)$. If the pressure at the center of the particle shown in Fig. 3.3 is denoted as p, then its average value on the two end faces that are perpendicular to the streamline are $p + \delta p_s$ and $p - \delta p_s$. Since the particle is "small," we can use a one-term Taylor series expansion for the pressure field (as was done in Chapter 2 for the pressure forces in static fluids) to obtain

$$\delta p_s \approx \frac{\partial p}{\partial s} \frac{\delta s}{2}$$

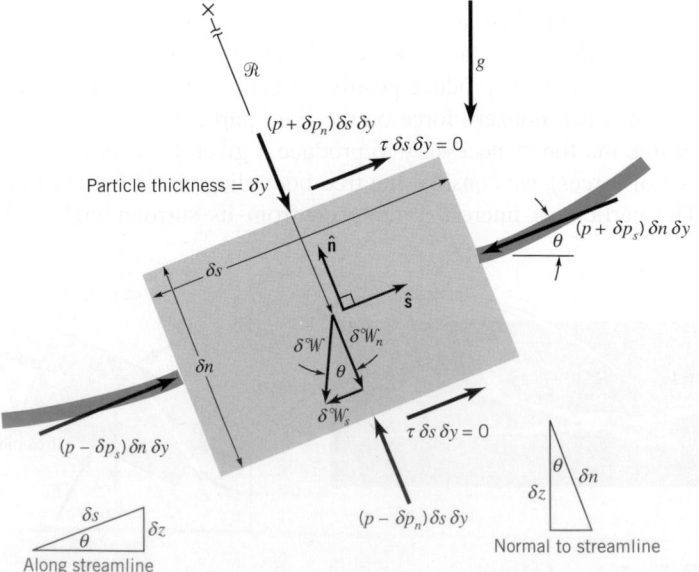

■ **FIGURE 3.3** Free-body diagram of a fluid particle for which the important forces are those due to pressure and gravity.

Thus, if δF_{ps} is the net pressure force on the particle in the streamline direction, it follows that

$$\delta F_{ps} = (p - \delta p_s)\,\delta n\,\delta y - (p + \delta p_s)\,\delta n\,\delta y = -2\,\delta p_s\,\delta n\,\delta y$$

$$= -\frac{\partial p}{\partial s}\,\delta s\,\delta n\,\delta y = -\frac{\partial p}{\partial s}\,\delta V\!\!\!/$$

Note that the actual level of the pressure, p, is not important. What produces a net pressure force is the fact that the pressure is not constant throughout the fluid. The nonzero pressure gradient, $\nabla p = \partial p/\partial s\,\hat{\mathbf{s}} + \partial p/\partial n\,\hat{\mathbf{n}}$, is what provides a net pressure force on the particle. Viscous forces, represented by $\tau\,\delta s\,\delta y$, are zero, since the fluid is inviscid.

Thus, the net force acting in the streamline direction on the particle shown in Fig. 3.3 is given by

$$\sum \delta F_s = \delta W_s + \delta F_{ps} = \left(-\gamma\sin\theta - \frac{\partial p}{\partial s}\right)\delta V\!\!\!/ \tag{3.3}$$

By combining Eqs. 3.2 and 3.3, we obtain the following equation of motion along the streamline direction:

$$-\gamma\sin\theta - \frac{\partial p}{\partial s} = \rho V\frac{\partial V}{\partial s} = \rho a_s \tag{3.4}$$

We have divided out the common particle volume factor, $\delta V\!\!\!/$, that appears in both the force and the acceleration portions of the equation. This is a representation of the fact that it is the fluid density (mass per unit volume), not the mass, per se, of the fluid particle that is important.

The physical interpretation of Eq. 3.4 is that a change in fluid particle speed is accomplished by the appropriate combination of pressure gradient and particle weight along the streamline. For fluid static situations this balance between pressure and gravity forces is such that no change in particle speed is produced—the right-hand side of Eq. 3.4 is zero, and the particle remains stationary. In a flowing fluid the pressure and weight forces do not necessarily balance—the force unbalance provides the appropriate acceleration and, hence, particle motion.

EXAMPLE 3.1   Pressure Variation along a Streamline

GIVEN   Consider the inviscid, incompressible, steady flow along the horizontal streamline A–B in front of the sphere of radius a, as shown in Fig. E3.1a. From a more advanced theory of flow past a sphere, the fluid velocity along this streamline is

$$V = V_0\left(1 + \frac{a^3}{x^3}\right)$$

as shown in Fig. E3.1b.

FIND   Determine the pressure variation along the streamline from point A far in front of the sphere ($x_A = -\infty$ and $V_A = V_0$) to point B on the sphere ($x_B = -a$ and $V_B = 0$).

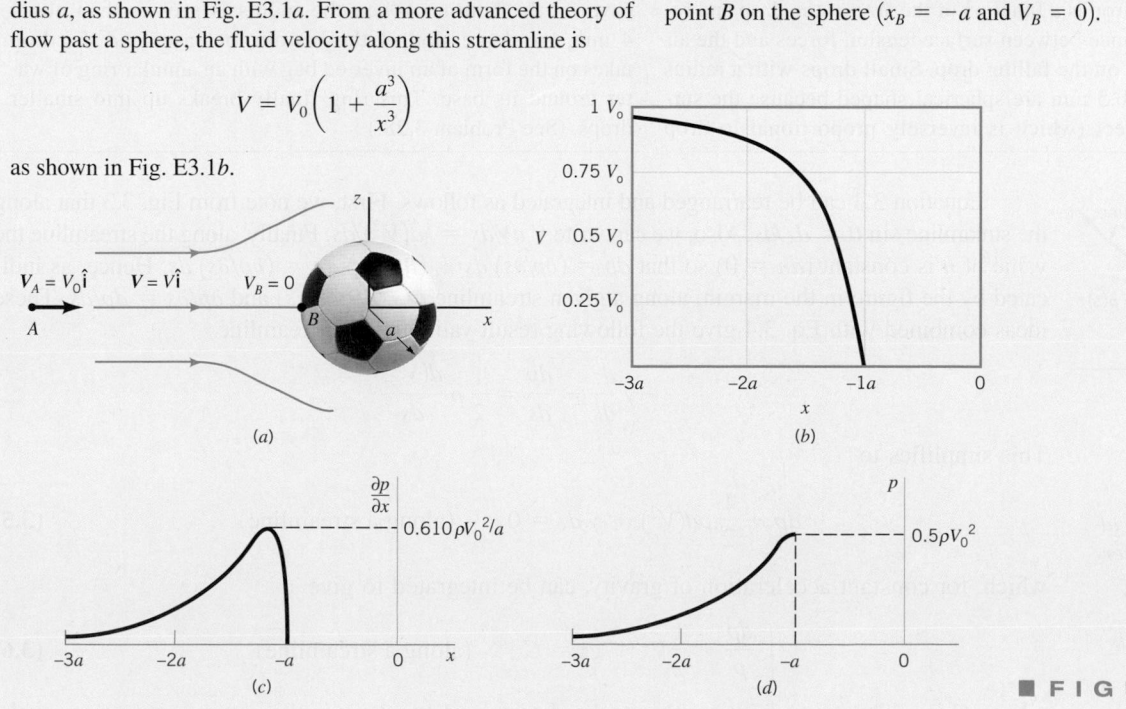

(a)

(b)

(c)

(d)

■ FIGURE E3.1

SOLUTION

Since the flow is steady and inviscid, Eq. 3.4 is valid. In addition, since the streamline is horizontal, $\sin \theta = \sin 0° = 0$ and the equation of motion along the streamline reduces to

$$\frac{\partial p}{\partial s} = -\rho V \frac{\partial V}{\partial s} \tag{1}$$

With the given velocity variation along the streamline, the acceleration term is

$$V \frac{\partial V}{\partial s} = V \frac{\partial V}{\partial x} = V_0 \left(1 + \frac{a^3}{x^3} \right) \left(-\frac{3V_0 a^3}{x^4} \right)$$

$$= -3V_0^2 \left(1 + \frac{a^3}{x^3} \right) \frac{a^3}{x^4}$$

where we have replaced s by x since the two coordinates are identical (within an additive constant) along streamline A–B. It follows that $V \, \partial V/\partial s < 0$ along the streamline. The fluid slows down from V_0 far ahead of the sphere to zero velocity on the "nose" of the sphere ($x = -a$).

Thus, according to Eq. 1, to produce the given motion the pressure gradient along the streamline is

$$\frac{\partial p}{\partial x} = \frac{3\rho a^3 V_0^2 (1 + a^3/x^3)}{x^4} \tag{2}$$

This variation is indicated in Fig. E3.1c. It is seen that the pressure increases in the direction of flow ($\partial p/\partial x > 0$) from point A to point B. The maximum pressure gradient ($0.610 \, \rho V_0^2/a$) occurs just slightly ahead of the sphere ($x = -1.205a$). It is the pressure gradient that slows the fluid down from $V_A = V_0$ to $V_B = 0$ as shown in Fig. E3.1b.

The pressure distribution along the streamline can be obtained by integrating Eq. 2 from $p = 0$ (gage) at $x = -\infty$ to pressure p at location x. The result, plotted in Fig. E3.1d, is

$$p = -\rho V_0^2 \left[\left(\frac{a}{x} \right)^3 + \frac{(a/x)^6}{2} \right] \tag{Ans}$$

COMMENT   The pressure at B, a stagnation point since $V_B = 0$, is the highest pressure along the streamline ($p_B = \rho V_0^2/2$). As shown in Chapter 9, this excess pressure on the front of the sphere (i.e., $p_B > 0$) contributes to the net drag force on the sphere. Note that the pressure gradient and pressure are directly proportional to the density of the fluid, a representation of the fact that the fluid inertia is proportional to its mass.

F l u i d s   i n   t h e   N e w s

Incorrect raindrop shape   The incorrect representation that raindrops are teardrop shaped is found nearly everywhere—from children's books, to weather maps on the *Weather Channel*. About the only time raindrops possess the typical teardrop shape is when they run down a windowpane. The actual shape of a falling raindrop is a function of the size of the drop and results from a balance between surface tension forces and the air pressure exerted on the falling drop. Small drops with a radius less than about 0.5 mm are spherical shaped because the surface tension effect (which is inversely proportional to drop size) wins over the increased pressure, $\rho V_0^2/2$, caused by the motion of the drop and exerted on its bottom. With increasing size, the drops fall faster and the increased pressure causes the drops to flatten. A 2-mm drop, for example, is flattened into a hamburger bun shape. Slightly larger drops are actually concave on the bottom. When the radius is greater than about 4 mm, the depression of the bottom increases and the drop takes on the form of an inverted bag with an annular ring of water around its base. This ring finally breaks up into smaller drops. (See Problem 3.28.)

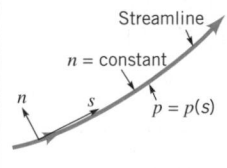

Equation 3.4 can be rearranged and integrated as follows. First, we note from Fig. 3.3 that along the streamline $\sin \theta = dz/ds$. Also, we can write $V \, dV/ds = \frac{1}{2} d(V^2)/ds$. Finally, along the streamline the value of n is constant ($dn = 0$) so that $dp = (\partial p/\partial s) \, ds + (\partial p/\partial n) \, dn = (\partial p/\partial s) \, ds$. Hence, as indicated by the figure in the margin, along a given streamline $p(s, n) = p(s)$ and $\partial p/\partial s = dp/ds$. These ideas combined with Eq. 3.4 give the following result valid along a streamline

$$-\gamma \frac{dz}{ds} - \frac{dp}{ds} = \frac{1}{2} \rho \frac{d(V^2)}{ds}$$

This simplifies to

For steady, inviscid flow the sum of certain pressure, velocity, and elevation effects is constant along a streamline.

$$dp + \frac{1}{2} \rho d(V^2) + \gamma \, dz = 0 \qquad \text{(along a streamline)} \tag{3.5}$$

which, for constant acceleration of gravity, can be integrated to give

$$\int \frac{dp}{\rho} + \frac{1}{2} V^2 + gz = C \qquad \text{(along a streamline)} \tag{3.6}$$

where C is a constant of integration to be determined by the conditions at some point on the streamline.

V3.2 Balancing ball

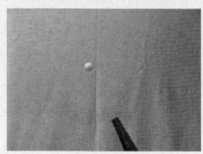

In general it is not possible to integrate the pressure term because the density may not be constant and, therefore, cannot be removed from under the integral sign. To carry out this integration we must know specifically how the density varies with pressure. This is not always easily determined. For example, for a perfect gas the density, pressure, and temperature are related according to $\rho = p/RT$, where R is the gas constant. To know how the density varies with pressure, we must also know the temperature variation. For now we will assume that the density and specific weight are constant (incompressible flow). The justification for this assumption and the consequences of compressibility will be considered further in Section 3.8.1 and more fully in Chapter 11.

With the additional assumption that the density remains constant (a very good assumption for liquids and also for gases if the speed is "not too high"), Eq. 3.6 assumes the following simple representation for steady, inviscid, incompressible flow.

$$p + \tfrac{1}{2}\rho V^2 + \gamma z = \text{constant along streamline} \qquad (3.7)$$

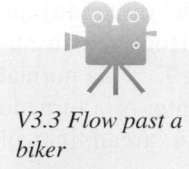

V3.3 Flow past a biker

This is the celebrated *Bernoulli equation*—a very powerful tool in fluid mechanics. In 1738 Daniel Bernoulli (1700–1782) published his *Hydrodynamics* in which an equivalent of this famous equation first appeared. To use it correctly we must constantly remember the basic assumptions used in its derivation: (1) viscous effects are assumed negligible, (2) the flow is assumed to be steady, (3) the flow is assumed to be incompressible, (4) the equation is applicable along a streamline. In the derivation of Eq. 3.7, we assume that the flow takes place in a plane (the x–z plane). In general, this equation is valid for both planar and nonplanar (three-dimensional) flows, provided it is applied along the streamline.

We will provide many examples to illustrate the correct use of the Bernoulli equation and will show how a violation of the basic assumptions used in the derivation of this equation can lead to erroneous conclusions. The constant of integration in the Bernoulli equation can be evaluated if sufficient information about the flow is known at one location along the streamline.

EXAMPLE 3.2   The Bernoulli Equation

GIVEN   Consider the flow of air around a bicyclist moving through still air with velocity V_0, as is shown in Fig. E3.2.

FIND   Determine the difference in the pressure between points (1) and (2).

SOLUTION

In a coordinate fixed to the ground, the flow is unsteady as the bicyclist rides by. However, in a coordinate system fixed to the bike, it appears as though the air is flowing steadily toward the bicyclist with speed V_0. Since use of the Bernoulli equation is restricted to steady flows, we select the coordinate system fixed to the bike. If the assumptions of Bernoulli's equation are valid (steady, incompressible, inviscid flow), Eq. 3.7 can be applied as follows along the streamline that passes through (1) and (2)

$$p_1 + \tfrac{1}{2}\rho V_1^2 + \gamma z_1 = p_2 + \tfrac{1}{2}\rho V_2^2 + \gamma z_2$$

We consider (1) to be in the free stream so that $V_1 = V_0$ and (2) to be at the tip of the bicyclist's nose and assume that $z_1 = z_2$ and $V_2 = 0$ (both of which, as is discussed in Section 3.4, are reasonable assumptions). It follows that the pressure at (2) is greater than that at (1) by an amount

$$p_2 - p_1 = \tfrac{1}{2}\rho V_1^2 = \tfrac{1}{2}\rho V_0^2 \qquad \text{(Ans)}$$

COMMENTS   A similar result was obtained in Example 3.1 by integrating the pressure gradient, which was known because

■ **FIGURE E3.2**

the velocity distribution along the streamline, $V(s)$, was known. The Bernoulli equation is a general integration of $\mathbf{F} = m\mathbf{a}$. To determine $p_2 - p_1$, knowledge of the detailed velocity distribution is not needed—only the "boundary conditions" at (1) and (2) are required. Of course, knowledge of the value of V along the streamline is needed to determine the pressure at points between (1) and (2). Note that if we measure $p_2 - p_1$ we can determine the speed, V_0. As discussed in Section 3.5, this is the principle upon which many velocity measuring devices are based.

If the bicyclist were accelerating or decelerating, the flow would be unsteady (i.e., $V_0 \neq$ constant) and the above analysis would be incorrect since Eq. 3.7 is restricted to steady flow.

The difference in fluid velocity between two points in a flow field, V_1 and V_2, can often be controlled by appropriate geometric constraints of the fluid. For example, a garden hose nozzle is designed to give a much higher velocity at the exit of the nozzle than at its entrance where it is attached to the hose. As is shown by the Bernoulli equation, the pressure within the hose must be larger than that at the exit (for constant elevation, an increase in velocity requires a decrease in pressure if Eq. 3.7 is valid). It is this pressure drop that accelerates the water through the nozzle. Similarly, an airfoil is designed so that the fluid velocity over its upper surface is greater (on the average) than that along its lower surface. From the Bernoulli equation, therefore, the average pressure on the lower surface is greater than that on the upper surface. A net upward force, the lift, results.

3.3   F = ma Normal to a Streamline

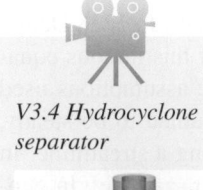

V3.4 Hydrocyclone separator

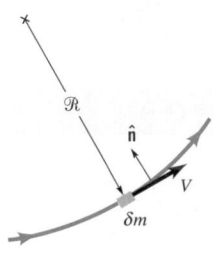

In this section we will consider application of Newton's second law in a direction normal to the streamline. In many flows the streamlines are relatively straight, the flow is essentially one-dimensional, and variations in parameters across streamlines (in the normal direction) can often be neglected when compared to the variations along the streamline. However, in numerous other situations valuable information can be obtained from considering $\mathbf{F} = m\mathbf{a}$ normal to the streamlines. For example, the devastating low-pressure region at the center of a tornado can be explained by applying Newton's second law across the nearly circular streamlines of the tornado.

We again consider the force balance on the fluid particle shown in Fig. 3.3 and the figure in the margin. This time, however, we consider components in the normal direction, $\hat{\mathbf{n}}$, and write Newton's second law in this direction as

$$\sum \delta F_n = \frac{\delta m \, V^2}{\mathcal{R}} = \frac{\rho \, \delta \mathbb{V} \, V^2}{\mathcal{R}} \tag{3.8}$$

where $\sum \delta F_n$ represents the sum of n components of all the forces acting on the particle and δm is particle mass. We assume the flow is steady with a normal acceleration $a_n = V^2/\mathcal{R}$, where $\mathcal{R}$ is the local radius of curvature of the streamlines. This acceleration is produced by the change in direction of the particle's velocity as it moves along a curved path.

We again assume that the only forces of importance are pressure and gravity. The component of the weight (gravity force) in the normal direction is

$$\delta \mathcal{W}_n = -\delta \mathcal{W} \cos \theta = -\gamma \, \delta \mathbb{V} \cos \theta$$

*To apply **F = ma** normal to streamlines, the normal components of force are needed.*

If the streamline is vertical at the point of interest, $\theta = 90°$, and there is no component of the particle weight normal to the direction of flow to contribute to its acceleration in that direction.

If the pressure at the center of the particle is p, then its values on the top and bottom of the particle are $p + \delta p_n$ and $p - \delta p_n$, where $\delta p_n = (\partial p/\partial n)(\delta n/2)$. Thus, if δF_{pn} is the net pressure force on the particle in the normal direction, it follows that

$$\delta F_{pn} = (p - \delta p_n)\,\delta s \, \delta y - (p + \delta p_n)\,\delta s \, \delta y = -2 \, \delta p_n \, \delta s \, \delta y$$

$$= -\frac{\partial p}{\partial n}\,\delta s \, \delta n \, \delta y = -\frac{\partial p}{\partial n}\,\delta \mathbb{V}$$

V3.5 Aircraft wing tip vortex

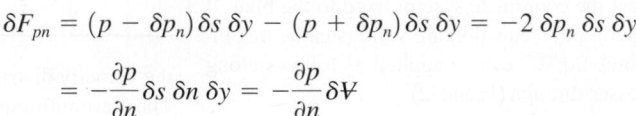

Hence, the net force acting in the normal direction on the particle shown in Fig 3.3 is given by

$$\sum \delta F_n = \delta \mathcal{W}_n + \delta F_{pn} = \left(-\gamma \cos \theta - \frac{\partial p}{\partial n} \right)\delta \mathbb{V} \tag{3.9}$$

By combining Eqs. 3.8 and 3.9 and using the fact that along a line normal to the streamline $\cos \theta = dz/dn$ (see Fig. 3.3), we obtain the following equation of motion along the normal direction

$$-\gamma \frac{dz}{dn} - \frac{\partial p}{\partial n} = \frac{\rho V^2}{\mathcal{R}} \tag{3.10a}$$

Weight and/or pressure can produce curved streamlines.

The physical interpretation of Eq. 3.10 is that a change in the direction of flow of a fluid particle (i.e., a curved path, $\mathcal{R} < \infty$) is accomplished by the appropriate combination of pressure gradient and particle weight normal to the streamline. A larger speed or density or a smaller radius of curvature of the motion requires a larger force unbalance to produce the motion. For example, if gravity is neglected (as is commonly done for gas flows) or if the flow is in a horizontal ($dz/dn = 0$) plane, Eq. 3.10 becomes

$$\frac{\partial p}{\partial n} = -\frac{\rho V^2}{\mathcal{R}} \tag{3.10b}$$

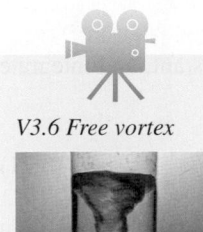

V3.6 Free vortex

This indicates that the pressure increases with distance away from the center of curvature ($\partial p/\partial n$ is negative since $\rho V^2/\mathcal{R}$ is positive—the positive n direction points toward the "inside" of the curved streamline). Thus, the pressure outside a tornado (typical atmospheric pressure) is larger than it is near the center of the tornado (where an often dangerously low partial vacuum may occur). This pressure difference is needed to balance the centrifugal acceleration associated with the curved streamlines of the fluid motion. (See Fig. E6.6a in Section 6.5.3.)

EXAMPLE 3.3   Pressure Variation Normal to a Streamline

GIVEN   Shown in Figs. E3.3a,b are two flow fields with circular streamlines. The velocity distributions are

$$V(r) = (V_0/r_0)r \quad \text{for case } (a)$$

and

$$V(r) = \frac{(V_0 r_0)}{r} \quad \text{for case } (b)$$

where V_0 is the velocity at $r = r_0$.

FIND   Determine the pressure distributions, $p = p(r)$, for each, given that $p = p_0$ at $r = r_0$.

SOLUTION

We assume the flows are steady, inviscid, and incompressible with streamlines in the horizontal plane ($dz/dn = 0$). Because the streamlines are circles, the coordinate n points in a direction opposite that of the radial coordinate, $\partial/\partial n = -\partial/\partial r$, and the radius of curvature is given by $\mathcal{R} = r$. Hence, Eq. 3.9 becomes

$$\frac{\partial p}{\partial r} = \frac{\rho V^2}{r}$$

For case (a) this gives

$$\frac{\partial p}{\partial r} = \rho(V_0/r_0)^2 r$$

whereas for case (b) it gives

$$\frac{\partial p}{\partial r} = \frac{\rho(V_0 r_0)^2}{r^3}$$

For either case the pressure increases as r increases since $\partial p/\partial r > 0$. Integration of these equations with respect to r, starting with a known pressure $p = p_0$ at $r = r_0$, gives

$$p - p_0 = (\rho V_0^2/2)[(r/r_0)^2 - 1] \quad \text{(Ans)}$$

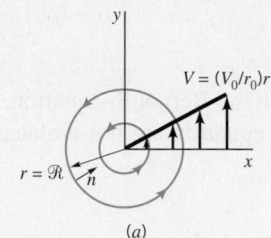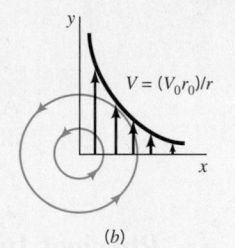

$$V = (V_0/r_0)r$$

$$r = \mathcal{R}$$

$$(a)$$

$$V = (V_0 r_0)/r$$

$$(b)$$

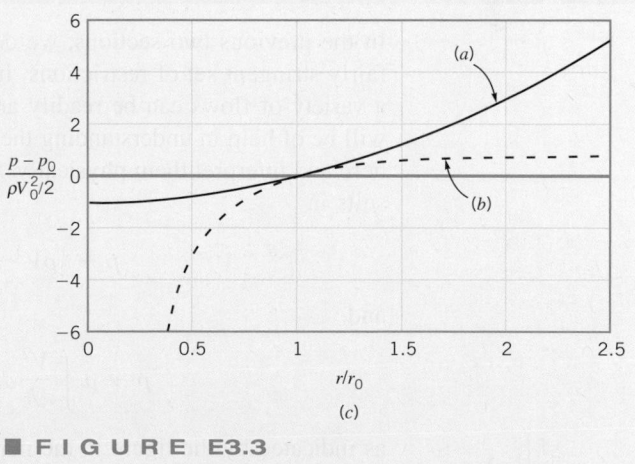

$$(c)$$

■ FIGURE E3.3

for case (a) and

$$p - p_0 = (\rho V_0^2/2)[1 - (r_0/r)^2] \quad \text{(Ans)}$$

for case (b). These pressure distributions are shown in Fig. E3.3c.

COMMENT   The pressure distributions needed to balance the centrifugal accelerations in cases (a) and (b) are not the same because the velocity distributions are different. In fact, for case (a) the pressure increases without bound as $r \to \infty$, whereas for case (b) the pressure approaches a finite value as $r \to \infty$. The streamline patterns are the same for each case, however.

Physically, case (a) represents rigid body rotation (as obtained in a can of water on a turntable after it has been "spun up") and case (b) represents a free vortex (an approximation to a tornado, a hurricane, or the swirl of water in a drain, the "bathtub vortex"). See Fig. E6.6 for an approximation of this type of flow.

The sum of pressure, elevation, and velocity effects is constant across streamlines.

If we multiply Eq. 3.10 by dn, use the fact that $\partial p/\partial n = dp/dn$ if s is constant, and integrate across the streamline (in the n direction) we obtain

$$\int \frac{dp}{\rho} + \int \frac{V^2}{\mathcal{R}} dn + gz = \text{constant across the streamline} \tag{3.11}$$

To complete the indicated integrations, we must know how the density varies with pressure and how the fluid speed and radius of curvature vary with n. For incompressible flow the density is constant and the integration involving the pressure term gives simply p/ρ. We are still left, however, with the integration of the second term in Eq. 3.11. Without knowing the n dependence in $V = V(s, n)$ and $\mathcal{R} = \mathcal{R}(s, n)$ this integration cannot be completed.

Thus, the final form of Newton's second law applied across the streamlines for steady, inviscid, incompressible flow is

$$\boxed{p + \rho \int \frac{V^2}{\mathcal{R}} dn + \gamma z = \text{constant across the streamline}} \tag{3.12}$$

As with the Bernoulli equation, we must be careful that the assumptions involved in the derivation of this equation are not violated when it is used.

3.4   Physical Interpretation

In the previous two sections, we developed the basic equations governing fluid motion under a fairly stringent set of restrictions. In spite of the numerous assumptions imposed on these flows, a variety of flows can be readily analyzed with them. A physical interpretation of the equations will be of help in understanding the processes involved. To this end, we rewrite Eqs. 3.7 and 3.12 here and interpret them physically. Application of $\mathbf{F} = m\mathbf{a}$ along and normal to the streamline results in

$$p + \tfrac{1}{2}\rho V^2 + \gamma z = \text{constant along the streamline} \tag{3.13}$$

and

$$p + \rho \int \frac{V^2}{\mathcal{R}} dn + \gamma z = \text{constant across the streamline} \tag{3.14}$$

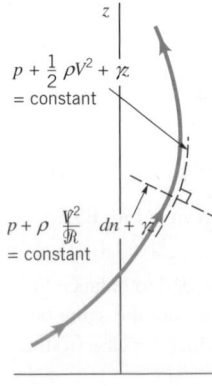

$p + \frac{1}{2}\rho V^2 + \gamma z$ = constant

$p + \rho \frac{V^2}{\mathcal{R}} dn + \gamma z$ = constant

as indicated by the figure in the margin.

The following basic assumptions were made to obtain these equations: The flow is steady and the fluid is inviscid and incompressible. In practice none of these assumptions is exactly true.

A violation of one or more of the above assumptions is a common cause for obtaining an incorrect match between the "real world" and solutions obtained by use of the Bernoulli equation. Fortunately, many "real-world" situations are adequately modeled by the use of Eqs. 3.13 and 3.14 because the flow is nearly steady and incompressible and the fluid behaves as if it were nearly inviscid.

The Bernoulli equation was obtained by integration of the equation of motion along the "natural" coordinate direction of the streamline. To produce an acceleration, there must be an unbalance of the resultant forces, of which only pressure and gravity were considered to be important. Thus, there are three processes involved in the flow—mass times acceleration (the $\rho V^2/2$ term), pressure (the p term), and weight (the γz term).

Integration of the equation of motion to give Eq. 3.13 actually corresponds to the work-energy principle often used in the study of dynamics [see any standard dynamics text (Ref. 1)]. This principle results from a general integration of the equations of motion for an object in a way very similar to that done for the fluid particle in Section 3.2. With certain assumptions, a statement of the work-energy principle may be written as follows:

> The work done on a particle by all forces acting on the particle is equal to the change of the kinetic energy of the particle.

The Bernoulli equation is a mathematical statement of this principle.

As the fluid particle moves, both gravity and pressure forces do work on the particle. Recall that the work done by a force is equal to the product of the distance the particle travels times the component of force in the direction of travel (i.e., work = $\mathbf{F} \cdot \mathbf{d}$). The terms γz and p in Eq. 3.13 are related to the work done by the weight and pressure forces, respectively. The remaining term, $\rho V^2/2$, is obviously related to the kinetic energy of the particle. In fact, an alternate method of deriving the Bernoulli equation is to use the first and second laws of thermodynamics (the energy and entropy equations), rather than Newton's second law. With the appropriate restrictions, the general energy equation reduces to the Bernoulli equation. This approach is discussed in Section 5.4.

The Bernoulli equation can be written in terms of heights called heads.

An alternate but equivalent form of the Bernoulli equation is obtained by dividing each term of Eq. 3.7 by the specific weight, γ, to obtain

$$\frac{p}{\gamma} + \frac{V^2}{2g} + z = \text{constant on a streamline}$$

Each of the terms in this equation has the units of energy per weight ($LF/F = L$) or length (meters) and represents a certain type of head.

The elevation term, z, is related to the potential energy of the particle and is called the *elevation head*. The pressure term, p/γ, is called the *pressure head* and represents the height of a column of the fluid that is needed to produce the pressure p. The velocity term, $V^2/2g$, is the *velocity head* and represents the vertical distance needed for the fluid to fall freely (neglecting friction) if it is to reach velocity V from rest. The Bernoulli equation states that the sum of the pressure head, the velocity head, and the elevation head is constant along a streamline.

EXAMPLE 3.4   Kinetic, Potential, and Pressure Energy

GIVEN Consider the flow of water from the syringe shown in Fig. E3.4(a). As indicated in Fig. E3.4b, a force, F, applied to the plunger will produce a pressure greater than atmospheric at point (1) within the syringe. The water flows from the needle, point (2), with relatively high velocity and coasts up to point (3) at the top of its trajectory.

FIND Discuss the energy of the fluid at points (1), (2), and (3) by using the Bernoulli equation.

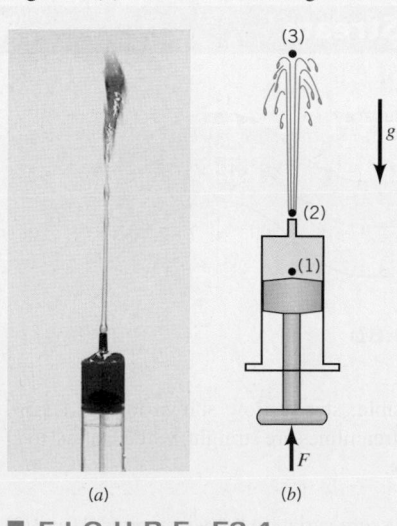

Point	Energy Type		
	Kinetic $\rho V^2/2$	**Potential** γz	**Pressure** p
1	Small	Zero	Large
2	Large	Small	Zero
3	Zero	Large	Zero

(a)   (b)

■ **FIGURE E3.4**

SOLUTION

If the assumptions (steady, inviscid, incompressible flow) of the Bernoulli equation are approximately valid, it then follows that the flow can be explained in terms of the partition of the total energy of the water. According to Eq. 3.13 the sum of the three types of energy (kinetic, potential, and pressure) or heads (velocity, elevation, and pressure) must remain constant. The table above indicates the relative magnitude of each of these energies at the three points shown in the figure.

The motion results in (or is due to) a change in the magnitude of each type of energy as the fluid flows from one location to another. An alternate way to consider this flow is as follows. The pressure gradient between (1) and (2) produces an acceleration to eject the water from the needle. Gravity acting on the particle between (2) and (3) produces a deceleration to cause the water to come to a momentary stop at the top of its flight.

COMMENT If friction (viscous) effects were important, there would be an energy loss between (1) and (3) and for the given p_1 the water would not be able to reach the height indicated in the figure. Such friction may arise in the needle (see Chapter 8 on pipe flow) or between the water stream and the surrounding air (see Chapter 9 on external flow).

F l u i d s   i n   t h e   N e w s

Armed with a water jet for hunting Archerfish, known for their ability to shoot down insects resting on foliage, are like submarine water pistols. With their snout sticking out of the water, they eject a high-speed water jet at their prey, knocking it onto the water surface where they snare it for their meal. The barrel of their water pistol is formed by placing their tongue against a groove in the roof of their mouth to form a tube. By snapping shut their gills, water is forced through the tube and directed with the tip of their tongue. The archerfish can produce a *pressure head* within their gills large enough so that the jet can reach 2 to 3 m. However, it is accurate to only about 1 m. Recent research has shown that archerfish are very adept at calculating where their prey will fall. Within 100 milliseconds (a reaction time twice as fast as a human's), the fish has extracted all the information needed to predict the point where the prey will hit the water. Without further visual cues it charges directly to that point. (See Problem 3.41.)

A net force is required to accelerate any mass. For steady flow the acceleration can be interpreted as arising from two distinct occurrences—a change in speed along the streamline and a change in direction if the streamline is not straight. Integration of the equation of motion along the streamline accounts for the change in speed (kinetic energy change) and results in the Bernoulli equation. Integration of the equation of motion normal to the streamline accounts for the centrifugal acceleration ($V^2/\mathcal{R}$) and results in Eq. 3.14.

The pressure variation across straight streamlines is hydrostatic.

When a fluid particle travels along a curved path, a net force directed toward the center of curvature is required. Under the assumptions valid for Eq. 3.14, this force may be either gravity or pressure, or a combination of both. In many instances the streamlines are nearly straight ($\mathcal{R} = \infty$) so that centrifugal effects are negligible and the pressure variation across the streamlines is merely hydrostatic (because of gravity alone), even though the fluid is in motion.

EXAMPLE 3.5   Pressure Variation in a Flowing Stream

GIVEN Water flows in a curved, undulating waterslide as shown in Fig. E3.5a. As an approximation to this flow, consider

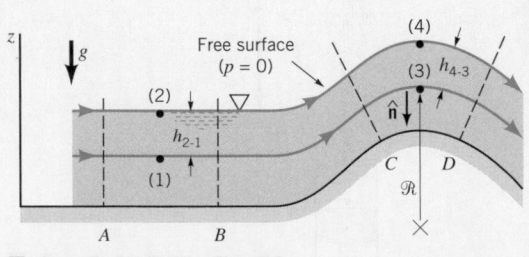

■ FIGURE E3.5b

the inviscid, incompressible, steady flow shown in Fig. E3.5b. From section A to B the streamlines are straight, while from C to D they follow circular paths.

FIND Describe the pressure variation between points (1) and (2) and points (3) and (4).

■ FIGURE E3.5a (Photo courtesy of Schlitterbahn® Waterparks.)

SOLUTION

With the above assumptions and the fact that $\mathcal{R} = \infty$ for the portion from A to B, Eq. 3.14 becomes

$$p + \gamma z = \text{constant}$$

The constant can be determined by evaluating the known variables at the two locations using $p_2 = 0$ (gage), $z_1 = 0$, and $z_2 = h_{2-1}$ to give

$$p_1 = p_2 + \gamma(z_2 - z_1) = p_2 + \gamma h_{2-1} \qquad \text{(Ans)}$$

Note that since the radius of curvature of the streamline is infinite, the pressure variation in the vertical direction is the same as if the fluid were stationary.

However, if we apply Eq. 3.14 between points (3) and (4) we obtain (using $dn = -dz$)

$$p_4 + \rho \int_{z_3}^{z_4} \frac{V^2}{\mathcal{R}}(-dz) + \gamma z_4 = p_3 + \gamma z_3$$

With $p_4 = 0$ and $z_4 - z_3 = h_{4-3}$ this becomes

$$p_3 = \gamma h_{4-3} - \rho \int_{z_3}^{z_4} \frac{V^2}{\mathcal{R}} dz \qquad \text{(Ans)}$$

To evaluate the integral, we must know the variation of V and $\mathcal{R}$ with z. Even without this detailed information we note that the integral has a positive value. Thus, the pressure at (3) is less than the hydrostatic value, γh_{4-3}, by an amount equal to $\rho \int_{z_3}^{z_4} (V^2/\mathcal{R}) \, dz$. This lower pressure, caused by the curved streamline, is necessary to accelerate the fluid around the curved path.

COMMENT   Note that we did not apply the Bernoulli equation (Eq. 3.13) across the streamlines from (1) to (2) or (3) to (4). Rather we used Eq. 3.14. As is discussed in Section 3.8, application of the Bernoulli equation across streamlines (rather than along them) may lead to serious errors.

3.5   Static, Stagnation, Dynamic, and Total Pressure

Each term in the Bernoulli equation can be interpreted as a form of pressure.

A useful concept associated with the Bernoulli equation deals with the stagnation and dynamic pressures. These pressures arise from the conversion of kinetic energy in a flowing fluid into a "pressure rise" as the fluid is brought to rest (as in Example 3.2). In this section we explore various results of this process. Each term of the Bernoulli equation, Eq. 3.13, has the dimensions of force per unit area—N/m². The first term, p, is the actual thermodynamic pressure of the fluid as it flows. To measure its value, one could move along with the fluid, thus being "static" relative to the moving fluid. Hence, it is normally termed the *static pressure*. Another way to measure the static pressure would be to drill a hole in a flat surface and fasten a piezometer tube as indicated by the location of point (3) in Fig. 3.4. As we saw in Example 3.5, the pressure in the flowing fluid at (1) is $p_1 = \gamma h_{3-1} + p_3$, the same as if the fluid were static. From the manometer considerations of Chapter 2, we know that $p_3 = \gamma h_{4-3}$. Thus, since $h_{3-1} + h_{4-3} = h$ it follows that $p_1 = \gamma h$.

The third term in Eq. 3.13, γz, is termed the *hydrostatic pressure,* in obvious regard to the hydrostatic pressure variation discussed in Chapter 2. It is not actually a pressure but does represent the change in pressure possible due to potential energy variations of the fluid as a result of elevation changes.

The second term in the Bernoulli equation, $\rho V^2/2$, is termed the *dynamic pressure*. Its interpretation can be seen in Fig. 3.4 by considering the pressure at the end of a small tube inserted into the flow and pointing upstream. After the initial transient motion has died out, the liquid will fill the tube to a height of H as shown. The fluid in the tube, including that at its tip, (2), will be stationary. That is, $V_2 = 0$, or point (2) is a *stagnation point*.

If we apply the Bernoulli equation between points (1) and (2), using $V_2 = 0$ and assuming that $z_1 = z_2$, we find that

$$p_2 = p_1 + \tfrac{1}{2}\rho V_1^2$$

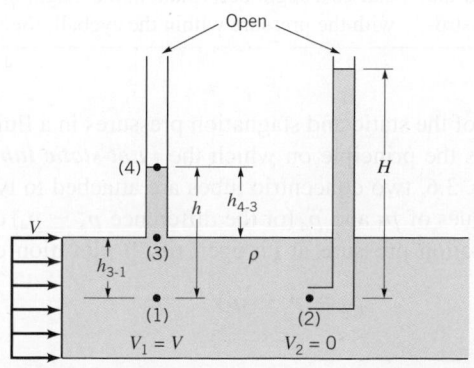

■ **FIGURE 3.4**   Measurement of static and stagnation pressures.

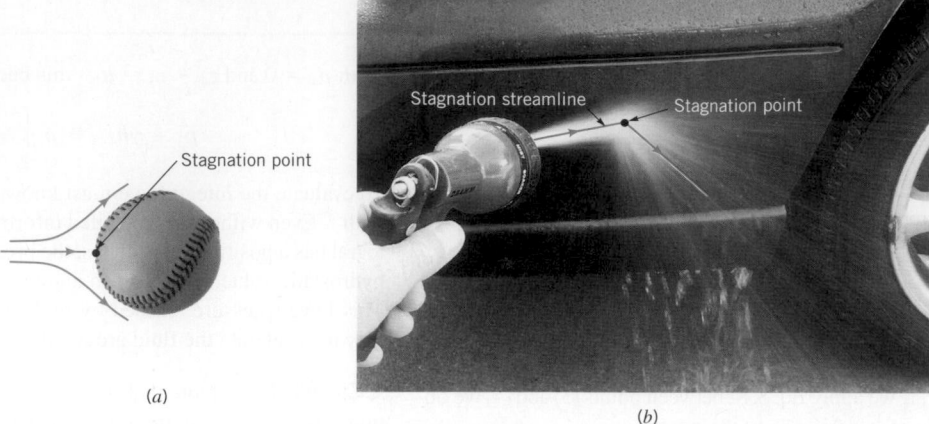

■ **F I G U R E 3.5** **Stagnation points.**

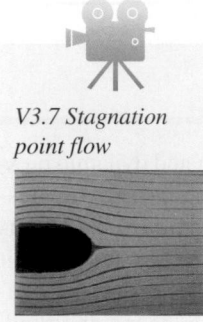

V3.7 Stagnation point flow

Hence, the pressure at the stagnation point is greater than the static pressure, p_1, by an amount $\rho V_1^2/2$, the dynamic pressure.

It can be shown that there is a stagnation point on any stationary body that is placed into a flowing fluid. Some of the fluid flows "over" and some "under" the object. The dividing line (or surface for two-dimensional flows) is termed the *stagnation streamline* and terminates at the stagnation point on the body. (See the photograph at the beginning of Chapter 3.) For symmetrical objects (such as a baseball) the stagnation point is clearly at the tip or front of the object as shown in Fig. 3.5a. For other flows such as a water jet against a car as shown in Fig. 3.5b, there is also a stagnation point on the car.

If elevation effects are neglected, the *stagnation pressure*, $p + \rho V^2/2$, is the largest pressure obtainable along a given streamline. It represents the conversion of all of the kinetic energy into a pressure rise. The sum of the static pressure, hydrostatic pressure, and dynamic pressure is termed the *total pressure, p_T*. The Bernoulli equation is a statement that the total pressure remains constant along a streamline. That is,

$$p + \tfrac{1}{2}\rho V^2 + \gamma z = p_T = \text{constant along a streamline} \tag{3.15}$$

Again, we must be careful that the assumptions used in the derivation of this equation are appropriate for the flow being considered.

F l u i d s   i n   t h e   N e w s

Pressurized eyes Our eyes need a certain amount of internal pressure in order to work properly, with the normal range being between 10 and 20 mm of mercury. The pressure is determined by a balance between the fluid entering and leaving the eye. If the pressure is above the normal level, damage may occur to the optic nerve where it leaves the eye, leading to a loss of the visual field termed glaucoma. Measurement of the pressure within the eye can be done by several different noninvasive types of instruments, all of which measure the slight deformation of the eyeball when a force is put on it. Some methods use a physical probe that makes contact with the front of the eye, applies a known force, and measures the deformation. One noncontact method uses a calibrated "puff" of air that is blown against the eye. The *stagnation pressure* resulting from the air blowing against the eyeball causes a slight deformation, the magnitude of which is correlated with the pressure within the eyeball. (See Problem 3.29.)

Knowledge of the values of the static and stagnation pressures in a fluid implies that the fluid speed can be calculated. This is the principle on which the *Pitot-static tube* is based [H. de Pitot (1695–1771)]. As shown in Fig. 3.6, two concentric tubes are attached to two pressure gages (or a differential gage) so that the values of p_3 and p_4 (or the difference $p_3 - p_4$) can be determined. The center tube measures the stagnation pressure at its open tip. If elevation changes are negligible,

$$p_3 = p + \tfrac{1}{2}\rho V^2$$

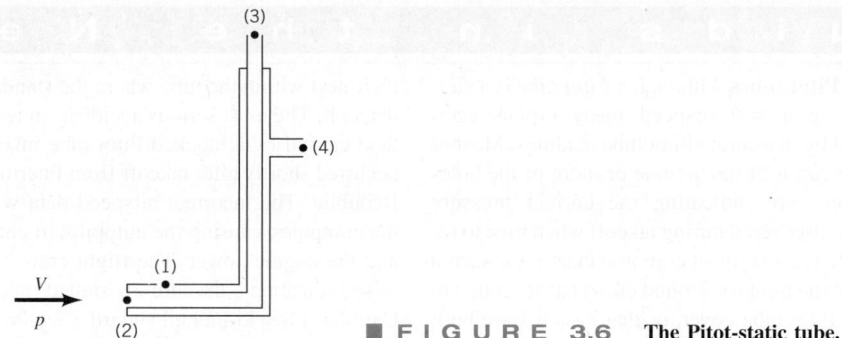

■ **F I G U R E 3.6**   The Pitot-static tube.

where p and V are the pressure and velocity of the fluid upstream of point (2). The outer tube is made with several small holes at an appropriate distance from the tip so that they measure the static pressure. If the effect of the elevation difference between (1) and (4) is negligible, then

$$p_4 = p_1 = p$$

Pitot-static tubes measure fluid velocity by converting velocity into pressure.

By combining these two equations we see that

$$p_3 - p_4 = \tfrac{1}{2}\rho V^2$$

which can be rearranged to give

$$V = \sqrt{2(p_3 - p_4)/\rho} \tag{3.16}$$

The actual shape and size of Pitot-static tubes vary considerably. A typical Pitot-static probe used to determine aircraft airspeed is shown in Fig. 3.7. (See Fig. E3.6a also.)

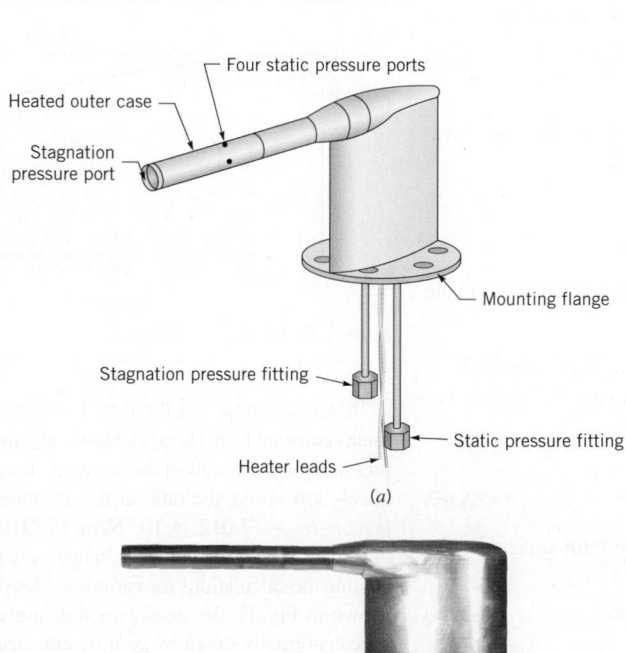

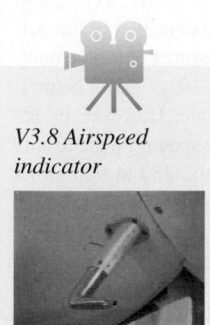

V3.8 Airspeed indicator

■ **F I G U R E 3.7**   **Airplane Pitot-static probe. (a) Schematic, (b) Photograph, (Photograph courtesy of SpaceAge Control, Inc.)**

F l u i d s   i n   t h e   N e w s

Bugged and plugged Pitot tubes Although a *Pitot tube* is a simple device for measuring aircraft airspeed, many airplane accidents have been caused by inaccurate Pitot tube readings. Most of these accidents are the result of having one or more of the holes blocked and, therefore, not indicating the correct pressure (speed). Usually this is discovered during takeoff when time to resolve the issue is short. The two most common causes for such a blockage are either that the pilot (or ground crew) has forgotten to remove the protective Pitot tube cover, or that insects have built

their nest within the tube where the standard visual check cannot detect it. The most serious accident (in terms of number of fatalities) caused by a blocked Pitot tube involved a Boeing 757 and occurred shortly after takeoff from Puerto Plata in the Dominican Republic. The incorrect airspeed data was automatically fed to the computer, causing the autopilot to change the angle of attack and the engine power. The flight crew became confused by the false indications, the aircraft stalled, and then plunged into the Caribbean Sea killing all aboard. (See Problem 3.30.)

EXAMPLE 3.6 Pitot-Static Tube

GIVEN An airplane flies 300 km/h at an elevation of 3000 m in a standard atmosphere as shown in Fig. E3.6*a*.

FIND Determine the pressure at point (1) far ahead of the airplane, the pressure at the stagnation point on the nose of the airplane, point (2), and the pressure difference indicated by a Pitot-static probe attached to the fuselage.

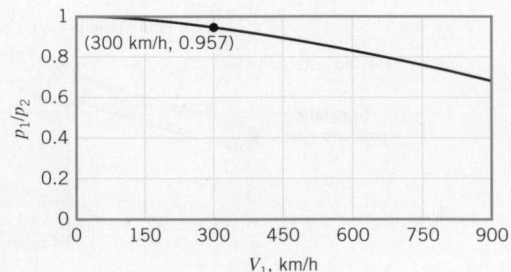

(2)

(1)
V_1 = 300 km/h

Pitot-static tube

■ **FIGURE E3.6*a*** **(Photo courtesy of Hawker Beechcraft.)**

SOLUTION

From Table C.1 we find that the static pressure at the altitude given is

$$p_1 = 7.012 \times 10^4 \, \text{N/m}^2 \, (\text{abs}) \qquad \text{(Ans)}$$

Also, the density is $\rho = 0.9093 \, \text{kg/m}^3$.

If the flow is steady, inviscid, and incompressible and elevation changes are neglected, Eq. 3.13 becomes

$$p_2 = p_1 + \frac{\rho V_1^2}{2}$$

With V_1 = 300 km/h = 83 m/s and V_2 = 0 (since the coordinate system is fixed to the airplane) we obtain

$$p_2 = 7.012 \times 10^4 \, \text{N/m}^2 + (0.9093 \, \text{kg/m}^3)(83^2 \, \text{m}^2/\text{s}^2)/2$$

$$= (7.012 \times 10^4 + 3132) \, \text{N/m}^2 \, (\text{abs})$$

Hence, in terms of gage pressure

$$p_2 = 3132 \, \text{N/m}^2 \qquad \text{(Ans)}$$

Thus, the pressure difference indicated by the Pitot-static tube is

$$p_2 - p_1 = \frac{\rho V_1^2}{2} = 3132 \, \text{N/m}^2 \qquad \text{(Ans)}$$

COMMENTS Note that it is very easy to obtain incorrect results by using improper units. Recall that $(\text{kg/m}^3)(\text{m}^2/\text{s}^2) = (\text{kg} \cdot \text{m/s}^2)/(\text{m}^2) = \text{N/m}^2$.

(300 km/h, 0.957)

p_1/p_2

V_1, km/h

■ **FIGURE E3.6*b***

It was assumed that the flow is incompressible—the density remains constant from (1) to (2). However, since $\rho = p/RT$, a change in pressure (or temperature) will cause a change in density. For this relatively low speed, the ratio of the absolute pressures is nearly unity [i.e., $p_1/p_2 = (7.012 \times 10^4 \, \text{N/m}^2)/(7.012 \times 10^4 + 3132 \, \text{N/m}^2)$ = 0.95], so that the density change is negligible. However, by repeating the calculations for various values of the speed, V_1, the results shown in Fig. E3.6*b* are obtained. Clearly at the 750 to 900 km/h speeds normally flown by commercial airliners, the pressure ratio is such that density changes are important. In such situations it is necessary to use compressible flow concepts to obtain accurate results. (See Section 3.8.1 and Chapter 11.)

The Pitot-static tube provides a simple, relatively inexpensive way to measure fluid speed. Its use depends on the ability to measure the static and stagnation pressures. Care is needed to obtain these values accurately. For example, an accurate measurement of static pressure requires that none of the fluid's kinetic energy be converted into a pressure rise at the point of

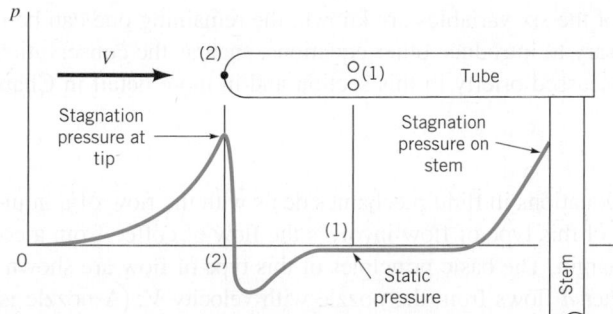

■ **FIGURE 3.8** **Incorrect and correct design of static pressure taps.**

■ **FIGURE 3.9** **Typical pressure distribution along a Pitot-static tube.**

Accurate measurement of static pressure requires great care.

measurement. This requires a smooth hole with no burrs or imperfections. As indicated in Fig. 3.8, such imperfections can cause the measured pressure to be greater or less than the actual static pressure.

Also, the pressure along the surface of an object varies from the stagnation pressure at its stagnation point to values that may be less than the free stream static pressure. A typical pressure variation for a Pitot-static tube is indicated in Fig. 3.9. Clearly it is important that the pressure taps be properly located to ensure that the pressure measured is actually the static pressure.

In practice it is often difficult to align the Pitot-static tube directly into the flow direction. Any misalignment will produce a nonsymmetrical flow field that may introduce errors. Typically, yaw angles up to 12 to 20° (depending on the particular probe design) give results that are less than 1% in error from the perfectly aligned results. Generally it is more difficult to measure static pressure than stagnation pressure.

One method of determining the flow direction and its speed (thus the velocity) is to use a directional-finding Pitot tube as is illustrated in Fig. 3.10. Three pressure taps are drilled into a small circular cylinder, fitted with small tubes, and connected to three pressure transducers. The cylinder is rotated until the pressures in the two side holes are equal, thus indicating that the center hole points directly upstream. The center tap then measures the stagnation pressure. The two side holes are located at a specific angle ($\beta = 29.5°$) so that they measure the static pressure. The speed is then obtained from $V = [2(p_2 - p_1)/\rho]^{1/2}$.

The above discussion is valid for incompressible flows. At high speeds, compressibility becomes important (the density is not constant) and other phenomena occur. Some of these ideas are discussed in Section 3.8, while others (such as shockwaves for supersonic Pitot-tube applications) are discussed in Chapter 11.

The concepts of static, dynamic, stagnation, and total pressure are useful in a variety of flow problems. These ideas are used more fully in the remainder of the book.

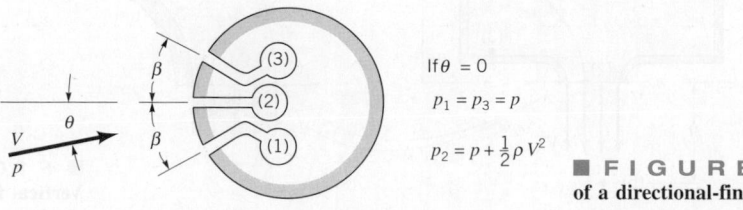

■ **FIGURE 3.10** **Cross section of a directional-finding Pitot-static tube.**

3.6 Examples of Use of the Bernoulli Equation

In this section we illustrate various additional applications of the Bernoulli equation. Between any two points, (1) and (2), on a streamline in steady, inviscid, incompressible flow the Bernoulli equation can be applied in the form

$$p_1 + \tfrac{1}{2}\rho V_1^2 + \gamma z_1 = p_2 + \tfrac{1}{2}\rho V_2^2 + \gamma z_2 \tag{3.17}$$

Obviously if five of the six variables are known, the remaining one can be determined. In many instances it is necessary to introduce other equations, such as the conservation of mass. Such considerations will be discussed briefly in this section and in more detail in Chapter 5.

3.6.1 Free Jets

One of the oldest equations in fluid mechanics deals with the flow of a liquid from a large reservoir. A modern version of this type of flow involves the flow of coffee from a coffee urn as indicated by the figure in the margin. The basic principles of this type of flow are shown in Fig. 3.11 where a jet of liquid of diameter d flows from the nozzle with velocity V. (A nozzle is a device shaped to accelerate a fluid.) Application of Eq. 3.17 between points (1) and (2) on the streamline shown gives

$$\gamma h = \tfrac{1}{2}\rho V^2$$

We have used the facts that $z_1 = h$, $z_2 = 0$, the reservoir is large ($V_1 \cong 0$) and open to the atmosphere ($p_1 = 0$ gage), and the fluid leaves as a *"free jet"* ($p_2 = 0$). Thus, we obtain

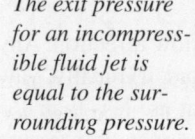

$$V = \sqrt{2\,\frac{\gamma h}{\rho}} = \sqrt{2gh} \tag{3.18}$$

which is the modern version of a result obtained in 1643 by Torricelli (1608–1647), an Italian physicist.

The exit pressure for an incompressible fluid jet is equal to the surrounding pressure.

The fact that the exit pressure equals the surrounding pressure ($p_2 = 0$) can be seen by applying $\mathbf{F} = m\mathbf{a}$, as given by Eq. 3.14, across the streamlines between (2) and (4). If the streamlines at the tip of the nozzle are straight ($\mathcal{R} = \infty$), it follows that $p_2 = p_4$. Since (4) is on the surface of the jet, in contact with the atmosphere, we have $p_4 = 0$. Thus, $p_2 = 0$ also. Since (2) is an arbitrary point in the exit plane of the nozzle, it follows that the pressure is atmospheric across this plane. Physically, since there is no component of the weight force or acceleration in the normal (horizontal) direction, the pressure is constant in that direction.

Once outside the nozzle, the stream continues to fall as a free jet with zero pressure throughout ($p_5 = 0$) and as seen by applying Eq. 3.17 between points (1) and (5), the speed increases according to

$$V = \sqrt{2g\,(h + H)}$$

where H is the distance the fluid has fallen outside the nozzle.

Equation 3.18 could also be obtained by writing the Bernoulli equation between points (3) and (4) using the fact that $z_4 = 0$, $z_3 = \ell$. Also, $V_3 = 0$ since it is far from the nozzle, and from hydrostatics, $p_3 = \gamma(h - \ell)$.

V3.9 Flow from a tank

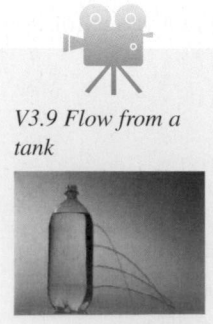

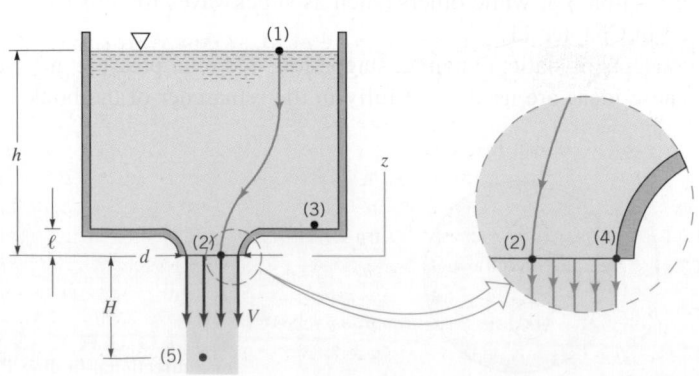

■ FIGURE 3.11
Vertical flow from a tank.

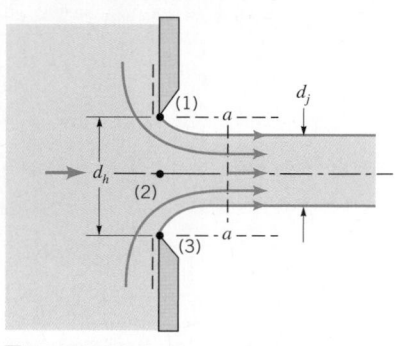

■ **F I G U R E   3.12**   **Horizontal flow from a tank.**

(a)   (b)

■ **F I G U R E   3.13**   **Vena contracta effect for a sharp-edged orifice.**

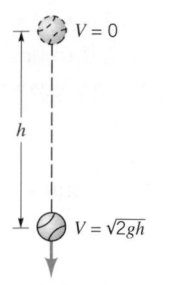

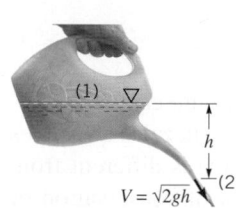

As learned in physics or dynamics and illustrated in the figure in the margin, any object dropped from rest that falls through a distance h in a vacuum will obtain the speed $V = \sqrt{2gh}$, the same as the water leaving the spout of the watering can shown in the figure in the margin. This is consistent with the fact that all of the particle's potential energy is converted to kinetic energy, provided viscous (friction) effects are negligible. In terms of heads, the elevation head at point (1) is converted into the velocity head at point (2). Recall that for the case shown in Fig. 3.11 the pressure is the same (atmospheric) at points (1) and (2).

For the horizontal nozzle of Fig. 3.12a, the velocity of the fluid at the centerline, V_2, will be slightly greater than that at the top, V_1, and slightly less than that at the bottom, V_3, due to the differences in elevation. In general, $d \ll h$ as shown in Fig. 3.12b and we can safely use the centerline velocity as a reasonable "average velocity."

If the exit is not a smooth, well-contoured nozzle, but rather a flat plate as shown in Fig. 3.13, the diameter of the jet, d_j, will be less than the diameter of the hole, d_h. This phenomenon, called a *vena contracta* effect, is a result of the inability of the fluid to turn the sharp 90° corner indicated by the dotted lines in the figure.

Since the streamlines in the exit plane are curved ($\mathcal{R} < \infty$), the pressure across them is not constant. It would take an infinite pressure gradient across the streamlines to cause the fluid to turn a "sharp" corner ($\mathcal{R} = 0$). The highest pressure occurs along the centerline at (2) and the lowest pressure, $p_1 = p_3 = 0$, is at the edge of the jet. Thus, the assumption of uniform velocity with straight streamlines and constant pressure is not valid at the exit plane. It is valid, however, in the plane of the vena contracta, section $a–a$. The uniform velocity assumption is valid at this section provided $d_j \ll h$, as is discussed for the flow from the nozzle shown in Fig. 3.12.

The diameter of a fluid jet is often smaller than that of the hole from which it flows.

The vena contracta effect is a function of the geometry of the outlet. Some typical configurations are shown in Fig. 3.14 along with typical values of the experimentally obtained *contraction coefficient*, $C_c = A_j/A_h$, where A_j and A_h are the areas of the jet at the vena contracta and the area of the hole, respectively.

F l u i d s   i n   t h e   N e w s

Cotton candy, glass wool, and steel wool Although cotton candy and glass wool insulation are made of entirely different materials and have entirely different uses, they are made by similar processes. Cotton candy, invented in 1897, consists of sugar fibers. Glass wool, invented in 1938, consists of glass fibers. In a cotton candy machine, sugar is melted and then forced by centrifugal action to flow through numerous tiny *orifices* in a spinning "bowl." Upon emerging, the thin streams of liquid sugar cool very quickly and become solid threads that are collected on a stick or cone. Making glass wool in-

sulation is somewhat more complex, but the basic process is similar. Liquid glass is forced through tiny orifices and emerges as very fine glass streams that quickly solidify. The resulting intertwined flexible fibers, glass wool, form an effective insulation material because the tiny air "cavities" between the fibers inhibit air motion. Although steel wool looks similar to cotton candy or glass wool, it is made by an entirely different process. Solid steel wires are drawn over special cutting blades which have grooves cut into them so that long, thin threads of steel are peeled off to form the matted steel wool.

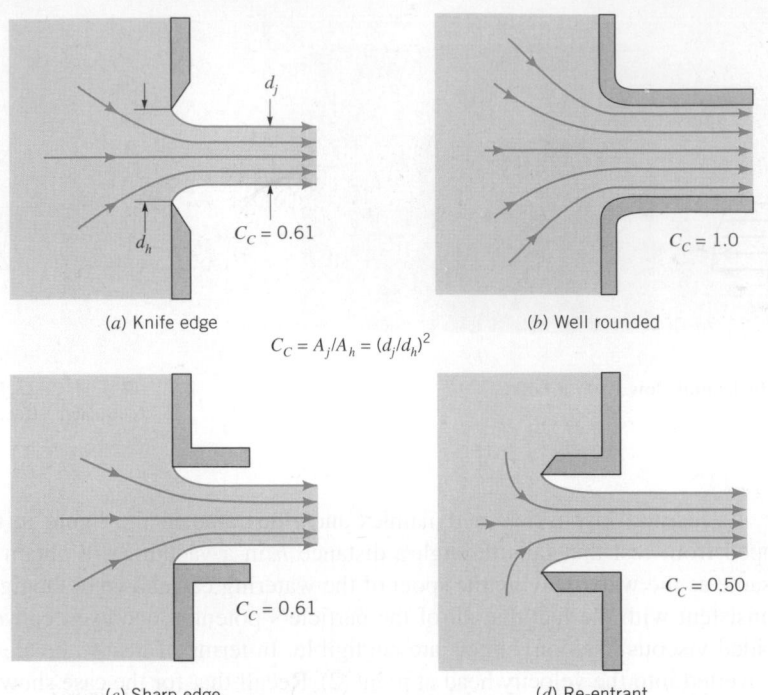

$$C_C = A_j/A_h = (d_j/d_h)^2$$

(a) Knife edge   (b) Well rounded

(c) Sharp edge   (d) Re-entrant

■ **FIGURE 3.14** **Typical flow patterns and contraction coefficients for various round exit configurations.** (*a*) **Knife edge,** (*b*) **Well rounded,** (*c*) **Sharp edge,** (*d*) **Re-entrant.**

3.6.2 Confined Flows

The continuity equation states that mass cannot be created or destroyed.

In many cases the fluid is physically constrained within a device so that its pressure cannot be prescribed a priori as was done for the free jet examples above. Such cases include nozzles and pipes of variable diameter for which the fluid velocity changes because the flow area is different from one section to another. For these situations it is necessary to use the concept of conservation of mass (the continuity equation) along with the Bernoulli equation. The derivation and use of this equation are discussed in detail in Chapters 4 and 5. For the needs of this chapter we can use a simplified form of the continuity equation obtained from the following intuitive arguments. Consider a fluid flowing through a fixed volume (such as a syringe) that has one inlet and one outlet as shown in Fig. 3.15a. If the flow is steady so that there is no additional accumulation of fluid within the volume, the rate at which the fluid flows into the volume must equal the rate at which it flows out of the volume (otherwise, mass would not be conserved).

The *mass flowrate* from an outlet, $\dot{m}$ (kg/s), is given by $\dot{m} = \rho Q$, where Q (m³/s) is the *volume flowrate*. If the outlet area is A and the fluid flows across this area (normal to the area) with an average velocity V, then the volume of the fluid crossing this area in a time interval δt is $VA\,\delta t$, equal to that in a volume of length $V\,\delta t$ and cross-sectional area A (see Fig. 3.15b). Hence, the volume flowrate (volume per unit time) is $Q = VA$. Thus, $\dot{m} = \rho VA$. To conserve mass, the inflow rate must equal the outflow rate. If the inlet is designated as (1) and the outlet as (2), it follows that $\dot{m}_1 = \dot{m}_2$. Thus, conservation of mass requires

$$\rho_1 A_1 V_1 = \rho_2 A_2 V_2$$

If the density remains constant, then $\rho_1 = \rho_2$, and the above becomes the *continuity equation* for incompressible flow

$$\boxed{A_1 V_1 = A_2 V_2, \text{ or } Q_1 = Q_2} \qquad \textbf{(3.19)}$$

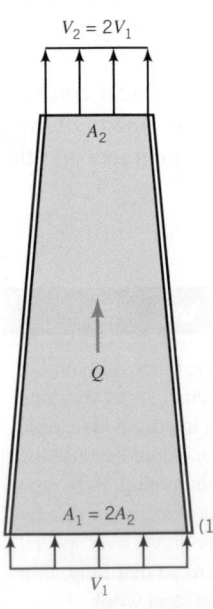

For example, if as shown by the figure in the margin the outlet flow area is one-half the size of the inlet flow area, it follows that the outlet velocity is twice that of the inlet velocity, since

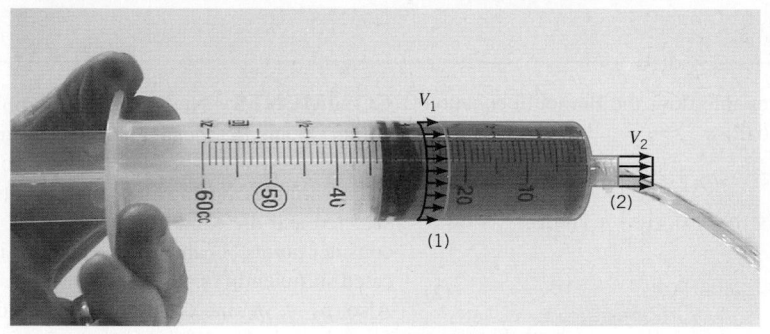

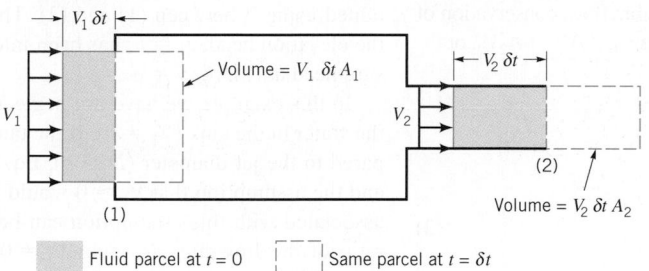

■ **FIGURE 3.15**   (*a*) **Flow through a syringe.** (*b*) **Steady flow into and out of a volume.**

$V_2 = A_1V_1/A_2 = 2V_1$. The use of the Bernoulli equation and the flowrate equation (continuity equation) is demonstrated by Example 3.7.

EXAMPLE 3.7   Flow from a Tank—Gravity

GIVEN   A stream of refreshing beverage of diameter $d = 0.01$ m flows steadily from the cooler of diameter $D = 0.20$ m as shown in Figs. E3.7*a* and *b*.

FIND   Determine the flowrate, Q, from the bottle into the cooler if the depth of beverage in the cooler is to remain constant at $h = 0.20$ m

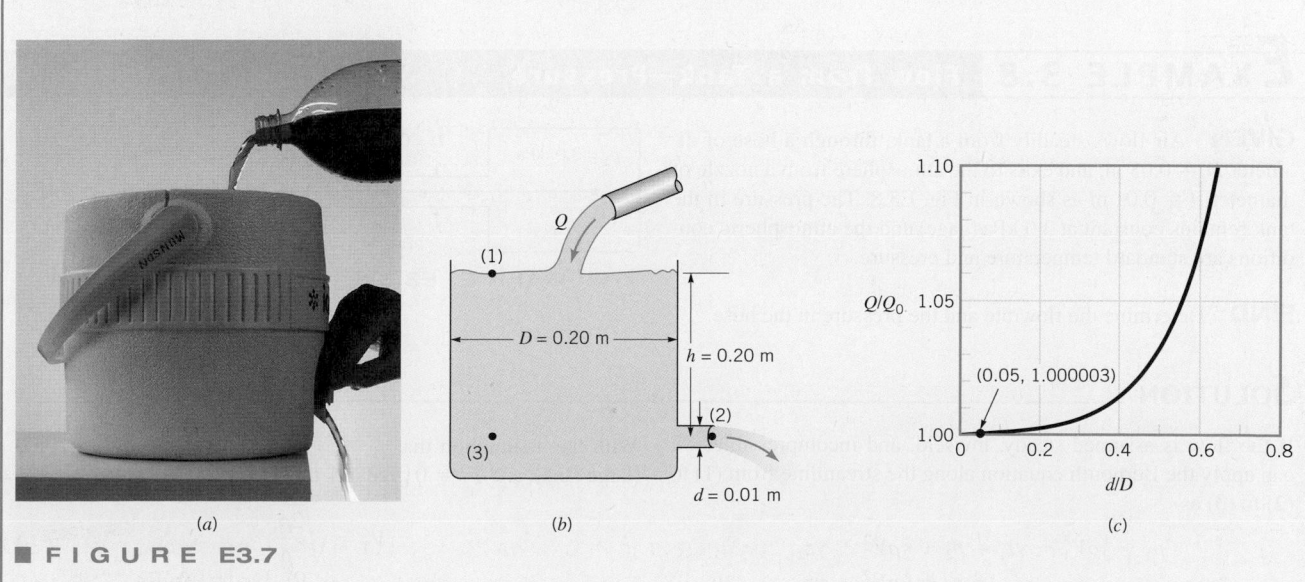

(*a*)   (*b*)   (*c*)

■ **FIGURE E3.7**

SOLUTION

For steady, inviscid, incompressible flow, the Bernoulli equation applied between points (1) and (2) is

$$p_1 + \tfrac{1}{2}\rho V_1^2 + \gamma z_1 = p_2 + \tfrac{1}{2}\rho V_2^2 + \gamma z_2 \qquad (1)$$

With the assumptions that $p_1 = p_2 = 0$, $z_1 = h$, and $z_2 = 0$, Eq. 1 becomes

$$\tfrac{1}{2}V_1^2 + gh = \tfrac{1}{2}V_2^2 \qquad (2)$$

Although the liquid level remains constant (h = constant), there is an average velocity, V_1, across section (1) because of the flow from the tank. From Eq. 3.19 for steady incompressible flow, conservation of mass requires $Q_1 = Q_2$, where $Q = AV$. Thus, $A_1V_1 = A_2V_2$, or

$$\frac{\pi}{4}D^2 V_1 = \frac{\pi}{4}d^2 V_2$$

Hence,

$$V_1 = \left(\frac{d}{D}\right)^2 V_2 \qquad (3)$$

Equations 1 and 3 can be combined to give

$$V_2 = \sqrt{\frac{2gh}{1 - (d/D)^4}} = \sqrt{\frac{2(9.81 \text{ m/s}^2)(0.20 \text{ m})}{1 - (0.01 \text{ m/0.20 m})^4}} = 1.98 \text{ m/s}$$

Thus,

$$Q = A_1V_1 = A_2V_2 = \frac{\pi}{4}(0.01 \text{ m})^2(1.98 \text{ m/s})$$

$$= 1.56 \times 10^{-4} \text{ m}^3/\text{s} \qquad \text{(Ans)}$$

COMMENTS Note that this problem was solved using points (1) and (2) located at the free surface and the exit of the pipe, respectively. Although this was convenient (because most of the variables are known at those points), other points could be selected and the same result would be obtained. For example, consider points (1) and (3) as indicated in Fig. E3.7b. At (3), located sufficiently far from the tank exit, $V_3 = 0$ and $z_3 = z_2 = 0$. Also, $p_3 = \gamma h$ since the pressure is hydrostatic sufficiently far from the exit. Use of this information in the Bernoulli equation applied between (1) and (3) gives the exact same result as obtained using it between (1) and (2). The only difference is that the elevation head, $z_1 = h$, has been interchanged with the pressure head at (3), $p_3/\gamma = h$.

In this example we have not neglected the kinetic energy of the water in the tank ($V_1 \neq 0$). If the tank diameter is large compared to the jet diameter ($D \gg d$), Eq. 3 indicates that $V_1 \ll V_2$ and the assumption that $V_1 \approx 0$ would be reasonable. The error associated with this assumption can be seen by calculating the ratio of the flowrate assuming $V_1 \neq 0$, denoted Q, to that assuming $V_1 = 0$, denoted Q_0. This ratio, written as

$$\frac{Q}{Q_0} = \frac{V_2}{V_2|_{D=\infty}} = \frac{\sqrt{2gh/[1 - (d/D)^4]}}{\sqrt{2gh}} = \frac{1}{\sqrt{1 - (d/D)^4}}$$

is plotted in Fig. E3.7c. With $0 < d/D < 0.4$ it follows that $1 < Q/Q_0 \lesssim 1.01$, and the error in assuming $V_1 = 0$ is less than 1%. For this example with $d/D = 0.01 \text{ m}/0.20 \text{ m} = 0.05$, it follows that $Q/Q_0 = 1.000003$. Thus, it is often reasonable to assume $V_1 = 0$.

The fact that a kinetic energy change is often accompanied by a change in pressure is shown by Example 3.8.

EXAMPLE 3.8   Flow from a Tank—Pressure

GIVEN Air flows steadily from a tank, through a hose of diameter $D = 0.03$ m, and exits to the atmosphere from a nozzle of diameter $d = 0.01$ m as shown in Fig. E3.8. The pressure in the tank remains constant at 3.0 kPa (gage) and the atmospheric conditions are standard temperature and pressure.

FIND Determine the flowrate and the pressure in the hose.

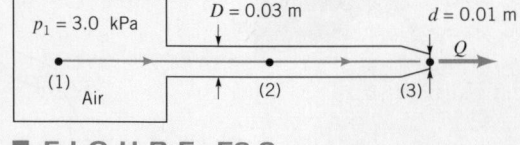

■ **FIGURE E3.8**

SOLUTION

If the flow is assumed steady, inviscid, and incompressible, we can apply the Bernoulli equation along the streamline from (1) to (2) to (3) as

$$p_1 + \tfrac{1}{2}\rho V_1^2 + \gamma z_1 = p_2 + \tfrac{1}{2}\rho V_2^2 + \gamma z_2$$
$$= p_3 + \tfrac{1}{2}\rho V_3^2 + \gamma z_3$$

With the assumption that $z_1 = z_2 = z_3$ (horizontal hose), $V_1 = 0$ (large tank), and $p_3 = 0$ (free jet), this becomes

$$V_3 = \sqrt{\frac{2p_1}{\rho}}$$

and

$$p_2 = p_1 - \tfrac{1}{2}\rho V_2^2 \qquad (1)$$

The density of the air in the tank is obtained from the perfect gas law, using standard absolute pressure and temperature, as

$$\rho = \frac{p_1}{RT_1}$$
$$= [(3.0 + 101) \text{ kN/m}^2]$$
$$\times \frac{10^3 \text{ N/kN}}{(286.9 \text{ N} \cdot \text{m/kg} \cdot \text{K})(15 + 273)\text{K}}$$
$$= 1.26 \text{ kg/m}^3$$

Thus, we find that

$$V_3 = \sqrt{\frac{2(3.0 \times 10^3 \text{ N/m}^2)}{1.26 \text{ kg/m}^3}} = 69.0 \text{ m/s}$$

or

$$Q = A_3 V_3 = \frac{\pi}{4} d^2 V_3 = \frac{\pi}{4}(0.01 \text{ m})^2 (69.0 \text{ m/s})$$
$$= 0.00542 \text{ m}^3/\text{s} \qquad \text{(Ans)}$$

The pressure within the hose can be obtained from Eq. 1 and the continuity equation (Eq. 3.19)

$$A_2 V_2 = A_3 V_3$$

Hence,

$$V_2 = A_3 V_3 / A_2 = \left(\frac{d}{D}\right)^2 V_3$$
$$= \left(\frac{0.01 \text{ m}}{0.03 \text{ m}}\right)^2 (69.0 \text{ m/s}) = 7.67 \text{ m/s}$$

and from Eq. 1

$$p_2 = 3.0 \times 10^3 \text{ N/m}^2 - \tfrac{1}{2}(1.26 \text{ kg/m}^3)(7.67 \text{ m/s})^2$$
$$= (3000 - 37.1)\text{N/m}^2 = 2963 \text{ N/m}^2 \qquad \text{(Ans)}$$

COMMENTS   Note that the value of V_3 is determined strictly by the value of p_1 (and the assumptions involved in the Bernoulli equation), independent of the "shape" of the nozzle. The pressure head within the tank, $p_1/\gamma = (3.0 \text{ kPa})/(9.81 \text{ m/s}^2)(1.26 \text{ kg/m}^3) = 243 \text{ m}$, is converted to the velocity head at the exit, $V_2^2/2g = (69.0 \text{ m/s})^2/(2 \times 9.81 \text{ m/s}^2) = 243 \text{ m}$. Although we used gage pressure in the Bernoulli equation ($p_3 = 0$), we had to use absolute pressure in the perfect gas law when calculating the density.

In the absence of viscous effects the pressure throughout the hose is constant and equal to p_2. Physically, the decreases in pressure from p_1 to p_2 to p_3 accelerate the air and increase its kinetic energy from zero in the tank to an intermediate value in the hose and finally to its maximum value at the nozzle exit. Since the air velocity in the nozzle exit is nine times that in the hose, most of the pressure drop occurs across the nozzle ($p_1 = 3000 \text{ N/m}^2, p_2 = 2963 \text{ N/m}^2$, and $p_3 = 0$).

Since the pressure change from (1) to (3) is not too great [i.e., in terms of absolute pressure $(p_1 - p_3)/p_1 = 3.0/101 = 0.03$], it follows from the perfect gas law that the density change is also not significant. Hence, the incompressibility assumption is reasonable for this problem. If the tank pressure were considerably larger or if viscous effects were important, the above results would be incorrect.

F l u i d s   i n   t h e   N e w s

Hi-tech inhaler The term inhaler often brings to mind a treatment for asthma or bronchitis. Work is underway to develop a family of inhalation devices that can do more than treat respiratory ailments. They will be able to deliver medication for diabetes and other conditions by spraying it to reach the bloodstream through the lungs. The concept is to make the spray droplets fine enough to penetrate to the lungs' tiny sacs, the alveoli, where exchanges between blood and the outside world take place. This is accomplished by use of a laser-machined *nozzle* containing an array of very fine holes that cause the liquid to divide into a mist of micron-scale droplets. The device fits the hand and accepts a disposable strip that contains the medicine solution sealed inside a blister of laminated plastic and the nozzle. An electrically actuated piston drives the liquid from its reservoir through the nozzle array and into the respiratory system. To take the medicine, the patient breathes through the device and a differential pressure transducer in the inhaler senses when the patient's breathing has reached the best condition for receiving the medication. At that point, the piston is automatically triggered.

In many situations the combined effects of kinetic energy, pressure, and gravity are important. Example 3.9 illustrates this.

*E*XAMPLE 3.9   Flow in a Variable Area Pipe

GIVEN   Water flows through a pipe reducer as is shown in Fig. E3.9. The static pressures at (1) and (2) are measured by the inverted U-tube manometer containing oil of specific gravity, SG, less than one.

FIND   Determine the manometer reading, h.

SOLUTION

With the assumptions of steady, inviscid, incompressible flow, the Bernoulli equation can be written as

$$p_1 + \tfrac{1}{2}\rho V_1^2 + \gamma z_1 = p_2 + \tfrac{1}{2}\rho V_2^2 + \gamma z_2$$

The continuity equation (Eq. 3.19) provides a second relationship between V_1 and V_2 if we assume the velocity profiles are uniform at those two locations and the fluid incompressible:

$$Q = A_1 V_1 = A_2 V_2$$

By combining these two equations we obtain

$$p_1 - p_2 = \gamma(z_2 - z_1) + \tfrac{1}{2}\rho V_2^2[1 - (A_2/A_1)^2] \quad \textbf{(1)}$$

This pressure difference is measured by the manometer and can be determined by using the pressure–depth ideas developed in Chapter 2. Thus,

$$p_1 - \gamma(z_2 - z_1) - \gamma\ell - \gamma h + SG\,\gamma h + \gamma\ell = p_2$$

or

$$p_1 - p_2 = \gamma(z_2 - z_1) + (1 - SG)\gamma h \quad \textbf{(2)}$$

As discussed in Chapter 2, this pressure difference is neither merely γh nor $\gamma(h + z_1 - z_2)$.

Equations 1 and 2 can be combined to give the desired result as follows:

$$(1 - SG)\gamma h = \frac{1}{2}\rho V_2^2\left[1 - \left(\frac{A_2}{A_1}\right)^2\right]$$

or since $V_2 = Q/A_2$

$$h = (Q/A_2)^2\,\frac{1 - (A_2/A_1)^2}{2g(1 - SG)} \quad \textbf{(Ans)}$$

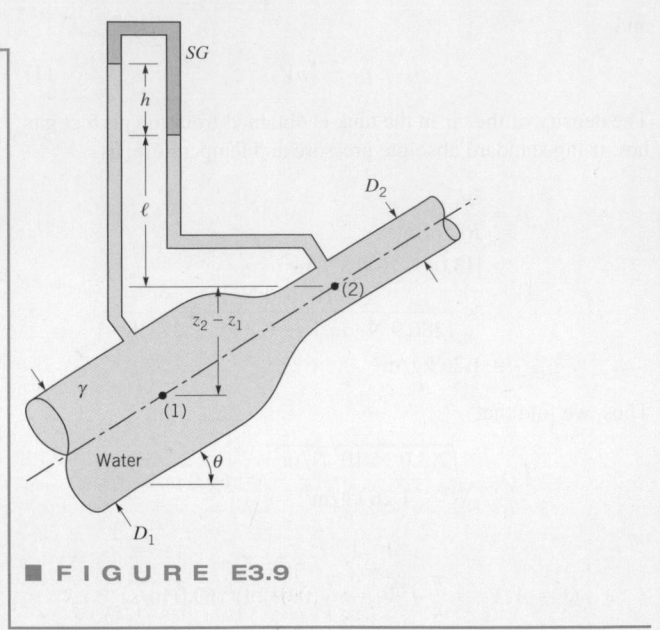

■ **FIGURE E3.9**

COMMENT The difference in elevation, $z_1 - z_2$, was not needed because the change in elevation term in the Bernoulli equation exactly cancels the elevation term in the manometer equation. However, the pressure difference, $p_1 - p_2$, depends on the angle θ, because of the elevation, $z_1 - z_2$, in Eq. 1. Thus, for a given flowrate, the pressure difference, $p_1 - p_2$, as measured by a pressure gage would vary with θ, but the manometer reading, h, would be independent of θ.

V3.10 Venturi channel

Cavitation occurs when the pressure is reduced to the vapor pressure.

In general, an increase in velocity is accompanied by a decrease in pressure. For example, the velocity of the air flowing over the top surface of an airplane wing is, on the average, faster than that flowing under the bottom surface. Thus, the net pressure force is greater on the bottom than on the top—the wing generates a lift.

If the differences in velocity are considerable, the differences in pressure can also be considerable. For flows of gases, this may introduce compressibility effects as discussed in Section 3.8 and Chapter 11. For flows of liquids, this may result in *cavitation,* a potentially dangerous situation that results when the liquid pressure is reduced to the vapor pressure and the liquid "boils."

As discussed in Chapter 1, the vapor pressure, p_v, is the pressure at which vapor bubbles form in a liquid. It is the pressure at which the liquid starts to boil. Obviously this pressure depends on the type of liquid and its temperature. For example, water, which boils at 100 °C (373 K) at standard atmospheric pressure, 101.3 kPa (abs), boils at 30 °C (303 K) if the pressure is 4.243 kPa (abs). That is, $p_v = 4.243$ kPa (abs) at 30 °C (303 K) and $p_v = 101.3$ kPa (abs) at 100 °C (373 K). (See Table B.1.)

One way to produce cavitation in a flowing liquid is noted from the Bernoulli equation. If the fluid velocity is increased (for example, by a reduction in flow area as shown in Fig. 3.16) the pressure will decrease. This pressure decrease (needed to accelerate the fluid through the constriction) can be large enough so that the pressure in the liquid is reduced to its vapor pressure. A simple example of cavitation can be demonstrated with an ordinary garden hose. If the hose is "kinked," a restriction in the flow area in some ways analogous to that shown in Fig. 3.16 will result. The water velocity through this restriction will be relatively large. With a sufficient amount of restriction the sound of the flowing water will change—a definite "hissing" sound is produced. This sound is a result of cavitation.

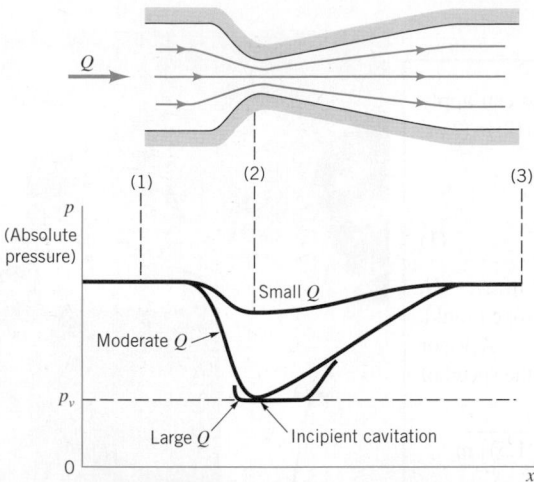

■ F I G U R E 3.16 Pressure variation and cavitation in a variable area pipe.

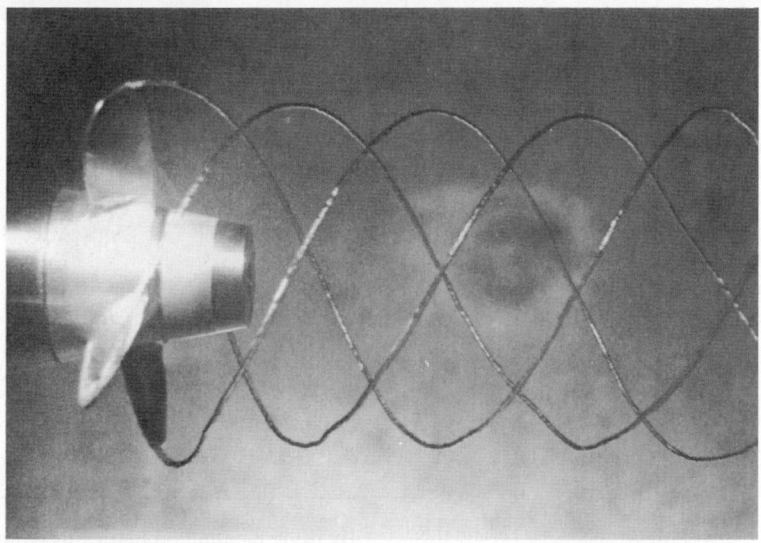

■ F I G U R E 3.17 Tip cavitation from a propeller. (Photograph courtesy of Garfield Thomas Water Tunnel, Pennsylvania State University.)

Cavitation can cause damage to equipment.

In such situations boiling occurs (though the temperature need not be high), vapor bubbles form, and then they collapse as the fluid moves into a region of higher pressure (lower velocity). This process can produce dynamic effects (imploding) that cause very large pressure transients in the vicinity of the bubbles. Pressures as large as 690 MPa are believed to occur. If the bubbles collapse close to a physical boundary they can, over a period of time, cause damage to the surface in the cavitation area. Tip cavitation from a propeller is shown in Fig. 3.17. In this case the high-speed rotation of the propeller produced a corresponding low pressure on the propeller. Obviously, proper design and use of equipment are needed to eliminate cavitation damage.

*E*XAMPLE 3.10   Siphon and Cavitation

GIVEN A liquid can be siphoned from a container as shown in Fig. E3.10a provided the end of the tube, point (3), is below the free surface in the container, point (1), and the maximum elevation of the tube, point (2), is "not too great." Consider water at 15° C (288 K) being siphoned from a large tank through a constant diameter hose as shown in Fig. E3.10b. The end of the siphon is 1.5 m below the bottom of the tank, and the atmospheric pressure is 101.3 kPa (abs).

FIND Determine the maximum height of the hill, H, over which the water can be siphoned without cavitation occurring.

SOLUTION

If the flow is steady, inviscid, and incompressible we can apply the Bernoulli equation along the streamline from (1) to (2) to (3) as follows:

$$p_1 + \tfrac{1}{2}\rho V_1^2 + \gamma z_1 = p_2 + \tfrac{1}{2}\rho V_2^2 + \gamma z_2$$
$$= p_3 + \tfrac{1}{2}\rho V_3^2 + \gamma z_3 \qquad (1)$$

With the tank bottom as the datum, we have $z_1 = 4.5$ m, $z_2 = H$, and $z_3 = -1.5$ m. Also, $V_1 = 0$ (large tank), $p_1 = 0$ (open tank), $p_3 = 0$ (free jet), and from the continuity equation $A_2 V_2 = A_3 V_3$, or because the hose is constant diameter, $V_2 = V_3$. Thus, the speed of the fluid in the hose is determined from Eq. 1 to be

$$V_3 = \sqrt{2g(z_1 - z_3)} = \sqrt{2(9.81 \text{ m/s}^2)[4.5 - (-1.5)]} \text{ m}$$
$$= 10.8 \text{ m/s} = V_2$$

Use of Eq. 1 between points (1) and (2) then gives the pressure p_2 at the top of the hill as

$$p_2 = p_1 + \tfrac{1}{2}\rho V_1^2 + \gamma z_1 - \tfrac{1}{2}\rho V_2^2 - \gamma z_2$$
$$= \gamma(z_1 - z_2) - \tfrac{1}{2}\rho V_2^2 \qquad (2)$$

From Table B.1, the vapor pressure of water at 15 °C (288 K) is 1.765 kPa (abs). Hence, for incipient cavitation the lowest pressure in the system will be $p = 1.765$ kPa (abs). Careful consideration of Eq. 2 and Fig. E3.10b will show that this lowest pressure will occur at the top of the hill. Since we have used gage pressure at point (1) ($p_1 = 0$), we must use gage pressure at point (2) also. Thus, $p_2 = 1.765 - 101.3 = -99.535$ kPa and Eq. 2 gives

$$99.535 \times 10^3 \text{ Pa}$$
$$= (9810 \text{ N/m}^3)(4.5 - H)\text{m} - \tfrac{1}{2}(1000 \text{ kg/m}^3)(10.8 \text{ m/s})^2$$

or

$$H = 8.70 \text{ m} \qquad \textbf{(Ans)}$$

For larger values of H, vapor bubbles will form at point (2) and the siphon action may stop.

COMMENTS Note that we could have used absolute pressure throughout ($p_2 = 1.765$ kPa (abs) and $p_1 = 101.3$ kPa (abs)) and obtained the same result. The lower the elevation of point (3), the larger the flowrate and, therefore, the smaller the value of H allowed.

We could also have used the Bernoulli equation between (2) and (3), with $V_2 = V_3$, to obtain the same value of H. In this case it would not have been necessary to determine V_2 by use of the Bernoulli equation between (1) and (3).

The above results are independent of the diameter and length of the hose (provided viscous effects are not important). Proper design of the hose (or pipe) is needed to ensure that it will not collapse due to the large pressure difference (vacuum) between the inside and outside of the hose.

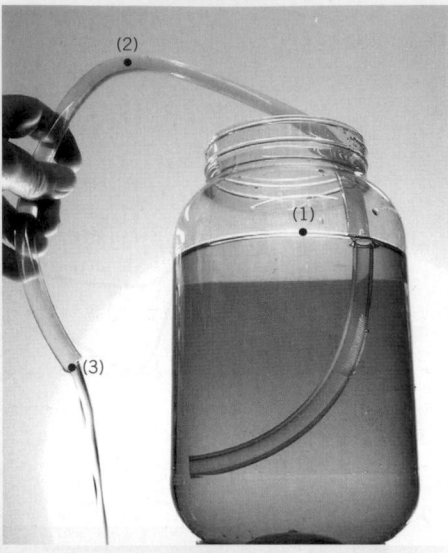

■ **FIGURE E3.10a**

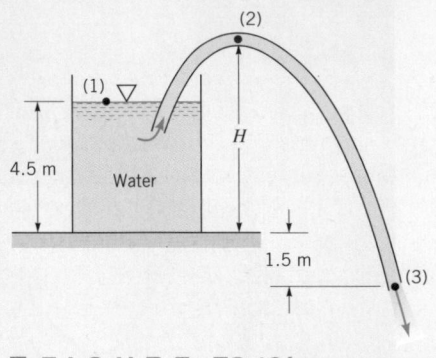

■ **FIGURE E3.10b**

By using the fluid properties listed in Table 1.5 and repeating the calculations for various fluids, the results shown in Fig. E3.10c are obtained. The value of H is a function of both the specific weight of the fluid, γ, and its vapor pressure, p_v.

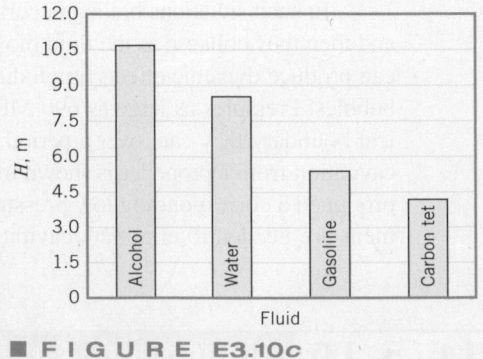

■ **FIGURE E3.10c**

3.6.3 Flowrate Measurement

Many types of devices using principles involved in the Bernoulli equation have been developed to measure fluid velocities and flowrates. The Pitot-static tube discussed in Section 3.5 is an example. Other examples discussed below include devices to measure flowrates in pipes and

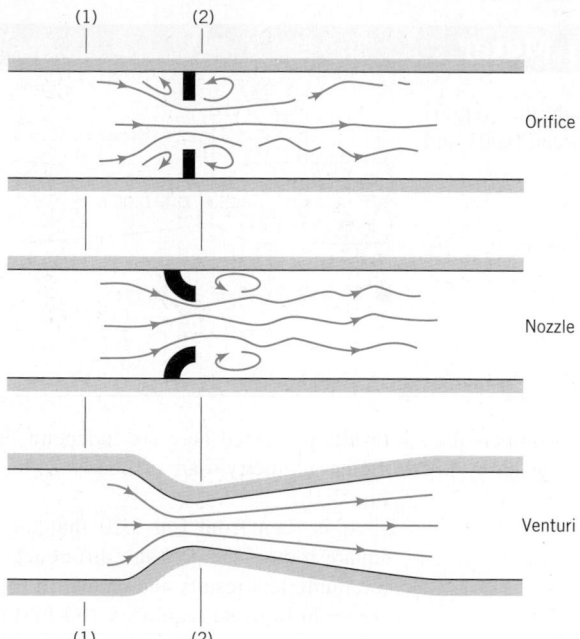

■ **F I G U R E   3.18**   Typical devices for measuring flowrate in pipes.

conduits and devices to measure flowrates in open channels. In this chapter we will consider "ideal" *flow meters*—those devoid of viscous, compressibility, and other "real-world" effects. Corrections for these effects are discussed in Chapters 8 and 10. Our goal here is to understand the basic operating principles of these simple flow meters.

An effective way to measure the flowrate through a pipe is to place some type of restriction within the pipe as shown in Fig. 3.18 and to measure the pressure difference between the low-velocity, high-pressure upstream section (1), and the high-velocity, low-pressure downstream section (2). Three commonly used types of flow meters are illustrated: the *orifice meter*, the *nozzle meter*, and the *Venturi meter*. The operation of each is based on the same physical principles—an increase in velocity causes a decrease in pressure. The difference between them is a matter of cost, accuracy, and how closely their actual operation obeys the idealized flow assumptions.

We assume the flow is horizontal ($z_1 = z_2$), steady, inviscid, and incompressible between points (1) and (2). The Bernoulli equation becomes

$$p_1 + \tfrac{1}{2}\rho V_1^2 = p_2 + \tfrac{1}{2}\rho V_2^2$$

(The effect of nonhorizontal flow can be incorporated easily by including the change in elevation, $z_1 - z_2$, in the Bernoulli equation.)

If we assume the velocity profiles are uniform at sections (1) and (2), the continuity equation (Eq. 3.19) can be written as

$$Q = A_1 V_1 = A_2 V_2$$

The flowrate varies as the square root of the pressure difference across the flow meter.

where A_2 is the small ($A_2 < A_1$) flow area at section (2). Combination of these two equations results in the following theoretical flowrate

$$Q = A_2 \sqrt{\frac{2(p_1 - p_2)}{\rho[1 - (A_2/A_1)^2]}} \tag{3.20}$$

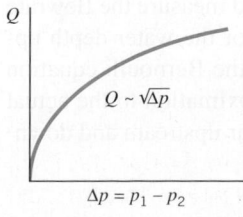

Thus, as shown by the figure in the margin, for a given flow geometry (A_1 and A_2) the flowrate can be determined if the pressure difference, $p_1 - p_2$, is measured. The actual measured flowrate, Q_{actual}, will be smaller than this theoretical result because of various differences between the "real world" and the assumptions used in the derivation of Eq. 3.20. These differences (which are quite consistent and may be as small as 1 to 2% or as large as 40%, depending on the geometry used) can be accounted for by using an empirically obtained discharge coefficient as discussed in Section 8.6.1.

EXAMPLE 3.11   Venturi Meter

GIVEN   Kerosene ($SG = 0.85$) flows through the Venturi meter shown in Fig. E3.11a with flowrates between 0.005 and 0.050 m³/s.

FIND   Determine the range in pressure difference, $p_1 - p_2$, needed to measure these flowrates.

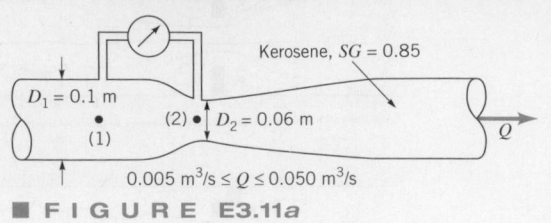

■ **FIGURE E3.11a**

SOLUTION

If the flow is assumed to be steady, inviscid, and incompressible, the relationship between flowrate and pressure is given by Eq. 3.20. This can be rearranged to give

$$p_1 - p_2 = \frac{Q^2 \rho [1 - (A_2/A_1)^2]}{2 A_2^2}$$

With the density of the flowing fluid

$$\rho = SG \, \rho_{H_2O} = 0.85(1000 \text{ kg/m}^3) = 850 \text{ kg/m}^3$$

and the area ratio

$$A_2/A_1 = (D_2/D_1)^2 = (0.06 \text{ m}/0.10 \text{ m})^2 = 0.36$$

the pressure difference for the smallest flowrate is

$$p_1 - p_2 = (0.005 \text{ m}^3/\text{s})^2 (850 \text{ kg/m}^3) \frac{(1 - 0.36^2)}{2 \left[(\pi/4)(0.06 \text{ m})^2 \right]^2}$$

$$= 1160 \text{ N/m}^2 = 1.16 \text{ kPa}$$

Likewise, the pressure difference for the largest flowrate is

$$p_1 - p_2 = (0.05)^2 (850) \frac{(1 - 0.36^2)}{2[(\pi/4)(0.06)^2]^2}$$

$$= 1.16 \times 10^5 \text{ N/m}^2 = 116 \text{ kPa}$$

Thus,

$$1.16 \text{ kPa} \leq p_1 - p_2 \leq 116 \text{ kPa} \qquad \textbf{(Ans)}$$

COMMENTS   These values represent the pressure differences for inviscid, steady, incompressible conditions. The ideal

results presented here are independent of the particular flow meter geometry—an orifice, nozzle, or Venturi meter (see Fig. 3.18).

It is seen from Eq. 3.20 that the flowrate varies as the square root of the pressure difference. Hence, as indicated by the numerical results and shown in Fig. E3.11b, a 10-fold increase in flowrate requires a 100-fold increase in pressure difference. This nonlinear relationship can cause difficulties when measuring flowrates over a wide range of values. Such measurements would require pressure transducers with a wide range of operation. An alternative is to use two flow meters in parallel—one for the larger and one for the smaller flowrate ranges.

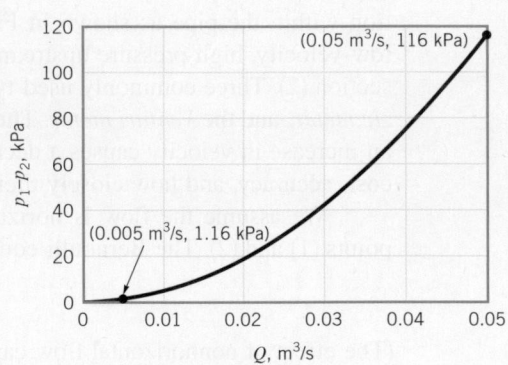

■ **FIGURE E3.11b**

Other flow meters based on the Bernoulli equation are used to measure flowrates in open channels such as flumes and irrigation ditches. Two of these devices, the *sluice gate* and the *sharp-crested weir*, are discussed below under the assumption of steady, inviscid, incompressible flow. These and other open-channel flow devices are discussed in more detail in Chapter 10.

Sluice gates like those shown in Fig. 3.19a are often used to regulate and measure the flowrate in open channels. As indicated in Fig. 3.19b, the flowrate, Q, is a function of the water depth upstream, z_1, the width of the gate, b, and the gate opening, a. Application of the Bernoulli equation and continuity equation between points (1) and (2) can provide a good approximation to the actual flowrate obtained. We assume the velocity profiles are uniform sufficiently far upstream and downstream of the gate.

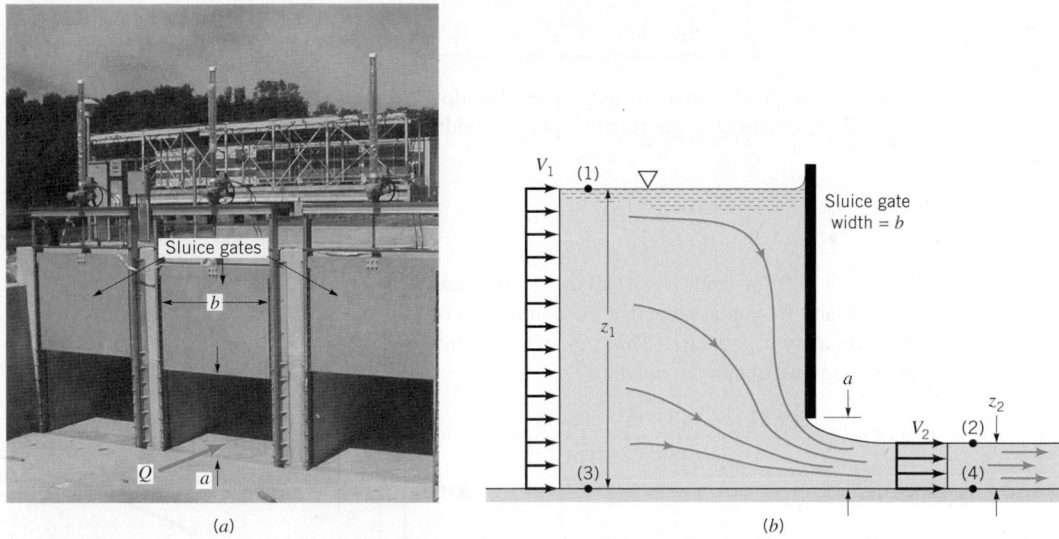

■ FIGURE 3.19   Sluice gate geometry. (Photograph courtesy of Plasti-Fab, Inc.)

Thus, we apply the Bernoulli equation between points on the free surfaces at (1) and (2) to give

$$p_1 + \tfrac{1}{2}\rho V_1^2 + \gamma z_1 = p_2 + \tfrac{1}{2}\rho V_2^2 + \gamma z_2$$

Also, if the gate is the same width as the channel so that $A_1 = bz_1$ and $A_2 = bz_2$, the continuity equation gives

$$Q = A_1 V_1 = bV_1 z_1 = A_2 V_2 = bV_2 z_2$$

With the fact that $p_1 = p_2 = 0$, these equations can be combined and rearranged to give the flowrate as

$$Q = z_2 b \sqrt{\frac{2g(z_1 - z_2)}{1 - (z_2/z_1)^2}} \qquad (3.21)$$

The flowrate under a sluice gate depends on the water depths on either side of the gate.

In the limit of $z_1 \gg z_2$ this result simply becomes

$$Q = z_2 b \sqrt{2gz_1}$$

This limiting result represents the fact that if the depth ratio, z_1/z_2, is large, the kinetic energy of the fluid upstream of the gate is negligible and the fluid velocity after it has fallen a distance $(z_1 - z_2) \approx z_1$ is approximately $V_2 = \sqrt{2gz_1}$.

The results of Eq. 3.21 could also be obtained by using the Bernoulli equation between points (3) and (4) and the fact that $p_3 = \gamma z_1$ and $p_4 = \gamma z_2$ since the streamlines at these sections are straight. In this formulation, rather than the potential energies at (1) and (2), we have the pressure contributions at (3) and (4).

The downstream depth, z_2, not the gate opening, a, was used to obtain the result of Eq. 3.21. As was discussed relative to flow from an orifice (Fig. 3.14), the fluid cannot turn a sharp 90° corner. A vena contracta results with a contraction coefficient, $C_c = z_2/a$, less than 1. Typically C_c is approximately 0.61 over the depth ratio range of $0 < a/z_1 < 0.2$. For larger values of a/z_1 the value of C_c increases rapidly.

***E*XAMPLE 3.12   Sluice Gate**

GIVEN Water flows under the sluice gate shown in Fig. E3.12a.   **FIND** Determine the approximate flowrate per unit width of the channel.

SOLUTION

Under the assumptions of steady, inviscid, incompressible flow, we can apply Eq. 3.21 to obtain Q/b, the flowrate per unit width, as

$$\frac{Q}{b} = z_2 \sqrt{\frac{2g(z_1 - z_2)}{1 - (z_2/z_1)^2}}$$

In this instance $z_1 = 5.0$ m and $a = 0.80$ m so the ratio $a/z_1 = 0.16 < 0.20$, and we can assume that the contraction coefficient is approximately $C_c = 0.61$. Thus, $z_2 = C_c a = 0.61$ (0.80 m) $= 0.488$ m and we obtain the flowrate

$$\frac{Q}{b} = (0.488 \text{ m}) \sqrt{\frac{2(9.81 \text{ m/s}^2)(5.0 \text{ m} - 0.488 \text{ m})}{1 - (0.488 \text{ m}/5.0 \text{ m})^2}}$$

$$= 4.61 \text{ m}^2/\text{s} \qquad \text{(Ans)}$$

COMMENT If we consider $z_1 \gg z_2$ and neglect the kinetic energy of the upstream fluid, we would have

$$\frac{Q}{b} = z_2 \sqrt{2gz_1} = 0.488 \text{ m} \sqrt{2(9.81 \text{ m/s}^2)(5.0 \text{ m})}$$

$$= 4.83 \text{ m}^2/\text{s}$$

In this case the difference in Q with or without including V_1 is not too significant because the depth ratio is fairly large $(z_1/z_2 = 5.0/0.488 = 10.2)$. Thus, it is often reasonable to neglect the kinetic energy upstream from the gate compared to that downstream of it.

By repeating the calculations for various flow depths, z_1, the results shown in Fig. E3.12b are obtained. Note that the

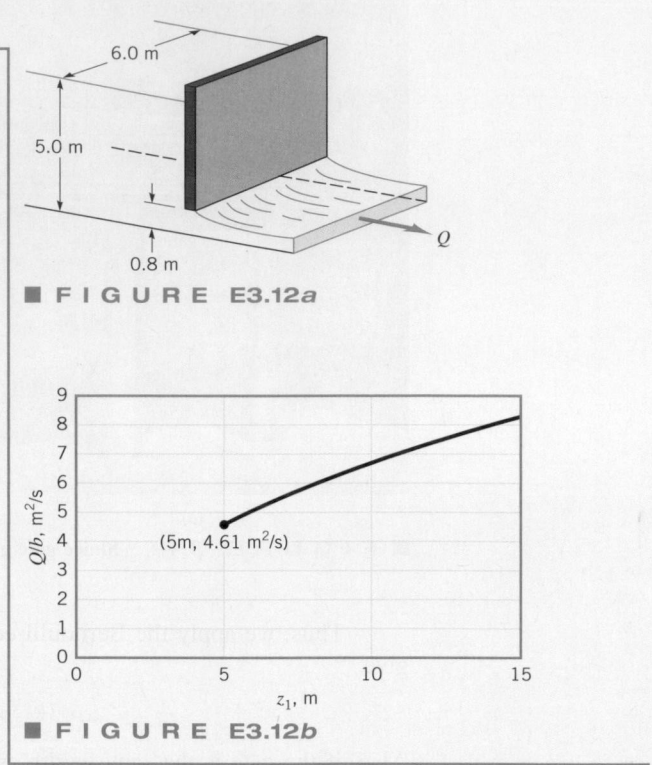

■ FIGURE E3.12a

■ FIGURE E3.12b

flowrate is not directly proportional to the flow depth. Thus, for example, if during flood conditions the upstream depth doubled from $z_1 = 5$ m to $z_1 = 10$ m, the flowrate per unit width of the channel would not double, but would increase only from 4.61 m²/s to 6.67 m²/s.

Another device used to measure flow in an open channel is a *weir*. A typical rectangular, sharp-crested weir is shown in Fig. 3.20. For such devices the flowrate of liquid over the top of the weir plate is dependent on the weir height, P_w, the width of the channel, b, and the head, H, of the water above the top of the weir. Application of the Bernoulli equation can provide a simple approximation of the flowrate expected for these situations, even though the actual flow is quite complex.

Between points (1) and (2) the pressure and gravitational fields cause the fluid to accelerate from velocity V_1 to velocity V_2. At (1) the pressure is $p_1 = \gamma h$, while at (2) the pressure is essentially atmospheric, $p_2 = 0$. Across the curved streamlines directly above the top of the weir plate (section a–a), the pressure changes from atmospheric on the top surface to some maximum value within the fluid stream and then to atmospheric again at the bottom surface. This distribution is indicated in Fig. 3.20. Such a pressure distribution, combined with the streamline curvature and gravity, produces a rather nonuniform velocity profile across this section. This velocity distribution can be obtained from experiments or a more advanced theory.

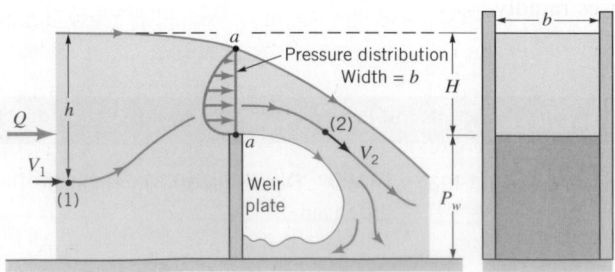

■ FIGURE 3.20
Rectangular, sharp-crested weir geometry.

For now, we will take a very simple approach and assume that the weir flow is similar in many respects to an orifice-type flow with a free streamline. In this instance we would expect the average velocity across the top of the weir to be proportional to $\sqrt{2gH}$ and the flow area for this rectangular weir to be proportional to Hb. Hence, it follows that

$$Q = C_1 Hb \sqrt{2gH} = C_1 b \sqrt{2g} \, H^{3/2}$$

where C_1 is a constant to be determined.

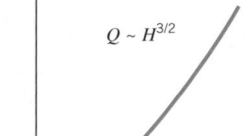

Simple use of the Bernoulli equation has provided a method to analyze the relatively complex flow over a weir. The correct functional dependence of Q on H has been obtained ($Q \sim H^{3/2}$, as indicated by the figure in the margin), but the value of the coefficient C_1 is unknown. Even a more advanced analysis cannot predict its value accurately. As is discussed in Chapter 10, experiments are used to determine the value of C_1.

EXAMPLE 3.13   Weir

GIVEN   Water flows over a triangular weir, as is shown in Fig. E3.13.

FIND   Based on a simple analysis using the Bernoulli equation, determine the dependence of the flowrate on the depth H. If the flowrate is Q_0 when $H = H_0$, estimate the flowrate when the depth is increased to $H = 3H_0$.

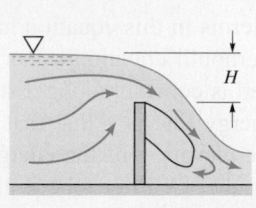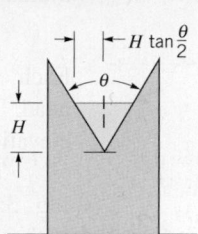

■ **FIGURE   E3.13**

SOLUTION

With the assumption that the flow is steady, inviscid, and incompressible, it is reasonable to assume from Eq. 3.18 that the average speed of the fluid over the triangular notch in the weir plate is proportional to $\sqrt{2gH}$. Also, the flow area for a depth of H is $H[H \tan (\theta/2)]$. The combination of these two ideas gives

$$Q = AV = H^2 \tan \frac{\theta}{2} (C_2 \sqrt{2gH}) = C_2 \tan \frac{\theta}{2} \sqrt{2g} \, H^{5/2} \quad \textbf{(Ans)}$$

where C_2 is an unknown constant to be determined experimentally.

Thus, an increase in the depth by a factor of three (from H_0 to $3H_0$) results in an increase of the flowrate by a factor of

$$\frac{Q_{3H_0}}{Q_{H_0}} = \frac{C_2 \tan(\theta/2) \sqrt{2g} \, (3H_0)^{5/2}}{C_2 \tan(\theta/2) \sqrt{2g} \, (H_0)^{5/2}}$$

$$= 15.6 \quad \textbf{(Ans)}$$

COMMENT   Note that for a triangular weir the flowrate is proportional to $H^{5/2}$, whereas for the rectangular weir discussed above, it is proportional to $H^{3/2}$. The triangular weir can be accurately used over a wide range of flowrates.

3.7   The Energy Line and the Hydraulic Grade Line

As was discussed in Section 3.4, the Bernoulli equation is actually an energy equation representing the partitioning of energy for an inviscid, incompressible, steady flow. The sum of the various energies of the fluid remains constant as the fluid flows from one section to another. A useful interpretation of the Bernoulli equation can be obtained through the use of the concepts of the *hydraulic grade line* (HGL) and the *energy line* (EL). These ideas represent a geometrical interpretation of a flow and can often be effectively used to better grasp the fundamental processes involved.

The hydraulic grade line and energy line are graphical forms of the Bernoulli equation.

For steady, inviscid, incompressible flow the total energy remains constant along a streamline. The concept of "head" was introduced by dividing each term in Eq. 3.7 by the specific weight, $\gamma = \rho g$, to give the Bernoulli equation in the following form

$$\frac{p}{\gamma} + \frac{V^2}{2g} + z = \text{constant on a streamline} = H \quad \textbf{(3.22)}$$

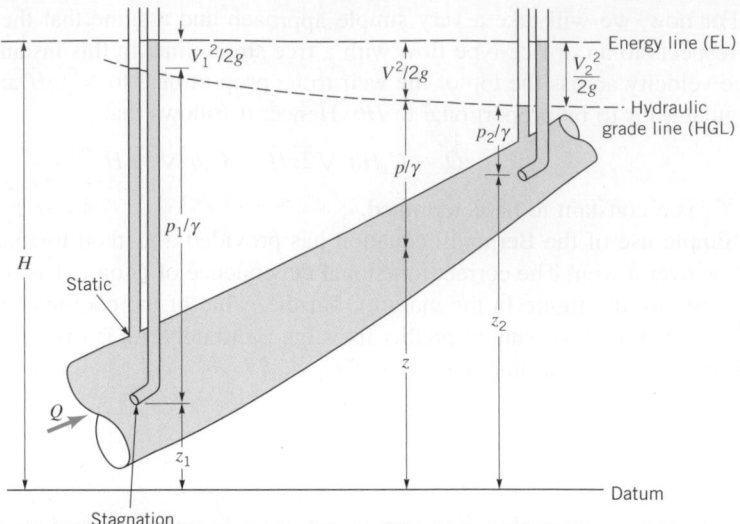

■ **F I G U R E   3.21**   **Representation of the energy line and the hydraulic grade line.**

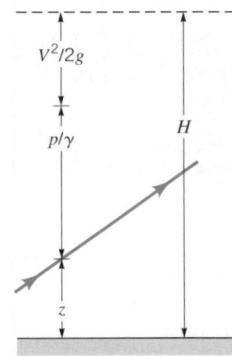

Each of the terms in this equation has the units of length (meters) and represents a certain type of head. The Bernoulli equation states that the sum of the pressure head, the velocity head, and the elevation head is constant along a streamline. This constant is called the *total head*, H.

The energy line is a line that represents the total head available to the fluid. As shown in Fig. 3.21, the elevation of the energy line can be obtained by measuring the stagnation pressure with a Pitot tube. (A Pitot tube is the portion of a Pitot-static tube that measures the stagnation pressure. See Section 3.5.) The stagnation point at the end of the Pitot tube provides a measurement of the total head (or energy) of the flow. The static pressure tap connected to the piezometer tube shown, on the other hand, measures the sum of the pressure head and the elevation head, $p/\gamma + z$. This sum is often called the *piezometric head*. The static pressure tap does not measure the velocity head.

According to Eq. 3.22, the total head remains constant along the streamline (provided the assumptions of the Bernoulli equation are valid). Thus, a Pitot tube at any other location in the flow will measure the same total head, as is shown in the figure. The elevation head, velocity head, and pressure head may vary along the streamline, however.

The locus of elevations provided by a series of Pitot tubes is termed the energy line, EL. The locus provided by a series of piezometer taps is termed the hydraulic grade line, HGL. Under the assumptions of the Bernoulli equation, the energy line is horizontal. If the fluid velocity changes along the streamline, the hydraulic grade line will not be horizontal. If viscous effects are important (as they often are in pipe flows), the total head does not remain constant due to a loss in energy as the fluid flows along its streamline. This means that the energy line is no longer horizontal. Such viscous effects are discussed in Chapters 5 and 8.

Under the assumptions of the Bernoulli equation, the energy line is horizontal.

The energy line and hydraulic grade line for flow from a large tank are shown in Fig. 3.22. If the flow is steady, incompressible, and inviscid, the energy line is horizontal and at the elevation of the liquid in the tank (since the fluid velocity in the tank and the pressure on the surface

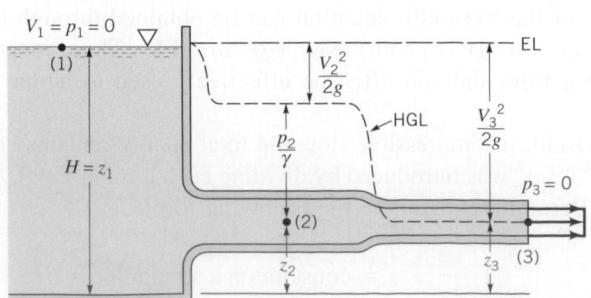

■ **F I G U R E   3.22**   **The energy line and hydraulic grade line for flow from a tank.**

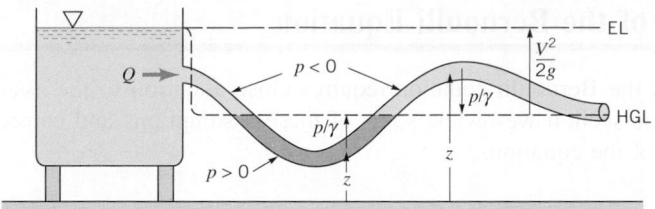

■ FIGURE 3.23
Use of the energy line and the hydraulic grade line.

are zero). The hydraulic grade line lies a distance of one velocity head, $V^2/2g$, below the energy line. Thus, a change in fluid velocity due to a change in the pipe diameter results in a change in the elevation of the hydraulic grade line. At the pipe outlet the pressure head is zero (gage) so the pipe elevation and the hydraulic grade line coincide.

For flow below (above) the hydraulic grade line, the pressure is positive (negative).

The distance from the pipe to the hydraulic grade line indicates the pressure within the pipe, as is shown in Fig. 3.23. If the pipe lies below the hydraulic grade line, the pressure within the pipe is positive (above atmospheric). If the pipe lies above the hydraulic grade line, the pressure is negative (below atmospheric). Thus, a scale drawing of a pipeline and the hydraulic grade line can be used to readily indicate regions of positive or negative pressure within a pipe.

EXAMPLE 3.14 Energy Line and Hydraulic Grade Line

GIVEN Water is siphoned from the tank shown in Fig. E3.14 through a hose of constant diameter. A small hole is found in the hose at location (1) as indicated.

FIND When the siphon is used, will water leak out of the hose, or will air leak into the hose, thereby possibly causing the siphon to malfunction?

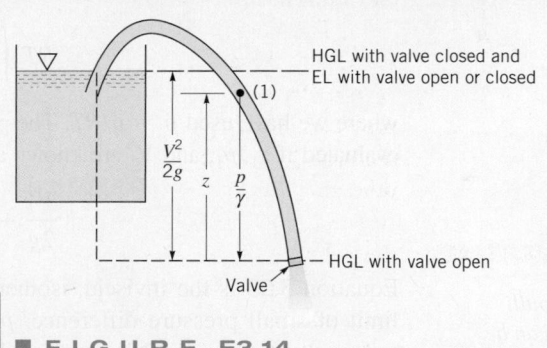

■ FIGURE E3.14

SOLUTION

Whether air will leak into or water will leak out of the hose depends on whether the pressure within the hose at (1) is less than or greater than atmospheric. Which happens can be easily determined by using the energy line and hydraulic grade line concepts. With the assumption of steady, incompressible, inviscid flow it follows that the total head is constant—thus, the energy line is horizontal.

Since the hose diameter is constant, it follows from the continuity equation (AV = constant) that the water velocity in the hose is constant throughout. Thus, the hydraulic grade line is a constant distance, $V^2/2g$, below the energy line as shown in Fig. E3.14. Since the pressure at the end of the hose is atmospheric, it follows that the hydraulic grade line is at the same elevation as the end of the hose outlet. The fluid within the hose at any point above the hydraulic grade line will be at less than atmospheric pressure.

Thus, air will leak into the hose through the hole at point (1). **(Ans)**

COMMENT In practice, viscous effects may be quite important, making this simple analysis (horizontal energy line) incorrect. However, if the hose is "not too small diameter," "not too long," the fluid "not too viscous," and the flowrate "not too large," the above result may be very accurate. If any of these assumptions are relaxed, a more detailed analysis is required (see Chapter 8). If the end of the hose were closed so that the flowrate were zero, the hydraulic grade line would coincide with the energy line ($V^2/2g = 0$ throughout), the pressure at (1) would be greater than atmospheric, and water would leak through the hole at (1).

The above discussion of the hydraulic grade line and the energy line is restricted to ideal situations involving inviscid, incompressible flows. Another restriction is that there are no "sources" or "sinks" of energy within the flow field. That is, there are no pumps or turbines involved. Alterations in the energy line and hydraulic grade line concepts due to these devices are discussed in Chapters 5 and 8.

3.8 Restrictions on Use of the Bernoulli Equation

Proper use of the Bernoulli equation requires close attention to the assumptions used in its derivation. In this section we review some of these assumptions and consider the consequences of incorrect use of the equation.

3.8.1 Compressibility Effects

One of the main assumptions is that the fluid is incompressible. Although this is reasonable for most liquid flows, it can, in certain instances, introduce considerable errors for gases.

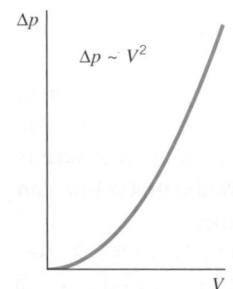

In the previous section, we saw that the stagnation pressure, p_{stag}, is greater than the static pressure, p_{static}, by an amount $\Delta p = p_{stag} - p_{static} = \rho V^2/2$, provided that the density remains constant. If this dynamic pressure is not too large compared with the static pressure, the density change between two points is not very large and the flow can be considered incompressible. However, since the dynamic pressure varies as V^2, the error associated with the assumption that a fluid is incompressible increases with the square of the velocity of the fluid, as indicated by the figure in the margin. To account for compressibility effects we must return to Eq. 3.6 and properly integrate the term $\int dp/\rho$ when ρ is not constant.

A simple, although specialized, case of compressible flow occurs when the temperature of a perfect gas remains constant along the streamline—isothermal flow. Thus, we consider $p = \rho RT$, where T is constant. (In general, p, ρ, and T will vary.) For steady, inviscid, isothermal flow, Eq. 3.6 becomes

$$RT \int \frac{dp}{p} + \frac{1}{2}V^2 + gz = \text{constant}$$

where we have used $\rho = p/RT$. The pressure term is easily integrated and the constant of integration evaluated if z_1, p_1, and V_1 are known at some location on the streamline. The result is

$$\frac{V_1^2}{2g} + z_1 + \frac{RT}{g}\ln\left(\frac{p_1}{p_2}\right) = \frac{V_2^2}{2g} + z_2 \tag{3.23}$$

The Bernoulli equation can be modified for compressible flows.

Equation 3.23 is the inviscid, isothermal analog of the incompressible Bernoulli equation. In the limit of small pressure difference, $p_1/p_2 = 1 + (p_1 - p_2)/p_2 = 1 + \varepsilon$, with $\varepsilon \ll 1$ and Eq. 3.23 reduces to the standard incompressible Bernoulli equation. This can be shown by use of the approximation $\ln(1 + \varepsilon) \approx \varepsilon$ for small ε. The use of Eq. 3.23 in practical applications is restricted by the inviscid flow assumption, since (as is discussed in Section 11.5) most isothermal flows are accompanied by viscous effects.

A much more common compressible flow condition is that of isentropic (constant entropy) flow of a perfect gas. Such flows are reversible adiabatic processes—"no friction or heat transfer"— and are closely approximated in many physical situations. As discussed fully in Chapter 11, for isentropic flow of a perfect gas the density and pressure are related by $p/\rho^k = C$, where k is the specific heat ratio and C is a constant. Hence, the $\int dp/\rho$ integral of Eq. 3.6 can be evaluated as follows. The density can be written in terms of the pressure as $\rho = p^{1/k}C^{-1/k}$ so that Eq. 3.6 becomes

$$C^{1/k} \int p^{-1/k}\, dp + \frac{1}{2}V^2 + gz = \text{constant}$$

The pressure term can be integrated between points (1) and (2) on the streamline and the constant C evaluated at either point ($C^{1/k} = p_1^{1/k}/\rho_1$ or $C^{1/k} = p_2^{1/k}/\rho_2$) to give the following:

$$C^{1/k}\int_{p1}^{p2} p^{-1/k}\, dp = C^{1/k}\left(\frac{k}{k-1}\right)[p_2^{(k-1)/k} - p_1^{(k-1)/k}]$$

$$= \left(\frac{k}{k-1}\right)\left(\frac{p_2}{\rho_2} - \frac{p_1}{\rho_1}\right)$$

Thus, the final form of Eq. 3.6 for compressible, isentropic, steady flow of a perfect gas is

$$\left(\frac{k}{k-1}\right)\frac{p_1}{\rho_1} + \frac{V_1^2}{2} + gz_1 = \left(\frac{k}{k-1}\right)\frac{p_2}{\rho_2} + \frac{V_2^2}{2} + gz_2 \tag{3.24}$$

The similarities between the results for compressible isentropic flow (Eq. 3.24) and incompressible isentropic flow (the Bernoulli equation, Eq. 3.7) are apparent. The only differences are the factors of $[k/(k-1)]$ that multiply the pressure terms and the fact that the densities are different ($\rho_1 \neq \rho_2$). In the limit of "low-speed flow" the two results are exactly the same, as is seen by the following.

We consider the stagnation point flow of Section 3.5 to illustrate the difference between the incompressible and compressible results. As is shown in Chapter 11, Eq. 3.24 can be written in dimensionless form as

$$\frac{p_2 - p_1}{p_1} = \left[\left(1 + \frac{k-1}{2}\,\mathrm{Ma}_1^2\right)^{k/k-1} - 1\right] \quad \text{(compressible)} \tag{3.25}$$

where (1) denotes the upstream conditions and (2) the stagnation conditions. We have assumed $z_1 = z_2$, $V_2 = 0$, and have denoted $\mathrm{Ma}_1 = V_1/c_1$ as the upstream *Mach number*—the ratio of the fluid velocity to the speed of sound, $c_1 = \sqrt{kRT_1}$.

A comparison between this compressible result and the incompressible result is perhaps most easily seen if we write the incompressible flow result in terms of the pressure ratio and the Mach number. Thus, we divide each term in the Bernoulli equation, $\rho V_1^2/2 + p_1 = p_2$, by p_1 and use the perfect gas law, $p_1 = \rho RT_1$, to obtain

$$\frac{p_2 - p_1}{p_1} = \frac{V_1^2}{2RT_1}$$

Since $\mathrm{Ma}_1 = V_1/\sqrt{kRT_1}$ this can be written as

$$\frac{p_2 - p_1}{p_1} = \frac{k\mathrm{Ma}_1^2}{2} \quad \text{(incompressible)} \tag{3.26}$$

Equations 3.25 and 3.26 are plotted in Fig. 3.24. In the low-speed limit of $\mathrm{Ma}_1 \to 0$, both of the results are the same. This can be seen by denoting $(k-1)\mathrm{Ma}_1^2/2 = \tilde{\varepsilon}$ and using the binomial expansion, $(1 + \tilde{\varepsilon})^n = 1 + n\tilde{\varepsilon} + n(n-1)\,\tilde{\varepsilon}^2/2 + \cdots$, where $n = k/(k-1)$, to write Eq. 3.25 as

$$\frac{p_2 - p_1}{p_1} = \frac{k\mathrm{Ma}_1^2}{2}\left(1 + \frac{1}{4}\,\mathrm{Ma}_1^2 + \frac{2-k}{24}\,\mathrm{Ma}_1^4 + \cdots\right) \quad \text{(compressible)}$$

For small Mach numbers the compressible and incompressible results are nearly the same.

For $\mathrm{Ma}_1 \ll 1$ this compressible flow result agrees with Eq. 3.26. The incompressible and compressible equations agree to within about 2% up to a Mach number of approximately $\mathrm{Ma}_1 = 0.3$. For larger Mach numbers the disagreement between the two results increases.

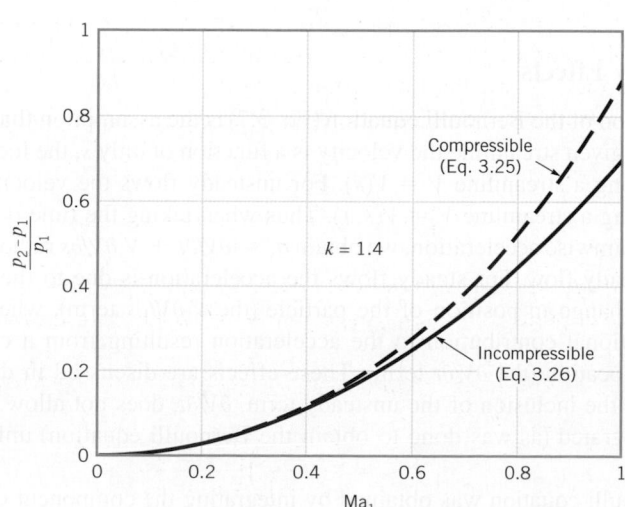

■ FIGURE 3.24   Pressure ratio as a function of Mach number for incompressible and compressible (isentropic) flow.

Thus, a "rule of thumb" is that the flow of a perfect gas may be considered as incompressible provided the Mach number is less than about 0.3. In standard air ($T_1 = 15\,°C$ (288 K), $c_1 = \sqrt{kRT_1} = 340$ m/s) this corresponds to a speed of $V_1 = Ma_1c_1 = 0.3(340$ m/s$) = 102$ m/s $= 367$ km/hr. At higher speeds, compressibility may become important.

EXAMPLE 3.15   Compressible Flow—Mach Number

GIVEN The jet shown in Fig. E3.15 flies at Mach 0.82 at an altitude of 10 km in a standard atmosphere.

FIND Determine the stagnation pressure on the leading edge of its wing if the flow is incompressible; and if the flow is compressible isentropic.

SOLUTION

From Tables 1.4 and C.1 we find that $p_1 = 26.5$ kPa (abs), $T_1 = -49.9\,°C$ $\rho = 0.414$ kg/m^3, and $k = 1.4$. Thus, if we assume incompressible flow, Eq. 3.26 gives

$$\frac{p_2 - p_1}{p_1} = \frac{kMa_1^2}{2} = 1.4\frac{(0.82)^2}{2} = 0.471$$

or

$$p_2 - p_1 = 0.471(26.5 \text{ kPa}) = 12.5 \text{ kPa} \qquad \textbf{(Ans)}$$

On the other hand, if we assume isentropic flow, Eq. 3.25 gives

$$\frac{p_2 - p_1}{p_1} = \left\{\left[1 + \frac{(1.4 - 1)}{2}(0.82)^2\right]^{1.4/(1.4-1)} - 1\right\}$$
$$= 0.555$$

or

$$p_2 - p_1 = 0.555(26.5 \text{ kPa}) = 14.7 \text{ kPa} \qquad \textbf{(Ans)}$$

COMMENT We see that at Mach 0.82 compressibility effects are of importance. The pressure (and, to a first approximation, the

■ **FIGURE E3.15** **(Photograph courtesy of Pure stock/superstock.)**

lift and drag on the airplane; see Chapter 9) is approximately $14.7/12.5 = 1.18$ times greater according to the compressible flow calculations. This may be very significant. As discussed in Chapter 11, for Mach numbers greater than 1 (supersonic flow) the differences between incompressible and compressible results are often not only quantitative but also qualitative.

Note that if the airplane were flying at Mach 0.30 (rather than 0.82) the corresponding values would be $p_2 - p_1 = 1.670$ kPa for incompressible flow and $p_2 - p_1 = 1.707$ kPa for compressible flow. The difference between these two results is about 2%.

3.8.2 Unsteady Effects

Another restriction of the Bernoulli equation (Eq. 3.7) is the assumption that the flow is steady. For such flows, on a given streamline the velocity is a function of only s, the location along the streamline. That is, along a streamline $V = V(s)$. For unsteady flows the velocity is also a function of time, so that along a streamline $V = V(s, t)$. Thus when taking the time derivative of the velocity to obtain the streamwise acceleration, we obtain $a_s = \partial V/\partial t + V\,\partial V/\partial s$ rather than just $a_s = V\,\partial V/\partial s$ as is true for steady flow. For steady flows the acceleration is due to the change in velocity resulting from a change in position of the particle (the $V\,\partial V/\partial s$ term), whereas for unsteady flow there is an additional contribution to the acceleration resulting from a change in velocity with time at a fixed location (the $\partial V/\partial t$ term). These effects are discussed in detail in Chapter 4. The net effect is that the inclusion of the unsteady term, $\partial V/\partial t$, does not allow the equation of motion to be easily integrated (as was done to obtain the Bernoulli equation) unless additional assumptions are made.

The Bernoulli equation can be modified for unsteady flows.

The Bernoulli equation was obtained by integrating the component of Newton's second law (Eq. 3.5) along the streamline. When integrated, the acceleration contribution to this equation, the

$\frac{1}{2}\rho d(V^2)$ term, gave rise to the kinetic energy term in the Bernoulli equation. If the steps leading to Eq. 3.5 are repeated with the inclusion of the unsteady effect ($\partial V/\partial t \neq 0$) the following is obtained:

$$\rho \frac{\partial V}{\partial t} ds + dp + \frac{1}{2}\rho d(V^2) + \gamma \, dz = 0 \qquad \text{(along a streamline)}$$

For incompressible flow this can be easily integrated between points (1) and (2) to give

$$p_1 + \frac{1}{2}\rho V_1^2 + \gamma z_1 = \rho \int_{s_1}^{s_2} \frac{\partial V}{\partial t} ds + p_2 + \frac{1}{2}\rho V_2^2 + \gamma z_2 \qquad \text{(along a streamline)} \qquad \textbf{(3.27)}$$

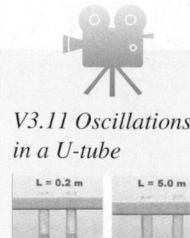

V3.11 Oscillations in a U-tube

L = 0.2 m   L = 5.0 m

Equation 3.27 is an unsteady form of the Bernoulli equation valid for unsteady, incompressible, inviscid flow. Except for the integral involving the local acceleration, $\partial V/\partial t$, it is identical to the steady Bernoulli equation. In general, it is not easy to evaluate this integral because the variation of $\partial V/\partial t$ along the streamline is not known. In some situations the concepts of "irrotational flow" and the "velocity potential" can be used to simplify this integral. These topics are discussed in Chapter 6.

EXAMPLE 3.16   Unsteady Flow—U-Tube

GIVEN   An incompressible, inviscid liquid is placed in a vertical, constant diameter U-tube as indicated in Fig. E3.16. When released from the nonequilibrium position shown, the liquid column will oscillate at a specific frequency.

FIND   Determine this frequency.

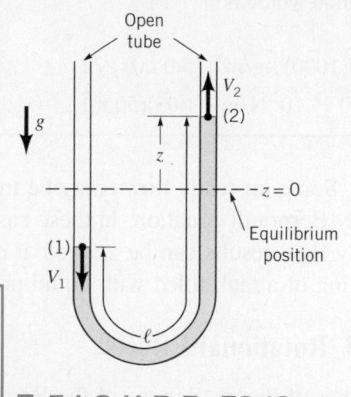

■ **FIGURE E3.16**

SOLUTION

The frequency of oscillation can be calculated by use of Eq. 3.27 as follows. Let points (1) and (2) be at the air–water interfaces of the two columns of the tube and $z = 0$ correspond to the equilibrium position of these interfaces. Hence, $p_1 = p_2 = 0$ and if $z_2 = z$, then $z_1 = -z$. In general, z is a function of time, $z = z(t)$. For a constant diameter tube, at any instant in time the fluid speed is constant throughout the tube, $V_1 = V_2 = V$, and the integral representing the unsteady effect in Eq. 3.27 can be written as

$$\int_{s_1}^{s_2} \frac{\partial V}{\partial t} ds = \frac{dV}{dt} \int_{s_1}^{s_2} ds = \ell \frac{dV}{dt}$$

where ℓ is the total length of the liquid column as shown in the figure. Thus, Eq. 3.27 can be written as

$$\gamma(-z) = \rho \ell \frac{dV}{dt} + \gamma z$$

Since $V = dz/dt$ and $\gamma = \rho g$, this can be written as the second-order differential equation describing simple harmonic motion

$$\frac{d^2 z}{dt^2} + \frac{2g}{\ell} z = 0$$

which has the solution $z(t) = C_1 \sin(\sqrt{2g/\ell}\,t) + C_2 \cos(\sqrt{2g/\ell}\,t)$. The values of the constants C_1 and C_2 depend on the initial state (velocity and position) of the liquid at $t = 0$. Thus, the liquid oscillates in the tube with a frequency

$$\omega = \sqrt{2g/\ell} \qquad \text{(Ans)}$$

COMMENT   This frequency depends on the length of the column and the acceleration of gravity (in a manner very similar to the oscillation of a pendulum). The period of this oscillation (the time required to complete an oscillation) is $t_0 = 2\pi \sqrt{\ell/2g}$.

In a few unsteady flow cases, the flow can be made steady by an appropriate selection of the coordinate system. Example 3.17 illustrates this.

EXAMPLE 3.17 Unsteady or Steady Flow

GIVEN A submarine moves through seawater ($SG = 1.03$) at a depth of 50 m with velocity $V_0 = 5.0$ m/s as shown in Fig. E3.17.

FIND Determine the pressure at the stagnation point (2).

SOLUTION

In a coordinate system fixed to the ground, the flow is unsteady. For example, the water velocity at (1) is zero with the submarine in its initial position, but at the instant when the nose, (2), reaches point (1) the velocity there becomes $\mathbf{V}_1 = -V_0\hat{\mathbf{i}}$. Thus, $\partial\mathbf{V}_1/\partial t \neq 0$ and the flow is unsteady. Application of the steady Bernoulli equation between (1) and (2) would give the incorrect result that "$p_1 = p_2 + \rho V_0^2/2$." According to this result the static pressure is greater than the stagnation pressure—an incorrect use of the Bernoulli equation.

We can either use an unsteady analysis for the flow (which is outside the scope of this text) or redefine the coordinate system so that it is fixed on the submarine, giving steady flow with respect to this system. The correct method would be

$$p_2 = \frac{\rho V_1^2}{2} + \gamma h = [(1.03)(1000) \text{ kg/m}^3] (5.0 \text{ m/s})^2/2$$
$$+ (9.80 \times 10^3 \text{ N/m}^3)(1.03)(50 \text{ m})$$

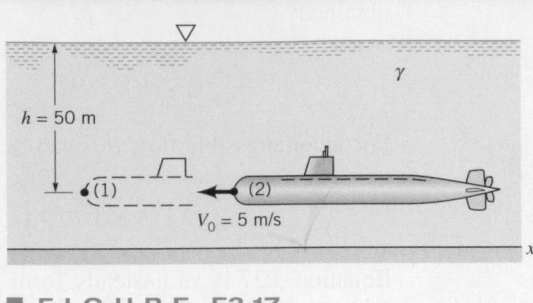

■ **FIGURE E3.17**

$$= (12,900 + 505,000) \text{ N/m}^2$$
$$= 518 \text{ kPa} \qquad \text{(Ans)}$$

similar to that discussed in Example 3.2.

COMMENT If the submarine were accelerating, $\partial V_0/\partial t \neq 0$, the flow would be unsteady in either of the above coordinate systems and we would be forced to use an unsteady form of the Bernoulli equation.

Some unsteady flows may be treated as "quasisteady" and solved approximately by using the steady Bernoulli equation. In these cases the unsteadiness is "not too great" (in some sense), and the steady flow results can be applied at each instant in time as though the flow were steady. The slow draining of a tank filled with liquid provides an example of this type of flow.

3.8.3 Rotational Effects

Care must be used in applying the Bernoulli equation across streamlines.

Another of the restrictions of the Bernoulli equation is that it is applicable along the streamline. Application of the Bernoulli equation across streamlines (i.e., from a point on one streamline to a point on another streamline) can lead to considerable errors, depending on the particular flow conditions involved. In general, the Bernoulli constant varies from streamline to streamline. However, under certain restrictions this constant is the same throughout the entire flow field. Example 3.18 illustrates this fact.

EXAMPLE 3.18 Use of Bernoulli Equation across Streamlines

GIVEN Consider the uniform flow in the channel shown in Fig. E3.18a. The liquid in the vertical piezometer tube is stationary.

FIND Discuss the use of the Bernoulli equation between points (1) and (2), points (3) and (4), and points (4) and (5).

SOLUTION

If the flow is steady, inviscid, and incompressible, Eq. 3.7 written between points (1) and (2) gives

$$p_1 + \tfrac{1}{2}\rho V_1^2 + \gamma z_1 = p_2 + \tfrac{1}{2}\rho V_2^2 + \gamma z_2$$
$$= \text{constant} = C_{12}$$

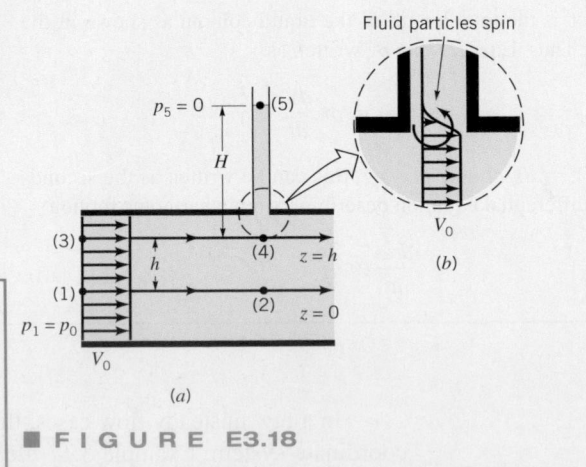

■ **FIGURE E3.18**

Since $V_1 = V_2 = V_0$ and $z_1 = z_2 = 0$, it follows that $p_1 = p_2 = p_0$ and the Bernoulli constant for this streamline, C_{12}, is given by

$$C_{12} = \tfrac{1}{2}\rho V_0^2 + p_0$$

Along the streamline from (3) to (4) we note that $V_3 = V_4 = V_0$ and $z_3 = z_4 = h$. As was shown in Example 3.5, application of $\mathbf{F} = m\mathbf{a}$ across the streamline (Eq. 3.12) gives $p_3 = p_1 - \gamma h$ because the streamlines are straight and horizontal. The above facts combined with the Bernoulli equation applied between (3) and (4) show that $p_3 = p_4$ and that the Bernoulli constant along this streamline is the same as that along the streamline between (1) and (2). That is, $C_{34} = C_{12}$, or

$$p_3 + \tfrac{1}{2}\rho V_3^2 + \gamma z_3 = p_4 + \tfrac{1}{2}\rho V_4^2 + \gamma z_4 = C_{34} = C_{12}$$

Similar reasoning shows that the Bernoulli constant is the same for any streamline in Fig. E3.18. Hence,

$$p + \tfrac{1}{2}\rho V^2 + \gamma z = \text{constant throughout the flow}$$

Again from Example 3.5 we recall that

$$p_4 = p_5 + \gamma H = \gamma H$$

If we apply the Bernoulli equation across streamlines from (4) to (5), we obtain the incorrect result "$H = p_4/\gamma + V_4^2/2g$." The correct result is $H = p_4/\gamma$.

From the above we see that we can apply the Bernoulli equation across streamlines (1)–(2) and (3)–(4) (i.e., $C_{12} = C_{34}$) but not across streamlines from (4) to (5). The reason for this is that while the flow in the channel is "irrotational," it is "rotational" between the flowing fluid in the channel and the stationary fluid in the piezometer tube. Because of the uniform velocity profile across the channel, it is seen that the fluid particles do not rotate or "spin" as they move. The flow is "irrotational." However, as seen in Fig. E3.18b, there is a very thin shear layer between (4) and (5) in which adjacent fluid particles interact and rotate or "spin." This produces a "rotational" flow. A more complete analysis would show that the Bernoulli equation cannot be applied across streamlines if the flow is "rotational" (see Chapter 6).

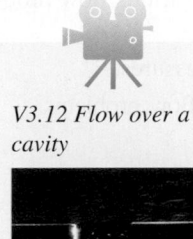

V3.12 Flow over a cavity

As is suggested by Example 3.18, if the flow is "irrotational" (i.e., the fluid particles do not "spin" as they move), it is appropriate to use the Bernoulli equation across streamlines. However, if the flow is "rotational" (fluid particles "spin"), use of the Bernoulli equation is restricted to flow along a streamline. The distinction between irrotational and rotational flow is often a very subtle and confusing one. These topics are discussed in more detail in Chapter 6. A thorough discussion can be found in more advanced texts (Ref. 3).

3.8.4 Other Restrictions

Another restriction on the Bernoulli equation is that the flow is inviscid. As is discussed in Section 3.4, the Bernoulli equation is actually a first integral of Newton's second law along a streamline. This general integration was possible because, in the absence of viscous effects, the fluid system considered was a conservative system. The total energy of the system remains constant. If viscous effects are important the system is nonconservative (dissipative) and energy losses occur. A more detailed analysis is needed for these cases. Such material is presented in Chapter 5.

The Bernoulli equation is not valid for flows that involve pumps or turbines.

The final basic restriction on use of the Bernoulli equation is that there are no mechanical devices (pumps or turbines) in the system between the two points along the streamline for which the equation is applied. These devices represent sources or sinks of energy. Since the Bernoulli equation is actually one form of the energy equation, it must be altered to include pumps or turbines, if these are present. The inclusion of pumps and turbines is covered in Chapters 5 and 12.

In this chapter we have spent considerable time investigating fluid dynamic situations governed by a relatively simple analysis for steady, inviscid, incompressible flows. Many flows can be adequately analyzed by use of these ideas. However, because of the rather severe restrictions imposed, many others cannot. An understanding of these basic ideas will provide a firm foundation for the remainder of the topics in this book.

3.9   Chapter Summary and Study Guide

In this chapter, several aspects of the steady flow of an inviscid, incompressible fluid are discussed. Newton's second law, $\mathbf{F} = m\mathbf{a}$, is applied to flows for which the only important forces are those due to pressure and gravity (weight)—viscous effects are assumed negligible. The result is the often-used Bernoulli equation, which provides a simple relationship among pressure, elevation, and velocity variations along a streamline. A similar but less often used equation is also obtained to describe the variations in these parameters normal to a streamline.

The concept of a stagnation point and the corresponding stagnation pressure is introduced as are the concepts of static, dynamic, and total pressure and their related heads.

Several applications of the Bernoulli equation are discussed. In some flow situations, such as the use of a Pitot-static tube to measure fluid velocity or the flow of a liquid as a free jet from a tank, a Bernoulli equation alone is sufficient for the analysis. In other instances, such as confined flows in tubes and flow meters, it is necessary to use both the Bernoulli equation and the continuity equation, which is a statement of the fact that mass is conserved as fluid flows.

The following checklist provides a study guide for this chapter. When your study of the entire chapter and end-of-chapter exercises has been completed, you should be able to

- write out meanings of the terms listed here in the margin and understand each of the related concepts. These terms are particularly important and are set in *italic, bold, and color* type in the text.

- explain the origin of the pressure, elevation, and velocity terms in the Bernoulli equation and how they are related to Newton's second law of motion.

- apply the Bernoulli equation to simple flow situations, including Pitot-static tubes, free jet flows, confined flows, and flow meters.

- use the concept of conservation of mass (the continuity equation) in conjunction with the Bernoulli equation to solve simple flow problems.

- apply Newton's second law across streamlines for appropriate steady, inviscid, incompressible flows.

- use the concepts of pressure, elevation, velocity, and total heads to solve various flow problems.

- explain and use the concepts of static, stagnation, dynamic, and total pressures.

- use the energy line and the hydraulic grade line concepts to solve various flow problems.

- explain the various restrictions on use of the Bernoulli equation.

Some of the important equations in this chapter are:

Streamwise and normal acceleration	$a_s = V \dfrac{\partial V}{\partial s}, \quad a_n = \dfrac{V^2}{\mathcal{R}}$	**(3.1)**
Force balance along a streamline for steady inviscid flow	$\displaystyle\int \dfrac{dp}{\rho} + \dfrac{1}{2} V^2 + gz = C \quad \text{(along a streamline)}$	**(3.6)**
The Bernoulli equation	$p + \tfrac{1}{2}\rho V^2 + \gamma z = \text{constant along streamline}$	**(3.7)**
Pressure gradient normal to streamline for inviscid flow in absence of gravity	$\dfrac{\partial p}{\partial n} = -\dfrac{\rho V^2}{\mathcal{R}}$	**(3.10b)**
Force balance normal to a streamline for steady, inviscid, incompressible flow	$p + \rho \displaystyle\int \dfrac{V^2}{\mathcal{R}}\, dn + \gamma z = \text{constant across the streamline}$	**(3.12)**
Velocity measurement for a Pitot-static tube	$V = \sqrt{2\,(p_3 - p_4)/\rho}$	**(3.16)**
Free jet	$V = \sqrt{2\,\dfrac{\gamma h}{\rho}} = \sqrt{2gh}$	**(3.18)**
Continuity equation	$A_1 V_1 = A_2 V_2, \text{ or } Q_1 = Q_2$	**(3.19)**
Flow meter equation	$Q = A_2 \sqrt{\dfrac{2(p_1 - p_2)}{\rho[1 - (A_2/A_1)^2]}}$	**(3.20)**
Sluice gate equation	$Q = z_2 b \sqrt{\dfrac{2g(z_1 - z_2)}{1 - (z_2/z_1)^2}}$	**(3.21)**
Total head	$\dfrac{p}{\gamma} + \dfrac{V^2}{2g} + z = \text{constant on a streamline} = H$	**(3.22)**

References

1. Riley, W. F., and Sturges, L. D., *Engineering Mechanics: Dynamics,* 2nd Ed., Wiley, New York, 1996.
2. Tipler, P. A., *Physics,* Worth, New York, 1982.
3. Panton, R. L., *Incompressible Flow,* Wiley, New York, 1984.

Review Problems

Go to Appendix G for a set of review problems with answers. Detailed solutions can be found in *Student Solution Manual and Study* *Guide for Fundamentals of Fluid Mechanics,* by Munson et al. (© 2009 John Wiley and Sons, Inc.).

Problems

Note: Unless otherwise indicated, use the values of fluid properties found in the tables on the inside of the front cover. Problems designated with an (*) are intended to be solved with the aid of a programmable calculator or a computer. Problems designated with a (†) are "open-ended" problems and require critical thinking in that to work them one must make various assumptions and provide the necessary data. There is not a unique answer to these problems.

Answers to the even-numbered problems are listed at the end of the book. Access to the videos that accompany problems can be obtained through the book's web site, www.wiley.com/college/munson. The lab-type problems can also be accessed on this web site.

Section 3.2 F = *m*a along a Streamline

3.1 Obtain a photograph/image of a situation which can be analyzed by use of the Bernoulli equation. Print this photo and write a brief paragraph that describes the situation involved.

3.2 Air flows steadily along a streamline from point (1) to point (2) with negligible viscous effects. The following conditions are measured: At point (1) $z_1 = 2$ m and $p_1 = 0$ kPa; at point (2) $z_2 = 10$ m, $p_2 = 20$ N/m^2, and $V_2 = 0$. Determine the velocity at point (1).

3.3 Water flows steadily through the variable area horizontal pipe shown in Fig. P3.3. The centerline velocity is given by $\mathbf{V} = 3(1 + 3x)\,\hat{\mathbf{i}}$ m/s, where x is in meters. Viscous effects are neglected. **(a)** Determine the pressure gradient, $\partial p/\partial x$, (as a function of x) needed to produce this flow. **(b)** If the pressure at section (1) is 345 kPa, determine the pressure at (2) by (i) integration of the pressure gradient obtained in **(a)**, (ii) application of the Bernoulli equation.

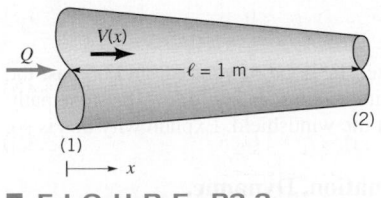

■ **FIGURE P3.3**

3.4 Repeat Problem 3.3 if the pipe is vertical with the flow down.

3.5 An incompressible fluid with density ρ flows steadily past the object shown in **Video V3.7** and Fig. P3.5. The fluid velocity along the horizontal dividing streamline $(-\infty \le x \le -a)$ is found to be $V = V_0(1 + a/x)$, where a is the radius of curvature of the front of the object and V_0 is the upstream velocity. **(a)** Determine the pressure gradient along this streamline. **(b)** If the upstream pressure is p_0, integrate the pressure gradient to obtain the pressure $p(x)$ for $-\infty \le x \le -a$. **(c)** Show from the result of part **(b)** that the pressure at the stagnation point $(x = -a)$ is $p_0 + \rho V_0^2/2$, as expected from the Bernoulli equation.

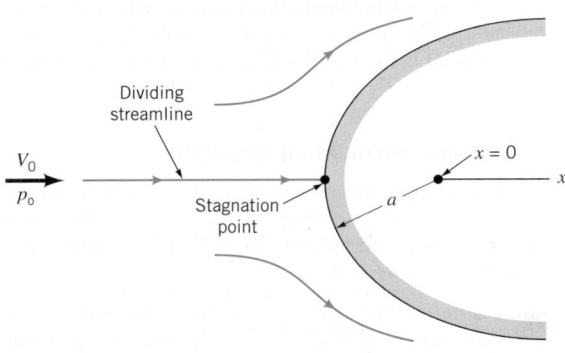

■ **FIGURE P3.5**

3.6 What pressure gradient along the streamline, dp/ds, is required to accelerate water in a horizontal pipe at a rate of 30 m/s^2?

3.7 A fluid with a specific weight of 15.7 kN/m^3 and negligible viscous effects flows in the pipe shown in Fig. P3.7. The pressures at points (1) and (2) are 19 kPa and 43 kPa, respectively. The velocities at points (1) and (2) are equal. Is the fluid accelerating uphill, downhill, or not accelerating? Explain.

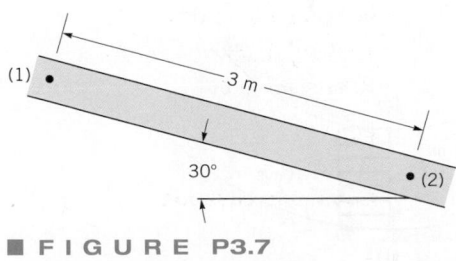

■ **FIGURE P3.7**

3.8 What pressure gradient along the streamline, dp/ds, is required to accelerate water upward in a vertical pipe at a rate of 9 m/s^2? What is the answer if the flow is downward?

3.9 Consider a compressible fluid for which the pressure and density are related by $p/\rho^n = C_0$, where n and C_0 are constants. Integrate the equation of motion along the streamline, Eq. 3.6, to obtain the "Bernoulli equation" for this compressible flow as $[n/(n-1)]p/\rho + V^2/2 + gz = \text{constant}$.

3.10 An incompressible fluid flows steadily past a circular cylinder as shown in Fig. P3.10. The fluid velocity along the dividing streamline $(-\infty \le x \le -a)$ is found to be $V = V_0(1 - a^2/x^2)$, where a is the radius of the cylinder and V_0 is the upstream velocity. (a) Determine the pressure gradient along this streamline. (b) If the upstream pressure is p_0, integrate the pressure gradient to obtain the pressure $p(x)$ for $-\infty \le x \le -a$. (c) Show from the result of part (b) that the pressure at the stagnation point $(x = -a)$ is $p_0 + \rho V_0^2/2$, as expected from the Bernoulli equation.

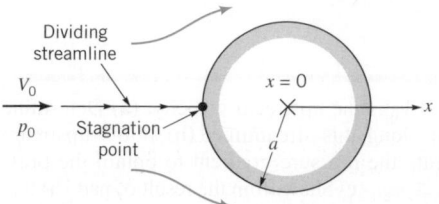

■ **FIGURE P3.10**

3.11 Consider a compressible liquid that has a constant bulk modulus. Integrate "$\mathbf{F} = m\mathbf{a}$" along a streamline to obtain the equivalent of the Bernoulli equation for this flow. Assume steady, inviscid flow.

Section 3.3 F = ma Normal to a Streamline

3.12 Obtain a photograph/image of a situation in which Newton's second law applied across the streamlines (as given by Eq. 3.12) is important. Print this photo and write a brief paragrph that describes the situation involved.

3.13 Air flows along a horizontal, curved streamline with a 6 m radius with a speed of 30 m/s. Determine the pressure gradient normal to the streamline.

3.14 Water flows around the vertical two-dimensional bend with circular streamlines and constant velocity as shown in Fig. P3.14. If the pressure is 40 kPa at point (1), determine the pressures at points (2) and (3). Assume that the velocity profile is uniform as indicated.

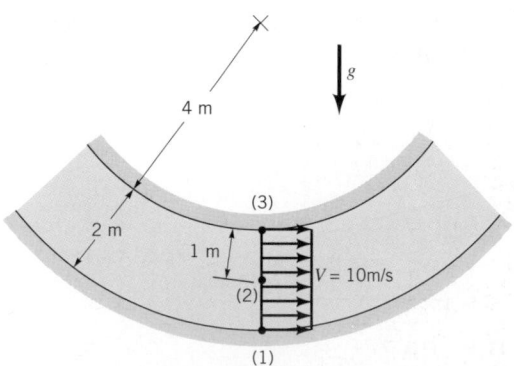

■ **FIGURE P3.14**

***3.15** Water flows around the vertical two-dimensional bend with circular streamlines as is shown in Fig. P3.15. The pressure at point (1) is measured to be $p_1 = 172$ kPa and the velocity across section a–a is as indicated in the table. Calculate and plot the pressure across section a–a of the channel [$p = p(z)$ for $0 \le z \le 0.6$ m].

z (m)	V (m/s)
0	0
0.06	2.4
0.12	4.4
0.18	6.0
0.24	5.9
0.30	4.8
0.36	2.5
0.42	1.9
0.48	1.1
0.54	0.6
0.60	0

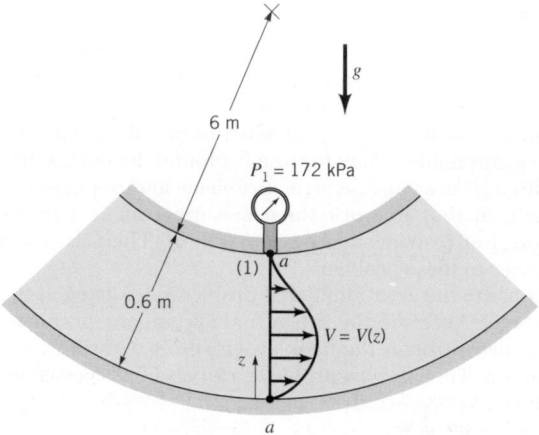

■ **FIGURE P3.15**

3.16 Water in a container and air in a tornado flow in horizontal circular streamlines of radius r and speed V as shown in Video V3.6 and Fig. P3.16. Determine the radial pressure gradient, $\partial p/\partial r$, needed for the following situations: (a) The fluid is water with $r = 7.6$ cm and $V = 0.24$ m/s. (b) The fluid is air with $r = 91$ m and $V = 322$ km/h.

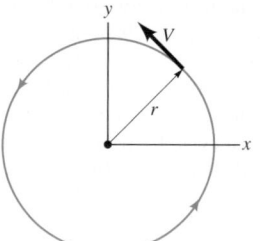

■ **FIGURE P3.16**

3.17 Air flows smoothly over the hood of your car and up past the windshield. However, a bug in the air does not follow the same path; it becomes splattered against the windshield. Explain why this is so.

Section 3.5 Static, Stagnation, Dynamic, and Total Pressure

3.18 Obtain a photograph/image of a situation in which the concept of the stagnation pressure is important. Print this photo and write a brief paragraph that describes the situation involved.

3.19 At a given point on a horizontal streamline in flowing air, the static pressure is -14 kPa (i.e., a vacuum) and the velocity is 46 m/s. Determine the pressure at a stagnation point on that streamline.

†3.20 Estimate the maximum pressure on the surface of your car when you wash it using a garden hose connected to your outside faucet. List all assumptions and show calculations.

3.21 When an airplane is flying 322 km/h at 2000 m altitude in a standard atmosphere, the air velocity at a certain point on the wing is 439 km/h relative to the airplane. **(a)** What suction pressure is developed on the wing at that point? **(b)** What is the pressure at the leading edge (a stagnation point) of the wing?

3.22 Some animals have learned to take advantage of Bernoulli effect without having read a fluid mechanics book. For example, a typical prairie dog burrow contains two entrances—a flat front door, and a mounded back door as shown in Fig. P3.22. When the wind blows with velocity V_0 across the front door, the average velocity across the back door is greater than V_0 because of the mound. Assume the air velocity across the back door is $1.07V_0$. For a wind velocity of 6 m/s, what pressure differences, $p_1 - p_2$, are generated to provide a fresh air flow within the burrow?

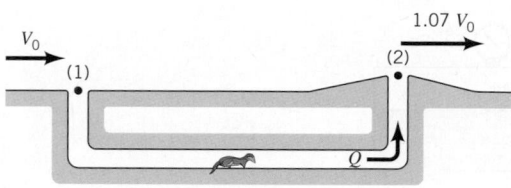

■ **FIGURE P3.22**

3.23 A loon is a diving bird equally at home "flying" in the air or water. What swimming velocity under water will produce a dynamic pressure equal to that when it flies in the air at 64 km/h?

3.24 A person thrusts his hand into the water while traveling 3 m/s in a motorboat. What is the maximum pressure on his hand?

3.25 A Pitot-static tube is used to measure the velocity of helium in a pipe. The temperature and pressure are 4 °C and 172 kPa (abs). A water manometer connected to the Pitot-static tube indicates a reading of 5.8 cm. Determine the helium velocity. Is it reasonable to consider the flow as incompressible? Explain.

3.26 An inviscid fluid flows steadily along the stagnation streamline shown in Fig. P3.26 and **Video V3.7**, starting with speed V_0 far upstream of the object. Upon leaving the stagnation point, point (1), the fluid speed along the surface of the object is assumed to be given by $V = 2 V_0 \sin \theta$, where θ is the angle indicated. At what angular position, θ_2, should a hole be drilled to give a pressure difference of $p_1 - p_2 = \rho V_0^2 / 2$? Gravity is negligible.

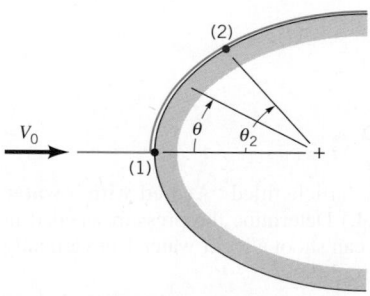

■ **FIGURE P3.26**

3.27 A water-filled manometer is connected to a Pitot-static tube to measure a nominal airspeed of 15 m/s. It is assumed that a change in the manometer reading of 0.05 mm can be detected. What is the minimum deviation from the 15 m/s airspeed that can be detected by this system? Repeat the problem if the nominal airspeed is 1.5 m/s.

3.28 (See Fluids in the News article titled "**Incorrect raindrop shape**," Section 3.2.) The speed, V, at which a raindrop falls is a function of its diameter, D, as shown in Fig. P3.28. For what sized raindrop will the stagnation pressure be equal to half the internal pressure caused by surface tension? Recall from Section 1.9 that the pressure inside a drop is $\Delta p = 4\sigma/D$ greater than the surrounding pressure, where σ is the surface tension.

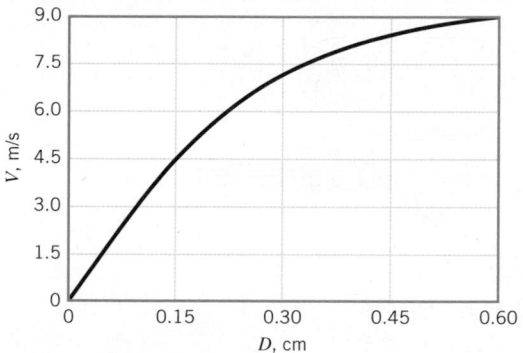

■ **FIGURE P3.28**

3.29 (See Fluids in the News article titled "**Pressurized eyes**," Section 3.5.) Determine the air velocity needed to produce a stagnation pressure equal to 10 mm of mercury.

3.30 (See Fluids in the News article titled "**Bugged and plugged Pitot tubes**," Section 3.5.) An airplane's Pitot tube used to indicate airspeed is partially plugged by an insect nest so that it measures 60% of the stagnation pressure rather than the actual stagnation pressure. If the airspeed indicator indicates that the plane is flying 240 km/h, what is the actual airspeed?

Section 3.6.1 Free Jets

3.31 Obtain a photograph/image of a situation in which the concept of a free jet is important. Print this photo and write a brief paragraph that describes the situation involved.

3.32 Water flows through a hole in the bottom of a large, open tank with a speed of 8 m/s. Determine the depth of water in the tank. Viscous effects are negligible.

3.33 Water flows from the faucet on the first floor of the building shown in Fig. P3.33 with a maximum velocity of 6 m/s. For steady

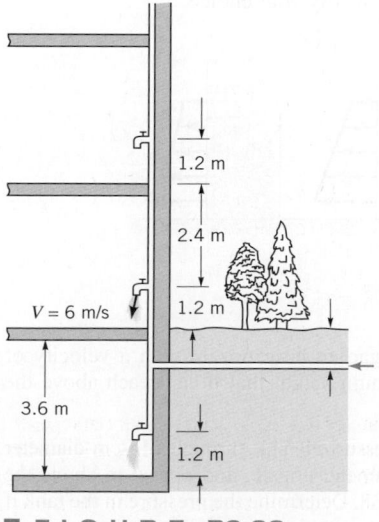

■ **FIGURE P3.33**

inviscid flow, determine the maximum water velocity from the basement faucet and from the faucet on the second floor (assume each floor is 3.6 m tall).

†**3.34** The "super soaker" water gun shown in Fig. P3.34 can shoot more than 9 m in the horizontal direction. Estimate the minimum pressure, p_1, needed in the chamber in order to accomplish this. List all assumptions and show all calculations.

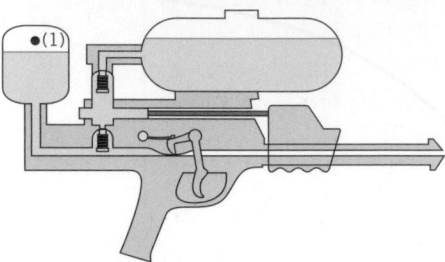

■ **F I G U R E P3.34**

3.35* An inviscid liquid drains from a large tank through a square duct of width b as shown in Fig. P3.35. The velocity of the fluid at the outlet is not precisely uniform because of the difference in elevation across the outlet. If $b \ll h$, this difference in velocity is negligible. For given b and h, determine v as a function of x and integrate the results to determine the average velocity, $V = Q/b^2$. Plot the velocity distribution, $v = v(x)$, across the outlet if $h = 1$ and $b = 0.1$, 0.2, 0.4, 0.6, 0.8, and 1.0 m. How small must b be if the centerline velocity, v at $x = b/2$, is to be within 3% of the average velocity?

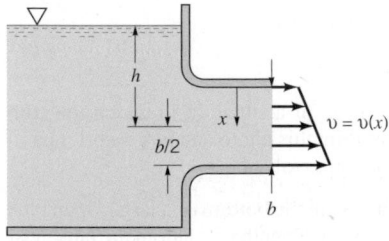

■ **F I G U R E P3.35**

3.36 Several holes are punched into a tin can as shown in Fig. P3.36. Which of the figures represents the variation of the water velocity as it leaves the holes? Justify your choice.

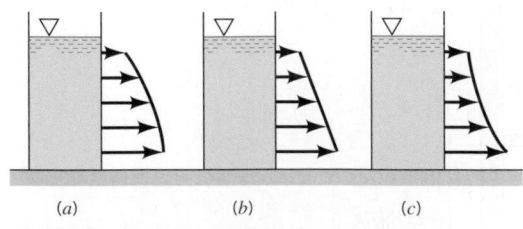

■ **F I G U R E P3.36**

3.37 Water flows from a garden hose nozzle with a velocity of 15 m/s. What is the maximum height that it can reach above the nozzle?

3.38 Water flows from a pressurized tank, through a 15-cm-diameter pipe, exits from a 5-cm-diameter nozzle, and rises 6 m above the nozzle as shown in Fig. P3.38. Determine the pressure in the tank if the flow is steady, frictionless, and incompressible.

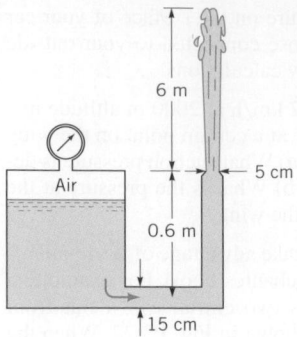

■ **F I G U R E P3.38**

3.39 An inviscid, incompressible liquid flows steadily from the large pressurized tank shown in Fig. P.3.39. The velocity at the exit is 12 m/s. Determine the specific gravity of the liquid in the tank.

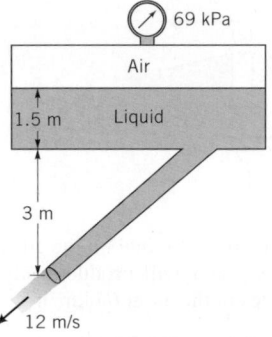

■ **F I G U R E P3.39**

3.40 Water flows from the tank shown in Fig. P3.40. If viscous effects are negligible, determine the value of h in terms of H and the specific gravity, SG, of the manometer fluid.

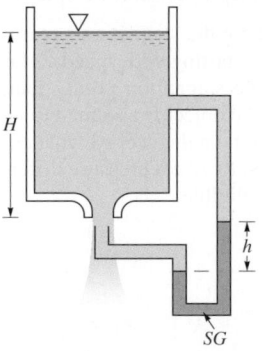

■ **F I G U R E P3.40**

3.41 (See Fluids in the News article titled "**Armed with a water jet for hunting,**" Section 3.4.) Determine the pressure needed in the gills of an archerfish if it can shoot a jet of water 1 m vertically upward. Assume steady, inviscid flow.

Section 3.6.2 Confined Flows (Also see Lab Problems 3.118 and 3.120.)

3.42 Obtain a photograph/image of a situation that involves a confined flow for which the Bernoulli and continuity equations are important. Print this photo and write a brief paragraph that describes the situation involved.

3.43 Air flows steadily through a horizontal 10-cm-diameter pipe and exits into the atmosphere through a 7.0-cm-diameter nozzle. The velocity at the nozzle exit is 46 m/s. Determine the pressure in the pipe if viscous effects are negligible.

3.44 A fire hose nozzle has a diameter of 2.86 cm. According to some fire codes, the nozzle must be capable of delivering at least 946 L/min. If the nozzle is attached to a 7.6-cm-diameter hose, what pressure must be maintained just upstream of the nozzle to deliver this flowrate?

3.45 Water flowing from the 1.9-cm-diameter outlet shown in **Video V8.14** and Fig. P3.45 rises 7 cm above the outlet. Determine the flowrate.

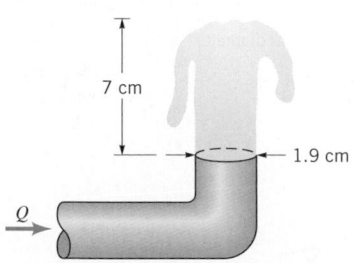

■ **F I G U R E P3.45**

3.46 Pop (with the same properties as water) flows from a 10-cm-diameter pop container that contains three holes as shown in Fig. P3.46 (see **Video 3.9**). The diameter of each fluid stream is 0.4 cm, and the distance between holes is 5 cm. If viscous effects are negligible and quasi-steady conditions are assumed, determine the time at which the pop stops draining from the top hole. Assume the pop surface is 5 cm above the top hole when $t = 0$. Compare your results with the time you measure from the video.

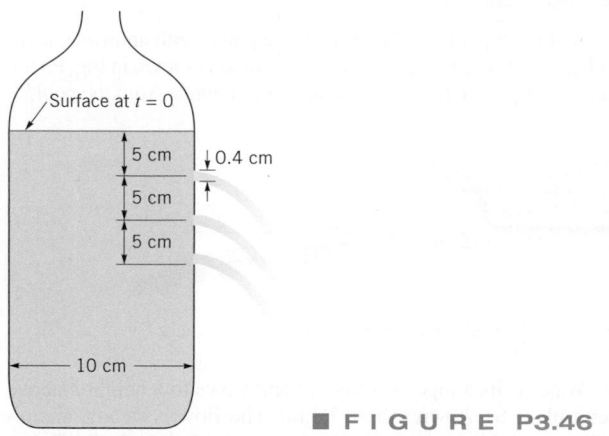

■ **F I G U R E P3.46**

3.47 Water (assumed inviscid and incompressible) flows steadily in the vertical variable-area pipe shown in Fig. P3.47. Determine the flowrate if the pressure in each of the gages reads 50 kPa.

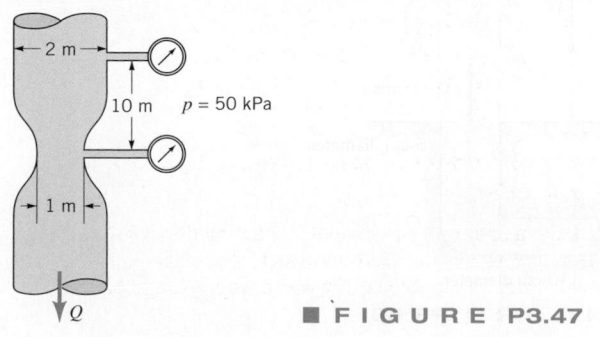

■ **F I G U R E P3.47**

3.48 Air is drawn into a wind tunnel used for testing automobiles as shown in Fig. P3.48. **(a)** Determine the manometer reading, h, when the velocity in the test section is 97 km/h. Note that there is a 2.5-cm column of oil on the water in the manometer. **(b)** Determine the difference between the stagnation pressure on the front of the automobile and the pressure in the test section.

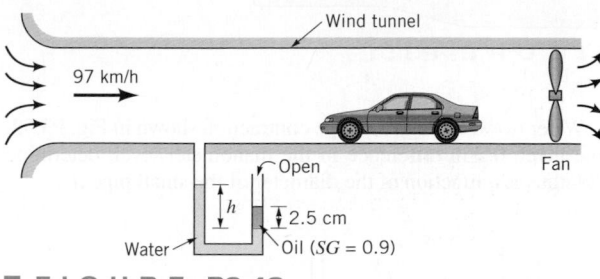

■ **F I G U R E P3.48**

3.49 Small-diameter, high-pressure liquid jets can be used to cut various materials as shown in Fig. P3.49. If viscous effects are negligible, estimate the pressure needed to produce a 0.10-mm-diameter water jet with a speed of 700 m/s. Determine the flowrate.

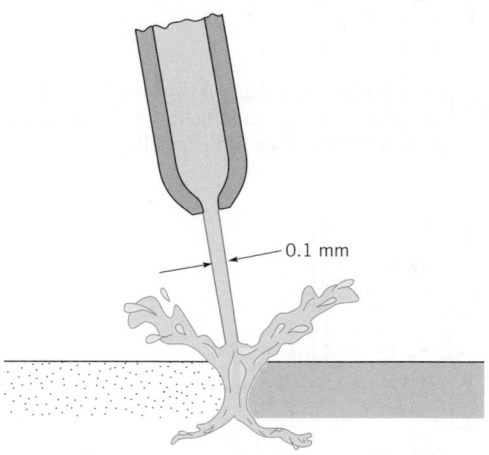

■ **F I G U R E P3.49**

3.50 Water (assumed inviscid and incompressible) flows steadily with a speed of 3 m/s from the large tank shown in Fig. P3.50. Determine the depth, H, of the layer of light liquid (specific weight = 7.85 kN/m^3) that covers the water in the tank.

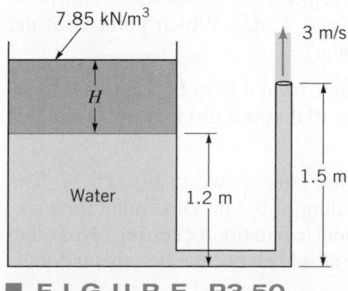

■ **F I G U R E P3.50**

3.51 Water flows through the pipe contraction shown in Fig. P3.51. For the given 0.2-m difference in manometer level, determine the flowrate as a function of the diameter of the small pipe, D.

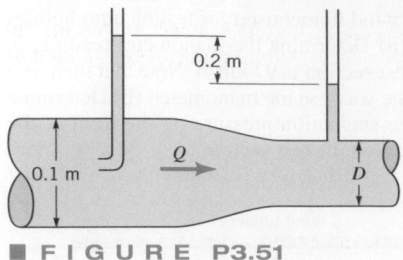

■ **FIGURE P3.51**

3.52 Water flows through the pipe contraction shown in Fig. P3.52. For the given 0.2-m difference in the manometer level, determine the flowrate as a function of the diameter of the small pipe, D.

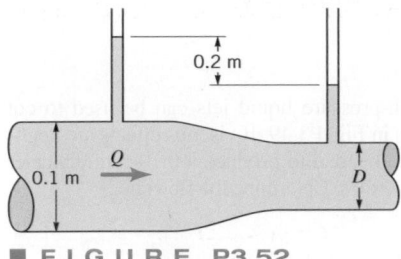

■ **FIGURE P3.52**

3.53 Water flows through the pipe contraction shown in Fig. P3.53. For the given 0.2-m difference in the manometer level, determine the flowrate as a function of the diameter of the small pipe, D.

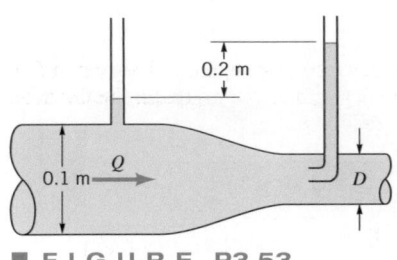

■ **FIGURE P3.53**

3.54 A 0.15-m-diameter pipe discharges into a 0.10-m-diameter pipe. Determine the velocity head in each pipe if they are carrying 0.12 m³/s of kerosene.

3.55 Carbon tetrachloride flows in a pipe of variable diameter with negligible viscous effects. At point A in the pipe the pressure and velocity are 138 kPa and 9 m/s, respectively. At location B the pressure and velocity are 159 kPa and 4 m/s. Which point is at the higher elevation and by how much?

3.56 The circular stream of water from a faucet is observed to taper from a diameter of 20 mm to 10 mm in a distance of 50 cm. Determine the flowrate.

3.57 Water is siphoned from the tank shown in Fig. P3.57. The water barometer indicates a reading of 9.2 m. Determine the maximum value of h allowed without cavitation occurring. Note that the pressure of the vapor in the closed end of the barometer equals the vapor pressure.

3.58 As shown in Fig. P3.58, water from a large reservoir flows without viscous effects through a siphon of diameter D and into a tank. It exits from a hole in the bottom of the tank as a stream of diameter d. The surface of the reservoir remains H above the bottom

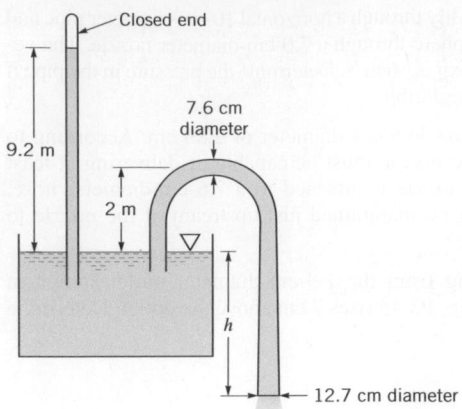

■ **FIGURE P3.57**

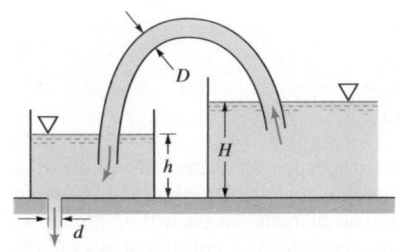

■ **FIGURE P3.58**

of the tank. For steady-state conditions, the water depth in the tank, h, is constant. Plot a graph of the depth ratio h/H as a function of the diameter ratio d/D.

3.59 A smooth plastic, 10-m-long garden hose with an inside diameter of 20 mm is used to drain a wading pool as is shown in Fig. P3.59. If viscous effects are neglected, what is the flowrate from the pool?

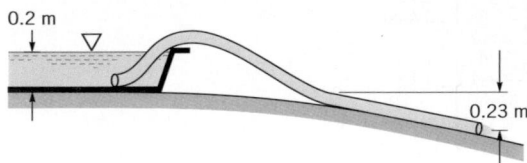

■ **FIGURE P3.59**

3.60 Water exits a pipe as a free jet and flows to a height h above the exit plane as shown in Fig. P3.60. The flow is steady, incompressible, and frictionless. **(a)** Determine the height h. **(b)** Determine the velocity and pressure at section (1).

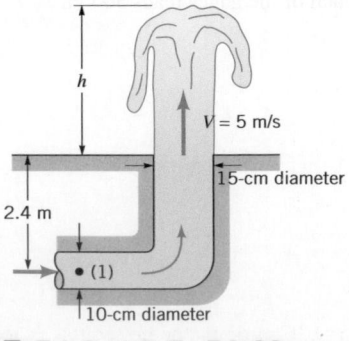

■ **FIGURE P3.60**

3.61 Water flows steadily from a large, closed tank as shown in Fig. P3.61. The deflection in the mercury manometer is 2.5 cm and viscous effects are negligible. **(a)** Determine the volume flowrate. **(b)** Determine the air pressure in the space above the surface of the water in the tank.

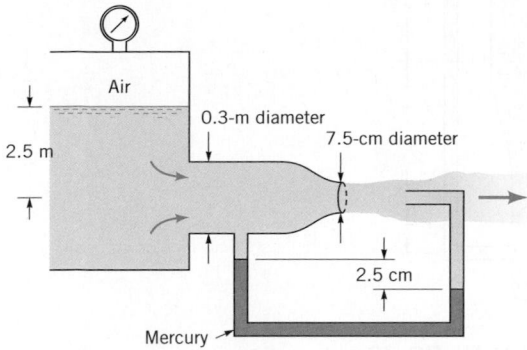

■ **F I G U R E   P3.61**

3.62 Blood ($SG = 1$) flows with a velocity of 0.5 m/s in an artery. It then enters an aneurysm in the artery (i.e., an area of weakened and stretched artery walls that cause a ballooning of the vessel) whose cross-sectional area is 1.8 times that of the artery. Determine the pressure difference between the blood in the aneurysm and that in the artery. Assume the flow is steady and inviscid.

3.63 Water flows steadily through the variable area pipe shown in Fig. P3.63 with negligible viscous effects. Determine the manometer reading, H, if the flowrate is 0.5 m³/s and the density of the manometer fluid is 600 kg/m³.

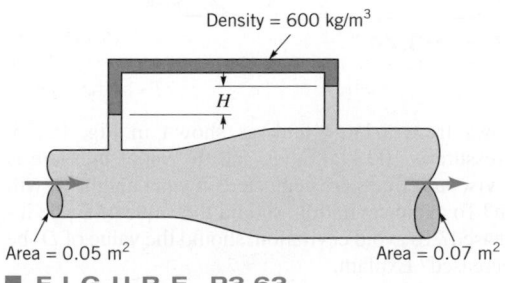

■ **F I G U R E   P3.63**

3.64 Water flows steadily with negligible viscous effects through the pipe shown in Fig. P3.64. It is known that the 10-cm-diameter section of thin-walled tubing will collapse if the pressure within it becomes less than 69 kPa below atmospheric pressure. Determine the maximum value that h can have without causing collapse of the tubing.

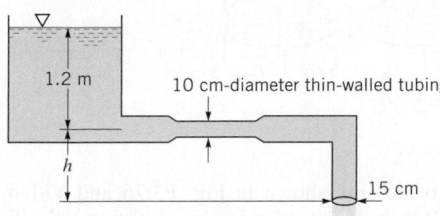

■ **F I G U R E   P3.64**

3.65 Helium flows through a 0.30-m-diameter horizontal pipe with a temperature of 20 °C and a pressure of 200 kPa (abs) at a rate

of 0.30 kg/s. If the pipe reduces to 0.25-m-diameter determine the pressure difference between these two sections. Assume incompressible, inviscid flow.

3.66 Water is pumped from a lake through an 20-cm pipe at a rate of 0.3 m³/s. If viscous effects are negligible, what is the pressure in the suction pipe (the pipe between the lake and the pump) at an elevation 2 m above the lake?

3.67 Air flows through a Venturi channel of rectangular cross section as shown in **Video V3.10** and Fig. P3.67. The constant width of the channel is 0.06 m and the height at the exit is 0.04 m. Compressibility and viscous effects are negligible. **(a)** Determine the flowrate when water is drawn up 0.10 m in a small tube attached to the static pressure tap at the throat where the channel height is 0.02 m. **(b)** Determine the channel height, h_2, at section (2) where, for the same flowrate as in part **(a)**, the water is drawn up 0.05 m. **(c)** Determine the pressure needed at section (1) to produce this flow.

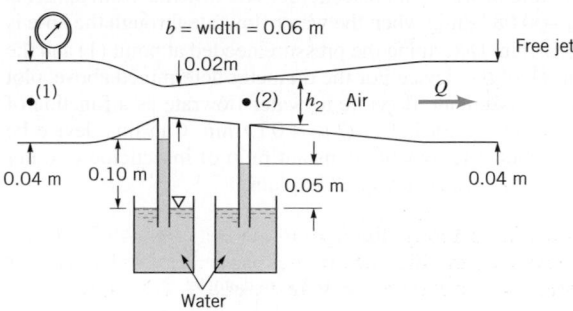

■ **F I G U R E   P3.67**

3.68 Water flows steadily from the large open tank shown in Fig. P3.68. If viscous effects are negligible, determine **(a)** the flowrate, Q, and **(b)** the manometer reading, h.

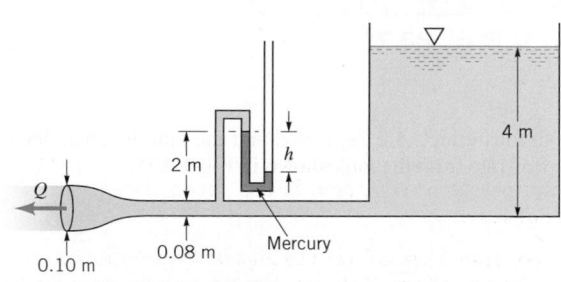

■ **F I G U R E   P3.68**

3.69 Water from a faucet fills a 0.5-L glass (volume = 5×10^{-4} m³) in 20 s. If the diameter of the jet leaving the faucet is 1.5 cm, what is the diameter of the jet when it strikes the water surface in the glass which is positioned 36 cm below the faucet?

3.70 Air flows steadily through a converging–diverging rectangular channel of constant width as shown in Fig. P3.70 and **Video V3.10**. The height of the channel at the exit and the exit velocity are H_0 and V_0, respectively. The channel is to be shaped so that the distance, d, that water is drawn up into tubes attached to static pressure taps along the channel wall is linear with distance along the channel. That is, $d = (d_{max}/L) x$, where L is the channel length and d_{max} is the maximum water depth (at the minimum channel height; $x = L$). Determine the height, $H(x)$, as a function of x and the other important parameters.

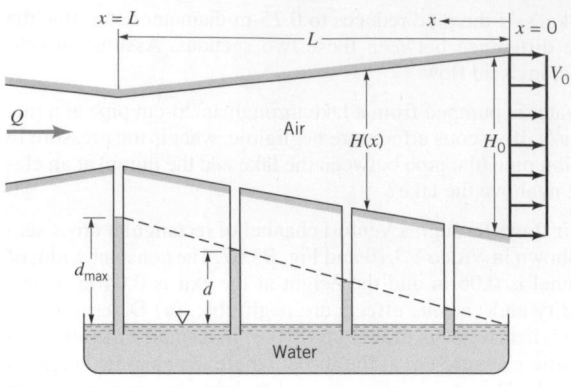

■ FIGURE P3.70

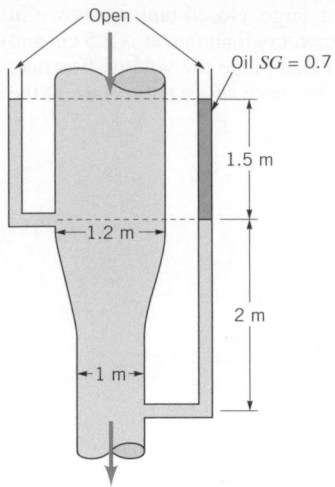

■ FIGURE P3.73

*3.71 The device shown in Fig. P3.71 is used to spray an appropriate mixture of water and insecticide. The flowrate from tank A is to be $Q_A = 0.08$ L/min when the water flowrate through the hose is $Q = 4.0$ L/min. Determine the pressure needed at point (1) and the diameter, D, of the device For the diameter determined above, plot the ratio of insecticide flowrate to water flowrate as a function of water flowrate, Q, for $0.4 \le Q \le 4.0$ L/min. Can this device be used to provide a reasonably constant ratio of insecticide to water regardless of the water flowrate? Explain.

3.74 Air at 27 °C and 101.3 kPa (abs) flows into the tank shown in Fig. P3.74. Determine the flowrate in m³/s, N/s, and kg/s. Assume incompressible flow.

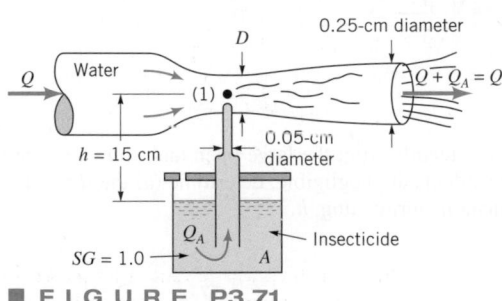

■ FIGURE P3.71

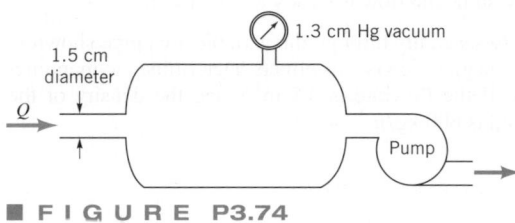

■ FIGURE P3.74

3.72 If viscous effects are neglected and the tank is large, determine the flowrate from the tank shown in Fig. P3.72.

3.75 Water flows from a large tank as shown in Fig. P3.75. Atmospheric pressure is 100 kPa (abs), and the vapor pressure is 11 kPa (abs). If viscous effects are neglected, at what height, h, will cavitation begin? To avoid cavitation, should the value of D_1 be increased or decreased? To avoid cavitation, should the value of D_2 be increased or decreased? Explain.

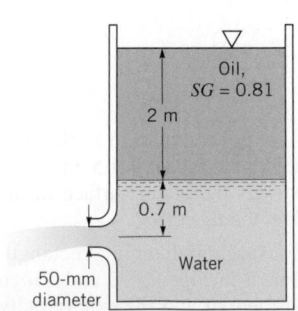

■ FIGURE P3.72

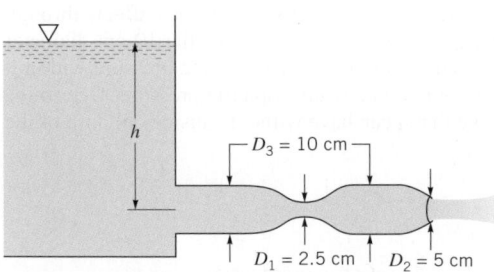

■ FIGURE P3.75

3.73 Water flows steadily downward in the pipe shown in Fig. 3.73 with negligible losses. Determine the flowrate.

3.76 Water flows into the sink shown in Fig. P3.76 and Video V5.1 at a rate of 8.0 L/min. If the drain is closed, the water will eventually flow through the overflow drain holes rather than over the edge of the sink. How many 1-cm-diameter drain holes are needed to ensure that the water does not overflow the sink? Neglect viscous effects.

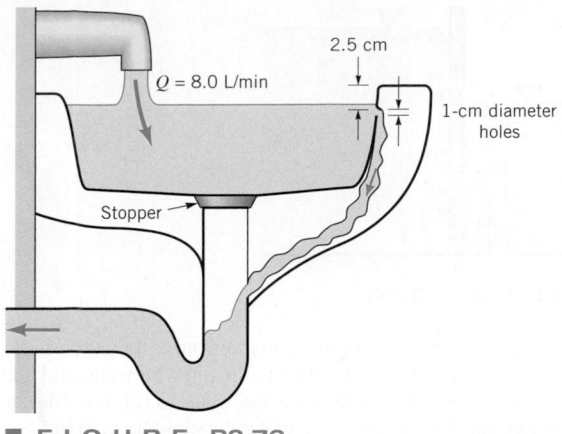

■ F I G U R E   P3.76

3.77 What pressure, p_1, is needed to produce a flowrate of 2.5×10^{-3} m³/s from the tank shown in Fig. P3.77?

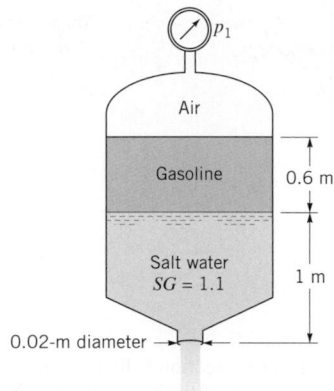

■ F I G U R E   P3.77

3.78 Water is siphoned from the tank shown in Fig. P3.78. Determine the flowrate from the tank and the pressures at points (1), (2), and (3) if viscous effects are negligible.

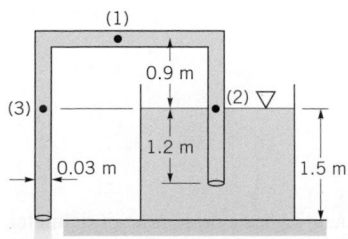

■ F I G U R E   P3.78

3.79 Water is siphoned from a large tank and discharges into the atmosphere through a 5-cm-diameter tube as shown in Fig. P3.79. The end of the tube is 1.0 m below the tank bottom, and viscous effects are negligible. **(a)** Determine the volume flowrate from the tank. **(b)** Determine the maximum height, H, over which the water can be siphoned without cavitation occurring. Atmospheric pressure is 101.3 kPa (abs), and the water vapor pressure is 1.8 kPa (abs).

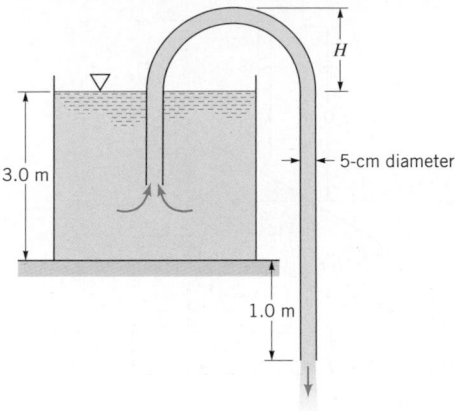

■ F I G U R E   P3.79

3.80 Determine the manometer reading, h, for the flow shown in Fig. P3.80.

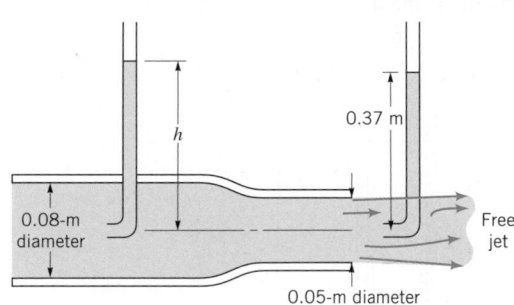

■ F I G U R E   P3.80

3.81 Air flows steadily through the variable area pipe shown in Fig. P3.81. Determine the flowrate if viscous and compressibility effects are negligible.

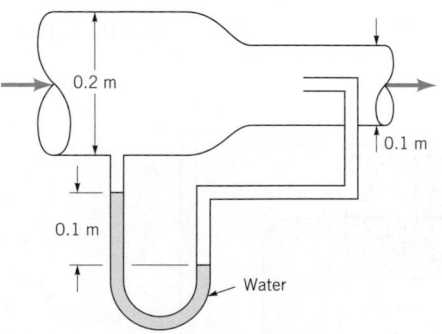

■ F I G U R E   P3.81

3.82 JP-4 fuel ($SG = 0.77$) flows through the Venturi meter shown in Fig. P3.82 with a velocity of 5 m/s in the 15-cm pipe. If viscous effects are negligible, determine the elevation, h, of the fuel in the open tube connected to the throat of the Venturi meter.

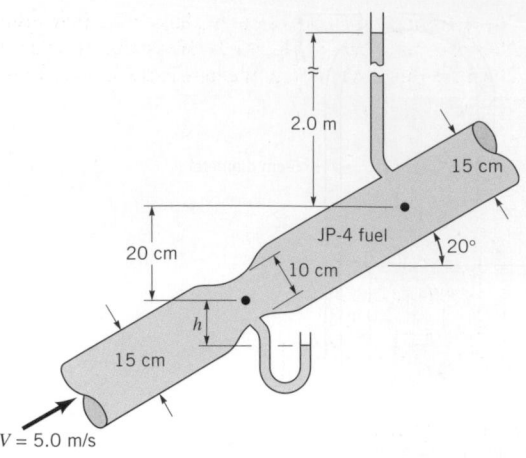

■ **FIGURE P3.82**

3.83 Repeat Problem 3.82 if the flowing fluid is water rather than JP-4 fuel.

3.84 Oil flows through the system shown in Fig. P3.84 with negligible losses. Determine the flowrate.

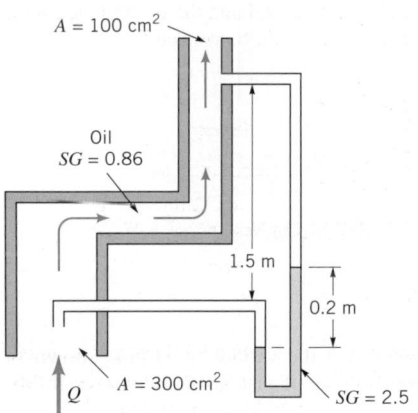

■ **FIGURE P3.84**

3.85 Water, considered an inviscid, incompressible fluid, flows steadily as shown in Fig. P3.85. Determine h.

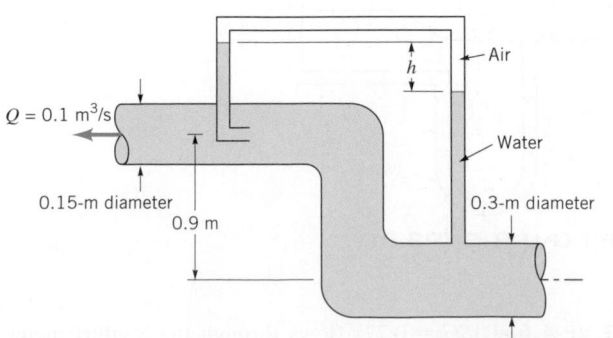

■ **FIGURE P3.85**

3.86 Determine the flowrate through the submerged orifice shown in Fig. P3.86 if the contraction coefficient is $C_c = 0.63$.

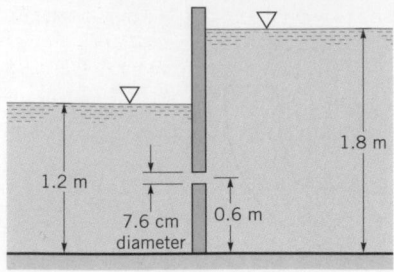

■ **FIGURE P3.86**

*3.87 An inexpensive timer is to be made from a funnel as indicated in Fig. P3.87. The funnel is filled to the top with water and the plug is removed at time $t = 0$ to allow the water to run out. Marks are to be placed on the wall of the funnel indicating the time in 15-s intervals, from 0 to 3 min (at which time the funnel becomes empty). If the funnel outlet has a diameter of $d = 0.25$ cm, draw to scale the funnel with the timing marks for funnels with angles of $\theta = 30, 45$, and 60°. Repeat the problem if the diameter is changed to 0.13 cm.

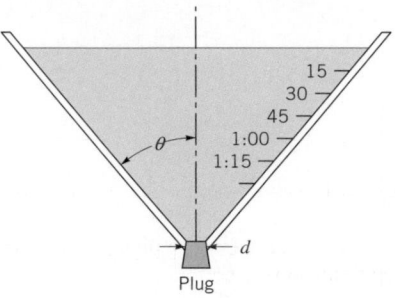

■ **FIGURE P3.87**

3.88 A long water trough of triangular cross section is formed from two planks as is shown in Fig. P3.88. A gap of 0.25 cm remains at the junction of the two planks. If the water depth initially was 0.6 m, how long a time does it take for the water depth to reduce to 0.3 m?

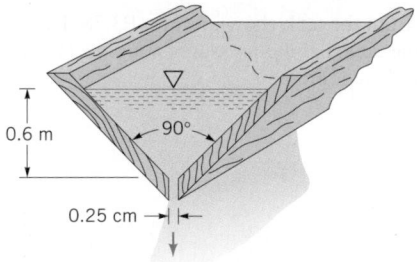

■ **FIGURE P3.88**

*3.89 A spherical tank of diameter D has a drain hole of diameter d at its bottom. A vent at the top of the tank maintains atmospheric pressure at the liquid surface within the tank. The flow is quasi-steady and inviscid and the tank is full of water initially. Determine the water depth as a function of time, $h = h(t)$, and plot graphs of $h(t)$ for tank diameters of 0.5, 1.5, 3.0, and 6 if $d = 2.5$ cm.

3.90 When the drain plug is pulled, water flows from a hole in the bottom of a large, open cylindrical tank. Show that if viscous effects are negligible and if the flow is assumed to be quasisteady, then it takes 3.41 times longer to empty the entire tank than it does to empty the first half of the tank. Explain why this is so.

*3.91 The surface area, A, of the pond shown in Fig. P3.91 varies with the water depth, h, as shown in the table. At time $t = 0$ a valve is opened and the pond is allowed to drain through a pipe of diameter D. If viscous effects are negligible and quasisteady conditions are assumed, plot the water depth as a function of time from when the valve is opened ($t = 0$) until the pond is drained for pipe diameters of $D = 0.15, 0.30, 0.45, 0.60, 0.75,$ and 0.9 m. Assume $h = 6$ m at $t = 0$.

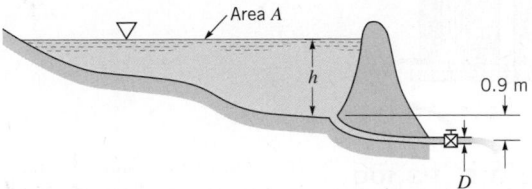

■ **FIGURE P3.91**

h (m)	A (m^2)
0	0
0.5	1200
1.0	2000
1.5	3200
2.0	3600
2.5	4500
3.0	6000
3.5	7300
4.0	9700
4.5	11300

3.92 Water flows through a horizontal branching pipe as shown in Fig. P3.92. Determine the pressure at section (3).

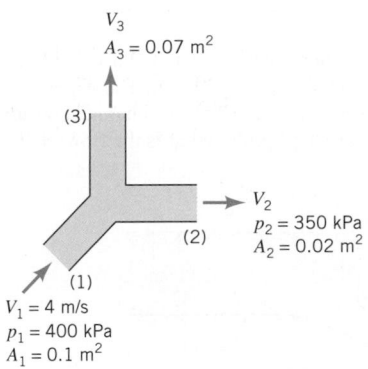

■ **FIGURE P3.92**

3.93 Water flows through the horizontal branching pipe shown in Fig. P3.93 at a rate of 0.3 m^3/s. If viscous effects are negligible, determine the water speed at section (2), the pressure at section (3), and the flowrate at section (4).

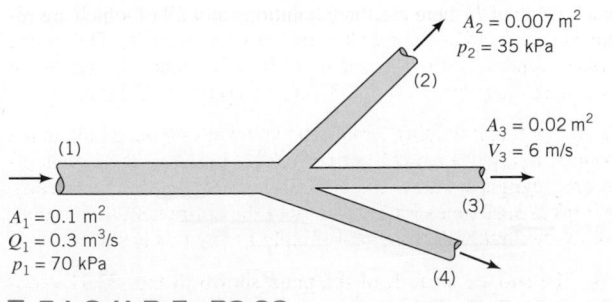

■ **FIGURE P3.93**

3.94 Water flows from a large tank through a large pipe that splits into two smaller pipes as shown in Fig. P3.94. If viscous effects are negligible, determine the flowrate from the tank and the pressure at point (1).

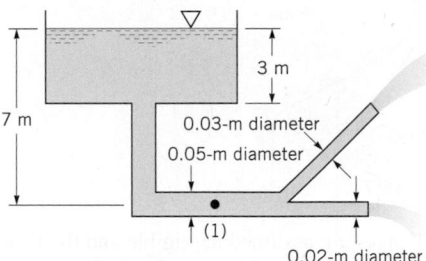

■ **FIGURE P3.94**

3.95 An air cushion vehicle is supported by forcing air into the chamber created by a skirt around the periphery of the vehicle as shown in Fig. P3.95. The air escapes through the 7.0 cm clearance between the lower end of the skirt and the ground (or water). Assume the vehicle weighs 50 kN and is essentially rectangular in shape, 9 by 20 m. The volume of the chamber is large enough so that the kinetic energy of the air within the chamber is negligible. Determine the flowrate, Q, needed to support the vehicle. If the ground clearance were reduced to 5 cm, what flowrate would be needed? If the vehicle weight were reduced to 25 kN and the ground clearance maintained at 7 cm, what flowrate would be needed?

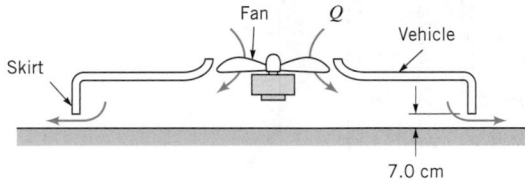

■ **FIGURE P3.95**

3.96 Water flows from the pipe shown in Fig. P3.96 as a free jet and strikes a circular flat plate. The flow geometry shown is axisymmetrical. Determine the flowrate and the manometer reading, H.

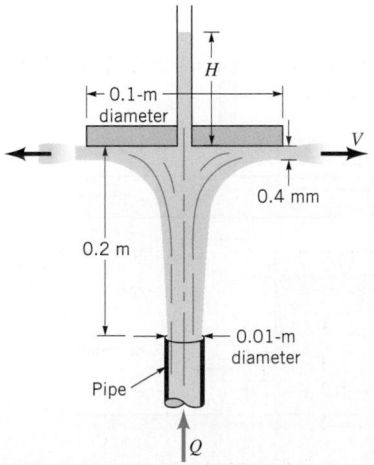

■ **FIGURE P3.96**

3.97 Air flows from a hole of diameter 0.03 m in a flat plate as shown in Fig. P3.97. A circular disk of diameter D is placed a distance h from the lower plate. The pressure in the tank is maintained at 1 kPa. Determine the flowrate as a function of h if viscous

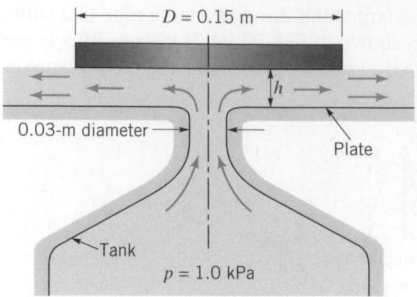

■ **FIGURE P3.97**

effects and elevation changes are assumed negligible and the flow exits radially from the circumference of the circular disk with uniform velocity.

3.98 A conical plug is used to regulate the air flow from the pipe shown in Fig. P3.98. The air leaves the edge of the cone with a uniform thickness of 0.02 m. If viscous effects are negligible and the flowrate is $0.50 \text{ m}^3/\text{s}$, determine the pressure within the pipe.

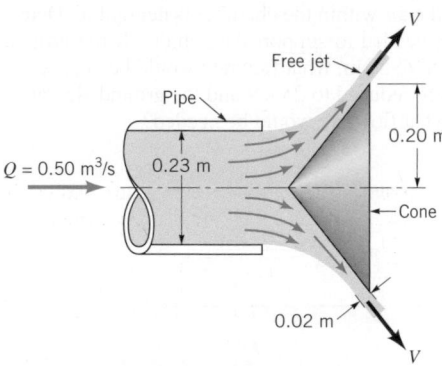

■ **FIGURE P3.98**

3.99 Water flows steadily from a nozzle into a large tank as shown in Fig. P3.99. The water then flows from the tank as a jet of diameter d. Determine the value of d if the water level in the tank remains constant. Viscous effects are negligible.

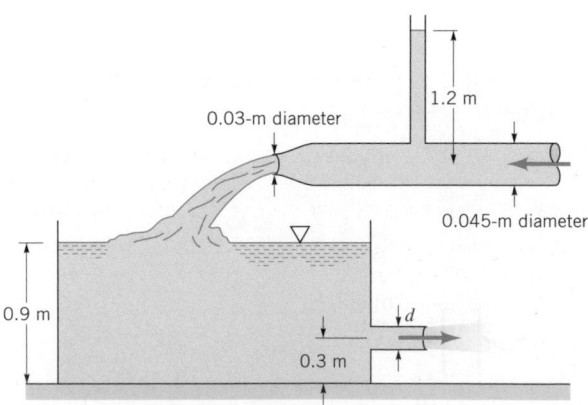

■ **FIGURE P3.99**

3.100 A small card is placed on top of a spool as shown in Fig. P3.100. It is not possible to blow the card off the spool by blowing air through the hole in the center of the spool. The harder one blows, the harder the card "sticks" to the spool. In fact, by blowing hard enough it is possible to keep the card against the

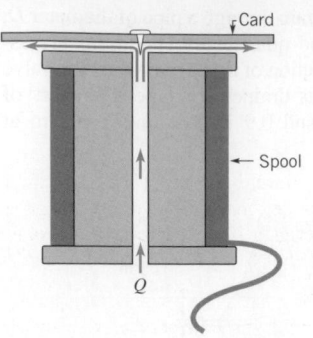

■ **FIGURE P3.100**

spool with the spool turned upside down. (*Note:* It may be necessary to use a thumb tack to prevent the card from sliding from the spool.) Explain this phenomenon.

3.101 Water flows down the sloping ramp shown in Fig. P3.101 with negligible viscous effects. The flow is uniform at sections (1) and (2). For the conditions given, show that three solutions for the downstream depth, h_2, are obtained by use of the Bernoulli and continuity equations. However, show that only two of these solutions are realistic. Determine these values.

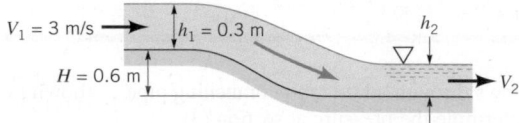

■ **FIGURE P3.101**

3.102 Water flows in a rectangular channel that is 2.0 m wide as shown in Fig. P3.102. The upstream depth is 70 mm. The water surface rises 40 mm as it passes over a portion where the channel bottom rises 10 mm. If viscous effects are negligible, what is the flowrate?

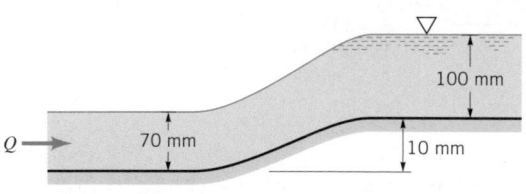

■ **FIGURE P3.102**

*3.103** Water flows up the ramp shown in Fig. P3.103 with negligible viscous losses. The upstream depth and velocity are maintained at $h_1 = 0.3$ m and $V_1 = 6$ m/s. Plot a graph of the downstream depth, h_2, as a function of the ramp height, H, for $0 \leq H \leq 2$ m. Note that for each value of H there are three solutions, not all of which are realistic.

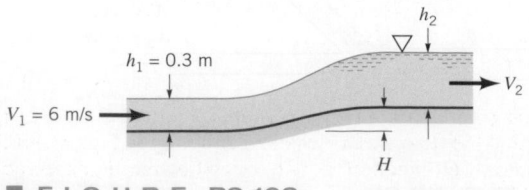

■ **FIGURE P3.103**

Section 3.6.3 Flowrate Measurement (Also see Lab Problems 3.119 and 3.121.)

3.104 Obtain a photograph/image of a situation that involves some type of flow meter. Print this photo and write a brief paragraph that describes the situation involved.

3.105 A Venturi meter with a minimum diameter of 7.0 cm is to be used to measure the flowrate of water through a 10-cm-diameter pipe. Determine the pressure difference indicated by the pressure gage attached to the flow meter if the flowrate is $1.4 \times 10^{-2} \, \text{m}^3/\text{s}$ and viscous effects are negligible.

3.106 Determine the flowrate through the Venturi meter shown in Fig. P3.106 if ideal conditions exist.

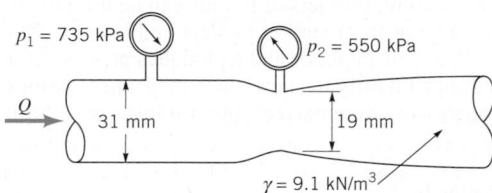

$p_1 = 735$ kPa $p_2 = 550$ kPa

Q 31 mm   19 mm

$\gamma = 9.1$ kN/m^3

■ **F I G U R E P3.106**

3.107 For what flowrate through the Venturi meter of Problem 3.106 will cavitation begin if $p_1 = 275$ kPa gage, atmospheric pressure is 101 kPa (abs), and the vapor pressure is 3.6 kPa (abs)?

3.108 What diameter orifice hole, d, is needed if under ideal conditions the flowrate through the orifice meter of Fig. P3.108 is to be 100 L/min of seawater with $p_1 - p_2 = 16.0$ kPa? The contraction coefficient is assumed to be 0.63.

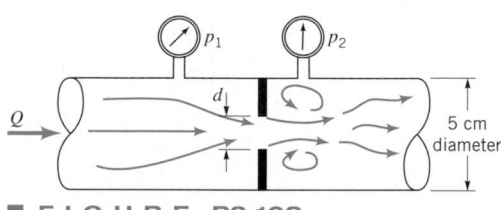

p_1 p_2

Q d 5 cm diameter

■ **F I G U R E P3.108**

3.109 Water flows over a weir plate (see **Video V10.13**) which has a parabolic opening as shown in Fig. P3.109. That is, the opening in the weir plate has a width $CH^{1/2}$, where C is a constant. Determine the functional dependence of the flowrate on the head, $Q = Q(H)$.

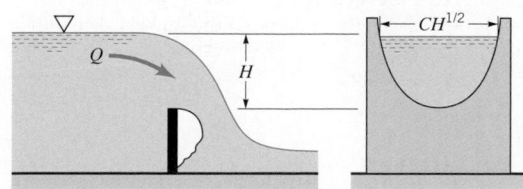

Q H $CH^{1/2}$

■ **F I G U R E P3.109**

3.110 A weir (see **Video V10.13**) of trapezoidal cross section is used to measure the flowrate in a channel as shown in Fig. P3.110. If the flowrate is Q_0 when $H = \ell/2$, what flowrate is expected when $H = \ell$?

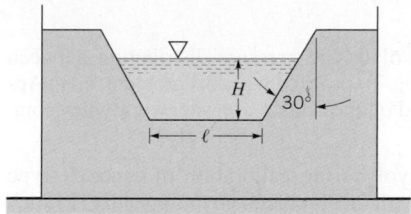

H 30° ℓ

■ **F I G U R E P3.110**

3.111 The flowrate in a water channel is sometimes determined by use of a device called a Venturi flume. As shown in Fig. P3.111, this device consists simply of a hump on the bottom of the channel. If the water surface dips a distance of 0.07 m for the conditions shown, what is the flowrate per width of the channel? Assume the velocity is uniform and viscous effects are negligible.

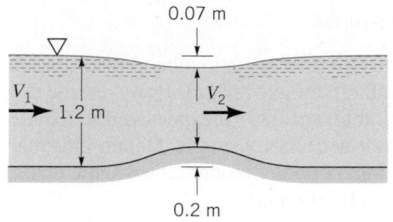

0.07 m

V_1 1.2 m V_2

0.2 m

■ **F I G U R E P3.111**

3.112 Water flows under the inclined sluice gate shown in Fig. P3.112. Determine the flowrate if the gate is 2.5 m wide.

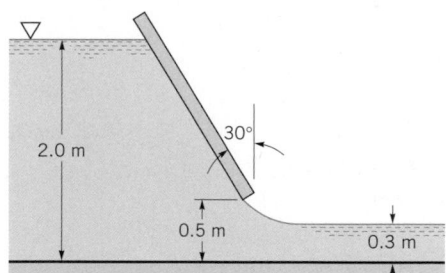

2.0 m   30°

0.5 m   0.3 m

■ **F I G U R E P3.112**

Section 3.7 The Energy Line and the Hydraulic Grade Line

3.113 Water flows in a vertical pipe of 0.15-m diameter at a rate of 0.2 m^3/s and a pressure of 200 kPa at an elevation of 25 m. Determine the velocity head and pressure head at elevations of 20 and 55 m.

3.114 Draw the energy line and the hydraulic grade line for the flow shown in Problem 3.78.

3.115 Draw the energy line and the hydraulic grade line for the flow of Problem 3.75.

3.116 Draw the energy line and hydraulic grade line for the flow shown in Problem 3.64.

Section 3.8 Restrictions on the Use of the Bernoulli Equation

3.117 Obtain a photograph/image of a flow in which it would not be appropriate to use the Bernoulli equation. Print this photo and write a brief paragraph that describes the situation involved.

■ **Lab Problems**

3.118 This problem involves the pressure distribution between two parallel circular plates. To proceed with this problem, go to Appendix H which is located on the book's web site, www.wiley.com/college/munson.

3.119 This problem involves the calibration of a nozzle-type flow meter. To proceed with this problem, go to Appendix H which is located on the book's web site, www.wiley.com/college/munson.

3.120 This problem involves the pressure distribution in a two-dimensional channel. To proceed with this problem, go to Appendix H which is located on the book's web site, www.wiley.com/college/munson.

3.121 This problem involves the determination of the flowrate under a sluice gate as a function of the water depth. To proceed with this problem, go to Appendix H which is located on the book's web site, www.wiley.com/college/munson.

■ **Life Long Learning Problems**

3.122 The concept of the use of a Pitot-static tube to measure the airspeed of an airplane is rather straightforward. However, the design and manufacture of reliable, accurate, inexpensive Pitot-static tube airspeed indicators is not necessarily simple. Obtain information about the design and construction of modern Pitot-static tubes. Summarize your findings in a brief report.

3.123 In recent years damage due to hurricanes has been significant, particularly in the southeastern United States. The low barometric pressure, high winds, and high tides generated by hurricanes can combine to cause considerable damage. According to some experts, in the coming years hurricane frequency may increase because of global warming. Obtain information about the fluid mechanics of hurricanes. Summarize your findings in a brief report.

3.124 Orifice, nozzle, or Venturi flow meters have been used for a long time to predict accurately the flowrate in pipes. However, recently there have been several new concepts suggested or used for such flowrate measurements. Obtain information about new methods to obtain pipe flowrate information. Summarize your findings in a brief report.

3.125 Ultra-high-pressure, thin jets of liquids can be used to cut various materials ranging from leather to steel and beyond. Obtain information about new methods and techniques proposed for liquid jet cutting and investigate how they may alter various manufacturing processes. Summarize your findings in a brief report.

■ **FE Exam Problems**

Sample FE (Fundamentals of Engineering) exam questions for fluid mechanics are provided on the book's web site, www.wiley.com/college/munson.

4 Fluid Kinematics

CHAPTER OPENING PHOTO: A vortex ring: The complex, three-dimensional structure of a smoke ring is indicated in this cross-sectional view. (Smoke in air.) [Photograph courtesy of R. H. Magarvey and C. S. MacLatchy (Ref. 4).]

Learning Objectives

After completing this chapter, you should be able to:

■ discuss the differences between the Eulerian and Lagrangian descriptions of fluid motion.

■ identify various flow characteristics based on the velocity field.

■ determine the streamline pattern and acceleration field given a velocity field.

■ discuss the differences between a system and control volume.

■ apply the Reynolds transport theorem and the material derivative.

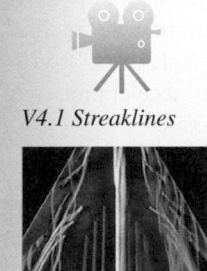

V4.1 Streaklines

In this chapter we will discuss various aspects of fluid motion without being concerned with the actual forces necessary to produce the motion. That is, we will consider the *kinematics* of the motion—the velocity and acceleration of the fluid, and the description and visualization of its motion. The analysis of the specific forces necessary to produce the motion (the *dynamics* of the motion) will be discussed in detail in the following chapters. A wide variety of useful information can be gained from a thorough understanding of fluid kinematics. Such an understanding of how to describe and observe fluid motion is an essential step to the complete understanding of fluid dynamics.

4.1 The Velocity Field

In general, fluids flow. That is, there is a net motion of molecules from one point in space to another point as a function of time. As is discussed in Chapter 1, a typical portion of fluid contains so many molecules that it becomes totally unrealistic (except in special cases) for us to attempt to

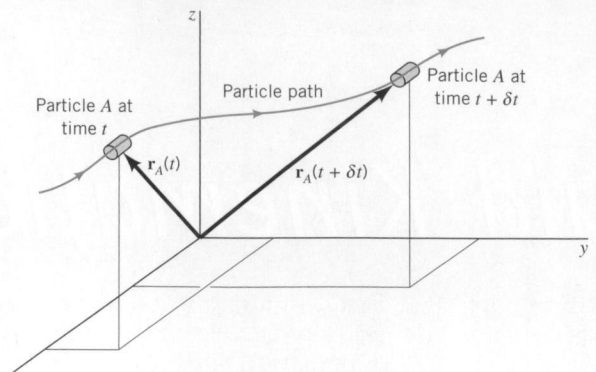

■ **FIGURE 4.1** **Particle location in terms of its position vector.**

V4.2 Velocity field

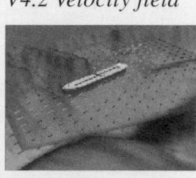

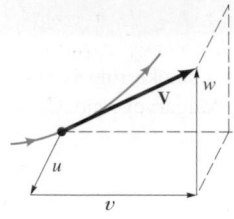

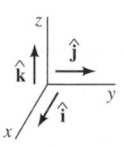

V4.3 Cylinder-velocity vectors

account for the motion of individual molecules. Rather, we employ the continuum hypothesis and consider fluids to be made up of fluid particles that interact with each other and with their surroundings. Each particle contains numerous molecules. Thus, we can describe the flow of a fluid in terms of the motion of fluid particles rather than individual molecules. This motion can be described in terms of the velocity and acceleration of the fluid particles.

The infinitesimal particles of a fluid are tightly packed together (as is implied by the continuum assumption). Thus, at a given instant in time, a description of any fluid property (such as density, pressure, velocity, and acceleration) may be given as a function of the fluid's location. This representation of fluid parameters as functions of the spatial coordinates is termed a *field representation* of the flow. Of course, the specific field representation may be different at different times, so that to describe a fluid flow we must determine the various parameters not only as a function of the spatial coordinates (x, y, z, for example) but also as a function of time, t. Thus, to completely specify the temperature, T, in a room we must specify the temperature field, $T = T(x, y, z, t)$, throughout the room (from floor to ceiling and wall to wall) at any time of the day or night.

Shown in the margin figure is one of the most important fluid variables, the *velocity field,*

$$\mathbf{V} = u(x, y, z, t)\hat{\mathbf{i}} + v(x, y, z, t)\hat{\mathbf{j}} + w(x, y, z, t)\hat{\mathbf{k}}$$

where u, v, and w are the x, y, and z components of the velocity vector. By definition, the velocity of a particle is the time rate of change of the position vector for that particle. As is illustrated in Fig. 4.1, the position of particle A relative to the coordinate system is given by its *position vector*, $\mathbf{r}_A$, which (if the particle is moving) is a function of time. The time derivative of this position gives the *velocity* of the particle, $d\mathbf{r}_A/dt = \mathbf{V}_A$. By writing the velocity for all of the particles we can obtain the field description of the velocity vector $\mathbf{V} = \mathbf{V}(x, y, z, t)$.

Since the velocity is a vector, it has both a direction and a magnitude. The magnitude of $\mathbf{V}$, denoted $V = |\mathbf{V}| = (u^2 + v^2 + w^2)^{1/2}$, is the speed of the fluid. (It is very common in practical situations to call V velocity rather than speed, i.e., "the velocity of the fluid is 12 m/s.") As is discussed in the next section, a change in velocity results in an acceleration. This acceleration may be due to a change in speed and/or direction.

F l u i d s   i n   t h e   N e w s

Follow those particles Superimpose two photographs of a bouncing ball taken a short time apart and draw an arrow between the two images of the ball. This arrow represents an approximation of the velocity (displacement/time) of the ball. The particle image velocimeter (PIV) uses this technique to provide the instantaneous *velocity field* for a given cross section of a flow. The flow being studied is seeded with numerous micron-sized particles which are small enough to follow the flow yet big enough to reflect enough light to be captured by the camera. The flow is illuminated with a light sheet from a double-pulsed laser. A digital camera captures both light pulses on the same image frame, allowing the movement of the particles to be tracked. By using appropriate computer software to carry out a pixel-by-pixel interrogation of the double image, it is possible to track the motion of the particles and determine the two components of velocity in the given cross section of the flow. By using two cameras in a stereoscopic arrangement it is possible to determine all three components of velocity. (See Problem 4.62.)

EXAMPLE 4.1   Velocity Field Representation

GIVEN A velocity field is given by $\mathbf{V} = (V_0/\ell)(-x\hat{\mathbf{i}} + y\hat{\mathbf{j}})$ where V_0 and ℓ are constants.

FIND At what location in the flow field is the speed equal to V_0? Make a sketch of the velocity field for $x \geq 0$ by drawing arrows representing the fluid velocity at representative locations.

SOLUTION

The x, y, and z components of the velocity are given by $u = -V_0 x/\ell$, $v = V_0 y/\ell$, and $w = 0$ so that the fluid speed, V, is

$$V = (u^2 + v^2 + w^2)^{1/2} = \frac{V_0}{\ell}(x^2 + y^2)^{1/2} \tag{1}$$

The speed is $V = V_0$ at any location on the circle of radius ℓ centered at the origin $[(x^2 + y^2)^{1/2} = \ell]$ as shown in Fig. E4.1a. **(Ans)**

The direction of the fluid velocity relative to the x axis is given in terms of $\theta = \arctan(v/u)$ as shown in Fig. E4.1b. For this flow

$$\tan \theta = \frac{v}{u} = \frac{V_0 y/\ell}{-V_0 x/\ell} = \frac{y}{-x}$$

Thus, along the x axis ($y = 0$) we see that $\tan \theta = 0$, so that $\theta = 0°$ or $\theta = 180°$. Similarly, along the y axis ($x = 0$) we obtain $\tan \theta = \pm\infty$ so that $\theta = 90°$ or $\theta = 270°$. Also, for $y = 0$ we find $\mathbf{V} = (-V_0 x/\ell)\hat{\mathbf{i}}$, while for $x = 0$ we have $\mathbf{V} = (V_0 y/\ell)\hat{\mathbf{j}}$,

indicating (if $V_0 > 0$) that the flow is directed away from the origin along the y axis and toward the origin along the x axis as shown in Fig. E4.1a.

By determining $\mathbf{V}$ and θ for other locations in the x–y plane, the velocity field can be sketched as shown in the figure. For example, on the line $y = x$ the velocity is at a 45° angle relative to the x axis ($\tan \theta = v/u = -y/x = -1$). At the origin $x = y = 0$ so that $\mathbf{V} = 0$. This point is a stagnation point. The farther from the origin the fluid is, the faster it is flowing (as seen from Eq. 1). By careful consideration of the velocity field it is possible to determine considerable information about the flow.

COMMENT The velocity field given in this example approximates the flow in the vicinity of the center of the sign shown in Fig. E4.1c. When wind blows against the sign, some air flows over the sign, some under it, producing a stagnation point as indicated.

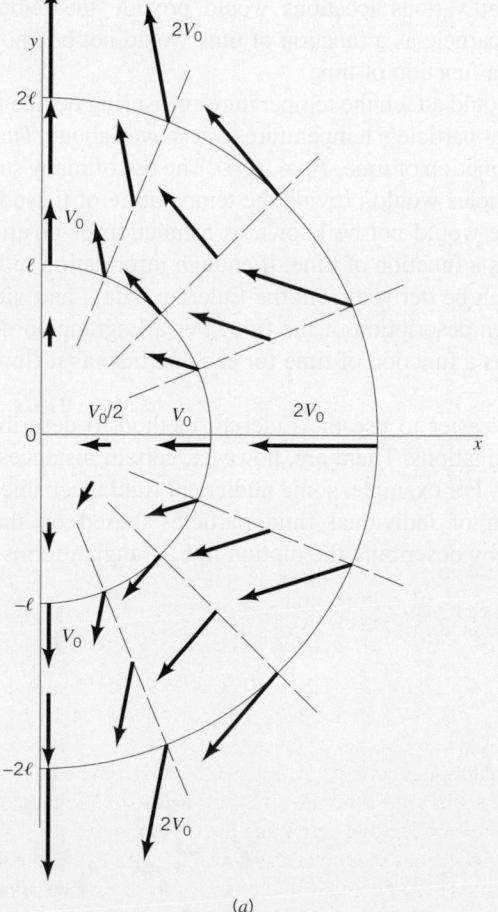

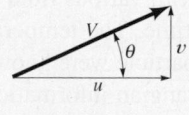

(b)

(c)

(a)

■ **FIGURE E4.1**

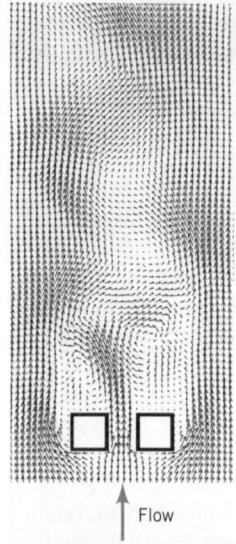

Flow

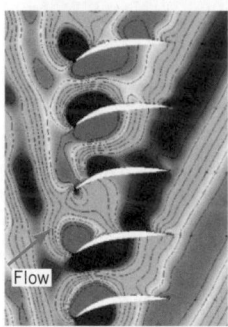

Flow

Either Eulerian or Lagrangian methods can be used to describe flow fields.

The figure in the margin shows the velocity field (i.e., velocity vectors) for flow past two square bars. It is possible to obtain much qualitative and quantitative information for complex flows by using plots such as this.

4.1.1 Eulerian and Lagrangian Flow Descriptions

There are two general approaches in analyzing fluid mechanics problems (or problems in other branches of the physical sciences, for that matter). The first method, called the *Eulerian method,* uses the field concept introduced above. In this case, the fluid motion is given by completely prescribing the necessary properties (pressure, density, velocity, etc.) as functions of space and time. From this method we obtain information about the flow in terms of what happens at fixed points in space as the fluid flows through those points.

A typical Eulerian representation of the flow is shown by the figure in the margin which involves flow past a row of turbine blades as occurs in a jet engine. The pressure field is indicated by using a contour plot showing lines of constant pressure, with grey shading indicating the intensity of the pressure.

The second method, called the *Lagrangian method,* involves following individual fluid particles as they move about and determining how the fluid properties associated with these particles change as a function of time. That is, the fluid particles are "tagged" or identified, and their properties determined as they move.

The difference between the two methods of analyzing fluid flow problems can be seen in the example of smoke discharging from a chimney, as is shown in Fig. 4.2. In the Eulerian method one may attach a temperature-measuring device to the top of the chimney (point 0) and record the temperature at that point as a function of time. At different times there are different fluid particles passing by the stationary device. Thus, one would obtain the temperature, T, for that location ($x = x_0, y = y_0,$ and $z = z_0$) as a function of time. That is, $T = T(x_0, y_0, z_0, t)$. The use of numerous temperature-measuring devices fixed at various locations would provide the temperature field, $T = T(x, y, z, t)$. The temperature of a particle as a function of time would not be known unless the location of the particle were known as a function of time.

In the Lagrangian method, one would attach the temperature-measuring device to a particular fluid particle (particle A) and record that particle's temperature as it moves about. Thus, one would obtain that particle's temperature as a function of time, $T_A = T_A(t)$. The use of many such measuring devices moving with various fluid particles would provide the temperature of these fluid particles as a function of time. The temperature would not be known as a function of position unless the location of each particle were known as a function of time. If enough information in Eulerian form is available, Lagrangian information can be derived from the Eulerian data—and vice versa.

Example 4.1 provides an Eulerian description of the flow. For a Lagrangian description we would need to determine the velocity as a function of time for each particle as it flows along from one point to another.

In fluid mechanics it is usually easier to use the Eulerian method to describe a flow—in either experimental or analytical investigations. There are, however, certain instances in which the Lagrangian method is more convenient. For example, some numerical fluid mechanics calculations are based on determining the motion of individual fluid particles (based on the appropriate interactions among the particles), thereby describing the motion in Lagrangian terms. Similarly, in

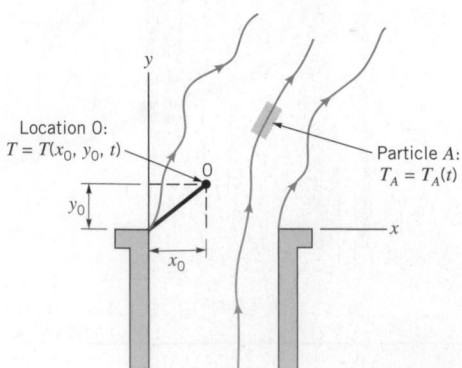

Location 0:
$T = T(x_0, y_0, t)$

Particle A:
$T_A = T_A(t)$

■ **FIGURE 4.2** **Eulerian and Lagrangian descriptions of temperature of a flowing fluid.**

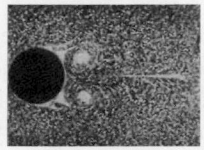

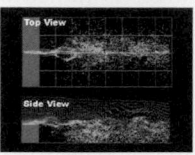

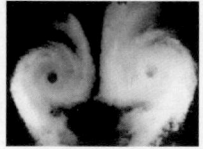

some experiments individual fluid particles are "tagged" and are followed throughout their motion, providing a Lagrangian description. Oceanographic measurements obtained from devices that flow with the ocean currents provide this information. Similarly, by using X-ray opaque dyes it is possible to trace blood flow in arteries and to obtain a Lagrangian description of the fluid motion. A Lagrangian description may also be useful in describing fluid machinery (such as pumps and turbines) in which fluid particles gain or lose energy as they move along their flow paths.

Another illustration of the difference between the Eulerian and Lagrangian descriptions can be seen in the following biological example. Each year thousands of birds migrate between their summer and winter habitats. Ornithologists study these migrations to obtain various types of important information. One set of data obtained is the rate at which birds pass a certain location on their migration route (birds per hour). This corresponds to an Eulerian description—"flowrate" at a given location as a function of time. Individual birds need not be followed to obtain this information. Another type of information is obtained by "tagging" certain birds with radio transmitters and following their motion along the migration route. This corresponds to a Lagrangian description—"position" of a given particle as a function of time.

4.1.2 One-, Two-, and Three-Dimensional Flows

Generally, a fluid flow is a rather complex three-dimensional, time-dependent phenomenon—$\mathbf{V} = \mathbf{V}(x, y, z, t) = u\hat{\mathbf{i}} + v\hat{\mathbf{j}} + w\hat{\mathbf{k}}$. In many situations, however, it is possible to make simplifying assumptions that allow a much easier understanding of the problem without sacrificing needed accuracy. One of these simplifications involves approximating a real flow as a simpler one- or two-dimensional flow.

In almost any flow situation, the velocity field actually contains all three velocity components (u, v, and w, for example). In many situations the *three-dimensional flow* characteristics are important in terms of the physical effects they produce. (See the photograph at the beginning of Chapter 4.) For these situations it is necessary to analyze the flow in its complete three-dimensional character. Neglect of one or two of the velocity components in these cases would lead to considerable misrepresentation of the effects produced by the actual flow.

The flow of air past an airplane wing provides an example of a complex three-dimensional flow. A feel for the three-dimensional structure of such flows can be obtained by studying Fig. 4.3, which is a photograph of the flow past a model wing; the flow has been made visible by using a flow visualization technique.

In many situations one of the velocity components may be small (in some sense) relative to the two other components. In situations of this kind it may be reasonable to neglect the smaller component and assume *two-dimensional flow*. That is, $\mathbf{V} = u\hat{\mathbf{i}} + v\hat{\mathbf{j}}$, where u and v are functions of x and y (and possibly time, t).

It is sometimes possible to further simplify a flow analysis by assuming that two of the velocity components are negligible, leaving the velocity field to be approximated as a *one-dimensional flow* field. That is, $\mathbf{V} = u\hat{\mathbf{i}}$. As we will learn from examples throughout the remainder of the book, although there are very few, if any, flows that are truly one-dimensional, there are

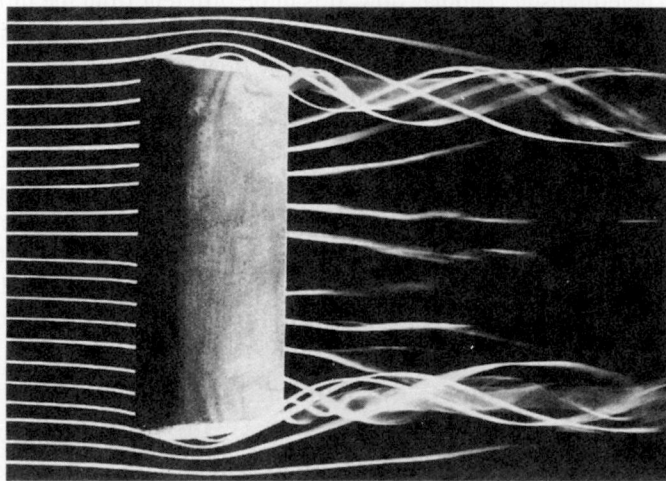

■ **FIGURE 4.3**
Flow visualization of the complex three-dimensional flow past a model wing. (Photograph by M. R. Head.)

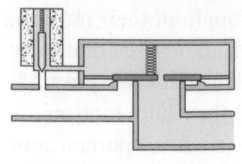

Solenoid off, valve closed

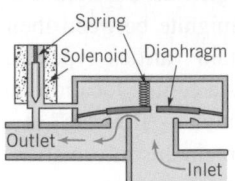

Solenoid on, valve open

V4.7 Flow types

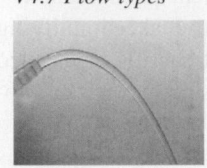

many flow fields for which the one-dimensional flow assumption provides a reasonable approximation. There are also many flow situations for which use of a one-dimensional flow field assumption will give completely erroneous results.

4.1.3 Steady and Unsteady Flows

In the previous discussion we have assumed *steady flow*—the velocity at a given point in space does not vary with time, $\partial \mathbf{V}/\partial t = 0$. In reality, almost all flows are unsteady in some sense. That is, the velocity does vary with time. It is not difficult to believe that *unsteady flows* are usually more difficult to analyze (and to investigate experimentally) than are steady flows. Hence, considerable simplicity often results if one can make the assumption of steady flow without compromising the usefulness of the results. Among the various types of unsteady flows are nonperiodic flow, periodic flow, and truly random flow. Whether or not unsteadiness of one or more of these types must be included in an analysis is not always immediately obvious.

An example of a nonperiodic, unsteady flow is that produced by turning off a faucet to stop the flow of water. Usually this unsteady flow process is quite mundane and the forces developed as a result of the unsteady effects need not be considered. However, if the water is turned off suddenly (as with the electrically operated valve in a dishwasher shown in the figure in the margin), the unsteady effects can become important [as in the "water hammer" effects made apparent by the loud banging of the pipes under such conditions (Ref. 1)].

In other flows the unsteady effects may be periodic, occurring time after time in basically the same manner. The periodic injection of the air–gasoline mixture into the cylinder of an automobile engine is such an example. The unsteady effects are quite regular and repeatable in a regular sequence. They are very important in the operation of the engine.

F l u i d s   i n   t h e   N e w s

New pulsed liquid-jet scalpel High-speed liquid-jet cutters are used for cutting a wide variety of materials such as leather goods, jigsaw puzzles, plastic, ceramic, and metal. Typically, compressed air is used to produce a continuous stream of water that is ejected from a tiny nozzle. As this stream impacts the material to be cut, a high pressure (the stagnation pressure) is produced on the surface of the material, thereby cutting the material. Such liquid-jet cutters work well in air, but are difficult to control if the jet must pass through a liquid as often happens in surgery. Researchers have developed a new pulsed jet cutting tool that may allow surgeons to perform microsurgery on tissues that are immersed in water. Rather than using a steady water jet, the system uses *unsteady flow.* A high-energy electrical discharge inside the nozzle momentarily raises the temperature of the microjet to approximately 10,000 °C. This creates a rapidly expanding vapor bubble in the nozzle and expels a tiny fluid jet from the nozzle. Each electrical discharge creates a single, brief jet, which makes a small cut in the material.

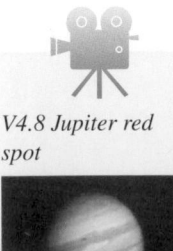

V4.8 Jupiter red spot

In many situations the unsteady character of a flow is quite random. That is, there is no repeatable sequence or regular variation to the unsteadiness. This behavior occurs in *turbulent flow* and is absent from *laminar flow.* The "smooth" flow of highly viscous syrup onto a pancake represents a "deterministic" laminar flow. It is quite different from the turbulent flow observed in the "irregular" splashing of water from a faucet onto the sink below it. The "irregular" gustiness of the wind represents another random turbulent flow. The differences between these types of flows are discussed in considerable detail in Chapters 8 and 9.

It must be understood that the definition of steady or unsteady flow pertains to the behavior of a fluid property as observed at a fixed point in space. For steady flow, the values of all fluid properties (velocity, temperature, density, etc.) at any fixed point are independent of time. However, the value of those properties for a given fluid particle may change with time as the particle flows along, even in steady flow. Thus, the temperature of the exhaust at the exit of a car's exhaust pipe may be constant for several hours, but the temperature of a fluid particle that left the exhaust pipe five minutes ago is lower now than it was when it left the pipe, even though the flow is steady.

4.1.4 Streamlines, Streaklines, and Pathlines

Although fluid motion can be quite complicated, there are various concepts that can be used to help in the visualization and analysis of flow fields. To this end we discuss the use of streamlines,

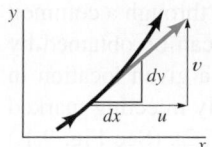

streaklines, and pathlines in flow analysis. The streamline is often used in analytical work while the streakline and pathline are often used in experimental work.

A *streamline* is a line that is everywhere tangent to the velocity field. If the flow is steady, nothing at a fixed point (including the velocity direction) changes with time, so the streamlines are fixed lines in space. (See the photograph at the beginning of Chapter 6.) For unsteady flows the streamlines may change shape with time. Streamlines are obtained analytically by integrating the equations defining lines tangent to the velocity field. As illustrated in the margin figure, for two-dimensional flows the slope of the streamline, dy/dx, must be equal to the tangent of the angle that the velocity vector makes with the x axis or

V4.9 Streamlines

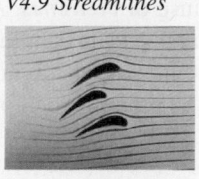

$$\frac{dy}{dx} = \frac{v}{u}. \tag{4.1}$$

If the velocity field is known as a function of x and y (and t if the flow is unsteady), this equation can be integrated to give the equation of the streamlines.

For unsteady flow there is no easy way to produce streamlines experimentally in the laboratory. As discussed below, the observation of dye, smoke, or some other tracer injected into a flow can provide useful information, but for unsteady flows it is not necessarily information about the streamlines.

EXAMPLE 4.2 Streamlines for a Given Velocity Field

GIVEN Consider the two-dimensional steady flow discussed in Example 4.1, $\mathbf{V} = (V_0/\ell)(-x\hat{\mathbf{i}} + y\hat{\mathbf{j}})$.

FIND Determine the streamlines for this flow.

SOLUTION

Since

$$u = (-V_0/\ell)x \text{ and } v = (V_0/\ell)y \tag{1}$$

it follows that streamlines are given by solution of the equation

$$\frac{dy}{dx} = \frac{v}{u} = \frac{(V_0/\ell)y}{-(V_0/\ell)x} = -\frac{y}{x}$$

in which variables can be separated and the equation integrated to give

$$\int \frac{dy}{y} = -\int \frac{dx}{x}$$

or

$$\ln y = -\ln x + \text{constant}$$

Thus, along the streamline

$$xy = C, \quad \text{where } C \text{ is a constant} \tag{Ans}$$

By using different values of the constant C, we can plot various lines in the x–y plane—the streamlines. The streamlines for $x \geq 0$ are plotted in Fig. E4.2. A comparison of this figure with Fig. E4.1a illustrates the fact that streamlines are lines tangent to the velocity field.

COMMENT Note that a flow is not completely specified by the shape of the streamlines alone. For example, the streamlines for the flow with $V_0/\ell = 10$ have the same shape as those for the flow with $V_0/\ell = -10$. However, the direction of the flow is opposite for these two cases. The arrows in Fig. E4.2 representing the flow direction are correct for $V_0/\ell = 10$ since, from Eq. 1, $u = -10x$ and $v = 10y$. That is, the flow is from right to left. For $V_0/\ell = -10$ the arrows are reversed. The flow is from left to right.

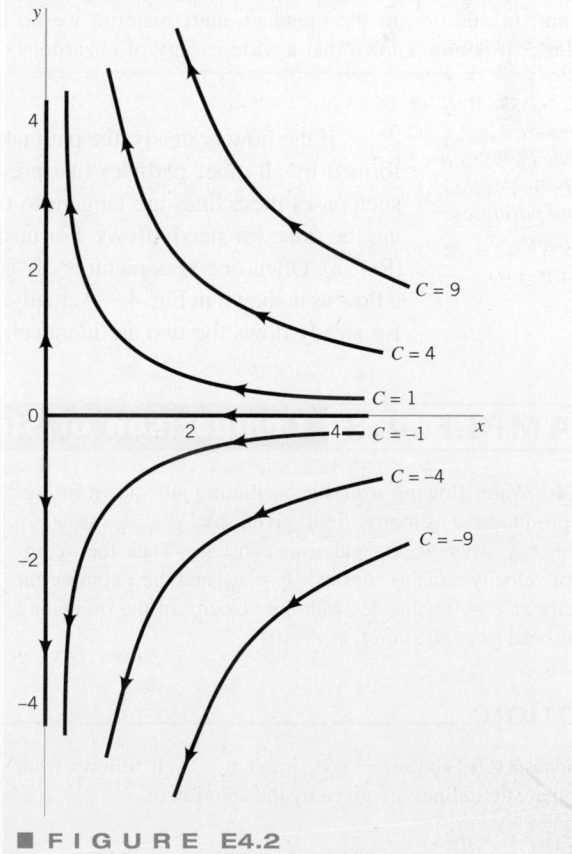

■ FIGURE E4.2

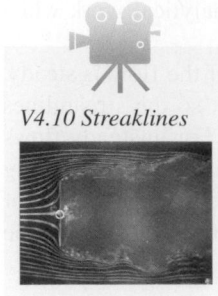

V4.10 Streaklines

A *streakline* consists of all particles in a flow that have previously passed through a common point. Streaklines are more of a laboratory tool than an analytical tool. They can be obtained by taking instantaneous photographs of marked particles that all passed through a given location in the flow field at some earlier time. Such a line can be produced by continuously injecting marked fluid (neutrally buoyant smoke in air, or dye in water) at a given location (Ref. 2). (See Fig. 9.1.) If the flow is steady, each successively injected particle follows precisely behind the previous one, forming a steady streakline that is exactly the same as the streamline through the injection point.

For unsteady flows, particles injected at the same point at different times need not follow the same path. An instantaneous photograph of the marked fluid would show the streakline at that instant, but it would not necessarily coincide with the streamline through the point of injection at that particular time nor with the streamline through the same injection point at a different time (see Example 4.3).

The third method used for visualizing and describing flows involves the use of *pathlines*. A pathline is the line traced out by a given particle as it flows from one point to another. The pathline is a Lagrangian concept that can be produced in the laboratory by marking a fluid particle (dying a small fluid element) and taking a time exposure photograph of its motion. (See the photograph at the beginning of Chapter 7.)

F l u i d s   i n   t h e   N e w s

Air bridge spanning the oceans It has long been known that large quantities of material are transported from one location to another by airborne dust particles. It is estimated that 2 billion metric tons (2×10^{12} kg) of dust are lifted into the atmosphere each year. Most of these particles settle out fairly rapidly, but significant amounts travel large distances. Scientists are beginning to understand the full impact of this phenomena—it is not only the tonnage transported, but the type of material transported that is significant. In addition to the mundane inert material we all term "dust," it is now known that a wide variety of hazardous materials and organisms are also carried along these literal *particle paths*. Satellite images reveal the amazing rate by which desert soils and other materials are transformed into airborne particles as a result of storms that produce strong winds. Once the tiny particles are aloft, they may travel thousands of miles, crossing the oceans and eventually being deposited on other continents. For the health and safety of all, it is important that we obtain a better understanding of the air bridges that span the oceans and also understand the ramification of such material transport.

For steady flow, streamlines, streaklines, and pathlines are the same.

If the flow is steady, the path taken by a marked particle (a pathline) will be the same as the line formed by all other particles that previously passed through the point of injection (a streakline). For such cases these lines are tangent to the velocity field. Hence, pathlines, streamlines, and streaklines are the same for steady flows. For unsteady flows none of these three types of lines need be the same (Ref. 3). Often one sees pictures of "streamlines" made visible by the injection of smoke or dye into a flow as is shown in Fig. 4.3. Actually, such pictures show streaklines rather than streamlines. However, for steady flows the two are identical; only the nomenclature is incorrectly used.

EXAMPLE 4.3   Comparison of Streamlines, Pathlines, and Streaklines

GIVEN Water flowing from the oscillating slit shown in Fig. E4.3a produces a velocity field given by $\mathbf{V} = u_0 \sin[\omega(t - y/v_0)]\hat{\mathbf{i}} + v_0\hat{\mathbf{j}}$, where u_0, v_0, and ω are constants. Thus, the y component of velocity remains constant ($v = v_0$) and the x component of velocity at $y = 0$ coincides with the velocity of the oscillating sprinkler head [$u = u_0 \sin(\omega t)$ at $y = 0$].

FIND (a) Determine the streamline that passes through the origin at $t = 0$; at $t = \pi/2\omega$. (b) Determine the pathline of the particle that was at the origin at $t = 0$; at $t = \pi/2$. (c) Discuss the shape of the streakline that passes through the origin.

SOLUTION

(a) Since $u = u_0 \sin[\omega(t - y/v_0)]$ and $v = v_0$ it follows from Eq. 4.1 that streamlines are given by the solution of

$$\frac{dy}{dx} = \frac{v}{u} = \frac{v_0}{u_0 \sin[\omega(t - y/v_0)]}$$

in which the variables can be separated and the equation integrated (for any given time t) to give

$$u_0 \int \sin\left[\omega\left(t - \frac{y}{v_0}\right)\right] dy = v_0 \int dx,$$

or

$$u_0(v_0/\omega) \cos\left[\omega\left(t - \frac{y}{v_0}\right)\right] = v_0 x + C \qquad (1)$$

where C is a constant. For the streamline at $t = 0$ that passes through the origin ($x = y = 0$), the value of C is obtained from Eq. 1 as $C = u_0 v_0/\omega$. Hence, the equation for this streamline is

$$x = \frac{u_0}{\omega}\left[\cos\left(\frac{\omega y}{v_0}\right) - 1\right] \qquad (2) \quad (Ans)$$

Similarly, for the streamline at $t = \pi/2\omega$ that passes through the origin, Eq. 1 gives $C = 0$. Thus, the equation for this streamline is

$$x = \frac{u_0}{\omega}\cos\left[\omega\left(\frac{\pi}{2\omega} - \frac{y}{v_0}\right)\right] = \frac{u_0}{\omega}\cos\left(\frac{\pi}{2} - \frac{\omega y}{v_0}\right)$$

or

$$x = \frac{u_0}{\omega}\sin\left(\frac{\omega y}{v_0}\right) \qquad (3) \quad (Ans)$$

COMMENT   These two streamlines, plotted in Fig. E4.3b, are not the same because the flow is unsteady. For example, at the origin ($x = y = 0$) the velocity is $\mathbf{V} = v_0\hat{\mathbf{j}}$ at $t = 0$ and $\mathbf{V} = u_0\hat{\mathbf{i}} + v_0\hat{\mathbf{j}}$ at $t = \pi/2\omega$. Thus, the angle of the streamline passing through the origin changes with time. Similarly, the shape of the entire streamline is a function of time.

(b)   The pathline of a particle (the location of the particle as a function of time) can be obtained from the velocity field and the definition of the velocity. Since $u = dx/dt$ and $v = dy/dt$ we obtain

$$\frac{dx}{dt} = u_0 \sin\left[\omega\left(t - \frac{y}{v_0}\right)\right] \quad \text{and} \quad \frac{dy}{dt} = v_0$$

The y equation can be integrated (since $v_0 = $ constant) to give the y coordinate of the pathline as

$$y = v_0 t + C_1 \qquad (4)$$

where C_1 is a constant. With this known $y = y(t)$ dependence, the x equation for the pathline becomes

$$\frac{dx}{dt} = u_0 \sin\left[\omega\left(t - \frac{v_0 t + C_1}{v_0}\right)\right] = -u_0 \sin\left(\frac{C_1\omega}{v_0}\right)$$

This can be integrated to give the x component of the pathline as

$$x = -\left[u_0 \sin\left(\frac{C_1\omega}{v_0}\right)\right]t + C_2 \qquad (5)$$

where C_2 is a constant. For the particle that was at the origin ($x = y = 0$) at time $t = 0$, Eqs. 4 and 5 give $C_1 = C_2 = 0$. Thus, the pathline is

$$x = 0 \quad \text{and} \quad y = v_0 t \qquad (6) \quad (Ans)$$

Similarly, for the particle that was at the origin at $t = \pi/2\omega$, Eqs. 4 and 5 give $C_1 = -\pi v_0/2\omega$ and $C_2 = -\pi u_0/2\omega$. Thus, the pathline for this particle is

$$x = u_0\left(t - \frac{\pi}{2\omega}\right) \quad \text{and} \quad y = v_0\left(t - \frac{\pi}{2\omega}\right) \qquad (7)$$

The pathline can be drawn by plotting the locus of $x(t)$, $y(t)$ values for $t \geq 0$ or by eliminating the parameter t from Eq. 7 to give

$$y = \frac{v_0}{u_0}x \qquad (8) \quad (Ans)$$

COMMENT   The pathlines given by Eqs. 6 and 8, shown in Fig. E4.3c, are straight lines from the origin (rays). The pathlines and streamlines do not coincide because the flow is unsteady.

(c)   The streakline through the origin at time $t = 0$ is the locus of particles at $t = 0$ that previously ($t < 0$) passed through the origin. The general shape of the streaklines can be seen as follows. Each particle that flows through the origin travels in a straight line (pathlines are rays from the origin), the slope of which lies between $\pm v_0/u_0$ as shown in Fig. E4.3d. Particles passing through the origin at different times are located on different rays from the origin and at different distances from the origin. The net result is that a stream of dye continually injected at the origin (a streakline) would have the shape shown in Fig. E4.3d. Because of the unsteadiness, the streakline will vary with time, although it will always have the oscillating, sinuous character shown.

COMMENT   Similar streaklines are given by the stream of water from a garden hose nozzle that oscillates back and forth in a direction normal to the axis of the nozzle.

In this example neither the streamlines, pathlines, nor streaklines coincide. If the flow were steady, all of these lines would be the same.

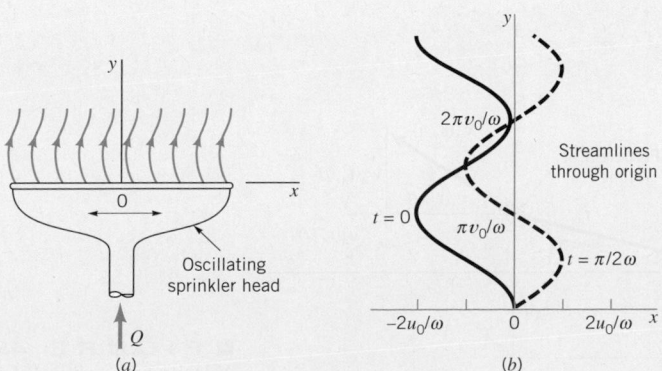

■ FIGURE E4.3(a), (b)

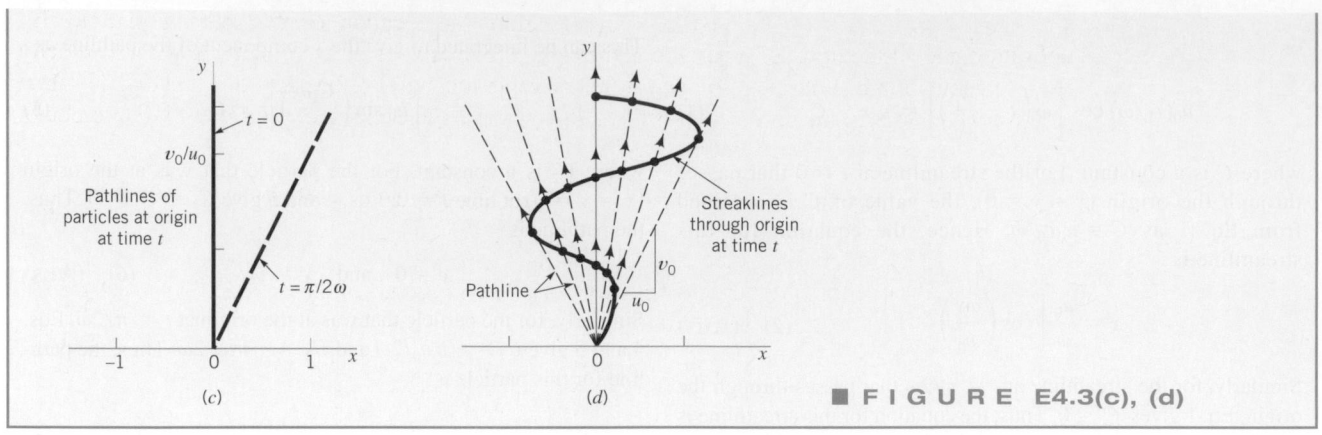

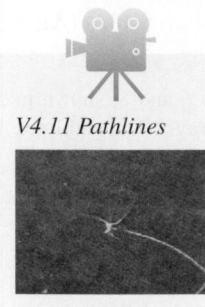

■ FIGURE E4.3(c), (d)

4.2 The Acceleration Field

V4.11 Pathlines

Acceleration is the time rate of change of velocity for a given particle.

As indicated in the previous section, we can describe fluid motion by either (1) following individual particles (Lagrangian description) or (2) remaining fixed in space and observing different particles as they pass by (Eulerian description). In either case, to apply Newton's second law ($\mathbf{F} = m\mathbf{a}$) we must be able to describe the particle acceleration in an appropriate fashion. For the infrequently used Lagrangian method, we describe the fluid acceleration just as is done in solid body dynamics—$\mathbf{a} = \mathbf{a}(t)$ for each particle. For the Eulerian description we describe the *acceleration field* as a function of position and time without actually following any particular particle. This is analogous to describing the flow in terms of the velocity field, $\mathbf{V} = \mathbf{V}(x, y, z, t)$, rather than the velocity for particular particles. In this section we will discuss how to obtain the acceleration field if the velocity field is known.

The acceleration of a particle is the time rate of change of its velocity. For unsteady flows the velocity at a given point in space (occupied by different particles) may vary with time, giving rise to a portion of the fluid acceleration. In addition, a fluid particle may experience an acceleration because its velocity changes as it flows from one point to another in space. For example, water flowing through a garden hose nozzle under steady conditions (constant number of liters per minute from the hose) will experience an acceleration as it changes from its relatively low velocity in the hose to its relatively high velocity at the tip of the nozzle.

4.2.1 The Material Derivative

Consider a fluid particle moving along its pathline as is shown in Fig. 4.4. In general, the particle's velocity, denoted $\mathbf{V}_A$ for particle A, is a function of its location and the time. That is,

$$\mathbf{V}_A = \mathbf{V}_A(\mathbf{r}_A, t) = \mathbf{V}_A[x_A(t), y_A(t), z_A(t), t]$$

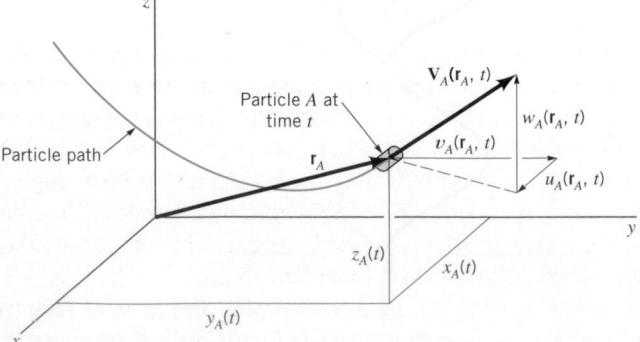

■ FIGURE 4.4
Velocity and position of particle A at time t.

where $x_A = x_A(t)$, $y_A = y_A(t)$, and $z_A = z_A(t)$ define the location of the moving particle. By definition, the acceleration of a particle is the time rate of change of its velocity. Since the velocity may be a function of both position and time, its value may change because of the change in time as well as a change in the particle's position. Thus, we use the chain rule of differentiation to obtain the acceleration of particle A, denoted $\mathbf{a}_A$, as

$$\mathbf{a}_A(t) = \frac{d\mathbf{V}_A}{dt} = \frac{\partial \mathbf{V}_A}{\partial t} + \frac{\partial \mathbf{V}_A}{\partial x}\frac{dx_A}{dt} + \frac{\partial \mathbf{V}_A}{\partial y}\frac{dy_A}{dt} + \frac{\partial \mathbf{V}_A}{\partial z}\frac{dz_A}{dt} \tag{4.2}$$

Using the fact that the particle velocity components are given by $u_A = dx_A/dt$, $v_A = dy_A/dt$, and $w_A = dz_A/dt$, Eq. 4.2 becomes

$$\mathbf{a}_A = \frac{\partial \mathbf{V}_A}{\partial t} + u_A\frac{\partial \mathbf{V}_A}{\partial x} + v_A\frac{\partial \mathbf{V}_A}{\partial y} + w_A\frac{\partial \mathbf{V}_A}{\partial z}$$

Since the above is valid for any particle, we can drop the reference to particle A and obtain the acceleration field from the velocity field as

$$\mathbf{a} = \frac{\partial \mathbf{V}}{\partial t} + u\frac{\partial \mathbf{V}}{\partial x} + v\frac{\partial \mathbf{V}}{\partial y} + w\frac{\partial \mathbf{V}}{\partial z} \tag{4.3}$$

This is a vector result whose scalar components can be written as

$$a_x = \frac{\partial u}{\partial t} + u\frac{\partial u}{\partial x} + v\frac{\partial u}{\partial y} + w\frac{\partial u}{\partial z}$$

$$a_y = \frac{\partial v}{\partial t} + u\frac{\partial v}{\partial x} + v\frac{\partial v}{\partial y} + w\frac{\partial v}{\partial z} \tag{4.4}$$

and

$$a_z = \frac{\partial w}{\partial t} + u\frac{\partial w}{\partial x} + v\frac{\partial w}{\partial y} + w\frac{\partial w}{\partial z}$$

where a_x, a_y, and a_z are the x, y, and z components of the acceleration.

The above result is often written in shorthand notation as

The material derivative is used to describe time rates of change for a given particle.

$$\mathbf{a} = \frac{D\mathbf{V}}{Dt}$$

where the operator

$$\frac{D(\ )}{Dt} \equiv \frac{\partial(\ )}{\partial t} + u\frac{\partial(\ )}{\partial x} + v\frac{\partial(\ )}{\partial y} + w\frac{\partial(\ )}{\partial z} \tag{4.5}$$

is termed the *material derivative* or *substantial derivative*. An often-used shorthand notation for the material derivative operator is

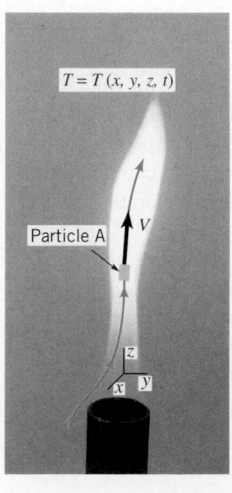

$$\frac{D(\ )}{Dt} = \frac{\partial(\ )}{\partial t} + (\mathbf{V} \cdot \nabla)(\ ) \tag{4.6}$$

The dot product of the velocity vector, $\mathbf{V}$, and the gradient operator, $\nabla(\ ) = \partial(\ )/\partial x\,\hat{\mathbf{i}} + \partial(\ )/\partial y\,\hat{\mathbf{j}} + \partial(\ )/\partial z\,\hat{\mathbf{k}}$ (a vector operator) provides a convenient notation for the spatial derivative terms appearing in the Cartesian coordinate representation of the material derivative. Note that the notation $\mathbf{V} \cdot \nabla$ represents the operator $\mathbf{V} \cdot \nabla(\ ) = u\partial(\ )/\partial x + v\partial(\ )/\partial y + w\partial(\ )/\partial z$.

The material derivative concept is very useful in analysis involving various fluid parameters, not just the acceleration. The material derivative of any variable is the rate at which that variable changes with time for a given particle (as seen by one moving along with the fluid—the Lagrangian description). For example, consider a temperature field $T = T(x, y, z, t)$ associated with a given flow, like the flame shown in the figure in the margin. It may be of interest to determine the time rate of change of temperature of a fluid particle (particle A) as it moves through this temperature field. If the velocity, $\mathbf{V} = \mathbf{V}(x, y, z, t)$, is known, we can apply the chain rule to determine the rate of change of temperature as

$$\frac{dT_A}{dt} = \frac{\partial T_A}{\partial t} + \frac{\partial T_A}{\partial x}\frac{dx_A}{dt} + \frac{\partial T_A}{\partial y}\frac{dy_A}{dt} + \frac{\partial T_A}{\partial z}\frac{dz_A}{dt}$$

This can be written as

$$\frac{DT}{Dt} = \frac{\partial T}{\partial t} + u\frac{\partial T}{\partial x} + v\frac{\partial T}{\partial y} + w\frac{\partial T}{\partial z} = \frac{\partial T}{\partial t} + \mathbf{V}\cdot\nabla T$$

As in the determination of the acceleration, the material derivative operator, $D(\ )/Dt$, appears.

EXAMPLE 4.4 — Acceleration along a Streamline

GIVEN An incompressible, inviscid fluid flows steadily past a ball of radius R, as shown in Fig. E4.4a. According to a more advanced analysis of the flow, the fluid velocity along streamline A–B is given by

$$\mathbf{V} = u(x)\hat{\mathbf{i}} = V_0\left(1 + \frac{R^3}{x^3}\right)\hat{\mathbf{i}}$$

where V_0 is the upstream velocity far ahead of the sphere.

FIND Determine the acceleration experienced by fluid particles as they flow along this streamline.

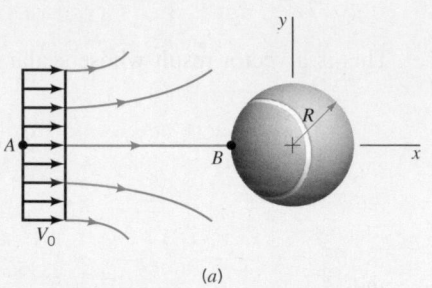

(a)

SOLUTION

Along streamline A–B there is only one component of velocity ($v = w = 0$) so that from Eq. 4.3

$$\mathbf{a} = \frac{\partial \mathbf{V}}{\partial t} + u\frac{\partial \mathbf{V}}{\partial x} = \left(\frac{\partial u}{\partial t} + u\frac{\partial u}{\partial x}\right)\hat{\mathbf{i}}$$

or

$$a_x = \frac{\partial u}{\partial t} + u\frac{\partial u}{\partial x}, \qquad a_y = 0, \qquad a_z = 0$$

Since the flow is steady the velocity at a given point in space does not change with time. Thus, $\partial u/\partial t = 0$. With the given velocity distribution along the streamline, the acceleration becomes

$$a_x = u\frac{\partial u}{\partial x} = V_0\left(1 + \frac{R^3}{x^3}\right)V_0[R^3(-3x^{-4})]$$

or

$$a_x = -3(V_0^2/R)\frac{1 + (R/x)^3}{(x/R)^4} \qquad \text{(Ans)}$$

COMMENTS Along streamline A–B ($-\infty \le x \le -R$ and $y = 0$) the acceleration has only an x component and it is negative (a deceleration). Thus, the fluid slows down from its upstream

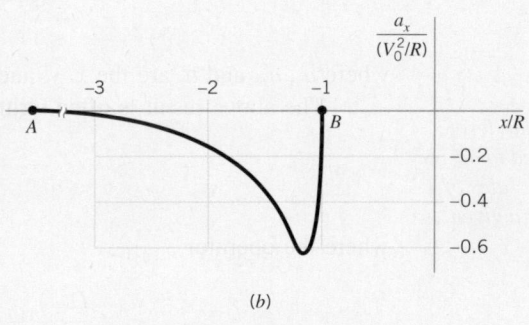

(b)

■ **FIGURE E4.4**

velocity of $\mathbf{V} = V_0\hat{\mathbf{i}}$ at $x = -\infty$ to its stagnation point velocity of $\mathbf{V} = 0$ at $x = -R$, the "nose" of the ball. The variation of a_x along streamline A–B is shown in Fig. E4.4b. It is the same result as is obtained in Example 3.1 by using the streamwise component of the acceleration, $a_x = V\,\partial V/\partial s$. The maximum deceleration occurs at $x = -1.205R$ and has a value of $a_{x,max} = -0.610\,V_0^2/R$. Note that this maximum deceleration increases with increasing velocity and decreasing size. As indicated in the following table, typical values of this deceleration can be quite large. For example, the $a_{x,max} = -1.2 \times 10^4$ m/s^2 value for a pitched baseball is a deceleration approximately 1500 times that of gravity.

Object	V_0 (m/s)	R (m)	$a_{x,max}$ (m/s²)
Rising weather balloon	0.30	1.2	−0.047
Soccer ball	6.00	0.24	−93.0
Baseball	25.00	0.037	-1.2×10^4
Tennis ball	30.00	0.032	-1.8×10^4
Golf ball	60.00	0.021	-1.1×10^5

In general, for fluid particles on streamlines other than $A–B$, all three components of the acceleration (a_x, a_y, and a_z) will be nonzero.

4.2.2 Unsteady Effects

The local derivative is a result of the unsteadiness of the flow.

As is seen from Eq. 4.5, the material derivative formula contains two types of terms—those involving the time derivative $[\partial(\ )/\partial t]$ and those involving spatial derivatives $[\partial(\ )/\partial x, \partial(\ )/\partial y,$ and $\partial(\ )/\partial z]$. The time derivative portions are denoted as the *local derivative*. They represent effects of the unsteadiness of the flow. If the parameter involved is the acceleration, that portion given by $\partial \mathbf{V}/\partial t$ is termed the ***local acceleration.*** For steady flow the time derivative is zero throughout the flow field $[\partial(\ )/\partial t \equiv 0]$, and the local effect vanishes. Physically, there is no change in flow parameters at a fixed point in space if the flow is steady. There may be a change of those parameters for a fluid particle as it moves about, however.

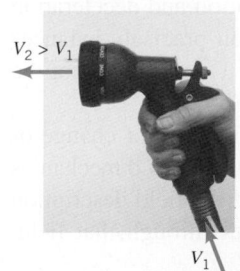

V4.12 Unsteady flow

If a flow is unsteady, its parameter values (velocity, temperature, density, etc.) at any location may change with time. For example, an unstirred ($\mathbf{V} = 0$) cup of coffee will cool down in time because of heat transfer to its surroundings. That is, $DT/Dt = \partial T/\partial t + \mathbf{V} \cdot \nabla T = \partial T/\partial t < 0$. Similarly, a fluid particle may have nonzero acceleration as a result of the unsteady effect of the flow. Consider flow in a constant diameter pipe as is shown in Fig. 4.5. The flow is assumed to be spatially uniform throughout the pipe. That is, $\mathbf{V} = V_0(t) \,\hat{\mathbf{i}}$ at all points in the pipe. The value of the acceleration depends on whether V_0 is being increased, $\partial V_0/\partial t > 0$, or decreased, $\partial V_0/\partial t < 0$. Unless V_0 is independent of time ($V_0 \equiv$ constant) there will be an acceleration, the local acceleration term. Thus, the acceleration field, $\mathbf{a} = \partial V_0/\partial t \,\hat{\mathbf{i}}$, is uniform throughout the entire flow, although it may vary with time ($\partial V_0/\partial t$ need not be constant). The acceleration due to the spatial variations of velocity ($u\,\partial u/\partial x, v\,\partial v/\partial y,$ etc.) vanishes automatically for this flow, since $\partial u/\partial x = 0$ and $v = w = 0$. That is,

$$\mathbf{a} = \frac{\partial \mathbf{V}}{\partial t} + u\frac{\partial \mathbf{V}}{\partial x} + v\frac{\partial \mathbf{V}}{\partial y} + w\frac{\partial \mathbf{V}}{\partial z} = \frac{\partial \mathbf{V}}{\partial t} = \frac{\partial V_0}{\partial t}\hat{\mathbf{i}}$$

4.2.3 Convective Effects

$V_2 > V_1$

The portion of the material derivative (Eq. 4.5) represented by the spatial derivatives is termed the *convective derivative*. It represents the fact that a flow property associated with a fluid particle may vary because of the motion of the particle from one point in space where the parameter has one value to another point in space where its value is different. For example, the water velocity at the inlet of the garden hose nozzle shown in the figure in the margin is different (both in direction and speed) than it is at the exit. This contribution to the time rate of change of the parameter for the particle can occur whether the flow is steady or unsteady.

V_1

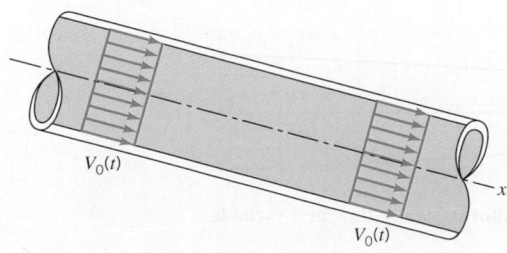

■ **FIGURE 4.5** **Uniform, unsteady flow in a constant diameter pipe.**

$V_0(t)$

x

$V_0(t)$

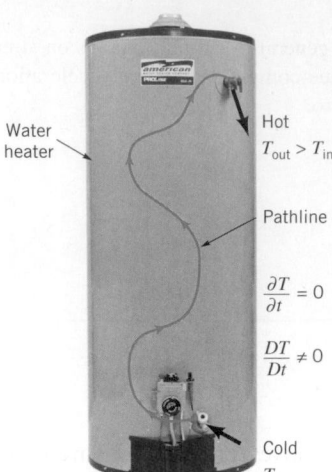

Water heater

Hot
$T_{out} > T_{in}$

Pathline

$\dfrac{\partial T}{\partial t} = 0$

$\dfrac{DT}{Dt} \neq 0$

Cold
T_{in}

■ **F I G U R E   4.6**   **Steady-state operation of a water heater. (Photo courtesy of American Water Heater Company.)**

The convective derivative is a result of the spatial variation of the flow.

It is due to the convection, or motion, of the particle through space in which there is a gradient $[\nabla(\ ) = \partial(\ )/\partial x\,\hat{\mathbf{i}} + \partial(\ )/\partial y\,\hat{\mathbf{j}} + \partial(\ )/\partial z\,\hat{\mathbf{k}}]$ in the parameter value. That portion of the acceleration given by the term $(\mathbf{V} \cdot \nabla)\mathbf{V}$ is termed the *convective acceleration*.

As is illustrated in Fig. 4.6, the temperature of a water particle changes as it flows through a water heater. The water entering the heater is always the same cold temperature and the water leaving the heater is always the same hot temperature. The flow is steady. However, the temperature, T, of each water particle increases as it passes through the heater—$T_{out} > T_{in}$. Thus, $DT/Dt \neq 0$ because of the convective term in the total derivative of the temperature. That is, $\partial T/\partial t = 0$, but $u\,\partial T/\partial x \neq 0$ (where x is directed along the streamline), since there is a nonzero temperature gradient along the streamline. A fluid particle traveling along this nonconstant temperature path ($\partial T/\partial x \neq 0$) at a specified speed ($u$) will have its temperature change with time at a rate of $DT/Dt = u\,\partial T/\partial x$ even though the flow is steady ($\partial T/\partial t = 0$).

The same types of processes are involved with fluid accelerations. Consider flow in a variable area pipe as shown in Fig. 4.7. It is assumed that the flow is steady and one-dimensional with velocity that increases and decreases in the flow direction as indicated. As the fluid flows from section (1) to section (2), its velocity increases from V_1 to V_2. Thus, even though $\partial \mathbf{V}/\partial t = 0$ (steady flow), fluid particles experience an acceleration given by $a_x = u\,\partial u/\partial x$ (convective acceleration). For $x_1 < x < x_2$, it is seen that $\partial u/\partial x > 0$ so that $a_x > 0$—the fluid accelerates. For $x_2 < x < x_3$, it is seen that $\partial u/\partial x < 0$ so that $a_x < 0$—the fluid decelerates. This acceleration and deceleration are shown in the figure in the margin. If $V_1 = V_3$, the amount of acceleration precisely balances the amount of deceleration even though the distances between x_2 and x_1 and x_3 and x_2 are not the same.

The concept of the material derivative can be used to determine the time rate of change of any parameter associated with a particle as it moves about. Its use is not restricted to fluid mechanics alone. The basic ingredients needed to use the material derivative concept are the field description of the parameter, $P = P(x, y, z, t)$, and the rate at which the particle moves through that field, $\mathbf{V} = \mathbf{V}(x, y, z, t)$.

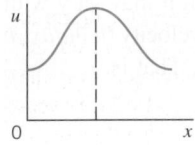

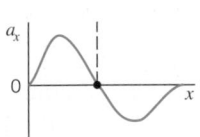

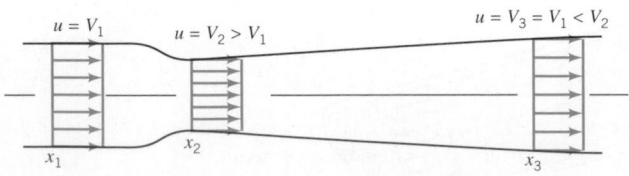

$u = V_1$

$u = V_2 > V_1$

$u = V_3 = V_1 < V_2$

x_1 x_2 x_3

■ **F I G U R E   4.7**   **Uniform, steady flow in a variable area pipe.**

EXAMPLE 4.5 | **Acceleration from a Given Velocity Field**

GIVEN   Consider the steady, two-dimensional flow field discussed in Example 4.2.

FIND   Determine the acceleration field for this flow.

SOLUTION

In general, the acceleration is given by

$$\mathbf{a} = \frac{D\mathbf{V}}{Dt} = \frac{\partial \mathbf{V}}{\partial t} + (\mathbf{V} \cdot \nabla)(\mathbf{V})$$

$$= \frac{\partial \mathbf{V}}{\partial t} + u \frac{\partial \mathbf{V}}{\partial x} + v \frac{\partial \mathbf{V}}{\partial y} + w \frac{\partial \mathbf{V}}{\partial z} \qquad (1)$$

where the velocity is given by $\mathbf{V} = (V_0/\ell)(-x\hat{\mathbf{i}} + y\hat{\mathbf{j}})$ so that $u = -(V_0/\ell)x$ and $v = (V_0/\ell)y$. For steady $[\partial(\ )/\partial t = 0]$, two-dimensional $[w = 0$ and $\partial(\ )/\partial z = 0]$ flow, Eq. 1 becomes

$$\mathbf{a} = u \frac{\partial \mathbf{V}}{\partial x} + v \frac{\partial \mathbf{V}}{\partial y}$$

$$= \left(u \frac{\partial u}{\partial x} + v \frac{\partial u}{\partial y}\right)\hat{\mathbf{i}} + \left(u \frac{\partial v}{\partial x} + v \frac{\partial v}{\partial y}\right)\hat{\mathbf{j}}$$

Hence, for this flow the acceleration is given by

$$\mathbf{a} = \left[\left(-\frac{V_0}{\ell}\right)(x)\left(-\frac{V_0}{\ell}\right) + \left(\frac{V_0}{\ell}\right)(y)(0)\right]\hat{\mathbf{i}}$$

$$+ \left[\left(-\frac{V_0}{\ell}\right)(x)(0) + \left(\frac{V_0}{\ell}\right)(y)\left(\frac{V_0}{\ell}\right)\right]\hat{\mathbf{j}}$$

or

$$a_x = \frac{V_0^2 x}{\ell^2}, \qquad a_y = \frac{V_0^2 y}{\ell^2} \qquad \textbf{(Ans)}$$

COMMENTS   The fluid experiences an acceleration in both the x and y directions. Since the flow is steady, there is no local acceleration—the fluid velocity at any given point is constant in time. However, there is a convective acceleration due to the change in velocity from one point on the particle's pathline to another. Recall that the velocity is a vector—it has both a magnitude and a direction. In this flow both the fluid speed (magnitude) and flow direction change with location (see Fig. E4.1a).

For this flow the magnitude of the acceleration is constant on circles centered at the origin, as is seen from the fact that

$$|\mathbf{a}| = (a_x^2 + a_y^2 + a_z^2)^{1/2} = \left(\frac{V_0}{\ell}\right)^2 (x^2 + y^2)^{1/2} \qquad (2)$$

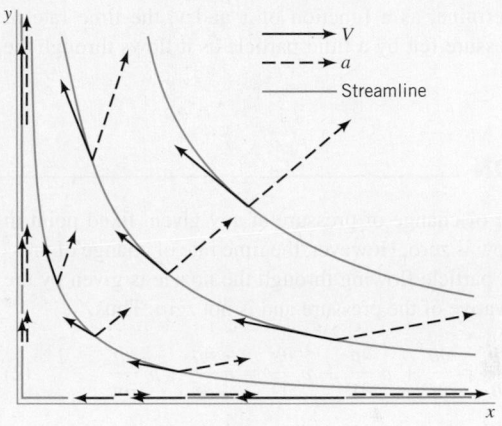

■ **F I G U R E   E4.5**

Also, the acceleration vector is oriented at an angle θ from the x axis, where

$$\tan \theta = \frac{a_y}{a_x} = \frac{y}{x}$$

This is the same angle as that formed by a ray from the origin to point (x, y). Thus, the acceleration is directed along rays from the origin and has a magnitude proportional to the distance from the origin. Typical acceleration vectors (from Eq. 2) and velocity vectors (from Example 4.1) are shown in Fig. E4.5 for the flow in the first quadrant. Note that **a** and **V** are not parallel except along the x and y axes (a fact that is responsible for the curved pathlines of the flow), and that both the acceleration and velocity are zero at the origin ($x = y = 0$). An infinitesimal fluid particle placed precisely at the origin will remain there, but its neighbors (no matter how close they are to the origin) will drift away.

EXAMPLE 4.6 | **The Material Derivative**

GIVEN   A fluid flows steadily through a two-dimensional nozzle of length ℓ as shown in Fig. E4.6a. The nozzle shape is given by

$$y/\ell = \pm 0.5/[1 + (x/\ell)]$$

If viscous and gravitational effects are negligible, the velocity field is approximately

$$u = V_0[1 + x/\ell], \quad v = -V_0 y/\ell \qquad (1)$$

and the pressure field is

$$p - p_0 = -(\rho V_0^2/2)[(x^2 + y^2)/\ell^2 + 2x/\ell]$$

where V_0 and p_0 are the velocity and pressure at the origin, $x = y = 0$. Note that the fluid speed increases as it flows through the nozzle. For example, along the center line $(y = 0)$, $V = V_0$ at $x = 0$ and $V = 2V_0$ at $x = \ell$.

FIND Determine, as a function of x and y, the time rate of change of pressure felt by a fluid particle as it flows through the nozzle.

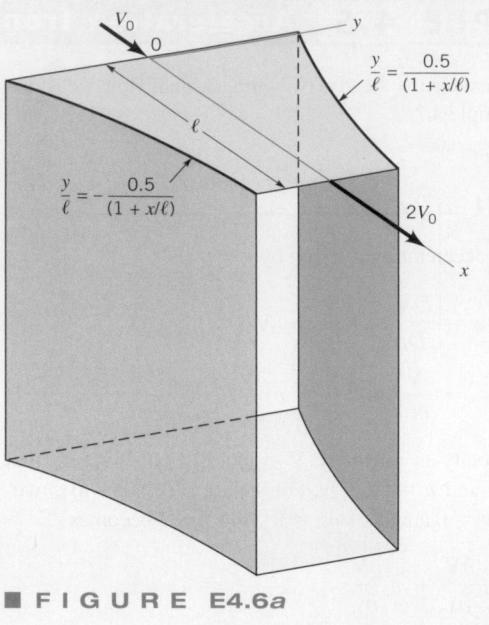

■ **FIGURE E4.6a**

SOLUTION

The time rate of change of pressure at any given, fixed point in this steady flow is zero. However, the time rate of change of pressure felt by a particle flowing through the nozzle is given by the material derivative of the pressure and is not zero. Thus,

$$\frac{Dp}{Dt} = \frac{\partial p}{\partial t} + u\frac{\partial p}{\partial x} + v\frac{\partial p}{\partial y} = u\frac{\partial p}{\partial x} + v\frac{\partial p}{\partial y} \qquad (2)$$

where the x- and y-components of the pressure gradient can be written as

$$\frac{\partial p}{\partial x} = -\frac{\rho V_0^2}{\ell}\left(\frac{x}{\ell} + 1\right) \qquad (3)$$

and

$$\frac{\partial p}{\partial y} = -\frac{\rho V_0^2}{\ell}\left(\frac{y}{\ell}\right) \qquad (4)$$

Therefore, by combining Eqs. (1), (2), (3), and (4) we obtain

$$\frac{Dp}{Dt} = V_0\left(1 + \frac{x}{\ell}\right)\left(-\frac{\rho V_0^2}{\ell}\right)\left(\frac{x}{\ell} + 1\right) + \left(-V_0\frac{y}{\ell}\right)\left(-\frac{\rho V_0^2}{\ell}\right)\left(\frac{y}{\ell}\right)$$

or

$$\frac{Dp}{Dt} = -\frac{\rho V_0^3}{\ell}\left[\left(\frac{x}{\ell} + 1\right)^2 - \left(\frac{y}{\ell}\right)^2\right] \qquad (5) \quad \textbf{(Ans)}$$

COMMENT Lines of constant pressure within the nozzle are indicated in Fig. E4.6b, along with some representative streamlines of the flow. Note that as a fluid particle flows along its streamline, it moves into areas of lower and lower pressure. Hence, even though the flow is steady, the time rate of change of the pressure for any given particle is negative. This can be verified from Eq. (5) which, when plotted in Fig. E4.6c, shows that for any point within the nozzle $Dp/Dt < 0$.

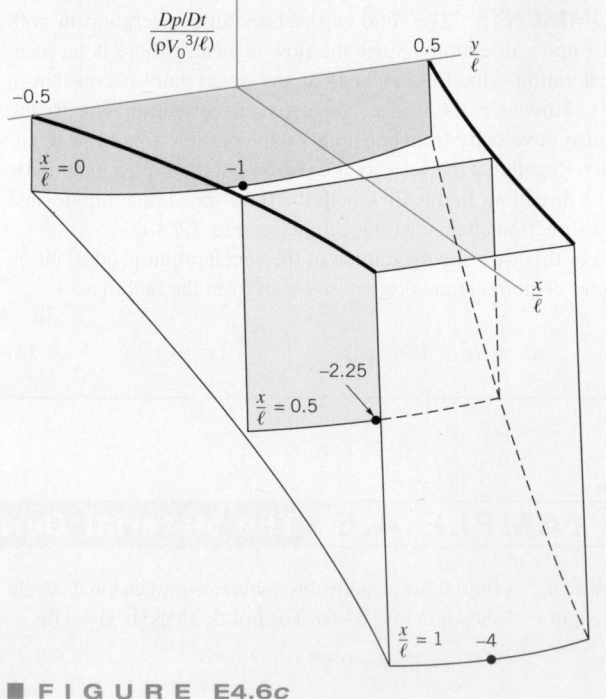

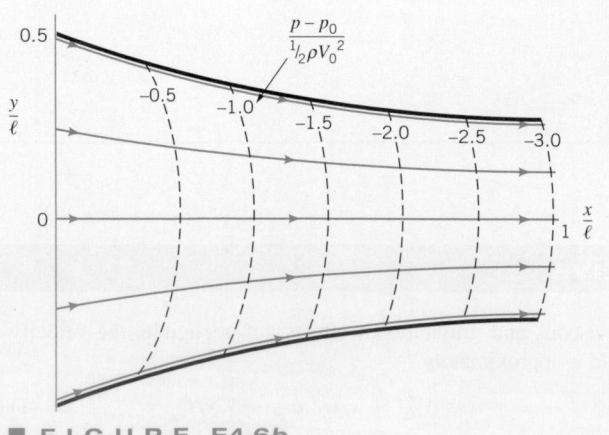

■ **FIGURE E4.6b**

■ **FIGURE E4.6c**

4.2.4 Streamline Coordinates

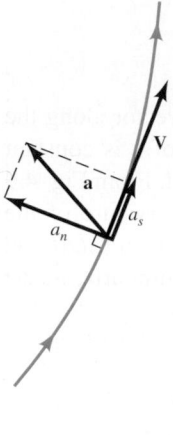

V4.13 Streamline coordinates

In many flow situations it is convenient to use a coordinate system defined in terms of the streamlines of the flow. An example for steady, two-dimensional flows is illustrated in Fig. 4.8. Such flows can be described either in terms of the usual x, y Cartesian coordinate system (or some other system such as the r, θ polar coordinate system) or the streamline coordinate system. In the streamline coordinate system the flow is described in terms of one coordinate along the streamlines, denoted s, and the second coordinate normal to the streamlines, denoted n. Unit vectors in these two directions are denoted by $\hat{s}$ and $\hat{n}$, as shown in the figure. Care is needed not to confuse the coordinate distance s (a scalar) with the unit vector along the streamline direction, $\hat{s}$.

The flow plane is therefore covered by an orthogonal curved net of coordinate lines. At any point the s and n directions are perpendicular, but the lines of constant s or constant n are not necessarily straight. Without knowing the actual velocity field (hence, the streamlines) it is not possible to construct this flow net. In many situations appropriate simplifying assumptions can be made so that this lack of information does not present an insurmountable difficulty. One of the major advantages of using the streamline coordinate system is that the velocity is always tangent to the s direction. That is,

$$\mathbf{V} = V\hat{s}$$

This allows simplifications in describing the fluid particle acceleration and in solving the equations governing the flow.

For steady, two-dimensional flow we can determine the acceleration as

$$\mathbf{a} = \frac{D\mathbf{V}}{Dt} = a_s\hat{s} + a_n\hat{n}$$

where a_s and a_n are the streamline and normal components of acceleration, respectively, as indicated by the figure in the margin. We use the material derivative because by definition the acceleration is the time rate of change of the velocity of a given particle as it moves about. If the streamlines

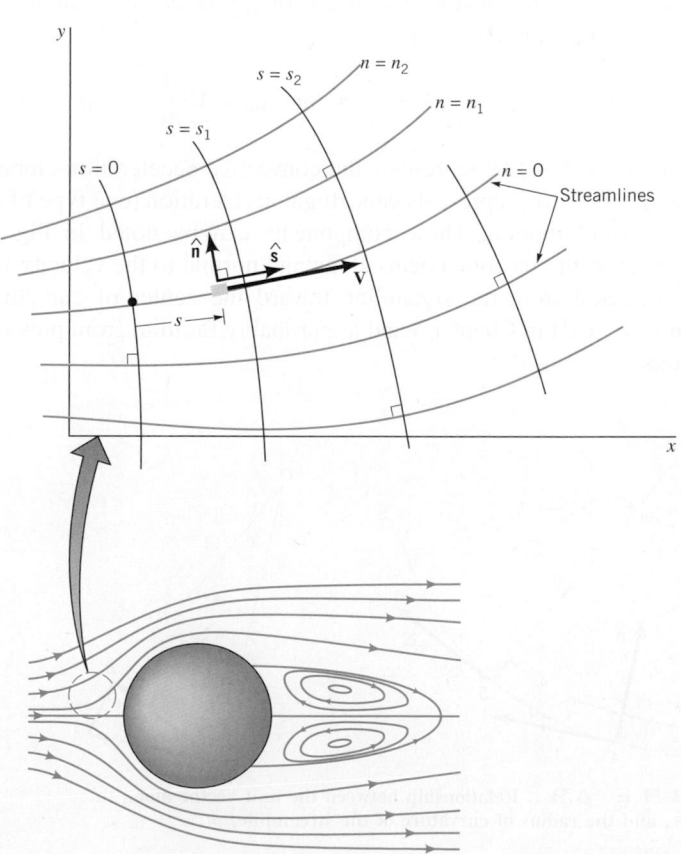

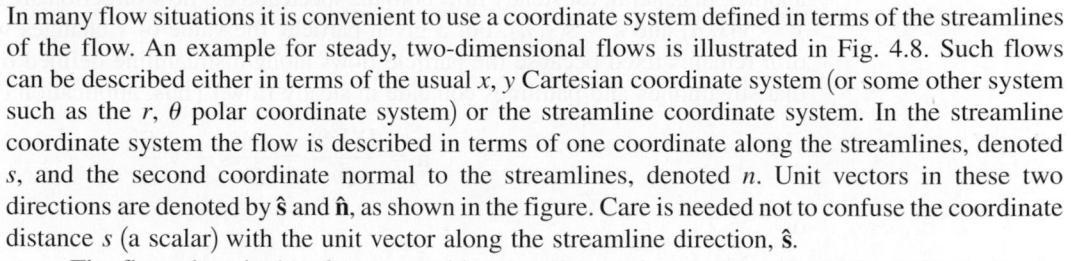

■ **FIGURE 4.8**
Streamline coordinate system for two-dimensional flow.

are curved, both the speed of the particle and its direction of flow may change from one point to another. In general, for steady flow both the speed and the flow direction are a function of location—$V = V(s, n)$ and $\hat{s} = \hat{s}(s, n)$. For a given particle, the value of s changes with time, but the value of n remains fixed because the particle flows along a streamline defined by n = constant. (Recall that streamlines and pathlines coincide in steady flow.) Thus, application of the chain rule gives

$$\mathbf{a} = \frac{D(V\hat{s})}{Dt} = \frac{DV}{Dt}\hat{s} + V\frac{D\hat{s}}{Dt}$$

or

$$\mathbf{a} = \left(\frac{\partial V}{\partial t} + \frac{\partial V}{\partial s}\frac{ds}{dt} + \frac{\partial V}{\partial n}\frac{dn}{dt}\right)\hat{s} + V\left(\frac{\partial \hat{s}}{\partial t} + \frac{\partial \hat{s}}{\partial s}\frac{ds}{dt} + \frac{\partial \hat{s}}{\partial n}\frac{dn}{dt}\right)$$

This can be simplified by using the fact that for steady flow nothing changes with time at a given point so that both $\partial V/\partial t$ and $\partial \hat{s}/\partial t$ are zero. Also, the velocity along the streamline is $V = ds/dt$ and the particle remains on its streamline (n = constant) so that $dn/dt = 0$. Hence,

$$\mathbf{a} = \left(V\frac{\partial V}{\partial s}\right)\hat{s} + V\left(V\frac{\partial \hat{s}}{\partial s}\right)$$

The quantity $\partial \hat{s}/\partial s$ represents the limit as $\delta s \to 0$ of the change in the unit vector along the streamline, $\delta\hat{s}$, per change in distance along the streamline, δs. The magnitude of $\hat{s}$ is constant ($|\hat{s}| = 1$; it is a unit vector), but its direction is variable if the streamlines are curved. From Fig. 4.9 it is seen that the magnitude of $\partial \hat{s}/\partial s$ is equal to the inverse of the radius of curvature of the streamline, $\mathcal{R}$, at the point in question. This follows because the two triangles shown (AOB and $A'O'B'$) are similar triangles so that $\delta s/\mathcal{R} = |\delta\hat{s}|/|\hat{s}| = |\delta\hat{s}|$, or $|\delta\hat{s}/\delta s| = 1/\mathcal{R}$. Similarly, in the limit $\delta s \to 0$, the direction of $\delta\hat{s}/\delta s$ is seen to be normal to the streamline. That is,

$$\frac{\partial \hat{s}}{\partial s} = \lim_{\delta s \to 0} \frac{\delta\hat{s}}{\delta s} = \frac{\hat{n}}{\mathcal{R}}$$

Hence, the acceleration for steady, two-dimensional flow can be written in terms of its streamwise and normal components in the form

$$\mathbf{a} = V\frac{\partial V}{\partial s}\hat{s} + \frac{V^2}{\mathcal{R}}\hat{n} \quad \text{or} \quad a_s = V\frac{\partial V}{\partial s}, \quad a_n = \frac{V^2}{\mathcal{R}} \tag{4.7}$$

The first term, $a_s = V \partial V/\partial s$, represents the convective acceleration along the streamline and the second term, $a_n = V^2/\mathcal{R}$, represents centrifugal acceleration (one type of convective acceleration) normal to the fluid motion. These components can be noted in Fig. E4.5 by resolving the acceleration vector into its components along and normal to the velocity vector. Note that the unit vector $\hat{n}$ is directed from the streamline toward the center of curvature. These forms of the acceleration were used in Chapter 3 and are probably familiar from previous dynamics or physics considerations.

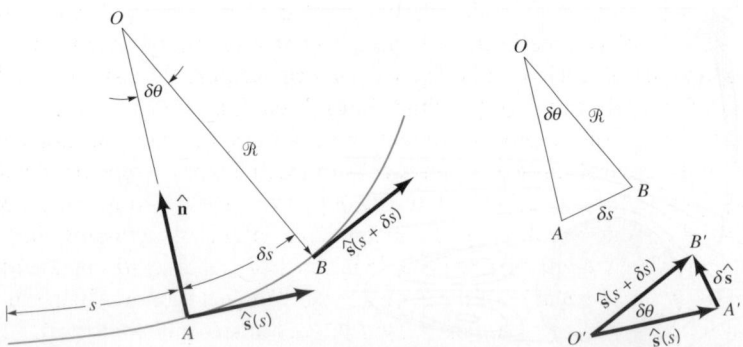

■ **FIGURE 4.9**   Relationship between the unit vector along the streamline, $\hat{s}$, and the radius of curvature of the streamline, $\mathcal{R}$.

4.3   Control Volume and System Representations

Both control volume and system concepts can be used to describe fluid flow.

As is discussed in Chapter 1, a fluid is a type of matter that is relatively free to move and interact with its surroundings. As with any matter, a fluid's behavior is governed by fundamental physical laws which are approximated by an appropriate set of equations. The application of laws such as the conservation of mass, Newton's laws of motion, and the laws of thermodynamics form the foundation of fluid mechanics analyses. There are various ways that these governing laws can be applied to a fluid, including the system approach and the control volume approach. By definition, a *system* is a collection of matter of fixed identity (always the same atoms or fluid particles), which may move, flow, and interact with its surroundings. A *control volume,* on the other hand, is a volume in space (a geometric entity, independent of mass) through which fluid may flow.

A system is a specific, identifiable quantity of matter. It may consist of a relatively large amount of mass (such as all of the air in the earth's atmosphere), or it may be an infinitesimal size (such as a single fluid particle). In any case, the molecules making up the system are "tagged" in some fashion (dyed red, either actually or only in your mind) so that they can be continually identified as they move about. The system may interact with its surroundings by various means (by the transfer of heat or the exertion of a pressure force, for example). It may continually change size and shape, but it always contains the same mass.

A mass of air drawn into an air compressor can be considered as a system. It changes shape and size (it is compressed), its temperature may change, and it is eventually expelled through the outlet of the compressor. The matter associated with the original air drawn into the compressor remains as a system, however. The behavior of this material could be investigated by applying the appropriate governing equations to this system.

One of the important concepts used in the study of statics and dynamics is that of the free-body diagram. That is, we identify an object, isolate it from its surroundings, replace its surroundings by the equivalent actions that they put on the object, and apply Newton's laws of motion. The body in such cases is our system—an identified portion of matter that we follow during its interactions with its surroundings. In fluid mechanics, it is often quite difficult to identify and keep track of a specific quantity of matter. A finite portion of a fluid contains an uncountable number of fluid particles that move about quite freely, unlike a solid that may deform but usually remains relatively easy to identify. For example, we cannot as easily follow a specific portion of water flowing in a river as we can follow a branch floating on its surface.

We may often be more interested in determining the forces put on a fan, airplane, or automobile by air flowing past the object than we are in the information obtained by following a given portion of the air (a system) as it flows along. Similarly, for the Space Shuttle launch vehicle shown in the margin, we may be more interested in determining the thrust produced than we are in the information obtained by following the highly complex, irregular path of the exhaust plume from the rocket engine nozzle. For these situations we often use the control volume approach. We identify a specific volume in space (a volume associated with the fan, airplane, or automobile, for example) and analyze the fluid flow within, through, or around that volume. In general, the control volume can be a moving volume, although for most situations considered in this book we will use only fixed, nondeformable control volumes. The matter within a control volume may change with time as the fluid flows through it. Similarly, the amount of mass within the volume may change with time. The control volume itself is a specific geometric entity, independent of the flowing fluid.

Examples of control volumes and *control surfaces* (the surface of the control volume) are shown in Fig. 4.10. For case (*a*), fluid flows through a pipe. The fixed control surface consists of the inside surface of the pipe, the outlet end at section (2), and a section across the pipe at (1). One portion of the control surface is a physical surface (the pipe), while the remainder is simply a surface in space (across the pipe). Fluid flows across part of the control surface, but not across all of it.

Another control volume is the rectangular volume surrounding the jet engine shown in Fig. 4.10*b*. If the airplane to which the engine is attached is sitting still on the runway, air flows through this control volume because of the action of the engine within it. The air that was within the engine itself at time $t = t_1$ (a system) has passed through the engine and is outside of the control volume at a later time $t = t_2$ as indicated. At this later time other air (a different system) is within the engine. If the airplane is moving, the control volume is fixed relative to an observer on the airplane, but it

(Photograph courtesy of NASA.)

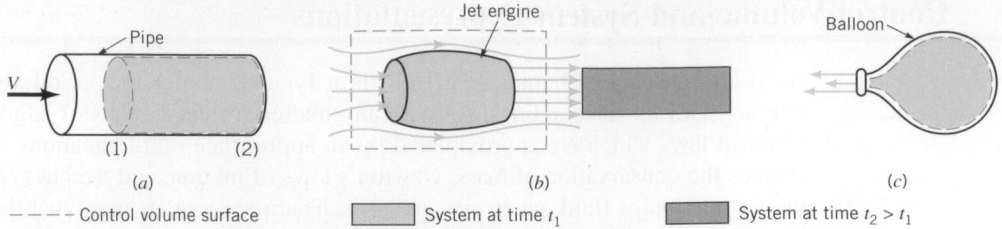

--- - Control volume surface   System at time t_1 System at time $t_2 > t_1$

■ **FIGURE 4.10**   Typical control volumes: (*a*) fixed control volume, (*b*) fixed or moving control volume, (*c*) deforming control volume.

is a moving control volume relative to an observer on the ground. In either situation air flows through and around the engine as indicated.

The deflating balloon shown in Fig. 4.10*c* provides an example of a deforming control volume. As time increases, the control volume (whose surface is the inner surface of the balloon) decreases in size. If we do not hold onto the balloon, it becomes a moving, deforming control volume as it darts about the room. The majority of the problems we will analyze can be solved by using a fixed, nondeforming control volume. In some instances, however, it will be advantageous, in fact necessary, to use a moving, deforming control volume.

In many ways the relationship between a system and a control volume is similar to the relationship between the Lagrangian and Eulerian flow description introduced in Section 4.1.1. In the system or Lagrangian description, we follow the fluid and observe its behavior as it moves about. In the control volume or Eulerian description we remain stationary and observe the fluid's behavior at a fixed location. (If a moving control volume is used, it virtually never moves with the system—the system flows through the control volume.) These ideas are discussed in more detail in the next section.

The governing laws of fluid motion are stated in terms of fluid systems, not control volumes.

All of the laws governing the motion of a fluid are stated in their basic form in terms of a system approach. For example, "the mass of a system remains constant," or "the time rate of change of momentum of a system is equal to the sum of all the forces acting on the system." Note the word system, not control volume, in these statements. To use the governing equations in a control volume approach to problem solving, we must rephrase the laws in an appropriate manner. To this end we introduce the Reynolds transport theorem in the following section.

4.4   The Reynolds Transport Theorem

We are sometimes interested in what happens to a particular part of the fluid as it moves about. Other times we may be interested in what effect the fluid has on a particular object or volume in space as fluid interacts with it. Thus, we need to describe the laws governing fluid motion using both system concepts (consider a given mass of the fluid) and control volume concepts (consider a given volume). To do this we need an analytical tool to shift from one representation to the other. The *Reynolds transport theorem* provides this tool.

All physical laws are stated in terms of various physical parameters. Velocity, acceleration, mass, temperature, and momentum are but a few of the more common parameters. Let *B* represent any of these (or other) fluid parameters and *b* represent the amount of that parameter per unit mass. That is,

$$B = mb$$

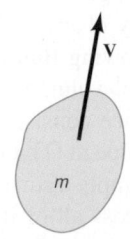

where *m* is the mass of the portion of fluid of interest. For example, as shown by the figure in the margin, if $B = m$, the mass, it follows that $b = 1$. The mass per unit mass is unity. If $B = mV^2/2$, the kinetic energy of the mass, then $b = V^2/2$, the kinetic energy per unit mass. The parameters *B* and *b* may be scalars or vectors. Thus, if $\mathbf{B} = m\mathbf{V}$, the momentum of the mass, then $\mathbf{b} = \mathbf{V}$. (The momentum per unit mass is the velocity.)

The parameter *B* is termed an *extensive property* and the parameter *b* is termed an *intensive property*. The value of *B* is directly proportional to the amount of the mass being considered, whereas the value of *b* is independent of the amount of mass. The amount of an extensive property that a system possesses at a given instant, B_{sys}, can be determined by adding up the amount associated with each fluid particle in the system. For infinitesimal fluid particles of size $\delta V\!\!\!\!-$ and mass $\rho \, \delta V\!\!\!\!-$,

B	$b = B/m$
m	1
$m\mathbf{V}$	$\mathbf{V}$
$\frac{1}{2}mV^2$	$\frac{1}{2}V^2$

this summation (in the limit of $\delta \mathcal{V} \to 0$) takes the form of an integration over all the particles in the system and can be written as

$$B_{\text{sys}} = \lim_{\delta \mathcal{V} \to 0} \sum_i b_i(\rho_i \, \delta \mathcal{V}_i) = \int_{\text{sys}} \rho b \, d\mathcal{V}$$

The limits of integration cover the entire system—a (usually) moving volume. We have used the fact that the amount of B in a fluid particle of mass $\rho \, \delta \mathcal{V}$ is given in terms of b by $\delta B = b\rho \, \delta \mathcal{V}$.

Most of the laws governing fluid motion involve the time rate of change of an extensive property of a fluid system—the rate at which the momentum of a system changes with time, the rate at which the mass of a system changes with time, and so on. Thus, we often encounter terms such as

$$\frac{dB_{\text{sys}}}{dt} = \frac{d\left(\displaystyle\int_{\text{sys}} \rho b \, d\mathcal{V} \right)}{dt} \tag{4.8}$$

To formulate the laws into a control volume approach, we must obtain an expression for the time rate of change of an extensive property within a control volume, B_{cv}, not within a system. This can be written as

$$\frac{dB_{\text{cv}}}{dt} = \frac{d\left(\displaystyle\int_{\text{cv}} \rho b \, d\mathcal{V} \right)}{dt} \tag{4.9}$$

Differences between control volume and system concepts are subtle but very important.

where the limits of integration, denoted by cv, cover the control volume of interest. Although Eqs. 4.8 and 4.9 may look very similar, the physical interpretation of each is quite different. Mathematically, the difference is represented by the difference in the limits of integration. Recall that the control volume is a volume in space (in most cases stationary, although if it moves it need not move with the system). On the other hand, the system is an identifiable collection of mass that moves with the fluid (indeed it is a specified portion of the fluid). We will learn that even for those instances when the control volume and the system momentarily occupy the same volume in space, the two quantities dB_{sys}/dt and dB_{cv}/dt need not be the same. The Reynolds transport theorem provides the relationship between the time rate of change of an extensive property for a system and that for a control volume—the relationship between Eqs. 4.8 and 4.9.

EXAMPLE 4.7   Time Rate of Change for a System and a Control Volume

GIVEN   Fluid flows from the fire extinguisher tank shown in Fig. E4.7a.

FIND   Discuss the differences between dB_{sys}/dt and dB_{cv}/dt if B represents mass.

SOLUTION

With $B = m$, the system mass, it follows that $b = 1$ and Eqs. 4.8 and 4.9 can be written as

$$\frac{dB_{\text{sys}}}{dt} \equiv \frac{dm_{\text{sys}}}{dt} = \frac{d\left(\displaystyle\int_{\text{sys}} \rho \, d\mathcal{V} \right)}{dt}$$

and

$$\frac{dB_{\text{cv}}}{dt} \equiv \frac{dm_{\text{cv}}}{dt} = \frac{d\left(\displaystyle\int_{\text{cv}} \rho \, d\mathcal{V} \right)}{dt}$$

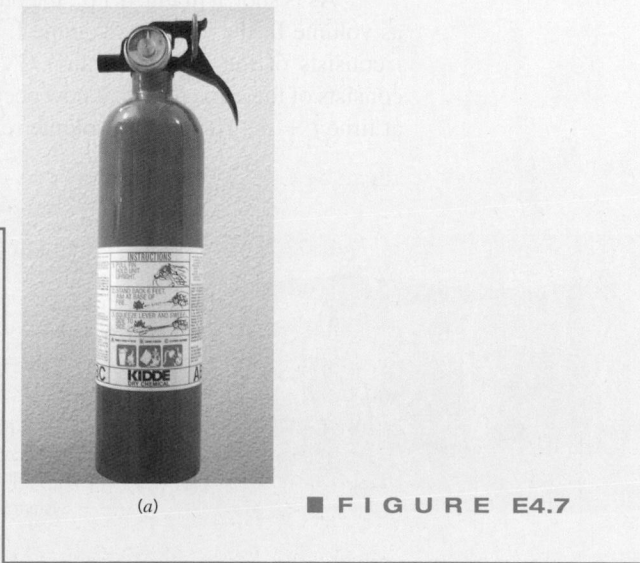

(a)

■ FIGURE E4.7

Physically these represent the time rate of change of mass within the system and the time rate of change of mass within the control volume, respectively. We choose our system to be the fluid within the tank at the time the valve was opened ($t = 0$) and the control volume to be the tank itself as shown in Fig. E4.7b. A short time after the valve is opened, part of the system has moved outside of the control volume as is shown in Fig. E4.7c. The control volume remains fixed. The limits of integration are fixed for the control volume; they are a function of time for the system.

Clearly, if mass is to be conserved (one of the basic laws governing fluid motion), the mass of the fluid in the system is constant, so that

$$\frac{d\left(\int_{\text{sys}} \rho \, d\forall\right)}{dt} = 0$$

On the other hand, it is equally clear that some of the fluid has left the control volume through the nozzle on the tank. Hence, the amount of mass within the tank (the control volume) decreases with time, or

$$\frac{d\left(\int_{\text{cv}} \rho \, d\forall\right)}{dt} < 0$$

The actual numerical value of the rate at which the mass in the control volume decreases will depend on the rate at which the fluid flows through the nozzle (i.e., the size of the nozzle and the speed and density of the fluid). Clearly the meanings of dB_{sys}/dt and dB_{cv}/dt are different. For this example, $dB_{\text{cv}}/dt < dB_{\text{sys}}/dt$. Other situations may have $dB_{\text{cv}}/dt \geq dB_{\text{sys}}/dt$.

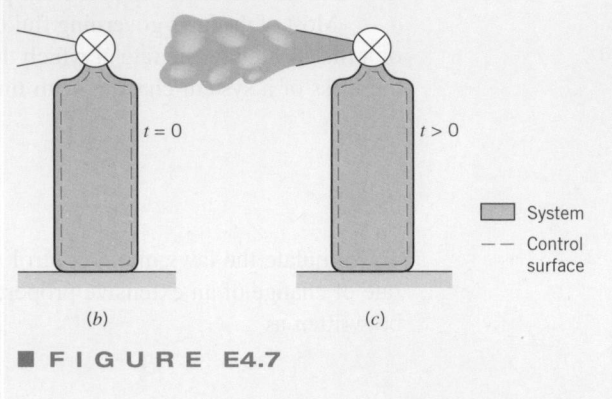

(b)   (c)

■ **FIGURE E4.7**

4.4.1 Derivation of the Reynolds Transport Theorem

A simple version of the Reynolds transport theorem relating system concepts to control volume concepts can be obtained easily for the one-dimensional flow through a fixed control volume such as the variable area duct section shown in Fig. 4.11a. We consider the control volume to be that stationary volume within the duct between sections (1) and (2) as indicated in Fig. 4.11b. The system that we consider is that fluid occupying the control volume at some initial time t. A short time later, at time $t + \delta t$, the system has moved slightly to the right. The fluid particles that coincided with section (2) of the control surface at time t have moved a distance $\delta\ell_2 = V_2 \, \delta t$ to the right, where V_2 is the velocity of the fluid as it passes section (2). Similarly, the fluid initially at section (1) has moved a distance $\delta\ell_1 = V_1 \, \delta t$, where V_1 is the fluid velocity at section (1). We assume the fluid flows across sections (1) and (2) in a direction normal to these surfaces and that V_1 and V_2 are constant across sections (1) and (2).

The moving system flows through the fixed control volume.

As is shown in Fig. 4.11c, the outflow from the control volume from time t to $t + \delta t$ is denoted as volume II, the inflow as volume I, and the control volume itself as CV. Thus, the system at time t consists of the fluid in section CV; that is, "SYS = CV" at time t. At time $t + \delta t$ the system consists of the same fluid that now occupies sections (CV − I) + II. That is, "SYS = CV − I + II" at time $t + \delta t$. The control volume remains as section CV for all time.

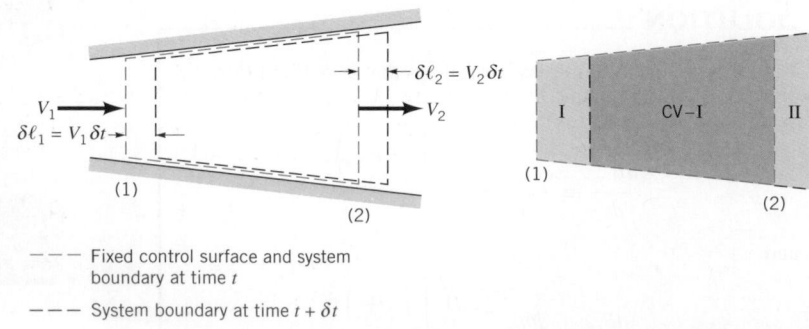

(a)   (b)   (c)

■ **FIGURE 4.11**   **Control volume and system for flow through a variable area pipe.**

If B is an extensive parameter of the system, then the value of it for the system at time t is

$$B_{\text{sys}}(t) = B_{\text{cv}}(t)$$

since the system and the fluid within the control volume coincide at this time. Its value at time $t + \delta t$ is

$$B_{\text{sys}}(t + \delta t) = B_{\text{cv}}(t + \delta t) - B_{\text{I}}(t + \delta t) + B_{\text{II}}(t + \delta t)$$

Thus, the change in the amount of B in the system in the time interval δt divided by this time interval is given by

$$\frac{\delta B_{\text{sys}}}{\delta t} = \frac{B_{\text{sys}}(t + \delta t) - B_{\text{sys}}(t)}{\delta t} = \frac{B_{\text{cv}}(t + \delta t) - B_{\text{I}}(t + \delta t) + B_{\text{II}}(t + \delta t) - B_{\text{sys}}(t)}{\delta t}$$

By using the fact that at the initial time t we have $B_{\text{sys}}(t) = B_{\text{cv}}(t)$, this ungainly expression may be rearranged as follows.

$$\frac{\delta B_{\text{sys}}}{\delta t} = \frac{B_{\text{cv}}(t + \delta t) - B_{\text{cv}}(t)}{\delta t} - \frac{B_{\text{I}}(t + \delta t)}{\delta t} + \frac{B_{\text{II}}(t + \delta t)}{\delta t} \tag{4.10}$$

The time rate of change of a system property is a Lagrangian concept.

In the limit $\delta t \rightarrow 0$, the left-hand side of Eq. 4.10 is equal to the time rate of change of B for the system and is denoted as DB_{sys}/Dt. We use the material derivative notation, $D(\ )/Dt$, to denote this time rate of change to emphasize the Lagrangian character of this term. (Recall from Section 4.2.1 that the material derivative, DP/Dt, of any quantity P represents the time rate of change of that quantity associated with a given fluid particle as it moves along.) Similarly, the quantity DB_{sys}/Dt represents the time rate of change of property B associated with a system (a given portion of fluid) as it moves along.

In the limit $\delta t \rightarrow 0$, the first term on the right-hand side of Eq. 4.10 is seen to be the time rate of change of the amount of B within the control volume

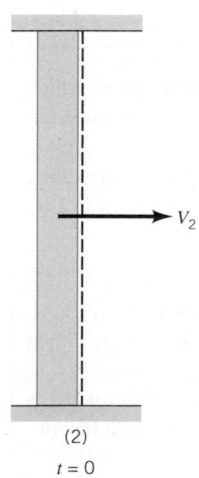

$$\lim_{\delta t \to 0} \frac{B_{\text{cv}}(t + \delta t) - B_{\text{cv}}(t)}{\delta t} = \frac{\partial B_{\text{cv}}}{\partial t} = \frac{\partial \left(\int_{\text{cv}} \rho b \, d\Psi \right)}{\partial t} \tag{4.11}$$

The third term on the right-hand side of Eq. 4.10 represents the rate at which the extensive parameter B flows from the control volume, across the control surface. As indicated by the figure in the margin, during the time interval from $t = 0$ to $t = \delta t$ the volume of fluid that flows across section (2) is given by $\delta\Psi_{\text{II}} = A_2 \delta\ell_2 = A_2(V_2\delta t)$. Thus, the amount of B within region II, the outflow region, is its amount per unit volume, ρb, times the volume

$$B_{\text{II}}(t + \delta t) = (\rho_2 b_2)(\delta\Psi_{\text{II}}) = \rho_2 b_2 A_2 V_2 \, \delta t$$

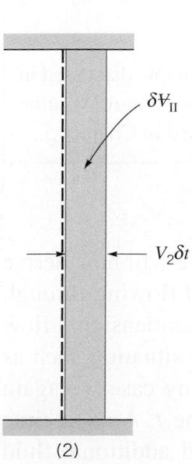

where b_2 and ρ_2 are the constant values of b and ρ across section (2). Thus, the rate at which this property flows from the control volume, $\dot{B}_{\text{out}}$, is given by

$$\dot{B}_{\text{out}} = \lim_{\delta t \to 0} \frac{B_{\text{II}}(t + \delta t)}{\delta t} = \rho_2 A_2 V_2 b_2 \tag{4.12}$$

Similarly, the inflow of B into the control volume across section (1) during the time interval δt corresponds to that in region I and is given by the amount per unit volume times the volume, $\delta\Psi_{\text{I}} = A_1 \delta\ell_1 = A_1(V_1 \delta t)$. Hence,

$$B_{\text{I}}(t + \delta t) = (\rho_1 b_1)(\delta\Psi_{\text{I}}) = \rho_1 b_1 A_1 V_1 \, \delta t$$

where b_1 and ρ_1 are the constant values of b and ρ across section (1). Thus, the rate of inflow of the property B into the control volume, $\dot{B}_{\text{in}}$, is given by

$$\dot{B}_{\text{in}} = \lim_{\delta t \to 0} \frac{B_{\text{I}}(t + \delta t)}{\delta t} = \rho_1 A_1 V_1 b_1 \tag{4.13}$$

If we combine Eqs. 4.10, 4.11, 4.12, and 4.13 we see that the relationship between the time rate of change of B for the system and that for the control volume is given by

$$\frac{DB_{sys}}{Dt} = \frac{\partial B_{cv}}{\delta t} + \dot{B}_{out} - \dot{B}_{in} \qquad (4.14)$$

or

$$\frac{DB_{sys}}{Dt} = \frac{\partial B_{cv}}{\partial t} + \rho_2 A_2 V_2 b_2 - \rho_1 A_1 V_1 b_1 \qquad (4.15)$$

The time derivative associated with a system may be different from that for a control volume.

This is a version of the Reynolds transport theorem valid under the restrictive assumptions associated with the flow shown in Fig. 4.11—fixed control volume with one inlet and one outlet having uniform properties (density, velocity, and the parameter b) across the inlet and outlet with the velocity normal to sections (1) and (2). Note that the time rate of change of B for the system (the left-hand side of Eq. 4.15 or the quantity in Eq. 4.8) is not necessarily the same as the rate of change of B within the control volume (the first term on the right-hand side of Eq. 4.15 or the quantity in Eq. 4.9). This is true because the inflow rate ($b_1 \rho_1 V_1 A_1$) and the outflow rate ($b_2 \rho_2 V_2 A_2$) of the property B for the control volume need not be the same.

EXAMPLE 4.8 ■ Use of the Reynolds Transport Theorem

GIVEN Consider again the flow from the fire extinguisher shown in Fig. E4.7. Let the extensive property of interest be the system mass ($B = m$, the system mass, or $b = 1$).

FIND Write the appropriate form of the Reynolds transport theorem for this flow.

SOLUTION

Again we take the control volume to be the fire extinguisher, and the system to be the fluid within it at time $t = 0$. For this case there is no inlet, section (1), across which the fluid flows into the control volume ($A_1 = 0$). There is, however, an outlet, section (2). Thus, the Reynolds transport theorem, Eq. 4.15, along with Eq. 4.9 with $b = 1$ can be written as

$$\frac{Dm_{sys}}{Dt} = \frac{\partial \left(\int_{cv} \rho \, d\Psi \right)}{\partial t} + \rho_2 A_2 V_2 \qquad (1) \quad \text{(Ans)}$$

COMMENT If we proceed one step further and use the basic law of conservation of mass, we may set the left-hand side of this equation equal to zero (the amount of mass in a system is constant) and rewrite Eq. 1 in the form

$$\frac{\partial \left(\int_{cv} \rho \, d\Psi \right)}{\partial t} = -\rho_2 A_2 V_2 \qquad (2)$$

The physical interpretation of this result is that the rate at which the mass in the tank decreases in time is equal in magnitude but opposite to the rate of flow of mass from the exit, $\rho_2 A_2 V_2$. Note the units for the two terms of Eq. 2 (kg/s). Note that if there were both an inlet and an outlet to the control volume shown in Fig. E4.7, Eq. 2 would become

$$\frac{\partial \left(\int_{cv} \rho \, d\Psi \right)}{\partial t} = \rho_1 A_1 V_1 - \rho_2 A_2 V_2 \qquad (3)$$

In addition, if the flow were steady, the left-hand side of Eq. 3 would be zero (the amount of mass in the control would be constant in time) and Eq. 3 would become

$$\rho_1 A_1 V_1 = \rho_2 A_2 V_2$$

This is one form of the conservation of mass principle discussed in Sect. 3.6.2—the mass flowrates into and out of the control volume are equal. Other more general forms are discussed in Chapter 5.

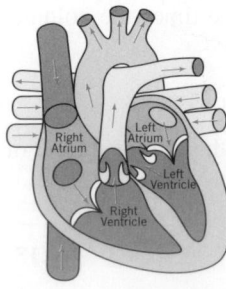

Equation 4.15 is a simplified version of the Reynolds transport theorem. We will now derive it for much more general conditions. A general, fixed control volume with fluid flowing through it is shown in Fig. 4.12. The flow field may be quite simple (as in the above one-dimensional flow considerations), or it may involve a quite complex, unsteady, three-dimensional situation such as the flow through a human heart as illustrated by the figure in the margin. In any case we again consider the system to be the fluid within the control volume at the initial time t. A short time later a portion of the fluid (region II) has exited from the control volume and additional fluid (region I, not part of the original system) has entered the control volume.

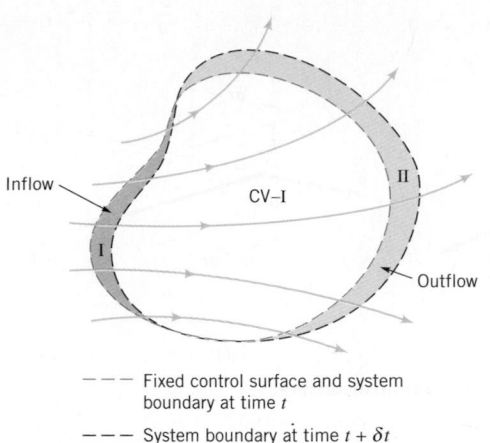

■ FIGURE 4.12 Control volume and system for flow through an arbitrary, fixed control volume.

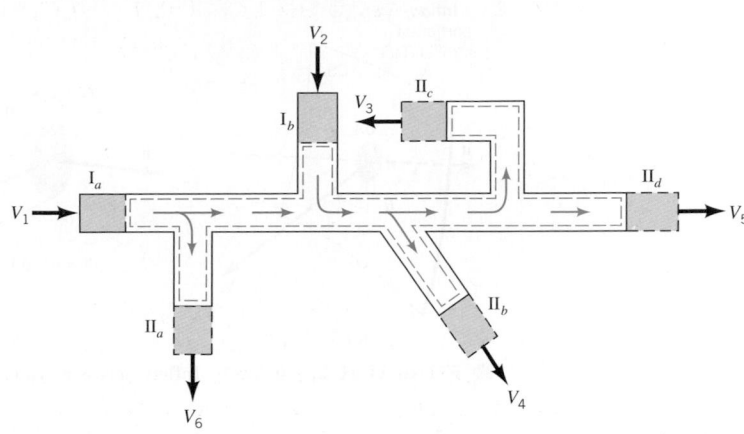

■ FIGURE 4.13 Typical control volume with more than one inlet and outlet.

The simplified Reynolds transport theorem can be easily generalized.

We consider an extensive fluid property B and seek to determine how the rate of change of B associated with the system is related to the rate of change of B within the control volume at any instant. By repeating the exact steps that we did for the simplified control volume shown in Fig. 4.11, we see that Eq. 4.14 is valid for the general case also, provided that we give the correct interpretation to the terms $\dot{B}_{out}$ and $\dot{B}_{in}$. In general, the control volume may contain more (or less) than one inlet and one outlet. A typical pipe system may contain several inlets and outlets as are shown in Fig. 4.13. In such instances we think of all inlets grouped together ($I = I_a + I_b + I_c + \cdots$) and all outlets grouped together ($II = II_a + II_b + II_c + \cdots$), at least conceptually.

The term $\dot{B}_{out}$ represents the net flowrate of the property B from the control volume. Its value can be thought of as arising from the addition (integration) of the contributions through each infinitesimal area element of size δA on the portion of the control surface dividing region II and the control volume. This surface is denoted CS_{out}. As is indicated in Fig. 4.14, in time δt the volume of fluid that passes across each area element is given by $\delta V = \delta \ell_n \, \delta A$, where $\delta \ell_n = \delta \ell \cos \theta$ is the height (normal to the base, δA) of the small volume element, and θ is the angle between the velocity vector and the outward pointing normal to the surface, $\hat{\mathbf{n}}$. Thus, since $\delta \ell = V \delta t$, the amount of the property B carried across the area element δA in the time interval δt is given by

$$\delta B = b\rho \, \delta V = b\rho (V \cos \theta \, \delta t) \, \delta A$$

The rate at which B is carried out of the control volume across the small area element δA, denoted $\delta \dot{B}_{out}$, is

$$\delta \dot{B}_{out} = \lim_{\delta t \to 0} \frac{\rho b \, \delta V}{\delta t} = \lim_{\delta t \to 0} \frac{(\rho b V \cos \theta \, \delta t) \, \delta A}{\delta t} = \rho b V \cos \theta \, \delta A$$

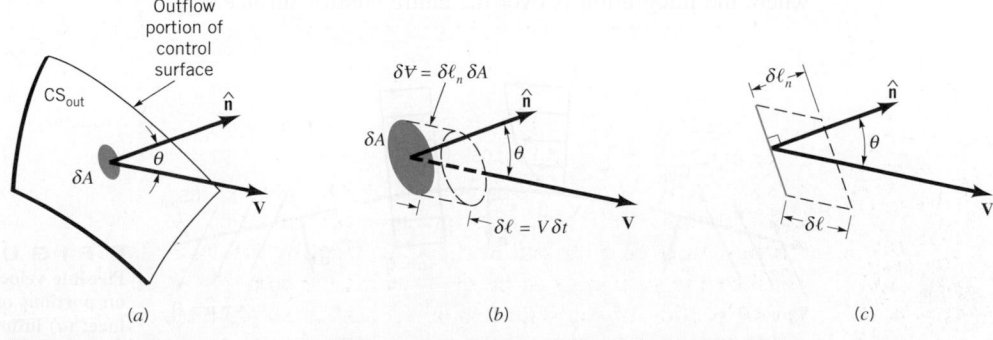

(a)   (b)   (c)

■ FIGURE 4.14 Outflow across a typical portion of the control surface.

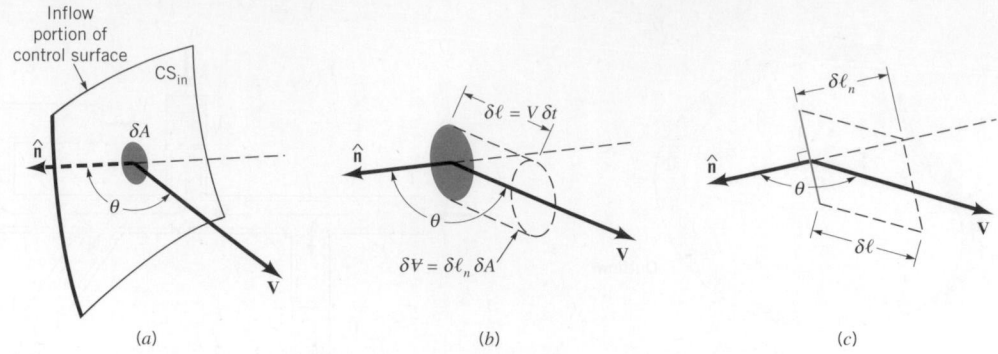

■ FIGURE 4.15   Inflow across a typical portion of the control surface.

By integrating over the entire outflow portion of the control surface, $\mathrm{CS_{out}}$, we obtain

$$\dot{B}_{\text{out}} = \int_{\mathrm{cs_{out}}} d\dot{B}_{\text{out}} = \int_{\mathrm{cs_{out}}} \rho b V \cos\theta \, dA$$

The quantity $V \cos\theta$ is the component of the velocity normal to the area element δA. From the definition of the dot product, this can be written as $V \cos\theta = \mathbf{V} \cdot \hat{\mathbf{n}}$. Hence, an alternate form of the outflow rate is

The flowrate of a parameter across the control surface is written in terms of a surface integral.

$$\dot{B}_{\text{out}} = \int_{\mathrm{cs_{out}}} \rho b \mathbf{V} \cdot \hat{\mathbf{n}} \, dA \tag{4.16}$$

In a similar fashion, by considering the inflow portion of the control surface, $\mathrm{CS_{in}}$, as shown in Fig. 4.15, we find that the inflow rate of B into the control volume is

$$\dot{B}_{\text{in}} = -\int_{\mathrm{cs_{in}}} \rho b V \cos\theta \, dA = -\int_{\mathrm{cs_{in}}} \rho b \mathbf{V} \cdot \hat{\mathbf{n}} \, dA \tag{4.17}$$

We use the standard notation that the unit normal vector to the control surface, $\hat{\mathbf{n}}$, points out from the control volume. Thus, as is shown in Fig. 4.16, $-90° < \theta < 90°$ for outflow regions (the normal component of $\mathbf{V}$ is positive; $\mathbf{V} \cdot \hat{\mathbf{n}} > 0$). For inflow regions $90° < \theta < 270°$ (the normal component of $\mathbf{V}$ is negative; $\mathbf{V} \cdot \hat{\mathbf{n}} < 0$). The value of $\cos\theta$ is, therefore, positive on the $\mathrm{CV_{out}}$ portions of the control surface and negative on the $\mathrm{CV_{in}}$ portions. Over the remainder of the control surface, there is no inflow or outflow, leading to $\mathbf{V} \cdot \hat{\mathbf{n}} = V \cos\theta = 0$ on those portions. On such portions either $V = 0$ (the fluid "sticks" to the surface) or $\cos\theta = 0$ (the fluid "slides" along the surface without crossing it) (see Fig. 4.16). Therefore, the net flux (flowrate) of parameter B across the entire control surface is

$$\dot{B}_{\text{out}} - \dot{B}_{\text{in}} = \int_{\mathrm{cs_{out}}} \rho b \mathbf{V} \cdot \hat{\mathbf{n}} \, dA - \left(-\int_{\mathrm{cs_{in}}} \rho b \mathbf{V} \cdot \hat{\mathbf{n}} \, dA \right)$$

$$= \int_{\mathrm{cs}} \rho b \mathbf{V} \cdot \hat{\mathbf{n}} \, dA \tag{4.18}$$

where the integration is over the entire control surface.

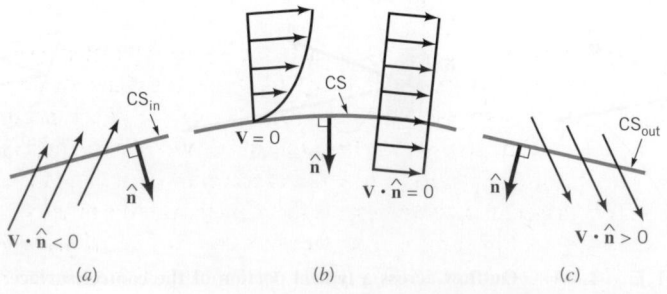

■ FIGURE 4.16
Possible velocity configurations on portions of the control surface: (*a*) inflow, (*b*) no flow across the surface, (*c*) outflow.

By combining Eqs. 4.14 and 4.18 we obtain

$$\frac{DB_{sys}}{Dt} = \frac{\partial B_{cv}}{\partial t} + \int_{cs} \rho b \mathbf{V} \cdot \hat{\mathbf{n}} \, dA$$

This can be written in a slightly different form by using $B_{cv} = \int_{cv} \rho b \, d\forall$ so that

$$\frac{DB_{sys}}{Dt} = \frac{\partial}{\partial t} \int_{cv} \rho b \, d\forall + \int_{cs} \rho b \, \mathbf{V} \cdot \hat{\mathbf{n}} \, dA \qquad (4.19)$$

Equation 4.19 is the general form of the Reynolds transport theorem for a fixed, nondeforming control volume. Its interpretation and use are discussed in the following sections.

4.4.2 Physical Interpretation

The Reynolds transport theorem as given in Eq. 4.19 is widely used in fluid mechanics (and other areas as well). At first it appears to be a rather formidable mathematical expression—perhaps one to be steered clear of if possible. However, a physical understanding of the concepts involved will show that it is a rather straightforward, relatively easy-to-use tool. Its purpose is to provide a link between control volume ideas and system ideas.

The left side of Eq. 4.19 is the time rate of change of an arbitrary extensive parameter of a system. This may represent the rate of change of mass, momentum, energy, or angular momentum of the system, depending on the choice of the parameter B.

Because the system is moving and the control volume is stationary, the time rate of change of the amount of B within the control volume is not necessarily equal to that of the system. The first term on the right side of Eq. 4.19 represents the rate of change of B within the control volume as the fluid flows through it. Recall that b is the amount of B per unit mass, so that $\rho b \, d\forall$ is the amount of B in a small volume $d\forall$. Thus, the time derivative of the integral of ρb throughout the control volume is the time rate of change of B within the control volume at a given time.

Control surface

The last term in Eq. 4.19 (an integral over the control surface) represents the net flowrate of the parameter B across the entire control surface. As illustrated by the figure in the margin, over a portion of the control surface this property is being carried out of the control volume ($\mathbf{V} \cdot \hat{\mathbf{n}} > 0$); over other portions it is being carried into the control volume ($\mathbf{V} \cdot \hat{\mathbf{n}} < 0$). Over the remainder of the control surface there is no transport of B across the surface since $b\mathbf{V} \cdot \hat{\mathbf{n}} = 0$, because either $b = 0$, $\mathbf{V} = 0$, or $\mathbf{V}$ is parallel to the surface at those locations. The mass flowrate through area element δA, given by $\rho \mathbf{V} \cdot \hat{\mathbf{n}} \, \delta A$, is positive for outflow (efflux) and negative for inflow (influx). Each fluid particle or fluid mass carries a certain amount of B with it, as given by the product of B per unit mass, b, and the mass. The rate at which this B is carried across the control surface is given by the area integral term of Eq. 4.19. This net rate across the entire control surface may be negative, zero, or positive depending on the particular situation involved.

4.4.3 Relationship to Material Derivative

In Section 4.2.1 we discussed the concept of the material derivative $D(\ )/Dt = \partial(\ )/\partial t + \mathbf{V} \cdot \nabla(\ ) = \partial(\ )/\partial t + u \, \partial(\ )/\partial x + v \, \partial(\ )/\partial y + w \, \partial(\ )/\partial z$. The physical interpretation of this derivative is that it provides the time rate of change of a fluid property (temperature, velocity, etc.) associated with a particular fluid particle as it flows. The value of that parameter for that particle may change because of unsteady effects [the $\partial(\ )/\partial t$ term] or because of effects associated with the particle's motion [the $\mathbf{V} \cdot \nabla(\ )$ term].

Careful consideration of Eq. 4.19 indicates the same type of physical interpretation for the Reynolds transport theorem. The term involving the time derivative of the control volume integral represents unsteady effects associated with the fact that values of the parameter within the control volume may change with time. For steady flow this effect vanishes—fluid flows through the control volume but the amount of any property, B, within the control volume is constant in time. The term involving the control surface integral represents the convective effects associated with the flow of the system across the fixed control surface. The sum of these two terms gives the rate of change of the parameter B for the system. This corresponds to the interpretation of the material derivative,

The Reynolds transport theorem is the integral counterpart of the material derivative.

$D(\ )/Dt = \partial(\ )/\partial t + \mathbf{V} \cdot \nabla(\ )$, in which the sum of the unsteady effect and the convective effect gives the rate of change of a parameter for a fluid particle. As is discussed in Section 4.2, the material derivative operator may be applied to scalars (such as temperature) or vectors (such as velocity). This is also true for the Reynolds transport theorem. The particular parameters of interest, B and b, may be scalars or vectors.

Thus, both the material derivative and the Reynolds transport theorem equations represent ways to transfer from the Lagrangian viewpoint (follow a particle or follow a system) to the Eulerian viewpoint (observe the fluid at a given location in space or observe what happens in the fixed control volume). The material derivative (Eq. 4.5) is essentially the infinitesimal (or derivative) equivalent of the finite size (or integral) Reynolds transport theorem (Eq. 4.19).

4.4.4 Steady Effects

Consider a steady flow $[\partial(\ )/\partial t \equiv 0]$ so that Eq. 4.19 reduces to

$$\frac{DB_{\text{sys}}}{Dt} = \int_{cs} \rho b \mathbf{V} \cdot \hat{\mathbf{n}}\, dA \tag{4.20}$$

In such cases if there is to be a change in the amount of B associated with the system (nonzero left-hand side), there must be a net difference in the rate that B flows into the control volume compared with the rate that it flows out of the control volume. That is, the integral of $\rho b \mathbf{V} \cdot \hat{\mathbf{n}}$ over the inflow portions of the control surface would not be equal and opposite to that over the outflow portions of the surface.

Consider steady flow through the "black box" control volume that is shown in Fig. 4.17. If the parameter B is the mass of the system, the left-hand side of Eq. 4.20 is zero (conservation of mass for the system as discussed in detail in Section 5.1). Hence, the flowrate of mass into the box must be the same as the flowrate of mass out of the box because the right-hand side of Eq. 4.20 represents the net flowrate through the control surface. On the other hand, assume the parameter B is the momentum of the system. The momentum of the system need not be constant. In fact, according to Newton's second law the time rate of change of the system momentum equals the net force, $\mathbf{F}$, acting on the system. In general, the left-hand side of Eq. 4.20 will therefore be nonzero. Thus, the right-hand side, which then represents the net flux of momentum across the control surface, will be nonzero. The flowrate of momentum into the control volume need not be the same as the flux of momentum from the control volume. We will investigate these concepts much more fully in Chapter 5. They are the basic principles describing the operation of such devices as jet or rocket engines like the one shown in the figure in the margin.

For steady flows the amount of the property B within the control volume does not change with time. The amount of the property associated with the system may or may not change with time, depending on the particular property considered and the flow situation involved. The difference between that associated with the control volume and that associated with the system is determined by the rate at which B is carried across the control surface—the term $\int_{cs} \rho b \mathbf{V} \cdot \hat{\mathbf{n}}\, dA$.

(Photograph courtesy of NASA.)

The Reynolds transport theorem involves both steady and unsteady effects.

4.4.5 Unsteady Effects

Consider unsteady flow $[\partial(\ )/\partial t \neq 0]$ so that all terms in Eq. 4.19 must be retained. When they are viewed from a control volume standpoint, the amount of parameter B within the system may change because the amount of B within the fixed control volume may change with time

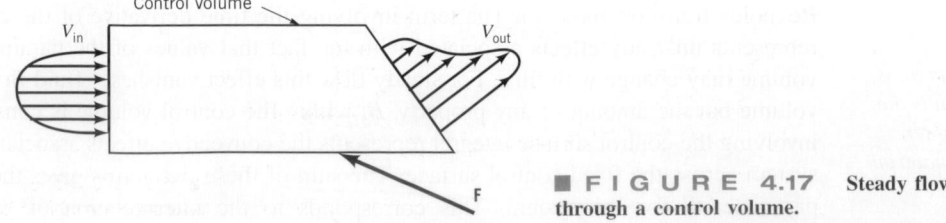

■ **FIGURE 4.17**   **Steady flow through a control volume.**

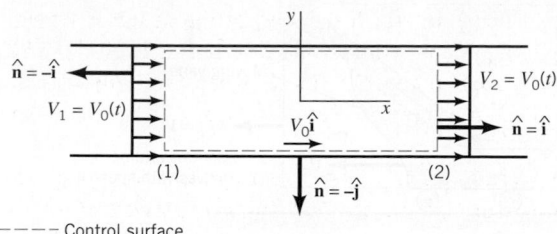

■ **F I G U R E   4.18**   **Unsteady flow through a constant diameter pipe.**

For some flow situations, certain portions of the Reynolds transport theorem are automatically zero.

[the $\partial(\int_{cv} \rho b\, d\Psi)/\partial t$ term] and because there may be a net nonzero flow of that parameter across the control surface (the $\int_{cs} \rho b\mathbf{V} \cdot \hat{\mathbf{n}}\, dA$ term).

For the special unsteady situations in which the rate of inflow of parameter B is exactly balanced by its rate of outflow, it follows that $\int_{cs} \rho b\mathbf{V} \cdot \hat{\mathbf{n}}\, dA = 0$, and Eq. 4.19 reduces to

$$\frac{DB_{sys}}{Dt} = \frac{\partial}{\partial t}\int_{cv} \rho b\, d\Psi \tag{4.21}$$

For such cases, any rate of change in the amount of B associated with the system is equal to the rate of change of B within the control volume. This can be illustrated by considering flow through a constant diameter pipe as is shown in Fig. 4.18. The control volume is as shown, and the system is the fluid within this volume at time t_0. We assume the flow is one-dimensional with $\mathbf{V} = V_0\hat{\mathbf{i}}$, where $V_0(t)$ is a function of time, and that the density is constant. At any instant in time, all particles in the system have the same velocity. We let $\mathbf{B}$ = system momentum = $m\mathbf{V} = mV_0\hat{\mathbf{i}}$, where m is the system mass, so that $\mathbf{b} = \mathbf{B}/m = \mathbf{V} = V_0\hat{\mathbf{i}}$, the fluid velocity. The magnitude of the momentum efflux across the outlet [section (2)] is the same as the magnitude of the momentum influx across the inlet [section (1)]. However, the sign of the efflux is opposite to that of the influx since $\mathbf{V} \cdot \hat{\mathbf{n}} > 0$ for the outflow and $\mathbf{V} \cdot \hat{\mathbf{n}} < 0$ for the inflow. Note that $\mathbf{V} \cdot \hat{\mathbf{n}} = 0$ along the sides of the control volume. Thus, with $\mathbf{V} \cdot \hat{\mathbf{n}} = -V_0$ on section (1), $\mathbf{V} \cdot \hat{\mathbf{n}} = V_0$ on section (2), and $A_1 = A_2$, we obtain

$$\int_{cs} \rho b\mathbf{V} \cdot \hat{\mathbf{n}}\, dA = \int_{cs} \rho(V_0\hat{\mathbf{i}})(\mathbf{V} \cdot \hat{\mathbf{n}})\, dA$$

$$= \int_{(1)} \rho(V_0\hat{\mathbf{i}})(-V_0)\, dA + \int_{(2)} \rho(V_0\hat{\mathbf{i}})(V_0)\, dA$$

$$= -\rho V_0^2 A_1\hat{\mathbf{i}} + \rho V_0^2 A_2\hat{\mathbf{i}} = 0$$

It is seen that for this special case Eq. 4.21 is valid. The rate at which the momentum of the system changes with time is the same as the rate of change of momentum within the control volume. If V_0 is constant in time, there is no rate of change of momentum of the system and for this special case each of the terms in the Reynolds transport theorem is zero by itself.

Consider the flow through a variable area pipe shown in Fig. 4.19. In such cases the fluid velocity is not the same at section (1) as it is at (2). Hence, the efflux of momentum from the control volume is not equal to the influx of momentum, so that the convective term in Eq. 4.20 [the integral of $\rho\mathbf{V}(\mathbf{V} \cdot \hat{\mathbf{n}})$ over the control surface] is not zero. These topics will be discussed in considerably more detail in Chapter 5.

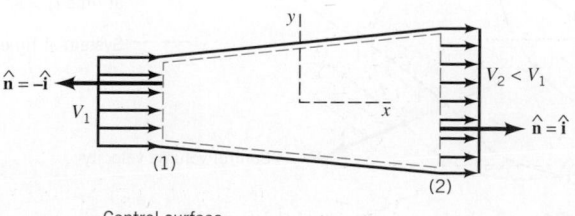

■ **F I G U R E   4.19**   **Flow through a variable area pipe.**

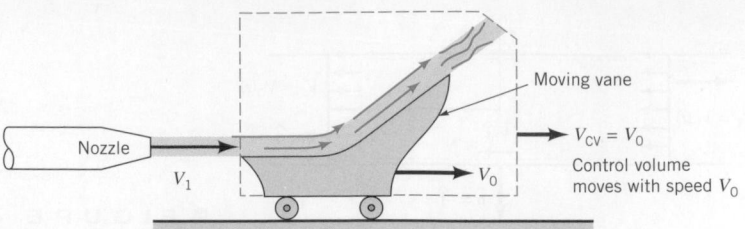

■ **FIGURE 4.20**   Example of a moving control volume.

4.4.6 Moving Control Volumes

For most problems in fluid mechanics, the control volume may be considered as a fixed volume through which the fluid flows. There are, however, situations for which the analysis is simplified if the control volume is allowed to move or deform. The most general situation would involve a control volume that moves, accelerates, and deforms. As one might expect, the use of these control volumes can become fairly complex.

A number of important problems can be most easily analyzed by using a nondeforming control volume that moves with a constant velocity. Such an example is shown in Fig. 4.20 in which a stream of water with velocity $\mathbf{V}_1$ strikes a vane that is moving with constant velocity $\mathbf{V}_0$. It may be of interest to determine the force, $\mathbf{F}$, that the water puts on the vane. Such problems frequently occur in turbines where a stream of fluid (water or steam, for example) strikes a series of blades that move past the nozzle. To analyze such problems it is advantageous to use a moving control volume. We will obtain the Reynolds transport theorem for such control volumes.

We consider a control volume that moves with a constant velocity as is shown in Fig. 4.21. The shape, size, and orientation of the control volume do not change with time. The control volume merely translates with a constant velocity, $\mathbf{V}_{cv}$, as shown. In general, the velocity of the control volume and the fluid are not the same, so that there is a flow of fluid through the moving control volume just as in the stationary control volume cases discussed in Section 4.4.2. The main difference between the fixed and the moving control volume cases is that it is the *relative velocity*, $\mathbf{W}$, that carries fluid across the moving control surface, whereas it is the *absolute velocity*, $\mathbf{V}$, that carries the fluid across the fixed control surface. The relative velocity is the fluid velocity relative to the moving control volume—the fluid velocity seen by an observer riding along on the control volume. The absolute velocity is the fluid velocity as seen by a stationary observer in a fixed coordinate system.

The absolute and relative velocities differ by an amount equal to the control volume velocity.

The difference between the absolute and relative velocities is the velocity of the control volume, $\mathbf{V}_{cv} = \mathbf{V} - \mathbf{W}$, or

$$\mathbf{V} = \mathbf{W} + \mathbf{V}_{cv} \tag{4.22}$$

Since the velocity is a vector, we must use vector addition as is shown in Fig. 4.22 to obtain the relative velocity if we know the absolute velocity and the velocity of the control volume. Thus, if the water leaves the nozzle in Fig. 4.20 with a velocity of $\mathbf{V}_1 = 30\hat{\mathbf{i}}$ m/s and the vane has a velocity of $\mathbf{V}_0 = 6\hat{\mathbf{i}}$ m/s (the same as the control volume), it appears to an observer riding on the vane that the water approaches the vane with a velocity of $\mathbf{W} = \mathbf{V} - \mathbf{V}_{cv} = 24\hat{\mathbf{i}}$ m/s. In general, the absolute

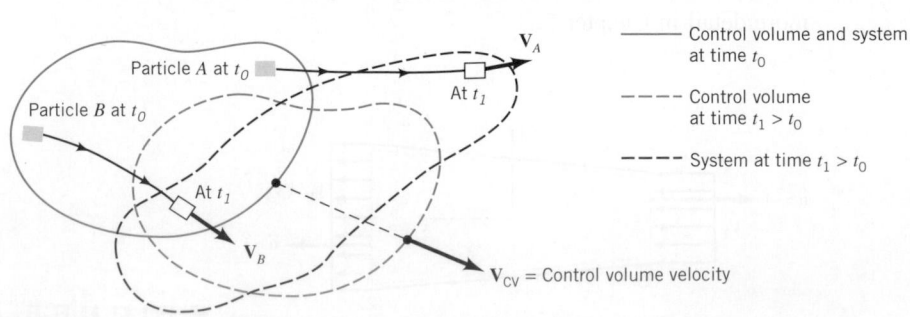

■ **FIGURE 4.21**   Typical moving control volume and system.

V_A = Absolute velocity of A

V_B

V_{CV}

W_A = Velocity of A relative to control volume

V_{CV}

W_B = Velocity of B relative to control volume

■ **F I G U R E   4.22**
Relationship between absolute and relative velocities.

velocity, **V**, and the control volume velocity, **V**$_{cv}$, will not be in the same direction so that the relative and absolute velocities will have different directions (see Fig. 4.22).

The Reynolds transport theorem for a moving, nondeforming control volume can be derived in the same manner that it was obtained for a fixed control volume. As is indicated in Fig. 4.23, the only difference that needs be considered is the fact that relative to the moving control volume the fluid velocity observed is the relative velocity, not the absolute velocity. An observer fixed to the moving control volume may or may not even know that he or she is moving relative to some fixed coordinate system. If we follow the derivation that led to Eq. 4.19 (the Reynolds transport theorem for a fixed control volume), we note that the corresponding result for a moving control volume can be obtained by simply replacing the absolute velocity, **V**, in that equation by the relative velocity, **W**. Thus, the Reynolds transport theorem for a control volume moving with constant velocity is given by

The Reynolds transport theorem for a moving control volume involves the relative velocity.

$$\frac{DB_{sys}}{Dt} = \frac{\partial}{\partial t} \int_{cv} \rho b \, d\forall + \int_{cs} \rho b \, \mathbf{W} \cdot \hat{\mathbf{n}} \, dA \qquad (4.23)$$

where the relative velocity is given by Eq. 4.22.

4.4.7 Selection of a Control Volume

Any volume in space can be considered as a control volume. It may be of finite size or it may be infinitesimal in size, depending on the type of analysis to be carried out. In most of our cases, the control volume will be a fixed, nondeforming volume. In some situations we will consider control volumes that move with constant velocity. In either case it is important that considerable thought go into the selection of the specific control volume to be used.

The selection of an appropriate control volume in fluid mechanics is very similar to the selection of an appropriate free-body diagram in dynamics or statics. In dynamics, we select the body in which we are interested, represent the object in a free-body diagram, and then apply the appropriate governing laws to that body. The ease of solving a given dynamics problem is often very dependent on the specific object that we select for use in our free-body diagram. Similarly, the ease of solving a given fluid mechanics problem is often very dependent on the choice of the control volume used. Only by practice can we develop skill at selecting the "best" control volume. None are "wrong," but some are "much better" than others.

Solution of a typical problem will involve determining parameters such as velocity, pressure, and force at some point in the flow field. It is usually best to ensure that this point is located on the control surface, not "buried" within the control volume. The unknown will then appear in the convective term (the surface integral) of the Reynolds transport theorem. If possible, the control

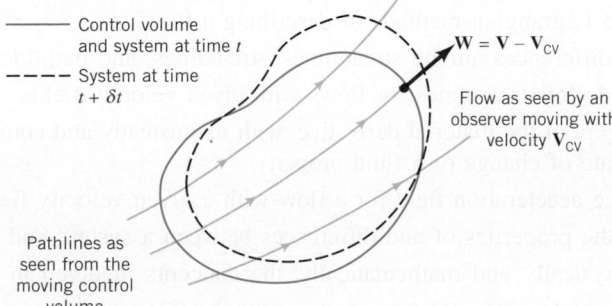

———— Control volume and system at time t

$\mathbf{W} = \mathbf{V} - \mathbf{V}_{CV}$

- - - - System at time $t + \delta t$

Flow as seen by an observer moving with velocity $\mathbf{V}_{CV}$

Pathlines as seen from the moving control volume

■ **F I G U R E   4.23**
Control volume and system as seen by an observer moving with the control volume.

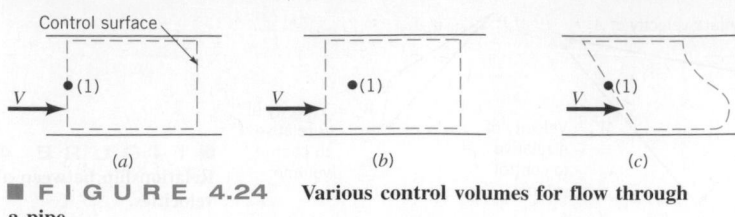

■ **F I G U R E   4.24** Various control volumes for flow through a pipe.

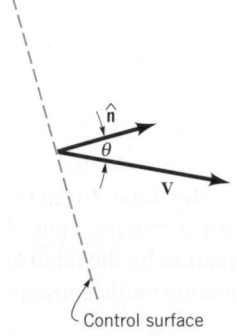

surface should be normal to the fluid velocity so that the angle θ ($\mathbf{V} \cdot \hat{\mathbf{n}} = V \cos \theta$ as shown by the figure in the margin) in the flux terms of Eq. 4.19 will be 0 or 180°. This will usually simplify the solution process.

Figure 4.24 illustrates three possible control volumes associated with flow through a pipe. If the problem is to determine the pressure at point (1), the selection of the control volume (a) is better than that of (b) because point (1) lies on the control surface. Similarly, control volume (a) is better than (c) because the flow is normal to the inlet and exit portions of the control volume. None of these control volumes are wrong—(a) will be easier to use. Proper control volume selection will become much clearer in Chapter 5 where the Reynolds transport theorem is used to transform the governing equations from the system formulation into the control volume formulation, and numerous examples using control volume ideas are discussed.

4.5   Chapter Summary and Study Guide

field representation
velocity field
Eulerian method
Lagrangian method
one-, two-, and three-
   dimensional flow
steady and unsteady
   flow
streamline
streakline
pathline
acceleration field
material derivative
local acceleration
convective acceleration
system
control volume
Reynolds transport
   theorem

This chapter considered several fundamental concepts of fluid kinematics. That is, various aspects of fluid motion are discussed without regard to the forces needed to produce this motion. The concepts of a field representation of a flow and the Eulerian and Lagrangian approaches to describing a flow are introduced, as are the concepts of velocity and acceleration fields.

The properties of one-, two-, or three-dimensional flows and steady or unsteady flows are introduced along with the concepts of streamlines, streaklines, and pathlines. Streamlines, which are lines tangent to the velocity field, are identical to streaklines and pathlines if the flow is steady. For unsteady flows, they need not be identical.

As a fluid particle moves about, its properties (i.e., velocity, density, temperature) may change. The rate of change of these properties can be obtained by using the material derivative, which involves both unsteady effects (time rate of change at a fixed location) and convective effects (time rate of change due to the motion of the particle from one location to another).

The concepts of a control volume and a system are introduced, and the Reynolds transport theorem is developed. By using these ideas, the analysis of flows can be carried out using a control volume (a volume, usually fixed, through which the fluid flows), whereas the governing principles are stated in terms of a system (a flowing portion of fluid).

The following checklist provides a study guide for this chapter. When your study of the entire chapter and end-of-chapter exercises has been completed you should be able to

■ write out meaning of the terms listed here in the margin and understand each of the related concepts. These terms are particularly important and are set in *italic, bold, and color* type in the text.

■ understand the concept of the field representation of a flow and the difference between Eulerian and Lagrangian methods of describing a flow.

■ explain the differences among streamlines, streaklines, and pathlines.

■ calculate and plot streamlines for flows with given velocity fields.

■ use the concept of the material derivative, with its unsteady and convective effects, to determine time rate of change of a fluid property.

■ determine the acceleration field for a flow with a given velocity field.

■ understand the properties of and differences between a system and a control volume.

■ interpret, physically and mathematically, the concepts involved in the Reynolds transport theorem.

Some of the important equations in this chapter are:

Equation for streamlines	$\dfrac{dy}{dx} = \dfrac{v}{u}$	**(4.1)**
Acceleration	$\mathbf{a} = \dfrac{\partial \mathbf{V}}{\partial t} + u\dfrac{\partial \mathbf{V}}{\partial x} + v\dfrac{\partial \mathbf{V}}{\partial y} + w\dfrac{\partial \mathbf{V}}{\partial z}$	**(4.3)**
Material derivative	$\dfrac{D(\ )}{Dt} = \dfrac{\partial(\ )}{\partial t} + (\mathbf{V} \cdot \nabla)(\ )$	**(4.6)**
Streamwise and normal components of acceleration	$a_s = V\dfrac{\partial V}{\partial s}, \qquad a_n = \dfrac{V^2}{\mathcal{R}}$	**(4.7)**
Reynolds transport theorem (restricted form)	$\dfrac{DB_{\text{sys}}}{Dt} = \dfrac{\partial B_{\text{cv}}}{\partial t} + \rho_2 A_2 V_2 b_2 - \rho_1 A_1 V_1 b_1$	**(4.15)**
Reynolds transport theorem (general form)	$\dfrac{DB_{\text{sys}}}{Dt} = \dfrac{\partial}{\partial t}\displaystyle\int_{\text{cv}} \rho b\, d\forall + \int_{\text{cs}} \rho b\, \mathbf{V} \cdot \hat{\mathbf{n}}\, dA$	**(4.19)**
Relative and absolute velocities	$\mathbf{V} = \mathbf{W} + \mathbf{V}_{\text{cv}}$	**(4.22)**

References

1. Streeter, V. L., and Wylie, E. B., *Fluid Mechanics*, 8th Ed., McGraw-Hill, New York, 1985.
2. Goldstein, R. J., *Fluid Mechanics Measurements*, Hemisphere, New York, 1983.
3. Homsy, G. M., et al., *Multimedia Fluid Mechanics* CD-ROM, 2nd Ed., Cambridge University Press, New York, 2007.
4. Magarvey, R. H., and MacLatchy, C. S., The Formation and Structure of Vortex Rings, *Canadian Journal of Physics*, Vol. 42, 1964.

Review Problems

Go to Appendix G for a set of review problems with answers. Detailed solutions can be found in *Student Solution Manual and Study Guide for Fundamentals of Fluid Mechanics*, by Munson et al. (© 2009 John Wiley and Sons, Inc.).

Problems

Note: Unless otherwise indicated, use the values of fluid properties found in the tables on the inside of the front cover. Problems designated with an (*) are intended to be solved with the aid of a programmable calculator or a computer. Problems designated with a (†) are "open-ended" problems and require critical thinking in that to work them one must make various assumptions and provide the necessary data. There is not a unique answer to these problems.

Answers to the even-numbered problems are listed at the end of the book. Access to the videos that accompany problems can be obtained through the book's web site, www.wiley.com/college/munson. The lab-type problems can also be accessed on this web site.

Section 4.1 The Velocity Field

4.1 Obtain a photograph/image that shows a flowing fluid. Print this photo and write a brief paragraph that describes the flow in terms of an Eulerian description; a Lagrangian description.

4.2 Obtain a photograph/image of a situation in which the unsteadiness of the flow is important. Print this photo and write a brief paragraph that describes the situation involved.

4.3 Obtain a photograph/image of a situation in which a fluid is flowing. Print this photo and draw in some lines to represent how you think some streamlines may look. Write a brief paragraph to describe the acceleration of a fluid particle as it flows along one of these streamlines.

4.4 The x- and y-components of a velocity field are given by $u = -(V_0/\ell)x$ and $v = -(V_0/\ell)y$, where V_0 and ℓ are constants. Make a sketch of the velocity field in the first quadrant $(x > 0, y > 0)$ by drawing arrows representing the fluid velocity at representative locations.

4.5 A two-dimensional velocity field is given by $u = 1 + y$ and $v = 1$. Determine the equation of the streamline that passes through the origin. On a graph, plot this streamline.

4.6 The velocity field of a flow is given by $\mathbf{V} = (5z - 3)\hat{\mathbf{i}} + (x + 4)\hat{\mathbf{j}} + 4y\hat{\mathbf{k}}$ m/s, where x, y, and z are in meters. Determine the fluid speed at the origin $(x = y = z = 0)$ and on the x axis $(y = z = 0)$.

4.7 A flow can be visualized by plotting the velocity field as velocity vectors at representative locations in the flow as shown in **Video V4.2** and Fig. E4.1. Consider the velocity field given in polar coordinates by $v_r = -10/r$, and $v_\theta = 10/r$. This flow approximates a fluid swirling into a sink as shown in Fig. P4.7. Plot the velocity field at locations given by $r = 1, 2$, and 3 with $\theta = 0, 30, 60$, and $90°$.

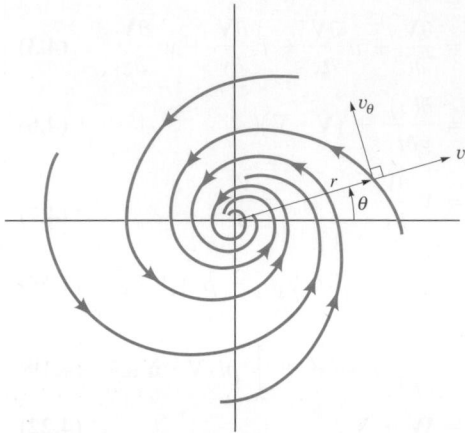

■ **FIGURE P4.7**

4.8 The velocity field of a flow is given by $\mathbf{V} = 20y/(x^2 + y^2)^{1/2}\hat{\mathbf{i}} - 20x/(x^2 + y^2)^{1/2}\hat{\mathbf{j}}$ m/s, where x and y are in meters. Determine the fluid speed at points along the x axis; along the y axis. What is the angle between the velocity vector and the x axis at points $(x, y) = (1.5, 0), (1.5, 1.5)$, and $(0, 1.5)$?

4.9 The components of a velocity field are given by $u = x + y$, $v = xy^3 + 16$, and $w = 0$. Determine the location of any stagnation points $(\mathbf{V} = 0)$ in the flow field.

4.10 The x and y components of velocity for a two-dimensional flow are $u = 2.0y$ m/s and $v = 1$ m/s, where y is in meters. Determine the equation for the streamlines and sketch representative streamlines in the upper half plane.

4.11 Show that the streamlines for a flow whose velocity components are $u = c(x^2 - y^2)$ and $v = -2cxy$, where c is a constant, are given by the equation $x^2y - y^3/3 =$ constant. At which point (points) is the flow parallel to the y axis? At which point (points) is the fluid stationary?

4.12 A velocity field is given by $\mathbf{V} = x\hat{\mathbf{i}} + x(x - 1)(y + 1)\hat{\mathbf{j}}$, where u and v are in m/s and x and y are in meters. Plot the streamline that passes through $x = 0$ and $y = 0$. Compare this streamline with the streakline through the origin.

4.13 From time $t = 0$ to $t = 5$ hr radioactive steam is released from a nuclear power plant accident located at $x = -1.6$ km and $y = 4.8$ km. The following wind conditions are expected: $\mathbf{V} = 16\hat{\mathbf{i}} - 8\hat{\mathbf{j}}$ km/hr for $0 < t < 3$ hr, $\mathbf{V} = 24\hat{\mathbf{i}} + 13\hat{\mathbf{j}}$ km/hr for $3 < t < 10$ hr, and $\mathbf{V} = 8\hat{\mathbf{i}}$ km/hr for $t > 10$ hr. Draw to scale the expected streakline of the steam for $t = 3, 10$, and 15 hr.

****4.14** Consider a ball thrown with initial speed V_0 at an angle of θ as shown in Fig. P4.14a. As discussed in beginning physics, if friction is negligible the path that the ball takes is given by

$$y = (\tan \theta)x - [g/(2 V_0^2 \cos^2 \theta)]x^2$$

That is, $y = c_1x + c_2x^2$, where c_1 and c_2 are constants. The path is a parabola. The pathline for a stream of water leaving a small nozzle is shown in Fig. P4.14b and **Video V4.12**. The coordinates for this water stream are given in the following table. **(a)** Use the given data to determine appropriate values for c_1 and c_2 in the above equation and, thus, show that these water particles also follow a parabolic pathline. **(b)** Use your values of c_1 and c_2 to determine the speed of the water, V_0, leaving the nozzle.

x (cm)	y (cm)
0	0
0.6	0.33
1.2	0.41
1.8	0.33
2.4	0.00
3.0	−0.51
3.5	−1.35
4.2	−2.29
4.8	−3.63

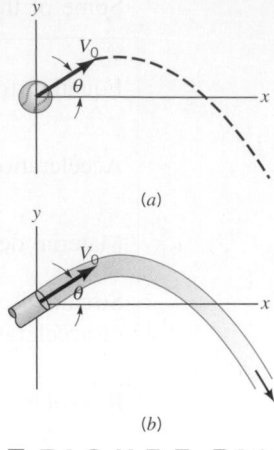

■ **FIGURE P4.14**

4.15 The x and y components of a velocity field are given by $u = x^2y$ and $v = -xy^2$. Determine the equation for the streamlines of this flow and compare it with those in Example 4.2. Is the flow in this problem the same as that in Example 4.2? Explain.

4.16 A flow in the x–y plane is given by the following velocity field: $u = 3$ and $v = 6$ m/s for $0 < t < 20$ s; $u = -4$ and $v = 0$ m/s for $20 < t < 40$ s. Dye is released at the origin $(x = y = 0)$ for $t \geq 0$. **(a)** Draw the pathlines at $t = 30$ s for two particles that were released from the origin—one released at $t = 0$ and the other released at $t = 20$ s. **(b)** On the same graph draw the streamlines at times $t = 10$ s and $t = 30$ s.

4.17 In addition to the customary horizontal velocity components of the air in the atmosphere (the "wind"), there often are vertical air currents (thermals) caused by buoyant effects due to uneven heating of the air as indicated in Fig. P4.17. Assume that the velocity field in a certain region is approximated by $u = u_0$, $v = v_0 (1 - y/h)$ for $0 < y < h$, and $u = u_0$, $v = 0$ for $y > h$. Plot the shape of the streamline that passes through the origin for values of $u_0/v_0 = 0.5, 1$, and 2.

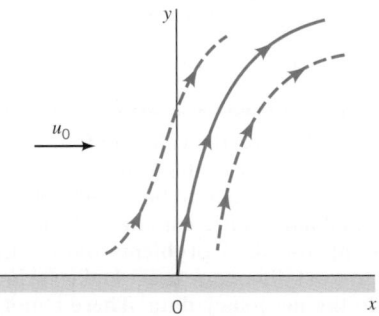

■ **FIGURE P4.17**

****4.18** Repeat Problem 4.17 using the same information except that $u = u_0y/h$ for $0 \leq y \leq h$ rather than $u = u_0$. Use values of $u_0/v_0 = 0, 0.1, 0.2, 0.4, 0.6, 0.8$, and 1.0.

4.19 As shown in **Video V4.6** and Fig. P4.19, a flying airplane produces swirling flow near the end of its wings. In certain circumstances this flow can be approximated by the velocity field $u = -Ky/(x^2 + y^2)$ and $v = Kx/(x^2 + y^2)$, where K is a constant depending on various parameters associated with the airplane (i.e., its weight, speed) and x and y are measured from the center of the swirl. **(a)** Show that for this flow the velocity is inversely proportional to the distance from the origin. That is, $V = K/(x^2 + y^2)^{1/2}$. **(b)** Show that the streamlines are circles.

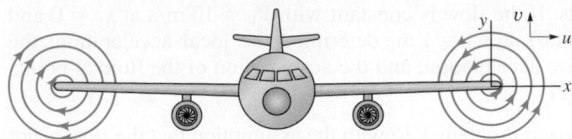

■ **FIGURE P4.19**

4.20 (See Fluids in the News article titled "Follow those particles," Section 4.1.) Two photographs of four particles in a flow past a sphere are superposed as shown in Fig. P4.20. The time interval between the photos is $\Delta t = 0.002$ s. The locations of the particles, as determined from the photos, are shown in the table. **(a)** Determine the fluid velocity for these particles. **(b)** Plot a graph to compare the results of part **(a)** with the theoretical velocity which is given by $V = V_0(1 + a^3/x^3)$, where a is the sphere radius and V_0 is the fluid speed far from the sphere.

Particle	x at $t = 0$ s (m)	x at $t = 0.002$ s (m)
1	−0.152	−0.146
2	−0.075	−0.070
3	−0.043	−0.039
4	−0.036	−0.034

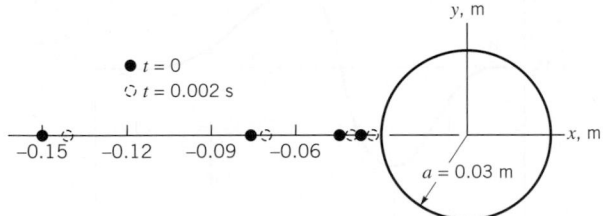

■ **FIGURE P4.20**

4.21 (See Fluids in the News article titled "Winds on Earth and Mars," Section 4.1.4.) A 3-m-diameter dust devil that rotates one revolution per second travels across the Martian surface (in the x-direction) with a speed of 1.5 m/s. Plot the pathline etched on the surface by a fluid particle 3 m from the center of the dust devil for time $0 \leq t \leq 3$ s. The particle position is given by the sum of that for a stationary swirl $[x = 3 \cos(2\pi t), y = 3 \sin(2\pi t)]$ and that for a uniform velocity ($x = 1.5t$, $y = $ constant), where x and y are in meters and t is in seconds.

Section 4.2 The Acceleration Field

4.22 The x- and y-components of a velocity field are given by $u = (V_0/\ell)x$ and $v = -(V_0/\ell)y$, where V_0 and ℓ are constants. Plot the streamlines for this flow and determine the acceleration field.

4.23 A velocity field is given by $u = cx^2$ and $v = cy^2$, where c is a constant. Determine the x and y components of the acceleration. At what point (points) in the flow field is the acceleration zero?

4.24 Determine the acceleration field for a three-dimensional flow with velocity components $u = -x$, $v = 4x^2y^2$, and $w = x - y$.

†4.25 Estimate the deceleration of a water particle in a raindrop as it strikes the sidewalk. List all assumptions and show all calculations.

4.26 The velocity of air in the diverging pipe shown in Fig. P4.26 is given by $V_1 = 1.2t$ m/s and $V_2 = 0.6t$ m/s, where t is in seconds. **(a)** Determine the local acceleration at points (1) and (2). **(b)** Is the average convective acceleration between these two points negative, zero, or positive? Explain.

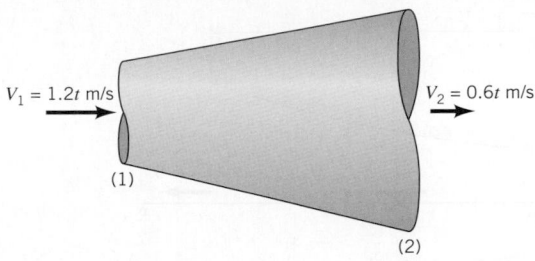

■ **FIGURE P4.26**

4.27 Water flows in a pipe so that its velocity triples every 20 s. At $t = 0$ it has $u = 5$ m/s. That is, $\mathbf{V} = u(t)\hat{\mathbf{i}} = 5\,(3^{t/20})\hat{\mathbf{i}}$ m/s. Determine the acceleration when $t = 0$, 10, and 20 s.

4.28 When a valve is opened, the velocity of water in a certain pipe is given by $u = 10(1 - e^{-t})$, $v = 0$, and $w = 0$, where u is in m/s and t is in seconds. Determine the maximum velocity and maximum acceleration of the water.

4.29 The velocity of the water in the pipe shown in Fig. P4.29 is given by $V_1 = 0.50t$ m/s and $V_2 = 1.0t$ m/s, where t is in seconds. Determine the local acceleration at points (1) and (2). Is the average convective acceleration between these two points negative, zero, or positive? Explain.

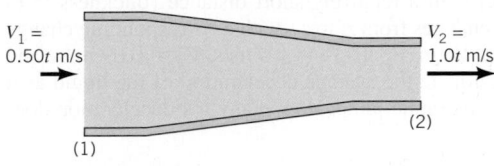

■ **FIGURE P4.29**

4.30 A shock wave is a very thin layer (thickness $= \ell$) in a high-speed (supersonic) gas flow across which the flow properties (velocity, density, pressure, etc.) change from state (1) to state (2) as shown in Fig. P4.30. If $V_1 = 549$ m/s, $V_2 = 213$ m/s, and $\ell = 2.5 \times 10^{-4}$ cm, estimate the average deceleration of the gas as it flows across the shock wave. How many g's deceleration does this represent?

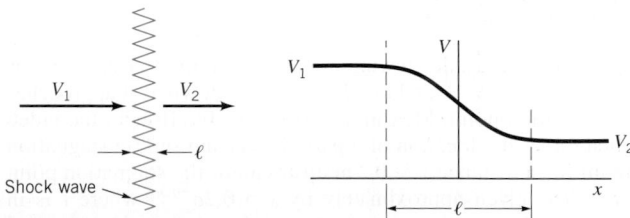

■ **FIGURE P4.30**

†4.31 Estimate the average acceleration of water as it travels through the nozzle on your garden hose. List all assumptions and show all calculations.

4.32 As a valve is opened, water flows through the diffuser shown in Fig. P4.32 at an increasing flowrate so that the velocity along the centerline is given by $\mathbf{V} = u\hat{\mathbf{i}} = V_0(1 - e^{-ct})\,(1 - x/\ell)\,\hat{\mathbf{i}}$, where u_0, c, and ℓ are constants. Determine the acceleration as a function of x and t. If $V_0 = 3$ m/s and $\ell = 1.5$ m, what value of c (other than $c = 0$) is needed to make the acceleration zero for any x at $t = 1$ s? Explain how the acceleration can be zero if the flowrate is increasing with time.

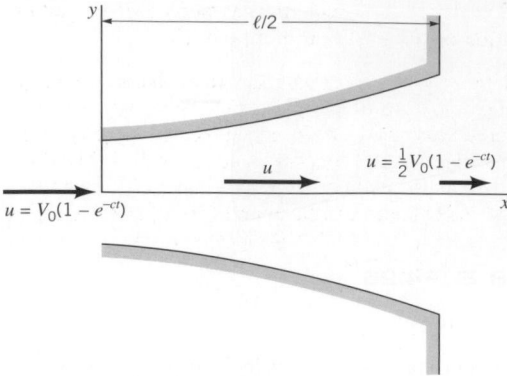

■ **F I G U R E   P4.32**

4.33 A fluid flows along the x axis with a velocity given by $\mathbf{V} = (x/t)\,\hat{\mathbf{i}}$, where x is in meters and t in seconds. (a) Plot the speed for $0 \le x \le 3$ m and $t = 3$ s. (b) Plot the speed for $x = 2.0$ m and $2 \le t \le 4$ s. (c) Determine the local and convective acceleration. (d) Show that the acceleration of any fluid particle in the flow is zero. (e) Explain physically how the velocity of a particle in this unsteady flow remains constant throughout its motion.

4.34 A hydraulic jump is a rather sudden change in depth of a liquid layer as it flows in an open channel as shown in Fig. P4.34 and **Video V10.12**. In a relatively short distance (thickness = ℓ) the liquid depth changes from z_1 to z_2, with a corresponding change in velocity from V_1 to V_2. If $V_1 = 0.4$ m/s, $V_2 = 0.09$ m/s, and $\ell = 0.01$ m, estimate the average deceleration of the liquid as it flows across the hydraulic jump. How many g's deceleration does this represent?

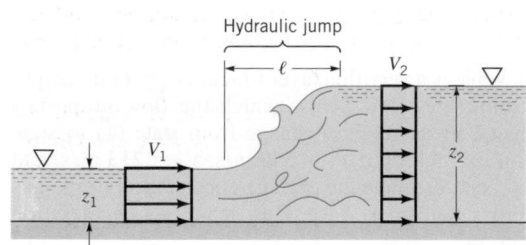

■ **F I G U R E   P4.34**

4.35 A fluid particle flowing along a stagnation streamline, as shown in **Video V4.9** and Fig. P4.35, slows down as it approaches the stagnation point. Measurements of the dye flow in the video indicate that the location of a particle starting on the stagnation streamline a distance $s = 0.2$ m upstream of the stagnation point at $t = 0$ is given approximately by $s = 0.2e^{-0.5t}$, where t is in seconds and s is in meters. (a) Determine the speed of a fluid particle as a function of time, $V_{\text{particle}}(t)$, as it flows along the streamline. (b) Determine the speed of the fluid as a function of position along the streamline, $V = V(s)$. (c) Determine the fluid acceleration along the streamline as a function of position, $a_s = a_s(s)$.

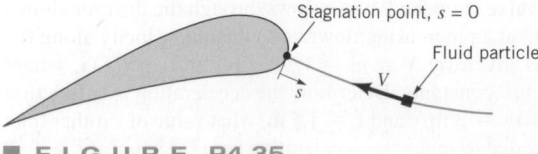

■ **F I G U R E   P4.35**

4.36 A nozzle is designed to accelerate the fluid from V_1 to V_2 in a linear fashion. That is, $V = ax + b$, where a and b are

constants. If the flow is constant with $V_1 = 10$ m/s at $x_1 = 0$ and $V_2 = 25$ m/s at $x_2 = 1$ m, determine the local acceleration, the convective acceleration, and the acceleration of the fluid at points (1) and (2).

4.37 Repeat Problem 4.36 with the assumption that the flow is not steady, but at the time when $V_1 = 10$ m/s and $V_2 = 25$ m/s, it is known that $\partial V_1/\partial t = 20$ m/s² and $\partial V_2/\partial t = 60$ m/s².

4.38 An incompressible fluid flows past a turbine blade as shown in Fig. P4.38a and **Video V4.9**. Far upstream and downstream of the blade the velocity is V_0. Measurements show that the velocity of the fluid along streamline A–F near the blade is as indicated in Fig. P4.38b. Sketch the streamwise component of acceleration, a_s, as a function of distance, s, along the streamline. Discuss the important characteristics of your result.

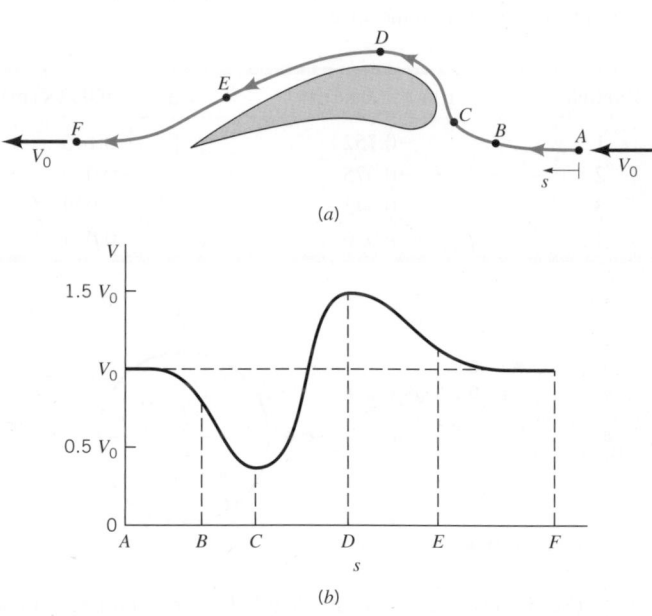

(a)

(b)

■ **F I G U R E   P4.38**

*4.39 Air flows steadily through a variable area pipe with a velocity of $\mathbf{V} = u(x)\hat{\mathbf{i}}$ m/s, where the approximate measured values of $u(x)$ are given in the table. Plot the acceleration as a function of x for $0 \le x \le 30.0$ cm. Plot the acceleration if the flowrate is increased by a factor of N (i.e., the values of u are increased by a factor of N) for $N = 2, 4, 10$.

x (cm)	u (m/s)	x (cm)	u (m/s)
0	3.0	17.5	6.1
2.5	3.1	20.0	5.3
5.0	4.0	22.5	4.1
7.5	6.1	25.0	3.6
10.0	8.6	27.5	3.1
12.5	8.7	30.0	3.0
15.0	7.9	33.5	3.0

*4.40 As is indicated in Fig. P4.40, the speed of exhaust in a car's exhaust pipe varies in time and distance because of the periodic nature of the engine's operation and the damping effect with distance from the engine. Assume that the speed is given by $V = V_0[1 + ae^{-bx} \sin(\omega t)]$, where $V_0 = 2.4$ m/s, $a = 0.05$, $b = 0.66$ m^{-1}, and $\omega = 50$ rad/s. Calculate and plot the fluid acceleration at $x = 0, 0.3, 0.6, 0.9, 1.2,$ and 1.5 m for $0 \le t \le \pi/25$ s.

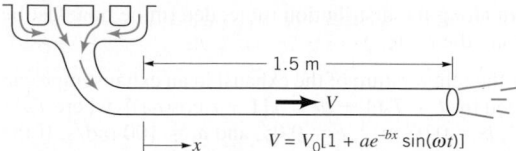

■ FIGURE P4.40

4.41 Water flows over the crest of a dam with speed V as shown in Fig. P4.41. Determine the speed if the magnitude of the normal acceleration at point (1) is to equal the acceleration of gravity, g.

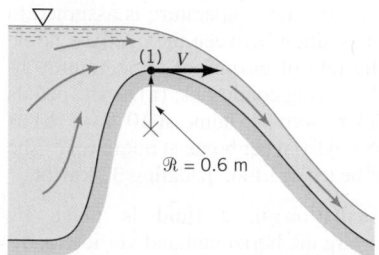

■ FIGURE P4.41

4.42 Assume that the streamlines for the wingtip vortices from an airplane (see Fig. P4.19 and **Video V4.6**) can be approximated by circles of radius r and that the speed is $V = K/r$, where K is a constant. Determine the streamline acceleration, a_s, and the normal acceleration, a_n, for this flow.

4.43 A fluid flows past a sphere with an upstream velocity of $V_0 = 40$ m/s as shown in Fig. P4.43. From a more advanced theory it is found that the speed of the fluid along the front part of the sphere is $V = \frac{3}{2}V_0 \sin \theta$. Determine the streamwise and normal components of acceleration at point A if the radius of the sphere is $a = 0.20$ m.

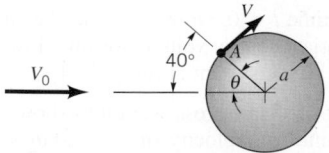

■ FIGURE P4.43

*4.44 For flow past a sphere as discussed in Problem 4.43, plot a graph of the streamwise acceleration, a_s, the normal acceleration, a_n, and the magnitude of the acceleration as a function of θ for $0 \le \theta \le 90°$ with $V_0 = 15$ m/s and $a = 0.03, 0.3$ and 3 m. Repeat for $V_0 = 1.5$ m/s. At what point is the acceleration a maximum; a minimum?

*4.45 The velocity components for steady flow through the nozzle shown in Fig. P4.45 are $u = -V_0 x/\ell$ and $v = V_0 \left[1 + (y/\ell)\right]$,

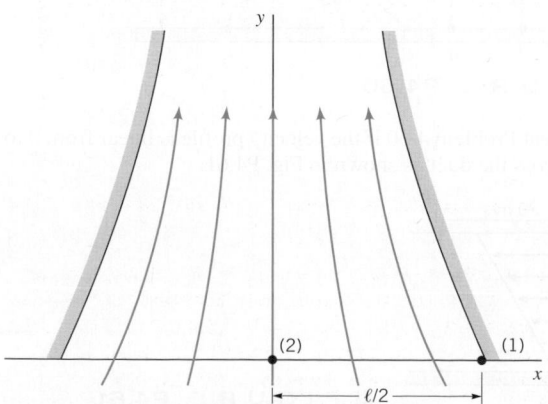

■ FIGURE P4.45

where V_0 and ℓ are constants. Determine the ratio of the magnitude of the acceleration at point (1) to that at point (2).

*4.46 A fluid flows past a circular cylinder of radius a with an upstream speed of V_0 as shown in Fig. P4.46. A more advanced theory indicates that if viscous effects are negligible, the velocity of the fluid along the surface of the cylinder is given by $V = 2V_0 \sin \theta$. Determine the streamline and normal components of acceleration on the surface of the cylinder as a function of V_0, a, and θ and plot graphs of a_s and a_n for $0 \le \theta \le 90°$ with $V_0 = 10$ m/s and $a = 0.01, 0.10, 1.0,$ and 10.0 m.

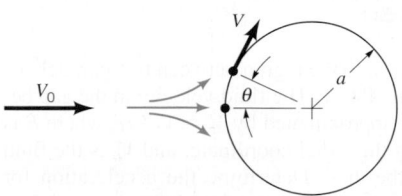

■ FIGURE P4.46

4.47 Determine the x and y components of acceleration for the flow given in Problem 4.11. If $c > 0$, is the particle at point $x = x_0 > 0$ and $y = 0$ accelerating or decelerating? Explain. Repeat if $x_0 < 0$.

4.48 When flood gates in a channel are opened, water flows along the channel downstream of the gates with an increasing speed given by $V = 1.2(1 + 0.1t)$ m/s, for $0 \le t \le 20$ s, where t is in seconds. For $t > 20$ s the speed is a constant $V = 3.6$ m/s. Consider a location in the curved channel where the radius of curvature of the streamlines is 15 m. For $t = 10$ s determine **(a)** the component of acceleration along the streamline, **(b)** the component of acceleration normal to the streamline, and **(c)** the net acceleration (magnitude and direction). Repeat for $t = 30$ s.

4.49 Water flows steadily through the funnel shown in Fig. P4.49. Throughout most of the funnel the flow is approximately radial (along rays from O) with a velocity of $V = c/r^2$, where r is the radial coordinate and c is a constant. If the velocity is 0.4 m/s when $r = 0.1$ m, determine the acceleration at points A and B.

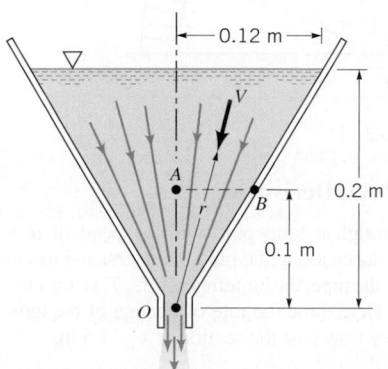

■ FIGURE P4.49

4.50 Water flows though the slit at the bottom of a two-dimensional water trough as shown in Fig. P4.50. Throughout most of the trough the flow is approximately radial (along rays from O) with a velocity of $V = c/r$, where r is the radial coordinate and c is a constant. If the velocity is 0.04 m/s when $r = 0.1$ m, determine the acceleration at points A and B.

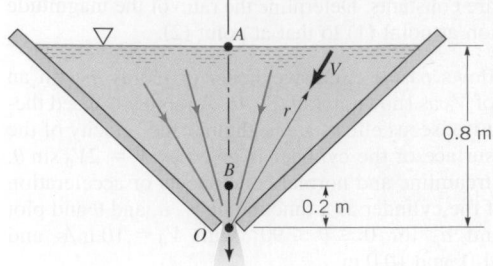

■ **FIGURE P4.50**

4.51 Air flows from a pipe into the region between two parallel circular disks as shown in Fig. P4.51. The fluid velocity in the gap between the disks is closely approximated by $V = V_0 R/r$, where R is the radius of the disk, r is the radial coordinate, and V_0 is the fluid velocity at the edge of the disk. Determine the acceleration for $r = 0.3, 0.6$, or 0.9 m if $V_0 = 1.5$ m/s and $R = 0.9$ m.

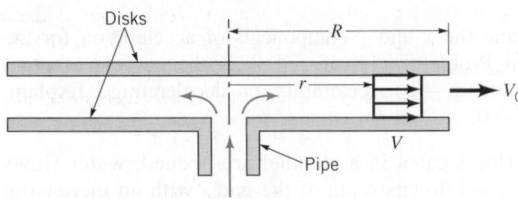

■ **FIGURE P4.51**

4.52 Air flows into a pipe from the region between a circular disk and a cone as shown in Fig. P4.52. The fluid velocity in the gap between the disk and the cone is closely approximated by $V = V_0 R^2/r^2$, where R is the radius of the disk, r is the radial coordinate, and V_0 is the fluid velocity at the edge of the disk. Determine the acceleration for $r = 0.2$ and 0.6 m if $V_0 = 1.5$ m/s and $R = 0.6$ m.

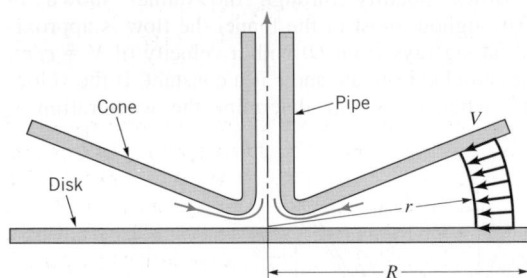

■ **FIGURE P4.52**

Section 4.2.1 The Material Derivative

4.53 Air flows steadily through a long pipe with a speed of $u = 15 + 0.5x$, where x is the distance along the pipe in meters, and u is in m/s. Due to heat transfer into the pipe, the air temperature, T, within the pipe is $T = 300 + 10x$ °C. Determine the rate of change of the temperature of air particles as they flow past the section at $x = 1.5$ m.

4.54 A company produces a perishable product in a factory located at $x = 0$ and sells the product along the distribution route $x > 0$. The selling price of the product, P, is a function of the length of time after it was produced, t, and the location at which it is sold, x. That is, $P = P(x, t)$. At a given location the price of the product decreases in time (it is perishable) according to $\partial P/\partial t = -8$ dollars/hr. In addition, because of shipping costs the price increases with distance from the factory according to $\partial P/\partial x = 0.1$ dollars/km. If the manufacturer wishes to sell the product for the same 100-dollar price anywhere along the distribution route, determine how fast he must travel along the route.

4.55 Assume the temperature of the exhaust in an exhaust pipe can be approximated by $T = T_0(1 + ae^{-bx})[1 + c\cos(\omega t)]$, where $T_0 = 100$ °C, $a = 3$, $b = 0.03$ m^{-1}, $c = 0.05$, and $\omega = 100$ rad/s. If the exhaust speed is a constant 3 m/s, determine the time rate of change of temperature of the fluid particles at $x = 0$ and $x = 4$ m when $t = 0$.

4.56 A bicyclist leaves from her home at 9 A.M. and rides to a beach 64 km away. Because of a breeze off the ocean, the temperature at the beach remains 15 °C throughout the day. At the cyclist's home the temperature increases linearly with time, going from 15 °C at 9 A.M. to 26 °C by 1 P.M. The temperature is assumed to vary linearly as a function of position between the cyclist's home and the beach. Determine the rate of change of temperature observed by the cyclist for the following conditions: **(a)** as she pedals 16 km/hr through a town 16 km from her home at 10 A.M.; **(b)** as she eats lunch at a rest stop 48 km from her home at noon; **(c)** as she arrives enthusiastically at the beach at 1 P.M., pedaling 32 km/hr.

4.57 The temperature distribution in a fluid is given by $T = 10x + 5y$, where x and y are the horizontal and vertical coordinates in meters and T is in degrees centigrade. Determine the time rate of change of temperature of a fluid particle traveling **(a)** horizontally with $u = 20$ m/s, $v = 0$ or **(b)** vertically with $u = 0$, $v = 20$ m/s.

Section 4.4 The Reynolds Transport Theorem

4.58 Obtain a photograph/image of a situation in which a fluid is flowing. Print this photo and draw a control volume through which the fluid flows. Write a brief paragraph that describes how the fluid flows into and out of this control volume.

4.59 The wind blows through the front door of a house with a speed of 2 m/s and exits with a speed of 1 m/s through two windows on the back of the house. Consider the system of interest for this flow to be the air within the house at time $t = 0$. Draw a simple sketch of the house and show an appropriate control volume for this flow. On the sketch, show the position of the system at time $t = 1$ s.

4.60 Water flows through a duct of square cross section as shown in Fig. P4.60 with a constant, uniform velocity of $V = 20$ m/s. Consider fluid particles that lie along line A–B at time $t = 0$. Determine the position of these particles, denoted by line A'–B', when $t = 0.20$ s. Use the volume of fluid in the region between lines A–B and A'–B' to determine the flowrate in the duct. Repeat the problem for fluid particles originally along line C–D; along line E–F. Compare your three answers.

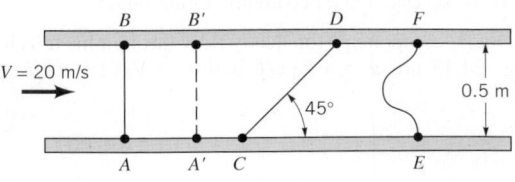

■ **FIGURE P4.60**

4.61 Repeat Problem 4.60 if the velocity profile is linear from 0 to 20 m/s across the duct as shown in Fig. P4.61.

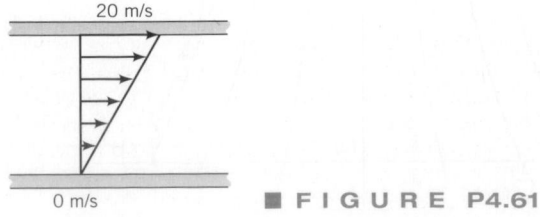

■ **FIGURE P4.61**

4.62 In the region just downstream of a sluice gate, the water may develop a reverse flow region as is indicated in Fig. P4.62 and **Video V10.9**. The velocity profile is assumed to consist of two uniform regions, one with velocity $V_a = 3$ m/s and the other with $V_b = 0.9$ m/s. Determine the net flowrate of water across the portion of the control surface at section (2) if the channel is 6 m wide.

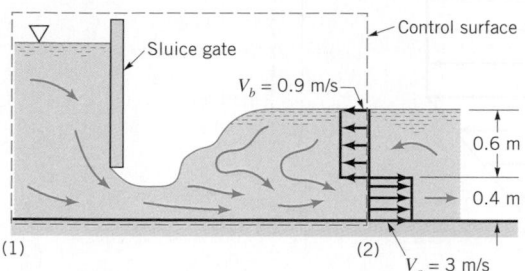

■ **FIGURE P4.62**

4.63 At time $t = 0$ the valve on an initially empty (perfect vacuum, $\rho = 0$) tank is opened and air rushes in. If the tank has a volume of V_0 and the density of air within the tank increases as $\rho = \rho_\infty(1 - e^{-bt})$, where b is a constant, determine the time rate of change of mass within the tank.

†4.64 From calculus, one obtains the following formula (Leibnitz rule) for the time derivative of an integral that contains time in both the integrand and the limits of the integration:

$$\frac{d}{dt}\int_{x_1(t)}^{x_2(t)} f(x, t)dx = \int_{x_1}^{x_2} \frac{\partial f}{\partial t}dx + f(x_2, t)\frac{dx_2}{dt} - f(x_1, t)\frac{dx_1}{dt}$$

Discuss how this formula is related to the time derivative of the total amount of a property in a system and to the Reynolds transport theorem.

4.65 Water enters the bend of a river with the uniform velocity profile shown in Fig. P4.65. At the end of the bend there is a region of separation or reverse flow. The fixed control volume $ABCD$ coincides with the system at time $t = 0$. Make a sketch to indicate (a) the system at time $t = 5$ s and (b) the fluid that has entered and exited the control volume in that time period.

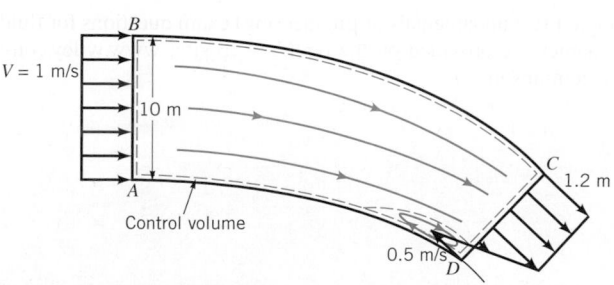

■ **FIGURE P4.65**

4.66 A layer of oil flows down a vertical plate as shown in Fig. P4.66 with a velocity of $\mathbf{V} = (V_0/h^2)(2hx - x^2)\hat{\mathbf{j}}$ where V_0 and h are constants. (a) Show that the fluid sticks to the plate and that the shear stress at the edge of the layer $(x = h)$ is zero. (b) Determine the flowrate across surface AB. Assume the width of the plate is b. (*Note:* The velocity profile for laminar flow in a pipe has a similar shape. See **Video V6.13**.)

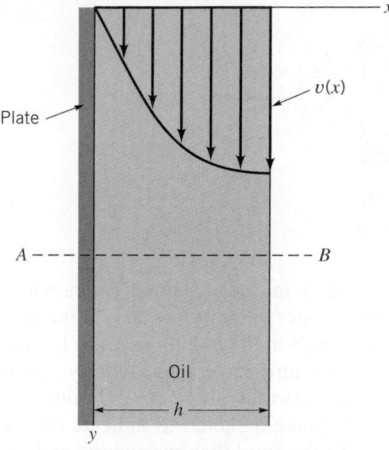

■ **FIGURE P4.66**

4.67 Water flows in the branching pipe shown in Fig. P4.67 with uniform velocity at each inlet and outlet. The fixed control volume indicated coincides with the system at time $t = 20$ s. Make a sketch to indicate (a) the boundary of the system at time $t = 20.1$ s, (b) the fluid that left the control volume during that 0.1-s interval, and (c) the fluid that entered the control volume during that time interval.

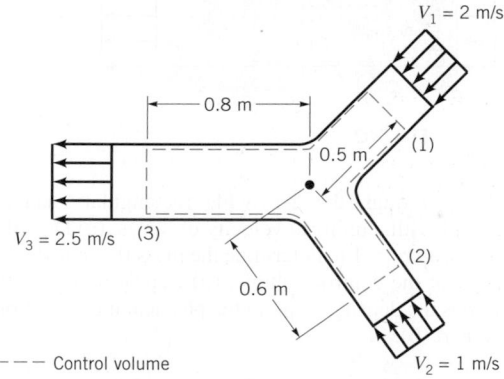

--- Control volume

■ **FIGURE P4.67**

4.68 Two plates are pulled in opposite directions with speeds of 0.3 m/s as shown in Fig. P4.68. The oil between the plates moves with a velocity given by $\mathbf{V} = 10\,y\hat{\mathbf{i}}$ m/s, where y is in meters. The fixed control volume $ABCD$ coincides with the system at time $t = 0$. Make a sketch to indicate (a) the system at time $t = 0.2$ s and (b) the fluid that has entered and exited the control volume in that time period.

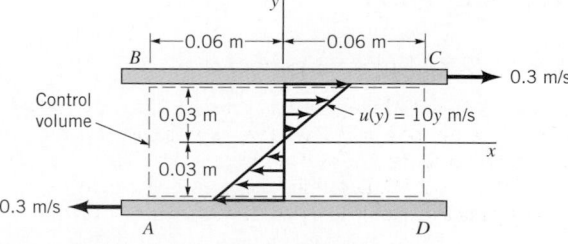

■ **FIGURE P4.68**

4.69 Water is squirted from a syringe with a speed of $V = 5$ m/s by pushing in the plunger with a speed of $V_p = 0.03$ m/s as shown in Fig. P4.69. The surface of the deforming control volume consists of the sides and end of the cylinder and the end of the plunger. The system consists of the water in the syringe at $t = 0$ when the plunger is at section (1) as shown. Make a sketch to indicate the control surface and the system when $t = 0.5$ s.

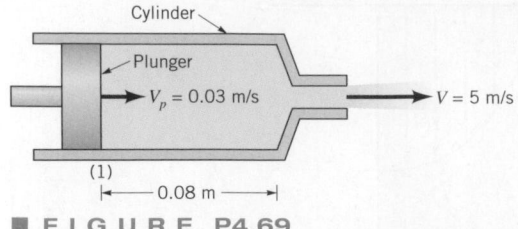

■ **F I G U R E P4.69**

4.70 Water enters a 5-m-wide, 1-m-deep channel as shown in Fig. P4.70. Across the inlet the water velocity is 6 m/s in the center portion of the channel and 1 m/s in the remainder of it. Farther downstream the water flows at a uniform 2 m/s velocity across the entire channel. The fixed control volume *ABCD* coincides with the system at time *t* = 0. Make a sketch to indicate **(a)** the system at time *t* = 0.5 s and **(b)** the fluid that has entered and exited the control volume in that time period.

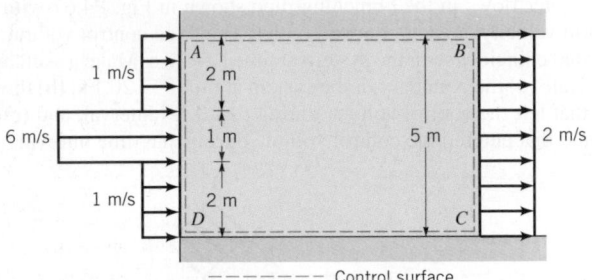

■ **F I G U R E P4.70**

4.71 Water flows through the 2-m-wide rectangular channel shown in Fig. P4.71 with a uniform velocity of 3 m/s. **(a)** Directly integrate Eq. 4.16 with *b* = 1 to determine the mass flowrate (kg/s) across section *CD* of the control volume. **(b)** Repeat part (a) with *b* = 1/*ρ*, where *ρ* is the density. Explain the physical interpretation of the answer to part **(b)**.

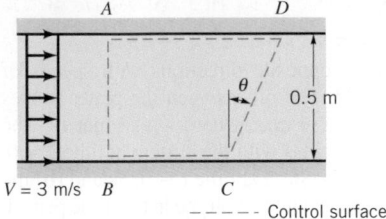

■ **F I G U R E P4.71**

4.72 The wind blows across a field with an approximate velocity profile as shown in Fig. P4.72. Use Eq. 4.16 with the parameter *b* equal to the velocity to determine the momentum flowrate across the vertical surface *A–B*, which is of unit depth into the paper.

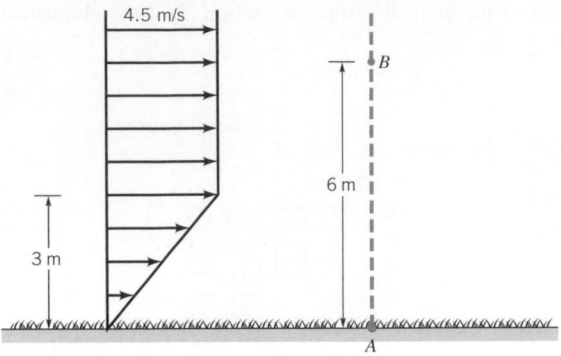

■ **F I G U R E P4.72**

■ **Life Long Learning Problems**

4.73 Even for the simplest flows it is often not be easy to visually represent various flow field quantities such as velocity, pressure, or temperature. For more complex flows, such as those involving three-dimensional or unsteady effects, it is extremely difficult to "show the data." However, with the use of computers and appropriate software, novel methods are being devised to more effectively illustrate the structure of a given flow. Obtain information about methods used to present complex flow data. Summarize your findings in a brief report.

4.74 For centuries people have obtained qualitative and quantitative information about various flow fields by observing the motion of objects or particles in a flow. For example, the speed of the current in a river can be approximated by timing how long it takes a stick to travel a certain distance. The swirling motion of a tornado can be observed by following debris moving within the tornado funnel. Recently various high-tech methods using lasers and minute particles seeded within the flow have been developed to measure velocity fields. Such techniques include the laser doppler anemometer (LDA), the particle image velocimeter (PIV), and others. Obtain information about new laser-based techniques for measuring velocity fields. Summarize your findings in a brief report.

■ **FE Exam Problems**

Sample FE (Fundamentals of Engineering) exam questions for fluid mechanics are provided on the book's web site, www.wiley.com/college/munson.

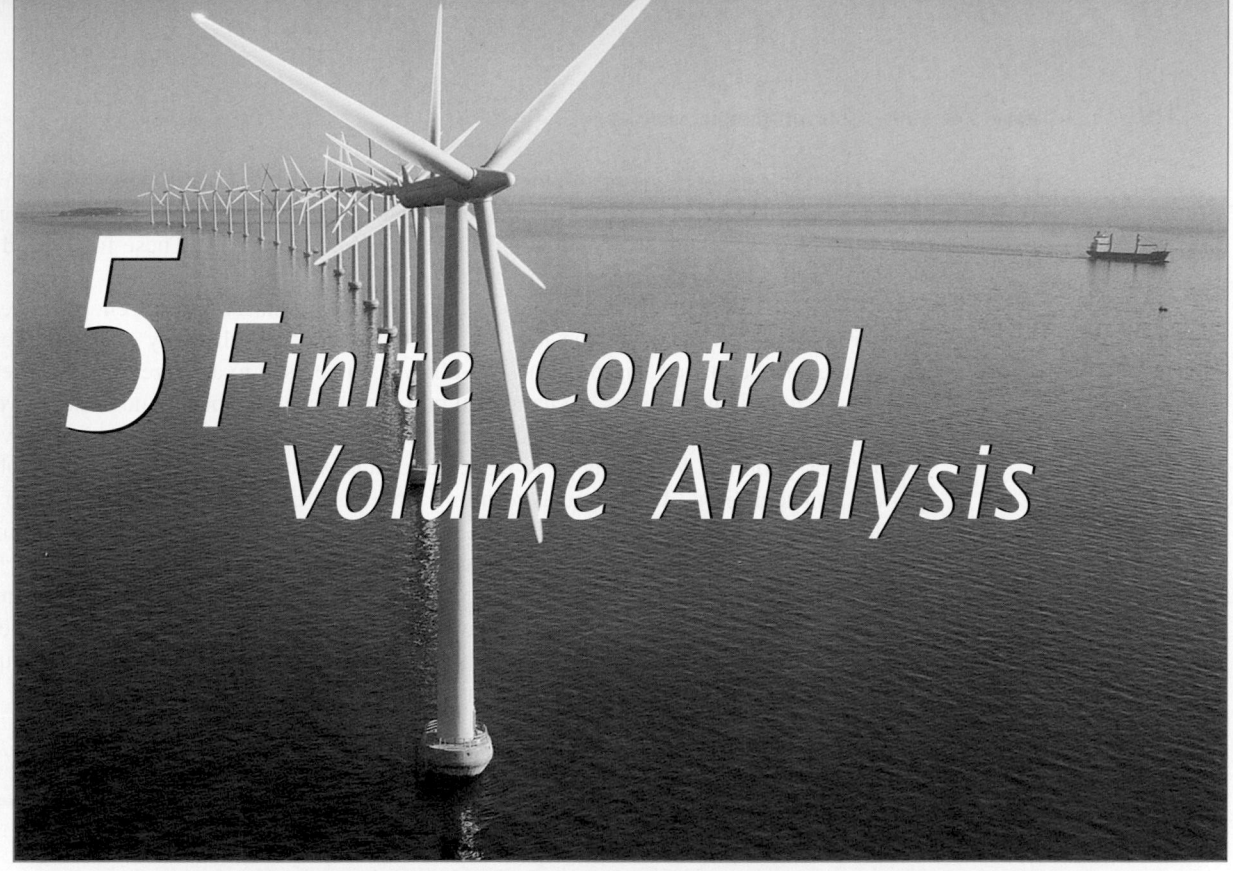

5 Finite Control Volume Analysis

CHAPTER OPENING PHOTO: Wind turbine farms (this is the Middelgrunden Offshore Wind Farm in Denmark) are becoming more common. Finite control volume analysis can be used to estimate the amount of energy transferred between the moving air and each turbine rotor. *(Photograph courtesy of Siemens Wind Power.)*

Learning Objectives

After completing this chapter, you should be able to:

- select an appropriate finite control volume to solve a fluid mechanics problem.
- apply conservation of mass and energy and Newton's second law of motion to the contents of a finite control volume to get important answers.
- know how velocity changes and energy transfers in fluid flows are related to forces and torques.
- understand why designing for minimum loss of energy in fluid flows is so important.

Many fluid mechanics problems can be solved by using control volume analysis.

To solve many practical problems in fluid mechanics, questions about the behavior of the contents of a finite region in space (a finite control volume) are answered. For example, we may be asked to estimate the maximum anchoring force required to hold a turbojet engine stationary during a test. Or we may be called on to design a propeller to move a boat both forward and backward. Or we may need to determine how much power it would take to move natural gas from one location to another many miles away.

The bases of finite control volume analysis are some fundamental laws of physics, namely, conservation of mass, Newton's second law of motion, and the first and second laws of thermodynamics. While some simplifying approximations are made for practicality, the engineering answers possible with the estimates of this powerful analysis method have proven valuable in numerous instances.

Conservation of mass is the key to tracking flowing fluid. How much enters and leaves a control volume can be ascertained.

187

Newton's second law of motion leads to the conclusion that forces can result from or cause changes in a flowing fluid's velocity magnitude and/or direction. Moment of force (torque) can result from or cause changes in a flowing fluid's moment of velocity. These forces and torques can be associated with work and power transfer.

The first law of thermodynamics is a statement of conservation of energy. The second law of thermodynamics identifies the loss of energy associated with every actual process. The mechanical energy equation based on these two laws can be used to analyze a large variety of steady, incompressible flows in terms of changes in pressure, elevation, speed, and of shaft work and loss.

Good judgment is required in defining the finite region in space, the control volume, used in solving a problem. What exactly to leave out of and what to leave in the control volume are important considerations. The formulas resulting from applying the fundamental laws to the contents of the control volume are easy to interpret physically and are not difficult to derive and use.

Because a finite region of space, a control volume, contains many fluid particles and even more molecules that make up each particle, the fluid properties and characteristics are often average values. In Chapter 6 an analysis of fluid flow based on what is happening to the contents of an infinitesimally small region of space or control volume through which numerous molecules simultaneously flow (what we might call a point in space) is considered.

5.1 Conservation of Mass—The Continuity Equation

5.1.1 Derivation of the Continuity Equation

A system is defined as a collection of unchanging contents, so the *conservation of mass* principle for a system is simply stated as

time rate of change of the system mass = 0

or

The amount of mass in a system is constant.

$$\frac{DM_{sys}}{Dt} = 0 \tag{5.1}$$

where the system mass, M_{sys}, is more generally expressed as

$$M_{sys} = \int_{sys} \rho \, d\forall \tag{5.2}$$

and the integration is over the volume of the system. In words, Eq. 5.2 states that the system mass is equal to the sum of all the density-volume element products for the contents of the system.

For a system and a fixed, nondeforming control volume that are coincident at an instant of time, as illustrated in Fig. 5.1, the Reynolds transport theorem (Eq. 4.19) with $B =$ mass and $b = 1$ allows us to state that

$$\frac{D}{Dt} \int_{sys} \rho \, d\forall = \frac{\partial}{\partial t} \int_{cv} \rho \, d\forall + \int_{cs} \rho \mathbf{V} \cdot \hat{\mathbf{n}} \, dA \tag{5.3}$$

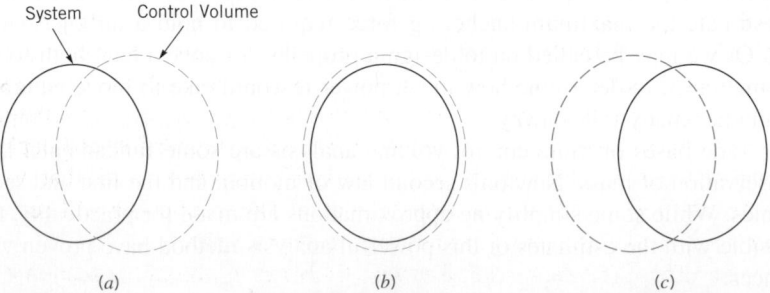

■ FIGURE 5.1 **System and control volume at three different instances of time.** (*a*) System and control volume at time $t - \delta t$. (*b*) System and control volume at time t, coincident condition. (*c*) System and control volume at time $t + \delta t$.

or

time rate of change of the mass of the coincident system	=	time rate of change of the mass of the contents of the coincident control volume	+	net rate of flow of mass through the control surface

In Eq. 5.3, we express the time rate of change of the system mass as the sum of two control volume quantities, the time rate of change of the mass of the contents of the control volume,

$$\frac{\partial}{\partial t} \int_{cv} \rho \, d\forall$$

and the net rate of mass flow through the control surface,

$$\int_{cs} \rho \mathbf{V} \cdot \hat{\mathbf{n}} \, dA$$

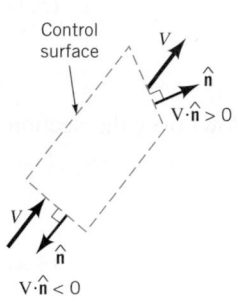

Control surface

$\mathbf{V} \cdot \hat{\mathbf{n}} > 0$

$\mathbf{V} \cdot \hat{\mathbf{n}} < 0$

When a flow is steady, all field properties (i.e., properties at any specified point) including density remain constant with time and the time rate of change of the mass of the contents of the control volume is zero. That is,

$$\frac{\partial}{\partial t} \int_{cv} \rho \, d\forall = 0$$

The integrand, $\mathbf{V} \cdot \hat{\mathbf{n}} \, dA$, in the mass flowrate integral represents the product of the component of velocity, $\mathbf{V}$, perpendicular to the small portion of control surface and the differential area, dA. Thus, $\mathbf{V} \cdot \hat{\mathbf{n}} \, dA$ is the volume flowrate through dA and $\rho \mathbf{V} \cdot \hat{\mathbf{n}} \, dA$ is the mass flowrate through dA. Furthermore, as shown in the sketch in the margin, the sign of the dot product $\mathbf{V} \cdot \hat{\mathbf{n}}$ is "+" for flow *out* of the control volume and "−" for flow *into* the control volume since $\hat{\mathbf{n}}$ is considered positive when it points out of the control volume. When all of the differential quantities, $\rho \mathbf{V} \cdot \hat{\mathbf{n}} \, dA$, are summed over the entire control surface, as indicated by the integral

$$\int_{cs} \rho \mathbf{V} \cdot \hat{\mathbf{n}} \, dA$$

the result is the net mass flowrate through the control surface, or

$$\int_{cs} \rho \mathbf{V} \cdot \hat{\mathbf{n}} \, dA = \sum \dot{m}_{\text{out}} - \sum \dot{m}_{\text{in}} \tag{5.4}$$

where $\dot{m}$ is the mass flowrate (kg/s). If the integral in Eq. 5.4 is positive, the net flow is out of the control volume; if the integral is negative, the net flow is into the control volume.

The control volume expression for conservation of mass, which is commonly called the *continuity equation*, for a fixed, nondeforming control volume is obtained by combining Eqs. 5.1, 5.2, and 5.3 to obtain

The continuity equation is a statement that mass is conserved.

$$\boxed{\frac{\partial}{\partial t} \int_{cv} \rho \, d\forall + \int_{cs} \rho \mathbf{V} \cdot \hat{\mathbf{n}} \, dA = 0} \tag{5.5}$$

In words, Eq. 5.5 states that to conserve mass the time rate of change of the mass of the contents of the control volume plus the net rate of mass flow through the control surface must equal zero. Actually, the same result could have been obtained more directly by equating the rates of mass flow into and out of the control volume to the rates of accumulation and depletion of mass within the control volume (see Section 3.6.2). It is reassuring, however, to see that the Reynolds transport theorem works for this simple-to-understand case. This confidence will serve us well as we develop control volume expressions for other important principles.

An often-used expression for *mass flowrate*, $\dot{m}$, through a section of control surface having area A is

$$\boxed{\dot{m} = \rho Q = \rho A V} \tag{5.6}$$

Mass flowrate equals the product of density and volume flowrate.

where ρ is the fluid density, Q is the volume flowrate (m³/s), and V is the component of fluid velocity perpendicular to area A. Since

$$\dot{m} = \int_A \rho \mathbf{V} \cdot \hat{\mathbf{n}} \, dA$$

application of Eq. 5.6 involves the use of *representative* or average values of fluid density, ρ, and fluid velocity, V. For incompressible flows, ρ is uniformly distributed over area A. For compressible flows, we will normally consider a uniformly distributed fluid density at each section of flow and allow density changes to occur only from section to section. The appropriate fluid velocity to use in Eq. 5.6 is the average value of the component of velocity normal to the section area involved. This average value, $\overline{V}$, defined as

$$\overline{V} = \frac{\int_A \rho \mathbf{V} \cdot \hat{\mathbf{n}} \, dA}{\rho A} \tag{5.7}$$

is shown in the figure in the margin.

If the velocity is considered uniformly distributed (one-dimensional flow) over the section area, A, then

$$\overline{V} = \frac{\int_A \rho \mathbf{V} \cdot \hat{\mathbf{n}} \, dA}{\rho A} = V \tag{5.8}$$

and the bar notation is not necessary (as in Example 5.1). When the flow is not uniformly distributed over the flow cross-sectional area, the bar notation reminds us that an average velocity is being used (as in Examples 5.2 and 5.4).

V5.1 Sink flow

5.1.2 Fixed, Nondeforming Control Volume

In many applications of fluid mechanics, an appropriate control volume to use is fixed and nondeforming. Several example problems that involve the continuity equation for fixed, nondeforming control volumes (Eq. 5.5) follow.

EXAMPLE 5.1 **Conservation of Mass—Steady, Incompressible Flow**

GIVEN Water flows steadily through a nozzle at the end of a fire hose as illustrated in Fig. E5.1a. According to local regulations, the nozzle exit velocity must be at least 20 m/s as shown in Fig. E5.1b.

FIND Determine the minimum pumping capacity, Q, required in m³/s.

■ **FIGURE E5.1a**

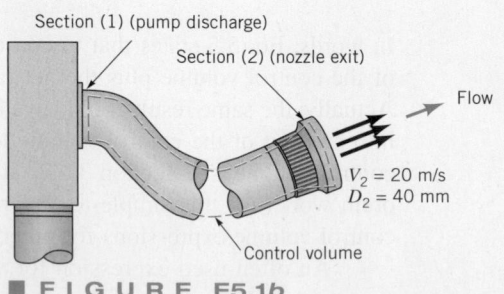

■ **FIGURE E5.1b**

SOLUTION

The pumping capacity sought is the volume flowrate delivered by the fire pump to the hose and nozzle. Since we desire knowledge about the pump discharge flowrate and we have information about the nozzle exit flowrate, we link these two flowrates with the control volume designated with the dashed line in Fig. E5.1b. This control volume contains, at any instant, water that is within the hose and nozzle from the pump discharge to the nozzle exit plane.

Equation 5.5 is applied to the contents of this control volume to give

$$\cancel{\frac{\partial}{\partial t} \int_{cv} \rho \, d\Psi}^{\text{0 (flow is steady)}} + \int_{cs} \rho \mathbf{V} \cdot \hat{\mathbf{n}} \, dA = 0 \qquad (1)$$

The time rate of change of the mass of the contents of this control volume is zero because the flow is steady. Because there is only one inflow [the pump discharge, section (1)] and one outflow [the nozzle exit, section (2)], Eq. (1) becomes

$$\rho_2 A_2 V_2 - \rho_1 A_1 V_1 = 0$$

so that with $\dot{m} = \rho A V$

$$\dot{m}_1 = \dot{m}_2 \qquad (2)$$

Because the mass flowrate is equal to the product of fluid density, ρ, and volume flowrate, Q (see Eq. 5.6), we obtain from Eq. 2

$$\rho_2 Q_2 = \rho_1 Q_1 \qquad (3)$$

Liquid flow at low speeds, as in this example, may be considered incompressible. Therefore

$$\rho_2 = \rho_1 \qquad (4)$$

and from Eqs. 3 and 4

$$Q_2 = Q_1 \qquad (5)$$

The pumping capacity is equal to the volume flowrate at the nozzle exit. If, for simplicity, the velocity distribution at the nozzle exit plane, section (2), is considered uniform (one-dimensional), then from Eq. 5

$$Q_1 = Q_2 = V_2 A_2$$

$$= V_2 \frac{\pi}{4} D_2^2 = (20 \text{ m/s}) \frac{\pi}{4} \left(\frac{40 \text{ mm}}{1000 \text{ mm/m}} \right)^2$$

$$= 0.0251 \text{ m}^3/\text{s} \qquad \textbf{(Ans)}$$

COMMENT By repeating the calculations for various values of the nozzle exit diameter, D_2, the results shown in Fig. E5.1c are obtained. The flowrate is proportional to the exit area, which varies as the diameter squared. Hence, if the diameter were doubled, the flowrate would increase by a factor of four, provided the exit velocity remained the same.

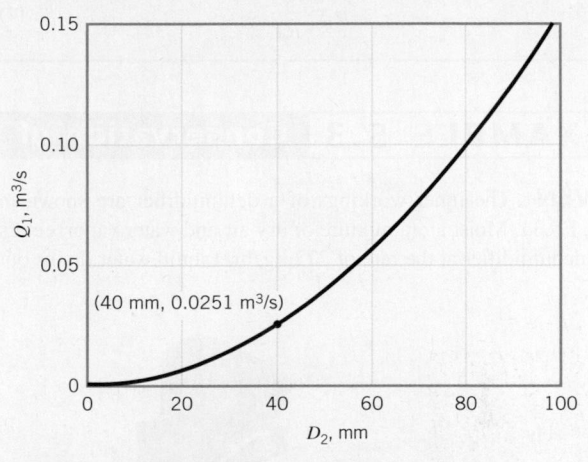

■ **FIGURE E5.1c**

EXAMPLE 5.2 Conservation of Mass—Steady, Compressible Flow

GIVEN Air flows steadily between two sections in a long, straight portion of 10 cm inside diameter pipe as indicated in Fig. E5.2. The uniformly distributed temperature and pressure at each section are given. The average air velocity (nonuniform velocity distribution) at section (2) is 305 m/s.

FIND Calculate the average air velocity at section (1).

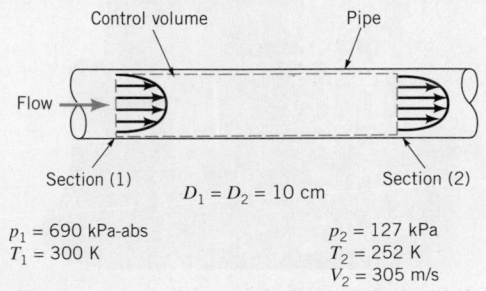

■ **FIGURE E5.2**

SOLUTION

The average fluid velocity at any section is that velocity which yields the section mass flowrate when multiplied by the section average fluid density and section area (Eq. 5.7). We relate the flows at sections (1) and (2) with the control volume designated with a dashed line in Fig. E5.2.

Equation 5.5 is applied to the contents of this control volume to obtain

$$\cancel{\frac{\partial}{\partial t} \int_{cv} \rho \, d\Psi}^{\text{0 (flow is steady)}} + \int_{cs} \rho \mathbf{V} \cdot \hat{\mathbf{n}} \, dA = 0$$

The time rate of change of the mass of the contents of this control volume is zero because the flow is steady. The control surface integral involves mass flowrates at sections (1) and (2) so that from Eq. 5.4 we get

$$\int_{cs} \rho \mathbf{V} \cdot \hat{\mathbf{n}} \, dA = \dot{m}_2 - \dot{m}_1 = 0$$

or

$$\dot{m}_1 = \dot{m}_2 \qquad (1)$$

and from Eqs. 1, 5.6, and 5.7 we obtain

$$\rho_1 A_1 \overline{V}_1 = \rho_2 A_2 \overline{V}_2 \qquad (2)$$

or since $A_1 = A_2$

$$\overline{V}_1 = \frac{\rho_2}{\rho_1} \overline{V}_2 \qquad (3)$$

Air at the pressures and temperatures involved in this example problem behaves like an ideal gas. The ideal gas equation of state (Eq. 1.8) is

$$\rho = \frac{p}{RT} \qquad (4)$$

Thus, combining Eqs. 3 and 4 we obtain

$$\overline{V}_1 = \frac{p_2 T_1 \overline{V}_2}{p_1 T_2}$$

$$= \frac{(127 \text{ kPa-abs})(300 \text{ K})(305 \text{ m/s})}{(690 \text{ kPa-abs})(252 \text{ K})} = 67 \text{ m/s} \qquad \textbf{(Ans)}$$

COMMENT We learn from this example that the continuity equation (Eq. 5.5) is valid for compressible as well as incompressible flows. Also, nonuniform velocity distributions can be handled with the average velocity concept. Significant average velocity changes can occur in pipe flow if the fluid is compressible.

EXAMPLE 5.3 Conservation of Mass—Two Fluids

GIVEN The inner workings of a dehumidifier are shown in Fig. E5.3a. Moist air (a mixture of dry air and water vapor) enters the dehumidifier at the rate of 272 kg/hr. Liquid water drains out of the dehumidifier at a rate of 1.36 kg/hr. A simplified sketch of the process is provided in Fig. E5.3b.

FIND Determine the mass flowrate of the dry air and the water vapor leaving the dehumidifier.

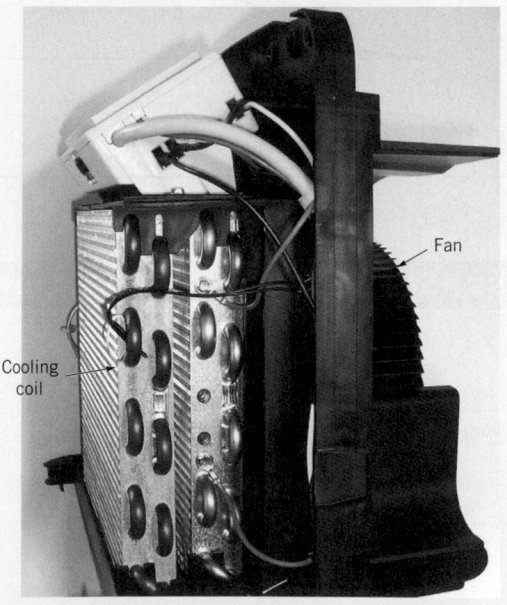

■ **FIGURE E5.3a**

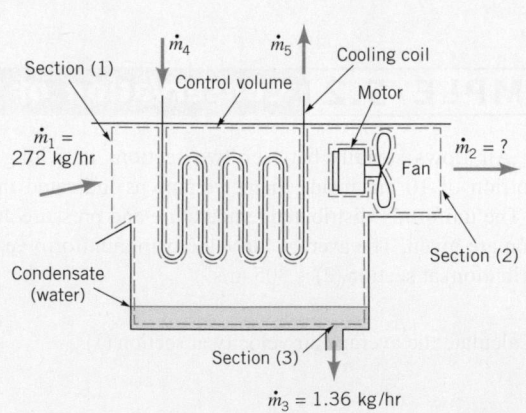

■ **FIGURE E5.3b**

SOLUTION

The unknown mass flowrate at section (2) is linked with the known flowrates at sections (1) and (3) with the control volume designated with a dashed line in Fig. E5.3b. The contents of the control volume are the air and water vapor mixture and the condensate (liquid water) in the dehumidifier at any instant.

Not included in the control volume are the fan and its motor, and the condenser coils and refrigerant. Even though the flow in the vicinity of the fan blade is unsteady, it is unsteady in a cyclical way. Thus, the flowrates at sections (1), (2), and (3) appear steady and the time rate of change of the mass of the contents of the control volume may be considered equal to zero on a time-average basis. The application of **Eqs. 5.4** and **5.5** to the control volume contents results in

$$\int_{cs} \rho \mathbf{V} \cdot \hat{\mathbf{n}} \, dA = -\dot{m}_1 + \dot{m}_2 + \dot{m}_3 = 0$$

or

$$\dot{m}_2 = \dot{m}_1 - \dot{m}_3 = 272 \text{ kg/hr} - 1.36 \text{ kg/hr}$$
$$= 270.64 \text{ kg/hr} \qquad \text{(Ans)}$$

COMMENT   Note that the continuity equation (Eq. 5.5) can be used when there is more than one stream of fluid flowing through the control volume.

The answer is the same with a control volume which includes the cooling coils to be within the control volume. The continuity equation becomes

$$\dot{m}_2 = \dot{m}_1 - \dot{m}_3 + \dot{m}_4 - \dot{m}_5 \qquad \text{(1)}$$

where $\dot{m}_4$ is the mass flowrate of the cooling fluid flowing into the control volume, and $\dot{m}_5$ is the flowrate out of the control volume through the cooling coil. Since the flow through the coils is steady, it follows that $\dot{m}_4 = \dot{m}_5$. Hence, Eq. 1 gives the same answer as obtained with the original control volume.

EXAMPLE 5.4   Conservation of Mass—Nonuniform Velocity Profile

GIVEN   Incompressible, laminar water flow develops in a straight pipe having radius R as indicated in Fig. E5.4a. At section (1), the velocity profile is uniform; the velocity is equal to a constant value U and is parallel to the pipe axis everywhere. At section (2), the velocity profile is axisymmetric and parabolic, with zero velocity at the pipe wall and a maximum value of u_{max} at the centerline.

FIND

(a)   How are U and u_{max} related?

(b)   How are the average velocity at section (2), $\overline{V}_2$, and u_{max} related?

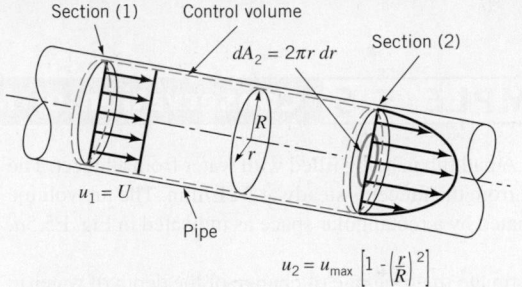

■ **FIGURE E5.4a**

SOLUTION

(a)   An appropriate control volume is sketched (dashed lines) in Fig. E5.4a. The application of Eq. 5.5 to the contents of this control volume yields

$$\frac{\partial}{\partial t}\int_{cv} \rho \, d\mathcal{V} + \int_{cs} \rho \mathbf{V} \cdot \hat{\mathbf{n}} \, dA = 0 \qquad \text{(1)}$$

overset: 0 (flow is steady)

At the inlet, section (1), the velocity is uniform with $V_1 = U$ so that

$$\int_{(1)} \rho \mathbf{V} \cdot \hat{\mathbf{n}} \, dA = -\rho_1 A_1 U \qquad \text{(2)}$$

At the outlet, section (2), the velocity is not uniform. However, the net flowrate through this section is the sum of flows through numerous small washer-shaped areas of size $dA_2 = 2\pi r \, dr$ as shown by the shaded area element in Fig. E5.4b. On each of

these infinitesimal areas the fluid velocity is denoted as u_2. Thus, in the limit of infinitesimal area elements, the summation is replaced by an integration and the outflow through section (2) is given by

$$\int_{(2)} \rho \mathbf{V} \cdot \hat{\mathbf{n}} \, dA = \rho_2 \int_0^R u_2 2\pi r \, dr \qquad \text{(3)}$$

By combining Eqs. 1, 2, and 3 we get

$$\rho_2 \int_0^R u_2 2\pi r \, dr - \rho_1 A_1 U = 0 \qquad \text{(4)}$$

Since the flow is considered incompressible, $\rho_1 = \rho_2$. The parabolic velocity relationship for flow through section (2) is used in Eq. 4 to yield

$$2\pi u_{max} \int_0^R \left[1 - \left(\frac{r}{R}\right)^2\right] r \, dr - A_1 U = 0 \qquad \text{(5)}$$

Integrating, we get from Eq. 5

$$2\pi u_{max} \left(\frac{r^2}{2} - \frac{r^4}{4R^2}\right)_0^R - \pi R^2 U = 0$$

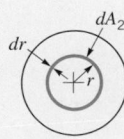

■ **FIGURE E5.4b**

or

$$u_{max} = 2U \qquad \text{(Ans)}$$

(b) Since this flow is incompressible, we conclude from Eq. 5.7 that U is the average velocity at all sections of the control volume. Thus, the average velocity at section (2), $\bar{V}_2$, is one-half the maximum velocity, u_{max}, there or

$$\bar{V}_2 = \frac{u_{max}}{2} \qquad \text{(Ans)}$$

COMMENT The relationship between the maximum velocity at section (2) and the average velocity is a function of the "shape" of the velocity profile. For the parabolic profile assumed in this example, the average velocity, $u_{max}/2$, is the actual "average" of the maximum velocity at section (2), $u_2 = u_{max}$, and the minimum velocity at that section, $u_2 = 0$. However, as shown in Fig. E5.4c, if the velocity profile is a different shape (non-parabolic), the average velocity is not necessarily one half of the maximum velocity.

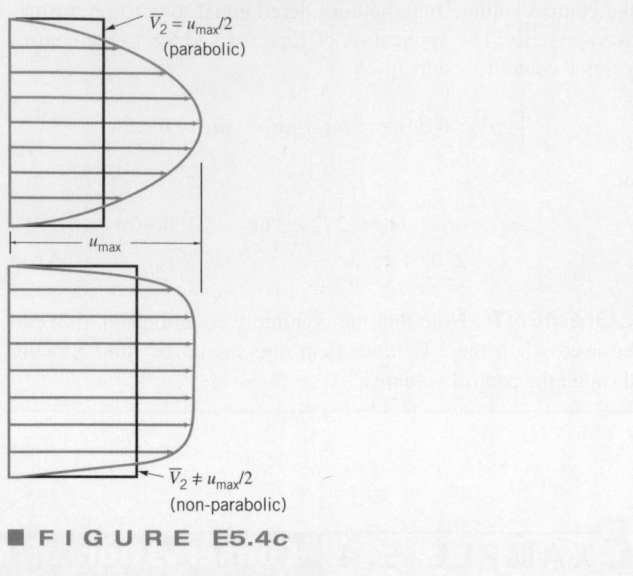

■ **FIGURE E5.4c**

EXAMPLE 5.5 │ Conservation of Mass—Unsteady Flow

GIVEN A bathtub is being filled with water from a faucet. The rate of flow from the faucet is steady at 34 L/min. The tub volume is approximated by a rectangular space as indicated in Fig. E5.5a.

FIND Estimate the time rate of change of the depth of water in the tub, $\partial h/\partial t$, in cm per minute at any instant.

SOLUTION

We use the fixed, nondeforming control volume outlined with a dashed line in Fig. E5.5a. This control volume includes in it, at any instant, the water accumulated in the tub, some of the water flowing from the faucet into the tub, and some air. Application of Eqs. 5.4 and 5.5 to these contents of the control volume results in

$$\frac{\partial}{\partial t} \int_{\substack{air \\ volume}} \rho_{air}\, d\mathcal{V}_{air} + \frac{\partial}{\partial t} \int_{\substack{water \\ volume}} \rho_{water}\, d\mathcal{V}_{water}$$
$$- \dot{m}_{water} + \dot{m}_{air} = 0 \quad \textbf{(1)}$$

Recall that the mass, dm, of fluid contained in a small volume $d\mathcal{V}$ is $dm = \rho\, d\mathcal{V}$. Hence, the two *integrals* in Eq. 1 represent the total amount of air and water in the control volume, and the sum of the first two *terms* is the time rate of change of mass within the control volume.

Note that the time rate of change of air mass and water mass are each not zero. Recognizing, however, that the air mass must be conserved, we know that the time rate of change of the mass of air in the control volume must be equal to the rate of air mass flow out of the control volume. For simplicity, we disregard any water evaporation that occurs. Thus, applying Eqs. 5.4 and 5.5 to the air only and to the water only, we obtain

$$\frac{\partial}{\partial t} \int_{\substack{air \\ volume}} \rho_{air}\, d\mathcal{V}_{air} + \dot{m}_{air} = 0$$

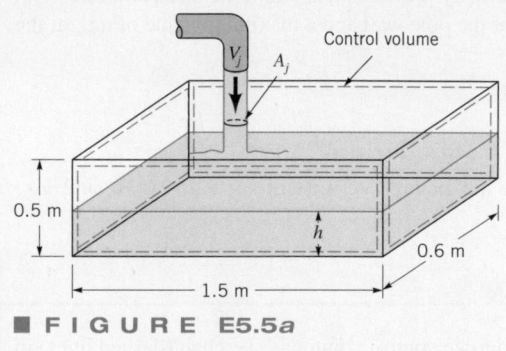

■ **FIGURE E5.5a**

for air, and

$$\frac{\partial}{\partial t} \int_{\substack{water \\ volume}} \rho_{water}\, d\mathcal{V}_{water} = \dot{m}_{water} \qquad \textbf{(2)}$$

for water. The volume of water in the control volume is given by

$$\int_{\substack{water \\ volume}} \rho_{water}\, dV_{water} = \rho_{water}\,[h(0.6\ \text{m})(1.5\ \text{m})$$
$$+ (0.5\ \text{m} - h)A_j] \qquad \textbf{(3)}$$

where A_j is the cross-sectional area of the water flowing from the faucet into the tub. Combining Eqs. 2 and 3, we obtain

$$\rho_{water}\,(0.9\ \text{m}^2 - A_j)\,\frac{\partial h}{\partial t} = \dot{m}_{water}$$

and, thus, since $\dot{m} = \rho Q$,

$$\frac{\partial h}{\partial t} = \frac{Q_{water}}{(0.9\ \text{m}^2 - A_j)}$$

For $A_j \ll 0.9$ m^2 we can conclude that

$$\frac{\partial h}{\partial t} = \frac{Q_{\text{water}}}{(0.9 \text{ m}^2)}$$

or

$$\frac{\partial h}{\partial t} = \frac{(34 \text{ L/min})(100 \text{ cm/m})}{(1000 \text{ L/m}^3)(0.9 \text{ m}^2)} = 3.78 \text{ cm/min} \quad \text{(Ans)}$$

COMMENT By repeating the calculations for the same flowrate but with various water jet diameters, D_j, the results shown in Fig. E5.5b are obtained. With the flowrate held constant, the value of $\partial h/\partial t$ is nearly independent of the jet diameter for values of the diameter less than about 25 cm.

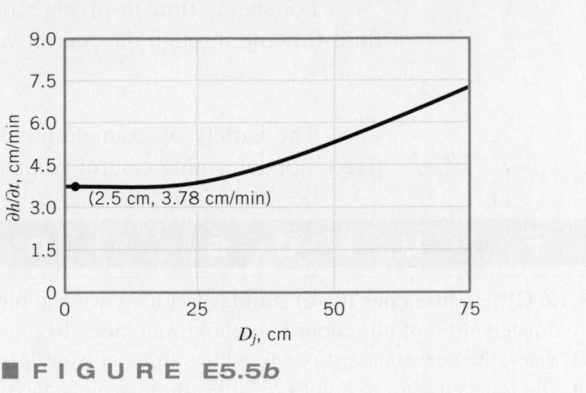

■ **FIGURE E5.5b**

The preceding example problems illustrate some important results of applying the conservation of mass principle to the contents of a fixed, nondeforming control volume. The dot product $\mathbf{V} \cdot \hat{\mathbf{n}}$ is "+" for flow out of the control volume and "−" for flow into the control volume. Thus, mass flowrate out of the control volume is "+" and mass flowrate in is "−." When the flow is steady, the time rate of change of the mass of the contents of the control volume

$$\frac{\partial}{\partial t} \int_{cv} \rho \, dV$$

The appropriate sign convention must be followed.

is zero and the net amount of mass flowrate, $\dot{m}$, through the control surface is therefore also zero

$$\sum \dot{m}_{\text{out}} - \sum \dot{m}_{\text{in}} = 0 \tag{5.9}$$

V5.2 Shop vac filter

If the steady flow is also incompressible, the net amount of volume flowrate, Q, through the control surface is also zero:

$$\sum Q_{\text{out}} - \sum Q_{\text{in}} = 0 \tag{5.10}$$

An unsteady, but cyclical flow can be considered steady on a time-average basis. When the flow is unsteady, the instantaneous time rate of change of the mass of the contents of the control volume is not necessarily zero and can be an important variable. When the value of

$$\frac{\partial}{\partial t} \int_{cv} \rho \, dV$$

is "+," the mass of the contents of the control volume is increasing. When it is "−," the mass of the contents of the control volume is decreasing.

When the flow is uniformly distributed over the opening in the control surface (one-dimensional flow),

$$\dot{m} = \rho A V$$

where V is the uniform value of the velocity component normal to the section area A. When the velocity is nonuniformly distributed over the opening in the control surface,

$$\dot{m} = \rho A \bar{V} \tag{5.11}$$

V5.3 Flow through a contraction

where $\bar{V}$ is the average value of the component of velocity normal to the section area A as defined by Eq. 5.7.

For steady flow involving only one stream of a specific fluid flowing through the control volume at sections (1) and (2),

$$\dot{m} = \rho_1 A_1 \bar{V}_1 = \rho_2 A_2 \bar{V}_2 \tag{5.12}$$

and for incompressible flow,

$$Q = A_1 \bar{V}_1 = A_2 \bar{V}_2 \tag{5.13}$$

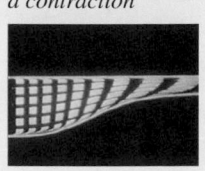

For steady flow involving more than one stream of a specific fluid or more than one specific fluid flowing through the control volume,

$$\sum \dot{m}_{\text{in}} = \sum \dot{m}_{\text{out}}$$

The variety of example problems solved above should give the correct impression that the fixed, nondeforming control volume is versatile and useful.

5.1.3 Moving, Nondeforming Control Volume

Some problems are most easily solved by using a moving control volume.

It is sometimes necessary to use a nondeforming control volume attached to a moving reference frame. Examples include control volumes containing a gas turbine engine on an aircraft in flight, the exhaust stack of a ship at sea, and the gasoline tank of an automobile passing by.

As discussed in Section 4.4.6, when a moving control volume is used, the fluid velocity relative to the moving control volume (relative velocity) is an important flow field variable. The relative velocity, **W**, is the fluid velocity seen by an observer moving with the control volume. The control volume velocity, $\mathbf{V}_{\text{cv}}$, is the velocity of the control volume as seen from a fixed coordinate system. The absolute velocity, **V**, is the fluid velocity seen by a stationary observer in a fixed coordinate system. These velocities are related to each other by the vector equation

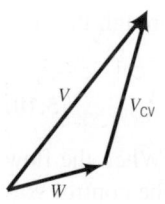

$$\mathbf{V} = \mathbf{W} + \mathbf{V}_{\text{cv}} \tag{5.14}$$

as illustrated by the figure in the margin. This is the same as Eq. 4.22, introduced earlier.

For a system and a moving, nondeforming control volume that are coincident at an instant of time, the Reynolds transport theorem (Eq. 4.23) for a moving control volume leads to

$$\frac{DM_{\text{sys}}}{Dt} = \frac{\partial}{\partial t} \int_{\text{cv}} \rho \, d\Psi + \int_{\text{cs}} \rho \mathbf{W} \cdot \hat{\mathbf{n}} \, dA \tag{5.15}$$

From Eqs. 5.1 and 5.15, we can get the control volume expression for conservation of mass (the continuity equation) for a moving, nondeforming control volume, namely,

$$\boxed{\frac{\partial}{\partial t} \int_{\text{cv}} \rho \, d\Psi + \int_{\text{cs}} \rho \mathbf{W} \cdot \hat{\mathbf{n}} \, dA = 0} \tag{5.16}$$

Some examples of the application of Eq. 5.16 follow.

EXAMPLE 5.6   Conservation of Mass—Compressible Flow with a Moving Control Volume

GIVEN An airplane moves forward at a speed of 971 km/hr as shown in Fig. E5.6a. The frontal intake area of the jet engine is 0.80 m² and the entering air density is 0.736 kg/m³. A stationary observer determines that relative to the earth, the jet engine exhaust gases move away from the engine with a speed of 1050 km/hr. The engine exhaust area is 0.558 m², and the exhaust gas density is 0.515 kg/m³.

FIND Estimate the mass flowrate of fuel into the engine in kg/hr.

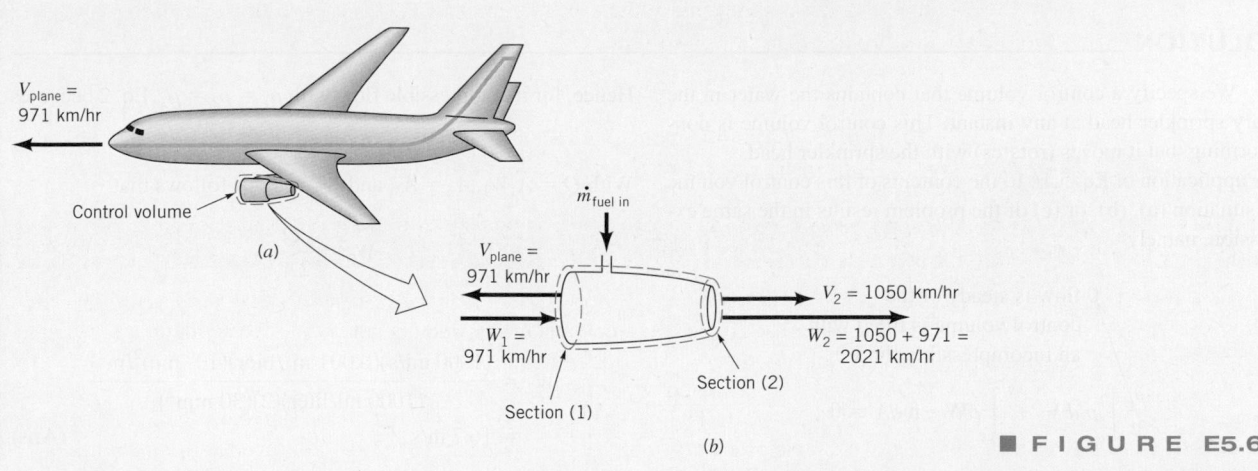

■ **FIGURE E5.6**

SOLUTION

The control volume, which moves with the airplane (see Fig. E5.6b), surrounds the engine and its contents and includes all fluids involved at an instant. The application of Eq. 5.16 to these contents of the control volume yields

$$\overset{\displaystyle 0 \text{ (flow relative to moving control}}{\overset{\text{volume is considered steady on a}}{\overset{\text{time-average basis)}}{\cancel{\frac{\partial}{\partial t} \int_{cv} \rho \, d\forall}} + \int_{cs} \rho \mathbf{W} \cdot \hat{\mathbf{n}} \, dA = 0}} \tag{1}$$

Assuming one-dimensional flow, we evaluate the surface integral in Eq. 1 and get

$$-\dot{m}_{\underset{\text{in}}{\text{fuel}}} - \rho_1 A_1 W_1 + \rho_2 A_2 W_2 = 0$$

or

$$\dot{m}_{\underset{\text{in}}{\text{fuel}}} = \rho_2 A_2 W_2 - \rho_1 A_1 W_1 \tag{2}$$

We consider the intake velocity, W_1, relative to the moving control volume, as being equal in magnitude to the speed of the airplane, 971 km/hr. The exhaust velocity, W_2, also needs to be measured relative to the moving control volume. Since a fixed

observer noted that the exhaust gases were moving away from the engine at a speed of 1050 km/hr, the speed of the exhaust gases relative to the moving control volume, W_2, is determined as follows by using Eq. 5.14

$$V_2 = W_2 + V_{\text{plane}}$$

or

$$W_2 = V_2 - V_{\text{plane}} = 1050 \text{ km/hr} - (-971 \text{ km/hr})$$
$$= 2021 \text{ km/hr}$$

and is shown in Fig. E5.6b.

From Eq. 2,

$$\dot{m}_{\underset{\text{in}}{\text{fuel}}} = (0.515 \text{ kg/m}^3)(0.558 \text{ m}^2)(2021 \text{ km/hr})(1000 \text{ m/km})$$
$$\quad - (0.736 \text{ kg/m}^3)(0.80 \text{ m}^2)(971 \text{ km/hr})(1000 \text{ m/km})$$
$$= (580,800 - 571,700) \text{ kg/hr}$$
$$\dot{m}_{\underset{\text{in}}{\text{fuel}}} = 9100 \text{ kg/hr} \tag{Ans}$$

COMMENT   Note that the fuel flowrate was obtained as the difference of two large, nearly equal numbers. Precise values of W_2 and W_1 are needed to obtain a modestly accurate value of $\dot{m}_{\text{fuel}}$.

EXAMPLE 5.7   Conservation of Mass—Relative Velocity

GIVEN   Water enters a rotating lawn sprinkler through its base at the steady rate of 1000 ml/s as sketched in Fig. E5.7. The exit area of each of the two nozzles is 30 mm².

FIND   Determine the average speed of the water leaving the nozzle, relative to the nozzle, if

(a) the rotary sprinkler head is stationary,

(b) the sprinkler head rotates at 600 rpm, and

(c) the sprinkler head accelerates from 0 to 600 rpm.

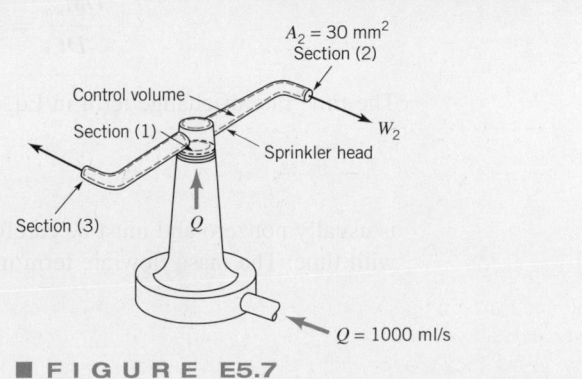

■ **FIGURE E5.7**

SOLUTION

(a) We specify a control volume that contains the water in the rotary sprinkler head at any instant. This control volume is non-deforming, but it moves (rotates) with the sprinkler head.

The application of Eq. 5.16 to the contents of this control volume for situation **(a)**, **(b)**, or **(c)** of the problem results in the same expression, namely

0 flow is steady or the control volume is filled with an incompressible fluid

$$\frac{\partial}{\partial t}\int_{cv}\rho\,d\mathcal{V} + \int_{cs}\rho\mathbf{W}\cdot\hat{\mathbf{n}}\,dA = 0$$

or

$$\sum\rho_{out}A_{out}W_{out} - \sum\rho_{in}A_{in}W_{in} = 0 \qquad (1)$$

The time rate of change of the mass of water in the control volume is zero because the flow is steady and the control volume is filled with water.

Because there is only one inflow [at the base of the rotating arm, section (1)] and two outflows [the two nozzles at the tips of the arm, sections (2) and (3), each have the same area and fluid velocity], Eq. 1 becomes

$$\rho_2 A_2 W_2 + \rho_3 A_3 W_3 - \rho_1 A_1 W_1 = 0 \qquad (2)$$

Hence, for incompressible flow with $\rho_1 = \rho_2 = \rho_3$, Eq. 2 becomes

$$A_2 W_2 + A_3 W_3 - A_1 W_1 = 0$$

With $Q = A_1 W_1$, $A_2 = A_3$, and $W_2 = W_3$ it follows that

$$W_2 = \frac{Q}{2A_2}$$

or

$$W_2 = \frac{(1000\ \text{ml/s})(0.001\ \text{m}^3/\text{liter})(10^6\ \text{mm}^2/\text{m}^2)}{(1000\ \text{ml/liter})(2)(30\ \text{mm}^2)}$$

$$= 16.7\ \text{m/s} \qquad \text{(Ans)}$$

(b), (c) The value of W_2 is independent of the speed of rotation of the sprinkler head and represents the average velocity of the water exiting from each nozzle with respect to the nozzle for cases **(a)**, **(b)**, and **(c)**.

COMMENT The velocity of water discharging from each nozzle, when viewed from a stationary reference (i.e., V_2), will vary as the rotation speed of the sprinkler head varies since from Eq. 5.14,

$$V_2 = W_2 - U$$

where $U = \omega R$ is the speed of the nozzle and ω and R are the angular velocity and radius of the sprinkler head, respectively.

When a moving, nondeforming control volume is used, the dot product sign convention used earlier for fixed, nondeforming control volume applications is still valid. Also, if the flow within the moving control volume is steady, or steady on a time-average basis, the time rate of change of the mass of the contents of the control volume is zero. Velocities seen from the control volume reference frame (relative velocities) must be used in the continuity equation. Relative and absolute velocities are related by a vector equation (Eq. 5.14), which also involves the control volume velocity.

5.1.4 Deforming Control Volume

Care is needed to ensure that absolute and relative velocities are used correctly.

Occasionally, a deforming control volume can simplify the solution of a problem. A deforming control volume involves changing volume size and control surface movement. Thus, the Reynolds transport theorem for a moving control volume can be used for this case, and Eqs. 4.23 and 5.1 lead to

$$\frac{DM_{sys}}{Dt} = \frac{\partial}{\partial t}\int_{cv}\rho\,d\mathcal{V} + \int_{cs}\rho\mathbf{W}\cdot\hat{\mathbf{n}}\,dA = 0 \qquad (5.17)$$

The time rate of change term in Eq. 5.17,

$$\frac{\partial}{\partial t}\int_{cv}\rho\,d\mathcal{V}$$

is usually nonzero and must be carefully evaluated because the extent of the control volume varies with time. The mass flowrate term in Eq. 5.17,

$$\int_{cs}\rho\mathbf{W}\cdot\hat{\mathbf{n}}\,dA$$

must be determined with the relative velocity, **W**, the velocity referenced to the control surface. Since the control volume is deforming, the control surface velocity is not necessarily uniform and identical to the control volume velocity, $\mathbf{V}_{cv}$, as was true for moving, nondeforming control volumes. For the deforming control volume,

> *The velocity of the surface of a deforming control volume is not the same at all points on the surface.*

$$\mathbf{V} = \mathbf{W} + \mathbf{V}_{cs} \tag{5.18}$$

where $\mathbf{V}_{cs}$ is the velocity of the control surface as seen by a fixed observer. The relative velocity, **W**, must be ascertained with care wherever fluid crosses the control surface. Two example problems that illustrate the use of the continuity equation for a deforming control volume, Eq. 5.17, follow.

EXAMPLE 5.8   Conservation of Mass—Deforming Control Volume

GIVEN A syringe (Fig. E5.8) is used to inoculate a cow. The plunger has a face area of 500 mm². The liquid in the syringe is to be injected steadily at a rate of 300 cm³/min. The leakage rate past the plunger is 0.10 times the volume flowrate out of the needle.

FIND With what speed should the plunger be advanced?

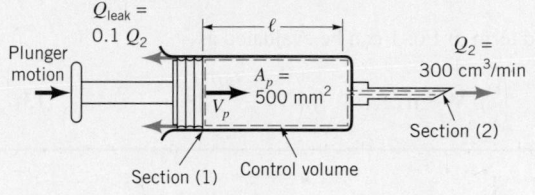

■ FIGURE E5.8

SOLUTION

The control volume selected for solving this problem is the deforming one illustrated in Fig. E5.8. Section (1) of the control surface moves with the plunger. The surface area of section (1), A_1, is considered equal to the circular area of the face of the plunger, A_p, although this is not strictly true, since leakage occurs. The difference is small, however. Thus,

$$A_1 = A_p \tag{1}$$

Liquid also leaves the needle through section (2), which involves fixed area A_2. The application of Eq. 5.17 to the contents of this control volume gives

$$\frac{\partial}{\partial t} \int_{cv} \rho \, d\mathcal{V} + \dot{m}_2 + \rho Q_{leak} = 0 \tag{2}$$

Even though Q_{leak} and the flow through section area A_2 are steady, the time rate of change of the mass of liquid in the shrinking control volume is not zero because the control volume is getting smaller. To evaluate the first term of Eq. 2, we note that

$$\int_{cv} \rho \, d\mathcal{V} = \rho(\ell A_1 + \mathcal{V}_{needle}) \tag{3}$$

where ℓ is the changing length of the control volume (see Fig. E5.8) and $\mathcal{V}_{needle}$ is the volume of the needle. From Eq. 3, we obtain

$$\frac{\partial}{\partial t} \int_{cv} \rho \, d\mathcal{V} = \rho A_1 \frac{\partial \ell}{\partial t} \tag{4}$$

Note that

$$-\frac{\partial \ell}{\partial t} = V_p \tag{5}$$

where V_p is the speed of the plunger sought in the problem statement. Combining Eqs. 2, 4, and 5 we obtain

$$-\rho A_1 V_p + \dot{m}_2 + \rho Q_{leak} = 0 \tag{6}$$

However, from Eq. 5.6, we see that

$$\dot{m}_2 = \rho Q_2 \tag{7}$$

and Eq. 6 becomes

$$-\rho A_1 V_p + \rho Q_2 + \rho Q_{leak} = 0 \tag{8}$$

Solving Eq. 8 for V_p yields

$$V_p = \frac{Q_2 + Q_{leak}}{A_1} \tag{9}$$

Since $Q_{leak} = 0.1 Q_2$, Eq. 9 becomes

$$V_p = \frac{Q_2 + 0.1 Q_2}{A_1} = \frac{1.1 Q_2}{A_1}$$

and

$$V_p = \frac{(1.1)(300 \text{ cm}^3/\text{min})}{(500 \text{ mm}^2)} \left(\frac{1000 \text{ mm}^3}{\text{cm}^3} \right)$$

$$= 660 \text{ mm/min} \tag{Ans}$$

EXAMPLE 5.9   Conservation of Mass—Deforming Control Volume

GIVEN Consider Example 5.5.

FIND Solve the problem of Example 5.5 using a deforming control volume that includes only the water accumulating in the bathtub.

SOLUTION

For this deforming control volume, Eq. 5.17 leads to

$$\frac{\partial}{\partial t}\int_{\substack{\text{water}\\ \text{volume}}} \rho\, d\text{V} + \int_{cs} \rho \mathbf{W}\cdot\hat{\mathbf{n}}\, dA = 0 \tag{1}$$

The first term of Eq. 1 can be evaluated as

$$\frac{\partial}{\partial t}\int_{\substack{\text{water}\\ \text{volume}}} \rho\, d\text{V} = \frac{\partial}{\partial t}\left[\rho h(0.6\text{ m})(1.5\text{ m})\right]$$

$$= \rho\,(0.9\text{ m}^2)\frac{\partial h}{\partial t} \tag{2}$$

The second term of Eq. 1 can be evaluated as

$$\int_{cs} \rho\,\mathbf{W}\cdot\hat{\mathbf{n}}\, dA = -\rho\left(V_j + \frac{\partial h}{\partial t}\right)A_j \tag{3}$$

where A_j and V_j are the cross-sectional area and velocity of the water flowing from the faucet into the tube. Thus, from Eqs. 1, 2, and 3 we obtain

$$\frac{\partial h}{\partial t} = \frac{V_j A_j}{(0.9\text{ m}^2 - A_j)} = \frac{Q_{\text{water}}}{(0.9\text{ m}^2 - A_j)}$$

or for $A_j \ll 0.9\text{ m}^2$

$$\frac{\partial h}{\partial t} = \frac{(34\text{ L/min})(100\text{ cm/m})}{(1000\text{ L/m}^3)(0.9\text{ m}^2)} = 3.78\text{ cm/min} \quad \textbf{(Ans)}$$

COMMENT Note that these results using a deforming control volume are the same as that obtained in Example 5.5 with a fixed control volume.

The conservation of mass principle is easily applied to the contents of a control volume. The appropriate selection of a specific kind of control volume (for example, fixed and nondeforming, moving and nondeforming, or deforming) can make the solution of a particular problem less complicated. In general, where fluid flows through the control surface, it is advisable to make the control surface perpendicular to the flow. In the sections ahead we learn that the conservation of mass principle is primarily used in combination with other important laws to solve problems.

5.2 Newton's Second Law—The Linear Momentum and Moment-of-Momentum Equations

5.2.1 Derivation of the Linear Momentum Equation

V5.4 Smokestack plume momentum

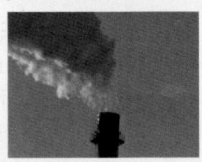

Newton's second law of motion for a system is

> time rate of change of the = sum of external forces
> linear momentum of the system acting on the system

Forces acting on a flowing fluid can change its velocity magnitude and/or direction.

Since momentum is mass times velocity, the momentum of a small particle of mass $\rho d\text{V}$ is $\mathbf{V}\rho d\text{V}$. Thus, the momentum of the entire system is $\int_{sys}\mathbf{V}\rho d\text{V}$ and Newton's law becomes

$$\frac{D}{Dt}\int_{sys} \mathbf{V}\rho\, d\text{V} = \sum \mathbf{F}_{sys} \tag{5.19}$$

Any reference or coordinate system for which this statement is true is called *inertial*. A fixed coordinate system is inertial. A coordinate system that moves in a straight line with constant velocity and is thus without acceleration is also inertial. We proceed to develop the control volume formula for this important law. When a control volume is coincident with a system at an instant of time, the forces acting on the system and the forces acting on the contents of the coincident control volume (see Fig. 5.2) are instantaneously identical, that is,

$$\sum \mathbf{F}_{sys} = \sum \mathbf{F}_{\substack{\text{contents of the}\\ \text{coincident control volume}}} \tag{5.20}$$

Furthermore, for a system and the contents of a coincident control volume that is fixed and nondeforming, the Reynolds transport theorem [Eq. 4.19 with b set equal to the velocity (i.e., momentum per unit mass), and B_{sys} being the system momentum] allows us to conclude that

$$\frac{D}{Dt}\int_{sys} \mathbf{V}\rho\, d\text{V} = \frac{\partial}{\partial t}\int_{cv} \mathbf{V}\rho\, d\text{V} + \int_{cs} \mathbf{V}\rho\mathbf{V}\cdot\hat{\mathbf{n}}\, dA \tag{5.21}$$

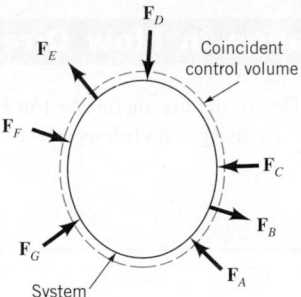

■ **FIGURE 5.2**   **External forces acting on system and coincident control volume.**

or

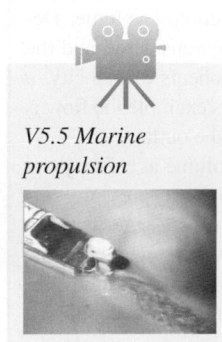

V5.5 Marine propulsion

| time rate of change of the linear momentum of the system | = | time rate of change of the linear momentum of the contents of the control volume | + | net rate of flow of linear momentum through the control surface |

Equation 5.21 states that the time rate of change of system linear momentum is expressed as the sum of the two control volume quantities: the time rate of change of the *linear momentum of the contents of the control volume*, and the net rate of *linear momentum flow through the control surface*. As particles of mass move into or out of a control volume through the control surface, they carry linear momentum in or out. Thus, linear momentum flow should seem no more unusual than mass flow.

For a control volume that is fixed (and thus inertial) and nondeforming, Eqs. 5.19, 5.20, and 5.21 provide an appropriate mathematical statement of Newton's second law of motion as

$$\frac{\partial}{\partial t} \int_{cv} \mathbf{V}\rho \, d\Psi + \int_{cs} \mathbf{V}\rho \mathbf{V} \cdot \hat{\mathbf{n}} \, dA = \sum \mathbf{F}_{\substack{\text{contents of the} \\ \text{control volume}}} \tag{5.22}$$

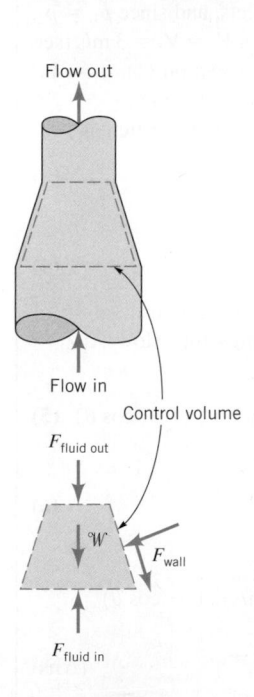

We call Eq. 5.22 the *linear momentum equation*.

In our application of the linear momentum equation, we initially confine ourselves to fixed, nondeforming control volumes for simplicity. Subsequently, we discuss the use of a moving but inertial, nondeforming control volume. We do not consider deforming control volumes and accelerating (noninertial) control volumes. If a control volume is noninertial, the acceleration components involved (for example, translation acceleration, Coriolis acceleration, and centrifugal acceleration) require consideration.

The forces involved in Eq. 5.22 are body and surface forces that act on what is contained in the control volume as shown in the sketch in the margin. The only body force we consider in this chapter is the one associated with the action of gravity. We experience this body force as weight, $\mathscr{W}$. The surface forces are basically exerted on the contents of the control volume by material just outside the control volume in contact with material just inside the control volume. For example, a wall in contact with fluid can exert a reaction surface force on the fluid it bounds. Similarly, fluid just outside the control volume can push on fluid just inside the control volume at a common interface, usually an opening in the control surface through which fluid flow occurs. An immersed object can resist fluid motion with surface forces.

The linear momentum terms in the momentum equation deserve careful explanation. We clarify their physical significance in the following sections.

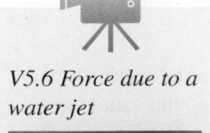

V5.6 Force due to a water jet

5.2.2   Application of the Linear Momentum Equation

The linear momentum equation for an inertial control volume is a vector equation (Eq. 5.22). In engineering applications, components of this vector equation resolved along orthogonal coordinates, for example, x, y, and z (rectangular coordinate system) or r, θ, and x (cylindrical coordinate system), will normally be used. A simple example involving steady, incompressible flow is considered first.

EXAMPLE 5.10 Linear Momentum—Change in Flow Direction

GIVEN As shown in Fig. E5.10a, a horizontal jet of water exits a nozzle with a uniform speed of $V_1 = 3$ m/s, strikes a vane, and is turned through an angle θ.

FIND Determine the anchoring force needed to hold the vane stationary if gravity and viscous effects are negligible.

SOLUTION

We select a control volume that includes the vane and a portion of the water (see Figs. E5.10b, c) and apply the linear momentum equation to this fixed control volume. The only portions of the control surface across which fluid flows are section (1) (the entrance) and section (2) (the exit). Hence, the x and z components of Eq. 5.22 become

$$\frac{\partial}{\partial t} \int_{cv} u\, \rho\, d\forall \overset{0\,\text{(flow is steady)}}{} + \int_{cs} u\, \rho\, \mathbf{V} \cdot \hat{\mathbf{n}}\, dA = \sum F_x$$

and

$$\frac{\partial}{\partial t} \int_{cv} w\, \rho\, d\forall \overset{0\,\text{(flow is steady)}}{} + \int_{cs} w\, \rho\, \mathbf{V} \cdot \hat{\mathbf{n}}\, dA = \sum F_z$$

or

$$u_2 \rho A_2 V_2 - u_1 \rho A_1 V_1 = \sum F_x \quad (1)$$

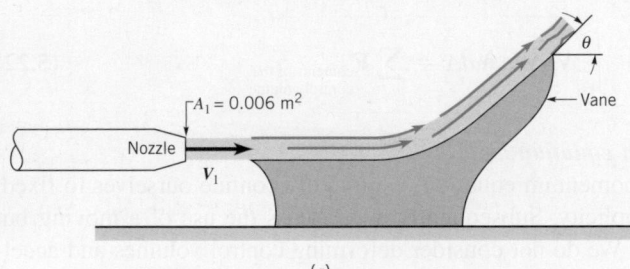

$A_1 = 0.006$ m^2

Nozzle

V_1

Vane

θ

(a)

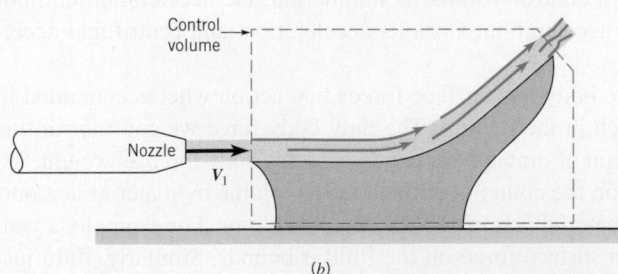

Control volume

Nozzle

V_1

(b)

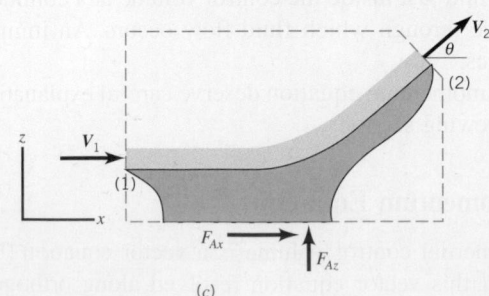

V_2

θ

(2)

z

V_1

(1)

x

F_{Ax}

F_{Az}

(c)

■ FIGURE E5.10

and

$$w_2 \rho A_2 V_2 - w_1 \rho A_1 V_1 = \sum F_z \quad (2)$$

where $\mathbf{V} = u\hat{\mathbf{i}} + w\hat{\mathbf{k}}$, and $\sum F_x$ and $\sum F_z$ are the net x and z components of force acting on the contents of the control volume. Depending on the particular flow situation being considered and the coordinate system chosen, the x and z components of velocity, u and w, can be positive, negative, or zero. In this example the flow is in the positive directions at both the inlet and the outlet.

The water enters and leaves the control volume as a free jet at atmospheric pressure. Hence, there is atmospheric pressure surrounding the entire control volume, and the net pressure force on the control volume surface is zero. If we neglect the weight of the water and vane, the only forces applied to the control volume contents are the horizontal and vertical components of the anchoring force, F_{Ax} and F_{Az}, respectively.

With negligible gravity and viscous effects, and since $p_1 = p_2$, the speed of the fluid remains constant so that $V_1 = V_2 = 3$ m/s (see the Bernoulli equation, Eq. 3.7). Hence, at section (1), $u_1 = V_1$, $w_1 = 0$, and at section (2), $u_2 = V_1 \cos\theta$, $w_2 = V_1 \sin\theta$.

By using this information, Eqs. 1 and 2 can be written as

$$V_1 \cos\theta\, \rho\, A_2 V_1 - V_1\, \rho\, A_1 V_1 = F_{Ax} \quad (3)$$

and

$$V_1 \sin\theta\, \rho\, A_2 V_1 - 0\, \rho\, A_1 V_1 = F_{Az} \quad (4)$$

Equations 3 and 4 can be simplified by using conservation of mass, which states that for this incompressible flow $A_1 V_1 = A_2 V_2$, or $A_1 = A_2$ since $V_1 = V_2$. Thus

$$F_{Ax} = -\rho A_1 V_1^2 + \rho A_1 V_1^2 \cos\theta = -\rho A_1 V_1^2 (1 - \cos\theta) \quad (5)$$

and

$$F_{Az} = \rho A_1 V_1^2 \sin\theta \quad (6)$$

With the given data we obtain

$$F_{Ax} = -(1000 \text{ kg/m}^3)(0.006 \text{ m}^2)(3 \text{ m/s})^2(1 - \cos\theta)$$
$$= -54(1 - \cos\theta) \text{ kg} \cdot \text{m/s}^2$$
$$= -54(1 - \cos\theta) \text{ N} \quad \text{(Ans)}$$

and

$$F_{Az} = (1000 \text{ kg/m}^3)(0.006 \text{ m}^2)(3 \text{ m/s})^2 \sin\theta$$
$$= 54 \sin\theta \text{ N} \quad \text{(Ans)}$$

COMMENTS The values of F_{Ax} and F_{Az} as a function of θ are shown in Fig. E5.10d. Note that if $\theta = 0$ (i.e., the vane does not turn the water), the anchoring force is zero. The inviscid fluid merely slides along the vane without putting any force on it. If $\theta = 90°$, then $F_{Ax} = -54$ N and $F_{Az} = 54$ N. It is necessary to push on the vane (and, hence, for the vane to push on the water)

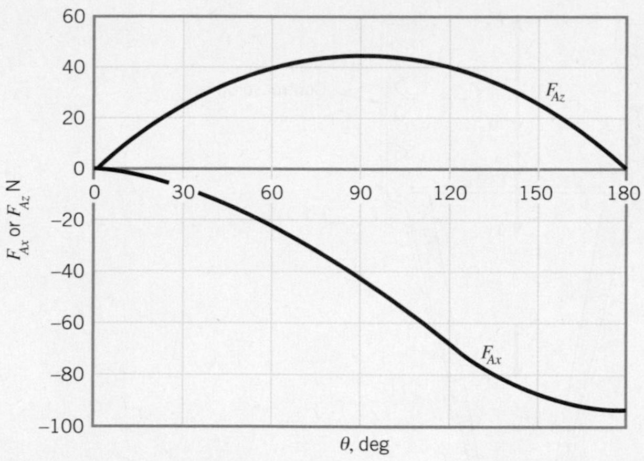

■ FIGURE E5.10d

to the left (F_{Ax} is negative) and up in order to change the direction of flow of the water from horizontal to vertical. This momentum change requires a force. If $\theta = 180°$, the water jet is turned back on itself. This requires no vertical force ($F_{Az} = 0$), but the horizontal force ($F_{Ax} = -108$ N) is two times that required if $\theta = 90°$. This horizontal fluid momentum change requires a horizontal force only.

Note that the anchoring force (Eqs. 5, 6) can be written in terms of the mass flowrate, $\dot{m} = \rho A_1 V_1$, as

$$F_{Ax} = -\dot{m} V_1 (1 - \cos \theta)$$

and

$$F_{Az} = \dot{m} V_1 \sin \theta$$

In this example exerting a force on a fluid flow resulted in a change in its direction only (i.e., change in its linear momentum).

F l u i d s   i n   t h e   N e w s

Where the plume goes Commercial airliners have wheel brakes very similar to those on highway vehicles. In fact, antilock brakes now found on most new cars were first developed for use on airplanes. However, when landing, the major braking force comes from the engine rather than the wheel brakes. Upon touchdown, a piece of engine cowling translates aft and blocker doors drop down, directing the engine airflow into a honeycomb structure called a cascade. The cascade reverses the direction of the high-speed engine exhausts by nearly 180° so that it flows forward. As predicted by the *momentum equation*, the air passing through the engine produces a substantial braking force—the reverse thrust. Designers must know the flow pattern of the exhaust plumes to eliminate potential problems. For example, the plumes of hot exhaust must be kept away from parts of the aircraft where repeated heating and cooling could cause premature fatigue. Also, the plumes must not re-enter the engine inlet, or blow debris from the runway in front of the engine, or envelop the vertical tail. (See Problem 5.67.)

---

*E*XAMPLE 5.11   Linear Momentum—Weight, Pressure, and Change in Speed

GIVEN   As shown in Fig. E5.11a, water flows through a nozzle attached to the end of a laboratory sink faucet with a flowrate of 0.6 liters/s. The nozzle inlet and exit diameters are 16 and 5 mm, respectively, and the nozzle axis is vertical. The mass of the nozzle is 0.1 kg. The pressure at section (1) is 464 kPa.

FIND   Determine the anchoring force required to hold the nozzle in place.

SOLUTION

The anchoring force sought is the reaction force between the faucet and nozzle threads. To evaluate this force we select a control volume that includes the entire nozzle and the water contained in the nozzle at an instant, as is indicated in Figs. E5.11a and E5.11b. All of the vertical forces acting on the contents of this control volume are identified in Fig. E5.11b. The action of atmospheric pressure cancels out in every direction and is not shown. Gage pressure forces do not cancel out in the vertical direction and are shown. Application of the vertical or z direction component of Eq. 5.22 to the contents of this control volume leads to

$$\frac{\partial}{\partial t} \int_{cv} w \rho \, dV + \int_{cs} w \rho \mathbf{V} \cdot \hat{\mathbf{n}} \, dA = F_A - \mathcal{W}_n - p_1 A_1$$
$$- \mathcal{W}_w + p_2 A_2 \quad (1)$$

(with "0 (flow is steady)" annotation over the first term)

where w is the z direction component of fluid velocity, and the various parameters are identified in the figure.

Note that the positive direction is considered "up" for the forces. We will use this same sign convention for the fluid velocity, w, in Eq. 1. In Eq. 1, the dot product, $\mathbf{V} \cdot \hat{\mathbf{n}}$, is "+" for flow out of the control volume and "−" for flow into the control volume. For this particular example

$$\mathbf{V} \cdot \hat{\mathbf{n}} \, dA = \pm |w| \, dA \quad (2)$$

with the "+" used for flow out of the control volume and "−" used for flow in. To evaluate the control surface integral in Eq. 1, we need to assume a distribution for fluid velocity, w, and fluid density, ρ. For simplicity, we assume that w is uniformly distributed or constant, with magnitudes of w_1 and w_2 over cross-sectional areas A_1 and A_2. Also, this flow is incompressible so the

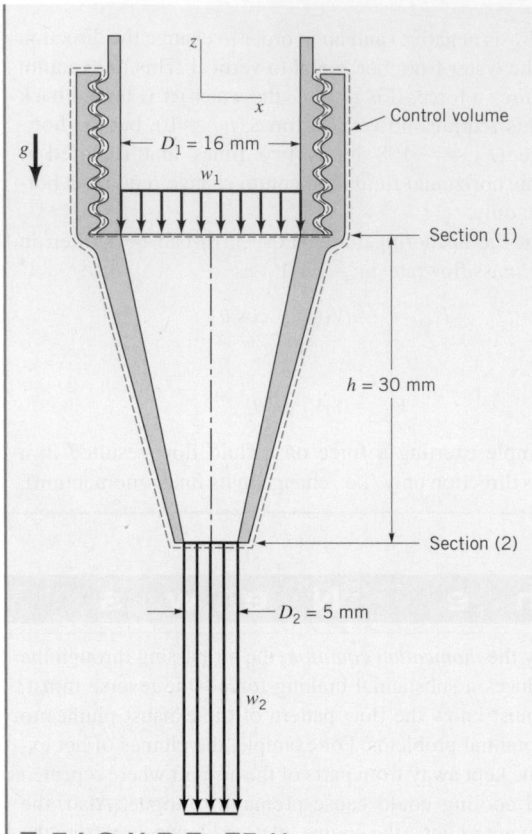

■ **FIGURE E5.11a**

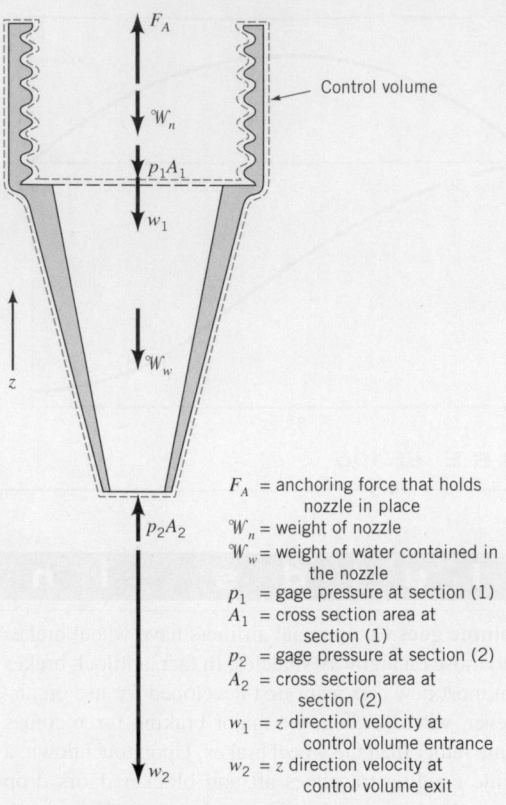

F_A = anchoring force that holds
 nozzle in place
$\mathcal{W}_n$ = weight of nozzle
$\mathcal{W}_w$ = weight of water contained in
 the nozzle
p_1 = gage pressure at section (1)
A_1 = cross section area at
 section (1)
p_2 = gage pressure at section (2)
A_2 = cross section area at
 section (2)
w_1 = z direction velocity at
 control volume entrance
w_2 = z direction velocity at
 control volume exit

■ **FIGURE E5.11b**

fluid density, ρ, is constant throughout. Proceeding further we obtain for Eq. 1

$$(-\dot{m}_1)(-w_1) + \dot{m}_2(-w_2)$$
$$= F_A - \mathcal{W}_n - p_1 A_1 - \mathcal{W}_w + p_2 A_2 \quad \text{(3)}$$

where $\dot{m} = \rho A V$ is the mass flowrate.

Note that $-w_1$ and $-w_2$ are used because both of these velocities are "down." Also, $-\dot{m}_1$ is used because it is associated with flow into the control volume. Similarly, $+\dot{m}_2$ is used because it is associated with flow out of the control volume. Solving Eq. 3 for the anchoring force, F_A, we obtain

$$F_A = \dot{m}_1 w_1 - \dot{m}_2 w_2 + \mathcal{W}_n + p_1 A_1 + \mathcal{W}_w - p_2 A_2 \quad \text{(4)}$$

From the conservation of mass equation, Eq. 5.12, we obtain

$$\dot{m}_1 = \dot{m}_2 = \dot{m} \quad \text{(5)}$$

which when combined with Eq. 4 gives

$$F_A = \dot{m}(w_1 - w_2) + \mathcal{W}_n + p_1 A_1 + \mathcal{W}_w - p_2 A_2 \quad \text{(6)}$$

It is instructive to note how the anchoring force is affected by the different actions involved. As expected, the nozzle weight, $\mathcal{W}_n$, the water weight, $\mathcal{W}_w$, and gage pressure force at section (1), $p_1 A_1$, all increase the anchoring force, while the gage pressure force at section (2), $p_2 A_2$, acts to decrease the anchoring force. The change in the vertical momentum flowrate, $\dot{m}(w_1 - w_2)$, will, in this instance, decrease the anchoring force because this change is negative ($w_2 > w_1$).

To complete this example we use quantities given in the problem statement to quantify the terms on the right-hand side of Eq. 6.

From Eq. 5.6,

$$\dot{m} = \rho w_1 A_1 = \rho Q$$
$$= (999 \text{ kg/m}^3)(0.6 \text{ liter/s})(10^{-3} \text{ m}^3/\text{liter})$$
$$= 0.599 \text{ kg/s} \quad \text{(7)}$$

and

$$w_1 = \frac{Q}{A_1} = \frac{Q}{\pi(D_1^2/4)}$$
$$= \frac{(0.6 \text{ liter/s})(10^{-3} \text{ m}^3/\text{liter})}{\pi(16 \text{ mm})^2/4(1000^2 \text{ mm}^2/\text{m}^2)} = 2.98 \text{ m/s} \quad \text{(8)}$$

Also from Eq. 5.6,

$$w_2 = \frac{Q}{A_2} = \frac{Q}{\pi(D_2^2/4)}$$
$$= \frac{(0.6 \text{ liter/s})(10^{-3} \text{ m}^3/\text{liter})}{\pi(5 \text{ mm})^2/4(1000^2 \text{ mm}^2/\text{m}^2)} = 30.6 \text{ m/s} \quad \text{(9)}$$

The weight of the nozzle, $\mathcal{W}_n$, can be obtained from the nozzle mass, m_n, with

$$\mathcal{W}_n = m_n g = (0.1 \text{ kg})(9.81 \text{ m/s}^2) = 0.981 \text{ N} \quad \text{(10)}$$

The weight of the water in the control volume, $\mathcal{W}_w$, can be obtained from the water density, ρ, and the volume of water, $\mathcal{V}_w$, in the truncated cone of height h. That is,

$$\mathcal{W}_w = \rho \mathcal{V}_w g$$

where

$$\mathcal{V}_w = \tfrac{1}{12}\pi h (D_1^2 + D_2^2 + D_1 D_2)$$

$$= \frac{1}{12}\pi \frac{(30 \text{ mm})}{(1000 \text{ mm/m})}$$

$$\times \left[\frac{(16 \text{ mm})^2 + (5 \text{ mm})^2 + (16 \text{ mm})(5 \text{ mm})}{(1000^2 \text{ mm}^2/\text{m}^2)} \right]$$

$$= 2.84 \times 10^{-6} \text{ m}^3$$

Thus,

$$\mathcal{W}_w = (999 \text{ kg/m}^3)(2.84 \times 10^{-6} \text{ m}^3)(9.81 \text{ m/s}^2)$$

$$= 0.0278 \text{ N} \tag{11}$$

The gage pressure at section (2), p_2, is zero since, as discussed in Section 3.6.1, when a subsonic flow discharges to the atmosphere as in the present situation, the discharge pressure is essentially atmospheric. The anchoring force, F_A, can now be determined from Eqs. 6 through 11 with

$$F_A = (0.599 \text{ kg/s})(2.98 \text{ m/s} - 30.6 \text{ m/s}) + 0.981 \text{ N}$$

$$+ (464 \text{ kPa})(1000 \text{ Pa/kPa}) \frac{\pi(16 \text{ mm})^2}{4(1000^2 \text{ mm}^2/\text{m}^2)}$$

$$+ 0.0278 \text{ N} - 0$$

or

$$F_A = -16.5 \text{ N} + 0.981 \text{ N} + 93.3 \text{ N} + 0.0278 \text{ N}$$

$$= 77.8 \text{ N} \tag{Ans}$$

Since the anchoring force, F_A, is positive, it acts upward in the z direction. The nozzle would be pushed off the pipe if it were not fastened securely.

COMMENT   The control volume selected above to solve problems such as these is not unique. The following is an alternate solution that involves two other control volumes—one containing only the nozzle and the other containing only the water in the nozzle. These control volumes are shown in Figs. E5.11c and E5.11d along with the vertical forces acting on the contents of each control volume. The new force involved, R_z, represents the interaction between the water and the conical inside surface of the nozzle. It includes the net pressure and viscous forces at this interface.

Application of Eq. 5.22 to the contents of the control volume of Fig. E5.11c leads to

$$F_A = \mathcal{W}_n + R_z - p_{atm}(A_1 - A_2) \tag{12}$$

The term $p_{atm}(A_1 - A_2)$ is the resultant force from the atmospheric pressure acting upon the exterior surface of the nozzle (i.e., that portion of the surface of the nozzle that is not in contact with the water). Recall that the pressure force on a curved surface (such as the exterior surface of the nozzle) is equal to the pressure times the projection of the surface area on a plane perpendicular to the axis of the nozzle. The projection of this area on a plane perpendicular to the z direction is $A_1 - A_2$. The effect of the atmospheric pressure on the internal area (between the nozzle and the water) is already included in R_z which represents the net force on this area.

Similarly, for the control volume of Fig. E5.11d we obtain

$$R_z = \dot{m}(w_1 - w_2) + \mathcal{W}_w + (p_1 + p_{atm})A_1 - (p_2 - p_{atm})A_2 \tag{13}$$

where p_1 and p_2 are gage pressures. From Eq. 13 it is clear that the value of R_z depends on the value of the atmospheric pressure, p_{atm}, since $A_1 \neq A_2$. That is, we must use absolute pressure, not gage pressure, to obtain the correct value of R_z. From Eq. 13 we can easily identify which forces acting on the flowing fluid change its velocity magnitude and thus linear momentum.

By combining Eqs. 12 and 13 we obtain the same result for F_A as before (Eq. 6):

$$F_A = \dot{m}(w_1 - w_2) + \mathcal{W}_n + p_1 A_1 + W_w - p_2 A_2$$

Note that although the force between the fluid and the nozzle wall, R_z, is a function of p_{atm}, the anchoring force, F_A, is not. That is, we were correct in using gage pressure when solving for F_A by means of the original control volume shown in Fig. E5.11b.

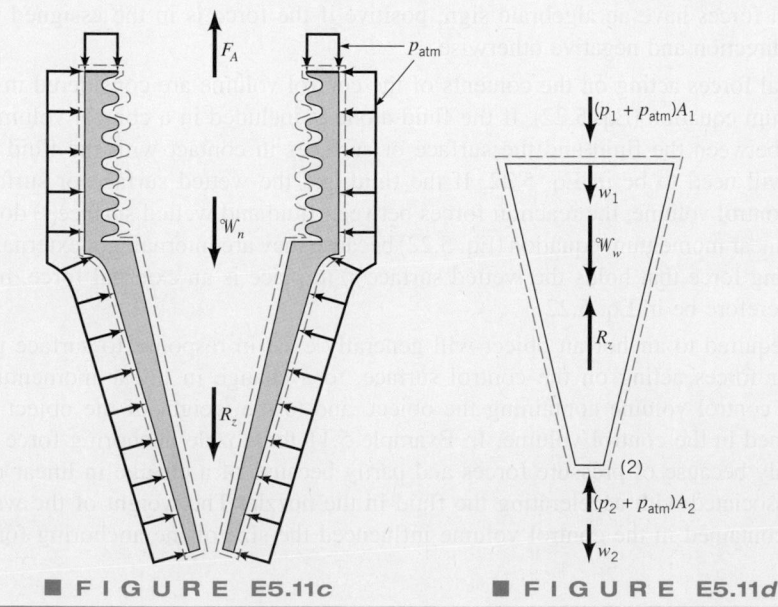

■ **FIGURE E5.11c**   ■ **FIGURE E5.11d**

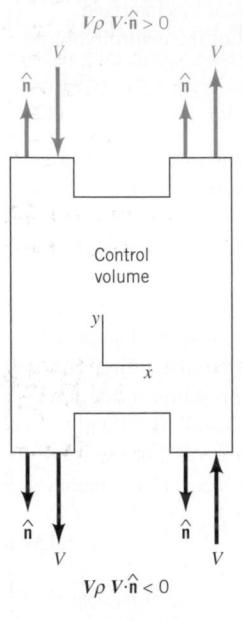

$V\rho \, V \cdot \hat{\mathbf{n}} > 0$

Control volume

$V\rho \, V \cdot \hat{\mathbf{n}} < 0$

V5.7 Running on water

A control volume diagram is similar to a free-body diagram.

Several important generalities about the application of the linear momentum equation (Eq. 5.22) are apparent in the example just considered.

1. When the flow is uniformly distributed over a section of the control surface where flow into or out of the control volume occurs, the integral operations are simplified. Thus, one-dimensional flows are easier to work with than flows involving nonuniform velocity distributions.

2. Linear momentum is directional; it can have components in as many as three orthogonal coordinate directions. Furthermore, along any one coordinate, the linear momentum of a fluid particle can be in the positive or negative direction and thus be considered as a positive or a negative quantity. In Example 5.11, only the linear momentum in the z direction was considered (all of it was in the negative z direction and was hence treated as being negative).

3. The flow of positive or negative linear momentum *into* a control volume involves a negative $\mathbf{V} \cdot \hat{\mathbf{n}}$ product. Momentum flow *out* of the control volume involves a positive $\mathbf{V} \cdot \hat{\mathbf{n}}$ product. The correct algebraic sign ($+$ or $-$) to assign to momentum flow ($V\rho \mathbf{V} \cdot \hat{\mathbf{n}} \, dA$) will depend on the sense of the velocity ($+$ in positive coordinate direction, $-$ in negative coordinate direction) and the $\mathbf{V} \cdot \hat{\mathbf{n}}$ product ($+$ for flow out of the control volume, $-$ for flow into the control volume). This is shown in the figure in the margin. In Example 5.11, the momentum flow into the control volume past section (1) was a positive ($+$) quantity while the momentum flow out of the control volume at section (2) was a negative ($-$) quantity.

4. The time rate of change of the linear momentum of the contents of a nondeforming control volume (i.e., $\partial/\partial t \int_{cv} \mathbf{V}\rho \, d\Psi$) is zero for steady flow. The momentum problems considered in this text all involve steady flow.

5. If the control surface is selected so that it is perpendicular to the flow where fluid enters or leaves the control volume, the surface force exerted at these locations by fluid outside the control volume on fluid inside will be due to pressure. Furthermore, when subsonic flow exits from a control volume into the atmosphere, atmospheric pressure prevails at the exit cross section. In Example 5.11, the flow was subsonic and so we set the exit flow pressure at the atmospheric level. The continuity equation (Eq. 5.12) allowed us to evaluate the fluid flow velocities w_1 and w_2 at sections (1) and (2).

6. The forces due to atmospheric pressure acting on the control surface may need consideration as indicated by Eq. 13 in Example 5.11 for the reaction force between the nozzle and the fluid. When calculating the anchoring force, F_A, the forces due to atmospheric pressure on the control surface cancel each other (for example, after combining Eqs. 12 and 13 the atmospheric pressure forces are no longer involved) and gage pressures may be used.

7. The external forces have an algebraic sign, positive if the force is in the assigned positive coordinate direction and negative otherwise.

8. Only external forces acting on the contents of the control volume are considered in the linear momentum equation (Eq. 5.22). If the fluid alone is included in a control volume, reaction forces between the fluid and the surface or surfaces in contact with the fluid [wetted surface(s)] will need to be in Eq. 5.22. If the fluid and the wetted surface or surfaces are within the control volume, the reaction forces between fluid and wetted surface(s) do not appear in the linear momentum equation (Eq. 5.22) because they are internal, not external forces. The anchoring force that holds the wetted surface(s) in place is an external force, however, and must therefore be in Eq. 5.22.

9. The force required to anchor an object will generally exist in response to surface pressure and/or shear forces acting on the control surface, to a change in linear momentum flow through the control volume containing the object, and to the weight of the object and the fluid contained in the control volume. In Example 5.11 the nozzle anchoring force was required mainly because of pressure forces and partly because of a change in linear momentum flow associated with accelerating the fluid in the nozzle. The weight of the water and the nozzle contained in the control volume influenced the size of the anchoring force only slightly.

Motorized surfboard When Bob Montgomery, a former professional surfer, started to design his motorized surfboard (called a jet board), he discovered that there were many engineering challenges to the design. The idea is to provide surfing to anyone, no matter where they live, near or far from the ocean. The rider stands on the device like a surfboard and steers it like a surfboard by shifting his/her body weight. A new, sleek, compact 45-horsepower (33.56 kW) engine and pump was designed to fit within the surfboard hull. Thrust is produced in response to the change in *linear momentum* of the water stream as it enters through the inlet passage and exits through an appropriately designed nozzle. Some of the fluid dynamic problems associated with designing the craft included one-way valves so that water does not get into the engine (at both the intake or exhaust ports), buoyancy, hydrodynamic lift, drag, thrust, and hull stability. (See Problem 5.68.)

To further demonstrate the use of the linear momentum equation (Eq. 5.22), we consider another one-dimensional flow example before moving on to other facets of this important equation.

EXAMPLE 5.12 | **Linear Momentum—Pressure and Change in Flow Direction**

GIVEN Water flows through a horizontal, 180° pipe bend as illustrated in Fig. E5.12a. The flow cross-sectional area is constant at a value of 0.01 m² through the bend. The magnitude of the flow velocity everywhere in the bend is axial and 15 m/s. The absolute pressures at the entrance and exit of the bend are 207 kPa (abs) and 165 kPa (abs), respectively.

FIND Calculate the horizontal (x and y) components of the anchoring force required to hold the bend in place.

SOLUTION _____

Since we want to evaluate components of the anchoring force to hold the pipe bend in place, an appropriate control volume (see dashed line in Fig. E5.12a) contains the bend and the water in the bend at an instant. The horizontal forces acting on the contents of this control volume are identified in Fig. E5.12b. Note that the weight of the water is vertical (in the negative z direction) and does not contribute to the x and y components of the anchoring force. All of the horizontal normal and tangential forces exerted on the fluid and the pipe bend are resolved and combined into the two resultant components, F_{Ax} and F_{Ay}. These two forces act on the control volume contents, and thus for the x direction, Eq. 5.22 leads to

$$\int_{cs} u\rho \mathbf{V} \cdot \hat{\mathbf{n}}\, dA = F_{Ax} \tag{1}$$

At sections (1) and (2), the flow is in the y direction and therefore $u = 0$ at both cross sections. There is no x direction momentum flow into or out of the control volume and we conclude from Eq. 1 that

$$F_{Ax} = 0 \tag{Ans}$$

For the y direction, we get from Eq. 5.22

$$\int_{cs} v\rho \mathbf{V} \cdot \hat{\mathbf{n}}\, dA = F_{Ay} + p_1 A_1 + p_2 A_2 \tag{2}$$

For one-dimensional flow, the surface integral in Eq. 2 is easy to evaluate and Eq. 2 becomes

$$(+v_1)(-\dot{m}_1) + (-v_2)(+\dot{m}_2) = F_{Ay} + p_1 A_1 + p_2 A_2 \tag{3}$$

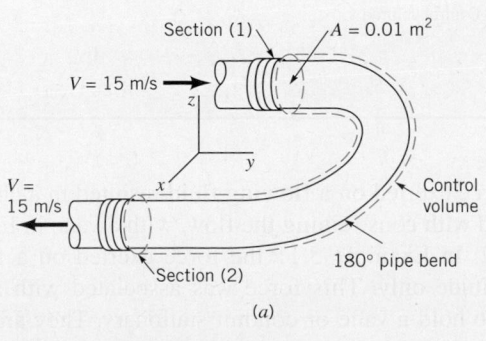

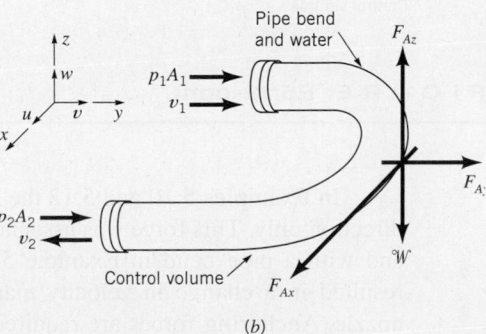

(a)   (b)

■ **FIGURE** E5.12

Note that the y component of velocity is positive at section (1) but is negative at section (2). Also, the mass flowrate term is negative at section (1) (flow in) and is positive at section (2) (flow out). From the continuity equation (Eq. 5.12), we get

$$\dot{m} = \dot{m}_1 = \dot{m}_2 \qquad (4)$$

and thus Eq. 3 can be written as

$$-\dot{m}(v_1 + v_2) = F_{Ay} + p_1A_1 + p_2A_2 \qquad (5)$$

Solving Eq. 5 for F_{Ay} we obtain

$$F_{Ay} = -\dot{m}(v_1 + v_2) - p_1A_1 - p_2A_2 \qquad (6)$$

From the given data we can calculate the mass flowrate, $\dot{m}$, from Eq. 5.6 as

$$\dot{m} = \rho_1A_1v_1 = (1000 \text{ kg/m}^3)(0.01 \text{ m}^2)(15 \text{ m/s})$$
$$= 150 \text{ kg/s}$$

For determining the anchoring force, F_{Ay}, the effects of atmospheric pressure cancel and thus gage pressures for p_1 and p_2 are appropriate. By substituting numerical values of variables into Eq. 6, and using the fact that $1 \text{ N} = 1 \text{ kg} \cdot \text{m/s}^2$ we get

$$F_{Ay} = -(150 \text{ kg/s})(15 \text{ m/s} + 15 \text{ m/s})$$
$$- (207 \text{ kPa (abs)} - 101.3 \text{ kPa (abs)})(0.01 \text{ m}^2)$$
$$- (165 \text{ kPa (abs)} - 101.3 \text{ kPa (abs)})(0.01 \text{ m}^2)$$

$$F_{Ay} = -4500 \text{ N} - 1057 \text{ N} - 637 \text{ N} = -6194 \text{ N} \quad \textbf{(Ans)}$$

The negative sign for F_{Ay} is interpreted as meaning that the y component of the anchoring force is actually in the negative y direction, not the positive y direction as originally indicated in Fig. E5.12b.

COMMENT As with Example 5.11, the anchoring force for the pipe bend is independent of the atmospheric pressure. However, the force that the bend puts on the fluid inside of it, R_y, depends on the atmospheric pressure. We can see this by using a control volume which surrounds only the fluid within the bend as shown in Fig. E5.12c. Application of the momentum equation to this situation gives

$$R_y = -\dot{m}(v_1 + v_2) - p_1A_1 - p_2A_2$$

where p_1 and p_2 must be in terms of absolute pressure because the force between the fluid and the pipe wall, R_y, is the complete pressure effect (i.e., absolute pressure). We see that forces exerted on the flowing fluid result in a change in its velocity direction (a change in linear momentum).

Thus, we obtain

$$R_y = -(150 \text{ kg/s})(15 \text{ m/s} + 15 \text{ m/s})$$
$$- (207 \text{ kPa (abs)})(0.01 \text{ m}^2)$$
$$- (165 \text{ kPa (abs)})(0.01 \text{ m}^2) \qquad (7)$$
$$= -8220 \text{ N}$$

We can use the control volume that includes just the pipe bend (without the fluid inside it) as shown in Fig. E5.12d to determine F_{Ay}, the anchoring force component in the y direction necessary to hold the bend stationary. The y component of the momentum equation applied to this control volume gives

$$F_{Ay} = R_y + p_{atm}(A_1 + A_2) \qquad (8)$$

where R_y is given by Eq. 7. The $p_{atm}(A_1 + A_2)$ term represents the net pressure force on the outside portion of the control volume. Recall that the pressure force on the inside of the bend is accounted for by R_y. By combining Eqs. 7 and 8 and using the fact that $p_{atm} = 101.3 \text{ kPa}$, we obtain

$$F_{Ay} = -8220 \text{ N} + 101.3 \text{ kPa} (0.01 \text{ m}^2 + 0.01 \text{ m}^2)$$
$$= -6194 \text{ N}$$

in agreement with the original answer obtained using the control volume of Fig. E5.12b.

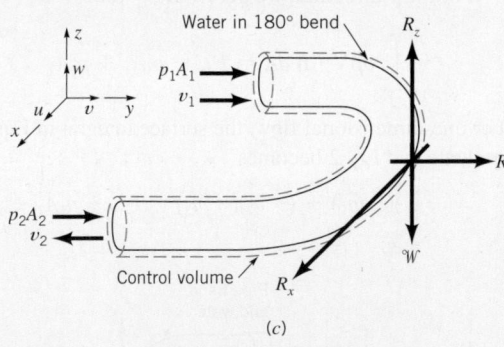

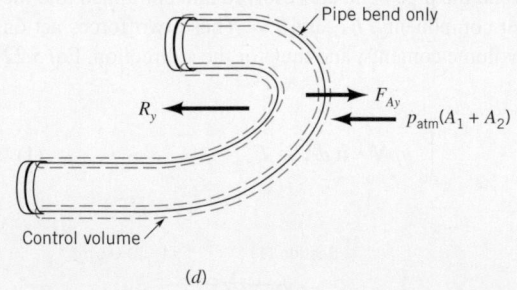

(c)   (d)

■ **F I G U R E E5.12 cont.**

V5.8 Fire hose

In Examples 5.10 and 5.12 the force exerted on a flowing fluid resulted in a change in flow direction only. This force was associated with constraining the flow, with a vane in Example 5.10, and with a pipe bend in Example 5.12. In Example 5.11 the force exerted on a flowing fluid resulted in a change in velocity magnitude only. This force was associated with a converging nozzle. Anchoring forces are required to hold a vane or conduit stationary. They are most easily estimated with a control volume that contains the vane or conduit and the flowing fluid involved. Alternately, two separate control volumes can be used, one containing the vane or conduit only and one containing the flowing fluid only.

EXAMPLE 5.13   Linear Momentum—Pressure, Change in Speed, and Friction

GIVEN   Air flows steadily between two cross sections in a long, straight portion of 10 cm inside diameter pipe as indicated in Fig. E5.13, where the uniformly distributed temperature and pressure at each cross section are given. If the average air velocity at section (2) is 305 m/s, we found in Example 5.2 that the average air velocity at section (1) must be 67 m/s. Assume uniform velocity distributions at sections (1) and (2).

FIND   Determine the frictional force exerted by the pipe wall on the air flow between sections (1) and (2).

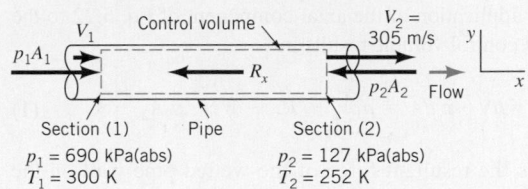

■ **FIGURE E5.13**

$p_1 = 690$ kPa(abs) $p_2 = 127$ kPa(abs)
$T_1 = 300$ K $T_2 = 252$ K

SOLUTION

The control volume of Example 5.2 is appropriate for this problem. The forces acting on the air between sections (1) and (2) are identified in Fig. E5.13. The weight of air is considered negligibly small. The reaction force between the wetted wall of the pipe and the flowing air, R_x, is the frictional force sought. Application of the axial component of Eq. 5.22 to this control volume yields

$$\int_{cs} u\rho \mathbf{V} \cdot \hat{\mathbf{n}}\, dA = -R_x + p_1 A_1 - p_2 A_2 \tag{1}$$

The positive x direction is set as being to the right. Furthermore, for uniform velocity distributions (one-dimensional flow), Eq. 1 becomes

$$(+u_1)(-\dot{m}_1) + (+u_2)(+\dot{m}_2) = -R_x + p_1 A_1 - p_2 A_2 \tag{2}$$

From conservation of mass (Eq. 5.12) we get

$$\dot{m} = \dot{m}_1 = \dot{m}_2 \tag{3}$$

so that Eq. 2 becomes

$$\dot{m}(u_2 - u_1) = -R_x + A_2(p_1 - p_2) \tag{4}$$

Solving Eq. 4 for R_x, we get

$$R_x = A_2(p_1 - p_2) - \dot{m}(u_2 - u_1) \tag{5}$$

The equation of state gives

$$\rho_2 = \frac{p_2}{RT_2} \tag{6}$$

and the equation for area A_2 is

$$A_2 = \frac{\pi D_2^2}{4} \tag{7}$$

Thus, from Eqs. 3, 6, and 7

$$\dot{m} = \left(\frac{p_2}{RT_2}\right)\left(\frac{\pi D_2^2}{4}\right) u_2$$

The gas constant, R, for air in SI units is

$$R = 286.9 \text{ J/kg} \cdot \text{k}$$

Hence, $\dot{m} = \dfrac{(127 \text{ kPa})\,(\text{abs})}{[286.9 \text{ J/kg} \cdot \text{k}]\,(252 \text{ k})}$

$$\times \frac{\pi (10 \text{ cm})^2}{4(10^4 \text{ cm}^2/\text{m}^2)}(305 \text{ m/s}) = 4.21 \text{ kg/s} \tag{8}$$

Thus, from Eqs. 5 and 8

$$R_x = \frac{\pi (10 \text{ cm})^2}{4\left(\dfrac{10^4 \text{ cm}^2}{\text{m}^2}\right)}(690 \text{ kPa (abs)} - 127 \text{ kPa (abs)})$$

$$- (4.21 \text{ kg/s})(305 \text{ m/s} - 67 \text{ m/s})$$

$$= 4422 \text{ N} - 1002 \text{ N}$$

or

$$R_x = 3420 \text{ N} \tag{Ans}$$

COMMENT   For this compressible flow, the pressure difference drives the motion which results in a frictional force, R_x, and an acceleration of the fluid (i.e., a velocity magnitude increase). For a similar incompressible pipe flow, a pressure difference results in fluid motion with a frictional force only (i.e., no change in velocity magnitude).

EXAMPLE 5.14   Linear Momentum—Weight, Pressure, Friction, and Nonuniform Velocity Profile

GIVEN   Consider the flow of Example 5.4 to be vertically upward.

FIND   Develop an expression for the fluid pressure drop that occurs between sections (1) and (2).

SOLUTION

A control volume (see dashed lines in Fig. E5.14) that includes only fluid from section (1) to section (2) is selected. The forces acting on the fluid in this control volume are identified in Fig. E5.14. The application of the axial component of Eq. 5.22 to the fluid in this control volume results in

$$\int_{cs} w\rho \mathbf{V} \cdot \hat{\mathbf{n}} \, dA = p_1 A_1 - R_z - \mathcal{W} - p_2 A_2 \tag{1}$$

where R_z is the resultant force of the wetted pipe wall on the fluid. Further, for uniform flow at section (1), and because the flow at section (2) is out of the control volume, Eq. 1 becomes

$$(+w_1)(-\dot{m}_1) + \int_{A_2} (+w_2)\rho(+w_2 \, dA_2) = p_1 A_1 - R_z \\ - \mathcal{W} - p_2 A_2 \tag{2}$$

The positive direction is considered up. The surface integral over the cross-sectional area at section (2), A_2, is evaluated by using the parabolic velocity profile obtained in Example 5.4, $w_2 = 2w_1[1 - (r/R)^2]$, as

$$\int_{A_2} w_2 \rho w_2 \, dA_2 = \rho \int_0^R w_2^2 \, 2\pi r \, dr$$

$$= 2\pi\rho \int_0^R (2w_1)^2 \left[1 - \left(\frac{r}{R}\right)^2\right]^2 r \, dr$$

or

$$\int_{A_2} w_2 \rho w_2 \, dA_2 = 4\pi\rho w_1^2 \frac{R^2}{3} \tag{3}$$

Combining Eqs. 2 and 3 we obtain

$$-w_1^2 \rho \pi R^2 + \tfrac{4}{3} w_1^2 \rho \pi R^2 = p_1 A_1 - R_z - \mathcal{W} - p_2 A_2 \tag{4}$$

Solving Eq. 4 for the pressure drop from section (1) to section (2), $p_1 - p_2$, we obtain

$$p_1 - p_2 = \frac{\rho w_1^2}{3} + \frac{R_z}{A_1} + \frac{\mathcal{W}}{A_1} \tag{Ans}$$

COMMENT   We see that the drop in pressure from section (1) to section (2) occurs because of the following:

1. The change in momentum flow between the two sections associated with going from a uniform velocity profile to a parabolic velocity profile, $\rho w_1^2/3$
2. Pipe wall friction, R_z.
3. The weight of the water column, $\mathcal{W}$; a hydrostatic pressure effect.

If the velocity profiles had been identically parabolic at sections (1) and (2), the momentum flowrate at each section would have

been identical, a condition we call "fully developed" flow. Then, the pressure drop, $p_1 - p_2$, would be due only to pipe wall friction and the weight of the water column. If in addition to being fully developed, the flow involved negligible weight effects (for example, horizontal flow of liquids or the flow of gases in any direction) the drop in pressure between any two sections, $p_1 - p_2$, would be a result of pipe wall friction only.

Note that although the average velocity is the same at section (1) as it is at section (2) $(\bar{V}_1 = \bar{V}_2 = w_1)$, the momentum flux across section (1) is not the same as it is across section (2). If it were, the left-hand side of Eq. (4) would be zero. For this nonuniform flow the momentum flux can be written in terms of the average velocity, $\bar{V}$, and the *momentum coefficient*, β, as

$$\beta = \frac{\int w\rho \mathbf{V} \cdot \hat{\mathbf{n}} \, dA}{\rho \bar{V}^2 A}$$

Hence the momentum flux can be written as

$$\int_{cs} w\rho \mathbf{V} \cdot \hat{\mathbf{n}} \, dA = -\beta_1 w_1^2 \rho \pi R^2 + \beta_2 w_1^2 \rho \pi R^2$$

where $\beta_1 = 1$ ($\beta = 1$ for uniform flow) and $\beta_2 = 4/3$ ($\beta > 1$ for any nonuniform flow).

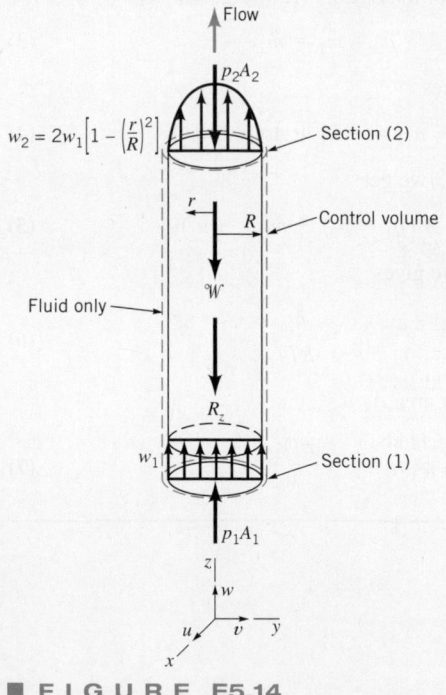

■ **FIGURE E5.14**

EXAMPLE 5.15   Linear Momentum—Thrust

GIVEN   A static thrust stand as sketched in Fig. E5.15 is to be designed for testing a jet engine. The following conditions are known for a typical test: Intake air velocity = 200 m/s; exhaust gas velocity = 500 m/s; intake cross-sectional area = 1 m²; intake

static pressure = −22.5 kPa = 78.5 kPa (abs); intake static temperature = 268 K; exhaust static pressure = 0 kPa = 101 kPa (abs).

FIND   Estimate the nominal anchoring force for which to design.

SOLUTION

The cylindrical control volume outlined with a dashed line in Fig. E5.15 is selected. The external forces acting in the axial direction are also shown. Application of the momentum equation (Eq. 5.22) to the contents of this control volume yields

$$\int_{cs} u\rho \mathbf{V} \cdot \hat{\mathbf{n}} \, dA = p_1 A_1 + F_{th} - p_2 A_2$$
$$- p_{atm}(A_1 - A_2) \qquad (1)$$

where the pressures are absolute. Thus, for one-dimensional flow, Eq. 1 becomes

$$(+u_1)(-\dot{m}_1) + (+u_2)(+\dot{m}_2) = (p_1 - p_{atm})A_1$$
$$- (p_2 - p_{atm})A_2 + F_{th} \qquad (2)$$

The positive direction is to the right. The conservation of mass equation (Eq. 5.12) leads to

$$\dot{m} = \dot{m}_1 = \rho_1 A_1 u_1 = \dot{m}_2 = \rho_2 A_2 u_2 \qquad (3)$$

Combining Eqs. 2 and 3 and using gage pressure we obtain

$$\dot{m}(u_2 - u_1) = p_1 A_1 - p_2 A_2 + F_{th} \qquad (4)$$

Solving Eq. 4 for the thrust force, F_{th}, we obtain

$$F_{th} = -p_1 A_1 + p_2 A_2 + \dot{m}(u_2 - u_1) \qquad (5)$$

We need to determine the mass flowrate, $\dot{m}$, to calculate F_{th}, and to calculate $\dot{m} = \rho_1 A_1 u_1$, we need ρ_1. From the ideal gas equation of state

$$\rho_1 = \frac{p_1}{RT_1} = \frac{(78.5 \text{ kPa})(1000 \text{ Pa/kPa})[1(\text{N/m}^2)/\text{Pa}]}{(286.9 \text{ J/kg} \cdot \text{K})(268 \text{ K})(1 \text{ N} \cdot \text{m/J})}$$
$$= 1.02 \text{ kg/m}^3$$

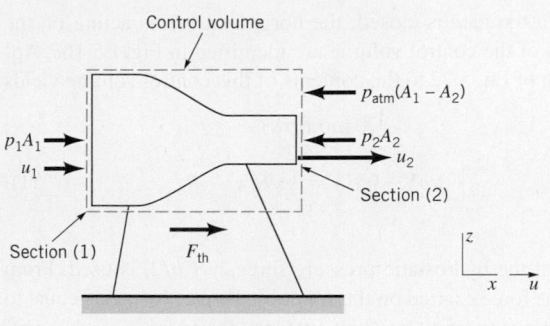

Control volume

■ **FIGURE E5.15**

Thus,

$$\dot{m} = \rho_1 A_1 u_1 = (1.02 \text{ kg/m}^3)(1 \text{ m}^2)(200 \text{ m/s})$$
$$= 204 \text{ kg/s} \qquad (6)$$

Finally, combining Eqs. 5 and 6 and substituting given data with $p_2 = 0$, we obtain

$$F_{th} = -(1 \text{ m}^2)(-22.5 \text{ kPa})(1000 \text{ Pa/kPa})[1(\text{N/m}^2)/\text{Pa}]$$
$$+ (204 \text{ kg/s})(500 \text{ m/s} - 200 \text{ m/s})[1 \text{ N}/(\text{kg} \cdot \text{m/s}^2)]$$

or

$$F_{th} = 22{,}500 \text{ N} + 61{,}200 \text{ N} = 83{,}700 \text{ N} \qquad \textbf{(Ans)}$$

COMMENT   The force of the thrust stand on the engine is directed toward the right. Conversely, the engine pushes to the left on the thrust stand (or aircraft).

F l u i d s   i n   t h e   N e w s

Bow thrusters In the past, large ships required the use of tugboats for precise maneuvering, especially when docking. Nowadays, most large ships (and many moderate to small ones as well) are equipped with bow thrusters to help steer in close quarters. The units consist of a mechanism (usually a ducted propeller mounted at right angles to the fore/aft axis of the ship) that takes water from one side of the bow and ejects it as a water jet on the other side. The *momentum flux* of this jet produces a starboard or port force

on the ship for maneuvering. Sometimes a second unit is installed in the stern. Initially used in the bows of ferries, these versatile control devices have became popular in offshore oil servicing boats, fishing vessels, and larger ocean-going craft. They permit unassisted maneuvering alongside of oilrigs, vessels, loading platforms, fishing nets, and docks. They also provide precise control at slow speeds through locks, narrow channels, and bridges, where the rudder becomes very ineffective. (See Problem 5.69.)

---

*E*XAMPLE 5.16   Linear Momentum—Nonuniform Pressure

GIVEN   A sluice gate across a channel of width b is shown in the closed and open positions in Figs. E5.16a and E5.16b.

FIND   Is the anchoring force required to hold the gate in place larger when the gate is closed or when it is open?

SOLUTION

We will answer this question by comparing expressions for the horizontal reaction force, R_x, between the gate and the water when the gate is closed and when the gate is open. The control

volume used in each case is indicated with dashed lines in Figs. E5.16a and E5.16b.

When the gate is closed, the horizontal forces acting on the contents of the control volume are identified in Fig. E5.16c. Application of Eq. 5.22 to the contents of this control volume yields

$$\int_{cs} u\rho \mathbf{V} \cdot \hat{\mathbf{n}} \, dA = \frac{1}{2}\gamma H^2 b - R_x \qquad \text{0 (no flow)} \tag{1}$$

Note that the hydrostatic pressure force, $\gamma H^2 b/2$, is used. From Eq. 1, the force exerted on the water by the gate (which is equal to the force necessary to hold the gate stationary) is

$$R_x = \frac{1}{2}\gamma H^2 b \tag{2}$$

which is equal in magnitude to the hydrostatic force exerted on the gate by the water.

When the gate is open, the horizontal forces acting on the contents of the control volume are shown in Fig. E5.16d. Application of Eq. 5.22 to the contents of this control volume leads to

$$\int_{cs} u\rho \mathbf{V} \cdot \hat{\mathbf{n}} \, dA = \frac{1}{2}\gamma H^2 b - R_x - \frac{1}{2}\gamma h^2 b - F_f \tag{3}$$

Note that because the water at sections (1) and (2) is flowing along straight, horizontal streamlines, the pressure distribution at those locations is hydrostatic, varying from zero at the free surface to γ times the water depth at the bottom of the channel (see Chapter 3, Section 3.4). Thus, the pressure forces at sections (1) and (2) (given by the pressure at the centroid times the area) are $\gamma H^2 b/2$ and $\gamma h^2 b/2$, respectively. Also, the frictional force between the channel bottom and the water is specified as F_f. The surface integral in Eq. 3 is nonzero only where there is flow across the control surface. With the assumption of uniform velocity distributions,

$$\int_{cs} u\rho \mathbf{V} \cdot \hat{\mathbf{n}} \, dA = (u_1)\rho(-u_1)Hb + (+u_2)\rho(+u_2)hb \tag{4}$$

Thus, Eqs. 3 and 4 combine to form

$$-\rho u_1^2 Hb + \rho u_2^2 hb = \frac{1}{2}\gamma H^2 b - R_x - \frac{1}{2}\gamma h^2 b - F_f \tag{5}$$

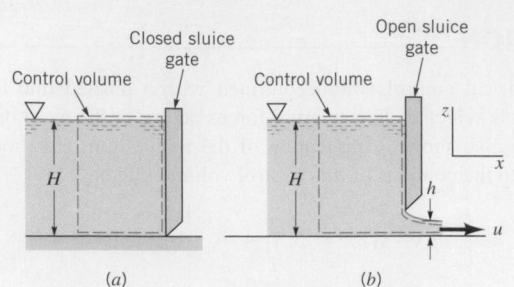

(a)   (b)

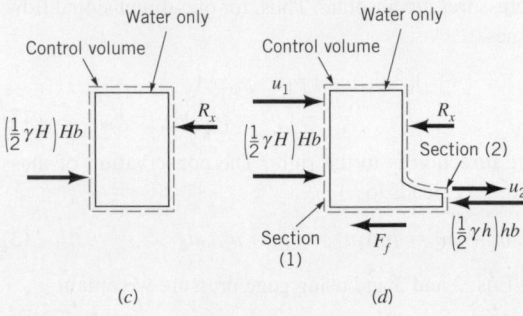

(c)   (d)

■ **FIGURE E5.16**

If $H \gg h$, the upstream velocity, u_1, is much less than u_2 so that the contribution of the incoming momentum flow to the control surface integral can be neglected and from Eq. 5 we obtain

$$R_x = \frac{1}{2}\gamma H^2 b - \frac{1}{2}\gamma h^2 b - F_f - \rho u_2^2 hb \tag{6}$$

By using the continuity equation, $\dot{m} = \rho b H u_1 = \rho b h u_2$, Eq. (6) can be rewritten as

$$R_x = \frac{1}{2}\gamma H^2 b - \frac{1}{2}\gamma h^2 b - F_f - \dot{m}(u_2 - u_1) \tag{7}$$

Hence, since $u_2 > u_1$, by comparing the expressions for R_x (Eqs. 2 and 7) we conclude that the reaction force between the gate and the water (and therefore the anchoring force required to hold the gate in place) is smaller when the gate is open than when it is closed. **(Ans)**

The linear momentum equation can be written for a moving control volume.

V5.9 Jelly fish

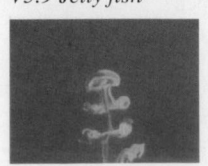

All of the linear momentum examples considered thus far have involved stationary and nondeforming control volumes which are thus inertial because there is no acceleration. A nondeforming control volume translating in a straight line at constant speed is also inertial because there is no acceleration. For a system and an inertial, moving, nondeforming control volume that are both coincident at an instant of time, the Reynolds transport theorem (Eq. 4.23) leads to

$$\frac{D}{Dt}\int_{sys} \mathbf{V}\rho \, d\Psi = \frac{\partial}{\partial t}\int_{cv} \mathbf{V}\rho \, d\Psi + \int_{cs} \mathbf{V}\rho \mathbf{W} \cdot \hat{\mathbf{n}} \, dA \tag{5.23}$$

When we combine Eq. 5.23 with Eqs. 5.19 and 5.20, we get

$$\frac{\partial}{\partial t}\int_{cv} \mathbf{V}\rho \, d\Psi + \int_{cs} \mathbf{V}\rho \mathbf{W} \cdot \hat{\mathbf{n}} \, dA = \sum \mathbf{F}_{\substack{\text{contents of the} \\ \text{control volume}}} \tag{5.24}$$

When the equation relating absolute, relative, and control volume velocities (Eq. 5.14) is used with Eq. 5.24, the result is

$$\frac{\partial}{\partial t}\int_{cv} (\mathbf{W} + \mathbf{V}_{cv})\rho \, d\Psi + \int_{cs} (\mathbf{W} + \mathbf{V}_{cv})\rho \mathbf{W} \cdot \hat{\mathbf{n}} \, dA = \sum \mathbf{F}_{\substack{\text{contents of the} \\ \text{control volume}}} \tag{5.25}$$

For a constant control volume velocity, $\mathbf{V}_{cv}$, and steady flow in the control volume reference frame,

$$\frac{\partial}{\partial t} \int_{cv} (\mathbf{W} + \mathbf{V}_{cv}) \rho \, d\mathcal{V} = 0 \tag{5.26}$$

Also, for this inertial, nondeforming control volume

$$\int_{cs} (\mathbf{W} + \mathbf{V}_{cv}) \rho \mathbf{W} \cdot \hat{\mathbf{n}} \, dA = \int_{cs} \mathbf{W} \rho \mathbf{W} \cdot \hat{\mathbf{n}} \, dA + \mathbf{V}_{cv} \int_{cs} \rho \mathbf{W} \cdot \hat{\mathbf{n}} \, dA \tag{5.27}$$

For steady flow (on an instantaneous or time-average basis), Eq. 5.15 gives

$$\int_{cs} \rho \mathbf{W} \cdot \hat{\mathbf{n}} \, dA = 0 \tag{5.28}$$

Combining Eqs. 5.25, 5.26, 5.27, and 5.28, we conclude that the linear momentum equation for an inertial, moving, nondeforming control volume that involves steady (instantaneous or time-average) flow is

The linear momentum equation for a moving control volume involves the relative velocity.

$$\int_{cs} \mathbf{W} \rho \mathbf{W} \cdot \hat{\mathbf{n}} \, dA = \sum \mathbf{F}_{\substack{\text{contents of the} \\ \text{control volume}}} \tag{5.29}$$

Example 5.17 illustrates the use of Eq. 5.29.

---

*E*XAMPLE 5.17   Linear Momentum—Moving Control Volume

GIVEN   A vane on wheels moves with constant velocity $\mathbf{V}_0$ when a stream of water having a nozzle exit velocity of $\mathbf{V}_1$ is turned 45° by the vane as indicated in Fig. E5.17a. Note that this is the same moving vane considered in Section 4.4.6 earlier. The speed of the water jet leaving the nozzle is 30 m/s,

and the vane is moving to the right with a constant speed of 6 m/s.

FIND   Determine the magnitude and direction of the force, $\mathbf{F}$, exerted by the stream of water on the vane surface.

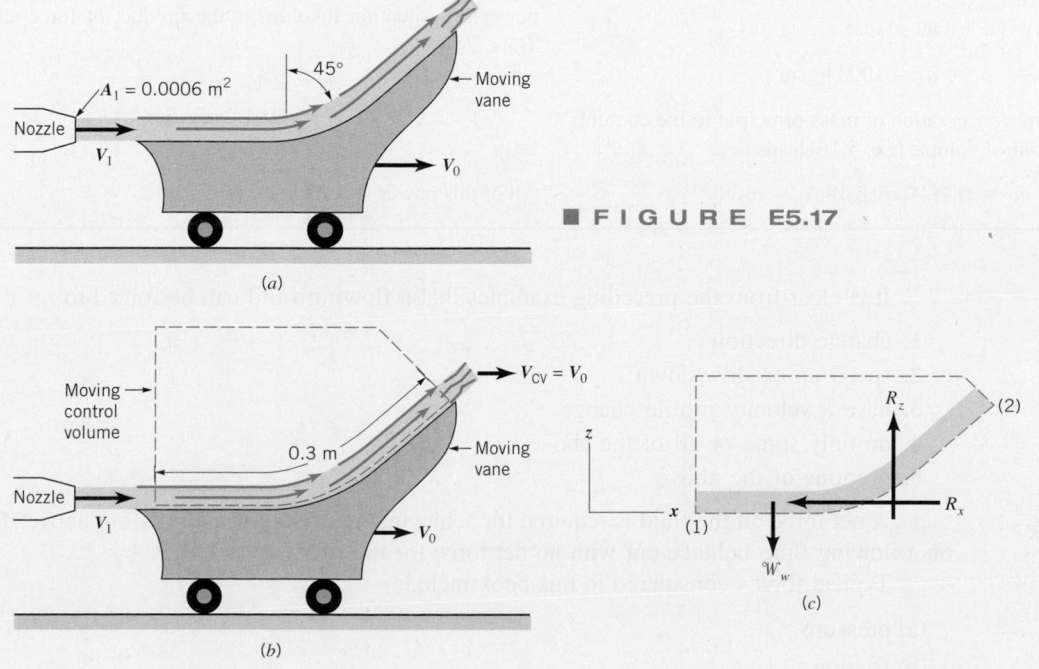

■ **FIGURE E5.17**

SOLUTION

To determine the magnitude and direction of the force, $\mathbf{F}$, exerted by the water on the vane, we apply Eq. 5.29 to the contents of the moving control volume shown in Fig. E5.17b. The forces acting on the contents of this control volume are indicated in Fig. E5.17c. Note that since the ambient pressure is atmospheric, all pressure forces cancel each other out. Equation 5.29 is applied to the contents of the moving control volume in component directions. For the x direction (positive to the right), we get

$$\int_{cs} W_x \rho \, \mathbf{W} \cdot \hat{\mathbf{n}} \, dA = -R_x$$

or

$$(+W_1)(-\dot{m}_1) + (+W_2 \cos 45°)(+\dot{m}_2) = -R_x \qquad (1)$$

where

$$\dot{m}_1 = \rho_1 W_1 A_1 \qquad \text{and} \qquad \dot{m}_2 = \rho_2 W_2 A_2.$$

For the vertical or z direction (positive up) we get

$$\int_{cs} W_z \rho \mathbf{W} \cdot \hat{\mathbf{n}} \, dA = R_z - \mathcal{W}_w$$

or

$$(+W_2 \sin 45°)(+\dot{m}_2) = R_z - \mathcal{W}_w \qquad (2)$$

We assume for simplicity that the water flow is frictionless and that the change in water elevation across the vane is negligible. Thus, from the Bernoulli equation (Eq. 3.7) we conclude that the speed of the water relative to the moving control volume, W, is constant or

$$W_1 = W_2$$

The relative speed of the stream of water entering the control volume, W_1, is

$$W_1 = V_1 - V_0 = 30 \text{ m/s} - 6 \text{ m/s} = 24 \text{ m/s} = W_2$$

The water density is constant so that

$$\rho_1 = \rho_2 = 1000 \text{ kg/m}^3$$

Application of the conservation of mass principle to the contents of the moving control volume (Eq. 5.16) leads to

$$\dot{m}_1 = \rho_1 W_1 A_1 = \rho_2 W_2 A_2 = \dot{m}_2$$

Combining results we get

$$R_x = \rho W_1^2 A_1 (1 - \cos 45°)$$

or

$$R_x = (1000 \text{ kg/m}^3)(24 \text{ m/s})^2(0.0006 \text{ m}^2)(1 - \cos 45°)$$
$$= 101.2 \text{ N}$$

Also,

$$R_z = \rho W_1^2(\sin 45°)A_1 + \mathcal{W}_w$$

where

$$\mathcal{W}_w = \rho g A_1 \ell$$

Thus,

$$R_z = (1000 \text{ kg/m}^3)(24 \text{ m/s})^2(\sin 45°)(0.0006 \text{ m}^2)$$
$$+ (9810 \text{ N/m}^3)(0.0006 \text{ m}^2)(0.3 \text{ m})$$
$$= 244.4 \text{ N} + 1.8 \text{ N} = 246 \text{ N}$$

Combining the components we get

$$R = \sqrt{R_x^2 + R_z^2} = [(101.2 \text{ N})^2 + (246 \text{ N})^2]^{1/2} = 266 \text{ N}$$

The angle of $\mathbf{R}$ from the x direction, α, is

$$\alpha = \tan^{-1}\frac{R_z}{R_x} = \tan^{-1}(246 \text{ N}/101.2 \text{ N}) = 67.6°$$

The force of the water on the vane is equal in magnitude but opposite in direction from $\mathbf{R}$; thus it points to the right and down at an angle of $67.6°$ from the x direction and is equal in magnitude to 266 N. **(Ans)**

COMMENT The force of the fluid on the vane in the x-direction, $R_x = 101.2$ N, is associated with x-direction motion of the vane at a constant speed of 6 m/s. Since the vane is not accelerating, this x-direction force is opposed mainly by a wheel friction force of the same magnitude. From basic physics we recall that the power this situation involves is the product of force and speed. Thus,

$$\mathcal{P} = R_x V_0$$
$$= (101.2 \text{ N})(6 \text{ m/s})$$
$$= 0.6 \text{ kW}$$

All of this power is consumed by friction.

It is clear from the preceding examples that a flowing fluid can be forced to

1. change direction
2. speed up or slow down
3. have a velocity profile change
4. do only some or all of the above
5. do none of the above

A net force on the fluid is required for achieving any or all of the first four above. The forces on a flowing fluid balance out with no net force for the fifth.

Typical forces considered in this book include

(a) pressure

(b) friction

(c) weight

and involve some type of constraint such as a vane, channel, or conduit to guide the flowing fluid. A flowing fluid can cause a vane, channel or conduit to move. When this happens, power is produced.

The selection of a control volume is an important matter. For determining anchoring forces, consider including fluid and its constraint in the control volume. For determining force between a fluid and its constraint, consider including only the fluid in the control volume.

5.2.3 Derivation of the Moment-of-Momentum Equation[2]

In many engineering problems, the moment of a force with respect to an axis, namely, *torque*, is important. Newton's second law of motion has already led to a useful relationship between forces and linear momentum flow. The linear momentum equation can also be used to solve problems involving torques. However, by forming the moment of the linear momentum and the resultant force associated with each particle of fluid with respect to a point in an inertial coordinate system, we will develop a *moment-of-momentum equation* that relates *torques* and *angular momentum flow* for the contents of a control volume. When torques are important, the moment-of-momentum equation is often more convenient to use than the linear momentum equation.

Application of Newton's second law of motion to a particle of fluid yields

The angular momentum equation is derived from Newton's second law.

$$\frac{D}{Dt}(\mathbf{V}\rho\,\delta\mathcal{V}) = \delta\mathbf{F}_{particle} \tag{5.30}$$

where $\mathbf{V}$ is the particle velocity measured in an inertial reference system, ρ is the particle density, $\delta\mathcal{V}$ is the infinitesimally small particle volume, and $\delta\mathbf{F}_{particle}$ is the resultant external force acting on the particle. If we form the moment of each side of Eq. 5.30 with respect to the origin of an inertial coordinate system, we obtain

$$\mathbf{r} \times \frac{D}{Dt}(\mathbf{V}\rho\,\delta\mathcal{V}) = \mathbf{r} \times \delta\mathbf{F}_{particle} \tag{5.31}$$

where $\mathbf{r}$ is the position vector from the origin of the inertial coordinate system to the fluid particle (Fig. 5.3). We note that

$$\frac{D}{Dt}[(\mathbf{r} \times \mathbf{V})\rho\,\delta\mathcal{V}] = \frac{D\mathbf{r}}{Dt} \times \mathbf{V}\rho\,\delta\mathcal{V} + \mathbf{r} \times \frac{D(\mathbf{V}\rho\,\delta\mathcal{V})}{Dt} \tag{5.32}$$

and

$$\frac{D\mathbf{r}}{Dt} = \mathbf{V} \tag{5.33}$$

Thus, since

$$\mathbf{V} \times \mathbf{V} = 0 \tag{5.34}$$

by combining Eqs. 5.31, 5.32, 5.33, and 5.34, we obtain the expression

$$\frac{D}{Dt}[(\mathbf{r} \times \mathbf{V})\rho\,\delta\mathcal{V}] = \mathbf{r} \times \delta\mathbf{F}_{particle} \tag{5.35}$$

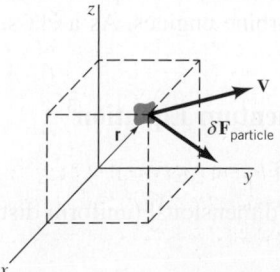

■ **FIGURE 5.3   Inertial coordinate system.**

Equation 5.35 is valid for every particle of a system. For a system (collection of fluid particles), we need to use the sum of both sides of Eq. 5.35 to obtain

$$\int_{\text{sys}} \frac{D}{Dt}[(\mathbf{r} \times \mathbf{V})\rho \, d\Psi] = \sum (\mathbf{r} \times \mathbf{F})_{\text{sys}} \tag{5.36}$$

where

$$\sum \mathbf{r} \times \delta\mathbf{F}_{\text{particle}} = \sum (\mathbf{r} \times \mathbf{F})_{\text{sys}} \tag{5.37}$$

We note that

$$\frac{D}{Dt}\int_{\text{sys}} (\mathbf{r} \times \mathbf{V})\rho \, d\Psi = \int_{\text{sys}} \frac{D}{Dt}[(\mathbf{r} \times \mathbf{V})\rho \, d\Psi] \tag{5.38}$$

since the sequential order of differentiation and integration can be reversed without consequence. (Recall that the material derivative, $D(\ )/Dt$, denotes the time derivative following a given system; see Section 4.2.1.) Thus, from Eqs. 5.36 and 5.38 we get

$$\frac{D}{Dt}\int_{\text{sys}} (\mathbf{r} \times \mathbf{V})\rho \, d\Psi = \sum (\mathbf{r} \times \mathbf{F})_{\text{sys}} \tag{5.39}$$

or

$$\begin{array}{c}\text{the time rate of change of the} \\ \text{moment-of-momentum of the system}\end{array} = \begin{array}{c}\text{sum of external torques} \\ \text{acting on the system}\end{array}$$

The sketch in the margin illustrates what torque, $\mathbf{T} = \mathbf{r} \times \mathbf{F}$, is. For a control volume that is instantaneously coincident with the system, the torques acting on the system and on the control volume contents will be identical:

$$\sum (\mathbf{r} \times \mathbf{F})_{\text{sys}} = \sum (\mathbf{r} \times \mathbf{F})_{\text{cv}} \tag{5.40}$$

Further, for the system and the contents of the coincident control volume that is fixed and nondeforming, the Reynolds transport theorem (Eq. 4.19) leads to

$$\frac{D}{Dt}\int_{\text{sys}} (\mathbf{r} \times \mathbf{V})\rho \, d\Psi = \frac{\partial}{\partial t}\int_{\text{cv}} (\mathbf{r} \times \mathbf{V})\rho \, d\Psi + \int_{\text{cs}} (\mathbf{r} \times \mathbf{V})\rho \mathbf{V} \cdot \hat{\mathbf{n}} \, dA \tag{5.41}$$

or

$$\begin{array}{c}\text{time rate of change} \\ \text{of the moment-of-} \\ \text{momentum of the} \\ \text{system}\end{array} = \begin{array}{c}\text{time rate of change} \\ \text{of the moment-of-} \\ \text{momentum of the} \\ \text{contents of the} \\ \text{control volume}\end{array} + \begin{array}{c}\text{net rate of flow} \\ \text{of the moment-of-} \\ \text{momentum through} \\ \text{the control surface}\end{array}$$

For a control volume that is fixed (and therefore inertial) and nondeforming, we combine Eqs. 5.39, 5.40, and 5.41 to obtain the moment-of-momentum equation:

$$\boxed{\frac{\partial}{\partial t}\int_{\text{cv}} (\mathbf{r} \times \mathbf{V})\rho \, d\Psi + \int_{\text{cs}} (\mathbf{r} \times \mathbf{V})\rho \mathbf{V} \cdot \hat{\mathbf{n}} \, dA = \sum (\mathbf{r} \times \mathbf{F})_{\substack{\text{contents of the} \\ \text{control volume}}}} \tag{5.42}$$

For a system, the rate of change of moment-of-momentum equals the net torque.

An important category of fluid mechanical problems that is readily solved with the help of the moment-of-momentum equation (Eq. 5.42) involves machines that rotate or tend to rotate around a single axis. Examples of these machines include rotary lawn sprinklers, ceiling fans, lawn mower blades, wind turbines, turbochargers, and gas turbine engines. As a class, these devices are often called turbomachines.

5.2.4 Application of the Moment-of-Momentum Equation[3]

We simplify our use of Eq. 5.42 in several ways:

1. We assume that flows considered are one-dimensional (uniform distributions of average velocity at any section).

---

[3]This section may be omitted, along with Sections 5.2.3 and 5.3.5, without loss of continuity in the text material. However, these sections are recommended for those interested in Chapter 12.

*V5.10 Rotating
lawn sprinkler*

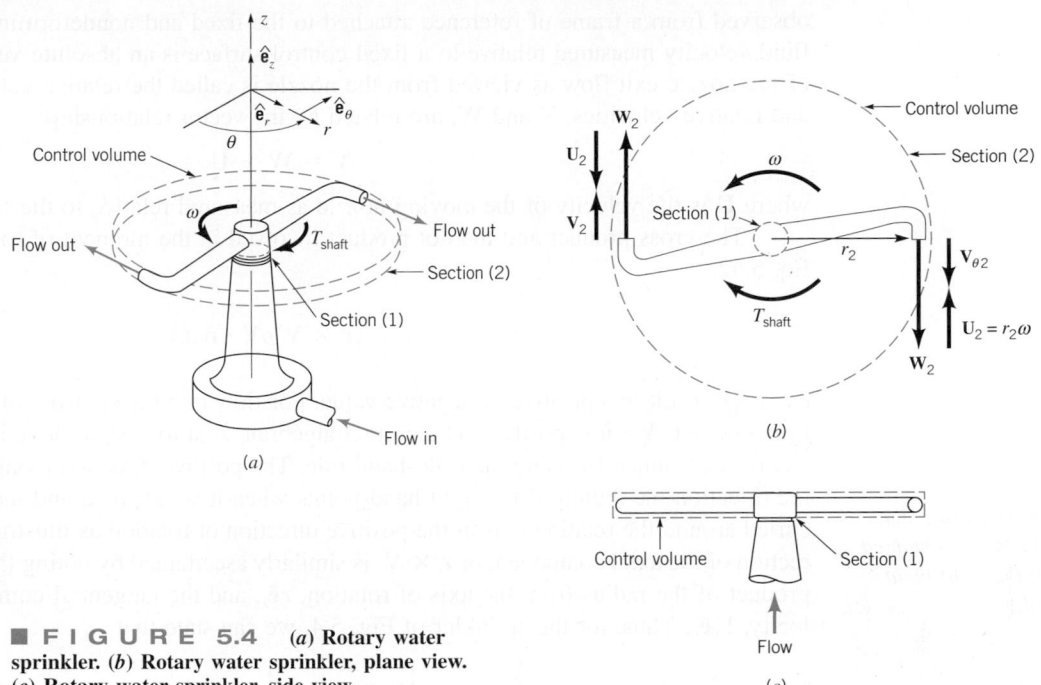

■ **F I G U R E   5.4**   (*a*) **Rotary water
sprinkler.** (*b*) **Rotary water sprinkler, plane view.**
(*c*) **Rotary water sprinkler, side view.**

2. We confine ourselves to steady or steady-in-the-mean cyclical flows. Thus,

$$\frac{\partial}{\partial t}\int_{cv}(\mathbf{r}\times\mathbf{V})\rho\,d\text{\textcrossedV} = 0$$

at any instant of time for steady flows or on a time-average basis for cyclical unsteady
flows.

3. We work only with the component of Eq. 5.42 resolved along the axis of rotation.

*Change in moment
of fluid velocity
around an axis can
result in torque and
rotation around
that same axis.*

Consider the rotating sprinkler sketched in Fig. 5.4. Because the direction and magnitude of the flow
through the sprinkler from the inlet [section (1)] to the outlet [section (2)] of the arm changes, the
water exerts a torque on the sprinkler head causing it to tend to rotate or to actually rotate in the direction shown, much like a turbine rotor. In applying the moment-of-momentum equation (Eq. 5.42)
to this flow situation, we elect to use the fixed and nondeforming control volume shown in Fig. 5.4.
This disk-shaped control volume contains within its boundaries the spinning or stationary sprinkler
head and the portion of the water flowing through the sprinkler contained in the control volume at
an instant. The control surface cuts through the sprinkler head's solid material so that the shaft torque
that resists motion can be clearly identified. When the sprinkler is rotating, the flow field in the stationary control volume is cyclical and unsteady, but steady in the mean. We proceed to use the axial component of the moment-of-momentum equation (Eq. 5.42) to analyze this flow.

The integrand of the moment-of-momentum flow term in Eq. 5.42,

$$\int_{cs}(\mathbf{r}\times\mathbf{V})\rho\mathbf{V}\cdot\hat{\mathbf{n}}\,dA$$

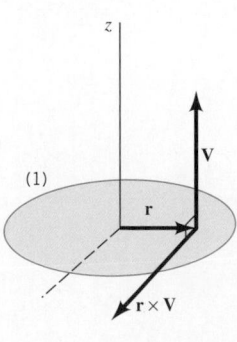

can be nonzero only where fluid is crossing the control surface. Everywhere else on the control
surface this term will be zero because $\mathbf{V}\cdot\hat{\mathbf{n}} = 0$. Water enters the control volume axially through
the hollow stem of the sprinkler at section (1). At this portion of the control surface, the component of $\mathbf{r}\times\mathbf{V}$ resolved along the axis of rotation is zero because as illustrated by the figure in the
margin, $\mathbf{r}\times\mathbf{V}$ lies in the plane of section (1), perpendicular to the axis of rotation. Thus, there is
no axial moment-of-momentum flow in at section (1). Water leaves the control volume through
each of the two nozzle openings at section (2). For the exiting flow, the magnitude of the axial
component of $\mathbf{r}\times\mathbf{V}$ is $r_2V_{\theta2}$, where r_2 is the radius from the axis of rotation to the nozzle centerline
and $V_{\theta2}$ is the value of the tangential component of the velocity of the flow exiting each nozzle as

observed from a frame of reference attached to the fixed and nondeforming control volume. The fluid velocity measured relative to a fixed control surface is an absolute velocity, $\mathbf{V}$. The velocity of the nozzle exit flow as viewed from the nozzle is called the relative velocity, $\mathbf{W}$. The absolute and relative velocities, $\mathbf{V}$ and $\mathbf{W}$, are related by the vector relationship

$$\mathbf{V} = \mathbf{W} + \mathbf{U} \tag{5.43}$$

where $\mathbf{U}$ is the velocity of the moving nozzle as measured relative to the fixed control surface.

The cross product and the dot product involved in the moment-of-momentum flow term of Eq. 5.42,

$$\int_{cs} (\mathbf{r} \times \mathbf{V})\rho \mathbf{V} \cdot \hat{\mathbf{n}}\, dA$$

can each result in a positive or negative value. For flow into the control volume, $\mathbf{V} \cdot \hat{\mathbf{n}}$ is negative. For flow out, $\mathbf{V} \cdot \hat{\mathbf{n}}$ is positive. The correct algebraic sign to assign the axis component of $\mathbf{r} \times \mathbf{V}$ can be ascertained by using the right-hand rule. The positive direction along the axis of rotation is the direction the thumb of the right hand points when it is extended and the remaining fingers are curled around the rotation axis in the positive direction of rotation as illustrated in Fig. 5.5. The direction of the axial component of $\mathbf{r} \times \mathbf{V}$ is similarly ascertained by noting the direction of the cross product of the radius from the axis of rotation, $r\hat{\mathbf{e}}_r$, and the tangential component of absolute velocity, $V_\theta\hat{\mathbf{e}}_\theta$. Thus, for the sprinkler of Fig. 5.4, we can state that

The algebraic sign of $\mathbf{r} \times \mathbf{V}$ is obtained by the right-hand rule.

$$\left[\int_{cs} (\mathbf{r} \times \mathbf{V})\rho \mathbf{V} \cdot \hat{\mathbf{n}}\, dA \right]_{axial} = (-r_2 V_{\theta 2})(+\dot{m}) \tag{5.44}$$

where, because of mass conservation, $\dot{m}$ is the total mass flowrate through both nozzles. As was demonstrated in Example 5.7, the mass flowrate is the same whether the sprinkler rotates or not. The correct algebraic sign of the axial component of $\mathbf{r} \times \mathbf{V}$ can be easily remembered in the following way: if $\mathbf{V}_\theta$ and $\mathbf{U}$ are in the same direction, use $+$; if $\mathbf{V}_\theta$ and $\mathbf{U}$ are in opposite directions, use $-$.

The torque term $[\sum (\mathbf{r} \times \mathbf{F})_{\text{contents of the control volume}}]$ of the moment-of-momentum equation (Eq. 5.42) is analyzed next. Confining ourselves to torques acting with respect to the axis of rotation only, we conclude that the shaft torque is important. The net torque with respect to the axis of rotation associated with normal forces exerted on the contents of the control volume will be very small if not zero. The net axial torque due to fluid tangential forces is also negligibly small for the control volume of Fig. 5.4. Thus, for the sprinkler of Fig. 5.4

$$\sum \left[(\mathbf{r} \times \mathbf{F})_{\substack{\text{contents of the} \\ \text{control volume}}} \right]_{axial} = \mathbf{T}_{shaft} \tag{5.45}$$

Note that we have entered T_{shaft} as a positive quantity in Eq. 5.45. This is equivalent to assuming that T_{shaft} is in the same direction as rotation.

For the sprinkler of Fig. 5.4, the axial component of the moment-of-momentum equation (Eq. 5.42) is, from Eqs. 5.44 and 5.45

$$-r_2 V_{\theta 2}\dot{m} = T_{shaft} \tag{5.46}$$

We interpret T_{shaft} being a negative quantity from Eq. 5.46 to mean that the shaft torque actually opposes the rotation of the sprinkler arms as shown in Fig. 5.4. The shaft torque, T_{shaft}, opposes rotation in all turbine devices.

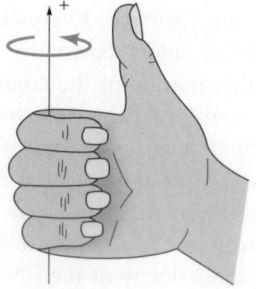

■ F I G U R E 5.5   Right-hand rule convention.

We could evaluate the *shaft power,* $\dot{W}_{shaft}$, associated with *shaft torque,* T_{shaft}, by forming the product of T_{shaft} and the rotational speed of the shaft, ω. [We use the notation that W = work, $(\,\cdot\,)$ = $d(\quad)/dt$, and thus $\dot{W}$ = power.] Thus, from Eq. 5.46 we get

$$\dot{W}_{shaft} = T_{shaft}\,\omega = -r_2 V_{\theta 2}\dot{m}\,\omega \tag{5.47}$$

Since $r_2\omega$ is the speed of each sprinkler nozzle, U, we can also state Eq. 5.47 in the form

$$\dot{W}_{shaft} = -U_2 V_{\theta 2}\dot{m} \tag{5.48}$$

Shaft work per unit mass, w_{shaft}, is equal to $\dot{W}_{shaft}/\dot{m}$. Dividing Eq. 5.48 by the mass flowrate, $\dot{m}$, we obtain

$$w_{shaft} = -U_2 V_{\theta 2} \tag{5.49}$$

Negative shaft work as in Eqs. 5.47, 5.48, and 5.49 is work out of the control volume, that is, work done by the fluid on the rotor and thus its shaft.

The principles associated with this sprinkler example can be extended to handle most simplified turbomachine flows. The fundamental technique is not difficult. However, the geometry of some turbomachine flows is quite complicated.

Example 5.18 further illustrates how the axial component of the moment-of-momentum equation (Eq. 5.46) can be used.

Power is equal to angular velocity times torque.

V5.11 Impulse-type lawn sprinkler

EXAMPLE 5.18 | Moment-of-Momentum—Torque

GIVEN   Water enters a rotating lawn sprinkler through its base at the steady rate of 1000 ml/s as sketched in Fig. E5.18a. The exit area of each of the two nozzles is 30 mm^2 and the flow leaving each nozzle is in the tangential direction. The radius from the axis of rotation to the centerline of each nozzle is 200 mm.

FIND   **(a)**   Determine the resisting torque required to hold the sprinkler head stationary.

(b)   Determine the resisting torque associated with the sprinkler rotating with a constant speed of 500 rev/min.

(c)   Determine the speed of the sprinkler if no resisting torque is applied.

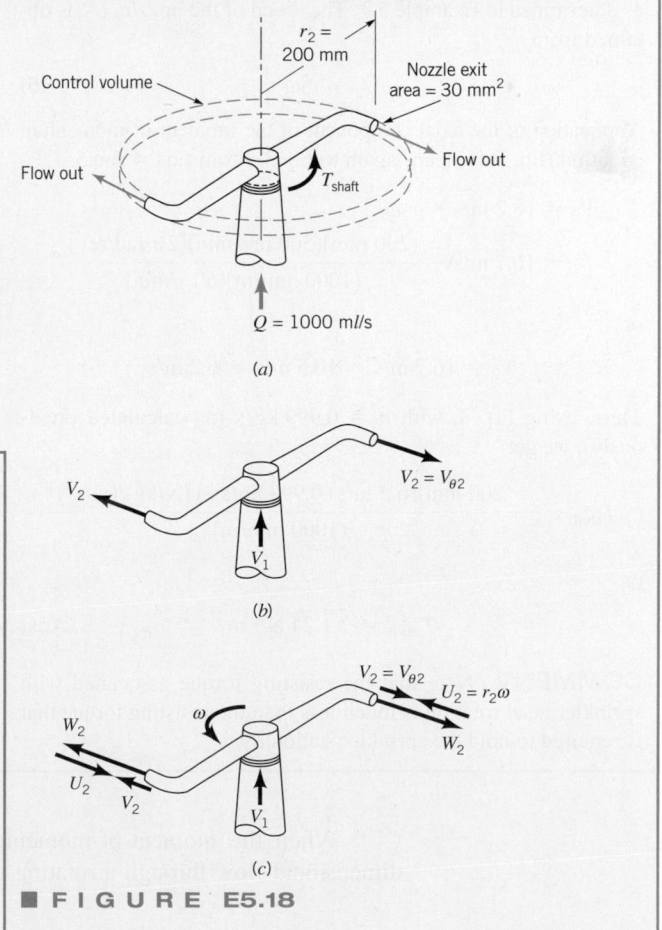

■ **F I G U R E   E5.18**

SOLUTION

To solve parts **(a)**, **(b)**, and **(c)** of this example we can use the same fixed and nondeforming, disk-shaped control volume illustrated in Fig. 5.4. As indicated in Fig. E5.18a, the only axial torque considered is the one resisting motion, T_{shaft}.

(a)   When the sprinkler head is held stationary as specified in part **(a)** of this example problem, the velocities of the fluid entering and leaving the control volume are shown in Fig. E5.18b. Equation 5.46 applies to the contents of this control volume. Thus,

$$T_{shaft} = -r_2 V_{\theta 2}\dot{m} \tag{1}$$

Since the control volume is fixed and nondeforming and the flow exiting from each nozzle is tangential,

$$V_{\theta 2} = V_2 \tag{2}$$

Equations 1 and 2 give

$$T_{shaft} = -r_2 V_2 \dot{m} \tag{3}$$

In Example 5.7, we ascertained that $V_2 = 16.7$ m/s. Thus, from Eq. 3 with

$$\dot{m} = Q\rho = \frac{(1000 \text{ ml/s})(10^{-3} \text{ m}^3/\text{liter})(999 \text{ kg/m}^3)}{(1000 \text{ ml/liter})}$$

$$= 0.999 \text{ kg/s}$$

we obtain

$$T_{\text{shaft}} = -\frac{(200 \text{ mm})(16.7 \text{ m/s})(0.999 \text{ kg/s})[1 (\text{N/kg})/(\text{m/s}^2)]}{(1000 \text{ mm/m})}$$

or

$$T_{\text{shaft}} = -3.34 \text{ N} \cdot \text{m} \qquad \text{(Ans)}$$

(b) When the sprinkler is rotating at a constant speed of 500 rpm, the flow field in the control volume is unsteady but cyclical. Thus, the flow field is steady in the mean. The velocities of the flow entering and leaving the control volume are as indicated in Fig. E5.18c. The absolute velocity of the fluid leaving each nozzle, V_2, is from Eq. 5.43,

$$V_2 = W_2 - U_2 \qquad (4)$$

where

$$W_2 = 16.7 \text{ m/s}$$

as determined in Example 5.7. The speed of the nozzle, U_2, is obtained from

$$U_2 = r_2\omega \qquad (5)$$

Application of the axial component of the moment-of-momentum equation (Eq. 5.46) leads again to Eq. 3. From Eqs. 4 and 5,

$$V_2 = 16.7 \text{ m/s} - r_2\omega$$

$$= 16.7 \text{ m/s} - \frac{(200 \text{ mm})(500 \text{ rev/min})(2\pi \text{ rad/rev})}{(1000 \text{ mm/m})(60 \text{ s/min})}$$

or

$$V_2 = 16.7 \text{ m/s} - 10.5 \text{ m/s} = 6.2 \text{ m/s}$$

Thus, using Eq. 3, with $\dot{m} = 0.999$ kg/s (as calculated previously), we get

$$T_{\text{shaft}} = -\frac{(200 \text{ mm})(6.2 \text{ m/s}) \, 0.999 \text{ kg/s} \, [1 (\text{N/kg})/(\text{m/s}^2)]}{(1000 \text{ mm/m})}$$

or

$$T_{\text{shaft}} = -1.24 \text{ N} \cdot \text{m} \qquad \text{(Ans)}$$

COMMENT Note that the resisting torque associated with sprinkler head rotation is much less than the resisting torque that is required to hold the sprinkler stationary.

(c) When no resisting torque is applied to the rotating sprinkler head, a maximum constant speed of rotation will occur as demonstrated below. Application of Eqs. 3, 4, and 5 to the contents of the control volume results in

$$T_{\text{shaft}} = -r_2(W_2 - r_2\omega)\dot{m} \qquad (6)$$

For no resisting torque, Eq. 6 yields

$$0 = -r_2(W_2 - r_2\omega)\dot{m}$$

Thus,

$$\omega = \frac{W_2}{r_2} \qquad (7)$$

In Example 5.4, we learned that the relative velocity of the fluid leaving each nozzle, W_2, is the same regardless of the speed of rotation of the sprinkler head, ω, as long as the mass flowrate of the fluid, $\dot{m}$, remains constant. Thus, by using Eq. 7 we obtain

$$\omega = \frac{W_2}{r_2} = \frac{(16.7 \text{ m/s})(1000 \text{ mm/m})}{(200 \text{ mm})} = 83.5 \text{ rad/s}$$

or

$$\omega = \frac{(83.5 \text{ rad/s})(60 \text{ s/min})}{2\pi \text{ rad/rev}} = 797 \text{ rpm} \qquad \text{(Ans)}$$

For this condition ($T_{\text{shaft}} = 0$), the water both enters and leaves the control volume with zero angular momentum.

COMMENT Note that forcing a change in direction of a flowing fluid, in this case with a sprinkler, resulted in rotary motion and a useful "sprinkling" of water over an area.

By repeating the calculations for various values of the angular velocity, ω, the results shown in Fig. E5.18d are obtained. It is seen that the magnitude of the resisting torque associated with rotation is less than the torque required to hold the rotor stationary. Even in the absence of a resisting torque, the rotor maximum speed is finite.

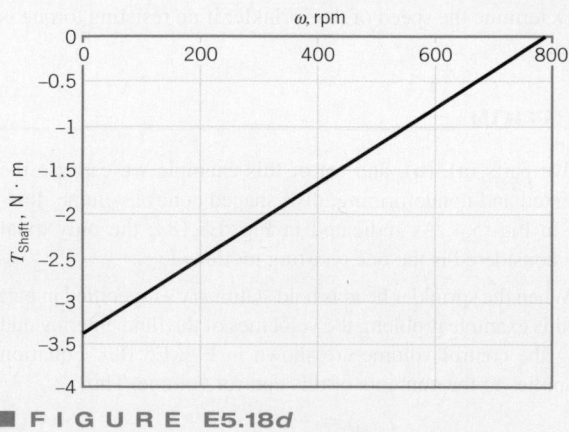

■ **F I G U R E   E5.18d**

When the moment-of-momentum equation (Eq. 5.42) is applied to a more general, one-dimensional flow through a rotating machine, we obtain

$$T_{\text{shaft}} = (-\dot{m}_{\text{in}})(\pm r_{\text{in}}V_{\theta\text{in}}) + \dot{m}_{\text{out}}(\pm r_{\text{out}}V_{\theta\text{out}}) \qquad (5.50)$$

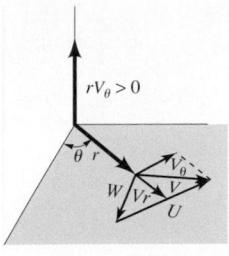

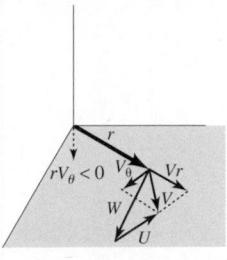

by applying the same kind of analysis used with the sprinkler of Fig. 5.4. The "−" is used with mass flowrate into the control volume, $\dot{m}_{in}$, and the "+" is used with mass flowrate out of the control volume, $\dot{m}_{out}$, to account for the sign of the dot product, $\mathbf{V} \cdot \hat{\mathbf{n}}$, involved. Whether "+" or "−" is used with the rV_θ product depends on the direction of $(\mathbf{r} \times \mathbf{V})_{axial}$. A simple way to determine the sign of the rV_θ product is to compare the direction of V_θ and the blade speed, U. As shown in the margin, if V_θ and U are in the same direction, then the rV_θ product is positive. If V_θ and U are in opposite directions, the rV_θ product is negative. The sign of the shaft torque is "+" if T_{shaft} is in the same direction along the axis of rotation as ω, and "−" otherwise.

The shaft power, $\dot{W}_{shaft}$, is related to shaft torque, T_{shaft}, by

$$\dot{W}_{shaft} = T_{shaft}\,\omega \tag{5.51}$$

Thus, using Eqs. 5.50 and 5.51 with a "+" sign for T_{shaft} in Eq. 5.50, we obtain

$$\dot{W}_{shaft} = (-\dot{m}_{in})(\pm r_{in}\omega V_{\theta in}) + \dot{m}_{out}(\pm r_{out}\omega V_{\theta out}) \tag{5.52}$$

or since $r\omega = U$

$$\boxed{\dot{W}_{shaft} = (-\dot{m}_{in})(\pm U_{in}V_{\theta in}) + \dot{m}_{out}(\pm U_{out}V_{\theta out})} \tag{5.53}$$

The "+" is used for the UV_θ product when U and V_θ are in the same direction; the "−" is used when U and V_θ are in opposite directions. Also, since $+T_{shaft}$ was used to obtain Eq. 5.53, when $\dot{W}_{shaft}$ is positive, power is into the fluid (for example, a pump), and when $\dot{W}_{shaft}$ is negative, power is out of the fluid (for example, a turbine).

When shaft torque and shaft rotation are in the same (opposite) direction, power is into (out of) the fluid.

The shaft work per unit mass, w_{shaft}, can be obtained from the shaft power, $\dot{W}_{shaft}$, by dividing Eq. 5.53 by the mass flowrate, $\dot{m}$. By conservation of mass,

$$\dot{m} = \dot{m}_{in} = \dot{m}_{out}$$

From Eq. 5.53, we obtain

$$\boxed{w_{shaft} = -(\pm U_{in}V_{\theta in}) + (\pm U_{out}V_{\theta out})} \tag{5.54}$$

The application of Eqs. 5.50, 5.53, and 5.54 is demonstrated in Example 5.19. More examples of the application of Eqs. 5.50, 5.53, and 5.54 are included in Chapter 12.

EXAMPLE 5.19 ▐ Moment-of-Momentum—Power

GIVEN An air fan has a bladed rotor of 30 cm outside diameter and 25 cm inside diameter as illustrated in Fig. E5.19a. The height of each rotor blade is constant at 2.5 cm from blade inlet to outlet. The flowrate is steady, on a time-average basis, at 6.5 m³/min and the absolute velocity of the air at blade inlet, $\mathbf{V}_1$, is radial. The blade discharge angle is 30° from the tangential direction. The rotor rotates at a constant speed of 1725 rpm.

FIND Estimate the power required to run the fan.

SOLUTION

We select a fixed and nondeforming control volume that includes the rotating blades and the fluid within the blade row at an instant, as shown with a dashed line in Fig. E5.19a. The flow within this control volume is cyclical, but steady in the mean. The only torque we consider is the driving shaft torque, T_{shaft}. This torque is provided by a motor. We assume that the entering and leaving flows are each represented by uniformly distributed velocities and flow properties. Since shaft power is sought, Eq. 5.53 is appropriate. Application of Eq. 5.53 to the contents of the control volume in Fig. E5.19 gives

$$\dot{W}_{shaft} = -\dot{m}_1(\pm U_1 \cancel{V_{\theta 1}})^{\,0\ (\mathbf{V}_1\ \text{is radial})} + \dot{m}_2(\pm U_2 V_{\theta 2}) \tag{1}$$

From Eq. 1 we see that to calculate fan power, we need mass flowrate, $\dot{m}$, rotor exit blade velocity, U_2, and fluid tangential velocity at blade exit, $V_{\theta 2}$. The mass flowrate, $\dot{m}$, is easily obtained from Eq. 5.6 as

$$\dot{m} = \rho Q = \frac{(1.2\ \text{kg/m}^3)(6.5\ \text{m}^3/\text{min})}{(60\ \text{s/min})}$$

$$= 0.13\ \text{kg/s} \tag{2}$$

The rotor exit blade speed, U_2, is

$$U_2 = r_2\omega = \frac{(15\ \text{cm})(1725\ \text{rpm})(2\pi\ \text{rad/rev})}{(100\ \text{cm/m})(60\ \text{s/min})}$$

$$= 27.1\ \text{m/s} \tag{3}$$

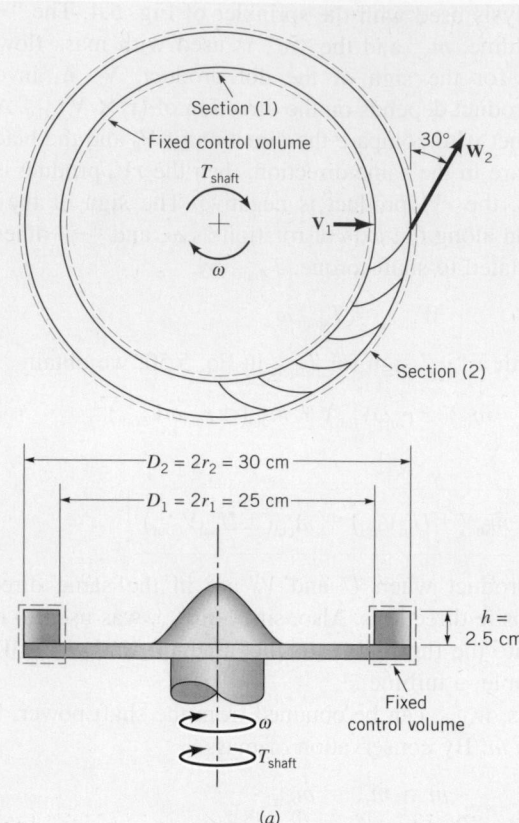

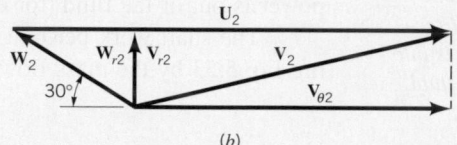

■ **FIGURE E5.19**

To determine the fluid tangential speed at the fan rotor exit, $V_{\theta 2}$, we use Eq. 5.43 to get

$$\mathbf{V_2 = W_2 + U_2} \tag{4}$$

The vector addition of Eq. 4 is shown in the form of a "velocity triangle" in Fig. E5.19b. From Fig. E5.19b, we can see that

$$V_{\theta 2} = U_2 - W_2 \cos 30° \tag{5}$$

To solve Eq. 5 for $V_{\theta 2}$ we need a value of W_2, in addition to the value of U_2 already determined (Eq. 3). To get W_2, we recognize that

$$W_2 \sin 30° = V_{r2} \tag{6}$$

where V_{r2} is the radial component of either $\mathbf{W_2}$ or $\mathbf{V_2}$. Also, using Eq. 5.6, we obtain

$$\dot{m} = \rho A_2 V_{r2} \tag{7}$$

or since

$$A_2 = 2\,\pi r_2 h \tag{8}$$

where h is the blade height, Eqs. 7 and 8 combine to form

$$\dot{m} = \rho 2\pi r_2 h V_{r2} \tag{9}$$

Taking Eqs. 6 and 9 together we get

$$W_2 = \frac{\dot{m}}{\rho 2\pi r_2 h \sin 30°} = \frac{\rho Q}{\rho 2\pi r_2 h \sin 30°} \tag{10}$$

$$= \frac{Q}{2\pi r_2 h \sin 30°}$$

Substituting known values into Eq. 10, we obtain

$$W_2 = \frac{(6.5 \text{ m}^3/\text{min})(100 \text{ cm/m})(100 \text{ cm/m})}{(60 \text{ s/min})2\pi(15 \text{ cm})(2.5 \text{ cm})\sin 30°}$$

$$= 9.2 \text{ m/s}$$

By using this value of W_2 in Eq. 5 we get

$$V_{\theta 2} = U_2 - W_2 \cos 30°$$

$$= 27.1 \text{ m/s} - (9.2 \text{ m/s})(0.866) = 19.13 \text{ m/s}$$

Equation 1 can now be used to obtain

$$\dot{W}_{\text{shaft}} = \dot{m}\, U_2 V_{\theta 2} = \frac{(0.13 \text{ kg/s})(27.1 \text{ m/s})(19.13 \text{ m/s})}{[1(\text{kg} \cdot \text{m/s}^2)/\text{N}]\,[1(\text{N} \cdot \text{m})/(\text{W} \cdot \text{s})]}$$

$$\dot{W}_{\text{shaft}} = 0.067 \text{ kW} \qquad \textbf{(Ans)}$$

COMMENT Note that the "+" was used with the $U_2 V_{\theta 2}$ product because U_2 and $V_{\theta 2}$ are in the same direction. This result, 0.067 kW, is the power that needs to be delivered through the fan shaft for the given conditions. Ideally, all of this power would go into the flowing air. However, because of fluid friction, only some of this power will produce useful effects (e.g., movement and pressure rise) on the air. How much useful effect depends on the efficiency of the energy transfer between the fan blades and the fluid.

5.3   First Law of Thermodynamics—The Energy Equation

5.3.1 Derivation of the Energy Equation

The *first law of thermodynamics* for a system is, in words

| time rate of
increase of the
total stored energy
of the system | = | net time rate of
energy addition by
heat transfer into
the system | + | net time rate of
energy addition by
work transfer into
the system |

In symbolic form, this statement is

$$\frac{D}{Dt} \int_{\text{sys}} e\rho \, d\Psi = \left(\sum \dot{Q}_{\text{in}} - \sum \dot{Q}_{\text{out}} \right)_{\text{sys}} + \left(\sum \dot{W}_{\text{in}} - \sum \dot{W}_{\text{out}} \right)_{\text{sys}}$$

The first law of thermodynamics is a statement of conservation of energy.

or

$$\frac{D}{Dt} \int_{\text{sys}} e\rho \, d\Psi = (\dot{Q}_{\underset{\text{in}}{\text{net}}} + \dot{W}_{\underset{\text{in}}{\text{net}}})_{\text{sys}} \tag{5.55}$$

Some of these variables deserve a brief explanation before proceeding further. The total stored energy per unit mass for each particle in the system, e, is related to the internal energy per unit mass, $\check{u}$, the kinetic energy per unit mass, $V^2/2$, and the potential energy per unit mass, gz, by the equation

$$e = \check{u} + \frac{V^2}{2} + gz \tag{5.56}$$

The net *rate of heat transfer* into the system is denoted with $\dot{Q}_{\text{net in}}$, and the net rate of work transfer into the system is labeled $\dot{W}_{\text{net in}}$. Heat transfer and work transfer are considered "+" going into the system and "−" coming out.

Equation 5.55 is valid for inertial and noninertial reference systems. We proceed to develop the control volume statement of the first law of thermodynamics. For the control volume that is coincident with the system at an instant of time

$$(\dot{Q}_{\underset{\text{in}}{\text{net}}} + \dot{W}_{\underset{\text{in}}{\text{net}}})_{\text{sys}} = (\dot{Q}_{\underset{\text{in}}{\text{net}}} + \dot{W}_{\underset{\text{in control volume}}{\text{net}}})_{\text{coincident}} \tag{5.57}$$

Furthermore, for the system and the contents of the coincident control volume that is fixed and nondeforming, the Reynolds transport theorem (Eq. 4.19 with the parameter b set equal to e) allows us to conclude that

$$\frac{D}{Dt} \int_{\text{sys}} e\rho \, d\Psi = \frac{\partial}{\partial t} \int_{\text{cv}} e\rho \, d\Psi + \int_{\text{cs}} e\rho \mathbf{V} \cdot \hat{\mathbf{n}} \, dA \tag{5.58}$$

or in words,

| the time rate
of increase
of the total
stored energy
of the system | = | the time rate of in-
crease of the total stored
energy of the contents
of the control volume | + | the net rate of flow
of the total stored energy
out of the control
volume through the
control surface |

Combining Eqs. 5.55, 5.57, and 5.58 we get the control volume formula for the first law of thermodynamics:

The energy equation involves stored energy and heat and work transfer.

$$\boxed{\frac{\partial}{\partial t} \int_{\text{cv}} e\rho \, d\Psi + \int_{\text{cs}} e\rho \mathbf{V} \cdot \hat{\mathbf{n}} \, dA = (\dot{Q}_{\underset{\text{in}}{\text{net}}} + \dot{W}_{\underset{\text{in}}{\text{net}}})_{\text{cv}}} \tag{5.59}$$

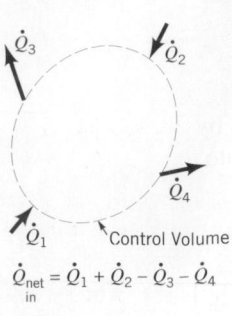

$\dot{Q}_{\text{net}} = \dot{Q}_1 + \dot{Q}_2 - \dot{Q}_3 - \dot{Q}_4$
in

The total stored energy per unit mass, e, in Eq. 5.59 is for fluid particles entering, leaving, and within the control volume. Further explanation of the heat transfer and work transfer involved in this equation follows.

The heat transfer rate, $\dot{Q}$, represents all of the ways in which energy is exchanged between the control volume contents and surroundings because of a temperature difference. Thus, radiation, conduction, and/or convection are possible. As shown by the figure in the margin, heat transfer into the control volume is considered positive, heat transfer out is negative. In many engineering applications, the process is *adiabatic*; the heat transfer rate, $\dot{Q}$, is zero. The net heat transfer rate, $\dot{Q}_{\text{net in}}$, can also be zero when $\sum \dot{Q}_{\text{in}} - \sum \dot{Q}_{\text{out}} = 0$.

The work transfer rate, $\dot{W}$, also called *power*, is positive when work is done on the contents of the control volume by the surroundings. Otherwise, it is considered negative. Work can be transferred across the control surface in several ways. In the following paragraphs, we consider some important forms of work transfer.

In many instances, work is transferred across the control surface by a moving shaft. In rotary devices such as turbines, fans, and propellers, a rotating shaft transfers work across that portion of the control surface that slices through the shaft. Even in reciprocating machines like positive displacement internal combustion engines and compressors that utilize piston-in-cylinder arrangements, a rotating crankshaft is used. Since work is the dot product of force and related displacement, rate of work (or power) is the dot product of force and related displacement per unit time. For a rotating shaft, the power transfer, $\dot{W}_{\text{shaft}}$, is related to the shaft torque that causes the rotation, T_{shaft}, and the angular velocity of the shaft, ω, by the relationship

$$\dot{W}_{\text{shaft}} = T_{\text{shaft}}\omega$$

When the control surface cuts through the shaft material, the shaft torque is exerted by shaft material at the control surface. To allow for consideration of problems involving more than one shaft we use the notation

$$\dot{W}_{\text{shaft}}_{\text{net in}} = \sum_{\text{in}} \dot{W}_{\text{shaft}} - \sum_{\text{out}} \dot{W}_{\text{shaft}} \tag{5.60}$$

Work transfer can also occur at the control surface when a force associated with fluid normal stress acts over a distance. Consider the simple pipe flow illustrated in Fig. 5.6 and the control volume shown. For this situation, the fluid normal stress, σ, is simply equal to the negative of fluid pressure, p, in all directions; that is,

$$\sigma = -p \tag{5.61}$$

This relationship can be used with varying amounts of approximation for many engineering problems (see Chapter 6).

The power transfer, $\dot{W}$, associated with a force **F** acting on an object moving with velocity **V** is given by the dot product $\mathbf{F} \cdot \mathbf{V}$. This is illustrated by the figure in the margin. Hence, the power transfer associated with normal stresses acting on a single fluid particle, $\delta\dot{W}_{\text{normal stress}}$, can be evaluated as the dot product of the normal stress force, $\delta\mathbf{F}_{\text{normal stress}}$, and the fluid particle velocity, **V**, as

$$\delta\dot{W}_{\text{normal stress}} = \delta\mathbf{F}_{\text{normal stress}} \cdot \mathbf{V}$$

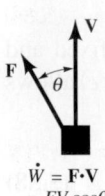

$\dot{W} = \mathbf{F} \cdot \mathbf{V}$
$= FV\cos\theta$

If the normal stress force is expressed as the product of local normal stress, $\sigma = -p$, and fluid particle surface area, $\hat{\mathbf{n}}\,\delta A$, the result is

$$\delta\dot{W}_{\text{normal stress}} = \sigma\hat{\mathbf{n}}\,\delta A \cdot \mathbf{V} = -p\hat{\mathbf{n}}\,\delta A \cdot \mathbf{V} = -p\mathbf{V} \cdot \hat{\mathbf{n}}\,\delta A$$

For all fluid particles on the control surface of Fig. 5.6 at the instant considered, power transfer due to fluid normal stress, $\dot{W}_{\text{normal stress}}$, is

$$\dot{W}_{\text{normal}}_{\text{stress}} = \int_{\text{cs}} \sigma\mathbf{V} \cdot \hat{\mathbf{n}}\,dA = \int_{\text{cs}} -p\mathbf{V} \cdot \hat{\mathbf{n}}\,dA \tag{5.62}$$

Note that the value of $\dot{W}_{\text{normal stress}}$ for particles on the wetted inside surface of the pipe is zero because $\mathbf{V} \cdot \hat{\mathbf{n}}$ is zero there. Thus, $\dot{W}_{\text{normal stress}}$ can be nonzero only where fluid enters and leaves the

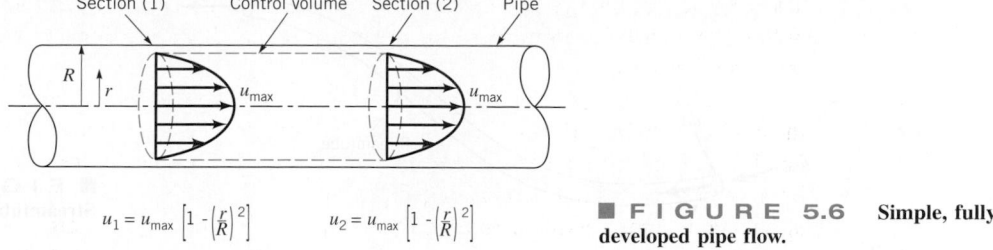

$$u_1 = u_{max}\left[1 - \left(\frac{r}{R}\right)^2\right] \qquad u_2 = u_{max}\left[1 - \left(\frac{r}{R}\right)^2\right]$$

■ **F I G U R E 5.6   Simple, fully developed pipe flow.**

control volume. Although only a simple pipe flow was considered, Eq. 5.62 is quite general and the control volume used in this example can serve as a general model for other cases.

Work transfer can also occur at the control surface because of tangential stress forces. Rotating shaft work is transferred by tangential stresses in the shaft material. For a fluid particle, shear stress force power, $\delta\dot{W}_{\text{tangential stress}}$, can be evaluated as the dot product of tangential stress force, $\delta\mathbf{F}_{\text{tangential stress}}$, and the fluid particle velocity, $\mathbf{V}$. That is,

$$\delta\dot{W}_{\text{tangential stress}} = \delta\mathbf{F}_{\text{tangential stress}} \cdot \mathbf{V}$$

Work is transferred by rotating shafts, normal stresses, and tangential stresses.

For the control volume of Fig. 5.6, the fluid particle velocity is zero everywhere on the wetted inside surface of the pipe. Thus, no tangential stress work is transferred across that portion of the control surface. Furthermore, if we select the control surface so that it is perpendicular to the fluid particle velocity, then the tangential stress force is also perpendicular to the velocity. Therefore, the tangential stress work transfer is zero on that part of the control surface. This is illustrated in the figure in the margin. Thus, in general, we select control volumes like the one of Fig. 5.6 and consider fluid tangential stress power transfer to be negligibly small.

Using the information we have developed about power, we can express the first law of thermodynamics for the contents of a control volume by combining Eqs. 5.59, 5.60, and 5.62 to obtain

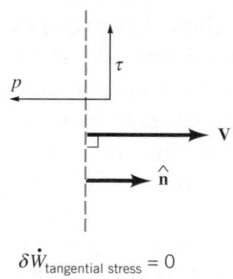

$$\frac{\partial}{\partial t}\int_{cv} e\rho\, d\forall + \int_{cs} e\rho\mathbf{V} \cdot \hat{\mathbf{n}}\, dA = \dot{Q}_{\substack{net \\ in}} + \dot{W}_{\substack{shaft \\ net\ in}} - \int_{cs} p\mathbf{V} \cdot \hat{\mathbf{n}}\, dA \qquad \textbf{(5.63)}$$

$\delta\dot{W}_{\text{tangential stress}} = 0$

When the equation for total stored energy (Eq. 5.56) is considered with Eq. 5.63, we obtain the *energy equation*:

$$\frac{\partial}{\partial t}\int_{cv} e\rho\, d\forall + \int_{cs}\left(\breve{u} + \frac{p}{\rho} + \frac{V^2}{2} + gz\right)\rho\mathbf{V} \cdot \hat{\mathbf{n}}\, dA = \dot{Q}_{\substack{net \\ in}} + \dot{W}_{\substack{shaft \\ net\ in}} \qquad \textbf{(5.64)}$$

5.3.2 Application of the Energy Equation

In Eq. 5.64, the term $\partial/\partial t\int_{cv} e\rho\, d\forall$ represents the time rate of change of the total stored energy, e, of the contents of the control volume. This term is zero when the flow is steady. This term is also zero in the mean when the flow is steady in the mean (cyclical).

In Eq. 5.64, the integrand of

$$\int_{cs}\left(\breve{u} + \frac{p}{\rho} + \frac{V^2}{2} + gz\right)\rho\mathbf{V} \cdot \hat{\mathbf{n}}\, dA$$

can be nonzero only where fluid crosses the control surface ($\mathbf{V} \cdot \hat{\mathbf{n}} \neq 0$). Otherwise, $\mathbf{V} \cdot \hat{\mathbf{n}}$ is zero and the integrand is zero for that portion of the control surface. If the properties within parentheses, $\breve{u}$, p/ρ, $V^2/2$, and gz, are all assumed to be uniformly distributed over the flow cross-sectional areas involved, the integration becomes simple and gives

$$\int_{cs}\left(\breve{u} + \frac{p}{\rho} + \frac{V^2}{2} + gz\right)\rho\mathbf{V} \cdot \hat{\mathbf{n}}\, dA = \sum_{\substack{flow \\ out}}\left(\breve{u} + \frac{p}{\rho} + \frac{V^2}{2} + gz\right)\dot{m}$$

$$- \sum_{\substack{flow \\ in}}\left(\breve{u} + \frac{p}{\rho} + \frac{V^2}{2} + gz\right)\dot{m} \qquad \textbf{(5.65)}$$

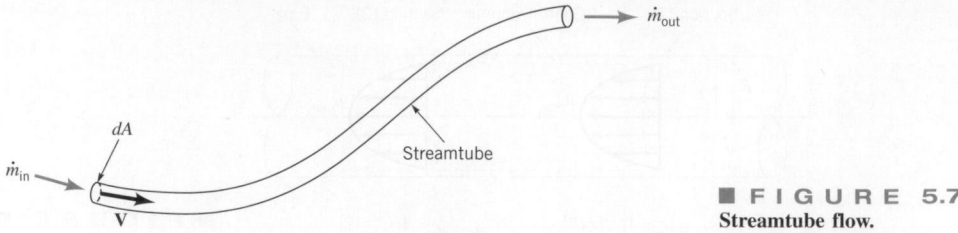

■ **FIGURE 5.7**
Streamtube flow.

Furthermore, if there is only one stream entering and leaving the control volume, then

$$\int_{cs} \left(\check{u} + \frac{p}{\rho} + \frac{V^2}{2} + gz \right) \rho \mathbf{V} \cdot \hat{\mathbf{n}} \, dA =$$

$$\left(\check{u} + \frac{p}{\rho} + \frac{V^2}{2} + gz \right)_{out} \dot{m}_{out} - \left(\check{u} + \frac{p}{\rho} + \frac{V^2}{2} + gz \right)_{in} \dot{m}_{in} \qquad \textbf{(5.66)}$$

Uniform flow as described above will occur in an infinitesimally small diameter streamtube as illustrated in Fig. 5.7. This kind of streamtube flow is representative of the steady flow of a particle of fluid along a pathline. We can also idealize actual conditions by disregarding nonuniformities in a finite cross section of flow. We call this one-dimensional flow and although such uniform flow rarely occurs in reality, the simplicity achieved with the one-dimensional approximation often justifies its use. More details about the effects of nonuniform distributions of velocities and other fluid flow variables are considered in Section 5.3.4 and in Chapters 8, 9, and 10.

If shaft work is involved, the flow must be unsteady, at least locally (see Refs. 1 and 2). The flow in any fluid machine that involves shaft work is unsteady within that machine. For example, the velocity and pressure at a fixed location near the rotating blades of a fan are unsteady. However, upstream and downstream of the machine, the flow may be steady. Most often shaft work is associated with flow that is unsteady in a recurring or cyclical way. On a time-average basis for flow that is one-dimensional, cyclical, and involves only one stream of fluid entering and leaving the control volume, Eq. 5.64 can be simplified with the help of **Eqs. 5.9** and 5.66 to form

$$\dot{m} \left[\check{u}_{out} - \check{u}_{in} + \left(\frac{p}{\rho} \right)_{out} - \left(\frac{p}{\rho} \right)_{in} + \frac{V_{out}^2 - V_{in}^2}{2} + g(z_{out} - z_{in}) \right] = \dot{Q}_{net \atop in} + \dot{W}_{shaft \atop net \, in} \qquad \textbf{(5.67)}$$

We call Eq. 5.67 the *one-dimensional energy equation for steady-in-the-mean flow*. Note that Eq. 5.67 is valid for incompressible and compressible flows. Often, the fluid property called *enthalpy*, $\check{h}$, where

$$\check{h} = \check{u} + \frac{p}{\rho} \qquad \textbf{(5.68)}$$

The energy equation is sometimes written in terms of enthalpy.

is used in Eq. 5.67. With enthalpy, the one-dimensional energy equation for steady-in-the-mean flow (Eq. 5.67) is

$$\dot{m} \left[\check{h}_{out} - \check{h}_{in} + \frac{V_{out}^2 - V_{in}^2}{2} + g(z_{out} - z_{in}) \right] = \dot{Q}_{net \atop in} + \dot{W}_{shaft \atop net \, in} \qquad \textbf{(5.69)}$$

Equation 5.69 is often used for solving compressible flow problems. Examples 5.20 and 5.21 illustrate how Eqs. 5.67 and 5.69 can be used.

EXAMPLE 5.20 Energy—Pump Power

GIVEN A pump delivers water at a steady rate of 1136 L/min as shown in Fig. E5.20. Just upstream of the pump [section (1)] where the pipe diameter is 9 cm, the pressure is 124 kPa. Just downstream of the pump [section (2)] where the pipe diameter is 2.5 cm, the pressure is 414 kPa. The change in water elevation across the pump is zero. The rise in internal energy of water, $\check{u}_2 - \check{u}_1$, associated with a temperature rise across the pump is 278 N · m/kg. The pumping process is considered to be adiabatic.

FIND Determine the power (kW) required by the pump.

SOLUTION

We include in our control volume the water contained in the pump between its entrance and exit sections. Application of Eq. 5.67 to the contents of this control volume on a time-average basis yields

0 (no elevation change)

$$\dot{m}\left[\breve{u}_2 - \breve{u}_1 + \left(\frac{p}{\rho}\right)_2 - \left(\frac{p}{\rho}\right)_1 + \frac{V_2^2 - V_1^2}{2} + g(z_2 - z_1)\right]$$

0 (adiabatic flow)

$$= \dot{Q}_{\underset{in}{net}} + \dot{W}_{\underset{net\ in}{shaft}} \tag{1}$$

We can solve directly for the power required by the pump, $\dot{W}_{shaft\ net\ in}$, from Eq. 1, after we first determine the mass flowrate, $\dot{m}$, the speed of flow into the pump, V_1, and the speed of the flow out of the pump, V_2. All other quantities in Eq. 1 are given in the problem statement. From Eq. 5.6, we get

$$\dot{m} = \rho Q = \frac{(1000\ kg/m^3)(1136\ L/min)}{(1000\ L/m^3)(60\ s/min)}$$

$$= 19\ kg/s \tag{2}$$

Also from Eq. 5.6,

$$V = \frac{Q}{A} = \frac{Q}{\pi D^2/4}$$

so

$$V_1 = \frac{Q}{A_1} = \frac{(1136\ L/min)4(100\ cm/m)^2}{(1000\ L/m^3)(60\ s/min)\pi(9\ cm)^2}$$

$$= 2.98\ m/s \tag{3}$$

and

$$V_2 = \frac{Q}{A_2} = \frac{(1136\ L/min)4\ (100\ cm/m)^2}{(1000\ L/m^3)(60\ s/min)\pi(2.5\ cm)^2}$$

$$= 38.57\ m/s \tag{4}$$

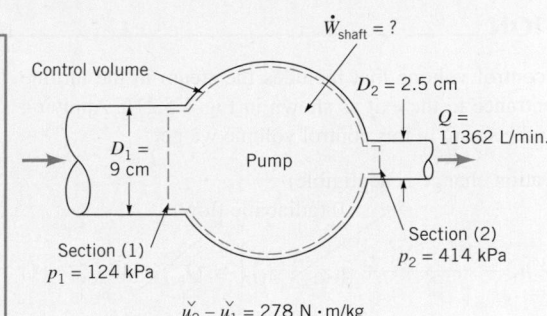

$\dot{W}_{shaft} = ?$

Control volume

$D_2 = 2.5$ cm

$D_1 = 9$ cm

Pump

$Q = 11362$ L/min.

Section (1)
$p_1 = 124$ kPa

Section (2)
$p_2 = 414$ kPa

$\breve{u}_2 - \breve{u}_1 = 278\ N \cdot m/kg$

■ **F I G U R E   E5.20**

Substituting the values of Eqs. 2, 3, and 4 and values from the problem statement into Eq. 1 we obtain

$$\dot{W}_{\underset{net\ in}{shaft}} = (19\ kg/s)\left[(278\ N \cdot m/kg)\right.$$

$$+ \frac{(414\ kPa)}{(1000\ kg/m^3)}$$

$$- \frac{(124\ kPa)}{(1000\ kg/m^3)}$$

$$\left.+ \frac{(38.57\ m/s)^2 - (2.98\ m/s)^2}{2[1\ kg/m/N \cdot s^2]}\right]$$

$$\times \frac{1}{[1\ N \cdot m/W \cdot s]} = 24.8\ kW \tag{Ans}$$

COMMENT   Of the total 24.8 kW, internal energy change accounts for 5.3 kW, the pressure rise accounts for 5.5 kW, and the kinetic energy increase accounts for 14.0 kW.

E XAMPLE 5.21 | Energy—Turbine Power per Unit Mass of Flow

GIVEN   A steam turbine generator unit used to produce electricity is shown in Fig. E5.21a. Assume the steam enters a turbine with a velocity of 30 m/s and enthalpy, h_1, of 3348 kJ/kg (see Fig. E5.21b). The steam leaves the turbine as a mixture of vapor and liquid having a velocity of 60 m/s and an enthalpy of 2550 kJ/kg. The flow through the turbine is adiabatic, and changes in elevation are negligible.

FIND   Determine the work output involved per unit mass of steam through-flow.

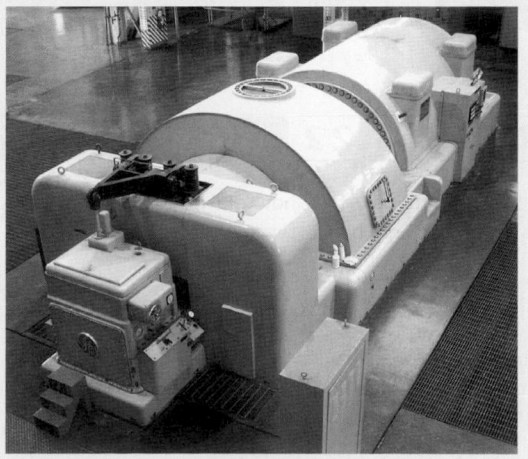

■ **F I G U R E   E5.21a**

SOLUTION

We use a control volume that includes the steam in the turbine from the entrance to the exit as shown in Fig. E5.21*b*. Applying Eq. 5.69 to the steam in this control volume we get

0 (elevation change is negligible)

0 (adiabatic flow)

$$\dot{m}\left[\check{h}_2 - \check{h}_1 + \frac{V_2^2 - V_1^2}{2} + g(z_2 - z_1)\right] = \dot{Q}_{\substack{net \\ in}} + \dot{W}_{\substack{shaft \\ net \ in}} \quad (1)$$

The work output per unit mass of steam through-flow, $w_{shaft \ net \ in}$, can be obtained by dividing Eq. 1 by the mass flow rate, $\dot{m}$, to obtain

$$w_{\substack{shaft \\ net \ in}} = \frac{\dot{W}_{\substack{shaft \\ net \ in}}}{\dot{m}} = \check{h}_2 - \check{h}_1 + \frac{V_2^2 - V_1^2}{2} \quad (2)$$

Since $w_{shaft \ net \ out} = -w_{shaft \ net \ in}$, we obtain

$$w_{\substack{shaft \\ net \ out}} = \check{h}_1 - \check{h}_2 + \frac{V_1^2 - V_2^2}{2}$$

or

$$w_{\substack{shaft \\ net \ out}} = 3348 \ kJ/kg - 2550 \ kJ/kg$$

$$+ \frac{[(30 \ m/s)^2 - (60 \ m/s)^2][1 \ J/(N \cdot m)]}{2[1 \ (kg \cdot m)/(N \cdot s^2)](1000 \ J/kJ)}$$

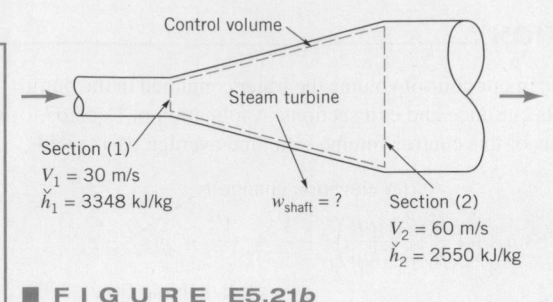

Control volume

Steam turbine

Section (1)
$V_1 = 30 \ m/s$
$\check{h}_1 = 3348 \ kJ/kg$

$w_{shaft} = ?$

Section (2)
$V_2 = 60 \ m/s$
$\check{h}_2 = 2550 \ kJ/kg$

■ **FIGURE E5.21*b***

Thus,

$$w_{\substack{shaft \\ net \ out}} = 3348 \ kJ/kg - 2550 \ kJ/kg - 1.35 \ kJ/kg$$

$$= 797 \ kJ/kg \qquad \text{(Ans)}$$

COMMENT   Note that in this particular example, the change in kinetic energy is small in comparison to the difference in enthalpy involved. This is often true in applications involving steam turbines. To determine the power output, $\dot{W}_{shaft}$, we must know the mass flowrate, $\dot{m}$.

V5.12 Pelton wheel turbine

If the flow is steady throughout, one-dimensional, and only one fluid stream is involved, then the shaft work is zero and the energy equation is

$$\dot{m}\left[\check{u}_{out} - \check{u}_{in} + \left(\frac{p}{\rho}\right)_{out} - \left(\frac{p}{\rho}\right)_{in} + \frac{V_{out}^2 - V_{in}^2}{2} + g(z_{out} - z_{in})\right] = \dot{Q}_{\substack{net \\ in}} \qquad \textbf{(5.70)}$$

We call Eq. 5.70 the *one-dimensional, steady flow energy equation*. This equation is valid for incompressible and compressible flows. For compressible flows, enthalpy is most often used in the one-dimensional, steady flow energy equation and, thus, we have

$$\dot{m}\left[\check{h}_{out} - \check{h}_{in} + \frac{V_{out}^2 - V_{in}^2}{2} + g(z_{out} - z_{in})\right] = \dot{Q}_{\substack{net \\ in}} \qquad \textbf{(5.71)}$$

An example of the application of Eq. 5.70 follows.

*E*XAMPLE 5.22   Energy—Temperature Change

GIVEN   The 128 m waterfall shown in Fig. E5.22*a* involves steady flow from one large body of water to another.

FIND   Determine the temperature change associated with this flow.

SOLUTION

To solve this problem we consider a control volume consisting of a small cross-sectional streamtube from the nearly motionless surface of the upper body of water to the nearly motionless surface of the lower body of water as is sketched in Fig. E5.22*b*. We need to determine $T_2 - T_1$. This temperature change is related to

the change of internal energy of the water, $\check{u}_2 - \check{u}_1$, by the relationship

$$T_2 - T_1 = \frac{\check{u}_2 - \check{u}_1}{\check{c}} \qquad \textbf{(1)}$$

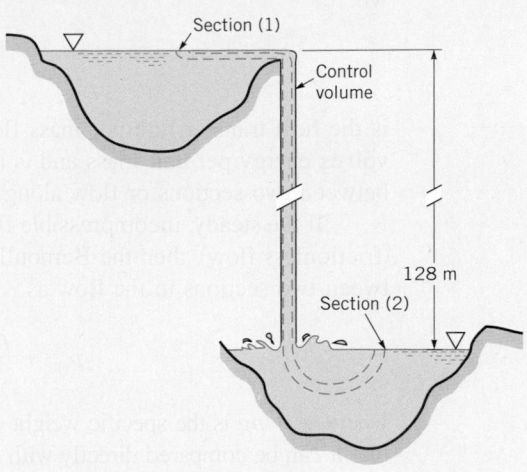

■ **F I G U R E E5.22b**

■ **F I G U R E E5.22a**
[Photograph of Akaka Falls (Hawaii) courtesy of Scott and Margaret Jones.]

where $\breve{c} = 4184$ J/kg · K is the specific heat of water. The application of Eq. 5.70 to the contents of this control volume leads to

$$\dot{m}\left[\breve{u}_2 - \breve{u}_1 + \left(\frac{p}{\rho}\right)_2 - \left(\frac{p}{\rho}\right)_1 + \frac{V_2^2 - V_1^2}{2} + g(z_2 - z_1)\right]$$

$$= \dot{Q}_{\text{net}\atop\text{in}} \tag{2}$$

We assume that the flow is adiabatic. Thus $\dot{Q}_{\text{net in}} = 0$. Also,

$$\left(\frac{p}{\rho}\right)_1 = \left(\frac{p}{\rho}\right)_2 \tag{3}$$

because the flow is incompressible and atmospheric pressure prevails at sections (1) and (2). Furthermore,

$$V_1 = V_2 = 0 \tag{4}$$

because the surface of each large body of water is considered motionless. Thus, Eqs. 1 through 4 combine to yield

$$T_2 - T_1 = \frac{g(z_1 - z_2)}{\breve{c}}$$

so that with

$$\breve{c} = [4184 \text{ J/kg} \cdot \text{K}]$$

$$T_2 - T_1 = \frac{(9.81 \text{m/s}^2)(128 \text{ m})}{[4184 \text{ J/kg} \cdot \text{K}][1 \text{ kg} \cdot \text{m/N} \cdot \text{s}^2]}$$

$$= 0.3 \text{ K} \tag{Ans}$$

COMMENT   Note that it takes a considerable change of potential energy to produce even a small increase in temperature.

A form of the energy equation that is most often used to solve incompressible flow problems is developed in the next section.

5.3.3 Comparison of the Energy Equation with the Bernoulli Equation

When the one-dimensional energy equation for steady-in-the-mean flow, Eq. 5.67, is applied to a flow that is steady, Eq. 5.67 becomes the one-dimensional, steady-flow energy equation, Eq. 5.70. The only difference between Eq. 5.67 and Eq. 5.70 is that shaft power, $\dot{W}_{\text{shaft net in}}$, is zero if the flow is steady throughout the control volume (fluid machines involve locally unsteady flow). If in addition to being steady, the flow is incompressible, we get from Eq. 5.70

$$\dot{m}\left[\breve{u}_{\text{out}} - \breve{u}_{\text{in}} + \frac{p_{\text{out}}}{\rho} - \frac{p_{\text{in}}}{\rho} + \frac{V_{\text{out}}^2 - V_{\text{in}}^2}{2} + g(z_{\text{out}} - z_{\text{in}})\right] = \dot{Q}_{\text{net}\atop\text{in}} \tag{5.72}$$

Dividing Eq. 5.72 by the mass flowrate, $\dot{m}$, and rearranging terms we obtain

$$\frac{p_{\text{out}}}{\rho} + \frac{V_{\text{out}}^2}{2} + gz_{\text{out}} = \frac{p_{\text{in}}}{\rho} + \frac{V_{\text{in}}^2}{2} + gz_{\text{in}} - (\breve{u}_{\text{out}} - \breve{u}_{\text{in}} - q_{\text{net}\atop\text{in}}) \tag{5.73}$$

where

$$q_{\underset{\text{in}}{\text{net}}} = \frac{\dot{Q}_{\text{net in}}}{\dot{m}}$$

is the heat transfer rate per mass flowrate, or heat transfer per unit mass. Note that Eq. 5.73 involves energy per unit mass and is applicable to one-dimensional flow of a single stream of fluid between two sections or flow along a streamline between two sections.

If the steady, incompressible flow we are considering also involves negligible viscous effects (frictionless flow), then the Bernoulli equation, Eq. 3.7, can be used to describe what happens between two sections in the flow as

$$p_{\text{out}} + \frac{\rho V_{\text{out}}^2}{2} + \gamma z_{\text{out}} = p_{\text{in}} + \frac{\rho V_{\text{in}}^2}{2} + \gamma z_{\text{in}} \tag{5.74}$$

where $\gamma = \rho g$ is the specific weight of the fluid. To get Eq. 5.74 in terms of energy per unit mass, so that it can be compared directly with Eq. 5.73, we divide Eq. 5.74 by density, ρ, and obtain

$$\frac{p_{\text{out}}}{\rho} + \frac{V_{\text{out}}^2}{2} + g z_{\text{out}} = \frac{p_{\text{in}}}{\rho} + \frac{V_{\text{in}}^2}{2} + g z_{\text{in}} \tag{5.75}$$

A comparison of Eqs. 5.73 and 5.75 prompts us to conclude that

$$\check{u}_{\text{out}} - \check{u}_{\text{in}} - q_{\underset{\text{in}}{\text{net}}} = 0 \tag{5.76}$$

when the steady incompressible flow is frictionless. For steady incompressible flow with friction, we learn from experience (second law of thermodynamics) that

$$\check{u}_{\text{out}} - \check{u}_{\text{in}} - q_{\underset{\text{in}}{\text{net}}} > 0 \tag{5.77}$$

In Eqs. 5.73 and 5.75, we can consider the combination of variables

$$\frac{p}{\rho} + \frac{V^2}{2} + g z$$

as equal to *useful* or *available energy*. Thus, from inspection of Eqs. 5.73 and 5.75, we can conclude that $\check{u}_{\text{out}} - \check{u}_{\text{in}} - q_{\text{net in}}$ represents the *loss* of useful or available energy that occurs in an incompressible fluid flow because of friction. In equation form we have

Minimizing loss is the central goal of fluid mechanical design.

$$\check{u}_{\text{out}} - \check{u}_{\text{in}} - q_{\underset{\text{in}}{\text{net}}} = \text{loss} \tag{5.78}$$

For a frictionless flow, Eqs. 5.73 and 5.75 tell us that loss equals zero.

It is often convenient to express Eq. 5.73 in terms of loss as

$$\frac{p_{\text{out}}}{\rho} + \frac{V_{\text{out}}^2}{2} + g z_{\text{out}} = \frac{p_{\text{in}}}{\rho} + \frac{V_{\text{in}}^2}{2} + g z_{\text{in}} - \text{loss} \tag{5.79}$$

An example of the application of Eq. 5.79 follows.

E**XAMPLE 5.23**   **Energy—Effect of Loss of Available Energy**

GIVEN As shown in Fig. E5.23a, air flows from a room through two different vent configurations: a cylindrical hole in the wall having a diameter of 120 mm and the same diameter cylindrical hole in the wall but with a well-rounded entrance. The room pressure is held constant at 1.0 kPa above atmospheric pressure. Both vents exhaust into the atmosphere. As discussed in Section 8.4.2, the loss in available energy associated with flow through the cylindrical vent from the room to the vent exit is $0.5 V_2^2/2$ where V_2 is the uniformly distributed exit velocity of air. The loss in available energy associated with flow through the rounded entrance vent from the room to the vent exit is $0.05 V_2^2/2$, where V_2 is the uniformly distributed exit velocity of air.

FIND Compare the volume flowrates associated with the two different vent configurations.

SOLUTION

We use the control volume for each vent sketched in Fig. E5.23a. What is sought is the flowrate, $Q = A_2 V_2$, where A_2 is the vent exit cross-sectional area, and V_2 is the uniformly distributed exit velocity. For both vents, application of Eq. 5.79 leads to

0 (no elevation change)

$$\frac{p_2}{\rho} + \frac{V_2^2}{2} + g\cancel{z_2} = \frac{p_1}{\rho} + \cancel{\frac{V_1^2}{2}} + g\cancel{z_1} - {}_1\text{loss}_2$$

0 ($V_1 \approx 0$)   (1)

where ${}_1\text{loss}_2$ is the loss between sections (1) and (2). Solving Eq. 1 for V_2 we get

$$V_2 = \sqrt{2\left[\left(\frac{p_1 - p_2}{\rho}\right) - {}_1\text{loss}_2\right]} \qquad (2)$$

Since

$$_1\text{loss}_2 = K_L \frac{V_2^2}{2} \qquad (3)$$

where K_L is the loss coefficient ($K_L = 0.5$ and 0.05 for the two vent configurations involved), we can combine Eqs. 2 and 3 to get

$$V_2 = \sqrt{2\left[\left(\frac{p_1 - p_2}{\rho}\right) - K_L \frac{V_2^2}{2}\right]} \qquad (4)$$

Solving Eq. 4 for V_2 we obtain

$$V_2 = \sqrt{\frac{p_1 - p_2}{\rho[(1 + K_L)/2]}} \qquad (5)$$

Therefore, for flowrate, Q, we obtain

$$Q = A_2 V_2 = \frac{\pi D_2^2}{4} \sqrt{\frac{p_1 - p_2}{\rho[(1 + K_L)/2]}} \qquad (6)$$

For the rounded entrance cylindrical vent, Eq. 6 gives

$$Q = \frac{\pi(120 \text{ mm})^2}{4(1000 \text{ mm/m})^2}$$
$$\times \sqrt{\frac{(1.0 \text{ kPa})(1000 \text{ Pa/kPa})[1(\text{N/m}^2)/(\text{Pa})]}{(1.23 \text{ kg/m}^3)[(1 + 0.05)/2][1(\text{N·s}^2)/(\text{kg·m})]}}$$

or

$$Q = 0.445 \text{ m}^3/\text{s} \qquad \text{(Ans)}$$

For the cylindrical vent, Eq. 6 gives us

$$Q = \frac{\pi(120 \text{ mm})^2}{4(1000 \text{ mm/m})^2}$$
$$\times \sqrt{\frac{(1.0 \text{ kPa})(1000 \text{ Pa/kPa})[1(\text{N/m}^2)/(\text{Pa})]}{(1.23 \text{ kg/m}^3)[(1 + 0.5)/2][1(\text{N·s}^2)/(\text{kg·m})]}}$$

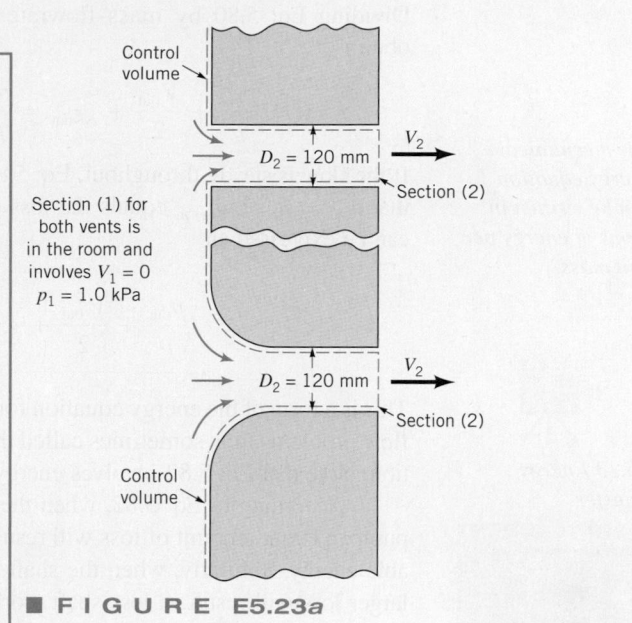

■ **FIGURE E5.23a**

or

$$Q = 0.372 \text{ m}^3/\text{s} \qquad \text{(Ans)}$$

COMMENT By repeating the calculations for various values of the loss coefficient, K_L, the results shown in Fig. E5.23b are obtained. Note that the rounded entrance vent allows the passage of more air than does the cylindrical vent because the loss associated with the rounded entrance vent is less than that for the cylindrical one. For this flow the pressure drop, $p_1 - p_2$, has two purposes: (1) overcome the loss associated with the flow, and (2) produce the kinetic energy at the exit. Even if there were no loss (i.e., $K_L = 0$), a pressure drop would be needed to accelerate the fluid through the vent.

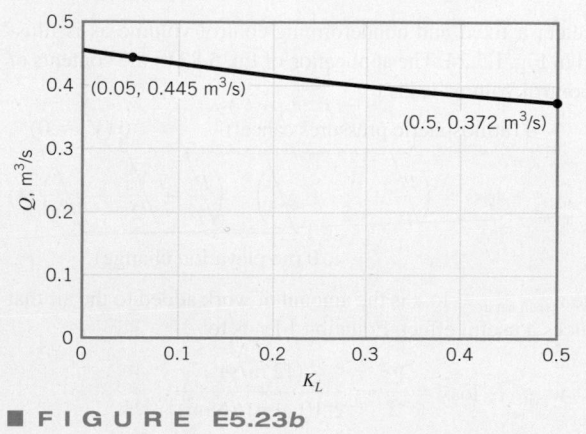

■ **FIGURE E5.23b**

An important group of fluid mechanics problems involves one-dimensional, incompressible, steady-in-the-mean flow with friction and shaft work. Included in this category are constant density flows through pumps, blowers, fans, and turbines. For this kind of flow, Eq. 5.67 becomes

$$\dot{m}\left[\check{u}_\text{out} - \check{u}_\text{in} + \frac{p_\text{out}}{\rho} - \frac{p_\text{in}}{\rho} + \frac{V_\text{out}^2 - V_\text{in}^2}{2} + g(z_\text{out} - z_\text{in})\right] = \dot{Q}_{\substack{\text{net} \\ \text{in}}} + \dot{W}_{\substack{\text{shaft} \\ \text{net in}}} \qquad (5.80)$$

Dividing Eq. 5.80 by mass flowrate and using the work per unit mass, $w_{\substack{shaft \\ net\ in}} = \dot{W}_{\substack{shaft \\ net\ in}} /\dot{m}$, we obtain

$$\frac{p_{out}}{\rho} + \frac{V_{out}^2}{2} + gz_{out} = \frac{p_{in}}{\rho} + \frac{V_{in}^2}{2} + gz_{in} + w_{\substack{shaft \\ net\ in}} - (\check{u}_{out} - \check{u}_{in} - q_{\substack{net \\ in}}) \qquad \textbf{(5.81)}$$

The mechanical energy equation can be written in terms of energy per unit mass.

If the flow is steady throughout, Eq. 5.81 becomes identical to Eq. 5.73, and the previous observation that $\check{u}_{out} - \check{u}_{in} - q_{net\ in}$ equals the loss of available energy is valid. Thus, we conclude that Eq. 5.81 can be expressed as

$$\frac{p_{out}}{\rho} + \frac{V_{out}^2}{2} + gz_{out} = \frac{p_{in}}{\rho} + \frac{V_{in}^2}{2} + gz_{in} + w_{\substack{shaft \\ net\ in}} - \text{loss} \qquad \textbf{(5.82)}$$

V5.13 Energy transfer

This is a form of the energy equation for steady-in-the-mean flow that is often used for incompressible flow problems. It is sometimes called the *mechanical energy* equation or the *extended Bernoulli* equation. Note that Eq. 5.82 involves energy per unit mass $(\text{N} \cdot \text{m} = \text{m}^2/\text{s}^2)$.

According to Eq. 5.82, when the shaft work is into the control volume, as for example with a pump, a larger amount of loss will result in more shaft work being required for the same rise in available energy. Similarly, when the shaft work is out of the control volume (for example, a turbine), a larger loss will result in less shaft work out for the same drop in available energy. Designers spend a great deal of effort on minimizing losses in fluid flow components. The following examples demonstrate why losses should be kept as small as possible in fluid systems.

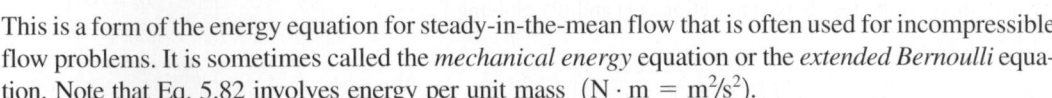

EXAMPLE 5.24 Energy—Fan Work and Efficiency

GIVEN An axial-flow ventilating fan driven by a motor that delivers 0.4 kW of power to the fan blades produces a 0.6-m-diameter axial stream of air having a speed of 12 m/s. The flow upstream of the fan involves negligible speed.

FIND Determine how much of the work to the air actually produces useful effects, that is, fluid motion and a rise in available energy. Estimate the fluid mechanical efficiency of this fan.

SOLUTION

We select a fixed and nondeforming control volume as is illustrated in Fig. E5.24. The application of Eq. 5.82 to the contents of this control volume leads to

0 (atmospheric pressures cancel)   0 ($V_1 \approx 0$)

$$w_{\substack{shaft \\ net\ in}} - \text{loss} = \left(\cancel{\frac{p_2}{\rho}} + \frac{V_2^2}{2} + \cancel{gz_2}\right) - \left(\cancel{\frac{p_1}{\rho}} + \cancel{\frac{V_1^2}{2}} + \cancel{gz_1}\right) \quad \textbf{(1)}$$

0 (no elevation change)

where $w_{shaft\ net\ in} - \text{loss}$ is the amount of work added to the air that produces a useful effect. Equation 1 leads to

$$w_{\substack{shaft \\ net\ in}} - \text{loss} = \frac{V_2^2}{2} = \frac{(12 \text{ m/s})^2}{2[1(\text{kg}\cdot\text{m})/(\text{N}\cdot\text{s}^2)]}$$
$$= 72.0 \text{ N}\cdot\text{m/kg} \qquad \textbf{(2)} \quad \textbf{(Ans)}$$

A reasonable estimate of *efficiency*, η, would be the ratio of amount of work that produces a useful effect, Eq. 2, to the amount of work delivered to the fan blades. That is

$$\eta = \frac{w_{\substack{shaft \\ net\ in}} - \text{loss}}{w_{\substack{shaft \\ net\ in}}} \qquad \textbf{(3)}$$

To calculate the efficiency, we need a value of $w_{shaft\ net\ in}$, which is related to the power delivered to the blades, $\dot{W}_{shaft\ net\ in}$. We note that

$$w_{\substack{shaft \\ net\ in}} = \frac{\dot{W}_{\substack{shaft \\ net\ in}}}{\dot{m}} \qquad \textbf{(4)}$$

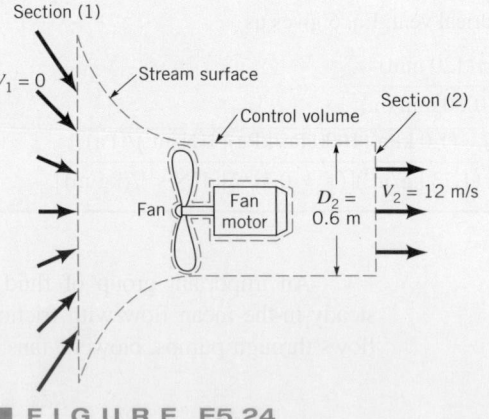

■ **FIGURE E5.24**

where the mass flowrate, $\dot{m}$, is (from Eq. 5.6)

$$\dot{m} = \rho A V = \rho \frac{\pi D_2^2}{4} V_2 \qquad (5)$$

For fluid density, ρ, we use 1.23 kg/m³ (standard air) and, thus, from Eqs. 4 and 5 we obtain

$$w_{\substack{\text{shaft} \\ \text{net in}}} = \frac{\dot{W}_{\substack{\text{shaft} \\ \text{net in}}}}{(\rho \pi D_2^2/4)V_2}$$

$$= \frac{(0.4\ \text{kW})[1000\ (\text{Nm})/(\text{skW})]}{(1.23\ \text{kg/m}^3)[(\pi)(0.6\ \text{m})^2/4](12\ \text{m/s})}$$

or

$$w_{\substack{\text{shaft} \\ \text{net in}}} = 95.8\ \text{N·m/kg} \qquad (6)$$

From Eqs. 2, 3, and 6 we obtain

$$\eta = \frac{72.0\ \text{N·m/kg}}{95.8\ \text{N·m/kg}} = 0.752 \qquad \text{(Ans)}$$

COMMENT   Note that only 75% of the power that was delivered to the air resulted in useful effects, and, thus, 25% of the shaft power is lost to air friction.

F l u i d s   i n   t h e   N e w s

Curtain of air An air curtain is produced by blowing air through a long rectangular nozzle to produce a high-velocity sheet of air, or a "curtain of air." This air curtain is typically directed over a doorway or opening as a replacement for a conventional door. The air curtain can be used for such things as keeping warm air from infiltrating dedicated cold spaces, preventing dust and other contaminates from entering a clean environment, and even just keeping insects out of the workplace, still allowing people to enter or exit. A disadvantage over conventional doors is the added *power requirements* to operate the air curtain, although the advantages can outweigh the disadvantage for various industrial applications. New applications for current air curtain designs continue to be developed. For example, the use of air curtains as a means of road tunnel fire security is currently being investigated. In such an application, the air curtain would act to isolate a portion of the tunnel where fire has broken out and not allow smoke and fumes to infiltrate the entire tunnel system. (See Problem 5.123.)

V5.14 Water plant aerator

If Eq. 5.82, which involves energy per unit mass, is multiplied by fluid density, ρ, we obtain

$$p_{\text{out}} + \frac{\rho V_{\text{out}}^2}{2} + \gamma z_{\text{out}} = p_{\text{in}} + \frac{\rho V_{\text{in}}^2}{2} + \gamma z_{\text{in}} + \rho w_{\substack{\text{shaft} \\ \text{net in}}} - \rho(\text{loss}) \qquad (5.83)$$

where $\gamma = \rho g$ is the specific weight of the fluid. Equation 5.83 involves *energy per unit volume* and the units involved are identical with those used for pressure $(\text{N} \cdot \text{m/m}^3 = \text{N/m}^2)$.

If Eq. 5.82 is divided by the acceleration of gravity, g, we get

$$\boxed{\frac{p_{\text{out}}}{\gamma} + \frac{V_{\text{out}}^2}{2g} + z_{\text{out}} = \frac{p_{\text{in}}}{\gamma} + \frac{V_{\text{in}}^2}{2g} + z_{\text{in}} + h_s - h_L} \qquad (5.84)$$

where

$$h_s = w_{\text{shaft net in}}/g = \frac{\dot{W}_{\substack{\text{shaft} \\ \text{net in}}}}{\dot{m}g} = \frac{\dot{W}_{\substack{\text{shaft} \\ \text{net in}}}}{\gamma Q} \qquad (5.85)$$

The energy equation written in terms of energy per unit weight involves heads.

is the *shaft work head* and $h_L = \text{loss}/g$ is the *head loss*. Equation 5.84 involves *energy per unit weight* $(\text{N} \cdot \text{m/N} = \text{m})$. In Section 3.7, we introduced the notion of "head," which is energy per unit weight. Units of length (for example, m) are used to quantify the amount of head involved. If a turbine is in the control volume, h_s is negative because it is associated with shaft work out of the control volume. For a pump in the control volume, h_s is positive because it is associated with shaft work into the control volume.

We can define a total head, H, as follows

$$H = \frac{p}{\gamma} + \frac{V^2}{2g} + z$$

Then Eq. 5.84 can be expressed as

$$H_{\text{out}} = H_{\text{in}} + h_s - h_L$$

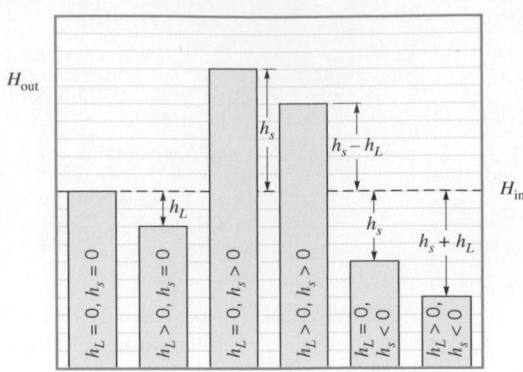

■ **F I G U R E 5.8** Total-head change in fluid flows.

Some important possible values of H_{out} in comparison to H_{in} are shown in Fig. 5.8. Note that h_L (head loss) always reduces the value of H_{out}, except in the ideal case when it is zero. Note also that h_L lessens the effect of shaft work that can be extracted from a fluid. When $h_L = 0$ (ideal condition) the shaft work head, h_s, and the change in total head are the same. This head change is sometimes called "ideal head change." The corresponding ideal shaft work head is the minimum required to achieve a desired effect. For work out, it is the maximum possible. Designers usually strive to minimize loss. In Chapter 12 we learn of one instance when minimum loss is sacrificed for survivability of fish coursing through a turbine rotor.

EXAMPLE 5.25   Energy—Head Loss and Power Loss

GIVEN   The pump shown in Fig. E5.25a adds 7.5 kilowatt to the water as it pumps water from the lower lake to the upper lake. The elevation difference between the lake surfaces is 9 m and the head loss is 4.5 m.

FIND   Determine

(a)   the flowrate and

(b)   the power loss associated with this flow.

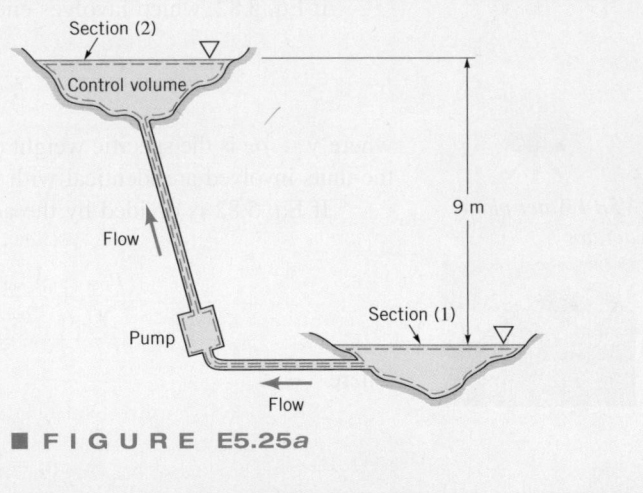

■ **F I G U R E E5.25a**

SOLUTION

(a)   The energy equation (Eq. 5.84) for this flow is

$$\frac{p_2}{\gamma} + \frac{V_2^2}{2g} + z_2 = \frac{p_1}{\gamma} + \frac{V_1^2}{2g} + z_1 + h_s - h_L \qquad (1)$$

where points 2 and 1 (corresponding to "out" and "in" in Eq. 5.84) are located on the lake surfaces. Thus, $p_2 = p_1 = 0$ and $V_2 = V_1 = 0$ so that Eq. 1 becomes

$$h_s = h_L + z_2 - z_1 \qquad (2)$$

where $z_2 = 9$ m, $z_1 = 0$, and $h_L = 4.5$ m. The pump head is obtained from Eq. 5.85 as

$$h_s = \dot{W}_{\text{shaft net in}}/\gamma Q$$
$$= (7.5 \text{ kW})(1 \text{ N} \cdot \text{m/W} \cdot \text{s})/(9810 \text{ N/m}^3) Q$$
$$= 0.76/Q$$

where h_s is in m when Q is in m³/s.

Hence, from Eq. 2,

$$0.76/Q = 4.5 \text{ m} + 9 \text{ m}$$

or

$$Q = 5.6 \times 10^{-2} \text{ m}^3/\text{s} \qquad (\text{Ans})$$

COMMENT   Note that in this example the purpose of the pump is to lift the water (a 9 m head) and overcome the head loss (a 4.5 m head); it does not, overall, alter the water's pressure or velocity.

(b)   The power lost due to friction can be obtained from Eq. 5.85 as

$$\dot{W}_{loss} = \gamma Q h_L = (9810 \ N/m^3)(5.6 \times 10^{-2} \ m^3/s)(4.5 \ m)$$
$$= 2472 \ N \cdot m/s \ (1 \ W \cdot s/ \ N \cdot m)$$
$$= 2.47 \ kW \qquad\qquad\qquad\qquad \textbf{(Ans)}$$

COMMENTS   The remaining 7.5 kW − 2.47 kW = 5.03 kW that the pump adds to the water is used to lift the water from the lower to the upper lake. This energy is not "lost," but it is stored as potential energy.

By repeating the calculations for various head losses, h_L, the results shown in Fig. E5.25b are obtained. Note that as the head loss increases, the flowrate decreases because an increasing portion of the 7.5 kW supplied by the pump is lost and, therefore, not available to lift the fluid to the higher elevation.

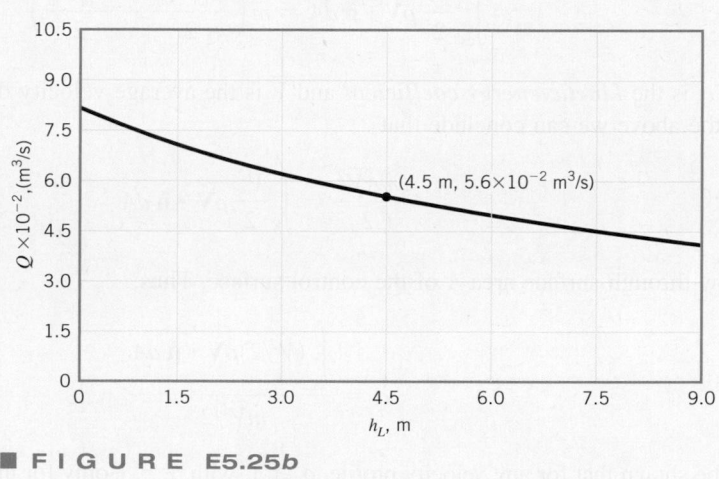

■ F I G U R E   E5.25b

A comparison of the energy equation and the Bernoulli equation has led to the concept of loss of available energy in incompressible fluid flows with friction. In Chapter 8, we discuss in detail some methods for estimating loss in incompressible flows with friction. In Section 5.4 and Chapter 11, we demonstrate that loss of available energy is also an important factor to consider in compressible flows with friction.

F l u i d s   i n   t h e   N e w s

Smart shocks Vehicle shock absorbers are dampers used to provide a smooth, controllable ride. When going over a bump, the relative motion between the tires and the vehicle body displaces a piston in the shock and forces a viscous fluid through a small orifice or channel. The viscosity of the fluid produces a *head loss* that dissipates energy to dampen the vertical motion. Current shocks use a fluid with fixed viscosity. However, recent technology has been developed that uses a synthetic oil with millions of tiny iron balls suspended in it. These tiny balls react to a magnetic field generated by an electric coil on the shock piston in a manner that changes the fluid viscosity, going anywhere from essentially no damping to a solid almost instantly. A computer adjusts the current to the coil to select the proper viscosity for the given conditions (i.e., wheel speed, vehicle speed, steering-wheel angle, lateral acceleration, brake application, and temperature). The goal of these adjustments is an optimally tuned shock that keeps the vehicle on a smooth, even keel while maximizing the contact of the tires with the pavement for any road conditions. (See Problem 5.107.)

5.3.4 Application of the Energy Equation to Nonuniform Flows

The forms of the energy equation discussed in Sections 5.3.2 and 5.3.3 are applicable to one-dimensional flows, flows that are approximated with uniform velocity distributions where fluid crosses the control surface.

If the velocity profile at any section where flow crosses the control surface is not uniform, inspection of the energy equation for a control volume, Eq. 5.64, suggests that the integral

$$\int_{cs} \frac{V^2}{2} \rho \mathbf{V} \cdot \hat{\mathbf{n}} \, dA$$

will require special attention. The other terms of Eq. 5.64 can be accounted for as already discussed in Sections 5.3.2 and 5.3.3.

For one stream of fluid entering and leaving the control volume, we can define the relationship

$$\int_{cs} \frac{V^2}{2} \rho \mathbf{V} \cdot \hat{\mathbf{n}} \, dA = \dot{m} \left(\frac{\alpha_{out} \overline{V}_{out}^2}{2} - \frac{\alpha_{in} \overline{V}_{in}^2}{2} \right)$$

where α is the *kinetic energy coefficient* and $\overline{V}$ is the average velocity defined earlier in Eq. 5.7. From the above we can conclude that

$$\frac{\dot{m} \alpha \overline{V}^2}{2} = \int_A \frac{V^2}{2} \rho \mathbf{V} \cdot \hat{\mathbf{n}} \, dA$$

for flow through surface area A of the control surface. Thus,

$$\alpha = \frac{\int_A (V^2/2) \rho \mathbf{V} \cdot \hat{\mathbf{n}} \, dA}{\dot{m} \overline{V}^2 / 2} \tag{5.86}$$

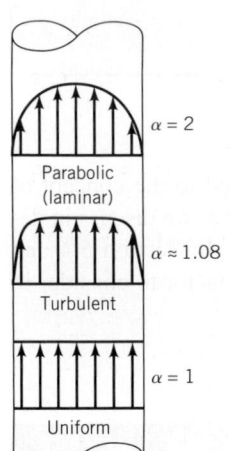

$\alpha = 2$

Parabolic (laminar)

$\alpha \approx 1.08$

Turbulent

$\alpha = 1$

Uniform

It can be shown that for any velocity profile, $\alpha \geq 1$, with $\alpha = 1$ only for uniform flow. Some typical velocity profile examples for flow in a conventional pipe are shown in the sketch in the margin. Therefore, for nonuniform velocity profiles, the energy equation on an energy per unit mass basis for the incompressible flow of one stream of fluid through a control volume that is steady in the mean is

$$\boxed{\frac{p_{out}}{\rho} + \frac{\alpha_{out} \overline{V}_{out}^2}{2} + gz_{out} = \frac{p_{in}}{\rho} + \frac{\alpha_{in} \overline{V}_{in}^2}{2} + gz_{in} + w_{\substack{shaft \\ net\ in}} - \text{loss}} \tag{5.87}$$

On an energy per unit volume basis we have

$$p_{out} + \frac{\rho \alpha_{out} \overline{V}_{out}^2}{2} + \gamma z_{out} = p_{in} + \frac{\rho \alpha_{in} \overline{V}_{in}^2}{2} + \gamma z_{in} + \rho w_{\substack{shaft \\ net\ in}} - \rho(\text{loss}) \tag{5.88}$$

and on an energy per unit weight or head basis we have

$$\boxed{\frac{p_{out}}{\gamma} + \frac{\alpha_{out} \overline{V}_{out}^2}{2g} + z_{out} = \frac{p_{in}}{\gamma} + \frac{\alpha_{in} \overline{V}_{in}^2}{2g} + z_{in} + \frac{w_{\substack{shaft \\ net\ in}}}{g} - h_L} \tag{5.89}$$

The following examples illustrate the use of the kinetic energy coefficient.

EXAMPLE 5.26   Energy—Effect of Nonuniform Velocity Profile

GIVEN The small fan shown in Fig. E5.26 moves air at a mass flowrate of 0.1 kg/min. Upstream of the fan, the pipe diameter is 60 mm, the flow is laminar, the velocity distribution is parabolic, and the kinetic energy coefficient, α_1, is equal to 2.0. Downstream of the fan, the pipe diameter is 30 mm, the flow is turbulent, the velocity profile is quite uniform, and the kinetic

energy coefficient, α_2, is equal to 1.08. The rise in static pressure across the fan is 0.1 kPa and the fan motor draws 0.14 W.

FIND Compare the value of loss calculated: **(a)** assuming uniform velocity distributions, **(b)** considering actual velocity distributions.

SOLUTION

Application of Eq. 5.87 to the contents of the control volume shown in Fig. E5.26 leads to

$$\frac{p_2}{\rho} + \frac{\alpha_2 \overline{V}_2^2}{2} + \cancel{gz_2}^{\,0 \text{ (change in } gz \text{ is negligible)}} = \frac{p_1}{\rho} + \frac{\alpha_1 \overline{V}_1^2}{2} + \cancel{gz_1}$$
$$- \text{loss} + w_{\substack{\text{shaft} \\ \text{net in}}} \qquad (1)$$

or solving Eq. 1 for loss we get

$$\text{loss} = w_{\substack{\text{shaft} \\ \text{net in}}} - \left(\frac{p_2 - p_1}{\rho}\right) + \frac{\alpha_1 \overline{V}_1^2}{2} - \frac{\alpha_2 \overline{V}_2^2}{2} \qquad (2)$$

To proceed further, we need values of $w_{\text{shaft net in}}$, $\overline{V}_1$, and $\overline{V}_2$. These quantities can be obtained as follows. For shaft work

$$w_{\substack{\text{shaft} \\ \text{net in}}} = \frac{\text{power to fan motor}}{\dot{m}}$$

or

$$w_{\substack{\text{shaft} \\ \text{net in}}} = \frac{(0.14 \text{ W})[(1 \text{ N} \cdot \text{m/s})/\text{W}]}{0.1 \text{ kg/min}} (60 \text{ s/min})$$
$$= 84.0 \text{ N} \cdot \text{m/kg} \qquad (3)$$

For the average velocity at section (1), $\overline{V}_1$, from Eq. 5.11 we obtain

$$\overline{V}_1 = \frac{\dot{m}}{\rho A_1}$$
$$= \frac{\dot{m}}{\rho(\pi D_1^2/4)} \qquad (4)$$
$$= \frac{(0.1 \text{ kg/min})(1 \text{ min/60 s})(1000 \text{ mm/m})^2}{(1.23 \text{ kg/m}^3)[\pi(60 \text{ mm})^2/4]}$$
$$= 0.479 \text{ m/s}$$

For the average velocity at section (2), $\overline{V}_2$,

$$\overline{V}_2 = \frac{(0.1 \text{ kg/min})(1 \text{ min/60 s})(1000 \text{ mm/m})^2}{(1.23 \text{ kg/m}^3)[\pi(30 \text{ mm})^2/4]}$$
$$= 1.92 \text{ m/s} \qquad (5)$$

(a) For the assumed uniform velocity profiles ($\alpha_1 = \alpha_2 = 1.0$), Eq. 2 yields

$$\text{loss} = w_{\substack{\text{shaft} \\ \text{net in}}} - \left(\frac{p_2 - p_1}{\rho}\right) + \frac{\overline{V}_1^2}{2} - \frac{\overline{V}_2^2}{2} \qquad (6)$$

Using Eqs. 3, 4, and 5 and the pressure rise given in the problem statement, Eq. 6 gives

$$\text{loss} = 84.0 \frac{\text{N} \cdot \text{m}}{\text{kg}} - \frac{(0.1 \text{ kPa})(1000 \text{ Pa/kPa})(1 \text{ N/m}^2/\text{Pa})}{1.23 \text{ kg/m}^3}$$
$$+ \frac{(0.479 \text{ m/s})^2}{2[1 \text{ (kg} \cdot \text{m)/(N} \cdot \text{s}^2)]} - \frac{(1.92 \text{ m/s})^2}{2[1 \text{ (kg} \cdot \text{m)/(N} \cdot \text{s}^2)]}$$

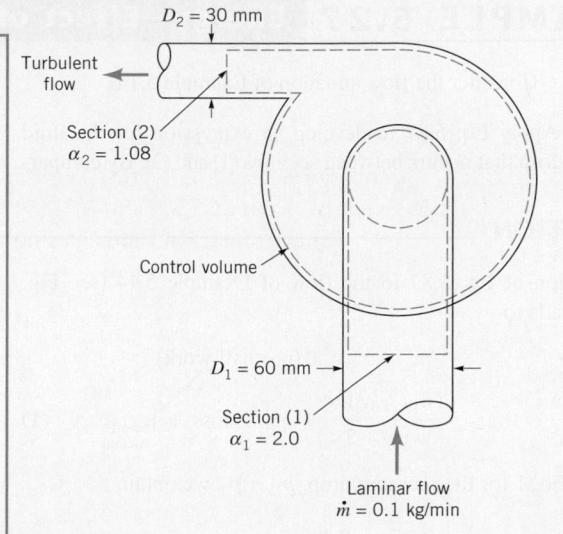

$D_2 = 30$ mm

Turbulent flow

Section (2)
$\alpha_2 = 1.08$

Control volume

$D_1 = 60$ mm

Section (1)
$\alpha_1 = 2.0$

Laminar flow
$\dot{m} = 0.1$ kg/min

■ **FIGURE E5.26**

or

$$\text{loss} = 84.0 \text{ N} \cdot \text{m/kg} - 81.3 \text{ N} \cdot \text{m/kg}$$
$$+ 0.115 \text{ N} \cdot \text{m/kg} - 1.84 \text{ N} \cdot \text{m/kg}$$
$$= 0.975 \text{ N} \cdot \text{m/kg} \qquad \textbf{(Ans)}$$

(b) For the actual velocity profiles ($\alpha_1 = 2$, $\alpha_2 = 1.08$), Eq. 1 gives

$$\text{loss} = w_{\substack{\text{shaft} \\ \text{net in}}} - \left(\frac{p_2 - p_1}{\rho}\right) + \alpha_1 \frac{\overline{V}_1^2}{2} - \alpha_2 \frac{\overline{V}_2^2}{2} \qquad (7)$$

If we use Eqs. 3, 4, and 5 and the given pressure rise, Eq. 7 yields

$$\text{loss} = 84 \text{ N} \cdot \text{m/kg} - \frac{(0.1 \text{ kPa})(1000 \text{ Pa/kPa})(1 \text{ N/m}^2/\text{Pa})}{1.23 \text{ kg/m}^3}$$
$$+ \frac{2(0.479 \text{ m/s})^2}{2[1 \text{ (kg} \cdot \text{m)/(N} \cdot \text{s}^2)]} - \frac{1.08(1.92 \text{ m/s})^2}{2[1 \text{ (kg} \cdot \text{m)/(N} \cdot \text{s}^2)]}$$

or

$$\text{loss} = 84.0 \text{ N} \cdot \text{m/kg} - 81.3 \text{ N} \cdot \text{m/kg}$$
$$+ 0.230 \text{ N} \cdot \text{m/kg} - 1.99 \text{ N} \cdot \text{m/kg}$$
$$= 0.940 \text{ N} \cdot \text{m/kg} \qquad \textbf{(Ans)}$$

COMMENT   The difference in loss calculated assuming uniform velocity profiles and actual velocity profiles is not large compared to $w_{\text{shaft net in}}$ for this fluid flow situation.

EXAMPLE 5.27   Energy—Effect of Nonuniform Velocity Profile

GIVEN   Consider the flow situation of Example 5.14.

FIND   Apply Eq. 5.87 to develop an expression for the fluid pressure drop that occurs between sections (1) and (2). By compar-ing the equation for pressure drop obtained presently with the re-sult of Example 5.14, obtain an expression for loss between sec-tions (1) and (2).

SOLUTION

Application of Eq. 5.87 to the flow of Example 5.14 (see Fig. E5.14) leads to

$$\frac{p_2}{\rho} + \frac{\alpha_2 \overline{w}_2^2}{2} + gz_2 = \frac{p_1}{\rho} + \frac{\alpha_1 \overline{w}_1^2}{2} + gz_1 - \text{loss} + \overset{\text{0 (no shaft work)}}{w_{\text{shaft}}} \tag{1}$$

Solving Eq. 1 for the pressure drop, $p_1 - p_2$, we obtain

$$p_1 - p_2 = \rho \left[\frac{\alpha_2 \overline{w}_2^2}{2} - \frac{\alpha_1 \overline{w}_1^2}{2} + g(z_2 - z_1) + \text{loss} \right] \tag{2}$$

Since the fluid velocity at section (1), w_1, is uniformly distributed over cross-sectional area A_1, the corresponding kinetic energy coefficient, α_1, is equal to 1.0. The kinetic energy coefficient at section (2), α_2, needs to be determined from the velocity profile distribution given in Example 5.14. Using Eq. 5.86 we get

$$\alpha_2 = \frac{\displaystyle\int_{A_2} \rho w_2^3 \, dA_2}{\dot{m} \overline{w}_2^2} \tag{3}$$

Substituting the parabolic velocity profile equation into Eq. 3 we obtain

$$\alpha_2 = \frac{\rho \displaystyle\int_0^R (2w_1)^3 [1 - (r/R)^2]^3 2\pi r \, dr}{(\rho A_2 \overline{w}_2) \overline{w}_2^2}$$

From conservation of mass, since $A_1 = A_2$

$$w_1 = \overline{w}_2 \tag{4}$$

Then, substituting Eq. 4 into Eq. 3, we obtain

$$\alpha_2 = \frac{\rho 8 \overline{w}_2^3 2\pi \displaystyle\int_0^R [1 - (r/R)^2]^3 r \, dr}{\rho \pi R^2 \overline{w}_2^3}$$

or

$$\alpha_2 = \frac{16}{R^2} \int_0^R [1 - 3(r/R)^2 + 3(r/R)^4 - (r/R)^6] r \, dr$$
$$= 2 \tag{5}$$

Now we combine Eqs. 2 and 5 to get

$$p_1 - p_2 = \rho \left[\frac{2.0 \overline{w}_2^2}{2} - \frac{1.0 \overline{w}_1^2}{2} + g(z_2 - z_1) + \text{loss} \right] \tag{6}$$

However, from conservation of mass $\overline{w}_2 = \overline{w}_1 = \overline{w}$ so that Eq. 6 becomes

$$p_1 - p_2 = \frac{\rho \overline{w}^2}{2} + \rho g(z_2 - z_1) + \rho(\text{loss}) \tag{7}$$

The term associated with change in elevation, $\rho g(z_2 - z_1)$, is equal to the weight per unit cross-sectional area, $\mathcal{W}/A$, of the water con-tained between sections (1) and (2) at any instant,

$$\rho g(z_2 - z_1) = \frac{\mathcal{W}}{A} \tag{8}$$

Thus, combining Eqs. 7 and 8 we get

$$p_1 - p_2 = \frac{\rho \overline{w}^2}{2} + \frac{\mathcal{W}}{A} + \rho(\text{loss}) \tag{9}$$

The pressure drop between sections (1) and (2) is due to:

1. The change in kinetic energy between sections (1) and (2) as-sociated with going from a uniform velocity profile to a par-abolic velocity profile.

2. The weight of the water column, that is, hydrostatic pressure effect.

3. Viscous loss.

Comparing Eq. 9 for pressure drop with the one obtained in Example 5.14 (i.e., the answer of Example 5.14) we obtain

$$\frac{\rho \overline{w}^2}{2} + \frac{\mathcal{W}}{A} + \rho(\text{loss}) = \frac{\rho \overline{w}^2}{3} + \frac{R_z}{A} + \frac{\mathcal{W}}{A} \tag{10}$$

or

$$\text{loss} = \frac{R_z}{\rho A} - \frac{\overline{w}^2}{6} \tag{Ans}$$

COMMENT   We conclude that while some of the pipe wall friction force, R_z, resulted in loss of available energy, a portion of this friction, $\rho A \overline{w}^2/6$, led to the velocity profile change.

5.3.5   Combination of the Energy Equation and the Moment-of-Momentum Equation[4]

If Eq. 5.82 is used for one-dimensional incompressible flow through a turbomachine, we can use Eq. 5.54, developed in Section 5.2.4 from the moment-of-momentum equation (Eq. 5.42), to evaluate

---

[4]This section may be omitted without loss of continuity in the text material. This section should not be considered without prior study of Sections 5.2.3 and 5.2.4. All of these sections are recommended for those interested in Chapter 12.

shaft work. This application of both Eqs. 5.54 and 5.82 allows us to ascertain the amount of loss that occurs in incompressible turbomachine flows as is demonstrated in Example 5.28.

EXAMPLE 5.28 | Energy—Fan Performance

GIVEN Consider the fan of Example 5.19.

FIND Show that only some of the shaft power into the air is converted into useful effects. Develop a meaningful efficiency equation and a practical means for estimating lost shaft energy.

SOLUTION

We use the same control volume used in Example 5.19. Application of Eq. 5.82 to the contents of this control volume yields

$$\frac{p_2}{\rho} + \frac{V_2^2}{2} + gz_2 = \frac{p_1}{\rho} + \frac{V_1^2}{2} + gz_1 + w_{\substack{\text{shaft} \\ \text{net in}}} - \text{loss} \qquad (1)$$

As in Example 5.26, we can see with Eq. 1 that a "useful effect" in this fan can be defined as

$$\text{useful effect} = w_{\substack{\text{shaft} \\ \text{net in}}} - \text{loss}$$

$$= \left(\frac{p_2}{\rho} + \frac{V_2^2}{2} + gz_2\right) - \left(\frac{p_1}{\rho} + \frac{V_1^2}{2} + gz_1\right) \qquad (2) \quad \text{(Ans)}$$

In other words, only a portion of the shaft work delivered to the air by the fan blades is used to increase the available energy of the air; the rest is lost because of fluid friction.

A meaningful efficiency equation involves the ratio of shaft work converted into a useful effect (Eq. 2) to shaft work into the air, $w_{\text{shaft net in}}$. Thus, we can express efficiency, η, as

$$\eta = \frac{w_{\substack{\text{shaft} \\ \text{net in}}} - \text{loss}}{w_{\substack{\text{shaft} \\ \text{net in}}}} \qquad (3)$$

However, when Eq. 5.54, which was developed from the moment-of-momentum equation (Eq. 5.42), is applied to the contents of the control volume of Fig. E5.19, we obtain

$$w_{\substack{\text{shaft} \\ \text{net in}}} = +U_2 V_{\theta 2} \qquad (4)$$

Combining Eqs. 2, 3, and 4, we obtain

$$\eta = \{[(p_2/\rho) + (V_2^2/2) + gz_2] - [(p_1/\rho) + (V_1^2/2) + gz_1]\}/U_2 V_{\theta 2} \qquad (5) \quad \text{(Ans)}$$

Equation 5 provides us with a practical means to evaluate the efficiency of the fan of Example 5.19.

Combining Eqs. 2 and 4, we obtain

$$\text{loss} = U_2 V_{\theta 2} - \left[\left(\frac{p_2}{\rho} + \frac{V_2^2}{2} + gz_2\right) - \left(\frac{p_1}{\rho} + \frac{V_1^2}{2} + gz_1\right)\right] \qquad (6) \quad \text{(Ans)}$$

COMMENT Equation 6 provides us with a useful method of evaluating the loss due to fluid friction in the fan of Example 5.19 in terms of fluid mechanical variables that can be measured.

5.4 Second Law of Thermodynamics—Irreversible Flow[5]

The second law of thermodynamics affords us with a means to formalize the inequality

$$\breve{u}_2 - \breve{u}_1 - q_{\substack{\text{net} \\ \text{in}}} \geq 0 \qquad (5.90)$$

The second law of thermodynamics formalizes the notion of loss.

for steady, incompressible, one-dimensional flow with friction (see Eq. 5.73). In this section we continue to develop the notion of loss of useful or available energy for flow with friction. Minimization of loss of available energy in any flow situation is of obvious engineering importance.

5.4.1 Semi-infinitesimal Control Volume Statement of the Energy Equation

If we apply the one-dimensional, steady flow energy equation, Eq. 5.70, to the contents of a control volume that is infinitesimally thin as illustrated in Fig 5.8, the result is

$$\dot{m}\left[d\breve{u} + d\left(\frac{p}{\rho}\right) + d\left(\frac{V^2}{2}\right) + g\,(dz)\right] = \delta\dot{Q}_{\substack{\text{net} \\ \text{in}}} \qquad (5.91)$$

---

[5]This entire section may be omitted without loss of continuity in the text material.

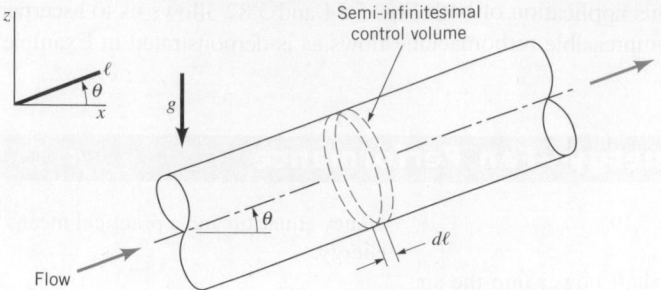

■ **FIGURE 5.9**   Semi-infinitesimal control volume.

For all pure substances including common engineering working fluids, such as air, water, oil, and gasoline, the following relationship is valid (see, for example, Ref. 3).

$$T \, ds = d\breve{u} + p \, d\left(\frac{1}{\rho}\right) \tag{5.92}$$

where T is the absolute temperature and s is the *entropy* per unit mass.

Combining Eqs. 5.91 and 5.92 we get

$$\dot{m}\left[T \, ds - p \, d\left(\frac{1}{\rho}\right) + d\left(\frac{p}{\rho}\right) + d\left(\frac{V^2}{2}\right) + g \, dz\right] = \delta\dot{Q}_{\substack{\text{net} \\ \text{in}}}$$

or, dividing through by $\dot{m}$ and letting $\delta q_{\substack{\text{net} \\ \text{in}}} = \delta\dot{Q}_{\substack{\text{net} \\ \text{in}}}/\dot{m}$, we obtain

$$\frac{dp}{\rho} + d\left(\frac{V^2}{2}\right) + g \, dz = -\left(T \, ds - \delta q_{\substack{\text{net} \\ \text{in}}}\right) \tag{5.93}$$

5.4.2 Semi-infinitesimal Control Volume Statement of the Second Law of Thermodynamics

A general statement of the second law of thermodynamics is

$$\frac{D}{Dt}\int_{\text{sys}} s\rho \, d\Psi \geq \sum\left(\frac{\delta\dot{Q}_{\substack{\text{net} \\ \text{in}}}}{T}\right)_{\text{sys}} \tag{5.94}$$

or in words,

The second law of thermodynamics involves entropy, heat transfer, and temperature.	the time rate of increase of the entropy of a system $\geq$ sum of the ratio of net heat transfer rate into system to absolute temperature for each particle of mass in the system receiving heat from surroundings

The right-hand side of Eq. 5.94 is identical for the system and control volume at the instant when system and control volume are coincident; thus,

$$\sum\left(\frac{\delta\dot{Q}_{\substack{\text{net} \\ \text{in}}}}{T}\right)_{\text{sys}} = \sum\left(\frac{\delta\dot{Q}_{\substack{\text{net} \\ \text{in}}}}{T}\right)_{\text{cv}} \tag{5.95}$$

With the help of the Reynolds transport theorem (Eq. 4.19) the system time derivative can be expressed for the contents of the coincident control volume that is fixed and nondeforming. Using Eq. 4.19, we obtain

$$\frac{D}{Dt}\int_{\text{sys}} s\rho \, d\Psi = \frac{\partial}{\partial t}\int_{\text{cv}} s\rho \, d\Psi + \int_{\text{cs}} s\rho\mathbf{V} \cdot \hat{\mathbf{n}} \, dA \tag{5.96}$$

For a fixed, nondeforming control volume, Eqs. 5.94, 5.95, and 5.96 combine to give

$$\frac{\partial}{\partial t} \int_{cv} s\rho \, d\!\!\!V + \int_{cs} s\rho \mathbf{V} \cdot \hat{\mathbf{n}} \, dA \geq \sum \left(\frac{\delta \dot{Q}_{\text{net}}^{\text{in}}}{T} \right)_{cv} \qquad (5.97)$$

At any instant for steady flow

$$\frac{\partial}{\partial t} \int_{cv} s\rho \, d\!\!\!V = 0 \qquad (5.98)$$

If the flow consists of only one stream through the control volume and if the properties are uniformly distributed (one-dimensional flow), Eqs. 5.97 and 5.98 lead to

$$\dot{m}(s_{\text{out}} - s_{\text{in}}) \geq \sum \frac{\delta \dot{Q}_{\text{net}}^{\text{in}}}{T} \qquad (5.99)$$

For the infinitesimally thin control volume of Fig. 5.8, Eq. 5.99 yields

$$\dot{m} \, ds \geq \sum \frac{\delta \dot{Q}_{\text{net}}^{\text{in}}}{T} \qquad (5.100)$$

If all of the fluid in the infinitesimally thin control volume is considered as being at a uniform temperature, T, then from Eq. 5.100 we get

$$T \, ds \geq \delta q_{\text{net}\atop \text{in}}$$

or

$$T \, ds - \delta q_{\text{net}\atop \text{in}} \geq 0 \qquad (5.101)$$

The equality is for any reversible (frictionless) process; the inequality is for all irreversible (friction) processes.

5.4.3 Combination of the Equations of the First and Second Laws of Thermodynamics

Combining Eqs. 5.93 and 5.101, we conclude that

$$-\left[\frac{dp}{\rho} + d\left(\frac{V^2}{2} \right) + g \, dz \right] \geq 0 \qquad (5.102)$$

The equality is for any steady, reversible (frictionless) flow, an important example being flow for which the Bernoulli equation (Eq. 3.7) is applicable. The inequality is for all steady, irreversible (friction) flows. The actual amount of the inequality has physical significance. It represents the extent of loss of useful or available energy which occurs because of irreversible flow phenomena including viscous effects. Thus, Eq. 5.102 can be expressed as

$$-\left[\frac{dp}{\rho} + d\left(\frac{V^2}{2} \right) + g \, dz \right] = \delta(\text{loss}) = \left(T \, ds - \delta q_{\text{net}\atop \text{in}} \right) \qquad (5.103)$$

The irreversible flow loss is zero for a frictionless flow and greater than zero for a flow with frictional effects. Note that when the flow is frictionless, Eq. 5.103 multiplied by density, ρ, is identical to Eq. 3.5. Thus, for steady frictionless flow, Newton's second law of motion (see Section 3.1) and the first and second laws of thermodynamics lead to the same differential equation,

$$\frac{dp}{\rho} + d\left(\frac{V^2}{2} \right) + g \, dz = 0 \qquad (5.104)$$

If some shaft work is involved, then the flow must be at least locally unsteady in a cyclical way and the appropriate form of the energy equation for the contents of an infinitesimally thin control volume can be developed starting with Eq. 5.67. The resulting equation is

$$-\left[\frac{dp}{\rho} + d\left(\frac{V^2}{2}\right) + g\,dz\right] = \delta(\text{loss}) - \delta w_{\substack{\text{shaft} \\ \text{net in}}} \tag{5.105}$$

Equations 5.103 and 5.105 are valid for incompressible and compressible flows. If we combine Eqs. 5.92 and 5.103, we obtain

$$d\breve{u} + pd\left(\frac{1}{\rho}\right) - \delta q_{\substack{\text{net} \\ \text{in}}} = \delta(\text{loss}) \tag{5.106}$$

For incompressible flow, $d(1/\rho) = 0$ and, thus, from Eq. 5.106,

$$d\breve{u} - \delta q_{\substack{\text{net} \\ \text{in}}} = \delta(\text{loss}) \tag{5.107}$$

Applying Eq. 5.107 to a finite control volume, we obtain

$$\breve{u}_{\text{out}} - \breve{u}_{\text{in}} - q_{\substack{\text{net} \\ \text{in}}} = \text{loss}$$

which is the same conclusion we reached earlier (see Eq. 5.78) for incompressible flows.

For compressible flow, $d(1/\rho) \neq 0$, and thus when we apply Eq. 5.106 to a finite control volume we obtain

$$\breve{u}_{\text{out}} - \breve{u}_{\text{in}} + \int_{\text{in}}^{\text{out}} pd\left(\frac{1}{\rho}\right) - q_{\substack{\text{net} \\ \text{in}}} = \text{loss} \tag{5.108}$$

indicating that $u_{\text{out}} - u_{\text{in}} - q_{\text{net in}}$ is not equal to loss.

5.4.4 Application of the Loss Form of the Energy Equation

Zero loss is associated with the Bernoulli equation.

Steady flow along a pathline in an incompressible and frictionless flow field provides a simple application of the loss form of the energy equation (Eq. 5.105). We start with Eq. 5.105 and integrate it term by term from one location on the pathline, section (1), to another one downstream, section (2). Note that because the flow is frictionless, loss = 0. Also, because the flow is steady throughout, $w_{\text{shaft net in}} = 0$. Since the flow is incompressible, the density is constant. The control volume in this case is an infinitesimally small diameter streamtube (Fig. 5.7). The resultant equation is

$$\frac{p_2}{\rho} + \frac{V_2^2}{2} + gz_2 = \frac{p_1}{\rho} + \frac{V_1^2}{2} + gz_1 \tag{5.109}$$

which is identical to the Bernoulli equation (Eq. 3.7) already discussed in Chapter 3.

If the frictionless and steady pathline flow of the fluid particle considered above was compressible, application of Eq. 5.105 would yield

$$\int_1^2 \frac{dp}{\rho} + \frac{V_2^2}{2} + gz_2 = \frac{V_1^2}{2} + gz_1 \tag{5.110}$$

To carry out the integration required, $\int_1^2 (dp/\rho)$, a relationship between fluid density, ρ, and pressure, p, must be known. If the frictionless compressible flow we are considering is adiabatic and involves the flow of an ideal gas, it is shown in Section 11.1 that

$$\frac{p}{\rho^k} = \text{constant} \tag{5.111}$$

where $k = c_p/c_v$ is the ratio of gas specific heats, c_p and c_v, which are properties of the fluid. Using Eq. 5.111 we get

$$\int_1^2 \frac{dp}{\rho} = \frac{k}{k-1}\left(\frac{p_2}{\rho_2} - \frac{p_1}{\rho_1}\right) \tag{5.112}$$

Thus, Eqs. 5.110 and 5.112 lead to

$$\frac{k}{k-1}\frac{p_2}{\rho_2} + \frac{V_2^2}{2} + gz_2 = \frac{k}{k-1}\frac{p_1}{\rho_1} + \frac{V_1^2}{2} + gz_1 \tag{5.113}$$

Note that this equation is identical to Eq. 3.24. An example application of Eqs. 5.109 and 5.113 follows.

EXAMPLE 5.29 Energy—Comparison of Compressible and Incompressible Flow

GIVEN Air steadily expands adiabatically and without friction from stagnation conditions of 690 kPa (abs) and 289 K to 101.3 kPa (abs).

FIND Determine the velocity of the expanded air assuming (a) incompressible flow, (b) compressible flow.

SOLUTION

(a) If the flow is considered incompressible, the Bernoulli equation, Eq. 5.109, can be applied to flow through an infinitesimal cross-sectional streamtube, like the one in Fig. 5.7, from the stagnation state (1) to the expanded state (2). From Eq. 5.109 we get

0 (1 is the stagnation state)

$$\frac{p_2}{\rho} + \frac{V_2^2}{2} + \cancel{gz_2} = \frac{p_1}{\rho} + \cancel{\frac{V_1^2}{2}} + \cancel{gz_1} \tag{1}$$

0 (changes in gz are negligible for air flow)

or

$$V_2 = \sqrt{2\left(\frac{p_1 - p_2}{\rho}\right)}$$

We can calculate the density at state (1) by assuming that air behaves like an ideal gas,

$$\rho = \frac{p_1}{RT_1} = \frac{(690 \text{ kPa (abs)})}{(286.9 \text{ J/kg} \cdot \text{K})(289 \text{ K})}$$
$$= 8.3 \text{ kg/m}^3 \tag{2}$$

Thus,

$$V_2 = \sqrt{\frac{2(690 \text{ kPa} - 101.3 \text{ kPa})}{(8.3 \text{ kg/m}^3)\,(1 \text{ N} \cdot \text{s}^2/\text{kg} \cdot \text{m})}}$$
$$= 377 \text{ m/s} \tag{Ans}$$

The assumption of incompressible flow is not valid in this case since for air a change from 690 kPa (abs) to 101.3 kPa (abs) would undoubtedly result in a significant density change.

(b) If the flow is considered compressible, Eq. 5.113 can be applied to the flow through an infinitesimal cross-sectional control volume, like the one in Fig. 5.7, from the stagnation state (1) to the expanded state (2). We obtain

0 (1 is the stagnation state)

$$\frac{k}{k-1}\frac{p_2}{\rho_2} + \frac{V_2^2}{2} + \cancel{gz_2} = \frac{k}{k-1}\frac{p_1}{\rho_1} + \cancel{\frac{V_1^2}{2}} + \cancel{gz_1} \tag{3}$$

0 (changes in gz are negligible for air flow)

or

$$V_2 = \sqrt{\frac{2k}{k-1}\left(\frac{p_1}{\rho_1} - \frac{p_2}{\rho_2}\right)} \tag{4}$$

Given in the problem statement are values of p_1 and p_2. A value of ρ_1 was calculated earlier (Eq. 2). To determine ρ_2 we need to make use of a property relationship for reversible (frictionless) and adiabatic flow of an ideal gas that is derived in Chapter 11; namely,

$$\frac{p}{\rho^k} = \text{constant} \tag{5}$$

where $k = 1.4$ for air. Solving Eq. 5 for ρ_2 we get

$$\rho_2 = \rho_1\left(\frac{p_2}{p_1}\right)^{1/k}$$

or

$$\rho_2 = (8.3 \text{ kg/m}^3)\left[\frac{101.3 \text{ kPa (abs)}}{690 \text{ kPa (abs)}}\right]^{1/1.4} = 2.1 \text{ kg/m}^3$$

Then, from Eq. 4, with $p_1 = 690$ kPa (abs) and $p_2 = 101.3$ kPa (abs),

$$V_2 = \sqrt{\frac{(2)(1.4)}{1.4-1}\left(\frac{690 \text{ kPa (abs)}}{8.3 \text{ kg/m}^3} - \frac{101.3 \text{ kPa (abs)}}{2.1 \text{ kg/m}^3}\right)}$$
$$= 494 \text{ (m}^2\text{/s}^2)^{1/2}$$

or

$$V_2 = 494 \text{ m/s} \tag{Ans}$$

COMMENT A considerable difference exists between the air velocities calculated assuming incompressible and compressible flow. In Section 3.8.1, a discussion of when a fluid flow may be appropriately considered incompressible is provided. Basically, when flow speed is less than a third of the speed of sound in the fluid involved, incompressible flow may be assumed with only a small error.

5.5 Chapter Summary and Study Guide

In this chapter the flow of a fluid is analyzed by using important principles including conservation of mass, Newton's second law of motion, and the first and second laws of thermodynamics as applied to control volumes. The Reynolds transport theorem is used to convert basic system-orientated laws into corresponding control volume formulations.

The continuity equation, a statement of the fact that mass is conserved, is obtained in a form that can be applied to any flow—steady or unsteady, incompressible or compressible. Simplified forms of the continuity equation enable tracking of fluid everywhere in a control volume, where it enters, where it leaves, and within. Mass or volume flowrates of fluid entering or leaving a control volume and rate of accumulation or depletion of fluid within a control volume can be estimated.

The linear momentum equation, a form of Newton's second law of motion applicable to flow of fluid through a control volume, is obtained and used to solve flow problems. Net force results from or causes changes in linear momentum (velocity magnitude and/or direction) of fluid flowing through a control volume. Work and power associated with force can be involved.

The moment-of-momentum equation, which involves the relationship between torque and changes in angular momentum, is obtained and used to solve flow problems dealing with turbines (energy extracted from a fluid) and pumps (energy supplied to a fluid).

The steady-state energy equation, obtained from the first law of thermodynamics (conservation of energy), is written in several forms. The first (Eq. 5.69) involves power terms. The second form (Eq. 5.82 or 5.84) is termed the mechanical energy equation or the extended Bernoulli equation. It consists of the Bernoulli equation with extra terms that account for energy losses due to friction in the flow, as well as terms accounting for the work of pumps or turbines in the flow.

The following checklist provides a study guide for this chapter. When your study of the entire chapter and end-of-chapter exercises has been completed you should be able to

conservation of mass
continuity equation
mass flowrate
linear momentum equation
moment-of-momentum equation
shaft power
shaft torque
first law of thermodynamics
heat transfer rate
energy equation
loss
shaft work head
head loss
kinetic energy coefficient

- ■ write out meanings of the terms listed here in the margin and understand each of the related concepts. These terms are particularly important and are set in *italic, bold, and color* type in the text.
- ■ select an appropriate control volume for a given problem and draw an accurately labeled control volume diagram.
- ■ use the continuity equation and a control volume to solve problems involving mass or volume flowrate.
- ■ use the linear momentum equation and a control volume, in conjunction with the continuity equation as necessary, to solve problems involving forces related to linear momentum change.
- ■ use the moment-of-momentum equation to solve problems involving torque and related work and power due to angular momentum change.
- ■ use the energy equation, in one of its appropriate forms, to solve problems involving losses due to friction (head loss) and energy input by pumps or extraction by turbines.
- ■ use the kinetic energy coefficient in the energy equation to account for nonuniform flows.

Some of the important equations in this chapter are given below.

Conservation of mass	$\dfrac{\partial}{\partial t} \displaystyle\int_{cv} \rho \, d\!\!\!V + \int_{cs} \rho \mathbf{V} \cdot \hat{\mathbf{n}} \, dA = 0$	(5.5)
Mass flowrate	$\dot{m} = \rho Q = \rho A V$	(5.6)
Average velocity	$\overline{V} = \dfrac{\displaystyle\int_{A} \rho \mathbf{V} \cdot \hat{\mathbf{n}} \, dA}{\rho A}$	(5.7)
Steady flow mass conservation	$\sum \dot{m}_{out} - \sum \dot{m}_{in} = 0$	(5.9)
Moving control volume mass conservation	$\dfrac{\partial}{\partial t} \displaystyle\int_{cv} \rho \, d\!\!\!V + \int_{cs} \rho \mathbf{W} \cdot \hat{\mathbf{n}} \, dA = 0$	(5.16)

Deforming control volume mass conservation	$$\frac{DM_{sys}}{Dt} = \frac{\partial}{\partial t}\int_{cv}\rho\,d\forall + \int_{cs}\rho\mathbf{W}\cdot\hat{\mathbf{n}}\,dA = 0$$	(5.17)
Force related to change in linear momentum	$$\frac{\partial}{\partial t}\int_{cv}\mathbf{V}\rho\,d\forall + \int_{cs}\mathbf{V}\rho\mathbf{V}\cdot\hat{\mathbf{n}}\,dA = \sum\mathbf{F}_{\substack{\text{contents of the}\\\text{control volume}}}$$	(5.22)
Moving control volume force related to change in linear momentum	$$\int_{cs}\mathbf{W}\rho\mathbf{W}\cdot\hat{\mathbf{n}}\,dA = \sum\mathbf{F}_{\substack{\text{contents of the}\\\text{control volume}}}$$	(5.29)
Vector addition of absolute and relative velocities	$$\mathbf{V} = \mathbf{W} + \mathbf{U}$$	(5.43)
Shaft torque from force	$$\sum\left[(\mathbf{r}\times\mathbf{F})_{\substack{\text{contents of the}\\\text{control volume}}}\right]_{axial} = \mathbf{T}_{shaft}$$	(5.45)
Shaft torque related to change in moment-of-momentum (angular momentum)	$$T_{shaft} = (-\dot{m}_{in})(\pm r_{in}V_{\theta in}) + \dot{m}_{out}(\pm r_{out}V_{\theta out})$$	(5.50)
Shaft power related to change in moment-of-momentum (angular momentum)	$$\dot{W}_{shaft} = (-\dot{m}_{in})(\pm U_{in}V_{\theta in}) + \dot{m}_{out}(\pm U_{out}V_{\theta out})$$	(5.53)
First law of thermodynamics (Conservation of energy)	$$\frac{\partial}{\partial t}\int_{cv}e\rho\,d\forall + \int_{cs}\left(\breve{u} + \frac{p}{\rho} + \frac{V^2}{2} + gz\right)\rho\mathbf{V}\cdot\hat{\mathbf{n}}\,dA = \dot{Q}_{\substack{net\\in}} + \dot{W}_{\substack{shaft\\net\,in}}$$	(5.64)
Conservation of power	$$\dot{m}\left[\breve{h}_{out} - \breve{h}_{in} + \frac{V_{out}^2 - V_{in}^2}{2} + g(z_{out} - z_{in})\right] = \dot{Q}_{\substack{net\\in}} + \dot{W}_{\substack{shaft\\net\,in}}$$	(5.69)
Conservation of mechanical energy	$$\frac{p_{out}}{\rho} + \frac{V_{out}^2}{2} + gz_{out} = \frac{p_{in}}{\rho} + \frac{V_{in}^2}{2} + gz_{in} + w_{\substack{shaft\\net\,in}} - loss$$	(5.82)

References

1. Eck, B., *Technische Stromungslehre*, Springer-Verlag, Berlin, Germany, 1957.
2. Dean, R. C., "On the Necessity of Unsteady Flow in Fluid Machines," *ASME Journal of Basic Engineering* 81D; 24–28, March 1959.
3. Moran, M. J., and Shapiro, H. N., *Fundamentals of Engineering Thermodynamics*, 6th Ed., Wiley, New York, 2008.

Review Problems

Go to Appendix G for a set of review problems with answers. Detailed solutions can be found in *Student Solution Manual and Study Guide for Fundamentals of Fluid Mechanics*, by Munson et al. (© 2009 John Wiley and Sons, Inc.).

Problems

Note: Unless otherwise indicated, use the values of fluid properties found in the tables on the inside of the front cover. Problems designated with an (*) are intended to be solved with the aid of a programmable calculator or a computer. Problems designated with a (†) are "open-ended" problems and require critical thinking in that to work them one must make various assumptions and provide the necessary data. There is not a unique answer to these problems.

Answers to the even-numbered problems are listed at the end of the book. Access to the videos that accompany problems can be obtained through the book's web site, www.wiley.com/college/munson. The lab-type problems can also be accessed on this web site.

Section 5.1.1 Derivation of the Continuity Equation

5.1 Explain why the mass of the contents of a system is constant with time.

5.2 Explain how the mass of the contents of a control volume can vary with time or not.

5.3 Explain the concept of a coincident control volume and system and why it is useful.

5.4 Obtain a photograph/image of a situation for which the conservation of mass law is important. Briefly describe the situation and its relevance.

Section 5.1.2 Fixed, Nondeforming Control Volume— Uniform Velocity Profile or Average Velocity.

5.5 Water enters a cylindrical tank through two pipes at rates of 950 and 400 L/min (see Fig. P5.5). If the level of the water in the tank remains constant, calculate the average velocity of the flow leaving the tank through an 20 cm inside-diameter pipe.

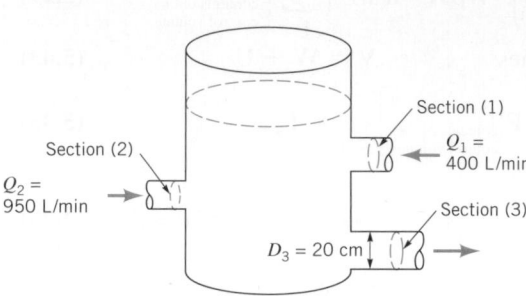

■ **FIGURE P5.5**

5.6 Water flows out through a set of thin, closely spaced blades as shown in Fig. 5.6 with a speed of $V = 3$ m/s around the entire circumference of the outlet. Determine the mass flowrate through the inlet pipe.

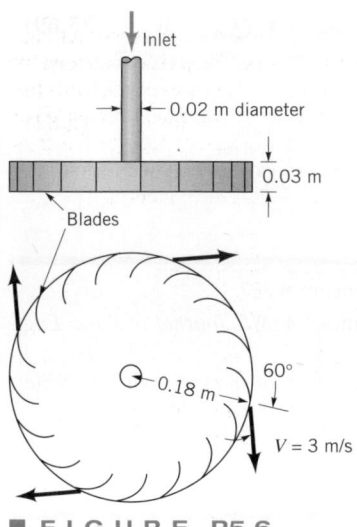

■ **FIGURE P5.6**

5.7 The pump shown in Fig. P5.7 produces a steady flow of 40 L/min through the nozzle. Determine the nozzle exit diameter, D_2, if the exit velocity is to be $V_2 = 30.0$ m/s.

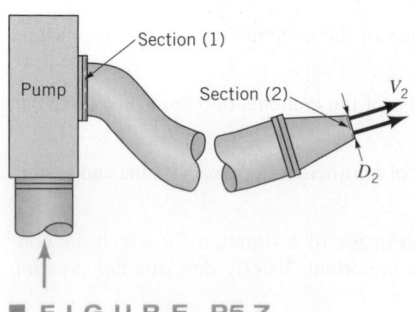

■ **FIGURE P5.7**

5.8 Water flows into a sink as shown in **Video V5.1** and Fig. P5.8 at the rate of 8 L/min. Determine the average velocity through each of the three 1-cm-diameter overflow holes if the drain is closed and the water level in the sink remains constant.

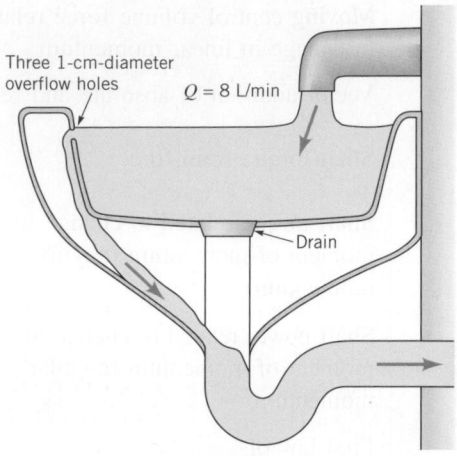

■ **FIGURE P5.8**

5.9 The wind blows through a 2 m $\times$ 3 m garage door opening with a speed of 1.5 m/s as shown in Fig. P5.9. Determine the average speed, V, of the air through the two 1.0 m $\times$ 1.2 m openings in the windows.

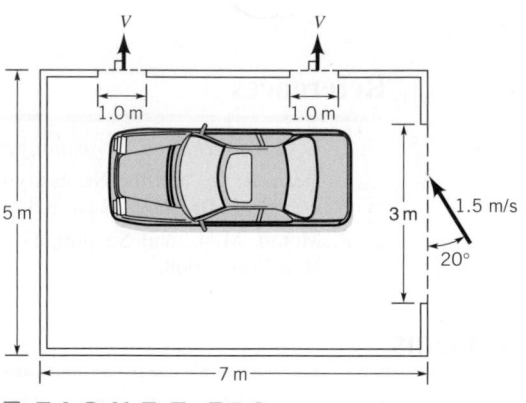

■ **FIGURE P5.9**

5.10 The human circulatory system consists of a complex branching pipe network ranging in diameter from the aorta (largest) to the capillaries (smallest). The average radii and the number of these vessels is shown in the table below. Does the average blood velocity increase, decrease, or remain constant as it travels from the aorta to the capillaries?

Vessel	Average Radius, mm	Number
Aorta	12.5	1
Arteries	2.0	159
Arterioles	0.03	1.4×10^7
Capillaries	0.006	3.9×10^9

5.11 Air flows steadily between two cross sections in a long, straight section of 0.1-m inside diameter pipe. The static temperature and pressure at each section are indicated in Fig. P5.11. If the average air velocity at section (1) is 205 m/s, determine the average air velocity at section (2).

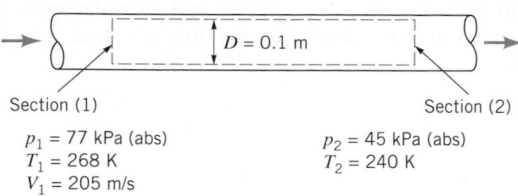

Section (1)   Section (2)

$p_1 = 77$ kPa (abs) $p_2 = 45$ kPa (abs)
$T_1 = 268$ K $T_2 = 240$ K
$V_1 = 205$ m/s

■ **FIGURE P5.11**

5.12 A hydraulic jump (see Video V10.10) is in place downstream from a spillway as indicated in Fig. P5.12. Upstream of the jump, the depth of the stream is 0.2 m and the average stream velocity is 5.0 m/s. Just downstream of the jump, the average stream velocity is 1 m/s. Calculate the depth of the stream, h, just downstream of the jump.

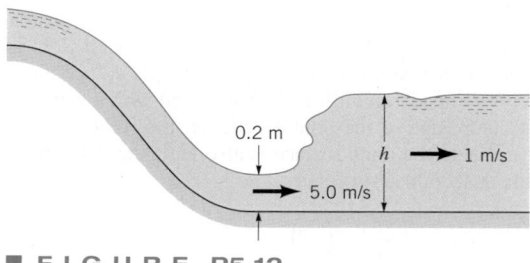

■ **FIGURE P5.12**

5.13 An evaporative cooling tower (see Fig. P5.13) is used to cool water from 38 to 27°C. Water enters the tower at a rate of 31.0 kg/s. Dry air (no water vapor) flows into the tower at a rate of 19 kg/s. If the rate of wet air flow out of the tower is 20 kg/s, determine the rate of water evaporation in kg/s and the rate of cooled water flow in kg/s.

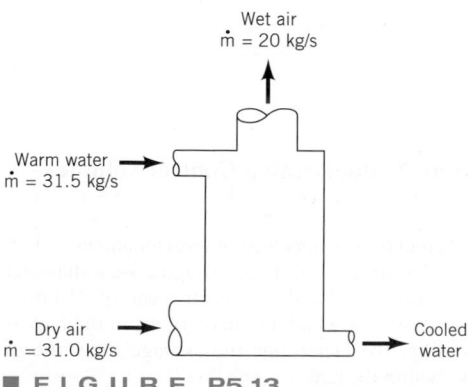

■ **FIGURE P5.13**

5.14 At cruise conditions, air flows into a jet engine at a steady rate of 30 kg/s. Fuel enters the engine at a steady rate of 0.3 kg/s. The average velocity of the exhaust gases is 500 m/s relative to the engine. If the engine exhaust effective cross-sectional area is 0.3 m², estimate the density of the exhaust gases in kg/m³.

5.15 Water at 0.1 m³/s and alcohol ($SG=0.8$) at 0.3 m³/s are mixed in a *y*-duct as shown in Fig. 5.15. What is the average density of the mixture of alcohol and water?

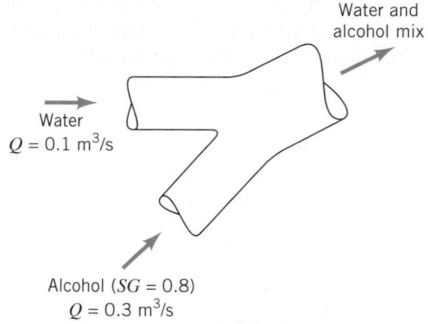

■ **FIGURE P5.15**

5.16 Freshwater flows steadily into an open 200 L drum initially filled with seawater. The freshwater mixes thoroughly with the seawater and the mixture overflows out of the drum. If the freshwater flowrate is 40 L/min, estimate the time in seconds required to decrease the difference between the density of the mixture and the density of freshwater by 50%.

Section 5.1.2 Fixed, Nondeforming Control Volume—Nonuniform Velocity Profile

5.17 A water jet pump (see Fig. P5.17) involves a jet cross-sectional area of 0.01 m², and a jet velocity of 30 m/s. The jet is surrounded by entrained water. The total cross-sectional area associated with the jet and entrained streams is 0.075 m². These two fluid streams leave the pump thoroughly mixed with an average velocity of 6 m/s through a cross-sectional area of 0.075 m². Determine the pumping rate (i.e., the entrained fluid flowrate) involved in liters/s.

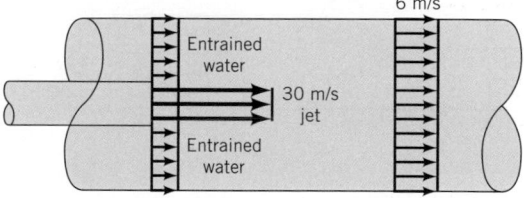

■ **FIGURE P5.17**

5.18 Two rivers merge to form a larger river as shown in Fig. P5.18. At a location downstream from the junction (before the two streams completely merge), the nonuniform velocity profile is as shown and the depth is 2.0 m. Determine the value of V.

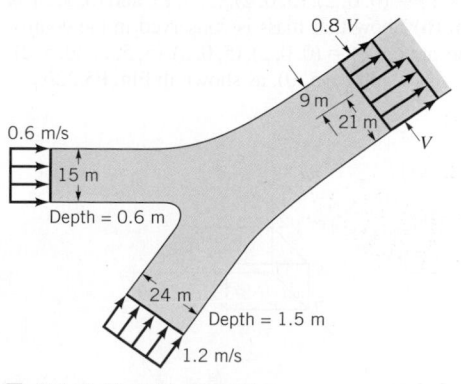

■ **FIGURE P5.18**

5.19 Various types of attachments can be used with the shop vac shown in **Video V5.2.** Two such attachments are shown in Fig. P5.19 —a nozzle and a brush. The flowrate is 0.03 m³/s. **(a)** Determine the average velocity through the nozzle entrance, V_n. **(b)** Assume the air enters the brush attachment in a radial direction all around the brush with a velocity profile that varies linearly from 0 to V_b along the length of the bristles as shown in the figure. Determine the value of V_b.

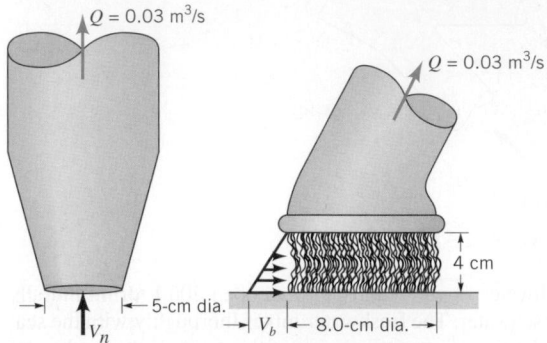

■ **FIGURE P5.19**

5.20 An appropriate turbulent pipe flow velocity profile is

$$\mathbf{V} = u_c\left(\frac{R-r}{R}\right)^{1/n}\hat{\mathbf{i}}$$

where u_c = centerline velocity, r = local radius, R = pipe radius, and $\hat{\mathbf{i}}$ = unit vector along pipe centerline. Determine the ratio of average velocity, $\bar{u}$, to centerline velocity, u_c, for **(a)** $n = 4$, **(b)** $n = 6$, **(c)** $n = 8$, **(d)** $n = 10$. Compare the different velocity profiles.

5.21 As shown in Fig. P5.21, at the entrance to a 1.0-m-wide channel the velocity distribution is uniform with a velocity V. Further downstream the velocity profile is given by $u = 4y - 2y^2$, where u is in m/s and y is in m. Determine the value of V.

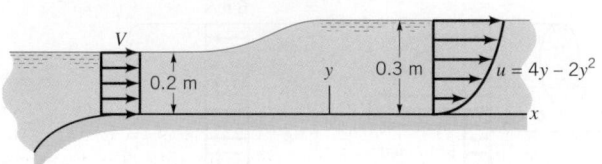

■ **FIGURE P5.21**

5.22 A water flow situation is described by the velocity field equation

$$\mathbf{V} = (3x + 2)\hat{\mathbf{i}} + (2y - 4)\hat{\mathbf{j}} - 5z\hat{\mathbf{k}} \text{ m/s}$$

where $x, y,$ and z are in meters. **(a)** Determine the mass flowrate through the rectangular area in the plane corresponding to $z = 2$ meters having corners at $(x, y, z) = (0, 0, 2), (5, 0, 2), (5, 5, 2),$ and $(0, 5, 2)$ as shown in Fig P5.22a. **(b)** Show that mass is conserved in the control volume having corners at $(x, y, z) = (0, 0, 2), (5, 0, 2), (5, 5, 2), (0, 5, 2), (0, 0, 0), (5, 0, 0), (5, 5, 0),$ and $(0, 5, 0)$, as shown in Fig. P5.22b.

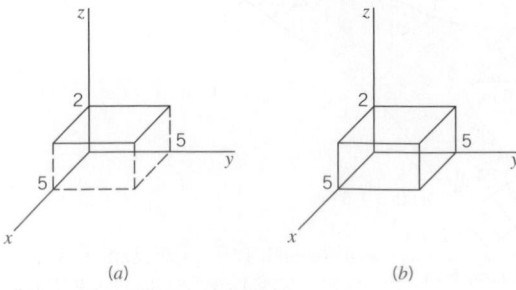

■ **FIGURE P5.22**

5.23 An incompressible flow velocity field (water) is given as

$$\mathbf{V} = -\frac{1}{r}\hat{\mathbf{e}}_r + \frac{1}{r}\hat{\mathbf{e}}_\theta \text{ m/s}$$

where r is in meters. **(a)** Calculate the mass flowrate through the cylindrical surface at $r = 1$ m from $z = 0$ to $z = 1$ m as shown in Fig.P5.23a. **(b)** Show that mass is conserved in the annular control volume from $r = 1$ m to $r = 2$ m and $z = 0$ to $z = 1$ m as shown in Fig. P5.23b.

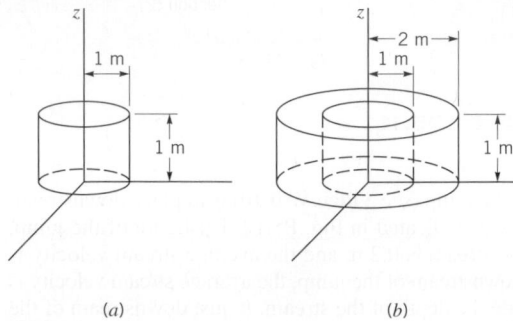

(a)   (b)

■ **FIGURE P5.23**

5.24 Flow of a viscous fluid over a flat plate surface results in the development of a region of reduced velocity adjacent to the wetted surface as depicted in Fig. P5.24. This region of reduced flow is called a boundary layer. At the leading edge of the plate, the velocity profile may be considered uniformly distributed with a value U. All along the outer edge of the boundary layer, the fluid velocity component parallel to the plate surface is also U. If the x direction velocity profile at section (2) is

$$\frac{u}{U} = \left(\frac{y}{\delta}\right)^{1/7}$$

develop an expression for the volume flowrate through the edge of the boundary layer from the leading edge to a location downstream at x where the boundary layer thickness is δ.

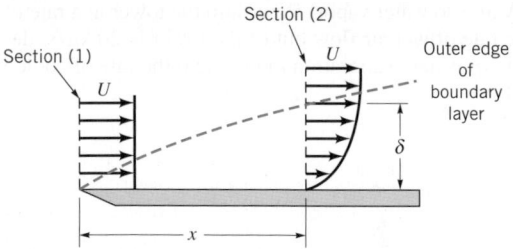

■ **FIGURE P5.24**

Section 5.1.2 Fixed, Nondeforming Control Volume— Unsteady Flow

5.25 Air at standard conditions enters the compressor shown in Fig. P5.25 at a rate of 0.3 m³/s. It leaves the tank through a 3-cm-diameter pipe with a density of 1.8 kg/m³ and a uniform speed of 210 m/s. **(a)** Determine the rate (kg/s) at which the mass of air in the tank is increasing or decreasing. **(b)** Determine the average time rate of change of air density within the tank.

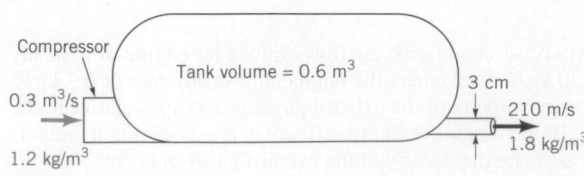

■ **FIGURE P5.25**

5.26 Estimate the time required to fill with water a cone-shaped container (see Fig. P5.26) 1.5 m high and 1.5 m across at the top if the filling rate is 76 L/min.

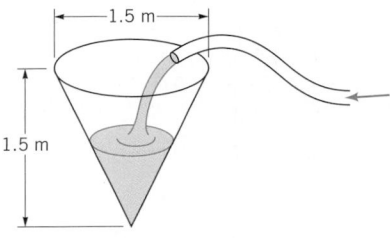

■ **F I G U R E P5.26**

†**5.27** Estimate the maximum flowrate of rainwater (during a heavy rain) that you would expect from the downspout connected to the gutters of your house. List all assumptions and show all calculations.

Section 5.1.3 Moving, Nondeforming Control Volume

5.28 For an automobile moving along a highway, describe the control volume you would use to estimate the flowrate of air across the radiator. Explain how you would estimate the velocity of that air.

Section 5.1.4 Deforming Control Volume

5.29 A hypodermic syringe (see Fig. P5.29) is used to apply a vaccine. If the plunger is moved forward at the steady rate of 20 mm/s and if vaccine leaks past the plunger at 0.1 of the volume flowrate out the needle opening, calculate the average velocity of the needle exit flow. The inside diameters of the syringe and the needle are 20 mm and 0.7 mm.

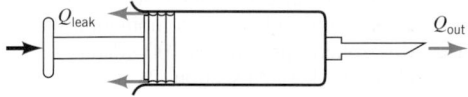

■ **F I G U R E P5.29**

5.30 The Hoover Dam (see **Video V2.4**) backs up the Colorado River and creates Lake Mead, which is approximately 185 km long and has a surface area of approximately 580 square kilometers. If during flood conditions the Colorado River flows into the lake at a rate of 1300 m³/s and the outflow from the dam is 227 m³/s, how many meters per 24-hour day will the lake level rise?

5.31 Storm sewer backup causes your basement to flood at the steady rate of 2.5 cm of depth per hour. The basement floor area is 139 m². What capacity (L/min) pump would you rent to **(a)** keep the water accumulated in your basement at a constant level until the storm sewer is blocked off, and **(b)** reduce the water accumulation in your basement at a rate of 75 mm/hr even while the backup problem exists?

5.32 (See Fluids in the News article **"New (6 lpf) standards,"** Section 5.1.2.) When a toilet is flushed, the water depth, h, in the tank as a function of time, t, is as given in the table. The size of the rectangular tank is 50 cm by 20 cm **(a)** Determine the volume of water used per flush, liters per flush. **(b)** Plot the flowrate for $0 \le t \le 6$ s.

t (s)	h (cm)
0	14.5
0.5	13.5
1.0	12.1
2.0	8.8
3.0	6.1
4.0	3.8
5.0	1.9
6.0	0

Section 5.2.1 Derivation of the Linear Momentum Equation

5.33 What is fluid linear momentum and the "flow" of linear momentum?

5.34 Explain the physical meaning of each of the terms of the linear momentum equation (Eq. 5.22).

5.35 What is an inertial control volume?

5.36 Distinguish between body and surface forces.

5.37 Obtain a photograph/image of a situation in which the linear momentum of a fluid changes during flow from one location to another. Explain briefly how force is involved.

Section 5.2.2 Application of the Linear Momentum Equation (Also see Lab Problems 5.140, 5.141, 5.142, and 5.143.)

5.38 A 10-mm diameter jet of water is deflected by a homogeneous rectangular block (15 mm by 200 mm by 100 mm) that weighs 6 N as shown in **Video V5.6** and Fig. P5.38. Determine the minimum volume flowrate needed to tip the block.

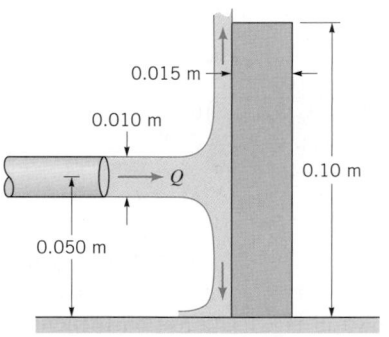

■ **F I G U R E P5.38**

5.39 Determine the anchoring force required to hold in place the conical nozzle attached to the end of the laboratory sink faucet shown in Fig. P5.39 when the water flowrate is 40 L/min. The nozzle weight is 1.0 N. The nozzle inlet and exit inside diameters are 1.5 and 0.5 cm, respectively. The nozzle axis is vertical and the axial distance between sections (1) and (2) is 3 cm. The pressure at section (1) is 470 kPa.

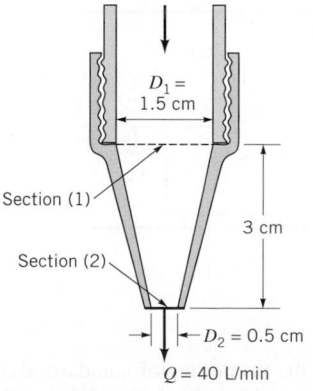

■ **F I G U R E P5.39**

5.40 Water flows through a horizontal, 180° pipe bend as is illustrated in Fig. P5.40. The flow cross section area is constant at a value of 9000 mm². The flow velocity everywhere in the bend is 15 m/s.

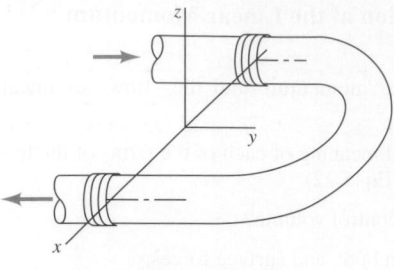

■ **F I G U R E   P5.40**

The pressures at the entrance and exit of the bend are 210 and 165 kPa, respectively. Calculate the horizontal (x and y) components of the anchoring force needed to hold the bend in place.

5.41 Water enters the horizontal, circular cross-sectional, sudden contraction nozzle sketched in Fig. P5.41 at section (1) with a uniformly distributed velocity of 7.5 m/s and a pressure of 500 kPa. The water exits from the nozzle into the atmosphere at section (2) where the uniformly distributed velocity is 30 m/s. Determine the axial component of the anchoring force required to hold the contraction in place.

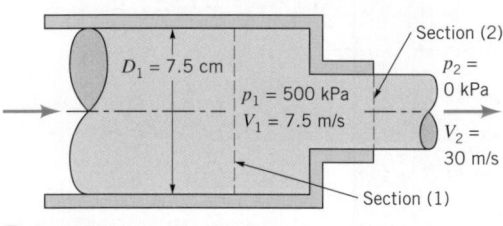

■ **F I G U R E   P5.41**

5.42 The four devices shown in Fig. P5.42 rest on frictionless wheels, are restricted to move in the x direction only, and are initially held stationary. The pressure at the inlets and outlets of each is atmospheric, and the flow is incompressible. The contents of each device is not known. When released, which devices will move to the right and which to the left? Explain.

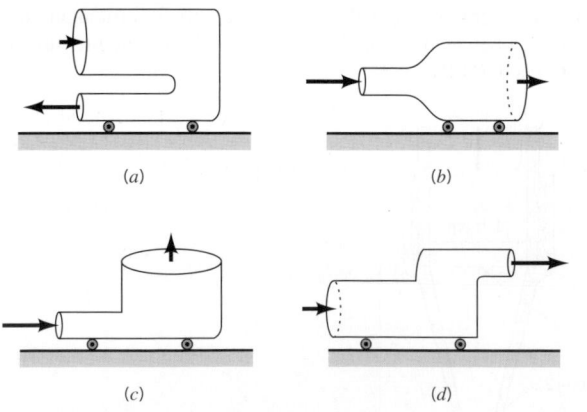

■ **F I G U R E   P5.42**

5.43 Exhaust (assumed to have the properties of standard air) leaves the 1.0-m-diameter chimney shown in **Video V5.4** and Fig. P5.43 with a speed of 2.0 m/s. Because of the wind, after a few diameters downstream the exhaust flows in a horizontal direction with the speed of the wind, 5.0 m/s. Determine the horizontal component of the force that the blowing wind puts on the exhaust gases.

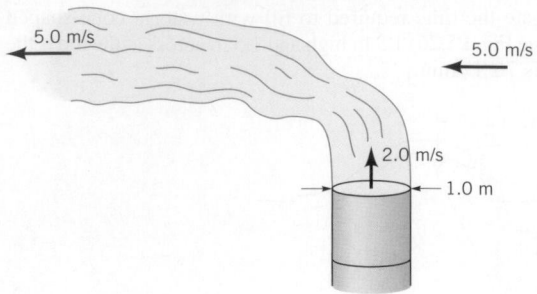

■ **F I G U R E   P5.43**

5.44 Air flows steadily between two cross sections in a long, straight section of 30-cm-inside diameter pipe. The static temperature and pressure at each section are indicated in Fig P5.44. If the average air velocity at section (2) is 320 m/s, determine the average air velocity at section (1). Determine the frictional force exerted by the pipe wall on the air flowing between sections (1) and (2). Assume uniform velocity distributions at each section.

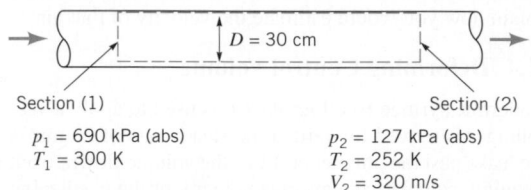

Section (1)
$p_1 = 690$ kPa (abs)
$T_1 = 300$ K

Section (2)
$p_2 = 127$ kPa (abs)
$T_2 = 252$ K
$V_2 = 320$ m/s

■ **F I G U R E   P5.44**

5.45 Determine the magnitude and direction of the anchoring force needed to hold the horizontal elbow and nozzle combination shown in Fig. P5.45 in place. Atmospheric pressure is 100 kPa(abs). The gage pressure at section (1) is 100 kPa. At section (2), the water exits to the atmosphere.

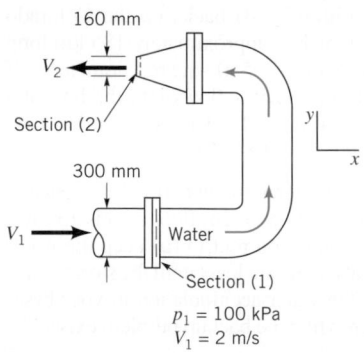

$p_1 = 100$ kPa
$V_1 = 2$ m/s

■ **F I G U R E   P5.45**

5.46 Water flows as two free jets from the tee attached to the pipe shown in Fig. P5.46. The exit speed is 15 m/s. If viscous effects and gravity are negligible, determine the x and y components of the force that the pipe exerts on the tee.

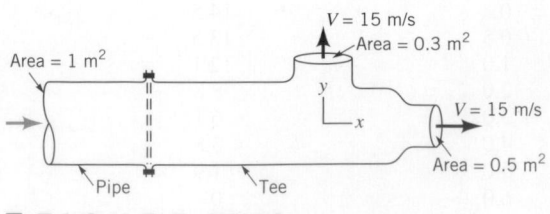

■ **F I G U R E   P5.46**

5.47 A converging elbow (see Fig. P5.47) turns water through an angle of 135° in a vertical plane. The flow cross section diameter is 400 mm at the elbow inlet, section (1), and 200 mm at the elbow outlet, section (2). The elbow flow passage volume is 0.2 m³ between sections (1) and (2). The water volume flowrate is 0.4 m³/s and the elbow inlet and outlet pressures are 150 kPa and 90 kPa. The elbow mass is 12 kg. Calculate the horizontal (x direction) and vertical (z direction) anchoring forces required to hold the elbow in place.

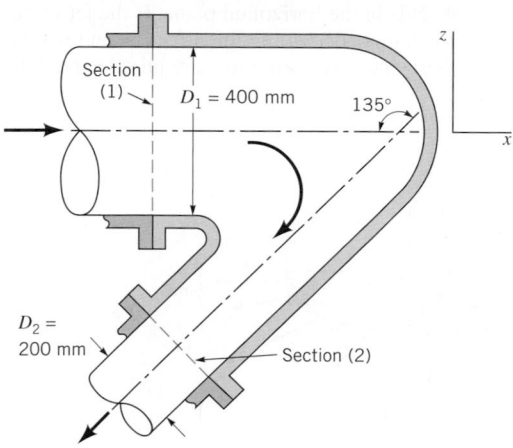

■ **FIGURE P5.47**

5.48 The hydraulic dredge shown in Fig. P5.48 is used to dredge sand from a river bottom. Estimate the thrust needed from the propeller to hold the boat stationary. Assume the specific gravity of the sand/water mixture is $SG = 1.2$.

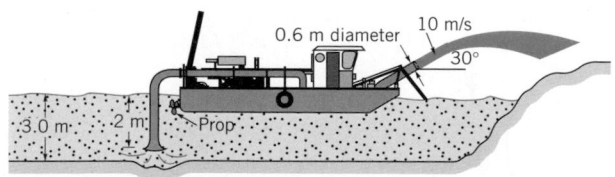

■ **FIGURE P5.48**

5.49 A static thrust stand is to be designed for testing a specific jet engine. Knowing the following conditions for a typical test,

$$\begin{aligned}
\text{intake air velocity} &= 210 \text{ m/s} \\
\text{exhaust gas velocity} &= 500 \text{ m/s} \\
\text{intake cross section area} &= 1 \text{ m}^2 \\
\text{intake static pressure} &= 79 \text{ kPa(abs)} \\
\text{intake static temperature} &= 267 \text{ K} \\
\text{exhaust gas pressure} &= 0 \text{ kPa}
\end{aligned}$$

estimate a nominal thrust to design for.

5.50 A horizontal, circular cross-sectional jet of air having a diameter of 15 cm strikes a conical deflector as shown in Fig. P5.50. A horizontal anchoring force of 22 N is required to hold the cone in

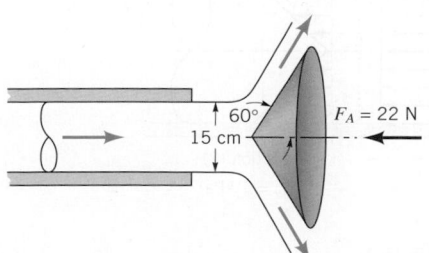

■ **FIGURE P5.50**

place. Estimate the nozzle flowrate in m³/s. The magnitude of the velocity of the air remains constant.

5.51 A vertical, circular cross-sectional jet of air strikes a conical deflector as indicated in Fig. P5.51. A vertical anchoring force of 0.1 N is required to hold the deflector in place. Determine the mass (kg) of the deflector. The magnitude of velocity of the air remains constant.

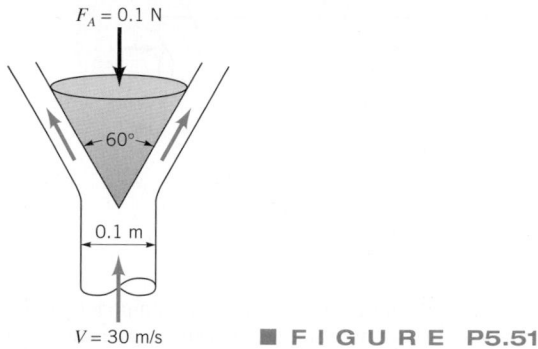

■ **FIGURE P5.51**

5.52 Water flows from a large tank into a dish as shown in Fig. P5.52. **(a)** If at the instant shown the tank and the water in it weigh W_1 N, what is the tension, T_1, in the cable supporting the tank? **(b)** If at the instant shown the dish and the water in it weigh W_2 N, what is the force, F_2, needed to support the dish?

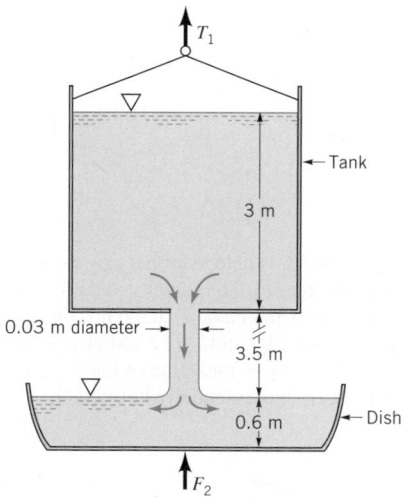

■ **FIGURE P5.52**

5.53 Two water jets of equal size and speed strike each other as shown in Fig. P5.53. Determine the speed, V, and direction, θ, of the resulting combined jet. Gravity is negligible.

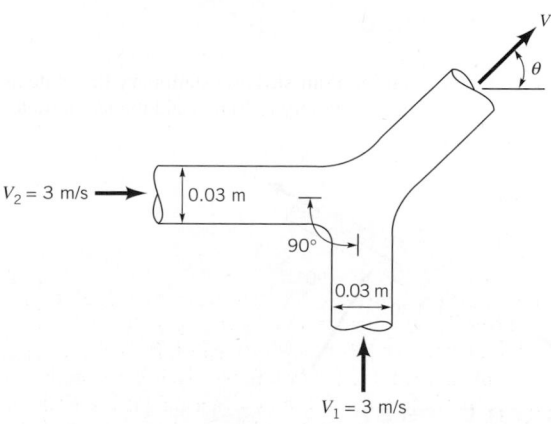

■ **FIGURE P5.53**

5.54 Assuming frictionless, incompressible, one-dimensional flow of water through the horizontal tee connection sketched in Fig. P5.54, estimate values of the x and y components of the force exerted by the tee on the water. Each pipe has an inside diameter of 1 m.

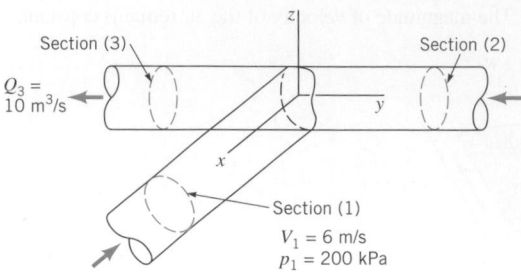

■ **FIGURE P5.54**

5.55 Determine the magnitude of the horizontal component of the anchoring force required to hold in place the sluice gate shown in Fig. 5.55. Compare this result with the size of the horizontal component of the anchoring force required to hold in place the sluice gate when it is closed and the depth of water upstream is 3 m.

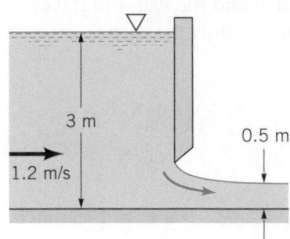

■ **FIGURE P5.55**

5.56 The rocket shown in Fig. P5.56. is held stationary by the horizontal force, F_x, and the vertical force, F_z. The velocity and pressure of the exhaust gas are 1500 m/s and 138 kPa (abs) at the nozzle exit, which has a cross section area of 390 cm². The exhaust mass flowrate is constant at 10 kg/s. Determine the value of the restraining force F_x. Assume the exhaust flow is essentially horizontal.

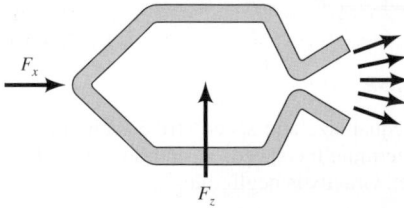

■ **FIGURE P5.56**

5.57 A horizontal circular jet of air strikes a stationary flat plate as indicated in Fig. 5.57. The jet velocity is 40 m/s and the jet diameter

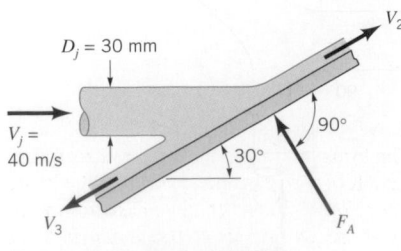

■ **FIGURE P5.57**

is 30 mm. If the air velocity magnitude remains constant as the air flows over the plate surface in the directions shown, determine: **(a)** the magnitude of F_A, the anchoring force required to hold the plate stationary; **(b)** the fraction of mass flow along the plate surface in each of the two directions shown; **(c)** the magnitude of F_A, the anchoring force required to allow the plate to move to the right at a constant speed of 10 m/s.

5.58 Water is sprayed radially outward over 180° as indicated in Fig. P5.58. The jet sheet is in the horizontal plane. If the jet velocity at the nozzle exit is 6 m/s, determine the direction and magnitude of the resultant horizontal anchoring force required to hold the nozzle in place.

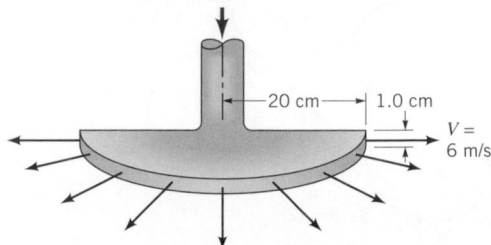

■ **FIGURE P5.58**

5.59 A sheet of water of uniform thickness ($h = 0.01$ m) flows from the device shown in Fig. P5.59. The water enters vertically through the inlet pipe and exits horizontally with a speed that varies linearly from 0 to 10 m/s along the 0.2-m length of the slit. Determine the y component of anchoring force necessary to hold this device stationary.

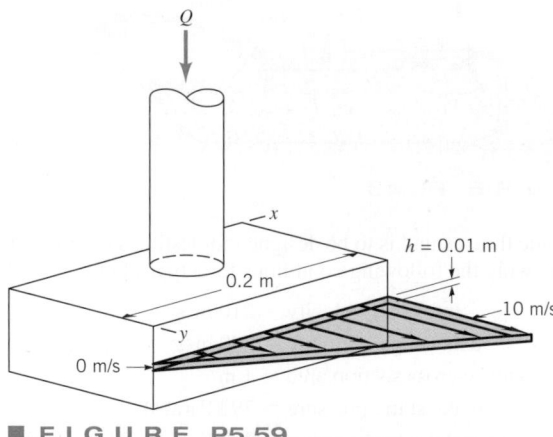

■ **FIGURE P5.59**

5.60 A variable mesh screen produces a linear and axisymmetric velocity profile as indicated in Fig. P5.60 in the air flow through a

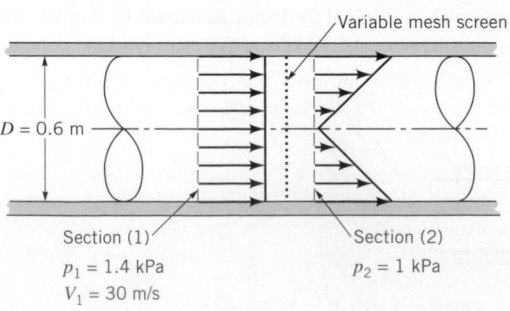

■ **FIGURE P5.60**

0.5-m-diameter circular cross section duct. The static pressures upstream and downstream of the screen are 1.4 and 1 kPa and are uniformly distributed over the flow cross section area. Neglecting the force exerted by the duct wall on the flowing air, calculate the screen drag force.

5.61 Water flows vertically upward in a circular cross-sectional pipe as shown in Fig. P5.61. At section (1), the velocity profile over the cross-sectional area is uniform. At section (2), the velocity profile is

$$\mathbf{V} = w_c \left(\frac{R - r}{R} \right)^{1/7} \hat{\mathbf{k}}$$

where $\mathbf{V}$ = local velocity vector, w_c = centerline velocity in the axial direction, R = pipe radius, and r = radius from pipe axis. Develop an expression for the fluid pressure drop that occurs between sections (1) and (2).

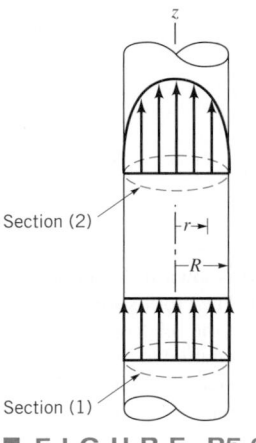

■ **FIGURE P5.61**

5.62 In a laminar pipe flow that is fully developed, the axial velocity profile is parabolic. That is,

$$u = u_c \left[1 - \left(\frac{r}{R} \right)^2 \right]$$

as is illustrated in Fig. P5.62. Compare the axial direction momentum flowrate calculated with the average velocity, $\bar{u}$, with the axial direction momentum flowrate calculated with the nonuniform velocity distribution taken into account.

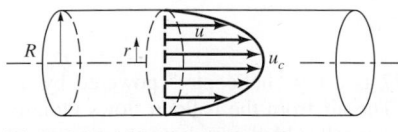

■ **FIGURE P5.62**

†**5.63** Water from a garden hose is sprayed against your car to rinse dirt from it. Estimate the force that the water exerts on the car. List all assumptions and show calculations.

5.64 A Pelton wheel vane directs a horizontal, circular cross-sectional jet of water symmetrically as indicated in Fig. P5.64 and **Video V5.6**. The jet leaves the nozzle with a velocity of 30 m/s. Determine the x direction component of anchoring force required to **(a)** hold the vane stationary, **(b)** confine the speed of the vane to a value of 3 m/s to the right. The fluid speed magnitude remains constant along the vane surface.

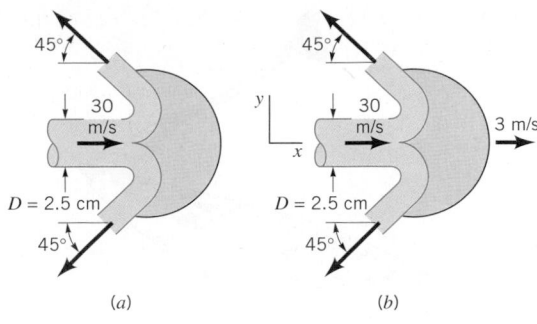

■ **FIGURE P5.64**

5.65 How much power is transferred to the moving vane of Problem 5.64?

5.66 The thrust developed to propel the jet ski shown in **Video V9.11** and Fig. P5.66 is a result of water pumped through the vehicle and exiting as a high-speed water jet. For the conditions shown in the figure, what flowrate is needed to produce a 1.3 kN thrust? Assume the inlet and outlet jets of water are free jets.

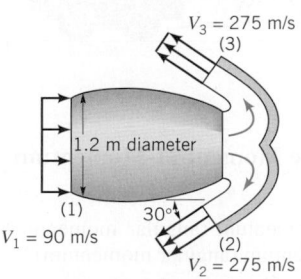

■ **FIGURE P5.66**

5.67 (See Fluids in the News article titled **"Where the plume goes,"** Section 5.2.2.) Air flows into the jet engine shown in Fig. P5.67 at a rate of 131 kg/s slugs/s and a speed of 90 m/s. Upon landing, the engine exhaust exits through the reverse thrust mechanism with a speed of 275 m/s in the direction indicated. Determine the reverse thrust applied by the engine to the airplane. Assume the inlet and exit pressures are atmospheric and that the mass flowrate of fuel is negligible compared to the air flowrate through the engine.

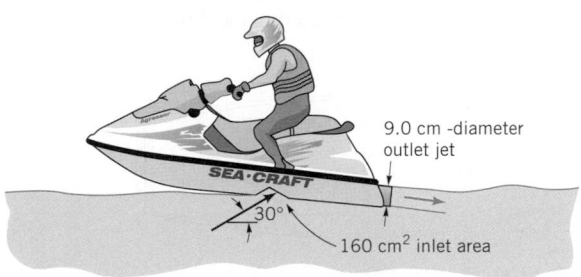

■ **FIGURE P5.67**

5.68 (See Fluids in the News article titled **"Motorized surfboard,"** Section 5.2.2.) The thrust to propel the powered surfboard shown in Fig. P5.68 is a result of water pumped through the board that exits as a high-speed 7-cm-diameter jet. Determine the flowrate and the velocity of the exiting jet if the thrust is to be 1.3 kN. Neglect the momentum of the water entering the pump.

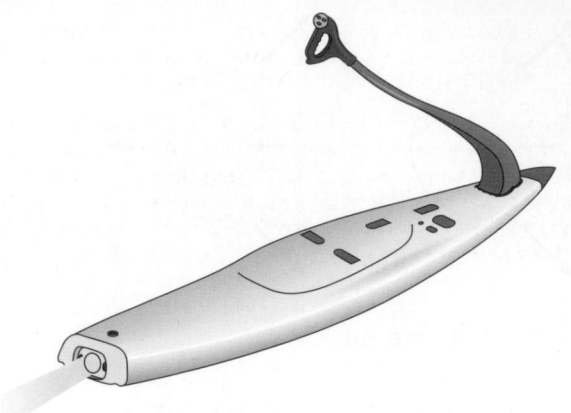

■ **F I G U R E   P5.68**

5.69 (See Fluids in the News article titled **"Bow thrusters,"** Section 5.2.2). The bow thruster on the boat shown in Fig. P5.69 is used to turn the boat. The thruster produces a 1-m-diameter jet of water with a velocity of 10 m/s. Determine the force produced by the thruster. Assume that the inlet and outlet pressures are zero and that the momentum of the water entering the thruster is negligible.

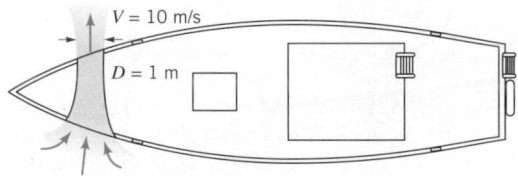

■ **F I G U R E   P5.69**

5.70 A snowplow mounted on a truck clears a path 3.5 m through heavy wet snow, as shown in Figure P5.70. The snow is 20 cm deep and its density is 160 kg/m³. The truck travels at 48 km/hr. The snow is discharged from the plow at an angle of 45° from the direction of travel and 45° above the horizontal, as shown in Figure P5.70. Estimate the force required to push the plow.

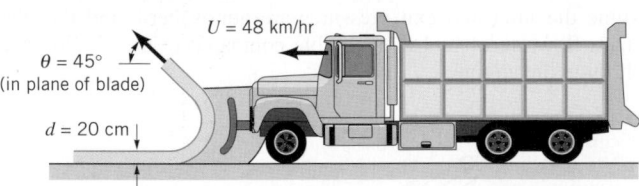

■ **F I G U R E   P5.70**

Section 5.2.3   Derivation of the Moment-of-Momentum Equation

5.71 What is fluid moment-of-momentum (angular momentum) and the "flow" of moment-of-momentum (angular momentum)?

5.72 Describe the orthogonal components of the moment-of-momentum equation (Eq. 5.42) and comment on the direction of each.

5.73 Describe a few examples (include photographs/images) of turbines where the force/torque of a flowing fluid leads to rotation of a shaft.

5.74 Describe a few examples (include photographs/images) of pumps where a fluid is forced to move by "blades" mounted on a rotating shaft.

Section 5.2.4   Application of the Moment-of-Momentum Equation

5.75 Water enters a rotating lawn sprinkler through its base at the steady rate of 60 L/min as shown in Fig. P5.75. The exit cross-sectional area of each of the two nozzles is 0.26 cm², and the flow leaving each nozzle is tangential. The radius from the axis of rotation to the centerline of each nozzle is 20 cm. (a) Determine the resisting torque required to hold the sprinkler head stationary. (b) Determine the resisting torque associated with the sprinkler rotating with a constant speed of 500 rev/min. (c) Determine the angular velocity of the sprinkler if no resisting torque is applied.

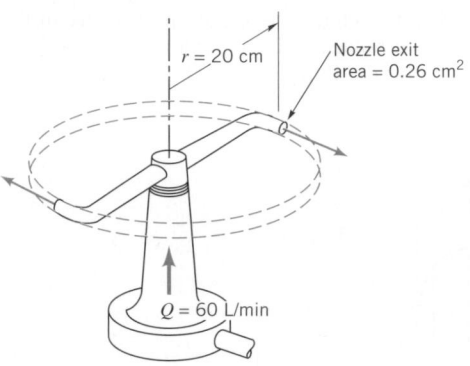

■ **F I G U R E   P5.75**

5.76 Five liters/s of water enter the rotor shown in **Video V5.10** and Fig. P5.76 along the axis of rotation. The cross-sectional area of each of the three nozzle exits normal to the relative velocity is 18 mm². How large is the resisting torque required to hold the rotor stationary? How fast will the rotor spin steadily if the resisting torque is reduced to zero and (a) $\theta = 0°$, (b) $\theta = 30°$, (c) $\theta = 60°$?

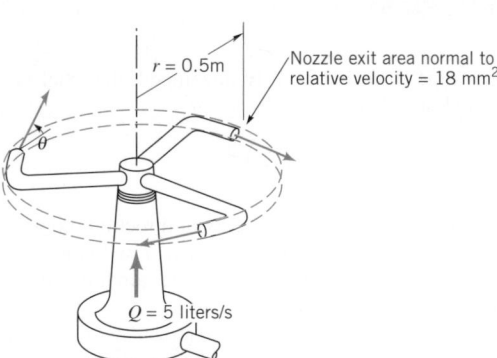

■ **F I G U R E   P5.76**

5.77 Shown in Fig. P5.77 is a toy "helicopter" powered by air escaping from a balloon. The air from the balloon flows radially through each of the three propeller blades and out through small nozzles at the tips of the blades. Explain physically how this flow can cause the rotation necessary to rotate the blades to produce the needed lifting force.

5.78 A simplified sketch of a hydraulic turbine runner is shown in Fig. P5.78. Relative to the rotating runner, water enters at section (1) (cylindrical cross section area A_1 at $r_1=1.5$ m) at an angle of 100° from the tangential direction and leaves at section (2) (cylindrical cross section area A_2 at $r_2=0.85$ m) at an angle of 50° from the tangential direction. The blade height at sections (1) and (2) is 0.45 m and the volume flowrate through the turbine is 30 m³/s. The runner speed is 130 rpm in the direction shown. Determine the shaft power developed.

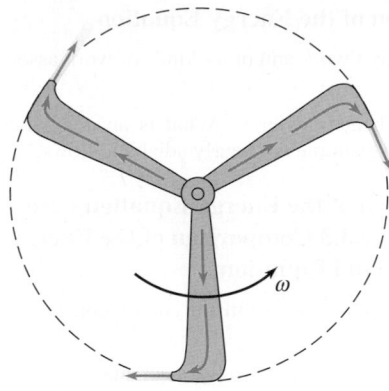

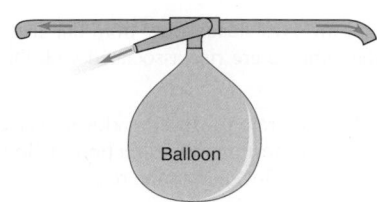

■ FIGURE P5.77

angle of 30° with the tangent to the rotor. The absolute exit velocity is directed radially inward. The angular speed of the rotor is 120 rpm. Find the power delivered to the shaft of the turbine.

5.80 Shown in Fig. P5.80 are front and side views of a centrifugal pump rotor or impeller. If the pump delivers 200 liters/s of water and the blade exit angle is 35° from the tangential direction, determine the power requirement associated with flow leaving at the blade angle. The flow entering the rotor blade row is essentially radial as viewed from a stationary frame.

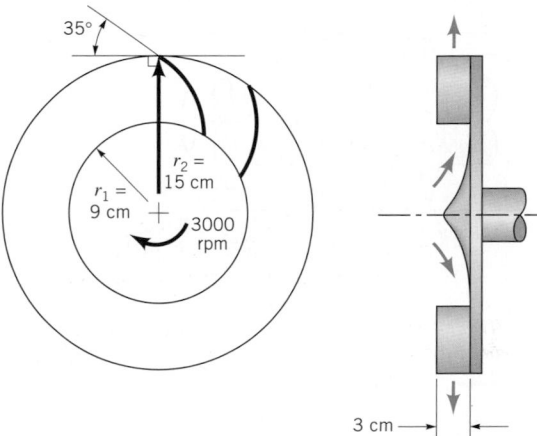

■ FIGURE P5.80

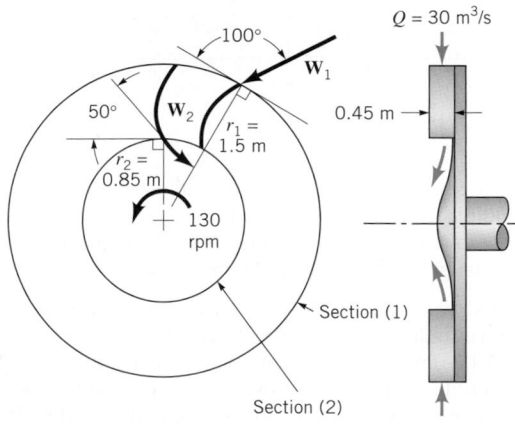

■ FIGURE P5.78

5.81 The velocity triangles for water flow through a radial pump rotor are as indicated in Fig. P5.81. **(a)** Determine the energy added to each unit mass (kg) of water as it flows through the rotor. **(b)** Sketch an appropriate blade section.

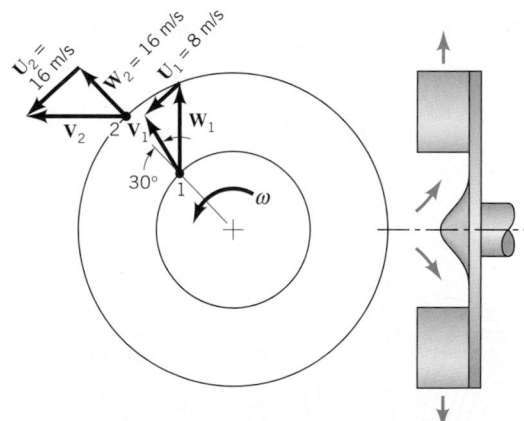

■ FIGURE P5.81

5.79 A water turbine with radial flow has the dimensions shown in Fig.P5.79.The absolute entering velocity is 15 m/s, and it makes an

5.82 An axial flow turbomachine rotor involves the upstream (1) and downstream (2) velocity triangles shown in Fig.P5.82. Is this turbomachine a turbine or a fan? Sketch an appropriate blade section and determine energy transferred per unit mass of fluid.

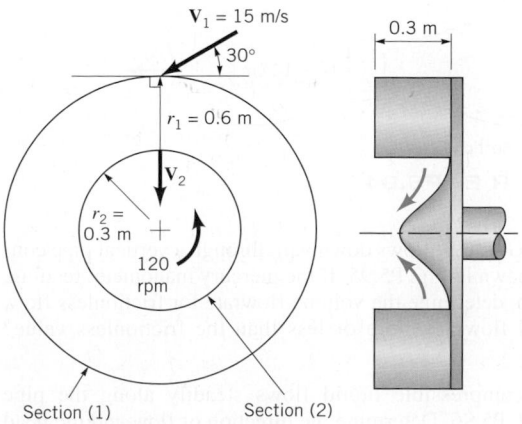

■ FIGURE P5.79

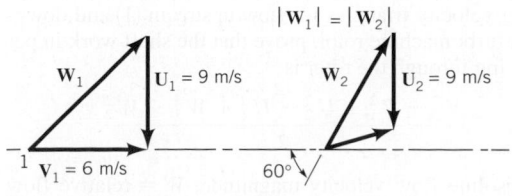

■ FIGURE P5.82

5.83 An axial flow gasoline pump (see Fig. P5.83) consists of a rotating row of blades (rotor) followed downstream by a stationary row of blades (stator). The gasoline enters the rotor axially (without any angular momentum) with an absolute velocity of 3 m/s. The rotor blade inlet and exit angles are 60° and 45° from the axial direction. The pump annulus passage cross-sectional area is constant. Consider the flow as being tangent to the blades involved. Sketch velocity triangles for flow just upstream and downstream of the rotor and just downstream of the stator where the flow is axial. How much energy is added to each kilogram of gasoline? Is this an actual or ideal amount?

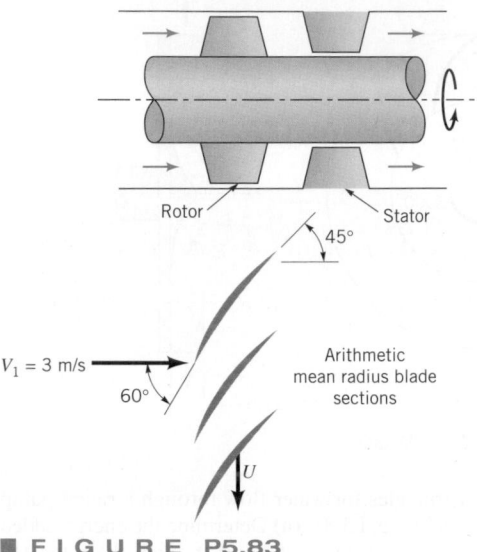

■ **F I G U R E   P5.83**

5.84 Sketch the velocity triangles for the flows entering and leaving the rotor of the turbine-type flow meter shown in Fig. P5.84. Show how rotor angular velocity is proportional to average fluid velocity.

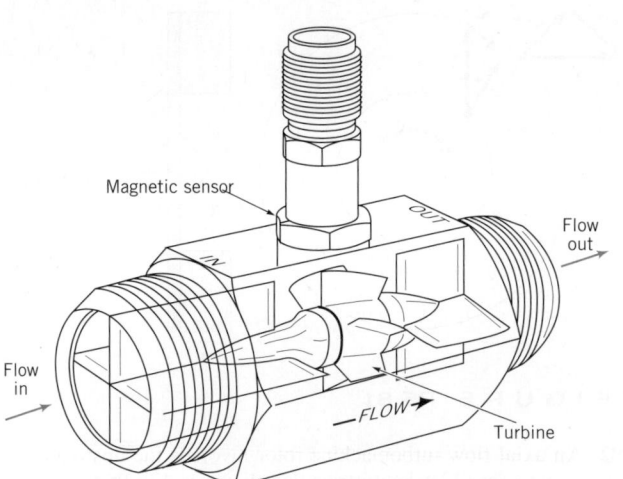

■ **F I G U R E   P5.84** **(Courtesy of EG&G Flow Technology, Inc.)**

5.85 By using velocity triangles for flow upstream (1) and downstream (2) of a turbomachine rotor, prove that the shaft work in per unit mass flowing through the rotor is

$$w_{\substack{\text{shaft} \\ \text{net in}}} = \frac{V_2^2 - V_1^2 + U_2^2 - U_1^2 + W_1^2 - W_2^2}{2}$$

where V = absolute flow velocity magnitude, W = relative flow velocity magnitude, and U = blade speed.

Section 5.3.1 Derivation of the Energy Equation

5.86 Distiguish between shaft work and other kinds of work associated with a flowing fluid.

5.87 Define briefly what heat transfer is. What is an adiabatic flow? Give several practical examples of nearly adiabatic flows.

Section 5.3.2 Application of the Energy Equation – No Shaft Work and Section 5.3.3 Comparison of the Energy Equation with the Bernoulli Equation

5.88 What is enthalpy and why is it useful for energy considerations in fluid mechanics?

5.89 Cite a few examples of evidence of loss of available energy in actual fluid flows. Why does loss occur?

5.90 Is zero heat transfer a necessary condition for application of the Bernoulli equation (Eq. 5.75)?

5.91 A 1000-m-high waterfall involves steady flow from one large body to another. Detemine the temperature rise associated with this flow.

5.92 A 30-m-wide river with a flowrate of 70 m³/s flows over a rock pile as shown in Fig. P5.92. Determine the direction of flow and the head loss associated with the flow across the rock pile.

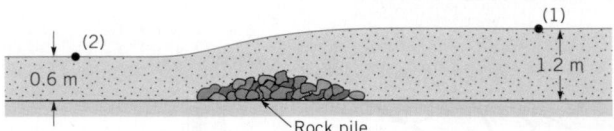

■ **F I G U R E   P5.92**

5.93 Air steadily expands adiabatically and without friction from stagnation conditions of 690 kPa (abs) and 290 K to a static pressure of 101 kPa (abs). Determine the velocity of the expanded air assuming: **(a)** incompressible flow; **(b)** compressible flow.

5.94 A horizontal Venturi flow meter consists of a converging-- diverging conduit as indicated in Fig. P5.94. The diameters of cross sections (1) and (2) are 15 and 10 cm. The velocity and static pressure are uniformly distributed at cross sections (1) and (2). Determine the volume flowrate (m³/s) through the meter if $p_1 - p_2 = 20$ kPa, the flowing fluid is oil ($\rho = 897$ kg/m³), and the loss per unit mass from (1) to (2) is negligibly small.

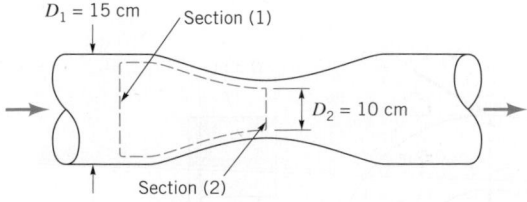

■ **F I G U R E   P5.94**

5.95 Oil ($SG = 0.9$) flows downward through a vertical pipe contraction as shown in Fig. P5.95. If the mercury manometer reading, h, is 100 mm, determine the volume flowrate for frictionless flow. Is the actual flowrate more or less than the frictionless value? Explain.

5.96 An incompressible liquid flows steadily along the pipe shown in Fig. P5.96. Determine the direction of flow and the head loss over the 6-m length of pipe.

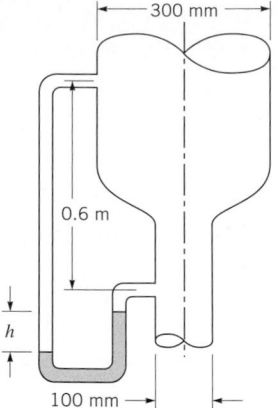

■ **F I G U R E P5.95**

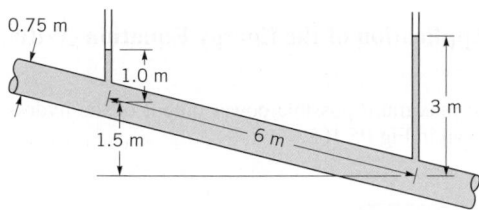

■ **F I G U R E P5.96**

5.97 Water flows through a vertical pipe, as is indicated in Fig. P5.97. Is the flow up or down in the pipe? Explain.

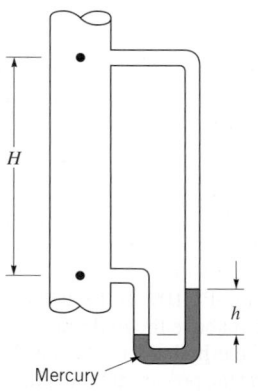

■ **F I G U R E P5.97**

5.98 A circular disk can be lifted up by blowing on it with the device shown in Fig. P5.98. Explain why this happens.

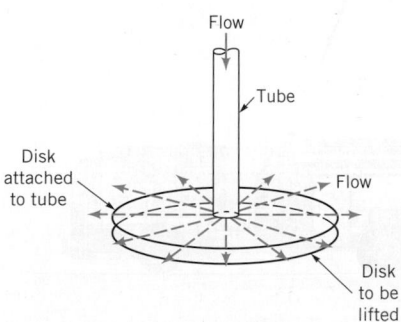

■ **F I G U R E P5.98**

5.99 A siphon is used to draw water at 20°C from a large container as indicated in Fig. P5.99. Does changing the elevation, h, of the siphon centerline above the water level in the tank vary the flowrate through the siphon? Explain. What is the maximum allowable value of h?

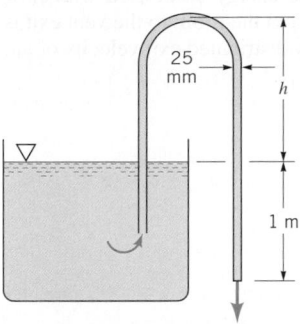

■ **F I G U R E P5.99**

5.100 A water siphon having a constant inside diameter of 7.5 cm is arranged as shown in Fig. P5.100. If the friction loss between A and B is $0.8V^2/2$, where V is the velocity of flow in the siphon, determine the flowrate involved.

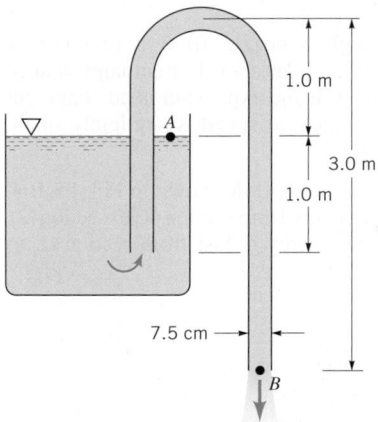

■ **F I G U R E P5.100**

5.101 Water flows through a valve (see Fig.P5.101) at the rate of 450 kg/s. The pressure just upstream of the valve is 620 kPa and the pressure drop across the valve is 345 kPa. The inside diameters of the valve inlet and exit pipes are 30 cm and 60 cm. If the flow through the valve occurs in a horizontal plane determine the loss in available energy across the valve.

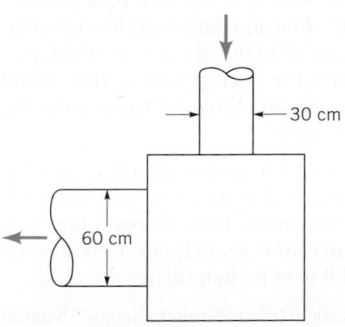

■ **F I G U R E P5.101**

5.102 Compare the volume flowrates associated with two different vent configurations, a cylindrical hole in the wall having a diameter of 10 cm and the same diameter cylindrical hole in the wall but with a well-rounded entrance (see Fig. P5.102). The room is held at a constant pressure of 10 kPa above atmospheric. Both vents exhaust into the atmosphere. The loss in available energy associated with flow through the cylindrical vent from the room to the vent exit is $0.5 V_2^2/2$, where V_2 is the uniformly distributed exit velocity of air. The loss in available energy associated with flow through the rounded entrance vent from the room to the vent exit is $0.05 V_2^2/s2$, where V_2 is the uniformly distributed exit velocity of air.

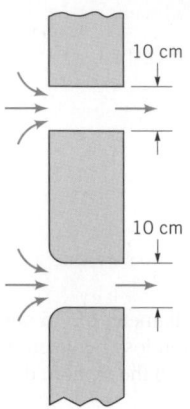

10 cm

10 cm

■ **FIGURE P5.102**

5.103 A gas expands through a nozzle from a pressure of 2068 kPa(abs) to a pressure of 34 kPa (abs). The enthalpy change involved, $\check{h}_1 - \check{h}_2$, is 350 kJ/kg. If the expansion is adiabatic but with frictional effects and the inlet gas speed is negligibly small, determine the exit gas velocity.

5.104 For the 180° elbow and nozzle flow shown in Fig. P5.104, determine the loss in available energy from section (1) to section (2). How much additional available energy is lost from section (2) to where the water comes to rest?

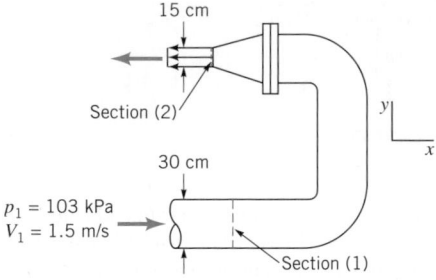

15 cm

Section (2)

30 cm

$p_1 = 103$ kPa
$V_1 = 1.5$ m/s

Section (1)

■ **FIGURE P5.104**

5.105 An automobile engine will work best when the back pressure at the interface of the exhaust manifold and the engine block is minimized. Show how reduction of losses in the exhaust manifold, piping, and muffler will also reduce the back pressure. How could losses in the exhaust system be reduced? What primarily limits the minimization of exhaust system losses?

†**5.106** Explain how, in terms of the loss of available energy involved, a home sink water faucet valve works to vary the flow from the shutoff condition to maximum flow. Explain how you would estimate the size of the overflow drain holes needed in the sink of **Video V5.1** (**Video V3.9** may be helpful).

5.107 (See Fluids in the News article titled "Smart shocks," Section 5.3.3.) A 890-N force applied to the end of the piston of the shock absorber shown in Fig. P5.107 causes the two ends of the shock absorber to move toward each other with a speed of 1.5 m/s. Determine the head loss associated with the flow of the oil through the channel. Neglect gravity and any friction force between the piston and cylinder walls.

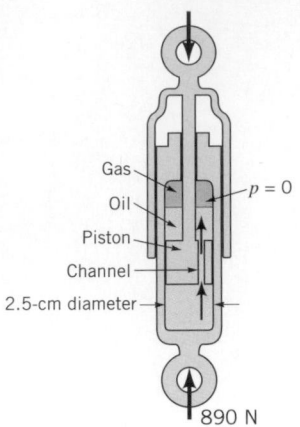

Gas
Oil
Piston
Channel
2.5-cm diameter
$p = 0$

890 N

■ **FIGURE P5.107**

Section 5.3.2 Application of the Energy Equation–With Shaft Work

5.108 What is the maximum possible power output of the hydroelectric turbine shown in Fig.P5.108?

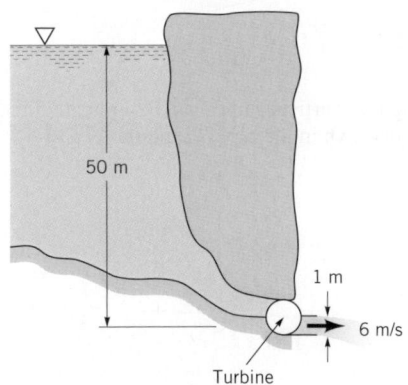

50 m

1 m

6 m/s

Turbine

■ **FIGURE P5.108**

5.109 The pumper truck shown in Fig. P5.109 is to deliver 0.04 m³/s to a maximum elevation of 18 m above the hydrant. The pressure at the 10 cm-diameter outlet of the hydrant is 70 kPa. If head losses are negligibly small, determine the power that the pump must add to the water.

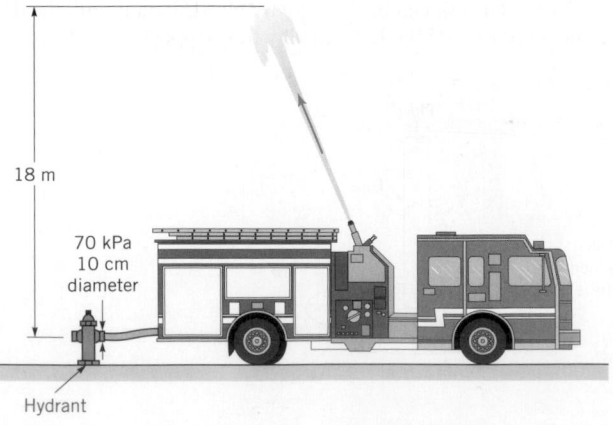

18 m

70 kPa
10 cm
diameter

Hydrant

■ **FIGURE P5.109**

5.110 The hydroelectric turbine shown in Fig. P5.110 passes 500 kL/s across a head of 180 m. What is the maximum amount of power output possible? Why will the actual amount be less?

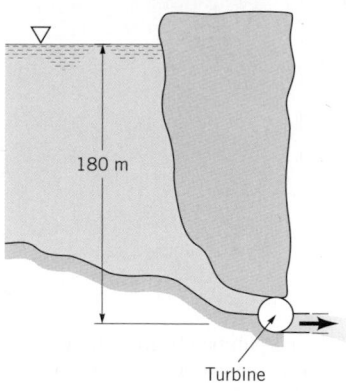

180 m

Turbine

■ **FIGURE P5.110**

5.111 A pump is to move water from a lake into a large, pressurized tank as shown in Fig. P5.111 at a rate of 4000 liters in 10 min or less. Will a pump that adds 2.2 kW to the water work for this purpose? Support your answer with appropriate calculations. Repeat the problem if the tank were pressurized to 300, rather than 200 kPa.

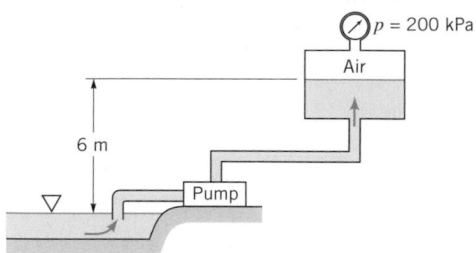

p = 200 kPa

Air

6 m

Pump

■ **FIGURE P5.111**

5.112 A hydraulic turbine is provided with 4.25 m³/s of water at 415 kPa. A vacuum gage in the turbine discharge 3 m below the turbine inlet centerline reads 250 mm Hg vacuum. If the turbine shaft output power is 1100 kW, calculate the power loss through the turbine. The supply and discharge pipe inside diameters are identically 80 mm.

5.113 Water is supplied at 4.50 m³/s and 415 kPa to a hydraulic turbine through a 1.0-m inside diameter inlet pipe as indicated in Fig. P5.113. The turbine discharge pipe has a 1.2-m inside diameter. The static pressure at section (2), 3 m below the turbine inlet, is 25 cm Hg vacuum. If the turbine develops 1.9 Mw, determine the power lost between sections (1) and (2).

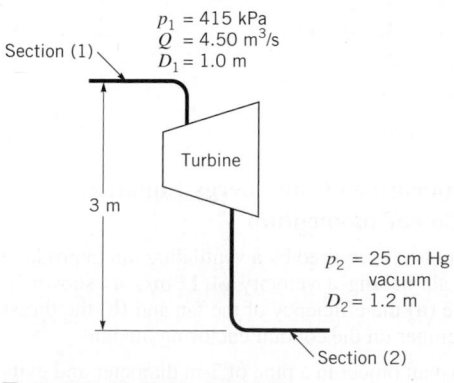

Section (1)

p_1 = 415 kPa
Q = 4.50 m³/s
D_1 = 1.0 m

Turbine

3 m

p_2 = 25 cm Hg vacuum
D_2 = 1.2 m

Section (2)

■ **FIGURE P5.113**

5.114 A centrifugal air compressor stage operates between an inlet stagnation pressure of 101.3 kPa (abs) and an exit stagnation pressure of 415 kPa(abs). The inlet stagnation temperature is 25 °C. If the loss of total pressure through the compressor stage associated with irreversible flow phenomena is 70 kPa, estimate the actual and ideal stagnation temperature rise through the compressor. Estimate the ratio of ideal to actual temperature rise to obtain an approximate value of the efficiency.

5.115 Water is pumped through a 10-cm-diameter pipe as shown in Fig. P5.115a. The pump characteristics (pump head versus flowrate) are given in Fig. P5.115b. Determine the flowrate if the head loss in the pipe is $h_L = 8V^2/2g$.

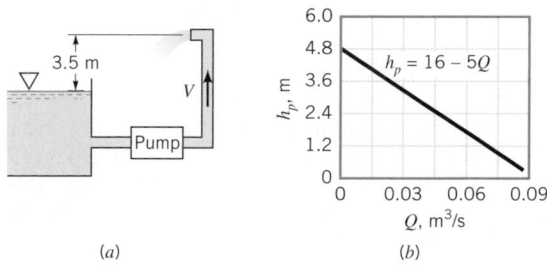

3.5 m

V

Pump

$h_p = 16 - 5Q$

(a)   (b)

■ **FIGURE P5.115**

5.116 Water is pumped from the large tank shown in Fig. P5.116. The head loss is known to be equal to $4V^2/2g$ and the pump head is $h_p = 20 - 4Q^2$, where h_p is in m when Q is in m³/s. Determine the flowrate.

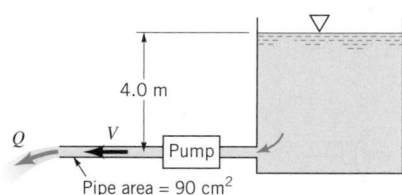

4.0 m

Q V Pump

Pipe area = 90 cm²

■ **FIGURE P5.116**

5.117 When a fan or pump is tested at the factory, head curves (head across the fan or pump versus volume flowrate) are often produced. A generic fan or pump head curve is shown in Fig. P5.117a. For any piping system, the drop in pressure or head involved because of loss can be estimated as a function of volume flowrate. A generic piping system loss curve is shown in Fig.P5.117b. When the pump or fan and piping system associated with the two curves of Fig.P5.117 are combined, what will the flowrate be? Why? How can the flowrate through this combined system be varied?

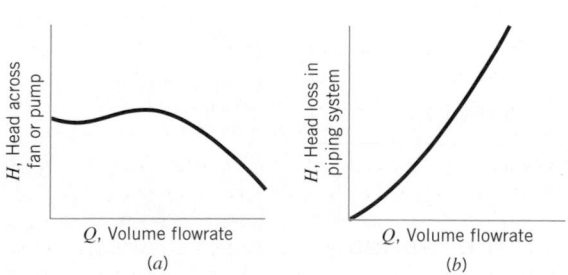

H, Head across fan or pump

Q, Volume flowrate

(a)

H, Head loss in piping system

Q, Volume flowrate

(b)

■ **FIGURE P5.117**

5.118 Water flows by gravity from one lake to another as sketched in Fig. P5.118 at the steady rate of 300 L/min. What is the loss in available energy associated with this flow? If this same amount of loss is associated with pumping the fluid from the lower lake to the higher one at the same flowrate, estimate the amount of pumping power required.

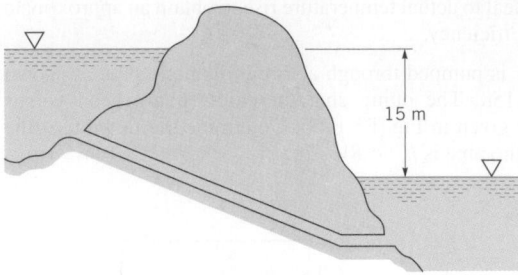

■ **F I G U R E   P5.118**

5.119 Water is pumped from a tank, point (1), to the top of a water plant aerator, point (2), as shown in **Video V5.14** and Fig. P5.119 at a rate of 0.10 m³/s (a) Determine the power that the pump adds to the water if the head loss from (1) to (2) where $V_2 = 0$ is 1.2 m. (b) Determine the head loss from (2) to the bottom of the aerator column, point (3), if the average velocity at (3) is $V_3 = 0.6$ m/s.

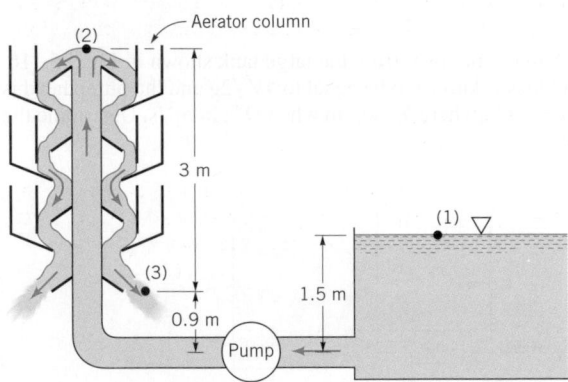

■ **F I G U R E   P5.119**

5.120 A liquid enters a fluid machine at section (1) and leaves at sections (2) and (3) as shown in Fig. P5.120. The density of the fluid is constant at 1000 kg/m³. All of the flow occurs in a horizontal plane and is frictionless and adiabatic. For the above-mentioned and additional conditions indicated in Fig. P5.120, determine the amount of shaft power involved.

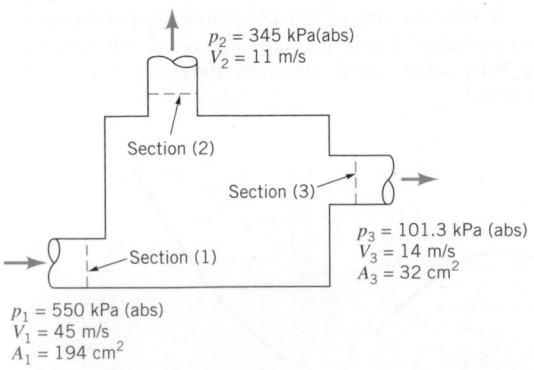

$p_2 = 345$ kPa(abs)
$V_2 = 11$ m/s

$p_3 = 101.3$ kPa (abs)
$V_3 = 14$ m/s
$A_3 = 32$ cm²

$p_1 = 550$ kPa (abs)
$V_1 = 45$ m/s
$A_1 = 194$ cm²

■ **F I G U R E   P5.120**

5.121 Water is to be moved from one large reservoir to another at a higher elevation as indicated in Fig. P5.121. The loss of available

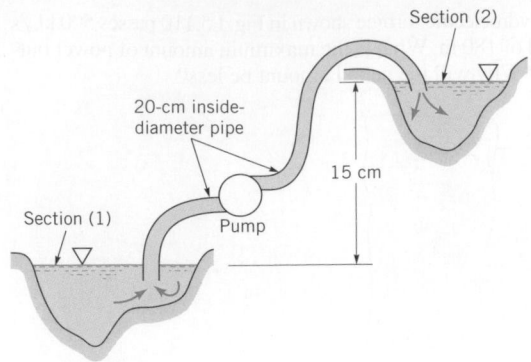

■ **F I G U R E   P5.121**

energy associated with 0.07 m³/s being pumped from sections (1) to (2) is loss = $61\overline{V}^2/2$ m²/s², where $\overline{V}$ is the average velocity of water in the 20 cm inside diameter piping involved. Determine the amount of shaft power required.

5.122 Water is to be pumped from the large tank shown in Fig. P5.122 with an exit velocity of 6 m/s. It was determined that the original pump (pump 1) that supplies 1 kW of power to the water did not produce the desired velocity. Hence, it is proposed that an additional pump (pump 2) be installed as indicated to increase the flowrate to the desired value. How much power must pump 2 add to the water? The head loss for this flow is $h_L = 250Q^2$, where h_L is in m when Q is in m³/s.

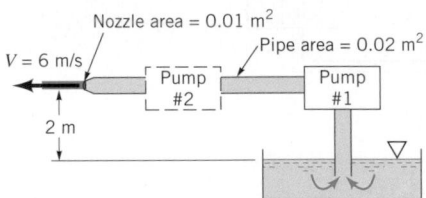

■ **F I G U R E   P5.122**

5.123 (See Fluids in the News article titled **"Curtain of air,"** Section 5.3.3.) The fan shown in Fig. P5.123 produces an air curtain to separate a loading dock from a cold storage room. The air curtain is a jet of air 3 m wide, 0.15 m thick moving with speed $V = 9$ m/s. The loss associated with this flow is loss = $K_L V^2/2$, where $K_L = 5$. How much power must the fan supply to the air to produce this flow?

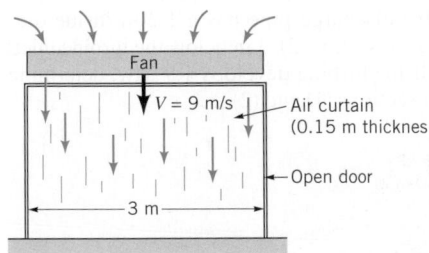

■ **F I G U R E   P5.123**

Section 5.3.2 Application of the Energy Equation— Combined with Linear momentum

5.124 If a 0.6 kW motor is required by a ventilating fan to produce a 60-cm stream of air having a velocity of 12 m/s as shown in Fig. P5.124, estimate **(a)** the efficiency of the fan and **(b)** the thrust of the supporting member on the conduit enclosing the fan.

5.125 Air flows past an object in a pipe of 2-m diameter and exits as a free jet as shown in Fig. P5.125. The velocity and pressure upstream are uniform at 10 m/s and 50 N/m², respectively. At the

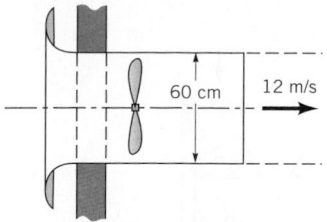

■ **F I G U R E P5.124**

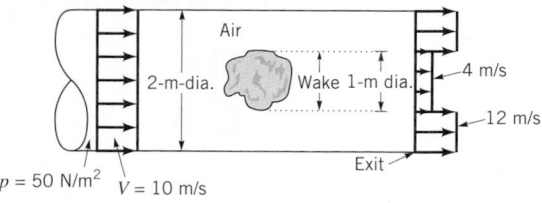

■ **F I G U R E P5.125**

pipe exit the velocity is nonuniform as indicated. The shear stress along the pipe wall is negligible. **(a)** Determine the head loss associated with a particle as it flows from the uniform velocity upstream of the object to a location in the wake at the exit plane of the pipe. **(b)** Determine the force that the air puts on the object.

5.126 Water flows through a 0.6-m-diameter pipe arranged horizontally in a circular arc as shown in Fig. P5.126. If the pipe discharges to the atmosphere ($p = 101.3$ kPa(abs)) determine the x and y components of the resultant force exerted by the water on the piping between sections (1) and (2). The steady flowrate is 85 m³/min. The loss in pressure due to fluid friction between sections (1) and (2) is 415 kPa.

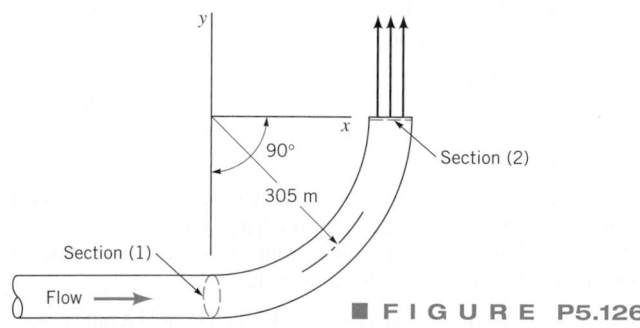

■ **F I G U R E P5.126**

5.127 Water flows steadily down the inclined pipe as indicated in Fig. P5.127. Determine the following: **(a)** the difference in pressure

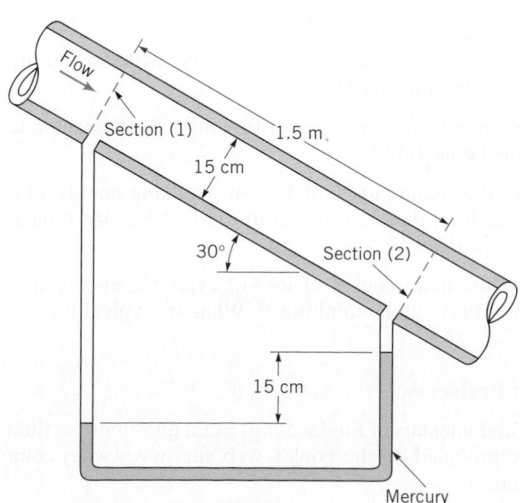

■ **F I G U R E P5.127**

$p_1 - p_2$, **(b)** the loss between sections (1) and (2), **(c)** the net axial force exerted by the pipe wall on the flowing water between sections (1) and (2).

5.128 Water flows steadily in a pipe and exits as a free jet through an end cap that contains a filter as shown in Fig. P5.128. The flow is in a horizontal plane. The axial component, R_y, of the anchoring force needed to keep the end cap stationary is 270 N. Determine the head loss for the flow through the end cap.

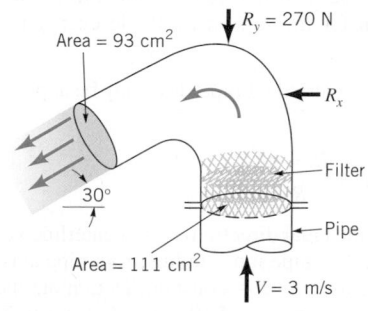

■ **F I G U R E P5.128**

5.129 When fluid flows through an abrupt expansion as indicated in Fig. P5.129, the loss in available energy across the expansion, loss$_{ex}$, is often expressed as

$$\text{loss}_{ex} = \left(1 - \frac{A_1}{A_2}\right)^2 \frac{V_1^2}{2}$$

where A_1 = cross-sectional area upstream of expansion, A_2 = cross-sectional area downstream of expansion, and V_1 = velocity of flow upstream of expansion. Derive this relationship.

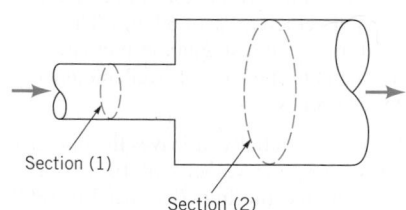

■ **F I G U R E P5.129**

5.130 Two water jets collide and form one homogeneous jet as shown in Fig. P5.130. **(a)** Determine the speed, V, and direction, θ, of the combined jet. **(b)** Determine the loss for a fluid particle flowing from (1) to (3), from (2) to (3). Gravity is negligible.

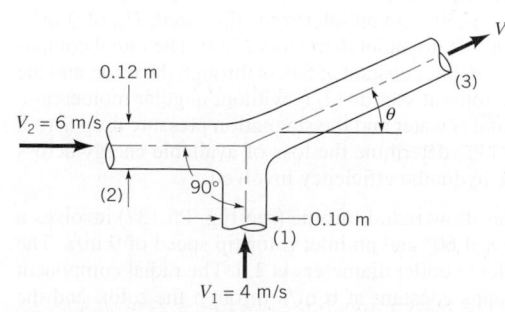

■ **F I G U R E P5.130**

Section 5.3.4 Application of the Energy Equation to Nonuniform Flows

5.131 Water flows vertically upward in a circular cross-sectional pipe. At section (1), the velocity profile over the cross-sectional area is uniform. At section (2), the velocity profile is

$$\mathbf{V} = w_c \left(\frac{R - r}{R} \right)^{1/7} \hat{\mathbf{k}}$$

where $\mathbf{V}$ = local velocity vector, w_c = centerline velocity in the axial direction, R = pipe inside radius, and, r = radius from pipe axis. Develop an expression for the loss in available energy between sections (1) and (2).

5.132 The velocity profile in a turbulent pipe flow may be approximated with the expression

$$\frac{u}{u_c} = \left(\frac{R - r}{R} \right)^{1/n}$$

where u = local velocity in the axial direction, u_c = centerline velocity in the axial direction, R = pipe inner radius from pipe axis, r = local radius from pipe axis, and n = constant. Determine the kinetic energy coefficient, α, for **(a)** $n = 5$, **(b)** $n = 6$, **(c)** $n = 7$, **(d)** $n = 8$, **(e)** $n = 9$, **(f)** $n = 10$.

5.133 A small fan moves air at a mass flowrate of 0.002 kg/s. Upstream of the fan, the pipe diameter is 6.4 cm, the flow is laminar, the velocity distribution is parabolic, and the kinetic energy coefficient, α_1, is equal to 2.0. Downstream of the fan, the pipe diameter is 2.5 cm, the flow is turbulent, the velocity profile is quite flat, and the kinetic energy coefficient, α_2, is equal to 1.08. If the rise in static pressure across the fan is 103 kPa and the fan shaft draws 0.18 W, compare the value of loss calculated: **(a)** assuming uniform velocity distributions, **(b)** considering actual velocity distributions.

Section 5.3.5 Combination of the Energy Equation and the Moment-of-Momentum Equation

5.134 Air enters a radial blower with zero angular momentum. It leaves with an absolute tangential velocity, V_θ, of 60 m/s. The rotor blade speed at rotor exit is 50 m/s. If the stagnation pressure rise across the rotor is 3.0 kPa, calculate the loss of available energy across the rotor and the rotor efficiency.

5.135 Water enters a pump impeller radially. It leaves the impeller with a tangential component of absolute velocity of 10 m/s. The impeller exit diameter is 60 mm, and the impeller speed is 1800 rpm. If the stagnation pressure rise across the impeller is 45 kPa, determine the loss of available energy across the impeller and the hydraulic efficiency of the pump.

5.136 Water enters an axial-flow turbine rotor with an absolute velocity tangential component, V_θ, of 5.0 m/s. The corresponding blade velocity, U, is 15 m/s. The water leaves the rotor blade row with no angular momentum. If the stagnation pressure drop across the turbine is 83 kPa, determine the hydraulic efficiency of the turbine.

5.137 An inward flow radial turbine (see Fig. P5.137) involves a nozzle angle, α_1, of 60° and an inlet rotor tip speed, U_1, of 9 m/s. The ratio of rotor inlet to outlet diameters is 2.0. The radial component of velocity remains constant at 6 m/s through the rotor, and the flow leaving the rotor at section (2) is without angular momentum. If the flowing fluid is water and the stagnation pressure drop across the rotor is 110 kPa, determine the loss of available energy across the rotor and the hydraulic efficiency involved.

5.138 An inward flow radial turbine (see Fig. P5.137) involves a nozzle angle, α_1, of 60° and an inlet rotor tip speed of 9 m/s. The ratio of rotor inlet to outlet diameters is 2.0. The radial component of velocity remains constant at 6 m/s through the rotor, and the

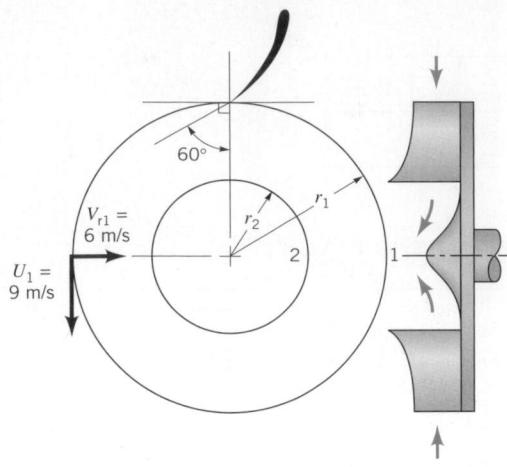

■ **FIGURE P5.137**

flow leaving the rotor at section (2) is without angular momentum. If the flowing fluid is air and the static pressure drop across the rotor is 70 Pa, determine the loss of available energy across the rotor and the rotor aerodynamic efficiency.

Section 5.4 Second Law of Thermodynamics— Irreversible Flow

5.139 Why do all actual fluid flows involve loss of available energy?

■ **Lab Problems**

5.140 This problem involves the force that a jet of air exerts on a flat plate as the air is deflected by the plate. To proceed with this problem, go to Appendix H which is located on the book's web site, www.wiley.com/college/munson.

5.141 This problem involves the pressure distribution produced on a flat plate that deflects a jet of air. To proceed with this problem, go to Appendix H which is located on the book's web site, www. wiley.com/college/munson.

5.142 This problem involves the force that a jet of water exerts on a vane when the vane turns the jet through a given angle. To proceed with this problem, go to Appendix H which is located on the book's web site, www.wiley.com/college/munson.

5.143 This problem involves the force needed to hold a pipe elbow stationary. To proceed with this problem, go to Appendix H which is located on the book's web site, www.wiley.com/college/munson.

■ **Life Long Learning Problems**

5.144 What are typical efficiencies associated with swimming and how can they be improved?

5.145 Explain how local ionization of flowing air can accelerate it. How can this be useful?

5.146 Discuss the main causes of loss of available energy in a turbo-pump and how they can be minimized. What are typical turbo-pump efficiencies?

5.147 Discuss the main causes of loss of available energy in a turbine and how they can be minimized. What are typical turbine efficiencies?

■ **FE Exam Problems**

Sample FE (Fundamentals of Engineering) exam questions for fluid mechanics are provided on the book's web site, www.wiley.com/college/munson.

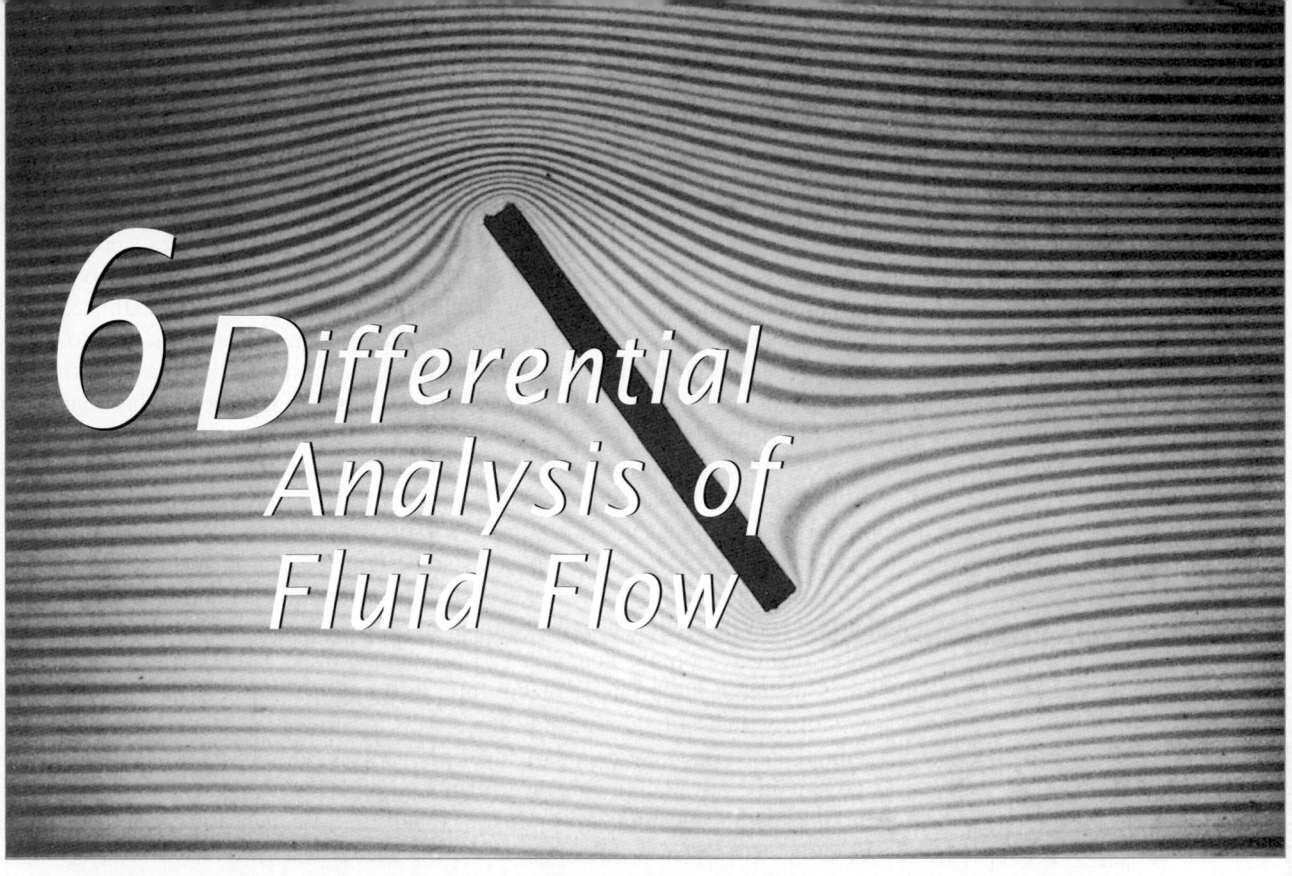

6 Differential Analysis of Fluid Flow

CHAPTER OPENING PHOTO: Flow past an inclined plate: The streamlines of a viscous fluid flowing slowly past a two-dimensional object placed between two closely spaced plates (a Hele-Shaw cell) approximate inviscid, irrotational (potential) flow. (Dye in water between glass plates spaced 1 mm apart.) (Photography courtesy of D. H. Peregrine.)

Learning Objectives

After completing this chapter, you should be able to:

- determine various kinematic elements of the flow given the velocity field.
- explain the conditions necessary for a velocity field to satisfy the continuity equation.
- apply the concepts of stream function and velocity potential.
- characterize simple potential flow fields.
- analyze certain types of flows using the Navier–Stokes equations.

In the previous chapter attention is focused on the use of finite control volumes for the solution of a variety of fluid mechanics problems. This approach is very practical and useful, since it does not generally require a detailed knowledge of the pressure and velocity variations within the control volume. Typically, we found that only conditions on the surface of the control volume were needed, and thus problems could be solved without a detailed knowledge of the flow field. Unfortunately, there are many situations that arise in which the details of the flow are important and the finite control volume approach will not yield the desired information. For example, we may need to know how the velocity varies over the cross section of a pipe, or how the pressure and shear stress vary along the surface of an airplane wing. In these circumstances we need to develop relationships that apply at a point, or at least in a very small infinitesimal region within a given flow field. This approach, which involves an *infinitesimal control volume*, as distinguished from a finite control volume, is commonly referred to as *differential analysis*, since (as we will soon discover) the governing equations are differential equations.

In this chapter we will provide an introduction to the differential equations that describe (in detail) the motion of fluids. Unfortunately, we will also find that these equations are rather complicated, non-linear partial differential equations that cannot be solved exactly except in a few cases, where simplifying assumptions are made. Thus, although differential analysis has the potential for supplying very detailed information about flow fields, this information is not easily extracted. Nevertheless, this approach provides a fundamental basis for the study of fluid mechanics. We do not want to be too discouraging at this point, since there are some exact solutions for laminar flow that can be obtained, and these have proved to be very useful. A few of these are included in this chapter. In addition, by making some simplifying assumptions many other analytical solutions can be obtained. For example, in some circumstances it may be reasonable to assume that the effect of viscosity is small and can be neglected. This rather drastic assumption greatly simplifies the analysis and provides the opportunity to obtain detailed solutions to a variety of complex flow problems. Some examples of these so-called *inviscid flow* solutions are also described in this chapter.

It is known that for certain types of flows the flow field can be conceptually divided into two regions—a very thin region near the boundaries of the system in which viscous effects are important, and a region away from the boundaries in which the flow is essentially inviscid. By making certain assumptions about the behavior of the fluid in the thin layer near the boundaries, and using the assumption of inviscid flow outside this layer, a large class of problems can be solved using differential analysis. These boundary layer problems are discussed in Chapter 9. Finally, it is to be noted that with the availability of powerful computers it is feasible to attempt to solve the differential equations using the techniques of numerical analysis. Although it is beyond the scope of this book to delve extensively into this approach, which is generally referred to as *computational fluid dynamics* (CFD), the reader should be aware of this approach to complex flow problems. CFD has become a common engineering tool and a brief introduction can be found in Appendix A. To introduce the power of CFD, two animations based on the numerical computations are provided as shown in the margin.

We begin our introduction to differential analysis by reviewing and extending some of the ideas associated with fluid kinematics that were introduced in Chapter 4. With this background the remainder of the chapter will be devoted to the derivation of the basic differential equations (which will be based on the principle of conservation of mass and Newton's second law of motion) and to some applications.

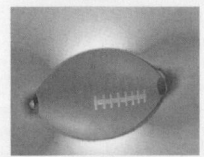

V6.1 Spinning football-velocity contours

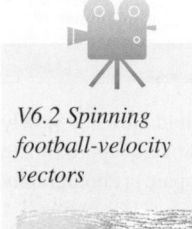

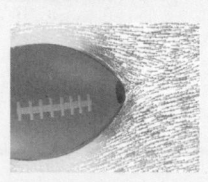

V6.2 Spinning football-velocity vectors

6.1 Fluid Element Kinematics

Fluid element motion consists of translation, linear deformation, rotation, and angular deformation.

In this section we will be concerned with the mathematical description of the motion of fluid elements moving in a flow field. A small fluid element in the shape of a cube which is initially in one position will move to another position during a short time interval δt as illustrated in Fig. 6.1. Because of the generally complex velocity variation within the field, we expect the element not only to translate from one position but also to have its volume changed (linear deformation), to rotate, and to undergo a change in shape (angular deformation). Although these movements and deformations occur simultaneously, we can consider each one separately as illustrated in Fig. 6.1. Since element motion and deformation are intimately related to the velocity and variation of velocity throughout the flow field, we will briefly review the manner in which velocity and acceleration fields can be described.

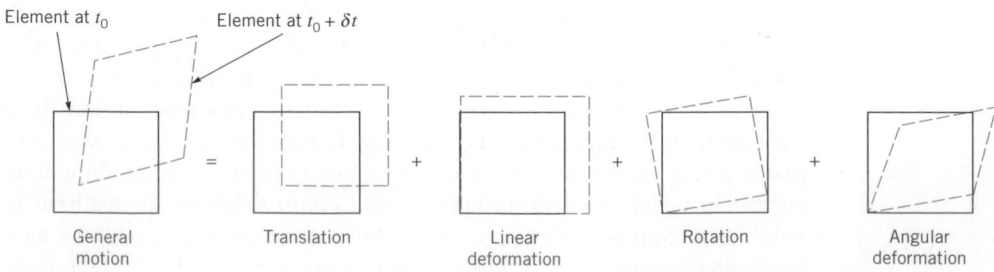

■ **FIGURE 6.1** Types of motion and deformation for a fluid element.

6.1.1 Velocity and Acceleration Fields Revisited

As discussed in detail in Section 4.1, the velocity field can be described by specifying the velocity **V** at all points, and at all times, within the flow field of interest. Thus, in terms of rectangular coordinates, the notation **V** (x, y, z, t) means that the velocity of a fluid particle depends on where it is located within the flow field (as determined by its coordinates, x, y, and z) and when it occupies the particular point (as determined by the time, t). As is pointed out in Section 4.1.1, this method of describing the fluid motion is called the Eulerian method. It is also convenient to express the velocity in terms of three rectangular components so that

$$\mathbf{V} = u\hat{\mathbf{i}} + v\hat{\mathbf{j}} + w\hat{\mathbf{k}} \tag{6.1}$$

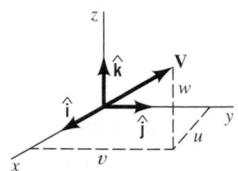

where u, v, and w are the velocity components in the x, y, and z directions, respectively, and $\hat{\mathbf{i}}$, $\hat{\mathbf{j}}$, and $\hat{\mathbf{k}}$ are the corresponding unit vectors, as shown by the figure in the margin. Of course, each of these components will, in general, be a function of x, y, z, and t. One of the goals of differential analysis is to determine how these velocity components specifically depend on x, y, z, and t for a particular problem.

With this description of the velocity field it was also shown in Section 4.2.1 that the acceleration of a fluid particle can be expressed as

$$\mathbf{a} = \frac{\partial \mathbf{V}}{\partial t} + u\frac{\partial \mathbf{V}}{\partial x} + v\frac{\partial \mathbf{V}}{\partial y} + w\frac{\partial \mathbf{V}}{\partial z} \tag{6.2}$$

and in component form:

$$a_x = \frac{\partial u}{\partial t} + u\frac{\partial u}{\partial x} + v\frac{\partial u}{\partial y} + w\frac{\partial u}{\partial z} \tag{6.3a}$$

$$a_y = \frac{\partial v}{\partial t} + u\frac{\partial v}{\partial x} + v\frac{\partial v}{\partial y} + w\frac{\partial v}{\partial z} \tag{6.3b}$$

$$a_z = \frac{\partial w}{\partial t} + u\frac{\partial w}{\partial x} + v\frac{\partial w}{\partial y} + w\frac{\partial w}{\partial z} \tag{6.3c}$$

The acceleration of a fluid particle is described using the concept of the material derivative.

The acceleration is also concisely expressed as

$$\mathbf{a} = \frac{D\mathbf{V}}{Dt} \tag{6.4}$$

where the operator

$$\frac{D(\ )}{Dt} = \frac{\partial(\ )}{\partial t} + u\frac{\partial(\ )}{\partial x} + v\frac{\partial(\ )}{\partial y} + w\frac{\partial(\ )}{\partial z} \tag{6.5}$$

is termed the *material derivative*, or *substantial derivative*. In vector notation

$$\frac{D(\ )}{Dt} = \frac{\partial(\ )}{\partial t} + (\mathbf{V} \cdot \nabla)(\ ) \tag{6.6}$$

where the gradient operator, $\nabla(\ )$, is

$$\nabla(\ ) = \frac{\partial(\ )}{\partial x}\hat{\mathbf{i}} + \frac{\partial(\ )}{\partial y}\hat{\mathbf{j}} + \frac{\partial(\ )}{\partial z}\hat{\mathbf{k}} \tag{6.7}$$

which was introduced in Chapter 2. As we will see in the following sections, the motion and deformation of a fluid element depend on the velocity field. The relationship between the motion and the forces causing the motion depends on the acceleration field.

6.1.2 Linear Motion and Deformation

The simplest type of motion that a fluid element can undergo is translation, as illustrated in Fig. 6.2. In a small time interval δt a particle located at point O will move to point O' as is illustrated in the figure. If all points in the element have the same velocity (which is only true if there are no velocity gradients), then the element will simply translate from one position to another. However,

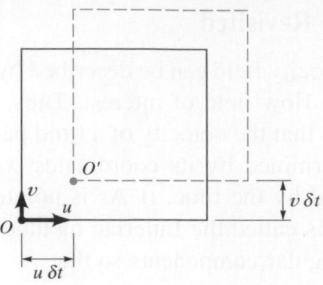

■ **FIGURE 6.2** Translation of a fluid element.

because of the presence of velocity gradients, the element will generally be deformed and rotated as it moves. For example, consider the effect of a single velocity gradient, $\partial u/\partial x$, on a small cube having sides δx, δy, and δz. As is shown in Fig. 6.3a, if the x component of velocity of O and B is u, then at nearby points A and C the x component of the velocity can be expressed as $u + (\partial u/\partial x)\,\delta x$. This difference in velocity causes a "stretching" of the volume element by an amount $(\partial u/\partial x)(\delta x)(\delta t)$ during the short time interval δt in which line OA stretches to OA' and BC to BC' (Fig. 6.3b). The corresponding change in the original volume, $\delta \mathcal{V} = \delta x\,\delta y\,\delta z$, would be

$$\text{Change in } \delta \mathcal{V} = \left(\frac{\partial u}{\partial x}\,\delta x\right)(\delta y\,\delta z)(\delta t)$$

and the *rate* at which the volume $\delta \mathcal{V}$ is changing *per unit volume* due to the gradient $\partial u/\partial x$ is

$$\frac{1}{\delta \mathcal{V}}\frac{d(\delta \mathcal{V})}{dt} = \lim_{\delta t \to 0}\left[\frac{(\partial u/\partial x)\,\delta t}{\delta t}\right] = \frac{\partial u}{\partial x} \tag{6.8}$$

The rate of volume change per unit volume is related to the velocity gradients.

If velocity gradients $\partial v/\partial y$ and $\partial w/\partial z$ are also present, then using a similar analysis it follows that, in the general case,

$$\frac{1}{\delta \mathcal{V}}\frac{d(\delta \mathcal{V})}{dt} = \frac{\partial u}{\partial x} + \frac{\partial v}{\partial y} + \frac{\partial w}{\partial z} = \boldsymbol{\nabla} \cdot \mathbf{V} \tag{6.9}$$

This rate of change of the volume per unit volume is called the *volumetric dilatation rate*. Thus, we see that the volume of a fluid may change as the element moves from one location to another in the flow field. However, for an *incompressible fluid* the volumetric dilatation rate is zero, since the element volume cannot change without a change in fluid density (the element mass must be conserved). Variations in the velocity in the direction of the velocity, as represented by the derivatives $\partial u/\partial x$, $\partial v/\partial y$, and $\partial w/\partial z$, simply cause a *linear deformation* of the element in the sense that the shape of the element does not change. Cross derivatives, such as $\partial u/\partial y$ and $\partial v/\partial x$, will cause the element to rotate and generally to undergo an *angular deformation*, which changes the shape of the element.

6.1.3 Angular Motion and Deformation

For simplicity we will consider motion in the x–y plane, but the results can be readily extended to the more general three dimensional case. The velocity variation that causes rotation and angular deformation is illustrated in Fig. 6.4a. In a short time interval δt the line segments OA and OB will

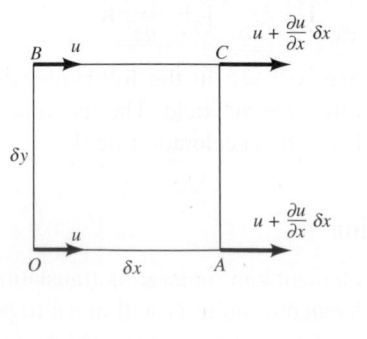

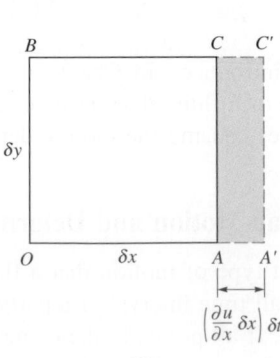

(a)   (b)

■ **FIGURE 6.3** Linear deformation of a fluid element.

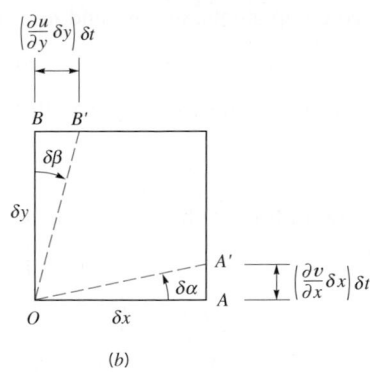

■ FIGURE 6.4
Angular motion and deformation of a fluid element.

rotate through the angles $\delta\alpha$ and $\delta\beta$ to the new positions OA' and OB', as is shown in Fig. 6.4b. The angular velocity of line OA, ω_{OA}, is

$$\omega_{OA} = \lim_{\delta t \to 0} \frac{\delta\alpha}{\delta t}$$

V6.3 Shear deformation

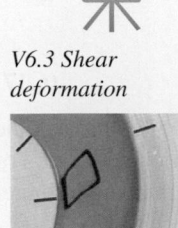

For small angles

$$\tan \delta\alpha \approx \delta\alpha = \frac{(\partial v/\partial x)\,\delta x\,\delta t}{\delta x} = \frac{\partial v}{\partial x}\,\delta t \qquad (6.10)$$

so that

$$\omega_{OA} = \lim_{\delta t \to 0} \left[\frac{(\partial v/\partial x)\,\delta t}{\delta t} \right] = \frac{\partial v}{\partial x}$$

Note that if $\partial v/\partial x$ is positive, ω_{OA} will be counterclockwise. Similarly, the angular velocity of the line OB is

$$\omega_{OB} = \lim_{\delta t \to 0} \frac{\delta\beta}{\delta t}$$

and

$$\tan \delta\beta \approx \delta\beta = \frac{(\partial u/\partial y)\,\delta y\,\delta t}{\delta y} = \frac{\partial u}{\partial y}\,\delta t \qquad (6.11)$$

so that

$$\omega_{OB} = \lim_{\delta t \to 0} \left[\frac{(\partial u/\partial y)\,\delta t}{\delta t} \right] = \frac{\partial u}{\partial y}$$

Rotation of fluid particles is related to certain velocity gradients in the flow field.

In this instance if $\partial u/\partial y$ is positive, ω_{OB} will be clockwise. The *rotation*, ω_z, of the element about the z axis is defined as the average of the angular velocities ω_{OA} and ω_{OB} of the two mutually perpendicular lines OA and OB.[1] Thus, if counterclockwise rotation is considered to be positive, it follows that

$$\omega_z = \frac{1}{2}\left(\frac{\partial v}{\partial x} - \frac{\partial u}{\partial y} \right) \qquad (6.12)$$

Rotation of the field element about the other two coordinate axes can be obtained in a similar manner with the result that for rotation about the x axis

$$\omega_x = \frac{1}{2}\left(\frac{\partial w}{\partial y} - \frac{\partial v}{\partial z} \right) \qquad (6.13)$$

and for rotation about the y axis

$$\omega_y = \frac{1}{2}\left(\frac{\partial u}{\partial z} - \frac{\partial w}{\partial x} \right) \qquad (6.14)$$

----
[1]With this definition ω_z can also be interpreted to be the angular velocity of the bisector of the angle between the lines OA and OB.

The three components, ω_x, ω_y, and ω_z can be combined to give the rotation vector, $\boldsymbol{\omega}$, in the form

$$\boldsymbol{\omega} = \omega_x \hat{\mathbf{i}} + \omega_y \hat{\mathbf{j}} + \omega_z \hat{\mathbf{k}} \qquad (6.15)$$

An examination of this result reveals that $\boldsymbol{\omega}$ is equal to one-half the curl of the velocity vector. That is,

$$\boldsymbol{\omega} = \tfrac{1}{2} \text{ curl } \mathbf{V} = \tfrac{1}{2} \boldsymbol{\nabla} \times \mathbf{V} \qquad (6.16)$$

since by definition of the vector operator $\boldsymbol{\nabla} \times \mathbf{V}$

$$\frac{1}{2} \boldsymbol{\nabla} \times \mathbf{V} = \frac{1}{2} \begin{vmatrix} \hat{\mathbf{i}} & \hat{\mathbf{j}} & \hat{\mathbf{k}} \\ \dfrac{\partial}{\partial x} & \dfrac{\partial}{\partial y} & \dfrac{\partial}{\partial z} \\ u & v & w \end{vmatrix}$$

$$= \frac{1}{2}\left(\frac{\partial w}{\partial y} - \frac{\partial v}{\partial z}\right)\hat{\mathbf{i}} + \frac{1}{2}\left(\frac{\partial u}{\partial z} - \frac{\partial w}{\partial x}\right)\hat{\mathbf{j}} + \frac{1}{2}\left(\frac{\partial v}{\partial x} - \frac{\partial u}{\partial y}\right)\hat{\mathbf{k}}$$

Vorticity in a flow field is related to fluid particle rotation.

The *vorticity*, $\boldsymbol{\zeta}$, is defined as a vector that is twice the rotation vector; that is,

$$\boldsymbol{\zeta} = 2\,\boldsymbol{\omega} = \boldsymbol{\nabla} \times \mathbf{V} \qquad (6.17)$$

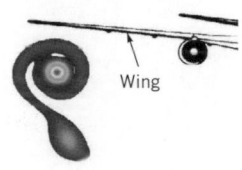
Wing

The use of the vorticity to describe the rotational characteristics of the fluid simply eliminates the $\left(\tfrac{1}{2}\right)$ factor associated with the rotation vector. The figure in the margin shows vorticity contours of the wing tip vortex flow shortly after an aircraft has passed. The lighter colors indicate stronger vorticity. (See also Fig. 4.3.)

We observe from Eq. 6.12 that the fluid element will rotate about the z axis as an *undeformed* block (i.e., $\omega_{OA} = -\omega_{OB}$) only when $\partial u/\partial y = -\partial v/\partial x$. Otherwise the rotation will be associated with an angular deformation. We also note from Eq. 6.12 that when $\partial u/\partial y = \partial v/\partial x$ the rotation around the z axis is zero. More generally if $\boldsymbol{\nabla} \times \mathbf{V} = 0$, then the rotation (and the vorticity) are zero, and flow fields for which this condition applies are termed *irrotational*. We will find in Section 6.4 that the condition of irrotationality often greatly simplifies the analysis of complex flow fields. However, it is probably not immediately obvious why some flow fields would be irrotational, and we will need to examine this concept more fully in Section 6.4.

EXAMPLE 6.1 | Vorticity

GIVEN For a certain two-dimensional flow field the velocity is given by the equation

$$\mathbf{V} = (x^2 - y^2)\hat{\mathbf{i}} - 2xy\hat{\mathbf{j}}$$

FIND Is this flow irrotational?

SOLUTION

For an irrotational flow the rotation vector, $\boldsymbol{\omega}$, having the components given by Eqs. 6.12, 6.13, and 6.14 must be zero. For the prescribed velocity field

$$u = x^2 - y^2 \qquad v = -2xy \qquad w = 0$$

and therefore

$$\omega_x = \frac{1}{2}\left(\frac{\partial w}{\partial y} - \frac{\partial v}{\partial z}\right) = 0$$

$$\omega_y = \frac{1}{2}\left(\frac{\partial u}{\partial z} - \frac{\partial w}{\partial x}\right) = 0$$

$$\omega_z = \frac{1}{2}\left(\frac{\partial v}{\partial x} - \frac{\partial u}{\partial y}\right) = \frac{1}{2}\left[(-2y) - (-2y)\right] = 0$$

Thus, the flow is irrotational. **(Ans)**

COMMENTS It is to be noted that for a two-dimensional flow field (where the flow is in the x–y plane) ω_x and ω_y will always be

zero, since by definition of two-dimensional flow u and v are not functions of z, and w is zero. In this instance the condition for irrotationality simply becomes $\omega_z = 0$ or $\partial v/\partial x = \partial u/\partial y$.

The streamlines for the steady, two-dimensional flow of this example are shown in Fig. E6.1. (Information about how to calculate

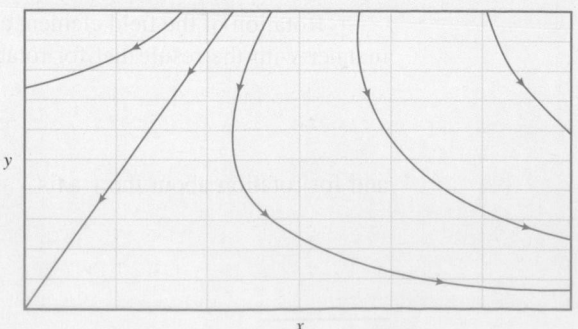

■ **FIGURE E6.1**

streamlines for a given velocity field is given in Sections 4.1.4 and 6.2.3.) It is noted that all of the streamlines (except for the one through the origin) are curved. However, because the flow is irrotational, there is no rotation of the fluid elements. That is, lines *OA* and *OB* of Fig. 6.4 rotate with the same speed but in opposite directions.

As shown by Eq. 6.17, the condition of irrotationality is equivalent to the fact that the vorticity, ζ, is zero or the curl of the velocity is zero.

In addition to the rotation associated with the derivatives $\partial u/\partial y$ and $\partial v/\partial x$, it is observed from Fig. 6.4*b* that these derivatives can cause the fluid element to undergo an *angular deformation*, which results in a change in shape of the element. The change in the original right angle formed by the lines *OA* and *OB* is termed the shearing strain, $\delta\gamma$, and from Fig. 6.4*b*

$$\delta\gamma = \delta\alpha + \delta\beta$$

where $\delta\gamma$ is considered to be positive if the original right angle is decreasing. The rate of change of $\delta\gamma$ is called the *rate of shearing strain* or the *rate of angular deformation* and is commonly denoted with the symbol $\dot{\gamma}$. The angles $\delta\alpha$ and $\delta\beta$ are related to the velocity gradients through Eqs. 6.10 and 6.11 so that

$$\dot{\gamma} = \lim_{\delta t \to 0} \frac{\delta\gamma}{\delta t} = \lim_{\delta t \to 0} \left[\frac{(\partial v/\partial x)\,\delta t + (\partial u/\partial y)\,\delta t}{\delta t} \right]$$

and, therefore,

$$\dot{\gamma} = \frac{\partial v}{\partial x} + \frac{\partial u}{\partial y} \tag{6.18}$$

As we will learn in Section 6.8, the rate of angular deformation is related to a corresponding shearing stress which causes the fluid element to change in shape. From Eq. 6.18 we note that if $\partial u/\partial y = -\partial v/\partial x$, the rate of angular deformation is zero, and this condition corresponds to the case in which the element is simply rotating as an undeformed block (Eq. 6.12). In the remainder of this chapter we will see how the various kinematical relationships developed in this section play an important role in the development and subsequent analysis of the differential equations that govern fluid motion.

6.2   Conservation of Mass

Conservation of mass requires that the mass of a system remain constant.

As is discussed in Section 5.1, conservation of mass requires that the mass, M, of a system remain constant as the system moves through the flow field. In equation form this principle is expressed as

$$\frac{DM_{\text{sys}}}{Dt} = 0$$

We found it convenient to use the control volume approach for fluid flow problems, with the control volume representation of the conservation of mass written as

$$\frac{\partial}{\partial t} \int_{\text{cv}} \rho\,d V + \int_{\text{cs}} \rho \mathbf{V} \cdot \hat{\mathbf{n}}\,dA = 0 \tag{6.19}$$

where the equation (commonly called the *continuity equation*) can be applied to a finite control volume (cv), which is bounded by a control surface (cs). The first integral on the left side of Eq. 6.19 represents the rate at which the mass within the control volume is changing, and the second integral represents the net rate at which mass is flowing out through the control surface (rate of mass outflow − rate of mass inflow). To obtain the differential form of the continuity equation, Eq. 6.19 is applied to an infinitesimal control volume.

6.2.1 Differential Form of Continuity Equation

We will take as our control volume the small, stationary cubical element shown in Fig. 6.5*a*. At the center of the element the fluid density is ρ and the velocity has components u, v, and w. Since the element is small, the volume integral in Eq. 6.19 can be expressed as

$$\frac{\partial}{\partial t} \int_{\text{cv}} \rho\,d V \approx \frac{\partial\rho}{\partial t}\,\delta x\,\delta y\,\delta z \tag{6.20}$$

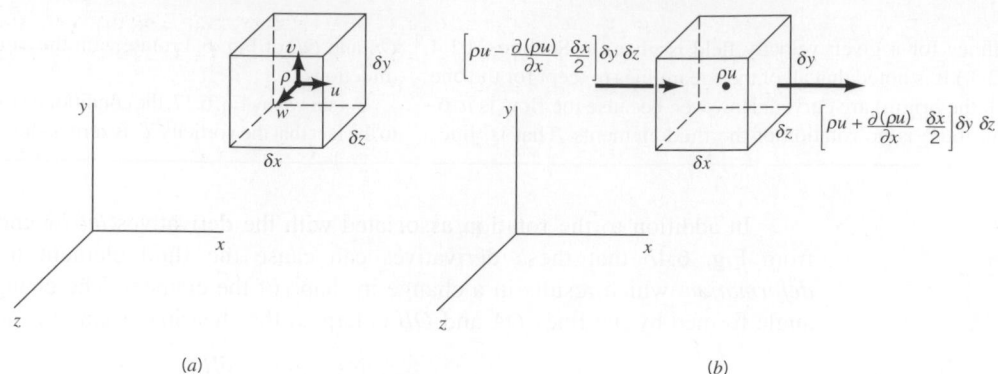

■ **F I G U R E 6.5**   **A differential element for the development of conservation of mass equation.**

The rate of mass flow through the surfaces of the element can be obtained by considering the flow in each of the coordinate directions separately. For example, in Fig. 6.5b flow in the x direction is depicted. If we let ρu represent the x component of the mass rate of flow per unit area at the center of the element, then on the right face

$$\rho u|_{x+(\delta x/2)} = \rho u + \frac{\partial(\rho u)}{\partial x}\frac{\delta x}{2} \qquad (6.21)$$

and on the left face

$$\rho u|_{x-(\delta x/2)} = \rho u - \frac{\partial(\rho u)}{\partial x}\frac{\delta x}{2} \qquad (6.22)$$

Note that we are really using a Taylor series expansion of ρu and neglecting higher order terms such as $(\delta x)^2$, $(\delta x)^3$, and so on. When the right-hand sides of Eqs. 6.21 and 6.22 are multiplied by the area $\delta y\,\delta z$, the rate at which mass is crossing the right and left sides of the element are obtained as is illustrated in Fig. 6.5b. When these two expressions are combined, the net rate of mass flowing from the element through the two surfaces can be expressed as

$$\begin{aligned}\text{Net rate of mass} \atop \text{outflow in } x \text{ direction} &= \left[\rho u + \frac{\partial(\rho u)}{\partial x}\frac{\delta x}{2}\right]\delta y\,\delta z \\ &\quad - \left[\rho u - \frac{\partial(\rho u)}{\partial x}\frac{\delta x}{2}\right]\delta y\,\delta z = \frac{\partial(\rho u)}{\partial x}\delta x\,\delta y\,\delta z \quad (6.23)\end{aligned}$$

For simplicity, only flow in the x direction has been considered in Fig. 6.5b, but, in general, there will also be flow in the y and z directions. An analysis similar to the one used for flow in the x direction shows that

$$\begin{matrix}\text{Net rate of mass} \\ \text{outflow in } y \text{ direction}\end{matrix} = \frac{\partial(\rho v)}{\partial y}\delta x\,\delta y\,\delta z \qquad (6.24)$$

and

$$\begin{matrix}\text{Net rate of mass} \\ \text{outflow in } z \text{ direction}\end{matrix} = \frac{\partial(\rho w)}{\partial z}\delta x\,\delta y\,\delta z \qquad (6.25)$$

Thus,

$$\begin{matrix}\text{Net rate of} \\ \text{mass outflow}\end{matrix} = \left[\frac{\partial(\rho u)}{\partial x} + \frac{\partial(\rho v)}{\partial y} + \frac{\partial(\rho w)}{\partial z}\right]\delta x\,\delta y\,\delta z \qquad (6.26)$$

From Eqs. 6.19, 6.20, and 6.26 it now follows that the differential equation for conservation of mass is

The continuity equation is one of the fundamental equations of fluid mechanics.

$$\boxed{\frac{\partial\rho}{\partial t} + \frac{\partial(\rho u)}{\partial x} + \frac{\partial(\rho v)}{\partial y} + \frac{\partial(\rho w)}{\partial z} = 0} \qquad (6.27)$$

As previously mentioned, this equation is also commonly referred to as the continuity equation.

The continuity equation is one of the fundamental equations of fluid mechanics and, as expressed in Eq. 6.27, is valid for steady or unsteady flow, and compressible or incompressible fluids. In vector notation, Eq. 6.27 can be written as

$$\frac{\partial \rho}{\partial t} + \nabla \cdot \rho \mathbf{V} = 0 \tag{6.28}$$

Two special cases are of particular interest. For *steady* flow of *compressible* fluids

$$\nabla \cdot \rho \mathbf{V} = 0$$

or

$$\frac{\partial(\rho u)}{\partial x} + \frac{\partial(\rho v)}{\partial y} + \frac{\partial(\rho w)}{\partial z} = 0 \tag{6.29}$$

For incompressible fluids the continuity equation reduces to a simple relationship involving certain velocity gradients.

This follows since by definition ρ is not a function of time for steady flow, but could be a function of position. For *incompressible* fluids the fluid density, ρ, is a constant throughout the flow field so that Eq. 6.28 becomes

$$\nabla \cdot \mathbf{V} = 0 \tag{6.30}$$

or

$$\frac{\partial u}{\partial x} + \frac{\partial v}{\partial y} + \frac{\partial w}{\partial z} = 0 \tag{6.31}$$

Equation 6.31 applies to both steady and unsteady flow of incompressible fluids. Note that Eq. 6.31 is the same as that obtained by setting the volumetric dilatation rate (Eq. 6.9) equal to zero. This result should not be surprising since both relationships are based on conservation of mass for incompressible fluids. However, the expression for the volumetric dilatation rate was developed from a system approach, whereas Eq. 6.31 was developed from a control volume approach. In the former case the deformation of a particular differential mass of fluid was studied, and in the latter case mass flow through a fixed differential volume was studied.

EXAMPLE 6.2   Continuity Equation

GIVEN   The velocity components for a certain incompressible, steady flow field are

$$u = x^2 + y^2 + z^2$$
$$v = xy + yz + z$$
$$w = \ ?$$

FIND   Determine the form of the z component, w, required to satisfy the continuity equation.

so that the required expression for $\partial w/\partial z$ is

$$\frac{\partial w}{\partial z} = -2x - (x + z) = -3x - z$$

Integration with respect to z yields

$$w = -3xz - \frac{z^2}{2} + f(x, y) \qquad \text{(Ans)}$$

SOLUTION

Any physically possible velocity distribution must for an incompressible fluid satisfy conservation of mass as expressed by the continuity equation

$$\frac{\partial u}{\partial x} + \frac{\partial v}{\partial y} + \frac{\partial w}{\partial z} = 0$$

For the given velocity distribution

$$\frac{\partial u}{\partial x} = 2x \quad \text{and} \quad \frac{\partial v}{\partial y} = x + z$$

COMMENT   The third velocity component cannot be explicitly determined since the function $f(x, y)$ can have any form and conservation of mass will still be satisfied. The specific form of this function will be governed by the flow field described by these velocity components—that is, some additional information is needed to completely determine w.

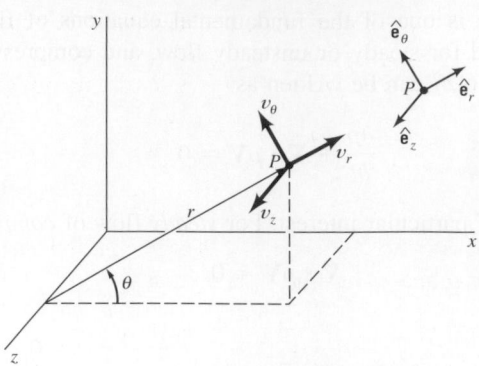

■ **FIGURE 6.6** **The representation of velocity components in cylindrical polar coordinates.**

6.2.2 Cylindrical Polar Coordinates

For some problems, velocity components expressed in cylindrical polar coordinates will be convenient.

For some problems it is more convenient to express the various differential relationships in cylindrical polar coordinates rather than Cartesian coordinates. As is shown in Fig. 6.6, with cylindrical coordinates a point is located by specifying the coordinates r, θ, and z. The coordinate r is the radial distance from the z axis, θ is the angle measured from a line parallel to the x axis (with counterclockwise taken as positive), and z is the coordinate along the z axis. The velocity components, as sketched in Fig. 6.6, are the radial velocity, v_r, the tangential velocity, v_θ, and the axial velocity, v_z. Thus, the velocity at some arbitrary point P can be expressed as

$$\mathbf{V} = v_r\hat{\mathbf{e}}_r + v_\theta\hat{\mathbf{e}}_\theta + v_z\hat{\mathbf{e}}_z \tag{6.32}$$

where $\hat{\mathbf{e}}_r$, $\hat{\mathbf{e}}_\theta$, and $\hat{\mathbf{e}}_z$ are the unit vectors in the r, θ, and z directions, respectively, as are illustrated in Fig. 6.6. The use of cylindrical coordinates is particularly convenient when the boundaries of the flow system are cylindrical. Several examples illustrating the use of cylindrical coordinates will be given in succeeding sections in this chapter.

The differential form of the continuity equation in cylindrical coordinates is

$$\frac{\partial\rho}{\partial t} + \frac{1}{r}\frac{\partial(r\rho v_r)}{\partial r} + \frac{1}{r}\frac{\partial(\rho v_\theta)}{\partial\theta} + \frac{\partial(\rho v_z)}{\partial z} = 0 \tag{6.33}$$

This equation can be derived by following the same procedure used in the preceding section (see Problem 6.20). For steady, compressible flow

$$\frac{1}{r}\frac{\partial(r\rho v_r)}{\partial r} + \frac{1}{r}\frac{\partial(\rho v_\theta)}{\partial\theta} + \frac{\partial(\rho v_z)}{\partial z} = 0 \tag{6.34}$$

For incompressible fluids (for steady or unsteady flow)

$$\frac{1}{r}\frac{\partial(rv_r)}{\partial r} + \frac{1}{r}\frac{\partial v_\theta}{\partial\theta} + \frac{\partial v_z}{\partial z} = 0 \tag{6.35}$$

6.2.3 The Stream Function

Steady, incompressible, plane, two-dimensional flow represents one of the simplest types of flow of practical importance. By plane, two-dimensional flow we mean that there are only two velocity components, such as u and v, when the flow is considered to be in the x–y plane. For this flow the continuity equation, Eq. 6.31, reduces to

$$\frac{\partial u}{\partial x} + \frac{\partial v}{\partial y} = 0 \tag{6.36}$$

We still have two variables, u and v, to deal with, but they must be related in a special way as indicated by Eq. 6.36. This equation suggests that if we define a function $\psi(x, y)$, called the *stream function,* which relates the velocities shown by the figure in the margin as

$$u = \frac{\partial \psi}{\partial y} \qquad v = -\frac{\partial \psi}{\partial x}$$

(6.37)

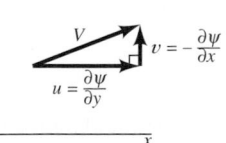

then the continuity equation is identically satisfied. This conclusion can be verified by simply substituting the expressions for u and v into Eq. 6.36 so that

$$\frac{\partial}{\partial x}\left(\frac{\partial \psi}{\partial y}\right) + \frac{\partial}{\partial y}\left(-\frac{\partial \psi}{\partial x}\right) = \frac{\partial^2 \psi}{\partial x \, \partial y} - \frac{\partial^2 \psi}{\partial y \, \partial x} = 0$$

Thus, whenever the velocity components are defined in terms of the stream function we know that conservation of mass will be satisfied. Of course, we still do not know what $\psi(x, y)$ is for a particular problem, but at least we have simplified the analysis by having to determine only one unknown function, $\psi(x, y)$, rather than the two functions, $u(x, y)$ and $v(x, y)$.

Another particular advantage of using the stream function is related to the fact that *lines along which ψ is constant are streamlines*. Recall from Section 4.1.4 that streamlines are lines in the flow field that are everywhere tangent to the velocities, as is illustrated in Fig. 6.7. It follows from the definition of the streamline that the slope at any point along a streamline is given by

$$\frac{dy}{dx} = \frac{v}{u}$$

The change in the value of ψ as we move from one point (x, y) to a nearby point $(x + dx, y + dy)$ is given by the relationship:

$$d\psi = \frac{\partial \psi}{\partial x} dx + \frac{\partial \psi}{\partial y} dy = -v \, dx + u \, dy$$

Along a line of constant ψ we have $d\psi = 0$ so that

$$-v \, dx + u \, dy = 0$$

and, therefore, along a line of constant ψ

$$\frac{dy}{dx} = \frac{v}{u}$$

which is the defining equation for a streamline. Thus, if we know the function $\psi(x, y)$ we can plot lines of constant ψ to provide the family of streamlines that are helpful in visualizing the pattern

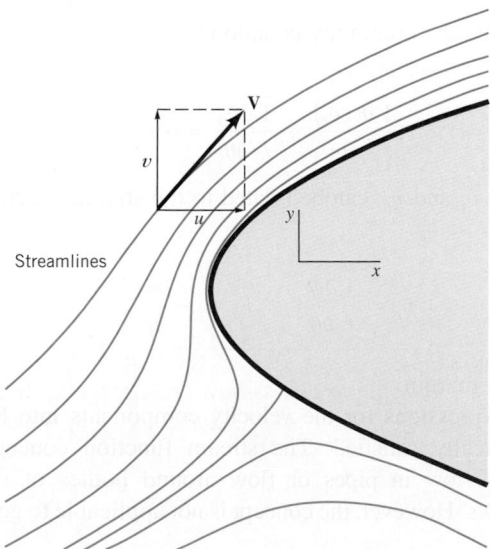

■ **FIGURE 6.7** **Velocity and velocity components along a streamline.**

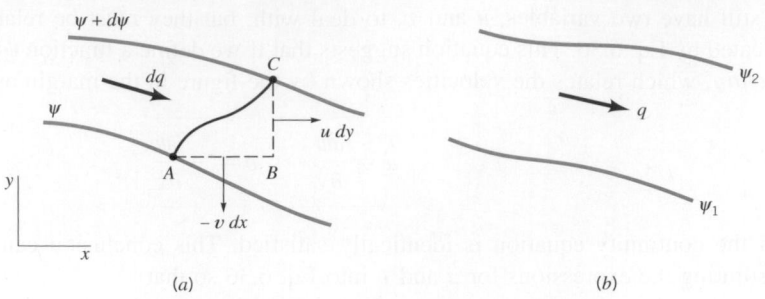

■ **FIGURE 6.8**   The flow between two streamlines.

of flow. There are an infinite number of streamlines that make up a particular flow field, since for each constant value assigned to ψ a streamline can be drawn.

The actual numerical value associated with a particular streamline is not of particular significance, but the change in the value of ψ is related to the volume rate of flow. Consider two closely spaced streamlines, shown in Fig. 6.8a. The lower streamline is designated ψ and the upper one $\psi + d\psi$. Let dq represent the volume rate of flow (per unit width perpendicular to the x–y plane) passing between the two streamlines. Note that flow never crosses streamlines, since by definition the velocity is tangent to the streamline. From conservation of mass we know that the inflow, dq, crossing the arbitrary surface AC of Fig. 6.8a must equal the net outflow through surfaces AB and BC. Thus,

> *The change in the value of the stream function is related to the volume rate of flow.*

$$dq = u\,dy - v\,dx$$

or in terms of the stream function

$$dq = \frac{\partial \psi}{\partial y}\,dy + \frac{\partial \psi}{\partial x}\,dx \tag{6.38}$$

The right-hand side of Eq. 6.38 is equal to $d\psi$ so that

$$dq = d\psi \tag{6.39}$$

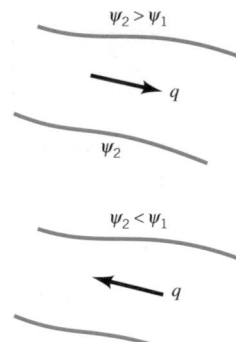

Thus, the volume rate of flow, q, between two streamlines such as ψ_1 and ψ_2 of Fig. 6.8b can be determined by integrating Eq. 6.39 to yield

$$q = \int_{\psi_1}^{\psi_2} d\psi = \psi_2 - \psi_1 \tag{6.40}$$

The relative value of ψ_2 with respect to ψ_1 determines the direction of flow, as shown by the figure in the margin.

In cylindrical coordinates the continuity equation (Eq. 6.35) for incompressible, plane, two-dimensional flow reduces to

$$\frac{1}{r}\frac{\partial (rv_r)}{\partial r} + \frac{1}{r}\frac{\partial v_\theta}{\partial \theta} = 0 \tag{6.41}$$

and the velocity components, v_r and v_θ, can be related to the stream function, $\psi(r, \theta)$, through the equations

$$\boxed{v_r = \frac{1}{r}\frac{\partial \psi}{\partial \theta} \qquad v_\theta = -\frac{\partial \psi}{\partial r}} \tag{6.42}$$

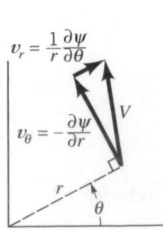

as shown by the figure in the margin.

Substitution of these expressions for the velocity components into Eq. 6.41 shows that the continuity equation is identically satisfied. The stream function concept can be extended to axisymmetric flows, such as flow in pipes or flow around bodies of revolution, and to two-dimensional compressible flows. However, the concept is not applicable to general three-dimensional flows.

EXAMPLE 6.3 | Stream Function

GIVEN The velocity components in a steady, incompressible, two-dimensional flow field are

$$u = 2y$$
$$v = 4x$$

FIND

(a) Determine the corresponding stream function and

(b) Show on a sketch several streamlines. Indicate the direction of flow along the streamlines.

SOLUTION

(a) From the definition of the stream function (Eqs. 6.37)

$$u = \frac{\partial \psi}{\partial y} = 2y$$

and

$$v = -\frac{\partial \psi}{\partial x} = 4x$$

The first of these equations can be integrated to give

$$\psi = y^2 + f_1(x)$$

where $f_1(x)$ is an arbitrary function of x. Similarly from the second equation

$$\psi = -2x^2 + f_2(y)$$

where $f_2(y)$ is an arbitrary function of y. It now follows that in order to satisfy both expressions for the stream function

$$\psi = -2x^2 + y^2 + C \qquad \text{(Ans)}$$

where C is an arbitrary constant.

COMMENT Since the velocities are related to the derivatives of the stream function, an arbitrary constant can always be added to the function, and the value of the constant is actually of no consequence. Usually, for simplicity, we set $C = 0$ so that for this particular example the simplest form for the stream function is

$$\psi = -2x^2 + y^2 \qquad \text{(1)} \quad \text{(Ans)}$$

Either answer indicated would be acceptable.

(b) Streamlines can now be determined by setting ψ = constant and plotting the resulting curve. With the above expression for ψ (with $C = 0$) the value of ψ at the origin is zero so that the equation of the streamline passing through the origin (the $\psi = 0$ streamline) is

$$0 = -2x^2 + y^2$$

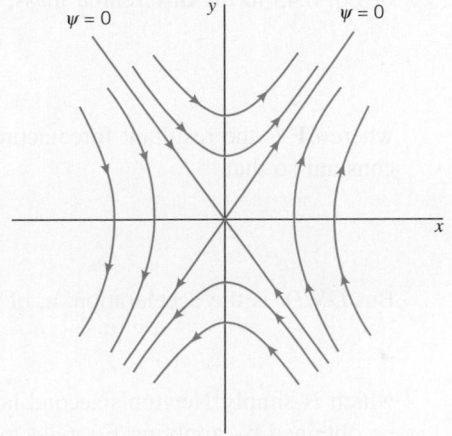

■ FIGURE E6.3

or

$$y = \pm\sqrt{2}x$$

Other streamlines can be obtained by setting ψ equal to various constants. It follows from Eq. 1 that the equations of these streamlines (for $\psi \neq 0$) can be expressed in the form

$$\frac{y^2}{\psi} - \frac{x^2}{\psi/2} = 1$$

which we recognize as the equation of a hyperbola. Thus, the streamlines are a family of hyperbolas with the $\psi = 0$ streamlines as asymptotes. Several of the streamlines are plotted in Fig. E6.3. Since the velocities can be calculated at any point, the direction of flow along a given streamline can be easily deduced. For example, $v = -\partial\psi/\partial x = 4x$ so that $v > 0$ if $x > 0$ and $v < 0$ if $x < 0$. The direction of flow is indicated on the figure.

6.3 Conservation of Linear Momentum

To develop the differential momentum equations we can start with the linear momentum equation

$$\mathbf{F} = \frac{D\mathbf{P}}{Dt}\bigg|_{sys} \qquad (6.43)$$

where $\mathbf{F}$ is the resultant force acting on a fluid mass, $\mathbf{P}$ is the linear momentum defined as

$$\mathbf{P} = \int_{sys} \mathbf{V}\,dm$$

and the operator $D(\ )/Dt$ is the material derivative (see Section 4.2.12). In the last chapter it was demonstrated how Eq. 6.43 in the form

$$\sum \mathbf{F}_{\substack{\text{contents of the} \\ \text{control volume}}} = \frac{\partial}{\partial t} \int_{cv} \mathbf{V} \rho \, d\mathcal{V} + \int_{cs} \mathbf{V} \rho \mathbf{V} \cdot \hat{\mathbf{n}} \, dA \qquad (6.44)$$

could be applied to a finite control volume to solve a variety of flow problems. To obtain the differential form of the linear momentum equation, we can either apply Eq. 6.43 to a differential system, consisting of a mass, δm, or apply Eq. 6.44 to an infinitesimal control volume, $\delta \mathcal{V}$, which initially bounds the mass δm. It is probably simpler to use the system approach since application of Eq. 6.43 to the differential mass, δm, yields

$$\delta \mathbf{F} = \frac{D(\mathbf{V} \, \delta m)}{Dt}$$

where $\delta \mathbf{F}$ is the resultant force acting on δm. Using this system approach δm can be treated as a constant so that

$$\delta \mathbf{F} = \delta m \frac{D\mathbf{V}}{Dt}$$

But $D\mathbf{V}/Dt$ is the acceleration, $\mathbf{a}$, of the element. Thus,

$$\delta \mathbf{F} = \delta m \, \mathbf{a} \qquad (6.45)$$

which is simply Newton's second law applied to the mass δ_m. This is the same result that would be obtained by applying Eq. 6.44 to an infinitesimal control volume (see Ref. 1). Before we can proceed, it is necessary to examine how the force $\delta \mathbf{F}$ can be most conveniently expressed.

6.3.1 Description of Forces Acting on the Differential Element

Both surface forces and body forces generally act on fluid particles.

In general, two types of forces need to be considered: *surface forces*, which act on the surface of the differential element, and *body forces*, which are distributed throughout the element. For our purpose, the only body force, $\delta \mathbf{F}_b$, of interest is the weight of the element, which can be expressed as

$$\delta \mathbf{F}_b = \delta m \, \mathbf{g} \qquad (6.46)$$

where $\mathbf{g}$ is the vector representation of the acceleration of gravity. In component form

$$\delta F_{bx} = \delta m \, g_x \qquad (6.47a)$$

$$\delta F_{by} = \delta m \, g_y \qquad (6.47b)$$

$$\delta F_{bz} = \delta m \, g_z \qquad (6.47c)$$

where g_x, g_y, and g_z are the components of the acceleration of gravity vector in the x, y, and z directions, respectively.

Surface forces act on the element as a result of its interaction with its surroundings. At any arbitrary location within a fluid mass, the force acting on a small area, δA, which lies in an arbitrary surface, can be represented by $\delta \mathbf{F}_s$, as is shown in Fig. 6.9. In general, $\delta \mathbf{F}_s$ will be inclined with respect to the surface. The force $\delta \mathbf{F}_s$ can be resolved into three components, δF_n, δF_1, and δF_2, where δF_n is normal to the area, δA, and δF_1 and δF_2 are parallel to the area and orthogonal to each other. The *normal stress*, σ_n, is defined as

$$\sigma_n = \lim_{\delta A \to 0} \frac{\delta F_n}{\delta A}$$

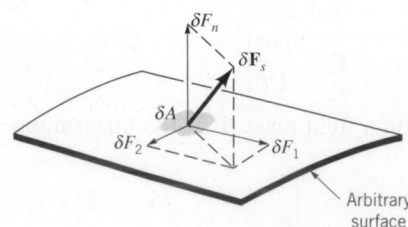

Arbitrary surface

■ **FIGURE 6.9** Components of force acting on an arbitrary differential area.

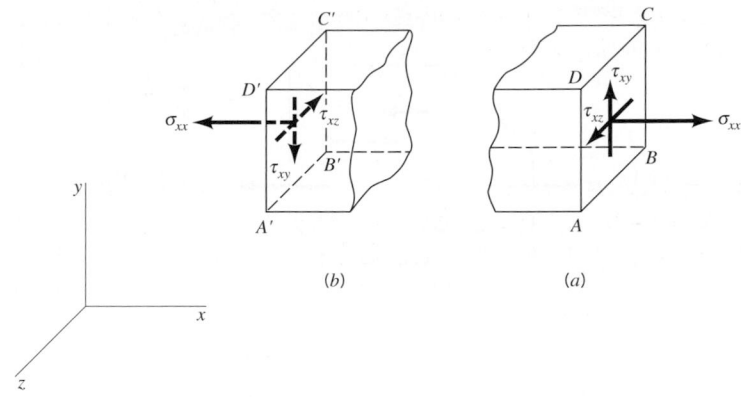

■ **F I G U R E   6.10**   **Double subscript notation for stresses.**

and the *shearing stresses* are defined as

$$\tau_1 = \lim_{\delta A \to 0} \frac{\delta F_1}{\delta A}$$

and

$$\tau_2 = \lim_{\delta A \to 0} \frac{\delta F_2}{\delta A}$$

We will use σ for normal stresses and τ for shearing stresses. The intensity of the force per unit area at a point in a body can thus be characterized by a normal stress and two shearing stresses, if the orientation of the area is specified. For purposes of analysis it is usually convenient to reference the area to the coordinate system. For example, for the rectangular coordinate system shown in Fig. 6.10 we choose to consider the stresses acting on planes parallel to the coordinate planes. On the plane *ABCD* of Fig. 6.10a, which is parallel to the y–z plane, the normal stress is denoted σ_{xx} and the shearing stresses are denoted as τ_{xy} and τ_{xz}. To easily identify the particular stress component we use a double subscript notation. The first subscript indicates the direction of the *normal* to the plane on which the stress acts, and the second subscript indicates the direction of the stress. Thus, normal stresses have repeated subscripts, whereas the subscripts for the shearing stresses are always different.

It is also necessary to establish a sign convention for the stresses. We define the positive direction for the stress as the positive coordinate direction on the surfaces for which the outward normal is in the positive coordinate direction. This is the case illustrated in Fig. 6.10a where the outward normal to the area *ABCD* is in the positive x direction. The positive directions for σ_{xx}, τ_{xy}, and τ_{xz} are as shown in Fig. 6.10a. If the outward normal points in the negative coordinate direction, as in Fig. 6.10b for the area $A'B'C'D'$, then the stresses are considered positive if directed in the negative coordinate directions. Thus, the stresses shown in Fig. 6.10b are considered to be positive when directed as shown. Note that positive normal stresses are tensile stresses; that is, they tend to "stretch" the material.

It should be emphasized that the state of stress at a point in a material is not completely defined by simply three components of a "stress vector." This follows, since any particular stress vector depends on the orientation of the plane passing through the point. However, it can be shown that the normal and shearing stresses acting on *any* plane passing through a point can be expressed in terms of the stresses acting on three orthogonal planes passing through the point (Ref. 2).

Surface forces can be expressed in terms of the shear and normal stresses.

We now can express the surface forces acting on a small cubical element of fluid in terms of the stresses acting on the faces of the element as shown in Fig. 6.11. It is expected that in general the stresses will vary from point to point within the flow field. Thus, through the use of Taylor series expansions we will express the stresses on the various faces in terms of the corresponding stresses at the center of the element of Fig. 6.11 and their gradients in the coordinate directions. For simplicity only the forces in the x direction are shown. Note that the stresses must be multiplied by the area on which they act to obtain the force. Summing all these forces in the x direction yields

$$\delta F_{sx} = \left(\frac{\partial \sigma_{xx}}{\partial x} + \frac{\partial \tau_{yx}}{\partial y} + \frac{\partial \tau_{zx}}{\partial z} \right) \delta x \, \delta y \, \delta z \qquad \text{(6.48a)}$$

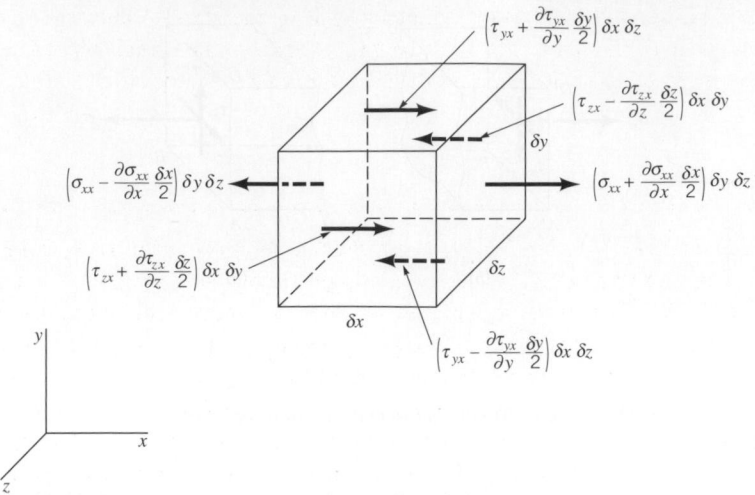

■ **F I G U R E   6.11**   **Surface forces in the x direction acting on a fluid element.**

for the resultant surface force in the x direction. In a similar manner the resultant surface forces in the y and z directions can be obtained and expressed as

$$\delta F_{sy} = \left(\frac{\partial \tau_{xy}}{\partial x} + \frac{\partial \sigma_{yy}}{\partial y} + \frac{\partial \tau_{zy}}{\partial z} \right) \delta x \, \delta y \, \delta z \tag{6.48b}$$

$$\delta F_{sz} = \left(\frac{\partial \tau_{xz}}{\partial x} + \frac{\partial \tau_{yz}}{\partial y} + \frac{\partial \sigma_{zz}}{\partial z} \right) \delta x \, \delta y \, \delta z \tag{6.48c}$$

The resultant surface force can now be expressed as

$$\delta \mathbf{F}_s = \delta F_{sx} \hat{\mathbf{i}} + \delta F_{sy} \hat{\mathbf{j}} + \delta F_{sz} \hat{\mathbf{k}} \tag{6.49}$$

and this force combined with the body force, $\delta \mathbf{F}_b$, yields the resultant force, $\delta \mathbf{F}$, acting on the differential mass, δm. That is, $\delta \mathbf{F} = \delta \mathbf{F}_s + \delta \mathbf{F}_b$.

6.3.2 Equations of Motion

The expressions for the body and surface forces can now be used in conjunction with Eq. 6.45 to develop the equations of motion. In component form Eq. 6.45 can be written as

$$\delta F_x = \delta m \, a_x$$

$$\delta F_y = \delta m \, a_y$$

$$\delta F_z = \delta m \, a_z$$

where $\delta m = \rho \, \delta x \, \delta y \, \delta z$, and the acceleration components are given by Eq. 6.3. It now follows (using Eqs. 6.47 and 6.48 for the forces on the element) that

The motion of a fluid is governed by a set of nonlinear differential equations.

$$\rho g_x + \frac{\partial \sigma_{xx}}{\partial x} + \frac{\partial \tau_{yx}}{\partial y} + \frac{\partial \tau_{zx}}{\partial z} = \rho \left(\frac{\partial u}{\partial t} + u \frac{\partial u}{\partial x} + v \frac{\partial u}{\partial y} + w \frac{\partial u}{\partial z} \right) \tag{6.50a}$$

$$\rho g_y + \frac{\partial \tau_{xy}}{\partial x} + \frac{\partial \sigma_{yy}}{\partial y} + \frac{\partial \tau_{zy}}{\partial z} = \rho \left(\frac{\partial v}{\partial t} + u \frac{\partial v}{\partial x} + v \frac{\partial v}{\partial y} + w \frac{\partial v}{\partial z} \right) \tag{6.50b}$$

$$\rho g_z + \frac{\partial \tau_{xz}}{\partial x} + \frac{\partial \tau_{yz}}{\partial y} + \frac{\partial \sigma_{zz}}{\partial z} = \rho \left(\frac{\partial w}{\partial t} + u \frac{\partial w}{\partial x} + v \frac{\partial w}{\partial y} + w \frac{\partial w}{\partial z} \right) \tag{6.50c}$$

where the element volume $\delta x \, \delta y \, \delta z$ cancels out.

Equations 6.50 are the general differential equations of motion for a fluid. In fact, they are applicable to any continuum (solid or fluid) in motion or at rest. However, before we can use the equations to solve specific problems, some additional information about the stresses must be obtained.

Otherwise, we will have more unknowns (all of the stresses and velocities and the density) than equations. It should not be too surprising that the differential analysis of fluid motion is complicated. We are attempting to describe, in detail, complex fluid motion.

6.4   Inviscid Flow

As is discussed in Section 1.6, shearing stresses develop in a moving fluid because of the viscosity of the fluid. We know that for some common fluids, such as air and water, the viscosity is small, and therefore it seems reasonable to assume that under some circumstances we may be able to simply neglect the effect of viscosity (and thus shearing stresses). Flow fields in which the shearing stresses are assumed to be negligible are said to be *inviscid, nonviscous,* or *frictionless.* These terms are used interchangeably. As is discussed in Section 2.1, for fluids in which there are no shearing stresses the normal stress at a point is independent of direction—that is, $\sigma_{xx} = \sigma_{yy} = \sigma_{zz}$. In this instance we define the pressure, p, as the negative of the normal stress so that

$$-p = \sigma_{xx} = \sigma_{yy} = \sigma_{zz}$$

The negative sign is used so that a *compressive* normal stress (which is what we expect in a fluid) will give a *positive* value for p.

In Chapter 3 the inviscid flow concept was used in the development of the Bernoulli equation, and numerous applications of this important equation were considered. In this section we will again consider the Bernoulli equation and will show how it can be derived from the general equations of motion for inviscid flow.

6.4.1 Euler's Equations of Motion

For an inviscid flow in which all the shearing stresses are zero, and the normal stresses are replaced by $-p$, the general equations of motion (Eqs. 6.50) reduce to

$$\rho g_x - \frac{\partial p}{\partial x} = \rho \left(\frac{\partial u}{\partial t} + u \frac{\partial u}{\partial x} + v \frac{\partial u}{\partial y} + w \frac{\partial u}{\partial z} \right) \tag{6.51a}$$

$$\rho g_y - \frac{\partial p}{\partial y} = \rho \left(\frac{\partial v}{\partial t} + u \frac{\partial v}{\partial x} + v \frac{\partial v}{\partial y} + w \frac{\partial v}{\partial z} \right) \tag{6.51b}$$

$$\rho g_z - \frac{\partial p}{\partial z} = \rho \left(\frac{\partial w}{\partial t} + u \frac{\partial w}{\partial x} + v \frac{\partial w}{\partial y} + w \frac{\partial w}{\partial z} \right) \tag{6.51c}$$

These equations are commonly referred to as *Euler's equations of motion,* named in honor of Leonhard Euler (1707–1783), a famous Swiss mathematician who pioneered work on the relationship between pressure and flow. In vector notation Euler's equations can be expressed as

Euler's equations of motion apply to an inviscid flow field.

$$\rho \mathbf{g} - \nabla p = \rho \left[\frac{\partial \mathbf{V}}{\partial t} + (\mathbf{V} \cdot \nabla)\mathbf{V} \right] \tag{6.52}$$

Although Eqs. 6.51 are considerably simpler than the general equations of motion, Eqs. 6.50, they are still not amenable to a general analytical solution that would allow us to determine the pressure and velocity at all points within an inviscid flow field. The main difficulty arises from the nonlinear velocity terms ($u \, \partial u/\partial x$, $v \, \partial u/\partial y$, etc.), which appear in the convective acceleration. Because of these terms, Euler's equations are nonlinear partial differential equations for which we do not have a general method of solving. However, under some circumstances we can use them to obtain useful information about inviscid flow fields. For example, as shown in the following section we can integrate Eq. 6.52 to obtain a relationship (the Bernoulli equation) between elevation, pressure, and velocity along a streamline.

6.4.2 The Bernoulli Equation

In Section 3.2 the Bernoulli equation was derived by a direct application of Newton's second law to a fluid particle moving along a streamline. In this section we will again derive this important

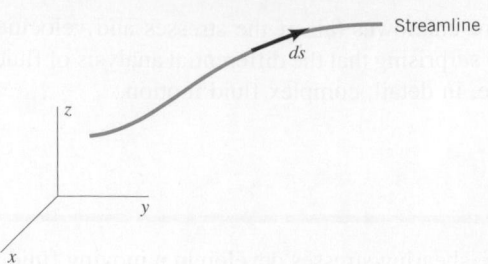

■ FIGURE 6.12 **The notation for differential length along a streamline.**

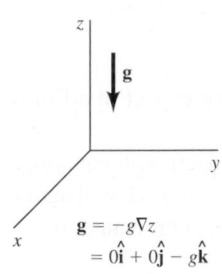

equation, starting from Euler's equations. Of course, we should obtain the same result since Euler's equations simply represent a statement of Newton's second law expressed in a general form that is useful for flow problems and maintains the restriction of zero viscosity. We will restrict our attention to steady flow so Euler's equation in vector form becomes

$$\rho \mathbf{g} - \boldsymbol{\nabla} p = \rho(\mathbf{V} \cdot \boldsymbol{\nabla})\mathbf{V} \tag{6.53}$$

We wish to integrate this differential equation along some arbitrary streamline (Fig. 6.12) and select the coordinate system with the z axis vertical (with "up" being positive) so that, as shown by the figure in the margin, the acceleration of gravity vector can be expressed as

$$\mathbf{g} = -g\boldsymbol{\nabla} z$$

where g is the magnitude of the acceleration of gravity vector. Also, it will be convenient to use the vector identity

$$(\mathbf{V} \cdot \boldsymbol{\nabla})\mathbf{V} = \tfrac{1}{2}\boldsymbol{\nabla}(\mathbf{V} \cdot \mathbf{V}) - \mathbf{V} \times (\boldsymbol{\nabla} \times \mathbf{V})$$

Equation 6.53 can now be written in the form

$$-\rho g\boldsymbol{\nabla} z - \boldsymbol{\nabla} p = \frac{\rho}{2}\boldsymbol{\nabla}(\mathbf{V} \cdot \mathbf{V}) - \rho\mathbf{V} \times (\boldsymbol{\nabla} \times \mathbf{V})$$

and this equation can be rearranged to yield

$$\frac{\boldsymbol{\nabla} p}{\rho} + \frac{1}{2}\boldsymbol{\nabla}(V^2) + g\boldsymbol{\nabla} z = \mathbf{V} \times (\boldsymbol{\nabla} \times \mathbf{V})$$

We next take the dot product of each term with a differential length $d\mathbf{s}$ along a streamline (Fig. 6.12). Thus,

$$\frac{\boldsymbol{\nabla} p}{\rho} \cdot d\mathbf{s} + \frac{1}{2}\boldsymbol{\nabla}(V^2) \cdot d\mathbf{s} + g\boldsymbol{\nabla} z \cdot d\mathbf{s} = [\mathbf{V} \times (\boldsymbol{\nabla} \times \mathbf{V})] \cdot d\mathbf{s} \tag{6.54}$$

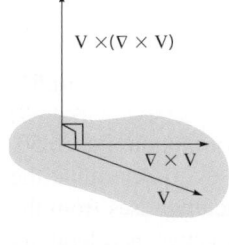

Since $d\mathbf{s}$ has a direction along the streamline, the vectors $d\mathbf{s}$ and $\mathbf{V}$ are parallel. However, as shown by the figure in the margin, the vector $\mathbf{V} \times (\boldsymbol{\nabla} \times \mathbf{V})$ is perpendicular to $\mathbf{V}$ (why?), so it follows that

$$[\mathbf{V} \times (\boldsymbol{\nabla} \times \mathbf{V})] \cdot d\mathbf{s} = 0$$

Recall also that the dot product of the gradient of a scalar and a differential length gives the differential change in the scalar in the direction of the differential length. That is, with $d\mathbf{s} = dx\,\hat{\mathbf{i}} + dy\,\hat{\mathbf{j}} + dz\,\hat{\mathbf{k}}$ we can write $\boldsymbol{\nabla} p \cdot d\mathbf{s} = (\partial p/\partial x)\,dx + (\partial p/\partial y)dy + (\partial p/\partial z)dz = dp$. Thus, Eq. 6.54 becomes

$$\frac{dp}{\rho} + \frac{1}{2}d(V^2) + g\,dz = 0 \tag{6.55}$$

Euler's equations can be arranged to give the relationship among pressure, velocity, and elevation for inviscid fluids

where the change in p, V, and z is along the streamline. Equation 6.55 can now be integrated to give

$$\int \frac{dp}{\rho} + \frac{V^2}{2} + gz = \text{constant} \tag{6.56}$$

which indicates that the sum of the three terms on the left side of the equation must remain a constant along a given streamline. Equation 6.56 is valid for both compressible and incompressible inviscid flows, but for compressible fluids the variation in ρ with p must be specified before the first term in Eq. 6.56 can be evaluated.

For inviscid, incompressible fluids (commonly called *ideal fluids*) Eq. 6.56 can be written as

$$\frac{p}{\rho} + \frac{V^2}{2} + gz = \text{constant along a streamline} \tag{6.57}$$

and this equation is the *Bernoulli equation* used extensively in Chapter 3. It is often convenient to write Eq. 6.57 between two points (1) and (2) along a streamline and to express the equation in the "head" form by dividing each term by g so that

$$\frac{p_1}{\gamma} + \frac{V_1^2}{2g} + z_1 = \frac{p_2}{\gamma} + \frac{V_2^2}{2g} + z_2 \tag{6.58}$$

It should be again emphasized that the Bernoulli equation, as expressed by Eqs. 6.57 and 6.58, is restricted to the following:

- inviscid flow
- incompressible flow
- steady flow
- flow along a streamline

You may want to go back and review some of the examples in Chapter 3 that illustrate the use of the Bernoulli equation.

6.4.3 Irrotational Flow

The vorticity is zero in an irrotational flow field.

If we make one additional assumption—that the flow is *irrotational*—the analysis of inviscid flow problems is further simplified. Recall from Section 6.1.3 that the rotation of a fluid element is equal to $\frac{1}{2}(\nabla \times \mathbf{V})$, and an irrotational flow field is one for which $\nabla \times \mathbf{V} = 0$ (i.e., the curl of velocity is zero). Since the vorticity, ζ, is defined as $\nabla \times \mathbf{V}$, it also follows that in an irrotational flow field the vorticity is zero. The concept of irrotationality may seem to be a rather strange condition for a flow field. Why would a flow field be irrotational? To answer this question we note that if $\frac{1}{2}(\nabla \times \mathbf{V}) = 0$, then each of the components of this vector, as are given by Eqs. 6.12, 6.13, and 6.14, must be equal to zero. Since these components include the various velocity gradients in the flow field, the condition of irrotationality imposes specific relationships among these velocity gradients. For example, for rotation about the z axis to be zero, it follows from Eq. 6.12 that

$$\omega_z = \frac{1}{2}\left(\frac{\partial v}{\partial x} - \frac{\partial u}{\partial y}\right) = 0$$

and, therefore,

$$\frac{\partial v}{\partial x} = \frac{\partial u}{\partial y} \tag{6.59}$$

Similarly from Eqs. 6.13 and 6.14

$$\frac{\partial w}{\partial y} = \frac{\partial v}{\partial z} \tag{6.60}$$

$$\frac{\partial u}{\partial z} = \frac{\partial w}{\partial x} \tag{6.61}$$

A general flow field would not satisfy these three equations. However, a uniform flow as is illustrated in Fig. 6.13 does. Since $u = U$ (a constant), $v = 0$, and $w = 0$, it follows that Eqs. 6.59, 6.60, and 6.61 are all satisfied. Therefore, a uniform flow field (in which there are no velocity gradients) is certainly an example of an irrotational flow.

Uniform flows by themselves are not very interesting. However, many interesting and important flow problems include uniform flow in some part of the flow field. Two examples are

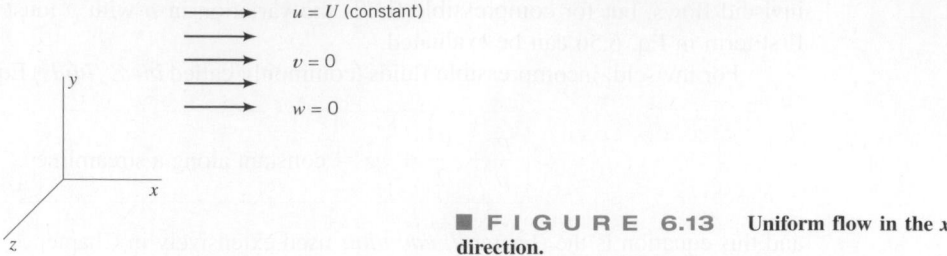

■ **F I G U R E   6.13**   **Uniform flow in the x direction.**

shown in Fig. 6.14. In Fig. 6.14*a* a solid body is placed in a uniform stream of fluid. Far away from the body the flow remains uniform, and in this far region the flow is irrotational. In Fig. 6.14*b*, flow from a large reservoir enters a pipe through a streamlined entrance where the velocity distribution is essentially uniform. Thus, at the entrance the flow is irrotational.

For an inviscid fluid there are no shearing stresses—the only forces acting on a fluid element are its weight and pressure forces. Since the weight acts through the element center of gravity, and the pressure acts in a direction normal to the element surface, neither of these forces can cause the element to rotate. Therefore, for an inviscid fluid, if some part of the flow field is irrotational, the fluid elements emanating from this region will not take on any rotation as they progress through the flow field. This phenomenon is illustrated in Fig. 6.14*a* in which fluid elements flowing far away from the body have irrotational motion, and as they flow around the body the motion remains irrotational except very near the boundary. Near the boundary the velocity changes rapidly from zero at the boundary (no-slip condition) to some relatively large value in a short distance from the boundary. This rapid change in velocity gives rise to a large velocity gradient normal to the boundary and produces significant shearing stresses, even though the viscosity is small. Of course if we had a truly inviscid fluid, the fluid would simply "slide" past the boundary and the flow would be irrotational everywhere. But this is not the case for real fluids, so we will typically have a layer (usually very thin) near any fixed surface in a moving stream in which shearing stresses are not negligible. This layer is called the *boundary layer*. Outside the boundary layer the flow can be treated as an irrotational flow. Another possible consequence of the boundary layer is that the main stream may "separate" from the surface and form a *wake* downstream from the body. (See the

Flow fields involving real fluids often include both regions of negligible shearing stresses and regions of significant shearing stresses.

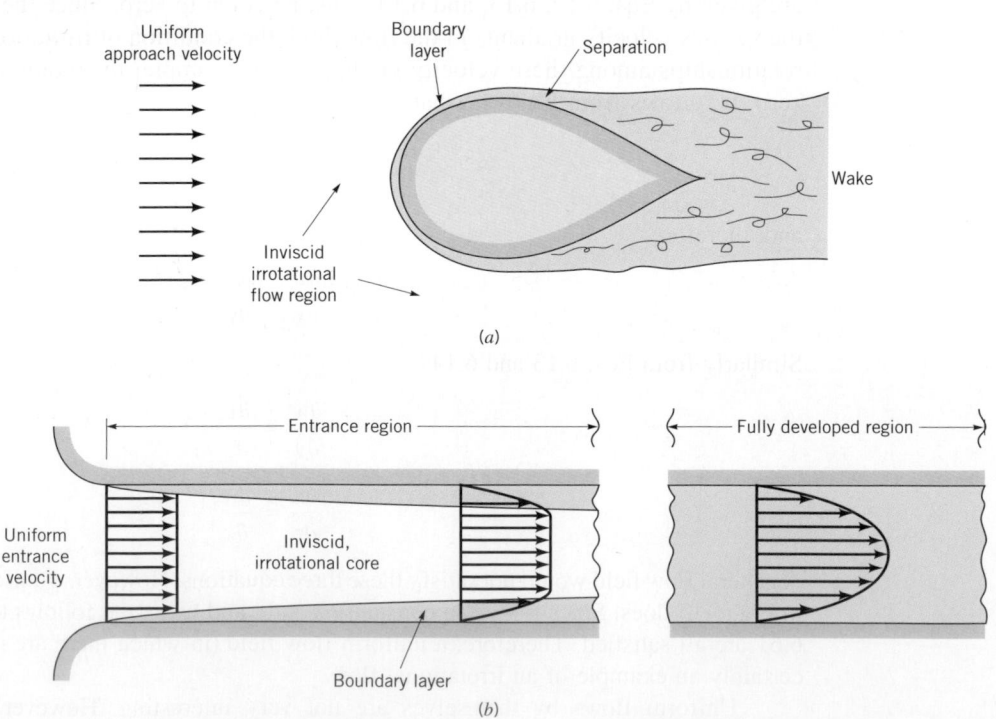

■ **F I G U R E   6.14**   **Various regions of flow: (*a*) around bodies; (*b*) through channels.**

photographs at the beginning of Chapters 7, 9, and 11.) The wake would include a region of slow, perhaps randomly moving fluid. To completely analyze this type of problem it is necessary to consider both the inviscid, irrotational flow outside the boundary layer, and the viscous, rotational flow within the boundary layer and to somehow "match" these two regions. This type of analysis is considered in Chapter 9.

As is illustrated in Fig. 6.14*b*, the flow in the entrance to a pipe may be uniform (if the entrance is streamlined), and thus will be irrotational. In the central core of the pipe the flow remains irrotational for some distance. However, a boundary layer will develop along the wall and grow in thickness until it fills the pipe. Thus, for this type of internal flow there will be an *entrance region* in which there is a central irrotational core, followed by a so-called *fully developed region* in which viscous forces are dominant. The concept of irrotationality is completely invalid in the fully developed region. This type of internal flow problem is considered in detail in Chapter 8.

The two preceding examples are intended to illustrate the possible applicability of irrotational flow to some "real fluid" flow problems and to indicate some limitations of the irrotationality concept. We proceed to develop some useful equations based on the assumptions of inviscid, incompressible, irrotational flow, with the admonition to use caution when applying the equations.

6.4.4 The Bernoulli Equation for Irrotational Flow

In the development of the Bernoulli equation in Section 6.4.2, Eq. 6.54 was integrated along a streamline. This restriction was imposed so the right side of the equation could be set equal to zero; that is,

$$[\mathbf{V} \times (\nabla \times \mathbf{V})] \cdot d\mathbf{s} = 0$$

(since $d\mathbf{s}$ is parallel to $\mathbf{V}$). However, for irrotational flow, $\nabla \times \mathbf{V} = 0$, so the right side of Eq. 6.54 is zero regardless of the direction of $d\mathbf{s}$. We can now follow the same procedure used to obtain Eq. 6.55, where the differential changes dp, $d(V^2)$, and dz can be taken in any direction. Integration of Eq. 6.55 again yields

$$\int \frac{dp}{\rho} + \frac{V^2}{2} + gz = \text{constant} \tag{6.62}$$

where for irrotational flow the constant is the same throughout the flow field. Thus, for incompressible, irrotational flow the Bernoulli equation can be written as

$$\boxed{\frac{p_1}{\gamma} + \frac{V_1^2}{2g} + z_1 = \frac{p_2}{\gamma} + \frac{V_2^2}{2g} + z_2} \tag{6.63}$$

The Bernoulli equation can be applied between any two points in an irrotational flow field.

between *any two points in the flow field*. Equation 6.63 is exactly the same form as Eq. 6.58 but is not limited to application along a streamline. However, Eq. 6.63 is restricted to

- inviscid flow
- steady flow
- incompressible flow
- irrotational flow

It may be worthwhile to review the use and misuse of the Bernoulli equation for rotational flow as is illustrated in Example 3.18.

6.4.5 The Velocity Potential

For an irrotational flow the velocity gradients are related through Eqs. 6.59, 6.60, and 6.61. It follows that in this case the velocity components can be expressed in terms of a scalar function $\phi(x, y, z, t)$ as

$$u = \frac{\partial \phi}{\partial x} \qquad v = \frac{\partial \phi}{\partial y} \qquad w = \frac{\partial \phi}{\partial z} \tag{6.64}$$

where ϕ is called the *velocity potential*. Direct substitution of these expressions for the velocity components into Eqs. 6.59, 6.60, and 6.61 will verify that a velocity field defined by Eqs. 6.64 is indeed irrotational. In vector form, Eqs. 6.64 can be written as

$$\boxed{\mathbf{V} = \nabla\phi} \tag{6.65}$$

so that for an irrotational flow the velocity is expressible as the gradient of a scalar function ϕ.

The velocity potential is a consequence of the irrotationality of the flow field, whereas the stream function is a consequence of conservation of mass (see Section 6.2.3). It is to be noted, however, that the velocity potential can be defined for a general three-dimensional flow, whereas the stream function is restricted to two-dimensional flows.

For an incompressible fluid we know from conservation of mass that

$$\nabla \cdot \mathbf{V} = 0$$

and therefore for incompressible, irrotational flow (with $\mathbf{V} = \nabla\phi$) it follows that

$$\nabla^2\phi = 0 \tag{6.66}$$

where $\nabla^2(\ ) = \nabla \cdot \nabla(\ )$ is the *Laplacian operator*. In Cartesian coordinates

$$\frac{\partial^2\phi}{\partial x^2} + \frac{\partial^2\phi}{\partial y^2} + \frac{\partial^2\phi}{\partial z^2} = 0$$

Inviscid, incompressible, irrotational flow fields are governed by Laplace's equation and are called potential flows.

This differential equation arises in many different areas of engineering and physics and is called *Laplace's equation*. Thus, inviscid, incompressible, irrotational flow fields are governed by Laplace's equation. This type of flow is commonly called a *potential flow*. To complete the mathematical formulation of a given problem, boundary conditions have to be specified. These are usually velocities specified on the boundaries of the flow field of interest. It follows that if the potential function can be determined, then the velocity at all points in the flow field can be determined from Eq. 6.64, and the pressure at all points can be determined from the Bernoulli equation (Eq. 6.63). Although the concept of the velocity potential is applicable to both steady and unsteady flow, we will confine our attention to steady flow.

Potential flows, governed by Eqs. 6.64 and 6.66, are irrotational flows. That is, the vorticity is zero throughout. If vorticity is present (e.g., boundary layer, wake), then the flow cannot be described by Laplace's equation. The figure in the margin illustrates a flow in which the vorticity is not zero in two regions—the separated region behind the bump and the boundary layer next to the solid surface. This is discussed in detail in Chapter 9.

For some problems it will be convenient to use cylindrical coordinates, r, θ, and z. In this coordinate system the gradient operator is

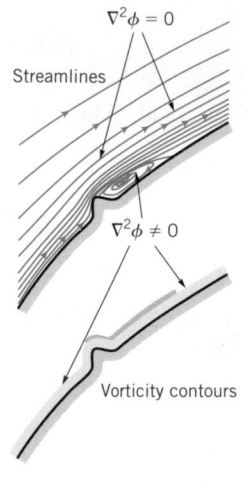

$\nabla^2\phi = 0$

Streamlines

$\nabla^2\phi \neq 0$

Vorticity contours

$$\nabla(\ ) = \frac{\partial(\ )}{\partial r}\hat{\mathbf{e}}_r + \frac{1}{r}\frac{\partial(\ )}{\partial\theta}\hat{\mathbf{e}}_\theta + \frac{\partial(\ )}{\partial z}\hat{\mathbf{e}}_z \tag{6.67}$$

so that

$$\nabla\phi = \frac{\partial\phi}{\partial r}\hat{\mathbf{e}}_r + \frac{1}{r}\frac{\partial\phi}{\partial\theta}\hat{\mathbf{e}}_\theta + \frac{\partial\phi}{\partial z}\hat{\mathbf{e}}_z \tag{6.68}$$

where $\phi = \phi(r, \theta, z)$. Since

$$\mathbf{V} = v_r\hat{\mathbf{e}}_r + v_\theta\hat{\mathbf{e}}_\theta + v_z\hat{\mathbf{e}}_z \tag{6.69}$$

it follows for an irrotational flow (with $\mathbf{V} = \nabla\phi$)

$$v_r = \frac{\partial\phi}{\partial r} \qquad v_\theta = \frac{1}{r}\frac{\partial\phi}{\partial\theta} \qquad v_z = \frac{\partial\phi}{\partial z} \tag{6.70}$$

Also, Laplace's equation in cylindrical coordinates is

$$\frac{1}{r}\frac{\partial}{\partial r}\left(r\frac{\partial\phi}{\partial r}\right) + \frac{1}{r^2}\frac{\partial^2\phi}{\partial\theta^2} + \frac{\partial^2\phi}{\partial z^2} = 0 \tag{6.71}$$

EXAMPLE 6.4 | Velocity Potential and Inviscid Flow Pressure

GIVEN The two-dimensional flow of a nonviscous, incompressible fluid in the vicinity of the 90° corner of Fig. E6.4a is described by the stream function

$$\psi = 2r^2 \sin 2\theta$$

where ψ has units of m^2/s when r is in meters. Assume the fluid density is $10^3 \ kg/m^3$ and the x–y plane is horizontal—

that is, there is no difference in elevation between points (1) and (2).

FIND

(a) Determine, if possible, the corresponding velocity potential.

(b) If the pressure at point (1) on the wall is 30 kPa, what is the pressure at point (2)?

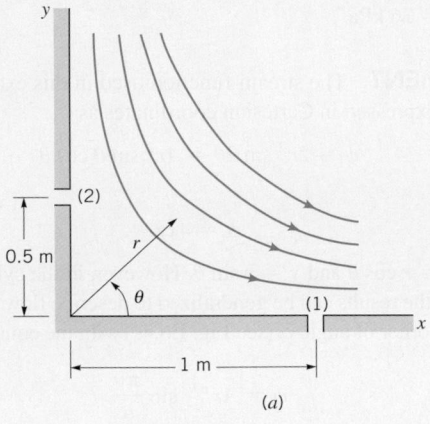

(a)

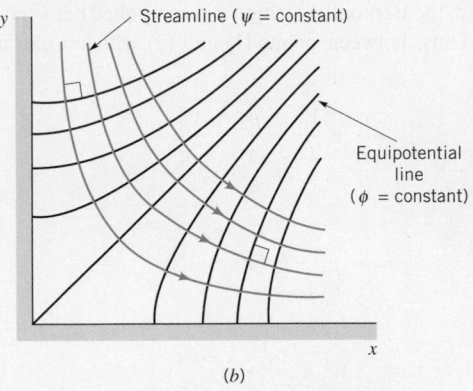

(b)

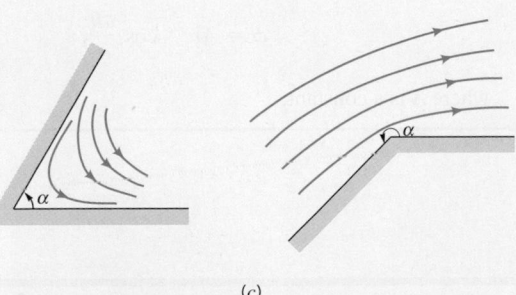

(c)

■ **FIGURE E6.4**

SOLUTION

(a) The radial and tangential velocity components can be obtained from the stream function as (see Eq. 6.42)

$$v_r = \frac{1}{r}\frac{\partial \psi}{\partial \theta} = 4r \cos 2\theta$$

and

$$v_\theta = -\frac{\partial \psi}{\partial r} = -4r \sin 2\theta$$

Since

$$v_r = \frac{\partial \phi}{\partial r}$$

it follows that

$$\frac{\partial \phi}{\partial r} = 4r \cos 2\theta$$

and therefore by integration

$$\phi = 2r^2 \cos 2\theta + f_1(\theta) \qquad (1)$$

where $f_1(\theta)$ is an arbitrary function of θ. Similarly

$$v_\theta = \frac{1}{r}\frac{\partial \phi}{\partial \theta} = -4r \sin 2\theta$$

and integration yields

$$\phi = 2r^2 \cos 2\theta + f_2(r) \qquad (2)$$

where $f_2(r)$ is an arbitrary function of r. To satisfy both Eqs. 1 and 2, the velocity potential must have the form

$$\phi = 2r^2 \cos 2\theta + C \qquad \text{(Ans)}$$

where C is an arbitrary constant. As is the case for stream functions, the specific value of C is not important, and it is customary to let $C = 0$ so that the velocity potential for this corner flow is

$$\phi = 2r^2 \cos 2\theta \qquad \text{(Ans)}$$

COMMENT In the statement of this problem it was implied by the wording "if possible" that we might not be able to find a corresponding velocity potential. The reason for this concern is that we can always define a stream function for two-dimensional flow, but the flow must be *irrotational* if there is a corresponding velocity potential. Thus, the fact that we were able to determine a velocity potential means that the flow is irrotational. Several streamlines and lines of constant ϕ are plotted in Fig. E6.4b. These two sets of lines are *orthogonal*. The reason why streamlines and lines of constant ϕ are always orthogonal is explained in Section 6.5.

(b) Since we have an irrotational flow of a nonviscous, incompressible fluid, the Bernoulli equation can be applied between any two points. Thus, between points (1) and (2) with no elevation change

$$\frac{p_1}{\gamma} + \frac{V_1^2}{2g} = \frac{p_2}{\gamma} + \frac{V_2^2}{2g}$$

or

$$p_2 = p_1 + \frac{\rho}{2}(V_1^2 - V_2^2) \tag{3}$$

Since

$$V^2 = v_r^2 + v_\theta^2$$

it follows that for any point within the flow field

$$V^2 = (4r\cos 2\theta)^2 + (-4r\sin 2\theta)^2$$
$$= 16r^2(\cos^2 2\theta + \sin^2 2\theta)$$
$$= 16r^2$$

This result indicates that the square of the velocity at any point depends only on the radial distance, r, to the point. Note that the constant, 16, has units of s^{-2}. Thus,

$$V_1^2 = (16\text{ s}^{-2})(1\text{ m})^2 = 16\text{ m}^2/\text{s}^2$$

and

$$V_2^2 = (16\text{ s}^{-2})(0.5\text{ m})^2 = 4\text{ m}^2/\text{s}^2$$

Substitution of these velocities into Eq. 3 gives

$$p_2 = 30 \times 10^3\text{ N/m}^2 + \frac{10^3\text{ kg/m}^3}{2}(16\text{ m}^2/\text{s}^2 - 4\text{ m}^2/\text{s}^2)$$

$$= 36\text{ kPa} \tag{Ans}$$

COMMENT The stream function used in this example could also be expressed in Cartesian coordinates as

$$\psi = 2r^2 \sin 2\theta = 4r^2 \sin\theta \cos\theta$$

or

$$\psi = 4xy$$

since $x = r\cos\theta$ and $y = r\sin\theta$. However, in the cylindrical polar form the results can be generalized to describe flow in the vicinity of a corner of angle α (see Fig. E6.4c) with the equations

$$\psi = Ar^{\pi/\alpha} \sin\frac{\pi\theta}{\alpha}$$

and

$$\phi = Ar^{\pi/\alpha} \cos\frac{\pi\theta}{\alpha}$$

where A is a constant.

6.5 Some Basic, Plane Potential Flows

For potential flow, basic solutions can be simply added to obtain more complicated solutions.

A major advantage of Laplace's equation is that it is a linear partial differential equation. Since it is linear, various solutions can be added to obtain other solutions—that is, if $\phi_1(x, y, z)$ and $\phi_2(x, y, z)$ are two solutions to Laplace's equation, then $\phi_3 = \phi_1 + \phi_2$ is also a solution. The practical implication of this result is that if we have certain basic solutions we can combine them to obtain more complicated and interesting solutions. In this section several basic velocity potentials, which describe some relatively simple flows, will be determined. In the next section these basic potentials will be combined to represent complicated flows.

For simplicity, only plane (two-dimensional) flows will be considered. In this case, by using Cartesian coordinates

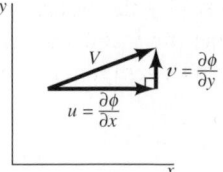

$$u = \frac{\partial\phi}{\partial x} \qquad v = \frac{\partial\phi}{\partial y} \tag{6.72}$$

or by using cylindrical coordinates

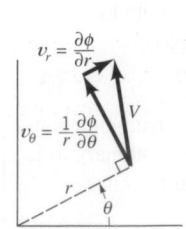

$$v_r = \frac{\partial\phi}{\partial r} \qquad v_\theta = \frac{1}{r}\frac{\partial\phi}{\partial\theta} \tag{6.73}$$

as shown by the figure in the margin. Since we can define a stream function for plane flow, we can also let

$$u = \frac{\partial\psi}{\partial y} \qquad v = -\frac{\partial\psi}{\partial x} \tag{6.74}$$

or

$$v_r = \frac{1}{r}\frac{\partial \psi}{\partial \theta} \qquad v_\theta = -\frac{\partial \psi}{\partial r} \qquad (6.75)$$

where the stream function was previously defined in Eqs. 6.37 and 6.42. We know that by defining the velocities in terms of the stream function, conservation of mass is identically satisfied. If we now impose the condition of irrotationality, it follows from Eq. 6.59 that

$$\frac{\partial u}{\partial y} = \frac{\partial v}{\partial x}$$

and in terms of the stream function

$$\frac{\partial}{\partial y}\left(\frac{\partial \psi}{\partial y}\right) = \frac{\partial}{\partial x}\left(-\frac{\partial \psi}{\partial x}\right)$$

or

$$\frac{\partial^2 \psi}{\partial x^2} + \frac{\partial^2 \psi}{\partial y^2} = 0$$

Thus, for a plane irrotational flow we can use either the velocity potential or the stream function—both must satisfy Laplace's equation in two dimensions. It is apparent from these results that the velocity potential and the stream function are somehow related. We have previously shown that lines of constant ψ are streamlines; that is,

$$\left.\frac{dy}{dx}\right|_{\text{along }\psi = \text{constant}} = \frac{v}{u} \qquad (6.76)$$

The change in ϕ as we move from one point (x, y) to a nearby point $(x + dx, y + dy)$ is given by the relationship

$$d\phi = \frac{\partial \phi}{\partial x}dx + \frac{\partial \phi}{\partial y}dy = u\,dx + v\,dy$$

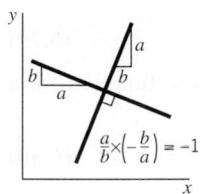

Along a line of constant ϕ we have $d\phi = 0$ so that

$$\left.\frac{dy}{dx}\right|_{\text{along }\phi = \text{constant}} = -\frac{u}{v} \qquad (6.77)$$

A comparison of Eqs. 6.76 and 6.77 shows that lines of constant ϕ (called *equipotential lines*) are orthogonal to lines of constant ψ (streamlines) at all points where they intersect. (Recall that two lines are orthogonal if the product of their slopes is -1, as illustrated by the figure in the margin.) For any potential flow field a *"flow net"* can be drawn that consists of a family of streamlines and equipotential lines. The flow net is useful in visualizing flow patterns and can be used to obtain graphical solutions by sketching in streamlines and equipotential lines and adjusting the lines until the lines are approximately orthogonal at all points where they intersect. An example of a flow net is shown in Fig. 6.15. Velocities can be estimated from the flow net, since the velocity is inversely proportional to the streamline spacing, as shown by the figure in the margin. Thus, for example, from Fig. 6.15 we can see that the velocity near the inside corner will be higher than the velocity along the outer part of the bend. (See the photographs at the beginning of Chapters 3 and 6.)

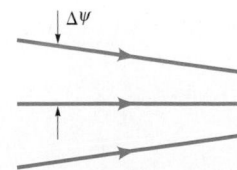

Streamwise acceleration

6.5.1 Uniform Flow

The simplest plane flow is one for which the streamlines are all straight and parallel, and the magnitude of the velocity is constant. This type of flow is called a *uniform flow*. For example, consider a uniform flow in the positive x direction as is illustrated in Fig. 6.16a. In this instance, $u = U$ and $v = 0$, and in terms of the velocity potential

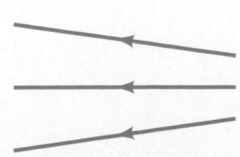

Streamwise deceleration

$$\frac{\partial \phi}{\partial x} = U \qquad \frac{\partial \phi}{\partial y} = 0$$

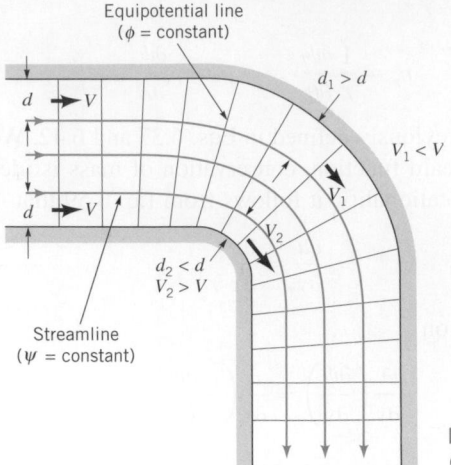

■ **FIGURE 6.15** **Flow net for a 90° bend. (From Ref. 3, used by permission.)**

These two equations can be integrated to yield

$$\phi = Ux + C$$

where C is an arbitrary constant, which can be set equal to zero. Thus, for a uniform flow in the positive x direction

$$\phi = Ux \tag{6.78}$$

The corresponding stream function can be obtained in a similar manner, since

$$\frac{\partial \psi}{\partial y} = U \qquad \frac{\partial \psi}{\partial x} = 0$$

and, therefore,

$$\psi = Uy \tag{6.79}$$

These results can be generalized to provide the velocity potential and stream function for a uniform flow at an angle α with the x axis, as in Fig. 6.16b. For this case

$$\phi = U(x \cos \alpha + y \sin \alpha) \tag{6.80}$$

and

$$\psi = U(y \cos \alpha - x \sin \alpha) \tag{6.81}$$

6.5.2 Source and Sink

Consider a fluid flowing radially outward from a line through the origin perpendicular to the x–y plane as is shown in Fig. 6.17. Let m be the volume rate of flow emanating from the line (per unit length), and therefore to satisfy conservation of mass

$$(2\pi r)v_r = m$$

or

$$v_r = \frac{m}{2\pi r}$$

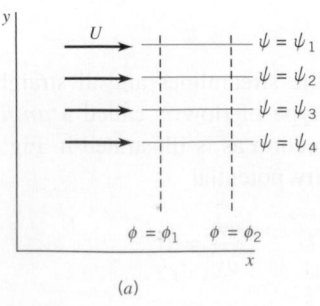

(a)

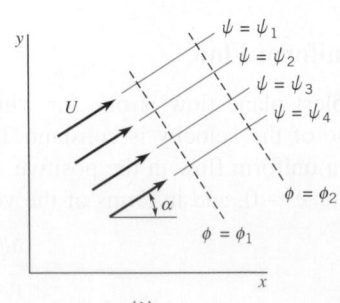

(b)

■ **FIGURE 6.16**
Uniform flow: (a) in the x direction; (b) in an arbitrary direction, α.

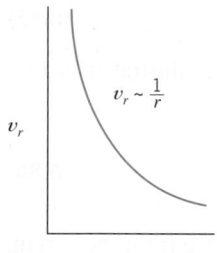

■ **F I G U R E   6.17**   **The streamline pattern for a source.**

A source or sink represents a purely radial flow.

Also, since the flow is a purely radial flow, $v_\theta = 0$, the corresponding velocity potential can be obtained by integrating the equations

$$\frac{\partial \phi}{\partial r} = \frac{m}{2\pi r} \qquad \frac{1}{r} \frac{\partial \phi}{\partial \theta} = 0$$

It follows that

$$\phi = \frac{m}{2\pi} \ln r \tag{6.82}$$

If m is positive, the flow is radially outward, and the flow is considered to be a *source* flow. If m is negative, the flow is toward the origin, and the flow is considered to be a *sink* flow. The flowrate, m, is the *strength* of the source or sink.

As shown by the figure in the margin, at the origin where $r = 0$ the velocity becomes infinite, which is of course physically impossible. Thus, sources and sinks do not really exist in real flow fields, and the line representing the source or sink is a mathematical *singularity* in the flow field. However, some real flows can be approximated at points away from the origin by using sources or sinks. Also, the velocity potential representing this hypothetical flow can be combined with other basic velocity potentials to approximately describe some real flow fields. This idea is further discussed in Section 6.6.

The stream function for the source can be obtained by integrating the relationships

$$v_r = \frac{1}{r} \frac{\partial \psi}{\partial \theta} = \frac{m}{2\pi r} \qquad v_\theta = -\frac{\partial \psi}{\partial r} = 0$$

to yield

$$\psi = \frac{m}{2\pi} \theta \tag{6.83}$$

It is apparent from Eq. 6.83 that the streamlines (lines of ψ = constant) are radial lines, and from Eq. 6.82 the equipotential lines (lines of ϕ = constant) are concentric circles centered at the origin.

$v_r \sim \dfrac{1}{r}$

v_r

r

EXAMPLE 6.5   Potential Flow—Sink

GIVEN   A nonviscous, incompressible fluid flows between wedge-shaped walls into a small opening as shown in Fig. E6.5. The velocity potential (in m^2/s), which approximately describes this flow is

$$\phi = -2 \ln r$$

FIND   Determine the volume rate of flow (per unit length) into the opening.

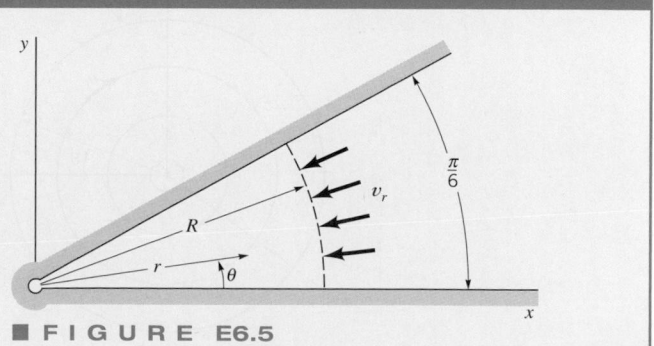

■ **F I G U R E   E6.5**

SOLUTION

The components of velocity are

$$v_r = \frac{\partial \phi}{\partial r} = -\frac{2}{r} \qquad v_\theta = \frac{1}{r}\frac{\partial \phi}{\partial \theta} = 0$$

which indicates we have a purely radial flow. The flowrate per unit width, q, crossing the arc of length $R\pi/6$ can thus be obtained by integrating the expression

$$q = \int_0^{\pi/6} v_r R\, d\theta = -\int_0^{\pi/6}\left(\frac{2}{R}\right)R\, d\theta$$

$$= -\frac{\pi}{3} = -1.05 \text{ m}^2/\text{s} \qquad \text{(Ans)}$$

COMMENT Note that the radius R is arbitrary since the flowrate crossing any curve between the two walls must be the same. The negative sign indicates that the flow is toward the opening, that is, in the negative radial direction.

6.5.3 Vortex

We next consider a flow field in which the streamlines are concentric circles—that is, we interchange the velocity potential and stream function for the source. Thus, let

$$\phi = K\theta \qquad (6.84)$$

A vortex represents a flow in which the streamlines are concentric circles.

and

$$\psi = -K \ln r \qquad (6.85)$$

where K is a constant. In this case the streamlines are concentric circles as are illustrated in Fig. 6.18, with $v_r = 0$ and

$$v_\theta = \frac{1}{r}\frac{\partial \phi}{\partial \theta} = -\frac{\partial \psi}{\partial r} = \frac{K}{r} \qquad (6.86)$$

This result indicates that the tangential velocity varies inversely with the distance from the origin, as shown by the figure in the margin, with a singularity occurring at $r = 0$ (where the velocity becomes infinite).

It may seem strange that this *vortex* motion is irrotational (and it is since the flow field is described by a velocity potential). However, it must be recalled that rotation refers to the orientation of a fluid element and not the path followed by the element. Thus, for an irrotational vortex, if a pair of small sticks were placed in the flow field at location A, as indicated in Fig. 6.19a, the sticks would rotate as they move to location B. One of the sticks, the one that is aligned along the streamline, would follow a circular path and rotate in a counterclockwise

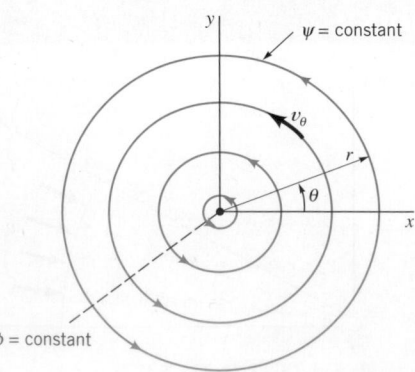

■ **FIGURE 6.18** The streamline pattern for a vortex.

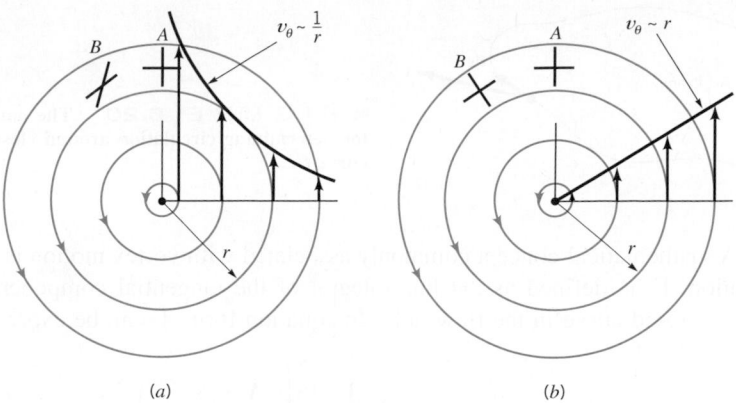

■ **F I G U R E 6.19**   **Motion of fluid element from *A* to *B*: (*a*) for irrotational (free) vortex; (*b*) for rotational (forced) vortex.**

direction. The other stick would rotate in a clockwise direction due to the nature of the flow field—that is, the part of the stick nearest the origin moves faster than the opposite end. Although both sticks are rotating, the average angular velocity of the two sticks is zero since the flow is irrotational.

If the fluid were rotating as a rigid body, such that $v_\theta = K_1 r$ where K_1 is a constant, then sticks similarly placed in the flow field would rotate as is illustrated in Fig. 6.19*b*. This type of vortex motion is *rotational* and cannot be described with a velocity potential. The rotational vortex is commonly called a *forced vortex*, whereas the irrotational vortex is usually called a *free vortex*. The swirling motion of the water as it drains from a bathtub is similar to that of a free vortex, whereas the motion of a liquid contained in a tank that is rotated about its axis with angular velocity ω corresponds to a forced vortex.

Vortex motion can be either rotational or irrotational.

A *combined vortex* is one with a forced vortex as a central core and a velocity distribution corresponding to that of a free vortex outside the core. Thus, for a combined vortex

$$v_\theta = \omega r \qquad r \leq r_0 \tag{6.87}$$

and

$$v_\theta = \frac{K}{r} \qquad r > r_0 \tag{6.88}$$

where K and ω are constants and r_0 corresponds to the radius of the central core. The pressure distribution in both the free and forced vortex was previously considered in Example 3.3. (See Fig. E6.6*a* for an approximation of this type of flow.)

F l u i d s   i n   t h e   N e w s

Some hurricane facts One of the most interesting, yet potentially devastating, naturally occurring fluid flow phenomenan is a hurricane. Broadly speaking a hurricane is a rotating mass of air circulating around a low pressure central core. In some respects the motion is similar to that of a *free vortex*. The Caribbean and Gulf of Mexico experience the most hurricanes, with the official hurricane season being from June 1 to November 30. Hurricanes are usually 300 to 400 miles (483 to 644 km) wide and are structured around a central eye in which the air is relatively calm. The eye is surrounded by an eye wall which is the region of strongest winds and precipitation. As one goes from the eye wall to the eye the wind speeds decrease sharply and within the eye the air is relatively calm and

clear of clouds. However, in the eye the pressure is at a minimum and may be 10% less than standard atmospheric pressure. This low pressure creates strong downdrafts of dry air from above. Hurricanes are classified into five categories based on their wind speeds:

Category one—74–95 mph (119–153 km/hr)

Category two—96–110 mph (154–177 km/hr)

Category three—111–130 mph (178–209 km/hr)

Category four—131–155 mph (210–249 km/hr)

Category five—greater than 155 mph (249 km/hr).

(See Problem 6.58.)

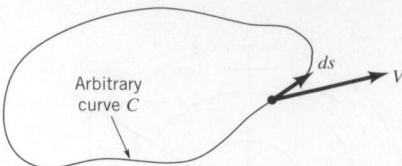

■ **FIGURE 6.20** The notation for determining circulation around closed curve *C*.

A mathematical concept commonly associated with vortex motion is that of *circulation.* The circulation, Γ, is defined as the line integral of the tangential component of the velocity taken around a closed curve in the flow field. In equation form, Γ can be expressed as

$$\Gamma = \oint_C \mathbf{V} \cdot d\mathbf{s} \qquad (6.89)$$

where the integral sign means that the integration is taken around a closed curve, *C*, in the counterclockwise direction, and *d***s** is a differential length along the curve as is illustrated in Fig. 6.20. For an irrotational flow, $\mathbf{V} = \nabla\phi$ so that $\mathbf{V} \cdot d\mathbf{s} = \nabla\phi \cdot d\mathbf{s} = d\phi$ and, therefore,

$$\Gamma = \oint_C d\phi = 0$$

This result indicates that for an irrotational flow the circulation will generally be zero. (Chapter 9 has further discussion of circulation in real flows.) However, if there are singularities enclosed within the curve the circulation may not be zero. For example, for the free vortex with $v_\theta = K/r$ the circulation around the circular path of radius *r* shown in Fig. 6.21 is

$$\Gamma = \int_0^{2\pi} \frac{K}{r}(r \, d\theta) = 2\pi K$$

The numerical value of the circulation may depend on the particular closed path considered.

which shows that the circulation is nonzero and the constant $K = \Gamma/2\pi$. However, for irrotational flows the circulation around any path that does not include a singular point will be zero. This can be easily confirmed for the closed path *ABCD* of Fig. 6.21 by evaluating the circulation around that path.

The velocity potential and stream function for the free vortex are commonly expressed in terms of the circulation as

$$\phi = \frac{\Gamma}{2\pi}\theta \qquad (6.90)$$

and

$$\psi = -\frac{\Gamma}{2\pi}\ln r \qquad (6.91)$$

V6.4 Vortex in a beaker

The concept of circulation is often useful when evaluating the forces developed on bodies immersed in moving fluids. This application will be considered in Section 6.6.3.

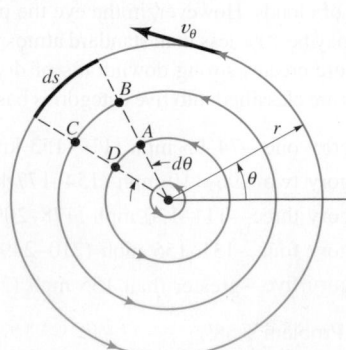

■ **FIGURE 6.21** Circulation around various paths in a free vortex.

EXAMPLE 6.6   Potential Flow—Free Vortex

GIVEN A liquid drains from a large tank through a small opening as illustrated in Fig. E6.6*a*. A vortex forms whose velocity distribution away from the tank opening can be approximated as that of a free vortex having a velocity potential

$$\phi = \frac{\Gamma}{2\pi}\theta$$

FIND Determine an expression relating the surface shape to the strength of the vortex as specified by the circulation Γ.

SOLUTION

Since the free vortex represents an irrotational flow field, the Bernoulli equation

$$\frac{p_1}{\gamma} + \frac{V_1^2}{2g} + z_1 = \frac{p_2}{\gamma} + \frac{V_2^2}{2g} + z_2$$

can be written between any two points. If the points are selected at the free surface, $p_1 = p_2 = 0$, so that

$$\frac{V_1^2}{2g} = z_s + \frac{V_2^2}{2g} \tag{1}$$

where the free surface elevation, z_s, is measured relative to a datum passing through point (1) as shown in Fig. E6.6*b*.

The velocity is given by the equation

$$v_\theta = \frac{1}{r}\frac{\partial\phi}{\partial\theta} = \frac{\Gamma}{2\pi r}$$

We note that far from the origin at point (1), $V_1 = v_\theta \approx 0$ so that Eq. 1 becomes

$$z_s = -\frac{\Gamma^2}{8\pi^2 r^2 g} \tag{Ans}$$

which is the desired equation for the surface profile.

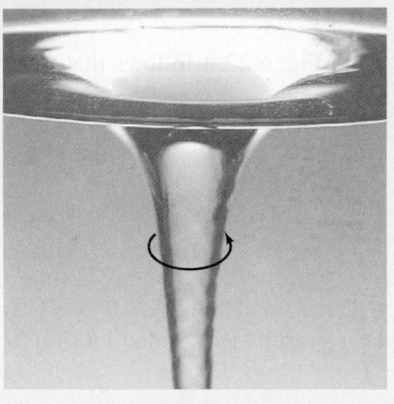

■ FIGURE E6.6*a*

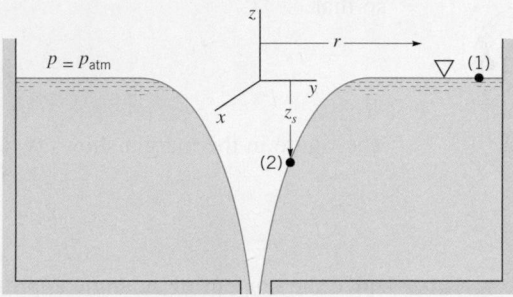

■ FIGURE E6.6*b*

COMMENT The negative sign indicates that the surface falls as the origin is approached as shown in Fig. E6.6. This solution is not valid very near the origin since the predicted velocity becomes excessively large as the origin is approached.

6.5.4 Doublet

A doublet is formed by an appropriate source–sink pair.

The final, basic potential flow to be considered is one that is formed by combining a source and sink in a special way. Consider the equal strength, source–sink pair of Fig. 6.22. The combined stream function for the pair is

$$\psi = -\frac{m}{2\pi}(\theta_1 - \theta_2)$$

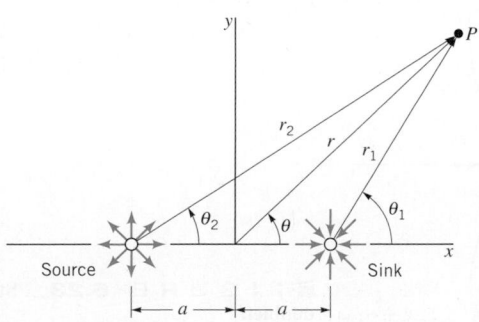

■ FIGURE 6.22   The combination of a source and sink of equal strength located along the *x* axis.

which can be rewritten as

$$\tan\left(-\frac{2\pi\psi}{m}\right) = \tan(\theta_1 - \theta_2) = \frac{\tan\theta_1 - \tan\theta_2}{1 + \tan\theta_1 \tan\theta_2} \tag{6.92}$$

From Fig. 6.22 it follows that

$$\tan\theta_1 = \frac{r\sin\theta}{r\cos\theta - a}$$

and

$$\tan\theta_2 = \frac{r\sin\theta}{r\cos\theta + a}$$

These results substituted into Eq. 6.92 give

$$\tan\left(-\frac{2\pi\psi}{m}\right) = \frac{2ar\sin\theta}{r^2 - a^2}$$

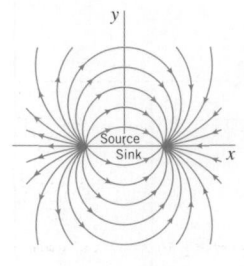

so that

$$\psi = -\frac{m}{2\pi}\tan^{-1}\left(\frac{2ar\sin\theta}{r^2 - a^2}\right) \tag{6.93}$$

The figure in the margin shows typical streamlines for this flow. For small values of the distance a

$$\psi = -\frac{m}{2\pi}\frac{2ar\sin\theta}{r^2 - a^2} = -\frac{mar\sin\theta}{\pi(r^2 - a^2)} \tag{6.94}$$

since the tangent of an angle approaches the value of the angle for small angles.

A doublet is formed by letting a source and sink approach one another.

The so-called *doublet* is formed by letting the source and sink approach one another $(a \rightarrow 0)$ while increasing the strength m $(m \rightarrow \infty)$ so that the product ma/π remains constant. In this case, since $r/(r^2 - a^2) \rightarrow 1/r$, Eq. 6.94 reduces to

$$\psi = -\frac{K\sin\theta}{r} \tag{6.95}$$

where K, a constant equal to ma/π, is called the *strength* of the doublet. The corresponding velocity potential for the doublet is

$$\phi = \frac{K\cos\theta}{r} \tag{6.96}$$

Plots of lines of constant ψ reveal that the streamlines for a doublet are circles through the origin tangent to the x axis as shown in Fig. 6.23. Just as sources and sinks are not physically realistic entities, neither are doublets. However, the doublet when combined with other basic potential flows

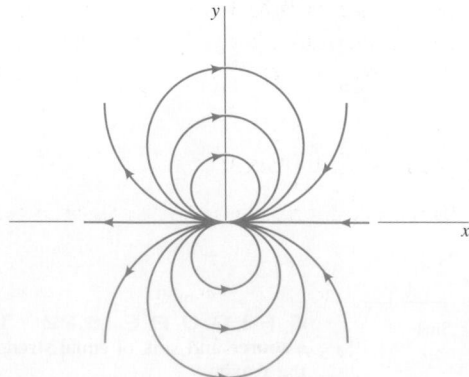

■ **F I G U R E   6.23**   Streamlines for a doublet.

■ TABLE 6.1
Summary of Basic, Plane Potential Flows

Description of Flow Field	Velocity Potential	Stream Function	Velocity Components[a]
Uniform flow at angle α with the x axis (see Fig. 6.16b)	$\phi = U(x \cos \alpha + y \sin \alpha)$	$\psi = U(y \cos \alpha - x \sin \alpha)$	$u = U \cos \alpha$ $v = U \sin \alpha$
Source or sink (see Fig. 6.17) $m > 0$ source $m < 0$ sink	$\phi = \dfrac{m}{2\pi} \ln r$	$\psi = \dfrac{m}{2\pi} \theta$	$v_r = \dfrac{m}{2\pi r}$ $v_\theta = 0$
Free vortex (see Fig. 6.18) $\Gamma > 0$ counterclockwise motion $\Gamma < 0$ clockwise motion	$\phi = \dfrac{\Gamma}{2\pi} \theta$	$\psi = -\dfrac{\Gamma}{2\pi} \ln r$	$v_r = 0$ $v_\theta = \dfrac{\Gamma}{2\pi r}$
Doublet (see Fig. 6.23)	$\phi = \dfrac{K \cos \theta}{r}$	$\psi = -\dfrac{K \sin \theta}{r}$	$v_r = -\dfrac{K \cos \theta}{r^2}$ $v_\theta = -\dfrac{K \sin \theta}{r^2}$

[a]Velocity components are related to the velocity potential and stream function through the relationships:

$$u = \frac{\partial \phi}{\partial x} = \frac{\partial \psi}{\partial y} \qquad v = \frac{\partial \phi}{\partial y} = -\frac{\partial \psi}{\partial x} \qquad v_r = \frac{\partial \phi}{\partial r} = \frac{1}{r}\frac{\partial \psi}{\partial \theta} \qquad v_\theta = \frac{1}{r}\frac{\partial \phi}{\partial \theta} = -\frac{\partial \psi}{\partial r}.$$

provides a useful representation of some flow fields of practical interest. For example, we will determine in Section 6.6.3 that the combination of a uniform flow and a doublet can be used to represent the flow around a circular cylinder. Table 6.1 provides a summary of the pertinent equations for the basic, plane potential flows considered in the preceding sections.

6.6   Superposition of Basic, Plane Potential Flows

As was discussed in the previous section, potential flows are governed by Laplace's equation, which is a linear partial differential equation. It therefore follows that the various basic velocity potentials and stream functions can be combined to form new potentials and stream functions. (Why is this true?) Whether such combinations yield useful results remains to be seen. It is to be noted that *any streamline in an inviscid flow field can be considered as a solid boundary*, since the conditions along a solid boundary and a streamline are the same—that is, there is no flow through the boundary or the streamline. Thus, if we can combine some of the basic velocity potentials or stream functions to yield a streamline that corresponds to a particular body shape of interest, that combination can be used to describe in detail the flow around that body. This method of solving some interesting flow problems, commonly called the *method of superposition*, is illustrated in the following three sections.

6.6.1 Source in a Uniform Stream—Half-Body

Flow around a half-body is obtained by the addition of a source to a uniform flow.

Consider the superposition of a source and a uniform flow as shown in Fig. 6.24a. The resulting stream function is

$$\psi = \psi_{\text{uniform flow}} + \psi_{\text{source}}$$

$$= Ur \sin \theta + \frac{m}{2\pi} \theta \qquad (6.97)$$

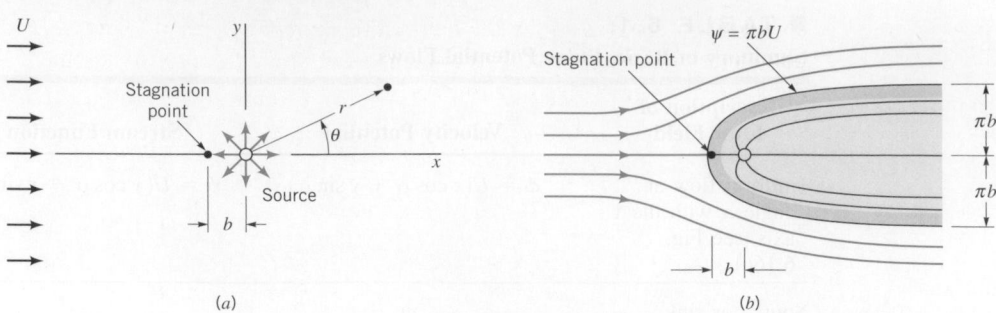

■ **F I G U R E   6.24**   The flow around a half-body: (*a*) superposition of a source and a uniform flow; (*b*) replacement of streamline $\psi = \pi b U$ with solid boundary to form half-body.

and the corresponding velocity potential is

$$\phi = Ur \cos \theta + \frac{m}{2\pi} \ln r \qquad (6.98)$$

It is clear that at some point along the negative x axis the velocity due to the source will just cancel that due to the uniform flow and a stagnation point will be created. For the source alone

$$v_r = \frac{m}{2\pi r}$$

so that the stagnation point will occur at $x = -b$ where

$$U = \frac{m}{2\pi b}$$

or

$$b = \frac{m}{2\pi U} \qquad (6.99)$$

The value of the stream function at the stagnation point can be obtained by evaluating ψ at $r = b$ and $\theta = \pi$, which yields from Eq. 6.97

$$\psi_{\text{stagnation}} = \frac{m}{2}$$

Since $m/2 = \pi b U$ (from Eq. 6.99) it follows that the equation of the streamline passing through the stagnation point is

$$\pi b U = Ur \sin \theta + bU\theta$$

or

$$r = \frac{b(\pi - \theta)}{\sin \theta} \qquad (6.100)$$

where θ can vary between 0 and 2π. A plot of this streamline is shown in Fig. 6.24*b*. If we replace this streamline with a solid boundary, as indicated in the figure, then it is clear that this combination of a uniform flow and a source can be used to describe the flow around a streamlined body placed in a uniform stream. The body is open at the downstream end, and thus is called a *half-body*. Other streamlines in the flow field can be obtained by setting $\psi = $ constant in Eq. 6.97 and plotting the resulting equation. A number of these streamlines are shown in Fig. 6.24*b*. Although the streamlines inside the body are shown, they are actually of no interest in this case, since we are concerned with the flow field outside the body. It should be noted that the singularity in the flow field (the source) occurs inside the body, and there are no singularities in the flow field of interest (outside the body).

The width of the half-body asymptotically approaches $2\pi b$. This follows from Eq. 6.100, which can be written as

$$y = b(\pi - \theta)$$

V6.5 Half-body

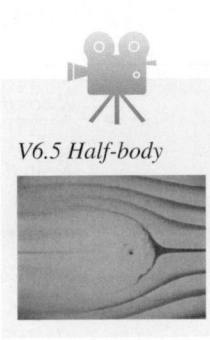

For inviscid flow, a streamline can be replaced by a solid boundary.

so that as $\theta \to 0$ or $\theta \to 2\pi$ the half-width approaches $\pm b\pi$. With the stream function (or velocity potential) known, the velocity components at any point can be obtained. For the half-body, using the stream function given by Eq. 6.97,

$$v_r = \frac{1}{r}\frac{\partial \psi}{\partial \theta} = U\cos\theta + \frac{m}{2\pi r}$$

and

$$v_\theta = -\frac{\partial \psi}{\partial r} = -U\sin\theta$$

Thus, the square of the magnitude of the velocity, V, at any point is

$$V^2 = v_r^2 + v_\theta^2 = U^2 + \frac{Um\cos\theta}{\pi r} + \left(\frac{m}{2\pi r}\right)^2$$

and since $b = m/2\pi U$

$$V^2 = U^2\left(1 + 2\frac{b}{r}\cos\theta + \frac{b^2}{r^2}\right) \tag{6.101}$$

With the velocity known, the pressure at any point can be determined from the Bernoulli equation, which can be written between any two points in the flow field since the flow is irrotational. Thus, applying the Bernoulli equation between a point far from the body, where the pressure is p_0 and the velocity is U, and some arbitrary point with pressure p and velocity V, it follows that

$$p_0 + \tfrac{1}{2}\rho U^2 = p + \tfrac{1}{2}\rho V^2 \tag{6.102}$$

where elevation changes have been neglected. Equation 6.101 can now be substituted into Eq. 6.102 to obtain the pressure at any point in terms of the reference pressure, p_0, and the upstream velocity, U.

   This relatively simple potential flow provides some useful information about the flow around the front part of a streamlined body, such as a bridge pier or strut placed in a uniform stream. An important point to be noted is that the velocity tangent to the surface of the body is not zero; that *For a potential flow* is, the fluid "slips" by the boundary. This result is a consequence of neglecting viscosity, the fluid *the fluid is allowed* property that causes real fluids to stick to the boundary, thus creating a "no-slip" condition. All *to slip past a fixed* potential flows differ from the flow of real fluids in this respect and do not accurately represent *solid boundary.* the velocity very near the boundary. However, outside this very thin boundary layer the velocity distribution will generally correspond to that predicted by potential flow theory if flow separation does not occur. (See Section 9.2.6.) Also, the pressure distribution along the surface will closely approximate that predicted from the potential flow theory, since the boundary layer is thin and there is little opportunity for the pressure to vary through the thin layer. In fact, as discussed in more detail in Chapter 9, the pressure distribution obtained from potential flow theory is used in conjunction with viscous flow theory to determine the nature of flow within the boundary layer.

EXAMPLE 6.7   Potential Flow—Half-body

GIVEN   A 64 km/hr wind blows toward a hill arising from a plain that can be approximated with the top section of a half-body as illustrated in Fig. E6.7a. The height of the hill approaches 60 m as shown. Assume an air density of 1.23 kg/m².

FIND

(a) What is the magnitude of the air velocity at a point on the hill directly above the origin [point (2)]?

(b) What is the elevation of point (2) above the plain and what is the difference in pressure between point (1) on the plain far from the hill and point (2)?

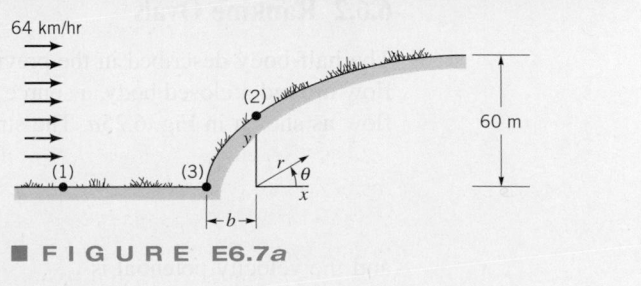

■ **FIGURE E6.7a**

SOLUTION

(a) The velocity is given by Eq. 6.101 as

$$V^2 = U^2 \left(1 + 2\frac{b}{r}\cos\theta + \frac{b^2}{r^2}\right)$$

At point (2), $\theta = \pi/2$, and since this point is on the surface (Eq. 6.100)

$$r = \frac{b(\pi - \theta)}{\sin\theta} = \frac{\pi b}{2} \qquad (1)$$

Thus,

$$V_2^2 = U^2\left[1 + \frac{b^2}{(\pi b/2)^2}\right]$$

$$= U^2\left(1 + \frac{4}{\pi^2}\right)$$

and the magnitude of the velocity at (2) for a 64 km/hr approaching wind is

$$V_2 = \left(1 + \frac{4}{\pi^2}\right)^{1/2}(64 \text{ km/hr}) = 76 \text{ km/hr} \qquad \textbf{(Ans)}$$

(b) The elevation at (2) above the plain is given by Eq. 1 as

$$y_2 = \frac{\pi b}{2}$$

Since the height of the hill approaches 60 m and this height is equal to πb, it follows that

$$y_2 = \frac{60 \text{ m}}{2} = 30 \text{ m} \qquad \textbf{(Ans)}$$

From the Bernoulli equation (with the y axis the vertical axis)

$$\frac{p_1}{\gamma} + \frac{V_1^2}{2g} + y_1 = \frac{p_2}{\gamma} + \frac{V_2^2}{2g} + y_2$$

so that

$$p_1 - p_2 = \frac{\rho}{2}(V_2^2 - V_1^2) + \gamma(y_2 - y_1)$$

and with

$$V_1 = (64 \text{ km/hr})\left(\frac{1000 \text{ m/km}}{3600 \text{ s/hr}}\right) = 17.8 \text{ m/s}$$

and

$$V_2 = (76 \text{ km/hr})\left(\frac{1000 \text{ m/km}}{3600 \text{ s/hr}}\right) = 21.1 \text{ m/s}$$

it follows that

$$p_1 - p_2 = \frac{(1.23 \text{ kg/m}^3)}{2}\left[(21.1 \text{ m/s})^2 - (17.8 \text{ m/s})^2\right]$$

$$+ (1.23 \text{ kg/m}^3)(9.81 \text{ m/s}^2)(30 \text{ m} - 0 \text{ m})$$

$$= 440.9 \text{ N/m}^2 = 0.44 \text{ kPa} \qquad \textbf{(Ans)}$$

COMMENTS This result indicates that the pressure on the hill at point (2) is slightly lower than the pressure on the plain at some distance from the base of the hill with a 0.37 kPa difference due to the elevation increase and a 0.07 kPa difference due to the velocity increase.

By repeating the calculations for various values of the upstream wind speed, U, the results shown in Fig. E6.7b are obtained. Note that as the wind speed increases, the pressure difference increases from the calm conditions of $p_1 - p_2 = 0.37$ kPa.

The maximum velocity along the hill surface does not occur at point (2) but farther up the hill at $\theta = 63°$. At this point $V_{\text{surface}} = 1.26U$. The minimum velocity ($V = 0$) and maximum pressure occur at point (3), the stagnation point.

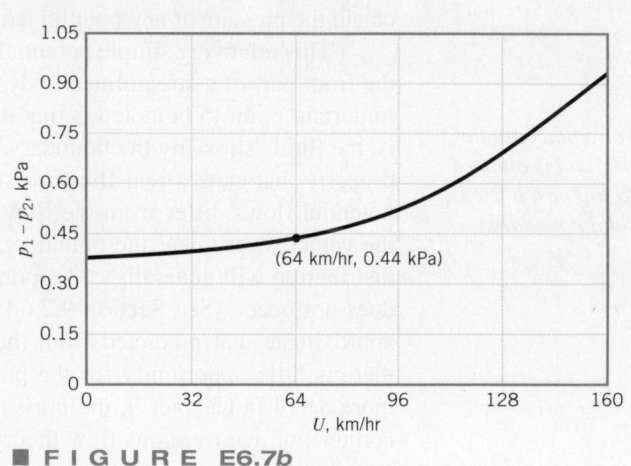

■ **FIGURE E6.7b**

6.6.2 Rankine Ovals

The half-body described in the previous section is a body that is "open" at one end. To study the flow around a closed body, a source and a sink of equal strength can be combined with a uniform flow as shown in Fig. 6.25a. The stream function for this combination is

$$\psi = Ur\sin\theta - \frac{m}{2\pi}(\theta_1 - \theta_2) \qquad \textbf{(6.103)}$$

and the velocity potential is

$$\phi = Ur\cos\theta - \frac{m}{2\pi}(\ln r_1 - \ln r_2) \qquad \textbf{(6.104)}$$

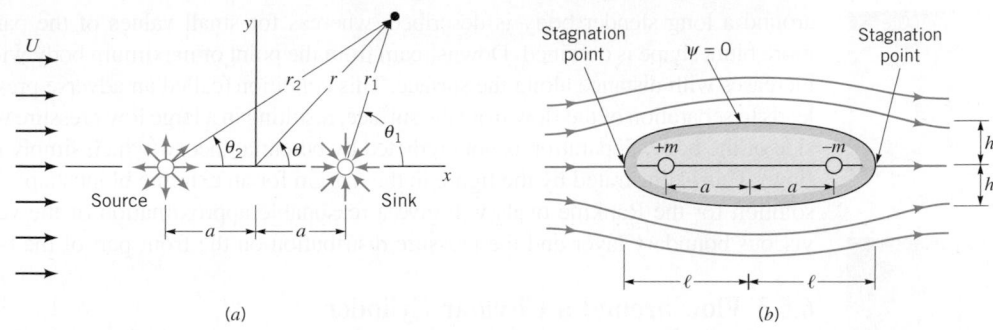

■ **FIGURE 6.25**   The flow around a Rankine oval: (*a*) superposition of source–sink pair and a uniform flow; (*b*) replacement of streamline $\psi = 0$ with solid boundary to form Rankine oval.

As discussed in Section 6.5.4, the stream function for the source–sink pair can be expressed as in Eq. 6.93 and, therefore, Eq. 6.103 can also be written as

$$\psi = Ur \sin \theta - \frac{m}{2\pi} \tan^{-1} \left(\frac{2ar \sin \theta}{r^2 - a^2} \right)$$

or

$$\psi = Uy - \frac{m}{2\pi} \tan^{-1} \left(\frac{2ay}{x^2 + y^2 - a^2} \right) \tag{6.105}$$

The corresponding streamlines for this flow field are obtained by setting ψ = constant. If several of these streamlines are plotted, it will be discovered that the streamline $\psi = 0$ forms a closed body as is illustrated in Fig. 6.25*b*. We can think of this streamline as forming the surface of a body of length 2ℓ and width $2h$ placed in a uniform stream. The streamlines inside the body are of no practical interest and are not shown. Note that since the body is closed, all of the flow emanating from the source flows into the sink. These bodies have an oval shape and are termed *Rankine ovals*.

Rankine ovals are formed by combining a source and sink with a uniform flow.

Stagnation points occur at the upstream and downstream ends of the body as are indicated in Fig. 6.25*b*. These points can be located by determining where along the *x* axis the velocity is zero. The stagnation points correspond to the points where the uniform velocity, the source velocity, and the sink velocity all combine to give a zero velocity. The locations of the stagnation points depend on the value of *a*, *m*, and *U*. The body half-length, ℓ (the value of $|x|$ that gives $\mathbf{V} = 0$ when $y = 0$), can be expressed as

$$\ell = \left(\frac{ma}{\pi U} + a^2 \right)^{1/2} \tag{6.106}$$

or

$$\frac{\ell}{a} = \left(\frac{m}{\pi Ua} + 1 \right)^{1/2} \tag{6.107}$$

The body half-width, *h*, can be obtained by determining the value of *y* where the *y* axis intersects the $\psi = 0$ streamline. Thus, from Eq. 6.105 with $\psi = 0$, $x = 0$, and $y = h$, it follows that

$$h = \frac{h^2 - a^2}{2a} \tan \frac{2\pi U h}{m} \tag{6.108}$$

or

$$\frac{h}{a} = \frac{1}{2} \left[\left(\frac{h}{a} \right)^2 - 1 \right] \tan \left[2 \left(\frac{\pi Ua}{m} \right) \frac{h}{a} \right] \tag{6.109}$$

Large *Ua/m*

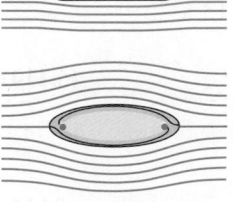

Small *Ua/m*

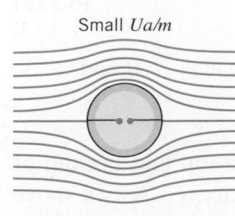

Equations 6.107 and 6.109 show that both ℓ/a and h/a are functions of the dimensionless parameter, $\pi Ua/m$. Although for a given value of Ua/m the corresponding value of ℓ/a can be determined directly from Eq. 6.107, h/a must be determined by a trial and error solution of Eq. 6.109.

A large variety of body shapes with different length to width ratios can be obtained by using different values of Ua/m, as shown by the figure in the margin. As this parameter becomes large, flow

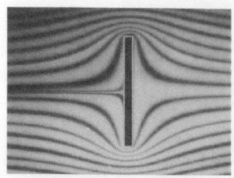

Potential Flow

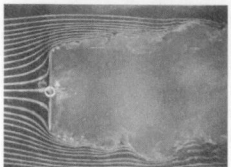

Viscous Flow

around a long slender body is described, whereas for small values of the parameter, flow around a more blunt shape is obtained. Downstream from the point of maximum body width the surface pressure increases with distance along the surface. This condition (called an adverse pressure gradient) typically leads to separation of the flow from the surface, resulting in a large low pressure wake on the downstream side of the body. Separation is not predicted by potential theory (which simply indicates a symmetrical flow). This is illustrated by the figure in the margin for an extreme blunt shape. Therefore, the potential solution for the Rankine ovals will give a reasonable approximation of the velocity outside the thin, viscous boundary layer and the pressure distribution on the front part of the body only.

6.6.3 Flow around a Circular Cylinder

A doublet combined with a uniform flow can be used to represent flow around a circular cylinder.

As was noted in the previous section, when the distance between the source–sink pair approaches zero, the shape of the Rankine oval becomes more blunt and in fact approaches a circular shape. Since the doublet described in Section 6.5.4 was developed by letting a source–sink pair approach one another, it might be expected that a uniform flow in the positive x direction combined with a doublet could be used to represent flow around a circular cylinder. This combination gives for the stream function

$$\psi = Ur \sin \theta - \frac{K \sin \theta}{r} \tag{6.110}$$

and for the velocity potential

$$\phi = Ur \cos \theta + \frac{K \cos \theta}{r} \tag{6.111}$$

V6.6 Circular cylinder

In order for the stream function to represent flow around a circular cylinder it is necessary that $\psi = $ constant for $r = a$, where a is the radius of the cylinder. Since Eq. 6.110 can be written as

$$\psi = \left(U - \frac{K}{r^2} \right) r \sin \theta$$

it follows that $\psi = 0$ for $r = a$ if

$$U - \frac{K}{a^2} = 0$$

which indicates that the doublet strength, K, must be equal to Ua^2. Thus, the stream function for flow around a circular cylinder can be expressed as

$$\psi = Ur \left(1 - \frac{a^2}{r^2} \right) \sin \theta \tag{6.112}$$

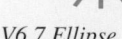

V6.7 Ellipse

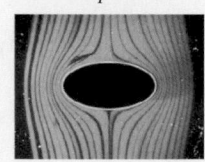

and the corresponding velocity potential is

$$\phi = Ur \left(1 + \frac{a^2}{r^2} \right) \cos \theta \tag{6.113}$$

A sketch of the streamlines for this flow field is shown in Fig. 6.26.

The velocity components can be obtained from either Eq. 6.112 or 6.113 as

$$v_r = \frac{\partial \phi}{\partial r} = \frac{1}{r} \frac{\partial \psi}{\partial \theta} = U \left(1 - \frac{a^2}{r^2} \right) \cos \theta \tag{6.114}$$

and

$$v_\theta = \frac{1}{r} \frac{\partial \phi}{\partial \theta} = -\frac{\partial \psi}{\partial r} = -U \left(1 + \frac{a^2}{r^2} \right) \sin \theta \tag{6.115}$$

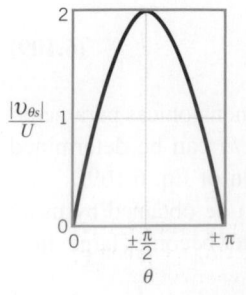

On the surface of the cylinder ($r = a$) it follows from Eq. 6.114 and 6.115 that $v_r = 0$ and

$$v_{\theta s} = -2U \sin \theta$$

As shown by the figure in the margin, the maximum velocity occurs at the top and bottom of the cylinder ($\theta = \pm \pi/2$) and has a magnitude of twice the upstream velocity, U. As we move away from the cylinder along the ray $\theta = \pi/2$ the velocity varies, as is illustrated in Fig. 6.26.

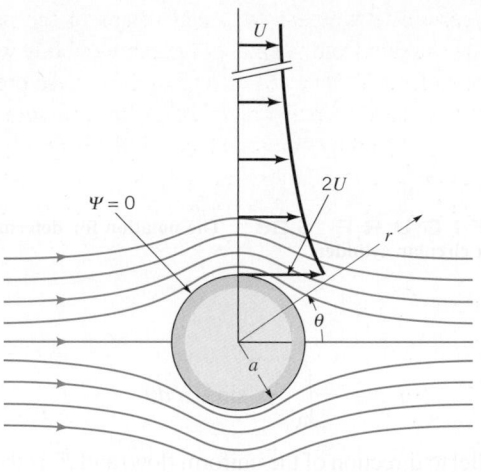

■ **FIGURE 6.26**   The flow around a circular cylinder.

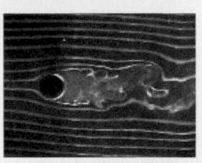

The pressure distribution on the cylinder surface is obtained from the Bernoulli equation written from a point far from the cylinder where the pressure is p_0 and the velocity is U so that

$$p_0 + \tfrac{1}{2}\rho U^2 = p_s + \tfrac{1}{2}\rho v_{\theta s}^2$$

where p_s is the surface pressure. Elevation changes are neglected. Since $v_{\theta s} = -2U \sin \theta$, the surface pressure can be expressed as

$$p_s = p_0 + \tfrac{1}{2}\rho U^2(1 - 4 \sin^2 \theta) \tag{6.116}$$

A comparison of this theoretical, symmetrical pressure distribution expressed in dimensionless form with a typical measured distribution is shown in Fig. 6.27. This figure clearly reveals that only on the upstream part of the cylinder is there approximate agreement between the potential flow and the experimental results. Because of the viscous boundary layer that develops on the cylinder, the main flow separates from the surface of the cylinder, leading to the large difference between the theoretical, frictionless fluid solution and the experimental results on the downstream side of the cylinder (see Chapter 9).

The resultant force (per unit length) developed on the cylinder can be determined by integrating the pressure over the surface. From Fig. 6.28 it can be seen that

$$F_x = -\int_0^{2\pi} p_s \cos \theta \, a \, d\theta \tag{6.117}$$

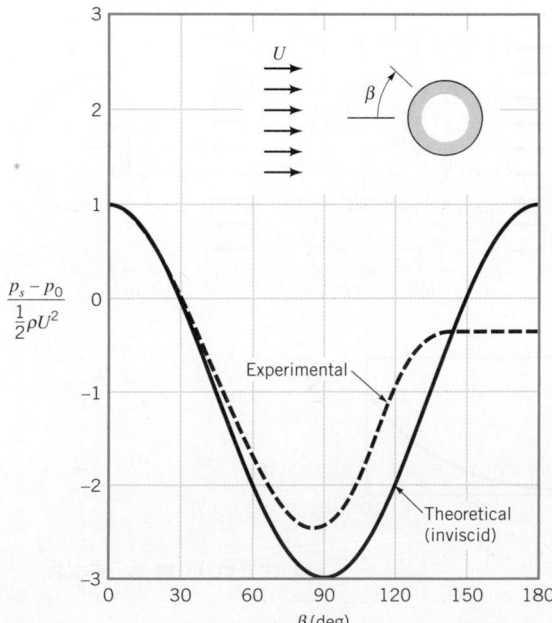

■ **FIGURE 6.27**   A comparison of theoretical (inviscid) pressure distribution on the surface of a circular cylinder with typical experimental distribution.

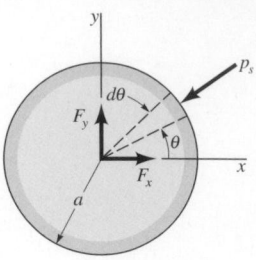

■ **FIGURE 6.28** The notation for determining lift and drag on a circular cylinder.

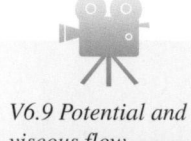

and

$$F_y = -\int_0^{2\pi} p_s \sin\theta \, a \, d\theta \tag{6.118}$$

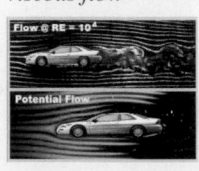

where F_x is the *drag* (force parallel to direction of the uniform flow) and F_y is the *lift* (force perpendicular to the direction of the uniform flow). Substitution for p_s from Eq. 6.116 into these two equations, and subsequent integration, reveals that $F_x = 0$ and $F_y = 0$ (Problem 6.73). These results indicate that both the drag and lift as predicted by potential theory for a fixed cylinder in a uniform stream are zero. Since the pressure distribution is symmetrical around the cylinder, this is not really a surprising result. However, we know from experience that there is a significant drag developed on a cylinder when it is placed in a moving fluid. This discrepancy is known as d'Alembert's paradox. The paradox is named after Jean le Rond d'Alembert (1717–1783), a French mathematician and philosopher, who first showed that the drag on bodies immersed in inviscid fluids is zero. It was not until the latter part of the nineteenth century and the early part of the twentieth century that the role viscosity plays in the steady fluid motion was understood and d'Alembert's paradox explained (see Section 9.1).

EXAMPLE 6.8 Potential Flow—Cylinder

GIVEN When a circular cylinder is placed in a uniform stream, a stagnation point is created on the cylinder as is shown in Fig. E6.8*a*. If a small hole is located at this point, the stagnation pressure, p_{stag}, can be measured and used to determine the approach velocity, U.

FIND

(a) Show how p_{stag} and U are related.

(b) If the cylinder is misaligned by an angle α (Figure E6.8*b*), but the measured pressure is still interpreted as the stagnation pressure, determine an expression for the ratio of the true velocity, U, to the predicted velocity, U'. Plot this ratio as a function of α for the range $-20° \le \alpha \le 20°$.

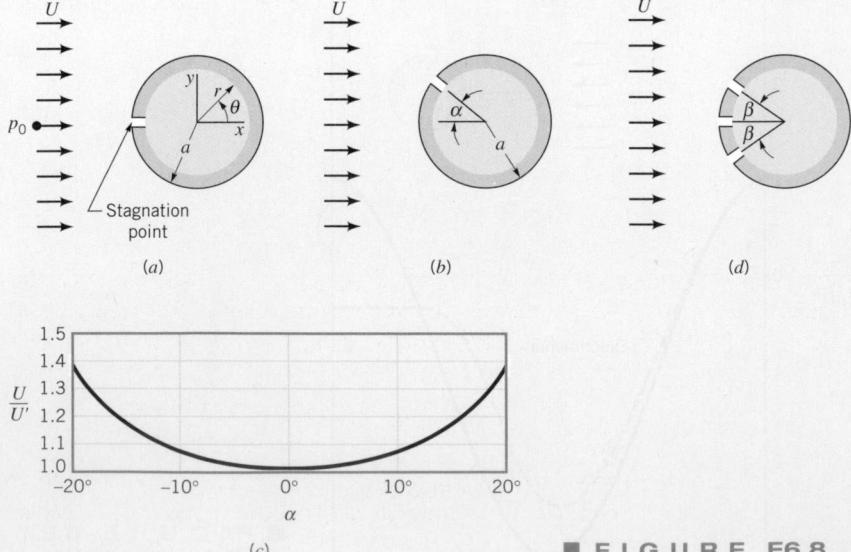

■ **FIGURE E6.8**

SOLUTION

(a) The velocity at the stagnation point is zero so the Bernoulli equation written between a point on the stagnation streamline upstream from the cylinder and the stagnation point gives

$$\frac{p_0}{\gamma} + \frac{U^2}{2g} = \frac{p_{\text{stag}}}{\gamma}$$

Thus,

$$U = \left[\frac{2}{\rho}(p_{\text{stag}} - p_0)\right]^{1/2} \qquad \textbf{(Ans)}$$

COMMENT A measurement of the difference between the pressure at the stagnation point and the upstream pressure can be used to measure the approach velocity. This is, of course, the same result that was obtained in Section 3.5 for Pitot-static tubes.

(b) If the direction of the fluid approaching the cylinder is not known precisely, it is possible that the cylinder is misaligned by some angle, α. In this instance the pressure actually measured, p_α, will be different from the stagnation pressure, but if the misalignment is not recognized the predicted approach velocity, U', would still be calculated as

$$U' = \left[\frac{2}{\rho}(p_\alpha - p_0)\right]^{1/2}$$

Thus,

$$\frac{U(\text{true})}{U'(\text{predicted})} = \left(\frac{p_{\text{stag}} - p_0}{p_\alpha - p_0}\right)^{1/2} \qquad \textbf{(1)}$$

The velocity on the surface of the cylinder, v_θ, where $r = a$, is obtained from Eq. 6.115 as

$$v_\theta = -2U \sin \theta$$

If we now write the Bernoulli equation between a point upstream of the cylinder and the point on the cylinder where $r = a$, $\theta = \alpha$, it follows that

$$p_0 + \frac{1}{2}\rho U^2 = p_\alpha + \frac{1}{2}\rho(-2U \sin \alpha)^2$$

and, therefore,

$$p_\alpha - p_0 = \tfrac{1}{2}\rho U^2(1 - 4 \sin^2\alpha) \qquad \textbf{(2)}$$

Since $p_{\text{stag}} - p_0 = \tfrac{1}{2}\rho U^2$ it follows from Eqs. 1 and 2 that

$$\frac{U(\text{true})}{U'(\text{predicted})} = (1 - 4 \sin^2\alpha)^{-1/2} \qquad \textbf{(Ans)}$$

This velocity ratio is plotted as a function of the misalignment angle α in Fig. E6.8c.

COMMENT It is clear from these results that significant errors can arise if the stagnation pressure tap is not aligned with the stagnation streamline. As is discussed in Section 3.5, if two additional, symmetrically located holes are drilled on the cylinder, as are illustrated in Fig. E6.8d, the correct orientation of the cylinder can be determined. The cylinder is rotated until the pressures in the two symmetrically placed holes are equal, thus indicating that the center hole coincides with the stagnation streamline. For $\beta = 30°$ the pressure at the two holes theoretically corresponds to the upstream pressure, p_0. With this orientation a measurement of the difference in pressure between the center hole and the side holes can be used to determine U.

An additional, interesting potential flow can be developed by adding a free vortex to the stream function or velocity potential for the flow around a cylinder. In this case

$$\psi = Ur\left(1 - \frac{a^2}{r^2}\right)\sin \theta - \frac{\Gamma}{2\pi}\ln r \qquad \textbf{(6.119)}$$

and

$$\phi = Ur\left(1 + \frac{a^2}{r^2}\right)\cos \theta + \frac{\Gamma}{2\pi}\theta \qquad \textbf{(6.120)}$$

where Γ is the circulation. We note that the circle $r = a$ will still be a streamline (and thus can be replaced with a solid cylinder), since the streamlines for the added free vortex are all circular. However, the tangential velocity, v_θ, on the surface of the cylinder $(r = a)$ now becomes

$$v_{\theta s} = -\frac{\partial \psi}{\partial r}\bigg|_{r=a} = -2U \sin \theta + \frac{\Gamma}{2\pi a} \qquad \textbf{(6.121)}$$

Flow around a rotating cylinder is approximated by the addition of a free vortex.

This type of flow field could be approximately created by placing a rotating cylinder in a uniform stream. Because of the presence of viscosity in any real fluid, the fluid in contact with the rotating cylinder would rotate with the same velocity as the cylinder, and the resulting flow field would resemble that developed by the combination of a uniform flow past a cylinder and a free vortex.

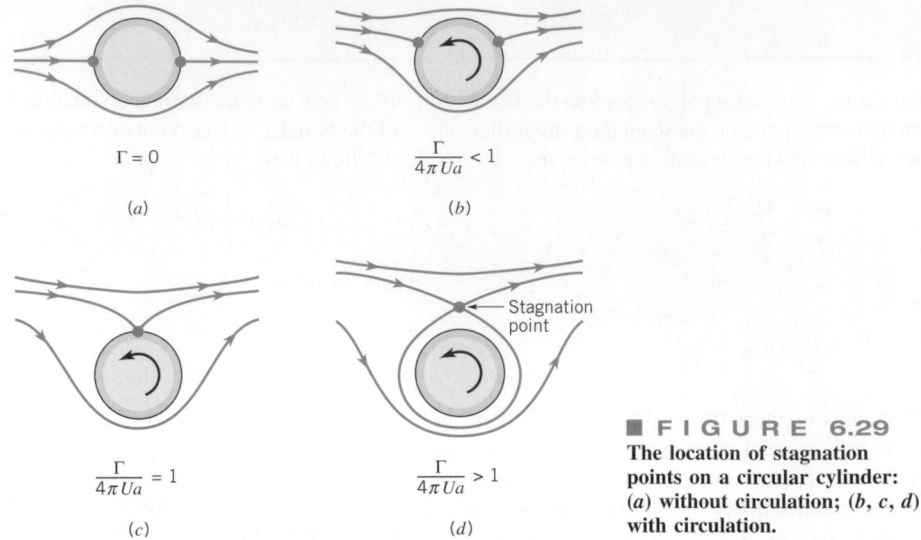

$\Gamma = 0$

(a)

$\dfrac{\Gamma}{4\pi Ua} < 1$

(b)

$\dfrac{\Gamma}{4\pi Ua} = 1$

(c)

$\dfrac{\Gamma}{4\pi Ua} > 1$

(d)

Stagnation point

■ **FIGURE 6.29**
The location of stagnation points on a circular cylinder: *(a)* **without circulation;** *(b, c, d)* **with circulation.**

A variety of streamline patterns can be developed, depending on the vortex strength, Γ. For example, from Eq. 6.121 we can determine the location of stagnation points on the surface of the cylinder. These points will occur at $\theta = \theta_{\text{stag}}$ where $v_\theta = 0$ and therefore from Eq. 6.121

$$\sin \theta_{\text{stag}} = \frac{\Gamma}{4\pi Ua} \tag{6.122}$$

If $\Gamma = 0$, then $\theta_{\text{stag}} = 0$ or π—that is, the stagnation points occur at the front and rear of the cylinder as are shown in Fig. 6.29a. However, for $-1 \le \Gamma/4\pi Ua \le 1$, the stagnation points will occur at some other location on the surface as illustrated in Figs. 6.29b,c. If the absolute value of the parameter $\Gamma/4\pi Ua$ exceeds 1, Eq. 6.122 cannot be satisfied, and the stagnation point is located away from the cylinder as shown in Fig. 6.29d.

The force per unit length developed on the cylinder can again be obtained by integrating the differential pressure forces around the circumference as in Eqs. 6.117 and 6.118. For the cylinder with circulation, the surface pressure, p_s, is obtained from the Bernoulli equation (with the surface velocity given by Eq. 6.121)

$$p_0 + \frac{1}{2}\rho U^2 = p_s + \frac{1}{2}\rho\left(-2U \sin\theta + \frac{\Gamma}{2\pi a}\right)^2$$

or

$$p_s = p_0 + \frac{1}{2}\rho U^2\left(1 - 4\sin^2\theta + \frac{2\Gamma \sin\theta}{\pi aU} - \frac{\Gamma^2}{4\pi^2 a^2 U^2}\right) \tag{6.123}$$

Equation 6.123 substituted into Eq. 6.117 for the drag, and integrated, again yields (Problem 6.74)

$$F_x = 0$$

Potential flow past a cylinder with circulation gives zero drag but non-zero lift.

That is, even for the rotating cylinder no force in the direction of the uniform flow is developed. However, use of Eq. 6.123 with the equation for the lift, F_y (Eq. 6.118), yields (Problem 6.74)

$$F_y = -\rho U\Gamma \tag{6.124}$$

Thus, for the cylinder with circulation, lift is developed equal to the product of the fluid density, the upstream velocity, and the circulation. The negative sign means that if U is positive (in the positive x direction) and Γ is positive (a free vortex with counterclockwise rotation), the direction of the F_y is downward.

Of course, if the cylinder is rotated in the clockwise direction ($\Gamma < 0$) the direction of F_y would be upward. This can be seen by studying the surface pressure distribution (Eq. 6.123), which is plotted in Fig. 6.30 for two situations. One has $\Gamma/4\pi Ua = 0$, which corresponds to no rotation of the cylinder. The other has $\Gamma/4\pi Ua = -0.25$, which corresponds to clockwise rotation of the cylinder. With no

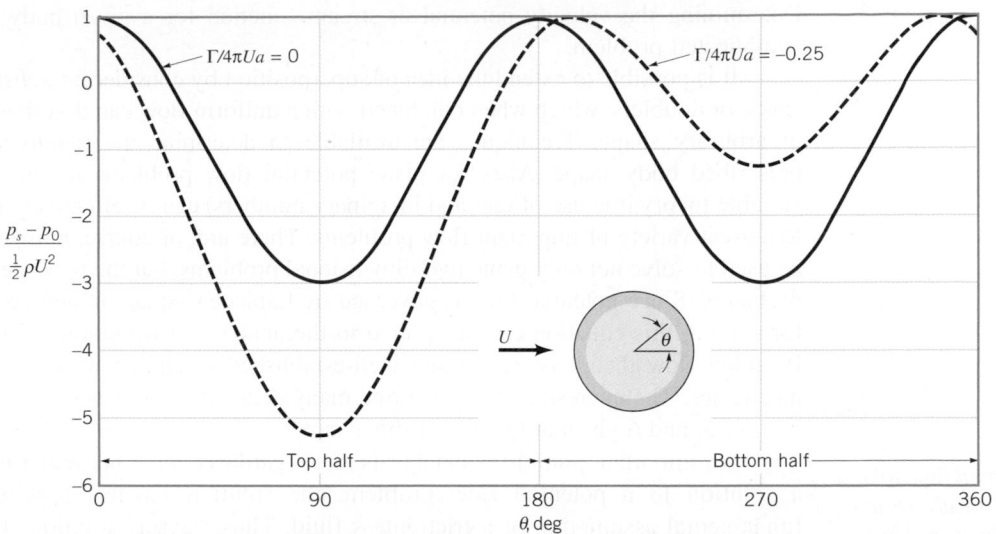

■ **FIGURE 6.30**   Pressure distribution on a circular cylinder with and without rotation.

rotation the flow is symmetrical both top to bottom and front to back on the cylinder. With rotation the flow is symmetrical front to back, but not top to bottom. In this case the two stagnation points [i.e., $(p_s - p_0)/(\rho U^2/2) = 1$] are located on the bottom of the cylinder and the average pressure on the top half of the cylinder is less than that on the bottom half. The result is an upward lift force. It is this force acting in a direction perpendicular to the direction of the approach velocity that causes baseballs and golf balls to curve when they spin as they are propelled through the air. The development of this lift on rotating bodies is called the *Magnus effect*. (See Section 9.4 for further comments.)

Although Eq. 6.124 was developed for a cylinder with circulation, it gives the lift per unit length for any two-dimensional object of any cross-sectional shape placed in a uniform, inviscid stream. The circulation is determined around any closed curve containing the body. The generalized equation relating lift to fluid density, velocity, and circulation is called the *Kutta–Joukowski law*, and is commonly used to determine the lift on airfoils (see Section 9.4.2 and Refs. 2–6).

F l u i d s   i n   t h e   N e w s

A sailing ship without sails A sphere or cylinder spinning about its axis when placed in an airstream develops a force at right angles to the direction of the airstream. This phenomenon is commonly referred to as the *Magnus effect* and is responsible for the curved paths of baseballs and golf balls. Another lesser-known application of the Magnus effect was proposed by a German physicist and engineer, Anton Flettner, in the 1920s. Flettner's idea was to use the Magnus effect to make a ship move. To demonstrate the practicality of the "rotor-ship" he purchased a sailing schooner and replaced the ship's masts and rigging with two vertical cylinders that were 50 feet (15 m) high and 9 feet (3 m) in diameter. The cylinders looked like smokestacks on the ship. Their spinning motion was developed by 45-hp (34 kW) motors. The combination of a wind and the rotating cylinders created a force (Magnus effect) to push the ship forward. The ship, named the *Baden Baden*, made a successful voyage across the Atlantic, arriving in New York Harbor on May 9, 1926. Although the feasibility of the rotor-ship was clearly demonstrated, it proved to be less efficient and practical than more conventional vessels and the idea was not pursued. (See Problem 6.72.)

6.7   Other Aspects of Potential Flow Analysis

In the preceding section the method of superposition of basic potentials has been used to obtain detailed descriptions of irrotational flow around certain body shapes immersed in a uniform stream. For the cases considered, two or more of the basic potentials were combined and the question is asked: What kind of flow does this combination represent? This approach is relatively simple and does not require the use of advanced mathematical techniques. It is, however, restrictive in its general applicability. It does not allow us to specify a priori the body shape and then determine the velocity potential or stream function that describes the flow around the particular body.

Determining the velocity potential or stream function for a given body shape is a much more complicated problem.

It is possible to extend the idea of superposition by considering a *distribution* of sources and sinks, or doublets, which when combined with a uniform flow can describe the flow around bodies of arbitrary shape. Techniques are available to determine the required distribution to give a prescribed body shape. Also, for plane potential flow problems it can be shown that complex variable theory (the use of real and imaginary numbers) can be effectively used to obtain solutions to a great variety of important flow problems. There are, of course, numerical techniques that can be used to solve not only plane two-dimensional problems, but the more general three-dimensional problems. Since potential flow is governed by Laplace's equation, any procedure that is available for solving this equation can be applied to the analysis of irrotational flow of frictionless fluids. Potential flow theory is an old and well-established discipline within the general field of fluid mechanics. The interested reader can find many detailed references on this subject, including Refs. 2, 3, 4, 5, and 6 given at the end of this chapter.

An important point to remember is that regardless of the particular technique used to obtain a solution to a potential flow problem, the solution remains approximate because of the fundamental assumption of a frictionless fluid. Thus, "exact" solutions based on potential flow theory represent, at best, only approximate solutions to real fluid problems. The applicability of potential flow theory to real fluid problems has been alluded to in a number of examples considered in the previous section. As a rule of thumb, potential flow theory will usually provide a reasonable approximation in those circumstances when we are dealing with a low viscosity fluid moving at a relatively high velocity, in regions of the flow field in which the flow is accelerating. Under these circumstances we generally find that the effect of viscosity is confined to the thin boundary layer that develops at a solid boundary. Outside the boundary layer the velocity distribution and the pressure distribution are closely approximated by the potential flow solution. However, in those regions of the flow field in which the flow is decelerating (for example, in the rearward portion of a bluff body or in the expanding region of a conduit), the pressure near a solid boundary will increase in the direction of flow. This so-called adverse pressure gradient can lead to flow separation, a phenomenon that causes dramatic changes in the flow field which are generally not accounted for by potential theory. However, as discussed in Chapter 9, in which boundary layer theory is developed, it is found that potential flow theory is used to obtain the appropriate pressure distribution that can then be combined with the viscous flow equations to obtain solutions near the boundary (and also to predict separation). The general differential equations that describe viscous fluid behavior and some simple solutions to these equations are considered in the remaining sections of this chapter.

Potential flow solutions are always approximate because the fluid is assumed to be frictionless.

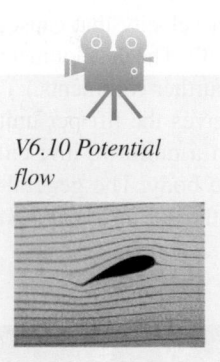

V6.10 Potential flow

6.8 Viscous Flow

To incorporate viscous effects into the differential analysis of fluid motion we must return to the previously derived general equations of motion, Eqs. 6.50. Since these equations include both stresses and velocities, there are more unknowns than equations, and therefore before proceeding it is necessary to establish a relationship between the stresses and velocities.

6.8.1 Stress–Deformation Relationships

For incompressible Newtonian fluids it is known that the stresses are linearly related to the rates of deformation and can be expressed in Cartesian coordinates as (for normal stresses)

$$\sigma_{xx} = -p + 2\mu \frac{\partial u}{\partial x} \tag{6.125a}$$

$$\sigma_{yy} = -p + 2\mu \frac{\partial v}{\partial y} \tag{6.125b}$$

$$\sigma_{zz} = -p + 2\mu \frac{\partial w}{\partial z} \tag{6.125c}$$

(for shearing stresses)

$$\tau_{xy} = \tau_{yx} = \mu\left(\frac{\partial u}{\partial y} + \frac{\partial v}{\partial x}\right) \tag{6.125d}$$

$$\tau_{yz} = \tau_{zy} = \mu\left(\frac{\partial v}{\partial z} + \frac{\partial w}{\partial y}\right) \tag{6.125e}$$

$$\tau_{zx} = \tau_{xz} = \mu\left(\frac{\partial w}{\partial x} + \frac{\partial u}{\partial z}\right) \tag{6.125f}$$

where p is the pressure, the negative of the average of the three normal stresses; that is, $-p = (\frac{1}{3})(\sigma_{xx} + \sigma_{yy} + \sigma_{zz})$. For viscous fluids in motion the normal stresses are not necessarily the same in different directions, thus, the need to define the pressure as the average of the three normal stresses. For fluids at rest, or frictionless fluids, the normal stresses are equal in all directions. (We have made use of this fact in the chapter on fluid statics and in developing the equations for inviscid flow.) Detailed discussions of the development of these stress–velocity gradient relationships can be found in Refs. 3, 7, and 8. An important point to note is that whereas for elastic solids the stresses are linearly related to the deformation (or strain), for Newtonian fluids the stresses are linearly related to the rate of deformation (or rate of strain).

For Newtonian fluids, stresses are linearly related to the rate of strain.

In cylindrical polar coordinates the stresses for incompressible Newtonian fluids are expressed as (for normal stresses)

$$\sigma_{rr} = -p + 2\mu\frac{\partial v_r}{\partial r} \tag{6.126a}$$

$$\sigma_{\theta\theta} = -p + 2\mu\left(\frac{1}{r}\frac{\partial v_\theta}{\partial \theta} + \frac{v_r}{r}\right) \tag{6.126b}$$

$$\sigma_{zz} = -p + 2\mu\frac{\partial v_z}{\partial z} \tag{6.126c}$$

(for shearing stresses)

$$\tau_{r\theta} = \tau_{\theta r} = \mu\left[r\frac{\partial}{\partial r}\left(\frac{v_\theta}{r}\right) + \frac{1}{r}\frac{\partial v_r}{\partial \theta}\right] \tag{6.126d}$$

$$\tau_{\theta z} = \tau_{z\theta} = \mu\left(\frac{\partial v_\theta}{\partial z} + \frac{1}{r}\frac{\partial v_z}{\partial \theta}\right) \tag{6.126e}$$

$$\tau_{zr} = \tau_{rz} = \mu\left(\frac{\partial v_r}{\partial z} + \frac{\partial v_z}{\partial r}\right) \tag{6.126f}$$

The double subscript has a meaning similar to that of stresses expressed in Cartesian coordinates—that is, the first subscript indicates the plane on which the stress acts, and the second subscript the direction. Thus, for example, σ_{rr} refers to a stress acting on a plane perpendicular to the radial direction and in the radial direction (thus a normal stress). Similarly, $\tau_{r\theta}$ refers to a stress acting on a plane perpendicular to the radial direction but in the tangential (θ direction) and is therefore a shearing stress.

6.8.2 The Navier–Stokes Equations

The stresses as defined in the preceding section can be substituted into the differential equations of motion (Eqs. 6.50) and simplified by using the continuity equation (Eq. 6.31) to obtain:

(x direction)

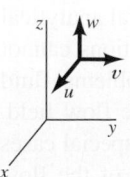

$$\rho\left(\frac{\partial u}{\partial t} + u\frac{\partial u}{\partial x} + v\frac{\partial u}{\partial y} + w\frac{\partial u}{\partial z}\right) = -\frac{\partial p}{\partial x} + \rho g_x + \mu\left(\frac{\partial^2 u}{\partial x^2} + \frac{\partial^2 u}{\partial y^2} + \frac{\partial^2 u}{\partial z^2}\right) \tag{6.127a}$$

(y direction)

$$\rho\left(\frac{\partial v}{\partial t} + u\frac{\partial v}{\partial x} + v\frac{\partial v}{\partial y} + w\frac{\partial v}{\partial z}\right) = -\frac{\partial p}{\partial y} + \rho g_y + \mu\left(\frac{\partial^2 v}{\partial x^2} + \frac{\partial^2 v}{\partial y^2} + \frac{\partial^2 v}{\partial z^2}\right) \tag{6.127b}$$

(z direction)

$$\rho\left(\frac{\partial w}{\partial t} + u\frac{\partial w}{\partial x} + v\frac{\partial w}{\partial y} + w\frac{\partial w}{\partial z}\right) = -\frac{\partial p}{\partial z} + \rho g_z + \mu\left(\frac{\partial^2 w}{\partial x^2} + \frac{\partial^2 w}{\partial y^2} + \frac{\partial^2 w}{\partial z^2}\right) \tag{6.127c}$$

where u, v, and w are the x, y, and z components of velocity as shown in the figure in the margin of the previous page. We have rearranged the equations so the acceleration terms are on the left side and the force terms are on the right. These equations are commonly called the *Navier–Stokes equations*, named in honor of the French mathematician L. M. H. Navier (1785–1836) and the English mechanician Sir G. G. Stokes (1819–1903), who were responsible for their formulation. These three equations of motion, when combined with the conservation of mass equation (Eq. 6.31), provide a complete mathematical description of the flow of incompressible Newtonian fluids. We have four equations and four unknowns (u, v, w, and p), and therefore the problem is "well-posed" in mathematical terms. Unfortunately, because of the general complexity of the Navier–Stokes equations (they are nonlinear, second-order, partial differential equations), they are not amenable to exact mathematical solutions except in a few instances. However, in those few instances in which solutions have been obtained and compared with experimental results, the results have been in close agreement. Thus, the Navier–Stokes equations are considered to be the governing differential equations of motion for incompressible Newtonian fluids.

The Navier–Stokes equations are the basic differential equations describing the flow of Newtonian fluids.

In terms of cylindrical polar coordinates (see the figure in the margin), the Navier–Stokes equations can be written as

(r direction)

$$\rho\left(\frac{\partial v_r}{\partial t} + v_r\frac{\partial v_r}{\partial r} + \frac{v_\theta}{r}\frac{\partial v_r}{\partial \theta} - \frac{v_\theta^2}{r} + v_z\frac{\partial v_r}{\partial z}\right)$$
$$= -\frac{\partial p}{\partial r} + \rho g_r + \mu\left[\frac{1}{r}\frac{\partial}{\partial r}\left(r\frac{\partial v_r}{\partial r}\right) - \frac{v_r}{r^2} + \frac{1}{r^2}\frac{\partial^2 v_r}{\partial \theta^2} - \frac{2}{r^2}\frac{\partial v_\theta}{\partial \theta} + \frac{\partial^2 v_r}{\partial z^2}\right] \tag{6.128a}$$

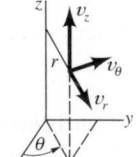

(θ direction)

$$\rho\left(\frac{\partial v_\theta}{\partial t} + v_r\frac{\partial v_\theta}{\partial r} + \frac{v_\theta}{r}\frac{\partial v_\theta}{\partial \theta} + \frac{v_r v_\theta}{r} + v_z\frac{\partial v_\theta}{\partial z}\right)$$
$$= -\frac{1}{r}\frac{\partial p}{\partial \theta} + \rho g_\theta + \mu\left[\frac{1}{r}\frac{\partial}{\partial r}\left(r\frac{\partial v_\theta}{\partial r}\right) - \frac{v_\theta}{r^2} + \frac{1}{r^2}\frac{\partial^2 v_\theta}{\partial \theta^2} + \frac{2}{r^2}\frac{\partial v_r}{\partial \theta} + \frac{\partial^2 v_\theta}{\partial z^2}\right] \tag{6.128b}$$

(z direction)

$$\rho\left(\frac{\partial v_z}{\partial t} + v_r\frac{\partial v_z}{\partial r} + \frac{v_\theta}{r}\frac{\partial v_z}{\partial \theta} + v_z\frac{\partial v_z}{\partial z}\right)$$
$$= -\frac{\partial p}{\partial z} + \rho g_z + \mu\left[\frac{1}{r}\frac{\partial}{\partial r}\left(r\frac{\partial v_z}{\partial r}\right) + \frac{1}{r^2}\frac{\partial^2 v_z}{\partial \theta^2} + \frac{\partial^2 v_z}{\partial z^2}\right] \tag{6.128c}$$

To provide a brief introduction to the use of the Navier–Stokes equations, a few of the simplest exact solutions are developed in the next section. Although these solutions will prove to be relatively simple, this is not the case in general. In fact, only a few other exact solutions have been obtained.

6.9 Some Simple Solutions for Laminar, Viscous, Incompressible Fluids

A principal difficulty in solving the Navier–Stokes equations is because of their nonlinearity arising from the convective acceleration terms (i.e., $u\,\partial u/\partial x$, $w\,\partial v/\partial z$, etc.). There are no general analytical schemes for solving nonlinear partial differential equations (e.g., superposition of solutions cannot be used), and each problem must be considered individually. For most practical flow problems, fluid particles do have accelerated motion as they move from one location to another in the flow field. Thus, the convective acceleration terms are usually important. However, there are a few special cases for which the convective acceleration vanishes because of the nature of the geometry of the flow system. In these cases exact solutions are often possible. The Navier–Stokes equations apply to both laminar and turbulent flow, but for turbulent flow each velocity component fluctuates randomly with respect to time and this added complication makes an analytical solution intractable. Thus, the exact solutions referred to are for laminar flows in which the velocity is either independent of time (steady flow) or dependent on time (unsteady flow) in a well-defined manner.

6.9.1 Steady, Laminar Flow between Fixed Parallel Plates

An exact solution can be obtained for steady laminar flow between fixed parallel plates.

We first consider flow between the two horizontal, infinite parallel plates of Fig. 6.31a. For this geometry the fluid particles move in the x direction parallel to the plates, and there is no velocity in the y or z direction—that is, $v = 0$ and $w = 0$. In this case it follows from the continuity equation (Eq. 6.31) that $\partial u/\partial x = 0$. Furthermore, there would be no variation of u in the z direction for infinite plates, and for steady flow $\partial u/\partial t = 0$ so that $u = u(y)$. If these conditions are used in the Navier–Stokes equations (Eqs. 6.127), they reduce to

$$0 = -\frac{\partial p}{\partial x} + \mu\left(\frac{\partial^2 u}{\partial y^2}\right) \tag{6.129}$$

$$0 = -\frac{\partial p}{\partial y} - \rho g \tag{6.130}$$

$$0 = -\frac{\partial p}{\partial z} \tag{6.131}$$

where we have set $g_x = 0$, $g_y = -g$, and $g_z = 0$. That is, the y axis points up. We see that for this particular problem the Navier–Stokes equations reduce to some rather simple equations.

Equations 6.130 and 6.131 can be integrated to yield

$$p = -\rho g y + f_1(x) \tag{6.132}$$

which shows that the pressure varies hydrostatically in the y direction. Equation 6.129, rewritten as

$$\frac{d^2 u}{dy^2} = \frac{1}{\mu}\frac{\partial p}{\partial x}$$

can be integrated to give

$$\frac{du}{dy} = \frac{1}{\mu}\left(\frac{\partial p}{\partial x}\right)y + c_1$$

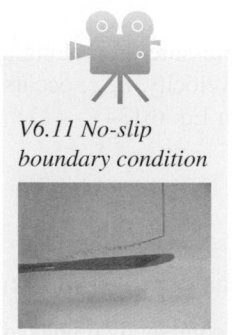

V6.11 No-slip boundary condition

and integrated again to yield

$$u = \frac{1}{2\mu}\left(\frac{\partial p}{\partial x}\right)y^2 + c_1 y + c_2 \tag{6.133}$$

Note that for this simple flow the pressure gradient, $\partial p/\partial x$, is treated as constant as far as the integration is concerned, since (as shown in Eq. 6.132) it is not a function of y. The two constants c_1 and c_2 must be determined from the boundary conditions. For example, if the two plates are

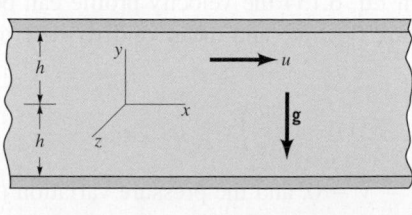
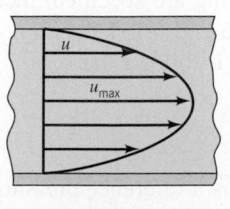

(a)   (b)

■ **F I G U R E   6.31**   The viscous flow between parallel plates: (*a*) coordinate system and notation used in analysis; (*b*) parabolic velocity distribution for flow between parallel fixed plates.

fixed, then $u = 0$ for $y = \pm h$ (because of the no-slip condition for viscous fluids). To satisfy this condition $c_1 = 0$ and

$$c_2 = -\frac{1}{2\mu}\left(\frac{\partial p}{\partial x}\right)h^2$$

Thus, the velocity distribution becomes

$$u = \frac{1}{2\mu}\left(\frac{\partial p}{\partial x}\right)(y^2 - h^2) \tag{6.134}$$

Equation 6.134 shows that the velocity profile between the two fixed plates is parabolic as illustrated in Fig. 6.31b.

The volume rate of flow, q, passing between the plates (for a unit width in the z direction) is obtained from the relationship

$$q = \int_{-h}^{h} u \, dy = \int_{-h}^{h} \frac{1}{2\mu}\left(\frac{\partial p}{\partial x}\right)(y^2 - h^2) \, dy$$

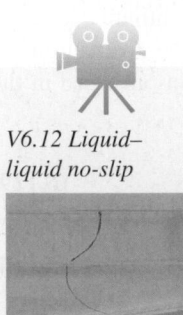

V6.12 Liquid–liquid no-slip

or

$$q = -\frac{2h^3}{3\mu}\left(\frac{\partial p}{\partial x}\right) \tag{6.135}$$

The pressure gradient $\partial p/\partial x$ is negative, since the pressure decreases in the direction of flow. If we let Δp represent the pressure *drop* between two points a distance ℓ apart, then

$$\frac{\Delta p}{\ell} = -\frac{\partial p}{\partial x}$$

and Eq. 6.135 can be expressed as

$$q = \frac{2h^3 \Delta p}{3\mu\ell} \tag{6.136}$$

The flow is proportional to the pressure gradient, inversely proportional to the viscosity, and strongly dependent $(\sim h^3)$ on the gap width. In terms of the mean velocity, V, where $V = q/2h$, Eq. 6.136 becomes

$$V = \frac{h^2 \Delta p}{3\mu\ell} \tag{6.137}$$

Equations 6.136 and 6.137 provide convenient relationships for relating the pressure drop along a parallel-plate channel and the rate of flow or mean velocity. The maximum velocity, u_{max}, occurs midway $(y = 0)$ between the two plates, as shown in Fig. 6.31b, so that from Eq. 6.134

$$u_{max} = -\frac{h^2}{2\mu}\left(\frac{\partial p}{\partial x}\right)$$

or

$$u_{max} = \tfrac{3}{2}V \tag{6.138}$$

The Navier–Stokes equations provide detailed flow characteristics for laminar flow between fixed parallel plates.

The details of the steady laminar flow between infinite parallel plates are completely predicted by this solution to the Navier–Stokes equations. For example, if the pressure gradient, viscosity, and plate spacing are specified, then from Eq. 6.134 the velocity profile can be determined, and from Eqs. 6.136 and 6.137 the corresponding flowrate and mean velocity determined. In addition, from Eq. 6.132 it follows that

$$f_1(x) = \left(\frac{\partial p}{\partial x}\right)x + p_0$$

where p_0 is a reference pressure at $x = y = 0$, and the pressure variation throughout the fluid can be obtained from

$$p = -\rho g y + \left(\frac{\partial p}{\partial x}\right)x + p_0 \tag{6.139}$$

For a given fluid and reference pressure, p_0, the pressure at any point can be predicted. This relatively simple example of an exact solution illustrates the detailed information about the flow field which can be obtained. The flow will be laminar if the Reynolds number, $\text{Re} = \rho V(2h)/\mu$, remains below about 1400. For flow with larger Reynolds numbers the flow becomes turbulent and the preceding analysis is not valid since the flow field is complex, three-dimensional, and unsteady.

F l u i d s   i n   t h e   N e w s

10 tons on 8 psi (9072 kg on 55 kPa) Place a golf ball on the end of a garden hose and then slowly turn the water on a small amount until the ball just barely lifts off the end of the hose, leaving a small gap between the ball and the hose. The ball is free to rotate. This is the idea behind the new "floating ball water fountains" developed in Finland. Massive, 10-ton (9072 kg), 6-ft (1.8 m) diameter stone spheres are supported by the pressure force of the water on the curved surface within a pedestal and rotate so easily that even a small child can change their direction of rotation. The key to the fountain design is the ability to grind and polish stone to an accuracy of a few thousandths of an inch. This allows the gap between the ball and its pedestal to be very small (on the order of 5/1000 in. or 13/1000 cm) and the water flowrate correspondingly small (on the order of 5 gallons per minute or 196/min). Due to the small gap, the flow in the gap is essentially that of *flow between parallel plates*. Although the sphere is very heavy, the pressure under the sphere within the pedestal needs to be only about 8 psi (55 kPa). (See Problem 6.88.)

6.9.2 Couette Flow

For a given flow geometry, the character and details of the flow are strongly dependent on the boundary conditions.

Another simple parallel-plate flow can be developed by fixing one plate and letting the other plate move with a constant velocity, U, as is illustrated in Fig. 6.32a. The Navier–Stokes equations reduce to the same form as those in the preceding section, and the solution for the pressure and velocity distribution are still given by Eqs. 6.132 and 6.133, respectively. However, for the moving plate problem the boundary conditions for the velocity are different. For this case we locate the origin of the coordinate system at the bottom plate and designate the distance between the two plates as b (see Fig. 6.32a). The two constants c_1 and c_2 in Eq. 6.133 can be determined from the boundary conditions, $u = 0$ at $y = 0$ and $u = U$ at $y = b$. It follows that

$$u = U\frac{y}{b} + \frac{1}{2\mu}\left(\frac{\partial p}{\partial x}\right)(y^2 - by) \tag{6.140}$$

or, in dimensionless form,

$$\frac{u}{U} = \frac{y}{b} - \frac{b^2}{2\mu U}\left(\frac{\partial p}{\partial x}\right)\left(\frac{y}{b}\right)\left(1 - \frac{y}{b}\right) \tag{6.141}$$

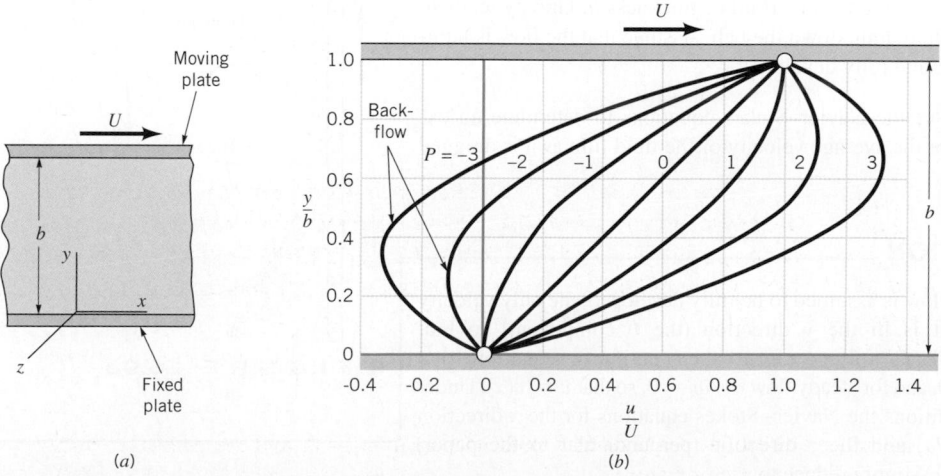

(a)   (b)

■ F I G U R E 6.32 The viscous flow between parallel plates with bottom plate fixed and upper plate moving (Couette flow): (*a*) coordinate system and notation used in analysis; (*b*) velocity distribution as a function of parameter, *P*, where $P = -(b^2/2\mu U)\,\partial p/\partial x$. (From Ref. 8, used by permission.)

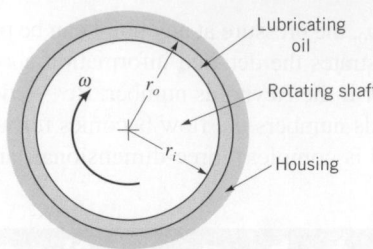

■ **FIGURE 6.33** **Flow in the narrow gap of a journal bearing.**

The actual velocity profile will depend on the dimensionless parameter

$$P = -\frac{b^2}{2\mu U}\left(\frac{\partial p}{\partial x}\right)$$

Flow between parallel plates with one plate fixed and the other moving is called Couette flow.

Several profiles are shown in Fig. 6.32b. This type of flow is called *Couette flow*.

The simplest type of Couette flow is one for which the pressure gradient is zero; that is, the fluid motion is caused by the fluid being dragged along by the moving boundary. In this case, with $\partial p/\partial x = 0$, Eq. 6.140 simply reduces to

$$u = U\frac{y}{b} \tag{6.142}$$

which indicates that the velocity varies linearly between the two plates as shown in Fig. 6.31b for $P = 0$. This situation would be approximated by the flow between closely spaced concentric cylinders in which one cylinder is fixed and the other cylinder rotates with a constant angular velocity, ω. As illustrated in Fig. 6.33, the flow in an unloaded journal bearing might be approximated by this simple Couette flow if the gap width is very small (i.e., $r_o - r_i \ll r_i$). In this case $U = r_i\omega$, $b = r_o - r_i$, and the shearing stress resisting the rotation of the shaft can be simply calculated as $\tau = \mu r_i\omega/(r_o - r_i)$. When the bearing is loaded (i.e., a force applied normal to the axis of rotation), the shaft will no longer remain concentric with the housing and the flow cannot be treated as flow between parallel boundaries. Such problems are dealt with in lubrication theory (see, for example, Ref. 9).

EXAMPLE 6.9 | Plane Couette Flow

GIVEN A wide moving belt passes through a container of a viscous liquid. The belt moves vertically upward with a constant velocity, V_0, as illustrated in Fig. E6.9a. Because of viscous forces the belt picks up a film of fluid of thickness h. Gravity tends to make the fluid drain down the belt. Assume that the flow is laminar, steady, and fully developed.

FIND Use the Navier–Stokes equations to determine an expression for the average velocity of the fluid film as it is dragged up the belt.

SOLUTION

Since the flow is assumed to be fully developed, the only velocity component is in the y direction (the v component) so that $u = w = 0$. It follows from the continuity equation that $\partial v/\partial y = 0$, and for steady flow $\partial v/\partial t = 0$, so that $v = v(x)$. Under these conditions the Navier–Stokes equations for the x direction (Eq. 6.127a) and the z direction (perpendicular to the paper) (Eq. 6.127c) simply reduce to

$$\frac{\partial p}{\partial x} = 0 \qquad \frac{\partial p}{\partial z} = 0$$

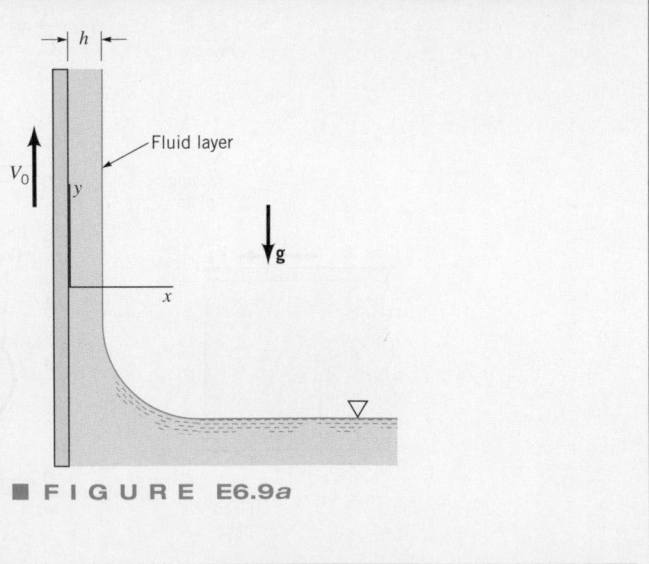

■ **FIGURE E6.9a**

This result indicates that the pressure does not vary over a horizontal plane, and since the pressure on the surface of the film $(x = h)$ is atmospheric, the pressure throughout the film must be atmospheric (or zero gage pressure). The equation of motion in the y direction (Eq. 6.127b) thus reduces to

$$0 = -\rho g + \mu \frac{d^2 v}{dx^2}$$

or

$$\frac{d^2 v}{dx^2} = \frac{\gamma}{\mu} \tag{1}$$

Integration of Eq. 1 yields

$$\frac{dv}{dx} = \frac{\gamma}{\mu} x + c_1 \tag{2}$$

On the film surface $(x = h)$ we assume the shearing stress is zero—that is, the drag of the air on the film is negligible. The shearing stress at the free surface (or any interior parallel surface) is designated as τ_{xy}, where from Eq. 6.125d

$$\tau_{xy} = \mu \left(\frac{dv}{dx} \right)$$

Thus, if $\tau_{xy} = 0$ at $x = h$, it follows from Eq. 2 that

$$c_1 = -\frac{\gamma h}{\mu}$$

A second integration of Eq. 2 gives the velocity distribution in the film as

$$v = \frac{\gamma}{2\mu} x^2 - \frac{\gamma h}{\mu} x + c_2$$

At the belt $(x = 0)$ the fluid velocity must match the belt velocity, V_0, so that

$$c_2 = V_0$$

and the velocity distribution is therefore

$$v = \frac{\gamma}{2\mu} x^2 - \frac{\gamma h}{\mu} x + V_0 \tag{3}$$

With the velocity distribution known we can determine the flowrate per unit width, q, from the relationship

$$q = \int_0^h v \, dx = \int_0^h \left(\frac{\gamma}{2\mu} x^2 - \frac{\gamma h}{\mu} x + V_0 \right) dx$$

and thus

$$q = V_0 h - \frac{\gamma h^3}{3\mu}$$

The average film velocity, V (where $q = Vh$), is therefore

$$V = V_0 - \frac{\gamma h^2}{3\mu} \tag{Ans}$$

COMMENT   Equation (3) can be written in dimensionless form as

$$\frac{v}{V_0} = c \left(\frac{x}{h} \right)^2 - 2c \left(\frac{x}{h} \right) + 1$$

where $c = \gamma h^2 / 2\mu V_0$. This velocity profile is shown in Fig. E6.9b. Note that even though the belt is moving upward, for $c > 1$ (e.g., for fluids with small enough viscosity or with a small enough belt speed) there are portions of the fluid that flow downward (as indicated by $v/V_0 < 0$).

It is interesting to note from this result that there will be a net upward flow of liquid (positive V) only if $V_0 > \gamma h^2/3\mu$. It takes a relatively large belt speed to lift a small viscosity fluid.

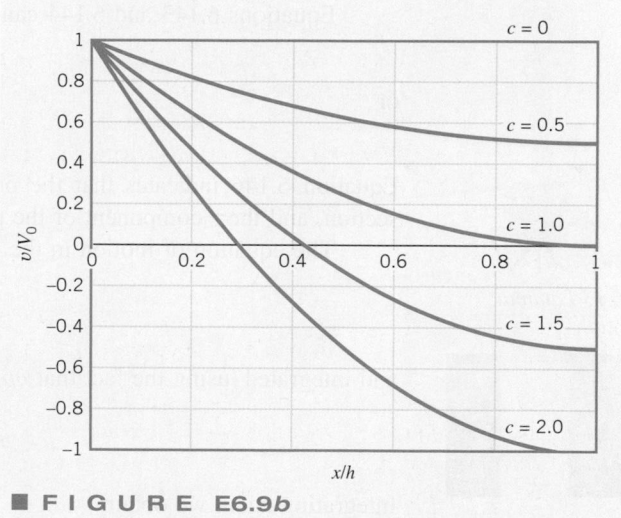

■ **FIGURE E6.9b**

6.9.3   Steady, Laminar Flow in Circular Tubes

An exact solution can be obtained for steady, incompressible, laminar flow in circular tubes.

Probably the best known exact solution to the Navier–Stokes equations is for steady, incompressible, laminar flow through a straight circular tube of constant cross section. This type of flow is commonly called *Hagen–Poiseuille flow*, or simply *Poiseuille flow*. It is named in honor of J. L. Poiseuille (1799–1869), a French physician, and G. H. L. Hagen (1797–1884), a German hydraulic engineer. Poiseuille was interested in blood flow through capillaries and deduced experimentally the resistance laws for laminar flow through circular tubes. Hagen's investigation of flow in tubes was also experimental. It was actually after the work of Hagen and Poiseuille that the theoretical results presented in this section were determined, but their names are commonly associated with the solution of this problem.

Consider the flow through a horizontal circular tube of radius R as is shown in Fig. 6.34a. Because of the cylindrical geometry it is convenient to use cylindrical coordinates. We assume that the flow is parallel to the walls so that $v_r = 0$ and $v_\theta = 0$, and from the continuity equation (6.34) $\partial v_z/\partial z = 0$. Also, for steady, axisymmetric flow, v_z is not a function of t or θ so the velocity, v_z,

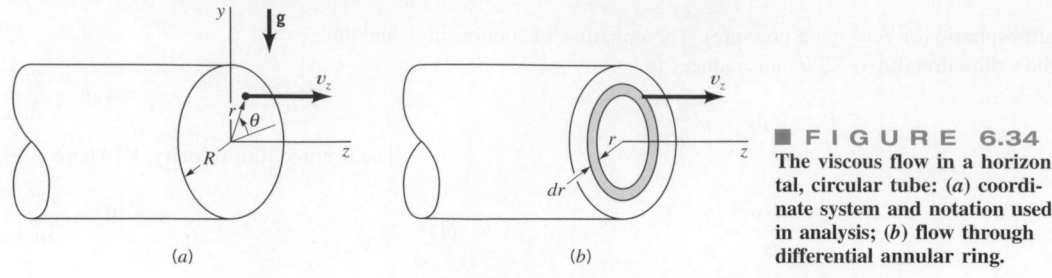

■ **FIGURE 6.34**
The viscous flow in a horizontal, circular tube: (*a*) coordinate system and notation used in analysis; (*b*) flow through differential annular ring.

is only a function of the radial position within the tube—that is, $v_z = v_z(r)$. Under these conditions the Navier–Stokes equations (Eqs. 6.128) reduce to

$$0 = -\rho g \sin\theta - \frac{\partial p}{\partial r} \tag{6.143}$$

$$0 = -\rho g \cos\theta - \frac{1}{r}\frac{\partial p}{\partial\theta} \tag{6.144}$$

$$0 = -\frac{\partial p}{\partial z} + \mu\left[\frac{1}{r}\frac{\partial}{\partial r}\left(r\frac{\partial v_z}{\partial r}\right)\right] \tag{6.145}$$

where we have used the relationships $g_r = -g\sin\theta$ and $g_\theta = -g\cos\theta$ (with θ measured from the horizontal plane).

Equations 6.143 and 6.144 can be integrated to give

$$p = -\rho g(r\sin\theta) + f_1(z)$$

or

$$p = -\rho gy + f_1(z) \tag{6.146}$$

Equation 6.146 indicates that the pressure is hydrostatically distributed at any particular cross section, and the z component of the pressure gradient, $\partial p/\partial z$, is not a function of r or θ.

The equation of motion in the z direction (Eq. 6.145) can be written in the form

$$\frac{1}{r}\frac{\partial}{\partial r}\left(r\frac{\partial v_z}{\partial r}\right) = \frac{1}{\mu}\frac{\partial p}{\partial z}$$

and integrated (using the fact that $\partial p/\partial z = $ constant) to give

$$r\frac{\partial v_z}{\partial r} = \frac{1}{2\mu}\left(\frac{\partial p}{\partial z}\right)r^2 + c_1$$

Integrating again we obtain

$$v_z = \frac{1}{4\mu}\left(\frac{\partial p}{\partial z}\right)r^2 + c_1\ln r + c_2 \tag{6.147}$$

Since we wish v_z to be finite at the center of the tube $(r = 0)$, it follows that $c_1 = 0$ [since $\ln(0) = -\infty$]. At the wall $(r = R)$ the velocity must be zero so that

$$c_2 = -\frac{1}{4\mu}\left(\frac{\partial p}{\partial z}\right)R^2$$

and the velocity distribution becomes

$$v_z = \frac{1}{4\mu}\left(\frac{\partial p}{\partial z}\right)(r^2 - R^2) \tag{6.148}$$

Thus, at any cross section the velocity distribution is parabolic.

To obtain a relationship between the volume rate of flow, Q, passing through the tube and the pressure gradient, we consider the flow through the differential, washer-shaped ring of Fig. 6.34*b*. Since v_z is constant on this ring, the volume rate of flow through the differential area $dA = (2\pi r)\,dr$ is

$$dQ = v_z(2\pi r)\,dr$$

V6.13 Laminar flow

The velocity distribution is parabolic for steady, laminar flow in circular tubes.

and therefore

$$Q = 2\pi \int_0^R v_z r \, dr \qquad (6.149)$$

Equation 6.148 for v_z can be substituted into Eq. 6.149, and the resulting equation integrated to yield

$$Q = -\frac{\pi R^4}{8\mu}\left(\frac{\partial p}{\partial z}\right) \qquad (6.150)$$

This relationship can be expressed in terms of the pressure *drop*, Δp, which occurs over a length, ℓ, along the tube, since

$$\frac{\Delta p}{\ell} = -\frac{\partial p}{\partial z}$$

and therefore

$$Q = \frac{\pi R^4 \Delta p}{8\mu\ell} \qquad (6.151)$$

Poiseuille's law relates pressure drop and flowrate for steady, laminar flow in circular tubes.

For a given pressure drop per unit length, the volume rate of flow is inversely proportional to the viscosity and proportional to the tube radius to the fourth power. A doubling of the tube radius produces a 16-fold increase in flow! Equation 6.151 is commonly called *Poiseuille's law*.

In terms of the mean velocity, V, where $V = Q/\pi R^2$, Eq. 6.151 becomes

$$V = \frac{R^2 \Delta p}{8\mu\ell} \qquad (6.152)$$

The maximum velocity v_{max} occurs at the center of the tube, where from Eq. 6.148

$$v_{max} = -\frac{R^2}{4\mu}\left(\frac{\partial p}{\partial z}\right) = \frac{R^2 \Delta p}{4\mu\ell} \qquad (6.153)$$

so that

$$v_{max} = 2V$$

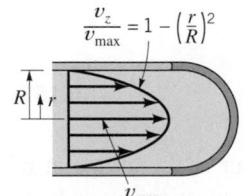

The velocity distribution, as shown by the figure in the margin, can be written in terms of v_{max} as

$$\frac{v_z}{v_{max}} = 1 - \left(\frac{r}{R}\right)^2 \qquad (6.154)$$

As was true for the similar case of flow between parallel plates (sometimes referred to as *plane Poiseuille flow*), a very detailed description of the pressure and velocity distribution in tube flow results from this solution to the Navier–Stokes equations. Numerous experiments performed to substantiate the theoretical results show that the theory and experiment are in agreement for the laminar flow of Newtonian fluids in circular tubes or pipes. In general, the flow remains laminar for Reynolds numbers, $Re = \rho V(2R)/\mu$, below 2100. Turbulent flow in tubes is considered in Chapter 8.

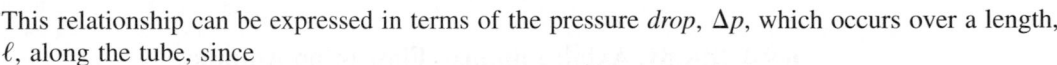

Fluids in the News

Poiseuille's law revisited Poiseuille's law governing *laminar flow* of fluids in tubes has an unusual history. It was developed in 1842 by a French physician, J. L. M. Poiseuille, who was interested in the flow of blood in capillaries. Poiseuille, through a series of carefully conducted experiments using water flowing through very small tubes, arrived at the formula, $Q = K\Delta p \, D^4/\ell$. In this formula Q is the flowrate, K an empirical constant, Δp the pressure drop over the length ℓ, and D the tube diameter. Another formula was given for the value of K as a function of the water temperature. It was not until the concept of viscosity was introduced at a later date that Poiseuille's law was derived mathematically and the constant K found to be equal to $\pi/8\mu$, where μ is the fluid viscosity. The experiments by Poiseuille have long been admired for their accuracy and completeness considering the laboratory instrumentation available in the mid nineteenth century.

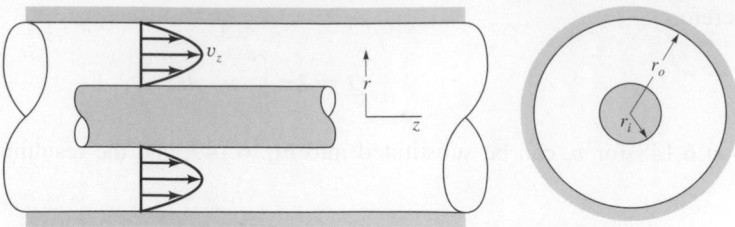

■ **F I G U R E   6.35**   The viscous flow through an annulus.

6.9.4 Steady, Axial, Laminar Flow in an Annulus

An exact solution can be obtained for axial flow in the annular space between two fixed, concentric cylinders.

The differential equations (Eqs. 6.143, 6.144, 6.145) used in the preceding section for flow in a tube also apply to the axial flow in the annular space between two fixed, concentric cylinders (Fig. 6.35). Equation 6.147 for the velocity distribution still applies, but for the stationary annulus the boundary conditions become $v_z = 0$ at $r = r_o$ and $v_z = 0$ for $r = r_i$. With these two conditions the constants c_1 and c_2 in Eq. 6.147 can be determined and the velocity distribution becomes

$$v_z = \frac{1}{4\mu}\left(\frac{\partial p}{\partial z}\right)\left[r^2 - r_o^2 + \frac{r_i^2 - r_o^2}{\ln(r_o/r_i)}\ln\frac{r}{r_o}\right] \tag{6.155}$$

The corresponding volume rate of flow is

$$Q = \int_{r_i}^{r_o} v_z(2\pi r)\,dr = -\frac{\pi}{8\mu}\left(\frac{\partial p}{\partial z}\right)\left[r_o^4 - r_i^4 - \frac{(r_o^2 - r_i^2)^2}{\ln(r_o/r_i)}\right]$$

or in terms of the pressure drop, Δp, in length ℓ of the annulus

$$Q = \frac{\pi \Delta p}{8\mu\ell}\left[r_o^4 - r_i^4 - \frac{(r_o^2 - r_i^2)^2}{\ln(r_o/r_i)}\right] \tag{6.156}$$

The velocity at any radial location within the annular space can be obtained from Eq. 6.155. The maximum velocity occurs at the radius $r = r_m$ where $\partial v_z/\partial r = 0$. Thus,

$$r_m = \left[\frac{r_o^2 - r_i^2}{2\ln(r_o/r_i)}\right]^{1/2} \tag{6.157}$$

An inspection of this result shows that the maximum velocity does not occur at the midpoint of the annular space, but rather it occurs nearer the inner cylinder. The specific location depends on r_o and r_i.

These results for flow through an annulus are valid only if the flow is laminar. A criterion based on the conventional Reynolds number (which is defined in terms of the tube diameter) cannot be directly applied to the annulus, since there are really "two" diameters involved. For tube cross sections other than simple circular tubes it is common practice to use an "effective" diameter, termed the *hydraulic diameter*, D_h, which is defined as

$$D_h = \frac{4 \times \text{cross-sectional area}}{\text{wetted perimeter}}$$

The wetted perimeter is the perimeter in contact with the fluid. For an annulus

$$D_h = \frac{4\pi(r_o^2 - r_i^2)}{2\pi(r_o + r_i)} = 2(r_o - r_i)$$

In terms of the hydraulic diameter, the Reynolds number is $\text{Re} = \rho D_h V/\mu$ (where $V = Q/$ cross-sectional area), and it is commonly assumed that if this Reynolds number remains below 2100 the flow will be laminar. A further discussion of the concept of the hydraulic diameter as it applies to other noncircular cross sections is given in Section 8.4.3.

EXAMPLE 6.10   Laminar Flow in an Annulus

GIVEN   A viscous liquid ($\rho = 1.18 \times 10^3$ kg/m^3; $\mu = 0.0045$ N·s/m^2) flows at a rate of 12 ml/s through a horizontal, 4-mm-diameter tube.

FIND   (a) Determine the pressure drop along a 1-m length of the tube which is far from the tube entrance so that the only component of velocity is parallel to the tube axis. (b) If a 2-mm-diameter rod is placed in the 4-mm-diameter tube to form a symmetric annulus, what is the pressure drop along a 1-m length if the flowrate remains the same as in part (a)?

SOLUTION

(a)   We first calculate the Reynolds number, Re, to determine whether or not the flow is laminar. With the diameter $D = 4$ mm $= 0.004$ m, the mean velocity is

$$V = \frac{Q}{(\pi/4)D^2} = \frac{(12 \text{ ml/s})(10^{-6} \text{ m}^3/\text{ml})}{(\pi/4)(0.004 \text{ m})^2}$$

$$= 0.955 \text{ m/s}$$

and, therefore,

$$\text{Re} = \frac{\rho V D}{\mu} = \frac{(1.18 \times 10^3 \text{ kg/m}^3)(0.955 \text{ m/s})(0.004 \text{ m})}{0.0045 \text{ N} \cdot \text{s/m}^2}$$

$$= 1000$$

Since the Reynolds number is well below the critical value of 2100 we can safely assume that the flow is laminar. Thus, we can apply Eq. 6.151, which gives for the pressure drop

$$\Delta p = \frac{8 \mu \ell Q}{\pi R^4}$$

$$= \frac{8(0.0045 \text{ N} \cdot \text{s/m}^2)(1 \text{ m})(12 \times 10^{-6} \text{ m}^3/\text{s})}{\pi (0.002 \text{ m})^4}$$

$$= 8.59 \text{ kPa} \qquad \textbf{(Ans)}$$

(b)   For flow in the annulus with an outer radius $r_o = 0.002$ m and an inner radius $r_i = 0.001$ m, the mean velocity is

$$V = \frac{Q}{\pi(r_o^2 - r_i^2)} = \frac{12 \times 10^{-6} \text{ m}^3/\text{s}}{(\pi)[(0.002 \text{ m})^2 - (0.001 \text{ m})^2]}$$

$$= 1.27 \text{ m/s}$$

and the Reynolds number [based on the hydraulic diameter, $D_h = 2(r_o - r_i) = 2(0.002 \text{ m} - 0.001 \text{ m}) = 0.002$ m] is

$$\text{Re} = \frac{\rho D_h V}{\mu}$$

$$= \frac{(1.18 \times 10^3 \text{ kg/m}^3)(0.002 \text{ m})(1.27 \text{ m/s})}{0.0045 \text{ N} \cdot \text{s/m}^2}$$

$$= 666$$

This value is also well below 2100 so the flow in the annulus should also be laminar. From Eq. 6.156,

$$\Delta p = \frac{8 \mu \ell Q}{\pi} \left[r_o^4 - r_i^4 - \frac{(r_o^2 - r_i^2)^2}{\ln(r_o/r_i)} \right]^{-1}$$

so that

$$\Delta p = \frac{8(0.0045 \text{ N} \cdot \text{s/m}^2)(1 \text{ m})(12 \times 10^{-6} \text{ m}^3/\text{s})}{\pi}$$

$$\times \left\{ (0.002 \text{ m})^4 - (0.001 \text{ m})^4 \right.$$

$$\left. - \frac{[(0.002 \text{ m})^2 - (0.001 \text{ m})^2]^2}{\ln(0.002 \text{ m}/0.001 \text{ m})} \right\}^{-1}$$

$$= 68.2 \text{ kPa} \qquad \textbf{(Ans)}$$

COMMENTS   The pressure drop in the annulus is much larger than that of the tube. This is not a surprising result, since to maintain the same flow in the annulus as that in the open tube, the average velocity must be larger (the cross-sectional area is smaller) and the pressure difference along the annulus must overcome the shearing stresses that develop along both an inner and an outer wall.

By repeating the calculations for various radius ratios, r_i/r_o, the results shown in Fig. E6.10 are obtained. It is seen that the pressure drop ratio, $\Delta p_{\text{annulus}}/\Delta p_{\text{tube}}$ (i.e., the pressure drop in the annulus compared to that in a tube with a radius equal to the outer radius of the annulus, r_o), is a strong function of the radius ratio. Even an annulus with a very small inner radius will have a pressure drop significantly larger than that of a tube. For example, if the inner radius is only 1/100 of the outer radius, $\Delta p_{\text{annulus}}/\Delta p_{\text{tube}} = 1.28$. As shown in the figure, for larger inner radii, the pressure drop ratio is much larger [i.e., $\Delta p_{\text{annulus}}/\Delta p_{\text{tube}} = 7.94$ for $r_i/r_o = 0.50$ as in part (b) of this example].

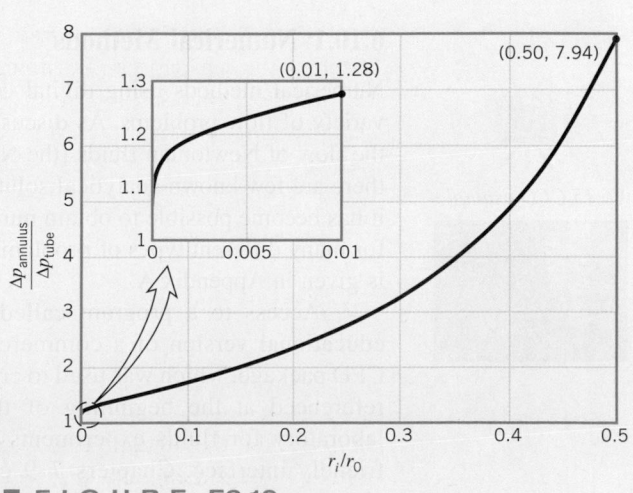

■ **FIGURE E6.10**

6.10   Other Aspects of Differential Analysis

In this chapter the basic differential equations that govern the flow of fluids have been developed. The Navier–Stokes equations, which can be compactly expressed in vector notation as

$$\rho\left(\frac{\partial \mathbf{V}}{\partial t} + \mathbf{V} \cdot \boldsymbol{\nabla}\mathbf{V}\right) = -\boldsymbol{\nabla}p + \rho\mathbf{g} + \mu\nabla^2\mathbf{V} \tag{6.158}$$

along with the continuity equation

$$\nabla \cdot \mathbf{V} = 0 \tag{6.159}$$

are the general equations of motion for incompressible Newtonian fluids. Although we have restricted our attention to incompressible fluids, these equations can be readily extended to include compressible fluids. It is well beyond the scope of this introductory text to consider in depth the variety of analytical and numerical techniques that can be used to obtain both exact and approximate solutions to the Navier–Stokes equations. Students, however, should be aware of the existence of these very general equations, which are frequently used as the basis for many advanced analyses of fluid motion. A few relatively simple solutions have been obtained and discussed in this chapter to indicate the type of detailed flow information that can be obtained by using differential analysis. However, it is hoped that the relative ease with which these solutions were obtained does not give the false impression that solutions to the Navier–Stokes equations are readily available. This is certainly not true, and as previously mentioned there are actually very few practical fluid flow problems that can be solved by using an exact analytical approach. In fact, there are no known analytical solutions to Eq. 6.158 for flow past any object such as a sphere, cube, or airplane.

Very few practical fluid flow problems can be solved using an exact analytical approach.

    Because of the difficulty in solving the Navier–Stokes equations, much attention has been given to various types of approximate solutions. For example, if the viscosity is set equal to zero, the Navier–Stokes equations reduce to Euler's equations. Thus, the frictionless fluid solutions discussed previously are actually approximate solutions to the Navier–Stokes equations. At the other extreme, for problems involving slowly moving fluids, viscous effects may be dominant and the nonlinear (convective) acceleration terms can be neglected. This assumption greatly simplifies the analysis, since the equations now become linear. There are numerous analytical solutions to these "*slow flow*" or "*creeping flow*" problems. Another broad class of approximate solutions is concerned with flow in the very thin boundary layer. L. Prandtl showed in 1904 how the Navier–Stokes equations could be simplified to study flow in boundary layers. Such "boundary layer solutions" play a very important role in the study of fluid mechanics. A further discussion of boundary layers is given in Chapter 9.

6.10.1   Numerical Methods

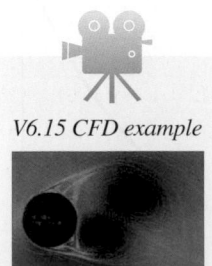

V6.15 CFD example

Numerical methods using digital computers are, of course, commonly utilized to solve a wide variety of flow problems. As discussed previously, although the differential equations that govern the flow of Newtonian fluids [the Navier–Stokes equations (6.158)] were derived many years ago, there are few known analytical solutions to them. With the advent of high-speed digital computers it has become possible to obtain numerical solutions to these (and other fluid mechanics) equations for many different types of problems. A brief introduction to computational fluid dynamics (CFD) is given in Appendix A.

    Access to a program called FlowLab is available with this textbook. FlowLab is an educational version of a commercial CFD program. The backbone of FlowLab is the Fluent CFD package, which was used to create the numerical animations of flow past a spinning football referenced at the beginning of the chapter (V6.1 and V6.2). FlowLab provides a virtual laboratory for fluids experiments that makes use of the power of CFD, but with a student-friendly interface. Chapters 7–9 contain fluids problems that require the use of FlowLab to obtain the solutions.

Fluids in the Academy Awards A computer science professor at Stanford University and his colleagues were awarded a Scientific and Technical Academy Award for applying the Navier–Stokes equations for use in Hollywood movies. These researchers make use of computational algorithms to numerically solve the Navier–Stokes equations (also termed computational fluid dynam-ics, or CFD) and simulate complex liquid flows. The realism of the simulations has found application in the entertainment industry. Movie producers have used the power of these numerical tools to simulate flows from ocean waves in "Pirates of the Caribbean" to lava flows in the final duel in "Star Wars: Revenge of the Sith." Therefore, even Hollywood has recognized the usefulness of CFD.

6.11 Chapter Summary and Study Guide

volumetric dilatation rate

vorticity

irrotational flow

continuity equation

stream function

Euler's equations of motion

ideal fluid

Bernoulli equation

velocity potential

potential flow

equipotential lines

flow net

uniform flow

source and sink

vortex

circulation

doublet

method of superposition

half-body

Rankine oval

Navier–Stokes equations

Couette flow

Poiseuille's law

Differential analysis of fluid flow is concerned with the development of concepts and techniques that can be used to provide a detailed, point by point, description of a flow field. Concepts related to the motion and deformation of a fluid element are introduced, including the Eulerian method for describing the velocity and acceleration of fluid particles. Linear deformation and angular deformation of a fluid element are described through the use of flow characteristics such as the volumetric dilatation rate, rate of angular deformation, and vorticity. The differential form of the conservation of mass equation (continuity equation) is derived in both rectangular and cylindrical polar coordinates.

Use of the stream function for the study of steady, incompressible, plane, two-dimensional flow is introduced. The general equations of motion are developed, and for inviscid flow these equations are reduced to the simpler Euler equations of motion. The Euler equations are integrated to give the Bernoulli equation, and the concept of irrotational flow is introduced. Use of the velocity potential for describing irrotational flow is considered in detail, and several basic velocity potentials are described, including those for a uniform flow, source or sink, vortex, and doublet. The technique of using various combinations of these basic velocity potentials, by superposition, to form new potentials is described. Flows around a half-body, a Rankine oval, and around a circular cylinder are obtained using this superposition technique.

Basic differential equations describing incompressible, viscous flow (the Navier–Stokes equations) are introduced. Several relatively simple solutions for steady, viscous, laminar flow between parallel plates and through circular tubes are included.

The following checklist provides a study guide for this chapter. When your study of the entire chapter and end-of-chapter exercises has been completed you should be able to

- write out meanings of the terms listed here in the margin and understand each of the related concepts. These terms are particularly important and are set in *italic bold, and color* type in the text.
- determine the acceleration of a fluid particle, given the equation for the velocity field.
- determine the volumetric dilatation rate, vorticity, and rate of angular deformation for a fluid element, given the equation for the velocity field.
- show that a given velocity field satisfies the continuity equation.
- use the concept of the stream function to describe a flow field.
- use the concept of the velocity potential to describe a flow field.
- use superposition of basic velocity potentials to describe simple potential flow fields.
- use the Navier–Stokes equations to determine the detailed flow characteristics of in-compressible, steady, laminar, viscous flow between parallel plates and through circular tubes.

Some of the important equations in this chapter are:

Acceleration of fluid particle	$\mathbf{a} = \dfrac{\partial \mathbf{V}}{\partial t} + u \dfrac{\partial \mathbf{V}}{\partial x} + v \dfrac{\partial \mathbf{V}}{\partial y} + w \dfrac{\partial \mathbf{V}}{\partial z}$	**(6.2)**
Vorticity	$\zeta = 2\,\boldsymbol{\omega} = \nabla \times \mathbf{V}$	**(6.17)**
Conservation of mass	$\dfrac{\partial \rho}{\partial t} + \dfrac{\partial(\rho u)}{\partial x} + \dfrac{\partial(\rho v)}{\partial y} + \dfrac{\partial(\rho w)}{\partial z} = 0$	**(6.27)**

Stream function $\qquad u = \dfrac{\partial \psi}{\partial y} \qquad v = -\dfrac{\partial \psi}{\partial x}$ (6.37)

Euler's equations of motion $\quad \rho g_x - \dfrac{\partial p}{\partial x} = \rho\left(\dfrac{\partial u}{\partial t} + u\dfrac{\partial u}{\partial x} + v\dfrac{\partial u}{\partial y} + w\dfrac{\partial u}{\partial z}\right)$ (6.51a)

$$\rho g_y - \dfrac{\partial p}{\partial y} = \rho\left(\dfrac{\partial v}{\partial t} + u\dfrac{\partial v}{\partial x} + v\dfrac{\partial v}{\partial y} + w\dfrac{\partial v}{\partial z}\right)$$ (6.51b)

$$\rho g_z - \dfrac{\partial p}{\partial z} = \rho\left(\dfrac{\partial w}{\partial t} + u\dfrac{\partial w}{\partial x} + v\dfrac{\partial w}{\partial y} + w\dfrac{\partial w}{\partial z}\right)$$ (6.51c)

Velocity potential $\qquad \mathbf{V} = \nabla \phi$ (6.65)

Laplace's equation $\qquad \nabla^2 \phi = 0$ (6.66)

Uniform potential flow $\quad \phi = U(x\cos\alpha + y\sin\alpha) \quad \psi = U(y\cos\alpha - x\sin\alpha) \quad u = U\cos\alpha$
$\qquad\qquad\qquad\qquad\qquad\qquad\qquad\qquad\qquad\qquad\qquad\qquad\qquad v = U\sin\alpha$

Source and sink $\qquad \phi = \dfrac{m}{2\pi}\ln r \qquad \psi = \dfrac{m}{2\pi}\theta \qquad v_r = \dfrac{m}{2\pi r}$
$\qquad\qquad\qquad\qquad\qquad\qquad\qquad\qquad\qquad\qquad\qquad v_\theta = 0$

Vortex $\qquad\qquad \phi = \dfrac{\Gamma}{2\pi}\theta \qquad \psi = -\dfrac{\Gamma}{2\pi}\ln r \qquad v_r = 0$
$\qquad\qquad\qquad\qquad\qquad\qquad\qquad\qquad\qquad\qquad\qquad v_\theta = \dfrac{\Gamma}{2\pi r}$

Doublet $\qquad\qquad \phi = \dfrac{K\cos\theta}{r} \qquad \psi = -\dfrac{K\sin\theta}{r} \qquad v_r = -\dfrac{K\cos\theta}{r^2}$
$\qquad\qquad\qquad\qquad\qquad\qquad\qquad\qquad\qquad\qquad\qquad v_\theta = -\dfrac{K\cos\theta}{r^2}$

The Navier–Stokes equations

(x direction)

$$\rho\left(\dfrac{\partial u}{\partial t} + u\dfrac{\partial u}{\partial x} + v\dfrac{\partial u}{\partial y} + w\dfrac{\partial u}{\partial z}\right) = -\dfrac{\partial p}{\partial x} + \rho g_x + \mu\left(\dfrac{\partial^2 u}{\partial x^2} + \dfrac{\partial^2 u}{\partial y^2} + \dfrac{\partial^2 u}{\partial z^2}\right)$$ (6.127a)

(y direction)

$$\rho\left(\dfrac{\partial v}{\partial t} + u\dfrac{\partial v}{\partial x} + v\dfrac{\partial v}{\partial y} + w\dfrac{\partial v}{\partial z}\right) = -\dfrac{\partial p}{\partial y} + \rho g_y + \mu\left(\dfrac{\partial^2 v}{\partial x^2} + \dfrac{\partial^2 v}{\partial y^2} + \dfrac{\partial^2 v}{\partial z^2}\right)$$ (6.127b)

(z direction)

$$\rho\left(\dfrac{\partial w}{\partial t} + u\dfrac{\partial w}{\partial x} + v\dfrac{\partial w}{\partial y} + w\dfrac{\partial w}{\partial z}\right) = -\dfrac{\partial p}{\partial z} + \rho g_z + \mu\left(\dfrac{\partial^2 w}{\partial x^2} + \dfrac{\partial^2 w}{\partial y^2} + \dfrac{\partial^2 w}{\partial z^2}\right)$$ (6.127c)

References

1. White, F. M., *Fluid Mechanics*, 5th Ed., McGraw-Hill, New York, 2003.
2. Streeter, V. L., *Fluid Dynamics*, McGraw-Hill, New York, 1948.
3. Rouse, H., *Advanced Mechanics of Fluids*, Wiley, New York, 1959.
4. Milne-Thomson, L. M., *Theoretical Hydrodynamics*, 4th Ed., Macmillan, New York, 1960.
5. Robertson, J. M., *Hydrodynamics in Theory and Application*, Prentice-Hall, Englewood Cliffs, N.J., 1965.
6. Panton, R. L., *Incompressible Flow*, 3rd Ed., Wiley, New York, 2005.
7. Li, W. H., and Lam, S. H., *Principles of Fluid Mechanics*, Addison-Wesley, Reading, Mass., 1964.
8. Schlichting, H., *Boundary-Layer Theory*, 8th Ed., McGraw-Hill, New York, 2000.
9. Fuller, D. D., *Theory and Practice of Lubrication for Engineers*, Wiley, New York, 1984.

Review Problems

Go to Appendix G for a set of review problems with answers. Detailed solutions can be found in *Student Solution Manual and Study Guide for Fundamentals of Fluid Mechanics*, by Munson et al. (© 2009 John Wiley and Sons, Inc.).

Problems

Section 6.1 Fluid Element Kinematics

6.1 Obtain a photograph/image of a situation in which a fluid is undergoing angular deformation. Print this photo and write a brief paragraph that describes the situation involved.

6.2 The velocity in a certain two-dimensional flow field is given by the equation

$$\mathbf{V} = 2xt\hat{\mathbf{i}} - 2yt\hat{\mathbf{j}}$$

where the velocity is in m/s when x, y, and t are in meters and seconds, respectively. Determine expressions for the local and convective components of acceleration in the x and y directions. What is the magnitude and direction of the velocity and the acceleration at the point $x = y = 2$ m at the time $t = 0$?

6.3 The velocity in a certain flow field is given by the equation

$$\mathbf{V} = x\hat{\mathbf{i}} + x^2z\hat{\mathbf{j}} + yz\hat{\mathbf{k}}$$

Determine the expressions for the three rectangular components of acceleration.

6.4 The three components of velocity in a flow field are given by

$$u = x^2 + y^2 + z^2$$
$$v = xy + yz + z^2$$
$$w = -3xz - z^2/2 + 4$$

(a) Determine the volumetric dilatation rate and interpret the results. **(b)** Determine an expression for the rotation vector. Is this an irrotational flow field?

6.5 Determine the vorticity field for the following velocity vector:

$$\mathbf{V} = (x^2 - y^2)\hat{\mathbf{i}} - 2xy\hat{\mathbf{j}}$$

6.6 Determine an expression for the vorticity of the flow field described by

$$\mathbf{V} = -xy^3\hat{\mathbf{i}} + y^4\hat{\mathbf{j}}$$

Is the flow irrotational?

6.7 A one-dimensional flow is described by the velocity field

$$u = ay + by^2$$
$$v = w = 0$$

where a and b are constants. Is the flow irrotational? For what combination of constants (if any) will the rate of angular deformation as given by Eq. 6.18 be zero?

6.8 For a certain incompressible, two-dimensional flow field the velocity component in the y direction is given by the equation

$$v = 3xy + x^2y$$

Determine the velocity component in the x direction so that the volumetric dilatation rate is zero.

6.9 An incompressible viscous fluid is placed between two large parallel plates as shown in Fig. P6.9. The bottom plate is fixed and the upper plate moves with a constant velocity, U. For these conditions the velocity distribution between the plates is linear and can be expressed as

$$u = U\frac{y}{b}$$

Determine: **(a)** the volumetric dilatation rate, **(b)** the rotation vector, **(c)** the vorticity, and **(d)** the rate of angular deformation.

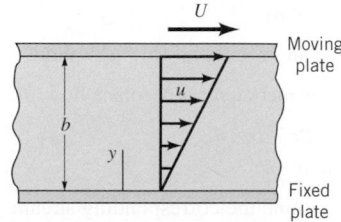

■ **F I G U R E P6.9**

6.10 A viscous fluid is contained in the space between concentric cylinders. The inner wall is fixed, and the outer wall rotates with an angular velocity ω. (See Fig. P6.10a and Video V6.3.) Assume that the velocity distribution in the gap is linear as illustrated in Fig. P6.10b. For the small rectangular element shown in Fig. P6.10b, determine the rate of change of the right angle γ due to the fluid motion. Express your answer in terms of r_0, r_i, and ω.

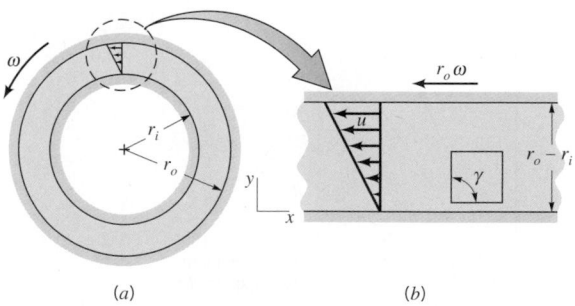

(a)   (b)

■ **F I G U R E P6.10**

Section 6.2 Conservation of Mass

6.11 Obtain a photograph/image of a situation in which streamlines indicate a feature of the flow field. Print this photo and write a brief paragraph that describes the situation involved.

6.12 Verify that the stream function in cylindrical coordinates satisfies the continuity equation.

6.13 For a certain incompressible flow field it is suggested that the velocity components are given by the equations

$$u = 2xy \quad v = -x^2y \quad w = 0$$

Is this a physically possible flow field? Explain.

6.14 The velocity components of an incompressible, two-dimensional velocity field are given by the equations

$$u = y^2 - x(1 + x)$$
$$v = y(2x + 1)$$

Show that the flow is irrotational and satisfies conservation of mass.

6.15 For each of the following stream functions, with units of m^2/s, determine the magnitude and the angle the velocity vector makes with the x axis at $x = 1$ m, $y = 2$ m. Locate any stagnation points in the flow field.
(a) $\psi = xy$
(b) $\psi = -2x^2 + y$

6.16 The stream function for an incompressible, two-dimensional flow field is

$$\psi = ay - by^3$$

where a and b are constants. Is this an irrotational flow? Explain.

6.17 The stream function for an incompressible, two-dimensional flow field is

$$\psi = ay^2 - bx$$

where a and b are constants. Is this an irrotational flow? Explain.

6.18 The velocity components for an incompressible, plane flow are

$$v_r = Ar^{-1} + Br^{-2} \cos \theta$$
$$v_\theta = Br^{-2} \sin \theta$$

where A and B are constants. Determine the corresponding stream function.

6.19 For a certain two-dimensional flow field

$$u = 0$$
$$v = V$$

(a) What are the corresponding radial and tangential velocity components? **(b)** Determine the corresponding stream function expressed in Cartesian coordinates and in cylindrical polar coordinates.

6.20 Make use of the control volume shown in Fig. P6.20 to derive the continuity equation in cylindrical coordinates (Eq. 6.33 in text).

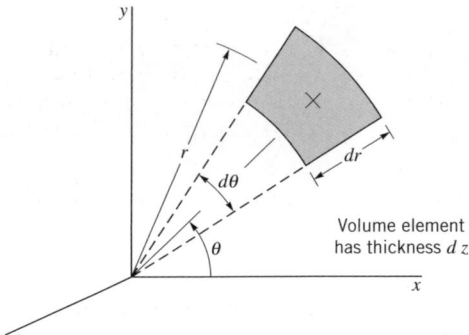

■ **FIGURE P6.20**

6.21 A two-dimensional, incompressible flow is given by $u = -y$ and $v = x$. Show that the streamline passing through the point $x = 10$ and $y = 0$ is a circle centered at the origin.

6.22 In a certain steady, two-dimensional flow field the fluid density varies linearly with respect to the coordinate x; that is, $\rho = Ax$ where A is a constant. If the x component of velocity u is given by the equation $u = y$, determine an expression for v.

6.23 In a two-dimensional, incompressible flow field, the x component of velocity is given by the equation $u = 2x$. **(a)** Determine the corresponding equation for the y component of velocity if $v = 0$ along the x axis. **(b)** For this flow field, what is the magnitude of the average velocity of the fluid crossing the surface OA of Fig. P6.23? Assume that the velocities are in meters per second when x and y are in meters.

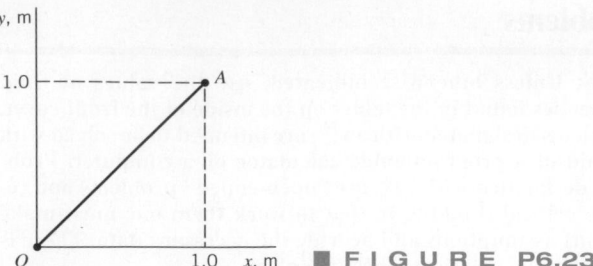

■ **FIGURE P6.23**

6.24 The radial velocity component in an incompressible, two-dimensional flow field $(v_z = 0)$ is

$$v_r = 2r + 3r^2 \sin \theta$$

Determine the corresponding tangential velocity component, v_θ, required to satisfy conservation of mass.

6.25 The stream function for an incompressible flow field is given by the equation

$$\psi = 3x^2y - y^3$$

where the stream function has the units of m^2/s with x and y in meters. **(a)** Sketch the streamline(s) passing through the origin. **(b)** Determine the rate of flow across the straight path AB shown in Fig. P6.25.

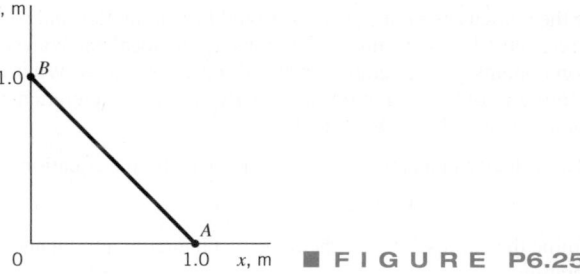

■ **FIGURE P6.25**

6.26 The streamlines in a certain incompressible, two-dimensional flow field are all concentric circles so that $v_r = 0$. Determine the stream function for **(a)** $v_\theta = Ar$ and for **(b)** $v_\theta = Ar^{-1}$, where A is a constant.

*6.27 The stream function for an incompressible, two-dimensional flow field is

$$\psi = 3x^2y + y$$

For this flow field, plot several streamlines.

6.28 Consider the incompressible, two-dimensional flow of a non-viscous fluid between the boundaries shown in Fig. P6.28. The velocity potential for this flow field is

$$\phi = x^2 - y^2$$

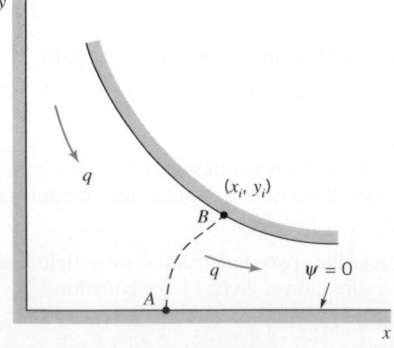

■ **FIGURE P6.28**

(a) Determine the corresponding stream function. **(b)** What is the relationship between the discharge, q, (per unit width normal to plane of paper) passing between the walls and the coordinates x_i, y_i of any point on the curved wall? Neglect body forces.

Section 6.3 Conservation of Linear Momentum

6.29 Obtain a photograph/image of a situation in which a fluid flow produces a force. Print this photo and write a brief paragraph that describes the situation involved.

Section 6.4 Inviscid Flow

6.30 Obtain a photograph/image of a situation in which all or part of a flow field could be approximated by assuming inviscid flow. Print this photo and write a brief paragraph that describes the situation involved.

6.31 Given the streamfunction for a flow as $\psi = 4x^2 - 4y^2$, show that the Bernoulli equation can be applied between any two points in the flow field.

6.32 A two-dimensional flow field for a nonviscous, incompressible fluid is described by the velocity components

$$u = U_0 + 2y$$
$$v = 0$$

where U_0 is a constant. If the pressure at the origin (Fig. P6.32) is p_0, determine an expression for the pressure at **(a)** point A, and **(b)** point B. Explain clearly how you obtained your answer. Assume that the units are consistent and body forces may be neglected.

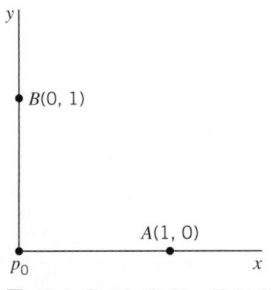

■ **FIGURE P6.32**

6.33 In a certain two-dimensional flow field, the velocity is constant with components $u = -1.2$ m/s and $v = -0.6$ m/s. Determine the corresponding stream function and velocity potential for this flow field. Sketch the equipotential line $\phi = 0$ which passes through the origin of the coordinate system.

6.34 The stream function for a given two-dimensional flow field is

$$\psi = 5x^2 y - (5/3)y^3$$

Determine the corresponding velocity potential.

6.35 Determine the stream function corresponding to the velocity potential

$$\phi = x^3 - 3xy^2$$

Sketch the streamline $\psi = 0$, which passes through the origin.

6.36 A certain flow field is described by the stream function

$$\psi = A\,\theta + B\,r\sin\theta$$

where A and B are positive constants. Determine the corresponding velocity potential and locate any stagnation points in this flow field.

6.37 It is known that the velocity distribution for two-dimensional flow of a viscous fluid between wide parallel plates (Fig. P6.37) is parabolic; that is,

$$u = U_c\left[1 - \left(\frac{y}{h}\right)^2\right]$$

with $v = 0$. Determine, if possible, the corresponding stream function and velocity potential.

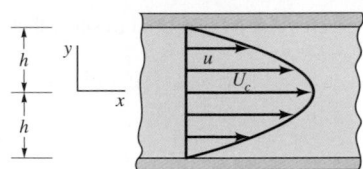

■ **FIGURE P6.37**

6.38 The velocity potential for a certain inviscid flow field is

$$\phi = -(9.8x^2 y - 3.28y^3)$$

where ϕ has the units of m^2/s when x and y are in meters. Determine the pressure difference (in kPa) between the points (0.3, 0.6) and (1.2, 1.2), where the coordinates are in meters, if the fluid is water and elevation changes are negligible.

6.39 The velocity potential for a flow is given by

$$\phi = \frac{a}{2}(x^2 - y^2)$$

where a is a constant. Determine the corresponding stream function and sketch the flow pattern.

6.40 The stream function for a two-dimensional, nonviscous, incompressible flow field is given by the expression

$$\psi = -0.6\,(x - y)$$

where the stream function has the units of m^2/s with x and y in meters. **(a)** Is the continuity equation satisfied? **(b)** Is the flow field irrotational? If so, determine the corresponding velocity potential. **(c)** Determine the pressure gradient in the horizontal x direction at the point $x = 0.6$ m, $y = 0.6$ m.

6.41 The velocity potential for a certain inviscid, incompressible flow field is given by the equation

$$\phi = 2x^2 y - (\tfrac{2}{3})y^3$$

where ϕ has the units of m^2/s when x and y are in meters. Determine the pressure at the point $x = 2$ m, $y = 2$ m if the pressure at $x = 1$ m, $y = 1$ m is 200 kPa. Elevation changes can be neglected, and the fluid is water.

6.42 A steady, uniform, incompressible, inviscid, two-dimensional flow makes an angle of 30° with the horizontal x axis. **(a)** Determine the velocity potential and the stream function for this flow. **(b)** Determine an expression for the pressure gradient in the vertical y direction. What is the physical interpretation of this result?

6.43 The streamlines for an incompressible, inviscid, two-dimensional flow field are all concentric circles, and the velocity varies directly with the distance from the common center of the streamlines; that is

$$v_\theta = Kr$$

where K is a constant. **(a)** For this *rotational* flow, determine, if possible, the stream function. **(b)** Can the pressure difference between the origin and any other point be determined from the Bernoulli equation? Explain.

6.44 The velocity potential

$$\phi = -k(x^2 - y^2) \qquad (k = \text{constant})$$

may be used to represent the flow against an infinite plane boundary, as illustrated in Fig. P6.44. For flow in the vicinity of a stagnation point, it is frequently assumed that the pressure gradient along the surface is of the form

$$\frac{\partial p}{\partial x} = Ax$$

where A is a constant. Use the given velocity potential to show that this is true.

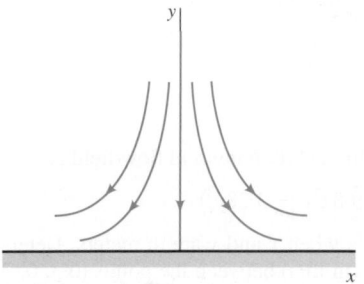

■ **FIGURE P6.44**

6.45 Water is flowing between wedge-shaped walls into a small opening as shown in Fig. P6.45. The velocity potential with units m²/s for this flow is $\phi = -2 \ln r$ with r in meters. Determine the pressure differential between points A and B.

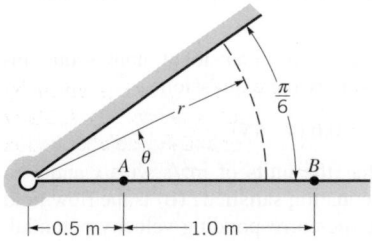

■ **FIGURE P6.45**

6.46 An ideal fluid flows between the inclined walls of a two-dimensional channel into a sink located at the origin (Fig. P6.46). The velocity potential for this flow field is

$$\phi = \frac{m}{2\pi} \ln r$$

where m is a constant. **(a)** Determine the corresponding stream function. Note that the value of the stream function along the wall OA is zero. **(b)** Determine the equation of the streamline passing through the point B, located at $x = 1$, $y = 4$.

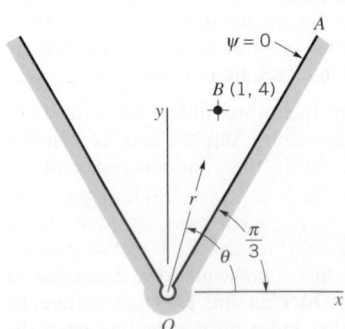

■ **FIGURE P6.46**

6.47 It is suggested that the velocity potential for the incompressible, nonviscous, two-dimensional flow along the wall shown in Fig. P6.47 is

$$\phi = r^{4/3} \cos \tfrac{4}{3}\theta$$

Is this a suitable velocity potential for flow along the wall? Explain.

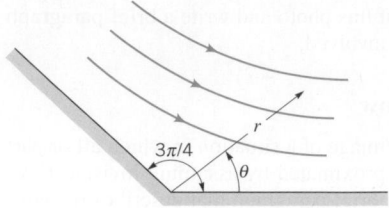

■ **FIGURE P6.47**

Section 6.5 Some Basic, Plane Potential Flows

6.48 Obtain a photograph/image of a situation which approximates one of the basic, plane potential flows. Print this photo and write a brief paragraph that describes the situation involved.

6.49 As illustrated in Fig. P6.49, a tornado can be approximated by a free vortex of strength Γ for $r > R_c$, where R_c is the radius of the core. Velocity measurements at points A and B indicate that $V_A = 38$ m/s and $V_B = 18$ m/s. Determine the distance from point A to the center of the tornado. Why can the free vortex model not be used to approximate the tornado throughout the flow field ($r \geq 0$)?

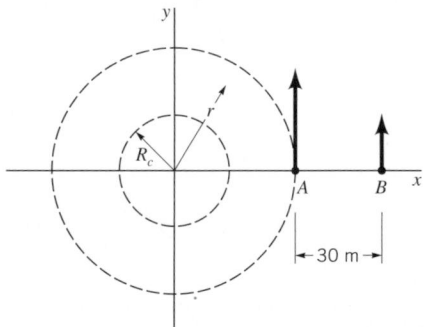

■ **FIGURE P6.49**

6.50 If the velocity field is given by $\mathbf{V} = ax\hat{\mathbf{i}} - ay\hat{\mathbf{j}}$, and a is a constant, find the circulation around the closed curve shown in Fig. P6.50.

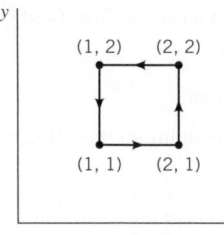

■ **FIGURE P6.50**

6.51 The streamlines in a particular two-dimensional flow field are all concentric circles, as shown in Fig. P6.51. The velocity is given by the equation $v_\theta = \omega r$ where ω is the angular velocity of the rotating mass of fluid. Determine the circulation around the path $ABCD$.

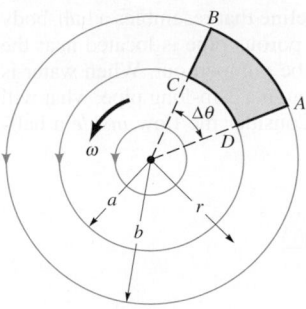

■ **F I G U R E   P6.51**

6.52 The motion of a liquid in an open tank is that of a combined vortex consisting of a forced vortex for $0 \leq r \leq 0.6$ m and a free vortex for $r > 0.6$ m. The velocity profile and the corresponding shape of the free surface are shown in Fig. P6.52. The free surface at the center of the tank is a depth h below the free surface at $r = \infty$. Determine the value of h. Note that $h = h_{\text{forced}} + h_{\text{free}}$, where h_{forced} and h_{free} are the corresponding depths for the forced vortex and the free vortex, respectively. (See Section 2.12.2 for further discussion regarding the forced vortex.)

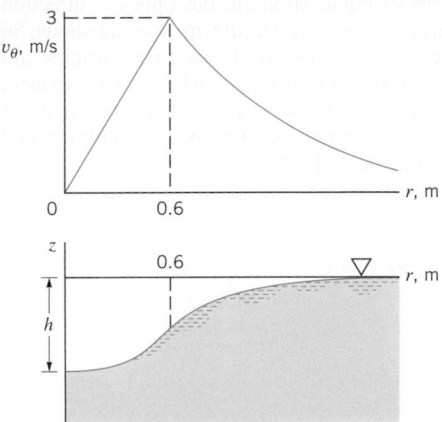

■ **F I G U R E   P6.52**

6.53 When water discharges from a tank through an opening in its bottom, a vortex may form with a curved surface profile, as shown in Fig. P6.53 and **Video V6.4**. Assume that the velocity distribution in the vortex is the same as that for a free vortex. At the same time the water is being discharged from the tank at point A, it is desired to discharge a small quantity of water through the pipe B. As the discharge through A is increased, the strength of the vortex, as indicated by its circulation, is increased. Determine the maximum strength that the vortex can have in order that no air is sucked in at B. Express your answer in terms of the circulation. Assume that the fluid level in the tank at a large distance from the opening at A remains constant and viscous effects are negligible.

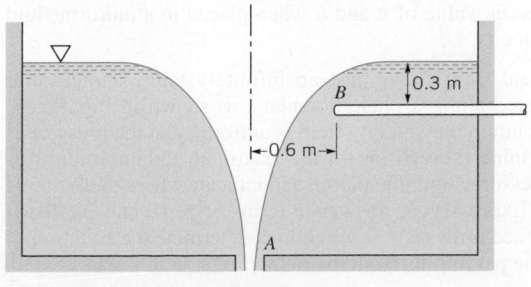

■ **F I G U R E   P6.53**

6.54 Water flows over a flat surface at 1.2 m/s, as shown in Fig. P6.54. A pump draws off water through a narrow slit at a volume rate of 0.01 m³/s per meter length of the slit. Assume that the fluid is incompressible and inviscid and can be represented by the combination of a uniform flow and a sink. Locate the stagnation point on the wall (point A) and determine the equation for the stagnation streamline. How far above the surface, H, must the fluid be so that it does not get sucked into the slit?

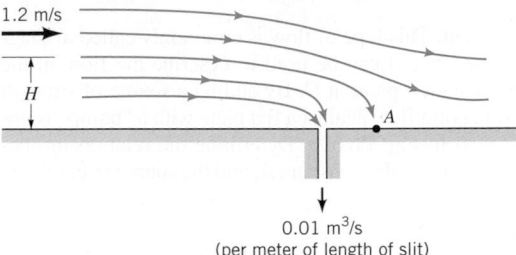

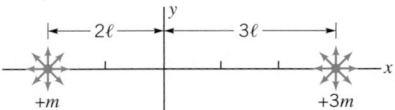

■ **F I G U R E   P6.54**

6.55 Two sources, one of strength m and the other with strength $3m$, are located on the x axis as shown in Fig. P6.55. Determine the location of the stagnation point in the flow produced by these sources.

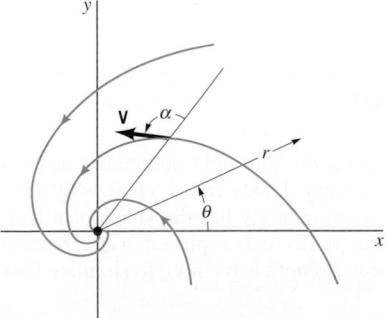

■ **F I G U R E   P6.55**

6.56 The velocity potential for a spiral vortex flow is given by $\phi = (\Gamma/2\pi) \theta - (m/2\pi) \ln r$, where Γ and m are constants. Show that the angle, α, between the velocity vector and the radial direction is constant throughout the flow field (see Fig. P6.56).

■ **F I G U R E   P6.56**

6.57 For a free vortex (see **Video V6.4**) determine an expression for the pressure gradient **(a)** along a streamline, and **(b)** normal to a streamline. Assume that the streamline is in a horizontal plane, and express your answer in terms of the circulation.

6.58 (See Fluids in the News article titled "Some hurricanes facts," Section 6.5.3.) Consider a category five hurricane that has a maximum wind speed of 260 km/hr at the eye wall, 16 km from the center of the hurricane. If the flow in the hurricane outside of the hurricane's eye is approximated as a free vortex, determine the wind speeds at locations 32 km, 48 km, and 64 km from the center of the storm.

Section 6.6 Superposition of Basic, Plane Potential Flows

6.59 Obtain a photograph/image of a situation that mimics the superposition of potential flows (see Ex. 6.7). Print this photo and write a brief paragraph that describes the situation involved.

6.60 Potential flow against a flat plate (Fig. P6.60a) can be described with the stream function

$$\psi = Axy$$

where A is a constant. This type of flow is commonly called a "stagnation point" flow since it can be used to describe the flow in the vicinity of the stagnation point at O. By adding a source of strength m at O, stagnation point flow against a flat plate with a "bump" is obtained as illustrated in Fig. P6.60b. Determine the relationship between the bump height, h, the constant, A, and the source strength, m.

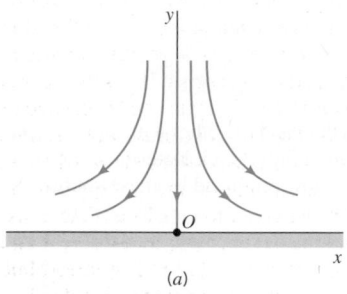

(a)

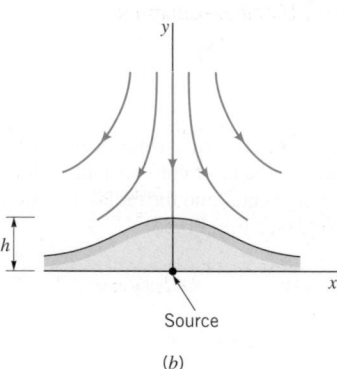

Source

(b)

■ **FIGURE P6.60**

6.61 The combination of a uniform flow and a source can be used to describe flow around a streamlined body called a half-body. (See Video V6.5.) Assume that a certain body has the shape of a half-body with a thickness of 0.5 m. If this body is placed in an airstream moving at 15 m/s, what source strength is required to simulate flow around the body?

6.62 A vehicle windshield is to be shaped as a portion of a half-body with the dimensions shown in Fig. P6.62. **(a)** Make a scale drawing of the windshield shape. **(b)** For a free stream velocity of 90 km/hr, determine the velocity of the air at points A and B.

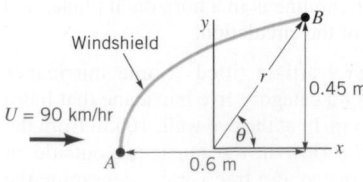

■ **FIGURE P6.62**

6.63 One end of a pond has a shoreline that resembles a half-body as shown in Fig. P6.63. A vertical porous pipe is located near the end of the pond so that water can be pumped out. When water is pumped at the rate of 0.08 m³/s through a 3-m-long pipe, what will be the velocity at point A? *Hint:* Consider the flow *inside* a half-body. (See Video V6.5.)

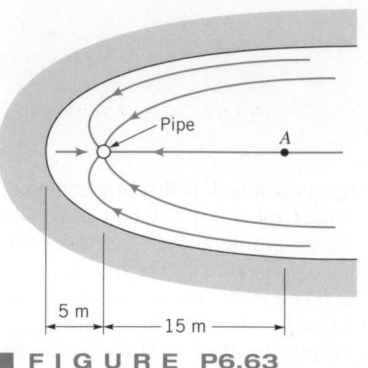

■ **FIGURE P6.63**

6.64 Two free vortices of equal strength, but opposite direction of rotation, are superimposed with a uniform flow as shown in Fig. P6.64. The stream functions for these two vortices are $\psi = -[\pm\Gamma/(2\pi)]\ln r$. **(a)** Develop an equation for the x-component of velocity, u, at point $P(x,y)$ in terms of Cartesian coordinates x and y. **(b)** Compute the x-component of velocity at point A and show that it depends on the ratio Γ/H.

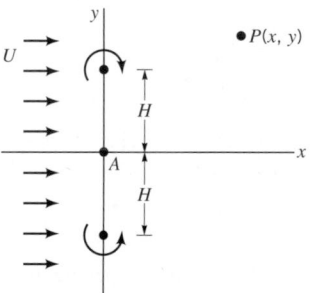

■ **FIGURE P6.64**

6.65 A Rankine oval is formed by combining a source–sink pair, each having a strength of 3.3 m²/s and separated by a distance of 3.6 m along the x axis, with a uniform velocity of 3 m/s (in the positive x direction). Determine the length and thickness of the oval.

***6.66** Make use of Eqs. 6.107 and 6.109 to construct a table showing how ℓ/a, h/a, and ℓ/h for Rankine ovals depend on the parameter $\pi Ua/m$. Plot ℓ/h versus $\pi Ua/m$ and describe how this plot could be used to obtain the required values of m and a for a Rankine oval having a specific value of ℓ and h when placed in a uniform fluid stream of velocity, U.

6.67 An ideal fluid flows past an infinitely long, semicircular "hump" located along a plane boundary, as shown in Fig. P6.67. Far from the hump the velocity field is uniform, and the pressure is p_0. **(a)** Determine expressions for the maximum and minimum values of the pressure along the hump, and indicate where these points are located. Express your answer in terms of ρ, U, and p_0. **(b)** If the solid surface is the $\psi = 0$ streamline, determine the equation of the streamline passing through the point $\theta = \pi/2$, $r = 2a$.

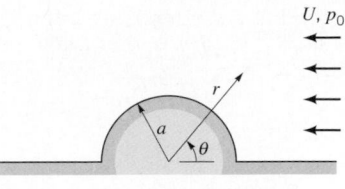

■ FIGURE P6.67

6.68 Water flows around a 2-m-diameter bridge pier with a velocity of 4 m/s. Estimate the force (per unit length) that the water exerts on the pier. Assume that the flow can be approximated as an ideal fluid flow around the front half of the cylinder, but due to flow separation (see **Video V6.8**), the average pressure on the rear half is constant and approximately equal to ½ the pressure at point *A* (see Fig. P6.68).

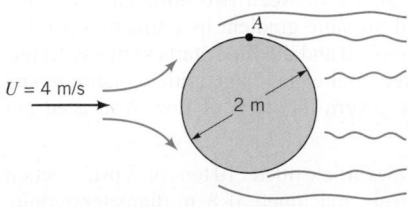

■ FIGURE P6.68

***6.69** Consider the steady potential flow around the circular cylinder shown in Fig. 6.26. On a plot show the variation of the magnitude of the dimensionless fluid velocity, V/U, along the positive *y* axis. At what distance, y/a (along the *y* axis), is the velocity within 1% of the free-stream velocity?

6.70 The velocity potential for a cylinder (Fig. P6.70) rotating in a uniform stream of fluid is

$$\phi = Ur\left(1 + \frac{a^2}{r^2}\right)\cos\theta + \frac{\Gamma}{2\pi}\theta$$

where Γ is the circulation. For what value of the circulation will the stagnation point be located at: **(a)** point *A*, **(b)** point *B*?

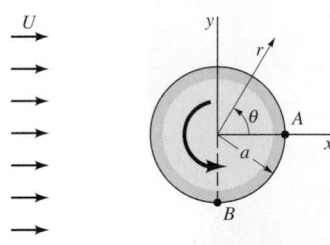

■ FIGURE P6.70

6.71 Show that for a rotating cylinder in a uniform flow, the following pressure ratio equation is true.

$$\frac{p_{top} - p_{bottom}}{p_{stagnation}} = \frac{8q}{U}$$

Here *U* is the velocity of the uniform flow and *q* is the surface speed of the rotating cylinder.

6.72 (See Fluids in the News article titled "A sailing ship without sails," Section 6.6.3.) Determine the magnitude of the total force developed by the two rotating cylinders on the Flettner "rotor-ship" due to the Magnus effect. Assume a wind speed relative to the ship of **(a)** 16 km/hr and **(b)** 48 km/hr. Each cylinder has a diameter of 3 m, a length of 15.0 m, and rotates at 750 rev/min. Use Eq. 6.124 and

calculate the circulation by assuming the air sticks to the rotating cylinders. *Note*: This calculated force is at right angles to the direction of the wind and it is the component of this force in the direction of motion of the ship that gives the propulsive thrust. Also, due to viscous effects, the actual propulsive thrust will be smaller than that calculated from Eq. 6.124 which is based on inviscid flow theory.

6.73 A fixed circular cylinder of infinite length is placed in a steady, uniform stream of an incompressible, nonviscous fluid. Assume that the flow is irrotational. Prove that the drag on the cylinder is zero. Neglect body forces.

6.74 Repeat Problem 6.73 for a rotating cylinder for which the stream function and velocity potential are given by Eqs. 6.119 and 6.120, respectively. Verify that the lift is not zero and can be expressed by Eq. 6.124.

6.75 At a certain point at the beach, the coast line makes a right-angle bend, as shown in Fig. P6.75a. The flow of salt water in this bend can be approximated by the potential flow of an incompressible fluid in a right-angle corner. **(a)** Show that the stream function for this flow is $\psi = A r^2 \sin 2\theta$, where *A* is a positive constant. **(b)** A fresh-water reservoir is located in the corner. The salt water is to be kept away from the reservoir to avoid any possible seepage of salt water into the fresh water (Fig. P6.75b). The fresh-water source can be approximated as a line source having a strength *m*, where *m* is the volume rate of flow (per unit length) emanating from the source. Determine *m* if the salt water is not to get closer than a distance *L* to the corner. *Hint*: Find the value of *m* (in terms of *A* and *L*) so that a stagnation point occurs at $y = L$. **(c)** The streamline passing through the stagnation point would represent the line dividing the fresh water from the salt water. Plot this streamline.

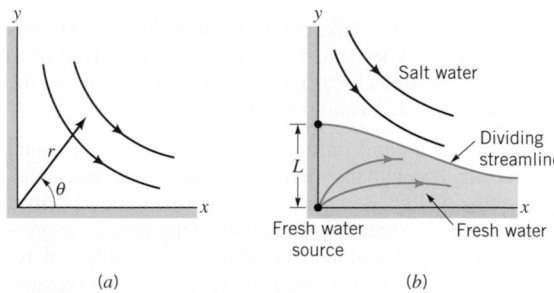

■ FIGURE P6.75

6.76 Typical inviscid flow solutions for flow around bodies indicate that the fluid flows smoothly around the body, even for blunt bodies as shown in **Video V6.10**. However, experience reveals that due to the presence of viscosity, the main flow may actually separate from the body creating a wake behind the body. As discussed in a later section (Section 9.2.6), whether or not separation takes place depends on the pressure gradient along the surface of the body, as calculated by inviscid flow theory. If the pressure decreases in the direction of flow (a *favorable* pressure gradient), no separation will occur. However, if the pressure increases in the direction of flow (an *adverse* pressure gradient), separation may occur. For the circular cylinder of Fig. P6.76 placed in a uniform stream with velocity, *U*,

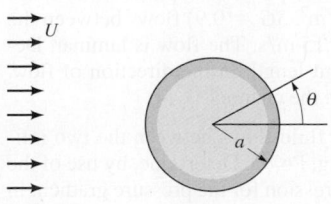

■ FIGURE P6.76

determine an expression for the pressure gradient in the direction of flow on the surface of the cylinder. For what range of values for the angle θ will an adverse pressure gradient occur?

Section 6.8 Viscous Flow

6.77 Obtain a photograph/image of a situation in which the cylindrical form of the Navier–Stokes equations would be appropriate for the solution. Print this photo and write a brief paragraph that describes the situation involved.

6.78 For a steady, two-dimensional, incompressible flow, the velocity is given by $\mathbf{V} = (ax - cy)\hat{\mathbf{i}} + (-ay + cx)\hat{\mathbf{j}}$, where a and c are constants. Show that this flow can be considered inviscid.

6.79 Determine the shearing stress for an incompressible Newtonian fluid with a velocity distribution of $\mathbf{V} = (3xy^2 - 4x^3)\hat{\mathbf{i}} + (12x^2y - y^3)\hat{\mathbf{j}}$.

6.80 The two-dimensional velocity field for an incompressible Newtonian fluid is described by the relationship

$$\mathbf{V} = (12xy^2 - 6x^3)\hat{\mathbf{i}} + (18x^2y - 4y^3)\hat{\mathbf{j}}$$

where the velocity has units of m/s when x and y are in meters. Determine the stresses σ_{xx}, σ_{yy}, and τ_{xy} at the point $x = 0.5$ m, $y = 1.0$ m if pressure at this point is 6 kPa and the fluid is glycerin at 20 °C. Show these stresses on a sketch.

6.81 For a two-dimensional incompressible flow in the $x - y$ plane show that the z component of the vorticity, ζ_z, varies in accordance with the equation

$$\frac{D\zeta_z}{Dt} = \nu\nabla^2\zeta_z$$

What is the physical interpretation of this equation for a nonviscous fluid? *Hint:* This *vorticity transport equation* can be derived from the Navier–Stokes equations by differentiating and eliminating the pressure between Eqs. 6.127a and 6.127b.

6.82 The velocity of a fluid particle moving along a horizontal streamline that coincides with the x axis in a plane, two-dimensional, incompressible flow field was experimentally found to be described by the equation $u = x^2$. Along this streamline determine an expression for (a) the rate of change of the v component of velocity with respect to y, (b) the acceleration of the particle, and (c) the pressure gradient in the x direction. The fluid is Newtonian.

Section 6.9.1 Steady, Laminar Flow between Fixed Parallel Plates

6.83 Obtain a photograph/image of a situation which can be approximated by one of the simple cases covered in Sec. 6.9. Print this photo and write a brief paragraph that describes the situation involved.

6.84 Oil ($\mu = 0.4$ N · s/m^2) flows between two fixed horizontal infinite parallel plates with a spacing of 5 mm. The flow is laminar and steady with a pressure gradient of -900 (N/m^2) per unit meter. Determine the volume flowrate per unit width and the shear stress on the upper plate.

6.85 Two fixed, horizontal, parallel plates are spaced 1 cm apart. A viscous liquid ($\mu = 0.4$ N · s/m^2, $SG = 0.9$) flows between the plates with a mean velocity of 0.15 m/s. The flow is laminar. Determine the pressure drop per unit length in the direction of flow. What is the maximum velocity in the channel?

6.86 A viscous, incompressible fluid flows between the two infinite, vertical, parallel plates of Fig. P6.86. Determine, by use of the Navier–Stokes equations, an expression for the pressure gradient in the direction of flow. Express your answer in terms of the mean velocity. Assume that the flow is laminar, steady, and uniform.

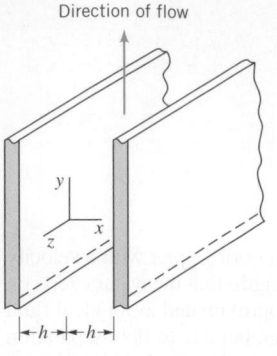

Direction of flow

y

x

z

$\leftarrow h \rightarrow \leftarrow h \rightarrow$

■ **F I G U R E   P6.86**

6.87 A fluid is initially at rest between two horizontal, infinite, parallel plates. A constant pressure gradient in a direction parallel to the plates is suddenly applied and the fluid starts to move. Determine the appropriate differential equation(s), initial condition, and boundary conditions that govern this type of flow. You need not solve the equation(s).

6.88 (See Fluids in the News article titled "10 tons on 8 psi," Section 6.9.1.) A massive, precisely machined, 1.8 m-diameter granite sphere rests upon a 1.2 m-diameter cylindrical pedestal as shown in Fig. P6.88. When the pump is turned on and the water pressure within the pedestal reaches 55 kPa, the sphere rises off the pedestal, creating a 1.3×10^{-2} cm gap through which the water flows. The sphere can then be rotated about any axis with minimal friction. (a) Estimate the pump flowrate, Q_0, required to accomplish this. Assume the flow in the gap between the sphere and the pedestal is essentially viscous flow between fixed, parallel plates. (b) Describe what would happen if the pump flowrate were increased to $2Q_0$.

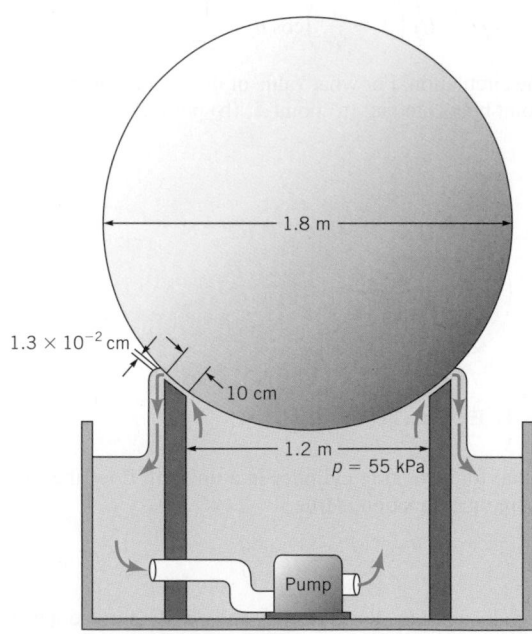

1.8 m

1.3×10^{-2} cm

10 cm

1.2 m

$p = 55$ kPa

Pump

■ **F I G U R E   P6.88**

Section 6.9.2 Couette Flow

6.89 Two horizontal, infinite, parallel plates are spaced a distance b apart. A viscous liquid is contained between the plates. The bottom plate is fixed, and the upper plate moves parallel to the bottom plate with a velocity U. Because of the no-slip boundary condition (see **Video V6.11**), the liquid motion is caused by the liquid being dragged along by the moving boundary. There is no pressure gradient in the direction of flow. Note that this is a so-called simple *Couette flow* discussed in Section 6.9.2. **(a)** Start with the Navier–Stokes equations and determine the velocity distribution between the plates. **(b)** Determine an expression for the flowrate passing between the plates (for a unit width). Express your answer in terms of b and U.

6.90 A layer of viscous liquid of constant thickness (no velocity perpendicular to plate) flows steadily down an infinite, inclined plane. Determine, by means of the Navier–Stokes equations, the relationship between the thickness of the layer and the discharge per unit width. The flow is laminar, and assume air resistance is negligible so that the shearing stress at the free surface is zero.

6.91 Due to the no-slip condition, as a solid is pulled out of a viscous liquid some of the liquid is also pulled along as described in Example 6.9 and shown in **Video V6.11**. Based on the results given in Example 6.9, show on a dimensionless plot the velocity distribution in the fluid film (v/V_0 vs. x/h) when the average film velocity, V, is 10% of the belt velocity, V_0.

6.92 An incompressible, viscous fluid is placed between horizontal, infinite, parallel plates as is shown in Fig. P6.92. The two plates move in opposite directions with constant velocities, U_1 and U_2, as shown. The pressure gradient in the x direction is zero, and the only body force is due to the fluid weight. Use the Navier–Stokes equations to derive an expression for the velocity distribution between the plates. Assume laminar flow.

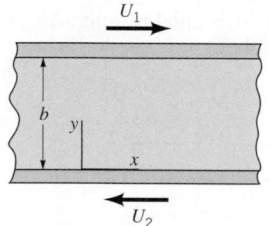

■ **F I G U R E   P6.92**

6.93 Two immiscible, incompressible, viscous fluids having the same densities but different viscosities are contained between two infinite, horizontal, parallel plates (Fig. P6.93). The bottom plate is fixed and the upper plate moves with a constant velocity U. Determine the velocity at the interface. Express your answer in terms of U, μ_1, and μ_2. The motion of the fluid is caused entirely by the movement of the upper plate; that is, there is no pressure gradient in the x direction. The fluid velocity and shearing stress are continuous across the interface between the two fluids. Assume laminar flow.

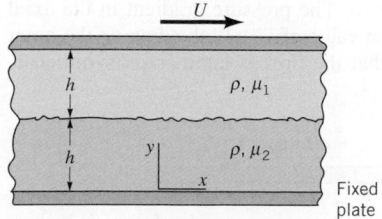

■ **F I G U R E   P6.93**

6.94 The viscous, incompressible flow between the parallel plates shown in Fig. P6.94 is caused by both the motion of the bottom plate and a pressure gradient, $\partial p/\partial x$. As noted in Section 6.9.2, an important dimensionless parameter for this type of problem is

$P = -(b^2/2\,\mu U)\,(\partial p/\partial x)$ where μ is the fluid viscosity. Make a plot of the dimensionless velocity distribution (similar to that shown in Fig. 6.32b) for $P = 3$. For this case where does the maximum velocity occur?

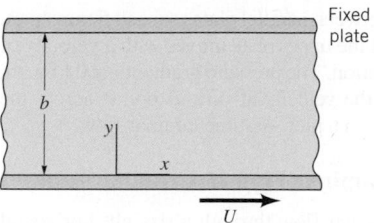

■ **F I G U R E   P6.94**

6.95 A viscous fluid (specific weight $= 1.26$ kN/m^3; viscosity $= 1.4$ N · s/m^2 is contained between two infinite, horizontal parallel plates as shown in Fig. P6.95. The fluid moves between the plates under the action of a pressure gradient, and the upper plate moves with a velocity U while the bottom plate is fixed. A U-tube manometer connected between two points along the bottom indicates a differential reading of 0.25 cm. If the upper plate moves with a velocity of 6×10^{-3} m/s, at what distance from the bottom plate does the maximum velocity in the gap between the two plates occur? Assume laminar flow.

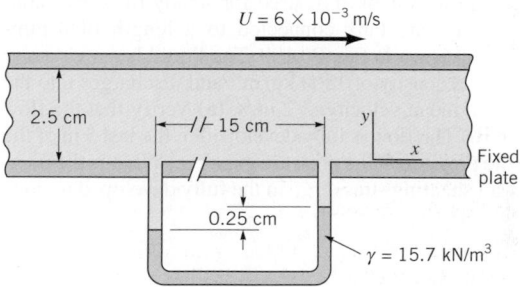

■ **F I G U R E   P6.95**

6.96 A vertical shaft passes through a bearing and is lubricated with an oil having a viscosity of 0.2 N · s/m^2 as shown in Fig. P6.96. Assume that the flow characteristics in the gap between the shaft and bearing are the same as those for laminar flow between infinite parallel plates with zero pressure gradient in the direction of flow. Estimate the torque required to overcome viscous resistance when the shaft is turning at 80 rev/min.

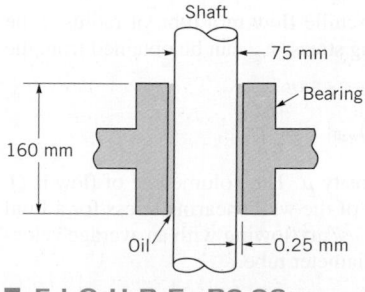

■ **F I G U R E   P6.96**

6.97 A viscous fluid is contained between two long concentric cylinders. The geometry of the system is such that the flow between the cylinders is approximately the same as the laminar flow between two infinite parallel plates. **(a)** Determine an expression for the torque required to rotate the outer cylinder with an angular velocity ω.

The inner cylinder is fixed. Express your answer in terms of the geometry of the system, the viscosity of the fluid, and the angular velocity. **(b)** For a small, rectangular element located at the fixed wall determine an expression for the rate of angular deformation of this element. (See Video V6.3 and Fig. P6.9.)

*6.98 Oil (SAE 30) flows between parallel plates spaced 5 mm apart. The bottom plate is fixed, but the upper plate moves with a velocity of 0.2 m/s in the positive x direction. The pressure gradient is 60 kPa/m, and it is negative. Compute the velocity at various points across the channel and show the results on a plot. Assume laminar flow.

Section 6.9.3 Steady, Laminar Flow in Circular Tubes

6.99 Consider a steady, laminar flow through a straight horizontal tube having the constant elliptical cross section given by the equation

$$\frac{x^2}{a^2} + \frac{y^2}{b^2} = 1$$

The streamlines are all straight and parallel. Investigate the possibility of using an equation for the z component of velocity of the form

$$w = A\left(1 - \frac{x^2}{a^2} - \frac{y^2}{b^2}\right)$$

as an exact solution to this problem. With this velocity distribution, what is the relationship between the pressure gradient along the tube and the volume flowrate through the tube?

6.100 A simple flow system to be used for steady flow tests consists of a constant head tank connected to a length of 4-mm-diameter tubing as shown in Fig. P6.100. The liquid has a viscosity of 0.015 N · s/m², a density of 1200 kg/m³, and discharges into the atmosphere with a mean velocity of 2 m/s. **(a)** Verify that the flow will be laminar. **(b)** The flow is fully developed in the last 3 m of the tube. What is the pressure at the pressure gage? **(c)** What is the magnitude of the wall shearing stress, τ_{rz}, in the fully developed region?

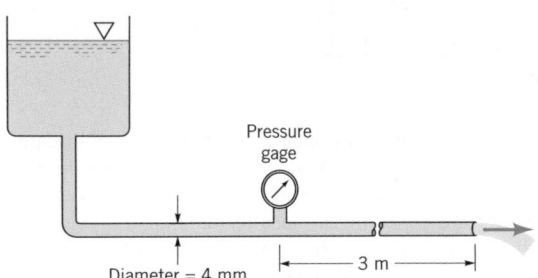

■ **FIGURE P6.100**

6.101 **(a)** Show that for Poiseuille flow in a tube of radius R the magnitude of the wall shearing stress, τ_{rz}, can be obtained from the relationship

$$|(\tau_{rz})_{\text{wall}}| = \frac{4\mu Q}{\pi R^3}$$

for a Newtonian fluid of viscosity μ. The volume rate of flow is Q. **(b)** Determine the magnitude of the wall shearing stress for a fluid having a viscosity of 0.004 N · s/m² flowing with an average velocity of 130 mm/s in a 2-mm-diameter tube.

6.102 An infinitely long, solid, vertical cylinder of radius R is located in an infinite mass of an incompressible fluid. Start with the Navier–Stokes equation in the θ direction and derive an expression for the velocity distribution for the steady flow case in which the cylinder is rotating about a fixed axis with a constant angular velocity ω. You need not consider body forces. Assume that the flow is axisymmetric and the fluid is at rest at infinity.

*6.103 As is shown by Eq. 6.150 the pressure gradient for laminar flow through a tube of constant radius is given by the expression

$$\frac{\partial p}{\partial z} = -\frac{8\mu Q}{\pi R^4}$$

For a tube whose radius is changing very gradually, such as the one illustrated in Fig. P6.103, it is expected that this equation can be used to approximate the pressure change along the tube if the actual radius, $R(z)$, is used at each cross section. The following measurements were obtained along a particular tube.

z/ℓ	0	0.1	0.2	0.3	0.4	0.5	0.6	0.7	0.8	0.9	1.0
$R(z)/R_o$	1.00	0.73	0.67	0.65	0.67	0.80	0.80	0.71	0.73	0.77	1.00

Compare the pressure drop over the length ℓ for this nonuniform tube with one having the constant radius R_o. *Hint:* To solve this problem you will need to numerically integrate the equation for the pressure gradient given above.

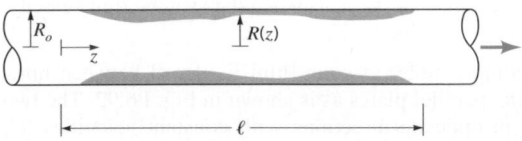

■ **FIGURE P6.103**

6.104 A liquid (viscosity = 0.002 N·s/m²; density = 1000 kg/m³) is forced through the circular tube shown in Fig. P6.104. A differential manometer is connected to the tube as shown to measure the pressure drop along the tube. When the differential reading, Δh, is 9 mm, what is the mean velocity in the tube?

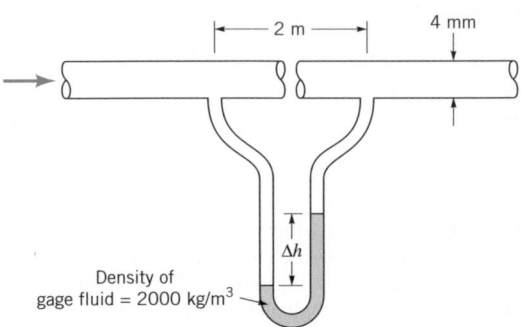

■ **FIGURE P6.104**

Section 6.9.4 Steady, Axial, Laminar Flow in an Annulus

6.105 An incompressible Newtonian fluid flows steadily between two infinitely long, concentric cylinders as shown in Fig. P6.105. The outer cylinder is fixed, but the inner cylinder moves with a longitudinal velocity V_0 as shown. The pressure gradient in the axial direction is $-\Delta p/\ell$. For what value of V_0 will the drag on the inner cylinder be zero? Assume that the flow is laminar, axisymmetric, and fully developed.

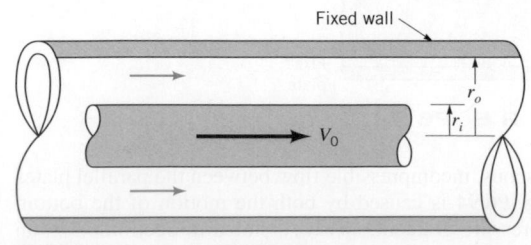

■ **FIGURE P6.105**

6.106 A viscous fluid is contained between two infinitely long, vertical, concentric cylinders. The outer cylinder has a radius r_o and rotates with an angular velocity ω. The inner cylinder is fixed and has a radius r_i. Make use of the Navier–Stokes equations to obtain an exact solution for the velocity distribution in the gap. Assume that the flow in the gap is axisymmetric (neither velocity nor pressure are functions of angular position θ within the gap) and that there are no velocity components other than the tangential component. The only body force is the weight.

6.107 For flow between concentric cylinders, with the outer cylinder rotating at an angular velocity ω and the inner cylinder fixed, it is commonly assumed that the tangential velocity (v_θ) distribution in the gap between the cylinders is linear. Based on the exact solution to this problem (see Problem 6.106) the velocity distribution in the gap is not linear. For an outer cylinder with radius $r_o = 5$ cm and an inner cylinder with radius $r_i = 4.5$ cm, show, with the aid of a plot, how the dimensionless velocity distribution, $v_\theta/r_o\omega$, varies with the dimensionless radial position, r/r_o, for the exact and approximate solutions.

6.108 A viscous liquid ($\mu = 0.6$ N $\cdot$ s/m^2, $\rho = 922.4$ kg/m^3) flows through the annular space between two horizontal, fixed, concentric cylinders. If the radius of the inner cylinder is 4 cm and the radius of the outer cylinder is 6.4 cm, what is the pressure drop along the axis of the annulus per meter when the volume flowrate is 4×10^{-3} m^3/s?

6.109 Show how Eq. 6.155 is obtained.

6.110 A wire of diameter d is stretched along the centerline of a pipe of diameter D. For a given pressure drop per unit length of pipe, by how much does the presence of the wire reduce the flowrate if **(a)** $d/D = 0.1$; **(b)** $d/D = 0.01$?

Section 6.10 **Other Aspects of Differential Analysis**

6.111 Obtain a photograph/image of a situation in which CFD has been used to solve a fluid flow problem. Print this photo and write a brief paragraph that describes the situation involved.

■ **Life Long Learning Problems**

6.112 What sometimes appear at first glance to be simple fluid flows can contain subtle, complex fluid mechanics. One such example is the stirring of tea leaves in a teacup. Obtain information about "Einstein's tea leaves" and investigate some of the complex fluid motions interacting with the leaves. Summarize your findings in a brief report.

6.113 Computational fluid dynamics (CFD) has moved from a research tool to a design tool for engineering. Initially, much of the work in CFD was focused in the aerospace industry, but now has expanded into other areas. Obtain information on what other industries (e.g., automotive) make use of CFD in their engineering design. Summarize your findings in a brief report.

■ **FE Exam Problems**

Sample FE (Fundamentals of Engineering) exam questions for fluid mechanics are provided on the book's web site, www.wiley.com/college/munson.

7 Dimensional Analysis, Similitude, and Modeling

Learning Objectives

After completing this chapter, you should be able to:

- apply the Buckingham pi theorem.
- develop a set of dimensionless variables for a given flow situation.
- discuss the use of dimensionless variables in data analysis.
- apply the concepts of modeling and similitude to develop prediction equations.

Experimentation and modeling are widely used techniques in fluid mechanics.

V7.1 Real and model flies

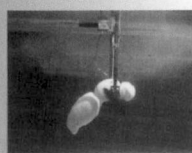

Although many practical engineering problems involving fluid mechanics can be solved by using the equations and analytical procedures described in the preceding chapters, there remain a large number of problems that rely on experimentally obtained data for their solution. In fact, it is probably fair to say that very few problems involving real fluids can be solved by analysis alone. The solution to many problems is achieved through the use of a combination of theoretical and numerical analysis and experimental data. Thus, engineers working on fluid mechanics problems should be familiar with the experimental approach to these problems so that they can interpret and make use of data obtained by others, such as might appear in handbooks, or be able to plan and execute the necessary experiments in their own laboratories. In this chapter we consider some techniques and ideas that are important in the planning and execution of experiments, as well as in understanding and correlating data that may have been obtained by other experimenters.

An obvious goal of any experiment is to make the results as widely applicable as possible. To achieve this end, the concept of *similitude* is often used so that measurements made on one system (for example, in the laboratory) can be used to describe the behavior of other similar systems (outside the laboratory). The laboratory systems are usually thought of as *models* and are used to study the phenomenon of interest under carefully controlled conditions. From these model studies, empirical formulations can be developed, or specific predictions of one or more characteristics of some other similar system can be made. To do this, it is necessary to establish the relationship between the laboratory model and the "other" system. In the following sections, we find out how this can be accomplished in a systematic manner.

F l u i d s   i n   t h e   N e w s

Model study of New Orleans levee breach caused by Hurricane Katrina Much of the devastation to New Orleans from Hurricane Katrina in 2005 was a result of flood waters that surged through a breach of the 17th Street Outfall Canal. To better understand why this occurred and to determine what can be done to prevent future occurrences, the U.S. Army Engineer Research and Development Center Coastal and Hydraulics Laboratory is conducting tests on a large (1:50 length scale) 15,000 square foot (1394 m²) hydraulic *model* that replicates 0.5 mile (0.8 km) of the canal surrounding the breach and more than a mile (or km) of the adjacent Lake Pontchartrain front. The objective of the study is to obtain information regarding the effect that waves had on the breaching of the canal and to investigate the surging water currents within the canals. The waves are generated by computer-controlled wave generators that can produce waves of varying heights, periods, and directions similar to the storm conditions that occurred during the hurricane. Data from the study will be used to calibrate and validate information that will be fed into various numerical model studies of the disaster.

7.1   Dimensional Analysis

To illustrate a typical fluid mechanics problem in which experimentation is required, consider the steady flow of an incompressible Newtonian fluid through a long, smooth-walled, horizontal, circular pipe. An important characteristic of this system, which would be of interest to an engineer designing a pipeline, is the pressure drop per unit length that develops along the pipe as a result of friction. Although this would appear to be a relatively simple flow problem, it cannot generally be solved analytically (even with the aid of large computers) without the use of experimental data.

The first step in the planning of an experiment to study this problem would be to decide on the factors, or variables, that will have an effect on the pressure drop per unit length, Δp_ℓ $[(\text{N/m}^2)/\text{m} = \text{N/m}^3]$. We expect the list to include the pipe diameter, D, the fluid density, ρ, fluid viscosity, μ, and the mean velocity, V, at which the fluid is flowing through the pipe. Thus, we can express this relationship as

$$\Delta p_\ell = f(D, \rho, \mu, V) \tag{7.1}$$

which simply indicates mathematically that we expect the pressure drop per unit length to be some function of the factors contained within the parentheses. At this point the nature of the function is unknown and the objective of the experiments to be performed is to determine the nature of this function.

It is important to develop a meaningful and systematic way to perform an experiment.

To perform the experiments in a meaningful and systematic manner, it would be necessary to change one of the variables, such as the velocity, while holding all others constant, and measure the corresponding pressure drop. This series of tests would yield data that could be represented graphically as is illustrated in Fig. 7.1a. It is to be noted that this plot would only be valid for the specific pipe and for the specific fluid used in the tests; this certainly does not give us the general formulation we are looking for. We could repeat the process by varying each of the other variables in turn, as is illustrated in Figs. 7.1b, 7.1c, and 7.1d. This approach to determining the functional relationship between the pressure drop and the various factors that influence it, although logical in concept, is fraught with difficulties. Some of the experiments would be hard to carry out—for example, to obtain the data illustrated in Fig. 7.1c it would be necessary to vary fluid density while holding viscosity constant. How would you do this? Finally, once we obtained the various curves shown in Figs. 7.1a, 7.1b, 7.1c, and 7.1d, how could we combine these data to obtain the desired general functional relationship between Δp_ℓ, D, ρ, μ, and V which would be valid for any similar pipe system?

Fortunately, there is a much simpler approach to this problem that will eliminate the difficulties described above. In the following sections we will show that rather than working with the

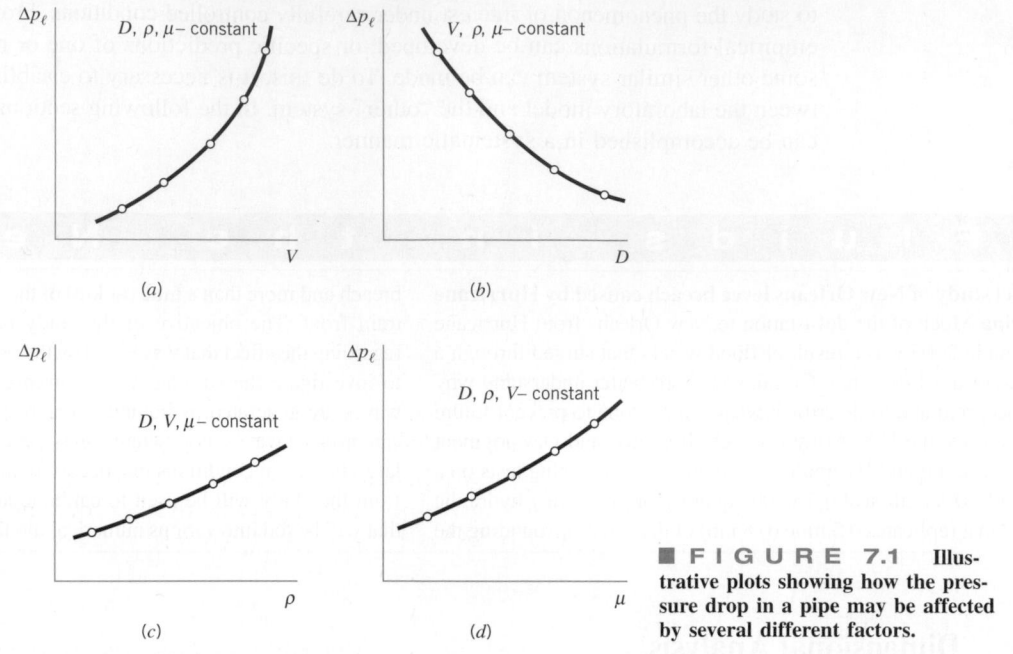

■ **FIGURE 7.1** Illustrative plots showing how the pressure drop in a pipe may be affected by several different factors.

Dimensionless products are important and useful in the planning, execution, and interpretation of experiments.

original list of variables, as described in Eq. 7.1, we can collect these into two nondimensional combinations of variables (called *dimensionless products* or *dimensionless groups*) so that

$$\frac{D\, \Delta p_\ell}{\rho V^2} = \phi\left(\frac{\rho V D}{\mu}\right)$$

(7.2)

Thus, instead of having to work with five variables, we now have only two. The necessary experiment would simply consist of varying the dimensionless product $\rho V D/\mu$ and determining the corresponding value of $D\, \Delta p_\ell/\rho V^2$. The results of the experiment could then be represented by a single, universal curve as is illustrated in Fig. 7.2. This curve would be valid for any combination of smooth-walled pipe and incompressible Newtonian fluid. To obtain this curve we could choose a pipe of convenient size and a fluid that is easy to work with. Note that we wouldn't have to use different pipe sizes or even different fluids. It is clear that the experiment would be much simpler, easier to do, and less expensive (which would certainly make an impression on your boss).

The basis for this simplification lies in a consideration of the dimensions of the variables involved. As was discussed in Chapter 1, a qualitative description of physical quantities can be given in terms of *basic dimensions* such as mass, M, length, L, and time, T.[1] Alternatively, we could use force, F, L, and T as basic dimensions, since from Newton's second law

$$F \doteq MLT^{-2}$$

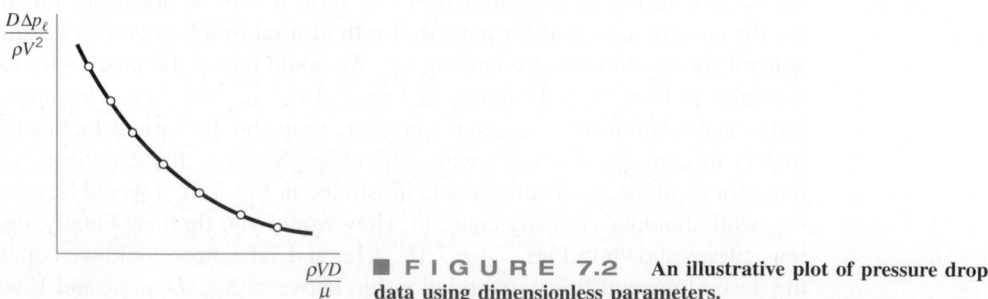

■ **FIGURE 7.2** An illustrative plot of pressure drop data using dimensionless parameters.

---

[1]As noted in Chapter 1, we will use T to represent the basic dimension of time, although T is also used for temperature in thermodynamic relationships (such as the ideal gas law).

(Recall from Chapter 1 that the notation $\doteq$ is used to indicate dimensional equality.) The dimensions of the variables in the pipe flow example are $\Delta p_\ell \doteq FL^{-3}$, $D \doteq L$, $\rho \doteq FL^{-4}T^2$, $\mu \doteq FL^{-2}T$, and $V \doteq LT^{-1}$. [Note that the pressure drop per unit length has the dimensions of $(F/L^2)/L = FL^{-3}$.] A quick check of the dimensions of the two groups that appear in Eq. 7.2 shows that they are in fact *dimensionless* products; that is,

$$\frac{D\,\Delta p_\ell}{\rho V^2} \doteq \frac{L(F/L^3)}{(FL^{-4}T^2)(LT^{-1})^2} \doteq F^0 L^0 T^0$$

and

$$\frac{\rho VD}{\mu} \doteq \frac{(FL^{-4}T^2)(LT^{-1})(L)}{(FL^{-2}T)} \doteq F^0 L^0 T^0$$

Not only have we reduced the number of variables from five to two, but the new groups are dimensionless combinations of variables, which means that the results presented in the form of Fig. 7.2 will be independent of the system of units we choose to use. This type of analysis is called *dimensional analysis*, and the basis for its application to a wide variety of problems is found in the *Buckingham pi theorem* described in the following section.

7.2   Buckingham Pi Theorem

A fundamental question we must answer is how many dimensionless products are required to replace the original list of variables? The answer to this question is supplied by the basic theorem of dimensional analysis that states the following:

> If an equation involving k variables is dimensionally homogeneous, it can be reduced to a relationship among $k - r$ independent dimensionless products, where r is the minimum number of reference dimensions required to describe the variables.

Dimensional analysis is based on the Buckingham pi theorem.

The dimensionless products are frequently referred to as "*pi terms*," and the theorem is called the *Buckingham pi theorem*.[2] Edgar Buckingham used the symbol Π to represent a dimensionless product, and this notation is commonly used. Although the pi theorem is a simple one, its proof is not so simple and we will not include it here. Many entire books have been devoted to the subject of similitude and dimensional analysis, and a number of these are listed at the end of this chapter (Refs. 1–15). Students interested in pursuing the subject in more depth (including the proof of the pi theorem) can refer to one of these books.

The pi theorem is based on the idea of dimensional homogeneity which was introduced in Chapter 1. Essentially we assume that for any physically meaningful equation involving k variables, such as

$$u_1 = f(u_2, u_3, \ldots, u_k)$$

the dimensions of the variable on the left side of the equal sign must be equal to the dimensions of any term that stands by itself on the right side of the equal sign. It then follows that we can rearrange the equation into a set of dimensionless products (pi terms) so that

$$\Pi_1 = \phi(\Pi_2, \Pi_3, \ldots, \Pi_{k-r})$$

where $\phi(\Pi_2, \Pi_3, \ldots, \Pi_{k-r})$ is a function of Π_2 through Π_{k-r}.

The required number of pi terms is fewer than the number of original variables by r, where r is determined by the minimum number of reference dimensions required to describe the original list of variables. Usually the reference dimensions required to describe the variables will be the basic dimensions M, L, and T or F, L, and T. However, in some instances perhaps only two dimensions, such as L and T, are required, or maybe just one, such as L. Also, in a few rare cases

---

[2]Although several early investigators, including Lord Rayleigh (1842–1919) in the nineteenth century, contributed to the development of dimensional analysis, Edgar Buckingham's (1867–1940) name is usually associated with the basic theorem. He stimulated interest in the subject in the United States through his publications during the early part of the twentieth century. See, for example, E. Buckingham, On Physically Similar Systems: Illustrations of the Use of Dimensional Equations, *Phys. Rev.*, 4 (1914), 345–376.

the variables may be described by some combination of basic dimensions, such as M/T^2 and L, and in this case r would be equal to two rather than three. Although the use of the pi theorem may appear to be a little mysterious and complicated, we will actually develop a simple, systematic procedure for developing the pi terms for a given problem.

7.3 Determination of Pi Terms

A dimensional analysis can be performed using a series of distinct steps.

Several methods can be used to form the dimensionless products, or pi terms, that arise in a dimensional analysis. Essentially we are looking for a method that will allow us to systematically form the pi terms so that we are sure that they are dimensionless and independent, and that we have the right number. The method we will describe in detail in this section is called the *method of repeating variables.*

It will be helpful to break the repeating variable method down into a series of distinct steps that can be followed for any given problem. With a little practice you will be able to readily complete a dimensional analysis for your problem.

Step 1 List all the variables that are involved in the problem. This step is the most difficult one and it is, of course, vitally important that all pertinent variables be included. Otherwise the dimensional analysis will not be correct! We are using the term "variable" to include any quantity, including dimensional and nondimensional constants, which play a role in the phenomenon under investigation. All such quantities should be included in the list of "variables" to be considered for the dimensional analysis. The determination of the variables must be accomplished by the experimenter's knowledge of the problem and the physical laws that govern the phenomenon. Typically the variables will include those that are necessary to describe the *geometry* of the system (such as a pipe diameter), to define any *fluid properties* (such as a fluid viscosity), and to indicate *external effects* that influence the system (such as a driving pressure drop per unit length). These general classes of variables are intended as broad categories that should be helpful in identifying variables. It is likely, however, that there will be variables that do not fit easily into one of these categories, and each problem needs to be carefully analyzed.

Since we wish to keep the number of variables to a minimum, so that we can minimize the amount of laboratory work, it is important that all variables be independent. For example, if in a certain problem the cross-sectional area of a pipe is an important variable, either the area or the pipe diameter could be used, but not both, since they are obviously not independent. Similarly, if both fluid density, ρ, and specific weight, γ, are important variables, we could list ρ and γ, or ρ and g (acceleration of gravity), or γ and g. However, it would be incorrect to use all three since $\gamma = \rho g$; that is, ρ, γ, and g are not independent. Note that although g would normally be constant in a given experiment, that fact is irrelevant as far as a dimensional analysis is concerned.

Step 2 Express each of the variables in terms of basic dimensions. For the typical fluid mechanics problem the basic dimensions will be either M, L, and T or F, L, and T. Dimensionally these two sets are related through Newton's second law ($\mathbf{F} = m\mathbf{a}$) so that $F \doteq MLT^{-2}$. For example, $\rho \doteq ML^{-3}$ or $\rho \doteq FL^{-4}T^2$. Thus, either set can be used. The basic dimensions for typical variables found in fluid mechanics problems are listed in Table 1.1 in Chapter 1.

Step 3 Determine the required number of pi terms. This can be accomplished by means of the Buckingham pi theorem, which indicates that the number of pi terms is equal to $k - r$, where k is the number of variables in the problem (which is determined from Step 1) and r is the number of reference dimensions required to describe these variables (which is determined from Step 2). The reference dimensions usually correspond to the basic dimensions and can be determined by an inspection of the dimensions of the variables obtained in Step 2. As previously noted, there may be occasions (usually rare) in which the basic dimensions appear in combinations so that the number of reference dimensions is less than the number of basic dimensions. This possibility is illustrated in Example 7.2.

Step 4 Select a number of repeating variables, where the number required is equal to the number of reference dimensions. Essentially what we are doing here is selecting from the original list of variables several of which can be combined with each of the remaining variables to form a pi term. All of the required reference dimensions must be included within the group of repeating variables, and each repeating variable must be dimensionally independent of the others (i.e., the dimensions of one repeating variable cannot be reproduced by some combination of products of powers of the remaining repeating variables). This means that the repeating variables cannot themselves be combined to form a dimensionless product.

For any given problem we usually are interested in determining how one particular variable is influenced by the other variables. We would consider this variable to be the dependent variable, and we would want this to appear in only one pi term. Thus, do *not* choose the dependent variable as one of the repeating variables, since the repeating variables will generally appear in more than one pi term.

Step 5   **Form a pi term by multiplying one of the nonrepeating variables by the product of the repeating variables, each raised to an exponent that will make the combination dimensionless.** Essentially each pi term will be of the form $u_i u_1^{a_i} u_2^{b_i} u_3^{c_i}$ where u_i is one of the nonrepeating variables; u_1, u_2, and u_3 are the repeating variables; and the exponents a_i, b_i, and c_i are determined so that the combination is dimensionless.

Step 6   **Repeat Step 5 for each of the remaining nonrepeating variables.** The resulting set of pi terms will correspond to the required number obtained from Step 3. If not, check your work—you have made a mistake!

Step 7   **Check all the resulting pi terms to make sure they are dimensionless.** It is easy to make a mistake in forming the pi terms. However, this can be checked by simply substituting the dimensions of the variables into the pi terms to confirm that they are all dimensionless. One good way to do this is to express the variables in terms of M, L, and T if the basic dimensions F, L, and T were used initially, or vice versa, and then check to make sure the pi terms are dimensionless.

Step 8   **Express the final form as a relationship among the pi terms, and think about what it means.** Typically the final form can be written as

$$\Pi_1 = \phi(\Pi_2, \Pi_3, \ldots, \Pi_{k-r})$$

where Π_1 would contain the dependent variable in the numerator. It should be emphasized that if you started out with the correct list of variables (and the other steps were completed correctly), then the relationship in terms of the pi terms can be used to describe the problem. You need only work with the pi terms—not with the individual variables. However, it should be clearly noted that this is as far as we can go with the dimensional analysis; that is, the actual functional relationship among the pi terms must be determined by experiment.

By using dimensional analysis, the original problem is simplified and defined with pi terms.

To illustrate these various steps we will again consider the problem discussed earlier in this chapter which was concerned with the steady flow of an incompressible Newtonian fluid through a long, smooth-walled, horizontal circular pipe. We are interested in the pressure drop per unit length, Δp_ℓ, along the pipe as illustrated by the figure in the margin. First (Step 1) we must list all of the pertinent variables that are involved based on the experimenter's knowledge of the problem. In this problem we assume that

$$\Delta p_\ell = f(D, \rho, \mu, V)$$

where D is the pipe diameter, ρ and μ are the fluid density and viscosity, respectively, and V is the mean velocity.

Next (Step 2) we express all the variables in terms of basic dimensions. Using F, L, and T as basic dimensions it follows that

$$\Delta p_\ell \doteq FL^{-3}$$
$$D \doteq L$$
$$\rho \doteq FL^{-4}T^2$$
$$\mu \doteq FL^{-2}T$$
$$V \doteq LT^{-1}$$

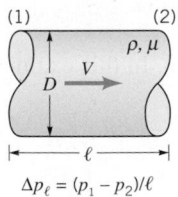

(1)   (2)

ρ, μ

D V

ℓ

$\Delta p_\ell = (p_1 - p_2)/\ell$

We could also use M, L, and T as basic dimensions if desired—the final result will be the same. Note that for density, which is a mass per unit volume (ML^{-3}), we have used the relationship $F \doteq MLT^{-2}$ to express the density in terms of F, L, and T. Do not mix the basic dimensions; that is, use either F, L, and T or M, L, and T.

We can now apply the pi theorem to determine the required number of pi terms (Step 3). An inspection of the dimensions of the variables from Step 2 reveals that all three basic dimensions are required to describe the variables. Since there are five ($k = 5$) variables (do not forget to count the dependent variable, Δp_ℓ) and three required reference dimensions ($r = 3$), then according to the pi theorem there will be ($5 - 3$), or two pi terms required.

The repeating variables to be used to form the pi terms (Step 4) need to be selected from the list D, ρ, μ, and V. Remember, we do not want to use the dependent variable as one of the repeating variables. Since three reference dimensions are required, we will need to select three repeating variables. Generally, we would try to select as repeating variables those that are the simplest, dimensionally. For example, if one of the variables has the dimension of a length, choose it as one of the repeating variables. In this example we will use D, V, and ρ as repeating variables. Note that these are dimensionally independent, since D is a length, V involves both length and time, and ρ involves force, length, and time. This means that we cannot form a dimensionless product from this set.

Special attention should be given to the selection of repeating variables as detailed in Step 4.

We are now ready to form the two pi terms (Step 5). Typically, we would start with the dependent variable and combine it with the repeating variables to form the first pi term; that is,

$$\Pi_1 = \Delta p_\ell D^a V^b \rho^c$$

Since this combination is to be dimensionless, it follows that

$$(FL^{-3})(L)^a(LT^{-1})^b(FL^{-4}T^2)^c \doteq F^0L^0T^0$$

The exponents, a, b, and c must be determined such that the resulting exponent for each of the basic dimensions—F, L, and T—must be zero (so that the resulting combination is dimensionless). Thus, we can write

$$1 + c = 0 \quad \text{(for } F\text{)}$$

$$-3 + a + b - 4c = 0 \quad \text{(for } L\text{)}$$

$$-b + 2c = 0 \quad \text{(for } T\text{)}$$

The solution of this system of algebraic equations gives the desired values for a, b, and c. It follows that $a = 1$, $b = -2$, $c = -1$ and, therefore,

$$\Pi_1 = \frac{\Delta p_\ell D}{\rho V^2}$$

The process is now repeated for the remaining nonrepeating variables (Step 6). In this example there is only one additional variable (μ) so that

$$\Pi_2 = \mu D^a V^b \rho^c$$

or

$$(FL^{-2}T)(L)^a(LT^{-1})^b(FL^{-4}T^2)^c \doteq F^0L^0T^0$$

and, therefore,

$$1 + c = 0 \quad \text{(for } F\text{)}$$

$$-2 + a + b - 4c = 0 \quad \text{(for } L\text{)}$$

$$1 - b + 2c = 0 \quad \text{(for } T\text{)}$$

Solving these equations simultaneously it follows that $a = -1$, $b = -1$, $c = -1$ so that

$$\Pi_2 = \frac{\mu}{DV\rho}$$

Note that we end up with the correct number of pi terms as determined from Step 3.

At this point stop and check to make sure the pi terms are actually dimensionless (Step 7). We will check using both *FLT* and *MLT* dimensions. Thus,

$$\Pi_1 = \frac{\Delta p_\ell D}{\rho V^2} \doteq \frac{(FL^{-3})(L)}{(FL^{-4}T^2)(LT^{-1})^2} \doteq F^0 L^0 T^0$$

$$\Pi_2 = \frac{\mu}{DV\rho} \doteq \frac{(FL^{-2}T)}{(L)(LT^{-1})(FL^{-4}T^2)} \doteq F^0 L^0 T^0$$

or alternatively,

$$\Pi_1 = \frac{\Delta p_\ell D}{\rho V^2} \doteq \frac{(ML^{-2}T^{-2})(L)}{(ML^{-3})(LT^{-1})^2} \doteq M^0 L^0 T^0$$

$$\Pi_2 = \frac{\mu}{DV\rho} \doteq \frac{(ML^{-1}T^{-1})}{(L)(LT^{-1})(ML^{-3})} \doteq M^0 L^0 T^0$$

Finally (Step 8), we can express the result of the dimensional analysis as

$$\frac{\Delta p_\ell D}{\rho V^2} = \tilde{\phi}\left(\frac{\mu}{DV\rho}\right)$$

This result indicates that this problem can be studied in terms of these two pi terms, rather than the original five variables we started with. The eight steps carried out to obtain this result are summarized by the figure in the margin.

Dimensional analysis will *not* provide the form of the function $\tilde{\phi}$. This can only be obtained from a suitable set of experiments. If desired, the pi terms can be rearranged; that is, the reciprocal of $\mu/DV\rho$ could be used, and of course the order in which we write the variables can be changed. Thus, for example, Π_2 could be expressed as

$$\Pi_2 = \frac{\rho VD}{\mu}$$

and the relationship between Π_1 and Π_2 as

$$\frac{D \, \Delta p_\ell}{\rho V^2} = \phi\left(\frac{\rho VD}{\mu}\right)$$

as shown by the figure in the margin.

This is the form we previously used in our initial discussion of this problem (Eq. 7.2). The dimensionless product $\rho VD/\mu$ is a very famous one in fluid mechanics—the Reynolds number. This number has been briefly alluded to in Chapters 1 and 6 and will be further discussed in Section 7.6.

To summarize, the steps to be followed in performing a dimensional analysis using the method of repeating variables are as follows:

Step 1 List all the variables that are involved in the problem.

Step 2 Express each of the variables in terms of basic dimensions.

Step 3 Determine the required number of pi terms.

Step 4 Select a number of repeating variables, where the number required is equal to the number of reference dimensions (usually the same as the number of basic dimensions).

Step 5 Form a pi term by multiplying one of the nonrepeating variables by the product of repeating variables each raised to an exponent that will make the combination dimensionless.

Step 6 Repeat Step 5 for each of the remaining nonrepeating variables.

Step 7 Check all the resulting pi terms to make sure they are dimensionless and independent.

Step 8 Express the final form as a relationship among the pi terms and think about what it means.

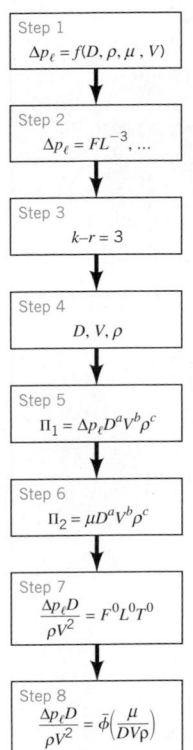

Step 1
$\Delta p_\ell = f(D, \rho, \mu, V)$

Step 2
$\Delta p_\ell = FL^{-3}, \ldots$

Step 3
$k-r = 3$

Step 4
D, V, ρ

Step 5
$\Pi_1 = \Delta p_\ell D^a V^b \rho^c$

Step 6
$\Pi_2 = \mu D^a V^b \rho^c$

Step 7
$\dfrac{\Delta p_\ell D}{\rho V^2} = F^0 L^0 T^0$

Step 8
$\dfrac{\Delta p_\ell D}{\rho V^2} = \tilde{\phi}\left(\dfrac{\mu}{DV\rho}\right)$

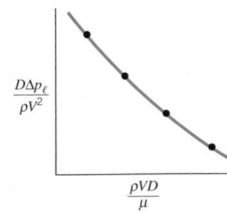

The method of repeating variables can be most easily carried out by following a step-by-step procedure.

EXAMPLE 7.1 Method of Repeating Variables

GIVEN A thin rectangular plate having a width w and a height h is located so that it is normal to a moving stream of fluid as shown in Fig. E7.1. Assume the drag, $\mathcal{D}$, that the fluid exerts on the plate is a function of w and h, the fluid viscosity and density, μ and ρ, respectively, and the velocity V of the fluid approaching the plate.

FIND Determine a suitable set of pi terms to study this problem experimentally.

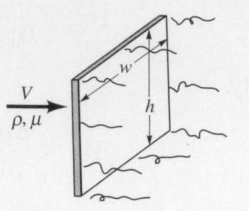

V7.2 Flow past a flat plate

SOLUTION

From the statement of the problem we can write

$$\mathcal{D} = f(w, h, \mu, \rho, V)$$

where this equation expresses the general functional relationship between the drag and the several variables that will affect it. The dimensions of the variables (using the *MLT* system) are

$$\mathcal{D} \doteq MLT^{-2}$$
$$w \doteq L$$
$$h \doteq L$$
$$\mu \doteq ML^{-1}T^{-1}$$
$$\rho \doteq ML^{-3}$$
$$V \doteq LT^{-1}$$

We see that all three basic dimensions are required to define the six variables so that the Buckingham pi theorem tells us that three pi terms will be needed (six variables minus three reference dimensions, $k - r = 6 - 3$).

We will next select three repeating variables such as w, V, and ρ. A quick inspection of these three reveals that they are dimensionally independent, since each one contains a basic dimension not included in the others. Note that it would be incorrect to use both w and h as repeating variables since they have the same dimensions.

Starting with the dependent variable, $\mathcal{D}$, the first pi term can be formed by combining $\mathcal{D}$ with the repeating variables such that

$$\Pi_1 = \mathcal{D}w^a V^b \rho^c$$

and in terms of dimensions

$$(MLT^{-2})(L)^a(LT^{-1})^b(ML^{-3})^c \doteq M^0L^0T^0$$

Thus, for Π_1 to be dimensionless it follows that

$$1 + c = 0 \qquad \text{(for } M\text{)}$$
$$1 + a + b - 3c = 0 \qquad \text{(for } L\text{)}$$
$$-2 - b = 0 \qquad \text{(for } T\text{)}$$

and, therefore, $a = -2$, $b = -2$, and $c = -1$. The pi term then becomes

$$\Pi_1 = \frac{\mathcal{D}}{w^2 V^2 \rho}$$

Next the procedure is repeated with the second nonrepeating variable, h, so that

$$\Pi_2 = hw^a V^b \rho^c$$

It follows that

$$(L)(L)^a(LT^{-1})^b(ML^{-3})^c \doteq M^0L^0T^0$$

and

$$c = 0 \qquad \text{(for } M\text{)}$$
$$1 + a + b - 3c = 0 \qquad \text{(for } L\text{)}$$
$$b = 0 \qquad \text{(for } T\text{)}$$

so that $a = -1$, $b = 0$, $c = 0$, and therefore

$$\Pi_2 = \frac{h}{w}$$

The remaining nonrepeating variable is μ so that

$$\Pi_3 = \mu w^a V^b \rho^c$$

with

$$(ML^{-1}T^{-1})(L)^a(LT^{-1})^b(ML^{-3})^c \doteq M^0L^0T^0$$

and, therefore,

$$1 + c = 0 \qquad \text{(for } M\text{)}$$
$$-1 + a + b - 3c = 0 \qquad \text{(for } L\text{)}$$
$$-1 - b = 0 \qquad \text{(for } T\text{)}$$

Solving for the exponents, we obtain $a = -1$, $b = -1$, $c = -1$ so that

$$\Pi_3 = \frac{\mu}{wV\rho}$$

Now that we have the three required pi terms we should check to make sure they are dimensionless. To make this check we use F, L, and T, which will also verify the correctness of the original dimensions used for the variables. Thus,

$$\Pi_1 = \frac{\mathcal{D}}{w^2 V^2 \rho} \doteq \frac{(F)}{(L)^2(LT^{-1})^2(FL^{-4}T^2)} \doteq F^0L^0T^0$$

$$\Pi_2 = \frac{h}{w} \doteq \frac{(L)}{(L)} \doteq F^0L^0T^0$$

$$\Pi_3 = \frac{\mu}{wV\rho} \doteq \frac{(FL^{-2}T)}{(L)(LT^{-1})(FL^{-4}T^2)} \doteq F^0L^0T^0$$

■ **FIGURE E7.1**

If these do not check, go back to the original list of variables and make sure you have the correct dimensions for each of the variables and then check the algebra you used to obtain the exponents a, b, and c.

Finally, we can express the results of the dimensional analysis in the form

$$\frac{\mathcal{D}}{w^2 V^2 \rho} = \tilde{\phi}\left(\frac{h}{w}, \frac{\mu}{wV\rho}\right) \qquad \text{(Ans)}$$

Since at this stage in the analysis the nature of the function $\tilde{\phi}$ is unknown, we could rearrange the pi terms if we so desire. For example, we could express the final result in the form

$$\frac{\mathcal{D}}{w^2 \rho V^2} = \phi\left(\frac{w}{h}, \frac{\rho V w}{\mu}\right) \qquad \text{(Ans)}$$

which would be more conventional, since the ratio of the plate width to height, w/h, is called the *aspect ratio*, and $\rho Vw/\mu$ is the Reynolds number.

COMMENT To proceed, it would be necessary to perform a set of experiments to determine the nature of the function ϕ, as discussed in Section 7.7.

7.4 Some Additional Comments about Dimensional Analysis

The preceding section provides a systematic approach for performing a dimensional analysis. Other methods could be used, although we think the method of repeating variables is the easiest for the beginning student to use. Pi terms can also be formed by inspection, as is discussed in Section 7.5. Regardless of the specific method used for the dimensional analysis, there are certain aspects of this important engineering tool that must seem a little baffling and mysterious to the student (and sometimes to the experienced investigator as well). In this section we will attempt to elaborate on some of the more subtle points that, based on our experience, can prove to be puzzling to students.

7.4.1 Selection of Variables

One of the most important, and difficult, steps in applying dimensional analysis to any given problem is the selection of the variables that are involved. As noted previously, for convenience we will use the term variable to indicate any quantity involved, including dimensional and nondimensional constants. There is no simple procedure whereby the variables can be easily identified. Generally, one must rely on a good understanding of the phenomenon involved and the governing physical laws. If extraneous variables are included, then too many pi terms appear in the final solution, and it may be difficult, time consuming, and expensive to eliminate these experimentally. If important variables are omitted, then an incorrect result will be obtained; and again, this may prove to be costly and difficult to ascertain. It is, therefore, imperative that sufficient time and attention be given to this first step in which the variables are determined.

Most engineering problems involve certain simplifying assumptions that have an influence on the variables to be considered. Usually we wish to keep the problem as simple as possible, perhaps even if some accuracy is sacrificed. A suitable balance between simplicity and accuracy is a desirable goal. How "accurate" the solution must be depends on the objective of the study; that is, we may be only concerned with general trends and, therefore, some variables that are thought to have only a minor influence in the problem may be neglected for simplicity.

It is often helpful to classify variables into three groups—geometry, material properties, and external effects.

For most engineering problems (including areas outside of fluid mechanics), pertinent variables can be classified into three general groups—geometry, material properties, and external effects.

Geometry. The geometric characteristics can usually be described by a series of lengths and angles. In most problems the geometry of the system plays an important role, and a sufficient number of geometric variables must be included to describe the system. These variables can usually be readily identified.

Material Properties. Since the response of a system to applied external effects such as forces, pressures, and changes in temperature is dependent on the nature of the materials involved in the system, the material properties that relate the external effects and the responses must be included as variables. For example, for Newtonian fluids the viscosity of the fluid is the property that relates the applied forces to the rates of deformation of the fluid. As the material behavior becomes more complex, such as would be true for non-Newtonian fluids, the determination of material properties becomes difficult, and this class of variables can be troublesome to identify.

External Effects. This terminology is used to denote any variable that produces, or tends to produce, a change in the system. For example, in structural mechanics, forces (either concentrated or distributed) applied to a system tend to change its geometry, and such forces would need to be considered as pertinent variables. For fluid mechanics, variables in this class would be related to pressures, velocities, or gravity.

The above general classes of variables are intended as broad categories that should be helpful in identifying variables. It is likely, however, that there will be important variables that do not fit easily into one of the above categories and each problem needs to be carefully analyzed.

Since we wish to keep the number of variables to a minimum, it is important that all variables are independent. For example, if in a given problem we know that the moment of inertia of the area of a circular plate is an important variable, we could list either the moment of inertia or the plate diameter as the pertinent variable. However, it would be unnecessary to include both moment of inertia and diameter, assuming that the diameter enters the problem only through the moment of inertia. In more general terms, if we have a problem in which the variables are

$$f(p, q, r, \ldots, u, v, w, \ldots) = 0 \tag{7.3}$$

and it is known that there is an additional relationship among some of the variables, for example,

$$q = f_1(u, v, w, \ldots) \tag{7.4}$$

then q is not required and can be omitted. Conversely, if it is known that the only way the variables $u, v, w, \ldots$ enter the problem is through the relationship expressed by Eq. 7.4, then the variables $u, v, w, \ldots$ can be replaced by the single variable q, therefore reducing the number of variables.

In summary, the following points should be considered in the selection of variables:

1. Clearly define the problem. What is the main variable of interest (the dependent variable)?
2. Consider the basic laws that govern the phenomenon. Even a crude theory that describes the essential aspects of the system may be helpful.
3. Start the variable selection process by grouping the variables into three broad classes: geometry, material properties, and external effects.
4. Consider other variables that may not fall into one of the above categories. For example, time will be an important variable if any of the variables are time dependent.
5. Be sure to include all quantities that enter the problem even though some of them may be held constant (e.g., the acceleration of gravity, g). For a dimensional analysis it is the dimensions of the quantities that are important—not specific values!
6. Make sure that all variables are independent. Look for relationships among subsets of the variables.

7.4.2 Determination of Reference Dimensions

For any given problem it is obviously desirable to reduce the number of pi terms to a minimum and, therefore, we wish to reduce the number of variables to a minimum; that is, we certainly do not want to include extraneous variables. It is also important to know how many reference dimensions are required to describe the variables. As we have seen in the preceding examples, F, L, and T appear to be a convenient set of basic dimensions for characterizing fluid-mechanical quantities. There is, however, really nothing "fundamental" about this set, and as previously noted M, L, and T would also be suitable. Actually any set of measurable quantities could be used as basic dimensions provided that the selected combination can be used to describe all secondary quantities. However, the use of *FLT* or *MLT* as basic dimensions is the simplest, and these dimensions can be used to describe fluid-mechanical phenomena. Of course, in some problems only one or two of these are required. In addition, we occasionally find that the number of reference dimensions needed to describe all variables is smaller than the number of basic dimensions. This point is illustrated in Example 7.2. Interesting discussions, both practical and philosophical, relative to the concept of basic dimensions can be found in the books by Huntley (Ref. 4) and by Isaacson and Isaacson (Ref. 12).

Typically, in fluid mechanics, the required number of reference dimensions is three, but in some problems only one or two are required.

EXAMPLE 7.2 | Determination of Pi Terms

GIVEN An open, cylindrical paint can having a diameter D is filled to a depth h with paint having a specific weight γ. The vertical deflection, δ, of the center of the bottom is a function of D, h, d, γ, and E, where d is the thickness of the bottom and E is the modulus of elasticity of the bottom material.

FIND Determine the functional relationship between the vertical deflection, δ, and the independent variables using dimensional analysis.

SOLUTION

From the statement of the problem

$$\delta = f(D, h, d, \gamma, E)$$

and the dimensions of the variables are

$$\delta \doteq L$$
$$D \doteq L$$
$$h \doteq L$$
$$d \doteq L$$
$$\gamma \doteq FL^{-3} \doteq ML^{-2}T^{-2}$$
$$E \doteq FL^{-2} \doteq ML^{-1}T^{-2}$$

where the dimensions have been expressed in terms of both the *FLT* and *MLT* systems.

We now apply the pi theorem to determine the required number of pi terms. First, let us use F, L, and T as our system of basic dimensions. There are six variables and two reference dimensions (F and L) required so that four pi terms are needed. For repeating variables, we can select D and γ so that

$$\Pi_1 = \delta \, D^a \gamma^b$$
$$(L)(L)^a(FL^{-3})^b \doteq F^0 L^0$$

and

$$1 + a - 3b = 0 \quad \text{(for } L)$$
$$b = 0 \quad \text{(for } F)$$

Therefore, $a = -1$, $b = 0$, and

$$\Pi_1 = \frac{\delta}{D}$$

Similarly,

$$\Pi_2 = h \, D^a \gamma^b$$

and following the same procedure as above, $a = -1$, $b = 0$ so that

$$\Pi_2 = \frac{h}{D}$$

The remaining two pi terms can be found using the same procedure, with the result

$$\Pi_3 = \frac{d}{D} \qquad \Pi_4 = \frac{E}{D\gamma}$$

■ **FIGURE E7.2**

Thus, this problem can be studied by using the relationship

$$\frac{\delta}{D} = \phi\left(\frac{h}{D}, \frac{d}{D}, \frac{E}{D\gamma}\right) \qquad \text{(Ans)}$$

COMMENTS Let us now solve the same problem using the *MLT* system. Although the number of variables is obviously the same, it would seem that there are three reference dimensions required, rather than two. If this were indeed true it would certainly be fortuitous, since we would reduce the number of required pi terms from four to three. Does this seem right? How can we reduce the number of required pi terms by simply using the *MLT* system of basic dimensions? The answer is that we cannot, and a closer look at the dimensions of the variables listed above reveals that actually only two reference dimensions, MT^{-2} and L, are required.

This is an example of the situation in which the number of reference dimensions differs from the number of basic dimensions. It does not happen very often and can be detected by looking at the dimensions of the variables (regardless of the systems used) and making sure how many reference dimensions are actually required to describe the variables. Once the number of reference dimensions has been determined, we can proceed as before. Since the number of repeating variables must equal the number of reference dimensions, it follows that two reference dimensions are still required and we could again use D and γ as repeating variables. The pi terms would be determined in the same manner. For example, the pi term containing E would be developed as

$$\Pi_4 = ED^a\gamma^b$$
$$(ML^{-1}T^{-2})(L)^a(ML^{-2}T^{-2})^b \doteq (MT^{-2})^0 L^0$$
$$1 + b = 0 \quad \text{(for } MT^{-2})$$
$$-1 + a - 2b = 0 \quad \text{(for } L)$$

and, therefore, $a = -1$, $b = -1$ so that

$$\Pi_4 = \frac{E}{D\gamma}$$

which is the same as Π_4 obtained using the *FLT* system. The other pi terms would be the same, and the final result is the same; that is,

$$\frac{\delta}{D} = \phi\left(\frac{h}{D}, \frac{d}{D}, \frac{E}{D\gamma}\right) \qquad \text{(Ans)}$$

This will always be true—you cannot affect the required number of pi terms by using M, L, and T instead of F, L, and T, or vice versa.

7.4.3 Uniqueness of Pi Terms

A little reflection on the process used to determine pi terms by the method of repeating variables reveals that the specific pi terms obtained depend on the somewhat arbitrary selection of repeating variables. For example, in the problem of studying the pressure drop in a pipe, we selected D, V, and ρ as repeating variables. This led to the formulation of the problem in terms of pi terms as

$$\frac{\Delta p_\ell D}{\rho V^2} = \phi\left(\frac{\rho VD}{\mu}\right) \qquad (7.5)$$

What if we had selected D, V, and μ as repeating variables? A quick check will reveal that the pi term involving Δp_ℓ becomes

$$\frac{\Delta p_\ell D^2}{V\mu}$$

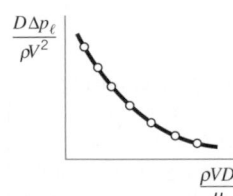

and the second pi term remains the same. Thus, we can express the final result as

$$\frac{\Delta p_\ell D^2}{V\mu} = \phi_1\left(\frac{\rho VD}{\mu}\right) \qquad (7.6)$$

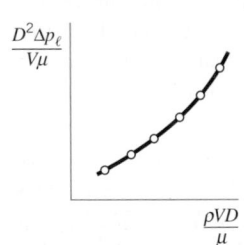

Both results are correct, and both would lead to the same final equation for Δp_ℓ. Note, however, that the functions ϕ and ϕ_1 in Eqs. 7.5 and 7.6 will be different because the dependent pi terms are different for the two relationships. As shown by the figure in the margin, the resulting graph of dimensionless data will be different for the two formulations. However, when extracting the physical variable, Δp_ℓ, from the two results, the values will be the same.

We can conclude from this illustration that there is *not* a unique set of pi terms which arises from a dimensional analysis. However, the required *number* of pi terms is fixed, and once a correct set is determined, all other possible sets can be developed from this set by combinations of products of powers of the original set. Thus, if we have a problem involving, say, three pi terms,

Once a correct set of pi terms is obtained, any other set can be obtained by manipulation of the original set.

$$\Pi_1 = \phi(\Pi_2, \Pi_3)$$

we could always form a new set from this one by combining the pi terms. For example, we could form a new pi term, Π_2', by letting

$$\Pi_2' = \Pi_2^a \, \Pi_3^b$$

where a and b are arbitrary exponents. Then the relationship could be expressed as

$$\Pi_1 = \phi_1(\Pi_2', \Pi_3)$$

or

$$\Pi_1 = \phi_2(\Pi_2, \Pi_2')$$

All of these would be correct. It should be emphasized, however, that the required number of pi terms cannot be reduced by this manipulation; only the form of the pi terms is altered. By using this technique we see that the pi terms in Eq. 7.6 could be obtained from those in Eq. 7.5; that is, we multiply Π_1 in Eq. 7.5 by Π_2 so that

$$\left(\frac{\Delta p_\ell D}{\rho V^2}\right)\left(\frac{\rho V D}{\mu}\right) = \frac{\Delta p_\ell D^2}{V\mu}$$

which is the Π_1 of Eq. 7.6.

There is no simple answer to the question: Which form for the pi terms is best? Usually our only guideline is to keep the pi terms as simple as possible. Also, it may be that certain pi terms will be easier to work with in actually performing experiments. The final choice remains an arbitrary one and generally will depend on the background and experience of the investigator. It should again be emphasized, however, that although there is no unique set of pi terms for a given problem, the *number* required is fixed in accordance with the pi theorem.

7.5   Determination of Pi Terms by Inspection

The method of repeating variables for forming pi terms has been presented in Section 7.3. This method provides a step-by-step procedure that if executed properly will provide a correct and complete set of pi terms. Although this method is simple and straightforward, it is rather tedious, particularly for problems in which large numbers of variables are involved. Since the only restrictions placed on the pi terms are that they be (1) correct in number, (2) dimensionless, and (3) independent, it is possible to simply form the pi terms by inspection, without resorting to the more formal procedure.

To illustrate this approach, we again consider the pressure drop per unit length along a smooth pipe. Regardless of the technique to be used, the starting point remains the same—determine the variables, which in this case are

$$\Delta p_\ell = f(D, \rho, \mu, V)$$

Next, the dimensions of the variables are listed:

$$\Delta p_\ell \doteq FL^{-3}$$

$$D \doteq L$$

$$\rho \doteq FL^{-4}T^2$$

$$\mu \doteq FL^{-2}T$$

$$V \doteq LT^{-1}$$

and subsequently the number of reference dimensions determined. The application of the pi theorem then tells us how many pi terms are required. In this problem, since there are five variables and three reference dimensions, two pi terms are needed. Thus, the required number of pi terms can be easily obtained. The determination of this number should always be done at the beginning of the analysis.

Pi terms can be formed by inspection by simply making use of the fact that each pi term must be dimensionless.

Once the number of pi terms is known, we can form each pi term by inspection, simply making use of the fact that each pi term must be dimensionless. We will always let Π_1 contain the dependent variable, which in this example is Δp_ℓ. Since this variable has the dimensions FL^{-3}, we need to combine it with other variables so that a nondimensional product will result. One possibility is to first divide Δp_ℓ by ρ so that

$$\frac{\Delta p_\ell}{\rho} \doteq \frac{(FL^{-3})}{(FL^{-4}T^2)} \doteq \frac{L}{T^2} \quad \text{(cancels } F\text{)}$$

The dependence on F has been eliminated, but $\Delta p_\ell/\rho$ is obviously not dimensionless. To eliminate the dependence on T, we can divide by V^2 so that

$$\left(\frac{\Delta p_\ell}{\rho}\right)\frac{1}{V^2} \doteq \left(\frac{L}{T^2}\right)\frac{1}{(LT^{-1})^2} \doteq \frac{1}{L} \quad \text{(cancels } T\text{)}$$

Finally, to make the combination dimensionless we multiply by D so that

$$\left(\frac{\Delta p_\ell}{\rho V^2}\right) D \doteq \left(\frac{1}{L}\right)(L) \doteq L^0 \qquad \text{(cancels } L\text{)}$$

Thus,

$$\Pi_1 = \frac{\Delta p_\ell D}{\rho V^2}$$

Next, we will form the second pi term by selecting the variable that was not used in Π_1, which in this case is μ. We simply combine μ with the other variables to make the combination dimensionless (but do not use Δp_ℓ in Π_2, since we want the dependent variable to appear only in Π_1). For example, divide μ by ρ (to eliminate F), then by V (to eliminate T), and finally by D (to eliminate L). Thus,

$$\Pi_2 = \frac{\mu}{\rho V D} \doteq \frac{(FL^{-2}T)}{(FL^{-4}T^2)(LT^{-1})(L)} \doteq F^0 L^0 T^0$$

and, therefore,

$$\frac{\Delta p_\ell D}{\rho V^2} = \phi\left(\frac{\mu}{\rho V D}\right)$$

which is, of course, the same result we obtained by using the method of repeating variables.

An additional concern, when one is forming pi terms by inspection, is to make certain that they are all independent. In the pipe flow example, Π_2 contains μ, which does not appear in Π_1, and therefore these two pi terms are obviously independent. In a more general case a pi term would not be independent of the others in a given problem if it can be formed by some combination of the others. For example, if Π_2 can be formed by a combination of say Π_3, Π_4, and Π_5 such as

$$\Pi_2 = \frac{\Pi_3^2 \, \Pi_4}{\Pi_5}$$

then Π_2 is not an independent pi term. We can ensure that each pi term is independent of those preceding it by incorporating a new variable in each pi term.

Although forming pi terms by inspection is essentially equivalent to the repeating variable method, it is less structured. With a little practice the pi terms can be readily formed by inspection, and this method offers an alternative to more formal procedures.

7.6 Common Dimensionless Groups in Fluid Mechanics

At the top of Table 7.1 is a list of variables that commonly arise in fluid mechanics problems. The list is obviously not exhaustive but does indicate a broad range of variables likely to be found in a typical problem. Fortunately, not all of these variables would be encountered in all problems. However, when combinations of these variables are present, it is standard practice to combine them into some of the common dimensionless groups (pi terms) given in Table 7.1. These combinations appear so frequently that special names are associated with them, as indicated in the table.

A useful physical interpretation can often be given to dimensionless groups.

It is also often possible to provide a physical interpretation to the dimensionless groups which can be helpful in assessing their influence in a particular application. For example, the Froude number is an index of the ratio of the force due to the acceleration of a fluid particle to the force due to gravity (weight). This can be demonstrated by considering a fluid particle moving along a streamline (Fig. 7.3). The magnitude of the component of inertia force F_I along the streamline can be expressed as $F_I = a_s m$, where a_s is the magnitude of the acceleration along the streamline for a particle having a mass m. From our study of particle motion along a curved path (see Section 3.1) we know that

$$a_s = \frac{dV_s}{dt} = V_s \frac{dV_s}{ds}$$

■ **TABLE 7.1**
Some Common Variables and Dimensionless Groups in Fluid Mechanics

Variables: Acceleration of gravity, g; Bulk modulus, E_v; Characteristic length, ℓ; Density, ρ; Frequency of oscillating flow, ω; Pressure, p (or Δp); Speed of sound, c; Surface tension, σ; Velocity, V; Viscosity, μ

Dimensionless Groups	Name	Interpretation (Index of Force Ratio Indicated)	Types of Applications
$\dfrac{\rho V \ell}{\mu}$	Reynolds number, Re	$\dfrac{\text{inertia force}}{\text{viscous force}}$	Generally of importance in all types of fluid dynamics problems
$\dfrac{V}{\sqrt{g\ell}}$	Froude number, Fr	$\dfrac{\text{inertia force}}{\text{gravitational force}}$	Flow with a free surface
$\dfrac{p}{\rho V^2}$	Euler number, Eu	$\dfrac{\text{pressure force}}{\text{inertia force}}$	Problems in which pressure, or pressure differences, are of interest
$\dfrac{\rho V^2}{E_v}$	Cauchy number,[a] Ca	$\dfrac{\text{inertia force}}{\text{compressibility force}}$	Flows in which the compressibility of the fluid is important
$\dfrac{V}{c}$	Mach number,[a] Ma	$\dfrac{\text{inertia force}}{\text{compressibility force}}$	Flows in which the compressibility of the fluid is important
$\dfrac{\omega \ell}{V}$	Strouhal number, St	$\dfrac{\text{inertia (local) force}}{\text{inertia (convective) force}}$	Unsteady flow with a characteristic frequency of oscillation
$\dfrac{\rho V^2 \ell}{\sigma}$	Weber number, We	$\dfrac{\text{inertia force}}{\text{surface tension force}}$	Problems in which surface tension is important

[a]The Cauchy number and the Mach number are related and either can be used as an index of the relative effects of inertia and compressibility. See accompanying discussion.

Special names along with physical interpretations are given to the most common dimensionless groups.

where s is measured along the streamline. If we write the velocity, V_s, and length, s, in dimensionless form, that is,

$$V_s^* = \frac{V_s}{V} \qquad s^* = \frac{s}{\ell}$$

where V and ℓ represent some characteristic velocity and length, respectively, then

$$a_s = \frac{V^2}{\ell} V_s^* \frac{dV_s^*}{ds^*}$$

and

$$F_I = \frac{V^2}{\ell} V_s^* \frac{dV_s^*}{ds^*} m$$

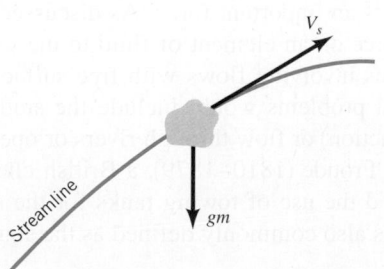

■ **FIGURE 7.3**   **The force of gravity acting on a fluid particle moving along a streamline.**

The magnitude of the weight of the particle, F_G, is $F_G = gm$, so the ratio of the inertia to the gravitational force is

$$\frac{F_I}{F_G} = \frac{V^2}{g\ell} V_s^* \frac{dV_s^*}{ds^*}$$

Thus, the force ratio F_I/F_G is proportional to $V^2/g\ell$, and the square root of this ratio, $V/\sqrt{g\ell}$, is called the *Froude number*. We see that a physical interpretation of the Froude number is that it is a measure of, or an index of, the relative importance of inertial forces acting on fluid particles to the weight of the particle. Note that the Froude number is not really *equal* to this force ratio, but is simply some type of average measure of the influence of these two forces. In a problem in which gravity (or weight) is not important, the Froude number would not appear as an important pi term. A similar interpretation in terms of indices of force ratios can be given to the other dimensionless groups, as indicated in Table 7.1, and a further discussion of the basis for this type of interpretation is given in the last section in this chapter. Some additional details about these important dimensionless groups are given below, and the types of application or problem in which they arise are briefly noted in the last column of Table 7.1.

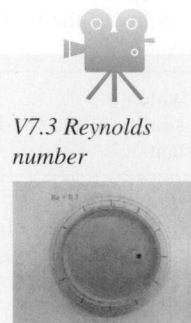

V7.3 Reynolds number

Reynolds Number. The Reynolds number is undoubtedly the most famous dimensionless parameter in fluid mechanics. It is named in honor of Osborne Reynolds (1842–1912), a British engineer who first demonstrated that this combination of variables could be used as a criterion to distinguish between laminar and turbulent flow. In most fluid flow problems there will be a characteristic length, ℓ, and a velocity, V, as well as the fluid properties of density, ρ, and viscosity, μ, which are relevant variables in the problem. Thus, with these variables the Reynolds number

$$Re = \frac{\rho V \ell}{\mu}$$

arises naturally from the dimensional analysis. The Reynolds number is a measure of the ratio of the inertia force on an element of fluid to the viscous force on an element. When these two types of forces are important in a given problem, the Reynolds number will play an important role. However, if the Reynolds number is very small ($Re \ll 1$), this is an indication that the viscous forces are dominant in the problem, and it may be possible to neglect the inertial effects; that is, the density of the fluid will not be an important variable. Flows at very small Reynolds numbers are commonly referred to as "creeping flows" as discussed in Section 6.10. Conversely, for large Reynolds number flows, viscous effects are small relative to inertial effects and for these cases it may be possible to neglect the effect of viscosity and consider the problem as one involving a "nonviscous" fluid. This type of problem is considered in detail in Sections 6.4 through 6.7. An example of the importance of the Reynolds number in determining the flow physics is shown in the figure in the margin for flow past a circular cylinder at two different Re values. This flow is discussed further in Chapter 9.

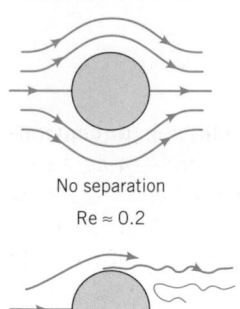

No separation
Re ≈ 0.2

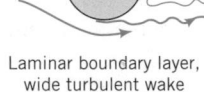

Laminar boundary layer,
wide turbulent wake
Re ≈ 20,000

Froude Number. The Froude number

$$Fr = \frac{V}{\sqrt{g\ell}}$$

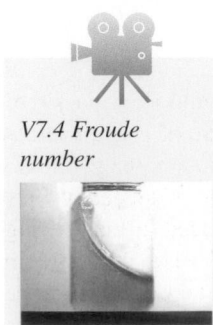

V7.4 Froude number

is distinguished from the other dimensionless groups in Table 7.1 in that it contains the acceleration of gravity, g. The acceleration of gravity becomes an important variable in a fluid dynamics problem in which the fluid weight is an important force. As discussed, the Froude number is a measure of the ratio of the inertia force on an element of fluid to the weight of the element. It will generally be important in problems involving flows with free surfaces since gravity principally affects this type of flow. Typical problems would include the study of the flow of water around ships (with the resulting wave action) or flow through rivers or open conduits. The Froude number is named in honor of William Froude (1810–1879), a British civil engineer, mathematician, and naval architect who pioneered the use of towing tanks for the study of ship design. It is to be noted that the Froude number is also commonly defined as the square of the Froude number listed in Table 7.1.

Euler Number.   The Euler number

$$\text{Eu} = \frac{p}{\rho V^2}$$

can be interpreted as a measure of the ratio of pressure forces to inertial forces, where p is some characteristic pressure in the flow field. Very often the Euler number is written in terms of a pressure difference, Δp, so that $\text{Eu} = \Delta p / \rho V^2$. Also, this combination expressed as $\Delta p / \frac{1}{2} \rho V^2$ is called the *pressure coefficient*. Some form of the Euler number would normally be used in problems in which pressure or the pressure difference between two points is an important variable. The Euler number is named in honor of Leonhard Euler (1707–1783), a famous Swiss mathematician who pioneered work on the relationship between pressure and flow. For problems in which cavitation is of concern, the dimensionless group $(p_r - p_v) / \frac{1}{2} \rho V^2$ is commonly used, where p_v is the vapor pressure and p_r is some reference pressure. Although this dimensionless group has the same form as the Euler number, it is generally referred to as the *cavitation number*.

Cauchy Number and Mach Number.   The Cauchy number

$$\text{Ca} = \frac{\rho V^2}{E_v}$$

The Mach number is a commonly used dimensionless parameter in compressible flow problems.

and the Mach number

$$\text{Ma} = \frac{V}{c}$$

are important dimensionless groups in problems in which fluid compressibility is a significant factor. Since the speed of sound, c, in a fluid is equal to $c = \sqrt{E_v / \rho}$ (see Section 1.7.3), it follows that

$$\text{Ma} = V \sqrt{\frac{\rho}{E_v}}$$

and the square of the Mach number

$$\text{Ma}^2 = \frac{\rho V^2}{E_v} = \text{Ca}$$

is equal to the Cauchy number. Thus, either number (but not both) may be used in problems in which fluid compressibility is important. Both numbers can be interpreted as representing an index of the ratio of inertial forces to compressibility forces. When the Mach number is relatively small (say, less than 0.3), the inertial forces induced by the fluid motion are not sufficiently large to cause a significant change in the fluid density, and in this case the compressibility of the fluid can be neglected. The Mach number is the more commonly used parameter in compressible flow problems, particularly in the fields of gas dynamics and aerodynamics. The Cauchy number is named in honor of Augustin Louis de Cauchy (1789–1857), a French engineer, mathematician, and hydrodynamicist. The Mach number is named in honor of Ernst Mach (1838–1916), an Austrian physicist and philosopher.

Strouhal Number.   The Strouhal number

$$\text{St} = \frac{\omega \ell}{V}$$

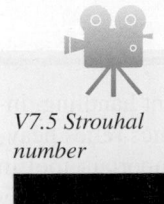

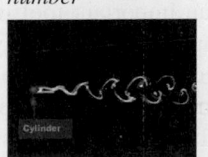

V7.5 Strouhal number

is a dimensionless parameter that is likely to be important in unsteady, oscillating flow problems in which the frequency of the oscillation is ω. It represents a measure of the ratio of inertial forces due to the unsteadiness of the flow (local acceleration) to the inertial forces due to changes in velocity from point to point in the flow field (convective acceleration). This type of unsteady flow may develop when a fluid flows past a solid body (such as a wire or cable) placed in the moving stream. For example, in a certain Reynolds number range, a periodic flow will develop downstream from a cylinder placed in a moving fluid due to a regular pattern of vortices that are shed from the body. (See the photograph at the beginning of this chapter and Fig. 9.21.) This system of vortices, called a *Kármán vortex trail* [named after Theodor von Kármán (1881–1963), a famous fluid mechanician], creates an oscillating flow at a discrete frequency, ω, such that the Strouhal number can be closely correlated with the Reynolds number. When the frequency is in the audible range, a sound can be heard and the bodies appear to "sing." In fact, the Strouhal number is named in honor of Vincenz Strouhal (1850–1922), who used this parameter in his study of "singing wires." The most dramatic evidence of this phenomenon occurred in 1940 with the collapse of the Tacoma Narrows bridge. The shedding frequency of the vortices coincided with the natural frequency of the bridge, thereby setting up a resonant condition that eventually led to the collapse of the bridge.

There are, of course, other types of oscillating flows. For example, blood flow in arteries is periodic and can be analyzed by breaking up the periodic motion into a series of harmonic components (Fourier series analysis), with each component having a frequency that is a multiple of the fundamental frequency, ω (the pulse rate). Rather than use the Strouhal number in this type of problem, a dimensionless group formed by the product of St and Re is used; that is

$$\text{St} \times \text{Re} = \frac{\rho \omega \ell^2}{\mu}$$

The square root of this dimensionless group is often referred to as the *frequency parameter*.

Weber Number. The Weber number

$$\text{We} = \frac{\rho V^2 \ell}{\sigma}$$

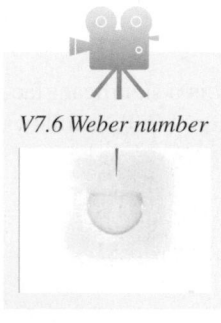

V7.6 Weber number

may be important in problems in which there is an interface between two fluids. In this situation the surface tension may play an important role in the phenomenon of interest. The Weber number can be thought of as an index of the inertial force to the surface tension force acting on a fluid element. Common examples of problems in which this parameter may be important include the flow of thin films of liquid, or the formation of droplets or bubbles. Clearly, not all problems involving flows with an interface will require the inclusion of surface tension. The flow of water in a river is not affected significantly by surface tension, since inertial and gravitational effects are dominant (We ≫ 1). However, as discussed in a later section, for river models (which may have small depths) caution is required so that surface tension does not become important in the model, whereas it is not important in the actual river. The Weber number is named after Moritz Weber (1871–1951), a German professor of naval mechanics who was instrumental in formalizing the general use of common dimensionless groups as a basis for similitude studies.

F l u i d s   i n   t h e   N e w s

Slip at the micro scale A goal in chemical and biological analyses is to miniaturize the experiment, which has many advantages including reduction in sample size. In recent years, there has been significant work on integrating these tests on a single microchip to form the "lab-on-a-chip" system. These devices are on the millimeter scale with complex passages for fluid flow on the micron scale (or smaller). While there are advantages to miniaturization, care must be taken in moving to smaller and smaller flow regimes, as you will eventually bump into the continuum assumption. To characterize this situation, a dimensionless number termed the Knudsen number, $\text{Kn} = \lambda/\ell$, is commonly employed. Here λ is the mean free path and ℓ is the characteristic length of the system. If Kn is smaller than 0.01, then the flow can be described by the Navier–Stokes equations with no-slip at the walls. For $0.01 < \text{Kn} < 0.3$, the same equations can be used, but there can be "slip" between the fluid and the wall so the boundary conditions need to be adjusted. For $\text{Kn} > 10$, the continuum assumption breaks down and the Navier–Stokes equations are no longer valid.

7.7 Correlation of Experimental Data

One of the most important uses of dimensional analysis is as an aid in the efficient handling, interpretation, and correlation of experimental data. Since the field of fluid mechanics relies heavily on empirical data, it is not surprising that dimensional analysis is such an important tool in this field. As noted previously, a dimensional analysis cannot provide a complete answer to any given problem, since the analysis only provides the dimensionless groups describing the phenomenon, and not the specific relationship among the groups. To determine this relationship, suitable experimental data must be obtained. The degree of difficulty involved in this process depends on the number of pi terms, and the nature of the experiments (How hard is it to obtain the measurements?). The simplest problems are obviously those involving the fewest pi terms, and the following sections indicate how the complexity of the analysis increases with the increasing number of pi terms.

7.7.1 Problems with One Pi Term

Application of the pi theorem indicates that if the number of variables minus the number of reference dimensions is equal to unity, then only *one* pi term is required to describe the phenomenon. The functional relationship that must exist for one pi term is

$$\Pi_1 = C$$

If only one pi term is involved in a problem, it must be equal to a constant.

where C is a constant. This is one situation in which a dimensional analysis reveals the specific form of the relationship and, as is illustrated by the following example, shows how the individual variables are related. The value of the constant, however, must still be determined by experiment.

EXAMPLE 7.3 | Flow with Only One Pi Term

GIVEN   As shown in Fig. E7.3, assume that the drag, $\mathcal{D}$, acting on a spherical particle that falls very slowly through a viscous fluid, is a function of the particle diameter, D, the particle velocity, V, and the fluid viscosity, μ.

FIND   Determine, with the aid of dimensional analysis, how the drag depends on the particle velocity.

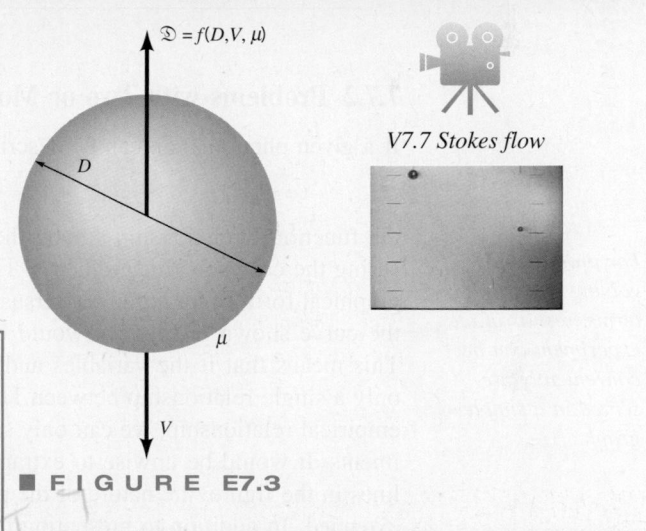

V7.7 Stokes flow

SOLUTION

From the information given, it follows that

$$\mathcal{D} = f(D, V, \mu)$$

and the dimensions of the variables are

$$\mathcal{D} \doteq F$$
$$D \doteq L$$
$$V \doteq LT^{-1}$$
$$\mu \doteq FL^{-2}T$$

We see that there are four variables and three reference dimensions (F, L, and T) required to describe the variables. Thus, according to the pi theorem, one pi term is required. This pi term can be easily formed by inspection and can be expressed as

$$\Pi_1 = \frac{\mathcal{D}}{\mu V D}$$

Because there is only one pi term, it follows that

$$\frac{\mathcal{D}}{\mu V D} = C$$

where C is a constant. Thus,

$$\mathcal{D} = C\mu V D$$

Thus, for a given particle and fluid, the drag varies directly with the velocity so that

$$\mathcal{D} \propto V \qquad \text{(Ans)}$$

■ **FIGURE E7.3**

COMMENTS   Actually, the dimensional analysis reveals that the drag not only varies directly with the velocity, but it also varies directly with the particle diameter and the fluid viscosity. We could not, however, predict the value of the drag, since the constant, C, is unknown. An experiment would have to be performed in which the drag and the corresponding velocity are measured for a given particle and fluid. Although in principle we would only have to run a single test, we would certainly want to repeat it several times to obtain a reliable value for C. It should be emphasized that once the value of C is determined it is not necessary to run similar tests by using different spherical particles and fluids; that is, C is a universal constant so long as the drag is a function only of particle diameter, velocity, and fluid viscosity.

An approximate solution to this problem can also be obtained theoretically, from which it is found that $C = 3\pi$ so that

$$\mathcal{D} = 3\pi\mu V D$$

This equation is commonly called *Stokes law* and is used in the study of the settling of particles. Our experiments would reveal that this result is only valid for small Reynolds numbers ($\rho V D/\mu \ll 1$). This follows, since in the original list of variables, we have neglected inertial effects (fluid density is not included as a variable). The inclusion of an additional variable would lead to another pi term so that there would be two pi terms rather than one.

Consider a free body diagram of a sphere in Stokes flow; there would be a buoyant force in the same direction as the drag in Fig. E7.3, as well as a weight force in the opposite direction. As shown above, the drag force is proportional to the product of the diameter and fall velocity, $\mathcal{D} \propto VD$. The weight and buoyant force are

proportional to the diameter cubed, W and $F_B \propto D^3$. Given equilibrium conditions, the force balance can be written as

$$\mathcal{D} = W - F_B$$

Based on the scaling laws for these terms, it follows that $VD \propto D^3$. Hence, the fall velocity will be proportional to the square of the diameter, $V \propto D^2$. Therefore, for two spheres, one having twice the diameter of the other, and falling through the same fluid, the sphere with the larger diameter will fall four times faster (see Video V7.7).

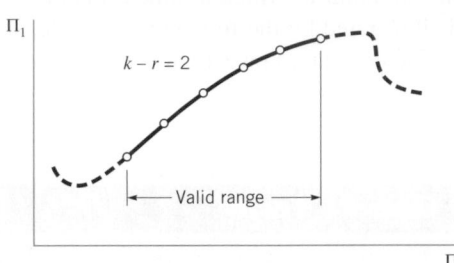

■ **FIGURE 7.4** The graphical presentation of data for problems involving two pi terms, with an illustration of the potential danger of extrapolation of data.

7.7.2 Problems with Two or More Pi Terms

If a given phenomenon can be described with two pi terms such that

$$\Pi_1 = \phi(\Pi_2)$$

For problems involving only two pi terms, results of an experiment can be conveniently presented in a simple graph.

the functional relationship among the variables can then be determined by varying Π_2 and measuring the corresponding values of Π_1. For this case the results can be conveniently presented in graphical form by plotting Π_1 versus Π_2 as is illustrated in Fig. 7.4. It should be emphasized that the curve shown in Fig. 7.4 would be a "universal" one for the particular phenomenon studied. This means that if the variables and the resulting dimensional analysis are correct, then there is only a single relationship between Π_1 and Π_2, as illustrated in Fig. 7.4. However, since this is an empirical relationship, we can only say that it is valid over the range of Π_2 covered by the experiments. It would be unwise to extrapolate beyond this range, since as illustrated with the dashed lines in the figure, the nature of the phenomenon could dramatically change as the range of Π_2 is extended. In addition to presenting the data graphically, it may be possible (and desirable) to obtain an empirical equation relating Π_1 and Π_2 by using a standard curve-fitting technique.

---

*E*XAMPLE 7.4   Dimensionless Correlation of Experimental Data

GIVEN The relationship between the pressure drop per unit length along a smooth-walled, horizontal pipe and the variables that affect the pressure drop is to be determined experimentally. In the laboratory the pressure drop was measured over a 1.5-cm length of smooth-walled pipe having an inside diameter of 1.25 m. The fluid used was water at 15 °C ($\mu = 112 \times 10^{-5}$ N · s/m^2, $\rho = 1000$ kg/m^3). Tests were run in which the velocity was varied and the corresponding pressure drop measured. The results of these tests are shown below:

Velocity (m/s)	Pressure drop for 1.5 m length (kPa)
0.36	0.304
0.6	0.76
0.89	1.48
1.78	5.0
3.41	15.8
5.2	32.7
7.2	58.4
8.83	83.6

Solution

The first step is to perform a dimensional analysis during the planning stage *before* the experiments are actually run. As was discussed in Section 7.3, we will assume that the pressure drop per unit length, Δp_ℓ, is a function of the pipe diameter, D, fluid

FIND Make use of these data to obtain a general relationship between the pressure drop per unit length and the other variables.

density, ρ, fluid viscosity, μ, and the velocity, V. Thus,

$$\Delta p_\ell = f(D, \rho, \mu, V)$$

and application of the pi theorem yields two pi terms

$$\Pi_1 = \frac{D\,\Delta p_\ell}{\rho V^2} \quad \text{and} \quad \Pi_2 = \frac{\rho V D}{\mu}$$

Hence,

$$\frac{D\,\Delta p_\ell}{\rho V^2} = \phi\left(\frac{\rho V D}{\mu}\right)$$

To determine the form of the relationship, we need to vary the Reynolds number, $\text{Re} = \rho V D / \mu$, and to measure the corresponding values of $D\,\Delta p_\ell / \rho V^2$. The Reynolds number could be varied by changing any one of the variables, ρ, V, D, or μ, or any combination of them. However, the simplest way to do this is to vary the velocity, since this will allow us to use the same fluid and pipe. Based on the data given, values for the two pi terms can be computed, with the result:

$D\,\Delta p_\ell / \rho V^2$	$\rho V D / \mu$
0.0195	4.01×10^3
0.0175	6.68×10^3
0.0155	9.97×10^3
0.0132	2.00×10^4
0.0113	3.81×10^4
0.0101	5.80×10^4
0.00939	8.00×10^4
0.00893	9.85×10^4

These are dimensionless groups so that their values are independent of the system of units used so long as a consistent system is used. For example, if the velocity is in m/s, then the diameter should be in m, not inches or km. Note that since the Reynolds numbers are all greater than 2100, the flow in the pipe is turbulent (see Section 8.1.1).

A plot of these two pi terms can now be made with the results shown in Fig. E7.4a. The correlation appears to be quite good, and if it was not, this would suggest that either we had large experimental measurement errors or that we had perhaps omitted an important variable. The curve shown in Fig. E7.4a represents the general relationship between the pressure drop and the other factors in the range of Reynolds numbers between 4.01×10^3 and 9.85×10^4. Thus, for this range of Reynolds numbers it is *not* necessary to repeat the tests for other pipe sizes or other fluids provided the assumed independent variables (D, ρ, μ, V) are the only important ones.

Since the relationship between Π_1 and Π_2 is nonlinear, it is not immediately obvious what form of empirical equation might be used to describe the relationship. If, however, the same data are

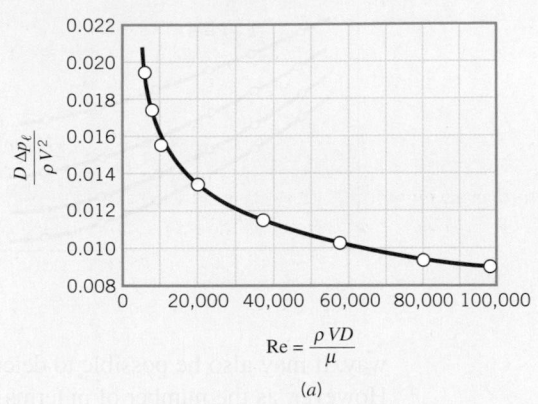

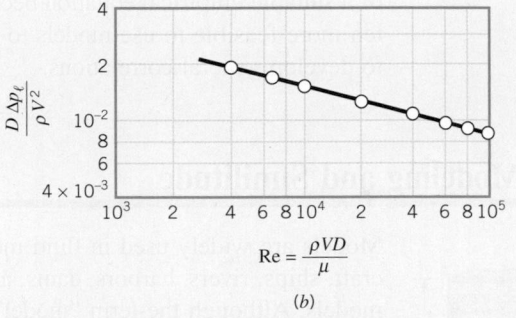

■ **F I G U R E E7.4**

plotted on a logarithmic graph paper, as is shown in Fig. E7.4b, the data form a straight line, suggesting that a suitable equation is of the form $\Pi_1 = A\Pi_2^n$ where A and n are empirical constants to be determined from the data by using a suitable curve-fitting technique, such as a nonlinear regression program. For the data given in this example, a good fit of the data is obtained with the equation

$$\Pi_1 = 0.150\,\Pi_2^{-0.25} \tag{Ans}$$

COMMENT In 1911, H. Blasius (1883–1970), a German fluid mechanician, established a similar empirical equation that is used widely for predicting the pressure drop in smooth pipes in the range $4 \times 10^3 < \text{Re} < 10^5$ (Ref. 16). This equation can be expressed in the form

$$\frac{D\,\Delta p_\ell}{\rho V^2} = 0.1582\left(\frac{\rho V D}{\mu}\right)^{-1/4}$$

The so-called Blasius formula is based on numerous experimental results of the type used in this example. Flow in pipes is discussed in more detail in the next chapter, where it is shown how pipe roughness (which introduces another variable) may affect the results given in this example (which is for smooth-walled pipes).

For problems involving more than two or three pi terms, it is often necessary to use a model to predict specific characteristics.

As the number of required pi terms increases, it becomes more difficult to display the results in a convenient graphical form and to determine a specific empirical equation that describes the phenomenon. For problems involving three pi terms

$$\Pi_1 = \phi(\Pi_2, \Pi_3)$$

it is still possible to show data correlations on simple graphs by plotting families of curves as illustrated in Fig. 7.5. This is an informative and useful way of representing the data in a general

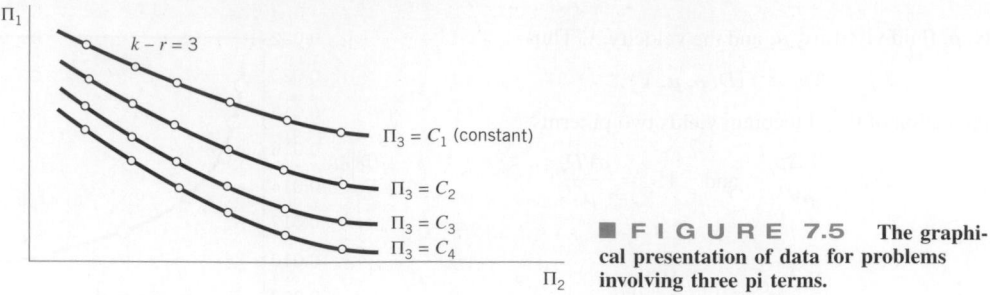

■ **FIGURE 7.5** The graphical presentation of data for problems involving three pi terms.

way. It may also be possible to determine a suitable empirical equation relating the three pi terms. However, as the number of pi terms continues to increase, corresponding to an increase in the general complexity of the problem of interest, both the graphical presentation and the determination of a suitable empirical equation become intractable. For these more complicated problems, it is often more feasible to use models to predict specific characteristics of the system rather than to try to develop general correlations.

7.8 Modeling and Similitude

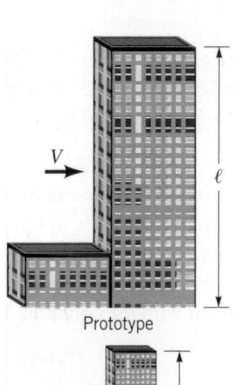

Prototype

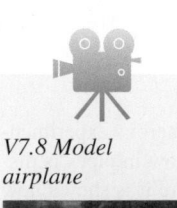

Model

V7.8 Model airplane

Models are widely used in fluid mechanics. Major engineering projects involving structures, aircraft, ships, rivers, harbors, dams, air and water pollution, and so on, frequently involve the use of models. Although the term "model" is used in many different contexts, the "engineering model" generally conforms to the following definition. *A **model** is a representation of a physical system that may be used to predict the behavior of the system in some desired respect.* The physical system for which the predictions are to be made is called the *prototype.* Although *mathematical* or *computer* models may also conform to this definition, our interest will be in physical models, that is, models that resemble the prototype but are generally of a different size, may involve different fluids, and often operate under different conditions (pressures, velocities, etc.). As shown by the figure in the margin, usually a model is smaller than the prototype. Therefore, it is more easily handled in the laboratory and less expensive to construct and operate than a large prototype (it should be noted that variables or pi terms without a subscript will refer to the prototype, whereas the subscript *m* will be used to designate the model variables or pi terms). Occasionally, if the prototype is very small, it may be advantageous to have a model that is larger than the prototype so that it can be more easily studied. For example, large models have been used to study the motion of red blood cells, which are approximately 8 μm in diameter. With the successful development of a valid model, it is possible to predict the behavior of the prototype under a certain set of conditions. We may also wish to examine a priori the effect of possible design changes that are proposed for a hydraulic structure or fluid-flow system. There is, of course, an inherent danger in the use of models in that predictions can be made that are in error and the error not detected until the prototype is found not to perform as predicted. It is, therefore, imperative that the model be properly designed and tested and that the results be interpreted correctly. In the following sections we will develop the procedures for designing models so that the model and prototype will behave in a similar fashion.

7.8.1 Theory of Models

The theory of models can be readily developed by using the principles of dimensional analysis. It has been shown that any given problem can be described in terms of a set of pi terms as

$$\Pi_1 = \phi(\Pi_2, \Pi_3, \ldots, \Pi_n) \tag{7.7}$$

In formulating this relationship, only a knowledge of the general nature of the physical phenomenon, and the variables involved, is required. Specific values for variables (size of components, fluid properties, and so on) are not needed to perform the dimensional analysis. Thus, Eq. 7.7 applies to any system that is governed by the same variables. If Eq. 7.7 describes the behavior of a particular prototype, a similar relationship can be written for a model of this prototype; that is,

$$\Pi_{1m} = \phi(\Pi_{2m}, \Pi_{3m}, \ldots, \Pi_{nm}) \tag{7.8}$$

where the form of the function will be the same as long as the same phenomenon is involved in both the prototype and the model. Variables, or pi terms, without a subscript will refer to the prototype, whereas the subscript m will be used to designate the model variables or pi terms.

The pi terms can be developed so that Π_1 contains the variable that is to be predicted from observations made on the model. Therefore, if the model is designed and operated under the following conditions,

$$\Pi_{2m} = \Pi_2$$
$$\Pi_{3m} = \Pi_3 \tag{7.9}$$
$$\vdots$$
$$\Pi_{nm} = \Pi_n$$

then with the presumption that the form of ϕ is the same for model and prototype, it follows that

$$\Pi_1 = \Pi_{1m} \tag{7.10}$$

The similarity requirements for a model can be readily obtained with the aid of dimensional analysis.

Equation 7.10 is the desired *prediction equation* and indicates that the measured value of Π_{1m} obtained with the model will be equal to the corresponding Π_1 for the prototype as long as the other pi terms are equal. The conditions specified by Eqs. 7.9 provide the *model design conditions*, also called *similarity requirements* or *modeling laws*.

As an example of the procedure, consider the problem of determining the drag, $\mathcal{D}$, on a thin rectangular plate ($w \times h$ in size) placed normal to a fluid with velocity, V, as shown by the figure in the margin. The dimensional analysis of this problem was performed in Example 7.1, where it was assumed that

$$\mathcal{D} = f(w, h, \mu, \rho, V)$$

Application of the pi theorem yielded

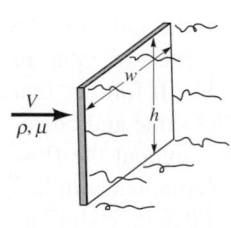

$$\frac{\mathcal{D}}{w^2 \rho V^2} = \phi\left(\frac{w}{h}, \frac{\rho V w}{\mu}\right) \tag{7.11}$$

We are now concerned with designing a model that could be used to predict the drag on a certain prototype (which presumably has a different size than the model). Since the relationship expressed by Eq. 7.11 applies to both prototype and model, Eq. 7.11 is assumed to govern the prototype, with a similar relationship

$$\frac{\mathcal{D}_m}{w_m^2 \rho_m V_m^2} = \phi\left(\frac{w_m}{h_m}, \frac{\rho_m V_m w_m}{\mu_m}\right) \tag{7.12}$$

for the model. The model design conditions, or similarity requirements, are therefore

$$\frac{w_m}{h_m} = \frac{w}{h} \qquad \frac{\rho_m V_m w_m}{\mu_m} = \frac{\rho V w}{\mu}$$

The size of the model is obtained from the first requirement which indicates that

$$w_m = \frac{h_m}{h} w \tag{7.13}$$

We are free to establish the height ratio h_m/h, but then the model plate width, w_m, is fixed in accordance with Eq. 7.13.

The second similarity requirement indicates that the model and prototype must be operated at the same Reynolds number. Thus, the required velocity for the model is obtained from the relationship

$$V_m = \frac{\mu_m}{\mu} \frac{\rho}{\rho_m} \frac{w}{w_m} V \tag{7.14}$$

Note that this model design requires not only geometric scaling, as specified by Eq. 7.13, but also the correct scaling of the velocity in accordance with Eq. 7.14. This result is typical of most model designs—there is more to the design than simply scaling the geometry!

With the foregoing similarity requirements satisfied, the prediction equation for the drag is

$$\frac{\mathcal{D}}{w^2 \rho V^2} = \frac{\mathcal{D}_m}{w_m^2 \rho_m V_m^2}$$

or

$$\mathcal{D} = \left(\frac{w}{w_m}\right)^2 \left(\frac{\rho}{\rho_m}\right)\left(\frac{V}{V_m}\right)^2 \mathcal{D}_m$$

Similarity between a model and a prototype is achieved by equating pi terms.

Thus, a measured drag on the model, $\mathcal{D}_m$, must be multiplied by the ratio of the square of the plate widths, the ratio of the fluid densities, and the ratio of the square of the velocities to obtain the predicted value of the prototype drag, $\mathcal{D}$.

Generally, as is illustrated in this example, to achieve similarity between model and prototype behavior, *all the corresponding pi terms must be equated between model and prototype.* Usually, one or more of these pi terms will involve ratios of important lengths (such as w/h in the foregoing example); that is, they are purely geometrical. Thus, when we equate the pi terms involving length ratios, we are requiring that there be complete *geometric similarity* between the model and prototype. This means that the model must be a scaled version of the prototype. Geometric scaling may extend to the finest features of the system, such as surface roughness, or small protuberances on a structure, since these kinds of geometric features may significantly influence the flow. Any deviation from complete geometric similarity for a model must be carefully considered. Sometimes complete geometric scaling may be difficult to achieve, particularly when dealing with surface roughness, since roughness is difficult to characterize and control.

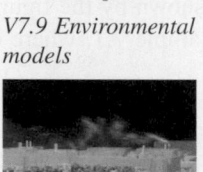

V7.9 Environmental models

Another group of typical pi terms (such as the Reynolds number in the foregoing example) involves force ratios as noted in Table 7.1. The equality of these pi terms requires the ratio of like forces in model and prototype to be the same. Thus, for flows in which the Reynolds numbers are equal, the ratio of viscous forces in model and prototype is equal to the ratio of inertia forces. If other pi terms are involved, such as the Froude number or Weber number, a similar conclusion can be drawn; that is, the equality of these pi terms requires the ratio of like forces in model and prototype to be the same. Thus, when these types of pi terms are equal in model and prototype, we have *dynamic similarity* between model and prototype. It follows that with both geometric and dynamic similarity the streamline patterns will be the same and corresponding velocity ratios (V_m/V) and acceleration ratios (a_m/a) are constant throughout the flow field. Thus, *kinematic similarity* exists between model and prototype. To have complete similarity between model and prototype, we must maintain geometric, kinematic, and dynamic similarity between the two systems. This will automatically follow if all the important variables are included in the dimensional analysis, and if all the similarity requirements based on the resulting pi terms are satisfied.

F l u i d s   i n   t h e   N e w s

Modeling parachutes in a water tunnel The first use of a parachute with a free-fall jump from an aircraft occurred in 1914, although parachute jumps from hot air balloons had occurred since the late 1700s. In more modern times parachutes are commonly used by the military, and for safety and sport. It is not surprising that there remains interest in the design and characteristics of parachutes, and researchers at the Worcester Polytechnic Institute have been studying various aspects of the aerodynamics associated with parachutes. An unusual part of their study is that they are using small-scale parachutes tested in a *water tunnel*. The *model parachutes* are reduced in size by a factor of 30 to 60 times. Various types of tests can be performed, ranging from the study of the velocity fields in the wake of the canopy with a steady free-stream velocity to the study of conditions during rapid deployment of the canopy. According to the researchers, the advantage of using water as the working fluid, rather than air, is that the velocities and deployment dynamics are slower than in the atmosphere, thus providing more time to collect detailed experimental data. (See Problem 7.47.)

EXAMPLE 7.5 Prediction of Prototype Performance from Model Data

GIVEN A long structural component of a bridge has an elliptical cross section shown in Fig. E7.5. It is known that when a steady wind blows past this type of bluff body, vortices may develop on the downwind side that are shed in a regular fashion at some definite frequency. Since these vortices can create harmful periodic forces acting on the structure, it is important to determine the shedding frequency. For the specific structure of interest, $D = 0.1$ m, $H = 0.3$ m, and a representative wind velocity is 50 km/hr. Standard air can be assumed. The shedding frequency is to be determined through the use of a small-scale model that is to be tested in a water tunnel. For the model $D_m = 20$ mm and the water temperature is 20 °C.

FIND Determine the model dimension, H_m, and the velocity at which the test should be performed. If the shedding frequency for the model is found to be 49.9 Hz, what is the corresponding frequency for the prototype?

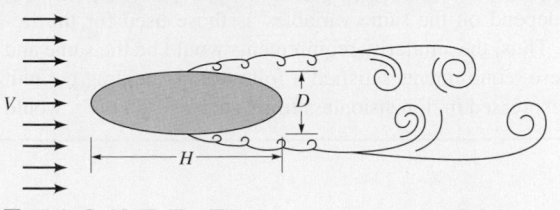

■ **FIGURE E7.5**

V7.10 Flow past an ellipse

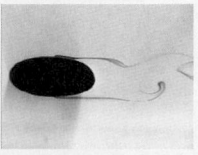

SOLUTION

We expect the shedding frequency, ω, to depend on the lengths D and H, the approach velocity, V, and the fluid density, ρ, and viscosity, μ. Thus,

$$\omega = f(D, H, V, \rho, \mu)$$

where

$$\omega \doteq T^{-1}$$
$$D \doteq L$$
$$H \doteq L$$
$$V \doteq LT^{-1}$$
$$\rho \doteq ML^{-3}$$
$$\mu \doteq ML^{-1}T^{-1}$$

Since there are six variables and three reference dimensions (MLT), three pi terms are required. Application of the pi theorem yields

$$\frac{\omega D}{V} = \phi\left(\frac{D}{H}, \frac{\rho V D}{\mu}\right)$$

We recognize the pi term on the left as the Strouhal number, and the dimensional analysis indicates that the Strouhal number is a function of the geometric parameter, D/H, and the Reynolds number. Thus, to maintain similarity between model and prototype

$$\frac{D_m}{H_m} = \frac{D}{H}$$

and

$$\frac{\rho_m V_m D_m}{\mu_m} = \frac{\rho V D}{\mu}$$

From the first similarity requirement

$$H_m = \frac{D_m}{D} H$$
$$= \frac{(20 \times 10^{-3} \text{ m})}{(0.1 \text{ m})} (0.3 \text{ m})$$
$$H_m = 60 \times 10^{-3} \text{ m} = 60 \text{ mm} \qquad \text{(Ans)}$$

The second similarity requirement indicates that the Reynolds number must be the same for model and prototype so that the model velocity must satisfy the condition

$$V_m = \frac{\mu_m}{\mu} \frac{\rho}{\rho_m} \frac{D}{D_m} V \qquad (1)$$

For air at standard conditions, $\mu = 1.79 \times 10^{-5}$ kg/m · s, $\rho = 1.23$ kg/m^3, and for water at 20 °C, $\mu = 1.00 \times 10^{-3}$ kg/m · s, $\rho = 998$ kg/m^3. The fluid velocity for the prototype is

$$V = \frac{(50 \times 10^3 \text{ m/hr})}{(3600 \text{ s/hr})} = 13.9 \text{ m/s}$$

The required velocity can now be calculated from Eq. 1 as

$$V_m = \frac{[1.00 \times 10^{-3} \text{ kg/(m · s)}] \, (1.23 \text{ kg/m}^3)}{[1.79 \times 10^{-5} \text{ kg/(m · s)}] \, (998 \text{ kg/m}^3)}$$
$$\times \frac{(0.1 \text{ m})}{(20 \times 10^{-3} \text{ m})} (13.9 \text{ m/s})$$
$$V_m = 4.79 \text{ m/s} \qquad \text{(Ans)}$$

This is a reasonable velocity that could be readily achieved in a water tunnel.

With the two similarity requirements satisfied, it follows that the Strouhal numbers for prototype and model will be the same so that

$$\frac{\omega D}{V} = \frac{\omega_m D_m}{V_m}$$

and the predicted prototype vortex shedding frequency is

$$\omega = \frac{V}{V_m} \frac{D_m}{D} \omega_m$$
$$= \frac{(13.9 \text{ m/s})}{(4.79 \text{ m/s})} \frac{(20 \times 10^{-3} \text{ m})}{(0.1 \text{ m})} (49.9 \text{ Hz})$$
$$\omega = 29.0 \text{ Hz} \qquad \text{(Ans)}$$

COMMENT This same model could also be used to predict the drag per unit length, $\mathcal{D}_\ell$ (N/m), on the prototype, since the drag would depend on the same variables as those used for the frequency. Thus, the similarity requirements would be the same and with these requirements satisfied it follows that the drag per unit length expressed in dimensionless form, such as $\mathcal{D}_\ell/D\rho V^2$, would be equal in model and prototype. The measured drag per unit length on the model could then be related to the corresponding drag per unit length on the prototype through the relationship

$$\mathcal{D}_\ell = \left(\frac{D}{D_m}\right)\left(\frac{\rho}{\rho_m}\right)\left(\frac{V}{V_m}\right)^2 \mathcal{D}_{\ell m}$$

7.8.2 Model Scales

It is clear from the preceding section that the ratio of like quantities for the model and prototype naturally arises from the similarity requirements. For example, if in a given problem there are two length variables ℓ_1 and ℓ_2, the resulting similarity requirement based on a pi term obtained from these two variables is

$$\frac{\ell_1}{\ell_2} = \frac{\ell_{1m}}{\ell_{2m}}$$

so that

$$\frac{\ell_{1m}}{\ell_1} = \frac{\ell_{2m}}{\ell_2}$$

The ratio of a model variable to the corresponding prototype variable is called the scale for that variable.

We define the ratio ℓ_{1m}/ℓ_1 or ℓ_{2m}/ℓ_2 as the *length scale*. For true models there will be only one length scale, and all lengths are fixed in accordance with this scale. There are, however, other scales such as the velocity scale, V_m/V, density scale, ρ_m/ρ, viscosity scale, μ_m/μ, and so on. In fact, we can define a scale for each of the variables in the problem. Thus, it is actually meaningless to talk about a "scale" of a model without specifying which scale.

We will designate the length scale as λ_ℓ, and other scales as λ_V, λ_ρ, λ_μ, and so on, where the subscript indicates the particular scale. Also, we will take the ratio of the model value to the prototype value as the scale (rather than the inverse). Length scales are often specified, for example, as 1 : 10 or as a $\frac{1}{10}$ scale model. The meaning of this specification is that the model is one-tenth the size of the prototype, and the tacit assumption is that all relevant lengths are scaled accordingly so the model is geometrically similar to the prototype.

F l u i d s   i n   t h e   N e w s

"Galloping Gertie" One of the most dramatic bridge collapses occurred in 1940 when the Tacoma Narrows bridge, located near Tacoma, Washington, failed due to aerodynamic instability. The bridge had been nicknamed "Galloping Gertie" due to its tendency to sway and move in high winds. On the fateful day of the collapse the wind speed was 65 km/hr. This particular combination of a high wind and the aeroelastic properties of the bridge created large oscillations leading to its failure. The bridge was replaced in 1950, and plans are underway to add a second bridge parallel to the existing structure. To determine possible wind interference effects due to two bridges in close proximity, wind tunnel tests were run in a 9 m × 9 m wind tunnel operated by the National Research Council of Canada. *Models* of the two side-by-side bridges, each having a length scale of 1 : 211, were tested under various wind conditions. Since the failure of the original Tacoma Narrows bridge, it is now common practice to use wind tunnel model studies during the design process to evaluate any bridge that is to be subjected to wind-induced vibrations. (See Problem 7.72.)

7.8.3 Practical Aspects of Using Models

Validation of Model Design. Most model studies involve simplifying assumptions with regard to the variables to be considered. Although the number of assumptions is frequently less stringent than that required for mathematical models, they nevertheless introduce some uncertainty in the model design. It is, therefore, desirable to check the design experimentally whenever possible. In some situations the purpose of the model is to predict the effects of certain proposed changes in a given prototype, and in this instance some actual prototype data may be available. The model can be designed, constructed, and tested, and the model prediction can be compared with these data. If the agreement is satisfactory, then the model can be changed in the desired manner, and the corresponding effect on the prototype can be predicted with increased confidence.

V7.11 Model of fish hatchery pond

Another useful and informative procedure is to run tests with a series of models of different sizes, where one of the models can be thought of as the prototype and the others as "models" of this prototype. With the models designed and operated on the basis of the proposed design, a necessary condition for the validity of the model design is that an accurate prediction be made between any pair of models, since one can always be considered as a model of the other. Although suitable agreement in validation tests of this type does not unequivocally indicate a correct model design (e.g., the length scales between laboratory models may be significantly different than required for actual prototype prediction), it is certainly true that if agreement between models cannot be achieved in these tests, there is no reason to expect that the same model design can be used to predict prototype behavior correctly.

Distorted Models.   Although the general idea behind establishing similarity requirements for models is straightforward (we simply equate pi terms), it is not always possible to satisfy all the known requirements. If one or more of the similarity requirements are not met, for example, if $\Pi_{2m} \neq \Pi_2$, then it follows that the prediction equation $\Pi_1 = \Pi_{1m}$ is not true; that is, $\Pi_1 \neq \Pi_{1m}$. Models for which one or more of the similarity requirements are not satisfied are called *distorted models*.

Models for which one or more similarity requirements are not satisfied are called distorted models.

Distorted models are rather commonplace, and they can arise for a variety of reasons. For example, perhaps a suitable fluid cannot be found for the model. The classic example of a distorted model occurs in the study of open channel or free-surface flows. Typically, in these problems both the Reynolds number, $\rho V \ell / \mu$, and the Froude number, $V / \sqrt{g\ell}$, are involved.

Froude number similarity requires

$$\frac{V_m}{\sqrt{g_m \ell_m}} = \frac{V}{\sqrt{g\ell}}$$

If the model and prototype are operated in the same gravitational field, then the required velocity scale is

$$\frac{V_m}{V} = \sqrt{\frac{\ell_m}{\ell}} = \sqrt{\lambda_\ell}$$

Reynolds number similarity requires

$$\frac{\rho_m V_m \ell_m}{\mu_m} = \frac{\rho V \ell}{\mu}$$

and the velocity scale is

$$\frac{V_m}{V} = \frac{\mu_m}{\mu} \frac{\rho}{\rho_m} \frac{\ell}{\ell_m}$$

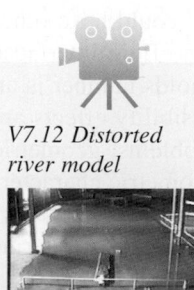

V7.12 Distorted river model

Since the velocity scale must be equal to the square root of the length scale, it follows that

$$\frac{\mu_m/\rho_m}{\mu/\rho} = \frac{\nu_m}{\nu} = (\lambda_\ell)^{3/2} \tag{7.15}$$

where the ratio μ/ρ is the kinematic viscosity, ν. Although in principle it may be possible to satisfy this design condition, it may be quite difficult, if not impossible, to find a suitable model fluid, particularly for small length scales. For problems involving rivers, spillways, and harbors, for which the prototype fluid is water, the models are also relatively large so that the only practical model fluid is water. However, in this case (with the kinematic viscosity scale equal to unity) Eq. 7.15 will not be satisfied, and a distorted model will result. Generally, hydraulic models of this type are distorted and are designed on the basis of the Froude number, with the Reynolds number different in model and prototype.

Distorted models can be successfully used, but the interpretation of results obtained with this type of model is obviously more difficult than the interpretation of results obtained with *true models* for which all similarity requirements are met. There are no general rules for handling distorted models, and essentially each problem must be considered on its own merits. The success of using distorted models depends to a large extent on the skill and experience of the investigator responsible for the design of the model and in the interpretation of experimental data obtained from the model. Distorted models are widely used, and additional information can be found in the references at the end of the chapter. References 14 and 15 contain detailed discussions of several practical examples of distorted fluid flow and hydraulic models.

F l u i d s   i n   t h e   N e w s

Old Man River in (large) miniature One of the world's largest scale models, a Mississippi River model, resides near Jackson, Mississippi. It is a detailed, complex model that covers many acres and replicates the 1,250,000 acre (5060 km²) Mississippi River basin. Built by the Army Corps of Engineers and used from 1943 to 1973, today it has mostly gone to ruin. As with many hydraulic models, this is a *distorted model*, with a horizontal scale of 1 to 2000 and a vertical scale of 1 to 100. One step along the model river corresponds to one mile (1.6 km) along the river. All essential river basin elements such as geological features, levees, and railroad embankments were sculpted by hand to match the actual contours. The main purpose of the model was to predict floods. This was done by supplying specific amounts of water at prescribed locations along the model and then measuring the water depths up and down the model river. Because of the length scale, there is a difference in the time taken by the corresponding model and prototype events. Although it takes days for the actual floodwaters to travel from Sioux City, Iowa, to Omaha, Nebraska, it would take only minutes for the simulated flow in the model.

7.9   Some Typical Model Studies

Models are used to investigate many different types of fluid mechanics problems, and it is difficult to characterize in a general way all necessary similarity requirements, since each problem is unique. We can, however, broadly classify many of the problems on the basis of the general nature of the flow and subsequently develop some general characteristics of model designs in each of these classifications. In the following sections we will consider models for the study of (1) flow through closed conduits, (2) flow around immersed bodies, and (3) flow with a free surface. Turbomachine models are considered in Chapter 12.

7.9.1 Flow through Closed Conduits

Geometric and Reynolds number similarity is usually required for models involving flow through closed conduits.

Common examples of this type of flow include pipe flow and flow through valves, fittings, and metering devices. Although the conduit cross sections are often circular, they could have other shapes as well and may contain expansions or contractions. Since there are no fluid interfaces or free surfaces, the dominant forces are inertial and viscous so that the Reynolds number is an important similarity parameter. For low Mach numbers (Ma < 0.3), compressibility effects are usually negligible for both the flow of liquids or gases. For this class of problems, geometric similarity between model and prototype must be maintained. Generally the geometric characteristics can be described by a series of length terms, $\ell_1, \ell_2, \ell_3, \ldots, \ell_i$, and ℓ, where ℓ is some particular length dimension for the system. Such a series of length terms leads to a set of pi terms of the form

$$\Pi_i = \frac{\ell_i}{\ell}$$

where $i = 1, 2, \ldots$, and so on. In addition to the basic geometry of the system, the roughness of the internal surface in contact with the fluid may be important. If the average height of surface roughness elements is defined as ε, then the pi term representing roughness will be ε/ℓ. This parameter indicates that for complete geometric similarity, surface roughness would also have to be scaled. Note that this implies that for length scales less than 1, the model surfaces should be smoother than those in the prototype since $\varepsilon_m = \lambda_\ell \varepsilon$. To further complicate matters, the pattern of roughness elements in model and prototype would have to be similar. These are conditions that are virtually impossible to satisfy exactly. Fortunately, in some problems the surface roughness plays a minor role and can be neglected. However, in other problems (such as turbulent flow through pipes) roughness can be very important.

It follows from this discussion that for flow in closed conduits at low Mach numbers, any dependent pi term (the one that contains the particular variable of interest, such as pressure drop) can be expressed as

$$\text{Dependent pi term} = \phi\left(\frac{\ell_i}{\ell}, \frac{\varepsilon}{\ell}, \frac{\rho V \ell}{\mu}\right) \qquad (7.16)$$

This is a general formulation for this type of problem. The first two pi terms of the right side of Eq. 7.16 lead to the requirement of geometric similarity so that

$$\frac{\ell_{im}}{\ell_m} = \frac{\ell_i}{\ell} \qquad \frac{\varepsilon_m}{\ell_m} = \frac{\varepsilon}{\ell}$$

or

$$\frac{\ell_{im}}{\ell_i} = \frac{\varepsilon_m}{\varepsilon} = \frac{\ell_m}{\ell} = \lambda_\ell$$

This result indicates that the investigator is free to choose a length scale, λ_ℓ, but once this scale is selected, all other pertinent lengths must be scaled in the same ratio.

The additional similarity requirement arises from the equality of Reynolds numbers

$$\frac{\rho_m V_m \ell_m}{\mu_m} = \frac{\rho V \ell}{\mu}$$

Accurate predictions of flow behavior require the correct scaling of velocities.

From this condition the velocity scale is established so that

$$\frac{V_m}{V} = \frac{\mu_m}{\mu} \frac{\rho}{\rho_m} \frac{\ell}{\ell_m} \qquad (7.17)$$

and the actual value of the velocity scale depends on the viscosity and density scales, as well as the length scale. Different fluids can be used in model and prototype. However, if the same fluid is used (with $\mu_m = \mu$ and $\rho_m = \rho$), then

$$\frac{V_m}{V} = \frac{\ell}{\ell_m}$$

Thus, $V_m = V/\lambda_\ell$, which indicates that the fluid velocity in the model will be larger than that in the prototype for any length scale less than 1. Since length scales are typically much less than unity, Reynolds number similarity may be difficult to achieve because of the large model velocities required.

With these similarity requirements satisfied, it follows that the dependent pi term will be equal in model and prototype. For example, if the dependent variable of interest is the pressure differential,[3] Δp, between two points along a closed conduit, then the dependent pi term could be expressed as

$$\Pi_1 = \frac{\Delta p}{\rho V^2}$$

The prototype pressure drop would then be obtained from the relationship

$$\Delta p = \frac{\rho}{\rho_m}\left(\frac{V}{V_m}\right)^2 \Delta p_m$$

so that from a measured pressure differential in the model, Δp_m, the corresponding pressure differential for the prototype could be predicted. Note that in general $\Delta p \neq \Delta p_m$.

---

[3]In some previous examples the pressure differential *per unit length*, Δp_ℓ, was used. This is appropriate for flow in long pipes or conduits in which the pressure would vary linearly with distance. However, in the more general situation the pressure may not vary linearly with position so that it is necessary to consider the pressure differential, Δp, as the dependent variable. In this case the distance between pressure taps is an additional variable (as well as the distance of one of the taps measured from some reference point within the flow system).

EXAMPLE 7.6 | Reynolds Number Similarity

GIVEN Model tests are to be performed to study the flow through a large check valve having a 0.6-m-diameter inlet and carrying water at a flowrate of 0.85 m³/s as shown in Fig. E7.6a. The working fluid in the model is water at the same temperature as that in the prototype. Complete geometric similarity exists between model and prototype, and the model inlet diameter is 7.5 cm.

FIND Determine the required flowrate in the model.

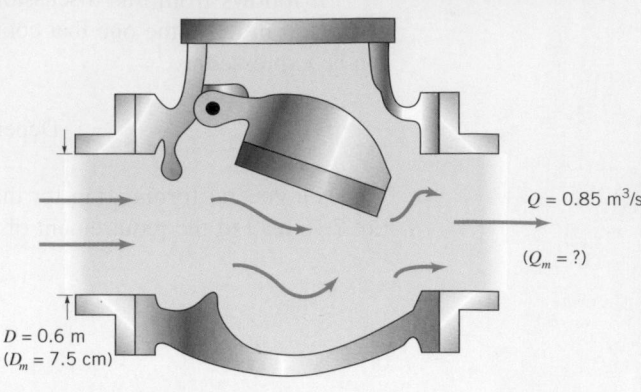

$Q = 0.85$ m³/s

$(Q_m = ?)$

$D = 0.6$ m
$(D_m = 7.5$ cm)

■ **FIGURE E7.6a**

SOLUTION

To ensure dynamic similarity, the model tests should be run so that

$$Re_m = Re$$

or

$$\frac{V_m D_m}{\nu_m} = \frac{VD}{\nu}$$

where V and D correspond to the inlet velocity and diameter, respectively. Since the same fluid is to be used in model and prototype, $\nu = \nu_m$, and therefore

$$\frac{V_m}{V} = \frac{D}{D_m}$$

The discharge, Q, is equal to VA, where A is the inlet area, so

$$\frac{Q_m}{Q} = \frac{V_m A_m}{VA} = \left(\frac{D}{D_m}\right)\frac{[(\pi/4)D_m^2]}{[(\pi/4)D^2]}$$

$$= \frac{D_m}{D}$$

and for the data given

$$Q_m = \frac{(7.5/100 \text{ m})}{(0.6 \text{ m})} 0.85 \text{ m}^3/\text{s}$$

$$Q_m = 0.106 \text{ m}^3/\text{s} \qquad \textbf{(Ans)}$$

COMMENT As indicated by the above analysis, to maintain Reynolds number similarity using the same fluid in model and prototype, the required velocity scale is inversely proportional to the length scale, that is, $V_m/V = (D_m/D)^{-1}$. This strong influence of the length scale on the velocity scale is shown in Fig. E7.6b.

For this particular example, $D_m/D = 0.125$, and the corresponding velocity scale is 8 (see Fig. E7.6b). Thus, with the prototype velocity equal to $V = (0.85 \text{ m}^3/\text{s})/(\pi/4)(0.6 \text{ m})^2 = 3$ m/s, the required model velocity is $V_m = 24$ m/s. Although this is a relatively large velocity, it could be attained in a laboratory facility. It is to be noted that if we tried to use a smaller model, say one with $D = 2.5$ cm, the required model velocity is 72 m/s, a very high velocity that would be difficult to achieve. These results are indicative of one of the difficulties encountered in maintaining Reynolds number similarity—the required model velocities may be impractical to obtain.

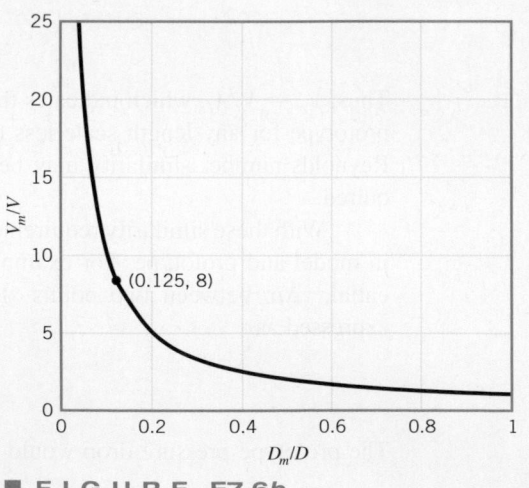

(0.125, 8)

V_m/V

D_m/D

■ **FIGURE E7.6b**

In some problems Reynolds number similarity may be relaxed.

Two additional points should be made with regard to modeling flows in closed conduits. First, for large Reynolds numbers, inertial forces are much larger than viscous forces, and in this case it may be possible to neglect viscous effects. The important practical consequence of this is that it would not be necessary to maintain Reynolds number similarity between model and prototype. However, *both* model and prototype would have to operate at large Reynolds numbers. Since we do not know, a priori, what a "large Reynolds number" is, the effect of Reynolds numbers would have to be determined from the model. This could be accomplished by varying the model Reynolds number to determine the range (if any) over which the dependent pi term ceases to be affected by changes in Reynolds number.

The second point relates to the possibility of cavitation in flow through closed conduits. For example, flow through the complex passages that may exist in valves may lead to local regions of high velocity (and thus low pressure), which can cause the fluid to cavitate. If the model is to be used to study cavitation phenomena, then the vapor pressure, p_v, becomes an important variable and an additional similarity requirement such as equality of the cavitation number $(p_r - p_v)/\frac{1}{2}\rho V^2$ is required, where p_r is some reference pressure. The use of models to study cavitation is complicated, since it is not fully understood how vapor bubbles form and grow. The initiation of bubbles seems to be influenced by the microscopic particles that exist in most liquids, and how this aspect of the problem influences model studies is not clear. Additional details can be found in Ref. 17.

7.9.2 Flow around Immersed Bodies

Geometric and Reynolds number similarity is usually required for models involving flow around bodies.

Models have been widely used to study the flow characteristics associated with bodies that are completely immersed in a moving fluid. Examples include flow around aircraft, automobiles, golf balls, and buildings. (These types of models are usually tested in wind tunnels as is illustrated in Fig. 7.6.) Modeling laws for these problems are similar to those described in the preceding section; that is, geometric and Reynolds number similarity is required. Since there are no fluid interfaces, surface tension (and therefore the Weber number) is not important. Also, gravity will not affect the flow patterns, so the Froude number need not be considered. The Mach number will be important for high-speed flows in which compressibility becomes an important factor, but for incompressible fluids (such as liquids or for gases at relatively low speeds) the Mach number can be omitted as a similarity requirement. In this case, a general formulation for these problems is

$$\text{Dependent pi term} = \phi\left(\frac{\ell_i}{\ell}, \frac{\varepsilon}{\ell}, \frac{\rho V \ell}{\mu}\right) \tag{7.18}$$

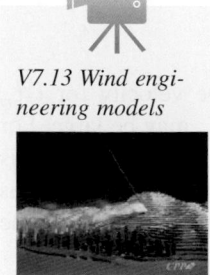

V7.13 Wind engineering models

where ℓ is some characteristic length of the system and ℓ_i represents other pertinent lengths, ε/ℓ is the relative roughness of the surface (or surfaces), and $\rho V \ell/\mu$ is the Reynolds number.

Frequently, the dependent variable of interest for this type of problem is the drag, $\mathcal{D}$, developed on the body, and in this situation the dependent pi term would usually be expressed in the form of a *drag coefficient*, C_D, where

$$C_D = \frac{\mathcal{D}}{\frac{1}{2}\rho V^2 \ell^2}$$

The numerical factor, $\frac{1}{2}$, is arbitrary but commonly included, and ℓ^2 is usually taken as some representative area of the object. Thus, drag studies can be undertaken with the formulation

$$\frac{\mathcal{D}}{\frac{1}{2}\rho V^2 \ell^2} = C_D = \phi\left(\frac{\ell_i}{\ell}, \frac{\varepsilon}{\ell}, \frac{\rho V \ell}{\mu}\right) \tag{7.19}$$

■ **FIGURE 7.6**   **Model of the National Bank of Commerce, San Antonio, Texas, for measurement of peak, rms, and mean pressure distributions. The model is located in a long-test-section, meteorological wind tunnel. (Photograph courtesy of Cermak Peterka Petersen, Inc.)**

It is clear from Eq. 7.19 that geometric similarity

$$\frac{\ell_{im}}{\ell_m} = \frac{\ell_i}{\ell} \qquad \frac{\varepsilon_m}{\ell_m} = \frac{\varepsilon}{\ell}$$

as well as Reynolds number similarity

$$\frac{\rho_m V_m \ell_m}{\mu_m} = \frac{\rho V \ell}{\mu}$$

must be maintained. If these conditions are met, then

For flow around bodies, drag is often the dependent variable of interest.

$$\frac{\mathcal{D}}{\frac{1}{2}\rho V^2 \ell^2} = \frac{\mathcal{D}_m}{\frac{1}{2}\rho_m V_m^2 \ell_m^2}$$

or

$$\mathcal{D} = \frac{\rho}{\rho_m}\left(\frac{V}{V_m}\right)^2 \left(\frac{\ell}{\ell_m}\right)^2 \mathcal{D}_m$$

Measurements of model drag, $\mathcal{D}_m$, can then be used to predict the corresponding drag, $\mathcal{D}$, on the prototype from this relationship.

As was discussed in the previous section, one of the common difficulties with models is related to the Reynolds number similarity requirement which establishes the model velocity as

$$V_m = \frac{\mu_m}{\mu}\frac{\rho}{\rho_m}\frac{\ell}{\ell_m}V \qquad\qquad (7.20)$$

or

$$V_m = \frac{\nu_m}{\nu}\frac{\ell}{\ell_m}V \qquad\qquad (7.21)$$

where ν_m/ν is the ratio of kinematic viscosities. If the same fluid is used for model and prototype so that $\nu_m = \nu$, then

$$V_m = \frac{\ell}{\ell_m}V$$

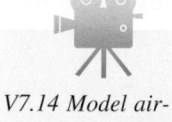

V7.14 Model airplane test in water

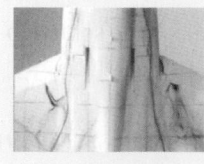

and, therefore, the required model velocity will be higher than the prototype velocity for ℓ/ℓ_m greater than 1. Since this ratio is often relatively large, the required value of V_m may be large. For example, for a $\frac{1}{10}$ length scale, and a prototype velocity of 80 km/h, the required model velocity is 800 km/h. This is a value that is unreasonably high to achieve with liquids, and for gas flows this would be in the range where compressibility would be important in the model (but not in the prototype).

As an alternative, we see from Eq. 7.21 that V_m could be reduced by using a different fluid in the model such that $\nu_m/\nu < 1$. For example, the ratio of the kinematic viscosity of water to that of air is approximately $\frac{1}{10}$, so that if the prototype fluid were air, tests might be run on the model using water. This would reduce the required model velocity, but it still may be difficult to achieve the necessary velocity in a suitable test facility, such as a water tunnel.

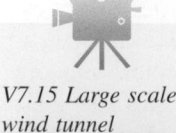

V7.15 Large scale wind tunnel

Another possibility for wind tunnel tests would be to increase the air pressure in the tunnel so that $\rho_m > \rho$, thus reducing the required model velocity as specified by Eq. 7.20. Fluid viscosity is not strongly influenced by pressure. Although pressurized tunnels have been used, they are obviously more complicated and expensive.

The required model velocity can also be reduced if the length scale is modest; that is, the model is relatively large. For wind tunnel testing, this requires a large test section which greatly increases the cost of the facility. However, large wind tunnels suitable for testing very large models (or prototypes) are in use. One such tunnel, located at the NASA Ames Research Center, Moffett Field, California, has a test section that is 12 m by 24 m and can accommodate test speeds to 555 km/h. Such a large and expensive test facility is obviously not feasible for university or industrial laboratories, so most model testing has to be accomplished with relatively small models.

EXAMPLE 7.7 Model Design Conditions and Predicted Prototype Performance

GIVEN The drag on the airplane shown in Fig. E7.7 cruising at 386 km/h in standard air is to be determined from tests on a 1:10 scale model placed in a pressurized wind tunnel. To minimize compressibility effects, the air speed in the wind tunnel is also to be 386 km/h.

FIND Determine

(a) the required air pressure in the tunnel (assuming the same air temperature for model and prototype) and

(b) the drag on the prototype corresponding to a measured force of 4 N on the model.

$V = 386$ km/h

■ **F I G U R E   E7.7**

SOLUTION

(a) From Eq. 7.19 it follows that drag can be predicted from a geometrically similar model if the Reynolds numbers in model and prototype are the same. Thus,

$$\frac{\rho_m V_m \ell_m}{\mu_m} = \frac{\rho V \ell}{\mu}$$

For this example, $V_m = V$ and $\ell_m / \ell = \frac{1}{10}$ so that

$$\frac{\rho_m}{\rho} = \frac{\mu_m}{\mu} \frac{V}{V_m} \frac{\ell}{\ell_m}$$

$$= \frac{\mu_m}{\mu}(1)(10)$$

and therefore

$$\frac{\rho_m}{\rho} = 10 \frac{\mu_m}{\mu}$$

This result shows that the same fluid with $\rho_m = \rho$ and $\mu_m = \mu$ cannot be used if Reynolds number similarity is to be maintained. One possibility is to pressurize the wind tunnel to increase the density of the air. We assume that an increase in pressure does not significantly change the viscosity so that the required increase in density is given by the relationship

$$\frac{\rho_m}{\rho} = 10$$

For an ideal gas, $p = \rho RT$ so that

$$\frac{p_m}{p} = \frac{\rho_m}{\rho}$$

for constant temperature ($T = T_m$). Therefore, the wind tunnel would need to be pressurized so that

$$\frac{p_m}{p} = 10$$

Since the prototype operates at standard atmospheric pressure, the required pressure in the wind tunnel is 10 atmospheres or

$$p_m = 10\,(101.3 \text{ kPa}) \text{ (abs)}$$
$$= 1013 \text{ kPa (abs)} \qquad \textbf{(Ans)}$$

COMMENT Thus, we see that a high pressure would be required and this could not be achieved easily or inexpensively. However, under these conditions, Reynolds similarity would be attained.

(b) The drag could be obtained from Eq. 7.19 so that

$$\frac{\mathcal{D}}{\frac{1}{2}\rho V^2 \ell^2} = \frac{\mathcal{D}_m}{\frac{1}{2}\rho_m V_m^2 \ell_m^2}$$

or

$$\mathcal{D} = \frac{\rho}{\rho_m}\left(\frac{V}{V_m}\right)^2 \left(\frac{\ell}{\ell_m}\right)^2 \mathcal{D}_m$$

$$= \left(\frac{1}{10}\right)(1)^2 (10)^2 \mathcal{D}_m$$

$$= 10 \mathcal{D}_m$$

Thus, for a drag of 4 N on the model the corresponding drag on the prototype is

$$\mathcal{D} = 40 \text{ N} \qquad \textbf{(Ans)}$$

V7.16 Wind tunnel train model

Fortunately, in many situations the flow characteristics are not strongly influenced by the Reynolds number over the operating range of interest. In these cases we can avoid the rather stringent similarity requirement of matching Reynolds numbers. To illustrate this point, consider the variation in the drag coefficient with the Reynolds number for a smooth sphere of diameter d placed in a uniform stream with approach velocity, V. Some typical data are shown in Fig. 7.7. We observe that for Reynolds numbers between approximately 10^3 and 2×10^5 the drag coefficient is relatively constant and does not strongly depend on the specific value of the Reynolds number. Thus, exact Reynolds number similarity is not required in this range. For other geometric shapes we would typically find that for high Reynolds numbers, inertial forces are dominant (rather than viscous forces), and the drag is essentially independent of the Reynolds number.

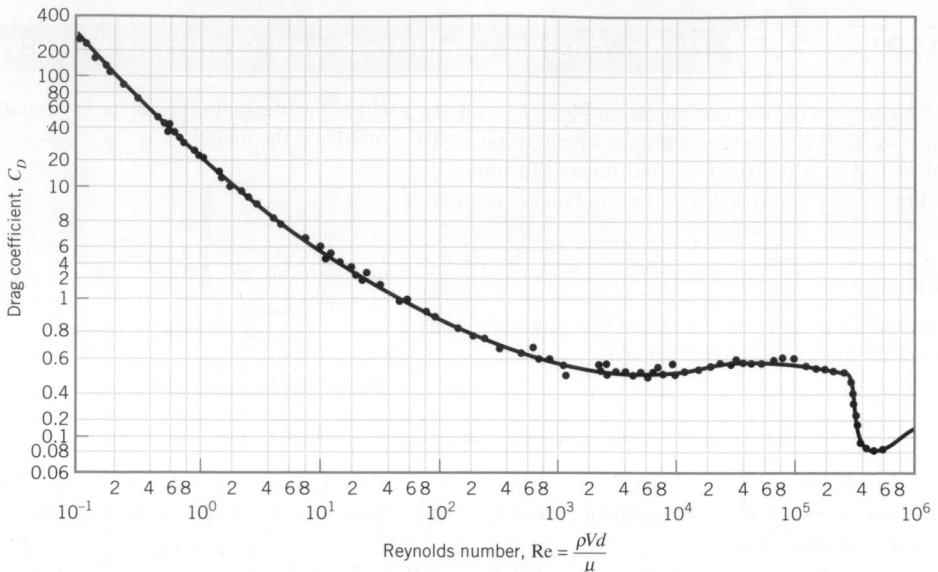

■ **FIGURE 7.7** The effect of Reynolds number on the drag coefficient, C_D, for a smooth sphere with $C_D = \mathcal{D}/\frac{1}{2}A\rho V^2$, where A is the projected area of sphere, $\pi d^2/4$. (Data from Ref. 16, used by permission.)

At high Reynolds numbers the drag is often essentially independent of the Reynolds number.

Another interesting point to note from Fig. 7.7 is the rather abrupt drop in the drag coefficient near a Reynolds number of 3×10^5. As is discussed in Section 9.3.3, this is due to a change in the flow conditions near the surface of the sphere. These changes are influenced by the surface roughness and, in fact, the drag coefficient for a sphere with a "rougher" surface will generally be less than that of the smooth sphere for high Reynolds number. For example, the dimples on a golf ball are used to reduce the drag over that which would occur for a smooth golf ball. Although this is undoubtedly of great interest to the avid golfer, it is also important to engineers responsible for fluid-flow models, since it does emphasize the potential importance of the surface roughness. However, for bodies that are sufficiently angular with sharp corners, the actual surface roughness is likely to play a secondary role compared with the main geometric features of the body.

One final note with regard to Fig. 7.7 concerns the interpretation of experimental data when plotting pi terms. For example, if ρ, μ, and d remain constant, then an increase in Re comes from an increase in V. Intuitively, it would seem in general that if V increases, the drag would increase. However, as shown in the figure, the drag coefficient generally decreases with increasing Re. When interpreting data, one needs to be aware if the variables are nondimensional. In this case, the physical drag force is proportional to the drag coefficient times the velocity squared. Thus, as shown by the figure in the margin, the drag force does, as expected, increase with increasing velocity. The exception occurs in the Reynolds number range $2 \times 10^5 < \text{Re} < 4 \times 10^5$ where the drag coefficient decreases dramatically with increasing Reynolds number (see Fig. 7.7). This phenomena is discussed in Section 9.3.

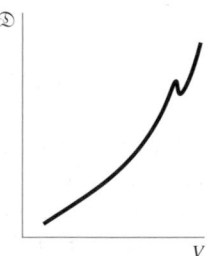

For problems involving high velocities in which the Mach number is greater than about 0.3, the influence of compressibility, and therefore the Mach number (or Cauchy number), becomes significant. In this case complete similarity requires not only geometric and Reynolds number similarity but also Mach number similarity so that

$$\frac{V_m}{c_m} = \frac{V}{c} \tag{7.22}$$

This similarity requirement, when combined with that for Reynolds number similarity (Eq. 7.21), yields

$$\frac{c}{c_m} = \frac{\nu}{\nu_m}\frac{\ell_m}{\ell} \tag{7.23}$$

Clearly the same fluid with $c = c_m$ and $\nu = \nu_m$ cannot be used in model and prototype unless the length scale is unity (which means that we are running tests on the prototype). In high-speed aerodynamics the prototype fluid is usually air, and it is difficult to satisfy Eq. 7.23 for reasonable length scales. Thus, models involving high-speed flows are often distorted with respect to Reynolds number similarity, but Mach number similarity is maintained.

7.9.3 Flow with a Free Surface

Froude number similarity is usually required for models involving free-surface flows.

Flows in canals, rivers, spillways, and stilling basins, as well as flow around ships, are all examples of flow phenomena involving a free surface. For this class of problems, both gravitational and inertial forces are important and, therefore, the Froude number becomes an important similarity parameter. Also, since there is a free surface with a liquid–air interface, forces due to surface tension may be significant, and the Weber number becomes another similarity parameter that needs to be considered along with the Reynolds number. Geometric variables will obviously still be important. Thus a general formulation for problems involving flow with a free surface can be expressed as

$$\text{Dependent pi term} = \phi\left(\frac{\ell_i}{\ell}, \frac{\varepsilon}{\ell}, \frac{\rho V \ell}{\mu}, \frac{V}{\sqrt{g\ell}}, \frac{\rho V^2 \ell}{\sigma}\right) \quad (7.24)$$

As discussed previously, ℓ is some characteristic length of the system, ℓ_i represents other pertinent lengths, and ε/ℓ is the relative roughness of the various surfaces. Since gravity is the driving force in these problems, Froude number similarity is definitely required so that

$$\frac{V_m}{\sqrt{g_m \ell_m}} = \frac{V}{\sqrt{g\ell}}$$

The model and prototype are expected to operate in the same gravitational field $(g_m = g)$, and therefore it follows that

$$\frac{V_m}{V} = \sqrt{\frac{\ell_m}{\ell}} = \sqrt{\lambda_\ell} \quad (7.25)$$

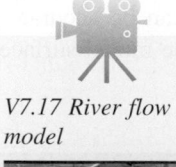

V7.17 River flow model

Thus, when models are designed on the basis of Froude number similarity, the velocity scale is determined by the square root of the length scale. As is discussed in Section 7.8.3, to simultaneously have Reynolds and Froude number similarity it is necessary that the kinematic viscosity scale be related to the length scale as

$$\frac{\nu_m}{\nu} = (\lambda_\ell)^{3/2} \quad (7.26)$$

The working fluid for the prototype is normally either freshwater or seawater and the length scale is small. Under these circumstances it is virtually impossible to satisfy Eq. 7.26, so models involving free-surface flows are usually distorted. The problem is further complicated if an attempt is made to model surface tension effects, since this requires the equality of Weber numbers, which leads to the condition

$$\frac{\sigma_m/\rho_m}{\sigma/\rho} = (\lambda_\ell)^2 \quad (7.27)$$

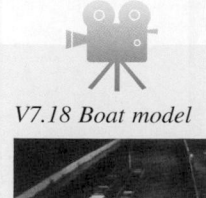

V7.18 Boat model

for the kinematic surface tension (σ/ρ). It is again evident that the same fluid cannot be used in model and prototype if we are to have similitude with respect to surface tension effects for $\lambda_\ell \neq 1$.

Fortunately, in many problems involving free-surface flows, both surface tension and viscous effects are small and consequently strict adherence to Weber and Reynolds number similarity is not required. Certainly, surface tension is not important in large hydraulic structures and rivers. Our only concern would be if in a model the depths were reduced to the point where surface tension becomes an important factor, whereas it is not in the prototype. This is of particular importance in the design of river models, since the length scales are typically small (so that the width of the model is reasonable), but with a small length scale the required model depth may be very small. To overcome this problem, different horizontal and vertical length scales are often used for river

■ **FIGURE 7.8** A scale hydraulic model (1:197) of the Guri Dam in Venezuela which is used to simulate the characteristics of the flow over and below the spillway and the erosion below the spillway. (Photograph courtesy of St. Anthony Falls Hydraulic Laboratory.)

models. Although this approach eliminates surface tension effects in the model, it introduces geometric distortion that must be accounted for empirically, usually by increasing the model surface roughness. It is important in these circumstances that verification tests with the model be performed (if possible) in which model data are compared with available prototype river flow data. Model roughness can be adjusted to give satisfactory agreement between model and prototype, and then the model subsequently used to predict the effect of proposed changes on river characteristics (such as velocity patterns or surface elevations).

V7.19 Dam model

For large hydraulic structures, such as dam spillways, the Reynolds numbers are large so that viscous forces are small in comparison to the forces due to gravity and inertia. In this case, Reynolds number similarity is not maintained and models are designed on the basis of Froude number similarity. Care must be taken to ensure that the model Reynolds numbers are also large, but they are not required to be equal to those of the prototype. This type of hydraulic model is usually made as large as possible so that the Reynolds number will be large. A spillway model is shown in Fig. 7.8. Also, for relatively large models the geometric features of the prototype can be accurately scaled, as well as surface roughness. Note that $\varepsilon_m = \lambda_\ell \varepsilon$, which indicates that the model surfaces must be smoother than the corresponding prototype surfaces for $\lambda_\ell < 1$.

F l u i d s   i n   t h e   N e w s

Ice engineering Various types of models have been studied in wind tunnels, water tunnels, and towing tanks for many years. But another type of facility is needed to study ice and ice-related problems. The U.S. Army Cold Regions Research and Engineering Laboratory has developed a unique complex that houses research facilities for studies related to the mechanical behavior of ice and ice–structure interactions. The laboratory contains three separate cold-rooms—a test basin, a flume, and a general research area. In the test basin, large-scale *model studies* of ice forces on structures such as dams, piers, ships, and offshore platforms can be performed. Ambient temperatures can be controlled as low as −20 °F (−29 °C or 244 K), and at this temperature a 2-mm per hour ice growth rate can be achieved. It is also possible to control the mechanical properties of the ice to properly match the physical scale of the model. Tests run in the recirculating flume can simulate river processes during ice formation. And in the large research area, scale models of lakes and rivers can be built and operated to model ice interactions with various types of engineering projects. (See Problem 7.73.)

*E*XAMPLE 7.8   Froude Number Similarity

GIVEN The spillway for the dam shown in Fig. E7.8*a* is 20 m wide and is designed to carry 125 m³/s at flood stage. A 1:15 model is constructed to study the flow characteristics through the spillway. The effects of surface tension and viscosity are to be neglected.

FIND

(a) Determine the required model width and flowrate.

(b) What operating time for the model corresponds to a 24-hr period in the prototype?

■ **FIGURE E7.8*a***

SOLUTION

The width, w_m, of the model spillway is obtained from the length scale, λ_ℓ, so that

$$\frac{w_m}{w} = \lambda_\ell = \frac{1}{15}$$

Therefore,

$$w_m = \frac{20 \text{ m}}{15} = 1.33 \text{ m} \quad \text{(Ans)}$$

Of course, all other geometric features (including surface roughness) of the spillway must be scaled in accordance with the same length scale.

With the neglect of surface tension and viscosity, Eq. 7.24 indicates that dynamic similarity will be achieved if the Froude numbers are equal between model and prototype. Thus,

$$\frac{V_m}{\sqrt{g_m \ell_m}} = \frac{V}{\sqrt{g\ell}}$$

and for $g_m = g$

$$\frac{V_m}{V} = \sqrt{\frac{\ell_m}{\ell}}$$

Since the flowrate is given by $Q = VA$, where A is an appropriate cross-sectional area, it follows that

$$\frac{Q_m}{Q} = \frac{V_m A_m}{VA} = \sqrt{\frac{\ell_m}{\ell}}\left(\frac{\ell_m}{\ell}\right)^2$$

$$= (\lambda_\ell)^{5/2}$$

where we have made use of the relationship $A_m/A = (\ell_m/\ell)^2$. For $\lambda_\ell = \frac{1}{15}$ and $Q = 125 \text{ m}^3/\text{s}$

$$Q_m = (\tfrac{1}{15})^{5/2}(125 \text{ m}^3/\text{s}) = 0.143 \text{ m}^3/\text{s} \quad \text{(Ans)}$$

The time scale can be obtained from the velocity scale, since the velocity is distance divided by time ($V = \ell/t$), and therefore

$$\frac{V}{V_m} = \frac{\ell}{t}\frac{t_m}{\ell_m}$$

or

$$\frac{t_m}{t} = \frac{V}{V_m}\frac{\ell_m}{\ell} = \sqrt{\frac{\ell_m}{\ell}} = \sqrt{\lambda_\ell}$$

This result indicates that time intervals in the model will be smaller than the corresponding intervals in the prototype if $\lambda_\ell < 1$. For $\lambda_\ell = \frac{1}{15}$ and a prototype time interval of 24 hr

$$t_m = \sqrt{\tfrac{1}{15}}(24 \text{ hr}) = 6.20 \text{ hr} \quad \text{(Ans)}$$

COMMENT As indicated by the above analysis, the time scale varies directly as the square root of the length scale. Thus, as shown in Fig. E7.8b, the model time interval, t_m, corresponding to a 24-hr prototype time interval can be varied by changing the length scale, λ_ℓ. The ability to scale times may be very useful, since it is possible to "speed up" events in the model which may occur over a relatively long time in the prototype. There is of course a practical limit to how small the length scale (and the corresponding time scale) can become. For example, if the length scale is too small then surface tension effects may become important in the model whereas they are not in the prototype. In such a case the present model design, based simply on Froude number similarity, would not be adequate.

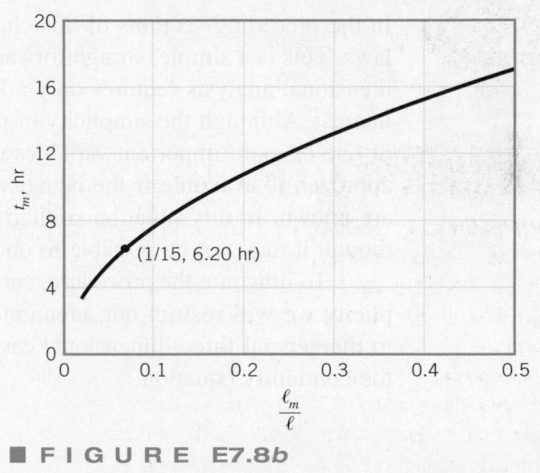

■ **FIGURE E7.8b**

V7.20 Testing of large yacht mode

There are, unfortunately, problems involving flow with a free surface in which viscous, inertial, and gravitational forces are all important. The drag on a ship as it moves through water is due to the viscous shearing stresses that develop along its hull, as well as a pressure-induced component of drag caused by both the shape of the hull and wave action. The shear drag is a function of the Reynolds number, whereas the pressure drag is a function of the Froude number. Since both Reynolds number and Froude number similarity cannot be simultaneously achieved by using water as the model fluid (which is the only practical fluid for ship models), some technique other than a straightforward model test must be employed. One common approach is to measure the total drag on a small, geometrically similar model as it is towed through a model basin at Froude numbers matching those of the prototype. The shear drag on the model is calculated using analytical techniques of the type described in Chapter 9. This calculated value is then subtracted from the total drag to obtain pressure drag, and using Froude number scaling the pressure drag on the prototype can then be predicted. The experimentally determined value can then be combined with a calculated value of the shear drag (again using analytical techniques) to provide the desired total drag

■ **FIGURE 7.9** Instrumented, small-waterplane-area, twin hull (SWATH) model suspended from a towing carriage. (Photograph courtesy of the U.S. Navy's David W. Taylor Research Center.)

on the ship. Ship models are widely used to study new designs, but the tests require extensive facilities (see Fig. 7.9).

It is clear from this brief discussion of various types of models involving free-surface flows that the design and use of such models requires considerable ingenuity, as well as a good understanding of the physical phenomena involved. This is generally true for most model studies. Modeling is both an art and a science. Motion picture producers make extensive use of model ships, fires, explosions, and the like. It is interesting to attempt to observe the flow differences between these distorted model flows and the real thing.

7.10 Similitude Based on Governing Differential Equations

Similarity laws can be directly developed from the equations governing the phenomenon of interest.

In the preceding sections of this chapter, dimensional analysis has been used to obtain similarity laws. This is a simple, straightforward approach to modeling, which is widely used. The use of dimensional analysis requires only a knowledge of the variables that influence the phenomenon of interest. Although the simplicity of this approach is attractive, it must be recognized that omission of one or more important variables may lead to serious errors in the model design. An alternative approach is available if the equations (usually differential equations) governing the phenomenon are known. In this situation similarity laws can be developed from the governing equations, even though it may not be possible to obtain analytic solutions to the equations.

To illustrate the procedure, consider the flow of an incompressible Newtonian fluid. For simplicity we will restrict our attention to two-dimensional flow, although the results are applicable to the general three-dimensional case. From Chapter 6 we know that the governing equations are the continuity equation

$$\frac{\partial u}{\partial x} + \frac{\partial v}{\partial y} = 0 \tag{7.28}$$

and the Navier–Stokes equations

$$\rho\left(\frac{\partial u}{\partial t} + u\frac{\partial u}{\partial x} + v\frac{\partial u}{\partial y}\right) = -\frac{\partial p}{\partial x} + \mu\left(\frac{\partial^2 u}{\partial x^2} + \frac{\partial^2 u}{\partial^2 y}\right) \tag{7.29}$$

$$\rho\left(\frac{\partial v}{\partial t} + u\frac{\partial v}{\partial x} + v\frac{\partial v}{\partial y}\right) = -\frac{\partial p}{\partial y} - \rho g + \mu\left(\frac{\partial^2 v}{\partial x^2} + \frac{\partial^2 v}{\partial y^2}\right) \tag{7.30}$$

where the y axis is vertical, so that the gravitational body force, ρg, only appears in the "y equation." To continue the mathematical description of the problem, boundary conditions are required. For example, velocities on all boundaries may be specified; that is, $u = u_B$ and $v = v_B$ at all boundary points $x = x_B$ and $y = y_B$. In some types of problems it may be necessary to specify the pressure over some part of the boundary. For time-dependent problems, initial conditions would also have to be provided, which means that the values of all dependent variables would be given at some time (usually taken at $t = 0$).

Once the governing equations, including boundary and initial conditions, are known, we are ready to proceed to develop similarity requirements. The next step is to define a new set of variables that are dimensionless. To do this we select a reference quantity for each type of variable. In this problem the variables are u, v, p, x, y, and t so we will need a reference velocity, V, a reference pressure, p_0, a reference length, ℓ, and a reference time, τ. These reference quantities should be parameters that appear in the problem. For example, ℓ may be a characteristic length of a body immersed in a fluid or the width of a channel through which a fluid is flowing. The velocity, V, may be the free-stream velocity or the inlet velocity. The new dimensionless (starred) variables can be expressed as

$$u^* = \frac{u}{V} \qquad v^* = \frac{v}{V} \qquad p^* = \frac{p}{p_0}$$

$$x^* = \frac{x}{\ell} \qquad y^* = \frac{y}{\ell} \qquad t^* = \frac{t}{\tau}$$

> *Each variable is made dimensionless by dividing by an appropriate reference quantity.*

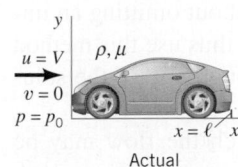

Actual

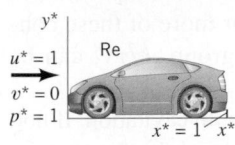

Dimensionless

as shown in the figure in the margin.

The governing equations can now be rewritten in terms of these new variables. For example,

$$\frac{\partial u}{\partial x} = \frac{\partial V u^*}{\partial x^*}\frac{\partial x^*}{\partial x} = \frac{V}{\ell}\frac{\partial u^*}{\partial x^*}$$

and

$$\frac{\partial^2 u}{\partial x^2} = \frac{V}{\ell}\frac{\partial}{\partial x^*}\left(\frac{\partial u^*}{\partial x^*}\right)\frac{\partial x^*}{\partial x} = \frac{V}{\ell^2}\frac{\partial^2 u^*}{\partial x^{*2}}$$

The other terms that appear in the equations can be expressed in a similar fashion. Thus, in terms of the new variables the governing equations become

$$\frac{\partial u^*}{\partial x^*} + \frac{\partial v^*}{\partial y^*} = 0 \tag{7.31}$$

and

$$\underbrace{\left[\frac{\rho V}{\tau}\right]}_{}\frac{\partial u^*}{\partial t^*} + \underbrace{\left[\frac{\rho V^2}{\ell}\right]}_{}\left(u^*\frac{\partial u^*}{\partial x^*} + v^*\frac{\partial u^*}{\partial y^*}\right) = -\left[\frac{p_0}{\ell}\right]\frac{\partial p^*}{\partial x^*} + \left[\frac{\mu V}{\ell^2}\right]\left(\frac{\partial^2 u^*}{\partial x^{*2}} + \frac{\partial^2 u^*}{\partial y^{*2}}\right) \tag{7.32}$$

$$\underbrace{\left[\frac{\rho V}{\tau}\right]}_{F_{I\ell}}\frac{\partial v^*}{\partial t^*} + \underbrace{\left[\frac{\rho V^2}{\ell}\right]}_{F_{Ic}}\left(u^*\frac{\partial v^*}{\partial x^*} + v^*\frac{\partial v^*}{\partial y^*}\right)$$

$$= -\underbrace{\left[\frac{p_0}{\ell}\right]}_{F_P}\frac{\partial p^*}{\partial y^*} - \underbrace{[\rho g]}_{F_G} + \underbrace{\left[\frac{\mu V}{\ell^2}\right]}_{F_V}\left(\frac{\partial^2 v^*}{\partial x^{*2}} + \frac{\partial^2 v^*}{\partial y^{*2}}\right) \tag{7.33}$$

The terms appearing in brackets contain the reference quantities and can be interpreted as indices of the various forces (per unit volume) that are involved. Thus, as is indicated in Eq. 7.33, $F_{I\ell}$ = inertia (local) force, F_{Ic} = inertia (convective) force, F_p = pressure force, F_G = gravitational force, and F_V = viscous force. As the final step in the nondimensionalization process, we will divide each term in Eqs. 7.32 and 7.33 by one of the bracketed quantities. Although any one of these quantities could be used, it is conventional to divide by the bracketed quantity $\rho V^2/\ell$ which is the index of the convective inertia force. The final nondimensional form then becomes

$$\left[\frac{\ell}{\tau V}\right]\frac{\partial u^*}{\partial t^*} + u^*\frac{\partial u^*}{\partial x^*} + v^*\frac{\partial u^*}{\partial y^*} = -\left[\frac{p_0}{\rho V^2}\right]\frac{\partial p^*}{\partial x^*} + \left[\frac{\mu}{\rho V \ell}\right]\left(\frac{\partial^2 u^*}{\partial x^{*2}} + \frac{\partial^2 u^*}{\partial y^{*2}}\right) \tag{7.34}$$

$$\left[\frac{\ell}{\tau V}\right]\frac{\partial v^*}{\partial t^*} + u^*\frac{\partial v^*}{\partial x^*} + v^*\frac{\partial v^*}{\partial y^*} = -\left[\frac{p_0}{\rho V^2}\right]\frac{\partial p^*}{\partial y^*} - \left[\frac{g\ell}{V^2}\right] + \left[\frac{\mu}{\rho V \ell}\right]\left(\frac{\partial^2 v^*}{\partial x^{*2}} + \frac{\partial^2 v^*}{\partial y^{*2}}\right) \tag{7.35}$$

We see that bracketed terms are the standard dimensionless groups (or their reciprocals) which were developed from dimensional analysis; that is, $\ell/\tau V$ is a form of the Strouhal number, $p_0/\rho V^2$

the Euler number, $g\ell/V^2$ the reciprocal of the square of the Froude number, and $\mu/\rho V\ell$ the reciprocal of the Reynolds number. From this analysis it is now clear how each of the dimensionless groups can be interpreted as the ratio of two forces, and how these groups arise naturally from the governing equations.

Governing equations expressed in terms of dimensionless variables lead to the appropriate dimensionless groups.

Although we really have not helped ourselves with regard to obtaining an analytical solution to these equations (they are still complicated and not amenable to an analytical solution), the dimensionless forms of the equations, Eqs. 7.31, 7.34, and 7.35, can be used to establish similarity requirements. From these equations it follows that if two systems are governed by these equations, then the solutions (in terms of u^*, v^*, p^*, x^*, y^*, and t^*) will be the same if the four parameters $\ell/\tau V$, $p_0/\rho V^2$, $V^2/g\ell$, and $\rho V\ell/\mu$ are equal for the two systems. The two systems will be dynamically similar. Of course, boundary and initial conditions expressed in dimensionless form must also be equal for the two systems, and this will require complete geometric similarity. These are the same similarity requirements that would be determined by a dimensional analysis if the same variables were considered. However, the advantage of working with the governing equations is that the variables appear naturally in the equations, and we do not have to worry about omitting an important one, provided the governing equations are correctly specified. We can thus use this method to deduce the conditions under which two solutions will be similar even though one of the solutions will most likely be obtained experimentally.

In the foregoing analysis we have considered a general case in which the flow may be unsteady, and both the actual pressure level, p_0, and the effect of gravity are important. A reduction in the number of similarity requirements can be achieved if one or more of these conditions is removed. For example, if the flow is steady the dimensionless group, $\ell/\tau V$, can be eliminated.

The actual pressure level will only be of importance if we are concerned with cavitation. If not, the flow patterns and the pressure differences will not depend on the pressure level. In this case, p_0 can be taken as ρV^2 (or $\frac{1}{2}\rho V^2$), and the Euler number can be eliminated as a similarity requirement. However, if we are concerned about cavitation (which will occur in the flow field if the pressure at certain points reaches the vapor pressure, p_v), then the actual pressure level is important. Usually, in this case, the characteristic pressure, p_0, is defined relative to the vapor pressure such that $p_0 = p_r - p_v$ where p_r is some reference pressure within the flow field. With p_0 defined in this manner, the similarity parameter $p_0/\rho V^2$ becomes $(p_r - p_v)/\rho V^2$. This parameter is frequently written as $(p_r - p_v)/\frac{1}{2}\rho V^2$, and in this form, as was noted previously in Section 7.6, is called the cavitation number. Thus we can conclude that if cavitation is not of concern we do not need a similarity parameter involving p_0, but if cavitation is to be modeled, then the cavitation number becomes an important similarity parameter.

The Froude number, which arises because of the inclusion of gravity, is important for problems in which there is a free surface. Examples of these types of problems include the study of rivers, flow through hydraulic structures such as spillways, and the drag on ships. In these situations the shape of the free surface is influenced by gravity, and therefore the Froude number becomes an important similarity parameter. However, if there are no free surfaces, the only effect of gravity is to superimpose a hydrostatic pressure distribution on the pressure distribution created by the fluid motion. The hydrostatic distribution can be eliminated from the governing equation (Eq. 7.30) by defining a new pressure, $p' = p - \rho gy$, and with this change the Froude number does not appear in the nondimensional governing equations.

We conclude from this discussion that for the steady flow of an incompressible fluid without free surfaces, dynamic and kinematic similarity will be achieved if (for geometrically similar systems) Reynolds number similarity exists. If free surfaces are involved, Froude number similarity must also be maintained. For free-surface flows we have tacitly assumed that surface tension is not important. We would find, however, that if surface tension is included, its effect would appear in the free-surface boundary condition, and the Weber number, $\rho V^2\ell/\sigma$, would become an additional similarity parameter. In addition, if the governing equations for compressible fluids are considered, the Mach number, V/c, would appear as an additional similarity parameter.

It is clear that all the common dimensionless groups that we previously developed by using dimensional analysis appear in the governing equations that describe fluid motion when these equations are expressed in terms of dimensionless variables. Thus, the use of the governing equations to obtain similarity laws provides an alternative to dimensional analysis. This approach has the advantage that the variables are known and the assumptions involved are clearly identified. In addition, a physical interpretation of the various dimensionless groups can often be obtained.

7.11   Chapter Summary and Study Guide

Many practical engineering problems involving fluid mechanics require experimental data for their solution. Thus, laboratory studies and experimentation play a significant role in this field. It is important to develop good procedures for the design of experiments so they can be efficiently completed with as broad applicability as possible. To achieve this end the concept of similitude is often used in which measurements made in the laboratory can be utilized for predicting the behavior of other similar systems. In this chapter, dimensional analysis is used for designing such experiments, as an aid for correlating experimental data, and as the basis for the design of physical models. As the name implies, dimensional analysis is based on a consideration of the dimensions required to describe the variables in a given problem. A discussion of the use of dimensions and the concept of dimensional homogeneity (which forms the basis for dimensional analysis) was included in Chapter 1.

Essentially, dimensional analysis simplifies a given problem described by a certain set of variables by reducing the number of variables that need to be considered. In addition to being fewer in number, the new variables are dimensionless products of the original variables. Typically these new dimensionless variables are much simpler to work with in performing the desired experiments. The Buckingham pi theorem, which forms the theoretical basis for dimensional analysis, is introduced. This theorem establishes the framework for reducing a given problem described in terms of a set of variables to a new set of fewer dimensionless variables. A simple method, called the repeating variable method, is described for actually forming the dimensionless variables (often called pi terms). Forming dimensionless variables by inspection is also considered. It is shown how the use of dimensionless variables can be of assistance in planning experiments and as an aid in correlating experimental data.

For problems in which there are a large number of variables, the use of physical models is described. Models are used to make specific predictions from laboratory tests rather than formulating a general relationship for the phenomenon of interest. The correct design of a model is obviously imperative for the accurate predictions of other similar, but usually larger, systems. It is shown how dimensional analysis can be used to establish a valid model design. An alternative approach for establishing similarity requirements using governing equations (usually differential equations) is presented.

The following checklist provides a study guide for this chapter. When your study of the entire chapter and end-of-chapter exercies has been completed you should be able to

similitude
dimensionless product
basic dimensions
pi term
Buckingham pi theorem
method of repeating
  variables
model
modeling laws
prototype
prediction equation
model design conditions
similarity requirements
modeling laws
length scale
distorted model
true model

- write out meanings of the terms listed here in the margin and understand each of the related concepts. These terms are particularly important and are set in *italic, bold, and color* type in the text.
- use the Buckingham pi theorem to determine the number of independent dimensionless variables needed for a given flow problem.
- form a set of dimensionless variables using the method of repeating variables.
- form a set of dimensionless variables by inspection.
- use dimensionless variables as an aid in interpreting and correlating experimental data.
- use dimensional analysis to establish a set of similarity requirements (and prediction equation) for a model to be used to predict the behavior of another similar system (the prototype).
- rewrite a given governing equation in a suitable nondimensional form and deduce similarity requirements from the nondimensional form of the equation.

Some of the important equations in this chapter are:

Reynolds number	$Re = \dfrac{\rho V \ell}{\mu}$
Froude number	$Fr = \dfrac{V}{\sqrt{g\ell}}$

Euler number	$\mathrm{Eu} = \dfrac{p}{\rho V^2}$
Cauchy number	$\mathrm{Ca} = \dfrac{\rho V^2}{E_v}$
Mach number	$\mathrm{Ma} = \dfrac{V}{c}$
Strouhal number	$\mathrm{St} = \dfrac{\omega \ell}{V}$
Weber number	$\mathrm{We} = \dfrac{\rho V^2 \ell}{\sigma}$

References

1. Bridgman, P. W., *Dimensional Analysis*, Yale University Press, New Haven, Conn., 1922.
2. Murphy, G., *Similitude in Engineering*, Ronald Press, New York, 1950.
3. Langhaar, H. L., *Dimensional Analysis and Theory of Models*, Wiley, New York, 1951.
4. Huntley, H. E., *Dimensional Analysis*, Macdonald, London, 1952.
5. Duncan, W. J., *Physical Similarity and Dimensional Analysis: An Elementary Treatise*, Edward Arnold, London, 1953.
6. Sedov, K. I., *Similarity and Dimensional Methods in Mechanics*, Academic Press, New York, 1959.
7. Ipsen, D. C., *Units, Dimensions, and Dimensionless Numbers*, McGraw-Hill, New York, 1960.
8. Kline, S. J., *Similitude and Approximation Theory*, McGraw-Hill, New York, 1965.
9. Skoglund, V. J., *Similitude—Theory and Applications*, International Textbook, Scranton, Pa., 1967.
10. Baker, W. E., Westline, P. S., and Dodge, F. T., *Similarity Methods in Engineering Dynamics—Theory and Practice of Scale Modeling*, Hayden (Spartan Books), Rochelle Park, N.J., 1973.
11. Taylor, E. S., *Dimensional Analysis for Engineers*, Clarendon Press, Oxford, 1974.
12. Isaacson, E. de St. Q., and Isaacson, M. de St. Q., *Dimensional Methods in Engineering and Physics*, Wiley, New York, 1975.
13. Schuring, D. J., *Scale Models in Engineering*, Pergamon Press, New York, 1977.
14. Yalin, M. S., *Theory of Hydraulic Models*, Macmillan, London, 1971.
15. Sharp, J. J., *Hydraulic Modeling*, Butterworth, London, 1981.
16. Schlichting, H., *Boundary-Layer Theory*, 7th Ed., McGraw-Hill, New York, 1979.
17. Knapp, R. T., Daily, J. W., and Hammitt, F. G., *Cavitation*, McGraw-Hill, New York, 1970.

Review Problems

Go to Appendix G for a set of review problems with answers. Detailed solutions can be found in *Student Solution Manual and Study* *Guide for Fundamentals of Fluid Mechanics*, by Munson et al. (© 2009 John Wiley and Sons, Inc.).

Problems

Note: Unless otherwise indicated, use the values of fluid properties found in the tables on the inside of the front cover. Problems designated with an (*) are intended to be solved with the aid of a programmable calculator or a computer. Problems designated with a (†) are "open-ended" problems and require critical thinking in that to work them one must make various assumptions and provide the necessary data. There is not a unique answer to these problems.

Answers to the even-numbered problems are listed at the end of the book. Access to the videos that accompany problems can be obtained through the book's web site, www.wiley.com/college/munson. The lab-type problems and FlowLab problems can also be accessed on this web site.

Section 7.1 Dimensional Analysis

7.1 Obtain a photograph/image of an experimental setup used to investigate some type of fluid flow phenomena. Print this photo and write a brief paragraph that describes the situation involved.

7.2 Verify the left-hand side of Eq. 7.2 is dimensionless using the MLT system.

7.3 The Reynolds number, $\rho VD/\mu$, is a very important parameter in fluid mechanics. Verify that the Reynolds number is dimensionless, using both the *FLT* system and the *MLT* system for basic dimensions, and determine its value for ethyl alcohol flowing at a velocity of 3 m/s through a 5-cm-diameter pipe.

7.4 What are the dimensions of acceleration of gravity, density, dynamic viscosity, kinematic viscosity, specific weight, and speed of sound in (a) the *FLT* system, and (b) the *MLT* system? Compare your results with those given in Table 1.1 in Chapter 1.

7.5 For the flow of a thin film of a liquid with a depth h and a free surface, two important dimensionless parameters are the Froude number, $V/\sqrt{gh}$, and the Weber number, $\rho V^2 h/\sigma$. Determine the value of these two parameters for glycerin (at 20 °C) flowing with a velocity of 0.7 m/s at a depth of 3 mm.

7.6 The Mach number for a body moving through a fluid with velocity V is defined as V/c, where c is the speed of sound in the fluid. This dimensionless parameter is usually considered to be important in fluid dynamics problems when its value exceeds 0.3. What would be the velocity of a body at a Mach number of 0.3 if the fluid is (a) air at standard atmospheric pressure and 20 °C, and (b) water at the same temperature and pressure?

Section 7.3 **Determination of Pi Terms**

7.7 Obtain a photograph/image of Osborne Reynolds, who developed the famous dimensionless quantity, the Reynolds number. Print this photo and write a brief paragraph about him.

7.8 The power, $\mathcal{P}$, required to run a pump that moves fluid within a piping system is dependent upon the volume flowrate, Q, density, ρ, impeller diameter, d, angular velocity, ω, and fluid viscosity, μ. Find the number of pi terms for this relationship.

7.9 For low speed flow over a flat plate, one measure of the boundary layer is the resulting thickness, δ, at a given downstream location. The boundary layer thickness is a function of the free stream velocity, V_∞, fluid density and viscosity ρ and μ, and the distance from the leading edge, x. Find the number of pi terms for this relationship.

7.10 The excess pressure inside a bubble (discussed in Chapter 1) is known to be dependent on bubble radius and surface tension. After finding the pi terms, determine the variation in excess pressure if we (a) double the radius and (b) double the surface tension.

7.11 It is known that the variation of pressure, Δp, within a static fluid is dependent upon the specific weight of the fluid and the elevation difference, Δz. Using dimensional analysis, find the form of the hydrostatic equation for pressure variation.

7.12 At a sudden contraction in a pipe the diameter changes from D_1 to D_2. The pressure drop, Δp, which develops across the contraction is a function of D_1 and D_2, as well as the velocity, V, in the larger pipe, and the fluid density, ρ, and viscosity, μ. Use D_1, V, and μ as repeating variables to determine a suitable set of dimensionless parameters. Why would it be incorrect to include the velocity in the smaller pipe as an additional variable?

7.13 Water sloshes back and forth in a tank as shown in Fig. P7.13. The frequency of sloshing, ω, is assumed to be a function of the acceleration of gravity, g, the average depth of the water, h, and the length of the tank, ℓ. Develop a suitable set of dimensionless parameters for this problem using g and ℓ as repeating variables.

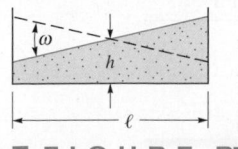

■ **FIGURE P7.13**

7.14 Assume that the power, $\mathcal{P}$, required to drive a fan is a function of the fan diameter, D, the fluid density, ρ, the rotational speed, ω, and the flowrate, Q. Use D, ω, and ρ as repeating variables to determine a suitable set of pi terms.

7.15 Assume that the flowrate, Q, of a gas from a smokestack is a function of the density of the ambient air, ρ_a, the density of the gas, ρ_g, within the stack, the acceleration of gravity, g, and the height and diameter of the stack, h and d, respectively. Use ρ_a, d, and g as repeating variables to develop a set of pi terms that could be used to describe this problem.

7.16 The pressure rise, Δp, across a pump can be expressed as

$$\Delta p = f(D, \rho, \omega, Q)$$

where D is the impeller diameter, ρ the fluid density, ω the rotational speed, and Q the flowrate. Determine a suitable set of dimensionless parameters.

7.17 A thin elastic wire is placed between rigid supports. A fluid flows past the wire, and it is desired to study the static deflection, δ, at the center of the wire due to the fluid drag. Assume that

$$\delta = f(\ell, d, \rho, \mu, V, E)$$

where ℓ is the wire length, d the wire diameter, ρ the fluid density, μ the fluid viscosity, V the fluid velocity, and E the modulus of elasticity of the wire material. Develop a suitable set of pi terms for this problem.

7.18 Because of surface tension, it is possible, with care, to support an object heavier than water on the water surface as shown in Fig. P7.18. (See **Video V1.9.**) The maximum thickness, h, of a square of material that can be supported is assumed to be a function of the length of the side of the square, ℓ, the density of the material, ρ, the acceleration of gravity, g, and the surface tension of the liquid, σ. Develop a suitable set of dimensionless parameters for this problem.

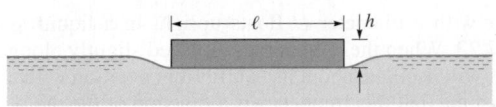

■ **FIGURE P7.18**

7.19 Under certain conditions, wind blowing past a rectangular speed limit sign can cause the sign to oscillate with a frequency ω. (See Fig. P7.19 and **Video V9.9.**) Assume that ω is a function of the sign width, b, sign height, h, wind velocity, V, air density, ρ, and an elastic constant, k, for the supporting pole. The constant, k, has dimensions of FL. Develop a suitable set of pi terms for this problem.

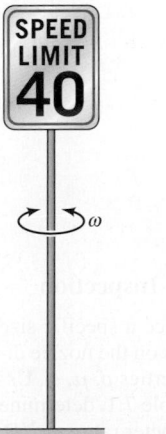

■ **FIGURE P7.19**

7.20 The height, h, that a liquid will rise in a capillary tube is a function of the tube diameter, D, the specific weight of the liquid, γ, and the surface tension, σ. Perform a dimensional analysis using both the *FLT* and *MLT* systems for basic dimensions. Note: the results should obviously be the same regardless of the system of dimensions used. If your analysis indicates otherwise, go back and check your work, giving particular attention to the required number of reference dimensions.

7.21 A cone and plate viscometer consists of a cone with a very small angle α which rotates above a flat surface as shown in Fig. P7.21. The torque, $\mathcal{T}$, required to rotate the cone at an angular velocity ω is a function of the radius, R, the cone angle, α, and the fluid viscosity, μ, in addition to ω. With the aid of dimensional analysis, determine how the torque will change if both the viscosity and angular velocity are doubled.

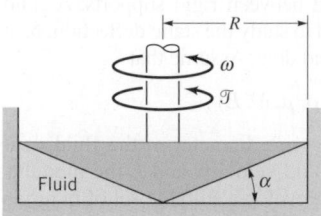

■ **FIGURE P7.21**

7.22 The pressure drop, Δp, along a straight pipe of diameter D has been experimentally studied, and it is observed that for laminar flow of a given fluid and pipe, the pressure drop varies directly with the distance, ℓ, between pressure taps. Assume that Δp is a function of D and ℓ, the velocity, V, and the fluid viscosity, μ. Use dimensional analysis to deduce how the pressure drop varies with pipe diameter.

7.23 A cylinder with a diameter D floats upright in a liquid as shown in Fig. P7.23. When the cylinder is displaced slightly along its vertical axis it will oscillate about its equilibrium position with a frequency, ω. Assume that this frequency is a function of the diameter, D, the mass of the cylinder, m, and the specific weight, γ, of the liquid. Determine, with the aid of dimensional analysis, how the frequency is related to these variables. If the mass of the cylinder were increased, would the frequency increase or decrease?

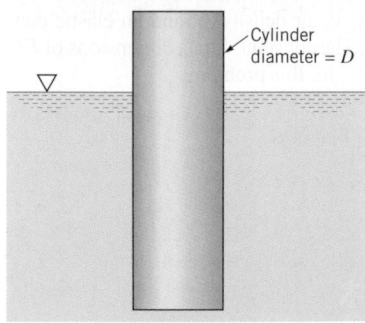

■ **FIGURE P7.23**

Section 7.5 Determination of Pi Terms by Inspection

7.24 A liquid spray nozzle is designed to produce a specific size droplet with diameter, d. The droplet size depends on the nozzle diameter, D, nozzle velocity, V, and the liquid properties ρ, μ, σ. Using the common dimensionless terms found in Table 7.1, determine the functional relationship for the dependent diameter ratio of d/D.

7.25 The velocity, c, at which pressure pulses travel through arteries (pulse-wave velocity) is a function of the artery diameter, D, and wall thickness, h, the density of blood, ρ, and the modulus of elasticity, E, of the arterial wall. Determine a set of nondimensional parameters that can be used to study experimentally the relationship

between the pulse-wave velocity and the variables listed. Form the nondimensional parameters by inspection.

7.26 As shown in Fig. P7.26 and **Video V5.6**, a jet of liquid directed against a block can tip over the block. Assume that the velocity, V, needed to tip over the block is a function of the fluid density, ρ, the diameter of the jet, D, the weight of the block, $\mathcal{W}$, the width of the block, b, and the distance, d, between the jet and the bottom of the block. (a) Determine a set of dimensionless parameters for this problem. Form the dimensionless parameters by inspection. (b) Use the momentum equation to determine an equation for V in terms of the other variables. (c) Compare the results of parts (a) and (b).

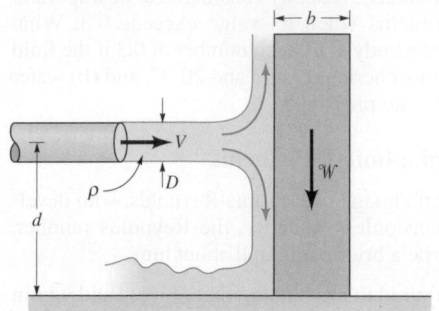

■ **FIGURE P7.26**

7.27 Assume that the drag, $\mathcal{D}$, on an aircraft flying at supersonic speeds is a function of its velocity, V, fluid density, ρ, speed of sound, c, and a series of lengths, $\ell_1, \ldots, \ell_i$, which describe the geometry of the aircraft. Develop a set of pi terms that could be used to investigate experimentally how the drag is affected by the various factors listed. Form the pi terms by inspection.

Section 7.7 Correlation of Experimental Data (Also See Lab Problems 7.82, 7.83, 7.84, and 7.85)

7.28 The measurement of pressure is typically an important task in fluids experiments. Obtain a photograph/image of a pressure measurement device. Print this photo and write a brief paragraph that describes its use.

***7.29** The pressure drop, Δp, over a certain length of horizontal pipe is assumed to be a function of the velocity, V, of the fluid in the pipe, the pipe diameter, D, and the fluid density and viscosity, ρ and μ. (a) Show that this flow can be described in dimensionless form as a "pressure coefficient," $C_p = \Delta p/(0.5\,\rho V^2)$ that depends on the Reynolds number, $\text{Re} = \rho VD/\mu$. (b) The following data were obtained in an experiment involving a fluid with $\rho = 1030$ kg/m³, $\mu = 0.1$ N · s/m², and $D = 0.03$ m. Plot a dimensionless graph and use a power law equation to determine the functional relationship between the pressure coefficient and the Reynolds number. (c) What are the limitations on the applicability of your equation obtained in part (b)?

V, m/s	Δp, kPa
1.0	9
3.0	34
5.0	52
6.0	61

***7.30** The pressure drop across a short hollowed plug placed in a circular tube through which a liquid is flowing (see Fig. P7.30) can be expressed as

$$\Delta p = f(\rho, V, D, d)$$

where ρ is the fluid density, and V is the mean velocity in the tube. Some experimental data obtained with $D = 0.06$ m, $\rho = 1030$ kg/m^3, and $V = 0.6$ m/s are given in the following table:

d (m)	0.018	0.026	0.031	0.046
Δp (kPa)	24	7.5	3	0.6

Plot the results of these tests, using suitable dimensionless parameters, on log–log graph paper. Use a standard curve-fitting technique to determine a general equation for Δp. What are the limits of applicability of the equation?

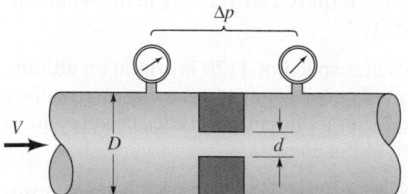

■ **FIGURE P7.30**

*7.31 Describe some everyday situations involving fluid flow and estimate the Reynolds numbers for them. Based on your results, do you think fluid inertia is important in most typical flow situations? Explain.

*7.32 As shown in Fig. 2.26, Fig. P7.32, and Video V2.10, a rectangular barge floats in a stable configuration provided the distance between the center of gravity, CG, of the object (boat and load) and the center of buoyancy, C, is less than a certain amount, H. If this distance is greater than H, the boat will tip over. Assume H is a function of the boat's width, b, length, ℓ, and draft, h. **(a)** Put this relationship into dimensionless form. **(b)** The results of a set of experiments with a model barge with a width of 1.0 m are shown in the table. Plot these data in dimensionless form and determine a power-law equation relating the dimensionless parameters.

ℓ, m	h, m	H, m
2.0	0.10	0.833
4.0	0.10	0.833
2.0	0.20	0.417
4.0	0.20	0.417
2.0	0.35	0.238
4.0	0.35	0.238

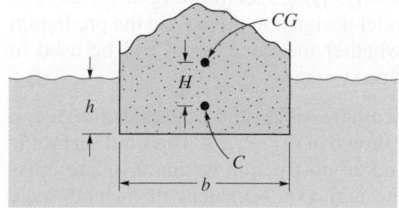

■ **FIGURE P7.32**

7.33 The time, t, it takes to pour a certain volume of liquid from a cylindrical container depends on several factors, including the viscosity of the liquid. (See Video V1.3.) Assume that for very viscous liquids the time it takes to pour out 2/3 of the initial volume depends on the initial liquid depth, ℓ, the cylinder diameter, D, the liquid viscosity, μ, and the liquid specific weight, γ. The data shown in the

following table were obtained in the laboratory. For these tests $\ell = 45$ mm, $D = 67$ mm, and $\gamma = 9.60$ kN/m^3. **(a)** Perform a dimensional analysis, and based on the data given, determine if variables used for this problem appear to be correct. Explain how you arrived at your answer. **(b)** If possible, determine an equation relating the pouring time and viscosity for the cylinder and liquids used in these tests. If it is not possible, indicate what additional information is needed.

μ (N·s/m^2)	11	17	39	61	107
t(s)	15	23	53	83	145

7.34 In order to maintain uniform flight, smaller birds must beat their wings faster than larger birds. It is suggested that the relationship between the wingbeat frequency, ω, beats per second, and the bird's wingspan, ℓ, is given by a power law relationship, $\omega \sim \ell^n$. **(a)** Use dimensional analysis with the assumption that the wingbeat frequency is a function of the wingspan, the specific weight of the bird, γ, the acceleration of gravity, g, and the density of the air, ρ_a, to determine the value of the exponent n. **(b)** Some typical data for various birds are given in the table below. Does this data support your result obtained in part (a)? Provide appropriate analysis to show how you arrived at your conclusion.

Bird	Wingspan, m	Wingbeat frequency, beats/s
purple martin	0.28	5.3
robin	0.36	4.3
mourning dove	0.46	3.2
crow	1.00	2.2
Canada goose	1.50	2.6
great blue heron	1.80	2.0

*7.35 The concentric cylinder device of the type shown in Fig. P7.35 is commonly used to measure the viscosity, μ, of liquids by relating the angle of twist, θ, of the inner cylinder to the angular velocity, ω, of the outer cylinder. Assume that

$$\theta = f(\omega, \mu, K, D_1, D_2, \ell)$$

where K depends on the suspending wire properties and has the dimensions FL. The following data were obtained in a series of tests for which $\mu = 0.5$ N·s/m^2, $K = 14$ N·m, $\ell = 0.3$ m, and D_1 and D_2 were constant.

θ (rad)	ω (rad/s)
0.89	0.30
1.50	0.50
2.51	0.82
3.05	1.05
4.28	1.43
5.52	1.86
6.40	2.14

Determine from these data, with the aid of dimensional analysis, the relationship between θ, ω, and μ for this particular apparatus. *Hint:* Plot the data using appropriate dimensionless parameters, and determine the equation of the resulting curve using a standard curve-fitting technique. The equation should satisfy the condition that $\theta = 0$ for $\omega = 0$.

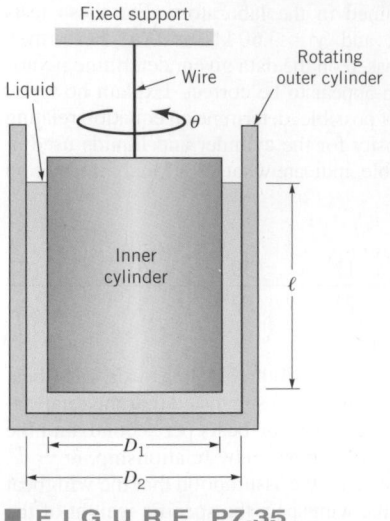

Fixed support

Liquid

Wire

Rotating outer cylinder

θ

Inner cylinder

ℓ

D_1

D_2

■ **F I G U R E   P7.35**

Section 7.8 Modeling and Similitude

7.36 Obtain a photograph/image of a prototype and the corresponding model that was used for testing. Print these photos and write a brief paragraph that describes the situation involved.

7.37 Air at 25 °C is to flow through a 0.6-m pipe at an average velocity of 2 m/s. What size pipe should be used to move water at 15 °C and average velocity of 1 m/s if Reynolds number similarity is enforced?

7.38 To test the aerodynamics of a new prototype automobile, a scale model will be tested in a wind tunnel. For dynamic similarity, it will be required to match Reynolds number between model and prototype. Assuming that you will be testing a one-tenth-scale model and both model and prototype will be exposed to standard air pressure, will it be better for the wind tunnel air to be colder or hotter than standard sea-level air temperature of 15 °C? Why?

7.39 You are to conduct wind tunnel testing of a new football design that has a smaller lace height than previous designs (see **Videos V6.1 and V6.2**). It is known that you will need to maintain Re and St similarity for the testing. Based on standard college quarterbacks, the prototype parameters are set at V = 64 km/hr and ω = 300 rpm. The prototype football has a 18-cm diameter. Due to instrumentation required to measure pressure and shear stress on the surface of the football, the model will require a length scale of 2:1 (the model will be larger than the prototype). Determine the required model freestream velocity and model angular velocity.

7.40 A model of a submarine, 1 : 15 scale, is to be tested at 55 m/s in a wind tunnel with standard sea-level air, while the prototype will be operated in seawater. Determine the speed of the prototype to ensure Reynolds number similarity.

7.41 SAE 30 oil at 15 °C is pumped through a 1-m-diameter pipeline at a rate of 400 L/s. A model of this pipeline is to be designed using a 7-cm-diameter pipe and water at 15 °C as the working fluid. To maintain Reynolds number similarity between these two systems, what fluid velocity will be required in the model?

7.42 The water velocity at a certain point along a 1 : 10 scale model of a dam spillway is 3 m/s. What is the corresponding prototype velocity if the model and prototype operate in accordance with Froude number similarity?

7.43 The drag characteristics of a torpedo are to be studied in a water tunnel using a 1 : 5 scale model. The tunnel operates with freshwater at 20 °C, whereas the prototype torpedo is to be used in

seawater at 15.6 °C. To correctly simulate the behavior of the prototype moving with a velocity of 30 m/s, what velocity is required in the water tunnel?

7.44 For a certain fluid-flow problem it is known that both the Froude number and the Weber number are important dimensionless parameters. If the problem is to be studied by using a 1 : 15 scale model, determine the required surface tension scale if the density scale is equal to 1. The model and prototype operate in the same gravitational field.

7.45 The fluid dynamic characteristics of an airplane flying 390 km/hr at 3000 m are to be investigated with the aid of a 1 : 20 scale model. If the model tests are to be performed in a wind tunnel using standard air, what is the required air velocity in the wind tunnel? Is this a realistic velocity?

7.46 If an airplane travels at a speed of 1120 km/hr at an altitude of 15 km, what is the required speed at an altitude of 8 km to satisfy Mach number similarity? Assume the air properties correspond to those for the U.S. standard atmosphere.

7.47 (See Fluids in the News article "Modeling parachutes in a water tunnel," Section 7.8.1.) Flow characteristics for a 9-m-diameter prototype parachute are to be determined by tests of a 0.3-m-diameter model parachute in a water tunnel. Some data collected with the model parachute indicate a drag of 75 N when the water velocity is 1 m/s. Use the model data to predict the drag on the prototype parachute falling through air at 3 m/s. Assume the drag to be a function of the velocity, V, the fluid density, ρ, and the parachute diameter, D.

7.48 The lift and drag developed on a hydrofoil are to be determined through wind tunnel tests using standard air. If full-scale tests are to be run, what is the required wind tunnel velocity corresponding to a hydrofoil velocity in seawater at 24 km/hr? Assume Reynolds number similarity is required.

7.49 A 1/50 scale model is to be used in a towing tank to study the water motion near the bottom of a shallow channel as a large barge passes over. (See **Video V7.16**.) Assume that the model is operated in accordance with the Froude number criteria for dynamic similitude. The prototype barge moves at a typical speed of 30 km/hr. **(a)** At what speed (in m/s) should the model be towed? **(b)** Near the bottom of the model channel a small particle is found to move 0.05 m in one second so that the fluid velocity at that point is approximately 0.05 m/s. Determine the velocity at the corresponding point in the prototype channel.

7.50 A solid sphere having a diameter d and specific weight γ_s is immersed in a liquid having a specific weight $\gamma_f(\gamma_f > \gamma_s)$ and then released. It is desired to use a model system to determine the maximum height, h, above the liquid surface that the sphere will rise upon release from a depth H. It can be assumed that the important liquid properties are the density, γ_f/g, specific weight, γ_f, and viscosity, μ_f. Establish the model design conditions and the prediction equation, and determine whether the same liquid can be used in both the model and prototype systems.

7.51 A thin layer of an incompressible fluid flows steadily over a horizontal smooth plate as shown in Fig. P7.51. The fluid surface is open to the atmosphere, and an obstruction having a square cross section is placed on the plate as shown. A model with a length scale

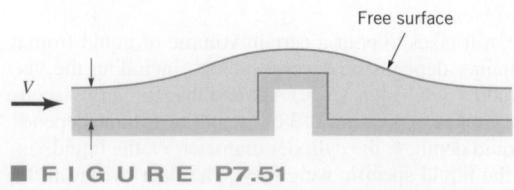

Free surface

V

■ **F I G U R E   P7.51**

of $\frac{1}{4}$ and a fluid density scale of 1.0 is to be designed to predict the depth of fluid, y, along the plate. Assume that inertial, gravitational, surface tension, and viscous effects are all important. What are the required viscosity and surface tension scales?

7.52 The drag on a 2-m-diameter satellite dish due to an 80-km/hr wind is to be determined through a wind tunnel test using a geometrically similar 0.4-m-diameter model dish. Assume standard air for both model and prototype. **(a)** At what air speed should the model test be run? **(b)** With all similarity conditions satisfied, the measured drag on the model was determined to be 170 N. What is the predicted drag on the prototype dish?

7.53 A large, rigid, rectangular billboard is supported by an elastic column as shown in Fig. P7.53. There is concern about the deflection, δ, of the top of the structure during a high wind of velocity V. A wind tunnel test is to be conducted with a 1 : 15 scale model. Assume the pertinent column variables are its length and cross-sectional dimensions, and the modulus of elasticity of the material used for the column. The only important "wind" variables are the air density and velocity. **(a)** Determine the model design conditions and the prediction equation for the deflection. **(b)** If the same structural materials are used for the model and prototype, and the wind tunnel operates under standard atmospheric conditions, what is the required wind tunnel velocity to match an 80 km/hr wind?

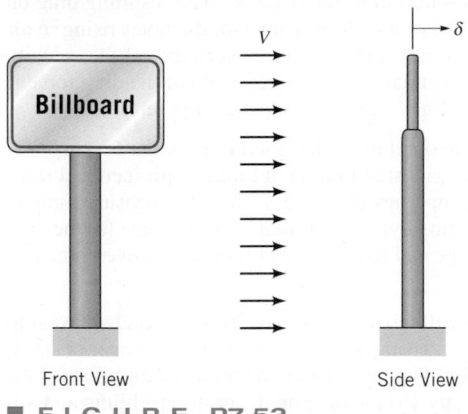

Front View   Side View

■ **F I G U R E   P7.53**

7.54 A thin flat plate having a diameter of 0.1 m is towed through a tank of oil ($\gamma = 8.3$ kN/m³) at a velocity of 1.5 m/s. The plane of the plate is perpendicular to the direction of motion, and the plate is submerged so that wave action is negligible. Under these conditions the drag on the plate is 6 N. If viscous effects are neglected, predict the drag on a geometrically similar, 0.6-m-diameter plate that is towed with a velocity of 1 m/s through water at 15 °C under conditions similar to those for the smaller plate.

7.55 For a certain model study involving a 1 : 5 scale model it is known that Froude number similarity must be maintained. The possibility of cavitation is also to be investigated, and it is assumed that the cavitation number must be the same for model and prototype. The prototype fluid is water at 30 °C, and the model fluid is water at 70 °C. If the prototype operates at an ambient pressure of 101 kPa (abs), what is the required ambient pressure for the model system?

7.56 A thin layer of particles rests on the bottom of a horizontal tube as shown in Fig. P7.56. When an incompressible fluid flows

■ **F I G U R E   P7.56**

through the tube, it is observed that at some critical velocity the particles will rise and be transported along the tube. A model is to be used to determine this critical velocity. Assume the critical velocity, V_c, to be a function of the pipe diameter, D, particle diameter, d, the fluid density, ρ, and viscosity, μ, the density of the particles, ρ_p, and the acceleration of gravity, g. **(a)** Determine the similarity requirements for the model, and the relationship between the critical velocity for model and prototype (the prediction equation). **(b)** For a length scale of $\frac{1}{2}$ and a fluid density scale of 1.0, what will be the critical velocity scale (assuming all similarity requirements are satisfied)?

7.57 The pressure rise, Δp, across a blast wave, as shown in Fig. P7.57 and **Video V11.7**, is assumed to be a function of the amount of energy released in the explosion, E, the air density, ρ, the speed of sound, c, and the distance from the blast, d. **(a)** Put this relationship in dimensionless form. **(b)** Consider two blasts: the prototype blast with energy release E and a model blast with 1/1000th the energy release ($E_m = 0.001\ E$). At what distance from the model blast will the pressure rise be the same as that at a distance of 1 km from the prototype blast?

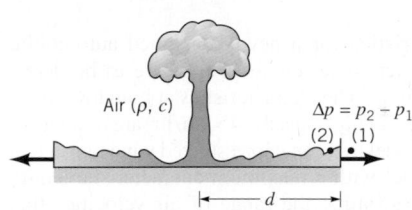

■ **F I G U R E   P7.57**

7.58 The drag, $\mathscr{D}$, on a sphere located in a pipe through which a fluid is flowing is to be determined experimentally (see Fig. P7.58). Assume that the drag is a function of the sphere diameter, d, the pipe diameter, D, the fluid velocity, V, and the fluid density, ρ. **(a)** What dimensionless parameters would you use for this problem? **(b)** Some experiments using water indicate that for $d = 0.5$ cm, $D = 1$ cm, and $V = 0.6$ m/s, the drag is 7×10^{-3} N. If possible, estimate the drag on a sphere located in a 0.6-m-diameter pipe through which water is flowing with a velocity of 2 m/s. The sphere diameter is such that geometric similarity is maintained. If it is not possible, explain why not.

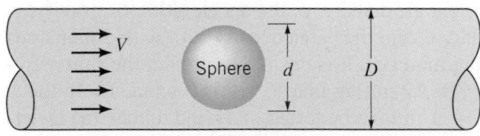

■ **F I G U R E   P7.58**

7.59 An incompressible fluid oscillates harmonically ($V = V_0 \sin \omega t$, where V is the velocity) with a frequency of 10 rad/s in a 10-cm-diameter pipe. A $\frac{1}{4}$ scale model is to be used to determine the pressure difference per unit length, Δp_ℓ (at any instant) along the pipe. Assume that

$$\Delta p_\ell = f(D, V_0, \omega, t, \mu, \rho)$$

where D is the pipe diameter, ω the frequency, t the time, μ the fluid viscosity, and ρ the fluid density. **(a)** Determine the similarity requirements for the model and the prediction equation for Δp_ℓ. **(b)** If the same fluid is used in the model and the prototype, at what frequency should the model operate?

7.60 As shown in Fig. P7.60, a "noisemaker" B is towed behind a minesweeper A to set off enemy acoustic mines such as at C. The drag force of the noisemaker is to be studied in a water tunnel at a ¼ scale model (model ¼ the size of the prototype). The drag force is assumed to be a function of the speed of the ship, the density and viscosity of the fluid, and the diameter of the noisemaker. **(a)** If the prototype towing speed in 3 m/s, determine the water velocity in the tunnel for the model tests. **(b)** If the model tests of part (a) produced a model drag of 900 N, determine the drag expected on the prototype.

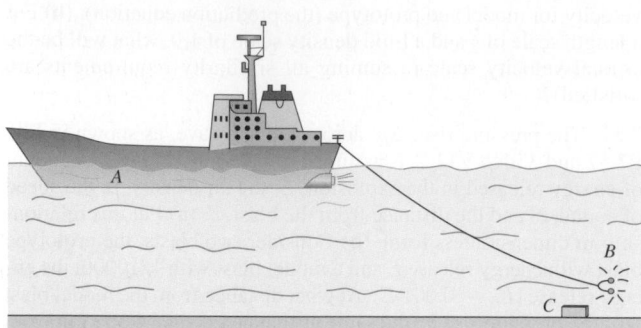

■ **F I G U R E   P7.60**

7.61 The drag characteristics for a newly designed automobile having a maximum characteristic length of 6 m are to be determined through a model study. The characteristics at both low speed (approximately 30 km/hr) and high speed (145 km/hr) are of interest. For a series of projected model tests, an unpressurized wind tunnel that will accommodate a model with a maximum characteristic length of 1 m is to be used. Determine the range of air velocities that would be required for the wind tunnel if Reynolds number similarity is desired. Are the velocities suitable? Explain.

7.62 The drag characteristics of an airplane are to be determined by model tests in a wind tunnel operated at an absolute pressure of 1300 kPa. If the prototype is to cruise in standard air at 385 km/hr, and the corresponding speed of the model is not to differ by more than 20% from this (so that compressibility effects may be ignored), what range of length scales may be used if Reynolds number similarity is to be maintained? Assume the viscosity of air is unaffected by pressure, and the temperature of air in the tunnel is equal to the temperature of the air in which the airplane will fly.

7.63 Wind blowing past a flag causes it to "flutter in the breeze." The frequency of this fluttering, ω, is assumed to be a function of the wind speed, V, the air density, ρ, the acceleration of gravity, g, the length of the flag, ℓ, and the "area density," ρ_A (with dimensions of ML^{-2}) of the flag material. It is desired to predict the flutter frequency of a large $\ell = 12$ m flag in a $V = 9$ m/s wind. To do this a model flag with $\ell = 1$ m is to be tested in a wind tunnel. **(a)** Determine the required area density of the model flag material if the large flag has $\rho_A = 1$ kg/m². **(b)** What wind tunnel velocity is required for testing the model? **(c)** If the model flag flutters at 6 Hz, predict the frequency for the large flag.

†7.64 If a large oil spill occurs from a tanker operating near a coastline, the time it would take for the oil to reach shore is of great concern. Design a model system that can be used to investigate this type of problem in the laboratory. Indicate all assumptions made in developing the design and discuss any difficulty that may arise in satisfying the similarity requirements arising from your model design.

7.65 The drag on a sphere moving in a fluid is known to be a function of the sphere diameter, the velocity, and the fluid viscosity and density. Laboratory tests on a 10-cm-diameter sphere were performed in a water tunnel and some model data are plotted in Fig. P7.65. For these tests the viscosity of the water was 1.0×10^{-3} N · s/m² and the water density was 998 kg/m³. Estimate the drag on an 2-m-diameter balloon moving in air at a velocity of 1 m/s. Assume

the air to have a viscosity of 1.8×10^{-5} N · s/m² and a density of 1.2 kg/m³.

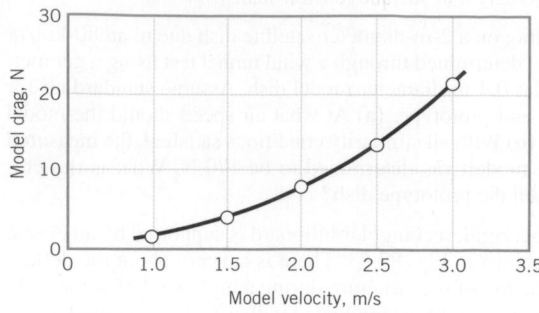

■ **F I G U R E   P7.65**

Section 7.9 Some Typical Model Studies

7.66 Obtain a photograph/image of a situation where a flow around an immersed body is being experimentally tested. Print this photo and write a brief paragraph that describes the situation involved.

7.67 Drag measurements were taken for a sphere, with a diameter of 5 cm, moving at 4 m/s in water at 20 °C. The resulting drag on the sphere was 10 N. For a balloon with 1-m diameter rising in air with standard temperature and pressure, determine **(a)** the velocity if Reynolds number similarity is enforced and **(b)** the drag force if the drag coefficient (Eq. 7.19) is the dependent pi term.

7.68 A prototype automobile is designed to travel at 65 km/hr. A model of this design is tested in a wind tunnel with identical standard sea-level air properties at a 1 : 5 scale. The measured model drag is 400 N, enforcing dynamic similarity. Determine **(a)** the drag force on the prototype and **(b)** the power required to overcome this drag. See Eq. 7.19.

7.69 A new blimp will move at 6 m/s in 20 °C air, and we want to predict the drag force. Using a 1 : 13-scale model in water at 20 °C and measuring a 2500-N drag force on the model, determine **(a)** the required water velocity, **(b)** the drag on the prototype blimp and, **(c)** the power that will be required to propel it through the air.

7.70 At a large fish hatchery the fish are reared in open, water-filled tanks. Each tank is approximately square in shape with curved corners, and the walls are smooth. To create motion in the tanks, water is supplied through a pipe at the edge of the tank. The water is drained from the tank through an opening at the center. (See **Video V7.9**.) A model with a length scale of 1 : 13 is to be used to determine the velocity, V, at various locations within the tank. Assume that $V = f(\ell, \ell_i, \rho, \mu, g, Q)$ where ℓ is some characteristic length such as the tank width, ℓ_i represents a series of other pertinent lengths, such as inlet pipe diameter, fluid depth, etc., ρ is the fluid density, μ is the fluid viscosity, g is the acceleration of gravity, and Q is the discharge through the tank. **(a)** Determine a suitable set of dimensionless parameters for this problem and the prediction equation for the velocity. If water is to be used for the model, can all of the similarity requirements be satisfied? Explain and support your answer with the necessary calculations. **(b)** If the flowrate into the full-sized tank is 950 L/min, determine the required value for the model discharge assuming Froude number similarity. What model depth will correspond to a depth of 80 cm in the full-sized tank?

7.71 Flow patterns that develop as winds blow past a vehicle, such as a train, are often studied in low-speed environmental (meteorological) wind tunnels. (See **Video V7.16**.) Typically, the air velocities in these tunnels are in the range of 0.1 m/s to 30 m/s. Consider a cross wind blowing past a train locomotive. Assume that the local wind velocity, V, is a function of the approaching wind velocity (at some distance from the locomotive), U, the locomotive length, ℓ, height, h, and width, b, the air density, ρ, and the air viscosity, μ. **(a)** Establish the similarity requirements and prediction equation for a model to be used in the wind tunnel to study the air velocity, V, around the locomotive. **(b)** If the model is to be used for cross winds gusting to $U = 25$ m/s, explain why it is not practical to maintain Reynolds number similarity for a typical length scale 1:50.

7.72 (See Fluids in the News article titled "Galloping Gertie," Section 7.8.2.) The Tacoma Narrows bridge failure is a dramatic example of the possible serious effects of wind-induced vibrations. As a fluid flows around a body, vortices may be created which are shed periodically creating an oscillating force on the body. If the frequency of the shedding vortices coincides with the natural frequency of the body, large displacements of the body can be induced as was the case with the Tacoma Narrows bridge. To illustrate this type of phenomenon, consider fluid flow past a circular cylinder. Assume the frequency, n, of the shedding vortices behind the cylinder is a function of the cylinder diameter, D, the fluid velocity, V, and the fluid kinematic viscosity, ν. **(a)** Determine a suitable set of dimensionless variables for this problem. One of the dimensionless variables should be the Strouhal number, nD/V. **(b)** Some results of experiments in which the shedding frequency of the vortices (in Hz) was measured, using a particular cylinder and Newtonian, incompressible fluid, are shown in Fig. P7.72. Is this a "universal curve" that can be used to predict the shedding frequency for any cylinder placed in any fluid? Explain. **(c)** A certain structural component in the form of a 2.5-cm-diameter, 4-m-long rod acts as a cantilever beam with a natural frequency of 19 Hz. Based on the data in Fig. P7.72, estimate the wind speed that may cause the rod to oscillate at its natural frequency. *Hint:* Use a trial and error solution.

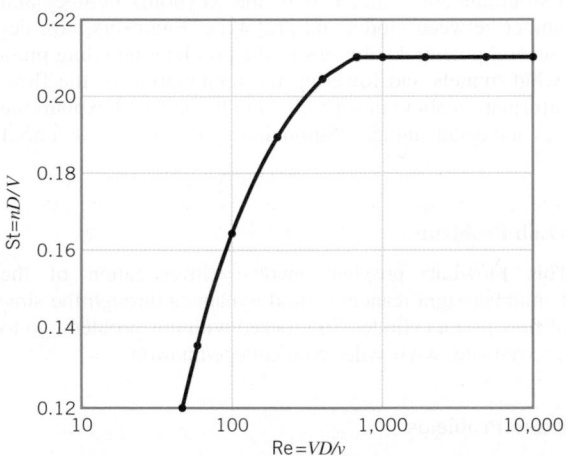

■ **F I G U R E P7.72**

7.73 (See Fluids in the News article titled "Ice engineering," Section 7.9.3.) A model study is to be developed to determine the force exerted on bridge piers due to floating chunks of ice in a river. The piers of interest have square cross sections. Assume that the force, R, is a function of the pier width, b, the depth of the ice, d, the velocity of the ice, V, the acceleration of gravity, g, the density of the ice, ρ_i, and a measure of the strength of the ice, E_i, where E_i has the dimensions FL^{-2}. **(a)** Based on these variables determine a suitable set of dimensionless variables for this problem. **(b)** The prototype conditions of interest include an ice thickness of 30 cm and an ice velocity of 2 m/s. What model ice thickness and velocity would be required if the length scale is to be 1/10? **(c)** If the model and prototype ice have the same density, can the model ice have the same strength properties as that of the prototype ice? Explain.

7.74 As illustrated in Video V7.9, models are commonly used to study the dispersion of a gaseous pollutant from an exhaust stack located near a building complex. Similarity requirements for the pollutant source involve the following independent variables: the stack gas speed, V, the wind speed, U, the density of the atmospheric air, ρ, the difference in densities between the air and the stack gas, $\rho - \rho_s$, the acceleration of gravity, g, the kinematic viscosity of the stack gas, ν_s, and the stack diameter, D. **(a)** Based on these variables, determine a suitable set of similarity requirements for modeling the pollutant source. **(b)** For this type of model a typical length scale might be 1:200. If the same fluids were used in model and prototype, would the similarity requirements be satisfied? Explain and support your answer with the necessary calculations.

7.75 River models are used to study many different types of flow situations. (See, for example, Video V7.12.) A certain small river has an average width and depth of 18 m and 1 m, respectively, and carries water at a flowrate of 20 m³/s. A model is to be designed based on Froude number similarity so that the discharge scale is 1/250. At what depth and flowrate would the model operate?

7.76 As winds blow past buildings, complex flow patterns can develop due to various factors such as flow separation and interactions between adjacent buildings. (See Video V7.13.) Assume that the local gage pressure, p, at a particular loaction on a building is a function of the air density, ρ, the wind speed, V, some characteristic length, ℓ, and all other pertinent lengths, ℓ_i, needed to characterize the geometry of the building or building complex. **(a)** Determine a suitable set of dimensionless parameters that can be used to study the pressure distribution. **(b)** An eight-story building that is 30 m tall is to be modeled in a wind tunnel. If a length scale of 1:300 is to be used, how tall should the model building be? **(c)** How will a measured pressure in the model be related to the corresponding prototype pressure? Assume the same air density in model and prototype. Based on the assumed variables, does the model wind speed have to be equal to the prototype wind speed? Explain.

Section 7.10 Similitude Based on Governing Differential Equations

7.77 Start with the two-dimensional continuity equation and the Navier–Stokes equations (Eqs. 7.28, 7.29, and 7.30) and verify the nondimensional forms of these equations (Eqs. 7.31, 7.34, and 7.35).

7.78 A viscous fluid is contained between wide, parallel plates spaced a distance h apart as shown in Fig. P7.78. The upper plate is fixed, and the bottom plate oscillates harmonically with a velocity amplitude U and frequency ω. The differential equation for the velocity distribution between the plates is

$$\rho \frac{\partial u}{\partial t} = \mu \frac{\partial^2 u}{\partial y^2}$$

where u is the velocity, t is time, and ρ and μ are fluid density and viscosity, respectively. Rewrite this equation in a suitable nondimensional form using h, U, and ω as reference parameters.

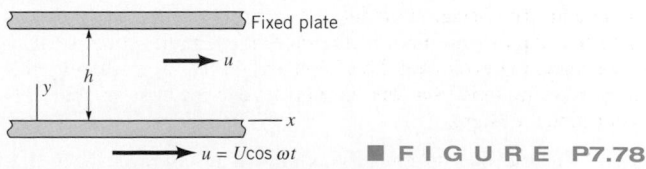

■ **F I G U R E P7.78**

7.79 The deflection of the cantilever beam of Fig. P7.79 is governed by the differential equation

$$EI \frac{d^2y}{dx^2} = P(x - \ell)$$

where E is the modulus of elasticity and I is the moment of inertia of the beam cross section. The boundary conditions are $y = 0$ at $x = 0$ and $dy/dx = 0$ at $x = 0$. **(a)** Rewrite the equation and boundary conditions in dimensionless form using the beam length, ℓ, as the reference length. **(b)** Based on the results of part **(a)**, what are the similarity requirements and the prediction equation for a model to predict deflections?

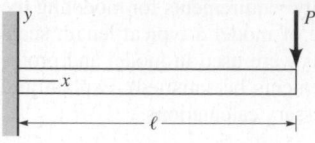

■ **F I G U R E   P7.79**

7.80 A liquid is contained in a pipe that is closed at one end as shown in Fig. P7.80. Initially the liquid is at rest, but if the end is suddenly opened the liquid starts to move. Assume the pressure p_1 remains constant. The differential equation that describes the resulting motion of the liquid is

$$\rho \frac{\partial v_z}{\partial t} = \frac{p_1}{\ell} + \mu \left(\frac{\partial^2 v_z}{\partial r^2} + \frac{1}{r} \frac{\partial v_z}{\partial r} \right)$$

where v_z is the velocity at any radial location, r, and t is time. Rewrite this equation in dimensionless form using the liquid density, ρ, the viscosity, μ, and the pipe radius, R, as reference parameters.

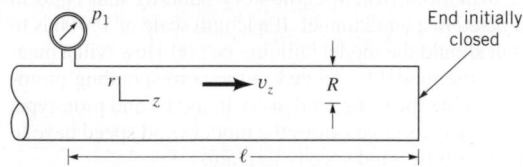

■ **F I G U R E   P7.80**

7.81 An incompressible fluid is contained between two infinite parallel plates as illustrated in Fig. P7.81. Under the influence of a harmonically varying pressure gradient in the x direction, the fluid oscillates harmonically with a frequency ω. The differential equation describing the fluid motion is

$$\rho \frac{\partial u}{\partial t} = X \cos \omega t + \mu \frac{\partial^2 u}{\partial y^2}$$

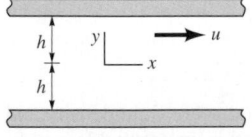

■ **F I G U R E   P7.81**

where X is the amplitude of the pressure gradient. Express this equation in nondimensional form using h and ω as reference parameters.

■ **Lab Problems**

7.82 This problem involves the time that it takes water to drain from two geometrically similar tanks. To proceed with this problem, go to the book's web site, www.wiley.com/college/munson.

7.83 This problem involves determining the frequency of vortex shedding from a circular cylinder as water flows past it. To proceed with this problem, go to the book's web site, www.wiley.com/college/munson.

7.84 This problem involves the determination of the head loss for flow through a valve. To proceed with this problem, go to the book's web site, www.wiley.com/college/munson.

7.85 This problem involves the calibration of a rotameter. To proceed with this problem, go to the book's web site, www.wiley.com/college/munson.

■ **Life Long Learning Problems**

7.86 Microfluidics is the study of fluid flow in fabricated devices at the micro scale. Advances in microfluidics have enhanced the ability of scientists and engineers to perform laboratory experiments using miniaturized devices known as a "lab-on-a-chip." Obtain information about a lab-on-a-chip device that is available commercially and investigate its capabilities. Summarize your findings in a brief report.

7.87 For some types of aerodynamic wind tunnel testing, it is difficult to simultaneously match both the Reynolds number and Mach number between model and prototype. Engineers have developed several potential solutions to the problem including pressurized wind tunnels and lowering the temperature of the flow. Obtain information about cryogenic wind tunnels and explain the advantages and disadvantages. Summarize your findings in a brief report.

■ **FlowLab Problems**

***7.88** This FlowLab problem involves investigation of the Reynolds number significance in fluid dynamics through the simulation of flow past a cylinder. To proceed with this problem, go to the book's web site, www.wiley.com/college/munson.

■ **FE Exam Problems**

Sample FE (Fundamental of Engineering) exam questions for fluid mechanics are provided on the book's web site, www.wiley.com/college/munson.

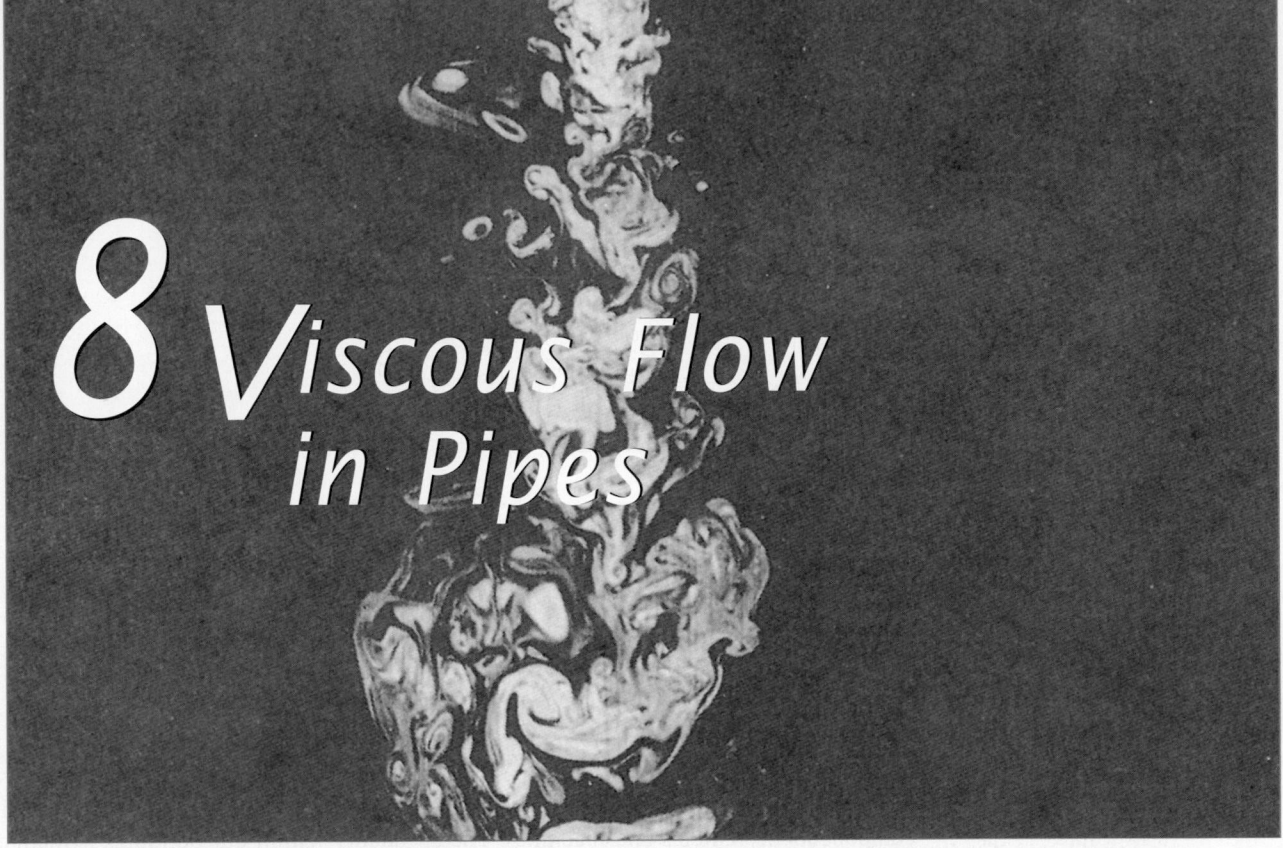

8 Viscous Flow in Pipes

CHAPTER OPENING PHOTO: Turbulent jet: The jet of water from the pipe is turbulent. The complex, irregular, unsteady structure typical of turbulent flows is apparent. (Laser-induced fluorescence of dye in water.) *(Photography by P. E. Dimotakis, R. C. Lye, and D. Z. Papantoniou.)*

Learning Objectives

V8.1 Turbulent jet

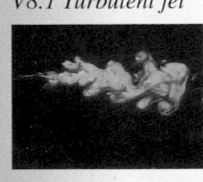

After completing this chapter, you should be able to:

- identify and understand various characteristics of the flow in pipes.
- discuss the main properties of laminar and turbulent pipe flow and appreciate their differences.
- calculate losses in straight portions of pipes as well as those in various pipe system components.
- apply appropriate equations and principles to analyze a variety of pipe flow situations.
- predict the flowrate in a pipe by use of common flowmeters.

In the previous chapters we have considered a variety of topics concerning the motion of fluids. The basic governing principles concerning mass, momentum, and energy were developed and applied, in conjunction with rather severe assumptions, to numerous flow situations. In this chapter we will apply the basic principles to a specific, important topic—the incompressible flow of viscous fluids in pipes and ducts.

Pipe flow is very important in our daily operations.

The transport of a fluid (liquid or gas) in a closed conduit (commonly called a *pipe* if it is of round cross section or a *duct* if it is not round) is extremely important in our daily operations. A brief consideration of the world around us will indicate that there is a wide variety of applications of pipe flow. Such applications range from the large, man-made Alaskan pipeline that carries crude oil almost 1290 km across Alaska, to the more complex (and certainly not less useful) natural systems of "pipes" that carry blood throughout our body and air into and out of our lungs. Other examples

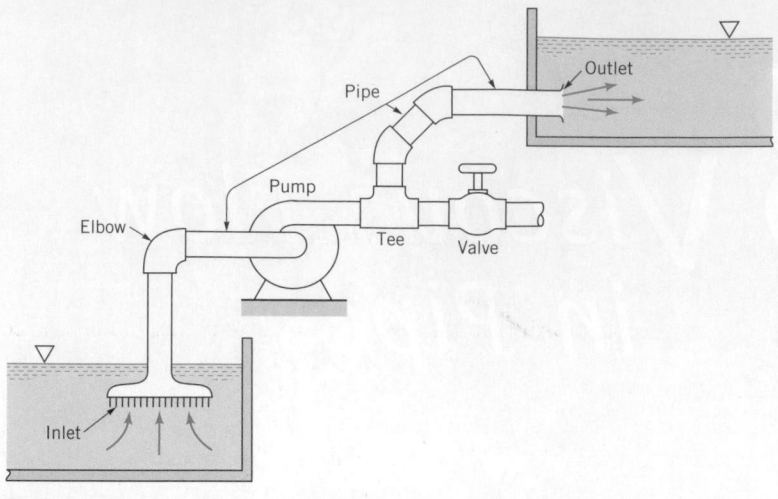

■ **FIGURE 8.1** Typical pipe system components.

include the water pipes in our homes and the distribution system that delivers the water from the city well to the house. Numerous hoses and pipes carry hydraulic fluid or other fluids to various components of vehicles and machines. The air quality within our buildings is maintained at comfortable levels by the distribution of conditioned (heated, cooled, humidified/dehumidified) air through a maze of pipes and ducts. Although all of these systems are different, the fluid mechanics principles governing the fluid motions are common. The purpose of this chapter is to understand the basic processes involved in such flows.

Some of the basic components of a typical *pipe system* are shown in Fig. 8.1. They include the pipes themselves (perhaps of more than one diameter), the various fittings used to connect the individual pipes to form the desired system, the flowrate control devices (valves), and the pumps or turbines that add energy to or remove energy from the fluid. Even the most simple pipe systems are actually quite complex when they are viewed in terms of rigorous analytical considerations. We will use an "exact" analysis of the simplest pipe flow topics (such as laminar flow in long, straight, constant diameter pipes) and dimensional analysis considerations combined with experimental results for the other pipe flow topics. Such an approach is not unusual in fluid mechanics investigations. When "real-world" effects are important (such as viscous effects in pipe flows), it is often difficult or "impossible" to use only theoretical methods to obtain the desired results. A judicious combination of experimental data with theoretical considerations and dimensional analysis often provides the desired results. The flow in pipes discussed in this chapter is an example of such an analysis.

8.1 General Characteristics of Pipe Flow

Before we apply the various governing equations to pipe flow examples, we will discuss some of the basic concepts of pipe flow. With these ground rules established we can then proceed to formulate and solve various important flow problems.

Although not all conduits used to transport fluid from one location to another are round in cross section, most of the common ones are. These include typical water pipes, hydraulic hoses, and other conduits that are designed to withstand a considerable pressure difference across their walls without undue distortion of their shape. Typical conduits of noncircular cross section include heating and air conditioning ducts that are often of rectangular cross section. Normally the pressure difference between the inside and outside of these ducts is relatively small. Most of the basic principles involved are independent of the cross-sectional shape, although the details of the flow may be dependent on it. Unless otherwise specified, we will assume that the conduit is round, although we will show how to account for other shapes.

The pipe is assumed to be completely full of the flowing fluid.

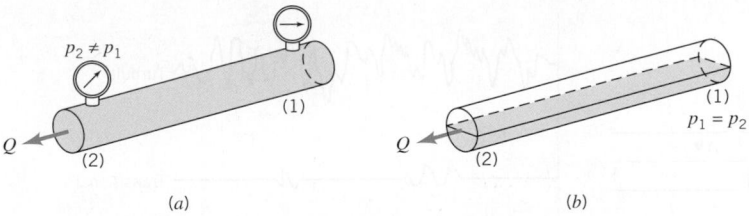

■ **F I G U R E 8.2** (*a*) **Pipe flow.** (*b*) **Open-channel flow.**

For all flows involved in this chapter, we assume that the pipe is completely filled with the fluid being transported as is shown in Fig. 8.2*a*. Thus, we will not consider a concrete pipe through which rainwater flows without completely filling the pipe, as is shown in Fig. 8.2*b*. Such flows, called open-channel flow, are treated in Chapter 10. The difference between open-channel flow and the pipe flow of this chapter is in the fundamental mechanism that drives the flow. For open-channel flow, gravity alone is the driving force—the water flows down a hill. For pipe flow, gravity may be important (the pipe need not be horizontal), but the main driving force is likely to be a pressure gradient along the pipe. If the pipe is not full, it is not possible to maintain this pressure difference, $p_1 - p_2$.

8.1.1 Laminar or Turbulent Flow

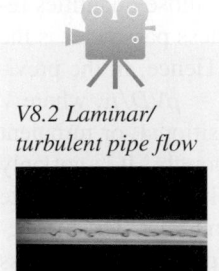

V8.2 Laminar/ turbulent pipe flow

The flow of a fluid in a pipe may be laminar flow or it may be turbulent flow. Osborne Reynolds (1842–1912), a British scientist and mathematician, was the first to distinguish the difference between these two classifications of flow by using a simple apparatus as shown by the figure in the margin, which is a sketch of Reynolds' dye experiment. Reynolds injected dye into a pipe in which water flowed due to gravity. The entrance region of the pipe is depicted in Fig. 8.3*a*. If water runs through a pipe of diameter D with an average velocity V, the following characteristics are observed by injecting neutrally buoyant dye as shown. For "small enough flowrates" the dye streak (a streakline) will remain as a well-defined line as it flows along, with only slight blurring due to molecular diffusion of the dye into the surrounding water. For a somewhat larger "intermediate flowrate" the dye streak fluctuates in time and space, and intermittent bursts of irregular behavior appear along the streak. On the other hand, for "large enough flowrates" the dye streak almost immediately becomes blurred and spreads across the entire pipe in a random fashion. These three characteristics, denoted as *laminar*, *transitional*, and *turbulent* flow, respectively, are illustrated in Fig. 8.3*b*.

A flow may be laminar, transitional, or turbulent.

The curves shown in Fig. 8.4 represent the x component of the velocity as a function of time at a point A in the flow. The random fluctuations of the turbulent flow (with the associated particle mixing) are what disperse the dye throughout the pipe and cause the blurred appearance illustrated in Fig. 8.3*b*. For laminar flow in a pipe there is only one component of velocity,

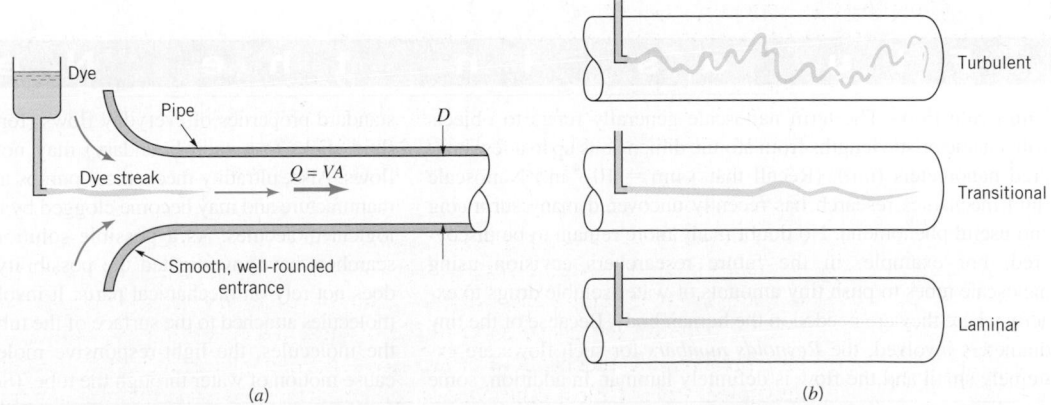

■ **F I G U R E 8.3** (*a*) **Experiment to illustrate type of flow.** (*b*) **Typical dye streaks.**

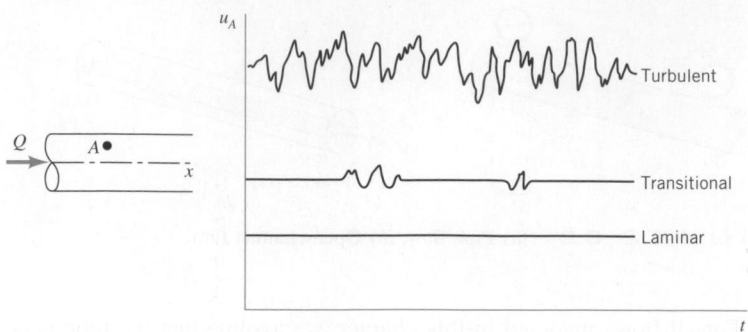

■ **FIGURE 8.4** **Time dependence of fluid velocity at a point.**

$\mathbf{V} = u\hat{\mathbf{i}}$. For turbulent flow the predominant component of velocity is also along the pipe, but it is unsteady (random) and accompanied by random components normal to the pipe axis, $\mathbf{V} = u\hat{\mathbf{i}} + v\hat{\mathbf{j}} + w\hat{\mathbf{k}}$. Such motion in a typical flow occurs too fast for our eyes to follow. Slow motion pictures of the flow can more clearly reveal the irregular, random, turbulent nature of the flow.

As was discussed in Chapter 7, we should not label dimensional quantities as being "large" or "small," such as "small enough flowrates" in the preceding paragraphs. Rather, the appropriate dimensionless quantity should be identified and the "small" or "large" character attached to it. A quantity is "large" or "small" only relative to a reference quantity. The ratio of those quantities results in a dimensionless quantity. For pipe flow the most important dimensionless parameter is the Reynolds number, Re—the ratio of the inertia to viscous effects in the flow. Hence, in the previous paragraph the term flowrate should be replaced by Reynolds number, $\text{Re} = \rho VD/\mu$, where V is the average velocity in the pipe. That is, the flow in a pipe is laminar, transitional, or turbulent provided the Reynolds number is "small enough," "intermediate," or "large enough." It is not only the fluid velocity that determines the character of the flow—its density, viscosity, and the pipe size are of equal importance. These parameters combine to produce the Reynolds number. The distinction between laminar and turbulent pipe flow and its dependence on an appropriate dimensionless quantity was first pointed out by Osborne Reynolds in 1883.

Pipe flow characteristics are dependent on the value of the Reynolds number.

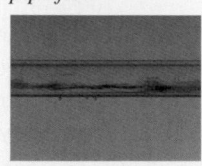

V8.3 Intermittent turbulent burst in pipe flow

The Reynolds number ranges for which laminar, transitional, or turbulent pipe flows are obtained cannot be precisely given. The actual transition from laminar to turbulent flow may take place at various Reynolds numbers, depending on how much the flow is disturbed by vibrations of the pipe, roughness of the entrance region, and the like. For general engineering purposes (i.e., without undue precautions to eliminate such disturbances), the following values are appropriate: The flow in a round pipe is laminar if the Reynolds number is less than approximately 2100. The flow in a round pipe is turbulent if the Reynolds number is greater than approximately 4000. For Reynolds numbers between these two limits, the flow may switch between laminar and turbulent conditions in an apparently random fashion (transitional flow).

F l u i d s   i n   t h e   N e w s

Nanoscale flows The term nanoscale generally refers to objects with characteristic lengths from atomic dimensions up to a few hundred nanometers (nm). (Recall that 1 nm $= 10^{-9}$ m.) Nanoscale fluid mechanics research has recently uncovered many surprising and useful phenomena. No doubt many more remain to be discovered. For example, in the future researchers envision using nanoscale tubes to push tiny amounts of water-soluble drugs to exactly where they are needed in the human body. Because of the tiny diameters involved, the *Reynolds numbers* for such flows are extremely small and the flow is definitely laminar. In addition, some standard properties of everyday flows (for example, the fact that a fluid sticks to a solid boundary) may not be valid for nanoscale flows. Also, ultratiny mechanical pumps and valves are difficult to manufacture and may become clogged by tiny particles such as biological molecules. As a possible solution to such problems, researchers have investigated the possibility of using a system that does not rely on mechanical parts. It involves using light-sensitive molecules attached to the surface of the tubes. By shining light onto the molecules, the light-responsive molecules attract water and cause motion of water through the tube. (See Problem 8.10.)

EXAMPLE 8.1   Laminar or Turbulent Flow

GIVEN   Water at a temperature of 10 °C flows through a pipe of diameter $D = 1.85$ cm and into a glass as shown in Fig. E8.1a.

FIND   Determine

(a)   the minimum time taken to fill a 0.355 L glass (volume = 355 cm^3) with water if the flow in the pipe is to be laminar. Repeat the calculations if the water temperature is 60 °C.

(b)   the maximum time taken to fill the glass if the flow is to be turbulent. Repeat the calculations if the water temperature is 60 °C.

SOLUTION

(a)   If the flow in the pipe is to remain laminar, the minimum time to fill the glass will occur if the Reynolds number is the maximum allowed for laminar flow, typically Re $= \rho VD/\mu = 2100$. Thus, $V = 2100\,\mu/\rho D$, where from Table B.1, $\rho = 1000$ kg/m^3 and $\mu = 1.307 \times 10^{-3}$ N $\cdot$ s/m^2 at 10 °C, while $\rho = 983.2$ kg/m^3 and $\mu = 4.665 \times 10^{-4}$ N $\cdot$ s/m^2 at 60 °C. Thus, the maximum average velocity for laminar flow in the pipe is

$$V = \frac{2100\mu}{\rho D} = \frac{2100(1.307 \times 10^{-3}\ \text{N} \cdot \text{s/m}^2)}{(1000\ \text{kg/m}^3)(1.85/100\ \text{m})}$$

$$= 0.148\ \text{N} \cdot \text{s/kg} = 0.148\ \text{m/s}$$

Similarly, $V = 0.054$ m/s at 60 °C. With $V\hspace{-0.9em}-$ = volume of glass and $V\hspace{-0.9em}- = Qt$ we obtain

$$t = \frac{V\hspace{-0.9em}-}{Q} = \frac{V\hspace{-0.9em}-}{(\pi/4)D^2 V} = \frac{4(3.55 \times 10^{-4}\ \text{m}^3)}{(\pi[1.85/100]^2\ \text{m}^2)(0.148\ \text{m/s})}$$

$$= 8.92\ \text{s at } T = 10\ °\text{C} \tag{Ans}$$

Similarly, $t = 24.4$ s at 60 °C. To maintain laminar flow, the less viscous hot water requires a lower flowrate than the cold water.

(b)   If the flow in the pipe is to be turbulent, the maximum time to fill the glass will occur if the Reynolds number is the minimum allowed for turbulent flow, Re $= 4000$. Thus, $V = 4000\mu/\rho D = 0.282$ m/s and

$$t = 4.67\ \text{s at } 10\ °\text{C} \tag{Ans}$$

Similarly, $V = 0.102$ m/s and $t = 13.0$ s at 60 °C.

COMMENTS   Note that because water is "not very viscous," the velocity must be "fairly small" to maintain laminar flow. In general, turbulent flows are encountered more often than laminar flows because of the relatively small viscosity of most common fluids (water, gasoline, air). By repeating the calculations at various water temperatures, T (i.e., with different densities and viscosities), the results shown in Fig. E8.1b are obtained. As the water temperature increases, the kinematic viscosity, $\nu = \mu/\rho$, decreases and the corresponding times to fill the glass increase as indicated. (Temperature effects on the viscosity of gases are the opposite; increase in temperature causes an increase in viscosity.)

■ **FIGURE E8.1a**

If the flowing fluid had been honey with a kinematic viscosity ($\nu = \mu/\rho$) 3000 times greater than that of water, the velocities given earlier would be increased by a factor of 3000 and the times reduced by the same factor. As shown in the following sections, the pressure needed to force a very viscous fluid through a pipe at such a high velocity may be unreasonably large.

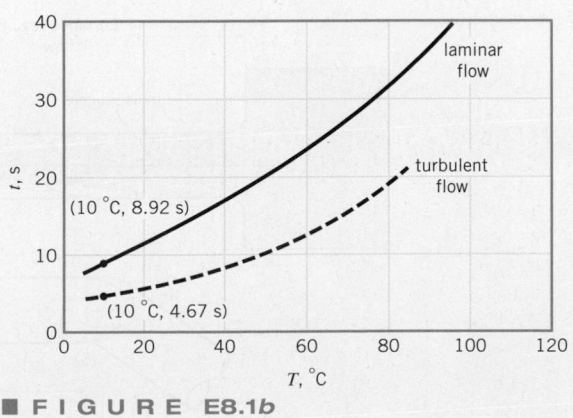

■ **FIGURE E8.1b**

8.1.2 Entrance Region and Fully Developed Flow

Any fluid flowing in a pipe had to enter the pipe at some location. The region of flow near where the fluid enters the pipe is termed the *entrance region* and is illustrated in Fig. 8.5. It may be the first few meters of a pipe connected to a tank or the initial portion of a long run of a hot air duct coming from a furnace.

As is shown in Fig. 8.5, the fluid typically enters the pipe with a nearly uniform velocity profile at section (1). As the fluid moves through the pipe, viscous effects cause it to stick to the pipe wall (the no-slip boundary condition). This is true whether the fluid is relatively inviscid air or a very viscous oil. Thus, a *boundary layer* in which viscous effects are important is produced along the pipe wall such that the initial velocity profile changes with distance along the pipe, x, until the fluid reaches the end of the entrance length, section (2), beyond which the velocity profile does not vary with x. The boundary layer has grown in thickness to completely fill the pipe. Viscous effects are of considerable importance within the boundary layer. For fluid outside the boundary layer [within the *inviscid core* surrounding the centerline from (1) to (2)], viscous effects are negligible.

The shape of the velocity profile in the pipe depends on whether the flow is laminar or turbulent, as does the length of the entrance region, ℓ_e. As with many other properties of pipe flow, the dimensionless **entrance length**, ℓ_e/D, correlates quite well with the Reynolds number. Typical entrance lengths are given by

$$\frac{\ell_e}{D} = 0.06 \, \mathrm{Re} \text{ for laminar flow} \tag{8.1}$$

The entrance length is a function of the Reynolds number.

and

$$\frac{\ell_e}{D} = 4.4 \, (\mathrm{Re})^{1/6} \text{ for turbulent flow} \tag{8.2}$$

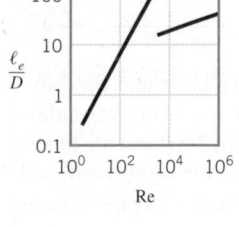

For very low Reynolds number flows the entrance length can be quite short ($\ell_e = 0.6D$ if Re = 10), whereas for large Reynolds number flows it may take a length equal to many pipe diameters before the end of the entrance region is reached ($\ell_e = 120D$ for Re = 2000). For many practical engineering problems, $10^4 < \mathrm{Re} < 10^5$ so that as shown by the figure in the margin, $20D < \ell_e < 30D$.

Calculation of the velocity profile and pressure distribution within the entrance region is quite complex. However, once the fluid reaches the end of the entrance region, section (2) of Fig. 8.5, the flow is simpler to describe because the velocity is a function of only the distance from the pipe centerline, r, and independent of x. This is true until the character of the pipe changes in some way, such as a change in diameter, or the fluid flows through a bend, valve, or some other component at section (3). The flow between (2) and (3) is termed *fully developed flow.* Beyond the interruption of the fully developed flow [at section (4)], the flow gradually begins its

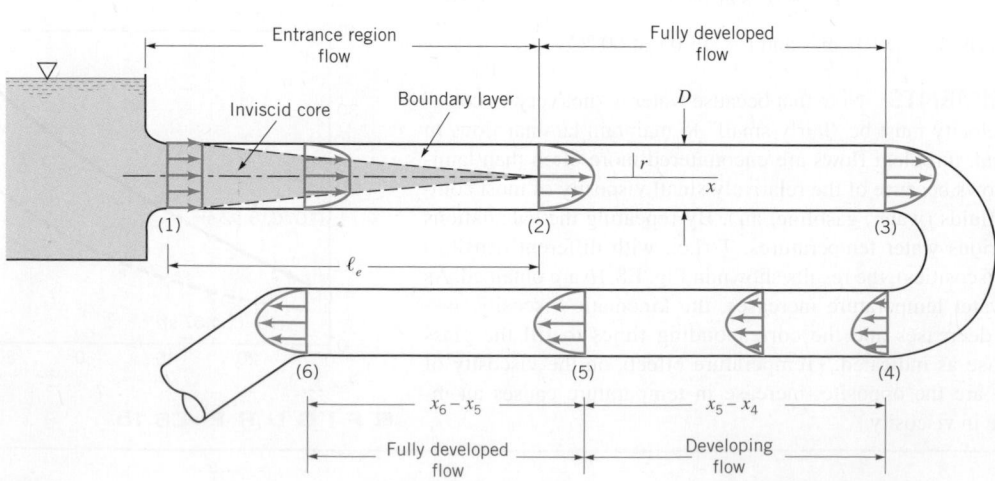

■ **F I G U R E 8.5**   **Entrance region, developing flow, and fully developed flow in a pipe system.**

return to its fully developed character [section (5)] and continues with this profile until the next pipe system component is reached [section (6)]. In many cases the pipe is long enough so that there is a considerable length of fully developed flow compared with the developing flow length $[(x_3 - x_2) \gg \ell_e$ and $(x_6 - x_5) \gg (x_5 - x_4)]$. In other cases the distances between one component (bend, tee, valve, etc.) of the pipe system and the next component is so short that fully developed flow is never achieved.

8.1.3 Pressure and Shear Stress

Fully developed steady flow in a constant diameter pipe may be driven by gravity and/or pressure forces. For horizontal pipe flow, gravity has no effect except for a hydrostatic pressure variation across the pipe, γD, that is usually negligible. It is the pressure difference, $\Delta p = p_1 - p_2$, between one section of the horizontal pipe and another which forces the fluid through the pipe. Viscous effects provide the restraining force that exactly balances the pressure force, thereby allowing the fluid to flow through the pipe with no acceleration. If viscous effects were absent in such flows, the pressure would be constant throughout the pipe, except for the hydrostatic variation.

In nonfully developed flow regions, such as the entrance region of a pipe, the fluid accelerates or decelerates as it flows (the velocity profile changes from a uniform profile at the entrance of the pipe to its fully developed profile at the end of the entrance region). Thus, in the entrance region there is a balance between pressure, viscous, and inertia (acceleration) forces. The result is a pressure distribution along the horizontal pipe as shown in Fig. 8.6. The magnitude of the pressure gradient, $\partial p / \partial x$, is larger in the entrance region than in the fully developed region, where it is a constant, $\partial p / \partial x = -\Delta p / \ell < 0$.

The fact that there is a nonzero pressure gradient along the horizontal pipe is a result of viscous effects. As is discussed in Chapter 3, if the viscosity were zero, the pressure would not vary with x. The need for the pressure drop can be viewed from two different standpoints. In terms of a force balance, the pressure force is needed to overcome the viscous forces generated. In terms of an energy balance, the work done by the pressure force is needed to overcome the viscous dissipation of energy throughout the fluid. If the pipe is not horizontal, the pressure gradient along it is due in part to the component of weight in that direction. As is discussed in Section 8.2.1, this contribution due to the weight either enhances or retards the flow, depending on whether the flow is downhill or uphill.

Laminar flow characteristics are different than those for turbulent flow.

The nature of the pipe flow is strongly dependent on whether the flow is laminar or turbulent. This is a direct consequence of the differences in the nature of the shear stress in laminar and turbulent flows. As is discussed in some detail in Section 8.3.3, the shear stress in laminar flow is a direct result of momentum transfer among the randomly moving molecules (a microscopic phenomenon). The shear stress in turbulent flow is largely a result of momentum transfer among the randomly moving, finite-sized fluid particles (a macroscopic phenomenon). The net result is that the physical properties of the shear stress are quite different for laminar flow than for turbulent flow.

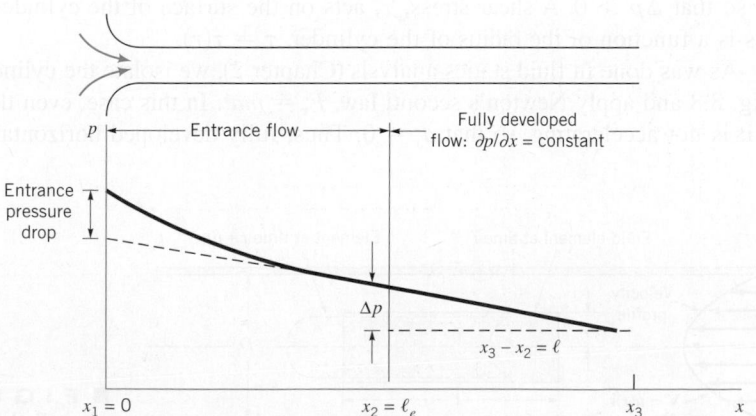

■ FIGURE 8.6   Pressure distribution along a horizontal pipe.

8.2 Fully Developed Laminar Flow

As is indicated in the previous section, the flow in long, straight, constant diameter sections of a pipe becomes fully developed. That is, the velocity profile is the same at any cross section of the pipe. Although this is true whether the flow is laminar or turbulent, the details of the velocity profile (and other flow properties) are quite different for these two types of flow. As will be seen in the remainder of this chapter, knowledge of the velocity profile can lead directly to other useful information such as pressure drop, head loss, flowrate, and the like. Thus, we begin by developing the equation for the velocity profile in fully developed laminar flow. If the flow is not fully developed, a theoretical analysis becomes much more complex and is outside the scope of this text. If the flow is turbulent, a rigorous theoretical analysis is as yet not possible.

Although most flows are turbulent rather than laminar, and many pipes are not long enough to allow the attainment of fully developed flow, a theoretical treatment and full understanding of fully developed laminar flow is of considerable importance. First, it represents one of the few theoretical viscous analyses that can be carried out "exactly" (within the framework of quite general assumptions) without using other ad hoc assumptions or approximations. An understanding of the method of analysis and the results obtained provides a foundation from which to carry out more complicated analyses. Second, there are many practical situations involving the use of fully developed laminar pipe flow.

There are numerous ways to derive important results pertaining to fully developed laminar flow. Three alternatives include: (1) from $\mathbf{F} = m\mathbf{a}$ applied directly to a fluid element, (2) from the Navier–Stokes equations of motion, and (3) from dimensional analysis methods.

8.2.1 From $\mathbf{F} = m\mathbf{a}$ Applied Directly to a Fluid Element

Steady, fully developed pipe flow experiences no acceleration.

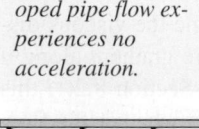
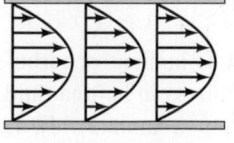

Velocity profiles

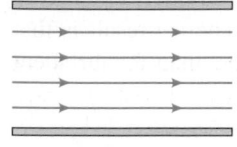

Streamlines

We consider the fluid element at time t as is shown in Fig. 8.7. It is a circular cylinder of fluid of length ℓ and radius r centered on the axis of a horizontal pipe of diameter D. Because the velocity is not uniform across the pipe, the initially flat ends of the cylinder of fluid at time t become distorted at time $t + \delta t$ when the fluid element has moved to its new location along the pipe as shown in the figure. If the flow is fully developed and steady, the distortion on each end of the fluid element is the same, and no part of the fluid experiences any acceleration as it flows, as shown by the figure in the margin. The local acceleration is zero ($\partial \mathbf{V}/\partial t = 0$) because the flow is steady, and the convective acceleration is zero ($\mathbf{V} \cdot \nabla \mathbf{V} = u\, \partial u/\partial x\, \hat{\mathbf{i}} = 0$) because the flow is fully developed. Thus, every part of the fluid merely flows along its streamline parallel to the pipe walls with constant velocity, although neighboring particles have slightly different velocities. The velocity varies from one pathline to the next. This velocity variation, combined with the fluid viscosity, produces the shear stress.

If gravitational effects are neglected, the pressure is constant across any vertical cross section of the pipe, although it varies along the pipe from one section to the next. Thus, if the pressure is $p = p_1$ at section (1), it is $p_2 = p_1 - \Delta p$ at section (2) where Δp is the pressure drop between sections (1) and (2). We anticipate the fact that the pressure decreases in the direction of flow so that $\Delta p > 0$. A shear stress, τ, acts on the surface of the cylinder of fluid. This viscous stress is a function of the radius of the cylinder, $\tau = \tau(r)$.

As was done in fluid statics analysis (Chapter 2), we isolate the cylinder of fluid as is shown in Fig. 8.8 and apply Newton's second law, $F_x = ma_x$. In this case, even though the fluid is moving, it is not accelerating, so that $a_x = 0$. Thus, fully developed horizontal pipe flow is merely a

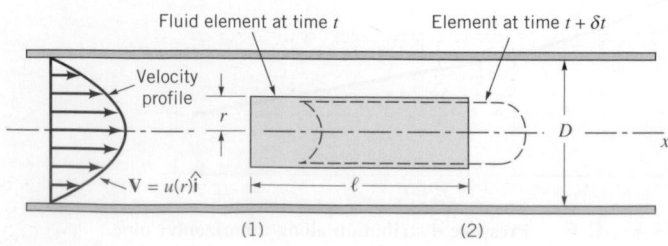

■ **FIGURE 8.7** Motion of a cylindrical fluid element within a pipe.

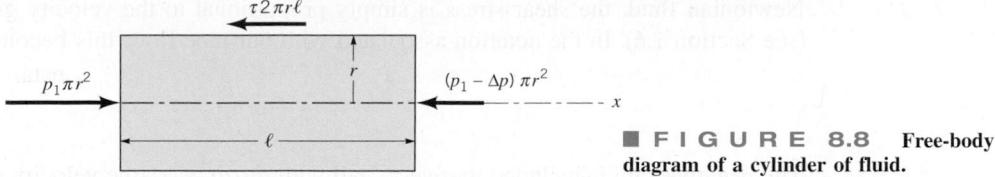

■ **FIGURE 8.8**   **Free-body diagram of a cylinder of fluid.**

balance between pressure and viscous forces—the pressure difference acting on the end of the cylinder of area πr^2, and the shear stress acting on the lateral surface of the cylinder of area $2\pi r\ell$. This force balance can be written as

$$(p_1)\pi r^2 - (p_1 - \Delta p)\pi r^2 - (\tau)2\pi r\ell = 0$$

which can be simplified to give

$$\frac{\Delta p}{\ell} = \frac{2\tau}{r} \tag{8.3}$$

Equation 8.3 represents the basic balance in forces needed to drive each fluid particle along the pipe with constant velocity. Since neither Δp nor ℓ are functions of the radial coordinate, r, it follows that $2\tau/r$ must also be independent of r. That is, $\tau = Cr$, where C is a constant. At $r = 0$ (the centerline of the pipe) there is no shear stress ($\tau = 0$). At $r = D/2$ (the pipe wall) the shear stress is a maximum, denoted τ_w, the *wall shear stress*. Hence, $C = 2\tau_w/D$ and the shear stress distribution throughout the pipe is a linear function of the radial coordinate

$$\tau = \frac{2\tau_w r}{D} \tag{8.4}$$

as is indicated in Fig. 8.9. The linear dependence of τ on r is a result of the pressure force being proportional to r^2 (the pressure acts on the end of the fluid cylinder; area $= \pi r^2$) and the shear force being proportional to r (the shear stress acts on the lateral sides of the cylinder; area $= 2\pi r\ell$). If the viscosity were zero there would be no shear stress, and the pressure would be constant throughout the horizontal pipe ($\Delta p = 0$). As is seen from Eqs. 8.3 and 8.4, the pressure drop and wall shear stress are related by

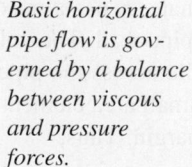

Basic horizontal pipe flow is governed by a balance between viscous and pressure forces.

$$\Delta p = \frac{4\ell\tau_w}{D} \tag{8.5}$$

A small shear stress can produce a large pressure difference if the pipe is relatively long ($\ell/D \gg 1$).

Although we are discussing laminar flow, a closer consideration of the assumptions involved in the derivation of Eqs. 8.3, 8.4, and 8.5 reveals that these equations are valid for both laminar and turbulent flow. To carry the analysis further we must prescribe how the shear stress is related to the velocity. This is the critical step that separates the analysis of laminar from that of turbulent flow—from being able to solve for the laminar flow properties and not being able to solve for the turbulent flow properties without additional ad hoc assumptions. As is discussed in Section 8.3, the shear stress dependence for turbulent flow is very complex. However, for laminar flow of a

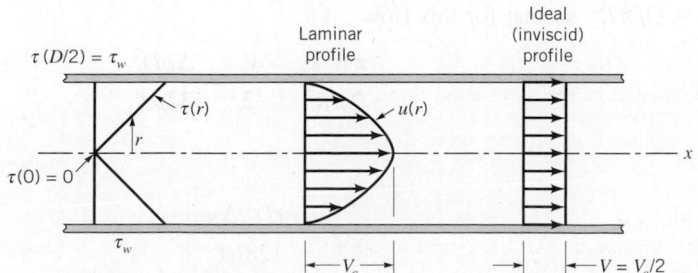

■ **FIGURE 8.9**
Shear stress distribution within the fluid in a pipe (laminar or turbulent flow) and typical velocity profiles.

Newtonian fluid, the shear stress is simply proportional to the velocity gradient, "$\tau = \mu\, du/dy$" (see Section 1.6). In the notation associated with our pipe flow, this becomes

$$\tau = -\mu \frac{du}{dr} \tag{8.6}$$

The negative sign is included to give $\tau > 0$ with $du/dr < 0$ (the velocity decreases from the pipe centerline to the pipe wall).

Equations 8.3 and 8.6 represent the two governing laws for fully developed laminar flow of a Newtonian fluid within a horizontal pipe. The one is Newton's second law of motion and the other is the definition of a Newtonian fluid. By combining these two equations we obtain

$$\frac{du}{dr} = -\left(\frac{\Delta p}{2\mu\ell}\right) r$$

which can be integrated to give the velocity profile as follows:

$$\int du = -\frac{\Delta p}{2\mu\ell} \int r\, dr$$

or

$$u = -\left(\frac{\Delta p}{4\mu\ell}\right) r^2 + C_1$$

where C_1 is a constant. Because the fluid is viscous it sticks to the pipe wall so that $u = 0$ at $r = D/2$. Thus, $C_1 = (\Delta p/16\mu\ell)D^2$. Hence, the velocity profile can be written as

$$u(r) = \left(\frac{\Delta p D^2}{16\mu\ell}\right)\left[1 - \left(\frac{2r}{D}\right)^2\right] = V_c\left[1 - \left(\frac{2r}{D}\right)^2\right] \tag{8.7}$$

where $V_c = \Delta p D^2/(16\mu\ell)$ is the centerline velocity. An alternative expression can be written by using the relationship between the wall shear stress and the pressure gradient (Eqs. 8.5 and 8.7) to give

Under certain restrictions the velocity profile in a pipe is parabolic.

$$u(r) = \frac{\tau_w D}{4\mu}\left[1 - \left(\frac{r}{R}\right)^2\right]$$

where $R = D/2$ is the pipe radius.

This velocity profile, plotted in Fig. 8.9, is parabolic in the radial coordinate, r, has a maximum velocity, V_c, at the pipe centerline, and a minimum velocity (zero) at the pipe wall. The volume flowrate through the pipe can be obtained by integrating the velocity profile across the pipe. Since the flow is axisymmetric about the centerline, the velocity is constant on small area elements consisting of rings of radius r and thickness dr as shown in the figure in the margin. Thus,

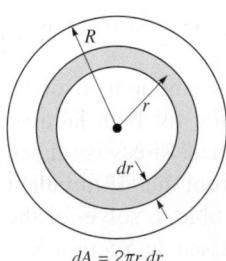

$dA = 2\pi r\, dr$

$$Q = \int u\, dA = \int_{r=0}^{r=R} u(r) 2\pi r\, dr = 2\pi V_c \int_0^R \left[1 - \left(\frac{r}{R}\right)^2\right] r\, dr$$

or

$$Q = \frac{\pi R^2 V_c}{2}$$

By definition, the average velocity is the flowrate divided by the cross-sectional area, $V = Q/A = Q/\pi R^2$, so that for this flow

$$V = \frac{\pi R^2 V_c}{2\pi R^2} = \frac{V_c}{2} = \frac{\Delta p D^2}{32\mu\ell} \tag{8.8}$$

and

$$Q = \frac{\pi D^4\, \Delta p}{128\mu\ell} \tag{8.9}$$

As is indicated in Eq. 8.8, the average velocity is one-half of the maximum velocity. In general, for velocity profiles of other shapes (such as for turbulent pipe flow), the average velocity is not merely the average of the maximum (V_c) and minimum (0) velocities as it is for the laminar parabolic profile. The two velocity profiles indicated in Fig. 8.9 provide the same flowrate—one is the fictitious ideal ($\mu = 0$) profile; the other is the actual laminar flow profile.

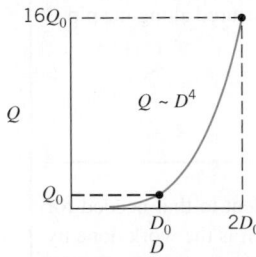

Poiseuille's law is valid for laminar flow only.

The above results confirm the following properties of laminar pipe flow. For a horizontal pipe the flowrate is (a) directly proportional to the pressure drop, (b) inversely proportional to the viscosity, (c) inversely proportional to the pipe length, and (d) proportional to the pipe diameter to the fourth power. With all other parameters fixed, an increase in diameter by a factor of 2 will increase the flowrate by a factor of $2^4 = 16$—the flowrate is very strongly dependent on pipe size. This dependence is shown by the figure in the margin. Likewise, a small error in pipe diameter can cause a relatively large error in flowrate. For example, a 2% error in diameter gives an 8% error in flowrate ($Q \sim D^4$ or $\delta Q \sim 4D^3 \delta D$, so that $\delta Q/Q = 4 \, \delta D/D$). This flow, the properties of which were first established experimentally by two independent workers, G. Hagen (1797–1884) in 1839 and J. Poiseuille (1799–1869) in 1840, is termed *Hagen–Poiseuille flow*. Equation 8.9 is commonly referred to as *Poiseuille's law*. Recall that all of these results are restricted to laminar flow (those with Reynolds numbers less than approximately 2100) in a horizontal pipe.

The adjustment necessary to account for nonhorizontal pipes, as shown in Fig. 8.10, can be easily included by replacing the pressure drop, Δp, by the combined effect of pressure and gravity, $\Delta p - \gamma \ell \sin \theta$, where θ is the angle between the pipe and the horizontal. (Note that $\theta > 0$ if the flow is uphill, while $\theta < 0$ if the flow is downhill.) This can be seen from the force balance in the x direction (along the pipe axis) on the cylinder of fluid shown in Fig. 8.10b. The method is exactly analogous to that used to obtain the Bernoulli equation (Eq. 3.6) when the streamline is not horizontal. The net force in the x direction is a combination of the pressure force in that direction, $\Delta p \pi r^2$, and the component of weight in that direction, $-\gamma \pi r^2 \ell \sin \theta$. The result is a slightly modified form of Eq. 8.3 given by

$$\frac{\Delta p - \gamma \ell \sin \theta}{\ell} = \frac{2\tau}{r} \qquad (8.10)$$

Thus, all of the results for the horizontal pipe are valid provided the pressure gradient is adjusted for the elevation term, that is, Δp is replaced by $\Delta p - \gamma \ell \sin \theta$ so that

$$V = \frac{(\Delta p - \gamma \ell \sin \theta)D^2}{32\mu\ell} \qquad (8.11)$$

and

$$Q = \frac{\pi(\Delta p - \gamma \ell \sin \theta)D^4}{128\mu\ell} \qquad (8.12)$$

It is seen that the driving force for pipe flow can be either a pressure drop in the flow direction, Δp, or the component of weight in the flow direction, $-\gamma \ell \sin \theta$. If the flow is downhill, gravity helps the flow (a smaller pressure drop is required; $\sin \theta < 0$). If the flow is uphill, gravity works against the flow (a larger pressure drop is required; $\sin \theta > 0$). Note that $\gamma \ell \sin \theta = \gamma \Delta z$ (where

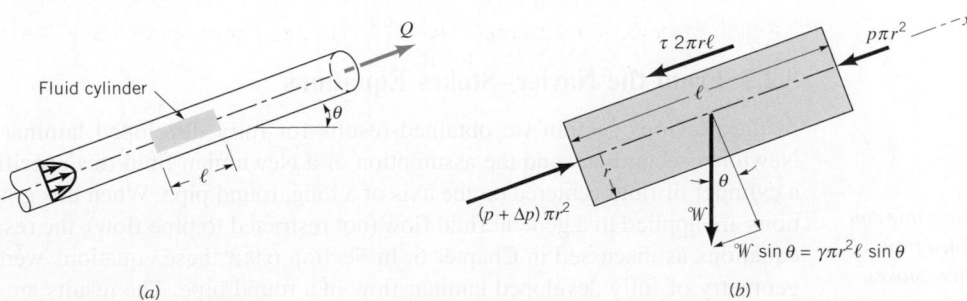

(a)   (b)

■ **FIGURE 8.10**   **Free-body diagram of a fluid cylinder for flow in a nonhorizontal pipe.**

Δz is the change in elevation) is a hydrostatic type pressure term. If there is no flow, $V = 0$ and $\Delta p = \gamma \ell \sin \theta = \gamma \Delta z$, as expected for fluid statics.

EXAMPLE 8.2   Laminar Pipe Flow

GIVEN An oil with a viscosity of $\mu = 0.40$ N · s/m^2 and density $\rho = 900$ kg/m^3 flows in a pipe of diameter $D = 0.020$ m.

FIND **(a)** What pressure drop, $p_1 - p_2$, is needed to produce a flowrate of $Q = 2.0 \times 10^{-5}$ m^3/s if the pipe is horizontal with $x_1 = 0$ and $x_2 = 10$ m?

(b) How steep a hill, θ, must the pipe be on if the oil is to flow through the pipe at the same rate as in part (a), but with $p_1 = p_2$?

(c) For the conditions of part (b), if $p_1 = 200$ kPa, what is the pressure at section $x_3 = 5$ m, where x is measured along the pipe?

SOLUTION

(a) If the Reynolds number is less than 2100 the flow is laminar and the equations derived in this section are valid. Since the average velocity is $V = Q/A = (2.0 \times 10^{-5}$ m^3/s)/$[\pi(0.020)^2$m$^2/4] = 0.0637$ m/s, the Reynolds number is Re $= \rho V D/\mu = 2.87 < 2100$. Hence, the flow is laminar and from Eq. 8.9 with $\ell = x_2 - x_1 = 10$ m, the pressure drop is

$$\Delta p = p_1 - p_2 = \frac{128 \mu \ell Q}{\pi D^4}$$

$$= \frac{128(0.40 \text{ N} \cdot \text{s/m}^2)(10.0 \text{ m})(2.0 \times 10^{-5} \text{ m}^3/\text{s})}{\pi(0.020 \text{ m})^4}$$

or

$$\Delta p = 20{,}400 \text{ N/m}^2 = 20.4 \text{ kPa} \qquad \text{(Ans)}$$

(b) If the pipe is on a hill of angle θ such that $\Delta p = p_1 - p_2 = 0$, Eq. 8.12 gives

$$\sin \theta = -\frac{128 \mu Q}{\pi \rho g D^4} \qquad (1)$$

or

$$\sin \theta = \frac{-128(0.40 \text{ N} \cdot \text{s/m}^2)(2.0 \times 10^{-5} \text{ m}^3/\text{s})}{\pi(900 \text{ kg/m}^3)(9.81 \text{ m/s}^2)(0.020 \text{ m})^4} \quad \text{(Ans)}$$

Thus, $\theta = -13.34°$.

COMMENT This checks with the previous horizontal result as is seen from the fact that a change in elevation of $\Delta z = \ell \sin \theta = (10 \text{ m}) \sin(-13.34°) = -2.31$ m is equivalent to a pressure change of $\Delta p = \rho g \Delta z = (900 \text{ kg/m}^3)(9.81 \text{ m/s}^2)$

$(2.31 \text{ m}) = 20{,}400$ N/m^2, which is equivalent to that needed for the horizontal pipe. For the horizontal pipe it is the work done by the pressure forces that overcomes the viscous dissipation. For the zero-pressure-drop pipe on the hill, it is the change in potential energy of the fluid "falling" down the hill that is converted to the energy lost by viscous dissipation. Note that if it is desired to increase the flowrate to $Q = 1.0 \times 10^{-4}$ m^3/s with $p_1 = p_2$, the value of θ given by Eq. 1 is $\sin \theta = -1.15$. Since the sine of an angle can not be greater than 1, this flow would not be possible. The weight of the fluid would not be large enough to offset the viscous force generated for the flowrate desired. A larger diameter pipe would be needed.

(c) With $p_1 = p_2$ the length of the pipe, ℓ, does not appear in the flowrate equation (Eq. 1). This is a statement of the fact that for such cases the pressure is constant all along the pipe (provided the pipe lies on a hill of constant slope). This can be seen by substituting the values of Q and θ from case (b) into Eq. 8.12 and noting that $\Delta p = 0$ for any ℓ. For example, $\Delta p = p_1 - p_3 = 0$ if $\ell = x_3 - x_1 = 5$ m. Thus, $p_1 = p_2 = p_3$ so that

$$p_3 = 200 \text{ kPa} \qquad \text{(Ans)}$$

COMMENT Note that if the fluid were gasoline ($\mu = 3.1 \times 10^{-4}$ N · s/m^2 and $\rho = 680$ kg/m^3), the Reynolds number would be Re $= 2790$, the flow would probably not be laminar, and use of Eqs. 8.9 and 8.12 would give incorrect results. Also note from Eq. 1 that the kinematic viscosity, $\nu = \mu/\rho$, is the important viscous parameter. This is a statement of the fact that with constant pressure along the pipe, it is the ratio of the viscous force ($\sim \mu$) to the weight force ($\sim \gamma = \rho g$) that determines the value of θ.

8.2.2 From the Navier–Stokes Equations

Poiseuille's law can be obtained from the Navier–Stokes equations.

In the previous section we obtained results for fully developed laminar pipe flow by applying Newton's second law and the assumption of a Newtonian fluid to a specific portion of the fluid—a cylinder of fluid centered on the axis of a long, round pipe. When this governing law and assumptions are applied to a general fluid flow (not restricted to pipe flow), the result is the Navier–Stokes equations as discussed in Chapter 6. In Section 6.9.3 these equations were solved for the specific geometry of fully developed laminar flow in a round pipe. The results are the same as those given in Eq. 8.7.

We will not repeat the detailed steps used to obtain the laminar pipe flow from the Navier–Stokes equations (see Section 6.9.3) but will indicate how the various assumptions used and steps applied in the derivation correlate with the analysis used in the previous section.

General motion of an incompressible Newtonian fluid is governed by the continuity equation (conservation of mass, Eq. 6.31) and the momentum equation (Eq. 6.127), which are rewritten here for convenience:

$$\nabla \cdot \mathbf{V} = 0 \tag{8.13}$$

$$\frac{\partial \mathbf{V}}{\partial t} + \mathbf{V} \cdot \nabla \mathbf{V} = -\frac{\nabla p}{\rho} + \mathbf{g} + \nu \nabla^2 \mathbf{V} \tag{8.14}$$

For steady, fully developed flow in a pipe, the velocity contains only an axial component, which is a function of only the radial coordinate [$\mathbf{V} = u(r)\hat{\mathbf{i}}$]. For such conditions, the left-hand side of the Eq. 8.14 is zero. This is equivalent to saying that the fluid experiences no acceleration as it flows along. The same constraint was used in the previous section when considering $\mathbf{F} = m\mathbf{a}$ for the fluid cylinder. Thus, with $\mathbf{g} = -g\hat{\mathbf{k}}$ the Navier–Stokes equations become

$$\nabla \cdot \mathbf{V} = 0$$

$$\nabla p + \rho g \hat{\mathbf{k}} = \mu \nabla^2 \mathbf{V} \tag{8.15}$$

The flow is governed by a balance of pressure, weight, and viscous forces in the flow direction, similar to that shown in Fig. 8.10 and Eq. 8.10. If the flow were not fully developed (as in an entrance region, for example), it would not be possible to simplify the Navier–Stokes equations to that form given in Eq. 8.15 (the nonlinear term $\mathbf{V} \cdot \nabla \mathbf{V}$ would not be zero), and the solution would be very difficult to obtain.

Because of the assumption that $\mathbf{V} = u(r)\hat{\mathbf{i}}$, the continuity equation, Eq. 8.13, is automatically satisfied. This conservation of mass condition was also automatically satisfied by the incompressible flow assumption in the derivation in the previous section. The fluid flows across one section of the pipe at the same rate that it flows across any other section (see Fig. 8.8).

When it is written in terms of polar coordinates (as was done in Section 6.9.3), the component of Eq. 8.15 along the pipe becomes

The governing differential equations can be simplified by appropriate assumptions.

$$\frac{\partial p}{\partial x} + \rho g \sin \theta = \mu \frac{1}{r} \frac{\partial}{\partial r}\left(r \frac{\partial u}{\partial r}\right) \tag{8.16}$$

Since the flow is fully developed, $u = u(r)$ and the right-hand side is a function of, at most, only r. The left-hand side is a function of, at most, only x. It was shown that this leads to the condition that the pressure gradient in the x direction is a constant—$\partial p/\partial x = -\Delta p/\ell$. The same condition was used in the derivation of the previous section (Eq. 8.3).

It is seen from Eq. 8.16 that the effect of a nonhorizontal pipe enters into the Navier–Stokes equations in the same manner as was discussed in the previous section. The pressure gradient in the flow direction is coupled with the effect of the weight in that direction to produce an effective pressure gradient of $-\Delta p/\ell + \rho g \sin \theta$.

The velocity profile is obtained by integration of Eq. 8.16. Since it is a second-order equation, two boundary conditions are needed—(1) the fluid sticks to the pipe wall (as was also done in Eq. 8.7) and (2) either of the equivalent forms that the velocity remains finite throughout the flow (in particular $u < \infty$ at $r = 0$) or, because of symmetry, that $\partial u/\partial r = 0$ at $r = 0$. In the derivation of the previous section, only one boundary condition (the no-slip condition at the wall) was needed because the equation integrated was a first-order equation. The other condition ($\partial u/\partial r = 0$ at $r = 0$) was automatically built into the analysis because of the fact that $\tau = -\mu\, du/dr$ and $\tau = 2\tau_w r/D = 0$ at $r = 0$.

The results obtained by either applying $\mathbf{F} = m\mathbf{a}$ to a fluid cylinder (Section 8.2.1) or solving the Navier–Stokes equations (Section 6.9.3) are exactly the same. Similarly, the basic assumptions regarding the flow structure are the same. This should not be surprising because the two methods are based on the same principle—Newton's second law. One is restricted to fully developed laminar pipe flow from the beginning (the drawing of the free-body diagram), and the other starts with the general governing equations (the Navier–Stokes equations) with the appropriate restrictions concerning fully developed laminar flow applied as the solution process progresses.

8.2.3 From Dimensional Analysis

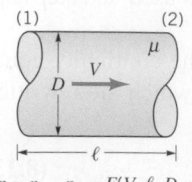

$\Delta p = p_1 - p_2 = F(V, \ell, D, \mu)$

Although fully developed laminar pipe flow is simple enough to allow the rather straightforward solutions discussed in the previous two sections, it may be worthwhile to consider this flow from a dimensional analysis standpoint. Thus, we assume that the pressure drop in the horizontal pipe, Δp, is a function of the average velocity of the fluid in the pipe, V, the length of the pipe, ℓ, the pipe diameter, D, and the viscosity of the fluid, μ, as shown by the figure in the margin. We have not included the density or the specific weight of the fluid as parameters because for such flows they are not important parameters. There is neither mass (density) times acceleration nor a component of weight (specific weight times volume) in the flow direction involved. Thus,

$$\Delta p = F(V, \ell, D, \mu)$$

There are five variables that can be described in terms of three reference dimensions (M, L, T). According to the results of dimensional analysis (Chapter 7), this flow can be described in terms of $k - r = 5 - 3 = 2$ dimensionless groups. One such representation is

$$\frac{D \, \Delta p}{\mu V} = \phi\left(\frac{\ell}{D}\right) \tag{8.17}$$

where $\phi(\ell/D)$ is an unknown function of the length to diameter ratio of the pipe.

Although this is as far as dimensional analysis can take us, it seems reasonable to impose a further assumption that the pressure drop is directly proportional to the pipe length. That is, it takes twice the pressure drop to force fluid through a pipe if its length is doubled. The only way that this can be true is if $\phi(\ell/D) = C\ell/D$, where C is a constant. Thus, Eq. 8.17 becomes

$$\frac{D \, \Delta p}{\mu V} = \frac{C\ell}{D}$$

which can be rewritten as

$$\frac{\Delta p}{\ell} = \frac{C\mu \, V}{D^2}$$

or

$$Q = AV = \frac{(\pi/4C) \, \Delta p D^4}{\mu \ell} \tag{8.18}$$

Dimensional analysis can be used to put pipe flow parameters into dimensionless form.

The basic functional dependence for laminar pipe flow given by Eq. 8.18 is the same as that obtained by the analysis of the two previous sections. The value of C must be determined by theory (as done in the previous two sections) or experiment. For a round pipe, $C = 32$. For ducts of other cross-sectional shapes, the value of C is different (see Section 8.4.3).

It is usually advantageous to describe a process in terms of dimensionless quantities. To this end we rewrite the pressure drop equation for laminar horizontal pipe flow, Eq. 8.8, as $\Delta p = 32\mu\ell V/D^2$ and divide both sides by the dynamic pressure, $\rho V^2/2$, to obtain the dimensionless form as

$$\frac{\Delta p}{\frac{1}{2}\rho V^2} = \frac{(32\mu\ell V/D^2)}{\frac{1}{2}\rho V^2} = 64\left(\frac{\mu}{\rho VD}\right)\left(\frac{\ell}{D}\right) = \frac{64}{\text{Re}}\left(\frac{\ell}{D}\right)$$

This is often written as

$$\Delta p = f\frac{\ell}{D}\frac{\rho V^2}{2}$$

where the dimensionless quantity

$$f = \Delta p(D/\ell)/(\rho V^2/2)$$

is termed the *friction factor,* or sometimes the *Darcy friction factor* [H. P. G. Darcy (1803–1858)]. (This parameter should not be confused with the less-used Fanning friction

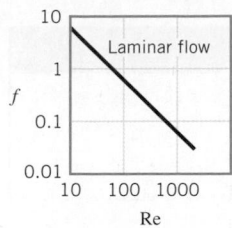

factor, which is defined to be $f/4$. In this text we will use only the Darcy friction factor.) Thus, the friction factor for laminar fully developed pipe flow is simply

$$f = \frac{64}{\text{Re}} \tag{8.19}$$

as shown by the figure in the margin.

By substituting the pressure drop in terms of the wall shear stress (Eq. 8.5), we obtain an alternate expression for the friction factor as a dimensionless wall shear stress

$$f = \frac{8\tau_w}{\rho V^2} \tag{8.20}$$

Knowledge of the friction factor will allow us to obtain a variety of information regarding pipe flow. For turbulent flow the dependence of the friction factor on the Reynolds number is much more complex than that given by Eq. 8.19 for laminar flow. This is discussed in detail in Section 8.4.

8.2.4 Energy Considerations

In the previous three sections we derived the basic laminar flow results from application of $\mathbf{F} = m\mathbf{a}$ or dimensional analysis considerations. It is equally important to understand the implications of energy considerations of such flows. To this end we consider the energy equation for incompressible, steady flow between two locations as is given in Eq. 5.89

$$\frac{p_1}{\gamma} + \alpha_1 \frac{V_1^2}{2g} + z_1 = \frac{p_2}{\gamma} + \alpha_2 \frac{V_2^2}{2g} + z_2 + h_L \tag{8.21}$$

Recall that the kinetic energy coefficients, α_1 and α_2, compensate for the fact that the velocity profile across the pipe is not uniform. For uniform velocity profiles, $\alpha = 1$, whereas for any nonuniform profile, $\alpha > 1$. The head loss term, h_L, accounts for any energy loss associated with the flow. This loss is a direct consequence of the viscous dissipation that occurs throughout the fluid in the pipe. For the ideal (inviscid) cases discussed in previous chapters, $\alpha_1 = \alpha_2 = 1$, $h_L = 0$, and the energy equation reduces to the familiar Bernoulli equation discussed in Chapter 3 (Eq. 3.7).

Even though the velocity profile in viscous pipe flow is not uniform, for fully developed flow it does not change from section (1) to section (2) so that $\alpha_1 = \alpha_2$. Thus, the kinetic energy is the same at any section ($\alpha_1 V_1^2/2 = \alpha_2 V_2^2/2$) and the energy equation becomes

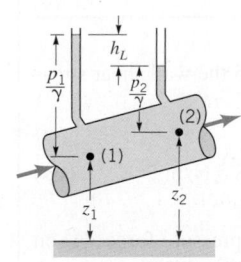

$$\left(\frac{p_1}{\gamma} + z_1 \right) - \left(\frac{p_2}{\gamma} + z_2 \right) = h_L \tag{8.22}$$

The energy dissipated by the viscous forces within the fluid is supplied by the excess work done by the pressure and gravity forces as shown by the figure in the margin.

A comparison of Eqs. 8.22 and 8.10 shows that the head loss is given by

$$h_L = \frac{2\tau \ell}{\gamma r}$$

(recall $p_1 = p_2 + \Delta p$ and $z_2 - z_1 = \ell \sin \theta$), which, by use of Eq. 8.4, can be rewritten in the form

The head loss in a pipe is a result of the viscous shear stress on the wall.

$$h_L = \frac{4\ell \tau_w}{\gamma D} \tag{8.23}$$

It is the shear stress at the wall (which is directly related to the viscosity and the shear stress throughout the fluid) that is responsible for the head loss. A closer consideration of the assumptions involved in the derivation of Eq. 8.23 will show that it is valid for both laminar and turbulent flow.

EXAMPLE 8.3 ■ Laminar Pipe Flow Properties

GIVEN The flowrate, Q, of corn syrup through the horizontal pipe shown in Fig. E8.3a is to be monitored by measuring the pressure difference between sections (1) and (2). It is proposed that $Q = K \Delta p$, where the calibration constant, K, is a function of temperature, T, because of the variation of the syrup's viscosity and density with temperature. These variations are given in Table E8.3.

FIND (a) Plot $K(T)$ versus T for $15 \, °C \leq T \leq 70 \, °C$. (b) Determine the wall shear stress and the pressure drop, $\Delta p = p_1 - p_2$, for $Q = 1.4 \times 10^{-2} \, m^3/s$ and $T = 38 \, °C$. (c) For the conditions of part (b), determine the net pressure force, $(\pi D^2/4) \Delta p$, and the net shear force, $\pi D \ell \tau_w$, on the fluid within the pipe between the sections (1) and (2).

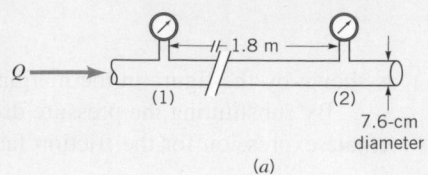

(a)

SOLUTION

(a) If the flow is laminar it follows from Eq. 8.9 that

$$Q = \frac{\pi D^4 \, \Delta p}{128 \mu \ell} = \frac{\pi (\frac{7.6}{100} \, m)^4 \, \Delta p}{128 \mu (1.8 \, m)}$$

or

$$Q = K \, \Delta p = \frac{4.5 \times 10^{-7}}{\mu} \, \Delta p \quad (1)$$

where the units on Q, Δp, and μ are m^3/s, N/m^2, and $N \cdot s/m^2$, respectively. Thus

$$K = \frac{4.5 \times 10^{-7}}{\mu} \quad \text{(Ans)}$$

where the units of K are $m^5/N \cdot s$. By using values of the viscosity from Table E8.3, the calibration curve shown in Fig. E8.3b is obtained. This result is valid only if the flow is laminar.

COMMENT As shown in Section 8.5, for turbulent flow the flowrate is not linearly related to the pressure drop so it would not be possible to have $Q = K \, \Delta p$. Note also that the value of K is independent of the syrup density (ρ was not used in the calculations) since laminar pipe flow is governed by pressure and viscous effects; inertia is not important.

(b) For $T = 38 \, °C$, the viscosity is $\mu = 1.8 \times 10^{-1} \, N \cdot s/m^2$ so that with a flowrate of $Q = 1.4 \times 10^{-2} \, m^3/s$ the pressure drop (according to Eq. 8.9) is

$$\Delta p = \frac{128 \mu \ell Q}{\pi D^4}$$

$$= \frac{128(1.8 \times 10^{-1} \, N \cdot s/m^2)(1.8 \, m)(1.4 \times 10^{-2} \, m^3/s)}{\pi (\frac{7.6}{100} \, m)^4}$$

$$= 5540 \, N/m^2 \quad \text{(Ans)}$$

provided the flow is laminar. For this case

$$V = \frac{Q}{A} = \frac{1.4 \times 10^{-2} \, m^3/s}{\frac{\pi}{4} (\frac{7.6}{100} \, m)^2} = 3.1 \, m/s$$

so that

$$Re = \frac{\rho V D}{\mu} = \frac{(1056 \, kg/m^3)(3.1 \, m/s)(\frac{7.6}{100} \, m)}{(1.8 \times 10^{-1} \, N \cdot s/m^2)}$$

$$= 1382 < 2100$$

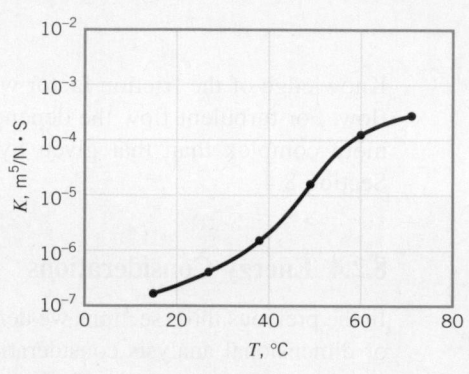

(b)

■ **FIGURE E8.3**

■ **TABLE E8.3**

$T \, (°C)$	$\rho \, (kg/m^3)$	$\mu \, (N \cdot s/m^2)$
15	1166	19.2×10^{-1}
27	1061	9.1×10^{-1}
38	1056	1.8×10^{-1}
49	1051	2.1×10^{-2}
60	1046	4.4×10^{-3}
71	1041	1.1×10^{-3}

Hence, the flow is laminar. From Eq. 8.5 the wall shear stress is

$$\tau_w = \frac{\Delta p D}{4 \ell} = \frac{(5540 \, N/m^2)(\frac{7.6}{100} \, m)}{4(1.8 \, m)} = 58.5 \, N/m^2 \quad \text{(Ans)}$$

(c) For the conditions of part (b), the net pressure force, F_p, on the fluid within the pipe between sections (1) and (2) is

$$F_p = \frac{\pi}{4} D^2 \, \Delta p = \frac{\pi}{4} \left(\frac{7.6}{100} \, m \right)^2 (5540 \, N/m^2) = 25 \, N \quad \text{(Ans)}$$

Similarly, the net viscous force, F_v, on that portion of the fluid is

$$F_v = 2\pi \left(\frac{D}{2} \right) \ell \tau_w$$

$$= 2\pi \left[\frac{7.6}{2(100)} \, m \right] (1.8 \, m)(58.5 \, N/m^2) = 25 \, N \quad \text{(Ans)}$$

COMMENT Note that the values of these two forces are the same. The net force is zero; there is no acceleration.

8.3   Fully Developed Turbulent Flow

In the previous section various properties of fully developed laminar pipe flow were discussed. Since turbulent pipe flow is actually more likely to occur than laminar flow in practical situations, it is necessary to obtain similar information for turbulent pipe flow. However, turbulent flow is a very complex process. Numerous persons have devoted considerable effort in attempting to understand the variety of baffling aspects of turbulence. Although a considerable amount of knowledge about the topic has been developed, the field of turbulent flow still remains the least understood area of fluid mechanics. In this book we can provide only some of the very basic ideas concerning turbulence. The interested reader should consult some of the many books available for further reading (Refs. 1–3).

8.3.1 Transition from Laminar to Turbulent Flow

Flows are classified as laminar or turbulent. For any flow geometry, there is one (or more) dimensionless parameter such that with this parameter value below a particular value the flow is laminar, whereas with the parameter value larger than a certain value the flow is turbulent. The important parameters involved (i.e., Reynolds number, Mach number) and their critical values depend on the specific flow situation involved. For example, flow in a pipe and flow along a flat plate (boundary layer flow, as is discussed in Section 9.2.4) can be laminar or turbulent, depending on the value of the Reynolds number involved. As a general rule for pipe flow, the value of the Reynolds number must be less than approximately 2100 for laminar flow and greater than approximately 4000 for turbulent flow. For flow along a flat plate the transition between laminar and turbulent flow occurs at a Reynolds number of approximately 500,000 (see Section 9.2.4), where the length term in the Reynolds number is the distance measured from the leading edge of the plate.

Consider a long section of pipe that is initially filled with a fluid at rest. As the valve is opened to start the flow, the flow velocity and, hence, the Reynolds number increase from zero (no flow) to their maximum steady-state flow values, as is shown in Fig. 8.11. Assume this transient process is slow enough so that unsteady effects are negligible (quasi-steady flow). For an initial time period the Reynolds number is small enough for laminar flow to occur. At some time the Reynolds number reaches 2100, and the flow begins its transition to turbulent conditions. Intermittent spots or bursts of turbulence appear. As the Reynolds number is increased, the entire flow field becomes turbulent. The flow remains turbulent as long as the Reynolds number exceeds approximately 4000.

Turbulent flows involve randomly fluctuating parameters.

A typical trace of the axial component of velocity measured at a given location in the flow, $u = u(t)$, is shown in Fig. 8.12. Its irregular, random nature is the distinguishing feature of turbulent flow. The character of many of the important properties of the flow (pressure drop, heat transfer, etc.) depends strongly on the existence and nature of the turbulent fluctuations or randomness

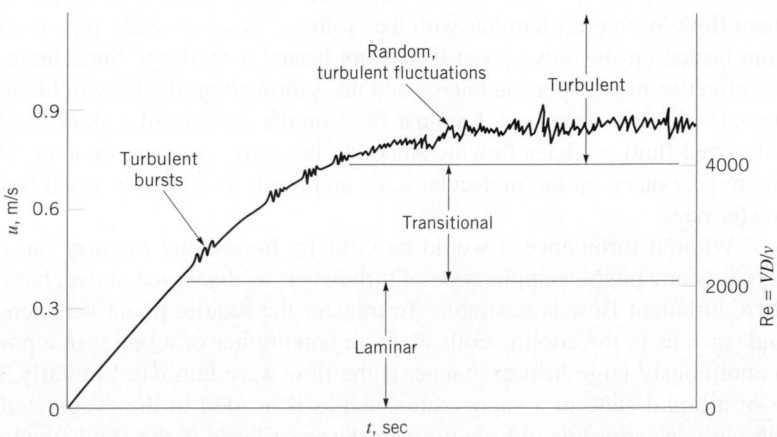

■ **F I G U R E   8.11**   **Transition from laminar to turbulent flow in a pipe.**

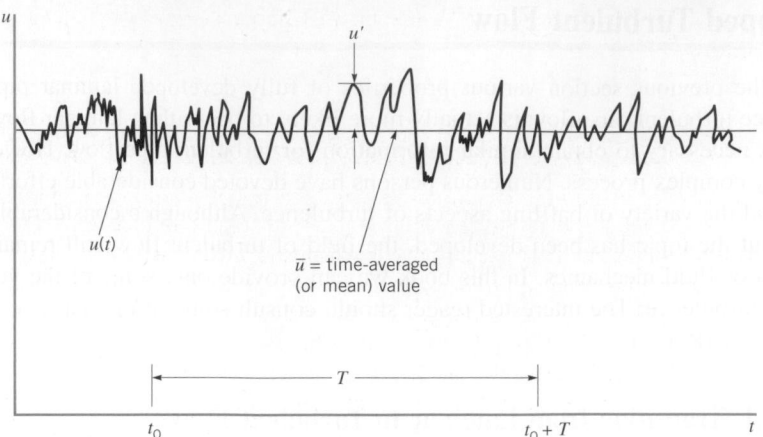

■ **F I G U R E   8.12**   The time-averaged, $\bar{u}$, and fluctuating, u', description of a parameter for turbulent flow.

indicated. In previous considerations involving inviscid flow, the Reynolds number is (strictly speaking) infinite (because the viscosity is zero), and the flow most surely would be turbulent. However, reasonable results were obtained by using the inviscid Bernoulli equation as the governing equation. The reason that such simplified inviscid analyses gave reasonable results is that viscous effects were not very important and the velocity used in the calculations was actually the time-averaged velocity, $\bar{u}$, indicated in Fig. 8.12. Calculation of the heat transfer, pressure drop, and many other parameters would not be possible without inclusion of the seemingly small, but very important, effects associated with the randomness of the flow.

Consider flow in a pan of water placed on a stove. With the stove turned off, the fluid is stationary. The initial sloshing has died out because of viscous dissipation within the water. With the stove turned on, a temperature gradient in the vertical direction, $\partial T/\partial z$, is produced. The water temperature is greatest near the pan bottom and decreases toward the top of the fluid layer. If the temperature difference is very small, the water will remain stationary, even though the water density is smallest near the bottom of the pan because of the decrease in density with an increase in temperature. A further increase in the temperature gradient will cause a buoyancy-driven instability that results in fluid motion—the light, warm water rises to the top, and the heavy, cold water sinks to the bottom. This slow, regular "turning over" increases the heat transfer from the pan to the water and promotes mixing within the pan. As the temperature gradient increases still further, the fluid motion becomes more vigorous and eventually turns into a chaotic, random, turbulent flow with considerable mixing, vaporization (boiling) and greatly increased heat transfer rate. The flow has progressed from a stationary fluid, to laminar flow, and finally to turbulent, multi-phase (liquid and vapor) flow.

Mixing processes and heat and mass transfer processes are considerably enhanced in turbulent flow compared to laminar flow. This is due to the macroscopic scale of the randomness in turbulent flow. We are all familiar with the "rolling," vigorous eddy type motion of the water in a pan being heated on the stove (even if it is not heated to boiling). Such finite-sized random mixing is very effective in transporting energy and mass throughout the flow field, thereby increasing the various rate processes involved. Laminar flow, on the other hand, can be thought of as very small but finite-sized fluid particles flowing smoothly in layers, one over another. The only randomness and mixing take place on the molecular scale and result in relatively small heat, mass, and momentum transfer rates.

Without turbulence it would be virtually impossible to carry out life as we now know it. Mixing is one positive application of turbulence, as discussed above, but there are other situations where turbulent flow is desirable. To transfer the required heat between a solid and an adjacent fluid (such as in the cooling coils of an air conditioner or a boiler of a power plant) would require an enormously large heat exchanger if the flow were laminar. Similarly, the required mass transfer of a liquid state to a vapor state (such as is needed in the evaporated cooling system associated with sweating) would require very large surfaces if the fluid flowing past the surface were

V8.4 Stirring color into paint

V8.5 Laminar and turbulent mixing

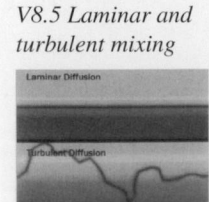

laminar rather than turbulent. As shown in Chapter 9, turbulence can also aid in delaying flow separation.

V8.6 Stirring cream into coffee

Turbulence is also of importance in the mixing of fluids. Smoke from a stack would continue for kilometers as a ribbon of pollutant without rapid dispersion within the surrounding air if the flow were laminar rather than turbulent. Under certain atmospheric conditions this is observed to occur. Although there is mixing on a molecular scale (laminar flow), it is several orders of magnitude slower and less effective than the mixing on a macroscopic scale (turbulent flow). It is considerably easier to mix cream into a cup of coffee (turbulent flow) than to thoroughly mix two colors of a viscous paint (laminar flow).

In other situations laminar (rather than turbulent) flow is desirable. The pressure drop in pipes (hence, the power requirements for pumping) can be considerably lower if the flow is laminar rather than turbulent. Fortunately, the blood flow through a person's arteries is normally laminar, except in the largest arteries with high blood flowrates. The aerodynamic drag on an airplane wing can be considerably smaller with laminar flow past it than with turbulent flow.

8.3.2 Turbulent Shear Stress

The fundamental difference between laminar and turbulent flow lies in the chaotic, random behavior of the various fluid parameters. Such variations occur in the three components of velocity, the pressure, the shear stress, the temperature, and any other variable that has a field description. Turbulent flow is characterized by random, three-dimensional vorticity (i.e., fluid particle rotation or spin; see Section 6.1.3). As is indicated in Fig. 8.12, such flows can be described in terms of their mean values (denoted with an overbar) on which are superimposed the fluctuations (denoted with a prime). Thus, if $u = u(x, y, z, t)$ is the x component of instantaneous velocity, then its time mean (or *time-average*) value, $\bar{u}$, is

Turbulent flow parameters can be described in terms of mean and fluctuating portions.

$$\bar{u} = \frac{1}{T} \int_{t_0}^{t_0+T} u(x, y, z, t)\, dt \tag{8.24}$$

where the time interval, T, is considerably longer than the period of the longest fluctuations, but considerably shorter than any unsteadiness of the average velocity. This is illustrated in Fig. 8.12.

The *fluctuating part* of the velocity, u', is that time-varying portion that differs from the average value

$$u = \bar{u} + u' \quad \text{or} \quad u' = u - \bar{u} \tag{8.25}$$

Clearly, the time average of the fluctuations is zero, since

$$\bar{u'} = \frac{1}{T} \int_{t_0}^{t_0+T} (u - \bar{u})\, dt = \frac{1}{T} \left(\int_{t_0}^{t_0+T} u\, dt - \bar{u} \int_{t_0}^{t_0+T} dt \right)$$

$$= \frac{1}{T} (T\bar{u} - T\bar{u}) = 0$$

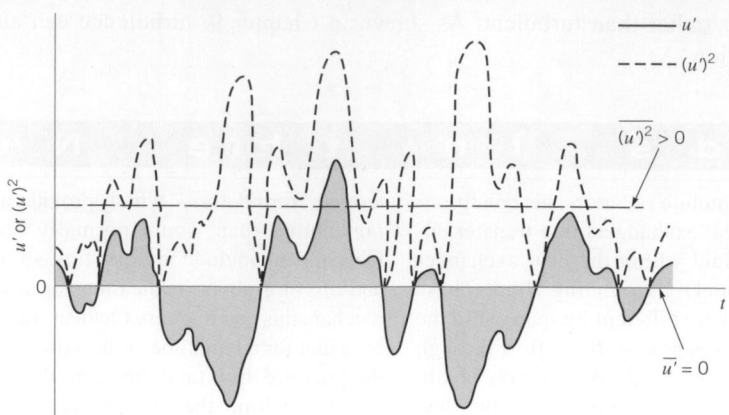

■ **F I G U R E 8.13** Average of the fluctuations and average of the square of the fluctuations.

The fluctuations are equally distributed on either side of the average. It is also clear, as is indicated in Fig. 8.13, that since the square of a fluctuation quantity cannot be negative $[(u')^2 \geq 0]$, its average value is positive. Thus,

$$\overline{(u')^2} = \frac{1}{T}\int_{t_0}^{t_0+T} (u')^2 \, dt > 0$$

On the other hand, it may be that the average of products of the fluctuations, such as $\overline{u'v'}$, are zero or nonzero (either positive or negative).

The structure and characteristics of turbulence may vary from one flow situation to another. For example, the *turbulence intensity* (or the level of the turbulence) may be larger in a very gusty wind than it is in a relatively steady (although turbulent) wind. The turbulence intensity, $\mathscr{I}$, is often defined as the square root of the mean square of the fluctuating velocity divided by the time-averaged velocity, or

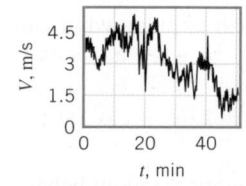

$$\mathscr{I} = \frac{\sqrt{\overline{(u')^2}}}{\bar{u}} = \frac{\left[\dfrac{1}{T}\displaystyle\int_{t_0}^{t_0+T} (u')^2 \, dt\right]^{1/2}}{\bar{u}}$$

The larger the turbulence intensity, the larger the fluctuations of the velocity (and other flow parameters). Well-designed wind tunnels have typical values of $\mathscr{I} \approx 0.01$, although with extreme care, values as low as $\mathscr{I} = 0.0002$ have been obtained. On the other hand, values of $\mathscr{I} \gtrsim 0.1$ are found for the flow in the atmosphere and rivers. A typical atmospheric wind speed graph is shown in the figure in the margin.

Another turbulence parameter that is different from one flow situation to another is the period of the fluctuations—the *time scale* of the fluctuations shown in Fig. 8.12. In many flows, such as the flow of water from a faucet, typical frequencies are on the order of 10, 100, or 1000 cycles per second (cps). For other flows, such as the Gulf Stream current in the Atlantic Ocean or flow of the atmosphere of Jupiter, characteristic random oscillations may have a period on the order of hours, days, or more.

The relationship between fluid motion and shear stress is very complex for turbulent flow.

It is tempting to extend the concept of viscous shear stress for laminar flow ($\tau = \mu \, du/dy$) to that of turbulent flow by replacing u, the instantaneous velocity, by $\bar{u}$, the time-averaged velocity. However, numerous experimental and theoretical studies have shown that such an approach leads to completely incorrect results. That is, $\tau \neq \mu \, d\bar{u}/dy$. A physical explanation for this behavior can be found in the concept of what produces a shear stress.

Laminar flow is modeled as fluid particles that flow smoothly along in layers, gliding past the slightly slower or faster ones on either side. As is discussed in Chapter 1, the fluid actually consists of numerous molecules darting about in an almost random fashion as is indicated in Fig. 8.14*a*. The motion is not entirely random—a slight bias in one direction produces the flowrate we associate

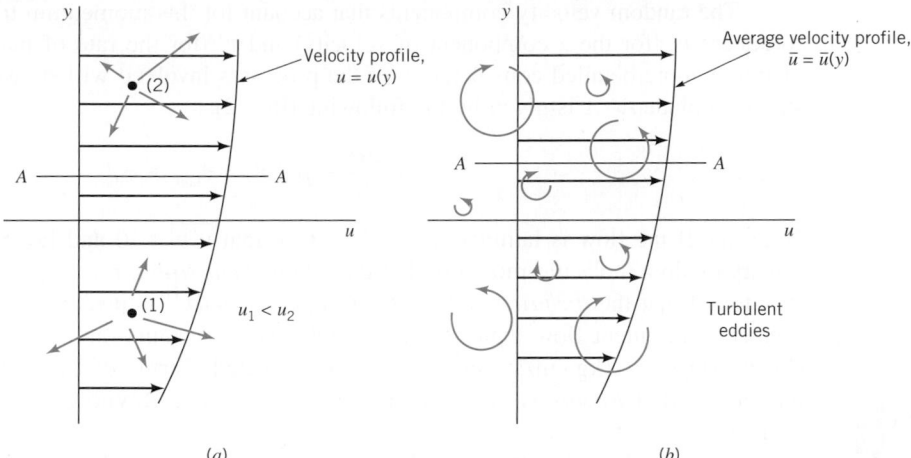

■ **F I G U R E   8.14**   (*a*) **Laminar flow shear stress caused by random motion of molecules.**
(*b*) **Turbulent flow as a series of random, three-dimensional eddies.**

with the motion of fluid particles, $\overline{u}$. As the molecules dart across a given plane (plane $A-A$, for example), the ones moving upward have come from an area of smaller average x component of velocity than the ones moving downward, which have come from an area of larger velocity.

The momentum flux in the x direction across plane $A-A$ gives rise to a drag (to the left) of the lower fluid on the upper fluid and an equal but opposite effect of the upper fluid on the lower fluid. The sluggish molecules moving upward across plane $A-A$ must be accelerated by the fluid above this plane. The rate of change of momentum in this process produces (on the macroscopic scale) a shear force. Similarly, the more energetic molecules moving down across plane $A-A$ must be slowed down by the fluid below that plane. This shear force is present only if there is a gradient in $u = u(y)$, otherwise the average x component of velocity (and momentum) of the upward and downward molecules is exactly the same. In addition, there are attractive forces between molecules. By combining these effects we obtain the well-known Newton viscosity law: $\tau = \mu\, du/dy$, where on a molecular basis μ is related to the mass and speed (temperature) of the random motion of the molecules.

Turbulent flow shear stress is larger than laminar flow shear stress because of the irregular, random motion.

Although the above random motion of the molecules is also present in turbulent flow, there is another factor that is generally more important. A simplistic way of thinking about turbulent flow is to consider it as consisting of a series of random, three-dimensional eddy type motions as is depicted (in one dimension only) in Fig. 8.14*b*. (See the photograph at the beginning of this chapter.) These eddies range in size from very small diameter (on the order of the size of a fluid particle) to fairly large diameter (on the order of the size of the object or flow geometry considered). They move about randomly, conveying mass with an average velocity $\overline{u} = \overline{u}(y)$. This eddy structure greatly promotes mixing within the fluid. It also greatly increases the transport of x momentum across plane $A-A$. That is, finite particles of fluid (not merely individual molecules as in laminar flow) are randomly transported across this plane, resulting in a relatively large (when compared with laminar flow) shear force. These particles vary in size but are much larger than molecules.

F l u i d s   i n   t h e   N e w s

Listen to the flowrate Sonar systems are designed to listen to transmitted and reflected sound waves in order to locate submerged objects. They have been used successfully for many years to detect and track underwater objects such as submarines and aquatic animals. Recently, sonar techniques have been refined so that they can be used to determine the flowrate in pipes. These new flow meters work for turbulent, not laminar, pipe flows because their operation depends strictly on the existence of *turbu-

lent eddies* within the flow. The flow meters contain a sonar-based array that listens to and interprets pressure fields generated by the turbulent motion in pipes. By listening to the pressure fields associated with the movement of the turbulent eddies, the device can determine the speed at which the eddies travel past an array of sensors. The flowrate is determined by using a calibration procedure which links the speed of the turbulent structures to the volumetric flowrate.

The random velocity components that account for this momentum transfer (hence, the shear force) are u' (for the x component of velocity) and v' (for the rate of mass transfer crossing the plane). A more detailed consideration of the processes involved will show that the apparent shear stress on plane A–A is given by the following (Ref. 2):

$$\tau = \mu\,\frac{d\bar{u}}{dy} - \rho\overline{u'v'} = \tau_{\text{lam}} + \tau_{\text{turb}} \tag{8.26}$$

The shear stress is the sum of a laminar portion and a turbulent portion.

Note that if the flow is laminar, $u' = v' = 0$, so that $\overline{u'v'} = 0$ and Eq. 8.26 reduces to the customary random molecule-motion-induced *laminar shear stress*, $\tau_{\text{lam}} = \mu\,d\bar{u}/dy$. For turbulent flow it is found that the *turbulent shear stress*, $\tau_{\text{turb}} = -\rho\overline{u'v'}$, is positive. Hence, the shear stress is greater in turbulent flow than in laminar flow. Note the units on τ_{turb} are (density)(velocity)^2 = $(\text{kg/m}^3)(\text{m/s})^2 = (\text{kg}\cdot\text{m/s}^2)/\text{m}^2 = \text{N/m}^2$, as expected. Terms of the form $-\rho\overline{u'v'}$ (or $-\rho\overline{v'w'}$, etc.) are called *Reynolds stresses* in honor of Osborne Reynolds who first discussed them in 1895.

It is seen from Eq. 8.26 that the shear stress in turbulent flow is not merely proportional to the gradient of the time-averaged velocity, $\bar{u}(y)$. It also contains a contribution due to the random fluctuations of the x and y components of velocity. The density is involved because of the momentum transfer of the fluid within the random eddies. Although the relative magnitude of τ_{lam} compared to τ_{turb} is a complex function dependent on the specific flow involved, typical measurements indicate the structure shown in Fig. 8.15a. (Recall from Eq. 8.4 that the shear stress is proportional to the distance from the centerline of the pipe.) In a very narrow region near the wall (the *viscous sublayer*), the laminar shear stress is dominant. Away from the wall (in the *outer layer*) the turbulent portion of the shear stress is dominant. The transition between these two regions occurs in the *overlap layer*. The corresponding typical velocity profile is shown in Fig. 8.15b.

The scale of the sketches shown in Fig. 8.15 is not necessarily correct. Typically the value of τ_{turb} is 100 to 1000 times greater than τ_{lam} in the outer region, while the converse is true in the viscous sublayer. A correct modeling of turbulent flow is strongly dependent on an accurate knowledge of τ_{turb}. This, in turn, requires an accurate knowledge of the fluctuations u' and v', or $\rho\overline{u'v'}$. As yet it is not possible to solve the governing equations (the Navier–Stokes equations) for these details of the flow, although numerical techniques (see Appendix A) using the largest and fastest computers available have produced important information about some of the characteristics of turbulence. Considerable effort has gone into the study of turbulence. Much remains to be learned. Perhaps studies in the new areas of chaos and fractal geometry will provide the tools for a better understanding of turbulence (see Section 8.3.5).

The vertical scale of Fig. 8.15 is also distorted. The viscous sublayer is usually a very thin layer adjacent to the wall. For example, for water flow in a 7.6-cm-diameter pipe with an average velocity of 3 m/s, the viscous sublayer is approximately 5×10^{-3} cm thick. Since the fluid motion within this thin layer is critical in terms of the overall flow (the no-slip condition and the wall shear stress occur in this layer), it is not surprising to find that turbulent pipe flow properties can be quite

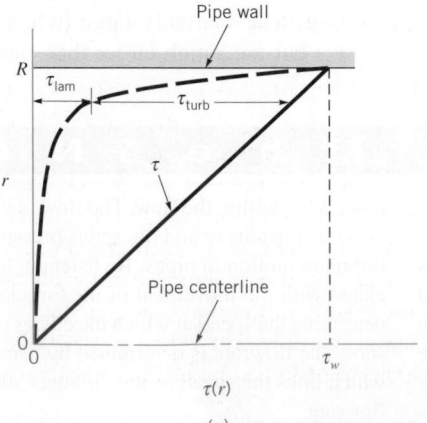

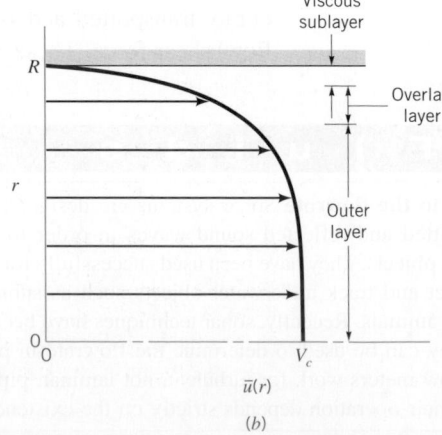

■ **F I G U R E 8.15**   Structure of turbulent flow in a pipe. (*a*) Shear stress. (*b*) Average velocity.

dependent on the roughness of the pipe wall, unlike laminar pipe flow which is independent of roughness. Small roughness elements (scratches, rust, sand or dirt particles, etc.) can easily disturb this viscous sublayer (see Section 8.4), thereby affecting the entire flow.

An alternate form for the shear stress for turbulent flow is given in terms of the *eddy viscosity*, η, where

$$\tau_{\text{turb}} = \eta \frac{d\bar{u}}{dy} \tag{8.27}$$

This extension of laminar flow terminology was introduced by J. Boussinesq, a French scientist, in 1877. Although the concept of an eddy viscosity is intriguing, in practice it is not an easy parameter to use. Unlike the absolute viscosity, μ, which is a known value for a given fluid, the eddy viscosity is a function of both the fluid and the flow conditions. That is, the eddy viscosity of water cannot be looked up in handbooks—its value changes from one turbulent flow condition to another and from one point in a turbulent flow to another.

The inability to accurately determine the Reynolds stress, $\overline{\rho u'v'}$, is equivalent to not knowing the eddy viscosity. Several semiempirical theories have been proposed (Ref. 3) to determine approximate values of η. L. Prandtl (1875–1953), a German physicist and aerodynamicist, proposed that the turbulent process could be viewed as the random transport of bundles of fluid particles over a certain distance, ℓ_m, the *mixing length*, from a region of one velocity to another region of a different velocity. By the use of some ad hoc assumptions and physical reasoning, it was concluded that the eddy viscosity was given by

> *Various ad hoc assumptions have been used to approximate turbulent shear stresses.*

$$\eta = \rho \ell_m^2 \left| \frac{d\bar{u}}{dy} \right|$$

Thus, the turbulent shear stress is

$$\tau_{\text{turb}} = \rho \ell_m^2 \left(\frac{d\bar{u}}{dy} \right)^2 \tag{8.28}$$

The problem is thus shifted to that of determining the mixing length, ℓ_m. Further considerations indicate that ℓ_m is not a constant throughout the flow field. Near a solid surface the turbulence is dependent on the distance from the surface. Thus, additional assumptions are made regarding how the mixing length varies throughout the flow.

The net result is that as yet there is no general, all-encompassing, useful model that can accurately predict the shear stress throughout a general incompressible, viscous turbulent flow. Without such information it is impossible to integrate the force balance equation to obtain the turbulent velocity profile and other useful information, as was done for laminar flow.

8.3.3 Turbulent Velocity Profile

Considerable information concerning turbulent velocity profiles has been obtained through the use of dimensional analysis, experimentation, numerical simulations, and semiempirical theoretical efforts. As is indicated in Fig. 8.15, fully developed turbulent flow in a pipe can be broken into three regions which are characterized by their distances from the wall: the viscous sublayer very near the pipe wall, the overlap region, and the outer turbulent layer throughout the center portion of the flow. Within the viscous sublayer the viscous shear stress is dominant compared with the turbulent (or Reynolds) stress, and the random, eddying nature of the flow is essentially absent. In the outer turbulent layer the Reynolds stress is dominant, and there is considerable mixing and randomness to the flow.

The character of the flow within these two regions is entirely different. For example, within the viscous sublayer the fluid viscosity is an important parameter; the density is unimportant. In the outer layer the opposite is true. By a careful use of dimensional analysis arguments for the flow in each layer and by a matching of the results in the common overlap layer, it has been possible to obtain the following conclusions about the turbulent velocity profile in a smooth pipe (Ref. 5).

In the viscous sublayer the velocity profile can be written in dimensionless form as

$$\frac{\bar{u}}{u^*} = \frac{yu^*}{\nu} \tag{8.29}$$

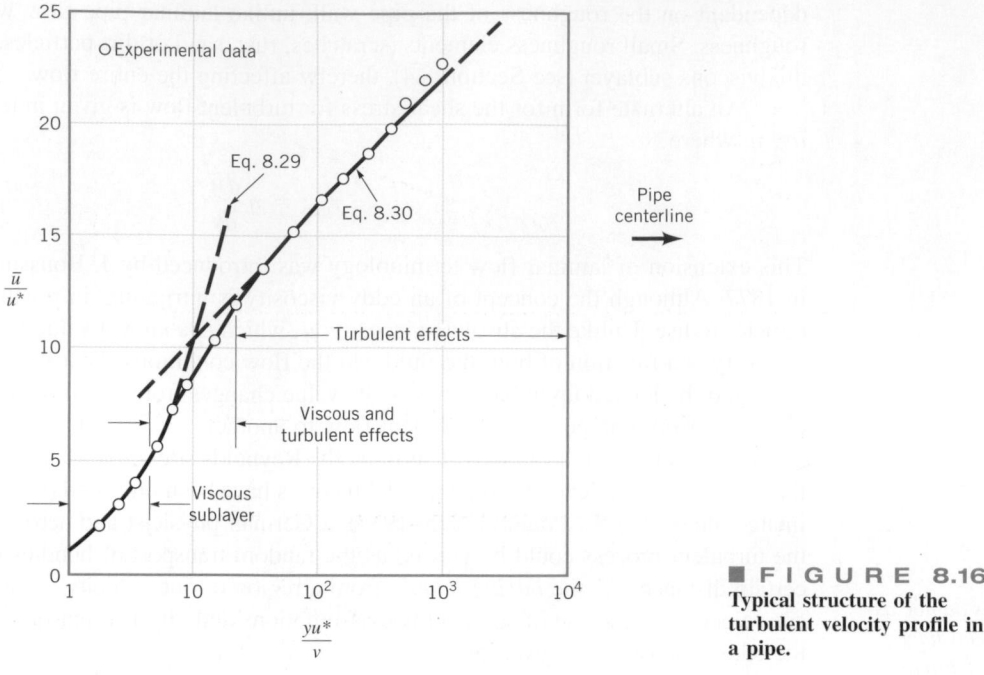

■ **FIGURE 8.16**
Typical structure of the turbulent velocity profile in a pipe.

where $y = R - r$ is the distance measured from the wall, $\bar{u}$ is the time-averaged x component of velocity, and $u^* = (\tau_w/\rho)^{1/2}$ is termed the *friction velocity*. Note that u^* is not an actual velocity of the fluid—it is merely a quantity that has dimensions of velocity. As is indicated in Fig. 8.16, Eq. 8.29 (commonly called the *law of the wall*) is valid very near the smooth wall, for $0 \leq yu^*/\nu \lesssim 5$.

Dimensional analysis arguments indicate that in the overlap region the velocity should vary as the logarithm of y. Thus, the following expression has been proposed:

A turbulent flow velocity profile can be divided into various regions.

$$\frac{\bar{u}}{u^*} = 2.5 \ln\left(\frac{yu^*}{\nu}\right) + 5.0 \tag{8.30}$$

where the constants 2.5 and 5.0 have been determined experimentally. As is indicated in Fig. 8.16, for regions not too close to the smooth wall, but not all the way out to the pipe center, Eq. 8.30 gives a reasonable correlation with the experimental data. Note that the horizontal scale is a logarithmic scale. This tends to exaggerate the size of the viscous sublayer relative to the remainder of the flow. As is shown in Example 8.4, the viscous sublayer is usually quite thin. Similar results can be obtained for turbulent flow past rough walls (Ref. 17).

A number of other correlations exist for the velocity profile in turbulent pipe flow. In the central region (the outer turbulent layer) the expression $(V_c - \bar{u})/u^* = 2.5 \ln(R/y)$, where V_c is the centerline velocity, is often suggested as a good correlation with experimental data. Another often-used (and relatively easy to use) correlation is the empirical *power-law velocity profile*

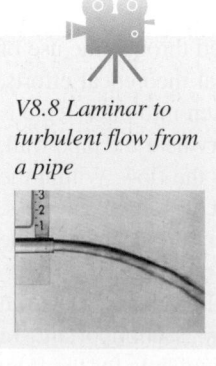

V8.8 Laminar to turbulent flow from a pipe

$$\frac{\bar{u}}{V_c} = \left(1 - \frac{r}{R}\right)^{1/n} \tag{8.31}$$

In this representation, the value of n is a function of the Reynolds number, as is indicated in Fig. 8.17. The one-seventh power-law velocity profile ($n = 7$) is often used as a reasonable approximation for many practical flows. Typical turbulent velocity profiles based on this power-law representation are shown in Fig. 8.18.

A closer examination of Eq. 8.31 shows that the power-law profile cannot be valid near the wall, since according to this equation the velocity gradient is infinite there. In addition, Eq. 8.31 cannot be precisely valid near the centerline because it does not give $d\bar{u}/dr = 0$ at $r = 0$. However, it does provide a reasonable approximation to the measured velocity profiles across most of the pipe.

Note from Fig. 8.18 that the turbulent profiles are much "flatter" than the laminar profile and that this flatness increases with Reynolds number (i.e., with n). Recall from Chapter 3 that

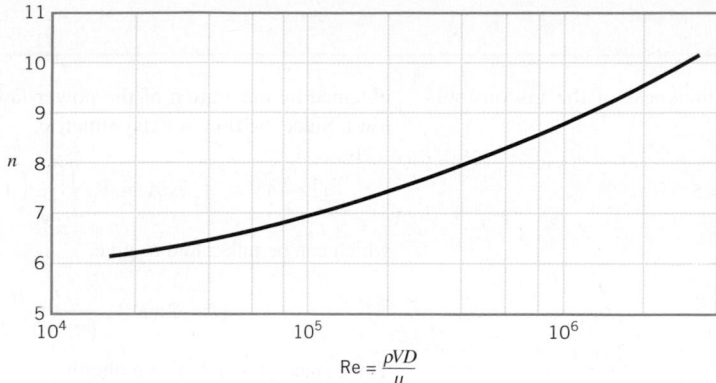

■ **F I G U R E 8.17**   Exponent, n, for power-law velocity profiles. (Adapted from Ref. 1.)

$$\mathrm{Re} = \frac{\rho V D}{\mu}$$

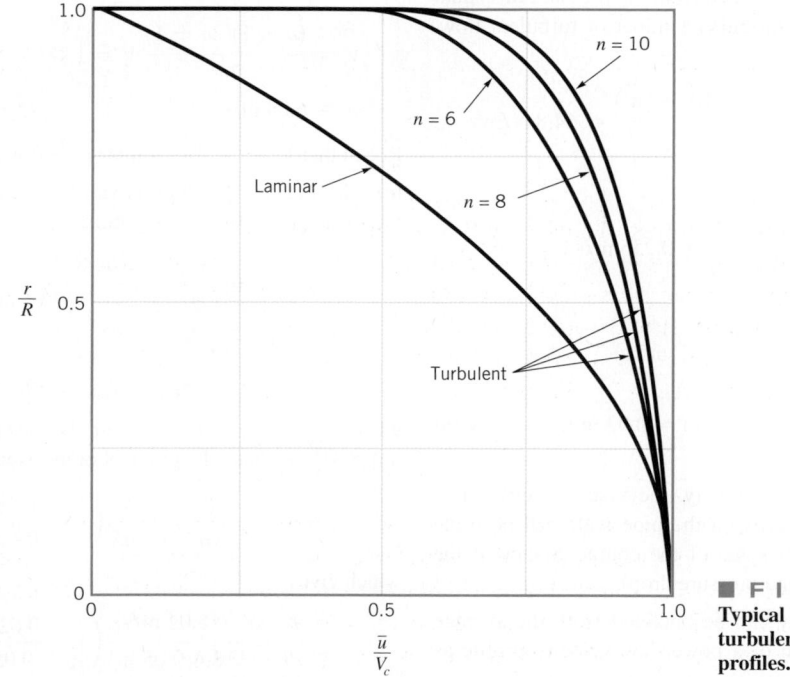

V8.9 Laminar/ turbulent velocity profiles

■ **F I G U R E 8.18** Typical laminar flow and turbulent flow velocity profiles.

reasonable approximate results are often obtained by using the inviscid Bernoulli equation and by assuming a fictitious uniform velocity profile. Since most flows are turbulent and turbulent flows tend to have nearly uniform velocity profiles, the usefulness of the Bernoulli equation and the uniform profile assumption is not unexpected. Of course, many properties of the flow cannot be accounted for without including viscous effects.

---

*E*XAMPLE 8.4   Turbulent Pipe Flow Properties

GIVEN Water at 20 °C ($\rho = 998$ kg/m^3 and $\nu = 1.004 \times 10^{-6}$ m^2/s) flows through a horizontal pipe of 0.1-m diameter with a flowrate of $Q = 4 \times 10^{-2}$ m^3/s and a pressure gradient of 2.59 kPa/m.

FIND (a) Determine the approximate thickness of the viscous sublayer.

(b) Determine the approximate centerline velocity, V_c.

(c) Determine the ratio of the turbulent to laminar shear stress, $\tau_{\mathrm{turb}}/\tau_{\mathrm{lam}}$, at a point midway between the centerline and the pipe wall (i.e., at $r = 0.025$ m).

SOLUTION

(a) According to Fig. 8.16, the thickness of the viscous sublayer, δ_s, is approximately

$$\frac{\delta_s u^*}{\nu} = 5$$

Therefore,

$$\delta_s = 5\frac{\nu}{u^*}$$

where

$$u^* = \left(\frac{\tau_w}{\rho}\right)^{1/2} \tag{1}$$

The wall shear stress can be obtained from the pressure drop data and Eq. 8.5, which is valid for either laminar or turbulent flow. Thus,

$$\tau_w = \frac{D\,\Delta p}{4\ell} = \frac{(0.1\text{ m})(2.59 \times 10^3\text{ N/m}^2)}{4(1\text{ m})} = 64.8\text{ N/m}^2$$

Hence, from Eq. 1 we obtain

$$u^* = \left(\frac{64.8\text{ N/m}^2}{998\text{ kg/m}^3}\right)^{1/2} = 0.255\text{ m/s}$$

so that

$$\delta_s = \frac{5(1.004 \times 10^{-6}\text{ m}^2/\text{s})}{0.255\text{ m/s}}$$
$$= 1.97 \times 10^{-5}\text{ m} \approx 0.02\text{ mm} \tag{Ans}$$

COMMENT As stated previously, the viscous sublayer is very thin. Minute imperfections on the pipe wall will protrude into this sublayer and affect some of the characteristics of the flow (i.e., wall shear stress and pressure drop).

(b) The centerline velocity can be obtained from the average velocity and the assumption of a power-law velocity profile as follows. For this flow with

$$V = \frac{Q}{A} = \frac{0.04\text{ m}^3/\text{s}}{\pi(0.1\text{ m})^2/4} = 5.09\text{ m/s}$$

the Reynolds number is

$$\text{Re} = \frac{VD}{\nu} = \frac{(5.09\text{ m/s})(0.1\text{ m})}{(1.004 \times 10^{-6}\text{ m}^2/\text{s})} = 5.07 \times 10^5$$

Thus, from Fig. 8.17, $n = 8.4$ so that

$$\frac{\bar{u}}{V_c} \approx \left(1 - \frac{r}{R}\right)^{1/8.4}$$

To determine the centerline velocity, V_c, we must know the relationship between V (the average velocity) and V_c. This can be obtained by integration of the power-law velocity profile as follows. Since the flow is axisymmetric,

$$Q = AV = \int \bar{u}\,dA = V_c \int_{r=0}^{r=R} \left(1 - \frac{r}{R}\right)^{1/n}(2\pi r)\,dr$$

which can be integrated to give

$$Q = 2\pi R^2 V_c \frac{n^2}{(n+1)(2n+1)}$$

Thus, since $Q = \pi R^2 V$, we obtain

$$\frac{V}{V_c} = \frac{2n^2}{(n+1)(2n+1)}$$

With $n = 8.4$ in the present case, this gives

$$V_c = \frac{(n+1)(2n+1)}{2n^2}V = 1.186V = 1.186\,(5.09\text{ m/s})$$
$$= 6.04\text{ m/s} \tag{Ans}$$

Recall that $V_c = 2V$ for laminar pipe flow.

(c) From Eq. 8.4, which is valid for laminar or turbulent flow, the shear stress at $r = 0.025$ m is

$$\tau = \frac{2\tau_w r}{D} = \frac{2(64.8\text{ N/m}^2)(0.025\text{ m})}{(0.1\text{ m})}$$

or

$$\tau = \tau_{\text{lam}} + \tau_{\text{turb}} = 32.4\text{ N/m}^2$$

where $\tau_{\text{lam}} = -\mu\,d\bar{u}/dr$. From the power-law velocity profile (Eq. 8.31) we obtain the gradient of the average velocity as

$$\frac{d\bar{u}}{dr} = -\frac{V_c}{nR}\left(1 - \frac{r}{R}\right)^{(1-n)/n}$$

which gives

$$\frac{d\bar{u}}{dr} = -\frac{(6.04\text{ m/s})}{8.4(0.05\text{ m})}\left(1 - \frac{0.025\text{ m}}{0.05\text{ m}}\right)^{(1-8.4)/8.4}$$
$$= -26.5/\text{s}$$

Thus,

$$\tau_{\text{lam}} = -\mu\frac{d\bar{u}}{dr} = -(\nu\rho)\frac{d\bar{u}}{dr}$$
$$= -(1.004 \times 10^{-6}\text{ m}^2/\text{s})(998\text{ kg/m}^3)(-26.5/\text{s})$$
$$= 0.0266\text{ N/m}^2$$

Thus, the ratio of turbulent to laminar shear stress is given by

$$\frac{\tau_{\text{turb}}}{\tau_{\text{lam}}} = \frac{\tau - \tau_{\text{lam}}}{\tau_{\text{lam}}} = \frac{32.4 - 0.0266}{0.0266} = 1220 \tag{Ans}$$

COMMENT As expected, most of the shear stress at this location in the turbulent flow is due to the turbulent shear stress.

The turbulent flow characteristics discussed in this section are not unique to turbulent flow in round pipes. Many of the characteristics introduced (i.e., the Reynolds stress, the viscous sublayer, the overlap layer, the outer layer, the general characteristics of the velocity profile, etc.) are found in other turbulent flows. In particular, turbulent pipe flow and turbulent flow past a solid wall (boundary layer flow) share many of these common traits. Such ideas are discussed more fully in Chapter 9.

8.3.4 Turbulence Modeling

Although it is not yet possible to theoretically predict the random, irregular details of turbulent flows, it would be useful to be able to predict the time-averaged flow fields (pressure, velocity, etc.) directly from the basic governing equations. To this end one can time average the governing Navier–Stokes equations (Eqs. 6.31 and 6.127) to obtain equations for the average velocity and pressure. However, because the Navier–Stokes equations are nonlinear, the resulting time-averaged differential equations contain not only the desired average pressure and velocity as variables, but also averages of products of the fluctuations—terms of the type that one tried to eliminate by averaging the equations! For example, the Reynolds stress $-\rho\overline{u'v'}$ (see Eq. 8.26) occurs in the time-averaged momentum equation.

Thus, it is not possible to merely average the basic differential equations and obtain governing equations involving only the desired averaged quantities. This is the reason for the variety of ad hoc assumptions that have been proposed to provide "closure" to the equations governing the average flow. That is, the set of governing equations must be a complete or closed set of equations—the same number of equation as unknowns.

Various attempts have been made to solve this closure problem (Refs. 1, 32). Such schemes involving the introduction of an eddy viscosity or the mixing length (as introduced in Section 8.3.2) are termed algebraic or zero-equation models. Other methods, which are beyond the scope of this book, include the one-equation model and the two-equation model. These turbulence models are based on the equation for the turbulence kinetic energy and require significant computer usage.

Turbulence modeling is an important and extremely difficult topic. Although considerable progress has been made, much remains to be done in this area.

8.3.5 Chaos and Turbulence

Chaos theory may eventually provide a deeper understanding of turbulence.

Chaos theory is a relatively new branch of mathematical physics that may provide insight into the complex nature of turbulence. This method combines mathematics and numerical (computer) techniques to provide a new way to analyze certain problems. Chaos theory, which is quite complex and is currently under development, involves the behavior of nonlinear dynamical systems and their response to initial and boundary conditions. The flow of a viscous fluid, which is governed by the nonlinear Navier–Stokes equations (Eq. 6.127), may be such a system.

To solve the Navier–Stokes equations for the velocity and pressure fields in a viscous flow, one must specify the particular flow geometry being considered (the boundary conditions) and the condition of the flow at some particular time (the initial conditions). If, as some researchers predict, the Navier–Stokes equations allow chaotic behavior, then the state of the flow at times after the initial time may be very, very sensitive to the initial conditions. A slight variation to the initial flow conditions may cause the flow at later times to be quite different than it would have been with the original, only slightly different initial conditions. When carried to the extreme, the flow may be "chaotic," "random," or perhaps (in current terminology), "turbulent."

The occurrence of such behavior would depend on the value of the Reynolds number. For example, it may be found that for sufficiently small Reynolds numbers the flow is not chaotic (i.e., it is laminar), while for large Reynolds numbers it is chaotic with turbulent characteristics.

Thus, with the advancement of chaos theory it may be found that the numerous ad hoc turbulence ideas mentioned in previous sections (i.e., eddy viscosity, mixing length, law of the wall, etc.) may not be needed. It may be that chaos theory can provide the turbulence properties and structure directly from the governing equations. As of now we must wait until this exciting topic is developed further. The interested reader is encouraged to consult Ref. 4 for a general introduction to chaos or Ref. 33 for additional material.

8.4 Dimensional Analysis of Pipe Flow

As noted previously, turbulent flow can be a very complex, difficult topic—one that as yet has defied a rigorous theoretical treatment. Thus, most turbulent pipe flow analyses are based on experimental data and semi-empirical formulas. These data are expressed conveniently in dimensionless form.

It is often necessary to determine the head loss, h_L, that occurs in a pipe flow so that the energy equation, Eq. 5.84, can be used in the analysis of pipe flow problems. As shown in Fig. 8.1, a typical pipe system usually consists of various lengths of straight pipe interspersed with various types of components (valves, elbows, etc.). The overall head loss for the pipe system consists of the head loss due to viscous effects in the straight pipes, termed the *major loss* and denoted $h_{L\,\text{major}}$, and the head loss in the various pipe components, termed the *minor loss* and denoted $h_{L\,\text{minor}}$. That is,

$$h_L = h_{L\,\text{major}} + h_{L\,\text{minor}}$$

The head loss designations of "major" and "minor" do not necessarily reflect the relative importance of each type of loss. For a pipe system that contains many components and a relatively short length of pipe, the minor loss may actually be larger than the major loss.

8.4.1 Major Losses

A dimensional analysis treatment of pipe flow provides the most convenient base from which to consider turbulent, fully developed pipe flow. An introduction to this topic was given in Section 8.3. As is discussed in Sections 8.2.1 and 8.2.4, the pressure drop and head loss in a pipe are dependent on the wall shear stress, τ_w, between the fluid and pipe surface. A fundamental difference between laminar and turbulent flow is that the shear stress for turbulent flow is a function of the density of the fluid, ρ. For laminar flow, the shear stress is independent of the density, leaving the viscosity, μ, as the only important fluid property.

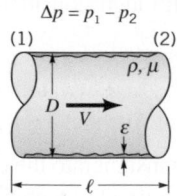

$\Delta p = p_1 - p_2$

Thus, as indicated by the figure in the margin, the pressure drop, Δp, for steady, incompressible turbulent flow in a horizontal round pipe of diameter D can be written in functional form as

$$\Delta p = F(V, D, \ell, \varepsilon, \mu, \rho) \tag{8.32}$$

where V is the average velocity, ℓ is the pipe length, and ε is a measure of the roughness of the pipe wall. It is clear that Δp should be a function of V, D, and ℓ. The dependence of Δp on the fluid properties μ and ρ is expected because of the dependence of τ on these parameters.

Turbulent pipe flow properties depend on the fluid density and the pipe roughness.

Although the pressure drop for laminar pipe flow is found to be independent of the roughness of the pipe, it is necessary to include this parameter when considering turbulent flow. As is discussed in Section 8.3.3 and illustrated in Fig. 8.19, for turbulent flow there is a relatively thin viscous sublayer formed in the fluid near the pipe wall. In many instances this layer is very thin; $\delta_s/D \ll 1$, where δ_s is the sublayer thickness. If a typical wall roughness element protrudes sufficiently far into (or even through) this layer, the structure and properties of the viscous sublayer (along with Δp and τ_w) will be different than if the wall were smooth. Thus, for turbulent flow the pressure drop is expected to be a function of the wall roughness. For laminar flow there is no thin viscous layer—viscous effects are important across the entire pipe. Thus, relatively small roughness elements have completely negligible effects on laminar pipe flow. Of course, for pipes with very large wall "roughness" ($\varepsilon/D \gtrsim 0.1$), such as that in corrugated pipes, the flowrate may be a function of the "roughness." We will consider only typical constant diameter pipes with relative roughnesses in the range $0 \leq \varepsilon/D \lesssim 0.05$. Analysis of flow in corrugated pipes does not fit into the standard constant diameter pipe category, although experimental results for such pipes are available (Ref. 30).

The list of parameters given in Eq. 8.32 is apparently a complete one. That is, experiments have shown that other parameters (such as surface tension, vapor pressure, etc.) do not affect the pressure drop for the conditions stated (steady, incompressible flow; round, horizontal pipe). Since there are seven variables ($k = 7$) which can be written in terms of the three reference dimensions MLT ($r = 3$), Eq. 8.32 can be written in dimensionless form in terms of $k - r = 4$ dimensionless groups. As was discussed in Section 7.9.1, one such representation is

$$\frac{\Delta p}{\frac{1}{2}\rho V^2} = \tilde{\phi}\left(\frac{\rho V D}{\mu}, \frac{\ell}{D}, \frac{\varepsilon}{D}\right)$$

This result differs from that used for laminar flow (see Eq. 8.17) in two ways. First, we have chosen to make the pressure dimensionless by dividing by the dynamic pressure, $\rho V^2/2$, rather than a characteristic viscous shear stress, $\mu V/D$. This convention was chosen in recognition of the fact that the shear stress for turbulent flow is normally dominated by τ_turb, which is a stronger function

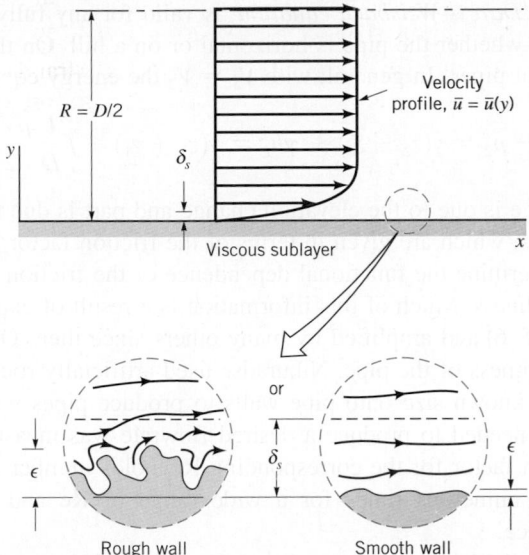

■ FIGURE 8.19   Flow in the viscous sublayer near rough and smooth walls.

of the density than it is of viscosity. Second, we have introduced two additional dimensionless parameters, the Reynolds number, $\text{Re} = \rho V D / \mu$, and the *relative roughness*, ε/D, which are not present in the laminar formulation because the two parameters ρ and ε are not important in fully developed laminar pipe flow.

As was done for laminar flow, the functional representation can be simplified by imposing the reasonable assumption that the pressure drop should be proportional to the pipe length. (Such a step is not within the realm of dimensional analysis. It is merely a logical assumption supported by experiments.) The only way that this can be true is if the ℓ/D dependence is factored out as

$$\frac{\Delta p}{\frac{1}{2}\rho V^2} = \frac{\ell}{D}\,\phi\!\left(\text{Re}, \frac{\varepsilon}{D}\right)$$

As was discussed in Section 8.2.3, the quantity $\Delta p D/(\ell\rho V^2/2)$ is termed the friction factor, f. Thus, for a horizontal pipe

$$\Delta p = f\frac{\ell}{D}\frac{\rho V^2}{2} \tag{8.33}$$

where

$$f = \phi\!\left(\text{Re}, \frac{\varepsilon}{D}\right)$$

For laminar fully developed flow, the value of f is simply $f = 64/\text{Re}$, independent of ε/D. For turbulent flow, the functional dependence of the friction factor on the Reynolds number and the relative roughness, $f = \phi(\text{Re}, \varepsilon/D)$, is a rather complex one that cannot, as yet, be obtained from a theoretical analysis. The results are obtained from an exhaustive set of experiments and usually presented in terms of a curve-fitting formula or the equivalent graphical form.

From Eq. 5.89 the energy equation for steady incompressible flow is

$$\frac{p_1}{\gamma} + \alpha_1\frac{V_1^2}{2g} + z_1 = \frac{p_2}{\gamma} + \alpha_2\frac{V_2^2}{2g} + z_2 + h_L$$

where h_L is the head loss between sections (1) and (2). With the assumption of a constant diameter ($D_1 = D_2$ so that $V_1 = V_2$), horizontal ($z_1 = z_2$) pipe with fully developed flow ($\alpha_1 = \alpha_2$), this becomes $\Delta p = p_1 - p_2 = \gamma h_L$, which can be combined with Eq. 8.33 to give

The major head loss in pipe flow is given in terms of the friction factor.

$$h_{L\,\text{major}} = f\frac{\ell}{D}\frac{V^2}{2g} \tag{8.34}$$

Equation 8.34, called the *Darcy–Weisbach equation*, is valid for any fully developed, steady, incompressible pipe flow—whether the pipe is horizontal or on a hill. On the other hand, Eq. 8.33 is valid only for horizontal pipes. In general, with $V_1 = V_2$ the energy equation gives

$$p_1 - p_2 = \gamma(z_2 - z_1) + \gamma h_L = \gamma(z_2 - z_1) + f\frac{\ell}{D}\frac{\rho V^2}{2}$$

Part of the pressure change is due to the elevation change and part is due to the head loss associated with frictional effects, which are given in terms of the friction factor, f.

It is not easy to determine the functional dependence of the friction factor on the Reynolds number and relative roughness. Much of this information is a result of experiments conducted by J. Nikuradse in 1933 (Ref. 6) and amplified by many others since then. One difficulty lies in the determination of the roughness of the pipe. Nikuradse used artificially roughened pipes produced by gluing sand grains of known size onto pipe walls to produce pipes with sandpaper-type surfaces. The pressure drop needed to produce a desired flowrate was measured and the data were converted into the friction factor for the corresponding Reynolds number and relative roughness. The tests were repeated numerous times for a wide range of Re and ε/D to determine the $f = \phi(\text{Re}, \varepsilon/D)$ dependence.

In commercially available pipes the roughness is not as uniform and well defined as in the artificially roughened pipes used by Nikuradse. However, it is possible to obtain a measure of the effective relative roughness of typical pipes and thus to obtain the friction factor. Typical roughness values for various pipe surfaces are given in Table 8.1. Figure 8.20 shows the functional dependence of f on Re and ε/D and is called the *Moody chart* in honor of L. F. Moody, who, along with C. F. Colebrook, correlated the original data of Nikuradse in terms of the relative roughness of commercially available pipe materials. It should be noted that the values of ε/D do not necessarily correspond to the actual values obtained by a microscopic determination of the average height of the roughness of the surface. They do, however, provide the correct correlation for $f = \phi(\text{Re}, \varepsilon/D)$.

It is important to observe that the values of relative roughness given pertain to new, clean pipes. After considerable use, most pipes (because of a buildup of corrosion or scale) may have a relative roughness that is considerably larger (perhaps by an order of magnitude) than that given. As shown by the figure in the margin, very old pipes may have enough scale buildup to not only alter the value of ε but also to change their effective diameter by a considerable amount.

The following characteristics are observed from the data of Fig. 8.20. For laminar flow, $f = 64/\text{Re}$, which is independent of relative roughness. For turbulent flows with very large Reynolds numbers, $f = \phi(\varepsilon/D)$, which, as shown by the figure in the margin, is independent of the Reynolds number. For such flows, commonly termed *completely turbulent flow* (or *wholly turbulent flow*), the laminar sublayer is so thin (its thickness decreases with increasing Re) that the surface roughness completely dominates the character of the flow near the wall. Hence, the pressure drop required is a

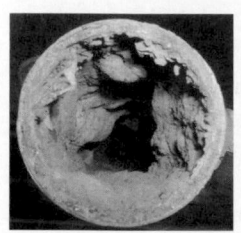

■ **TABLE 8.1**

Equivalent Roughness for New Pipes [From Moody (Ref. 7) and Colebrook (Ref. 8)]

| | Equivalent Roughness, ε |
Pipe	Millimeters
Riveted steel	0.9–9.0
Concrete	0.3–3.0
Wood stave	0.18–0.9
Cast iron	0.26
Galvanized iron	0.15
Commercial steel or wrought iron	0.045
Drawn tubing	0.0015
Plastic, glass	0.0 (smooth)

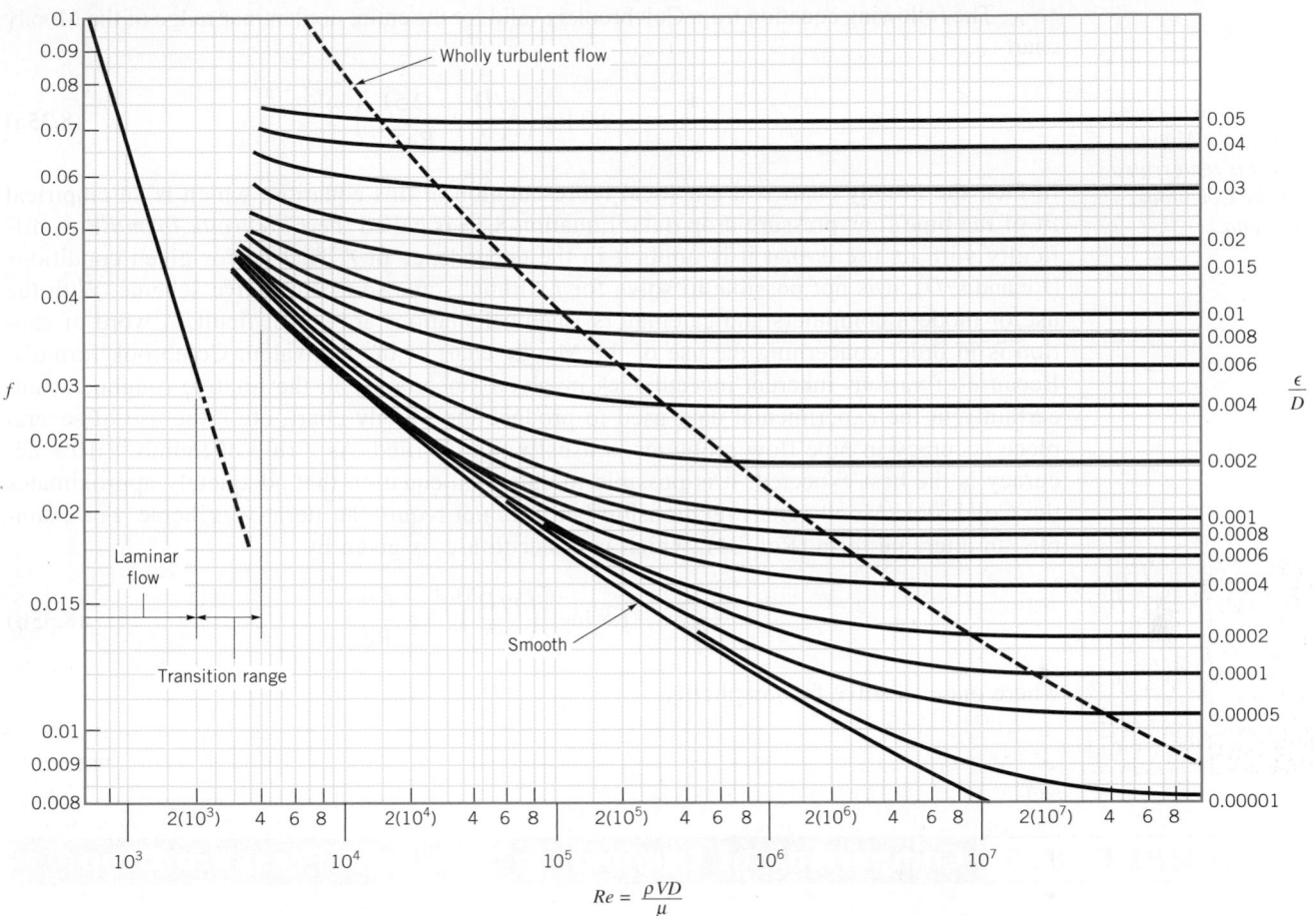

■ **F I G U R E 8.20**   **Friction factor as a function of Reynolds number and relative roughness for round pipes—the Moody chart. (Data from Ref. 7 with permission.)**

result of an inertia-dominated turbulent shear stress rather than the viscosity-dominated laminar shear stress normally found in the viscous sublayer. For flows with moderate values of Re, the friction factor is indeed dependent on both the Reynolds number and relative roughness—$f = \phi(\text{Re}, \varepsilon/D)$. The gap in the figure for which no values of f are given (the $2100 < \text{Re} < 4000$ range) is a result of the fact that the flow in this transition range may be laminar or turbulent (or an unsteady mix of both) depending on the specific circumstances involved.

For any pipe, even smooth ones, the head loss is not zero.

Note that even for smooth pipes ($\varepsilon = 0$) the friction factor is not zero. That is, there is a head loss in any pipe, no matter how smooth the surface is made. This is a result of the no-slip boundary condition that requires any fluid to stick to any solid surface it flows over. There is always some microscopic surface roughness that produces the no-slip behavior (and thus $f \neq 0$) on the molecular level, even when the roughness is considerably less than the viscous sublayer thickness. Such pipes are called *hydraulically smooth*.

Various investigators have attempted to obtain an analytical expression for $f = \phi(\text{Re}, \varepsilon/D)$. Note that the Moody chart covers an extremely wide range in flow parameters. The nonlaminar region covers more than four orders of magnitude in Reynolds number—from $\text{Re} = 4 \times 10^3$ to $\text{Re} = 10^8$. Obviously, for a given pipe and fluid, typical values of the average velocity do not cover this range. However, because of the large variety in pipes (D), fluids (ρ and μ), and velocities (V), such a wide range in Re is needed to accommodate nearly all applications of pipe flow. In many cases the particular pipe flow of interest is confined to a relatively small region of the Moody chart, and simple semiempirical expressions can be developed for those conditions. For example, a company that manufactures cast iron water pipes with diameters between 5 and 30 cm may use a simple equation valid for their conditions only. The Moody chart, on the other hand, is universally valid for all steady, fully developed, incompressible pipe flows.

The following equation from Colebrook is valid for the entire nonlaminar range of the Moody chart

The turbulent portion of the Moody chart is represented by the Colebrook formula.

$$\frac{1}{\sqrt{f}} = -2.0 \log\left(\frac{\varepsilon/D}{3.7} + \frac{2.51}{\text{Re}\sqrt{f}}\right) \tag{8.35a}$$

In fact, the Moody chart is a graphical representation of this equation, which is an empirical fit of the pipe flow pressure drop data. Equation 8.35 is called the *Colebrook formula*. A difficulty with its use is that it is implicit in the dependence of *f*. That is, for given conditions (Re and ε/D), it is not possible to solve for *f* without some sort of iterative scheme. With the use of modern computers and calculators, such calculations are not difficult. A word of caution is in order concerning the use of the Moody chart or the equivalent Colebrook formula. Because of various inherent inaccuracies involved (uncertainty in the relative roughness, uncertainty in the experimental data used to produce the Moody chart, etc.), the use of several place accuracy in pipe flow problems is usually not justified. As a rule of thumb, a 10% accuracy is the best expected. It is possible to obtain an equation that adequately approximates the Colebrook/Moody chart relationship but does not require an iterative scheme. For example, an alternate form (Ref. 34), which is easier to use, is given by

$$\frac{1}{\sqrt{f}} = -1.8 \log\left[\left(\frac{\varepsilon/D}{3.7}\right)^{1.11} + \frac{6.9}{\text{Re}}\right] \tag{8.35b}$$

where one can solve for *f* explicitly.

EXAMPLE 8.5 ■ Comparison of Laminar or Turbulent Pressure Drop

GIVEN Air under standard conditions flows through a 4.0-mm-diameter drawn tubing with an average velocity of $V = 50$ m/s. For such conditions the flow would normally be turbulent. However, if precautions are taken to eliminate disturbances to the flow (the entrance to the tube is very smooth, the air is dust free, the tube does not vibrate, etc.), it may be possible to maintain laminar flow.

FIND (a) Determine the pressure drop in a 0.1-m section of the tube if the flow is laminar.

(b) Repeat the calculations if the flow is turbulent.

SOLUTION

Under standard temperature and pressure conditions the density and viscosity are $\rho = 1.23$ kg/m³ and $\mu = 1.79 \times 10^{-5}$ N·s/m². Thus, the Reynolds number is

$$\text{Re} = \frac{\rho VD}{\mu} = \frac{(1.23 \text{ kg/m}^3)(50 \text{ m/s})(0.004 \text{ m})}{1.79 \times 10^{-5} \text{ N·s/m}^2} = 13,700$$

which would normally indicate turbulent flow.

(a) If the flow were laminar, then $f = 64/\text{Re} = 64/13,700 = 0.00467$ and the pressure drop in a 0.1-m-long horizontal section of the pipe would be

$$\Delta p = f\frac{\ell}{D}\frac{1}{2}\rho V^2$$

$$= (0.00467)\frac{(0.1 \text{ m})}{(0.004 \text{ m})}\frac{1}{2}(1.23 \text{ kg/m}^3)(50 \text{ m/s})^2$$

or

$$\Delta p = 0.179 \text{ kPa} \qquad \text{(Ans)}$$

COMMENT Note that the same result is obtained from Eq. 8.8:

$$\Delta p = \frac{32\mu\ell}{D^2}V$$

$$= \frac{32(1.79 \times 10^{-5} \text{ N·s/m}^2)(0.1 \text{ m})(50 \text{ m/s})}{(0.004 \text{ m})^2}$$

$$= 179 \text{ N/m}^2$$

(b) If the flow were turbulent, then $f = \phi(\text{Re}, \varepsilon/D)$, where from Table 8.1, $\varepsilon = 0.0015$ mm so that $\varepsilon/D = 0.0015$ mm/4.0 mm $= 0.000375$. From the Moody chart with Re $= 1.37 \times 10^4$ and $\varepsilon/D = 0.000375$ we obtain $f = 0.028$. Thus, the pressure drop in this case would be approximately

$$\Delta p = f\frac{\ell}{D}\frac{1}{2}\rho V^2 = (0.028)\frac{(0.1 \text{ m})}{(0.004 \text{ m})}\frac{1}{2}(1.23 \text{ kg/m}^3)(50 \text{ m/s})^2$$

or

$$\Delta p = 1.076 \text{ kPa} \qquad \text{(Ans)}$$

COMMENT A considerable savings in effort to force the fluid through the pipe could be realized (0.179 kPa rather than 1.076 kPa) if the flow could be maintained as laminar flow at this Reynolds number. In general this is very difficult to do, although laminar flow in pipes has been maintained up to Re ≈ 100,000 in rare instances.

An alternate method to determine the friction factor for the turbulent flow would be to use the Colebrook formula, Eq. 8.35a. Thus,

$$\frac{1}{\sqrt{f}} = -2.0 \log\left(\frac{\varepsilon/D}{3.7} + \frac{2.51}{\text{Re}\sqrt{f}}\right) = -2.0 \log\left(\frac{0.000375}{3.7} + \frac{2.51}{1.37 \times 10^4 \sqrt{f}}\right)$$

or

$$\frac{1}{\sqrt{f}} = -2.0 \log\left(1.01 \times 10^{-4} + \frac{1.83 \times 10^{-4}}{\sqrt{f}}\right) \qquad (1)$$

By using a root-finding technique on a computer or calculator, the solution to Eq. 1 is determined to be $f = 0.0291$, in agreement (within the accuracy of reading the graph) with the Moody chart method of $f = 0.028$.

Eq. 8.35b provides an alternate form to the Colebrook formula that can be used to solve for the friction factor directly.

$$\frac{1}{\sqrt{f}} = -1.8 \log\left[\left(\frac{\varepsilon/D}{3.7}\right)^{1.11} + \frac{6.9}{\text{Re}}\right] = -1.8 \log\left[\left(\frac{0.000375}{3.7}\right)^{1.11} + \frac{6.9}{1.37 \times 10^4}\right]$$

$$= 0.0289$$

This agrees with the Colebrook formula and Moody chart values obtained above.

Numerous other empirical formulas can be found in the literature (Ref. 5) for portions of the Moody chart. For example, an often-used equation, commonly referred to as the Blasius formula, for turbulent flow in smooth pipes ($\varepsilon/D = 0$) with Re $< 10^5$ is

$$f = \frac{0.316}{\text{Re}^{1/4}}$$

For our case this gives

$$f = 0.316(13,700)^{-0.25} = 0.0292$$

which is in agreement with the previous results. Note that the value of f is relatively insensitive to ε/D for this particular situation. Whether the tube was smooth glass ($\varepsilon/D = 0$) or the drawn tubing ($\varepsilon/D = 0.000375$) would not make much difference in the pressure drop. For this flow, an increase in relative roughness by a factor of 30 to $\varepsilon/D = 0.0113$ (equivalent to a commercial steel surface; see Table 8.1) would give $f = 0.043$. This would represent an increase in pressure drop and head loss by a factor of $0.043/0.0291 = 1.48$ compared with that for the original drawn tubing.

The pressure drop of 1.076 kPa in a length of 0.1 m of pipe corresponds to a change in absolute pressure [assuming $p = 101$ kPa (abs) at $x = 0$] of approximately $1.076/101 = 0.0107$, or about 1%. Thus, the incompressible flow assumption on which the above calculations (and all of the formulas in this chapter) are based is reasonable. However, if the pipe were 2-m long the pressure drop would be 21.5 kPa, approximately 20% of the original pressure. In this case the density would not be approximately constant along the pipe, and a compressible flow analysis would be needed. Such considerations are discussed in Chapter 11.

8.4.2 Minor Losses

As discussed in the previous section, the head loss in long, straight sections of pipe, the major losses, can be calculated by use of the friction factor obtained from either the Moody chart or the Colebrook equation. Most pipe systems, however, consist of considerably more than straight pipes. These additional components (valves, bends, tees, and the like) add to the overall head loss of the system. Such losses are generally termed *minor losses*, with the corresponding head loss denoted $h_{L\,\text{minor}}$. In this section we indicate how to determine the various minor losses that commonly occur in pipe systems.

The head loss associated with flow through a valve is a common minor loss. The purpose of a valve is to provide a means to regulate the flowrate. This is accomplished by changing the geometry of the system (i.e., closing or opening the valve alters the flow pattern through the valve), which in turn alters the losses associated with the flow through the valve. The flow resistance or head loss through the valve may be a significant portion of the resistance in the system. In fact, with the valve closed, the resistance to the flow is infinite—the fluid cannot flow. Such minor losses may be very important indeed. With the valve wide open the extra resistance due to the presence of the valve may or may not be negligible.

The flow pattern through a typical component such as a valve is shown in Fig. 8.21. It is not difficult to realize that a theoretical analysis to predict the details of such flows to obtain the head loss for these components is not, as yet, possible. Thus, the head loss information for essentially all components is given in dimensionless form and based on experimental data. The most common method used to determine these head losses or pressure drops is to specify the *loss coefficient*, K_L, which is defined as

Losses due to pipe system components are given in terms of loss coefficients.

$$K_L = \frac{h_{L\,\text{minor}}}{(V^2/2g)} = \frac{\Delta p}{\frac{1}{2}\rho V^2}$$

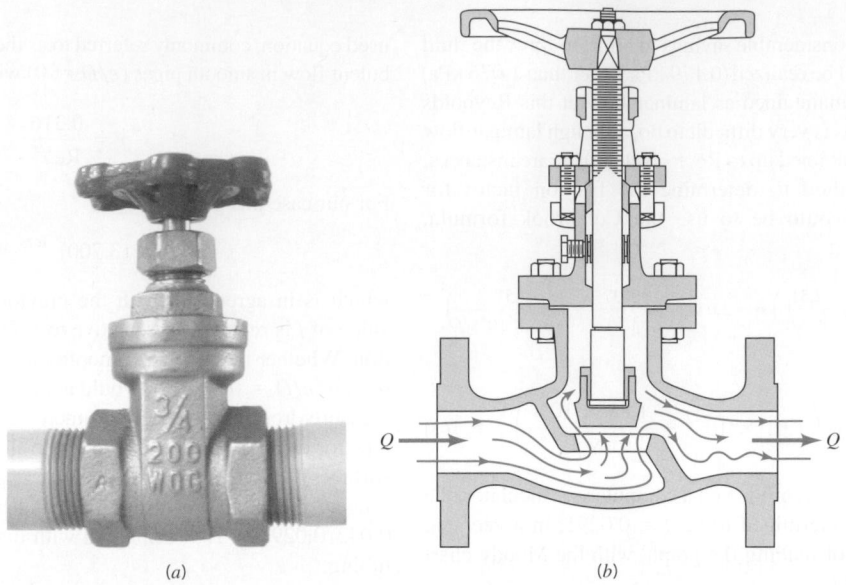

■ **FIGURE 8.21** **Flow through a valve.**

so that

$$\Delta p = K_L \tfrac{1}{2}\rho V^2$$

or

$$h_{L\,\text{minor}} = K_L \frac{V^2}{2g} \tag{8.36}$$

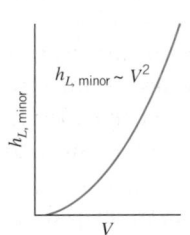

The pressure drop across a component that has a loss coefficient of $K_L = 1$ is equal to the dynamic pressure, $\rho V^2/2$. As shown by Eq. 8.36 and the figure in the margin, for a given value of K_L the head loss is proportional to the square of the velocity.

The actual value of K_L is strongly dependent on the geometry of the component considered. It may also be dependent on the fluid properties. That is,

$$K_L = \phi(\text{geometry, Re})$$

where $\text{Re} = \rho V D/\mu$ is the pipe Reynolds number. For many practical applications the Reynolds number is large enough so that the flow through the component is dominated by inertia effects, with viscous effects being of secondary importance. This is true because of the relatively large accelerations and decelerations experienced by the fluid as it flows along a rather curved, variable area (perhaps even torturous) path through the component (see Fig. 8.21). In a flow that is dominated by inertia effects rather than viscous effects, it is usually found that pressure drops and head losses correlate directly with the dynamic pressure. This is the reason why the friction factor for very large Reynolds number, fully developed pipe flow is independent of the Reynolds number. The same condition is found to be true for flow through pipe components. Thus, in most cases of practical interest the loss coefficients for components are a function of geometry only, $K_L = \phi(\text{geometry})$.

For most flows the loss coefficient is independent of the Reynolds number.

Minor losses are sometimes given in terms of an *equivalent length*, ℓ_{eq}. In this terminology, the head loss through a component is given in terms of the equivalent length of pipe that would produce the same head loss as the component. That is,

$$h_{L\,\text{minor}} = K_L \frac{V^2}{2g} = f \frac{\ell_{\text{eq}}}{D} \frac{V^2}{2g}$$

or

$$\ell_{\text{eq}} = \frac{K_L D}{f}$$

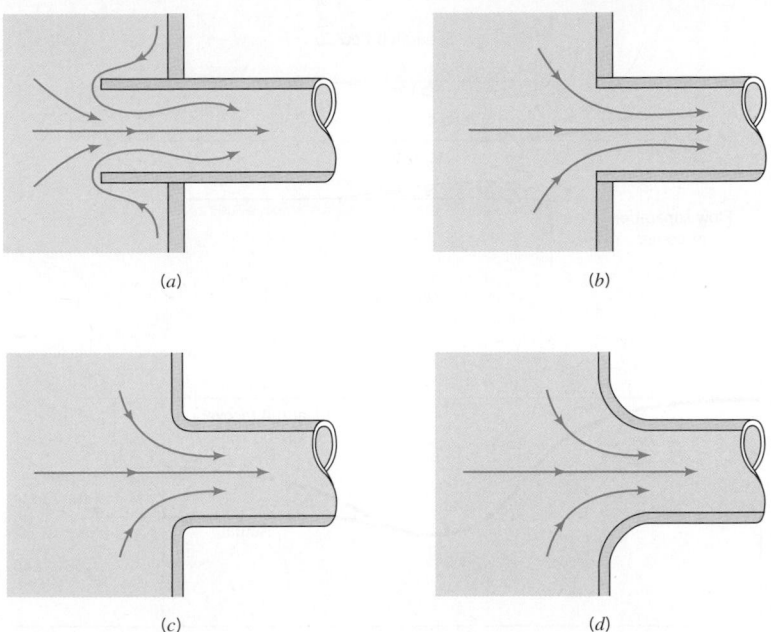

(a)

(b)

(c)

(d)

■ **F I G U R E   8.22**   **Entrance flow conditions and loss coefficient** (Refs. 28, 29). (*a*) Reentrant, $K_L = 0.8$, (*b*) sharp-edged, $K_L = 0.5$, (*c*) slightly rounded, $K_L = 0.2$ (see Fig. 8.24), (*d*) well-rounded, $K_L = 0.04$ (see Fig. 8.24).

where D and f are based on the pipe containing the component. The head loss of the pipe system is the same as that produced in a straight pipe whose length is equal to the pipes of the original system plus the sum of the additional equivalent lengths of all of the components of the system. Most pipe flow analyses, including those in this book, use the loss coefficient method rather than the equivalent length method to determine the minor losses.

Many pipe systems contain various transition sections in which the pipe diameter changes from one size to another. Such changes may occur abruptly or rather smoothly through some type of area change section. Any change in flow area contributes losses that are not accounted for in the fully developed head loss calculation (the friction factor). The extreme cases involve flow into a pipe from a reservoir (an entrance) or out of a pipe into a reservoir (an exit).

A fluid may flow from a reservoir into a pipe through any number of differently shaped entrance regions as are sketched in Fig. 8.22. Each geometry has an associated loss coefficient. A typical flow pattern for flow entering a pipe through a square-edged entrance is sketched in Fig. 8.23. As was discussed in Chapter 3, a vena contracta region may result because the fluid cannot turn a sharp right-angle corner. The flow is said to separate from the sharp corner. The maximum velocity at section (2) is greater than that in the pipe at section (3), and the pressure there is lower. If this high-speed fluid could slow down efficiently, the kinetic energy could be converted into pressure (the Bernoulli effect), and the ideal pressure distribution indicated in Fig. 8.23 would result. The head loss for the entrance would be essentially zero.

Minor head losses are often a result of the dissipation of kinetic energy.

Such is not the case. Although a fluid may be accelerated very efficiently, it is very difficult to slow down (decelerate) a fluid efficiently. Thus, the extra kinetic energy of the fluid at section (2) is partially lost because of viscous dissipation, so that the pressure does not return to the ideal value. An entrance head loss (pressure drop) is produced as is indicated in Fig. 8.23. The majority of this loss is due to inertia effects that are eventually dissipated by the shear stresses within the fluid. Only a small portion of the loss is due to the wall shear stress within the entrance region. The net effect is that the loss coefficient for a square-edged entrance is approximately $K_L = 0.50$. One-half of a velocity head is lost as the fluid enters the pipe. If the pipe protrudes into the tank (a reentrant entrance) as is shown in Fig. 8.22a, the losses are even greater.

An obvious way to reduce the entrance loss is to round the entrance region as is shown in Fig. 8.22c, thereby reducing or eliminating the vena contracta effect. Typical values for the loss coefficient for entrances with various amounts of rounding of the lip are shown in Fig. 8.24. A significant reduction in K_L can be obtained with only slight rounding.

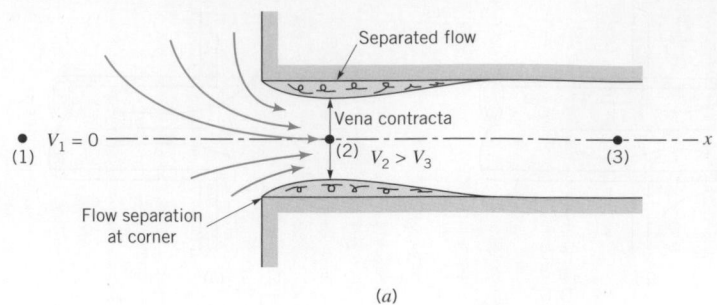

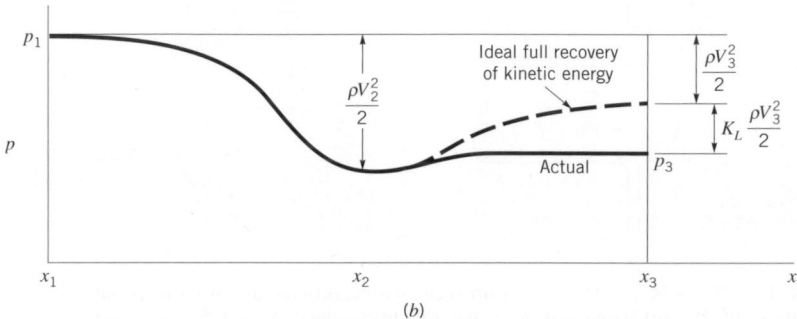

■ FIGURE 8.23 **Flow pattern and pressure distribution for a sharp-edged entrance.**

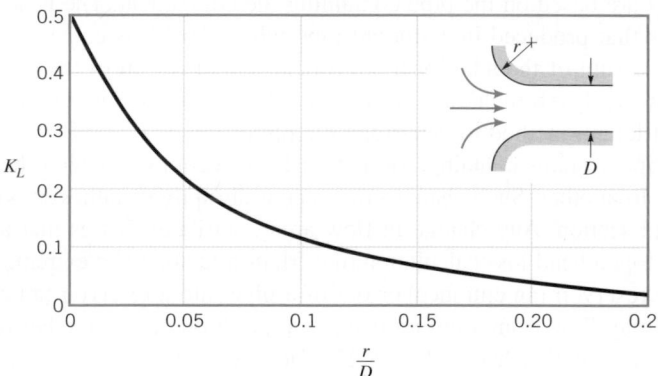

■ FIGURE 8.24
Entrance loss coefficient as a function of rounding of the inlet edge (Ref. 9).

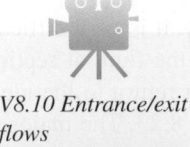

V8.10 Entrance/exit flows

A head loss (the exit loss) is also produced when a fluid flows from a pipe into a tank as is shown in Fig. 8.25. In these cases the entire kinetic energy of the exiting fluid (velocity V_1) is dissipated through viscous effects as the stream of fluid mixes with the fluid in the tank and eventually comes to rest ($V_2 = 0$). The exit loss from points (1) and (2) is therefore equivalent to one velocity head, or $K_L = 1$.

Losses also occur because of a change in pipe diameter as is shown in Figs. 8.26 and 8.27. The sharp-edged entrance and exit flows discussed in the previous paragraphs are limiting cases of this type of flow with either $A_1/A_2 = \infty$, or $A_1/A_2 = 0$, respectively. The loss coefficient for a sudden contraction, $K_L = h_L/(V_2^2/2g)$, is a function of the area ratio, A_2/A_1, as is shown in Fig. 8.26. The value of K_L changes gradually from one extreme of a sharp-edged entrance ($A_2/A_1 = 0$ with $K_L = 0.50$) to the other extreme of no area change ($A_2/A_1 = 1$ with $K_L = 0$).

In many ways, the flow in a sudden expansion is similar to exit flow. As is indicated in Fig. 8.28, the fluid leaves the smaller pipe and initially forms a jet-type structure as it enters the larger pipe. Within a few diameters downstream of the expansion, the jet becomes dispersed across the pipe, and fully developed flow becomes established again. In this process [between sections (2) and (3)] a portion of the kinetic energy of the fluid is dissipated as a result of viscous effects. A square-edged exit is the limiting case with $A_1/A_2 = 0$.

A sudden expansion is one of the few components (perhaps the only one) for which the loss coefficient can be obtained by means of a simple analysis. To do this we consider the continuity

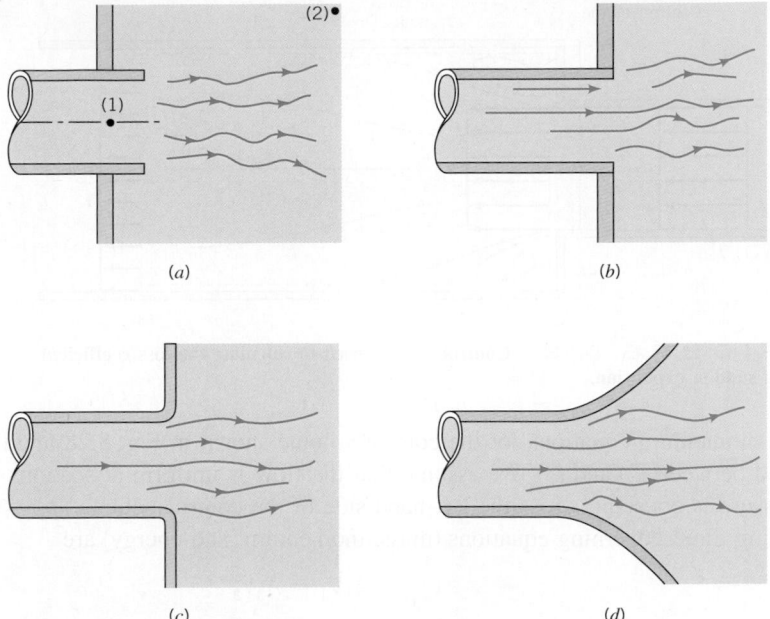

■ FIGURE 8.25 **Exit flow conditions and loss coefficient.**
(*a*) Reentrant, $K_L = 1.0$, (*b*) sharp-edged, $K_L = 1.0$, (*c*) slightly rounded, $K_L = 1.0$,
(*d*) well-rounded, $K_L = 1.0$.

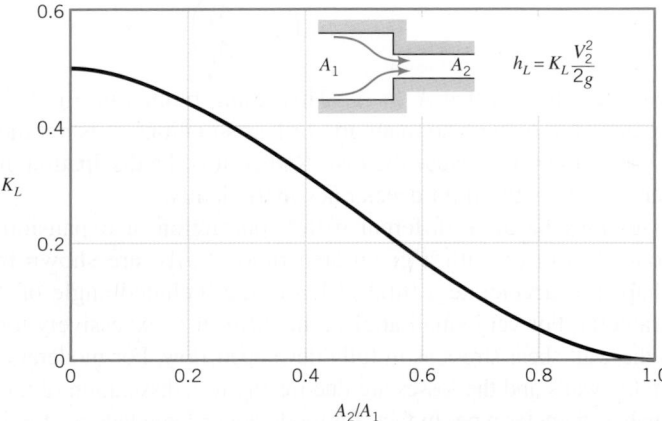

■ FIGURE 8.26
Loss coefficient for a sudden contraction (Ref. 10).

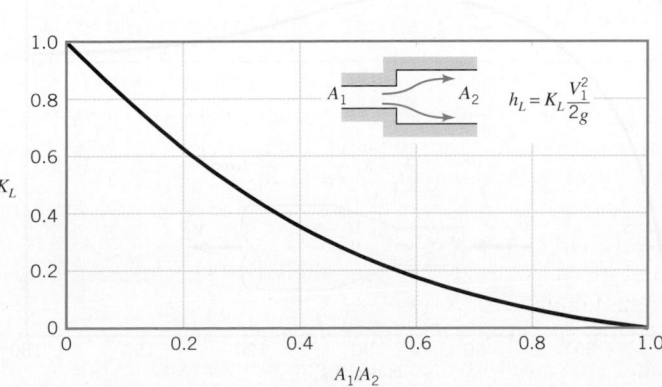

■ FIGURE 8.27
Loss coefficient for a sudden expansion (Ref. 10).

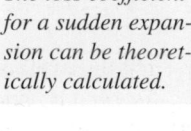

■ **FIGURE 8.28** Control volume used to calculate the loss coefficient for a sudden expansion.

and momentum equations for the control volume shown in Fig. 8.28 and the energy equation applied between (2) and (3). We assume that the flow is uniform at sections (1), (2), and (3) and the pressure is constant across the left-hand side of the control volume ($p_a = p_b = p_c = p_1$). The resulting three governing equations (mass, momentum, and energy) are

$$A_1 V_1 = A_3 V_3$$

$$p_1 A_3 - p_3 A_3 = \rho A_3 V_3 (V_3 - V_1)$$

The loss coefficient for a sudden expansion can be theoretically calculated.

and

$$\frac{p_1}{\gamma} + \frac{V_1^2}{2g} = \frac{p_3}{\gamma} + \frac{V_3^2}{2g} + h_L$$

These can be rearranged to give the loss coefficient, $K_L = h_L/(V_1^2/2g)$, as

$$K_L = \left(1 - \frac{A_1}{A_2}\right)^2$$

where we have used the fact that $A_2 = A_3$. This result, plotted in Fig. 8.27, is in good agreement with experimental data. As with so many minor loss situations, it is not the viscous effects directly (i.e., the wall shear stress) that cause the loss. Rather, it is the dissipation of kinetic energy (another type of viscous effect) as the fluid decelerates inefficiently.

The losses may be quite different if the contraction or expansion is gradual. Typical results for a conical *diffuser* with a given area ratio, A_2/A_1, are shown in Fig. 8.29. (A diffuser is a device shaped to decelerate a fluid.) Clearly the included angle of the diffuser, θ, is a very important parameter. For very small angles, the diffuser is excessively long and most of the head loss is due to the wall shear stress as in fully developed flow. For moderate or large angles, the flow separates from the walls and the losses are due mainly to a dissipation of the kinetic energy of the jet leaving the smaller diameter pipe. In fact, for moderate or large values of θ (i.e., $\theta > 35°$ for the case

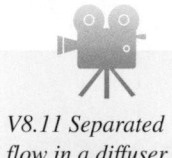

V8.11 Separated flow in a diffuser

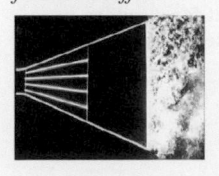

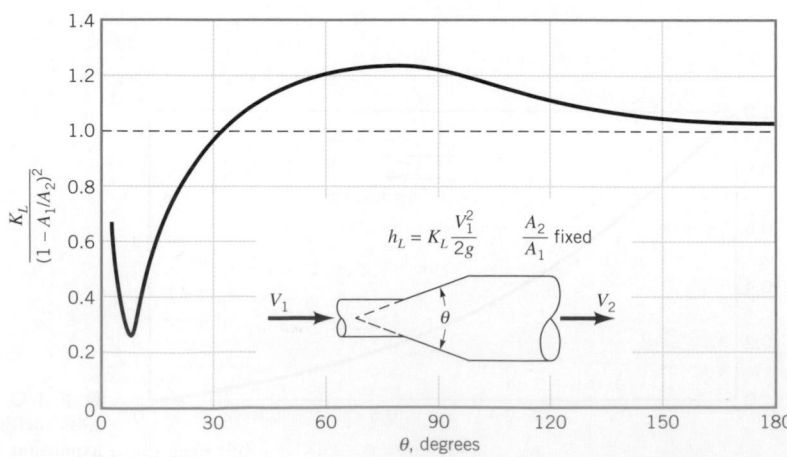

■ **FIGURE 8.29** Loss coefficient for a typical conical diffuser (Ref. 5).

shown in Fig. 8.29), the conical diffuser is, perhaps unexpectedly, less efficient than a sharp-edged expansion which has $K_L = (1 - A_1/A_2)^2$. There is an optimum angle ($\theta \approx 8°$ for the case illustrated) for which the loss coefficient is a minimum. The relatively small value of θ for the minimum K_L results in a long diffuser and is an indication of the fact that it is difficult to efficiently decelerate a fluid.

It must be noted that the conditions indicated in Fig. 8.29 represent typical results only. Flow through a diffuser is very complicated and may be strongly dependent on the area ratio A_2/A_1, specific details of the geometry, and the Reynolds number. The data are often presented in terms of a *pressure recovery coefficient*, $C_p = (p_2 - p_1)/(\rho V_1^2/2)$, which is the ratio of the static pressure rise across the diffuser to the inlet dynamic pressure. Considerable effort has gone into understanding this important topic (Refs. 11, 12).

Flow in a conical contraction (a nozzle; reverse the flow direction shown in Fig. 8.29) is less complex than that in a conical expansion. Typical loss coefficients based on the downstream (high-speed) velocity can be quite small, ranging from $K_L = 0.02$ for $\theta = 30°$, to $K_L = 0.07$ for $\theta = 60°$, for example. It is relatively easy to accelerate a fluid efficiently.

Bends in pipes produce a greater head loss than if the pipe were straight. The losses are due to the separated region of flow near the inside of the bend (especially if the bend is sharp) and the swirling secondary flow that occurs because of the imbalance of centripetal forces as a result of the curvature of the pipe centerline. These effects and the associated values of K_L for large Reynolds number flows through a 90° bend are shown in Fig. 8.30. The friction loss due to the axial length of the pipe bend must be calculated and added to that given by the loss coefficient of Fig. 8.30.

For situations in which space is limited, a flow direction change is often accomplished by use of miter bends, as is shown in Fig. 8.31, rather than smooth bends. The considerable losses in such bends can be reduced by the use of carefully designed guide vanes that help direct the flow with less unwanted swirl and disturbances.

Another important category of pipe system components is that of commercially available pipe fittings such as elbows, tees, reducers, valves, and filters. The values of K_L for such components depend strongly on the shape of the component and only very weakly on the Reynolds number for typical large Re flows. Thus, the loss coefficient for a 90° elbow depends on whether the pipe joints are threaded or flanged but is, within the accuracy of the data, fairly independent of the pipe diameter, flow rate, or fluid properties (the Reynolds number effect). Typical values of K_L for such components are given in Table 8.2. These typical components are designed more for ease of manufacturing and costs than for reduction of the head losses that they produce. The flowrate from a faucet in a typical house is sufficient whether the value of K_L for an elbow is the typical $K_L = 1.5$, or it is reduced to $K_L = 0.2$ by use of a more expensive long-radius, gradual bend (Fig. 8.30).

V8.12 Car exhaust system

Extensive tables are available for loss coefficients of standard pipe components.

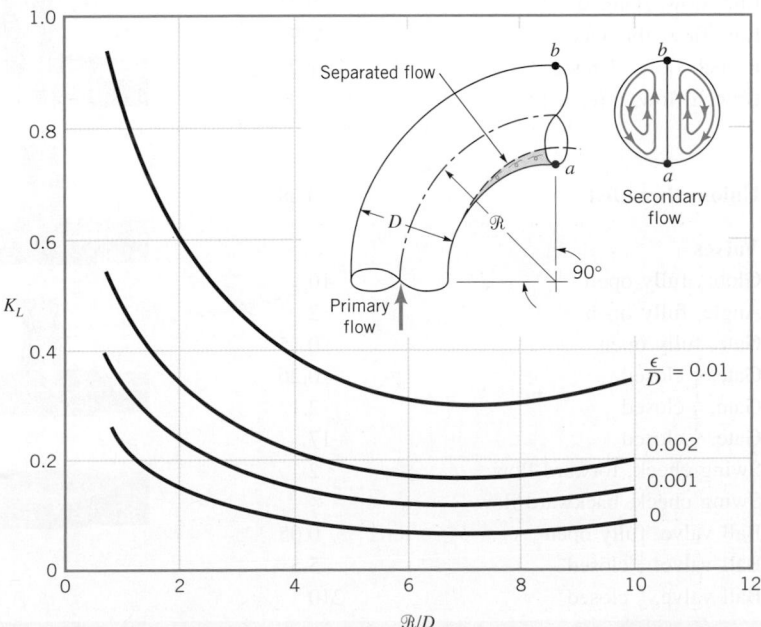

■ **FIGURE 8.30** Character of the flow in a 90° bend and the associated loss coefficient (Ref. 5).

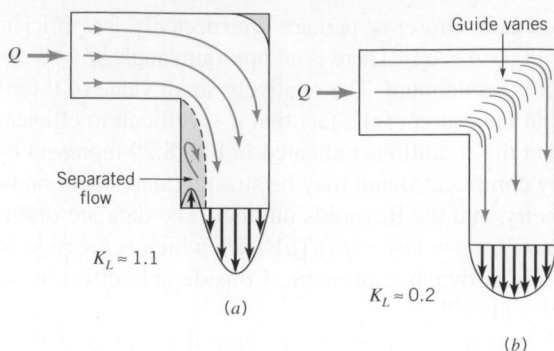

■ **FIGURE 8.31** Character of the flow in a 90° mitered bend and the associated loss coefficient: (a) without guide vanes, (b) with guide vanes.

■ **TABLE 8.2**

Loss Coefficients for Pipe Components $\left(h_L = K_L \dfrac{V^2}{2g} \right)$ (Data from Refs. 5, 10, 27)

Component	K_L
a. Elbows	
Regular 90°, flanged	0.3
Regular 90°, threaded	1.5
Long radius 90°, flanged	0.2
Long radius 90°, threaded	0.7
Long radius 45°, flanged	0.2
Regular 45°, threaded	0.4
b. 180° return bends	
180° return bend, flanged	0.2
180° return bend, threaded	1.5
c. Tees	
Line flow, flanged	0.2
Line flow, threaded	0.9
Branch flow, flanged	1.0
Branch flow, threaded	2.0
d. Union, threaded	0.08
*e. Valves	
Globe, fully open	10
Angle, fully open	2
Gate, fully open	0.15
Gate, $\frac{1}{4}$ closed	0.26
Gate, $\frac{1}{2}$ closed	2.1
Gate, $\frac{3}{4}$ closed	17
Swing check, forward flow	2
Swing check, backward flow	∞
Ball valve, fully open	0.05
Ball valve, $\frac{1}{3}$ closed	5.5
Ball valve, $\frac{2}{3}$ closed	210

*See Fig. 8.32 for typical valve geometry.

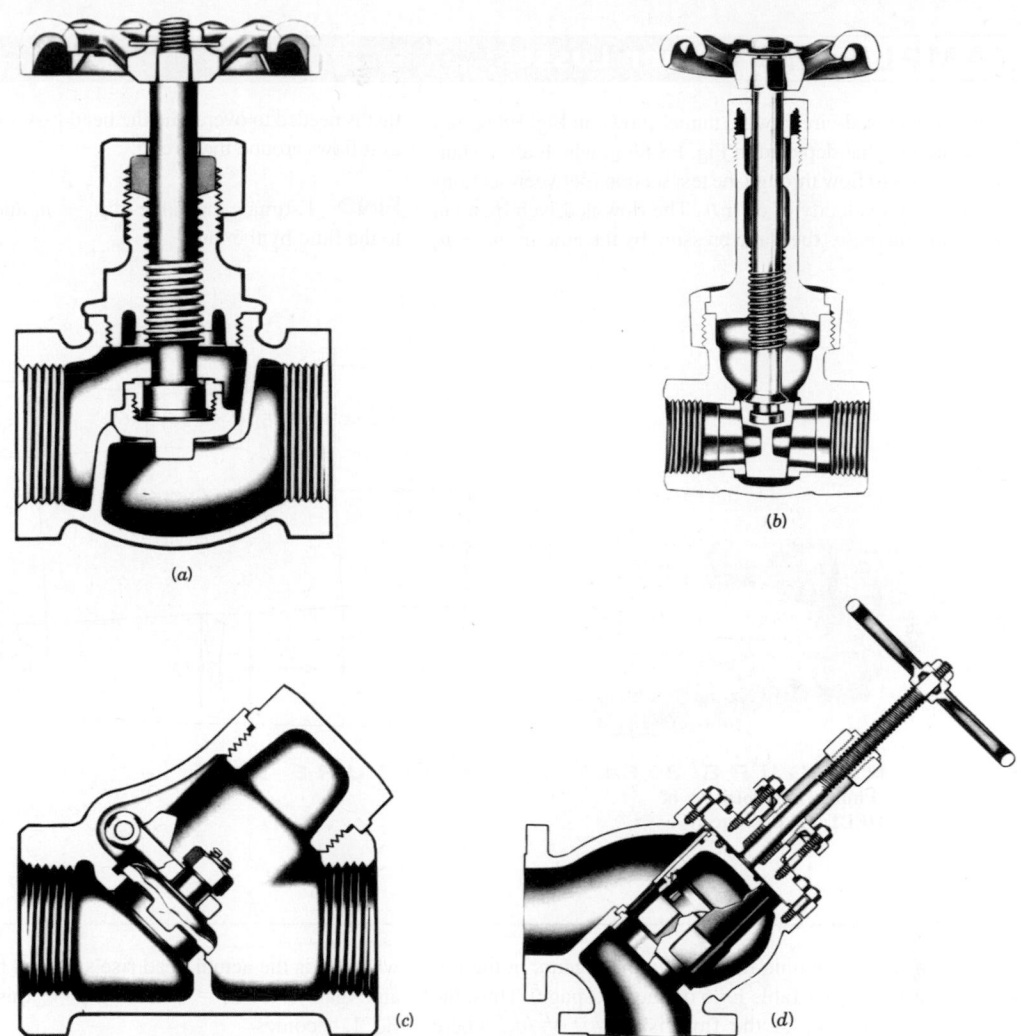

■ **FIGURE 8.32**   Internal structure of various valves: (*a*) globe valve, (*b*) gate valve, (*c*) swing check valve, (*d*) stop check valve. (Courtesy of Crane Co., Valve Division.)

Valves control the flowrate by providing a means to adjust the overall system loss coefficient to the desired value. When the valve is closed, the value of K_L is infinite and no fluid flows. Opening of the valve reduces K_L, producing the desired flowrate. Typical cross sections of various types of valves are shown in Fig. 8.32. Some valves (such as the conventional globe valve) are designed for general use, providing convenient control between the extremes of fully closed and fully open. Others (such as a needle valve) are designed to provide very fine control of the flowrate. The check valve provides a diode type operation that allows fluid to flow in one direction only.

Loss coefficients for typical valves are given in Table 8.2. As with many system components, the head loss in valves is mainly a result of the dissipation of kinetic energy of a high-speed portion of the flow. This high speed, V_3, is illustrated in Fig. 8.33.

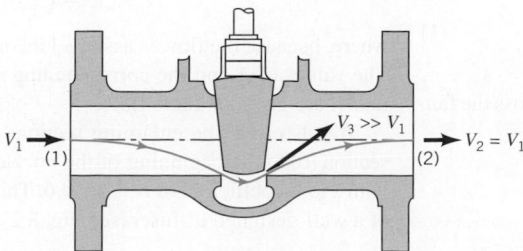

■ **FIGURE 8.33**   Head loss in a valve is due to dissipation of the kinetic energy of the large-velocity fluid near the valve seat.

EXAMPLE 8.6 Minor Losses

GIVEN The closed-circuit wind tunnel shown in Fig. E8.6a is a smaller version of that depicted in Fig. E8.6b in which air at standard conditions is to flow through the test section [between sections (5) and (6)] with a velocity of 60 m/s. The flow is driven by a fan that essentially increases the static pressure by the amount $p_1 - p_9$

that is needed to overcome the head losses experienced by the fluid as it flows around the circuit.

FIND Estimate the value of $p_1 - p_9$ and the kilowatts supplied to the fluid by the fan.

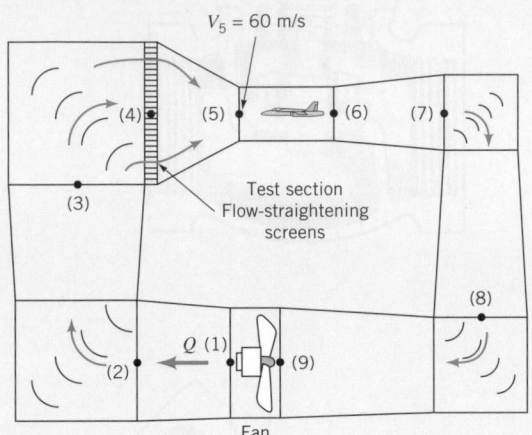

■ **FIGURE E8.6b**

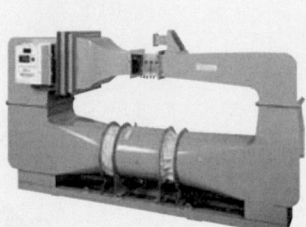

■ **FIGURE E8.6a**
(Photograph courtesy of DELTALAB.France.)

SOLUTION

The maximum velocity within the wind tunnel occurs in the test section (smallest area; see Table E8.6 on the next page). Thus, the maximum Mach number of the flow is $Ma_5 = V_5/c_5$, where $V_5 = 60$ m/s and from Eq. 1.20 the speed of sound is $c_5 = (kRT_5)^{1/2} = \{1.4(286.9 \text{ J/kg} \cdot \text{K})(273 + 15)\text{K}\}^{1/2} = 340$ m/s. Thus, $Ma_5 = 60/340 = 0.176$. As was indicated in Chapter 3 and discussed fully in Chapter 11, most flows can be considered as incompressible if the Mach number is less than about 0.3. Hence, we can use the incompressible formulas for this problem.

The purpose of the fan in the wind tunnel is to provide the necessary energy to overcome the net head loss experienced by the air as it flows around the circuit. This can be found from the energy equation between points (1) and (9) as

$$\frac{p_1}{\gamma} + \frac{V_1^2}{2g} + z_1 = \frac{p_9}{\gamma} + \frac{V_9^2}{2g} + z_9 + h_{L_{1-9}}$$

where $h_{L_{1-9}}$ is the total head loss from (1) to (9). With $z_1 = z_9$ and $V_1 = V_9$ this gives

$$\frac{p_1}{\gamma} - \frac{p_9}{\gamma} = h_{L_{1-9}} \qquad (1)$$

Similarly, by writing the energy equation (Eq. 5.84) across the fan, from (9) to (1), we obtain

$$\frac{p_9}{\gamma} + \frac{V_9^2}{2g} + z_9 + h_p = \frac{p_1}{\gamma} + \frac{V_1^2}{2g} + z_1$$

where h_p is the actual head rise supplied by the pump (fan) to the air. Again since $z_9 = z_1$ and $V_9 = V_1$ this, when combined with Eq. 1, becomes

$$h_p = \frac{(p_1 - p_9)}{\gamma} = h_{L_{1-9}}$$

The actual power supplied to the air (kilowatts, $\mathscr{P}_a$) is obtained from the fan head by

$$\mathscr{P}_a = \gamma Q h_p = \gamma A_5 V_5 h_p = \gamma A_5 V_5 h_{L_{1-9}} \qquad (2)$$

Thus, the power that the fan must supply to the air depends on the head loss associated with the flow through the wind tunnel. To obtain a reasonable, approximate answer we make the following assumptions. We treat each of the four turning corners as a mitered bend with guide vanes so that from Fig. 8.31 $K_{L_{\text{corner}}} = 0.2$. Thus, for each corner

$$h_{L_{\text{corner}}} = K_L \frac{V^2}{2g} = 0.2 \frac{V^2}{2g}$$

where, because the flow is assumed incompressible, $V = V_5 A_5/A$. The values of A and the corresponding velocities throughout the tunnel are given in Table E8.6.

We also treat the enlarging sections from the end of the test section (6) to the beginning of the nozzle (4) as a conical diffuser with a loss coefficient of $K_{L_{\text{dif}}} = 0.6$. This value is larger than that of a well-designed diffuser (see Fig. 8.29, for example). Since the

■ **TABLE E8.6**

Location	Area (m²)	Velocity (m/s)
1	2.0	11.1
2	2.6	8.7
3	3.3	6.9
4	3.3	6.9
5	0.4	60.0
6	0.4	60.0
7	0.9	24.0
8	1.7	13.5
9	2.0	11.1

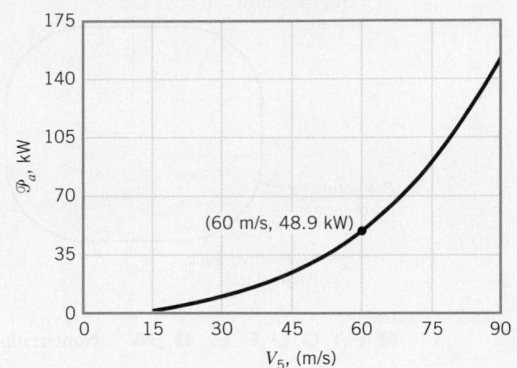

■ **F I G U R E   E8.6c**

wind tunnel diffuser is interrupted by the four turning corners and the fan, it may not be possible to obtain a smaller value of $K_{L_{\text{dif}}}$ for this situation. Thus,

$$h_{L_{\text{dif}}} = K_{L_{\text{dif}}} \frac{V_6^2}{2g} = 0.6 \frac{V_6^2}{2g}$$

The loss coefficients for the conical nozzle between section (4) and (5) and the flow-straightening screens are assumed to be $K_{L_{\text{noz}}} = 0.2$ and $K_{L_{\text{scr}}} = 4.0$ (Ref. 13), respectively. We neglect the head loss in the relatively short test section.

Thus, the total head loss is

$$h_{L_{1-9}} = h_{L_{\text{corner7}}} + h_{L_{\text{corner8}}} + h_{L_{\text{corner2}}} + h_{L_{\text{corner3}}}$$
$$+ h_{L_{\text{dif}}} + h_{L_{\text{noz}}} + h_{L_{\text{scr}}}$$

or

$$h_{L_{1-9}} = [0.2(V_7^2 + V_8^2 + V_2^2 + V_3^2)$$
$$+ 0.6V_6^2 + 0.2V_5^2 + 4.0V_4^2]/2g$$
$$= [0.2(24.0^2 + 13.5^2 + 8.7^2 + 6.9^2) + 0.6(61)^2$$
$$+ 0.2(61)^2 + 4.0(6.9)^2]\ \text{m}^2/\text{s}^2/[2(9.81\ \text{m/s}^2)]$$

or

$$h_{L_{1-9}} = 170\ \text{m}$$

Hence, from Eq. 1 we obtain the pressure rise across the fan as

$$p_1 - p_9 = \gamma h_{L_{1-9}} = (12\ \text{N/m}^3)(170\ \text{m})$$
$$= 2040\ \text{N/m}^2 = 2.04\ \text{kPa} \qquad \textbf{(Ans)}$$

From Eq. 2 we obtain the power added to the fluid as

$$\mathcal{P}_a = (12\ \text{N/m}^3)(0.4\ \text{m}^2)(60\ \text{m/s})(170\ \text{m})$$
$$= 48.9\ \text{kN} \cdot \text{m/s}$$

or

$$\mathcal{P}_a = \frac{48.9\ \text{kN} \cdot \text{m/s}}{1\ (\text{N} \cdot \text{m/s})/\text{W}} = 48.9\ \text{kW} \qquad \textbf{(Ans)}$$

COMMENTS   By repeating the calculations with various test section velocities, V_5, the results shown in Fig. E8.6c are obtained. Since the head loss varies as V_5^2 and the power varies as head loss times V_5, it follows that the power varies as the cube of the velocity. Thus, doubling the wind tunnel speed requires an eightfold increase in power.

With a closed-return wind tunnel of this type, all of the power required to maintain the flow is dissipated through viscous effects, with the energy remaining within the closed tunnel. If heat transfer across the tunnel walls is negligible, the air temperature within the tunnel will increase in time. For steady-state operations of such tunnels, it is often necessary to provide some means of cooling to maintain the temperature at acceptable levels.

It should be noted that the actual size of the motor that powers the fan must be greater than the calculated 48.9 kW because the fan is not 100% efficient. The power calculated above is that needed by the fluid to overcome losses in the tunnel, excluding those in the fan. If the fan were 60% efficient, it would require a shaft power of $\mathcal{P} = 48.9\ \text{kW}/(0.60) = 81.5\ \text{kW}$ to run the fan. Determination of fan (or pump) efficiencies can be a complex problem that depends on the specific geometry of the fan. Introductory material about fan performance is presented in Chapter 12; additional material can be found in various references (Refs. 14, 15, 16, for example).

It should also be noted that the above results are only approximate. Clever, careful design of the various components (corners, diffuser, etc.) may lead to improved (i.e., lower) values of the various loss coefficients, and hence lower power requirements. Since h_L is proportional to V^2, the components with the larger V tend to have the larger head loss. Thus, even though $K_L = 0.2$ for each of the four corners, the head loss for corner (7) is $(V_7/V_3)^2 = (24/6.9)^2 = 12.2$ times greater than it is for corner (3).

8.4.3 Noncircular Conduits

Many of the conduits that are used for conveying fluids are not circular in cross section. Although the details of the flows in such conduits depend on the exact cross-sectional shape, many round pipe results can be carried over, with slight modification, to flow in conduits of other shapes.

Theoretical results can be obtained for fully developed laminar flow in noncircular ducts, although the detailed mathematics often becomes rather cumbersome. For an arbitrary

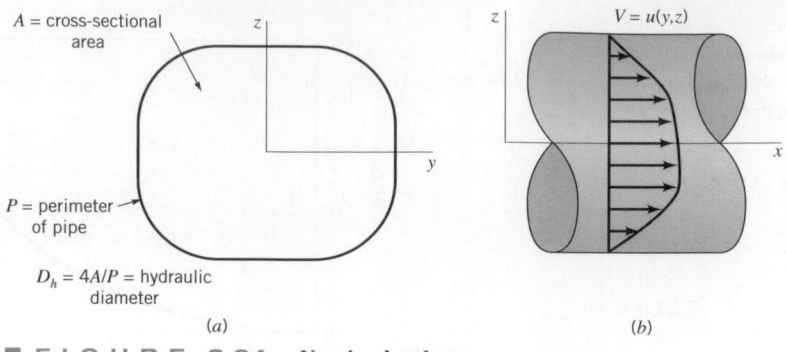

■ **FIGURE 8.34** **Noncircular duct.**

cross section, as is shown in Fig. 8.34, the velocity profile is a function of both y and z $[\mathbf{V} = u(y, z)\hat{\mathbf{i}}]$. This means that the governing equation from which the velocity profile is obtained (either the Navier–Stokes equations of motion or a force balance equation similar to that used for circular pipes, Eq. 8.6) is a partial differential equation rather than an ordinary differential equation. Although the equation is linear (for fully developed flow the convective acceleration is zero), its solution is not as straightforward as for round pipes. Typically the velocity profile is given in terms of an infinite series representation (Ref. 17).

Practical, easy-to-use results can be obtained as follows. Regardless of the cross-sectional shape, there are no inertia effects in fully developed laminar pipe flow. Thus, the friction factor can be written as $f = C/\text{Re}_h$, where the constant C depends on the particular shape of the duct, and Re_h is the Reynolds number, $\text{Re}_h = \rho V D_h / \mu$, based on the hydraulic diameter. The *hydraulic diameter* defined as $D_h = 4A/P$ is four times the ratio of the cross-sectional flow area divided by the wetted perimeter, P, of the pipe as is illustrated in Fig. 8.34. It represents a characteristic length that defines the size of a cross section of a specified shape. The factor of 4 is included in the definition of D_h so that for round pipes the diameter and hydraulic diameter are equal $[D_h = 4A/P = 4(\pi D^2/4)/(\pi D) = D]$. The hydraulic diameter is also used in the definition of the friction factor, $h_L = f(\ell/D_h)V^2/2g$, and the relative roughness, ε/D_h.

The values of $C = f\,\text{Re}_h$ for laminar flow have been obtained from theory and/or experiment for various shapes. Typical values are given in Table 8.3 along with the hydraulic diameter. Note

> *The hydraulic diameter is used for noncircular duct calculations.*

■ **TABLE 8.3**
Friction Factors for Laminar Flow in Noncircular Ducts (Data from Ref. 18)

Shape	Parameter	$C = f\,\text{Re}_h$
I. Concentric Annulus $D_h = D_2 - D_1$	D_1/D_2	
	0.0001	71.8
	0.01	80.1
	0.1	89.4
	0.6	95.6
	1.00	96.0
II. Rectangle $D_h = \dfrac{2ab}{a+b}$	a/b	
	0	96.0
	0.05	89.9
	0.10	84.7
	0.25	72.9
	0.50	62.2
	0.75	57.9
	1.00	56.9

that the value of C is relatively insensitive to the shape of the conduit. Unless the cross section is very "thin" in some sense, the value of C is not too different from its circular pipe value, $C = 64$. Once the friction factor is obtained, the calculations for noncircular conduits are identical to those for round pipes.

The Moody chart, developed for round pipes, can also be used for noncircular ducts.

Calculations for fully developed turbulent flow in ducts of noncircular cross section are usually carried out by using the Moody chart data for round pipes with the diameter replaced by the hydraulic diameter and the Reynolds number based on the hydraulic diameter. Such calculations are usually accurate to within about 15%. If greater accuracy is needed, a more detailed analysis based on the specific geometry of interest is needed.

EXAMPLE 8.7 | Noncircular Conduit

GIVEN Air at a temperature of 50 °C and standard pressure flows from a furnace through a 20-cm-diameter pipe with an average velocity of 3 m/s. It then passes through a transition section similar to the one shown in Fig. E8.7 and into a square duct whose side is of length a. The pipe and duct surfaces are smooth ($\varepsilon = 0$). The head loss per meter is to be the same for the pipe and the duct.

FIND Determine the duct size, a.

SOLUTION

We first determine the head loss per meter for the pipe, $h_L/\ell = (f/D) V^2/2g$, and then size the square duct to give the same value. For the given pressure and temperature we obtain (from Table B.2) $\nu = 1.76 \times 10^{-5} \text{ m}^2/\text{s}$ so that

$$\text{Re} = \frac{VD}{\nu} = \frac{(3 \text{ m/s})(\frac{20}{100}\text{ m})}{1.76 \times 10^{-5} \text{ m}^2/\text{s}} = 34{,}100$$

With this Reynolds number and with $\varepsilon/D = 0$ we obtain the friction factor from Fig. 8.20 as $f = 0.022$ so that

$$\frac{h_L}{\ell} = \frac{0.022}{(\frac{20}{100}\text{ m})} \frac{(3 \text{ m/s})^2}{2(9.81 \text{ m/s}^2)} = 0.0505$$

Thus, for the square duct we must have

$$\frac{h_L}{\ell} = \frac{f}{D_h} \frac{V_s^2}{2g} = 0.0505 \qquad (1)$$

where

$$D_h = 4A/P = 4a^2/4a = a \quad \text{and}$$

$$V_s = \frac{Q}{A} = \frac{\frac{\pi}{4}\left(\frac{20}{100}\text{ m}\right)^2 (3 \text{ m/s})}{a^2} = \frac{0.09}{a^2} \qquad (2)$$

is the velocity in the duct.

By combining Eqs. 1 and 2 we obtain

$$0.0505 = \frac{f}{a} \frac{(0.09/a^2)^2}{2(9.81)}$$

or

$$a = 0.38 f^{1/5} \qquad (3)$$

■ **FIGURE E8.7**

where a is in meters. Similarly, the Reynolds number based on the hydraulic diameter is

$$\text{Re}_h = \frac{V_s D_h}{\nu} = \frac{(0.09/a^2)a}{1.76 \times 10^{-5}} = \frac{5.1 \times 10^3}{a} \qquad (4)$$

We have three unknowns (a, f, and Re_h) and three equations—Eqs. 3, 4, and either in graphical form the Moody chart (Fig. 8.20) or the Colebrook equation (Eq. 8.35a).

If we use the Moody chart, we can use a trial and error solution as follows. As an initial attempt, assume the friction factor for the duct is the same as for the pipe. That is, assume $f = 0.022$. From Eq. 3 we obtain $a = 0.18 \text{ m}$, while from Eq. 4 we have $\text{Re}_h = 2.88 \times 10^4$. From Fig. 8.20, with this Reynolds number and the given smooth duct we obtain $f = 0.023$, which does not quite agree with the assumed value of f. Hence, we do not have the solution. We try again, using the latest calculated value of $f = 0.023$ as our guess. The calculations are repeated until the guessed value of f agrees with the value obtained from Fig. 8.20. The final result (after only two iterations) is $f = 0.023$, $\text{Re}_h = 2.85 \times 10^4$, and

$$a = 0.18 \text{ m} = 18 \text{ cm} \qquad \text{(Ans)}$$

COMMENTS Alternatively, we can use the Colebrook equation (rather than the Moody chart) to obtain the solution as follows. For a smooth pipe ($\varepsilon/D_h = 0$) the Colebrook equation, Eq. 8.35a, becomes

$$\frac{1}{\sqrt{f}} = -2.0 \log\left(\frac{\varepsilon/D_h}{3.7} + \frac{2.51}{Re_h \sqrt{f}}\right)$$

$$= -2.0 \log\left(\frac{2.51}{Re_h \sqrt{f}}\right) \qquad (5)$$

where from Eq. 3,

$$f = 122.3\, a^5 \qquad (6)$$

If we combine Eqs. 4, 5, and 6 and simplify, Eq. 7 is obtained for a.

$$0.09\, a^{-5/2} = -2 \log(4.5 \times 10^{-5}\, a^{-3/2}) \qquad (7)$$

By using a root-finding technique on a computer or calculator, the solution to Eq. 7 is determined to be $a = 0.18$ m, in agreement (given the accuracy of reading the Moody chart) with that obtained by the trial and error method given above.

Note that the length of the side of the equivalent square duct is $a/D = 18/20 = 0.90$, or approximately 90% of the diameter of the equivalent duct. It can be shown that this value, 90%, is a very good approximation for any pipe flow—laminar or turbulent. The cross-sectional area of the duct ($A = a^2 = 324$ cm^2) is greater than that of the round pipe ($A = \pi D^2/4 = 314$ cm^2). Also, it takes less material to form the round pipe (perimeter $= \pi D = 63$ cm) than the square duct (perimeter $= 4a = 72$ cm). Circles are very efficient shapes.

8.5   Pipe Flow Examples

Pipe systems may contain a single pipe with components or multiple interconnected pipes.

In the previous sections of this chapter, we discussed concepts concerning flow in pipes and ducts. The purpose of this section is to apply these ideas to the solutions of various practical problems. The application of the pertinent equations is straightforward, with rather simple calculations that give answers to problems of engineering importance. The main idea involved is to apply the energy equation between appropriate locations within the flow system, with the head loss written in terms of the friction factor and the minor loss coefficients. We will consider two classes of pipe systems: those containing a single pipe (whose length may be interrupted by various components), and those containing multiple pipes in parallel, series, or network configurations.

F l u i d s   i n   t h e   N e w s

New hi-tech fountains Ancient Egyptians used fountains in their palaces for decorative and cooling purposes. Current use of fountains continues, but with a hi-tech flair. Although the basic fountain still consists of a typical *pipe system* (i.e., pump, pipe, regulating valve, nozzle, filter, and basin), recent use of computer-controlled devices has led to the design of innovative fountains with special effects. For example, by using several rows of multiple nozzles, it is possible to program and activate control valves to produce water jets that resemble symbols, letters, or the time of day. Other fountains use specially designed nozzles to produce coherent, laminar streams of water that look like glass rods flying through the air. By using fast-acting control valves in a synchronized manner it is possible to produce mesmerizing three-dimensional patterns of water droplets. The possibilities are nearly limitless. With the initial artistic design of the fountain established, the initial engineering design (i.e., the capacity and pressure requirements of the nozzles and the size of the pipes and pumps) can be carried out. It is often necessary to modify the artistic and/or engineering aspects of the design in order to obtain a functional, pleasing fountain. (See Problem 8.64.)

8.5.1   Single Pipes

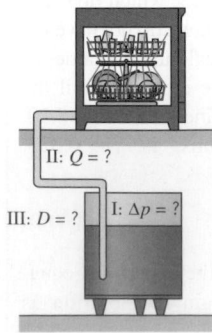

The nature of the solution process for pipe flow problems can depend strongly on which of the various parameters are independent parameters (the "given") and which is the dependent parameter (the "determine"). The three most common types of problems are shown in Table 8.4 in terms of the parameters involved. We assume the pipe system is defined in terms of the length of pipe sections used and the number of elbows, bends, and valves needed to convey the fluid between the desired locations. In all instances we assume the fluid properties are given.

In a Type I problem we specify the desired flowrate or average velocity and determine the necessary pressure difference or head loss. For example, if a flowrate of 7.6 L/min is required for a dishwasher that is connected to the water heater by a given pipe system as shown by the figure in the margin, what pressure is needed in the water heater?

In a Type II problem we specify the applied driving pressure (or, alternatively, the head loss) and determine the flowrate. For example, how many liters/min of hot water are supplied to the dishwasher if the pressure within the water heater is 414 kPa and the pipe system details (length, diameter, roughness of the pipe; number of elbows; etc.) are specified?

■ **T A B L E   8 . 4**
Pipe Flow Types

Pipe flow problems can be categorized by what parameters are given and what is to be calculated.

Variable	Type I	Type II	Type III
a. Fluid			
Density	Given	Given	Given
Viscosity	Given	Given	Given
b. Pipe			
Diameter	Given	Given	Determine
Length	Given	Given	Given
Roughness	Given	Given	Given
c. Flow			
Flowrate or Average Velocity	Given	Determine	Given
d. Pressure			
Pressure Drop or Head Loss	Determine	Given	Given

In a Type III problem we specify the pressure drop and the flowrate and determine the diameter of the pipe needed. For example, what diameter of pipe is needed between the water heater and dishwasher if the pressure in the water heater is 414 kPa (determined by the city water system) and the flowrate is to be not less than 7.6 L/min (determined by the manufacturer)?

Several examples of these types of problems follow.

EXAMPLE 8.8   Type I, Determine Pressure Drop

GIVEN   Water at 15 °C flows from the basement to the second floor through the 1.9 cm (1.9×10^{-2} m)-diameter copper pipe (a drawn tubing) at a rate of $Q = 45$ L/min $= 7.5 \times 10^{-4}$ m³/s and exits through a faucet of diameter 1.3 cm as shown in Fig. E8.8a.

FIND   Determine the pressure at point (1) if

(a)   all losses are neglected,

(b)   the only losses included are major losses, or

(c)   all losses are included.

*S*OLUTION

Since the fluid velocity in the pipe is given by $V_1 = Q/A_1 = Q/(\pi D^2/4) = (7.5 \times 10^{-4}$ m³/s$)/[\pi(1.9 \times 10^{-2}$ m$)^2/4] = 2.65$ m/s, and the fluid properties are $\rho = 998.2$ kg/m³ and $\mu = 1.002 \times 10^{-3}$ N·s/m² (see Table B.1), it follows that Re $= \rho V D/\mu = (998.2$ kg/m³$)(2.65$ m/s$)(1.9 \times 10^{-2}$ m$)/1.002 \times 10^{-3}$ N·s/m² $= 50,100$. Thus, the flow is turbulent. The governing equation for either case (a), (b), or (c) is the energy equation given by Eq. 8.21,

$$\frac{p_1}{\gamma} + \alpha_1 \frac{V_1^2}{2g} + z_1 = \frac{p_2}{\gamma} + \alpha_2 \frac{V_2^2}{2g} + z_2 + h_L$$

where $z_1 = 0$, $z_2 = 6$ m, $p_2 = 0$ (free jet), $\gamma = \rho g = 9.79$ kN/m³, and the outlet velocity is $V_2 = Q/A_2 = (7.5 \times 10^{-4}$ m³/s$)/[\pi(1.3$ cm$/100)^2$ m²$/4] = 5.65$ m/s. We assume that the kinetic energy coefficients α_1 and α_2 are unity. This is reasonable because turbulent velocity profiles are nearly uniform across the pipe. Thus,

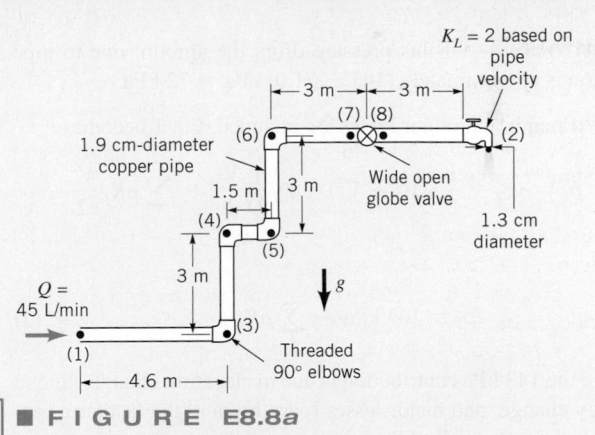

■ **F I G U R E   E8.8a**

$$p_1 = \gamma z_2 + \tfrac{1}{2}\rho(V_2^2 - V_1^2) + \gamma h_L \qquad (1)$$

where the head loss is different for each of the three cases.

(a)   If all losses are neglected ($h_L = 0$), Eq. 1 gives

$$p_1 = (9.79\ \text{kN/m}^3)(6\ \text{m})$$
$$+ \frac{998.2\ \text{kg/m}^3}{2}\left[\left(5.65\ \frac{\text{m}}{\text{s}}\right)^2 - \left(2.65\ \frac{\text{m}}{\text{s}}\right)^2\right]$$
$$= (58,740 + 12,428)\ \text{N/m}^2 = 71.0\ \text{kN/m}^2$$

or

$$p_1 = 71.0\ \text{kPa} \qquad \text{(Ans)}$$

COMMENT Note that for this pressure drop, the amount due to elevation change (the hydrostatic effect) is $\gamma(z_2 - z_1) = 59$ kPa and the amount due to the increase in kinetic energy is $\rho(V_2^2 - V_1^2)/2 = 12.0$ kPa.

(b) If the only losses included are the major losses, the head loss is

$$h_L = f \frac{\ell}{D} \frac{V_1^2}{2g}$$

From Table 8.1 the roughness for a 1.9-cm-diameter copper pipe (drawn tubing) is $\varepsilon = 1.5 \times 10^{-6}$ m so that $\varepsilon/D = 8 \times 10^{-5}$. With this ε/D and the calculated Reynolds number (Re = 50,100), the value of f is obtained from the Moody chart as $f = 0.0215$. Note that the Colebrook equation (Eq. 8.35) would give the same value of f. Hence, with the total length of the pipe as $\ell = (4.6 + 1.5 + 3 + 3 + 6)$ m = 18.1 m and the elevation and kinetic energy portions the same as for part (a), Eq. 1 gives

$$p_1 = \gamma z_2 + \frac{1}{2} \rho(V_2^2 - V_1^2) + \rho f \frac{\ell}{D} \frac{V_1^2}{2}$$

$$= (58{,}740 + 12{,}428) \text{ N/m}^2$$

$$+ (998.2 \text{ kg/m}^3)(0.0215)\left(\frac{18.1 \text{ m}}{1.9 \times 10^{-2} \text{ m}}\right)\frac{(2.65 \text{ m/s})^2}{2}$$

$$= (58{,}740 + 12{,}428 + 71{,}786) \text{ N/m}^2 = 143 \text{ k} \cdot \text{N/m}^2$$

or

$$p_1 = 143 \text{ kPa} \qquad \textbf{(Ans)}$$

COMMENT Of this pressure drop, the amount due to pipe friction is approximately $(143 - 71.0)$ kPa = 72 kPa.

(c) If major and minor losses are included, Eq. 1 becomes

$$p_1 = \gamma z_2 + \frac{1}{2} \rho(V_2^2 - V_1^2) + f\gamma \frac{\ell}{D} \frac{V_1^2}{2g} + \sum \rho K_L \frac{V^2}{2}$$

or

$$p_1 = 143 \text{ kPa} + \sum \rho K_L \frac{V^2}{2} \qquad (2)$$

where the 143 kPa contribution is due to elevation change, kinetic energy change, and major losses [part (b)], and the last term represents the sum of all of the minor losses. The loss coefficients of the components ($K_L = 1.5$ for each elbow and $K_L = 10$ for the wide-open globe valve) are given in Table 8.2 (except for the loss coefficient of the faucet, which is given in Fig. E8.8a as $K_L = 2$). Thus,

$$\sum \rho K_L \frac{V^2}{2} = (998.2 \text{ kg/m}^3)\frac{(2.65 \text{ m/s})^2}{2}[10 + 4(1.5) + 2]$$

$$= 63.0 \text{ kN/m}^2$$

or

$$\sum \rho K_L \frac{V^2}{2} = 63.0 \text{ kPa} \qquad (3)$$

Note that we did not include an entrance or exit loss because points (1) and (2) are located within the fluid streams, not within an at-

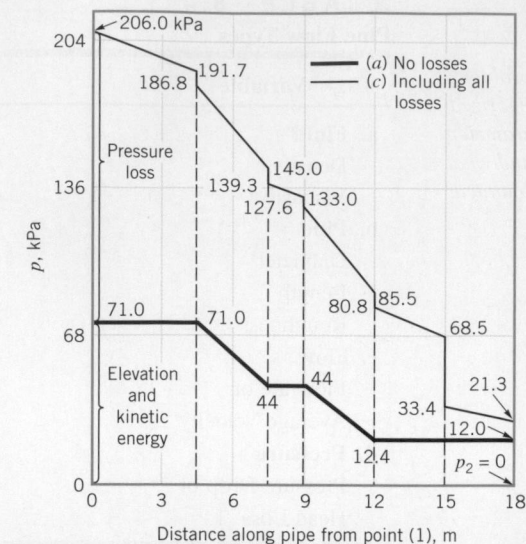

■ **FIGURE E8.8b**

taching reservoir where the kinetic energy is zero. Thus, by combining Eqs. 2 and 3 we obtain the entire pressure drop as

$$p_1 = (143 + 63.0) \text{ kPa} = 206.0 \text{ kPa} \qquad \textbf{(Ans)}$$

This pressure drop calculated by including all losses should be the most realistic answer of the three cases considered.

COMMENTS More detailed calculations will show that the pressure distribution along the pipe is as illustrated in Fig. E8.8b for cases (a) and (c)—neglecting all losses or including all losses. Note that not all of the pressure drop, $p_1 - p_2$, is a "pressure loss." The pressure change due to the elevation and velocity changes is completely reversible. The portion due to the major and minor losses is irreversible.

This flow can be illustrated in terms of the energy line and hydraulic grade line concepts introduced in Section 3.7. As is shown in Fig. E8.8c, for case (a) there are no losses and the energy line (EL) is horizontal, one velocity head ($V^2/2g$) above the hydraulic grade line (HGL), which is one pressure head (γz) above the pipe itself. For cases (b) or (c) the energy line is not horizontal. Each bit of friction in the pipe or loss in a component reduces the available

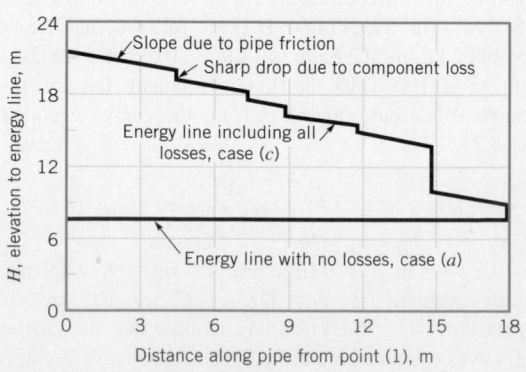

■ **FIGURE E8.8c**

energy, thereby lowering the energy line. Thus, for case (a) the total head remains constant throughout the flow with a value of

$$H = \frac{p_1}{\gamma} + \frac{V_1^2}{2g} + z_1 = \frac{(71.0 \text{ kN/m}^2)}{(9.79 \text{ kN/m}^3)} + \frac{(2.65 \text{ m/s})^2}{2(9.81 \text{ m/s}^2)} + 0$$

$$= 7.6 \text{ m}$$

$$= \frac{p_2}{\gamma} + \frac{V_2^2}{2g} + z_2 = \frac{p_3}{\gamma} + \frac{V_3^3}{2g} + z_3 = \cdots$$

For case (c) the energy line starts at

$$H_1 = \frac{p_1}{\gamma} + \frac{V_1^2}{2g} + z_1$$

$$= \frac{(206.0)\text{kN/m}^2}{(9.79 \text{ kN/m}^3)} + \frac{(2.65 \text{ m/s})^2}{2(9.81 \text{ m/s}^2)} + 0 = 21.4 \text{ m}$$

and falls to a final value of

$$H_2 = \frac{p_2}{\gamma} + \frac{V_2^2}{2g} + z_2 = 0 + \frac{(5.65 \text{ m/s})^2}{2(9.81 \text{ m/s}^2)} + 6 \text{ m}$$

$$= 7.6 \text{ m}$$

The elevation of the energy line can be calculated at any point along the pipe. For example, at point (7), 15 m from point (1),

$$H_7 = \frac{p_7}{\gamma} + \frac{V_7^2}{2g} + z_7$$

$$= \frac{(68.5) \text{ kN/m}^2}{(9.79 \text{ kN/m}^3)} + \frac{(2.65 \text{ m/s})^2}{2(9.81 \text{ m/s}^2)} + 6 \text{ m}$$

$$= 13.4 \text{ m}$$

The head loss per meter of pipe is the same all along the pipe. That is,

$$\frac{h_L}{\ell} = f \frac{V^2}{2gD} = \frac{0.0215(2.65 \text{ m/s})^2}{2(9.81 \text{ m/s}^2)(1.9 \times 10^{-2} \text{ m})} = 0.405$$

Thus, the energy line is a set of straight line segments of the same slope separated by steps whose height equals the head loss of the minor component at that location. As is seen from Fig. E8.8c, the globe valve produces the largest of all the minor losses.

Although the governing pipe flow equations are quite simple, they can provide very reasonable results for a variety of applications, as is shown in the next example.

EXAMPLE 8.9 | Type I, Determine Head Loss

GIVEN As shown in Fig. E8.9a, crude oil at 60 °C with $\gamma = 8.4 \text{ kN/m}^3$ and $\mu = 3.8 \times 10^{-3} \text{ N} \cdot \text{s/m}^2$ (about four times the viscosity of water) is pumped across Alaska through the Alaskan pipeline, a 1280-km-long, 1.2-m-diameter steel pipe, at a maximum rate of $Q = 2.0 \text{ ML/day} = 3.3 \text{ m}^3/\text{s}$.

FIND Determine the kilowatt needed for the pumps that drive this large system.

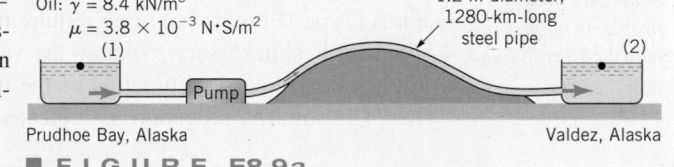

Oil: $\gamma = 8.4 \text{ kN/m}^3$
$\mu = 3.8 \times 10^{-3} \text{ N} \cdot \text{S/m}^2$

1.2-m-diameter, 1280-km-long steel pipe

(1) Pump (2)

Prudhoe Bay, Alaska   Valdez, Alaska

■ **FIGURE E8.9a**

SOLUTION

From the energy equation (Eq. 8.21) we obtain

$$\frac{p_1}{\gamma} + \frac{V_1^2}{2g} + z_1 + h_p = \frac{p_2}{\gamma} + \frac{V_2^2}{2g} + z_2 + h_L$$

where points (1) and (2) represent locations within the large holding tanks at either end of the line and h_p is the head provided to the oil by the pumps. We assume that $z_1 = z_2$ (pumped from sea level to sea level), $p_1 = p_2 = V_1 = V_2 = 0$ (large, open tanks) and $h_L = (f\ell/D)V^2/2g$. Minor losses are negligible because of the large length-to-diameter ratio of the relatively straight, uninterrupted pipe; $\ell/D = (1280 \text{ km}) \times (1000 \text{ m/km})/(1.2 \text{ m}) = 1.06 \times 10^6$. Thus,

$$h_p = h_L = f \frac{\ell}{D} \frac{V^2}{2g}$$

where $V = Q/A = (3.3 \text{ m}^3/\text{s})/[\pi(1.2 \text{ m})^2/4] = 2.9 \text{ m/s}$. From Fig. 8.20 or Eq. 8.35, $f = 0.0125$ since $\varepsilon/D = (4.5 \times 10^{-5}\text{m})/(1.2 \text{ m})$

$= 0.0000375$ (see Table 8.1) and Re $= \rho VD/\mu = [(8.4 \times 10^3/9.81)$ kg/m$^3]$ $(2.9 \text{ m/s})(1.2 \text{ m})/(3.8 \times 10^{-3} \text{ N} \cdot \text{s/m}^2) = 7.84 \times 10^5$. Thus,

$$h_p = 0.0125(1.06 \times 10^6) \frac{(2.9 \text{ m/s})^2}{2(9.81 \text{ m/s}^2)} = 5680 \text{ m}$$

and the actual power supplied to the fluid, $\mathcal{P}_a$, is

$$\mathcal{P}_a = \gamma Q h_p = (8.4 \text{ kN/m}^3)(3.3 \text{ m}^3/\text{s})(5680 \text{ m})$$

$$= 1.57 \times 10^8 \text{ m} \cdot \text{N/s} \left(\frac{1 \text{ W}}{1 \text{ N} \cdot \text{m/s}}\right)$$

$$= 157 \text{ MW} \qquad \text{(Ans)}$$

COMMENTS There are many reasons why it is not practical to drive this flow with a single pump of this size. First, there are no pumps this large! Second, the pressure at the pump outlet would need to be $p = \gamma h_L = (8.4 \text{ kN/m}^3)(5680 \text{ m}) = 48 \text{ mPa}$. No practical 1.2-m-diameter pipe would withstand this pressure. An equally unfeasible alternative would be to place the holding tank at the beginning of the pipe on top of a hill of height $h_L = 5680$ m and let gravity force the oil through the 1280 km pipe! How much power would it take to lift the oil to the top of the hill?

To produce the desired flow, the actual system contains 12 pumping stations positioned at strategic locations along the pipeline. Each station contains four pumps, three of which operate at any one time (the fourth is in reserve in case of emergency). Each pump is driven by a 10-MW motor, thereby producing a total kilowatts of $\mathscr{P} = 12$ stations (3 pump/station) (10 MW/pump) = 362 MW. If we assume that the pump/motor combination is approximately 60% efficient, there is a total of 0.60 (362) MW = 217 MW available to drive the fluid. This number compares favorably with the 157-MW answer calculated above.

The assumption of a 60 °C oil temperature may not seem reasonable for flow across Alaska. Note, however, that the oil is warm when it is pumped from the ground and that the 157 MW needed to pump the oil is dissipated as a head loss (and therefore a temperature rise) along the pipe. However, if the oil temperature were 21 °C rather than 60 °C, the viscosity would be approximately 7.6×10^{-3} N · s/m² (twice as large), but the friction factor would only increase from $f = 0.0125$ at 60 °C (Re $= 7.84 \times 10^5$) to $f = 0.0140$ at 21 °C (Re $= 3.92 \times 10^5$). This doubling of

viscosity would result in only an 11% increase in power (from 157 to 174 MW). Because of the large Reynolds numbers involved, the shear stress is due mostly to the turbulent nature of the flow. That is, the value of Re for this flow is large enough (on the relatively flat part of the Moody chart) so that f is nearly independent of Re (or viscosity).

By repeating the calculations for various values of the pipe diameter, D, the results shown in Fig. E8.9b are obtained. Clearly the required pump power, $\mathscr{P}_a$, is a strong function of the pipe diameter, with $\mathscr{P}_a \sim D^{-4}$ if the friction factor is constant. The actual 1.2-m-diameter pipe used represents a compromise between using smaller diameter pipes which are less expensive to make but require considerably more pump power, and larger diameter pipes which require less pump power but are very expensive to make and maintain.

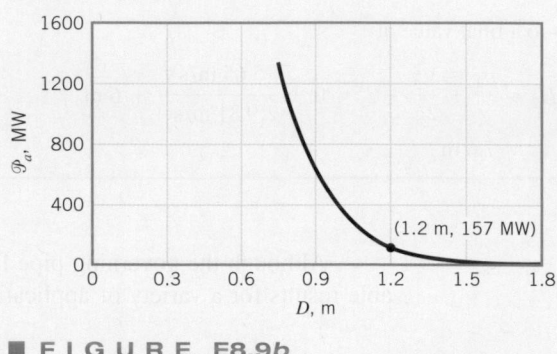

■ **FIGURE E8.9b**

Some pipe flow problems require a trial-and-error solution technique.

Pipe flow problems in which it is desired to determine the flowrate for a given set of conditions (Type II problems) often require trial-and-error or numerical root-finding techniques. This is because it is necessary to know the value of the friction factor to carry out the calculations, but the friction factor is a function of the unknown velocity (flowrate) in terms of the Reynolds number. The solution procedure is indicated in Example 8.10.

*E*XAMPLE 8.10 | Type II, Determine Flowrate

GIVEN Air at a temperature of 38 °C and standard pressure flows from a clothes dryer. According to the appliance manufacturer, the 10-cm-diameter galvanized iron vent on the clothes dryer is not to contain more than 6 m of pipe and four 90° elbows.

FIND Under these conditions determine the air flowrate if the pressure at the start of the vent pipe, directly downstream of the dryer fan, is 0.5 cm of water.

SOLUTION

Application of the energy equation (Eq. 8.21) between the beginning of the vent pipe, point (1), and the exit of the pipe, point (2), gives

$$\frac{p_1}{\gamma} + \frac{V_1^2}{2g} + z_1 = \frac{p_2}{\gamma} + \frac{V_2^2}{2g} + z_2 + f\frac{\ell}{D}\frac{V^2}{2g} + \sum K_L \frac{V^2}{2g} \quad (1)$$

where K_L for each elbow is assumed to be 1.5. In addition, $V_1 = V_2$ and $z_1 = z_2$. (The change in elevation is often negligible for gas flows.) Also, $p_2 = 0$, and $p_1/\gamma_{H_2O} = 0.5$ cm or

$$p_1 = (0.5 \text{ cm})\left(\frac{1 \text{ m}}{100 \text{ cm}}\right)(9800 \text{ N/m}^3) = 49 \text{ N/m}^2$$

Thus, with $\gamma = 11.05$ N/m³ (see Table B.2) and $V_2 = V$ (the air velocity in the pipe), Eq. 1 becomes

$$\frac{(49 \text{ N/m}^2)}{(11.05 \text{ N/m}^3)} = \left[f\frac{(6 \text{ m})}{(\frac{10}{100} \text{ m})} + 4(1.5)\right]\frac{V^2}{2(9.81 \text{ m/s}^2)}$$

or

$$87 = (6.0 + 60 f)V^2 \quad (2)$$

where V is in m/s.

The value of f is dependent on Re, which is dependent on V, an unknown. However, from Table B.2, $\nu = 1.66 \times 10^{-5}$ m²/s and we obtain

$$\text{Re} = \frac{VD}{\nu} = \frac{\left(\frac{10}{100}\text{ m}\right) V}{1.66 \times 10^{-5}\text{ m}^2/\text{s}}$$

or

$$\text{Re} = 6020\, V \tag{3}$$

where again V is in m/s.

Also, since $\varepsilon/D = (1.5 \times 10^{-4}\text{ m})/(10/100\text{ m}) = 0.0015$ (see Table 8.1 for the value of ε), we know which particular curve of the Moody chart is pertinent to this flow. Thus, we have three relationships (Eqs. 2, 3, and the $\varepsilon/D = 0.0015$ curve of Fig. 8.20) from which we can solve for the three unknowns f, Re, and V. This is done easily by an iterative scheme as follows.

It is usually simplest to assume a value of f, calculate V from Eq. 2, calculate Re from Eq. 3, and look up the appropriate value of f in the Moody chart for this value of Re. If the assumed f and the new f do not agree, the assumed answer is not correct—we do not have the solution to the three equations. Although values of either f, V, or Re could be assumed as starting values, it is usually simplest to assume a value of f because the correct value often lies on the relatively flat portion of the Moody chart for which f is quite insensitive to Re.

Thus, we assume $f = 0.022$, approximately the large Re limit for the given relative roughness. From Eq. 2 we obtain

$$V = \left[\frac{87}{6.0 + 60(0.022)}\right]^{1/2} = 3.5\text{ m/s}$$

and from Eq. 3

$$\text{Re} = 6020(3.5) = 21{,}100$$

With this Re and ε/D, Fig. 8.20 gives $f = 0.029$, which is not equal to the assumed solution $f = 0.022$ (although it is close!). We try again, this time with the newly obtained value of $f = 0.029$, which gives $V = 3.4$ m/s and Re $= 20{,}500$. With these values, Fig. 8.20 gives $f = 0.029$, which agrees with the assumed value. Thus, the solution is $V = 3.4$ m/s, or

$$Q = AV = \frac{\pi}{4}\left(\tfrac{10}{100}\text{ m}\right)^2 (3.4\text{ m/s}) = 0.03\text{ m}^3/\text{s} \quad \textbf{(Ans)}$$

COMMENTS Note that the need for the iteration scheme is because one of the equations, $f = \phi(\text{Re}, \varepsilon/D)$, is in graphical form (the Moody chart). If the dependence of f on Re and ε/D is known in equation form, this graphical dependency is eliminated, and the solution technique may be easier. Such is the case if the flow is laminar so that the friction factor is simply $f = 64/\text{Re}$. For turbulent flow, we can use the Colebrook equation rather than the Moody chart. Thus, we keep Eqs. 2 and 3 and use the Colebrook equation (Eq. 8.35a) with $\varepsilon/D = 0.0015$ to give

$$\frac{1}{\sqrt{f}} = -2.0 \log\left(\frac{\varepsilon/D}{3.7} + \frac{2.51}{\text{Re}\sqrt{f}}\right)$$

$$= -2.0 \log\left(4.05 \times 10^{-4} + \frac{2.51}{\text{Re}\sqrt{f}}\right) \tag{4}$$

From Eq. 2 we have $V = [87/(6.0 + 60 f)]^{1/2}$, which can be combined with Eq. 3 to give

$$\text{Re} = \frac{56{,}150}{\sqrt{6.0 + 60 f}} \tag{5}$$

The combination of Eqs. 4 and 5 provides a single equation for the determination of f

$$\frac{1}{\sqrt{f}} = -2.0 \log\Bigg(4.05 \times 10^{-4}$$
$$+\, 4.47 \times 10^{-5}\sqrt{60 + \frac{6.0}{f}}\Bigg) \tag{6}$$

By using a root-finding technique on a computer or calculator, the solution to this equation is determined to be $f = 0.029$, in agreement with the above solution which used the Moody chart.

Note that unlike the Alaskan pipeline example (Example 8.9) in which we assumed minor losses are negligible, minor losses are of importance in this example because of the relatively small length-to-diameter ratio: $\ell/D = 6/(10/100) = 60$. The ratio of minor to major losses in this case is $K_L/(f\ell/D) = 6.0/[0.029\,(60)] = 3.45$. The elbows and entrance produce considerably more loss than the pipe itself.

EXAMPLE 8.11 | Type II, Determine Flowrate

GIVEN The turbine shown in Fig. E8.11 extracts 35 kW from the water flowing through it. The 0.3-m-diameter, 90-m-long pipe is assumed to have a friction factor of 0.02. Minor losses are negligible.

FIND Determine the flowrate through the pipe and turbine.

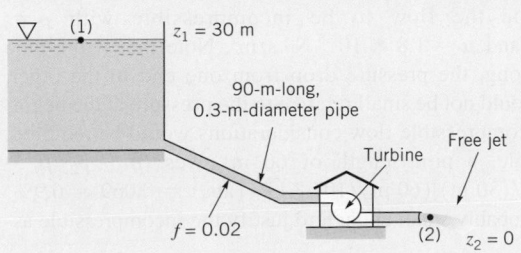

■ FIGURE E8.11

SOLUTION

The energy equation (Eq. 8.21) can be applied between the surface of the lake [point (1)] and the outlet of the pipe as

$$\frac{p_1}{\gamma} + \frac{V_1^2}{2g} + z_1 = \frac{p_2}{\gamma} + \frac{V_2^2}{2g} + z_2 + h_L + h_T \quad (1)$$

where h_T is the turbine head, $p_1 = V_1 = p_2 = z_2 = 0$, $z_1 = 30$ m. and $V_2 = V$, the fluid velocity in the pipe. The head loss is given by

$$h_L = f\frac{\ell}{D}\frac{V^2}{2g} = 0.02\frac{(90 \text{ m})}{(0.3 \text{ m})}\frac{V^2}{2(9.81 \text{ m/s}^2)} = 0.03V^2 \text{ m}$$

where V is in m/s. Also, the turbine head is

$$h_T = \frac{\mathcal{P}_a}{\gamma Q} = \frac{\mathcal{P}_a}{\gamma(\pi/4)D^2V}$$

$$= \frac{(35 \text{ kW})[(1 \text{ N} \cdot \text{m/s})/\text{W}]}{(9800 \text{ N/m}^3)[(\pi/4)(0.3 \text{ m})^2V]} = \frac{51}{V} \text{ m}$$

Thus, Eq. 1 can be written as

$$30 = \frac{V^2}{2(9.81)} + 0.30V^2 + \frac{51}{V}$$

or

$$0.35V^3 - 30V + 51 = 0 \quad (2)$$

where V is in m/s. The velocity of the water in the pipe is found as the solution of Eq. 2. Surprisingly, there are two real, positive roots: $V = 1.8$ m/s or $V = 8.2$ m/s. The third root is negative ($V = -10$ m/s) and has no physical meaning for this flow. Thus, the two acceptable flowrates are

$$Q = \frac{\pi}{4}D^2V = \frac{\pi}{4}(0.3 \text{ m})^2(1.8 \text{ m/s}) = 0.13 \text{ m}^3/\text{s} \quad \textbf{(Ans)}$$

or

$$Q = \frac{\pi}{4}(0.3 \text{ m})^2(8.2 \text{ m/s}) = 0.58 \text{ m}^3/\text{s} \quad \textbf{(Ans)}$$

COMMENTS Either of these two flowrates gives the same power, $\mathcal{P}_a = \gamma Q h_T$. The reason for two possible solutions can be seen from the following. With the low flowrate ($Q = 0.13$ m³/s), we obtain the head loss and turbine head as $h_L = 1$ m and $h_T = 28$ m. Because of the relatively low velocity there is a relatively small head loss and, therefore, a large head available for the turbine. With the large flowrate ($Q = 0.58$ m³/s), we find $h_L = 20$ m and $h_T = 6.2$ m. The high-speed flow in the pipe produces a relatively large loss due to friction, leaving a relatively small head for the turbine. However, in either case the product of the turbine head times the flowrate is the same. That is, the power extracted ($\mathcal{P}_a = \gamma Q h_T$) is identical for each case. Although either flowrate will allow the extraction of 35 kW from the water, the details of the design of the turbine itself will depend strongly on which flowrate is to be used. Such information can be found in Chapter 12 and various references about turbomachines (Refs. 14, 19, 20).

If the friction factor were not given, the solution to the problem would be much more lengthy. A trial-and-error solution similar to that in Example 8.10 would be required along with the solution of a cubic equation.

In pipe flow problems for which the diameter is the unknown (Type III), an iterative or numerical root-finding technique is required. This is, again, because the friction factor is a function of the diameter—through both the Reynolds number and the relative roughness. Thus, neither $\text{Re} = \rho VD/\mu = 4\rho Q/\pi\mu D$ nor ε/D are known unless D is known. Examples 8.12 and 8.13 illustrate this.

EXAMPLE 8.12   Type III without Minor Losses, Determine Diameter

GIVEN Air at standard temperature and pressure flows through a horizontal, galvanized iron pipe ($\varepsilon = 1.5 \times 10^{-4}$ m) at a rate of 5.6×10^{-2} m³/s. The pressure drop is to be no more than 3.5 kPa per 30 m of pipe.

FIND Determine the minimum pipe diameter.

SOLUTION

We assume the flow to be incompressible with $\rho = 1.2$ kg/m³ and $\mu = 1.8 \times 10^{-5}$ N · s/m². Note that if the pipe were too long, the pressure drop from one end to the other, $p_1 - p_2$, would not be small relative to the pressure at the beginning, and compressible flow considerations would be required. For example, a pipe length of 60 m gives $(p_1 - p_2)/p_1 = [(3.5 \text{ kPa})/(30 \text{ m})](60 \text{ m})/101.3 \text{ kPa (abs)} = 0.069 = 6.9\%$, which is probably small enough to justify the incompressible assumption.

With $z_1 = z_2$ and $V_1 = V_2$ the energy equation (Eq. 8.21) becomes

$$p_1 = p_2 + f\frac{\ell}{D}\frac{\rho V^2}{2} \quad (1)$$

where $V = Q/A = 4Q/(\pi D^2) = 4(5.6 \times 10^{-2} \text{ m}^3/\text{s})/\pi D^2$, or

$$V = \frac{0.07}{D^2}$$

where D is in meters. Thus, with $p_1 - p_2 = 3.5$ kPa and $\ell = 30$ m, Eq. 1 becomes

$$p_1 - p_2 = 3.5 \text{ kN/m}^2$$

$$= f\frac{(30 \text{ m})}{D}(1.2 \text{ kg/m}^3)\frac{1}{2}\left(\frac{0.07 \text{ m}}{D^2 \text{ s}}\right)^2$$

or

$$D = 0.12 f^{1/5} \qquad (2)$$

where D is in meters. Also $\text{Re} = \rho VD/\mu = (1.2 \text{ kg/m}^3)$ $[(0.07/D^2) \text{ m/s}]D/(1.8 \times 10^{-5} \text{ N} \cdot \text{s/m}^2)$, or

$$\text{Re} = \frac{4.6 \times 10^3}{D} \qquad (3)$$

and

$$\frac{\varepsilon}{D} = \frac{1.5 \times 10^{-4}}{D} \qquad (4)$$

Thus, we have four equations (Eqs. 2, 3, 4, and either the Moody chart or the Colebrook equation) and four unknowns (f, D, ε/D, and Re) from which the solution can be obtained by trial-and-error methods.

If we use the Moody chart, it is probably easiest to assume a value of f, use Eqs. 2, 3, and 4 to calculate D, Re, and ε/D, and then compare the assumed f with that from the Moody chart. If they do not agree, try again. Thus, we assume $f = 0.02$, a typical value, and obtain $D = 0.12(0.02)^{1/5} = 5.4 \times 10^{-2}$ m, which gives $\varepsilon/D = 1.5 \times 10^{-4}/5.4 \times 10^{-2}$ m $= 2.8 \times 10^{-3}$ and $\text{Re} = 4.6 \times 10^3/5.4 \times 10^{-2} = 8.5 \times 10^4$. From the Moody chart we obtain $f = 0.027$ for these values of ε/D and Re. Since this is not the same as our assumed value of f, we try again. With $f = 0.027$, we obtain $D = 5.8 \times 10^{-2}$ m, $\varepsilon/D = 2.6 \times 10^{-3}$, and $\text{Re} = 8.0 \times 10^4$, which in turn give $f = 0.027$, in agreement with the assumed value. Thus, the diameter of the pipe should be

$$D = 5.8 \times 10^{-2} \text{ m} \qquad \text{(Ans)}$$

COMMENT   If we use the Colebrook equation (Eq. 8.35a) with $\varepsilon/D = 1.5 \times 10^{-4}/0.12 f^{1/5} = 0.00125/f^{1/5}$ and $\text{Re} = 4.6 \times 10^3/0.12 f^{1/5} = 3.8 \times 10^4/f^{1/5}$, we obtain

$$\frac{1}{\sqrt{f}} = -2.0 \log\left(\frac{\varepsilon/D}{3.7} + \frac{2.51}{\text{Re}\sqrt{f}}\right)$$

or

$$\frac{1}{\sqrt{f}} = -2.0 \log\left(\frac{3.38 \times 10^{-4}}{f^{1/5}} + \frac{6.60 \times 10^{-5}}{f^{3/10}}\right)$$

By using a root-finding technique on a computer or calculator, the solution to this equation is determined to be $f = 0.027$, and hence $D = 5.8 \times 10^{-2}$ m, in agreement with the Moody chart method.

By repeating the calculations for various values of the flowrate, Q, the results shown in Fig. E8.12 are obtained. Although an increase in flowrate requires a larger diameter pipe (for the given pressure drop), the increase in diameter is minimal. For example, if the flowrate is doubled from 0.03 m³/s to 0.06 m³/s, the diameter increases from 0.05 m to 0.06 m.

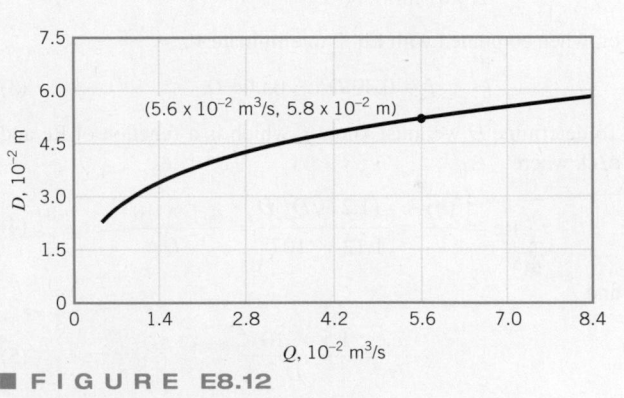

■ **FIGURE E8.12**

In the previous example we only had to consider major losses. In some instances the inclusion of major and minor losses can cause a slightly more lengthy solution procedure, even though the governing equations are essentially the same. This is illustrated in Example 8.13.

*E*XAMPLE 8.13   Type III with Minor Losses, Determine Diameter

GIVEN   Water at 15 °C ($\nu = 1.12 \times 10^{-6}$ m²/s, see Table 1.3) is to flow from reservoir A to reservoir B through a pipe of length 520 m and roughness 1.5×10^{-4} m at a rate of $Q = 1$ m³/s as shown in Fig. E8.13a. The system contains a sharp-edged entrance and four flanged 45° elbows.

FIND   Determine the pipe diameter needed.

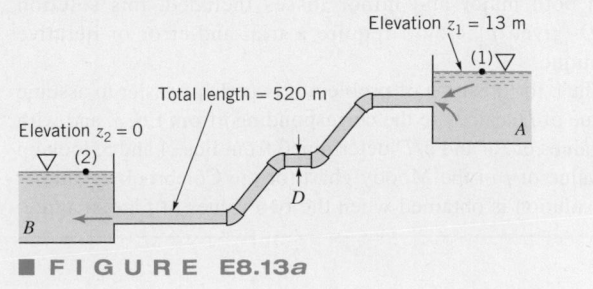

■ **FIGURE E8.13a**

SOLUTION

The energy equation (Eq. 8.21) can be applied between two points on the surfaces of the reservoirs ($p_1 = p_2 = V_1 = V_2 = z_2 = 0$) as follows:

$$\frac{p_1}{\gamma} + \frac{V_1^2}{2g} + z_1 = \frac{p_2}{\gamma} + \frac{V_2^2}{2g} + z_2 + h_L$$

or

$$z_1 = \frac{V^2}{2g}\left(f\frac{\ell}{D} + \sum K_L\right) \qquad (1)$$

where $V = Q/A = 4Q/\pi D^2 = 4(1 \text{ m}^3/\text{s})/\pi D^2$, or

$$V = \frac{1.27}{D^2} \qquad (2)$$

is the velocity within the pipe. (Note that the units on V and D are m/s and m, respectively.) The loss coefficients are obtained from Table 8.2 and Figs. 8.22 and 8.25 as $K_{L_{\text{ent}}} = 0.5$, $K_{L_{\text{elbow}}} = 0.2$, and $K_{L_{\text{exit}}} = 1$. Thus, Eq. 1 can be written as

$$13 \text{ m} = \frac{V^2}{2(9.81 \text{ m/s}^2)}\left\{\frac{520}{D}f + [4(0.2) + 0.5 + 1]\right\}$$

or, when combined with Eq. 2 to eliminate V,

$$f = 0.30\,D^5 - 0.004\,D \qquad (3)$$

To determine D we must know f, which is a function of Re and ε/D, where

$$\text{Re} = \frac{VD}{\nu} = \frac{[(1.27)/D^2]D}{1.12 \times 10^{-6}} = \frac{1.1 \times 10^6}{D} \qquad (4)$$

and

$$\frac{\varepsilon}{D} = \frac{1.5 \times 10^{-4}}{D} \qquad (5)$$

where D is in meters. Again, we have four equations (Eqs. 3, 4, 5, and the Moody chart or the Colebrook equation) for the four unknowns D, f, Re, and ε/D.

Consider the solution by using the Moody chart. Although it is often easiest to assume a value of f and make calculations to determine if the assumed value is the correct one, with the inclusion of minor losses this may not be the simplest method. For example, if we assume $f = 0.02$ and calculate D from Eq. 3, we would have to solve a fifth-order equation. With only major losses (see Example 8.12), the term proportional to D in Eq. 3 is absent, and it is easy to solve for D if f is given. With both major and minor losses included, this solution for D (given f) would require a trial-and-error or iterative technique.

Thus, for this type of problem it is perhaps easier to assume a value of D, calculate the corresponding f from Eq. 3, and with the values of Re and ε/D determined from Eqs. 4 and 5, look up the value of f in the Moody chart (or the Colebrook equation). The solution is obtained when the two values of f are in agreement. A few rounds of calculation will reveal that the solution is given by

$$D \approx 0.5 \text{ m} \qquad \text{(Ans)}$$

COMMENTS   Alternatively, we can use the Colebrook equation rather than the Moody chart to solve for D. This is easily done by using the Colebrook equation (Eq. 8.35a) with f as a function of D obtained from Eq. 3 and Re and ε/D as functions of D from Eqs. 4 and 5. The resulting single equation for D can be solved by using a root-finding technique on a computer or calculator to obtain $D = 0.5$ m. This agrees with the solution obtained using the Moody chart.

By repeating the calculations for various pipe lengths, ℓ, the results shown in Fig. E8.13b are obtained. As the pipe length increases it is necessary, because of the increased friction, to increase the pipe diameter to maintain the same flowrate.

It is interesting to attempt to solve this example if all losses are neglected so that Eq. 1 becomes $z_1 = 0$. Clearly from Fig. E8.13a, $z_1 = 13$ m. Obviously something is wrong. A fluid cannot flow from one elevation, beginning with zero pressure and velocity, and end up at a lower elevation with zero pressure and velocity unless energy is removed (i.e., a head loss or a turbine) somewhere between the two locations. If the pipe is short (negligible friction) and the minor losses are negligible, there is still the kinetic energy of the fluid as it leaves the pipe and enters the reservoir. After the fluid meanders around in the reservoir for some time, this kinetic energy is lost and the fluid is stationary. No matter how small the viscosity is, the exit loss cannot be neglected. The same result can be seen if the energy equation is written from the free surface of the upstream tank to the exit plane of the pipe, at which point the kinetic energy is still available to the fluid. In either case the energy equation becomes $z_1 = V^2/2g$ in agreement with the inviscid results of Chapter 3 (the Bernoulli equation).

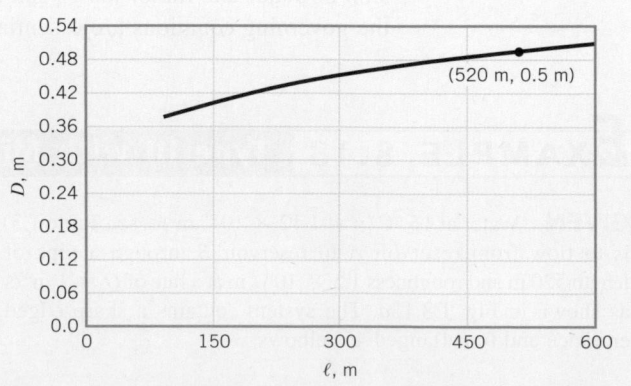

■ **F I G U R E   E8.13b**

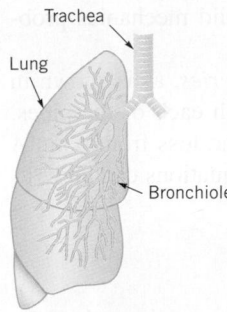

Trachea

Lung

Bronchiole

8.5.2 Multiple Pipe Systems

In many pipe systems there is more than one pipe involved. The complex system of tubes in our lungs (beginning as shown by the figure in the margin, with the relatively large-diameter trachea and ending in tens of thousands of minute bronchioles after numerous branchings) and the maze of pipes in a city's water distribution system are typical of such systems. The governing mechanisms for the flow in *multiple pipe systems* are the same as for the single pipe systems discussed in this chapter. However, because of the numerous unknowns involved, additional complexities may arise in solving for the flow in multiple pipe systems. Some of these complexities are discussed in this section.

F l u i d s   i n   t h e   N e w s

Deepwater pipeline Pipelines used to transport oil and gas are commonplace. But south of New Orleans, in deep waters of the Gulf of Mexico, a not-so-common *multiple pipe system* is being built. The new so-called Mardi Gras system of pipes is being laid in water depths of 4300 to 7300 feet (1310 to 2230 meters). It will transport oil and gas from five deepwater fields with the interesting names of Holstein, Mad Dog, Thunder Horse, Atlantis, and Na Kika. The deepwater pipelines will connect with lines at intermediate water depths to transport the oil and gas to shallow-water fixed platforms and shore. The steel pipe used is 28 inches (71 cm) in diameter with a wall thickness of 1 1/8 in (3.5 cm). The thick-walled pipe is needed to withstand the large external pressure which is about 3250 psi (22.4 MPa) at a depth of 7300 ft (2225 m). The pipe is installed in 240-ft (73 m) sections from a vessel the size of a large football stadium. Upon completion, the deepwater pipeline system will have a total length of more than 450 miles (724 km) and the capability of transporting more than 1 million barrels (159 ML) of oil per day and 1.5 billion cubic feet (4.2×10^7 m^3) of gas per day. (See Problem 8.113.)

The simplest multiple pipe systems can be classified into series or parallel flows, as are shown in Fig. 8.35. The nomenclature is similar to that used in electrical circuits. Indeed, an analogy between fluid and electrical circuits is often made as follows. In a simple electrical circuit, there is a balance between the voltage (e), current (i), and resistance (R) as given by Ohm's law: $e = iR$. In a fluid circuit there is a balance between the pressure drop (Δp), the flowrate or velocity (Q or V), and the flow resistance as given in terms of the friction factor and minor loss coefficients (f and K_L). For a simple flow $[\Delta p = f(\ell/D)(\rho V^2/2)]$, it follows that $\Delta p = Q^2 \tilde{R}$, where $\tilde{R}$, a measure of the resistance to the flow, is proportional to f.

The main differences between the solution methods used to solve electrical circuit problems and those for fluid circuit problems lie in the fact that Ohm's law is a linear equation (doubling the voltage doubles the current), while the fluid equations are generally nonlinear (doubling the pressure drop does not double the flowrate unless the flow is laminar). Thus, although some of the

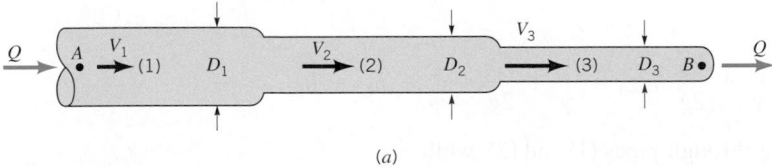

(a)

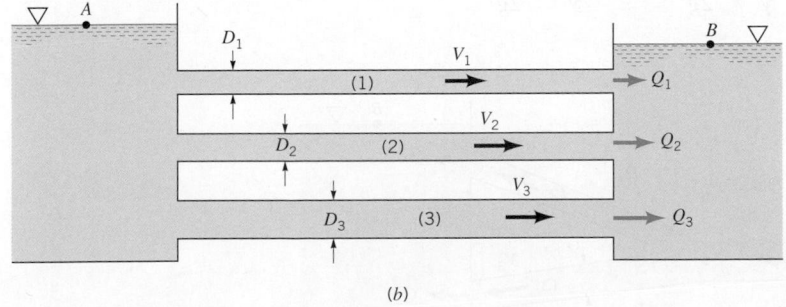

(b)

■ **F I G U R E   8.35**   (*a*) Series and (*b*) parallel pipe systems.

standard electrical engineering methods can be carried over to help solve fluid mechanics problems, others cannot.

One of the simplest multiple pipe systems is that containing pipes in *series*, as is shown in Fig. 8.35a. Every fluid particle that passes through the system passes through each of the pipes. Thus, the flowrate (but not the velocity) is the same in each pipe, and the head loss from point A to point B is the sum of the head losses in each of the pipes. The governing equations can be written as follows:

$$Q_1 = Q_2 = Q_3$$

and

$$h_{L_{A-B}} = h_{L_1} + h_{L_2} + h_{L_3}$$

Series and parallel pipe systems are often encountered.

where the subscripts refer to each of the pipes. In general, the friction factors will be different for each pipe because the Reynolds numbers ($Re_i = \rho V_i D_i / \mu$) and the relative roughnesses (ε_i / D_i) will be different. If the flowrate is given, it is a straightforward calculation to determine the head loss or pressure drop (Type I problem). If the pressure drop is given and the flowrate is to be calculated (Type II problem), an iteration scheme is needed. In this situation none of the friction factors, f_i, are known, so the calculations may involve more trial-and-error attempts than for corresponding single pipe systems. The same is true for problems in which the pipe diameter (or diameters) is to be determined (Type III problems).

Another common multiple pipe system contains pipes in *parallel*, as is shown in Fig. 8.35b. In this system a fluid particle traveling from A to B may take any of the paths available, with the total flowrate equal to the sum of the flowrates in each pipe. However, by writing the energy equation between points A and B it is found that the head loss experienced by any fluid particle traveling between these locations is the same, independent of the path taken. Thus, the governing equations for parallel pipes are

$$Q = Q_1 + Q_2 + Q_3$$

and

$$h_{L_1} = h_{L_2} = h_{L_3}$$

Again, the method of solution of these equations depends on what information is given and what is to be calculated.

Another type of multiple pipe system called a *loop* is shown in Fig. 8.36. In this case the flowrate through pipe (1) equals the sum of the flowrates through pipes (2) and (3), or $Q_1 = Q_2 + Q_3$. As can be seen by writing the energy equation between the surfaces of each reservoir, the head loss for pipe (2) must equal that for pipe (3), even though the pipe sizes and flowrates may be different for each. That is,

$$\frac{p_A}{\gamma} + \frac{V_A^2}{2g} + z_A = \frac{p_B}{\gamma} + \frac{V_B^2}{2g} + z_B + h_{L_1} + h_{L_2}$$

for a fluid particle traveling through pipes (1) and (2), while

$$\frac{p_A}{\gamma} + \frac{V_A^2}{2g} + z_A = \frac{p_B}{\gamma} + \frac{V_B^2}{2g} + z_B + h_{L_1} + h_{L_3}$$

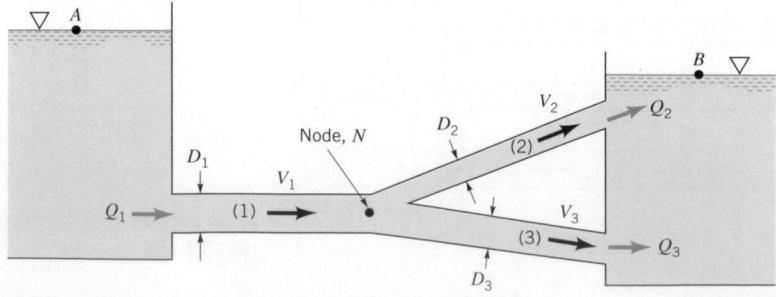

■ **FIGURE 8.36**   **Multiple pipe loop system.**

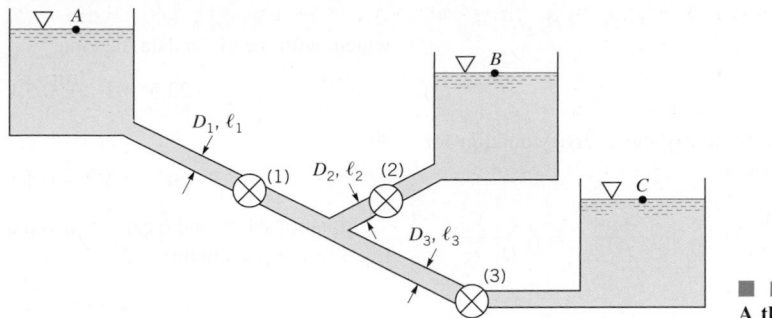

■ FIGURE 8.37
A three-reservoir system.

for fluid that travels through pipes (1) and (3). These can be combined to give $h_{L_2} = h_{L_3}$. This is a statement of the fact that fluid particles that travel through pipe (2) and particles that travel through pipe (3) all originate from common conditions at the junction (or node, N) of the pipes and all end up at the same final conditions.

The flow in a relatively simple looking multiple pipe system may be more complex than it appears initially. The branching system termed the *three-reservoir problem* shown in Fig. 8.37 is such a system. Three reservoirs at known elevations are connected together with three pipes of known properties (lengths, diameters, and roughnesses). The problem is to determine the flowrates into or out of the reservoirs. If valve (1) were closed, the fluid would flow from reservoir B to C, and the flowrate could be easily calculated. Similar calculations could be carried out if valves (2) or (3) were closed with the others open.

With all valves open, however, it is not necessarily obvious which direction the fluid flows. For the conditions indicated in Fig. 8.37, it is clear that fluid flows from reservoir A because the other two reservoir levels are lower. Whether the fluid flows into or out of reservoir B depends on the elevation of reservoirs B and C and the properties (length, diameter, roughness) of the three pipes. In general, the flow direction is not obvious, and the solution process must include the determination of this direction. This is illustrated in Example 8.14.

For some pipe systems, the direction of flow is not known a priori.

EXAMPLE 8.14   Three-Reservoir, Multiple-Pipe System

GIVEN Three reservoirs are connected by three pipes as are shown in Fig. E8.14. For simplicity we assume that the diameter of each pipe is 0.3 m, the friction factor for each is 0.02, and because of the large length-to-diameter ratio, minor losses are negligible.

FIND Determine the flowrate into or out of each reservoir.

SOLUTION

It is not obvious which direction the fluid flows in pipe (2). However, we assume that it flows out of reservoir B, write the governing equations for this case, and check our assumption. The continuity equation requires that $Q_1 + Q_2 = Q_3$, which, since the diameters are the same for each pipe, becomes simply

$$V_1 + V_2 = V_3 \tag{1}$$

The energy equation for the fluid that flows from A to C in pipes (1) and (3) can be written as

$$\frac{p_A}{\gamma} + \frac{V_A^2}{2g} + z_A = \frac{p_C}{\gamma} + \frac{V_C^2}{2g} + z_C + f_1 \frac{\ell_1}{D_1} \frac{V_1^2}{2g} + f_3 \frac{\ell_3}{D_3} \frac{V_3^2}{2g}$$

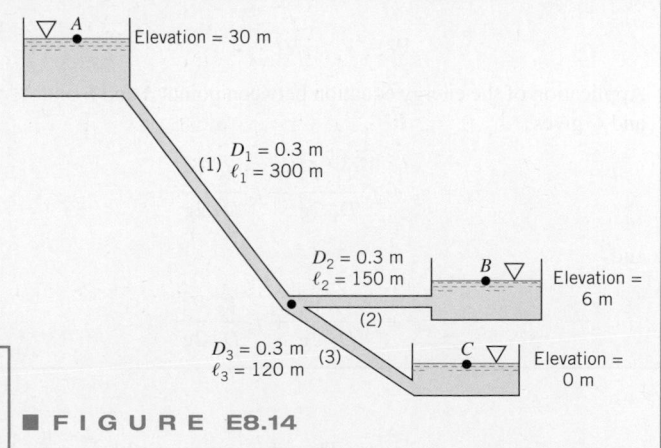

■ FIGURE E8.14

By using the fact that $p_A = p_C = V_A = V_C = z_C = 0$, this becomes

$$z_A = f_1 \frac{\ell_1}{D_1} \frac{V_1^2}{2g} + f_3 \frac{\ell_3}{D_3} \frac{V_3^2}{2g}$$

For the given conditions of this problem we obtain

$$30 \text{ m} = \frac{0.02}{2(9.81 \text{ m/s}^2)} \frac{1}{(0.3 \text{ m})} \left[(300 \text{ m})V_1^2 + (120 \text{ m})V_3^2 \right]$$

or

$$29.4 = V_1^2 + 0.4 V_3^2 \tag{2}$$

where V_1 and V_3 are in m/s. Similarly the energy equation for fluid flowing from B and C is

$$\frac{p_B}{\gamma} + \frac{V_B^2}{2g} + z_B = \frac{p_C}{\gamma} + \frac{V_C^2}{2g} + z_C + f_2 \frac{\ell_2}{D_2} \frac{V_2^2}{2g} + f_3 \frac{\ell_3}{D_3} \frac{V_3^2}{2g}$$

or

$$z_B = f_2 \frac{\ell_2}{D_2} \frac{V_2^2}{2g} + f_3 \frac{\ell_3}{D_3} \frac{V_3^2}{2g}$$

For the given conditions this can be written as

$$11.8 = V_2^2 + 1.25 V_3^2 \tag{3}$$

Equations 1, 2, and 3 (in terms of the three unknowns V_1, V_2, and V_3) are the governing equations for this flow, provided the fluid flows from reservoir B. It turns out, however, that there is no solution for these equations with positive, real values of the velocities. Although these equations do not appear to be complicated, there is no simple way to solve them directly. Thus, a trial-and-error solution is suggested. This can be accomplished as follows. Assume a value of $V_1 > 0$, calculate V_3 from Eq. 2, and then V_2 from Eq. 3. It is found that the resulting V_1, V_2, V_3 trio does not satisfy Eq. 1 for any value of V_1 assumed. There is no solution to Eqs. 1, 2, and 3 with real, positive values of V_1, V_2, and V_3. Thus, our original assumption of flow out of reservoir B must be incorrect.

To obtain the solution, assume the fluid flows into reservoirs B and C and out of A. For this case the continuity equation becomes

$$Q_1 = Q_2 + Q_3$$

or

$$V_1 = V_2 + V_3 \tag{4}$$

Application of the energy equation between points A and B and A and C gives

$$z_A = z_B + f_1 \frac{\ell_1}{D_1} \frac{V_1^2}{2g} + f_2 \frac{\ell_2}{D_2} \frac{V_2^2}{2g}$$

and

$$z_A = z_C + f_1 \frac{\ell_1}{D_1} \frac{V_1^2}{2g} + f_3 \frac{\ell_3}{D_3} \frac{V_3^2}{2g}$$

which, with the given data, become

$$23.5 = V_1^2 + 0.5 \, V_2^2 \tag{5}$$

and

$$29.4 = V_1^2 + 0.4 \, V_3^2 \tag{6}$$

Equations 4, 5, and 6 can be solved as follows. By subtracting Eq. 5 from 6 we obtain

$$V_3 = \sqrt{14.8 + 1.25 V_2^2}$$

Thus, Eq. 5 can be written as

$$23.5 = (V_2 + V_3)^2 + 0.5 V_2^2$$
$$= (V_2 + \sqrt{14.8 + 1.25 V_2^2})^2 + 0.5 V_2^2$$

or

$$2 V_2 \sqrt{14.8 + 1.25 V_2^2} = 8.7 - 2.75 V_2^2 \tag{7}$$

which, upon squaring both sides, can be written as

$$V_2^4 - 41.8 \, V_2^2 + 29.6 = 0$$

By using the quadratic formula we can solve for V_2^2 to obtain either $V_2^2 = 41$ or $V_2^2 = 0.72$. Thus, either $V_2 = 6.40$ m/s or $V_2 = 0.85$ m/s. The value $V_2 = 6.40$ m/s is not a root of the original equations. It is an extra root introduced by squaring Eq. 7, which with $V_2 = 21.3$ becomes "$104 = -104$." Thus, $V_2 = 0.85$ m/s and from Eq. 5, $V_1 = 4.8$ m/s. The corresponding flowrates are

$$Q_1 = A_1 V_1 = \frac{\pi}{4} D_1^2 V_1 = \frac{\pi}{4} (0.3 \text{ m})^2 (4.8 \text{ m/s})$$
$$= 0.34 \text{ m}^3/\text{s from } A \tag{Ans}$$

$$Q_2 = A_2 V_2 = \frac{\pi}{4} D_2^2 V_2 = \frac{\pi}{4} (0.3 \text{ m})^2 (0.85 \text{ m/s})$$
$$= 0.06 \text{ m}^3/\text{s into } B \tag{Ans}$$

and

$$Q_3 = Q_1 - Q_2 = (0.34 - 0.06) \text{ m}^3/\text{s}$$
$$= 0.28 \text{ m}^3/\text{s into } C \tag{Ans}$$

Note the slight differences in the governing equations depending on the direction of the flow in pipe (2)—compare Eqs. 1, 2, and 3 with Eqs. 4, 5, and 6.

COMMENT If the friction factors were not given, a trial-and-error procedure similar to that needed for Type II problems (see Section 8.5.1) would be required.

The ultimate in multiple pipe systems is a *network* of pipes such as that shown in Fig. 8.38. Networks like these often occur in city water distribution systems and other systems that may have multiple "inlets" and "outlets." The direction of flow in the various pipes is by no means obvious—in fact, it may vary in time, depending on how the system is used from time to time.

The solution for pipe network problems is often carried out by use of node and loop equations similar in many ways to that done in electrical circuits. For example, the continuity equation requires that for each *node* (the junction of two or more pipes) the net flowrate is zero. What flows into a node must flow out at the same rate. In addition, the net pressure difference completely around a *loop* (starting at one location in a pipe and returning to that location) must be zero. By combining these ideas with the usual head loss and pipe flow equations, the flow throughout the entire network can

Pipe network problems can be solved using node and loop concepts.

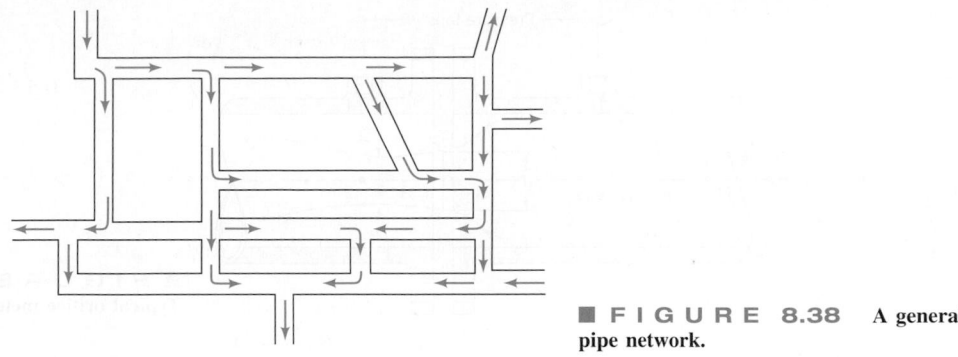

■ **FIGURE 8.38** A general pipe network.

be obtained. Of course, trial-and-error solutions are usually required because the direction of flow and the friction factors may not be known. Such a solution procedure using matrix techniques is ideally suited for computer use (Refs. 21, 22).

8.6 Pipe Flowrate Measurement

It is often necessary to determine experimentally the flowrate in a pipe. In Chapter 3 we introduced various types of flow-measuring devices (Venturi meter, nozzle meter, orifice meter, etc.) and discussed their operation under the assumption that viscous effects were not important. In this section we will indicate how to account for the ever-present viscous effects in these flow meters. We will also indicate other types of commonly used flow meters.

8.6.1 Pipe Flowrate Meters

Orifice, nozzle and Venturi meters involve the concept "high velocity gives low pressure."

Three of the most common devices used to measure the instantaneous flowrate in pipes are the orifice meter, the nozzle meter, and the Venturi meter. As was discussed in Section 3.6.3, each of these meters operates on the principle that a decrease in flow area in a pipe causes an increase in velocity that is accompanied by a decrease in pressure. Correlation of the pressure difference with the velocity provides a means of measuring the flowrate. In the absence of viscous effects and under the assumption of a horizontal pipe, application of the Bernoulli equation (Eq. 3.7) between points (1) and (2) shown in Fig. 8.39 gave

$$Q_{\text{ideal}} = A_2 V_2 = A_2 \sqrt{\frac{2(p_1 - p_2)}{\rho(1 - \beta^4)}} \tag{8.37}$$

where $\beta = D_2/D_1$. Based on the results of the previous sections of this chapter, we anticipate that there is a head loss between (1) and (2) so that the governing equations become

$$Q = A_1 V_1 = A_2 V_2$$

and

$$\frac{p_1}{\gamma} + \frac{V_1^2}{2g} = \frac{p_2}{\gamma} + \frac{V_2^2}{2g} + h_L$$

The ideal situation has $h_L = 0$ and results in Eq. 8.37. The difficulty in including the head loss is that there is no accurate expression for it. The net result is that empirical coefficients are used in the flowrate equations to account for the complex real-world effects brought on by the nonzero viscosity. The coefficients are discussed in this section.

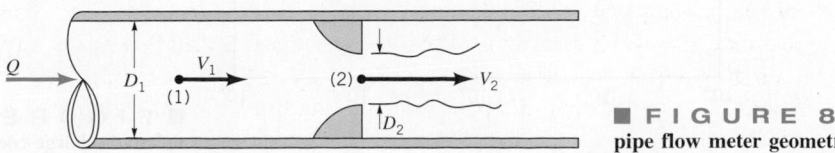

■ **FIGURE 8.39** Typical pipe flow meter geometry.

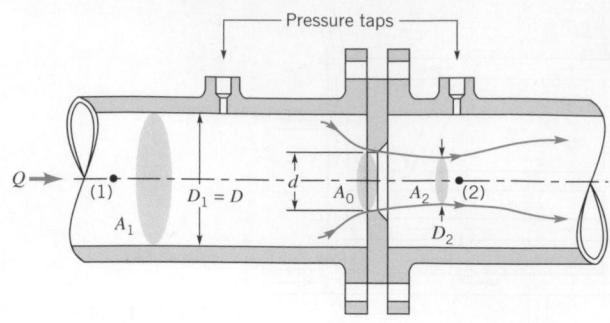

■ **FIGURE 8.40**
Typical orifice meter construction.

An orifice discharge coefficient is used to account for non-ideal effects.

A typical *orifice meter* is constructed by inserting between two flanges of a pipe a flat plate with a hole, as shown in Fig. 8.40. The pressure at point (2) within the vena contracta is less than that at point (1). Nonideal effects occur for two reasons. First, the vena contracta area, A_2, is less than the area of the hole, A_o, by an unknown amount. Thus, $A_2 = C_c A_o$, where C_c is the contraction coefficient ($C_c < 1$). Second, the swirling flow and turbulent motion near the orifice plate introduce a head loss that cannot be calculated theoretically. Thus, an *orifice discharge coefficient, C_o,* is used to take these effects into account. That is,

$$Q = C_o Q_{ideal} = C_o A_o \sqrt{\frac{2(p_1 - p_2)}{\rho(1 - \beta^4)}} \tag{8.38}$$

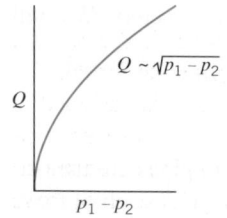

where $A_o = \pi d^2/4$ is the area of the hole in the orifice plate. The value of C_o is a function of $\beta = d/D$ and the Reynolds number $Re = \rho VD/\mu$, where $V = Q/A_1$. Typical values of C_o are given in Fig. 8.41. As shown by Eq. 8.38 and the figure in the margin, for a given value of C_o, the flowrate is proportional to the square root of the pressure difference. Note that the value of C_o depends on the specific construction of the orifice meter (i.e., the placement of the pressure taps, whether the orifice plate edge is square or beveled, etc.). Very precise conditions governing the construction of standard orifice meters have been established to provide the greatest accuracy possible (Refs. 23, 24).

Another type of pipe flow meter that is based on the same principles used in the orifice meter is the *nozzle meter,* three variations of which are shown in Fig. 8.42. This device uses a contoured nozzle (typically placed between flanges of pipe sections) rather than a simple (and less expensive) plate with a hole as in an orifice meter. The resulting flow pattern for the nozzle meter is closer to ideal than the orifice meter flow. There is only a slight vena contracta and the secondary

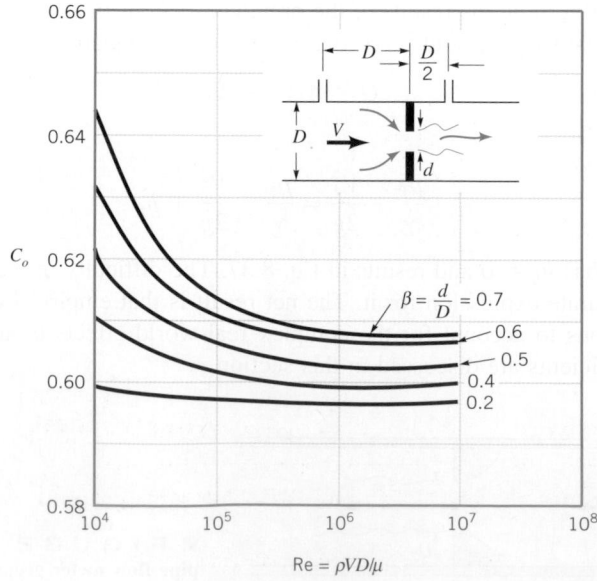

■ **FIGURE 8.41** **Orifice meter discharge coefficient (Ref. 24).**

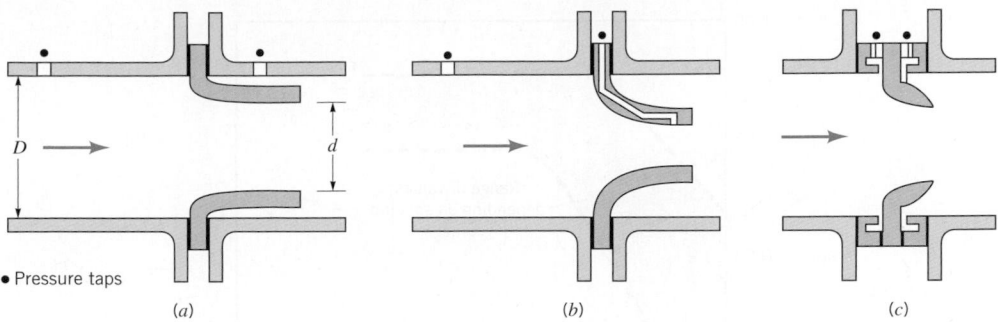

● Pressure taps

(a)   (b)   (c)

■ **FIGURE 8.42**   Typical nozzle meter construction.

The nozzle meter is more efficient than the orifice meter.

flow separation is less severe, but there still are viscous effects. These are accounted for by use of the *nozzle discharge coefficient*, C_n, where

$$Q = C_n Q_{\text{ideal}} = C_n A_n \sqrt{\frac{2(p_1 - p_2)}{\rho(1 - \beta^4)}} \qquad \textbf{(8.39)}$$

with $A_n = \pi d^2/4$. As with the orifice meter, the value of C_n is a function of the diameter ratio, $\beta = d/D$, and the Reynolds number, $\text{Re} = \rho V D/\mu$. Typical values obtained from experiments are shown in Fig. 8.43. Again, precise values of C_n depend on the specific details of the nozzle design. Accepted standards have been adopted (Ref. 24). Note that $C_n > C_o$; the nozzle meter is more efficient (less energy dissipated) than the orifice meter.

The most precise and most expensive of the three obstruction-type flow meters is the *Venturi meter* shown in Fig. 8.44 [G. B. Venturi (1746–1822)]. Although the operating principle for this device is the same as for the orifice or nozzle meters, the geometry of the Venturi meter is designed to reduce head losses to a minimum. This is accomplished by providing a relatively streamlined contraction (which eliminates separation ahead of the throat) and a very gradual expansion downstream of the throat (which eliminates separation in this decelerating portion of the device). Most of the head loss that occurs in a well-designed Venturi meter is due to friction losses along the walls rather than losses associated with separated flows and the inefficient mixing motion that accompanies such flow.

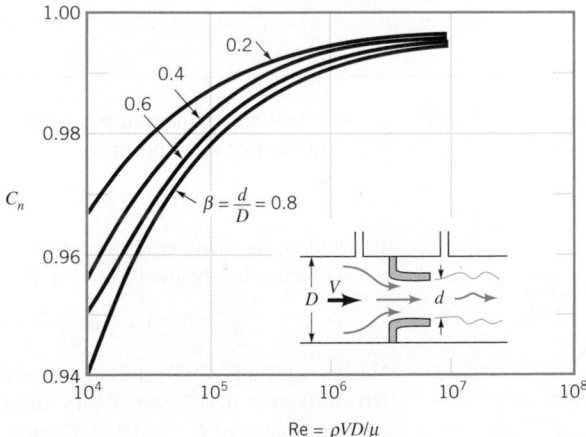

■ **FIGURE 8.43**   Nozzle meter discharge coefficient (Ref. 24).

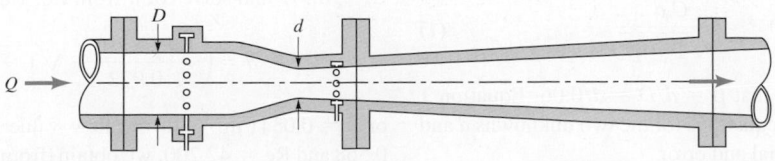

■ **FIGURE 8.44**   Typical Venturi meter construction.

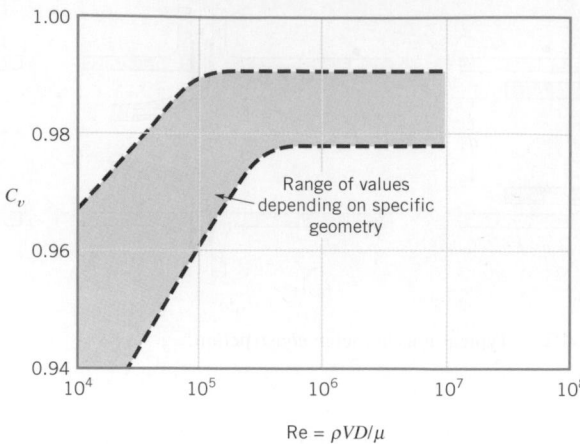

$$\text{Re} = \rho VD/\mu$$

■ FIGURE 8.45 Venturi meter discharge coefficient (Ref. 23).

Thus, the flowrate through a Venturi meter is given by

$$Q = C_v Q_{\text{ideal}} = CA_T \sqrt{\frac{2(p_1 - p_2)}{\rho(1 - \beta^4)}} \tag{8.40}$$

The Venturi discharge coefficient is a function of the specific geometry of the meter.

where $A_T = \pi d^2/4$ is the throat area. The range of values of C_v, the *Venturi discharge coefficient*, is given in Fig. 8.45. The throat-to-pipe diameter ratio ($\beta = d/D$), the Reynolds number, and the shape of the converging and diverging sections of the meter are among the parameters that affect the value of C_v.

Again, the precise values of C_n, C_o, and C_v depend on the specific geometry of the devices used. Considerable information concerning the design, use, and installation of standard flow meters can be found in various books (Refs. 23, 24, 25, 26, 31).

EXAMPLE 8.15 Nozzle Flow Meter

GIVEN Ethyl alcohol flows through a pipe of diameter $D = 60$ mm in a refinery. The pressure drop across the nozzle meter used to measure the flowrate is to be $\Delta p = 4.0$ kPa when the flowrate is $Q = 0.003$ m³/s.

FIND Determine the diameter, d, of the nozzle.

SOLUTION

From Table 1.3 the properties of ethyl alcohol are $\rho = 789$ kg/m³ and $\mu = 1.19 \times 10^{-3}$ N·s/m². Thus,

$$\text{Re} = \frac{\rho VD}{\mu} = \frac{4\rho Q}{\pi D \mu}$$

$$= \frac{4(789 \text{ kg/m}^3)(0.003 \text{ m}^3/\text{s})}{\pi(0.06 \text{ m})(1.19 \times 10^{-3} \text{ N·s/m}^2)} = 42{,}200$$

From Eq. 8.39 the flowrate through the nozzle is

$$Q = 0.003 \text{ m}^3/\text{s} = C_n \frac{\pi}{4} d^2 \sqrt{\frac{2(4 \times 10^3 \text{ N/m}^2)}{789 \text{ kg/m}^3(1 - \beta^4)}}$$

or

$$1.20 \times 10^{-3} = \frac{C_n d^2}{\sqrt{1 - \beta^4}} \tag{1}$$

where d is in meters. Note that $\beta = d/D = d/0.06$. Equation 1 and Fig. 8.43 represent two equations for the two unknowns d and C_n that must be solved by trial and error.

As a first approximation we assume that the flow is ideal, or $C_n = 1.0$, so that Eq. 1 becomes

$$d = (1.20 \times 10^{-3} \sqrt{1 - \beta^4})^{1/2} \tag{2}$$

In addition, for many cases $1 - \beta^4 \approx 1$, so that an approximate value of d can be obtained from Eq. 2 as

$$d = (1.20 \times 10^{-3})^{1/2} = 0.0346 \text{ m}$$

Hence, with an initial guess of $d = 0.0346$ m or $\beta = d/D = 0.0346/0.06 = 0.577$, we obtain from Fig. 8.43 (using Re = 42,200) a value of $C_n = 0.972$. Clearly this does not agree with our initial assumption of $C_n = 1.0$. Thus, we do not have the solution to Eq. 1 and Fig. 8.43. Next we assume $\beta = 0.577$ and $C_n = 0.972$ and solve for d from Eq. 1 to obtain

$$d = \left(\frac{1.20 \times 10^{-3}}{0.972} \sqrt{1 - 0.577^4}\right)^{1/2}$$

or $d = 0.0341$ m. With the new value of $\beta = 0.0341/0.060 = 0.568$ and Re = 42,200, we obtain (from Fig. 8.43) $C_n \approx 0.972$ in agreement with the assumed value. Thus,

$$d = 34.1 \text{ mm} \qquad \text{(Ans)}$$

COMMENTS   If numerous cases are to be investigated, it may be much easier to replace the discharge coefficient data of Fig. 8.43 by the equivalent equation, $C_n = \phi(\beta, \text{Re})$, and use a computer to iterate for the answer. Such equations are available in the literature (Ref. 24). This would be similar to using the Colebrook equation rather than the Moody chart for pipe friction problems.

By repeating the calculations, the nozzle diameters, d, needed for the same flowrate and pressure drop but with different fluids are shown in Fig. E8.15. The diameter is a function of the fluid viscosity because the nozzle coefficient, C_n, is a function of the Reynolds number (see Fig. 8.43). In addition, the diameter is a function of the density because of this Reynolds number effect and, perhaps more importantly, because the density is involved directly in the flowrate equation, Eq. 8.39. These factors all combine to produce the results shown in the figure.

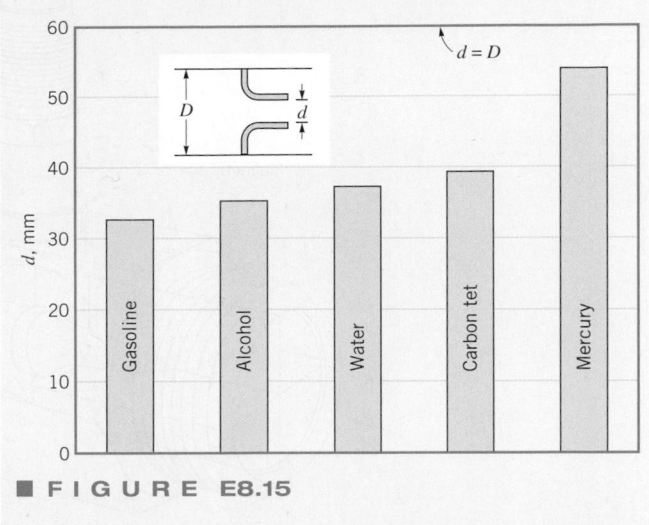

■ **FIGURE E8.15**

There are many types of flow meters.

Numerous other devices are used to measure the flowrate in pipes. Many of these devices use principles other than the high-speed/low-pressure concept of the orifice, nozzle, and Venturi meters.

A quite common, accurate, and relatively inexpensive flow meter is the *rotameter*, or variable area meter as is shown in Fig. 8.46. In this device a float is contained within a tapered, transparent metering tube that is attached vertically to the pipeline. As fluid flows through the meter (entering at the bottom), the float will rise within the tapered tube and reach an equilibrium height that is a function of the flowrate. This height corresponds to an equilibrium condition for which the net force on the float (buoyancy, float weight, fluid drag) is zero. A calibration scale in the tube provides the relationship between the float position and the flowrate.

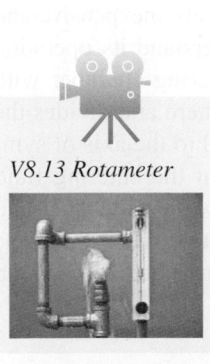

V8.13 Rotameter

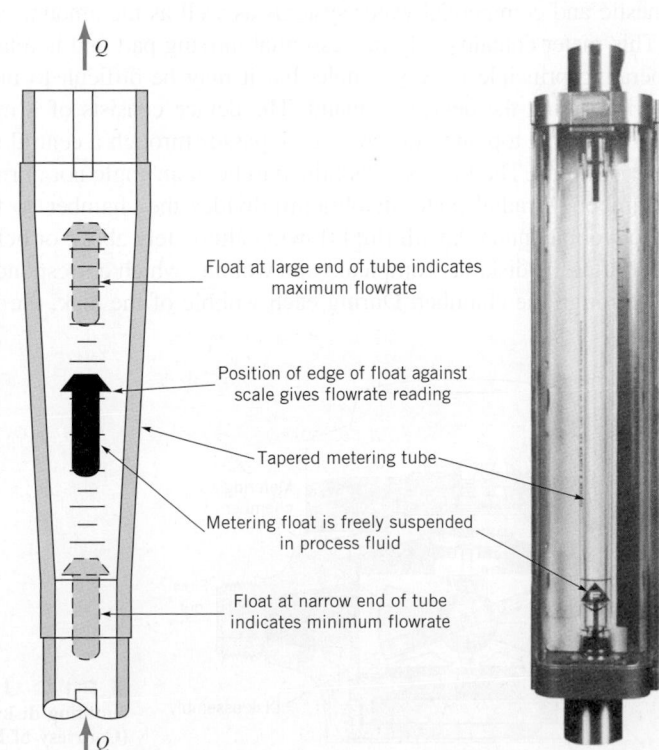

Float at large end of tube indicates maximum flowrate

Position of edge of float against scale gives flowrate reading

Tapered metering tube

Metering float is freely suspended in process fluid

Float at narrow end of tube indicates minimum flowrate

■ **FIGURE 8.46**
Rotameter-type flow meter.
(Courtesy of Fischer & Porter Co.)

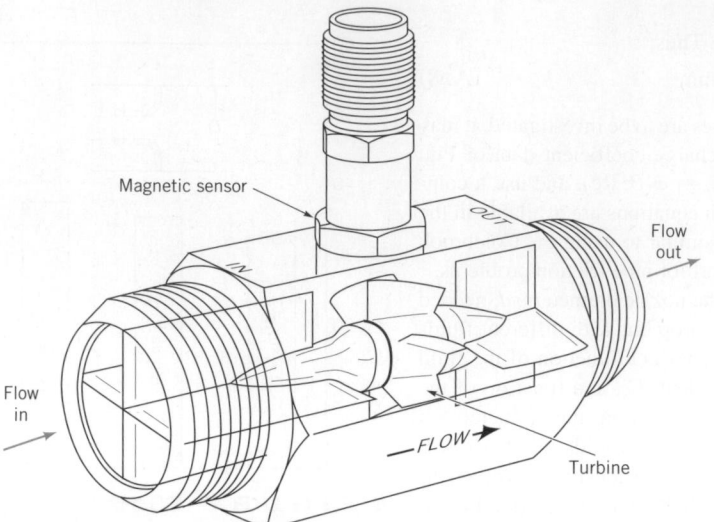

■ **FIGURE 8.47**
Turbine-type flow meter.
(Courtesy of E G & G Flow
Technology, Inc.)

Another useful pipe flowrate meter is a *turbine meter* as is shown in Fig. 8.47. A small, freely rotating propeller or turbine within the turbine meter rotates with an angular velocity that is a function of (nearly proportional to) the average fluid velocity in the pipe. This angular velocity is picked up magnetically and calibrated to provide a very accurate measure of the flowrate through the meter.

8.6.2 Volume Flow Meters

Volume flow meters measure volume rather than volume flowrate.

In many instances it is necessary to know the amount (volume or mass) of fluid that has passed through a pipe during a given time period, rather than the instantaneous flowrate. For example, we are interested in how many liters of gasoline are pumped into the tank in our car rather than the rate at which it flows into the tank. There are numerous quantity-measuring devices that provide such information.

The *nutating disk meter* shown in Fig. 8.48 is widely used to measure the net amount of water used in domestic and commercial water systems as well as the amount of gasoline delivered to your gas tank. This meter contains only one essential moving part and is relatively inexpensive and accurate. Its operating principle is very simple, but it may be difficult to understand its operation without actually inspecting the device firsthand. The device consists of a metering chamber with spherical sides and conical top and bottom. A disk passes through a central sphere and divides the chamber into two portions. The disk is constrained to be at an angle not normal to the axis of symmetry of the chamber. A radial plate (diaphragm) divides the chamber so that the entering fluid causes the disk to wobble (nutate), with fluid flowing alternately above or below the disk. The fluid exits the chamber after the disk has completed one wobble, which corresponds to a specific volume of fluid passing through the chamber. During each wobble of the disk, the pin attached to the tip

V8.14 Water meter

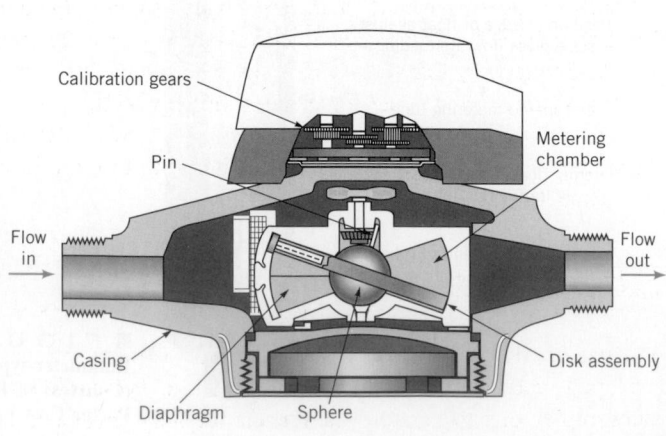

■ **FIGURE 8.48**
Nutating disk flow meter.
(Courtesy of Badger Meter,
Inc.)

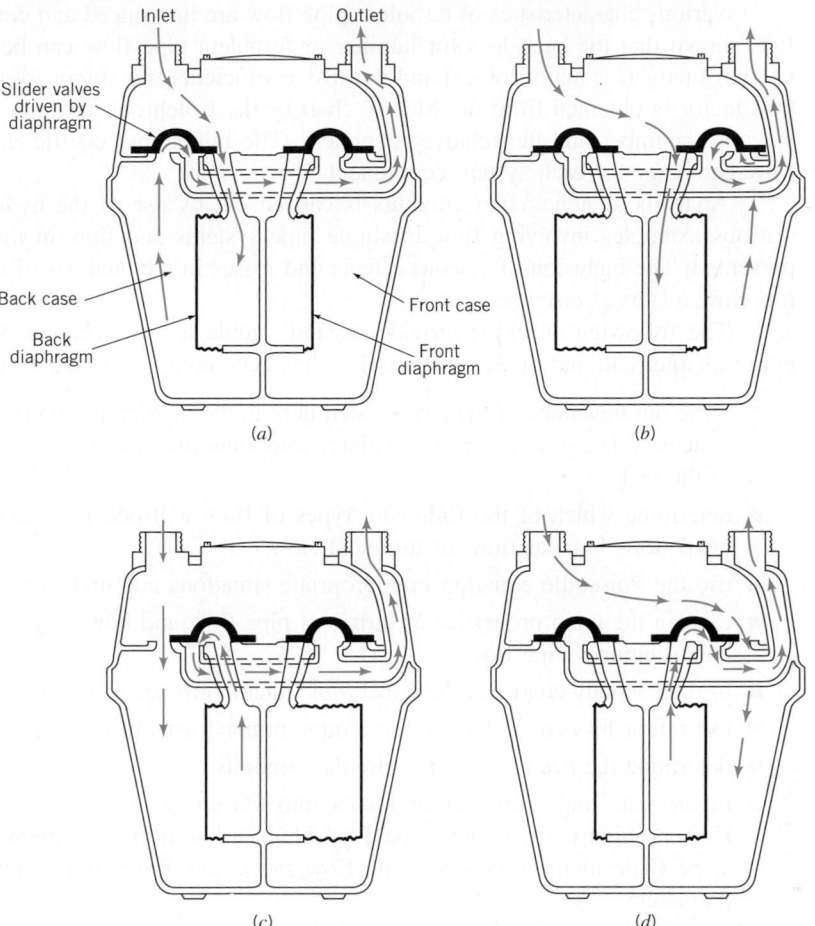

Inlet   Outlet

Slider valves
driven by
diaphragm

Back case

Front case

Back
diaphragm

Front
diaphragm

(a)   (b)

(c)   (d)

■ **F I G U R E   8.49**   **Bellows-type flow meter. (Courtesy of BTR—Rockwell Gas Products). (*a*) Back case emptying, back diaphragm filling. (*b*) Front diaphragm filling, front case emptying. (*c*) Back case filling, back diaphragm emptying. (*d*) Front diaphragm emptying, front case filling.**

The nutating disk meter has only one moving part; the bellows meter has a complex set of moving parts.

of the center sphere, normal to the disk, completes one circle. The volume of fluid that has passed through the meter can be obtained by counting the number of revolutions completed.

Another quantity-measuring device that is used for gas flow measurements is the *bellows meter* as shown in Fig. 8.49. It contains a set of bellows that alternately fill and empty as a result of the pressure of the gas and the motion of a set of inlet and outlet valves. The common household natural gas meter is of this type. For each cycle [(*a*) through (*d*)] a known volume of gas passes through the meter.

The nutating disk meter (water meter) is an example of extreme simplicity—one cleverly designed moving part. The bellows meter (gas meter), on the other hand, is relatively complex—it contains many moving, interconnected parts. This difference is dictated by the application involved. One measures a common, safe-to-handle, relatively high-pressure liquid, whereas the other measures a relatively dangerous, low-pressure gas. Each device does its intended job very well.

There are numerous devices used to measure fluid flow, only a few of which have been discussed here. The reader is encouraged to review the literature to gain familiarity with other useful, clever devices (Refs. 25, 26).

8.7   Chapter Summary and Study Guide

This chapter discussed the flow of a viscous fluid in a pipe. General characteristics of laminar, turbulent, fully developed, and entrance flows are considered. Poiseuille's equation is obtained to describe the relationship among the various parameters for fully developed laminar flow.

Various characteristics of turbulent pipe flow are introduced and contrasted to laminar flow. It is shown that the head loss for laminar or turbulent pipe flow can be written in terms of the friction factor (for major losses) and the loss coefficients (for minor losses). In general, the friction factor is obtained from the Moody chart or the Colebrook formula and is a function of the Reynolds number and the relative roughness. The minor loss coefficients are a function of the flow geometry for each system component.

Analysis of noncircular conduits is carried out by use of the hydraulic diameter concept. Various examples involving flow in single pipe systems and flow in multiple pipe systems are presented. The inclusion of viscous effects and losses in the analysis of orifice, nozzle, and Venturi flow meters is discussed.

The following checklist provides a study guide for this chapter. When your study of the entire chapter and end-of-chapter exercises has been completed you should be able to

■ write out meanings of the terms listed here in the margin and understand each of the related concepts. These terms are particularly important and are set in *italic, bold, and color* type in the text.

■ determine which of the following types of flow will occur: entrance flow, or fully developed flow; laminar flow, or turbulent flow.

■ use the Poiseuille equation in appropriate situations and understand its limitations.

■ explain the main properties of turbulent pipe flow and how they are different from or similar to laminar pipe flow.

■ use the Moody chart and the Colebrook equation to determine major losses in pipe systems.

■ use minor loss coefficients to determine minor losses in pipe systems.

■ determine the head loss in noncircular conduits.

■ incorporate major and minor losses into the energy equation to solve a variety of pipe flow problems, including Type I problems (determine the pressure drop or head loss), Type II problems (determine the flow rate), and Type III problems (determine the pipe diameter).

■ solve problems involving multiple pipe systems.

■ determine the flowrate through orifice, nozzle, and Venturi flowmeters as a function of the pressure drop across the meter.

Some of the important equations in this chapter are given below.

Entrance length	$\dfrac{\ell_e}{D} = 0.06\ \mathrm{Re}$ for laminar flow	**(8.1)**
	$\dfrac{\ell_e}{D} = 4.4\ (\mathrm{Re})^{1/6}$ for turbulent flow	**(8.2)**
Pressure drop for fully developed laminar pipe flow	$\Delta p = \dfrac{4\ell\tau_w}{D}$	**(8.5)**
Velocity profile for fully developed laminar pipe flow	$u(r) = \left(\dfrac{\Delta p D^2}{16\mu\ell}\right)\left[1 - \left(\dfrac{2r}{D}\right)^2\right] = V_c\left[1 - \left(\dfrac{2r}{D}\right)^2\right]$	**(8.7)**
Volume flowrate for fully developed laminar pipe flow	$Q = \dfrac{\pi D^4\ \Delta p}{128\mu\ell}$	**(8.9)**
Friction factor for fully developed laminar pipe flow	$f = \dfrac{64}{\mathrm{Re}}$	**(8.19)**
Pressure drop for a horizontal pipe	$\Delta p = f\dfrac{\ell}{D}\dfrac{\rho V^2}{2}$	**(8.33)**
Head loss due to major losses	$h_{L\,\mathrm{major}} = f\dfrac{\ell}{D}\dfrac{V^2}{2g}$	**(8.34)**

Colebrook formula	$\dfrac{1}{\sqrt{f}} = -2.0 \log \left(\dfrac{\varepsilon/D}{3.7} + \dfrac{2.51}{\text{Re}\sqrt{f}} \right)$	**(8.35a)**
Explicit alternative to Colebrook formula	$\dfrac{1}{\sqrt{f}} = -1.8 \log \left[\left(\dfrac{\varepsilon/D}{3.7} \right)^{1.11} + \dfrac{6.9}{\text{Re}} \right]$	**(8.35b)**
Head loss due to minor losses	$h_{L\,\text{minor}} = K_L \dfrac{V^2}{2g}$	**(8.36)**
Volume flowrate for orifice, nozzle, or Venturi meter	$Q = C_i A_i \sqrt{\dfrac{2(p_1 - p_2)}{\rho(1 - \beta^4)}}$	**(8.38, 8.39, 8.40)**

References

1. Hinze, J. O., *Turbulence*, 2nd Ed., McGraw-Hill, New York, 1975.
2. Panton, R. L., *Incompressible Flow*, 3rd Ed., Wiley, New York, 2005.
3. Schlichting, H., *Boundary Layer Theory*, 8th Ed., McGraw-Hill, New York, 2000.
4. Gleick, J., *Chaos: Making a New Science*, Viking Penguin, New York, 1987.
5. White, F. M., *Fluid Mechanics*, 6th Ed., McGraw-Hill, New York, 2008.
6. Nikuradse, J., "Stomungsgesetz in Rauhen Rohren," *VDI-Forschungsch*, No. 361, 1933; or see NACA Tech Memo 1922.
7. Moody, L. F., "Friction Factors for Pipe Flow," *Transactions of the ASME*, Vol. 66, 1944.
8. Colebrook, C. F., "Turbulent Flow in Pipes with Particular Reference to the Transition Between the Smooth and Rough Pipe Laws," *Journal of the Institute of Civil Engineers London*, Vol. 11, 1939.
9. *ASHRAE Handbook of Fundamentals*, ASHRAE, Atlanta, 1981.
10. Streeter, V. L., ed., *Handbook of Fluid Dynamics*, McGraw-Hill, New York, 1961.
11. Sovran, G., and Klomp, E. D., "Experimentally Determined Optimum Geometries for Rectilinear Diffusers with Rectangular, Conical, or Annular Cross Sections," in *Fluid Mechanics of Internal Flow*, Sovran, G., ed., Elsevier, Amsterdam, 1967.
12. Runstadler, P. W., "Diffuser Data Book," Technical Note 186, Creare, Inc., Hanover, NH, 1975.
13. Laws, E. M., and Livesey, J. L., "Flow Through Screens," *Annual Review of Fluid Mechanics*, Vol. 10, Annual Reviews, Inc., Palo Alto, CA, 1978.
14. Balje, O. E., *Turbomachines: A Guide to Design, Selection and Theory*, Wiley, New York, 1981.
15. Wallis, R. A., *Axial Flow Fans and Ducts*, Wiley, New York, 1983.
16. Karassick, I. J. et al., *Pump Handbook*, 2nd Ed., McGraw-Hill, New York, 1985.
17. White, F. M., *Viscous Fluid Flow*, 3rd Ed., McGraw-Hill, New York, 2006.
18. Olson, R. M., *Essentials of Engineering Fluid Mechanics*, 4th Ed., Harper & Row, New York, 1980.
19. Dixon, S. L., *Fluid Mechanics of Turbomachinery*, 3rd Ed., Pergamon, Oxford, 1978.
20. Finnemore, E. J., and Franzini, J. R., *Fluid Mechanics*, 10th Ed., McGraw-Hill, New York, 2002.
21. Streeter, V. L., and Wylie, E. B., *Fluid Mechanics*, 8th Ed., McGraw-Hill, New York, 1985.
22. Jeppson, R. W., *Analysis of Flow in Pipe Networks*, Ann Arbor Science Publishers, Ann Arbor, Mich., 1976.
23. Bean, H. S., ed., *Fluid Meters: Their Theory and Application*, 6th Ed., American Society of Mechanical Engineers, New York, 1971.
24. "Measurement of Fluid Flow by Means of Orifice Plates, Nozzles, and Venturi Tubes Inserted in Circular Cross Section Conduits Running Full," Int. Organ. Stand. Rep. DIS-5167, Geneva, 1976.
25. Goldstein, R. J., ed., *Flow Mechanics Measurements*, 2nd Ed., Taylor and Francis, Philadelphia, 1996.
26. Benedict, R. P., *Measurement of Temperature, Pressure, and Flow*, 2nd Ed., Wiley, New York, 1977.
27. Hydraulic Institute, *Engineering Data Book*, 1st Ed., Cleveland Hydraulic Institute, 1979.
28. Harris, C. W., *University of Washington Engineering Experimental Station Bulletin*, 48, 1928.
29. Hamilton, J. B., *University of Washington Engineering Experimental Station Bulletin*, 51, 1929.
30. Miller, D. S., *Internal Flow Systems*, 2nd Ed., BHRA, Cranfield, UK, 1990.
31. Spitzer, D. W., ed., *Flow Measurement: Practical Guides for Measurement and Control*, Instrument Society of America, Research Triangle Park, North Carolina, 1991.
32. Wilcox, D. C., *Turbulence Modeling for CFD*, DCW Industries, Inc., La Canada, California, 1994.
33. Mullin, T., ed., *The Nature of Chaos*, Oxford University Press, Oxford, 1993.
34. Haaland, S.E., "Simple and Explicit Formulas for the Friction-Factor in Turbulent Pipe Flow," *Transactions of the ASME, Journal of Fluids Engineering*, Vol. 105, 1983.

Review Problems

Go to Appendix G for a set of review problems with answers. Detailed solutions can be found in *Student Solution Manual and Study* *Guide for Fundamentals of Fluid Mechanics*, by Munson et al. (© 2009 John Wiley and Sons, Inc.).

Problems

Note: Unless otherwise indicated use the values of fluid properties found in the tables on the inside of the front cover. Problems designated with an (*) are intended to be solved with the aid of a programmable calculator or a computer. Problems designated with a (†) are "open-ended" problems and require critical thinking in that to work them one must make various assumptions and provide the necessary data. There is not a unique answer to these problems.

Answers to the even-numbered problems are listed at the end of the book. Access to the videos that accompany problems can be obtained through the book's web site, www.wiley.com/ college/munson. The lab-type problems and FlowLab problems can also be accessed on this web site.

Section 8.1 General Characteristics of Pipe Flow (Also see Lab Problem 8.130.)

8.1 Obtain a photograph/image of a piping system that would likely contain "pipe flow" and not "open channel flow." Print this photo and write a brief paragraph that describes the situation involved.

8.2 Water flows through a 15-m pipe with a 1.3-cm diameter at 20 L/min. What fraction of this pipe can be considered an entrance region?

8.3 Rainwater runoff from a parking lot flows through a 0.9-m-diameter pipe, completely filling it. Whether flow in a pipe is laminar or turbulent depends on the value of the Reynolds number. (See Video V8.2.) Would you expect the flow to be laminar or turbulent? Support your answer with appropriate calculations.

8.4 Blue and yellow streams of paint at 15 °C (each with a density of 825 kg/m³ and a viscosity 1000 times greater than water) enter a pipe with an average velocity of 1.2 m/s as shown in Fig. P8.4. Would you expect the paint to exit the pipe as green paint or separate streams of blue and yellow paint? Explain. Repeat the problem if the paint were "thinned" so that it is only 10 times more viscous than water. Assume the density remains the same.

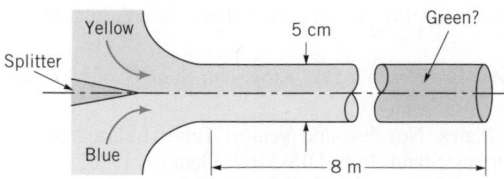

■ **FIGURE P8.4**

8.5 Air at 90 °C flows at standard atmospheric pressure in a pipe at a rate of 0.4 N/s. Determine the minimum diameter allowed if the flow is to be laminar.

8.6 To cool a given room it is necessary to supply 0.1 m³/s of air through a 20-cm-diameter pipe. Approximately how long is the entrance length in this pipe?

8.7 A long small-diameter tube is to be used as a viscometer by measuring the flowrate through the tube as a function of the pressure drop along the tube. The calibration constant, $K = Q/\Delta p$, is calculated by assuming the flow is laminar. For tubes of diameter 0.5, 1.0, and 2.0 mm, determine the maximum flowrate allowed (in cm³/s) if the fluid is **(a)** 20 °C water, or **(b)** standard air.

8.8 Carbon dioxide at 20 °C and a pressure of 550 kPa (abs) flows in a pipe at a rate of 0.04 N/s. Determine the maximum diameter allowed if the flow is to be turbulent.

8.9 The pressure distribution measured along a straight, horizontal portion of a 50-mm-diameter pipe attached to a tank is shown in the table below. Approximately how long is the entrance length? In the fully developed portion of the flow, what is the value of the wall shear stress?

x (m) (±0.01 m)	p (mm H₂O) (±5 mm)
0 (tank exit)	520
0.5	427
1.0	351
1.5	288
2.0	236
2.5	188
3.0	145
3.5	109
4.0	73
4.5	36
5.0 (pipe exit)	0

8.10 (See Fluids in the News article titled "Nanoscale flows," Section 8.1.1.) **(a)** Water flows in a tube that has a diameter of $D = 0.1$ m. Determine the Reynolds number if the average velocity is 10 diameters per second. **(b)** Repeat the calculations if the tube is a nanoscale tube with a diameter of $D = 100$ nm.

Section 8.2 Fully Developed Laminar Flow

8.11 Obtain a photograph/image of a piping system that contains both entrance region flow and fully developed flow. Print this photo and write a brief paragraph that describes the situation involved.

8.12 For fully developed laminar pipe flow in a circular pipe, the velocity profile is given by $u(r) = 2 (1 - r^2/R^2)$ in m/s, where R is the inner radius of the pipe. Assuming that the pipe diameter is 4 cm, find the maximum and average velocities in the pipe as well as the volume flow rate.

8.13 The wall shear stress in a fully developed flow portion of a 30-cm-diameter pipe carrying water is 90 N/m². Determine the pressure gradient, $\partial p/\partial x$, where x is in the flow direction, if the pipe is **(a)** horizontal, **(b)** vertical with flow up, or **(c)** vertical with flow down.

8.14 The pressure drop needed to force water through a horizontal 2.5-cm-diameter pipe is 4 kPa, for every 3.6 m length of pipe. Determine the shear stress on the pipe wall. Determine the shear stress at distances 1 and 1.3 cm away from the pipe wall.

8.15 Repeat Problem 8.14 if the pipe is on a 20° hill. Is the flow up or down the hill? Explain.

8.16 Water flows in a constant diameter pipe with the following conditions measured: At section (a) $p_a = 223$ kPa and $z_a = 15$ m; at section (b) $p_b = 205$ kPa and $z_b = 20$ m. Is the flow from (a) to (b) or from (b) to (a)? Explain.

***8.17** Some fluids behave as a non-Newtonian power-law fluid characterized by $\tau = -C(du/dr)^n$, where $n = 1, 3, 5$, and so on, and C is a constant. (If $n = 1$, the fluid is the customary Newtonian fluid.) **(a)** For flow in a round pipe of a diameter D, integrate the force balance equation (Eq. 8.3) to obtain the velocity profile

$$u(r) = \frac{-n}{(n+1)} \left(\frac{\Delta p}{2\ell C} \right)^{1/n} \left[r^{(n+1)/n} - \left(\frac{D}{2} \right)^{(n+1)/n} \right]$$

(b) Plot the dimensionless velocity profile u/V_c, where V_c is the centerline velocity (at $r = 0$), as a function of the dimensionless radial coordinate $r/(D/2)$, where D is the pipe diameter. Consider values of $n = 1, 3, 5$, and 7.

8.18 For laminar flow in a round pipe of diameter D, at what distance from the centerline is the actual velocity equal to the average velocity?

8.19 Water at 20 °C flows through a horizontal 1-mm-diameter tube to which are attached two pressure taps a distance 1 m apart. **(a)** What is the maximum pressure drop allowed if the flow is to be laminar? **(b)** Assume the manufacturing tolerance on the tube diameter is $D = 1.0 \pm 0.1$ mm. Given this uncertainty in the tube diameter, what is the maximum pressure drop allowed if it must be assured that the flow is laminar?

8.20 Glycerin at 20 °C flows upward in a vertical 75-mm-diameter pipe with a centerline velocity of 1.0 m/s. Determine the head loss and pressure drop in a 10-m length of the pipe.

8.21 Determine the magnitude of the velocity gradient at points 10, 20, and 30 mm from the pipe wall for the flow in Problem 8.20.

8.22 A large artery in a person's body can be approximated by a tube of diameter 9 mm and length 0.35 m. Also assume that blood has a viscosity of approximately 4×10^{-3} N · s/m², a specific gravity of 1.0, and that the pressure at the beginning of the artery is equivalent to 120 mm Hg. If the flow were steady (it is not) with $V = 0.2$ m/s, determine the pressure at the end of the artery if it is oriented **(a)** vertically up (flow up) or **(b)** horizontal.

8.23 At time $t = 0$ the level of water in tank A shown in Fig. P8.23 is 0.6 m above that in tank B. Plot the elevation of the water in tank A as a function of time until the free surfaces in both tanks are at the same elevation. Assume quasisteady conditions—that is, the steady pipe flow equations are assumed valid at any time, even though the flowrate does change (slowly) in time. Neglect minor losses. *Note:* Verify and use the fact that the flow is laminar.

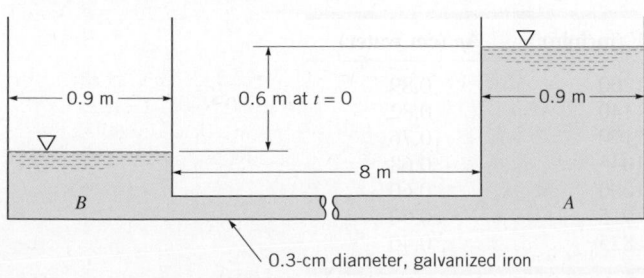

■ **FIGURE P8.23**

8.24 A fluid flows through a horizontal 0.3-cm-diameter pipe. When the Reynolds number is 1500, the head loss over a 6-m length of the pipe is 2 m. Determine the fluid velocity.

8.25 A viscous fluid flows in a 0.10-m-diameter pipe such that its velocity measured 0.012 m away from the pipe wall is 0.8 m/s. If the flow is laminar, determine the centerline velocity and the flowrate.

8.26 Oil flows through the horizontal pipe shown in Fig. P8.26 under laminar conditions. All sections are the same diameter except one. Which section of the pipe (A, B, C, D, or E) is slightly smaller in diameter than the others? Explain.

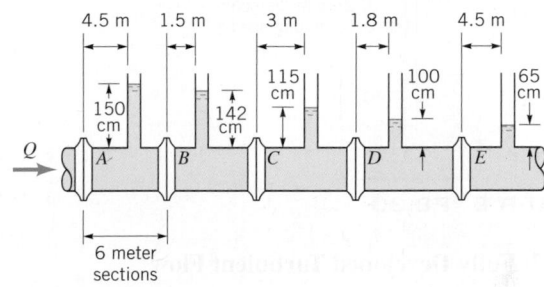

■ **FIGURE P8.26**

8.27 Asphalt at 50 °C, considered to be a Newtonian fluid with a viscosity 80,000 times that of water and a specific gravity of 1.09, flows through a pipe of diameter 5 cm. If the pressure gradient is 35 kPa/m determine the flowrate assuming the pipe is **(a)** horizontal; **(b)** vertical with flow up.

8.28 Oil of $SG = 0.87$ and a kinematic viscosity $\nu = 2.2 \times 10^{-4}$ m²/s flows through the vertical pipe shown in Fig. P8.28 at a rate of 4×10^{-4} m³/s. Determine the manometer reading, h.

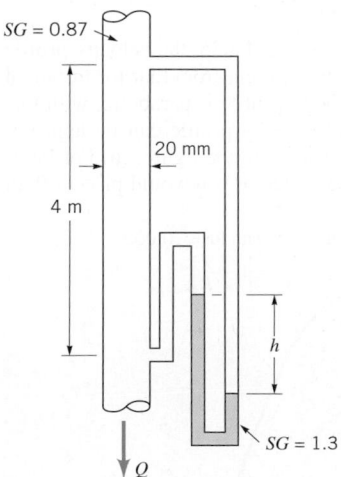

■ **FIGURE P8.28**

8.29 Determine the manometer reading, h, for Problem 8.28 if the flow is up rather than down the pipe. *Note:* The manometer reading will be reversed.

8.30 A liquid with $SG = 0.96$, $\mu = 9.2 \times 10^{-4}$ N · s/m², and vapor pressure $p_v = 1.2 \times 10^4$ N/m²(abs) is drawn into the syringe as is indicated in Fig. P8.30. What is the maximum flowrate if cavitation is not to occur in the syringe?

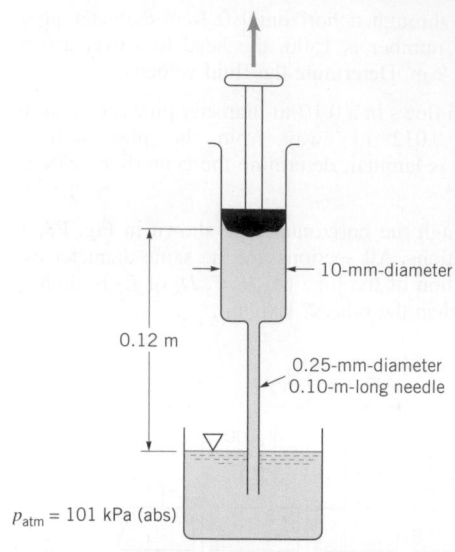

$p_{atm} = 101$ kPa (abs)

■ **F I G U R E   P8.30**

Section 8.3   Fully Developed Turbulent Flow

8.31   Obtain a photograph/image of a "turbulator." (See Fluids in the News article titled "Smaller heat exchangers" in Section 8.3.1.) Print this photo and write a brief paragraph that describes its use.

8.32   For oil ($SG = 0.86$, $\mu = 0.025$ Ns/m^2) flow of 0.3 m^3/s through a round pipe with diameter of 500 mm, determine the Reynolds number. Is the flow laminar or turbulent?

8.33   For air at a pressure of 200 kPa (abs) and temperature of 15 °C, determine the maximum laminar volume flowrate for flow through a 2.0-cm-diameter tube.

8.34   Show that the power-law approximation for the velocity profile in turbulent pipe flow (Eq. 8.31) cannot be accurate at the centerline or at the pipe wall because the velocity gradients at these locations are not correct. Explain.

8.35   As shown in **Video V8.9** and Fig. P8.35, the velocity profile for laminar flow in a pipe is quite different from that for turbulent flow. With laminar flow the velocity profile is parabolic; with turbulent flow at Re = 10,000 the velocity profile can be approximated by the power-law profile shown in the figure. **(a)** For laminar flow, determine at what radial location you would place a Pitot

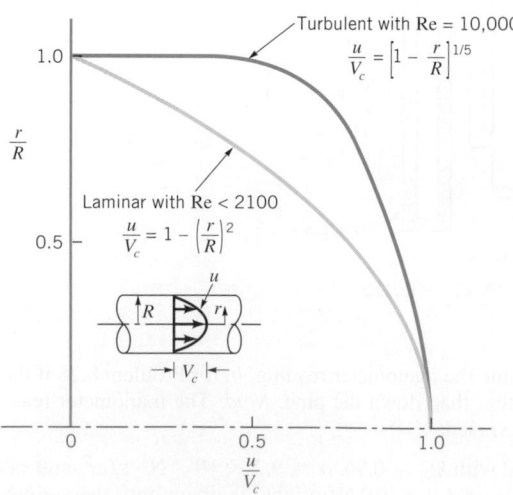

■ **F I G U R E   P8.35**

tube if it is to measure the average velocity in the pipe. **(b)** Repeat part **(a)** for turbulent flow with Re = 10,000.

8.36   The kinetic energy coefficient, α, is defined in Eq. 5.86. Show that its value for a power-law turbulent velocity profile (Eq. 8.31) is given by $\alpha = (n + 1)^3(2n + 1)^3/[4n^4(n + 3)(2n + 3)]$.

8.37   When soup is stirred in a bowl, there is considerable turbulence in the resulting motion (see **Video V8.7**). From a very simplistic standpoint, this turbulence consists of numerous intertwined swirls, each involving a characteristic diameter and velocity. As time goes by, the smaller swirls (the fine scale structure) die out relatively quickly, leaving the large swirls that continue for quite some time. Explain why this is to be expected.

8.38   Determine the thickness of the viscous sublayer in a smooth 20-cm-diameter pipe if the Reynolds number is 25,000.

8.39   Water at 15 °C flows through a 15-cm-diameter pipe with an average velocity of 4.5 m/s. Approximately what is the height of the largest roughness element allowed if this pipe is to be classified as smooth?

Section 8.4.1   Major Losses (Also see Lab Problem 8.126.)

8.40   Obtain photographs/images for round pipes of different materials. Print these photos and write a brief paragraph that describes the different pipes.

8.41   A person with no experience in fluid mechanics wants to estimate the friction factor for 2.5-cm-diameter galvanized iron pipe at a Reynolds number of 8,000. They stumble across the simple equation of $f = 64/Re$ and use this to calculate the friction factor. Explain the problem with this approach and estimate their error.

8.42   Water flows through a horizontal plastic pipe with a diameter of 0.2 m at a velocity of 10 cm/s. Determine the pressure drop per meter of pipe using the Moody chart.

8.43   For Problem 8.42, calculate the power lost to the friction per meter of pipe.

8.44   Oil ($SG = 0.9$), with a kinematic viscosity of 6.5×10^{-4} m^2/s, flows in a 8-cm-diameter pipe at 2.8×10^{-4} m^3/s. Determine the head loss per unit length of this flow.

8.45   Water flows through a 15-cm-diameter horizontal pipe at a rate of 0.06 m^3/s and a pressure drop of 1 kPa per meter of pipe. Determine the friction factor.

8.46   Water flows downward through a vertical 10-mm-diameter galvanized iron pipe with an average velocity of 5.0 m/s and exits as a free jet. There is a small hole in the pipe 4 m above the outlet. Will water leak out of the pipe through this hole, or will air enter into the pipe through the hole? Repeat the problem if the average velocity is 0.5 m/s.

8.47   Air at standard conditions flows through an 20-cm-diameter, 4.5-m-long, straight duct with the velocity versus pressure drop data indicated in the following table. Determine the average friction factor over this range of data.

V (m/min)	Δp (cm water)
1200	0.89
1140	0.82
1100	0.76
1045	0.68
1000	0.60
915	0.50
825	0.40

8.48 Water flows through a horizontal 60-mm-diameter galvanized iron pipe at a rate of 0.02 m³/s. If the pressure drop is 135 kPa per 10 m of pipe, do you think this pipe is **(a)** a new pipe, **(b)** an old pipe with a somewhat increased roughness due to aging, or **(c)** a very old pipe that is partially clogged by deposits? Justify your answer.

8.49 Water flows at a rate of 40 liters per minute in a new horizontal 2-cm-diameter galvanized iron pipe. Determine the pressure gradient, $\Delta p/\ell$, along the pipe.

8.50 Two equal length, horizontal pipes, one with a diameter of 2.5 cm, the other with a diameter of 5 cm, are made of the same material and carry the same fluid at the same flow rate. Which pipe produces the larger head loss? Justify your answer.

†8.51 A 15-cm-diameter water main in your town has become very rough due to rust and corrosion. It has been suggested that the flowrate through this pipe can be increased by inserting a smooth plastic liner into the pipe. Although the new diameter will be smaller, the pipe will be smoother. Will such a procedure produce a greater flowrate? List all assumptions and show all calculations.

8.52 Blood (assume $\mu = 2.15 \times 10^{-3}$ N · s/m², $SG = 1.0$) flows through an artery in the neck of a giraffe from its heart to its head at a rate of 7×10^{-6} m³/s. Assume the length is 3 m and the diameter is 0.5 cm. If the pressure at the beginning of the artery (outlet of the heart) is equivalent to 21 cm Hg, determine the pressure at the end of the artery when the head is **(a)** 2.4 m above the heart, or **(b)** 1.8 m below the heart. Assume steady flow. How much of this pressure difference is due to elevation effects, and how much is due to frictional effects?

8.53 A 40-m-long, 12-mm-diameter pipe with a friction factor of 0.020 is used to siphon 30 °C water from a tank as shown in Fig. P8.53. Determine the maximum value of h allowed if there is to be no cavitation within the hose. Neglect minor losses.

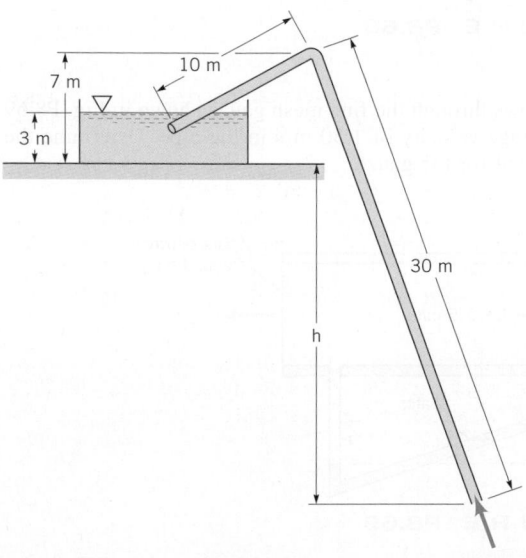

■ **F I G U R E   P8.53**

8.54 Gasoline flows in a smooth pipe of 40-mm diameter at a rate of 0.001 m³/s. If it were possible to prevent turbulence from occurring, what would be the ratio of the head loss for the actual turbulent flow compared to that if it were laminar flow?

8.55 A 1-m-diameter duct is used to carry ventilating air into a vehicular tunnel at a rate of 255 m³/min. Tests show that the pressure drop is 4 cm of water per 460 m of duct. What is the value of the friction factor for this duct and the approximate size of the equivalent roughness of the surface of the duct?

Section 8.4.2 Minor Losses (Also see Lab Problem 8.131.)

8.56 Obtain photographs/images of various pipe components that would cause minor losses in the system. Print these photos and write a brief paragraph that discusses these components.

8.57 An optional method of stating minor losses from pipe components is to express the loss in terms of equivalent length; the head loss from the component is quoted as the length of straight pipe with the same diameter that would generate an equivalent loss. Develop an equation for the equivalent length, ℓ_{eq}.

8.58 Given 90° threaded elbows used in conjunction with copper pipe (drawn tubing) of 2-cm diameter, convert the loss for a single elbow to equivalent length of copper pipe for wholly turbulent flow.

8.59 Based on Problem 8.57, develop a graph to predict equivalent length, ℓ_{eq}, as a function of pipe diameter for a 45° threaded elbow connecting copper piping (drawn tubing) for wholly turbulent flow.

8.60 A regular 90° threaded elbow is used to connect two straight portions of 10-cm-diameter galvanized iron pipe. **(a)** If the flow is assumed to be wholly turbulent, determine the equivalent length of straight pipe for this elbow. **(b)** Does a pipe fitting such as this elbow have a significant or negligible effect on the flow? Explain.

8.61 To conserve water and energy, a "flow reducer" is installed in the shower head as shown in Fig. P8.61. If the pressure at point (1) remains constant and all losses except for that in the "flow reducer" are neglected, determine the value of the loss coefficient (based on the velocity in the pipe) of the "flow reducer" if its presence is to reduce the flowrate by a factor of 2. Neglect gravity.

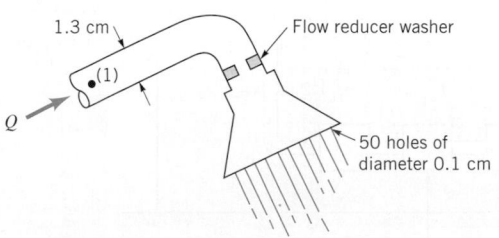

■ **F I G U R E   P8.61**

8.62 Water flows at a rate of 0.040 m³/s in a 0.12-m-diameter pipe that contains a sudden contraction to a 0.06-m-diameter pipe. Determine the pressure drop across the contraction section. How much of this pressure difference is due to losses and how much is due to kinetic energy changes?

8.63 A sign like the one shown in Fig. P8.63 is often attached to the side of a jet engine as a warning to airport workers. Based on **Video V8.10** or Figs. 8.22 and 8.25, explain why the danger areas (indicated in color) are the shape they are.

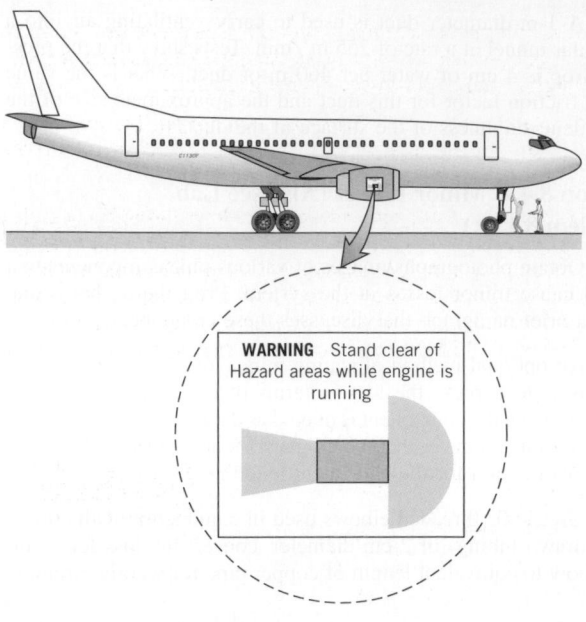

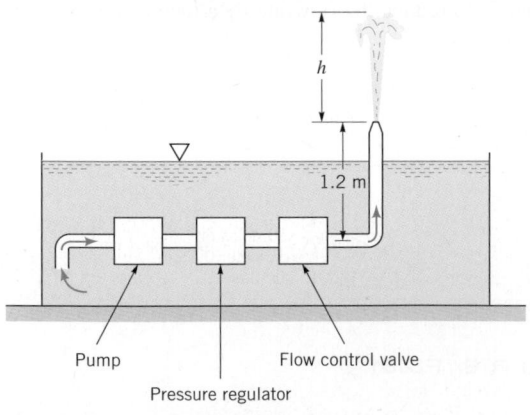

■ **F I G U R E   P8.63**

8.64 (See Fluids in the News article titled "New hi-tech fountains," Section 8.5.) The fountain shown in Fig. P8.64 is designed to provide a stream of water that rises $h = 3$ m to $h = 6$ m above the nozzle exit in a periodic fashion. To do this the water from the pool enters a pump, passes through a pressure regulator that maintains a constant pressure ahead of the flow control valve. The valve is electronically adjusted to provide the desired water height. With $h = 3$ m the loss coefficient for the valve is $K_L = 50$. Determine the valve loss coefficient needed for $h = 6$ m. All losses except for the flow control valve are negligible. The area of the pipe is 5 times the area of the exit nozzle.

■ **F I G U R E   P8.64**

*8.65** Water flows from a large open tank through a sharp-edged entrance and into a galvanized iron pipe of length 100 m and diameter 10 mm. The water exits the pipe as a free jet at a distance h below the free surface of the tank. Plot a log–log graph of the flowrate, Q, as a function of h for $0.1 \le h \le 10$ m.

8.66 Air flows through the mitered bend shown in Fig. P8.66 at a rate of 0.1 m³/s. To help straighten the flow after the bend, a set of 0.6-cm-diameter drinking straws is placed in the pipe as shown.

Estimate the extra pressure drop between points (1) and (2) caused by these straws.

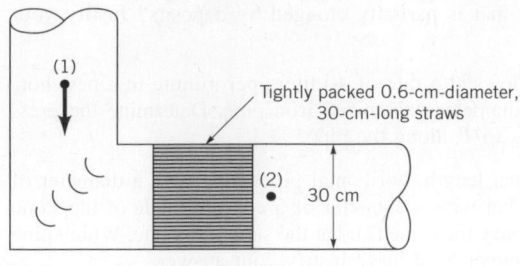

■ **F I G U R E   P8.66**

8.67 Repeat Problem 8.66 if the straws are replaced by a piece of porous foam rubber that has a loss coefficient equal to 5.4.

8.68 As shown in Fig. P8.68, water flows from one tank to another through a short pipe whose length is n times the pipe diameter. Head losses occur in the pipe and at the entrance and exit. (See Video V8.10.) Determine the maximum value of n if the major loss is to be no more than 10% of the minor loss and the friction factor is 0.02.

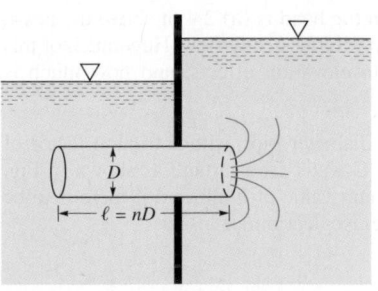

■ **F I G U R E   P8.68**

8.69 Air flows through the fine mesh gauze shown in Fig. P8.69 with an average velocity of 1.50 m/s in the pipe. Determine the loss coefficient for the gauze.

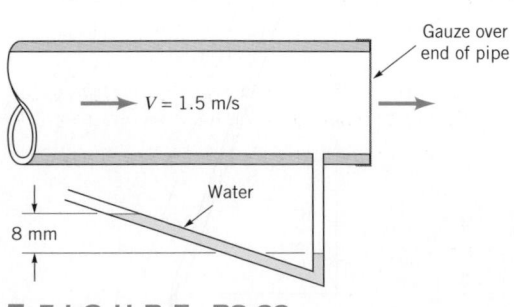

■ **F I G U R E   P8.69**

8.70 Water flows steadily through the 2-cm-diameter galvanized iron pipe system shown in Video V8.14 and Fig. P8.70 at a rate of 5.6×10^{-4} m³/s. Your boss suggests that friction losses in the straight pipe sections are negligible compared to losses in the threaded elbows and fittings of the system. Do you agree or disagree with your boss? Support your answer with appropriate calculations.

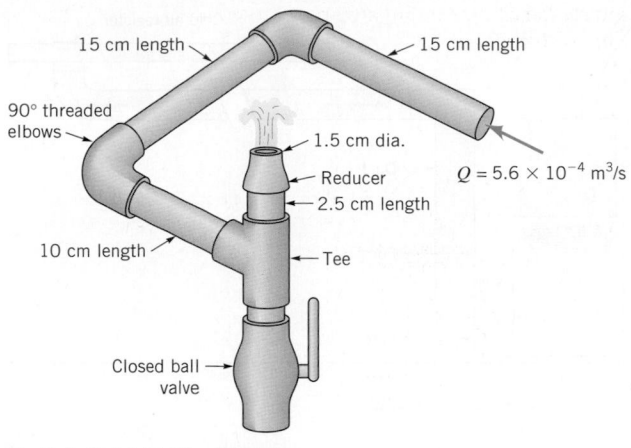

15 cm length — 15 cm length

90° threaded elbows

1.5 cm dia.

Reducer

2.5 cm length

$Q = 5.6 \times 10^{-4}$ m^3/s

10 cm length

Tee

Closed ball valve

■ **F I G U R E   P8.70**

Section 8.4.3 **Noncircular Conduits**

8.71 Obtain a photograph/image of a noncircular duct. Print this photo and write a brief paragraph that describes the situation involved.

8.72 Given two rectangular ducts with equal cross-sectional area, but different aspect ratios (width/height) of 2 and 4, which will have the greater frictional losses? Explain your answer.

8.73 Air at standard temperature and pressure flows at a rate of 0.2 m^3/s through a horizontal, galvanized iron duct that has a rectangular cross-sectional shape of 30 cm by 15 cm. Estimate the pressure drop per 60 m of duct.

8.74 Air flows through a rectangular galvanized iron duct of size 0.30 m by 0.15 m at a rate of 0.068 m^3/s. Determine the head loss in 12 m of this duct.

8.75 Air at standard conditions flows through a horizontal 0.3 m by 0.5 m rectangular wooden duct at a rate of 140 m^3/min. Determine the head loss, pressure drop, and power supplied by the fan to overcome the flow resistance in 150 m of the duct.

Section 8.5.1 **Single Pipes—Determine Pressure Drop**

8.76 Assume a car's exhaust system can be approximated as 4 m of 0.05-m-diameter cast-iron pipe with the equivalent of six 90° flanged elbows and a muffler. (See **Video V8.12.**) The muffler acts as a resistor with a loss coefficient of $K_L = 8.5$. Determine the pressure at the beginning of the exhaust system if the flowrate is 3.0×10^{-3} m^3/s, the temperature is 120 °C, and the exhaust has the same properties as air.

8.77 The pressure at section (2) shown in Fig. P8.77 is not to fall below 415 kPa when the flowrate from the tank varies from 0 to

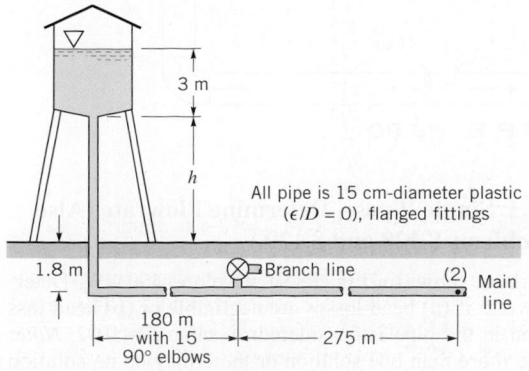

3 m

h

1.8 m

All pipe is 15 cm-diameter plastic ($\epsilon/D = 0$), flanged fittings

Branch line

(2) Main line

180 m with 15 90° elbows

275 m

■ **F I G U R E   P8.77**

0.03 m^3/s and the branch line is shut off. Determine the minimum height, h, of the water tank under the assumption that **(a)** minor losses are negligible, **(b)** minor losses are not negligible.

8.78 Repeat Problem 8.77 with the assumption that the branch line is open so that half of the flow from the tank goes into the branch, and half continues in the main line.

8.79 The exhaust from your car's engine flows through a complex pipe system as shown in Fig. P8.79 and **Video V8.12.** Assume that the pressure drop through this system is Δp_1 when the engine is idling at 1000 rpm at a stop sign. Estimate the pressure drop (in terms of Δp_1) with the engine at 3000 rpm when you are driving on the highway. List all the assumptions that you made to arrive at your answer.

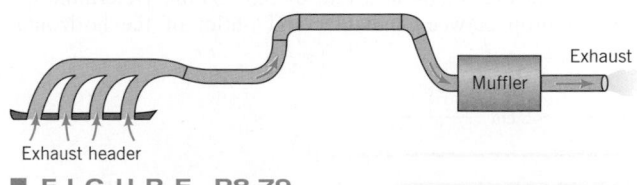

Exhaust

Muffler

Exhaust header

■ **F I G U R E   P8.79**

8.80 According to fire regulations in a town, the pressure drop in a commercial steel horizontal pipe must not exceed 0.2 kPa per m of pipe for flowrates up to 2 kL/min. If the water temperature is above 10° C, can a 15-cm-diameter pipe be used?

8.81 As shown in **Video V8.14** and Fig. P8.81, water "bubbles up" 8 cm above the exit of the vertical pipe attached to three horizontal pipe segments. The total length of the 2-cm-diameter galvanized iron pipe between point (1) and the exit is 55 cm. Determine the pressure needed at point (1) to produce this flow.

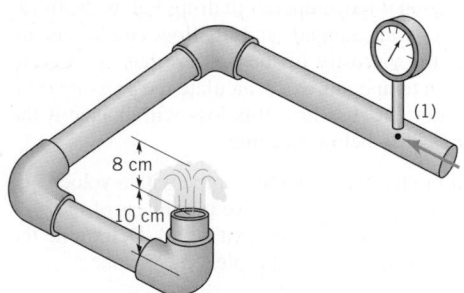

(1)

8 cm

10 cm

■ **F I G U R E   P8.81**

8.82 Water at 10 °C is pumped from a lake as shown in Fig. P8.82. If the flowrate is 0.011 m^3/s, what is the maximum length inlet pipe, ℓ, that can be used without cavitation occurring?

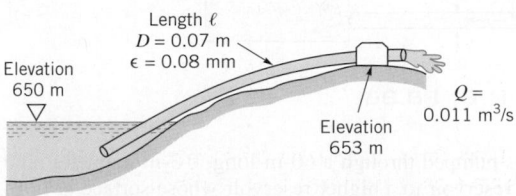

Length ℓ
$D = 0.07$ m
$\epsilon = 0.08$ mm

Elevation 650 m

Elevation 653 m

$Q = 0.011$ m^3/s

■ **F I G U R E   P8.82**

8.83 Water flows through the pipe system shown in Fig. P8.83 at a rate of 8.5×10^{-3} m^3/s. The pipe diameter is 5 cm, and its roughness is 0.005 cm. The loss coefficient for each of the five filters is 6.0, and all other minor losses are negligible. Determine the power added to the water by the pump if the pressure immediately before the pump is to be the same as that immediately after the last filter. The length of the pipe between these two locations is 25 m.

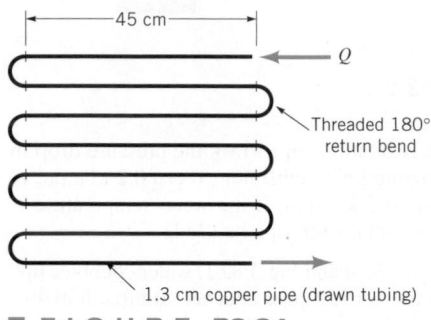

■ **F I G U R E   P8.83**

8.84 Water at 5 °C flows through the coils of the heat exchanger as shown in Fig. P8.84 at a rate of 3.5 L/min. Determine the pressure drop between the inlet and outlet of the horizontal device.

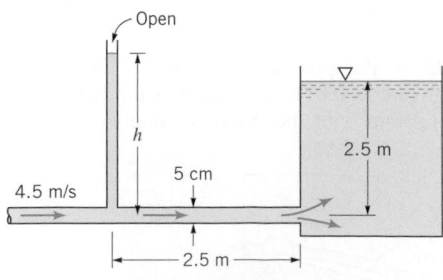

■ **F I G U R E   P8.84**

8.85 For the flow in Problem 8.84, ethylene glycol is added to the water for freeze protection if the temperature drops below the freezing point. The density is unchanged, and all flow conditions are the same except that the viscosity of the mixture has changed to 0.01 Ns/m^2 at the given temperature. Recalculate the pressure drop between inlet and outlet. Discuss how this loss will change if the fluid temperature does drop below freezing.

8.86 Water flows through a 5-cm-diameter pipe with a velocity of 4.5 m/s as shown in Fig. P8.86. The relative roughness of the pipe is 0.004, and the loss coefficient for the exit is 1.0. Determine the height, h, to which the water rises in the piezometer tube.

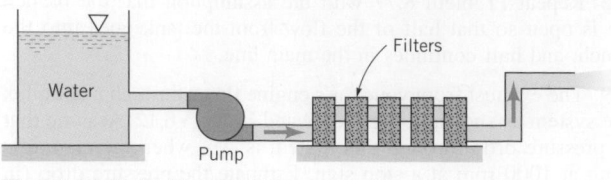

■ **F I G U R E   P8.86**

8.87 Water is pumped through a 60-m-long, 0.3-m-diameter pipe from a lower reservoir to a higher reservoir whose surface is 10 m above the lower one. The sum of the minor loss coefficients for the system is $K_L = 14.5$. When the pump adds 40 kW to the water the flowrate is $0.20 \text{ m}^3/\text{s}$. Determine the pipe roughness.

†**8.88** Estimate the pressure drop associated with the air flow from the cold air register in your room to the furnace (see Figure P8.88). List all assumptions and show all calculations.

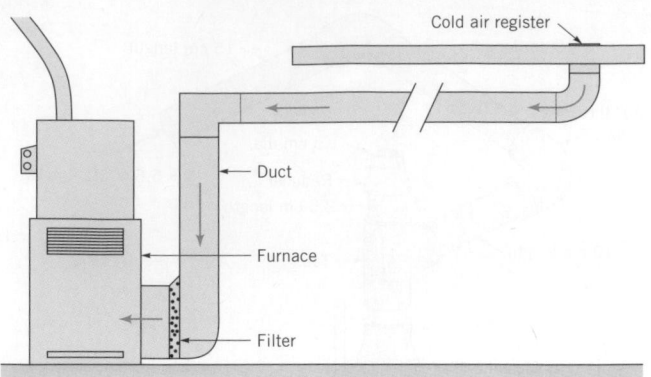

■ **F I G U R E   P8.88**

8.89 As shown in Fig. P8.89, a standard household water meter is incorporated into a lawn irrigation system to measure the volume of water applied to the lawn. Note that these meters measure volume, not volume flowrate. (See **Video V8.14.**) With an upstream pressure of $p_1 = 345 \text{ kPa}$ the meter registered that 3.4 m^3 of water was delivered to the lawn during an "on" cycle. Estimate the upstream pressure, p_1, needed if it is desired to have 4 m^3 delivered during an "on" cycle. List any assumptions needed to arrive at your answer.

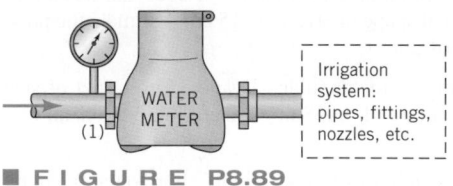

■ **F I G U R E   P8.89**

8.90 A fan is to produce a constant air speed of 40 m/s throughout the pipe loop shown in Fig. P8.90. The 3-m-diameter pipes are smooth, and each of the four 90° elbows has a loss coefficient of 0.30. Determine the power that the fan adds to the air.

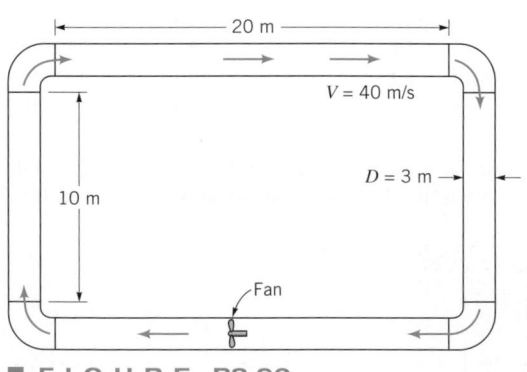

■ **F I G U R E   P8.90**

Section 8.5.1 **Single Pipes—Determine Flowrate (Also see Lab Problems 8.128 and 8.129.)**

8.91 The turbine shown in Fig. P8.91 develops 400 kW. Determine the flowrate if **(a)** head losses are negligible or **(b)** head loss due to friction in the pipe is considered. Assume $f = 0.02$. *Note:* There may be more than one solution or there may be no solution to this problem.

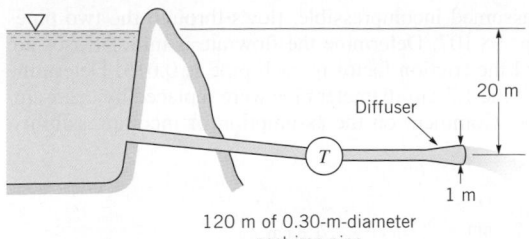

■ FIGURE P8.91

*8.92 In some locations with very "hard" water, a scale can build up on the walls of pipes to such an extent that not only does the roughness increases with time, but the diameter significantly decreases with time. Consider a case for which the roughness and diameter vary as $\varepsilon = 0.02 + 0.01t$ mm, $D = 50 \ (1 - 0.02t)$ mm, where t is in years. Plot the flowrate as a function of time for $t = 0$ to $t = 10$ years if the pressure drop per 12 m of horizontal pipe remains constant at $\Delta p = 1.3$ kPa.

8.93 Water flows from the nozzle attached to the spray tank shown in Fig. P8.93. Determine the flowrate if the loss coefficient for the nozzle (based on upstream conditions) is 0.75 and the friction factor for the rough hose is 0.11.

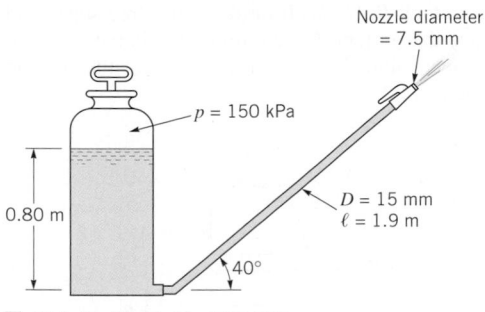

■ FIGURE P8.93

8.94 When the pump shown in Fig. P8.94 adds 150 Watts to the flowing water, the pressures indicated by the two gages are equal. Determine the flowrate.

Length of pipe between gages = 18 m
Pipe diameter = 0.03 m
Pipe friction factor = 0.03
Filter loss coefficient = 12

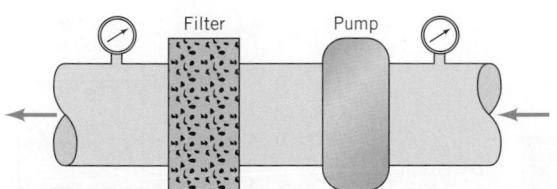

■ FIGURE P8.94

8.95 Water is pumped between two large open tanks as shown in Fig. P8.95. If the pump adds 50 kW of power to the fluid, what is

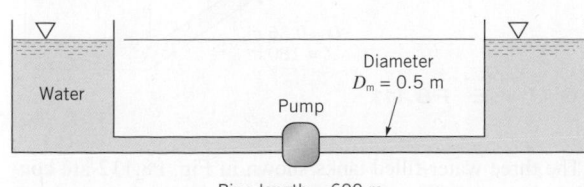

■ FIGURE P8.95

the flowrate passing between the tanks? Assume the friction factor to be equal to 0.02 and minor losses to be negligible.

†8.96 Gasoline is unloaded from the tanker truck shown in Fig. P8.96 through a 10-cm-diameter rough-surfaced hose. This is a "gravity dump" with no pump to enhance the flowrate. It is claimed that the 33 kL capacity truck can be unloaded in 28 minutes. Do you agree with this claim? Support your answer with appropriate calculations.

■ FIGURE P8.96

8.97 The pump shown in Fig. P8.97 delivers a head of 80 m to the water. Determine the power that the pump adds to the water. The difference in elevation of the two ponds is 60 m.

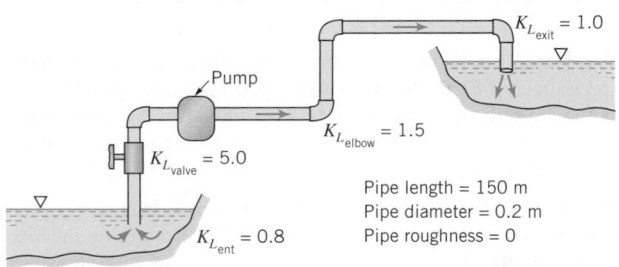

■ FIGURE P8.97

8.98 Water flows through two sections of the vertical pipe shown in Fig. P8.98. The bellows connection cannot support any force in the vertical direction. The 0.12-m-diameter pipe weighs 3 N/m, and the friction factor is assumed to be 0.02. At what velocity will the force, F, required to hold the pipe be zero?

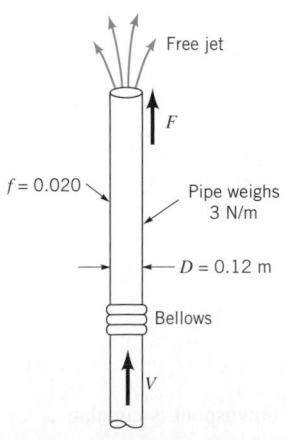

■ FIGURE P8.98

8.99 Water is circulated from a large tank, through a filter, and back to the tank as shown in Fig. P8.99. The power added to the water by the pump is 270 N · m/s. Determine the flowrate through the filter.

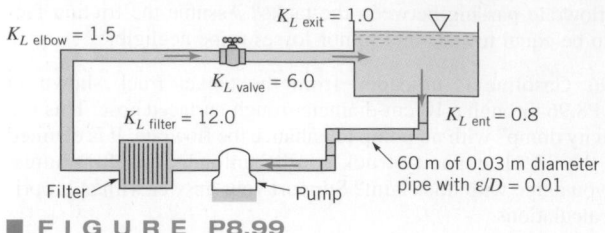

$K_{L\ elbow} = 1.5$

$K_{L\ exit} = 1.0$

$K_{L\ valve} = 6.0$

$K_{L\ filter} = 12.0$

$K_{L\ ent} = 0.8$

Filter

Pump

60 m of 0.03 m diameter pipe with $\varepsilon/D = 0.01$

■ **FIGURE P8.99**

Section 8.5.1 Single Pipes—Determine Diameter

8.100 A certain process requires 6.5×10^{-2} m³/s of water to be delivered at a pressure of 205 kPa. This water comes from a large-diameter supply main in which the pressure remains at 415 kPa. If the galvanized iron pipe connecting the two locations is 60 m long and contains six threaded 90° elbows, determine the pipe diameter. Elevation differences are negligible.

8.101 Water is pumped between two large open reservoirs through 1.5 km of smooth pipe. The water surfaces in the two reservoirs are at the same elevation. When the pump adds 20 kW to the water the flowrate is 1 m³/s. If minor losses are negligible, determine the pipe diameter.

8.102 Determine the diameter of a steel pipe that is to carry 8 kL/min of gasoline with a pressure drop of 12 kPa per 30 m of horizontal pipe.

8.103 Water is to be moved from a large, closed tank in which the air pressure is 140 kPa into a large, open tank through 610 m of smooth pipe at the rate of 0.01 m³/s. The fluid level in the open tank is 45 m below that in the closed tank. Determine the required diameter of the pipe. Neglect minor losses.

8.104 Rainwater flows through the galvanized iron downspout shown in Fig. P8.104 at a rate of 0.006 m³/s. Determine the size of the downspout cross section if it is a rectangle with an aspect ratio of 1.7 to 1 and it is completely filled with water. Neglect the velocity of the water in the gutter at the free surface and the head loss associated with the elbow.

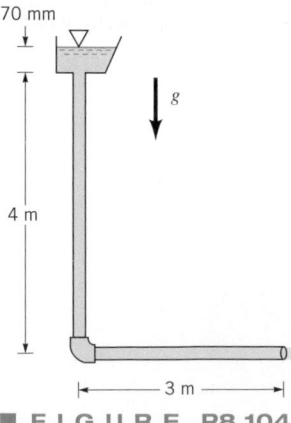

70 mm

g

4 m

3 m

■ **FIGURE P8.104**

*8.105 Repeat Problem 8.104 if the downspout is circular.

Section 8.5.2 Multiple Pipe Systems

8.106 Obtain a photograph/image of a multiple pipe system with series of parallel flows. Print this photo and write a brief paragraph that describes the situation involved.

8.107 Air, assumed incompressible, flows through the two pipes shown in Fig. P8.107. Determine the flowrate if minor losses are neglected and the friction factor in each pipe is 0.015. Determine the flowrate if the 1.3-cm-diameter pipe were replaced by a 2.5-cm-diameter pipe. Comment on the assumption of incompressibility.

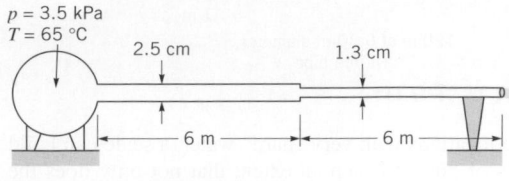

$p = 3.5$ kPa
$T = 65\ °C$

2.5 cm

1.3 cm

6 m

6 m

■ **FIGURE P8.107**

*8.108 Repeat Problem 8.107 if the pipes are galvanized iron and the friction factors are not known a priori.

†8.109 Estimate the power that the human heart must impart to the blood to pump it through the two carotid arteries from the heart to the brain. List all assumptions and show all calculations.

8.110 The flowrate between tank A and tank B shown in Fig. P8.110 is to be increased by 30% (i.e., from Q to $1.30Q$) by the addition of a second pipe (indicated by the dotted lines) running from node C to tank B. If the elevation of the free surface in tank A is 8 m above that in tank B, determine the diameter, D, of this new pipe. Neglect minor losses and assume that the friction factor for each pipe is 0.02.

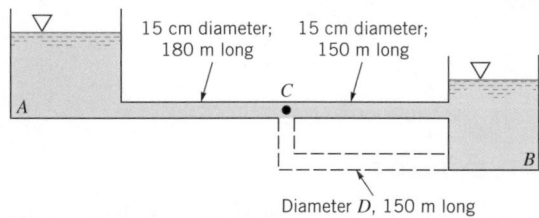

15 cm diameter; 180 m long

15 cm diameter; 150 m long

A

C

B

Diameter D, 150 m long

■ **FIGURE P8.110**

8.111 The three tanks shown in Fig. P8.111 are connected by pipes with friction factors of 0.03 for each pipe. Determine the water velocity in each pipe. Neglect minor losses.

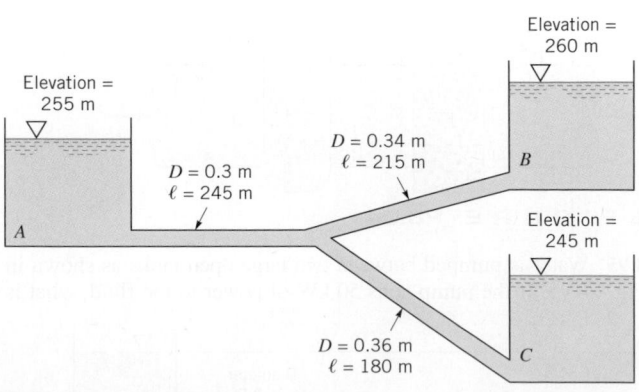

Elevation = 255 m

Elevation = 260 m

$D = 0.3$ m
$\ell = 245$ m

$D = 0.34$ m
$\ell = 215$ m

B

A

Elevation = 245 m

$D = 0.36$ m
$\ell = 180$ m

C

■ **FIGURE P8.111**

8.112 The three water-filled tanks shown in Fig. P8.112 are connected by pipes as indicated. If minor losses are neglected, determine the flowrate in each pipe.

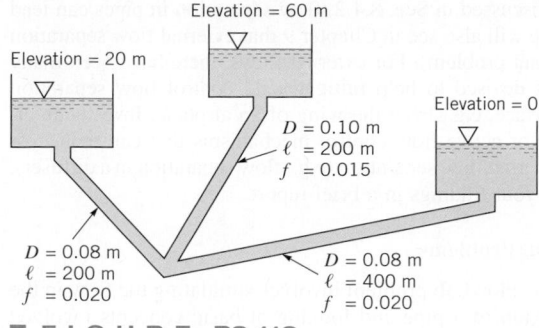

■ FIGURE P8.112

8.113 (See Fluids in the News article titled "Deepwater pipeline," Section 8.5.2.) Five oil fields, each producing an output of Q barrels per day, are connected to the 70-cm-diameter "main line pipe" ($A–B–C$) by 40-cm-diameter "lateral pipes" as shown in Fig. P8.113. The friction factor is the same for each of the pipes and elevation effects are negligible. (a) For section $A–B$ determine the ratio of the pressure drop per kilometer in the main line pipe to that in the lateral pipes. (b) Repeat the calculations for section $B–C$.

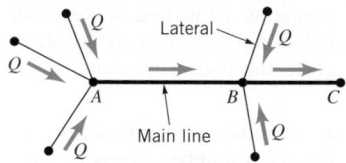

■ FIGURE P8.113

†8.114 As shown in Fig. P8.114, cold water ($T = 10$ °C) flows from the water meter to either the shower or the hot water heater. In the hot water heater it is heated to a temperature of 65 °C. Thus, with equal amounts of hot and cold water, the shower is at a comfortable 40 °C. However, when the dishwasher is turned on, the shower water becomes too cold. Indicate how you would predict this new shower temperature (assume the shower faucet is not adjusted). State any assumptions needed in your analysis.

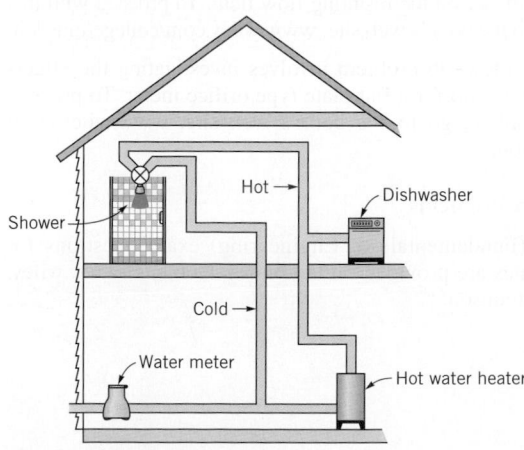

■ FIGURE P8.114

Section 8.6 Pipe Flowrate Measurement (Also see Lab Problem 8.127.)

8.115 Obtain a photograph/image of a flowrate measurement device. Print this photo and write a brief paragraph that describes the measurement range of the device.

8.116 A 5-cm-diameter orifice plate is inserted in a 8-cm-diameter pipe. If the water flowrate through the pipe is 0.03 m³/s, determine the pressure difference indicated by a manometer attached to the flow meter.

8.117 Air to ventilate an underground mine flows through a large 2-m-diameter pipe. A crude flowrate meter is constructed by placing a sheet metal "washer" between two sections of the pipe. Estimate the flowrate if the hole in the sheet metal has a diameter of 1.6 m and the pressure difference across the sheet metal is 8.0 mm of water.

8.118 Water flows through a 40-mm-diameter nozzle meter in a 75-mm-diameter pipe at a rate of 0.015 m³/s. Determine the pressure difference across the nozzle if the temperature is (a) 10 °C, or (b) 80 °C.

8.119 Air at 90 °C and 415 kPa (abs) flows in a 10-cm-diameter pipe at a rate of 2.3 N/s. Determine the pressure at the 5-cm-diameter throat of a Venturi meter placed in the pipe.

8.120 A 6.5-cm-diameter flow nozzle is installed in a 10-cm-diameter pipe that carries water at 70 °C. If the air–water manometer used to measure the pressure difference across the meter indicates a reading of 1 m, determine the flowrate.

8.121 A 0.064-m-diameter nozzle meter is installed in a 0.097 m-diameter pipe that carries water at 60 °C. If the inverted air–water U-tube manometer used to measure the pressure difference across the meter indicates a reading of 1 m, determine the flowrate.

8.122 Water flows through the Venturi meter shown in Fig. P8.122. The specific gravity of the manometer fluid is 1.52. Determine the flowrate.

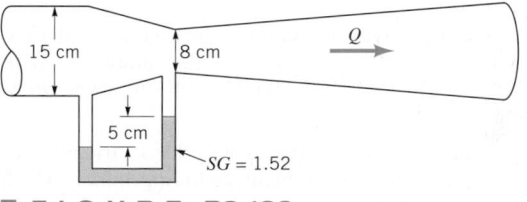

■ FIGURE P8.122

8.123 Water flows through the orifice meter shown in Fig. P8.123 at a rate of 3×10^{-3} m³/s. If $d = 0.03$ m, determine the value of h.

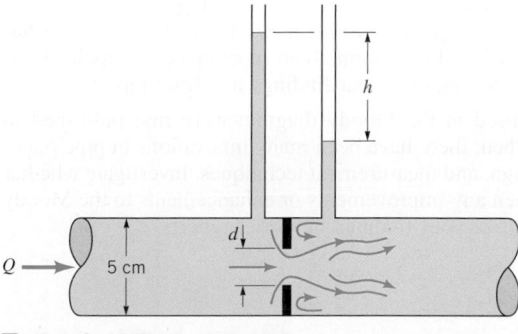

■ FIGURE P8.123

8.124 Water flows through the orifice meter shown in Fig. P8.123 such that $h = 0.5$ m with $d = 4$ cm. Determine the flowrate.

8.125 The scale reading on the rotameter shown in Fig. P8.125 and **Video V8.14** (also see Fig. 8.46) is directly proportional to the volumetric flowrate. With a scale reading of 2.6 the water bubbles up approximately 8 cm. How far will it bubble up if the scale reading is 5.0?

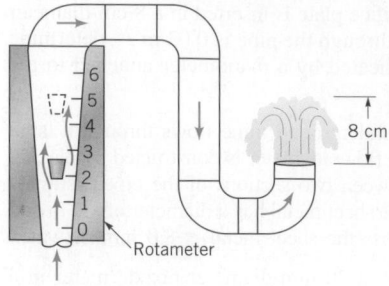

■ **F I G U R E   P8.125**

■ **Lab Problems**

8.126 This problem involves the determination of the friction factor in a pipe for laminar and transitional flow conditions. To proceed with this problem, go to Appendix H which is located on the book's web site, www.wiley.com/college/munson.

8.127 This problem involves the calibration of an orifice meter and a Venturi meter. To proceed with this problem, go to Appendix H which is located on the book's web site, www.wiley.com/college/munson.

8.128 This problem involves the flow of water from a tank and through a pipe system. To proceed with this problem, go to Appendix H which is located on the book's web site, www.wiley.com/college/munson.

8.129 This problem involves the flow of water pumped from a tank and through a pipe system. To proceed with this problem, go to Appendix H which is located on the book's web site, www.wiley.com/college/munson.

8.130 This problem involves the pressure distribution in the entrance region of a pipe. To proceed with this problem, go to Appendix H which is located on the book's web site, www.wiley.com/college/munson.

8.131 This problem involves the power loss due to friction in a coiled pipe. To proceed with this problem, go to Appendix H which is located on the book's web site, www.wiley.com/college/munson.

■ **Life Long Learning Problems**

8.132 The field of bioengineering has undergone significant growth in recent years. Some universities have undergraduate and graduate programs in this field. Bioengineering applies engineering principles to help solve problems in the medical field for human health. Obtain information about bioengineering applications in blood flow. Summarize your findings in a brief report.

8.133 Data used in the Moody diagram were first published in 1944. Since then, there have been many innovations in pipe material, pipe design, and measurement techniques. Investigate whether there have been any improvements or enhancements to the Moody chart. Summarize your findings in a brief report.

8.134 As discussed in Sec. 8.4.2, flow separation in pipes can lead to losses (we will also see in Chapter 9 that external flow separation is a significant problem). For external flows, there have been many mechanisms devised to help mitigate and control flow separation from the surface, e.g., from the wing of an airplane. Investigate either passive or active flow control mechanisms that can reduce or eliminate internal flow separation (e.g., flow separation in a diffuser). Summarize your findings in a brief report.

■ **FlowLab Problems**

***8.135** This FlowLab problem involves simulating the flow in the entrance region of a pipe and looking at basic concepts involved with the flow regime. To proceed with this problem, go to the book's web site, www.wiley.com/college/munson.

***8.136** This FlowLab problem involves investigation of the centerline pressure distribution along a pipe. To proceed with this problem, go to the book's web site, www.wiley.com/college/munson.

***8.137** This FlowLab problem involves conducting a parametric study to see how Reynolds number affects the entrance length of a pipe. To proceed with this problem, go to the book's web site, www.wiley.com/college/munson.

***8.138** This FlowLab problem involves investigation of pressure drop in the entrance region of a pipe as a function of Reynolds number as well as comparing simulation results to analytic values. To proceed with this problem, go to the book's web site, www.wiley.com/college/munson.

***8.139** This FlowLab problem involves the simulation of fully developed pipe flow and how the Reynolds number affects the wall friction. To proceed with this problem, go to the book's web site, www.wiley.com/college/munson.

***8.140** This FlowLab problem involves conducting a parametric study on the effects of a sudden pipe expansion on the overall pressure drop in a pipe. To proceed with this problem, go to the book's web site, www.wiley.com/college/munson.

***8.141** This FlowLab problem involves investigation of effects of the pipe expansion ratio on flow separation. To proceed with this problem, go to the book's web site, www.wiley.com/college/munson.

***8.142** This FlowLab problem involves investigation of geometric effects of a diffuser on the resulting flow field. To proceed with this problem, go to the book's web site, www.wiley.com/college/munson.

***8.143** This FlowLab problem involves investigating the effects of the diameter ratio for a flat plate type orifice meter. To proceed with this problem, go to the book's web site, www.wiley.com/college/munson.

■ **FE Exam Problems**

Sample FE (Fundamentals of Engineering) exam questions for fluid mechanics are provided on the book's web site, www.wiley.com/college/munson.

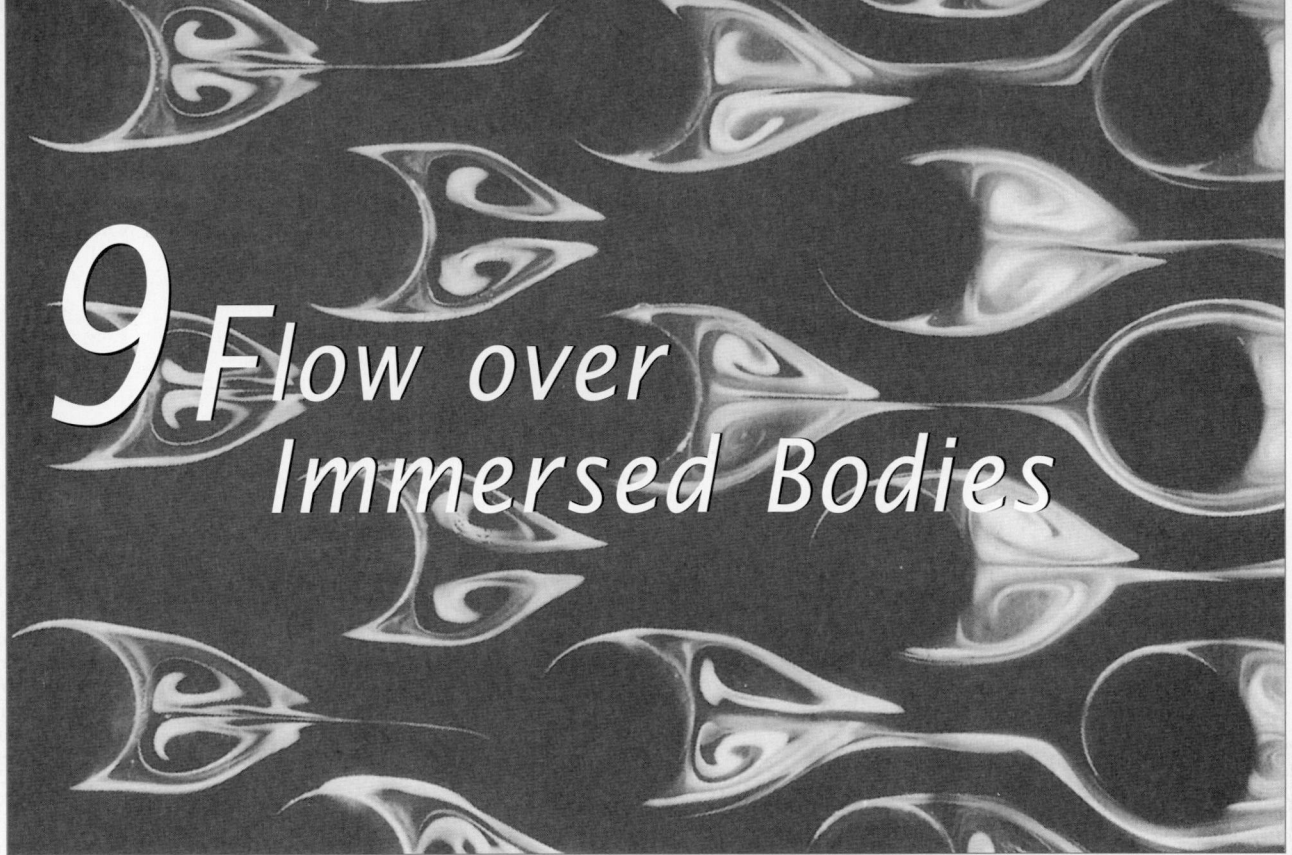

9 Flow over Immersed Bodies

Learning Objectives

After completing this chapter, you should be able to:

- identify and discuss the features of external flow.
- explain the fundamental characteristics of a boundary layer, including laminar, transitional, and turbulent regimes.
- calculate boundary layer paremeters for flow past a flat plate.
- provide a description of boundary layer separation.
- calculate the lift and drag forces for various objects.

In this chapter we consider various aspects of the flow over bodies that are immersed in a fluid. Examples include the flow of air around airplanes, automobiles, and falling snowflakes, or the flow of water around submarines and fish. In these situations the object is completely surrounded by the fluid and the flows are termed *external flows.*

External flows involving air are often termed aerodynamics in response to the important external flows produced when an object such as an airplane flies through the atmosphere. Although this field of external flows is extremely important, there are many other examples that are of equal importance. The fluid force (lift and drag) on surface vehicles (cars, trucks, bicycles) has become a very important topic. By correctly designing cars and trucks, it has become possible to greatly decrease the fuel consumption and improve the handling characteristics of the vehicle. Similar efforts have resulted in improved ships, whether they are surface vessels (surrounded by two fluids, air and water) or submersible vessels (surrounded completely by water).

Other applications of external flows involve objects that are not completely surrounded by fluid, although they are placed in some external-type flow. For example, the proper design of a

Many practical situations involve flow past objects.

461

(a)

(b)

■ **FIGURE 9.1** (a) Flow past a full-sized streamlined vehicle in the GM aerodynamics laboratory wind tunnel, an 18-ft (5.5-m) by 34-ft (10-m) test section facility driven by a 4000-hp (3-MW), 43-ft (13-m) diameter fan. (Photograph courtesy of General Motors Corporation.) (b) Predicted streamlines for flow past a Formula 1 race car as obtained by using computational fluid dynamics techniques. (Courtesy of Ansys, Inc.)

building (whether it is your house or a tall skyscraper) must include consideration of the various wind effects involved.

As with other areas of fluid mechanics, various approaches (theoretical, numerical and experimental) are used to obtain information on the fluid forces developed by external flows. Theoretical (i.e., analytical) techniques can provide some of the needed information about such flows. However, because of the complexities of the governing equations and the complexities of the geometry of the objects involved, the amount of information obtained from purely theoretical methods is limited.

Much of the information about external flows comes from experiments carried out, for the most part, on scale models of the actual objects. Such testing includes the obvious wind tunnel testing of model airplanes, buildings, and even entire cities. In some instances the actual device, not a model, is tested in wind tunnels. Figure 9.1a shows a test of a vehicle in a wind tunnel. Better performance of cars, bikes, skiers, and numerous other objects has resulted from testing in wind tunnels. The use of water tunnels and towing tanks also provides useful information about the flow around ships and other objects. With advancement in computational fluid dynamics, or CFD, numerical methods are also capable of predicting external flows past objects. Figure 9.1b shows streamlines around a Formula 1 car as predicted by CFD. Appendix A provides an introduction to CFD.

In this chapter we consider characteristics of external flow past a variety of objects. We investigate the qualitative aspects of such flows and learn how to determine the various forces on objects surrounded by a moving liquid.

9.1 General External Flow Characteristics

For external flows it is usually easiest to use a coordinate system fixed to the object.

A body immersed in a moving fluid experiences a resultant force due to the interaction between the body and the fluid surrounding it. In some instances (such as an airplane flying through still air) the fluid far from the body is stationary and the body moves through the fluid with velocity U. In other instances (such as the wind blowing past a building) the body is stationary and the fluid flows past the body with velocity U. In any case, we can fix the coordinate system in the body and treat the situation as fluid flowing past a stationary body with velocity U, the *upstream velocity*. For the purposes of this book, we will assume that the upstream velocity is constant in both time and location. That is, there is a uniform, constant velocity fluid flowing past the object. In actual situations this is often not true. For example, the wind blowing past a smokestack is nearly always turbulent and gusty (unsteady) and probably not of uniform velocity from the top to the bottom of the stack. Usually the unsteadiness and nonuniformity are of minor importance.

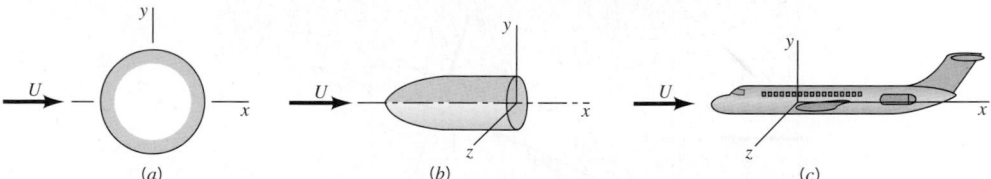

■ **FIGURE 9.2** **Flow classification:** (*a*) **two-dimensional,** (*b*) **axisymmetric,** (*c*) **three-dimensional.**

V9.1 Space shuttle landing

Even with a steady, uniform upstream flow, the flow in the vicinity of an object may be unsteady. Examples of this type of behavior include the flutter that is sometimes found in the flow past airfoils (wings), the regular oscillation of telephone wires that "sing" in a wind, and the irregular turbulent fluctuations in the wake regions behind bodies.

The structure of an external flow and the ease with which the flow can be described and analyzed often depend on the nature of the body in the flow. Three general categories of bodies are shown in Fig. 9.2. They include (a) two-dimensional objects (infinitely long and of constant cross-sectional size and shape), (b) axisymmetric bodies (formed by rotating their cross-sectional shape about the axis of symmetry), and (c) three-dimensional bodies that may or may not possess a line or plane of symmetry. In practice there can be no truly two-dimensional bodies—nothing extends to infinity. However, many objects are sufficiently long so that the end effects are negligibly small.

Another classification of body shape can be made depending on whether the body is streamlined or blunt. The flow characteristics depend strongly on the amount of streamlining present. In general, *streamlined bodies* (i.e., airfoils, racing cars, etc.) have little effect on the surrounding fluid, compared with the effect that *blunt bodies* (i.e., parachutes, buildings, etc.) have on the fluid. Usually, but not always, it is easier to force a streamlined body through a fluid than it is to force a similar-sized blunt body at the same velocity. There are important exceptions to this basic rule.

9.1.1 Lift and Drag Concepts

When any body moves through a fluid, an interaction between the body and the fluid occurs; this effect can be given in terms of the forces at the fluid–body interface. These forces can be described in terms of the stresses—wall shear stresses, τ_w, due to viscous effects and normal stresses due to the pressure, p. Typical shear stress and pressure distributions are shown in Figs. 9.3*a* and 9.3*b*. Both τ_w and p vary in magnitude and direction along the surface.

It is often useful to know the detailed distribution of shear stress and pressure over the surface of the body, although such information is difficult to obtain. Many times, however, only the

A body interacts with the surrounding fluid through pressure and shear stresses.

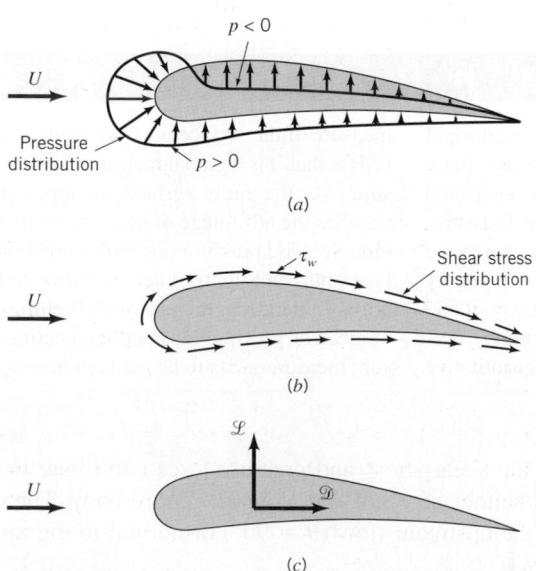

■ **FIGURE 9.3** **Forces from the surrounding fluid on a two-dimensional object:** (*a*) **pressure force,** (*b*) **viscous force,** (*c*) **resultant force (lift and drag).**

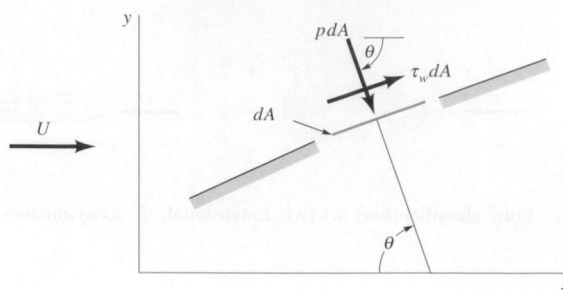

Pressure and shear forces on a small element of the surface of a body.

integrated or resultant effects of these distributions are needed. The resultant force in the direction of the upstream velocity is termed the *drag,* $\mathscr{D}$, and the resultant force normal to the upstream velocity is termed the *lift,* $\mathscr{L}$, as is indicated in Fig. 9.3c. For some three-dimensional bodies there may also be a side force that is perpendicular to the plane containing $\mathscr{D}$ and $\mathscr{L}$.

The resultant of the shear stress and pressure distributions can be obtained by integrating the effect of these two quantities on the body surface as is indicated in Fig. 9.4. The x and y components of the fluid force on the small area element dA are

$$dF_x = (p\, dA) \cos \theta + (\tau_w\, dA) \sin \theta$$

and

$$dF_y = -(p\, dA) \sin \theta + (\tau_w\, dA) \cos \theta$$

Thus, the net x and y components of the force on the object are

$$\mathscr{D} = \int dF_x = \int p \cos \theta\, dA + \int \tau_w \sin \theta\, dA \qquad (9.1)$$

and

$$\mathscr{L} = \int dF_y = -\int p \sin \theta\, dA + \int \tau_w \cos \theta\, dA \qquad (9.2)$$

Lift and drag on a section of a body depend on the orientation of the surface.

Of course, to carry out the integrations and determine the lift and drag, we must know the body shape (i.e., θ as a function of location along the body) and the distribution of τ_w and p along the surface. These distributions are often extremely difficult to obtain, either experimentally or theoretically. The pressure distribution can be obtained experimentally by use of a series of static pressure taps along the body surface. On the other hand, it is usually quite difficult to measure the wall shear stress distribution.

F l u i d s   i n   t h e   N e w s

Pressure-sensitive paint For many years, the conventional method for measuring *surface pressure* has been to use static pressure taps consisting of small holes on the surface connected by hoses from the holes to a pressure measuring device. Pressure-sensitive paint (PSP) is now gaining acceptance as an alternative to the static surface pressure ports. The PSP material is typically a luminescent compound that is sensitive to the pressure on it and can be excited by an appropriate light which is captured by special video imaging equipment. Thus, it provides a quantitative measure of the surface pressure. One of the biggest advantages of PSP is that it is a global measurement technique, measuring pressure over the entire surface, as opposed to discrete points. PSP also has the advantage of being nonintrusive to the flow field. Although static pressure port holes are small, they do alter the surface and can slightly alter the flow, thus affecting downstream ports. In addition, the use of PSP eliminates the need for a large number of pressure taps and connecting tubes. This allows pressure measurements to be made in less time and at a lower cost.

It is seen that both the shear stress and pressure force contribute to the lift and drag, since for an arbitrary body θ is neither zero nor 90° along the entire body. The exception is a flat plate aligned either parallel to the upstream flow ($\theta = 90°$) or normal to the upstream flow ($\theta = 0$) as is discussed in Example 9.1.

EXAMPLE 9.1 Drag from Pressure and Shear Stress Distributions

GIVEN Air at standard conditions flows past a flat plate as is indicated in Fig. E9.1. In case (a) the plate is parallel to the upstream flow, and in case (b) it is perpendicular to the upstream flow. The pressure and shear stress distributions on the surface are as indicated (obtained either by experiment or theory).

FIND Determine the lift and drag on the plate.

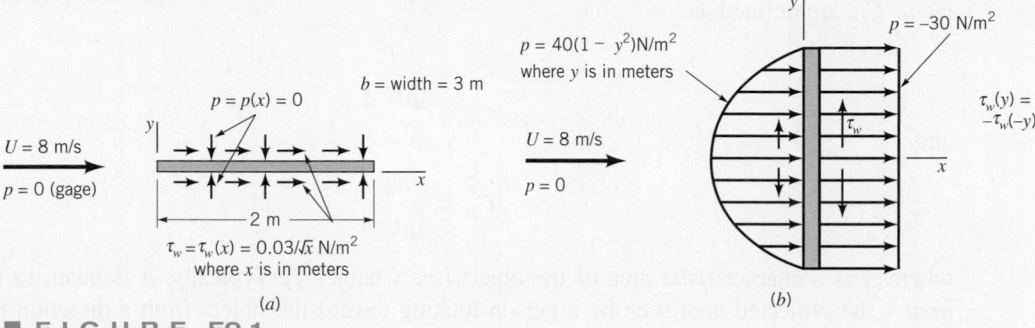

FIGURE E9.1

SOLUTION

For either orientation of the plate, the lift and drag are obtained from Eqs. 9.1 and 9.2. With the plate parallel to the upstream flow we have $\theta = 90°$ on the top surface and $\theta = 270°$ on the bottom surface so that the lift and drag are given by

$$\mathscr{L} = -\int_{\text{top}} p \, dA + \int_{\text{bottom}} p \, dA = 0$$

and

$$\mathscr{D} = \int_{\text{top}} \tau_w \, dA + \int_{\text{bottom}} \tau_w \, dA = 2\int_{\text{top}} \tau_w \, dA \qquad (1)$$

where we have used the fact that because of symmetry the shear stress distribution is the same on the top and the bottom surfaces, as is the pressure also [whether we use gage ($p = 0$) or absolute ($p = p_{\text{atm}}$) pressure]. There is no lift generated—the plate does not know up from down. With the given shear stress distribution, Eq. 1 gives

$$\mathscr{D} = 2\int_{x=0}^{2\,\text{m}} \left(\frac{0.03}{x^{1/2}} \, \text{N/m}^2\right)(3\,\text{m}) \, dx$$

or

$$\mathscr{D} = 0.509 \, \text{N} \qquad \text{(Ans)}$$

With the plate perpendicular to the upstream flow, we have $\theta = 0°$ on the front and $\theta = 180°$ on the back. Thus, from Eqs. 9.1 and 9.2

$$\mathscr{L} = \int_{\text{front}} \tau_w \, dA - \int_{\text{back}} \tau_w \, dA = 0$$

and

$$\mathscr{D} = \int_{\text{front}} p \, dA - \int_{\text{back}} p \, dA$$

Again there is no lift because the pressure forces act parallel to the upstream flow (in the direction of $\mathscr{D}$ not $\mathscr{L}$) and the shear stress is

symmetrical about the center of the plate. With the given relatively large pressure on the front of the plate (the center of the plate is a stagnation point) and the negative pressure (less than the upstream pressure) on the back of the plate, we obtain the following drag

$$\mathscr{D} = \int_{y=-1}^{1\,\text{m}} \left[40(1 - y^2)\text{N/m}^2 - (-30 \, \text{N/m}^2)\right](3\,\text{m}) \, dy$$

or

$$\mathscr{D} = 2.50 \, \text{N} \qquad \text{(Ans)}$$

COMMENTS Clearly there are two mechanisms responsible for the drag. On the ultimately streamlined body (a zero thickness flat plate parallel to the flow) the drag is entirely due to the shear stress at the surface and, in this example, is relatively small. For the ultimately blunted body (a flat plate normal to the upstream flow) the drag is entirely due to the pressure difference between the front and back portions of the object and, in this example, is relatively large.

If the flat plate were oriented at an arbitrary angle relative to the upstream flow as indicated in Fig. E9.1c, there would be both a lift and a drag, each of which would be dependent on both the shear stress and the pressure. Both the pressure and shear stress distributions would be different for the top and bottom surfaces.

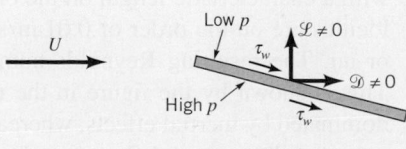

FIGURE E9.1 *(Continued)*

Although Eqs. 9.1 and 9.2 are valid for any body, the difficulty in their use lies in obtaining the appropriate shear stress and pressure distributions on the body surface. Considerable effort has gone into determining these quantities, but because of the various complexities involved, such information is available only for certain simple situations.

Without detailed information concerning the shear stress and pressure distributions on a body, Eqs. 9.1 and 9.2 cannot be used. The widely used alternative is to define dimensionless lift and drag coefficients and determine their approximate values by means of either a simplified analysis, some numerical technique, or an appropriate experiment. The *lift coefficient, C_L,* and *drag coefficient, C_D,* are defined as

Lift coefficients and drag coefficients are dimensionless forms of lift and drag.

$$C_L = \frac{\mathscr{L}}{\frac{1}{2}\rho U^2 A}$$

and

$$C_D = \frac{\mathscr{D}}{\frac{1}{2}\rho U^2 A}$$

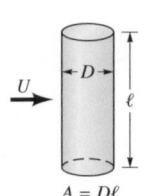

$A = D\ell$

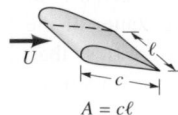

$A = c\ell$

where A is a characteristic area of the object (see Chapter 7). Typically, A is taken to be *frontal area*—the projected area seen by a person looking toward the object from a direction parallel to the upstream velocity, U, as indicated by the figure in the margin. It would be the area of the shadow of the object projected onto a screen normal to the upstream velocity as formed by a light shining along the upstream flow. In other situations A is taken to be the *planform area*—the projected area seen by an observer looking toward the object from a direction normal to the upstream velocity (i.e., from "above" it). Obviously, which characteristic area is used in the definition of the lift and drag coefficients must be clearly stated.

9.1.2 Characteristics of Flow Past an Object

External flows past objects encompass an extremely wide variety of fluid mechanics phenomena. Clearly the character of the flow field is a function of the shape of the body. Flows past relatively simple geometric shapes (i.e., a sphere or circular cylinder) are expected to have less complex flow fields than flows past a complex shape such as an airplane or a tree. However, even the simplest-shaped objects produce rather complex flows.

For a given-shaped object, the characteristics of the flow depend very strongly on various parameters such as size, orientation, speed, and fluid properties. As is discussed in Chapter 7, according to dimensional analysis arguments, the character of the flow should depend on the various dimensionless parameters involved. For typical external flows the most important of these parameters are the Reynolds number, $\text{Re} = \rho U \ell / \mu = U \ell / \nu$, the Mach number, $\text{Ma} = U/c$, and for flows with a free surface (i.e., flows with an interface between two fluids, such as the flow past a surface ship), the Froude number, $\text{Fr} = U/\sqrt{g\ell}$. (Recall that ℓ is some characteristic length of the object and c is the speed of sound.)

The character of flow past an object is dependent on the value of the Reynolds number.

For the present, we consider how the external flow and its associated lift and drag vary as a function of Reynolds number. Recall that the Reynolds number represents the ratio of inertial effects to viscous effects. In the absence of all viscous effects ($\mu = 0$), the Reynolds number is infinite. On the other hand, in the absence of all inertial effects (negligible mass or $\rho = 0$), the Reynolds number is zero. Clearly, any actual flow will have a Reynolds number between (but not including) these two extremes. The nature of the flow past a body depends strongly on whether $\text{Re} \gg 1$ or $\text{Re} \ll 1$.

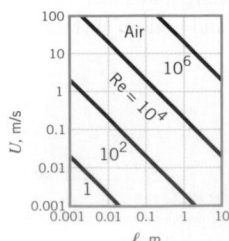

Most external flows with which we are familiar are associated with moderately sized objects with a characteristic length on the order of $0.01\text{ m} < \ell < 10\text{ m}$. In addition, typical upstream velocities are on the order of $0.01\text{ m/s} < U < 100\text{ m/s}$ and the fluids involved are typically water or air. The resulting Reynolds number range for such flows is approximately $10 < \text{Re} < 10^9$. This is shown by the figure in the margin for air. As a rule of thumb, flows with $\text{Re} > 100$ are dominated by inertial effects, whereas flows with $\text{Re} < 1$ are dominated by viscous effects. Hence, most familiar external flows are dominated by inertia.

On the other hand, there are many external flows in which the Reynolds number is considerably less than 1, indicating in some sense that viscous forces are more important than inertial forces. The gradual settling of small particles of dirt in a lake or stream is governed by low Reynolds number flow principles because of the small diameter of the particles and their small settling speed. Similarly, the Reynolds number for objects moving through large viscosity oils is small because μ is large. The general differences between small and large Reynolds number flow past stream-lined and blunt objects can be illustrated by considering flows past two objects—one a flat plate parallel to the upstream velocity and the other a circular cylinder.

For low Reynolds number flows, viscous effects are felt far from the object.

Flows past three flat plates of length ℓ with Re $= \rho U\ell/\mu = 0.1$, 10, and 10^7 are shown in Fig. 9.5. If the Reynolds number is small, the viscous effects are relatively strong and the plate affects the uniform upstream flow far ahead, above, below, and behind the plate. To reach that portion of the flow field where the velocity has been altered by less than 1% of its undisturbed value (i.e., $U - u < 0.01\,U$) we must travel relatively far from the plate. In low Reynolds number flows the viscous effects are felt far from the object in all directions.

As the Reynolds number is increased (by increasing U, for example), the region in which viscous effects are important becomes smaller in all directions except downstream, as is shown in

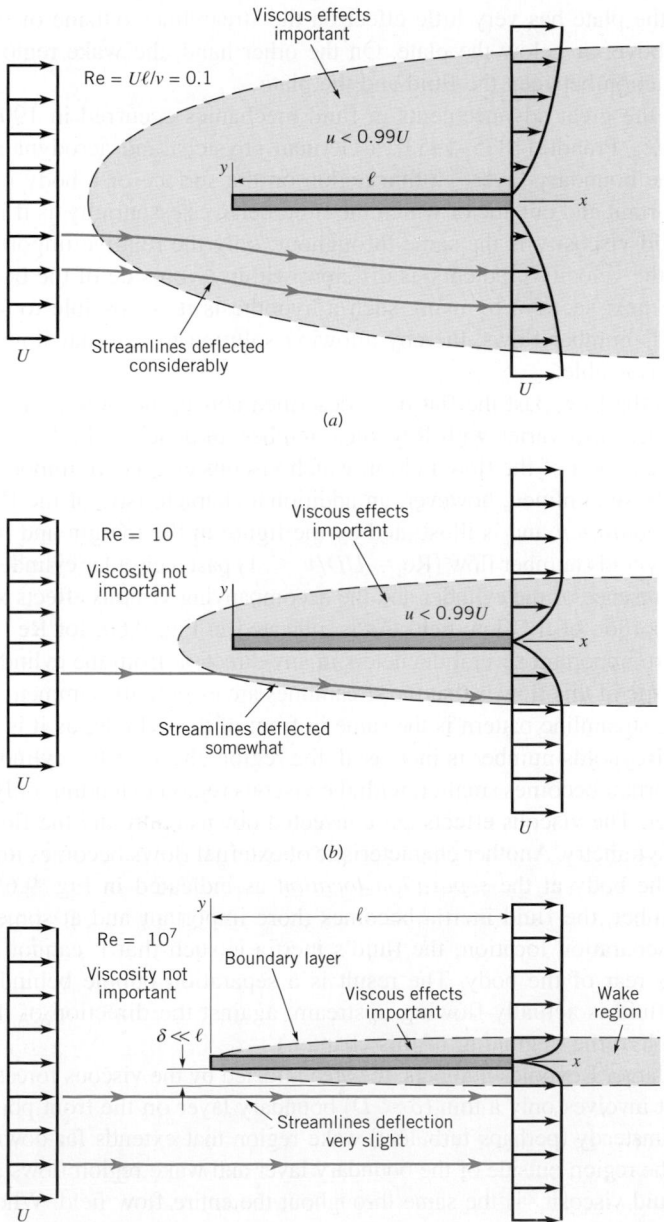

■ **FIGURE 9.5**   **Character of the steady, viscous flow past a flat plate parallel to the upstream velocity: (a) low Reynolds number flow, (b) moderate Reynolds number flow, (c) large Reynolds number flow.**

Fig. 9.5*b*. One does not need to travel very far ahead, above, or below the plate to reach areas in which the viscous effects of the plate are not felt. The streamlines are displaced from their original uniform upstream conditions, but the displacement is not as great as for the Re = 0.1 situation shown in Fig. 9.5*a*.

If the Reynolds number is large (but not infinite), the flow is dominated by inertial effects and the viscous effects are negligible everywhere except in a region very close to the plate and in the relatively thin *wake region* behind the plate, as shown in Fig. 9.5*c*. Since the fluid viscosity is not zero (Re < ∞), it follows that the fluid must stick to the solid surface (the no-slip boundary condition). There is a thin *boundary layer* region of thickness $\delta = \delta(x) \ll \ell$ (i.e., thin relative to the length of the plate) next to the plate in which the fluid velocity changes from the upstream value of $u = U$ to zero velocity on the plate. The thickness of this layer increases in the direction of flow, starting from zero at the forward or leading edge of the plate. The flow within the boundary layer may be laminar or turbulent, depending on various parameters involved.

The streamlines of the flow outside of the boundary layer are nearly parallel to the plate. As we will see in the next section, the slight displacement of the external streamlines that are outside of the boundary layer is due to the thickening of the boundary layer in the direction of flow. The existence of the plate has very little effect on the streamlines outside of the boundary layer—either ahead, above, or below the plate. On the other hand, the wake region is due entirely to the viscous interaction between the fluid and the plate.

Thin boundary layers may develop in large Reynolds number flows.

One of the great advancements in fluid mechanics occurred in 1904 as a result of the insight of Ludwig Prandtl (1875–1953), a German physicist and aerodynamicist. He conceived of the idea of the boundary layer—a thin region on the surface of a body in which viscous effects are very important and outside of which the fluid behaves essentially as if it were inviscid. Clearly the actual fluid viscosity is the same throughout; only the relative importance of the viscous effects (due to the velocity gradients) is different within or outside of the boundary layer. As is discussed in the next section, by using such a hypothesis it is possible to simplify the analysis of large Reynolds number flows, thereby allowing solution to external flow problems that are otherwise still unsolvable.

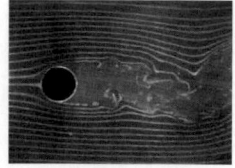

As with the flow past the flat plate described above, the flow past a blunt object (such as a circular cylinder) also varies with Reynolds number. In general, the larger the Reynolds number, the smaller the region of the flow field in which viscous effects are important. For objects that are not sufficiently streamlined, however, an additional characteristic of the flow is observed. This is termed *flow separation* and is illustrated by the figure in the margin and in Fig. 9.6.

Low Reynolds number flow (Re = UD/ν < 1) past a circular cylinder is characterized by the fact that the presence of the cylinder and the accompanying viscous effects are felt throughout a relatively large portion of the flow field. As is indicated in Fig. 9.6*a*, for Re = UD/ν = 0.1, the viscous effects are important several diameters in any direction from the cylinder. A somewhat surprising characteristic of this flow is that the streamlines are essentially symmetric about the center of the cylinder—the streamline pattern is the same in front of the cylinder as it is behind the cylinder.

As the Reynolds number is increased, the region ahead of the cylinder in which viscous effects are important becomes smaller, with the viscous region extending only a short distance ahead of the cylinder. The viscous effects are convected downstream and the flow loses its upstream to downstream symmetry. Another characteristic of external flows becomes important—the flow separates from the body at the *separation location* as indicated in Fig. 9.6*b*. With the increase in Reynolds number, the fluid inertia becomes more important and at some location on the body, denoted the separation location, the fluid's inertia is such that it cannot follow the curved path around to the rear of the body. The result is a separation bubble behind the cylinder in which some of the fluid is actually flowing upstream, against the direction of the upstream flow. (See the photograph at the beginning of this chapter.)

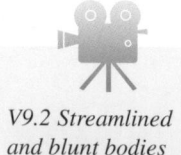

V9.2 Streamlined and blunt bodies

At still larger Reynolds numbers, the area affected by the viscous forces is forced farther downstream until it involves only a thin ($\delta \ll D$) boundary layer on the front portion of the cylinder and an irregular, unsteady (perhaps turbulent) wake region that extends far downstream of the cylinder. The fluid in the region outside of the boundary layer and wake region flows as if it were inviscid. Of course, the fluid viscosity is the same throughout the entire flow field. Whether viscous effects are important or not depends on which region of the flow field we consider. The velocity gradients within the boundary layer and wake regions are much larger than those in the remainder of the flow field.

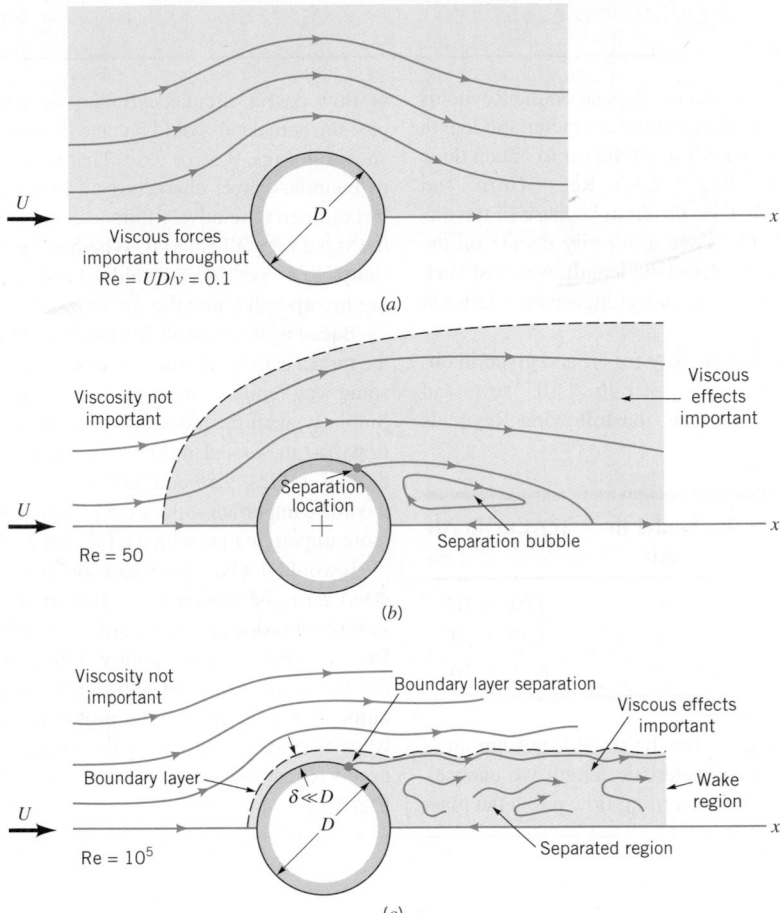

■ **FIGURE 9.6**   Character of the steady, viscous flow past a circular cylinder: (*a*) low Reynolds number flow, (*b*) moderate Reynolds number flow, (*c*) large Reynolds number flow.

Since the shear stress (i.e., viscous effect) is the product of the fluid viscosity and the velocity gradient, it follows that viscous effects are confined to the boundary layer and wake regions.

Most familiar flows involve large Reynolds numbers.

The characteristics described in Figs. 9.5 and 9.6 for flow past a flat plate and a circular cylinder are typical of flows past streamlined and blunt bodies, respectively. The nature of the flow depends strongly on the Reynolds number. (See Ref. 31 for many examples illustrating this behavior.) Most familiar flows are similar to the large Reynolds number flows depicted in Figs. 9.5*c* and 9.6*c*, rather than the low Reynolds number flow situations. (See the photograph at the beginning of Chapters 7 and 11.) In the remainder of this chapter we will investigate more thoroughly these ideas and determine how to calculate the forces on immersed bodies.

---

*E*XAMPLE 9.2   Characteristics of Flow Past Objects

GIVEN   It is desired to experimentally determine the various characteristics of flow past a car as shown in Fig E9.2. The following tests could be carried out: (a) $U = 20$ mm/s flow of glycerin past a scale model that is 34-mm tall, 100-mm long, and 40-mm wide, (b) $U = 20$ mm/s air flow past the same scale model, or (c) $U = 25$ m/s air flow past the actual car, which is 1.7-m tall, 5-m long, and 2-m wide.

FIND   Would the flow characteristics for these three situations be similar? Explain.

■ **FIGURE E9.2**

SOLUTION

The characteristics of flow past an object depend on the Reynolds number. For this instance we could pick the characteristic length to be the height, h, width, b, or length, ℓ, of the car to obtain three possible Reynolds numbers, $Re_h = Uh/\nu$, $Re_b = Ub/\nu$, and $Re_\ell = U\ell/\nu$. These numbers will be different because of the different values of h, b, and ℓ. Once we arbitrarily decide on the length we wish to use as the characteristic length, we must stick with it for all calculations when using comparisons between model and prototype.

With the values of kinematic viscosity for air and glycerin obtained from Tables 1.4 and 1.3 as $\nu_{air} = 1.46 \times 10^{-5}$ m^2/s and $\nu_{glycerin} = 1.19 \times 10^{-3}$ m^2/s, we obtain the following Reynolds numbers for the flows described.

Reynolds Number	(a) Model in Glycerin	(b) Model in Air	(c) Car in Air
Re_h	0.571	46.6	2.91×10^6
Re_b	0.672	54.8	3.42×10^6
Re_ℓ	1.68	137.0	8.56×10^6

Clearly, the Reynolds numbers for the three flows are quite different (regardless of which characteristic length we choose). Based on the previous discussion concerning flow past a flat plate or flow past a circular cylinder, we would expect that the flow past the actual car would behave in some way similar to the flows shown in Figs. 9.5c or 9.6c. That is, we would expect some type of boundary layer characteristic in which viscous effects would be confined to relatively thin layers near the surface of the car and the wake region behind it. Whether the car would act more like a flat plate or a cylinder would depend on the amount of streamlining incorporated into the car's design.

Because of the small Reynolds number involved, the flow past the model car in glycerin would be dominated by viscous effects, in some way reminiscent of the flows depicted in Figs. 9.5a or 9.6a. Similarly, with the moderate Reynolds number involved for the air flow past the model, a flow with characteristics similar to those indicated in Figs. 9.5b and 9.6b would be expected. Viscous effects would be important—not as important as with the glycerin flow, but more important than with the full-sized car.

It would not be a wise decision to expect the flow past the full-sized car to be similar to the flow past either of the models. The same conclusions result regardless of whether we use Re_h, Re_b, or Re_ℓ. As is indicated in Chapter 7, the flows past the model car and the full-sized prototype will not be similar unless the Reynolds numbers for the model and prototype are the same. It is not always an easy task to ensure this condition. One (expensive) solution is to test full-sized prototypes in very large wind tunnels (see Fig. 9.1).

9.2 Boundary Layer Characteristics

As was discussed in the previous section, it is often possible to treat flow past an object as a combination of viscous flow in the boundary layer and inviscid flow elsewhere. If the Reynolds number is large enough, viscous effects are important only in the boundary layer regions near the object (and in the wake region behind the object). The boundary layer is needed to allow for the no-slip boundary condition that requires the fluid to cling to any solid surface that it flows past. Outside of the boundary layer the velocity gradients normal to the flow are relatively small, and the fluid acts as if it were inviscid, even though the viscosity is not zero. A necessary condition for this structure of the flow is that the Reynolds number be large.

Large Reynolds number flow fields may be divided into viscous and inviscid regions.

9.2.1 Boundary Layer Structure and Thickness on a Flat Plate

There can be a wide variety in the size of a boundary layer and the structure of the flow within it. Part of this variation is due to the shape of the object on which the boundary layer forms. In this section we consider the simplest situation, one in which the boundary layer is formed on an infinitely long flat plate along which flows a viscous, incompressible fluid as is shown in Fig. 9.7. If the surface were curved (i.e., a circular cylinder or an airfoil), the boundary layer structure would be more complex. Such flows are discussed in Section 9.2.6.

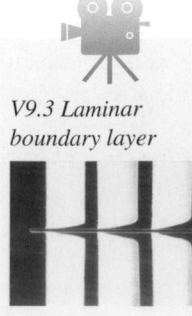

V9.3 Laminar boundary layer

If the Reynolds number is sufficiently large, only the fluid in a relatively thin boundary layer on the plate will feel the effect of the plate. That is, except in the region next to the plate the flow velocity will be essentially $\mathbf{V} = U\hat{\mathbf{i}}$, the upstream velocity. For the infinitely long flat plate extending from $x = 0$ to $x = \infty$, it is not obvious how to define the Reynolds number because there is no characteristic length. The plate has no thickness and is not of finite length!

For a finite length plate, it is clear that the plate length, ℓ, can be used as the characteristic length. For an infinitely long plate we use x, the coordinate distance along the plate from the leading edge, as the characteristic length and define the Reynolds number as $Re_x = Ux/\nu$. Thus, for

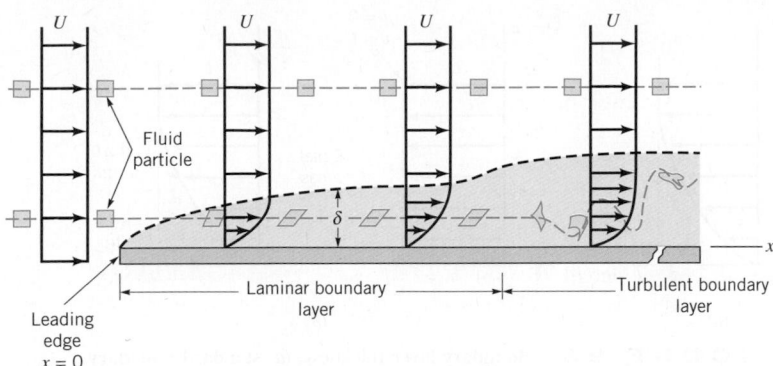

■ FIGURE 9.7   **Distortion of a fluid particle as it flows within the boundary layer.**

Fluid particles within the boundary layer experience viscous effects.

V9.4 Laminar/turbulent transition

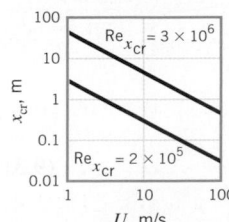

any fluid or upstream velocity the Reynolds number will be sufficiently large for boundary layer type flow (i.e., Fig. 9.5*c*) if the plate is long enough. Physically, this means that the flow situations illustrated in Fig. 9.5 could be thought of as occurring on the same plate, but should be viewed by looking at longer portions of the plate as we step away from the plate to see the flows in Fig. 9.5*a*, 9.5*b*, and 9.5*c*, respectively.

If the plate is sufficiently long, the Reynolds number Re = $U\ell/\nu$ is sufficiently large so that the flow takes on its boundary layer character (except very near the leading edge). The details of the flow field near the leading edge are lost to our eyes because we are standing so far from the plate that we cannot make out these details. On this scale (Fig. 9.5*c*) the plate has negligible effect on the fluid ahead of the plate. The presence of the plate is felt only in the relatively thin boundary layer and wake regions. As previously noted, Prandtl in 1904 was the first to hypothesize such a concept. It has become one of the major turning points in fluid mechanics analysis.

A better appreciation of the structure of the boundary layer flow can be obtained by considering what happens to a fluid particle that flows into the boundary layer. As is indicated in Fig. 9.7, a small rectangular particle retains its original shape as it flows in the uniform flow outside of the boundary layer. Once it enters the boundary layer, the particle begins to distort because of the velocity gradient within the boundary layer—the top of the particle has a larger speed than its bottom. The fluid particles do not rotate as they flow along outside the boundary layer, but they begin to rotate once they pass through the fictitious boundary layer surface and enter the world of viscous flow. The flow is said to be irrotational outside the boundary layer and rotational within the boundary layer. (In terms of the kinematics of fluid particles as is discussed in Section 6.1, the flow outside the boundary layer has zero vorticity, and the flow within the boundary layer has nonzero vorticity.)

At some distance downstream from the leading edge, the boundary layer flow becomes turbulent and the fluid particles become greatly distorted because of the random, irregular nature of the turbulence. One of the distinguishing features of turbulent flow is the occurrence of irregular mixing of fluid particles that range in size from the smallest fluid particles up to those comparable in size with the object of interest. For laminar flow, mixing occurs only on the molecular scale. This molecular scale is orders of magnitude smaller in size than typical size scales for turbulent flow mixing. The transition from a *laminar boundary layer* to a *turbulent boundary layer* occurs at a critical value of the Reynolds number, Re_{xcr}, on the order of 2×10^5 to 3×10^6, depending on the roughness of the surface and the amount of turbulence in the upstream flow, as is discussed in Section 9.2.4. As shown by the figure in the margin, the location along the plate where the flow becomes turbulent, x_{cr}, moves towards the leading edge as the free-stream velocity increases.

The purpose of the boundary layer is to allow the fluid to change its velocity from the upstream value of U to zero on the surface. Thus, $\mathbf{V} = 0$ at $y = 0$ and $\mathbf{V} \approx U\hat{\mathbf{i}}$ at the edge of the boundary layer, with the velocity profile, $u = u(x, y)$ bridging the boundary layer thickness. This boundary layer characteristic occurs in a variety of flow situations, not just on flat plates. For example, boundary layers form on the surfaces of cars, in the water running down the gutter of the street, and in the atmosphere as the wind blows across the surface of the earth (land or water).

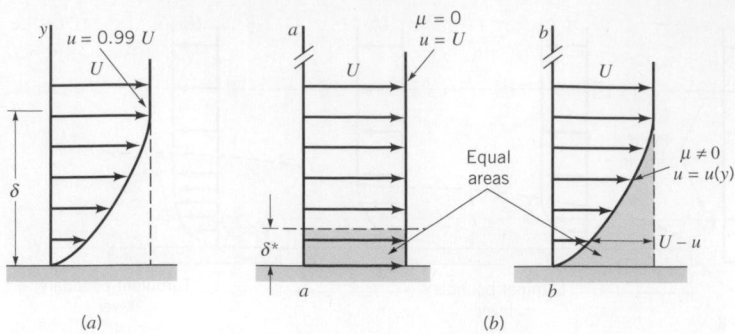

■ **FIGURE 9.8** Boundary layer thickness: (*a*) standard boundary layer thickness, (*b*) boundary layer displacement thickness.

In actuality (both mathematically and physically), there is no sharp "edge" to the boundary layer; that is, $u \to U$ as we get farther from the plate. We define the *boundary layer thickness, δ,* as that distance from the plate at which the fluid velocity is within some arbitrary value of the upstream velocity. Typically, as indicated in Fig. 9.8*a,*

$$\delta = y \quad \text{where} \quad u = 0.99U$$

To remove this arbitrariness (i.e., what is so special about 99%; why not 98%?), the following definitions are introduced. Shown in Fig. 9.8*b* are two velocity profiles for flow past a flat plate—one if there were no viscosity (a uniform profile) and the other if there are viscosity and zero slip at the wall (the boundary layer profile). Because of the velocity deficit, $U - u$, within the boundary layer, the flowrate across section *b–b* is less than that across section *a–a*. However, if we displace the plate at section *a–a* by an appropriate amount δ^*, the *boundary layer displacement thickness*, the flowrates across each section will be identical. This is true if

The boundary layer displacement thickness is defined in terms of volumetric flowrate.

$$\delta^* b U = \int_0^\infty (U - u) b \, dy$$

where *b* is the plate width. Thus,

$$\delta^* = \int_0^\infty \left(1 - \frac{u}{U}\right) dy \tag{9.3}$$

The displacement thickness represents the amount that the thickness of the body must be increased so that the fictitious uniform inviscid flow has the same mass flowrate properties as the actual viscous flow. It represents the outward displacement of the streamlines caused by the viscous effects on the plate. This idea allows us to simulate the presence that the boundary layer has on the flow outside of the boundary layer by adding the displacement thickness to the actual wall and treating the flow over the thickened body as an inviscid flow. The displacement thickness concept is illustrated in Example 9.3.

EXAMPLE 9.3 — Boundary Layer Displacement Thickness

GIVEN Air flowing into a 0.6-m-square duct with a uniform velocity of 3 m/s forms a boundary layer on the walls as shown in Fig. E9.3a. The fluid within the core region (outside the boundary layers) flows as if it were inviscid. From advanced calculations it is determined that for this flow the boundary layer displacement thickness is given by

$$\delta^* = 0.004(x)^{1/2} \quad (1)$$

where δ^* and x are in meters.

FIND Determine the velocity $U = U(x)$ of the air within the duct but outside of the boundary layer.

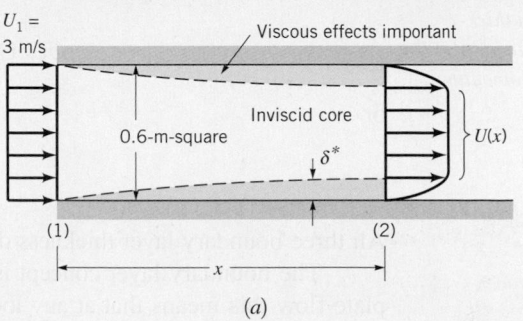

(a)

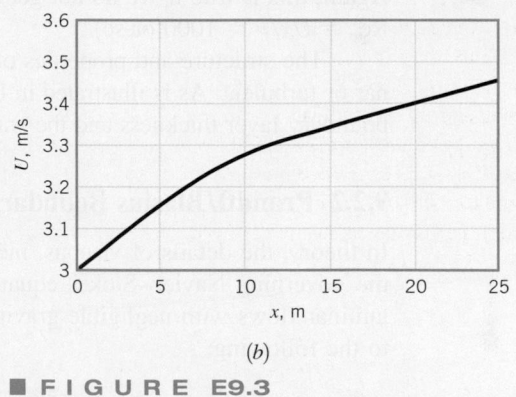

(b)

■ **FIGURE E9.3**

SOLUTION

If we assume incompressible flow (a reasonable assumption because of the low velocities involved), it follows that the volume flowrate across any section of the duct is equal to that at the entrance (i.e., $Q_1 = Q_2$). That is,

$$U_1 A_1 = 3 \text{ m/s} (0.6 \text{ m})^2 = 1.08 \text{ m}^3/\text{s} = \int_{(2)} u \, dA$$

According to the definition of the displacement thickness, δ^*, the flowrate across section (2) is the same as that for a uniform flow with velocity U through a duct whose walls have been moved inward by δ^*. That is,

$$1.08 \text{ m}^3/\text{s} = \int_{(2)} u \, dA = U(0.6 \text{ m} - 2\delta^*)^2 \quad (2)$$

By combining Eqs. 1 and 2 we obtain

$$1.08 \text{ m}^3/\text{s} = U(0.6 - 0.008x^{1/2})^2$$

or

$$U = \frac{1.08}{(0.6 - 0.008x^{1/2})^2} \text{ m/s} \quad \text{(Ans)}$$

COMMENTS Note that U increases in the downstream direction. For example, as shown in Fig. E9.3b, $U = 3.44$ m/s at $x = 25$ m. The viscous effects that cause the fluid to stick to the walls of the duct reduce the effective size of the duct, thereby (from conservation of mass principles) causing the fluid to accelerate. The pressure drop necessary to do this can be obtained by using the Bernoulli equation (Eq. 3.7) along the inviscid streamlines from section (1) to (2). (Recall that this equation is not valid for viscous flows within the boundary layer. It is, how-

ever, valid for the inviscid flow outside the boundary layer.) Thus,

$$p_1 + \tfrac{1}{2}\rho U_1^2 = p + \tfrac{1}{2}\rho U^2$$

Hence, with $\rho = 1.20$ kg/m³ and $p_1 = 0$ we obtain

$$p = \frac{1}{2}\rho (U_1^2 - U^2)$$

$$= \frac{1}{2}(1.20 \text{ kg/m}^3)$$

$$\times \left[(3 \text{ m/s})^2 - \frac{(1.08)^2}{(0.6 - 0.008x^{1/2})^4} \text{ m}^2/\text{s}^2 \right]$$

or

$$p = 0.6 \left[9 - \frac{1.17}{(0.6 - 0.008x^{1/2})^4} \right]$$

For example, $p = -2.12$ N/m² at $x = 30$ m.

If it were desired to maintain a constant velocity along the centerline of this entrance region of the duct, the walls could be displaced outward by an amount equal to the boundary layer displacement thickness, δ^*.

Another boundary layer thickness definition, the *boundary layer momentum thickness*, Θ, is often used when determining the drag on an object. Again because of the velocity deficit, $U - u$, in the boundary layer, the momentum flux across section b–b in Fig. 9.8 is less than that across section a–a. This deficit in momentum flux for the actual boundary layer flow on a plate of width b is given by

$$\int \rho u (U - u)\, dA = \rho b \int_0^\infty u(U - u)\, dy$$

The boundary layer momentum thickness is defined in terms of momentum flux.

which by definition is the momentum flux in a layer of uniform speed U and thickness Θ. That is,

$$\rho b U^2 \Theta = \rho b \int_0^\infty u(U - u)\, dy$$

or

$$\Theta = \int_0^\infty \frac{u}{U}\left(1 - \frac{u}{U}\right) dy \tag{9.4}$$

All three boundary layer thickness definitions, δ, δ^*, and Θ, are of use in boundary layer analyses.

The boundary layer concept is based on the fact that the boundary layer is thin. For the flat plate flow this means that at any location x along the plate, $\delta \ll x$. Similarly, $\delta^* \ll x$ and $\Theta \ll x$. Again, this is true if we do not get too close to the leading edge of the plate (i.e., not closer than $\text{Re}_x = Ux/\nu = 1000$ or so).

The structure and properties of the boundary layer flow depend on whether the flow is laminar or turbulent. As is illustrated in Fig. 9.9 and discussed in Sections 9.2.2 through 9.2.5, both the boundary layer thickness and the wall shear stress are different in these two regimes.

9.2.2 Prandtl/Blasius Boundary Layer Solution

In theory, the details of viscous, incompressible flow past any object can be obtained by solving the governing Navier–Stokes equations discussed in Section 6.8.2. For steady, two-dimensional laminar flows with negligible gravitational effects, these equations (Eqs. 6.127a, b, and c) reduce to the following:

$$u \frac{\partial u}{\partial x} + v \frac{\partial u}{\partial y} = -\frac{1}{\rho} \frac{\partial p}{\partial x} + \nu \left(\frac{\partial^2 u}{\partial x^2} + \frac{\partial^2 u}{\partial y^2} \right) \tag{9.5}$$

$$u \frac{\partial v}{\partial x} + v \frac{\partial v}{\partial y} = -\frac{1}{\rho} \frac{\partial p}{\partial y} + \nu \left(\frac{\partial^2 v}{\partial x^2} + \frac{\partial^2 v}{\partial y^2} \right) \tag{9.6}$$

which express Newton's second law. In addition, the conservation of mass equation, Eq. 6.31, for incompressible flow is

$$\frac{\partial u}{\partial x} + \frac{\partial v}{\partial y} = 0 \tag{9.7}$$

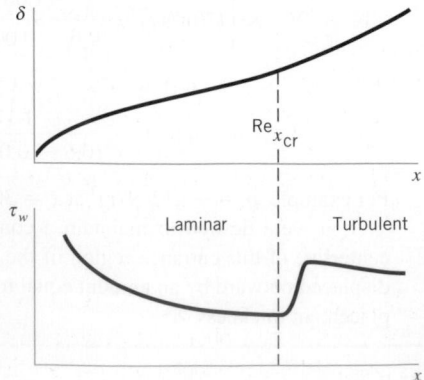

■ **FIGURE 9.9** **Typical characteristics of boundary layer thickness and wall shear stress for laminar and turbulent boundary layers.**

The appropriate boundary conditions are that the fluid velocity far from the body is the upstream velocity and that the fluid sticks to the solid body surfaces. Although the mathematical problem is well-posed, no one has obtained an analytical solution to these equations for flow past any shaped body! Currently much work is being done to obtain numerical solutions to these governing equations for many flow geometries.

By using boundary layer concepts introduced in the previous sections, Prandtl was able to impose certain approximations (valid for large Reynolds number flows), and thereby to simplify the governing equations. In 1908, H. Blasius (1883–1970), one of Prandtl's students, was able to solve these simplified equations for the boundary layer flow past a flat plate parallel to the flow. A brief outline of this technique and the results are presented below. Additional details may be found in the literature (Refs. 1–3).

Since the boundary layer is thin, it is expected that the component of velocity normal to the plate is much smaller than that parallel to the plate and that the rate of change of any parameter across the boundary layer should be much greater than that along the flow direction. That is,

$$v \ll u \quad \text{and} \quad \frac{\partial}{\partial x} \ll \frac{\partial}{\partial y}$$

Physically, the flow is primarily parallel to the plate and any fluid property is convected downstream much more quickly than it is diffused across the streamlines.

With these assumptions it can be shown that the governing equations (Eqs. 9.5, 9.6, and 9.7) reduce to the following boundary layer equations:

$$\frac{\partial u}{\partial x} + \frac{\partial v}{\partial y} = 0 \tag{9.8}$$

$$u \frac{\partial u}{\partial x} + v \frac{\partial u}{\partial y} = \nu \frac{\partial^2 u}{\partial y^2} \tag{9.9}$$

The Navier–Stokes equations can be simplified for boundary layer flow analysis.

Although both these boundary layer equations and the original Navier–Stokes equations are non-linear partial differential equations, there are considerable differences between them. For one, the y momentum equation has been eliminated, leaving only the original, unaltered continuity equation and a modified x momentum equation. One of the variables, the pressure, has been eliminated, leaving only the x and y components of velocity as unknowns. For boundary layer flow over a flat plate the pressure is constant throughout the fluid. The flow represents a balance between viscous and inertial effects, with pressure playing no role.

As shown by the figure in the margin, the boundary conditions for the governing boundary layer equations are that the fluid sticks to the plate

$$u = v = 0 \quad \text{on} \quad y = 0 \tag{9.10}$$

and that outside of the boundary layer the flow is the uniform upstream flow $u = U$. That is,

$$u \to U \quad \text{as} \quad y \to \infty \tag{9.11}$$

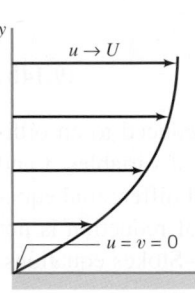

Mathematically, the upstream velocity is approached asymptotically as one moves away from the plate. Physically, the flow velocity is within 1% of the upstream velocity at a distance of δ from the plate.

In mathematical terms, the Navier–Stokes equations (Eqs. 9.5 and 9.6) and the continuity equation (Eq. 9.7) are elliptic equations, whereas the equations for boundary layer flow (Eqs. 9.8 and 9.9) are parabolic equations. The nature of the solutions to these two sets of equations, therefore, is different. Physically, this fact translates to the idea that what happens downstream of a given location in a boundary layer cannot affect what happens upstream of that point. That is, whether the plate shown in Fig. 9.5c ends with length ℓ or is extended to length 2ℓ, the flow within the first segment of length ℓ will be the same. In addition, the presence of the plate has no effect on the flow ahead of the plate. On the other hand, ellipticity allows flow information to propagate in all directions, including upstream.

In general, the solutions of nonlinear partial differential equations (such as the boundary layer equations, Eqs. 9.8 and 9.9) are extremely difficult to obtain. However, by applying a clever coordinate transformation and change of variables, Blasius reduced the partial differential equations to an ordinary differential equation that he was able to solve. A brief description of this process is given below. Additional details can be found in standard books dealing with boundary layer flow (Refs. 1, 2).

It can be argued that in dimensionless form the boundary layer velocity profiles on a flat plate should be similar regardless of the location along the plate. That is,

$$\frac{u}{U} = g\left(\frac{y}{\delta}\right)$$

where $g(y/\delta)$ is an unknown function to be determined. In addition, by applying an order of magnitude analysis of the forces acting on fluid within the boundary layer, it can be shown that the boundary layer thickness grows as the square root of x and inversely proportional to the square root of U. That is,

$$\delta \sim \left(\frac{\nu x}{U}\right)^{1/2}$$

Such a conclusion results from a balance between viscous and inertial forces within the boundary layer and from the fact that the velocity varies much more rapidly in the direction across the boundary layer than along it.

The boundary layer equations can be written in terms of a similarity variable.

Thus, we introduce the dimensionless *similarity variable* $\eta = (U/\nu x)^{1/2}y$ and the stream function $\psi = (\nu x U)^{1/2} f(\eta)$, where $f = f(\eta)$ is an unknown function. Recall from Section 6.2.3 that the velocity components for two-dimensional flow are given in terms of the stream function as $u = \partial\psi/\partial y$ and $v = -\partial\psi/\partial x$, which for this flow become

$$u = Uf'(\eta) \tag{9.12}$$

and

$$v = \left(\frac{\nu U}{4x}\right)^{1/2} (\eta f' - f) \tag{9.13}$$

with the notation $(\ )' = d/d\eta$. We substitute Eqs. 9.12 and 9.13 into the governing equations, Eqs. 9.8 and 9.9, to obtain (after considerable manipulation) the following nonlinear, third-order ordinary differential equation:

$$2f''' + ff'' = 0 \tag{9.14a}$$

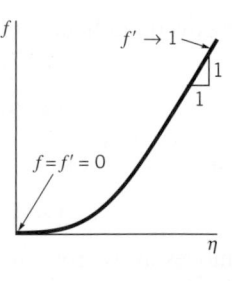

As shown by the figure in the margin, the boundary conditions given in Eqs. 9.10 and 9.11 can be written as

$$f = f' = 0 \text{ at } \eta = 0 \quad \text{and} \quad f' \to 1 \text{ as } \eta \to \infty \tag{9.14b}$$

The original partial differential equation and boundary conditions have been reduced to an ordinary differential equation by use of the similarity variable η. The two independent variables, x and y, were combined into the similarity variable in a fashion that reduced the partial differential equation (and boundary conditions) to an ordinary differential equation. This type of reduction is not generally possible. For example, this method does not work on the full Navier–Stokes equations, although it does on the boundary layer equations (Eqs. 9.8 and 9.9).

Although there is no known analytical solution to Eq. 9.14, it is relatively easy to integrate this equation on a computer. The dimensionless boundary layer profile, $u/U = f'(\eta)$, obtained by numerical solution of Eq. 9.14 (termed the Blasius solution), is sketched in Fig. 9.10a and is tabulated in Table 9.1. The velocity profiles at different x locations are similar in that there is only one curve necessary to describe the velocity at any point in the boundary layer. Because the similarity variable η contains both x and y, it is seen from Fig. 9.10b that the actual velocity profiles are a function of both x and y. The profile at location x_1 is the same as that at x_2 except that the y coordinate is stretched by a factor of $(x_2/x_1)^{1/2}$.

From the solution it is found that $u/U \approx 0.99$ when $\eta = 5.0$. Thus,

$$\delta = 5\sqrt{\frac{\nu x}{U}} \tag{9.15}$$

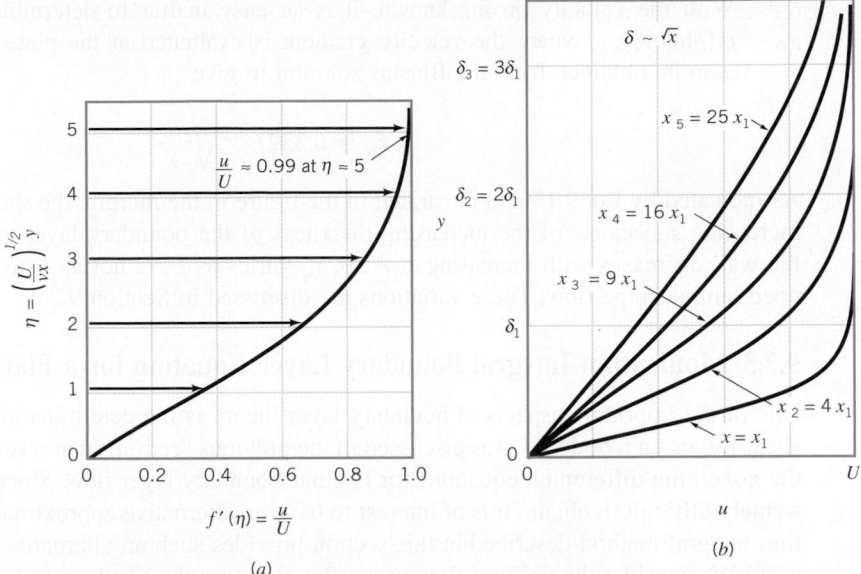

■ **FIGURE 9.10** Blasius boundary layer profile: (*a*) boundary layer profile in dimensionless form using the similarity variable η, (*b*) similar boundary layer profiles at different locations along the flat plate.

or

$$\frac{\delta}{x} = \frac{5}{\sqrt{Re_x}}$$

where $Re_x = Ux/\nu$. It can also be shown that the displacement and momentum thicknesses are given by

$$\frac{\delta^*}{x} = \frac{1.721}{\sqrt{Re_x}} \tag{9.16}$$

and

$$\frac{\Theta}{x} = \frac{0.664}{\sqrt{Re_x}} \tag{9.17}$$

As postulated, the boundary layer is thin provided that Re_x is large (i.e., $\delta/x \to 0$ as $Re_x \to \infty$).

For large Reynolds numbers the boundary layer is relatively thin.

■ **TABLE 9.1**

Laminar Flow along a Flat Plate (the Blasius Solution)

$\eta = y(U/\nu x)^{1/2}$	$f'(\eta) = u/U$	η	$f'(\eta)$
0	0	3.6	0.9233
0.4	0.1328	4.0	0.9555
0.8	0.2647	4.4	0.9759
1.2	0.3938	4.8	0.9878
1.6	0.5168	5.0	0.9916
2.0	0.6298	5.2	0.9943
2.4	0.7290	5.6	0.9975
2.8	0.8115	6.0	0.9990
3.2	0.8761	∞	1.0000

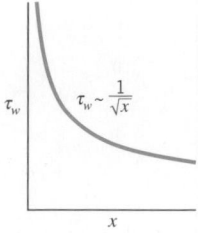

With the velocity profile known, it is an easy matter to determine the wall shear stress, $\tau_w = \mu(\partial u/\partial y)_{y=0}$, where the velocity gradient is evaluated at the plate. The value of $\partial u/\partial y$ at $y = 0$ can be obtained from the Blasius solution to give

$$\tau_w = 0.332 U^{3/2} \sqrt{\frac{\rho\mu}{x}} \tag{9.18}$$

As indicated by Eq. 9.18 and illustrated in the figure in the margin, the shear stress decreases with increasing x because of the increasing thickness of the boundary layer—the velocity gradient at the wall decreases with increasing x. Also, τ_w varies as $U^{3/2}$, not as U as it does for fully developed laminar pipe flow. These variations are discussed in Section 9.2.3.

9.2.3 Momentum Integral Boundary Layer Equation for a Flat Plate

One of the important aspects of boundary layer theory is the determination of the drag caused by shear forces on a body. As was discussed in the previous section, such results can be obtained from the governing differential equations for laminar boundary layer flow. Since these solutions are extremely difficult to obtain, it is of interest to have an alternative approximate method. The momentum integral method described in this section provides such an alternative.

We consider the uniform flow past a flat plate and the fixed control volume as shown in Fig. 9.11. In agreement with advanced theory and experiment, we assume that the pressure is constant throughout the flow field. The flow entering the control volume at the leading edge of the plate [section (1)] is uniform, while the velocity of the flow exiting the control volume [section (2)] varies from the upstream velocity at the edge of the boundary layer to zero velocity on the plate.

The fluid adjacent to the plate makes up the lower portion of the control surface. The upper surface coincides with the streamline just outside the edge of the boundary layer at section (2). It need not (in fact, does not) coincide with the edge of the boundary layer except at section (2). If we apply the x component of the momentum equation (Eq. 5.22) to the steady flow of fluid within this control volume we obtain

$$\sum F_x = \rho \int_{(1)} u\mathbf{V} \cdot \hat{\mathbf{n}} \, dA + \rho \int_{(2)} u\mathbf{V} \cdot \hat{\mathbf{n}} \, dA$$

where for a plate of width b

$$\sum F_x = -\mathcal{D} = -\int_{\text{plate}} \tau_w \, dA = -b \int_{\text{plate}} \tau_w \, dx \tag{9.19}$$

and $\mathcal{D}$ is the drag that the plate exerts on the fluid. Note that the net force caused by the uniform pressure distribution does not contribute to this flow. Since the plate is solid and the upper surface of the control volume is a streamline, there is no flow through these areas. Thus,

$$-\mathcal{D} = \rho \int_{(1)} U(-U) \, dA + \rho \int_{(2)} u^2 \, dA$$

The drag on a flat plate depends on the velocity profile within the boundary layer.

or

$$\mathcal{D} = \rho U^2 bh - \rho b \int_0^\delta u^2 \, dy \tag{9.20}$$

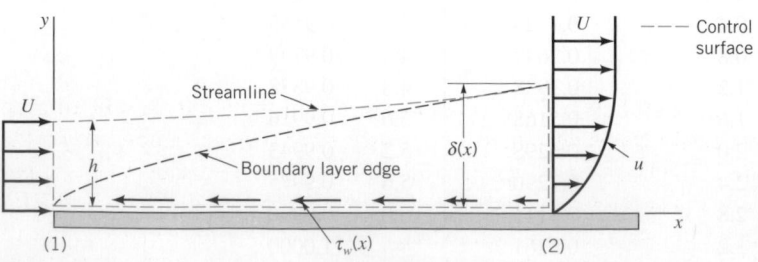

■ **FIGURE 9.11** **Control volume used in the derivation of the momentum integral equation for boundary layer flow.**

Although the height h is not known, it is known that for conservation of mass the flowrate through section (1) must equal that through section (2), or

$$Uh = \int_0^\delta u \, dy$$

which can be written as

$$\rho U^2 bh = \rho b \int_0^\delta Uu \, dy \qquad (9.21)$$

Thus, by combining Eqs. 9.20 and 9.21 we obtain the drag in terms of the deficit of momentum flux across the outlet of the control volume as

$$\mathcal{D} = \rho b \int_0^\delta u(U - u) \, dy \qquad (9.22)$$

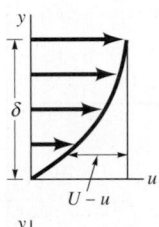

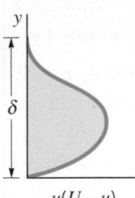

The idea of a momentum deficit is illustrated in the figure in the margin. If the flow were inviscid, the drag would be zero, since we would have $u \equiv U$ and the right-hand side of Eq. 9.22 would be zero. (This is consistent with the fact that $\tau_w = 0$ if $\mu = 0$.) Equation 9.22 points out the important fact that boundary layer flow on a flat plate is governed by a balance between shear drag (the left-hand side of Eq. 9.22) and a decrease in the momentum of the fluid (the right-hand side of Eq. 9.22). As x increases, δ increases and the drag increases. The thickening of the boundary layer is necessary to overcome the drag of the viscous shear stress on the plate. This is contrary to horizontal fully developed pipe flow in which the momentum of the fluid remains constant and the shear force is overcome by the pressure gradient along the pipe.

The development of Eq. 9.22 and its use was first put forth in 1921 by T. von Kármán (1881–1963), a Hungarian/German aerodynamicist. By comparing Eqs. 9.22 and 9.4 we see that the drag can be written in terms of the momentum thickness, Θ, as

$$\mathcal{D} = \rho b U^2 \, \Theta \qquad (9.23)$$

Note that this equation is valid for laminar or turbulent flows.

The shear stress distribution can be obtained from Eq. 9.23 by differentiating both sides with respect to x to obtain

$$\frac{d\mathcal{D}}{dx} = \rho b U^2 \frac{d\Theta}{dx} \qquad (9.24)$$

The increase in drag per length of the plate, $d\mathcal{D}/dx$, occurs at the expense of an increase of the momentum boundary layer thickness, which represents a decrease in the momentum of the fluid.

Since $d\mathcal{D} = \tau_w \, b \, dx$ (see Eq. 9.19) it follows that

$$\frac{d\mathcal{D}}{dx} = b\tau_w \qquad (9.25)$$

Hence, by combining Eqs. 9.24 and 9.25 we obtain the *momentum integral equation* for the boundary layer flow on a flat plate

$$\tau_w = \rho U^2 \frac{d\Theta}{dx} \qquad (9.26)$$

The usefulness of this relationship lies in the ability to obtain approximate boundary layer results easily by using rather crude assumptions. For example, if we knew the detailed velocity profile in the boundary layer (i.e., the Blasius solution discussed in the previous section), we could evaluate either the right-hand side of Eq. 9.23 to obtain the drag, or the right-hand side of Eq. 9.26 to obtain the shear stress. Fortunately, even a rather crude guess at the velocity profile will allow us to obtain reasonable drag and shear stress results from Eq. 9.26. This method is illustrated in Example 9.4.

EXAMPLE 9.4 | Momentum Integral Boundary Layer Equation

GIVEN Consider the laminar flow of an incompressible fluid past a flat plate at $y = 0$. The boundary layer velocity profile is approximated as $u = Uy/\delta$ for $0 \leq y \leq \delta$ and $u = U$ for $y > \delta$, as is shown in Fig. E9.4.

FIND Determine the shear stress by using the momentum integral equation. Compare these results with the Blasius results given by Eq. 9.18.

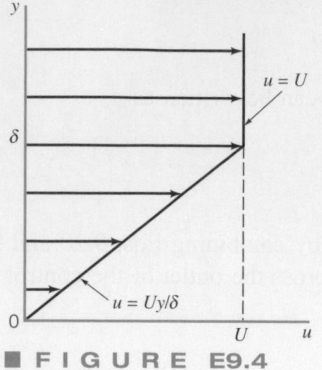

■ **FIGURE E9.4**

SOLUTION

From Eq. 9.26 the shear stress is given by

$$\tau_w = \rho U^2 \frac{d\Theta}{dx} \qquad (1)$$

while for laminar flow we know that $\tau_w = \mu(\partial u/\partial y)_{y=0}$. For the assumed profile we have

$$\tau_w = \mu \frac{U}{\delta} \qquad (2)$$

and from Eq. 9.4

$$\Theta = \int_0^\infty \frac{u}{U}\left(1 - \frac{u}{U}\right) dy = \int_0^\delta \frac{u}{U}\left(1 - \frac{u}{U}\right) dy$$

$$= \int_0^\delta \left(\frac{y}{\delta}\right)\left(1 - \frac{y}{\delta}\right) dy$$

or

$$\Theta = \frac{\delta}{6} \qquad (3)$$

Note that as yet we do not know the value of δ (but suspect that it should be a function of x).

By combining Eqs. 1, 2, and 3 we obtain the following differential equation for δ:

$$\frac{\mu U}{\delta} = \frac{\rho U^2}{6} \frac{d\delta}{dx}$$

or

$$\delta \, d\delta = \frac{6\mu}{\rho U} dx$$

This can be integrated from the leading edge of the plate, $x = 0$ (where $\delta = 0$) to an arbitrary location x where the boundary layer thickness is δ. The result is

$$\frac{\delta^2}{2} = \frac{6\mu}{\rho U} x$$

or

$$\delta = 3.46 \sqrt{\frac{\nu x}{U}} \qquad (4)$$

Note that this approximate result (i.e., the velocity profile is not actually the simple straight line we assumed) compares favorably with the (much more laborious to obtain) Blasius result given by Eq. 9.15.

The wall shear stress can also be obtained by combining Eqs. 1, 3, and 4 to give

$$\tau_w = 0.289 U^{3/2} \sqrt{\frac{\rho\mu}{x}} \qquad \textbf{(Ans)}$$

Again this approximate result is close (within 13%) to the Blasius value of τ_w given by Eq. 9.18.

As is illustrated in Example 9.4, the momentum integral equation, Eq. 9.26, can be used along with an assumed velocity profile to obtain reasonable, approximate boundary layer results. The accuracy of these results depends on how closely the shape of the assumed velocity profile approximates the actual profile.

Thus, we consider a general velocity profile

$$\frac{u}{U} = g(Y) \quad \text{for} \quad 0 \leq Y \leq 1$$

and

Approximate velocity profiles are used in the momentum integral equation.

$$\frac{u}{U} = 1 \quad \text{for} \quad Y > 1$$

where the dimensionless coordinate $Y = y/\delta$ varies from 0 to 1 across the boundary layer. The dimensionless function $g(Y)$ can be any shape we choose, although it should be a reasonable

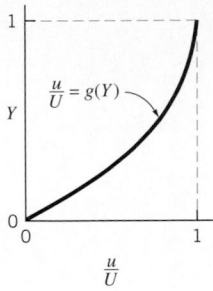

approximation to the boundary layer profile, as shown by the figure in the margin. In particular, it should certainly satisfy the boundary conditions $u = 0$ at $y = 0$ and $u = U$ at $y = \delta$. That is,

$$g(0) = 0 \quad \text{and} \quad g(1) = 1$$

The linear function $g(Y) = Y$ used in Example 9.4 is one such possible profile. Other conditions, such as $dg/dY = 0$ at $Y = 1$ (i.e., $\partial u/\partial y = 0$ at $y = \delta$), could also be incorporated into the function $g(Y)$ to more closely approximate the actual profile.

For a given $g(Y)$, the drag can be determined from Eq. 9.22 as

$$\mathscr{D} = \rho b \int_0^{\delta} u(U - u)\, dy = \rho b U^2 \delta \int_0^1 g(Y)[1 - g(Y)]\, dY$$

or

$$\mathscr{D} = \rho b U^2 \delta C_1 \tag{9.27}$$

where the dimensionless constant C_1 has the value

$$C_1 = \int_0^1 g(Y)[1 - g(Y)]\, dY$$

Also, the wall shear stress can be written as

$$\tau_w = \mu \left.\frac{\partial u}{\partial y}\right|_{y=0} = \frac{\mu U}{\delta} \left.\frac{dg}{dY}\right|_{Y=0} = \frac{\mu U}{\delta} C_2 \tag{9.28}$$

where the dimensionless constant C_2 has the value

$$C_2 = \left.\frac{dg}{dY}\right|_{Y=0}$$

By combining Eqs. 9.25, 9.27, and 9.28 we obtain

$$\delta\, d\delta = \frac{\mu C_2}{\rho U C_1}\, dx$$

which can be integrated from $\delta = 0$ at $x = 0$ to give

$$\delta = \sqrt{\frac{2\nu C_2 x}{U C_1}}$$

or

$$\frac{\delta}{x} = \frac{\sqrt{2 C_2/C_1}}{\sqrt{\text{Re}_x}} \tag{9.29}$$

By substituting this expression back into Eqs. 9.28 we obtain

$$\tau_w = \sqrt{\frac{C_1 C_2}{2}}\, U^{3/2} \sqrt{\frac{\rho\mu}{x}} \tag{9.30}$$

Approximate boundary layer results are obtained from the momentum integral equation.

To use Eqs. 9.29 and 9.30 we must determine the values of C_1 and C_2. Several assumed velocity profiles and the resulting values of δ are given in Fig. 9.12 and Table 9.2. The more closely the assumed shape approximates the actual (i.e., Blasius) profile, the more accurate the final results. For any assumed profile shape, the functional dependence of δ and τ_w on the physical parameters ρ, μ, U, and x is the same. Only the constants are different. That is, $\delta \sim (\mu x/\rho U)^{1/2}$ or $\delta \text{Re}_x^{1/2}/x = $ constant, and $\tau_w \sim (\rho \mu U^3/x)^{1/2}$, where $\text{Re}_x = \rho U x/\mu$.

It is often convenient to use the dimensionless *local friction coefficient*, c_f, defined as

$$c_f = \frac{\tau_w}{\frac{1}{2}\rho U^2} \tag{9.31}$$

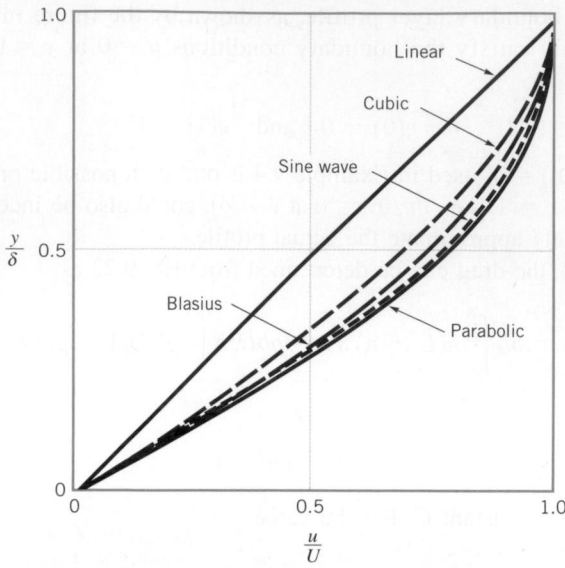

■ **FIGURE 9.12** Typical approximate boundary layer profiles used in the momentum integral equation.

to express the wall shear stress. From Eq. 9.30 we obtain the approximate value

$$c_f = \sqrt{2C_1 C_2} \sqrt{\frac{\mu}{\rho U x}} = \frac{\sqrt{2C_1 C_2}}{\sqrt{Re_x}}$$

while the Blasius solution result is given by

$$c_f = \frac{0.664}{\sqrt{Re_x}} \tag{9.32}$$

These results are also indicated in Table 9.2.

For a flat plate of length ℓ and width b, the net friction drag, $\mathscr{D}_f$, can be expressed in terms of the *friction drag coefficient*, C_{Df}, as

$$C_{Df} = \frac{\mathscr{D}_f}{\frac{1}{2}\rho U^2 b \ell} = \frac{b \int_0^\ell \tau_w \, dx}{\frac{1}{2}\rho U^2 b \ell}$$

The friction drag coefficient is an integral of the local friction coefficient.

or

$$C_{Df} = \frac{1}{\ell} \int_0^\ell c_f \, dx \tag{9.33}$$

■ **TABLE 9.2**

Flat Plate Momentum Integral Results for Various Assumed Laminar Flow Velocity Profiles

Profile Character	$\delta Re_x^{1/2}/x$	$c_f Re_x^{1/2}$	$C_{Df} Re_\ell^{1/2}$
a. Blasius solution	5.00	0.664	1.328
b. Linear $u/U = y/\delta$	3.46	0.578	1.156
c. Parabolic $u/U = 2y/\delta - (y/\delta)^2$	5.48	0.730	1.460
d. Cubic $u/U = 3(y/\delta)/2 - (y/\delta)^3/2$	4.64	0.646	1.292
e. Sine wave $u/U = \sin[\pi(y/\delta)/2]$	4.79	0.655	1.310

We use the above approximate value of $c_f = (2C_1C_2\mu/\rho Ux)^{1/2}$ to obtain

$$C_{Df} = \frac{\sqrt{8C_1C_2}}{\sqrt{\text{Re}_\ell}}$$

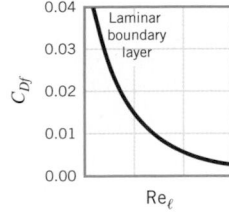

where $\text{Re}_\ell = U\ell/\nu$ is the Reynolds number based on the plate length. The corresponding value obtained from the Blasius solution (Eq. 9.32) and shown by the figure in the margin gives

$$C_{Df} = \frac{1.328}{\sqrt{\text{Re}_\ell}}$$

These results are also indicated in Table 9.2.

The momentum integral boundary layer method provides a relatively simple technique to obtain useful boundary layer results. As is discussed in Sections 9.2.5 and 9.2.6, this technique can be extended to boundary layer flows on curved surfaces (where the pressure and fluid velocity at the edge of the boundary layer are not constant) and to turbulent flows.

9.2.4 Transition from Laminar to Turbulent Flow

The boundary layer on a flat plate will become turbulent if the plate is long enough.

The analytical results given in Table 9.2 are restricted to laminar boundary layer flows along a flat plate with zero pressure gradient. They agree quite well with experimental results up to the point where the boundary layer flow becomes turbulent, which will occur for any free-stream velocity and any fluid provided the plate is long enough. This is true because the parameter that governs the *transition* to turbulent flow is the Reynolds number—in this case the Reynolds number based on the distance from the leading edge of the plate, $\text{Re}_x = Ux/\nu$.

The value of the Reynolds number at the transition location is a rather complex function of various parameters involved, including the roughness of the surface, the curvature of the surface (for example, a flat plate or a sphere), and some measure of the disturbances in the flow outside the boundary layer. On a flat plate with a sharp leading edge in a typical airstream, the transition takes place at a distance x from the leading edge given by $\text{Re}_{xcr} = 2 \times 10^5$ to 3×10^6. Unless otherwise stated, we will use $\text{Re}_{xcr} = 5 \times 10^5$ in our calculations.

The actual transition from laminar to turbulent boundary layer flow may occur over a region of the plate, not at a specific single location. This occurs, in part, because of the spottiness of the transition. Typically, the transition begins at random locations on the plate in the vicinity of $\text{Re}_x = \text{Re}_{xcr}$. These spots grow rapidly as they are convected downstream until the entire width of the plate is covered with turbulent flow. The photo shown in Fig. 9.13 illustrates this transition process.

The complex process of transition from laminar to turbulent flow involves the instability of the flow field. Small disturbances imposed on the boundary layer flow (i.e., from a vibration of the plate, a roughness of the surface, or a "wiggle" in the flow past the plate) will either grow (instability) or decay (stability), depending on where the disturbance is introduced into the flow. If these disturbances occur at a location with $\text{Re}_x < \text{Re}_{xcr}$ they will die out, and the boundary layer will return to laminar flow at that location. Disturbances imposed at a location with $\text{Re}_x > \text{Re}_{xcr}$ will grow and transform the boundary layer flow downstream of this location into turbulence. The study of the initiation, growth, and structure of these turbulent bursts or spots is an active area of fluid mechanics research.

V9.5 Transition on flat plate

■ **FIGURE 9.13**
Turbulent spots and the transition from laminar to turbulent boundary layer flow on a flat plate. Flow from left to right. (Photograph courtesy of B. Cantwell, Stanford University.)

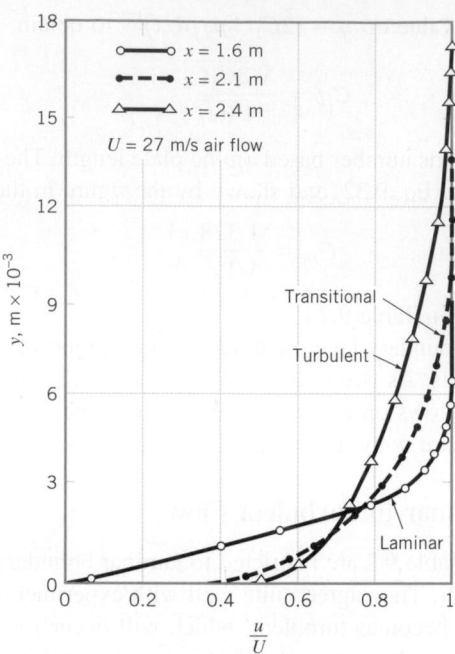

■ **FIGURE 9.14** Typical boundary layer profiles on a flat plate for laminar, transitional, and turbulent flow (Ref. 1).

Transition from laminar to turbulent flow also involves a noticeable change in the shape of the boundary layer velocity profile. Typical profiles obtained in the neighborhood of the transition location are indicated in Fig. 9.14. The turbulent profiles are flatter, have a larger velocity gradient at the wall, and produce a larger boundary layer thickness than do the laminar profiles.

EXAMPLE 9.5 | Boundary Layer Transition

GIVEN A fluid flows steadily past a flat plate with a velocity of $U = 3$ m/s.

FIND At approximately what location will the boundary layer become turbulent, and how thick is the boundary layer at that point if the fluid is (a) water at 15 °C, (b) standard air, or (c) glycerin at 20 °C?

SOLUTION

For any fluid, the laminar boundary layer thickness is found from Eq. 9.15 as

$$\delta = 5 \sqrt{\frac{\nu x}{U}}$$

The boundary layer remains laminar up to

$$x_{cr} = \frac{\nu \mathrm{Re}_{xcr}}{U}$$

Thus, if we assume $\mathrm{Re}_{xcr} = 5 \times 10^5$ we obtain

$$x_{cr} = \frac{5 \times 10^5}{3 \text{ m/s}} \nu = 1.7 \times 10^5 \, \nu$$

and

$$\delta_{cr} \equiv \delta|_{x=x_{cr}} = 5 \left[\frac{\nu}{3} \left(1.7 \times 10^5 \, \nu \right) \right]^{1/2} = 1190 \, \nu$$

■ **TABLE E9.5**

Fluid	ν (m²/s)	x_{cr} (m)	δ_{cr} (m)
a. Water	1.12×10^{-6}	0.190	1.3×10^{-3}
b. Air	1.46×10^{-5}	2.482	0.017
c. Glycerin	1.19×10^{-3}	202.3	1.42

(Ans)

where ν is in m²/s and x_{cr} and δ_{cr} are in meters. The values of the kinematic viscosity obtained from Tables 1.3 and 1.4 are listed in Table E9.5 along with the corresponding x_{cr} and δ_{cr}.

COMMENT Laminar flow can be maintained on a longer portion of the plate if the viscosity is increased. However, the boundary layer flow eventually becomes turbulent, provided the plate is long enough. Similarly, the boundary layer thickness is greater if the viscosity is increased.

9.2.5 Turbulent Boundary Layer Flow

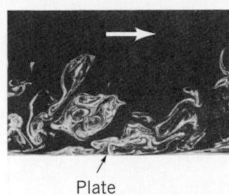

Plate

The structure of turbulent boundary layer flow is very complex, random, and irregular. It shares many of the characteristics described for turbulent pipe flow in Section 8.3. In particular, the velocity at any given location in the flow is unsteady in a random fashion. The flow can be thought of as a jumbled mix of intertwined eddies (or swirls) of different sizes (diameters and angular velocities). The figure in the margin shows a laser-induced fluorescence visualization of a turbulent boundary layer on a flat plate (side view). The various fluid quantities involved (i.e., mass, momentum, energy) are convected downstream in the free-stream direction as in a laminar boundary layer. For turbulent flow they are also convected across the boundary layer (in the direction perpendicular to the plate) by the random transport of finite-sized fluid particles associated with the turbulent eddies. There is considerable mixing involved with these finite-sized eddies—considerably more than is associated with the mixing found in laminar flow where it is confined to the molecular scale. Although there is considerable random motion of fluid particles perpendicular to the plate, there is very little net transfer of mass across the boundary layer—the largest flowrate by far is parallel to the plate.

Random transport of finite-sized fluid particles occurs within turbulent boundary layers.

There is, however, a considerable net transfer of x component of momentum perpendicular to the plate because of the random motion of the particles. Fluid particles moving toward the plate (in the negative y direction) have some of their excess momentum (they come from areas of higher velocity) removed by the plate. Conversely, particles moving away from the plate (in the positive y direction) gain momentum from the fluid (they come from areas of lower velocity). The net result is that the plate acts as a momentum sink, continually extracting momentum from the fluid. For laminar flows, such cross-stream transfer of these properties takes place solely on the molecular scale. For turbulent flow the randomness is associated with fluid particle mixing. Consequently, the shear force for turbulent boundary layer flow is considerably greater than it is for laminar boundary layer flow (see Section 8.3.2).

There are no "exact" solutions for turbulent boundary layer flow. As is discussed in Section 9.2.2, it is possible to solve the Prandtl boundary layer equations for laminar flow past a flat plate to obtain the Blasius solution (which is "exact" within the framework of the assumptions involved in the boundary layer equations). Since there is no precise expression for the shear stress in turbulent flow (see Section 8.3), solutions are not available for turbulent flow. However, considerable headway has been made in obtaining numerical (computer) solutions for turbulent flow by using approximate shear stress relationships. Also, progress is being made in the area of direct, full numerical integration of the basic governing equations, the Navier–Stokes equations.

Approximate turbulent boundary layer results can also be obtained by use of the momentum integral equation, Eq. 9.26, which is valid for either laminar or turbulent flow. What is needed for the use of this equation are reasonable approximations to the velocity profile $u = U\, g(Y)$, where $Y = y/\delta$ and u is the time-averaged velocity (the overbar notation, $\bar{u}$, of Section 8.3.2 has been dropped for convenience), and a functional relationship describing the wall shear stress. For laminar flow the wall shear stress was used as $\tau_w = \mu(\partial u/\partial y)_{y=0}$. In theory, such a technique should work for turbulent boundary layers also. However, as is discussed in Section 8.3, the details of the velocity gradient at the wall are not well understood for turbulent flow. Thus, it is necessary to use some empirical relationship for the wall shear stress. This is illustrated in Example 9.6.

EXAMPLE 9.6   Turbulent Boundary Layer Properties

GIVEN Consider turbulent flow of an incompressible fluid past a flat plate. The boundary layer velocity profile is assumed to be $u/U = (y/\delta)^{1/7} = Y^{1/7}$ for $Y = y/\delta \leq 1$ and $u = U$ for $Y > 1$ as shown in Fig. E9.6. This is a reasonable approximation of experimentally observed profiles, except very near the plate where this formula gives $\partial u/\partial y = \infty$ at $y = 0$. Note the differences between the assumed turbulent profile and the laminar profile. Also assume that the shear stress agrees with the experimentally determined formula:

$$\tau_w = 0.0225\rho U^2 \left(\frac{\nu}{U\delta}\right)^{1/4} \tag{1}$$

FIND Determine the boundary layer thicknesses δ, δ^*, and Θ and the wall shear stress, τ_w, as a function of x. Determine the friction drag coefficient, C_{Df}.

SOLUTION

Whether the flow is laminar or turbulent, it is true that the drag force is accounted for by a reduction in the momentum of the fluid flowing past the plate. The shear is obtained from Eq. 9.26 in terms of the rate at which the momentum boundary layer thickness, Θ, increases with distance along the plate as

$$\tau_w = \rho U^2 \frac{d\Theta}{dx}$$

For the assumed velocity profile, the boundary layer momentum thickness is obtained from Eq. 9.4 as

$$\Theta = \int_0^\infty \frac{u}{U}\left(1 - \frac{u}{U}\right) dy = \delta \int_0^1 \frac{u}{U}\left(1 - \frac{u}{U}\right) dY$$

or by integration

$$\Theta = \delta \int_0^1 Y^{1/7}\left(1 - Y^{1/7}\right) dY = \frac{7}{72}\delta \qquad (2)$$

where δ is an unknown function of x. By combining the assumed shear force dependence (Eq. 1) with Eq. 2, we obtain the following differential equation for δ:

$$0.0225\rho U^2\left(\frac{\nu}{U\delta}\right)^{1/4} = \frac{7}{72}\rho U^2 \frac{d\delta}{dx}$$

or

$$\delta^{1/4}\, d\delta = 0.231\left(\frac{\nu}{U}\right)^{1/4} dx$$

This can be integrated from $\delta = 0$ at $x = 0$ to obtain

$$\delta = 0.370\left(\frac{\nu}{U}\right)^{1/5} x^{4/5} \qquad (3) \quad \text{(Ans)}$$

or in dimensionless form

$$\frac{\delta}{x} = \frac{0.370}{\text{Re}_x^{1/5}}$$

Strictly speaking, the boundary layer near the leading edge of the plate is laminar, not turbulent, and the precise boundary condition should be the matching of the initial turbulent boundary layer thickness (at the transition location) with the thickness of the laminar boundary layer at that point. In practice, however, the laminar boundary layer often exists over a relatively short portion of the plate, and the error associated with starting the turbulent boundary layer with $\delta = 0$ at $x = 0$ can be negligible.

The displacement thickness, δ^*, and the momentum thickness, Θ, can be obtained from Eqs. 9.3 and 9.4 by integrating as follows:

$$\delta^* = \int_0^\infty \left(1 - \frac{u}{U}\right) dy = \delta \int_0^1 \left(1 - \frac{u}{U}\right) dY$$

$$= \delta \int_0^1 \left(1 - Y^{1/7}\right) dY = \frac{\delta}{8}$$

Thus, by combining this with Eq. 3 we obtain

$$\delta^* = 0.0463\left(\frac{\nu}{U}\right)^{1/5} x^{4/5} \qquad \text{(Ans)}$$

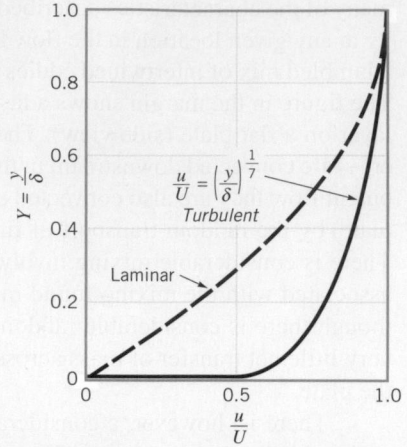

■ **FIGURE E9.6**

Similarly, from Eq. 2,

$$\Theta = \tfrac{7}{72}\delta = 0.0360\left(\frac{\nu}{U}\right)^{1/5} x^{4/5} \qquad (4) \quad \text{(Ans)}$$

The functional dependence for δ, δ^*, and Θ is the same; only the constants of proportionality are different. Typically, $\Theta < \delta^* < \delta$.

By combining Eqs. 1 and 3, we obtain the following result for the wall shear stress

$$\tau_w = 0.0225\rho U^2\left[\frac{\nu}{U(0.370)(\nu/U)^{1/5}x^{4/5}}\right]^{1/4}$$

$$= \frac{0.0288\rho U^2}{\text{Re}_x^{1/5}} \qquad \text{(Ans)}$$

This can be integrated over the length of the plate to obtain the friction drag on one side of the plate, $\mathcal{D}_f$, as

$$\mathcal{D}_f = \int_0^\ell b\tau_w\, dx = b(0.0288\rho U^2)\int_0^\ell \left(\frac{\nu}{Ux}\right)^{1/5} dx$$

or

$$\mathcal{D}_f = 0.0360\rho U^2 \frac{A}{\text{Re}_\ell^{1/5}}$$

where $A = b\ell$ is the area of the plate. (This result can also be obtained by combining Eq. 9.23 and the expression for the momentum thickness given in Eq. 4.) The corresponding friction drag coefficient, C_{Df}, is

$$C_{Df} = \frac{\mathcal{D}_f}{\tfrac{1}{2}\rho U^2 A} = \frac{0.0720}{\text{Re}_\ell^{1/5}} \qquad \text{(Ans)}$$

COMMENT Note that for the turbulent boundary layer flow the boundary layer thickness increases with x as $\delta \sim x^{4/5}$ and the shear stress decreases as $\tau_w \sim x^{-1/5}$. For laminar flow these dependencies are $x^{1/2}$ and $x^{-1/2}$, respectively. The random character of the turbulent flow causes a different structure of the flow.

Obviously the results presented in this example are valid only in the range of validity of the original data—the assumed velocity profile and shear stress. This range covers smooth flat plates with $5 \times 10^5 < \text{Re}_\ell < 10^7$.

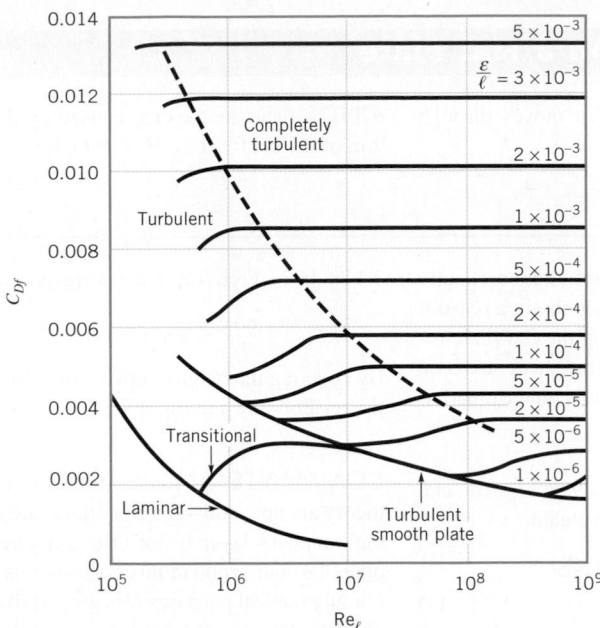

■ **F I G U R E   9.15** **Friction drag coefficient for a flat plate parallel to the upstream flow (Ref. 18, with permission).**

In general, the drag coefficient for a flat plate of length ℓ is a function of the Reynolds number, Re_ℓ, and the relative roughness, ε/ℓ. The results of numerous experiments covering a wide range of the parameters of interest are shown in Fig. 9.15. For laminar boundary layer flow the drag coefficient is a function of only the Reynolds number—surface roughness is not important. This is similar to laminar flow in a pipe. However, for turbulent flow, the surface roughness does affect the shear stress and, hence, the drag coefficient. This is similar to turbulent pipe flow in which the surface roughness may protrude into or through the viscous sublayer next to the wall and alter the flow in this thin, but very important, layer (see Section 8.4.1). Values of the roughness, ε, for different materials can be obtained from Table 8.1.

The flat plate drag coefficient is a function of relative roughness and Reynolds number.

The drag coefficient diagram of Fig. 9.15 (boundary layer flow) shares many characteristics in common with the familiar Moody diagram (pipe flow) of Fig. 8.23, even though the mechanisms governing the flow are quite different. Fully developed horizontal pipe flow is governed by a balance between pressure forces and viscous forces. The fluid inertia remains constant throughout the flow. Boundary layer flow on a horizontal flat plate is governed by a balance between inertia effects and viscous forces. The pressure remains constant throughout the flow. (As is discussed in Section 9.2.6, for boundary layer flow on curved surfaces, the pressure is not constant.)

It is often convenient to have an equation for the drag coefficient as a function of the Reynolds number and relative roughness rather than the graphical representation given in Fig. 9.15. Although there is not one equation valid for the entire $Re_\ell - \varepsilon/\ell$ range, the equations presented in Table 9.3 do work well for the conditions indicated.

■ **TABLE 9.3**

Empirical Equations for the Flat Plate Drag Coefficient (Ref. 1)

Equation	Flow Conditions
$C_{Df} = 1.328/(Re_\ell)^{0.5}$	Laminar flow
$C_{Df} = 0.455/(\log Re_\ell)^{2.58} - 1700/Re_\ell$	Transitional with $Re_{xcr} = 5 \times 10^5$
$C_{Df} = 0.455/(\log Re_\ell)^{2.58}$	Turbulent, smooth plate
$C_{Df} = [1.89 - 1.62 \log(\varepsilon/\ell)]^{-2.5}$	Completely turbulent

EXAMPLE 9.7 | Drag on a Flat Plate

GIVEN The water ski shown in Fig. E9.7a moves through 20 °C water with a velocity U.

FIND Estimate the drag caused by the shear stress on the bottom of the ski for $0 < U < 9$ m/s.

SOLUTION

Clearly the ski is not a flat plate, and it is not aligned exactly parallel to the upstream flow. However, we can obtain a reasonable approximation to the shear force by using the flat plate results. That is, the friction drag, $\mathcal{D}_f$, caused by the shear stress on the bottom of the ski (the wall shear stress) can be determined as

$$\mathcal{D}_f = \tfrac{1}{2}\rho U^2 \ell b C_{Df}$$

With $A = \ell b = 1$ m $\times$ 0.15 m $= 0.15$ m^2, $\rho = 998.2$ kg/m^3, and $\mu = 1.002 \times 10^{-3}$ N$\cdot$s/m^2 (see Table B.1) we obtain

$$\mathcal{D}_f = \tfrac{1}{2}(998.2 \text{ kg/m}^3)(0.15 \text{ m}^2)U^2 C_{Df}$$
$$= 75\, U^2 C_{Df} \tag{1}$$

where $\mathcal{D}_f$ and U are in N and m/s, respectively.

The friction coefficient, C_{Df}, can be obtained from Fig. 9.15 or from the appropriate equations given in Table 9.3. As we will see, for this problem, much of the flow lies within the transition regime where both the laminar and turbulent portions of the boundary layer flow occupy comparable lengths of the plate. We choose to use the values of C_{Df} from the table.

For the given conditions we obtain

$$\text{Re}_\ell = \frac{\rho U \ell}{\mu} = \frac{(998.2 \text{ kg/m}^3)(1 \text{ m})U}{1.002 \times 10^{-3} \text{ N}\cdot\text{s/m}^2} = 9.96 \times 10^5\, U$$

where U is in m/s. With $U = 3$ m/s, or $\text{Re}_\ell = 3 \times 10^6$, we obtain from Table 9.3 $C_{Df} = 0.455/(\log \text{Re}_\ell)^{2.58} - 1700/\text{Re}_\ell =$

3.1×10^{-3}. From Eq. 1 the corresponding drag is

$$\mathcal{D}_f = \tfrac{1}{2} \times 998.2 \times (3)^2 \times (0.15) \times (3.1 \times 10^{-3}) = 2.1 \text{ N}$$

By covering the range of upstream velocities of interest we obtain the results shown in Fig. E9.7b. **(Ans)**

COMMENTS If Re $\lesssim 1000$, the results of boundary layer theory are not valid—inertia effects are not dominant enough and the boundary layer is not thin compared with the length of the plate. For our problem this corresponds to $U = 1.0 \times 10^{-3}$ m/s. For all practical purposes U is greater than this value, and the flow past the ski is of the boundary layer type.

The approximate location of the transition from laminar to turbulent boundary layer flow as defined by $\text{Re}_{cr} = \rho U x_{cr}/\mu = 4.5 \times 10^5$ is indicated in Fig. E9.7b. Up to $U = 0.4$ m/s the entire boundary layer is laminar. The fraction of the boundary layer that is laminar decreases as U increases until only the front 0.05 m is laminar when $U = 9$ m/s.

For anyone who has water skied, it is clear that it can require considerably more force to be pulled along at 9 m/s than the 2×22 N $= 44$ N (two skis) indicated in Fig. E9.7b. As is discussed in Section 9.3, the total drag on an object such as a water ski consists of more than just the friction drag. Other components, including pressure drag and wave-making drag, add considerably to the total resistance.

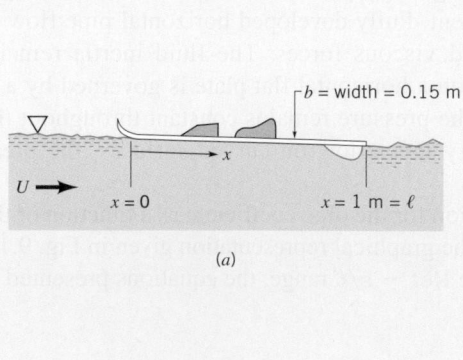

(a)

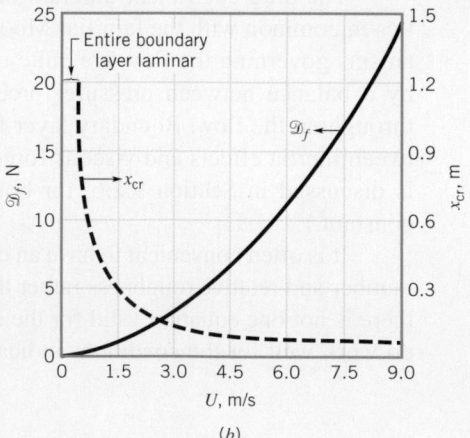

(b)

■ **FIGURE E9.7**

9.2.6 Effects of Pressure Gradient

The boundary layer discussions in the previous parts of Section 9.2 have dealt with flow along a flat plate in which the pressure is constant throughout the fluid. In general, when a fluid flows past an object other than a flat plate, the pressure field is not uniform. As shown in Fig. 9.6, if the Reynolds number is large, relatively thin boundary layers will develop along the surfaces. Within

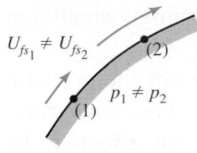

The free-stream velocity on a curved surface is not constant.

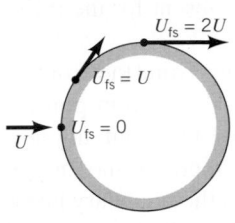

these layers the component of the pressure gradient in the streamwise direction (i.e., along the body surface) is not zero, although the pressure gradient normal to the surface is negligibly small. That is, if we were to measure the pressure while moving across the boundary layer from the body to the boundary layer edge, we would find that the pressure is essentially constant. However, the pressure does vary in the direction along the body surface if the body is curved, as shown by the figure in the margin. The variation in the *free-stream velocity, U_{fs}*, the fluid velocity at the edge of the boundary layer, is the cause of the pressure gradient in this direction. The characteristics of the entire flow (both within and outside of the boundary layer) are often highly dependent on the pressure gradient effects on the fluid within the boundary layer.

For a flat plate parallel to the upstream flow, the upstream velocity (that far ahead of the plate) and the free-stream velocity (that at the edge of the boundary layer) are equal—$U = U_{fs}$. This is a consequence of the negligible thickness of the plate. For bodies of nonzero thickness, these two velocities are different. This can be seen in the flow past a circular cylinder of diameter D. The upstream velocity and pressure are U and p_0, respectively. If the fluid were completely inviscid ($\mu = 0$), the Reynolds number would be infinite ($Re = \rho UD/\mu = \infty$) and the streamlines would be symmetrical, as are shown in Fig. 9.16a. The fluid velocity along the surface would vary from $U_{fs} = 0$ at the very front and rear of the cylinder (points A and F are stagnation points) to a maximum of $U_{fs} = 2U$ at the top and bottom of the cylinder (point C). This is also indicated in the figure in the margin. The pressure on the surface of the cylinder would be symmetrical about the vertical midplane of the cylinder, reaching a maximum value of $p_0 + \rho U^2/2$ (the stagnation pressure) at both the front and back of the cylinder, and a minimum of $p_0 - 3\rho U^2/2$ at the top and bottom of the cylinder. The pressure and free-stream velocity distributions are shown in Figs. 9.16b and 9.16c. These characteristics can be obtained from potential flow analysis of Section 6.6.3.

Because of the absence of viscosity (therefore, $\tau_w = 0$) and the symmetry of the pressure distribution for inviscid flow past a circular cylinder, it is clear that the drag on the cylinder is zero. Although it is not obvious, it can be shown that the drag is zero for any object that does not produce a lift (symmetrical or not) in an inviscid fluid (Ref. 4). Based on experimental evidence, however, we know that there must be a net drag. Clearly, since there is no purely inviscid fluid, the reason for the observed drag must lie on the shoulders of the viscous effects.

To test this hypothesis, we could conduct an experiment by measuring the drag on an object (such as a circular cylinder) in a series of fluids with decreasing values of viscosity. To our initial surprise we would find that no matter how small we make the viscosity (provided it is not precisely zero) we would measure a finite drag, essentially independent of the value of μ. As was noted in Section 6.6.3, this leads to what has been termed *d'Alembert's paradox*—the drag on an

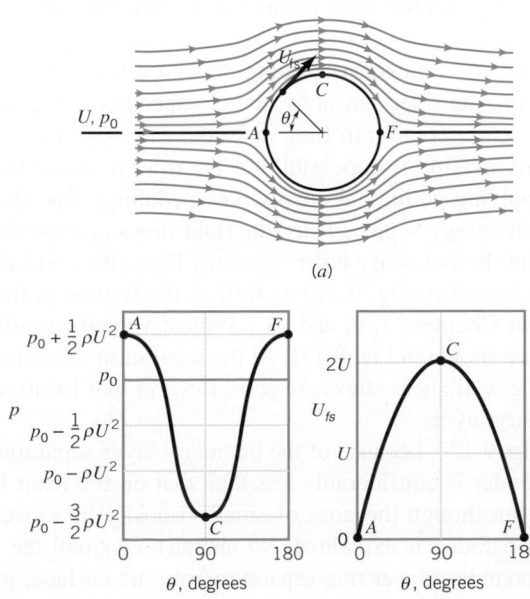

■ **FIGURE 9.16** Inviscid flow past a circular cylinder: (*a*) streamlines for the flow if there were no viscous effects, (*b*) pressure distribution on the cylinder's surface, (*c*) free-stream velocity on the cylinder's surface.

object in an inviscid fluid is zero, but the drag on an object in a fluid with vanishingly small (but nonzero) viscosity is not zero.

The reason for the above paradox can be described in terms of the effect of the pressure gradient on boundary layer flow. Consider large Reynolds number flow of a real (viscous) fluid past a circular cylinder. As was discussed in Section 9.1.2, we expect the viscous effects to be confined to thin boundary layers near the surface. This allows the fluid to stick ($\mathbf{V} = 0$) to the surface—a necessary condition for any fluid, provided $\mu \neq 0$. The basic idea of boundary layer theory is that the boundary layer is thin enough so that it does not greatly disturb the flow outside the boundary layer. Based on this reasoning, for large Reynolds numbers the flow throughout most of the flow field would be expected to be as is indicated in Fig. 9.16a, the inviscid flow field.

The pressure gradient in the external flow is imposed throughout the boundary layer fluid.

The pressure distribution indicated in Fig. 9.16b is imposed on the boundary layer flow along the surface of the cylinder. In fact, there is negligible pressure variation across the thin boundary layer so that the pressure within the boundary layer is that given by the inviscid flow field. This pressure distribution along the cylinder is such that the stationary fluid at the nose of the cylinder ($U_{fs} = 0$ at $\theta = 0$) is accelerated to its maximum velocity ($U_{fs} = 2U$ at $\theta = 90°$) and then is decelerated back to zero velocity at the rear of the cylinder ($U_{fs} = 0$ at $\theta = 180°$). This is accomplished by a balance between pressure and inertia effects; viscous effects are absent for the inviscid flow outside the boundary layer.

Physically, in the absence of viscous effects, a fluid particle traveling from the front to the back of the cylinder coasts down the "pressure hill" from $\theta = 0$ to $\theta = 90°$ (from point A to C in Fig. 9.16b) and then back up the hill to $\theta = 180°$ (from point C to F) without any loss of energy. There is an exchange between kinetic and pressure energy, but there are no energy losses. The same pressure distribution is imposed on the viscous fluid within the boundary layer. The decrease in pressure in the direction of flow along the front half of the cylinder is termed a *favorable pressure gradient*. The increase in pressure in the direction of flow along the rear half of the cylinder is termed an ***adverse pressure gradient.***

Consider a fluid particle within the boundary layer indicated in Fig. 9.17a. In its attempt to flow from A to F it experiences the same pressure distribution as the particles in the free stream immediately outside the boundary layer—the inviscid flow field pressure. However, because of the viscous effects involved, the particle in the boundary layer experiences a loss of energy as it flows along. This loss means that the particle does not have enough energy to coast all of the way up the pressure hill (from C to F) and to reach point F at the rear of the cylinder. This kinetic energy deficit is seen in the velocity profile detail at point C, shown in Fig. 9.17a. Because of friction, the boundary layer fluid cannot travel from the front to the rear of the cylinder. (This conclusion can also be obtained from the concept that due to viscous effects the particle at C does not have enough momentum to allow it to coast up the pressure hill to F.)

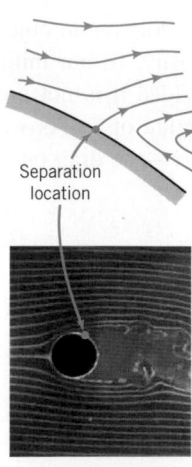

Separation location

The situation is similar to a bicyclist coasting down a hill and up the other side of the valley. If there were no friction, the rider starting with zero speed could reach the same height from which he or she started. Clearly friction (rolling resistance, aerodynamic drag, etc.) causes a loss of energy (and momentum), making it impossible for the rider to reach the height from which he or she started without supplying additional energy (i.e., pedaling). The fluid within the boundary layer does not have such an energy supply. Thus, the fluid flows against the increasing pressure as far as it can, at which point the boundary layer separates from (lifts off) the surface. This *boundary layer separation* is indicated in Fig. 9.17a as well as the figures in the margin. (See the photograph at the beginning of Chapters 7, 9, and 11.) Typical velocity profiles at representative locations along the surface are shown in Fig. 9.17b. At the separation location (profile D), the velocity gradient at the wall and the wall shear stress are zero. Beyond that location (from D to E) there is reverse flow in the boundary layer.

V9.6 Snow drifts

As is indicated in Fig. 9.17c, because of the boundary layer separation, the average pressure on the rear half of the cylinder is considerably less than that on the front half. Thus, a large pressure drag is developed, even though (because of small viscosity) the viscous shear drag may be quite small. D'Alembert's paradox is explained. No matter how small the viscosity, provided it is not zero, there will be a boundary layer that separates from the surface, giving a drag that is, for the most part, independent of the value of μ.

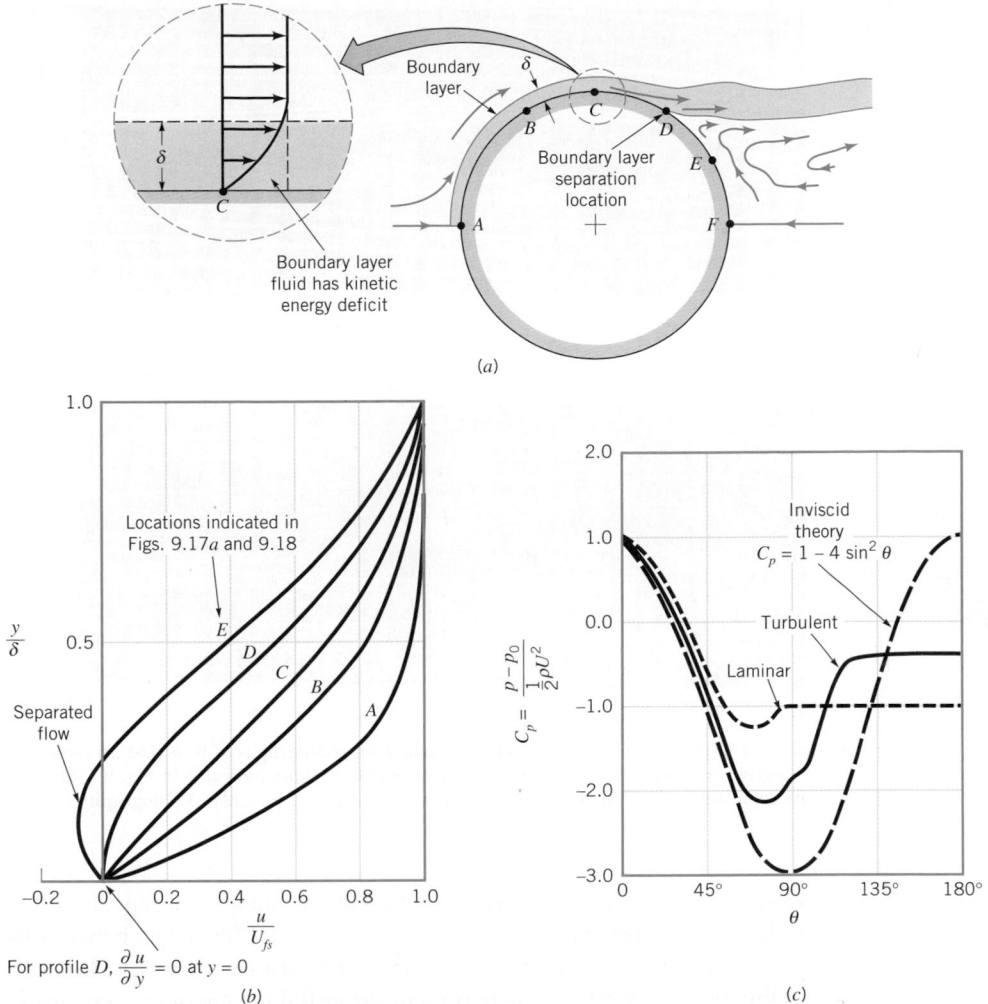

■ FIGURE 9.17 Boundary layer characteristics on a circular cylinder: (*a*) boundary layer separation location, (*b*) typical boundary layer velocity profiles at various locations on the cylinder, (*c*) surface pressure distributions for inviscid flow and boundary layer flow.

F l u i d s   i n   t h e   N e w s

Increasing truck mpg A large portion of the aerodynamic drag on semis (tractor-trailer rigs) is a result of the low pressure on the flat back end of the trailer. Researchers have recently developed a drag-reducing attachment that could reduce fuel costs on these big rigs by 10 percent. The device consists of a set of flat plates (attached to the rear of the trailer) that fold out into a box shape, thereby making the originally flat rear of the trailer a somewhat more "aerodynamic" shape. Based on thorough wind tunnel testing and actual tests conducted with a prototype design used in a series of cross-country runs, it is estimated that trucks using the device could save approximately $6,000 a year in fuel costs.

Viscous effects within the boundary layer cause boundary layer separation.

The location of separation, the width of the wake region behind the object, and the pressure distribution on the surface depend on the nature of the boundary layer flow. Compared with a laminar boundary layer, a turbulent boundary layer flow has more kinetic energy and momentum associated with it because: (1) as is indicated in Fig. E9.6, the velocity profile is fuller, more nearly like the ideal uniform profile, and (2) there can be considerable energy associated with the swirling, random components of the velocity that do not appear in the time-averaged *x* component of velocity. Thus, as is indicated in Fig. 9.17*c*, the turbulent boundary layer can flow farther around the cylinder (farther up the pressure hill) before it separates than can the laminar boundary layer.

The structure of the flow field past a circular cylinder is completely different for a zero viscosity fluid than it is for a viscous fluid, no matter how small the viscosity is, provided it is not

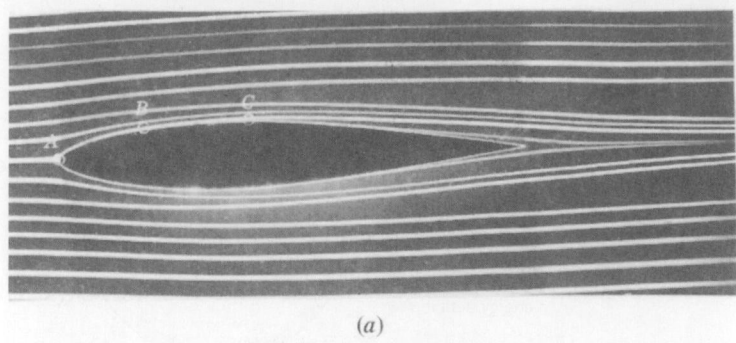

(a)

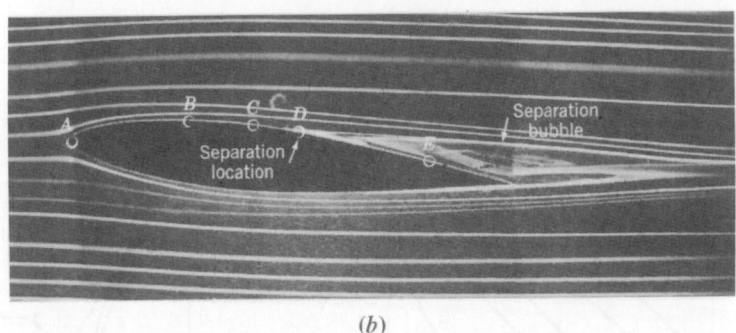

(b)

■ FIGURE 9.18 Flow visualization photographs of flow past an airfoil (the boundary layer velocity profiles for the points indicated are similar to those indicated in Fig. 9.17b): (a) zero angle of attack, no separation, (b) 5° angle of attack, flow separation. Dye in water. (Photograph courtesy of ONERA, France.)

zero. This is due to boundary layer separation. Similar concepts hold for other shaped bodies as well. The flow past an airfoil at zero *angle of attack* (the angle between the upstream flow and the axis of the object) is shown in Fig. 9.18a; flow past the same airfoil at a 5° angle of attack is shown in Fig. 9.18b. Over the front portion of the airfoil the pressure decreases in the direction of flow—a favorable pressure gradient. Over the rear portion the pressure increases in the direction of flow—an adverse pressure gradient. The boundary layer velocity profiles at representative locations are similar to those indicated in Fig. 9.17b for flow past a circular cylinder. If the adverse pressure gradient is not too great (because the body is not too "thick" in some sense), the boundary layer fluid can flow into the slightly increasing pressure region (i.e., from C to the trailing edge in Fig. 9.18a) without separating from the surface. However, if the pressure gradient is too adverse (because the angle of attack is too large), the boundary layer will separate from the surface as indicated in Fig. 9.18b. Such situations can lead to the catastrophic loss of lift called *stall*, which is discussed in Section 9.4.

Streamlined bodies generally have no separated flow.

Streamlined bodies are generally those designed to eliminate (or at least to reduce) the effects of separation, whereas nonstreamlined bodies generally have relatively large drag due to the low pressure in the separated regions (the wake). Although the boundary layer may be quite thin, it can appreciably alter the entire flow field because of boundary layer separation. These ideas are discussed in Section 9.3.

9.2.7 Momentum Integral Boundary Layer Equation with Nonzero Pressure Gradient

The boundary layer results discussed in Sections 9.2.2 and 9.2.3 are valid only for boundary layers with zero pressure gradients. They correspond to the velocity profile labeled C in Fig. 9.17b. Boundary layer characteristics for flows with nonzero pressure gradients can be obtained from nonlinear, partial differential boundary layer equations similar to Eqs. 9.8 and 9.9, provided the pressure gradient is appropriately accounted for. Such an approach is beyond the scope of this book (Refs. 1, 2).

An alternative approach is to extend the momentum integral boundary layer equation technique (Section 9.2.3) so that it is applicable for flows with nonzero pressure gradients. The momentum integral equation for boundary layer flows with zero pressure gradient, Eq. 9.26, is a statement of the balance between the shear force on the plate (represented by τ_w) and rate of change of momentum of the fluid within the boundary layer [represented by $\rho U^2 \, (d\Theta/dx)$]. For such flows the free-stream velocity is constant ($U_{fs} = U$). If the free-stream velocity is not constant [$U_{fs} = U_{fs}(x)$, where x is the distance measured along the curved body], the pressure will not be constant. This follows from the Bernoulli equation with negligible gravitational effects, since $p + \rho U_{fs}^2/2$ is constant along the streamlines outside the boundary layer. Thus,

$$\frac{dp}{dx} = -\rho U_{fs} \frac{dU_{fs}}{dx} \tag{9.34}$$

For a given body the free-stream velocity and the corresponding pressure gradient on the surface can be obtained from inviscid flow techniques (potential flow) discussed in Section 6.7. (This is how the circular cylinder results of Fig. 9.16 were obtained.)

Flow in a boundary layer with nonzero pressure gradient is very similar to that shown in Fig. 9.11, except that the upstream velocity, U, is replaced by the free-stream velocity, $U_{fs}(x)$, and the pressures at sections (1) and (2) are not necessarily equal. By using the x component of the momentum equation (Eq. 5.22) with the appropriate shear forces and pressure forces acting on the control surface indicated in Fig. 9.11, the following integral momentum equation for boundary layer flows is obtained:

Pressure gradient effects can be included in the momentum integral equation.

$$\tau_w = \rho \frac{d}{dx}(U_{fs}^2 \, \Theta) + \rho \delta^* \, U_{fs} \frac{dU_{fs}}{dx} \tag{9.35}$$

The derivation of this equation is similar to that of the corresponding equation for constant-pressure boundary layer flow, Eq. 9.26, although the inclusion of the pressure gradient effect brings in additional terms (Refs. 1, 2, 3). For example, both the boundary layer momentum thickness, Θ, and the displacement thickness, δ^*, are involved.

Equation 9.35, the general momentum integral equation for two-dimensional boundary layer flow, represents a balance between viscous forces (represented by τ_w), pressure forces (represented by $\rho U_{fs} \, dU_{fs}/dx = -dp/dx$), and the fluid momentum (represented by Θ, the boundary layer momentum thickness). In the special case of a flat plate, $U_{fs} = U = $ constant, and Eq. 9.35 reduces to Eq. 9.26.

Equation 9.35 can be used to obtain boundary layer information in a manner similar to that done for the flat plate boundary layer (Section 9.2.3). That is, for a given body shape the free-stream velocity, U_{fs}, is determined, and a family of approximate boundary layer profiles is assumed. Equation 9.35 is then used to provide information about the boundary layer thickness, wall shear stress, and other properties of interest. The details of this technique are not within the scope of this book (Refs. 1, 3).

9.3   Drag

As was discussed in Section 9.1, any object moving through a fluid will experience a drag, $\mathcal{D}$—a net force in the direction of flow due to the pressure and shear forces on the surface of the object. This net force, a combination of flow direction components of the normal and tangential forces on the body, can be determined by use of Eqs. 9.1 and 9.2, provided the distributions of pressure, p, and wall shear stress, τ_w, are known. Only in very rare instances can these distributions be determined analytically. The boundary layer flow past a flat plate parallel to the upstream flow as is discussed in Section 9.2 is one such case. Current advances in computational fluid dynamics, CFD, (i.e., the use of computers to solve the governing equations of the flow field) have provided encouraging results for more complex shapes. However, much work in this area remains.

Most of the information pertaining to drag on objects is a result of numerous experiments with wind tunnels, water tunnels, towing tanks, and other ingenious devices that are used to measure the drag on scale models. As was discussed in Chapter 7, these data can be put into dimensionless form and the results can be appropriately ratioed for prototype calculations. Typically, the result for a given-shaped object is a drag coefficient, C_D, where

$$C_D = \frac{\mathscr{D}}{\frac{1}{2}\rho U^2 A}$$

(9.36)

and C_D is a function of other dimensionless parameters such as Reynolds number, Re, Mach number, Ma, Froude number, Fr, and relative roughness of the surface, ε/ℓ. That is,

$$C_D = \phi(\text{shape, Re, Ma, Fr, } \varepsilon/\ell)$$

The character of C_D as a function of these parameters is discussed in this section.

9.3.1 Friction Drag

Friction drag, $\mathscr{D}_f$, is that part of the drag that is due directly to the shear stress, τ_w, on the object. It is a function of not only the magnitude of the wall shear stress, but also of the orientation of the surface on which it acts. This is indicated by the factor $\tau_w \sin \theta$ in Eq. 9.1. If the surface is parallel to the upstream velocity, the entire shear force contributes directly to the drag. This is true for the flat plate parallel to the flow as was discussed in Section 9.2. If the surface is perpendicular to the upstream velocity, the shear stress contributes nothing to the drag. Such is the case for a flat plate normal to the upstream velocity as was discussed in Section 9.1.

In general, the surface of a body will contain portions parallel to and normal to the upstream flow, as well as any direction in between. A circular cylinder is such a body. Because the viscosity of most common fluids is small, the contribution of the shear force to the overall drag on a body is often quite small. Such a statement should be worded in dimensionless terms. That is, because the Reynolds number of most familiar flows is quite large, the percent of the drag caused directly by the shear stress is often quite small. For highly streamlined bodies or for low Reynolds number flow, however, most of the drag may be due to friction drag.

The friction drag on a flat plate of width b and length ℓ oriented parallel to the upstream flow can be calculated from

$$\mathscr{D}_f = \frac{1}{2}\rho U^2 b\ell C_{Df}$$

Friction (viscous) drag is the drag produced by viscous shear stresses.

where C_{Df} is the friction drag coefficient. The value of C_{Df}, given as a function of Reynolds number, $\text{Re}_\ell = \rho U\ell/\mu$, and relative surface roughness, ε/ℓ, in Fig. 9.15 and Table 9.3, is a result of boundary layer analysis and experiments (see Section 9.2). Typical values of roughness, ε, for various surfaces are given in Table 8.1. As with the pipe flow discussed in Chapter 8, the flow is divided into two distinct categories—laminar or turbulent, with a transitional regime connecting them. The drag coefficient (and, hence, the drag) is not a function of the plate roughness if the flow is laminar. However, for turbulent flow the roughness does considerably affect the value of C_{Df}. As with pipe flow, this dependence is a result of the surface roughness elements protruding into or through the laminar sublayer (see Section 8.3).

Most objects are not flat plates parallel to the flow; instead, they are curved surfaces along which the pressure varies. As was discussed in Section 9.2.6, this means that the boundary layer character, including the velocity gradient at the wall, is different for most objects from that for a flat plate. This can be seen in the change of shape of the boundary layer profile along the cylinder in Fig. 9.17b.

The precise determination of the shear stress along the surface of a curved body is quite difficult to obtain. Although approximate results can be obtained by a variety of techniques (Refs. 1, 2), these are outside the scope of this text. As is shown by the following example, if the shear stress is known, its contribution to the drag can be determined.

E**XAMPLE 9.8** ■ **Drag Coefficient Based on Friction Drag**

GIVEN A viscous, incompressible fluid flows past the circular cylinder shown in Fig. E9.8a. According to a more advanced theory of boundary layer flow, the boundary layer remains attached to the cylinder up to the separation location at $\theta \approx 108.8°$, with the dimensionless wall shear stress as is indicated in Fig. E9.8b (Ref. 1). The shear stress on the cylinder in the wake region, $108.8 < \theta < 180°$, is negligible.

FIND Determine C_{Df}, the drag coefficient for the cylinder based on the friction drag only.

SOLUTION

The friction drag, $\mathscr{D}_f$, can be determined from Eq. 9.1 as

$$\mathscr{D}_f = \int \tau_w \sin\theta \, dA = 2\left(\frac{D}{2}\right) b \int_0^\pi \tau_w \sin\theta \, d\theta$$

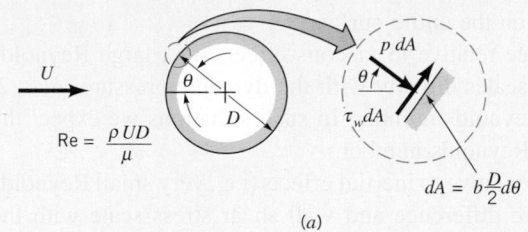

$$\text{Re} = \frac{\rho UD}{\mu}$$

$$dA = b\frac{D}{2}d\theta$$

(a)

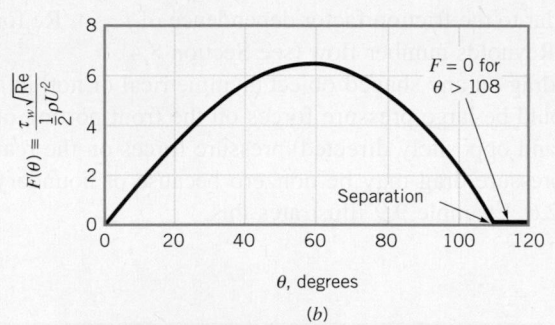

θ, degrees

(b)

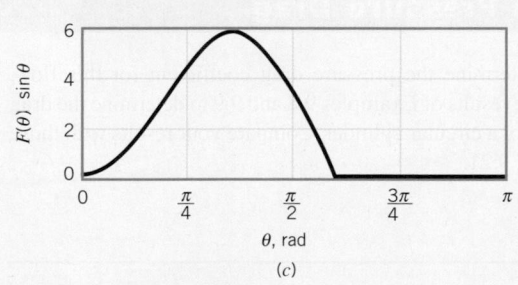

θ, rad

(c)

■ **FIGURE E9.8**

where b is the length of the cylinder. Note that θ is in radians (not degrees) to ensure the proper dimensions of $dA = 2\,(D/2)\,b\,d\theta$. Thus,

$$C_{Df} = \frac{\mathscr{D}_f}{\frac{1}{2}\rho U^2 bD} = \frac{2}{\rho U^2} \int_0^\pi \tau_w \sin\theta \, d\theta$$

This can be put into dimensionless form by using the dimensionless shear stress parameter, $F(\theta) = \tau_w \sqrt{\text{Re}}/(\rho U^2/2)$, given in Fig. E9.8b as follows:

$$C_{Df} = \int_0^\pi \frac{\tau_w}{\frac{1}{2}\rho U^2} \sin\theta \, d\theta = \frac{1}{\sqrt{\text{Re}}} \int_0^\pi \frac{\tau_w \sqrt{\text{Re}}}{\frac{1}{2}\rho U^2} \sin\theta \, d\theta$$

where Re $= \rho UD/\mu$. Thus,

$$C_{Df} = \frac{1}{\sqrt{\text{Re}}} \int_0^\pi F(\theta) \sin\theta \, d\theta \tag{1}$$

The function $F(\theta) \sin\theta$, obtained from Fig. E9.8b, is plotted in Fig. E9.8c. The necessary integration to obtain C_{Df} from Eq. 1 can be done by an appropriate numerical technique or by an approximate graphical method to determine the area under the given curve.

The result is $\int_0^\pi F(\theta) \sin\theta \, d\theta = 5.93$, or

$$C_{Df} = \frac{5.93}{\sqrt{\text{Re}}} \tag{Ans}$$

COMMENTS Note that the total drag must include both the shear stress (friction) drag and the pressure drag. As we will see in Example 9.9, for the circular cylinder most of the drag is due to the pressure force.

The above friction drag result is valid only if the boundary layer flow on the cylinder is laminar. As is discussed in Section 9.3.3, for a smooth cylinder this means that Re $= \rho UD/\mu < 3 \times 10^5$. It is also valid only for flows that have a Reynolds number sufficiently large to ensure the boundary layer structure to the flow. For the cylinder, this means Re > 100.

9.3.2 Pressure Drag

Pressure (form) drag is the drag produced by normal stresses.

Pressure drag, $\mathscr{D}_p$, is that part of the drag that is due directly to the pressure, p, on an object. It is often referred to as *form drag* because of its strong dependency on the shape or form of the object. Pressure drag is a function of the magnitude of the pressure and the orientation of the surface element on which the pressure force acts. For example, the pressure force on either side of a flat plate parallel to the flow may be very large, but it does not contribute to the drag because it acts in the direction normal to the upstream velocity. On the other hand, the pressure force on a flat plate normal to the flow provides the entire drag.

As previously noted, for most bodies, there are portions of the surface that are parallel to the upstream velocity, others normal to the upstream velocity, and the majority of which are at some angle in between, as shown by the figure in the margin. The pressure drag can be obtained from Eq. 9.1 provided a detailed description of the pressure distribution and the body shape is given. That is,

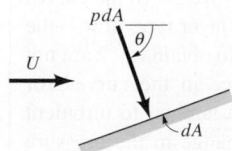

$$\mathscr{D}_p = \int p \cos\theta \, dA$$

which can be rewritten in terms of the *pressure drag coefficient*, C_{Dp}, as

$$C_{Dp} = \frac{\mathcal{D}_p}{\frac{1}{2}\rho U^2 A} = \frac{\int p \cos\theta\, dA}{\frac{1}{2}\rho U^2 A} = \frac{\int C_p \cos\theta\, dA}{A} \tag{9.37}$$

The pressure coefficient is a dimensionless form of the pressure.

Here $C_p = (p - p_0)/(\rho U^2/2)$ is the *pressure coefficient*, where p_0 is a reference pressure. The level of the reference pressure does not influence the drag directly because the net pressure force on a body is zero if the pressure is constant (i.e., p_0) on the entire surface.

For flows in which inertial effects are large relative to viscous effects (i.e., large Reynolds number flows), the pressure difference, $p - p_0$, scales directly with the dynamic pressure, $\rho U^2/2$, and the pressure coefficient is independent of Reynolds number. In such situations we expect the drag coefficient to be relatively independent of Reynolds number.

For flows in which viscous effects are large relative to inertial effects (i.e., very small Reynolds number flows), it is found that both the pressure difference and wall shear stress scale with the characteristic viscous stress, $\mu U/\ell$, where ℓ is a characteristic length. In such situations we expect the drag coefficient to be proportional to 1/Re. That is, $C_D \sim \mathcal{D}/(\rho U^2/2) \sim (\mu U/\ell)/(\rho U^2/2) \sim \mu/\rho U\ell = 1/\text{Re}$. These characteristics are similar to the friction factor dependence of $f \sim 1/\text{Re}$ for laminar pipe flow and $f \sim$ constant for large Reynolds number flow (see Section 8.4).

If the viscosity were zero, the pressure drag on any shaped object (symmetrical or not) in a steady flow would be zero. There perhaps would be large pressure forces on the front portion of the object, but there would be equally large (and oppositely directed) pressure forces on the rear portion. If the viscosity is not zero, the net pressure drag may be nonzero because of boundary layer separation as is discussed in Section 9.2.6. Example 9.9 illustrates this.

EXAMPLE 9.9 Drag Coefficient Based on Pressure Drag

GIVEN A viscous, incompressible fluid flows past the circular cylinder shown in Fig. E9.8a. The pressure coefficient on the surface of the cylinder (as determined from experimental measurements) is as indicated in Fig. E9.9a.

FIND Determine the pressure drag coefficient for this flow. Combine the results of Examples 9.8 and 9.9 to determine the drag coefficient for a circular cylinder. Compare your results with those given in Fig. 9.21.

SOLUTION

The pressure (form) drag coefficient, C_{Dp}, can be determined from Eq. 9.37 as

$$C_{Dp} = \frac{1}{A}\int C_p \cos\theta\, dA = \frac{1}{bD}\int_0^{2\pi} C_p \cos\theta\, b\left(\frac{D}{2}\right) d\theta$$

or because of symmetry

$$C_{Dp} = \int_0^\pi C_p \cos\theta\, d\theta$$

where b and D are the length and diameter of the cylinder. To obtain C_{Dp}, we must integrate the $C_p \cos\theta$ function from $\theta = 0$ to $\theta = \pi$ radians. Again, this can be done by some numerical integration scheme or by determining the area under the curve shown in Fig. E9.9b. The result is

$$C_{Dp} = 1.17 \tag{1} \text{ (Ans)}$$

Note that the positive pressure on the front portion of the cylinder $(0 \leq \theta \leq 30°)$ and the negative pressure (less than the upstream value) on the rear portion $(90 \leq \theta \leq 180°)$ produce positive contributions to the drag. The negative pressure on the front portion of the cylinder $(30 < \theta < 90°)$ reduces the drag by pulling on the cylinder in the upstream direction. The positive area under the $C_p \cos\theta$ curve is greater than the negative area—there is a net pressure drag. In the absence of viscosity, these two contributions would be equal—there would be no pressure (or friction) drag.

The net drag on the cylinder is the sum of friction and pressure drag. Thus, from Eq. 1 of Example 9.8 and Eq. 1 of this example, we obtain the drag coefficient

$$C_D = C_{Df} + C_{Dp} = \frac{5.93}{\sqrt{\text{Re}}} + 1.17 \tag{2} \text{ (Ans)}$$

This result is compared with the standard experimental value (obtained from Fig. 9.21) in Fig. E9.9c. The agreement is very good over a wide range of Reynolds numbers. For Re < 10 the curves diverge because the flow is not a boundary layer type flow—the shear stress and pressure distributions used to obtain Eq. 2 are not valid in this range. The drastic divergence in the curves for Re $> 3 \times 10^5$ is due to the change from a laminar to turbulent boundary layer, with the corresponding change in the pressure distribution. This is discussed in Section 9.3.3.

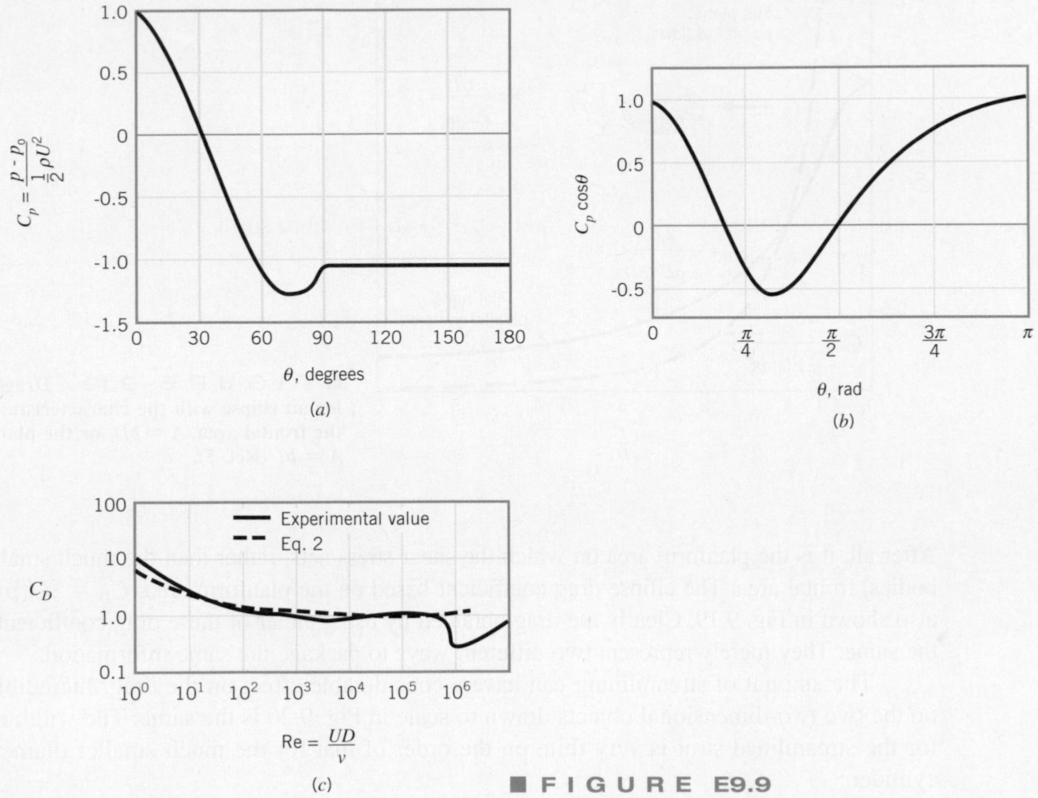

$$C_p = \frac{p - p_o}{\frac{1}{2}\rho U^2}$$

(a)

θ, degrees

$C_p \cos\theta$

(b)

θ, rad

$$Re = \frac{UD}{\nu}$$

C_D

— Experimental value
- - Eq. 2

(c)

■ **FIGURE E9.9**

COMMENT   It is of interest to compare the friction drag to the total drag on the cylinder. That is,

$$\frac{\mathscr{D}_f}{\mathscr{D}} = \frac{C_{Df}}{C_D} = \frac{5.93/\sqrt{Re}}{(5.93/\sqrt{Re}) + 1.17} = \frac{1}{1 + 0.197\sqrt{Re}}$$

For $Re = 10^3$, 10^4, and 10^5 this ratio is 0.138, 0.0483, and 0.0158, respectively. Most of the drag on the blunt cylinder is pressure drag—a result of the boundary layer separation.

9.3.3 Drag Coefficient Data and Examples

As was discussed in previous sections, the net drag is produced by both pressure and shear stress effects. In most instances these two effects are considered together, and an overall drag coefficient, C_D, as defined in Eq. 9.36 is used. There is an abundance of such drag coefficient data available in the literature. This information covers incompressible and compressible viscous flows past objects of almost any shape of interest—both man-made and natural objects. In this section we consider a small portion of this information for representative situations. Additional data can be obtained from various sources (Refs. 5, 6).

V9.7 Skydiving practice

The drag coefficient may be based on the frontal area or the planform area.

Shape Dependence.   Clearly the drag coefficient for an object depends on the shape of the object, with shapes ranging from those that are streamlined to those that are blunt. The drag on an ellipse with aspect ratio ℓ/D, where D and ℓ are the thickness and length parallel to the flow, illustrates this dependence. The drag coefficient $C_D = \mathscr{D}/(\rho U^2 bD/2)$, based on the frontal area, $A = bD$, where b is the length normal to the flow, is as shown in Fig. 9.19. The more blunt the body, the larger the drag coefficient. With $\ell/D = 0$ (i.e., a flat plate normal to the flow) we obtain the flat plate value of $C_D = 1.9$. With $\ell/D = 1$ the corresponding value for a circular cylinder is obtained. As ℓ/D becomes larger the value of C_D decreases.

For very large aspect ratios ($\ell/D \to \infty$) the ellipse behaves as a flat plate parallel to the flow. For such cases, the friction drag is greater than the pressure drag, and the value of C_D based on the frontal area, $A = bD$, would increase with increasing ℓ/D. (This occurs for larger ℓ/D values than those shown in the figure.) For such extremely thin bodies (i.e., an ellipse with $\ell/D \to \infty$, a flat plate, or very thin airfoils) it is customary to use the planform area, $A = b\ell$, in defining the drag coefficient.

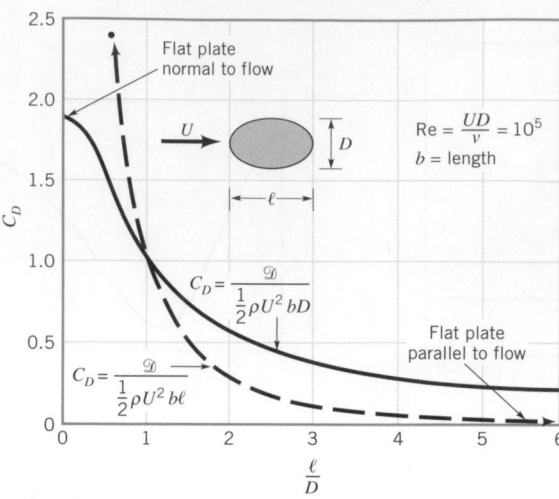

■ **F I G U R E 9.19**   Drag coefficient for an ellipse with the characteristic area either the frontal area, $A = bD$, or the planform area, $A = b\ell$ (Ref. 5).

After all, it is the planform area on which the shear stress acts, rather than the much smaller (for thin bodies) frontal area. The ellipse drag coefficient based on the planform area, $C_D = \mathcal{D}/(\rho U^2 b\ell/2)$, is also shown in Fig. 9.19. Clearly the drag obtained by using either of these drag coefficients would be the same. They merely represent two different ways to package the same information.

The amount of streamlining can have a considerable effect on the drag. Incredibly, the drag on the two two-dimensional objects drawn to scale in Fig. 9.20 is the same. The width of the wake for the streamlined strut is very thin, on the order of that for the much smaller diameter circular cylinder.

Reynolds Number Dependence. Another parameter on which the drag coefficient can be very dependent is the Reynolds number. The main categories of Reynolds number dependence are (1) very low Reynolds number flow, (2) moderate Reynolds number flow (laminar boundary layer), and (3) very large Reynolds number flow (turbulent boundary layer). Examples of these three situations are discussed below.

Low Reynolds number flows (Re < 1) are governed by a balance between viscous and pressure forces. Inertia effects are negligibly small. In such instances the drag on a three-dimensional body is expected to be a function of the upstream velocity, U, the body size, ℓ, and the viscosity, μ. Thus, for a small grain of sand settling in a lake (see margin figure)

$$\mathcal{D} = f(U, \ell, \mu)$$

From dimensional considerations (see Section 7.7.1)

$$\mathcal{D} = C\mu\ell U \tag{9.38}$$

where the value of the constant C depends on the shape of the body. If we put Eq. 9.38 into dimensionless form using the standard definition of the drag coefficient, we obtain

$$C_D = \frac{\mathcal{D}}{\frac{1}{2}\rho U^2 \ell^2} = \frac{2C\mu\ell U}{\rho U^2 \ell^2} = \frac{2C}{\text{Re}}$$

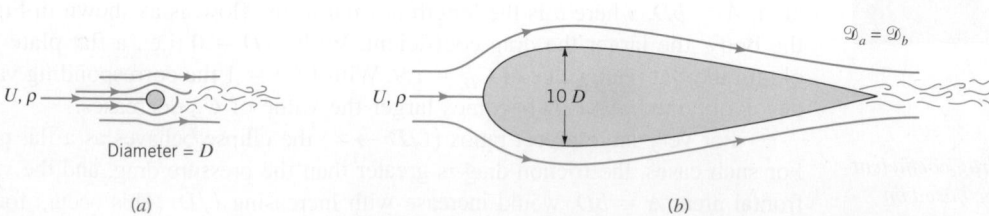

■ **F I G U R E 9.20**   Two objects of considerably different size that have the same drag force: (a) circular cylinder $C_D = 1.2$; (b) streamlined strut $C_D = 0.12$.

■ **TABLE 9.4**
Low Reynolds Number Drag Coefficients (Ref. 7) (Re = $\rho UD/\mu$, $A = \pi D^2/4$)

	$C_D = \mathcal{D}/(\rho U^2 A/2)$		
Object	(for Re $\lesssim$ 1)	Object	C_D
a. Circular disk normal to flow 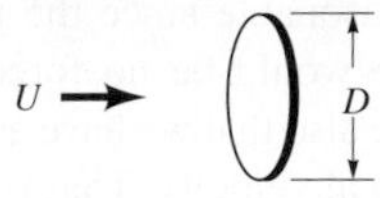	20.4/Re	c. Sphere 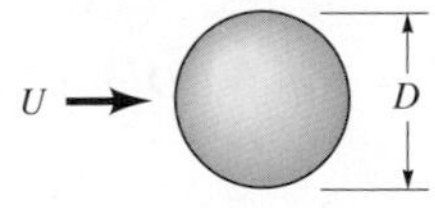	24.0/Re
b. Circular disk parallel to flow 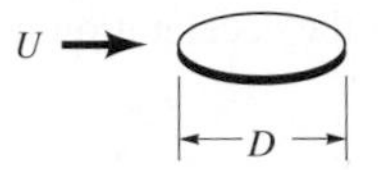	13.6/Re	d. Hemisphere 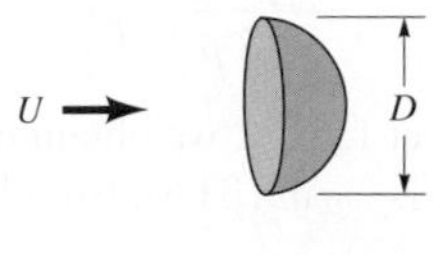	22.2/Re

where Re = $\rho U\ell/\mu$. The use of the dynamic pressure, $\rho U^2/2$, in the definition of the drag coefficient is somewhat misleading in the case of creeping flows (Re < 1) because it introduces the fluid density, which is not an important parameter for such flows (inertia is not important). Use of this standard drag coefficient definition gives the 1/Re dependence for small Re drag coefficients.

Typical values of C_D for low Reynolds number flows past a variety of objects are given in Table 9.4. It is of interest that the drag on a disk normal to the flow is only 1.5 times greater than that on a disk parallel to the flow. For large Reynolds number flows this ratio is considerably larger (see Example 9.1). Streamlining (i.e., making the body slender) can produce a considerable drag reduction for large Reynolds number flows; for very small Reynolds number flows it can actually increase the drag because of an increase in the area on which shear forces act. For most objects, the low Reynolds number flow results are valid up to a Reynolds number of about 1.

For very small Reynolds number flows, the drag coefficient varies inversely with the Reynolds number.

EXAMPLE 9.10 ■ Low Reynolds Number Flow Drag

GIVEN A small grain of sand, diameter $D = 0.10$ mm and specific gravity $SG = 2.3$, settles to the bottom of a lake after having been stirred up by a passing boat.

FIND Determine how fast it falls through the still water.

SOLUTION

A free-body diagram of the particle (relative to the moving particle) is shown in Fig. E9.10a. The particle moves downward with a constant velocity U that is governed by a balance between the weight of the particle, $\mathcal{W}$, the buoyancy force of the surrounding water, F_B, and the drag of the water on the particle, $\mathcal{D}$.

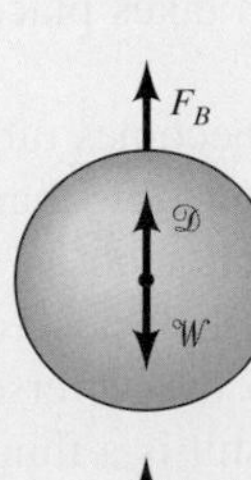

From the free-body diagram, we obtain

$$\mathcal{W} = \mathcal{D} + F_B$$

where

$$\mathcal{W} = \gamma_{sand}\,\forall = SG\,\gamma_{H_2O}\,\frac{\pi}{6}D^3 \tag{1}$$

and

$$F_B = \gamma_{H_2O}\,\forall = \gamma_{H_2O}\,\frac{\pi}{6}D^3 \tag{2}$$

We assume (because of the smallness of the object) that the flow will be creeping flow (Re < 1) with $C_D = 24/\text{Re}$ (see Table 9.4) so that

$$\mathcal{D} = \frac{1}{2}\rho_{H_2O}U^2\,\frac{\pi}{4}D^2C_D = \frac{1}{2}\rho_{H_2O}U^2\,\frac{\pi}{4}D^2\left(\frac{24}{\rho_{H_2O}UD/\mu_{H_2O}}\right)$$

■ **FIGURE E9.10a**

or

$$\mathcal{D} = 3\pi\mu_{H_2O}UD \tag{3}$$

We must eventually check to determine if this assumption (Re < 1) is valid or not. Equation 3 is called Stokes's law in honor of G. G. Stokes (1819–1903), a British mathematician and physicist. By combining Eqs. 1, 2, and 3, we obtain

$$SG\,\gamma_{H_2O}\frac{\pi}{6}D^3 = 3\pi\mu_{H_2O}UD + \gamma_{H_2O}\frac{\pi}{6}D^3$$

or, since $\gamma = \rho g$,

$$U = \frac{(SG - 1)\rho_{H_2O}\,gD^2}{18\,\mu} \tag{4}$$

From Table 1.6 for water at 15.6 °C we obtain $\rho_{H_2O} = 999$ kg/m^3 and $\mu_{H_2O} = 1.12 \times 10^{-3}$ N·s/m^2. Thus, from Eq. 4 we obtain

$$U = \frac{(2.3 - 1)(999 \text{ kg/m}^3)(9.81 \text{ m/s}^2)(0.10 \times 10^{-3} \text{ m})^2}{18(1.12 \times 10^{-3} \text{ N·s/m}^2)}$$

or

$$U = 6.32 \times 10^{-3} \text{ m/s} \tag{Ans}$$

Since

$$\text{Re} = \frac{\rho DU}{\mu} = \frac{(999 \text{ kg/m}^3)(0.10 \times 10^{-3} \text{ m})(0.00632 \text{ m/s})}{1.12 \times 10^{-3} \text{ N·s/m}^2}$$
$$= 0.564$$

we see that Re < 1, and the form of the drag coefficient used is valid.

COMMENTS By repeating the calculations for various particle diameters, D, the results shown in Fig. E9.10b are obtained. Note that very small particles fall extremely slowly. Thus, it can take considerable time for silt to settle to the bottom of a river or lake.

Note that if the density of the particle were the same as the surrounding water (i.e., $SG = 1$), from Eq. 4 we would obtain $U = 0$. This is reasonable since the particle would be neutrally buoyant and there would be no force to overcome the motion-induced drag. Note also that we have assumed that the particle falls at its steady terminal velocity. That is, we have neglected the acceleration of the particle from rest to its terminal velocity. Since the terminal velocity is small, this acceleration time is quite small. For faster objects (such as a free-falling sky diver) it may be important to consider the acceleration portion of the fall.

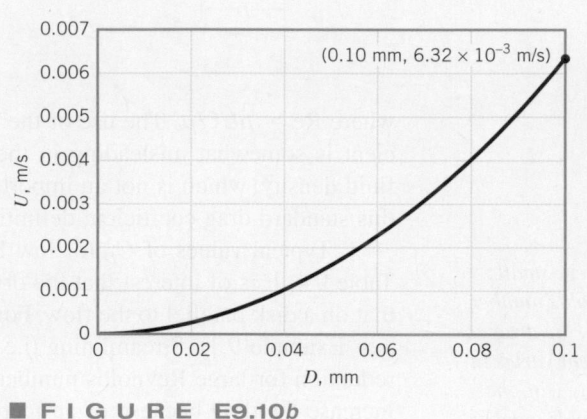

■ **FIGURE E9.10b**

Flow past a cylinder can take on a variety of different structures.

Moderate Reynolds number flows tend to take on a boundary layer flow structure. For such flows past streamlined bodies, the drag coefficient tends to decrease slightly with Reynolds number. The $C_D \sim \text{Re}^{-1/2}$ dependence for a laminar boundary layer on a flat plate (see Table 9.3) is such an example. Moderate Reynolds number flows past blunt bodies generally produce drag coefficients that are relatively constant. The C_D values for the spheres and circular cylinders shown in Fig. 9.21a indicate this character in the range $10^3 < \text{Re} < 10^5$.

The structure of the flow field at selected Reynolds numbers indicated in Fig. 9.21a is shown in Fig. 9.21b. For a given object there is a wide variety of flow situations, depending on the Reynolds number involved. The curious reader is strongly encouraged to study the many beautiful photographs and videos of these (and other) flow situations found in Refs. 8 and 31. (See also the photograph at the beginning of Chapter 7.)

For many shapes there is a sudden change in the character of the drag coefficient when the boundary layer becomes turbulent. This is illustrated in Fig. 9.15 for the flat plate and in Fig. 9.21 for the sphere and the circular cylinder. The Reynolds number at which this transition takes place is a function of the shape of the body.

V9.8 Karman vortex street

For streamlined bodies, the drag coefficient increases when the boundary layer becomes turbulent because most of the drag is due to the shear force, which is greater for turbulent flow than for laminar flow. On the other hand, the drag coefficient for a relatively blunt object, such as a cylinder or sphere, actually decreases when the boundary layer becomes turbulent. As is discussed in Section 9.2.6, a turbulent boundary layer can travel further along the surface into the adverse pressure gradient on the rear portion of the cylinder before separation occurs. The result is a thinner wake and smaller pressure drag for turbulent boundary layer flow. This is indicated in Fig. 9.21

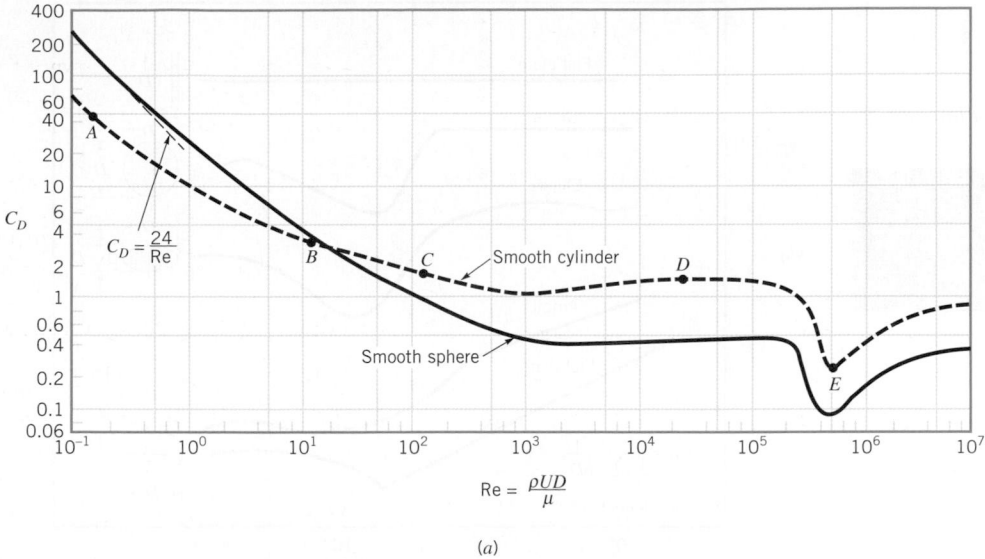

(a)

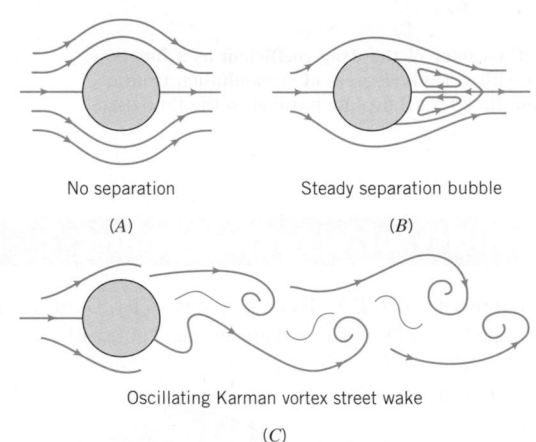

No separation

(A)

Steady separation bubble

(B)

Oscillating Karman vortex street wake

(C)

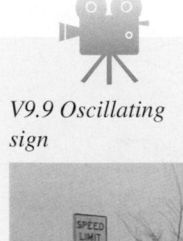

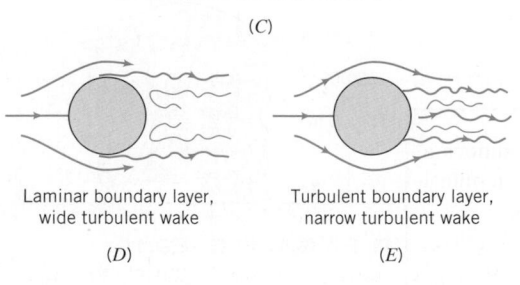

Laminar boundary layer, wide turbulent wake

(D)

Turbulent boundary layer, narrow turbulent wake

(E)

(b)

■ **F I G U R E 9.21** (a) Drag coefficient as a function of Reynolds number for a smooth circular cylinder and a smooth sphere. (b) Typical flow patterns for flow past a circular cylinder at various Reynolds numbers as indicated in (a).

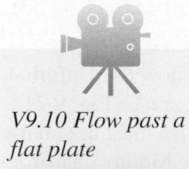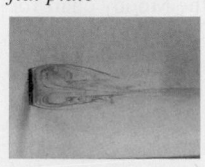

by the sudden decrease in C_D for $10^5 < \text{Re} < 10^6$. In a portion of this range the actual drag (not just the drag coefficient) decreases with increasing speed. It would be very difficult to control the steady flight of such an object in this range—an increase in velocity requires a decrease in thrust (drag). In all other Reynolds number ranges the drag increases with an increase in the upstream velocity (even though C_D may decrease with Re).

For extremely blunt bodies, like a flat plate perpendicular to the flow, the flow separates at the edge of the plate regardless of the nature of the boundary layer flow. Thus, the drag coefficient shows very little dependence on the Reynolds number.

The drag coefficients for a series of two-dimensional bodies of varying bluntness are given as a function of Reynolds number in Fig. 9.22. The characteristics described above are evident.

V9.11 Flow past an ellipse

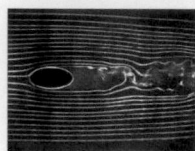

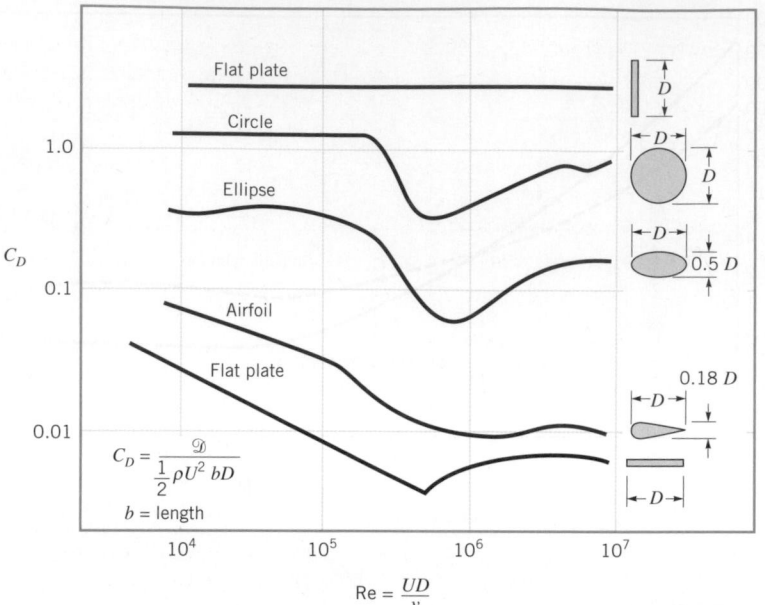

■ **FIGURE 9.22** **Character of the drag coefficient as a function of Reynolds number for objects with various degrees of streamlining, from a flat plate normal to the upstream flow to a flat plate parallel to the flow (two-dimensional flow) (Ref. 5).**

EXAMPLE 9.11 Terminal Velocity of a Falling Object

GIVEN Hail is produced by the repeated rising and falling of ice particles in the updraft of a thunderstorm, as is indicated in Fig. E9.11a. When the hail becomes large enough, the aerodynamic drag from the updraft can no longer support the weight of the hail, and it falls from the storm cloud.

FIND Estimate the velocity, U, of the updraft needed to make $D = 4$-cm-diameter (i.e., "golf ball-sized") hail.

SOLUTION

As is discussed in Example 9.10, for steady-state conditions a force balance on an object falling through a fluid at its terminal velocity, U, gives

$$\mathcal{W} = \mathcal{D} + F_B$$

where $F_B = \gamma_{air} \mathcal{V}$ is the buoyant force of the air on the particle, $\mathcal{W} = \gamma_{ice} \mathcal{V}$ is the particle weight, and $\mathcal{D}$ is the aerodynamic drag. This equation can be rewritten as

$$\tfrac{1}{2}\rho_{air}U^2 \frac{\pi}{4} D^2 C_D = \mathcal{W} - F_B \quad (1)$$

With $\mathcal{V} = \pi D^3/6$ and since $\gamma_{ice} \gg \gamma_{air}$ (i.e., $\mathcal{W} \gg F_B$), Eq. 1 can be simplified to

$$U = \left(\frac{4}{3}\frac{\rho_{ice}}{\rho_{air}}\frac{gD}{C_D}\right)^{1/2} \quad (2)$$

By using $\rho_{ice} = 948 \text{ kg/m}^3$, $\rho_{air} = 1.23 \text{ kg/m}^3$, and $D = 4 \text{ cm} = 0.04 \text{ m}$, Eq. 2 becomes

$$U = \left[\frac{4(948 \text{ kg/m}^3)(9.81 \text{ m/s}^2)(0.04 \text{ m})}{3(1.23 \text{ kg/m}^3)C_D}\right]^{1/2}$$

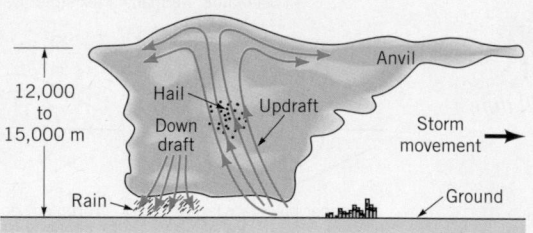

■ **FIGURE E9.11a**

or

$$U = \frac{20.08}{\sqrt{C_D}} \quad (3)$$

where U is in m/s. To determine U, we must know C_D. Unfortunately, C_D is a function of the Reynolds number (see Fig. 9.21), which is not known unless U is known. Thus, we must use an iterative technique similar to that done with the Moody chart for certain types of pipe flow problems (see Section 8.5).

From Fig. 9.21 we expect that C_D is on the order of 0.5. Thus, we assume $C_D = 0.5$ and from Eq. 3 obtain

$$U = \frac{20.08}{\sqrt{0.5}} = 28.4 \text{ m/s}$$

The corresponding Reynolds number (assuming $v = 1.46 \times 10^{-5}\,\text{m}^2/\text{s}$) is

$$\text{Re} = \frac{UD}{\nu} = \frac{28.4\,\text{m/s}\,(0.04\,\text{m})}{1.46 \times 10^{-5}\,\text{m}^2/\text{s}} = 7.78 \times 10^4$$

For this value of Re we obtain from Fig. 9.21, $C_D = 0.5$. Thus, our assumed value of $C_D = 0.5$ was correct. The corresponding value of U is

$$U = 28.4\,\text{m/s} = 102.2\,\text{km/hr} \qquad \textbf{(Ans)}$$

COMMENTS   By repeating the calculations for various altitudes, z, above sea level (using the properties of the U.S. Standard Atmosphere given in Appendix C), the results shown in Fig. E9.11b are obtained. Because of the decrease in density with altitude, the hail falls even faster through the upper portions of the storm than when it hits the ground.

Clearly, an airplane flying through such an updraft would feel its effects (even if it were able to dodge the hail). As seen from Eq. 2, the larger the hail, the stronger the necessary updraft.

Hailstones greater than 15 cm in diameter have been reported. In reality, a hailstone is seldom spherical and often not smooth. However, the calculated updraft velocities are in agreement with measured values.

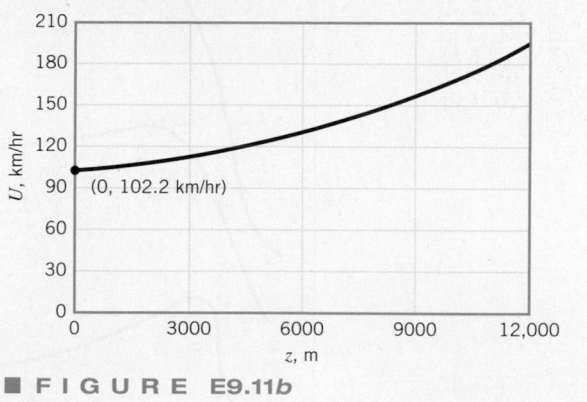

■ **FIGURE   E9.11b**

Compressibility Effects.   The above discussion is restricted to incompressible flows. If the velocity of the object is sufficiently large, compressibility effects become important and the drag coefficient becomes a function of the Mach number, $\text{Ma} = U/c$, where c is the speed of sound in the fluid. The introduction of Mach number effects complicates matters because the drag coefficient for a given object is then a function of both Reynolds number and Mach number— $C_D = \phi(\text{Re}, \text{Ma})$. The Mach number and Reynolds number effects are often closely connected because both are directly proportional to the upstream velocity. For example, both Re and Ma increase with increasing flight speed of an airplane. The changes in C_D due to a change in U are due to changes in both Re and Ma.

The precise dependence of the drag coefficient on Re and Ma is generally quite complex (Ref. 13). However, the following simplifications are often justified. For low Mach numbers, the drag coefficient is essentially independent of Ma as is indicated in Fig. 9.23. For this situation, if $\text{Ma} < 0.5$ or so, compressibility effects are unimportant. On the other hand, for larger Mach number flows, the drag coefficient can be strongly dependent on Ma, with only secondary Reynolds number effects.

For most objects, values of C_D increase dramatically in the vicinity of $\text{Ma} = 1$ (i.e., sonic flow). This change in character, indicated by Fig. 9.24, is due to the existence of shock waves as

The drag coefficient is usually independent of Mach number for Mach numbers up to approximately 0.5.

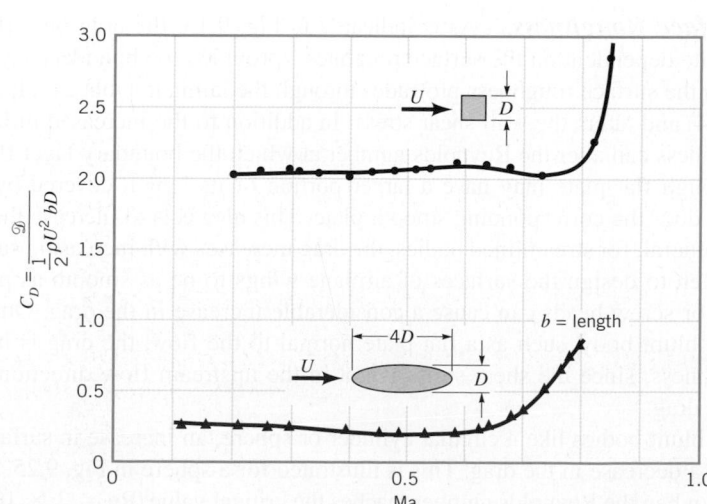

■ **FIGURE 9.23**
Drag coefficient as a function of Mach number for two-dimensional objects in subsonic flow (Ref. 5).

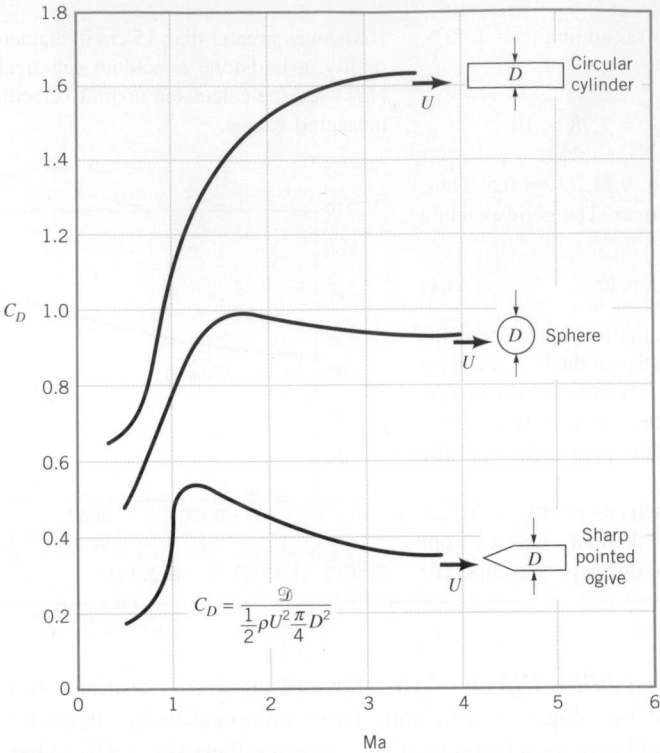

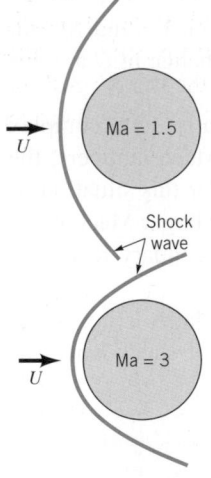

*Depending on the
body shape, an in-
crease in surface
roughness may in-
crease or decrease
drag.*

indicated by the figure in the margin. Shock waves are extremely narrow regions in the flow field
across which the flow parameters change in a nearly discontinuous manner, which are discussed
in Chapter 11. Shock waves, which cannot exist in subsonic flows, provide a mechanism for the
generation of drag that is not present in the relatively low-speed subsonic flows. (See the photograph at the beginning of Chapter 11.)

The character of the drag coefficient as a function of Mach number is different for blunt bodies than for sharp bodies. As is shown in Fig. 9.24, sharp-pointed bodies develop their maximum
drag coefficient in the vicinity of Ma = 1 (sonic flow), whereas the drag coefficient for blunt bodies increases with Ma far above Ma = 1. This behavior is due to the nature of the shock wave
structure and the accompanying flow separation. The leading edges of wings for subsonic aircraft
are usually quite rounded and blunt, while those of supersonic aircraft tend to be quite pointed and
sharp. More information on these important topics can be found in standard texts about compressible flow and aerodynamics (Refs. 9, 10, 29).

Surface Roughness. As is indicated in Fig. 9.15, the drag on a flat plate parallel to the
flow is quite dependent on the surface roughness, provided the boundary layer flow is turbulent. In
such cases the surface roughness protrudes through the laminar sublayer adjacent to the surface (see
Section 8.4) and alters the wall shear stress. In addition to the increased turbulent shear stress, surface roughness can alter the Reynolds number at which the boundary layer flow becomes turbulent.
Thus, a rough flat plate may have a larger portion of its length covered by a turbulent boundary
layer than does the corresponding smooth plate. This also acts to increase the net drag on the plate.

In general, for streamlined bodies, the drag increases with increasing surface roughness. Great
care is taken to design the surfaces of airplane wings to be as smooth as possible, since protruding rivets or screw heads can cause a considerable increase in the drag. On the other hand, for an
extremely blunt body, such as a flat plate normal to the flow, the drag is independent of the surface roughness, since the shear stress is not in the upstream flow direction and contributes nothing to the drag.

For blunt bodies like a circular cylinder or sphere, an increase in surface roughness can actually cause a decrease in the drag. This is illustrated for a sphere in Fig. 9.25. As is discussed in Section 9.2.6, when the Reynolds number reaches the critical value (Re = 3×10^5 for a smooth sphere),

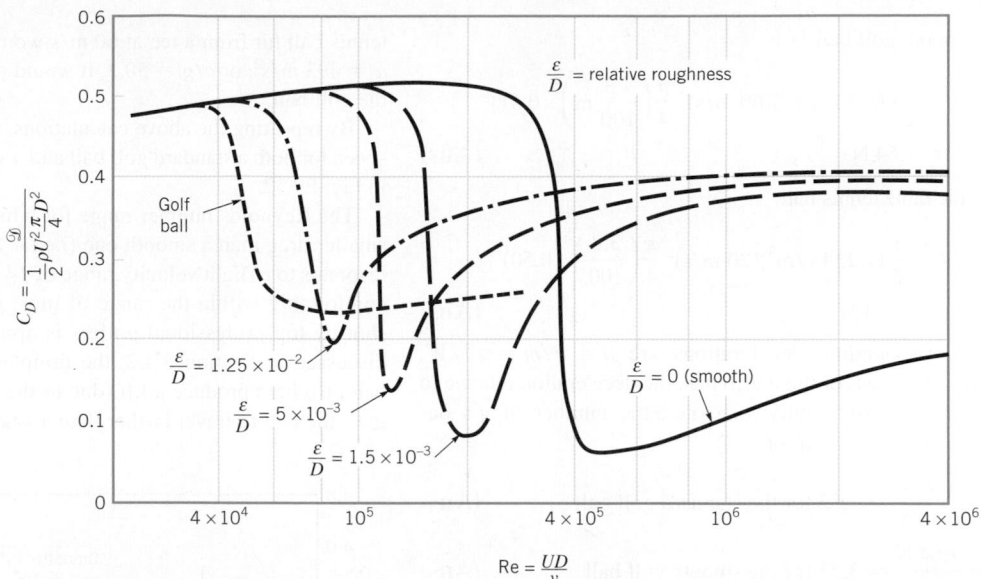

■ **FIGURE 9.25** The effect of surface roughness on the drag coefficient of a sphere in the Reynolds number range for which the laminar boundary layer becomes turbulent (Ref. 5).

the boundary layer becomes turbulent and the wake region behind the sphere becomes considerably narrower than if it were laminar (see Fig. 9.17). The result is a considerable drop in pressure drag with a slight increase in friction drag, combining to give a smaller overall drag (and C_D).

Surface roughness can cause the boundary layer to become turbulent.

The boundary layer can be tripped into turbulence at a smaller Reynolds number by using a rough-surfaced sphere. For example, the critical Reynolds number for a golf ball is approximately Re = 4×10^4. In the range $4 \times 10^4 <$ Re $< 4 \times 10^5$, the drag on the standard rough (i.e., dimpled) golf ball is considerably less ($C_{Drough}/C_{Dsmooth} \approx 0.25/0.5 = 0.5$) than for the smooth ball. As is shown in Example 9.12, this is precisely the Reynolds number range for well-hit golf balls—hence, a reason for dimples on golf balls. The Reynolds number range for well-hit table tennis balls is less than Re = 4×10^4. Thus, table tennis balls are smooth.

EXAMPLE 9.12 Effect of Surface Roughness

GIVEN A well-hit golf ball (diameter D = 4.3 cm, weight $\mathcal{W}$ = 0.44 N) can travel at U = 60 m/s as it leaves the tee. A well-hit table tennis ball (diameter D = 3.8 cm, weight $\mathcal{W}$ = 0.025 N) can travel at U = 20 m/s as it leaves the paddle.

FIND Determine the drag on a standard golf ball, a smooth golf ball, and a table tennis ball for the conditions given. Also determine the deceleration of each ball for these conditions.

SOLUTION

For either ball, the drag can be obtained from

$$\mathcal{D} = \frac{1}{2}\rho U^2 \frac{\pi}{4} D^2 C_D \qquad (1)$$

where the drag coefficient, C_D, is given in Fig. 9.25 as a function of the Reynolds number and surface roughness. For the golf ball in standard air

$$Re = \frac{UD}{\nu} = \frac{(60 \text{ m/s})(4.3/100 \text{ m})}{1.46 \times 10^{-5} \text{ m}^2/\text{s}} = 1.79 \times 10^5$$

while for the table tennis ball

$$Re = \frac{UD}{\nu} = \frac{(20 \text{ m/s})(3.8/100 \text{ m})}{1.46 \times 10^{-5} \text{ m}^2/\text{s}} = 5.2 \times 10^4$$

The corresponding drag coefficients are C_D = 0.25 for the standard golf ball, C_D = 0.51 for the smooth golf ball, and C_D = 0.50 for the table tennis ball. Hence, from Eq. 1 for the standard golf ball

$$\mathcal{D} = \frac{1}{2}(1.23 \text{ kg/m}^3)(60 \text{ m/s})^2 \frac{\pi}{4}\left(\frac{4.3}{100} \text{ m}\right)^2 (0.25)$$

$$= 0.8 \text{ N} \qquad \text{(Ans)}$$

for the smooth golf ball

$$\mathcal{D} = \frac{1}{2}(1.23 \text{ kg/m}^3)(60 \text{ m/s})^2 \frac{\pi}{4}\left(\frac{4.3}{100} \text{ m}\right)^2 (0.51)$$

$$= 1.64 \text{ N} \qquad \text{(Ans)}$$

and for the table tennis ball

$$\mathcal{D} = \frac{1}{2}(1.23 \text{ kg/m}^3)(20 \text{ m/s})^2 \frac{\pi}{4}\left(\frac{3.8}{100}\right)^2 (0.50)$$

$$= 0.14 \text{ N} \qquad \text{(Ans)}$$

The corresponding decelerations are $a = \mathcal{D}/m = g\mathcal{D}/\mathcal{W}$, where m is the mass of the ball. Thus, the deceleration relative to the acceleration of gravity, a/g (i.e., the number of g's deceleration) is $a/g = \mathcal{D}/\mathcal{W}$ or

$$\frac{a}{g} = \frac{0.8 \text{ N}}{0.44 \text{ N}} = 1.82 \text{ for the standard golf ball} \qquad \text{(Ans)}$$

$$\frac{a}{g} = \frac{1.64 \text{ N}}{0.44 \text{ N}} = 3.73 \text{ for the smooth golf ball} \qquad \text{(Ans)}$$

$$\frac{a}{g} = \frac{0.14 \text{ N}}{0.025 \text{ N}} = 5.6 \text{ for the table tennis ball} \qquad \text{(Ans)}$$

COMMENTS   Note that there is a considerably smaller deceleration for the rough golf ball than for the smooth one. Because of its much larger drag-to-mass ratio, the table tennis ball slows down relatively quickly and does not travel as far as the golf ball. Note that with $U = 20$ m/s the standard golf ball has a drag of $\mathcal{D} = 0.09$ N and a deceleration of $a/g = 0.205$, considerably less than the $a/g = 5.6$ of the table tennis ball. Conversely, a table

tennis ball hit from a tee at 60 m/s would decelerate at a rate of $a = 493 \text{ m/s}^2$, or $a/g = 50.3$. It would not travel nearly as far as the golf ball.

By repeating the above calculations, the drag as a function of speed for both a standard golf ball and a smooth golf ball is shown in Fig. E9.12.

The Reynolds number range for which a rough golf ball has smaller drag than a smooth one (i.e., 4×10^4 to 3.6×10^5) corresponds to a flight velocity range of $14 < U < 122$ m/s. This is comfortably within the range of most golfers. (The fastest tee shot by top professional golfers is approximately 85 m/s.) As discussed in Section 9.4.2, the dimples (roughness) on a golf ball also help produce a lift (due to the spin of the ball) that allows the ball to travel farther than a smooth ball.

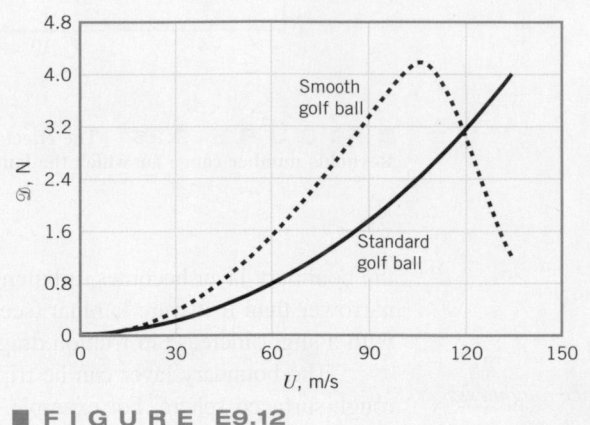

■ **F I G U R E   E9.12**

V9.12 Jet ski

The drag coefficient for surface ships is a function of the Froude number.

Froude Number Effects.   Another parameter on which the drag coefficient may be strongly dependent is the Froude number, $\text{Fr} = U/\sqrt{g\ell}$. As is discussed in Chapter 10, the Froude number is a ratio of the free-stream speed to a typical wave speed on the interface of two fluids, such as the surface of the ocean. An object moving on the surface, such as a ship, often produces waves that require a source of energy to generate. This energy comes from the ship and is manifest as a drag. [Recall that the rate of energy production (power) equals speed times force.] The nature of the waves produced often depends on the Froude number of the flow and the shape of the object—the waves generated by a water skier "plowing" through the water at a low speed (low Fr) are different than those generated by the skier "planing" along the surface at high speed (large Fr).

Thus, the drag coefficient for surface ships is a function of Reynolds number (viscous effects) and Froude number (wave-making effects); $C_D = \phi(\text{Re}, \text{Fr})$. As was discussed in Chapter 7, it is often quite difficult to run model tests under conditions similar to those of the prototype (i.e., same Re and Fr for surface ships). Fortunately, the viscous and wave effects can often be separated, with the total drag being the sum of the drag of these individual effects. A detailed account of this important topic can be found in standard texts (Ref. 11).

As is indicated in Fig. 9.26, the wave-making drag, $\mathcal{D}_w$, can be a complex function of the Froude number and the body shape. The rather "wiggly" dependence of the wave drag coefficient,

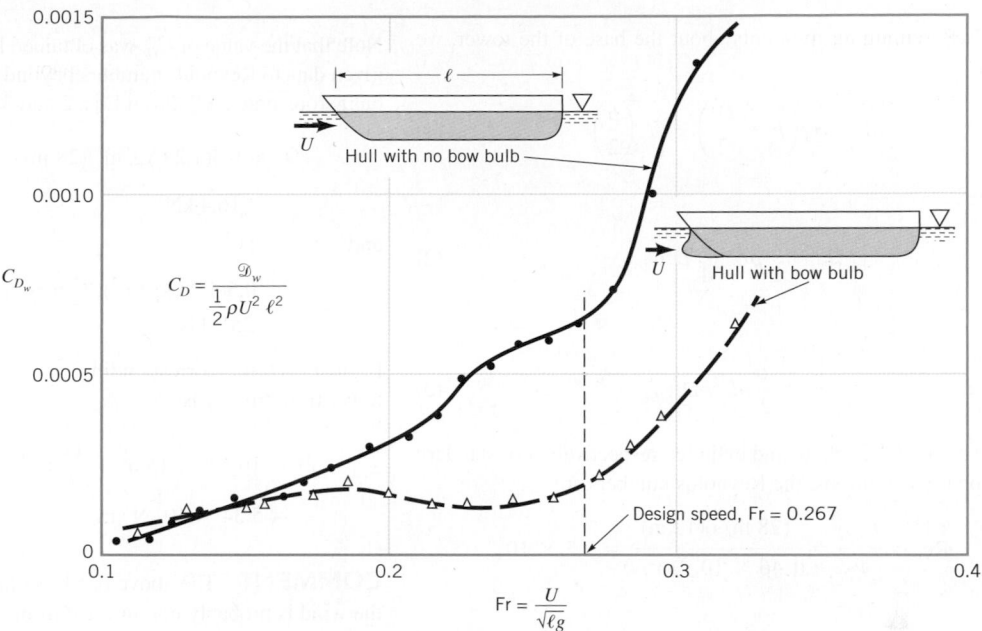

■ FIGURE 9.26 Typical drag coefficient data as a function of Froude number and hull characteristics for that portion of the drag due to the generation of waves (adapted from Ref. 25).

$C_{Dw} = \mathcal{D}_w/(\rho U^2 \ell^2/2)$, on the Froude number shown is typical. It results from the fact that the structure of the waves produced by the hull is a strong function of the ship speed or, in dimensionless form, the Froude number. This wave structure is also a function of the body shape. For example, the bow wave, which is often the major contributor to the wave drag, can be reduced by use of an appropriately designed bulb on the bow, as is indicated in Fig. 9.26. In this instance the streamlined body (hull without a bulb) has more drag than the less streamlined one.

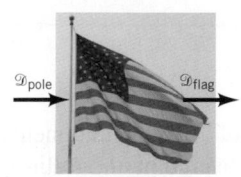

The drag on a complex body can be approximated as the sum of the drag on its parts.

Composite Body Drag. Approximate drag calculations for a complex body can often be obtained by treating the body as a composite collection of its various parts. For example, the total force on a flag pole because of the wind (see the figure in the margin) can be approximated by adding the aerodynamic drag produced by the various components involved—the drag on the flag and the drag on the pole. In some cases considerable care must be taken in such an approach because of the interactions between the various parts. It may not be correct to merely add the drag of the components to obtain the drag of the entire object, although such approximations are often reasonable.

EXAMPLE 9.13 Drag on a Composite Body

GIVEN A 100 km/hr (i.e., 28 m/s) wind blows past the water tower shown in Fig. E9.13a.

FIND Estimate the moment (torque), M, needed at the base to keep the tower from tipping over.

SOLUTION

We treat the water tower as a sphere resting on a circular cylinder and assume that the total drag is the sum of the drag from these parts. The free-body diagram of the tower is shown in Fig.

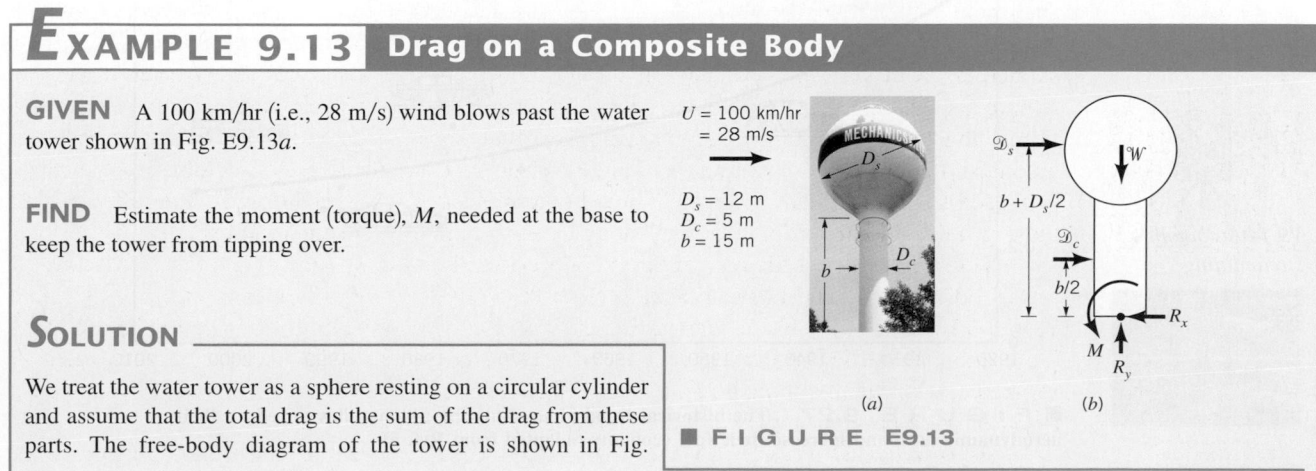

$U = 100$ km/hr
$= 28$ m/s

$D_s = 12$ m
$D_c = 5$ m
$b = 15$ m

(a)   (b)

■ FIGURE E9.13

E9.13*b*. By summing moments about the base of the tower, we obtain

$$M = \mathscr{D}_s\left(b + \frac{D_s}{2}\right) + \mathscr{D}_c\left(\frac{b}{2}\right) \quad (1)$$

where

$$\mathscr{D}_s = \frac{1}{2}\rho U^2 \frac{\pi}{4}D_s^2 C_{Ds} \quad (2)$$

and

$$\mathscr{D}_c = \frac{1}{2}\rho U^2 b D_c C_{Dc} \quad (3)$$

are the drag on the sphere and cylinder, respectively. For standard atmospheric conditions, the Reynolds numbers are

$$\mathrm{Re}_s = \frac{UD_s}{\nu} = \frac{(28\text{ m/s})(12\text{ m})}{1.46 \times 10^{-5}\text{ m}^2/\text{s}} = 2.3 \times 10^7$$

and

$$\mathrm{Re}_c = \frac{UD_c}{\nu} = \frac{(28\text{ m/s})(5\text{ m})}{1.46 \times 10^{-5}\text{ m}^2/\text{s}} = 9.6 \times 10^6$$

The corresponding drag coefficients, C_{Ds} and C_{Dc}, can be approximated from Fig. 9.21 as

$$C_{Ds} \approx 0.3 \quad \text{and} \quad C_{Dc} \approx 0.7$$

Note that the value of C_{Ds} was obtained by an extrapolation of the given data to Reynolds numbers beyond those given (a potentially dangerous practice!). From Eqs. 2 and 3 we obtain

$$\mathscr{D}_s = 0.5(1.23\text{ kg/m}^3)(28\text{ m/s})^2\frac{\pi}{4}(12\text{ m})^2(0.3)$$

$$= 16.4\text{ kN}$$

and

$$\mathscr{D}_c = 0.5(1.23\text{ kg/m}^3)(28\text{ m/s})^2(15\text{ m} \times 5\text{ m})(0.7)$$

$$= 25.3\text{ kN}$$

From Eq. 1 the corresponding moment needed to prevent the tower from tipping is

$$M = 16.4\text{ kN}\left(15\text{ m} + \frac{12}{2}\text{ m}\right) + 25.3\text{ kN}\left(\frac{15}{2}\text{ m}\right)$$

$$= 5.34 \times 10^5\text{ N} \cdot \text{m} \quad \text{(Ans)}$$

COMMENT The above result is only an estimate because (a) the wind is probably not uniform from the top of the tower to the ground, (b) the tower is not exactly a combination of a smooth sphere and a circular cylinder, (c) the cylinder is not of infinite length, (d) there will be some interaction between the flow past the cylinder and that past the sphere so that the net drag is not exactly the sum of the two, and (e) a drag coefficient value was obtained by extrapolation of the given data. However, such approximate results are often quite accurate.

V9.13 Drag on a truck

The aerodynamic drag on automobiles provides an example of the use of adding component drag forces. The power required to move a car along a level street is used to overcome the rolling resistance and the aerodynamic drag. For speeds above approximately 50 km/hr, the aerodynamic drag becomes a significant contribution to the net propulsive force needed. The contribution of the drag due to various portions of car (i.e., front end, windshield, roof, rear end, windshield peak, rear roof/trunk, and cowl) have been determined by numerous model and full-sized tests as well as by

V9.14 Automobile streamlining

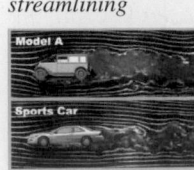

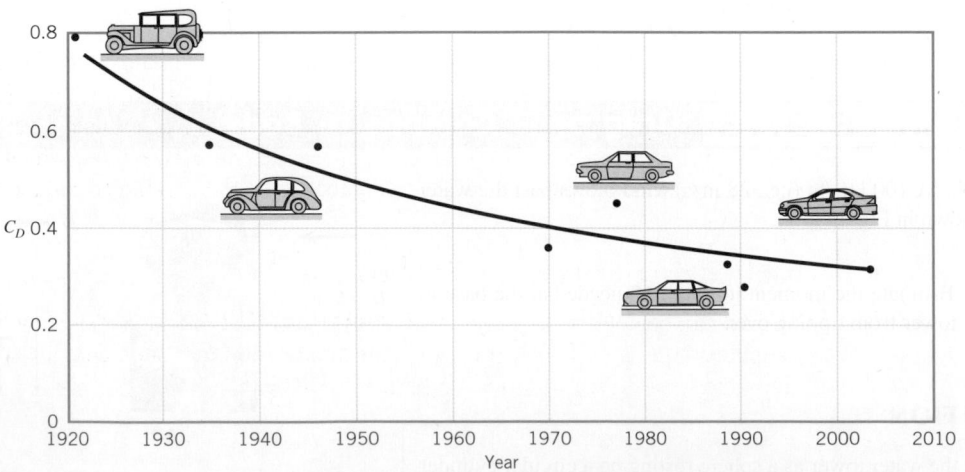

■ **F I G U R E   9.27**   The historical trend of streamlining automobiles to reduce their aerodynamic drag and increase their fuel economy (adapted from Ref. 5).

numerical calculations. As a result it is possible to predict the aerodynamic drag on cars of a wide variety of body styles.

As is indicated in Fig. 9.27, the drag coefficient for cars has decreased rather continuously over the years. This reduction is a result of careful design of the shape and the details (such as window molding, rear view mirrors, etc.). An additional reduction in drag has been accomplished by a reduction of the projected area. The net result is a considerable increase in the gas mileage, especially at highway speeds. Considerable additional information about the aerodynamics of road vehicles can be found in the literature (Ref. 30).

F l u i d s   i n   t h e   N e w s

At 10,240 mpg (4352 km/L) it doesn't cost much to "fill 'er up" Typical gas consumption for a Formula 1 racer, a sports car, and a sedan is approximately 2 mpg (0.85 km/L), 15 mpg (6.4 km/L), and 30 mpg (12.8 km/L), respectively. Thus, just how did the winning entry in the 2002 Shell Eco-Marathon achieve an incredible 10,240 mpg (4352 km/L)? To be sure, this vehicle is not as fast as a Formula 1 racer (although the rules require it to average at least 15 mph or 24 km/hr) and it can't carry as large a load as your family sedan can (the vehicle has barely enough room for the driver).

However, by using a number of clever engineering design considerations, this amazing fuel efficiency was obtained. The type (and number) of tires, the appropriate engine power and weight, the specific chassis design, and the design of the body shell are all important and interrelated considerations. To reduce *drag*, the aerodynamic shape of the high-efficiency vehicle was given special attention through theoretical considerations and wind tunnel model testing. The result is an amazing vehicle that can travel a long distance without hearing the usual "fill 'er up." (See Problem 9.90.)

The effect of several important parameters (shape, Re, Ma, Fr, and roughness) on the drag coefficient for various objects has been discussed in this section. As stated previously, drag coefficient information for a very wide range of objects is available in the literature. Some of this information is given in Figs. 9.28, 9.29, and 9.30 below for a variety of two- and three-dimensional, natural and man-made objects. Recall that a drag coefficient of unity is equivalent to the drag produced by the dynamic pressure acting on an area of size A. That is, $\mathcal{D} = \frac{1}{2}\rho U^2 A C_D = \frac{1}{2}\rho U^2 A$ if $C_D = 1$. Typical nonstreamlined objects have drag coefficients on this order.

9.4 Lift

As is indicated in Section 9.1, any object moving through a fluid will experience a net force of the fluid on the object. For objects symmetrical perpendicular to the upstream flow, this force will be in the direction of the free stream—a drag, $\mathcal{D}$. If the object is not symmetrical (or if it does not produce a symmetrical flow field, such as the flow around a rotating sphere), there may also be a force normal to the free stream—a lift, $\mathcal{L}$. Considerable effort has been put forth to understand the various properties of the generation of lift. Some objects, such as an airfoil, are designed to generate lift. Other objects are designed to reduce the lift generated. For example, the lift on a car tends to reduce the contact force between the wheels and the ground, causing reduction in traction and cornering ability. It is desirable to reduce this lift.

9.4.1 Surface Pressure Distribution

The lift can be determined from Eq. 9.2 if the distributions of pressure and wall shear stress around the entire body are known. As is indicated in Section 9.1, such data are usually not known. Typically, the lift is given in terms of the lift coefficient,

The lift coefficient is a dimensionless form of the lift.

$$C_L = \frac{\mathcal{L}}{\frac{1}{2}\rho U^2 A}$$

(9.39)

Shape	Reference area A (b = length)	Drag coefficient $C_D = \dfrac{\mathcal{D}}{\frac{1}{2}\rho U^2 A}$	Reynolds number $\text{Re} = \rho U D/\mu$
Square rod with rounded corners	$A = bD$	$\begin{array}{c\|c} R/D & C_D \\ \hline 0 & 2.2 \\ 0.02 & 2.0 \\ 0.17 & 1.2 \\ 0.33 & 1.0 \end{array}$	$\text{Re} = 10^5$
Rounded equilateral triangle	$A = bD$	$\begin{array}{c\|c\|c} R/D & \rightarrow~C_D & \leftarrow~C_D \\ \hline 0 & 1.4 & 2.1 \\ 0.02 & 1.2 & 2.0 \\ 0.08 & 1.3 & 1.9 \\ 0.25 & 1.1 & 1.3 \end{array}$	$\text{Re} = 10^5$
Semicircular shell	$A = bD$	$\rightarrow$ 2.3 $\leftarrow$ 1.1	$\text{Re} = 2 \times 10^4$
Semicircular cylinder	$A = bD$	$\rightarrow$ 2.15 $\leftarrow$ 1.15	$\text{Re} > 10^4$
T-beam	$A = bD$	$\rightarrow$ 1.80 $\leftarrow$ 1.65	$\text{Re} > 10^4$
I-beam	$A = bD$	2.05	$\text{Re} > 10^4$
Angle	$A = bD$	$\rightarrow$ 1.98 $\leftarrow$ 1.82	$\text{Re} > 10^4$
Hexagon	$A = bD$	1.0	$\text{Re} > 10^4$
Rectangle	$A = bD$	$\begin{array}{c\|c} \ell/D & C_D \\ \hline \leq 0.1 & 1.9 \\ 0.5 & 2.5 \\ 0.65 & 2.9 \\ 1.0 & 2.2 \\ 2.0 & 1.6 \\ 3.0 & 1.3 \end{array}$	$\text{Re} = 10^5$

■ **F I G U R E   9.28**   **Typical drag coefficients for regular two-dimensional objects (Refs. 5, 6).**

which is obtained from experiments, advanced analysis, or numerical considerations. The lift coefficient is a function of the appropriate dimensionless parameters and, as the drag coefficient, can be written as

$$C_L = \phi(\text{shape, Re, Ma, Fr, } \varepsilon/\ell)$$

The lift coefficient is a function of other dimensionless parameters.

The Froude number, Fr, is important only if there is a free surface present, as with an underwater "wing" used to support a high-speed hydrofoil surface ship. Often the surface roughness, ε, is relatively unimportant in terms of lift—it has more of an effect on the drag. The Mach number, Ma, is of importance for relatively high-speed subsonic and supersonic flows (i.e., Ma > 0.8), and the Reynolds number effect is often not great. The most important parameter that affects the lift coefficient is the shape of the object. Considerable effort has gone into designing optimally shaped lift-producing devices. We will emphasize the effect of the shape on lift—the effects of the other dimensionless parameters can be found in the literature (Refs. 13, 14, 29).

Shape	Reference area A	Drag coefficient C_D	Reynolds number $Re = \rho UD/\mu$
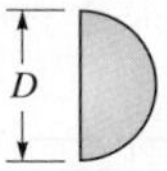 Solid hemisphere	$A = \frac{\pi}{4}D^2$	→ 1.17 ← 0.42	$Re > 10^4$
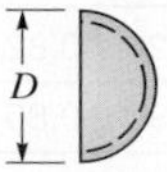 Hollow hemisphere	$A = \frac{\pi}{4}D^2$	→ 1.42 ← 0.38	$Re > 10^4$
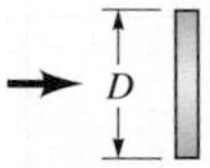 Thin disk	$A = \frac{\pi}{4}D^2$	1.1	$Re > 10^3$
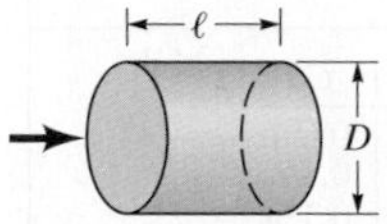 Circular rod parallel to flow	$A = \frac{\pi}{4}D^2$	ℓ/D: 0.5→1.1, 1.0→0.93, 2.0→0.83, 4.0→0.85	$Re > 10^5$
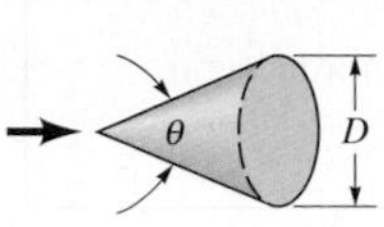 Cone	$A = \frac{\pi}{4}D^2$	θ: 10→0.30, 30→0.55, 60→0.80, 90→1.15	$Re > 10^4$
Cube	$A = D^2$	1.05	$Re > 10^4$
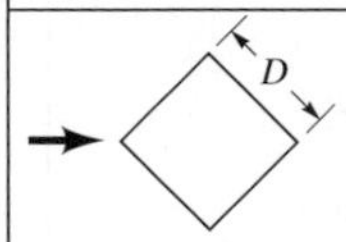 Cube	$A = D^2$	0.80	$Re > 10^4$
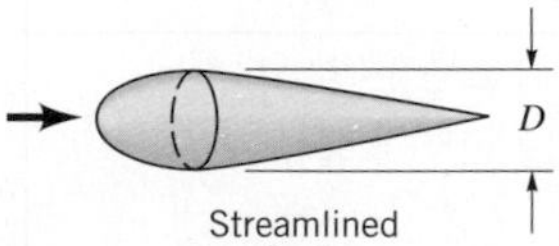 Streamlined body	$A = \frac{\pi}{4}D^2$	0.04	$Re > 10^5$

■ **FIGURE 9.29** **Typical drag coefficients for regular three-dimensional objects (Ref. 5).**

Usually most lift comes from pressure forces, not viscous forces.

Most common lift-generating devices (i.e., airfoils, fans, spoilers on cars, etc.) operate in the large Reynolds number range in which the flow has a boundary layer character, with viscous effects confined to the boundary layers and wake regions. For such cases the wall shear stress, τ_w, contributes little to the lift. Most of the lift comes from the surface pressure distribution. A typical pressure distribution on a moving car is shown in Fig. 9.31. The distribution, for the most part, is consistent with simple Bernoulli equation analysis. Locations with high-speed flow (i.e., over the roof and hood) have low pressure, while locations with low-speed flow (i.e., on the grill and windshield) have high pressure. It is easy to believe that the integrated effect of this pressure distribution would provide a net upward force.

For objects operating in very low Reynolds number regimes (i.e., Re < 1), viscous effects are important, and the contribution of the shear stress to the lift may be as important as that of the pressure. Such situations include the flight of minute insects and the swimming of microscopic organisms. The relative importance of τ_w and p in the generation of lift in a typical large Reynolds number flow is shown in Example 9.14.

Shape	Reference area	Drag coefficient C_D			
Parachute	Frontal area $A = \frac{\pi}{4}D^2$	1.4			
Porous parabolic dish	Frontal area $A = \frac{\pi}{4}D^2$	Porosity	0	0.2	0.5
		$\longrightarrow$	1.42	1.20	0.82
		$\longleftarrow$	0.95	0.90	0.80
		Porosity = open area/total area			
Average person	Standing	$C_D A = 0.8 \text{ m}^2$			
	Sitting	$C_D A = 0.6 \text{ m}^2$			
	Crouching	$C_D A = 0.2 \text{ m}^2$			
Fluttering flag	$A = \ell D$	ℓ/D	C_D		
		1	0.07		
		2	0.12		
		3	0.15		
Empire State Building	Frontal area	1.4			
Six-car passenger train	Frontal area	1.8			
Bikes Upright commuter	$A = 0.5 \text{ m}^2$	1.1			
Racing	$A = 0.4 \text{ m}^2$	0.88			
Drafting	$A = 0.4 \text{ m}^2$	0.50			
Streamlined	$A = 0.46 \text{ m}^2$	0.12			
Tractor-trailer trucks Standard	Frontal area	0.96			
With fairing	Frontal area	0.76			
With fairing and gap seal	Frontal area	0.70			
Tree $U = 10$ m/s $U = 20$ m/s $U = 30$ m/s	Frontal area	0.43 0.26 0.20			
Dolphin	Wetted area	0.0036 at Re = 6×10^6 (flat plate has $C_{Df} = 0.0031$)			
Large birds	Frontal area	0.40			

■ **F I G U R E   9.30**   **Typical drag coefficients for objects of interest (Refs. 5, 6, 15, 20).**

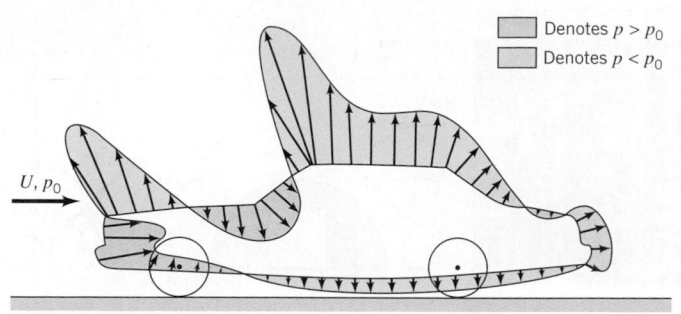

Denotes $p > p_0$
Denotes $p < p_0$

U, p_0

■ **FIGURE 9.31**   Pressure distribution on the surface of an automobile.

EXAMPLE 9.14   Lift from Pressure and Shear Stress Distributions

GIVEN   When a uniform wind of velocity U blows past the semicircular building shown in Fig. E9.14a,b, the wall shear stress and pressure distributions on the outside of the building are as given previously in Figs. E9.8b and E9.9a, respectively.

FIND   If the pressure in the building is atmospheric (i.e., the value, p_0, far from the building), determine the lift coefficient and the lift on the roof.

SOLUTION

From Eq. 9.2 we obtain the lift as

$$\mathcal{L} = -\int p \sin \theta \, dA + \int \tau_w \cos \theta \, dA \tag{1}$$

As is indicated in Fig. E9.14b, we assume that on the inside of the building the pressure is uniform, $p = p_0$, and that there is no shear stress. Thus, Eq. 1 can be written as

$$\mathcal{L} = -\int_0^\pi (p - p_0) \sin \theta \, b\left(\frac{D}{2}\right) d\theta$$
$$+ \int_0^\pi \tau_w \cos \theta \, b\left(\frac{D}{2}\right) d\theta$$

or

$$\mathcal{L} = \frac{bD}{2}\left[-\int_0^\pi (p - p_0) \sin \theta \, d\theta + \int_0^\pi \tau_w \cos \theta \, d\theta\right] \tag{2}$$

where b and D are the length and diameter of the building, respectively, and $dA = b(D/2)d\theta$. Equation 2 can be put into dimensionless form by using the dynamic pressure, $\rho U^2/2$, planform area, $A = bD$, and dimensionless shear stress

$$F(\theta) = \tau_w(\mathrm{Re})^{1/2}/(\rho U^2/2)$$

to give

$$\mathcal{L} = \frac{1}{2}\rho U^2 A \left[-\frac{1}{2}\int_0^\pi \frac{(p - p_0)}{\frac{1}{2}\rho U^2} \sin \theta \, d\theta\right.$$
$$\left.+ \frac{1}{2\sqrt{\mathrm{Re}}}\int_0^\pi F(\theta) \cos \theta \, d\theta\right] \tag{3}$$

From the data in Figs. E9.8b and E9.9a, the values of the two integrals in Eq. 3 can be obtained by determining the area under the

curves of $[(p - p_0)/(\rho U^2/2)] \sin \theta$ versus θ and $F(\theta) \cos \theta$ versus θ plotted in Figs. E9.14c and E9.14d. The results are

$$\int_0^\pi \frac{(p - p_0)}{\frac{1}{2}\rho U^2} \sin \theta \, d\theta = -1.76$$

and

$$\int_0^\pi F(\theta) \cos \theta \, d\theta = 3.92$$

Thus, the lift is

$$\mathcal{L} = \frac{1}{2}\rho U^2 A \left[\left(-\frac{1}{2}\right)(-1.76) + \frac{1}{2\sqrt{\mathrm{Re}}}(3.92)\right]$$

or

$$\mathcal{L} = \left(0.88 + \frac{1.96}{\sqrt{\mathrm{Re}}}\right)\left(\frac{1}{2}\rho U^2 A\right) \tag{Ans}$$

and

$$C_L = \frac{\mathcal{L}}{\frac{1}{2}\rho U^2 A} = 0.88 + \frac{1.96}{\sqrt{\mathrm{Re}}} \tag{4} (Ans)$$

COMMENTS   Consider a typical situation with $D = 6$ m, $U = 10$ m/s, $b = 15$ m, and standard atmospheric conditions ($\rho = 1.23$ kg/m³ and $\nu = 1.46 \times 10^{-5}$ m²/s), which gives a Reynolds number of

$$\mathrm{Re} = \frac{UD}{\nu} = \frac{(10 \text{ m/s})(6 \text{ m})}{1.46 \times 10^{-5} \text{ m}^2/\text{s}} = 4.1 \times 10^6$$

Hence, the lift coefficient is

$$C_L = 0.88 + \frac{1.96}{(4.1 \times 10^6)^{1/2}} = 0.88 + 0.001 = 0.881$$

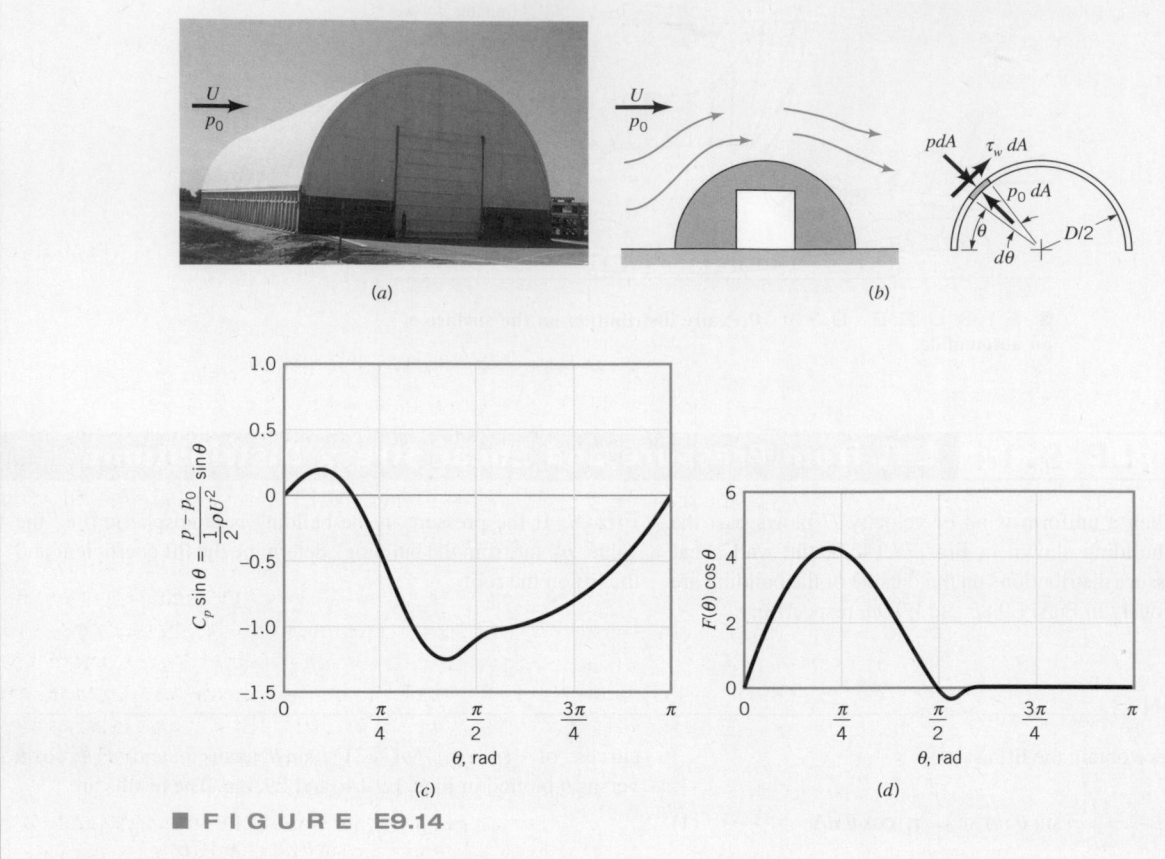

(a)

(b)

$$C_p \sin\theta = \frac{p - p_0}{\frac{1}{2}\rho U^2}\sin\theta$$

(c)

$F(\theta)\cos\theta$

(d)

■ **F I G U R E   E9.14**

Note that the pressure contribution to the lift coefficient is 0.88 whereas that due to the wall shear stress is only $1.96/(\mathrm{Re}^{1/2}) = 0.001$. The Reynolds number dependency of C_L is quite minor. The lift is pressure dominated. Recall from Example 9.9 that this is also true for the drag on a similar shape.

From Eq. 4 with $A = 6\ \mathrm{m} \times 15\ \mathrm{m} = 90\ \mathrm{m}^2$, we obtain the lift for the assumed conditions as

$$\mathscr{L} = \tfrac{1}{2}\rho U^2 A C_L = \tfrac{1}{2}(1.23\ \mathrm{kg/m^3})(10\ \mathrm{m/s})^2(90\ \mathrm{m}^2)(0.881)$$

or

$$\mathscr{L} = 4876\ \mathrm{N}$$

There is a considerable tendency for the building to lift off the ground. Clearly this is due to the object being nonsymmetrical. The lift force on a complete circular cylinder is zero, although the fluid forces do tend to pull the upper and lower halves apart.

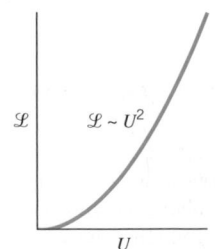

A typical device designed to produce lift does so by generating a pressure distribution that is different on the top and bottom surfaces. For large Reynolds number flows these pressure distributions are usually directly proportional to the dynamic pressure, $\rho U^2/2$, with viscous effects being of secondary importance. Hence, as indicated by the figure in the margin, for a given airfoil the lift is proportional to the square of the airspeed. Two airfoils used to produce lift are indicated in Fig. 9.32. Clearly the symmetrical one cannot produce lift unless the angle of attack, α, is nonzero. Because of the asymmetry of the nonsymmetric airfoil, the pressure distributions on the upper and lower surfaces are different, and a lift is produced even with $\alpha = 0$. Of course, there will be a certain value of α (less than zero for this case) for which the lift is zero. For this situation, the pressure distributions on the upper and lower surfaces are different, but their resultant (integrated) pressure forces will be equal and opposite.

Since most airfoils are thin, it is customary to use the planform area, $A = bc$, in the definition of the lift coefficient. Here b is the length of the airfoil and c is the *chord length*—the length from the leading edge to the trailing edge as indicated in Fig. 9.32. Typical lift coefficients so defined are on the order of unity. That is, the lift force is on the order of the dynamic pressure times the planform area of the wing, $\mathscr{L} \approx (\rho U^2/2)A$. The *wing loading*, defined as the average lift per unit area of the wing, $\mathscr{L}/A$, therefore, increases with speed. For example, the wing loading of the

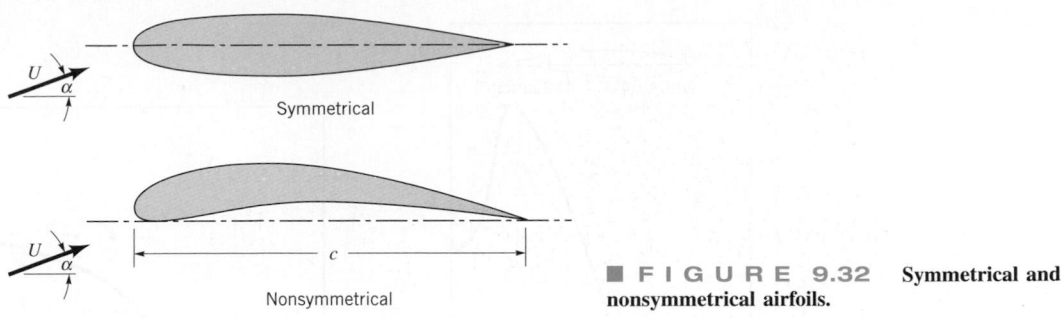

■ **F I G U R E 9.32** **Symmetrical and nonsymmetrical airfoils.**

1903 Wright Flyer aircraft was 72 N/m², while for the present-day Boeing 747 aircraft it is 7200 N/m². The wing loading for a bumble bee is approximately 48 N/m² (Ref. 15).

Typical lift and drag coefficient data as a function of angle of attack, α, and *aspect ratio*, $\mathcal{A}$, are indicated in Figs. 9.33a and 9.33b. The aspect ratio is defined as the ratio of the square of the wing length to the planform area, $\mathcal{A} = b^2/A$. If the chord length, c, is constant along the length of the wing (a rectangular planform wing), this reduces to $\mathcal{A} = b/c$.

In general, the lift coefficient increases and the drag coefficient decreases with an increase in aspect ratio. Long wings are more efficient because their wing tip losses are relatively more minor than for short wings. The increase in drag due to the finite length ($\mathcal{A} < \infty$) of the wing is often termed induced drag. It is due to the interaction of the complex swirling flow structure near the wing tips (see Fig. 9.37) and the free stream (Ref. 13). High-performance soaring airplanes and highly efficient soaring birds (i.e., the albatross and sea gull) have long, narrow wings. Such wings, however, have considerable inertia that inhibits rapid maneuvers. Thus, highly maneuverable fighter or acrobatic airplanes and birds (i.e., the falcon) have small-aspect-ratio wings.

Although viscous effects and the wall shear stress contribute little to the direct generation of lift, they play an extremely important role in the design and use of lifting devices. This is because of the viscosity-induced boundary layer separation that can occur on nonstreamlined bodies such as airfoils that have too large an angle of attack (see Fig. 9.18). As is indicated in Fig. 9.33, up to a certain point, the lift coefficient increases rather steadily with the angle of attack. If α is too large, the boundary layer on the upper surface separates, the flow over the wing develops a wide, turbulent wake region, the lift decreases, and the drag increases. This condition, as indicated by the figures in the margin, is termed *stall.* Such conditions are extremely dangerous if they occur while the airplane is flying at a low altitude where there is not sufficient time and altitude to recover from the stall.

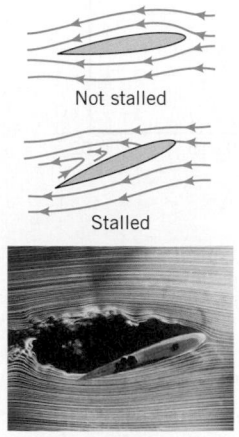

Not stalled

Stalled

At large angles of attack the boundary layer separates and the wing stalls.

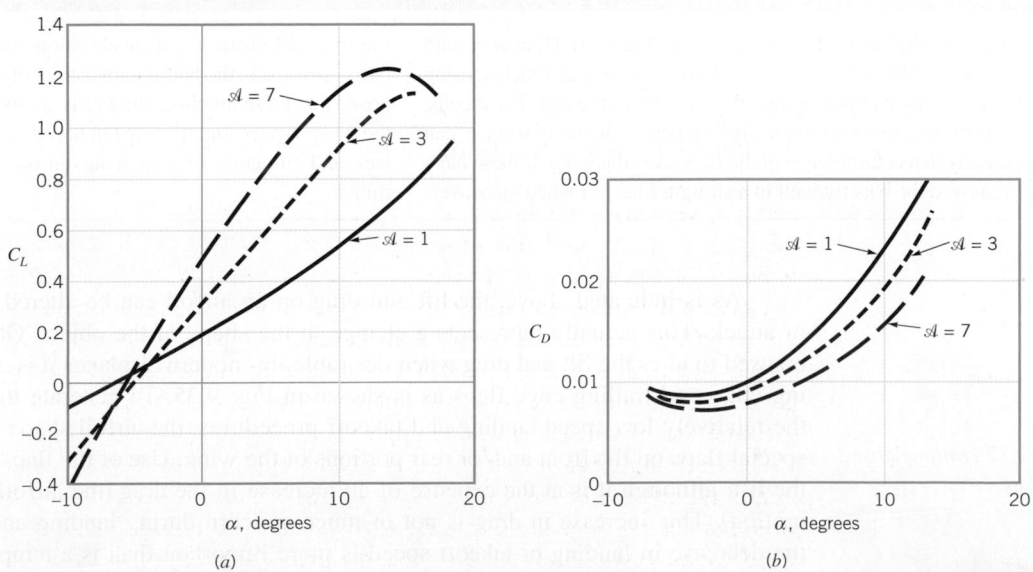

■ **F I G U R E 9.33** **Typical lift and drag coefficient data as a function of angle of attack and the aspect ratio of the airfoil: (a) lift coefficient, (b) drag coefficient.**

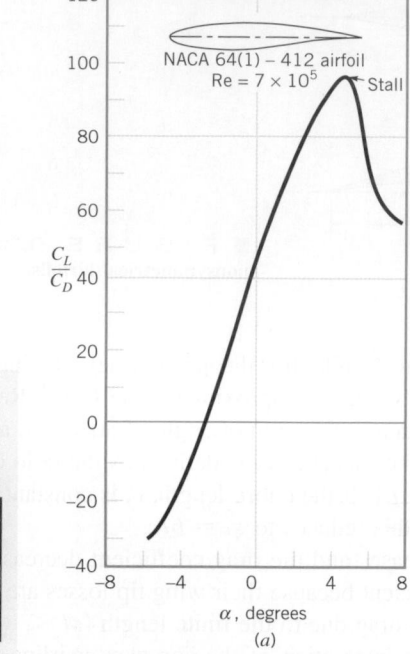

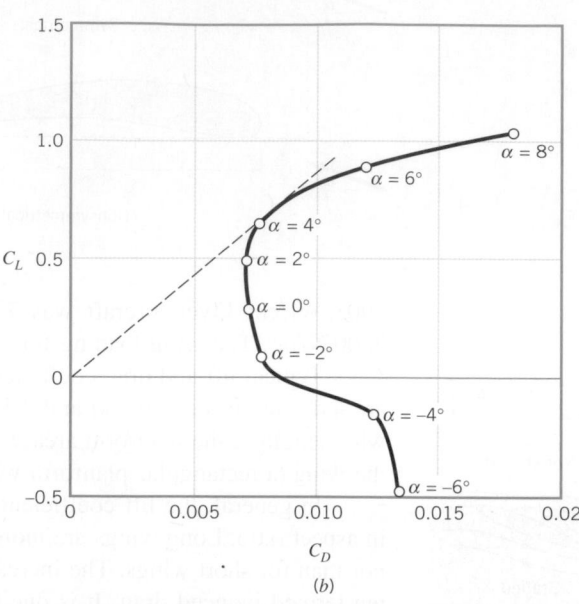

V9.15 Stalled airfoil

■ FIGURE 9.34 Two representations of the same lift and drag data for a typical airfoil: (*a*) lift-to-drag ratio as a function of angle of attack, with the onset of boundary layer separation on the upper surface indicated by the occurrence of stall, (*b*) the lift and drag polar diagram with the angle of attack indicated (Ref. 27).

V9.16 Bat flying

In many lift-generating devices the important quantity is the ratio of the lift to drag developed, $\mathscr{L}/\mathscr{D} = C_L/C_D$. Such information is often presented in terms of C_L/C_D versus α, as is shown in Fig. 9.34*a*, or in a *lift-drag polar* of C_L versus C_D with α as a parameter, as is shown in Fig. 9.34*b*. The most efficient angle of attack (i.e., largest C_L/C_D) can be found by drawing a line tangent to the $C_L - C_D$ curve from the origin, as is shown in Fig. 9.34*b*. High-performance airfoils generate lift that is perhaps 100 or more times greater than their drag. This translates into the fact that in still air they can glide a horizontal distance of 100 m for each 1 m drop in altitude.

F l u i d s   i n   t h e   N e w s

Bats feel turbulence Researchers have discovered that at certain locations on the wings of bats, there are special touch-sensing cells with a tiny hair poking out of the center of the cell. These cells, which are very sensitive to air flowing across the wing surface, can apparently detect turbulence in the flow over the wing. If these hairs are removed the bats fly well in a straight line, but when maneuvering to avoid obstacles, their elevation control is erratic. When the hairs grow back, the bats regain their complete flying skills. It is proposed that these touch-sensing cells are used to detect turbulence on the wing surface and thereby tell bats when to adjust the angle of attack and curvature of their wings in order to avoid stalling out in midair.

V9.17 Trailing edge flap

As is indicated above, the lift and drag on an airfoil can be altered by changing the angle of attack. This actually represents a change in the shape of the object. Other shape changes can be used to alter the lift and drag when desirable. In modern airplanes it is common to utilize leading edge and trailing edge flaps as is shown in Fig. 9.35. To generate the necessary lift during the relatively low-speed landing and takeoff procedures, the airfoil shape is altered by extending special flaps on the front and/or rear portions of the wing. Use of the flaps considerably enhances the lift, although it is at the expense of an increase in the drag (the airfoil is in a "dirty" configuration). This increase in drag is not of much concern during landing and takeoff operations—the decrease in landing or takeoff speed is more important than is a temporary increase in drag. During normal flight with the flaps retracted (the "clean" configuration), the drag is relatively small, and the needed lift force is achieved with the smaller lift coefficient and the larger dynamic pressure (higher speed).

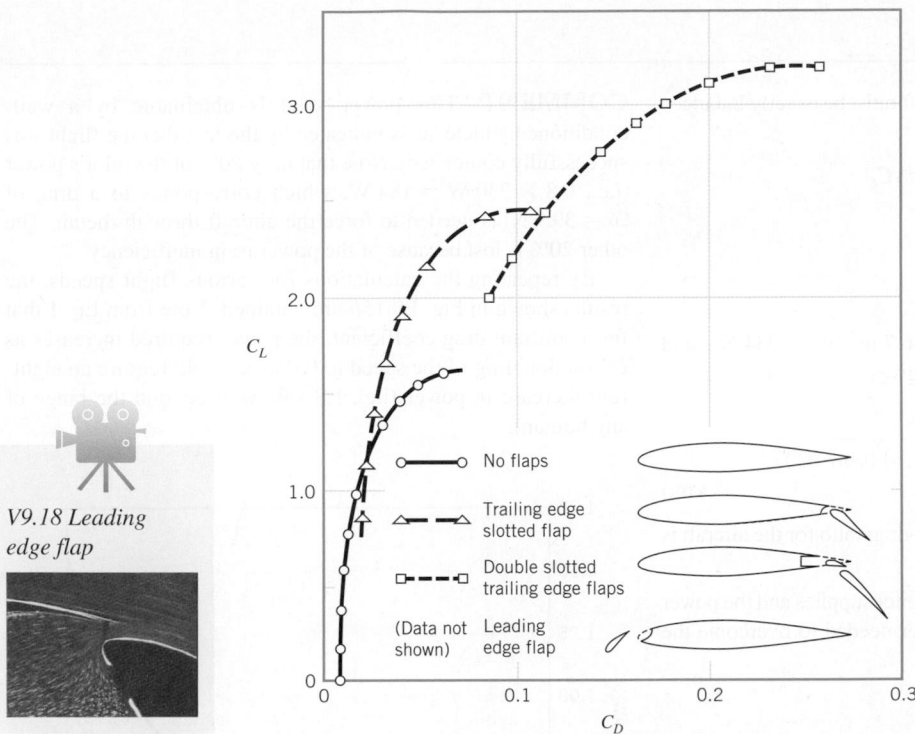

V9.18 Leading edge flap

■ **FIGURE 9.35** Typical lift and drag alterations possible with the use of various types of flap designs (Ref. 21).

Fluids in the News

Learning from nature For hundreds of years humans looked toward nature, particularly birds, for insight about flying. However, all early airplanes that closely mimicked birds proved to be unsuccessful. Only after much experimenting with rigid (or at least nonflapping) wings did human flight become possible. Recently, however, engineers have been turning to living systems—birds, insects, and other biological models—in an attempt to produce breakthroughs in aircraft design. Perhaps it is possible that nature's basic design concepts can be applied to airplane systems. For example, by morphing and rotating their wings in three dimensions, birds have remarkable maneuverability that to date has no technological parallel. Birds can control the airflow over their wings by moving the feathers on their wingtips and the leading edges of their wings, providing designs that are more efficient than the flaps and rigid, pivoting tail surfaces of current aircraft (Ref. 15). On a smaller scale, understanding the mechanism by which insects dynamically manage unstable flow to generate lift may provide insight into the development of microscale air vehicles. With new hi-tech materials, computers, and automatic controls, aircraft of the future may mimic nature more than was once thought possible. (See Problem 9.110.)

A wide variety of lift and drag information for airfoils can be found in standard aerodynamics books (Ref. 13, 14, 29).

EXAMPLE 9.15 Lift and Power for Human Powered Flight

GIVEN In 1977 the *Gossamer Condor*, shown in Fig. E9.15a, won the Kremer prize by being the first human-powered aircraft to complete a prescribed figure-of-eight course around two turning points 0.8 km apart (Ref. 22). The following data pertain to this aircraft:

$$\text{flight speed} = U = 4.6 \text{ m/s}$$

$$\text{wing size} = b = 29 \text{ m}, c = 2.3 \text{ m (average)}$$

$$\text{weight (including pilot)} = \mathcal{W} = 934 \text{ N}$$

$$\text{drag coefficient} = C_D = 0.046 \text{ (based on planform area)}$$

$$\text{power train efficiency} = \eta$$

$$= \text{power to overcome drag/pilot power} = 0.8$$

FIND Determine

(a) the lift coefficient, C_L, and

(b) the power, $\mathcal{P}$, required by the pilot.

■ **FIGURE E9.15a**
(Photograph copyright © Don Monroe.)

Solution

(a) For steady flight conditions the lift must be exactly balanced by the weight, or

$$W = \mathscr{L} = \tfrac{1}{2}\rho U^2 A C_L$$

Thus,

$$C_L = \frac{2W}{\rho U^2 A}$$

where $A = bc = 29 \text{ m} \times 2.3 \text{ m} = 66.7 \text{ m}^2$, $W = 934 \text{ N}$, and $\rho = 1.23 \text{ kg/m}^3$ for standard air. This gives

$$C_L = \frac{2(934 \text{ N})}{(1.23 \text{ kg/m}^3)(4.6 \text{ m/s})^2(66.7 \text{ m}^2)}$$

$$= 1.08 \qquad \textbf{(Ans)}$$

a reasonable number. The overall lift-to-drag ratio for the aircraft is $C_L/C_D = 1.08/0.046 = 23.5$.

(b) The product of the power that the pilot supplies and the power train efficiency equals the useful power needed to overcome the drag, $\mathscr{D}$. That is,

$$\eta \mathscr{P} = \mathscr{D} U$$

where

$$\mathscr{D} = \tfrac{1}{2}\rho U^2 A C_D$$

Thus,

$$\mathscr{P} = \frac{\mathscr{D}U}{\eta} = \frac{\tfrac{1}{2}\rho U^2 A C_D U}{\eta} = \frac{\rho A C_D U^3}{2\eta} \qquad \textbf{(1)}$$

or

$$\mathscr{P} = \frac{(1.23 \text{ kg/m}^3)(66.7 \text{ m}^2)(0.046)(4.6 \text{ m/s})^3}{2(0.8)}$$

$$\mathscr{P} = 230 \text{ W} \qquad \textbf{(Ans)}$$

COMMENT This power level is obtainable by a well-conditioned athlete (as is indicated by the fact that the flight was successfully completed). Note that only 80% of the pilot's power (i.e., $0.8 \times 230 \text{ W} = 184 \text{ W}$, which corresponds to a drag of $\mathscr{D} = 39.9 \text{ N}$) is needed to force the aircraft through the air. The other 20% is lost because of the power train inefficiency.

By repeating the calculations for various flight speeds, the results shown in Fig. E9.15b are obtained. Note from Eq. 1 that for a constant drag coefficient, the power required increases as U^3—a doubling of the speed to 9.0 m/s would require an eight-fold increase in power (i.e., 1.8 kW, well beyond the range of any human).

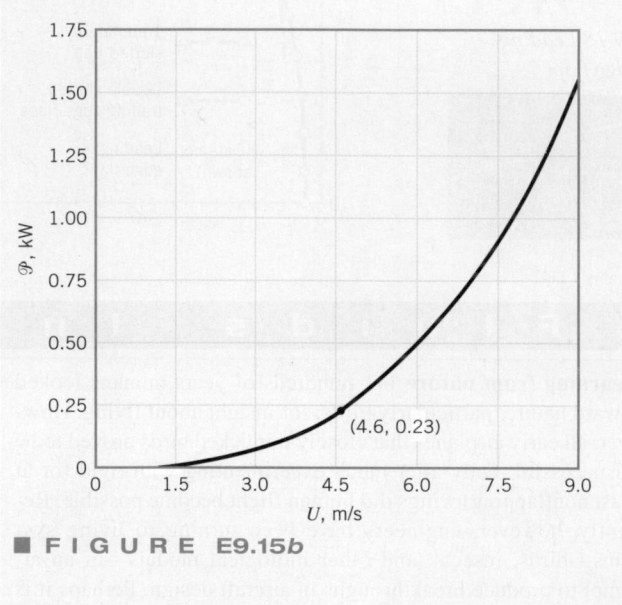

■ **FIGURE E9.15b**

9.4.2 Circulation

Inviscid flow analysis can be used to obtain ideal flow past airfoils.

Since viscous effects are of minor importance in the generation of lift, it should be possible to calculate the lift force on an airfoil by integrating the pressure distribution obtained from the equations governing inviscid flow past the airfoil. That is, the potential flow theory discussed in Chapter 6 should provide a method to determine the lift. Although the details are beyond the scope of this book, the following is found from such calculations (Ref. 4).

The calculation of the inviscid flow past a two-dimensional airfoil gives a flow field as indicated in Fig. 9.36. The predicted flow field past an airfoil with no lift (i.e., a symmetrical airfoil at zero angle of attack, Fig. 9.36a) appears to be quite accurate (except for the absence of thin boundary layer regions). However, as is indicated in Fig. 9.36b, the calculated flow past the same airfoil at a nonzero angle of attack (but one small enough so that boundary layer separation would not occur) is not proper near the trailing edge. In addition, the calculated lift for a nonzero angle of attack is zero—in conflict with the known fact that such airfoils produce lift.

In reality, the flow should pass smoothly over the top surface as is indicated in Fig. 9.36c, without the strange behavior indicated near the trailing edge in Fig. 9.36b. As is shown in Fig. 9.36d, the unrealistic flow situation can be corrected by adding an appropriate clockwise swirling flow around the airfoil. The results are twofold: (1) The unrealistic behavior near the trailing edge is eliminated (i.e.,

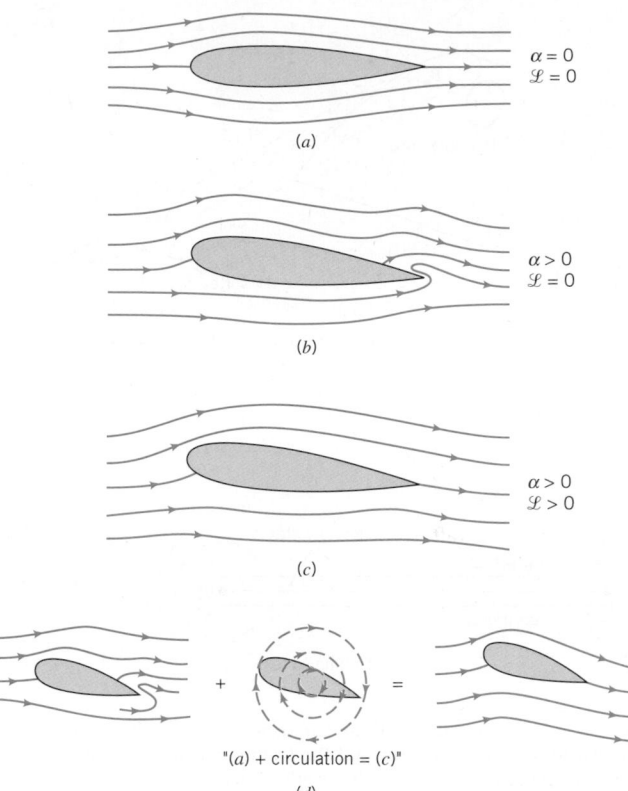

$\alpha = 0$
$\mathscr{L} = 0$

(a)

$\alpha > 0$
$\mathscr{L} = 0$

(b)

$\alpha > 0$
$\mathscr{L} > 0$

(c)

+   =

"(a) + circulation = (c)"

(d)

■ FIGURE 9.36   Inviscid flow past an airfoil: (a) symmetrical flow past the symmetrical airfoil at a zero angle of attack; (b) same airfoil at a nonzero angle of attack—no lift, flow near trailing edge not realistic; (c) same conditions as for (b) except circulation has been added to the flow—nonzero lift, realistic flow; (d) superposition of flows to produce the final flow past the airfoil.

the flow pattern of Fig. 9.36b is changed to that of Fig. 9.36c), and (2) the average velocity on the upper surface of the airfoil is increased while that on the lower surface is decreased. From the Bernoulli equation concepts (i.e., $p/\gamma + V^2/2g + z = $ constant), the average pressure on the upper surface is decreased and that on the lower surface is increased. The net effect is to change the original zero lift condition to that of a lift-producing airfoil.

The addition of the clockwise swirl is termed the addition of *circulation*. The amount of swirl (circulation) needed to have the flow leave the trailing edge smoothly is a function of the airfoil size and shape and can be calculated from potential flow (inviscid) theory (see Section 6.6.3 and Ref. 29). Although the addition of circulation to make the flow field physically realistic may seem artificial, it has well-founded mathematical and physical grounds. For example, consider the flow past a finite length airfoil, as is indicated in Fig. 9.37. For lift-generating conditions the average pressure on the lower surface is greater than that on the upper surface. Near the tips of the wing this pressure difference will cause some of the fluid to attempt to migrate from the lower to the upper surface, as is indicated in Fig. 9.37b. At the same time, this fluid is swept downstream, forming a *trailing vortex* (swirl) from each wing tip (see Fig. 4.3). It is speculated that the reason some birds migrate in vee-formation is to take advantage of the updraft produced by the trailing vortex of the preceding bird. [It is calculated that for a given expenditure of energy, a flock of 25 birds flying in vee-formation could travel 70% farther than if each bird were to fly separately (Ref. 15).]

The trailing vortices from the right and left wing tips are connected by the *bound vortex* along the length of the wing. It is this vortex that generates the circulation that produces the lift. The combined vortex system (the bound vortex and the trailing vortices) is termed a horseshoe vortex. The strength of the trailing vortices (which is equal to the strength of the bound vortex) is proportional to the lift generated. Large aircraft (for example, a Boeing 747) can generate very strong trailing vortices that persist for a long time before viscous effects and instability mechanisms finally cause them to die out. Such vortices are strong enough to flip smaller aircraft out of control if they follow too closely behind the large aircraft. The figure in the margin clearly shows a trailing vortex produced during a wake vortex study in which an airplane flew through a column of smoke.

V9.19 Wing tip vortices

(Photograph courtesy of NASA.)

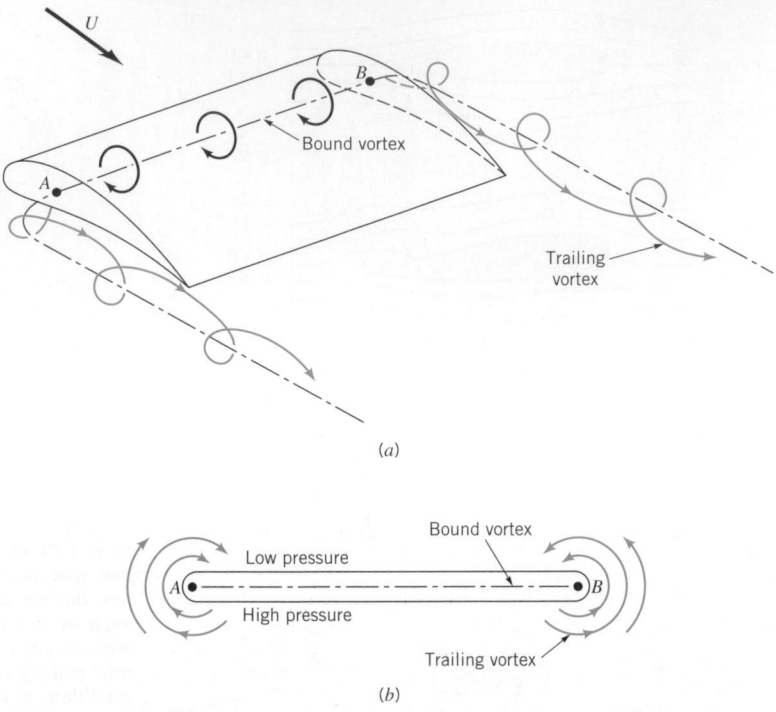

■ **FIGURE 9.37** **Flow past a finite length wing:** (*a*) **the horseshoe vortex system produced by the bound vortex and the trailing vortices;** (*b*) **the leakage of air around the wing tips produces the trailing vortices.**

As is indicated above, the generation of lift is directly related to the production of a swirl or vortex flow around the object. A nonsymmetric airfoil, by design, generates its own prescribed amount of swirl and lift. A symmetric object like a circular cylinder or sphere, which normally provides no lift, can generate swirl and lift if it rotates.

As is discussed in Section 6.6.3, the inviscid flow past a circular cylinder has the symmetrical flow pattern indicated in Fig. 9.38*a*. By symmetry the lift and drag are zero. However, if the cylinder is rotated about its axis in a stationary real $(\mu \neq 0)$ fluid, the rotation will drag some of the fluid around, producing circulation about the cylinder as in Fig. 9.38*b*. When this circulation is combined with an ideal, uniform upstream flow, the flow pattern indicated in Fig. 9.38*c* is obtained. The flow is no longer symmetrical about the horizontal plane through the center of the cylinder; the average pressure is greater on the lower half of the cylinder than on the upper half, and a lift is generated. This effect is called the *Magnus effect,* after Heinrich Magnus (1802–1870), a German chemist and physicist who first investigated this phenomenon. A similar lift is generated on a rotating sphere. It accounts for the various types of pitches in baseball (i.e., curve ball, floater, sinker, etc.), the ability of a soccer player to hook the ball, and the hook or slice of a golf ball.

A spinning sphere or cylinder can generate lift.

Typical lift and drag coefficients for a smooth, spinning sphere are shown in Fig. 9.39. Although the drag coefficient is fairly independent of the rate of rotation, the lift coefficient is strongly

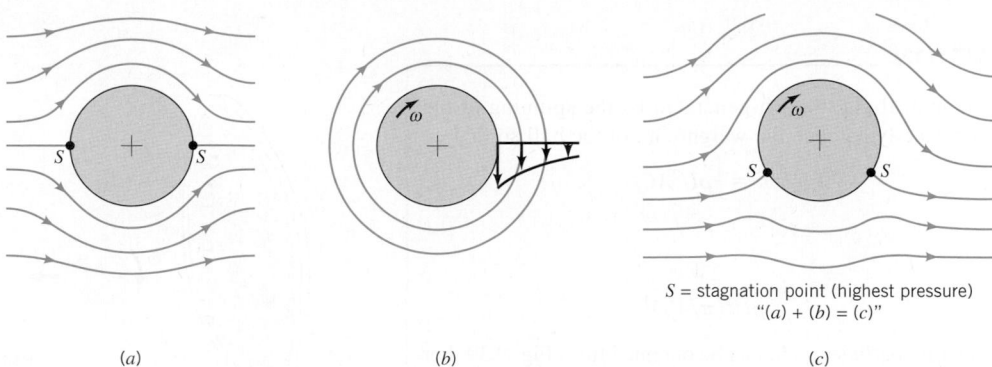

S = stagnation point (highest pressure)
"(a) + (b) = (c)"

(a)   (b)   (c)

■ **FIGURE 9.38**   **Inviscid flow past a circular cylinder:** (a) **uniform upstream flow without circulation,** (b) **free vortex at the center of the cylinder,** (c) **combination of free vortex and uniform flow past a circular cylinder giving nonsymmetric flow and a lift.**

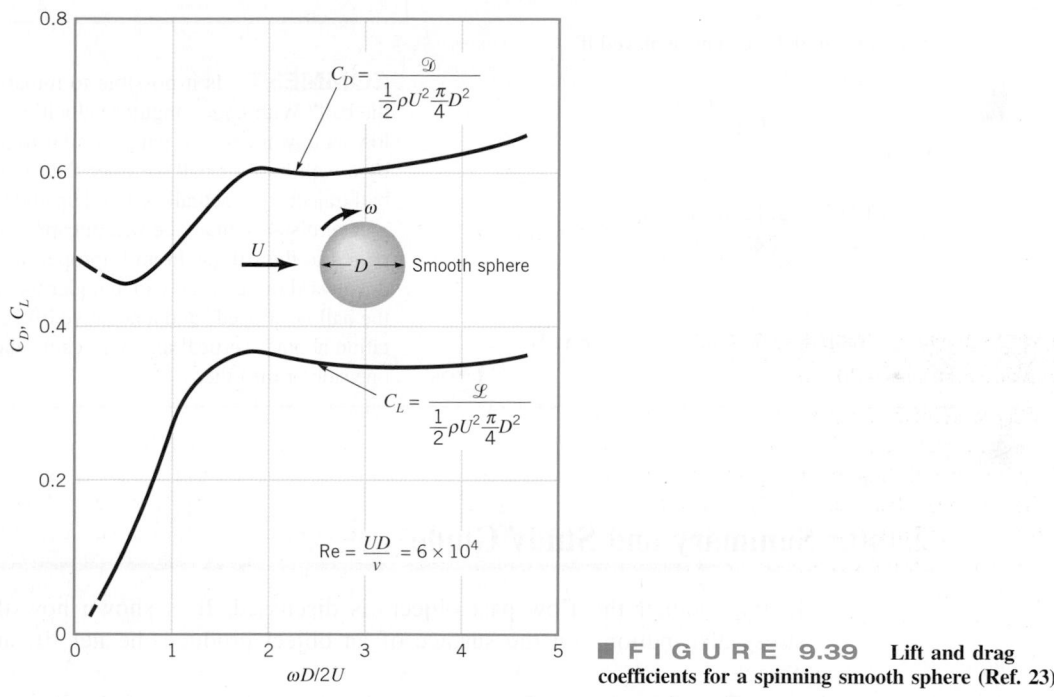

■ **FIGURE 9.39**   **Lift and drag coefficients for a spinning smooth sphere (Ref. 23).**

dependent on it. In addition (although not indicated in the figure), both C_L and C_D are dependent on the roughness of the surface. As was discussed in Section 9.3, in a certain Reynolds number range an increase in surface roughness actually decreases the drag coefficient. Similarly, an increase in surface roughness can increase the lift coefficient because the roughness helps drag more fluid around the sphere increasing the circulation for a given angular velocity. Thus, a rotating, rough golf ball travels farther than a smooth one because the drag is less and the lift is greater. However, do not expect a severely roughed up (cut) ball to work better—extensive testing has gone into obtaining the optimum surface roughness for golf balls.

A dimpled golf ball has less drag and more lift than a smooth one.

EXAMPLE 9.16   Lift on a Rotating Sphere

GIVEN   A table tennis ball weighing 2.45×10^{-2} N with diameter $D = 3.8 \times 10^{-2}$ m is hit at a velocity of $U = 12$ m/s with a back spin of angular velocity ω as is shown in Fig. E9.16.

FIND   What is the value of ω if the ball is to travel on a horizontal path, not dropping due to the acceleration of gravity?

SOLUTION

For horizontal flight, the lift generated by the spinning of the ball must exactly balance the weight, $\mathcal{W}$, of the ball so that

$$\mathcal{W} = \mathcal{L} = \tfrac{1}{2}\rho U^2 A C_L$$

or

$$C_L = \frac{2\mathcal{W}}{\rho U^2 (\pi/4) D^2}$$

where the lift coefficient, C_L, can be obtained from Fig. 9.39. For standard atmospheric conditions with $\rho = 1.23 \text{ kg/m}^3$ we obtain

$$C_L = \frac{2(2.45 \times 10^{-2}\text{ N})}{(1.23 \text{ kg/m}^3)(12 \text{ m/s})^2(\pi/4)(3.8 \times 10^{-2}\text{ m})^2}$$

$$= 0.244$$

which, according to Fig. 9.39, can be achieved if

$$\frac{\omega D}{2U} = 0.9$$

or

$$\omega = \frac{2U(0.9)}{D} = \frac{2(12 \text{ m/s})(0.9)}{3.8 \times 10^{-2}\text{ m}} = 568 \text{ rad/s}$$

Thus,

$$\omega = (568 \text{ rad/s})(60 \text{ s/min})(1 \text{ rev}/2\pi \text{ rad})$$

$$= 5420 \text{ rpm} \qquad \text{(Ans)}$$

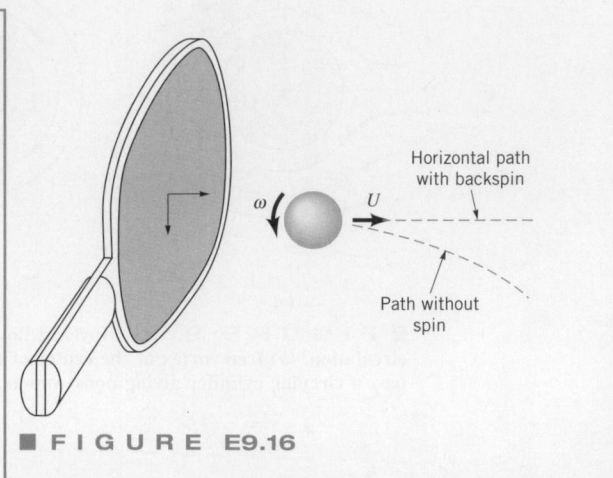

Horizontal path with backspin

Path without spin

■ **F I G U R E E9.16**

COMMENT   Is it possible to impart this angular velocity to the ball? With larger angular velocities the ball will rise and follow an upward curved path. Similar trajectories can be produced by a well-hit golf ball—rather than falling like a rock, the golf ball trajectory is actually curved up and the spinning ball travels a greater distance than one without spin. However, if topspin is imparted to the ball (as in an improper tee shot) the ball will curve downward more quickly than under the action of gravity alone—the ball is "topped" and a negative lift is generated. Similarly, rotation about a vertical axis will cause the ball to hook or slice to one side or the other.

9.5   Chapter Summary and Study Guide

drag
lift
lift coefficient
drag coefficient
wake region
boundary layer
laminar boundary layer
turbulent boundary layer
boundary layer thickness
transition
free-stream velocity
favorable pressure
  gradient
adverse pressure
  gradient
boundary layer
  separation
friction drag
pressure drag
stall
circulation
Magnus effect

In this chapter the flow past objects is discussed. It is shown how the pressure and shear stress distributions on the surface of an object produce the net lift and drag forces on the object.

The character of flow past an object is a function of the Reynolds number. For large Reynolds number flows a thin boundary layer forms on the surface. Properties of this boundary layer flow are discussed. These include the boundary layer thickness, whether the flow is laminar or turbulent, and the wall shear stress exerted on the object. In addition, boundary layer separation and its relationship to the pressure gradient are considered.

The drag, which contains portions due to friction (viscous) effects and pressure effects, is written in terms of the dimensionless drag coefficient. It is shown how the drag coefficient is a function of shape, with objects ranging from very blunt to very streamlined. Other parameters affecting the drag coefficient include the Reynolds number, Froude number, Mach number, and surface roughness.

The lift is written in terms of the dimensionless lift coefficient, which is strongly dependent on the shape of the object. Variation of the lift coefficient with shape is illustrated by the variation of an airfoil's lift coefficient with angle of attack.

The following checklist provides a study guide for this chapter. When your study of the entire chapter and end-of-chapter exercises has been completed you should be able to

■ write out meanings of the terms listed here in the margin and understand each of the related concepts. These terms are particularly important and are set in *italic, bold, and color* type in the text.

- determine the lift and drag on an object from the given pressure and shear stress distributions on the object.

- for flow past a flat plate, calculate the boundary layer thickness, the wall shear stress, the friction drag, and determine whether the flow is laminar or turbulent.

- explain the concept of the pressure gradient and its relationship to boundary layer separation.

- for a given object, obtain the drag coefficient from appropriate tables, figures, or equations and calculate the drag on the object.

- explain why golf balls have dimples.

- for a given object, obtain the lift coefficient from appropriate figures and calculate the lift on the object.

Some of the important equations in this chapter are:

Lift coefficient and drag coefficient
$$C_L = \frac{\mathscr{L}}{\frac{1}{2}\rho U^2 A}, \quad C_D = \frac{\mathscr{D}}{\frac{1}{2}\rho U^2 A} \qquad \text{(9.39), (9.36)}$$

Boundary layer displacement thickness
$$\delta^* = \int_0^\infty \left(1 - \frac{u}{U}\right) dy \qquad \text{(9.3)}$$

Boundary layer momentum thickness
$$\Theta = \int_0^\infty \frac{u}{U}\left(1 - \frac{u}{U}\right) dy \qquad \text{(9.4)}$$

Blasius boundary layer thickness, displacement thickness, and momentum thickness for flat plate
$$\frac{\delta}{x} = \frac{5}{\sqrt{\text{Re}_x}}, \quad \frac{\delta^*}{x} = \frac{1.721}{\sqrt{\text{Re}_x}}, \quad \frac{\Theta}{x} = \frac{0.664}{\sqrt{\text{Re}_x}} \qquad \text{(9.15), (9.16), (9.17)}$$

Blasius wall shear stress for flat plate
$$\tau_w = 0.332 U^{3/2} \sqrt{\frac{\rho\mu}{x}} \qquad \text{(9.18)}$$

Drag on flat plate
$$\mathscr{D} = \rho b U^2 \Theta \qquad \text{(9.23)}$$

Blasius wall friction coefficient and friction drag coefficient for flat plate
$$c_f = \frac{0.664}{\sqrt{\text{Re}_x}}, \quad C_{Df} = \frac{1.328}{\sqrt{\text{Re}_\ell}} \qquad \text{(9.32)}$$

References

1. Schlichting, H., *Boundary Layer Theory*, 8th Ed., McGraw-Hill, New York, 2000.
2. Rosenhead, L., *Laminar Boundary Layers*, Oxford University Press, London, 1963.
3. White, F. M., *Viscous Fluid Flow*, 3rd Ed., McGraw-Hill, New York, 2005.
4. Currie, I. G., *Fundamental Mechanics of Fluids*, McGraw-Hill, New York, 1974.
5. Blevins, R. D., *Applied Fluid Dynamics Handbook*, Van Nostrand Reinhold, New York, 1984.
6. Hoerner, S. F., *Fluid-Dynamic Drag*, published by the author, Library of Congress No. 64,19666, 1965.
7. Happel, J., *Low Reynolds Number Hydrodynamics*, Prentice-Hall, Englewood Cliffs, NJ, 1965.
8. Van Dyke, M., *An Album of Fluid Motion*, Parabolic Press, Stanford, Calif., 1982.
9. Thompson, P. A., *Compressible-Fluid Dynamics*, McGraw-Hill, New York, 1972.
10. Zucrow, M. J., and Hoffman, J. D., *Gas Dynamics, Vol. I*, Wiley, New York, 1976.
11. Clayton, B. R., and Bishop, R. E. D., *Mechanics of Marine Vehicles*, Gulf Publishing Co., Houston, 1982.
12. *CRC Handbook of Tables for Applied Engineering Science*, 2nd Ed., CRC Press, Boca Raton, Florida, 1973.
13. Shevell, R. S., *Fundamentals of Flight*, 2nd Ed., Prentice-Hall, Englewood Cliffs, NJ, 1989.
14. Kuethe, A. M., and Chow, C. Y., *Foundations of Aerodynamics, Bases of Aerodynamics Design*, 4th Ed., Wiley, New York, 1986.

15. Vogel, J., *Life in Moving Fluids*, 2nd Ed., Willard Grant Press, Boston, 1994.

16. Kreider, J. F., *Principles of Fluid Mechanics*, Allyn and Bacon, Newton, Mass., 1985.

17. Dobrodzicki, G. A., Flow Visualization in the National Aeronautical Establishment's Water Tunnel, National Research Council of Canada, Aeronautical Report LR-557, 1972.

18. White, F. M., *Fluid Mechanics*, 6th Ed., McGraw-Hill, New York, 2008.

19. Vennard, J. K., and Street, R. L., *Elementary Fluid Mechanics*, 7th Ed., Wiley, New York, 1995.

20. Gross, A. C., Kyle, C. R., and Malewicki, D. J., The Aerodynamics of Human Powered Land Vehicles, *Scientific American*, Vol. 249, No. 6, 1983.

21. Abbott, I. H., and Von Doenhoff, A. E., *Theory of Wing Sections*, Dover Publications, New York, 1959.

22. MacReady, P. B., "Flight on 0.33 Horsepower: The Gossamer Condor," *Proc. AIAA 14th Annual Meeting* (Paper No. 78-308), Washington, DC, 1978.

23. Goldstein, S., *Modern Developments in Fluid Dynamics*, Oxford Press, London, 1938.

24. Achenbach, E., Distribution of Local Pressure and Skin Friction around a Circular Cylinder in Cross-Flow up to Re = 5×10^6, *Journal of Fluid Mechanics*, Vol. 34, Pt. 4, 1968.

25. Inui, T., Wave-Making Resistance of Ships, *Transactions of the Society of Naval Architects and Marine Engineers*, Vol. 70, 1962.

26. Sovran, G., et al. (ed.), *Aerodynamic Drag Mechanisms of Bluff Bodies and Road Vehicles*, Plenum Press, New York, 1978.

27. Abbott, I. H., von Doenhoff, A. E., and Stivers, L. S., Summary of Airfoil Data, NACA Report No. 824, Langley Field, Va., 1945.

28. Society of Automotive Engineers Report HSJ1566, "Aerodynamic Flow Visualization Techniques and Procedures," 1986.

29. Anderson, J. D., *Fundamentals of Aerodynamics*, 4th Ed., McGraw-Hill, New York, 2007.

30. Hucho, W. H., *Aerodynamics of Road Vehicles*, Butterworth–Heinemann, 1987.

31. Homsy, G. M., et al., *Multimedia Fluid Mechanics*, 2nd Ed., CD-ROM, Cambridge University Press, New York, 2008.

Review Problems

Go to Appendix G for a set of review problems with answers. Detailed solutions can be found in *Student Solution Manual and Study Guide for Fundamentals of Fluid Mechanics*, by Munson et al. (© 2009 John Wiley and Sons, Inc.).

Problems

Note: Unless otherwise indicated use the values of fluid properties found in the tables on the inside of the front cover. Problems designated with an (*) are intended to be solved with the aid of a programmable calculator or a computer. Problems designated with a (†) are "open ended" problems and require critical thinking in that to work them one must make various assumptions and provide the necessary data. There is not a unique answer to these problems.

Answers to the even-numbered problems are listed at the end of the book. Access to the videos that accompany problems can be obtained through the book's web site, www.wiley.com/college/munson. The lab-type problems and FlowLab problems can also be accessed on this web site.

Section 9.1 General External Flow Characteristics

9.1 Obtain photographs/images of external flow objects that are exposed to both a low Reynolds number and high Reynolds number. Print these photos and write a brief paragraph that describes the situations involved.

9.2 A thin square is oriented perpendicular to the upstream velocity in a uniform flow. The average pressure on the front side of the square is 0.7 times the stagnation pressure and the average pressure on the back side is a vacuum (i.e., less than the free stream pressure) with a magnitude 0.4 times the stagnation pressure. Determine the drag coefficient for this square.

9.3 A small 15-mm-long fish swims with a speed of 20 mm/s. Would a boundary layer type flow be developed along the sides of the fish? Explain.

9.4 The average pressure and shear stress acting on the surface of the 1-m-square flat plate are as indicated in Fig. P9.4. Determine the lift and drag generated. Determine the lift and drag if the shear stress is neglected. Compare these two sets of results.

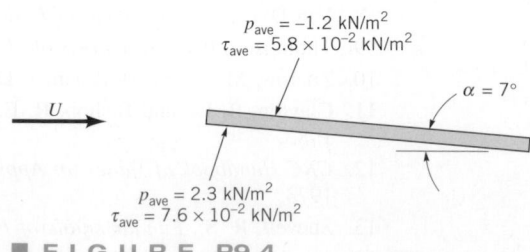

■ **FIGURE P9.4**

*9.5 The pressure distribution on the 1-m-diameter circular disk in Fig. P9.5 is given in the table. Determine the drag on the disk.

r (m)	p (kN/m^2)
0	4.34
0.05	4.28
0.10	4.06
0.15	3.72
0.20	3.10
0.25	2.78
0.30	2.37
0.35	1.89
0.40	1.41
0.45	0.74
0.50	0.0

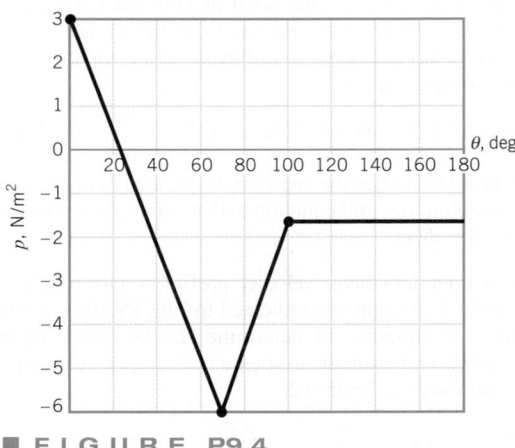

■ **FIGURE P9.5**

9.6 When you walk through still air at a rate of 1 m/s, would you expect the character of the air flow around you to be most like that depicted in Fig. 9.6a, b, or c? Explain.

9.7 A 0.10-m-diameter circular cylinder moves through air with a speed U. The pressure distribution on the cylinder's surface is approximated by the three straight line segments shown in Fig. P9.7. Determine the drag coefficient on the cylinder. Neglect shear forces.

■ **FIGURE P9.4**

9.8 Typical values of the Reynolds number for various animals moving through air or water are listed below. For which cases is inertia of the fluid important? For which cases do viscous effects dominate? For which cases would the flow be laminar; turbulent? Explain.

Animal	Speed	Re
(a) large whale	10 m/s	300,000,000
(b) flying duck	20 m/s	300,000
(c) large dragonfly	7 m/s	30,000
(d) invertebrate larva	1 mm/s	0.3
(e) bacterium	0.01 mm/s	0.00003

†9.9 Estimate the Reynolds numbers associated with the following objects moving through water: (a) a kayak, (b) a minnow, (c) a submarine, (d) a grain of sand settling to the bottom, (e) you swimming.

Section 9.2 Boundary Layer Characteristics (Also see Lab Problems 9.112 and 9.113.)

9.10 Obtain a photograph/image of an object that can be approximated as flow past a flat plate, in which you could use equations from Section 9.2 to approximate the boundary layer characteristics. Print this photo and write a brief paragraph that describes the situation involved.

9.11 Discuss any differences in boundary layers between internal flows (e.g., pipe flow) and external flows.

9.12 Water flows past a flat plate that is oriented parallel to the flow with an upstream velocity of 0.5 m/s. Determine the approximate location downstream from the leading edge where the boundary layer becomes turbulent. What is the boundary layer thickness at this location?

9.13 A viscous fluid flows past a flat plate such that the boundary layer thickness at a distance 1.3 m from the leading edge is 12 mm. Determine the boundary layer thickness at distances of 0.20, 2.0, and 20 m from the leading edge. Assume laminar flow.

9.14 If the upstream velocity of the flow in Problem 9.13 is $U = 1.5$ m/s, determine the kinematic viscosity of the fluid.

9.15 Water flows past a flat plate with an upstream velocity of $U = 0.02$ m/s. Determine the water velocity a distance of 10 mm from the plate at distances of $x = 1.5$ m and $x = 15$ m from the leading edge.

9.16 Approximately how fast can the wind blow past a 1-cm-diameter twig if viscous effects are to be of importance throughout the entire flow field (i.e., Re < 1)? Explain. Repeat for a 0.01-cm-diameter hair and a 2-m-diameter smokestack.

9.17 As is indicated in Table 9.2, the laminar boundary layer results obtained from the momentum integral equation are relatively insensitive to the shape of the assumed velocity profile. Consider the profile given by $u = U$ for $y > \delta$, and $u = U\{1 - [(y - \delta)/\delta]^2\}^{1/2}$ for $y \le \delta$ as shown in Fig. P9.17. Note that this satisfies the conditions $u = 0$ at $y = 0$ and $u = U$ at $y = \delta$. However, show that such a profile produces meaningless results when used with the momentum integral equation. Explain.

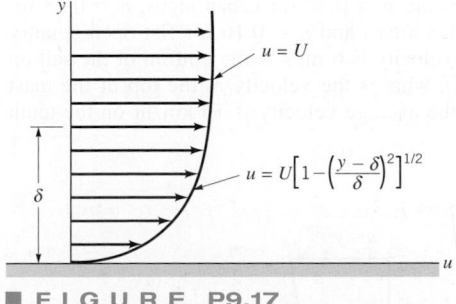

■ **FIGURE P9.17**

9.18 If a high-school student who has completed a first course in physics asked you to explain the idea of a boundary layer, what would you tell the student?

9.19 Because of the velocity deficit, $U - u$, in the boundary layer, the streamlines for flow past a flat plate are not exactly parallel to the plate. This deviation can be determined by use of the displacement thickness, δ^*. For air blowing past the flat plate shown in Fig. P9.19, plot the streamline $A-B$ that passes through the edge of the boundary layer $(y = \delta_B$ at $x = \ell)$ at point B. That is, plot $y = y(x)$ for streamline $A-B$. Assume laminar boundary layer flow.

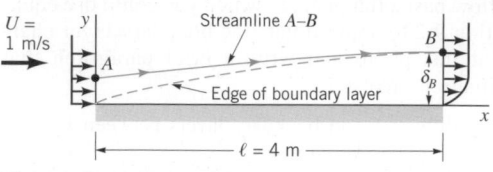

■ **F I G U R E** **P9.19**

9.20 Air enters a square duct through a 0.5-m opening as is shown in Fig. P9.20. Because the boundary layer displacement thickness increases in the direction of flow, it is necessary to increase the cross-sectional size of the duct if a constant $U = 1$ m/s velocity is to be maintained outside the boundary layer. Plot a graph of the duct size, d, as a function of x for $0 \le x \le 3$ m if U is to remain constant. Assume laminar flow.

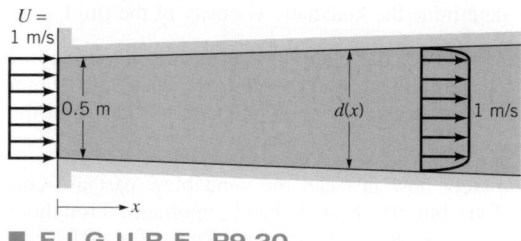

■ **F I G U R E** **P9.20**

9.21 A smooth, flat plate of length $\ell = 6$ m and width $b = 4$ m is placed in water with an upstream velocity of $U = 0.5$ m/s. Determine the boundary layer thickness and the wall shear stress at the center and the trailing edge of the plate. Assume a laminar boundary layer.

9.22 An atmospheric boundary layer is formed when the wind blows over the earth's surface. Typically, such velocity profiles can be written as a power law: $u = ay^n$, where the constants a and n depend on the roughness of the terrain. As is indicated in Fig. P9.22, typical values are $n = 0.40$ for urban areas, $n = 0.28$ for woodland or suburban areas, and $n = 0.16$ for flat open country (Ref. 23). **(a)** If the velocity is 6 m/s at the bottom of the sail on your boat $(y = 1$ m), what is the velocity at the top of the mast $(y = 10$ m)? **(b)** If the average velocity is 16 km/hr on the tenth

floor of an urban building, what is the average velocity on the sixtieth floor?

9.23 It is relatively easy to design an efficient nozzle to accelerate a fluid. Conversely, it is very difficult to build an efficient diffuser to decelerate a fluid without boundary layer separation and its subsequent inefficient flow behavior. Use the ideas of favorable and adverse pressure gradients to explain these facts.

9.24 A 30-story office building (each story is 4 m tall) is built in a suburban industrial park. Plot the dynamic pressure, $\rho u^2/2$, as a function of elevation if the wind blows at hurricane strength (120 km/hr) at the top of the building. Use the atmospheric boundary layer information of Problem 9.22.

9.25 Show that for any function $f = f(\eta)$ the velocity components u and v determined by Eqs. 9.12 and 9.13 satisfy the incompressible continuity equation, Eq. 9.8.

*9.26 Integrate the Blasius equation (Eq. 9.14) numerically to determine the boundary layer profile for laminar flow past a flat plate. Compare your results with those of Table 9.1.

9.27 An airplane flies at a speed of 645 km/hr at an altitude of 3000 m. If the boundary layers on the wing surfaces behave as those on a flat plate, estimate the extent of laminar boundary layer flow along the wing. Assume a transitional Reynolds number of $Re_{xcr} = 5 \times 10^5$. If the airplane maintains its 645 km/hr speed but descends to sea-level elevation, will the portion of the wing covered by a laminar boundary layer increase or decrease compared with its value at 3000 m? Explain.

†9.28 If the boundary layer on the hood of your car behaves as one on a flat plate, estimate how far from the front edge of the hood the boundary layer becomes turbulent. How thick is the boundary layer at this location?

9.29 A laminar boundary layer velocity profile is approximated by $u/U = [2 - (y/\delta)](y/\delta)$ for $y \le \delta$, and $u = U$ for $y > \delta$. **(a)** Show that this profile satisfies the appropriate boundary conditions. **(b)** Use the momentum integral equation to determine the boundary layer thickness, $\delta = \delta(x)$.

9.30 A laminar boundary layer velocity profile is approximated by the two straight-line segments indicated in Fig. P9.30. Use the momentum integral equation to determine the boundary layer thickness, $\delta = \delta(x)$, and wall shear stress, $\tau_w = \tau_w(x)$. Compare these results with those in Table 9.2.

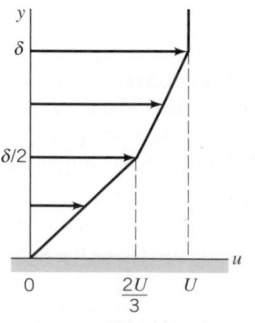

■ **F I G U R E** **P9.30**

*9.31 For a fluid of specific gravity $SG = 0.86$ flowing past a flat plate with an upstream velocity of $U = 5$ m/s, the wall shear stress on a flat plate was determined to be as indicated in the table below. Use the momentum integral equation to determine the boundary

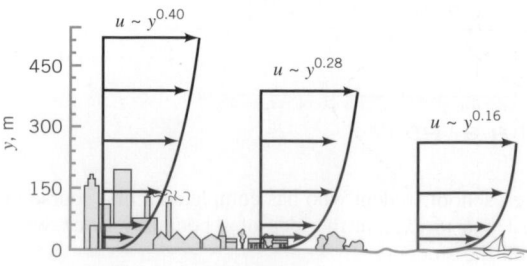

■ **F I G U R E** **P9.22**

layer momentum thickness, $\Theta = \Theta(x)$. Assume $\Theta = 0$ at the leading edge, $x = 0$.

x (m)	τ_w (N/m²)
0	—
0.2	13.4
0.4	9.25
0.6	7.68
0.8	6.51
1.0	5.89
1.2	6.57
1.4	6.75
1.6	6.23
1.8	5.92
2.0	5.26

Section 9.3 Drag

9.32 Obtain a photograph/image of an everyday item in which drag plays a key role. Print this photo and write a brief paragraph that describes the situation involved.

9.33 Should a canoe paddle be made rough to get a "better grip on the water" for paddling purposes? Explain.

9.34 Define the purpose of "streamlining" a body.

9.35 Water flows over two flat plates with the same laminar free-stream velocity. Both plates have the same width, but Plate #2 is twice as long as Plate #1. What is the relationship between the drag force for these two plates?

9.36 Fluid flows past a flat plate with a drag force $\mathcal{D}_1$. If the free-stream velocity is doubled, will the new drag force, $\mathcal{D}_2$, be larger or smaller than $\mathcal{D}_1$ and by what amount?

9.37 A model is placed in an air flow with a given velocity and then placed in water flow with the same velocity. If the drag coefficients are the same between these two cases, how do the drag forces compare between the two fluids?

9.38 The drag coefficient for a newly designed hybrid car is predicted to be 0.21. The cross-sectional area of the car is 3 m². Determine the aerodynamic drag on the car when it is driven through still air at 90 km/hr.

9.39 A 5-m-diameter parachute of a new design is to be used to transport a load from flight altitude to the ground with an average vertical speed of 3 m/s. The total weight of the load and parachute is 200 N. Determine the approximate drag coefficient for the parachute.

9.40 A 80-km/hr wind blows against an outdoor movie screen that is 20 m wide and 6 m tall. Estimate the wind force on the screen.

9.41 The aerodynamic drag on a car depends on the "shape" of the car. For example, the car shown in Fig. P9.41 has a drag coefficient of 0.36 with the windows and roof closed. With the windows and roof open, the drag coefficient increases to 0.45.

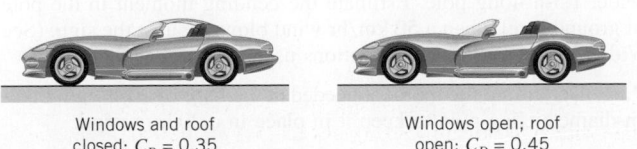

Windows and roof closed: $C_D = 0.35$ Windows open; roof open: $C_D = 0.45$

■ F I G U R E   P9.41

With the windows and roof open, at what speed is the amount of power needed to overcome aerodynamic drag the same as it is at 105 km/hr with the windows and roof closed? Assume the frontal area remains the same. Recall that power is force times velocity.

9.42 A rider on a bike with the combined mass of 100 kg attains a terminal speed of 15 m/s on a 12% slope. Assuming that the only forces affecting the speed are the weight and the drag, calculate the drag coefficient. The frontal area is 0.9 m². Speculate whether the rider is in the upright or racing position.

9.43 A baseball is thrown by a pitcher at 150 km/hr through standard air. The diameter of the baseball is 7 cm. Estimate the drag force on the baseball.

9.44 A logging boat is towing a log that is 2 m in diameter and 8 m long at 4 m/s through water. Estimate the power required if the axis of the log is parallel to the tow direction.

9.45 A sphere of diameter D and density ρ_s falls at a steady rate through a liquid of density ρ and viscosity μ. If the Reynolds number, $\text{Re} = \rho D U / \mu$, is less than 1, show that the viscosity can be determined from $\mu = g D^2 (\rho_s - \rho) / 18\, U$.

9.46 The square, flat plate shown in Fig. P9.46a is cut into four equal-sized pieces and arranged as shown in Fig. P9.46b. Determine the ratio of the drag on the original plate [case **(a)**] to the drag on the plates in the configuration shown in **(b)**. Assume laminar boundary flow. Explain your answer physically.

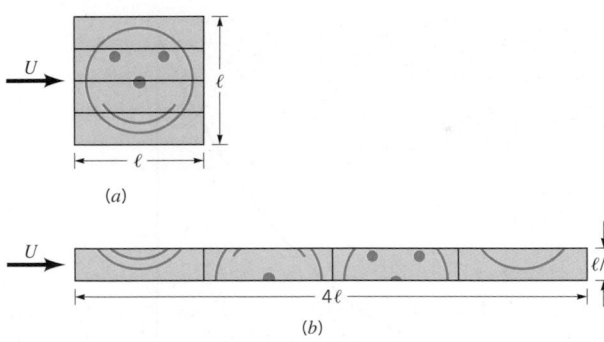

(a)

(b)

■ F I G U R E   P9.46

9.47 If the drag on one side of a flat plate parallel to the upstream flow is $\mathcal{D}$ when the upstream velocity is U, what will the drag be when the upstream velocity is $2U$; or $U/2$? Assume laminar flow.

9.48 Water flows past a triangular flat plate oriented parallel to the free stream as shown in Fig. P9.48. Integrate the wall shear stress over the plate to determine the friction drag on one side of the plate. Assume laminar boundary layer flow.

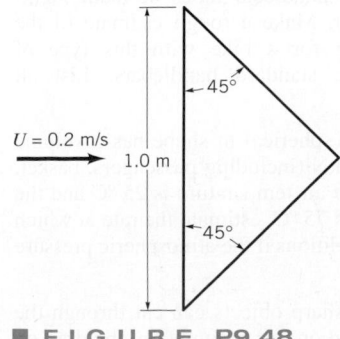

$U = 0.2$ m/s   1.0 m   45°   45°

■ F I G U R E   P9.48

9.49 For small Reynolds number flows the drag coefficient of an object is given by a constant divided by the Reynolds number (see Table 9.4). Thus, as the Reynolds number tends to zero, the drag coefficient becomes infinitely large. Does this mean that for small velocities (hence, small Reynolds numbers) the drag is very large? Explain.

9.50 A rectangular car-top carrier of 0.5-m height, 1.5-m length (front to back), and 1.3-m width is attached to the top of a car. Estimate the additional power required to drive the car with the carrier at 100 km/hr through still air compared with the power required to driving only the car at 100 km/hr.

9.51 As shown in **Video V9.2** and Fig. P9.51a, a kayak is a relatively streamlined object. As a first approximation in calculating the drag on a kayak, assume that the kayak acts as if it were a smooth, flat plate 5 m long and 0.5 m wide. Determine the drag as a function of speed and compare your results with the measured values given in Fig. P9.51b. Comment on reasons why the two sets of values may differ.

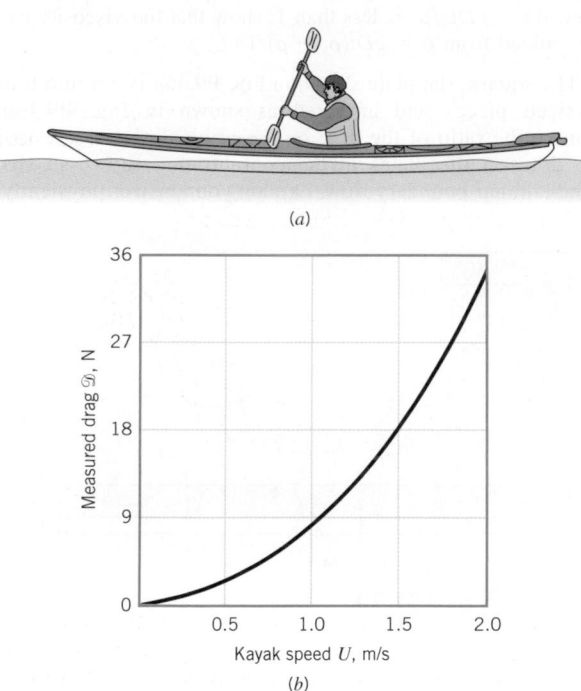

(a)

(b)

■ **F I G U R E P9.51**

9.52 A 38.1-mm-diameter, 0.0245-N table tennis ball is released from the bottom of a swimming pool. With what velocity does it rise to the surface? Assume it has reached its terminal velocity.

9.53 To reduce aerodynamic drag on a bicycle, it is proposed that the cross-sectional shape of the handlebar tubes be made "teardrop" shape rather than circular. Make a rough estimate of the reduction in aerodynamic drag for a bike with this type of handlebars compared with the standard handlebars. List all assumptions.

9.54 A hot air balloon roughly spherical in shape has a volume of 2000 m³ and a weight of 2.2 kN (including passengers, basket, ballon fabric, etc.). If the outside air temperature is 25 °C and the temperature within the balloon is 75 °C, estimate the rate at which it will rise under steady state conditions if the atmospheric pressure is 101.3 kPa.

9.55 It is often assumed that "sharp objects can cut through the air better than blunt ones." Based on this assumption, the drag on

the object shown in Fig. P9.55 should be less when the wind blows from right to left than when it blows from left to right. Experiments show that the opposite is true. Explain.

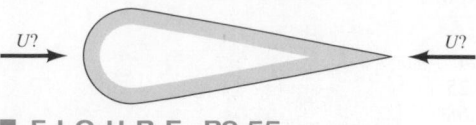

■ **F I G U R E P9.55**

*****9.56** The device shown in Fig. P9.56 is to be designed to measure the wall shear stress as air flows over the smooth surface with an upstream velocity U. It is proposed that τ_w can be obtained by measuring the bending moment, M, at the base [point (1)] of the support that holds the small surface element which is free from contact with the surrounding surface. Plot a graph of M as a function of U for $5 \le U \le 50$ m/s, with $\ell = 2, 3, 4$, and 5 m.

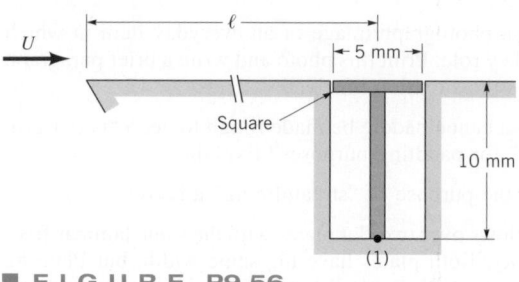

■ **F I G U R E P9.56**

9.57 A 12-mm-diameter cable is strung between a series of poles that are 50 m apart. Determine the horizontal force this cable puts on each pole if the wind velocity is 30 m/s.

9.58 How fast do small water droplets of 0.06 μm (6×10^{-8} m) diameter fall through the air under standard sea-level conditions? Assume the drops do not evaporate. Repeat the problem for standard conditions at 5000-m altitude.

9.59 A strong wind can blow a golf ball off the tee by pivoting it about point 1 as shown in Fig. P9.59. Determine the wind speed necessary to do this.

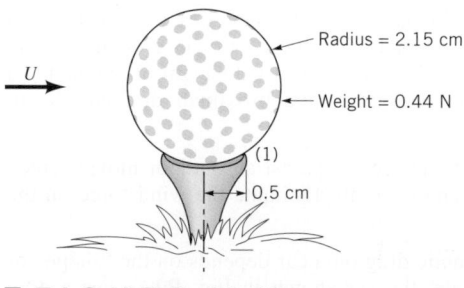

■ **F I G U R E P9.59**

9.60 A 55-cm by 85-cm speed limit sign is supported on a 8-cm wide, 1.5-m-long pole. Estimate the bending moment in the pole at ground level when a 50 km/hr wind blows against the sign. (See **Video V9.9.**) List any assumptions used in your calculations.

9.61 Determine the moment needed at the base of 20-m-tall, 0.12-m-diameter flag pole to keep it in place in a 20 m/s wind.

9.62 Repeat Problem 9.61 if a 2-m by 2.5-m flag is attached to the top of the pole. See Fig. 9.30 for drag coefficient data for flags.

†9.63 During a flash flood, water rushes over a road as shown in Fig. P9.63 with a speed of 20 km/hr. Estimate the maximum water depth, h, that would allow a car to pass without being swept away. List all assumptions and show all calculations.

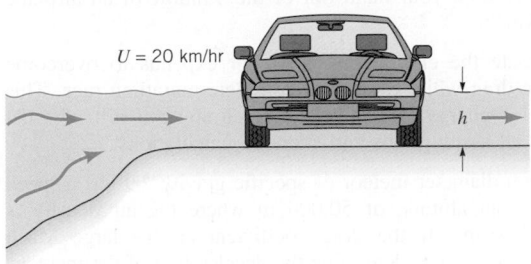

$U = 20$ km/hr

h

■ FIGURE P9.63

9.64 How much more power is required to pedal a bicycle at 25 km/hr into a 30-km/hr head-wind than at 25 km/hr through still air? Assume a frontal area of 0.4 m^2 and a drag coefficient of $C_D = 0.88$.

†9.65 Estimate the wind velocity necessary to knock over a 45-N garbage can that is 1 m tall and 0.5 m in diameter. List your assumptions.

9.66 On a day without any wind, your car consumes x liters of gasoline when you drive at a constant speed, U, from point A to point B and back to point A. Assume that you repeat the journey, driving at the same speed, on another day when there is a steady wind blowing from B to A. Would you expect your fuel consumption to be less than, equal to, or greater than x liters for this windy round-trip? Support your answer with appropriate analysis.

9.67 The structure shown in Fig. P9.67 consists of three cylindrical support posts to which an elliptical flat-plate sign is attached. Estimate the drag on the structure when a 80 km/hr wind blows against it.

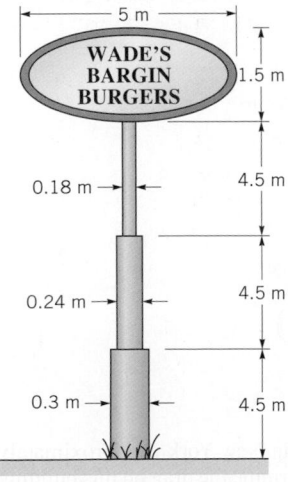

5 m

WADE'S
BARGIN
BURGERS

1.5 m

0.18 m

4.5 m

0.24 m

4.5 m

0.3 m

4.5 m

■ FIGURE P9.67

9.68 As shown in Video V9.13 and Fig. P9.68, the aerodynamic drag on a truck can be reduced by the use of appropriate air deflectors. A reduction in drag coefficient from $C_D = 0.96$ to $C_D = 0.70$ corresponds to a reduction of how many horsepower needed at a highway speed of 105 km/hr?

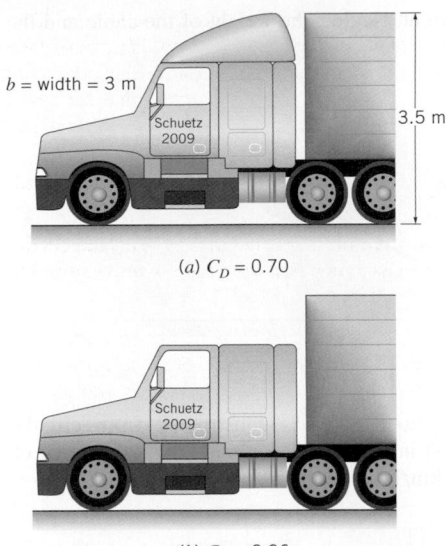

$b = $ width $= 3$ m

3.5 m

Schuetz
2009

(a) $C_D = 0.70$

Schuetz
2009

(b) $C_D = 0.96$

■ FIGURE P9.68

9.69 As shown in Video V9.7 and Fig. P9.69, a vertical wind tunnel can be used for skydiving practice. Estimate the vertical wind speed needed if a 670 N person is to be able to "float" motionless when the person (a) curls up as in a crouching position or (b) lies flat. See Fig. 9.30 for appropriate drag coefficient data.

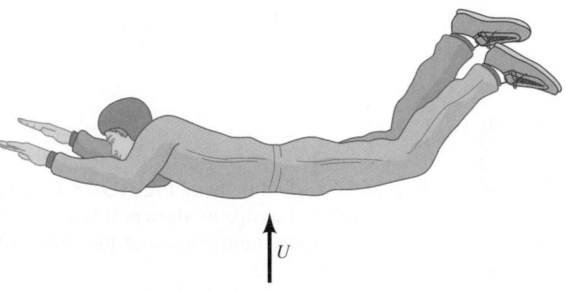

U

■ FIGURE P9.69

*9.70 The helium-filled balloon shown in Fig. P9.70 is to be used as a wind speed indicator. The specific weight of the helium is $\gamma = 1.7$ N/m^3, the weight of the balloon material is 0.9 N, and the weight of the anchoring cable is negligible. Plot a graph of θ as a function of U for $1.5 \leq U \leq 80$ km/hr. Would this be an effective device over the range of U indicated? Explain.

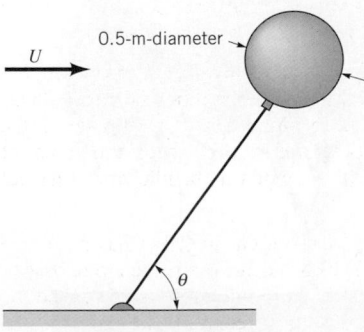

0.5-m-diameter

U

θ

■ FIGURE P9.70

9.71 A 0.30-m-diameter cork ball ($SG = 0.21$) is tied to an object on the bottom of a river as is shown in Fig. P9.71. Estimate the speed of the river current. Neglect the weight of the cable and the drag on it.

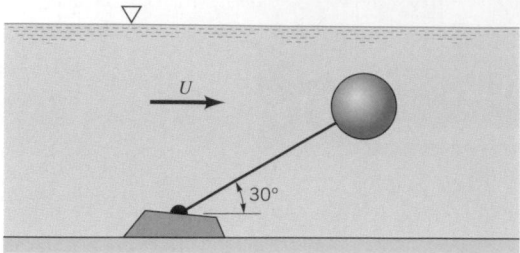

■ **F I G U R E   P9.71**

9.72 A shortwave radio antenna is constructed from circular tubing, as is illustrated in Fig. P9.72. Estimate the wind force on the antenna in a 100 km/hr wind.

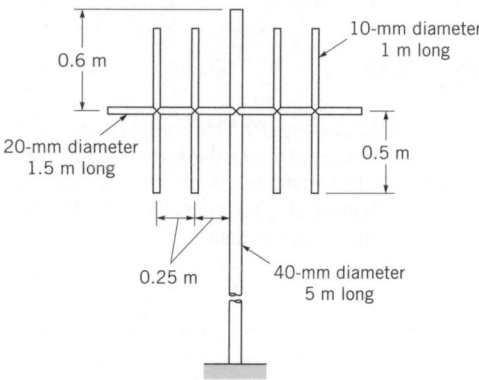

■ **F I G U R E   P9.72**

9.73 The large, newly planted tree shown in Fig. P9.73 is kept from tipping over in a wind by use of a rope as shown. It is assumed that the sandy soil cannot support any moment about the center of

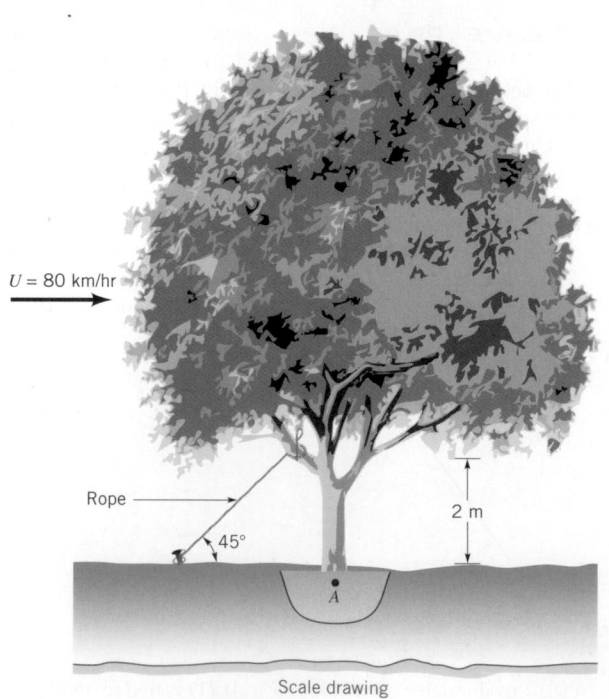

■ **F I G U R E   P9.73**

the soil ball, point A. Estimate the tension in the rope if the wind is 80 km/hr. See Fig. 9.30 for drag coefficient data.

9.74 Estimate the wind force on your hand when you hold it out of your car window while driving 90 km/hr. Repeat your calculations if you were to hold your hand out of the window of an airplane flying 900 km/hr.

†9.75 Estimate the energy that a runner expends to overcome aerodynamic drag while running a complete marathon race. This expenditure of energy is equivalent to climbing a hill of what height? List all assumptions and show all calculations.

9.76 A 2-mm-diameter meteor of specific gravity 2.9 has a speed of 6 km/s at an altitude of 50,000 m where the air density is 1.03×10^{-3} kg/m³. If the drag coefficient at this large Mach number condition is 1.5, determine the deceleration of the meteor.

9.77 Air flows past two equal sized spheres (one rough, one smooth) that are attached to the arm of a balance as is indicated in Fig. P9.77. With $U = 0$ the beam is balanced. What is the minimum air velocity for which the balance arm will rotate clockwise?

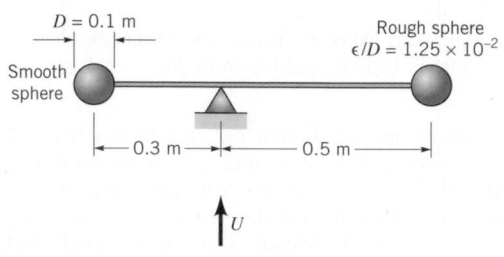

■ **F I G U R E   P9.77**

9.78 A 5-cm-diameter sphere weighing 0.6 N is suspended by the jet of air shown in Fig. P9.78 and **Video V3.2**. The drag coefficient for the sphere is 0.5. Determine the reading on the pressure gage if friction and gravity effects can be neglected for the flow between the pressure gage and the nozzle exit.

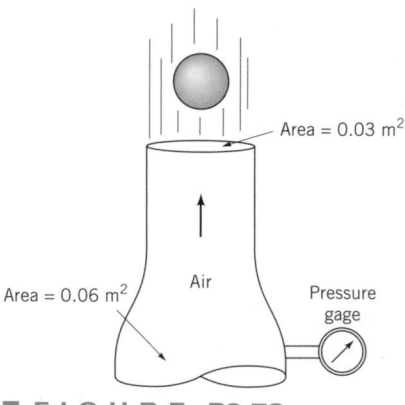

■ **F I G U R E   P9.78**

9.79 The United Nations Building in New York is approximately 87.5-m wide and 154-m tall. **(a)** Determine the drag on this building if the drag coefficient is 1.3 and the wind speed is a uniform 20 m/s. **(b)** Repeat your calculations if the velocity profile against the building is a typical profile for an urban area (see Problem 9.22) and the wind speed halfway up the building is 20 m/s.

9.80 A regulation football is 17 cm in diameter and weighs 4 N. If its drag coefficient is $C_D = 0.2$, determine its deceleration if it has a speed of 6 m/s at the top of its trajectory.

9.81 An airplane tows a banner that is $b = 0.8$ m tall and $\ell = 25$ m long at a speed of 150 km/hr. If the drag coefficient based on the area $b\ell$ is $C_D = 0.06$, estimate the power required to tow the banner. Compare the drag force on the banner with that on a rigid flat plate of the same size. Which has the larger drag force and why?

†9.82 Skydivers often join together to form patterns during the free-fall portion of their jump. The current *Guiness Book of World Records* record is 297 skydivers joined hand-to-hand. Given that they can't all jump from the same airplane at the same time, describe how they manage to get together (see *Video V9.7*). Use appropriate fluid mechanics equations and principles in your answer.

9.83 The paint stirrer shown in Fig. P9.83 consists of two circular disks attached to the end of a thin rod that rotates at 80 rpm. The specific gravity of the paint is $SG = 1.1$ and its viscosity is $\mu = 96 \times 10^{-2}$ N · s/m^2. Estimate the power required to drive the mixer if the induced motion of the liquid is neglected.

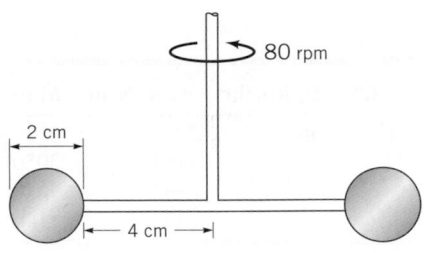

■ **F I G U R E P9.83**

†9.84 If the wind becomes strong enough, it is "impossible" to paddle a canoe into the wind. Estimate the wind speed at which this will happen. List all assumptions and show all calculations.

9.85 A fishnet consists of 0.25-cm-diameter strings tied into squares 10 cm per side. Estimate the force needed to tow a 4.5-m by 9.0-m section of this net through seawater at 1.5 m/s.

9.86 As indicated in Fig. P9.86, the orientation of leaves on a tree is a function of the wind speed, with the tree becoming "more streamlined" as the wind increases. The resulting drag coefficient for the tree (based on the frontal area of the tree, HW) as a function of Reynolds number (based on the leaf length, L) is approximated as shown. Consider a tree with leaves of length $L = 0.1$ m. What wind speed will produce a drag on the tree that is 6 times greater than the drag on the tree in a 5 m/s wind?

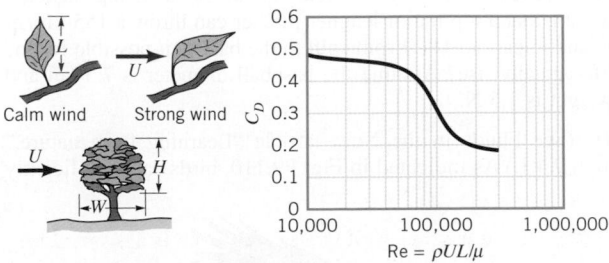

■ **F I G U R E P9.86**

9.87 The blimp shown in Fig. P9.87 is used at various athletic events. It is 40 m long and has a maximum diameter of 10 m. If its drag coefficient (based on the frontal area) is 0.060, estimate the power required to propel it **(a)** at its 56 km/hr cruising speed, or **(b)** at its maximum 90 km/hr speed.

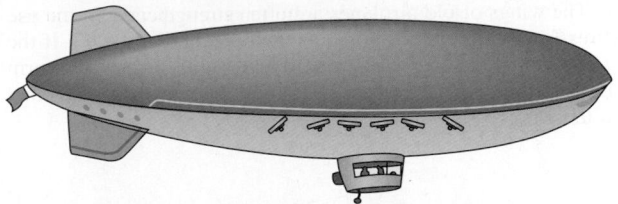

■ **F I G U R E P9.87**

9.88 Show that for level flight at a given speed, the power required to overcome aerodynamic drag decreases as the altitude increases. Assume that the drag coefficient remains constant. This is one reason why airlines fly at high altitudes.

9.89 (See Fluids in the News article "Dimpled baseball bats," Section 9.3.3.) How fast must a 9-cm-diameter, dimpled baseball bat move through the air in order to take advantage of drag reduction produced by the dimples on the bat. Although there are differences, assume the bat (a cylinder) acts the same as a golf ball in terms of how the dimples affect the transition from a laminar to a turbulent boundary layer.

9.90 (See Fluids in the News article "At 10,240 mpg it doesn't cost much to 'fill 'er up,' " Section 9.3.3.) **(a)** Determine the power it takes to overcome aerodynamic drag on a small (0.6 m^2 cross section), streamlined ($C_D = 0.12$) vehicle traveling 24 km/hr. **(b)** Compare the power calculated in part **(a)** with that for a large (3 m^2 cross-sectional area), nonstreamlined ($C_D = 0.48$) SUV traveling 105 km/hr on the interstate.

Section 9.4 Lift

9.91 Obtain a photograph/image of a device, other than an aircraft wing, that creates lift. Print this photo and write a brief paragraph that describes the situation involved.

9.92 A rectangular wing with an aspect ratio of 6 is to generate 4.5 kN of lift when it flies at a speed of 60 m/s. Determine the length of the wing if its lift coefficient is 1.0.

9.93 Explain why aircraft and birds take off and land into the wind.

9.94 A Piper Cub airplane has a gross weight of 7.8 kN, a cruising speed of 185 km/hr, and a wing area of 17 m^2. Determine the lift coefficient of this airplane for these conditions.

9.95 A light aircraft with a wing area of 20 m^2 and a weight of 9 kN has a lift coefficient of 0.40 and a drag coefficient of 0.05. Determine the power required to maintain level flight.

9.96 As shown in *Video V9.19* and Fig. P9.96, a spoiler is used on race cars to produce a negative lift, thereby giving a better tractive force. The lift coefficient for the airfoil shown is $C_L = 1.1$, and the coefficient of friction between the wheels and the pavement is 0.6. At a speed of 320 km/hr, by how much would use of the spoiler increase the maximum tractive force that could be generated between the wheels and ground? Assume the air speed past the spoiler equals the car speed and that the airfoil acts directly over the drive wheels.

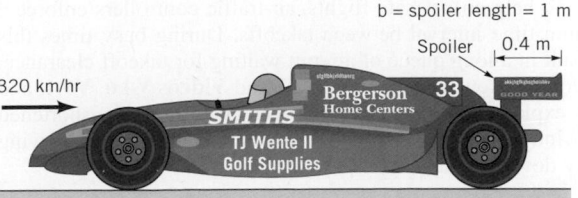

■ **F I G U R E P9.96**

9.97 The wings of old airplanes are often strengthened by the use of wires that provided cross-bracing as shown in Fig. P9.97. If the drag coefficient for the wings was 0.020 (based on the planform area), determine the ratio of the drag from the wire bracing to that from the wings.

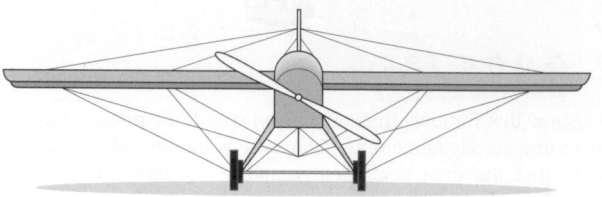

Speed: 110 km/hr
Wing area: 14 m²
Wire: length = 50 m
          diameter = 0.1 cm

■ **F I G U R E   P9.97**

9.98 A wing generates a lift $\mathscr{L}$ when moving through sea-level air with a velocity U. How fast must the wing move through the air at an altitude of 10,000 m with the same lift coefficient if it is to generate the same lift?

9.99 Air blows over the flat-bottomed, two-dimensional object shown in Fig. P9.99. The shape of the object, $y = y(x)$, and the fluid speed along the surface, $u = u(x)$, are given in the table. Determine the lift coefficient for this object.

x(% c)	y(% c)	u/U
0	0	0
2.5	3.72	0.971
5.0	5.30	1.232
7.5	6.48	1.273
10	7.43	1.271
20	9.92	1.276
30	11.14	1.295
40	11.49	1.307
50	10.45	1.308
60	9.11	1.195
70	6.46	1.065
80	3.62	0.945
90	1.26	0.856
100	0	0.807

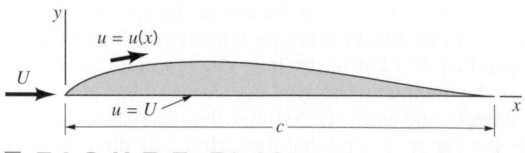

■ **F I G U R E   P9.99**

9.100 To help ensure safe flights, air-traffic controllers enforce a minimum time interval between takeoffs. During busy times this can result in a long queue of aircraft waiting for takeoff clearance. Based on the flow shown in Fig. 9.37 and **Videos V4.6, V9.1,** and **V9.19,** explain why the interval between takeoffs can be shortened if the wind has a cross-runway component (as opposed to blowing directly down the runway).

9.101 A Boeing 747 aircraft weighing 2.6 MN when loaded with fuel and 100 passengers takes off with an airspeed of 225 km/hr. With the same configuration (i.e., angle of attack, flap settings, etc.), what is its takeoff speed if it is loaded with 372 passengers? Assume each passenger with luggage weighs 890 N.

9.102 Show that for unpowered flight (for which the lift, drag, and weight forces are in equilibrium) the glide slope angle, θ, is given by $\tan \theta = C_D/C_L$.

9.103 If the lift coefficient for a Boeing 777 aircraft is 15 times greater than its drag coefficient, can it glide from an altitude of 9000 m to an airport 130 km away if it loses power from its engines? Explain. (See Problem 9.102.)

9.104 On its final approach to the airport, an airplane flies on a flight path that is 3.0° relative to the horizontal. What lift-to-drag ratio is needed if the airplane is to land with its engines idled back to zero power? (See Problem 9.102.)

9.105 Over the years there has been a dramatic increase in the flight speed (U) and altitude (h), weight ($\mathscr{W}$), and wing loading ($\mathscr{W}/A$ = weight divided by wing area) of aircraft. Use the data given in the table below to determine the lift coefficient for each of the aircraft listed.

Aircraft	Year	$\mathscr{W}$, kN	U, km/hr	$\mathscr{W}/A$, N/m²	h, m
Wright Flyer	1903	3.3	56	72	0
Douglas DC-3	1935	111.2	290	1197	3050
Douglas DC-6	1947	467	507	3447	4570
Boeing 747	1970	3600	917	7182	9150

9.106 The landing speed of an airplane such as the Space Shuttle is dependent on the air density. (See **Video V9.1.**) By what percent must the landing speed be increased on a day when the temperature is 45 °C compared to a day when it is 10 °C? Assume that the atmospheric pressure remains constant.

9.107 Commercial airliners normally cruise at relatively high altitudes (9000 to 10,700 m). Discuss how flying at this high altitude (rather than 3000 m, for example) can save fuel costs.

9.108 A pitcher can pitch a "curve ball" by putting sufficient spin on the ball when it is thrown. A ball that has absolutely no spin will follow a "straight" path. A ball that is pitched with a very small amount of spin (on the order of one revolution during its flight between the pitcher's mound and home plate) is termed a knuckle ball. A ball pitched this way tends to "jump around" and "zig-zag" back and forth. Explain this phenomenon. Note: A baseball has seams.

9.109 For many years, hitters have claimed that some baseball pitchers have the ability to actually throw a rising fastball. Assuming that a top major leaguer pitcher can throw a 155-km/hr pitch and impart an 1800-rpm spin to the ball, is it possible for the ball to actually rise? Assume the baseball diameter is 7.5 cm and its weight is 1.5 N.

9.110 (See Fluids in the News article "Learning from nature," Section 9.4.1.) As indicated in Fig. P9.110, birds can significantly

■ **F I G U R E   P9.110**

alter their body shape and increase their planform area, *A*, by spreading their wing and tail feathers, thereby reducing their flight speed. If during landing the planform area is increased by 50% and the lift coefficient increased by 30% while all other parameters are held constant, by what percent is the flight speed reduced?

9.111 (See Fluids in the News article "Why winglets?," Section 9.4.2.) It is estimated that by installing appropriately designed winglets on a certain airplane the drag coefficient will be reduced by 5%. For the same engine thrust, by what percent will the aircraft speed be increased by use of the winglets?

■ **Lab Problems**

9.112 This problem involves measuring the boundary layer profile on a flat plate. To proceed with this problem, go to Appendix H which is located on the book's web site, www.wiley.com/college/munson.

9.113 This problem involves measuring the pressure distribution on a circular cylinder. To proceed with this problem, go to Appendix H which is located on the book's web site, www.wiley.com/college/munson.

■ **Life Long Learning Problems**

9.114 One of the "Fluids in the News" articles in this chapter discusses pressure-sensitive paint—a new technique of measuring surface pressure. There have been other advances in fluid measurement techniques, particularly in velocity measurements. One such technique is particle image velocimetry, or PIV. Obtain information about PIV and its advantages. Summarize your findings in a brief report.

9.115 For typical aircraft flying at cruise conditions, it is advantageous to have as much laminar flow over the wing as possible since there is an increase in friction drag once the flow becomes turbulent. Various techniques have been developed to help promote laminar flow over the wing, both in airfoil geometry configurations as well as active flow control mechanisms. Obtain information on one of these techniques. Summarize your findings in a brief report.

9.116 We have seen in this chapter that streamlining an automobile can help to reduce the drag coefficient. One of the methods of reducing the drag has been to reduce the projected area. However, it is difficult for some road vehicles, such as a tractor-trailer, to reduce this projected area due to the storage volume needed to haul the required load. Over the years, work has been done to help minimize some of the drag on this type of vehicle. Obtain information on a method that has been developed to reduce drag on a tractor-trailer. Summarize your findings in a brief report.

■ **FlowLab Problems**

*9.117 This FlowLab problem involves simulation of flow past an airfoil and investigation of the surface pressure distribution as a function of angle of attack. To proceed with this problem, go to the book's web site, www.wiley.com/college/munson.

*9.118 This FlowLab problem involves investigation of the effects of angle-of-attack on lift and drag for flow past an airfoil. To proceed with this problem, go to the book's web site, www.wiley.com/college/munson.

*9.119 This FlowLab problem involves simulating the effects of altitude on the lift and drag of an airfoil. To proceed with this problem, go to the book's web site, www.wiley.com/college/munson.

*9.120 This FlowLab problem involves comparison between inviscid and viscous flows past an airfoil. To proceed with this problem, go to the book's web site, www.wiley.com/college/munson.

*9.121 This FlowLab problem involves simulating the pressure distribution for flow past a cylinder and investigating the differences between inviscid and viscous flows. To proceed with this problem, go to the book's web site, www.wiley.com/college/munson.

*9.122 This FlowLab problem involves comparing CFD predictions and theoretical values of the drag coefficient of flow past a cylinder. To proceed with this problem, go to the book's web site, www.wiley.com/college/munson.

*9.123 This FlowLab problem involves simulating the unsteady flow past a cylinder. To proceed with this problem, go to the book's web site, www.wiley.com/college/munson.

■ **FE Exam Problems**

Sample FE (Fundamentals of Engineering) exam questions for fluid mechanics are provided on the book's web site, www.wiley.com/college/munson.

10 Open-Channel Flow

CHAPTER OPENING PHOTO: **Hydraulic jump:** Under certain conditions, when water flows in an open channel, even if it has constant geometry, the depth of the water may increase considerably over a short distance along the channel. This phenomenon is termed a hydraulic jump (water flow from left to right).

Learning Objectives

After completing this chapter, you should be able to:

- discuss the general characteristics of open-channel flow.
- use a specific energy diagram.
- apply appropriate equations to analyze open-channel flow with uniform depth.
- calculate key properties of a hydraulic jump.
- determine flowrates based on open-channel flow-measuring devices.

Open-channel flow involves the flow of a liquid in a channel or conduit that is not completely filled. A free surface exists between the flowing fluid (usually water) and fluid above it (usually the atmosphere). The main driving force for such flows is the fluid weight—gravity forces the fluid to flow downhill. Most open-channel flow results are based on correlations obtained from model and full-scale experiments. Additional information can be gained from various analytical and numerical efforts.

Open-channel flows are essential to the world as we know it. The natural drainage of water through the numerous creek and river systems is a complex example of open-channel flow. Although the flow geometry for these systems is extremely complex, the resulting flow properties are of considerable economic, ecological, and recreational importance. Other examples of open-channel flows include the flow of rainwater in the gutters of our houses; the flow in canals, drainage ditches, sewers, and gutters along roads; the flow of small rivulets and sheets of water across fields or parking lots; and the flow in the chutes of water rides in amusement parks.

Open-channel flow involves the existence of a free surface which can distort into various shapes. Thus, a brief introduction into the properties and characteristics of surface waves is included.

The purpose of this chapter is to investigate the concepts of open-channel flow. Because of the amount and variety of material available, only a brief introduction to the topic can be presented. Further information can be obtained from the references indicated.

V10.1 Off-shore oil drilling platform.

10.1   General Characteristics of Open-Channel Flow

Open-channel flow can have a variety of characteristics.

Uniform flow

Rapidly varying flow (photograph courtesy of Stillwater Sciences).

In our study of pipe flow (Chapter 8), we found that there are many ways to classify a flow—developing, fully developed, laminar, turbulent, and so on. For open-channel flow, the existence of a free surface allows additional types of flow. The extra freedom that allows the fluid to select its free-surface location and configuration (because it does not completely fill a pipe or conduit) allows important phenomena in open-channel flow that cannot occur in pipe flow. Some of the classifications of the flows are described below.

The manner in which the fluid depth, y, varies with time, t, and distance along the channel, x, is used to partially classify a flow. For example, the flow is *unsteady* or *steady* depending on whether the depth at a given location does or does not change with time. Some unsteady flows can be viewed as steady flows if the reference frame of the observer is changed. For example, a tidal bore (difference it water level) moving up a river is unsteady to an observer standing on the bank, but steady to an observer moving along the bank with the speed of the wave front of the bore. Other flows are unsteady regardless of the reference frame used. The complex, time-dependent, wind-generated waves on a lake are in this category. In this book we will consider only steady open-channel flows.

An open-channel flow is classified as *uniform flow* (UF) if the depth of flow does not vary along the channel ($dy/dx = 0$). Conversely, it is *nonuniform flow* or *varied flow* if the depth varies with distance ($dy/dx \neq 0$). Nonuniform flows are further classified as *rapidly varying flow* (RVF) if the flow depth changes considerably over a relatively short distance; $dy/dx \sim 1$. *Gradually varying flows* (GVF) are those in which the flow depth changes slowly with distance along the channel; $dy/dx \ll 1$. Examples of these types of flow are illustrated in Fig. 10.1 and the photographs in the margin. The relative importance of the various types of forces involved (pressure, weight, shear, inertia) is different for the different types of flows.

As for any flow geometry, open-channel flow may be *laminar*, *transitional*, or *turbulent*, depending on various conditions involved. Which type of flow occurs depends on the Reynolds number, $\text{Re} = \rho V R_h/\mu$, where V is the average velocity of the fluid and R_h is the hydraulic radius of the channel (see Section 10.4). A general rule is that open-channel flow is laminar if $\text{Re} < 500$, turbulent if $\text{Re} > 12{,}500$, and transitional otherwise. The values of these dividing Reynolds numbers are only approximate—a precise knowledge of the channel geometry is necessary to obtain specific values. Since most open-channel flows involve water (which has a fairly small viscosity) and have relatively large characteristic lengths, it is rare to have laminar open-channel flows. For example, flow of $10\,°C$ water ($\nu = 1.3 \times 10^{-6}\,\text{m}^2/\text{s}$) with an average velocity of $V = 0.3\,\text{m/s}$ in a river with a hydraulic radius of $R_h = 3\,\text{m}$ has $\text{Re} = V R_h/\nu = 6.9 \times 10^5$. The flow is turbulent. However, flow of a thin sheet of water down a driveway with an average velocity of $V = 0.08\,\text{m/s}$ such that $R_h = 6 \times 10^{-3}\,\text{m}$ (in such cases the hydraulic radius is approximately equal to the fluid depth; see Section 10.4) has $\text{Re} = 369$. The flow is laminar.

In some cases *stratified flows* are important. In such situations layers of two or more fluids of different densities flow in a channel. A layer of oil on water is one example of this type of flow. All of the open-channel flows considered in this book are *homogeneous flows*. That is, the fluid has uniform properties throughout.

Open-channel flows involve a free surface that can deform from its undisturbed relatively flat configuration to form waves. Such waves move across the surface at speeds that depend on

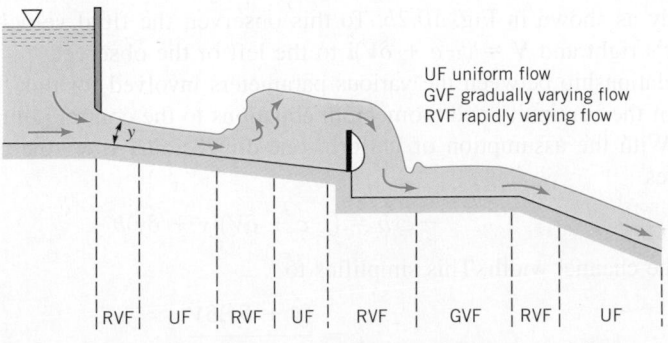

UF uniform flow
GVF gradually varying flow
RVF rapidly varying flow

■ F I G U R E   10.1   Classification of open-channel flow.

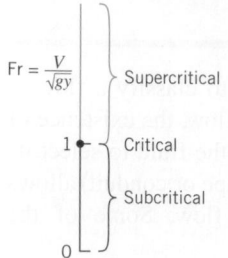

$$\mathrm{Fr} = \frac{V}{\sqrt{gy}}$$

Supercritical

1 — Critical

Subcritical

0

their size (height, length) and properties of the channel (depth, fluid velocity, etc.). The character of an open-channel flow may depend strongly on how fast the fluid is flowing relative to how fast a typical wave moves relative to the fluid. The dimensionless parameter that describes this behavior is termed the *Froude number*, $\mathrm{Fr} = V/(g\ell)^{1/2}$, where ℓ is an appropriate characteristic length of the flow. This dimensionless parameter was introduced in Chapter 7 and is discussed more fully in Section 10.2. As shown by the figure in the margin, the special case of a flow with a Froude number of unity, $\mathrm{Fr} = 1$, is termed a *critical flow*. If the Froude number is less than 1, the flow is *subcritical* (or *tranquil*). A flow with the Froude number greater than 1 is termed *supercritical* (or *rapid*).

10.2 Surface Waves

The distinguishing feature of flows involving a free surface (as in open-channel flows) is the opportunity for the free surface to distort into various shapes. The surface of a lake or the ocean is seldom "smooth as a mirror." It is usually distorted into ever-changing patterns associated with surface waves as shown in the photos in the margin. Some of these waves are very high, some barely ripple the surface; some waves are very long (the distance between wave crests), some are short; some are breaking waves that form whitecaps, others are quite smooth. Although a general study of this wave motion is beyond the scope of this book, an understanding of certain fundamental properties of simple waves is necessary for open-channel flow considerations. The interested reader is encouraged to use some of the excellent references available for further study about wave motion (Refs. 1, 2, 3).

F l u i d s   i n   t h e   N e w s

Rogue Waves There is a long history of stories concerning giant rogue ocean *waves* that come out of nowhere and capsize ships. The movie *Poseidon* (2006) is based on such an event. Although these giant, freakish waves were long considered fictional, recent satellite observations and computer simulations prove that, although rare, they are real. Such waves are single, sharply-peaked mounds of water that travel rapidly across an otherwise relatively calm ocean. Although most ships are designed to withstand waves up to 15 meters high, satellite measurements and data from offshore oil platforms indicate that such rogue waves can reach a height of 30 meters. Although researchers still do not understand the formation of these large rogue waves, there are several suggestions as to how ordinary smaller waves can be focused into one spot to produce a giant wave. Additional theoretical calculations and wave tank experiments are needed to adequately grasp the nature of such waves. Perhaps it will eventually be possible to predict the occurrence of these destructive waves, thereby reducing the loss of ships and life because of them.

10.2.1 Wave Speed

V10.2 Filling your car's gas tank.

Consider the situation illustrated in Fig. 10.2a in which a single elementary wave of small height, δy, is produced on the surface of a channel by suddenly moving the initially stationary end wall with speed δV. The water in the channel was stationary at the initial time, $t = 0$. A stationary observer will observe a single wave move down the channel with a *wave speed* c, with no fluid motion ahead of the wave and a fluid velocity of δV behind the wave. The motion is unsteady for such an observer. For an observer moving along the channel with speed c, the flow will appear steady as shown in Fig. 10.2b. To this observer, the fluid velocity will be $\mathbf{V} = -c\hat{\mathbf{i}}$ on the observer's right and $\mathbf{V} = (-c + \delta V)\hat{\mathbf{i}}$ to the left of the observer.

The relationship between the various parameters involved for this flow can be obtained by application of the continuity and momentum equations to the control volume shown in Fig. 10.2b as follows. With the assumption of uniform one-dimensional flow, the continuity equation (Eq. 5.12) becomes

$$-cyb = (-c + \delta V)(y + \delta y)b$$

where b is the channel width. This simplifies to

$$c = \frac{(y + \delta y)\delta V}{\delta y}$$

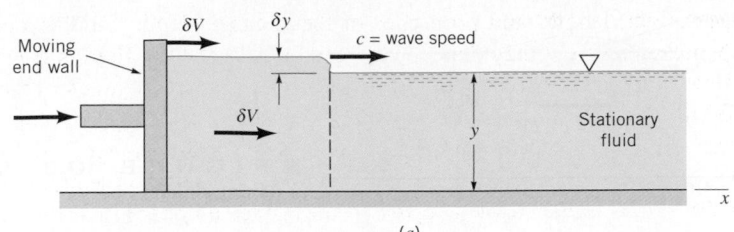

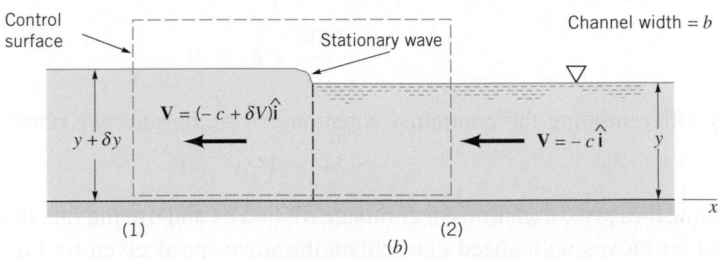

FIGURE 10.2 (a) Production of a single elementary wave in a channel as seen by a stationary observer. (b) Wave as seen by an observer moving with a speed equal to the wave speed.

or in the limit of small amplitude waves with $\delta y \ll y$

$$c = y \frac{\delta V}{\delta y} \tag{10.1}$$

Similarly, the momentum equation (Eq. 5.22) is

$$\tfrac{1}{2}\gamma y^2 b - \tfrac{1}{2}\gamma(y + \delta y)^2 b = \rho b c y[(c - \delta V) - c]$$

where we have written the mass flowrate as $\dot{m} = \rho b c y$ and have assumed that the pressure variation is hydrostatic within the fluid. That is, the pressure forces on the channel cross sections (1) and (2) are $F_1 = \gamma y_{c1} A_1 = \gamma(y + \delta y)^2 b/2$ and $F_2 = \gamma y_{c2} A_2 = \gamma y^2 b/2$, respectively. If we again impose the assumption of small amplitude waves [i.e., $(\delta y)^2 \ll y\,\delta y$], the momentum equation reduces to

The wave speed can be obtained from the continuity and momentum equations.

$$\frac{\delta V}{\delta y} = \frac{g}{c} \tag{10.2}$$

Combination of Eqs. 10.1 and 10.2 gives the wave speed

$$c = \sqrt{gy} \tag{10.3}$$

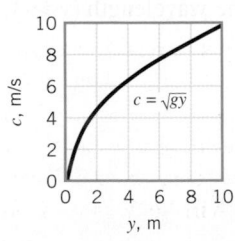

as indicated by the figure in the margin.

The speed of a small amplitude solitary wave as is indicated in Fig. 10.2 is proportional to the square root of the fluid depth, y, and independent of the wave amplitude, δy. The fluid density is not an important parameter, although the acceleration of gravity is. This is a result of the fact that such wave motion is a balance between inertial effects (proportional to ρ) and weight or hydrostatic pressure effects (proportional to $\gamma = \rho g$). A ratio of these forces eliminates the common factor ρ but retains g. For very small waves (like those produced by insects on water as shown in the photograph on the cover of the book), Eq. 10.3 is not valid because the effects of surface tension are significant.

The wave speed can also be calculated by using the energy and continuity equations rather than the momentum and continuity equations as is done above. A simple wave on the surface is shown in Fig. 10.3. As seen by an observer moving with the wave speed, c, the flow is steady. Since the pressure is constant at any point on the free surface, the Bernoulli equation for this frictionless flow is simply

$$\frac{V^2}{2g} + y = \text{constant}$$

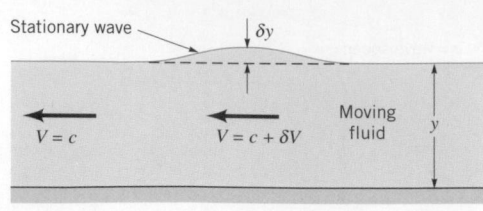

■ F I G U R E 10.3   **Stationary simple wave in a flowing fluid.**

V10.3 Water strider

or by differentiating

$$\frac{V\,\delta V}{g} + \delta y = 0$$

Also, by differentiating the continuity equation, $Vy = $ constant, we obtain

$$y\,\delta V + V\,\delta y = 0$$

We combine these two equations to eliminate δV and δy and use the fact that $V = c$ for this situation (the observer moves with speed c) to obtain the wave speed given by Eq. 10.3.

The above results are restricted to waves of small amplitude because we have assumed one-dimensional flow. That is, $\delta y/y \ll 1$. More advanced analysis and experiments show that the wave speed for finite-sized solitary waves exceeds that given by Eq. 10.3. To a first approximation, one obtains (Ref. 4)

$$c \approx \sqrt{gy}\left(1 + \frac{\delta y}{y}\right)^{1/2}$$

As indicated by the figure in the margin, the larger the amplitude, the faster the wave travels.

A more general description of wave motion can be obtained by considering continuous (not solitary) waves of sinusoidal shape as is shown in Fig. 10.4. By combining waves of various wavelengths, λ, and amplitudes, δy, it is possible to describe very complex surface patterns found in nature, such as the wind-driven waves on a lake. Mathematically, such a process consists of using a Fourier series (each term of the series represented by a wave of different wavelength and amplitude) to represent an arbitrary function (the free-surface shape).

A more advanced analysis of such sinusoidal surface waves of small amplitude shows that the wave speed varies with both the wavelength and fluid depth as (Ref. 1)

V10.4 Sinusoidal waves

$$c = \left[\frac{g\lambda}{2\pi}\tanh\left(\frac{2\pi y}{\lambda}\right)\right]^{1/2} \tag{10.4}$$

where $\tanh(2\pi y/\lambda)$ is the hyperbolic tangent of the argument $2\pi y/\lambda$. The result is plotted in Fig. 10.5. For conditions for which the water depth is much greater than the wavelength ($y \gg \lambda$, as in the ocean), the wave speed is independent of y and given by

$$c = \sqrt{\frac{g\lambda}{2\pi}}$$

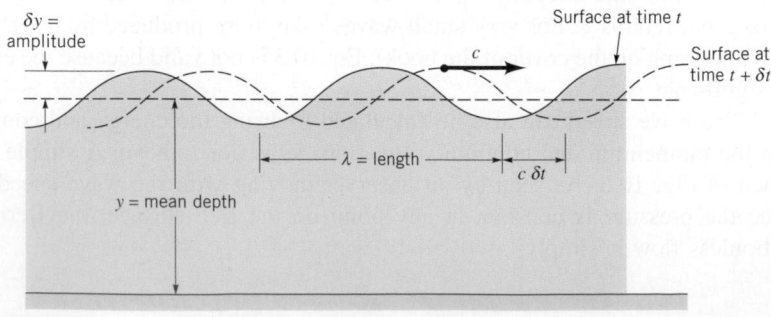

■ F I G U R E 10.4   **Sinusoidal surface wave.**

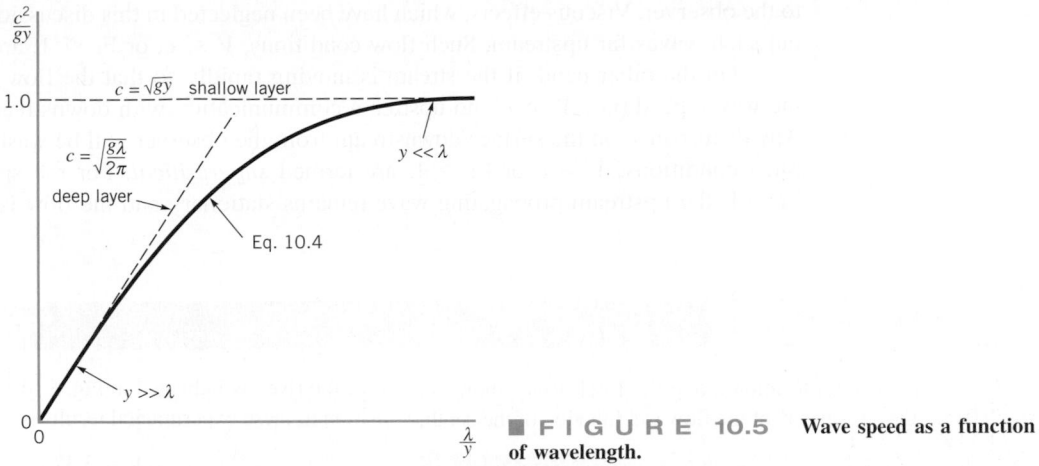

■ FIGURE 10.5 Wave speed as a function of wavelength.

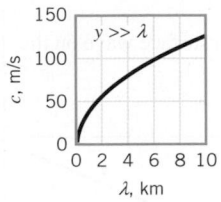

This result, shown in the figure in the margin, follows from Eq. 10.4, since $\tanh(2\pi y/\lambda) \to 1$ as $y/\lambda \to \infty$. Note that waves with very long wavelengths [e.g., waves created by a tsunami ("tidal wave") with wavelengths on the order of several kilometers] travel very rapidly. On the other hand, if the fluid layer is shallow ($y \ll \lambda$, as often happens in open channels), the wave speed is given by $c = (gy)^{1/2}$, as derived for the solitary wave in Fig. 10.2. This result also follows from Eq. 10.4, since $\tanh(2\pi y/\lambda) \to 2\pi y/\lambda$ as $y/\lambda \to 0$. These two limiting cases are shown in Fig. 10.5. For moderate depth layers ($y \sim \lambda$), the results are given by the complete Eq. 10.4. Note that for a given fluid depth, the long wave travels fastest. Hence, for our purposes we will consider the wave speed to be this limiting situation, $c = (gy)^{1/2}$.

F l u i d s   i n   t h e   N e w s

Tsunami, the nonstorm wave A tsunami, often miscalled a "tidal wave," is a wave produced by a disturbance (for example, an earthquake, volcanic eruption, or meteorite impact) that vertically displaces the water column. Tsunamis are characterized as shallow-water waves, with long periods, very long wavelengths, and extremely large wave speeds. For example, the waves of the great December 2005, Indian Ocean tsunami traveled with speeds to 500–1000 m/s. Typically, these waves were of small amplitude in deep water far from land. Satellite radar measured the wave height less than 1 m in these areas. However, as the waves approached shore and moved into shallower water, they slowed down considerably and reached heights up to 30 m. Because the rate at which a wave loses its energy is inversely related to its wavelength, tsunamis, with their wavelengths on the order of 100 km, not only travel at high speeds, they also travel great distances with minimal energy loss. The furthest reported death from the Indian Ocean tsunami occurred approximately 8000 km from the epicenter of the earthquake that produced it. (See Problem 10.14.)

10.2.2 Froude Number Effects

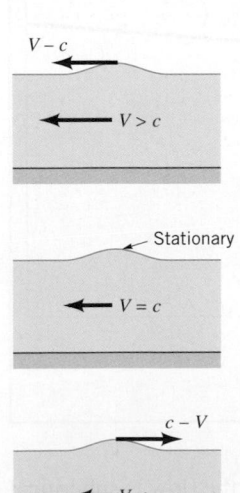

Consider an elementary wave traveling on the surface of a fluid, as is shown in the figure in the margin and Fig. 10.2a. If the fluid layer is stationary, the wave moves to the right with speed c relative to the fluid and the stationary observer. If the fluid is flowing to the left with velocity $V < c$, the wave (which travels with speed c relative to the fluid) will travel to the right with a speed of $c - V$ relative to a fixed observer. If the fluid flows to the left with $V = c$, the wave will remain stationary, but if $V > c$ the wave will be washed to the left with speed $V - c$.

The above ideas can be expressed in dimensionless form by use of the Froude number, $\mathrm{Fr} = V/(gy)^{1/2}$, where we take the characteristic length to be the fluid depth, y. Thus, the Froude number, $\mathrm{Fr} = V/(gy)^{1/2} = V/c$, is the ratio of the fluid velocity to the wave speed.

The following characteristics are observed when a wave is produced on the surface of a moving stream, as happens when a rock is thrown into a river. If the stream is not flowing, the wave spreads equally in all directions. If the stream is nearly stationary or moving in a tranquil manner (i.e., $V < c$), the wave can move upstream. Upstream locations are said to be in hydraulic communication with the downstream locations. That is, an observer upstream of a disturbance can tell that there has been a disturbance on the surface because that disturbance can propagate upstream to the observer. Viscous effects, which have been neglected in this discussion, will eventually damp out such waves far upstream. Such flow conditions, $V < c$, or Fr < 1, are termed *subcritical*.

On the other hand, if the stream is moving rapidly so that the flow velocity is greater than the wave speed (i.e., $V > c$), no upstream communication with downstream locations is possible. Any disturbance on the surface downstream from the observer will be washed farther downstream. Such conditions, $V > c$ or Fr > 1, are termed *supercritical*. For the special case of $V = c$ or Fr $= 1$, the upstream propagating wave remains stationary and the flow is termed *critical*.

*E*XAMPLE 10.1

GIVEN At a certain location along the Rock River shown in Fig. E10.1a, the velocity, V, of the flow is a function of the depth, y, of the river as indicated in Fig. E10.1b. A reasonable approximation to these experimental results is

$$V = 3.37\, y^{2/3} \qquad (1)$$

where V is in m/s and y is in m.

FIND For what range of water depth will a surface wave on the river be able to travel upstream?

■ **FIGURE E10.1a**

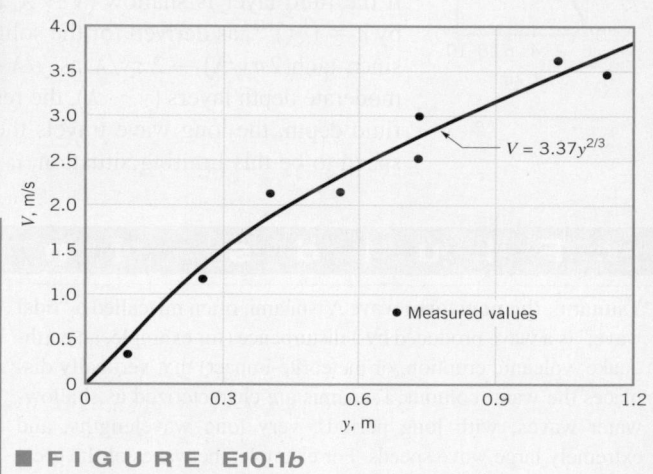

■ **FIGURE E10.1b**

SOLUTION

While the river travels to the left with speed V, the surface wave travels upstream (to the right) with speed $c = (g\, y)^{1/2}$ relative to the water (not relative to the ground). Hence relative to the stationary ground, the wave travels to the right with speed

$$c - V = (g\, y)^{1/2} - 3.37\, y^{2/3}$$
$$= (9.81 \text{ m/s}^2\, y)^{1/2} - 3.37\, y^{2/3} \qquad (2)$$

For the wave to travel upstream, $c - V > 0$ so that from Eq. 2,

$$(9.81\, y)^{1/2} > 3.37\, y^{2/3}$$

or

$$y < 0.64 \text{ m} \qquad \text{(Ans)}$$

COMMENT As shown above, if the river depth is less than 0.64 m, its velocity is less than the wave speed and the wave can travel upstream. This is consistent with the fact that if a wave is to travel upstream, the flow must be subcritical (i.e., Fr $= V/c < 1$). For this flow

$$\text{Fr} = V/c = (3.37\, y^{2/3})/(g\, y)^{1/2}$$
$$= 3.37\, y^{1/6}/(9.81 \text{ m/s}^2)^{1/2}$$
$$= 1.08\, y^{1/6}$$

This result is plotted in Fig. E10.1c. Note that in agreement with the above answer, for $y < 0.64$ the flow is subcritical; the wave can travel upstream.

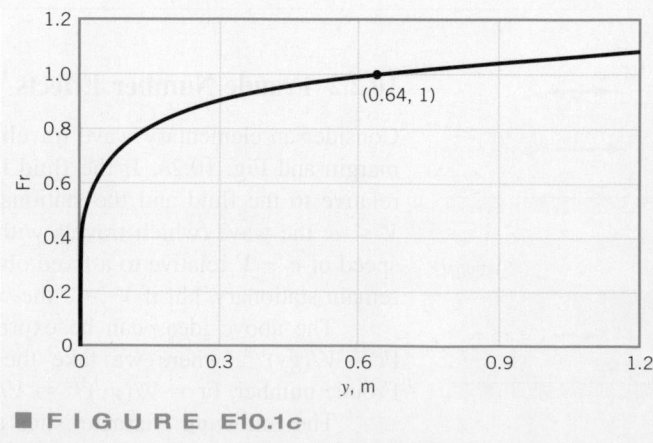

■ **FIGURE E10.1c**

The character of an open-channel flow may depend strongly on whether the flow is subcritical or supercritical. The characteristics of the flow may be completely opposite for subcritical flow than for supercritical flow. For example, as is discussed in Section 10.3, a "bump" on the bottom

V10.5 Bicycle through a puddle

of a river (such as a submerged log) may cause the surface of the river to dip below the level it would have had if the log were not there, or it may cause the surface level to rise above its undisturbed level. Which situation will happen depends on the value of Fr. Similarly, for supercritical flows it is possible to produce steplike discontinuities in the fluid depth (called a hydraulic jump; see Section 10.6.1). For subcritical flows, however, changes in depth must be smooth and continuous. Certain open-channel flows, such as the broad-crested weir (Section 10.6.3), depend on the existence of critical flow conditions for their operation.

As strange as it may seem, there exist many similarities between the open-channel flow of a liquid and the compressible flow of a gas. The governing dimensionless parameter in each case is the fluid velocity, V, divided by a wave speed, the surface wave speed for open-channel flow or sound wave speed for compressible flow. Many of the differences between subcritical (Fr < 1) and supercritical (Fr > 1) open-channel flows have analogs in subsonic (Ma < 1) and supersonic (Ma > 1) compressible gas flow, where Ma is the Mach number. Some of these similarities are discussed in this chapter and in Chapter 11.

10.3   Energy Considerations

The slope of the bottom of most open channels is very small; the bottom is nearly horizontal.

A typical segment of an open-channel flow is shown in Fig. 10.6. The slope of the channel bottom (or *bottom slope*), $S_0 = (z_1 - z_2)/\ell$, is assumed constant over the segment shown. The fluid depths and velocities are y_1, y_2, V_1, and V_2 as indicated. Note that the fluid depth is measured in the vertical direction and the distance x is horizontal. For most open-channel flows the value of S_0 is very small (the bottom is nearly horizontal). For example, the Mississippi River drops a distance of 450 m in its 3780 km length to give an average value of $S_0 = 1.2 \times 10^{-4}$. In such circumstances the values of x and y are often taken as the distance along the channel bottom and the depth normal to the bottom, with negligibly small differences introduced by the two coordinate schemes.

With the assumption of a uniform velocity profile across any section of the channel, the one-dimensional energy equation for this flow (Eq. 5.84) becomes

$$\frac{p_1}{\gamma} + \frac{V_1^2}{2g} + z_1 = \frac{p_2}{\gamma} + \frac{V_2^2}{2g} + z_2 + h_L \tag{10.5}$$

where h_L is the head loss due to viscous effects between sections (1) and (2) and $z_1 - z_2 = S_0\ell$. Since the pressure is essentially hydrostatic at any cross section, we find that $p_1/\gamma = y_1$ and $p_2/\gamma = y_2$ so that Eq. 10.5 becomes

$$y_1 + \frac{V_1^2}{2g} + S_0\ell = y_2 + \frac{V_2^2}{2g} + h_L \tag{10.6}$$

One of the difficulties of analyzing open-channel flow, similar to that discussed in Chapter 8 for pipe flow, is associated with the determination of the head loss in terms of other physical parameters. Without getting into such details at present, we write the head loss in terms of the slope of the energy line, $S_f = h_L/\ell$ (often termed the *friction slope*), as indicated in Fig. 10.6. Recall from

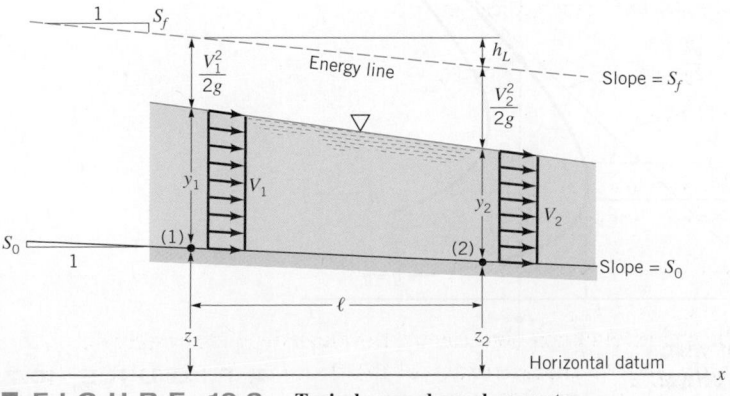

■ **FIGURE 10.6**   **Typical open-channel geometry.**

Chapter 3 that the energy line is located a distance z (the elevation from some datum to the channel bottom) plus the pressure head (p/γ) plus the velocity head ($V^2/2g$) above the datum. Therefore, Eq. 10.6 can be written as

$$y_1 - y_2 = \frac{(V_2^2 - V_1^2)}{2g} + (S_f - S_0)\ell \tag{10.7}$$

If there is no head loss, the energy line is horizontal ($S_f = 0$), and the total energy of the flow is free to shift between kinetic energy and potential energy in a conservative fashion. In the specific instance of a horizontal channel bottom ($S_0 = 0$) and negligible head loss ($S_f = 0$), Eq. 10.7 simply becomes

$$y_1 - y_2 = \frac{(V_2^2 - V_1^2)}{2g}$$

10.3.1 Specific Energy

The concept of the *specific energy* or specific head, E, defined as

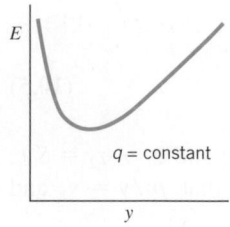

The specific energy is the sum of potential energy and kinetic energy (per unit weight).

$$E = y + \frac{V^2}{2g} \tag{10.8}$$

is often useful in open-channel flow considerations. The energy equation, Eq. 10.7, can be written in terms of E as

$$E_1 = E_2 + (S_f - S_0)\ell \tag{10.9}$$

If head losses are negligible, then $S_f = 0$ so that $(S_f - S_0)\ell = -S_0\ell = z_2 - z_1$ and the sum of the specific energy and the elevation of the channel bottom remains constant (i.e., $E_1 + z_1 = E_2 + z_2$, a statement of the Bernoulli equation).

If we consider a simple channel whose cross-sectional shape is a rectangle of width b, the specific energy can be written in terms of the flowrate per unit width, $q = Q/b = Vyb/b = Vy$, as

$$E = y + \frac{q^2}{2gy^2} \tag{10.10}$$

which is illustrated by the figure in the margin.

For a given channel of constant width, the value of q remains constant along the channel, although the depth, y, may vary. To gain insight into the flow processes involved, we consider the *specific energy diagram,* a graph of $E = E(y)$, with q fixed, as shown in Fig. 10.7. The relationship between the flow depth, y, and the velocity head, $V^2/2g$, as given by Eq. 10.8 is indicated in the figure.

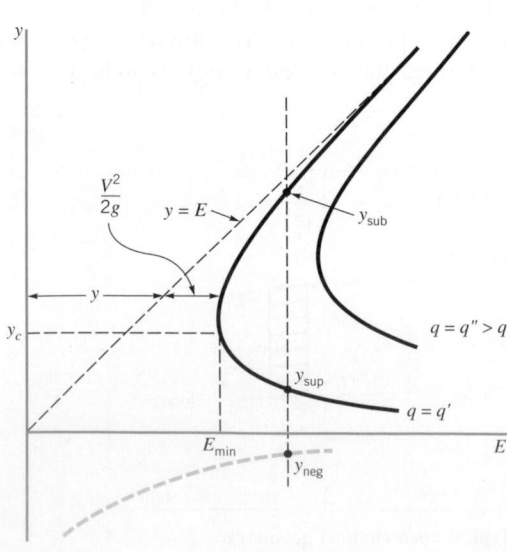

■ **FIGURE 10.7**   **Specific energy diagram.**

For a given value
of specific energy, a
flow may have al-
ternate depths.

For given q and E, Eq. 10.10 is a cubic equation $[y^3 - Ey^2 + (q^2/2g) = 0]$ with three solutions, y_{sup}, y_{sub}, and y_{neg}. If the specific energy is large enough (i.e., $E > E_{min}$, where E_{min} is a function of q), two of the solutions are positive and the other, y_{neg}, is negative. The negative root, represented by the curved dashed line in Fig. 10.7, has no physical meaning and can be ignored. Thus, for a given flowrate and specific energy there are two possible depths, unless the vertical line from the E axis does not intersect the specific energy curve corresponding to the value of q given (i.e., $E < E_{min}$). These two depths are termed *alternate depths*.

For large values of E the upper and lower branches of the specific energy diagram (y_{sub} and y_{sup}) approach $y = E$ and $y = 0$, respectively. These limits correspond to a very deep channel flowing very slowly ($E = y + V^2/2g \rightarrow y$ as $y \rightarrow \infty$ with $q = Vy$ fixed), or a very high-speed flow in a shallow channel ($E = y + V^2/2g \rightarrow V^2/2g$ as $y \rightarrow 0$).

As is indicated in Fig. 10.7, $y_{sup} < y_{sub}$. Thus, since $q = Vy$ is constant along the curve, it follows that $V_{sup} > V_{sub}$, where the subscripts "sub" and "sup" on the velocities correspond to the depths so labeled. The specific energy diagram consists of two portions divided by the E_{min} "nose" of the curve. We will show that the flow conditions at this location correspond to critical conditions (Fr = 1), those on the upper portion of the curve correspond to subcritical conditions (hence, the "sub" subscript), and those on the lower portion of the curve correspond to supercritical conditions (hence, the "sup" subscript).

To determine the value of E_{min}, we use Eq. 10.10 and set $dE/dy = 0$ to obtain

$$\frac{dE}{dy} = 1 - \frac{q^2}{gy^3} = 0$$

or

$$y_c = \left(\frac{q^2}{g}\right)^{1/3} \tag{10.11}$$

where the subscript "c" denotes conditions at E_{min}. By substituting this back into Eq. 10.10 we obtain

$$E_{min} = \frac{3y_c}{2}$$

By combining Eq. 10.11 and $V_c = q/y_c$, we obtain

$$V_c = \frac{q}{y_c} = \frac{(y_c^{3/2}g^{1/2})}{y_c} = \sqrt{gy_c}$$

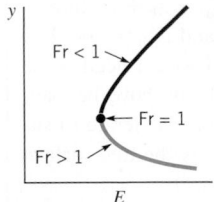

or $Fr_c \equiv V_c/(gy_c)^{1/2} = 1$. Thus, critical conditions (Fr = 1) occur at the location of E_{min}. Since the layer is deeper and the velocity smaller for the upper part of the specific energy diagram (compared with the conditions at E_{min}), such flows are subcritical (Fr < 1). Conversely, flows for the lower part of the diagram are supercritical. This is shown by the figure in the margin. Thus, for a given flowrate, q, if $E > E_{min}$ there are two possible depths of flow, one subcritical and the other supercritical.

It is often possible to determine various characteristics of a flow by considering the specific energy diagram. Example 10.2 illustrates this for a situation in which the channel bottom elevation is not constant.

EXAMPLE 10.2 Specific Energy Diagram—Quantitative

GIVEN Water flows up a 0.15-m-tall ramp in a constant width rectangular channel at a rate $q = 0.5$ m²/s as is shown in Fig. E10.2a. (For now disregard the "bump.") The upstream depth is 0.7 m and viscous effects are negligible.

FIND Determine the elevation of the water surface downstream of the ramp, $y_2 + z_2$.

SOLUTION

With $S_0\ell = z_1 - z_2$ and $h_L = 0$, conservation of energy (Eq. 10.6 which, under these conditions, is actually the Bernoulli equation) requires that

$$y_1 + \frac{V_1^2}{2g} + z_1 = y_2 + \frac{V_2^2}{2g} + z_2$$

For the conditions given ($z_1 = 0$, $z_2 = 0.15$ m, $y_1 = 0.7$ m, and $V_1 = q/y_1 = 0.8$ m/s), this becomes

$$0.58 = y_2 + \frac{V_2^2}{19.6} \qquad (1)$$

where V_2 and y_2 are in m/s and meters, respectively. The continuity equation provides the second equation

$$y_2 V_2 = y_1 V_1$$

or

$$y_2 V_2 = 0.5 \text{ m}^2/\text{s} \qquad (2)$$

Equations 1 and 2 can be combined to give

$$y_2^3 - 0.58 y_2^2 + 0.013 = 0$$

which has solutions

$$y_2 = 0.53 \text{ m}, \qquad y_2 = 0.18 \text{ m}, \quad \text{or} \quad y_2 = -0.13 \text{ m}$$

Note that two of these solutions are physically realistic, but the negative solution is meaningless. This is consistent with the previous discussions concerning the specific energy (recall the three roots indicated in Fig. 10.7). The corresponding elevations of the free surface are either

$$y_2 + z_2 = 0.53 \text{ m} + 0.15 \text{ m} = 0.68 \text{ m}$$

or

$$y_2 + z_2 = 0.18 \text{ m} + 0.15 \text{ m} = 0.33 \text{ m}$$

The question is which of these two flows is to be expected? This can be answered by use of the specific energy diagram obtained from Eq. 10.10, which for this problem is

$$E = y + \frac{0.013}{y^2}$$

where E and y are in meters. The diagram is shown in Fig. E10.2b. The upstream condition corresponds to subcritical flow; the downstream condition is either subcritical or supercritical, corresponding to points 2 or 2'. Note that since $E_1 = E_2 + (z_2 - z_1) = E_2 + 0.15$ m, it follows that the downstream conditions are located 0.15 m to the left of the upstream conditions on the diagram.

With a constant width channel, the value of q remains the same for any location along the channel. That is, all points for the flow from (1) to (2) or (2') must lie along the $q = 0.5$ m²/s curve shown. Any deviation from this curve would imply either a change in q or a relaxation of the one-dimensional flow assumption. To stay on the curve and go from (1) around the critical point (point c) to point (2') would require a reduction in

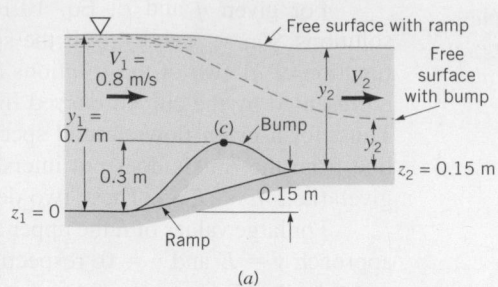

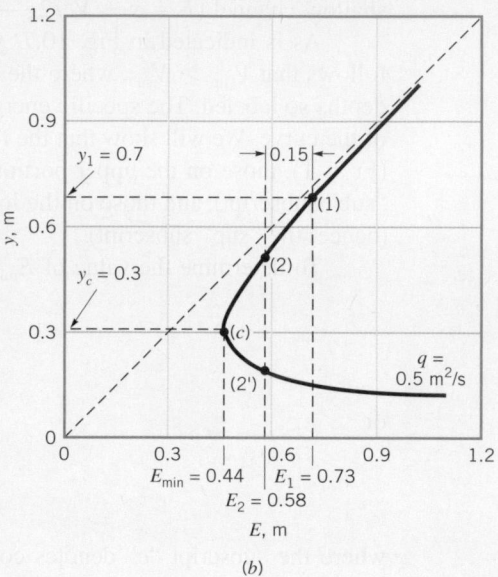

■ **FIGURE E10.2**

specific energy to E_{min}. As is seen from Fig. E10.2a, this would require a specified elevation (bump) in the channel bottom so that critical conditions would occur above this bump. The height of this bump can be obtained from the energy equation (Eq. 10.9) written between points (1) and (c) with $S_f = 0$ (no viscous effects) and $S_0\ell = z_1 - z_c$. That is, $E_1 = E_{min} - z_1 + z_c$. In particular, since $E_1 = y_1 + 0.013/y_1^2 = 0.73$ m and $E_{min} = 3y_c/2 = 3(q^2/g)^{1/3}/2 = 0.44$ m, the top of this bump would need to be $z_c - z_1 = E_1 - E_{min} = 0.73$ m $- 0.44$ m $= 0.3$ m above the channel bottom at section (1). The flow could then accelerate to supercritical conditions ($Fr_{2'} > 1$) as is shown by the free surface represented by the dashed line in Fig. E10.2a.

Since the actual elevation change (a ramp) shown in Fig. E10.2a does not contain a bump, the downstream conditions will correspond to the subcritical flow denoted by (2), not the supercritical condition (2'). Without a bump on the channel bottom, the state (2') is inaccessible from the upstream condition state (1). Such considerations are often termed the *accessibility of flow regimes*. Thus, the surface elevation is

$$y_2 + z_2 = 0.68 \text{ m} \qquad \text{(Ans)}$$

Note that since $y_1 + z_1 = 0.7$ m and $y_2 + z_2 = 0.6$ m, the elevation of the free surface decreases as it goes across the ramp.

COMMENT If the flow conditions upstream of the ramp were supercritical, the free-surface elevation and fluid depth would increase as the fluid flows up the ramp. This is indicated in Fig. E10.2c along with the corresponding specific energy diagram, as is shown in Fig. E10.2d. For this case the flow starts at (1) on the lower (supercritical) branch of the specific energy curve and ends at (2) on the same branch with $y_2 > y_1$. Since both y and z increase from (1) to (2), the surface elevation, $y + z$, also increases. Thus, flow up a ramp is different for subcritical than it is for supercritical conditions.

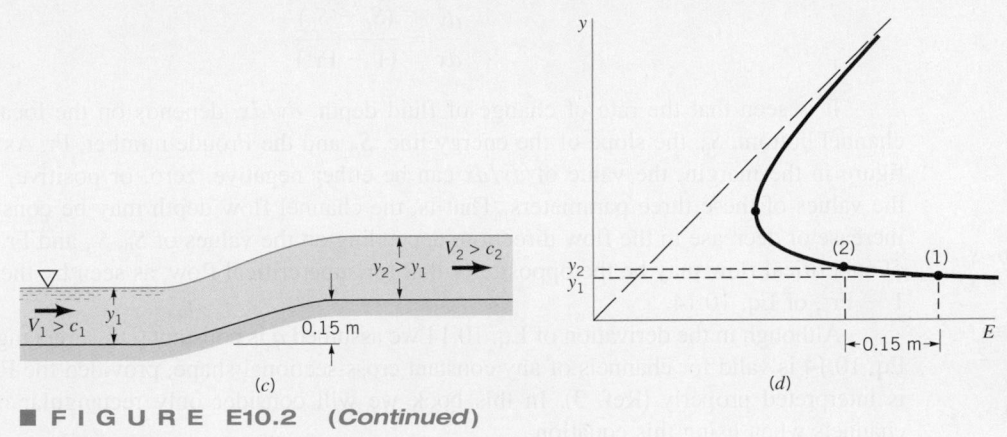

■ **F I G U R E   E10.2**   (*Continued*)

10.3.2 Channel Depth Variations

By using the concepts of the specific energy and critical flow conditions (Fr $= 1$), it is possible to determine how the depth of a flow in an open channel changes with distance along the channel. In some situations the depth change is very rapid so that the value of dy/dx is of the order of 1. Complex effects involving two- or three-dimensional flow phenomena are often involved in such flows.

In this section we consider only gradually varying flows. For such flows, $dy/dx \ll 1$ and it is reasonable to impose the one-dimensional velocity assumption. At any section the total head is $H = V^2/2g + y + z$ and the energy equation (Eq. 10.5) becomes

$$H_1 = H_2 + h_L$$

where h_L is the head loss between sections (1) and (2).

As is discussed in the previous section, the slope of the energy line is $dH/dx = dh_L/dx = S_f$ and the slope of the channel bottom is $dz/dx = S_0$. Thus, since

$$\frac{dH}{dx} = \frac{d}{dx}\left(\frac{V^2}{2g} + y + z\right) = \frac{V}{g}\frac{dV}{dx} + \frac{dy}{dx} + \frac{dz}{dx}$$

we obtain

$$\frac{dh_L}{dx} = \frac{V}{g}\frac{dV}{dx} + \frac{dy}{dx} + S_0$$

or

$$\frac{V}{g}\frac{dV}{dx} + \frac{dy}{dx} = S_f - S_0 \tag{10.12}$$

For a given flowrate per unit width, q, in a rectangular channel of constant width b, we have $V = q/y$ or by differentiation

$$\frac{dV}{dx} = -\frac{q}{y^2}\frac{dy}{dx} = -\frac{V}{y}\frac{dy}{dx}$$

so that the kinetic energy term in Eq. 10.12 becomes

$$\frac{V}{g}\frac{dV}{dx} = -\frac{V^2}{gy}\frac{dy}{dx} = -\text{Fr}^2\frac{dy}{dx} \qquad (10.13)$$

where $\text{Fr} = V/(gy)^{1/2}$ is the local Froude number of the flow. Substituting Eq. 10.13 into Eq. 10.12 and simplifying gives

$$\frac{dy}{dx} = \frac{(S_f - S_0)}{(1 - \text{Fr}^2)} \qquad (10.14)$$

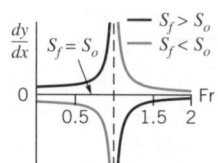

It is seen that the rate of change of fluid depth, dy/dx, depends on the local slope of the channel bottom, S_0, the slope of the energy line, S_f, and the Froude number, Fr. As shown by the figure in the margin, the value of dy/dx can be either negative, zero, or positive, depending on the values of these three parameters. That is, the channel flow depth may be constant or it may increase or decrease in the flow direction, depending on the values of S_0, S_f, and Fr. The behavior of subcritical flow may be the opposite of that for supercritical flow, as seen by the denominator, $1 - \text{Fr}^2$, of Eq. 10.14.

Although in the derivation of Eq. 10.14 we assumed q is constant (i.e., a rectangular channel), Eq. 10.14 is valid for channels of any constant cross-sectional shape, provided the Froude number is interpreted properly (Ref. 3). In this book we will consider only rectangular cross-sectional channels when using this equation.

10.4 Uniform Depth Channel Flow

V10.6 Merging channels

Many channels are designed to carry fluid at a uniform depth all along their length. Irrigation canals are frequently of uniform depth and cross section for considerable lengths. Natural channels such as rivers and creeks are seldom of uniform shape, although a reasonable approximation to the flowrate in such channels can often be obtained by assuming uniform flow. In this section we will discuss various aspects of such flows.

Uniform depth flow $(dy/dx = 0)$ can be accomplished by adjusting the bottom slope, S_0, so that it precisely equals the slope of the energy line, S_f. That is, $S_0 = S_f$. This can be seen from Eq. 10.14. From an energy point of view, uniform depth flow is achieved by a balance between the potential energy lost by the fluid as it coasts downhill and the energy that is dissipated by viscous effects (head loss) associated with shear stresses throughout the fluid. Similar conclusions can be reached from a force balance analysis as discussed in the following section.

10.4.1 Uniform Flow Approximations

We consider fluid flowing in an open channel of constant cross-sectional size and shape such that the depth of flow remains constant as is indicated in Fig. 10.8. The area of the section is A and the *wetted perimeter* (i.e., the length of the perimeter of the cross section in contact with the fluid) is P. The interaction between the fluid and the atmosphere at the free surface is assumed negligible so that this portion of the perimeter is not included in the definition of the wetted perimeter.

The wall shear stress acts on the wetted perimeter of the channel.

Since the fluid must adhere to the solid surfaces, the actual velocity distribution in an open channel is not uniform. Some typical velocity profiles measured in channels of various shapes are indicated in Fig. 10.9a. The maximum velocity is often found somewhat below the free surface,

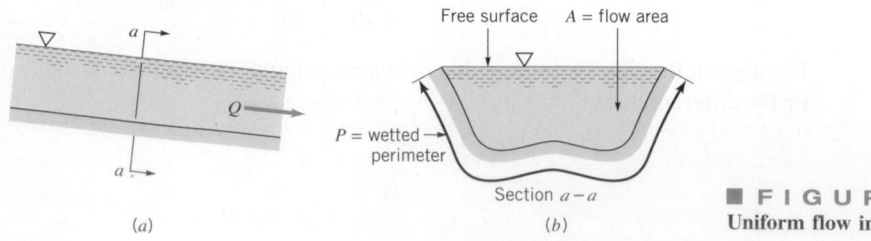

(a)   Section a–a   (b)

■ FIGURE 10.8
Uniform flow in an open channel.

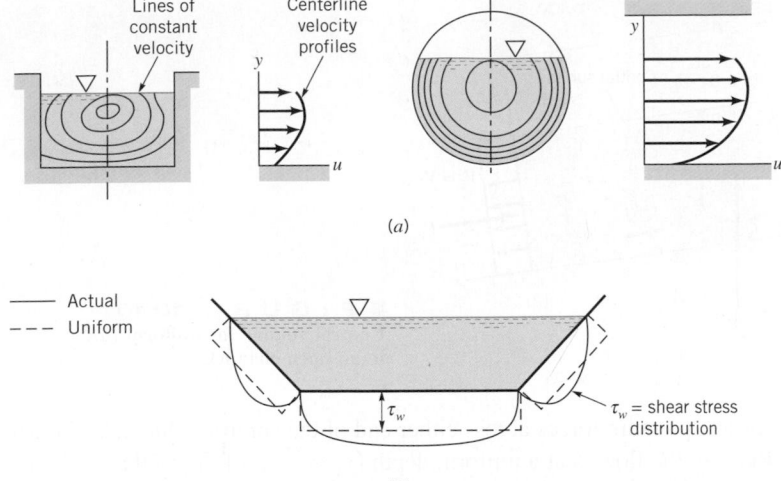

■ **FIGURE 10.9**   **Typical velocity and shear stress distributions in an open channel: (a) velocity distribution throughout the cross section, (b) shear stress distribution on the wetted perimeter.**

and the fluid velocity is zero on the wetted perimeter, where a wall shear stress, τ_w, is developed. This shear stress is seldom uniform along the wetted perimeter, with typical variations as are indicated in Fig. 10.9b.

Fortunately, reasonable analytical results can be obtained by assuming a uniform velocity profile, V, and a constant wall shear stress, τ_w. Similar assumptions were made for pipe flow situations (Chapter 8), with the friction factor being used to obtain the head loss.

F l u i d s   i n   t h e   N e w s

Plumbing the Everglades Because of all of the economic development that has occurred in southern Florida, the natural drainage pattern of that area has been greatly altered during the past century. Previously there was a vast network of surface flow southward from the Orlando area, to Lake Okeechobee, through the Everglades, and out to the Gulf of Mexico. Currently a vast amount of freshwater from Lake Okeechobee and surrounding waterways (1.7 billion gallons (or, 6.4 billion L) per day) is sluiced into the ocean for flood control, bypassing the Everglades. A new long-term Comprehensive Everglades Restoration Plan is being implemented to restore, preserve, and protect the south Florida ecosystem. Included in the plan are the use of numerous aquifer-storage-and-recovery systems that will recharge the ecosystem. In addition, surface water reservoirs using artificial wetlands will clean agricultural runoff. In an attempt to improve the historical flow from north to south, old levees will be removed, parts of the Tamiami Trail causeway will be altered, and stored water will be redirected through miles of new pipes and rebuilt *canals*. Strictly speaking, the Everglades will not be "restored." However, by 2030, 1.6 million acres (6,475 square km) of national parkland will have cleaner water and more of it. (See Problem 10.77.)

10.4.2 The Chezy and Manning Equations

The basic equations used to determine the uniform flowrate in open channels were derived many years ago. Continual refinements have taken place to obtain better values of the empirical coefficients involved. The result is a semiempirical equation that provides reasonable engineering results. A more refined analysis is perhaps not warranted because of the complexity and uncertainty of the flow geometry (i.e., channel shape and the irregular makeup of the wetted perimeter, particularly for natural channels).

Under the assumptions of steady uniform flow, the x component of the momentum equation (Eq. 5.22) applied to the control volume indicated in Fig. 10.10 simply reduces to

$$\Sigma F_x = \rho Q(V_2 - V_1) = 0$$

For steady, uniform depth flow in an open channel there is no fluid acceleration.

since $V_1 = V_2$. There is no acceleration of the fluid, and the momentum flux across section (1) is the same as that across section (2). The flow is governed by a simple balance between the forces in the direction of the flow. Thus, $\Sigma F_x = 0$, or

$$F_1 - F_2 - \tau_w P\ell + \mathcal{W} \sin \theta = 0 \qquad \textbf{(10.15)}$$

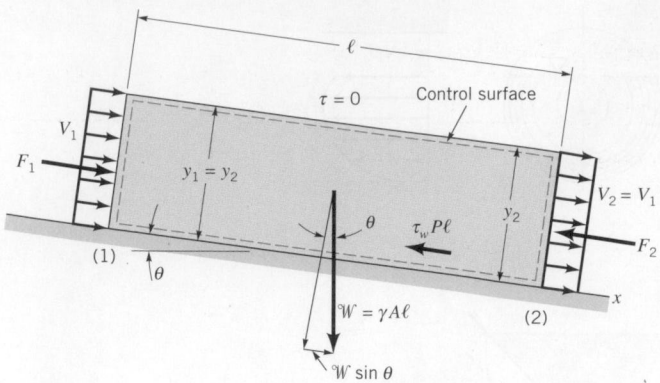

■ **FIGURE 10.10**
Control volume for uniform flow in an open channel.

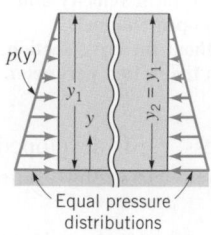

Equal pressure distributions

where F_1 and F_2 are the hydrostatic pressure forces across either end of the control volume, as shown by the figure in the margin. Because the flow is at a uniform depth ($y_1 = y_2$), it follows that $F_1 = F_2$ so that these two forces do not contribute to the force balance. The term $\mathcal{W}\sin\theta$ is the component of the fluid weight that acts down the slope, and $\tau_w P\ell$ is the shear force on the fluid, acting up the slope as a result of the interaction of the water and the channel's wetted perimeter. Thus, Eq. 10.15 becomes

$$\tau_w = \frac{\mathcal{W}\sin\theta}{P\ell} = \frac{\mathcal{W} S_0}{P\ell}$$

where we have used the approximation that $\sin\theta \approx \tan\theta = S_0$, since the bottom slope is typically very small (i.e., $S_0 \ll 1$). Since $\mathcal{W} = \gamma A\ell$ and the *hydraulic radius* is defined as $R_h = A/P$, the force balance equation becomes

$$\tau_w = \frac{\gamma A\ell S_0}{P\ell} = \gamma R_h S_0 \tag{10.16}$$

For uniform depth, channel flow is governed by a balance between friction and weight.

Most open-channel flows are turbulent rather than laminar. In fact, typical Reynolds numbers are quite large, well above the transitional value and into the wholly turbulent regime. As was discussed in Chapter 8, and shown by the figure in the margin, for very large Reynolds number pipe flows (wholly turbulent flows), the friction factor, f, is found to be independent of Reynolds number, dependent only on the relative roughness, ε/D, of the pipe surface. For such cases, the wall shear stress is proportional to the dynamic pressure, $\rho V^2/2$, and independent of the viscosity. That is,

$$\tau_w = K\rho \frac{V^2}{2}$$

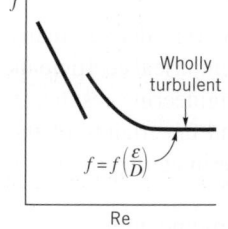

where K is a constant dependent upon the roughness of the pipe.

It is not unreasonable that similar shear stress dependencies occur for the large Reynolds number open-channel flows. In such situations, Eq. 10.16 becomes

$$K\rho \frac{V^2}{2} = \gamma R_h S_0$$

or

$$V = C\sqrt{R_h S_0} \tag{10.17}$$

where the constant C is termed the Chezy coefficient and Eq. 10.17 is termed the *Chezy equation.* This equation, one of the oldest in the area of fluid mechanics, was developed in 1768 by A. Chezy (1718–1798), a French engineer who designed a canal for the Paris water supply. The value of the Chezy coefficient, which must be determined by experiments, is not dimensionless but has the dimensions of (length)$^{1/2}$ per time (i.e., the square root of the units of acceleration).

From a series of experiments it was found that the slope dependence of Eq. 10.17 ($V \sim S_0^{1/2}$) is reasonable, but that the dependence on the hydraulic radius is more nearly $V \sim R_h^{2/3}$ rather than $V \sim R_h^{1/2}$. In 1889, R. Manning (1816–1897), an Irish engineer, developed the following somewhat modified equation for open-channel flow to more accurately describe the R_h dependence:

$$V = \frac{R_h^{2/3} S_0^{1/2}}{n} \tag{10.18}$$

■ **TABLE 10.1**
Values of the Manning Coefficient, *n* (Ref. 6)

Wetted Perimeter	*n*	Wetted Perimeter	*n*
A. Natural channels		**D. Artificially lined channels**	
Clean and straight	0.030	Glass	0.010
Sluggish with deep pools	0.040	Brass	0.011
Major rivers	0.035	Steel, smooth	0.012
		Steel, painted	0.014
B. Floodplains		Steel, riveted	0.015
Pasture, farmland	0.035	Cast iron	0.013
Light brush	0.050	Concrete, finished	0.012
Heavy brush	0.075	Concrete, unfinished	0.014
Trees	0.15	Planed wood	0.012
		Clay tile	0.014
C. Excavated earth channels		Brickwork	0.015
Clean	0.022	Asphalt	0.016
Gravelly	0.025	Corrugated metal	0.022
Weedy	0.030	Rubble masonry	0.025
Stony, cobbles	0.035		

Equation 10.18 is termed the *Manning equation*, and the parameter *n* is the *Manning resistance coefficient*. Its value is dependent on the surface material of the channel's wetted perimeter and is obtained from experiments. It is not dimensionless, having the units of $s/m^{1/3}$.

As is discussed in Chapter 7, any correlation should be expressed in dimensionless form, with the coefficients that appear being dimensionless coefficients, such as the friction factor for pipe flow or the drag coefficient for flow past objects. Thus, Eq. 10.18 should be expressed in dimensionless form. Unfortunately, the Manning equation is so widely used and has been used for so long that it will continue to be used in its dimensional form with a coefficient, *n*, that is not dimensionless. The values of *n* found in the literature (such as Table 10.1) were developed for SI units.

The Manning equation is used to obtain the velocity or flowrate in an open channel.

Thus, uniform flow in an open channel is obtained from the Manning equation written as

$$V = \frac{\kappa}{n} R_h^{2/3} S_0^{1/2}$$ **(10.19)**

and

$$Q = \frac{\kappa}{n} A R_h^{2/3} S_0^{1/2}$$ **(10.20)**

where $\kappa = 1$ if SI units are used. Thus, by using R_h in meters, A in m^2, and $\kappa = 1$, the average velocity is m/s and the flowrate m^3/s.

Typical values of the Manning coefficient are indicated in Table 10.1. As expected, the rougher the wetted perimeter, the larger the value of *n*. For example, the roughness of floodplain surfaces increases from pasture to brush to tree conditions. So does the corresponding value of the Manning coefficient. Thus, for a given depth of flooding, the flowrate varies with floodplain roughness as indicated by the figure in the margin.

Precise values of *n* are often difficult to obtain. Except for artificially lined channel surfaces like those found in new canals or flumes, the channel surface structure may be quite complex and variable. There are various methods used to obtain a reasonable estimate of the value of *n* for a given situation (Ref. 5). For the purpose of this book, the values from Table 10.1 are sufficient. Note that the error in *Q* is directly proportional to the error in *n*. A 10%

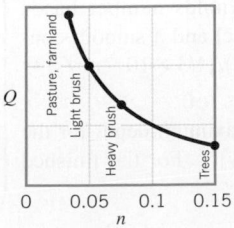

error in the value of n produces a 10% error in the flowrate. Considerable effort has been put forth to obtain the best estimate of n, with extensive tables of values covering a wide variety of surfaces (Ref. 7). It should be noted that the values of n given in Table 10.1 are valid only for water as the flowing fluid.

Both the friction factor for pipe flow and the Manning coefficient for channel flow are parameters that relate the wall shear stress to the makeup of the bounding surface. Thus, various results are available that describe n in terms of the equivalent pipe friction factor, f, and the surface roughness, ε (Ref. 8). For our purposes we will use the values of n from Table 10.1.

V10.7 Uniform channel flow

10.4.3 Uniform Depth Examples

A variety of interesting and useful results can be obtained from the Manning equation. The following examples illustrate some of the typical considerations.

The main parameters involved in uniform depth open-channel flow are the size and shape of the channel cross section (A, R_h), the slope of the channel bottom (S_0), the character of the material lining the channel bottom and walls (n), and the average velocity or flowrate $(V$ or $Q)$. Although the Manning equation is a rather simple equation, the ease of using it depends in part on which variables are given and which are to be determined.

Determination of the flowrate of a given channel with flow at a given depth (often termed the *normal flowrate* for *normal depth*, sometimes denoted y_n) is obtained from a straightforward calculation as is shown in Example 10.3.

EXAMPLE 10.3 | Uniform Flow, Determine Flow Rate

GIVEN Water flows in the canal of trapezoidal cross section shown in Fig. E10.3a. The bottom drops 0.4 m per 305 m of length. The canal is lined with new finished concrete.

FIND Determine

(a) the flowrate and

(b) the Froude number for this flow.

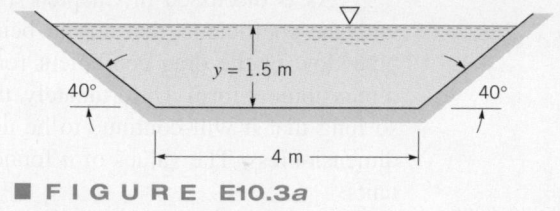

■ **FIGURE E10.3a**

SOLUTION

(a) From Eq. 10.20,

$$Q = \frac{1}{n} A R_h^{2/3} S_0^{1/2} \qquad (1)$$

where we have used $\kappa = 1$, since the dimensions are given in SI units. For a depth of $y = 1.5$ m, the flow area is

$$A = 4 \text{ m} (1.5 \text{ m}) + 1.5 \text{ m} \left(\frac{1.5}{\tan 40^\circ} \text{ m} \right) = 8.7 \text{ m}^2$$

so that with a wetted perimeter of $P = 4 \text{ m} + 2(1.5/\sin 40^\circ \text{ m}) = 8.6$ m, the hydraulic radius is determined to be $R_h = A/P = 1$ m. Note that even though the channel is quite wide (the free-surface width is 7 m), the hydraulic radius is only 1 m, which is less than the depth.

Thus, with $S_0 = 0.4 \text{ m}/305 \text{ m} = 0.0013$, Eq. 1 becomes

$$Q = \frac{1}{n} (8.7 \text{ m}^2)(1 \text{ m})^{2/3}(0.0013)^{1/2} = \frac{0.31}{n}$$

where Q is in m³/s.

From Table 10.1, we obtain $n = 0.012$ for the finished concrete. Thus,

$$Q = \frac{0.31}{0.012} = 25.8 \text{ m}^3/\text{s} \qquad \text{(Ans)}$$

COMMENT The corresponding average velocity, $V = Q/A$, is 3 m/s. It does not take a very steep slope ($S_0 = 0.0013$ or $\theta = \tan^{-1}(0.0013) = 0.074^\circ$) for this velocity.

By repeating the calculations for various surface types (i.e., various Manning coefficient values), the results shown in Fig. E10.3b are obtained. Note that the increased roughness causes a decrease in the flowrate. This is an indication that for the turbulent flows involved, the wall shear stress increases with surface roughness. [For water at 10 °C, the Reynolds number based on the 1 m hydraulic radius of the channel and a smooth concrete surface is Re $= R_h V/\nu = 1$ m $(3 \text{ m/s})/(1.3 \times 10^{-6} \text{ m}^2/\text{s}) = 2.31 \times 10^6$, well into the turbulent regime.]

(b) The Froude number based on the maximum depth for the flow can be determined from Fr $= V/(gy)^{1/2}$. For the finished

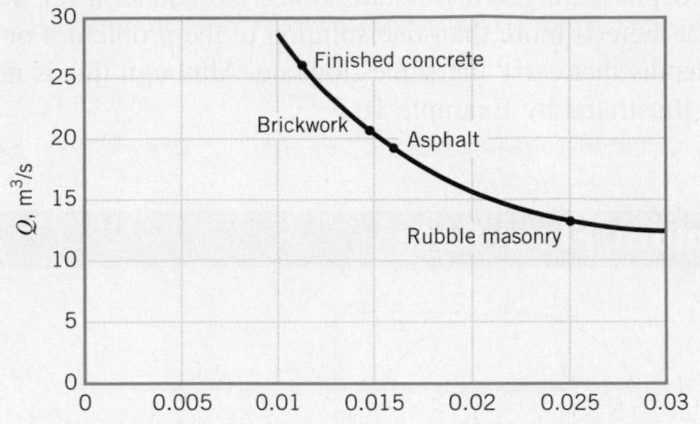

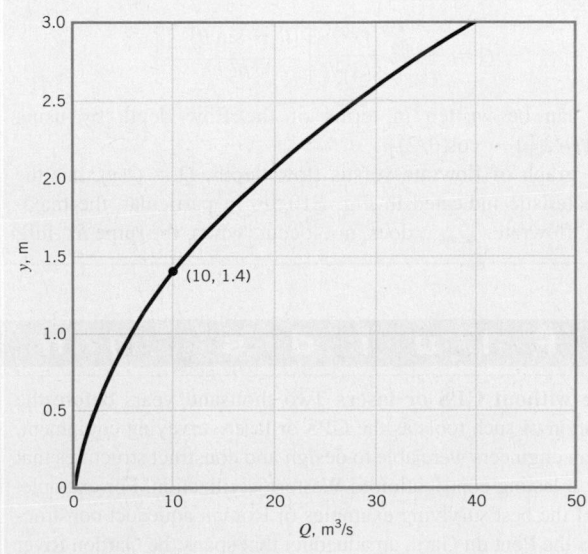

■ FIGURE E10.3b

concrete case,

$$Fr = \frac{3 \text{ m/s}}{(9.81 \text{ m/s}^2 \times 1.5 \text{ m})^{1/2}} = 0.8 \qquad \textbf{(Ans)}$$

The flow is subcritical.

In some instances a trial-and-error or iteration method must be used to solve for the dependent variable. This is often encountered when the flowrate, channel slope, and channel material are given, and the flow depth is to be determined as illustrated in the following examples.

EXAMPLE 10.4   Uniform Flow, Determine Flow Depth

GIVEN Water flows in the channel shown in Fig. E10.3a at a rate of $Q = 10.0 \text{ m}^3/\text{s}$. The canal lining is weedy.

FIND Determine the depth of the flow.

SOLUTION

In this instance neither the flow area nor the hydraulic radius are known, although they can be written in terms of the depth, y. Hence, the bottom width is 4 m and the area is

$$A = y\left(\frac{y}{\tan 40°}\right) + 4y = 1.19y^2 + 4y$$

where A and y are in square meters and meters, respectively. Also, the wetted perimeter is

$$P = 4 + 2\left(\frac{y}{\sin 40°}\right) = 3.11y + 4$$

so that

$$R_h = \frac{A}{P} = \frac{1.19y^2 + 4y}{3.11y + 4}$$

where R_h and y are in meters. Thus, with $n = 0.030$ (from Table 10.1), Eq. 10.20 can be written as

$$Q = 10 = \frac{\kappa}{n} A R_h^{2/3} S_0^{1/2}$$

$$= \frac{1.0}{0.030}(1.19y^2 + 4y)\left(\frac{1.19y^2 + 4y}{3.11y + 4}\right)^{2/3}$$

$$\times (0.0014)^{1/2}$$

which can be rearranged into the form

$$(1.19y^2 + 4y)^5 - 515(3.11y + 4)^2 = 0 \qquad \textbf{(1)}$$

where y is in meters. The solution of Eq. 1 can be easily obtained by use of a simple rootfinding numerical technique or by trial-

and-error methods. The only physically meaningful root of Eq. 1 (i.e., a positive, real number) gives the solution for the normal flow depth at this flowrate as

$$y = 1.40 \text{ m} \qquad \textbf{(Ans)}$$

COMMENT By repeating the calculations for various flowrates, the results shown in Fig. E10.4 are obtained. Note that the water depth is not linearly related to the flowrate. That is, if the flowrate is doubled, the depth is not doubled.

■ FIGURE E10.4

In Example 10.4 we found the flow depth for a given flowrate. Since the equation for this depth is a nonlinear equation, it may be that there is more than one solution to the problem. For a given channel there may be two or more depths that carry the same flowrate. Although this is not normally so, it can and does happen, as is illustrated by Example 10.5.

EXAMPLE 10.5 | Uniform Flow, Maximum Flow Rate

GIVEN Water flows in a round pipe of diameter D at a depth of $0 \le y \le D$, as is shown in Fig. E10.5a. The pipe is laid on a constant slope of S_0, and the Manning coefficient is n.

FIND (a) At what depth does the maximum flowrate occur?

(b) Show that for certain flowrates there are two depths possible with the same flowrate. Explain this behavior.

SOLUTION

(a) According to the Manning equation (Eq. 10.20) the flowrate is

$$Q = \frac{\kappa}{n} A R_h^{2/3} S_0^{1/2} \tag{1}$$

where S_0, n, and κ are constants for this problem. From geometry it can be shown that

$$A = \frac{D^2}{8} (\theta - \sin \theta)$$

where θ, the angle indicated in Fig. E10.5a, is in radians. Similarly, the wetted perimeter is

$$P = \frac{D\theta}{2}$$

so that the hydraulic radius is

$$R_h = \frac{A}{P} = \frac{D(\theta - \sin \theta)}{4\theta}$$

Therefore, Eq. 1 becomes

$$Q = \frac{\kappa}{n} S_0^{1/2} \frac{D^{8/3}}{8(4)^{2/3}} \left[\frac{(\theta - \sin \theta)^{5/3}}{\theta^{2/3}} \right]$$

This can be written in terms of the flow depth by using $y = (D/2)[1 - \cos(\theta/2)]$.

A graph of flowrate versus flow depth, $Q = Q(y)$, has the characteristic indicated in Fig. E10.5b. In particular, the maximum flowrate, Q_{max}, does not occur when the pipe is full;

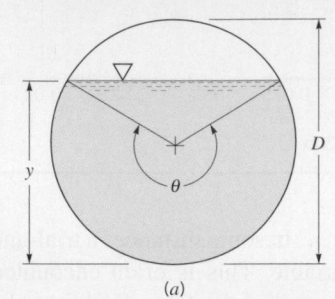

(a)

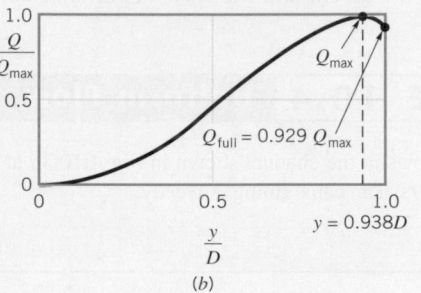

(b)

■ **FIGURE E10.5**

$Q_{full} = 0.929 Q_{max}$. It occurs when $y = 0.938D$, or $\theta = 5.28$ rad $= 303°$. Thus,

$$Q = Q_{max} \text{ when } y = 0.938D \tag{Ans}$$

(b) For any $0.929 < Q/Q_{max} < 1$ there are two possible depths that give the same Q. The reason for this behavior can be seen by considering the gain in flow area, A, compared to the increase in wetted perimeter, P, for $y \approx D$. The flow area increase for an increase in y is very slight in this region, whereas the increase in wetted perimeter, and hence the increase in shear force holding back the fluid, is relatively large. The net result is a decrease in flowrate as the depth increases.

COMMENT For most practical problems, the slight difference between the maximum flowrate and full pipe flowrates is negligible, particularly in light of the usual inaccuracy of the value of n.

F l u i d s   i n   t h e   N e w s

Done without GPS or lasers Two thousand years before the invention of such tools as the GPS or laser surveying equipment, Roman engineers were able to design and construct structures that made a lasting contribution to Western civilization. For example, one of the best surviving examples of Roman aqueduct construction is the Pont du Gard, an aqueduct that spans the Gardon River near Nîmes, France. This aqueduct is part of a circuitous, 50 km long open channel that transported water to Rome from a spring located 20 km from Rome. The spring is only 14.6 m above the point of delivery, giving an average *bottom slope* of only 3×10^{-4}. It is obvious that to carry out such a project, the Roman understanding of hydraulics, surveying, and construction was well advanced. (See Problem 10.59.)

For many open channels, the surface roughness varies across the channel.

In many man-made channels and in most natural channels, the surface roughness (and hence the Manning coefficient) varies along the wetted perimeter of the channel. A drainage ditch, for example, may have a rocky bottom surface with concrete side walls to prevent erosion. Thus, the effective n will be different for shallow depths than for deep depths of flow. Similarly, a river channel may have one value of n appropriate for its normal channel and another very different value of n during its flood stage when a portion of the flow occurs across fields or through floodplain woods. An ice-covered channel usually has a different value of n for the ice than for the remainder of the wetted perimeter (Ref. 7). (Strictly speaking, such ice-covered channels are not "open" channels, although analysis of their flow is often based on open-channel flow equations. This is acceptable, since the ice cover is often thin enough so that it represents a fixed boundary in terms of the shear stress resistance, but it cannot support a significant pressure differential as in pipe flow situations.)

A variety of methods has been used to determine an appropriate value of the effective roughness of channels that contain subsections with different values of n. Which method gives the most accurate, easy-to-use results is not firmly established, since the results are nearly the same for each method (Ref. 5). A reasonable approximation is to divide the channel cross section into N subsections, each with its own wetted perimeter, P_i, area, A_i, and Manning coefficient, n_i. The P_i values do not include the imaginary boundaries between the different subsections. The total flowrate is assumed to be the sum of the flowrates through each section. This technique is illustrated by Example 10.7.

---

*E*XAMPLE 10.6   **Uniform Flow, Variable Roughness**

GIVEN   Water flows along the drainage canal having the properties shown in Fig. E10.6a. The bottom slope is $S_0 = 0.3 \text{ m}/150 \text{ m} = 0.002$.

FIND   Estimate the flowrate when the depth is $y = 0.24 \text{ m} + 0.18 \text{ m} = 0.42 \text{ m}$.

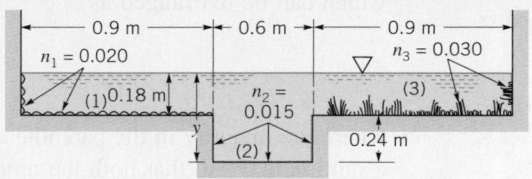

■ **FIGURE   E10.6a**

SOLUTION

We divide the cross section into three subsections as is indicated in Fig. E10.6a and write the flowrate as $Q = Q_1 + Q_2 + Q_3$, where for each section

$$Q_i = \frac{1}{n_i} A_i R_{h_i}^{2/3} S_0^{1/2}$$

The appropriate values of A_i, P_i, R_{hi}, and n_i are listed in Table E10.6. Note that the imaginary portions of the perimeters between sections (denoted by the vertical dashed lines in Fig. E10.6a) are not included in the P_i. That is, for section (2)

$$A_2 = 0.6 \text{ m} (0.24 + 0.18) \text{ m} = 0.25 \text{ m}^2$$

and

$$P_2 = 0.6 \text{ m} + 2(0.24 \text{ m}) = 1 \text{ m}$$

so that

$$R_{h_2} = \frac{A_2}{P_2} = \frac{0.25 \text{ m}^2}{1 \text{ m}} = 0.25 \text{ m}$$

Thus, the total flowrate is

$$Q = Q_1 + Q_2 + Q_3 = 1(0.002)^{1/2}$$
$$\times \left[\frac{(0.16 \text{ m}^2)(0.16 \text{ m})^{2/3}}{0.020} + \frac{(0.25 \text{ m}^2)(0.25 \text{ m})^{2/3}}{0.015} \right.$$
$$\left. + \frac{(0.16 \text{ m}^2)(0.16 \text{ m})^{2/3}}{0.030} \right]$$

or

$$Q = 0.47 \text{ m}^3/\text{s} \qquad \text{(Ans)}$$

COMMENTS   If the entire channel cross section were considered as one flow area, then $A = A_1 + A_2 + A_3 = 0.6 \text{ m}^2$ and $P = P_1 + P_2 + P_3 = 3 \text{ m}$, or $R_h = A/P = 0.6 \text{ m}^2/3 \text{ m} = 0.2 \text{ m}$. The flowrate is given by Eq. 10.20, which can be written as

$$Q = \frac{1}{n_{\text{eff}}} A R_h^{2/3} S_0^{1/2}$$

■ **TABLE E10.6**

i	A_i (m²)	P_i (m)	R_{hi} (m)	n_i
1	0.16	1	0.16	0.020
2	0.25	1	0.25	0.015
3	0.16	1	0.16	0.030

where n_{eff} is the effective value of n for this channel. With $Q = 0.47 \text{ m}^3/\text{s}$ as determined above, the value of n_{eff} is found to be

$$n_{eff} = \frac{1 A R_h^{2/3} S_0^{1/2}}{Q}$$

$$= \frac{1(0.6)(0.2)^{2/3}(0.002)^{1/2}}{0.47} = 0.019$$

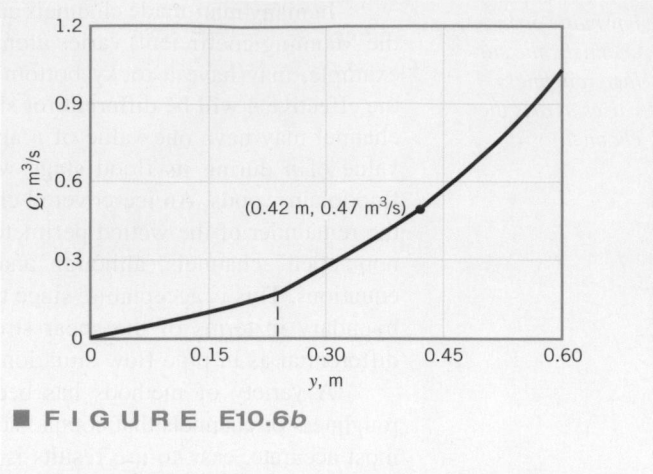

As expected, the effective roughness (Manning n) is between the minimum ($n_2 = 0.015$) and maximum ($n_3 = 0.030$) values for the individual subsections.

By repeating the calculations for various depths, y, the results shown in Fig. E10.6b are obtained. Note that there are two distinct portions of the graph—one when the water is contained entirely within the main, center channel ($y < 0.24$ m); the other when the water overflows into the side portions of the channel ($y > 0.24$ m).

■ **FIGURE E10.6***b*

One type of problem often encountered in open-channel flows is that of determining the *best hydraulic cross section* defined as the section of the minimum area for a given flowrate, Q, slope, S_0, and roughness coefficient, n. By using $R_h = A/P$ we can write Eq. 10.20 as

For a given flow-rate, the channel of minimum area is denoted as the best hydraulic cross section.

$$Q = \frac{\kappa}{n} A \left(\frac{A}{P}\right)^{2/3} S_0^{1/2} = \frac{\kappa}{n} \frac{A^{5/3} S_0^{1/2}}{P^{2/3}}$$

which can be rearranged as

$$A = \left(\frac{nQ}{\kappa S_0^{1/2}}\right)^{3/5} P^{2/5}$$

where the quantity in the parentheses is a constant. Thus, a channel with minimum A is one with a minimum P, so that both the amount of excavation needed and the amount of material to line the surface are minimized by the best hydraulic cross section.

The best hydraulic cross section possible is that of a semicircular channel. No other shape has as small a wetted perimeter for a given area. It is often desired to determine the best shape for a class of cross sections. The results (given here without proof) for rectangular, trapezoidal (with 60° sides), and triangular shapes are shown in Fig. 10.11. For example, the best hydraulic cross section for a rectangle is one whose depth is half its width; for a triangle it is a 90° triangle.

10.5 Gradually Varied Flow

In many situations the flow in an open channel is not of uniform depth ($y =$ constant) along the channel. This can occur because of several reasons: The bottom slope is not constant, the cross-sectional shape and area vary in the flow direction, or there is some obstruction across a portion of the channel. Such flows are classified as gradually varying flows if $dy/dx \ll 1$.

If the bottom slope and the energy line slope are not equal, the flow depth will vary along the channel, either increasing or decreasing in the flow direction. In such cases $dy/dx \neq 0$, $dV/dx \neq 0$, and the right-hand side of Eq. 10.10 is not zero. Physically, the difference between the

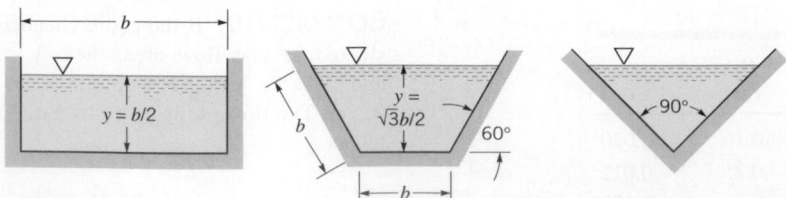

■ **FIGURE 10.11** Best hydraulic cross sections for a rectangle, a 60° trapezoid, and a triangle.

component of weight and the shear forces in the direction of flow produces a change in the fluid momentum that requires a change in velocity and, from continuity considerations, a change in depth. Whether the depth increases or decreases depends on various parameters of the flow, with a variety of surface profile configurations [flow depth as a function of distance, $y = y(x)$] possible (Refs. 5, 9).

10.6   Rapidly Varied Flow

In many open channels, flow depth changes occur over a relatively short distance so that $dy/dx \sim 1$. Such *rapidly varied flow* conditions are often quite complex and difficult to analyze in a precise fashion. Fortunately, many useful approximate results can be obtained by using a simple one-dimensional model along with appropriate experimentally determined coefficients when necessary. In this section we discuss several of these flows.

Some rapidly varied flows occur in constant area channels for reasons that are not immediately obvious. The hydraulic jump is one such case. As is indicated in Fig. 10.12, the flow may change from a relatively shallow, high-speed condition into a relatively deep, low-speed condition within a horizontal distance of just a few channel depths. Other rapidly varied flows may be due to a sudden change in the channel geometry such as the flow in an expansion or contraction section of a channel as is indicated in Fig. 10.13.

In such situations the flow field is often two- or three-dimensional in character. There may be regions of flow separation, flow reversal, or unsteady oscillations of the free surface. For the purpose of some analyses, these complexities can be neglected and a simplified analysis can be undertaken. In other cases, however, it is the complex details of the flow that are the most important property of the flow; any analysis must include their effects. The scouring of a river bottom in the neighborhood of a bridge pier, as is indicated in Fig. 10.14, is such an example. A one- or two-dimensional model of this flow would not be sufficient to describe the complex structure of the flow that is responsible for the erosion near the foot of the bridge pier.

Many open-channel flow-measuring devices are based on principles associated with rapidly varied flows. Among these devices are broad-crested weirs, sharp-crested weirs, critical flow flumes, and sluice gates. The operation of such devices is discussed in the following sections.

In many cases the flow depth may change significantly in a short distance.

V10.8 Erosion in a channel

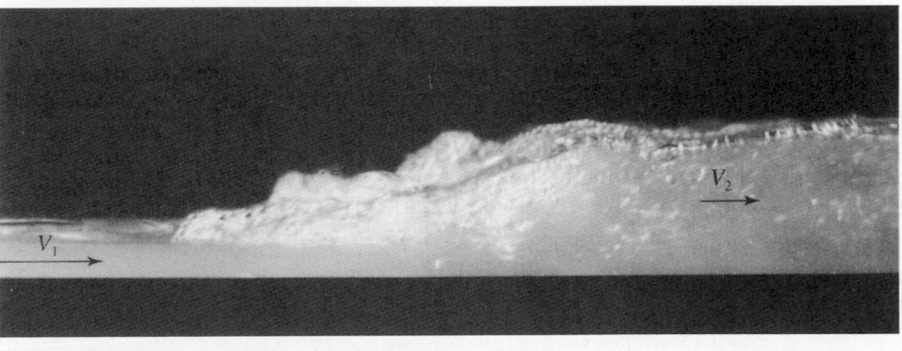

■ **FIGURE 10.12**   **Hydraulic jump.**

V10.9 Bridge pier scouring

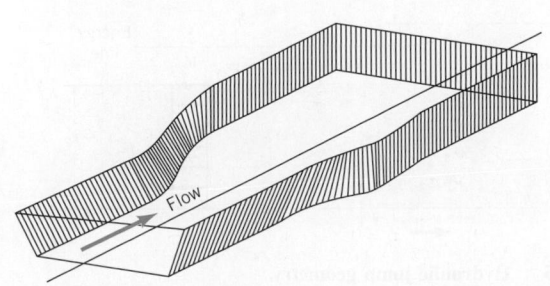

■ **FIGURE 10.13**   **Rapidly varied flow may occur in a channel transition section.**

V10.10 Big Sioux River bridge collapse

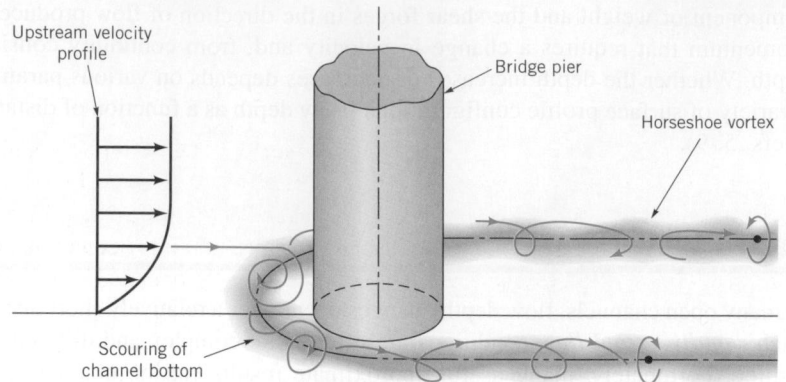

Upstream velocity profile

Bridge pier

Horseshoe vortex

Scouring of channel bottom

■ **F I G U R E 10.14**　The complex three-dimensional flow structure around a bridge pier.

10.6.1 The Hydraulic Jump

Observations of flows in open channels show that under certain conditions it is possible that the fluid depth will change very rapidly over a short length of the channel without any change in the channel configuration. Such changes in depth can be approximated as a discontinuity in the free-surface elevation ($dy/dx = \infty$). For reasons discussed below, this step change in depth is always from a shallow to a deeper depth—always a step up, never a step down.

A hydraulic jump is a steplike increase in fluid depth in an open channel.

Physically, this near discontinuity, called a *hydraulic jump*, may result when there is a conflict between the upstream and downstream influences that control a particular section (or reach) of a channel. For example, a sluice gate may require that the conditions at the upstream portion of the channel (downstream of the gate) be supercritical flow, while obstructions in the channel on the downstream end of the reach may require that the flow be subcritical. The hydraulic jump provides the mechanism (a nearly discontinuous one at that) to make the transition between the two types of flow.

The simplest type of hydraulic jump occurs in a horizontal, rectangular channel as is indicated in Fig. 10.15. Although the flow within the jump itself is extremely complex and agitated, it is reasonable to assume that the flow at sections (1) and (2) is nearly uniform, steady, and one-dimensional. In addition, we neglect any wall shear stresses, τ_w, within the relatively short segment between these two sections. Under these conditions the x component of the momentum equation (Eq. 5.22) for the control volume indicated can be written as

$$F_1 - F_2 = \rho Q(V_2 - V_1) = \rho V_1 y_1 b(V_2 - V_1)$$

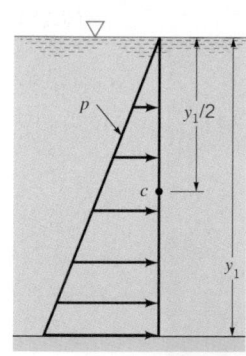

where, as indicated by the figure in the margin, the pressure force at either section is hydrostatic. That is, $F_1 = p_{c1}A_1 = \gamma y_1^2 b/2$ and $F_2 = p_{c2}A_2 = \gamma y_2^2 b/2$, where $p_{c1} = \gamma y_1/2$ and $p_{c2} = \gamma y_2/2$ are the pressures at the centroids of the channel cross sections and b is the channel width. Thus, the momentum equation becomes

$$\frac{y_1^2}{2} - \frac{y_2^2}{2} = \frac{V_1 y_1}{g}(V_2 - V_1) \tag{10.21}$$

In addition to the momentum equation, we have the conservation of mass equation (Eq. 5.12)

$$y_1 b V_1 = y_2 b V_2 = Q \tag{10.22}$$

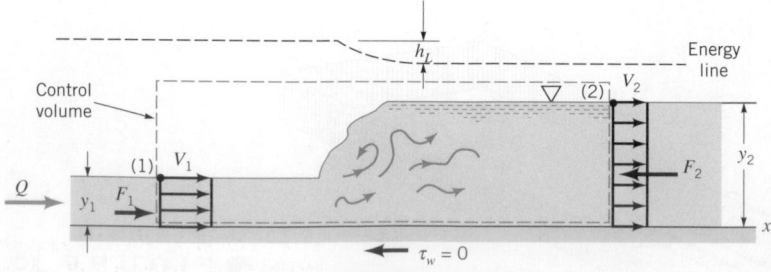

Energy line

Control volume

(2) V_2

(1) V_1

Q

y_1 F_1

h_L

F_2

y_2

x

$\tau_w = 0$

■ **F I G U R E 10.15**　Hydraulic jump geometry.

V10.11 Hydraulic jump in a river

and the energy equation (Eq. 5.84)

$$y_1 + \frac{V_1^2}{2g} = y_2 + \frac{V_2^2}{2g} + h_L \tag{10.23}$$

The head loss, h_L, in Eq. 10.23 is due to the violent turbulent mixing and dissipation that occur within the jump itself. We have neglected any head loss due to wall shear stresses.

Clearly Eqs. 10.21, 10.22, and 10.23 have a solution $y_1 = y_2$, $V_1 = V_2$, and $h_L = 0$. This represents the trivial case of no jump. Since these are nonlinear equations, it may be possible that more than one solution exists. The other solutions can be obtained as follows. By combining Eqs. 10.21 and 10.22 to eliminate V_2 we obtain

$$\frac{y_1^2}{2} - \frac{y_2^2}{2} = \frac{V_1 y_1}{g}\left(\frac{V_1 y_1}{y_2} - V_1\right) = \frac{V_1^2 y_1}{g y_2}(y_1 - y_2)$$

which can be simplified by factoring out a common nonzero factor $y_1 - y_2$ from each side to give

$$\left(\frac{y_2}{y_1}\right)^2 + \left(\frac{y_2}{y_1}\right) - 2\,\mathrm{Fr}_1^2 = 0$$

where $\mathrm{Fr}_1 = V_1/\sqrt{gy_1}$ is the upstream Froude number. By using the quadratic formula we obtain

$$\frac{y_2}{y_1} = \frac{1}{2}(-1 \pm \sqrt{1 + 8\mathrm{Fr}_1^2})$$

The depth ratio across a hydraulic jump depends on the Froude number only.

Clearly the solution with the minus sign is not possible (it would give a negative y_2/y_1). Thus,

$$\frac{y_2}{y_1} = \frac{1}{2}(-1 + \sqrt{1 + 8\mathrm{Fr}_1^2}) \tag{10.24}$$

This depth ratio, y_2/y_1, across the hydraulic jump is shown as a function of the upstream Froude number in Fig. 10.16. The portion of the curve for $\mathrm{Fr}_1 < 1$ is dashed in recognition of the fact that to have a hydraulic jump the flow must be supercritical. That is, the solution as given by Eq. 10.24 must be restricted to $\mathrm{Fr}_1 \geq 1$, for which $y_2/y_1 \geq 1$. This can be shown by consideration of the energy equation, Eq. 10.23, as follows. The dimensionless head loss, h_L/y_1, can be obtained from Eq. 10.23 as

$$\frac{h_L}{y_1} = 1 - \frac{y_2}{y_1} + \frac{\mathrm{Fr}_1^2}{2}\left[1 - \left(\frac{y_1}{y_2}\right)^2\right] \tag{10.25}$$

where, for given values of Fr_1, the values of y_2/y_1 are obtained from Eq. 10.24. As is indicated in Fig. 10.16, the head loss is negative if $\mathrm{Fr}_1 < 1$. Since negative head losses violate the second law

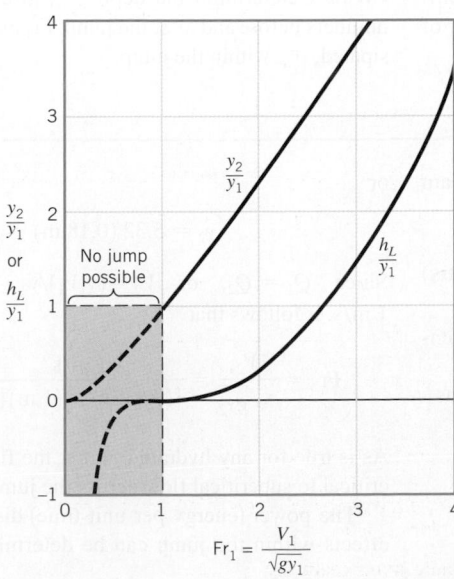

■ **F I G U R E   10.16**   Depth ratio and dimensionless head loss across a hydraulic jump as a function of upstream Froude number.

of thermodynamics (viscous effects dissipate energy, they cannot create energy; see Section 5.3), it is not possible to produce a hydraulic jump with $Fr_1 < 1$. The head loss across the jump is indicated by the lowering of the energy line shown in Fig. 10.15.

A flow must be supercritical (Froude number > 1) to produce the discontinuity called a hydraulic jump. This is analogous to the compressible flow ideas discussed in Chapter 11 in which it is shown that the flow of a gas must be supersonic (Mach number > 1) to produce the discontinuity called a normal shock wave. However, the fact that a flow is supercritical (or supersonic) does not guarantee the production of a hydraulic jump (or shock wave). The trivial solution $y_1 = y_2$ and $V_1 = V_2$ is also possible.

The fact that there is an energy loss across a hydraulic jump is useful in many situations. For example, the relatively large amount of energy contained in the fluid flowing down the spillway of a dam like that shown in the figure in the margin could cause damage to the channel below the dam. By placing suitable flow control objects in the channel downstream of the spillway, it is possible (if the flow is supercritical) to produce a hydraulic jump on the apron of the spillway and thereby dissipate a considerable portion of the energy of the flow. That is, the dam spillway produces supercritical flow, and the channel downstream of the dam requires subcritical flow. The resulting hydraulic jump provides the means to change the character of the flow.

(Photograph courtesy of U.S. Army Corps of Engineers.)

Hydraulic jumps dissipate energy.

F l u i d s   i n   t h e   N e w s

Grand Canyon rapids building Virtually all of the rapids in the Grand Canyon were formed by rock debris carried into the Colorado River from side canyons. Severe storms wash large amounts of sediment into the river, building debris fans that narrow the river. This debris forms crude dams which back up the river to form quiet pools above the rapids. Water exiting the pool through the narrowed channel can reach supercritical conditions and produce *hydraulic jumps* downstream. Since the configuration of the jumps is a function of the flowrate, the difficulty in running the rapids can change from day to day. Also, rapids change over the years as debris is added to or removed from the rapids. For example, Crystal Rapid, one of the notorious rafting stretches of the river, changed very little between the first photos of 1890 and those of 1966. However, a debris flow from a severe winter storm in 1966 greatly constricted the river. Within a few minutes the configuration of Crystal Rapid was completely changed. The new, immature rapid was again drastically changed by a flood in 1983. While Crystal Rapid is now considered full grown, it will undoubtedly change again, perhaps in 100 or 1000 years. (See Problem 10.100.)

---

*E*XAMPLE 10.7 | Hydraulic Jump

GIVEN Water on the horizontal apron of the 30-m-wide spillway shown in Fig. E10.7a has a depth of 0.18 m and a velocity of 5.5 m/s.

FIND Determine the depth, y_2, after the jump, the Froude numbers before and after the jump, Fr_1 and Fr_2, and the power dissipated, $\mathcal{P}_d$, within the jump.

SOLUTION

Conditions across the jump are determined by the upstream Froude number

$$Fr_1 = \frac{V_1}{\sqrt{gy_1}} = \frac{5.5 \text{ m/s}}{[(9.81 \text{ m/s}^2)(0.18 \text{ m})]^{1/2}} = 4.10 \quad \text{(Ans)}$$

Thus, the upstream flow is supercritical, and it is possible to generate a hydraulic jump as sketched.

From Eq. 10.24 we obtain the depth ratio across the jump as

$$\frac{y_2}{y_1} = \frac{1}{2}\left(-1 + \sqrt{1 + 8 \, Fr_1^2}\right)$$

$$= \frac{1}{2}\left[-1 + \sqrt{1 + 8(4.10)^2}\right] = 5.32$$

or

$$y_2 = 5.32 \, (0.18 \text{ m}) = 1 \text{ m} \quad \text{(Ans)}$$

Since $Q_1 = Q_2$, or $V_2 = (y_1 V_1)/y_2 = 0.18 \text{ m} (5.5 \text{ m/s})/1 \text{ m} = 1 \text{ m/s}$, it follows that

$$Fr_2 = \frac{V_2}{\sqrt{gy_2}} = \frac{1 \text{ m/s}}{[(9.81 \text{ m/s}^2)(1 \text{ m})]^{1/2}} = 0.32 \quad \text{(Ans)}$$

As is true for any hydraulic jump, the flow changes from supercritical to subcritical flow across the jump.

The power (energy per unit time) dissipated, $\mathcal{P}_d$, by viscous effects within the jump can be determined from the head loss as (see Eq. 5.85)

$$\mathcal{P}_d = \gamma Q h_L = \gamma b y_1 V_1 h_L \qquad (1)$$

where h_L is obtained from Eqs. 10.23 or 10.25 as

$$h_L = \left(y_1 + \frac{V_1^2}{2g}\right) - \left(y_2 + \frac{V_2^2}{2g}\right) = \left[0.18\ \text{m} + \frac{(5.5\ \text{m/s})^2}{2(9.81\ \text{m/s}^2)}\right]$$
$$- \left[1\ \text{m} + \frac{(1\ \text{m/s})^2}{2(9.8\ \text{m/s}^2)}\right]$$

or

$$h_L = 0.67\ \text{m}$$

Thus, from Eq. 1,

$$\mathcal{P}_d = (9800\ \text{N/m}^3)(30\ \text{m})(0.18\ \text{m})(5.5\ \text{m/s})(0.67\ \text{m})$$
$$= 2 \times 10^5\ \text{N} \cdot \text{m/s}$$

or

$$\mathcal{P}_d = \frac{2 \times 10^5\ \text{N} \cdot \text{m/s}}{[1\ \text{N} \cdot \text{m/s/W}]} = 2 \times 10^5\ \text{W} \qquad \textbf{(Ans)}$$

COMMENTS This power, which is dissipated within the highly turbulent motion of the jump, is converted into an increase in water temperature, T. That is, $T_2 > T_1$. Although the power dissipated is considerable, the difference in temperature is not great because the flowrate is quite large.

By repeating the calculations for the given flowrate $Q_1 = A_1 V_1 = b_1 y_1 V_1 = 30\ \text{m}\ (0.18\ \text{m})(5.5\ \text{m/s}) = 30\ \text{m}^3/\text{s}$ but with various upstream depths, y_1, the results shown in Fig. E10.7b are obtained. Note that a slight change in water depth can produce a considerable change in energy dissipated. Also, if $y_1 > 0.5$ m the flow is subcritical (Fr$_1 < 1$) and no hydraulic jump can occur.

The hydraulic jump flow process can be illustrated by use of the specific energy concept introduced in Section 10.3 as follows. Equation 10.23 can be written in terms of the specific energy, $E = y + V^2/2g$, as $E_1 = E_2 + h_L$, where $E_1 = y_1 + V_1^2/2g = 1.72$ m and $E_2 = y_2 + V_2^2/2g = 1.05$ m. As is discussed in Section 10.3, the specific energy diagram for this flow can be obtained by using $V = q/y$, where

$$q = q_1 = q_2 = \frac{Q}{b} = y_1 V_1 = 0.18\ \text{m}\ (5.5\ \text{m/s})$$
$$= 1\ \text{m}^2/\text{s}$$

Thus,

$$E = y + \frac{q^2}{2gy^2} = y + \frac{(1\ \text{m}^2/\text{s})^2}{2(9.81\ \text{m/s}^2)y^2} = y + \frac{0.05}{y^2}$$

where y and E are in meters. The resulting specific energy diagram is shown in Fig. E10.7c. Because of the head loss across

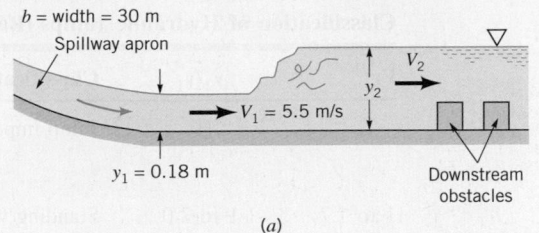

(a)

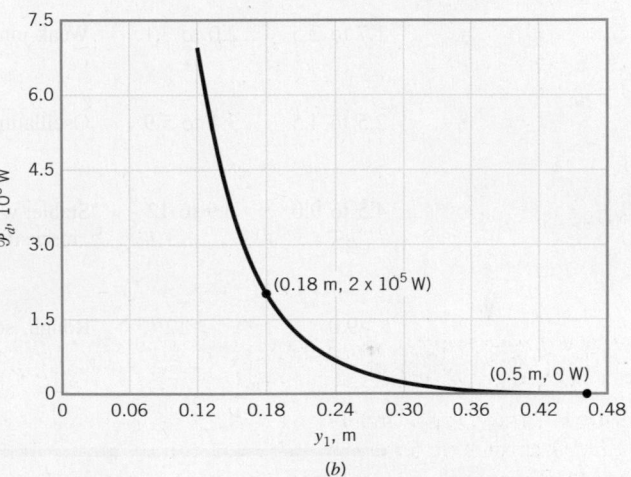

(b)

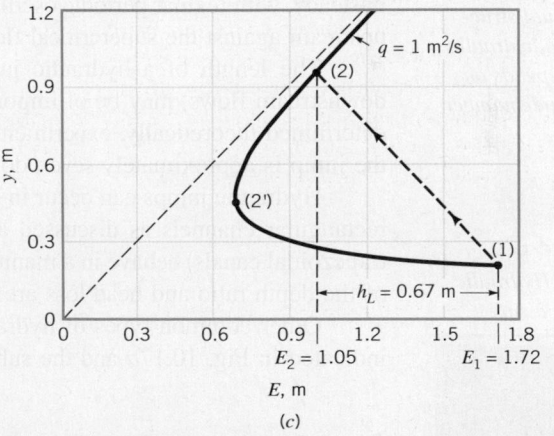

(c)

■ **FIGURE E10.7**

the jump, the upstream and downstream values of E are different. In going from state (1) to state (2) the fluid does not proceed along the specific energy curve and pass through the critical condition at state 2′. Rather, it jumps from (1) to (2) as is represented by the dashed line in the figure. From a one-dimensional consideration, the jump is a discontinuity. In actuality, the jump is a complex three-dimensional flow incapable of being represented on the one-dimensional specific energy diagram.

The actual structure of a hydraulic jump is a complex function of Fr$_1$, even though the depth ratio and head loss are given quite accurately by a simple one-dimensional flow analysis (Eqs. 10.24 and 10.25). A detailed investigation of the flow indicates that there are essentially five types of surface and jump conditions. The classification of these jumps is indicated in Table 10.2, along with sketches of the structure of the jump. For flows that are barely supercritical, the jump is more like a standing wave, without a nearly step change in depth. In some Froude number ranges the jump is

■ **TABLE 10.2**
Classification of Hydraulic Jumps (Ref. 12)

Fr_1	y_2/y_1	Classification	Sketch
<1	1	Jump impossible	
1 to 1.7	1 to 2.0	Standing wave or undulant jump	
1.7 to 2.5	2.0 to 3.1	Weak jump	
2.5 to 4.5	3.1 to 5.9	Oscillating jump	
4.5 to 9.0	5.9 to 12	Stable, well-balanced steady jump; insensitive to downstream conditions	
>9.0	>12	Rough, somewhat intermittent strong jump	

The actual structure of a hydraulic jump depends on the Froude number.

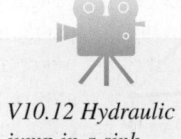

V10.12 Hydraulic jump in a sink

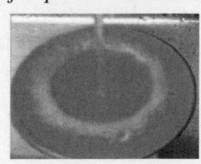

unsteady, with regular periodic oscillations traveling downstream. (Recall that the wave cannot travel upstream against the supercritical flow.)

The length of a hydraulic jump (the distance between the nearly uniform upstream and downstream flows) may be of importance in the design of channels. Although its value cannot be determined theoretically, experimental results indicate that over a wide range of Froude numbers, the jump is approximately seven downstream depths long (Ref. 5).

Hydraulic jumps can occur in a variety of channel flow configurations, not just in horizontal, rectangular channels as discussed above. Jumps in nonrectangular channels (i.e., circular pipes, trapezoidal canals) behave in a manner quite like those in rectangular channels, although the details of the depth ratio and head loss are somewhat different from jumps in rectangular channels.

Other common types of hydraulic jumps include those that occur in sloping channels as is indicated in Fig. 10.17a and the submerged hydraulic jumps that can occur just downstream of a

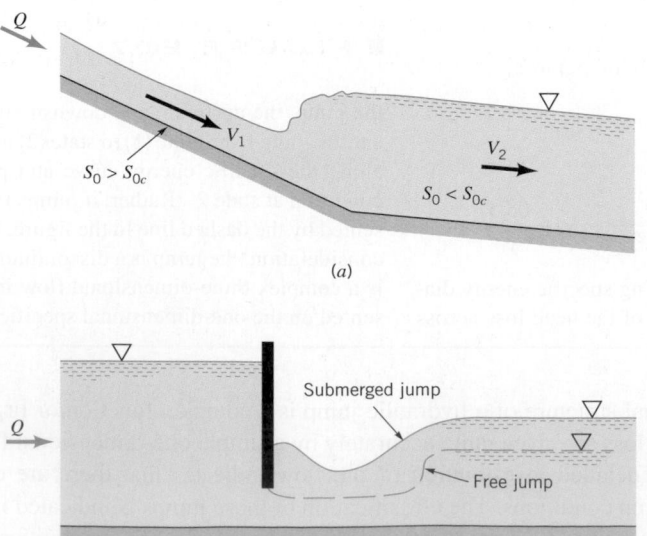

■ **FIGURE 10.17**
Hydraulic jump variations: (a) jump caused by a change in channel slope, (b) submerged jump.

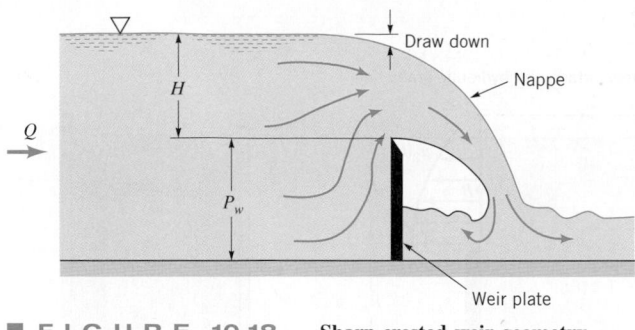

■ FIGURE 10.18 Sharp-crested weir geometry.

sluice gate as is indicated in Fig. 10.17b. Details of these and other jumps can be found in standard open-channel flow references (Refs. 3 and 5).

10.6.2 Sharp-Crested Weirs

A weir is an obstruction on a channel bottom over which the fluid must flow. It provides a convenient method of determining the flowrate in an open channel in terms of a single depth measurement. A *sharp-crested weir* is essentially a vertical sharp-edged flat plate placed across the channel in a way such that the fluid must flow across the sharp edge and drop into the pool downstream of the weir plate, as is shown in Fig. 10.18. The specific shape of the flow area in the plane of the weir plate is used to designate the type of weir. Typical shapes include the rectangular weir, the triangular weir, and the trapezoidal weir, as indicated in Fig. 10.19.

A sharp-crested weir can be used to determine the flowrate.

The complex nature of the flow over a weir makes it impossible to obtain precise analytical expressions for the flow as a function of other parameters, such as the weir height, P_w, *weir head,* H, the fluid depth upstream, and the geometry of the weir plate (angle θ for triangular weirs or aspect ratio, b/H, for rectangular weirs). The flow structure is far from one-dimensional, with a variety of interesting flow phenomena obtained.

The main mechanisms governing flow over a weir are gravity and inertia. From a highly simplified point of view, gravity accelerates the fluid from its free-surface elevation upstream of the weir to larger velocity as it flows down the hill formed by the nappe. Although viscous and surface tension effects are usually of secondary importance, such effects cannot be entirely neglected. Generally, appropriate experimentally determined coefficients are used to account for these effects.

As a first approximation, we assume that the velocity profile upstream of the weir plate is uniform and that the pressure within the nappe is atmospheric. In addition, we assume that the fluid flows horizontally over the weir plate with a nonuniform velocity profile, as indicated in Fig. 10.20. With $p_B = 0$ the Bernoulli equation for flow along the arbitrary streamline $A-B$ indicated can be written as

$$\frac{p_A}{\gamma} + \frac{V_1^2}{2g} + z_A = (H + P_w - h) + \frac{u_2^2}{2g} \tag{10.26}$$

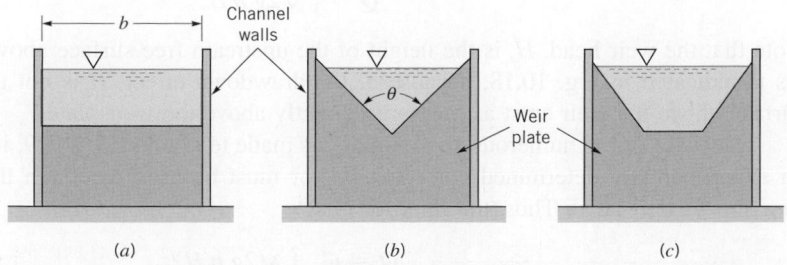

■ FIGURE 10.19 Sharp-crested weir plate geometry: (a) rectangular, (b) triangular, (c) trapezoidal.

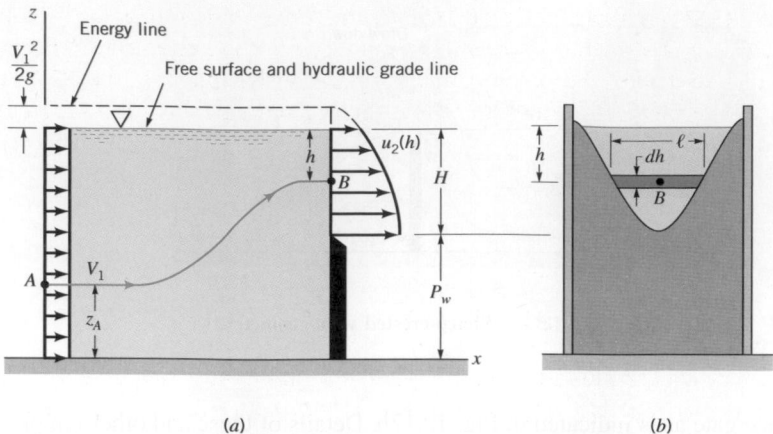

(a)   (b)

■ **FIGURE 10.20**   **Assumed flow structure over a weir.**

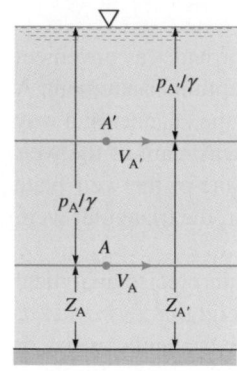

where h is the distance that point B is below the free surface. We do not know the location of point A from which came the fluid that passes over the weir at point B. However, since the total head for any particle along the vertical section (1) is the same, $z_A + p_A/\gamma + V_1^2/2g = H + P_w + V_1^2/2g$, the specific location of A (i.e., A or A' shown in the figure in the margin) is not needed, and the velocity of the fluid over the weir plate is obtained from Eq. 10.26 as

$$u_2 = \sqrt{2g\left(h + \frac{V_1^2}{2g}\right)}$$

The flowrate can be calculated from

$$Q = \int_{(2)} u_2 \, dA = \int_{h=0}^{h=H} u_2 \ell \, dh \tag{10.27}$$

where $\ell = \ell(h)$ is the cross-channel width of a strip of the weir area, as is indicated in Fig. 10.20b. For a rectangular weir ℓ is constant. For other weirs, such as triangular or circular weirs, the value of ℓ is known as a function of h.

For a rectangular weir, $\ell = b$, and the flowrate becomes

$$Q = \sqrt{2g}\, b \int_0^H \left(h + \frac{V_1^2}{2g}\right)^{1/2} dh$$

or

$$Q = \frac{2}{3}\sqrt{2g}\, b \left[\left(H + \frac{V_1^2}{2g}\right)^{3/2} - \left(\frac{V_1^2}{2g}\right)^{3/2}\right] \tag{10.28}$$

Equation 10.28 is a rather cumbersome expression that can be simplified by using the fact that with $P_w \gg H$ (as often happens in practical situations) the upstream velocity is negligibly small. That is, $V_1^2/2g \ll H$ and Eq. 10.28 simplifies to the basic rectangular weir equation

$$Q = \tfrac{2}{3}\sqrt{2g}\, b\, H^{3/2} \tag{10.29}$$

Note that the weir head, H, is the height of the upstream free surface above the crest of the weir. As is indicated in Fig. 10.18, because of the drawdown effect, H is not the distance of the free surface above the weir crest as measured directly above the weir plate.

A weir coefficient is used to account for nonideal conditions excluded in the simplified analysis.

Because of the numerous approximations made to obtain Eq. 10.29, it is not unexpected that an experimentally determined correction factor must be used to obtain the actual flowrate as a function of weir head. Thus, the final form is

$$Q = C_{wr}\tfrac{2}{3}\sqrt{2g}\, b\, H^{3/2} \tag{10.30}$$

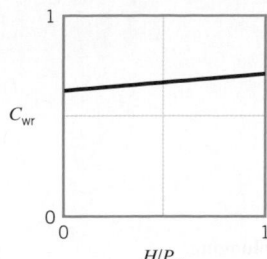

where C_{wr} is the rectangular weir coefficient. From dimensional analysis arguments, it is expected that C_{wr} is a function of Reynolds number (viscous effects), Weber number (surface tension effects), and H/P_w (geometry). In most practical situations, the Reynolds and Weber number effects are negligible, and the following correlation, shown in the figure in the margin, can be used (Refs. 4, 7):

$$C_{wr} = 0.611 + 0.075\left(\frac{H}{P_w}\right) \tag{10.31}$$

More precise values of C_{wr} can be found in the literature, if needed (Refs. 3, 14).

The triangular sharp-crested weir is often used for flow measurements, particularly for measuring flowrates over a wide range of values. For small flowrates, the head, H, for a rectangular weir would be very small and the flowrate could not be measured accurately. However, with the triangular weir, the flow width decreases as H decreases so that even for small flowrates, reasonable heads are developed. Accurate results can be obtained over a wide range of Q.

The triangular weir equation can be obtained from Eq. 10.27 by using

$$\ell = 2(H - h)\tan\left(\frac{\theta}{2}\right)$$

V10.13 Triangular weir

where θ is the angle of the V-notch (see Figs. 10.19 and 10.20). After carrying out the integration and again neglecting the upstream velocity ($V_1^2/2g \ll H$), we obtain

$$Q = \frac{8}{15}\tan\left(\frac{\theta}{2}\right)\sqrt{2g}\,H^{5/2}$$

An experimentally determined triangular weir coefficient, C_{wt}, is used to account for the real-world effects neglected in the analysis so that

$$Q = C_{wt}\frac{8}{15}\tan\left(\frac{\theta}{2}\right)\sqrt{2g}\,H^{5/2} \tag{10.32}$$

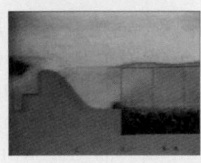

V10.14 Low-head dam

Typical values of C_{wt} for triangular weirs are in the range of 0.58 to 0.62, as is shown in Fig. 10.21. Note that although C_{wt} and θ are dimensionless, the value of C_{wt} is given as a function of the weir head, H, which is a dimensional quantity. Although using dimensional parameters is not recommended (see the dimensional analysis discussion in Chapter 7), such parameters are often used for open-channel flow.

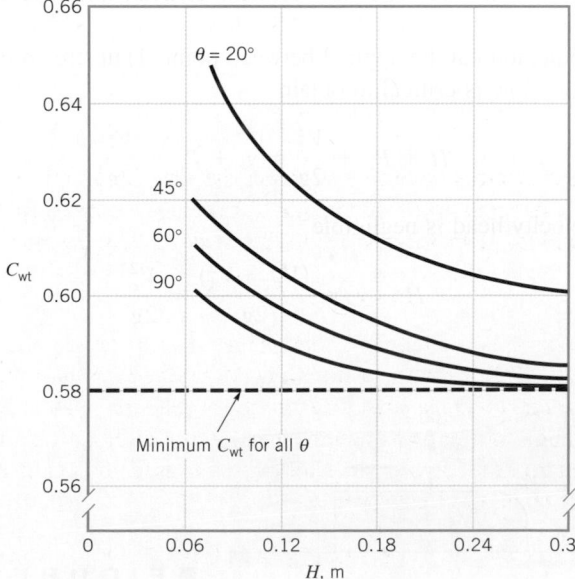

■ **FIGURE 10.21** Weir coefficient for triangular sharp-crested weirs (adapted from Ref. 10).

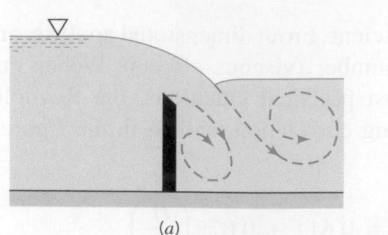 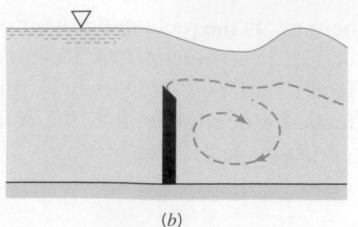

■ **F I G U R E 10.22** Flow conditions over a weir without a free nappe: (*a*) plunging nappe, (*b*) submerged nappe.

Flowrate over a weir depends on whether the nappe is free or submerged.

The above results for sharp-crested weirs are valid provided the area under the nappe is ventilated to atmospheric pressure. Although this is not a problem for triangular weirs, for rectangular weirs it is sometimes necessary to provide ventilation tubes to ensure atmospheric pressure in this region. In addition, depending on downstream conditions, it is possible to obtain submerged weir operation, as is indicated in Fig. 10.22. Clearly the flowrate will be different for these situations than that given by Eqs. 10.30 and 10.32.

10.6.3 Broad-Crested Weirs

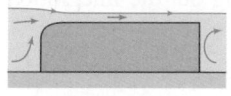

$H/L_w = 0.08$

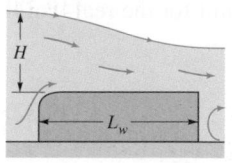

$H/L_w = 0.50$

A *broad-crested weir* is a structure in an open channel that has a horizontal crest above which the fluid pressure may be considered hydrostatic. A typical configuration is shown in Fig. 10.23. Generally, to ensure proper operation, these weirs are restricted to the range $0.08 < H/L_w < 0.50$. These conditions are drawn to scale in the figure in the margin. For long weir blocks (H/L_w less than 0.08), head losses across the weir cannot be neglected. On the other hand, for short weir blocks (H/L_w greater than 0.50) the streamlines of the flow over the weir block are not horizontal. Although broad-crested weirs can be used in channels of any cross-sectional shape, we restrict our attention to rectangular channels.

The operation of a broad-crested weir is based on the fact that nearly uniform critical flow is achieved in the short reach above the weir block. (If $H/L_w < 0.08$, viscous effects are important, and the flow is subcritical over the weir.) If the kinetic energy of the upstream flow is negligible, then $V_1^2/2g \ll y_1$ and the upstream specific energy is $E_1 = V_1^2/2g + y_1 \approx y_1$. Observations show that as the flow passes over the weir block, it accelerates and reaches critical conditions, $y_2 = y_c$ and $\mathrm{Fr}_2 = 1$ (i.e., $V_2 = c_2$), corresponding to the nose of the specific energy curve (see Fig. 10.7). The flow does not accelerate to supercritical conditions ($\mathrm{Fr}_2 > 1$). To do so would require the ability of the downstream fluid to communicate with the upstream fluid to let it know that there is an end of the weir block. Since waves cannot propagate upstream against a critical flow, this information cannot be transmitted. The flow remains critical, not supercritical, across the weir block.

The Bernoulli equation can be applied between point (1) upstream of the weir and point (2) over the weir where the flow is critical to obtain

$$H + P_w + \frac{V_1^2}{2g} = y_c + P_w + \frac{V_c^2}{2g}$$

or, if the upstream velocity head is negligible

$$H - y_c = \frac{(V_c^2 - V_1^2)}{2g} = \frac{V_c^2}{2g}$$

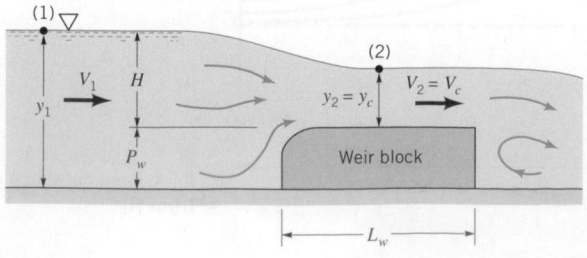

■ **F I G U R E 10.23** Broad-crested weir geometry.

However, since $V_2 = V_c = (gy_c)^{1/2}$, we find that $V_c^2 = gy_c$ so that we obtain

$$H - y_c = \frac{y_c}{2}$$

The broad-crested weir is governed by critical flow across the weir block.

or

$$y_c = \frac{2H}{3}$$

Thus, the flowrate is

$$Q = by_2V_2 = by_cV_c = by_c(gy_c)^{1/2} = b\sqrt{g}\, y_c^{3/2}$$

or

$$Q = b\sqrt{g}\left(\frac{2}{3}\right)^{3/2} H^{3/2}$$

Again an empirical weir coefficient is used to account for the various real-world effects not included in the above simplified analysis. That is

$$Q = C_{wb}\, b\sqrt{g}\left(\frac{2}{3}\right)^{3/2} H^{3/2} \tag{10.33}$$

where approximate values of C_{wb}, the broad-crested weir coefficient shown in the figure in the margin, can be obtained from the equation (Ref. 6)

$$C_{wb} = 1.125\left(\frac{1 + H/P_w}{2 + H/P_w}\right)^{1/2} \tag{10.34}$$

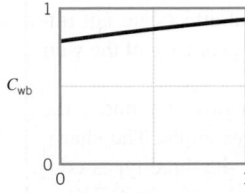

C_{wb} vs H/P_w

EXAMPLE 10.8   Sharp-Crested and Broad-Crested Weirs

GIVEN Water flows in a rectangular channel of width $b = 2$ m with flowrates between $Q_{min} = 0.02$ m^3/s and $Q_{max} = 0.60$ m^3/s. This flowrate is to be measured by using either (a) a rectangular sharp-crested weir, (b) a triangular sharp-crested weir with $\theta = 90°$, or (c) a broad-crested weir. In all cases the bottom of the flow area over the weir is a distance $P_w = 1$ m above the channel bottom.

FIND Plot a graph of $Q = Q(H)$ for each weir and comment on which weir would be best for this application.

SOLUTION

(a) For the rectangular weir with $P_w = 1$ m, Eqs. 10.30 and 10.31 give

$$Q = C_{wr}\frac{2}{3}\sqrt{2g}\, bH^{3/2}$$

$$= \left(0.611 + 0.075\frac{H}{P_w}\right)\frac{2}{3}\sqrt{2g}\, bH^{3/2}$$

Thus,

$$Q = (0.611 + 0.075H)\frac{2}{3}\sqrt{2(9.81 \text{ m/s}^2)}\,(2 \text{ m})\, H^{3/2}$$

or

$$Q = 5.91(0.611 + 0.075H)H^{3/2} \tag{1}$$

where H and Q are in meters and m^3/s, respectively. The results from Eq. 1 are plotted in Fig. E10.8.

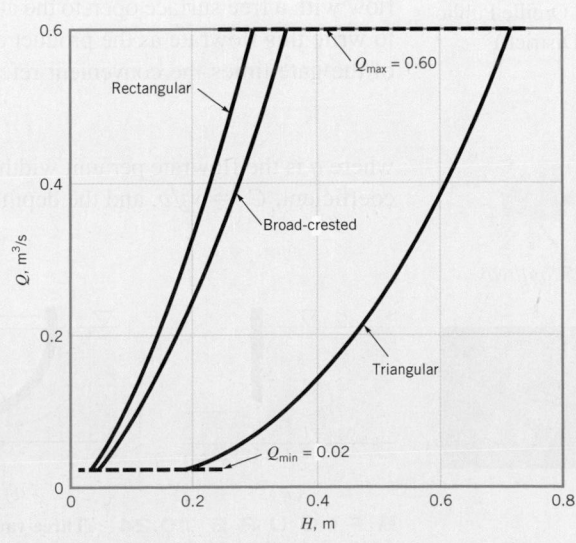

■ **FIGURE E10.8**

(b) Similarly, for the triangular weir, Eq. 10.32 gives

$$Q = C_{wt} \frac{8}{15} \tan\left(\frac{\theta}{2}\right) \sqrt{2g}\, H^{5/2}$$

$$= C_{wt} \frac{8}{15} \tan(45°) \sqrt{2(9.81 \text{ m/s}^2)}\, H^{5/2}$$

or

$$Q = 2.36 C_{wt}\, H^{5/2} \qquad (2)$$

where H and Q are in meters and m³/s and C_{wt} is obtained from Fig. 10.21. For example, with $H = 0.20$ m, we find $C_{wt} = 0.60$, or $Q = 2.36\,(0.60)(0.20)^{5/2} = 0.0253$ m³/s. The triangular weir results are also plotted in Fig. E10.8.

(c) For the broad-crested weir, Eqs. 10.28 and 10.29 give

$$Q = C_{wb}\, b \sqrt{g} \left(\frac{2}{3}\right)^{3/2} H^{3/2}$$

$$= 1.125 \left(\frac{1 + H/P_w}{2 + H/P_w}\right)^{1/2} b \sqrt{g} \left(\frac{2}{3}\right)^{3/2} H^{3/2}$$

Thus, with $P_w = 1$ m

$$Q = 1.125 \left(\frac{1 + H}{2 + H}\right)^{1/2} (2 \text{ m}) \sqrt{9.81 \text{ m/s}^2} \left(\frac{2}{3}\right)^{3/2} H^{3/2}$$

or

$$Q = 3.84 \left(\frac{1 + H}{2 + H}\right)^{1/2} H^{3/2} \qquad (3)$$

where, again, H and Q are in meters and m³/s. This result is also plotted in Fig. E10.8.

COMMENTS Although it appears as though any of the three weirs would work well for the upper portion of the flowrate range, neither the rectangular nor the broad-crested weir would be very accurate for small flowrates near $Q = Q_{min}$ because of the small head, H, at these conditions. The triangular weir, however, would allow reasonably large values of H at the lowest flowrates. The corresponding heads with $Q = Q_{min} = 0.02$ m³/s for rectangular, triangular, and broad-crested weirs are 0.0312, 0.182, and 0.0375 m, respectively.

In addition, as discussed in this section, for proper operation the broad-crested weir geometry is restricted to $0.08 < H/L_w < 0.50$, where L_w is the weir block length. From Eq. 3 with $Q_{max} = 0.60$ m³/s, we obtain $H_{max} = 0.349$. Thus, we must have $L_w > H_{max}/0.5 = 0.698$ m to maintain proper critical flow conditions at the largest flowrate in the channel. However, with $Q = Q_{min} = 0.02$ m³/s, we obtain $H_{min} = 0.0375$ m. Thus, we must have $L_w < H_{min}/0.08 = 0.469$ m to ensure that frictional effects are not important. Clearly, these two constraints on the geometry of the weir block, L_w, are incompatible.

A broad-crested weir will not function properly under the wide range of flowrates considered in this example. The sharp-crested triangular weir would be the best of the three types considered, provided the channel can handle the $H_{max} = 0.719$-m head.

10.6.4 Underflow Gates

(Photograph courtesy of Pend Oreille Public Utility District.)

A variety of *underflow gate* structures is available for flowrate control at the crest of an overflow spillway (as shown by the figure in the margin), or at the entrance of an irrigation canal or river from a lake. Three types are illustrated in Fig. 10.24. Each has certain advantages and disadvantages in terms of costs of construction, ease of use, and the like, although the basic fluid mechanics involved are the same in all instances.

The flow under a gate is said to be free outflow when the fluid issues as a jet of supercritical flow with a free surface open to the atmosphere as shown in Fig. 10.24. In such cases it is customary to write this flowrate as the product of the distance, a, between the channel bottom and the bottom of the gate times the convenient reference velocity $(2gy_1)^{1/2}$. That is,

$$q = C_d a \sqrt{2gy_1} \qquad (10.35)$$

where q is the flowrate per unit width. The discharge coefficient, C_d, is a function of the contraction coefficient, $C_c = y_2/a$, and the depth ratio y_1/a. Typical values of the discharge coefficient for free

V10.15 Spillway gate

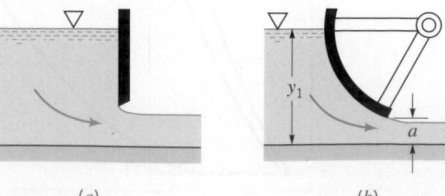

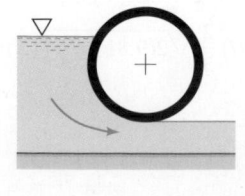

(a)　　　　　(b)　　　　　(c)

■ **FIGURE 10.24** Three variations of underflow gates: (*a*) vertical gate, (*b*) radial gate, (*c*) drum gate.

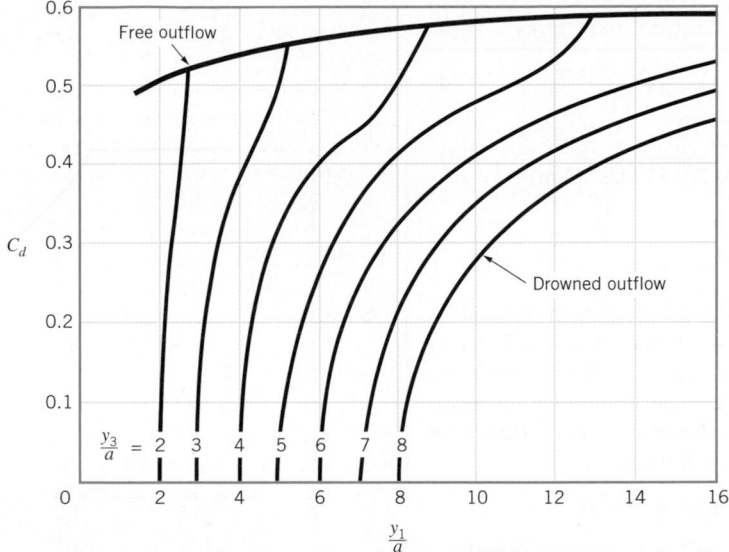

■ **F I G U R E   10.25** **Typical discharge coefficients for underflow gates (Ref. 3).**

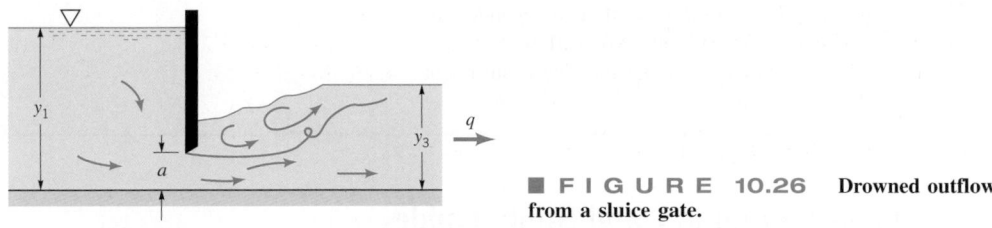

■ **F I G U R E   10.26** **Drowned outflow from a sluice gate.**

The flowrate from an underflow gate depends on whether the outlet is free or drowned.

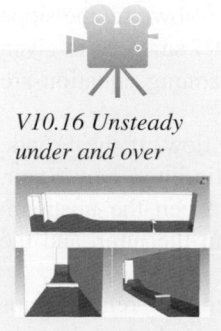

V10.16 Unsteady under and over

outflow (or free discharge) from a vertical sluice gate are on the order of 0.55 to 0.60 as indicated by the top line in Fig. 10.25 (Ref. 3).

As indicated in Fig. 10.26, in certain situations the depth downstream of the gate is controlled by some downstream obstacle and the jet of water issuing from under the gate is overlaid by a mass of water that is quite turbulent.

The flowrate for a submerged (or drowned) gate can be obtained from the same equation that is used for free outflow (Eq. 10.35), provided the discharge coefficient is modified appropriately. Typical values of C_d for drowned outflow cases are indicated as the series of lower curves in Fig. 10.25. Consider flow for a given gate and upstream conditions (i.e., given y_1/a) corresponding to a vertical line in the figure. With $y_3/a = y_1/a$ (i.e., $y_3 = y_1$) there is no head to drive the flow so that $C_d = 0$ and the fluid is stationary. For a given upstream depth (y_1/a fixed), the value of C_d increases with decreasing y_3/a until the maximum value of C_d is reached. This maximum corresponds to the free discharge conditions and is represented by the free outflow line so labeled in Fig. 10.25. For values of y_3/a that give C_d values between zero and its maximum, the jet from the gate is overlaid (drowned) by the downstream water and the flowrate is therefore reduced when compared with a free discharge situation. Similar results are obtained for the radial gate and drum gate.

E**XAMPLE   10.9**   **Sluice Gate**

GIVEN   Water flows under the sluice gate shown in Fig. E10.9. The channel width is $b = 6$ m, the upstream depth is $y_1 = 2$ m, and the gate is $a = 0.3$ m off the channel bottom.

FIND   Plot a graph of flowrate, Q, as a function of y_3.

SOLUTION

From Eq. 10.35 we have

$$Q = bq = baC_d \sqrt{2gy_1}$$
$$= 6 \text{ m } (0.3 \text{ m}) \, C_d \sqrt{2(9.81 \text{ m/s}^2)(2 \text{ m})}$$

or

$$Q = 11.0 C_d \text{ m}^2/\text{s} \qquad (1)$$

The value of C_d is obtained from Fig. 10.25 along the vertical line $y_1/a = 2 \text{ m}/0.3 \text{ m} = 6.7$. For $y_3 = 2 \text{ m}$ (i.e., $y_3/a = 6.7 = y_1/a$) we obtain $C_d = 0$, indicating that there is no flow when there is no head difference across the gate. The value of C_d increases as y_3/a decreases, reaching a maximum of $C_d = 0.56$ when $y_3/a = 3.2$. Thus, with $y_3 = 3.2a = 0.96 \text{ m}$

$$Q = 11.0 \, (0.56) \text{ m}^2/\text{s} = 6 \text{ m}^3/\text{s}$$

The flowrate for $0.96 \text{ m} \leq y_3 \leq 2 \text{ m}$ is obtained from Eq. 1 and the C_d values of Fig. 10.24 with the results as indicated in Fig. E10.9.

COMMENT For $y_3 < 0.96 \text{ m}$ the flowrate is independent of y_3, and the outflow is a free (not submerged) outflow. For such cases the inertia of the water flowing under the gate is sufficient to produce free outflow even with $y_3 > a$.

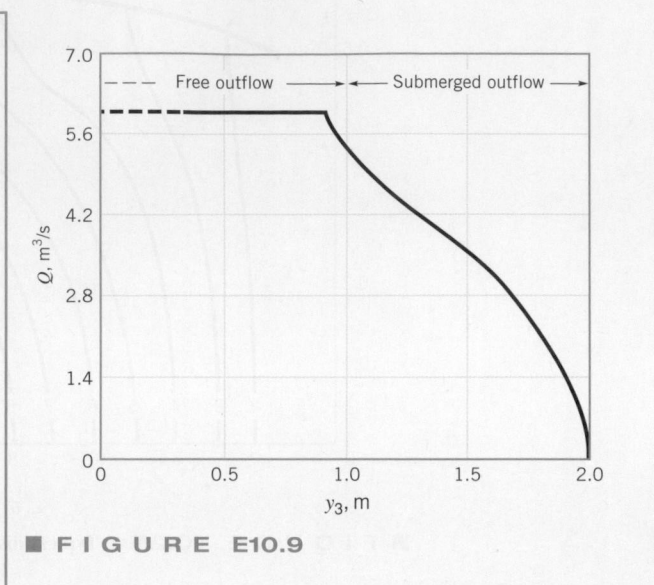

■ **FIGURE E10.9**

10.7 Chapter Summary and Study Guide

open-channel flow
Froude number
critical flow
subcritical flow
supercritical flow
wave speed
specific energy
specific energy diagram
uniform depth flow
wetted perimeter
hydraulic radius
Chezy equation
Manning equation
Manning coefficient
rapidly varied flow
hydraulic jump
sharp-crested weir
weir head
broad-crested weir
underflow gate

This chapter discussed various aspects of flows in an open channel. A typical open-channel flow is driven by the component of gravity in the direction of flow. The character of such flows can be a strong function of the Froude number, which is a ratio of the fluid speed to the free-surface wave speed. The specific energy diagram is used to provide insight into the flow processes involved in open-channel flow.

Uniform depth channel flow is achieved by a balance between the potential energy lost by the fluid as it coasts downhill and the energy dissipated by viscous effects. Alternately, it represents a balance between weight and friction forces. The relationship among the flowrate, the slope of the channel, the geometry of the channel, and the roughness of the channel surfaces is given by the Manning equation. Values of the Manning coefficient used in the Manning equation are dependent on the surface material roughness.

The hydraulic jump is an example of nonuniform depth open-channel flow. If the Froude number of a flow is greater than one, the flow is supercritical, and a hydraulic jump may occur. The momentum and mass equations are used to obtain the relationship between the upstream Froude number and the depth ratio across the jump. The energy dissipated in the jump and the head loss can then be determined by use of the energy equation.

The use of weirs to measure the flowrate in an open channel is discussed. The relationships between the flowrate and the weir head are given for both sharp-crested and broad-crested weirs.

The following checklist provides a study guide for this chapter. When your study of the entire chapter and end-of-chapter exercises has been completed you should be able to

■ write out meanings of the terms listed here in the margin and understand each of the related concepts. These terms are particularly important and are set in *italic, bold, and color* type in the text.

■ determine the Froude number for a given flow and explain the concepts of subcritical, critical, and supercritical flows.

■ plot and interpret the specific energy diagram for a given flow.

- use the Manning equation to analyze uniform depth flow in an open channel.
- calculate properties such as the depth ratio and the head loss for a hydraulic jump.
- determine the flowrates over sharp-crested weirs, broad-crested weirs, and under underflow gates.

Some of the important equations in this chapter are:

Froude number	$Fr = V/(gy)^{1/2}$	
Wave speed	$c = \sqrt{gy}$	(10.3)
Specific energy	$E = y + \dfrac{V^2}{2g}$	(10.8)
Manning equation	$V = \dfrac{\kappa}{n} R_h^{2/3} S_0^{1/2}$	(10.19)
Hydraulic jump depth ratio	$\dfrac{y_2}{y_1} = \dfrac{1}{2}(-1 + \sqrt{1 + 8Fr_1^2})$	(10.24)
Hydraulic jump head loss	$\dfrac{h_L}{y_1} = 1 - \dfrac{y_2}{y_1} + \dfrac{Fr_1^2}{2}\left[1 - \left(\dfrac{y_1}{y_2}\right)^2\right]$	(10.25)
Rectangular sharp-crested weir	$Q = C_{wr}\dfrac{2}{3}\sqrt{2g}\, b\, H^{3/2}$	(10.30)
Triangular sharp-crested weir	$Q = C_{wt}\dfrac{8}{15}\tan\left(\dfrac{\theta}{2}\right)\sqrt{2g}\, H^{5/2}$	(10.32)
Broad-crested weir	$Q = C_{wb}\, b\, \sqrt{g}\left(\dfrac{2}{3}\right)^{3/2} H^{3/2}$	(10.33)
Underflow gate	$q = C_d a\sqrt{2gy_1}$	(10.35)

References

1. Currie, C. G., and Currie, I. G., *Fundamental Mechanics of Fluids*, Third Edition, Marcel Dekker, New York, 2003.
2. Stoker, J. J., *Water Waves*, Interscience, New York, 1957.
3. Henderson, F. M., *Open Channel Flow*, Macmillan, New York, 1966.
4. Rouse, H., *Elementary Fluid Mechanics*, Wiley, New York, 1946.
5. French, R. H., *Open Channel Hydraulics*, McGraw-Hill, New York, 1992.
6. Chow, V. T., *Open Channel Hydraulics*, McGraw-Hill, New York, 1959.
7. Blevins, R. D., *Applied Fluid Dynamics Handbook*, Van Nostrand Reinhold, New York, 1984.
8. Daugherty, R. L., and Franzini, J. B., *Fluid Mechanics with Engineering Applications*, McGraw-Hill, New York, 1977.
9. Vennard, J. K., and Street, R. L., *Elementary Fluid Mechanics*, Seventh Edition Wiley, New York, 1995.
10. Lenz, A. T., "Viscosity and Surface Tension Effects on V-Notch Weir Coefficients," *Transactions of the American Society of Chemical Engineers*, Vol. 108, 759–820, 1943.
11. White, F. M., *Fluid Mechanics*, 5th Ed., McGraw-Hill, New York, 2003.
12. U.S. Bureau of Reclamation, Research Studies on Stilling Basins, Energy Dissipators, and Associated Appurtenances, Hydraulic Lab Report Hyd.-399, June 1, 1955.
13. Wallet, A., and Ruellan, F., *Houille Blanche*, Vol. 5, 1950.
14. Spitzer, D. W., ed., *Flow Measurement: Practical Guides for Measurement and Control*, Instrument Society of America, Research Triangle Park, NC, 1991.

Review Problems

Go to Appendix G for a set of review problems with answers. Detailed solutions can be found in *Student Solution Manual and Study Guide for Fundamentals of Fluid Mechanics*, by Munson et al. (© 2009 John Wiley and Sons, Inc.).

Problems

Note: Unless otherwise indicated, use the values of fluid properties found in the tables on the inside of the front cover. Problems designated with an (*) are intended to be solved with the aid of a programmable calculator or a computer. Problems designated with a (†) are "open-ended" problems and require critical thinking in that to work them one must make various assumptions and provide the necessary data. There is not a unique answer to these problems.

Answers to the even-numbered problems are listed at the end of the book. Access to the videos that accompany problems can be obtained through the book's web site, www.wiley.com/college/munson. The lab-type problems can also be accessed on this web site.

Section 10.2 Surface Waves

10.1 Obtain a photograph/image of surface waves. Print this photo and write a brief paragraph that describes the similarities and differences between these waves and those depicted in Fig. 10.4.

10.2 On a distant planet small amplitude waves travel across a 1-m-deep pond with a speed of 5 m/s. Determine the acceleration of gravity on the surface of that planet.

10.3 The flowrate in a 15-m-wide, 0.6-m-deep river is $Q = 5$ m³/s. Is the flow subcritical or supercritical?

10.4 The flowrate per unit width in a wide channel is $q = 2.3$ m²/s. Is the flow subcritical or supercritical if the depth is **(a)** 0.2 m, **(b)** 0.8 m, or **(c)** 2.5 m?

10.5 A rectangular channel 3 m wide carries 10 m³/s at a depth of 2 m. Is the flow subcritical or supercritical? For the same flowrate, what depth will give critical flow?

10.6 Consider waves made by dropping objects (one after another from a fixed location) into a stream of depth y that is moving with speed V as shown in Fig. P10.6 (see **Video V10.5**). The circular wave crests that are produced travel with speed $c = (gy)^{1/2}$ relative to the moving water. Thus, as the circular waves are washed downstream, their diameters increase and the center of each circle is fixed relative to the moving water. **(a)** Show that if the flow is supercritical, lines tangent to the waves generate a wedge of half-angle $\alpha/2 = \arcsin(1/\text{Fr})$, where $\text{Fr} = V/(gy)^{1/2}$ is the Froude number. **(b)** Discuss what happens to the wave pattern when the flow is subcritical, $\text{Fr} < 1$.

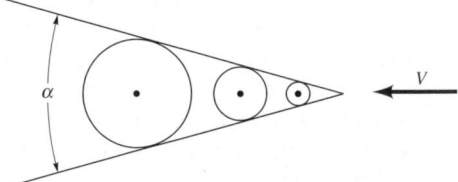

■ **F I G U R E P10.6**

10.7 Waves on the surface of a tank are observed to travel at a speed of 2 m/s. How fast would these waves travel if **(a)** the tank were in an elevator accelerating downward at a rate of 4 m/s², **(b)** the tank accelerates horizontally at a rate of 9.81 m/s², **(c)** the tank were aboard the orbiting Space Shuttle? Explain.

10.8 In flowing from section (1) to section (2) along an open channel, the water depth decreases by a factor of two and the Froude number changes from a subcritical value of 0.5 to a supercritical value of 3.0. Determine the channel width at (2) if it is 4 m wide at (1).

10.9 Observations at a shallow sandy beach show that even though the waves several hundred yards out from the shore are not parallel to the beach, the waves often "break" on the beach nearly parallel to the shore as indicated in Fig. P10.9. Explain this behavior based on the wave speed $c = (gy)^{1/2}$.

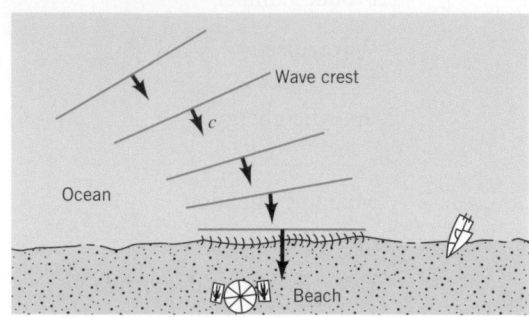

■ **F I G U R E P10.9**

†10.10 Explain, physically, why surface tension increases the speed of surface waves.

10.11 Often when an earthquake shifts a segment of the ocean floor, a relatively small amplitude wave of very long wavelength is produced. Such waves go unnoticed as they move across the open ocean; only when they approach the shore do they become dangerous (a tsunami or "tidal wave"). Determine the wave speed if the wavelength, λ, is 1800 m and the ocean depth is 4500 m.

10.12 A bicyclist rides through a 8-cm-deep puddle of water as shown in **Video V10.5** and Fig. P10.12. If the angle made by the V-shaped wave pattern produced by the front wheel is observed to be 40°, estimate the speed of the bike through the puddle. *Hint:* Make a sketch of the current location of the bike wheel relative to where it was Δt seconds ago. Also indicate on this sketch the current location of the wave that the wheel made Δt seconds ago. Recall that the wave moves radially outward in all directions with speed c relative to the stationary water.

■ **F I G U R E P10.12**

10.13 Determine the minimum depth in a 3-m-wide rectangular channel if the flow is to be subcritical with a flowrate of $Q = 60$ m³/s.

10.14 (See Fluids in the News article titled "Tsunami, the nonstorm wave," Section 10.2.1.) An earthquake causes a shift in the ocean floor that produces a tsunami with a wavelength of 100 km. How fast will this wave travel across the ocean surface if the ocean depth is 3000 m?

Section 10.3 Energy Considerations

10.15 Water flows in a 10-m-wide open channel with a flowrate of 5 m³/s. Determine the two possible depths if the specific energy of the flow is $E = 0.6$ m.

10.16 Water flows in a rectangular channel with a flowrate per unit width of $q = 2.5$ m²/s. Plot the specific energy diagram for this flow. Determine the two possible depths of flow if $E = 2.5$ m.

10.17 Water flows radially outward on a horizontal round disk as shown in **Video V10.12** and Fig. P10.17. **(a)** Show that the specific energy can be written in terms of the flowrate, Q, the radial distance from the axis of symmetry, r, and the fluid depth, y, as

$$E = y + \left(\frac{Q}{2\pi r}\right)^2 \frac{1}{2gy^2}$$

(b) For a constant flowrate, sketch the specific energy diagram. Recall Fig. 10.7, but note that for the present case r is a variable. Explain the important characteristics of your sketch. **(c)** Based on the results of Part **(b)**, show that the water depth increases in the flow direction if the flow is subcritical, but that it decreases in the flow direction if the flow is supercritical.

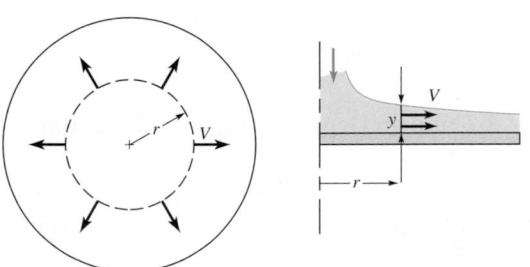

■ **FIGURE** **P10.17**

10.18 Water flows in a 3-m-wide rectangular channel with a flowrate of 6 m³/s. Plot the specific energy diagram for this flow. Determine the two possible flowrates when the specific energy is 2 m.

10.19 Water flows in a rectangular channel at a rate of $q = 2$ m³s/m. When a Pitot tube is placed in the stream, water in the tube rises to a level of 1.5 m above the channel bottom. Determine the two possible flow depths in the channel. Illustrate this flow on a specific energy diagram.

10.20 Water flows in a 1.5-m-wide rectangular channel with a flowrate of $Q = 0.8$ m³/s and an upstream depth of $y_1 = 1$ m as is shown in Fig. P10.20. Determine the flow depth and the surface elevation at section (2).

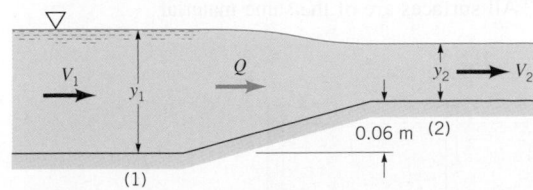

■ **FIGURE** **P10.20**

10.21 Repeat Problem 10.20 if the upstream depth is $y_1 = 0.2$ m.

*10.22 Water flows over the bump in the bottom of the rectangular channel shown in Fig. P10.22 with a flowrate per unit width of

$q = 4$ m²/s. The channel bottom contour is given by $z_B = 0.2e^{-x^2}$, where z_B and x are in meters. The water depth far upstream of the bump is $y_1 = 2$ m. Plot a graph of the water depth, $y = y(x)$, and the surface elevation, $z = z(x)$, for -4 m $\leq x \leq 4$ m. Assume one-dimensional flow.

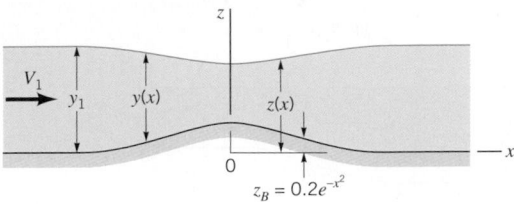

■ **FIGURE** **P10.22**

*10.23 Repeat Problem 10.22 if the upstream depth is 0.4 m.

10.24 Water in a rectangular channel flows into a gradual contraction section as is indicated in Fig. P10.24. If the flowrate is $Q = 0.7$ m³/s and the upstream depth is $y_1 = 0.6$ m, determine the downstream depth, y_2.

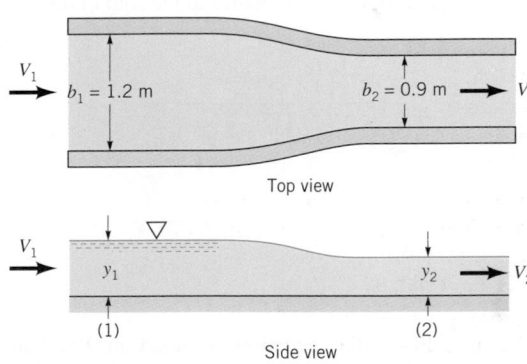

Top view

Side view

■ **FIGURE** **P10.24**

10.25 Sketch the specific energy diagram for the flow of Problem 10.24 and indicate its important characteristics. Note that $q_1 \neq q_2$.

10.26 Repeat Problem 10.24 if the upstream depth is $y_1 = 0.2$ m. Assume that there are no losses between sections (1) and (2).

10.27 Water flows in a rectangular channel with a flowrate per unit width of $q = 1.5$ m²/s and a depth of 0.5 m at section (1). The head loss between sections (1) and (2) is 0.03 m. Plot the specific energy diagram for this flow and locate states (1) and (2) on this diagram. Is it possible to have a head loss of 0.06 m? Explain.

10.28 Water flows in a horizontal rectangular channel with a flowrate per unit width of $q = 1$ m²/s and a depth of 0.3 m at the downstream section (2). The head loss between section (1) upstream and section (2) is 0.06 m. Plot the specific energy diagram for this flow and locate states (1) and (2) on this diagram.

10.29 Water flows in a horizontal, rectangular channel with an initial depth of 1 m and an initial velocity of 4 m/s. Determine the depth downstream if losses are negligible. Note that there may be more than one solution.

10.30 A smooth transition section connects two rectangular channels as shown in Fig. P10.30. The channel width increases from 1.8 to 2 m and the water surface elevation is the same in each channel. If the upstream depth of flow is 1 m, determine h, the amount the channel bed needs to be raised across the transition section to maintain the same surface elevation.

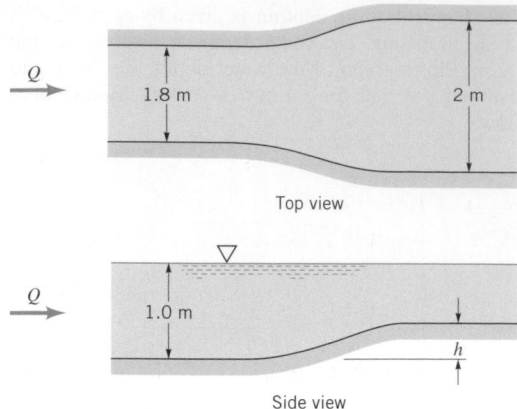

■ **F I G U R E P10.30**

10.31 Water flows over a bump of height $h = h(x)$ on the bottom of a wide rectangular channel as is indicated in Fig. P10.31. If energy losses are negligible, show that the slope of the water surface is given by $dy/dx = -(dh/dx)/[1 - (V^2/gy)]$, where $V = V(x)$ and $y = y(x)$ are the local velocity and depth of flow. Comment on the sign (i.e., <0, $= 0$, or >0) of dy/dx relative to the sign of dh/dx.

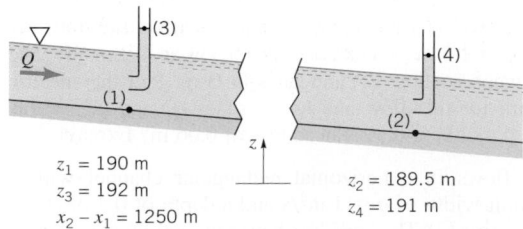

■ **F I G U R E P10.31**

10.32 Integrate the differential equation obtained in Problem 10.31 to determine the draw-down distance, $\ell = \ell(x)$, indicated in Fig. P10.31. Comment on your results.

10.33 Water flows in the river shown in Fig. P10.33 with a uniform bottom slope. The total head at each section is measured by using Pitot tubes as indicated. Determine the value of dy/dx at the location where the Froude number is 0.357.

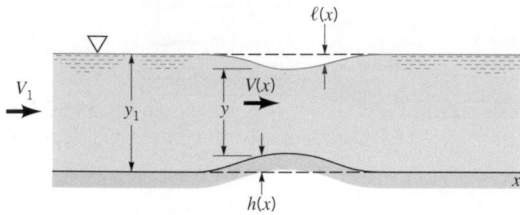

$z_1 = 190$ m
$z_3 = 192$ m
$x_2 - x_1 = 1250$ m

$z_2 = 189.5$ m
$z_4 = 191$ m

■ **F I G U R E P10.33**

10.34 Repeat Problem 10.33 if the Froude number is 2.75.

10.35 Water flows in a horizontal rectangular channel at a depth of 0.2 m and a velocity of 2.4 m/s. Determine the two possible depths at a location slightly downstream. Viscous effects between the water and the channel surface are negligible.

Section 10.4.2 The Manning Equation

10.36 Water flows in a 5-m-wide channel with a speed of 2 m/s and a depth of 1 m. The channel bottom slopes at a rate of 1 m per 1000 m. Determine the Manning coefficient for this channel.

10.37 Fluid properties such as viscosity or density do not appear in the Manning equation (Eq. 10.20). Does this mean that this equation is valid for any open-channel flow such as that involving mercury, water, oil, or molasses? Explain.

10.38 The following data are taken from measurements on Indian Fork Creek: $A = 26$ m^2, $P = 16$ m, and $S_0 = 0.02$ m/62 m. Determine the average shear stress on the wetted perimeter of this channel.

10.39 The following data are obtained for a particular reach of the Provo River in Utah: $A = 17$ m^2, free-surface width $= 15$ m, average depth $= 1$ m, $R_h = 1$ m, $V = 2$ m/s, length of reach $= 35$ m, and elevation drop of reach $= 0.3$ m. Determine (**a**) the average shear stress on the wetted perimeter, (**b**) the Manning coefficient, n, and (**c**) the Froude number of the flow.

10.40 At a particular location the cross section of the Columbia River is as indicated in Fig. P10.40. If on a day without wind it takes 5 min to float 1 km along the river, which drops 0.14 m in that distance, determine the value of the Manning coefficient, n.

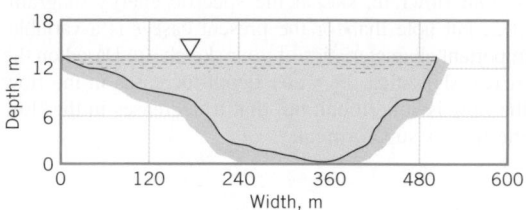

■ **F I G U R E P10.40**

Section 10.4.3 Uniform Depth Examples—Determine Flowrate

10.41 A 2-m-diameter pipe made of finished concrete lies on a slope of 1 m elevation change per 1000 m horizontal distance. Determine the flowrate when the pipe is half full.

10.42 Rainwater flows down a street whose cross section is shown in Fig. P10.42. The street is on a hill at an angle of 2°. Determine the maximum flowrate possible if the water is not to overflow onto the sidewalk.

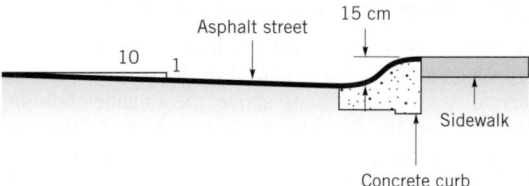

■ **F I G U R E P10.42**

10.43 By what percent is the flowrate reduced in the rectangular channel shown in Fig. P10.43 because of the addition of the thin center board? All surfaces are of the same material.

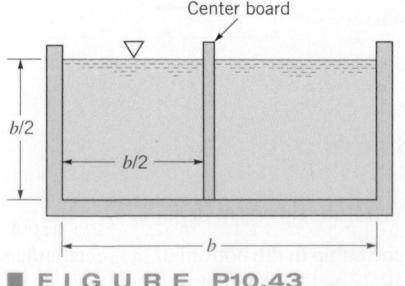

■ **F I G U R E P10.43**

10.44 The great Kings River flume in Fresno County, California, was used from 1890 to 1923 to carry logs from an elevation of 1400 m where trees were cut to an elevation of 90 m at the railhead. The flume was 85 km long, constructed of wood, and had a V-cross section as indicated in Fig. P10.44. It is claimed that logs would travel the length of the flume in 15 hours. Do you agree with this claim? Provide appropriate calculations to support your answer.

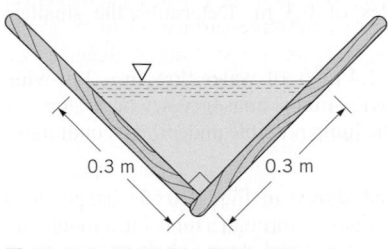

■ **F I G U R E P10.44**

10.45 Water flows in a channel as shown in Fig. P10.45. The velocity is 1.2 m/s when the channel is half full with depth d. Determine the velocity when the channel is completely full, depth $2d$.

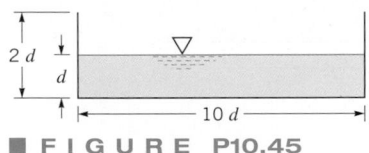

■ **F I G U R E P10.45**

10.46 A trapezoidal channel with a bottom width of 3.0 m and sides with a slope of 2 : 1 (horizontal:vertical) is lined with fine gravel ($n = 0.020$) and is to carry 10 m³/s. Can this channel be built with a slope of $S_0 = 0.00010$ if it is necessary to keep the velocity below 0.75 m/s to prevent scouring of the bottom? Explain.

10.47 Water flows in a 2-m-diameter finished concrete pipe so that it is completely full and the pressure is constant all along the pipe. If the slope is $S_0 = 0.005$, determine the flowrate by using open-channel flow methods. Compare this result with that obtained by using pipe flow methods of Chapter 8.

10.48 Water flows in a weedy earthen channel at a rate of 30 m³/s. What flowrate can be expected if the weeds are removed and the depth remains constant?

10.49 A round concrete storm sewer pipe used to carry rainfall runoff from a parking lot is designed to be half full when the rainfall rate is a steady 2.5 cm/hr. Will this pipe be able to handle the flow from a 5-cm/hr rainfall without water backing up into the parking lot? Support your answer with appropriate calculations.

10.50 A 3-m-wide rectangular channel is built to bypass a dam so that fish can swim upstream during their migration. During normal conditions when the water depth is 1.2 m, the water velocity is 1.5 m/s. Determine the velocity during a flood when the water depth is 2.4 m.

†**10.51** Overnight a thin layer of ice forms on the surface of a river. Estimate the percent reduction in flowrate caused by this condition. List all assumptions and show all calculations.

* **10.52** Water flows in the painted steel rectangular channel with rounded corners shown in Fig. P10.52. The bottom slope is 0.3 m/60 m. Plot a graph of flowrate as a function of water depth for $0 \le y \le 0.3$ m with corner radii of $r = 0$, 0.06, 0.12, 0.18, 0.24, and 0.3 m.

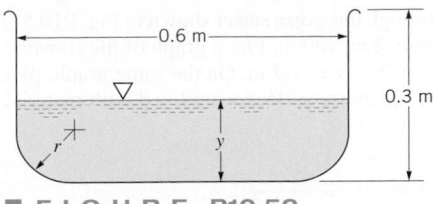

■ **F I G U R E P10.52**

* **10.53** The cross section of a long tunnel carrying water through a mountain is as indicated in Fig. P10.53. Plot a graph of flowrate as a function of water depth, y, for $0 \le y \le 5$ m. The slope is 0.4 m/km and the surface of the tunnel is rough rock (equivalent to rubble masonry). At what depth is the flowrate maximum? Explain.

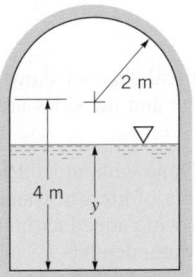

■ **F I G U R E P10.53**

10.54 The smooth concrete-lined channel shown in Fig. P10.54 is built on a slope of 2 m/km. Determine the flowrate if the depth is $y = 1.5$ m.

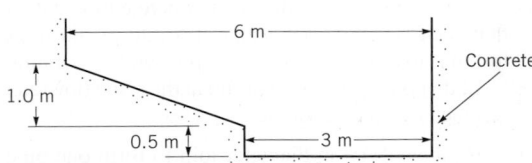

■ **F I G U R E P10.54**

* **10.55** At a given location, under normal conditions a river flows with a Manning coefficient of 0.030 and a cross section as indicated in Fig. P10.55a. During flood conditions at this location, the river has a Manning coefficient of 0.040 (because of trees and brush in the floodplain) and a cross section as shown in Fig. P10.55b. Determine the ratio of the flowrate during flood conditions to that during normal conditions.

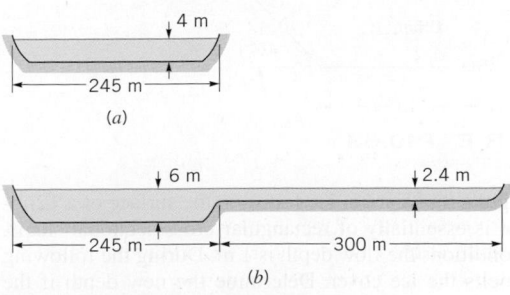

■ **F I G U R E P10.55**

10.56 Repeat Problem 10.54 if the surfaces are smooth concrete as is indicated, except for the diagonal surface, which is gravelly with $n = 0.025$.

*10.57 Water flows through the storm sewer shown in Fig. P10.57. The slope of the bottom is 2 m/400 m. Plot a graph of the flowrate as a function of depth for $0 \le y \le 1.7$ m. On the same graph, plot the flowrate expected if the entire surface were lined with material similar to that of a clay tile.

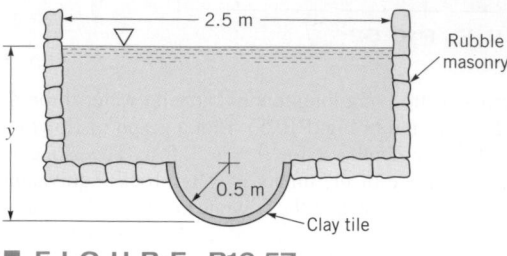

■ **F I G U R E   P10.57**

10.58 Determine the flowrate for the symmetrical channel shown in Fig. P10.80 if the bottom is smooth concrete and the sides are weedy. The bottom slope is $S_0 = 0.001$.

10.59 (See Fluids in the News article titled "Done without a GPS or lasers," Section 10.4.3.) Determine the number of liters of water delivered per day by a rubble masonry, 1.2-m-wide aqueduct laid on an average slope of 14.6 m per 50 km if the water depth is 1.8 m.

Section 10.4.3 Uniform Depth Examples—Determine Depth or Size

10.60 Water flows in a rectangular, finished concrete channel at a rate of 2 m³/s. The bottom slope is 0.001. Determine the channel width if the water depth is to be equal to its width.

10.61 An old, rough-surfaced, 2-m-diameter concrete pipe with a Manning coefficient of 0.025 carries water at a rate of 5.0 m³/s when it is half full. It is to be replaced by a new pipe with a Manning coefficient of 0.012 that is also to flow half full at the same flowrate. Determine the diameter of the new pipe.

10.62 Four sewer pipes of 0.5-m diameter join to form one pipe of diameter D. If the Manning coefficient, n, and the slope are the same for all of the pipes, and if each pipe flows half-full, determine D.

10.63 The flowrate in the clay-lined channel ($n = 0.025$) shown in Fig. P10.63 is to be 8.5 m³/s. To prevent erosion of the sides, the velocity must not exceed 1.5 m/s. For this maximum velocity, determine the width of the bottom, b, and the slope, S_0.

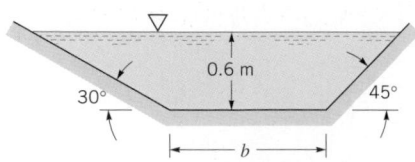

■ **F I G U R E   P10.63**

10.64 Overnight a thin layer of ice forms on the surface of a 12-m-wide river that is essentially of rectangular cross-sectional shape. Under these conditions the flow depth is 1 m. During the following day the sun melts the ice cover. Determine the new depth if the flowrate remains the same and the surface roughness of the ice is essentially the same as that for the bottom and sides of the river.

10.65 A rectangular, unfinished concrete channel of 9-m-width is laid on a slope of 1.5 m/km. Determine the flow depth and Froude number of the flow if the flowrate is 11 m³/s.

10.66 An engineer is to design a channel lined with planed wood to carry water at a flowrate of 2 m³/s on a slope of 10 m/800 m. The channel cross section can be either a 90° triangle or a rectangle with a cross section twice as wide as its depth. Which would require less wood and by what percent?

10.67 A circular finished concrete culvert is to carry a discharge of 1.4 m³/s on a slope of 0.0010. It is to flow not more than half-full. The culvert pipes are available from the manufacture with diameters that are multiples of 0.3 m. Determine the smallest suitable culvert diameter.

10.68 At what depth will 1.4 m³/s of water flow in a 2-m-wide rectangular channel lined with rubble masonry set on a slope of 0.3 m in 150 m? Is a hydraulic jump possible under these conditions? Explain.

10.69 The rectangular canal shown in Fig. P10.69 changes to a round pipe of diameter D as it passes through a tunnel in a mountain. Determine D if the surface material and slope remain the same and the round pipe is to flow completely full.

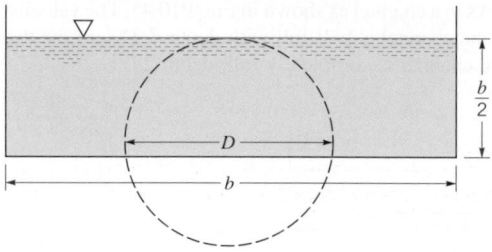

■ **F I G U R E   P10.69**

10.70 The flowrate through the trapezoidal canal shown in Fig. P10.70 is Q. If it is desired to double the flowrate to $2Q$ without changing the depth, determine the additional width, L, needed. The bottom slope, surface material, and the slope of the walls are to remain the same.

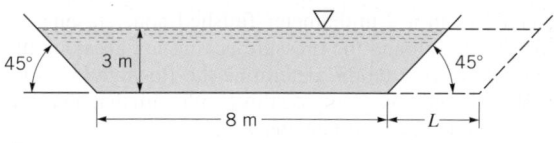

■ **F I G U R E   P10.70**

10.71 When the channel of triangular cross section shown in Fig. P10.71 was new, a flowrate of Q caused the water to reach $L = 2$ m up the side as indicated. After considerable use, the walls of the channel became rougher and the Manning coefficient, n, doubled. Determine the new value of L if the flowrate stayed the same.

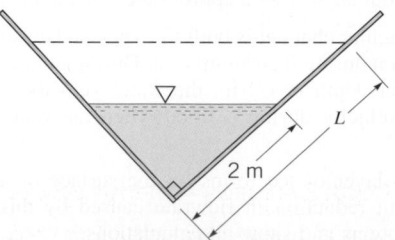

■ **F I G U R E   P10.71**

10.72 A smooth steel water slide at an amusement park is of semicircular cross section with a diameter of 1.0 m. The slide descends a vertical distance of 10 m in its 130 m length. If pumps supply water to the slide at a rate of 0.2 m³/s, determine the depth of flow. Neglect the effects of the curves and bends of the slide.

10.73 Two canals join to form a larger canal as shown in **Video V10.6** and Fig. P10.73. Each of the three rectangular canals is lined with the same material and has the same bottom slope. The water depth in each is to be 2 m. Determine the width of the merged canal, b. Explain physically (i.e., without using any equations) why it is expected that the width of the merged canal is less than the combined widths of the two original canals (i.e., $b < 4\text{ m} + 8\text{ m} = 12\text{ m}$).

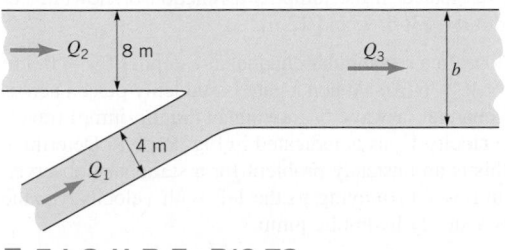

■ **F I G U R E   P10.73**

10.74 Water flows uniformly at a depth of 1 m in a channel that is 5 m wide as shown in Fig. P10.74. Further downstream the channel cross section changes to that of a square of width and height b. Determine the value of b if the two portions of this channel are made of the same material and are constructed with the same bottom slope.

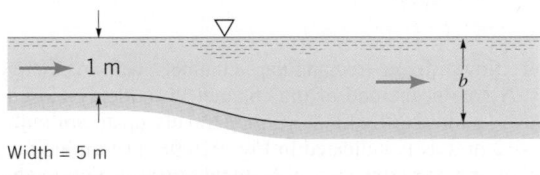

Width = 5 m

■ **F I G U R E   P10.74**

10.75 Determine the flow depth for the channel shown in Fig. P10.54 if the flowrate is 15 m³/s.

10.76 Rainwater runoff from a 60-m by 150-m parking lot is to drain through a circular concrete pipe that is laid on a slope of 0.5 m/km. Determine the pipe diameter if it is to be full with a steady rainfall of 4 cm/hr.

10.77 (See Fluids in the News article titled "Plumbing the Everglades," Section 10.4.1.) The canal shown in Fig. P10.77 is to be widened so that it can carry twice the amount of water. Determine the additional width, L, required if all other parameters (i.e., flow depth, bottom slope, surface material, side slope) are to remain the same.

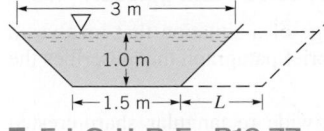

■ **F I G U R E   P10.77**

Section 10.4.3 Uniform Depth Examples—Determine Slope

10.78 Water flows 1 m deep in a 2-m-wide finished concrete channel. Determine the slope if the flowrate is 3 m³/s.

10.79 Water flows in the channel shown in Fig. P10.79 at a rate of 2.5 m³/s. Determine the minimum slope that this channel can have so that the water does not overflow the sides. The Manning coefficient for this channel is $n = 0.014$.

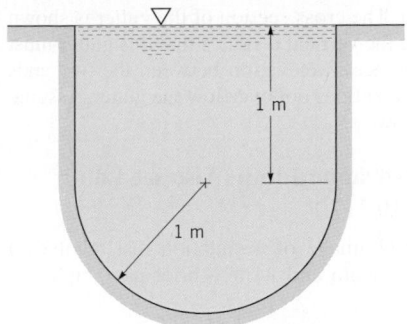

■ **F I G U R E   P10.79**

10.80 To prevent weeds from growing in a clean earthen-lined canal, it is recommended that the velocity be no less than 0.8 m/s. For the symmetrical canal shown in Fig. P10.80, determine the minimum slope needed.

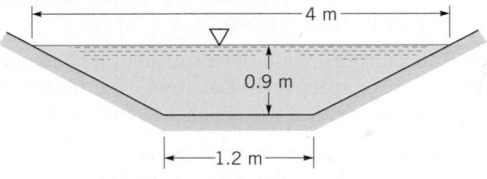

■ **F I G U R E   P10.80**

10.81 The smooth, concrete-lined, symmetrical channel shown in **Video V10.7** and Fig. P10.80 carries water from the silt-laden Colorado River. If the velocity must be 1.2 m/s to prevent the silt from settling out (and eventually clogging the channel), determine the minimum slope needed.

10.82 The symmetrical channel shown in Fig. P10.80 is dug in sandy loam soil with $n = 0.020$. For such surface material it is recommended that to prevent scouring of the surface the average velocity be no more than 0.5 m/s. Determine the maximum slope allowed.

10.83 The depth downstream of a sluice gate in a rectangular wooden channel of width 5 m is 0.60 m. If the flowrate is 18 m³/s, determine the channel slope needed to maintain this depth. Will the depth increase or decrease in the flow direction if the slope is (**a**) 0.02; (**b**) 0.01?

10.84 Water in a painted steel rectangular channel of width $b = 0.3$ m and depth y is to flow at critical conditions, Fr = 1. Plot a graph of the critical slope, S_{0c}, as a function of y for $0.02 \leq y \leq 1.5$ m. What is the maximum slope allowed if critical flow is not to occur regardless of the depth?

10.85 A 15-m-long aluminum gutter (Manning coefficient $n = 0.011$) on a section of a roof is to handle a flowrate of 4×10^{-3} m³/s

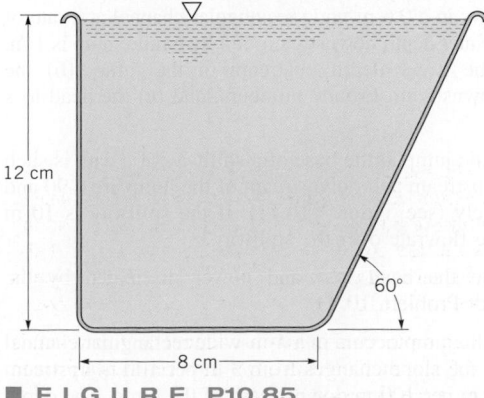

■ **F I G U R E   P10.85**

during a heavy rain storm. The cross section of the gutter is shown in Fig. P10.85. Determine the vertical distance that this gutter must be pitched (i.e., the difference in elevation between the two ends of the gutter) so that the water does not overflow the gutter. Assume uniform depth channel flow.

Section 10.6.1 The Hydraulic Jump (Also see Lab Problems 10.116 and 10.117.)

10.86 Obtain a photograph/image of a situation that involves a hydraulic jump. Print this photo and write a brief paragraph that describes the flow.

10.87 Water flows upstream of a hydraulic jump with a depth of 0.5 m and a velocity of 6 m/s. Determine the depth of the water downstream of the jump.

10.88 A 0.6-m standing wave is produced at the bottom of the rectangular channel in an amusement park water ride. If the water depth upstream of the wave is estimated to be 0.5 m, determine how fast the boat is traveling when it passes through this standing wave (hydraulic jump) for its final "splash."

10.89 The water depths upstream and downstream of a hydraulic jump are 0.3 and 1.2 m, respectively. Determine the upstream velocity and the power dissipated if the channel is 50 m wide.

10.90 Under appropriate conditions, water flowing from a faucet, onto a flat plate, and over the edge of the plate can produce a circular hydraulic jump as shown in Fig. P10.90 and **Video V10.12**. Consider a situation where a jump forms 8 cm from the center of the plate with depths upstream and downstream of the jump of 0.13 cm and 0.5 cm, respectively. Determine the flowrate from the faucet.

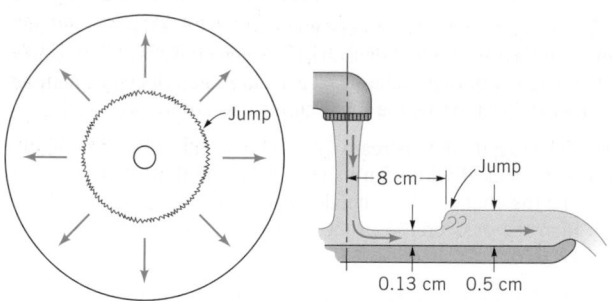

■ **FIGURE P10.90**

10.91 Show that the Froude number downstream of a hydraulic jump in a rectangular channel is $(y_1/y_2)^{3/2}$ times the Froude number upstream of the jump, where (1) and (2) denote the upstream and downstream conditions, respectively.

10.92 Water flows in a 0.6-m-wide rectangular channel at a rate of 0.3 m³/s. If the water depth downstream of a hydraulic jump is 1 m, determine **(a)** the water depth upstream of the jump, **(b)** the upstream and downstream Froude numbers, and **(c)** the head loss across the jump.

10.93 A hydraulic jump at the base of a spillway of a dam is such that the depths upstream and downstream of the jump are 0.90 and 3.6 m, respectively (see **Video V10.11**). If the spillway is 10 m wide, what is the flowrate over the spillway?

10.94 Determine the head loss and power dissipated by the hydraulic jump of Problem 10.93.

10.95 A hydraulic jump occurs in a 4-m-wide rectangular channel at a point where the slope changes from 3 m per 100 m upstream of the jump to h m per 100 m downstream of the jump. The depth

and velocity of the uniform flow upstream of the jump are 0.5 m and 8 m/s, respectively. Determine the value of h if the flow downstream of the jump is to be uniform flow.

10.96 At a given location in a 4-m-wide rectangular channel the flowrate is 25 m³/s and the depth is 1.0 m. Is this location upstream or downstream of the hydraulic jump that occurs in this channel? Explain.

*10.97 A rectangular channel of width b is to carry water at flowrates from $1 \le Q \le 17$ m³/s. The water depth upstream of the hydraulic jump that occurs (if one does occur) is to remain 0.5 m for all cases. Plot the power dissipated in the jump as a function of flowrate for channels of width $b = 3, 6, 9$, and 12 m.

10.98 Water flows in a rectangular channel at a depth of $y = 0.3$ m and a velocity of $V = 6$ m/s. When a gate is suddenly placed across the end of the channel, a wave (a moving hydraulic jump) travels upstream with velocity V_w as is indicated in Fig. P10.98. Determine V_w. Note that this is an unsteady problem for a stationary observer. However, for an observer moving to the left with velocity V_w, the flow appears as a steady hydraulic jump.

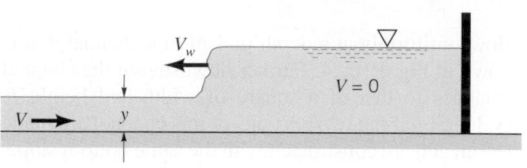

■ **FIGURE P10.98**

10.99 Water flows in a rectangular channel with velocity $V = 6$ m/s. A gate at the end of the channel is suddenly closed so that a wave (a moving hydraulic jump) travels upstream with velocity $V_w = 2$ m/s as is indicated in Fig. P10.98. Determine the depths ahead of and behind the wave. Note that this is an unsteady problem for a stationary observer. However, for an observer moving to the left with velocity V_w, the flow appears as a steady hydraulic jump.

10.100 (See Fluids in the News article titled "Grand Canyon rapids building," Section 10.6.1.) During the flood of 1983, a large hydraulic jump formed at "Crystal Hole" rapid on the Colorado River. People rafting the river at that time report "entering the rapid at almost 50 km/hr, hitting a 6-m-tall wall of water, and exiting at about 15 km/hr." Is this information (i.e., upstream and downstream velocities and change in depth) consistent with the principles of a hydraulic jump? Show calculations to support your answer.

Section 10.6.2,3 Sharp-Crested and Broad-Crested Weirs (Also see Lab Problems 10.114 and 10.115.)

10.101 Obtain a photograph/image of a situation that involves a weir. Print this photo and write a brief paragraph that describes the flow.

10.102 Water flows over a 2-m-wide rectangular sharp-crested weir. Determine the flowrate if the weir head is 0.1 m and the channel depth is 1 m.

10.103 Water flows over a 1.5-m-wide, rectangular sharp-crested weir that is $P_w = 1.4$ m tall. If the depth upstream is 1.5 m, determine the flowrate.

10.104 A rectangular sharp-crested weir is used to measure the flowrate in a channel of width 3 m. It is desired to have the channel flow depth be 2 m when the flowrate is 1.4 m³/s. Determine the height, P_w, of the weir plate.

10.105 Water flows from a storage tank, over two triangular weirs, and into two irrigation channels as shown in Video V10.13 and Fig. P10.105. The head for each weir is 0.12 m, and the flowrate in the channel fed by the 90°-V-notch weir is to be twice the flowrate in the other channel. Determine the angle θ for the second weir.

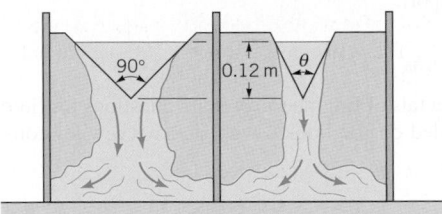

■ **FIGURE** **P10.105**

10.106 Rain water from a parking lot flows into a 8100-m² retention pond. After a heavy rain when there is no more inflow into the pond, the rectangular weir shown in Fig. P10.106 at the outlet of the pond has a head of $H = 0.2$ m. **(a)** Determine the rate at which the level of the water in the pond decreases, dH/dt, at this condition. **(b)** Determine how long it will take to reduce the pond level by 0.17 meter; that is, to $H = 0.03$ m.

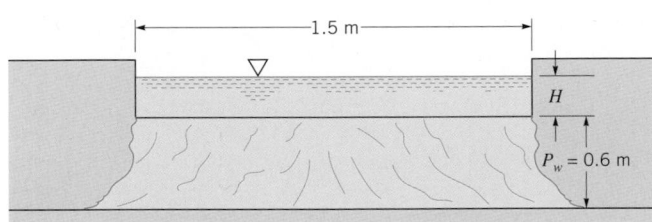

■ **FIGURE** **P10.106**

10.107 A basin at a water treatment plant is 20-m long, 3-m wide, and 1.5-m deep. Water flows from the basin over a 1-m-long, rectangular weir whose crest is 1.2 m above the bottom of the basin. Estimate how long it will take for the depth of the water in the basin to change from 1.4 m to 1.3 m if there is no flow into the basin.

10.108 Water flows over a sharp-crested triangular weir with $\theta = 90°$. The head range covered is $0.06 \le H \le 0.3$ m and the accuracy in the measurement of the head, H, is $\delta H = \pm 0.003$ m. Plot a graph of the percent error expected in Q as a function of Q.

10.109 **(a)** The rectangular sharp-crested weir shown in Fig. P10.109a is used to maintain a relatively constant depth in the channel upstream of the weir. How much deeper will the water be upstream of the weir during a flood when the flowrate is 1.3 m³/s compared to normal conditions when the flowrate is 0.8 m³/s? Assume the weir coefficient remains constant at $C_{wr} = 0.62$.

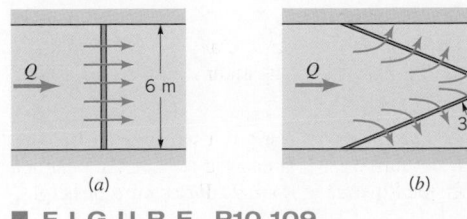

■ **FIGURE** **P10.109**

(b) Repeat the calculations if the weir of part **(a)** is replaced by a rectangular sharp-crested "duck bill" weir which is oriented at an angle of 30° relative to the channel centerline as shown in Fig. P10.109b. The weir coefficient remains the same.

10.110 Water flows in a rectangular channel of width $b = 6$ m at a rate of 3 m³/s. The flowrate is to be measured by using either a rectangular weir of height $P_w = 1$ m or a triangular $(\theta = 90°)$ sharp-crested weir. Determine the head, H, necessary. If measurement of the head is accurate to only ± 0.012 m, determine the accuracy of the measured flowrate expected for each of the weirs. Which weir would be the most accurate? Explain.

Section 10.6.4 Underflow Gates

10.111 Water flows under a sluice gate in a 20-m-wide finished concrete channel as is shown in Fig. P10.111. Determine the flowrate. If the slope of the channel is 1 m/60 m, will the water depth increase or decrease downstream of the gate? Assume $C_c = y_2/a = 0.65$. Explain.

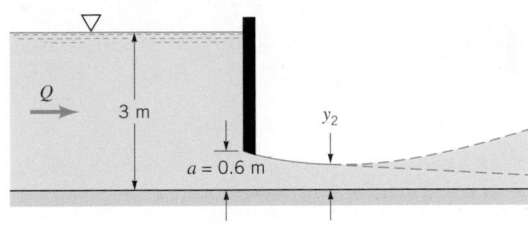

■ **FIGURE** **P10.111**

10.112 Water flows under a sluice gate in a channel of 3-m width. If the upstream depth remains constant at 1.5 m, plot a graph of flowrate as a function of the distance between the gate and the channel bottom as the gate is slowly opened. Assume free outflow.

10.113 A water-level regulator (not shown) maintains a depth of 2.0 m downstream from a 10-m-wide drum gate as shown in Fig. P10.113. Plot a graph of flowrate, Q, as a function of water depth upstream of the gate, y_1, for $2.0 \le y_1 \le 5.0$ m.

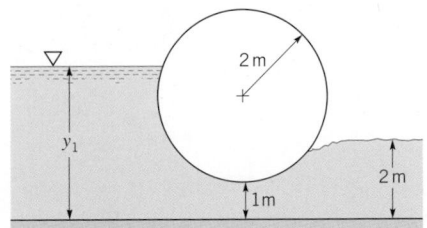

■ **FIGURE** **P10.113**

■ **Lab Problems**

10.114 This problem involves the calibration of a triangular weir. To proceed with this problem, go to Appendix H which is located on the book's web site, www.wiley.com/college/munson.

10.115 This problem involves the calibration of a rectangular weir. To proceed with this problem, go to Appendix H which is located on the book's web site, www.wiley.com/college/munson.

10.116 This problem involves the depth ratio across a hydraulic jump. To proceed with this problem, go to Appendix H which is located on the book's web site, www.wiley.com/college/munson.

10.117 This problem involves the head loss across a hydraulic jump. To proceed with this problem, go to Appendix H which is located on the book's web site, www.wiley.com/college/munson.

■ **Life Long Learning Problems**

10.118 With the increased usage of low-lying coastal areas and the possible rise in ocean levels because of global warming, the potential for widespread damage from tsunamis (i.e., "tidal waves") is increasing. Obtain information about new and improved methods available to predict the occurrence of these damaging waves and how to better use coastal areas so that massive loss of life and property does not occur. Summarize your findings in a brief report.

10.119 Recent photographs from NASA's Mars Orbiter Camera on the Mars Global Surveyor provide new evidence that water may still flow on the surface of Mars. Obtain information about the possibility of current or past open-channel flows on Mars and other planets or their satellites. Summarize your findings in a brief report.

10.120 Hydraulic jumps are normally associated with water flowing in rivers, gullies, and other such relatively high-speed open channels. However, recently, hydraulic jumps have been used in various manufacturing processes involving fluids other than water (such as liquid metal solder) in relatively small-scale flows. Obtain information about new manufacturing processes that involve hydraulic jumps as an integral part of the process. Summarize your findings in a brief report.

■ **FE Exam Problems**

Sample FE (Fundamentals of Engineering) exam questions for fluid mechanics are provided on the book's web site, www.wiley.com/college/munson.

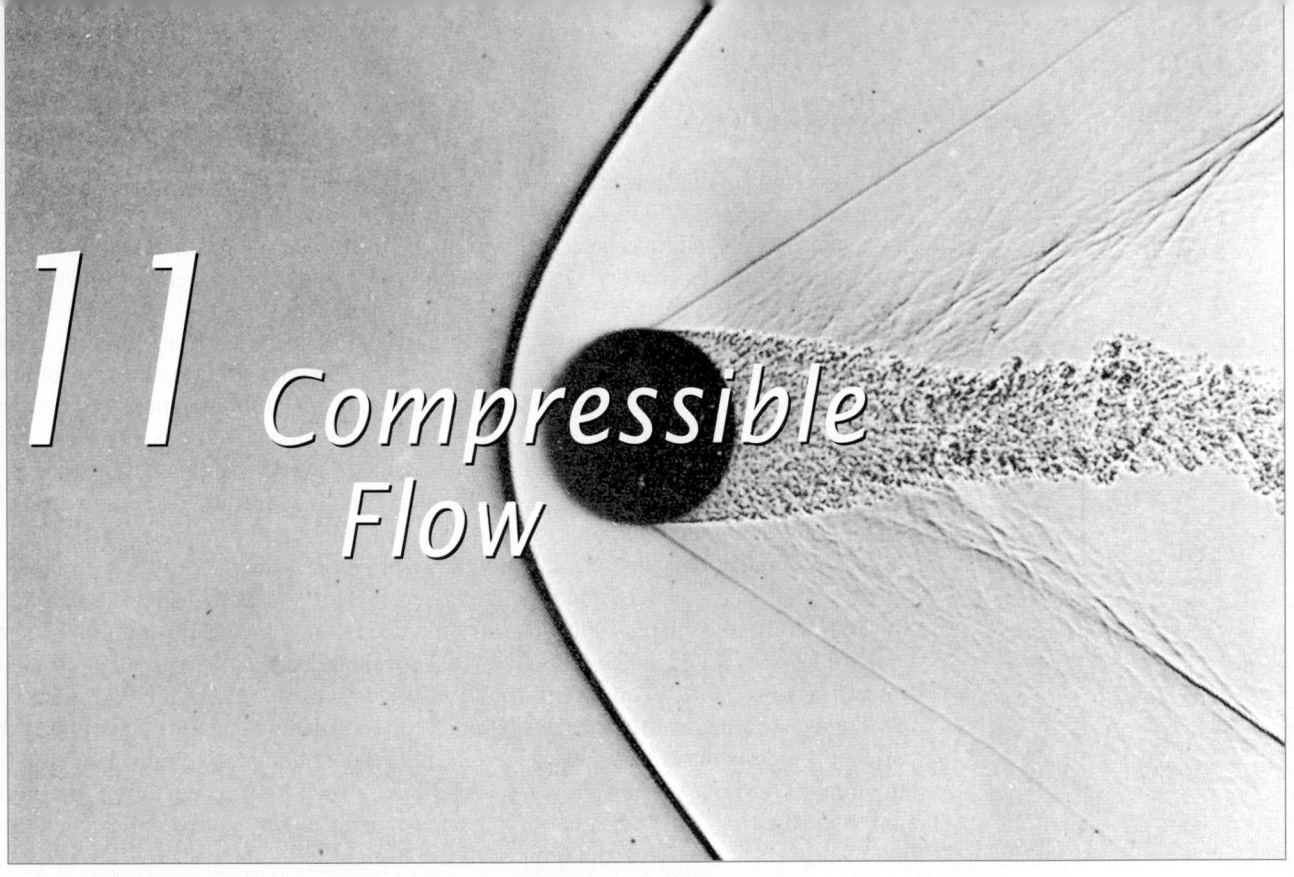

11 Compressible Flow

CHAPTER OPENING PHOTO: Flow past a sphere at Mach 1.53: An object moving through a fluid at supersonic speed (Mach number greater than one) creates a shock wave (a discontinuity in flow conditions shown by the dark curved line), which is heard as a sonic boom as the object passes overhead. The turbulent wake is also shown (shadowgraph technique used in air). *(Photograph courtesy of A. C. Charters.)*

Learning Objectives

After completing this chapter, you should be able to:

- distinguish between incompressible and compressible flows, and know when the approximations associated with assuming fluid incompressibility are acceptable.
- understand some important features of different categories of compressible flows of ideal gases.
- explain speed of sound and Mach number and their practical significance.
- solve useful problems involving isentropic and nonisentropic flows including flows across normal shock waves.
- appreciate the compelling similarities between compressible flows of gases and open channel flows of liquids.
- move on to understanding more advanced concepts about compressible flows.

Most first courses in fluid mechanics concentrate on constant density (incompressible) flows. In earlier chapters of this book, we mainly considered incompressible flow behavior. In a few instances, variable density (compressible) flow effects were covered briefly. The notion of an incompressible fluid is convenient because when constant density and constant (including zero) viscosity are assumed, problem solutions are greatly simplified. Also, fluid incompressibility allows us to build on the Bernoulli equation as was done, for example, in Chapter 5. Preceding examples should have convinced us that nearly incompressible flows are common in everyday experiences.

579

Any study of fluid mechanics would, however, be incomplete without a brief introduction to compressible flow behavior. Fluid compressibility is a very important consideration in numerous engineering applications of fluid mechanics. For example, the measurement of high-speed flow velocities requires compressible flow theory. The flows in gas turbine engine components are generally compressible. Many aircraft fly fast enough to involve compressible flow.

The variation of fluid density for compressible flows requires attention to density and other fluid property relationships. The fluid equation of state, often unimportant for incompressible flows, is vital in the analysis of compressible flows. Also, temperature variations for compressible flows are usually significant and thus the energy equation is important. Curious phenomena can occur with compressible flows. For example, with compressible flows we can have fluid acceleration because of friction, fluid deceleration in a converging duct, fluid temperature decrease with heating, and the formation of abrupt discontinuities in flows across which fluid properties change appreciably.

For simplicity, in this introductory study of compressibility effects we mainly consider the steady, one-dimensional, constant (including zero) viscosity, compressible flow of an ideal gas. We limit our study to compressibility due to high speed flow. In this chapter, one-dimensional flow refers to flow involving uniform distributions of fluid properties over any flow cross-sectional area. Both frictionless ($\mu = 0$) and frictional ($\mu \neq 0$) compressible flows are considered. If the change in volume associated with a change of pressure is considered a measure of compressibility, our experience suggests that gases and vapors are much more compressible than liquids. We focus our attention on the compressible flow of a gas because such flows occur often. We limit our discussion to ideal gases, since the equation of state for an ideal gas is uncomplicated, yet representative of actual gases at pressures and temperatures of engineering interest, and because the flow trends associated with an ideal gas are generally applicable to other compressible fluids.

An excellent film about compressible flow is available (see Ref. 1). This resource is a useful supplement to the material covered in this chapter.

11.1   Ideal Gas Relationships

V11.1 Lighter flame

Before we can proceed to develop *compressible flow* equations, we need to become more familiar with the fluid we will work with, the ideal gas. Specifically, we must learn how to evaluate ideal gas property changes. The equation of state for an *ideal gas* is

$$\rho = \frac{p}{RT} \tag{11.1}$$

We have already discussed fluid pressure, p, density, ρ, and temperature, T, in earlier chapters. The gas constant, R, represents a constant for each distinct ideal gas or mixture of ideal gases, where

$$R = \frac{\lambda}{M_{\text{gas}}} \tag{11.2}$$

We consider ideal gas flows only.

With this notation, λ is the universal gas constant and M_{gas} is the molecular weight of the ideal gas or gas mixture. Listed in Table 1.4 is values of the gas constants of some commonly used gases. Knowing the pressure and temperature of a gas, we can estimate its density. Nonideal gas state equations are beyond the scope of this text, and those interested in this topic are directed to texts on engineering thermodynamics, for example, Ref. 2. Note that the trends of ideal gas flows are generally good indicators of what nonideal gas flow behavior is like.

For an ideal gas, *internal energy*, $\breve{u}$, is part of the stored energy of the gas as explained in Section 5.3 and is considered to be a function of temperature only (Ref. 2). Thus, the ideal gas specific heat at constant volume, c_v, can be expressed as

$$c_v = \left(\frac{\partial \breve{u}}{\partial T}\right)_v = \frac{d\breve{u}}{dT} \tag{11.3}$$

where the subscript v on the partial derivative refers to differentiation at constant specific volume, $v = 1/\rho$. From Eq. 11.3 we conclude that for a particular ideal gas, c_v is a function of temperature only. Equation 11.3 can be rearranged to yield

$$d\breve{u} = c_v \, dT$$

Thus,

$$\check{u}_2 - \check{u}_1 = \int_{T_1}^{T_2} c_v \, dT \tag{11.4}$$

Equation 11.4 is useful because it allows us to evaluate the change in internal energy, $\check{u}_2 - \check{u}_1$, associated with ideal gas flow from section (1) to section (2) in a flow. For simplicity, we can assume that c_v is constant for a particular ideal gas and obtain from Eq. 11.4

$$\boxed{\check{u}_2 - \check{u}_1 = c_v(T_2 - T_1)} \tag{11.5}$$

Actually, c_v for a particular gas varies with temperature (see Ref. 2). However, for moderate changes in temperature, the constant c_v assumption is reasonable.

The fluid property *enthalpy*, $\check{h}$, is defined as

$$\check{h} = \check{u} + \frac{p}{\rho} \tag{11.6}$$

It combines internal energy, $\check{u}$, and pressure energy, p/ρ, and is useful when dealing with the energy equation (Eq. 5.69). For an ideal gas, we have already stated that

$$\check{u} = \check{u}(T)$$

From the equation of state (Eq. 11.1)

$$\frac{p}{\rho} = RT$$

Thus, it follows that

$$\check{h} = \check{h}(T)$$

Since for an ideal gas, enthalpy is a function of temperature only, the ideal gas specific heat at constant pressure, c_p, can be expressed as

$$c_p = \left(\frac{\partial \check{h}}{\partial T}\right)_p = \frac{d\check{h}}{dT} \tag{11.7}$$

where the subscript p on the partial derivative refers to differentiation at constant pressure, and c_p is a function of temperature only. The rearrangement of Eq. 11.7 leads to

$$d\check{h} = c_p \, dT$$

and

$$\check{h}_2 - \check{h}_1 = \int_{T_1}^{T_2} c_p \, dT \tag{11.8}$$

Equation 11.8 is useful because it allows us to evaluate the change in enthalpy, $\check{h}_2 - \check{h}_1$, associated with ideal gas flow from section (1) to section (2) in a flow. For simplicity, we can assume that c_p is constant for a specific ideal gas and obtain from Eq. 11.8

$$\boxed{\check{h}_2 - \check{h}_1 = c_p(T_2 - T_1)} \tag{11.9}$$

As is true for c_v, the value of c_p for a given gas varies with temperature. Nevertheless, for moderate changes in temperature, the constant c_p assumption is reasonable.

From Eqs. 11.5 and 11.9 we see that changes in internal energy and enthalpy are related to changes in temperature by values of c_v and c_p. We turn our attention now to developing useful relationships for determining c_v and c_p. Combining Eqs. 11.6 and 11.1 we get

For moderate temperature changes, specific heat values can be considered constant.

$$\check{h} = \check{u} + RT \tag{11.10}$$

Differentiating Eq. 11.10 leads to

$$d\breve{h} = d\breve{u} + R\,dT$$

or

$$\frac{d\breve{h}}{dT} = \frac{d\breve{u}}{dT} + R \tag{11.11}$$

From Eqs. 11.3, 11.7, and 11.11 we conclude that

$$\boxed{c_p - c_v = R} \tag{11.12}$$

Equation 11.12 indicates that the difference between c_p and c_v is constant for each ideal gas regardless of temperature. Also $c_p > c_v$. If the *specific heat ratio*, k, is defined as

$$k = \frac{c_p}{c_v} \tag{11.13}$$

then combining Eqs. 11.12 and 11.13 leads to

$$\boxed{c_p = \frac{Rk}{k-1}} \tag{11.14}$$

and

$$\boxed{c_v = \frac{R}{k-1}} \tag{11.15}$$

The gas constant is related to the specific heat values.

Actually, c_p, c_v, and k are all somewhat temperature dependent for any ideal gas. We will assume constant values for these variables in this book. Values of k and R for some commonly used gases at nominal temperatures are listed in Table 1.4. These tabulated values can be used with Eqs. 11.13 and 11.14 to determine the values of c_p and c_v. Example 11.1 demonstrates how internal energy and enthalpy changes can be calculated for a flowing ideal gas having constant c_p and c_v.

EXAMPLE 11.1   Internal Energy, Enthalpy, and Density for an Ideal Gas

GIVEN Air flows steadily between two sections in a long straight portion of 10-cm-diameter pipe as is indicated in Fig. E11.1. The uniformly distributed temperature and pressure at each section are $T_1 = 300$ K, $p_1 = 690$ kPa (abs), and $T_2 = 252$ K, $p_2 = 127$ kPa (abs).

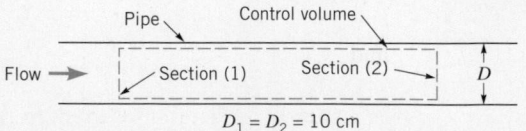

■ **FIGURE E11.1**

SOLUTION

(a) Assuming air behaves as an ideal gas, we can use Eq. 11.5 to evaluate the change in internal energy between sections (1) and (2). Thus

$$\breve{u}_2 - \breve{u}_1 = c_v(T_2 - T_1) \tag{1}$$

From Eq. 11.15 we have

$$c_v = \frac{R}{k-1} \tag{2}$$

and from Table 1.4, $R = 286.9$ J/(kg · K) and $k = 1.4$. Throughout this book, we use the nominal values of k for common gases

FIND Calculate the (a) change in internal energy between sections (1) and (2), (b) change in enthalpy between sections (1) and (2), and (c) change in density between sections (1) and (2).

listed in Table 1.4 and consider these values as being representative. From Eq. 2 we obtain

$$c_v = \frac{286.9}{(1.4 - 1)} \text{ J/(kg · K)}$$

$$= 717 \text{ J/(kg · K)} \tag{3}$$

Combining Eqs. 1 and 3 yields

$$\breve{u}_2 - \breve{u}_1 = c_v(T_2 - T_1) = 717 \text{ J/(kg} \cdot \text{K)}$$
$$\times (252 \text{ K} - 300 \text{ K})$$
$$= -34{,}416 \text{ J/kg} \qquad \textbf{(Ans)}$$

(b) For enthalpy change we use Eq. 11.9. Thus

$$\breve{h}_2 - \breve{h}_1 = c_p(T_2 - T_1) \qquad \textbf{(4)}$$

where since $k = c_p/c_v$ we obtain

$$c_p = kc_v = (1.4)[717 \text{ J/(kg} \cdot \text{K)}]$$
$$= 1004 \text{ J/(kg} \cdot \text{K)} \qquad \textbf{(5)}$$

From Eqs. 4 and 5 we obtain

$$\breve{h}_2 - \breve{h}_1 = c_p(T_2 - T_1) = 1004 \text{ J/(kg} \cdot \text{K)}$$
$$\times (252 \text{ K} - 300 \text{ K})$$
$$= -48{,}192 \text{ J/kg} \qquad \textbf{(Ans)}$$

(c) For density change we use the ideal gas equation of state (Eq. 11.1) to get

$$\rho_2 - \rho_1 = \frac{p_2}{RT_2} - \frac{p_1}{RT_1} = \frac{1}{R}\left(\frac{p_2}{T_2} - \frac{p_1}{T_1}\right) \qquad \textbf{(6)}$$

Using the pressures and temperatures given in the problem statement we calculate from Eq. 6

$$\rho_2 - \rho_1 = \frac{1}{286.9 \text{ J/kg} \cdot \text{K}}$$
$$\times \left[\frac{127 \times 10^3 \text{ Pa (abs)}}{252 \text{ K}}\right.$$
$$\left. - \frac{690 \times 10^3 \text{ Pa (abs)}}{300 \text{ K}}\right]$$

or

$$\rho_2 - \rho_1 = -6.26 \text{ kg/m}^3 \qquad \textbf{(Ans)}$$

COMMENT   This is a significant change in density when compared with the upstream density

$$\rho_1 = \frac{p_1}{RT_1} = \frac{[690 \times 10^3 \text{ Pa (abs)}]}{[286.9 \text{ J/kg} \cdot \text{K})(300 \text{ K})]}$$
$$= 8.02 \text{ kg/m}^3$$

Compressibility effects are important for this flow.

For compressible flows, changes in the thermodynamic property *entropy*, s, are important. For any pure substance including ideal gases, the "first $T \, ds$ equation" is (see Ref. 2)

$$T \, ds = d\breve{u} + pd\left(\frac{1}{\rho}\right) \qquad \textbf{(11.16)}$$

Changes in entropy are important because they are related to loss of available energy.

where T is absolute temperature, s is entropy, $\breve{u}$ is internal energy, p is absolute pressure, and ρ is density. Differentiating Eq. 11.6 leads to

$$d\breve{h} = d\breve{u} + pd\left(\frac{1}{\rho}\right) + \left(\frac{1}{\rho}\right)dp \qquad \textbf{(11.17)}$$

By combining Eqs. 11.16 and 11.17, we obtain

$$T \, ds = d\breve{h} - \left(\frac{1}{\rho}\right)dp \qquad \textbf{(11.18)}$$

Equation 11.18 is often referred to as the "second $T \, ds$ equation." For an ideal gas, Eqs. 11.1, 11.3, and 11.16 can be combined to yield

$$ds = c_v\frac{dT}{T} + \frac{R}{1/\rho}d\left(\frac{1}{\rho}\right) \qquad \textbf{(11.19)}$$

and Eqs. 11.1, 11.7, and 11.18 can be combined to yield

$$ds = c_p\frac{dT}{T} - R\frac{dp}{p} \qquad \textbf{(11.20)}$$

If c_p and c_v are assumed to be constant for a given gas, Eqs. 11.19 and 11.20 can be integrated to get

$$s_2 - s_1 = c_v\ln\frac{T_2}{T_1} + R\ln\frac{\rho_1}{\rho_2} \qquad \textbf{(11.21)}$$

Changes in entropy are related to changes in temperature, pressure, and density.

and

$$s_2 - s_1 = c_p \ln \frac{T_2}{T_1} - R \ln \frac{p_2}{p_1} \qquad (11.22)$$

Equations 11.21 and 11.22 allow us to calculate the change of entropy of an ideal gas flowing from one section to another with constant specific heat values (c_p and c_v).

EXAMPLE 11.2   Entropy for an Ideal Gas

GIVEN   Consider the air flow of Example 11.1.

FIND   Calculate the change in entropy, $s_2 - s_1$, between sections (1) and (2).

SOLUTION

Assuming that the flowing air in Fig. E11.1 behaves as an ideal gas, we can calculate the entropy change between sections by using either Eq. 11.21 or Eq. 11.22. We use both to demonstrate that the same result is obtained either way.

From Eq. 11.21,

$$s_2 - s_1 = c_v \ln \frac{T_2}{T_1} + R \ln \frac{\rho_1}{\rho_2} \qquad (1)$$

To evaluate $s_2 - s_1$ from Eq. 1 we need the density ratio, ρ_1/ρ_2, which can be obtained from the ideal gas equation of state (Eq. 11.1) as

$$\frac{\rho_1}{\rho_2} = \left(\frac{p_1}{T_1}\right)\left(\frac{T_2}{p_2}\right) \qquad (2)$$

and thus from Eqs. 1 and 2,

$$s_2 - s_1 = c_v \ln \frac{T_2}{T_1} + R \ln \left[\left(\frac{p_1}{T_1}\right)\left(\frac{T_2}{p_2}\right)\right] \qquad (3)$$

By substituting values already identified in the Example 11.1 problem statement and solution into Eq. 3 with

$$\left(\frac{p_1}{T_1}\right)\left(\frac{T_2}{p_2}\right) = \left(\frac{690 \times 10^3 \, \text{Pa (abs)}}{300 \, \text{K}}\right)\left(\frac{252 \, \text{K}}{127 \times 10^3 \, \text{Pa (abs)}}\right)$$

$$= 4.56$$

we get

$$s_2 - s_1 = [717 \, \text{J/kg} \cdot \text{K}] \ln \left(\frac{252 \, \text{K}}{300 \, \text{K}}\right)$$
$$+ [286.9 \, \text{J/kg} \cdot \text{K}] \ln 4.56$$

or

$$s_2 - s_1 = 310 \, \text{J/(kg} \cdot \text{K)} \qquad \text{(Ans)}$$

From Eq. 11.22,

$$s_2 - s_1 = c_p \ln \frac{T_2}{T_1} - R \ln \frac{p_2}{p_1} \qquad (4)$$

By substituting known values into Eq. 4 we obtain

$$s_2 - s_1 = [1004 \, (\text{J/kg} \cdot \text{K})] \ln \left(\frac{252 \, \text{K}}{300 \, \text{K}}\right)$$

$$- [286.9 \, \text{J/(kg} \cdot \text{K)}] \ln \left(\frac{127 \times 10^3 \, \text{Pa (abs)}}{690 \times 10^3 \, \text{Pa (abs)}}\right)$$

or

$$s_2 - s_1 = 310 \, \text{J/(kg} \cdot \text{K)} \qquad \text{(Ans)}$$

COMMENT   As anticipated, both Eqs. 11.21 and 11.22 yield the same result for the entropy change, $s_2 - s_1$.

Note that since the ideal gas equation of state was used in the derivation of the entropy difference equations, both the pressures and temperatures used must be absolute.

If internal energy, enthalpy, and entropy changes for ideal gas flow with variable specific heats are desired, Eqs. 11.4, 11.8, and 11.19 or 11.20 must be used as explained in Ref. 2. Detailed tables (see, for example, Ref. 3) are available for variable specific heat calculations.

The second law of thermodynamics requires that the *adiabatic* and frictionless flow of any fluid results in $ds = 0$ or $s_2 - s_1 = 0$. Constant entropy flow is called *isentropic* flow. For the isentropic flow of an ideal gas with constant c_p and c_v, we get from Eqs. 11.21 and 11.22

$$c_v \ln \frac{T_2}{T_1} + R \ln \frac{\rho_1}{\rho_2} = c_p \ln \frac{T_2}{T_1} - R \ln \frac{p_2}{p_1} = 0 \qquad (11.23)$$

By combining Eq. 11.23 with Eqs. 11.14 and 11.15 we obtain

$$\left(\frac{T_2}{T_1}\right)^{k/(k-1)} = \left(\frac{\rho_2}{\rho_1}\right)^k = \left(\frac{p_2}{p_1}\right) \tag{11.24}$$

which is a useful relationship between temperature, density, and pressure for the isentropic flow of an ideal gas. From Eq. 11.24 we can conclude that

$$\boxed{\frac{p}{\rho^k} = \text{constant}} \tag{11.25}$$

for an ideal gas with constant c_p and c_v flowing isentropically, a result already used without proof earlier in Chapters 1, 3, and 5.

F l u i d s   i n   t h e   N e w s

Hilsch tube (Ranque vortex tube) Years ago (around 1930) a French physics student (George Ranque) discovered that appreciably warmer and colder portions of *rapidly swirling air flow* could be separated in a simple apparatus consisting of a tube open at both ends into which was introduced, somewhere in between the two openings, swirling air at *high pressure*. Warmer air near the outer portion of the swirling air flowed out one open end of the tube through a simple valve and colder air near the inner portion of the swirling air flowed out the opposite end of the tube. Rudolph Hilsch, a German physicist, improved on this discovery (ca. 1947). Hot air temperatures of 260 °F (127 °C) and cold air temperatures of −50 °F (−46 °C) have been claimed in an optimized version of this apparatus. Thus far the inefficiency of the process has prevented it from being widely adopted. (See Problems 11.79.)

11.2   Mach Number and Speed of Sound

The *Mach number*, Ma, was introduced in Chapters 1 and 7 as a dimensionless measure of compressibility in a fluid flow. In this and subsequent sections, we develop some useful relationships involving the Mach number. The Mach number is defined as the ratio of the value of the local flow velocity, V, to the local *speed of sound*, c. In other words,

Mach number is the ratio of local flow and sound speeds.

$$\text{Ma} = \frac{V}{c}$$

What we perceive as sound generally consists of weak pressure pulses that move through air with a Mach number of one. When our ear drums respond to a succession of moving pressure pulses, we hear sounds.

To better understand the notion of speed of sound, we analyze the one-dimensional fluid mechanics of an infinitesimally thin, weak pressure pulse moving at the speed of sound through a fluid at rest (see Fig. 11.1a). Ahead of the pressure pulse, the fluid velocity is zero and the fluid pressure and density are p and ρ. Behind the pressure pulse, the fluid velocity has changed by an amount δV, and the pressure and density of the fluid have also changed by amounts δp and $\delta \rho$. We select an infinitesimally thin control volume that moves with the pressure pulse as is sketched

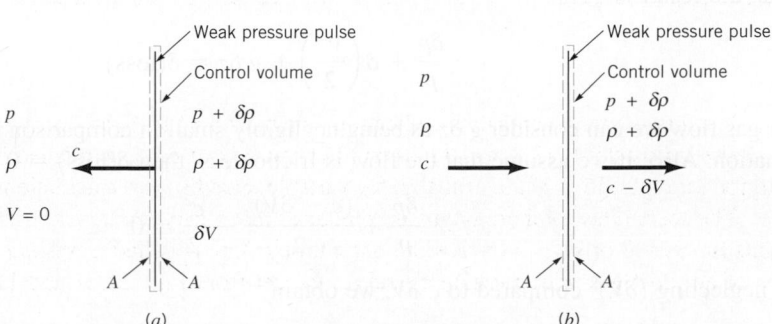

■ **FIGURE   11.1**   (a) **Weak pressure pulse moving through a fluid at rest.** (b) **The flow relative to a control volume containing a weak pressure pulse.**

in Fig. 11.1a. The speed of the weak pressure pulse is considered constant and in one direction only; thus, our control volume is inertial.

For an observer moving with this control volume (Fig. 11.1b), it appears as if fluid is entering the control volume through surface area A with speed c at pressure p and density ρ and leaving the control volume through surface area A with speed $c - \delta V$, pressure $p + \delta p$, and density $\rho + \delta \rho$. When the continuity equation (Eq. 5.16) is applied to the flow through this control volume, the result is

$$\rho A c = (\rho + \delta \rho) A (c - \delta V) \tag{11.26}$$

or

$$\rho c = \rho c - \rho\, \delta V + c\, \delta \rho - (\delta \rho)(\delta V) \tag{11.27}$$

Since $(\delta \rho)(\delta V)$ is much smaller than the other terms in Eq. 11.27, we drop it from further consideration and keep

$$\rho\, \delta V = c\, \delta \rho \tag{11.28}$$

The linear momentum equation (Eq. 5.29) can also be applied to the flow through the control volume of Fig. 11.1b. The result is

$$-c \rho c A + (c - \delta V)(\rho + \delta \rho)(c - \delta V)A = pA - (p + \delta p)A \tag{11.29}$$

Note that any frictional forces are considered as being negligibly small. We again neglect higher order terms [such as $(\delta V)^2$ compared to $c\, \delta V$, for example] and combine Eqs. 11.26 and 11.29 to get

$$-c \rho c A + (c - \delta V)\rho c A = -\delta p A$$

or

$$\rho \delta V = \frac{\delta p}{c} \tag{11.30}$$

From Eqs. 11.28 (continuity) and 11.30 (linear momentum) we obtain

$$c^2 = \frac{\delta p}{\delta \rho}$$

or

$$c = \sqrt{\frac{\delta p}{\delta \rho}} \tag{11.31}$$

This expression for the speed of sound results from application of the conservation of mass and conservation of linear momentum principles to the flow through the control volume of Fig. 11.1b. These principles were similarly used in Section 10.2.1 to obtain an expression for the speed of surface waves traveling on the surface of fluid in a channel.

The conservation of energy principle can also be applied to the flow through the control volume of Fig. 11.1b. If the energy equation (Eq. 5.103) is used for the flow through this control volume, the result is

$$\frac{\delta p}{\rho} + \delta\left(\frac{V^2}{2}\right) + g\, \delta z = \delta(\text{loss}) \tag{11.32}$$

For gas flow we can consider $g\, \delta z$ as being negligibly small in comparison to the other terms in the equation. Also, if we assume that the flow is frictionless, then $\delta(\text{loss}) = 0$ and Eq. 11.32 becomes

$$\frac{\delta p}{\rho} + \frac{(c - \delta V)^2}{2} - \frac{c^2}{2} = 0$$

or, neglecting $(\delta V)^2$ compared to $c\, \delta V$, we obtain

$$\rho\, \delta V = \frac{\delta p}{c} \tag{11.33}$$

By combining Eqs. 11.28 (continuity) and 11.33 (energy) we again find that

$$c = \sqrt{\frac{\delta p}{\delta \rho}}$$

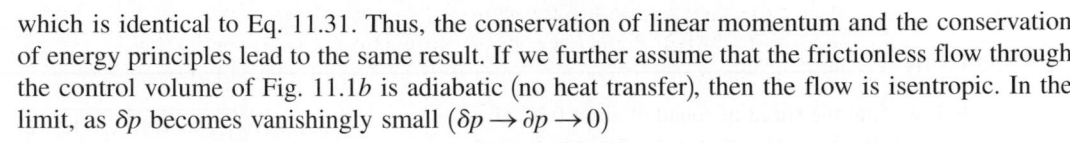

which is identical to Eq. 11.31. Thus, the conservation of linear momentum and the conservation of energy principles lead to the same result. If we further assume that the frictionless flow through the control volume of Fig. 11.1b is adiabatic (no heat transfer), then the flow is isentropic. In the limit, as δp becomes vanishingly small ($\delta p \to \partial p \to 0$)

$$c = \sqrt{\left(\frac{\partial p}{\partial \rho}\right)_s} \tag{11.34}$$

where the subscript s is used to designate that the partial differentiation occurs at constant entropy.

Equation 11.34 suggests to us that we can calculate the speed of sound by determining the partial derivative of pressure with respect to density at constant entropy. For the isentropic flow of an ideal gas (with constant c_p and c_v), we learned earlier (Eq. 11.25) that

$$p = (\text{constant})(\rho^k)$$

and thus

$$\left(\frac{\partial p}{\partial \rho}\right)_s = (\text{constant}) k\rho^{k-1} = \frac{p}{\rho^k} k\rho^{k-1} = \frac{p}{\rho} k = RTk \tag{11.35}$$

Thus, for an ideal gas

$$c = \sqrt{RTk} \tag{11.36}$$

From Eq. 11.36 and the charts in the margin we conclude that for a given temperature, the speed of sound, c, in hydrogen and in helium, is higher than in air.

More generally, the bulk modulus of elasticity, E_v, of any fluid including liquids is defined as (see Section 1.7.1)

$$E_v = \frac{dp}{d\rho/\rho} = \rho\left(\frac{\partial p}{\partial \rho}\right)_s \tag{11.37}$$

Thus, in general, from Eqs. 11.34 and 11.37,

$$c = \sqrt{\frac{E_v}{\rho}} \tag{11.38}$$

Values of the speed of sound are tabulated in Table B.1 for water and in Table B.2 for air. From experience we know that air is more easily compressed than water. Note from the values of c in Tables B.1 and B.2 and the graph in the margin that the speed of sound in air is much less than it is in water. From Eq. 11.37, we can conclude that if a fluid is truly incompressible, its bulk modulus would be infinitely large, as would be the speed of sound in that fluid. Thus, an incompressible flow must be considered an idealized approximation of reality.

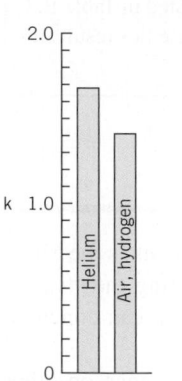

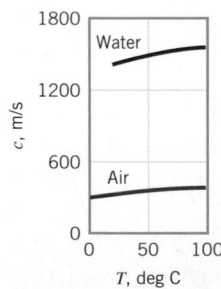

Speed of sound is larger in fluids that are more difficult to compress.

F l u i d s   i n   t h e   N e w s

Sonification The normal human ear is capable of detecting even very subtle sound patterns produced by *sound waves.* Most of us can distinguish the bark of a dog from the meow of a cat or the roar of a lion, or identify a person's voice on the telephone before they identify who is calling. The number of "things" we can identify from subtle sound patterns is enormous. Combine this ability with the power of computers to transform the information from sensor transducers into variations in pitch, rhythm, and volume and you have *sonification,* the representation of data in the form of sound. With this emerging technology, pathologists may soon learn to "hear" abnormalities in tissue samples, engineers may "hear" flaws in gas turbine engine blades being inspected, and scientists may "hear" a desired attribute in a newly invented material. Perhaps the concept of hearing the trends in data sets may become as commonplace as seeing them. Analysts may listen to the stock market and make decisions. Of course, none of this can happen in a vacuum.

EXAMPLE 11.3 Speed of Sound

GIVEN Consider the data in Table B.2.

FIND Verify the speed of sound for air at 0 °C.

SOLUTION

In Table B.2, we find the speed of sound of air at 0 °C given as 331.4 m/s. Assuming that air behaves as an ideal gas, we can calculate the speed of sound from Eq. 11.36 as

$$c = \sqrt{RTk} \qquad (1)$$

The value of the gas constant is obtained from Table 1.4 as

$$R = 286.9 \ J/(kg \cdot K)$$

and the specific heat ratio is listed in Table B.2 as

$$k = 1.401$$

By substituting values of R, k, and T into Eq. 1 we obtain

$$c = \sqrt{[(286.9) \ J/(kg \cdot K)](273.15 \ K)(1.401)}$$
$$= 331.4 \ (J/kg)^{1/2}$$

Thus, since $1 \ J/kg = 1 \ N \cdot m/kg = 1 \ (kg \cdot m/s^2) \cdot m/kg = 1 \ (m/s)^2$, we obtain

$$c = 331.4 \ m/s \qquad \text{(Ans)}$$

COMMENT The value of the speed of sound calculated with Eq. 11.36 agrees very well with the value of c listed in Table B.2. The ideal gas approximation does not compromise this result significantly.

11.3 Categories of Compressible Flow

Compressibility effects are more important at higher Mach numbers.

In Section 3.8.1, we learned that the effects of compressibility become more significant as the Mach number increases. For example, the error associated with using $\rho V^2/2$ in calculating the *stagnation pressure* of an ideal gas increases at larger Mach numbers. From Fig. 3.24 we can conclude that incompressible flows can only occur at low Mach numbers.

Experience has also demonstrated that compressibility can have a large influence on other important flow variables. For example, in Fig. 11.2 the variation of drag coefficient with Reynolds

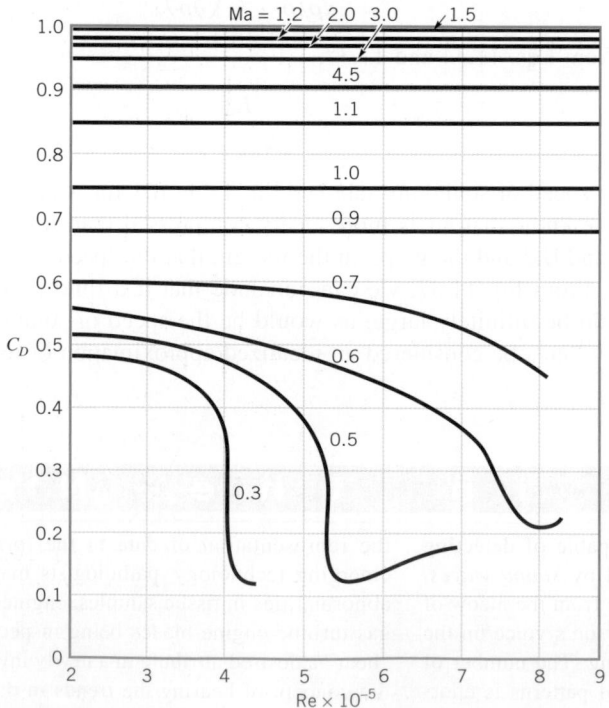

■ FIGURE 11.2 The variation of the drag coefficient of a sphere with Reynolds number and Mach number. (Adapted from Fig. 1.8 in Ref. 1 of Chapter 9.)

number and Mach number is shown for air flow over a sphere. Compressibility effects can be of considerable importance.

To further illustrate some curious features of compressible flow, a simplified example is considered. Imagine the emission of weak pressure pulses from a point source. These pressure waves are spherical and expand radially outward from the point source at the speed of sound, c. If a pressure wave is emitted at different times, t_{wave}, we can determine where several waves will be at a common instant of time, t, by using the relationship

$$r = (t - t_{wave})c$$

The wave pattern from a moving source is not symmetrical.

where r is the radius of the sphere-shaped wave emitted at time $= t_{wave}$. For a stationary point source, the symmetrical wave pattern shown in Fig. 11.3a is involved.

When the point source moves to the left with a constant velocity, V, the wave pattern is no longer symmetrical. In Figs. 11.3b, 11.3c, and 11.3d are illustrated the wave patterns at $t = 3$ s for different values of V. Also shown with a "+" are the positions of the moving point source at values of time, t, equal to 0 s, 1 s, 2 s, and 3 s. Knowing where the point source has been at different instances is important because it indicates to us where the different waves originated.

From the pressure wave patterns of Fig. 11.3, we can draw some useful conclusions. Before doing this we should recognize that if instead of moving the point source to the left, we held the point source stationary and moved the fluid to the right with velocity V, the resulting pressure wave patterns would be identical to those indicated in Fig. 11.3.

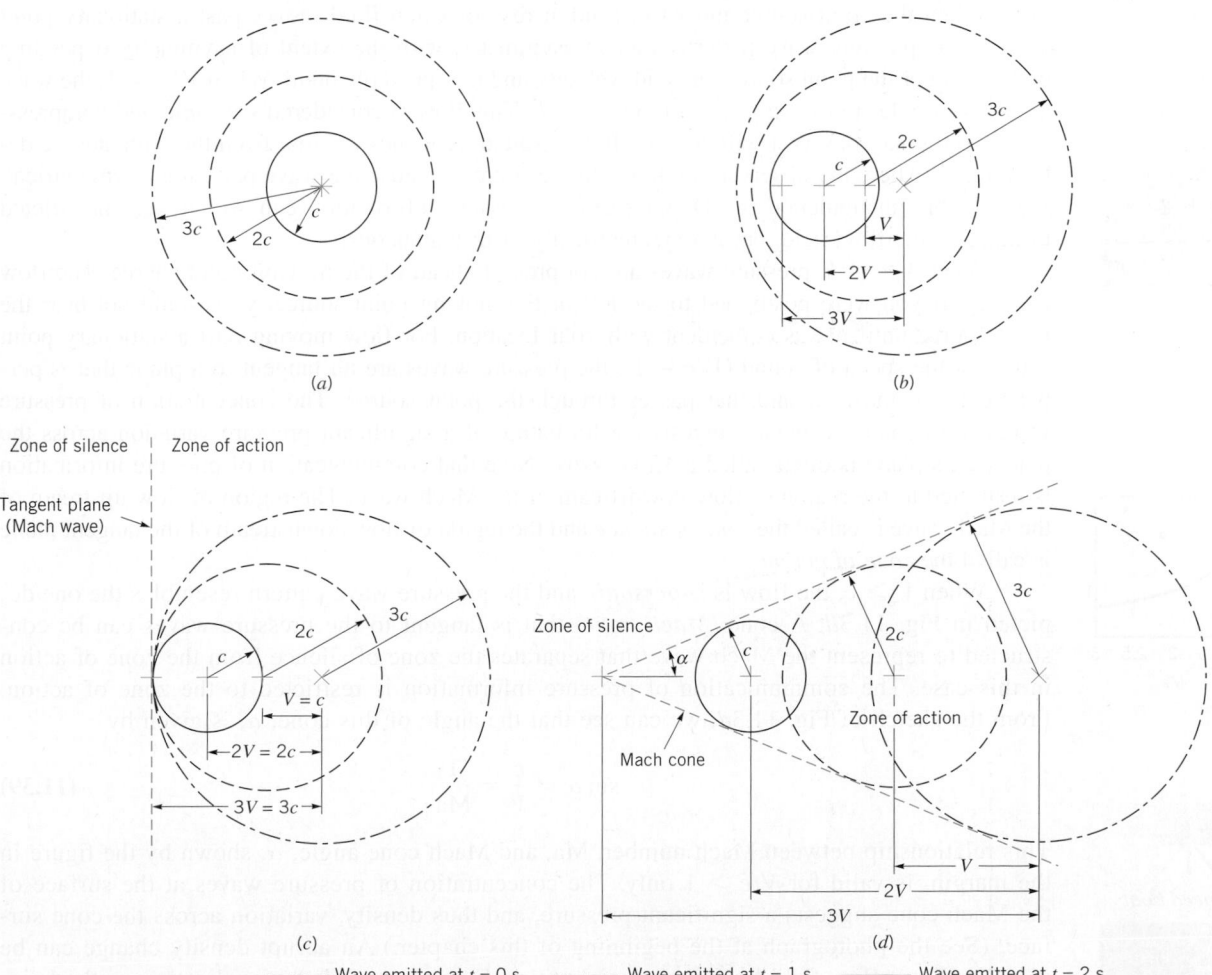

- - · - - Wave emitted at $t = 0$ s - - - - Wave emitted at $t = 1$ s —— Wave emitted at $t = 2$ s

✕ Source at $t = 0$ s + Source at $t = 1$, 2, or 3 s

■ **FIGURE   11.3**   (a) **Pressure waves at $t = 3$ s, $V = 0$; (b) pressure waves at $t = 3$ s, $V < c$; (c) pressure waves at $t = 3$ s, $V = c$; (d) pressure waves at $t = 3$ s, $V > c$.**

When the point source and the fluid are stationary, the pressure wave pattern is symmetrical (Fig. 11.3a) and an observer anywhere in the pressure field would hear the same sound frequency from the point source. When the velocity of the point source (or the fluid) is very small in comparison with the speed of sound, the pressure wave pattern will still be nearly symmetrical. The speed of sound in an incompressible fluid is infinitely large. Thus, the stationary point source and stationary fluid situation are representative of incompressible flows. For truly incompressible flows, the communication of pressure information throughout the flow field is unrestricted and instantaneous ($c = \infty$).

Pistol shrimp confound blast detectors Authorities are on the trail of fishermen in Southeast Asia and along Africa's east coast who illegally blast coral reefs to rubble to increase their catch. Researchers at Hong Kong University of Science and Technology have developed a method of using underwater microphones (hydrophones) to pick up the noise from such blasts. One complicating factor in the development of such a system is the noise produced by the claw-clicking pistol shrimp that live on the reefs. The third right appendage of the 2-in. (5-cm)-long pistol shrimp is adapted into a huge claw with a moveable finger that can be snapped shut with so much force that the resulting *sound waves* kill or stun nearby prey. When near the hydrophones, the shrimp can generate short-range shock waves that are bigger than the signal from a distant blast. By recognizing the differences between the signatures of the sound from an explosion and that of the pistol shrimp "blast," the scientists can differentiate between the two and pinpoint the location of the illegal blasts.

V11.2 Jet noise

When the point source moves in fluid at rest (or when fluid moves past a stationary point source), the pressure wave patterns vary in asymmetry, with the extent of asymmetry depending on the ratio of the point source (or fluid) velocity and the speed of sound. When $V/c < 1$, the wave pattern is similar to the one shown in Fig. 11.3b. This flow is considered *subsonic* and compressible. A stationary observer will hear a different sound frequency coming from the point source depending on where the observer is relative to the source because the wave pattern is asymmetrical. We call this phenomenon the Doppler effect. Pressure information can still travel unrestricted throughout the flow field, but not symmetrically or instantaneously.

When $V/c = 1$, pressure waves are not present ahead of the moving point source. The flow is *sonic*. If you were positioned to the left of the moving point source, you would not hear the point source until it was coincident with your location. For flow moving past a stationary point source at the speed of sound ($V/c = 1$), the pressure waves are all tangent to a plane that is perpendicular to the flow and that passes through the point source. The concentration of pressure waves in this tangent plane suggests the formation of a significant pressure variation across the plane. This plane is often called a *Mach wave*. Note that communication of pressure information is restricted to the region of flow downstream of the Mach wave. The region of flow upstream of the Mach wave is called the *zone of silence* and the region of flow downstream of the tangent plane is called the *zone of action*.

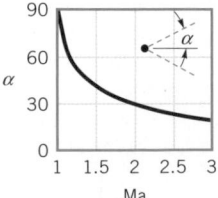

When $V > c$, the flow is *supersonic* and the pressure wave pattern resembles the one depicted in Fig. 11.3d. A cone (*Mach cone*) that is tangent to the pressure waves can be constructed to represent the Mach wave that separates the zone of silence from the zone of action in this case. The communication of pressure information is restricted to the zone of action. From the sketch of Fig. 11.3d, we can see that the angle of this cone, α, is given by

$$\sin \alpha = \frac{c}{V} = \frac{1}{\mathrm{Ma}} \tag{11.39}$$

V11.3 Speed boat

This relationship between Mach number, Ma, and Mach cone angle, α, shown by the figure in the margin, is valid for $V/c > 1$ only. The concentration of pressure waves at the surface of the Mach cone suggests a significant pressure, and thus density, variation across the cone surface. (See the photograph at the beginning of this chapter.) An abrupt density change can be visualized in a flow field by using special optics. Examples of flow visualization methods include the schlieren, shadowgraph, and interferometer techniques (see Ref. 4). A schlieren photo of a flow for which $V > c$ is shown in Fig. 11.4. The air flow through the row of compressor blade airfoils is as shown with the arrow. The flow enters supersonically ($\mathrm{Ma}_1 = 1.14$) and

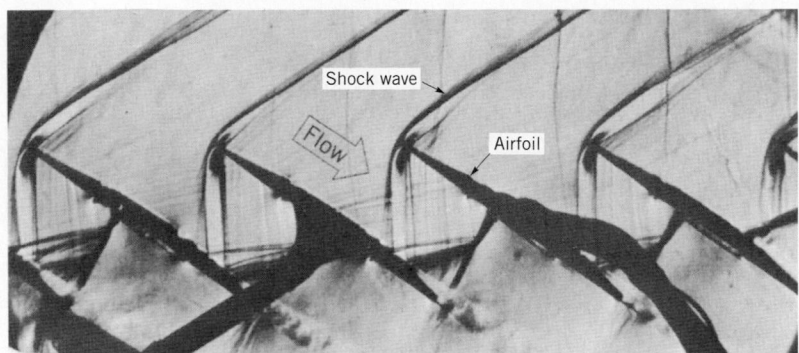

■ FIGURE 11.4   **The Schlieren visualization of flow (supersonic to subsonic) through a row of compressor airfoils. (Photograph provided by Dr. Hans Starken, Germany.)**

V11.4 Compressible flow visualization

Abrupt changes in fluid properties can occur in supersonic flows.

leaves subsonically (Ma$_2$ = 0.86). The center two airfoils have pressure tap hoses connected to them. Regions of significant changes in fluid density appear in the supersonic portion of the flow. Also, the region of separated flow on each airfoil is visible.

This discussion about pressure wave patterns suggests the following categories of fluid flow:

1. Incompressible flow: Ma $\leq$ 0.3. Unrestricted, nearly symmetrical and instantaneous pressure communication.

2. Compressible subsonic flow: 0.3 < Ma < 1.0. Unrestricted but noticeably asymmetrical pressure communication.

3. Compressible supersonic flow: Ma $\geq$ 1.0. Formation of Mach wave; pressure communication restricted to zone of action.

In addition to the above-mentioned categories of flows, two other regimes are commonly referred to: namely, *transonic flows* (0.9 $\leq$ Ma $\leq$ 1.2) and *hypersonic flows* (Ma > 5). Modern aircraft are mainly powered by gas turbine engines that involve transonic flows. When a space shuttle reenters the earth's atmosphere, the flow is hypersonic. Future aircraft may be expected to operate from subsonic to hypersonic flow conditions.

F l u i d s   i n   t h e   N e w s

Supersonic and compressible flows in gas turbines Modern gas turbine engines commonly involve compressor and turbine blades that are moving so fast that the fluid flows over the blades are locally *supersonic*. Density varies considerably in these flows so they are also considered to be *compressible*. *Shock waves* can form when these supersonic flows are sufficiently decelerated. Shocks formed at blade leading edges or on blade surfaces can interact with other blades and shocks and seriously affect blade aerodynamic and structural performance. It is possible to have supersonic flows past blades near the outer diameter of a rotor with *subsonic flows* near the inner diameter of the same rotor. These rotors are considered to be *transonic* in their operation. Very large aero gas turbines can involve thrust levels exceeding 100,000 lb (445 kN). Two of these engines are sufficient to carry over 350 passengers halfway around the world at high subsonic speed. (See Problem 11.80.)

*E*XAMPLE 11.4   Mach Cone

GIVEN   An aircraft cruising at 1000-m elevation, z, above you moves past in a flyby. It is moving with a Mach number equal to 1.5 and the ambient temperature is 20 °C.

FIND   How many seconds after the plane passes overhead do you expect to wait before you hear the aircraft?

S*OLUTION*

Since the aircraft is moving supersonically (Ma > 1), we can imagine a Mach cone originating from the forward tip of the craft as is illustrated in Fig. E11.4*a*. A photograph of this phenomenon is shown in Fig. E11.4*b*. When the surface of the cone reaches the

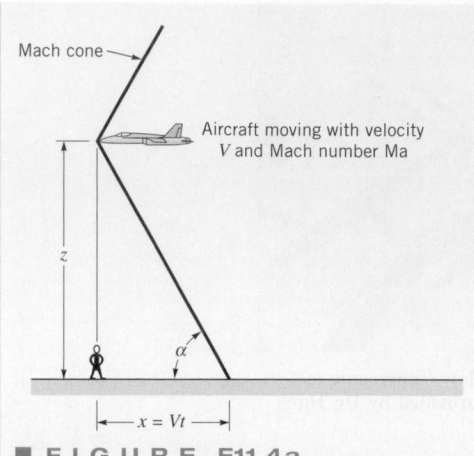

■ **FIGURE E11.4a**

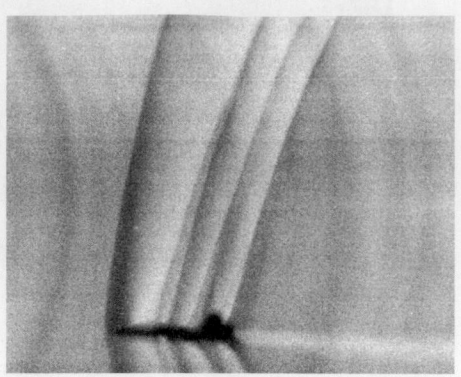

■ **FIGURE E11.4b** NASA
**Schlieren photograph of shock waves from a
T-38 aircraft at Mach 1.1, 13,000 feet (3962 m).**

observer, the "sound" of the aircraft is perceived. The angle α in Fig. E11.4 is related to the elevation of the plane, z, and the ground distance, x, by

$$\alpha = \tan^{-1} \frac{z}{x} = \tan^{-1} \frac{1000}{Vt} \qquad (1)$$

Also, assuming negligible change of Mach number with elevation, we can use Eq. 11.39 to relate Mach number to the angle α. Thus,

$$Ma = \frac{1}{\sin \alpha} \qquad (2)$$

Combining Eqs. 1 and 2 we obtain

$$Ma = \frac{1}{\sin\left[\tan^{-1}(1000/Vt)\right]} \qquad (3)$$

The speed of the aircraft can be related to the Mach number with

$$V = (Ma)c \qquad (4)$$

where c is the speed of sound. From Table B.2, $c = 343.3$ m/s. Using Ma = 1.5, we get from Eqs. 3 and 4

$$1.5 = \frac{1}{\sin\left\{\tan^{-1}\left[\dfrac{1000 \text{ m}}{(1.5)(343.3 \text{ m/s})t}\right]\right\}}$$

or

$$t = 2.17 \text{ s} \qquad \text{(Ans)}$$

COMMENT By repeating the calculations for various values of Mach number, Ma, the results shown in Fig. E11.4c are obtained. Note that for subsonic flight (Ma < 1) there is no delay since the sound travels faster than the aircraft. You can hear a subsonic aircraft approaching.

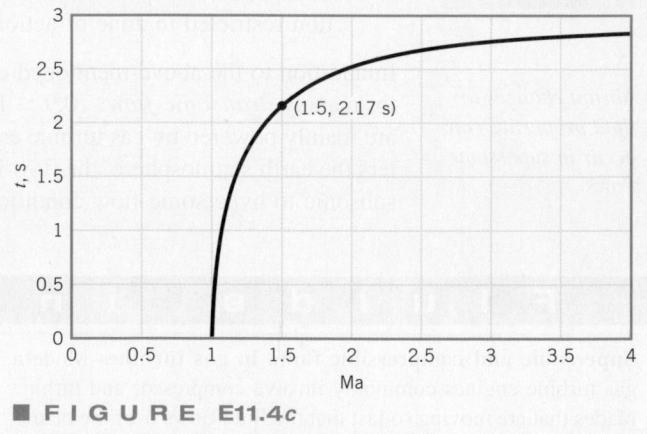

■ **FIGURE E11.4c**

11.4 Isentropic Flow of an Ideal Gas

In this section, we consider in further detail the steady, one-dimensional, isentropic flow of an ideal gas with constant specific heat values (c_p and c_v). Because the flow is steady throughout, shaft work cannot be involved. Also, as explained earlier, the one-dimensionality of flows we discuss in this chapter implies velocity and fluid property changes in the streamwise direction only. We consider flows through finite control volumes with uniformly distributed velocities and fluid properties at each section of flow. Much of what we develop can also apply to the flow of a fluid particle along its pathline.

An important class of isentropic flow involves no heat transfer and zero friction.

Isentropic flow involves constant entropy and was discussed earlier in Section 11.1, where we learned that adiabatic and frictionless (reversible) flow is one form of isentropic flow. Some ideal gas relationships for isentropic flows were developed in Section 11.1. An isentropic flow is not achievable with actual fluids because of friction. Nonetheless, the study of isentropic flow trends is useful because it helps us to gain an understanding of actual compressible flow phenomena including choked flow, shock waves, acceleration from subsonic to supersonic flow, and deceleration from supersonic to subsonic flow.

11.4.1 Effect of Variations in Flow Cross-Sectional Area

When fluid flows steadily through a conduit that has a flow cross-sectional area that varies with axial distance, the conservation of mass (continuity) equation

$$\dot{m} = \rho A V = \text{constant} \tag{11.40}$$

can be used to relate the flow rates at different sections. For incompressible flow, the fluid density remains constant and the flow velocity from section to section varies inversely with cross-sectional area. However, when the flow is compressible, density, cross-sectional area, and flow velocity can all vary from section to section. We proceed to determine how fluid density and flow velocity change with axial location in a variable area duct when the fluid is an ideal gas and the flow through the duct is steady and isentropic.

In Chapter 3, Newton's second law was applied to the inviscid (frictionless) and steady flow of a fluid particle. For the streamwise direction, the result (Eq. 3.5) for either compressible or incompressible flows is

$$dp + \tfrac{1}{2}\rho\, d(V^2) + \gamma\, dz = 0 \tag{11.41}$$

The frictionless flow from section to section through a finite control volume is also governed by Eq. 11.41, if the flow is one-dimensional, because every particle of fluid involved will have the same experience. For ideal gas flow, the potential energy difference term, $\gamma\, dz$, can be dropped because of its small size in comparison to the other terms, namely, dp and $d(V^2)$. Thus, an appropriate equation of motion in the streamwise direction for the steady, one-dimensional, and isentropic (adiabatic and frictionless) flow of an ideal gas is obtained from Eq. 11.41 as

$$\frac{dp}{\rho V^2} = -\frac{dV}{V} \tag{11.42}$$

If we form the logarithm of both sides of the continuity equation (Eq. 11.40), the result is

$$\ln \rho + \ln A + \ln V = \text{constant} \tag{11.43}$$

Density, cross-sectional area, and velocity may all vary for a compressible flow.

Differentiating Eq. 11.43 we get

$$\frac{d\rho}{\rho} + \frac{dA}{A} + \frac{dV}{V} = 0$$

or

$$-\frac{dV}{V} = \frac{d\rho}{\rho} + \frac{dA}{A} \tag{11.44}$$

Now we combine Eqs. 11.42 and 11.44 to obtain

$$\frac{dp}{\rho V^2}\left(1 - \frac{V^2}{dp/d\rho}\right) = \frac{dA}{A} \tag{11.45}$$

Since the flow being considered is isentropic, the speed of sound is related to variations of pressure with density by Eq. 11.34, repeated here for convenience as

$$c = \sqrt{\left(\frac{\partial p}{\partial \rho}\right)_s}$$

Equation 11.34, combined with the definition of Mach number

$$\text{Ma} = \frac{V}{c} \tag{11.46}$$

and Eq. 11.45 yields

$$\frac{dp}{\rho V^2}(1 - \text{Ma}^2) = \frac{dA}{A} \tag{11.47}$$

Equations 11.42 and 11.47 merge to form

$$\frac{dV}{V} = -\frac{dA}{A}\frac{1}{(1 - Ma^2)} \tag{11.48}$$

We can use Eq. 11.48 to conclude that when the flow is subsonic (Ma < 1), velocity and section area changes are in opposite directions. In other words, the area increase associated with subsonic flow through a diverging duct like the one shown in Fig. 11.5a is accompanied by a velocity decrease. Subsonic flow through a converging duct (see Fig. 11.5b) involves an increase of velocity. These trends are consistent with incompressible flow behavior, which we described earlier in this book, for instance, in Chapters 3 and 8.

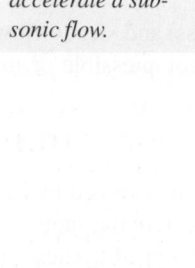

A converging duct will decelerate a supersonic flow and accelerate a subsonic flow.

Equation 11.48 also serves to show us that when the flow is supersonic (Ma > 1), velocity and area changes are in the same direction. A diverging duct (Fig. 11.5a) will accelerate a supersonic flow. A converging duct (Fig. 11.5b) will decelerate a supersonic flow. These trends are the opposite of what happens for incompressible and subsonic compressible flows.

To better understand why subsonic and supersonic duct flows are so different, we combine Eqs. 11.44 and 11.48 to form

$$\frac{d\rho}{\rho} = \frac{dA}{A}\frac{Ma^2}{(1 - Ma^2)} \tag{11.49}$$

Using Eq. 11.49, we can conclude that for subsonic flows (Ma < 1), density and area changes are in the same direction, whereas for supersonic flows (Ma > 1), density and area changes are in opposite directions. Since ρAV must remain constant (Eq. 11.40), when the duct diverges and the flow is subsonic, density and area both increase and thus flow velocity must decrease. However, for supersonic flow through a diverging duct, when the area increases, the density decreases enough so that the flow velocity has to increase to keep ρAV constant.

By rearranging Eq. 11.48, we can obtain

$$\frac{dA}{dV} = -\frac{A}{V}(1 - Ma^2) \tag{11.50}$$

Equation 11.50 gives us some insight into what happens when Ma = 1. For Ma = 1, Eq. 11.50 requires that $dA/dV = 0$. This result suggests that the area associated with Ma = 1 is either a minimum or a maximum amount.

A *converging–diverging duct* (Fig. 11.6a and margin photograph) involves a minimum area. If the flow entering such a duct were subsonic, Eq. 11.48 discloses that the fluid velocity would increase in the converging portion of the duct, and achievement of a sonic condition (Ma = 1) at the minimum area location appears possible. If the flow entering the converging–diverging duct is supersonic, Eq. 11.48 states that the fluid velocity would decrease in the converging portion of the duct and the sonic condition at the minimum area is possible.

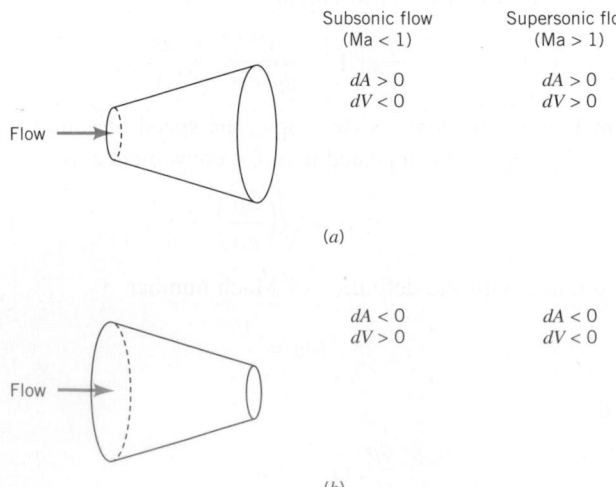

Subsonic flow
(Ma < 1)

Supersonic flow
(Ma > 1)

dA > 0
dV < 0

dA > 0
dV > 0

Flow

(a)

dA < 0
dV > 0

dA < 0
dV < 0

Flow

(b)

■ **FIGURE 11.5** (a) A diverging duct. (b) A converging duct.

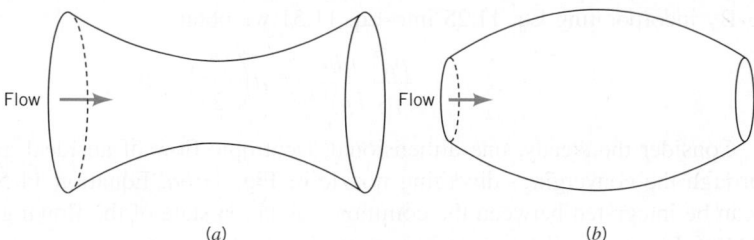

(a)   (b)

■ **F I G U R E 11.6**   (*a*) **A converging–diverging duct. (***b***) A diverging–converging duct.**

A diverging–converging duct (Fig. 11.6*b*), on the other hand, would involve a maximum area. If the flow entering this duct were subsonic, the fluid velocity would decrease in the diverging portion of the duct and the sonic condition could not be attained at the maximum area location. For supersonic flow in the diverging portion of the duct, the fluid velocity would increase and thus Ma = 1 at the maximum area is again impossible.

A converging–diverging duct is required to accelerate a flow from subsonic to supersonic flow conditions.

For the steady isentropic flow of an ideal gas, we conclude that the sonic condition (Ma = 1) can be attained in a converging–diverging duct at the minimum area location. This minimum area location is often called the ***throat*** of the converging–diverging duct. Furthermore, to achieve supersonic flow from a subsonic state in a duct, a converging–diverging area variation is necessary. For this reason, we often refer to such a duct as a *converging–diverging nozzle*. Note that a converging–diverging duct can also decelerate a supersonic flow to subsonic conditions. Thus, a converging–diverging duct can be a nozzle or a diffuser depending on whether the flow in the converging portion of the duct is subsonic or supersonic. A supersonic wind tunnel test section is generally preceded by a converging–diverging nozzle and followed by a converging–diverging diffuser (see Ref. 1). Further details about steady, isentropic, ideal gas flow through a converging–diverging duct are discussed in the next section.

11.4.2 Converging–Diverging Duct Flow

In the preceding section, we discussed the variation of density and velocity of the steady isentropic flow of an ideal gas through a variable area duct. We proceed now to develop equations that help us determine how other important flow properties vary in these flows.

It is convenient to use the stagnation state of the fluid as a reference state for compressible flow calculations. The stagnation state is associated with zero flow velocity and an entropy value that corresponds to the entropy of the flowing fluid. The subscript 0 is used to designate the stagnation state. Thus, stagnation temperature and pressure are T_0 and p_0. For example, if the fluid flowing through the converging–diverging duct of Fig. 11.6*a* were drawn isentropically from the atmosphere, the atmospheric pressure and temperature would represent the stagnation state of the flowing fluid. The stagnation state can also be achieved by isentropically decelerating a flow to zero velocity. This can be accomplished with a diverging duct for subsonic flows or a converging–diverging duct for supersonic flows. Also, as discussed earlier in Chapter 3, an approximately isentropic deceleration can be accomplished with a Pitot-static tube (see Fig. 3.6). It is thus possible to measure, with only a small amount of uncertainty, values of stagnation pressure, p_0, and stagnation temperature, T_0, of a flowing fluid.

In Section 11.1, we demonstrated that for the isentropic flow of an ideal gas (see Eq. 11.25)

$$\frac{p}{\rho^k} = \text{constant} = \frac{p_0}{\rho_0^k}$$

The streamwise equation of motion for steady and frictionless flow (Eq. 11.41) can be expressed for an ideal gas as

$$\frac{dp}{\rho} + d\left(\frac{V^2}{2}\right) = 0 \qquad (11.51)$$

since the potential energy term, $\gamma \, dz$, can be considered as being negligibly small in comparison with the other terms involved.

By incorporating Eq. 11.25 into Eq. 11.51 we obtain

$$\frac{p_0^{1/k}}{\rho_0} \frac{dp}{(p)^{1/k}} + d\left(\frac{V^2}{2}\right) = 0 \tag{11.52}$$

Consider the steady, one-dimensional, isentropic flow of an ideal gas with constant c_p and c_v through the converging–diverging nozzle of Fig. 11.6a. Equation 11.52 is valid for this flow and can be integrated between the common stagnation state of the flowing fluid to the state of the gas at any location in the converging–diverging duct to give

$$\frac{k}{k-1}\left(\frac{p_0}{\rho_0} - \frac{p}{\rho}\right) - \frac{V^2}{2} = 0 \tag{11.53}$$

By using the ideal gas equation of state (Eq. 11.1) with Eq. 11.53 we obtain

$$\frac{kR}{k-1}(T_0 - T) - \frac{V^2}{2} = 0 \tag{11.54}$$

It is of interest to note that combining Eqs. 11.14 and 11.54 leads to

$$c_p(T_0 - T) - \frac{V^2}{2} = 0$$

For isentropic flows the temperature, pressure, and density ratios are functions of the Mach number.

which, when merged with Eq. 11.9, results in

$$\check{h}_0 - \left(\check{h} + \frac{V^2}{2}\right) = 0 \tag{11.55}$$

where $\check{h}_0$ is the stagnation enthalpy. If the steady flow energy equation (Eq. 5.69) is applied to the flow situation we are presently considering, the resulting equation will be identical to Eq. 11.55. Further, we conclude that the stagnation enthalpy is constant. The conservation of momentum and energy principles lead to the same equation (Eq. 11.55) for steady isentropic flows.

The definition of Mach number (Eq. 11.46) and the speed of sound relationship for ideal gases (Eq. 11.36) can be combined with Eq. 11.54 to yield

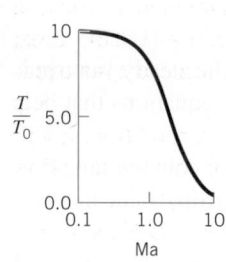

$$\boxed{\frac{T}{T_0} = \frac{1}{1 + [(k-1)/2]\text{Ma}^2}} \tag{11.56}$$

which is graphed in the margin for air. With Eq. 11.56 we can calculate the temperature of an ideal gas anywhere in the converging–diverging duct of Fig. 11.6a if the flow is steady, one-dimensional, and isentropic, provided we know the value of the local Mach number and the stagnation temperature.

We can also develop an equation for pressure variation. Since $p/\rho = RT$, then

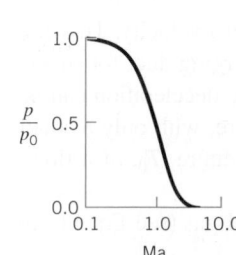

$$\left(\frac{p}{p_0}\right)\left(\frac{\rho_0}{\rho}\right) = \frac{T}{T_0} \tag{11.57}$$

From Eqs. 11.57 and 11.25 we obtain

$$\left(\frac{p}{p_0}\right) = \left(\frac{T}{T_0}\right)^{k/(k-1)} \tag{11.58}$$

Combining Eqs. 11.58 and 11.56 leads to

$$\boxed{\frac{p}{p_0} = \left\{\frac{1}{1 + [(k-1)/2]\text{Ma}^2}\right\}^{k/(k-1)}} \tag{11.59}$$

For density variation we consolidate Eqs. 11.56, 11.57, and 11.59 to get

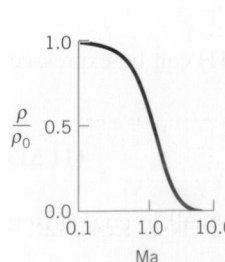

$$\boxed{\frac{\rho}{\rho_0} = \left\{\frac{1}{1 + [(k-1)/2]\text{Ma}^2}\right\}^{1/(k-1)}} \tag{11.60}$$

These relationships are graphed in the margin for air.

A very useful means of keeping track of the states of an isentropic flow of an ideal gas involves a *temperature–entropy (T–s) diagram,* as is shown in Fig. 11.7. Experience has shown (see, for example, Refs. 2 and 3) that lines of constant pressure are generally as are sketched in Fig. 11.7. An isentropic flow is confined to a vertical line on a T–s diagram. The vertical line in Fig. 11.7 is representative of flow between the stagnation state and any state within the converging–diverging nozzle. Equation 11.56 shows that fluid temperature decreases with an increase in Mach number. Thus, the lower temperature levels on a T–s diagram correspond to higher Mach numbers. Equation 11.59 suggests that fluid pressure also decreases with an increase in Mach number. Thus, lower fluid temperatures and pressures are associated with higher Mach numbers in our isentropic converging–diverging duct example.

One way to produce flow through a converging–diverging duct like the one in Fig. 11.6a is to connect the downstream end of the duct to a vacuum pump. When the pressure at the downstream end of the duct (the back pressure) is decreased slightly, air will flow from the atmosphere through the duct and vacuum pump. Neglecting friction and heat transfer and considering the air to act as an ideal gas, Eqs. 11.56, 11.59, and 11.60 and a T–s diagram can be used to describe steady flow through the converging–diverging duct.

If the pressure in the duct is only slightly less than atmospheric pressure, we predict with Eq. 11.59 that the Mach number levels in the duct will be low. Thus, with Eq. 11.60 we conclude that the variation of fluid density in the duct is also small. The continuity equation (Eq. 11.40) leads us to state that there is a small amount of fluid flow acceleration in the converging portion of the duct followed by flow deceleration in the diverging portion of the duct. We considered this type of flow when we discussed the Venturi meter in Section 3.6.3. The T–s diagram for this flow is sketched in Fig. 11.8.

We next consider what happens when the back pressure is lowered further. Since the flow starts from rest upstream of the converging portion of the duct of Fig. 11.6a, Eqs. 11.48 and 11.50 reveal to us that flow up to the nozzle throat can be accelerated to a maximum allowable Mach number of 1 at the throat. Thus, when the duct back pressure is lowered sufficiently, the Mach number at the throat of the duct will be 1. Any further decrease of the back pressure will not affect the flow in the converging portion of the duct because, as is discussed in Section 11.3, information about pressure cannot move upstream when Ma = 1. When Ma = 1 at the throat of the converging–diverging duct, we have a condition called *choked flow.* Some useful equations for choked flow are developed below.

Choked flow occurs when the Mach number is 1.0 at the minimum cross-sectional area.

We have already used the stagnation state for which Ma = 0 as a reference condition. It will prove helpful to us to use the state associated with Ma = 1 and the same entropy level as the flowing fluid as another reference condition we shall call the *critical state,* denoted ( )*.

The ratio of pressure at the converging–diverging duct throat for choked flow, p^*, to stagnation pressure, p_0, is referred to as the *critical pressure ratio.* By substituting Ma = 1 into Eq. 11.59 we obtain

$$\frac{p^*}{p_0} = \left(\frac{2}{k+1}\right)^{k/(k-1)} \tag{11.61}$$

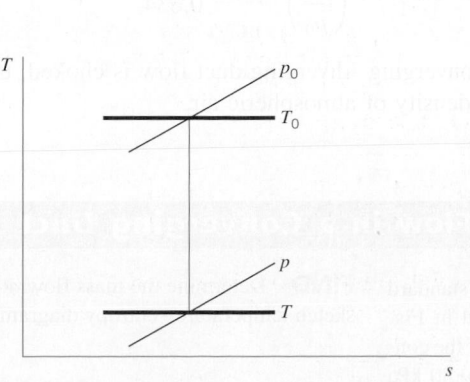

■ **FIGURE 11.7** The (T–s) diagram relating stagnation and static states.

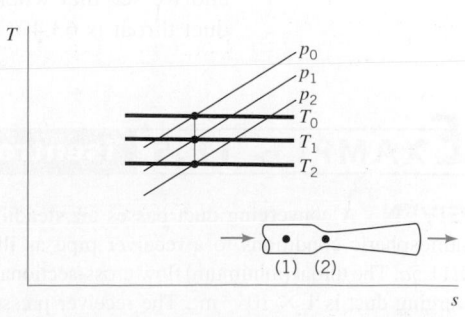

■ **FIGURE 11.8** The T–s diagram for Venturi meter flow.

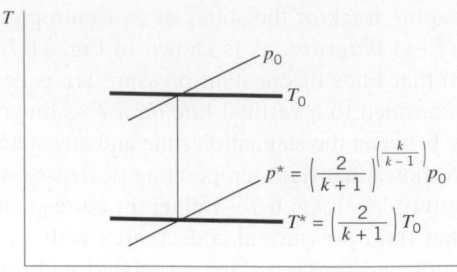

■ **FIGURE 11.9**   The relationship between the stagnation and critical states.

For $k = 1.4$, the nominal value of k for air, Eq. 11.61 yields

$$\left(\frac{p^*}{p_0}\right)_{k=1.4} = 0.528 \tag{11.62}$$

Because the stagnation pressure for our converging–diverging duct example is the atmospheric pressure, p_{atm}, the throat pressure for choked air flow is, from Eq. 11.62

$$p^*_{k=1.4} = 0.528 p_{\text{atm}}$$

We can get a relationship for the critical temperature ratio, T^*/T_0, by substituting Ma = 1 into Eq. 11.56. Thus,

$$\frac{T^*}{T_0} = \frac{2}{k+1} \tag{11.63}$$

or for $k = 1.4$

$$\left(\frac{T^*}{T_0}\right)_{k=1.4} = 0.833 \tag{11.64}$$

For the duct of Fig. 11.6a, Eq. 11.64 yields

$$T^*_{k=1.4} = 0.833 T_{\text{atm}}$$

The stagnation and critical states are at the same entropy level.

The stagnation and critical pressures and temperatures are shown on the T–s diagram of Fig. 11.9.

When we combine the ideal gas equation of state (Eq. 11.1) with Eqs. 11.61 and 11.63, for Ma = 1 we get

$$\frac{\rho^*}{\rho_0} = \left(\frac{p^*}{T^*}\right)\left(\frac{T_0}{p_0}\right) = \left(\frac{2}{k+1}\right)^{k/(k-1)}\left(\frac{k+1}{2}\right) = \left(\frac{2}{k+1}\right)^{1/(k-1)} \tag{11.65}$$

For air ($k = 1.4$), Eq. 11.65 leads to

$$\left(\frac{\rho^*}{\rho_0}\right)_{k=1.4} = 0.634 \tag{11.66}$$

and we see that when the converging–diverging duct flow is choked, the density of the air at the duct throat is 63.4% of the density of atmospheric air.

EXAMPLE 11.5   Isentropic Flow in a Converging Duct

GIVEN   A converging duct passes air steadily from standard atmospheric conditions to a receiver pipe as illustrated in Fig. E11.5a. The throat (minimum) flow cross-sectional area of the converging duct is 1×10^{-4} m^2. The receiver pressure is (a) 80 kPa (abs), (b) 40 kPa (abs).

FIND   Determine the mass flowrate through the duct and sketch temperature–entropy diagrams for situations (a) and (b).

SOLUTION

To determine the mass flowrate through the converging duct we use Eq. 11.40. Thus,

$$\dot{m} = \rho A V = \text{constant}$$

or in terms of the given throat area, A_{th},

$$\dot{m} = \rho_{th} A_{th} V_{th} \qquad (1)$$

We assume that the flow through the converging duct is isentropic and that the air behaves as an ideal gas with constant c_p and c_v. Then, from Eq. 11.60

$$\frac{\rho_{th}}{\rho_0} = \left\{ \frac{1}{1 + [(k-1)/2]\text{Ma}_{th}^2} \right\}^{1/(k-1)} \qquad (2)$$

The stagnation density, ρ_0, for the standard atmosphere is 1.23 kg/m^3 and the specific heat ratio is 1.4. To determine the throat Mach number, Ma_{th}, we can use Eq. 11.59,

$$\frac{p_{th}}{p_0} = \left\{ \frac{1}{1 + [(k-1)/2]\text{Ma}_{th}^2} \right\}^{k/(k-1)} \qquad (3)$$

The critical pressure, p^*, is obtained from Eq. 11.62 as

$$p^* = 0.528 p_0 = 0.528 p_{atm}$$
$$= (0.528)[101 \text{ kPa (abs)}] = 53.3 \text{ kPa (abs)}$$

If the receiver pressure, p_{re}, is greater than or equal to p^*, then $p_{th} = p_{re}$. If $p_{re} < p^*$, then $p_{th} = p^*$ and the flow is choked. With p_{th}, p_0, and k known, Ma_{th} can be obtained from Eq. 3, and ρ_{th} can be determined from Eq. 2.

The flow velocity at the throat can be obtained from Eqs. 11.36 and 11.46 as

$$V_{th} = \text{Ma}_{th} c_{th} = \text{Ma}_{th} \sqrt{RT_{th} k} \qquad (4)$$

The value of temperature at the throat, T_{th}, can be calculated from Eq. 11.56,

$$\frac{T_{th}}{T_0} = \frac{1}{1 + [(k-1)/2]\text{Ma}_{th}^2} \qquad (5)$$

Since the flow through the converging duct is assumed to be isentropic, the stagnation temperature is considered constant at the standard atmosphere value of $T_0 = 15 \text{ K} + 273 \text{ K} = 288 \text{ K}$. Note that absolute pressures and temperatures are used.

(a)   For $p_{re} = 80 \text{ kPa (abs)} > 53.3 \text{ kPa (abs)} = p^*$, we have $p_{th} = 80 \text{ kPa (abs)}$. Then from Eq. 3

$$\frac{80 \text{ kPa (abs)}}{101 \text{ kPa (abs)}} = \left\{ \frac{1}{1 + [(1.4-1)/2]\text{Ma}_{th}^2} \right\}^{1.4/(1.4-1)}$$

or

$$\text{Ma}_{th} = 0.587$$

From Eq. 2

$$\frac{\rho_{th}}{1.23 \text{ kg/m}^3} = \left\{ \frac{1}{1 + [(1.4-1)/2](0.587)^2} \right\}^{1/(1.4-1)}$$

or

$$\rho_{th} = 1.04 \text{ kg/m}^3$$

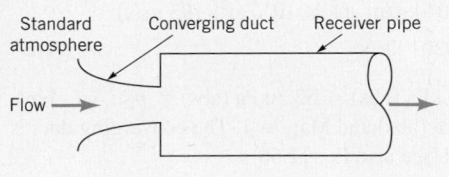

(a)

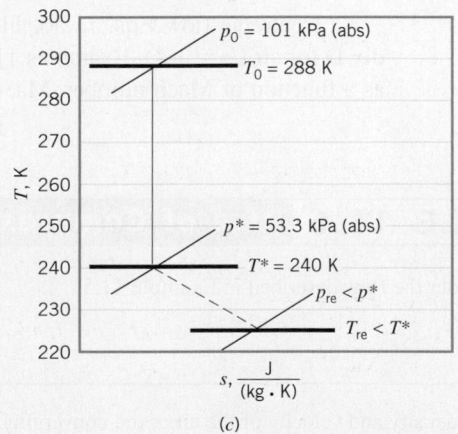

(b)

(c)

■ **FIGURE E11.5**

From Eq. 5

$$\frac{T_{th}}{288 \text{ K}} = \frac{1}{1 + [(1.4-1)/2](0.587)^2}$$

or

$$T_{th} = 269 \text{ K}$$

Substituting $\text{Ma}_{th} = 0.587$ and $T_{th} = 269 \text{ K}$ into Eq. 4 we obtain

$$V_{th} = 0.587 \sqrt{[286.9 \text{ J/(kg} \cdot \text{K)}](269 \text{ K})(1.4)}$$
$$= 193 \text{ (J/kg)}^{1/2}$$

Thus,   since $1 \text{ J/kg} = 1 \text{ N} \cdot \text{m/kg} = 1 \text{ (kg} \cdot \text{m/s}^2) \cdot \text{m/kg} = \text{(m/s)}^2$, we obtain

$$V_{th} = 193 \text{ m/s}$$

Finally from Eq. 1 we have

$$\dot{m} = (1.04 \text{ kg/m}^3)(1 \times 10^{-4} \text{ m}^2)(193 \text{ m/s})$$
$$= 0.0201 \text{ kg/s} \qquad \text{(Ans)}$$

(b) For $p_{re} = 40 \text{ kPa (abs)} < 53.3 \text{ kPa (abs)} = p^*$, we have $p_{th} = p^* = 53.3 \text{ kPa (abs)}$ and $\text{Ma}_{th} = 1$. The converging duct is choked. From Eq. 2 (see also Eq. 11.66)

$$\frac{\rho_{th}}{1.23 \text{ kg/m}^3} = \left\{ \frac{1}{1 + [(1.4 - 1)/2](1)^2} \right\}^{1/(1.4 - 1)}$$

or

$$\rho_{th} = 0.780 \text{ kg/m}^3$$

From Eq. 5 (see also Eq. 11.64),

$$\frac{T_{th}}{288 \text{ K}} = \frac{1}{1 + [(1.4 - 1)/2](1)^2}$$

or

$$T_{th} = 240 \text{ K}$$

From Eq. 4,

$$V_{th} = (1) \sqrt{[286.9 \text{ J/(kg} \cdot \text{K)}](240 \text{ K})(1.4)}$$
$$= 310 \text{ (J/kg)}^{1/2} = 310 \text{ m/s}$$

since $1 \text{ J/kg} = 1 \text{ N} \cdot \text{m/kg} = 1 (\text{kg} \cdot \text{m/s}^2) \cdot \text{m/kg} = (\text{m/s})^2$. Finally from Eq. 1

$$\dot{m} = (0.780 \text{ kg/m}^3)(1 \times 10^{-4} \text{ m}^2)(310 \text{ m/s})$$
$$= 0.0242 \text{ kg/s} \qquad \text{(Ans)}$$

From the values of throat temperature and throat pressure calculated above for flow situations (a) and (b), we can construct the temperature–entropy diagram shown in Fig. E11.5b.

COMMENT Note that the flow from standard atmosphere to the receiver for receiver pressure, p_{re}, greater than or equal to the critical pressure, p^*, is isentropic. When the receiver pressure is less than the critical pressure as in situation (b) above, what is the flow like downstream from the exit of the converging duct? Experience suggests that this flow, when $p_{re} < p^*$, is three-dimensional and nonisentropic and involves a drop in pressure from p_{th} to p_{re}, a drop in temperature, and an increase of entropy as are indicated in Fig. E11.5c.

Isentropic flow Eqs. 11.56, 11.59, and 11.60 have been used to construct Fig. D.1 in Appendix D for air ($k = 1.4$). Examples 11.6 and 11.7 illustrate how these graphs of T/T_0, p/p_0, and ρ/ρ_0 as a function of Mach number, Ma, can be used to solve compressible flow problems.

EXAMPLE 11.6 Use of Compressible Flow Graphs in Solving Problems

GIVEN Consider the flow described in Example 11.5.

FIND Solve Example 11.5 using Fig. D.1 of Appendix D.

SOLUTION

We still need the density and velocity of the air at the converging duct throat to solve for mass flowrate from

$$\dot{m} = \rho_{th} A_{th} V_{th} \qquad (1)$$

(a) Since the receiver pressure, $p_{re} = 80 \text{ kPa (abs)}$, is greater than the critical pressure, $p^* = 53.3 \text{ kPa (abs)}$, the throat pressure, p_{th}, is equal to the receiver pressure. Thus

$$\frac{p_{th}}{p_0} = \frac{80 \text{ kPa (abs)}}{101 \text{ kPa (abs)}} = 0.792$$

From Fig. D.1, for $p/p_0 = 0.79$, we get from the graph

$$\text{Ma}_{th} = 0.59$$

$$\frac{T_{th}}{T_0} = 0.94 \qquad (2)$$

$$\frac{\rho_{th}}{\rho_0} = 0.85 \qquad (3)$$

Thus, from Eqs. 2 and 3

$$T_{th} = (0.94)(288 \text{ K}) = 271 \text{ K}$$

and

$$\rho_{th} = (0.85)(1.23 \text{ kg/m}^3) = 1.04 \text{ kg/m}^3$$

Furthermore, using Eqs. 11.36 and 11.46 we get

$$V_{th} = \text{Ma}_{th} \sqrt{RT_{th}k}$$
$$= (0.59) \sqrt{[286.9 \text{ J/(kg} \cdot \text{K)}](269 \text{ K})(1.4)}$$
$$= 194 \text{ (J/kg)}^{1/2} = 194 \text{ m/s}$$

since $1 \text{ J/kg} = 1 \text{ N} \cdot \text{m/kg} = 1 (\text{kg} \cdot \text{m/s}^2) \cdot \text{m/kg} = (\text{m/s})^2$. Finally, from Eq. 1

$$\dot{m} = (1.04 \text{ kg/m}^3)(1 \times 10^{-4} \text{ m}^2)(194 \text{ m/s})$$
$$= 0.0202 \text{ kg/s} \qquad \text{(Ans)}$$

(b) For $p_{re} = 40$ kPa (abs) < 53.3 kPa (abs) $= p^*$, the throat pressure is equal to 53.3 kPa (abs) and the duct is choked with $Ma_{th} = 1$. From Fig. D.1, for Ma = 1 we get

$$\frac{T_{th}}{T_0} = 0.83 \tag{4}$$

and

$$\frac{\rho_{th}}{\rho_0} = 0.64 \tag{5}$$

From Eqs. 4 and 5 we obtain

$$T_{th} = (0.83)(288 \text{ K}) = 240 \text{ K}$$

and

$$\rho_{th} = (0.64)(1.23 \text{ kg/m}^3) = 0.79 \text{ kg/m}^3$$

Also, from Eqs. 11.36 and 11.46 we conclude that

$$\begin{aligned} V_{th} &= Ma_{th} \sqrt{RT_{th}k} \\ &= (1) \sqrt{[286.9 \text{ J/(kg} \cdot \text{K)}](240 \text{ K})(1.4)} \\ &= 310 \text{ (J/kg)}^{1/2} = 310 \text{ m/s} \end{aligned}$$

Then, from Eq. 1

$$\begin{aligned} \dot{m} &= (0.79 \text{ kg/m}^3)(1 \times 10^{-4} \text{ m}^2)(310 \text{ m/s}) \\ &= 0.024 \text{ kg/s} \end{aligned} \tag{Ans}$$

COMMENT   The values from Fig. D.1 resulted in answers for mass flowrate that are close to those using the ideal gas equations (see Example 11.5).

The temperature–entropy diagrams remain the same as those provided in the solution of Example 11.5.

EXAMPLE 11.7   Static to Stagnation Pressure Ratio

GIVEN   The static pressure to stagnation pressure ratio at a point in a flow stream is measured with a Pitot-static tube (see Fig. 3.6) as being equal to 0.82. The stagnation temperature of the fluid is 20 °C.

FIND   Determine the flow velocity if the fluid is (a) air, (b) helium.

SOLUTION

We consider both air and helium, flowing as described above, to act as ideal gases with constant specific heats. Then, we can use any of the ideal gas relationships developed in this chapter. To determine the flow velocity, we can combine Eqs. 11.36 and 11.46 to obtain

$$V = Ma \sqrt{RTk} \tag{1}$$

By knowing the value of static to stagnation pressure ratio, p/p_0, and the specific heat ratio we can obtain the corresponding Mach number from Eq. 11.59, or for air, from Fig. D.1. Figure D.1 cannot be used for helium, since k for helium is 1.66 and Fig. D.1 is for $k = 1.4$ only. With Mach number, specific heat ratio, and stagnation temperature known, the value of static temperature can be subsequently ascertained from Eq. 11.56 (or Fig. D.1 for air).

(a)   For air, $p/p_0 = 0.82$; thus from Fig. D.1,

$$Ma = 0.54 \tag{2}$$

and

$$\frac{T}{T_0} = 0.94 \tag{3}$$

Then, from Eq. 3

$$T = (0.94)[(20 + 274)\text{K}] = 294 \text{ K} \tag{4}$$

and using Eqs. 1, 2, and 4 we get

$$\begin{aligned} V &= (0.54) \sqrt{[286.9 \text{ J/(kg} \cdot \text{K)}](294 \text{ K})(1.4)} \\ &= 186 \text{ m/s} \end{aligned} \tag{Ans}$$

(b)   For helium, $p/p_0 = 0.82$ and $k = 1.66$. By substituting these values into Eq. 11.59 we get

$$0.82 = \left\{ \frac{1}{1 + [(1.66 - 1)/2] \, Ma^2} \right\}^{1.66/(1.66 - 1)}$$

or

$$Ma = 0.499$$

From Eq. 11.56 we obtain

$$\frac{T}{T_0} = \frac{1}{1 + [(k - 1)/2]Ma^2}$$

Thus,

$$\begin{aligned} T &= \left\{ \frac{1}{1 + [(1.66 - 1)/2](0.499)^2} \right\}[(20 + 274)\text{K}] \\ &= 272 \text{ K} \end{aligned}$$

From Eq. 1 we obtain

$$\begin{aligned} V &= (0.499) \sqrt{(286.9 \text{ J/kg} \cdot \text{K})(272 \text{ K})(1.66)} \\ &= 180 \text{ m/s} \end{aligned} \tag{Ans}$$

COMMENT Note that the isentropic flow equations and Fig. D.1 for $k = 1.4$ were used presently to describe fluid particle isentropic flow along a pathline in a stagnation process. Even though these equations and graph were developed for one-dimensional duct flows, they can be used for frictionless, adiabatic pathline flows also.

Furthermore, while the Mach numbers calculated above are of similar size for the air and helium flows, the flow speed is much larger for helium than for air because the speed of sound in helium is much larger than it is in air.

Also included in Fig. D.1 is a graph of the ratio of local area, A, to critical area, A^*, for different values of local Mach number. The importance of this area ratio is clarified below.

For choked flow through the converging–diverging duct of Fig. 11.6a, the conservation of mass equation (Eq. 11.40) yields

$$\rho A V = \rho^* A^* V^*$$

or

$$\frac{A}{A^*} = \left(\frac{\rho^*}{\rho}\right)\left(\frac{V^*}{V}\right) \tag{11.67}$$

From Eqs. 11.36 and 11.46, we obtain

$$V^* = \sqrt{RT^*k} \tag{11.68}$$

and

$$V = \text{Ma}\,\sqrt{RTk} \tag{11.69}$$

By combining Eqs. 11.67, 11.68, and 11.69 we get

$$\frac{A}{A^*} = \frac{1}{\text{Ma}}\left(\frac{\rho^*}{\rho_0}\right)\left(\frac{\rho_0}{\rho}\right)\sqrt{\frac{(T^*/T_0)}{(T/T_0)}} \tag{11.70}$$

The incorporation of Eqs. 11.56, 11.60, 11.63, 11.65, and 11.70 results in

$$\frac{A}{A^*} = \frac{1}{\text{Ma}}\left\{\frac{1 + [(k - 1)/2]\text{Ma}^2}{1 + [(k - 1)/2]}\right\}^{(k+1)/[2(k-1)]} \tag{11.71}$$

The ratio of flow area to the critical area is a useful concept for isentropic duct flow.

Equation 11.71 was used to generate the values of A/A^* for air ($k = 1.4$) in Fig. D.1. These values of A/A^* are graphed as a function of Mach number in Fig. 11.10. As is demonstrated in the following examples, whether or not the critical area, A^*, is physically present in the flow, the area ratio, A/A^*, is still a useful concept for the isentropic flow of an ideal gas through a converging–diverging duct.

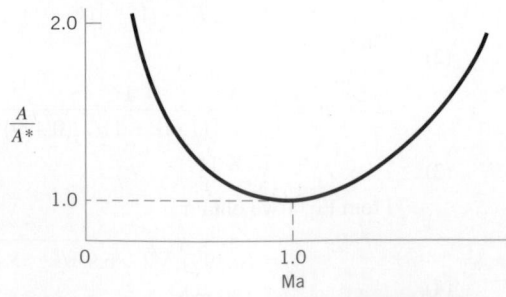

■ **FIGURE 11.10** The variation of area ratio with Mach number for isentropic flow of an ideal gas ($k = 1.4$, linear coordinate scales).

EXAMPLE 11.8 | Isentropic Choked Flow in a Converging–Diverging Duct with Subsonic Entry

GIVEN Air enters subsonically from standard atmosphere and flows isentropically through a choked converging–diverging duct having a circular cross-sectional area, A, that varies with axial distance from the throat, x, according to the formula

$$A = 0.1 + x^2$$

where A is in square meters and x is in meters. The duct extends from $x = -0.5$ m to $x = +0.5$ m.

FIND For this flow situation, sketch the side view of the duct and graph the variation of Mach number, static temperature to stagnation temperature ratio, T/T_0, and static pressure to stagnation pressure ratio, p/p_0, through the duct from $x = -0.5$ m to $x = +0.5$ m. Also show the possible fluid states at $x = -0.5$ m, 0 m, and $+0.5$ m using temperature–entropy coordinates.

SOLUTION

The side view of the converging–diverging duct is a graph of radius r from the duct axis as a function of axial distance. For a circular flow cross section we have

$$A = \pi r^2 \tag{1}$$

where

$$A = 0.1 + x^2 \tag{2}$$

Thus, combining Eqs. 1 and 2, we have

$$r = \left(\frac{0.1 + x^2}{\pi}\right)^{1/2} \tag{3}$$

and a graph of radius as a function of axial distance can be easily constructed (see Fig. E11.8a).

Since the converging–diverging duct in this example is choked, the throat area is also the critical area, A^*. From Eq. 2 we see that

$$A^* = 0.1 \text{ m}^2 \tag{4}$$

For any axial location, from Eqs. 2 and 4 we get

$$\frac{A}{A^*} = \frac{0.1 + x^2}{0.1} \tag{5}$$

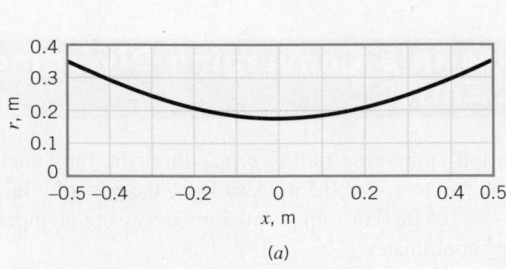

(a)

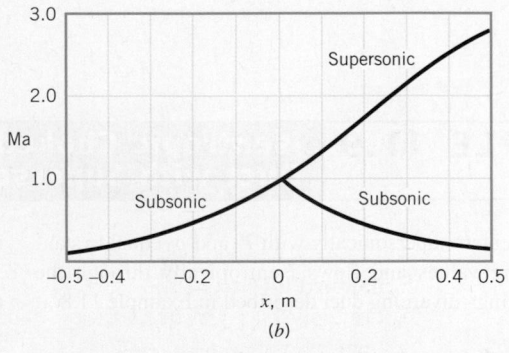

(b)

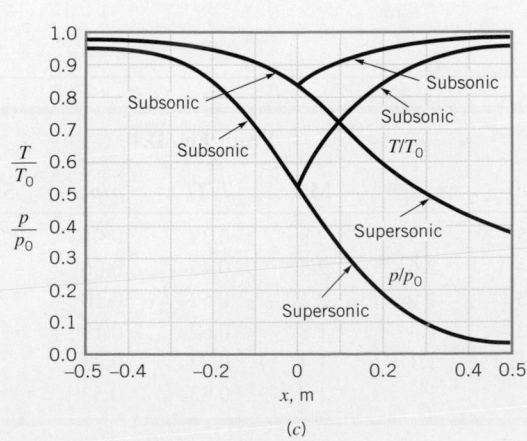

(c)

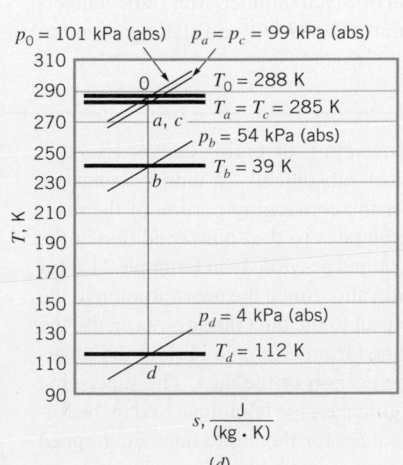

(d)

■ **FIGURE E11.8**

Values of A/A^* from Eq. 5 can be used in Eq. 11.71 to calculate corresponding values of Mach number, Ma. For air with $k = 1.4$, we could enter Fig. D.1 with values of A/A^* and read off values of the Mach number. With values of Mach number ascertained, we could use Eqs. 11.56 and 11.59 to calculate related values of T/T_0 and p/p_0. For air with $k = 1.4$, Fig. D.1 could be entered with A/A^* or Ma to get values of T/T_0 and p/p_0. To solve this example, we elect to use values from Fig. D.1.

The following table was constructed by using Eqs. 3 and 5 and Fig. D.1.

With the air entering the choked converging–diverging duct subsonically, only one isentropic solution exists for the converging portion of the duct. This solution involves an accelerating flow that becomes sonic (Ma = 1) at the throat of the passage. Two isentropic flow solutions are possible for the diverging portion of the duct—one subsonic, the other supersonic. If the pressure ratio, p/p_0, is set at 0.98 at $x = +0.5$ m (the outlet), the subsonic flow will occur. Alternatively, if p/p_0 is set at 0.04 at $x = +0.5$ m, the supersonic flow field will exist. These conditions are illustrated in Fig. E11.8. An unchoked subsonic flow through the converging–diverging duct of this example is discussed in Example 11.10. Choked flows involving flows other than the two isentropic flows in the diverging portion of the duct of this example are discussed after Example 11.10.

COMMENT   Note that if the diverging portion of this duct is extended, larger values of A/A^* and Ma are achieved. From Fig. D1, note that further increases of A/A^* result in smaller changes of Ma after A/A^* values of about 10. The ratio of p/p_0

x (m)	From Eq. 3, r (m)	From Eq. 5, A/A^*	From Fig. D.1 Ma	T/T_0	p/p_0	State
Subsonic Solution						
−0.5	0.334	3.5	0.17	0.99	0.98	a
−0.4	0.288	2.6	0.23	0.99	0.97	
−0.3	0.246	1.9	0.32	0.98	0.93	
−0.2	0.211	1.4	0.47	0.96	0.86	
−0.1	0.187	1.1	0.69	0.91	0.73	
0	0.178	1	1.00	0.83	0.53	b
+0.1	0.187	1.1	0.69	0.91	0.73	
+0.2	0.211	1.4	0.47	0.96	0.86	
+0.3	0.246	1.9	0.32	0.98	0.93	
+0.4	0.288	2.6	0.23	0.99	0.97	
+0.5	0.344	3.5	0.17	0.99	0.98	c
Supersonic Solution						
+0.1	0.187	1.1	1.37	0.73	0.33	
+0.2	0.211	1.4	1.76	0.62	0.18	
+0.3	0.246	1.9	2.14	0.52	0.10	
+0.4	0.288	2.6	2.48	0.45	0.06	
+0.5	0.334	3.5	2.80	0.39	0.04	d

becomes vanishingly small, suggesting a practical limit to the expansion.

EXAMPLE 11.9   Isentropic Choked Flow in a Converging–Diverging Duct with Supersonic Entry

GIVEN   Air enters supersonically with T_0 and p_0 equal to standard atmosphere values and flows isentropically through the choked converging–diverging duct described in Example 11.8.

FIND   Graph the variation of Mach number, Ma, static temperature to stagnation temperature ratio, T/T_0, and static pressure to stagnation pressure ratio, p/p_0, through the duct from $x = -0.5$ m to $x = +0.5$ m. Also show the possible fluid states at $x = -0.5$ m, 0 m, and $+0.5$ m by using temperature–entropy coordinates.

SOLUTION

With the air entering the converging–diverging duct of Example 11.8 supersonically instead of subsonically, a unique isentropic flow solution is obtained for the converging portion of the duct. Now, however, the flow decelerates to the sonic condition at the throat. The two solutions obtained previously in Example 11.8 for the diverging portion are still valid. Since the area variation in the duct is symmetrical with respect to the duct throat, we can use the supersonic flow values obtained from Example 11.8 for the supersonic flow in the converging portion of the duct. The supersonic flow solution for the converging passage is summarized in the following table. The solution values for the entire duct are graphed in Fig. E11.9.

x (m)	A/A^*	From Fig. D.1 Ma	T/T_0	p/p_0	State
−0.5	3.5	2.8	0.39	0.04	e
−0.4	2.6	2.5	0.45	0.06	
−0.3	1.9	2.1	0.52	0.10	
−0.2	1.4	1.8	0.62	0.18	
−0.1	1.1	1.4	0.73	0.33	
0	1	1.0	0.83	0.53	b

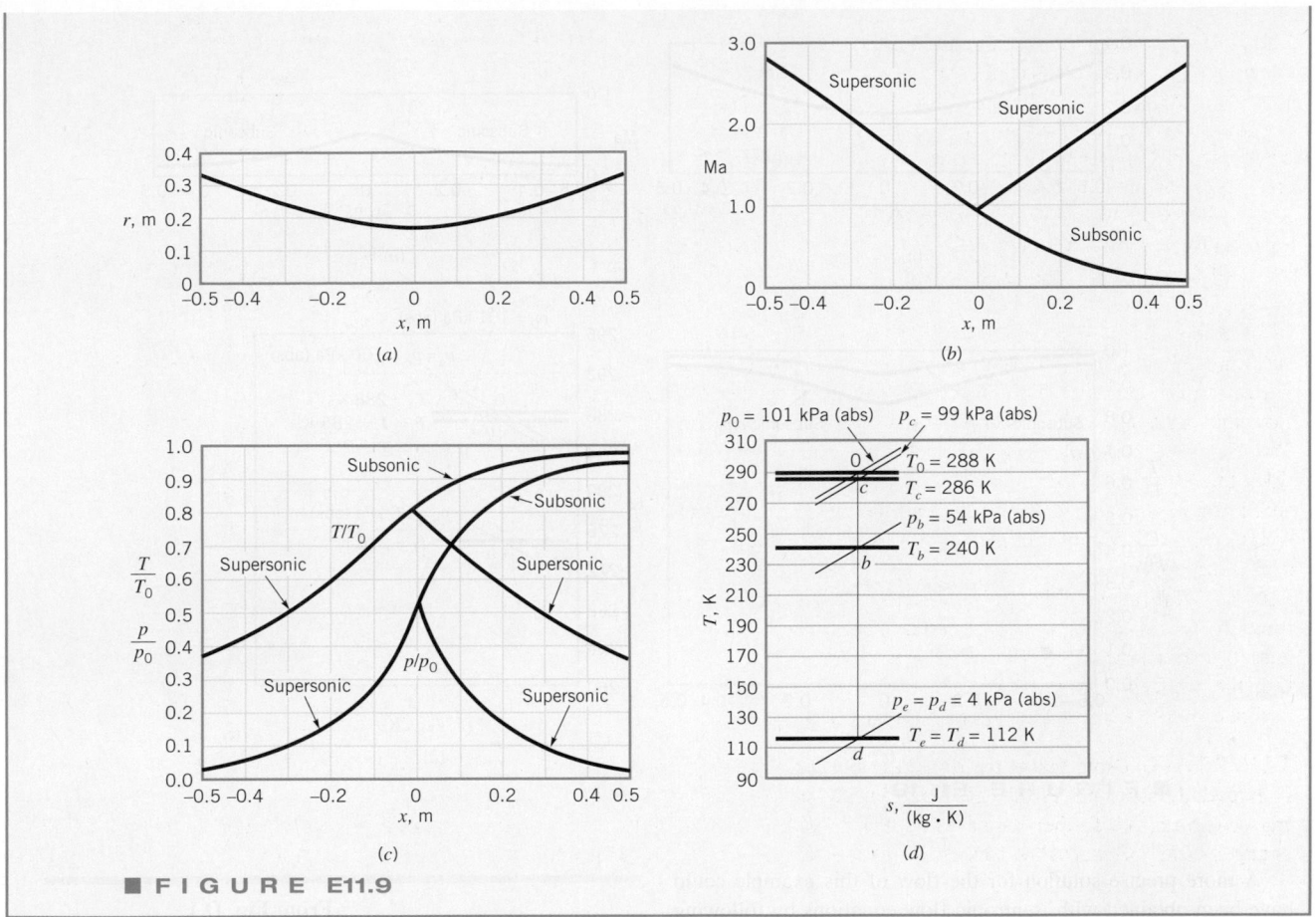

■ **FIGURE E11.9**

EXAMPLE 11.10 — Isentropic Unchoked Flow in a Converging–Diverging Duct

GIVEN Air flows subsonically and isentropically through the converging–diverging duct of Example 11.8.

FIND Graph the variation of Mach number, Ma, static temperature to stagnation temperature ratio, T/T_0, and the static pressure to stagnation pressure ratio, p/p_0, through the duct from $x = -0.5$ m to $x = +0.5$ m for Ma = 0.48 at $x = 0$ m. Show the corresponding temperature–entropy diagram.

SOLUTION

Since for this example, Ma = 0.48 at $x = 0$ m, the isentropic flow through the converging–diverging duct will be entirely subsonic and not choked. For air ($k = 1.4$) flowing isentropically through the duct, we can use Fig. D.1 for flow field quantities. Entering Fig. D.1 with Ma = 0.48 we read off $p/p_0 = 0.85$, $T/T_0 = 0.96$, and $A/A^* = 1.4$. Even though the duct flow is not choked in this example and A^* does not therefore exist physically, it still represents a valid reference. For a given isentropic flow, p_0, T_0, and A^* are constants. Since A at $x = 0$ m is equal to 0.10 m² (from Eq. 2 of Example 11.8),

A^* for this example is

$$A^* = \frac{A}{(A/A^*)} = \frac{0.10 \text{ m}^2}{1.4} = 0.07 \text{ m}^2 \qquad (1)$$

With known values of duct area at different axial locations, we can calculate corresponding area ratios, A/A^*, knowing $A^* = 0.07$ m². Having values of the area ratio, we can use Fig. D.1 and obtain related values of Ma, T/T_0, and p/p_0. The following table summarizes flow quantities obtained in this manner. The results are graphed in Fig. E11.10.

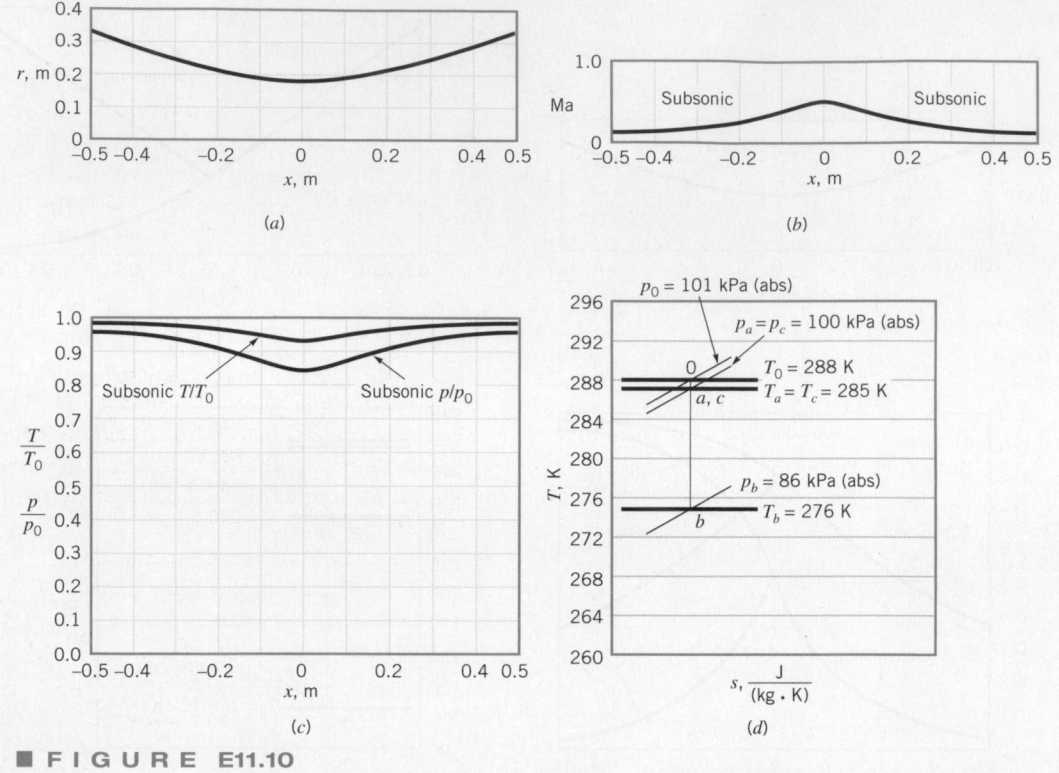

■ **F I G U R E   E11.10**

A more precise solution for the flow of this example could have been obtained with isentropic flow equations by following the steps outlined below.

1. Use Eq. 11.59 to get p/p_0 at $x = 0$ knowing k and Ma = 0.48.
2. From Eq. 11.71, obtain value of A/A^* at $x = 0$ knowing k and Ma.
3. Determine A^* knowing A and A/A^* at $x = 0$.
4. Determine A/A^* at different axial locations, x.
5. Use Eq. 11.71 and A/A^* from step 4 above to get values of Mach numbers at different axial locations.
6. Use Eqs. 11.56 and 11.59 and Ma from step 5 above to obtain T/T_0 and p/p_0 at different axial locations, x.

COMMENT There are an infinite number of subsonic, isentropic flow solutions for the converging–diverging duct considered in this example (one for any given Ma < 1 at $x = 0$).

	Calculated,	From Fig. D.1			
x (m)	A/A^*	Ma	T/T_0	p/p_0	State
−0.5	5.0	0.12	0.99	0.99	a
−0.4	3.7	0.16	0.99	0.98	
−0.3	2.7	0.23	0.99	0.96	
−0.2	2.0	0.31	0.98	0.94	
−0.1	1.6	0.40	0.97	0.89	
0	1.4	0.48	0.96	0.85	b
+0.1	1.6	0.40	0.97	0.89	
+0.2	2.0	0.31	0.98	0.94	
+0.3	2.7	0.23	0.99	0.96	
+0.4	3.7	0.16	0.99	0.98	
+0.5	5.0	0.12	0.99	0.99	c

F l u i d s   i n   t h e   N e w s

Liquid knife A supersonic stream of liquid nitrogen is capable of cutting through engineering materials like steel and concrete. Originally developed at the Idaho National Engineering Laboratory for cutting open barrels of waste products, this technology is now more widely available. The fast moving nitrogen enters the cracks and crevices of the material being cut then expands rapidly and breaks up the solid material it has penetrated. After doing its work, the nitrogen gas simply becomes part of the atmosphere which is mostly nitrogen already. This technology is also useful for stripping coatings even from delicate surfaces.

The isentropic flow behavior for the converging–diverging duct discussed in Examples 11.8, 11.9, and 11.10 is summarized in the area ratio–Mach number graphs sketched in Fig. 11.11. The points a, b, and c represent states at axial distance $x = -0.5$ m, 0 m, and $+0.5$ m. In Fig. 11.11a, the isentropic flow through the converging–diverging duct is subsonic without choking at the throat. This situation was discussed in Example 11.10. Figure 11.11b represents subsonic to subsonic choked flow (Example 11.8) and Fig. 11.11c is for subsonic to supersonic choked flow (also Example 11.8). The states in Fig. 11.11d are related to the supersonic to supersonic choked flow of Example 11.9; the states in Fig. 11.11e are for the supersonic to subsonic choked flow of Example 11.9. Not covered by an example but also possible are the isentropic flow states a, b, and c shown in Fig. 11.11f for supersonic to supersonic flow without choking. These six categories generally represent the possible kinds of isentropic, ideal gas flow through a converging–diverging duct.

For a given stagnation state (i.e., T_0 and p_0 fixed), ideal gas ($k =$ constant), and converging–diverging duct geometry, an infinite number of isentropic subsonic to subsonic (not choked) and supersonic to supersonic (not choked) flow solutions exist. In contrast, the isentropic subsonic to supersonic (choked), subsonic to subsonic (choked), supersonic to subsonic (choked), and supersonic to supersonic (choked) flow solutions are each unique. The above-mentioned isentropic

A variety of flow situations can occur for flow in a converging–diverging duct.

V11.5 Rocket engine start-up

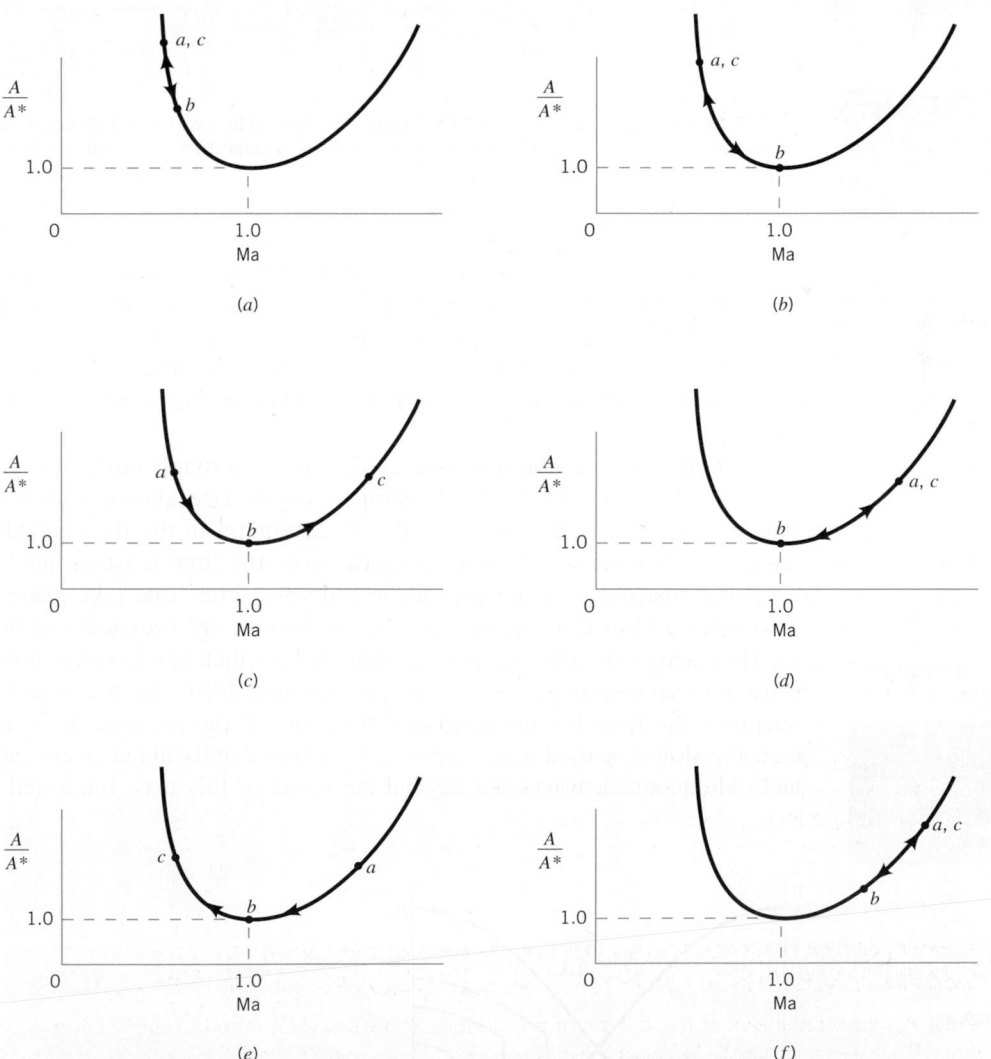

■ **FIGURE 11.11**   (*a*) Subsonic to subsonic isentropic flow (not choked). (*b*) Subsonic to subsonic isentropic flow (choked). (*c*) Subsonic to supersonic isentropic flow (choked). (*d*) Supersonic to supersonic isentropic flow (choked). (*e*) Supersonic to subsonic isentropic flow (choked). (*f*) Supersonic to supersonic isentropic flow (not choked).

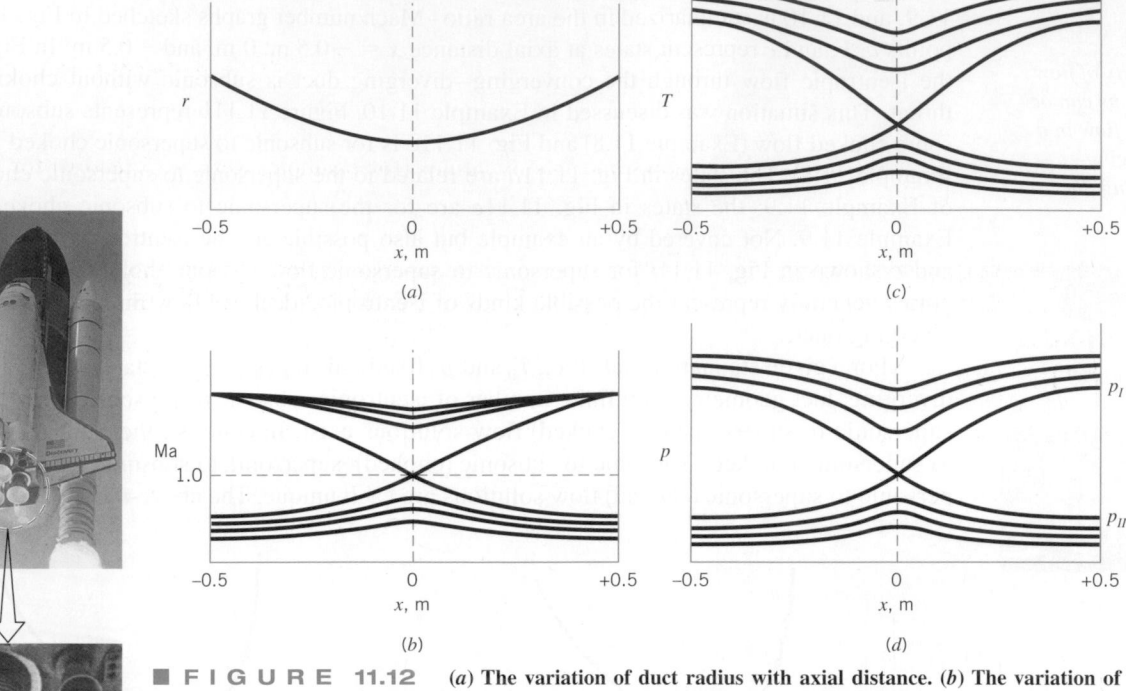

■ **FIGURE 11.12** (*a*) **The variation of duct radius with axial distance.** (*b*) **The variation of Mach number with axial distance.** (*c*) **The variation of temperature with axial distance.** (*d*) **The variation of pressure with axial distance.**

Photographs courtesy of NASA.

V11.6 Supersonic nozzle flow

flow solutions are represented in Fig. 11.12. When the pressure at $x = +0.5$ (exit) is greater than or equal to p_I indicated in Fig. 11.12*d*, an isentropic flow is possible. When the pressure at $x = +0.5$ is equal to or less than p_{II}, isentropic flows in the duct are possible. However, when the exit pressure is less than p_I and greater than p_{III} as indicated in Fig. 11.13, isentropic flows are no longer possible in the duct. Determination of the value of p_{III} is discussed in Example 11.19.

Some possible nonisentropic choked flows through our converging–diverging duct are represented in Fig. 11.13. Each abrupt pressure rise shown within and at the exit of the flow passage occurs across a very thin discontinuity in the flow called a *normal shock wave*. Except for flow across the normal shock wave, the flow is isentropic. The nonisentropic flow equations that describe the changes in fluid properties that take place across a normal shock wave are developed in Section 11.5.3. The less abrupt pressure rise or drop that occurs after the flow leaves the duct is nonisentropic and attributable to three-dimensional *oblique shock waves* or **expansion waves** (see margin photograph). If the pressure rises downstream of the duct exit, the flow is considered *overexpanded*. If the pressure drops downstream of the duct exit, the flow is called *underexpanded*. Further details about over- and underexpanded flows and oblique shock waves are beyond the scope of this text. Interested readers are referred to

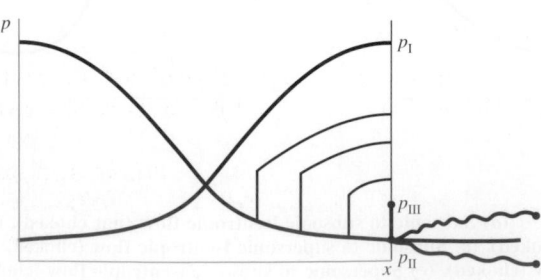

■ **FIGURE 11.13** **Shock formation in converging–diverging duct flows.**

Constant area duct

Fluid flow ⟶

■ **F I G U R E   11.14   Constant area duct flow.**

texts on compressible flows and gas dynamics (for example, Refs. 4, 5, and 6) for additional material on this subject.

F l u i d s   i n   t h e   N e w s

Rocket nozzles To develop the massive thrust needed for Space Shuttle liftoff, the gas leaving the rocket *nozzles* must be moving *supersonically*. For this to happen, the nozzle flow path must first *converge*, then *diverge*. Entering the nozzle at very high pressure and temperature, the gas accelerates in the converging portion of the nozzle until the flow *chokes* at the nozzle *throat*. Downstream of the throat, the gas further accelerates in the diverging portion of the nozzle (area ratio of 77.5 to 1), finally exiting into the atmos-

phere supersonically. At launch, the static pressure of the gas flowing from the nozzle exit is less than atmospheric and so the flow is *overexpanded*. At higher elevations where the atmospheric pressure is much less than at launch level, the static pressure of the gas flowing from the nozzle exit is greater than atmospheric and so now the flow is *underexpanded*, the result being expansion or divergence of the exhaust gas as it exits into the atmosphere. (See Problem 11.48.)

11.4.3 Constant Area Duct Flow

For steady, one-dimensional, isentropic flow of an ideal gas through a constant area duct (see Fig. 11.14), Eq. 11.50 suggests that $dV = 0$ or that flow velocity remains constant. With the energy equation (Eq. 5.69) we can conclude that since flow velocity is constant, the fluid enthalpy and thus temperature are also constant for this flow. This information and Eqs. 11.36 and 11.46 indicate that the Mach number is constant for this flow also. This being the case, Eqs. 11.59 and 11.60 tell us that fluid pressure and density also remain unchanged. Thus, we see that a steady, one-dimensional, isentropic flow of an ideal gas does not involve varying velocity or fluid properties unless the flow cross-sectional area changes.

In Section 11.5 we discuss nonisentropic, steady, one-dimensional flows of an ideal gas through a constant area duct and also a normal shock wave. We learn that friction and/or heat transfer can also accelerate or decelerate a fluid.

11.5   Nonisentropic Flow of an Ideal Gas

Fanno flow involves wall friction with no heat transfer and constant cross-sectional area.

Actual fluid flows are generally nonisentropic. An important example of *nonisentropic flow* involves adiabatic (no heat transfer) flow with friction. Flows with heat transfer (diabatic flows) are generally nonisentropic also. In this section we consider the adiabatic flow of an ideal gas through a constant area duct with friction. This kind of flow is often referred to as *Fanno flow*. We also analyze the diabatic flow of an ideal gas through a constant area duct without friction (*Rayleigh flow*). The concepts associated with Fanno and Rayleigh flows lead to further discussion of normal shock waves.

11.5.1 Adiabatic Constant Area Duct Flow with Friction (Fanno Flow)

Consider the steady, one-dimensional, and adiabatic flow of an ideal gas through the constant area duct shown in Fig. 11.15. This is Fanno flow. For the control volume indicated, the energy equation (Eq. 5.69) leads to

0(negligibly small for gas flow)   0(flow is adiabatic)   0(flow is steady throughout)

$$\dot{m}\left[\check{h}_2 - \check{h}_1 + \frac{V_2^2 - V_1^2}{2} + g(z_2 - z_1)\right] = \dot{Q}_{\substack{net \\ in.}} + \dot{W}_{\substack{shaft \\ net\ in}}$$

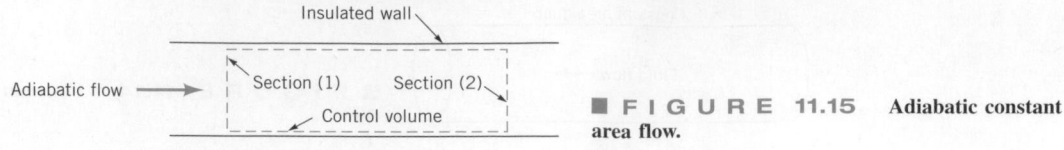

■ FIGURE 11.15 Adiabatic constant area flow.

or

$$\check{h} + \frac{V^2}{2} = \check{h}_0 = \text{constant} \tag{11.72}$$

where h_0 is the stagnation enthalpy. For an ideal gas we gather from Eq. 11.9 that

$$\check{h} - \check{h}_0 = c_p(T - T_0) \tag{11.73}$$

so that by combining Eqs. 11.72 and 11.73 we get

$$T + \frac{V^2}{2c_p} = T_0 = \text{constant}$$

or

$$T + \frac{(\rho V)^2}{2c_p\rho^2} = T_0 = \text{constant} \tag{11.74}$$

By substituting the ideal gas equation of state (Eq. 11.1) into Eq. 11.74 we obtain

$$T + \frac{(\rho V)^2 T^2}{2c_p(p^2/R^2)} = T_0 = \text{constant} \tag{11.75}$$

From the continuity equation (Eq. 11.40) we can conclude that the density–velocity product, ρV, is constant for a given Fanno flow since the area, A, is constant. Also, for a particular Fanno flow, the stagnation temperature, T_0, is fixed. Thus, Eq. 11.75 allows us to calculate values of fluid temperature corresponding to values of fluid pressure in the Fanno flow. We postpone our discussion of how pressure is determined until later.

As with earlier discussions in this chapter, it is helpful to describe Fanno flow with a temperature–entropy diagram. From the second $T\,ds$ relationship, an expression for entropy variation was already derived (Eq. 11.22). If the temperature, T_1, pressure, p_1, and entropy, s_1, at the entrance of the Fanno flow duct are considered as reference values, then Eq. 11.22 yields

$$s - s_1 = c_p \ln \frac{T}{T_1} - R \ln \frac{p}{p_1} \tag{11.76}$$

Entropy increases in Fanno flows because of wall friction.

Equations 11.75 and 11.76 taken together result in a curve with $T–s$ coordinates as is illustrated in Fig. 11.16. This curve involves a given gas (c_p and R) with fixed values of stagnation temperature, density–velocity product, and inlet temperature, pressure, and entropy. Curves like the one sketched in Fig. 11.16 are called Fanno lines.

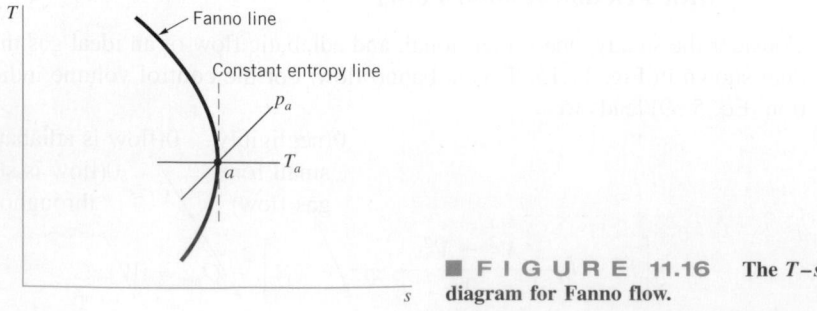

■ FIGURE 11.16 The $T–s$ diagram for Fanno flow.

EXAMPLE 11.11   Compressible Flow with Friction (Fanno Flow)

GIVEN Air ($k = 1.4$) enters [section (1)] an insulated, constant cross-sectional area duct with the following properties:

$$T_0 = 288 \text{ K}$$
$$T_1 = 286 \text{ K}$$
$$p_1 = 99 \text{ kPa (abs)}$$

FIND For Fanno flow, determine corresponding values of fluid temperature and entropy change for various values of downstream pressures and plot the related Fanno line.

SOLUTION

To plot the Fanno line we can use Eq. 11.75

$$T + \frac{(\rho V)^2 T^2}{2 c_p p^2 / R^2} = T_0 = \text{constant} \qquad (1)$$

and Eq. 11.76

$$s - s_1 = c_p \ln \frac{T}{T_1} - R \ln \frac{p}{p_1} \qquad (2)$$

to construct a table of values of temperature and entropy change corresponding to different levels of pressure in the Fanno flow.

We need values of the ideal gas constant and the specific heat at constant pressure to use in Eqs. 1 and 2. From Table 1.4 we read for air

$$R = 286.9 \text{ J/kg} \cdot \text{K}$$

From Eq. 11.14 we obtain

$$c_p = \frac{Rk}{k-1} \qquad (3)$$

or

$$c_p = \frac{(286.9 \text{ J/kg} \cdot \text{K})(1.4)}{1.4 - 1}$$
$$= 1004 \text{ J/kg} \cdot \text{K}$$

From Eqs. 11.1 and 11.69 we obtain

$$\rho V = \frac{p}{RT} \text{Ma} \sqrt{RTk}$$

and ρV is constant for this flow

$$\rho V = \rho_1 V_1 = \frac{p_1}{RT_1} \text{Ma}_1 \sqrt{RT_1 k} \qquad (4)$$

But

$$\frac{T_1}{T_0} = \frac{286 \text{ K}}{288 \text{ K}} = 0.993$$

and from Eq. 11.56

$$\text{Ma}_1 = \sqrt{\left(\frac{1}{0.993} - 1\right) / .02} = 0.2$$

Thus, with

$$\sqrt{RT_1 k} = \sqrt{(1.4)(286.9 \text{ J/kg} \cdot \text{K})(286 \text{ K})}$$
$$= 339 \text{ m/s}$$

Eq. 4 becomes

$$\rho V = \frac{99 \times 10^3 \text{ Pa } 0.2(339 \text{ m/s})}{(286.9 \text{ J/kg} \cdot \text{K})(286 \text{ K})}$$

or

$$\rho V = 81.8 \text{ kg/(m}^2 \cdot \text{s)}$$

For $p = 48$ kPa (abs) we have from Eq. 1

$$T + \frac{[81.8 \text{ kg/(m}^2 \cdot \text{s)}]^2 T^2}{2[1004 \text{ J/(kg} \cdot \text{K)}] \dfrac{(48 \times 10^3 \text{ Pa})^2}{[286.9 \text{ J/(kg} \cdot \text{K)}]^2}}$$
$$= 288 \text{ K}$$

or

$$0.12 \times 10^{-3} \left[(\text{kg} \cdot \text{m/s}^2)/(\text{N} \cdot \text{K})\right] T^2 + T - 288 \text{ K} = 0$$

Thus, since $1 \text{ N} = 1 \text{ kg} \cdot \text{m/s}^2$ we obtain

$$0.12 \times 10^{-3} T^2 + T - 288 = 0$$

Hence,

$$T = 278.7 \text{ K} \qquad \textbf{(Ans)}$$

where T is in K.

From Eq. 2, we obtain

$$s - s_1 = (1004 \text{ J/kg} \cdot \text{K}) \ln\left(\frac{278.7 \text{ K}}{286 \text{ K}}\right)$$
$$- (286.9 \text{ J/kg} \cdot \text{K}) \ln\left(\frac{48 \times 10^3 \text{ Pa (abs)}}{99 \times 10^3 \text{ Pa (abs)}}\right)$$

or

$$s - s_1 = 181.7 \text{ J/(kg} \cdot \text{K)} \qquad \textbf{(Ans)}$$

Proceeding as outlined above, we construct the table of values shown below and graphed as the Fanno line in Fig. E11.11. The

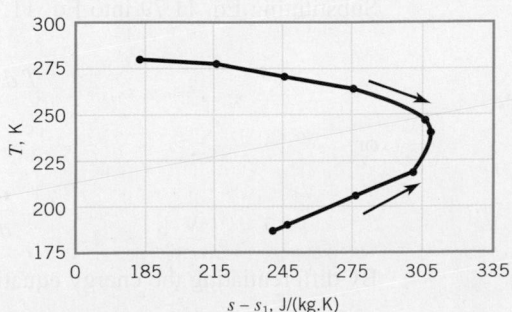

■ **FIGURE E11.11**

maximum entropy difference occurs at a pressure of 18 kPa (abs) and a temperature of 239.4 K.

COMMENT Note that for Fanno flow the entropy must increase in the direction of flow. Hence, this flow can proceed either from subsonic conditions upstream to a sonic condition (Ma = 1) downstream or from supersonic conditions upstream to a sonic condition downstream. The arrows in Fig. 11.11 indicate in which direction a Fanno flow can proceed.

p kPa (abs)	T (K)	$s - s_1$ [J/(kg · K)]
48	278.7	181.7
41	275.6	215.7
34	270.6	251.0
28	264.0	282.0
21	249.3	307.0
18	239.4	310.5
14	220.0	297.8
12	206.8	279.9
10	190.0	247.0
9.6	185.6	235.3

We can learn more about Fanno lines by further analyzing the equations that describe the physics involved. For example, the second $T\,ds$ equation (Eq. 11.18) is

$$T\,ds = d\check{h} - \frac{dp}{\rho} \tag{11.18}$$

For an ideal gas

$$d\check{h} = c_p\,dT \tag{11.7}$$

and

$$\rho = \frac{p}{RT} \tag{11.1}$$

or

$$\frac{dp}{p} = \frac{d\rho}{\rho} + \frac{dT}{T} \tag{11.77}$$

Fanno flow properties can be obtained from the second T ds equation combined with the continuity and energy equations.

Thus, consolidating Eqs. 11.1, 11.7, 11.18, and 11.77 we obtain

$$T\,ds = c_p\,dT - RT\left(\frac{d\rho}{\rho} + \frac{dT}{T}\right) \tag{11.78}$$

Also, from the continuity equation (Eq. 11.40), we get for Fanno flow ρV = constant, or

$$\frac{d\rho}{\rho} = -\frac{dV}{V} \tag{11.79}$$

Substituting Eq. 11.79 into Eq. 11.78 yields

$$T\,ds = c_p\,dT - RT\left(-\frac{dV}{V} + \frac{dT}{T}\right)$$

or

$$\frac{ds}{dT} = \frac{c_p}{T} - R\left(-\frac{1}{V}\frac{dV}{dT} + \frac{1}{T}\right) \tag{11.80}$$

By differentiating the energy equation (11.74) obtained earlier, we obtain

$$\frac{dV}{dT} = -\frac{c_p}{V} \tag{11.81}$$

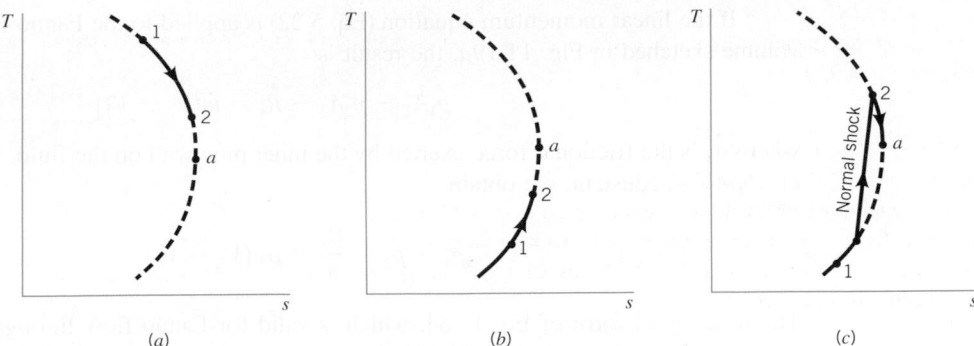

■ **FIGURE 11.17** *(a) Subsonic Fanno flow. (b) Supersonic Fanno flow. (c) Normal shock occurrence in Fanno flow.*

which, when substituted into Eq. 11.80, results in

$$\frac{ds}{dT} = \frac{c_p}{T} - R\left(\frac{c_p}{V^2} + \frac{1}{T}\right) \tag{11.82}$$

The Fanno line in Fig. 11.16 goes through a state (labeled state *a*) for which $ds/dT = 0$. At this state, we can conclude from Eqs. 11.14 and 11.82 that

$$V_a = \sqrt{RT_a k} \tag{11.83}$$

However, by comparing Eqs. 11.83 and 11.36 we see that the Mach number at state *a* is 1. Since the stagnation temperature is the same for all points on the Fanno line [see energy equation (Eq. 11.74)], the temperature at point *a* is the critical temperature, T^*, for the entire Fanno line. Thus, Fanno flow corresponding to the portion of the Fanno line above the critical temperature must be subsonic, and Fanno flow on the line below T^* must be supersonic.

The second law of thermodynamics states that, based on all past experience, entropy can only remain constant or increase for adiabatic flows. For Fanno flow to be consistent with the second law of thermodynamics, flow can only proceed along the Fanno line toward state *a*, the critical state. The critical state may or may not be reached by the flow. If it is, the Fanno flow is *choked*. Some examples of Fanno flow behavior are summarized in Fig. 11.17. A case involving subsonic Fanno flow that is accelerated by friction to a higher Mach number without choking is illustrated in Fig. 11.17*a*. A supersonic flow that is decelerated by friction to a lower Mach number without choking is illustrated in Fig. 11.17*b*. In Fig. 11.17*c*, an abrupt change from supersonic to subsonic flow in the Fanno duct is represented. This sudden deceleration occurs across a standing *normal shock wave* that is described in more detail in Section 11.5.3.

Friction accelerates a subsonic Fanno flow.

The qualitative aspects of Fanno flow that we have already discussed are summarized in Table 11.1 and Fig. 11.18. To quantify Fanno flow behavior we need to combine a relationship that represents the linear momentum law with the set of equations already derived in this chapter.

■ **TABLE 11.1**
Summary of Fanno Flow Behavior

Parameter	Flow	
	Subsonic Flow	**Supersonic Flow**
Stagnation temperature	Constant	Constant
Ma	Increases (maximum is 1)	Decreases (minimum is 1)
Friction	Accelerates flow	Decelerates flow
Pressure	Decreases	Increases
Temperature	Decreases	Increases

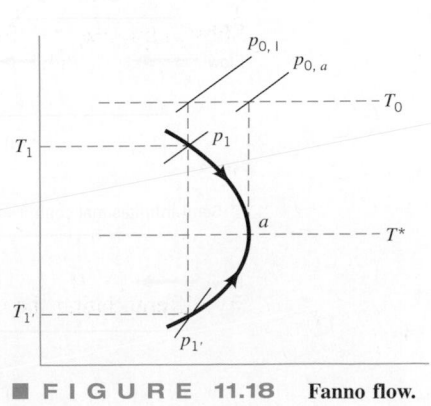

■ **FIGURE 11.18** **Fanno flow.**

If the linear momentum equation (Eq. 5.22) is applied to the Fanno flow through the control volume sketched in Fig. 11.19a, the result is

$$p_1 A_1 - p_2 A_2 - R_x = \dot{m}(V_2 - V_1)$$

where R_x is the frictional force exerted by the inner pipe wall on the fluid. Since $A_1 = A_2 = A$ and $\dot{m} = \rho A V =$ constant, we obtain

$$p_1 - p_2 - \frac{R_x}{A} = \rho V (V_2 - V_1) \tag{11.84}$$

The differential form of Eq. 11.84, which is valid for Fanno flow through the semi-infinitesimal control volume shown in Fig. 11.19b, is

$$-dp - \frac{\tau_w \pi D \, dx}{A} = \rho V \, dV \tag{11.85}$$

The wall shear stress, τ_w, is related to the wall friction factor, f, by Eq. 8.20 as

Friction forces in Fanno flow are given in terms of the friction factor.

$$f = \frac{8 \tau_w}{\rho V^2} \tag{11.86}$$

By substituting Eq. 11.86 and $A = \pi D^2 / 4$ into Eq. 11.85, we obtain

$$-dp - f \rho \frac{V^2}{2} \frac{dx}{D} = \rho V \, dV \tag{11.87}$$

or

$$\frac{dp}{p} + \frac{f}{p} \frac{\rho V^2}{2} \frac{dx}{D} + \frac{\rho}{p} \frac{d(V^2)}{2} = 0 \tag{11.88}$$

Combining the ideal gas equation of state (Eq. 11.1), the ideal gas speed-of-sound equation (Eq. 11.36), and the Mach number definition (Eq. 11.46) with Eq. 11.88 leads to

$$\frac{dp}{p} + \frac{fk}{2} \mathrm{Ma}^2 \frac{dx}{D} + k \frac{\mathrm{Ma}^2}{2} \frac{d(V^2)}{V^2} = 0 \tag{11.89}$$

Since $V = \mathrm{Ma} \, c = \mathrm{Ma} \sqrt{RTk}$, then

$$V^2 = \mathrm{Ma}^2 RTk$$

or

$$\frac{d(V^2)}{V^2} = \frac{d(\mathrm{Ma}^2)}{\mathrm{Ma}^2} + \frac{dT}{T} \tag{11.90}$$

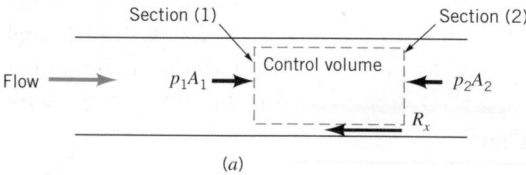

(a)

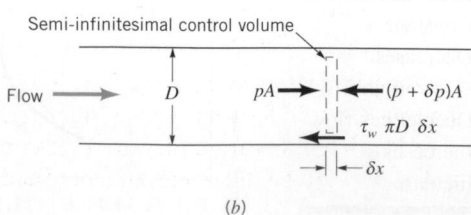

(b)

■ **F I G U R E 11.19** (a) Finite control volume. (b) Semi-infinitesimal control volume.

The application of the energy equation (Eq. 5.69) to Fanno flow gave Eq. 11.74. If Eq. 11.74 is differentiated and divided by temperature, the result is

$$\frac{dT}{T} + \frac{d(V^2)}{2c_p T} = 0 \tag{11.91}$$

Substituting Eqs. 11.14, 11.36, and 11.46 into Eq. 11.91 yields

$$\frac{dT}{T} + \frac{k-1}{2} Ma^2 \frac{d(V^2)}{V^2} = 0 \tag{11.92}$$

which can be combined with Eq. 11.90 to form

$$\frac{d(V^2)}{V^2} = \frac{d(Ma^2)/Ma^2}{1 + [(k-1)/2]Ma^2} \tag{11.93}$$

We can merge Eqs. 11.77, 11.79, and 11.90 to get

$$\frac{dp}{p} = \frac{1}{2}\frac{d(V^2)}{V^2} - \frac{d(Ma^2)}{Ma^2} \tag{11.94}$$

Consolidating Eqs. 11.94 and 11.89 leads to

$$\frac{1}{2}(1 + kMa^2)\frac{d(V^2)}{V^2} - \frac{d(Ma^2)}{Ma^2} + \frac{fk}{2}Ma^2\frac{dx}{D} = 0 \tag{11.95}$$

Finally, incorporating Eq. 11.93 into Eq. 11.95 yields

$$\frac{(1 - Ma^2)\,d(Ma^2)}{\{1 + [(k-1)/2]Ma^2\}kMa^4} = f\frac{dx}{D} \tag{11.96}$$

Equation 11.96 can be integrated from one section to another in a Fanno flow duct. We elect to use the critical (*) state as a reference and to integrate Eq. 11.96 from an upstream state to the critical state. Thus

$$\int_{Ma}^{Ma^*=1}\frac{(1 - Ma^2)\,d(Ma^2)}{\{1 + [(k-1)/2]\,Ma^2\}kMa^4} = \int_{\ell}^{\ell*} f\frac{dx}{D} \tag{11.97}$$

where ℓ is length measured from an arbitrary but fixed upstream reference location to a section in the Fanno flow. For an approximate solution, we can assume that the friction factor is constant at an average value over the integration length, $\ell* - \ell$. We also consider a constant value of k. Thus, we obtain from Eq. 11.97

$$\boxed{\frac{1}{k}\frac{(1 - Ma^2)}{Ma^2} + \frac{k+1}{2k}\ln\left\{\frac{[(k+1)/2]Ma^2}{1 + [(k-1)/2]Ma^2}\right\} = \frac{f(\ell* - \ell)}{D}} \tag{11.98}$$

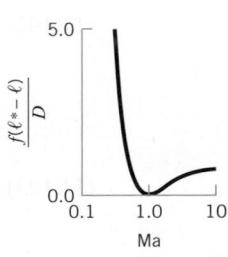

For a given gas, values of $f(\ell* - \ell)/D$ can be tabulated as a function of Mach number for Fanno flow. For example, values of $f(\ell* - \ell)/D$ for air ($k = 1.4$) Fanno flow are graphed as a function of Mach number in Fig. D.2 in Appendix D and in the figure in the margin. Note that the critical state does not have to exist in the actual Fanno flow being considered, since for any two sections in a given Fanno flow

$$\boxed{\frac{f(\ell* - \ell_2)}{D} - \frac{f(\ell* - \ell_1)}{D} = \frac{f}{D}(\ell_1 - \ell_2)} \tag{11.99}$$

The sketch in Fig. 11.20 illustrates the physical meaning of Eq. 11.99.

For a given Fanno flow (constant specific heat ratio, duct diameter, and friction factor) the length of duct required to change the Mach number from Ma_1 to Ma_2 can be determined from Eqs. 11.98 and 11.99 or a graph such as Fig. D.2. To get the values of other fluid properties in the Fanno flow field we need to develop more equations.

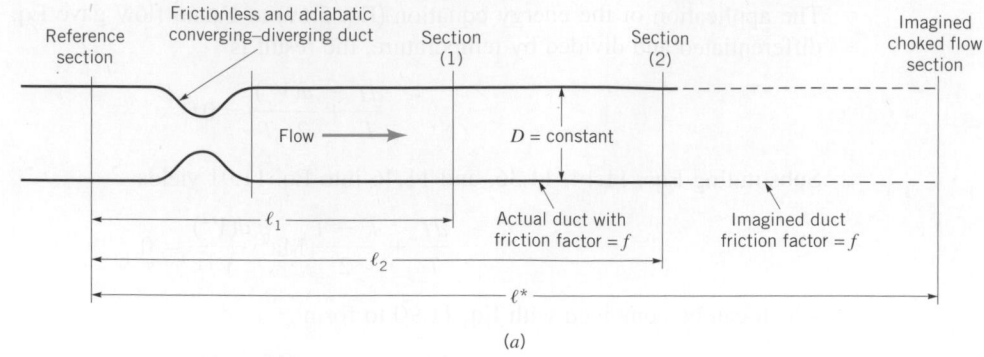

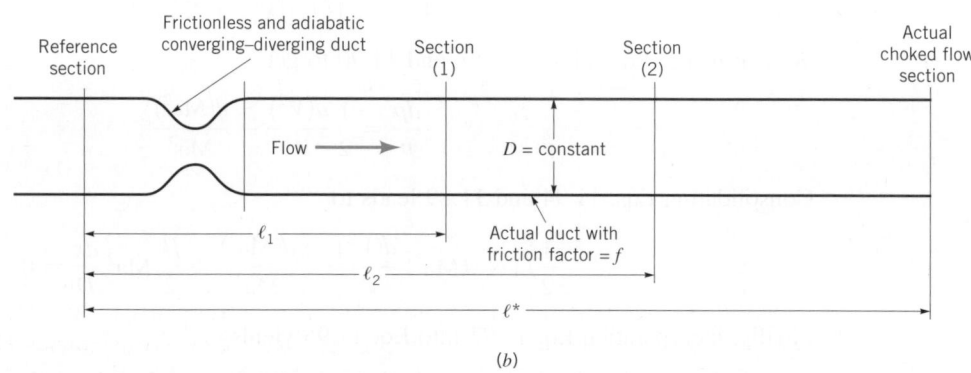

■ **F I G U R E 11.20** *(a) Unchoked Fanno flow. (b) Choked Fanno flow.*

For Fanno flow, the length of duct needed to produce a given change in Mach number can be determined.

By consolidating Eqs. 11.90 and 11.92 we obtain

$$\frac{dT}{T} = -\frac{(k-1)}{2\{1 + [(k-1)/2]\text{Ma}^2\}} d(\text{Ma}^2) \tag{11.100}$$

Integrating Eq. 11.100 from any state upstream in a Fanno flow to the critical (*) state leads to

$$\frac{T}{T^*} = \frac{(k+1)/2}{1 + [(k-1)/2]\text{Ma}^2} \tag{11.101}$$

Equations 11.68 and 11.69 allow us to write

$$\frac{V}{V^*} = \frac{\text{Ma}\ \sqrt{RTk}}{\sqrt{RT^*k}} = \text{Ma}\sqrt{\frac{T}{T^*}} \tag{11.102}$$

Substituting Eq. 11.101 into Eq. 11.102 yields

$$\frac{V}{V^*} = \left\{ \frac{[(k+1)/2]\text{Ma}^2}{1 + [(k-1)/2]\text{Ma}^2} \right\}^{1/2} \tag{11.103}$$

Equations 11.101 and 11.103 are graphed in the margin for air.

From the continuity equation (Eq. 11.40) we get for Fanno flow

$$\frac{\rho}{\rho^*} = \frac{V^*}{V} \tag{11.104}$$

Combining 11.104 and 11.103 results in

$$\frac{\rho}{\rho^*} = \left\{ \frac{1 + [(k-1)/2]\text{Ma}^2}{[(k+1)/2]\text{Ma}^2} \right\}^{1/2} \tag{11.105}$$

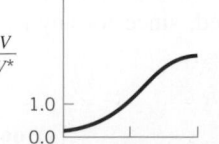

The ideal gas equation of state (Eq. 11.1) leads to

$$\frac{p}{p^*} = \frac{\rho}{\rho^*}\frac{T}{T^*}$$ (11.106)

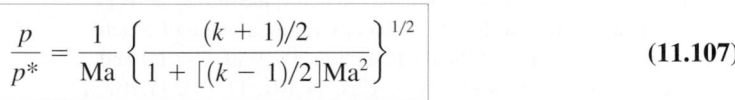

and merging Eqs. 11.106, 11.105, and 11.101 gives

$$\boxed{\frac{p}{p^*} = \frac{1}{\text{Ma}}\left\{\frac{(k+1)/2}{1 + [(k-1)/2]\text{Ma}^2}\right\}^{1/2}}$$ (11.107)

This relationship is graphed in the margin for air.

Finally, the stagnation pressure ratio can be written as

$$\frac{p_0}{p_0^*} = \left(\frac{p_0}{p}\right)\left(\frac{p}{p^*}\right)\left(\frac{p^*}{p_0^*}\right)$$ (11.108)

For Fanno flow, thermodynamic and flow properties can be calculated as a function of Mach number.

which by use of Eqs. 11.59 and 11.107 yields

$$\boxed{\frac{p_0}{p_0^*} = \frac{1}{\text{Ma}}\left[\left(\frac{2}{k+1}\right)\left(1 + \frac{k-1}{2}\text{Ma}^2\right)\right]^{[(k+1)/2(k-1)]}}$$ (11.109)

Values of $f(\ell^* - \ell)/D$, T/T^*, V/V^*, p/p^*, and p_0/p_0^* for Fanno flow of air $(k = 1.4)$ are graphed as a function of Mach number (using Eqs. 11.99, 11.101, 11.103, 11.107, and 11.109) in Fig. D.2 of Appendix D. The usefulness of Fig. D.2 is illustrated in Examples 11.12, 11.13, and 11.14. See Ref. 7 for additional compressible internal flow material.

EXAMPLE 11.12   Choked Fanno Flow

GIVEN Standard atmospheric air $[T_0 = 288\text{ K}, p_0 = 101$ kPa (abs)$]$ is drawn steadily through a frictionless, adiabatic converging nozzle into an adiabatic, constant area duct as shown in Fig. E11.12a. The duct is 2 m long and has an inside diameter of 0.1 m. The average friction factor for the duct is estimated as being equal to 0.02.

FIND What is the maximum mass flowrate through the duct? For this maximum flowrate, determine the values of static temperature, static pressure, stagnation temperature, stagnation pressure, and velocity at the inlet [section (1)] and exit [section (2)] of the constant area duct. Sketch a temperature–entropy diagram for this flow.

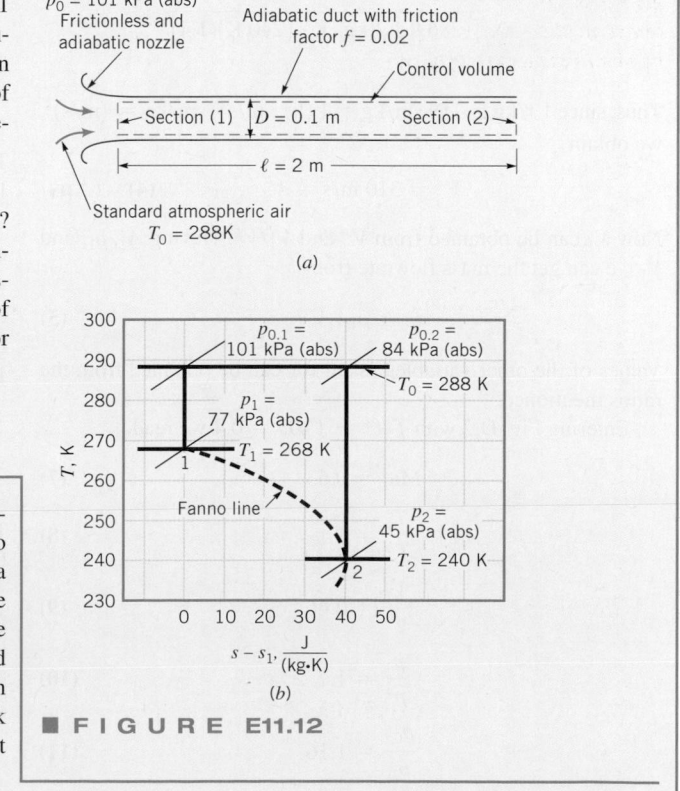

■ **FIGURE E11.12**

SOLUTION

We consider the flow through the converging nozzle to be isentropic and the flow through the constant area duct to be Fanno flow. A decrease in the pressure at the exit of the constant area duct (back pressure) causes the mass flowrate through the nozzle and the duct to increase. The flow throughout is subsonic. The maximum flowrate will occur when the back pressure is lowered to the extent that the constant area duct chokes and the Mach number at the duct exit is equal to 1. Any further decrease of back pressure will not affect the flowrate through the nozzle–duct combination.

For the maximum flowrate condition, the constant area duct must be choked, and

$$\frac{f(\ell^* - \ell_1)}{D} = \frac{f(\ell_2 - \ell_1)}{D} = \frac{(0.02)(2 \text{ m})}{(0.1 \text{ m})} = 0.4 \qquad (1)$$

With $k = 1.4$ for air and the above calculated value of $f(\ell^* - \ell_1)/D = 0.4$, we could use Eq. 11.98 to determine a value of Mach number at the entrance of the duct [section (1)]. With $k = 1.4$ and Ma_1 known, we could then rely on Eqs. 11.101, 11.103, 11.107, and 11.109 to obtain values of T_1/T^*, V_1/V^*, p_1/p^*, and $p_{0,1}/p_0^*$. Alternatively, for air ($k = 1.4$), we can use Fig. D.2 with $f(\ell^* - \ell_1)/D = 0.4$ and read off values of Ma_1, T_1/T^*, V_1/V^*, p_1/p^*, and $p_{0,1}/p_0^*$.

The pipe entrance Mach number, Ma_1, also represents the Mach number at the throat (and exit) of the isentropic, converging nozzle. Thus, the isentropic flow equations of Section 11.4 or Fig. D.1 can be used with Ma_1. We use Fig. D.1 in this example.

With Ma_1 known, we can enter Fig. D.1 and get values of T_1/T_0, p_1/p_0, and ρ_1/ρ_0. Through the isentropic nozzle, the values of T_0, p_0, and ρ_0 are each constant, and thus T_1, p_1, and ρ_1 can be readily obtained.

Since T_0 also remains constant through the constant area duct (see Eq. 11.75), we can use Eq. 11.63 to get T^*. Thus,

$$\frac{T^*}{T_0} = \frac{2}{k + 1} = \frac{2}{1.4 + 1} = 0.8333 \qquad (2)$$

Since $T_0 = 288$ K, we get from Eq. 2,

$$T^* = (0.8333)(288 \text{ K}) = 240 \text{ K} = T_2 \qquad (3) \quad \textbf{(Ans)}$$

With T^* known, we can calculate V^* from Eq. 11.36 as

$$\begin{aligned} V^* &= \sqrt{RT^*k} \\ &= \sqrt{[(286.9 \text{ J})/(\text{kg} \cdot \text{K})](240 \text{ K})(1.4)} \\ &= 310 \ (\text{J/kg})^{1/2} \end{aligned}$$

Thus, since $1 \text{ J/kg} = 1 \text{ N} \cdot \text{m/kg} = 1 \ (\text{kg} \cdot \text{m/s}^2) \cdot \text{m/kg} = (\text{m/s})^2$, we obtain

$$V^* = 310 \text{ m/s} = V_2 \qquad (4) \quad \textbf{(Ans)}$$

Now V_1 can be obtained from V^* and V_1/V^*. Having A_1, ρ_1, and V_1 we can get the mass flowrate from

$$\dot{m} = \rho_1 A_1 V_1 \qquad (5)$$

Values of the other variables asked for can be obtained from the ratios mentioned.

Entering Fig. D.2 with $f(\ell^* - \ell)/D = 0.4$ we read

$$\text{Ma}_1 = 0.63 \qquad (7)$$

$$\frac{T_1}{T^*} = 1.1 \qquad (8)$$

$$\frac{V_1}{V^*} = 0.66 \qquad (9)$$

$$\frac{p_1}{p^*} = 1.7 \qquad (10)$$

$$\frac{p_{0,1}}{p_0^*} = 1.16 \qquad (11)$$

Entering Fig. D.1 with $\text{Ma}_1 = 0.63$ we read

$$\frac{T_1}{T_0} = 0.93 \qquad (12)$$

$$\frac{p_1}{p_{0,1}} = 0.76 \qquad (13)$$

$$\frac{\rho_1}{\rho_{0,1}} = 0.83 \qquad (14)$$

Thus, from Eqs. 4 and 9 we obtain

$$V_1 = (0.66)(310 \text{ m/s}) = 205 \text{ m/s} \qquad \textbf{(Ans)}$$

From Eq. 14 we get

$$\rho_1 = 0.83\rho_{0,1} = (0.83)(1.23 \text{ kg/m}^3) = 1.02 \text{ kg/m}^3$$

and from Eq. 5 we conclude that

$$\begin{aligned} \dot{m} &= (1.02 \text{ kg/m}^3)\left[\frac{\pi(0.1 \text{ m})^2}{4}\right](206 \text{ m/s}) \\ &= 1.65 \text{ kg/s} \end{aligned} \qquad \textbf{(Ans)}$$

From Eq. 12, it follows that

$$T_1 = (0.93)(288 \text{ K}) = 268 \text{ K} \qquad \textbf{(Ans)}$$

Equation 13 yields

$$p_1 = (0.76)[101 \text{ kPa (abs)}] = 77 \text{ kPa (abs)} \qquad \textbf{(Ans)}$$

The stagnation temperature, T_0, remains constant through this adiabatic flow at a value of

$$T_{0,1} = T_{0,2} = 288 \text{ K} \qquad \textbf{(Ans)}$$

The stagnation pressure, p_0, at the entrance of the constant area duct is the same as the constant value of stagnation pressure through the isentropic nozzle. Thus

$$p_{0,1} = 101 \text{ kPa (abs)} \qquad \textbf{(Ans)}$$

To obtain the duct exit pressure ($p_2 = p^*$) we can use Eqs. 10 and 13. Thus,

$$\begin{aligned} p_2 &= \left(\frac{p^*}{p_1}\right)\left(\frac{p_1}{p_{0,1}}\right)(p_{0,1}) = \left(\frac{1}{1.7}\right)(0.76)[101 \text{ kPa (abs)}] \\ &= 45 \text{ kPa (abs)} \end{aligned} \qquad \textbf{(Ans)}$$

For the duct exit stagnation pressure ($p_{0,2} = p_0^*$) we can use Eq. 11 as

$$\begin{aligned} p_{0,2} &= \left(\frac{p_0^*}{p_{0,1}}\right)(p_{0,1}) = \left(\frac{1}{1.16}\right)[101 \text{ kPa (abs)}] \\ &= 87.1 \text{ kPa (abs)} \end{aligned} \qquad \textbf{(Ans)}$$

The stagnation pressure, p_0, decreases in a Fanno flow because of friction.

COMMENT   Use of graphs such as Figs. D.1 and D.2 illustrates the solution of a problem involving Fanno flow. The T–s diagram for this flow is shown in Fig. E.11.12b, where the entropy difference, $s_2 - s_1$, is obtained from Eq. 11.22.

EXAMPLE 11.13   Effect of Duct Length on Choked Fanno Flow

GIVEN   The duct in Example 11.12 is shortened by 50%, but the duct discharge pressure is maintained at the choked flow value for Example 11.12, namely,

$$p_d = 45 \text{ kPa (abs)}$$

FIND   Will shortening the duct cause the mass flowrate through the duct to increase or decrease? Assume that the average friction factor for the duct remains constant at a value of $f = 0.02$.

SOLUTION

We guess that the shortened duct will still choke and check our assumption by comparing p_d with p^*. If $p_d < p^*$, the flow is choked; if not, another assumption has to be made. For choked flow we can calculate the mass flowrate just as we did for Example 11.12. For unchoked flow, we will have to devise another strategy.

For choked flow

$$\frac{f(\ell^* - \ell_1)}{D} = \frac{(0.02)(1 \text{ m})}{0.1 \text{ m}} = 0.2$$

and from Fig. D.2, we read the values $Ma_1 = 0.70$ and $p_1/p^* = 1.5$. With $Ma_1 = 0.70$, we use Fig. D.1 and get

$$\frac{p_1}{p_0} = 0.72$$

Now the duct exit pressure ($p_2 = p^*$) can be obtained from

$$p_2 = p^* = \left(\frac{p^*}{p_1}\right)\left(\frac{p_1}{p_{0,1}}\right)(p_{0,1})$$

$$= \left(\frac{1}{1.5}\right)(0.72)[101 \text{ kPa (abs)}] = 48.5 \text{ kPa (abs)}$$

and we see that $p_d < p^*$. Our assumption of choked flow is justified. The pressure at the exit plane is greater than the surrounding pressure outside the duct exit. The final drop of pressure from 48.5 kPa (abs) to 45 kPa (abs) involves complicated three-dimensional flow downstream of the exit.

To determine the mass flowrate we use

$$\dot{m} = \rho_1 A_1 V_1 \tag{1}$$

The density at section (1) is obtained from

$$\frac{\rho_1}{\rho_{0,1}} = 0.79 \tag{2}$$

which is read in Fig. D.1 for $Ma_1 = 0.7$. Thus,

$$\rho_1 = (0.79)(1.23 \text{ kg/m}^3) = 0.97 \text{ kg/m}^3 \tag{3}$$

We get V_1 from

$$\frac{V_1}{V^*} = 0.73 \tag{4}$$

from Fig. D.2 for $Ma_1 = 0.7$. The value of V^* is the same as it was in Example 11.12, namely,

$$V^* = 310 \text{ m/s} \tag{5}$$

Thus, from Eqs. 4 and 5 we obtain

$$V_1 = (0.73)(310) = 226 \text{ m/s} \tag{6}$$

and from Eqs. 1, 3, and 6 we get

$$\dot{m} = (0.97 \text{ kg/m}^3)\left[\frac{\pi(0.1 \text{m})^2}{4}\right](226 \text{ m/s})$$

$$= 1.73 \text{ kg/s} \qquad \textbf{(Ans)}$$

The mass flowrate associated with a shortened tube is larger than the mass flowrate for the longer tube, $\dot{m} = 1.65$ kg/s. This trend is general for subsonic Fanno flow.

COMMENT   For the same upstream stagnation state and downstream pressure, the mass flowrate for the Fanno flow will decrease with increase in length of duct for subsonic flow. Equivalently, if the length of the duct remains the same but the wall friction is increased, the mass flowrate will decrease.

EXAMPLE 11.14   Unchoked Fanno Flow

GIVEN   The same flowrate obtained in Example 11.12 ($\dot{m} = 1.65$ kg/s) is desired through the shortened duct of Example 11.13 ($\ell_2 - \ell_1 = 1$ m). Assume f remains constant at a value of 0.02.

FIND   Determine the Mach number at the exit of the duct, M_2, and the back pressure, p_2, required.

SOLUTION

Since the mass flowrate of Example 11.12 is desired, the Mach number and other properties at the entrance of the constant area duct remain at the values determined in Example 11.12. Thus,

from Example 11.12, $Ma_1 = 0.63$ and from Fig. D.2

$$\frac{f(\ell^* - \ell_1)}{D} = 0.4$$

For this example,

$$\frac{f(\ell_2 - \ell_1)}{D} = \frac{f(\ell^* - \ell_1)}{D} - \frac{f(\ell^* - \ell_2)}{D}$$

or

$$\frac{(0.02)(1 \text{ m})}{0.1 \text{ m}} = 0.4 - \frac{f(\ell^* - \ell_2)}{D}$$

so that

$$\frac{f(\ell^* - \ell_2)}{D} = 0.2 \qquad \text{(1)}$$

By using the value from Eq. 1 and Fig. D.2, we get

$$\text{Ma}_2 = 0.70 \qquad \text{(Ans)}$$

and

$$\frac{p_2}{p^*} = 1.5 \qquad \text{(2)}$$

We obtain p_2 from

$$p_2 = \left(\frac{p_2}{p^*}\right)\left(\frac{p^*}{p_1}\right)\left(\frac{p_1}{p_{0,1}}\right)(p_{0,1})$$

where p_2/p^* is given in Eq. 2 and p^*/p_1, $p_1/p_{0,1}$, and $p_{0,1}$ are the same as they were in Example 11.12. Thus,

$$p_2 = (1.5)\left(\frac{1}{1.7}\right)(0.76)[101 \text{ kPa (abs)}]$$

$$= 68.0 \text{ kPa (abs)} \qquad \text{(Ans)}$$

COMMENT   A larger back pressure [68.0 kPa(abs)] than the one associated with choked flow through a Fanno duct [45 kPa(abs)] will maintain the same flowrate through a shorter Fanno duct with the same friction coefficient. The flow through the shorter duct is not choked. It would not be possible to maintain the same flowrate through a Fanno duct longer than the choked one with the same friction coefficient, regardless of what back pressure is used.

11.5.2 Frictionless Constant Area Duct Flow with Heat Transfer (Rayleigh Flow)

Rayleigh flow involves heat transfer with no wall friction and constant cross-sectional area.

Consider the steady, one-dimensional, and frictionless flow of an ideal gas through the constant area duct with heat transfer illustrated in Fig. 11.21. This is *Rayleigh flow*. Application of the linear momentum equation (Eq. 5.22) to the Rayleigh flow through the finite control volume sketched in Fig. 11.21 results in

$$p_1 A_1 + \dot{m} V_1 = p_2 A_2 + \dot{m} V_2 + \overset{0\text{(frictionless flow)}}{\cancel{R_x}}$$

or

$$p + \frac{(\rho V)^2}{\rho} = \text{constant} \qquad \text{(11.110)}$$

Use of the ideal gas equation of state (Eq. 11.1) in Eq. 11.110 leads to

$$p + \frac{(\rho V)^2 RT}{p} = \text{constant} \qquad \text{(11.111)}$$

Since the flow cross-sectional area remains constant for Rayleigh flow, from the continuity equation (Eq. 11.40) we conclude that

$$\rho V = \text{constant}$$

For a given Rayleigh flow, the constant in Eq. 11.111, the density–velocity product, ρV, and the ideal gas constant are all fixed. Thus, Eq. 11.111 can be used to determine values of fluid temperature corresponding to the local pressure in a Rayleigh flow.

To construct a temperature–entropy diagram for a given Rayleigh flow, we can use Eq. 11.76, which was developed earlier from the second $T\,ds$ relationship. Equations 11.111 and 11.76 can be solved simultaneously to obtain the curve sketched in Fig. 11.22. Curves like the one in Fig. 11.22 are called *Rayleigh lines*.

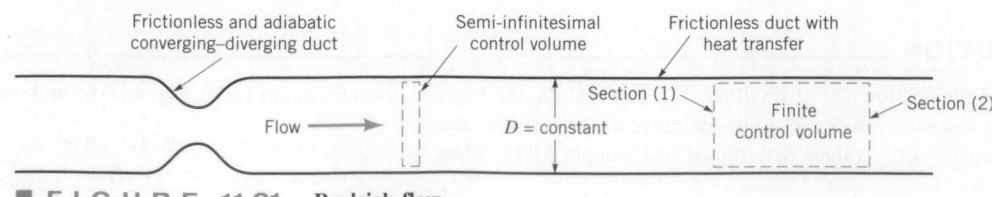

■ **FIGURE 11.21**   Rayleigh flow.

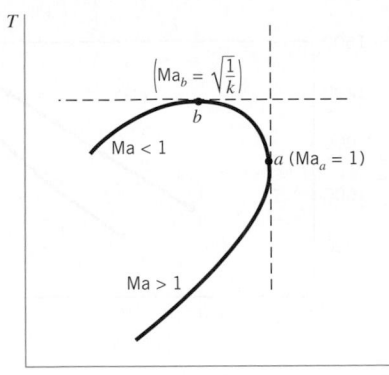

■ **FIGURE 11.22** Rayleigh line.

EXAMPLE 11.15 — Frictionless, Constant Area Compressible Flow with Heat Transfer (Rayleigh Flow)

GIVEN Air ($k = 1.4$) enters [section (1)] a frictionless, constant flow cross-sectional area duct with the following properties (the same as in Example 11.11):

$$T_0 = 288 \text{ K}$$
$$T_1 = 286 \text{ K}$$
$$p_1 = 99 \text{ kPa (abs)}$$

FIND For Rayleigh flow, determine corresponding values of fluid temperature and entropy change for various levels of downstream pressure and plot the related Rayleigh line.

SOLUTION

To plot the Rayleigh line asked for, use Eq. 11.111

$$p + \frac{(\rho V)^2 \, RT}{p} = \text{constant} \qquad (1)$$

and Eq. 11.76

$$s - s_1 = c_p \ln \frac{T}{T_1} - R \ln \frac{p}{p_1} \qquad (2)$$

to construct a table of values of temperature and entropy change corresponding to different levels of pressure downstream in a Rayleigh flow.

Use the value of ideal gas constant for air from Table 1.4

$$R = (286.9 \text{ J})/(\text{kg} \cdot \text{K})$$

and the value of specific heat at constant pressure for air from Example 11.11, namely,

$$c_p = 1004 \text{ J}/(\text{kg} \cdot \text{K})$$

Also, from Example 11.11, $\rho V = 81.8 \text{ kg}/(\text{m}^2 \cdot \text{s})$. For the given inlet [section (1)] conditions, we get

$$\frac{RT_1}{p_1} = \frac{[286.9 \text{ J}/(\text{kg} \cdot \text{K})](286 \text{ K})}{99 \times 10^3 \text{ Pa (abs)}}$$
$$= 0.8 \text{ m}^3/\text{kg}$$

Thus, from Eq. 1 we get

$$p + \frac{(\rho V)^2 \, RT}{p} = 99 \text{ kPa (abs)} + [81.8 \text{ kg}/(\text{m}^2 \cdot \text{s})]^2 (0.8 \text{ m}^3/\text{kg})$$
$$= 99 \text{ kPa (abs)} + 5353 \text{ kg}/(\text{m} \cdot \text{s}^2) = \text{constant}$$

or, since, $1 \text{ kg}/(\text{m} \cdot \text{s}^2) = \text{N/m}^2$,

$$p + \frac{(\rho V)^2 \, RT}{p} = 99 \times 10^3 \text{ N/m}^2 \text{ (abs)} + [5353 \text{ N/m}^2 \text{(abs)}]$$
$$= 104 \text{ kPa (abs)} = \text{constant} \qquad (3)$$

With the downstream pressure of $p = 93 \text{ kPa (abs)}$, we can obtain the downstream temperature by using Eq. 3 with the fact that

$$\frac{(\rho V)^2 R}{p} = \frac{[81.8 \text{ kg}/(\text{m}^2 \cdot \text{s})]^2 [286.9 \text{ J}/(\text{kg} \cdot \text{K})]}{93 \times 10^3 \text{ Pa (abs)}}$$
$$= 20.6 \text{ (N/m}^2)/\text{K}$$

Hence, from Eq. 3,

$$93 \times 10^3 \text{ Pa (abs)} + [20.6 \text{ (N/m}^2)/\text{K}] \, T = 104 \times 10^3 \text{ Pa (abs)}$$

or

$$T = 534 \text{ K}$$

From Eq. 2 with the downstream pressure $p = 93 \times 10^3 \text{ Pa (abs)}$ and temperature $T = 534 \text{ K}$ we get

$$s - s_1 = [1004 \text{ J}/(\text{kg} \cdot \text{K})] \ln\left(\frac{534 \text{ K}}{286 \text{ K}}\right)$$
$$- [286.9 \text{ J}/(\text{kg} \cdot \text{K})] \ln\left(\frac{93000 \text{ Pa (abs)}}{99000 \text{ Pa (abs)}}\right)$$

$$s - s_1 = 645 \text{ J}/(\text{kg} \cdot \text{K})$$

By proceeding as outlined above, we can construct the table of values shown below and graph the Rayleigh line of Fig. E11.15.

p kPa (abs)	T (K)	$s - s_1$ [J/(kg · K)]
93	534	645
86	807	1082
79	1028	1349
72	1199	1530
62	1356	1697
55	1404	1766
52	1409	1786
51.5	1409	1789
48	1400	1802
43	1366	1809
41	1346	1808
38	1306	1800
34	1240	1799
31	1179	1755
28	1109	1723
14	656	1395
7	354	974

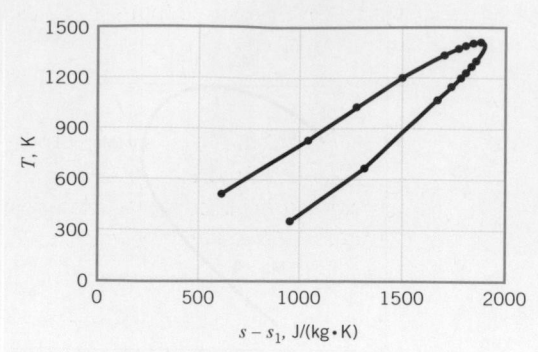

■ **FIGURE E11.15**

COMMENT   Depending on whether the flow is being heated or cooled, it can proceed in either direction along the curve.

At point a on the Rayleigh line of Fig. 11.22, $ds/dT = 0$. To determine the physical importance of point a, we analyze further some of the governing equations. By differentiating the linear momentum equation for Rayleigh flow (Eq. 11.110) we obtain

$$dp = -\rho V \, dV$$

or

$$\frac{dp}{\rho} = -V \, dV \tag{11.112}$$

Combining Eq. 11.112 with the second $T \, ds$ equation (Eq. 11.18) leads to

$$T \, ds = d\breve{h} + V \, dV \tag{11.113}$$

For an ideal gas (Eq. 11.7) $d\breve{h} = c_p \, dT$. Thus, substituting Eq. 11.7 into Eq. 11.113 gives

$$T \, ds = c_p \, dT + V \, dV$$

or

$$\frac{ds}{dT} = \frac{c_p}{T} + \frac{V}{T}\frac{dV}{dT} \tag{11.114}$$

Consolidation of Eqs. 11.114, 11.112 (linear momentum), 11.1, 11.77 (differentiated equation of state), and 11.79 (continuity) leads to

$$\frac{ds}{dT} = \frac{c_p}{T} + \frac{V}{T}\frac{1}{[(T/V) - (V/R)]} \tag{11.115}$$

The maximum entropy state on the Rayleigh line corresponds to sonic conditions.

Hence, at state a where $ds/dT = 0$, Eq. 11.115 reveals that

$$V_a = \sqrt{RT_a k} \tag{11.116}$$

Comparison of Eqs. 11.116 and 11.36 tells us that the Mach number at state a is equal to 1,

$$\text{Ma}_a = 1 \tag{11.117}$$

At point b on the Rayleigh line of Fig. 11.22, $dT/ds = 0$. From Eq. 11.115 we get

$$\frac{dT}{ds} = \frac{1}{ds/dT} = \frac{1}{(c_p/T) + (V/T)[(T/V) - (V/R)]^{-1}}$$

which for $dT/ds = 0$ (point b) gives

$$\mathrm{Ma}_b = \sqrt{\frac{1}{k}} \tag{11.118}$$

The flow at point b is subsonic ($\mathrm{Ma}_b < 1.0$). Recall that $k > 1$ for any gas.

To learn more about Rayleigh flow, we need to consider the energy equation in addition to the equations already used. Application of the energy equation (Eq. 5.69) to the Rayleigh flow through the finite control volume of Fig. 11.21 yields

$$\dot{m}\left[\check{h}_2 - \check{h}_1 + \frac{V_2^2 - V_1^2}{2} + g(z_2 - z_1)\right] = \dot{Q}_{\substack{\text{net} \\ \text{in}}} + \dot{W}_{\substack{\text{shaft} \\ \text{net in}}}$$

with annotations: the $g(z_2 - z_1)$ term marked "0(negligibly small for gas flow)" and the $\dot{W}_{\text{shaft net in}}$ term marked "0(flow is steady throughout)".

or in differential form for Rayleigh flow through the semi-infinitesimal control volume of Fig. 11.21

$$d\check{h} + V\,dV = \delta q \tag{11.119}$$

where δq is the heat transfer per unit mass of fluid in the semi-infinitesimal control volume.

By using $d\check{h} = c_p\,dT = Rk\,dT/(k-1)$ in Eq. 11.119, we obtain

$$\frac{dV}{V} = \frac{\delta q}{c_p T}\left[\frac{V}{T}\frac{dT}{dV} + \frac{V^2(k-1)}{kRT}\right]^{-1} \tag{11.120}$$

Thus, by combining Eqs. 11.36 (ideal gas speed of sound), 11.46 (Mach number), 11.1 and 11.77 (ideal gas equation of state), 11.79 (continuity), and 11.112 (linear momentum) with Eq. 11.120 (energy) we get

$$\frac{dV}{V} = \frac{\delta q}{c_p T}\frac{1}{(1 - \mathrm{Ma}^2)} \tag{11.121}$$

With the help of Eq. 11.121, we see clearly that when the Rayleigh flow is subsonic ($\mathrm{Ma} < 1$), fluid heating ($\delta q > 0$) increases fluid velocity while fluid cooling ($\delta q < 0$) decreases fluid velocity. When Rayleigh flow is supersonic ($\mathrm{Ma} > 1$), fluid heating decreases fluid velocity and fluid cooling increases fluid velocity.

The second law of thermodynamics states that, based on experience, entropy increases with heating and decreases with cooling. With this additional insight provided by the conservation of energy principle and the second law of thermodynamics, we can say more about the Rayleigh line in Fig. 11.22. A summary of the qualitative aspects of Rayleigh flow is outlined in Table 11.2 and Fig. 11.23. Along the upper portion of the line, which includes point b, the flow is subsonic. Heating the fluid results in flow acceleration to a maximum Mach number of 1 at point a. Note that between points b and a along the Rayleigh line, heating the fluid results in a temperature decrease and cooling the fluid leads to a temperature increase. This trend is not surprising if we consider the stagnation temperature and fluid velocity changes that occur between points a and b when the fluid is heated or cooled. Along the lower portion of the Rayleigh curve the flow is supersonic. Rayleigh flows may or may not be choked. The amount of heating or cooling involved determines what will happen in a specific instance. As with Fanno flows, an abrupt deceleration from supersonic flow to subsonic flow across a normal shock wave can also occur in Rayleigh flows.

Fluid temperature reduction can accompany heating a subsonic Rayleigh flow.

■ **TABLE 11.2**
Summary of Rayleigh Flow Characteristics

	Heating	Cooling
$\mathrm{Ma} < 1$	Acceleration	Deceleration
$\mathrm{Ma} > 1$	Deceleration	Acceleration

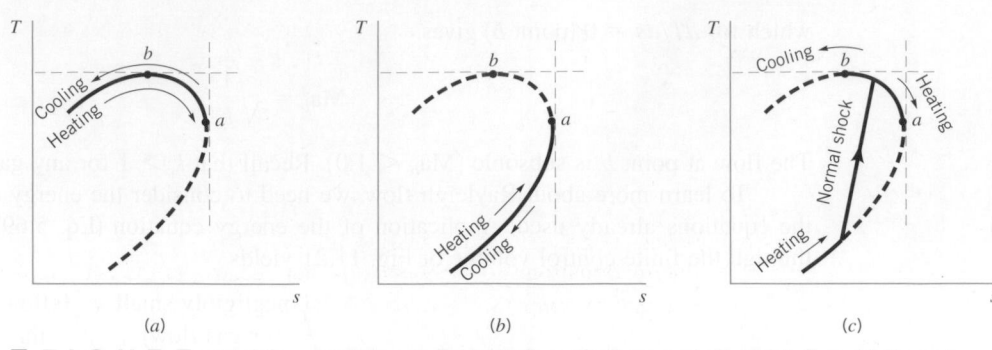

■ **F I G U R E 11.23** (*a*) **Subsonic Rayleigh flow.** (*b*) **Supersonic Rayleigh flow.** (*c*) **Normal shock in a Rayleigh flow.**

To quantify Rayleigh flow behavior we need to develop appropriate forms of the governing equations. We elect to use the state of the Rayleigh flow fluid at point *a* of Fig. 11.22 as the reference state. As shown earlier, the Mach number at point *a* is 1. Even though the Rayleigh flow being considered may not choke and state *a* is not achieved by the flow, this reference state is useful.

If we apply the linear momentum equation (Eq. 11.110) to Rayleigh flow between any upstream section and the section, actual or imagined, where state *a* is attained, we get

$$p + \rho V^2 = p_a + \rho_a V_a^2$$

or

$$\frac{p}{p_a} + \frac{\rho V^2}{p_a} = 1 + \frac{\rho_a}{p_a} V_a^2 \tag{11.122}$$

By substituting the ideal gas equation of state (Eq. 11.1) into Eq. 11.122 and making use of the ideal gas speed-of-sound equation (Eq. 11.36) and the definition of Mach number (Eq. 11.46), we obtain

$$\boxed{\frac{p}{p_a} = \frac{1 + k}{1 + k\mathrm{Ma}^2}} \tag{11.123}$$

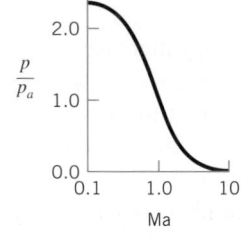

This relationship is graphed in the margin for air.

From the ideal gas equation of state (Eq. 11.1) we conclude that

$$\frac{T}{T_a} = \frac{p}{p_a} \frac{\rho_a}{\rho} \tag{11.124}$$

Conservation of mass (Eq. 11.40) with constant *A* gives

$$\frac{\rho_a}{\rho} = \frac{V}{V_a} \tag{11.125}$$

which when combined with Eqs. 11.36 (ideal gas speed of sound) and 11.46 (Mach number definition) gives

$$\frac{\rho_a}{\rho} = \mathrm{Ma} \sqrt{\frac{T}{T_a}} \tag{11.126}$$

Combining Eqs. 11.124 and 11.126 leads to

$$\frac{T}{T_a} = \left(\frac{p}{p_a} \mathrm{Ma} \right)^2 \tag{11.127}$$

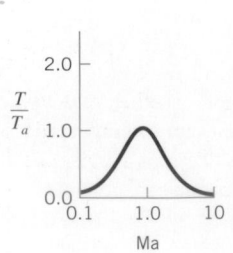

which when combined with Eq. 11.123 gives

$$\boxed{\frac{T}{T_a} = \left[\frac{(1 + k)\mathrm{Ma}}{1 + k\mathrm{Ma}^2} \right]^2} \tag{11.128}$$

This relationship is graphed in the margin on the previous page for air.

From Eqs. 11.125, 11.126, and 11.128 we see that

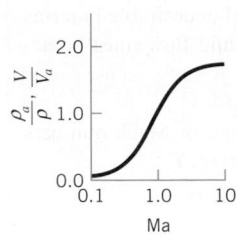

$$\frac{\rho_a}{\rho} = \frac{V}{V_a} = \text{Ma}\left[\frac{(1+k)\text{Ma}}{1+k\text{Ma}^2}\right] \tag{11.129}$$

This relationship is graphed in the margin for air.

The energy equation (Eq. 5.69) tells us that because of the heat transfer involved in Rayleigh flows, the stagnation temperature varies. We note that

$$\frac{T_0}{T_{0,a}} = \left(\frac{T_0}{T}\right)\left(\frac{T}{T_a}\right)\left(\frac{T_a}{T_{0,a}}\right) \tag{11.130}$$

Unlike Fanno flow, the stagnation temperature in Rayleigh flow varies.

We can use Eq. 11.56 (developed earlier for steady, isentropic, ideal gas flow) to evaluate T_0/T and T_a/T_{0a} because these two temperature ratios, by definition of the stagnation state, involve isentropic processes. Equation 11.128 can be used for T/T_a. Thus, consolidating Eqs. 11.130, 11.56, and 11.128 we obtain

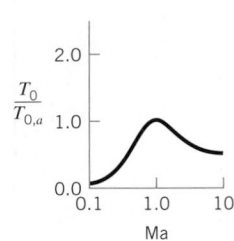

$$\frac{T_0}{T_{0,a}} = \frac{2(k+1)\text{Ma}^2\left(1 + \dfrac{k-1}{2}\text{Ma}^2\right)}{(1+k\text{Ma}^2)^2} \tag{11.131}$$

This relationship is graphed in the margin for air.

Finally, we observe that

$$\frac{p_0}{p_{0,a}} = \left(\frac{p_0}{p}\right)\left(\frac{p}{p_a}\right)\left(\frac{p_a}{p_{0,a}}\right) \tag{11.132}$$

We can use Eq. 11.59 developed earlier for steady, isentropic, ideal gas flow to evaluate p_0/p and $p_a/p_{0,a}$ because these two pressure ratios, by definition, involve isentropic processes. Equation 11.123 can be used for p/p_a. Together, Eqs. 11.59, 11.123, and 11.132 give

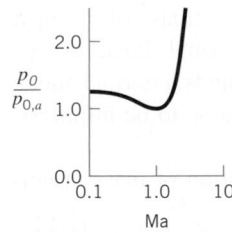

$$\frac{p_0}{p_{0,a}} = \frac{(1+k)}{(1+k\text{Ma}^2)}\left[\left(\frac{2}{k+1}\right)\left(1 + \frac{k-1}{2}\text{Ma}^2\right)\right]^{k/(k-1)} \tag{11.133}$$

This relationship is graphed in the margin for air.

Values of p/p_a, T/T_a, ρ/ρ_a or V/V_a, $T_0/T_{0,a}$, and $p_0/p_{0,a}$ are graphed in Fig. D.3 of Appendix D as a function of Mach number for Rayleigh flow of air ($k = 1.4$). The values in Fig. D.3 were calculated from Eqs. 11.123, 11.128, 11.129, 11.131, and 11.133. The usefulness of Fig. D.3 is illustrated in Example 11.16.

See Ref. 7 for a more advanced treatment of internal flows with heat transfer.

EXAMPLE 11.16 Effect of Mach Number and Heating/Cooling for Rayleigh Flow

GIVEN The information in Table 11.2 shows us that subsonic Rayleigh flow accelerates when heated and decelerates when cooled. Supersonic Rayleigh flow behaves just opposite to subsonic Rayleigh flow; it decelerates when heated and accelerates when cooled.

FIND Using Fig. D.3 for air ($k = 1.4$), state whether velocity, Mach number, static temperature, stagnation temperature, static pressure, and stagnation pressure increase or decrease as subsonic and supersonic Rayleigh flow is (a) heated, (b) cooled.

SOLUTION

Acceleration occurs when V/V_a in Fig. D.3 increases. For deceleration, V/V_a decreases. From Fig. D.3 and Table 11.2 the following chart can be constructed.

From the Rayleigh flow trends summarized in the table above, we note that heating affects Rayleigh flows much like friction affects Fanno flows. Heating and friction both accelerate subsonic flows and decelerate supersonic flows. More importantly, both

heating and friction cause the stagnation pressure to decrease. Since stagnation pressure loss is considered undesirable in terms of fluid mechanical efficiency, heating a fluid flow must be accomplished with this loss in mind.

COMMENT   Note that for a small range of Mach numbers cooling actually results in a rise in temperature, T.

	Heating		Cooling	
	Subsonic	**Supersonic**	**Subsonic**	**Supersonic**
V	Increase	Decrease	Decrease	Increase
Ma	Increase	Decrease	Decrease	Increase
T	Increase for $0 \leq \text{Ma} \leq \sqrt{1/k}$ Decrease for $\sqrt{1/k} \leq \text{Ma} \leq 1$	Increase	Decrease for $0 \leq \text{Ma} \leq \sqrt{1/k}$ Increase for $\sqrt{1/k} \leq \text{Ma} \leq 1$	Decrease
T_0	Increase	Increase	Decrease	Decrease
p	Decrease	Increase	Increase	Decrease
p_0	Decrease	Decrease	Increase	Increase

11.5.3 Normal Shock Waves

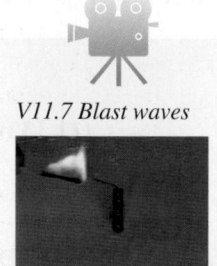

V11.7 Blast waves

As mentioned earlier, normal shock waves can occur in supersonic flows through converging–diverging and constant area ducts. Past experience suggests that normal shock waves involve deceleration from a supersonic flow to a subsonic flow, a pressure rise, and an increase of entropy. To develop the equations that verify this observed behavior of flows across a normal shock, we apply first principles to the flow through a control volume that completely surrounds a normal shock wave (see Fig. 11.24). We consider the normal shock and thus the control volume to be infinitesimally thin and stationary.

For steady flow through the control volume of Fig. 11.24, the conservation of mass principle yields

$$\rho V = \text{constant} \tag{11.134}$$

because the flow cross-sectional area remains essentially constant within the infinitesimal thickness of the normal shock. Note that Eq. 11.134 is identical to the continuity equation used for Fanno and Rayleigh flows considered earlier.

Normal shock waves are assumed to be infinitesimally thin discontinuities.

The friction force acting on the contents of the infinitesimally thin control volume surrounding the normal shock is considered to be negligibly small. Also for ideal gas flow, the effect of gravity is neglected. Thus, the linear momentum equation (Eq. 5.22) describing steady gas flow through the control volume of Fig. 11.24 is

$$p + \rho V^2 = \text{constant}$$

or for an ideal gas for which $p = \rho RT$,

$$p + \frac{(\rho V)^2 RT}{p} = \text{constant} \tag{11.135}$$

Equation 11.135 is the same as the linear momentum equation for Rayleigh flow, which was derived earlier (Eq. 11.111).

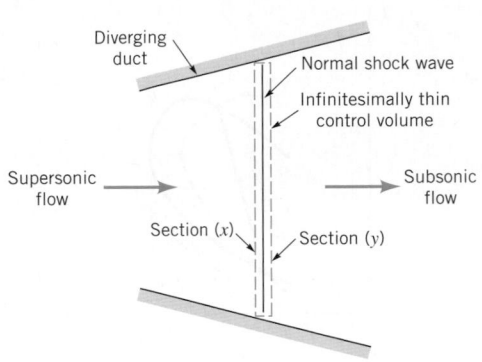

■ **F I G U R E   11.24   Normal shock control volume.**

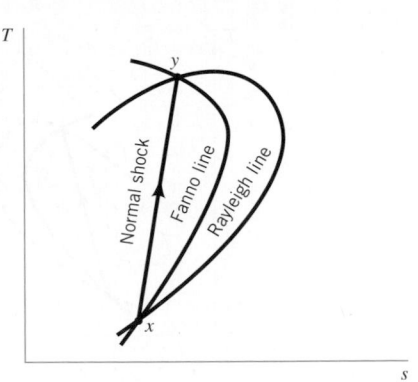

■ **F I G U R E   11.25   The relationship between a normal shock and Fanno and Rayleigh lines.**

For the control volume containing the normal shock, no shaft work is involved and the heat transfer is assumed negligible. Thus, the energy equation (Eq. 5.69) can be applied to steady gas flow through the control volume of Fig. 11.24 to obtain

$$\check{h} + \frac{V^2}{2} = \check{h}_0 = \text{constant}$$

or, for an ideal gas, since $\check{h} - \check{h}_0 = c_p(T - T_0)$ and $p = \rho R T$

$$T + \frac{(\rho V)^2 T^2}{2c_p(p^2/R^2)} = T_0 = \text{constant} \tag{11.136}$$

Equation 11.136 is identical to the energy equation for Fanno flow analyzed earlier (Eq. 11.75).

The $T\,ds$ relationship previously used for ideal gas flow (Eq. 11.22) is valid for the flow through the normal shock (Fig. 11.24) because it (Eq. 11.22) is an ideal gas property relationship.

The energy equation for Fanno flow and the momentum equation for Rayleigh flow are valid for flow across normal shocks.

From the analyses in the previous paragraphs, it is apparent that the steady flow of an ideal gas across a normal shock is governed by some of the same equations used for describing both Fanno and Rayleigh flows (energy equation for Fanno flows and momentum equation for Rayleigh flow). Thus, for a given density–velocity product (ρV), gas (R, k), and conditions at the inlet of the normal shock (T_x, p_x, and s_x), the conditions downstream of the shock (state y) will be on both a Fanno line and a Rayleigh line that pass through the inlet state (state x), as is illustrated in Fig. 11.25. To conform with common practice we designate the states upstream and downstream of the normal shock with x and y instead of numerals 1 and 2. The Fanno and Rayleigh lines describe more of the flow field than just in the vicinity of the normal shock when Fanno and Rayleigh flows are actually involved (solid lines in Figs. 11.26a and 11.26b). Otherwise, these lines (dashed lines in Figs. 11.26a, 11.26b, and 11.26c) are useful mainly as a way to better visualize how the governing equations combine to yield a solution to the normal shock flow problem.

The second law of thermodynamics requires that entropy must increase across a normal shock wave. This law and sketches of the Fanno line and Rayleigh line intersections, like those of Figs. 11.25 and 11.26, persuade us to conclude that flow across a normal shock can only proceed from supersonic to subsonic flow. Similarly, in open-channel flows (see Chapter 10) the flow across a hydraulic jump proceeds from supercritical to subcritical conditions.

Since the states upstream and downstream of a normal shock wave are represented by the supersonic and subsonic intersections of actual and/or imagined Fanno and Rayleigh lines, we should be able to use equations developed earlier for Fanno and Rayleigh flows to quantify normal shock flow. For example, for the Rayleigh line of Fig. 11.26b

$$\frac{p_y}{p_x} = \left(\frac{p_y}{p_a}\right)\left(\frac{p_a}{p_x}\right) \tag{11.137}$$

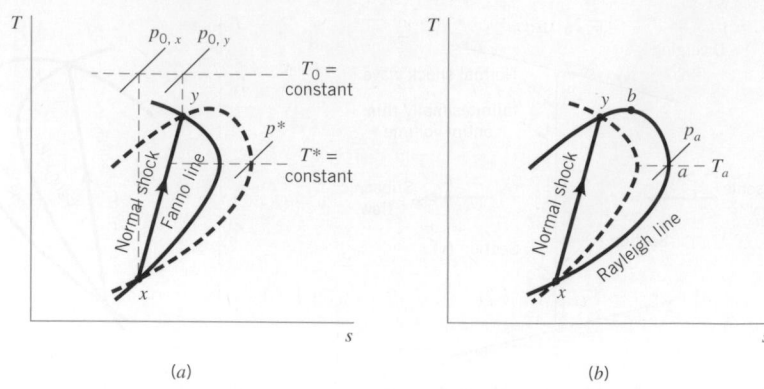

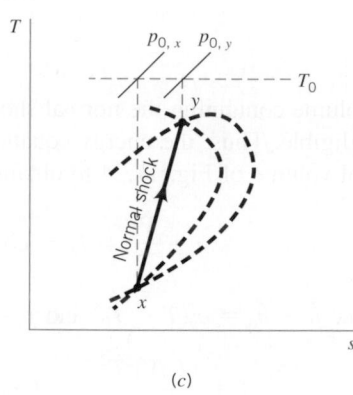

■ **F I G U R E 11.26** (*a*) The normal shock in a Fanno flow. (*b*) The normal shock in a Rayleigh flow. (*c*) The normal shock in a frictionless and adiabatic flow.

But from Eq. 11.123 for Rayleigh flow we get

$$\frac{p_y}{p_a} = \frac{1 + k}{1 + k\mathrm{Ma}_y^2} \tag{11.138}$$

and

$$\frac{p_x}{p_a} = \frac{1 + k}{1 + k\mathrm{Ma}_x^2} \tag{11.139}$$

Thus, by combining Eqs. 11.137, 11.138, and 11.139 we get

$$\boxed{\frac{p_y}{p_x} = \frac{1 + k\mathrm{Ma}_x^2}{1 + k\mathrm{Ma}_y^2}} \tag{11.140}$$

Equation 11.140 can also be derived starting with

$$\frac{p_y}{p_x} = \left(\frac{p_y}{p^*}\right)\left(\frac{p^*}{p_x}\right)$$

and using the Fanno flow equation (Eq. 11.107)

$$\frac{p}{p^*} = \frac{1}{\mathrm{Ma}}\left\{\frac{(k+1)/2}{1 + [(k-1)/2]\mathrm{Ma}^2}\right\}^{1/2}$$

As might be expected, Eq. 11.140 can be obtained directly from the linear momentum equation

$$p_x + \rho_x V_x^2 = p_y + \rho_y V_y^2$$

Ratios of thermodynamic properties across a normal shock are functions of the Mach numbers.

since $\rho V^2/p = V^2/RT = kV^2/RTk = k\,\mathrm{Ma}^2$.

For the Fanno flow of Fig. 11.26*a*,

$$\frac{T_y}{T_x} = \left(\frac{T_y}{T^*}\right)\left(\frac{T^*}{T_x}\right) \tag{11.141}$$

From Eq. 11.101 for Fanno flow we get

$$\frac{T_y}{T^*} = \frac{(k+1)/2}{1 + [(k-1)/2]\text{Ma}_y^2}$$

(11.142)

and

$$\frac{T_x}{T^*} = \frac{(k+1)/2}{1 + [(k-1)/2]\text{Ma}_x^2}$$

(11.143)

A consolidation of Eqs. 11.141, 11.142, and 11.143 gives

$$\boxed{\frac{T_y}{T_x} = \frac{1 + [(k-1)/2]\text{Ma}_x^2}{1 + [(k-1)/2]\text{Ma}_y^2}}$$

(11.144)

We seek next to develop an equation that will allow us to determine the Mach number downstream of the normal shock, Ma_y, when the Mach number upstream of the normal shock, Ma_x, is known. From the ideal gas equation of state (Eq. 11.1), we can form

$$\frac{p_y}{p_x} = \left(\frac{T_y}{T_x}\right)\left(\frac{\rho_y}{\rho_x}\right)$$

(11.145)

Using the continuity equation

$$\rho_x V_x = \rho_y V_y$$

with Eq. 11.145 we obtain

$$\frac{p_y}{p_x} = \left(\frac{T_y}{T_x}\right)\left(\frac{V_x}{V_y}\right)$$

(11.146)

When combined with the Mach number definition (Eq. 11.46) and the ideal gas speed-of-sound equation (Eq. 11.36), Eq. 11.146 becomes

$$\frac{p_y}{p_x} = \left(\frac{T_y}{T_x}\right)^{1/2}\left(\frac{\text{Ma}_x}{\text{Ma}_y}\right)$$

(11.147)

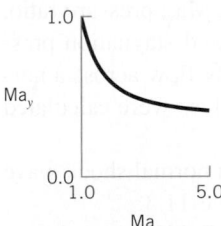

Thus, Eqs. 11.147 and 11.144 lead to

$$\frac{p_y}{p_x} = \left\{\frac{1 + [(k-1)/2]\text{Ma}_x^2}{1 + [(k-1)/2]\text{Ma}_y^2}\right\}^{1/2}\frac{\text{Ma}_x}{\text{Ma}_y}$$

(11.148)

which can be merged with Eq. 11.140 to yield

$$\boxed{\text{Ma}_y^2 = \frac{\text{Ma}_x^2 + [2/(k-1)]}{[2k/(k-1)]\text{Ma}_x^2 - 1}}$$

(11.149)

The flow changes from supersonic to subsonic across a normal shock.

This relationship is graphed in the margin for air.

Thus, we can use Eq. 11.149 to calculate values of Mach number downstream of a normal shock from a known Mach number upstream of the shock. As suggested by Fig. 11.26, to have a normal shock we must have $\text{Ma}_x > 1$. From Eq. 11.149 we find that $\text{Ma}_y < 1$.

If we combine Eqs. 11.149 and 11.140, we get

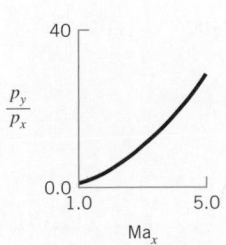

$$\boxed{\frac{p_y}{p_x} = \frac{2k}{k+1}\text{Ma}_x^2 - \frac{k-1}{k+1}}$$

(11.150)

This relationship is graphed in the margin for air.

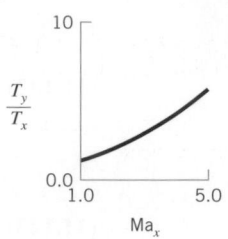

This equation allows us to calculate the pressure ratio across a normal shock from a known upstream Mach number. Similarly, taking Eqs. 11.149 and 11.144 together we obtain

$$\frac{T_y}{T_x} = \frac{\{1 + [(k-1)/2]\text{Ma}_x^2\}\{[2k/(k-1)]\text{Ma}_x^2 - 1\}}{\{(k+1)^2/[2(k-1)]\}\text{Ma}_x^2} \tag{11.151}$$

This relationship is graphed in the margin for air.

From the continuity equation (Eq. 11.40), we have for flow across a normal shock

$$\frac{\rho_y}{\rho_x} = \frac{V_x}{V_y} \tag{11.152}$$

and from the ideal gas equation of state (Eq. 11.1)

$$\frac{\rho_y}{\rho_x} = \left(\frac{p_y}{p_x}\right)\left(\frac{T_x}{T_y}\right) \tag{11.153}$$

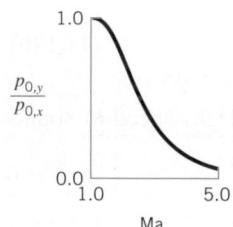

Thus, by combining Eqs. 11.152, 11.153, 11.150, and 11.151, we get

$$\frac{\rho_y}{\rho_x} = \frac{V_x}{V_y} = \frac{(k+1)\text{Ma}_x^2}{(k-1)\text{Ma}_x^2 + 2} \tag{11.154}$$

This relationship is graphed in the margin for air.

The stagnation pressure ratio across the shock can be determined by combining

$$\frac{p_{0,y}}{p_{0,x}} = \left(\frac{p_{0,y}}{p_y}\right)\left(\frac{p_y}{p_x}\right)\left(\frac{p_x}{p_{0,x}}\right) \tag{11.155}$$

with Eqs. 11.59, 11.149, and 11.150 to get

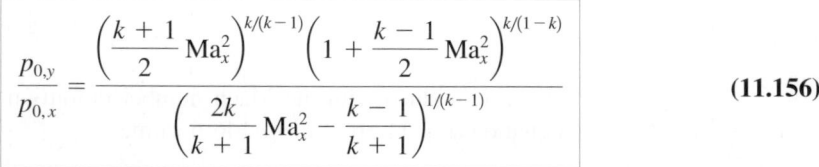

$$\frac{p_{0,y}}{p_{0,x}} = \frac{\left(\frac{k+1}{2}\text{Ma}_x^2\right)^{k/(k-1)}\left(1 + \frac{k-1}{2}\text{Ma}_x^2\right)^{k/(1-k)}}{\left(\frac{2k}{k+1}\text{Ma}_x^2 - \frac{k-1}{k+1}\right)^{1/(k-1)}} \tag{11.156}$$

This relationship is graphed in the margin for air.

Figure D.4 in Appendix D graphs values of downstream Mach numbers, Ma_y, pressure ratio, p_y/p_x, temperature ratio, T_y/T_x, density ratio, ρ_y/ρ_x, or velocity ratio, V_x/V_y, and stagnation pressure ratio, $p_{0,y}/p_{0,x}$, as a function of upstream Mach number, Ma_x, for the steady flow across a normal shock wave of an ideal gas having a specific heat ratio $k = 1.4$. These values were calculated from Eqs. 11.149, 11.150, 11.151, 11.154, and 11.156.

Important trends associated with the steady flow of an ideal gas across a normal shock wave can be determined by studying Fig. D.4. These trends are summarized in Table 11.3.

Examples 11.17 and 11.18 illustrate how Fig. D.4 can be used to solve fluid flow problems involving normal shock waves.

Across a normal shock the values of some parameters increase, some remain constant, and some decrease.

■ **TABLE 11.3**
Summary of Normal Shock Wave Characteristics

Variable	Change Across Normal Shock Wave
Mach number	Decrease
Static pressure	Increase
Stagnation pressure	Decrease
Static temperature	Increase
Stagnation temperature	Constant
Density	Increase
Velocity	Decrease

EXAMPLE 11.17   Stagnation Pressure Drop across a Normal Shock

GIVEN   Designers involved with fluid mechanics work hard at minimizing loss of available energy in their designs. Adiabatic, frictionless flows involve no loss in available energy. Entropy remains constant for these idealized flows. Adiabatic flows with friction involve available energy loss and entropy increase. Generally, larger entropy increases imply larger losses.

FIND   For normal shocks, show that the stagnation pressure drop (and thus loss) is larger for higher Mach numbers.

SOLUTION

We assume that air $(k = 1.4)$ behaves as a typical gas and use Fig. D.4 to respond to the above-stated requirements. Since

$$1 - \frac{p_{0,y}}{p_{0,x}} = \frac{p_{0,x} - p_{0,y}}{p_{0,x}}$$

we can construct the following table with values of $p_{0,y}/p_{0,x}$ from Fig. D.4.

COMMENT   When the Mach number of the flow entering the shock is low, say $Ma_x = 1.2$, the flow across the shock is nearly isentropic and the loss in stagnation pressure is small. However, as shown in Fig. E11.17, at larger Mach numbers, the entropy change across the normal shock rises dramatically and the stagna-

tion pressure drop across the shock is appreciable. If a shock occurs at $Ma_x = 2.5$, only about 50% of the upstream stagnation pressure is recovered.

In devices where supersonic flows occur, for example, high-performance aircraft engine inlet ducts and high-speed wind tunnels, designers attempt to prevent shock formation, or if shocks must occur, they design the flow path so that shocks are positioned where they are weak (small Mach number).

Of interest also is the static pressure rise that occurs across a normal shock. These static pressure ratios, p_y/p_x, obtained from Fig. D.4 are shown in the table for a few Mach numbers. For a developing boundary layer, any pressure rise in the flow direction is considered as an adverse pressure gradient that can possibly cause flow separation (see Section 9.2.6). Thus, shock–boundary layer interactions are of great concern to designers of high-speed flow devices.

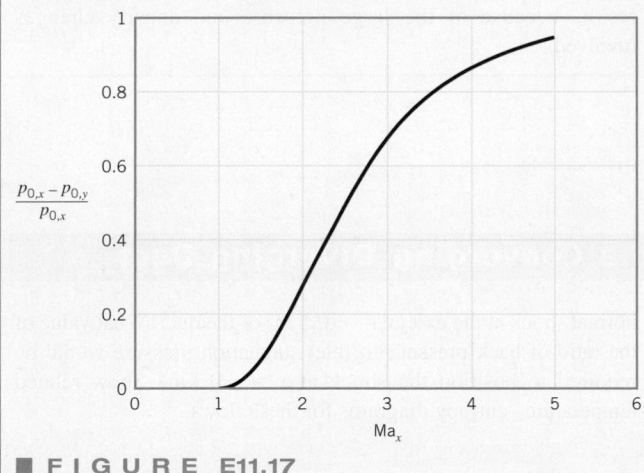

■ FIGURE E11.17

Ma_x	$p_{0,y}/p_{0,x}$	$\dfrac{p_{0,x} - p_{0,y}}{p_{0,x}}$	Ma_x	p_y/p_x
1.0	1.0	0	1.0	1.0
1.2	0.99	0.01	1.2	1.5
1.5	0.93	0.07	1.5	2.5
2.0	0.72	0.28	2.0	4.5
2.5	0.50	0.50	3.0	10
3.0	0.33	0.67	4.0	18
3.5	0.21	0.79	5.0	29
4.0	0.14	0.86		
5.0	0.06	0.94		

EXAMPLE 11.18   Supersonic Flow Pitot Tube

GIVEN   A total pressure probe is inserted into a supersonic air flow. A shock wave forms just upstream of the impact hole and head as illustrated in Fig. E11.18. The probe measures a total pressure of 414 kPa (abs). The stagnation temperature at the probe head is 555 K. The static pressure upstream of the shock is measured with a wall tap to be 82 kPa (abs).

FIND   Determine the Mach number and velocity of the flow.

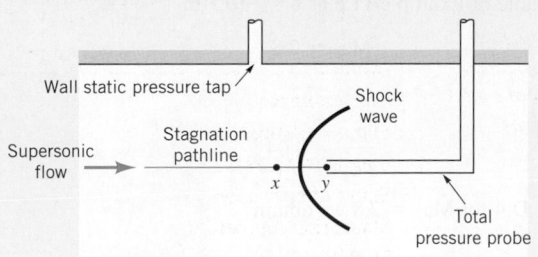

■ FIGURE E11.18

SOLUTION

We assume that the flow along the stagnation pathline is isentropic except across the shock. Also, the shock is treated as a normal shock. Thus, in terms of the data we have

$$\frac{p_{0,y}}{p_x} = \left(\frac{p_{0,y}}{p_{0,x}}\right)\left(\frac{p_{0,x}}{p_x}\right) \tag{1}$$

where $p_{0,y}$ is the stagnation pressure measured by the probe, and p_x is the static pressure measured by the wall tap. The stagnation pressure upstream of the shock, $p_{0,x}$, is not measured.

Combining Eqs. 1, 11.156, and 11.59 we obtain

$$\frac{p_{0,y}}{p_x} = \frac{\{[(k+1)/2]\mathrm{Ma}_x^2\}^{k/(k-1)}}{\{[2k/(k+1)]\mathrm{Ma}_x^2 - [(k-1)/(k+1)]\}^{1/(k-1)}} \tag{2}$$

which is called the *Rayleigh Pitot-tube formula*. Values of $p_{0,y}/p_x$ from Eq. 2 are considered important enough to be included in Fig. D.4 for $k = 1.4$. Thus, for $k = 1.4$ and

$$\frac{p_{0,y}}{p_x} = \frac{414 \text{ kPa (abs)}}{82 \text{ kPa (abs)}} = 5$$

we use Fig. D.4 (or Eq. 2) to ascertain that

$$\mathrm{Ma}_x = 1.9 \qquad \text{(Ans)}$$

To determine the flow velocity we need to know the static temperature upstream of the shock, since Eqs. 11.36 and 11.46 can be used to yield

$$V_x = \mathrm{Ma}_x c_x = \mathrm{Ma}_x \sqrt{RT_x k} \tag{3}$$

The stagnation temperature downstream of the shock was measured and found to be

$$T_{0,y} = 555 \text{ K}$$

Since the stagnation temperature remains constant across a normal shock (see Eq. 11.136),

$$T_{0,x} = T_{0,y} = 555 \text{ K}$$

For the isentropic flow upstream of the shock, Eq. 11.56 or Fig. D.1 can be used. For $\mathrm{Ma}_x = 1.9$,

$$\frac{T_x}{T_{0,x}} = 0.59$$

or

$$T_x = (0.59)(555 \text{ K}) = 327 \text{ K}$$

With Eq. 3 we obtain

$$V_x = 1.87 \sqrt{[286.9 \text{ J/(kg} \cdot \text{K)}](327 \text{ K})(1.4)}$$
$$= 678 \text{ m/s} \qquad \text{(Ans)}$$

COMMENT Application of the incompressible flow Pitot tube results (see Section 3.5) would give highly inaccurate results because of the large pressure and density changes involved.

EXAMPLE 11.19 ■ Normal Shock in a Converging–Diverging Duct

GIVEN Consider the converging–diverging duct of Example 11.8.

FIND Determine the ratio of back pressure to inlet stagnation pressure, $p_{\mathrm{III}}/p_{0,x}$ (see Fig. 11.13), that will result in a standing normal shock at the exit ($x = +0.5$ m) of the duct. What value of the ratio of back pressure to inlet stagnation pressure would be required to position the shock at $x = +0.3$ m? Show related temperature–entropy diagrams for these flows.

SOLUTION

For supersonic, isentropic flow through the nozzle to just upstream of the standing normal shock at the duct exit, we have from the table of Example 11.8 at $x = +0.5$ m

$$\mathrm{Ma}_x = 2.8$$

and

$$\frac{p_x}{p_{0,x}} = 0.04$$

From Fig. D.4 for $\mathrm{Ma}_x = 2.8$ we obtain

$$\frac{p_y}{p_x} = 9.0$$

Thus,

$$\frac{p_y}{p_{0,x}} = \left(\frac{p_y}{p_x}\right)\left(\frac{p_x}{p_{0,x}}\right) = (9.0)(0.04)$$

$$= 0.36 = \frac{p_{\mathrm{III}}}{p_{0,x}} \qquad \text{(Ans)}$$

When the ratio of duct back pressure to inlet stagnation pressure, $p_{\mathrm{III}}/p_{0,x}$, is set equal to 0.36, the air will accelerate through the converging–diverging duct to a Mach number of 2.8 at the duct exit. The air will subsequently decelerate to a subsonic flow across a normal shock at the duct exit. The stagnation pressure ratio across the normal shock, $p_{0,y}/p_{0,x}$, is 0.38 (Fig. D.4 for

$Ma_x = 2.8$). A considerable amount of available energy is lost across the shock.

For a normal shock at $x = +0.3$ m, we note from the table of Example 11.8 that $Ma_x = 2.14$ and

$$\frac{p_x}{p_{0,x}} = 0.10 \tag{1}$$

From Fig. D.4 for $Ma_x = 2.14$ we obtain $p_y/p_x = 5.2$, $Ma_y = 0.56$, and

$$\frac{p_{0,y}}{p_{0,x}} = 0.66 \tag{2}$$

From Fig. D.1 for $Ma_y = 0.56$ we get

$$\frac{A_y}{A^*} = 1.24 \tag{3}$$

For $x = +0.3$ m, the ratio of duct exit area to local area (A_2/A_y) is, using the area equation from Example 11.8,

$$\frac{A_2}{A_y} = \frac{0.1 + (0.5)^2}{0.1 + (0.3)^2} = 1.842 \tag{4}$$

Using Eqs. 3 and 4 we get

$$\frac{A_2}{A^*} = \left(\frac{A_y}{A^*}\right)\left(\frac{A_2}{A_y}\right) = (1.24)(1.842) = 2.28$$

Note that for the isentropic flow upstream of the shock, $A^* = 0.10$ m^2 (the actual throat area), while for the isentropic flow downstream of the shock, $A^* = A_2/2.28 = 0.35$ m$^2/2.28 = 0.15$ m^2. With $A_2/A^* = 2.28$ we use Fig. D.1 and find $Ma_2 = 0.26$ and

$$\frac{p_2}{p_{0,y}} = 0.95 \tag{5}$$

Combining Eqs. 2 and 5 we obtain

$$\frac{p_2}{p_{0,x}} = \left(\frac{p_2}{p_{0,y}}\right)\left(\frac{p_{0,y}}{p_{0,x}}\right) = (0.95)(0.66) = 0.63 \quad \textbf{(Ans)}$$

When the back pressure, p_2, is set equal to 0.63 times the inlet stagnation pressure, $p_{0,x}$, the normal shock will be positioned at $x = +0.3$ m. The corresponding $T-s$ diagrams are shown in Figs. E11.19a and E11.19b.

COMMENT   Note that $p_2/p_{0,x} = 0.63$ is less than the value of this ratio for subsonic isentropic flow through the converging–diverging duct, $p_2/p_0 = 0.98$ (from Example 11.8) and is larger than $p_{III}/p_{0,x} = 0.36$, for duct flow with a normal shock at the exit (see Fig. 11.13). Also the stagnation pressure ratio with the shock at $x = +0.3$ m, $p_{0,y}/p_{0,x} = 0.66$, is much greater than the stagnation pressure ratio, 0.38, when the shock occurs at the exit $(x = +0.5$ m) of the duct.

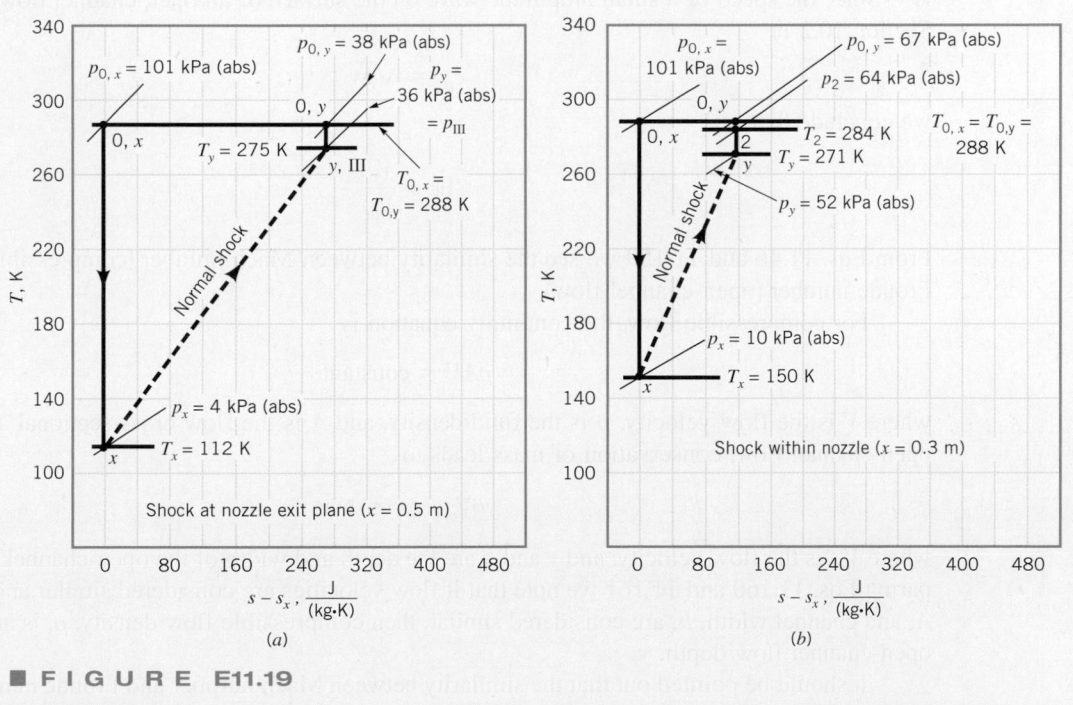

■ **FIGURE E11.19**

11.6   Analogy between Compressible and Open-Channel Flows

During a first course in fluid mechanics, students rarely study both open-channel flows (Chapter 10) and compressible flows. This is unfortunate because these two kinds of flows are strikingly similar in several ways. Furthermore, the analogy between open-channel and compressible flows is useful because important two-dimensional compressible flow phenomena can be simply and inexpensively demonstrated with a shallow, open-channel flow field in a *ripple tank* or *water table*.

The propagation of weak pressure pulses (sound waves) in a compressible flow can be considered to be comparable to the movement of small amplitude waves on the surface of an open-channel flow. In each case—two-dimensional compressible flow and open-channel flow—the influence of flow velocity on wave pattern is similar. When the flow velocity is less than the wave speed, wave fronts can move upstream of the wave source and the flow is subsonic (compressible flow) or subcritical (open-channel flow). When the flow velocity is equal to the wave speed, wave fronts cannot move upstream of the wave source and the flow is sonic (compressible flow) or critical (open-channel flow). When the flow velocity is greater than the wave speed, the flow is supersonic (compressible flow) or supercritical (open-channel flow). Normal shocks can occur in supersonic compressible flows. Hydraulic jumps can occur in supercritical open-channel flows. Comparison of the characteristics of normal shocks (Section 11.5.3) and hydraulic jumps (Section 10.6.1) suggests a strong resemblance and thus analogy between the two phenomena.

Compressible gas flows and open-channel liquid flows are strikingly similar in several ways.

For compressible flows a meaningful dimensionless variable is the Mach number, where

$$\text{Ma} = \frac{V}{c} \tag{11.46}$$

In open-channel flows, an important dimensionless variable is the Froude number, where

$$\text{Fr} = \frac{V_{oc}}{\sqrt{gy}} \tag{11.157}$$

The velocity of the channel flow is V_{oc}, the acceleration of gravity is g, and the depth of the flow is y. Since the speed of a small amplitude wave on the surface of an open-channel flow, c_{oc}, is (see Section 10.2.1)

$$c_{oc} = \sqrt{gy} \tag{11.158}$$

we conclude that

$$\text{Fr} = \frac{V_{oc}}{c_{oc}} \tag{11.159}$$

From Eqs. 11.46 and 11.159 we see the similarity between Mach number (compressible flow) and Froude number (open-channel flow).

For compressible flow, the continuity equation is

$$\rho A V = \text{constant} \tag{11.160}$$

where V is the flow velocity, ρ is the fluid density, and A is the flow cross-sectional area. For an open-channel flow, conservation of mass leads to

$$ybV_{oc} = \text{constant} \tag{11.161}$$

where V_{oc} is the flow velocity, and y and b are the depth and width of the open-channel flow. Comparing Eqs. 11.160 and 11.161 we note that if flow velocities are considered similar and flow area, A, and channel width, b, are considered similar, then compressible flow density, ρ, is analogous to open-channel flow depth, y.

It should be pointed out that the similarity between Mach number and Froude number is generally not exact. If compressible flow and open-channel flow velocities are considered to be similar, then it follows that for Mach number and Froude number similarity the wave speeds c and c_{oc} must also be similar.

From the development of the equation for the speed of sound in an ideal gas (see Eqs. 11.34 and 11.35) we have for the compressible flow

$$c = \sqrt{(\text{constant}) \, k\rho^{k-1}} \tag{11.162}$$

From Eqs. 11.162 and 11.158, we see that if y is to be similar to ρ as suggested by comparing Eq. 11.160 and 11.161, then k should be equal to 2. Typically $k = 1.4$ or 1.67, not 2. This limitation to exactness is, however, usually not serious enough to compromise the benefits of the analogy between compressible and open-channel flows.

11.7   Two-Dimensional Compressible Flow

A brief introduction to two-dimensional compressible flow is included here for those who are interested. We begin with a consideration of supersonic flow over a wall with a small change of direction as sketched in Fig. 11.27.

We apply the component of the linear momentum equation (Eq. 5.22) parallel to the Mach wave to the flow across the Mach wave. (See Eq. 11.39 for the definition of a Mach wave.) The result is that the component of velocity parallel to the Mach wave is constant across the Mach wave. That is, $V_{t1} = V_{t2}$. Thus, from the simple velocity triangle construction indicated in Fig. 11.27, we conclude that the flow accelerates because of the change in direction of the flow. If several changes in wall direction are involved as shown in Fig. 11.28, then the supersonic flow accelerates (expands) because of the changes in flow direction across the Mach waves (also called *expansion* waves). Each Mach wave makes an appropriately smaller angle α with the upstream wall because of the increase in Mach number that occurs with each direction change (see Section 11.3). A rounded expansion corner may be considered as a series of infinitesimal changes in direction. Conversely, even sharp corners are actually rounded when viewed on a small enough scale. Thus, expansion fans as illustrated in Fig. 11.29 are commonly used for supersonic flow around a "sharp" corner. If the flow across the Mach waves is considered to be isentropic, then Eq. 11.42 suggests that the increase in flow speed is accompanied by a decrease in static pressure.

Supersonic flows accelerate across expansion Mach waves.

When the change in supersonic flow direction involves the change in wall orientation sketched in Fig. 11.30, compression rather than expansion occurs. The flow decelerates and the static pressure increases across the Mach wave. For several changes in wall direction, as indicated in Fig. 11.31, several Mach waves occur, each at an appropriately larger angle α with the upstream wall. A rounded compression corner may be considered as a series of infinitesimal changes in

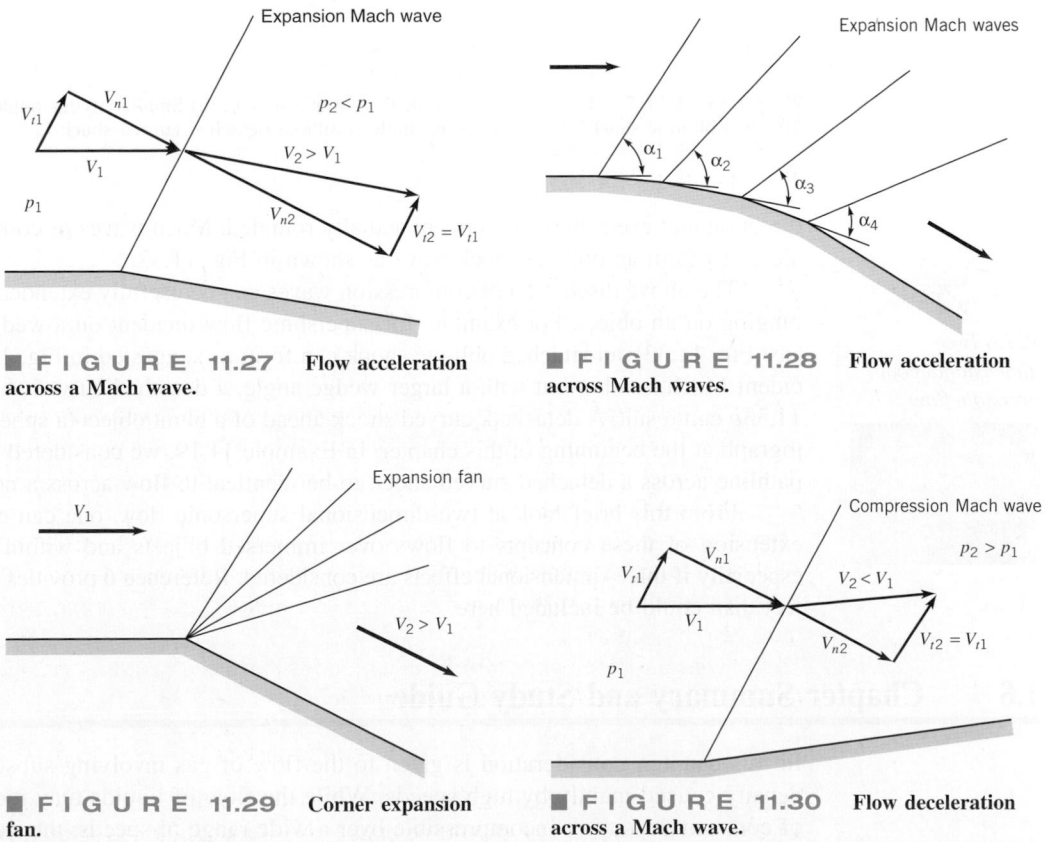

■ **F I G U R E   11.27**   **Flow acceleration across a Mach wave.**

■ **F I G U R E   11.28**   **Flow acceleration across Mach waves.**

■ **F I G U R E   11.29**   **Corner expansion fan.**

■ **F I G U R E   11.30**   **Flow deceleration across a Mach wave.**

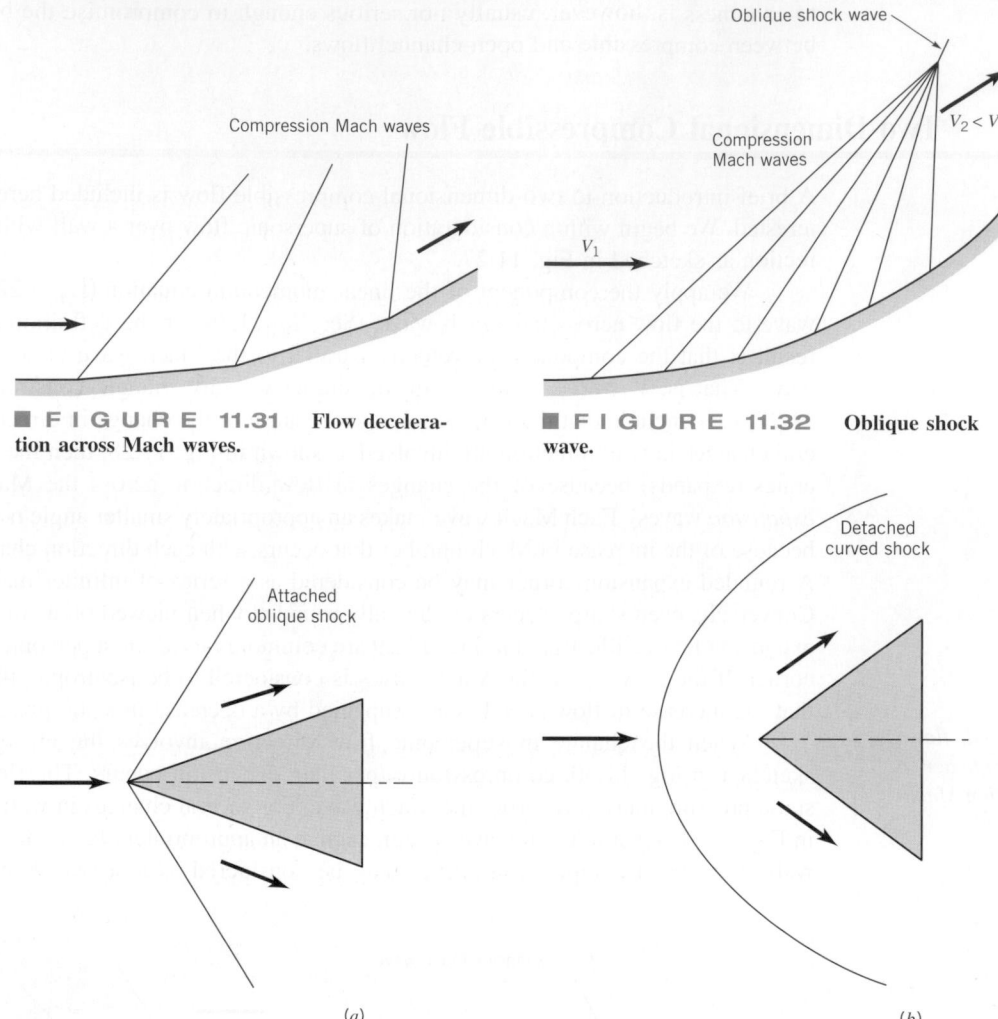

■ **FIGURE 11.31** **Flow deceleration across Mach waves.**

■ **FIGURE 11.32** **Oblique shock wave.**

(a)

(b)

■ **FIGURE 11.33** **Supersonic flow over a wedge:** (*a*) **Smaller wedge angle results in attached oblique shock.** (*b*) **Large wedge angle results in detached curved shock.**

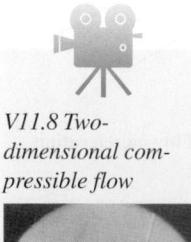

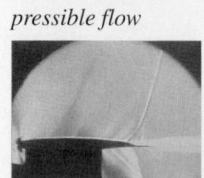

V11.8 Two-dimensional compressible flow

direction and even sharp corners are actually rounded. Mach waves or compression waves can co-alesce to form an oblique shock wave as shown in Fig. 11.32.

The above discussion of compression waves can be usefully extended to supersonic flow im-pinging on an object. For example, for supersonic flow incident on a wedge-shaped leading edge (see Fig. 11.33), an attached oblique shock can form as suggested in Fig. 11.33*a*. For the same in-cident Mach number but with a larger wedge angle, a detached curved shock as sketched in Fig. 11.33*b* can result. A detached, curved shock ahead of a blunt object (a sphere) is shown in the pho-tograph at the beginning of this chapter. In Example 11.19, we considered flow along a stagnation pathline across a detached curved shock to be identical to flow across a normal shock wave.

From this brief look at two-dimensional supersonic flow, one can easily conclude that the extension of these concepts to flows over immersed objects and within ducts can be exciting, especially if three-dimensional effects are considered. Reference 6 provides much more on this sub-ject than could be included here.

11.8 Chapter Summary and Study Guide

In this chapter, consideration is given to the flow of gas involving substantial changes in fluid density caused mainly by high speeds. While the flow of liquids may most often be considered of constant density or incompressible over a wide range of speeds, the flow of gases and vapors may involve substantial fluid density changes at higher speeds. At lower speeds, gas and vapor density changes are not appreciable and so these flows may be treated as incompressible.

Since fluid density and other fluid property changes are significant in compressible flows, property relationships are important. An ideal gas, with well-defined fluid property relationships, is used as an approximation of an actual gas. This profound simplification still allows useful conclusions to be made about compressible flows.

The Mach number is a key variable in compressible flow theory. Most easily understood as the ratio of the local speed of flow and the speed of sound in the flowing fluid, it is a measure of the extent to which the flow is compressible or not. It is used to define categories of compressible flows which range from subsonic (Mach number less than 1) to supersonic (Mach number greater than 1). The speed of sound in a truly incompressible fluid is infinite so the Mach numbers associated with liquid flows are generally low.

The notion of an isentropic or constant entropy flow is introduced. The most important isentropic flow is one that is adiabatic (no heat transfer to or from the flowing fluid) and frictionless (zero viscosity). This simplification, like the one associated with approximating real gases with an ideal gas, leads to useful results including trends associated with accelerating and decelerating flows through converging, diverging, and converging–diverging flow paths. Phenomena including flow choking, acceleration in a diverging passage, deceleration in a converging passage, and the achievement of supersonic flows are discussed.

Three major nonisentropic compressible flows considered in this chapter are Fanno flows, Rayleigh flows, and flows across normal shock waves. Unusual outcomes include the conclusions that friction can accelerate a subsonic Fanno flow, heating can result in fluid temperature reduction in a subsonic Rayleigh flow, and a flow can decelerate from supersonic flow to subsonic flow across a very small distance. The value of temperature–entropy ($T–s$) diagramming of flows to better understand them is demonstrated.

Numerous formulas describing a variety of ideal gas compressible flows are derived. These formulas can be easily solved with computers. However, to provide the learner with a better grasp of the details of a compressible flow process, a graphical approach, albeit approximate, is used.

The striking analogy between compressible and open-channel flows leads to a brief discussion of the usefulness of a ripple tank or water table to simulate compressible flows.

Expansion and compression Mach waves associated with two-dimensional compressible flows are introduced as is the formation of oblique shock waves from compression Mach waves.

The following checklist provides a study guide for this chapter. When your study of the entire chapter and end-of-chapter exercises is completed you should be able to

- write out the meanings of the terms listed here in the margin and understand each of the related concepts. These terms are particularly important and are set in *italic, **bold,** and color* type in the text.
- estimate the change in ideal gas properties in a compressible flow.
- calculate Mach number value for a specific compressible flow.
- estimate when a flow may be considered incompressible and when it must be considered compressible to preserve accuracy.
- estimate details of isentropic flows of an ideal gas though converging, diverging, and converging–diverging passages.
- estimate details of nonisentropic Fanno and Rayleigh flows and flows across normal shock waves.
- explain the analogy between compressible and open-channel flows.

Some of the important equations in this chapter are:

Ideal gas equation of state	$\rho = \dfrac{p}{RT}$	**(11.1)**
Internal energy change	$\breve{u}_2 - \breve{u}_1 = c_v(T_2 - T_1)$	**(11.5)**
Enthalpy	$\breve{h} = \breve{u} + \dfrac{p}{\rho}$	**(11.6)**

Enthalpy change	$\breve{h}_2 - \breve{h}_1 = c_p(T_2 - T_1)$	**(11.9)**
Specific heat difference	$c_p - c_v = R$	**(11.12)**
Specific heat ratio	$k = \dfrac{c_p}{c_v}$	**(11.13)**
Specific heat at constant pressure	$c_p = \dfrac{Rk}{k-1}$	**(11.14)**
Specific heat at constant volume	$c_v = \dfrac{R}{k-1}$	**(11.15)**
First $T\,ds$ equation	$T\,ds = d\breve{u} + p\,d\left(\dfrac{1}{\rho}\right)$	**(11.16)**
Second $T\,ds$ equation	$T\,ds = d\breve{h} - \left(\dfrac{1}{\rho}\right)dp$	**(11.18)**
Entropy change	$s_2 - s_1 = c_v \ln \dfrac{T_2}{T_1} + R \ln \dfrac{\rho_1}{\rho_2}$	**(11.21)**
Entropy change	$s_2 - s_1 = c_p \ln \dfrac{T_2}{T_1} - R \ln \dfrac{p_2}{p_1}$	**(11.22)**
Isentropic flow	$\dfrac{p}{\rho^k} = \text{constant}$	**(11.25)**
Speed of sound	$c = \sqrt{\left(\dfrac{\partial p}{\partial \rho}\right)_s}$	**(11.34)**
Speed of sound in gas	$c = \sqrt{RTk}$	**(11.36)**
Speed of sound in liquid	$c = \sqrt{\dfrac{E_v}{\rho}}$	**(11.38)**
Mach cone angle	$\sin \alpha = \dfrac{c}{V} = \dfrac{1}{\text{Ma}}$	**(11.39)**
Mach number	$\text{Ma} = \dfrac{V}{c}$	**(11.46)**
Isentropic flow	$\dfrac{dV}{V} = -\dfrac{dA}{A}\dfrac{1}{(1 - \text{Ma}^2)}$	**(11.48)**
Isentropic flow	$\dfrac{d\rho}{\rho} = \dfrac{dA}{A}\dfrac{\text{Ma}^2}{(1 - \text{Ma}^2)}$	**(11.49)**
Isentropic flow	$\dfrac{T}{T_0} = \dfrac{1}{1 + [(k-1)/2]\text{Ma}^2}$	**(11.56)**
Isentropic flow	$\dfrac{p}{p_0} = \left\{\dfrac{1}{1 + [(k-1)/2]\text{Ma}^2}\right\}^{k/(k-1)}$	**(11.59)**
Isentropic flow	$\dfrac{\rho}{\rho_0} = \left\{\dfrac{1}{1 + [(k-1)/2]\text{Ma}^2}\right\}^{1/(k-1)}$	**(11.60)**
Isentropic flow-critical pressure ratio	$\dfrac{p^*}{p_0} = \left(\dfrac{2}{k+1}\right)^{k/(k-1)}$	**(11.61)**
Isentropic flow-critical temperature ratio	$\dfrac{T^*}{T_0} = \dfrac{2}{k+1}$	**(11.63)**

Isentropic flow	$\dfrac{A}{A^*} = \dfrac{1}{\mathrm{Ma}} \left\{ \dfrac{1 + [(k-1)/2]\mathrm{Ma}^2}{1 + [(k-1)/2]} \right\}^{(k+1)/[2(k-1)]}$	**(11.71)**
Fanno flow	$\dfrac{1}{k}\dfrac{(1-\mathrm{Ma}^2)}{\mathrm{Ma}^2} + \dfrac{k+1}{2k}\ln\left\{ \dfrac{[(k+1)/2]\mathrm{Ma}^2}{1 + [(k-1)/2]\mathrm{Ma}^2} \right\} = \dfrac{f(\ell^* - \ell)}{D}$	**(11.98)**
Fanno flow	$\dfrac{T}{T^*} = \dfrac{(k+1)/2}{1 + [(k-1)/2]\mathrm{Ma}^2}$	**(11.101)**
Fanno flow	$\dfrac{V}{V^*} = \left\{ \dfrac{[(k+1)/2]\mathrm{Ma}^2}{1 + [(k-1)/2]\mathrm{Ma}^2} \right\}^{1/2}$	**(11.103)**
Fanno flow	$\dfrac{p}{p^*} = \dfrac{1}{\mathrm{Ma}} \left\{ \dfrac{(k+1)/2}{1 + [(k-1)/2]\mathrm{Ma}^2} \right\}^{1/2}$	**(11.107)**
Fanno flow	$\dfrac{p_0}{p_0^*} = \dfrac{1}{\mathrm{Ma}} \left[\left(\dfrac{2}{k+1}\right)\left(1 + \dfrac{k-1}{2}\mathrm{Ma}^2\right) \right]^{[(k+1)/2(k-1)]}$	**(11.109)**
Rayleigh flow	$\dfrac{p}{p_a} = \dfrac{1+k}{1+k\mathrm{Ma}^2}$	**(11.123)**
Rayleigh flow	$\dfrac{T}{T_a} = \left[\dfrac{(1+k)\mathrm{Ma}}{1+k\mathrm{Ma}^2} \right]^2$	**(11.128)**
Rayleigh flow	$\dfrac{\rho_a}{\rho} = \dfrac{V}{V_a} = \mathrm{Ma}\left[\dfrac{(1+k)\mathrm{Ma}}{1+k\mathrm{Ma}^2} \right]$	**(11.129)**
Rayleigh flow	$\dfrac{T_0}{T_{0,a}} = \dfrac{2(k+1)\mathrm{Ma}^2\left(1 + \dfrac{k-1}{2}\mathrm{Ma}^2\right)}{(1+k\mathrm{Ma}^2)^2}$	**(11.131)**
Rayleigh flow	$\dfrac{p_0}{p_{0,a}} = \dfrac{(1+k)}{(1+k\mathrm{Ma}^2)}\left[\left(\dfrac{2}{k+1}\right)\left(1 + \dfrac{k-1}{2}\mathrm{Ma}^2\right) \right]^{k/(k-1)}$	**(11.133)**
Normal shock	$\mathrm{Ma}_y^2 = \dfrac{\mathrm{Ma}_x^2 + [2/(k-1)]}{[2k/(k-1)]\mathrm{Ma}_x^2 - 1}$	**(11.149)**
Normal shock	$\dfrac{p_y}{p_x} = \dfrac{2k}{k+1}\mathrm{Ma}_x^2 - \dfrac{k-1}{k+1}$	**(11.150)**
Normal shock	$\dfrac{T_y}{T_x} = \dfrac{\{1 + [(k-1)/2]\mathrm{Ma}_x^2\}\{[2k/(k-1)]\mathrm{Ma}_x^2 - 1\}}{\{(k+1)^2/[2(k-1)]\}\mathrm{Ma}_x^2}$	**(11.151)**
Normal shock	$\dfrac{\rho_y}{\rho_x} = \dfrac{V_x}{V_y} = \dfrac{(k+1)\mathrm{Ma}_x^2}{(k-1)\mathrm{Ma}_x^2 + 2}$	**(11.154)**
Normal shock	$\dfrac{p_{0,y}}{p_{0,x}} = \dfrac{\left(\dfrac{k+1}{2}\mathrm{Ma}_x^2\right)^{k/(k-1)}\left(1 + \dfrac{k-1}{2}\mathrm{Ma}_x^2\right)^{k/(1-k)}}{\left(\dfrac{2k}{k+1}\mathrm{Ma}_x^2 - \dfrac{k-1}{k+1}\right)^{1/(k-1)}}$	**(11.156)**

References

1. Coles, D., "Channel Flow of a Compressible Fluid," Summary description of film in *Illustrated Experiments in Fluid Mechanics, The NCFMF Book of Film Notes,* MIT Press, Cambridge, Mass., 1972.

2. Moran, M. J., and Shapiro, H. N., *Fundamentals of Engineering Thermodynamics,* 6th Ed., Wiley, New York, 2008.

3. Keenan, J. H., Chao, J., and Kaye, J., *Gas Tables,* 2nd Ed., Wiley, New York, 1980.

4. Shapiro, A. H., *The Dynamics and Thermodynamics of Compressible Fluid Flow,* Vol. 1, Wiley, New York, 1953.

5. Liepmann, H. W., and Roshko, A., *Elements of Gasdynamics,* Dover Publications, 2002.

6. Anderson, J. D., Jr., *Modern Compressible Flow with Historical Perspective,* 3rd Ed., McGraw-Hill, New York, 2003.

7. Greitzer, E. M., Tan, C. S., and Graf, M. B., *Internal Flow Concepts and Applications,* Cambridge University Press, U. K., 2004.

Review Problems

Go to Appendix G for a set of review problems with answers. Detailed solutions can be found in *Student Solution Manual and Study Guide for Fundamentals of Fluid Mechanics,* by Munson et al. (© 2009 John Wiley and Sons, Inc.).

Problems

Note: Unless otherwise indicated, use the values of fluid properties found in the tables on the inside of the front cover. Problems designated with an (*) **are intended to be solved with the aid of a programmable calculator or a computer. If** $k = 1.4$ **the figures of Appendix D can be used to simplify a problem solution. Problems designated with a** (†) **are "open-ended" problems and require critical thinking in that to work them one must make various assumptions and provide the necessary data. There is not a unique answer to these problems.**

Answers to the even-numbered problems are listed at the end of the book. Access to the videos that accompany problems can be obtained through the book's web site, www.wiley.com/college/munson.

Section 11.1 Ideal Gas Relationships

11.1 Distinguish between flow of an ideal gas and inviscid flow of a fluid.

11.2 Compare the density of standard air listed in Table 1.4 with the value of standard air calculated with the ideal gas equation of state, and comment on what you discover.

11.3 Two kilograms mass of air is heated in a closed, rigid container from 25 °C, 103 kPa (abs) to 260 °C. Estimate the final pressure of the air and the entropy rise involved.

11.4 Air flows steadily between two sections in a duct. At section (1), the temperature and pressure are $T_1 = 80$ °C, $p_1 = 301$ kPa (abs), and at section (2), the temperature and pressure are $T_2 = 180$ °C, $p_2 = 181$ kPa (abs). Calculate the **(a)** change in internal energy between sections (1) and (2), **(b)** change in enthalpy between sections (1) and (2), **(c)** change in density between sections (1) and (2), **(d)** change in entropy between sections (1) and (2). How would you estimate the loss of available energy between the two sections of this flow?

11.5 Does the entropy change during the process of Example 11.2 indicate a loss of available energy by the flowing fluid?

11.6 As demonstrated in **Video V11.1**, fluid density differences in a flow may be seen with the help of a schlieren optical system. Discuss what variables affect fluid density and the different ways in which a variable density flow can be achieved.

11.7 Describe briefly how a schlieren optical visualization system (**Videos V11.1** and **V11.4**, also Fig. 11.4) works. How else might density changes in a fluid flow be made visible to the eye?

11.8 Explain why the Bernoulli equation (Eq. 3.7) cannot be accurately used for compressible flows.

11.9 Air at 101.3 kPa (abs) and 20 °C is compressed adiabatically by a centrifugal compressor to a pressure of 690 kPa (abs). What is the minimum temperature rise possible? Explain.

11.10 Methane is compressed adiabatically from 100 kPa (abs) and 25 °C to 200 kPa (abs). What is the minimum compressor exit temperature possible? Explain.

11.11 Air expands adiabatically through a turbine from a pressure and temperature of 1240 kPa (abs), 615 K to a pressure of 101.3 kPa (abs). If the actual temperature change is 85% of the ideal temperature change, determine the actual temperature of the expanded air and the actual enthalpy and entropy differences across the turbine.

11.12 An expression for the value of c_p for carbon dioxide as a function of temperature is

$$c_p = 1538 - \frac{3.44 \times 10^5}{T} + \frac{4.14 \times 10^6}{T^2}$$

where c_p is in J/(kg · K) and T is in kelvin. Compare the change in enthalpy of carbon dioxide using the constant value of c_p (see Table 1.4) with the change in enthalpy of carbon dioxide using the expression above, for $T_2 - T_1$ equal to **(a)** 5 K, **(b)** 555 K, and **(c)** 1666 K. Set $T_1 = 300$ K.

11.13 Are the flows shown in **Videos V11.1** and **V11.4** compressible? Do they involve high-speed flow velocities? Discuss.

Section 11.2 Mach Number and Speed of Sound

11.14 Confirm the speed of sound for air at 20 °C listed in Table B.2.

11.15 From Table B.1 we can conclude that the speed of sound in water at 20 °C is 1481 m/s. Is this value of c consistent with the value of bulk modulus, E_v, listed in Table 1.3?

11.16 If the observed speed of sound in steel is 5300 m/s, determine the bulk modulus of elasticity of steel in N/m^3. The density of steel is nominally 7790 kg/m^3. How does your value of E_v for steel compare with E_v for water at 15.6 °C? Compare the speeds of sound in steel, water, and air at standard atmospheric pressure and 15 °C and comment on what you observe.

11.17 Using information provided in Table C.1, develop a table of speed of sound in m/s as a function of elevation for U.S. standard atmosphere.

11.18 Determine the Mach number of a car moving in standard air at a speed of **(a)** 40 km/hr, **(b)** 90 km/hr, and **(c)** 160 km/hr.

†11.19 Estimate the Mach number levels associated with space shuttle main engine nozzle exit flows at launch (see Video V11.3).

Section 11.3 **Categories of Compressible Flow**

11.20 Obtain a photograph/image showing visualisation of flow phenomena caused by an object moving through a fluid at a Mach number exceeding 1.0. Explain what is happening, and identify zones of silence and of action.

11.21 Cite one specific and actual example each of a hypersonic flow, a supersonic flow, a transonic flow, and a compressible subsonic flow.

11.22 At a given instant of time, two pressure waves, each moving at the speed of sound, emitted by a point source moving with constant velocity in a fluid at rest are shown in Fig. P11.23. Determine the Mach number involved and indicate with a sketch the instantaneous location of the point source.

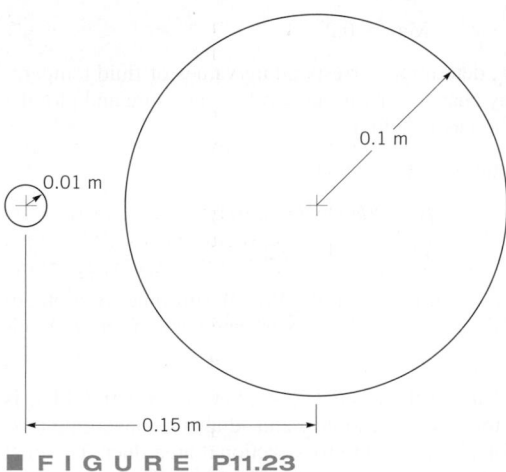

■ **FIGURE** **P11.23**

11.23 At a given instant of time, two pressure waves, each moving at the speed of sound, emitted by a point source moving with constant velocity in a fluid at rest, are shown in Fig. P11.24. Determine the Mach number involved and indicate with a sketch the instantaneous location of the point source.

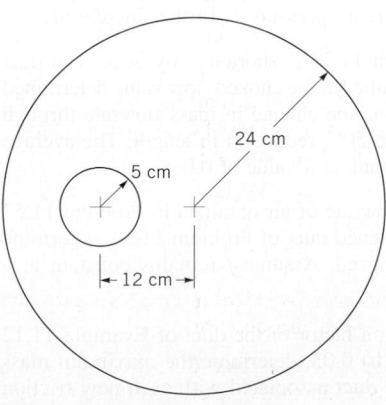

■ **FIGURE** **P11.24**

11.24 Sound waves are very small amplitude pressure pulses that travel at the "speed of sound." Do very large amplitude waves such as a blast wave caused by an explosion (see Video V11.7) travel less than, equal to, or greater than the speed of sound? Explain.

11.25 How would you estimate the distance between you and an approaching storm front involving lightning and thunder?

11.26 If a person inhales helium and then talks, his or her voice sounds like "Donald Duck." Explain why this happens.

11.27 If a high-performance aircraft is able to cruise at a Mach number of 3.0 at an altitude of 24,000 m, how fast is this in **(a)** km/hr, **(b)** m/s?

11.28 At the seashore, you observe a high-speed aircraft moving overhead at an elevation of 3000 m. You hear the plane 8 s after it passes directly overhead. Using a nominal air temperature of 5 °C, estimate the Mach number and speed of the aircraft.

11.29 Explain how you could vary the Mach number but not the Reynolds number in air flow past a sphere. For a constant Reynolds number of 300,000, estimate how much the drag coefficient will increase as the Mach number is increased from 0.3 to 1.0.

Section 11.4 **Isentropic Flow of an Ideal Gas**

11.30 Obtain photographs/images of converging/diverging nozzles used to achieve supersonic flows, and briefly explain each application.

11.31 Obtain photographs/images of supersonic diffusers used to decelerate supersonic flows to subsonic flows, and briefly explain each application.

11.32 Starting with the enthalpy form of the energy equation (Eq. 5.69), show that for isentropic flows, the stagnation temperature remains constant. Why is this important?

11.33 Explain how fluid pressure varies with cross-sectional area change for the isentropic flow of an ideal gas when the flow is **(a)** subsonic, **(b)** supersonic.

11.34 For any ideal gas, prove that the slope of constant pressure lines on a temperature–entropy diagram is positive and that higher pressure lines are above lower pressure lines. Why is this important?

11.35 Air flows steadily and isentropically from standard atmospheric conditions to a receiver pipe through a converging duct. The cross-sectional area of the throat of the converging duct is 5×10^{-3} m^2. Determine the mass flowrate through the duct if the receiver pressure is **(a)** 70 kPa (abs), **(b)** 35 kPa (abs). Sketch temperature–entropy diagrams for situations **(a)** and **(b)**. Verify results obtained with values from the appropriate graph in Appendix D with calculations involving ideal gas equations. Is condensation of water vapor a concern? Explain.

11.36 Determine the static pressure to stagnation pressure ratio associated with the following motion in standard air: **(a)** a runner moving at the rate of 16 km/hr, **(b)** a cyclist moving at the rate of 65 km/hr, **(c)** a car moving at the rate of 105 km/hr, **(d)** an airplane moving at the rate of 800 km/hr.

11.37 The static pressure to stagnation pressure ratio at a point in a gas flow field is measured with a Pitot-static probe as being equal to 0.6. The stagnation temperature of the gas is 20 °C. Determine the flow speed in m/s and the Mach number if the gas is air. What error would be associated with assuming that the flow is incompressible?

11.38 The stagnation pressure and temperature of air flowing past a probe are 120 kPa (abs) and 100 °C, respectively. The air pressure is 80 kPa (abs). Determine the air speed and the Mach number considering the flow to be (a) incompressible, (b) compressible.

11.39 The stagnation pressure indicated by a Pitot tube mounted on an airplane in flight is 45 kPa (abs). If the aircraft is cruising in standard atmosphere at an altitude of 10,000 m, determine the speed and Mach number involved.

†**11.40** Estimate the stagnation pressure level necessary at the entrance of a space shuttle main engine nozzle to achieve the over-expansion condition shown in Video V11.5.

*11.41 An ideal gas enters subsonically and flows isentropically through a choked converging–diverging duct having a circular cross-sectional area A that varies with axial distance from the throat, x, according to the formula

$$A = 0.01 + x^2$$

where A is in square meters and x is in meters. For this flow situation, sketch the side view of the duct and graph the variation of Mach number, static temperature to stagnation temperature ratio, T/T_0, and static pressure to stagnation pressure ratio, p/p_0, through the duct from $x = -0.2$ m to $x = +0.2$ m. Also show the possible fluid states at $x = -0.2$ m, 0 m, and $+0.2$ m using temperature–entropy coordinates. Consider the gas as being helium (use $0.051 \le$ Ma ≤ 5.193). Sketch on your pressure variation graph the nonisentropic paths that would occur with over- and underexpanded duct exit flows (see Video V11.6) and explain when they will occur. When will isentropic supersonic duct exit flow occur?

*11.42 An ideal gas enters supersonically and flows isentropically through the choked converging–diverging duct described in Problem 11.41. Graph the variation of Ma, T/T_0, and p/p_0 from the entrance to the exit sections of the duct for helium (use $0.051 \le$ Ma ≤ 5.193). Show the possible fluid states at $x = -0.2$ m, 0 m, and $+0.2$ m using temperature–entropy coordinates. Sketch on your pressure variation graph the nonisentropic paths that would occur with over- and underexpanded duct exit flows (see Video V11.6) and explain when they will occur. When will isentropic supersonic duct exit flow occur?

11.43 An ideal gas flows subsonically and isentropically through the converging–diverging duct described in Problem 11.41. Graph the variation of Ma, T/T_0, and p/p_0 from the entrance to the exit sections of the duct for air. The value of p/p_0 is 0.6708 at $x = 0$ m. Sketch important states on a T–s diagram.

11.44 An ideal gas is to flow isentropically from a large tank where the air is maintained at a temperature and pressure of 15 °C and 550 kPa (abs) to standard atmospheric discharge conditions. Describe in general terms the kind of duct involved and determine the duct exit Mach number and velocity in m/s if the gas is air.

11.45 An ideal gas flows isentropically through a converging–diverging nozzle. At a section in the converging portion of the nozzle, $A_1 = 0.1$ m^2 $p_1 = 600$ kPa (abs), $T_1 = 20$ °C, and Ma$_1 = 0.6$. For section (2) in the diverging part of the nozzle, determine A_2, p_2, and T_2 if Ma$_2 = 3.0$ and the gas is air.

11.46 Upstream of the throat of an isentropic converging–diverging nozzle at section (1), $V_1 = 150$ m/s, $p_1 = 100$ kPa (abs), and $T_1 = 20$ °C. If the discharge flow is supersonic and the throat area is 0.1 m^2, determine the mass flowrate in kg/s for the flow of air.

11.47 The flow blockage associated with the use of an intrusive probe can be important. Determine the percentage increase in section velocity corresponding to a 0.5% reduction in flow area due to probe blockage for air flow if the section area is 1.0 m^2, $T_0 = 20$ °C, and the unblocked flow Mach numbers are (a) Ma $= 0.2$, (b) Ma $= 0.8$, (c) Ma $= 1.5$, (d) Ma $= 30$.

11.48 (See Fluids in the News article titled "Rocket nozzles," Section 11.4.2.) Comment on the practical limits of area ratio for the diverging portion of a converging–diverging nozzle designed to achieve supersonic exit flow.

Section 11.5.1 Adiabatic Constant Area Duct Flow with Friction (Fanno Flow)

11.49 Cite an example of an actual subsonic flow of practical importance that can be approximated with a Fanno flow.

11.50 An ideal gas enters [section (1)] an insulated, constant cross-sectional area duct with the following properties:

$$T_0 = 293 \text{ K}$$

$$p_0 = 101 \text{ kPa (abs)}$$

$$\text{Ma}_1 = 0.2$$

For Fanno flow, determine corresponding values of fluid temperature and entropy change for various levels of pressure and plot the Fanno line if the gas is helium.

11.51 For Fanno flow, prove that

$$\frac{dV}{V} = \frac{fk(\text{Ma}^2/2)(dx/D)}{1 - \text{Ma}^2}$$

and in so doing show that when the flow is subsonic, friction accelerates the fluid, and when the flow is supersonic, friction decelerates the fluid.

11.52 Standard atmospheric air ($T_0 = 15$ °C, $p_0 = 101.3$ kPa) is drawn steadily through a frictionless and adiabatic converging nozzle into an adiabatic, constant cross-sectional area duct. The duct is 3 m long and has an inside diameter of 0.15 m. The average friction factor for the duct may be estimated as being equal to 0.03. What is the maximum mass flowrate in kg/s through the duct? For this maximum flowrate, determine the values of static temperature, static pressure, stagnation temperature, stagnation pressure, and velocity at the inlet [section (1)] and exit [section (2)] of the constant area duct. Sketch a temperature–entropy diagram for this flow.

11.53 The upstream pressure of a Fanno flow venting to the atmosphere is increased until the flow chokes. What will happen to the flowrate when the upstream pressure is further increased?

11.54 The duct in Problem 11.52 is shortened by 50%. The duct discharge pressure is maintained at the choked flow value determined in Problem 11.52. Determine the change in mass flowrate through the duct associated with the 50% reduction in length. The average friction factor remains constant at a value of 0.03.

11.55 If the same mass flowrate of air obtained in Problem 11.52 is desired through the shortened duct of Problem 11.54, determine the back pressure, p_2, required. Assume f remains constant at a value of 0.03.

11.56 If the average friction factor of the duct of Example 11.12 is changed to (a) 0.01 or (b) 0.03, determine the maximum mass flowrate of air through the duct associated with each new friction factor; compare with the maximum mass flowrate value of Example 11.12.

11.57 Air flows adiabatically between two sections in a constant area pipe. At upstream section (1), $p_{0,1} = 690$ kPa (abs), $T_{0,1} = 330$ K, and $Ma_1 = 0.5$. At downstream section (2), the flow is choked. Estimate the magnitude of the force per unit cross-sectional area exerted by the inside wall of the pipe on the fluid between sections (1) and (2).

Section 11.5.2 Frictionless Constant Area Duct Flow with Heat Transfer (Rayleigh Flow)

11.58 Cite an example of an actual subsonic flow of practical importance that may be approximated with a Rayleigh flow.

11.59 Standard atmospheric air [$T_0 = 288$ K, $p_0 = 101$ kPa(abs)] is drawn steadily through an isentropic converging nozzle into a frictionless diabatic ($q = 500$ kJ/kg) constant area duct. For maximum flow, determine the values of static temperature, static pressure, stagnation temperature, stagnation pressure, and flow velocity at the inlet [section (1)] and exit [section (2)] of the constant area duct. Sketch a temperature–entropy diagram for this flow.

11.60 Air enters a 0.15-m inside diameter duct with $p_1 = 140$ kPa (abs), $T_1 = 30\,°C$, and $V_1 = 60$ m/s. What frictionless heat addition rate in J/s is necessary for an exit gas temperature $T_2 = 815\,°C$? Determine p_2, V_2, and Ma_2 also.

11.61 Air enters a length of constant area pipe with $p_1 = 200$ kPa (abs), $T_1 = 500$ K, and $V_1 = 400$ m/s. If 500 kJ/kg of energy is removed from the air by frictionless heat transfer between sections (1) and (2), determine p_2, T_2, and V_2. Sketch a temperature–entropy diagram for the flow between sections (1) and (2).

11.62 Describe what happens to a Fanno flow when heat transfer is allowed to occur. Is this the same as a Rayleigh flow with friction considered?

Section 11.5.3 Normal Shock Waves

11.63 Obtain a photograph/image of a normal shock wave and explain briefly the situation involved.

11.64 The Mach number and stagnation pressure of air are 2.0 and 200 kPa (abs) just upstream of a normal shock. Estimate the stagnation pressure loss across the shock.

11.65 The stagnation pressure ratio across a normal shock in an air flow is 0.6. Estimate the Mach number of the flow entering the shock.

11.66 Just upstream of a normal shock in an air flow, Ma = 3.0, $T = 330$ K, and $p = 210$ kPa (abs). Estimate values of Ma, T_0, T, p_0, p, and V downstream of the shock.

11.67 A total pressure probe like the one shown in **Video V3.8** is inserted into a supersonic air flow. A shock wave forms just upstream of the impact hole. The probe measures a total pressure of 500 kPa (abs). The stagnation temperature at the probe head is 500 K. The static pressure upstream of the shock is measured with a wall tap to be 100 kPa (abs). From these data, estimate the Mach number and velocity of the flow.

11.68 The Pitot tube on a supersonic aircraft (see **Video V3.8**) cruising at an altitude of 9000 m senses a stagnation pressure of 82 kPa (abs). If the atmosphere is considered standard, determine the airspeed and Mach number of the aircraft. A shock wave is present just upstream of the probe impact hole.

11.69 An aircraft cruises at a Mach number of 2.0 at an altitude of 15 km. Inlet air is decelerated to a Mach number of 0.4 at the engine compressor inlet. A normal shock occurs in the inlet diffuser upstream of the compressor inlet at a section where the Mach number is 1.2. For isentropic diffusion, except across the shock, and for standard atmosphere, determine the stagnation temperature and pressure of the air entering the engine compressor.

11.70 Determine, for the air flow through the frictionless and adiabatic converging–diverging duct of Example 11.8, the ratio of duct exit pressure to duct inlet stagnation pressure that will result in a standing normal shock at: **(a)** $x = +0.1$ m, **(b)** $x = +0.2$ m, **(c)** $x = +0.4$ m. How large is the stagnation pressure loss in each case?

11.71 A normal shock is positioned in the diverging portion of a frictionless, adiabatic, converging–diverging air flow duct where the cross-sectional area is 0.01 m^2 and the local Mach number is 2.0. Upstream of the shock, $p_0 = 1380$ kPa (abs) and $T_0 = 110$ K. If the duct exit area is 14×10^{-3} m^2, determine the exit area temperature and pressure and the duct mass flowrate.

11.72 Supersonic air flow enters an adiabatic, constant area (inside diameter = 0.5 m) 9-m-long pipe with $Ma_1 = 3.0$. The pipe friction factor is estimated to be 0.02. What ratio of pipe exit pressure to pipe inlet stagnation pressure would result in a normal shock wave standing at **(a)** $x = 1.5$ m, or **(b)** $x = 3$ m, where x is the distance downstream from the pipe entrance? Determine also the duct exit Mach number and sketch the temperature–entropy diagram for each situation.

11.73 Supersonic air flow enters an adiabatic, constant area pipe (inside diameter = 0.1 m) with $Ma_1 = 2.0$. The pipe friction factor is 0.02. If a standing normal shock is located right at the pipe exit, and the Mach number just upstream of the shock is 1.2, determine the length of the pipe.

11.74 Air enters a frictionless, constant area duct with $Ma_1 = 2.0$, $T_{0,1} = 15\,°C$, and $p_{0,1} = 101.3$ kPa (abs). The air is decelerated by heating until a normal shock wave occurs where the local Mach number is 1.5. Downstream of the normal shock, the subsonic flow is accelerated with heating until it chokes at the duct exit. Determine the static temperature and pressure, the stagnation temperature and pressure, and the fluid velocity at the duct entrance, just upstream and downstream of the normal shock, and at the duct exit. Sketch the temperature–entropy diagram for this flow.

11.75 Air enters a frictionless, constant area duct with Ma = 2.5, $T_0 = 20\,°C$, and $p_0 = 101$ kPa (abs). The gas is decelerated by heating until a normal shock occurs where the local Mach number is 1.3. Downstream of the shock, the subsonic flow is accelerated with heating until it exits with a Mach number of 0.9. Determine the static temperature and pressure, the stagnation temperature and pressure, and the fluid velocity at the duct entrance, just upstream and downstream of the normal shock, and at the duct exit. Sketch the temperature–entropy diagram for this flow.

■ **Life Long Learning Problems**

11.76 Is there a limit to how fast an object can move through the atmosphere? Explain.

11.77 Discuss the similarities between hydraulic jumps in open-channel flow and shock waves in compressible flow. Explain how this knowledge can be useful.

11.78 Estimate the surface temperature associated with the reentry of the Space Shuttle into the earth's atmosphere. Why is knowing this important?

11.79 [See Fluids in the News article titled "Hilsch tube (Ranque vortex tube)," Section 11.1.] Explain why a Hilsch tube works and cite some high and low gas temperatures actually achieved. What is the most important limitation of a Hilsch tube and how can it be overcome?

11.80 [See Fluids in the News article titled "Supersonic and compressible flows in gas turbines," Section 11.3.] Using typical physical dimensions and rotation speeds of manufactured gas turbine rotors, consider the possibility that supersonic fluid velocities relative to blade surfaces are possible. How do designers use this knowledge?

11.81 Develop useful equations describing the constant temperature (isothermal) flow of an ideal gas through a constant cross section area pipe. What important practical flow situations would these equations be useful for? How are real gas effects estimated?

■ **FE Exam Problems**

Sample FE (Fundamentals of Engineering) exam questions for fluid mechanics are provided on the book's web site, www.wiley. com/college/munson.

12 Turbomachines

CHAPTER OPENING PHOTO: A mixed-flow, transonic compressor stage. (Photograph courtesy of Concepts NREC.)

Learning Objectives

After completing this chapter, you should be able to:

- explain how and why a turbomachine works.
- know the basic differences between a turbine and a pump.
- recognize the importance of minimizing loss in a turbomachine.
- select an appropriate class of turbomachine for a particular application.
- understand why turbomachine blades are shaped like they are.
- appreciate the basic fundamentals of sensibly scaling turbomachines that are larger or smaller than a prototype.
- move on to more advanced engineering work involving the fluid mechanics of turbomachinery (e.g., design, development, research).

In previous chapters we often used generic "black boxes" to represent fluid machines such as pumps or turbines. The purpose of this chapter is to understand the fluid mechanics of these devices when they are turbomachines.

Pumps and turbines (often turbomachines) occur in a wide variety of configurations. In general, pumps add energy to the fluid—they do work on the fluid to move and/or increase the pressure of the fluid; turbines extract energy from the fluid—the fluid does work on them. The term "pump" will be used to generically refer to all pumping machines, including *pumps*, *fans*, *blowers*, and *compressors*.

Turbomachines are dynamic fluid machines that add (for pumps) or extract (for turbines) flow energy.

Turbomachines involve a collection of blades, buckets, flow channels, or passages arranged around an axis of rotation to form a rotor. A fluid that is moving can force rotation and produce

645

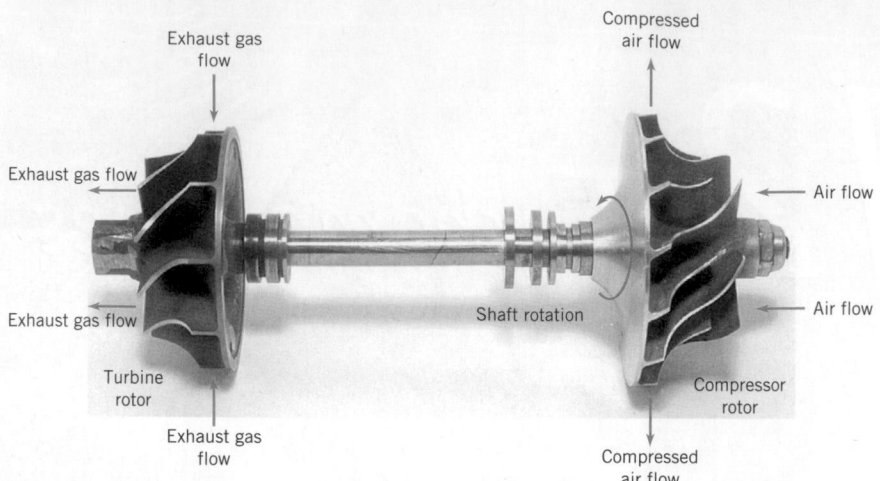

■ **F I G U R E   12.1**   **Automotive turbocharger turbine and compressor rotors. (Photograph courtesy of Concepts NREC.)**

shaft power. In this case we have a turbine. On the other hand, we can exert a shaft torque, typically with a motor, and by using blades, flow channels, or passages force the fluid to move. In this case we have a pump. In Fig. 12.1 are shown the turbine and compressor (pump) rotors of an automobile turbocharger. Examples of turbomachine-type pumps include simple window fans, propellers on ships or airplanes, squirrel-cage fans on home furnaces, axial-flow water pumps used in deep wells, and compressors in automobile turbochargers. Examples of turbines include the turbine portion of gas turbine engines on aircraft, steam turbines used to drive generators at electrical generation stations, and the small, high-speed air turbines that power dentist drills.

Turbomachines serve in an enormous array of applications in our daily lives and thus play an important role in modern society. These machines can have a high power density (large power transfer per size), relatively few moving parts, and reasonable efficiency. The following sections provide an introduction to the fluid mechanics of these important machines. References 1–3 are a few examples of the many books that offer much more knowledge about turbomachines.

12.1   Introduction

Turbomachines involve the related parameters of force, torque, work, and power.

(Photograph courtesy of Mid American Energy.)

Turbomachines are mechanical devices that either extract energy from a fluid (turbine) or add energy to a fluid (pump) as a result of dynamic interactions between the device and the fluid. While the actual design and construction of these devices often require considerable insight and effort, their basic operating principles are quite simple.

Using a food blender to make a fruit drink is an example of turbo-pump action. The blender blades are forced to rotate around an axis by a motor. The moving blades pulverize fruit and ice and mix them with a base liquid to form a "smoothie."

Conversely, the dynamic effect of the wind blowing past the sail on a boat creates pressure differences on the sail. The wind force on the moving sail in the direction of the boat's motion provides power to propel the boat. The sail and boat act as a machine extracting energy from the air. Turbine blades are like sails. See, for example, the enormous wind turbine blades in the figure in the margin.

The fluid involved can be either a gas (as with a window fan or a gas turbine engine) or a liquid (as with the water pump on a car or a turbine at a hydroelectric power plant). While the basic operating principles are the same whether the fluid is a liquid or a gas, important differences in the fluid dynamics involved can occur. For example, cavitation may be an important design consideration when liquids are involved if the pressure at any point within the flow is reduced to the vapor pressure. Compressibility effects may be important when gases are involved if the Mach number becomes large enough.

Many turbomachines contain some type of housing or casing that surrounds the rotating blades or rotor, thus forming an internal flow passageway through which the fluid flows (see Fig. 12.2).

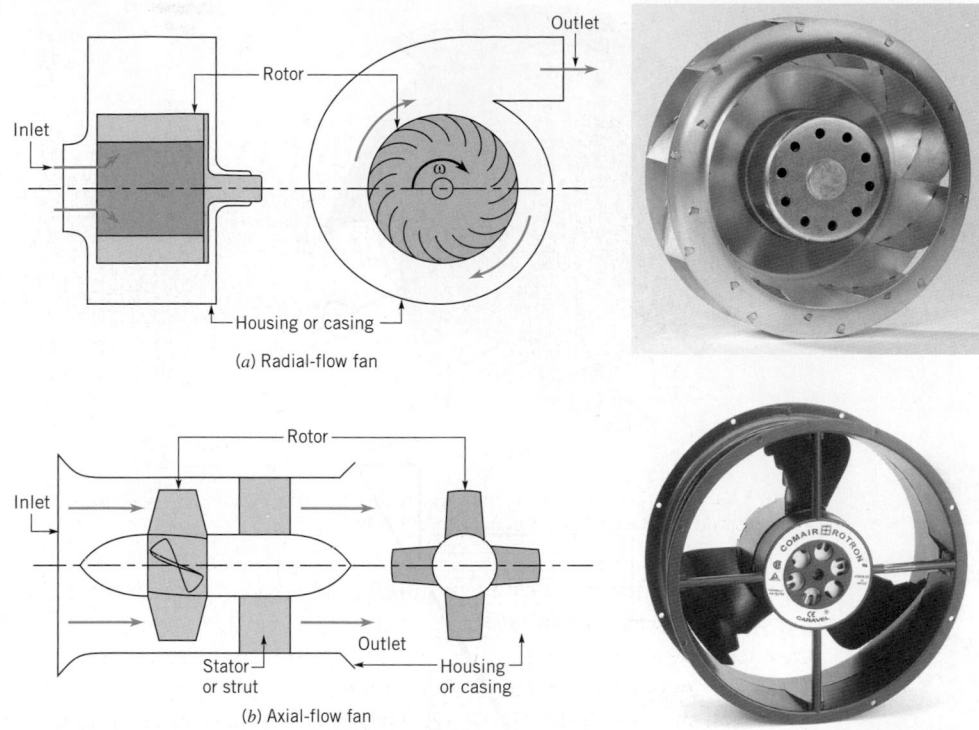

■ **F I G U R E   12.2**   (*a*) **A radial-flow turbomachine, (*b*) an axial-flow turbomachine. (Photographs courtesy of Comair Rotron, Inc.)**

A group of blades moving with or against a lift force is the essence of a turbomachine.

Others, such as a windmill or a window fan, are unducted. Some turbomachines include stationary blades or vanes in addition to rotor blades. These stationary vanes can be arranged to accelerate the flow and thus serve as nozzles. Or, these vanes can be set to diffuse the flow and act as diffusers.

Turbomachines are classified as *axial-flow*, *mixed-flow*, or *radial-flow* machines depending on the predominant direction of the fluid motion relative to the rotor's axis as the fluid passes the blades (see Fig. 12.2). For an axial-flow machine the fluid maintains a significant axial-flow direction component from the inlet to outlet of the rotor. For a radial-flow machine the flow across the blades involves a substantial radial-flow component at the rotor inlet, exit, or both. In other machines, designated as mixed-flow machines, there may be significant radial- and axial-flow velocity components for the flow through the rotor row. Each type of machine has advantages and disadvantages for different applications and in terms of fluid-mechanical performance.

12.2   Basic Energy Considerations

An understanding of the work transfer in turbomachines can be obtained by considering the basic operation of a household fan (pump) and a windmill (turbine). Although the actual flows in such devices are very complex (i.e., three-dimensional and unsteady), the essential phenomena can be illustrated by use of simplified flow considerations and velocity triangles.

Consider a fan blade driven at constant angular velocity, ω, by a motor as is shown in Fig. 12.3*a*. We denote the blade speed as $U = \omega r$, where r is the radial distance from the axis of the fan. The absolute fluid velocity (that seen by a person sitting stationary at the table on which the fan rests) is denoted **V**, and the relative velocity (that seen by a person riding on the fan blade) is denoted **W**. As shown by the figure in the margin, the actual (absolute) fluid velocity is the vector sum of the relative velocity and the blade velocity

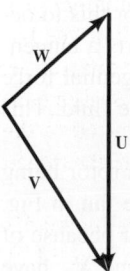

$$\mathbf{V} = \mathbf{W} + \mathbf{U} \tag{12.1}$$

A simplified sketch of the fluid velocity as it "enters" and "exits" the fan at radius r is shown in Fig. 12.3*b*. The shaded surface labeled $a-b-c-d$ is a portion of the cylindrical surface (including a "slice" through the blade) shown in Fig. 12.3*a*. We assume for simplicity that the

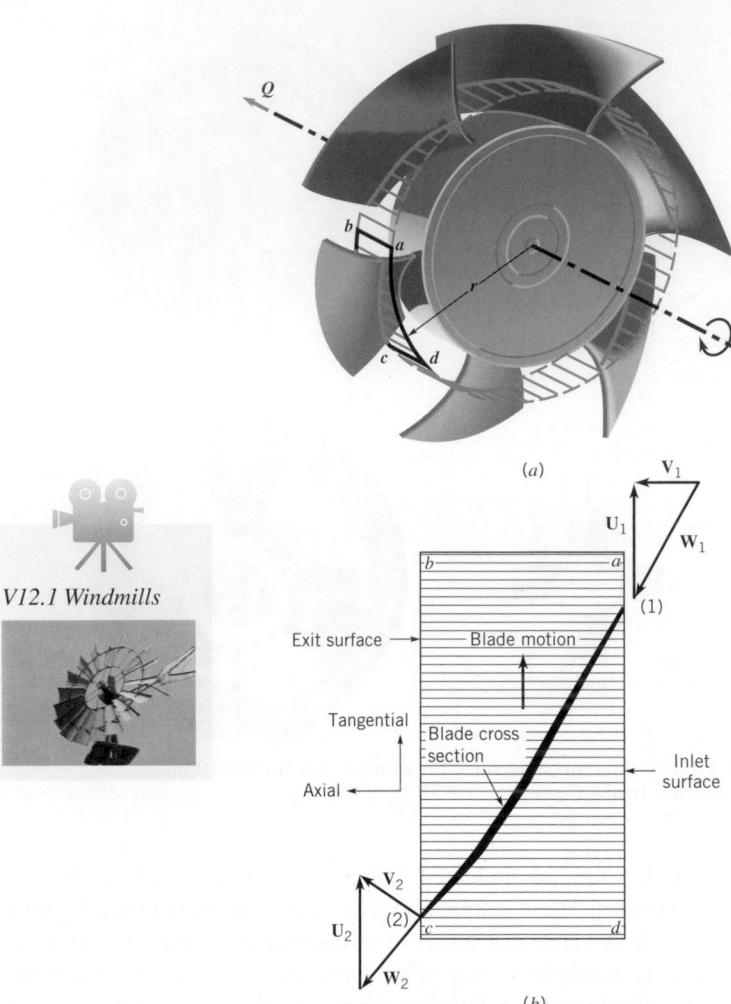

V12.1 Windmills

■ **FIGURE 12.3** **Idealized flow through a fan:** (*a*) **fan blade geometry;** (*b*) **absolute velocity, V; relative velocity, W; and blade velocity, U at the inlet and exit of the fan blade section.**

flow moves smoothly along the blade so that relative to the moving blade the velocity is parallel to the leading and trailing edges (points 1 and 2) of the blade. For now we assume that the fluid enters and leaves the fan at the same distance from the axis of rotation; thus, $U_1 = U_2 = \omega r$. In actual turbomachines, the entering and leaving flows are not necessarily tangent to the blades, and the fluid pathlines can involve changes in radius. These considerations are important at design and off-design operating conditions. Interested readers are referred to Refs. 1, 2, and 3 for more information about these aspects of turbomachine flows.

With this information we can construct the *velocity triangles* shown in Fig. 12.3*b*. Note that this view is from the top of the fan, looking radially down toward the axis of rotation. The motion of the blade is up; the motion of the incoming air is assumed to be directed along the axis of rotation. The important concept to grasp from this sketch is that the fan blade (because of its shape and motion) "pushes" the fluid, causing it to change direction. The absolute velocity vector, **V**, is turned during its flow across the blade from section (1) to section (2). Initially the fluid had no component of absolute velocity in the direction of the motion of the blade, the θ (or tangential) direction. When the fluid leaves the blade, this tangential component of absolute velocity is nonzero. For this to occur, the blade must push on the fluid in the tangential direction. That is, the blade exerts a tangential force component on the fluid in the direction of the motion of the blade. This tangential force component and the blade motion are in the same direction—the blade does work on the fluid. This device is a pump.

When blades move because of the fluid force, we have a turbine; when blades are forced to move fluid, we have a pump.

On the other hand, consider the windmill shown in Fig. 12.4*a*. Rather than the rotor being driven by a motor, the blades move in the direction of the lift force (compared to the fan in Fig. 12.3) exerted on each blade by the wind blowing through the rotor. We again note that because of the blade shape and motion, the absolute velocity vectors at sections (1) and (2), $\mathbf{V}_1$ and $\mathbf{V}_2$, have

(a)

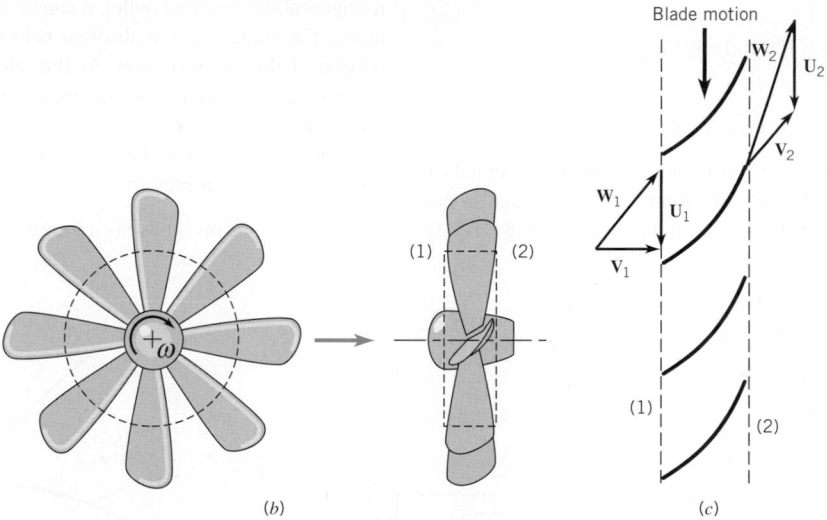

(b)   (c)

■ **FIGURE 12.4**   **Idealized flow through a windmill:** (a) **windmill;** (b) **windmill blade geometry;** (c) **absolute velocity, V; relative velocity, W; and blade velocity, U; at the inlet and exit of the windmill blade section.**

different directions. For this to happen, the blades must have pushed up on the fluid—opposite to the direction of blade motion. Alternatively, because of equal and opposite forces (action/reaction) the fluid must have pushed on the blades in the direction of their motion—the fluid does work on the blades. This extraction of energy from the fluid is the purpose of a turbine.

These examples involve work transfer to or from a flowing fluid in two axial-flow turbomachines. Similar concepts hold for other turbomachines including mixed-flow and radial-flow configurations.

<h1>Fluids in the News</h1>

Current from currents The use of large, efficient wind *turbines* to generate electrical power is becoming more commonplace throughout the world. "Wind farms" containing numerous turbines located at sites that have proper wind conditions can produce a significant amount of electrical power. Recently, researchers in the United States, the United Kingdom and Canada have been investigating the possibility of harvesting the power of ocean currents and tides by using current turbines that function much like wind turbines. Rather than being driven by wind, they derive energy from ocean currents that occur at many locations in the 70% of the earth's surface that is water. Clearly, a 4-knot (2.5 m/s) tidal current is not as fast as a 40-mph (70 km/hr) wind driving a wind turbine. However, since turbine power output is proportional to the fluid density, and since seawater is more than 800 times as dense as air, significant power can be extracted from slow, but massive, ocean currents. One promising configuration involves *blades* twisted in a helical pattern. This technology may provide electrical power that is both ecologically and economically sound. (See Problem 12.6.)

EXAMPLE 12.1 Basic Difference between a Pump and a Turbine

GIVEN The rotor shown in Fig. E12.1a rotates at a constant angular velocity of $\omega = 100$ rad/s. Although the fluid initially approaches the rotor in an axial direction, the flow across the blades is primarily outward (see Fig. 12.2a). Measurements indicate that the absolute velocity at the inlet and outlet are $V_1 = 12$ m/s and $V_2 = 15$ m/s, respectively.

FIND Is this device a pump or a turbine?

SOLUTION

To answer this question, we need to know if the tangential component of the force of the blade on the fluid is in the direction of the blade motion (a pump) or opposite to it (a turbine). We assume that the blades are tangent to the incoming relative velocity and that the relative flow leaving the rotor is tangent to the blades as shown in Fig. E12.1b. We can also calculate the inlet and outlet blade speeds as

$$U_1 = \omega r_1 = (100 \text{ rad/s})(0.1 \text{ m}) = 10 \text{ m/s}$$

and

$$U_2 = \omega r_2 = (100 \text{ rad/s})(0.2 \text{ m}) = 20 \text{ m/s}$$

With the known, absolute fluid velocity and blade velocity at the inlet, we can draw the velocity triangle (the graphical representation of Eq. 12.1) at that location as shown in Fig. E12.1c.

Note that we have assumed that the absolute flow at the blade row inlet is radial (i.e., the direction of $\mathbf{V}_1$ is radial). At the outlet we know the blade velocity, $\mathbf{U}_2$, the outlet speed, V_2, and the relative velocity direction, β_2 (because of the blade geometry). Therefore, we can graphically (or trigonometrically) construct the outlet velocity triangle as shown in the figure. By comparing the velocity triangles at the inlet and outlet, it can be seen that as the fluid flows across the blade row, the absolute velocity vector turns in the direction of the blade motion. At the inlet there is no component of absolute velocity in the direction of rotation; at the outlet this component is not zero. That is, the blade pushes and turns the fluid in the direction of the blade motion, thereby doing work on the fluid, adding energy to it.

This device is a pump. (Ans)

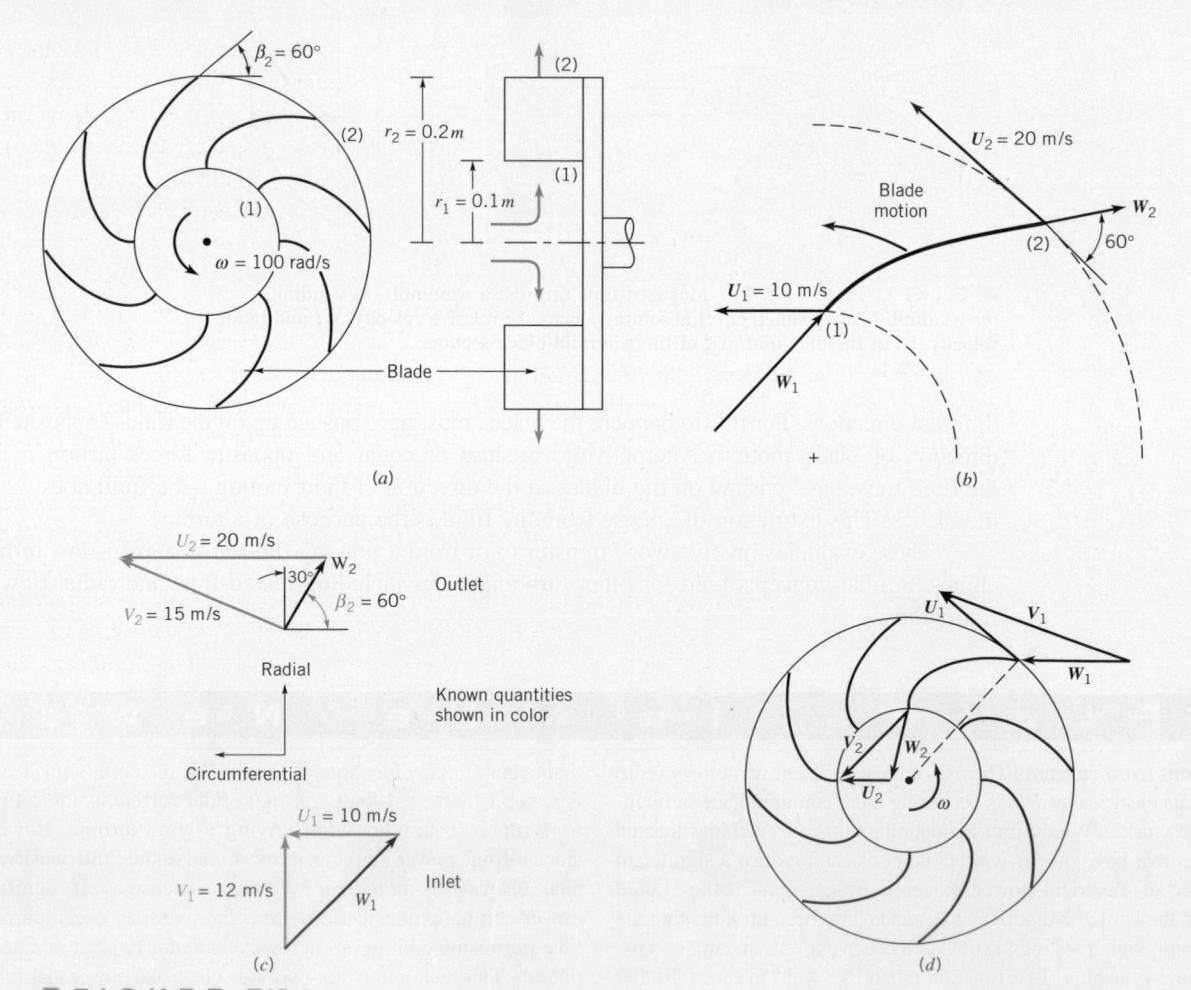

■ **FIGURE E12.1**

COMMENT   On the other hand, by reversing the direction of flow from larger to smaller radii, this device can become a radial-flow turbine. In this case (Fig. E12.1*d*) the flow direction is reversed (compared to that in Figs. E12.1*a*, *b*, and *c*) and the velocity triangles are as indicated. Stationary vanes around the perimeter of the rotor would be needed to achieve $\mathbf{V}_1$ as shown. Note that the component of the absolute velocity, $\mathbf{V}$, in the di-rection of the blade motion is smaller at the outlet than at the inlet. The blade must push against the fluid in the direction opposite the motion of the blade to cause this. Hence (by equal and opposite forces), the fluid pushes against the blade in the direction of blade motion, thereby doing work on the blade. There is a transfer of work from the fluid to the blade—a turbine operation.

12.3   Basic Angular Momentum Considerations

In the previous section we indicated how work transfer to or from a fluid flowing through a pump or a turbine occurs by interaction between moving rotor blades and the fluid. Since all of these turbomachines involve the rotation of an impeller or a rotor about a central axis, it is appropriate to discuss their performance in terms of torque and angular momentum.

Recall that work can be written as force times distance or as torque times angular displacement. Hence, if the shaft torque (the torque that the shaft applies to the rotor) and the rotation of the rotor are in the same direction, energy is transferred from the shaft to the rotor and from the rotor to the fluid—the machine is a pump. Conversely, if the torque exerted by the shaft on the rotor is opposite to the direction of rotation, the energy transfer is from the fluid to the rotor—a turbine. The amount of shaft torque (and hence shaft work) can be obtained from the moment-of-momentum equation derived formally in Section 5.2.3 and discussed as follows.

When shaft torque and rotation are in the same direction, we have a pump; otherwise we have a turbine.

Consider a fluid particle traveling outward through the rotor in the radial-flow machine shown in Figs. E12.1*a*, *b*, and *c*. For now, assume that the particle enters the rotor with a radial velocity only (i.e., no "swirl"). After being acted upon by the rotor blades during its passage from the inlet [section (1)] to the outlet [section (2)], this particle exits with radial (r) and circumferential (θ) components of velocity. Thus, the particle enters with no angular momentum about the rotor axis of rotation but leaves with nonzero angular momentum about that axis. (Recall that the axial component of angular momentum for a particle is its mass times the distance from the axis times the θ component of absolute velocity.)

V12.2 Self-propelled lawn sprinkler

A similar experience can occur at the neighborhood playground. Consider yourself as a particle and a merry-go-round as a rotor. Walk from the center to the edge of the spinning merry-go-round and note the forces involved. The merry-go-round does work on you—there is a "sideward force" on you. Another person must apply a torque (and power) to the merry-go-round to maintain a constant angular velocity, otherwise the angular momentum of the system (you and the merry-go-round) is conserved and the angular velocity decreases as you increase your distance from the axis of rotation. (Similarly, if the motor driving a pump is turned off, the pump will obviously slow down and stop.) Your friend is the motor supplying energy to the rotor that is transferred to you. Is the amount of energy your friend expends to keep the angular velocity constant dependent upon what path you follow along the merry-go-round (i.e., the blade shape); on how fast and in what direction you walk off the edge (i.e., the exit velocity); on how much you weigh (i.e., the density of the fluid)? What happens if you walk from the outside edge toward the center of the rotating merry-go-round? Recall that the opposite of a pump is a turbine.

In a turbomachine a series of particles (a continuum) passes through the rotor. Thus, the moment-of-momentum equation applied to a control volume as derived in Section 5.2.3 is valid. For steady flow (or for turbomachine rotors with steady-in-the-mean or steady-on-average cyclical flow), Eq. 5.42 gives

$$\sum (\mathbf{r} \times \mathbf{F}) = \int_{cs} (\mathbf{r} \times \mathbf{V})\, \rho \mathbf{V} \cdot \hat{\mathbf{n}}\, dA$$

Recall that the left-hand side of this equation represents the sum of the external torques (moments) acting on the contents of the control volume, and the right-hand side is the net rate of flow of moment-of-momentum (*angular momentum*) through the control surface.

The axial component of this equation applied to the one-dimensional simplification of flow through a turbomachine rotor with section (1) as the inlet and section (2) as the outlet results in

$$T_{shaft} = -\dot{m}_1(r_1 V_{\theta 1}) + \dot{m}_2(r_2 V_{\theta 2}) \tag{12.2}$$

where T_{shaft} is the *shaft torque* applied to the contents of the control volume. The "−" is associated with mass flowrate into the control volume and the " + " is used with the outflow. The sign of the V_θ component depends on the direction of V_θ and the blade motion, U. If V_θ and U are in the same direction, then V_θ is positive. The sign of the torque exerted by the shaft on the rotor, T_{shaft}, is positive if T_{shaft} is in the same direction as rotation, and negative otherwise.

As seen from Eq. 12.2, the shaft torque is directly proportional to the mass flowrate, $\dot{m} = \rho Q$. (It takes considerably more torque and power to pump water than to pump air with the same volume flowrate.) The torque also depends on the tangential component of the absolute velocity, V_θ. Equation 12.2 is often called the *Euler turbomachine equation*.

Also recall that the *shaft power*, $\dot{W}_{shaft}$, is related to the shaft torque and angular velocity by

The Euler turbomachine equation is the axial component of the moment-of-momentum equation.

$$\dot{W}_{shaft} = T_{shaft}\,\omega \tag{12.3}$$

By combining Eqs. 12.2 and 12.3 and using the fact that $U = \omega r$, we obtain

$$\dot{W}_{shaft} = -\dot{m}_1(U_1 V_{\theta 1}) + \dot{m}_2(U_2 V_{\theta 2}) \tag{12.4}$$

Again, the value of V_θ is positive when V_θ and U are in the same direction and negative otherwise. Also, $\dot{W}_{shaft}$ is positive when the shaft torque and ω are in the same direction and negative otherwise. Thus, $\dot{W}_{shaft}$ is positive when power is supplied to the contents of the control volume (pumps) and negative otherwise (turbines). This outcome is consistent with the sign convention involving the work term in the energy equation considered in Chapter 5 (see Eq. 5.67).

Finally, in terms of work per unit mass, $w_{shaft} = \dot{W}_{shaft}/\dot{m}$, we obtain

$$w_{shaft} = -U_1 V_{\theta 1} + U_2 V_{\theta 2} \tag{12.5}$$

where we have used the fact that by conservation of mass, $\dot{m}_1 = \dot{m}_2$. Equations 12.3, 12.4, and 12.5 are the basic governing equations for pumps or turbines whether the machines are radial-, mixed-, or axial-flow devices and for compressible and incompressible flows. Note that neither the axial nor the radial component of velocity enter into the specific work (work per unit mass) equation. [In the above merry-go-round example the amount of work your friend does is independent of how fast you jump "up" (axially) or "out" (radially) as you exit. The only thing that counts is your θ component of velocity.]

Another useful but more laborious form of Eq. 12.5 can be obtained by writing the right-hand side in a slightly different form based on the velocity triangles at the entrance or exit as shown generically in Fig. 12.5. The velocity component V_x is the generic through-flow component of velocity and it can be axial, radial, or in-between depending on the rotor configuration. From the large right triangle we note that

$$V^2 = V_\theta^2 + V_x^2$$

or

$$V_x^2 = V^2 - V_\theta^2 \tag{12.6}$$

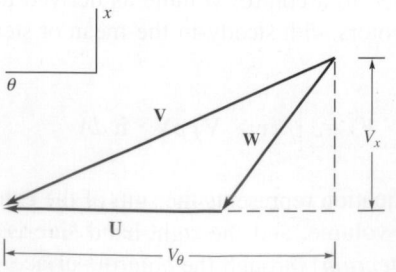

■ **F I G U R E 12.5** Velocity triangle: V = absolute velocity, W = relative velocity, U = blade velocity.

From the small right triangle we note that

$$V_x^2 + (V_\theta - U)^2 = W^2 \tag{12.7}$$

By combining Eqs. 12.6 and 12.7 we obtain

$$V_\theta U = \frac{V^2 + U^2 - W^2}{2}$$

which when written for the inlet and exit and combined with Eq. 12.5 gives

$$w_{\text{shaft}} = \frac{V_2^2 - V_1^2 + U_2^2 - U_1^2 - (W_2^2 - W_1^2)}{2} \tag{12.8}$$

Turbomachine work is related to changes in absolute, relative, and blade velocities.

Thus, the power and the shaft work per unit mass can be obtained from the speed of the blade, U, the absolute fluid speed, V, and the fluid speed relative to the blade, W. This is an alternative to using fewer components of the velocity as suggested by Eq. 12.5. Equation 12.8 contains more terms than Eq. 12.5; however, it is an important concept equation because it shows how the work transfer is related to absolute, relative, and blade velocity changes. Because of the general nature of the velocity triangle in Fig. 12.5, Eq. 12.8 is applicable for axial-, radial-, and mixed-flow rotors.

F l u i d s   i n   t h e   N e w s

1948 Buick Dynaflow started it Prior to 1948 almost all cars had manual transmissions which required the use of a clutch pedal to shift gears. The 1948 Buick Dynaflow was the first automatic transmission to use the hydraulic *torque* converter and was the model for present-day automatic transmissions. Currently, in the U.S. over 84% of the cars have automatic transmissions. The torque converter replaces the clutch found on manual shift vehicles and allows the engine to continue running when the vehicle comes to a stop. In principle, but certainly not in detail or complexity, op-

eration of a torque converter is similar to blowing air from a *fan* onto another fan which is unplugged. One can hold the *blade* of the unplugged fan and keep it from turning, but as soon as it is let go, it will begin to speed up until it comes close to the speed of the powered fan. The torque converter uses transmission *fluid* (not air) and consists of a *pump* (the powered fan) driven by the engine drive shaft, a *turbine* (the unplugged fan) connected to the input shaft of the transmission, and a *stator* (absent in the fan model) to efficiently direct the flow between the pump and turbine.

12.4   The Centrifugal Pump

One of the most common radial-flow turbomachines is the *centrifugal pump*. This type of pump has two main components: an *impeller* attached to a rotating shaft, and a stationary *casing, housing*, or *volute* enclosing the impeller. The impeller consists of a number of blades (usually curved), also sometimes called *vanes*, arranged in a regular pattern around the shaft. A sketch showing the essential features of a centrifugal pump is shown in Fig. 12.6. As the impeller rotates, fluid is sucked in through the *eye* of the casing and flows radially outward. Energy is added to the fluid by the rotating blades, and both pressure and absolute velocity are increased as the fluid flows from the eye to the periphery of the blades. For the simplest type of centrifugal pump, the

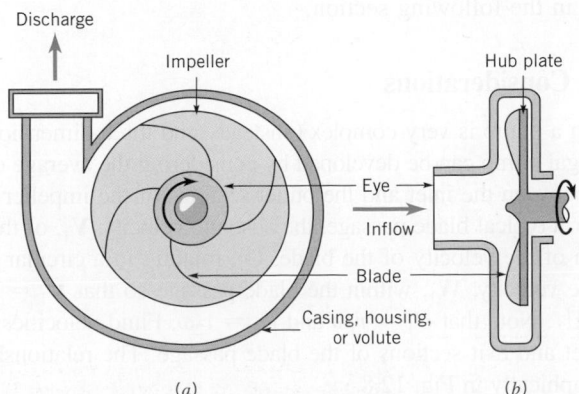

■ **FIGURE 12.6**   Schematic diagram of basic elements of a centrifugal pump.

(a)   *(b)*

■ **FIGURE 12.7**
(*a*) Open impeller, (*b*) enclosed
or shrouded impeller.
(Courtesy of Ingersoll-Dresser
Pump Company.)

fluid discharges directly into a volute-shaped casing. The casing shape is designed to reduce the velocity as the fluid leaves the impeller, and this decrease in kinetic energy is converted into an increase in pressure. The volute-shaped casing, with its increasing area in the direction of flow, is used to produce an essentially uniform velocity distribution as the fluid moves around the casing into the discharge opening. For large centrifugal pumps, a different design is often used in which diffuser guide vanes surround the impeller. The diffuser vanes decelerate the flow as the fluid is directed into the pump casing. This type of centrifugal pump is referred to as a *diffuser* pump.

Impellers are generally of two types. For one configuration the blades are arranged on a hub or backing plate and are open on the other (casing or shroud) side. A typical *open impeller* is shown in Fig. 12.7a. For the second type of impeller, called an *enclosed* or *shrouded* impeller, the blades are covered on both hub and shroud ends as shown in Fig. 12.7b.

Pump impellers can also be *single* or *double suction*. For the single-suction impeller the fluid enters through the eye on only one side of the impeller, whereas for the double-suction impeller the fluid enters the impeller along its axis from both sides. The double-suction arrangement reduces end thrust on the shaft, and also, since the net inlet flow area is larger, inlet velocities are reduced.

Pumps can be *single* or *multistage*. For a single-stage pump, only one impeller is mounted on the shaft, whereas for multistage pumps, several impellers are mounted on the same shaft. The stages operate in series, that is, the discharge from the first stage flows into the eye of the second stage, the discharge from the second stage flows into the eye of the third stage, and so on. The flowrate is the same through all stages, but each stage develops an additional pressure rise. Thus, a very large discharge pressure, or head, can be developed by a multistage pump.

Centrifugal pumps involve radially outward flows.

Centrifugal pumps come in a variety of arrangements (open or shrouded impellers, volute or diffuser casings, single- or double-suction, single- or multistage), but the basic operating principle remains the same. Work is done on the fluid by the rotating blades (centrifugal action and tangential blade force acting on the fluid over a distance), creating a large increase in kinetic energy of the fluid flowing through the impeller. This kinetic energy is converted into an increase in pressure as the fluid flows from the impeller into the casing enclosing the impeller. A simplified theory describing the behavior of the centrifugal pump was introduced in the previous section and is expanded in the following section.

12.4.1 Theoretical Considerations

Although flow through a pump is very complex (unsteady and three-dimensional), the basic theory of operation of a centrifugal pump can be developed by considering the average one-dimensional flow of the fluid as it passes between the inlet and the outlet sections of the impeller as the blades rotate. As shown in Fig. 12.8, for a typical blade passage, the absolute velocity, $\mathbf{V}_1$, of the fluid entering the passage is the vector sum of the velocity of the blade, $\mathbf{U}_1$, rotating in a circular path with angular velocity ω, and the relative velocity, $\mathbf{W}_1$, within the blade passage so that $\mathbf{V}_1 = \mathbf{W}_1 + \mathbf{U}_1$. Similarly, at the exit $\mathbf{V}_2 = \mathbf{W}_2 + \mathbf{U}_2$. Note that $U_1 = r_1\omega$ and $U_2 = r_2\omega$. Fluid velocities are taken to be average velocities over the inlet and exit sections of the blade passage. The relationship between the various velocities is shown graphically in Fig. 12.8.

As discussed in Section 12.3, the moment-of-momentum equation indicates that the shaft torque, T_{shaft}, required to rotate the pump impeller is given by equation Eq. 12.2 applied to a pump with $\dot{m}_1 = \dot{m}_2 = \dot{m}$. That is,

$$T_{shaft} = \dot{m}(r_2 V_{\theta 2} - r_1 V_{\theta 1})$$ (12.9)

or

$$T_{shaft} = \rho Q(r_2 V_{\theta 2} - r_1 V_{\theta 1})$$ (12.10)

where $V_{\theta 1}$ and $V_{\theta 2}$ are the tangential components of the absolute velocities, $\mathbf{V}_1$ and $\mathbf{V}_2$ (see Figs. 12.8b,c).

For a rotating shaft, the power transferred, $\dot{W}_{shaft}$, is given by

$$\dot{W}_{shaft} = T_{shaft}\omega$$

and therefore from Eq. 12.10

$$\dot{W}_{shaft} = \rho Q\omega(r_2 V_{\theta 2} - r_1 V_{\theta 1})$$

Since $U_1 = r_1\omega$ and $U_2 = r_2\omega$ we obtain

$$\dot{W}_{shaft} = \rho Q(U_2 V_{\theta 2} - U_1 V_{\theta 1})$$ (12.11)

Equation 12.11 shows how the power supplied to the shaft of the pump is transferred to the flowing fluid. It also follows that the shaft power per unit mass of flowing fluid is

$$w_{shaft} = \frac{\dot{W}_{shaft}}{\rho Q} = U_2 V_{\theta 2} - U_1 V_{\theta 1}$$ (12.12)

For incompressible pump flow, we get from Eq. 5.82

$$w_{shaft} = \left(\frac{p_{out}}{\rho} + \frac{V_{out}^2}{2} + gz_{out}\right) - \left(\frac{p_{in}}{\rho} + \frac{V_{in}^2}{2} + gz_{in}\right) + loss$$

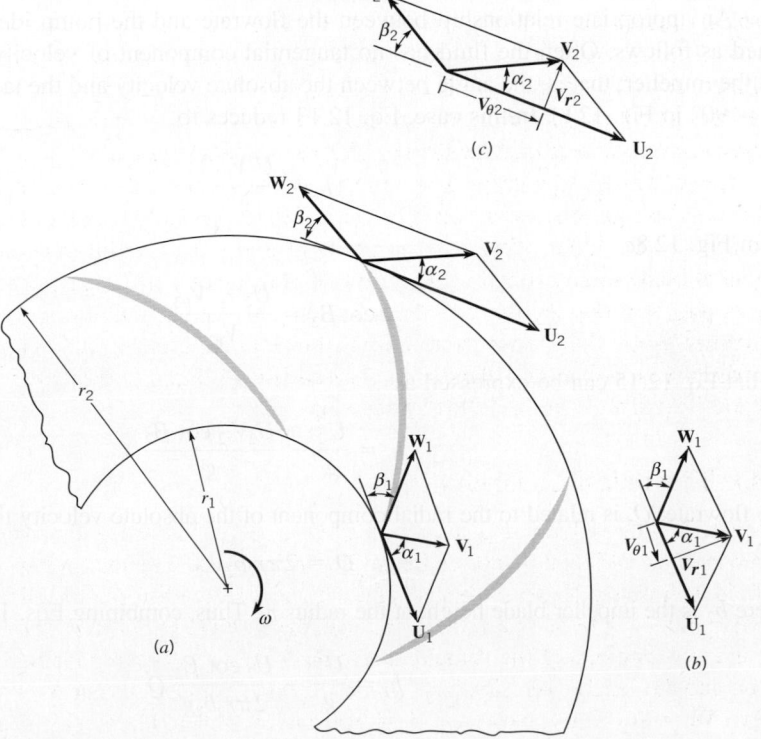

■ **FIGURE 12.8**   Velocity diagrams at the inlet and exit of a centrifugal pump impeller.

Combining Eq. 12.12 with this, we get

$$U_2 V_{\theta 2} - U_1 V_{\theta 1} = \left(\frac{p_{\text{out}}}{\rho} + \frac{V_{\text{out}}^2}{2} + g z_{\text{out}} \right) - \left(\frac{p_{\text{in}}}{\rho} + \frac{V_{\text{in}}^2}{2} + g z_{\text{in}} \right) + loss$$

Dividing both sides of this equation by the acceleration of gravity, g, we obtain

$$\frac{U_2 V_{\theta 2} - U_1 V_{\theta 1}}{g} = H_{\text{out}} - H_{\text{in}} + h_L$$

where H is total head defined by

$$H = \frac{p}{\rho g} + \frac{V^2}{2g} + z$$

and h_L is head loss.

From this equation we see that $(U_2 V_{\theta 2} - U_1 V_{\theta 1})/g$ is the shaft work head added to the fluid by the pump. Head loss, h_L, reduces the actual head rise, $H_{\text{out}} - H_{\text{in}}$, achieved by the fluid. Thus, the ideal head rise possible, h_i, is

$$h_i = \frac{U_2 V_{\theta 2} - U_1 V_{\theta 1}}{g} \tag{12.13}$$

The pump actual head rise is less than the pump ideal head rise by an amount equal to the head loss in the pump.

The actual head rise, $H_{\text{out}} - H_{\text{in}} = h_a$, is always less than the ideal head rise, h_i, by an amount equal to the head loss, h_L, in the pump. Some additional insight into the meaning of Eq. 12.13 can be obtained by using the following alternate version (see Eq. 12.8).

$$h_i = \frac{1}{2g} [(V_2^2 - V_1^2) + (U_2^2 - U_1^2) + (W_1^2 - W_2^2)] \tag{12.14}$$

A detailed examination of the physical interpretation of Eq. 12.14 would reveal the following. The first term in brackets on the right-hand side represents the increase in the kinetic energy of the fluid, and the other two terms represent the pressure head rise that develops across the impeller due to the centrifugal effect, $U_2^2 - U_1^2$, and the diffusion of relative flow in the blade passages, $W_1^2 - W_2^2$.

An appropriate relationship between the flowrate and the pump ideal head rise can be obtained as follows. Often the fluid has no tangential component of velocity $V_{\theta 1}$, or *swirl*, as it enters the impeller; that is, the angle between the absolute velocity and the tangential direction is 90° ($\alpha_1 = 90°$ in Fig. 12.8). In this case, Eq. 12.13 reduces to

$$h_i = \frac{U_2 V_{\theta 2}}{g} \tag{12.15}$$

From Fig. 12.8c

$$\cot \beta_2 = \frac{U_2 - V_{\theta 2}}{V_{r2}}$$

so that Eq. 12.15 can be expressed as

$$h_i = \frac{U_2^2}{g} - \frac{U_2 V_{r2} \cot \beta_2}{g} \tag{12.16}$$

The flowrate, Q, is related to the radial component of the absolute velocity through the equation

$$Q = 2\pi r_2 b_2 V_{r2} \tag{12.17}$$

where b_2 is the impeller blade height at the radius r_2. Thus, combining Eqs. 12.16 and 12.17 yields

$$h_i = \frac{U_2^2}{g} - \frac{U_2 \cot \beta_2}{2\pi r_2 b_2 g} Q \tag{12.18}$$

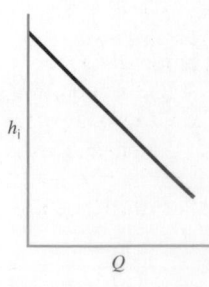

This equation is graphed in the margin and shows that the ideal or maximum head rise for a centrifugal pump varies linearly with Q for a given blade geometry and angular velocity. For actual pumps, the blade angle β_2 falls in the range of 15°–35°, with a normal range of 20° < β_2 < 25°, and with 15° < β_1 < 50° (Ref. 10). Blades with β_2 < 90° are called *backward curved*, whereas blades with β_2 > 90° are called *forward curved*. Pumps are not usually designed with forward curved vanes since such pumps tend to suffer unstable flow conditions.

EXAMPLE 12.2 | Centrifugal Pump Performance Based on Inlet/Outlet Velocities

GIVEN   Water is pumped at the rate of 5300 L/min through a centrifugal pump operating at a speed of 1750 rpm. The impeller has a uniform blade height, b, of 5 cm with r_1 = 4 cm and r_2 = 18 cm, and the exit blade angle β_2 is 23° (see Fig. 12.8). Assume ideal flow conditions and that the tangential velocity component, $V_{\theta1}$, of the water entering the blade is zero (α_1 = 90°).

FIND   Determine (a) the tangential velocity component, $V_{\theta2}$, at the exit, (b) the ideal head rise, h_i, and (c) the power, $\dot{W}_{shaft}$, transferred to the fluid. Discuss the difference between ideal and actual head rise. Is the power, $\dot{W}_{shaft}$, ideal or actual? Explain.

SOLUTION

(a)   At the exit the velocity diagram is as shown in Fig. 12.8c, where $\mathbf{V}_2$ is the absolute velocity of the fluid, $\mathbf{W}_2$ is the relative velocity, and $\mathbf{U}_2$ is the tip velocity of the impeller with

$$U_2 = r_2\omega = (18/100 \text{ m})(2\pi \text{ rad/rev})\frac{(1750 \text{ rpm})}{(60 \text{ s/min})}$$

$$= 33 \text{ m/s}$$

Since the flowrate is given, it follows that

$$Q = 2\pi r_2 b_2 V_{r2}$$

or

$$V_{r2} = \frac{Q}{2\pi r_2 b_2}$$

$$= \frac{5300 \text{ L/min}}{(1000 \text{ L/m}^3)(60 \text{ s/min})(2\pi)(18/100 \text{ m})(5/100 \text{ m})}$$

$$= 1.6 \text{ m/s}$$

From Fig. 12.8c we see that

$$\cot \beta_2 = \frac{U_2 - V_{\theta2}}{V_{r2}}$$

so that

$$V_{\theta2} = U_2 - V_{r2} \cot \beta_2$$

$$= (33 - 1.6 \cot 23°) \text{ m/s}$$

$$= 29 \text{ m/s} \qquad\text{(Ans)}$$

(b)   From Eq. 12.15 the ideal head rise is given by

$$h_i = \frac{U_2 V_{\theta2}}{g} = \frac{(33 \text{ m/s})(29 \text{ m/s})}{9.81 \text{ m/s}^2}$$

$$= 98 \text{ m} \qquad\text{(Ans)}$$

Alternatively, from Eq. 12.16, the ideal head rise is

$$h_i = \frac{U_2^2}{g} - \frac{U_2 V_{r2} \cot \beta_2}{g}$$

$$= \frac{(33 \text{ m/s})^2}{9.81 \text{ m/s}^2} - \frac{(33 \text{ m/s})(1.6 \text{ m/s}) \cot 23°}{9.81 \text{ m/s}^2}$$

$$= 98 \text{ m} \qquad\text{(Ans)}$$

(c)   From Eq. 12.11, with $V_{\theta1}$ = 0, the power transferred to the fluid is given by the equation

$$\dot{W}_{shaft} = \rho Q U_2 V_{\theta2}$$

$$= \frac{(1000 \text{ kg/m}^3)(5300 \text{ L/min})(33 \text{ m/s})(29 \text{ m/s})}{[1(\text{kg} \cdot \text{m/s}^2)/\text{N}](1000 \text{ L/m}^3)(60 \text{ s/min})}$$

$$= (84{,}500 \text{ N} \cdot \text{m/s})(1 \text{ W/N} \cdot \text{m/s}) = 84.5 \text{ kW} \qquad\text{(Ans)}$$

Note that the ideal head rise and the power transferred to the fluid are related through the relationship

$$\dot{W}_{shaft} = \rho g Q h_i$$

COMMENT   It should be emphasized that results given in the previous equation involve the ideal head rise. The actual head-rise performance characteristics of a pump are usually determined by experimental measurements obtained in a testing laboratory. The actual head rise is always less than the ideal head rise for a specific flowrate because of the loss of available energy associated with actual flows. Also, it is important to note that even if actual values of U_2 and V_{r2} are used in Eq. 12.16, the ideal head rise is calculated. The only idealization used in this example problem is that the exit flow angle is identical to the blade angle at the exit. If the actual exit flow angle was made available in this example, it could have been used in Eq. 12.16 to calculate the ideal head rise.

The pump power, $\dot{W}_{shaft}$, is the actual power required to achieve a blade speed of 33 m/s, a flowrate of 5300 L/min, and the tangential velocity, $V_{\theta2}$, associated with this example. If pump losses could somehow be reduced to zero (every pump designer's dream), the actual and ideal head rise would have been identical at 98 m. As is, the ideal head rise is 98 m and the actual head rise something less.

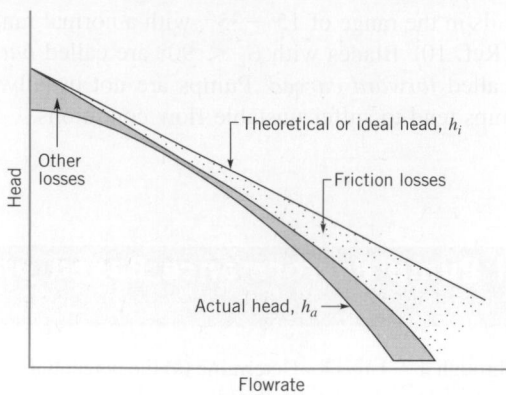

■ **FIGURE 12.9** **Effect of losses on the pump head–flowrate curve.**

Ideal and actual head rise levels differ by the head loss.

Figure 12.9 shows the ideal head versus flowrate curve (Eq. 12.18) for a centrifugal pump with backward curved vanes ($\beta_2 < 90°$). Since there are simplifying assumptions (i.e., zero losses) associated with the equation for h_i, we would expect that the actual rise in head of fluid, h_a, would be less than the ideal head rise, and this is indeed the case. As shown in Fig. 12.9, the h_a versus Q curve lies below the ideal head-rise curve and shows a nonlinear variation with Q. The differences between the two curves (as represented by the shaded areas between the curves) arise from several sources. These differences include losses due to fluid skin friction in the blade passages, which vary as Q^2, and other losses due to such factors as flow separation, impeller blade-casing clearance flows, and other three-dimensional flow effects. Near the design flowrate, some of these other losses are minimized.

Centrifugal pump design is a highly developed field, with much known about pump theory and design procedures (see, for example, Refs. 4–6). However, due to the general complexity of flow through a centrifugal pump, the actual performance of the pump cannot be accurately predicted on a completely theoretical basis as indicated by the data of Fig. 12.9. Actual pump performance is determined experimentally through tests on the pump. From these tests, pump characteristics are determined and presented as *pump performance curves.* It is this information that is most helpful to the engineer responsible for incorporating pumps into a given flow system.

12.4.2 Pump Performance Characteristics

The actual head rise, h_a, gained by fluid flowing through a pump can be determined with an experimental arrangement of the type shown in Fig. 12.10, using the energy equation (Eq. 5.84 with $h_a = h_s - h_L$ where h_s is the shaft work head and is identical to h_i, and h_L is the pump head loss)

$$h_a = \frac{p_2 - p_1}{\gamma} + z_2 - z_1 + \frac{V_2^2 - V_1^2}{2g} \tag{12.19}$$

with sections (1) and (2) at the pump inlet and exit, respectively. Typically, the differences in elevations and velocities are small so that

$$h_a \approx \frac{p_2 - p_1}{\gamma} \tag{12.20}$$

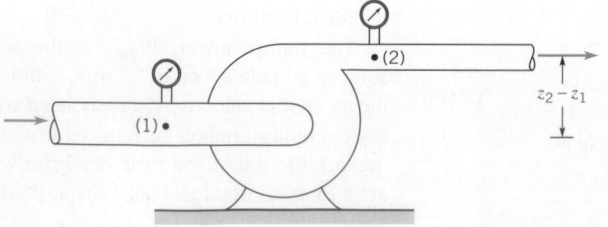

■ **FIGURE 12.10** **Typical experimental arrangement for determining the head rise gained by a fluid flowing through a pump.**

The power, $\mathcal{P}_f$, gained by the fluid is given by the equation

$$\mathcal{P}_f = \gamma Q h_a \qquad \text{(12.21)}$$

and this quantity, expressed in terms of kilowatts. Thus,

$$\mathcal{P}_f = \gamma Q h_a \qquad \text{(12.22)}$$

with γ expressed in N/m³, Q in m³/s, and h_a in m. Note that if the pumped fluid is not water, the γ appearing in Eq. 12.22 must be the specific weight of the fluid moving through the pump.

In addition to the head or power added to the fluid, the *overall efficiency, η*, is of interest, where

$$\eta = \frac{\text{power gained by the fluid}}{\text{shaft power driving the pump}} = \frac{\mathcal{P}_f}{\dot{W}_{\text{shaft}}} \qquad \text{(12.23)}$$

Pump overall efficiency is the ratio of power actually gained by the fluid to the shaft power supplied.

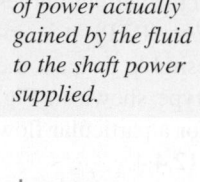
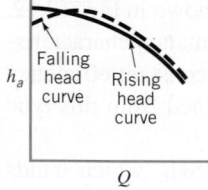

h_a | Falling head curve | Rising head curve

Q

The overall pump efficiency is affected by the *hydraulic losses* in the pump, as previously discussed, and in addition, by the *mechanical losses* in the bearings and seals. There may also be some power loss due to leakage of the fluid between the back surface of the impeller hub plate and the casing, or through other pump components. This leakage contribution to the overall efficiency is called the *volumetric loss*. Thus, the overall efficiency arises from three sources, the *hydraulic efficiency, η_h*, the *mechanical efficiency, η_m*, and the *volumetric efficiency, η_v*, so that $\eta = \eta_h \eta_m \eta_v$.

Performance characteristics for a given pump geometry and operating speed are usually given in the form of plots of h_a, η, and power versus Q (commonly referred to as *capacity*) as illustrated in Fig. 12.11. Actually, only two curves are needed since h_a, η, and power are related through Eq. 12.23. For convenience, all three curves are usually provided. Note that for the pump characterized by the data of Fig. 12.11, the head curve continuously rises as the flowrate decreases, and in this case the pump is said to have a *rising head* curve. As shown by the figure in the margin, pumps may also have $h_a - Q$ curves that initially rise as Q is decreased from the design value and then fall with a continued decrease in Q. These pumps have a *falling head* curve. The head developed by the pump at zero discharge is called the shutoff head, and it represents the rise in pressure head across the pump with the discharge valve closed. Since there is no flow with the valve closed, the related efficiency is zero, and the power supplied by the pump (power at $Q = 0$) is simply dissipated as heat. Although centrifugal pumps can be operated for short periods of time with the discharge valve closed, damage will occur due to overheating and large mechanical stress with any extended operation with the valve closed.

As can be seen from Fig. 12.11, as the discharge is increased from zero the power increases, with a subsequent fall as the maximum discharge is approached. As previously noted, with h_a and power known, the efficiency can be calculated. As shown in Fig. 12.11, the efficiency is a function

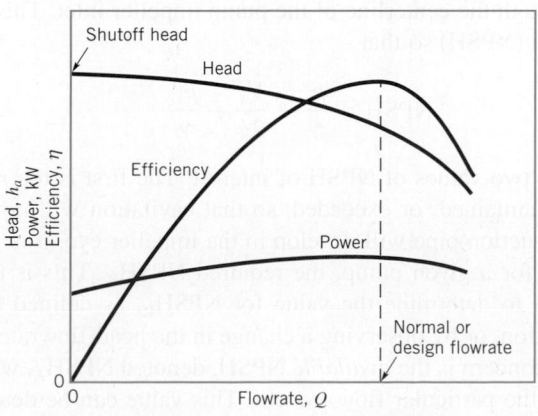

■ F I G U R E 12.11 **Typical performance characteristics for a centrifugal pump of a given size operating at a constant impeller speed.**

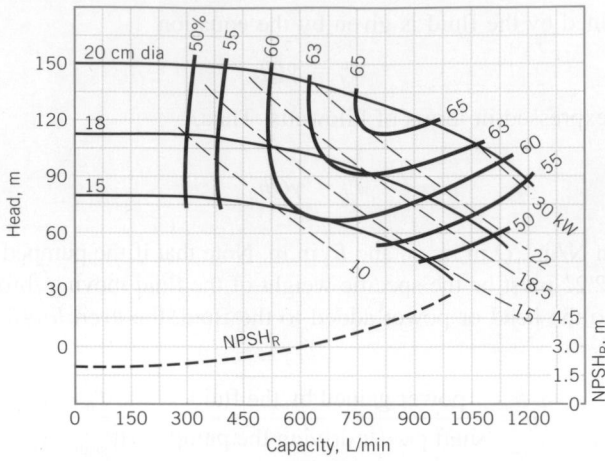

■ **F I G U R E 12.12** Performance curves for a two-stage centrifugal pump operating at 3500 rpm. Data given for three different impeller diameters.

of the flowrate and reaches a maximum value at some particular value of the flowrate, commonly referred to as the *normal* or *design* flowrate or capacity for the pump. The points on the various curves corresponding to the maximum efficiency are denoted as the *best efficiency points* (BEP). It is apparent that when selecting a pump for a particular application, it is usually desirable to have the pump operate near its maximum efficiency. Thus, performance curves of the type shown in Fig. 12.11 are very important to the engineer responsible for the selection of pumps for a particular flow system. Matching the pump to a particular flow system is discussed in Section 12.4.4.

Pump performance characteristics are also presented in charts of the type shown in Fig. 12.12. Since impellers with different diameters may be used in a given casing, performance characteristics for several impeller diameters can be provided with corresponding lines of constant efficiency and power as illustrated in Fig. 12.12. Thus, the same information can be obtained from this type of graph as from the curves shown in Fig. 12.11.

It is to be noted that an additional curve is given in Fig. 12.12, labeled NPSH$_R$, which stands for *required net positive suction head*. As discussed in the following section, the significance of this curve is related to conditions on the suction side of the pump, which must also be carefully considered when selecting and positioning a pump.

12.4.3 Net Positive Suction Head (NPSH)

Cavitation, which may occur when pumping a liquid, is usually avoided.

On the suction side of a pump, low pressures are commonly encountered, with the concomitant possibility of cavitation occurring within the pump. As discussed in Section 1.8, cavitation occurs when the liquid pressure at a given location is reduced to the vapor pressure of the liquid. When this occurs, vapor bubbles form (the liquid starts to "boil"); this phenomenon can cause a loss in efficiency as well as structural damage to the pump. To characterize the potential for cavitation, the difference between the total head on the suction side, near the pump impeller inlet, $p_s/\gamma + V_s^2/2g$, and the liquid vapor pressure head, p_v/γ, is used. The position reference for the elevation head passes through the centerline of the pump impeller inlet. This difference is called the net positive suction head (NPSH) so that

$$\text{NPSH} = \frac{p_s}{\gamma} + \frac{V_s^2}{2g} - \frac{p_v}{\gamma} \tag{12.24}$$

There are actually two values of NPSH of interest. The first is the *required* NPSH, denoted NPSH$_R$, that must be maintained, or exceeded, so that cavitation will not occur. Since pressures lower than those in the suction pipe will develop in the impeller eye, it is usually necessary to determine experimentally, for a given pump, the required NPSH$_R$. This is the curve shown in Fig. 12.12. Pumps are tested to determine the value for NPSH$_R$, as defined by Eq. 12.24, by either directly detecting cavitation, or by observing a change in the head-flowrate curve (Ref. 7). The second value for NPSH of concern is the *available* NPSH, denoted NPSH$_A$, which represents the head that actually occurs for the particular flow system. This value can be determined experimentally, or calculated if the system parameters are known. For example, a typical flow system is shown in

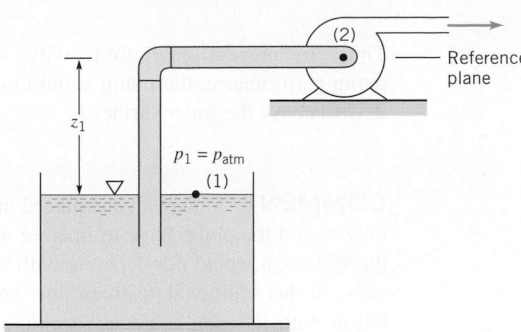

■ **FIGURE 12.13   Schematic of a pump installation in which the pump must lift fluid from one level to another.**

Fig. 12.13. The energy equation applied between the free liquid surface, where the pressure is atmospheric, p_{atm}, and a point on the suction side of the pump near the impeller inlet yields

$$\frac{p_{atm}}{\gamma} - z_1 = \frac{p_s}{\gamma} + \frac{V_s^2}{2g} + \sum h_L$$

where Σh_L represents head losses between the free surface and the pump impeller inlet. Thus, the head available at the pump impeller inlet is

$$\frac{p_s}{\gamma} + \frac{V_s^2}{2g} = \frac{p_{atm}}{\gamma} - z_1 - \sum h_L$$

so that

$$\text{NPSH}_A = \frac{p_{atm}}{\gamma} - z_1 - \sum h_L - \frac{p_v}{\gamma} \qquad (12.25)$$

For this calculation, absolute pressures are normally used since the vapor pressure is usually specified as an absolute pressure. For proper pump operation it is necessary that

$$\text{NPSH}_A \geq \text{NPSH}_R$$

For proper pump operation, the available net positive suction head must be greater than the required net positive suction head.

It is noted from Eq. 12.25 that as the height of the pump impeller above the fluid surface, z_1, is increased, the NPSH_A is decreased. Therefore, there is some critical value for z_1 above which the pump cannot operate without cavitation. The specific value depends on the head losses and the value of the vapor pressure. It is further noted that if the supply tank or reservoir is *above* the pump, z_1 will be negative in Eq. 12.25, and the NPSH_A will increase as this height is increased.

---

*E*XAMPLE 12.3   **Net Positive Suction Head**

GIVEN   A centrifugal pump is to be placed above a large, open water tank, as shown in Fig. 12.13, and is to pump water at a rate of $1.4 \times 10^{-2}\,\text{m}^3/\text{s}$. At this flowrate the required net positive suction head, NPSH_R, is 4.5 cm, as specified by the pump manufacturer. The water temperature is 30 °C and atmospheric pressure is 101.3 kPa. Assume that the major head loss between the tank and the pump inlet is due to filter at the pipe inlet having a minor loss

coefficient $K_L = 20$. Other losses can be neglected. The pipe on the suction side of the pump has a diameter of 10 cm.

FIND   Determine the maximum height, z_1, that the pump can be located above the water surface without cavitation. If you were required to place a valve in the flow path would you place it upstream or downstream of the pump? Why?

SOLUTION _____

From Eq. 12.25 the available net positive suction head, NPSH_A, is given by the equation

$$\text{NPSH}_A = \frac{p_{atm}}{\gamma} - z_1 - \sum h_L - \frac{p_v}{\gamma}$$

and the maximum value for z_1 will occur when $\text{NPSH}_A = \text{NPSH}_R$. Thus,

$$(z_1)_{max} = \frac{p_{atm}}{\gamma} - \sum h_L - \frac{p_v}{\gamma} - \text{NPSH}_R \qquad (1)$$

Since the only head loss to be considered is the loss

$$\sum h_L = K_L \frac{V^2}{2g}$$

with

$$V = \frac{Q}{A} = \frac{1.4 \times 10^{-2} \text{ m}^3/\text{s}}{(\pi/4)(10/100 \text{ cm})^2} = 1.8 \text{ m/s}$$

it follows that

$$\sum h_L = \frac{(20)(1.8 \text{ m/s})^2}{2(9.81 \text{ m/s}^2)} = 3 \text{ m}$$

From Table 8.1 the water vapor pressure at 30 °C is 4.243 kPa and $\gamma = 9.765 \text{ kN/m}^3$. Equation (1) can now be written as

$$(z_1)_{max} = \frac{(101.3 \text{ kPa})}{9.765 \text{ kN/m}^3} - 3 \text{ m}$$

$$- \frac{(4.243 \times 10^3 \text{ kPa})}{9.765 \text{ kN/m}^2} - 4.5 \text{ m}$$

$$= 2.37 \text{ m} \qquad \text{(Ans)}$$

Thus, to prevent cavitation, with its accompanying poor pump performance, the pump should not be located higher than 2.37 m above the water surface.

COMMENT If the valve is placed upstream of the pump, not only would the pump have to operate with an additional loss in the system, it would now operate with a lower inlet pressure because of this additional upstream loss and could now suffer cavitation with its usually negative consequences. If the valve is placed downstream of the pump, the pump would need to operate with more loss in the system and with higher back pressure than without the valve. Depending on the stability of the pump at higher back pressures, this could be inconsequential or important. Usually, pumps are stable even with higher back pressures. So, placing the valve on the downstream side of the pump is normally the better choice.

12.4.4 System Characteristics and Pump Selection

The system equation relates the actual head gained by the fluid to the flowrate.

A typical flow system in which a pump is used is shown in Fig. 12.14. The energy equation applied between points (1) and (2) indicates that

$$h_a = z_2 - z_1 + \sum h_L \qquad (12.26)$$

where h_a is the actual head gained by the fluid from the pump, and $\sum h_L$ represents all friction losses in the pipe and minor losses for pipe fittings and valves. From our study of pipe flow, we know that typically h_L varies approximately as the flowrate squared; that is, $h_L \propto Q^2$ (see Section 8.4). Thus, Eq. 12.26 can be written in the form

$$h_a = z_2 - z_1 + KQ^2 \qquad (12.27)$$

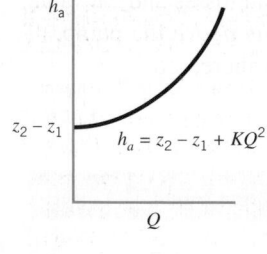

where K depends on the pipe sizes and lengths, friction factors, and minor loss coefficients. Equation 12.27, which is shown in the figure in the margin, is the *system equation* and shows how the actual head gained by the fluid from the pump is related to the system parameters. In this case the parameters include the change in elevation head, $z_2 - z_1$, and the losses due to friction as expressed by KQ^2. Each flow system has its own specific system equation. If the flow is laminar, the frictional losses will be proportional to Q rather than Q^2 (see Section 8.2).

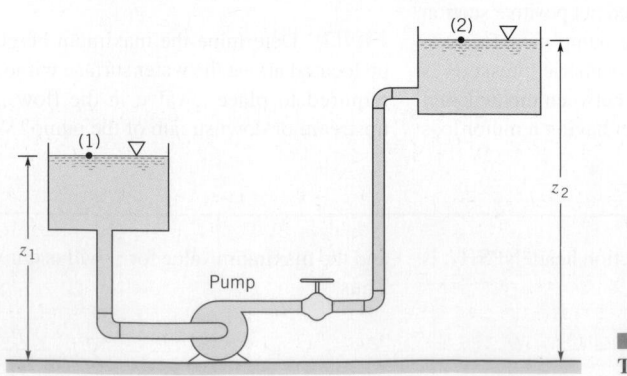

■ **FIGURE 12.14**
Typical flow system.

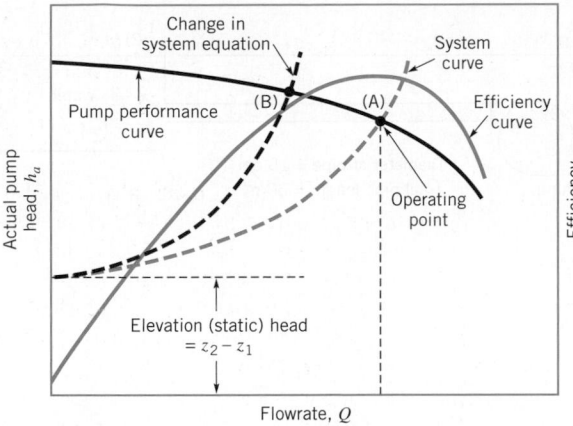

■ **F I G U R E** 12.15 **Utilization of the system curve and the pump performance curve to obtain the operating point for the system.**

The intersection of the pump performance curve and the system curve is the operating point.

There is also a unique relationship between the actual pump head gained by the fluid and the flowrate, which is governed by the pump design (as indicated by the pump performance curve). To select a pump for a particular application, it is necessary to utilize both the *system curve*, as determined by the system equation, and the pump performance curve. If both curves are plotted on the same graph, as illustrated in Fig. 12.15, their intersection (point A) represents the operating point for the system. That is, this point gives the head and flowrate that satisfies both the system equation and the pump equation. On the same graph the pump efficiency is shown. Ideally, we want the operating point to be near the best efficiency point (BEP) for the pump. For a given pump, it is clear that as the system equation changes, the operating point will shift. For example, if the pipe friction increases due to pipe wall fouling, the system curve changes, resulting in the operating point A shifting to point B in Fig. 12.15 with a reduction in flowrate and efficiency. The following example shows how the system and pump characteristics can be used to decide if a particular pump is suitable for a given application.

F l u i d s   i n   t h e   N e w s

Space Shuttle fuel pumps The fuel *pump* of your car engine is vital to its operation. Similarly, the fuels (liquid hydrogen and oxygen) of each Space Shuttle main engine (there are three per shuttle) rely on multistage *turbopumps* to get from storage tanks to main combustors. High pressures are utilized throughout the pumps to avoid cavitation. The pumps, some *centrifugal* and some *axial*, are driven by axial-flow, *multistage* turbines. Pump speeds are as high as 35,360 rpm. The liquid oxygen is pumped from 100 to 7420 psia (0.7 to 51 MPa), the liquid hydrogen from 30 to 6515 psia (0.21 to 50 MPa). Liquid hydrogen and oxygen flowrates of about 17,200 gpm (65 kL/min) and 6100 gpm (23 kL/min), respectively, are achieved. These pumps could empty your home swimming pool in seconds. The hydrogen goes from $-423\ °F\ (-252.7\ °C)$ in storage to $+6000\ °F\ (3315.5\ °C)$ in the combustion chamber!

E XAMPLE 12.4 　 Use of Pump Performance Curves

GIVEN　Water is to be pumped from one large, open tank to a second large, open tank as shown in Fig. E12.4a. The pipe diameter throughout is 15 cm and the total length of the pipe between the pipe entrance and exit is 60 m. Minor loss coefficients for the entrance, exit, and the elbow are shown, and the friction factor for the pipe can be assumed constant and equal to 0.02. A certain centrifugal pump having the performance characteristics shown in Fig. E12.4b is suggested as a good pump for this flow system.

FIND　With this pump, what would be the flowrate between the tanks? Do you think this pump would be a good choice?

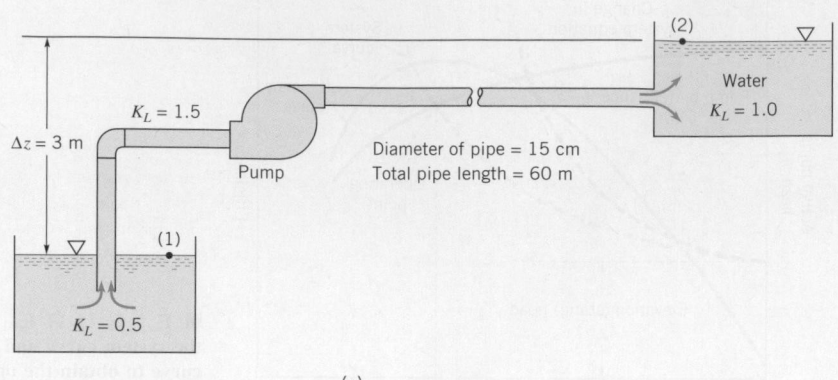

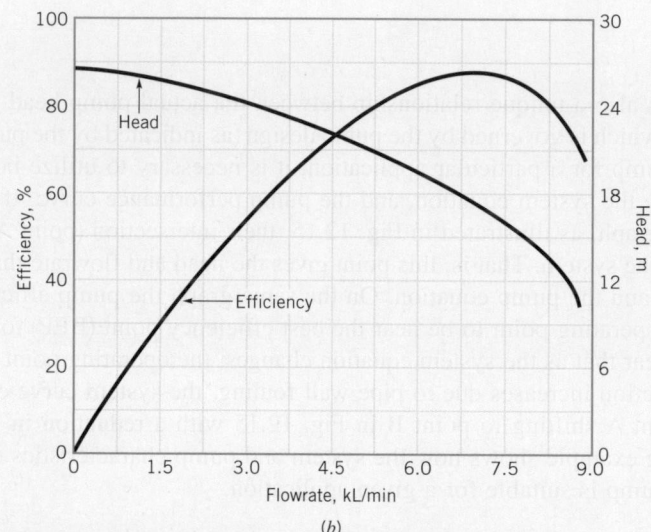

(a)

(b)

■ **F I G U R E E12.4a, b**

SOLUTION

Application of the energy equation between the two free surfaces, points (1) and (2) as indicated, gives

$$\frac{p_1}{\gamma} + \frac{V_1^2}{2g} + z_1 + h_a = \frac{p_2}{\gamma} + \frac{V_2^2}{2g} + z_2$$

$$+ f\frac{\ell}{D}\frac{V^2}{2g} + \mathbf{a}\ K_L\frac{V^2}{2g} \qquad (1)$$

Thus, with $p_1 = p_2 = 0$, $V_1 = V_2 = 0$, $\Delta z = z_2 - z_1 = 3$ m, $f = 0.02$, $D = 15/100$ m, and $\ell = 60$ m, Eq. 1 becomes

$$h_a = 3 + \left[0.02\frac{(60\ m)}{(15/100\ m)} \right.$$

$$\left. + (0.5 + 1.5 + 1.0) \right]\frac{V^2}{2(9.81\ m/s^2)} \qquad (2)$$

where the given minor loss coefficients have been used. Since

$$V = \frac{Q}{A} = \frac{Q(m^3/s)}{(\pi/4)(15/100\ m)^2}$$

Eq. 2 can be expressed as

$$h_a = 3 + 1795\ Q^2 \qquad (3)$$

where Q is in m³/s, or with Q in liters per minute

$$h_a = 3 + 5 \times 10^{-7}Q^2 \qquad (4)$$

Equation 3 or 4 represents the system equation for this particular flow system and reveals how much actual head the fluid will need to gain from the pump to maintain a certain flowrate. Performance data shown in Fig. E12.4b indicate the actual head the fluid will gain from this particular pump when it operates at a certain flowrate. Thus, when Eq. 4 is plotted on the same graph with performance data, the intersection of the two curves represents the operating point for the pump and the system. This combination is shown in Fig. E12.4c with the intersection (as obtained graphically) occurring at

$$Q = 6\ kL/min \qquad \text{(Ans)}$$

with the corresponding actual head gained equal to 21 m.

Another concern is whether the pump is operating efficiently at the operating point. As can be seen from Fig. E12.4c, although this is not peak efficiency, which is about 86%, it is close (about 84%). Thus, this pump would be a satisfactory choice, assuming the 6 kL/min flowrate is at or near the desired flowrate.

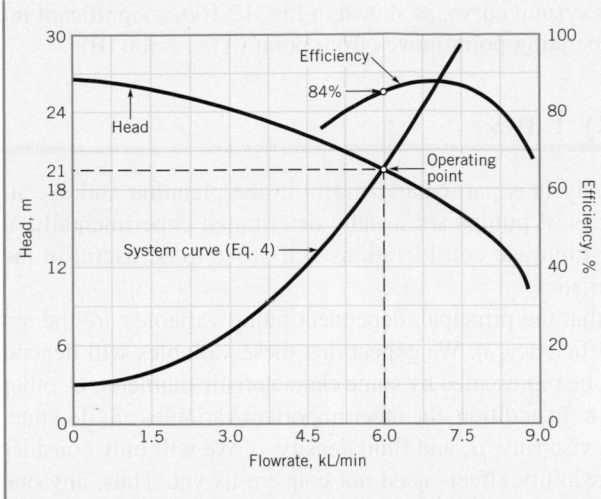

■ **FIGURE E12.4c (Continued)**

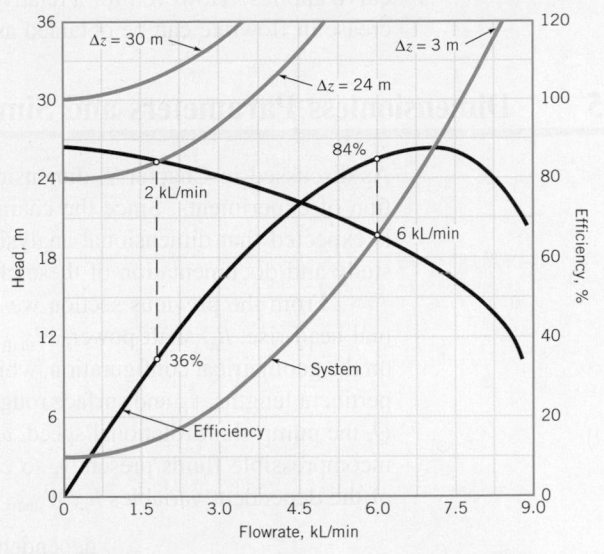

■ **FIGURE E12.4d**

The amount of pump head needed at the pump shaft is 21 m/0.84 = 25 m. The power needed to drive the pump is

$$\dot{W}_{\text{shaft}} = \frac{\gamma Q h_a}{\eta}$$

$$= \frac{(9800 \text{ N/m}^3)[(6 \text{ kL/min})/(1000 \text{ L/m}^3)(60 \text{ s/min})](21 \text{ m})}{0.84}$$

$$= 24{,}500 \text{ N} \cdot \text{m/s} = 24.5 \text{ kW}$$

COMMENT By repeating the calculations for $\Delta z = z_2 - z_1 = 24$ m and 30 m (rather than the given 3 m), the results shown in Fig. E12.4d are obtained. Although the given pump could be used with $\Delta z = 24$ m (provided that the 2 kL/min flowrate produced is acceptable), it would not be an ideal pump for this application since its efficiency would be only 36 percent. Energy could

be saved by using a different pump with a performance curve that more nearly matches the new system requirements (i.e., higher efficiency at the operating condition). On the other hand, the given pump would not work at all for $\Delta z = 30$ m since its maximum head ($h_a = 26.8$ m when $Q = 0$) is not enough to lift the water 30 m, let alone overcome head losses. This is shown in Fig. E12.4d by the fact that for $\Delta z = 30$ m the system curve and the pump performance curve do not intersect.

Note that head loss within the pump itself was accounted for with the pump efficiency, η. Thus, $h_s = h_a/\eta$, where h_s is the pump shaft work head and h_a is the actual head rise experienced by the flowing fluid.

For two pumps in series, add heads; for two in parallel, add flowrates.

Pumps can be arranged in series or in parallel to provide for additional head or flow capacity. When two pumps are placed in *series*, the resulting pump performance curve is obtained by adding heads at the same flowrate. As illustrated in Fig. 12.16a, for two identical pumps in series, both the actual head gained by the fluid and the flowrate are increased, but neither will be doubled if the system curve remains the same. The operating point is at (A) for one pump and moves to (B) for two pumps in series. For two identical pumps in *parallel*, the combined performance curve is obtained by adding flowrates at the same head, as shown in Fig. 12.16b. As illustrated, the flowrate for the system will not be doubled with the addition of two pumps in parallel (if the same system

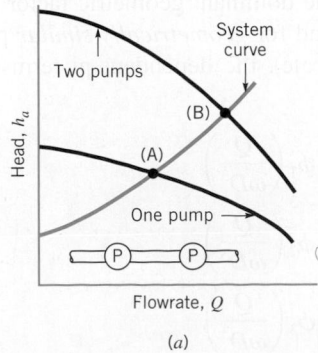

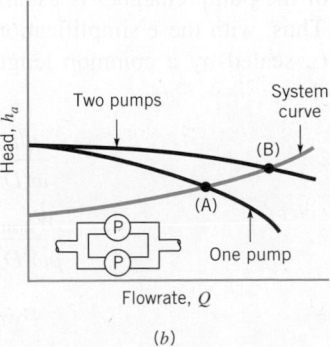

■ **FIGURE 12.16**
Effect of operating pumps in (a) series and (b) in parallel.

curve applies). However, for a relatively flat system curve, as shown in Fig. 12.16b, a significant increase in flowrate can be obtained as the operating point moves from point (A) to point (B).

12.5 Dimensionless Parameters and Similarity Laws

As discussed in Chapter 7, dimensional analysis is particularly useful in the planning and execution of experiments. Since the characteristics of pumps are usually determined experimentally, it is expected that dimensional analysis and similitude considerations will prove to be useful in the study and documentation of these characteristics.

From the previous section we know that the principal, dependent pump variables are the actual head rise, h_a, shaft power, $\dot{W}_{shaft}$, and efficiency, η. We expect that these variables will depend on the geometrical configuration, which can be represented by some characteristic diameter, D, other pertinent lengths, ℓ_i, and surface roughness, ε. In addition, the other important variables are flowrate, Q, the pump shaft rotational speed, ω, fluid viscosity, μ, and fluid density, ρ. We will only consider incompressible fluids presently, so compressibility effects need not concern us yet. Thus, any one of the dependent variables h_a, $\dot{W}_{shaft}$, and η can be expressed as

$$\text{dependent variable} = f(D, \ell_i, \varepsilon, Q, \omega, \mu, \rho)$$

and a straightforward application of dimensional analysis leads to

$$\text{dependent pi term} = \phi\left(\frac{\ell_i}{D}, \frac{\varepsilon}{D}, \frac{Q}{\omega D^3}, \frac{\rho\omega D^2}{\mu}\right) \tag{12.28}$$

Dimensionless pi terms and similarity laws are important pump considerations.

The dependent pi term involving the head is usually expressed as $C_H = gh_a/\omega^2 D^2$, where gh_a is the actual head rise in terms of energy per unit mass, rather than simply h_a, which is energy per unit weight. This dimensionless parameter is called the *head rise coefficient*. The dependent pi term involving the shaft power is expressed as $C_{\mathcal{P}} = \dot{W}_{shaft}/\rho\omega^3 D^5$, and this standard dimensionless parameter is termed the *power coefficient*. The rotational speed, ω, which appears in these dimensionless groups is expressed in rad/s. The final dependent pi term is the efficiency, η, which is already dimensionless. Thus, in terms of dimensionless parameters the performance characteristics are expressed as

$$C_H = \frac{gh_a}{\omega^2 D^2} = \phi_1\left(\frac{\ell_i}{D}, \frac{\varepsilon}{D}, \frac{Q}{\omega D^3}, \frac{\rho\omega D^2}{\mu}\right)$$

$$C_{\mathcal{P}} = \frac{\dot{W}_{shaft}}{\rho\omega^3 D^5} = \phi_2\left(\frac{\ell_i}{D}, \frac{\varepsilon}{D}, \frac{Q}{\omega D^3}, \frac{\rho\omega D^2}{\mu}\right)$$

$$\eta = \frac{\rho g Q h_a}{\dot{W}_{shaft}} = \phi_3\left(\frac{\ell_i}{D}, \frac{\varepsilon}{D}, \frac{Q}{\omega D^3}, \frac{\rho\omega D^2}{\mu}\right)$$

The last pi term in each of the above equations is a form of Reynolds number that represents the relative influence of viscous effects. When the pump flow involves high Reynolds numbers, as is usually the case, experience has shown that the effect of the Reynolds number can be neglected. For simplicity, the relative roughness, ε/D, can also be neglected in pumps since the highly irregular shape of the pump chamber is usually the dominant geometric factor rather than the surface roughness. Thus, with these simplifications and for *geometrically similar* pumps (all pertinent dimensions, ℓ_i, scaled by a common length scale), the dependent pi terms are functions of only $Q/\omega D^3$, so that

$$\frac{gh_a}{\omega^2 D^2} = \phi_1\left(\frac{Q}{\omega D^3}\right) \tag{12.29}$$

$$\frac{\dot{W}_{shaft}}{\rho\omega^3 D^5} = \phi_2\left(\frac{Q}{\omega D^3}\right) \tag{12.30}$$

$$\eta = \phi_3\left(\frac{Q}{\omega D^3}\right) \tag{12.31}$$

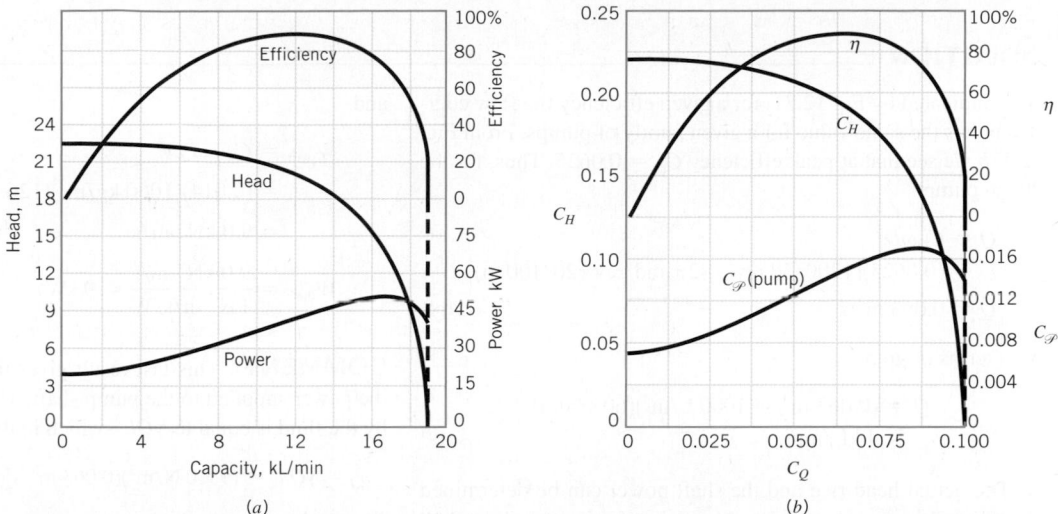

■ **FIGURE 12.17**   Typical performance data for a centrifugal pump: (*a*) characteristic curves for a 12-in. centrifugal pump operating at 1000 rpm, (*b*) dimensionless characteristic curves. (Data adapted from Ref. 8, used by permission.)

The dimensionless parameter $C_Q = Q/\omega D^3$ is called the *flow coefficient*. These three equations provide the desired similarity relationships among a family of geometrically similar pumps. If two pumps from the family are operated at the same value of flow coefficient

$$\left(\frac{Q}{\omega D^3}\right)_1 = \left(\frac{Q}{\omega D^3}\right)_2 \tag{12.32}$$

it then follows that

$$\left(\frac{gh_a}{\omega^2 D^2}\right)_1 = \left(\frac{gh_a}{\omega^2 D^2}\right)_2 \tag{12.33}$$

$$\left(\frac{\dot{W}_{\text{shaft}}}{\rho\omega^3 D^5}\right)_1 = \left(\frac{\dot{W}_{\text{shaft}}}{\rho\omega^3 D^5}\right)_2 \tag{12.34}$$

$$\eta_1 = \eta_2 \tag{12.35}$$

where the subscripts 1 and 2 refer to any two pumps from the family of geometrically similar pumps.

Pump scaling laws relate geometrically similar pumps.

With these so-called *pump scaling laws* it is possible to experimentally determine the performance characteristics of one pump in the laboratory and then use these data to predict the corresponding characteristics for other pumps within the family under different operating conditions. Figure 12.17*a* shows some typical curves obtained for a centrifugal pump. Figure 12.17*b* shows the results plotted in terms of the dimensionless coefficients, C_Q, C_H, $C_\mathscr{P}$, and η. From these curves the performance of different-sized, geometrically similar pumps can be predicted, as can the effect of changing speeds on the performance of the pump from which the curves were obtained. It is to be noted that the efficiency, η, is related to the other coefficients through the relationship $\eta = C_Q C_H C_\mathscr{P}^{-1}$. This follows directly from the definition of η.

*E*XAMPLE 12.5   Use of Pump Scaling Laws

GIVEN   An 20-cm-diameter centrifugal pump operating at 1200 rpm is geometrically similar to the 30-cm-diameter pump having the performance characteristics of Figs. 12.17*a* and 12.17*b* while operating at 1000 rpm. The working fluid is water at 15 °C.

FIND   For peak efficiency, predict the discharge, actual head rise, and shaft power for this smaller pump.

SOLUTION

As is indicated by Eq. 12.31, for a given efficiency the flow coefficient has the same value for a given family of pumps. From Fig. 12.17b we see that at peak efficiency $C_Q = 0.0625$. Thus, for the 20 cm pump

$$Q = C_Q \omega D^3$$
$$= (0.0625)(1200/60 \text{ rev/s})(2\pi \text{ rad/rev})(20/100 \text{ m})^3$$
$$Q = 0.063 \text{ m}^3/\text{s} \qquad \text{(Ans)}$$

or in terms of gpm

$$Q = (0.063 \text{ m}^3/\text{s})(1000 \text{ L/m}^3)(60 \text{ s/min})$$
$$= 4 \text{ kL/min} \qquad \text{(Ans)}$$

The actual head rise and the shaft power can be determined in a similar manner since at peak efficiency $C_H = 0.19$ and $C_\mathscr{P} = 0.014$, so that with $\omega = 1200 \text{ rev/min}(1 \text{ min/60 s})$ $(2\pi \text{ rad/rev}) = 126 \text{ rad/s}$

$$h_a = \frac{C_H \omega^2 D^2}{g} = \frac{(0.19)(126 \text{ rad/s})^2(20/100 \text{ m})^2}{9.81 \text{ m/s}^2} = 12.3 \text{ m} \quad \text{(Ans)}$$

and

$$\dot{W}_{\text{shaft}} = C_\mathscr{P} \rho \omega^3 D^5$$
$$= (0.014)(1000 \text{ kg/m}^3)(126 \text{ rad/s})^3(20/100 \text{ m})^5$$
$$= 9.0 \text{ kN} \cdot \text{m/s}$$
$$\dot{W}_{\text{shaft}} = \frac{9.0 \text{ kN} \cdot \text{m/s}}{1 \text{ N} \cdot \text{m/s/W}} = 9 \text{ kW} \qquad \text{(Ans)}$$

COMMENT   This last result gives the shaft power, which is the power supplied to the pump shaft. The power actually gained by the fluid is equal to $\gamma Q h_a$, which in this example is

$$\mathscr{P}_f = \gamma Q h_a = (9800 \text{ N/m}^3)(0.063 \text{ m}^3/\text{s})(12.3 \text{ m}) = 7.6 \text{ kN} \cdot \text{m/s}$$

Thus, the efficiency, η, is

$$\eta = \frac{\mathscr{P}_f}{\dot{W}_{\text{shaft}}} = \frac{7.6}{9} = 84\%$$

which checks with the efficiency curve of Fig. 12.17b.

12.5.1 Special Pump Scaling Laws

Effects of changes in pump operating speed and impeller diameter are often of interest.

Two special cases related to pump similitude commonly arise. In the first case we are interested in how a change in the operating speed, ω, for a *given pump*, affects pump characteristics. It follows from Eq. 12.32 that for the same flow coefficient (and therefore the same efficiency) with $D_1 = D_2$ (the same pump)

$$\frac{Q_1}{Q_2} = \frac{\omega_1}{\omega_2} \qquad (12.36)$$

The subscripts 1 and 2 now refer to the same pump operating at two different speeds at the same flow coefficient. Also, from Eqs. 12.33 and 12.34 it follows that

$$\frac{h_{a1}}{h_{a2}} = \frac{\omega_1^2}{\omega_2^2} \qquad (12.37)$$

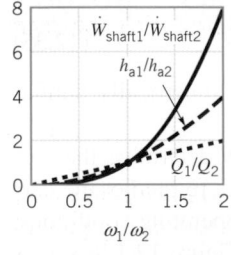

and

$$\frac{\dot{W}_{\text{shaft1}}}{\dot{W}_{\text{shaft2}}} = \frac{\omega_1^3}{\omega_2^3} \qquad (12.38)$$

Thus, for a given pump operating at a given flow coefficient, the flow varies directly with speed, the head varies as the speed squared, and the power varies as the speed cubed. These effects of angular velocity variation are illustrated in the sketch in the margin. These scaling laws are useful in estimating the effect of changing pump speed when some data are available from a pump test obtained by operating the pump at a particular speed.

In the second special case we are interested in how a change in the impeller diameter, D, of a geometrically similar family of pumps, operating at a *given speed*, affects pump characteristics. As before, it follows from Eq. 12.32 that for the same flow coefficient with $\omega_1 = \omega_2$

$$\frac{Q_1}{Q_2} = \frac{D_1^3}{D_2^3} \qquad (12.39)$$

Similarly, from Eqs. 12.33 and 12.34

$$\frac{h_{a1}}{h_{a2}} = \frac{D_1^2}{D_2^2} \qquad (12.40)$$

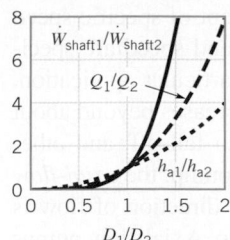

Pump affinity laws relate the same pump at different speeds or geometrically similar pumps at the same speed.

and

$$\frac{\dot{W}_{shaft1}}{\dot{W}_{shaft2}} = \frac{D_1^5}{D_2^5} \tag{12.41}$$

Thus, for a family of geometrically similar pumps operating at a given speed and the same flow coefficient, the flow varies as the diameter cubed, the head varies as the diameter squared, and the power varies as the diameter raised to the fifth power. These strong effects of diameter variation are illustrated in the sketch in the margin. These scaling relationships are based on the condition that, as the impeller diameter is changed, all other important geometric variables are properly scaled to maintain geometric similarity. This type of geometric scaling is not always possible due to practical difficulties associated with manufacturing the pumps. It is common practice for manufacturers to put impellers of different diameters in the same pump casing. In this case, complete geometric similarity is not maintained, and the scaling relationships expressed in Eqs. 12.39, 12.40, and 12.41 will not, in general, be valid. However, experience has shown that if the impeller diameter change is not too large, less than about 20%, these scaling relationships can still be used to estimate the effect of a change in the impeller diameter. The pump similarity laws expressed by Eqs. 12.36 through 12.41 are sometimes referred to as the *pump affinity laws*.

The effects of viscosity and surface roughness have been neglected in the foregoing similarity relationships. However, it has been found that as the pump size decreases these effects more significantly influence efficiency because of smaller clearances and blade size. An approximate, empirical relationship to estimate the influence of diminishing size on efficiency is (Ref. 9)

$$\frac{1 - \eta_2}{1 - \eta_1} \approx \left(\frac{D_1}{D_2}\right)^{1/5} \tag{12.42}$$

In general, it is to be expected that the similarity laws will not be very accurate if tests on a model pump with water are used to predict the performance of a prototype pump with a highly viscous fluid, such as oil, because at the much smaller Reynolds number associated with the oil flow, the fluid physics involved is different from the higher Reynolds number flow associated with water.

12.5.2 Specific Speed

A useful pi term can be obtained by eliminating diameter D between the flow coefficient and the head rise coefficient. This is accomplished by raising the flow coefficient to an appropriate exponent (1/2) and dividing this result by the head coefficient raised to another appropriate exponent (3/4) so that

$$\frac{(Q/\omega D^3)^{1/2}}{(gh_a/\omega^2 D^2)^{3/4}} = \frac{\omega\sqrt{Q}}{(gh_a)^{3/4}} = N_s \tag{12.43}$$

The dimensionless parameter N_s is called the *specific speed*. Specific speed varies with flow coefficient just as the other coefficients and efficiency discussed earlier do. However, for any pump it is customary to specify a value of specific speed at the flow coefficient corresponding to peak efficiency only. For pumps with low Q and high h_a, the specific speed is low compared to a pump with high Q and low h_a. Centrifugal pumps typically are low-capacity, high-head pumps, and therefore have low specific speeds.

Specific speed as defined by Eq. 12.43 is dimensionless, and therefore independent of the system of units used in its evaluation as long as a consistent unit system is used. However, in the United States a modified, dimensional form of specific speed, N_{sd}, is commonly used, where

$$N_{sd} = \frac{\omega(\text{rpm})\sqrt{Q(\text{gallons/min})}}{[h_a(\text{ft})]^{3/4}} \tag{12.44}$$

In this case N_{sd} is said to be expressed in *U.S. customary units*. Typical values of N_{sd} are in the range $500 < N_{sd} < 4000$ for centrifugal pumps. Both N_s and N_{sd} have the same physical meaning, but their magnitudes will differ by a constant conversion factor $(N_{sd} = 2733\,N_s)$ when ω in Eq. 12.43 is expressed in rad/s.

Each family or class of pumps has a particular range of values of specific speed associated with it. Thus, pumps that have low-capacity, high-head characteristics will have specific speeds that are

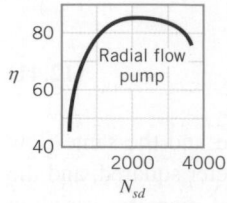

smaller than pumps that have high-capacity, low-head characteristics. The concept of specific speed is very useful to engineers and designers, since if the required head, flowrate, and speed are specified, it is possible to select an appropriate (most efficient) type of pump for a particular application. For example, as shown by the figure in the margin, as the specific speed, N_{sd}, increases beyond about 2000 the peak efficiency, η of the purely radial-flow centrifugal pump starts to fall off, and other types of more efficient pump design are preferred. In addition to the centrifugal pump, the *axial-flow* pump is widely used. As discussed in Section 12.6, in an axial-flow pump the direction of flow is primarily parallel to the rotating shaft rather than radial as in the centrifugal pump. Axial-flow pumps are essentially high-capacity, low-head pumps, and therefore have large specific speeds ($N_{sd} > 9000$) compared to centrifugal pumps. *Mixed-flow* pumps combine features of both radial-flow and axial-flow pumps and have intermediate values of specific speed. Figure 12.18 illustrates how the specific speed changes as the configuration of the pump changes from centrifugal or radial to axial.

12.5.3 Suction Specific Speed

With an analysis similar to that used to obtain the specific speed pi term, the *suction specific speed*, S_s, can be expressed as

$$S_s = \frac{\omega\sqrt{Q}}{[g(\text{NPSH}_R)]^{3/4}} \tag{12.45}$$

where h_a in Eq. 12.43 has been replaced by the required net positive suction head (NPSH_R). This dimensionless parameter is useful in determining the required operating conditions on the suction side of the pump. As was true for the specific speed, N_s, the value for S_s commonly used is for peak efficiency. For a family of geometrically similar pumps, S_s should have a fixed value. If this value is known, then the NPSH_R can be estimated for other pumps within the same family operating at different values of ω and Q.

Specific speed may be used to approximate what general pump geometry (axial, mixed or radial) to use for maximum efficiency.

As noted for N_s, the suction specific speed as defined by Eq. 12.45 is also dimensionless, and the value for S_s is independent of the system of units used. However, as was the case for specific speed, in the United States a modified dimensional form for the suction specific speed, designated as S_{sd}, is commonly used, where

$$S_{sd} = \frac{\omega(\text{rpm})\sqrt{Q(\text{gallons/min})}}{[\text{NPSH}_R\,(\text{ft})]^{3/4}} \tag{12.46}$$

For double-suction pumps the discharge, Q, in Eq. 12.46 is one-half the total discharge.

Typical values for S_{sd} fall in the range of 7000 to 12,000 (Ref. 11). If S_{sd} is specified, Eq. 12.46 can be used to estimate the NPSH_R for a given set of operating conditions. However, this calculation would generally only provide an approximate value for the NPSH_R, and the actual determination of the NPSH_R for a particular pump should be made through a direct measurement whenever possible. Note that $S_{sd} = 2733\,S_s$, with ω expressed in rad/s in Eq. 12.45.

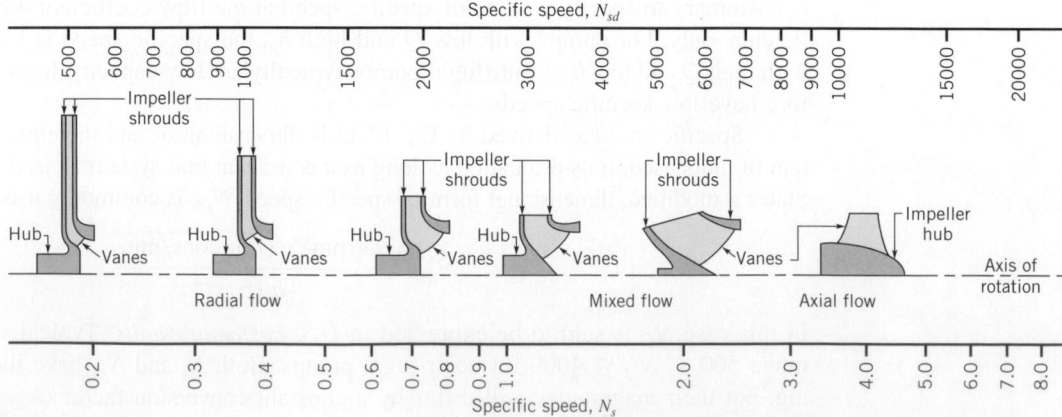

■ **FIGURE 12.18** Variation in specific speed at maximum efficiency with type of pump. (Adapted from Ref. 10, used with permission.)

12.6   Axial-Flow and Mixed-Flow Pumps

Axial-flow pumps often have alternating rows of stator blades and rotor blades.

As noted previously, centrifugal pumps are radial-flow machines that operate most efficiently for applications requiring high heads at relatively low flowrates. This head–flowrate combination typically yields specific speeds (N_s) that are less than approximately 1.5. For many applications, such as those associated with drainage and irrigation, high flowrates at low heads are required and centrifugal pumps are not suitable. In this case, axial-flow pumps are commonly used. This type of pump consists essentially of a propeller confined within a cylindrical casing. Axial-flow pumps are often called *propeller pumps*. For this type of pump the flow is primarily in the axial direction (parallel to the axis of rotation of the shaft), as opposed to the radial flow found in the centrifugal pump. Whereas the head developed by a centrifugal pump includes a contribution due to centrifugal action, the head developed by an axial-flow pump is due primarily to the tangential force exerted by the rotor blades on the fluid. A schematic of an axial-flow pump arranged for vertical operation is shown in Fig. 12.19. The rotor is connected to a motor through a shaft, and as it rotates (usually at a relatively high speed) the fluid is sucked in through the inlet. Typically the fluid discharges through a row of fixed stator (guide) vanes used to straighten the flow leaving the rotor. Some axial-flow pumps also have inlet guide vanes upstream of the rotor row, and some are multistage in which pairs (*stages*) of rotating blades (*rotor blades*) and fixed vanes (*stator blades*) are arranged in series. Axial-flow pumps usually have specific speeds (N_s) in excess of 3.3.

The definitions and broad concepts that were developed for centrifugal pumps are also applicable to axial-flow pumps. The actual flow characteristics, however, are quite different. In Fig. 12.20 typical head, power, and efficiency characteristics are compared for a centrifugal pump and an axial-flow pump. It is noted that at design capacity (maximum efficiency) the head and power are the same for the two pumps selected. But as the flowrate decreases, the power input to the centrifugal pump falls to 134 kW at shutoff, whereas for the axial-flow pump the power input increases to 388 kW at shutoff. This characteristic of the axial-flow pump can cause overloading of the drive motor if the flowrate is reduced significantly from the design capacity. It is also noted that the head curve for the axial-flow pump is much steeper than that for the centrifugal pump. Thus, with axial-flow pumps there will be a large change in head with a small change in the flowrate, whereas for the centrifugal pump, with its relatively flat head curve, there will be only a small change in head with large changes in the flowrate. It is further observed from Fig. 12.20 that, except at design capacity, the efficiency of the axial-flow pump is lower than that of the centrifugal pump. To improve operating characteristics, some axial-flow pumps are constructed with adjustable blades.

For applications requiring specific speeds intermediate to those for centrifugal and axial-flow pumps, mixed-flow pumps have been developed that operate efficiently in the specific speed range $1.5 < N_s < 3.3$. As the name implies, the flow in a mixed-flow pump has both a radial and an axial component. Figure 12.21 shows some typical data for centrifugal, mixed-flow, and axial-flow pumps, each operating with the same flowrate. These data indicate that as we proceed from the centrifugal pump to the mixed-flow pump to the axial-flow pump, the specific speed increases, the

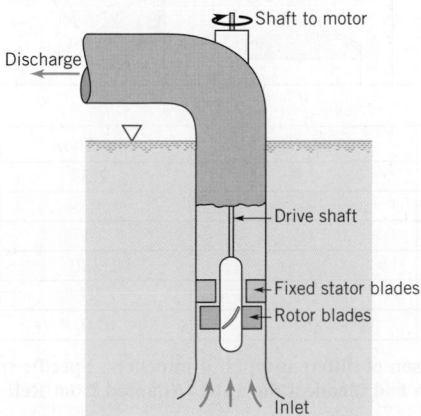

■ **FIGURE 12.19**   Schematic diagram of an axial-flow pump arranged for vertical operation.

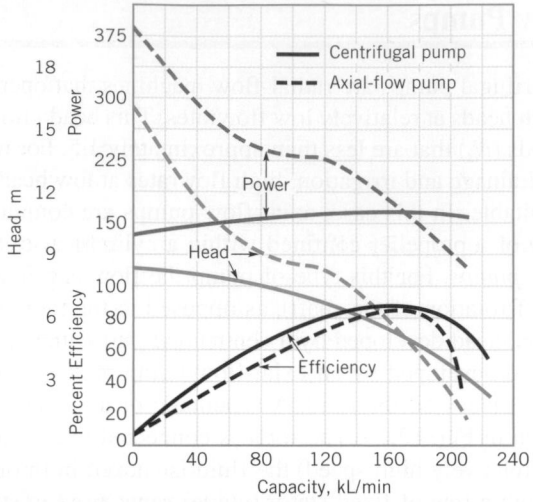

■ **F I G U R E 12.20** Comparison of performance characteristics for a centrifugal pump and an axial-flow pump, each rated 159 L/min at a 5-m head. (Data adapted from Ref. 12, used with permission.)

head decreases, the speed increases, the impeller diameter decreases, and the eye diameter increases. These general trends are commonly found when these three types of pumps are compared.

The dimensionless parameters and scaling relationships developed in the previous sections apply to all three types of pumps—centrifugal, mixed-flow, and axial-flow—since the dimensional analysis used is not restricted to a particular type of pump. Additional information about pumps can be found in Refs. 4, 7, 9, 12, and 13.

F l u i d s   i n   t h e   N e w s

Mechanical heart assist devices As with any pump, the human heart can suffer various malfunctions and problems during its useful life. Recent developments in artificial heart technology may be able to provide help to those whose pumps have broken down beyond repair. One of the more promising techniques is use of a left-ventricular assist device (LVAD), which supplements a diseased heart. Rather than replacing a diseased heart, an LVAD pump is implanted alongside the heart and works in parallel with the cardiovascular system to assist the pumping function of the heart's left ventricle. (The left ventricle supplies oxygenated blood to the entire body and performs about 80% of the heart's work.) Some LVADs are *centrifugal or axial flow pumps* that provide a continuous flow of blood. The continuous flow may take some adjustment on the part of patients, who do not hear a pulse or a heartbeat. Despite advances in artificial heart technology, it is probably still several years before fully implantable, quiet, and reliable devices will be considered for widespread use.

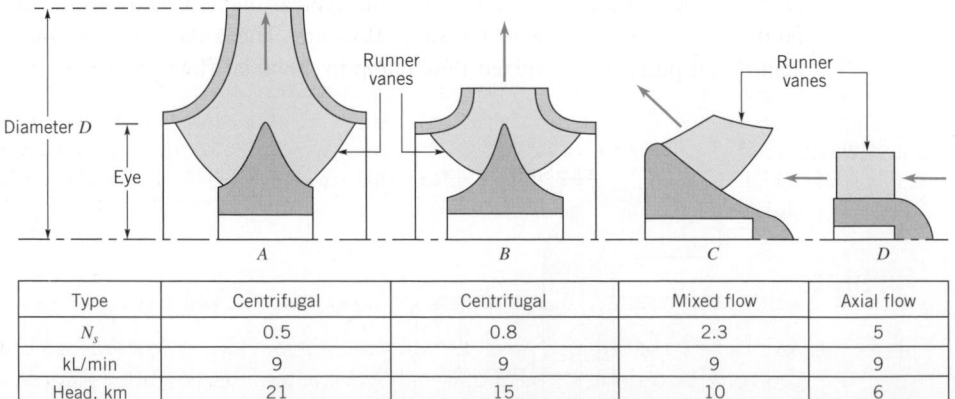

Type	Centrifugal	Centrifugal	Mixed flow	Axial flow
N_s	0.5	0.8	2.3	5
kL/min	9	9	9	9
Head, km	21	15	10	6
Rpm	870	1,160	1,750	2,600
D, cm	48	30	25	18
D_{eye}/D	0.5	0.7	0.9	1.0

■ **F I G U R E 12.21** Comparison of different types of impellers. Specific speed for centrifugal pumps based on single suction and identical flowrate. (Adapted from Ref. 12, used with permission.)

12.7   Fans

Fans are used to pump air and other gases and vapors.

When the fluid to be moved is air, or some other gas or vapor, *fans* are commonly used. Types of fans vary from the small fan used for cooling desktop computers to large fans used in many industrial applications such as ventilating of large buildings. Fans typically operate at relatively low rotation speeds and are capable of moving large volumes of gas. Although the fluid of interest is a gas, the change in gas density through the fan does not usually exceed 7%, which for air represents a change in pressure of only about 7 kPa (Ref. 14). Thus, in dealing with fans, the gas density is treated as a constant, and the flow analysis is based on incompressible flow concepts. Because of the low pressure rise involved, fans are often constructed of lightweight sheet metal. Fans are also called *blowers*, *boosters*, and *exhausters* depending on the location within the system; that is, blowers are located at the system entrance, exhausters are at the system exit, and boosters are located at some intermediate position within the system. Turbomachines used to produce larger changes in gas density and pressure than possible with fans are called *compressors* (see Section 12.9.1).

As is the case for pumps, fan designs include centrifugal (radial-flow) fans, as well as mixed-flow and axial-flow (propeller) fans, and the analysis of fan performance closely follows that previously described for pumps. The shapes of typical performance curves for centrifugal and axial-flow fans are quite similar to those shown in Fig. 12.20 for centrifugal and axial-flow pumps. However, fan head-rise data are often given in terms of pressure rise, either static or total, rather than the more conventional head rise commonly used for pumps.

Scaling relationships for fans are the same as those developed for pumps, that is, Eqs. 12.32 through 12.35 apply to fans as well as pumps. As noted above, for fans it is common to replace the head, h_a, in Eq. 12.33 with pressure head, $p_a/\rho g$, so that Eq. 12.33 becomes

$$\left(\frac{p_a}{\rho\omega^2 D^2}\right)_1 = \left(\frac{p_a}{\rho\omega^2 D^2}\right)_2 \tag{12.47}$$

where, as before, the subscripts 1 and 2 refer to any two fans from the family of geometrically similar fans. Equations 12.47, 12.32 and 12.34, are called the *fan laws* and can be used to scale performance characteristics between members of a family of geometrically similar fans. Additional information about fans can be found in Refs. 14–17.

F l u i d s   i n   t h e   N e w s

Hi-tech ceiling fans Energy savings of up to 25% can be realized if thermostats in air-conditioned homes are raised by a few degrees. This can be accomplished by using ceiling *fans* and taking advantage of the increased sensible cooling brought on by air moving over skin. If the energy used to run the fans can be reduced, additional energy savings can be realized. Most ceiling fans use flat, fixed pitch, nonaerodynamic *blades* with uniform chord length. Because the tip of a paddle moves through air faster than its root does, airflow over such fan blades is lowest near the hub and highest at the tip. By making the fan blade more propeller-like, it is possible to have a more uniform, efficient distribution. However, since ceiling fans are restricted by law to operate at less than 200 rpm, ordinary airplane propeller design is not appropriate. After considerable design effort, a highly efficient ceiling fan capable of delivering the same airflow as the conventional design with only half the power has been successfully developed and marketed. The fan blades are based on the slowly turning prop used in the *Gossamer Albatross*, the human-powered aircraft that flew across the English Channel in 1979. (See Problem 12.58.)

12.8   Turbines

As discussed in Section 12.2, turbines are devices that extract energy from a flowing fluid. The geometry of turbines is such that the fluid exerts a torque on the rotor in the direction of its rotation. The shaft power generated is available to drive generators or other devices.

In the following sections we discuss mainly the operation of hydraulic turbines (those for which the working fluid is water) and to a lesser extent gas and steam turbines (those for which the density of the working fluid may be much different at the inlet than at the outlet).

Although there are numerous ingenious hydraulic turbine designs, most of these turbines can be classified into two basic types—*impulse turbines* and *reaction turbines*. (Reaction is related to

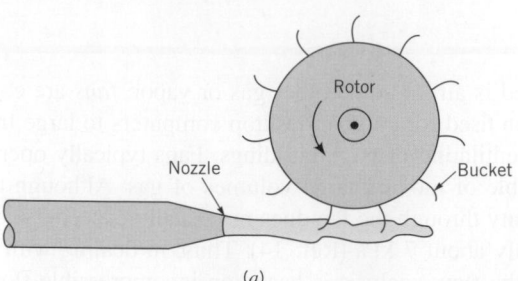

(a)

(b)

■ **F I G U R E 12.22** *(a)* **Schematic diagram of a Pelton wheel turbine,** *(b)* **photograph of a Pelton wheel turbine. (Courtesy of Voith Hydro, York, PA.)**

The two basic types of hydraulic turbines are impulse and reaction.

the ratio of static pressure drop that occurs across the rotor to static pressure drop across the turbine stage, with larger rotor pressure drop corresponding to larger reaction.) For hydraulic impulse turbines, the pressure drop across the rotor is zero; all of the pressure drop across the turbine stage occurs in the nozzle row. The *Pelton wheel* shown in Fig. 12.22 is a classical example of an impulse turbine. In these machines the total head of the incoming fluid (the sum of the pressure head, velocity head, and elevation head) is converted into a large velocity head at the exit of the supply nozzle (or nozzles if a multiple nozzle configuration is used). Both the pressure drop across the bucket (blade) and the change in relative speed (i.e., fluid speed relative to the moving bucket) of the fluid across the bucket are negligible. The space surrounding the rotor is not completely filled with fluid. It is the impulse of the individual jets of fluid striking the buckets that generates the torque.

For reaction turbines, on the other hand, the rotor is surrounded by a casing (or volute), which is completely filled with the working fluid. There is both a pressure drop and a fluid relative speed change across the rotor. As shown for the radial-inflow turbine in Fig 12.23, guide vanes act as nozzles to accelerate the flow and turn it in the appropriate direction as the fluid enters the rotor. Thus, part of the pressure drop occurs across the guide vanes and part occurs across the rotor. In many respects the operation of a reaction turbine is similar to that of a pump "flowing backward," although such oversimplification can be quite misleading.

Both impulse and reaction turbines can be analyzed using the moment-of-momentum principles discussed in Section 12.3. In general, impulse turbines are high-head, low-flowrate devices, while reaction turbines are low-head, high-flowrate devices.

12.8.1 Impulse Turbines

Although there are various types of impulse turbine designs, perhaps the easiest to understand is the Pelton wheel (see Fig. 12.24). Lester Pelton (1829–1908), an American mining engineer during the

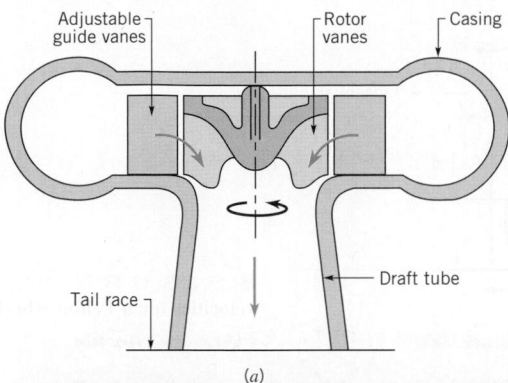

Adjustable guide vanes — Rotor vanes — Casing

Tail race — Draft tube

(a)

(b)

■ **F I G U R E   12.23** (*a*) **Schematic diagram of a reaction turbine,**
(*b*) **photograph of a reaction turbine. (Courtesy of Voith Hydro, York, PA.)**

California gold-mining days, is responsible for many of the still-used features of this type of turbine. It is most efficient when operated with a large head (for example, a water source from a lake located significantly above the turbine nozzle), which is converted into a relatively large velocity at the exit of the nozzle. Among the many design considerations for such a turbine are the head loss that occurs in the pipe (the penstock) transporting the water to the turbine, the design of the nozzle, and the design of the buckets on the rotor.

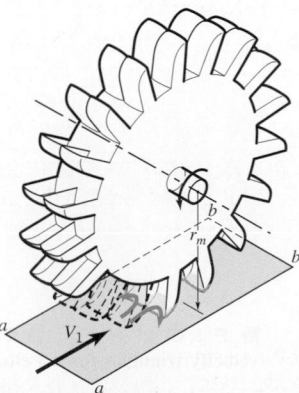

■ **F I G U R E   12.24** **Details of Pelton wheel turbine bucket.**

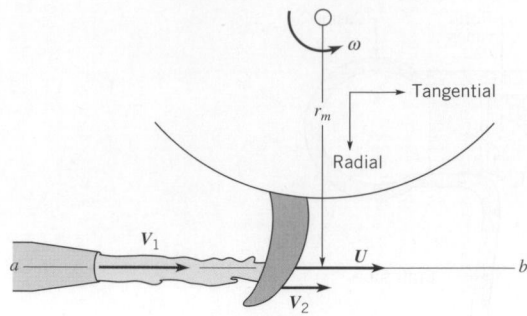

■ **F I G U R E   12.25**   **Ideal fluid velocities for a Pelton wheel turbine.**

As shown in Fig. 12.24, a high-speed jet of water strikes the Pelton wheel buckets and is deflected. The water enters and leaves the control volume surrounding the wheel as free jets (atmospheric pressure). In addition, a person riding on the bucket would note that the speed of the water does not change as it slides across the buckets (assuming viscous effects are negligible). That is, the magnitude of the relative velocity does not change, but its direction does. The change in direction of the velocity of the fluid jet causes a torque on the rotor, resulting in a power output from the turbine.

Design of the optimum, complex shape of the buckets to obtain maximum power output is a very difficult matter. Ideally, the fluid enters and leaves the control volume shown in Fig. 12.25 with no radial component of velocity. (In practice there often is a small but negligible radial component.) In addition, the buckets would ideally turn the relative velocity vector through a 180° turn, but physical constraints dictate that β, the angle of the exit edge of the blade, is less than 180°. Thus, the fluid leaves with an axial component of velocity as shown in Fig. 12.26.

The inlet and exit velocity triangles at the arithmetic mean radius, r_m, are assumed to be as shown in Fig. 12.27. To calculate the torque and power, we must know the tangential components of the absolute velocities at the inlet and exit. (Recall from the discussion in Section 12.3 that neither the radial nor the axial components of velocity enter into the torque or power equations.) From Fig. 12.27 we see that

$$V_{\theta 1} = V_1 = W_1 + U \tag{12.48}$$

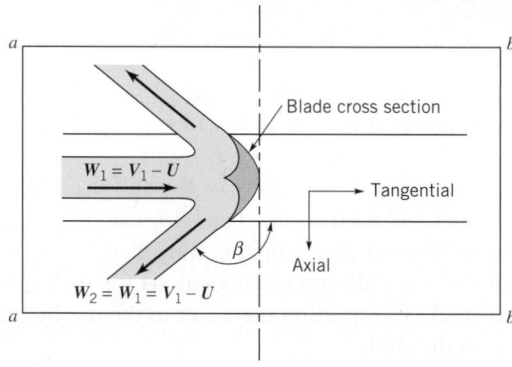

■ **F I G U R E   12.26**   **Flow as viewed by an observer riding on the Pelton wheel—relative velocities.**

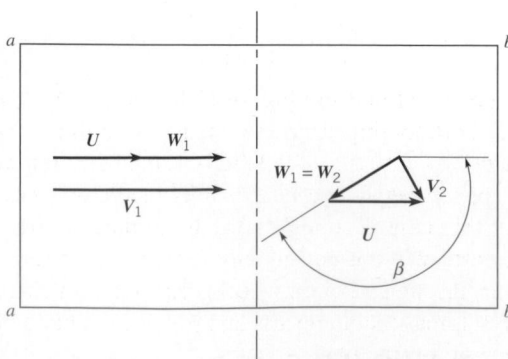

■ **F I G U R E   12.27**   **Inlet and exit velocity triangles for a Pelton wheel turbine.**

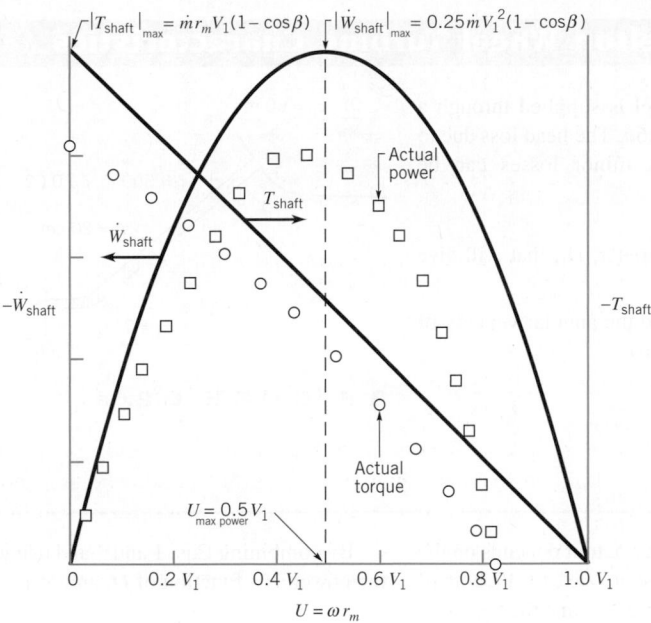

■ **FIGURE 12.28** Typical theoretical and experimental power and torque for a Pelton wheel turbine as a function of bucket speed.

and

$$V_{\theta 2} = W_2 \cos \beta + U \tag{12.49}$$

In Pelton wheel analyses, we assume the relative speed of the fluid is constant (no friction).

Thus, with the assumption that $W_1 = W_2$ (i.e., the relative speed of the fluid does not change as it is deflected by the buckets), we can combine Eqs. 12.48 and 12.49 to obtain

$$V_{\theta 2} - V_{\theta 1} = (U - V_1)(1 - \cos \beta) \tag{12.50}$$

This change in tangential component of velocity combined with the torque and power equations developed in Section 12.3 (i.e., Eqs. 12.2 and 12.4) gives

$$T_{\text{shaft}} = \dot{m} r_m (U - V_1)(1 - \cos \beta)$$

where $\dot{m} = \rho Q$ is the mass flowrate through the turbine. Since $U = \omega R_m$, it follows that

$$\dot{W}_{\text{shaft}} = T_{\text{shaft}} \omega = \dot{m} U (U - V_1)(1 - \cos \beta) \tag{12.51}$$

These results are plotted in Fig. 12.28 along with typical experimental results. Note that $V_1 > U$ (i.e., the jet impacts the bucket), and $\dot{W}_{\text{shaft}} < 0$ (i.e., the turbine extracts power from the fluid).

Several interesting points can be noted from the above results. First, the power is a function of β. However, a typical value of $\beta = 165°$ (rather than the optimum 180°) results in a relatively small (less than 2%) reduction in power since $1 - \cos 165° = 1.966$, compared to $1 - \cos 180° = 2$. Second, although the torque is maximum when the wheel is stopped ($U = 0$), there is no power under this condition—to extract power one needs force and motion. On the other hand, the power output is a maximum when

$$U|_{\text{max power}} = \frac{V_1}{2} \tag{12.52}$$

This can be shown by using Eq. 12.51 and solving for U that gives $d\dot{W}_{\text{shaft}}/dU = 0$. A bucket speed of one-half the speed of the fluid coming from the nozzle gives the maximum power. Third, the maximum speed occurs when $T_{\text{shaft}} = 0$ (i.e., the load is completely removed from the turbine, as would happen if the shaft connecting the turbine to the generator were to break and frictional torques were negligible). For this case $U = \omega R = V_1$, the turbine is "free wheeling," and the water simply passes across the rotor without putting any force on the buckets.

Although the actual flow through a Pelton wheel is considerably more complex than assumed in the above simplified analysis, reasonable results and trends are obtained by this simple application of the moment-of-momentum principle.

EXAMPLE 12.6 Pelton Wheel Turbine Characteristics

GIVEN Water to drive a Pelton wheel is supplied through a pipe from a lake as indicated in Fig. E12.6a. The head loss due to friction in the pipe is important, but minor losses can be neglected.

FIND (a) Determine the nozzle diameter, D_1, that will give the maximum power output.

(b) Determine the maximum power and the angular velocity of the rotor at the conditions found in part (a).

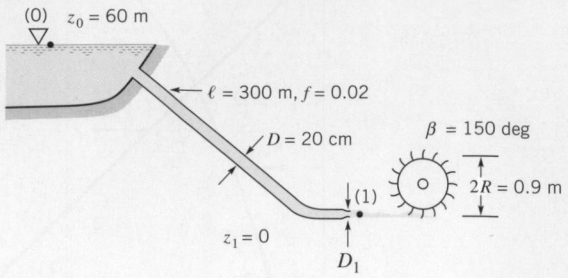

■ **FIGURE E12.6a**

SOLUTION

(a) As indicated by Eq. 12.51, the power output depends on the flowrate, $Q = \dot{m}/\rho$, and the jet speed at the nozzle exit, V_1, both of which depend on the diameter of the nozzle, D_1, and the head loss associated with the supply pipe. That is

$$\dot{W}_{\text{shaft}} = \rho Q U(U - V_1)(1 - \cos\beta) \quad (1)$$

The nozzle exit speed, V_1, can be obtained by applying the energy equation (Eq. 5.85) between a point on the lake surface (where $V_0 = p_0 = 0$) and the nozzle outlet (where $z_1 = p_1 = 0$) to give

$$z_0 = \frac{V_1^2}{2g} + h_L \quad (2)$$

where the head loss is given in terms of the friction factor, f, as (see Eq. 8.34)

$$h_L = f \frac{\ell}{D} \frac{V^2}{2g}$$

The speed, V, of the fluid in the pipe of diameter D is obtained from the continuity equation

$$V = \frac{A_1 V_1}{A} = \left(\frac{D_1}{D}\right)^2 V_1$$

We have neglected minor losses associated with the pipe entrance and the nozzle. With the given data, Eq. 2 becomes

$$z_0 = \left[1 + f \frac{\ell}{D}\left(\frac{D_1}{D}\right)^4\right] \frac{V_1^2}{2g} \quad (3)$$

or

$$V_1 = \left[\frac{2g z_0}{1 + f \frac{\ell}{D}\left(\frac{D_1}{D}\right)^4}\right]^{1/2}$$

$$= \left[\frac{2(9.81 \text{ m/s}^2)(60 \text{ m})}{1 + 0.02\left(\frac{300 \text{ m}}{20/100 \text{ m}}\right)\left(\frac{D_1}{20/100}\right)^4}\right]^{1/2}$$

$$= \frac{34.3}{\sqrt{1 + 18{,}750 \, D_1^4}} \quad (4)$$

where D_1 is in meters.

By combining Eqs. 1 and 4 and using $Q = \pi D_1^2 V_1/4$ we obtain the power as a function of D_1 and U as

$$\dot{W}_{\text{shaft}} = \frac{5.02 \times 10^4 \, U D_1^2}{\sqrt{1 + 18{,}750 \, D_1^4}}\left[U - \frac{34.3}{\sqrt{1 + 18{,}750 \, D_1^4}}\right] \quad (5)$$

where U is in meters per second and $\dot{W}_{\text{shaft}}$ is in N · m/s. These results are plotted as a function of U for various values of D_1 in Fig. 12.6b.

As shown by Eq. 12.52, the maximum power (in terms of its variation with U) occurs when $U = V_1/2$, which, when used with Eqs. 4 and 5, gives

$$\dot{W}_{\text{shaft}} = -\frac{15 \times 10^6 \, D_1^2}{(1 + 18{,}750 \, D_1^4)^{3/2}} \quad (6)$$

The maximum power possible occurs when $d\dot{W}_{\text{shaft}}/dD_1 = 0$, which according to Eq. 6 can be found as

$$\frac{d\dot{W}_{\text{shaft}}}{dD_1} = -15 \times 10^6 \left[\frac{2 \, D_1}{(1 + 18{,}750 \, D_1^4)^{3/2}}\right.$$

$$\left. - \left(\frac{3}{2}\right)\frac{4(18{,}750) \, D_1^5}{(1 + 18{,}750 \, D_1^4)^{5/2}}\right] = 0$$

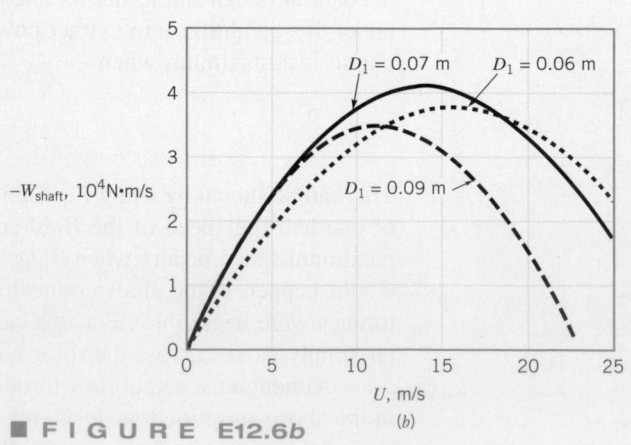

■ **FIGURE E12.6b**

(b)

or

$$37{,}250\, D_1^4 = 1$$

Thus, the nozzle diameter for maximum power output is

$$D_1 = 0.07 \text{ m} \qquad \text{(Ans)}$$

(b)   The corresponding maximum power can be determined from Eq. 6 as

$$\dot{W}_{\text{shaft}} = -\frac{15 \times 10^6 \,(0.07)^2}{[1 + 18{,}750(0.07)^4]^{3/2}} = -4.2 \times 10^4 \text{ N} \cdot \text{m/s}$$

or

$$\dot{W}_{\text{shaft}} = -4.2 \times 10^4 \text{ N} \cdot \text{m/s} \times \frac{1 \text{ W}}{1 \text{ N} \cdot \text{m/s}}$$
$$= -42 \text{ kW} \qquad \text{(Ans)}$$

The rotor speed at the maximum power condition can be obtained from

$$U = \omega R = \frac{V_1}{2}$$

where V_1 is given by Eq. 4. Thus,

$$\omega = \frac{V_1}{2R} = \frac{\dfrac{34.3}{\sqrt{1 + 18{,}750(0.07)^4}} \text{ m/s}}{2\left(\dfrac{0.9}{2} \text{ m}\right)}$$

$$= 31.6 \text{ rad/s} \times 1 \text{ rev}/2\pi \text{ rad} \times 60 \text{ s/min}$$
$$= 302 \text{ rpm} \qquad \text{(Ans)}$$

COMMENT   The reason that an optimum diameter nozzle exists can be explained as follows. A larger diameter nozzle will allow a larger flowrate, but will produce a smaller jet velocity because of the head loss within the supply side. A smaller diameter nozzle will reduce the flowrate but will produce a larger jet velocity. Since the power depends on a product combination of flowrate and jet velocity (see Eq. 1), there is an optimum-diameter nozzle that gives the maximum power.

These results can be generalized (i.e., without regard to the specific parameter values of this problem) by considering Eqs. 1 and 3 and the condition that $U = V_1/2$ to obtain

$$\dot{W}_{\text{shaft}\,|U = V_1/2} = -\frac{\pi}{16}\, \rho(1 - \cos \beta)$$
$$\times (2gz_0)^{3/2}\, D_1^2 \Big/ \left(1 + f\frac{\ell}{D^5} D_1^4\right)^{3/2}$$

By setting $d\dot{W}_{\text{shaft}}/dD_1 = 0$, it can be shown (see Problem 12.67) that the maximum power occurs when

$$D_1 = D \Big/ \left(2f\frac{\ell}{D}\right)^{1/4}$$

which gives the same results obtained earlier for the specific parameters of the example problem. Note that the optimum condition depends only on the friction factor and the length-to-diameter ratio of the supply pipe. What happens if the supply pipe is frictionless or of essentially zero length?

In previous chapters we mainly treated turbines (and pumps) as "black boxes" in the flow that removed (or added) energy to the fluid. We treated these devices as objects that removed a certain shaft work head from or added a certain shaft work head to the fluid. The relationship between the shaft work head and the power output as described by the moment-of-momentum considerations is illustrated in Example 12.7.

EXAMPLE 12.7   **Maximum Power Output for a Pelton Wheel Turbine**

GIVEN   Water flows through the Pelton wheel turbine shown in Fig. 12.24. For simplicity we assume that the water is turned 180° by the blade.

FIND   Show, based on the energy equation (Eq. 5.84), that the maximum power output occurs when the absolute velocity of the fluid exiting the turbine is zero.

SOLUTION

As indicated by Eq. 12.51, the shaft power of the turbine is given by

$$\dot{W}_{\text{shaft}} = \rho Q U (U - V_1)(1 - \cos \beta)$$
$$= 2\rho Q(U^2 - V_1 U) \qquad (1)$$

For this impulse turbine with $\beta = 180°$, the velocity triangles simplify into the diagram types shown in Fig. E12.7. Three possibilities are indicated:

(a)   the exit absolute velocity, $\mathbf{V}_2$, is directed back toward the nozzle,

(b)   the absolute velocity at the exit is zero, or

(c)   the exiting stream flows in the direction of the incoming stream.

According to Eq. 12.52, the maximum power occurs when $U = V_1/2$, which corresponds to the situation shown in Fig. E12.7b, that is, $U = V_1/2 = W_1$. If viscous effects are negligible, then $W_1 = W_2$ and we have $U = W_2$, which gives

$$V_2 = 0 \qquad \text{(Ans)}$$

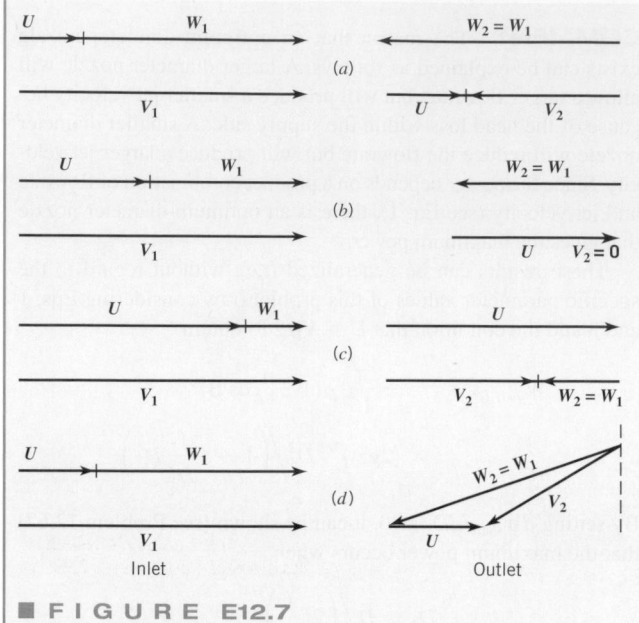

■ F I G U R E E12.7

If we consider the energy equation (Eq. 5.84) for flow across the rotor we have

$$\frac{p_1}{\gamma} + \frac{V_1^2}{2g} + z_1 + h_S = \frac{p_2}{\gamma} + \frac{V_2^2}{2g} + z_2 + h_L$$

where h_S is the shaft work head. This simplifies to

$$h_S = \frac{V_2^2 - V_1^2}{2g} + h_L \tag{2}$$

since $p_1 = p_2$ and $z_1 = z_2$. Note that the impulse turbine obtains its energy from a reduction in the velocity head. The largest shaft work head possible (and therefore the largest power) occurs when all of the kinetic energy available is extracted by the turbine, giving

$$V_2 = 0 \qquad\qquad \textbf{(Ans)}$$

This is consistent with the maximum power condition represented by Fig. E12.7b.

COMMENT As indicated by Eq. 1, if the exit absolute velocity is not in the plane of the rotor (i.e., $\beta < 180°$), there is a reduction in the power available (by a factor of $1 - \cos\beta$). This is also supported by the energy equation, Eq. 2, as follows. For $\beta < 180°$ the inlet and exit velocity triangles are as shown in Fig. E12.7d. Regardless of the bucket speed, U, it is not possible to reduce the value of V_2 to zero—there is always a component in the axial direction. Thus, according to Eq. 2, the turbine cannot extract the entire velocity head; the exiting fluid has some kinetic energy left in it.

A second type of impulse turbine that is widely used (most often with air as the working fluid) is indicated in Fig. 12.29. A circumferential series of fluid jets strikes the rotating blades which, as with the Pelton wheel, alter both the direction and magnitude of the absolute velocity. As with the Pelton wheel, the inlet and exit pressures (i.e., on either side of the rotor) are equal, and the magnitude of the relative velocity is unchanged as the fluid slides across the blades (if frictional effects are negligible).

Dentist drill turbines are usually of the impulse class.

Typical inlet and exit velocity triangles (absolute, relative, and blade velocities) are shown in Fig. 12.30. As discussed in Section 12.2, in order for the absolute velocity of the fluid to be changed

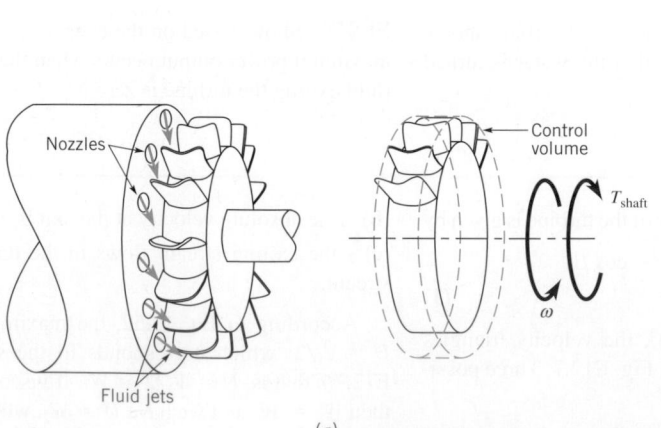

(a)

(b)

■ F I G U R E 12.29 **A multinozzle, non-Pelton wheel impulse turbine commonly used with air as the working fluid.**

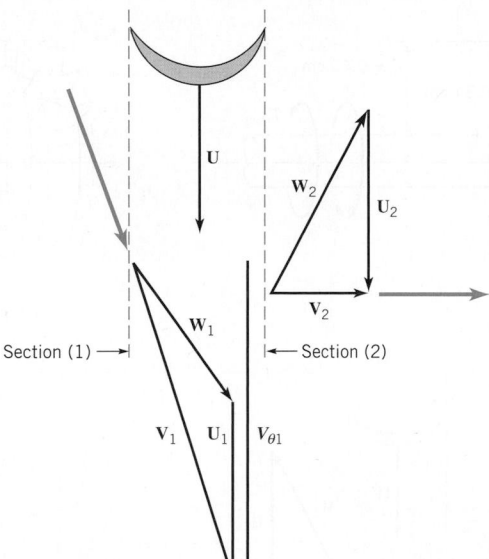

■ **F I G U R E 12.30** Inlet and exit velocity triangles for the impulse turbine shown in Fig. 12.29.

as indicated during its passage across the blade, the blade must push on the fluid in the direction opposite of the blade motion. Hence, the fluid pushes on the blade in the direction of the blade's motion—the fluid does work on the blade (a turbine).

EXAMPLE 12.8 | Non-Pelton Wheel Impulse Turbine (Dental Drill)

GIVEN An air turbine used to drive the high-speed drill used by your dentist is shown in Figs. 12.29 and E12.8a. Air exiting from the upstream nozzle holes forces the turbine blades to move in the direction shown. The turbine rotor speed is 300,000 rpm, the tangential component of velocity out of the nozzle is twice the blade speed, and the tangential component of the absolute velocity out of the rotor is zero.

FIND Estimate the shaft energy per unit mass of air flowing through the turbine.

SOLUTION

We use the fixed, nondeforming control volume that includes the turbine rotor and the fluid in the rotor blade passages at an instant of time (see Fig. E12.8b). The only torque acting on this control volume is the shaft torque. For simplicity we analyze this problem using an arithmetic mean radius, r_m, where

$$r_m = \frac{1}{2}(r_0 + r_i)$$

A sketch of the velocity triangles at the rotor entrance and exit is shown in Fig. E12.8c.

Application of Eq. 12.5 (a form of the moment-of-momentum equation) gives

$$w_{\text{shaft}} = -U_1 V_{\theta 1} + U_2 V_{\theta 2} \tag{1}$$

where w_{shaft} is shaft energy per unit of mass flowing through the turbine. From the problem statement, $V_{\theta 1} = 2U$ and $V_{\theta 2} = 0$, where

$$U = \omega r_m = (300{,}000 \text{ rev/min})(1 \text{ min/60 s})(2\pi \text{ rad/rev})$$
$$\times (0.43 \text{ cm} + 0.34 \text{ cm})/2(100 \text{ cm/m}) \tag{2}$$
$$= 121 \text{ m/s}$$

is the mean-radius blade velocity. Thus, Eq. (1) becomes

$$\begin{aligned}
w_{\text{shaft}} &= -U_1 V_{\theta 1} = -2U^2 = -2(121 \text{ m/s})^2 \\
&= -293 \times 10^3 \text{ m}^2/\text{s}^2 \\
&= (-29.3 \times 10^3 \text{ m}^2/\text{s}^2)/(1(\text{m} \cdot \text{kg})/(\text{N} \cdot \text{s}^2)) \\
&= -29 \text{ kN} \cdot \text{m/kg} \qquad \textbf{(Ans)}
\end{aligned}$$

COMMENT For each kg of air passing through the turbine there is 29 kN · m of energy available at the shaft to drive the drill. However, because of fluid friction, the actual amount of energy given up by each kg of air will be greater than the amount available at the shaft. How much greater depends on the efficiency of the fluid-mechanical energy transfer between the fluid and the turbine blades.

Recall that the shaft power, $\dot{W}_{\text{shaft}}$, is given by

$$\dot{W}_{\text{shaft}} = \dot{m} w_{\text{shaft}}$$

Hence, to determine the power we need to know the mass flowrate, $\dot{m}$, which depends on the size and number of the nozzles. Although the energy per unit mass is large (i.e., 29 kN · m/kg), the flowrate is small, so the power is not "large."

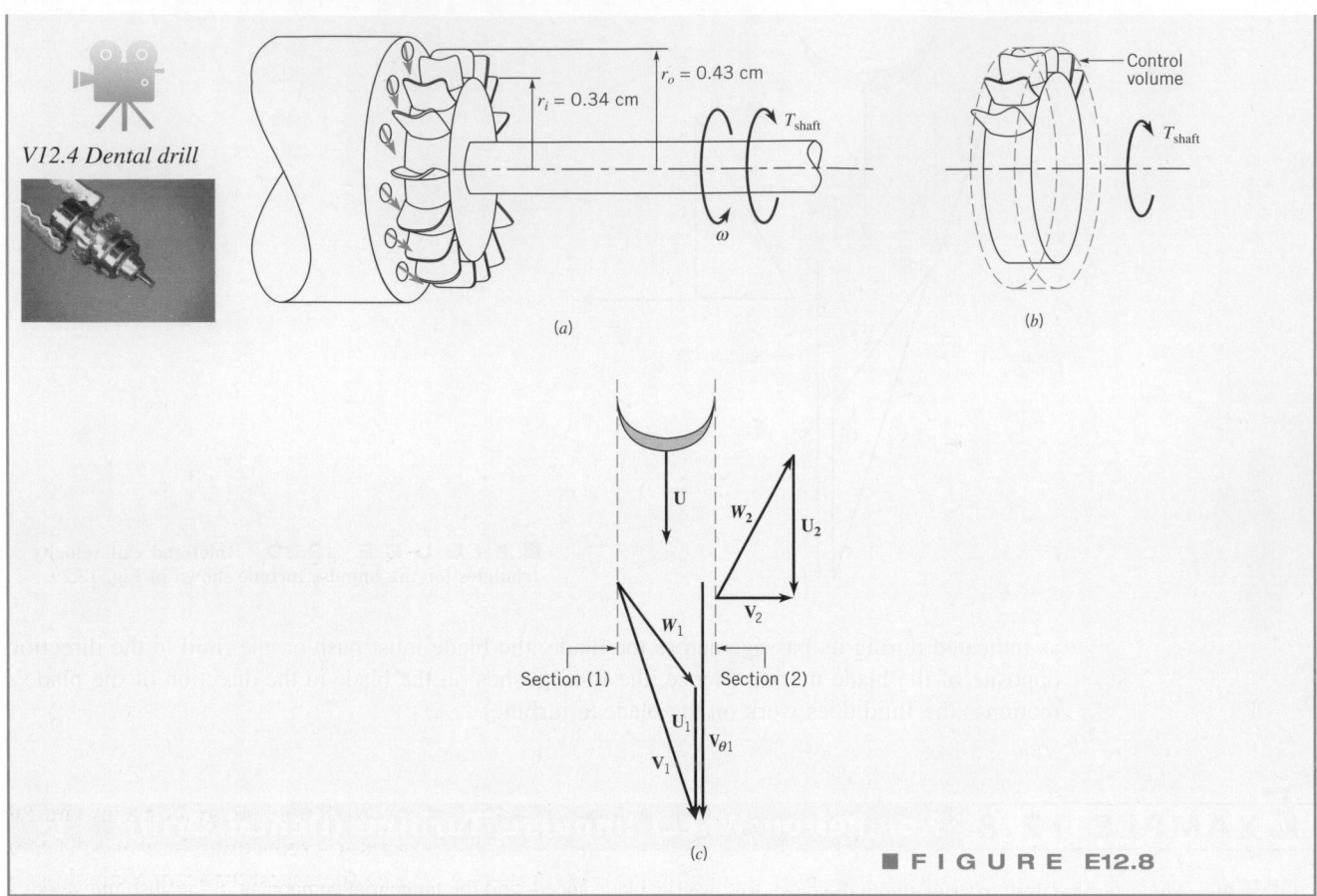

V12.4 Dental drill

$r_o = 0.43$ cm

$r_i = 0.34$ cm

T_{shaft}

ω

Control volume

T_{shaft}

(a)

(b)

U

W_2

U_2

V_2

W_1

Section (1)

Section (2)

U_1

$V_{\theta 1}$

V_1

(c)

■ **FIGURE E12.8**

12.8.2 Reaction Turbines

Reaction turbines are best suited for higher flowrate and lower head situations.

As indicated in the previous section, impulse turbines are best suited (i.e., most efficient) for lower flowrate and higher head operations. Reaction turbines, on the other hand, are best suited for higher flowrate and lower head situations such as are often encountered in hydroelectric power plants associated with a dammed river, for example.

In a reaction turbine the working fluid completely fills the passageways through which it flows (unlike an impulse turbine, which contains one or more individual unconfined jets of fluid). The angular momentum, pressure, and velocity of the fluid decrease as it flows through the turbine rotor—the turbine rotor extracts energy from the fluid.

As with pumps, turbines are manufactured in a variety of configurations—radial-flow, mixed-flow, and axial-flow. Typical radial- and mixed-flow hydraulic turbines are called *Francis turbines*, named after James B. Francis, an American engineer. At very low heads the most efficient type of turbine is the axial-flow or propeller turbine. The *Kaplan turbine*, named after Victor Kaplan, a German professor, is an efficient axial-flow hydraulic turbine with adjustable blades. Cross sections of these different turbine types are shown in Fig. 12.31.

As shown in Fig. 12.31*a*, flow across the rotor blades of a radial-inflow turbine has a major component in the radial direction. Inlet guide vanes (which may be adjusted to allow optimum performance) direct the water into the rotor with a tangential component of velocity. The absolute velocity of the water leaving the rotor is essentially without tangential velocity. Hence, the rotor decreases the angular momentum of the fluid, the fluid exerts a torque on the rotor in the direction of rotation, and the rotor extracts energy from the fluid. The Euler turbomachine equation (Eq. 12.2) and the corresponding power equation (Eq. 12.4) are equally valid for this turbine as they are for the centrifugal pump discussed in Section 12.4.

As shown in Fig. 12.31*b*, for an axial-flow Kaplan turbine, the fluid flows through the inlet guide vanes and achieves a tangential velocity in a vortex (swirl) motion before it reaches the

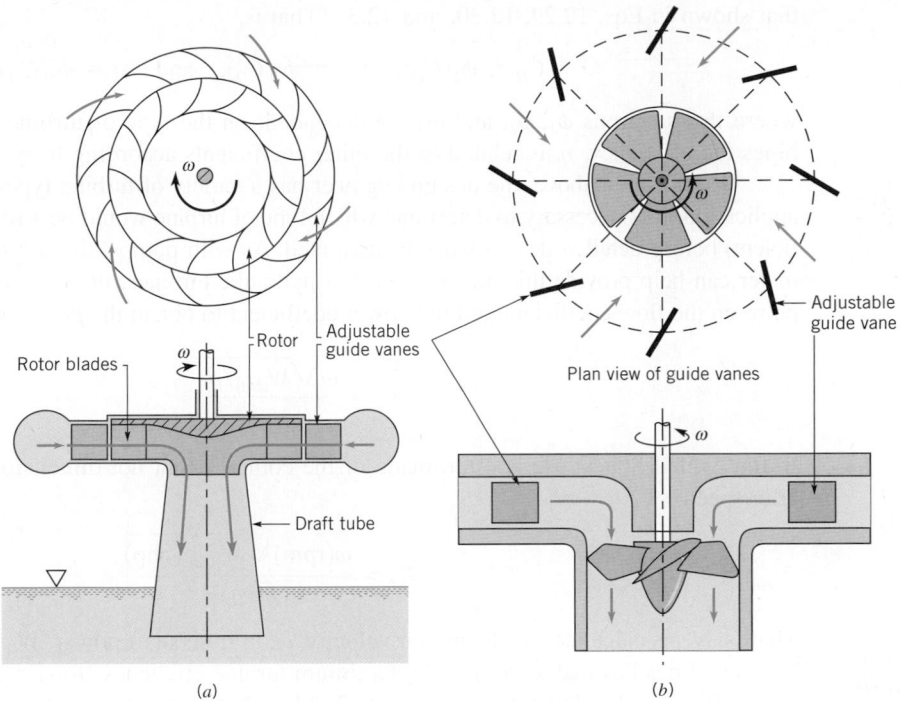

■ FIGURE 12.31   (*a*) **Typical radial-flow Francis turbine,** (*b*) **typical axial-flow Kaplan turbine.**

rotor. Flow across the rotor contains a major axial component. Both the inlet guide vanes and the turbine blades can be adjusted by changing their setting angles to produce the best match (optimum output) for the specific operating conditions. For example, the operating head available may change from season to season and/or the flowrate through the rotor may vary.

<table>
<tr><td colspan="2" align="center">**F l u i d s   i n   t h e   N e w s**</td></tr>
<tr><td>

Fish friendly hydraulic turbine Based on data about what actually kills fish as they pass through *hydraulic turbines*, Concepts NREC produced a rotor design that allows a larger flow passage, a more uniform pressure distribution, lower levels of shear stress,

</td><td>

and other acceptable trade offs between efficiency and fish survivability. Tests and projections suggest that the fish friendly turbine design will achieve 90 percent efficiency, with fish survivability increased from 60% to 98%.

</td></tr>
</table>

Pumps and turbines are often thought of as the "inverse" of each other. Pumps add energy to the fluid; turbines remove energy. The propeller on an outboard motor (a pump) and the propeller on a Kaplan turbine are in some ways geometrically similar, but they perform opposite tasks. Similar comparisons can be made for centrifugal pumps and Francis turbines. In fact, some large turbomachines at hydroelectric power plants are designed to be run as turbines during high-power demand periods (i.e., during the day) and as pumps to resupply the upstream reservoir from the downstream reservoir during low-demand times (i.e., at night). Thus, a pump type often has its corresponding turbine type. However, is it possible to have the "inverse" of a Pelton wheel turbine—an impulse pump?

Actual head available for a turbine, h_a, is always greater than shaft work head, h_s, because of head loss, h_L, in the turbine.

As with pumps, incompressible flow turbine performance is often specified in terms of appropriate dimensionless parameters. The flow coefficient, $C_Q = Q/\omega D^3$, the head coefficient, $C_H = gh_a/\omega^2 D^2$, and the power coefficient, $C_{\mathscr{P}} = \dot{W}_{\text{shaft}}/\rho\omega^3 D^5$, are defined in the same way for pumps and turbines. On the other hand, turbine efficiency, η, is the inverse of pump efficiency. That is, the efficiency is the ratio of the shaft power output to the power available in the flowing fluid, or

$$\eta = \frac{\dot{W}_{\text{shaft}}}{\rho g Q h_a}$$

For geometrically similar turbines and for negligible Reynolds number and surface roughness difference effects, the relationships between the dimensionless parameters are given functionally by that shown in Eqs. 12.29, 12.30, and 12.31. That is,

$$C_H = \phi_1(C_Q), \quad C_{\mathcal{P}} = \phi_2(C_Q), \quad \text{and} \quad \eta = \phi_3(C_Q)$$

where the functions ϕ_1, ϕ_2, and ϕ_3 are dependent on the type of turbine involved. Also, for turbines the efficiency, η, is related to the other coefficients according to $\eta = C_{\mathcal{P}}/C_H C_Q$.

As indicated above, the design engineer has a variety of turbine types available for any given application. It is necessary to determine which type of turbine would best fit the job (i.e., be most efficient) before detailed design work is attempted. As with pumps, the use of a specific speed parameter can help provide this information. For hydraulic turbines, the rotor diameter D is eliminated between the flow coefficient and the power coefficient to obtain the *power specific speed*, N_s', where

$$N_s' = \frac{\omega \sqrt{\dot{W}_{\text{shaft}}/\rho}}{(gh_a)^{5/4}} \tag{12.53}$$

In the United States, use is often made of the common, but not dimensionless, definition of specific speed

$$N_{sd}' = \frac{\omega(\text{rpm})\sqrt{\dot{W}_{\text{shaft}}(\text{bhp})}}{[h_a(\text{ft})]^{5/4}}$$

Specific speed may be used to approximate what kind of turbine geometry (axial to radial) would operate most efficiently.

That is, N_{sd}' is calculated with angular velocity, ω, in rpm; shaft power, $\dot{W}_{\text{shaft}}$, in brake horsepower; and actual head available, h_a, in feet. Optimum turbine efficiency (for large turbines) as a function of specific speed is indicated in Fig. 12.32. Also shown are representative rotor and casing cross sections. Note that impulse turbines are best at low specific speeds; that is, when operating with large heads and small flowrate. The other extreme is axial-flow turbines, which are the most efficient type if the head is low and if the flowrate is large. For intermediate values of specific speeds, radial- and mixed-flow turbines offer the best performance.

The data shown in Fig. 12.32 are meant only to provide a guide for turbine-type selection. The actual turbine efficiency for a given turbine depends very strongly on the detailed design of the turbine. Considerable analysis, testing, and experience are needed to produce an efficient turbine. However, the data of Fig. 12.32 are representative. Much additional information can be found in the literature.

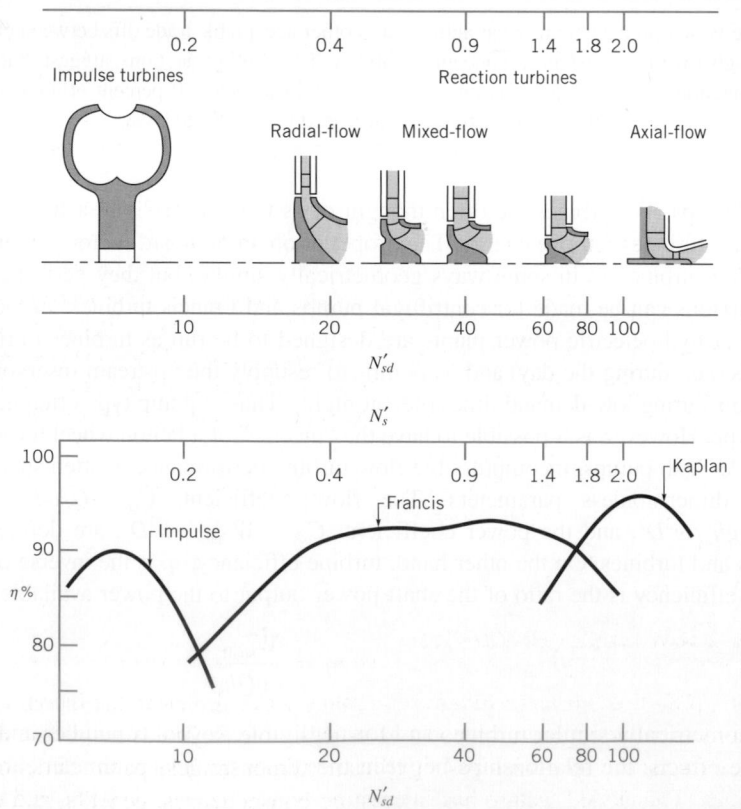

■ **F I G U R E 12.32**
Typical turbine cross sections and maximum efficiencies as a function of specific speed.

F l u i d s   i n   t h e   N e w s

Cavitation damage in hydraulic turbines The occurrence of *cavitation* in hydraulic *pumps* seem to be an obvious possibility since low suction *pressures* are expected. Cavitation damage can also occur in hydraulic *turbines* even though they do not seem obviously prone to this kind of problem. Local *acceleration* of liquid over *blade* surfaces can be sufficient to result in local pressures low enough to cause fluid vaporization or cavitation.

Further along the flow path, the fluid can decelerate rapidly enough with accompanying increase in local pressure to make cavitation bubbles collapse with enough intensity to cause blade surface damage in the form of material erosion. Over time, this erosion can be severe enough to require blade repair or replacement which is very expensive. (See Problem 12.80.)

EXAMPLE 12.9   Use of Specific Speed to Select Turbine Type

GIVEN A hydraulic turbine is to operate at an angular velocity of 6 rev/s, a flowrate of 0.28 m³/s, and a head of 6 m.

FIND What type of turbine should be selected? Explain.

SOLUTION

The most efficient type of turbine to use can be obtained by calculating the specific speed, N'_s, and using the information of Fig. 12.32. For the rotor speed we get

$$\omega = 6 \text{ rev/s} \times 2\pi \text{ rad/rev} = 37.7 \text{ rad/s}$$

To estimate the shaft power, we assume all of the available head is converted into power and multiply this amount by an assumed efficiency (94%).

$$\dot{W}_{\text{shaft}} = \gamma Q z \eta = (9800 \text{ N/m}^3)(0.28 \text{ m}^3/\text{s})\left[\frac{6 \text{ m}(0.94)}{1 \text{ N} \cdot \text{m/s} \cdot \text{W}}\right]$$

$$\dot{W}_{\text{shaft}} = 15.5 \text{ kW}$$

Thus for this turbine,

$$N'_s = \frac{\omega \sqrt{\dot{W}_{\text{shaft}}/s}}{(gh_a)^{5/4}} = \frac{(37.7 \text{ rad/s})\sqrt{\frac{15.5 \text{ kW}}{999 \text{ kg/m}^3}}}{(9.81 \text{ m/s}^2 \times 6 \text{ m})^{5/4}} = 0.9$$

According to the information of Fig. 12.32,

> A mixed-flow Francis turbine would probably give the highest efficiency and an assumed efficiency of 0.94 is appropriate.   **(Ans)**

COMMENT What would happen if we wished to use a Pelton wheel for this application? Note that with only a 6-m head, the maximum jet velocity, V_1, obtainable (neglecting viscous effects) would be

$$V_1 = \sqrt{2 \, gz} = \sqrt{2 \times 9.81 \text{ m/s}^2 \times 6 \text{ m}} = 10.8 \text{ m/s}$$

As shown by Eq. 12.52, for maximum efficiency of a Pelton wheel the jet velocity is ideally two times the blade velocity. Thus, $V_1 = 2\omega R$, or the wheel diameter, $D = 2R$, is

$$D = \frac{V_1}{\omega} = \frac{10.8 \text{ m/s}}{(6 \text{ rev/s} \times 2\pi \text{ rad/rev})} = 0.29 \text{ m}$$

To obtain a flowrate of $Q = 0.28 \text{ m}^3/\text{s}$ at a velocity of $V_1 = 10.8 \text{ m/s}$, the jet diameter, d_1, must be given by

$$Q = \frac{\pi}{4} d_1^2 V_1$$

or

$$d_1 = \left[\frac{4Q}{\pi V_1}\right]^{1/2} = \left[\frac{4(0.28 \text{ m}^3/\text{s})}{\pi(10.8 \text{ m/s})}\right]^{1/2} = 0.18 \text{ m}$$

A Pelton wheel with a diameter of $D = 0.29 \text{ m}$ supplied with water through a nozzle of diameter $d_1 = 0.18 \text{ m}$ is not a practical design. Typically $d_1 \ll D$ (see Fig. 12.22). By using multiple jets it would be possible to reduce the jet diameter. However, even with 8 jets, the jet diameter would be 0.06 m, which is still too large (relative to the wheel diameter) to be practical. Hence, the above calculations reinforce the results presented in Fig. 12.32—a Pelton wheel would not be practical for this application. If the flowrate were considerably smaller, the specific speed could be reduced to the range where a Pelton wheel would be the type to use (rather than a mixed-flow reaction turbine).

12.9   Compressible Flow Turbomachines

Compressible flow turbomachines are in many ways similar to the incompressible flow pumps and turbines described in previous portions of this chapter. The main difference is that the density of the fluid (a gas or vapor) changes significantly from the inlet to the outlet of the compressible flow machines. This added feature has interesting consequences, benefits, and complications.

Compressors are pumps that add energy to the fluid, causing a significant pressure rise and a corresponding significant increase in density. Compressible flow turbines, on the other hand, remove energy from the fluid, causing a lower pressure and a smaller density at the outlet than at the inlet. The information provided earlier about basic energy considerations (Section 12.2) and basic angular momentum considerations (Section 12.3) is directly applicable to these turbomachines in the ways demonstrated earlier.

As discussed in Chapter 11, compressible flow study requires an understanding of the principles of thermodynamics. Similarly, an in-depth analysis of compressible flow turbo-machines requires use of various thermodynamic concepts. In this section we provide only a brief discussion of some of the general properties of compressors and compressible flow turbines. The interested reader is encouraged to read some of the excellent references available for further information (e.g., Refs. 1–3, 18–20).

12.9.1 Compressors

Turbocompressors operate with the continuous compression of gas flowing through the device. Since there is a significant pressure and density increase, there is also a considerable temperature increase.

Radial-flow (or centrifugal) compressors are essentially centrifugal pumps (see Section 12.4) that use a gas (rather than a liquid) as the working fluid. They are typically high pressure rise, low flowrate, and axially compact turbomachines. A photograph of the rotor of a centrifugal compressor rotor is shown in Fig. 12.33.

The amount of compression is typically given in terms of the *total pressure ratio*, $PR = p_{02}/p_{01}$, where the pressures are absolute. Thus, a radial flow compressor with $PR = 3.0$ can compress standard atmospheric air from 101.3 kPa to 3.0×101.3 kPa $= 304$ kPa.

Higher pressure ratios can be obtained by using *multiple stage* devices in which flow from the outlet of the preceding stage proceeds to the inlet of the following stage. If each stage has the same pressure ratio, PR, the overall pressure ratio after n stages is PR^n. Thus, as shown by the figure in the margin, a four-stage compressor with individual stage $PR = 2.0$ can compress standard air from $p_{0\,in} = 101.3$ kPa to $p_{0\,out} = 2^4 \times 101.3 = 1620$ kPa. Adiabatic (i.e., no heat transfer) compression of a gas causes an increase in temperature and requires more work than isothermal (constant temperature) compression of a gas. An interstage cooler (i.e., an intercooler heat exchanger) as shown in Fig. 12.34 can be used to reduce the compressed gas temperature and thus the work required.

Multistaging is common in high-pressure ratio compressors.

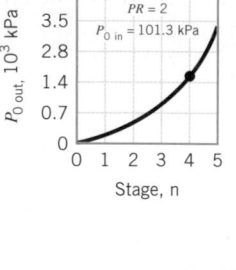

Relative to centrifugal water pumps, radial compressors of comparable size rotate at much higher speeds. It is not uncommon for the rotor blade exit speed and the speed of the absolute flow leaving the impeller to be greater than the speed of sound. That such large speeds are necessary for compressors can be seen by noting that the large pressure rise designed for is related to the differences of several squared speeds (see Eq. 12.14).

The axial-flow compressor is the other widely used configuration. This type of turbomachine has a lower pressure rise per stage, a higher flowrate, and is more radially compact than a centrifugal compressor. As shown in Fig. 12.35, axial-flow compressors usually consist of several stages, with each stage containing a rotor/stator row pair. For an 11-stage compressor, a compression ratio of $PR = 1.2$ per stage gives an overall pressure ratio of $p_{02}/p_{01} = 1.2^{11} = 7.4$. As the gas

■ FIGURE 12.33 Centrifugal compressor rotor. (Photograph courtesy of concepts NREC.)

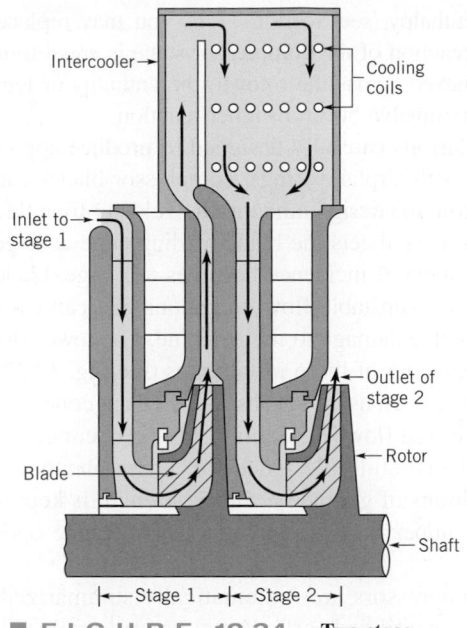

■ F I G U R E   12.34   Two-stage centrifugal compressor with an intercooler.

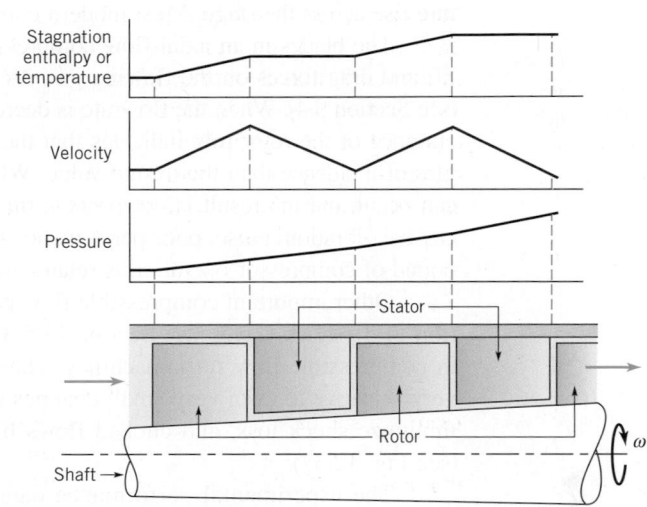

■ F I G U R E   12.35   Enthalpy, velocity, and pressure distribution in an axial-flow compressor.

Axial-flow compressor multistaging requires less space than centrifugal compressors.

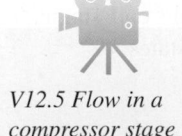

V12.5 Flow in a compressor stage

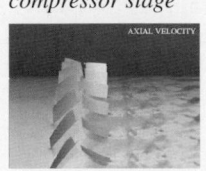

is compressed and its density increases, a smaller annulus cross-sectional area is required and the flow channel size decreases from the inlet to the outlet of the compressor. The typical jet aircraft engine uses an axial-flow compressor as one of its main components (see Fig. 12.36 and Ref. 21).

An axial-flow compressor can include a set of *inlet guide vanes* upstream of the first rotor row. These guide vanes optimize the size of the relative velocity into the first rotor row by directing the flow away from the axial direction. *Rotor blades* push on the gas in the direction of blade motion and to the rear, adding energy (like in an axial-pump) and moving the gas through the compressor. The *stator blade* rows act as diffusers, turning the fluid back toward the axial direction and increasing the static pressure. The stator blades cannot add energy to the fluid because they are stationary. Typical pressure, velocity, and enthalpy distributions along the axial direction are shown in Fig. 12.35. [If you are

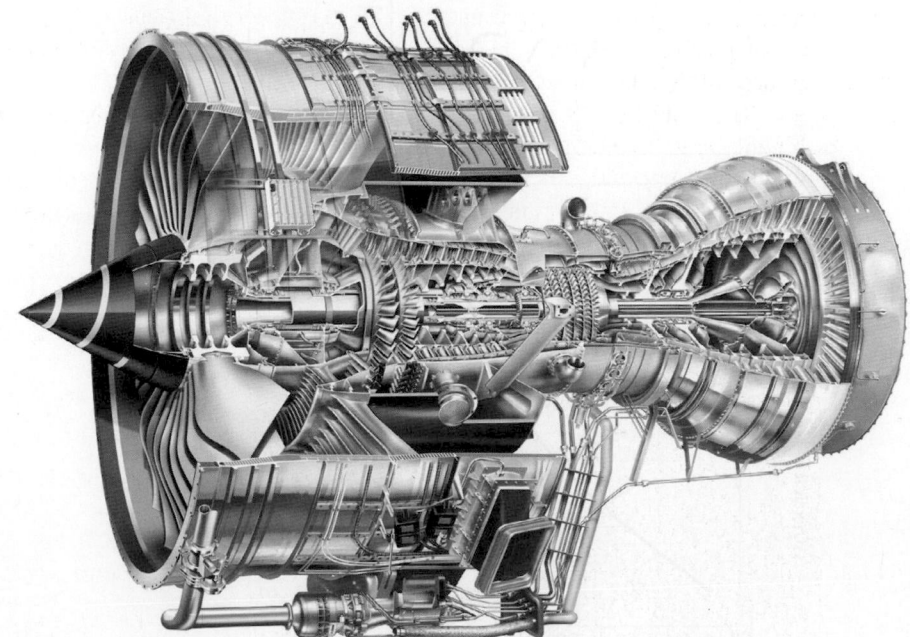

■ F I G U R E   12.36   Rolls-Royce Trent 900 three-shaft propulsion system.
(Courtesy of Rolls-Royce plc.)

Compressor blades can stall, and unstable flow conditions can subsequently occur.

not familiar with the thermodynamic concept of enthalpy (see Section 11.1), you may replace "enthalpy" by temperature as an approximation.] The reaction of the compressor stage is equal to the ratio of the rise in static enthalpy or temperature achieved across the rotor to the enthalpy or temperature rise across the stage. Most modern compressors involve 50% or higher reaction.

The blades in an axial-flow compressor are airfoils carefully designed to produce appropriate lift and drag forces on the flowing gas. As occurs with airplane wings, compressor blades can stall (see Section 9.4). When the flowrate is decreased from the design amount, the velocity triangle at the entrance of the rotor row indicates that the relative flow meets the blade leading edge at larger angles of incidence than the design value. When the angle of incidence becomes too large, blade stall can occur and the result is *compressor surge* or *stall*—unstable flow conditions that can cause excessive vibration, noise, poor performance, and possible damage to the machine. The lower flowrate bound of compressor operation is related to the beginning of these instabilities (see Fig. 12.37).

Other important compressible flow phenomena such as variations of the Mach cone (see Section 11.3), shock waves (see Section 11.5.3), and choked flow (see Section 11.4.2) occur commonly in compressible flow turbomachines. They must be carefully designed for. These phenomena are very sensitive to even very small changes or variations of geometry. Shock strength is kept low to minimize shock loss, and choked flows limit the upper flowrate boundary of machine operation (see Fig. 12.37).

The experimental performance data for compressors are systematically summarized with parameters prompted by dimensional analysis. As mentioned earlier, total pressure ratio, p_{02}/p_{01}, is used instead of the head-rise coefficient associated with pumps, blowers, and fans.

Either isentropic or polytropic efficiencies are used to characterize compressor performance. A detailed explanation of these efficiencies is beyond the scope of this text. Those interested in learning more about these parameters should study any of several available books on turbomachines (for example, Refs. 2 and 3). Basically, each of these compressor efficiencies involves a ratio of ideal work to actual work required to accomplish the compression. The isentropic efficiency involves a ratio of the ideal work required with an adiabatic and frictionless (no loss) compression process to the actual work required to achieve the same total pressure rise. The polytropic efficiency involves a ratio of the ideal work required to achieve the actual end state of the compression with a polytropic and frictionless process between the actual beginning and end stagnation states across the compressor and the actual work involved between these same states.

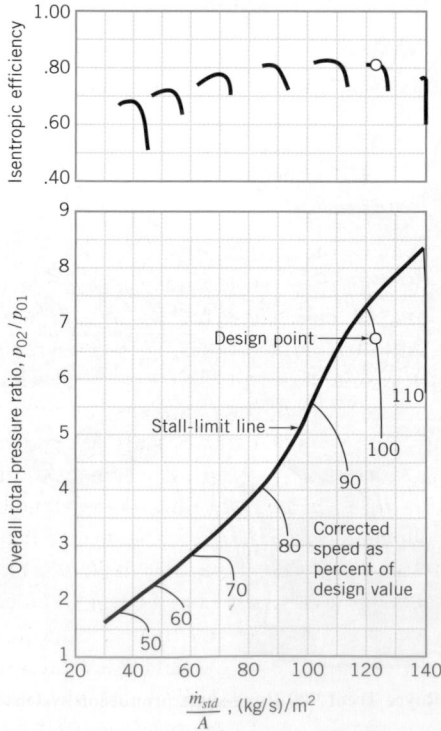

■ **FIGURE 12.37** Performance characteristics of an axial-flow compressor (Adapted from Ref. 19.).

The flow parameter commonly used for compressors is based on the following dimensionless grouping from dimensional analysis

$$\frac{R\dot{m}\sqrt{kT_{01}}}{D^2 p_{01}}$$

where R is the gas constant, $\dot{m}$ the mass flowrate, k the specific heat ratio, T_{01} the stagnation temperature at the compressor inlet, D a characteristic length, and p_{01} the stagnation pressure at the compressor inlet.

To account for variations in test conditions, the following strategy is employed. We set

$$\left(\frac{R\dot{m}\sqrt{kT_{01}}}{D^2 p_{01}}\right)_{\text{test}} = \left(\frac{R\dot{m}\sqrt{kT_{01}}}{D^2 p_{01}}\right)_{\text{std}}$$

where the subscript "test" refers to a specific test condition and "std" refers to the standard atmosphere ($p_0 = 101.3$ kPa (abs), $T_0 = 288$ K) condition. When we consider a given compressor operating on a given working fluid (so that R, k, and D are constant), the above equation reduces to

$$\dot{m}_{\text{std}} = \frac{\dot{m}_{\text{test}}\sqrt{T_{01\,\text{test}}/T_{0\,\text{std}}}}{p_{01\,\text{test}}/p_{0\,\text{std}}} \tag{12.54}$$

In essence, $\dot{m}_{\text{std}}$ is the compressor-test mass flowrate "corrected" to the standard atmosphere inlet condition. *The corrected compressor mass flowrate, $\dot{m}_{\text{std}}$, is used instead of flow coefficient.* Often, $\dot{m}_{\text{std}}$ is divided by A, the frontal area of the compressor flow path.

While for pumps, blowers, and fans, rotor speed was accounted for in the flow coefficient, it is not in the corrected mass flowrate derived above. Thus, for compressors, rotor speed needs to be accounted for with an additional group. This dimensionless group is

$$\frac{ND}{\sqrt{kRT_{01}}}$$

For the same compressor operating on the same gas, we eliminate D, k and R and, as with corrected mass flowrate, obtain a corrected speed, N_{std}, where

$$N_{\text{std}} = \frac{N}{\sqrt{T_{01}/T_{\text{std}}}} \tag{12.55}$$

Often, the percentage of the corrected speed design value is used.

An example of how compressor performance data are typically summarized is shown in Fig. 12.37.

12.9.2 Compressible Flow Turbines

Turbines that use a gas or vapor as the working fluid are in many respects similar to hydraulic turbines (see Section 12.8). Compressible flow turbines may be impulse or reaction turbines, and mixed-, radial-, or axial-flow turbines. The fact that the gas may expand (compressible flow) in coursing through the turbine can introduce some important phenomena that do not occur in hydraulic turbines. (*Note:* It is tempting to label turbines that use a gas as the working fluid as gas turbines. However, the terminology "gas turbine" is commonly used to denote a *gas turbine engine*, as employed, for example, for aircraft propulsion or stationary power generation. As shown in Fig. 12.36, these engines typically contain a compressor, combustion chamber, and turbine.)

A gas turbine engine generally consists of a compressor, a combustor, and a turbine.

Although for compressible flow turbines the axial-flow type is common, the radial-inflow type is also used for various purposes. As shown in Fig. 12.33, the turbine that drives the typical automobile turbocharger compressor is a radial-inflow type. The main advantages of the radial-inflow turbine are: (1) It is robust and durable, (2) it is axially compact, and (3) it can be relatively inexpensive. A radial-flow turbine usually has a lower efficiency than an axial-flow turbine, but lower initial costs may be the compelling incentive in choosing a radial-flow turbine over an axial-flow one.

Axial-flow turbines are widely used compressible flow turbines. Steam engines used in electrical generating plants and marine propulsion and the turbines used in gas turbine engines are usually of the axial-flow type. Often they are multistage turbomachines, although single-stage compressible turbines are also produced. They may be either an impulse type or a reaction type. With compressible flow turbines, the ratio of static enthalpy or temperature drop across the rotor to this drop across the stage, rather than the ratio of static pressure differences, is used to determine reaction. Strict impulse (zero pressure drop) turbines have slightly negative reaction; the static enthalpy or temperature actually increases across the rotor. Zero-reaction turbines involve no change of static enthalpy or temperature across the rotor but do involve a slight pressure drop.

A two-stage, axial-flow impulse turbine is shown in Fig. 12.38a. The gas accelerates through the supply nozzles, has some of its energy removed by the first-stage rotor blades, accelerates again through the second-stage nozzle row, and has additional energy removed by the second-stage rotor blades. As shown in Fig. 12.38b, the static pressure remains constant across the rotor rows. Across the second-stage nozzle row, the static pressure decreases, absolute velocity increases, and the stagnation enthalpy (temperature) is constant. Flow across the second rotor is similar to flow across the first rotor. Since the working fluid is a gas, the significant decrease in static pressure across the turbine results in a significant decrease in density—the flow is compressible. Hence, more detailed analysis of this flow must incorporate various compressible flow concepts developed in Chapter 11. Interesting phenomena such as shock waves and choking due to sonic conditions at the "throat" of the flow passage between blades can occur because of compressibility effects. The interested reader is encouraged to consult the various references available (e.g., Refs. 2, 3, 20) for fascinating applications of compressible flow principles in turbines.

The rotor and nozzle blades in a three-stage, axial-flow reaction turbine are shown in Fig. 12.39a. The axial variations of pressure and velocity are shown in Fig. 12.39c. Both the stationary and rotor blade (passages) act as flow-accelerating nozzles. That is, the static pressure and enthalpy (temperature) decrease in the direction of flow for both the fixed and the rotating blade rows. This distinguishes the reaction turbine from the impulse turbine (see Fig. 12.38b). Energy is removed from the fluid by the rotors only (the stagnation enthalpy or temperature is constant across the adiabatic flow stators).

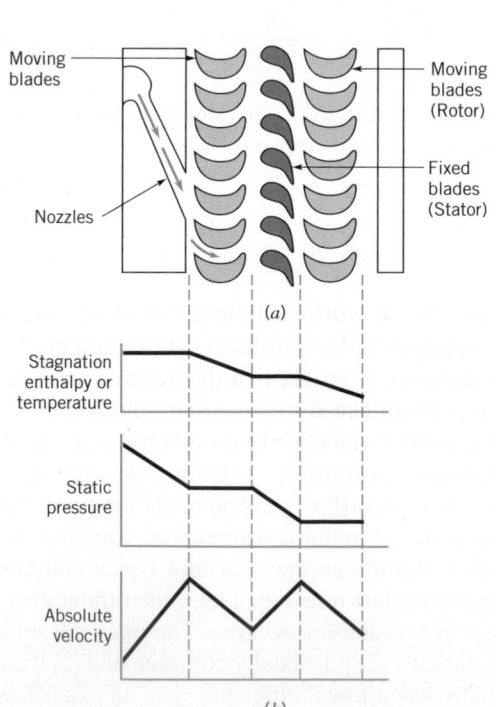

■ **F I G U R E 12.38** Enthalpy, velocity, and pressure distribution in two-stage impulse turbine.

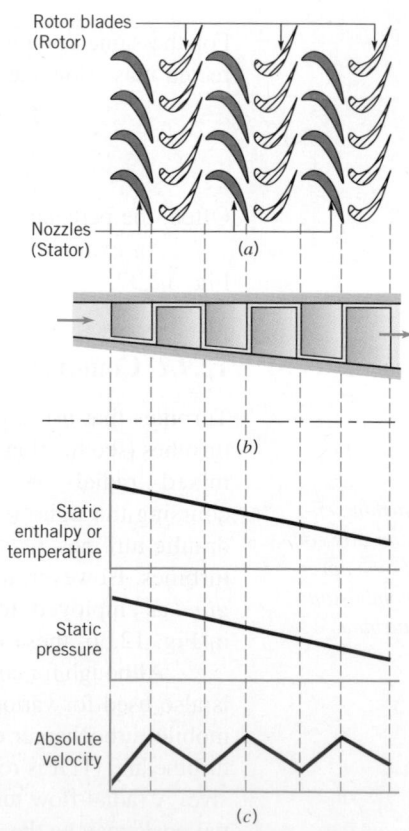

■ **F I G U R E 12.39** Enthalpy, pressure, and velocity distribution in a three-stage reaction turbine.

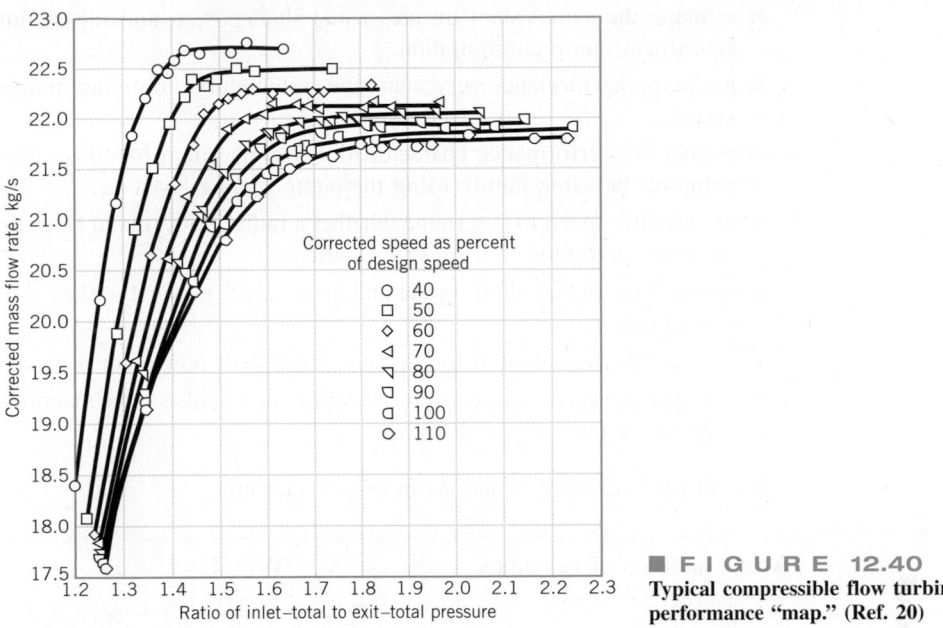

■ FIGURE 12.40
Typical compressible flow turbine performance "map." (Ref. 20)

Because of the reduction of static pressure in the downstream direction, the gas expands, and the flow passage area must increase from the inlet to the outlet of this turbine. This is seen in Fig. 12.39b.

Performance data for compressible flow turbines are summarized with the help of parameters derived from dimensional analysis. Isentropic and polytropic efficiencies (see Refs. 2, 3, and 20) are commonly used as are inlet-to-outlet total pressure ratios (p_{01}/p_{02}), corrected rotor speed (see Eq. 12.55), and corrected mass flowrate (see Eq. 12.54). In Fig. 12.40 is shown a compressible flow turbine performance "map."

Turbine perfor-mance maps are used to display complex turbine characteristics.

12.10   Chapter Summary and Study Guide

turbomachine

axial-, mixed-, and radial-flow

velocity triangle

angular momentum

shaft torque

Euler turboma-chine equation

shaft power

centrifugal pump

pump performance curve

overall efficiency

system equation

head rise coefficient

power coefficient

flow coefficient

pump scaling laws

specific speed

impulse turbine

reaction turbine

Pelton wheel

Various aspects of turbomachine flow are considered in this chapter. The connection between fluid angular momentum change and shaft torque is key to understanding how turbo-pumps and turbines operate.

The shaft torque associated with change in the axial component of angular momentum of a fluid as it flows through a pump or turbine is described in terms of the inlet and outlet velocity triangles diagrams. Such diagrams indicate the relationship among absolute, relative, and blade velocities.

Performance characteristics for centrifugal pumps are discussed. Standard dimensionless pump parameters, similarity laws, and the concept of specific speed are presented for use in pump analysis. How to use pump performance curves and the system curve for proper pump selection is presented. A brief discussion of axial-flow and mixed-flow pumps is given.

An analysis of impulse turbines is provided, with emphasis on the Pelton wheel turbine. For impulse turbines there is negligible pressure difference across the blade; the torque is a result of the change in direction of the fluid jet striking the blade. Radial-flow and axial-flow reaction turbines are also briefly discussed.

The following checklist provides a study guide for this chapter. When your study of the entire chapter and end-of-chapter exercises has been completed you should be able to

■ write out meanings of the terms listed here in the margin and understand each of the related concepts. These terms are particularly important and are set in *italic, bold, and color* type in the text.

■ draw appropriate velocity triangles for flows entering and leaving given pump or turbine configurations.

- estimate the actual shaft torque, actual shaft power, and ideal pump head rise for a given centrifugal pump configuration.
- use pump performance curves and the system curve to predict pump performance in a given system.
- predict the performance characteristics for one pump based on the performance of another pump of the same family using the pump scaling laws.
- use specific speed to determine whether a radial flow, mixed flow, or axial flow pump would be most appropriate for a given situation.
- estimate the actual shaft torque and actual shaft power for flow through an impulse turbine configuration.
- estimate the actual shaft torque and actual shaft power for a given reaction turbine.
- use specific speed to determine whether an impulse or a reaction turbine would be most appropriate for a given situation.

Some of the important equations in this chapter are:

Vector addition of velocities	$\mathbf{V} = \mathbf{W} + \mathbf{U}$	(12.1)
Shaft torque	$T_{shaft} = -\dot{m}_1(r_1 V_{\theta 1}) + \dot{m}_2(r_2 V_{\theta 2})$	(12.2)
Shaft power	$\dot{W}_{shaft} = T_{shaft}\, \omega$	(12.3)
Shaft power	$\dot{W}_{shaft} = -\dot{m}_1(U_1 V_{\theta 1}) + \dot{m}_2(U_2 V_{\theta 2})$	(12.4)
Shaft work	$w_{shaft} = \dfrac{V_2^2 - V_1^2 + U_2^2 - U_1^2 - (W_2^2 - W_1^2)}{2}$	(12.8)
Pump ideal head rise		
Pump actual head rise	$h_a = \dfrac{p_2 - p_1}{\gamma} + z_2 - z_1 + \dfrac{V_2^2 - V_1^2}{2g}$	(12.19)
Pump similarity relationship	$\dfrac{g h_a}{\omega^2 D^2} = \phi_1\left(\dfrac{Q}{\omega D^3}\right)$	(12.29)
Pump similarity relationship	$\dfrac{\dot{W}_{shaft}}{\rho \omega^3 D^5} = \phi_2\left(\dfrac{Q}{\omega D^3}\right)$	(12.30)
Pump similarity relationship	$\eta = \phi_3\left(\dfrac{Q}{\omega D^3}\right)$	(12.31)
Pump scaling law	$\left(\dfrac{Q}{\omega D^3}\right)_1 = \left(\dfrac{Q}{\omega D^3}\right)_2$	(12.32)
Pump scaling law	$\left(\dfrac{g h_a}{\omega^2 D^2}\right)_1 = \left(\dfrac{g h_a}{\omega^2 D^2}\right)_2$	(12.33)
Pump scaling law	$\left(\dfrac{\dot{W}_{shaft}}{\rho \omega^3 D^5}\right)_1 = \left(\dfrac{\dot{W}_{shaft}}{\rho \omega^3 D^5}\right)_2$	(12.34)
Pump scaling law	$\eta_1 = \eta_2$	(12.35)
Specific speed (pumps)	$N_s = \dfrac{\omega \sqrt{Q}}{(g h_a)^{3/4}}$	(12.43)
Suction specific speed	$S_s = \dfrac{\omega \sqrt{Q}}{[g(\mathrm{NPSH_R})]^{3/4}}$	(12.45)
Specific speed (turbines)	$N_s' = \dfrac{\omega \sqrt{\dot{W}_{shaft}/s}}{(g h_a)^{5/4}}$	(12.53)

Corrected compressor mass flowrate	$\dot{m}_{std} = \dfrac{\dot{m}_{test}\sqrt{T_{01\,test}/T_{0\,std}}}{p_{01\,test}/p_{0\,std}}$	**(12.54)**
Corrected compressor speed	$N_{std} = \dfrac{N}{\sqrt{T_{01}/T_{std}}}$	**(12.55)**

References

1. Cumpsty, N. A., *Jet Propulsion*, 2nd Ed., Cambridge University Press, Cambridge, UK, 2003.
2. Saravanamuttoo, H. I. H., Rogers, G. F. C., and Cohen, H., *Gas Turbine Theory*, 5th Ed., Prentice-Hall, Saddle River, New Jersey, 2001.
3. Wilson, D. G., and Korakianitis, T., *The Design of High-Efficiency Turbomachinery and Gas Turbines*, 2nd Ed., Prentice-Hall, Saddle River, New Jersey, 1998.
4. Stepanoff, H. J., *Centrifugal and Axial Flow Pumps*, 2nd Ed., Wiley, New York, 1957.
5. Wislicenus, G. F., *Preliminary Design of Turbopumps and Related Machinery*, NASA Reference Publication 1170, 1986.
6. Neumann, B., *The Interaction Between Geometry and Performance of a Centrifugal Pump*, Mechanical Engineering Publications Limited, London, 1991.
7. Garay, P. N., *Pump Application Desk Book*, Fairmont Press, Lilburn, Georgia, 1990.
8. Rouse, H., *Elementary Mechanics of Fluids*, Wiley, New York, 1946.
9. Moody, L. F., and Zowski, T., "Hydraulic Machinery," in *Handbook of Applied Hydraulics*, 3rd Ed., by C. V. Davis and K. E. Sorensen, McGraw-Hill, New York, 1969.
10. Hydraulic Institute, *Hydraulic Institute Standards*, 14th Ed., Hydraulic Institute, Cleveland, Ohio, 1983.
11. Heald, C. C., ed., *Cameron Hydraulic Data*, 17th Ed., Ingersoll-Rand, Woodcliff Lake, New Jersey, 1988.
12. Kristal, F. A., and Annett, F. A., *Pumps: Types, Selection, Installation, Operation, and Maintenance*, McGraw-Hill, New York, 1953.
13. Karassick, I. J., et al., *Pump Handbook*, McGraw-Hill, New York, 1985.
14. Stepanoff, A. J., *Turboblowers*, Wiley, New York, 1955.
15. Berry, C. H., *Flow and Fan Principles of Moving Air Through Ducts*, Industrial Press, New York, 1954.
16. Wallis, R. A., *Axial Flow Fans and Ducts*, Wiley, New York, 1983.
17. Reason, J., "Fans," *Power*, Vol. 127, No. 9, 103–128, 1983.
18. Cumpsty, N. A., *Compressor Aerodynamics*, Longman Scientific & Technical, Essex, UK, and John Wiley & Sons, Inc., New York, 1989.
19. Johnson, I. A., and Bullock, R. D., eds., *Aerodynamic Design of Axial-Flow Compressors*, NASA SP-36, National Aeronautics and Space Administration, Washington, 1965.
20. Glassman, A. J., ed., *Turbine Design and Application*, Vol. 3, NASA SP-290, National Aeronautics and Space Administration, Washington, 1975.
21. Saeed Farokhi, *Aircraft Propulsion*, Wiley, New York, 2009.

Review Problems

Go to Appendix G for a set of review problems with answers. Detailed solutions can be found in *Student Solution Manual and Study* *Guide for Fundamentals of Fluid Mechanics*, by Munson et al. (© 2009 John Wiley and Sons, Inc.).

Problems

Notes: Unless otherwise indicated, use the values of fluid properties found in the tables on the inside of the front cover. Problems designated with a (†) are "open-ended" problems and require critical thinking in that to work them one must make various assumptions and provide the necessary data. There is not a unique answer to these problems.

Answers to the even-numbered problems are listed at the end of the book. Access to the videos that accompany problems can be obtained through the book's web site, www.wiley.com/college/munson.

Section 12.1 Introduction and Section 12.2 Basic Energy Considerations

12.1 Obtain a photograph/image of the blades of an actual axial-flow turbomachine. Briefly explain how and why the machine works and whether it is a "pump" or a "turbine."

12.2 Obtain a photograph/image of the blades of an actual radial-flow turbomachine. Briefly explain how and why the machine works and whether it is a "pump" or a "turbine."

12.3 List ten examples of turbomachines you have encountered.

12.4 The rotor shown in Fig. P12.4 rotates clockwise. Assume that the fluid enters in the radial direction and the relative velocity is tangent to the blades and remains constant across the entire rotor. Is the device a pump or a turbine? Explain.

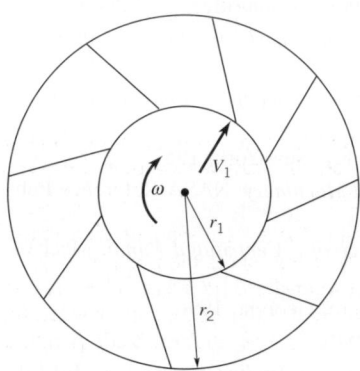

■ **FIGURE P12.4**

12.5 Obtain a schematic of a hydraulic turbine system, and briefly explain the main elements of how potential energy is converted to produce electricity.

12.6 (See Fluids in the News article titled "Current from currents," Section 12.2.) What is the Betz limit associated with wind turbines and why does it exist?

12.7 Would a turbine rotor that is forced to rotate in a fluid by applying a torque to the shaft move that fluid? Explain. Comment on the impact of rotation direction.

Section 12.3 Basic Angular Momentum Considerations

12.8 Identify typical units for the variables work per unit mass and power in the SI system of units.

12.9 Obtain a schematic of a torque converter, and briefly explain how it works.

12.10 Water flows through a rotating sprinkler arm as shown in Fig. P12.10 and Video V12.2. Estimate the minimum water pressure necessary for an angular velocity of 150 rpm. Is this a turbine or a pump?

12.11 Water is supplied to a dishwasher through the manifold shown in Fig. P12.11. Determine the rotational speed of the manifold if bearing friction and air resistance are neglected. The total flowrate of 9 L/min is divided evenly among the six outlets, each of which produces a 4/5-cm-diameter stream.

12.12 Water flows axially up the shaft and out through the two sprinkler arms as sketched in Fig. P12.10 and as shown in Video V12.2. With the help of the moment-of-momentum equation explain why only at a threshold amount of water flow, the sprinkler arms begin to rotate. What happens when the flowrate increases above this threshold amount? If the exit nozzle could be varied, what would happen for a set flowrate above the threshold amount, when the angle is increased to 90°? Decreased to 0°?

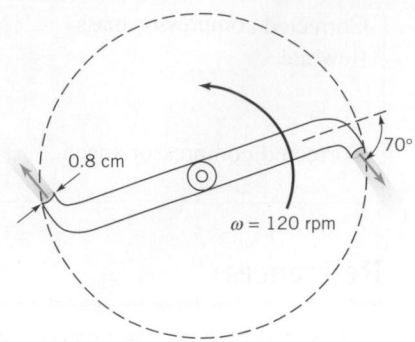

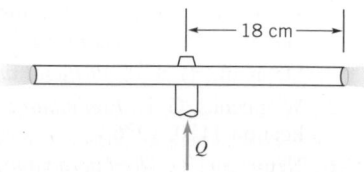

■ **FIGURE P12.10**

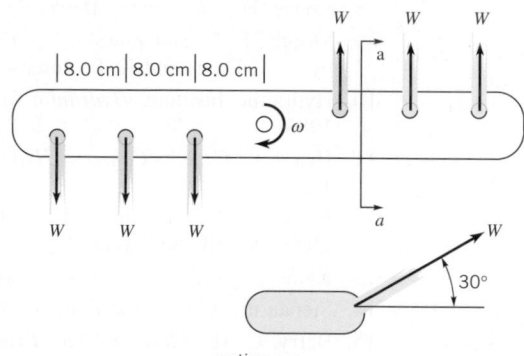

section a-a

■ **FIGURE P12.11**

12.13 At a given radial location, a 24 km/hr wind against a windmill (see Video V12.1) results in the upstream (1) and downstream (2) velocity triangles shown in Fig. P12.13. Sketch an appropriate blade section at that radial location and determine the energy transferred per unit mass of fluid.

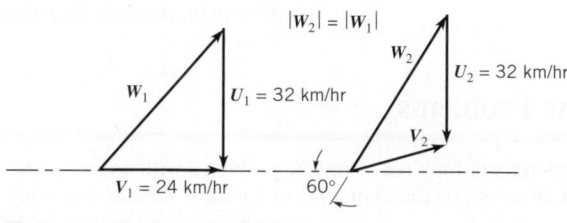

■ **FIGURE P12.13**

12.14 Sketch how you would arrange four 8-cm-wide by 30-cm-long thin but rigid strips of sheet metal on a hub to create a windmill like the one shown in Video V12.1. Discuss, with the help of velocity triangles, how you would arrange each blade on the hub and how you would orient your windmill in the wind.

12.15 Sketched in Fig. P12.15 are the upstream [section (1)] and downstream [section (2)] velocity triangles at the arithmetic mean radius for flow through an axial-flow turbomachine rotor. The axial

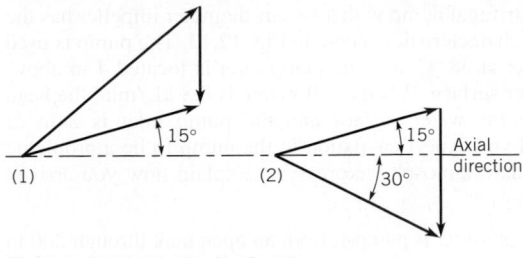

■ **F I G U R E P12.15**

component of velocity is 15 m/s at sections (1) and (2). **(a)** Label each velocity vector appropriately. Use **V** for absolute velocity, **W** for relative velocity, and **U** for blade velocity. **(b)** Are you dealing with a turbine or a fan? **(c)** Calculate the work per unit mass involved. **(d)** Sketch a reasonable blade section. Do you think that the actual blade exit angle will need to be less or greater than 15°? Why?

12.16 Shown in Fig. P12.16 is a toy "helicopter" powered by air escaping from a balloon. The air from the balloon flows radially through each of the three propeller blades and out small nozzles at the tips of the blades. The nozzles (along with the rotating propeller blades) are tilted at a small angle as indicated. Sketch the velocity triangle (i.e., blade, absolute, and relative velocities) for the flow from the nozzles. Explain why this toy tends to move upward. Is this a turbine? Pump?

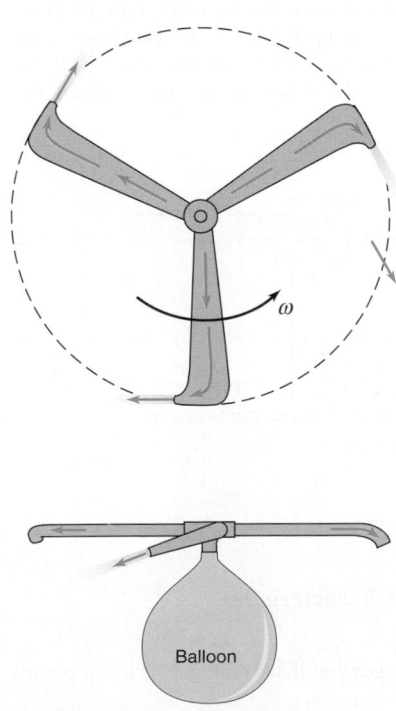

■ **F I G U R E P12.16**

Section 12.4 The Centrifugal Pump and Section 12.4.1 Theoretical Considerations

12.17 Obtain photographs/images of a variety of centrifugal pump rotors. Does the predominant direction of flow through the rotor make sense? Explain.

12.18 The radial component of velocity of water leaving the centrifugal pump sketched in Fig. P12.18 is 14 m/s. The magnitude

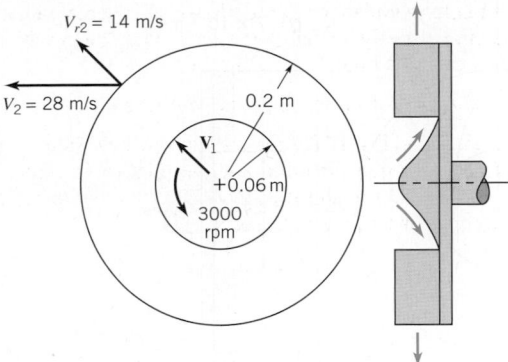

■ **F I G U R E P12.18**

of the absolute velocity at the pump exit is 28 m/s. The fluid enters the pump rotor radially. Calculate the shaft work required per unit mass flowing through the pump.

12.19 A centrifugal water pump having an impeller diameter of 0.5 m operates at 900 rpm. The water enters the pump parallel to the pump shaft. If the exit blade angle, β_2 (see Fig. 12.8), is 25°, determine the shaft power required to turn the impeller when the flow through the pump is 0.16 m³/s. The uniform blade height is 50 mm.

12.20 A centrifugal pump impeller is rotating at 1200 rpm in the direction shown in Fig. P12.20. The flow enters parallel to the axis of rotation and leaves at an angle of 30° to the radial direction. The absolute exit velocity, V_2, is 28 m/s. **(a)** Draw the velocity triangle for the impeller exit flow. **(b)** Estimate the torque necessary to turn the impeller if the fluid is water. What will the impeller rotation speed become if the shaft breaks?

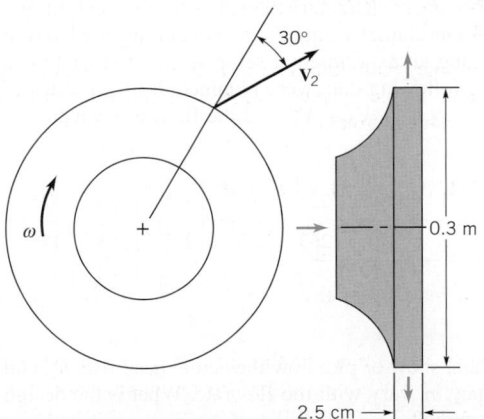

■ **F I G U R E P12.20**

12.21 Discuss the main simplifying assumptions associated with Eq. 12.13 and explain why actual head rise is always less than ideal head rise. Discuss how ideal head rise is head "added" to the fluid and actual head rise is head "gained" by the fluid. Can Eq. 12.13 be used for a turbine? Explain in terms of actual and ideal changes in head.

12.22 A centrifugal radial water pump has the dimensions shown in Fig. P12.22. The volume rate of flow is 7×10^3 m³/s and the absolute inlet velocity is directed radially outward. The angular velocity of the impeller is 960 rpm. The exit velocity as seen from a coordinate system attached to the impeller can be assumed to be tangent to the vane at its trailing edge. Calculate the power required to drive the pump.

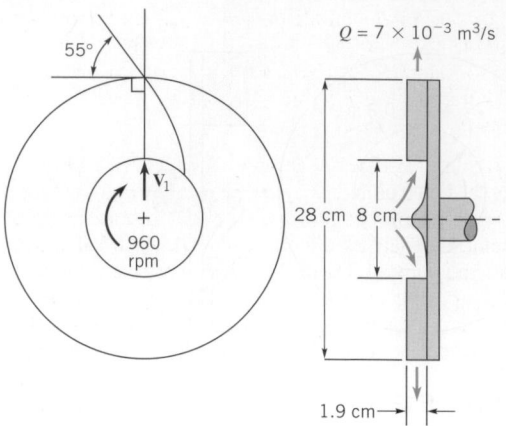

■ **F I G U R E   P12.22**

Section 12.4.2 Pump Performance Characteristics

12.23 Water is pumped with a centrifugal pump, and measurements made on the pump indicate that for a flowrate of 1 kL/min the required input power is 4.5 kW. For a pump efficiency of 62%, what is the actual head rise of the water being pumped?

12.24 The performance characteristics of a certain centrifugal pump are determined from an experimental setup similar to that shown in Fig. 12.10. When the flowrate of a liquid ($SG = 0.9$) through the pump is 0.5 kL/min, the pressure gage at (1) indicates a vacuum of 95 mm of mercury and the pressure gage at (2) indicates a pressure of 80 kPa. The diameter of the pipe at the inlet is 110 mm and at the exit it is 55 mm. If $z_2 - z_1 = 0.5$ m, what is the actual head rise across the pump? Explain how you would estimate the pump motor power requirement.

12.25 The performance characteristics of a certain centrifugal pump having a 23-cm-diameter impeller and operating at 1750 rpm are determined using an experimental setup similar to that shown in Fig. 12.10. The following data were obtained during a series of tests in which $z_2 - z_1 = 0$, $V_2 = V_1$, and the fluid was water.

Q (L/min)	75	150	225	300	375	450	525
$p_2 - p_1$ (kPa)	277	276	263	250	231	208	178
Power input (kW)	1.2	1.7	2.0	2.2	2.4	2.6	3.0

Based on these data, show or plot how the actual head rise, h_a, and the pump efficiency, η, vary with the flowrate. What is the design flowrate for this pump?

12.26 It is sometimes useful to have $h_a - Q$ pump performance curves expressed in the form of an equation. Fit the $h_a - Q$ data given in Problem 12.25 to an equation of the form $h_a = h_o - kQ^2$ and compare the values of h_a determined from the equation with the experimentally determined values. (*Hint*: Plot h_a versus Q^2 and use the method of least squares to fit the data to the equation.)

Section 12.4.3 Net Positive Suction Head (NPSH)

12.27 Obtain a photograph/image of cavitation damage to a centrifugal pump rotor. Is the damage where you expect it to occur? Explain.

12.28 In Example 12.3, how will the maximum height, z_1, that the pump can be located above the water surface change if the water temperature is decreased to 5 °C?

12.29 A centrifugal pump with a 18-cm-diameter impeller has the performance characteristics shown in Fig. 12.12. The pump is used to pump water at 38 °C, and the pump inlet is located 4 m above the open water surface. When the flowrate is 0.8 kL/min, the head loss between the water surface and the pump inlet is 2 m of water. Would you expect cavitation in the pump to be a problem? Assume standard atmospheric pressure. Explain how you arrived at your answer.

12.30 Water at 40 °C is pumped from an open tank through 200 m of 50-mm-diameter smooth horizontal pipe as shown in Fig. P12.30 and discharges into the atmosphere with a velocity of 3 m/s. Minor losses are negligible. **(a)** If the efficiency of the pump is 70%, how much power is being supplied to the pump? **(b)** What is the $NPSH_A$ at the pump inlet? Neglect losses in the short section of pipe connecting the pump to the tank. Assume standard atmospheric pressure.

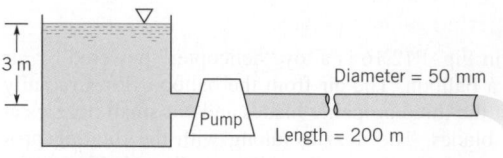

■ **F I G U R E   P12.30**

12.31 The centrifugal pump shown in Fig. P12.31 is not self-priming. That is, if the water is drained from the pump and pipe as shown in Fig. P12.31(*a*), the pump will not draw the water into the pump and start pumping when the pump is turned on. However, if the pump is primed [i.e., filled with water as in Fig. P12.31(*b*)], the pump does start pumping water when turned on. Explain this behavior.

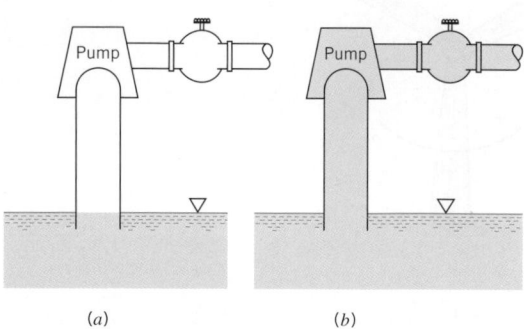

(*a*)   (*b*)

■ **F I G U R E   P12.31**

Section 12.4.4 System Characteristics and Pump Selection

12.32 Contrast the advantages and disadvantages of using pumps in parallel and in series.

12.33 Owing to fouling of the pipe wall, the friction factor for the pipe of Example 12.4 increases from 0.02 to 0.03. Determine the new flowrate, assuming all other conditions remain the same. What is the pump efficiency at this new flowrate? Explain how a line valve could be used to vary the flowrate through the pipe of Example 12.4. Would it be better to place the valve upstream or downstream of the pump? Why?

12.34 A centrifugal pump having a head-capacity relationship given by the equation $h_a = 54 - 1.2 \times 10^{-5}Q^2$, with h_a in meters when Q is in L/min, is to be used with a system similar to that shown in

Fig. 12.14. For $z_2 - z_1 = 15$ m, what is the expected flowrate if the total length of constant diameter pipe is 180 m and the fluid is water? Assume the pipe diameter to be 10 cm and the friction factor to be equal to 0.02. Neglect all minor losses.

12.35 A centrifugal pump having a 15-cm-diameter impeller and the characteristics shown in Fig. 12.12 is to be used to pump gasoline through 1200 m of commercial steel 8-cm-diameter pipe. The pipe connects two reservoirs having open surfaces at the same elevation. Determine the flowrate. Do you think this pump is a good choice? Explain.

12.36 Determine the new flowrate for the system described in Problem 12.35 if the pipe diameter is increased from 8 cm to 10 cm. Is this pump still a good choice? Explain.

12.37 A centrifugal pump having the characteristics shown in Example 12.4 is used to pump water between two large open tanks through 30 m of 20-cm-diameter pipe. The pipeline contains 4 regular flanged 90° elbows, a check valve, and a fully open globe valve. Other minor losses are negligible. Assume the friction factor $f = 0.02$ for the 30 m section of pipe. If the static head (difference in height of fluid surfaces in the two tanks) is 9 m, what is the expected flowrate? Do you think this pump is a good choice? Explain.

12.38 In a chemical processing plant a liquid is pumped from an open tank, through a 0.1-m-diameter vertical pipe, and into another open tank as shown in Fig. P12.38(*a*). A valve is located in the pipe, and the minor loss coefficient for the valve as a function of the valve setting is shown in Fig. P12.38(*b*). The pump head-capacity relationship is given by the equation $h_a = 52.0 - 1.01 \times 10^3 Q^2$ with h_a in meters when Q is in m³/s. Assume the friction

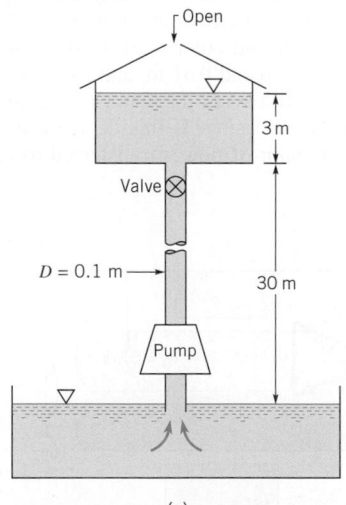

(*a*)

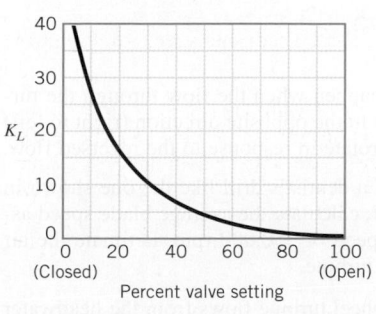

(*b*)

■ **F I G U R E P12.38**

factor $f = 0.02$ for the pipe, and all minor losses, except for the valve, are negligible. The fluid levels in the two tanks can be assumed to remain constant. **(a)** Determine the flowrate with the valve wide open. **(b)** Determine the required valve setting (percent open) to reduce the flowrate by 50%.

†12.39 Water is pumped between the two tanks described in Example 12.4 once a day, 365 days a year, with each pumping period lasting two hours. The water levels in the two tanks remain essentially constant. Estimate the annual cost of the electrical power needed to operate the pump if it were located in your city. You will have to make a reasonable estimate for the efficiency of the motor used to drive the pump. Due to aging, it can be expected that the overall resistance of the system will increase with time. If the operating point shown in Fig. E12.4*c* changes to a point where the flowrate has been reduced to 4 kL/min, what will be the new annual cost of operating the pump? Assume that the cost of electrical power remains the same.

Section 12.5 Dimensionless Parameters and Similarity Laws

12.40 Obtain photographs/images of a series of production pump rotors that suggest they are geometrically similar though different in feature size.

12.41 What is the rationale for operating two geometrically similar pumps differing in feature size at the same flow coefficient?

12.42 A centrifugal pump having an impeller diameter of 1 m is to be constructed so that it will supply a head rise of 200 m at a flowrate of 4.1 m³/s of water when operating at a speed of 1200 rpm. To study the characteristics of this pump, a 1/5 scale, geometrically similar model operated at the same speed is to be tested in the laboratory. Determine the required model discharge and head rise. Assume that both model and prototype operate with the same efficiency (and therefore the same flow coefficient).

12.43 A centrifugal pump with a 30-cm-diameter impeller requires a power input of 45 kW when the flowrate is 12 kL/min against a 18-m head. The impeller is changed to one with a 25 cm diameter. Determine the expected flowrate, head, and input power if the pump speed remains the same.

12.44 Do the head-flowrate data shown in Fig. 12.12 appear to follow the similarity laws as expressed by Eqs. 12.39 and 12.40? Explain.

12.45 A centrifugal pump has the performance characteristics of the pump with the 15-cm-diameter impeller described in Fig. 12.12. Note that the pump in this figure is operating at 3500 rpm. What is the expected head gained if the speed of this pump is reduced to 2800 rpm while operating at peak efficiency?

12.46 A centrifugal pump provides a flowrate of 2 kL/min when operating at 1750 rpm against a 30-m head. Determine the pump's flowrate and developed head if the pump speed is increased to 3500 rpm.

12.47 Explain how Fig. 12.18 was constructed from test data. Why is this use of specific speed important? Illustrate with a specific example.

12.48 Use the data given in Problem 12.25 and plot the dimensionless coefficients C_H, $C_{\mathscr{P}}$, η versus C_Q for this pump. Calculate a meaningful value of specific speed, discuss its usefulness, and compare the result with data of Fig. 12.18.

12.49 In a certain application a pump is required to deliver 19 kL/min against a 90-m head when operating at 1200 rpm. What type of pump would you recommend?

Section 12.6 **Axial-Flow and Mixed-Flow Pumps**

12.50 Obtain photographs/images of a variety of axial-flow and mixed-flow pump rotors. Explain any unusual features.

12.51 (See Fluids in the News Article titled "Mechanical heart assist devices" Section 12.6.) Obtain photographs/images of blood flow pumps that are turbomachines.

12.52 Explain how a marine propeller and an axial-flow pump are similar in the main effect they produce.

12.53 A certain axial-flow pump has a specific speed of $N_S = 5.0$. If the pump is expected to deliver 11 kL/min when operating against a 4.5-m head, at what speed (rpm) should the pump be run?

12.54 A certain pump is known to have a capacity of 3 m³/s when operating at a speed of 60 rad/s against a head of 20 m. Based on the information in Fig. 12.18, would you recommend a radial-flow, mixed-flow, or axial-flow pump?

12.55 Fuel oil (sp. wt = 7.5 kN/m³, viscosity = 9.6 × 10⁻⁴ N · s/m²) is pumped through the piping system of Fig. P12.55 with a velocity of 1.4 m/s. The pressure 60 m upstream from the pump is 35 kPa. Pipe losses downstream from the pump are negligible, but minor losses are not (minor loss coefficients are given on the figure). **(a)** For a pipe diameter of 5 cm with a relative roughness $\varepsilon/D = 0.001$, determine the head that must be added by the pump. **(b)** For a pump operating speed of 1750 rpm, what type of pump (radial-flow, mixed-flow, or axial-flow) would you recommend for this application?

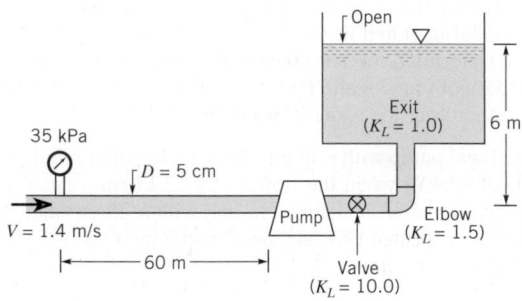

■ **F I G U R E   P12.55**

12.56 The axial-flow pump shown in Fig. 12.19 is designed to move 19 kL/min of water over a head rise of 1.5 m of water. Estimate the motor power requirement and the $U_2 V_{\theta 2}$ needed to achieve this flowrate on a continuous basis. Comment on any cautions associated with where the pump is placed vertically in the pipe.

Section 12.7 **Fans**

12.57 Obtain photographs/images of a variety of fan rotors and categorize them as axial-flow, radial-flow, or mixed-flow fans. Note any unusual features.

12.58 (See Fluids in the News Article titled "High-tech ceiling fans," Section 12.7.) Explain why reversing the direction of rotation of a ceiling fan results in airflow in the opposite direction.

12.59 For the fan of both Examples 5.19 and 5.28 discuss what fluid flow properties you would need to measure to estimate fan efficiency.

Section 12.8 **Turbines**

12.60 Obtain photographs/images of very small and very large turbine rotors, and explain briefly where each is used.

12.61 Consider the Pelton wheel turbine illustrated in Figs. 12.24, 12.25, 12.26, and 12.27. This kind of turbine is used to drive the oscillating sprinkler shown in **Video V12.3.** Explain how this kind of sprinkler is started, and subsequently operated at constant oscillating speed. What is the physical significance of the zero torque condition with the Pelton wheel rotating?

12.62 A small Pelton wheel is used to power an oscillating lawn sprinkler as shown in **Video V12.3** and Fig. P12.62. The arithmetic mean radius of the turbine is 2.5 cm, and the exit angle of the blade is 135° relative to the blade motion. Water is supplied through a single 0.5-cm-diameter nozzle at a speed of 15 m/s. Determine the flowrate, the maximum torque developed, and the maximum power developed by this turbine.

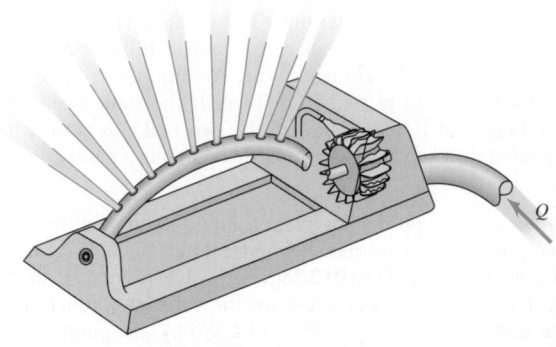

■ **F I G U R E   P12.62**

12.63 The single-stage, axial-flow turbomachine shown in Fig. P12.63 involves water flow at a volumetric flowrate of 9 m³/s. The rotor revolves at 600 rpm. The inner and outer radii of the annular flow path through the stage are 0.46 and 0.61 m, and $\beta_2 = 60°$. The flow entering the rotor row and leaving the stator row is axial when viewed from the stationary casing. Is this device a turbine or a pump? Estimate the amount of power transferred to or from the fluid.

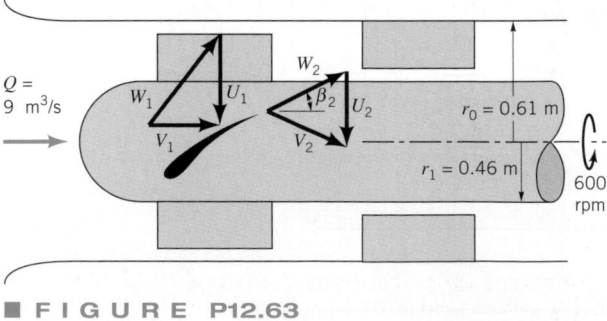

■ **F I G U R E   P12.63**

12.64 Describe what will happen when the flow through the turbomachine of Fig. P12.63 is in the opposite direction (right to left) and the shaft is freed up to rotate in response to the reversed flow.

12.65 For an air turbine of a dentist's drill like the one shown in Fig. E12.8 and **Video V12.4,** calculate the average blade speed associated with a rotational speed of 350,000 rpm. Estimate the air pressure needed to run this turbine.

12.66 Water for a Pelton wheel turbine flows from the headwater and through the penstock as shown in Fig. P12.66. The effective friction factor for the penstock, control valves, and the like is 0.032

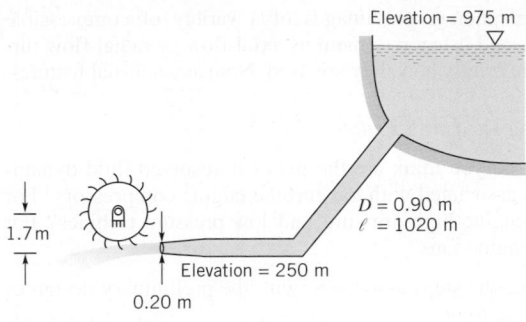

■ **F I G U R E P12.66**

and the diameter of the jet is 0.20 m. Determine the maximum power output.

12.67 Water to run a Pelton wheel is supplied by a penstock of length ℓ and diameter D with a friction factor f. If the only losses associated with the flow in the penstock are due to pipe friction, show that the maximum power output of the turbine occurs when the nozzle diameter, D_1, is given by $D_1 = D/(2f\,\ell/D)^{1/4}$.

12.68 A hydraulic turbine operating at 180 rpm with a head of 30 meters develops 15 MW. Estimate the power if the same turbine were to operate under a head of 15 m.

12.69 Draft tubes as shown in Fig. P12.69 are often installed at the exit of Kaplan and Francis turbines. Explain why such draft tubes are advantageous.

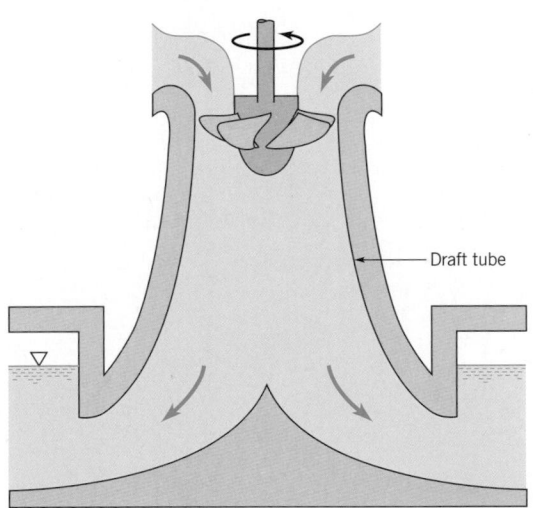

■ **F I G U R E P12.69**

12.70 Turbines are to be designed to develop 22 MW while operating under a head of 20 m and an angular velocity of 60 rpm. What type of turbine is best suited for this purpose? Estimate the flowrate needed.

12.71 Show how you would estimate the relationship between feature size and power production for a wind turbine like the one shown in **Video V12.1.**

12.72 Test data for the small Francis turbine shown in Fig. P12.72 is given in the table below. The test was run at a constant 10 m head just upstream of the turbine. The Prony brake on the turbine output shaft was adjusted to give various angular velocities, and the force on the brake arm, F, was recorded. Use the given data to plot curves of torque as a function of angular velocity and turbine efficiency as a function of angular velocity.

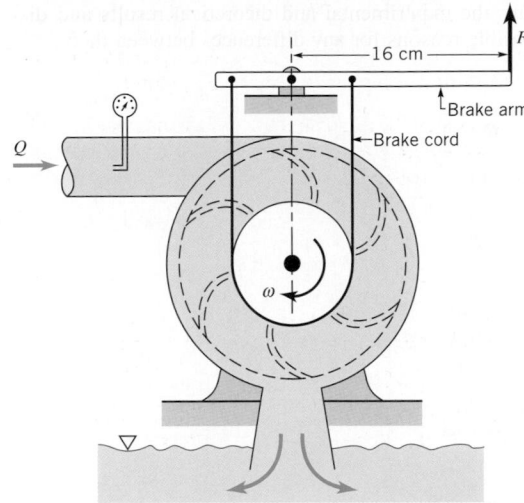

ω (rpm)	$Q(10^{-3} m^3/s)$	F (N)
0	3.6	12
1000	3.6	11
1500	3.6	10
1870	3.5	8.5
2170	3.3	6.6
2350	2.7	4.0
2580	2.2	1.5
2710	2.0	0.4

■ **F I G U R E P12.72**

†12.73 It is possible to generate power by using the water from your garden hose to drive a small Pelton wheel turbine (see **Video V12.3**). Provide a preliminary design of such a turbine and estimate the power output expected. List all assumptions and show calculations.

12.74 The device shown in Fig. P12.74 is used to investigate the power produced by a Pelton wheel turbine. Water supplied at a constant flowrate issues from a nozzle and strikes the turbine buckets as indicated. The angular velocity, ω, of the turbine wheel is varied by adjusting the tension on the Prony brake spring, thereby varying the torque, T_{shaft}, applied to the output shaft. This torque can be determined from the measured force, R, needed to keep the brake arm stationary as $T_{shaft} = F\ell$, where ℓ is the moment arm of the brake force.

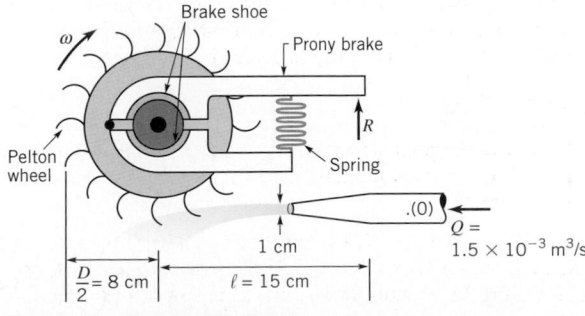

■ **F I G U R E P12.74**

Experimentally determined values of ω and R are shown in the following table. Use these results to plot a graph of torque as a function of the angular velocity. On another graph plot the power output, $\dot{W}_{shaft} = T_{shaft}\,\omega$, as a function of the angular velocity. On each of these graphs plot the theoretical curves for this turbine, assuming 100% efficiency.

Compare the experimental and theoretical results and discuss some possible reasons for any differences between them.

ω (rpm)	R (N)
0	11
360	8.5
450	8.2
600	7.5
700	7.0
940	5.2
1120	4.0
1480	0.7

Section 12.9 Compressible Flow Turbomachines

12.75 Obtain photographs/images of a variety of turbo-compressor rotors and categorize them as axial-flow or radial-flow compressors. Explain briefly how they are used. Note any unusual features.

12.76 Obtain photographs/images of a variety of compressible flow turbines and categorize them as axial-flow or radial-flow turbines. Explain briefly how they are used. Note any unusual features.

■ **Life Long Learning Problems**

12.77 What do you think are the major unresolved fluid dynamics problems associated with gas turbine engine compressors? For gas turbine engine high-pressure and low-pressure turbines? For gas turbine engine fans?

12.78 Outline the steps associated with the preliminary design of a turbomachine rotor.

12.79 What are current efficiencies achieved by the following categories of turbomachines? **(a)** Wind turbines; **(b)** hydraulic turbines; **(c)** power plant steam turbines; **(d)** aircraft gas turbine engines; **(e)** natural gas pipeline compressors; **(f)** home vacuum cleaner blowers; **(g)** laptop computer cooling fan; **(h)** irrigation pumps; **(i)** dentist drill air turbines. What is being done to improve these devices?

12.80 (See Fluids in the News Article titled "Cavitation damage in hydraulic turbines," Section 12.8.2.) How is cavitation and, more importantly, the damage it can cause detected in hydraulic turbines? How can this damage be minimized?

Appendix A
Computational Fluid Dynamics and FlowLab

A.1   Introduction

Numerical methods using digital computers are, of course, commonly utilized to solve a wide variety of flow problems. As discussed in Chapter 6, although the differential equations that govern the flow of Newtonian fluids [the Navier–Stokes equations (Eq. 6.127)] were derived many years ago, there are few known analytical solutions to them. However, with the advent of high-speed digital computers it has become possible to obtain approximate numerical solutions to these (and other fluid mechanics) equations for a wide variety of circumstances.

Computational fluid dynamics (CFD) involves replacing the partial differential equations with discretized algebraic equations that approximate the partial differential equations. These equations are then numerically solved to obtain flow field values at the discrete points in space and/or time. Since the Navier–Stokes equations are valid everywhere in the flow field of the fluid continuum, an analytical solution to these equations provides the solution for an infinite number of points in the flow. However, analytical solutions are available for only a limited number of simplified flow geometries. To overcome this limitation, the governing equations can be discretized and put in algebraic form for the computer to solve. The CFD simulation solves for the relevant flow variables only at the discrete points, which make up the grid or mesh of the solution (discussed in more detail below). Interpolation schemes are used to obtain values at non-grid point locations.

CFD can be thought of as a numerical experiment. In a typical fluids experiment, an experimental model is built, measurements of the flow interacting with that model are taken, and the results are analyzed. In CFD, the building of the model is replaced with the formulation of the governing equations and the development of the numerical algorithm. The process of obtaining measurements is replaced with running an algorithm on the computer to simulate the flow interaction. Of course, the analysis of results is common ground to both techniques.

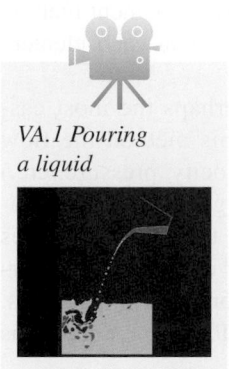

VA.1 Pouring a liquid

CFD can be classified as a subdiscipline to the study of fluid dynamics. However, it should be pointed out that a thorough coverage of CFD topics is well beyond the scope of this textbook. This appendix highlights some of the more important topics in CFD, but is only intended as a brief introduction. The topics include discretization of the governing equations, grid generation, boundary conditions, application of CFD, and some representative examples. Also included is a section on FlowLab, which is the educational CFD software incorporated with this textbook. FlowLab offers the reader the opportunity to begin using CFD to solve flow problems as well as to reinforce concepts covered in the textbook. For more information, go to the book's website, www.wiley.com/college/munson, to access the FlowLab problems, tutorials, and users guide.

A.2   Discretization

The process of *discretization* involves developing a set of algebraic equations (based on discrete points in the flow domain) to be used in place of the partial differential equations. Of the various discretization techniques available for the numerical solution of the governing differential equations, the following three types are most common: (1) the finite difference method, (2) the finite element (or finite volume) method, and (3) the boundary element method. In each of these methods, the continuous flow field (i.e., velocity or pressure as a function of space and time) is described in terms of discrete (rather than continuous) values at prescribed locations. Through this technique the differential equations are replaced by a set of algebraic equations that can be solved on the computer.

701

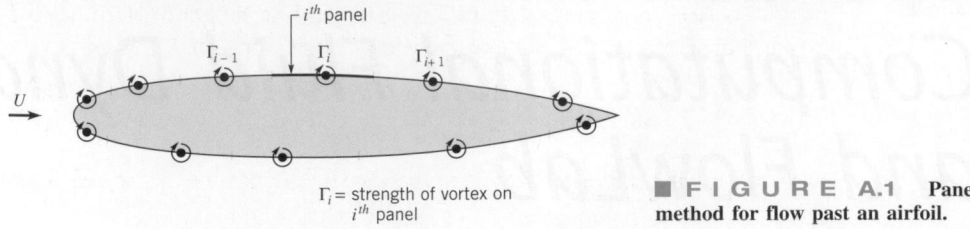

■ **FIGURE A.1**   **Panel method for flow past an airfoil.**

For the *finite element* (or *finite volume*) method, the flow field is broken into a set of small fluid elements (usually triangular areas if the flow is two-dimensional, or small volume elements if the flow is three-dimensional). The conservation equations (i.e., conservation of mass, momentum, and energy) are written in an appropriate form for each element, and the set of resulting algebraic equations for the flow field is solved numerically. The number, size, and shape of elements are dictated in part by the particular flow geometry and flow conditions for the problem at hand. As the number of elements increases (as is necessary for flows with complex boundaries), the number of simultaneous algebraic equations that must be solved increases rapidly. Problems involving one million (or more) grid cells are not uncommon in today's CFD community, particularly for complex three-dimensional geometries. Further information about this method can be found in Refs. 1 and 2.

For the *boundary element method*, the boundary of the flow field (not the entire flow field as in the finite element method) is broken into discrete segments (Ref. 3) and appropriate singularities such as sources, sinks, doublets, and vortices are distributed on these boundary elements. The strengths and type of the singularities are chosen so that the appropriate boundary conditions of the flow are obtained on the boundary elements. For points in the flow field not on the boundary, the flow is calculated by adding the contributions from the various singularities on the boundary. Although the details of this method are rather mathematically sophisticated, it may (depending on the particular problem) require less computational time and space than the finite element method. Typical boundary elements and their associated singularities (vortices) for two-dimensional flow past an airfoil are shown in Fig. A.1. Such use of the boundary element method in aerodynamics is often termed the *panel method* in recognition of the fact that each element plays the role of a panel on the airfoil surface (Ref. 4).

The *finite difference method* for computational fluid dynamics is perhaps the most easily understood and widely used of the three methods listed above. For this method the flow field is dissected into a set of grid points and the continuous functions (velocity, pressure, etc.) are approximated by discrete values of these functions calculated at the grid points. Derivatives of the functions are approximated by using the differences between the function values at local grid points divided by the grid spacing. The standard method for converting the partial differential equations to algebraic equations is through the use of Taylor series expansions. (See Ref. 5.) For example, assume a standard rectangular grid is applied to a flow domain as shown in Fig. A.2.

This grid stencil shows five grid points in $x-y$ space with the center point being labeled as i, j. This index notation is used as subscripts on variables to signify location. For example, $u_{i+1,j}$ is the u component of velocity at the first point to the right of the center point i, j. The grid spacing in the i and j directions is given as Δx and Δy, respectively.

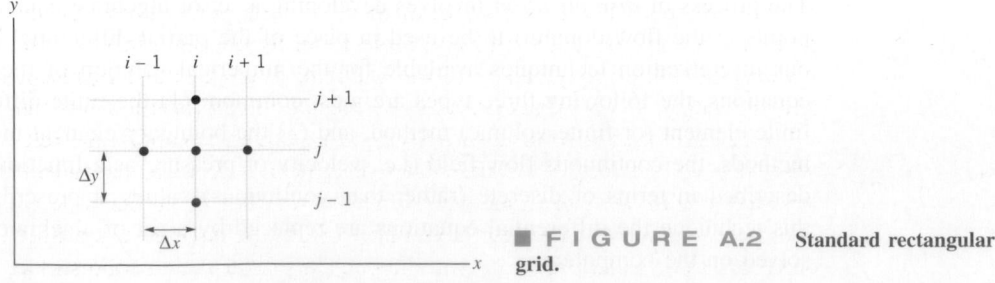

■ **FIGURE A.2**   **Standard rectangular grid.**

To find an algebraic approximation to a first derivative term such as $\partial u / \partial x$ at the i, j grid point, consider a Taylor series expansion written for u at $i + 1$ as

$$u_{i+1,j} = u_{i,j} + \underline{\left(\frac{\partial u}{\partial x}\right)_{i,j} \frac{\Delta x}{1!}} + \left(\frac{\partial^2 u}{\partial x^2}\right)_{i,j} \frac{(\Delta x)^2}{2!} + \left(\frac{\partial^3 u}{\partial x^3}\right)_{i,j} \frac{(\Delta x)^3}{3!} + \cdots \qquad \text{(A.1)}$$

Solving for the underlined term in the above equation results in the following:

$$\left(\frac{\partial u}{\partial x}\right)_{i,j} = \frac{u_{i+1,j} - u_{i,j}}{\Delta x} + O(\Delta x) \qquad \text{(A.2)}$$

where $O(\Delta x)$ contains higher order terms proportional to $\Delta x, (\Delta x)^2$, and so forth. Equation A.2 represents a forward difference equation to approximate the first derivative using values at $i + 1, j$ and i, j along with the grid spacing in the x direction. Obviously in solving for the $\partial u / \partial x$ term we have ignored higher order terms such as the second and third derivatives present in Eq. A.1. This process is termed *truncation* of the Taylor series expansion. The lowest order term that was truncated included $(\Delta x)^2$. Notice that the first derivative term contains Δx. When solving for the first derivative, all terms on the right-hand side were divided by Δx. Therefore, the term $O(\Delta x)$ signifies that this equation has error of "order (Δx)," which is due to the neglected terms in the Taylor series and is called truncation error. Hence, the forward difference is termed first-order accurate.

Thus, we can transform a partial derivative into an algebraic expression involving values of the variable at neighboring grid points. This method of using the Taylor series expansions to obtain discrete algebraic equations is called the finite difference method. Similar procedures can be used to develop approximations termed backward difference and central difference representations of the first derivative. The central difference makes use of both the left and right points (i.e., $i - 1, j$ and $i + 1, j$) and is second-order accurate. In addition, finite difference equations can be developed for the other spatial directions (i.e., $\partial u / \partial y$) as well as for second derivatives ($\partial^2 u / \partial x^2$), which are also contained in the Navier–Stokes equations (see Ref. 5 for details).

Applying this method to all terms in the governing equations transfers the differential equations into a set of algebraic equations involving the physical variables at the grid points (i.e., $u_{i,j}, p_{i,j}$ for $i = 1, 2, 3, \ldots$ and $j = 1, 2, 3, \ldots$, etc.). This set of equations is then solved by appropriate numerical techniques. The larger the number of grid points used, the larger the number of equations that must be solved.

A student of CFD should realize that the discretization of the continuum governing equations involves the use of algebraic equations that are an approximation to the original partial differential equation. Along with this approximation comes some amount of error. This type of error is termed truncation error because the Taylor series expansion used to represent a derivative is "truncated" at some reasonable point and the higher order terms are ignored. The truncation errors tend to zero as the grid is refined by making Δx and Δy smaller, so grid refinement is one method of reducing this type of error. Another type of unavoidable numerical error is the so-called round-off error. This type of error is due to the limit of the computer on the number of digits it can retain in memory. Engineering students can run into round-off errors from their calculators if they plug values into the equations at an early stage of the solution process. Fortunately, for most CFD cases, if the algorithm is setup properly, round-off errors are usually negligible.

A.3   Grids

CFD computations using the finite difference method provide the flow field at discrete points in the flow domain. The arrangement of these discrete points is termed the *grid* or the *mesh*. The type of grid developed for a given problem can have a significant impact on the numerical simulation, including the accuracy of the solution. The grid must represent the geometry correctly and accurately, since an error in this representation can have a significant effect on the solution.

The grid must also have sufficient grid resolution to capture the relevant flow physics, otherwise they will be lost. This particular requirement is problem dependent. For example, if a flow field has small-scale structures, the grid resolution must be sufficient to capture these structures. It is usually necessary to increase the number of grid points (i.e., use a finer mesh) where large gradients are to be expected, such as in the boundary layer near a solid surface. The same can also be

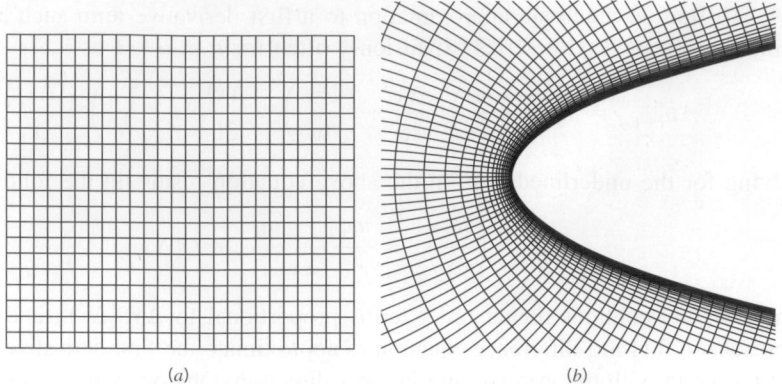

■ **F I G U R E   A.3**   Structured grids. (*a*) Rectangular grid.
(*b*) Grid around a parabolic surface.

said for the temporal resolution. The time step, Δt, used for unsteady flows must be smaller than the smallest time scale of the flow features being investigated.

Generally, the types of grids fall into two categories: structured and unstructured, depending on whether or not there exists a systematic pattern of connectivity of the grid points with their neighbors. As the name implies, a *structured grid* has some type of regular, coherent structure to the mesh layout that can be defined mathematically. The simplest structured grid is a uniform rectangular grid, as shown in Fig. A.3*a*. However, structured grids are not restricted to rectangular geometries. Fig. A.3*b* shows a structured grid wrapped around a parabolic surface. Notice that grid points are clustered near the surface (i.e., grid spacing in normal direction increases as one moves away from the surface) to help capture the steep flow gradients found in the boundary layer region. This type of variable grid spacing is used wherever there is a need to increase grid resolution and is termed grid stretching.

For the *unstructured grid*, the grid cell arrangement is irregular and has no systematic pattern. The grid cell geometry usually consists of various-sized triangles for two-dimensional problems and tetrahedrals for three-dimensional grids. An example of an unstructured grid is shown in Fig. A.4. Unlike structured grids, for an unstructured grid each grid cell and the connection information to neighboring cells is defined separately. This produces an increase in the computer code complexity as well as a significant computer storage requirement. The advantage to an unstructured grid is that it can be applied to complex geometries, where structured grids would have severe difficulty. The finite difference method is restricted to structured grids whereas the finite volume (or finite element) method can use either structured or unstructured grids.

Other grids include hybrid, moving, and adaptive grids. A grid that uses a combination of grid elements (rectangles, triangles, etc.) is termed a *hybrid grid*. As the name implies, the *moving grid*

VA.2 Dynamic grid

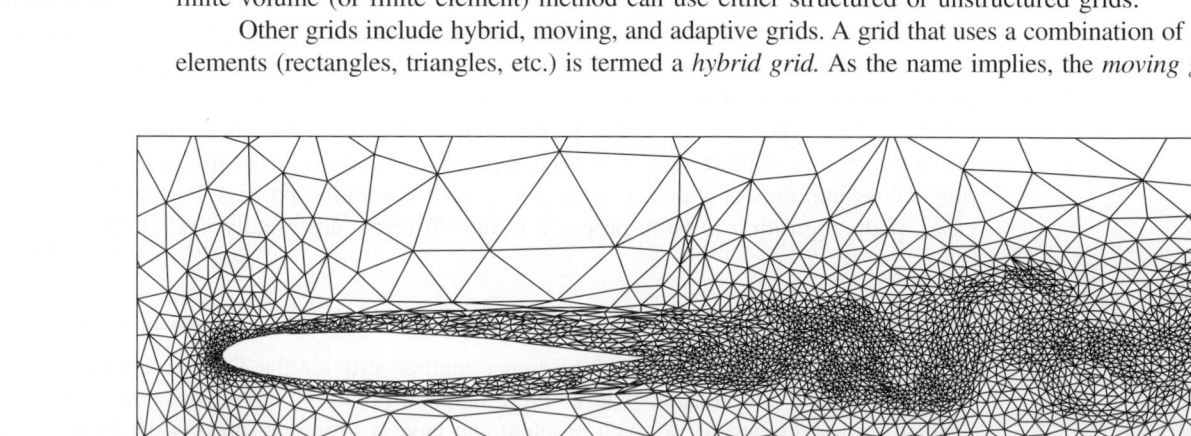

■ **F I G U R E   A.4**   Anisotropic adaptive mesh for the calculation of viscous flow over a NACA 0012 airfoil at a Reynolds number of 10,000, Mach number of 0.755, and angle of attack of 1.5°. (From CFD Laboratory, Concordia University, Montreal, Canada. Used by permission.)

is helpful for flows involving a time-dependent geometry. If, for example, the problem involves simulating the flow within a pumping heart or the flow around a flapping wing, a mesh that moves with the geometry is desired. The nature of the adaptive grid lies in its ability to literally adapt itself during the simulation. For this type of grid, while the CFD code is trying to reach a converged solution, the grid will adapt itself to place additional grid resources in regions of high flow gradients. Such a grid is particularly useful when a new problem arises and the user is not quite sure where to refine the grid due to high flow gradients.

A.4   Boundary Conditions

The same governing equations, the Navier–Stokes equations (Eq. 6.127), are valid for all incompressible Newtonian fluid flow problems. Thus, if the same equations are solved for all types of problems, how is it possible to achieve different solutions for different types of flows involving different flow geometries? The answer lies in the *boundary conditions* of the problem. The boundary conditions are what allow the governing equations to differentiate between different flow fields (for example, flow past an automobile and flow past a person running) and produce a solution unique to the given flow geometry.

It is critical to specify the correct boundary conditions so that the CFD simulation is a well-posed problem and is an accurate representation of the physical problem. Poorly defined boundary conditions can ultimately affect the accuracy of the solution. One of the most common boundary conditions used for simulation of viscous flow is the no-slip condition, as discussed in Section 1.6. Thus, for example, for two-dimensional external or internal flows, the x and y components of velocity (u and v) are set to zero at the stationary wall to satisfy the no-slip condition. Other boundary conditions that must be appropriately specified involve inlets, outlets, far-field, wall gradients, etc. It is important to not only select the correct physical boundary condition for the problem, but also to correctly implement this boundary condition into the numerical simulation.

A.5   Basic Representative Examples

A very simple one-dimensional example of the finite difference technique is presented in the following example.

EXAMPLE A.1   Flow from a Tank

A viscous oil flows from a large, open tank and through a long, small-diameter pipe as shown in Fig. EA.1a. At time $t = 0$ the fluid depth is H. Use a finite difference technique to determine the liquid depth as a function of time, $h = h(t)$. Compare this result with the exact solution of the governing equation.

SOLUTION

Although this is an unsteady flow (i.e., the deeper the oil, the faster it flows from the tank) we assume that the flow is "quasisteady" and apply steady flow equations as follows.

As shown by Eq. 6.152, the mean velocity, V, for steady laminar flow in a round pipe of diameter D is given by

$$V = \frac{D^2 \Delta p}{32 \mu \ell} \tag{1}$$

where Δp is the pressure drop over the length ℓ. For this problem the pressure at the bottom of the tank (the inlet of the pipe) is γh and that at the pipe exit is zero. Hence, $\Delta p = \gamma h$ and Eq. 1 becomes

$$V = \frac{D^2 \gamma h}{32 \mu \ell} \tag{2}$$

Conservation of mass requires that the flowrate from the tank, $Q = \pi D^2 V / 4$, is related to the rate of change of depth of oil in the tank, dh/dt, by

$$Q = -\frac{\pi}{4} D_T^2 \frac{dh}{dt}$$

where D_T is the tank diameter. Thus,

$$\frac{\pi}{4} D^2 V = -\frac{\pi}{4} D_T^2 \frac{dh}{dt}$$

or

$$V = -\left(\frac{D_T}{D}\right)^2 \frac{dh}{dt} \tag{3}$$

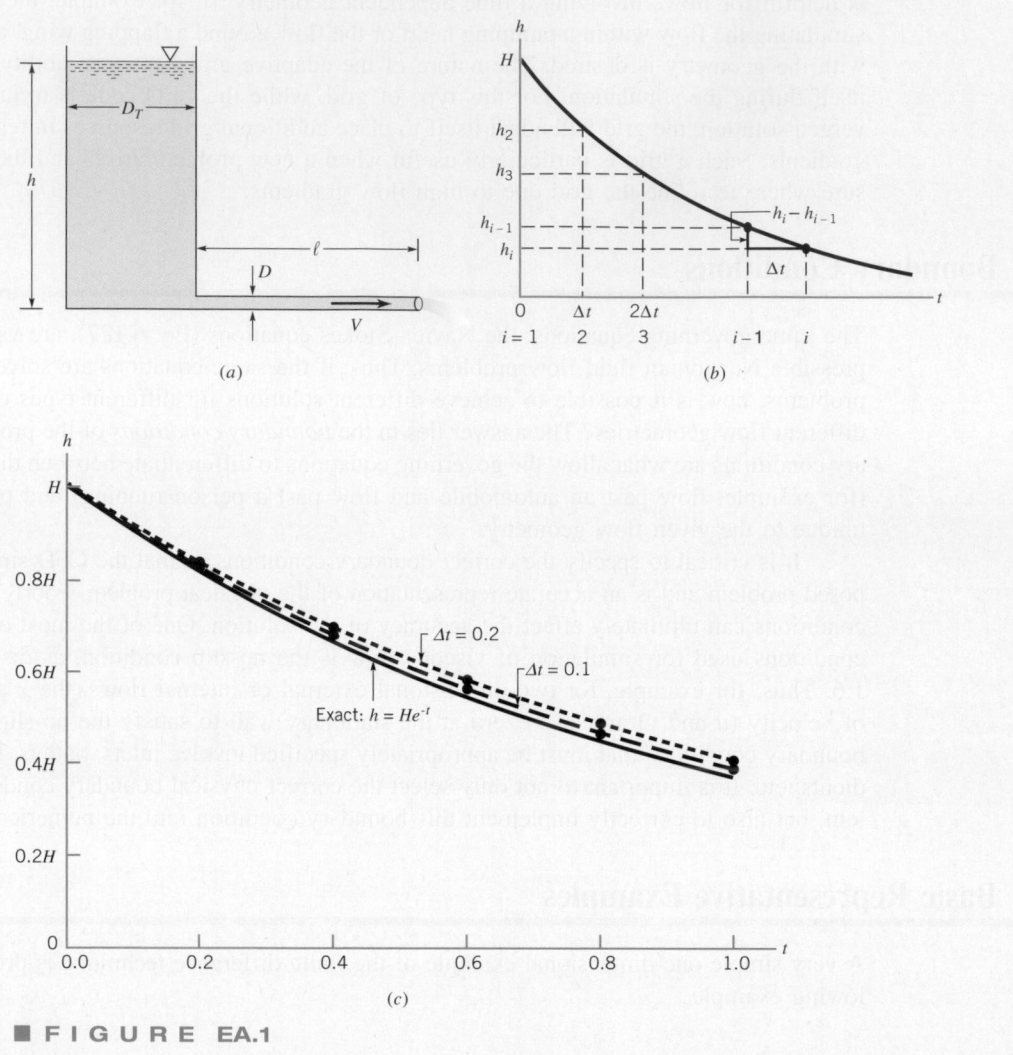

■ **F I G U R E   EA.1**

By combining Eqs. 2 and 3 we obtain

$$\frac{D^2\gamma h}{32\mu\ell} = -\left(\frac{D_T}{D}\right)^2\frac{dh}{dt}$$

or

$$\frac{dh}{dt} = -Ch$$

where $C = \gamma D^4/32\mu\ell D_T^2$ is a constant. For simplicity we assume the conditions are such that $C = 1$. Thus, we must solve

$$\frac{dh}{dt} = -h \quad \text{with} \quad h = H \text{ at } t = 0 \tag{4}$$

The exact solution to Eq. 4 is obtained by separating the variables and integrating to obtain

$$h = He^{-t} \tag{5}$$

However, assume this solution was not known. The following finite difference technique can be used to obtain an approximate solution.

As shown in Fig. EA.1b, we select discrete points (nodes or grid points) in time and approximate the time derivative of h by the expression

$$\left.\frac{dh}{dt}\right|_{t=t_i} \approx \frac{h_i - h_{i-1}}{\Delta t} \tag{6}$$

where Δt is the time step between the different node points on the time axis and h_i and h_{i-1} are the approximate values of h at nodes i and $i - 1$. Equation 6 is called the backward-difference approximation to dh/dt. We are free to select whatever value of Δt that we wish. (Although we do not need to space the nodes at equal distances, it is often convenient to do so.) Since the governing equation (Eq. 4) is an ordinary differential equation, the "grid" for the finite difference method is a one-dimensional grid as shown in Fig. EA.1b rather than a two-dimensional grid (which occurs for partial differential equations) as shown in Fig. EA.2b, or a three-dimensional grid.

Thus, for each value of $i = 2, 3, 4, \ldots$ we can approximate the governing equation, Eq. 4, as

$$\frac{h_i - h_{i-1}}{\Delta t} = -h_i$$

or

$$h_i = \frac{h_{i-1}}{(1 + \Delta t)} \quad (7)$$

We cannot use Eq. 7 for $i = 1$ since it would involve the non-existing h_0. Rather we use the initial condition (Eq. 4), which gives

$$h_1 = H$$

The result is the following set of N algebraic equations for the N approximate values of h at times $t_1 = 0, t_2 = \Delta t, \ldots, t_N = (N-1)\Delta t$.

$$h_1 = H$$

$$h_2 = h_1/(1 + \Delta t)$$

$$h_3 = h_2/(1 + \Delta t)$$

$$\vdots \qquad \vdots$$

$$h_N = h_{N-1}/(1 + \Delta t)$$

For most problems the corresponding equations would be more complicated than those just given, and a computer would be used to solve for the h_i. For this problem the solution is simply

$$h_2 = H/(1 + \Delta t)$$

$$h_3 = H/(1 + \Delta t)^2$$

$$\vdots \qquad \vdots$$

or in general

$$h_i = H/(1 + \Delta t)^{i-1}$$

The results for $0 < t < 1$ are shown in Fig. EA.1c. Tabulated values of the depth for $t = 1$ are listed in the table below.

Δt	i for $t = 1$	h_i for $t = 1$
0.2	6	0.4019H
0.1	11	0.3855H
0.01	101	0.3697H
0.001	1001	0.3681H
Exact (Eq. 5)	—	0.3678H

It is seen that the approximate results compare quite favorably with the exact solution given by Eq. 5. It is expected that the finite difference results would more closely approximate the exact results as Δt is decreased since in the limit of $\Delta t \to 0$ the finite difference approximation for the derivatives (Eq. 6) approaches the actual definition of the derivative.

For most CFD problems the governing equations to be solved are partial differential equations [rather than an ordinary differential equation as in the above example (Eq. A.1)] and the finite difference method becomes considerably more involved. The following example illustrates some of the concepts involved.

EXAMPLE A.2   Flow Past a Cylinder

Consider steady, incompressible flow of an inviscid fluid past a circular cylinder as shown in Fig. EA.2a. The stream function, ψ, for this flow is governed by the Laplace equation (see Section 6.5)

$$\frac{\partial^2 \psi}{\partial x^2} + \frac{\partial^2 \psi}{\partial y^2} = 0 \quad (1)$$

The exact analytical solution is given in Section 6.6.3.

Describe a simple finite difference technique that can be used to solve this problem.

SOLUTION

The first step is to define a flow domain and set up an appropriate grid for the finite difference scheme. Since we expect the flow field to be symmetrical both above and below and in front of and behind the cylinder, we consider only one-quarter of the entire flow domain as indicated in Fig. EA.2b. We locate the upper boundary and right-hand boundary far enough from the cylinder so that we expect the flow to be essentially uniform at these locations. It is not always clear how far from the object these boundaries must be located. If they are not far enough, the solution obtained will be incorrect because we have imposed artificial, uniform flow conditions at a location where the actual flow is not uniform. If these boundaries are farther than necessary from the object, the flow domain will be larger than necessary and excessive computer time and storage will be required. Experience in solving such problems is invaluable!

Once the flow domain has been selected, an appropriate grid is imposed on this domain (see Fig. EA.2b). Various grid structures can be used. If the grid is too coarse, the numerical solution may not be capable of capturing the fine scale structure of the actual flow field. If the grid is too fine, excessive computer time and

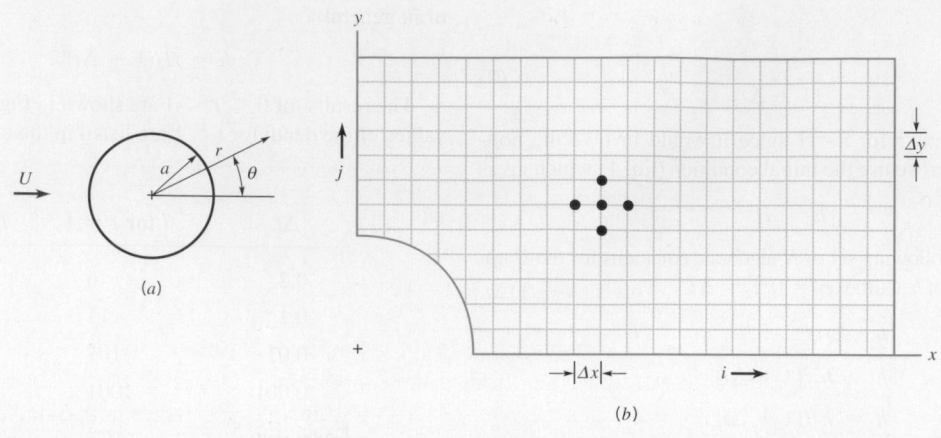

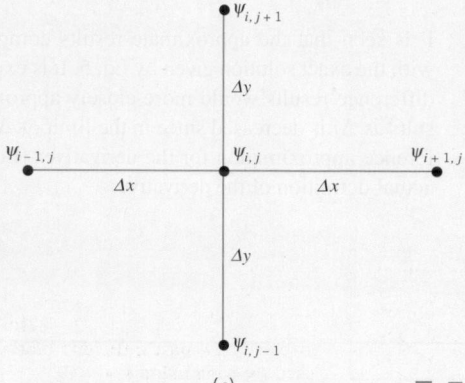

■ **FIGURE EA.2**

storage may be required. Considerable work has gone into forming appropriate grids (Ref. 6). We consider a grid that is uniformly spaced in the x and y directions, as shown in Fig. EA.2b.

As shown in Eq. 6.112, the exact solution to Eq. 1 (in terms of polar coordinates r, θ rather than Cartesian coordinates x, y) is $\psi = Ur(1 - a^2/r^2) \sin \theta$. The finite difference solution approximates these stream function values at a discrete (finite) number of locations (the grid points) as $\psi_{i,j}$, where the i and j indices refer to the corresponding x_i and y_j locations.

The derivatives of ψ can be approximated as follows:

$$\frac{\partial \psi}{\partial x} \approx \frac{1}{\Delta x} (\psi_{i+1,j} - \psi_{i,j})$$

and

$$\frac{\partial \psi}{\partial y} \approx \frac{1}{\Delta y} (\psi_{i,j+1} - \psi_{i,j})$$

This particular approximation is called a forward-difference approximation. Other approximations are possible. By similar reasoning, it is possible to show that the second derivatives of ψ can be written as follows:

$$\frac{\partial^2 \psi}{\partial x^2} \approx \frac{1}{(\Delta x)^2} (\psi_{i+1,j} - 2\psi_{i,j} + \psi_{i-1,j}) \tag{2}$$

and

$$\frac{\partial^2 \psi}{\partial y^2} \approx \frac{1}{(\Delta y)^2} (\psi_{i,j+1} - 2\psi_{i,j} + \psi_{i,j-1}) \tag{3}$$

Thus, by combining Eqs. 1, 2, and 3 we obtain

$$\frac{\partial^2 \psi}{\partial x^2} + \frac{\partial^2 \psi}{\partial y^2} \approx \frac{1}{(\Delta x)^2} (\psi_{i+1,j} + \psi_{i-1,j}) + \frac{1}{(\Delta y)^2} (\psi_{i,j+1} + \psi_{i,j-1}) - 2 \left(\frac{1}{(\Delta x)^2} + \frac{1}{(\Delta y)^2} \right) \psi_{i,j} = 0 \tag{4}$$

Equation 4 can be solved for the stream function at x_i and y_j to give

$$\psi_{i,j} = \frac{1}{2[(\Delta x)^2 + (\Delta y)^2]} \left[(\Delta y)^2 (\psi_{i+1,j} + \psi_{i-1,j}) + (\Delta x)^2 (\psi_{i,j+1} + \psi_{i,j-1}) \right] \tag{5}$$

Note that the value of $\psi_{i,j}$ depends on the values of the stream function at neighboring grid points on either side and above and below the point of interest (see Eq. 5 and Fig. EA. 2c).

To solve the problem (either exactly or by the finite difference technique) it is necessary to specify boundary conditions for points located on the boundary of the flow domain (see Section 6.6.3). For example, we may specify that $\psi = 0$ on the lower boundary of the domain (see Fig. EA.2b) and $\psi = C$, a constant, on the upper boundary of the domain. Appropriate boundary conditions on the two vertical ends of the flow domain can also be specified. Thus, for points interior to the boundary Eq. 5 is valid; similar equations or specified values of $\psi_{i,j}$ are valid for boundary points. The result is an equal number of equations and unknowns, $\psi_{i,j}$, one for every grid point. For this problem, these equations represent a set of linear algebraic equations for $\psi_{i,j}$, the solution of which provides the finite difference approximation for the stream function at discrete grid points in the flow field. Streamlines (lines of constant ψ) can be obtained by interpolating values of $\psi_{i,j}$ between the grid points and "connecting the dots" of $\psi = $ constant. The velocity field can be obtained from the derivatives of the stream function according to Eq. 6.74. That is,

$$u = \frac{\partial \psi}{\partial y} \approx \frac{1}{\Delta y}(\psi_{i,j+1} - \psi_{i,j})$$

and

$$v = -\frac{\partial \psi}{\partial x} \approx -\frac{1}{\Delta x}(\psi_{i+1,j} - \psi_{i,j})$$

Further details of the finite difference technique can be found in standard references on the topic (Refs. 5, 7, 8). Also, see the completely solved viscous flow CFD problem in Section A6.

The preceding two examples are rather simple because the governing equations are not too complex. A finite difference solution of the more complicated, nonlinear Navier–Stokes equation (Eq. 6.127) requires considerably more effort and insight and larger and faster computers. A typical finite difference grid for a more complex flow, the flow past a turbine blade, is shown in Fig. A.5. Note that the mesh is much finer in regions where large gradients are to be expected (i.e., near the leading and trailing edges of the blade) and more coarse away from the blade.

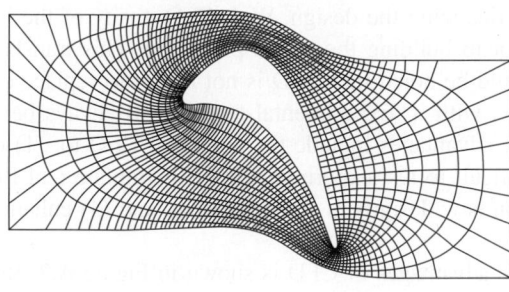

■ FIGURE A.5 Finite difference grid for flow past a turbine blade. (From Ref. 9, used by permission.)

A.6 Methodology

In general, most applications of CFD take the same basic approach. Some of the differences include problem complexity, available computer resources, available expertise in CFD, and whether a commercially available CFD package is used, or a problem-specific CFD algorithm is developed. In today's market, there are many commercial CFD codes available to solve a wide variety of problems. However, if the intent is to conduct a thorough investigation of a specific fluid flow problem such as in a research environment, it is possible that taking the time to develop a problem-specific algorithm may be most efficient in the long run. The features common to most CFD applications can be summarized in the flow chart shown in Fig. A.6. A complete, detailed CFD solution for a viscous flow obtained by using the steps summarized in the flow chart can be accessed from the book's website at www.wiley.com/college/munson.

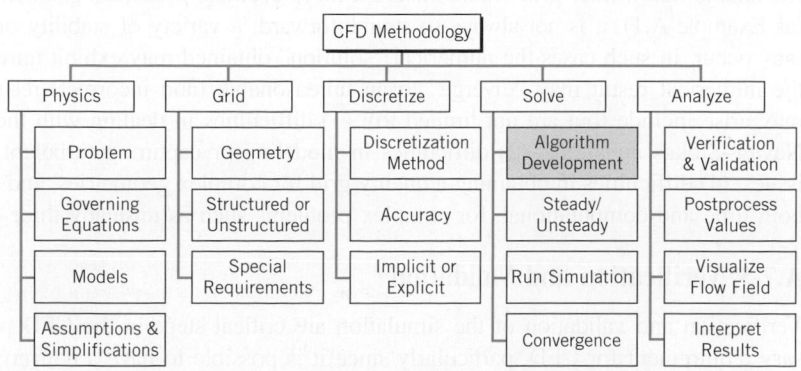

■ FIGURE A.6 Flow chart of general CFD methodology.

The Algorithm Development box is grayed because this step is required only when developing your own CFD code. When using a commercial CFD code, this step is not necessary. This chart represents a generalized methodology to CFD. There are other more complex components that are hidden in the above steps, which are beyond the scope of a brief introduction to CFD.

A.7   Application of CFD

In the early stages of CFD, research and development was primarily driven by the aerospace industry. Today, CFD is still used as a research tool, but it also has found a place in industry as a design tool. There is now a wide variety of industries that make at least some use of CFD, including automotive, industrial, HVAC, naval, civil, chemical, biological, and others. Industries are using CFD as an added engineering tool that complements the experimental and theoretical work in fluid dynamics.

A.7.1 Advantages of CFD

There are many advantages to using CFD for simulation of fluid flow. One of the most important advantages is the realizable savings in time and cost for engineering design. In the past, coming up with a new engineering design meant somewhat of a trial-and-error method of building and testing multiple prototypes prior to finalizing the design. With CFD, many of the issues dealing with fluid flow can be flushed out prior to building the actual prototype. This translates to a significant savings in time and cost. It should be noted that CFD is not meant to replace experimental testing, but rather to work in conjunction with it. Experimental testing will always be a necessary component of engineering design. Other advantages include the ability of CFD to: (1) obtain flow information in regions that would be difficult to test experimentally, (2) simulate real flow conditions, (3) conduct large parametric tests on new designs in a shorter time, and (4) enhance visualization of complex flow phenomena.

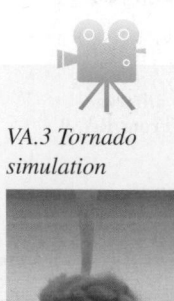

VA.3 Tornado simulation

A good example of the advantages of CFD is shown in Figure A.7. Researchers use a type of CFD approach called "large-eddy simulation" or LES to simulate the fluid dynamics of a tornado as it encounters a debris field and begins to pick up sand-sized particles. A full animation of this tornado simulation can be accessed by visiting the book website. The motivation for this work is to investigate whether there are significant differences in the fluid mechanics when debris particles are present. Historically it has been difficult to get comprehensive experimental data throughout a tornado so CFD is helping to shine some light on the complex fluid dynamics involved in such a flow.

A.7.2 Difficulties in CFD

One of the key points that a beginning CFD student should understand is that one cannot treat the computer as a "magic black box" when performing flow simulations. It is quite possible to obtain a fully converged solution for the CFD simulation, but this is no guarantee that the results are physically correct. This is why it is important to have a good understanding of the flow physics and how they are modeled. Any numerical technique (including those discussed above), no matter how simple in concept, contains many hidden subtleties and potential problems. For example, it may seem reasonable that a finer grid would ensure a more accurate numerical solution. While this may be true (as Example A.1), it is not always so straightforward; a variety of stability or convergence problems may occur. In such cases the numerical "solution" obtained may exhibit unreasonable oscillations or the numerical result may "diverge" to an unreasonable (and incorrect) result. Other problems that may arise include (but are not limited to): (1) difficulties in dealing with the nonlinear terms of the Navier–Stokes equations, (2) difficulties in modeling or capturing turbulent flows, (3) convergence issues, (4) difficulties in obtaining a quality grid for complex geometries, and (5) managing resources, both time and computational, for complex problems such as unsteady three-dimensional flows.

A.7.3 Verification and Validation

Verification and validation of the simulation are critical steps in the CFD process. This is a necessary requirement for CFD, particularly since it is possible to have a converged solution that is non-physical. Figure A.8 shows the streamlines for viscous flow past a circular cylinder at a given instant

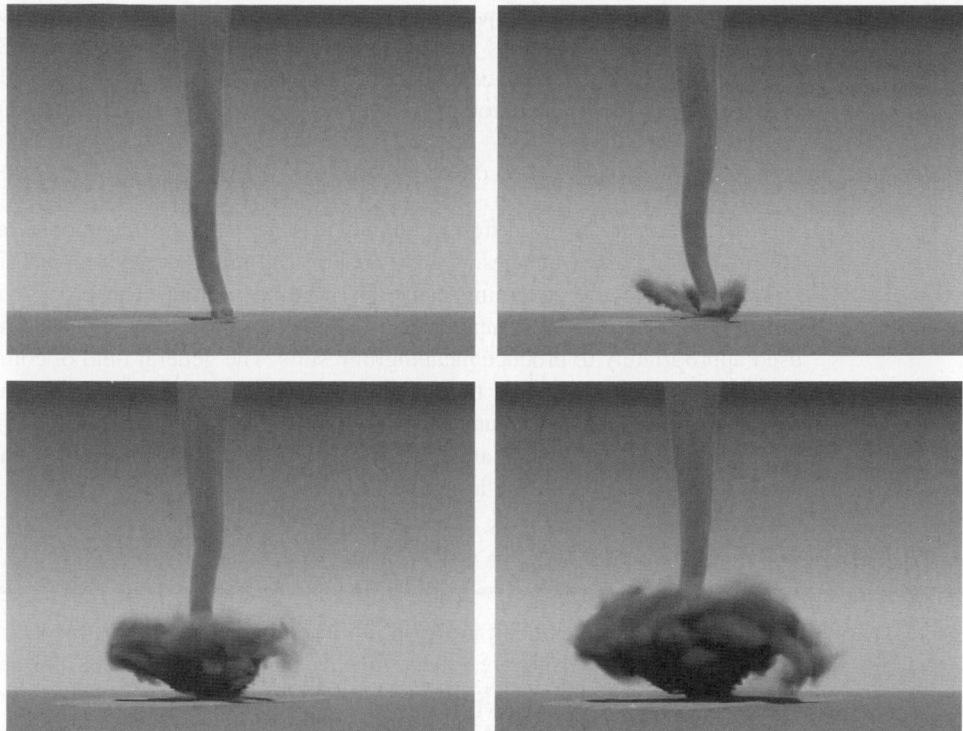

■ FIGURE A.7   **Results from a large-eddy simulation showing the visual appearance of the debris and funnel cloud from a simulated medium swirl F3-F4 tornado. The funnel cloud is translating at 15 m/s and is ingesting 1-mm-diameter "sand" from the surface as it encounters a debris field. Please visit the book website to access a full animation of this tornado simulation. (Photographs and animation courtesy of Dr. David Lewellen, Ref. 10, and Paul Lewellen, West Virginia University.)**

after it was impulsively started from rest. The lower half of the figure represents the results of a finite difference calculation; the upper half of the figure represents the photograph from an experiment of the same flow situation. It is clear that the numerical and experimental results agree quite well. For any CFD simulation, there are several levels of testing that need to be accomplished before one can have confidence in the solution. The most important verification to be performed is grid convergence testing. In its simplest form, it consists of proving that further refinement of the grid (i.e., increasing the number of grid points) does not alter the final solution. When this has been achieved, you have a grid-independent solution. Other verification factors that need to be investigated include the suitability

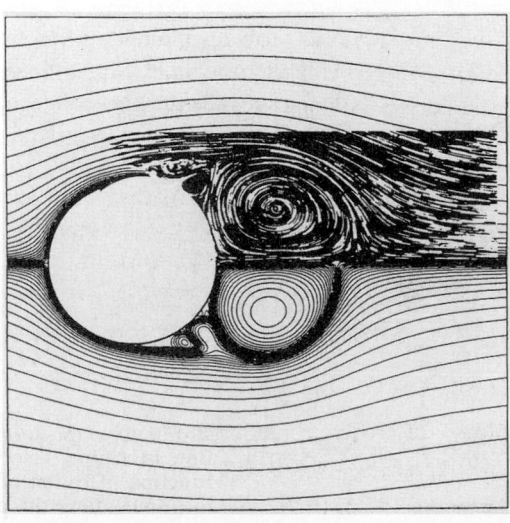

■ FIGURE A.8   **Streamlines for flow past a circular cylinder at a short time after the flow was impulsively started. The upper half is a photograph from a flow visualization experiment. The lower half is from a finite difference calculation. (See the photograph at the beginning of Chapter 9.) (From Ref. 9, used by permission.)**

of the convergence criterion, whether the time step is adequate for the time scale of the problem, and comparison of CFD solutions to existing data, at least for baseline cases. Even when using a commercial CFD code that has been validated on many problems in the past, the CFD practitioner still needs to verify the results through such measures as grid-dependence testing.

A.7.4 Summary

In CFD, there are many different numerical schemes, grid techniques, etc. They all have their advantages and disadvantages. A great deal of care must be used in obtaining approximate numerical solutions to the governing equations of fluid motion. The process is not as simple as the often-heard "just let the computer do it." Remember that CFD is a tool and as such needs to be used appropriately to produce meaningful results. The general field of computational fluid dynamics, in which computers and numerical analysis are combined to solve fluid flow problems, represents an extremely important subject area in advanced fluid mechanics. Considerable progress has been made in the past relatively few years, but much remains to be done. The reader is encouraged to consult some of the available literature.

A.8   FlowLab

The authors of this textbook are working in collaboration with Fluent, Inc., the largest provider of commercial CFD software (www.fluent.com), to offer students the opportunity to use a new CFD tool called FlowLab. FlowLab is designed to be a virtual fluids laboratory to help enhance the educational experience in fluids courses. It uses computational fluid dynamics to help the student grasp various concepts in fluid dynamics and introduces the student to the use of CFD in solving fluid flow problems. Go to the book's website at www.wiley.com/college/munson to access FlowLab resources for this textbook.

The motivation behind incorporating FlowLab with a fundamental fluid mechanics textbook is twofold: (1) expose the student to computational fluid dynamics and (2) offer a mechanism for students to conduct experiments in fluid dynamics, numerically in this case. This educational software allows students to reinforce basic concepts covered in class, conduct parametric studies to gain a better understanding of the interaction between geometry, fluid properties, and flow conditions, and provides the student a visualization tool for various flow phenomena.

One of the strengths of FlowLab is the ease-of-use. The CFD simulations are based on previously developed templates which allow the user to start using CFD to solve flow problems without requiring an extensive background in the subject. FlowLab provides the student the opportunity to focus on the results of the simulation rather than the development of the simulation. Typical results showing the developing velocity profile in the entrance region of a pipe are shown in the solution window of Fig. A.9.

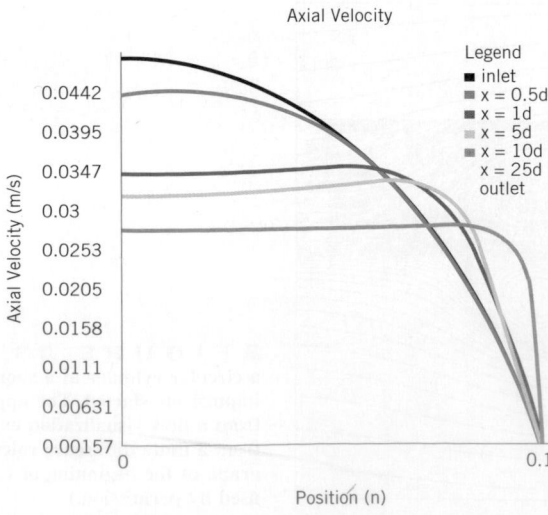

■ **F I G U R E   A.2**   **Entrance flow in a pipe. Velocity profiles as a function of radial position for various locations along the pipe length.**

Problems have been developed that take advantage of the FlowLab capability of this textbook. Go to the book's website, www.wiley.com/college/munson, to access these problems (contained in Chapters 7, 8, and 9) as well as a basic tutorial on using FlowLab. The course instructor can provide information on accessing the FlowLab software. The book's website also has a brief example using FlowLab.

References

1. Baker, A. J., *Finite Element Computational Fluid Mechanics*, McGraw-Hill, New York, 1983.
2. Carey, G. F., and Oden, J. T., *Finite Elements: Fluid Mechanics*, Prentice-Hall, Englewood Cliffs, N.J., 1986.
3. Brebbia, C. A., and Dominguez, J., *Boundary Elements: An Introductory Course*, McGraw-Hill, New York, 1989.
4. Moran, J., *An Introduction to Theoretical and Computational Aerodynamics*, Wiley, New York, 1984.
5. Anderson, J.D., *Computational Fluid Dynamics: The Basics with Applications*, McGraw-Hill, New York, 1995.
6. Thompson, J. F., Warsi, Z. U. A., and Mastin, C. W., *Numerical Grid Generation: Foundations and Applications*, North-Holland, New York, 1985.
7. Peyret, R., and Taylor, T. D., *Computational Methods for Fluid Flow*, Springer-Verlag, New York, 1983.
8. Tannehill, J. C., Anderson, D. A., and Pletcher, R. H., *Computational Fluid Mechanics and Heat Transfer*, 2nd Ed., Taylor and Francis, Washington, D.C., 1997.
9. Hall, E. J., and Pletcher, R. H., *Simulation of Time Dependent, Compressible Viscous Flow Using Central and Upwind-Biased Finite-Difference Techniques*, Technical Report HTL-52, CFD-22, College of Engineering, Iowa State University, 1990.
10. Lewellen, D. C., Gong, B., and Lewellen, W. S., *Effects of Debris on Near-Surface Tornado Dynamics*, 22nd Conference on Severe Local Storms, Paper 15.5, American Meteorological Society, 2004.

Appendix B
Physical Properties of Fluids

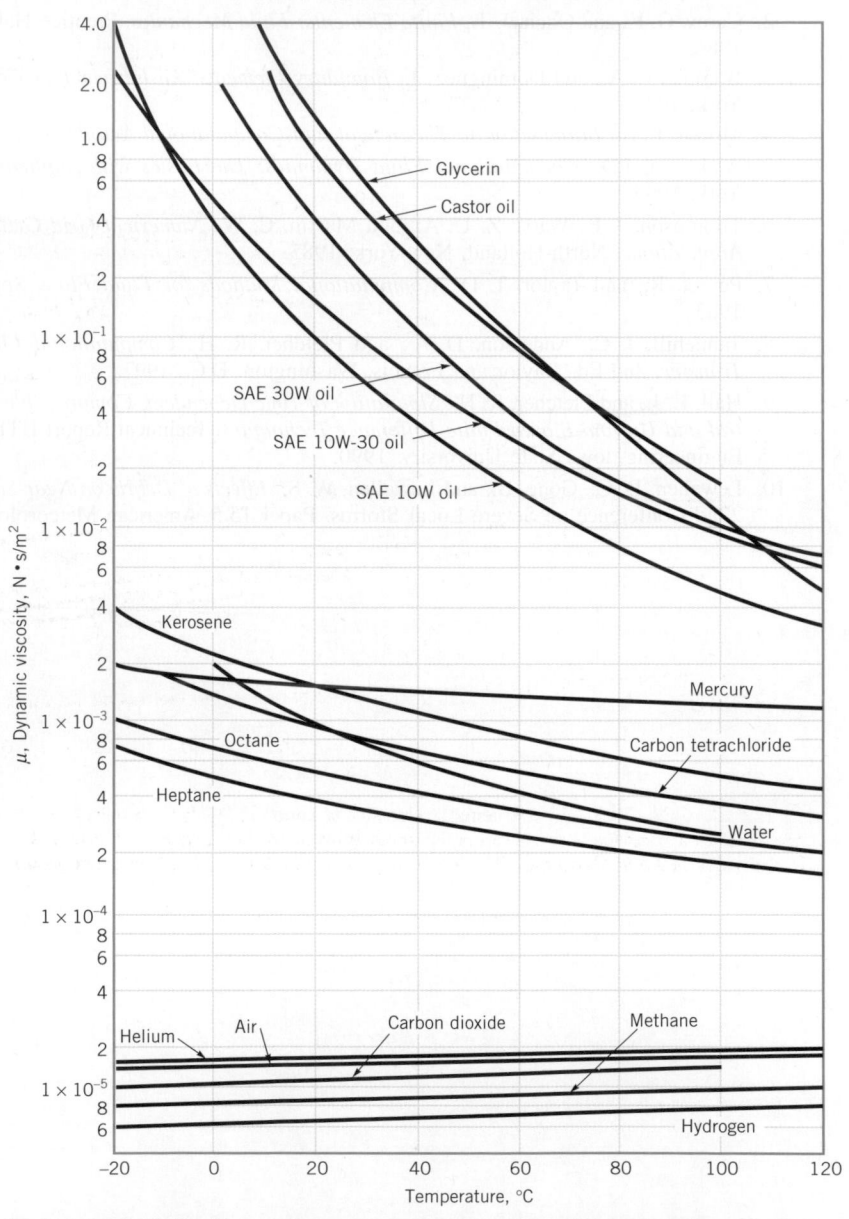

■ FIGURE B.1   Dynamic (absolute) viscosity of common fluids as a function of temperature. (Curves from R. W. Fox and A. T. McDonald, *Introduction to Fluid Mechanics*, 3rd Ed., Wiley, New York, 1985. Used by permission.)

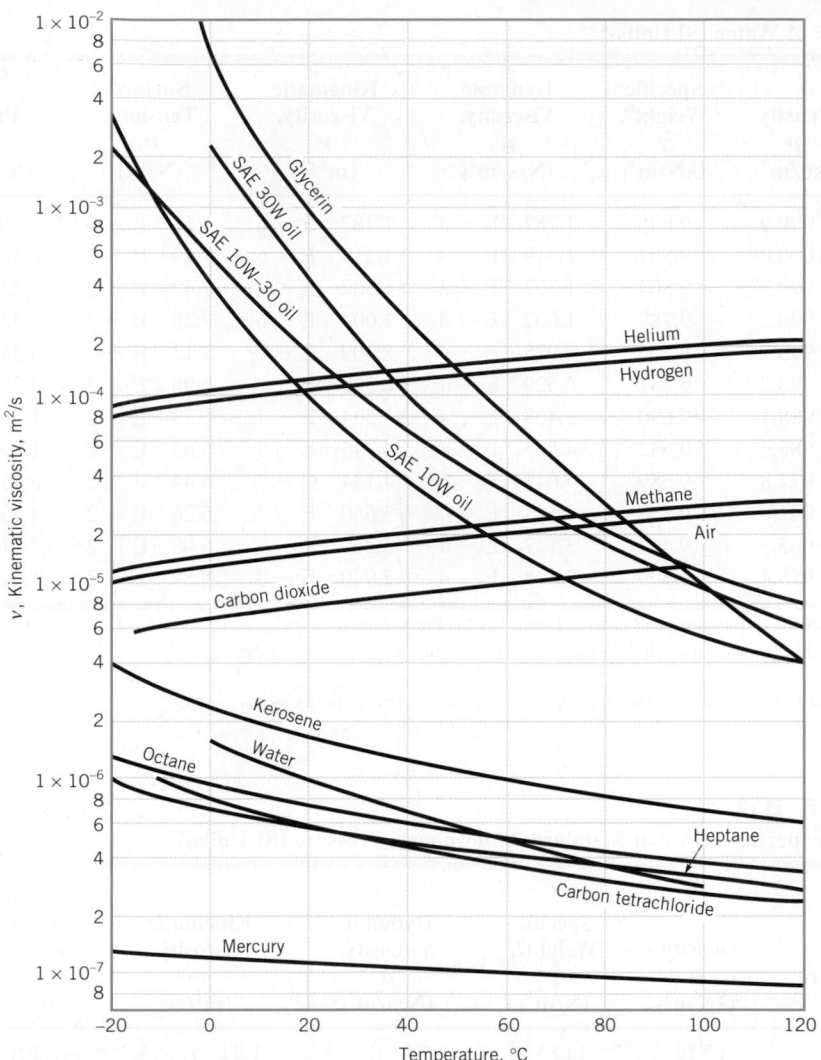

■ **FIGURE B.2**   Kinematic viscosity of common fluids (at atmospheric pressure) as a function of temperature. (Curves from R. W. Fox and A. T. McDonald, *Introduction to Fluid Mechanics*, 3rd Ed., Wiley, New York, 1985. Used by permission.)

■ **TABLE B.1**
Physical Properties of Water (SI Units)[a]

Temperature (°C)	Density, ρ (kg/m³)	Specific Weight[b], γ (kN/m³)	Dynamic Viscosity, μ (N·s/m²)	Kinematic Viscosity, ν (m²/s)	Surface Tension[c], σ (N/m)	Vapor Pressure, p_v [N/m²(abs)]	Speed of Sound[d], c (m/s)
0	999.9	9.806	1.787 E − 3	1.787 E − 6	7.56 E − 2	6.105 E + 2	1403
5	1000.0	9.807	1.519 E − 3	1.519 E − 6	7.49 E − 2	8.722 E + 2	1427
10	999.7	9.804	1.307 E − 3	1.307 E − 6	7.42 E − 2	1.228 E + 3	1447
20	998.2	9.789	1.002 E − 3	1.004 E − 6	7.28 E − 2	2.338 E + 3	1481
30	995.7	9.765	7.975 E − 4	8.009 E − 7	7.12 E − 2	4.243 E + 3	1507
40	992.2	9.731	6.529 E − 4	6.580 E − 7	6.96 E − 2	7.376 E + 3	1526
50	988.1	9.690	5.468 E − 4	5.534 E − 7	6.79 E − 2	1.233 E + 4	1541
60	983.2	9.642	4.665 E − 4	4.745 E − 7	6.62 E − 2	1.992 E + 4	1552
70	977.8	9.589	4.042 E − 4	4.134 E − 7	6.44 E − 2	3.116 E + 4	1555
80	971.8	9.530	3.547 E − 4	3.650 E − 7	6.26 E − 2	4.734 E + 4	1555
90	965.3	9.467	3.147 E − 4	3.260 E − 7	6.08 E − 2	7.010 E + 4	1550
100	958.4	9.399	2.818 E − 4	2.940 E − 7	5.89 E − 2	1.013 E + 5	1543

[a]Based on data from *Handbook of Chemistry and Physics*, 69th Ed., CRC Press, 1988.
[b]Density and specific weight are related through the equation $\gamma = \rho g$. For this table, $g = 9.807$ m/s².
[c]In contact with air.
[d]From R. D. Blevins, *Applied Fluid Dynamics Handbook*, Van Nostrand Reinhold Co., Inc., New York, 1984.

■ **TABLE B.2**
Physical Properties of Air at Standard Atmospheric Pressure (SI Units)[a]

Temperature (°C)	Density, ρ (kg/m³)	Specific Weight[b], γ (N/m³)	Dynamic Viscosity, μ (N·s/m²)	Kinematic Viscosity, ν (m²/s)	Specific Heat Ratio, k (—)	Speed of Sound, c (m/s)
−40	1.514	14.85	1.57 E − 5	1.04 E − 5	1.401	306.2
−20	1.395	13.68	1.63 E − 5	1.17 E − 5	1.401	319.1
0	1.292	12.67	1.71 E − 5	1.32 E − 5	1.401	331.4
5	1.269	12.45	1.73 E − 5	1.36 E − 5	1.401	334.4
10	1.247	12.23	1.76 E − 5	1.41 E − 5	1.401	337.4
15	1.225	12.01	1.80 E − 5	1.47 E − 5	1.401	340.4
20	1.204	11.81	1.82 E − 5	1.51 E − 5	1.401	343.3
25	1.184	11.61	1.85 E − 5	1.56 E − 5	1.401	346.3
30	1.165	11.43	1.86 E − 5	1.60 E − 5	1.400	349.1
40	1.127	11.05	1.87 E − 5	1.66 E − 5	1.400	354.7
50	1.109	10.88	1.95 E − 5	1.76 E − 5	1.400	360.3
60	1.060	10.40	1.97 E − 5	1.86 E − 5	1.399	365.7
70	1.029	10.09	2.03 E − 5	1.97 E − 5	1.399	371.2
80	0.9996	9.803	2.07 E − 5	2.07 E − 5	1.399	376.6
90	0.9721	9.533	2.14 E − 5	2.20 E − 5	1.398	381.7
100	0.9461	9.278	2.17 E − 5	2.29 E − 5	1.397	386.9
200	0.7461	7.317	2.53 E − 5	3.39 E − 5	1.390	434.5
300	0.6159	6.040	2.98 E − 5	4.84 E − 5	1.379	476.3
400	0.5243	5.142	3.32 E − 5	6.34 E − 5	1.368	514.1
500	0.4565	4.477	3.64 E − 5	7.97 E − 5	1.357	548.8
1000	0.2772	2.719	5.04 E − 5	1.82 E − 4	1.321	694.8

[a]Based on data from R. D. Blevins, *Applied Fluid Dynamics Handbook*, Van Nostrand Reinhold Co., Inc., New York, 1984.
[b]Density and specific weight are related through the equation $\gamma = \rho g$. For this table $g = 9.807$ m/s².

Appendix C
Properties of the U.S. Standard Atmosphere

■ **TABLE C.1**
Properties of the U.S. Standard Atmosphere (SI Units)[a]

Altitude (m)	Temperature (°C)	Acceleration of Gravity, g (m/s²)	Pressure, p [N/m²(abs)]		Density, ρ (kg/m³)		Dynamic Viscosity, μ (N·s/m²)	
−1,000	21.50	9.810	1.139	E + 5	1.347	E + 0	1.821	E − 5
0	15.00	9.807	1.013	E + 5	1.225	E + 0	1.789	E − 5
1,000	8.50	9.804	8.988	E + 4	1.112	E + 0	1.758	E − 5
2,000	2.00	9.801	7.950	E + 4	1.007	E + 0	1.726	E − 5
3,000	−4.49	9.797	7.012	E + 4	9.093	E − 1	1.694	E − 5
4,000	−10.98	9.794	6.166	E + 4	8.194	E − 1	1.661	E − 5
5,000	−17.47	9.791	5.405	E + 4	7.364	E − 1	1.628	E − 5
6,000	−23.96	9.788	4.722	E + 4	6.601	E − 1	1.595	E − 5
7,000	−30.45	9.785	4.111	E + 4	5.900	E − 1	1.561	E − 5
8,000	−36.94	9.782	3.565	E + 4	5.258	E − 1	1.527	E − 5
9,000	−43.42	9.779	3.080	E + 4	4.671	E − 1	1.493	E − 5
10,000	−49.90	9.776	2.650	E + 4	4.135	E − 1	1.458	E − 5
15,000	−56.50	9.761	1.211	E + 4	1.948	E − 1	1.422	E − 5
20,000	−56.50	9.745	5.529	E + 3	8.891	E − 2	1.422	E − 5
25,000	−51.60	9.730	2.549	E + 3	4.008	E − 2	1.448	E − 5
30,000	−46.64	9.715	1.197	E + 3	1.841	E − 2	1.475	E − 5
40,000	−22.80	9.684	2.871	E + 2	3.996	E − 3	1.601	E − 5
50,000	−2.50	9.654	7.978	E + 1	1.027	E − 3	1.704	E − 5
60,000	−26.13	9.624	2.196	E + 1	3.097	E − 4	1.584	E − 5
70,000	−53.57	9.594	5.221	E + 0	8.283	E − 5	1.438	E − 5
80,000	−74.51	9.564	1.052	E + 0	1.846	E − 5	1.321	E − 5

[a]Data abridged from *U.S. Standard Atmosphere*, 1976, U.S. Government Printing Office, Washington, D.C.

Appendix D
Compressible Flow Graphs for an Ideal Gas (k = 1.4)

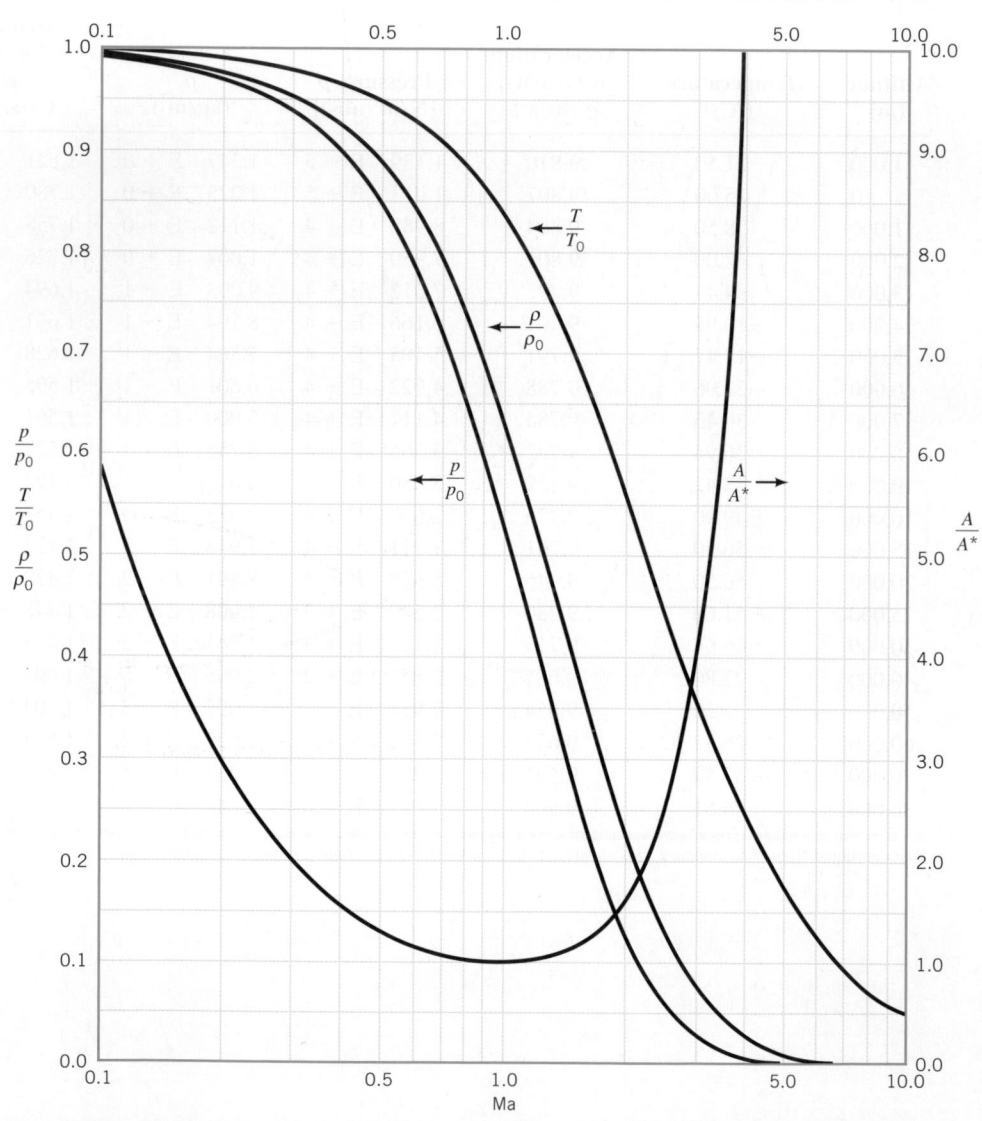

■ FIGURE D.1 Isentropic flow of an ideal gas with k = 1.4. (Graph provided by Dr. Bruce A. Reichert.)

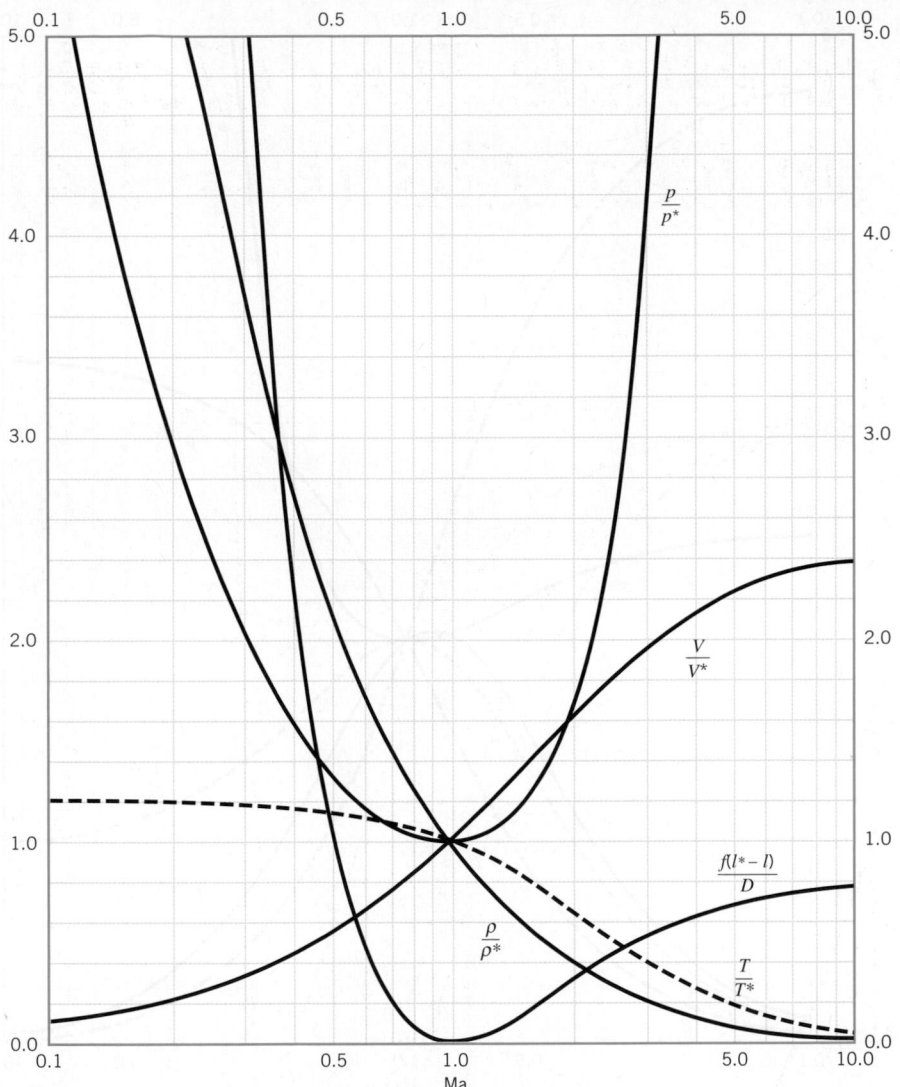

■ **FIGURE D.2** Fanno flow of an ideal gas with $k = 1.4$. (Graph provided by Dr. Bruce A. Reichert.)

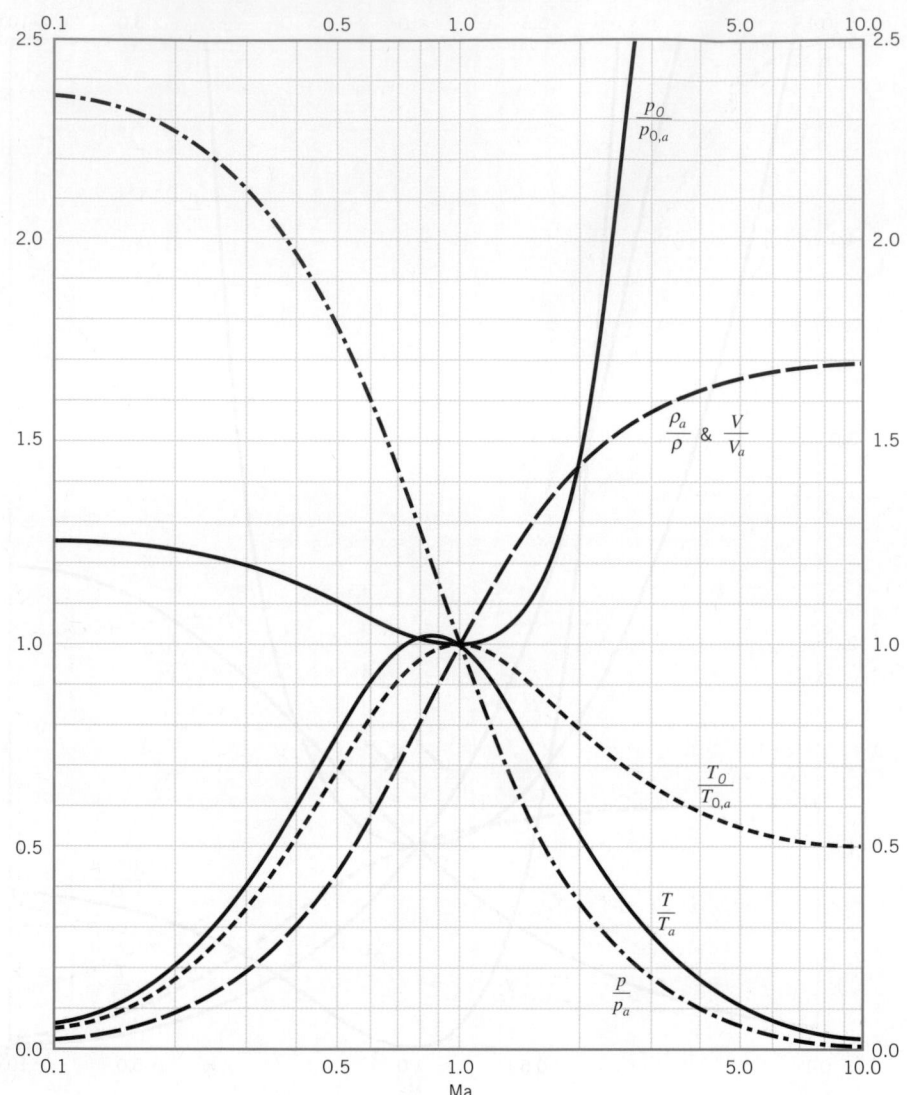

■ **FIGURE D.3** Rayleigh flow of an ideal gas with $k = 1.4$. (Graph provided by Dr. Bruce A. Reichert.)

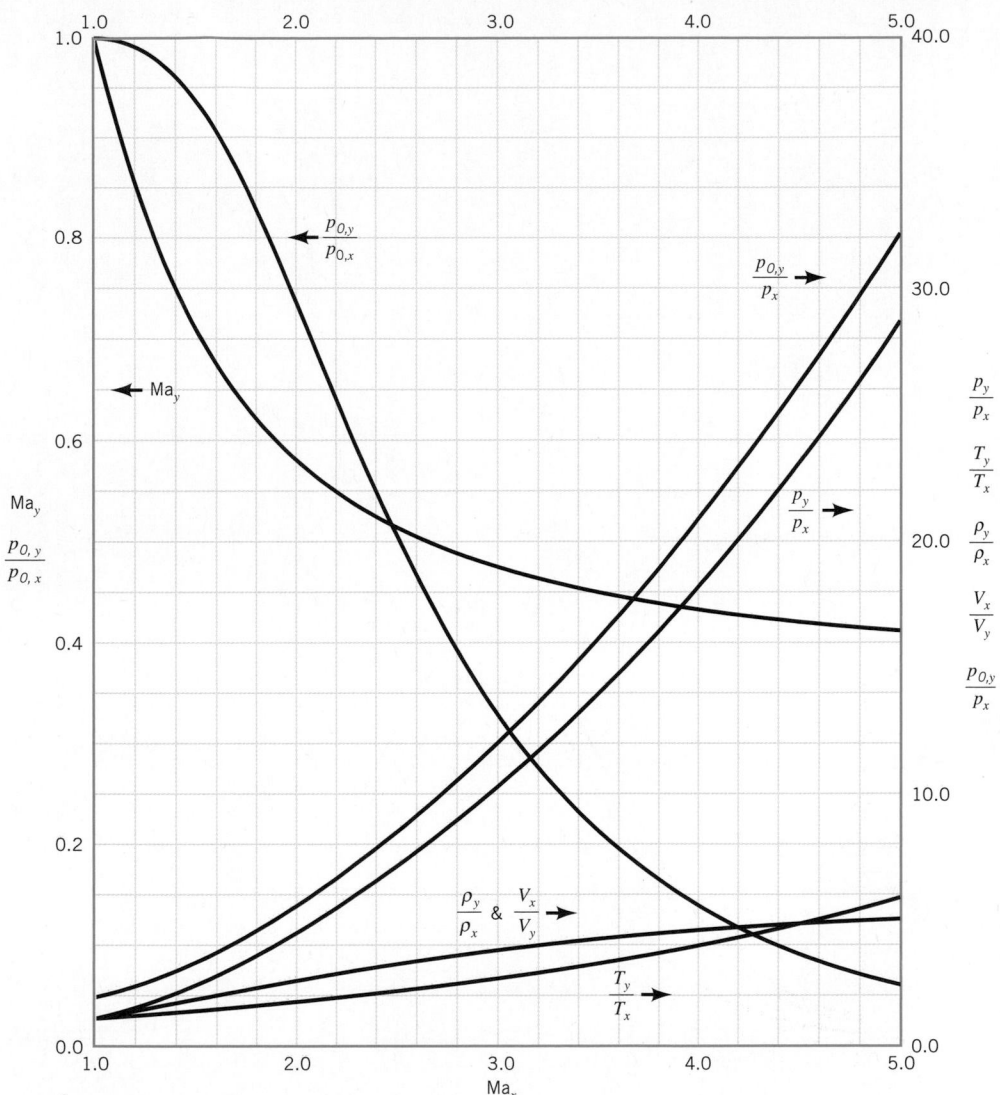

■ **FIGURE D.4**   Normal shock flow of an ideal gas with $k = 1.4$. (Graph provided by Dr. Bruce A. Reichert.)

Online Appendix List

Appendix E: **Comprehensive Table of Conversion Factors**
See book web site, www.wiley.com/go/global/munson, for this material.

Appendix F: **Video Library**
See book web site, www.wiley.com/go/global/munson, for this material.

Appendix G: **Review Problems**
See book web site, www.wiley.com/go/global/munson, for this material.

Appendix H: **Laboratory Problems**
See book web site, www.wiley.com/go/global/munson, for this material.

Appendix I: **CFD Driven Cavity Example**
See book web site, www.wiley.com/go/global/munson, for this material.

Appendix J: **FlowLab Tutorial and User's Guide**
See book web site, www.wiley.com/go/global/munson, for this material.

Appendix K: **FlowLab Problems**
See book web site, www.wiley.com/go/global/munson, for this material.

Answers to Selected Even-Numbered Homework Problems

Chapter 1

1.2   (a) L^3; (b) LT^{-2}; (c) $FL^{-1}T^2$; L^4; ML^2T^{-2}
1.6   (a) L^2T^{-2}; (b) $M^2L^{-3}T^{-3}$; (c) $M^0L^0T^0$
1.8 LT^{-1}, $F^0L^0T^0$, LT^{-1}
1.10 Yes
1.12 1/2, −1/2
1.14 No, no
1.16 1.24
1.18 0.250 m^3
1.20 (a) 3581 N; (b) 365 kg, 597 N
1.22 0.775; 7.60 kN/m^3
1.24 5×10^{-4} m^3; No
1.26 4.464 N
1.28 991.5 kg/m^3
1.32 64 °C
1.36 680 kPa (abs)
1.40 $893\ \dfrac{1}{\text{s}}$
1.42 $0.727\ \dfrac{\text{N} \cdot \text{s}}{\text{m}^2}$
1.44 23.7; 2.55×10^{-2}
1.46 $5.6\ \dfrac{\text{N} \cdot \text{s}}{\text{m}^2}$
1.48 15,000 (water); 752 (air)
1.50 1.43×10^{-6} kg/(m · s · K$^{1/2}$); S=107 K
1.52 D=1.767×10^{-6} N · s/m^2; B=1.870×10^3 K; 5.76×10^{-4} N · s/m^2
1.54 $\dfrac{V_1}{V} = 2c/(c+1)$
1.56 3.33×10^{-5} m
1.58 0.0883 m/s
1.62 1.27 N · m; $23.9\ \dfrac{\text{N} \cdot \text{m}}{\text{s}}$
1.64 $\mathcal{T} = 2\pi R_i^3 \ell \mu w (R_o - R_i)$
1.70 28500 kPa
1.72 (a) 343 m/s; (b) 1010 m/s; (c) 446 m/s
1.74 1171 kPa
1.76 2.17 kg/m^3; 156 °C
1.82 3000 m
1.84 13 kPa (abs)
1.86 5.81 kPa (abs)
1.88 0.060 N/m
1.90 12.2 Pa
1.92 (a) 0.292/D; (b) 0.292 cm
1.94 3.00 mm
1.96 0.5 cm; 10 cm

Chapter 2

2.2   59.2 kPa
2.4   (a) 16.0 kPa; 9.31 kPa; (b) no
2.6   50.5 MPa
2.8   0.552 kPa
2.12 70 kPa (abs)
2.14 (a) 58.8 kPa; (b) 442 mm Hg
2.16 (a) 59.2 kPa (abs); (b) 49.7 Kpa (abs); (c) 60.8 kPa (abs)
2.18 12.1 kPa; 0.195 kg/m^3
2.20 74.6 cm Hg
2.22 60 kPa
2.24 −8.82 kN/m^2
2.26 (a) 0.759 m; 0.759 m (without vapor pressure); (b) 10.1 m; 10.3 m (without vapor pressure); (c) 12.3 m; 13.0 m (without vapor pressure)
2.28 32 kPa
2.30 −3.32 kPa
2.32 1.5 kPa
2.34 $h = (p_1 - p_2)/(\gamma_2 - \gamma_1)$
2.36 0.040 m
2.38 94.9 kPa
2.40 0.449 m
2.42 27.3 kN/m^2
2.44 28.4 deg
2.46 0.083 m (down)
2.48 $\ell = [d + 11.31 - (d^2 + 18.61d + 128)^{1/2}]/2$
2.50 pkPa = 0.286 θ
2.52 889 N
2.54 92.4 kN; 0.0723 m along gate below centroid
2.56 639 kN; 0.990 m above bottom
2.58 107 kPa
2.64 143 kN
2.66 (a) 16.2 m; (b) No
2.68 1.60 m
2.70 3.55 m
2.72 (a) 2.11 m; (b) 941 kN
2.74 78.5 kN; 2.03 m below free surface
2.76 134 N · m

2.78 14.5 m; 18.4 m; 21.6 m; 24.4 m; 26.8 m

2.84 87.3 kN

2.86 34.5×10^9 N acting 124 m up from base of dam

2.88 64.4 kN

2.90 60.8 kN; 0.100 m below center of tank end wall

2.92 14.3 kN, 3.74 kN

2.94 485 kN

2.98 (a) 34.2×10^5 N; (b) 11.5×10^5 N

2.100 233 kN

2.104 2480 kg

2.106 2.13 kN up; 1.99 m to right of A

2.108 89.5

2.110 -2.63 m/s^2

2.112 (a) 3.3 kPa; (b) 255.8 N

2.114 37.4 deg

2.116 $h = a\ell/g$

2.118 2.52 m

2.120 28.8 kPa

Chapter 3

3.2 13.7 m/s

3.4 (a) $27000\,(1 + 3x) + 9800$ N/m^3;
(b) 287.3 kPa

3.6 -30.0 kPa/m

3.8 $-18.8 \dfrac{\text{kPa}}{\text{m}}$; 0.8 kPa/m

3.10 (a) $-2\rho a^2 V_0^2\,[1 - (a/x)^2]/x^3$
(b) $p_0 + \rho V_0^2\,[(a/x)^2 - (a/x)^4/2]$

3.14 12.0 kPa; -20.1 kPa

3.16 (a) 758 N/m^3; (b) 107 N/m^3

3.22 3.21 N/m^2

3.24 4.50 kPa

3.26 30 deg

3.28 4 mm

3.30 310 km/h

3.32 3.26 m

3.36 Tank (a)

3.38 58.8 kPa

3.40 $h = 0$

3.44 296.5 kPa

3.46 10.7 s

3.48 (a) 0.068 m; (b) 448.34 N/m^2

3.50 0.95 m

3.52 $0.0156\,D^2/[(0.1)^4 - D^4]^{1/2}$ m^3/s
when $D \sim$ m

3.54 2.35 m; 11.9 m

3.56 2.54×10^{-4} m^3/s

3.58 $h/H = 1/[1 + (d/D)^4]$

3.60 (a) 1.27 m; (b) 11.25 m/s, -27.3 kPa

3.62 86.3 Pa

3.64 0.42 m

3.66 -65.2 kPa

3.68 (a) 0.0696 m^3/s; (b) 0.574 m

3.70 $H/H_0 = 1/(1 + cx/L)^{1/2}$, where
$c = 2\gamma_{\text{H}_2\text{O}}\,d_{\text{max}}/\rho V_0^2$

3.72 0.0132 m^3/s

3.74 $9.54 \times 10^{-3}\dfrac{\text{m}^3}{\text{s}}$, 0.01 kg/s; 0.0981 N/s

3.76 4

3.78 $3.83\ 10^{-3}\dfrac{\text{m}^3}{\text{s}}$, -23.52 kPa, -14.7kPa, -14.7 kPa

3.80 0.37 m

3.82 2.98 m

3.84 0.0274 m^3/s

3.86 9.8×10^{-3} m^3/s

3.88 36.2 s

3.92 404.5 kPa

3.94 9.10×10^{-3} m^3/s; 57.9 kPa

3.96 2.00×10^{-4} m^3/s; 0.129 m

3.98 155 N/m^2

3.102 0.174 m^3/s

3.106 6.10×10^{-3} m^3/s

3.108 2.5 cm

3.110 $3.46\,Q_0$

3.112 4.5 m^3/s

Chapter 4

4.6 5 m/s; $(x^2 + 8x + 25)^{1/2}$ m/s

4.8 20 m/s for any x, y; -90 deg, -45 deg, 0 deg

4.10 $y^2 = x + c$

4.12 $y = e^{(x^2/2 - x)} - 1$

4.14 (a) 0.712, -8.50; (b) 1.69 ft/s

4.18 $x/h = (u_0/v_0)[\ln(h/(h - y) - y/h]$

4.20 3 m/s, 2.5 m/s, 2 m/s, 1 m/s

4.22 $a_x = (V_0/\ell)^2 x$; $a_y = (V_0/\ell)^2 y$

4.24 $a_x = x$; $a_y = 8x^2 y^2 (4x^2 y - 1)$; $a_z = -(x + 4x^2 y^2)$

4.26 (a) 1.2 m/s^2, 0.6 m/s^2; (b) negative

4.28 10 m/s; 10 m/s^2

4.30 $-5.12 \times 10^{10}\dfrac{\text{m}}{\text{s}^2}$; -5.22×10^9

4.32 $V_0(1 - x/\ell)[ce^{-ct} - (V_0/\ell)(1 - e^{-ct})^2]$;
$c = 0.490$ s^{-1}

4.34 -7.6 m/s^2; 0.77

4.36 $(225x + 150)$ m/s^2; 0; $(225x + 150)$ m/s^2;
150 m/s^2; 375 m/s^2

4.42 0, K/r^3

4.44 $(9V_0^2/4a)\sin\theta\cos\theta$; $(9V_0^2/4a)\sin^2\theta$; 0 deg; 90 deg

4.46 $(4V_0^2/a)\sin\theta\cos\theta$; $(4V_0^2/a)\sin^2\theta$

4.48 (a) 0.12 m/s^2, 0; (b) 0.38 m/s^2, 0.86 m/s^2;
(c) 0.4 m/s^2, 0.86 m/s^2

4.50 3.13×10^{-5} m/s^2; 2.00×10^{-3} m/s^2

4.52 7.64×10^3 m/s^2; 7.5 m/s^2

4.54 80 km/hr

4.56 (a) 2.5 °/hr; (b) 1.25 °/hr; (c) -10 °/hr

4.60 5.0 m^3/s

4.62 3.96 m^3/s

4.66 $2V_0 hb/3$

4.72 29.9 $\hat{\mathbf{i}}$ kg · m/s^2

Chapter 5

5.6 51 kg/s

5.8 0.56 m/s

5.10 decrease

5.12   1 m
5.14   0.202 kg/m^3
5.16   208 s
5.18   0.86 m/s
5.20   (a) 0.711; (b) 0.791; (c) 0.837; (d) 0.866
5.22 25×10^4 kg/s
5.24 $(7/8)\, U\ell\delta$
5.26   11.6 min
5.30   0.16 m/day
5.32   14.5 L/flush
5.38 2.66×10^{-4} m^3/s
5.40   0; 7420 N
5.42   motion to right (a), (b), (c); motion to left (d)
5.44   70.1 m/s; 30 kN
5.46   72,000 N; 67,400 N
5.48   29.4 kN
5.50   0.8 m^3/s
5.52   (a) $W_1 - 42$ N; (b) $W_2 + 62$ N
5.54 -185 kN; $+45.8$ kN
5.56   16.4 kN
5.58   144 N
5.60   51.4 N
5.62   3/4
5.64   (a) 754 N; (b) 611 N
5.66   0.112 m^3/s
5.68   0.071 m^3/s; 18.4 m/s
5.70   34 kN
5.76   (a) 231 N · m; 185 rad/s; (b) 200 N · m;
           160 rad/s; (c) 116 N · m; 92.5 rad/s
5.78   12.8 MW
5.80   348 kW
5.82   turbine; -3.3 J/kg
5.90   No
5.92   right to left; 0.009 m
5.94   0.06 m^3/s
5.96   right to left; 0.5 m
5.100 0.021 m^3/s
5.102 0.82 m^3/s; 0.97 m^3/s
5.104 86 N · m/kg; 18 N · m/kg
5.108 2.22 MW
5.110 8.82×10^8 W
5.112 930 kW
5.114 168 k; 148 k; 0.88
5.116 0.087 m^3/s
5.118 147.2 N · m/kg; 14.72 W
5.120 22.74 kW
5.122 1.78 kW
5.124 (a) 49%; (b) 48.84 N
5.126 139 kN; 35.7 kN
5.128 0.8 m
5.130 (a) 4.29 m/s; 17.2° (b) 558 N · m/s
5.132 (a) 1.11; (b) 1.08; (c) 1.06; (d) 1.05; (e) 1.04;
           (f) 1.03
5.134 561 N · m/kg; 81%
5.136 90.4%
5.138 17.3 J/kg; 84%

Chapter 6

6.2 $2x$, $4xt^2$; $-2y$, $4yt^2$; $V = 0$; $\mathbf{a} = 4\hat{\mathbf{i}} - 4\hat{\mathbf{j}}$ m/s^2;
          $a = 5.66$ m/s^2
6.4   (a) 0; (b) $\omega = -(y/2 + z)\hat{\mathbf{i}} + (5z/2)\hat{\mathbf{j}} - (y/2)\hat{\mathbf{k}}$;
          No
6.6 $\zeta = 3xy^2\,\hat{k}$; No
6.8 $u = -\dfrac{3}{2}x^2 + \dfrac{x^3}{3} + f(y)$
6.10 $\dot{\gamma} = \dfrac{-r_0\omega}{r_0 - r_i}$
6.16 No
6.18 $\psi = A\theta + Br^{-1}\sin\theta + C$
6.22 $v = \dfrac{-y^2}{2x} + f(x)$
6.24 $v_\theta = -4r\theta - 9r^2\cos\theta + f(r)$
6.26 (a) $\psi = \dfrac{-Ar^2}{2} + C$; (b) $\psi = -A\ln r + C$
6.28 (a) $\psi = 2xy$; (b) $q = 2x_iy_i$
6.32 (a) $p_A = p_0$; (b) $p_B = p_0$
6.34 $\phi = (5/3)x^3 - 5xy^2 + C$
6.36 $\phi = A\ln r + Br\cos\theta + C$; $\theta = \pi$, $r = A/B$
6.38 388 kPa
6.40 (a) Yes; (b) Yes, $\phi = 0.6(x + y) + C$; (c) 0
6.42 (a) $\phi = U(0.866x + 0.500y)$; $\psi = U(0.866y - 0.500x)$; (b) $\dfrac{\partial p}{\partial y} = -\gamma$
6.46 (a) $\psi = m\left(\dfrac{\theta}{2\pi} - \dfrac{1}{6}\right)$; (b) $\theta = 1.33$ rad
6.50 $\Gamma = 0$
6.52 0.92 m
6.54 $y = \dfrac{m}{2\pi U}\theta$, $H = 8.33 \times 10^{-3}$ m
6.58 130 km/hr; 86.7 km/hr; 65 km/hr
6.60 $h^2 = \dfrac{m}{2\pi A}$
6.62 (b) $V_A = 0$, $V_B = 119$ km/hr
6.64 $u_A = U - (\Gamma/H)/\pi$
6.68 $F_x = 18.6$ N/m
6.70 (a) $\Gamma = 0$; (b) $\Gamma = -4\pi Ua$
6.72 (a) $F_y = 182$ kN; (b) $F_y = 546$ kN
6.76 $\partial p_s/\partial\theta = 4\rho U^2\sin\theta\cos\theta$; θ falls in range
          of $\pm90°$
6.80 $\sigma_{xx} = -5.98$ kPa, $\sigma_{yy} = -6.02$ kPa, $\tau_{xy} = 45.0$ Pa
6.82 (a) $\partial v/\partial y = -2x$; (b) $\mathbf{a} = 2x^3\hat{\mathbf{i}}$; (c) $\partial p/\partial x = 2\mu - 2\rho x^3$
6.84 $q/\ell = 2.3 \times 10^{-5}$ m^2/s, $\tau = 2.25$ N/m^2
6.86 $\dfrac{\partial p}{\partial y} = \dfrac{-3\mu V}{h^2} - \rho g$
6.88 20.4 L/min
6.90 $q = (\rho gh^3\sin\alpha)/3\mu$
6.92 $u = [(U_1 + U_2)/b]y - U_2$
6.94 $y/b = \frac{1}{3}$
6.96 0.355 N · m
6.100 (a) Re $= 640 < 2100$; (b) 180 kPa; (c) 60.0 N/m^2

6.102 $v_\theta = R^2 \omega / r$

6.104 $V = 1.10 \times 10^{-2}$ m/s

6.106 $v_\theta = \dfrac{r\omega}{(1 - r_i^2/r_o^2)}\left[1 - \dfrac{r_i^2}{r^2}\right]$

6.108 $\dfrac{\Delta p}{\ell} = 6.4$ k$\dfrac{\text{N}}{\text{m}^2}$ per m

6.110 (a) 42.6%; (b) 21.7%

Chapter 7

7.6   (a) 103 m/s; (b) 444 m/s

7.8 $k - r = 3$ pi terms

7.10 (a) Δp is halved; (b) Δp is doubled

7.12 $\Delta p D_1/(V\mu) = \phi(D_2/D_1, \rho D_1 V/\mu)$

7.14 $\mathcal{P}/(\rho D^5 \omega^3) = \phi(Q/D^3\omega)$

7.16 $\Delta p/(D^2\rho\omega^2) = \phi(Q/D^3\omega)$

7.18 $h/\ell = \phi(\sigma/\ell^2 g\rho)$

7.20 $h/D = \phi(\sigma/\gamma D^2)$

7.22 $\Delta p \propto 1/D^2$ (for a given velocity)

7.24 $d/D = \phi(\rho VD/\mu, \sigma/\rho V^2 D)$

7.26 (a) $VD\sqrt{\rho/\mathcal{W}} = \phi(b/d, d/D)$; (b) $V = \sqrt{2\mathcal{W}b/\pi\rho dD^2}$

7.30 $\Delta p/(\rho V^2) = 0.508\,(D/d)^{4.03}$

7.32 (a) $H/b = \phi(h/b, \ell/b)$;
(b) $H/b = 0.0833\,(h/b)^{-1.00}$

7.34 (a) $n = -1/2$; (b) Yes

7.38 Colder

7.40 0.29 m/s

7.42 9.49 m/s

7.44 $\sigma_m/\sigma = 4.44 \times 10^{-3}$

7.46 1170 km/hr

7.48 299.5 km/hr

7.50 $h/d = \phi(H/d, \gamma_s/\gamma_f, \mu_f g^{1/2}/(\gamma_f d^{3/2}))$

7.52 (a) 400 km/hr; (b) 170 N

7.54 113.5 N

7.56 (a) $d_m/D_m = d/D$, $\rho_m/\rho_{pm} = \rho/\rho_p$, $g_m d_m^3 \rho_m^2/\mu_m^2 = gd^3\rho^2/\mu^2$, $\rho V_c D/\mu = \rho_m V_{cm} D_m/\mu_m$; (b) $V_{cm}/V_c = 0.707$

7.58 (a) $\mathcal{D}/\rho V^2 D^2 = \phi(d/D)$; (b) 280 N

7.60 (a) 12 m/s; (b) 900 N

7.62 0.0647 to 0.0971

7.68 (a) 400 N; (b) 7,220 W

7.70 (a) $V\ell^2/Q = \phi(\ell_i/\ell, Q^2/\ell^5 g, \rho Q/\ell\mu)$; no
(b) 1.56 L/min; 6.15 cm

7.72 (a) $nD/V = \phi(VD/\nu)$; (b) Yes; (c) 2.26 m/s

7.74 (a) $V_m/U_m = V/U$, $V_m D_m/\nu_{sm} = VD/\nu_s$, $V_m^2/g_m D_m = V^2/g D$, $(\rho - \rho_s)_m/\rho_m = (\rho - \rho_s)/\rho$; (b) no

7.76 (a) $p/\rho V^2 = \phi(\ell/\ell_i)$; (b) 0.333 ft; (c) $p = (V/V_m^2)\,p_m$, No

7.78 $(\rho\omega h^2/\mu)\,\partial u^*/\partial t^* = \partial^2 u^*/\partial y^{*2}$

7.80 $\partial v_z^*/\partial t^* = p_1\rho R^3/(\ell^2\mu) + \partial^2 v_z^*/\partial r^{*2} + (1/r^*)\,\partial v_z^*/\partial r^*$

Chapter 8

8.2   0.32 m

8.4   blue and yellow streams; green

8.6   5.22 m

8.8   0.0883 m

8.10 (a) 8.93×10^4; (b) 8.93×10^{-8}

8.12 $V_{\max} = 2$ m/s, $V = 1$ m/s, $Q = 1.3 \times 10^{-3}$ m^3/s

8.14 7.23 N/m^2, 1.67 N/m^2; 0 N/m^2

8.16 from (b) to (a)

8.18 0.354 D

8.20 3.43 m, 166 kPa

8.22 (a) 12.42 kPa; (b) 15.85 kPa

8.24 0.68 m/s

8.26 D

8.28 18.5 m

8.30 8.88×10^{-8} m^3/s

8.32 26,300, turbulent

8.38 7.3×10^{-4} m

8.42 0.65 Pa/m

8.44 0.019 m/m

8.46 water out if $V = 5$ m/s; air in if $V = 0.5$ m/s

8.48 (b)

8.50 smaller pipe

8.52 (a) 1.46 kN/m^2; (a) 42.6 kN/m^2

8.54 21.0

8.58 2.6 m

8.60 (a) 7.14 m

8.62 133 kPa; 39.7 kPa from losses and 93.0 kPa from change in kinetic energy

8.64 13.2

8.66 9.54 N/m^2

8.68 9

8.70 disagree

8.72 aspect ratio equal to 4

8.74 0.188 m

8.76 14.8 N/m^2

8.78 (a) 47 m; (b) 47.8 m

8.80 No

8.82 50.0 m

8.84 2.115 kPa

8.86 5.01 m

8.90 379 kW

8.94 1.7×10^{-3} m^3/s

8.98 1.78 m/s

8.100 0.135 m

8.102 0.863 m

8.104 0.031 m by 0.053 m

8.108 2.19×10^{-3} m^3/s; 9.33×10^{-3} m^3/s

8.110 0.12 m

8.112 0.0284 m^3/s; 0.0143 m^3/s; 0.0141 m^3/s

8.116 267.6 kPa

8.118 64.8 kPa

8.120 0.016 m^3/s

8.122 3.6×10^{-3} m^3/s

8.124 3.17×10^{-3} m^2/s

Chapter 9

9.2 1.1

9.4 3.45 kN, 0.560 kN; 3.47 kN, 0.427 kN

9.6 Fig. 9.6c

9.12 1.12 m, 7.92×10^{-3} m

9.14 6.65×10^{-6} m^2/s

9.16 2.30×10^{-3} m/s; 0.145 m/s; 8.1×10^{-6} m/s

9.22 (a) 8.67 m/s; (b) 32.76 km/hr

9.30 $\delta = 4.12\sqrt{\nu x/U}$, $\tau_w = 4\mu U/(3\delta)$

9.36 Larger by a factor of 4

9.38 240.2 N

9.40 42 kN

9.42 1.4, upright

9.44 85.4 kW

9.46 2

9.48 0.0296 N

9.50 11.2 kW

9.52 1.06 m/s

9.54 5.70 m/s

9.58 1.10×10^{-7} m/s, 1.20×10^{-7} m/s

9.60 230 N · m

9.62 8,950 N · m

9.64 280.4 kW

9.66 greater

9.68 41.8 kW

9.72 180 N

9.76 7220 m/s^2

9.78 457.7 N/m^2

9.80 0.24 m/s^2

9.86 13.9 m/s

9.90 (a) 13.2W; (b) 21.9 kW

9.92 2.04 m

9.94 0.282

9.96 1286 N

9.98 1.72 $U_{\text{sea level}}$

9.104 19.1

9.106 6.01% increase

9.110 28.4%

Chapter 10

10.2 25 m/s^2

10.4 (a) supercritical; (b) supercritical; (c) subcritical

10.8 1.89 m

10.12 2.60 m/s

10.14 616 km/hr

10.16 2.45 m; 0.388 m

10.18 0.12 m; 1.68 m

10.20 0.956 m; 1.016 m

10.24 0.59 m

10.26 0.054 m

10.30 0.1 m

10.34 $- 6.1 \times 10^{-5}$

10.36 0.0126

10.38 5.14 N/m^2

10.40 0.0125

10.42 0.25 m^3/s

10.44 Yes

10.46 Yes

10.48 40.9 m^3/s

10.50 1.86 m/s

10.54 18.2 m^3/s

10.56 17.3 m^3/s

10.58 3.23 m^3/s

10.60 1.19 m

10.62 0.840 m

10.64 0.77 m

10.66 Same

10.68 0.71 m; Not possible

10.70 8.77 m

10.72 0.114 m

10.74 2.21 m

10.76 0.52 m

10.78 0.000816

10.80 0.000676

10.82 0.000218

10.84 0.00766

10.88 4.16 m/s

10.90 2.22×10^{-4} m^3/s

10.92 (a) 0.05 m; (b) 14.3, 0.160; (c) 4.15 m

10.94 1.51 m, 12,500 kW

10.96 upstream

10.98 1.32 m/s

10.100 Yes

10.102 0.116 m^3/s

10.104 1.60 m

10.106 -0.112 m/hr; 5.53 hr

10.110 $H_r = 0.41$ m; $H_t = 1.369$ m; triangular

Chapter 11

11.4 (a) 71,700 J/kg; (b) 100,000 J/kg; (c) -1.58 kg/m^3; (d) 396 J/kg · K

11.10 351 K

11.12 (a) 4.1 kJ/kg; 2230 J/kg; (b) 4.54×10^5 J/kg; 5.02×10^5 J/kg; (c) 1.36×10^6 J/kg; 1.93×10^6 J/kg

11.16 $E_{v_{\text{steel}}} = 2.19 \times 10^{11}$ N/m^2, $c_{\text{water}} = 1470$ m/s, $c_{\text{air}} = 340$ m/s

11.18 (a) 0.0327; (b) 0.0735; (c) 0.131

11.22 Ma = 1.67

11.28 Ma = 2.22; V = 735 m/s

11.36 (a) 0.99988; (b) 0.998; (c) 0.9949; (d) 0.700

11.38 (a) V = 267 m/s; Ma = 0.725; (b) V = 285 m/s; Ma = 0.78

11.44 Ma$_{\text{exit}}$ = 1.8; V$_{\text{exit}}$ = 482 m/s

11.46 $\dot{m} = 26.5$ kg/s

11.52 $\dot{m} = 3.43$ kg/s
$P_1 = 81$ kPa (abs); $T_1 = 271.2$ K; $V_1 = 186.3$ m/s
$P_2 = 43.5$ kPa (abs); $T_2 = 240$ K; $V_2 = 310.5$ m/s

11.54   10.5%

11.56 $\dot{m}_{f=0.01} = 1.70$ kg/s; $\dot{m}_{f=0.02} = 1.65$ kg/s;
$\dot{m}_{f=0.03} = 1.52$ kg/s

11.60 $\dot{Q}_{net \atop in} = 1.4$ MJ/s; $P_2 = 119.3$ kPa (abs);
$V_2 = 264$ m/s; $Ma_2 = 0.40$

11.64   56 kPa

11.66 $Ma = 0.475$; $T = 891$ K; $T_0 = 928$ K;
$P_0 = 2520$ kPa (abs); $P = 2163$ kPa (abs);
$V = 284$ m/s

11.68 $Ma = 1.25$; $V = 3.80$ m/s

11.70   (a) 0.94; (b) 0.8; (c) 0.47; $P_{01} - P_{02} = 50$ kPa

11.72   (a) 0.213; $Ma_2 = 0.58$
(b) 0.9; $Ma_2 = 0.89$

11.74 $P_1 = 13.2$ kPa (abs); $V_1 = 509$ m/s; $T_1 = 161.3$ K
$P_2 = 36.7$ kPa (abs); $V_2 = 351$ m/s; $T_2 = 304$ K

Chapter 12

12.4   pump

12.8   SI, (N · m)/kg or J/kg; J/s or watt (W)

12.10   530 N/m²; turbine

12.16   turbine and pump

12.18 1.52×10^3 J/kg

12.20   (b) 1.2 kJ; 0 rpm

12.22   1.36 kW

12.24   11.6 m

12.28   increase to 2.74 m

12.30   (a) 2.07 kW; (b) 12.6 m

12.34   1388 L/min

12.36   964 L/min; no

12.38   (a) 0.0529 m³/s; (b) 13%

12.42   0.0328 m³/s; 8 m

12.44   Yes

12.46   4 kL/min; 120 m

12.48   671 at maximum efficiency

12.54   mixed-flow

12.56   5819 W; 18.4 m²/s²

12.62 2.95×10^{-4} m³/s; 0.19 N · m; 28.3 W

12.66   23,200 kW

12.68   5.27 MW

12.70   Francis; 112.2 m³/s

Index

Index of Fluids Phenomena Videos

Available on www.wiley.com/go/global/munson
Use the registration code included with this new text to access the videos.

V1.1
Mt. St. Helens
Eruption

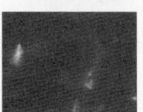
V1.2
E coli swimming

V1.3
Viscous fluids

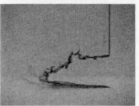
V1.4
No-slip condition

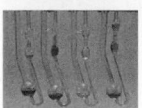
V1.5
Capillary tube vis-
cometer

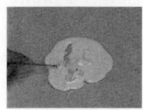
V1.6
Non-Newtonian
behavior

V1.7
Water balloon

V1.8
As fast as a speeding
bullet

V1.9
Floating razor blade

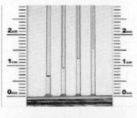
V1.10
Capillary rise

V2.1
Pressure on a car

V2.2
Blood pressure meas-
urement

V2.3
Bourdon gage

V2.4
Hoover dam

V2.5
Pop bottle

V2.6
Atmospheric
buoyancy

V2.7
Cartesian Diver

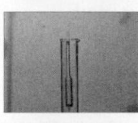
V2.8
Hydrometer

V2.9
Stability of a floating
cube

V2.10
Stability of a model
barge

V3.1
Streamlines past an
airfoil

V3.2
Balancing ball

V3.3
Flow past a biker

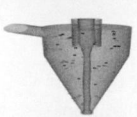
V3.4
Hydrocyclone sepa-
rator

V3.5
Aircraft wing tip
vortex

V3.6
Free vortex

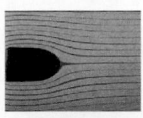
V3.7
Stagnation point flow

V3.8
Airspeed indicator

V3.9
Flow from a tank

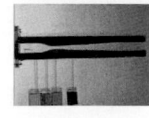
V3.10
Venturi channel

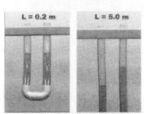
V3.11
Oscillations in a
U-tube

V3.12
Flow over a cavity

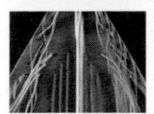
V4.1
Streaklines

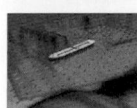
V4.2
Velocity field

V4.3
Cylinder-velocity
vectors

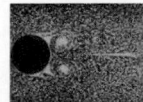
V4.4
Follow the particles
(experiment)

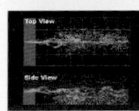
V4.5
Follow the particles
(computer)

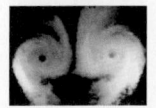
V4.6
Flow past a wing

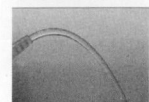
V4.7
Flow types

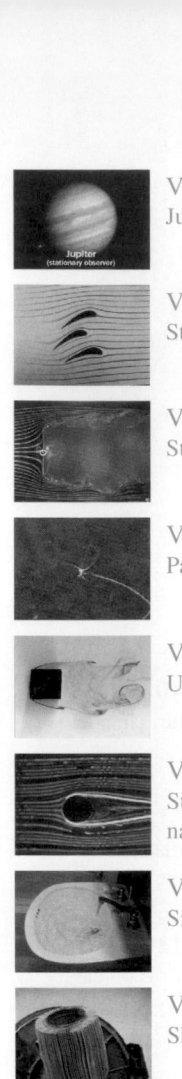

V4.8
Jupiter red spot

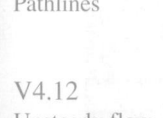

V4.9
Streamlines

V4.10
Streaklines

V4.11
Pathlines

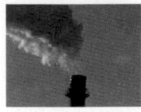

V4.12
Unsteady flow

V4.13
Streamline coordinates

V5.1
Sink flow

V5.2
Shop vac filter

V5.3
Flow through a contraction

V5.4
Smokestack plume momentum

V5.5
Marine propulsion

V5.6
Force due to a water jet

V5.7
Running on water

V5.8
Fire hose

V5.9
Jelly fish

V5.10
Rotating lawn sprinkler

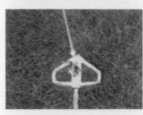

V5.11
Impulse-type lawn sprinkler

V5.12
Pelton wheel turbine

V5.13
Energy transfer

V5.14
Water plant aerator

V6.1
Spinning football-velocity contours

V6.2
Spinning football-velocity vectors

V6.3
Shear deformation

V6.4
Vortex in a beaker

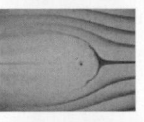

V6.5
Half-body

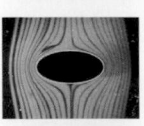

V6.6
Circular cylinder

V6.7
Ellipse

V6.8
Circular cylinder with separation

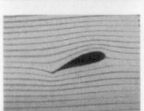

V6.9
Potential and viscous flow

V6.10
Potential flow

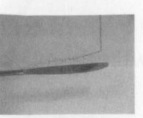

V6.11
No-slip boundary condition

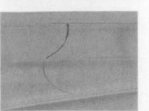

V6.12
Liquid–liquid no-slip

V6.13
Laminar flow

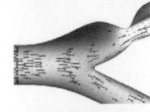

V6.14
Complex pipe flow

V6.15
CFD example

V7.1
Real and model flies

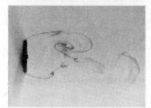

V7.2
Flow past a flat plate

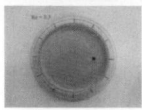

V7.3
Reynolds number

V7.4
Froude number

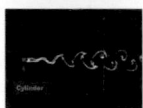

V7.5
Strouhal number

V7.6
Weber number

V7.7
Stokes flow

V7.8
Model airplane

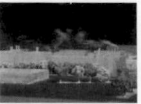

V7.9
Environmental models

V7.10
Flow past an ellipse

V7.11
Model of fish hatchery pond

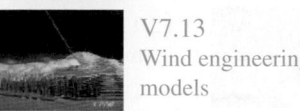
V7.12
Distorted river model

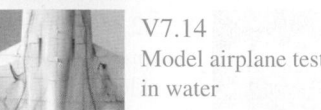
V7.13
Wind engineering models

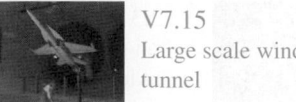
V7.14
Model airplane test in water

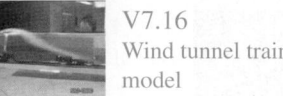
V7.15
Large scale wind tunnel

V7.16
Wind tunnel train model

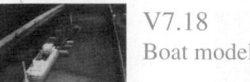
V7.17
River flow model

V7.18
Boat model

V7.19
Dam model

V7.20
Testing of large yacht mode

V8.1
Turbulent jet

V8.2
Laminar/turbulent pipe flow

V8.3
Intermittent turbulent burst in pipe flow

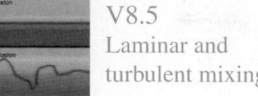
V8.4
Stirring color into paint

V8.5
Laminar and turbulent mixing

V8.6
Stirring cream into coffee

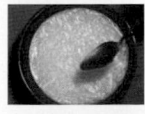
V8.7
Turbulence in a bowl

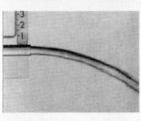
V8.8
Laminar to turbulent flow from a pipe

V8.9
Laminar/turbulent velocity profiles

V8.10
Entrance/exit flows

V8.11
Separated flow in a diffuser

V8.12
Car exhaust system

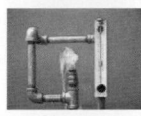
V8.13
Rotameter

V8.14
Water meter

V9.1
Space shuttle landing

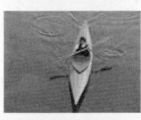
V9.2
Streamlined and blunt bodies

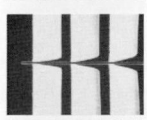
V9.3
Laminar boundary layer

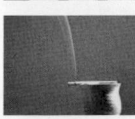
V9.4
Laminar/turbulent transition

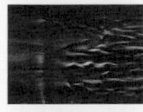
V9.5
Transition on flat plate

V9.6
Snow drifts

V9.7
Skydiving practice

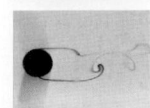
V9.8
Karman vortex street

V9.9
Oscillating sign

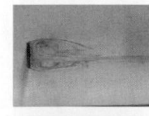
V9.10
Flow past a flat plate

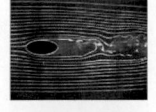
V9.11
Flow past an ellipse

V9.12
Jet ski

V9.13
Drag on a truck

V9.14
Automobile streamlining

V9.15
Stalled airfoil

V9.16
Bat flying

V9.17
Trailing edge flap

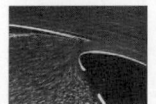
V9.18
Leading edge flap

V9.19
Wing tip vortices

V10.1
Off-shore oil drilling platform.

V10.2
Filling your car's gas tank.

V10.3
Water strider

V10.4
Sinusoidal waves

V10.5
Bicycle through a puddle

V10.6
Merging channels

V10.7
Uniform channel flow

V10.8
Erosion in a channel

V10.9
Bridge pier scouring

V10.10
Big Sioux River bridge collapse

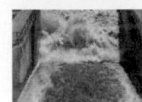
V10.11
Hydraulic jump in a river

V10.12
Hydraulic jump in a sink

V10.13
Triangular weir

V10.14
Low-head dam

V10.15
Spillway gate

V10.16
Unsteady under and over

V11.1
Lighter flame

V11.2
Jet noise

V11.3
Speed boat

V11.4
Compressible flow visualization

V11.5
Rocket engine start-up

V11.6
Supersonic nozzle flow

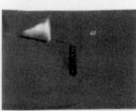
V11.7
Blast waves

V11.8
Two-dimensional compressible flow

V12.1
Windmills

V12.2
Self-propelled lawn sprinkler

V12.3
Pelton wheel lawn sprinkler

V12.4
Dental drill

V12.5
Flow in a compressor stage

VA.1
Pouring a liquid

VA.2
Dynamic grid

VA.3
Tornado simulation

■ TABLE 1.3
Conversion Factors from BG and EE Units to SI Units[a]

	To Convert from	to	Multiply by
Acceleration	ft/s^2	m/s^2	3.048 E − 1
Area	ft^2	m^2	9.290 E − 2
Density	lbm/ft^3	kg/m^3	1.602 E + 1
	slugs/ft^3	kg/m^3	5.154 E + 2
Energy	Btu	J	1.055 E + 3
	ft · lb	J	1.356
Force	lb	N	4.448
Length	ft	m	3.048 E − 1
	in.	m	2.540 E − 2
	mile	m	1.609 E + 3
Mass	lbm	kg	4.536 E − 1
	slug	kg	1.459 E + 1
Power	ft · lb/s	W	1.356
	hp	W	7.457 E + 2
Pressure	in. Hg (60 °F)	N/m^2	3.377 E + 3
	lb/ft^2 (psf)	N/m^2	4.788 E + 1
	lb/in.^{2} (psi)	N/m^2	6.895 E + 3
Specific weight	lb/ft^3	N/m^3	1.571 E + 2
Temperature	°F	°C	$T_C = (5/9)(T_F − 32°)$
	°R	K	5.556 E − 1
Velocity	ft/s	m/s	3.048 E − 1
	mi/hr (mph)	m/s	4.470 E − 1
Viscosity (dynamic)	lb · s/ft^2	N · s/m^2	4.788 E + 1
Viscosity (kinematic)	ft^2/s	m^2/s	9.290 E − 2
Volume flowrate	ft^3/s	m^3/s	2.832 E − 2
	gal/min (gpm)	m^3/s	6.309 E − 5

[a]If more than four-place accuracy is desired, refer to Appendix E.

Conversion Factors from SI Units to BG and EE Units[a]

	To Convert from	to	Multiply by
Acceleration	m/s^2	ft/s^2	3.281
Area	m^2	ft^2	1.076 E + 1
Density	kg/m^3	lbm/ft^3	6.243 E − 2
	kg/m^3	slugs/ft^3	1.940 E − 3
Energy	J	Btu	9.478 E − 4
	J	ft · lb	7.376 E − 1
Force	N	lb	2.248 E − 1
Length	m	ft	3.281
	m	in.	3.937 E + 1
	m	mile	6.214 E − 4
Mass	kg	lbm	2.205
	kg	slug	6.852 E − 2
Power	W	ft · lb/s	7.376 E − 1
	W	hp	1.341 E − 3
Pressure	N/m^2	in. Hg (60 °F)	2.961 E − 4
	N/m^2	lb/ft^2 (psf)	2.089 E − 2
	N/m^2	lb/in.^{2} (psi)	1.450 E − 4
Specific weight	N/m^3	lb/ft^3	6.366 E − 3
Temperature	°C	°F	$T_F = 1.8\,T_C + 32°$
	K	°R	1.800
Velocity	m/s	ft/s	3.281
	m/s	mi/hr (mph)	2.237
Viscosity (dynamic)	N · s/m^2	lb · s/ft^2	2.089 E − 2
Viscosity (kinematic)	m^2/s	ft^2/s	1.076 E + 1
Volume flowrate	m^3/s	ft^3/s	3.531 E + 1
	m^3/s	gal/min (gpm)	1.585 E + 4

[a]If more than four-place accuracy is desired, refer to Appendix E.